The Birds of Wales

Adar Cymru

The Birds of Wales
Adar Cymru

EDITED BY
**Rhion Pritchard,
Julian Hughes,
Ian M. Spence,
Bob Haycock
and Anne Brenchley**

**Rare and scarce species: Jon Green and Reg Thorpe
Photographic editor: Ben Porter
Cover and frontispiece illustrations: Philip Snow**

**Published on behalf of the
Welsh Ornithological Society**

LIVERPOOL UNIVERSITY PRESS

First published 2021 by
Liverpool University Press
4 Cambridge Street
Liverpool
L69 7ZU

British Library Cataloguing-in-Publication data
A British Library CIP record is available

ISBN 978-1-800-85972-2

Typeset by Carnegie Book Production, Lancaster
Printed and bound by Gomer

This book is dedicated to all the birdwatchers, past and present, whose records have added to our understanding of birds in Wales, and to the next generation of birdwatchers, whose recording and passion should secure their future.

Yr wylan deg ar lanw, dioer,
Unlliw ag eiry neu wenlloer,
Dilwch yw dy degwch di,
Darn fal haul, dyrnfol heli.
Ysgafn ar don eigion wyd,
Esgudfalch edn bysgodfwyd.

Fair gull on the rising tide, in truth,
Colour of snow or the white moon,
Unblemished is your beauty,
A piece of the sun, glove of the brine.
Lightly you go on the ocean's waves,
Nimble and proud, fish-eating bird.

Dafydd ap Gwilym, *fl.* 1340–70,
from *Yr Wylan* ('The Gull')

Translation by Rhion Pritchard

Contents

Preface

We could not be more pleased to be asked to write the Preface to this follow up to *Birds in Wales*, some quarter of a century since our original work was published as part of the Poyser series. It is, we feel, a timely and auspicious moment for the Welsh Ornithological Society to undertake this task—not only to define the considerable changes to the Welsh avifauna over the past 27 years, but also since Wales is on the cusp of the major agricultural and environmental changes that result from having recently left the European Union. Rising temperatures will also play a part.

As we indicated in 1994, the agricultural land use changes since, in particular, the Second World War have had a substantial and detrimental impact on bird populations in Wales. It is a very sad testament that our concerns are now even greater for many of the species that are regarded as special for Wales.

Breeding waders continue to be a particular worry. Golden Plovers, of which there were more than 250–300 breeding pairs in the 1970s, are now much diminished in range and numbers, with fewer than 20 pairs. Lapwings are declining rapidly and all the main colonies are confined to nature reserves. Curlews are 69% below the 1995 level indicated by the BTO/RSPB/JNCC Breeding Bird Survey. Only around 70 breeding pairs of Redshank remain, of which 40% are at a single site, RSPB Ynys-hir in Ceredigion. Those much-loved specialities of our hanging oakwoods, Pied Flycatchers and Wood Warblers, are declining. Seabirds such as Kittiwakes and Fulmars appear to be in trouble. Among farmland bird species, Turtle Dove and Corn Bunting have already gone, while Yellowhammers are 65% below their 1995 level and Linnets down by 25%.

It is not just breeding birds that are struggling. Recent results from the long-term Wetland Bird Survey reveal alerts (decreases greater than 25%) for many of the wintering species on the Dee Estuary in Flintshire/Wirral, the Severn Estuary in East Glamorgan and Gwent, and the Burry Inlet, Gower/Carmarthenshire, although not on Traeth Lafan, Caernarfonshire/Anglesey. In part at least, this appears to be due to climate change, since many birds from the Fennoscandian and Russian populations now winter farther east in response to warming temperatures.

Much has changed in ornithological recording in recent years. The quality of binoculars, telescopes and digital cameras available to the ornithologist of today would have been beyond our wildest dreams when we first ventured into the field many years ago! Even more dramatic have been the opportunities for innovative research through satellite telemetry, by fitting tiny GPS tags to birds. This has added greatly to our understanding of the breeding biology and movements of many Welsh species, not least, for example, the foraging excursions of Manx Shearwaters from Skomer in Pembrokeshire and Bardsey in Caernarfonshire, the year-round movements of Greenland White-fronted Geese from the small and declining flock on the Dyfi Estuary and the habitat requirements of Curlews on the Migneint in northwest Wales. The ground-breaking BTO Cuckoo tracking programme has revealed that the adults leave the UK much earlier than previously thought. Satellite tagging of Hen Harriers, including some from Welsh nests, has highlighted the horrific scale of persecution of this magnificent raptor on driven grouse moors throughout the United Kingdom. Through the wonders of DNA analysis, the challenging identification of closely related species can now be clarified by diligent collection of a feather or a faecal sample.

Since 1994 there has been a very welcome growth in interest about birds in Wales. This is illustrated by the marked increase in support for three relevant organisations. The membership of the Welsh Ornithological Society, which had hovered around 200 members since the society's inception in 1988, has risen to more than 350 by 2020. BTO membership in Wales, which in 1994 was around 500 members, now numbers over 1,000. RSPB membership in Wales numbers over 58,000 in 2020, up by more than 4,000 since 2016; the UK membership rose by 40% between 1994 and 2020. The stimulation of interest has, without doubt, been furthered by the establishment of educational work at visitor centres on Welsh nature reserves, among which are the Wildlife Trust Centre at the Teifi Marshes in Pembrokeshire, WWT Llanelli Wetland Centre in Carmarthenshire, RSPB Conwy in Denbighshire, RSPB Lake Vyrnwy in Montgomeryshire, and Newport Wetlands in Gwent. This expanding interest is reflected in the increased voluntary effort dedicated to the vital work of monitoring bird populations. For breeding birds, for example, the coverage in Wales of the Breeding Bird Survey increased from 122 squares in 1994 to 324 in 2019, providing a detailed and very valuable measure of long-term trends.

One of the greatest conservation success stories in Wales must surely be the struggle to maintain Red Kite as a UK and Welsh breeding species, an epic that covered virtually the whole of the 20th century and existed in one guise or another since 1903, with many vicissitudes along the way. It is hard to imagine that the breeding population was as low as two pairs in the 1930s, when they are now so numerous that monitoring the whole population annually is no longer feasible or necessary. One other species that merits this accolade is the Manx Shearwater, subject of a pioneering study by Ronald Lockley on Skokholm, Pembrokeshire, in the 1930s. It has been followed by a galaxy of other ground-breaking seabird studies on Skokholm and Skomer, which hold a high percentage of the world population of this species.

While recent years have seen, in general, a move away from concentrating on single species in favour of a broader habitat-based philosophy, there are still many instances where the former approach is beneficial. Priority species action on a UK scale played its part in work carried out in Wales. The EU 'Urgent Action for the Bittern' plan 1996–2000 stimulated work by the RSPB on freshwater habitat creation in Anglesey and elsewhere. As a result, it is gratifying that the Bittern is now firmly re-established as a Welsh breeding species. A network of wetlands has been created and restored by the WWT at Penclacwydd, by NRW at Newport Wetlands, and by the RSPB at Cors Ddyga and Valley Wetlands on Anglesey, and at Burton Mere Wetlands in Flintshire/Wirral. Species such as Avocet, Bearded Tit and Marsh Harrier are now well established, and most wetland birds have increased their range.

Work on other species has also proved very beneficial, none more so than the indefatigable long-term and ongoing Chough study by the Cross and Stratford team, which has identified the best management practices for this species and provided artificial nesting sites in vertiginous locations. It is encouraging that Choughs have fared well here in modern times, but the evidence of declines inland make it even more important that detailed ringing and monitoring continues. A success story for a species that also deserves mention is the wardening scheme for the Little Tern colony at Gronant, organised by Denbighshire Countryside Services. This has resulted in an increasing population, from 52 pairs in 1992 to 161 pairs in 2019, helping to secure its future as a Welsh breeding species.

There is a mixed message on woodland birds. Marked declines of Willow Tits and Lesser Spotted Woodpeckers, and to a lesser extent Pied Flycatchers and Wood Warblers, in broadleaved woodlands contrast with an increase in species such as Siskins and Crossbills in maturing conifer plantations. Species of broadleaved woodland will benefit from the EU-LIFE project (2018–25) that aims to restore the favourable conservation status of temperate Celtic Rainforests by ridding them of invasive species such as *Rhododendron ponticum* and finding ways of improving management, such as changing how the woods are grazed. The announcement of the creation of a Welsh National Forest in 2020 to provide a "connected ecological network of woodland" running through "the length and breadth of Wales" should provide a welcome boost to woodland birds. Important ongoing research on the enigmatic Hawfinch at the two main populations around Dolgellau, Meirionnydd, and the Wye Valley, Gwent/Gloucestershire, is investigating nest success and productivity and year-round diet.

While there is some encouragement from efforts to maintain, or increase, populations of a few species, there are disappointing outcomes for many lowland farmland species. Corn Bunting and Turtle Dove have gone completely, and there is the sad possibility that they could be joined by Tree Sparrow, Grey Partridge, Western Yellow Wagtail and Yellowhammer in coming years. Disappointingly, agri-environment schemes in Wales have had limited benefits for this suite of species.

Upland farmed areas are also notably impoverished, although there is one species which has benefited from a collaborative venture between interested parties, both voluntary and statutory, over a wide area; a style of approach which must surely be the way forward for bird conservation in Wales. The Welsh Black Grouse Recovery Project, which commenced in 1999, was designed to arrest a serious decline and involved targeted habitat improvements through collaborative work between the RSPB, Countryside Council for Wales (predecessor of NRW), rural and forestry interests, and estates. The results are encouraging in that the number of lekking males has gone up from 216 in 1997 to 328 in 2018, but it is worrying that Black Grouse are increasingly localised, with a marked contraction of range. On the habitat front, active blanket bogs in North Wales have been aided by European Union LIFE money that restored the wet peat and removed plantation forestry and invasive non-native trees on two Special Areas of Conservation.

Roger Lovegrove, Iolo Williams and Graham Williams
© Julian Hughes

Among much that is generally encouraging and positive, there is no escaping the clear message that all is far from well with the overall situation for Welsh birds. That is perhaps epitomised by the plight of Curlew, recognised now as the UK's highest conservation priority bird species. In 2018 concerns for its future led to the formation of Gylfinir Cymru, a collaborative initiative made up of 16 organisations from government, conservation, farming and game management with a shared determination to stop the decline of this iconic bird. Much work is already underway through RSPB Cymru's Curlew LIFE project in upland Conwy and by the Shropshire and Welsh Marches Recovery Project for Curlew Country. The will and intention are clearly there for the massive partnership effort that will be needed to halt the decline.

The future well-being of Welsh bird populations now rests firmly on impending legislation. We cannot overstate the significance of the Agriculture (Wales) Act when it replaces the existing Common Agriculture Policy farm support scheme. The Welsh Government has a unique opportunity, under the 'public goods model', to enact the environmental measures so urgently needed to halt decades of decline. We fervently urge that it takes this opportunity to do so. Also of significance is the continued development of landscape-scale initiatives involving partnerships of interested parties, local communities and land managers, to connect existing protected areas, as in the Wildlife Trusts Living Landscape Partnerships and the RSPB's Futurescapes Programme.

If we may permit ourselves the luxury of predicting changes in the Welsh avifauna over the next quarter of a century, we must take account of rising temperatures. Thus, we anticipate there will be breeding colonists from the south. Cattle and Great White Egrets and Spoonbills have already nested close to the Welsh border, so these seem a safe prediction, but they could very well be joined by Little Bitterns, Glossy Ibises, Purple Herons and Black-winged Stilts. Further expansions of Dartford and Cetti's Warbler, Quail and Hobby are very likely as suitable habitats shift northwards. Colonists from the east could add Rose-coloured Starlings, Blyth's Reed Warblers and even Pallid Harriers to our list of breeding species. Last, but by no means least, we envisage that there are likely to be successful re-establishments of Golden and White-tailed Eagles to grace Welsh skies.

Roger Lovegrove, Graham Williams and Iolo Williams

Introduction

The first full avifauna for Wales was published in 1994. This was *Birds in Wales*, by Roger Lovegrove, Graham Williams and Iolo Williams, who have encouraged us in our own efforts. For most species, this volume included records up to 1991. In 2002, the Welsh Ornithological Society published an update volume, *Birds in Wales 1992–2000* by Jon Green. During the early years of the 21st century, a great many changes in the fortunes of our birds were documented in *Welsh Bird Reports*, in papers in the journal *Birds in Wales* and elsewhere. Over the same period, the volume and quality of data collected to monitor bird populations at a national level has increased greatly. Some species, such as Little Egret, bred in Wales for the first time and quickly became established as regular breeders. Populations of some formerly common breeding birds, such as Curlew, declined rapidly, raising fears of possible extinction in Wales unless the trend can be reversed. Because of these changes, several people felt that the time had come for a new full avifauna of Wales. Following many informal discussions, in November 2018 a proposal was made to the Council of the Welsh Ornithological Society. The publication of a new avifauna was considered a good way to mark the thirtieth anniversary of WOS.

Publishing the avifauna online, rather than as a book, was considered. However, there can be no guarantee that information published only on the web will remain available in the long term. An avifauna of this type remains of value for many years after publication. In gathering material, we have made much use of books published well over a century ago, so it is reasonable to hope that someone 100 years from now will find that this work remains useful. In view of this, we agreed that a printed book would be most appropriate.

Council agreed to the suggestion, and we are very grateful for this, and for its support during the process of turning the proposal into reality. It was decided that, where possible, the species accounts would be written by people with specialist knowledge of those species in Wales. The names of many of the authors will be well known to anyone with any interest in Welsh, or indeed British, ornithology, and the book has benefited greatly from their expertise and willingness to be involved. The photographs also include examples of the work of some of our best bird photographers.

Liverpool University Press agreed to publish the volume, providing that WOS could raise half the publication cost. This was achieved by offering species for sponsorship by individuals or groups. The response proved to be excellent, and we are very grateful for the very generous support received. During the work on the book, it became clear that there was much more new information available about many species than we had anticipated initially. We decided that the number of pages in the volume would have to be increased, resulting in a corresponding increase in costs. Fortunately, the response from individuals prepared to support the book by sponsoring species, together with grants from LERC Wales (Local Environmental Records Centres Wales) and Natural Resources Wales, enabled us to raise the additional funding required. Ideally, this book would have been bilingual, or at least partly bilingual. Unfortunately, constraints of time, money and the sheer size of the volume made that impossible.

As we recognise in the dedication, the volume has only been made possible by the thousands of observers who have taken the trouble to submit their records, either directly to county recorders or, recently, through BirdTrack. A great deal of new information has become available from recently published county avifaunas and other sources. We have been greatly aided by the wealth of data collected through monitoring schemes run or monitored by the British Trust for Ornithology, and made publicly accessible online. Older material is also now more easily accessible than it was previously. Many books and periodicals from the 19th century and earlier times are available in digital format on the web, and this enabled us to add to the historical information available for a number of species.

A little under two years may seem a long time to spend writing a book, but is actually quite a short period for producing a country avifauna of this kind. The second year of our work was dominated by the global coronavirus pandemic. This affected work on the book, for example by preventing access to libraries. However, with travel temporarily restricted and few alternative distractions, it did at least ensure that we completed the work on time. Many people found solace in nature during lockdown, and much has been written about the need for a green recovery, to rebuild society and the economy in a sustainable way. History will tell whether these hopes are justified, but having documented the changes in the avifauna of Wales, we are all too aware of the risks to nature of carrying on with business as usual.

We certainly learned a great deal in researching, writing and editing this volume, and we hope that readers find it interesting and useful. If it contributes to the conservation of birds in Wales, the time and effort will have been amply rewarded

Rhion Pritchard
Julian Hughes
Ian M. Spence
Bob Haycock
Anne Brenchley

Acknowledgements

We are indebted to a huge number of individuals and organisations who have provided data and information, and made suggestions, responded to our many questions, and encouraged us in our efforts. Our thanks go to:

The authors of species accounts, without whom there would have been no book, and the past and present county bird recorders who have collated records over many decades.

The authors of the species accounts were:

Ian Beggs
Tim Birkhead
Katharine Bowgen
Anne Brenchley
Nigel Brown
Peter Coffey
Henry Cook
Heather Corfield
Tony Cross
Heather Crump
Peter Dare
Richard Dobbins
Steve Dodd
Annette Fayet
Tony Fox
Geoff Gibbs
Jon Green
Mick Green
Bob Harris
Ian Hawkins
Annie Haycock
Bob Haycock
Jane Hodges
Julian Hughes
Marc Hughes
Daniel Jenkins-Jones
Paddy Jenks
Chris M. Jones
Chris R. Jones
Kelvin Jones
Andrew King
Jerry Lewis
Patrick Lindley
Carl Mitchell
Wayne Morris
David Norman
Keith Offord
David Parker
Chris Perrins
Rhion Pritchard
Stephen Roberts
Robin Sandham
Mike Shewring
David Smith
Ian M. Spence
Ben Stammers
Adrienne Stratford
Steve Sutcliffe
Steph Tyler
Colin Wells
David Winnard

The photographers who have kindly donated their fantastic images:

Mike Alexander
Anglesey Archives
Alan Clewes
Tim Collier
Tony Cross
Richard Crossen
Andy Davis
Glaslyn Osprey Project
Gary Eisenhauer
Tommy Evans
John Freeman
Bob Garrett
John Hawkins
Annie Haycock
Bob Haycock
Peter Howlett
Julian Hughes
Martin Jones
Kev Joynes
Elfyn Lewis
Tate Lloyd
John Marsh
Hugh Miles
Jeremy Moore
Keith Offord
Carlton Parry
Tony Pope
Ben Porter
Steve Porter
Fausto Riccioni
Colin Richards
Dan Rouse
Rob Sandham
Alan Saunders
David Saunders Collection
Richard Smith
Steve Stansfield
Michael Steciuk
Richard Stonier
Norman West
Tony White
Steve Wilce
Tom Wright
Steve Young

Philip Snow for his superb illustrations for the cover and frontispiece.

The Welsh Ornithological Society was not able to fund its half of the publication costs from existing resources. We are very grateful to everyone who has sponsored species. Their names are shown in green at the bottom of the relevant species accounts:

All sponsors who provided a minimum of £200 per species (in bold);

All individual sponsors who provided a minimum of £50 per species;

The two corporate sponsors: LERC Wales (the Local Environmental Records Centres Wales) and Natural Resources Wales;

Thanks also to Beryl Moore for a donation of £500 in memory of Derek Moore, previously Chairman of the Welsh Ornithological Society.

The British Trust for Ornithology (BTO) deserves particular thanks and support, for curating huge amounts of data gathered by volunteers and making it publicly accessible. We have drawn extensively on the population monitoring schemes that are run or administered by the BTO, including the Breeding Bird Survey, the Wetland Bird Survey, the British and Irish Ringing Scheme and the Heronries Census. We would especially like to thank the following staff who responded so promptly and patiently to our queries: Dawn Balmer, Teresa Frost, Simon Gillings, Bridget Griffin, Sarah Harris and Ian Woodward.

For provision of data: Bardsey Bird and Field Observatory, Skokholm Bird Observatory, Local Environmental Records Centres (LERC) Wales, Grahame Madge at the Met Office, Matty Murphy at NRW, Aisling May at Cofnod, Martin Scott at HiDef Aerial Surveying Ltd, Ilka Win at JNCC, Chris Wynne at North Wales Wildlife Trust, and Ian Sims and Simon Wotton at RSPB. We owe a special debt of gratitude to current and former county bird recorders, who have gathered, ordered and published records in annual reports, since the 1950s in some cases.

For access to archives and unpublished works: Dawn Balmer, Mark Eaton and Mark Holling (on behalf of the Rare Breeding Birds Panel), Richard Farmer, Verena Keller (on behalf of EBBA2), John Lloyd, Graham Rees and Adrian Rogers.

For commenting extensively on the draft manuscript and providing additional information from county bird clubs and recording groups: Graham Appleton, Kane Brides, Phil Bristow, Nigel Brown, Jim Dustow, Mick Green, Mike Haigh, Gary Harper, Annie Haycock, Rob Hunt, Clive Hurford, Pete Jennings, Russell Jones, Andrew King, Jerry Lewis, Carl Mitchell, Ian Newton, David Parker, David Saunders, Ken Smith, Darryl Spittle, Rob Taylor, Reg Thorpe, Steph Tyler, Colin Wells, Arfon Williams, Graham Williams and Iolo Williams. For authoring the section on weather's influence on birds in Wales: David Lee.

For help with information about particular species or places: Richard Broughton, Dan Brown, Richard Brown, Jim Clark, Richard Clarke, Martin Clift, Heather Coates, Rob Cockbain, Peter Coffey, Ken Croft, Steve Culley, Tom Dalrymple, Andrew Dixon, Steve Dodd, Julian Driver, Giselle Eagle, Emyr Evans, Richard Facey, Tony Fox, Chris Griffiths, Mike Haigh, Kevin Hemsley, Lauren Hough, Gareth Howells, Brian Iddon, Jill Jackson, Paddy Jenks, Graham Jones, Huw Jones, Kelvin Jones, Martin Jones, Hugh Knott, Dewi Lewis, Jerry Lewis and Llangorse Ringing Group,

Patrick Lindley, John Lloyd, Graham McElwaine, Merseyside Ringing Group, Allen Moore, Greg Morgan, Lisa Morgan, Wayne Morris, Matty Murphy, Sarah Perkins, Rhian Pierce, Ivor Rees, Colin Richards, John Lawton Roberts, James Roden, Robin Sandham, Elwyn Sharps, Ian Sims, Jez Smith, Brian Southern, Steve Stansfield, Reg Thorpe, Steph Tyler, Nathan Wilkie, Matt Wood, Ian Woodward and Sylwia Zbijewska.

The maps of Wales and a few of those of Britain were prepared using DMAP, developed by Dr Alan J. Morton, with additional help from Lucy Frontani and Heather Crump.

The maps of ringing recoveries were prepared using QGIS from data provided by the BTO. All ringers who allowed us to use their data are thanked for their cooperation, including Sean Gray, Louise Samson, Kevin Scott, Manx Ringing Group and Tees Ringing Group. Geoff Radford is also thanked for his help in the initial stages of using the mapping software.

Thanks also to staff at the NRW library and the RSPB Cymru office in Bangor, and to Geoff Gibbs. In particular, we would like to thank Reg Thorpe for his detailed research into historic records of scarce and rare species and Trevor Payne for proof-reading and improving our work. We would also like to thank the Council of the Welsh Ornithological Society for their support throughout the project, in particular Daniel Jenkins-Jones, Giles Pepler and Dan Rouse.

Major data sources used in *The Birds of Wales*

The Breeding Bird Survey (BBS) is run by the British Trust for Ornithology (BTO) and is jointly funded by the BTO, the Royal Society for the Protection of Birds (RSPB) and the Joint Nature Conservation Committee (JNCC)*. © copyright and database right 2020.

The Wetland Bird Survey (WeBS) data is from *Waterbirds in the UK 2018/19.* WeBS is a partnership jointly funded by the BTO, RSPB and JNCC*, in association with WWT, with fieldwork conducted by volunteers. © copyright and database right 2020.

The BTO Ringing Scheme is funded by a partnership of the British Trust for Ornithology, the JNCC*, the National Parks and Wildlife Service (Ireland) and the ringers themselves.

Many of the seabird counts are based on data extracted from the Seabird Monitoring Programme Database (https://app.bto.org/seabirds). Data have been provided to the SMP by the generous contributions of nature conservation and research organisations, and many volunteers throughout Britain and Ireland.

Information about rare breeding species was provided by the Rare Breeding Birds Panel (RBBP). The Panel is funded by the JNCC* and the RSPB, with additional financial contributions from the BTO.

BirdTrack records sightings of birds throughout Britain and Ireland, through a partnership between the BTO, the RSPB, Birdwatch Ireland, the Scottish Ornithologists' Club and the Welsh Ornithological Society.

*JNCC involvement in the partnerships listed is on behalf of the statutory nature conservation bodies: the Department of Agriculture, Environment and Rural Affairs Northern Ireland; Natural England; Natural Resources Wales; and NatureScot.

1. The land, seas and birds of Wales

Wales is relatively temperate, without the continental winds of the east coast of Britain. It has altitudinal variation, but the highest summits are lower than those in Scotland. Its coastline is extensive but its seas are shallower than those that surround western Scotland or Ireland. While watered by Atlantic weather fronts, it is less exposed to the most severe winter storms than other west-facing coasts in neighbouring nations. From that description, Wales may sound rather average, but within its 20,735km^2 of land, 15,000km^2 of sea and intertidal, and 2,740km of coast is considerable natural diversity, as geography and land use combine to influence the birds of Wales. The nation comprises 9% of the UK's land area and is home to just over three million people, 5% of the UK population. As context, Wales is less than one-third the area of Scotland but with more than twice the population density.

Habitats for birds

Ask anyone who knows Wales to describe its landscape and it is more than likely that they will, unprompted, mention the hills and the sea. The natural and semi-natural habitats of Wales are founded on geology and fashioned by ice, rainfall, wind and sea. These physical processes continue to shape a varied landform of mountains and lowlands and a rugged coastline, interspersed with long shallow beaches and bays. The wet, uncultivated uplands have played an important role in its political and social history. The Industrial Revolution and developments in agriculture have transformed land use and influenced its wildlife. The natural and semi-natural habitats of Wales are founded on geology and rainfall.

The geology is complex and includes Pre-Cambrian and Cambrian outcrops in the northwest and southwest that are among the oldest rocks in the world, formed more than 500 million years ago. Volcanic rocks in Snowdonia are largely responsible for the area's rugged appearance, while much of mid-Wales is formed from Silurian sandstones and mudstones, 420–435 million years old. The rocks of the South Wales Valleys west to southern Pembrokeshire are younger, from the Carboniferous period, comprising a lower layer of limestone, overlaid by coarse sandstone, mudstone and the coal seams that came to define the area globally in the 19th century. At this time, 300–350 million years ago, shallow tropical seas extended over much of what is now North and South Wales, depositing limestone that would later form steep cliffs on which seabirds would nest. The most recent rocks are those from the Permian Period, 250–300 million years ago, that underlie the Vale of Clwyd and those of the Jurassic that form the Vale of Glamorgan. Across this complex geology, successive ice sheets ebbed and flowed. The Last Glacial Period, which ended 11,700 years ago, covered almost all of Wales save for the very far south. It is known by British geologists as the Devensian, after the River Dee (Deva) in North Wales. The ice had a substantial effect on the land surface, especially in the way that river basin catchments would form, bringing water to almost every square metre of Wales.

A little under 31% of terrestrial Wales is classed as semi-natural habitat, comprised primarily of unimproved grassland, heathland, blanket bog, semi-natural woodland and intertidal habitats (Natural Resources Wales 2019). This is now far from evenly spread: only 17% of the lowlands are semi-natural habitat, compared to 84% of the uplands (Natural Resources Wales 2016). While Wales is perceived as mountainous, only a small proportion is above the altitude of the natural treeline. Centuries of grazing and burning make the treeline very difficult to detect, however (Ingrouille 1995). The mountains of Wales do not support the specialist bird species, such as Ptarmigan, associated with montane habitats farther north, but those in Snowdonia hold some of the most southerly populations in Britain of Ring Ouzel and Twite and provided a refuge from persecution for populations of species such as Raven and Peregrine during the 19th century. The Carneddau, Caernarfonshire, is the largest area of land over 1,000m in the UK outside Scotland. At slightly lower altitude, the moorland plateaux that form the spine of Wales, from Denbighshire to the Brecon Beacons, provide extensive grass and heathland that are home to birds such as Hen Harrier, Merlin and Red Grouse. Grazing, over

Cwm Idwal, Caernarfonshire, is a spectacular result of glaciation and the oldest National Nature Reserve in Wales © Jeremy Moore

High rainfall in the Welsh mountains creates the extensive river network and the humid conditions for the 'Celtic Rainforest', such as at Coed Ganllwyd, Meirionnydd © Mike Alexander

many centuries, has prevented the natural vegetation succession of these high, remote areas, which has in some places created a tundra-like habitat on which Golden Plover and Dunlin nest. Around the fringes of the moorland is *ffridd* (or *coedcae*), a diverse mixture of heath, bracken, scattered trees, scrub, bog and rock, at altitudes between 100m and 450m. *Ffridd* has the highest mean habitat diversity/km^2 when compared to coastal, lowland and upland landscapes (Blackstock *et al.* 2010) and is valuable for birds such as Whinchat, Yellowhammer, Ring Ouzel and Cuckoo.

For more than 2,000 years, successive generations of people and their grazing animals have reduced the natural woodland and created open habitat (Birks 1988). During 1871–1913, standing timber in Wales covered just *c.*4% of the land, likely to have been the lowest point since the end of the Ice Age, and felling during the First World War was more severe in Wales than in some other parts of Britain (Linnard 1982). Forest area trebled in the century that followed and woodland now covers around 15% of Wales. Almost all of that increase was conifer plantations and much of the remaining native woodland is highly fragmented across the landscape. The woodland type of highest conservation importance is Western Atlantic Oakwood, for which Wales is responsible for half the global resource. A critical habitat for rare lichens and bryophytes, it also supports high densities of Wood Warbler and Pied Flycatcher, and to a lesser extent Redstart, the classic trio of the Celtic Rainforest.

These habitats, especially in the west, are watered by the high rainfall that sweeps in from the Atlantic Ocean. The rain gives rise to a 24,000km network of streams and rivers (Natural Resources Wales 2016). Waterlogged pools in the uplands provide a nesting habitat for birds such as Teal, while Dipper, Grey Wagtail and Red-breasted Merganser are found in the more wooded upper and middle reaches of the fast-flowing rivers and streams. The middle sections are also home to Goosander, Kingfisher and, where winter floods erode sandy banksides, Sand Martins. Rivers pass through some of the 558 lakes in Wales, including 150 reservoirs that were created to supply drinking water. The tributaries ultimately combine to form 35 main river channels that enter the sea through estuaries (Natural Resources Wales 2016). These support huge waterbird populations, especially in winter when they are generally free of ice compared to those, farther east, in Europe. The two largest estuaries, the Dee and Severn, each hold up to 100,000 waterbirds outside the breeding season and are particularly important for species such as Pintail and Pink-footed Goose, Shelduck and Dunlin. Smaller populations of wintering wildfowl occur on the western coasts, but some of these sites are important, in a British context, for birds from the East Canadian High Arctic population of Pale-bellied Brent Goose and of Greenland White-fronted Goose.

The lower sections of many rivers, including the floodplains, were heavily modified as settlements developed around them. Aside from the coal towns of Wrexham and the South Wales Valleys, the main conurbations of Wales are around estuaries. More than 80% of the resident human population live on 10% of the land area, particularly in Swansea, Cardiff and Newport in the south, and the Deeside towns in the northeast. Development over several centuries put pressure on adjacent semi-natural habitats, especially those that were viewed as unproductive, such as sand dune systems and saltmarsh in which species such as Little Tern and Redshank nested. Urbanisation did produce some avian winners, however, as Swift and House Martin expanded their range and eventually abandoned ancient woodland and sea cliffs, respectively, for a more metropolitan lifestyle. More recently, Herring Gull and Lesser Black-backed Gull have taken advantage of rooftops and a ready supply of human food waste, to follow them into town.

Many coastal Herring Gull colonies are far smaller now than during the last century, but the sea cliffs and islands remain of huge importance for seabird populations. Grassholm, Pembrokeshire, is the fourth-largest Gannet colony in the world, and Wales also holds at least 10% of the UK's breeding Little Tern and Cormorant populations, which feed within a few kilometres of their nesting colonies. Almost all of our nesting seabirds feed within the 12 nautical miles of inshore waters, an area that makes up 43% of Welsh territory.

Also associated with the rugged coastal slopes and sea cliffs is the Chough, the bird for which Wales holds a particular UK

In many places, such as southwest of Strumble Head, Pembrokeshire, intensive farming extends to the top of the steep clifftops © Jeremy Moore

responsibility. The Chough has made a welcome expansion of its range into South Wales in recent years, but farther north the availability of suitable short-grazed pasture for foraging may have been reduced both inland and around the coast. Farming is the dominant land use in Wales, accounting for 88% of the surface area. It has created features that have benefited birds such as Skylark and Swallow in a highly modified landscape. Grassland makes up nearly two-thirds of land-cover in Wales (more than 1 million ha), but less than 20% of this is semi-natural. The other 80% is periodically ploughed and reseeded, with rye-grass or clover mixes that produce no seeds or nectar, for sheep in the hills and dairy cattle in the valleys. This pasture, described as 'improved' in terms of its agricultural productivity, provides limited opportunities for breeding birds, although it can provide foraging for species such as Ring Ouzel and Curlew in the breeding season and Golden Plover and Woodcock in winter. Arable now forms only 5% of Welsh farmland, is concentrated in the drier eastern valleys and has become more intensively managed. Birds associated with that habitat, such as Grey Partridge, Tree Sparrow and Corn Bunting, have retracted to the point of extinction in Wales.

Weather patterns in Wales

As a broad statement it can be said that Wales has an essentially maritime climate, characterised by weather that is often cloudy, wet and windy but mild. This is very true between October and March, when depressions moving in from the Atlantic are at their most frequent and vigorous. These weather systems, powered by the polar jet stream, bring large amounts of precipitation that is often magnified over high ground, particularly in Snowdonia, where it exceeds 3,000mm annually.

River levels are very sensitive to the frequency of Atlantic depressions during the winter. They can rise rapidly after a close sequence of storms, causing extensive flooding in lowland valleys and producing temporary wetlands that provide additional wintering areas for wildfowl. Gales in the Irish Sea bring many seabirds close to shore, especially during August and September, as those departing their breeding grounds around Britain and farther north make their way into the Atlantic. Favoured headlands become hotspots for seeing species that otherwise spend much of their time out at sea, particularly when birds pushed into Liverpool Bay and Cardigan Bay find their way to the open ocean as storms move away.

Prevailing southwesterly winds help to drive the North Atlantic Current, which brings an offshoot of the Gulf Stream towards Britain and Ireland. This ensures that sea temperatures remain in double figures in the Celtic Sea through the winter. With winds predominantly from the southwest or west, coastal areas have no risk of frost or snow for much of the time in most winters. Many bird species have developed strategies to exploit these mild conditions and over-winter here. Occasionally the sequence of depressions is broken, and winds arrive from the north or east, but even when this occurs the coastal areas still offer more amenable feeding conditions and a greater chance of survival when inland areas become less hospitable.

If a cold spell lasts more than a few days, the outflows from meltwater in the mountains can reduce the sea surface temperature by several degrees, especially in relatively shallow waters such as the Bristol Channel and Liverpool Bay. When this occurs, the maritime influence in these areas is much reduced and becomes restricted to the far west coasts. Species that need open water or soft ground will move out of the Low Countries, eastern and northern Britain, and find suitable locations on Anglesey, in Pembrokeshire or on Llŷn, Caernarfonshire. In more extreme conditions, these birds cross to Ireland. Waders most sensitive to frozen ground, such as Lapwing and Golden Plover, move first, but other waders and wildfowl follow if the cold spell intensifies. Conservation management that provides habitat and refuge areas on the Dee and on Anglesey may be important in shaping wintering habits in relation to freezing weather.

The coastal zones around Cardigan Bay have protection from high ground to the east, but eventually very cold air can reach even western coasts, when a severe outbreak of cold weather from central and eastern Europe persists. This happens rarely, but occurs when cold spells become entrenched over Britain and Ireland, such as in 1947 and 1963. These particularly bad winters caused very high mortality for many species.

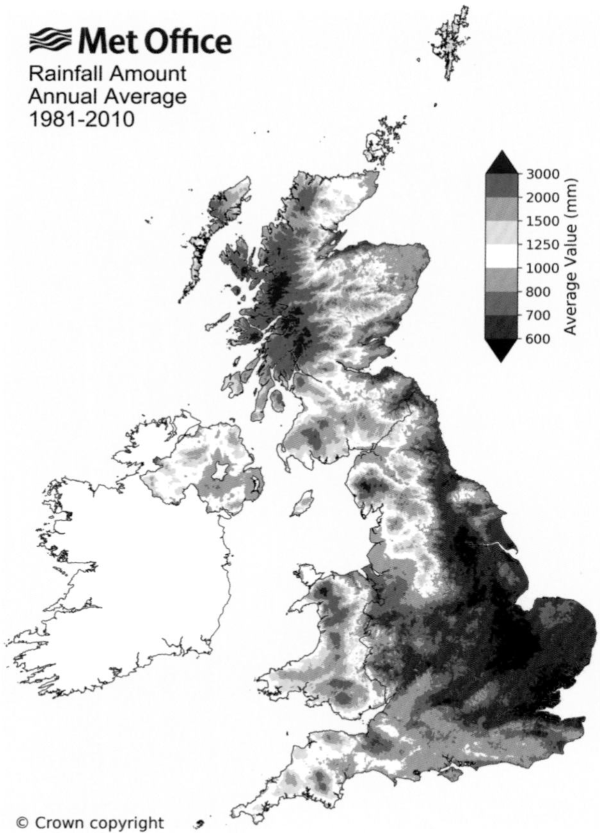

Figure 1. Average rainfall (mm) in the UK, 1981–2010. This shows the substantial difference that altitude and aspect make across relatively short distances, particularly in North Wales (reproduced with the kind permission of The Met Office).

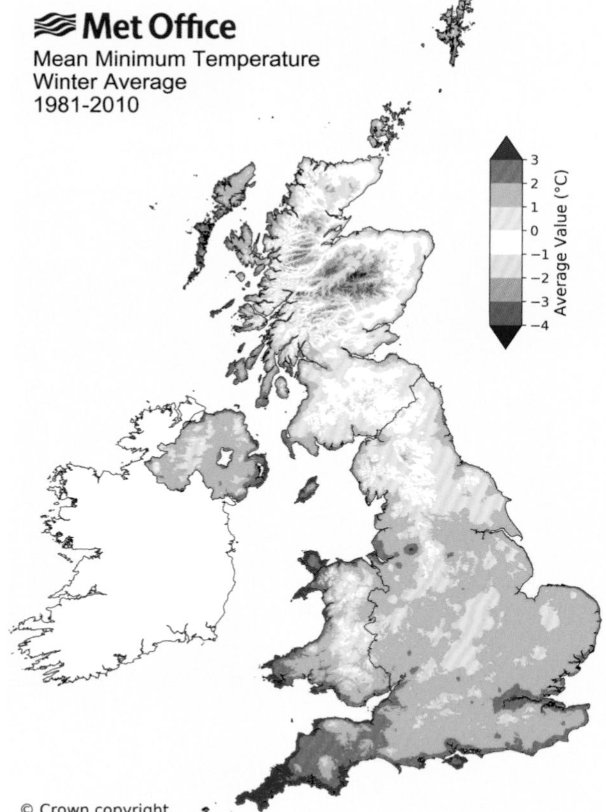

Figure 2. Mean minimum average winter temperature (°C) 1981–2010. This shows that the Welsh coast is consistently milder than most of the rest of the UK, providing a refuge for birds from farther north and east (reproduced with the kind permission of The Met Office).

It is not only the length of interruption in the predominantly westerly winds or the severity of any cold spell, but also the timing that determines the effects on the birds. A cold spell of similar duration and intensity is likely to have a greater effect in February than December, as this heightens the stresses on wild populations, highlighting the 'hunger gap' with which all wildlife has to cope during late winter and early spring. Birds fleeing bad weather from the Continent move west but, struggling for food or water, can become concentrated on west-coast headlands or islands. One such mass-mortality event resulted in the deaths of hundreds of birds on Skomer in January 1962, all of which were 30–50% underweight (Harris 1962). The worst of the 1947 winter was in February and March. There were also notable adverse effects from late snow in South Wales in 1983 and very heavy snowfall in North Wales in March 2013. Spells of cold weather in late winter or early spring can be just as severe as in midwinter, and even a short outbreak can have a significant impact, as occurred in late February and early March 2018.

Recent colonists, such as Dartford Warbler and Cetti's Warbler, that do not have a significant migration strategy, can be lost on a local or regional scale following prolonged cold weather, but species with the ability to respond by moving, such as Little Egret, suffer less. It remains to be seen how species expected to colonise Wales, such as Great White Egret and Cattle Egret, manage the occasional severe winters that are inevitable, despite a general climate drift to higher average temperatures.

Autumn and winter bird movements in Wales may be the result of adverse conditions elsewhere, as birds move ahead of, or around, poor conditions at migration periods. Strong winds can bring flocks of strong-flying wildfowl, such as Pink-footed Geese, that have strayed off course. The quick onset of a Scandinavian winter will bring a rapid influx of Snipe and Woodcock to Britain, and eventually to Wales. Other waders are also affected by the weather to the east of Britain. Numbers of Curlew Sandpiper and Little Stint moving through Wales in September are, at least partially, linked to wind patterns over Scandinavia and the North Sea.

Spring and early summer can be a complete contrast from year to year, so species that breed in Wales must cope with a great variety of weather conditions during this most important period. The months of April, May and June are, on average, the sunniest and driest of the year, with often some quite lengthy spells of fine weather. However, cold and wet conditions that hark back to late winter can feature strongly in some years; and snow often occurs in April above 500m. Such conditions can occur right through to the end of May and so cut deep into the heart of the breeding season. When these occur, even if only for a few days in succession, the exposed nests of upland birds, such as Black Grouse and Hen Harrier, can be chilled and waterlogged. Adults can struggle to forage for their chicks, resulting in the loss of otherwise-healthy broods. Heavy summer thunderstorms, that can wash out riverside nests, do occur, but the intense summer thunderstorms, characteristic of the Midlands of England, are a much rarer occurrence over Wales, especially in the west.

The canopied oak woodland is something of an enclosed habitat, often in deep and steep valleys, where all but the most seriously inclement weather in mid-spring is kept somewhat at bay. However, for the young chicks of the single-brooded Pied Flycatcher, which need a constant supply of insect food, a few days of cold, wet weather can wipe out a whole year's breeding cohort. Swifts can ride out spells of cold weather that restrict food supply, and even in poor summers can fledge young late in their season. However, an extended period of cold and wet or hot and dry weather can have a greater impact on species that are usually multi-brooded, such as Swallows. Breeding species have been able to survive well with the historic frequency of 'good' and 'bad' seasons, but any change in this balance can have a deleterious effect. This is explored in Chapter 6.

Male Black Grouse lek in spring irrespective of the weather, but their chicks are sensitive to high rainfall in late June, which may reduce the availability of invertebrates © John Hawkins

Important sites for birds

Sites designated by government as important for their bird interest, and thus assuming some degree of protection from development or damage, are based on criteria designed to assist the most vulnerable species, including those that migrate and congregate during or outside the breeding season. All of these are Sites of Special Scientific Interest (SSSIs), but some have additional protection either as Special Protection Areas (SPAs), designated under the EU 'Birds Directive', or wetlands of international importance, designated under the Ramsar Convention on Wetlands. In the marine environment, not shown in Fig. 3, are 139 Marine Protected Areas (MPAs), which cover 69% of Welsh inshore waters and 50% of all Welsh waters (to the median line with adjacent territories). These comprise 15 Special Areas of Conservation (SACs), one Marine Conservation Zone and other coastal SSSIs (Natural Resources Wales 2020). Additionally, the Irish Front SPA (18,000ha), 36km to the northwest of Anglesey, has the third-largest marine aggregation of breeding Manx Shearwater identified in the UK. More than 12,000 Manx Shearwaters are estimated to use the area during the breeding season, including birds tracked from colonies in Wales, Northern Ireland and Devon.

There are no recent data on the condition and quality of the habitat within most of these SSSIs and the foraging areas for seabirds are not features of most of the designated marine areas. Despite these deficiencies, which are a major handicap to achieving the recovery of species, protected areas are a cornerstone of conservation effort in Wales.

An estimated 37% of the UK Redstart population is found in Welsh woodland © Bob Garrett

Figure 3. Ramsar sites, Special Protection Areas, and other SSSIs, including rivers, designated for their bird interest. Marine Protected Areas are not shown. Contains Natural Resources Wales information © Natural Resources Wales and Database Right. All rights reserved. Contains Ordnance Survey Data. Ordnance Survey Licence number 100019741. Crown Copyright and Database Right.

Special Protection Area	Area (ha)	Qualifying species (breeding, non-breeding or passage)
Anglesey Terns/Morwenoliaid Ynys Môn	101,931	Arctic Tern, Common Tern, Roseate Tern, Sandwich Tern
Bae Caerfyrddin/Carmarthen Bay	33,450	Common Scoter
Berwyn	24,268	Hen Harrier, Merlin, Peregrine, Red Kite
Burry Inlet	6,673	Curlew, Dunlin, Grey Plover, Knot, Oystercatcher, Pintail, Redshank, Shelduck, Shoveler, Teal, Turnstone, Wigeon
Castlemartin Coast	1,114	Chough
Craig yr Aderyn (Bird's Rock)	89	Chough
The Dee Estuary*	14,295	Bar-tailed Godwit, Black-tailed Godwit, Common Tern, Curlew, Dunlin, Grey Plover, Knot, Little Tern, Oystercatcher, Pintail, Redshank, Sandwich Tern, Shelduck, Teal
Dyfi Estuary/Aber Dyfi	2,057	Greenland White-fronted Goose
Elenydd-Mallaen	30,008	Merlin, Red Kite
Glannau Aberdaron ac Ynys Enlli/ Aberdaron Coast and Bardsey Island	33,942	Chough, Manx Shearwater
Glannau Ynys Gybi/Holy Island Coast	604	Chough
Grassholm	1,774	Gannet
Liverpool Bay/Bae Lerpwl*	252,177	Common Scoter, Common Tern, Little Gull, Little Tern, Red-throated Diver
Migneint-Arenig-Dduallt	19,968	Hen Harrier, Merlin, Peregrine
Mynydd Cilan, Trwyn y Wylfa ac Ynysoedd Sant Tudwal	373	Chough
Northern Cardigan Bay/Gogledd Bae Ceredigion	82,704	Red-throated Diver
Ramsey and St Davids Peninsula Coast	831	Chough
Severn Estuary*	24,488	Bewick's Swan, Dunlin, European White-fronted Goose, Gadwall, Redshank, Shelduck
Skomer, Skokholm and the seas off Pembrokeshire/Sgomer, Sgogwm a Moroedd Penfro	166,801	Chough, Lesser Black-backed Gull, Manx Shearwater, Puffin, Short-eared Owl, Storm Petrel
Traeth Lafan/Lavan Sands, Conway Bay	2,703	Curlew, Great Crested Grebe, Oystercatcher, Red-breasted Merganser, Redshank
Ynys Seiriol/Puffin Island	31	Cormorant

Table 1. Special Protection Areas designated under the EU 'Birds Directive' in Wales (those marked * are shared with England).

The birds of Wales

The geology, soils and climate create the right conditions for the very diverse suite of breeding and wintering birds that make Wales a very special place. For some species it is of particular importance. Foremost among these is Chough, of which 79% of the UK breeding population and 94% of the wintering population occur in Wales. The difference in numbers between the seasons is a result of Wales holding a greater number of pre-breeding Choughs than other parts of the UK. Chough has been associated with the cliffs and grasslands of Wales for centuries, and formerly with the northeast and south coasts of England.

Welsh woodlands support at least 50% of the UK's populations of Pied Flycatcher, Hawfinch and Goshawk, along with a high proportion of Wood Warbler, Redstart and Nuthatch (Hughes *et al.* 2020). A visit to a broadleaved woodland in late April or May can be magical, with the repeated whistle of a Nuthatch, the tacking call of a Redstart, the falling cascade of a Willow Warbler and the pulsating trill of a Wood Warbler, likened to a spinning coin on a marble slab. The short song and call of Hawfinch is almost anonymous by comparison, but Wales has become more important for the UK's breeding population, as the species has disappeared from much of England. While perhaps overlooked in a few places, it maintains only two strongholds in Wales, in the Wye Valley and Meirionnydd. Willow Tit has suffered one of the greatest declines of any UK bird species in recent decades. While this decline has occurred in Wales too, around 20% of its British population is now found here, mostly in a band from north Pembrokeshire to Montgomeryshire.

The west coast islands, particularly those in Pembrokeshire, hold over half of the global population of breeding Manx Shearwaters, around 2% of the global population of breeding Sandwich Tern at Cemlyn on Anglesey, and the largest Arctic Tern colony in the UK, on The Skerries, Anglesey. The Little Tern colony at Gronant, Flintshire, has consistently been one of the largest in the UK in recent years. Wales may also have become relatively more important for terns and auks, following the reduction of other colonies in Scotland and the North Sea. The reverse is true of Red Kite, which is recovering well across the UK, thanks to re-establishment programmes in the other nations, leading to the Welsh contribution to the population becoming proportionately smaller. Nonetheless, our population remains significant in global terms, since the global range of this Near Threatened species is largely restricted to western Europe.

Welsh estuaries are important for wildfowl and waders, especially as refuge areas during cold conditions farther east in Britain. The Dee Estuary, Flintshire/Wirral, holds up to 155,000 non-breeding waterbirds, making it the fourth most important wetland in Britain, while the Severn Estuary supports 87,000 and the Burry Inlet, Gower/Carmarthenshire, over 42,000. The Dyfi Estuary, Ceredigion/Meirionnydd/Montgomeryshire, is the only UK site outside Scotland that is designated as a Special Protection Area for Greenland White-fronted Goose. Sites in North Wales hold an increasing proportion of the East Canadian High Arctic population of Pale-bellied Brent Goose, while Wales also holds around 10% of the biogeographic population of wintering Pintail. The seas around Wales can also be important outside the breeding season, especially for Common Scoter. Welsh waters can hold more than 50% of the UK's non-breeding population in some winters. New monitoring techniques are revealing larger numbers than previously appreciated, potentially of global significance.

Breeding species	% of UK population	Wintering species	% of UK population
Chough	79	Chough	94
Hawfinch	70	Common Scoter	55
Pied Flycatcher	68–76	Goshawk	46
Manx Shearwater	57[1]	Red Kite	41
Goshawk	50	Brambling	33
Wood Warbler	44	Great Grey Shrike	27
Redstart	37	Raven	25
Red Kite	31	Nuthatch	23
Garden Warbler	27	Willow Tit	20
Nuthatch	24	Guillemot	19

Table 2. Species for which Wales holds a significant proportion of the UK population. Breeding figures based on dedicated surveys (Chough, Manx Shearwater), an estimate made in this book (Hawfinch) and extrapolation in Hughes *et al.* 2020 (all others). Winter figures based on abundance estimates in occupied 10-km squares recorded during the *Britain and Ireland Atlas 2007–11* (from Bladwell *et al.* 2018), except for Chough (Cross *et al.* 2020).

[1] Manx Shearwater is a minimum and is likely to be revised upon completion of the Seabirds Count (2015–21) survey.

Rare visitors have piqued the interest of birdwatchers for more than a century, and while sites in Wales are not so well positioned as, for example, Fair Isle and Cape Clear to attract ultra-rarities, the coastal bird observatories have played an important role in the study of migration, of both abundant and rare species. Nonetheless, Wales recorded the first British records of 16 species: Yellow-billed Cuckoo, Little Whimbrel, Grey-tailed Tattler, Ring-billed Gull, American Royal Tern, Lesser Crested Tern, Nutcracker, Western Bonelli's Warbler, Swainson's Thrush, Grey Catbird, Moussier's Redstart, Olive-backed Pipit, Blackburnian Warbler, Yellow Warbler, Summer Tanager and Indigo Bunting. In addition, Britain's first Glaucous-winged Gull also visited Wales, having been seen previously in neighbouring Gloucestershire, the first British record of *Motacilla alba personata* (Masked Wagtail) was in Wales, as was the first, and to date only, breeding record of Iberian Chiffchaff in Britain.

These special birds, and the habitats in which they live, attract birdwatchers from elsewhere into Wales. Precious little assessment has been made of the role that birds play in the Welsh tourist economy, but surveys show that Welsh beaches and countryside are important reasons to visit. A single nature reserve, RSPB South Stack, Anglesey, supported 92 full-time equivalent jobs and £3.8 million of visitor spend into the local economy (RSPB 2019c).

This brief appetiser has, we hope, given you a taste for the birds of Wales and their habitats, and will encourage you to dip into the species accounts for a more satisfying repast.

Julian Hughes, with weather information by David Lee

2. Bird recording

This book, and all of our knowledge of the past history of birds in Wales, is based on people recording birds, even if some were not specifically aiming to do so. The earliest records come from archaeological excavations, some dating back over 240,000 years. Other records date from the period of the last glaciation, from about 40,000–120,000 years ago. At its peak, the ice sheet covered the whole of Wales, except for a small area of the Glamorgan and the south Pembrokeshire coast. There can have been very few birds that were able to survive these conditions, but as the ice withdrew, more were recorded. Archaeological excavations in Wales have recorded a suite of birds that existed here at this time, including Black Grouse, Lesser White-fronted Goose, Great Bustard, Cory's Shearwater, Red Kite and Chough (Pritchard 2020). Wildfowl remains such as Greylag Goose, Mallard, Wigeon and Teal provided evidence of contemporary diet in Britain (Yalden and Albarella 2009). The most productive sites in Wales for bird bones have been caves along the south coast, mainly on the Gower Peninsula and in Pembrokeshire. At Port Eynon cave on the Gower, for example, deposits dating from around 6,000–9,000 years ago produced the bones of 43 bird species. Several sites from the Roman period have also produced a good number of bird bones, including the legionary fortress of Caerleon, Gwent, while excavations at medieval castles have added valuable information.

Written records of birds in Wales began with references in Welsh poetry, some dating back to the 8th century or earlier. In the earliest poetry, the birds usually mentioned were Ravens and eagles, feasting on the flesh of the fallen after a battle. Gradually the emphasis changed, and later poetry had references to many species of bird, particularly that written in the 14th century. There were also references to birds in prose texts. Giraldus Cambrensis travelled around Wales in 1188, with the Archbishop of Canterbury, to recruit for the Third Crusade. His account of the journey, the *Itinerarium Cambriae*, has several references to birds, including what must be one of the earliest documented bird-identification disputes, when the party engaged in an argument about whether a bird heard between Caernarfon and Bangor was an oriole or a woodpecker. The texts of the Welsh laws also have references to birds, particularly relating to falconry, in manuscripts dating from the 13th century.

The antiquary John Leland was commissioned by Henry VIII to collect information on the antiquities of England and Wales. He wrote an account of his travels and noted a little information about birds, including a story of an eagle that nested at Dinas Brân, Llangollen, Denbighshire. The first work to do more than mention selective bird species was George Owen's *The Description of Penbrokshire* [*sic*] of 1603, which had a list of the county's birds with comments on their status. This included the only reference to Crane breeding in Wales.

The botanist William Johnson visited Wales in 1639. His account, published in 1641, also mentioned eagles in Snowdonia. He was followed by two noted naturalists, John Ray and Francis Willughby, who visited Wales in 1658 and 1662. Their second journey, described in Ray's *Third Itinerary*, provided some valuable information on birds. One of the most accomplished naturalists of the next generation was Edward Llwyd (1660–1709), who visited every county in Wales, collected a great deal of information from correspondents and contributed to the revision of the Welsh section of Camden's *Britannia*. The revised version, edited by Edmund Gibson and published in 1695, included a number of Llwyd's bird records. If the book he had intended to write, *A Natural History of Wales*, had seen the light of day, it would probably have given us much more information about the birds of Wales in this period.

Thomas Pennant (1726–98) included a number of bird records from Wales in his volume *British Zoology*, published in parts between 1776 and 1777, particularly birds recorded in the area around his home in Flintshire. Pennant published his *Tour in Wales* in 1778, followed by *A Journey to Snowdon*, part one in 1781 and part two in 1783, and these were later published together as *Tours in Wales* (1810). His focus was mainly on antiquities and he included less information on birds than in his *A Tour in Scotland*, but it was interesting information, nonetheless. Pennant was followed by many other travellers to Wales in subsequent years, particularly when the Napoleonic Wars made the Grand Tour of Europe impossible. Many travellers had little interest in birds or simply repeated what previous authors had said, but some accounts contained valuable information, such as W. Bingley's *A Tour Round North Wales* (1800) and Richard Fenton's *A Historical Tour Through Pembrokeshire* (1811). The *Swansea Guide* of 1802, published anonymously but written by J.P. Oldsworth and others, was a new type of book, a guide for travellers through Glamorgan and Monmouthshire. Like several later guides of this kind, it included a section on the birds of the area.

The 1830s saw the publication of several works that included sections on birds, such as John Williams's *Faunula Grustensis* (1830), on the area of the Conwy Valley around Llanrwst, Caernarfonshire/Denbighshire, and Robert Williams's *The History and Antiquities of the town of Aberconwy and its Neighbourhood* (1835), on the area around Conwy, Caernarfonshire. T.C. Eyton published 'An attempt to ascertain the Fauna of Shropshire and North Wales' in *Annals of Natural History* in 1838. This contained much valuable material, although the comments on the status of some species may have been more accurate for Shropshire than for North Wales. Lewis Weston Dillwyn's *Materials for a Fauna and Flora of Swansea and the neighbourhood* (1848) included a useful section on the birds of the area, as did Doddridge-Knight's 'An account of Newton Nottage', published in *Archaeologia Cambrensis* in 1853. In 1858, *Llandeilo-Vawr and its Neighbourhood* by William Davies (Gwilym Teilo) included a section on the birds of this part of Carmarthenshire. These sources were, however, the exception, and by the middle of the 19th century much of Wales still held few ornithologists. In 1865, A.G. More wrote 'On the Distribution of Birds in Great Britain during the Nesting-season' in *The Ibis* but found it difficult to obtain information from many parts of Wales. Some useful information was, however, published in *The Zoologist* in this period, such as James Tracy's *Catalogue of birds taken in Pembrokeshire* in 1850, and Thomas Dix's notes on birds in Carmarthenshire in 1865 and in Pembrokeshire in 1866 and 1869.

There was a distinct increase in ornithological activity in the last years of the 19th century, mainly in the southern part of Wales. This period saw the publication of the first county avifaunas. In 1882, Edward Cambridge Philips produced *The Birds of Breconshire*, a reprint of a paper published in several parts in *The Zoologist* in 1881 and 1882. He later greatly expanded this in the revised version published in 1899. In 1894, *The Birds of Pembrokeshire and its Islands*, by the Rev. Murray A. Mathew, was published. The Rev. Digby Seys Whitlock Nicholl of Llantwit Major produced a list of the birds of Glamorgan in a series of notes in *The Zoologist* in 1889. Several natural history societies were established in the second half of the 19th century, with the Cardiff Naturalists' Society, founded in 1867, being particularly active. The society published annual *Transactions* from 1868 to 1986, which included much of ornithological interest, including papers on birds outside the borders of Glamorgan. It also produced a county avifauna, *The Birds of Glamorgan*, in 1900, with a revised edition in the *Transactions* for 1925. Sir J.T. Dillwyn-Llewelyn, grandson of Lewis Weston Dillwyn, produced bird lists for Swansea in 1900 and 1925, but these were not published. Mid-Wales had far fewer ornithologists than the southeast, but Professor J.H. Salter, who held the Chair of Botany at the University College of Wales, Aberystwyth, wrote a good deal on the birds of northern Ceredigion between 1895 and 1904 and again after he returned to the area in 1923. These included *A list of the birds of Aberystwyth and neighbourhood* (1900).

Little information was published about the birds of North Wales between Eyton's work in the 1830s and the final years of the 19th century, although some notes appeared in *The Zoologist*, for example, by John Cordeaux in 1866. W.H. Dobie's 'Birds of West Cheshire, Denbighshire and Flintshire' appeared as a paper in the *Proceedings of the Chester Society of Natural Science* (1894). Observers such as Oliver Aplin and Thomas Coward visited Llŷn and Anglesey respectively, around the turn of the 20th century, and reported their findings in *The Zoologist*. The first comprehensive account of the birds of North Wales, including Montgomeryshire, was published in 1907 as part of *The Vertebrate Fauna of North Wales* written by H. Edward Forrest. It was based on information collected from a large number of correspondents. The late 19th century, and the period up to the First World War, was the heyday of large-scale game shooting. The unpublished records of large estates, which maintained detailed gamebags and often records of predators shot, are particularly valuable.

In 1919, Forrest wrote *A Handbook to the Vertebrate Fauna of North Wales*, updating his 1907 volume. Apart from this work, the period of the First World War and up to the early 1930s was fairly unproductive for Welsh ornithology. Records for parts of Wales were published in the Caradog and Severn Field Club's annual *Record of Bare Facts* in the 1920s and 1930s. The first bird observatory in Britain was established by Ronald Lockley on Skokholm, Pembrokeshire, in 1933.

In the foreword to his book *The Vertebrate Fauna of North Wales* in 1907, Forrest suggested that someone should write a similar volume for South Wales. Although Geoffrey Ingram (1883–1971) and Harry Morrey Salmon (1891–1985) started work on such a volume, it was never published, but they did produce seven county avifaunas, starting with *Birds of Glamorgan* in 1936 (in *Glamorgan County History*, Volume 1) and *The Birds of Monmouthshire,* published in 1939 in *The Transactions of the Cardiff Naturalists' Society.* The only work of note published during the Second World War was Eric Hardy's *The Birds of the Liverpool Area* (1941). Despite the title, it included a good deal of information on birds in northeast Wales.

Prior to the Second World War, ornithology in Wales had depended on the efforts of a comparatively small number of pioneers, their contribution described and celebrated by G.A. Williams (2013). The period after the war saw a great increase in ornithological activity across the UK and the number of birdwatchers increased rapidly. A greater interest in rare birds resulted in the establishment of the British Birds Rarities Committee (BBRC) in 1959, to determine whether records of rare species were acceptable, reported annually in *British Birds*.

County and island bird reports

The Cardiff Naturalists' Society formed an Ornithological Section after 1945 (Ballance 2020). Several other local ornithological societies were formed in the post-war years with the aim of sharing knowledge and producing bird reports. A network of county recorders developed, taking responsibility for bird recording within each county. Annual bird reports were produced by Skokholm Bird Observatory from 1936 to 1976, except during 1941–45. The production of county reports began with one for Montgomeryshire in 1947. The *Cambrian Bird Report* began in 1953–54, initially covering Anglesey, Caernarfonshire and Denbighshire, but later changed to include Meirionnydd and exclude Denbighshire.

More county bird reports were established in the 1960s, including the *Glamorgan Bird Report* in 1962. The more interesting bird records from Ceredigion, Pembrokeshire (excluding Skokholm) and Carmarthenshire, appeared regularly in *Nature in Wales* during 1955–67, after which more comprehensive five-year reports were produced for the three counties in a Dyfed Bird Report by the West Wales Naturalists' Trust. Separate reports for each county were published from 1981. In recent years, reports have been produced for all Welsh counties except Radnorshire, although selected records from that county have been provided for the *Welsh Bird Report*. The Wildlife Trust of South and West Wales publishes bird reports for Skomer and Skokholm and the Bardsey Bird and Field Observatory produces an annual report, in recent years called *Bardsey's Wildlife*.

Welsh Bird Report

At a meeting in Aberystwyth in 1967, which included representatives of all the regional organisations concerned with birds in Wales, it was agreed to produce a *Welsh Bird Report,* based on records in regional reports. The first report, for 1967, compiled by Peter Davis and Peter Hope Jones, was published in the journal *Nature in Wales.* This journal was published by the West Wales Field Society/Naturalists' Trust from 1955–81 and then by the National Museum of Wales until 1987. The Welsh Records Advisory Group (WRAG) was founded in 1972, to consider records of species not covered by the BBRC, but that were rare or scarce in Wales. As a

The second bird observatory in Wales was founded in 1953, on Bardsey, Caernarfonshire © Ben Porter

result of a change in editorial policy, owing to financial problems, *Nature in Wales* ceased to publish the *Welsh Bird Report* after 1977.

The Welsh Ornithological Society (WOS) was founded in 1989, following a conference in Aberystwyth in 1988. WOS resurrected the *Welsh Bird Report*, initially publishing a report covering the period 1978–87, followed by annual reports that have continued to the present. In 1994, the Welsh Records Panel (WRP) was formed to replace WRAG as the body assessing claims of scarce birds in Wales, defined as those recorded five times or fewer each year on average. Decisions are published in the report *Scarce and rare birds in Wales*, originally as a printed report and more recently on the WOS website. In 2019, the WRP changed its name to the Welsh Birds Rarities Committee (WBRC).

County and other avifaunas

In 1949, Ronald Lockley, Geoffrey Ingram and Harry Morrey Salmon wrote *The Birds of Pembrokeshire.* The indefatigable Ingram and Salmon went on to produce three more county avifaunas in the 1950s: Carmarthenshire (1954), Radnorshire (1955) and Breconshire (1957). In the north, Paul Whalley published a list of birds seen in Anglesey and Caernarfonshire, with notes on their status, in 1954.

The Birds of Monmouthshire by P.N. Humphreys, a revised version of the original work by Ingram and Salmon, was published in 1963. In 1966 Ingram and Salmon, this time with William Condry, wrote yet another county avifauna, *The Birds of Cardiganshire*. The centenary of the Cardiff Naturalists' Society, 1967, was marked by the publication of a new *Birds of Glamorgan*, by A. Heathcote, D. Griffin and H. Morrey Salmon. Very little had been published on the birds of North Wales since H.E. Forrest's volumes in the early 20th century, but a checklist of *The Birds of Flintshire* by Birch *et al.* (1968) was published by the Flintshire Ornithological Society. In 1974 Peter Hope Jones produced an avifauna for Meirionnydd, *Birds of Merioneth*, following this with *Birds of Caernarvonshire*, co-written with Peter Dare, in 1976. *Birds of Breconshire* by M.E. Massey was also published in 1976, and P.N. Ferns *et al.* produced a new avifauna for Gwent, published by the Gwent Ornithological Society in 1977. In 1983, Peter Hope Jones and John Lawton Roberts wrote a checklist of Denbighshire's birds in *Nature in Wales*, and Martin Peers produced *The Birds of Radnorshire and Mid-Powys* in 1985. *Birds of Breconshire* by Martin Peers and Michael Shrubb was published in 1990, and *Birds of Pembrokeshire* by Jack Donovan and Graham Rees in 1994. The first full avifauna for Wales, *Birds in Wales*, by Roger Lovegrove, Graham Williams and Iolo Williams, was published by Poyser in 1994. An update to this volume, *Birds in Wales 1992–2000* by Jon Green, was published by the Welsh Ornithological Society in 2002.

Birds of Glamorgan, by Clive Hurford and Peter Lansdown, appeared in 1995, and *The Birds of Caernarfonshire* by John Barnes was published in 1997. The authors of *Birds in Wales* (1994) had commented that there were several gaps in coverage of Welsh counties by avifaunas, with nothing comprehensive for Anglesey or Montgomeryshire since Forrest (1919). These gaps were filled in the early 21st century, with the publication of *Birds of Anglesey* by Peter Hope Jones and Paul Whalley in 2004 and *The Birds of Montgomeryshire* by Brayton Holt and Graham Williams in 2008. Updated avifaunas were produced for several other counties: *Birds of Ceredigion* by Hywel Roderick and Peter Davis in 2010, *Birds of Meirionnydd* by Rhion Pritchard in 2012, *The Birds of Radnorshire* by Peter Jennings in 2014 and *Birds of Caernarfonshire* by Rhion Pritchard in 2017. There have also been avifaunas for some of the Welsh islands. *The Birds of Bardsey* by Peter Roberts was published in 1985, *Birds of Skokholm* by Michael Betts in 1992 and *The natural history of Skokholm Island* by G.V.F. Thompson in 2008. On the mainland, David Gilmore's *The Birds of Cardiff* was published in 2006.

There still remain a number of gaps in the ornithological literature. No modern county avifauna has been published for Denbighshire or Flintshire, nor for Carmarthenshire since 1954, although the manuscript of an unpublished avifauna for the latter county, by John Lloyd *et al.*, has kindly been made available to assist research for this book. The publication of this work would be most welcome. Breconshire and Glamorgan have been among the counties best served in terms of avifaunas over the past 120 years, but the period since the last avifauna for these counties means that a new version would be invaluable.

Bird surveys

A number of surveys of individual species were undertaken in the first half of the 20th century, but the only one to be repeated annually was the BTO Heronries Census, which began in 1928 and continues today. Regular counts of wildfowl began at a number of sites in Britain, in 1951/52, and from winter 1955/56 there were counts on the Severn Estuary and at Aber Ogwen, Caernarfonshire. Counts of waders were added when the Birds of Estuaries Enquiry began in 1969–70, and there have been regular counts of waterbirds at many Welsh sites since. In 1993, full integration was achieved with the launch of the Wetland Bird Survey (WeBS), a joint BTO, WWT, RSPB and JNCC scheme, which is now managed primarily by the BTO. Initially, counts were only undertaken at coastal sites at high tide, but inland waters were soon added, and winter 1992/93 saw the beginning of the WeBS Low Tide Count scheme. Much use is made of data from WeBS counts in this book.

In 1961, the BTO introduced the Common Birds Census (CBC), commissioned by the Nature Conservancy to monitor population trends among widespread breeding birds. The initial emphasis

Sea cliffs, such as at Stack Rocks in Pembrokeshire, bustle with thousands of nesting seabirds during May and June © Mike Alexander

was on farmland, but woodland plots were added by 1964. This enabled population trends to be assessed for the UK, but the number of CBC plots covered in Wales was very small compared to England, and few plots in Wales were covered annually for long periods. In the early 1980s, the Nature Conservancy Council (NCC) and the RSPB carried out upland bird surveys across large parts of Wales, which provided important baseline information at a time of major changes in land use, but these have not been repeated.

An ambitious project, Operation Seafarer, to survey the coastal colonies of seabirds, was undertaken in 1969–70 by the recently formed Seabird Group and organised by David Saunders. Repeat surveys were undertaken in 1985–88 (the Seabird Colony Register) and in 1998–2002 (Seabird 2000). This is being updated through Seabirds Count (2015–21). These surveys achieved near-complete coverage in Wales, although until the most recent census, counts for Manx Shearwater and Storm Petrel were probably inaccurate, because of the difficulty of surveying these species, and there was limited effort to count roof-nesting gulls.

In 1994, the BTO, RSPB and the JNCC, on behalf of the statutory conservation agencies, introduced the Breeding Bird Survey (BBS) to replace and improve the CBC's monitoring of breeding birds, based on two visits in spring to survey randomly selected 1-km squares. This involved a much larger number of volunteers than the CBC. The number of squares surveyed in Wales has made it possible to calculate population trends for up to 60 bird species in Wales, plus Grey Squirrel and Rabbit. Much use is made of population trends derived from the BBS in this book.

Many breeding species are too scarce to be covered adequately by the BBS. Some, such as Chough, Black Grouse, Ring Ouzel and several raptor species, have been covered by the Statutory Conservation Agencies and RSPB Annual Breeding Bird Surveys (SCARABBS), which constitute a programme of regular single-species surveys co-ordinated by the RSPB. Most single-species surveys are UK-wide, but in 2012–13 the BTO conducted a survey of breeding Stonechat, Whinchat and Wheatear solely in Wales (Henderson *et al.* 2017). Other, mainly RSPB and NCC, surveys conducted in the late 1970s and 1980s covered all the species in the main upland and woodland areas, and the lengths of the Wye and Severn rivers. Woodland Bird Surveys, undertaken in 1965–72, 1973–80 and 1981–88, were repeated in 2003–04, in a major collaboration by the RSPB and the BTO (Amar *et al.* 2006). Several individuals have carried out their own surveys that have been reported in papers in *Birds in Wales* (*Welsh Birds* until 2010) and have provided valuable information for the species accounts in this volume.

Bird atlases

The publication of *The Atlas of Breeding Birds in Britain and Ireland*, based on fieldwork organised by the BTO and the Irish Wildbird Conservancy during 1968–72 (Sharrock 1976), provided the first overview of the distribution of birds in Wales at 10-km square level (a 10-km square has an area of 100km^2). Wintering birds were covered by *The Atlas of Wintering Birds in Britain and Ireland*, with fieldwork carried out during 1981–84 (Lack 1986). A second breeding-season atlas, *The New Atlas of Breeding Birds in Britain and Ireland: 1988–1991* (Gibbons *et al.* 1993), included maps of relative abundance as well as distribution. However, differences in methodology between these two atlases mean that any comparisons must be undertaken with care.

A development in the 1980s was the production of county atlases at tetrad level. A tetrad is a 2km by 2km square, with an area of 4km^2. A tetrad atlas thus provides a much more detailed picture of bird distribution than the 10-km squares used for the Britain and Ireland atlases. *The Gwent Atlas of Breeding Birds* by Tyler *et al.*, based on tetrads visited during fieldwork between 1981 and 1985, was published in 1987. Similar fieldwork during 1984–89 led to the publication by the Gower Ornithological Society of *An Atlas of Breeding Birds in West Glamorgan* by D.K. Thomas in 1992. Distribution maps from this survey for the whole of Glamorgan were published in Hurford and Lansdown (1995). Tetrads in Pembrokeshire were surveyed during 1984–88, with the results appearing in Donovan and Rees (1994). Just under half the tetrads in Breconshire were surveyed during 1988–90, with the results summarised in the species accounts in Peers and Shrubb (1990).

Second tetrad atlases in Gwent, East Glamorgan and Pembrokeshire enabled comparison with earlier periods. The second Gwent atlas was based on fieldwork between 1998 and 2003, and the results published in *The Birds of Gwent* (Venables *et al.* 2008). The results of the second Pembrokeshire atlas, undertaken during 2003–07, appeared in the *Atlas of Breeding Birds in Pembrokeshire 2003–07* (Rees *et al.* 2009), now published with updated information at pembsavifauna.co.uk. A second tetrad atlas of Eastern Glamorgan, in 2007–11, was also published online at eastglamorganbirdatlas.org.uk.

A new Britain and Ireland atlas, covering both breeding and wintering birds at 10-km square level, was published as *Bird Atlas 2007–11: the breeding and wintering birds of Britain and Ireland* (Balmer *et al.* 2013). The breeding codes used were the same as those used in the 1968–72 *Atlas*, enabling direct comparison. Some areas took advantage of the fieldwork and the BTO's online data-entry system to produce their own tetrad atlases, including the East Glamorgan atlas mentioned above. The authors of the *Birds in Wales* (1994) regretted that the efforts to produce tetrad atlases in some of the southern counties had not been emulated farther north. This was remedied by the publication of *The Breeding Birds of North Wales* (Brenchley *et al.* 2013), a tetrad atlas of the five North Wales counties, based on fieldwork between 2008 and 2012.

Some gaps remain however, particularly in mid-Wales. There has never been a tetrad atlas of Carmarthenshire, Ceredigion, Radnorshire or Montgomeryshire. Spence (2016) suggested that an all-Wales tetrad-level atlas, with the fieldwork coinciding with that for the next Britain and Ireland atlas, would be feasible.

Bird ringing and tracking

Although the ringing scheme in Britain was set up in 1909, there was little activity in Wales in the early years. Ronald Lockley was ringing Manx Shearwaters, and possibly other seabirds, on Skokholm from at least 1929 and set up a small Heligoland trap there in 1933. The number ringed there was initially limited by the cost of rings, but an appeal for funds led to the construction of a larger trap in 1935 and an increased level of ringing (Lockley 1936). During 1936–76, 284,355 birds of 164 species (including over 170,000 Manx Shearwaters) were ringed at Skokholm Bird Observatory (Betts 1992). In parallel, Orielton Duck Decoy in Pembrokeshire, used to catch birds for the table during 1871–1918, was re-opened in 1934 to catch waterbirds for ringing. By 1960, when ringing ceased, almost 12,500 ducks of six species had been caught (Saunders 2019). The number of birds ringed in Wales began to increase after the Second World War, and grew when ringing began on Bardsey in 1953. By the end of 2019, 293,967 birds of 196 species had been ringed there (*Bardsey's Wildlife* 2020). Skokholm lost its status as a bird observatory in 1976, when bird ringing lapsed due to the preference of the island's owner, although a warden remained in residence. It regained its status as an official bird observatory in 2014. Many ringers work as individuals, but there are also around ten ringing groups in Wales. Some have operated for many years, such as the Merseyside Ringing Group founded in 1954, which includes Flintshire and Denbighshire in the area it covers. The SCAN Ringing Group, founded in 1973, uses cannon nets to catch waders on the north coast of Wales, mainly on Traeth Lafan, Caernarfonshire. Over one million birds were ringed in Wales and reported to the BTO Ringing and Nest Recording Scheme during 2000–19.

More recently, other methods of tracking the movements of birds have been developed, including the use of geolocators and satellite tags. Geolocators record light levels every few minutes and if the bird can be recaptured and the device retrieved, the data are used to calculate the locations that the bird has visited. Satellite tags record the position of the bird using GPS or similar systems. The smallest types store the data, so that the bird has to be recaught and the device removed, but larger devices can transmit data in real time. The use of these devices in Wales has already produced much fascinating information that is used in a number of species accounts in this book, such as that of the Cuckoo.

Migration studies have been undertaken on several Welsh islands, including on Bardsey, Caernarfonshire, where the lighthouse beam attracted thousands of birds on some nights until 2014, when it was replaced by red LED lights © Hugh Miles

Online databases

An important development in bird recording has been the creation of databases that allow observers to enter their records online, rather than sending record cards or lists to county recorders. BirdTrack, established in 2004, is a partnership between the BTO, the RSPB, Birdwatch Ireland, the Scottish Ornithologists' Club and WOS. County recorders can access the records for their area, leading to BirdTrack records becoming one of the primary sources of records for the production of county bird reports. Another development has been the creation of the Local Environmental Records Centres (LERCs), which gather records of all taxa, including birds. Wales is covered by four LERCs: Cofnod in North Wales, the West Wales Biodiversity Information Centre (WWBIC), the Biodiversity Information Service for Powys and Brecon Beacons National Park (BIS), and the South East Wales Biodiversity Records Centre (SEWBReC). Aderyn, LERC Wales' Biodiversity Information and Reporting Database, is a joint project between these four LERCs. Both BirdTrack and LERC Wales have apps that enable observers to enter records on their smartphones.

In the 27 years since the publication of *Birds in Wales* in 1994, far more information has been published on the birds of Wales than in any comparable period previously. However, some areas and some species remain little known. Perhaps the publication of this volume will inspire others to get into the field and add to our knowledge of the birds of Wales.

Rhion Pritchard

Online portals and smartphone apps, such as BirdTrack supported by the Welsh Ornithological Society, enable birdwatchers to contribute to large citizen science projects © Ben Porter

3. Bird conservation

Before the 19th century, there was limited written evidence that birds were noticed in Wales. They must surely have been noticed, of course, but to many people they were perhaps merely a background to the challenges of surviving daily rural life. Aside from poets and writers, if people thought about birds at all, it was when they were considered to be a problem. People, recovering from a series of plague epidemics after 1348, then faced serious malnutrition following a series of disastrous crop harvests during the Tudor period. This led to granivorous birds being seen as competitors. Their control became a requirement under the Preservation of Grain Act 1532 and was accelerated by a series of vermin Acts from 1566 onwards. Parishes had to raise levies to pay bounties for birds and mammals killed. Although few Welsh parish records remain, there is no reason to believe that killing was any less here than elsewhere in Britain. In Hanmer, Denbighshire, for example, churchwardens paid bounties for over 150,000 House Sparrows between 1789 and 1835 (Lovegrove 2007), although the impact on populations is impossible to know.

The development of game shooting as a leisure activity of the rich accelerated from the 1840s, as a result of improvements to shotguns and increased mobility with the expansion of the railway network. Game laws provided protection to some birds and mammals that were hunted, primarily to provide for closed seasons and to restrict poaching. The resultant growth in gamekeeping resulted in predators being wiped out on many estates. However, with 75% of Welsh land being above 300m, and a history of land tenure that worked against the creation of large estates away from the Marches, the levels of predator control did not have the same degree of impact on bird populations as in England and Scotland. The remoter areas of Wales probably provided refuges for species decimated elsewhere. Nonetheless, there were local impacts. For example, the gamekeepers on Penrhyn Estate, between Penmachno and Dolwyddelan in Caernarfonshire, killed 1,988 Kestrels, 735 Sparrowhawks, 228 Merlins and 135 Buzzards during 1874–1902 (Lovegrove 2007).

The Age of Enlightenment across Europe resulted in a developing interest in nature, although much of this amateur science involved netting or shooting birds for their skins and collecting their eggs. A small cadre of people took a more studious approach, some combining their observations with romantic prose, which had the effect for some readers of popularising nature. Major social changes during the Victorian era saw the start of middle-class concerns about animal welfare and nature, resulting in the formation of the RSPCA in 1824 and the RSPB in 1889. Local groups formed, such as the Cardiff Naturalists' Society in 1867 and the South Pembrokeshire Naturalists' Field Club in 1879, bringing together people with a common interest. Other groups, based in Chester, Shrewsbury and Hereford, organised excursions to the wildlife-rich areas in Wales. From these societies, bird conservation grew slowly. An early example of these groups campaigning for wildlife, told in *Birds in Wales* (1994), involved a national outcry when, in 1890, members of the Cardiff Naturalists' Society found a company from the Royal Navy shooting the seabirds on Grassholm, Pembrokeshire. The RSPCA ultimately took a successful private prosecution against those involved. Protection efforts by individuals and small organisations before the Second World War, grew into government and civil society action from the 1950s, as birds and the factors affecting them grew in public consciousness.

Protecting birds from exploitation

The late 19th century saw the first attempts to protect Red Kites in Wales, initially by landowners in north Breconshire. Following a century of persecution, Red Kites were on the brink of extirpation in Britain, with only small populations remaining in the upper Tywi Valley, Carmarthenshire, and the upper Wye Valley in Breconshire/Radnorshire. A protection scheme was formalised by the British Ornithologists' Club (BOC) from 1904, with local people paid to guard several nests from egg- and skin-collectors, who could secure a good price for a stuffed Kite. Farmers and gamekeepers were paid to protect the birds, but that did not prevent them being shot, trapped and poisoned, as part of widespread and indiscriminate 'vermin' control. The RSPB had taken over organising nest protection from the BOC by 1923. The Red Kite population reached its nadir in the early 1930s, with just two pairs nesting successfully each year, but it gained a respite owing to constraints on travel caused by the Second World War. Protection and monitoring became better co-ordinated from 1950 with the creation of a new Kite Committee, whose work was taken over by the RSPB and Nature Conservancy from 1970, and from 1996 by the Welsh Kite Trust. Action for Red Kites in Wales was arguably the longest-running conservation scheme in the world, and such was its success that intervention is now rarely required (Lovegrove 1990).

The recovery of Red Kite in Wales, from its low point in the 1930s to an estimated 2,500 pairs today, is a celebrated conservation success © Tony Cross

The RSPB's involvement in Wales had started some years before the Red Kite nest protection, as this example shows:

> In the summer of 1909 information reached the Society that Pole-Traps were still being used in Wales and Scotland [having been outlawed in 1904]; and for some days the Inspector tramped the Welsh hills in search of them—an unknown country, made none the easier by mountain mists—deriving what help he could from a scattered Welsh-speaking population. Two convictions, one in Carmarthen and one in Pembrokeshire resulted; and owing to the ignorance of the law among the people in general, the Society decided to issue the Pole-Trap Act in Welsh (RSPB 1909).

The RSPB stepped up its activity in 1911, when its scheme of 'Watchers' was extended to Wales. A population of 'hundreds' of pairs of Roseate Terns on Ynys Llanddwyn, Anglesey, had been targeted by egg-collectors, prompting the landowner, the Honourable F.G. Wynn, to ask the RSPB to protect the birds on the island. Three slate-boat pilots, who worked for the Caernarfon Harbour Company and lived in a row of whitewashed cottages on the island, were appointed as Watchers in 1912. They were a significant proportion of the 22 Watchers employed by the Society across Britain and Ireland that year.

> The colony is well worth full protection. It is placed on a spot easily accessible and also ideal for observation by genuine naturalists. It was hard to calculate the numbers of birds, but I should say 150 pairs was somewhere near the mark. It is a splendid little colony, and in its present position is most fascinating to watch. (Blathwayt 1923)

Despite these efforts, the Roseate Tern colony abandoned Ynys Llanddwyn and moved back to The Skerries in the 1920s, following changes to the vegetation and the expansion of Cormorants and gulls (Jones and Whalley 2004). RSPB involvement on the island continued, however, with the Society taking on the grazing lease (£15 per year) from 1935. It erected fencing to protect nesting birds from the huge number of tourists for whom the landscape and history of the island have always been a draw. Mrs Jones, the widow of one of the original Watchers, continued to be employed by the RSPB, assisted by her son Sam, a former policeman. Her determination to protect the birds led to a stand-off with Newborough villagers, who organised a procession to 'occupy' the island on August Bank Holiday 1938, even drawing coverage in *The Daily Mail*.

Changing public attitudes, stronger laws and improved enforcement reduced the threats from egg-collectors from the 1960s, but the actions of a hardcore of 'eggers' continued to require nest protection and joint RSPB/police operations in Wales through the 1970s and 1980s. Red Kites, Choughs and Peregrines were targeted by criminals, the last of the three by gangs that could procure good money for falcons that could be used in hunting. Direct nest protection of birds is now rarely necessary in Wales. Measures are in place for a few colonial species, but are now used primarily to protect them from human disturbance, notably at Wales' sole colonies of Sandwich Tern at Cemlyn, Anglesey, and Little Tern at Gronant, Flintshire.

Mrs Jones was an RSPB 'Watcher' on Llanddwyn, Anglesey, through the 1920s and 1930s © Isle of Anglesey archives

Reserved for birds: land for nature

Buying land is the ultimate means to protect nature from threat, and to manage improvements that cannot be sustained by market forces in the wider landscape. Such sanctuaries could be viewed as a symbol of failure to manage adequately for nature across the wider countryside, but there are numerous species whose status would be poorer without reserves, and in Wales many seabirds nest almost exclusively on land protected in this way.

As with protection efforts, it was private landowners who instigated the first measures to protect their land for birds. J.J. Neale, for example, banned landings and photography on Skomer, Pembrokeshire, in order to safeguard the island's wildlife soon after he acquired the lease in 1904. Just weeks after it was formed in 1895, the National Trust received its first land donation, at Dinas Oleu, Meirionnydd, although this was primarily for public enjoyment rather than for its wildlife interest.

It was a passionate naturalist, Ronald (R.M.) Lockley, who—at the age of 24—initiated the concept of nature reserves in Wales. Taking on the lease of Skokholm, Pembrokeshire, from Dale Castle Estate in 1927, he studied Manx Shearwaters, Puffins and Storm Petrels. The RSPB paid the lease of £26 a year, in lieu of a salary, requiring Lockley to generate an income from the sale of wild Rabbits caught on the island as part of a failed eradication attempt aimed at improving the grazing for livestock. Lockley's attempt to farm Chinchillas for their fur was also aborted. In 1933, he established the first Bird Observatory in Britain on Skokholm (Saunders and Sutcliffe 2020). With the agreement of the individual landowners, the RSPB also paid for Watchers on Skomer, Grassholm and Ramsey to prevent tour-boats landing, and in the early 1930s, Newport steel magnate Leonard Whitehead offered to buy Ramsey for the RSPB on the condition that Lockley become Supervisory Watcher for all the islands. The deal collapsed because Mrs Lemon—the driving force behind the RSPB for its first 50 years and long-time chair of its Watchers Committee—was unhappy that Lockley was running boat services for tourists to the islands, permitting landing in contravention of the annual rent paid to him to maintain the sanctuary.

R.M. Lockley established the first Bird Observatory in Britain, on Skokholm, Pembrokeshire, in 1933 and pioneered studies of seabirds, including Manx Shearwater © David Saunders collection

Whitehead supported the creation of the Pembrokeshire Bird Protection Society, to which the RSPB contributed £100 per year, with Lockley as its first honorary Secretary and Superintendent Watcher. From 1938, it took over watching duties for the entire coast including all the islands, and would evolve into the West Wales Field Society (WWFS) in 1945, the West Wales Naturalists' Trust in 1962 and merge with the Glamorgan Wildlife Trust to become the Wildlife Trust of South and West Wales from 2002. Meanwhile, 30 years after the Ramsey deal had fallen through, the RSPB leased the island as a nature reserve in the 1960s and 1970s, although following a subsequent break, it would be 1993 before it finally had the opportunity to buy the island.

The West Wales Field Society took on the lease of Skokholm in 1948, when Lockley no longer wanted it, and sub-let it to the predecessor of the Field Studies Council, which appointed a warden, Peter Conder, who would later go on to lead the RSPB from 1963 to 1976.

The WWFS attempted to purchase Skomer in the early 1950s, but neither it nor the RSPB (of which Ronald Lockley was by now a trustee) could raise the £10,000 required. Following an appeal, the WWFS met the purchase price in 1958 and sold it to the Nature Conservancy for £6,000 the following year. The NCC leased it back to the WWFS on a 42-year term and built a house for the warden, the first of whom was David Saunders.

The Nature Conservancy had been created in 1949 under the National Parks and Access to the Countryside Act, and Skomer was just one site that it acquired and designated as a National Nature Reserve (NNR). By 2019, 76 NNRs had been designated in Wales, of which 58 are partly or wholly managed by government, now through Natural Resources Wales. Many are designated for wildlife reasons other than birds, but other sites with important bird components include Aber Valley in Caernarfonshire, Traeth Bach in Meirionnydd, Berwyn in Denbighshire/Meirionnydd/Montgomeryshire, the Dyfi Estuary in Meirionnydd/Ceredigion, Cors Caron in Ceredigion, Kenfig in East Glamorgan and Newport Wetlands in Gwent.

The Second World War brought many scientific and conservation efforts to a temporary halt and several parts of the Welsh coast were requisitioned by the military, including the Great Orme, Caernarfonshire, and an extensive part of the former Stackpole Estate, Pembrokeshire. While many were returned to civilian ownership after the war, some remain under Ministry of Defence stewardship. Training needs, while not always conducive to sympathetic conservation interests, mean that such places have largely resisted the intensification of land management that occurred outside the fence. Later in the 20th century, the MoD would work with nature conservation recording groups to help inform the management of its land.

Having bought the then 7,000-strong gannetry on Grassholm in 1948—its first land purchase in Wales—the RSPB maintained its presence on Ynys Llanddwyn, but there was no sign of the tern colony being re-established. Mrs Jones had died and the RSPB questioned its continued involvement on the island. Philip Brown, now in charge of Watchers and Sanctuaries at the RSPB, wrote to wildlife illustrator Charles Tunnicliffe who had moved to a nearby cottage, 'Shorelands', on the Cefni Estuary, hoping that he would take a role in protecting the site. With the Bank Holiday occupation still in recent memory, the artist rebuffed him: "My dear Philip, you must not ask me to do any policing of this area. I don't mind drawing for the R.S.P.B. but I must refuse to snoop for it" (Tunnicliffe 1949). In 1954 the RSPB surrendered the lease of Llanddwyn to the Nature Conservancy and it became part of the newly designated Newborough Warren NNR.

In 1965, the National Trust launched Project Neptune, an ambitious fundraising plan to acquire, or protect, under covenant, extensive stretches of coastline for public access and nature conservation. The very first Neptune purchase was in Wales, at Whiteford Burrows, Gower, followed by extensive additional areas of the Gower coast in 1967. Other nature sites followed into Neptune as land and funds became available, including Cemlyn Bay on Anglesey in 1968, the Marloes Peninsula in Pembrokeshire in 1981 and Taf Estuary saltmarsh in Carmarthenshire in 1997. Many coastal sites were acquired primarily for landscape reasons and to facilitate public access, especially in the early years, but many proved valuable for their wildlife too.

In 1969, the RSPB's first-ever fundraising reserve appeal to members enabled land to be purchased at Gwenffrwd-Dinas, Carmarthenshire, primarily for its national importance for breeding Red Kites, and at Ynys-hir, Ceredigion, where William Condry was appointed the first warden. Not only did Bill Condry ensure good protection and management of the site on the south side of the Dyfi Estuary, his talks and writing promoted the wonders of Welsh wildlife to a wider audience. Like all nature reserve acquisitions in Wales until the 1990s, the focus was on protecting important places at risk of change or development. Active management of nature sites was in its infancy, trialled on a few wetland sites such as Minsmere, Suffolk.

Through the 1960s and 1970s, development pressures put many UK estuaries at risk, with port expansion and associated dredging, land reclamation and tidal barriers among the threats. Development pressures on the Dee, specifically the threat of a tidal barrage in 1972, led to the creation of the Dee Estuary Conservation Group, an alliance of organisations with an interest in its habitats. Just prior to that in 1970, at the request of the Merseyside Ringing Group, British Steel declared the cooling lagoons within their Shotton Works and surrounding reedbeds a nature reserve. This enabled the Group to construct the rafts that became home to one of only four colonies of Common Tern in Wales.

Almost 50ha of Dee saltmarsh was lost in the mid-1970s, when work started on an oil-fired power station at Connah's Quay, the first in Britain built to use new North Sea production. Fortunately, the power station superintendent, Keith Steward, was an RSPB member and saw an opportunity to create new coastal habitat around the works, advised by the RSPB and supported by the Deeside Naturalists' Society, which built the hides and organised bird monitoring. Lagoons were created in 1976, drawing on experience from habitat creation work at Minsmere, Suffolk, and Titchwell, Norfolk. Wildfowling was prohibited at Connah's Quay, with a no-shooting zone extended north to Flint Castle by landowner Delyn District Council.

From 1979, a series of land purchases was made by the RSPB to protect the Dee. The first acquisition, from British Steel, was conducted in strict secrecy, much to the anger of the Dee Wildfowlers Club, which lost a large area of its shooting rights. In 1986 the RSPB purchased Inner Marsh Farm, waterlogged arable land straddling the national boundary, from Cheshire County Council, to which was added adjoining farmland acquired from the Welsh Development Agency in 2007. This combined area has been transformed into Burton Mere Wetlands. The RSPB acquired Oakenholt Marsh in 1988, which strengthened the Deeside Naturalists' Society's adjacent reserve, and at about the same time leased land at the Point of Ayr. Almost the whole of the Dee Estuary in Wales is now managed by the RSPB, save for a small area around Mostyn Docks, and thus is largely protected from development.

The mid-1970s also saw areas of the west coast brought under the management of conservation charities, including part of the Stackpole Estate, Pembrokeshire, by the National Trust in 1976 and the first lease to the RSPB at South Stack, Anglesey, in 1977. Later purchases enabled extensive areas of lowland heath to be managed there, as well as the seabird cliffs. The Bardsey Island Trust was set up in 1979 to manage the island, following a public appeal and campaign to buy the island from Lord Cowdray.

Protecting land from development is one thing, but to maintain the wildlife interest—and to make the habitats resilient to external threats—requires positive management, using the best available evidence and on a scale sufficient to be sustainable. In many cases, that involves commercial leases and tenancy arrangements with local farmers, but by the 1990s, more innovative measures were being trialled with rare-breed livestock and growing expertise from the likes of Pori Natur a Threftadaeth (PoNT). Much of this is relatively small scale, but at Lake Vyrnwy—where the RSPB entered a nature reserve monitoring agreement with landowner Severn Trent Water (now Hafren Dyfrdwy) in 1977—conservation is on a landscape scale. RSPB Cymru has managed the farming operation there since 1996, and since 2000 it has been the largest organic farm in England and Wales, with over 3,000 head of sheep, Welsh Black cattle and hill ponies grazing its 4,500ha.

By the 1990s, the network of most important terrestrial sites had been notified as Sites of Special Scientific Interest (SSSIs) and

Susan Cowdy played a major role in the development of Bardsey Bird & Field Observatory, studies of Chough and the public appeal to buy the island © Hugh Miles

Newport Wetlands nature reserve was created in the late 1990s and designated as a National Nature Reserve in 2008 © Bob Haycock

the first Special Protection Areas had been designated under the EU Birds Directive. Conservation charities then focused attention on habitat-creation opportunities, since the law should now protect the best areas from gross loss. The RSPB took an early step with the purchase of fields in the Cefni Valley, Anglesey, from 1994, with the intention of restoring a mosaic of wet grassland and reedbed. This was part of a UK strategy to provide a refuge for Bitterns away from the east coast of England, where most sites were at long-term risk of sea-level rise, and to increase the connectivity of wetlands on Anglesey. Although Bitterns were regular in the winter at RSPB Malltraeth Marsh (now Cors Ddyga) from 1996, it was another 20 years before they nested, coinciding with the first nesting Marsh Harriers, while Lapwing numbers were restored with the aid of an electrified anti-predator fence.

Land also came under conservation management through new avenues, particularly as a result of planning decisions. In 1993, 'land'—created from material from the A55 road tunnel dumped on intertidal habitats in the Conwy Estuary—was formed into a nature reserve by the RSPB under a lease with The Crown Estate (Hughes and Money 2016). It demonstrated that simple wetland habitats could be created with relative speed, a model that has subsequently been followed on a larger scale across the UK at former brownfield sites.

A much larger wetland was created in 1999 in the Gwent Levels, as legal compensation for the loss of the Taff-Ely Estuary SSSI, East Glamorgan, during construction of the Cardiff Bay barrage (Dalrymple 2020). The reedbed and grassland at Newport Wetlands and the lagoons and scrub at RSPB Conwy by no means replace the intertidal habitats destroyed by development, but have provided both managed breeding habitat for a range of wetland species, including Bittern, Bearded Tit and Avocet at the former site, and facilities to enthuse and educate future generations of birdwatchers.

Having finally acquired Ramsey in 1993, some 60 years after initial discussions, the RSPB set about restoring its former seabird colonies. First to go was the feral cat population, rehomed on the mainland with the help of the Cats Protection League and the RSPCA, and in 2000, Brown Rats were eradicated from the island and the offshore islets. It became the second Welsh island whose seabirds benefited from such intervention, after the RSPB and Countryside Council for Wales (CCW) had cleared Ynys Seiriol/Puffin Island, Anglesey, of rats in 1998–99. Death duties forced the Dale Castle Estate to sell Skokholm after 260 years of ownership, and the Wildlife Trust of South and West Wales was able to purchase the island after raising £650,000 in 12 months, and later a further £110,000 to renovate the buildings and quay and a further £100,000 for the lighthouse. In 2014 it re-established the Bird Observatory on the island that had ceased operation in 1976.

As we entered the 21st century, opportunities for large-scale nature reserve acquisition in Wales were diminishing, but mindful of sea-level rise, conservation land managers were increasingly assessing potential to create new coastal habitat, either in its own right, or to protect areas farther inland from the sea. The topography of Wales limits the opportunities for this compared to England's east coast, but the National Trust has undertaken a small-scale managed realignment at Cwm Ivy Marsh, Gower, where land, claimed from the sea in the 17th century, was allowed to revert to saltmarsh through the deliberate breach of the sea wall.

Campaigns for stronger legislation

The scale of killing of wild birds in the 19th century, to protect game interests and to furnish the plumage trade, as well as extensive trapping for the market in caged birds, led the fledgling animal welfare and conservation organisations to demand greater legal protection for birds. The *Sea Birds Preservation Act 1869* was the first nature protection law in the world, a result of extensive lobbying by the Association for the Protection of Seabirds (Barclay-Smith 1959). It was followed by the Wild Birds Protection Act 1896, which gave powers to County Councils to protect birds, their eggs or particular habitats. The RSPB lobbied councils to introduce the necessary Orders, but even where these were in place, local government resources were rarely available to enforce the legislation, so the RSPB provided Watchers wherever it could. The Protection of Animals Act 1911 made it illegal to use poisons and poisoned baits in situations that put wild birds at risk. This and the Pole-trap Act 1904 were the first national legislation aimed at protecting birds of prey. The RSPB's founding campaign resulted in the Importation of Plumage (Prohibition) Act 1921, although Etta Lemon wrote in the RSPB annual report that "it is impossible to say that the Act is a wholly satisfactory one" and it would take several years more of amendments before the sale of feathers was completely banned (Boase 2018).

Landmark legislation came after the Second World War, with the 1949 Act that created the Nature Conservancy, National Nature Reserves and Sites of Special Scientific Interest. The passing of the Protection of Birds Act 1954, "bringing to an end well over thirty years of miserable procrastination" (Anon. 1954), for the first time made it illegal to kill or injure (most) wild birds, their nests and eggs. This was a private member's bill promoted by Lady Tweedsmuir—the longest bill on record—rather than government-sponsored legislation (Evans 1997). The 1954 Act also allowed for the creation of sanctuaries, under which additional protection was given to all birds and their eggs at some sites. Sanctuaries in the Burry Inlet, Gower/Carmarthenshire, created in 1969, and Cleddau Estuary, Pembrokeshire, created in 1970, remain in place as Areas of Special Protection under the Wildlife and Countryside Act 1981 (not to be confused with Special Protection Areas below).

The 1981 Act (as amended) remains the backbone of wild bird legislation in Britain, securing more complete protection for all wild birds across Britain, as required by the European Community Directive on the Conservation of Wild Birds 1979. It also introduced protection measures for a broader range of species and greatly strengthened the protection afforded to SSSIs. The 'Birds Directive' required member states to designate Special Protection Areas (SPA) for migratory species and specific rare or vulnerable birds. In Wales, 18 SPAs have been designated in-country, covering almost 510,000ha. Another three—the Dee and Severn estuaries, and Liverpool Bay—cross the English boundary. SPA designation has provided an additional tier of protection by constraining built developments, since it requires compensatory habitat to be created if the SPA is damaged. It also places a requirement on government to improve the status of the species for which they are designated.

Subsequent changes came through the 'Habitats Regulations' (more properly known as the Conservation (Natural Habitats &c) Regulations 1994), which transposed the European Union Habitats Directive into UK law and required the designation of Special Areas of Conservation (SAC). Together with the SPAs, these were known as "Natura 2000" sites. Wales would go on to receive extensive EU-LIFE funding, focused on conservation action within Natura 2000 sites.

Designation as SPA, underpinned by the domestic SSSI legislation, has been important in protecting some important habitats from gross destruction, since owners, local authorities and utility companies are legally bound to consult with the Government's nature conservation adviser before sanctioning work. All SSSIs have plans designed to achieve favourable management of the species and habitats featured, but that does not mean the plans are followed or favourable conditions are achieved. The status of 55% of feature species in Natura 2000 sites in Wales is unfavourable, although the situation is better for the 43 bird species covered, with 86% in favourable condition. However, 75% of SAC habitats in Wales are in unfavourable condition, and some habitats are in a far worse state: for example, 92% of grassland SAC features are in unfavourable condition (Natural Resources Wales 2016). For SSSIs outside the Natura 2000 network, there has been no thorough assessment for 15 years (Countryside Council for Wales 2006).

In the 21st century came the Countryside and Rights of Way Act 2000 and the Natural Environment and Rural Communities Act 2006, which for the first time included a duty for public bodies to protect and enhance biodiversity. Creation of the Welsh Government, through a successive series of Acts following a referendum in 1997, transferred management of most environmental and agricultural legislation and policy from Westminster to the Senedd Cymru/Welsh Parliament. This allowed pursuit of a different agenda to other parts of the UK, although to date, the principal nature protection legislation has remained unchanged. The Environment (Wales) Act 2016 introduced some administrative changes to nature conservation, and while the Well-being of Future Generations (Wales) Act 2015 was heralded as world leading, to date it has made minimal material difference to the status of birds or other wildlife.

This section does not pretend to be anything other than a cursory summary of the laws protecting birds and their sites in Wales. While, during 1973–2019, the European Union was the primary driver of legislation, it was public campaigning—and the hard work of individual MPs, peers (and latterly Members of the Senedd)—that earned the additional protection and the closure of a number of loopholes.

Campaigning for better land and marine management

Strong bird-protection laws and well-managed nature reserves have been critical to nature conservation, but have not prevented significant declines in birds and other wildlife over the last century. Technological "improvements" to farming and fishing accelerated during and after the Second World War, without full appreciation of their impact on nature. A wake-up call came when it became evident, during the 1960s, that organochlorine chemicals used on crops were leading to the deaths of huge numbers of small birds, and in turn, predatory birds at the top of the food web. The impact of these pesticides was probably less significant in Wales as a whole than in England, since livestock farming is more dominant here, but it was the first time that high-quality science—conducted by both government and the growing non-governmental sector—was used to inform and persuade decision-makers on bird conservation. Organisations such as the RSPB, reasserting its roots as a campaign organisation for the first time since its early days, backed this with emotional public advocacy.

Now, of course, charities such as the RSPB and Wildlife Trusts Wales are at the forefront of civic society's influence on public policy, but limited resources constrain their ambition to speak up for nature. Built development is a very obvious cause of destruction to habitats, and during the 1960s and 1970s, estuaries were under particular threat. Even by the 1980s, with improved legislation, intertidal habitats were at real risk. In 1985–86, for example, the RSPB commented on 140 developments across the UK relating to estuaries, prompting a major public campaign to highlight the threats. On Welsh estuaries, proposals included a barrage, port development, dredging and road proposals on the Dee, Flintshire; a marina on the Mawddach, Meirionnydd; and barrages on the Severn in Gwent, Loughor in Carmarthenshire, Usk in Gwent and Taff/Ely confluence in East Glamorgan.

The last of these was the subject of a major campaign in the 1990s by the RSPB and Friends of the Earth, who took the case to the European Court of Justice. The ECJ authorised the flooding of the estuaries by a Cardiff Bay barrage, so long as 'compensation' was provided. Intertidal compensation proved unachievable in the vicinity of two major cities. As an alternative, the Government created Newport Wetlands Reserve on the Gwent Levels, which Friends of the Earth director Tony Juniper (later chair of Natural England) described as "like knocking down the Tower of Pisa and building a cinema and calling it compensation" (BBC News 1999).

The Gwent Levels were to be the subject of another major battle 20 years later, when the Welsh Government proposed a Relief Road to reduce congestion on the M4. The route backed by the planning inspector crossed seven SSSIs and farmland on which Cranes had started to nest. A 'Save the Levels' public campaign, backed by all the major conservation NGOs and Sophie Howe, the Future Generations Commissioner appointed by the Welsh Government, argued for more sustainable transport options. In June 2019, First Minister Mark Drakeford announced that the road would not be built.

To counter the intensification of farming encouraged by public subsidies, the RSPB and others lobbied for agri-environment schemes under which farmers would be paid to take action that benefits wildlife. In response, government introduced the first Environmentally Sensitive Areas from 1987 and the Countryside Stewardship Scheme from 1991, with add-on schemes for moorland and organic conversion in 1995. Political devolution provided an opportunity to develop a Welsh approach to agri-environment schemes, with Tir Gofal during 1999–2011, which was replaced by Glastir.

Priority species and habitats

Such has been the paucity of financial resources available to conservation organisations, both charities and statutory, that for most of their history, bird conservation priorities were defined by the enthusiasm of ornithologists. That is not a criticism, for without it we would probably have lost Red Kite as a breeding species from Britain, and many more of Wales' seabird colonies too. However, as monies from taxpayers or donors were deployed, an evidence-led approach became more desirable. From the 1960s, this was done at an international level, through a series of

Chough is a priority species in Wales, an amber-listed bird of conservation concern © Bob Haycock

global and regional IUCN Red Lists of Threatened Species, which evaluate the risk of extinction of all taxonomic groups.

A milestone for bird conservation came when the RSPB and the Nature Conservancy Council produced a Red Data List of birds in Britain (Batten *et al.* 1990), which was published as a book available to the growing community of birders and volunteer ornithologists. This assessment of the risk of extinction set the agenda for the next decade, although inevitably it focused on species that were relatively rare and did not include some of Wales' most rapidly declining species, simply because monitoring schemes, primarily the BTO Common Birds Census (1962–2000), did not adequately represent Welsh habitats.

A new assessment of birds in the UK and its Crown Dependencies came six years later (Gibbons *et al.* 1996). It introduced criteria that were considered to represent better the conservation priorities, and a Red–Amber–Green listing that has become widely understood by ecologists, politicians, birders and the wider public. The UK assessment has been repeated three times to date, most recently in 2015. A similar approach has been adopted in Wales, initiated by Thorpe and Young (2002) and most recently by Johnstone and Bladwell (2016).

A criteria-led approach to nature conservation priorities was also adopted by the UK Government as its response to the Convention on Biological Diversity (CBD), signed in Rio de Janeiro in 1992. While government was proud to flaunt its Biodiversity Action Plan (UKBAP) as a world first, it had been stimulated by a partnership of six environmental NGOs that formed a 'Biodiversity Challenge group'. Led by the RSPB, the coalition convinced ministers and officials that clear action plans would focus resources and effort on the species and habitats under greatest threat (Wynne 1993). For the first time, government and third-sector groups worked to a shared nature recovery agenda, with timescales, targets and plans. Action plans for the most-threatened species and habitats were set out to aid recovery. Reports produced every three-to-five years showed how the BAP was contributing to the UK's progress on significantly reducing biodiversity loss, as called for by the CBD.

Political devolution changed the dynamic of policy-making in the UK, especially for nature conservation, that was a devolved power in all three Celtic administrations. The new Welsh Minister for Environment signed up to a UK framework to maintain the momentum (UK Biodiversity Partnership 2007), but ever-tightening resources at the Countryside Council for Wales and its successor body, Natural Resources Wales, starved the initiative of the governmental co-ordination and support required to be successful, especially since the NGO community in Wales is relatively small.

Section 7 of the Environment (Wales) Act 2016 requires the Welsh Government to identify species and habitats of principal importance, "for the purpose of maintaining and enhancing biodiversity". For convenience, the initial list adopted was that devised under Section 42 of the Natural Environment and Rural Communities Act 2016, and included 51 birds. The populations of fewer than half of the species on the interim Section 7 list are stable or increasing.

Restoring the populations of declining species takes effort and resources. However, as the Red List of bird species grew longer, funding for nature conservation by government was reduced (Hayhow *et al.* 2019). The language of 'natural resources' presented nature for its essential ecosystem services, such as fresh water and climate regulation. Politicians—and some ecologists—in Wales use this language to argue for environmental

Colonies of seabirds, such as Gannet, have been the subject of long-term conservation effort. Conservation charities also campaign to protect their maritime feeding areas © Ben Porter

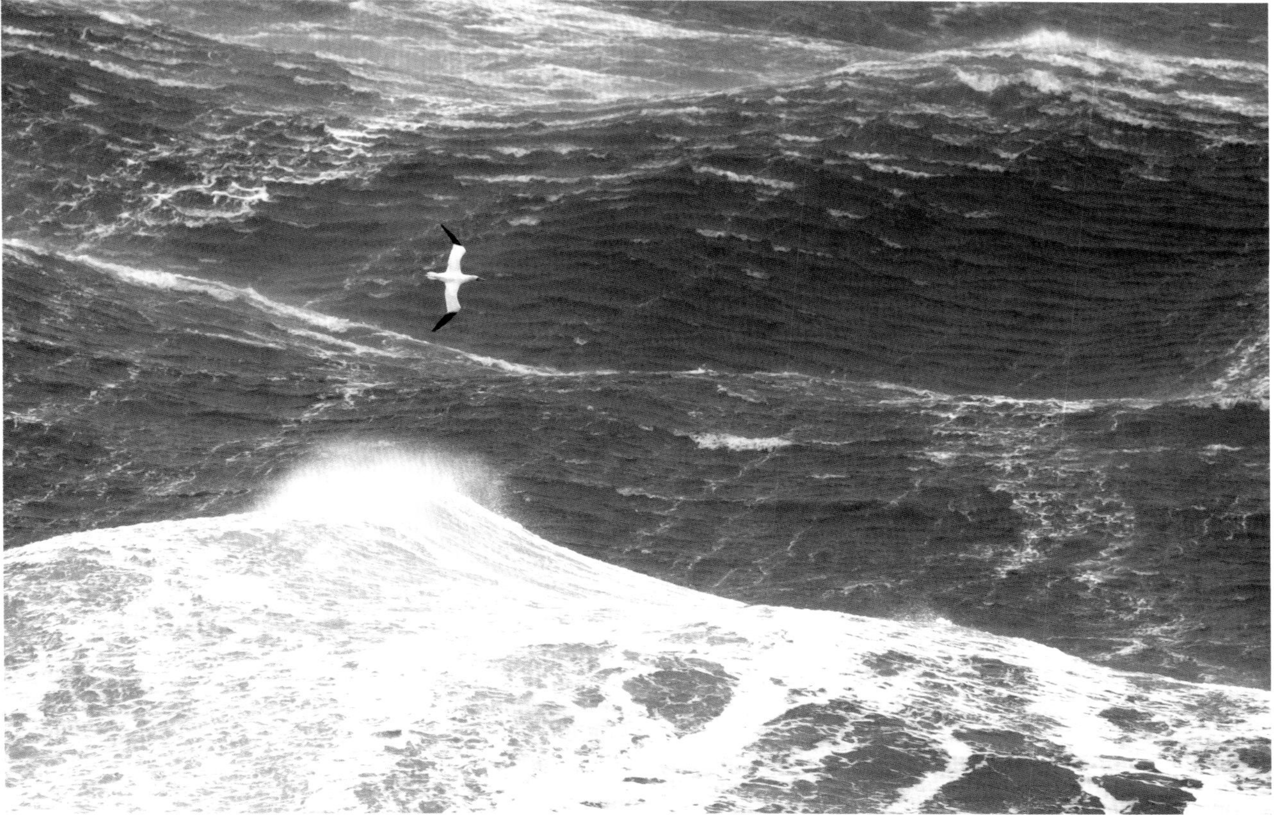

improvement. An Ecosystems Approach is enshrined in the Environment (Wales) Act 2016, through seven principles for the Sustainable Management of Natural Resources (SMNR). But while landscape-scale habitat recovery is essential, there is a real risk that the fine-grain needs of individual species will be ignored or lost. Landscape habitats cannot be divorced from their constituent elements: saving or creating a woodland will not, of itself, make Lesser Spotted Woodpeckers widespread again, and restoring blanket bogs in order to prevent further release of greenhouse gases will not alone secure a future for Curlew in Wales.

The nature conservation community in Wales

The Society for the Promotion of Nature Reserves was formed in 1912, initially to identify a list of Britain's finest wildlife sites for protection. The list, published in 1915 and known as the 'Rothschild reserves' after banker and entomologist Charles Rothschild, who masterminded the initiative, was produced through a survey sent to SPNR members and local natural history societies. It comprised 284 sites in Britain and Ireland, of which 21 were in Wales, including several for their bird interest: Ynys Llanddwyn, Puffin Island and The Skerries on Anglesey, Llandudno Sand Hills in Caernarfonshire and Worm's Head in Gower. Rothschild's intention was to fund the purchase of these sites, declare them nature reserves and hand their management to the National Trust under special conditions. But buying the sites proved complex, the National Trust went cool on the idea and the Government refused to get involved. In the end, the SPNR acquired only a handful of sites, but it did strongly influence the vision for National Parks and National Nature Reserves developed during and immediately after the Second World War. In the meantime, Naturalists' Trusts were set up in several English counties and sought co-ordination of their work, a role which the SPNR assumed from 1959.

The West Wales Field Society was described by E.M. (Max) Nicholson in 1945 as a "Naturalists' Trust in all but name" (per David Saunders), even though its formal affiliation did not come until later. Naturalists' Trusts were set up in Glamorgan (1961), West Wales (1962), North Wales (1963), Monmouthshire (1963), Brecknockshire (1964), Montgomeryshire (1982) and Radnorshire (1987), while the SPNR underwent several identity changes to become the Royal Society of Wildlife Trusts in 2004. Aside from a handful of salaried wardens, much Welsh trust activity was run by volunteers until the 1980s, and most staff teams remain small, focusing on managing 230 nature reserves and education work. Trust mergers in Wales during the first two decades of the 21st century reduced the total to five, with a small central team based in Cardiff that supports the Trusts and advocates nature policy to the Welsh Government, with marine conservation an area of particular strength. Birds have always been a focus for some Wildlife Trusts in Wales, perhaps because of the development of one of the early Trusts from the Pembrokeshire Bird Protection Society of the 1930s. Bird groups are associated with several county trusts *in lieu* of county bird clubs.

Wales became the last of the three Celtic nations to have a dedicated RSPB officer, with the appointment of Roger Lovegrove in 1970. He was given £200 to set up an office close to his home in Newtown and recruited Graham Williams as assistant regional officer in 1973. In the mid-1980s, Dr Stephanie Tyler joined the team as conservation officer and Iolo Williams as species officer. All four would contribute greatly to bird conservation in Wales. As well as promoting the RSPB through public events, such as film shows, their focus was on protecting threatened species and planning casework. Management of the RSPB's reserves in Wales would be devolved away from its Bedfordshire headquarters only later. As the challenges facing birds grew, so did support for the RSPB, and membership in Wales now exceeds 58,000 people. Since the 1990s, when the offices moved from Newtown to sites in Cardiff and Bangor, RSPB policy expertise in Wales has broadened to include agriculture, forestry, planning casework and marine issues.

As well as resurrecting the annual *Welsh Bird Report,* following its creation in 1989, the Welsh Ornithological Society also published an annual journal, *Welsh Birds,* from 1995, that was renamed *Birds in Wales* in 2010. This provided an outlet for both amateur and professional ornithological studies to be shared with the ornithological community of Wales. WOS has also organised a Conference annually since 1990, in recent years with the support of RSPB Cymru and BTO Cymru. It has campaigned on several issues during its third decade, including promoting a ban on

The Welsh Ring Ouzel population has contracted north into Snowdonia National Park during the last century © Ben Porter

hunting Greenland White-fronted Geese in Wales. The British Trust for Ornithology in Wales had been entirely run by volunteers, organising surveys, finding volunteers for monitoring schemes and successfully achieving the necessary tetrad coverage for the three breeding and two wintering atlases of Britain and Ireland. Recognising the need to grow both its supporter and science bases in Wales, the BTO now has both development and research staff in Wales, based at Bangor University.

Government nature conservation efforts were stepped up following assent of the National Parks and Access to the Countryside Act 1949, which created the Nature Conservancy, as well as Snowdonia, Brecon Beacons and Pembrokeshire Coast National Parks. It also created Areas of Outstanding National Beauty in Anglesey (1969), Wye Valley (1971), Gower (1956), Llŷn (1956), and the Clwydian Hills (1985), extended to the Dee Valley in 2011. The National Parks and AONBs are home to several priority Welsh bird species, including a good proportion of Wales' populations of Chough, Ring Ouzel, Whinchat, Pied Flycatcher, Wood Warbler and an assemblage of seabirds. For much of the Parks' 70-year history, most birds have fared little better within than outside National Parks. This is despite one of their two functions being to "conserve and enhance the natural beauty, wildlife and cultural heritage" and the Welsh Government's assertion that they should be exemplars of the sustainable management of natural resources (Welsh Government 2018b). There are signs of change, such as Snowdonia National Park's work on Sessile Oak woodland, under the EU-funded Celtic Rainforest Wales LIFE Project, launched in 2018, and Brecon Beacons Nature Recovery Action Plan that has adopted the 'Lawton Principles' (see below).

The Nature Conservancy was superseded by the Nature Conservancy Council in 1973, but the job remained the same: to notify Sites of Special Scientific Interest, manage National Nature Reserves and provide nature conservation advice to government. The NCC was broken into four country agencies in 1991, to the outrage of environmental NGOs. In Wales its functions were combined with those of the Countryside Commission to form the Countryside Council for Wales. The Joint Nature Conservation Committee (JNCC) was created to advise the UK and devolved governments on its international commitments, including the Natura 2000 protected site network. CCW governance was transferred from Westminster on the formation of the Welsh Government in 1998. With 500 staff and a budget of £45 million, CCW also had responsibility for public access to the countryside and implementing the Welsh Government's commitment to a Wales Coast Path, which opened in 2012. CCW had strong ecological capability and, within the small conservation community, worked closely with nature-oriented NGOs, in part to build their capacity within Wales.

As the Welsh Government entered its second decade, it sought a more joined-up approach to deliver its environmental and land management goals. Leading conservation thinker, Morgan Parry—founder of WWF Cymru and chair of CCW—saw opportunity when he wrote: "*A Living Wales* looks for a new contract between environmental managers and regulators, industry and commerce, marine stakeholders, landowners and the public. In doing this, Wales can reflect its deep historical and cultural links to the natural environment. We should find a distinctively Welsh approach, based on the best scientific evidence and latest thinking from around the world. We could be the first country to put it into practice" (Parry 2010).

The Government's chosen solution was to merge CCW with Environment Agency Wales and Forestry Commission Wales in 2013, with a staff of 2,000, an annual budget of £120 million and a landholding that comprised 7% of the land area. The merger, creating Natural Resources Wales, was presented by politicians as an opportunity to save money (£158 million over the first ten years). However, the cuts came unevenly across the new organisation, with political pressure to prioritise resources for flood defences, and then a procurement scandal involving timber sales. These incidents squeezed the resources available for nature conservation and, critics argued, reduced the body's ecological capability. Ultimately, NRW's reduced resources meant that all the promises and commitments made by politicians could not be delivered by an organisation that underwent a 32% cut in funding for its regulatory duties during 2013–20 (Rose 2020).

The Lawton Principles identified the need for more wildlife sites, and for existing sites to be bigger, better managed and joined up through habitat creation initiatives (Lawton *et al.* 2010). Although written for England, the principles have been adopted far more widely. *State of Nature* (Hayhow *et al.* 2019) showed there has been no significant improvement in the condition of nature since the 2016 report described the UK as "among the most nature-depleted countries in the world". Wales, along with the UK as a whole, failed to meet its 2020 international and national biodiversity targets. Does the public and political will exist to adopt the Lawton Principles in Wales? Will the resources be available to properly deliver the vision? And what will the outcomes be for birds in Wales? It will be down to the authors of a future avifauna of Wales to judge the success of bird conservation in the 21st century.

Julian Hughes

4. Bird trends

Due to short life-spans and faulty memories, humans have a poor conception of how much of the natural world has been degraded by our actions because our 'baseline' shifts with every generation, and sometimes even in an individual. What we see as pristine nature would be seen by our ancestors as hopelessly degraded, and what we see as degraded our children will view as 'natural'.

Jeremy Hance (2009)

It is only in recent decades that changes in populations of birds in Wales have been measured, thanks to the foresight of those who set up surveys to monitor wintering waterbirds from the 1960s and widespread terrestrial breeding birds from 1994. In ecological terms, of course, this is a narrow window, but they are the best data available. Accounts by naturalists in the 19th and early 20th centuries provide a flavour of the changes that have occurred since the late Victorian era, coinciding with the early days of the nature conservation movement described in Chapter 3. It is tempting to discount the earlier evidence and focus solely on the 50 years for which we have robust data. However, *State of Nature 2019* (Hayhow *et al.* 2019) emphasised that a great deal of biodiversity loss occurred before 1970. Among birds, this was particularly true of wetland species that were lost to habitat drainage and of predatory species, particularly raptors, which were trapped, shot and poisoned.

Changes since 1900

People inevitably judge the state of nature today by their own lifetime experience. Even where data are available, perceptions of the scale of change are often less than the reality (e.g. Papworth *et al.* 2009, Jones *et al.* 2020). An 80-year-old birdwatcher in 2020 will consider that many raptors are much more common than in their childhood and lament the absence of large overwintering flocks of granivorous songbirds. A 20-year-old birdwatcher, now, cannot imagine struggling to see a Sparrowhawk, but will not have seen a Corn Bunting in Wales.

Bringing together the information for this book has allowed an assessment of change to be made over a longer period, using descriptions in avifaunas written around the turn of the 20th century, even though information about abundance was rarely quantified. Each species has been assigned a tiered assessment of its status (Abundant, Locally Common, Scare, Rare or Absent) for each of the breeding and non-breeding seasons around 1900 and in 2020. In addition, it was noted where a species had already been lost as a breeder prior to 1900. Around 40% of species were considered to have a similar status now as in 1900. Around one-third of breeding species, and 44% of species outside the breeding season, have reduced in abundance by one or more category ('impoverished' in Fig. 4). The status of only 25% of breeding species, and 13% of non-breeding species, has improved over the last 120 years.

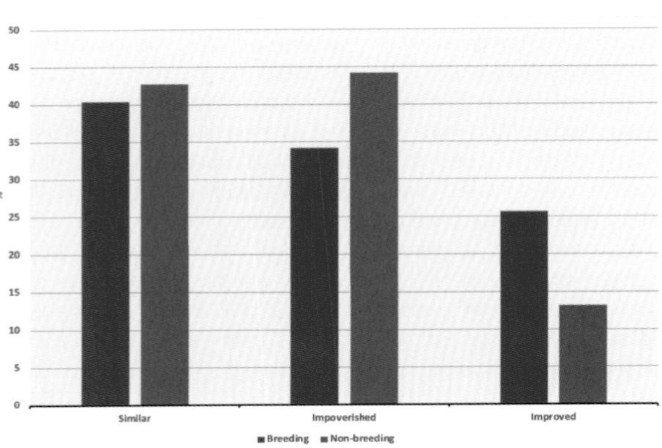

Figure 4. Status of all species in Wales during (n=176) and outside (n=197) the breeding season in 2020 compared to 1900, based on accounts in this book. Rare visitors are not included.

While the number of species with a more impoverished status is higher outside the breeding season (which includes passage and winter visitors as well as residents), the magnitude of the losses is greater for breeding species. For example, of the 49 breeding species that were considered abundant in 1900, 16 are now locally common, three are scarce and one (Corncrake) has been lost entirely as a breeding species from Wales. Of the 50 that were locally common in 1900, nine are now scarce, five are rare and two (Turtle Dove and Corn Bunting) have been lost. By contrast, 25 species that did not breed in 1900 now do so, including one (Collared Dove) that is abundant in Wales, having made a spectacular range expansion through Europe during the 20th century.

The list of species lost from Wales at some point during the last 200 years numbers a minimum of 16, although Honey-buzzard, Hen Harrier, Marsh Harrier, Goshawk and Bittern have all returned subsequently, the last of those for its second time. The Goshawk's return was a result of releases and escapes from falconry, though these are believed mainly to be of the physically larger subspecies *Accipiter gentilis buteoides* from northern Scandinavia, rather than

	<1875	1875–1900	1901–1969	1970–2019
Extirpated	Crane (*c.*1600) Bittern (*c.*1850) Goshawk (*c.*1800)	Marsh Harrier (1882)	Hen Harrier (1910–58) Montagu's Harrier (1969) Wryneck (1906) Red-backed Shrike (1952) Cirl Bunting (1968)	Turtle Dove (2011) Corncrake (*c.* 1973) Woodlark (*c.*1980) Nightingale (1981) Corn Bunting (2007)
Colonised		Tufted Duck (1892) Hawfinch (*c.*1900) Siskin (1899) Cirl Bunting (1875)	Garganey (1940) Gadwall (1969) Pochard (1905) Red-breasted Merganser (1953) Collared Dove (1961) [Black-necked Grebe 1904–57] [Common Gull 1960–80] Sandwich Tern (1915) Fulmar (*c.*1940) [Bearded Tit 1967–86]	Greylag Goose (1970) [Pintail 1988–2004] Eider (1997) Goosander (1970) Little Ringed Plover (1970) Avocet (2003) Little Egret (1996) Mediterranean Gull (2010) Honey-buzzard (1991) Osprey (2004) Bearded Tit (2005) Cetti's Warbler (1985) Dartford Warbler (1998) Firecrest (1977)
Returned			[Bittern 1947–84] Goshawk (1969)	Marsh Harrier (2014) Bittern (2016)

Table 3. Native species (and year) extirpated, colonised and re-established, excluding those that nested fewer than five times or only sporadically (e.g. Hoopoe, Black Redstart, Marsh Harrier between 1882 and 2014). Populations in square brackets were not sustained beyond the period of colonisation. It is recognised that colonising species prior to 1875 may be under-represented, but figures are based on known records.

the nominate *A.g. gentilis* that was rendered extinct in Wales around 1800. Western Yellow Wagtail could easily join the list of losses in the near future, with as few as 20 breeding pairs in recent years.

In numerical terms, the nine species lost, apparently permanently, since 1900 are counter-balanced by 21 colonists that have become established. Breeding by Black-necked Grebe, Common Gull and Pintail was localised and short-lived during the 20th century, and nesting by Crane in southeast Wales during 2015–19 is not included in the table, as it is not yet clear whether the reintroduction in Somerset, from which the adults originate, is self-sustaining.

Changes since 1970

The supreme effort of volunteers, co-ordinated in Wales by the British Trust for Ornithology, has enabled three breeding atlases and two winter atlases to measure changes in distribution in Britain and Ireland over a 40-year period. Among the species that have shown the greatest gains are raptors recovering from historical persecution, and species for which active conservation programmes have been undertaken, several associated with lowland wetlands and others that are not native.

Species gains	% change	Species losses	% change
Canada Goose	600	Corncrake*	-98
Siskin	597	Turtle Dove*	-96
Red Kite	468	Woodlark*	-95
Peregrine	324	Grey Partridge	-78
Red-legged Partridge	280	Woodcock	-78
Reed Warbler	238	W. Yellow Wagtail	-72
Quail	182	Black Grouse	-68
Hen Harrier	179	Redshank	-67
Lesser Whitethroat	124	Tree Sparrow	-66
Little Grebe	100	Short-eared Owl	-62
Tufted Duck	84	Long-eared Owl	-57
Great Crested Grebe	83	Black-headed Gull	-52
Lesser Black-backed Gull	63	Ring Ouzel	-50
Chough	50	Lapwing	-46
Rock Dove (Feral Pigeon)	49	Snipe	-46
Stonechat	43	Red Grouse	-45
Mute Swan	37	Little Owl	-44
Herring Gull	33	Willow Tit	-41
Collared Dove	33	Curlew	-39
Shelduck	32	Merlin	-35

Table 4. Species showing the greatest gains and losses in occupancy of 10-km squares in Wales during the breeding seasons between 1968–72 and 2008–11 (data provided by the British Trust for Ornithology). Those marked* are now extinct as a regular breeding species in Wales.

Twelve of the species experiencing the greatest losses in range are associated with lowland farmland or upland farmed habitats, and as noted above, the three with the greatest losses are now effectively extinct as Welsh breeding species, along with Nightingale and Corn Bunting. The list also includes woodland species, such as Willow Tit and Woodcock, that have declined rapidly in Wales, as across Britain. The marked decline in Black-headed Gull, however, appears to be a Welsh phenomenon that is not evident in the rest of the UK. There were 33 species recorded with breeding codes during 2008–11 that were absent in 1968–72. Half of these were introductions of non-native species, though only populations of Barnacle Goose and Mandarin Duck have become established.

Native species		Introduced species	
Species	**10-km squares**	**Species**	**10-km squares**
Cetti's Warbler	47	Mallard (domestic)	34
Little Egret	27	Mandarin Duck*	24
Dartford Warbler	19	Greylag Goose (domestic)	12
Osprey	12	Muscovy Duck	10
Honey-buzzard	11	Indian Peafowl	6
Firecrest	9	Black Swan	5
Eider	7	Barnacle Goose*	4
Marsh Harrier	5	Ruddy Duck*	4
hybrid Carrion x Hooded Crow	5	Ring-necked Parakeet	3
Hooded Crow	3	Reeves's Pheasant	3
Mediterranean Gull	2	Helmeted Guineafowl	2
Black Redstart	2	Wood Duck	1
Bearded Tit	2	Black Duck	1
Black-necked Grebe	1	Lady Amherst's Pheasant*	1
Whooper Swan	1	Eagle Owl sp.	1
Pintail	1	Falcated Duck	1
Avocet	1	South African Shelduck	1
		Chiloé Wigeon	1
		Zebra Finch	1

Table 5. Number of 10-km squares occupied during 2007–11 by species that had not bred in Wales in 1968–72 (excluding vagrants that held territory). Species marked * established breeding populations between the two periods, although Ruddy Duck and Lady Amherst's Pheasant no longer breed in Wales.

Barnacle Goose is also the waterbird species that has shown the biggest increase in abundance outside the breeding season since 1970. Colour-ringing has demonstrated that these are primarily from a naturalised population that commutes seasonally between Cumbria and West Wales. Of the species monitored by the BTO/RSPB/JNCC Wetland Bird Survey and its predecessor schemes since 1970, only one wader species, Black-tailed Godwit, is among the ten showing the greatest population increase, whereas waders comprise seven of the top ten declining species.

Gains	% change	Losses	% change
Barnacle Goose	10,800	Bewick's Swan	-88
Black-tailed Godwit	10,100	Bar-tailed Godwit	-85
Greylag Goose	9,900	Pochard	-83
Brent Goose	9,900	White-fronted Goose	-69
Gadwall	4,800	Grey Plover	-64
Canada Goose	1,920	Dunlin	-60
Goosander	600	Ringed Plover	-58
Mute Swan	431	Knot	-41
Shelduck	292	Turnstone	-38
Shoveler	285	Golden Plover	-28

Table 6. Non-breeding waterbird species that recorded the greatest increases and declines in abundance in Wales, winter 1970/71 to winter 2018/19 (from Frost *et al.* 2020).

Changes since 1995

While the breeding season and winter atlases provided a periodic snapshot of all bird species in Wales, changes in abundance have been measured only for a much smaller group of species. The BTO/JNCC/RSPB Breeding Bird Survey is currently able to produce trends for 60 species in Wales, and is used to produce the wild-bird indicator for Wales that combines the trends for all species monitored annually. The trend for all species monitored has been fairly stable for 25 years but it masks major changes. For example, a suite of 29 woodland species has increased since around 2006 but most of these are widespread and abundant, whereas many of the more specialist woodland birds, not included in the index, are in decline. On lowland farmland, an index of 13 species associated with lowland farmland fell by 28% and that of 20 species of upland farmed habitat fell by 20% during the same period (Bladwell *et al.* 2018).

Gains	**% change**	**Losses**	**% change**
Red Kite	413	Swift	-72
Canada Goose	359	Greenfinch	-71
Stonechat	191	Curlew	-69
Great Spotted Woodpecker	189	Starling	-65
Blackcap	143	Yellowhammer	-64
Goldfinch	104	Rook	-58
House Sparrow	92	Goldcrest	-54
Siskin	84	Wheatear	-48
Stock Dove	54	Magpie	-43
Chiffchaff	54	Chaffinch	-38

Table 7. Breeding species monitored by the Breeding Bird Survey that recorded the greatest increases and declines in abundance, 1995–2018 (from Harris *et al.* 2020).

The success of the Red Kite's recovery in Wales is reflected in its position at the top of the rankings, having increased by over 400% since 1995. It is also encouraging that Stock Dove and House Sparrow are among the winners, the latter in stark contrast to the UK as a whole. However, the number of widespread and abundant species that increased is almost outweighed by those that have declined, including four that are dependent on the farmed landscape: Curlew, Starling, Yellowhammer and Rook. Only two summer migrants are among the ten species showing the greatest declines, but the scale of decline for Swift and the speed of decline for Wheatear, most of which has occurred in just five years (2013–18), are shocking. Of critical significance is that these trends relate only to the most abundant species. Many of those that have suffered the biggest declines, including Lapwing, Marsh Tit and Kestrel, are already too scarce for BBS annual trends to be produced in Wales. Generally, common species have become more abundant and the less common have become rarer.

Birds of Conservation Concern Wales

The multiple sources of monitoring data are brought together every few years in an assessment of the conservation status of birds in Wales. This assessment uses systematic criteria to produce Red, Amber and Green lists, shown in the table at the start of each species account in this book. Changes in the criteria between each assessment mean that the figures are not directly comparable, but in each review since 2002, declines in abundance and/or range resulted in a net increase in the number of species on the Red List. The most recent assessment, in 2016, found that birds of coastal, farmland and upland habitats had the greatest proportions (70–93%) of their total on the Red and Amber lists, while around half the birds of lowland wetland, woodland and urban habitats (50–52%) were on the Green List.

	2002	**2010**	**2016**
Red	27	46	55
Amber	68	99	89
Green	125	67	69

Table 8. Numbers of species in each category in *Birds of Conservation Concern Wales* (from Thorpe and Young 2002, Johnstone *et al.* 2010a and Johnstone and Bladwell 2016).

What level of nature recovery should we be striving for? There is no prescriptive answer, and it seems an academic question when the Red List lengthens and even half of the most abundant species are in decline. To avoid Shifting Baseline Syndrome, *Birds of Conservation Concern Wales* includes a criterion of Historic Decline and uses data available from the longest period available (up to 46 years at the time of the 2016 assessment). The information in this book judges the state of birds in an even longer timeframe, of more than a century. As the next two chapters discuss, it will require substantial changes in land management, and mitigation against climate change, to achieve recovery of some of our most rapidly declining species.

Julian Hughes

5. What has changed for birds in Wales?

Land management before the mid-20th century

Around 15,000 years ago, as the glaciers of the last Ice Age melted, western Europe's fauna and flora were typical of today's northern tundra: open landscapes with low or sparse vegetation. Woodland then became established in the valleys and, from around 10,000 years ago, in the uplands. The first trees were birch, and after another 1,500 years, oak and, in some places, pine (Rhind and Jones 2003). Around the same time, humans used caves in Wales as bases from which to hunt birds and mammals. Periods of climatic variation resulted in fairly rapid changes to wildlife, including the extinction of some of the large vertebrates, such as Irish Elk, and later Reindeer and Wild Horse.

The Late Glacial and Mesolithic periods marked the start of the modern avifauna of Britain. Archaeological finds indicate that the small human population was already consuming wild birds as food. Such consumption is likely to have had a negligible effect on the populations of these species. The pressures on bird populations from then until perhaps 1,000 years ago were mainly natural, and changes were relatively slow once the rapid climate change of the Late Glacial was over.

As the climate warmed, extensive woodland grew quickly. This pattern dominated for the next few thousand years, except at the very highest elevations in northwest Wales: the natural tree line is usually held to be about 650m above sea level (Newton 2020). From the arrival of settled farming around 6,000 years ago, woodlands began to give way to agriculture, as the human population increased. Sir John Wynn, writing of the start of the 15th century, wrote that "in those dayes, the countrey of Nantconway was not onely wooded, but alsoe all Carnarvon, Merioneth, and Denbigh shires seemed to be but one forrest haveing few inhabitants" (Wynn 1770). From the start of the Neolithic Period, about 5,000 years ago, farming became more widespread, "initially a form of shifting cultivation in which forests were cleared locally and, for several years, crops were grown in the clearings" (Rhind and Jones 2003). Given its topography, the northwest was probably the last part of Wales to face the axe. Peat bogs started to form around this time, partly due to climatic change, but hastened by human clearance of woodland, the landscape kept open by both wild mammals and early livestock (Moore 1973).

From this point, the fortunes of the birds of Wales have been shaped by humans. Each species account in this book provides a historic overview, where this has been recorded (see Chapter 2). Roman and Anglo-Norman invaders drove roads through the woodlands of Wales and expanded farming, in the case of the latter through a network of landed estates. By the 12th century even the remotest parts of Snowdonia started to be systematically cleared (Rhind and Jones 2003), although John Leland described the heights of the main range (Craig Eryri) as being forested in 1536–39 (Toulmin Smith 1906). The next significant changes to topography occurred from the late 18th century through to the mid-20th century, when rivers and wetlands were managed on a scale that had not been possible hitherto. Wetlands were drained and embankments in the lower reaches of rivers were constructed to protect farmland and settlements from flooding. In some places, sea walls ('cobs') were built to allow land to be claimed from the sea for farming. Later, lowland wetlands were drained for industrial development, especially along the South Wales coast. Suitable areas for wetland birds, both breeding and wintering species, greatly diminished during this period.

Until the middle of the 19th century, much of Britain's farming was an integrated rotation of mixed arable and livestock, usually with a year or two of fallow to enable the fertility of the soil to be restored. Farming was small scale, the land mostly leased from large estates. While Wales was dominated by livestock, particularly in the mountains, a substantial amount of land in Wales was tilled. Although there are no bird data from this period, it may well have been the heyday for many of our 'farmland birds', particularly those in the lowlands. Small-scale, mixed farming with only natural inputs (farmyard manure) and weedy leys, would have provided food for birds throughout the year, and some species would have been attracted to seeds spilt from hayricks around farmyards.

The repeal of the Corn Laws in 1846 and expansion of mining, industry and the railway network, enabled more rapid change, including bulk ships that could import wheat from abroad, and centralised flour mills in the counties that grew most grain. Livestock farms benefited from cheaper grain, and so had no need to grow their own. Changes in the landscape were not immediate, and farming in much of Wales remained a mix of livestock and arable. In 1875, 41–70% of the land area of Anglesey, Denbighshire, Flintshire, Montgomery, Ceredigion, Gower and East Glamorgan was arable and ley (Shrubb 2003b). For several centuries, livestock were moved seasonally between lowlands and uplands in a system known as 'transhumance' and reflected in the farming residences of *hafod* (the summer dwelling) and *hendre* (permanent farmhouse). Sheep, in particular, would initially have been shepherded, but in time grew acclimatised to their "heft", a trait passed down the generations. By the First World War, the movements were made over much longer distances by rail, and then by road, and few farmers would rear all their stock from birth to the abattoir.

Significant social change was underway, as the great landed estates that had existed in some form since the late 15th century were sold off by owners who wanted to consolidate their holdings in England. In 1887, 90% of farmland in Wales was tenanted; by the 1970s that had fallen to 38% and there was a much greater level of ownership than in England (Davies 1974). The development of road transport made it more efficient to transport liquid milk from rural areas to urban markets, so dairy herds expanded farther away from the cities. The mechanisation of farming from the 1920s, the increased marketing and use of chemical and nitrogen inputs from the 1930s, the practice of ploughing and re-seeding pasture from the 1940s, and the growing of silage instead of hay-making from the 1960s, all combined to intensify agricultural production.

Farming had expanded at the cost of woodland. In 1919, woodland in Wales almost certainly covered less than 4% of the land area, a proportion that had been falling since at least the Middle Ages, and probably longer (Linnard 1982, Dyda *et al.* 2009). While management was evidently not always sustainable, woodlands were shaped by their use for grazing livestock, harvesting timber, coppicing for charcoal, hunting animals and gathering fruits and fungi. The trees were used as a source of hurdles, poles and tan bark. Bracken, leaf litter and branches were collected for animal bedding and fodder. The First World War highlighted the military and industrial need for a domestic supply of quick-growing timber. The creation of the Forestry Commission in 1919 led to a significant increase in planting, mostly of non-native conifers. This accelerated after the Second World War.

Management for shooting, principally Red Grouse and Black Grouse, but increasingly for released Pheasants, developed during the second half of the 19th century and reached its peak in Wales just before the First World War. It was achieved through heavy levels of persecution that affected populations of many birds of prey and corvids. During the same period, egg-collectors raided the nests of some of the rarest birds, including Red Kite. In the decade following the First World War, many of the largest family estates in Wales were sold and broken up. It is estimated that at least one-quarter of the land in Wales was sold by estates, mostly to their tenants, in 1918–22 (Davies 1974). For this reason, and because most moors were climatically less suitable than those elsewhere, driven grouse shooting declined in Wales compared to England and Scotland. However, these land sales opened the door to afforestation and higher numbers of sheep in the uplands.

Causes of change since 1950

As the authors of *Birds in Wales* (1994) describe in the preface to this book, changes in farming practices, driven by government policies, consumer demand and technological efficiency, lay behind the biggest changes to birdlife in Wales during the last 70 years. These changes were limited neither to Wales nor to birds. Burns *et al.* (2016) showed that intensive management of agricultural land was the primary driver of declines in abundance of 400 species of various taxa in the UK during 1970–2012. Climate change was the second most significant factor in declines, but it was the primary driver of population increases during that period.

Inspired by the approach of Burns *et al.* (2016), we reviewed the species accounts written for *The Birds of Wales* to identify the themes cited as the causes of change since around 1950. Agriculture has had, by far, the biggest influence on the birds of Wales, accounting for 41% of the references to declines, and only a handful of increases. Climate change is the second-biggest driver of decline (17%) and, like farming, is a broad heading for multiple causes of change. Collectively, the planting, management and operations relating to conifer afforestation have had a similar impact to climate change, but the changes have differed over time. There are limitations to this, but we suspect that they are unlikely to change the overall picture for the following reasons:

(1) account authors have cited many studies that have investigated declines in populations, but were not expected to be exhaustive in their reviews;
(2) limited resources have been directed to understanding the ecology or drivers of change of declining species, so there are fewer studies of populations that have been stable or have increased; and
(3) worrying numbers of species have declined in the last 20 years, on which very little research has been undertaken, in Wales or elsewhere.

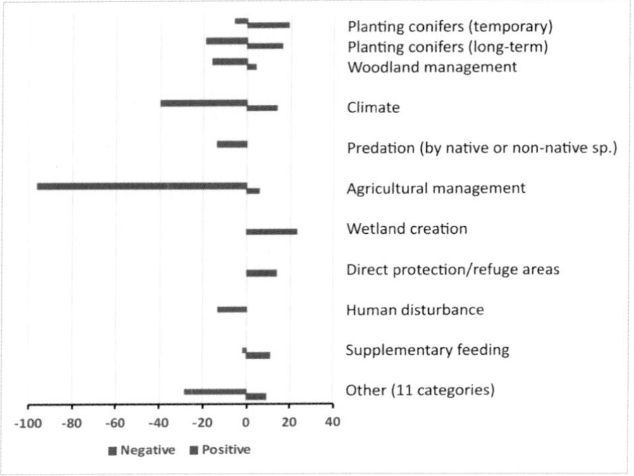

Figure 5. Factors cited as evidence of population change in Wales, based on species accounts within *The Birds of Wales* (n=127).

Before exploring the individual factors, it is worth placing the impacts of these changes in Wales in a wider context. Globally, a lack of long-term data for most wildlife taxa led to the development of a Biodiversity Intactness Index (BII), which described the state of each country's wildlife habitats for which data were available (Steffen *et al.* 2015). Of a maximum possible score of 100%, the authors considered that ecosystems may no longer reliably meet society's needs below a score of 90%. The UK's BII score, of 81%, placed it 189th of the 218 countries ranked. Wales was the second-best of the four UK nations, with a BII of 51%, but at 224th of 240 territories ranked, it was in the bottom 10% of the global league (Natural History Museum 2020).

Agriculture

Some 88% of Wales is farmed, a higher proportion than the UK average (70%), and so it is not surprising that management for agriculture has had a significant effect on the nation's birds. Around half of the farmland is permanent grassland and another 22% is rough grazing. Around 12% is grass ley that is less than five years old and 8% is cropped, of which half is non-cereal crops, such as Oil-seed Rape. The remainder is crops grown to feed livestock (such as Maize), cereal, horticulture and potatoes. Cereal accounts for a little over 2% of Welsh farmland, predominantly in the southeast and the Marches, but with some in Pembrokeshire, where potatoes are also prevalent (Armstrong 2016).

The numbers of sheep in Wales increased three-fold between 1950 and 1999 © Ben Porter

Around 79% of the pasture lies in the Less Favoured Area (LFA), a term used to describe places where land, climate and cultivation conditions are poor for farming. Around 10% of Welsh farmland is common land, used by multiple farmers who have grazing rights, and much of this is also in the LFA. Wales is grazed by 9.6 million sheep, primarily for their meat, and just under 1.1 million cattle, the majority of which are bred for dairy production.

The Second World War was a social and economic landmark, as well as a military one, which had major consequences for the birdlife of Wales. Demands for food security, in light of agricultural recessions of previous decades, led to the UK Government passing the Agriculture Act 1947. It guaranteed prices for crops and livestock and provided grants for land 'improvements' such as land drainage, ploughing old grassland, fencing and liming of fields. Grants and production subsidies, including headage payments per animal, continued under the Common Agricultural Policy (CAP) once the UK had joined the EEC in 1973. Thus, for more than 70 years public subsidies have warped the economics of farming and encouraged management that has had a dramatic effect on birds. The policies set a floor below which commodity prices would not fall, and this has made it difficult to influence changes that would benefit nature.

Since 1970, changes in populations of many farmland birds in the UK have been measured, although it was the mid-1990s before such data were available separately for Wales. The UK farmland bird index (comprising Grey Partridge, Woodpigeon, Stock Dove, Turtle Dove, Lapwing, Kestrel, Jackdaw, Rook, Whitethroat, Skylark, Starling, Tree Sparrow, Western Yellow Wagtail, Greenfinch, Goldfinch, Linnet, Reed Bunting, Yellowhammer, Corn Bunting) declined by 54% between 1970 and 2019, and while the

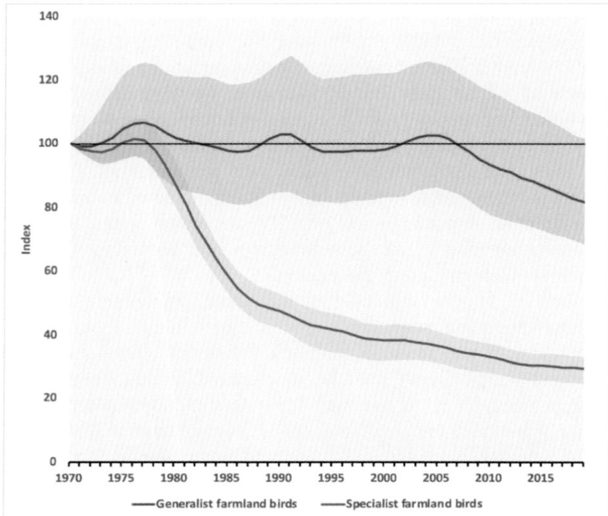

Figure 6. UK indicator of farmland birds, 1970–2019. The shaded areas show the 97.5% confidence limits (DEFRA 2020).

most rapid decline occurred during 1975–85, it continued to fall in the subsequent 30 years.

The figures for Wales look no better. It is impossible to know what an index for Wales would have looked like prior to 1995, but there is a strong indication from distribution changes between the *Britain and Ireland Breeding Atlas 1968–72* and that of 2008–11. More than half of species that underwent range contractions in Wales during this period (Table 4, Chapter 4) are farmland birds, including Corncrake (-98%), Turtle Dove (-96%) and Grey Partridge (-78%), while the Corn Bunting no longer breeds in Wales.

The Wales index for birds of farmland habitats remained relatively stable during 1995–2005 but fell by 24% during 2006–16 (Bladwell *et al.* 2018). Populations of species associated with lowland farmland fell even further, by 30% over the same period. The species listed above, and others such as Lapwing, were already too rare by 1995 to be included in the index. While farming intensification has, in general, had a negative impact on birds in Wales, this is not true for all species. For example, pasture managed for livestock provides additional foraging habitat for wintering wildfowl such as Whooper Swan, Wigeon and Pink-footed Geese. In the uplands, some level of grazing is necessary to produce patches of short grass for species such as Chough, Wheatear, Ring Ouzel, Curlew and Golden Plover to forage. The extent and intensity of grazing is crucial: high stocking densities over a large area remove the diversity of vegetation or longer swards in which the last three species hide their nests.

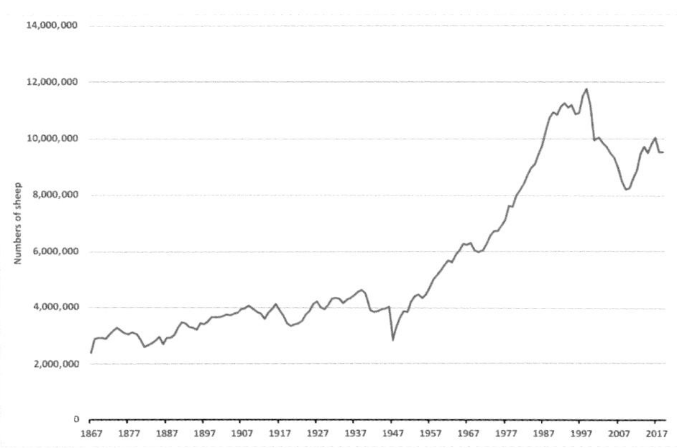

Figure 7. Number of sheep and lambs in Wales, 1867–2019 (Welsh Government 2019a).

No single factor caused declines in birds of the uplands and enclosed farmland, but grazing is a common element to many. Sheep numbers doubled between 1867 and 1939, doubled again between 1948 and 1970, and had doubled again by 2000. Numbers fell during the foot-and-mouth disease outbreak in 2001 and numbered 9.5 million in 2019 (Welsh Government 2019a). Nonetheless, sheep densities in Wales are the highest in Britain, at 240–460 animals/km^2 across Denbighshire, Ceredigion and the Cambrian Mountains (APHA 2019). The increases in sheep numbers and grazing densities were only made possible through large-scale habitat change, starting with the drainage of marshy low ground and peaty uplands. Wetland species such as Shoveler, Teal, Lapwing, Snipe, Redshank, Golden Plover and Curlew lost significant breeding habitat as a result of drainage for grazing (and forestry), while drainage of damp habitat in lowland fields is likely to have affected species such as Song Thrush, Western Yellow Wagtail and Reed Bunting. Drained land could support a higher density of sheep and became accessible to machinery that, over time, could plough, lime and reseed the grassland. Within several decades, large areas of rough grazing were transformed from permanent grassland, with a diverse topography and mix of flora, to rye-grass leys. These support a far lower abundance of invertebrates and less cover for small mammals and nesting passerines, which had previously supported Hen Harrier, Merlin, Barn Owl and Short-eared Owl.

However, the changes are more nuanced than simply the number of sheep. In the 1980s and 1990s, most 'light lambs', reared on semi-natural grassland and moor, were exported to southern Europe, a market that has now all but disappeared. The market for lamb increasingly demands animals that weigh more than 15kg, which require more nutrition than can be acquired from rough grazing. Farmers have cross-bred their animals with heavier breeds reared on 'improved' pasture on the lower slopes, and in some parts of Wales grazing pressure has been deliberately reduced to create a mosaic of grassy patches among Heather or Purple Moor-grass.

The other major change that resulted from increases in livestock densities was the growth in silage to provide animal feed in winter. Unlike hay, which is cut once late in the summer, silage is grown with high nitrogen inputs and cut while green. In lowland Wales, a minimum of three cuts each year is typical, starting in late April or early May, which has made large areas of Wales unsuitable for field-nesting species such as Skylark, which no longer have the opportunity to raise a brood before mowing. It is likely that breeding Curlews, which spread from upland into lowland Wales in the first half of the 20th century, have since been largely lost in the lowlands because of poor breeding productivity in silage fields, which look attractive to breeding adults at the start of the breeding season. As well as the cutting regime, which would also have been the final nail in the coffin for Corncrake, silage fields support fewer seed-eating or insectivorous birds, such as Linnet, Reed Bunting and Twite, than the hay meadows they replaced.

Silage, grown as winter feed for livestock, is cut several times between April and August, and so has made large areas of farmland unsuitable for ground-nesting birds © Annie Haycock

As pressure to maximise production has increased, so 'marginal' land of wildlife value has been squeezed. That is particularly the case for *ffridd* (or *coedcae*), the rocky knolls, patchy scrub and steep slopes that connect the enclosed grazing land and open moor. This complex mosaic of diverse vegetation, especially in North and mid-Wales and the Brecon Beacons (Blackstock *et al.* 2010), is important for birds such as Whinchat, Cuckoo, Yellowhammer and Ring Ouzel, and a wealth of fungi, lichens and invertebrates. It has, however, become fragmented by conversion to more intensively managed grassland, by burning or abandonment, and in some places being targeted for forestry.

Running a hill farm in Wales is challenging and unprofitable without state support, and many farmers have either intensified their management or abandoned it altogether. A lack of grazing on nutrient-enriched soils can mean the pasture soon becomes unsuitable for foraging species such as Curlew, Ring Ouzel and Golden Plover, and becomes attractive for alternative land use such as forestry. A decline in Choughs breeding inland in North and mid-Wales may be related to the abundance or availability of food. Reduced grazing, or the absence of sheep dung that provides a food source for invertebrates, may make some higher-elevation land less suitable for Chough, but the reseeding of valley slopes has made adjacent areas less suitable too.

Cattle numbers increased by 50% to 0.9 million between 1867 and 1945, and by another 70% to 1.55 million animals by the mid-1970s. Much of the growth was for dairy production, which intensified through selective breeding, automated milking, land drainage, inputs of nitrogen fertilisers and a switch from hay to silage for winter feeding. Overproduction of dairy products led to an EU milk quota during 1983–2015, which limited the volume that each farm could produce; as a result, the number of animals

Ffridd/coedcae, a distinctive and diverse habitat on slopes below moorland, is valuable for upland birds but at risk of agricultural abandonment or intensification © Robin Sandham

fell to around one million in Wales. Dairy cattle, and the associated intensive management, are concentrated in Pembrokeshire, Carmarthenshire and the Vale of Clwyd in Denbighshire. Although there have been few studies of the effects of dairy cattle on bird populations in Britain, it is evident that birds such as Skylark and Green Woodpecker have been lost from areas where dairy predominates. The fast-growing grass, fewer damp hollows and lower abundance of invertebrates, make it less suitable for ground-nesting birds, such as Meadow Pipit, Western Yellow Wagtail, Redshank and Curlew, and where birds do live among high densities of cattle, their nests are at risk from trampling. In drier parts of Wales, Maize is grown as a fodder crop for cattle that spend the winter indoors, and this provides little of the seed that can be accessed by wintering passerines such as Yellowhammer.

Most of the remaining beef production is on Anglesey and Llŷn, Caernarfonshire (APHA 2020), but there is now little in the uplands. Welsh Black cattle would, over several hundred years, have grazed on the *ffridd* but changing trends, cultural as much as economic, mean that fewer farmers are prepared to risk putting their valued cows on marginal land, where the vegetation provides less nutrition than on lowland rye-grass. Unlike sheep that nibble with their teeth, cattle pull tufts of vegetation out of the ground with the tongue, thus leaving tussocks of grass that are used by insects and small mammals. Providing they are not stocked at too high a density and veterinary medicines are not over-used, cattle can produce a more bird-friendly landscape, through their dung and grazing of scrub and herbs, than sheep grazing.

The other two major livestock trends have been the virtual disappearance of pig rearing, from 250,000 animals until the 1930s to around 25,000 since 2008 (over two-thirds of which were in just 30 holdings), and a doubling in the number of poultry since the Second World War (Welsh Government 2019a). Concern has grown over the last decade at the increase in intensive poultry units approved by Powys County Council (156 during 2015–19). The phosphate-rich manure is cleaned out every two to three months and spread onto fields, from which it is washed into streams and rivers, killing invertebrate eggs and larvae and using up oxygen, causing fish to suffocate. This has knock-on consequences for fish-eating birds and insectivorous species that live on or close to the water. Airborne ammonia from the sheds reduces plant diversity, particularly sensitive lichens in old-growth native woodland.

While Wales is dominated by livestock production, the mineral soils and drier conditions east of the Cambrian Mountains hold a greater proportion of arable farms, though these comprise only 1% of the UK total. Driven by public subsidies after the Second World War, hedges were removed and almost every corner of available fields was brought into production. Organochlorine pesticides had a dramatic effect on small granivorous birds and predators up the food chain. Although this was less pervasive in Wales than in more arable-dominated parts of England, it did have consequences for falcons that feed on small birds. A Merlin egg, laid in Radnorshire in the 1970s, contained the highest concentration of polychlorinated biphenyl, a persistent organic pollutant, ever discovered at that time. While organochlorines were progressively banned from the 1960s, the application of other chemicals, particularly herbicides and fungicides, has increased (Garthwaite *et al.* 2019), with impacts directly on the food chain for birds. Grey Partridge and Turtle Dove are functionally extinct as breeding birds in Wales as a result of changes in farming practice, including herbicides that reduced the plants that support the invertebrates on which their chicks depended. Fungicides and growth regulators are sprayed during April to June, at the height of the breeding season for

Grazing by beef cattle, at moderate densities, can create a diverse sward that benefits ground-nesting waders © Ben Porter

farmland birds, with adverse consequences for species that nest within fields, such as Skylark. There are also downstream impacts away from the fields, which can affect species dependent on aquatic invertebrates, such as Dipper. Rodenticides, particularly second-generation anti-coagulants (SGAR), are found in the livers of predatory birds, and may affect their body condition or hunting capability. Monitoring found no reduction in the proportion of dead Barn Owls in Britain (87%) containing residues in 2018, compared to 2006 when restrictions were tightened, supposedly to lower the risk of poisoning (Shore *et al.* 2019). SGARs were also found in 100% of Red Kites analysed in England and Wales in 2017 and 2018 (Walker *et al.* 2019), and have been found in other raptors, such as Kestrel.

The lack of seeds available in the farmed landscape in late winter, referred to as the 'hungry gap', is a particular problem for small passerines such as finches and buntings. While some, such as Chaffinch, Goldfinch and Greenfinch, have taken advantage of supplementary food provided in gardens, species such as Yellowhammer and Reed Bunting are less willing to move into gardens, even in rural areas. Cereal crops and rye-grass, deliberately left for foraging birds, produced seed in late winter and were used by both species, although for reasons unknown this did not translate into an increase in local population size of either species (Johnstone *et al.* 2019).

Greater specialisation of farms has meant that few livestock farmers are now self-sufficient. Most, for example, graze animals on their land for only part of the rearing cycle and buy in winter fodder grown elsewhere. Mixed farms of livestock and arable are no longer the norm in Wales, and so the mosaic of fields used for different produce has been lost from the landscape. Whole regions are now dominated by a single farming type: sheep in northwest and mid-Wales, dairy in the northeast and southwest, potatoes in Pembrokeshire, and arable in the east and south.

There are, of course, examples of excellent practice around Wales, where nature-friendly farmers are working together to protect habitats and showcase regenerative farming. Some, such as the farmer-led Pontbren Project in Montgomeryshire, have been underway for more than 20 years, demonstrating the value of long-term planning. Others are more recent, enabled by the Welsh Government's Sustainable Management Scheme. These should inform future support for farming, and if scaled-up, could achieve real change for birds.

Agriculture employs 53,500 people on farms, around 3.6% of the workforce, but accounts for less than 0.6% of the Welsh economy (Wiseall 2018). Incomes from livestock, in particular, provide a relatively poor and highly erratic return for farmers. Cattle and sheep farm income averaged £18,900 in the LFA and £17,100 outside it in 2018–19, half the level of five years previously, while dairy farm income fell 43% in just one year between 2017–18 and 2018–19. Around 20% of farms are loss-making and another 20% make less than £10,000 annually. A fundamental problem is the flawed belief that spending more on agricultural inputs will result in increased profit, but beyond a certain point, that has been shown not to be the case: "farmers need to learn to farm with nature… Farm business mindset needs to move from high productivity at all costs to a balance of farming with natural assets and careful management for our countryside" (NFFN 2020). In 2018–19, for example, the average farm in Wales made no money from what was grown or reared. Around 90% of income came from the taxpayer, in the form of agri-environment schemes (18%) and Basic Farm Payment (72%), and the remaining 10% from other business interests (Welsh Government 2019b). Without public subsidy, sheep and beef cattle production makes no economic sense for 95% of farms.

High densities of sheep grazing on a monoculture of rye-grass prevents the natural regeneration of scrub and trees © Ben Porter

The economics of farming in Wales have worked neither for farming families nor birds over recent decades. Many small farms, which contributed to the mosaic of habitats, have been subsumed into larger management units. Around 4,500 holdings have been de-registered, which is not the sign of a successful industry even with such heavy public subsidy. Agri-environment schemes (AES) have attempted to redress some of the negative impacts of agricultural intensification. Tir Gofal, the AES that ran from 1999 to 2013, was most successful for birds in woodland, scrub and hedgerows, but failed to deliver detectable benefits for many priority species in arable and grassland fields, in particular for waders, and on wet grassland (Dadam and Siriwardena 2019). Overall, few differences in bird abundance were evident between land within Welsh agri-environment schemes and land outside it (Macdonald *et al.* 2019). As AES are voluntary and are a small proportion of income from the public purse, there has been little incentive for many farmers to do more than the basics required. They have also failed to target the species and habitats in greatest need, because insufficient investment has been made in giving farmers good wildlife advice. In addition, farmers have complained about a lack of flexibility in prescriptions. The Welsh Government is proposing that future payments be based on outcomes delivered, not just actions undertaken. It is funding projects in the Elan Valley and North Wales that incorporate this approach.

Climate change

The climate has shaped the geology and natural history of the land and sea, and has always been subject to change. However, this change has quickened since the Industrial Revolution. The planet's average surface temperature has risen 1.14°C since the late 19th century, and the amount of carbon dioxide in the atmosphere is higher than at any time in at least 800,000 years (https://climate.nasa.gov/evidence).

'The Greenhouse Effect' and 'Global Warming' have increased dramatically in public and political consciousness since 1990, when the UK Prime Minister told the second World Climate Conference in Geneva that "our ability to come together to stop or limit damage to the world's environment will be perhaps the greatest test of how far we can act as a world community… We shall need statesmanship of a rare order" (Thatcher 1990). Ecologists contend that we still await that statesmanship, and in the 30 subsequent years, birdwatchers have witnessed changes that illustrate climate change to be in the present, not only the future, tense. Thorough reviews of the science (e.g. Dunn and Møller 2019) showed that significant changes for the birds of Wales are underway.

Just some of the examples included in this book are:

- the breakdown in synchrony between hatching of chicks and the availability of invertebrate food (e.g. Pied Flycatcher, Blue Tit),
- the northward shift of sandeel species beyond the reach of coastal seabird colonies (e.g. evident in the North Sea for Kittiwake and Sandwich Tern),
- 'short-stopping' by waders and wildfowl that means Arctic species no longer winter so far west (e.g. Bewick's Swan, Pochard) and
- sea-level rise that squeezes the availability of intertidal habitats (for waders and wildfowl) and shoreline breeding habitats (for terns).

The shifting distribution of breeding species is also evident. Between 1968–72 and 2007–11, the northern limits of British breeding species moved north by an average of 32km (Gillings *et al.* 2015). Even within Wales, the local differences in Willow Warbler densities are already noticeable between south and north, especially in the lowlands. At a broader scale, the population trends of European migrant species are demonstrably related to climate trends on their breeding grounds—positively for short-distance migrants, but negatively for trans-Saharan migrants. However, land-cover change is a greater driver of their trends when they are away from the breeding grounds (Howard *et al.* 2020). Species that breed in the Arctic may already have been more affected by the changing climate, which is more acute at higher latitudes: the area above 64°N has warmed 3.8 times faster than the global mean in the last 30 years (https://data.giss.nasa.gov/gistemp).

At home, less-frequent cold episodes have benefited the over-winter survival of resident, small-bodied species, such as Wren. Other changes may be occurring but not so detectable: does climate explain latitude shifts of Ring Ouzel and are Merlin simply running out of elevated moors on which to breed?

Woodlands

Numerous species accounts in *The Birds of Wales* refer to the expansion in conifer forest during the 20th century. Approximately 15% of Wales (306,000ha) is now forested, just under half of which is comprised of native tree species. Approximately 30% of woodland is classed as ancient, in existence since at least 1600 and holding the greatest species diversity, including the suite of woodland specialist birds. However, this classification by Welsh Government includes land where ancient trees were felled and replaced with non-native conifers. Planting of conifers in Wales accelerated after the Second World War, aided by grants and mechanisation. Since 2001, the area of conifers has decreased by 18,000ha, with new planting now at its lowest level for a century and around 2,000ha of felled plantations restocked each year. During that period, the area of broadleaved woodland has increased by 35,000ha (Warren-Thomas and Henderson 2017).

Cumulatively, the planting of conifers benefited many bird species in Wales, but for more than half of species assessed, the benefits lasted only a decade or two before the tree canopy closed. Birds such as Black Grouse had a short-lived boom in the 1960s and 1970s as grazing animals were fenced out to enable trees to become established. However, the drainage of vast areas of peaty soils, including blanket bog, that preceded planting had far greater negative long-term consequences for adjacent habitats. The consequences were evident downstream through the acidification of watercourses with negative impacts for species such as Dipper. Predators of moorland ground-nesting birds, such as Fox and Carrion Crow, found shelter in these plantations, which coincided with a reduction in gamekeeping, as driven grouse shooting was abandoned. Ground-nesting birds, such as Curlew, Golden Plover and Hen Harrier, faced a triple whammy: smaller areas of open habitat, of lower quality with less heather or fewer wet flushes rich in invertebrates, and surrounded by trees that supported an increase in generalist predators.

The bird species that benefited from planting in the long term were primarily abundant lowland species, such as Blackbird and Great Spotted Woodpecker, which were able to extend their distribution into previously unsuitable open habitats. Conifer specialists such as Siskin and Crossbill greatly increased their range and abundance in Wales, from being rare breeding species to locally common. The expansion of conifer plantations also increased carrying capacity for a number of predatory species, including Goshawk, Raven and Long-eared Owl. The management of conifer plantations makes a critical difference to their potential value for birds. Rotationally clear-felled areas, for example, have become the primary nesting habitat for Nightjars in Wales and attract Tree Pipits for a short period, but have not been inhabited by Woodlarks as they are on the sandy soils in East Anglia. Clear-fells have not, generally, attracted species such as Meadow Pipit and Black Grouse, that did well during the initial planting on moorland, as the wet flushes had long since been drained.

Native broadleaved woodlands are extremely important for some birds, including Pied Flycatcher, Wood Warbler and Hawfinch, all of which occur in higher densities in Wales than elsewhere in Britain. Only a small proportion, approximately 5%, of Welsh woodland is designated for nature conservation purposes, and only a minority of this is managed well for nature. The proportion of woodland Special Areas of Conservation (SACs) in favourable condition declined from 26% in 2002–06 to 21% in 2007–12 (Warren-Thomas and Henderson 2017). Some woods have had their ground flora degraded by grazing, but where livestock have been fenced out, the understorey has often become dominated by a dense shrub layer of Bramble, Holly and invasive non-native species such as *Rhododendron ponticum*. This reduced a woodland's suitability for specialists such as Wood Warbler and Pied Flycatcher. In addition, high levels of pollution from airborne

Plantations of conifers, such as at Pant Mawr, Breconshire, have resulted in significant changes to birdlife in the Welsh uplands © Colin Richards

nitrogen oxide and ammonia have encouraged nettles, rather than a more diverse and specialist woodland flora.

Episodic tree diseases have had a significant impact on the quality and quantity of woodland. Dutch Elm Disease, caused by two introduced fungi, was present in all Welsh counties by 1975. It spread rapidly from southeast Wales after 1969, accelerated by its importation via a saw-mill in mid-Wales, although away from southeast Wales the proportion of trees lost was less than in southern England (Jones 1981). This may have provided short-term benefits for birds such as Willow Tit and Lesser Spotted Woodpecker, which require rotting wood in which to feed or nest (Gibbons *et al.* 1993), but the long-term result was less woodland connectivity in the countryside, as many of the felled trees were not replaced. More recently an introduced Asian fungus is causing significant dieback of European Ash trees, and by May 2020 had been confirmed in 98% of 10-km squares in Wales (Forest Research 2020). Since European Ash is a significant component of many of the older native woodlands in Wales, including Sessile Oak woods, it is expected to have a dramatic effect on these ecosystems, but the effects on birds are likely to vary by species.

Game management

Agriculture and conifer afforestation have had the most significant effect on birds in Wales, but other land-management decisions have also had local effects. Game management is not practised on the scale of other parts of Britain, but can be equally intensive where it does occur. A single moor is managed for driven grouse shooting and a handful for walked-up grouse shooting, so burning on peat soils and the use of medicated grit have been less of a concern than in other parts of Britain. Similarly, deer stalking has never been a widespread practice in Wales. Approximately 4% of woodlands are managed for pheasant shooting (GWCT data), but there is concern that the associated management has become more intensive. There is growing evidence about the impact of high densities of released, non-native Pheasants and Red-legged Partridges on plants, reptiles and amphibians, as well on the invertebrates and natural seeds that are eaten by wild birds (Mason *et al.* 2020). Some marginal farms in Wales have been developed as game-shooting enterprises and are believed to release far greater numbers of Pheasants than syndicated shoots operated as a weekend leisure activity by farmers or woodland owners.

Illegal killing and egg-collecting

Stronger legislation and increased penalties since 1954 have gradually reduced the killing of birds, and the theft of eggs for collections and of nestlings for falconry. Using the number of confirmed incidents as an index of persecution, the illegal killing of raptors in Wales was half the level in 2000–09 that it had been in 1990–99, but levels in 2010–19 were similar to the previous decade (Hughes *et al.* in prep). The use of poison, in particular, appears to have become more prolific, which is a concerning trend, given both the risks to multiple birds and mammals from a single bait. Its use in the open has been illegal for more than a century. Peregrine, Buzzard and Red Kite are the three raptor species that most frequently fall victim to illegal persecution. While this has not prevented the recovery of their populations, there is some evidence that the increase in Peregrines was slowed in South Wales, where interference and poisoning related to pigeon racing was a problem and may remain so locally. While trends in persecution have generally declined, this is not uniform across Wales, and the substantial increase in Breconshire is notable. Satellite tagging has provided a greater insight into the causes of mortality of Hen Harriers and, albeit from a small sample size,

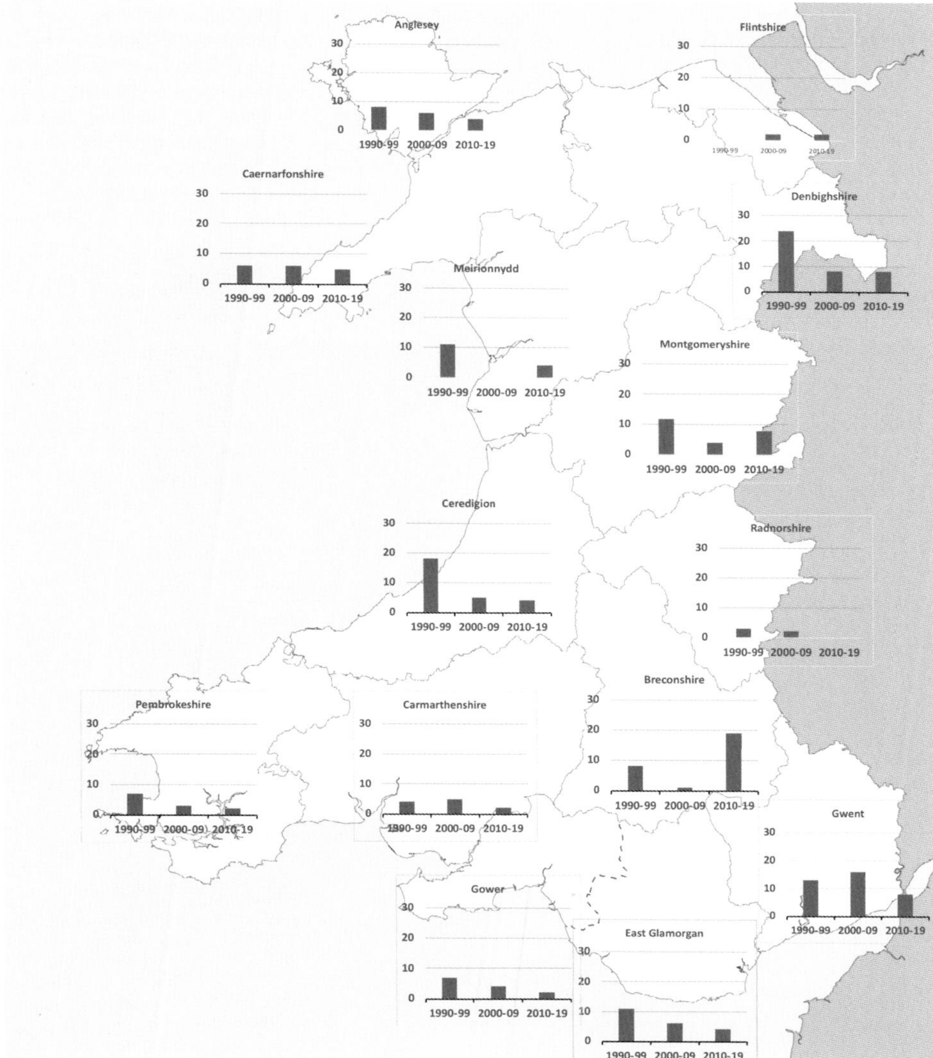

Figure 8. Number of confirmed raptor persecution incidents each decade, by county, 1990–2019 (from Hughes *et al.* in prep).

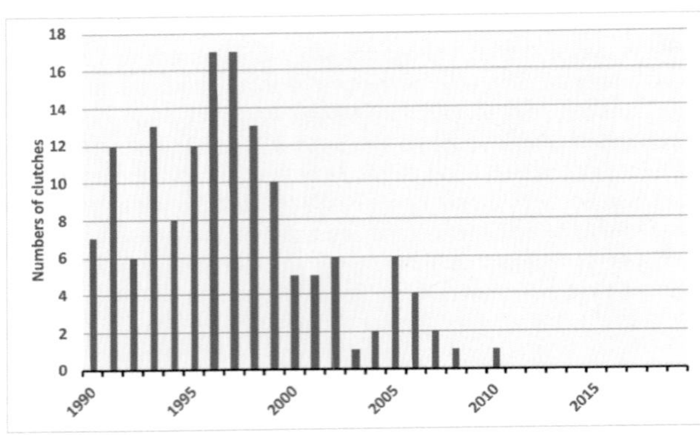

Figure 9. Number of clutches of Schedule 1 species known to have been taken illegally from nests in Wales, 1990–2019 (RSPB data).

suggests that illegal persecution contributes to low survival rates in Wales in their first year.

The decline in egg-collecting does appear to be genuine and this practice is hopefully on its final steps to becoming historic. During 1960–89, when the Red Kite population was far smaller than today, 72 clutches were known to have been taken in Wales, equating to 8.5% of all nesting attempts, and this was assessed to have slowed down the rate of population growth by 0.6% each year (Bibby *et al.* 1990). As a relatively rare bird, the Chough was also targeted by egg-collectors, even in the 1990s, when 30 clutches were known to have been lost, but such incidents have also declined and there have been no known clutch losses since 2007 (RSPB data).

Predation

Predation is a more visible threat to breeding success or survival than many other causes of change, such as climate and land use. Several reviews have assessed the impact of predation by native species on the conservation of birds in the UK (e.g. Gibbons *et al.* 2007). The consensus is that it has not been the primary cause of decline for most species, including most songbirds. However, as semi-natural habitats have been eroded and human activity has squeezed birds into a smaller area, predation of nests has become a more significant factor for reduced populations of a small number of ground-nesting species. Effective removal of Brown Rats, described in Chapter 3, has enabled the recovery of several important seabird islands. Maintaining good biosecurity to prevent Brown Rats from reaching these and other important colonies is critical, since mainland nesting opportunities for these birds are very limited.

Predation of eggs or chicks is cited in 11 of the species accounts: five breeding terns and five ground-nesting birds of moorland or wet grassland. The other affected species that occurs here, only outside the breeding season, is the globally threatened Balearic Shearwater.

Terns' ability to reduce the risk of predation and disturbance at colonies by using multiple sites has been greatly constrained by the lack of choice of alternatives, as a result of changing coastal morphology, predation by Foxes, rats and corvids, and greater recreational use of beaches by people. There are now only five regularly occupied tern colonies in Wales, all in the north and dependent on wardening and electric fencing, so these birds' eggs are very much in one basket. Single events such as Otter predation at Cemlyn, Anglesey, in 2017 and disturbance by Peregrine that prevented settling on The Skerries, Anglesey, in 2020, can jeopardise a whole season.

The other five species—Black Grouse, Red Grouse, Lapwing, Curlew and Redshank—have been well studied. Declines are a result of multiple factors, but predation becomes more significant in a declining population, and its significance can increase in combination with other changes, such as wet weather during the chick-rearing stage for grouse. For all these species, predation pressure has increased, at least in part, as a result of the planting of trees on previously open habitats, which has enabled Foxes, Carrion Crows and Goshawks to live at densities that would not have otherwise been feasible.

As a semi-colonial species, Lapwing can be assisted by the construction of extensive (and expensive) electrified fencing, which has proven effective in keeping out mammals such as Fox and Badger. These have been deployed at several sites in Wales, including the RSPB reserves at Ynys-hir in Ceredigion, Cors Ddyga in Anglesey and Burton Mere Wetlands in Flintshire/Wirral. However, this is not a solution for Lapwing outside nature reserves or for the other species, which do not nest so closely together. Lethal control of Carrion Crows and mammals is currently the only option available, but it does not always sit comfortably with conservation organisations, is time intensive, costly and unlikely to be a long-term solution. In its fragmented

A combination of land-use changes has led to Curlew becoming one of the breeding species at greatest risk of extinction in Wales © Tony Pope

Estuaries are important for non-breeding waterbirds, and many, such as the Mawddach Estuary in Meirionnydd, are designated as Sites of Special Scientific Interest © Jeremy Moore

and heavily managed landscape, Britain has the highest densities of some generalist predators in Europe. It may be that previously high levels of predator control, associated with game management, enabled ground-nesting birds to increase to artificially high numbers in a landscape that is 'naturally' sub-optimal. In the short to medium term, it seems likely that the targeted control of widespread generalist predators may be necessary, in tandem with habitat restoration (McMahon *et al.* 2020). In the long term, maintaining these ground-nesting bird species will depend on evolving the landscape to make it less attractive and productive for these predators. One area requiring greater investigation is the role of released gamebirds in supporting these predators, as suggested by Pringle *et al.* (2019).

Protected areas

While attempts have been made to prevent declines of birds on farmland and forestry through regulation, woodland management standards and agri-environment schemes, for a selected suite of species, the network of protected areas has made a real difference. While by no means perfect, they provide measurable conservation benefits for birds (e.g. Donald *et al.* 2007) as they give government conservation agencies some level of influence and control over activities undertaken. This is effective when dealing with direct threats, such as built development or disturbance through activities such as wildfowling, especially where combined with targeted management for the designated species, such as on the Dee Estuary. However, protected-area status has been less effective in tackling more fundamental change. As one example, 63% of Welsh SSSIs partly or entirely exceed the ammonia Critical Level for their qualifying habitat or species features (Rowe *et al.* 2019).

There are no recent published data on the condition of protected areas in Wales, nor on the achievement of site management objectives. In 2013, the condition of 55% of species features on Natura 2000 sites in Wales was unfavourable (Natural Resources Wales 2016). There also remain significant gaps in the SPA network for some key species, such as Curlew and Chough. For SSSIs outside the Natura 2000 network, there is no recent published information on habitat condition.

Protected areas are particularly effective for those species that aggregate over a relatively small area, such as nesting seabirds and roosting wildfowl, but have not prevented declines in more widespread species, nor deterioration in conditions outside nature-reserve boundaries. Several initiatives, some funded by the EU-LIFE programme, have restored a small proportion of the lowland wetlands lost in Wales. Quarrying for sand and rock has created after-use habitat for waterbirds, often providing accessible nature close to urban areas. Others have been purposely created, such as RSPB Cors Ddyga in Anglesey, RSPB Burton Mere Wetlands and Newport Wetlands in Gwent, or have been brought into improved management such as RSPB Valley Wetlands, Anglesey. These have benefited species such as Gadwall, Teal, Water Rail and Reed Warbler, and rarer species including Bittern, Marsh Harrier and Bearded Tit. These wetlands also provide the opportunity for pioneering species, such as Great White Egret and Avocet, to gain a foothold in Wales. However, even these refuges are not immune from the management of the wider farmed landscape. Several fen nature reserves suffer from eutrophication caused by nutrients running into watercourses from upstream farms, and invasive non-native species, such as Signal Crayfish and Himalayan Balsam, do not stop their spread at the boundaries of a nature reserve.

Marine environment

Breeding seabirds nest on our coasts and islands, but some forage beyond Welsh waters in the breeding season, and almost all move west and south outside the breeding season. Tracking has shown that the Irish Sea Front SPA, a food-rich area about 36km northwest of Anglesey, hosts the third-largest marine aggregation of breeding Manx Shearwaters in the UK (Kober *et al.* 2012), while Manx Shearwaters breeding on Bardsey, Caernarfonshire, travel as far as western Scotland to feed (Porter and Stansfield 2018). Breeding seabird populations in Scotland and on the North Sea coast of England, whose diet is dominated by sandeels, have struggled in recent decades, as these fish have followed their zooplankton prey north, beyond the reach of coastal-nesting seabirds. In the North Sea, the zooplankton *Calanus finmarchicus* (the main diet of sandeels) has declined by 70% since the 1960s, as the sea surface temperature has warmed by 1°C (Edwards *et al.* 2020). However, in the Irish Sea several seabird species have been able to switch to *Clupeids* (herring and sprat) that favour warmer waters and are more abundant here than in the North Sea. Also, feeding conditions for seabirds may be better than in the heavily fished North Sea and English Channel, although the absence of a commercial fishery in the Irish Sea means that precious little data have been collected on the trophic fisheries here. Cardigan Bay is likely to be of particular importance for foraging seabirds, especially in late summer as auks and Manx Shearwaters fledge from the islands, but only the seas off Meirionnydd and around Bardsey have any formal designation for seabirds, and there is no management of protected areas aimed at seabird conservation away from their nest sites.

An absence of industrial fishing in the Irish Sea has meant that this has not been the same threat to our breeding seabirds as in the North Sea, but this may have bred a degree of complacency. Fishery management measures in Marine Protected Areas in the Irish Sea have not even been proposed (Chaniotis *et al.* 2018). The availability of suitable fish is a factor in the declines of Kittiwake, Lesser Black-backed Gull, Puffin and Fulmar, based on studies elsewhere in the UK, but Welsh data are lacking. Mortality of birds caught in nets or on longlines is cited as a problem elsewhere in their European range for populations of four species that occur in Wales outside the breeding species: Scaup, Velvet Scoter, Long-tailed Duck and Balearic Shearwater. In addition, there are concerns about the effects of Fulmar 'bycatch' from longlining off the Scottish coast, but there has been no research into whether the declining Welsh breeding population is affected. Fixed fishing gear set close to shore can be a problem for species that feed in shallow waters, especially when it is damaged or not removed (so called 'ghost gear'). Crustaceans and fish can become entangled in gill and trammel nets, and so can birds that are attracted by the rotting animals. Nets can continue to catch birds, un-noticed, for several years, and in an experiment off Pembrokeshire, two 90m monofilament nylon nets caught three Shag between 43 and 70 days (Kaiser *et al.* 1996). Globally, it is estimated that there is 4.4km of 'ghost' fishing gear per square kilometre of fishing grounds (Gilman *et al.* 2016), though no estimates appear to have been made for Wales.

Oil pollution has been a major risk for seabirds since the construction of the first multi-hold tankers in the 1880s. Tackling pollution became the RSPB's next significant campaign after the feather trade, and in 1931 it prosecuted the owners of the *MV Ben Robinson* for causing an eight-mile oil slick near Skokholm. The company was fined £25, approximately £1,700 at 2020 prices. The destruction of shipping during the Second World War led to significant oil pollution in the North Atlantic and around the European coasts, and it is now estimated that Pembrokeshire's Guillemots declined by *c.*95% as a result. The grounding of the *Torrey Canyon* off Cornwall in March 1967 and, off Pembrokeshire, the *Christos Betos* in autumn 1978 and the *Sea Empress* in February 1996 affected breeding populations of gulls and auks, and wintering scoters and divers over a shorter period. Oil can still be a localised problem and remains a major risk off West Wales, where Milford Haven is the largest oil and gas port in the UK.

Feeding garden birds

Supplementary feeding of birds in gardens has grown hugely since the 1970s, into a multi-million-pound industry, providing many people with a daily connection with nature (Orros and Fellowes 2015). Species likely to have benefited from feeding, by increasing their overwinter survival or adult survival during the breeding season, include Blue Tit, Great Tit, Goldfinch and Siskin, and the increase in wintering Blackcaps in Britain may only have been possible because of bird feeding. However, it is not without its risks. The transfer of the disease Trichomonosis from pigeons to other species is believed to have occurred around shared feeding stations, resulting in a recent crash in the Greenfinch population and potentially in Chaffinch numbers too. The potential impact on our rare and localised Hawfinch populations is unknown.

Furthermore, the support given by garden feeding to Blue Tits and Great Tits has been suggested as a contributory factor in increased competition with declining populations of Marsh Tit and Willow Tit (Richard Broughton *in litt.*). Similarly, feeding may have boosted numbers of Great Spotted Woodpecker, which can be both a competitor and nestling-predator of other hole-nesting species. These serve as a cautionary reminder about the unwitting harm that an otherwise-positive human action could cause to bird populations.

Other drivers of change

Most of the remaining factors mentioned in the accounts involved just a small number of species. Human disturbance is the most frequently cited, and the effects are typically local, but population-level data are lacking. Coastal terns, Shelduck, Ringed Plover and Oystercatcher are the species most likely to be affected in the breeding season, along with Common Sandpipers in the uplands and, outside the breeding season, birds such as Turnstone, Sanderling and Wigeon. Several species are threatened by overhunting outside the UK and a number, including Mallard and Pochard, by the use of lead shot over wetlands, even though this has been illegal in Wales since 2002. A contrary success story was the ban on the use by anglers of lead weights up to 1oz (25g) in 1988, which led to a subsequent increase in Mute Swan populations. Loss of nest sites, as a result of changes to buildings, is cited as a factor in the decline of Swift, House Martin and Barn Owl, though it is difficult to know the significance of the loss, compared with reductions in the biomass of aerial invertebrates and of small mammals in rough grazing.

These factors describe, briefly, some of the reasons behind the changes in bird communities in Wales. It feels, writing in 2020, that we are at a tipping point for the way in which these factors are managed, and the next chapter will explore whether we can learn from the changes of the last century.

Julian Hughes

6. Future prospects

We shan't save all we should like to—but we shall save a great deal more than if we never tried.

Sir Peter Scott

It will come as little surprise that this chapter will be dominated by climate change. When *Birds in Wales* was written in 1994, there was barely a mention of the changing climate. Understanding of its likely impact on human life or wildlife was probably no higher among most ecologists and birdwatchers than among the wider public, but in the subsequent quarter of a century the accelerating pace of its effects has become evident. Even as this book was being drafted, Britain witnessed its joint hottest ever summer in 2018, the highest ever temperature in 2019 and its wettest ever February in 2020 (Met Office data).

Shifting distributions

Many of the species accounts in this book refer to distribution modelling undertaken by Huntley *et al.* (2007). They used three bioclimatic variables (mean temperature of the coldest month, the number of days when the temperature was over 5°C, and a ratio of evapotranspiration) and the same Hadley CM3 climate predictions that were used in UK and international climate impact assessments. The modelling showed the expected change in the European range of breeding species between 1960–90 and 2070–2100. Of the 168 bird species that bred in Wales regularly during 1960–90, 33% were expected to lose some (23%) or all (10%) of their range here. Most of the 16 species expected to be lost completely to Wales are upland breeders, such as Red Grouse, Black Grouse, Golden Plover, Dunlin, Merlin, Short-eared Owl, Ring Ouzel and Twite. Other species are expected to shift northwards over the coming century: 22 are predicted to breed in Wales by 2070–2100 that did not breed regularly in 1960–90. Some of these, such as Little Egret, Marsh Harrier, Hobby, Dartford Warbler and Bearded Tit, have already arrived, while the climatic conditions would be suitable for some former breeders, including Woodlark and Cirl Bunting. A few species could also arrive that do not currently breed in Britain, including Middle Spotted Woodpecker, Bluethroat, Zitting Cisticola, Melodious Warbler, Short-toed Treecreeper and Serin. Other factors, both natural and anthropogenic, also affect species' distributions. Avocet and Mediterranean Gull now breed here in small numbers, for example, which was not forecast by this climate modelling. Nonetheless, it did provide a window into the future.

It would be easy to dismiss the last 70 years of nature conservation as being ineffective in the face of such dramatic changes, but the core architecture of nature protection is as relevant, and arguably more essential, than ever. The network of protected areas, for example, is "needed to facilitate waterbird distribution change in response to climate warming in the Western Palearctic" (Gaget *et al.* 2020).

Melodious Warbler, the population of which has increased in France in recent decades, is expected to breed in southern Britain later this century © Steve Stansfield

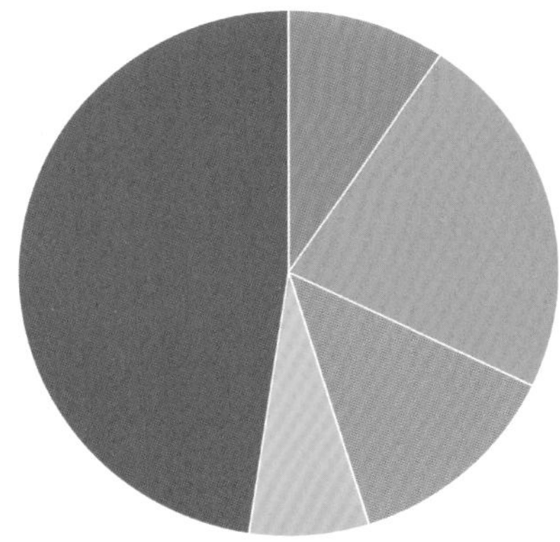

Figure 10. Projected change in status of native Welsh breeding birds between 1960–90 and 2070–2100, based on Huntley *et al.* (2007).

Climate changes

The *European Breeding Bird Atlas 2: Distribution, Abundance and Change* (Keller *et al.* 2020) shows that distributional changes are already underway, but the direction of travel for some species is not as expected. Western Yellow Wagtail and Nightingale are two examples of species that are predicted to spread north, but the recent trend is for their ranges to retract to the southeast. This illustrates that either the effects of climate change are more complex than can be described by the three variables used in the models, or that other factors are currently over-riding that of climate. In the 15 years since Huntley *et al.* (2007), studies have provided detail on the mechanisms that are driving those changes. While the climate change headline is expressed as global temperature, associated storm events or changes in precipitation can have a more immediate effect on birds. Current medium emissions scenarios for Wales in 2050 indicate a 4–6°C increase in mean summer temperatures, compared to 1961–90, 20–50% reduction in summer rainfall and 10–30% increase in winter rainfall (Met Office Hadley Centre 2019). These averages are dramatic, but the extremes may be more damaging. Just one or two weather events can, at least, have a short-term effect on resident birds with short lives. Multi-brooded species, such as Stonechat and Wren, can bounce back quickly from cold winter snaps, but for other species, climate change is not the only pressure that, in combination, could affect their populations. Chough, for example, may already be under stress from a decline in food availability associated with changes in grazing, but April storms during the early stage of nesting or long, dry summers that bake the soil hard could exacerbate the pressure. Any significant variation in the frequencies of summer wet and dry spells would be a problem. Heavy summer thunderstorms, characteristic of central England and continental Europe, may occur more frequently in Wales, making river valleys more prone to summer flooding. Conservation efforts over the coming decades will need to focus on measures to mitigate the effects of a changing climate, *and* to compensate for the losses of habitat that will occur. It will be critical that land management supports the changing distribution of birds and their food chain, while providing conditions that welcome natural arrivals from the south.

The impacts of climate change on migratory species are likely to be even greater than on our resident birds, since their effects will differ at the multiple sites on which they depend along the flyway. In general, species breeding at more northern latitudes will be more negatively affected as they simply run out of suitable land to nest and resources on which to feed. The loss of stopover sites, due to sea-level rise or other climatic effects, could reduce survival of both summer and winter visitors to Wales. Some species, such as Cuckoo, depend on the availability of food at a critical time in one part of their migration route, and this may not change in synchrony with the availability of food at a different point in their journey. Species such as Red-flanked Bluetail and Blyth's Reed Warbler have shown rapid recent westward extensions in their range, and who knows where this will take them in another

Changing climate is already affecting farming, river systems and woodlands in Wales. Habitat restoration, such as in the Dyfi Valley, Montgomeryshire, should help to increase natural resilience © Jeremy Moore

decade or two? Similarly, the migration strategy of species such as Yellow-browed Warbler is changing, and perhaps it will become a common passage migrant, or even winter regularly in Wales.

Woodland expansion

There is increasing global recognition that we are living in a 'climate emergency', even if the extent of action is as yet insufficient to limit the effects below the high-impact scenarios. But the political and economic responses to this emergency could also have significant repercussions for birds in Wales. Politicians have been quick to reach for the carbon-gathering properties of trees as a solution to climate change. The Welsh Government has committed to 100,000ha of new woodland between 2010 and 2030, which would represent a 25% increase on the current extent (Welsh Government 2009). There is no doubt that woodland cover is below its historic levels and that woodland birds would benefit from these increases in time. Wood Warbler and Pied Flycatcher need space for expansion, as the suitability of their southern and lower-altitude habitats becomes squeezed. Enabling the natural regeneration of Sessile Oak woodland up valley sides in northwest Wales would benefit these species, providing the understorey remains open, but it will take at least a century to achieve the right conditions, even if the start was made now. The projections suggest that both species may be lost from much of Wales before then.

The species accounts tell a repeated story from the 20th century of bogs and moorlands in upland Wales being ploughed and planted with non-native spruce and fir. This aided the uphill spread of some common birds, but caused loss of habitat for upland birds and indirect effects from increased predation. The mantra of 'the right tree in the right place' is conservationists' shorthand to encourage planting of broadleaved trees for long-term carbon sequestration, at locations that do not cause damage to wildlife or compromise the role of existing habitats, such as peat soils, to store carbon. This requires robust data about wildlife, fine-scale climate modelling and good spatial planning.

The challenge comes with working out where these trees should go. The areas frequently selected are those considered to be of lower value for farming. This 'marginal land' is often that to which our dwindling populations of birds, such as Curlew, has already been pushed by a century of increased farming intensity. Much work is going into modelling future land-use scenarios and understanding the consequences for the release and sequestration of greenhouse gases. On average, for example, Sitka Spruce lock away only 61% of the carbon dioxide achieved by native broadleaved trees that are not harvested, and the benefits are even lower if trees are planted on peat soils. The highest levels of carbon sequestration are on mineral soils that are currently used for arable or improved grassland. Therefore, there are some major decisions to be made about whether tree planting is deployed as a 'dash for carbon' or, as the UK Committee on Climate Change favours, accounting for a range of other needs, including nature conservation. The changing climate will, itself, reduce the area of Wales in which Sessile Oak, for example, could grow. Currently, around 12% of Wales is suitable for Sessile Oak, but that would reduce to 9% by 2050, with large areas of the south and east becoming less suitable (Environmental Systems 2019).

There is a real fear that political imperatives to be seen to be tackling the climate crisis will over-ride ecological sense and evidence, and that some of the mistakes of the 1950s and 1960s will be repeated. Barely any woodland is currently being planted in Wales—just 10% of the Welsh Government target for 2030, set in 2009, had been achieved by 2018–19 (Woodland Trust data). As the target date gets closer, politicians may feel a need to short-cut their way to more trees. A robust spatial plan that identifies where the greatest benefits will accrue for wildlife, climate and people would provide greater confidence in the future and direct the planting (or natural regeneration) of trees to the right places. However, it requires the Welsh Government to give clear signals about the type of woodland and the need for habitat connectivity. It must commit to its long-term management, and not be tempted by the short-term attraction of timber production. The scale, location and type of trees planted in the next 30 years, and their management over subsequent generations, will make a huge difference to the resilience of woodland birds to climate change and the maintenance of important birds of open habitats, such as Curlew.

Western Atlantic Oak woods in Wales are of international importance for biodiversity, but in need of management and expansion to respond to a changing climate © Ben Porter

Peat restoration

Peat soils are a major store of carbon dioxide, and account for 17% of the land surface of Wales (RSPB 2019a). While most peat soils are in the uplands, lowland mires, stillwaters and intertidal habitats also lock in carbon. Releasing carbon by allowing these habitats to dry out accelerates climate change. Planting trees on peat soils would accelerate its release, especially if softwoods such as conifers were grown and then felled. Even planting native trees such as Scots Pine and birch on peat had a negative impact on carbon stocks over 40 years, measured in a Scottish study (Friggens *et al.* 2020). The best route to protecting these peat soils is to restore their function through re-wetting. Raising the water table helps to lock in carbon and encourages the growth of *Sphagnum* mosses. These boggy pools, interspersed with a mix of vegetation, can provide nesting and feeding areas for moorland species such as Curlew, Teal and Golden Plover. While individual projects have been undertaken, Wales currently has no strategy to restore these important habitats, in contrast to Scotland, whose government has committed to £250 million to restore 250,000ha of blanket bog by 2030.

Drainage, overgrazing and burning have damaged peatlands. Work is now underway to restore some of these blanket bogs, such as in the Brecon Beacons © Colin Richards

Peatland restoration is one 'nature solution' to broader societal challenges, such as climate change and the need for clean water. Globally, 37% of the Paris climate goal, which committed to limiting temperature increase to less than 2°C of pre-industrial levels, could be achieved through such natural solutions (IUCN 2020). Wales is well-placed to achieve such a 'win–win' for nature and people.

Renewable energy

Achieving the remainder of the climate goals will need to come from transformational changes in the way that we reduce and, ultimately, cease greenhouse gas emissions. Switching energy production from coal, oil and gas to renewable sources will be a major factor, and Wales' diverse natural resources make it an obvious place to develop such technologies. This switch has potential to affect Wales' birds too.

On land, wind turbines and solar arrays have already been installed in some parts of Wales. Like post-war forestry plantations, these have frequently been sited on so-called 'marginal land' that is valuable for birds. Several solar arrays have been constructed on wet grassland in or adjacent to SSSIs, such as in the Gwent Levels and around the upper Dee Estuary in Flintshire, which held nesting Lapwings and wintering Whooper Swans respectively. It seems likely that the growth area for energy production over the coming decades will be in the marine environment. Tidal lagoons have been proposed for the coasts of North and South Wales, on a scale never previously seen, but with significant implications for feeding waders and non-breeding seabirds such as Common Scoter.

Extensions and additional large offshore wind farms are planned between North Wales and the Isle of Man, and as the southern North Sea becomes increasingly crowded with turbines, the search area is being extended to the west coast and the Celtic Sea. There is concern about population level impacts for several gull species, including Kittiwake, from the cumulative effect of blade collisions in the North Sea, as well as the direct loss of foraging areas. Less studied are indirect ecosystem effects that, cumulatively, may also be significant. Examples include the effect of noise from piling that can be heard by herrings and sprats up to 80km away. The rocks used to protect the base of turbines can provide a refuge for species such as Cod, but may drive out smaller fish that are prey for both fish and seabirds. Also, the 'wind-wake' effect may cause up-welling and down-welling in the water column, across an area ten times larger than a wind farm's footprint, affecting the transport of nutrients that are the base of the marine food chain (Perrow 2019). There is an urgent need to understand the potential effects of these wind farms on the availability of the right fish of the right size, within a few kilometres of shore, that are critical for Welsh populations of Kittiwakes, Puffins, Razorbills, Guillemots and terns. Anchored, floating wind farms are expected to be the next step in offshore wind technology. With no pile-driving, these are cheaper to deploy, have no aural impact on *Clupeids*, and if directed to deep-water sites that are less used by seabirds for foraging or migration, they could limit the cumulative impact of wind power on seabird populations. However, the capability to install a wind turbine in almost any marine location reinforces the urgent need for a spatial plan for the seas off Wales. The Wales National Marine Plan (Welsh Government 2019e) has little to tie development into biodiversity targets. On the plus side, a reduction in use of oil and gas should reduce the risk of maritime oil spills around the Welsh coast, although it is by no means the only pollutant, and there is no real understanding of the role that micro-plastics may have on the food chains in either the marine or terrestrial environments.

Finding solutions to plan the construction and operation of renewable energy plants is challenging when a lack of understanding of marine ecology makes it difficult to model what will happen when such large energy projects are undertaken. Tidal stream technology is suggested by some as the next step in renewable energy generation, but a test area proposed for developers to trial turbines in a 35km^2 area off the west coast of Anglesey has the potential to cause the loss of 60% of breeding Guillemots and 98% of breeding Razorbills on the sea cliffs at nearby RSPB South Stack (Menter Môn Morlais Limited 2020). It illustrates how a 'green' solution must take account of all the environmental impacts, or some of the nation's special birdlife could be sacrificed on the altar of climate-change mitigation.

Agricultural change

Achieving 'net zero' greenhouse-gas emissions is just one of many challenges facing the future of farming. The industry accounts for *c.*14% of Welsh emissions (Welsh Government 2019d). As described in the previous chapter, agriculture was the single-biggest factor that affected birdlife in Wales in the 20th century, and much of farming is uneconomic without public support. The future of farming in Wales will be driven by public policy on agriculture and trade, the technological push of international agri-business, climate change, and the response of individual owners and tenants to all three. Withdrawal from the European Union in 2020 has triggered a major review of how £250 million in taxpayer support for farming in Wales should be spent each year. The decisions were, at the time of writing, yet to be made, but are likely to be a major determinant of the fortunes of Wales' birds over the coming decades. The principle of public money delivering public goods, including wildlife, has gained political acceptance, so that farmers can be paid to deliver outcomes for nature, such as peat bogs, species-rich hay meadows or Black Grouse. Whether saving nature becomes a core purpose of land management will depend on counter-pressures from the main farming unions and the willingness of the government in power to listen to different voices, when critical decisions need to be made. The changing market for lamb, especially if opportunities to export to the EU are less, could result in significant de-stocking in the uplands, and abandonment of many farms. Nature conservation may, then, prove to be the most valid reason to maintain some grazing animals in the uplands to manage for open-ground species such as Curlew, if the alternative is forestry. In the valleys and lowlands, if action for nature continues to be voluntary, such as through agri-environment schemes, it will remain on the periphery of most busy farmers' radar and the recovery of farmland birds is much less likely.

The responses of farmers to a changing climate are equally hard to predict. Milder winters could, for example, extend the season for a different suite of crops to be grown. The expansion of silage has already transformed the farmed landscape in the lowlands, and with it the wild plants, invertebrates and nesting birds that used to live in the meadows it replaced. Currently, silage in the uplands tends to be taken in a single cut, usually in June, which can enable field-nesting birds to fledge or, for species such as Curlew, to leave the nest prior to mowing. But an earlier growing season, combined with new breeds of silage grass, could bring forward cutting to earlier in the season and so reduce further the area of land available to nesting birds. Treatment of livestock with pharmaceuticals can lead to significant reductions in fly larvae in animal dung, on which Choughs feed (Gilbert *et al.* 2019). These treatments are now used more frequently by farmers in western Scotland in response to elevated levels of parasites associated with higher temperatures. These are just two examples, and there will doubtless be many more adaptive changes in farming management that have unintended consequences for the ecological food chain, on top of which birds frequently sit.

Variable levels of grazing produce a diverse landscape of pasture, *ffridd*, woodland and moors © Jeremy Moore

Farming will also come under pressure to mitigate its own contribution to causing climate change. If it continues at current levels whilst other parts of society, particularly energy generation, reduce their greenhouse gas emissions, it is likely to be one of the biggest emitters by 2050. Nature-based solutions, including restoration of peatland and permanent grassland, planting hedgerows and spring-sown crops, would all directly benefit nature, as well as locking up carbon in the soil. Reductions in red-meat consumption may yet prove to be the biggest driver of change. Ultimately, the fate of many bird species will depend on how much money is invested in these changes, along with the choices made between biodiversity, a dash for carbon or food production, and the long-term security that will enable individual land managers to make decisions that can be sustained.

Invasive and problem species

Invasive non-native species (INNS) or problematic native species have been a relatively small factor in the declines of biodiversity in Britain to date, and the direct effects on birds have been negligible (Burns *et al.* 2016). However, increasing parasitic loads in cattle, mentioned above, is just one example of nature becoming more susceptible to pathogens as a result of changing climate. Tree disease is another example of the potential landscape effects. Dutch Elm Disease resulted in short-term benefits, but long-term losses, for woodland birds such as Lesser Spotted Woodpecker in the late 20th century (Ken Smith *in litt.*).

Woodland managers in Wales are currently dealing with *Phytophthora ramorum* in larch trees and *Hymenoscyphus fraxineus* in Ash trees. *P. ramorum* has had a significant impact on larch plantations in southern Wales since 2009, and by 2018 had spread as far north as Coed Gwydir, Caernarfonshire. While the impact of tree disease is primarily viewed as economic (Ash Dieback alone is forecast to cost £15 billion over the next century), the potential for *P. ramorum* to affect native oak woodland would have huge long-term consequences for specialist woodland species. The response to disease can be as dramatic, in the short-term, as the disease itself. Clear-felling provides a temporary boost for nesting Nightjars, for example, but a loss of nesting and hunting habitat for Honey-buzzards.

It is estimated that 36–48 new invasive species will become established in Britain by 2040, and their effects on wildlife and the economy could be huge. The response by public authorities is currently insufficient to deal with the forecast impact. A UK parliamentary inquiry said that "the UK is very good at horizon scanning for new invasive species, but is poor at taking action" (EAC 2019). Dealing with the problem once it has arrived is expensive and can be controversial (see Ruddy Duck), but failing to deal with the pathways of introduction could lead to transformative changes for birds and their habitats in Wales.

Public opinion and education play a considerable part in the acceptability of actions to tackle invasive species once they have arrived. The same is true of native species, such as Fox and Carrion Crow, which threaten the nesting attempts of waders and terns. Conservation has to balance what is ecologically necessary in the short term with the views of people whose values consider the life of a crow or a Fox equal to that of a Roseate Tern or a Curlew. This is a real dilemma for organisations that depend on money from donations or taxpayers, and yearn for a more intact landscape, where lethal predator control is unnecessary. It also fails to deal with the crux of the problem: why generalist predators are more abundant in Britain than the rest of Europe. There has been precious little funding to investigate this, although some studies point to factors such as the release of a high biomass of non-native gamebirds such as Pheasants (Pringle *et al.* 2019). Resolving both the science and the public acceptability surrounding these issues will be important if the full management toolkit is to be available to land managers responsible for rare birds in tandem with landscape-scale habitat restoration. On its own, killing native predators would be a never-ending task, sucking up huge resources that could be better directed to more sustainable solutions.

Conservation recovery

Having read this far through the introduction, it would be tempting to think that too much hope is already lost. As described in Chapter 3, the nature conservation movement developed from tiny beginnings to deliver real change on the ground and in the halls of policy-making. Conservation does work. At a global level, ten bird species went extinct during 1993–2020, but intervention prevented the extinction of up to 32 others (Bolam *et al.* 2020). During that period, Corn Bunting was lost as a regular breeding species in Wales, and the increasing number of species that are Red-listed here is a reminder that the direction of travel for many more is downward. However, the accounts in the book also tell of successful recovery, where the causes of decline have been tackled and populations have been made more resilient by improvements in the area and quality of habitat. Black Grouse would probably have been lost were it not for actions started in the 1990s. Lapwings would now have been lost as a breeding species from most counties and all lowland habitats in Wales, were it not for management interventions on lowland grassland nature reserves. Wetland creation in Anglesey and southeast Wales has enabled the return of Bittern and Marsh Harrier as breeding birds, and a suite of other plants, invertebrates and other fauna. This has been

Diseases and pathogens in trees, such as Larch and Ash, could dramatically transform the Welsh landscape, and are expected to increase as the climate warms © Ben Porter

A reduction in grazing can quickly result in regeneration of heather and trees, although in many places such as above the Tal y Llyn pass, Meirionnydd, non-native conifers also spread without intervention © Mike Alexander

achieved mostly by conservation charities and a small number of other private landowners, and to some extent by government. Restoring wetlands is relatively quick to achieve, and the response by birds is rapid compared to restoration of many other habitats.

The problem is one of scale and finances. In many cases, the causes of decline are known and the techniques to recovery have been trialled, or could be, if the money were available. The resources for nature conservation are dwarfed by the money from government and business that contribute to nature's decline. The proportion of the £250 million paid to Welsh farmers each year that delivers for nature is small, and public-sector expenditure on biodiversity conservation in the UK, as a proportion of GDP, fell by 42% between 2008–09 and 2018–19 (Hayhow *et al.* 2019). This will only change through political will and public support. The choices to be made following withdrawal from the European Union, especially for fisheries management, farm funding and the network of protected areas, make the immediate future one of uncertainty but also possibility.

The need to restore habitats on a large scale is the biggest change in nature conservation ambition in the last 20 years. The 'Lawton Principles', summarised as 'bigger, better and more connected' (Lawton *et al.* 2010), were written for England, but apply to the ecological network in every country. Nature reserves with dedicated management will, surely, continue to play a valuable role in securing bird populations, especially of the rarest species, but on their own are not enough. Increased conservation efforts may flatten the curve, but it will require transformational change to make production and consumption by humans genuinely sustainable (Leclère *et al.* 2020).

'Rewilding' is one approach that has found some popular resonance. Definitions vary, but it is essentially about restoring ecosystem function at a landscape scale, either by letting nature take its course or with a degree of continued management that is far less intensive than exploitative land use or many nature reserves. It is not without controversy, especially among upland farmers who see it as a threat to their livelihood, land values and culture. Additional concerns come from proposals to reintroduce apex predators or keystone species, such as Golden Eagle and European Beaver, whether or not associated with rewilding habitat projects. It is early days for rewilding, with several ambitious projects underway in Scotland, but it has had a faltering start in Wales. There is no 'right' menu for rewilding, and there is no doubt that it has significant potential to restore ecosystem function in many parts of Wales. Not all ecologists are convinced by the popular attraction of rewilding, however. It would require a reduction, perhaps initially a complete removal, of grazing herbivores to allow the vegetation to recover. There is no space here to describe the complex range of likely outcomes, but they are likely to vary greatly. Where there is a nearby seed source, the vegetation would probably become more diverse quite rapidly. There could be fairly immediate negative responses for birds that require short grass swards, such as Chough and Lapwing, but the growth of scrub could benefit species such as Whinchat and Cuckoo, as hillsides could return to *ffridd*-like conditions. Without the reintroduction of herbivores at some point, succession will lead to climax woodland in these places, although that is likely to be a mix of native and non-native species, rather than a romantic notion of wildwood.

In other places, change would be much slower in the absence of grazing. On the higher, thin soils of Pumlumon, Ceredigion/Montgomeryshire, for example, there is minimal scrub growth, perhaps because there is no seed source. Here, 12 years of zero grazing has led to the recovery of rare plant species, and successful nesting by Golden Plover and Merlin. With relatively few deer and a moderate climate, rewilding in Wales may well be more straightforward and rapid than in Scotland. The barriers are not ecological but psychological: "the contrast in attitudes to wild nature between Britain and the Continent is striking. The British public seem to have a mistrust of land 'out of control', and a fear of 'dangerous' animals" (Newton 2020). Or is it the landowners and farming unions who fear loss of control, in the same way that urban authorities have had an unwarranted devotion to 'tidiness'?

Rewilding can be cheaper than any form of subsidised land use and can also support more people economically, mainly through wildlife tourism, but it will not be appropriate everywhere and involves accepting the loss of species currently of high conservation priority, including Curlew, Chough and Lapwing in those areas. It may be better to consider rewilding as one end of a continuum, alongside regenerative farming that aims to increase biodiversity, improve soils and enhance ecosystem services. Nature-friendly farming is still viewed as outside the mainstream by many farming leaders, but it may become more widely supported if, for example, the alternative is farm abandonment, and if future payments from the public purse are linked to outcomes such as birds, pollinators and the right trees in the right place.

The future

Where possible, each species account in *The Birds of Wales* makes a prediction of its future prospects. Some will be accurate, others will miss the mark, because there are things about bird ecology that are not understood or have not even been studied. There may be factors that we have not even spotted. Making predictions may be seen as an act of brave folly at a time of so much uncertainty, and we applaud those authors who have attempted to describe them. It will be for the authors of the next national avifauna to judge how successful we were.

These accounts mark a point in time and there is still opportunity and hope that the next century can be more positive than the last. The blueprint for recovery exists. The nature-based solutions to some societal problems are known. The areas of greatest importance for Wales' birds have been identified. The political decisions taken in Cardiff, Westminster and globally will, more than those at any other time in history, determine the future abundance and diversity of Wales' birds.

Julian Hughes

If we make the right decisions at this critical moment, we can safeguard our planet's ecosystems, its extraordinary biodiversity and all its inhabitants… What happens next is up to every one of us.

Sir David Attenborough

7. Introduction to the species accounts

The concept behind a systematic list seems simple enough: to give an account of the fortunes of all the bird species that have been recorded in Wales. In practice, a great many decisions have to be made. First of all, there is no universal agreement on what constitutes a species and what a subspecies. Several lists of the bird species of the world are available, and these differ in their approach. We have followed IOC World Bird List 10.2 (Gill *et al.* 2020).

Decisions also need to be taken on which species are to be included in the systematic list. For this we have followed the Welsh List for 2021, available at https://birdsin.wales/welsh-list-2021-2/. A validated list of the birds of Wales is a very recent enterprise, first published by the Welsh Ornithological Society in 2017. It includes all those species that are considered to have occurred as 'wild', including those that have been introduced or become re-established and whose populations are now self-sustaining. Each of the 451 species on the Welsh List is assigned to a category, shown in the status box at the head of each species account.

- Category A: Birds recorded in an apparently natural state at least once since 1 January 1950. The great majority (436) of the species listed are in this category.
- Category B: Birds recorded in an apparently natural state at least once between 1 January 1800 and 31 December 1949. Seven species are listed in this category, of which three also now occur in Category C, which are included in the total above.
- Category C: 16 introduced species that have established self-sustaining populations, either in Wales or in an adjacent territory from which birds in Wales have originated. There are several subcategories:

C1	Naturalised introduction, split into C1(1) (introduced prior to 1950, e.g. Canada Goose) and C1(2) (introduced since 1950, e.g. Mandarin Duck)
C2	An established population resulting from introductions but also occurs in an apparently natural state, e.g. Gadwall
C3	Naturalised re-establishments of a species formerly breeding in Wales, e.g. Goshawk
C4	Naturalised feral species, e.g. Feral Pigeon
C5	Vagrant naturalised species, from an established population elsewhere in Britain or beyond, e.g. Egyptian Goose
C6	Former naturalised species, population now no longer self-sustaining or extinct.

Species placed in the other two categories are not part of the Welsh List and are not included in the main accounts:

- Category D: Species for which there is reasonable doubt that it has occurred in a natural state.
- Category E: Species introduced, escaped from captivity or transported by human agency, with no self-sustaining population.

Some species in these categories considered of particular interest are included in Appendix 1. A species may be allocated to more than one category; for example, Ruddy Shelduck is in Category B and in Category E.

Collecting the information for this book provided an opportunity to give a historical account of the development of the Welsh List since 1900. As the 20th century dawned, 280 species had been recorded in Wales, based on current criteria of valid records. Almost all of these were resident or regular visitors, and this core of the list has changed very little. The growth has come almost entirely from the increased recording of scarce and rare visitors. In each of the first four decades of the 20th century, the list increased at an average of four new species each decade. The growth in birdwatching, ringing and recording after the Second World War,

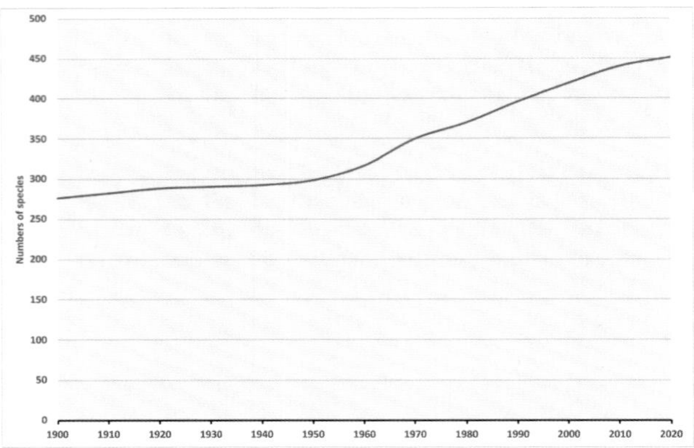

Figure 11. Numbers of species on the Welsh List at the start of each decade, based on the first dated records shown in the species accounts.

described in Chapter 2, accelerated the speed at which new species were added. Skokholm, in Pembrokeshire, and Bardsey, in Caernarfonshire, have added 27 and 37 species respectively since the establishment of a bird observatory on each island.

The number of bird species recorded in Wales increased to 300 in April 1948 when an Olive-backed Pipit was found on Skokholm and surpassed 400 species in 1990, with a Lanceolated Warbler on Bardsey. The growth in numbers came from a great increase in birdwatching, as people had more money and more leisure time, together with easier travel and more nature reserves to visit. The availability of better equipment, particularly binoculars, telescopes and cameras, also greatly aided the identification of new species. In the subsequent three decades, some additions to the Welsh List would come at the hands of taxonomists, as the use of mtDNA sequencing improved understanding of differences and relationships between species.

Almost half (46%) of the species on the Welsh List are rare visitors (Welsh Ornithological Society 2021). The remainder occur regularly, as breeders, non-breeders and passage migrants, although some species—Wigeon and Teal, for example—can be classified as all three, hence the total in Table 9 is greater than the number of species on the List.

Regular breeder	154
Former regular breeder	13
Sporadic breeder	8
Non-breeding (passage/winter)	203
Rare and scarce	211

Table 9. Native species on the Welsh List (categories A–C), classified according to their occurrence.

Period covered

The accounts feature records of birds in Wales from the earliest remains found during archaeological excavations, to 31 December 2019. The information about rare and scarce species is based on records accepted up to October 2020, with the exception of a small number removed subsequently by a review of pre-1950 records (Naylor 2021) and the addition of those accepted by the BBRC (Stoddart and Hudson 2021). A few records from 2020 are mentioned, by exception, where this was thought to be relevant.

Names

The English names follow those of the BOU's British List, scientific names are those used in the IOC World Bird List 10.2 (Gill *et al.* 2020) and the Welsh names are those recommended by *Adar y*

Byd (Cymdeithas Edward Llwyd a Cymdeithas Ted Breeze Jones 2015). These are the conventions adopted by the Welsh List (2021). Note that some bird subspecies do not have generally accepted English or Welsh names. Species other than birds referenced in the accounts are listed in Appendix 2.

Ageing of birds

Following the convention adopted by WOS and the BBRC, we describe the age of birds as first calendar-year, second calendar-year, etc. In lists, this is abbreviated to 1CY or 2CY. A bird is in its first calendar-year from the time of hatching to 31 December that year, in its second calendar-year from 1 January to 31 December the following year, etc. This is only presented for species where the age of the bird can be told by plumage characteristics in the hand, or in the field, such as gulls.

Conservation status

The box at the start of each species account includes the global International Union for Conservation of Nature (IUCN) assessment of the extinction risk for that species (BirdLife International 2020), and the assessments for Europe (BirdLife International 2015) and Britain (Stanbury *et al.* 2017), where applicable. The assessments, in order of gravity, are: RE: Regionally Extinct; CR: Critically Endangered; EN: Endangered; VU: Vulnerable; NT: Near Threatened; and LC: Least Concern. Where an assessment of extinction risk for the British breeding and non-breeding population differed in gravity, we show the highest level, qualified by B (breeding population) or NB (non-breeding population).

The other assessment shown is the conservation status in Birds of Conservation Concern (BoCC). The population status of birds regularly found in the UK, the Channel Islands and the Isle of Man is reviewed regularly to provide an assessment of conservation priorities. Species are placed in one of three lists—Red, Amber and Green—based on quantitative criteria, with Red indicating the highest level of concern. The assessment from the UK BoCC 2015 is shown. There have also been reviews of Birds of Conservation Concern specifically for Wales. The first of these was in 2002, and there have been others in 2010 and 2016. All three assessments for Wales are presented here. The criteria used for the most-recent BoCC assessment in Wales are given in Johnstone and Bladwell (2016).

Distribution

The global distribution of each species is outlined at the start of each species account. The distribution of subspecies is not normally described unless that subspecies has been recorded in Wales.

Inclusion of records

Before a species can be included on the Welsh List, it must first be included on the British List. The decision on adding a new species to the British List is the responsibility of the British Ornithologists' Union Records Committee (BOURC). It was important to use some criteria to determine whether records of birds were sufficiently reliable to be included in the accounts. For the rarest species, the British Birds Rarities Committee (BBRC) has assessed records since 1958. The Welsh Records Advisory Group (WRAG) was set up in 1972 to advise on the acceptance of records of species not rare enough in Britain to be dealt with by the BBRC, but still rare or scarce in Wales. The WRAG was later replaced by the Welsh Records Panel (WRP), which has now changed its name to the Welsh Birds Rarities Committee (WBRC). The lists of species considered by the BBRC and the WBRC have changed considerably over the years. For example, Little Egret records were originally considered by the BBRC, then were moved to the WRP/WBRC list and later dropped from this list, too, as the species became more common in Wales. As a matter of policy, records of species considered by these committees are only included here if a description, or a photograph, was submitted to the relevant committee, which accepted it as a valid record. All records up to 31 December 2019 that had been accepted by the BBRC or WBRC by 31 October 2020 are included. The status box shows, where appropriate, which committee is currently assessing records of the species, and also notes whether the species is on the list of the Rare Breeding Birds Panel (RBBP), which collects records of rarer breeding species. Not all birdwatchers submit their records, and consequently a great deal of interesting information posted in blogs on the web and elsewhere cannot be used in this book.

Older records, from the period before these groups were established, have had to be assessed separately. For this book, the WBRC set up a Historical Records Group, headed by Reg Thorpe, to assess these. As a result, some older records previously published have been deemed insufficiently well documented and are not included here (Green 2020a). A few previously published records of commoner birds have also not been included, for example some exceptional counts, because they are no longer considered reliable. This has usually been based on the advice of the local County Recorder and other local ornithologists. For a small number of species, only records since 1967 are included in the statistics, because it has proven difficult to produce accurate figures with certainty before the start of reporting by the WRAG.

Recording areas

Most ornithological works in Wales until fairly recent times used the historic counties of Wales as the recording units. These 13 counties were the basic units of local government in Wales for centuries, until 1974 when local government was reorganised. Wales was then divided into eight larger units: Clwyd, Gwynedd, Powys, Dyfed, West Glamorgan, Mid Glamorgan, South Glamorgan and Gwent. Although these units were sometimes used for bird recording, the authors of *Birds in Wales* (1994) used the historic counties to maintain consistency with past records. In 1996, local government in Wales was again reorganised, and a system of smaller unitary authorities was established. Some of these bore the same names as the historic counties, but the boundaries were often different. These changes had the potential to cause considerable problems for bird recording, so following the 1996 reorganisation, the Welsh Ornithological Society, in consultation with local bird clubs and recorders, switched to the Watsonian vice-county system as the basis of bird recording in Wales. This system was first proposed by

Figure 12. Counties in Wales used for bird-recording. Note the zones that differ from the 'old' counties: in (a) records are maintained by Denbighshire (VC50) although administratively it is in Powys; those from (b) and (c), while historically in Breconshire, are maintained by recorders in East Glamorgan and Gwent respectively; area (d), while historically in Gwent, is part of the East Glamorgan recording area and follows the modern boundary of the Cardiff and Newport unitary authorities.

the botanist Lyall Watson in the 1850s and is similar to the pre-1974 historic counties, so retains a degree of continuity (Dandy 1969).

However, there are some differences between the Watsonian vice-counties and the historic administrative counties. The vice-counties essentially followed the county boundaries as they were in the mid-19th century, but any detached portions of counties were included in the vice-county surrounding them in order to simplify the boundaries. The differences between the pre-1974 county, the post-1996 county and the Watsonian vice-county are perhaps best seen in Flintshire. The pre-1974 county of Flintshire had a large detached part to the southeast of the main area. In the vice-county system, this detached area became part of Denbighshire. It is also not included in the post-1996 unitary authority of Flintshire, instead forming part of the county borough of Wrexham. A map showing the vice-county boundaries can be seen at cucaera.co.uk/grp. The Welsh Ornithological Society decided to use the name 'Gwent' for the vice-county of Monmouth to prevent confusion with the new unitary authority of Monmouth, which covers only part of the vice-county, and this has been followed here. It also uses 'Ceredigion' rather than 'Cardiganshire' and 'Meirionnydd' rather than 'Merionethshire'. It was also decided at this time to split the vice-county of Glamorgan into two parts for bird-recording purposes: Gower and East Glamorgan. Note that 'Gower' here refers to the western part of the vice-county of Glamorgan, corresponding to the old unitary authority of West Glamorgan, not merely to the Gower Peninsula. The vice-county remains Glamorgan, however. In some areas, local arrangements mean that the standard boundaries of the vice-counties are not followed.

Population estimates

Population estimates for breeding birds in Wales have been derived from several sources:

(a) from dedicated surveys, such as the Seabird Monitoring Programme, the BTO Heronries Census and periodic surveys conducted as part of the Statutory Conservation Agency and RSPB Annual Breeding Bird Scheme (SCARABBS), such as for Ring Ouzel and Twite;

(b) information submitted to the Rare Breeding Birds Panel (RBBP) or published in bird reports; and

(c) for more abundant species, except those that breed colonially or hunt nocturnally, the estimates produced by Hughes *et al.* (2020). These used Timed Tetrad Visit (TTV) abundance data from fieldwork for the *Britain and Ireland Atlas 2007–11* to estimate the proportion of the British population found in Wales. The reliability of these estimates is dependent on those made by the Avian Population Estimates Panel for Britain (e.g. Musgrove *et al.* 2013, Woodward *et al.* 2020), on which they are based.

Readers may be tempted to compare population estimates in this book with those published in its predecessors, *Birds in Wales* (1994, 2002). The population estimates that are derived from Hughes *et al.* (2020) should not be directly compared with previous estimates, however. For example, the population estimates cited in *Birds in Wales* (2002) used data from the *Britain and Ireland Atlas 1988–91* that were based only on the percentage of occupied squares. Unlike the estimates by Hughes *et al.* (2020), those did not allow for differences in relative abundance between Wales and other parts of Britain.

Hughes *et al.* (2020) highlighted the importance of participation in the Breeding Bird Survey (BBS) and atlas fieldwork in order to obtain more data that can be used to assess population trends. Future calculations will be even more reliable if there is an increase in the numbers of BBS surveys conducted in Wales and if an all-Wales tetrad atlas, with as many Timed Tetrad Visits as possible, could be achieved.

Sources of information

The species accounts draw on a wide range of sources, including the *Welsh Bird Report* and county bird reports, national, regional and county atlases, and country and county avifaunas, described in Chapter 2. The *Welsh Bird Report* and county bird reports are considered as basic sources for records, and we have not referenced every record taken from them. However, we do give a reference if the opinion of the author (or authors) is cited. References are generally given for records taken from other sources. Data in charts are based on records in Welsh Bird Reports or from bird observatory records in BirdTrack unless otherwise stated.

Much use is made of data from the BTO/JNCC/RSPB Breeding Bird Survey (BBS), and graphs showing trends for Wales are included where available (Harris *et al.* 2020). We have not made extensive use of earlier data from the Common Birds Census (CBC) as few of the plots covered were in Wales and therefore trends shown for Britain may not apply to Wales. Much information on waterbirds outside the breeding season from the Wetland Bird Survey (WeBS) is available, and where appropriate, we have included graphs showing trends in Wales and five-year average counts at the most important sites. For the two estuaries shared by Wales and England, the Dee and the Severn, counts from both the Welsh side and the whole wetland are shown where available ('na' denotes that counts were not available separately for the Welsh side for some years).

Information on seabird numbers comes from a series of co-ordinated censuses of breeding seabirds, from Operation Seafarer (1969–70) to the recent Seabirds Count 2015–21, and from regular counts at colonies as part of the Seabird Monitoring Programme (SMP) run by the JNCC. Other surveys of specific species or groups of species have also been used. Information from the BTO online ringing report (Robinson *et al.* 2020) and from the *BTO Migration Atlas* (Wernham *et al.* 2002), together with data received from the BTO, has also been used for many species. The ringing maps, which use data supplied by the BTO with permission of ringers, show exchanges with Wales from outside Britain, unless otherwise stated.

Extensive use has also been made of BirdTrack records, which include historical records, notably those from the bird observatories on Bardsey, Caernarfonshire, and Skokholm, Pembrokeshire. For a number of species, these have been used to generate graphs showing the number of passage migrants recorded at these observatories. For rarer breeding species, the reports of the Rare Breeding Birds Panel (RBBP) have been of great value, together with additional information provided by the Panel. As well as works already published, we were able to obtain information from *European Breeding Bird Atlas 2*, prior to its publication in December 2020.

Abbreviations used in the species accounts

BBRC	British Birds Rarities Committee
BBS	Breeding Bird Survey
BOURC	British Ornithologists' Union Records Committee
BTO	British Trust for Ornithology
CBC	Common Birds Census
CCW	Countryside Council for Wales (now part of NRW)
IUCN	International Union for Conservation of Nature
JNCC	Joint Nature Conservation Committee
NRW	Natural Resources Wales
RBBP	Rare Breeding Birds Panel
RSPB	Royal Society for the Protection of Birds
WBRC	Welsh Birds Rarities Committee (formerly the Welsh Records Panel)
WeBS	Wetland Bird Survey
WWT	The Wildfowl & Wetlands Trust

Abbreviations of references for data sources

We thought that it would be more convenient for readers if the works most frequently used in references were presented as titles or abbreviated titles rather than conventional citations by author and date. These are as follows:

Quoted as:	**Reference (full reference in Bibliography)**
BirdLife International (2015)	European Red List and supplementary data, from: http://datazone.birdlife.org/info/euroredlist
BirdLife International (2020)	Species information from the BirdLife datazone: http://datazone.birdlife.org/home
Birds in England (2005)	Brown, A., and Grice, P. (2005)
Birds in Ireland (1989)	Hutchinson, C. (1989)
Birds in Wales (1994)	Lovegrove, R., *et al.* (1994)
Birds in Wales (2002)	Green, J. (2002)
The Birds of Scotland (2007)	Forrester, R., and Andrews, I. (2007)
Britain and Ireland Atlas 1968–72	Sharrock, J.T.R. (1976)
Britain and Ireland Winter Atlas 1981–84	Lack, P.C. (1986)
Britain and Ireland Atlas 1988–91	Gibbons, D.W., *et al.* (1993)
Britain and Ireland Atlas 2007–11	Balmer, D.E., *et al.* (2013)
BTO Migration Atlas (2002)	Wernham, C.V., *et al.* (2002)
BWP (Birds of the Western Palearctic)	Cramp, S., and Simmons, K.E.L. 1977–1994
East Glamorgan Atlas 1984–89 and *East Glamorgan Atlas 2008–11*	eastglamorganbirdatlas.org.uk
European Atlas (1997)	Hagemeijer, W.J.M., and Blair, M.J. (1997)
European Atlas (2020)	Keller, V., *et al.* (2020)
Gwent Atlas 1981–85	Tyler, S., *et al.* (1987)
Gwent Atlas 1998–2003	Venables, W.A., *et al.* (2008)
HBW (Handbook of the Birds of the World)	del Hoyo, J., *et al.* (eds.) (1992 onwards) Online version: birdsoftheworld.org/bow/home
Historical Atlas 1875–1900	Holloway, S. (1996)
IOC	Gill, F., *et al.* (2020)
North Wales Atlas 2008–12	Brenchley, A., *et al.* (2013)
PECBMS 2020	Pan European Common Bird Monitoring Scheme https://pecbms.info/trends-and-indicators/species-trends
Pembrokeshire Atlas 1984–88	Donovan, J., and Rees, G. (1994)
Pembrokeshire Atlas 2003–07	Rees, G., *et al.* (2009) Some figures have been updated/corrected from the online *Pembrokeshire Avifauna* (2020).
West Glamorgan Atlas 1984–89	Thomas, D.K. (1992b)

Photographs

All the photographs have been taken in Wales. Where images were taken at the nest of species listed on Schedule 1 of the Wildlife and Countryside Act 1981, the photographer was licensed by the relevant statutory conservation agency.

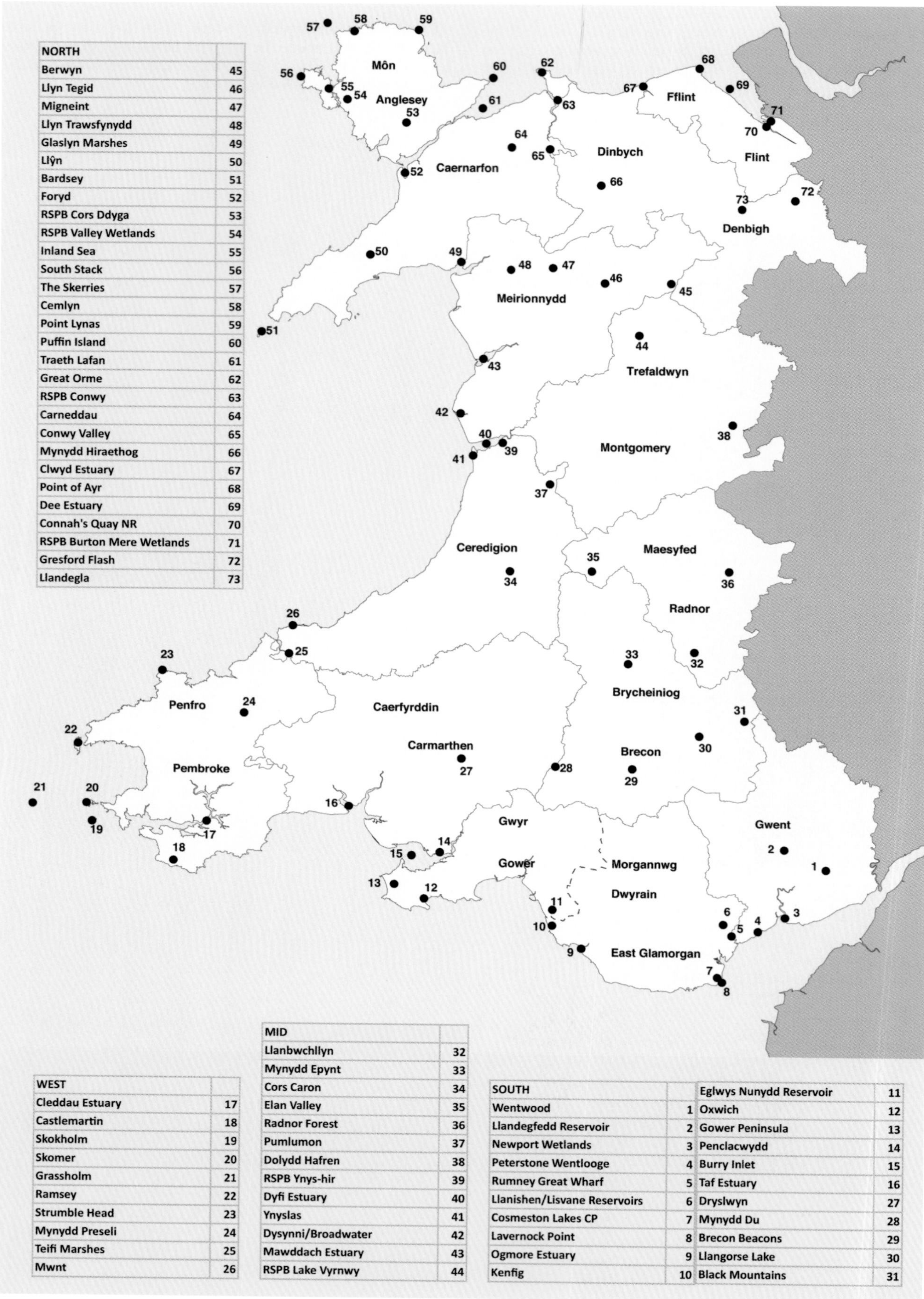

NORTH	
Berwyn	45
Llyn Tegid	46
Migneint	47
Llyn Trawsfynydd	48
Glaslyn Marshes	49
Llŷn	50
Bardsey	51
Foryd	52
RSPB Cors Ddyga	53
RSPB Valley Wetlands	54
Inland Sea	55
South Stack	56
The Skerries	57
Cemlyn	58
Point Lynas	59
Puffin Island	60
Traeth Lafan	61
Great Orme	62
RSPB Conwy	63
Carneddau	64
Conwy Valley	65
Mynydd Hiraethog	66
Clwyd Estuary	67
Point of Ayr	68
Dee Estuary	69
Connah's Quay NR	70
RSPB Burton Mere Wetlands	71
Gresford Flash	72
Llandegla	73

WEST	
Cleddau Estuary	17
Castlemartin	18
Skokholm	19
Skomer	20
Grassholm	21
Ramsey	22
Strumble Head	23
Mynydd Preseli	24
Teifi Marshes	25
Mwnt	26

MID	
Llanbwchllyn	32
Mynydd Epynt	33
Cors Caron	34
Elan Valley	35
Radnor Forest	36
Pumlumon	37
Dolydd Hafren	38
RSPB Ynys-hir	39
Dyfi Estuary	40
Ynyslas	41
Dysynni/Broadwater	42
Mawddach Estuary	43
RSPB Lake Vyrnwy	44

SOUTH			
		Eglwys Nunydd Reservoir	11
Wentwood	1	Oxwich	12
Llandegfedd Reservoir	2	Gower Peninsula	13
Newport Wetlands	3	Penclacwydd	14
Peterstone Wentlooge	4	Burry Inlet	15
Rumney Great Wharf	5	Taf Estuary	16
Llanishen/Lisvane Reservoirs	6	Dryslwyn	27
Cosmeston Lakes CP	7	Mynydd Du	28
Lavernock Point	8	Brecon Beacons	29
Ogmore Estuary	9	Llangorse Lake	30
Kenfig	10	Black Mountains	31

Principal birdwatching sites mentioned in the text.

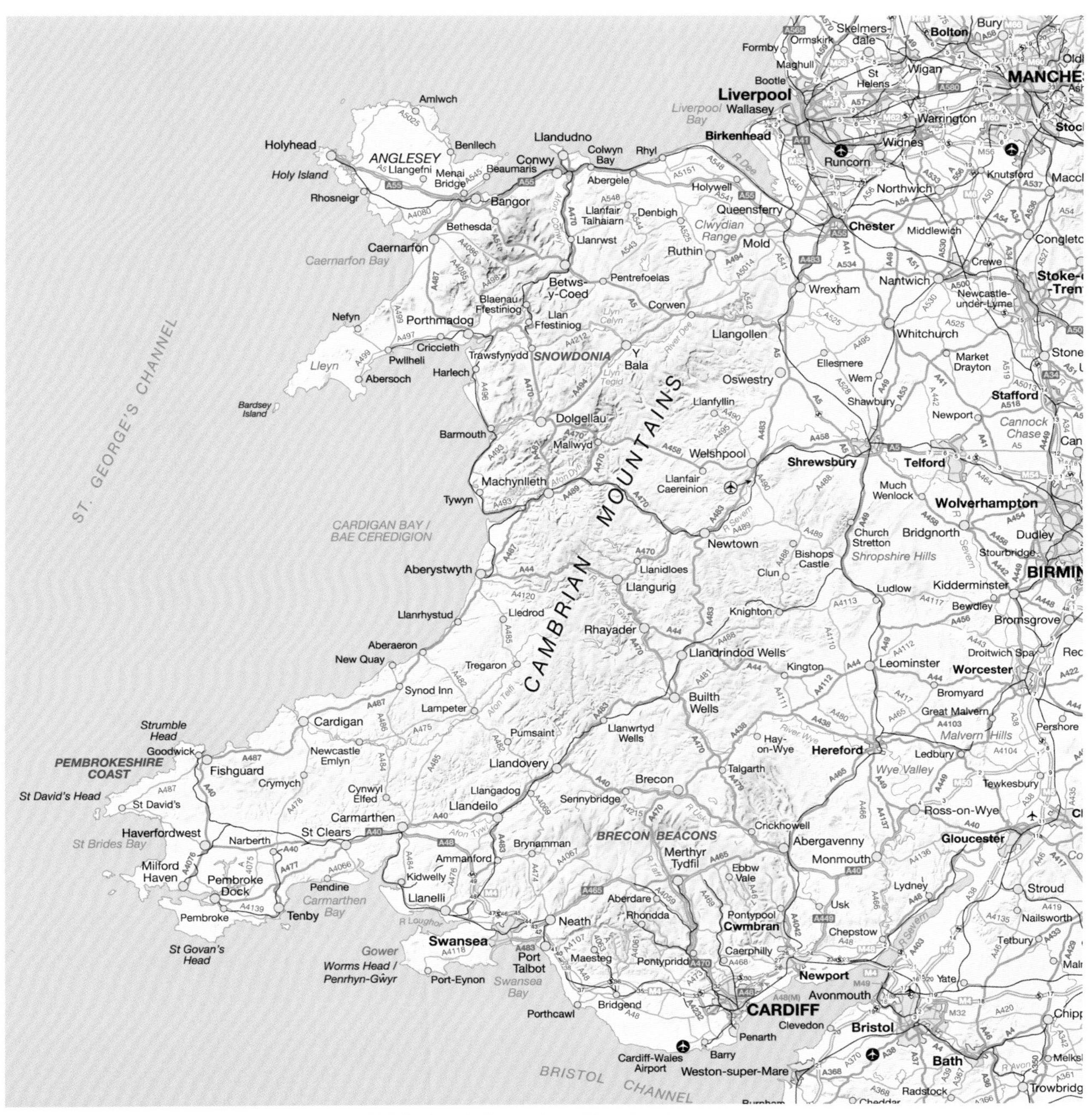

Public domain image from **Mapsland.com.**

SPECIES ACCOUNTS

Black Grouse *Lyrurus tetrix* Grugiar Ddu

Welsh List Category	IUCN Red List			Birds of Conservation Concern			
	Global	Europe	GB	UK	Wales		
A	LC	LC	VU	Red	2002	2010	2016

The Black Grouse occurs in a broad geographic band from Europe through temperate Russia and Central Asia to northeast China, reaching the Sea of Japan (*HBW*). The global breeding population was estimated to be several million (Storch 2000), with the highest densities in Europe found in Fennoscandia. Six subspecies are recognised, with the one occurring in Britain, *L.t. britannicus,* being the only one that is geographically isolated.

In northern Europe, the Black Grouse is typically associated with transitional habitats where woodland or forest meets open heath, bog, fen, steppe or poorly cultivated areas. In Britain, particularly Scotland and Wales, the close proximity of trees, particularly conifers, seems to be an essential habitat requirement. Newly planted open stands are favoured, particularly where there is a healthy understorey of food plants such as heather and bilberry, but this is only available until the tree canopy has closed.

Black Grouse are the stylish thespians of the bird world: males gather at communal display sites or leks, predominantly at first light. Most prefer a single lekking site within their breeding territory, but some have several lek sites that they occupy on a daily, rotational basis. Höglund *et al.* (2002) suggested that many males at a lek are related but the risk of in-breeding is significantly reduced by the species' reproductive strategy, whereby juvenile females disperse farther from their natal areas than young males.

Thomas Evans, vicar of Llanberis, reported the species as abundant in that part of Caernarfonshire in the late 17th century (Emery 1985), while in 1781 Pennant described it as abundant around Ffestiniog, Meirionnydd, but already rare in the Clwydian Hills, Denbighshire/Flintshire (Pennant 1810). It appears to have been lost from many areas in the 19th century, perhaps as a result of overshooting or a series of poor breeding years. Forrest (1907) documented its extirpation in many locations and stated that by the early 20th century, records in North Wales were sporadic but that it remained locally common in Montgomeryshire. It was also described as rare in Carmarthenshire by Davies (1858), but formerly had been plentiful on the Black Mountain (Y Mynydd Ddu). It was regular in Glamorgan as far south as Penllergaer, Gower, at the end of the 18th century, but was rare in the county by the end of the 19th century (Hurford and Lansdown 1995).

The *Historical Atlas 1875–1900* found evidence of extant populations only in Gwent, Radnorshire and Montgomeryshire. However, gamebags show that small numbers were shot in Radnorshire during 1860–1900, from Maesllwch in the south to Rhayader in the north (Jennings 2014). An average of 168 was shot on the Cawdor Estate, Carmarthenshire, during 1888–95, and *Birds in Wales* (1994) reported an increase in Breconshire during 1870–1900. Records from Coed Dias Estate, Gwent/Breconshire,

Male Black Grouse © Keith Offord

show that 22 were shot during 1902/03–1911/12, but the species did not feature subsequently.

The natural range of Black Grouse is clouded by releases for shooting in many areas. Forrest (1907) wrote of numerous unsuccessful attempts to (re)establish Black Grouse at sites across North Wales through the 19th century, as far back as the 1830s on the Berwyn, Meirionnydd (Eyton 1838). Fletcher (2001), assessing the diaries of the Duke of Newcastle, stated that Black Grouse were translocated from the Isle of Arran, western Scotland, to the Hafod Estate, Ceredigion, in 1833. There is no evidence that Black Grouse occurred naturally on Anglesey, but several unsuccessful introduction attempts were made in the late 1920s, and similar efforts were made in several other counties between the wars. Such releases probably account for the handful of records in Pembrokeshire, the last of which was in 1971.

Records in Caernarfonshire in the 20th century were relatively sparse until after the Second World War, when forest-planting at Coed Gwydyr resulted in the establishment of a population that numbered 20 males at three leks in the mid-1980s (Nigel Brown pers. comm.). They probably also occurred around Beddgelert, Llanberis and Mynydd Llandygai until the late 1960s. Jones and Dare (1976) described them as scarce, and Barnes (1997) reported just a handful in the county. Afforestation also boosted the population in Meirionnydd during 1940–85, with some evidence of range expansion continuing in the county in the early 1980s (Pritchard 2012). In Montgomeryshire, a review of Forestry Commission landholdings suggested they were spreading in Coedydd Hafren, Dyfi and Dyfnant (Cadman 1949). Numbers fell sharply after 1975 and the species retreated to just a few sites in the county, principally Cwm Nant yr Eira, Lake Vyrnwy and ultimately Llanbrynmair, where planting of conifers on the moorland started in the 1960s and continued into the 1980s (Holt and Williams 2008). There is uncertainty about the status of Black Grouse in Ceredigion prior to the Second World War, with sporadic sightings, a few gamebag records and claims of releases. As elsewhere they benefited from young forestry in the 1950s and 1960s and were well-established in the Tywi Forest in the south and Myherin Forest in the north, as well as Cors Fochno in the 1970s. Numbers contracted rapidly in the 1980s and by the early 21st century, just a handful remained around Pwllpeirian (Roderick and Davis 2010).

Ingram and Salmon (1954) considered that northwest Carmarthenshire was the main stronghold in South Wales, with populations also well-established around Llandovery and Llanfynydd in the southeast. In the north, 20–30 were shot at a lek on Mynydd Llanybydder in 1959. Numbers there recovered to at least 20 in 1969, fell during the 1970s and 1980s, and only a few leks existed in the county by the end of the latter decade. In neighbouring Breconshire, the species was reportedly not uncommon in the western half of the county (Massey 1976a), but there were only eight records during 1980–89 and it had probably died out by the end of the decade. Evidence from Gwent is fragmentary but Humphreys (1963) was of the view that it had been common in the hills in the 19th century and was still occasionally seen in the Black Mountains (Y Mynyddoed Duon) in the 1950s. As the range contracted from the south and west, the last county records in South Wales were: near Llanygynwyd, East Glamorgan, in 1936; Gaer Mountain, Gwent, in 1977; Rheola Forest, Gower, in 1984; The Whimble, Radnorshire, in June 1987; and near Brechfa, Carmarthenshire, in 1999.

County	1986	1992	1995	1997	2002	2005
Caernarfonshire	7	10	3	2		
Carmarthenshire	4	1	0	0		
Ceredigion	13	13	7	7		
Denbighshire	96	65	59	64		
Meirionnydd	126	113	77	41		
Montgomeryshire	18	17	18	17		
Total	**264**	**219**	**164**	**131**	**243**	**213**

Black Grouse: Lekking males in Wales, 1986–2005. The two later surveys produced a population estimate for Wales but not for each county.

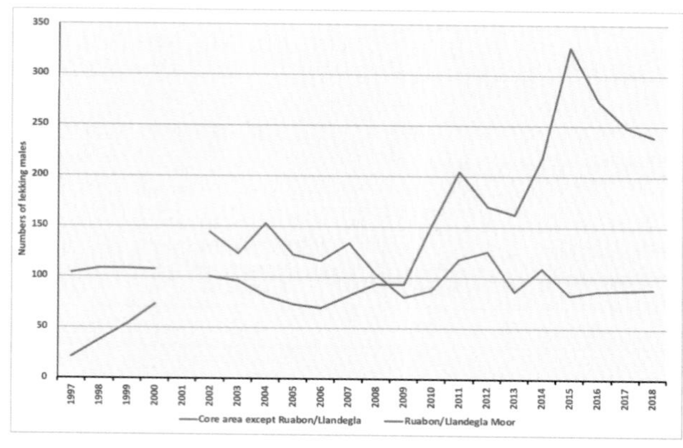

Black Grouse: Lekking males within core area, 1997–2018. No counts were undertaken in 2001 due to the outbreak of foot-and-mouth disease (RSPB data).

The first all-Wales Black Grouse survey recorded 264 lekking males at 91 leks in 1986 (Grove *et al.* 1988). Surveys were repeated three times in the 1990s, recording a 50% decline in numbers over just 11 years. The UK Black Grouse population was estimated at 5,078 displaying males in 2005 (Sim *et al.* 2008), a 22% decline from 1995/96 (Hancock *et al.* 1999). That survey showed the Welsh population had increased from 153 to 213 males, the only part of Britain in which an increase was recorded. However, these surveys showed extensive contraction in range from south to north and west to east and Black Grouse were mostly confined to a handful of upland sites in Meirionnydd, Denbighshire and Flintshire, with isolated and declining populations in Caernarfonshire and Montgomeryshire.

Wide-scale habitat loss and deterioration has occurred at a landscape level, including degradation of scrub and heather moorland and the maturation of non-native conifers, driven by land-use policies. There is growing evidence that climate change may also negatively influence Black Grouse in parts of its global range (Ludwig *et al.* 2006). Modelling suggests that northwest Scotland may be the only part of Britain that is climatically suitable for the species by the end of this century (Huntley *et al.* 2007). In the UK, although declines are strongly correlated with changes to land management (Pearce-Higgins *et al.* 2007), local extinctions between 1995–96 and 2005 were mostly at lower altitudes, a trend consistent with the potential impacts of climate change (Pearce-Higgins *et al.* 2016). Black Grouse chicks are sensitive to high rainfall in late June, which may reduce the availability of invertebrates. This forces chicks to forage more widely, use more energy and possibly feed in sparser vegetation, where they are more vulnerable to predators.

The severity of its decline prompted a Welsh Black Grouse recovery programme by the RSPB from 1997, which led to large-scale habitat management at eight key sites in the core range across North Wales that held over 80% of the remaining population. Habitat management was targeted at improving fledging rates in a 1.5km radius around the lek, the home range of males. That created a mosaic of moorland vegetation, known to be important for breeding success (Calladine *et al.* 2002), by varying its structure and composition within forests and on surrounding land through rotational cutting and burning. Non-native conifers were selectively thinned on woodland edges adjacent to moorland and within the forest (Lindley *et al.* 2003), with Foxes and Carrion Crows controlled at some sites in parallel.

Within this area, the number of lekking males stabilised during 1997–2008 and increased from 2009, but funding for management varied between sites. In 2015, 411 displaying males were in the core area—the highest Welsh total for at least 30 years—but fell to 328 in 2018. The recovery was driven by numbers on Ruabon and Llandegla Moors, Denbighshire, which accounted for 25% of the core-area population in 1997–99 and 74% in 2016–18. Much of this is managed for driven shooting of Red Grouse. In 2018, numbers there—along with Moel Famau, Denbighshire/Flintshire—constituted 82% of the Welsh population. Outside this core management area, numbers have continued to decline and the range has contracted further, with regional extinctions at Llanbrynmair, Montgomeryshire, and the Rhinogau, Meirionnydd; a small population around Lake Vyrnwy is believed to be the most southerly population in Britain. The species is Red-listed in Wales owing to the 67% decline in its range between 1968–72 and 2008–11 (Johnstone and Bladwell 2016).

Pearce-Higgins *et al.* (2019) showed that targeted conservation action, such as rotational vegetation cutting and burning, woodland management and legal control of predators, can mitigate the negative effects of climate change, particularly high June rainfall that has a detrimental effect on productivity. Such targeted intervention may increase the resilience of populations in the south of its range to climate change. There is an urgent need to revitalise targeted conservation activity in its core area of Wales to enable this species to adapt and survive. Consideration is being given to reintroducing Black Grouse to Radnorshire, but there is a long way to go before the Black Grouse population in Wales can be considered as secure and sustainable.

Patrick Lindley

Sponsored by Clwyd Bird Recording Group

Red Grouse — *Lagopus lagopus* — Grugiar Goch

Welsh List Category	IUCN Red List			Birds of Conservation Concern			
	Global	Europe	GB	UK	Wales		
					2002	2010	2016
A	LC	VU	LC	Amber			

For many years, the Red Grouse was considered the only bird species endemic to Britain and Ireland. It is now recognised as an endemic subspecies, *L.l. scoticus*, of Willow Ptarmigan, a species that has a very wide distribution through most of northern Eurasia and northern North America, from the Arctic tundra to the edge of the boreal forest (*HBW*). Wales was always the most southerly extent of the subspecies' natural range, although Red Grouse were introduced to Exmoor and Dartmoor in southwest England from 1915–16 (*Birds in England* 2005). The Red Grouse is a bird of moorland, feeding largely on heather, although it does also eat other plants. It prefers a mixture of young heather to provide fresh shoots for eating and older heather to provide cover. There are no ringing recoveries involving Wales, but it is normally very sedentary: 95% of ringed birds were recovered less than 1km from the ringing site, although females move farther than males (*BTO Migration Atlas* 2002). There can be hard-weather movements, particularly in response to heavy snowfall. Forrest (1907) related that during frozen conditions around 1898, "some hundreds"

Red Grouse © John Hawkins

came to Mynydd Rhiw in western Llŷn, Caernarfonshire, and many flew on to Bardsey. A few stayed and bred around Rhiw.

Remains of Red/Willow Grouse, dating either to just before or just after the Last Glacial Maximum, 22,000 years ago, have been discovered at two caves in Pembrokeshire: Hoyle's Mouth and Little Hoyle (Eastham 2016). The earliest written records date from the 17th century. George Owen (1603) listed it as breeding in Pembrokeshire and Thomas Evans, vicar of Llanberis, refers to "grouse in abundance" in that part of Caernarfonshire in a letter to Edward Llwyd in 1693. The earliest reference to shooting Red Grouse was apparently by Pennant in 1781, who commented that "a few black and red grouse have escaped the rage of shooters" in the Clwydian Hills of northeast Wales (Pennant 1810). He recorded larger numbers, including "multitudes of red grouse and a few black", in the bog and heath above the Vale of Mawddwy, Meirionnydd. Three were shot on the Gogerddan Estate, Ceredigion, in August 1804 and there was a reference to grouse shooting by Walford (1818), who stated that an inn at Llandybïe, Carmarthenshire, was much frequented in the grouse-shooting season.

The expansion of the railway network in Wales from about 1850 led to grouse shooting on a much larger scale, assisted by the increased availability of breech-loading shotguns, which made large bags possible. There are two types of grouse shooting: driven shooting involves a team of beaters driving birds towards a line of guns, positioned in butts, whereas 'walked-up' shooting involves walking in an extended line with gun dogs to flush the grouse. From being a recreational activity for landowners and their friends, grouse shooting became a commercial enterprise and, as early as 1868, Abraham Feetham, Britain's first commercial shoot manager, offered grouse shooting for "up to four guns" at Hafod, Ceredigion (*Fieldsports* October 2018). Driven shooting requires high densities of Red Grouse, achieved by rotational heather burning, restricted grazing and the killing of predators. For example, Lord Penrhyn spent large sums of money on drainage and erecting butts on his 60,000ha moorland estate in Caernarfonshire, which stretched from Penmachno to Dolwyddelan, and engaged an experienced Scottish head gamekeeper, Mr Foster, to manage it. Predators shot on this estate during 1874–1902 include 1,988 Kestrels, 1,538 Carrion Crows and 228 Merlins (Lovegrove 2007). In Wales, driven grouse shooting was largely restricted to the northeast; shooting elsewhere was mainly walked-up.

Red Grouse seem always to have been most abundant in northeast Wales, particularly on Ruabon Mountain and the Berwyn range. Forrest (1907) commented that there were "few finer grouse moors in the kingdom" than those on the Berwyn, and large numbers were shot on several estates there. On Rhug Estate near Corwen, Meirionnydd, 4,354 Red Grouse were shot between 1835 and 1858, an average of 198 each year. Considerably higher totals were shot there in later years, over 1,000 in some years between 1870 and the late 1930s, including 1,854 in 1932–33 (Pritchard 2012). An average of 464 were shot annually on the Llymystyn Estate on the southern Berwyn, Montgomeryshire, during 1866–1951, and an average of 696 annually on Nantyr Estate, Denbighshire, during 1877–1952. Larger numbers were shot on Ruabon Estate, Denbighshire, with an average of 4,658 each year during 1900–13, peaking at 7,142 in 1912 and including a bag of 1,774 in a single day, both records for Wales (Roberts 2010). The neighbouring Bodidris Estate on Llandegla Moor, Denbighshire, had an average annual bag of 2,111 in 1905–13 (Roberts 2010), while 4,500 were shot on the nearby Llantysilio Estate in 1912 (Hardy 1941).

Farther west, around Arenig-Penmachno, Meirionnydd/Caernarfonshire, 4,800 were shot in 1912 (Hardy 1941) and the annual bag on moorland around Llyn Conwy was 2,000–3,000 birds before the First World War (North *et al.* 1949). On Mynydd Hiraethog, Denbighshire, up to 200 had been shot in a day (Hardy 1941). In Montgomeryshire, the average bag on Llanbrynmair Moor during 1906–52 was 701, with a peak of 2,103 in 1913, and at Lake Vyrnwy was 1,600 during 1890–1914 (Holt and Williams 2008). Numbers were lower in Breconshire, although there were 'keepered moors on all the main hill ranges until 1914; Phillips (1899) mentioned a maximum of only 70 birds shot. In Radnorshire, the Maesllwch Estate shot an average annual bag of 207 during 1868–1935, with a peak of 611 in 1904 (Jennings 2014). Gwent held the most southerly shooting moors within the Red Grouse's natural range and numbers shot there were generally fairly low. However, Blaenavon Moor shot an average of 410 each season, with a maximum of 834 in 1905 (Venables *et al.* 2008). The species was widely distributed in the north of Glamorgan for much of the 19th century, but much suitable habitat was lost to slag dumping, urbanisation and later afforestation (Hurford and Lansdown 1995).

In western counties, the population always seems to have been lower. There is no record of Red Grouse breeding on Anglesey, where the first record was not until 1905, and the few subsequent records on the island may have been birds wandering from Caernarfonshire (Jones and Whalley 2004). Caernarfonshire, west of the Migneint, had a number of 'keepered moors, as far west as Cefn Du in the Snowdon range. Red Grouse were recorded even farther west on Yr Eifl in Llŷn, by Pennant in the 18th century and by several authors in the early 20th century. In Meirionnydd, Red Grouse were found as far west as the Rhinogau, though in lower densities than farther east. Ceredigion had several 'keepered moors, such as Gogerddan Estate where 110 were shot during 1804–56 and 1,447 during 1881–90. Smaller numbers were shot on estates in the south of the county (Roderick and Davis 2010). The species seems to have been less common in Carmarthenshire, despite the early reference to grouse shooting around Llandybïe in 1818. Davies (1858) described it as common on Mynydd Du, "but not so plentiful as in former years", and Barker (1905) mentioned a few localities in the east and northwest of the county, but gave no further details. Mathew (1894) thought it extremely doubtful that a single Red Grouse was left on Mynydd Preseli, Pembrokeshire, and noted that the last of which he was aware had been shot in 1885. No more were reported in Pembrokeshire until up to three on Mynydd Preseli in December 1952; one was on Ramsey on 15 October 1975 but there has been no subsequent record in the county.

In its core areas numbers remained high between the wars, particularly in the 1930s, but there were very steep declines in Britain and Ireland in the 1940s, which Watson and Moss (2008) attributed to a wider factor, such as climate, rather than local management. Numbers recovered in some areas of Britain in the decade after 1945, but not to any great extent in Wales. There were some 40 productive grouse moors in Wales between 1850 and 1939, but by the 1990s shooting was conducted, even once a season, on fewer than six (Shrubb 1997). The number shot/km² decreased by 63% in Wales during 1920–50 and, while there was no significant trend during 1950–80, declined by a further 83% during 1980–2010 (Robertson *et al.* 2017a). Driven grouse shooting was abandoned as no longer viable in many places, and there were only half a dozen or so 'keepered moors in Wales by 1990 (*Birds in Wales* 1994). Counts on Ruabon Mountain, Denbighshire, during 1978–2005 showed a large decline in Red

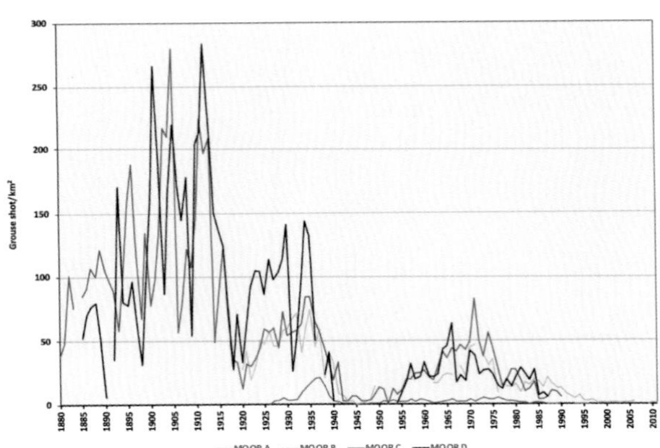

Red Grouse: Numbers shot/km² on four moors now in the Berwyn Special Protection Area, 1880–2010. The moors are Ruabon, Palé, Llanarmon and Lake Vyrnwy, but at the request of some owners, the location of each trend is not identified (from Warren and Baines 2012).

Grouse numbers, which correlated strongly with a decline in shooting bags there during 1970–93 (Roberts 2010). There were ten grouse shoots on Berwyn in 1994: two driven and five walked-up, with no shooting on two and no data for another; only one had a full-time gamekeeper and there were part-time gamekeepers on another four (Walker and Kenmir 1994).

By the late 1990s, driven-grouse-moor management had effectively ceased on Berwyn, except on Palé Moor, where Foxes and some corvids were controlled during 1996–2000, in an attempt to increase Red Grouse numbers sufficiently to enable sport shooting (Warren and Baines 2012). Red Grouse numbers failed to respond, primarily because of the tick-borne Louping Ill virus; management ceased there in 2000 (Game Conservancy Trust 2001). Ruabon Estate had been leased out for shooting of released Red-legged Partridges during 2005–15, but was taken back in hand by the owner to concentrate on Red Grouse, of which 2,000 were counted in autumn 2016 (Saffery Champness 2016). It is now the only grouse moor in Wales with a full-time gamekeeper. In Gwent, habitat management and a moratorium on shooting over Blaenavon Moor resulted in an increase in Red Grouse from 67 in 2011 to over 100 by 2016. A small number are now hunted there by falconers under licence from the local authority.

The *Britain and Ireland Atlas 1968–72* showed Red Grouse to be fairly widespread in Wales, though absent from Anglesey, Llŷn, Pembrokeshire and the south coast. By 1988–91 there were losses in all parts of Wales, and further losses were evident by the 2007–11 *Atlas*, with the species now absent from almost the entire southwest quarter of Wales. Abundance was relatively low everywhere and had reduced since 1988–91. Local atlases confirmed this pattern of decline in finer detail. Birds were recorded in 10% of tetrads in the *Gwent Atlas 1981–85*, but in only 7% in 1998–2003. In East Glamorgan, Red Grouse were found in just four tetrads in 1984–89 and not recorded during the breeding season in 2008–11, though they were recorded in three tetrads in winter. The *North Wales Atlas 2008–12* recorded the species in 7% of tetrads, restricted to a few upland areas, mainly in Meirionnydd, where it was found in 10% of the county's tetrads, and in Denbighshire. Numbers have declined in Breconshire, but small populations persist in the Cambrian and the Black Mountains, where small numbers were shot until the 1990s and on occasional shooting days subsequently. In the Brecon Beacons, seven territories were found in 2km^2 on Mynyddau Llangatwg/Llangynidr in 2018, and a count in the central Beacons estimated a population of 20 pairs.

The population in Wales is too small for population trends to be generated, but in its Welsh stronghold, Y Berwyn SPA, numbers encountered on upland breeding bird surveys declined by 54% between 1983–85 and 2002, the density falling from 1.09/km^2 to 0.50/km^2 (Warren and Baines 2012).

Red Grouse is a Red-listed bird of conservation concern in Wales owing to the historic decline in its breeding population and its assessment as Vulnerable in Europe (Johnstone and Bladwell 2016). Restoring its numbers to previous levels is not a priority for nature conservation, since these were the result of intensive management for shooting. Grouse at high densities suffer from a number of ailments, including *strongylosis*, caused by the nematode worm *Trichostrongylus tenuis,* and Louping Ill virus, transmitted between host animals by ticks. Once rotational burning of a heather moor ceases, tall heather predominates and grouse numbers remain at a low level, although inappropriate burning can make matters worse. Overgrazing can lead to the replacement of heather by grasses, which can lead to the spread of Bracken. Some moors have been completely lost to afforestation. An increase in numbers of generalist predators, particularly Foxes and corvids, is also likely to be a factor.

Many populations away from the core range in Wales are small and isolated and therefore at risk of dying out. Sample surveys by the RSPB in winter 1991 estimated the population in Wales to be 4,800 individuals, of which nearly 40% were on Berwyn (Williams *et al.* 1991). The Welsh breeding population in 2016 was estimated to be 835 (490–1,450) pairs (Hughes *et al.* 2020), of which a minimum of 210 territorial males were on moorland at RSPB Lake Vyrnwy and 65 on adjacent land in 2015. Its future in Wales does not look bright, particularly as modelling suggests that Wales will no longer have a suitable climate for Red Grouse by the end of the 21st century (Huntley *et al.* 2007).

Rhion Pritchard

Sponsored in memory of Zlatina Lawton-Roberts, 1949–2020

Red-legged Partridge *Alectoris rufa* Petrisen Goesgoch

Welsh List Category	IUCN Red List	
	Global	Europe
C1(1), E*	LC	NT

The Red-legged Partridge is a native of southwest Europe, with the largest populations in Spain and southwest France, where it favours dry hilly land with scattered small bushes (*HBW*), but is undergoing a marked decline (*European Atlas* 2020). The population in Wales only occurs because of releases for shooting and may well not be self-sustaining, at least in much of the western half of the country. Roderick and Davis (2010), for example, stated that there was no indication that birds were becoming established in the wild in Ceredigion. It is generally sedentary, although there was a surprising record of one on the summit of Yr Wyddfa, Caernarfonshire, in July 2008 and presumably the same bird returned there in July 2009. The only ringing recovery involving Wales was one hit by a car near Mitchell Troy, Gwent, in 1981, close to where it had been ringed six months previously.

Attempts to introduce this species to Britain began as far back as 1673 (Potts 2012), but it did not become established until large numbers were released in Suffolk in the 1770s (*Birds in England* 2005), from which the core of the present naturalised population is thought to derive (Lever 2009). The first recorded introduction into Wales was on the Dunraven Estate, East Glamorgan, *c.*1850 (Hurford and Lansdown 1995). On Anglesey, several shot at Aberffraw and Llanerchymedd during 1860–70 were thought to have come from an estate at Lligwy where they had been introduced some years earlier, but the population died out by 1869 (Jones and Whalley 2004). Later in the 19th century, there were releases in Meirionnydd, Pembrokeshire and Glamorgan, but although there were records of successful breeding, several of these populations expired too; that in Glamorgan was thought to have died out by 1935 (Parslow 1973).

Releases for shooting continued in the early 20th century but involved only small numbers by the 1960s when the National Gamebag Census index began. The numbers released across the UK in 2011 were almost 200 times greater than in 1961 (GWCT undated [a]) and increased further to 10 million birds by 2016 (Aebischer 2019). There is little public information on numbers released by estates in Wales, but over 5,000 were released by Dulas Estate, Anglesey, in 1999 and 10,000 onto Tregoyd Estate, Breconshire, in August 2011. High counts around releases in Denbighshire include 500 at Tainant in September 2000, and 300 at Ruabon Moor and 200 at Newtown Mountain in October 2008. Around 3% of UK releases are in Wales, based on those reported on the DEFRA poultry register (Bicknell *et al.* 2010); it is estimated that 480,000 Red-legged Partridges were released in Wales in 2016. The numbers breeding are far smaller, estimated to be 1,400 (940–1,850) pairs in 2016 (Hughes *et al.* 2020). Not all are shot, however. Many are undoubtedly predated and others are killed by road traffic; a flock of at least 50 that flew out to sea from Point Lynas, Anglesey, in October 2017, landed on the water and were swept out to sea.

The distribution shown by the successive Britain and Ireland breeding and wintering atlases is the best guide to releases over the last 50 years. In 1968–72 they were almost entirely confined to eastern Wales, and that remained the case in 1981–84, except for three 10-km squares on Anglesey and a handful elsewhere. The *Britain and Ireland Atlas 1988–91* showed a trebling of occupied 10-km squares since winter 1981–84, mainly in the aforementioned areas. By the 2007–11 *Atlas*, the number of 10-km squares had increased by a further 70% to *c*.30% of those in Wales. Its distribution is based primarily on business decisions rather than habitat availability and is concentrated in Gwent, Glamorgan, Pembrokeshire and in western Llŷn, Caernarfonshire. The *Gwent Atlas 1981–85* found the species in 21% of tetrads, but by 1998–2003 only 11% of tetrads were occupied, whereas in Pembrokeshire they were found in only 1% of tetrads in 1984–88 but in 5% in 2003–07.

There is little research to assess the effect of Red-legged Partridges on native wildlife but the benefits of local habitat management may be outweighed by the negative impacts associated with releases and shooting (Mason *et al.* 2020). Releases of gamebirds boost numbers of some avian predators (five crow species and Buzzard) at a UK scale, with potential consequences for declining ground-nesting species, such as Curlew (Pringle *et al.* 2019). On Ruabon Moor, Denbighshire, an increase in numbers of corvids and generalist raptors was thought to have resulted from the introduction of Red-legged Partridges for shooting (Roberts 2010). Watson *et al.* (2007) concluded that, ultimately, shooting of released non-native species such as Red-legged Partridges would not threaten the conservation status of the native Grey Partridge, which in any case has already vanished from large areas of Wales. From the 1970s the closely related Chukar (*A. chukar*) and Chukar hybrids, which can interbreed with Red-legged Partridge, were also released for shooting. The release of Chukar and hybrids was banned from 1992 (Potts 2012).

The distribution of the Red-legged Partridge in Britain coincides with the 19°C isotherm for average maximum temperatures in July (Tapper 1999), although Howells (1962) stated that it is restricted to areas where the average annual rainfall was less than 890mm; by either criterion, large parts of Wales are unsuitable. Climate change modelling indicates that Wales will become even less suitable for Red-legged Partridge through the 21st century (Huntley *et al.* 2007).

Rhion Pritchard

Grey Partridge *Perdix perdix* **Petrisen**

Welsh List Category	IUCN Red List			Birds of Conservation Concern			
	Global	Europe	GB	UK	Wales		
A, E*	LC	LC	VU	Red	2002	2010	2016

The Grey Partridge has an extensive distribution through temperate Europe and Central Asia, as far south as Iran. It was originally a steppe species, but is now mainly a bird of farmland, particularly arable (Potts 2012). Most populations are resident but can become partially migratory in severe winters (*European Atlas* 1997). It feeds mainly on seeds, but chicks require an insect diet. Anglesey and Pembrokeshire are suboptimal for the species and the rest of Wales is unsuitable (Aebischer 2009).

There are few early records of the Grey Partridge from Wales, though Owen (1603) listed it as breeding in Pembrokeshire. During the 19th century, it appeared to have been common over most of Wales, except on the highest ground. Numbers increased in the second half of the century as a result of more-intensive game preservation.

Record numbers were shot on several estates in winter 1887/88, such as 1,019 at Rhug, Meirionnydd, and 1,052 at Voelas, Denbighshire. In 1889, 1,273 were shot at Stackpole Court, Pembrokeshire (Matheson 1960), and in 1898, 1,157 at Maesllwch Estate, Radnorshire (Jennings 2014). The Powys and Lymore estates in Montgomeryshire recorded an average of 823 shot each year during 1890–99, with a peak of 1,713 in 1896/97. On one estate on Anglesey, over 2,000 were frequently shot in a season during 1875–1900, with the annual bag seldom below 1,000. Around the turn of the 20th century, it was described as being "fairly plentiful" in southern Pembrokeshire, where large numbers were shot (Mathew 1894), and as common in all suitable country throughout North Wales, being especially abundant on Anglesey (Forrest 1907).

Even in the early years of the 20th century, some authors considered that the species was declining in parts of Wales. Grey Partridges were said to have become noticeably scarcer in parts of Meirionnydd, Ceredigion and Carmarthenshire following the First World War (*Birds in Wales* 1994), perhaps because of a reduction in management for gamebird shooting. The number of gamekeepers in Britain peaked, at 23,000, in 1914, but fell during the war and never returned to former levels (Potts 2012). Arable farming also began to decline in Wales in this period. There was a steep decline in Grey Partridge numbers following the Second World War as a result of the introduction of selective herbicides that killed many plants that host the insects on which chicks depend. Chick survival rates fell from an average of 45% to under 30% during 1952–62 and by the 1980s, the abundance of insects in cereals had fallen by at least 75% (GWCT undated [c]).

Gamebags indicate that Grey Partridge numbers declined by 98% in Britain between 1914 and the early 21st century (Potts 2012), with much of that occurring from the 1960s; the population declined by 92% between 1967 and 2017 (Massimino *et al.* 2019). Its range contracted substantially in Wales during the latter period too. The *Britain and Ireland Atlas 1968–72* showed the Grey Partridge to be widely distributed in lowland areas, found in most 10-km squares in eastern Wales, Anglesey and Llŷn, but absent from most of Meirionnydd and with a much patchier distribution in southwest Wales. The *Britain and Ireland Atlas 1988–91* showed extensive losses over most of Wales, though it appeared to be holding its ground better in the extreme east and on Anglesey. In western areas of Wales, there were only a few scattered records and, by 2007–11, atlas fieldwork showed further widespread losses, even in the east.

County tetrad atlases tell a similar story. Grey Partridge was recorded in 46% of tetrads in the *Gwent Atlas 1981–85* but in just 8% of tetrads in 1998–2003, and the number with confirmed breeding fell from 59 to seven, an 83% reduction. The *East Glamorgan Bird Atlas 1984–89* recorded it in 11% of tetrads but confirmed breeding in only 3% and by 2008–11 there was no confirmed breeding in the area, and some evidence in just two tetrads (<1%). In Pembrokeshire, the population was already at a low level in 1984–88, occurring in just 2% of tetrads; by 2003–07 it was found in only 1% of tetrads and it was considered "doubtful that a self-sustaining population [still] existed" in the county (Rees *et al.* 2009). The *North Wales Atlas 2008–12* found it in just 2% of tetrads, almost all on Anglesey, eastern Flintshire and Denbighshire.

The size of broods recorded by birdwatchers in Wales fell from an average of 12–14 birds in the early 1980s to 2–4 in the late 1990s (Young 2001); although the age of these broods was not known, it is a strong indication that reduced breeding success drove the decline. There is little information on the number of Grey Partridges released for shooting in Wales, though it is presumed to be small compared to releases of Red-legged Partridge. Dulas Estate on Anglesey, for example, released 500 Greys and 5,000 Red-legs in 1999. Where Red-legged Partridges are released in high numbers, this can reduce numbers of wild Grey Partridges, partly because shooters are liable to shoot them by mistake (Potts 1978). Released Grey Partridges have a poor survival rate

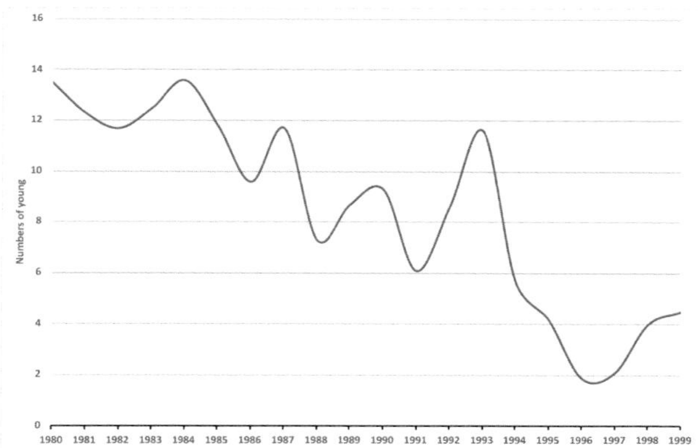

Grey Partridge: Average number of chicks/brood (of variable age) in Wales (from Young 2001).

and poorer breeding performance than wild birds (Parish and Sotherton 2007, Buner *et al.* 2011), so releases are unlikely to be effective in boosting the wild population. The species appears to be extinct in Meirionnydd, where none has been recorded since 1991 except for two seen at 400m in the Rhinogau on 24 September 2015 that had perhaps been released with Red-legged Partridges. One at Bwlch, Breconshire, in March 2014 was the first county record for ten years and three at Ffairfach in October 2018 were the first reported in Carmarthenshire since 2000.

There were 22 at Nash Point, East Glamorgan, in September 1999 and 11 pairs at Queensferry/Saltney, Flintshire, in 2000, but there has been no subsequent count in Wales higher than 15. In some counties, recent records are thought mostly to be of released birds, and there is no doubt that the wild population is in serious trouble. Habitat management and predator control has increased numbers on some estates in England (Potts 2012), but this may be less applicable to Wales. It is a Red-listed species of conservation concern in Wales owing to its historic decline, the decline in its UK population and the 78% decline in its range in Wales between 1968–72 and 2007–11, as well as a 67% decline in its wintering range (Johnstone and Bladwell 2016). It seems unlikely that the Grey Partridge will ever be widespread in Wales again.

Rhion Pritchard

Quail *Coturnix coturnix* Sofliar

Welsh List Category	IUCN Red List			Birds of Conservation Concern			
	Global	Europe	GB	UK	Wales		
A	LC	LC	LC	Amber	2002	2010	2016

Wales is on the northwestern edge of the breeding range of the Quail, which extends over most of Europe, except the far north, and eastwards to central Russia and western China, with other subspecies breeding in Africa, Cape Verde and the Azores. Most birds breeding in Europe are thought to winter in Africa south of the Sahara, mainly in the Sahel. It breeds in open habitats, including cultivated crops, and prefers dense herbage less than 1m tall (*HBW*), hence most Quail are detected by song, usually around dawn and dusk, and confirmed breeding evidence is very difficult to obtain. It is by no means purely a lowland species, and birds have been recorded singing at 570m at RSPB Lake Vyrnwy, Montgomeryshire. Grassland dominated by Purple Moor-grass is the favoured habitat in the uplands.

Quail are very mobile within a single breeding season. Most European breeders probably first reproduce in northern Africa and the Mediterranean basin in March or April, then migrate to central and northern Europe to breed for a second time in late May–July. Remarkably, Quail arriving in Wales during the summer can include reproductively active birds that hatched just a few weeks earlier (Guyomarc'h *et al.* 1998). Some males will call for a few days and move north if they do not attract a mate (Balmer *et al.* 2013), so a record of a singing male may not indicate breeding.

The number of Quail recorded varies greatly from year to year, with large numbers in some 'Quail years'. The reasons are not fully understood, but this seems to be linked to warm dry springs and southeasterly winds (Balmer *et al.* 2013). There is little information on the species in Wales before the 19th century, though Owen (1603) mentioned it as breeding in Pembrokeshire, and *Birds in Wales* (1994) quoted Pennant as saying that it bred in Caernarfonshire some time before 1776. A nest and eggs were found near Conwy in 1796 (Forrest 1907), and in Gower there were breeding pairs in Swansea and on the Peninsula about 1818 (Hurford and Lansdown 1995). Moreau (1951) noted "a marked and progressive decline of Quail in Britain in the first half of the nineteenth century" and reported that, from questionnaires in a study of Quail in Britain by A.G. More in 1865, Pembrokeshire was the only Welsh county where it was said to be regular. 1870 was an exceptional 'Quail year', when "considerable numbers" appeared in Flintshire, Denbighshire and southwest Meirionnydd, with many nests found (Forrest 1907). Quail were also abundant in Ceredigion and north Pembrokeshire that year, with many nests, though only one or two were reported in the south of the county

Quail © Steve Wilce

Quail: Approximate numbers of birds recorded each year, 1990–2019.

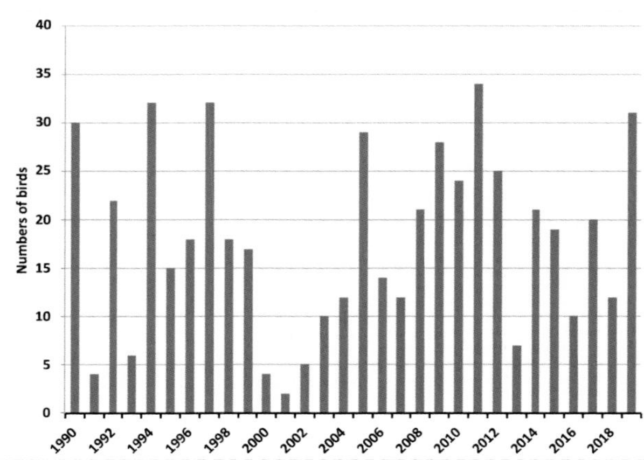

(Dix 1870). Dix knew of 330 having been killed by 18 shooters, but thought that the total number shot must have been four or five times greater. Most were shot in September, including 53 on the Gogerddan Estate near Aberystwyth between 1 September and 5 November (Roderick and Davis 2010). There were several other good years for Quail in the late 19th century, notably 1893, but not on the same scale, apparently, as 1870. Salter (1900) described it as an annual visitor to Ceredigion, shot most years at Gogerddan, but Roderick and Davis (2010) pointed out that the Gogerddan game books do not support this claim.

Even in some good Quail years in Britain, few birds reached Wales. There were several such years in England in the 1940s, but none was recorded in Wales in 1944 and few in 1947 other than in Montgomeryshire. Moreau (1956) noted that "Wales, which had in the past so few Quail, shared in the 1953 influx", with birds in Radnorshire, Denbighshire, Glamorgan, Caernarfonshire and Anglesey, but none in Pembrokeshire. 1970 was a good year, and 1977 produced an above-average number of records in Wales, though it was unremarkable elsewhere in Britain. 1989 was a genuine 'Quail year', with 237 records in Wales, the great majority in southern counties and only 33 farther north than Ceredigion and Radnorshire (Tew 1990). Pembrokeshire had the largest number: 80, including 50 on a single farm. There were 42 in Radnorshire and 34 in Ceredigion in 1989, although Roderick and Davis (2010) stated that at least 50 were recorded in Ceredigion that year. Since 1989 there has been no real influx year in Wales, with numbers ranging from just two birds in 2001 to 34 in 2011, including 11 in Radnorshire. The latter year was, across Britain, the biggest Quail year for decades, with more birds recorded than in 1989, but the largest numbers were in central and southern Scotland and northern England (Holling *et al.* 2013). Records are mainly between May and early October, peaking in June, though there have been a few winter records, particularly in December with a few in January.

It is an Amber-listed species of conservation concern in Wales owing to its status as a rare breeder (Johnstone and Bladwell 2016). With a huge global population, albeit declining, Wales is not the focus of this species' conservation effort, and its future status in Wales is likely to depend largely on conditions elsewhere.

Rhion Pritchard

Pheasant *Phasianus colchicus* Ffesant

Welsh List Category	IUCN Red List	
	Global	Europe
C1(1)	LC	LC

The natural range of the Pheasant extends from around the Caspian Sea discontinuously east to the Sea of Japan (*HBW*). The populations in Europe have been introduced, and with significant levels of hybridisation (Hill and Robertson 1988), individual subspecies—of which 30 are recognised—cannot be identified. The Pheasant is essentially a bird of woodland edge, favouring about 30% woodland and 70% farmland, with as much shrubby edge as possible (Woodburn and Robertson 1990).

It is not certain when the Pheasant was introduced into Britain. The few Pheasant remains found at Roman-era archaeological sites in England do not prove that it had become established in the wild in this period. In Wales, the earliest archaeological record of the species from Hen Domen, Montgomeryshire, is thought to be of the nominate subspecies *P.c. colchicus* from the Transcaucasian region and is probably of early Norman date (Yalden and Albarella 2009). There are several references to Pheasants in Welsh poetry of the 15th and 16th centuries, both as prey for falconry and as a food item (Williams 2014). A 16th-century poem by Gruffydd Hiraethog suggested that the Pheasants mentioned were the first ever seen in Meirionnydd. George Owen (1603) reported that Sir Thomas Perrot brought the first Pheasants to Pembrokeshire, from Ireland, in the 1580s, but that they were not doing well and only a few remained. It was apparently still a rare bird in some areas, even in the early 19th century. Isaac Foulkes in *Cymru Fu* (1862) printed a story about a great commotion in the Llandrillo area of Meirionnydd, when an animal thought to be a kind of winged viper appeared in 1812. Eventually the beast was identified as a cock Pheasant that had wandered from the Wynnstay Estate. On most estates the Grey Partridge appeared to have been more important at that time, but from the 1830s the Pheasant rapidly became a popular target for shooters. The proportion of Pheasants increased in gamebags on estates in Wales during the 19th century (Matheson 1963). For example, at Gogerddan in Ceredigion, Pheasants formed less than 1% of the bag during 1805–14, but 10% in 1835–44 and 13% in 1855–64, while at Chirk, Denbighshire, they constituted over one-third of the total game shot during 1827–36. Pheasants became much more important to game shooting in the second half of the 19th century and the early years of the 20th century. At Stackpole, Pembrokeshire, Pheasants formed 21% of the bag in 1868–77, but 75% by 1908–17.

Mathew (1894) commented that "the Pheasant thrives remarkably well in Pembrokeshire, not only in the preserves, but in the wild unpreserved districts in the county" and reported that the ring-necked form *P.c. torquatus*, native to China, was the predominant variety. In North Wales, Forrest (1907) described it as "more or less common in places where it is preserved, but only up to a moderate altitude". Around the same period, it was said to be plentiful in Glamorgan where it was preserved as game, but rare elsewhere (Hurford and Lansdown 1995). In Gwent, the largest releases for shooting were on Hendre Estate, where 6,000–7,000 were reared annually during the 1890s and early 20th century, and around 5,000 were shot in some seasons (Venables *et al.* 2008). Numbers fell markedly in Wales during the First World War, with no releases and many gamekeepers in the armed forces. *Birds in Wales* (1994) stated that game rearing was never undertaken on such a large scale in Wales again, although on some estates the number of Pheasants shot peaked in the 1930s, such as 4,219 birds shot in 1935/36 on Rhug Estate, Meirionnydd (Pritchard 2012). In some counties, large-scale afforestation of the uplands allowed the species to colonise higher altitudes. The Second World War led to a further decrease in releases in Wales, and by the 1950s, Pheasants were rare in some areas. Ingram and Salmon (1954) stated that there were relatively few wild Pheasants in Carmarthenshire, for example.

In the 1960s, the development of artificial rear-and-release methods led to a substantial increase in the number of Pheasants in Britain (Tapper 1992), partly in response to a decline in stocks of Grey Partridge that resulted from agricultural intensification during the 1950s and 1960s (GWCT undated [b]). In 2009, two million Pheasants were registered in Wales on DEFRA's poultry register, about 5.4% of the total in Britain and five times the number of Red-legged Partridge (Bicknell *et al.* 2010). Applying that proportion to UK estimates by Aebischer (2019) suggests that around 2.54 million are released and 0.75 million are shot in Wales each year. The proportion of released Pheasants shot each season was relatively stable until 1990 but has since declined

Pheasant: Smoothed BBS index for Wales, 1995–2018. The shaded area indicates the 85% confidence limits.

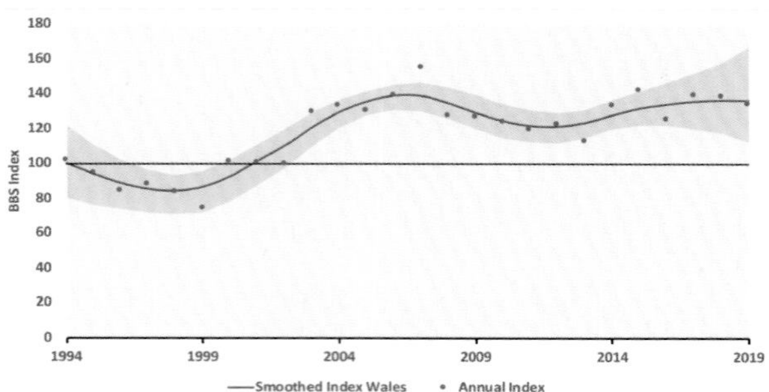

steeply (Robertson *et al.* 2017b). The numbers released increased nine-fold between 1961 and 2011, but now only 32% of these are shot (Aebischer 2019). Of the remainder, an estimated 35% are predated/scavenged, 13% are killed on roads or die of disease, and 16% survive to the following season (Mason *et al.* 2020).

While releases result in increased numbers for shooting, the number of 'wild-bred' Pheasants has declined (Robertson *et al.* 2017b). The Welsh breeding population in 2016 was estimated to be 165,000 (160,000–175,000) females (Hughes *et al.* 2020). The *Britain and Ireland Atlas 1968–72* showed the Pheasant to be widely distributed in Wales but with two noticeable gaps, in northern Glamorgan and western Meirionnydd. The 1988–91 *Atlas* showed little net change, but the 2007–11 *Atlas* showed gains in areas from which the species had been largely absent previously, particularly in Glamorgan. The *East Glamorgan Atlas 2008–11* recorded breeding codes for Pheasant in 26% of tetrads, whereas it was in 63% of tetrads in Pembrokeshire in 2003–07 (*Pembrokeshire Avifauna* 2020). The *North Wales Atlas 2008–12* found birds in 60% of tetrads, with most tetrads occupied in Flintshire, Denbighshire and Anglesey, but only 35% of tetrads in Meirionnydd occupied. The BBS showed that Pheasants increased by 46% in Wales between 1995 and 2018 (Harris *et al.* 2020).

Numbers in Wales would be much lower without annual releases. Pheasant and partridge densities are relatively low, 0–50 birds/km^2, across much of Wales, but 100–200 birds/km^2 in parts of North Wales and 1,000–1,500 birds/km^2 in southeast Montgomeryshire are among the highest anywhere in Britain (AHVLA 2013). Densities can be extremely high within a mile or two of woodlands where birds are released for shooting. These appear to have very low breeding success, and in some western areas the population may not be self-sustaining.

Only about 4% of woodland in Wales is managed for shooting Pheasants, far less than the 28% of English woodlands (GWCT 2018), but locally the effects may be considerable. In a scored review, the negative ecological impacts of released gamebirds outweighed the positive in five of six themes. The positive effect was on habitat management, supplementary feeding and legal lethal predator control, whereas negative effects were the direct impact of released birds on vegetation and other animals (such as reptiles and amphibians), illegal persecution of protected species, shooting practices (such as lead shot), disease transmission and their role as a source of supplementary food for predators (Mason *et al.* 2020). Large-scale releases have "fundamentally changed the community structure of wild birds in the UK" (Mason *et al.* 2020): non-native Pheasants contribute more biomass and use more energy than any other breeding bird species except Woodpigeon (Blackburn and Gaston 2018). Based on estimates by Mason *et al.* (2020), Pheasants in Wales account for *c.*2,373 tonnes of biomass. These are food for predatory and scavenging birds and mammals, and have boosted the numbers of five crow species and Buzzard in the UK (Pringle *et al.* 2019). There is an urgent need to test whether Pheasant releases enhance the local abundance of mammal and avian predators, and consequentially increase predation-related pressure on ground-nesting birds such as Curlew.

Rhion Pritchard

Brent Goose *Branta bernicla* Gŵydd Ddu

Welsh List Category	IUCN Red List			Birds of Conservation Concern			
	Global	Europe	GB	UK	Wales		
					2002	2010	2016
A	LC	LC	LC	Amber			

There are three well-defined subspecies of Brent Goose: two winter regularly in Wales, Dark-bellied *B.b. bernicla* primarily in the south and Pale-bellied Brent Geese *B.b. hrota* mostly in the northwest, while Black Brant *B.b. nigricans* is a rare visitor. All breed in the high Arctic on lowland tundra, usually near the coast, and winter in estuaries and sheltered bays farther south, as far as northwest Spain in Europe but to northern Mexico and Japan on other continents.

Most early records in Wales referred only to 'Brent Geese' without specifying the subspecies, though the majority were likely to have been Dark-bellied, which was the more common form in Wales until an increase in numbers of Pale-bellied Brent Geese in the 1990s.

Large flocks were sometimes found in the Dee Estuary, Flintshire/Wirral, in the late 19th and early 20th century, such as 200 there in February 1888 (Dobie 1894), and in the Conwy Estuary, Caernarfonshire/Denbighshire, almost every winter (Forrest 1907), but these were said to have virtually disappeared by 1924 (Jones and Dare 1976). A correspondent of Forrest also reported that it was common on the Dyfi Estuary, Ceredigion/Meirionnydd. Mathew (1894) described it as "sometimes abundant" in Pembrokeshire, while Ingram and Salmon (1954) said that there was little doubt that it had formerly been a regular visitor in fair numbers to the estuaries of Carmarthenshire, but that recent records had been scant. The status of the species in Glamorgan at the end of the 19th century was said to have been unclear (Hurford and Lansdown 1995). There are no early records from Gwent, where the species was first recorded in 1928, although a flock of 70 was at Goldcliff in February 1929. Brent Goose populations declined during the 1930s when their winter food plant, Common Eelgrass, was affected by disease, although hunting on the birds' breeding grounds may have been a more important factor (Ward 2004). In 1990/91–1994/1995, almost 95% of Brent Geese identified to subspecies in Wales were Dark-bellied, whereas in 2014/15–2018/19 the equivalent proportion was 40% (Frost *et al.* 2020).

Dark-bellied Brent Goose *B. bernicla bernicla* Gŵydd Ddu Fol-dywyll

The Dark-bellied Brent Goose breeds in the Russian high Arctic east of Novaya Zemlya. Those from the western part of its breeding range winter on North Sea coasts, from Denmark southwest through The Netherlands and England, to western France. The Thames Estuary, Chichester Harbour and The Wash are the most important sites in England. The global population was estimated at 15,000 in winter 1955/56, a low point in the population following a decline in the 1930s. This was followed by a dramatic increase, peaking at 329,000 in 1991, but it subsequently declined to 250,000, around which it has apparently stabilised (Fox and Leafloor 2018).

Small numbers wintered regularly in the Burry Inlet by 1925. A mean winter peak of 15 in the late 1960s increased to 35 in the early 1970s, followed by a dramatic increase from the mid-1970s (Hurford and Lansdown 1995). While averages were lower, peak counts in each decade were 352 in January 1977, 1,410 in November 1988, 1,525 in January 1992, 1,680 in December 2005 and 1,627 in December 2012.

Numbers at other sites in Wales are usually much lower, though large flocks sometimes occur, such as 400 in Carmarthen Bay in March 2000 and 200 there in November 2012. In the north, some Dark-bellied birds are usually among the Pale-bellied flock at the western end of the Menai Strait, where the highest counts on each coast have been 78 at Foryd Bay, Caernarfonshire, in November 2005 and 62 at the Braint Estuary, Anglesey, in November 1973. It is rare inland, even in counties with an extensive coastline such as Anglesey. Only 14 have been recorded in Radnorshire, mostly singles, although five were on the Royal Welsh Showground at Llanelwedd from 30 November to 12 December 1991 and one was shot illegally at Aberedw in October 2014.

A Medium WeBS Alert was issued in 2019 following a 36% long-term (25 year) decline in Wales, a greater reduction than in

Site	County	74/75 to 78/79	79/80 to 83/84	84/85 to 88/89	89/90 to 93/94	94/95 to 98/99	99/00 to 03/04	04/05 to 08/09	09/10 to 13/14	14/15 to 18/19
Burry Inlet	Gower/Carmarthenshire	140	366	721	924	993	1140	846	754	749
Foryd Bay	Caernarfonshire	0	0	3	20	24	12	46	22	24
Carmarthen Bay	Carmarthenshire	0	0	0	14	10	1	16	89	12

Dark-bellied Brent Goose: Five-year average peak counts during Wetland Bird Surveys between 1974/75 and 2018/19 at sites in Wales averaging 40 or more at least once. International importance threshold: 2,100; Great Britain importance threshold: 980.

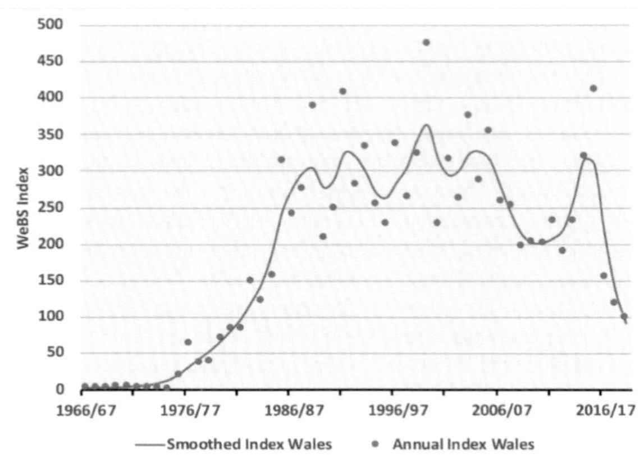

Dark-bellied Brent Goose: Wetland Bird Survey Indices, 1966/67 to 2018/19.

England. This was driven by declines on the Burry Inlet, which was a site of British importance until 2006/07. Being on the south-western fringe of its wintering range, this may be an early sign of climate-related short-stopping in the population.

Pale-bellied Brent Goose *B. bernicla hrota* Gŵydd Ddu Canada

This subspecies breeds in northeast Canada, northern Greenland and Svalbard, and winters in eastern North America and northwest Europe. There are four distinct populations, but only one visits Wales: the 'East Canadian High Arctic Pale-bellied Brent Goose' undertakes the longest migration of any goose that occurs in western Europe, crossing the Greenland ice cap and stopping in Iceland before crossing the North Atlantic to winter mainly in Ireland (Robinson *et al.* 2004a).

An extensive colour-ringing scheme that started in Ireland in 2001, which also involved ringing in Canada and Iceland, showed that Strangford Lough, Northern Ireland, holds over 75% of the population in late autumn. As winter progresses, birds disperse around the coast of Ireland where numbers gradually increased between the early 1960s and late 1980s (*Birds in Ireland* 1989). This population was almost entirely confined to Ireland until the 1990s, but was seen in coastal Pembrokeshire during September and December in some years, which were perhaps migrants over-shooting Ireland. The highest count there during that period was 48 in Jack Sound on 8 October 1949 (Donovan and Rees 1994). There were very few records from elsewhere in Wales before the 1980s and it was not recorded on Wetland Bird Surveys anywhere in the country until 1981/82. Small flocks began to appear on Anglesey and at Foryd Bay, Caernarfonshire, with a maximum of nine during the 1980s. Numbers increased from winter 1990/91, and on Anglesey there were 46 at the Inland Sea/Alaw Estuary in February 1996 and 63 in March 1997, 262 at Traeth Melynog in December 2005, and across the Menai Strait in Caernarfonshire, 440 in Foryd Bay in November 2012. Numbers have now levelled off, though 823 at Cymyran-Beddmanarch Bay, Anglesey, in November 2019 was the highest-ever count in Wales.

On Anglesey, those on the Cymyran Strait interchange with those on the Inland Sea, while those at Traeth Melynog and Foryd Bay interchange between the two shores of the western part of the Menai Strait. A smaller population, which reached a high of 97 in March 2012, moves between Anglesey and Caernarfonshire at the eastern part of the Menai Strait (Arnold 2006). Numbers also increased in the Dee Estuary from 2004/05, but the birds are usually on the English side, with fewer in Wales; 95 at Point of Ayr in January 2006 is the highest Flintshire count. Pale-bellied Brent Geese are usually less numerous farther south, but counts on the Cleddau Estuary, Pembrokeshire, increased after 2008/09, to an average of 67 during 2014/15–2018/19. They were recorded in the Burry Inlet in 47 of the 53 years during 1967–2019, almost always on the Gower side of the bay, and on the south shore of Gower at Blackpill, where there were 220 on 5 May 2013 and 189 on 27 April 2012.

Site	County	74/75 to 78/79	79/80 to 83/84	84/85 to 88/89	89/90 to 93/94	94/95 to 98/99	99/00 to 03/04	04/05 to 08/09	09/10 to 13/14	14/15 to 18/19
Traeth Melynog	Anglesey	0	0	0	0	13	117	253	279	346
Inland Sea and Alaw Estuary	Anglesey	0	2	3	15	36	79	121	256	281
Cymyran Strait	Anglesey							0	217	255
Foryd Bay	Caernarfonshire	0	0	1	4	15	37	138	243	196

Pale-bellied Brent Goose: Five-year average peak counts during Wetland Bird Surveys between 1974/75 and 2018/19 at sites in Wales averaging 200 or more at least once. International importance threshold: 400; Great Britain importance threshold: 16.

Pale-bellied Brent Goose © Ben Porter

A few hundred Pale-bellied Brent Geese regularly winter in the Channel Islands, northwest France and Galicia in northwest Spain (Robinson *et al.* 2004a) and it is likely that some of the larger counts in South and West Wales relate to this population. Examples include 190 off Strumble Head, Pembrokeshire, on 4 November 2005, *c.*200 past Saundersfoot and Stackpole, Pembrokeshire, on 26 April 2012, 220 at Blackpill, Gower, on 5 May 2013 and 144 in the Burry Inlet in March 2015. A flock on the Gwent Levels in September/October 2008 was only the tenth record for the county and resulted in a maximum count of 65 at Peterstone on 25 September. A few are sometimes recorded inland, such as four at Llangorse Lake, Breconshire, in January 2002.

Several Pale-bellied Brent Geese seen in Wales were

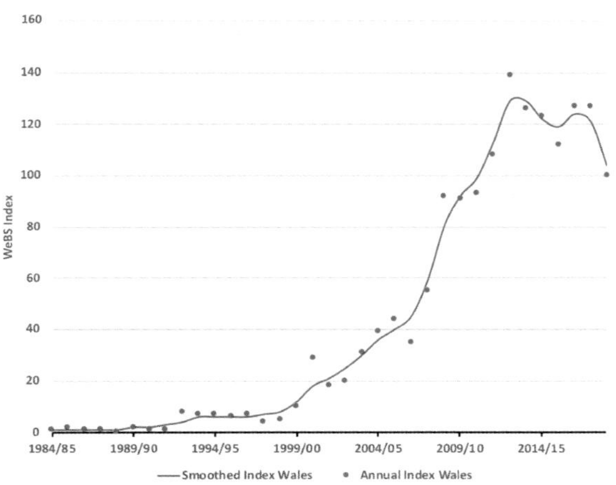

Pale-bellied Brent Goose: Wetland Bird Survey index for Wales, 1984/85 to 2018/19.

colour-ringed on their breeding grounds on Axel Heiberg Island in the Canadian Arctic. Once in Wales, some individuals are very site-faithful while others have been recorded crossing between Ireland and Wales more than once in a single winter (Graham McElwaine *in litt.*). The flyway population was estimated at 38,000–40,000 in 2017 (Graham McElwaine *in litt.*), a significant increase on the estimate of fewer than 6,000 in 1946–56 (Ruttledge and Hall Watt 1958). The spread to Wales was a consequence of this increase and although numbers are still much lower than in Ireland, Traeth Melynog is heading towards qualification as internationally important.

The nuanced effects of climate change on birds are illustrated by a study in northeast Canada that showed Pale-bellied Brent Geese produce more young in warmer years, but the mortality rate of females also increases. It is believed that females face more risk from predators while on their nests, whereas in colder years more nests fail early in incubation (Cleasby *et al.* 2017). It is unclear whether the increase in productivity will outweigh the increase in adult mortality and so influence the numbers that visit Wales in winter.

Black Brant *B. bernicla nigricans*
Gŵydd Ddu Siberia

This subspecies breeds on both sides of the Bering Strait, from the Taymyr Peninsula to Chukotka in Russia and in Alaska and northwest Canada. It winters in eastern China and Baja California, Mexico (*HBW*). Its range is believed to be expanding westwards, potentially within the breeding range of Dark-bellied Brent Goose in the Russian high Arctic (Tony Fox pers. comm.). It is an annual visitor to Britain, mainly to eastern and southern England. There were 402 British records to the end of 2018, numbers having increased from an average of less than one each year before 1979 to 14 each year during 2010–18 (White and Kehoe 2020a). Records in Wales are assessed by the WBRC and only two have been accepted. The first was one with a flock of Pale-bellied Brent Geese in Beddmanarch Bay, Anglesey, from 23 to 30 March 1997; the second was at Wernffrwd and Llanrhidian on the Gower side of the Burry Inlet from 26 November 2003 to 2 January 2004.

Rhion Pritchard

Red-breasted Goose *Branta ruficollis* Gŵydd Frongoch

Welsh List Category	IUCN Red List		Rarity recording
	Global	Europe	
A, E	VU	NT	BBRC

This migratory species breeds on the Taymyr, Gydan and Yamal peninsulas in northern Russia and winters around the Black Sea and a few other sites in eastern Europe (*HBW*). Of 90 assumed genuine wild individuals recorded in Britain to the end of 2019 (Holt *et al.* 2020), two were in Wales: an immature shot in Milford Haven, Pembrokeshire, on 18 January 1935 (Witherby *et al.* 1940) and one at Camlad Meadows, Montgomeryshire, on 4 March 1950 in the company of 650 White-fronted Geese and two Barnacle Geese. No others have been accepted as wild individuals but several have occurred that are considered to have escaped from captivity (see Appendix 1).

Jon Green and Robin Sandham

Canada Goose *Branta canadensis* Gŵydd Canada

Welsh List Category	IUCN Red List	
	Global	Europe
C1(1), E*	LC	LC

The native range of Canada Goose is in North America, where it breeds over most of Canada and Alaska, as well as the northern part of the USA. A population of 1,000–5,000 pairs breeds in western Greenland (BirdLife International 2015). Wild individuals of two subspecies, Todd's Canada Goose *B.c. interior* from south-central Canada and Parvipes Canada Goose *B.c. parvipes* from central Canada, have been recorded elsewhere in Britain, but not as being of wild origin in Wales. Birds introduced to Britain show characteristics of several subspecies.

This species was introduced into England in 1665, becoming part of Charles II's wildfowl collection in St James's Park, London and later became popular in collections, but there were few breeding records until the latter part of the 19th century (*Birds in England* 2005). In Wales, Dillwyn (1848) described the species as "nearly naturalised, and not unfrequently absents itself from the lake at Penllergare [*sic*]", north of Swansea, Gower. Five, of which one was shot, were in the Rhymney Estuary, East Glamorgan, in 1866 (Ingram and Salmon 1939). In North Wales, Forrest (1907) noted that semi-wild birds were sometimes seen on pools and marshes near the English border, generally in winter, and that it bred regularly on the wetlands around Ellesmere, Shropshire. Canada Geese were introduced to Leighton Park near Welshpool, Montgomeryshire, in 1908, but the species remained scarce in Wales in the first half of the 20th century. There were only three records in Glamorgan during 1900–50 (Hurford and Lansdown 1995), no modern record in Gwent prior to 1960 and no definite record in Radnorshire until 1959, when there was a breeding pair, thought likely to have been on Llyn Penybont (Jennings 2014).

By the early 1950s, Canada Geese were causing agricultural damage in some areas of England and the Wildfowl Trust (now WWT) suggested translocating birds to other areas, an operation with which the Wildfowlers' Association of Great Britain and Ireland (now BASC) assisted (Jones and Whalley 2004). On Anglesey, birds were released at Valley Wetlands in 1950 and elsewhere in 1954 and 1956, in Ceredigion at Cors Caron in 1954 (Roderick and Davis 2010), and in Pembrokeshire at Boulston in 1955 and Orielton in 1957 (Donovan and Rees 1994). Further introductions followed in the 1960s, for example in Gwent in 1960 and 1962 (Venables *et al.* 2008). Some introductions were more successful than others; the species failed to become established at this time in Gwent or Ceredigion but increased rapidly on Anglesey, with 200 on Llyn Traffwll in 1965 and over 600 there in 1970. A population that became established in southern Llŷn, Caernarfonshire, by about 1965 presumably derived from the Anglesey birds.

The *Britain and Ireland Atlas 1968–72* showed that the species had a patchy distribution in Wales, with records in

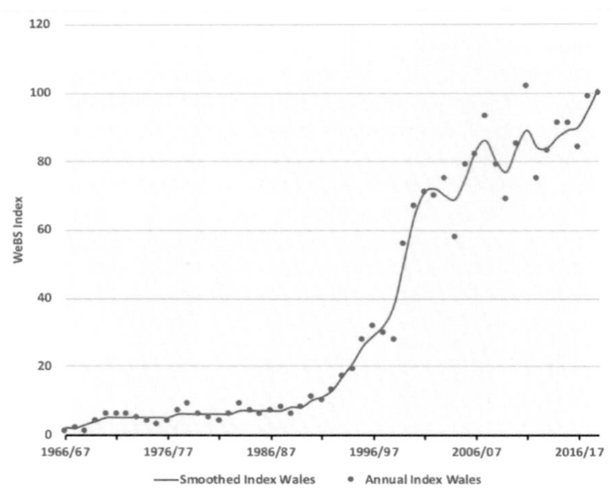

Canada Goose: Wetland Bird Survey indices, 1966/67 to 2018/19.

eastern mid-Wales, where it had probably spread from the English Midlands, and populations on Anglesey, Caernarfonshire and southern Pembrokeshire. Elsewhere, Canada Geese were largely absent. The *Britain and Ireland Atlas 1988–91* showed a considerable increase, including a spread westward from border counties into Ceredigion, Flintshire and Denbighshire, but it was still largely absent from Glamorgan and Meirionnydd. A dramatic extension in range was evident by the 2007–11 *Atlas*, with birds breeding in all parts of Wales and absent from only a few 10-km squares. While nest sites are generally close to water, they can occur in other habitats, such as on a steep heather-clad cliff at RSPB Lake Vyrnwy, Montgomeryshire, in 2015 and amid a seabird colony on Ynys Gwylan, Caernarfonshire, in 2014. The Welsh population in 2018 was estimated to be 8,850 pairs (Hughes *et al.* 2020).

Outside the breeding season, the WeBS index for Wales shows a fairly stable population through the 1970s and 1980s and a remarkable increase between 1990/91 and 2000/01, after which the increase continued at a slower rate. The BBS index for Wales increased by 359% between 1995 and 2018, the largest change for any species for which a Welsh trend can be calculated, except Red Kite (Harris *et al.* 2020). Possible reasons for the increase include the creation of new wetlands and a decrease in shooting. Greylag Goose shows a broadly similar but less pronounced trend.

Although the Welsh population is not migratory, there are considerable movements: the largest gatherings of Canada Geese tend to occur at the end of the breeding season, although the locations vary. High counts include *c.*7,000 on the Dee Estuary at Flint in September 2018 and 3,319 on the Dyfi Estuary in December 2008. The Teifi, Nevern and Cleddau estuaries in Pembrokeshire, the Tywi Valley in Carmarthenshire, the Glaslyn/Dwyryd estuaries in Meirionnydd/Caernarfonshire and Dolydd Hafren, Montgomeryshire, have held over 1,000 at times.

Most ringing takes place during moult in late June and early July, when Canada Geese are mainly flightless. The moulting herd at Llangorse Lake, Breconshire, increased from 28 in 1981 to over 600 in 2006; numbers fell to around 400 as some moved to nearby Talybont Reservoir, then increased again to 600 in 2017/18. In excess of 5,000 have been ringed at Llangorse since 1979, with 250–300 ringed annually in 2003–16 and 200 each year thereafter. In the early years, this revealed interchange with the Beauly and Moray Firths, Scotland: two juveniles ringed at Llangorse were later recaptured while moulting as adults in the Firth, an adult ringed in July 1983 was recaptured in the Moray Firth in July 1984 having apparently changed its moulting area, and one ringed in the Beauly Firth in July 1980 was shot at Montgomery in 1983. There has been no subsequent record of such movements, possibly due to reduced ringing activity at the Scottish site, but there were subsequent exchanges with central Scotland in 2007, Highland Region in 2011, and Perth and Kinross in 2014. Movements between Llangorse Lake and North Wales are unusual, however, with just two birds moving to northwestern counties. The vast majority of the 700 recoveries from Llangorse Ringing Group have been to the north and northeast, with many moving into Herefordshire from August, before they spread across the Midlands and northwest England, and some to Yorkshire. In more recent years, recoveries showed a movement southeast down the Usk Valley into Gwent, though only seven recoveries came from Chew Valley Lake, Somerset, where an annual goose roundup also takes place, suggesting that the Severn Estuary may be a barrier or a diversion to onward movement. By contrast, Canada Geese from Chew move predominantly to the southwest. Another site with which Canada Geese from southeast Wales exchange is Lake Windermere, Cumbria: six ringed there visited Llangorse during 2013–19 and an additional five were shot near Monmouth, Gwent (Jerry Lewis *in. litt.*)

The Canada Goose was formerly confined, in Wales, to the lowlands but in recent years it has spread to the uplands. Breeding pairs are now found on many mountain lakes, and nesting has occurred on Yr Wyddfa's Llyn Glas, Caernarfonshire, at an altitude of 530m and at Llyn Fyrddon Fach in Ceredigion at 540m. Concern has been expressed that their droppings could change the ecology of these nutrient-poor lakes. Owen *et al.* (2006) considered Canada Goose to be of medium risk to native biodiversity, noting that negative impacts were little understood, but that there was evidence of local problems through grazing, damage to reedbeds and the enrichment of wetlands with nutrients from their droppings. Subject to certain conditions, Canada Geese can be shot throughout the year under General Licences. In Wales they can be shot only to prevent serious damage to agriculture, but in England they can also be killed for reasons of public health and safety, and nature conservation. No information is collected on the numbers shot or eggs destroyed.

Prospects for the introduced Canada Goose in Wales seem good unless there is a large-scale effort to reduce the population. The growth in the population in Wales slowed in the second decade of the 21st century but has clearly yet to reach its capacity. Rehfisch *et al.* (2010) commented that if the species continued to increase at over 9% each year and with much apparently suitable habitat still unused, Canada Geese would start to have a more dramatic effect on Britain's biodiversity.

Rhion Pritchard

Site	**County**	**74/75 to 78/79**	**79/80 to 83/84**	**84/85 to 88/89**	**89/90 to 93/94**	**94/95 to 98/99**	**99/00 to 03/04**	**04/05 to 08/09**	**09/10 to 13/14**	**14/15 to 18/19**
Dee Estuary: Welsh side (whole estuary)	Flintshire	na (0)	na (9)	na (23)	120 (368)	386 (794)	658 (1,875)	550 (2,370)	1,107 (2,529)	1,900 (2,900)
Dyfi Estuary	Ceredigion/ Meirionnydd	0	0	19	223	909	2,337	2,773	2,280	1,653
Severn Estuary: Welsh side (whole estuary)	Gwent/East Glamorgan	0 (51)	0 (74)	0 (299)	0 (346)	4 (310)	80 (365)	181 (607)	309 (753)	373 (1,128)
Teifi Estuary	Pembrokeshire/ Ceredigion	0	0	0	4	107	273	423	805	762
Nevern Estuary	Pembrokeshire	0	0	0	0	0	4	345	636	739
Cleddau Estuary	Pembrokeshire	89	109	176	125	389	922	739	605	455

Canada Goose: Five-year average peak counts during Wetland Bird Surveys between 1974/75 and 2018/19 at sites in Wales averaging 600 or more at least once. No site thresholds have been determined for Canada Goose as it is not native to Britain.

Barnacle Goose *Branta leucopsis* Gŵydd Wyran

Welsh List Category	IUCN Red List			Birds of Conservation Concern			
	Global	Europe	GB	UK	Wales		
A, C5, E	LC	LC	LC	Amber	2002	2010	2016

This goose has very localised breeding and wintering grounds. It breeds on crags and rocky outcrops on the tundra of east Greenland, Svalbard and northwest Russia, particularly Novaya Zemlya, and winters on lowland meadows near the coast, mainly in Britain, Ireland and around the southern North Sea (*HBW*). Most are highly faithful to wintering sites (*The Birds of Scotland* 2007).

Birds breeding in Greenland spend the winter along western coasts of Scotland and Ireland, with the most recent flyway census recording 72,162 in Scotland and Ireland in 2018, of which almost half were on Islay, although this was 10% down from a peak of 80,670 in 2013 (Mitchell and Hall 2020). Birds breeding in Svalbard spend the winter almost entirely around the Solway Firth, Cumbria/Dumfriesshire, numbering about 40,000 in recent years. To the east, birds from Russia have extended their breeding range through the Baltic Sea and winter mainly along the southern North Sea coast, with the largest numbers in The Netherlands; these mix with a sedentary population extending from Belgium to southwest Denmark. An adult ringed in Greenland in July 1963 was shot at Rhandirmwyn, Carmarthenshire, in November 1965; *Birds in Wales* (1994) erroneously stated that it was shot at Pontrhydfendigaid, Ceredigion.

The Barnacle Goose's status in Wales is, however, complicated by the presence of a growing naturalised breeding population, thought to number 1,450 pairs in Britain in 2012–15 (Woodward *et al.* 2020). These are presumed, mostly, to originate from birds that escaped from collections. The 2007–11 *Atlas* found that this population had expanded its distribution by 88% since 1988–91, mostly in eastern and southern England. There is a small naturalised breeding population in Wales.

Barnacle Geese bones found at Little Hoyle Cave at Penally, Pembrokeshire, have been dated to about 22,800 years ago (Green and Walker 1991) and were thought to represent a local breeding population during a period of severe cold in the last Glacial period. Historic records all relate to a wintering population and suggest that, in Wales, it was most common in the north. Thomas Pennant, in his *Tours in Wales*, mentioned that a gale accompanied by intense frost killed a large number of birds in February 1776, whose bodies—including those of Barnacle Geese—were strewn along the shore near Criccieth, Caernarfonshire. Williams (1835) noted that "the last well authenticated visit of these birds was to the Avon ganol [Caernarfonshire/Denbighshire] in 1812; where they came in great numbers: several were caught from the effect of famine". A large flock was in wheat fields at Llandrillo Bay, Caernarfonshire, during a severe winter about 1816 (Dobie 1894), though this may be the same as the 1812 record. A small flock was recorded feeding on the Afon Conwy, Caernarfonshire/Denbighshire, in winter 1894.

Until the 1870s, Barnacle Geese were said to be present in their thousands in the Dee Estuary, Flintshire (Williams 1977). Forrest (1907) wrote that although several authors mentioned the species as a common winter visitor to the north coast, recent records had been few, and he suspected that previous authors had confused this species with Brent Goose. However, Reynardson (1887) specifically said that the geese on the Dee were not Brent Geese and wrote of "a huge blue-looking mass that covered acres of ground" some years previously, reporting that the geese fed on the Wirral side of the estuary. Dockray (1910) lamented "gone are the thousands of Barnacles, once the geese of the river.'

The species was less common farther south in the 19th century and in the first half of the 20th century. Salter (1900) was told by Sir Pryse Pryse of Gogerddan that Barnacle Geese occurred on the Dyfi Estuary, Ceredigion/Meirionnydd, particularly during winter 1854/55, but there was no further record for Ceredigion until March 1948 (Roderick and Davis 2010). It was a winter visitor to Pembrokeshire, generally arriving at Goodwick Sands about 1 October and often found in flocks of Brent Geese (Mathew 1894).

The first record for Carmarthenshire was not until January 1931, when birds were on the Tywi Estuary near Ferryside, the first dated record for Gower was of five off Whiteford Burrows in December 1933 and the first in Gwent in October 1970.

Prior to the 1960s, counts in Wales were small, although a flock of *c.*30 was on the Gwendraeth Estuary, Carmarthenshire, in December 1945. Numbers began to increase from the 1960s, including a flock of about 50 that settled on Ramsey, Pembrokeshire, in October/November 1968, although it was not known whether they remained over the winter. From 1981 a few wintered on Skomer, Pembrokeshire, the flock peaking at 130 in 1988/89, and commuting between the island and Marloes Mere, usually from early January. Donovan and Rees (1994) suggested that in 1989 the same birds had been at Bittell Reservoir, Worcestershire, during the early part of the winter and speculated that the presence of a Dutch-ringed bird in the English Midlands that year indicated that Siberia was the likely origin. No more than eight were recorded on Skomer in 1990 and the flock did not return subsequently. Other records during winter 1988/89 included 60 at Bronydd, Radnorshire, from December to February, and *c.*50 on the upper Taf Estuary, Carmarthenshire, in January, which were thought to be from this Skomer flock (*Birds in Wales* 1994). In Glamorgan, Barnacle Geese were recorded in nearly half the years between 1963 and 1992, with a total of 80 individuals, all between 21 October and 29 March (Hurford and Lansdown 1995), but in the period since, the only Barnacle Geese in the county believed to have been of wild origin were 16 at Rumney Great Wharf on 6 December 1998.

An increase in numbers wintering on the Dyfi Estuary was noted from the early 1990s, reaching 126 at the end of 2004, 274 in November 2007, 380 in December 2012 and 608 in November 2019. These were usually in Ceredigion but have also been recorded in Montgomeryshire and Meirionnydd. Although earlier authors (e.g. Roderick and Davis 2010) had speculated that these were wild birds, Barnacle Geese fitted with colour rings on the Dyfi in November 2013 were re-sighted at Talsarnau, Meirionnydd, in January and then at Derwent Water in Cumbria in March and April, demonstrating that they were from the naturalised breeding population (Dodd 2017). Wetland Bird Surveys at the Dyfi Estuary and non-WeBS summer counts on Derwent Water showed a strong correlation, at least up to 2009.

Elsewhere, Llyn Traffwll, Anglesey, held an average of 12 each winter during 2014/15–2018/19, there were 67 in Carmarthen Bay in January 2016 and 196 in the Burry Inlet, Gower/Carmarthenshire, in March 2014, although the species was not recorded at this site

Barnacle Goose: Wetland Bird Survey indices, 1966/67 to 2018/19.

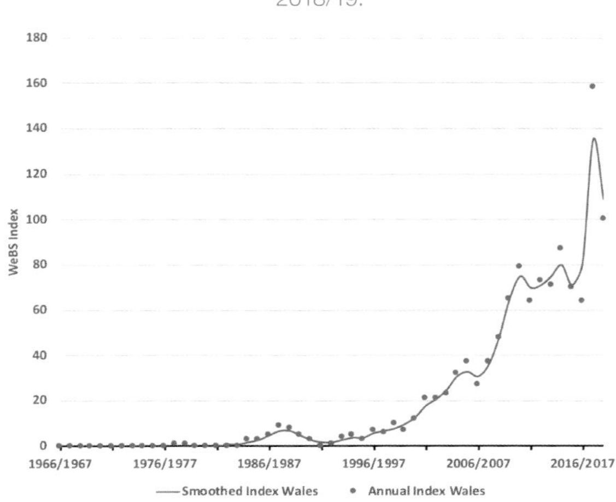

in most years. Up to 250 are recorded on the Severn Estuary and up to 30 on the Dee Estuary each winter, though few are on the Welsh side.

A breeding population on Cardigan Island, Ceredigion, became established in 2003 and does not move far, spending the winter on the nearby Teifi Estuary, Pembrokeshire/Ceredigion. The wintering flock numbered up to 40 birds until December 2010, when an additional 105 birds, believed to have come from the Dyfi Estuary, arrived during cold weather. Subsequent counts on the Teifi have regularly been over 100 (Richard Dobbins *in litt.*) and reached 130 in September 2019. The Cardigan Island breeding population numbered 40 nests by 2019. Barnacle Geese first bred at Llangorse Lake, Breconshire, in 2005, and breeding numbers grew rapidly to 44 adults rearing ten juveniles in 2019 (Waldron 2020); it is not known where those birds winter. In East Glamorgan, a small feral flock, originally pinioned, was at Roath Park Lake, Cardiff, from the late 1990s until 2017. Other records in the county since 1998, including 14 on 3 January 2017 and a county record 19 at Lavernock Point on 29 June 2018, are believed to have originated from Cardigan Island (Phil Bristow *in litt.*). Pairs have also occasionally bred at RSPB Ynys-hir, Ceredigion, and on Anglesey, where there have also been records of hybridisation with Canada Goose. Since a considerable number of birds from the English naturalised population winter in Wales, and a breeding population has now become established in Wales, all records are now regarded as Category C (introduced) unless a marked bird proves otherwise.

Rhion Pritchard

Greylag Goose *Anser anser* Gŵydd Lwyd

Welsh List Category	IUCN Red List			Birds of Conservation Concern
	Global	Europe	GB	UK
B, C1(1), E, E*	LC	LC	LC	Amber

The Greylag Goose breeds over a wide area, from Iceland across northern and central Europe and through temperate Asia as far as the Pacific Ocean. Most populations are migratory, wintering to the south of their breeding range (*HBW*); those breeding in Iceland winter mainly in Scotland and northern England. A sedentary population breeds in northwest Scotland, mainly in the Outer Hebrides. It has become re-established elsewhere in Britain south of the Great Glen since around 1930, initially using eggs taken from South Uist to Stranraer, Galloway (*The Birds of Scotland* 2007). Further translocations by the Wildfowlers' Association of Great Britain and Ireland (now BASC, the British Association for Shooting and Conservation) and individual wildfowling clubs followed in other parts of Britain, using birds from Scotland. Mitchell *et al.* (2012) made a convincing argument that Greylags breeding in Britain can no longer be regarded as two separate populations and that as the species formerly bred in England, it should be considered a re-establishment, but there is no written evidence of historic breeding in Wales.

Greylag Geese recorded in Wales prior to 1930, when birds were first translocated away from the Hebrides, were wintering birds, presumably from Iceland. Forrest (1907) commented that this was the rarest of the grey geese in North Wales, although "the older naturalists were apt to assume that most of the 'Grey Geese' seen in winter belonged to this species". He quoted Inchbald as stating that it "would seem to appear occasionally on the Conway River [Caernarfonshire/Denbighshire]" and was a scarce visitor around this time to the Dyfi Estuary, Ceredigion/Meirionnydd, where the highest count was nine in December 1892. Forrest knew of only two records from the Dee Estuary, where the species was said to be very rare. In Pembrokeshire there was apparently no record until two were shot at "Pont Clew" [*sic*] in January 1911 (Lockley *et al.* 1949), which may refer to Portclew near Freshwater East. It was described as occasional in Carmarthenshire (Barker 1905), although there was a subsequent record of *c.*50 on the Pembrey-Cydweli marshes in November 1922. In Gower, wild Greylags were sometimes recorded among White-fronted Geese at Margam Moors (Hurford and Lansdown 1995). Phillips (1899) stated that it had been shot occasionally in Breconshire, but this was doubted by later authors, and there were no early records from Radnorshire or Montgomeryshire. Several pairs in Cardiff's Roath Park, East Glamorgan, early in the 20th century were presumed to be of feral origin (Ingram and Salmon 1920 cited in *Birds in Wales* 1994).

By the late 1940s and 1950s larger counts were being recorded; it is impossible to know whether these were of Icelandic origin or from the naturalised population elsewhere in Britain. On Anglesey, 170 were on the Cefni Estuary in March 1948, with records of 100 there in 1950 and 1951. By 1954, flocks of 50–60 were regular on the island (Jones and Whalley 2004). Greylag Geese were released on Anglesey from 1960, eggs and young having been taken from the Loch Insh area in the central Highlands, where adults were causing damage to crops. Anglesey proved to have plenty of suitable habitat and a population quickly became established, with breeding confirmed at Llyn Dinam in 1970. The *Britain and Ireland Atlas 1968–72* included just two records in Wales, both on Anglesey, one of confirmed breeding, presumably at Llyn Dinam, and one of possible breeding. There were 136 at Llyn Traffwll in January 1973 and 540 there in October 1978. By the early 1990s the breeding population on the island was estimated to be over 150 pairs (Jones and Whalley 2004). Owen *et al.* (1986) calculated that the population increased at an average rate of 13% each year across Britain between the late 1970s and 1986; Austin *et al.* (2007) calculated the annual increase to be 9.4% during 1991–2000.

By the 1988–91 *Atlas*, breeding was widespread on Anglesey and there were nesting records in Denbighshire, Carmarthenshire, on the Dyfi Estuary, Ceredigion/Meirionnydd, and in two 10-km squares in southeast Llŷn, Caernarfonshire. By 1993, two adjoining populations were established in Carmarthenshire, on the Taf, Tywi and Gwendraeth estuaries and in the southeast at Penclacwydd-Machynys (*Birds in Wales* 2002). The first confirmed breeding record for Gwent was at Llanwern in 2000, where numbers increased to seven pairs in 2008, and from which Newport Wetlands Reserve was colonised. The species now breeds regularly there, and also bred at Magor Marsh in 2016 and Peterstone Golf Course in 2013–15. Breeding in Pembrokeshire occurred for the first time in 2006, but is not recorded annually. The *2007–11 Atlas* showed Greylags to be widespread in Wales, with the highest relative abundance in Gower, Carmarthenshire and northern counties, but still quite scarce in much of central and southeastern Wales. Breeding was confirmed in three tetrads in East Glamorgan in 2008–11. A pair at Llyn Gwyn in 2010 and

Greylag Goose: Wetland Bird Survey indices, 1966/67 to 2018/19.

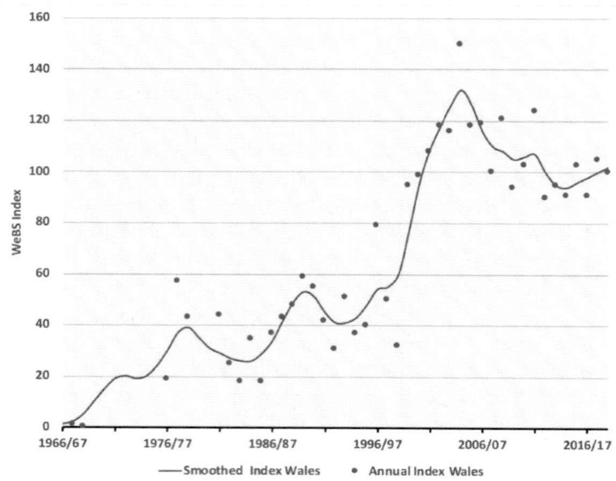

Site	County	74/75 to 78/79	79/80 to 83/84	84/85 to 88/89	89/90 to 93/94	94/95 to 98/99	99/00 to 03/04	04/05 to 08/09	09/10 to 13/14	14/15 to 18/19
RSPB Valley Wetlands inc. Llyn Traffwll	Anglesey	294	210	322	274	615	867	783	796	884
Traeth Lafan	Caernarfonshire/ Anglesey	0	0	0	38	50	793	352	345	636
Dee Estuary: Welsh side (whole estuary)	Flintshire	na (0)	na (0)	na (0)	1 (6)	3 (21)	13 (89)	57 (292)	181 (444)	311 (631)
RSPB Cors Ddyga	Anglesey	60	60	92	1	68	18	66	282	376
Llyn Coron	Anglesey	176	116	220	169	209	168	341	385	232
Llyn Alaw	Anglesey	145	257	292	433	367	416	445	264	91

Greylag Goose: Five-year average peak counts during Wetland Bird Surveys between 1974/75 and 2018/19 at sites in Wales averaging 300 or more at least once. The mobility of Greylag Geese on Anglesey means that these are likely to involve a substantial number of the same birds. Great Britain importance threshold: 1,400.

2011 were the first confirmed breeding records in Radnorshire (Jennings 2014). It has now bred in all Welsh counties, though is only abundant in south Carmarthenshire, Anglesey, Denbighshire and Flintshire. In 2016 the Welsh population was estimated to be 1,650 (965–2,050) pairs (Hughes *et al.* 2020).

In Scotland, Greylag Geese nest on heather moorland near lochs in parts of northern Scotland. Similar habitat is available in North Wales but naturalised birds nest near well-vegetated lowland lakes, with grassland nearby, sometimes along slow-flowing rivers and in urban parks. Islands in lakes are particularly favoured for nesting as they provide some protection from mammalian predators. While most breeding in Wales occurs in the lowlands, a few have been recorded in heather close to moorland pools at 400–550m (Jim Dustow *in litt.*). Greylags do not usually breed until they are three years old so the population is likely to include a considerable number of non-breeding birds (Nilsson *et al.* 1997). Ringing recoveries involving Wales show movements within Britain but none farther afield. Those covering the greatest distances were birds that bred in East Glamorgan and Gwent that flew around 500km north in late summer each year to Hogganfield Loch, Glasgow, and returned south as soon as the primary wing moult was complete (WWT/JNCC 2010). Other northerly movements from a colour-ringing project of birds breeding at Llanwern, Gwent, included birds seen subsequently in Dumfries and Galloway and Lancashire, although the majority of movements were along the South Wales coast as far west as Carmarthenshire and to the east into Gloucestershire (Richard Clarke *in litt.*).

Outside the breeding season, the highest counts usually come from Anglesey, where the birds move frequently between lakes. Counts of 1,000 were made on Llyn Maelog on 25 September 1998 and Malltraeth in October 2008; it is unclear whether this relates to RSPB Cors Ddyga or to Malltraeth Cob/Cefni Estuary. The largest flock recorded in Wales was 1,160 on Llyn Traffwll in July 2013. The Anglesey birds may also have been responsible for peak counts across the Menai Strait at Aber Ogwen, Caernarfonshire, where there were 1,010 in September 2001 and 1,037 in September 2002. In northeast Wales 702 were on the Dee Estuary, Flintshire, in November 2018, and 650 at Gresford Flash, Denbighshire, in November 2017. Greylags were rarely recorded in Pembrokeshire prior to 2000, but numbers increased from 2011/12, notably on the Cleddau Estuary, where up to 200 have been recorded. Numbers have also increased in the Tywi Valley, Carmarthenshire; 407 were at Dryslwyn/Cilsan in January 2019.

The apparent reduction in numbers measured by the WeBS index since 2004/05 may be a result of more birds feeding on farmland away from count sites, but in the longer term, Wales is expected to be climatically less suitable for Greylag Geese, particularly in the south (Huntley *et al.* 2007).

Five Greylag Geese at Dryslwyn, Carmarthenshire, in December 1983 marked with yellow neckbands originated in (then) East Germany, but it is not clear whether they were of the east European subspecies *A.a. rubrirostris* (Gary Harper *in litt.*). A party of nine with Bewick's Swans at Olway Meadows, Gwent, on 13 February 1993 were also thought to be *A.a. rubrirostris* (Venables *et al.* 2008), but some Greylag Geese introduced to England were of this subspecies (*Birds in England* 2005), so may not have been wild birds.

Rhion Pritchard

Taiga/Tundra Bean Goose *Anser fabalis/serrirostris* Gŵydd Lafur y Taiga/Twndra

These two species are treated consecutively below, even though they do not follow each other in the systematic list (IOC). Their taxonomic status remains a subject of debate given that reproductive isolation is incomplete and both taxa are known to hybridise (Ottenburghs *et al.* 2020). This introduction covers both species, as the identity of most early records cannot be determined. Both are scarce and irregular winter visitors to Wales.

Bones of Bean Geese dating from the Ipswichian interglacial period (115,000–130,000 years ago) have been found in Bacon Hole, Gower, suggesting that it may have once been more widespread. However, historic records are clouded by confusion between this and other grey geese, particularly Pink-footed Goose. Forrest (1907) described the Bean Goose as being "the most frequent wild goose in Wales", but recognised the identification problems and, based on more recent trends, it was surely always a scarce visitor. Bean Geese occurred, perhaps annually, on the Dee Estuary, Flintshire/Wirral, but in small numbers. Larger groups included 15 on the Dee on 22 January 1908, a flock on the Mawddach Estuary, Meirionnydd, in winter 1881, 7–8 near Stone Hall, Pembrokeshire, in 1880 and seven at Pencelli, Breconshire, in January 1894. There was also a single record from Radnorshire, shot at Dolyhir in December 1903.

From 1920, there was a decline in Welsh records, but it is unclear whether this was due to improved identification or a real change in status; some records may have been feral or escaped birds. The only regular wintering birds were to be found among

Total	1950–59	1960–69	1970–79	1980–89	1990–99	2000–09	2010–19
76	1	1	25	27	8	2	12

Taiga and Tundra Bean Goose: Presumed genuine wild individuals recorded in Wales by decade, since 1950.

the White-fronted Goose flock on the Afon Tywi at Dryslwyn, Carmarthenshire, where there were singles in January 1973 and January–March 1974, up to five in January/February 1976, two from December 1981 to February 1982, and singles in December 1982, 1984 and 1987. Other records were sporadic, mostly of singles and small groups, but larger groups included 12 on Dowrog Common, Pembrokeshire, on 22 April 1949, four at Weobley, Gower, in January 1976, nine at Hook, Pembrokeshire, on 2 November 1978 and 14 at Blackpill, Gower, on 7 November 1982.

All records considered to be genuine wild Tundra and Taiga Bean Geese occurred between November and April, the vast majority being recorded in southwestern counties. Prior to 'Bean Goose' being recognised as two species, the form was not recorded.

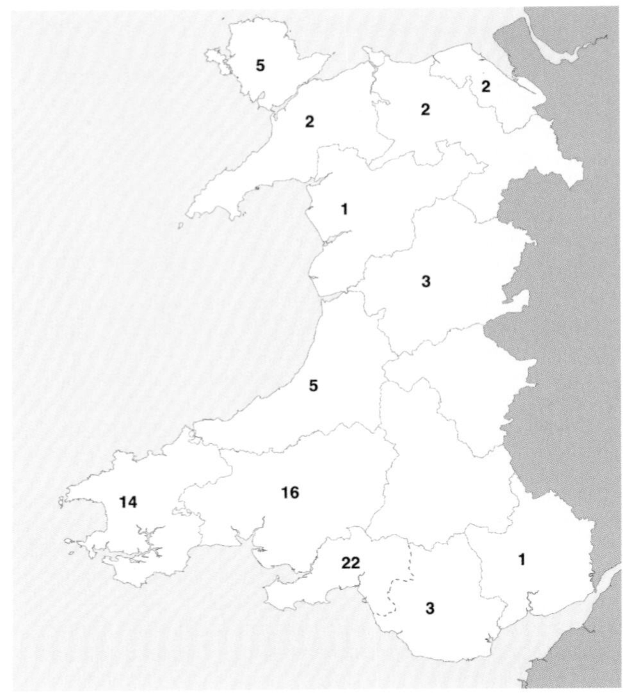

Taiga and Tundra Bean Goose: Numbers of birds from each recording area, 1950–2019. One was seen in both Gwent and East Glamorgan.

Taiga Bean Goose *Anser fabalis* Gŵydd Lafur y Taiga

Welsh List Category	IUCN Red List			Birds of Conservation Concern	Rarity recording
	Global	Europe	GB	UK	
A, E	LC	LC	CR	Amber	WBRC

This species breeds in Scandinavia and across Russia as far as the Pacific Ocean. The nominate subspecies winters mainly in southern Sweden and to a lesser extent in northern Germany and Poland, with small numbers in Denmark, Falkirk in Scotland and Norfolk in England. It is very rare in Wales, with just eight birds on four occasions:

- one at RSPB Ynys-hir, Ceredigion, on 8–11 April 1982;
- one at Ogmore, East Glamorgan, on 22 January 1996 moved to Nedern, Gwent, on 31 January and then to nearby Mathern on 10–18 February;
- one at Ffairfach, Carmarthenshire, from 24 December 1996 to 7 January 1997 and
- five at St Davids, Pembrokeshire, on 27 December 1997.

Breeding numbers are declining and its European wintering range is contracting as the population winters closer to its breeding areas. Thus, its occurrence in Wales is likely to become even less frequent.

Jon Green and Robin Sandham

Tundra Bean Goose *Anser serrirostris* Gŵydd Lafur y Twndra

Welsh List Category	IUCN Red List			Birds of Conservation Concern	Rarity recording
	Global	Europe	GB	UK	
A, E	LC	LC	VU	Amber	WBRC

As the name suggests, this species breeds on tundra, from Norway eastwards through Siberia. Birds of the subspecies *A.s. rossicus* breed west of the Taymyr Peninsula and winter in eastern and central Europe. The European winter population numbers *c.*300,000 and during severe cold weather, birds move west into The Netherlands and eastern England, but it is very rare in Wales, with only 12 individuals recorded:

- one at Cors Fochno, Ceredigion, on 23 January 1955;
- one at Dryslwyn, Carmarthenshire, on 18 to 28 December 1987;
- one at Kenfig, East Glamorgan, on 5 January 2002;
- three at Llyn Coron, Anglesey, on 13 to 19 November 2011, one of which remained until 27 November;
- two at Shotwick, Flintshire, on 4 January 2012;
- three at Dolydd Hafren, Montgomeryshire, from 28 January to 5 February 2012 and
- one at Aber Ogwen, Caernarfonshire, on 1 February to 2 March 2012.

Jon Green and Robin Sandham

Pink-footed Goose *Anser brachyrhynchus* Gŵydd Droetbinc

Welsh List Category	IUCN Red List			Birds of Conservation Concern			
	Global	Europe	GB	UK	Wales		
A, E	LC	LC	LC	Amber	2002	2010	2016

There are two discrete populations of Pink-footed Goose. One breeds in eastern Greenland and Iceland and winters in Britain, the majority of these birds wintering in Scotland, northwest and eastern England. The other breeds in Svalbard and winters on southern North Sea coasts, from Belgium to Denmark (*HBW*). Numbers of both have increased substantially in Britain in recent decades, from 48,000 to 172,000 during 1960–87 (Fox *et al.* 1989); the population reached a record 536,000 by 2015 (Mitchell 2016). The Icelandic breeding population has expanded its numbers and range towards the coast, which may be a recolonisation following historic persecution in the lowlands. The availability of high-quality grazing in the farmed landscape in Britain, improved protection and sanctuary areas have all contributed to the increase (Tony Fox *in litt.*).

Historic records reveal confusion between this species and Bean Goose, making the authenticity of some older records doubtful. Pink-footed Geese were reported on several occasions in South Wales during the 19th century, but few reliable records exist. A large flock was at Llanwern, Gwent, from October to December 1899 (*Birds in Wales* 1994). In the early 20th century, Pink-footed Geese occurred occasionally on the Dyfi Estuary, Ceredigion/Meirionnydd, and in the Menai Strait, Caernarfonshire/Anglesey. In 1901–02 small flocks were recorded inland at Llandrillo, Meirionnydd, where on 15 January 1901 one was shot from a flock of nine. During the 1950s and 1960s, a small flock regularly visited the lower Clwyd Valley, Denbighshire/Flintshire, but numbers dwindled by the early 1970s. They remained scarce in South Wales, with just a few double-digit counts, including three flocks at Peterstone Pill, Gwent, in December 1935, 20 in Breconshire on 5 December 1945 and 12 at Peterstone, Gwent, on 2 November 1975.

There is little evidence that Pink-footed Geese occurred on the Dee Estuary, Flintshire/Wirral, before the mid-19th century, but it is now one of the main UK wintering sites. In the late 19th century, around 1,000 were on the estuary and nearby fields, congregating at high tide on saltings now occupied by a Ministry of Defence rifle range and Deeside Industrial Estate. They also fed in fields at Sealand and Saltney, Flintshire, and at low tide roosted on sandbanks in the middle estuary (Coward 1910). By the 1930s the flocks had moved to the potato fields of south Lancashire as a result of disturbance due to the spread of industrialisation (Raines 1962) and being regularly harried by wildfowlers and punt-gunners. Leonard Brooke, a Dee punt-gunner, bagged 21 with one shot, although this was exceptional (Dockray 1910).

Williams (1949) wrote that "The Dee Pink-foots of Coward's Fauna are now little more than a memory of the past. But who can tell? Geese are capricious birds, and one day they may return in spite of industrialism. It seems hardly likely Pink-foots will ever return to the Dee in their former numbers". That proved unduly pessimistic, although it was not until winter 2005/06 when 250 wintered off Parkgate, Wirral, that the geese returned to the Dee Estuary in large numbers. Counts increased steadily to a record 14,659 in the latter part of winter 2018/19, although the record on the Welsh side was 8,000 in 2015. The Dee Estuary is now internationally important for Pink-footed Goose, with the five-year average of 9,648 substantially greater than 1% (5,400) of the northwest European population (Frost *et al.* 2020). However, it has yet to be included as a qualifying feature of the estuary's Special Protection Area.

Small numbers arrive from mid-September, increasing as autumn and winter progress. The main night-time roost is on the RSPB nature reserve between Bagillt, Flintshire, and the outer saltmarsh on the Wales/England border. These provide an undisturbed sanctuary when shooting occurs at Burton and Heswall marshes. Some graze on saltmarsh on the Wirral side of the estuary, at Parkgate and the northern edge of Burton Marsh; they also use White Sands, Flintshire, when there is no wildfowling.

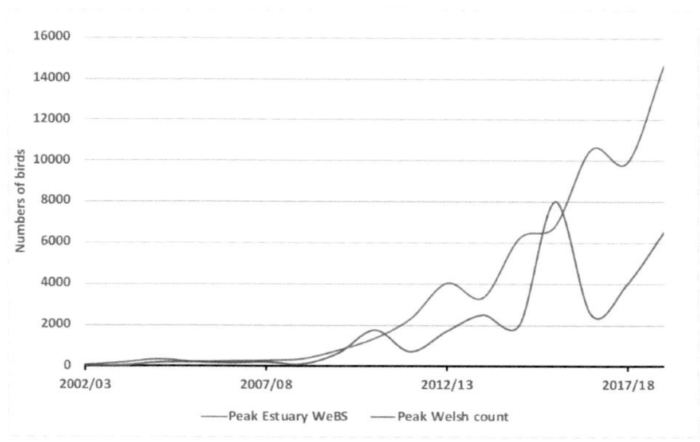

Pink-footed Goose: Peak annual counts on the Dee Estuary, 2002/03 to 2018/19. Not all Welsh counts are from the Wetland Bird Survey, hence the peak in 2015 was higher than the all-estuary total.

During winters 2014/15–2017/18, up to 5,000 occasionally visited RSPB Oakenholt Marsh and Flint Marsh on the Welsh side of the main channel.

The birds' use of stubble, winter barley/wheat and grass pasture varies depending on availability and level of disturbance. They generally move to and from these areas at dawn and dusk, although they will feed at night during a full moon. The principal feeding areas are on the English side of the estuary, as far inland as Marbury, Cheshire, 38km to the southeast (Anne Brenchley pers. comm.). The main feeding sites in Flintshire are Shotwick Fields, RSPB Burton Mere Wetlands and to a lesser extent the Sealand area. Despite objections from conservation groups, a solar farm was constructed on Shotwick Fields in 2016, for which planning conditions require an adjacent area to be managed for goose and swan foraging and without shooting for 25 years. The neighbouring RSPB Burton Mere Wetlands, with wet grassland and spring-sown barley followed by winter stubble, provides excellent conditions for the geese; up to 3,000 used these during winters 2010/11–2018/19.

With the end of the shooting season below the high-water mark on 20 February, the geese cease to fly long distances inland to feed, but use undisturbed sheep-grazed saltmarsh that straddles the national boundary. Numbers vary but counts on the Welsh part (White Sands/Taylors) at that time of year include 8,000 on 28 February 2016 and 6,500 on 8 March 2019. Pink-footed Geese are comparatively scarce in the lower estuary, although 45 were at Point of Ayr, Flintshire from 23 January to 9 March 2003.

Away from the Dee Estuary, there has not been a significant increase in Welsh records, and many records from South and mid-Wales are singles thought to be escapees from wildfowl collections. Flocks have been recorded during cold weather seeking unfrozen ground. Such conditions in early 1979 brought an estimated 3,500 Pink-footed Geese to North Wales, presumably from Lancashire, including 1,200 at Sealand, Flintshire, on 23 February. The following month there were 500 at Llyn Alaw, Anglesey, and in Denbighshire 500 over Gresford/Ruabon and 350 at Bodelwyddan. There were 316 on the Dyfi Estuary on 2 February and smaller numbers farther south including 16 at Marloes Mere, Pembrokeshire, on 8 January, 60 over Ffair Rhos, Ceredigion, on 26 January and 60 at Llanfarian, Ceredigion, in February.

Other cold-weather movements in the modern period were restricted to mid- and North Wales, notably in 1981, 1987, 2004 and 2018. Aside from 42 at Ynyslas, Ceredigion, on 27 December 1981, the influx that month was concentrated on Anglesey, where over 500 were present (*Birds in Wales* 1994). In early 1987 there were 63 at Aber Ogwen, Caernarfonshire, on 18 January,

300 (probably of this species) over Menai Bridge, Anglesey, on 26 February, and 150 over Llandudno, Caernarfonshire, on 27 February. On 26 January 2004, 178 were on the Dyfi Estuary during cold weather, including a neck-collared bird that had been seen earlier that winter in Scotland and Norfolk. There were also 200 at Towyn, Denbighshire, on 26 February 2004. Strong northerly gales in January 2018 brought a flock of 250 over Rhiwbryfdir, Meirionnydd, on 6 January, 140 over Lake Vyrnwy, Montgomeryshire, on 7 January, and several flocks of over 40 in Ceredigion over those two days. The following month, 120 were at Llyn Alaw, Anglesey, the largest group there since 136 in the winters of 1995/96 and 1996/97.

Recoveries of Pink-footed Geese ringed in Iceland have been made in Denbighshire, Flintshire and Ceredigion, the greatest distance being one ringed on 1 August 1953 and recovered 1,581km to the southeast in Ceredigion on 14 February 1954. There were also recoveries in Wales of Pink-footed Geese ringed in England and Scotland in the 1950s and 1960s, including from Wainfleet in Lincolnshire to Pistyll in Caernarfonshire; from Methven in Perth and Kinross to Shotton in Flintshire; and from the Lammermuir Hills in Borders to Trefnant in Denbighshire. More recently, birds with GPS tags fitted at Martin Mere, Lancashire, have been tracked to the Dee Estuary (WWT/Orsted).

The growth in numbers on the Dee Estuary during the second decade of the 21st century is spectacular, greatly aided by the creation of the large sanctuary area by the RSPB. The carrying capacity of the estuary and its environs is not known, but depends partly on the cropping and grazing choices made by farmers around the estuary and the extent of shooting over parts of the estuary not managed for the geese.

Colin Wells

White-fronted Goose *Anser albifrons* Gŵydd Dalcenwen

Welsh List Category	IUCN Red List			Birds of Conservation Concern			
	Global	Europe	GB	UK	Wales		
					2002	2010	2016
A	LC	LC	CR	Red			

There are five subspecies of White-fronted Goose, all of which breed on the Arctic tundra and two of which occur in Wales. The Greenland form, *A.a. flavirostris*, is a localised winter visitor to Anglesey and mid-Wales, while the nominate European subspecies is an increasingly infrequent visitor (*HBW*).

Greenland White-fronted Goose *A. albifrons flavirostris* Gŵydd Dalcenwen yr Ynys Las

The subspecies breeds in western Greenland between 64°N and 73°N and migrates via Iceland to winter at a small number of traditional haunts in Ireland, Scotland and, in smaller numbers, Wales. Its winter distribution reflects the climatic conditions required for the formation of oceanic blanket bog, which was almost certainly its traditional habitat outside the breeding season before agriculture transformed the landscape. Most leave Iceland in October, arriving at the two primary wintering sites in Wales—the Dyfi Estuary, Ceredigion, and on Anglesey—later that month. The majority now remain until late March, although in the 1960s and 1970s spring departure occurred in late April (Fox and Stroud 1985). Birds may occur elsewhere, usually singly or in small flocks at coastal localities on passage in March/April and late September/October, but occasionally in midwinter.

As the Greenland White-fronted Goose was first described only in 1948, records before this time cannot be ascribed to subspecies with certainty. However, the timing of movements in Wales drew previous reviews (e.g. *Birds in Wales* 1994, Fox and Stroud 1985) to conclude that it was formerly more numerous, with perhaps as many as 1,100 in the early 20th century.

White-fronted Geese have wintered in Ceredigion, on the Dyfi Estuary and Cors Fochno, since the early 20th century, and from their habits and arrival date, it seems likely that these were from Greenland. Birds fed on upland lakes and bogs in the north of the county, for example on Bugeilyn and Llyn Craig y Pistyll to the north of Pumlumon and on Cors Fochno, but it was not clear where they roosted. Small numbers were also recorded on upland bogs in Breconshire in the first half of the 20th century, suggestive of Greenland birds. Contemporary observers suggested that both subspecies occurred on the Dyfi Estuary in the 1940s and 1950s. In March 1951, 220 Greenland White-fronted Geese were on Cors Fochno and in January 1953, 350–500 were on the Dyfi. Numbers subsequently declined on the Dyfi, to 140 in 1970, 130 in 1971 and 36 in 1975.

Elsewhere in Ceredigion, White-fronted Geese wintered on Cors Caron from at least the late 19th century until the late 1970s. Their provenance was confirmed by a bird ringed at Ikamiut, west Greenland, in July 1946 and shot near Tregaron in January 1947. In January 1940, 250–300 were recorded, increasing to 340 in March 1954, 400 in winter 1957/58, and peaking at 550–600 in autumn 1962. However, the hard winter of 1962/63 caused high mortality and dispersal, and fewer than 200 returned after the thaw. Only 117 were counted in 1963/64, falling to 85 in January 1967 and only 3–4 during 1975–79. It has never wintered regularly in Pembrokeshire although 'many' White-fronted Geese that appeared during the severe winter of 1962/63 may have included Greenland birds displaced from Cors Caron alongside European birds from farther east. In Carmarthenshire, Greenland birds were occasionally identified among a larger flock of European birds at Dryslwyn: six on 31 January 1971, for example.

Greenland White-fronted Geese wintered at a few sites in North Wales, such as in the Porthmadog area, Caernarfonshire, in the early 1900s and at Ystumllyn, Caernarfonshire, in that period and in the 1940s, and up to 30 wintered there during 1958–68. On Anglesey, Charles Tunnicliffe's illustrations of White-fronted Geese made in winter 1947/48 suggest the flocks of up to 40 he recorded in the Cefni Valley during December and January were of the European subspecies (Tunnicliffe 1992). Small numbers of Greenland birds were thought to have wintered on Anglesey in the mid-20th century, increasing to 200 in the early 1970s, after which they were replaced or joined by birds of the European subspecies. In Montgomeryshire, small flocks of White-fronted Geese, subsequently confirmed as Greenland birds, were recorded on lakes above Caersws and on upland mires and lakes above Llanbrynmair in the 1960s and 1970s. They fed near Llyn Mawr, but ranged widely and may have moved to the Dyfi Estuary or Cors Caron in hard weather; the peak count was 57 in October 1976, but numbers declined to 24 in March 1984.

Since the mid-1970s, the two principal areas in Wales have been the Dyfi Estuary and western Anglesey. From the early 1970s, those on the Dyfi Estuary fed regularly in agricultural fields on the south shore and remained in the valley to roost. There was a simultaneous decrease in the use of the upland lakes and bogs. Cors Caron ceased to be a regular wintering site, although 39 were present in December 1982 and five were at Pont Einion, Tregaron, in November 1997. After a period of recovery in the 1980s, aligned with an increase in the global population, numbers on the Dyfi peaked at 167 in 1998, but declined to 17 in winter 2018/19, the lowest count since regular counting began in 1959. It has been included as a feature of the Dyfi Estuary Special Protection Area since 2001.

On Anglesey, nine were identified with 24 European birds at Llyn Bodgylched in December 1980 and 24 were at Llyn Traffwll in December 1987. More recent records of Greenland birds have

Greenland White-fronted Goose © John Hawkins

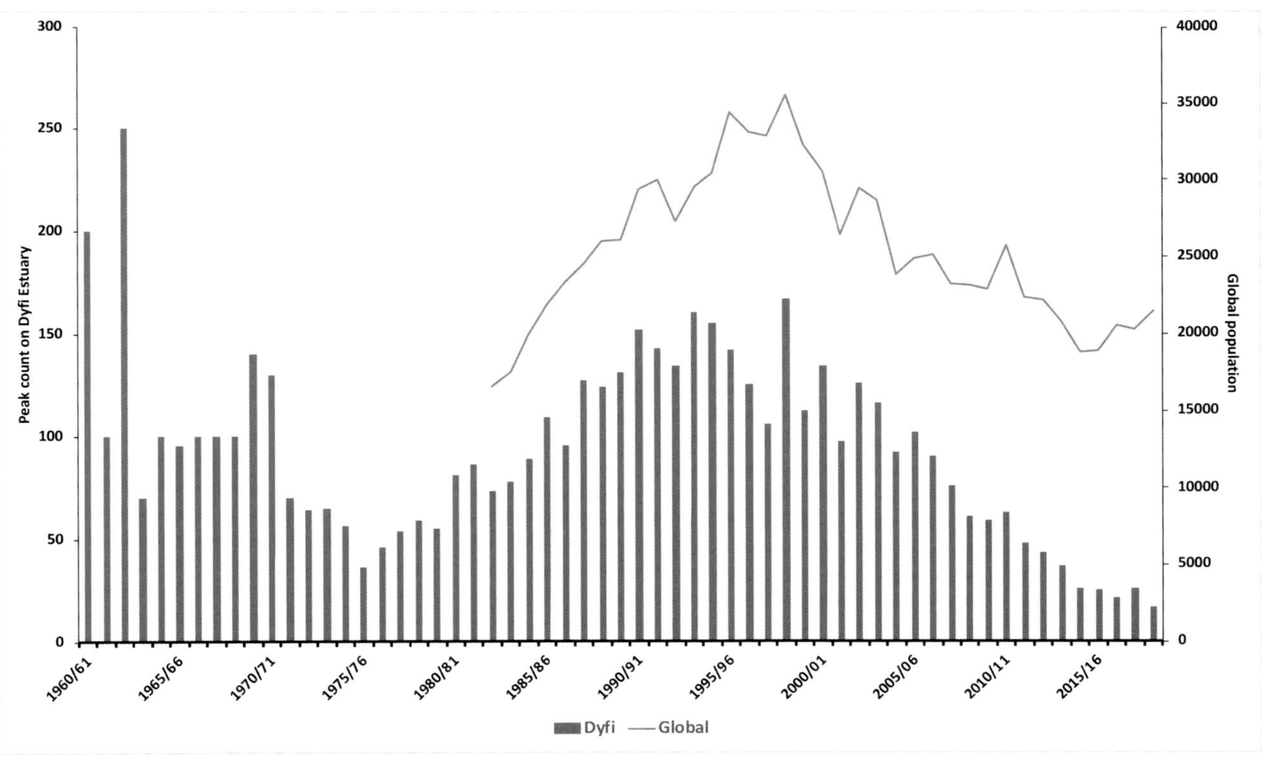

Greenland White-fronted Goose: Peak counts on the Dyfi Estuary, 1960/61 to 2018/19 (bars) and the global population 1982–2018 (line). Data courtesy of RSPB Cymru and the Greenland White-fronted Goose Study Group.

been scattered across the island at sites such as Llyn Coron, Llyn Alaw and Llyn Padrig. The highest total on Anglesey was 61 in 1995/96 (*Birds in Wales* 2002). Up to 19 wintered regularly, using a number of sites, including RSPB Cors Ddyga, in the late 2010s.

It is a rare species in the south, with just three records in Gwent, for example, though this includes a family party of two adults and four first-calendar-year birds near West Usk Lighthouse, St Brides Wentlooge, and then at Newport Wetlands on 10 November 2016. There were records from Marloes Mere, Pembrokeshire, in eight years during 2000–19, with a maximum of 16 in winter 2009/10. These were presumed to involve overshooting Irish-wintering birds during autumn migration. Greenland birds are also occasionally seen off Bardsey, Caernarfonshire, during migration. They are rare in eastern counties, although there were three at Dolydd Hafren, Montgomeryshire, from 4 March to 2 April 2010, five at Lleweni, Denbighshire, in February/March 2015 and nine at Talacre, Flintshire, from November 2017 to March 2018.

In December 2016, two adult females on the Dyfi Estuary were fitted with GPS tags (Mitchell *et al.* 2018), one of which remained on the estuary all winter with the majority of the flock. Telemetry data confirmed their use of agricultural fields throughout the winter, with some feeding on the saltmarsh, especially in March. The other, with its mate, flew to Co. Wexford, Ireland, eight days later and returned to the Dyfi Estuary in mid-March 2017 with six associates. Both tagged geese were tracked to staging areas in southern Iceland and then breeding grounds in west Greenland (between 67.8°N and 68.2°N). One returned to the Dyfi Estuary in the following two winters and the other spent both winters on Coll, Scotland, confirming connectivity and interchange between wintering sites.

Until recently, Greenland White-fronted Geese were legal quarry in Wales, although a voluntary shooting ban was in place, and largely observed by local wildfowling clubs, on the Dyfi Estuary from 1972. On Anglesey, 29 were shot between 1998 and 2010 (Stroud 2011). The subspecies was removed from Schedule 2 of the Wildlife and Countryside Act 1981 (the 'quarry list') in Wales in April 2020.

It is a Red-listed species of conservation concern in Wales owing to a decline in its population in the long term and because over half the population occurs at a single site (Johnstone and Bladwell 2016). Recent declines in the global population are driven by reduced breeding success since the mid-1990s, caused at least in part by increased incidence of snow and lower spring temperatures in west Greenland, but it is not yet evident whether this is consistent with climate change. The geese were subject to considerable hunting pressure in Iceland, where some 3,000 were shot annually during autumn migration, an unsustainable level once annual reproductive success began to decline. Hunting in Iceland was stopped in 2006 and although the population was slow to respond, global population counts since 2016/17 suggest that the decline has halted (Fox *et al.* 2019b). Within Wales, the provision of undisturbed sanctuary areas and high-quality grazing are crucial to the geese being in good condition ahead of their pre-breeding migration. While that is in place around the Dyfi Estuary, the difficulty of locating the birds on Anglesey and understanding their favoured feeding and roost areas remains a gap in conservation provision.

Greenland White-fronted Goose: Distribution of records in Wales, 2000–19 (n=67 1-km squares, excluding records from the Dyfi Estuary, marked with a black asterisk). Where repeat observations were made within a winter, only the maximum number for the duration is shown. The key shows the relative sizes of the dots, not the actual sizes.

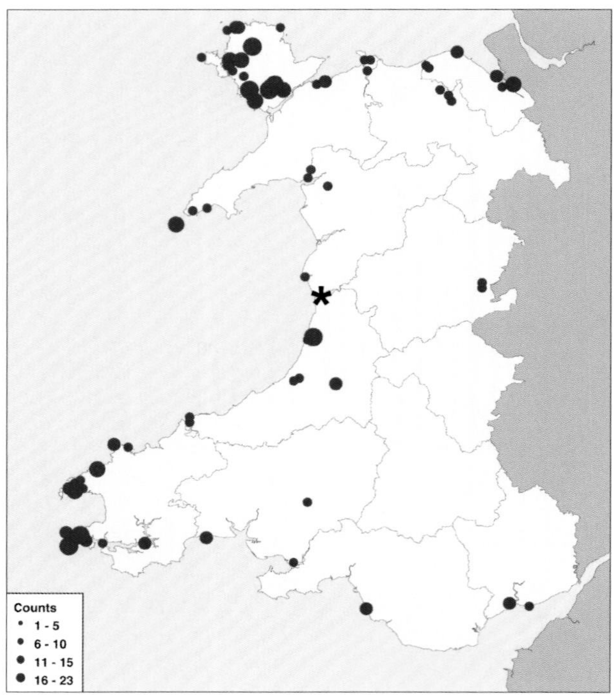

European White-fronted Goose *A. albifrons albifrons*
Gŵydd Dalcenwen Ewrop

This subspecies, formerly described as 'Russian', breeds in Arctic Russia and migrates to winter in northwest Europe, with 6,000–8,000 regularly in England and Wales in the early 1960s. The main wintering areas were Slimbridge, Gloucestershire, which held up to 7,600 in 1969/70, and the valley floodlands and grazing marshes of Montgomeryshire/Shropshire. Despite increases in the global population, more now remain in continental Europe to winter and smaller numbers reach England, an average 1,724 at ten regularly used sites during 2014/15–2018/19. Even fewer make it to Wales: most sightings since 2000 involve singles or small groups staying for short periods, mostly in coastal counties, although larger flocks are sometimes seen during cold weather.

From the 1880s, the Tywi floodplain at Dryslwyn, Carmarthenshire, attracted large flocks. In the early 20th century, White-fronted Geese were recorded regularly on Anglesey and sometimes in Montgomeryshire, in the upper Severn Valley, principally around Leighton. On the south Carmarthenshire coast, 200–300 were regular visitors to the Gwendraeth Estuary, but the planting of Pembrey Forest and the construction of an airfield in the late 1930s destroyed their main feeding area; 300–400 were present in December 1946 despite the disturbance. During 1948–55 numbers there reduced to 100–150, coincident with the increase at Dryslwyn, after which the flock abandoned the estuary (Lloyd *et al.* 2015).

By the 1940s, over 500 wintered annually in the upper Severn Valley, which may have included birds displaced from the Mersey Estuary, Cheshire. The flock, which was relatively mobile, increased to 1,500–2,500 in the early 1960s, but a rapid decline followed: to 1,400 in 1968/69, 750 in 1969/70, 500 in 1970/71, and 150 by 1976/77. Fewer than 50 were recorded in the 1980s, after which the flock ceased to regularly winter there. On Anglesey, 400 were recorded at Malltraeth in March 1951, although it is not clear to which population these belonged.

In Gower, flocks of 2,000–2,500 were recorded at Margam Moors during the early 1940s, but declined when a railway line and steelworks were built. In 1946/47, fewer than 500 were recorded there and by the 1950s the flock had gone (Hurford and Lansdown 1995). Numbers increased at Dryslwyn around this time, from 500 in December 1953 to around 1,000 each winter during 1964–73, and peaked at 2,500 in January 1971. In Flintshire, birds wintered regularly on the Dee Estuary, mainly on Burton Marsh, where 400 were counted in January 1950. Birds that fed farther up the Dee Valley, around Rossett and Pulford in Denbighshire in the mid-20th century, probably roosted on the Mersey Estuary (*Birds in Wales* 1994).

A regular flock of European birds became established around the Dyfi Estuary, Ceredigion, from the late 1940s until the 1970s. For example, a mixed flock of 155 European and Greenland birds at Cors Fochno was recorded in early 1954. Numbers on the estuary peaked at 200–300 between January and March, with exceptionally up to 600 in some winters, presumably a result of cold-weather movements. An adult female shot in January 1963 at Rhydyfelin, Ceredigion, had been ringed in The Netherlands in

March 1962. By the 1960s, however, numbers of European birds around the Dyfi had declined to around 100 each winter and by the 1970s they had become far less frequent.

In the 1970s and 1980s, records of European birds became more frequent on Anglesey, with a peak of 72 at Llyn Bodgylched in January 1973 and up to 40 at other sites, including Llyn Coron and Llyn Alaw, in the 1980s. European White-fronted Geese are occasional visitors to Gwent, mostly to the Levels and the Usk Valley, probably involving birds from Slimbridge. In the 1960s, counts included 150 over Llanfoist in December 1967, 1,000 over Abergavenny on 7 March 1969 and 200 over Caldicot on 30 April 1969. A flock of 29 was at Wentlooge on 7 January 1982, and numbers have subsequently been much smaller in Gwent: since 2000, there have been ten records in the county, involving 37 birds, of which 25 were in 2000 itself.

Cold weather would historically have brought an influx to Wales, such as in southeast Wales during winter 1962/63, including 200 at Peterstone, Gwent, on 3 March 1963. A flock of 1,100 at Dryslwyn in January 1979 may have moved there from Slimbridge, Gloucestershire. On the Dyfi Estuary there were 39 in February 1979, 54 in December 1982 and 164 in December 1987. More recent weather-related arrivals have been far more modest in number, such as 12 at Llyn Coron, Anglesey, in November 2011.

By the mid-1990s, numbers at Dryslwyn rarely reached 100, but since 12 were recorded there in 1996/97 no site in Wales has regularly held the subspecies. Notable flocks since 2000 include 25 on the Conwy Estuary, Caernarfonshire/Denbighshire, in winter 2003/04, 30 at Morfa Nefyn, Caernarfonshire, on 5 November 2006, 15 on the Dyfi Estuary and, concurrently, 16 at Llanon, Ceredigion, in October 2016, and 27 at RSPB Valley Wetlands, Anglesey, in January 2019. It is now less than annual in most counties, with fewer than 50 birds spread across half a dozen sites each year.

The general contraction eastward is thought to be the main reason that major wintering sites in Wales were abandoned. Numbers wintering in Britain were relatively stable through the 1950s and early 1960s, increased to more than 10,000 during 1967–71, but have since declined substantially. More now winter in The Netherlands and north Germany, where feeding conditions have improved, winters have become milder and hunting pressure has reduced. This is a classic case of short-stopping, where birds benefit from wintering closer to their breeding grounds. Wales is at the western end of the flyway and was one of the first places to see a decline. Few European White-fronted Geese are now seen in Wales, and that is likely to be the case for the foreseeable future.

Carl Mitchell and Mick Green

Sponsored by Tony Fox

Lesser White-fronted Goose *Anser erythropus* Gŵydd Dalcenwen Fechan

Welsh List Category	IUCN Red List		Rarity recording
	Global	Europe	
A, E	VU	EN	BBRC

This is the smaller cousin of the White-fronted Goose and breeds in the far north of Scandinavia and east through Siberia. It underwent a massive population decline in Fennoscandia, from more than 10,000 birds in the early 20th century to "functionally extinct" (Holt *et al.* 2020), and is now one of the most severely threatened Arctic-breeding birds (Marchant and Musgrove 2011). The global population is now quite small, estimated at 24,000–40,000 individuals, as a result of habitat fragmentation, drainage of wetlands for agriculture and high levels of hunting on its migration route and wintering grounds (BirdLife International 2020).

Although bones were identified as belonging to this species during excavations at Coygan Cave, Laugharne, Carmarthenshire (E. Walker *in litt.*, cited in *Birds in Wales* 1994), there have been only two records in the modern era: an adult female shot on the Dyfi Estuary, Ceredigion/Meirionnydd, on 11 February 1955, from a flock of 180 White-fronted Geese and a few Pink-footed Geese; and one at Dryslwyn Meadows on 7 March 1971, where a large flock of European White-fronted Geese had been present all winter. The number of British records annually declined from three in the 1950s and 1960s and two in the 1970s, to an average of one every two years since 1990 (Holt *et al.* 2020). One from a reintroduction scheme in Finland was recorded in Wales in 1991 (see Appendix 1).

Rhion Pritchard

Mute Swan *Cygnus olor* Alarch Dof

Welsh List Category	IUCN Red List			Birds of Conservation Concern			
	Global	Europe	GB	UK	Wales		
A, C2	LC	LC	LC	Amber	2002	2010	2016

The Mute Swan has a somewhat patchy distribution across Europe and Asia, from Ireland to eastern China, in lowland freshwater habitats, on estuaries and sheltered coasts. Naturalised populations have become established in several other parts of the world. The European population is mainly resident, but elsewhere wild populations are largely migratory (*HBW*). Adult breeding pairs usually remain in their territories to moult, if they have bred successfully. Non-breeders and failed breeders move to moulting sites between May and September (*BTO Migration Atlas* 2002). Many from Wirral, Cheshire and Shropshire have appeared at moulting sites in Wales (Jones *et al.* 1995). A few move farther: several ringed in Ireland have been found in Wales, and one ringed as a juvenile near Builth Wells, Breconshire, in September 2001 was found dead at Strathclyde Country Park, Lanarkshire, in February 2005. Colour-ringing shows that immature birds disperse from their natal sites, but usually not very far.

Habitat in much of Wales is unsuitable for Mute Swans, which seldom breed above 300m. There are no records of remains at archaeological sites in Wales, but there is a record from 1304 when a millpond was constructed at Caernarfon Castle with an island for a swan's nest (cited by Condry 1981, without further reference). In the medieval period, the species occurred as domesticated or semi-domesticated: Williams (2014) considered it likely that someone claimed ownership of every Mute Swan in Wales during this period, when the young were considered a delicacy. Breeding pairs were not kept in captivity, but some juveniles would be collected and kept in pens to be fattened for the table. A list of the resources available to the kitchen at Plas Isaf, Llanrwst,

Caernarfonshire, in a poem by Tudur Aled (*c.*1465–1525), included 120 Swans (Williams 2014) and there were many other references to Mute Swan in Welsh poetry, particularly in the 16th century. Edward Llwyd (1660–1709), quoted in Fenton (1811), mentioned up to 80 kept on two small lakes at Pembrey, Carmarthenshire, by the Lord of the Manor.

It appears that the Mute Swan gradually reverted to be a wild bird in Britain. Mathew (1894) said that the main site for the species in Pembrokeshire was at Stackpole Court, where there were nearly 100 birds "in what may be considered a wild state". Forrest (1907) said that the species was found in a domesticated state on many lakes and pools in North Wales, but was seldom found as a truly wild bird, though it bred "in a more or less wild state" near Pwllheli, Caernarfonshire, and occurred at Whixall Moss, Shropshire, just over the Denbighshire border.

	1955–56	1978	1983	1990
Estimated number of birds	780	590	700	840
Change from previous survey	-	-24%	+19%	+20%

Mute Swan: Estimated number, including non-breeders, and change in Wales during dedicated surveys (Campbell 1960, Ogilvie 1981, Ogilvie 1986, Delany *et al.* 1992).

The first full census of Mute Swans, in 1955–56, found breeding pairs in all Welsh counties and a total of 115 nesting pairs (Campbell 1960). Numbers ranged from 17 pairs in Gwent and 14 pairs in Carmarthenshire, to two pairs each in Ceredigion and Radnorshire, though coverage was incomplete in many counties. Numbers fell following the severe 1962/63 winter and had not recovered by the next survey in 1978, when a population of 218 breeding birds (109 pairs) was estimated for Wales (Ogilvie 1981). The highest numbers were in Gwent and Glamorgan, each with 14 pairs, but coverage in some counties was incomplete, particularly in Flintshire, where no breeding birds were recorded and coverage was described as "very poor". Another survey in 1983 (Ogilvie 1986) also had poor coverage in some counties and the number of breeding pairs in Wales could not be calculated, but there was evidence of an increase in the northwest. The 1990 survey (Delany *et al.* 1992) had good coverage in most counties, but eastern Clwyd (Flintshire and eastern parts of Denbighshire) was not covered. This estimated the total at 113 pairs, with another 45 pairs holding territory that were not thought to have bred. The highest counts were in Gwent, with 17 breeding pairs, and Montgomeryshire with 12 pairs. The most recent survey, in 2002, estimated an increase of 23% in the British population, but the method did not allow the calculation of a total for Wales (Ward *et al.* 2007). The Welsh population in 2016 was estimated to be 485 (400–585) pairs (Hughes *et al.* 2020).

Site	County	74/75 to 78/79	79/80 to 83/84	84/85 to 88/89	89/90 to 93/94	94/95 to 98/99	99/00 to 03/04	04/05 to 08/09	09/10 to 13/14	14/15 to 18/19
Severn Estuary: Welsh side (whole estuary)	Gwent/East Glamorgan	1 (5)	0 (18)	1 (37)	3 (127)	5 (235)	104 (305)	125 (412)	170 (363)	270 (441)
Llangorse Lake	Breconshire	16	12	10	15	74	91	67	150	150
Traeth Lafan	Caernarfonshire/ Anglesey	3	2	24	10	15	152	148	104	112
Roath Park Lake	East Glamorgan				0	5	35	128	122	92
Dee Estuary: Welsh side (whole estuary)	Flintshire	na (4)	na (5)	na (17)	8 (54)	36 (91)	77 (120)	62 (100)	88 (98)	90 (92)

Mute Swan: Five-year average peak counts during Wetland Bird Surveys between 1974/75 and 2018/19 at sites in Wales averaging 100 or more at least once. Threshold for sites of British and international importance: 500.

In the late 1970s, it became clear that in some parts of Britain birds were dying as a result of poisoning, having swallowed lead from anglers' weights and shotgun pellets. This effect was most serious in areas where coarse fishing was popular and may have been less detrimental in western Wales where angling is predominantly for game species. Lead weights were responsible for 4,000 Mute Swan deaths each year in Britain and a 15% decline in the population during 1956–78 (Wood *et al.* 2019). The use of most sizes of lead weight for fishing was banned in England and Wales in 1987 (Delany *et al.* 1992) and the use of lead shot in shotgun cartridges over wetlands was banned in Wales in 2001 (Ward *et al.* 2007), although compliance studies show that some 70% or more of wildfowl sold were illegally shot with lead ammunition (e.g. Cromie *et al.* 2010). The proportion of Mute Swans dying of lead poisoning fell from 34% to 6% following the introduction of the regulations (Wood *et al.* 2019).

The *Britain and Ireland Atlas 2007–11* showed extensive gains in range in most parts of Wales since 1988–91, although birds remained absent from higher ground. Relative abundance was highest in eastern Wales, around the southern coastal strip and on Anglesey, but was lower than much of England and Ireland. Restrictions on lead weights accounted for 82% of the change in Mute Swan population from 1987 to 2017, whereas factors such as food supply on arable fields, river habitat quality and winter air temperature had little effect. Following a rapid increase in the late 1980s and 1990s, the Mute Swan population in Britain has remained relatively stable in the 21st century.

Outside the breeding season, numbers peak at different sites at different times of year. Numbers on the Dee Estuary, for example, usually peak in winter, while the congregation at Traeth Lafan, usually at Aber Ogwen in Caernarfonshire, is mainly of non-breeding birds during moult. The annual maximum there is usually between July and September. Numbers on Llangorse Lake, Breconshire, peak in September in most years.

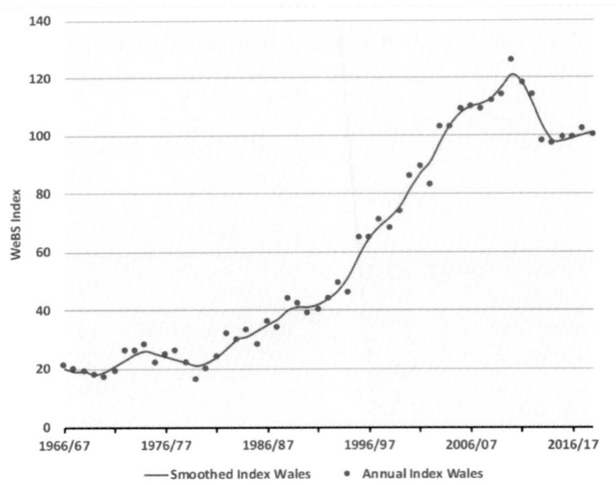

Mute Swan: Wetland Bird Survey indices, 1966/67 to 2018/19.

The WeBS index for Mute Swan in Wales has increased considerably since the early 1990s, though there have been local changes as flocks have switched favoured sites. For example, in East Glamorgan a non-breeding flock of up to 150 at Roath Park Lake, Cardiff, declined after the mid-1960s (*Birds in Wales* 1994), increased again from about 2003 to a peak of 154 in August 2008, and declined therafter. Elsewhere in the county, Knap Boating Lake held 209 in January 2003 and there were 278 at Cardiff Bay in August 2019, which was the largest count in Wales. The Welsh side of the Severn Estuary held 345 in August 2014.

In the long term, southern Wales is projected to become climatically less suitable for the species (Huntley *et al.* 2007). Mute Swans have been helped by the creation of new wetlands, from large-scale projects such as Newport Wetlands Reserve, Gwent, and RSPB Cors Ddyga, Anglesey, to smaller farm and urban ponds, where birds sometimes get human assistance in the form of artificial nests and fencing of the nest area against dogs.

Rhion Pritchard

Sponsored by The Laspen Trust

Bewick's Swan *Cygnus columbianus* Alarch Bewick

Welsh List Category	IUCN Red List			Birds of Conservation Concern			
	Global	Europe	GB	UK	Wales		
A	LC	EN	CR	Amber	2002	2010	2016

Bewick's Swan, or Tundra Swan as it is known in North America, breeds by shallow pools, lakes and rivers on tundra close to water in the Arctic Circle (*HBW*). Birds of the *C.c. bewickii* subspecies breed across northern Siberia; those from the west of their range spend the winter on grasslands in western Europe, where the median air temperature is 5.5°C (Nuijten *et al.* 2020). Many of the ringing recoveries in Wales are birds ringed at the Wildfowl & Wetlands Trust reserves at Slimbridge, Gloucestershire, or Martin Mere, Lancashire. However, a bird ringed northeast of Arkhangelsk, Russia, in August 1996 was identified by a numbered collar on Llyn Alaw, Anglesey, on 26 October that year, having previously been seen on the Estonia/Russia border and in The Netherlands. From Anglesey, it went on to spend the winter on Lough Neagh, Northern Ireland. Another, at Rheidol gravel pits, Ceredigion, in March 1997 had been ringed as an adult on the Pechora Delta, Russia, during the previous summer.

Bewick's and Whooper Swan were recognised as separate species in 1830. Bewick's Swan may have been less common in Wales in the late 19th than in the 20th century: there was only one record in Glamorgan prior to 1921, the first record in Gwent was in 1909 and in Carmarthenshire in 1935. Mathew (1894) said it "cannot be considered rare in Pembrokeshire" and reported several records, mostly of birds that had been shot. Flocks of up to 50 during the severe winter of 1890/91 were thought to be this species. Salter (1900) noted several records in Ceredigion in this period, including 42 on Llyn Conach in January 1894 and 40–50 on the Dyfi Estuary in February 1923. Forrest (1907) described it as "not uncommon" in North Wales, particularly on the western estuaries during severe weather, but said that it was very rare inland, with no records from Montgomeryshire.

Numbers increased in the second half of the 20th century, with Anglesey emerging as a stronghold, particularly Malltraeth Cob and surrounding farmland, where there were up to 35 in 1946, 42 in 1951 and 79 in January 1955. Numbers remained high on the island in the early 1980s, with up to 85 present, but fewer visited in subsequent winters. Some large flocks were also recorded on the Dee Estuary, such as 60 at Sealand, Flintshire, in February 1979. None was recorded in Radnorshire until 1962, after which it became regular in the county until 1999. Counts peaked at 44 in the Glasbury/Clyro/Bronydd area in January 1992, but it has become much scarcer there since 2000. They were even less regular in Montgomeryshire, which also had its first record in 1962, but 41 were at Morfa Dyfi in March 1996. There were also records in Breconshire from the early 1960s, with 34 at Brechfa Pool in December 1973. Numbers increased in Carmarthenshire from the 1940s, with 15–18 on the Afon Tywi in winter 1946/47. The highest count in Pembrokeshire was during cold weather early in 1985, when at least 49 were in the county, including 32 at Hendre Eynon in January. In Gwent, there had been only one record prior to 1960, after which it was recorded annually, with a regular wintering flock in the Usk Valley peaking at over 60 in the 1980s, and 70 at Nedern in January 1983.

From the late 1960s to the early 1990s, there was a considerable spring passage through western counties in late February and early March, thought to be birds returning from Ireland, where up to 2,500 wintered in the 1970s (*Birds in Ireland* 1989). The largest numbers were usually around the Dyfi Estuary, Ceredigion/Meirionnydd, over which two flocks totalling about 200 flew on 3 March 1991. Farther north, 104 were on the sea off Llandanwg, Meirionnydd, on 3 March 1968 and smaller numbers were recorded over Caernarfonshire and Anglesey. Around 50 flew east over Berthlwyd, East Glamorgan, in March 1987. A smaller autumn passage was noted from mid-October to early November, with the largest count recorded in Wales being a flock of *c.*320 over Brynsiencyn, Anglesey, on 4 November 1968.

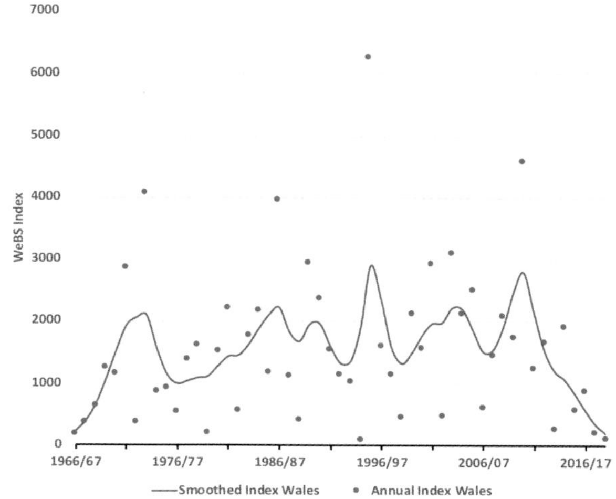

Bewick's Swan: Wetland Bird Survey indices, 1966/67 to 2018/19.

Few Bewick's Swans now winter in Ireland: just 21 were found during the 2015 International Swan Census (Crowe *et al.* 2015). As a result there is no longer passage through Wales, and numbers on its former stronghold of Anglesey are also much reduced. They were recorded on the island in only one year during 2005–19. There have been no confirmed records in Caernarfonshire since 1997 and it had become scarce in Ceredigion after the late 1990s (Roderick and Davis 2010). In recent years, records have been annual or almost annual only in Gwent and Flintshire. The flock on the Dee Estuary is the largest in Wales, though not always on the Welsh side of the border, and as elsewhere has been declining. There were 109 at Shotwick Fields in December 2010 and 46 in January 2015, but only three in winter 2018/19.

The increase in records in Wales during the 1970s reflected increased counts elsewhere in Britain (*Birds in England* 2005), as the population wintering in northwest Europe increased from *c.*10,000 in the mid-1970s to 29,000 in 1995. This was followed by a steep decline, and by the end of the 2010s fewer than ten were recorded in Wales each winter. The distribution of the wintering population shifted east at a rate of 13km/year during 1970–2017 and the time spent by birds in the wintering areas is now 38 days less than in 1989 (Nuijten *et al.* 2020). The numbers wintering in Germany have increased during that period (Beekman *et al.* 2019). The breeding and wintering population in northwest Europe is projected to be 50–79% lower in 2038 than it was in 2000 (BirdLife International 2020). It is a Red-listed species of conservation concern in Wales owing to the 55% decline in its range between 1981–84 and 2007–11 (Johnstone and Bladwell 2016). It seems that this species, which has already become scarce in Wales, will soon cease to be a regular winter visitor.

Rhion Pritchard

Whooper Swan *Cygnus cygnus* Alarch y Gogledd

Welsh List Category	IUCN Red List			Birds of Conservation Concern			
	Global	Europe	GB	UK	Wales		
					2002	2010	2016
A	LC	LC	EN	Amber			

Whooper Swans © Ben Porter

The Whooper Swan, of which there are five discrete populations, breeds on islands in, or along the banks of, shallow freshwater pools, lakes, slow-flowing rivers and marshes, birch forest and shrub tundra across northern Europe and Asia. It generally avoids open tundra, selecting riparian habitats with abundant emergent vegetation. Almost all those that breed in Iceland spend the winter in Britain and Ireland, while a separate population breeds in Fennoscandia (*HBW*). In Britain, Whooper Swan bred on Orkney until the late 18th century and re-established a small breeding population in Scotland in the 20th century. Around 30 pairs nest in Britain each year, around half in Shetland, but there are small numbers elsewhere in Scotland, and in Northern Ireland and Norfolk (Eaton *et al.* 2020).

Whooper Swans undertake the longest sea crossing of any European swan, migrating up to 1,400km between Britain/Ireland and Iceland. The majority arrive from Iceland between mid-October and mid-November and depart in March and April. Ring resightings show that movements occur between the Icelandic and Fennoscandian populations, with swans ringed in Iceland seen in continental Europe and several individuals ringed in Europe re-sighted in Britain (Hall *et al.* 2016). Individuals summer in Wales occasionally, often due to injury.

Forrest (1907) described this species as "by no means uncommon" in North Wales, and most frequent on Anglesey. He knew of no records in Montgomeryshire but recorded it regularly, though not annually, in the Glaslyn Valley, Meirionnydd/Caernarfonshire, and cited Williams' *Llandudno Guide 1861* as saying that it was more numerous in the lower Conwy Valley, Caernarfonshire/Denbighshire, prior to drainage of the Afon Ganol. Whooper Swan probably occurred regularly at Cors Caron, Ceredigion, in the late 19th century (Salter 1895), but there were no records during 1897–1950. It occurred in increasing numbers subsequently, with a flock of 30–40 annually through the 1970s.

Flocks of up to 20 were recorded annually on Anglesey lakes in the late 19th century, a pattern that continued through the first half of the 20th century, although Jones and Whalley (2004) considered that it was never so numerous as Bewick's Swan. Numbers on the island increased in the 1960s and 1970s, with a roving flock of 30–40 visiting large waterbodies such as Llyn Hafodol, Llyn Cefni, Llyn Llywenan and Llyn Alaw. Counts continued to increase in the 1980s, with up to 60 on Llyn Alaw in January 1988, and a herd of 50–60 has grazed regularly on pasture between Llangefni and Llanerchymedd since the 1990s.

The Glaslyn flock, more recently favouring the Meirionnydd side of the valley, was probably present most winters through the 20th century, but increased in size from 20–30 in 1986/87 to 50–80 in the late 2010s. There are no early records in Carmarthenshire, where the first dated sighting was at Llandovery in 1955, and

Site	County	74/75 to 78/79	79/80 to 83/84	84/85 to 88/89	89/90 to 93/94	94/95 to 98/99	99/00 to 03/04	04/05 to 08/09	09/10 to 13/14	14/15 to 18/19
Glaslyn Valley	Meirionnydd/ Caernarfonshire			46	46	39	55	63	46	60
Tywi Valley	Carmarthenshire					4	15	25	33	56
Dee Estuary: Welsh side (whole estuary)	Flintshire	na (0)	na (0)	na (0)	0 (3)	1 (4)	3 (8)	32 (32)	57 (58)	31 (42)
Cors Caron	Ceredigion	34	19	8		3	10	14	11	11
Llyn Alaw	Anglesey	14	23	33	30	17	6	3	3	2

Whooper Swan: Five-year average peak counts during Wetland Bird Surveys between 1974/75 and 2018/19 at sites in Wales averaging 30 or more at least once. International importance threshold: 340; Great Britain importance threshold: 160.

records were sporadic through the remainder of the 20th century (Lloyd *et al.* 2015). Whooper Swans have become an annual winter visitor to the Tywi Valley around Cilsan Bridge since 1994, which marked the first appearance of a neck-collared bird of Icelandic origin that returned annually until 2005.

Whooper Swan was first recorded in Radnorshire in 1962, and from 1980/81 small numbers wintered in the lower Wye floodplain, but Jennings (2014) observed that the flock was diminishing and stayed for an ever-shorter period each winter. It was similarly sporadic in Breconshire until the mid-1970s, since when 10–20 have been recorded annually. Kenfig Marsh, East Glamorgan, was a regular wintering site for up to a dozen from the 1960s, but numbers declined through the 1980s and they ceased to visit regularly by the mid-1990s. From the mid-1970s, 20–30 wintered between Aberhafesp and Caersws in the Severn Valley, Montgomeryshire, but counts have declined in recent years.

While there are a handful of regular sites for Whooper Swan in Wales, numbers at each vary, perhaps with changes to the suitability of foraging habitat. At all sites, they are associated with Mute Swans and forage predominantly on pasture and arable farmland. The largest wintering flocks are now in the Tywi Valley, where there was a marked increase during 2012–19 and a record 80 on 2 February 2019, and in the Glaslyn Valley, where a peak count of 82 was recorded in February 2010. On the Dee Estuary, they usually occur on the sheep-grazed saltmarsh at White Sands and Burton Marsh and on fields created as mitigation for the construction of a solar energy farm on their favoured fields at Shotwick in 2016. At night, these birds roost at RSPB Burton Mere Wetlands.

Whooper Swan occurs principally as a passage migrant in other counties such as Pembrokeshire, where the first dated record was at Cosheston in January 1933, and Gwent, where the first record was at Caldicot Moor in February 1960. In counties without a regular wintering site, small numbers are usually seen annually but only for a few days, although larger counts occurred during periods of hard weather in Scotland and northern England.

The WeBS index for wintering Whooper Swan in Britain increased four-fold between 1985/86 and 2017/18. The drivers behind this increase are not clear but it may be that changes in

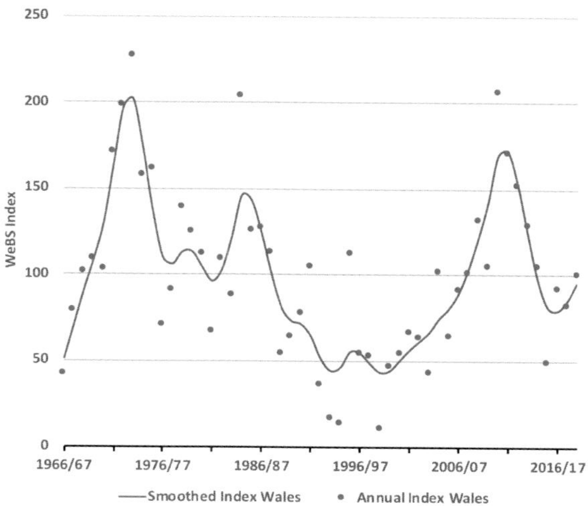

Whooper Swan: Wetland Bird Survey indices, 1966/67 to 2018/19.

survival rates and perhaps an interchange with the northwest European population are responsible, as well as improvements in breeding success (Hall *et al.* 2016). However, the trend for Wales has fluctuated more dramatically, falling by 56% during 2011/12–2016/17 and thus triggering a High WeBS Alert (Woodward *et al.* 2019).

The wintering Whooper Swan population in Britain is estimated to be 16,100 birds (Frost *et al.* 2019a). January counts, featured in *Welsh Bird Reports* in 2014/15–2018/19, suggest that only *c.*300 of these winter in Wales, representing less than 2% of the British population. Since this species is not a habitat specialist in winter, the future of the Whooper Swan in Wales is likely to be guided by its breeding success in Iceland and future climatic trends that determine how far south they spend the winter.

Patrick Lindley

Sponsored by John Hickerton

Egyptian Goose *Alopochen aegyptiaca* Gŵydd yr Aifft

Welsh List Category	IUCN Red List
	Global
C5, E	LC

Despite its name, the Egyptian Goose is not associated particularly with northeast Africa but breeds on wetlands across most of sub-Saharan Africa (*HBW*). It was introduced to England in the late 17th century, and by the mid-19th century many estates in England had flocks of full-winged birds. A naturalised population became established, mainly in Norfolk and north Suffolk, although Rutland Water also attracted significant numbers. The *Britain and Ireland Atlas 2007–11* showed a westward spread from the core area compared to 1988–91, but this has been slow relative to some other introduced wildfowl, perhaps because of low breeding productivity. Egyptian Geese breed early in the year and goslings retain their down for over a month. Consequently, breeding success is currently low but warmer, dry springs could increase survival of young (*Birds in England* 2005). It is included on Category C5 of the Welsh List on the basis that there is a naturalised population in England from which Welsh birds could derive. Most movements in Britain have been less than 50km and the *BTO Migration Atlas* (2002) considered that records outside the main breeding range were likely to have escaped from local collections. However, one with a Belgian ring at Penclacwydd, Carmarthenshire, on 20–23 April 2019 illustrates that naturalised Egyptian Geese are capable of long-distance travel.

Forrest (1907) noted that escaped birds were sometimes encountered in North Wales, but cited Brockholes saying that one shot on the Dee marshes, presumably in Flintshire, in November 1870 was from a group of four that were wary and appeared to be wild. Several records in Glamorgan, including six during 1860–89 and seven at Port Talbot on 20 September 1906, were considered all to be probably of captive origin (Hurford and Lansdown 1995). The first record in Gwent was in 1971, with four further records up to 2007 and all singles except for two at Llanwern in April 1983. Venables *et al.* (2008) considered these to have been birds from

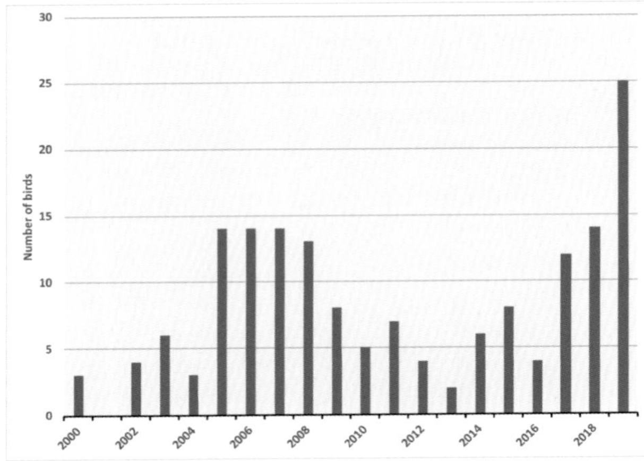

Egyptian Goose: Approximate numbers of individuals recorded in Wales, 2000–19.

England. There have been around a dozen subsequent records in the county, including two at Newport Wetlands Reserve for much of 2016 and 2017, one there in February/March 2018, three in April 2018 and one in May 2019.

In Carmarthenshire, an unringed bird that frequented the Tywi Estuary, from 1 September 1991 until it was shot on 5 January 1992, was admitted as the first record for the county; previous records were considered to have been escaped birds. There have been several other subsequent records of 1–2 in the county. In Pembrokeshire, four records since the first in 2007 were believed to be from the English population (*Pembrokeshire Avifauna* 2020) and two earlier records were considered of captive origin, as were five records in Ceredigion between 1991 and 2007 (Roderick and Davis 2010). In Meirionnydd it was first recorded in 1993, and there were several subsequent records around the Dysynni and Glaslyn/Dwyryd estuaries; three at the latter site from June to August 2002 were also seen in Caernarfonshire (Pritchard 2012). On Anglesey, sightings in 1989 and the 1990s were considered by Jones and Whalley (2004) to be escapes. The only Radnorshire record, on the River Wye from May to September 2008, was considered to be a wanderer from England (Jennings 2014) and the first Montgomeryshire record was two at Llyn Coed-y-Dinas on 22 January 2013.

Egyptian Goose has been recorded annually in Wales since 2002, but there is no strong trend. The largest groups were six at Rhuddlan, Flintshire, in August 2005, later also seen at Ridleywood, Denbighshire, and a Welsh record of eight at Cemaes Bay and later in Holyhead harbour, Anglesey, on 8–9 July 2005. The size of these groups suggested that they were from the naturalised English population, rather than local escapes. There seems no reason why the species should not eventually become established as a breeding bird in Wales. That seems mostly likely to occur via the northeast, where sightings have become more frequent. A pair bred just a few miles across the border in Burton, Wirral, in 2020.

Rhion Pritchard

Sheldduck *Tadorna tadorna* Hwyaden yr Eithin

Welsh List Category	IUCN Red List			Birds of Conservation Concern			
	Global	Europe	GB	UK	Wales		
					2002	2010	2016
A	LC	LC	EN	Amber			

Shelduck © Bob Garrett

This large goose-like ground-nesting duck is widely distributed within the Palearctic, from Ireland to northeast China (*HBW*). In the northwest of its range, the Shelduck occupies the shallow, muddy shores of estuaries and brackish lagoons in countries bordering the Atlantic Ocean, North and Baltic Seas, though a small but increasing proportion of the UK population is on inland freshwaters. The Shelduck requires foraging habitat of high biological productivity. Its diet in northwest Europe is dominated by the mud snail *Peringia ulvae*, which is typically 'sieved' from, or just below, the surface of slightly 'sloppy' mud, and supplemented by other molluscs and small crustaceans (Viain *et al.* 2011). The male defends its breeding territory, including foraging areas, and its brood from potential predators. It usually nests in burrows and holes under thick, thorny vegetation that provides concealment from potential predators, but also nests in tree cavities (*BWP*) and has been known to nest in (or under) artificial structures including buildings (Balmer *et al.* 2013). Nest sites can be up to 1km from foraging areas, so ducklings can face a trek from the nest to the relative safety of the water, during which obstacles such as roads, fences, ditches and banks have to be negotiated and predators avoided.

A key feature of its life cycle is the annual moult migration. Third-calendar-year pre-breeders depart from breeding areas for the moult grounds from mid-June, followed by post-breeding adults and those that failed to breed. Peak passage is during the first half of July. Some adults, which postpone their annual moult until later in the season, remain on the breeding grounds to 'mind' crèches of ducklings that form within 2–3 weeks of hatching. First-calendar-year Shelducks disperse from breeding areas during the autumn.

Almost all of the northwest European population migrates to moult in the remote mudflats of the Wadden Sea, along the coast of southeast Denmark, northwest Germany and The Netherlands. However, a small proportion of British and Irish breeding birds moult locally in large estuaries and bays, including Bridgwater Bay in Somerset and the Mersey Estuary, Cheshire. The latter site holds over 10,000 in July and August (*BTO Migration Atlas* 2002; Green *et al.* 2019). The majority of overseas recoveries of British-ringed Shelduck and 13 of the 14 overseas ringing exchanges with Wales come, not surprisingly, from the Wadden Sea (Robinson *et al.* 2020). Research on the Severn Estuary showed that Shelducks sometimes undertake significant movements within Wales. Birds that were dye-marked to monitor their use of intertidal habitats around Cardiff and Newport, were reported from Gower and Caernarfonshire, although most remained on the Severn Estuary (Scragg 2016).

In Wales, the Shelduck is a locally common breeding resident in coastal areas, a more widespread winter visitor and scarce but annual passage migrant inland. It decreased locally during the 19th century (*Birds in Wales* 1994) and increased during the 20th century, though this trend was not geographically uniform. In 1992, a pilot survey estimated the Welsh breeding population to be 500–800 pairs, of which around 75% were in North Wales (Delany 1992). By an extrapolated method, the population was estimated to be 1,250 (725–2,250) pairs in 2016, around 16% of the British total (Hughes *et al.* 2020). The breeding population in Britain decreased by 28% and in Ireland by 13% (to *c.*958 pairs) between 1988–91 and 2016 (Green *et al.* 2019).

The *North Wales Atlas 2008–12* recorded Shelduck in 13% of tetrads, some of which were in the Conwy and Clwyd valleys, away from the open coast. Occupation of 10-km squares in the region had increased by 42% since 1968–72. In Caernarfonshire, the breeding population is centred around the Conwy Estuary, Traeth Lafan and the Menai Strait, but aside from the area between Pwllheli and Abersoch, it is largely absent from the more exposed, rocky coastline of Llŷn, although up to ten pairs have bred on Bardsey (Pritchard 2017). In Meirionnydd, the Shelduck has always been a fairly common breeding species in coastal areas such as Traeth Bach, the Mawddach and Dyfi estuaries. On Anglesey, it has bred for many years in the Aberffraw, Newborough Warren and Red Wharf Bay dune systems and at Penmon. Several pairs were recorded on Puffin Island in 1902, where one nest was found in a Rabbit burrow at the top of 35m cliffs (Jones and Whalley 2004), but there is no such recent evidence (Stephen Dodd pers. comm.). In 1985, inland pairs were at Cors Ddyga and nests have also been recorded at Llyn Traffwll. The Anglesey breeding population was estimated to be 40–70 pairs in 1986, which was considered to be lower than previously as a result of human disturbance in coastal areas (Jones and Whalley 2004). In northeast Wales, crèches of 57 ducklings on the Afon Conwy, Caernarfon/Denbighshire, and 60 independent juveniles on the Clwyd Estuary, Denbighshire/Flintshire, in 2016 indicate that populations remain in both catchments. There was previously a significant population along the Flintshire coast of the Dee Estuary (Delany 1992), and while it was still widespread there in 2008–12, there is no recent information on its size.

In Ceredigion, the Shelduck has never been particularly common as a breeding species. It bred in Ynyslas dunes at the end of the 19th century, where fishermen would catch broods in nets (Roderick and Davis 2010), and around the Dyfi and Teifi estuaries. Today, the Shelduck breeds at Ynyslas and Cors Fochno on the Dyfi Estuary, on the Teifi Estuary, and since 1987—when a bird was found occupying an artificial hole created for Manx Shearwaters—on Cardigan Island. In Pembrokeshire, Mathew (1894) noted that "a pair or two nest on sand hills below Milford Haven" but otherwise breeding Shelducks were uncommon. By the 1940s, Shelduck bred annually near Dale, in the upper reaches of the Cleddau Estuary and on the Nevern and Teifi estuaries (Lockley *et al.* 1949). In the early 1990s, the county breeding population was estimated to be 50 pairs (Donovan and Rees 1994), which has since been increased to 60–70 pairs, with breeding confirmed in 5% of tetrads in 2003–07, including on Skomer and Skokholm (*Pembrokeshire Avifauna* 2020).

In Carmarthenshire, the Shelduck was numerous in sand hills near Laugharne Marsh [Witchett Pool] in 1869 and it was described as "common on the coast" in the early 20th century (Barker 1905). It continued to breed in the extensive dune systems around Carmarthen Bay, the Taf/Tywi/Gwendraeth estuaries, on the north shore of the Burry Inlet and on the Loughor Estuary in the early 1990s. Most breeding records are from around Llanelli, originally from the area now covered by Pembrey Forest, but latterly from Penclacwydd, where at least 20 pairs bred in 2011, a total that has not been matched subsequently. There is a single inland breeding record in Carmarthenshire, at Dinefwr Ponds, in July 1984 (Lloyd *et al.* 2015). *Birds in Wales* (1994) reported a steady decline in Carmarthenshire and in neighbouring Glamorgan, as a result of increased disturbance in their coastal habitats.

In East Glamorgan, the species was very common in the late 1890s and in the first quarter of the 20th century, it bred on Flat Holm and along the coast between the Rhymney Estuary and Barry and Sully islands. Several pairs also nested in the Aberthaw area, at Merthyr Mawr and Kenfig (Hurford and Lansdown 1995). The breeding population had declined by the 1960s, and by the 1980s it was largely restricted to the area around Cardiff and on Flat Holm. The *Britain and Ireland Atlas 2007–11* confirmed breeding around the Ogmore, Taff/Ely and Rhymney estuaries and around Barry. In Gower, it bred historically at Margam and Baglan Bay but is now to be found only in small numbers around Swansea Bay and on the Burry Inlet. In Gwent, Shelduck breeds on the coast, estuaries and upper tidal reaches of the major rivers and has bred on inland freshwaters such as Wentwood Reservoir. There was no significant change in the distribution or size of the breeding population, estimated at 100–200 pairs, between the *Gwent Atlas 1981–85* and 1998–2003 *Atlas*. Inland, the Shelduck is recorded in small numbers as a scarce winter visitor and passage migrant in Radnorshire (Jennings 2014) and Breconshire (Peers and Shrubb 1990), while in Montgomeryshire, breeding occurs around Dyfi Junction, and in some years at Dolydd Hafren and Llyn Coed-y-Dinas.

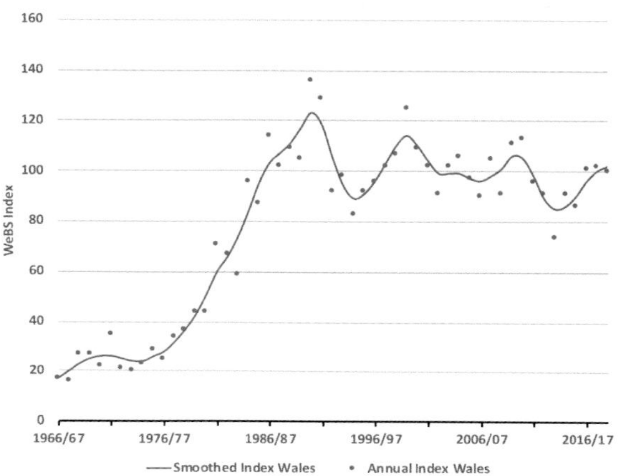

Shelduck: Wetland Bird Survey indices, 1966/67 to 2018/19.

The winter distribution of Shelduck in Wales is similar to that of the breeding population. Numbers increase through the autumn and early winter months when birds return from their annual moult. The WeBS index for Wales increased five-fold between 1973 and 1990, since when numbers have fallen by 20%. The UK index shows a similar pattern, albeit with a greater decline in the last 30 years.

The two most important Welsh sites for Shelduck outside the breeding season are the Dee and Severn estuaries, where the species is a qualifying feature of the Special Protection Areas. Shelduck numbers had been increasing on the Severn but the average count over the last decade has fallen by half, though less markedly on the Welsh side. Bridgwater Bay, on the southern shore, is the largest moulting area outside the Wadden Sea and may be used by some Welsh breeding birds. In East Glamorgan, the mouth of the Afon Rhymney is an important wintering area (Natural England and Countryside Council for Wales 2009), as was the Taff/Ely confluence prior to construction of the Cardiff barrage. In the north, the Dee forms part of a complex of estuaries that also includes the neighbouring Mersey Estuary, which may be used by some Welsh breeding birds during moult. A Medium WeBS Alert has been issued for the Dee Estuary on account of a decline of 27% over 25 years (Woodward *et al.* 2019), but numbers on the Welsh side have increased and it now holds over 20% of the estuary population compared to 10% in the early 1990s.

Declines in non-breeding numbers have been observed elsewhere in Wales such as on the Cleddau Estuary, which regularly held over 1,000 Shelduck during the 1980s and 1990s, but numbers have since fallen below the threshold of British

Site	County	74/75 to 78/79	79/80 to 83/84	84/85 to 88/89	89/90 to 93/94	94/95 to 98/99	99/00 to 03/04	04/05 to 08/09	09/10 to 13/14	14/15 to 18/19
Dee Estuary: Welsh side (whole estuary)	Flintshire	na (2,648)	na (5,358)	na (5,567)	793 (6,630)	888 (7,725)	1,879 (10,748)	1,435 (10,771)	1,743 (8,681)	2,102 (9,338)
Severn Estuary: Welsh side (whole estuary)	Gwent/East Glamorgan	195 (1,770)	678 (1,315)	2,274 (1,590)	1,944 (3,107)	1,499 (3,522)	1,449 (2,702)	1,437 (4,142)	1,813 (3,825)	1,144 (2,102)
Burry Inlet	Gower/ Carmarthenshire	715	1,118	1,463	1,400	959	1,080	732	790	882
Traeth Lafan	Caernarfonshire/ Anglesey	197	403	448	437	271	491	400	616	546
Cleddau Estuary	Pembrokeshire	290	515	1,434	891	1,014	715	656	581	387

Shelduck: Five-year average peak counts during Wetland Bird Surveys between 1974/75 and 2018/19 at sites in Wales averaging 600 or more at least once. International importance threshold: 2,500; Great Britain importance threshold: 470.

importance. Wintering numbers also declined in the Burry Inlet, where Shelduck is a qualifying feature of the Special Protection Area, from the late 1990s but have stabilised more recently.

It is an Amber-listed species of conservation concern in Wales, owing to the 33% decline in its winter range between 1981–84 and 2007–11 and because Wales holds 2% of the European breeding and 7% of the wintering population (Johnstone and Bladwell 2016). The reasons for the Wales-wide increase and subsequent decline in the wintering population are not clear, although external factors such as a shift in the locations of moult grounds may have played a part. Although a common wintering and breeding species in much of lowland coastal Wales, it is vulnerable to a range of pressures, such as loss of foraging habitat to coastal development and disturbance from recreation and other human activities. Breeding success rates are generally low, primarily a consequence of the high levels of predation of ducklings by mammals and birds. A combination of local and global factors is likely to influence the future of the Shelduck as a wintering and breeding species in Wales.

Jane Hodges

Sponsored by Deeside Naturalists' Society

Ruddy Shelduck *Tadorna ferruginea* **Hwyaden Goch yr Eithin**

Welsh List Category	IUCN Red List		Rarity recording
	Global	Europe	
B, E	LC	LC	BBRC, RBBP

The native range of Ruddy Shelduck extends from southeast Europe to northeast China, with populations in northwest Africa and Ethiopia (*HBW*). It was formerly more widespread in the western part of its range, though birds have been reintroduced in some countries. Most populations are dispersive or nomadic, although some are migratory. Their preferred habitat is brackish or freshwater lakes in open, sparsely vegetated areas (*European Atlas* 1997).

The British Ornithologists' Union Records Committee (BOURC) reviewed all British records of Ruddy Shelduck and no record after 1892 was considered acceptable (Harrop 2002). In the summer of that year an influx of birds to Ireland, Norway, Sweden, Iceland, Greenland and Britain was accepted as being of wild origin. This included one from Wales that was shot from a small flock of (presumed) Ruddy Shelduck near St Davids, Pembrokeshire, in July. It is the only record of a wild bird in Wales, although Forrest (1907) reported one on the Afon Conwy in April "about 1894", which could possibly have been part of the 1892 influx, as could a bird that was shot in Glamorgan "some years before 1899" (Hurford and Lansdown 1995). On the basis of the St Davids record, Ruddy Shelduck is on category B of the Welsh List, recorded in an apparently natural state prior to 1950, but not subsequently. More recent sightings are in Category E (see Appendix 1).

Rhion Pritchard

Sponsored by Tadorna Tours

Mandarin Duck *Aix galericulata* **Hwyaden Mandarin**

Welsh List Category	IUCN Red List
	Global
C1(2)	LC

Mandarin Duck is native to eastern Russia, northeast China, Taiwan and Japan, where forest clearance and wetland drainage plus the capture of birds for local consumption and export to ornamental gardens in Europe have greatly reduced their number. The population in Japan has made a recovery, to an estimated 13,340 pairs (Brazil 1991), but the wider Asian population of around 65,000 individuals may still be decreasing (van Kleunen and Lemaire 2014, BirdLife International 2020).

Mandarin Duck was introduced to Britain from China shortly before 1745 and first reported to have bred in the wild in 1866 (Lever 2009). The majority are in south, central and eastern England, but it has also occurred in northern England, Devon, Scotland and Wales for at least four decades (*Britain and Ireland Winter Atlas 1981–84*). Breeding populations became established in five other European countries and the USA during the second half of the 20th century, through escapes or releases of captive birds (van Kleunen and Lemaire 2014).

The species was not recorded in Wales by the *Britain and Ireland Atlas 1968–72*, although Davies (1988) believed that birds were overlooked due to their shy nature and preference for well-vegetated waters. *Birds in Wales* (1994) reported a failed nesting attempt near Abergele, Denbighshire, in 1974 and noted their presence on the Severn/Vyrnwy rivers from the early 1980s. By the time of the *Britain and Ireland Winter Atlas 1981–84*, it was recorded in one 10-km square each in Gower and East Glamorgan, but the four 10-km squares with confirmed breeding during the 1988–91 *Atlas* were in mid- and North Wales. A suggestion that breeding occurred at RSPB Ynys-hir, Ceredigion, in 1988 was speculative (Russell Jones pers. comm.).

Mandarin Duck first bred in East Glamorgan at Kenfig Hill in 1985, and in Gower at Aberdulais Falls in 1990 (Hurford and Lansdown 1995); pairs have subsequently bred in the Neath Valley, Gower. In Gwent, the first was seen at Llandegfedd Reservoir in 1987. The first possible breeding in the county was in 1999, when a brood was on the River Wye (though may have originated on the English bank). A nest with eggs by the Monmouth and Brecon Canal in 2003 was the first confirmed breeding in Gwent (Venables *et al.* 2008). There was an apparent decline in Montgomeryshire during the late 1990s, but new breeding populations became established in Caernarfonshire, Meirionnydd and Denbighshire.

During the last 20 years the species has consolidated its range in Wales. It breeds regularly in Meirionnydd (ten pairs in 2014), Radnorshire (15–20 pairs in 2014), Denbighshire, in the middle Dee and Clwyd valleys (*c.* ten pairs in 2018). It is also regular in Montgomeryshire, where Mandarins breed on the Afon Vyrnwy and most, if not all of its tributaries, notably the Banwy, Cain and Tanat. In the south, it breeds annually in Breconshire (ten pairs in 2018) and in Gwent (20–25 pairs), mainly on the River Monnow but also on other rivers, lakes and reservoirs such as Llandegfedd Reservoir (Tyler 2010, 2015). Small numbers also breed on the Afonau Elwy and Dulas, Denbighshire, on the Afon Glaslyn, Meirionnydd/Caernarfonshire, and it may have bred in the Conwy Valley in 2019. Two broods were on Ddol Reservoir, Flintshire, in 2018. A juvenile was at Dinefwr Ponds, Carmarthenshire, on 12 July 1997, following a series of records along the Afon Tywi that summer, but the first confirmed breeding in the county was not until 2020, near Glanaman. There are no breeding records on Anglesey, nor in Pembrokeshire, although there have been occasional long-staying individuals.

In their native range, Mandarin Ducks migrate from northern China to lowland east China and southern Japan in winter. Naturalised birds in Europe appear to have lost their instinct to migrate, but there is seasonal dispersal in late September. In autumn and winter, birds flock on rivers and lakes. The largest count recorded in Wales was an exceptional 195, of which two-thirds were adult males, on Garreg-ddu Reservoir, Radnorshire, on 10 November 2020. Other large counts include a pre-breeding gathering of 67 there in January/February 2017, 70 on the River Wye above Llyswen, Breconshire, on 10 September 2018, 48 on Wynnstay Lake, Ruabon, Denbighshire, on 26 May 2019 and 41 at Meifod, Montgomeryshire, on 19 November 2017. At dusk birds leave these gatherings and fly to nearby rivers to feed.

Despite their mainly sedentary nature, there have been ringing recoveries in Britain of Mandarin Ducks ringed in The Netherlands and even one from Berkshire to European Russia. However, the only ringing recovery involving Wales was a male found dead at Monmouth, Gwent, on 28 January 2015, which had been ringed as an adult at Parkend, Gloucestershire, on 18 September 2011. The post-breeding population in the Forest of Dean, Gloucestershire, is around 200 birds and has remained fairly constant since around 2000, suggesting that Mandarin Ducks have reached their carrying capacity there (Richard Baatsen pers. comm.). This is a likely source of birds in mid-Wales.

Birds in Wales (1994) suggested that the Welsh population was fewer than 30 pairs and *Birds in Wales* (2002) estimated it to be fewer than 20 pairs, but its secretive nature means that records do not reflect its real status. In 2016 the estimated breeding population in Wales was 485 pairs (Hughes *et al.* 2020). While this may be too high, it is not inconceivable, given that the UK WeBS index of non-breeding Mandarin Ducks doubled between 1999/2000 and 2018–19 (Harris *et al.* 2020). Numbers would be expected to have increased more rapidly in areas of colonisation than at the core of its range. Given the expanding population, it would seem likely that Mandarin Ducks could yet extend their range across much of Wales, given the availability of suitable habitat, at least up to a couple of hundred metres above sea level.

Steph Tyler

Garganey *Spatula querquedula* Hwyaden Addfain

Welsh List Category	IUCN Red List			Birds of Conservation Concern				Rarity recording
	Global	Europe	GB	UK	Wales			
A	LC	LC	CR	Amber	2002	2010	2016	RBBP

The Garganey is unique among our ducks in being a summer visitor. Wales is on the western fringe of an extensive breeding range that extends through Europe and temperate Asia, as far as the Pacific Ocean. The largest populations in Europe are in Russia, Belarus and Ukraine, breeding in open areas with shallow water and dense shore vegetation. Most of the European population winters on rivers and lakes in sub-Saharan Africa, probably flying non-stop across the Mediterranean and the Sahara (*European Atlas* 1997). Around 120 breeding pairs are reported in the UK annually with a 40% increase since 2009, although only *c.*15% of these are confirmed (Eaton *et al.* 2020). There are no ringing recoveries involving Wales, but recoveries of birds ringed elsewhere in Britain suggest an easterly migration route through Italy.

The densely vegetated habitat favoured by breeding Garganey, together with the preference of ducklings to stay in cover, makes it very difficult to confirm breeding, particularly since female and young Garganey are easily confused with Teal. It is likely that the species has bred in Wales more frequently than published records suggest.

Early authors did not mention the possibility of breeding. Forrest (1907) knew of about seven records in North Wales, mainly on Anglesey and the Dee Estuary, and could only add one more in his 1919 volume. There were two 19th-century records from Cors Caron, Ceredigion, while Mathew (1894) described it as a rare summer visitor to Pembrokeshire. In Carmarthenshire, Dillwyn (1848) recorded it from Llanelli Marsh, and Barker (1905) knew of two other records. There were six records in Glamorgan and a few records from other counties in the 19th century.

There have been nine, possibly ten, records of confirmed breeding in Wales. Parslow (1973) said that Garganey bred in Wales in 1936 but gave no further information. This may have been the record mentioned by Jones and Whalley (2004), who stated that the species was thought to have bred on Anglesey in 1936 but that the evidence was not conclusive. The first properly documented record of breeding came in May 1940, when a female and at least one duckling were seen at Kenfig Pool, East Glamorgan. A pair bred successfully at Llyn Llywenan, Anglesey, in 1951 and *Birds in Wales* (1994) stated that at least two pairs bred successfully there in 1952. Broods were at Magor Marsh, Gwent, in June 1965 and May 1969; a pair bred at Cors Caron, Ceredigion, in 1968; and a nest with eggs was found at Llyn Dinam, Anglesey, in 1980. There was no further record of confirmed breeding until a pair bred on flooded ditches at Aberleri, Ceredigion, in 1997 and 1999. Breeding has not been confirmed in Wales since, but there have been several records of probable breeding.

Pairs can be present in suitable habitat for a lengthy period during the breeding season and sometimes show behaviour suggestive of breeding. Nesting was suspected at Llyn Ystumllyn, Caernarfonshire, in 1946 and at several sites on Anglesey in 1951–52, including 4–5 pairs at Valley Wetlands in 1951. It was also suspected at Shotton Pools, Flintshire, in 1959 and probably in 1960, when the behaviour of a female suggested breeding and six immature or eclipse birds were seen in July. At Kenfig, East Glamorgan, breeding was suspected in 1962 and again in 1984, when two males, a female and four juveniles were together on 2 August. Breeding behaviour was also witnessed at Llyn Llywenan in 1992, at RSPB Ynys-hir, Ceredigion, in 2007 and at a site in Breconshire in 2008. There has been no real evidence of breeding subsequently, although six juveniles were at Penclacwydd, Carmarthenshire, in September 2012 and a pair possibly bred on the Wirral side of RSPB Burton Mere Wetlands in 2019. Its status as a rare breeder accounts for its Amber-listing in *Birds of Conservation Concern Wales* (Johnstone and Bladwell 2016).

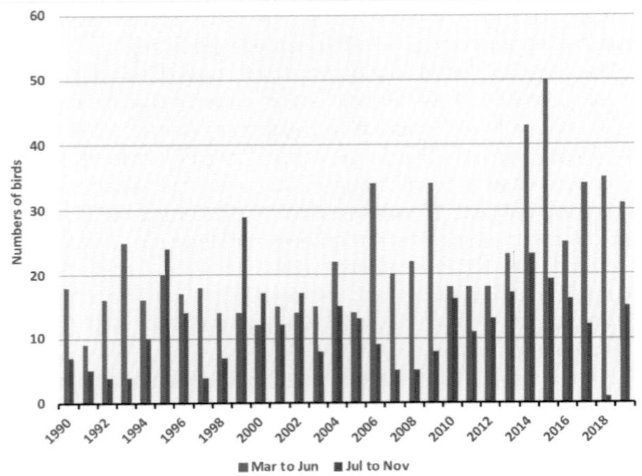

Garganey: Total numbers of individuals in Wales, March to June and July to November for 1990–2019.

Most Garganey in Wales are birds on passage, from mid-March to early June, with peak numbers in April and May. Most counties have had March records during 2000–19, the earliest being at Kenfig, in both East Glamorgan and Gower, on 12 March 2015. Most are singles or pairs, but there are several records of multiple arrivals on spring passage, including nine at Malltraeth Cob Pool, Anglesey, on 12 March 1948; eight at Witchett Pool, Carmarthenshire, from 28 March to 1 April 1969; and nine at Llanfairfechan, Caernarfonshire, on 13 March 1993. Spring passage in 1969 included more than 20 across Glamorgan (Hurford and Lansdown 1995). During 2010–19, numbers recorded in Wales in spring have been higher than for many years, with an estimated 43 in 2014 and 50 in 2015.

Numbers are usually lower on autumn passage, from late July to October, with a peak in August and September. There were November records from six counties during 2000–19, the latest being on Skomer, Pembrokeshire, on 28 November 2005. There was an average of 12 each autumn during 1990–2019, the highest seasonal total being 29 in 1999. The highest autumn totals at a single site were six at Newport Wetlands Reserve, Gwent, in 2000, and two records previously mentioned: six possible juveniles at Shotton, Flintshire, in July 1960, and six juveniles at Penclacwydd, Carmarthenshire, in September 2012.

A small number of historic records were in winter: two over-wintered at Roath Park Lake, Cardiff, in 1929/30 and there were a number of other records in February. Such records have been fewer in recent years, with only two in the 21st century, at Point of Ayr, Flintshire, in December 2006 and at Penclacwydd from 1 February to 19 April 2007.

Rhion Pritchard

Male Garganey © Jerry Moore

Blue-winged Teal *Spatula discors* Corhwyaden Asgell-las

Welsh List Category	IUCN Red List	Rarity recording
	Global	
A, E	LC	BBRC

This duck breeds across North America, and winters from the southern USA to northern South America (*HBW*). A total of 282 were recorded in Britain up to 2019, on average five a year over the last three decades (Holt *et al.* 2020). It is a very rare visitor to Wales with nine records involving ten individuals:

- male shot at Hollard Arms, Anglesey, in 1919;
- female/immature at Skokholm, Pembrokeshire, on 17 September 1960;
- male at Cemlyn, Anglesey, on 13–16 March 1963;
- pair at Shotwick, Flintshire, on 8–9 June 1983;
- male at RSPB Point of Ayr, Flintshire, on 2–12 October 1997;
- female at Penclacwydd, Carmarthenshire, on 12–26 March 2000;
- female at Malltraeth Cob, Anglesey, on 12–24 April 2005;
- male at Sandy Water Park, Carmarthenshire, from 20 March to 11 April 2011 and then at Penclacwydd from 16 April to 1 May 2011 and
- male at RSPB Burton Mere Wetlands, Flintshire/Wirral, on 19–20 April 2013.

Jon Green and Robin Sandham

Shoveler *Spatula clypeata* Hwyaden Lydanbig

Welsh List Category	IUCN Red List			Birds of Conservation Concern				Rarity recording
	Global	Europe	GB	UK	Wales			
A	LC	LC	LC	Amber	2002	2010	2016	RBBP

The Shoveler has a very extensive range, across most of northern North America and northern Eurasia except the high Arctic (*HBW*). It is found on a wide variety of shallow freshwater wetlands, preferably well-vegetated lakes and marshes. The large spatulate bill is adapted for filter feeding on zooplankton, particularly small crustaceans (Cabot 2009). Over most of its range it is highly migratory, though it is found year-round in parts of Europe (*HBW*). Around 1,200 pairs breed in the UK each year (Eaton *et al.* 2020).

Ringing recoveries involving Wales suggest that some breeding birds from northeastern Europe spend the winter here: two ringed as nestlings on Engure Lake, Latvia, were shot on Anglesey in their first autumn, one in October 1973 and the other in December 1983. Nestlings ringed in Estonia in June 1973 and June 1978 were subsequently shot during their first winter in Carmarthenshire and Pembrokeshire. Birds ringed in Sweden, Denmark and The Netherlands have also been recovered in Wales. Shovelers recovered at Orielton, Pembrokeshire, include birds ringed on the Volga and Pechora rivers in the western Siberian Plain, Russia (Donovan and Rees 1994). There is no information on movements by Shovelers that breed in Wales, but recoveries of those ringed elsewhere in Britain suggest that many move south in winter to France, the Iberian Peninsula and North Africa, while a proportion of the population may be resident (*BTO Migration Atlas* 2002).

The species was sometimes called the "Spoonbill" by early authors. Such a reference in the *Swansea Guide* of 1802 was thought by Hurford and Lansdown (1995) to refer to Shoveler. It was a very rare breeder in Britain during the early 1800s, and in 1850 was only known to breed in Norfolk. Uncontrolled wildfowling was thought to be mainly responsible for its scarcity, both as a breeding bird and in winter (*Historical Atlas 1875–1900*). There was a phase of expansion in western Europe, including in Britain and Ireland, in the early 20th century (Parslow 1973), when it was chiefly a winter visitor in small numbers to Wales, but it became established as a breeding species on Anglesey, particularly on Llyn Llywenan, where "a flourishing colony" was strictly protected (Forrest 1907). It bred or probably bred on several other Anglesey lakes at this time, but the only breeding pair reported by Forrest in mainland North Wales was on Llyn Mynyllod, Meirionnydd, in 1896. The Shoveler also bred regularly in Glamorgan, with several pairs annually at Morfa Pools, Gower, in the 1890s (Hurford and Lansdown 1995). A few pairs bred in Pembrokeshire: Lockley *et al.* (1949) stated that three-quarter-grown young were shot at Dowrog in 1904 and a nest and eggs found at Angle in 1931.

A possible breeding male was recorded at Cydweli Burrows, Carmarthenshire, in 1897, but the first confirmed breeding in the county was in 1929, when a nest was found there. An estimated 6–8 pairs bred until 1939, when observations ceased owing to wartime activity (Lloyd *et al.* 2015). Numbers breeding in Wales had apparently increased by the 1940s. Bruce Campbell visited Ystumllyn, Caernarfonshire, in 1946 and commented "The shoveler nests quite freely—Mr Speake has found up to a dozen nests in a good season" (North *et al.* 1949), but this was prior to the site being partially drained in 1955. Some 4–5 pairs still bred at Cei Cydweli in 1950, the last time the site was visited before it too was drained (Lloyd *et al.* 2015). Breeding was first recorded in Ceredigion in 1949, at Cors Caron, and in Montgomeryshire at an un-named site in 1954. In Flintshire, it bred irregularly on the coast from Point of Ayr to Ffynnongroyw, at Llyn Helyg and Shotton, with two pairs breeding at the last site in 1959 (Birch *et al.* 1968). The first confirmed breeding in Gwent was in 1966 and the first confirmed breeding in Breconshire was in 1973.

It is difficult to estimate the size or trend of the breeding population in Wales. Nests are usually well hidden in dense vegetation, and confirmation of breeding depends mainly on seeing broods. Lovegrove *et al.* (1980) thought that fewer than ten pairs bred in Wales in most years. A survey of Anglesey in 1986 confirmed six pairs breeding, 13 pairs probably breeding and nine pairs possibly breeding (RSPB 1986). *Birds in Wales* (1994) estimated 40 pairs breeding in Wales, while *Birds in Wales* (2002) estimated the population at below 90 pairs. During 2015–18, the Rare Breeding Birds Panel reported 45–59 pairs in Wales, including nine pairs confirmed breeding in two of those years, the highest reported to the RBBP since the panel began collecting data on Shoveler in 2006. In that four-year period, breeding was confirmed in all counties except East Glamorgan, Ceredigion, Montgomeryshire and Radnorshire, the latter being the only one with no breeding records ever (RBBP data). Up to five pairs breed on the Flintshire part of RSPB Burton Mere Wetlands, assisted by the installation of a fence to exclude mammalian predators. The Welsh breeding population is estimated to be 75 pairs (Hughes *et al.* 2020).

Larger numbers pass through Wales in autumn or spend the winter here, with peaks in December and February. There are also suggestions of a spring passage in March. Numbers have increased since the late 19th century; for example, 140 Shovelers were caught for eating at the Orielton Decoy, Pembrokeshire, during 1877–1918, but when the decoy was used for ringing in the 1930s, 129 were caught in winter 1938/39 alone (Saunders 2019) and at least 300 in December 1955 (Lockley 1961). In the late 1950s the Burry Inlet was the most important site for the species in Wales, with a mean winter peak of 400, although this declined to 60 by the end of the 1960s (Hurford and Lansdown 1995), before rising again later in the century. There were 200 at Point of Ayr in January 1967, up to 200 at Kenfig Pool, East Glamorgan, in the early 1970s and 309 at Llyn Traffwll in November 1976. On the

Site	County	74/75 to 78/79	79/80 to 83/84	84/85 to 88/89	89/90 to 93/94	94/95 to 98/99	99/00 to 03/04	04/05 to 08/09	09/10 to 13/14	14/15 to 18/19
Severn Estuary: Welsh side (whole estuary)	Gwent/East Glamorgan	61 (160)	11 (129)	74 (167)	50 (89)	89 (212)	178 (334)	349 (561)	340 (488)	257 (486)
RSPB Valley Wetlands inc. Llyn Traffwll	Anglesey	293	207	95	98	137	310	342	332	269
Dee Estuary: Welsh side (whole estuary)	Flintshire	33 (na)	25 (na)	16 (na)	13 (59)	46 (94)	35 (97)	21 (97)	22 (128)	90 (183)
RSPB Cors Ddyga	Anglesey	22	0	8	2	91	151	175	131	101
Burry Inlet	Gower/ Carmarthenshire	30	46	236	192	614	318	223	220	91

Shoveler: Five-year average peak counts during Wetland Bird Surveys between 1974/75 and 2018/19 at sites in Wales averaging 150 or more at least once. International importance threshold: 650; Great Britain importance threshold: 190.

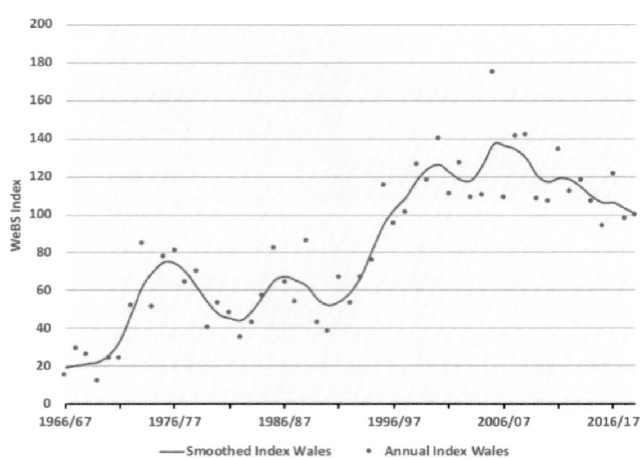

Shoveler: Wetland Bird Survey indices, 1966/67 to 2018/19.

Cardigan Bay coast, 50–100 wintered annually at Pwllheli harbour, Caernarfonshire, in the 1950s and 1960s, but it has seldom been recorded there since the construction of a marina. There were 100 in the Glaslyn/Dwyryd Estuary, Meirionnydd/Caernarfonshire, in December 1989 and 103 in the Dyfi Estuary, Ceredigion/Meirionnydd, in March 1976.

RSPB Valley Wetlands and the Dee and Severn estuaries are of British importance for Shoveler. There were 608 on the Welsh side of the Severn Estuary in January 2008 and 401 in December 2011, although recent counts have been lower. A slight downward trend has been evident at RSPB Valley Wetlands since the turn of the century, when there was a peak of 464 in January 2001. The Burry Inlet was designated as a Special Protection Area in 1992, with Shoveler as a qualifying feature. Counts there have fallen by 85% since the late 1990s, when the mean peak winter count was 1,034 (*Birds in Wales* 2002). Inland, good numbers have occurred at Llangorse Lake, Breconshire, in some winters, such as 81 in January/February 2006, but peak winter counts of 40 are more typical.

The estimated 75 pairs of Shoveler breeding in Wales are a small fraction of the total breeding in Britain, estimated at 1,196 pairs during 2014–18 (Eaton *et al.* 2020). Wintering numbers in Britain are estimated at 19,000 (Frost *et al.* 2019a), of which *c.*1,000–1,200 are in Wales. The fact that the species is very local in Wales, with significant numbers restricted to a few key sites, makes it very important that these sites are maintained in good condition and are regularly monitored. It is an Amber-listed species of conservation concern in Wales, owing to its status as a rare breeder and because 4% of the flyway population winters in Wales (Johnstone and Bladwell 2016). The species has suffered from the drainage of key sites in the past, but in recent years has benefited from the creation of sites such as RSPB Cors Ddyga. Its prospects in Wales may be improving as a breeding species, but recent declines in wintering numbers suggest that this may not be the case outside the breeding season.

Rhion Pritchard

Gadwall *Mareca strepera* Hwyaden Lwyd

Welsh List Category	IUCN Red List			Birds of Conservation Concern			
	Global	Europe	GB	UK	Wales		
					2002	2010	2016
A, C2	LC	LC	LC	Amber	2002	2010	2016

The Gadwall breeds across North America and Eurasia in a temperate band from the Atlantic coast to northern Japan and winters to the south, as far as the Horn of Africa, although the species is found year-round in ice-free parts of western Europe and the coastal USA (*HBW*). It favours highly productive, nutrient-rich freshwater marshes and lakes that support dense emergent and submerged vegetation (BirdLife International 2020). Many of the highly biologically productive wetlands occupied by breeding Gadwalls are also important outside the breeding season. The restricted nature of such habitats in Wales helps to explain its relative scarcity and patchy distribution. The species' dramatic range extension over the last 200 years may have originally been the result of drought in the steppe wetlands of southeast Europe and western Asia, but it has consolidated its distribution in Europe in the last 50 years, with expansion in the north, south and west of Europe. Breeding densities have increased six-fold in The Netherlands, for example (*European Atlas* 2020).

Gadwalls were released by wildfowlers in Britain in the 1930s and again in the 1950s and 1960s, but it is not clear what contribution these birds made to the establishment of a breeding population (Fox 1988). Britain holds one of the largest wintering populations in Europe (BirdLife International 2015). It is probable that many are resident in Wales throughout the year, although some may migrate out of Wales. Although extensive ringing recoveries are lacking, it is assumed that the obvious increase in numbers in autumn reflects the arrival of birds from Iceland and continental Europe (Fox and Mitchell 1988). One ringed as a nestling in Estonia in June 1977 was recovered at Connah's Quay, Flintshire, on 3 December the same year, while one shot near Builth Wells, Breconshire, in October 1966 had been ringed as a fledged first-calendar-year near Seville, Spain, earlier that year.

Gadwalls were a very rare winter visitor to most parts of Wales prior to 1920, almost unheard of in North Wales (Forrest 1919) and Breconshire (Ingram and Salmon 1957). Even by 1949 there had been only 21 records in Pembrokeshire (Lockley *et al.* 1949). Although several were said to have summered at a coastal marsh in Caernarfonshire—potentially Ystumllyn—as early as 1924, the species was only first proven to breed in Wales in 1969, when a brood was seen at Kenfig, East Glamorgan, not at Margam as stated in *Birds in Wales* (1994). A brood was seen at nearby Eglwys Nunydd Reservoir, Gower, in 1971, and a nest with three eggs was found at Rhosgoch, Radnorshire, in 1973.

There is strong evidence that Gadwalls bred in 1975 at Morfa Harlech, Meirionnydd, where birds were recorded in spring annually until 1984, and in Caernarfonshire, where 1–2 pairs bred on Afonwen Pools in the late 1980s. No subsequent breeding activity has been suspected in either county. Following an "unprecedented influx" to Anglesey in 1976, breeding was confirmed on Llyn Pen-y-Parc and suspected at two other lakes (Jones and Whalley 2004). A pair bred at Llwyngwair Mill Pond, Pembrokeshire, in 1981, which Donovan and Rees (1994) considered to have been released birds. The development of the National Wetlands Centre at Penclacwydd from 1990 created nesting habitat in Carmarthenshire and a modest breeding population rapidly became established, with around a dozen broods seen each year. They also occur regularly in good numbers on Anglesey, such as at Llyn Cefni, RSPB Valley Wetlands (29 males present in spring 2018) and RSPB Cors Ddyga (25 males), although the number of breeding attempts was likely to be lower than suggested by these counts. In Gwent, breeding was strongly suspected at Greenmoor Pool in the early 1990s (Venables *et al.* 2008) and has occurred regularly since 2001 at Newport Wetlands Reserve (Goldcliff) and since 2006 at Llanwern steelworks.

In Gower, Gadwall has bred sporadically at Staffal Haegr Pond (the maximum being four pairs in 2015), Llanrhidian and Oxwich Marsh; single pairs bred at Kenfig in 2016 and 2018–19, the first in East Glamorgan since 1969. Gadwall summer most years at Llangorse Lake, Breconshire, and probably breed there sporadically. They also summer regularly at sites in Pembrokeshire and in the northeast. In Flintshire, broods were seen at Shotton in five years during 2010–19, and in Denbighshire, 1–2 broods occur most years at RSPB Conwy. From being a very rare species, Gadwall are now regular but localised breeding birds in Wales. In 2007–11, 7.3% of the Gadwall's British range was in Wales, but abundance

Site	County	74/75 to 78/79	79/80 to 83/84	84/85 to 88/89	89/90 to 93/94	94/95 to 98/99	99/00 to 03/04	04/05 to 08/09	09/10 to 13/14	14/15 to 18/19
Severn Estuary: Welsh side (whole estuary)	Gwent/East Glamorgan	0 (121)	0 (260)	0 (286)	2 (322)	0 (255)	41 (277)	76 (243)	79 (207)	69 (191)
Burry Inlet	Gower/ Carmarthenshire	0	0	0	43	38	40	32	42	164
RSPB Valley Wetlands inc. Llyn Traffwll	Anglesey	10	14	39	60	61	135	109	115	142
Dee Estuary: Welsh side (whole estuary)	Flintshire	na (0)	na (5)	na (8)	2 (16)	18 (25)	26 (30)	22 (29)	10 (45)	109 (127)
RSPB Cors Ddyga	Anglesey	0	0	0	0	3	65	82	69	62

Gadwall: Five-year average peak counts during Wetland Bird Surveys between 1974/75 and 2018/19 at sites in Wales averaging 70 or more at least once. International importance threshold: 1,200; Great Britain importance threshold: 310.

was lower than elsewhere. The population in 2016 was estimated to be 62 pairs (Hughes *et al.* 2020) but is likely to fluctuate between 50–100 pairs. The densely vegetated nature of its wetland habitat often precludes the effective monitoring of confirmed breeding.

Ahead of colonisation of Wales by breeding birds, numbers wintering in Britain increased dramatically from the 1960s, thought to be a result of the proliferation of shallow inland waters such as gravel pits and lowland reservoirs (Fox and Salmon 1989). Wales' topography means that there are fewer such opportunities than in central and eastern England. The species had become commonplace throughout much of Wales by the 1990s, but its winter distribution is far from even, with concentrations confined to no more than 10–15 sites.

Newly created wetlands attract Gadwall as they mature. The Burry Inlet has, since the creation of the National Wetlands Centre at Penclacwydd, become the single most important Welsh site, the maximum count being 275 in December 2017. More than 100 have occured regularly at Newport Wetlands Reserve and up to 80 at RSPB Conwy. Peak counts at Anglesey sites were an average of 450 during 2014/15 to 2018/19, including 166 on Llyn Cefni in October 2019. Numbers have increased significantly on the Welsh side of the Dee Estuary, Flintshire, in the last decade, from an average annual peak of ten in 2009/10–2013/14 to 109 birds in 2014/15–2018/19. Elsewhere, Kenfig in East Glamorgan held a county record of 88 in January 2019 and there were 82 on Penrhyngwyn Pond, Carmarthenshire, in October 2019. Llangorse Lake consistently holds 30–80 each winter. Away from these sites, however, Gadwall remains uncommon, limited by a lack of suitable habitat.

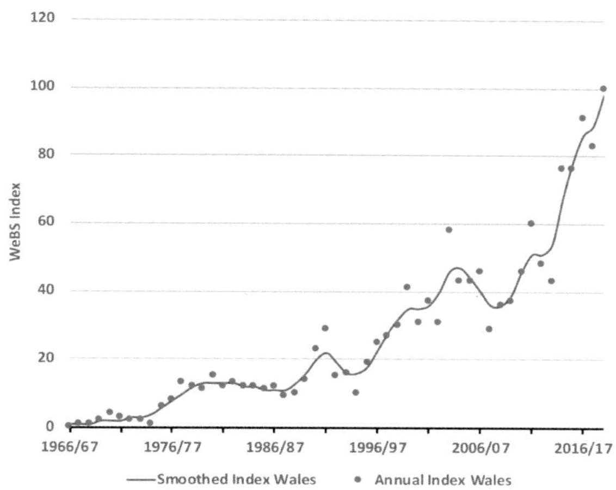

Gadwall: Wetland Bird Survey indices, 1966/67 to 2018/19.

A sustained increase in the WeBS index for Wales from the 1970s to the early 2000s has continued, reaching an all-time high in 2018/19. It seems unlikely that more than 1,000 Gadwall winter in Wales annually, but this represents a substantial increase and consolidation in recent years. Further geographic expansion will rely upon provision of lowland base-rich wetlands and lakes, which will always be geologically limited. However, existing wetlands have shown an ability to hold greater numbers in some years, which may suggest a current under-used capacity.

The Gadwall's kleptoparasitic nature enables it to obtain the submerged parts of aquatic plants that form its diet, brought to the surface by up-ending species such as Mute Swan and diving herbivores such as Coot (e.g. Amat and Soriguer 1984). This trait has enabled the Gadwall to subsist on waterbodies that would otherwise be unsuitable. It probably benefits more than most dabbling duck species from the eutrophication of waterbodies, as nutrient and waste-water runoff encourages growth of submerged vegetation, in extreme cases leading to the development of filamentous algal blooms. North American dietary studies show that Gadwall will eat these algae, which are not considered highly nutritional for other waterbird species (e.g. McKnight and Hepp 1998), although this has not yet been demonstrated in Europe.

Thanks to these features of its modest ecological success, there is no reason to think that the increase in Gadwall abundance will cease in Europe, or within Wales in particular. The immediate future looks secure for the species, so long as sites are safeguarded, wetland creation continues and sensitive management is in place.

Tony Fox

Falcated Duck *Mareca falcata* Hwyaden Grymanblu

Welsh List Category	IUCN Red List	Rarity recording
	Global	
A	NT	BBRC

This species breeds in eastern Siberia, northern Mongolia, China and Japan, making a relatively short-distance migration to the south in winter (*HBW*). Although the population is larger than previously thought, it is threatened by overhunting for food, mainly in China (BirdLife International 2020). It is a very rare vagrant to Britain, with seven records up to 2019 (Holt *et al.* 2020). The only Welsh record was a male at Llyn Traffwll, Anglesey, on 17–29 May 2008.

Julian Hughes

Wigeon *Mareca penelope* Chwiwell

Welsh List Category	IUCN Red List			Birds of Conservation Concern			
	Global	Europe	GB	UK	Wales		
A	LC	LC	NT (B)	Amber	2002	2010	2016

The Wigeon breeds over a wide area of northern Europe and Asia, from Iceland across Fennoscandia and Siberia to the Pacific. It breeds around shallow freshwater marshes, lakes and lagoons surrounded by scattered trees or open forest. It is migratory, wintering to the south of its breeding range, as far as the Sahel and Nile Valley in Africa (*HBW*). Britain is on the southern edge of the Wigeon's breeding range and was only colonised during the 19th century, during a period of cooler climate. Small numbers breed in Scotland and in northern and eastern England. The Scottish population was estimated at 240–400 breeding pairs in *The Birds of Scotland* (2007), but the average number reported across the UK is lower: 211 breeding pairs during 2014–18 (Eaton *et al.* 2020). Ringing recoveries involving Wales show that birds wintering here include individuals that breed in Iceland, Fennoscandia and Russia. Those from northeast Europe form the majority of our wintering population, based on birds ringed in The Netherlands and neighbouring countries that were presumed to have been on passage from that region.

There are very few records of confirmed breeding in Wales. Summering birds, including pairs, are regularly recorded, some with a damaged wing that have been unable to migrate. The use of breeding atlas definitions will over-estimate Wigeon's status, since the presence of a pair in suitable habitat equates to probable breeding, but only a few records include behaviour suggestive of breeding. Salter mentioned a report of breeding at Cors Caron, Ceredigion, prior to 1892 (Roderick and Davis 2010), but the earliest record of confirmed breeding was a brood recorded by Thomas Ruddy on Llyn Mynyllod near Llandrillo, Meirionnydd, in 1898, one of which was shot for examination to confirm the identification. Two pairs were seen at this site in April 1902, a young bird was shot there on 30 September 1904 and a brood was reported in 1934. A pair with four ducklings was reported on the Gower coast between Blackpill and West Cross in 1937, but a record of apparent breeding on the Gwent Levels between 1965 and 1968 was discounted by Lovegrove *et al.* (1980) for lack of hard evidence. A pair that raised six ducklings at Kenfig, East Glamorgan, in 1978 is the last confirmed breeding record in Wales. There have been more records of summering birds since the early 1990s, including eight at Newport Wetlands Reserve, Gwent, in summer 2006 and three pairs at RSPB Cors Ddyga, Anglesey, in 2007 and 2018. Behaviour suggestive of breeding included the female of a pair at RSPB Ynys-hir, Ceredigion, in early June 2005 that disappeared for about two weeks, and a female showing "furtive behaviour" at Cors Geirch, Caernarfonshire, in April 2010.

Wales holds good numbers of Wigeon in winter. No site comes anywhere near the numbers required for international importance but the Severn, Dee and Cleddau estuaries are of British importance. A few arrive during August and numbers build through September, but their presence at some sites is transitional as birds move farther south or west to winter. Most have left by late March.

It is possible that greater numbers wintered in Wales in the late 19th century but authors from this period gave few figures. Forrest (1907) described the Wigeon as an abundant winter visitor to flat coasts in North Wales, sometimes appearing in "vast flocks". Mackworth Praed stated that 9,000–10,000 were usually present at Orielton, Pembrokeshire, prior to the 1930s (Donovan and Rees 1994); *Birds in Wales* (1994) mentioned up to 12,000 there in the late 19th century, the highest site total recorded in Wales. Wigeon was the most numerous duck caught for eating at the Orielton Decoy in the late 19th century, peaking at 1,439 caught in winter 1893/94 but numbers fell in the early 20th century. When the decoy was operated for ringing from 1934/35, just 702 were caught in the 22 years to 1960/61, fewer than Teals or Mallards in the same period (Saunders 2019). Some birds pass through Wales, illustrated by a bird ringed at Orielton in October 1938, that was shot in Portugal the following January.

Numbers on the Welsh part of the Dee Estuary increased at the start of the 21st century and have now overtaken counts on the Welsh part of the Severn Estuary. Numbers have stabilised on the Cleddau Estuary and Foryd following periods of increase, whereas counts on the Dyfi Estuary and the Burry Inlet have declined, particularly at the latter, which was the most important site in Wales for this species in the 1980s; winter counts there peaked at 9,200 in January 1987. Several other sites, such as Traeth Lafan in Caernarfonshire/Anglesey and Traeth Melynog on Anglesey also hold 1,000 or more birds on average, and the Inland Sea/Alaw Estuary and Traeth Bach on Anglesey did so until recently. The Dysynni Estuary, Meirionnydd, had peak counts of over 2,000 in several years during 1995–2002, but subsequent counts have been much lower. Cold weather can result in increased numbers, as birds move from their usual winter areas farther east, and such influxes can be significant, such as 6,000–10,000 at Llyn Coron, Anglesey, in February 1955 and 9,835 in the Burry Inlet in October 1987. No inland site has regularly held large numbers, although numbers have increased markedly at Llangorse Lake, Breconshire, since 2009/10. The average peak count there was 340 in 2014/15–2018/19, and the highest WeBS count was 424 in January 2018.

Site	County	74/75 to 78/79	79/80 to 83/84	84/85 to 88/89	89/90 to 93/94	94/95 to 98/99	99/00 to 03/04	04/05 to 08/09	09/10 to 13/14	14/15 to 18/19
Severn Estuary: Welsh side (whole estuary)	Gwent/East Glamorgan	27 (4,210)	31 (7,289)	532 (8,429)	444 (3,853)	528 (6,564)	1,227 (6,191)	2,354 (8,249)	2,421 (7,667)	2,035 (7,881)
Dee Estuary: Welsh side (whole estuary)	Flintshire	na (377)	na (1,220)	na (2,278)	27 (3,047)	141 (4,526)	1,287 (4,323)	1,237 (3,839)	995 (3,002)	2,886 (7,371)
Cleddau Estuary	Pembrokeshire	1,052	1,188	3,867	3,044	3,255	4,122	8,022	6,357	6,748
Foryd	Caernarfonshire	818	523	340	2,102	2,868	3,320	2,478	2,633	3,178
Dyfi Estuary	Ceredigion/ Meirionnydd	1,594	1,701	3,880	4,056	3,622	2,434	2,993	2,426	2,554
Burry Inlet	Gower/ Carmarthenshire	2,881	3,184	6,771	1,933	2,915	2,509	2,133	1,221	1,450

Wigeon: Five-year average peak counts during Wetland Bird Surveys between 1974/75 and 2018/19 at sites in Wales averaging 3,000 or more at least once. International importance threshold: 14,000; Great Britain importance threshold: 4,500.

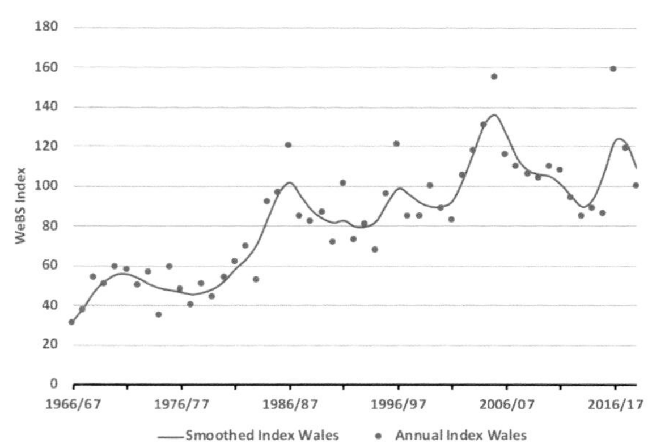

Wigeon: Wetland Bird Survey indices, 1966/67 to 2018/19.

The Wigeon is a popular quarry species for wildfowlers and while there is no evidence that numbers shot in Wales have a significant effect on its population status, it does influence their behaviour and distribution. Numbers on the Dee Estuary increased after the RSPB created sanctuary areas with no wildfowling from 1979 (Hirons and Thomas 1993). Disturbance from recreational activities also had a significant effect on Wigeon numbers and distribution at Llangorse in the early 1980s (Tuite *et al.* 1983). It is an Amber-listed species of conservation concern in Wales, owing to its rare breeding status and because Wales holds 3% of the flyway population (Johnstone and Bladwell 2016). Wales can be important as a refuge if there is severe weather farther north and east, so suitable habitat needs to be maintained in good condition and protected from excessive recreational disturbance.

Rhion Pritchard

An overwintering Wigeon flock with Yr Wyddfa in the distance © Ben Porter

American Wigeon — *Mareca americana* — Chwiwell America

Welsh List Category	IUCN Red List	Rarity recording
	Global	
A	LC	WBRC

This species breeds widely across northern North America and winters in the USA, Central and northern South America. The annual number of British records increased to 15–20 during 1990–2018, with the majority seen in Scotland and northern England (White and Kehoe 2020a). It is a rare visitor to Wales and shows no sign of becoming more frequent. Several individuals are presumed to have returned annually and visited the same or neighbouring sites in successive years.

The first Welsh record was a male at Llyn Llywenan, Anglesey, on 21–23 June 1910, but it was another 60 years before the second occurred, a first-calendar-year male at Kenfig, East Glamorgan, between 19 October and 2 November 1975. This was followed by a male at Llyn Bodgylched, Anglesey, in January–February 1977, which returned there in January–March 1979. A female was at RSPB Ynys-hir, Ceredigion, in November 1981 and a first-calendar-year male spent October–November 1985 at Kenfig.

A male, initially on the Conwy Estuary, Caernarfonshire/Denbighshire, from 20 December 1994 to 12 February 1995, was thought to have returned annually to winter in North Wales until 2006, often arriving with Wigeons in late summer and moving with the flock. During those 12 winters it also visited the Foryd, Llanfairfechan and Morfa Madryn in Caernarfonshire, the Braint Estuary on Anglesey, the Clwyd Estuary in Denbighshire, RSPB Burton Mere Wetlands, Flintshire/Wirral, and was last seen at Connah's Quay nature reserve, Flintshire, on 31 January 2006.

A male spent successive winters among the Wigeon flock in the Cleddau Estuary, Pembrokeshire, from 2003/04 to 2007/08, initially at Cresswell and West Williamston and later downstream at Angle Bay. Other records during the 1990s were a female at Dingestow, Gwent, in August–September 1995; a male at

Total	Pre-1950	1950–59	1960–69	1970–79	1980–89	1990–99	2000–09	2010–19
23	1	0	0	2	2	6	7	5

American Wigeon: Individuals recorded in Wales by decade.

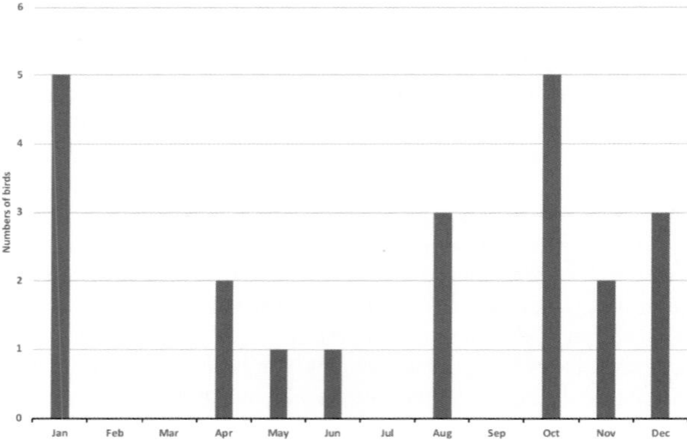

American Wigeon: Totals by month of arrival, 1910–2019.

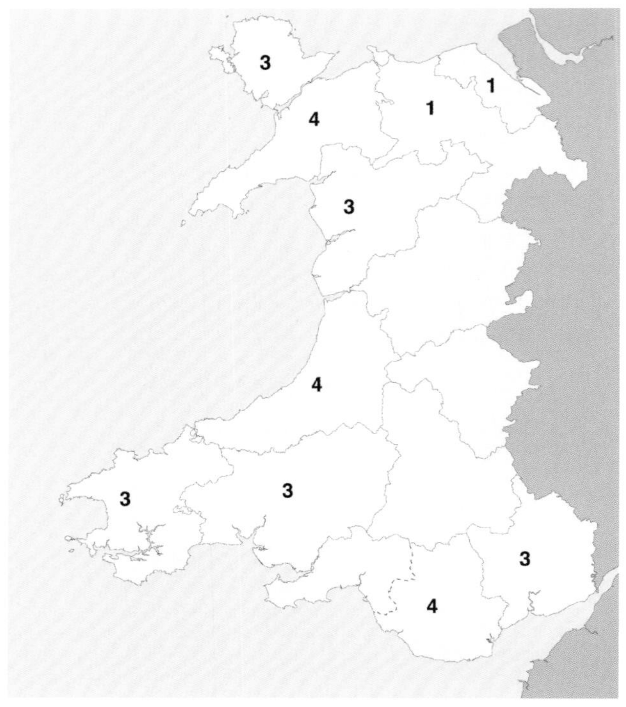

American Wigeon: Numbers of birds in each recording area, 1910–2019. Some birds were seen in more than one recording area.

Penclacwydd, Carmarthenshire, in October 1998; a first-calendar-year male at Peterstone Wentlooge, Gwent, from 31 October to 27 November 1999 that was also at Goldcliff Pill, Gwent, on 28 November; and a male at Cors Caron, Ceredigion, between December 1999 and March 2000.

Porthmadog, Meirionnydd/Caernarfonshire, hosted three in six winters: a male in August–September 2000, a female from December 2001 to February 2002 and a different male in January 2007. Farther south, a male was in Cardiff Bay, East Glamorgan, in April 2002 and a female at Cors Caron in August 2004. A male summered in Cardiff Bay from May to September 2006 but it was seven years before the next new bird, a male at Newport Wetlands Reserve (Uskmouth), Gwent, in January 2014. There have been four subsequent records, at Penclacwydd in October 2016, on the Dyfi Estuary in Ceredigion in November 2016, on the Gann Estuary in Pembrokeshire in January–March 2017 and at Penclacwydd in April 2019.

Jon Green and Robin Sandham

Mallard *Anas platyrhynchos* Hwyaden Wyllt

Welsh List Category	IUCN Red List			Birds of Conservation Concern			
	Global	Europe	GB	UK	Wales		
A, C2, C4, E*	LC	LC	NT (B)	Amber	2002	2010	2016

Many children's first experience of birds is feeding the Mallards on an urban pond. It is the most numerous and widespread of the world's ducks, found over most of Eurasia and North America and introduced to other parts of the world. Northernmost populations migrate south for the winter, while those in temperate areas are usually resident, though there are often local movements. This is a very adaptable and opportunistic species, breeding on almost any type of wetland, providing it has fairly shallow water and some cover (*HBW*). Mallards have nested at up to 535m in Caernarfonshire (Pritchard 2017) and 500m in Ceredigion (Roderick and Davis 2010). They also breed on the larger Welsh islands, though heavy predation, mainly by large gulls, means that in many years few or none of the young fledge.

The Mallard appears to have been a common species in Britain for a very long time, with bones found at many archaeological sites. Early authors, who often called this species the Wild Duck, described it as common, although Mathew (1894) concluded, based on numbers shot, that it must have been far more abundant in Pembrokeshire 50 years earlier. Phillips (1899) considered that Mallards were increasing in Breconshire around the turn of the 20th century, probably as a result of the release of captive-bred birds for shooting, and Forrest (1907) described it as by far the most numerous duck in North Wales. He saw small "flappers" (juveniles on the point of fledging) by a lake at over 455m on Berwyn, near Bala, Meirionnydd, and described it as particularly numerous along the west coast of Anglesey, where hundreds were shot every winter.

Greater numbers of Mallard are shot annually by wildfowlers than any other duck species, including a considerable number that were reared and released for the purpose. Harradine (1985) estimated the number of Mallards shot each year in the UK at 600,000–700,000, of which *c.*500,000 birds were reared on estates and inland waters and mostly shot early in the season. The number shot each year had grown to 940,000 in 2016 (Aebischer 2019), but there are no data specifically for Wales.

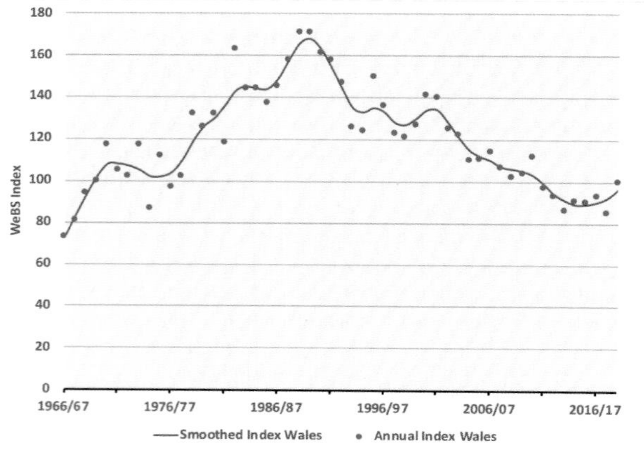

Mallard: Wetland Bird Survey indices, 1966/67 to 2018/19.

The *Britain and Ireland Atlas 1968–72* recorded Mallard in every 10-km square in Wales, except in a small area of south Ceredigion and the uplands of Glamorgan. The *Gwent Atlas 1981–85* and *East Glamorgan Atlas 1984–89* showed that it was absent from large areas of land above 200m in the coalfields, where industrial pollution and heavy shooting were thought to be factors. Densities on the Gwent Levels and in the larger river valleys were much higher. Anglesey seems always to have held a particularly large population: Lovegrove *et al.* (1980) calculated an average breeding density of 2–3 pairs/km^2 on the island, giving a population of 1,500 to 2,000 breeding pairs, from an estimated total for Wales of 4,000–5,000 pairs at that time.

The *Britain and Ireland Atlas 1988–91* showed gains in south Ceredigion and the Glamorgan uplands but losses elsewhere. *Birds in Wales* (1994) considered the 1980 estimate of the Welsh population to be low and suggested 7,000–8,000 pairs. The *Britain and Ireland Atlas 2007–11* showed a few gains since 1988–91 with lower abundance relative to many areas of England but higher in parts of eastern Wales and on Anglesey. The Welsh population in 2018 was estimated to be 4,600–11,000 pairs (Hughes *et al.* 2020).

In counties where there have been two breeding atlases at tetrad level, the results show an increase in distribution. The *Gwent Atlas* recorded it in 66% of tetrads in 1981–85 and 76% in 1998–2003, with significant colonisation of the western valleys. The *East Glamorgan Atlas 1984–89* recorded breeding codes for Mallard in 31% of tetrads and 41% by 2008–11, while in Pembrokeshire, 38% of tetrads were occupied in 1984–88, increasing to 48% in 2003–07. In Gower, the *West Glamorgan Atlas 1984–89* found Mallard in only 29% of tetrads and the author commented that it was surprisingly uncommon. Mallard were present in 23% of tetrads visited in Breconshire in 1988–90, while the *North Wales Atlas 2008–12* found the species in 50% of tetrads.

Numbers in autumn and winter are augmented by arrivals from northeast Europe and Russia. Most ring recoveries are from birds that have been shot. The largest number of recoveries in Wales of birds ringed outside Britain and Ireland come from Denmark and The Netherlands, presumably on passage from farther east. An adult male ringed at Penrhos Country Park, Anglesey, in March 1965 was shot in the Arkhangelsk area, Russia, in August 1966, while an adult male ringed in the Murmansk area of Russia in July 1969 was shot at Whitland, Carmarthenshire, in December 1971. Four birds ringed on Anglesey in autumn have been recovered in Iceland. Some sites show a peak in autumn, particularly October, after which the birds move on.

Based on the proportion of the British population recorded in Wales in 2008–11, Wales is estimated to have held *c.*78,000 individuals outside the breeding season in 2016 (Woodward *et al.* 2020). There has been a marked decline in numbers on the Dyfi Estuary, which regularly held counts of over 1,000 in the 1980s and early 1990s, for example 1,505 in September 1980 and 1,719 in winter 1992/93. Although individual lakes on Anglesey are not among the top sites in Wales, collectively these waterbodies can hold good numbers, although an unknown proportion have been released for shooting. Over 1,000 were regularly on Llyn Alaw in the 1980s and early 1990s. Similar counts sometimes occurred on Llyn Llywenan in the first decade of the 21st century, but none has reached the total of almost 5,000 at Llyn Llywenan in November 1968. Elsewhere in Wales, there have been few site counts over 1,000 in recent decades: 1,500 on Skomer, Pembrokeshire, in October 1981, was exceptional and probably included passage migrants. There were 2,498 at Undy, Gwent, in October 2012, 2,115 at Connah's Quay nature reserve, Flintshire, in August 2015, and 1,027 in the Burry Inlet in August 2017.

Site	County	74/75 to 78/79	79/80 to 83/84	84/85 to 88/89	89/90 to 93/94	94/95 to 98/99	99/00 to 03/04	04/05 to 08/09	09/10 to 13/14	14/15 to 18/19
Severn Estuary: Welsh side (whole estuary)	Gwent/East Glamorgan	336 (3,112)	702 (3,062)	885 (3,529)	692 (3,647)	674 (2,581)	1,005 (2,913)	1,138 (3,389)	1,153 (3,046)	765 (2,379)
Dee Estuary: Welsh side (whole estuary)	Flintshire	na (1,501)	na (3,168)	na (4,448)	804 (2,660)	831 (1,705)	745 (1,642)	495 (1,126)	500 (1,030)	585 (1,275)
Burry Inlet	Gower/ Carmarthenshire	143	160	385	355	408	428	351	468	603
Traeth Bach	Meirionnydd	556	322	636	368	301	240	192	452	389
Traeth Lafan	Caernarfonshire/ Anglesey	289	457	651	591	568	628	374	310	376
Dyfi Estuary	Ceredigion/ Meirionnydd	875	1,322	1,043	1,035	642	571	489	494	211
Llyn Llywenan	Anglesey	31	20	51	229	168	692	885	128	160

Mallard: Five-year average peak counts during Wetland Bird Surveys between 1974/75 and 2018/19 at sites in Wales averaging 600 or more at least once. International importance threshold: 20,000; Great Britain importance threshold: 6,700.

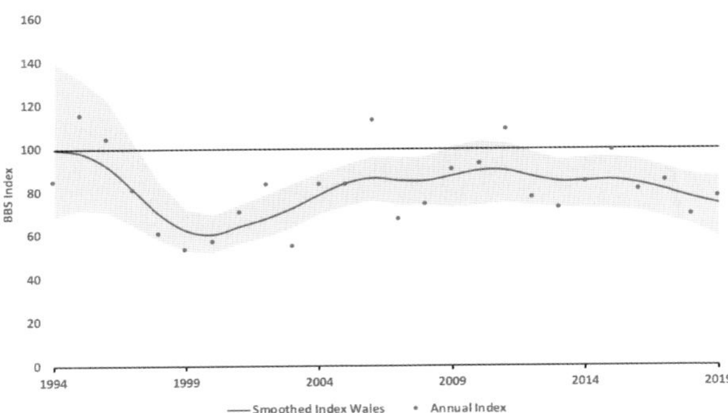

Mallard: Breeding Bird Survey indices, 1994–2019. The red line shows the smoothed index, 1995–2018. The shaded area indicates the 85% confidence limits.

Mallard numbers increased significantly both during and outside the breeding season in the 1970s and 1980s, but have subsequently fallen across the UK. This is reflected in the trends in Wales: the BBS index decreased by 21% during 1995–2018 (Harris *et al.* 2020) and the WeBS index fell 43% between 1989/90 and 2018/19 (Frost *et al.* 2020). Its Amber conservation status in Wales is due to the decline in wintering numbers (Johnstone and Bladwell 2016), although the trend is complicated by the number of birds released for shooting (Cabot 2009). There is no clear reason for the decline but there is a strong relationship with ingestion of lead shot (Green and Pain 2016). Recoveries of birds ringed in Britain and Ireland in winter show there has been a reduction in the number of winter visitors from continental Europe, which made up over 50% of all recoveries in the 1950s and 1960s but fewer than 25% in the 1980s and 1990s (*BTO Migration Atlas* 2002). On current trends, we may expect to see smaller flocks of wintering Mallards, but it should remain a widespread species for future generations to meet in urban parks.

Rhion Pritchard

Sponsored by Jos and Ed Turner

Black Duck *Anas rubripes* Hwyaden Ddu America

Welsh List Category	IUCN Red List	Rarity recording
	Global	
A	LC	BBRC

Black Ducks breed in eastern North America from Labrador to North Carolina and west to Manitoba. Most are resident or disperse locally in winter, but those breeding in the north migrate to the coast or lakes in the eastern USA. There were 41 records in Britain up to 2018 (Holt *et al.* 2019) and no additional records in 2019. It is a very rare visitor to Wales, having been recorded four times. The first was a male at Abergwyngregyn, Caernarfonshire, from 11 February 1979 until 29 January 1985 that mated with female Mallards and produced hybrid young: three were seen in autumn 1980 and in 1981. Mating was observed in 1982 and 1983 and broods of hybrid young were seen in September 1983 and autumn 1984. This was only the second series of breeding attempts involving an American Black Duck in Britain, following records involving a female with a male Mallard on the Isles of Scilly during 1978–83.

The three subsequent Welsh records were a male at Heathfield gravel pits, Pembrokeshire, from 20 January to 14 March 2001, a female at Marloes Mere, Pembrokeshire, from 16 March to 19 May 2008 and an immature male on the Conwy Estuary, Caernarfonshire/Denbighshire, on 7–9 April 2010.

Jon Green and Robin Sandham

Pintail *Anas acuta* Hwyaden Lostfain

Welsh List Category	IUCN Red List			Birds of Conservation Concern			
	Global	Europe	GB	UK	Wales		
A	LC	LC	CR (B)	Amber	2002	2010	2016

The Pintail breeds across the northern hemisphere, in both Arctic and temperate zones of North America, Europe and Asia, farther north than any other dabbling duck (*BTO Migration Atlas* 2002). It nests beside shallow freshwater marshes, small lakes and rivers, and is omnivorous, taking a wide variety of plant and animal food in shallow water up to 30cm deep. In estuaries, the Pintail feeds on small *Hydrobia* molluscs and seeds of saltmarsh plants, including glassworts and Annual Sea-blite, and in freshwater habitat, on seeds, tubers, rhizomes, insects, molluscs and crustaceans (*BWP*).

It is highly migratory, flying south to winter in lower latitudes, from the temperate zone to the tropics, reaching beyond the equator in East Africa. Pintails that winter in Britain and Ireland, or which pass through on migration, come from widely dispersed breeding grounds, ranging from Iceland, Fennoscandia, the Baltic states and Russia up to 80° East (*BTO Migration Atlas* 2002). Some passage migrants move much farther south, reaching major wintering areas in West Africa, such as the Senegal Delta (Trolliet and Girard 2006).

Around 25–30 pairs breed in Britain, mainly in Scotland (Orkney, North Uist and Tiree), where it is on the southern edge of its breeding range and thought to be declining. There are also small numbers in East Anglia and Kent (Eaton *et al.* 2020) and it has bred on several occasions in Wales. Two broods were recorded on Skomer, Pembrokeshire, in 1988 and three broods in 1989, when a pair also nested near Pont ar Sais in the Gwili Valley, Carmarthenshire. On Skomer, 1–2 pairs bred in 1990–93, but no young fledged, and an attempt on Skokholm in 1993 also failed. On Skomer, four pairs fledged seven young in 1994, two pairs fledged two young in 1995, but both pairs failed in 1996. In 1997 a male Pintail showing features of Mallard paired with a female Pintail, fledging a hybrid young. The last confirmed breeding record was in 2002, when a female Pintail was seen with six young on Skomer, but none survived. A female Pintail paired with a male Mallard in 2007 and produced a brood of 13 chicks but none of these survived. The low survival rate of chicks on the island was due to predation by large gulls. *Birds in Wales* (1994) cast doubt on the wild origin of the 1988 and 1989 breeding records, due to releases on the Severn Estuary in the late 1980s, but gave no reason to connect the two. Elsewhere, a pair of Pintail probably bred on Anglesey in 2004 and there have been occasional records of summering pairs and individuals in other counties.

The Pintail is widespread during the winter, being found on sheltered coasts and estuaries, grazing marshes and river flood-plains, primarily on the western coasts of Wales and England. Peak numbers in Wales tend to occur in January, three months later than in Scotland, indicating the southward movement through the autumn. Pintails ringed in Estonia and France have been recovered in Wales and birds ringed here have been recovered in The Netherlands and Iceland. There have been intra-UK recoveries too: one ringed at Slimbridge, Gloucestershire, on 19 January 2015 was shot at Greenfield, Flintshire, on 5 November 2015.

Early authors indicated that it occurred in small numbers around the North Wales coast during the late 19th and early 20th century, but was rare elsewhere in Wales except during hard weather. That status changed significantly in the second half of the 20th century and Wales is now very important for wintering Pintail: the population in Britain is estimated to be 19,500 (Woodward *et al.* 2020) and, in the five years to 2018/19, average peak counts at sites in Wales, or straddling the border, totalled 9,713 birds, 50% of the total (Frost *et al.* 2020).

The Dee Estuary, Flintshire/Wirral, has supported Pintails since the late 19th century. Flocks of a few hundred were reported until the 1930s, after which there was a large increase (Williams 1977). The estuary has consistently supported internationally important numbers since at least the 1950s and is currently the most important wintering site in Britain. It held about half of the British wintering population with at least 5,000 in the early 1950s, but numbers gradually declined and by the mid-1960s peak winter counts were 1,200–2,000. Numbers increased above 4,000 in the 1970s and further during the 1980s, and peaked at 11,945 in 1989/90. Numbers have since fluctuated and the highest recent count was almost 8,000 in November 2018 (Frost *et al.* 2020). The Pintail is a qualifying feature of the Special Protection Area, with the 2018/19 total constituting 13.3% of the northwest European population.

The Dee population is concentrated on saltmarshes on the English side of the estuary, feeding at the saltmarsh edge as the tide rises, lifting seed and making it available to the birds. These marshes are accreting, colonised by pioneer plants, and have a high tidal range, which may be key features of the estuary's attraction to the species. As the tide falls, most move to the low-tide channel at Bagillt Bank, Flintshire. Low numbers are, therefore, recorded by high-tide Wetland Bird Surveys in Flintshire, but low-tide counts include 8,000 at Bagillt Bank in October 1984, 6,120 in October 2001 and 4,000 in November 2015, 6,000 at Flint in October 2005 and 5,000 at RSPB Oakenholt Marsh in October 2015. It is entirely possible that most of the estuary population spends at least part of each tidal cycle in Wales, making the whole estuary important to them. Much of the feeding and low-tide areas have been acquired by the RSPB from 1979, where the birds are largely safe from disturbance. Little is known about the behaviour of the Pintail at night.

Upstream, the Dee Flood Meadows between Holt and Worthenbury are used by Pintails when floodwaters are receding, creating ideal conditions for feeding on stubbles and grassland (Colin Wells pers. comm.). Numbers there are of British importance

Site	County	74/75 to 78/79	79/80 to 83/84	84/85 to 88/89	89/90 to 93/94	94/95 to 98/99	99/00 to 03/04	04/05 to 08/09	09/10 to 13/14	14/15 to 18/19
Dee Estuary: Welsh side (whole estuary)	Flintshire	na (4,639)	na (6,788)	na (7,137)	1045 (8,605)	402 (5,505)	1,162 (4,982)	865 (5,187)	245 (3,299)	575 (5,499)
Burry Inlet	Gower/ Carmarthenshire	784	1,287	1,874	1,712	2,249	3,285	4,073	1,456	1,570
Severn Estuary: Welsh side (whole estuary)	Gwent/East Glamorgan	9 (344)	117 (306)	259 (448)	360 (598)	355 (584)	684 (883)	409 (816)	237 (501)	315 (786)
Dee Flood Meadows	Denbighshire/ Cheshire					325	812	498	826	565
Traeth Bach	Meirionnydd	87	61	98	124	185	142	243	261	391
Malltraeth Cob	Anglesey	94	151	305	422	215	207	303	315	124

Pintail: Five-year average peak counts during Wetland Bird Surveys between 1974/75 and 2018/19 at sites in Wales averaging 300 or more at least once. International importance threshold: 600; Great Britain importance threshold: 200.

annually and have been of international importance in half of the five-year periods in the early 21st century; the peak count was 1,920 in February 2016.

The Pintail is a qualifying feature of the Burry Inlet Special Protection Area, Gower/Carmarthenshire, which is the fourth most important site in the UK for the species, although annual peak counts have fallen below 500 since 2016/17. Numbers had grown steadily to peaks of 5,772 in 2003/04 and 6,224 in 2007/08. Although the average declined subsequently, large counts still occur, such as 9,170 at Llanrhidian, Gower, in January 2014. It favours the extensive grazed saltmarshes on the south shore between Whiteford Point and Penclawdd, Gower. The high tidal range, over 10m on some spring tides, could be a factor in creating attractive conditions for Pintail. There are few Pintail on the north shore, where much of the shoreline is mud and sand, and even where there is saltmarsh in the upper Loughor Estuary, only a few are recorded (Lyndon Jeffery and Bob Howells pers. comm.). Smaller numbers are on the Severn, mainly in the Rhymney Estuary, East Glamorgan, and along the shoreline of the Gwent Levels, probably due to the current paucity of saltmarsh that was long ago claimed for agriculture.

Elsewhere, numbers on the Dyfi Estuary, Ceredigion/ Meirionnydd, have fluctuated between 100 and 300 since 1974, returning to the top of that range since 2015/16, while sites around Anglesey and north Caernarfonshire, such as Traeth Melynog, Foryd Bay and Traeth Lafan, showed increases from the 1990s. The reason is not known, but could be linked to better site protection/management and reduced disturbance. These estuaries can also receive influxes caused by hard weather farther east. No inland site, with the exception of the Dee Flood Meadows, Denbighshire/ Cheshire, and the Severn/Vyrnwy confluence, Montgomeryshire, regularly holds more than a handful of Pintail, though they are frequent at Llangorse Lake, Breconshire, where the highest count was 17 in December 2002.

It is an Amber-listed species of conservation concern in Wales, owing to the significant proportion of the flyway population that occurs here and the localised nature of its distribution (Johnstone and Bladwell 2016). Given the importance to wintering Pintail of Welsh estuaries, the Dee and Burry Inlet in particular,

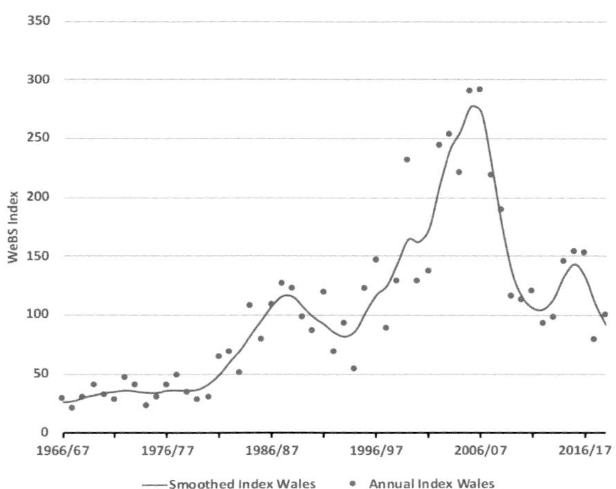

Pintail: Wetland Bird Survey indices, 1966/67 to 2018/19.

it is significant that declines are taking place across all the main sites. A WeBS High Alert over the medium term (ten years) was triggered by a decline of 52% in Wales, greater than in any other UK country. High Alerts have also been issued for the Dee Estuary SPA, owing to a 51% long-term decline and the Burry Inlet SPA, following a 61% medium-term decline (Woodward *et al.* 2019). It is not possible to link the decline with any change in management at either site (pers. obs., Lyndon Jeffery pers. comm.). The general reduction in the population appears to be caused by a decline in breeding success, such as that recorded in Finland (BirdLife International 2020). Climate change is resulting in more wildfowl short-stopping, able to winter in continental Europe and no longer needing to move west to Britain in normal conditions. Given this, our best course of action to conserve the Pintail in Wales is to ensure that all the main sites continue to be legislatively protected, well managed, and valued by their local communities.

David Parker

Sponsored by David Parker

Teal *Anas crecca* Corhwyaden

Welsh List Category	IUCN Red List			Birds of Conservation Concern			
	Global	Europe	GB	UK	Wales		
					2002	2010	2016
A	LC	LC	LC	Amber	2002	2010	2016

The Teal is one of the most abundant Old World ducks, found over most of temperate and subarctic Eurasia (*HBW*). Breeding densities are highest in the north of its range; Finland, for example, holds the largest breeding population in Europe outside Russia (BirdLife International 2015). It is highly migratory, wintering as far south as the Sahel and Nile Valley in Africa, and India and Southeast Asia. It breeds on shallow freshwater lakes and marshes with abundant fringe vegetation (*HBW*), favouring moorland pools, bogs and patterned mires in the uplands of Britain and Ireland. The greatest numbers nest in the north and west, although substantial numbers also breed in well-vegetated wetlands in lowland Britain (Fox 1986a). Ringing recoveries involving Wales suggest that birds wintering here are mainly from the east, including Denmark, Sweden, Finland, Latvia and Russia. Teal breeding in Iceland are known to winter in Britain and Ireland, but there are no exchanges between Wales and Iceland (Robinson *et al.* 2020). Most Teal breeding in Britain and Ireland are thought to winter here, though movements to France and Spain have been recorded, particularly in cold weather (Cabot 2009). Teal are a popular quarry species for wildfowlers, which is reflected in the fact that 97% of all ring recoveries in Britain and Ireland are from birds that were shot. Hunting was thought to have reduced populations in Europe in the 19th century, but it is uncertain whether hunting is now a significant limiting factor on population size (Guillemain and Elmberg 2014).

The Teal is quite common in the archaeological record in Britain. In Wales it was recorded in a Late Glacial context at Cat Hole cave, Gower (Yalden and Albarella 2009) and in Roman deposits at Caerwent and Caerleon, Gwent. Comments by 19th- and early-20th-century authors suggest that breeding was widespread, but they gave little indication of numbers involved. It is likely that the Teal was more numerous during that period than now, as several former breeding sites have been destroyed by drainage, afforestation and built development. Teal occasionally bred on lakes in North Wales (Eyton 1838) and "used to breed on the Avon Ganol [Caernarfonshire/Denbighshire] before the drainage, and that river was the resort of many hundreds of these, and other kinds of the duck tribe" (Williams 1835).

At the turn of the 20th century the Teal bred in all counties of North Wales including Montgomeryshire, though not numerously. In Caernarfonshire it bred on moorland bogs and tarns, including in Cwm Eigiau at over 490m and possibly on Ffynnon Llugwy at nearly 550m (Forrest 1907). In Ceredigion, four nests were found on Cors Caron in 1895 (Salter 1900), though the first successful breeding confirmed in the county was not until 1948 (Roderick and Davis 2010). Mathew (1894) considered it "extremely likely that a few pairs of Teal may nest annually in suitable places" in Pembrokeshire, but had no proof. Records in Glamorgan were also poorly documented prior to the late 1950s but it had bred in fair numbers at Morfa Pools, Gower, with small numbers at other sites (Hurford and Lansdown 1995). There was no confirmed breeding in Gwent until 1943 and none in Flintshire since 1955.

Ceredigion is probably the most important Welsh county for breeding Teal, with a population estimated at 40 pairs in *Birds in Wales* (1994) and 80–110 pairs by Roderick and Davis (2010). Regularly used sites include RSPB Ynys-hir, where numbers peaked at 18 pairs in 2003, but fell back to eight pairs in 2019. At Cors Caron, the breeding population was estimated at 20 pairs in the 1970s and 50 pairs in 2005 (Roderick and Davis 2010), and at Cors Fochno, 13–18 pairs bred in 1980–83 (Fox 1986a). In mid-Wales, breeding numbers have been in long-term decline: in Radnorshire, Jennings (2014) estimated a county population of 10–15 pairs in most years since 2000, while Peers and Shrubb (1990) thought that the Breconshire population did not exceed ten pairs, with 3–5 pairs more probable.

Farther north, *Birds in Wales* (1994) estimated that 20–25 pairs bred in Meirionnydd, although the number is now likely to be no more than ten pairs (pers. obs.). Surveys of the Migneint found five pairs in 1976 (RSPB), 4–6 pairs in 2007 and three pairs in 2008. The Caernarfonshire breeding population is unlikely to be more than five pairs and the same is true of Denbighshire, where a survey of Mynydd Hiraethog found only four pairs in 1975–78 (RSPB). Anglesey is the Teal stronghold in North Wales: an RSPB survey in 1986 found 3–20 breeding pairs, but wetland restoration has subsequently created more habitat: the average number of males during the breeding season in 2015–19 was 13 at RSPB Valley Wetlands and two at RSPB Cors Ddyga.

Numbers breeding in southern Wales are probably lower, although Hurford and Lansdown (1995) estimated the breeding population in Glamorgan at up to 20 pairs. In Gower, the *West Glamorgan Atlas 1984–89* recorded probable breeding in six tetrads and considered it annual at Oxwich Marsh and Crymlyn Bog, while the *East Glamorgan Atlas 2008–11* found probable breeding Teal in six tetrads. In Gwent, breeding is probably more sporadic; a pair at Newport Wetlands Reserve in 2009 was only the sixth confirmed record in the county. Breeding is occasionally recorded in Carmarthenshire and Pembrokeshire, including on the islands (Lloyd *et al.* 2015, Donovan and Rees 1994).

It is a difficult species to prove breeding unless a brood is seen. Adults early in the breeding season may be late-staying birds from winter, nests are in dense vegetation, young are prone to remain in cover and the presence of males later in the breeding season does not necessarily indicate breeding. Breeding records in the *Britain and Ireland Atlas 1968–72* were concentrated in northern Britain and sparsely distributed in Wales; Caernarfonshire and Ceredigion held the majority of 10-km squares with confirmed breeding. The *Britain and Ireland Atlas 1988–91* showed severe losses, particularly in Caernarfonshire, that were only partly balanced by gains on Anglesey. Breeding occurred in nearly all Welsh counties during 2000–19, although only 2–3 pairs were confirmed in some years. Estimates made in the last 40 years suggest fewer than 100 pairs of Teal in Wales (Lovegrove *et al.* 1980, *Birds in Wales* 2002), although *Birds in Wales* (1994) suggested as many as 125–150 breeding pairs. Between 1968–72 and 2007–11, there was a 30% contraction in range in Wales, part of a broader shift north and west in Britain, although there were some gains near the west coast. The Welsh population was estimated to be 60–250 pairs in 2016 (Hughes *et al.* 2020) and, based on the information above, is likely to be *c.*100–150 pairs.

Far greater numbers winter in Wales than breed, though numbers counted vary considerably, depending on the severity of the weather. Teal are particularly vulnerable to cold winter conditions and will move in search of ice-free conditions (Ridgill and

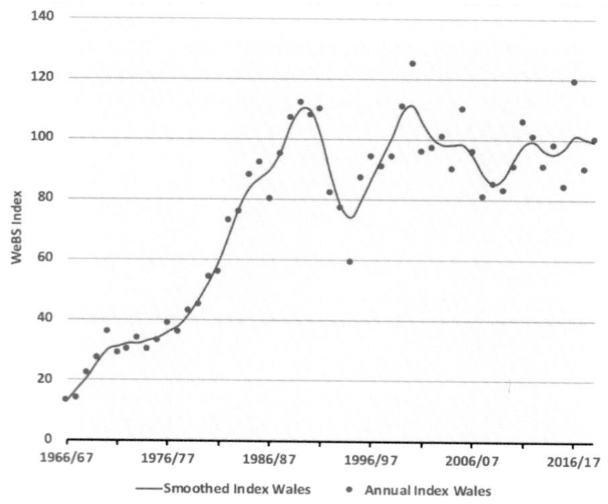

Teal: Wetland Bird Survey indices, 1966/67 to 2018/19.

Site	County	74/75 to 78/79	79/80 to 83/84	84/85 to 88/89	89/90 to 93/94	94/95 to 98/99	99/00 to 03/04	04/05 to 08/09	09/10 to 1 3/14	14/15 to 18/19
Dee Estuary: Welsh side (whole estuary)	Flintshire	na (685)	na (2,189)	na (4,167)	2,767 (7,116)	1,709 (5,259)	1,585 (4,988)	1,150 (2,861)	1,018 (4,252)	2,343 (6,231)
Severn Estuary: Welsh side (whole estuary)	Gwent/East Glamorgan	286 (816)	253 (1,449)	1,436 (2,015)	883 (2,730)	586 (3,633)	906 (4,215)	901 (4,626)	1,222 (5,627)	800 (5,028)
Cleddau Estuary	Pembrokeshire	435	1,051	2,721	2,290	2,234	2,142	1,851	2,568	2,393
RSPB Cors Ddyga	Anglesey	22	20	410	55	716	855	1,377	1,585	1,335
Dyfi Estuary	Ceredigion/ Meirionnydd	345	740	651	469	361	822	1,153	1,076	914
Burry Inlet	Gower/ Carmarthenshire	1,241	1,376	2,028	826	1,351	1,111	804	472	499

Teal: Five-year average peak counts during Wetland Bird Surveys between 1974/75 and 2018/19 at sites in Wales averaging 1,000 or more at least once. International importance threshold: 5,000; Great Britain importance threshold: 4,300.

Fox 1990). Following a large increase during the 1970s and 1980s, the WeBS index for Wales has since fluctuated, but remained stable. Numbers usually peak in December and January (Frost *et al.* 2020).

It was one of the most numerous species at the Orielton Decoy, Pembrokeshire, during 1868–1960, with a peak count of 2,300 in 1935 (Saunders 2019). The Dee Estuary Special Protection Area (SPA), for which Teal is a qualifying feature, held a peak of 10,715 birds in December 1991, though the highest count on the Welsh side was 6,651 in November 2018. Numbers there have declined by 25% over the long term (25 years), triggering a Medium Alert (Woodward *et al.* 2019), and have also fallen on the Burry Inlet SPA, which regularly held over 1,000 birds until 2006/07 and peaked at 3,655 in December 1984. Other large counts include 2,938 at Llyn Alaw, Anglesey, in December 1990 and 3,813 on the Cleddau Estuary in January 2017. The increase at RSPB Cors Ddyga, reaching 2,751 in December 2017, illustrates the response of Teal to wetland creation. Numbers have also increased on the Severn Estuary, a site of international importance, but Teal has yet to be included as a qualifying feature of the SPA.

The Teal's Amber status in Wales is based on a moderate long-term decline in its Welsh breeding range and because Wales holds 4% of the flyway population (Johnstone and Bladwell 2016). It is likely that breeding numbers have also declined, though its decline would have been greater had it not been for wetland creation since the 1990s. At Cors Fochno, blocking ditches gave rise to deep linear pools, with abundant emergent vegetation and was followed by a three- to four-fold increase in the number of nesting Teal (Fox 1986a). Wales holds only a tiny proportion of the European population, but its breeding habitat needs more attention to avoid a further decline in numbers. A dedicated survey to obtain a more accurate total for the number of breeding pairs would be a valuable first step.

Rhion Pritchard

Green-winged Teal *Anas carolinensis* Corhwyaden Asgellwerdd

Welsh List Category	IUCN Red List Global	Rarity recording
A	LC	WBRC

The Green-winged Teal breeds on lakes across Alaska, Canada and the northern USA and winters in the southern USA, Mexico and the Caribbean (*HBW*). It was formerly considered a subspecies of Teal (IOC). The numbers recorded in Britain increased towards the end of the 20th century, with mean annual totals doubling each decade, from three in 1960–69 to 45 in 2000–09, and have since remained stable (White and Kehoe 2020a). It is a scarce visitor to Wales, though the frequency of sightings has also increased here. All records were males, though it is quite likely that females occur undetected owing to their similarity to female Teal.

The first Welsh record was of two at Cors Caron, Ceredigion, on 11 December 1968. Several records are assumed to have been individuals returning to the same area in successive years. Welsh records show a western bias, with Pembrokeshire and Ceredigion having more than any other counties. In Pembrokeshire, one on Skokholm and Marloes in April–May 1996 is presumed to be the same as one on Skomer the following year; one on Skomer in November 2000 then wintered annually on the Cleddau Estuary and adjacent freshwaters until 2005, when it was joined by a second male. Another male, first seen at Marloes Mere in May 2009, occurred annually in the county until February 2015.

One at Penclacwydd, Carmarthenshire, in May 2007 was seen almost annually there until 2010. In Ceredigion, one wintered on the Dyfi Estuary almost annually between winter 2002/03 and 2008/09 and there were birds at Cors Caron in March 2002, October 2003 and April 2006, and at Llyn Eiddwen in April 2004.

In North Wales, two were at Morfa Madryn/Aber Ogwen, Caernarfonshire, in February–March 2000 and one wintered at RSPB Inner Marsh Farm, Flintshire/Wirral, between November 2003 and February 2004, returning from October 2004 to March 2005. On Anglesey, one at RSPB Cors Ddyga from November 2006 was seen in subsequent winters at Llyn Parc Mawr in January 2007, Llyn Coron in February 2009, and at Llyn Coron and Llyn Padrig in January–February 2010. In Gwent, there were single birds at Newport Wetlands Reserve (Goldcliff) in January 2009 and in October 2010, which was also seen at Peterstone in March/April 2011. One at Kenfig in March/April 2014 was seen in

Total	Pre-1950	1950–59	1960–69	1970–79	1980–89	1990–99	2000–09	2010–19
46	0	0	2	2	5	6	21	10

Green-winged Teal: Individuals recorded in Wales by decade.

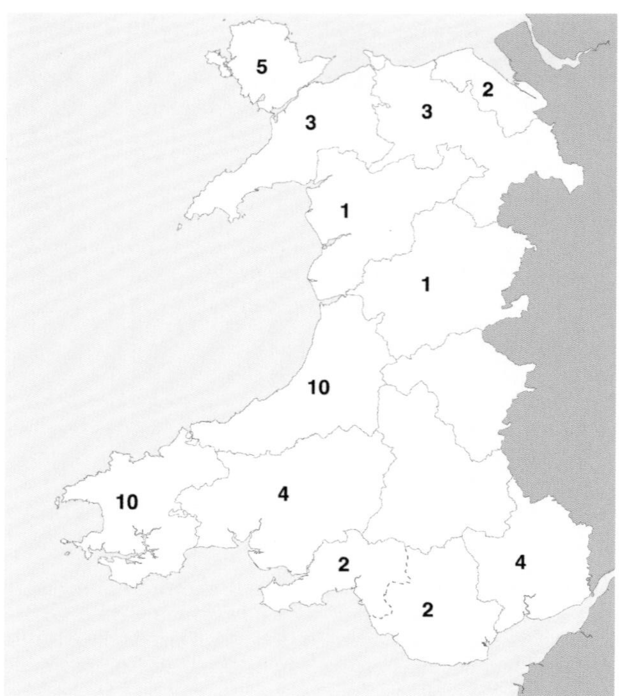

Green-winged Teal: Numbers of birds in each recording area, 1968–2019. One was seen in both East Glamorgan and Gower.

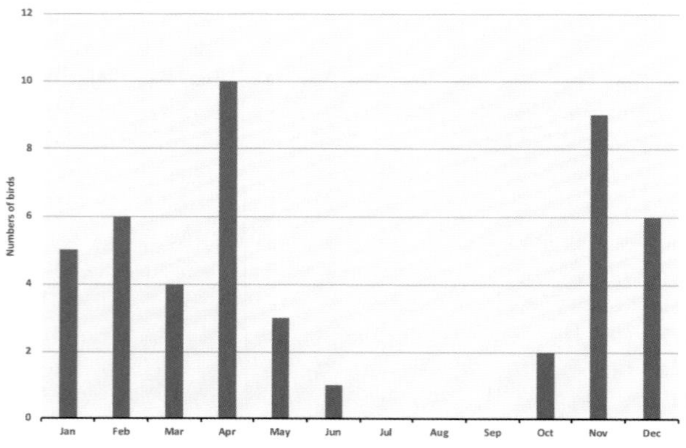

Green-winged Teal: Totals by month of arrival, 1968–2019.

both East Glamorgan and Gower. The only record from an inland county was at Llyn Coed y Dinas, Montgomeryshire, in April 1997.

Jon Green and Robin Sandham

Red-crested Pochard *Netta rufina* Hwyaden Gribgoch

Welsh List Category	IUCN Red List		Rarity recording
	Global	Europe	
C5, E	LC	LC	RBBP

The native range of Red-crested Pochard, although fragmented, extends from Spain to western Mongolia, but the bulk of the population is east of the Black Sea. It is a partial migrant, wintering to the south of its breeding range. It feeds by both diving and dabbling and prefers deep, large lakes and lagoons of fresh or brackish water, with abundant riparian vegetation (*HBW*). The closest population to Wales, numbering about 1,500 pairs, is on the Ebro Delta in northeast Spain (*HBW*). At least some early records in Britain are thought likely to have been vagrants. Pyman (1959) concluded that such occurrence was increasing, but recent Welsh records probably originate from a naturalised population in England. The first record of breeding in the wild in Britain was in 1937, when a pair bred in Lincolnshire that were thought to have escaped from the wildfowl collection at Woburn Park, Bedfordshire (Lever 2009). In 2018, there were 73 pairs breeding in England, but none has bred in Wales or Scotland (Eaton *et al.* 2020). The site with the largest population is Cotswold Water Park, Gloucestershire/ Wiltshire, which held 26 pairs in 2018, although the numbers outside the breeding season have fallen by 40% since the peak count of 432 in November 2011.

Birds in Wales (1994) reported 34 individuals in Wales up to 1991, the majority in Glamorgan. There were a further 28 records in Wales between 1992 and 2000, including seven in Glamorgan in 2000 (*Birds in Wales* 2002), and during 2001–18, about 45 individuals were reported. It occurred every year in Wales during 1989–2011, but there have been several subsequent years with no records, and the annual average during 2010–19 was just 1–2 birds. This coincides with the decline in numbers at Cotswold Water Park.

All counties in Wales have recorded a few Red-crested Pochards and it may be under-recorded if observers assume birds are likely to have escaped from collections. There are few early records of the species in Wales. Mathew (1894) mentions a female shot at Stackpole, Pembrokeshire, which was presented to the British Museum, but did not give a date. The first record in East Glamorgan was of two males at Llanishen Reservoir in December 1940, but it became more regular in the county from 1959, with 19 recorded up 1994. Hurford and Lansdown (1995) considered that although captive origin could not be ruled out for any of these birds, some were likely to originate from elsewhere in western Europe. There have been 13 records (16 birds) in Gwent, from the first record in 1972, and Anglesey recorded at least six individuals after the first in 1962 (Jones and Whalley 2004).

Most records in Wales are of 1–3 birds, though four were together in Gwent in 1995, initially at Llandegfedd Reservoir in July–August, then at Ynysfro Reservoir until October. Four were on Llangorse Lake, Breconshire, in November 2013. Several pairs have been reported, including one that displayed at Roath Park Lake, East Glamorgan, in April 2002. In Denbighshire, a female and four recently fledged young were at RSPB Conwy in August 2009, but had not bred on site.

This species could well be added to the list of breeding birds in Wales in the future. It is likely that many of those in Wales come from the English naturalised population or have escaped from captivity. Unless a marked bird from Europe is seen, it will be impossible to prove an indigenous origin.

Rhion Pritchard

Pochard *Aythya ferina* Hwyaden Bengoch

Welsh List Category	IUCN Red List			Birds of Conservation Concern				Rarity recording
	Global	Europe	GB	UK	Wales			
A	VU	VU	EN	Red	2002	2010	2016	RBBP

This omnivorous duck breeds in an almost continuous broad band across Eurasia as far east as the Kamchatka Peninsula and south to Turkmenistan. It winters to the south, as far as the Sahel and Nile Valley in Africa, and from India east to Japan (*HBW*). It favours relatively nutrient-rich freshwater bodies, usually with dense emergent and submergent vegetation (*BWP*). The acidic and nutrient-poor nature of many upland lakes in Wales restricts its breeding to lowland habitats below 300m. This is not strictly mirrored in its wintering habitat, when larger numbers gather on open water, including at slightly higher altitudes on more acidic waters that are shunned during the breeding season, suggesting differences in seasonal diet. Around 760 pairs breed annually in the UK, a three-fold increase since 1986–90, about 70% of which are in east and southeast England (Eaton *et al.* 2020).

The Pochard probably did not breed in Wales prior to 1900 (*Historical Atlas 1875–1900*), but its cryptic nature means that it may have gone unrecorded. Although breeding was suspected at Llyn Pen-y-Parc, Anglesey, in 1904 and pairs were present on upland lakes in Meirionnydd in spring 1901 and 1902 (Forrest 1907), the first confirmed breeding was a brood on Llyn Tegid, Meirionnydd, in 1905 (Bolam 1913) and a pair or two may have bred there in 1906. Two pairs nested on Whixall Moss in 1915, but it is unclear whether these were on the Denbighshire or Shropshire side of the border. Single pairs also bred at Llyn Maelog, Anglesey, in 1925, and at Witchett Pool, Carmarthenshire, in 1937. Breeding was claimed in Caernarfonshire by Hardy (1941), but without any detail. Such records remained sporadic until the late 1940s, after which breeding occured more frequently, particularly on Anglesey, where it had been suspected in earlier decades. The only Pembrokeshire breeding record was at Orielton in 1959. Following probable breeding at Cors Ddyga in 1951 and a handful of breeding records in the late 1960s, there were 15 pairs nesting on Anglesey by the end of the 1970s (Lovegrove *et al.* 1980). Breeding occurred again at Witchett Pool in 1971 and irregularly there up to 1992. A pair bred at Oxwich Marsh, Gower, in 1975 and annually thereafter, peaking at five breeding pairs in 1979. Elsewhere, breeding has been sporadic, but was confirmed at several sites in Carmarthenshire, at Eglwys Nunydd Reservoir, Gower, in 1950 and at Orielton, Pembrokeshire, in 1959.

Up to 40 pairs may have attempted to breed annually on Anglesey during the 21st century: the average number of confirmed, probable and possible breeding pairs in 2006–18 was 26, with more than 30 pairs present in several years (Eaton *et al.* 2020). A greater area of suitable breeding habitat exists on Anglesey than elsewhere, so it is likely to remain the core Welsh breeding area. Up to 25 pairs bred at Penclacwydd, Carmarthenshire, in the late 1990s, but numbers were subsequently lower, with a maximum of 11 broods in 2013 and 3–6 broods in most years during 2008–19. During the latter period up to five broods were reported each year from Newport Wetlands Reserve and Llanwern, Gwent. Breeding is extremely irregular away from these few favoured sites. The Pochard nests late in the season so, as well as being cryptic, broods are often present after general breeding bird surveys have been completed. Consequently, breeding may be under-recorded. However, it is unlikely that the Welsh breeding population greatly exceeds 50 pairs in most years, and habitat constraints means that this is unlikely to increase.

Wetland Bird Surveys show relatively stable numbers in Wales during July–October, when moulting migrant males mix with the small breeding population. Numbers rapidly increase by November to wintering levels that are maintained until March, after which birds disperse back to continental breeding areas (Frost *et al.* 2020). Ringing recoveries indicate that Pochards have less discrete northeast/southwest orientated flyways between breeding and winter quarters than many other European ducks. A large proportion of those wintering in western Europe originate from the southwest Asia flyway and particularly from the Ob river catchment, Russia, while others come from northwest and central Europe (Folliot *et al.* 2018). Pochards make considerable movements during the winter, which tend to be more complex than other common species in distance and direction (Keller *et al.* 2009). Some Pochards that breed in Britain move to France and Spain (*BTO Migration Atlas* 2002), so Welsh breeding birds do not necessarily winter locally, although there is no direct evidence for this. Birds recovered in Wales during the winter included one from Russia and several that had been ringed as chicks in Latvia.

The winter distribution of the Pochard in Wales is more widespread than its breeding range; it can occur on less eutrophic waters in winter than summer and was reported from marine waters around Anglesey by Forrest (1907). It seems to have been scarce at the turn of the century, with contemporary authors mentioning maximum counts of a few dozen in Pembrokeshire and North Wales (Mathew 1894, Forrest 1907). Numbers increased prior to the Second World War, with several hundred wintering at Kenfig Pool in the 1930s (Hurford and Lansdown 1995). A large influx in winter 1961/62 associated with cold weather in eastern Europe resulted in an exceptional count of 280 on the Dyfi Estuary, Ceredigion/Meirionnydd, and in 1962/63 there were 600 on the

Site	County	74/75 to 78/79	79/80 to 83/84	84/85 to 88/89	89/90 to 93/94	94/95 to 98/99	99/00 to 03/04	04/05 to 08/09	09/10 to 13/14	14/15 to 18/19
Severn Estuary: Welsh side (whole estuary)	Gwent/East Glamorgan	0 (212)	144 (810)	606 (1,742)	523 (1,626)	266 (1,467)	316 (1,044)	16 (680)	47 (512)	15 (254)
RSPB Valley Wetlands inc. Llyn Traffwll	Anglesey	127	91	158	226	164	197	125	172	80
Kenfig Pool	East Glamorgan	100	163	130	111	373	98	150	161	46
Eglwys Nunydd Reservoir	Gower	182	234	164	162	243	61	157	162	45
Cosmeston Lakes CP	East Glamorgan		4	2	119	52	150	344	133	40

Pochard: Five-year average peak counts during Wetland Bird Surveys between 1974/75 and 2018/19 at sites in Wales averaging 150 or more at least once. International importance threshold: 2,000; Great Britain importance threshold: 230.

Taff/Ely estuary, East Glamorgan, when lakes and reservoirs in Cardiff froze. In the 1960s, up to 800 wintered regularly at Eglwys Nunydd Reservoir, Gower, with 870 there in February 1964, and 550 were on Talybont Reservoir, Breconshire, in November 1965. Through that period, until the early 1990s, it seems likely that over 5,000 Pochard wintered annually in Wales. During that period, *c.*500 wintered in the Rhymney Estuary, East Glamorgan, and the site peak, of 1,000 in February 1987, is probably the highest Welsh count. Pochards no longer use the estuary, perhaps owing to improvements in water quality from sewer outfalls discharging into the Severn Estuary (Phil Bristow *in. litt.*). Numbers in North Wales were smaller, but several hundred wintered on Anglesey lakes, with the highest counts being on Llyn Cefni: 350 in November/December 1974 and 470 in December 1982.

The Pochard's winter range contracted by 27% between the *Britain and Ireland Winter Atlas 1981–84* and that of 2007–11, with losses from 10-km squares across mid- and North Wales and a few gains in the southeast. The Welsh side of the Severn Estuary is no longer as important as it was during the late 1980s, when an average of 1,831 were recorded between 1985/86 and 1989/90, although the estuary as a whole remains of British importance. Numbers in Wales have been declining since at least 1966, but this has accelerated in the last 25 years, and particularly in the last decade. The number of sites in Wales holding over 20 Pochards at least once during the year fell from 21 in 2010 to fewer than ten by the end of the decade, and in 2017 no site recorded a three-figure count for the first time. The most significant declines are evident in western counties. For example, Bosherston Lakes, a small eutrophic lake system in Pembrokeshire, regularly attracted 200–300 Pochards each winter in the 1970s and 1980s, but now it is a rarity there, even though the abundance of the lake's well-monitored submerged macrophyte vegetation has not reduced (Haycock and Hinton 2010). RSPB Valley Wetlands is now the only site in Wales to maintain a five-year average peak annual count of more than 50 birds.

Johnstone and Bladwell (2016) included Pochard on the Red List of highest conservation concern in Wales, based on the decrease in its long-term wintering abundance—now at 88%—and its global status. The decline has triggered WeBS High Alerts for Wales for the long, medium and short term. The Welsh index fell by 68% in just five years, the steepest rate of any UK country (Woodward *et al.* 2019), reflecting changes across western Europe. Following a long period of increase from the late 1940s, most likely the result of habitat creation and modest eutrophication (Fox and Salmon 1988), and a period of stability during 1991–2000, Pochard numbers have declined at an alarmingly rapid rate across Europe (Folliot *et al.* 2018). The precise causes are not clear but most experts consider that changes in water quality at breeding sites are among the most significant drivers, especially the hyper-eutrophication of nutrient-rich lakes, because of agricultural and urban runoff (Fox *et al.* 2016). Particularly conspicuous declines in eastern Europe, from where at least some Welsh wintering birds originate, were associated with adverse changes in freshwater fish aquaculture that followed the end of communist rule. Declines in Black-headed Gulls, within whose colonies Pochard nest and derive considerable benefit (Pöysä *et al.* 2019), may also have affected the major part of the European breeding population. An increase in introduced predators such as Raccoon Dog and American Mink (e.g. Väänänen *et al.* 2009) in

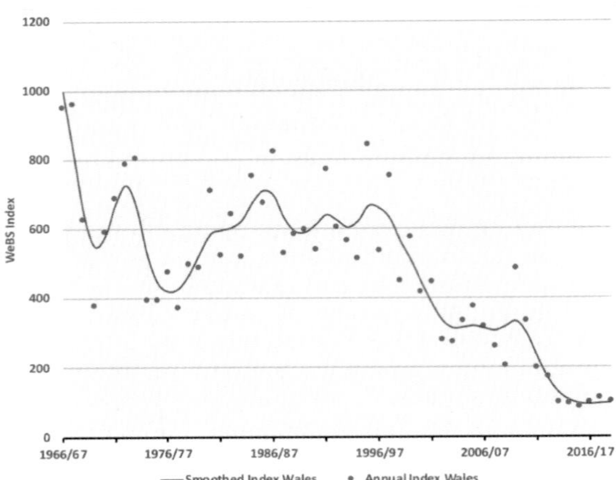

Pochard: Wetland Bird Survey indices, 1966/67 to 2018/19.

the core breeding area may have increased adult female mortality, since there are recent signs of a major decline in the ratio of females to males. Reduced survival of females could contribute to their declining reproductive success (Brides *et al.* 2017, Fox and Christensen 2018). Recent lake restorations show that control of runoff, combined with fish stock management, can restore turbid hyper-eutrophic lakes to their former clear-water ecological state, with great benefit to submerged macrophyte growth and thus to breeding Pochards (Fox *et al.* 2019a). Breeding numbers in the UK, France and The Netherlands have been stable or increasing in recent years (Fox *et al.* 2016), so such issues are not affecting the small Welsh breeding population.

We know little about how changes to annual survival may affect the rate of population change. The species has been a popular quarry species in some countries, although the annual bag has declined in proportion to its abundance in France (Guillemain *et al.* 2016) and the UK, where the estimated number shot fell from 2,400 in 2004 to 370 in 2016 (Aebischer 2019); there are no estimates specifically for Wales. Among the commoner duck species, the Pochard is thought especially susceptible to lead poisoning (Green and Pain 2016).

Although Wales makes a relatively modest contribution to the overall flyway population, the sites utilised still require effective protection. Most key breeding and wintering areas are in sympathetic conservation management, but it is critical to ensure that their clear-water nutrient-rich status is maintained, which is often beyond the control of conservation managers. Given the low production of ducklings compared to the number of potentially breeding females, it would be useful to understand better the precise conditions at key sites, in order to enhance reproductive success everywhere. Marking and telemetry of these birds would identify where Welsh breeding birds and their offspring reside post-breeding. However, the declines in wintering Pochards in Wales reflect declines in its global population, so activities here will not address the challenges facing the flyway population as a whole.

Tony Fox

Ferruginous Duck *Aythya nyroca* Hwyaden Lygadwen

Welsh List Category	IUCN Red List		Rarity recording
	Global	Europe	
A	NT	LC	BBRC

There are four populations of this species, which breeds discontinuously from northwest Africa to China and winters in the Sahel zone of Africa, the Arabian Peninsula and South and East Asia (*HBW*). The trend in each population is unclear, but the European population declined rapidly as a result of loss and deterioration of the well-vegetated shallow pools used for breeding due to drainage, water abstraction and eutrophication, and the introduction of non-native Grass Carp, Wels Catfish and Muskrat (BirdLife International 2020). It occurs in Britain primarily during the winter months, often accompanying Pochards, and is presumed to originate from the population that breeds in southeast Europe (Vinicombe 2000). There have been almost 700 records in Britain since 1950, averaging around nine each year during 1990–2019 (Holt *et al.* 2020). The decline in its European population has led to it becoming an increasingly rare visitor to Wales. It is commonly kept in captivity and it is likely that at least some recent records relate to escaped birds.

Over one-quarter of the Welsh total was recorded before 1950. The first Welsh record was a female or first-calendar-year bird at Boultibrooke, Radnorshire, from late 1859 to March 1860. Several historic records are now considered doubtful as wild birds, but a party of seven (of which one was shot) near Machynlleth, Montgomeryshire, in April 1906 remains an accepted record (Naylor 2021). Numbers recorded in Wales increased until the 1980s, but have since declined dramatically, with just four records since 2000, undoubtedly linked to declines in Europe. The distribution of records shows a southeast bias, with one-third of the total in East Glamorgan. Lisvane Reservoir has hosted five different birds and Kenfig four, including two males on 28 April 1999. Six have occurred in Gwent, including pairs at Llanvihangel Gobion on 27 December 1982 and at Wentwood Reservoir on 19 March 1987.

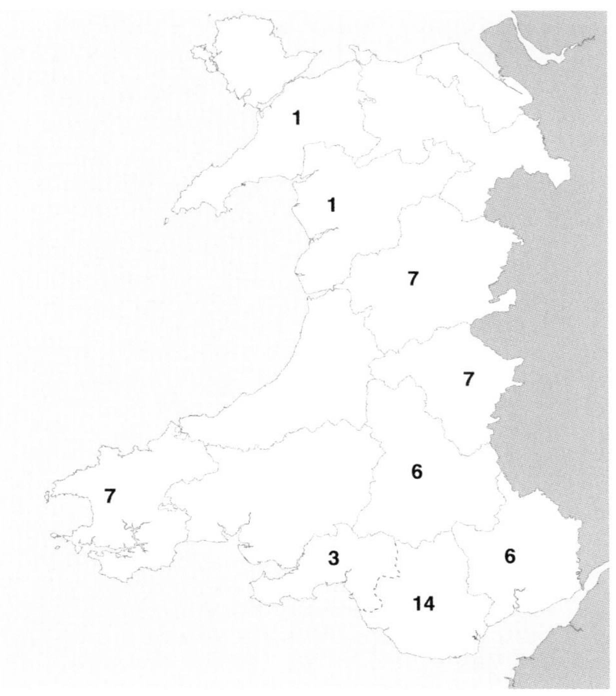

Ferruginous Duck: Numbers of birds from each recording area, 1992–2019.

Total	Pre-1950	1950–59	1960–69	1970–79	1980–89	1990–99	2000–09	2010–19
47	16	5	2	7	9	4	2	2

Ferruginous Duck: Individuals recorded in Wales by decade.

A scheme in northern Germany aimed at restoring Ferruginous Duck to part of its former range, released several hundred captive-bred juveniles at Steinhuder Meer, Lower Saxony, during 2011–16. A ringed bird from this scheme was seen at sites in Radnorshire and Breconshire, including Llan Bwch-llyn Lake and Llangorse Lake between 7 March and 27 April 2018. It was in Breconshire again from 6 October to 7 December 2019, with a visit to Llyn Heilyn, Radnorshire, on 15–16 October. Released birds from this scheme are ringed (most also bear yellow plastic rings) but wild-reared young are not. Future occurrence of Ferruginous Duck in Wales may well depend on this re-establishment scheme, if the eastern European population continues to decline. In the short term at least, this will be a very rare visitor.

Jon Green and Robin Sandham

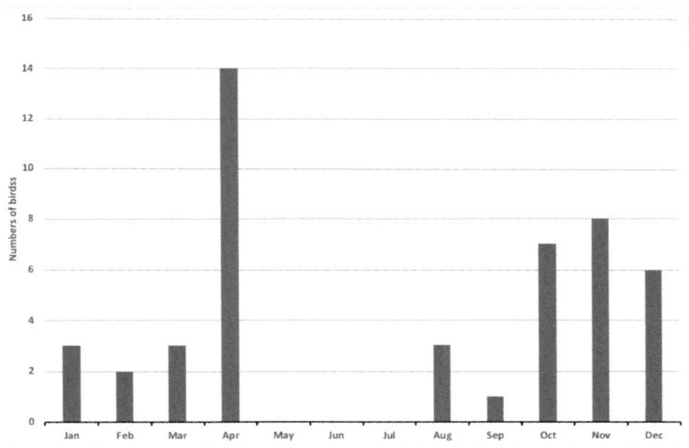

Ferruginous Duck: Totals by month of arrival, 1900–2019.

Ring-necked Duck *Aythya collaris* Hwyaden Dorchog

Welsh List Category	IUCN Red List	Rarity recording
	Global	
A	LC	WBRC

Total	Pre-1950	1950–59	1960–69	1970–79	1980–89	1990–99	2000–09	2010–19
52	0	0	2	2	8	7	19	14

Ring-necked Duck: Individuals recorded in Wales by decade.

This species breeds mostly in subarctic Canada and winters in the USA and Central America (*HBW*). There is a westerly bias to British records, notably in Scotland and southwest England, but it is a rare visitor to Wales. The numbers recorded annually in Britain increased from the 1970s to an annual mean of 32 during 2000–09, but subsequently fell by over 30% in 2010–18 (White and Kehoe 2020a).

The first Welsh record was a male at Bosherston, Pembrokeshire, between 12 February and 8 March 1967. The second was shot at Llangorse Lake, Breconshire, on 26 December 1967. It had been ringed in New Brunswick, Canada, on 7 September that year, 4,505km to the west. It is likely that several records relate to the same individual returning to the same area in successive years, hence there is uncertainty about the precise number of birds involved. For example, a male and female visited southeast Wales and the Bristol area annually during 2000–07 including Newport Wetlands Reserve, Gwent, and in East Glamorgan, Lisvane Reservoir, Kenfig and Cosmeston Lakes; it is possible that more than two birds were involved. Other suspected movements within Wales include a female seen at Llyn Blaenmelindwr and Llyn Rhosrhydd, Ceredigion, in November/December 2009, which was also at Llyn Pencarreg, Carmarthenshire, in February–March 2010 and returned there from October 2010 to January 2011, before moving within the county to Talley at the end of January. A female at Llangorse, Breconshire, on 11 December 2017 moved to Penclacwydd from 12 December 2017 to 6 April 2018. A male at Llyn Caer Euni, Meirionnydd, in October–November 2018 summered at Llyn Brân, Denbighshire, in March–July 2019, before returning to Llyn Caer Euni in October–November and then moving to Llyn Tegid, Meirionnydd, on 17 December 2019.

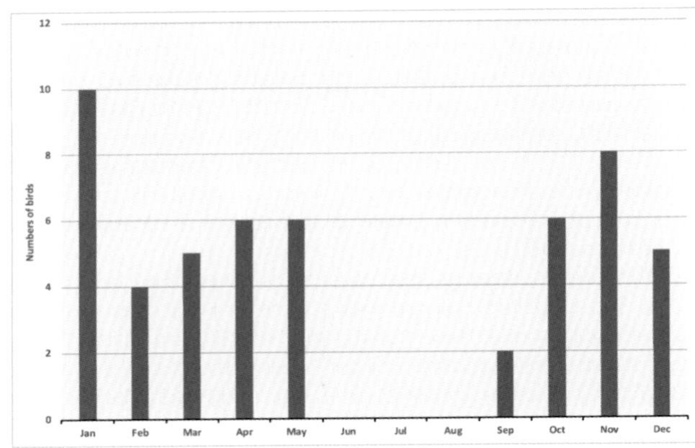

Ring-necked Duck: Totals by month of arrival, 1967–2019.

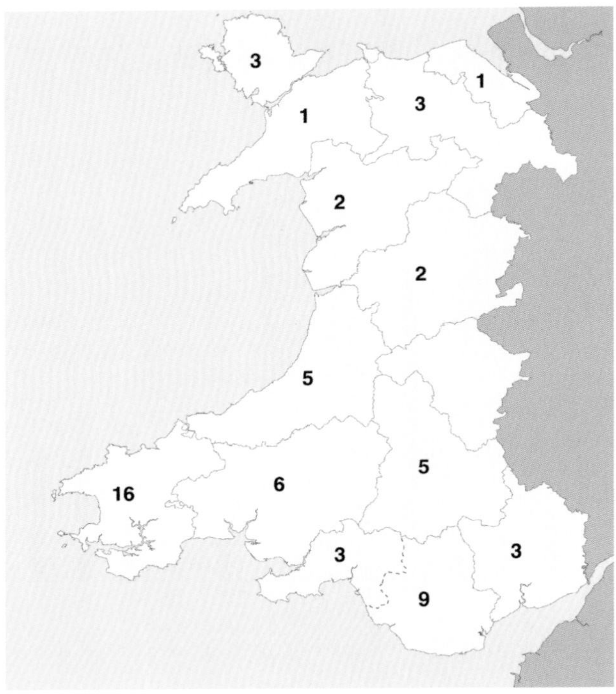

Ring-necked Duck: Numbers of birds from each recording area, 1967–2019. Some birds were seen in more than one recording area.

The majority of records have been males, although at least 22 females have been recorded. Most records have come from the southern half of Wales. Those from Pembrokeshire include a male and three females at Heathfield gravel pits from January to March 2001, Wales' only multiple record, and six different individuals at Bosherston. The eight in East Glamorgan include three different individuals at Kenfig Pool and three at Lisvane Reservoir.

Jon Green and Robin Sandham

Tufted Duck *Aythya fuligula* Hwyaden Gopog

Welsh List Category	IUCN Red List			Birds of Conservation Concern			
	Global	Europe	GB	UK	Wales		
					2002	2010	2016
A	LC	LC	LC	Green			

The Tufted Duck is the most numerous and widespread diving duck in the world, with a population estimated at 2.6–2.9 million individuals (BirdLife International 2020), breeding across most of temperate Eurasia, from Iceland to Kamchatka. The west European population is largely resident, with only short-distance movements, but those that breed farther east are migratory, wintering around the Mediterranean, in Africa, India and China. It breeds on freshwater lakes, reservoirs, ponds and along wide, sluggish rivers, and on islands where these are available. It feeds largely by diving, with molluscs usually a major item in its diet, although it also takes crustaceans, aquatic insects and aquatic plants (*HBW*). The global population is considered to be stable (BirdLife International 2020).

Tufted Ducks prefer calcium-rich and eutrophic waterbodies of over 1ha for breeding (Cabot 2009), so many Welsh lakes are unsuitable. Most breeding lakes have densely vegetated banks that provide protection from predators and make nests difficult to find. They breed later than most duck species, many broods not appearing until July or even early August, so breeding can easily be missed by fieldwork carried out in May and June.

There is little information on the movements of the Welsh breeding population in winter and comparatively few ringing recoveries in Wales from elsewhere. However, a duckling ringed at Rezekne, Latvia, in June 1984 was shot at Brecon in December 1985 and an adult ringed at Cosmeston Lakes, East Glamorgan, in November 2010 was shot on the River Ob near Nyalinskoye, Russia, in May 2015. The Icelandic breeding population winters mainly in Ireland (*BTO Migration Atlas* 2002), but some evidently reach Wales: a duckling ringed at Mývatn in July 1947 was recovered at Porthmadog, Caernarfonshire, in December the same year. Tufted Ducks wintering in northwest Europe exhibit alloheimy of the sexes, whereby females migrate farther south than males. Males outnumber females in Britain by a ratio of 1.4:1 (*BTO Migration Atlas* 2002), though there appear to be no data specifically for Wales. Around 5,000 Tufted Ducks are shot in the UK each year, with no real change over the previous 12 years (Aebischer 2019).

The Tufted Duck was recorded as a winter visitor in Wales as far back as February 1776, when Thomas Pennant listed it among the casualties from a storm-induced 'wreck' on the shore near Criccieth, Caernarfonshire. In 1849, it was first recorded breeding in Britain (*Historical Atlas 1875–1900*) and rapidly expanded in number and range following the introduction of Wild Bird Protection Acts that established a close season for shooting ducks. The Tufted Duck benefited from gravel pits that flooded after mineral extraction. A further surge during the 1950s was driven by the spread of the Zebra Mussel and Jenkins' Spire Snail, both non-native invasive species (Cabot 2009). The Zebra Mussel was introduced to Britain in the 1820s and became an important food item for Tufted Duck in Britain, although it has been recorded only at a limited number of sites in Wales (NBN Atlas undated).

The Tufted Duck only became established as a breeding species in Wales from the end of the 19th century. The first confirmed breeding record seems to be a brood at Llyn Maelog, Anglesey, on 1 August 1892, although Forrest (1907) felt that the species, though by no means plentiful, had escaped notice in many localities in North Wales. Mathew (1894) speculated that wintering birds might sometimes nest in Pembrokeshire but had no evidence. Breeding was confirmed in 1906 on Llyn Tegid, Meirionnydd, where 1–2 pairs hatched young that did not reach maturity (Bolam 1913). There were no further records of breeding in Meirionnydd for over 25 years, but there was a slow increase on Anglesey, where a pair bred annually on Llyn Bodgylched (Forrest 1919) and a brood was seen at Cors Goch in 1925. The first record of breeding in South Wales came in 1922, when two pairs nested and a brood of seven young was seen at Morfa Pools, Gower. Breeding continued there in subsequent years, prior to the habitat being destroyed by construction of Port Talbot steelworks (Hurford and Lansdown 1995). The first breeding record for Breconshire was at Llangorse Lake in 1930, and it was also reported to have bred in Radnorshire before 1939 (Jennings 2014). Breeding was recorded again in Meirionnydd in 1933, when two pairs bred near Llyn Trawsfynydd. In Caernarfonshire, Bruce Campbell reported that the species nested on a small dam above Tremadog a few years before 1946, and there was also a nesting attempt in Llŷn in the 1940s, probably at Ystumllyn (Pritchard 2017). A female and two males at Witchett Pool, Carmarthenshire, in May 1950, were thought to have possibly bred (Ingram and Salmon 1954) and breeding occurred in Montgomeryshire from the 1950s, particularly at Lymore (Holt and Williams 2008). A pair bred at Shotton, Flintshire, in 1959, but breeding was not recorded in Gwent until 1984. In Pembrokeshire, although a wild bird bred with one of captive origin in 1988, the first breeding involving a wild pair was at Marloes Mere in 1996. In Ceredigion, a pair almost certainly bred at Llyn Frongoch in 1995, but proof of breeding came only in 2000, when small young were seen at five sites in the county.

The *Britain and Ireland Atlas 1968–72* showed a patchy breeding distribution in Wales, occupying 21% of 10-km squares, with most records from Anglesey, Gower and eastern areas of mid-Wales, while it was absent from Pembrokeshire and much of

Site	County	74/75 to 78/79	79/80 to 83/84	84/85 to 88/89	89/90 to 93/94	94/95 to 98/99	99/00 to 03/04	04/05 to 08/09	09/10 to 13/14	14/15 to 18/19
Severn Estuary: Welsh side (whole estuary)	Gwent/East Glamorgan	8 (228)	36 (465)	1,079 (1,335)	517 (828)	337 (664)	188 (624)	126 (548)	228 (786)	306 (871)
Llangorse Lake	Breconshire	80	14	29	28	186	215	208	303	314
Llyn Cefni	Anglesey	141	191	155	165	239	283	221	278	286
Cosmeston Lakes CP	East Glamorgan		2	2	20	21	87	206	293	273
RSPB Valley Wetlands inc. Llyn Traffwll	Anglesey	156	99	122	146	195	315	252	303	228
Llyn Alaw	Anglesey	166	75	93	170	509	424	473	402	111

Tufted Duck: Five-year average peak counts during Wetland Bird Surveys between 1974/75 and 2018/19 at sites in Wales averaging 250 or more at least once. International importance threshold: 8,900; Great Britain importance threshold: 1,300.

Carmarthenshire. The *Britain and Ireland Atlas 1988–91* showed an increase in occupation to 30% of squares, with Anglesey still the most important area. A survey of the island in 1980 counted 31 broods and estimated the population to be 40–60 pairs. A repeat survey in 1986 confirmed 37 breeding pairs and recorded a further 32 probable breeding and 11 possible pairs (RSPB 1986). A survey of Llŷn by the RSPB in 1986–87, recorded three probable and two possible breeding pairs. The *Britain and Ireland Atlas 2007–11* showed an extension in breeding range to 39% of 10-km squares, with gains particularly in Gwent, Flintshire and in some areas of the southwest. Gwent is a good example of an increase being assisted by the creation of new wetlands: the *Gwent Atlas 1981–85* confirmed breeding in only one tetrad, but the *Gwent Atlas 1998–2003* confirmed breeding in 17 tetrads, with the county population estimated at 20–30 pairs.

This is mainly a lowland breeder in Wales, as elsewhere in Britain and Ireland. Cabot (2009) stated that it is never found breeding above 400m, but it has been recorded breeding on Llyn Brân on Mynydd Hiraethog, Denbighshire, at an altitude of 437m (*North Wales Atlas 2008–12*). Many pairs in Wales breed on smaller ponds, including those in urban parks, where it may become very tame. The Tufted Duck, like other waterbirds, has benefited from the creation of new wetlands, such as Penclacwydd, Carmarthenshire, and Newport Wetlands Reserve, Gwent, which have had up to 19 and 16 pairs respectively. There were 10–12 pairs at Llyn Coed y Dinas, Montgomeryshire, in 2005 and 23 broods at RSPB Valley Wetlands, Anglesey, in 2014. One or two pairs have bred on Skomer, Pembrokeshire, in some years, though usually with little success. It is suspected that the recent cessation of nesting at Brechfa Pool, Breconshire, is a result of the lake's infestation with the non-native plant, New Zealand Pygmyweed (Andrew King *in. litt.*). As a breeding species it appears to be doing well in Wales, although numbers are not well monitored. Lovegrove *et al.* (1980) estimated the breeding population to be about 80 pairs and *Birds in Wales* (1994) suggested 100–120 pairs. Based on densities during the 2007–11 *Atlas*, Hughes *et al.* (2020) estimated it to be 765–1,350 pairs in 2016, but while it has increased, even the lower estimate may be on the high side.

The Tufted Duck is more widespread and abundant outside the breeding season. Eyton (1838) described it as being mainly coastal in North Wales in winter and while it is still sometimes to be found on saltwater in estuaries or bays, particularly in cold weather, it generally winters on lakes and slow-flowing rivers. Numbers in the early part of the 20th century were low, but by the 1930s flocks of several hundred wintered regularly at Kenfig Pool, East Glamorgan, though numbers there fell to *c.*20 by 1966 (Hurford and Lansdown 1995). There was a large increase on Anglesey, where there were over 400 on Llyn Coron in November 1953 and 500 on Llyn Cefni in January 1957. Hundreds wintered on reservoirs in Glamorgan in the 1960s and 1970s, including *c.*450 at Llanishen in the cold winter of 1962/63 and 480 in January 1977. Flocks were recorded moving between Kenfig, East Glamorgan, and Eglwys Nunydd Reservoir, Gower, where there were 430 immediately after flooding in January 1966. Kenfig Pool held counts as high as 265 in October 1991 but numbers have been much lower in recent years. Other large counts include 1,336 on the Welsh side of the Severn Estuary in February 1986, 1,323 there in February 1987, and 900 in Cardiff Bay, East Glamorgan, in January 2010. Llyn Alaw also holds good numbers, particularly between August and October: 748 were there in September 1994, 885 in August 2004 and 856 in October 2013. Numbers have slowly increased at Llangorse Lake over recent decades, with 506 in November 2019, a record count for Breconshire.

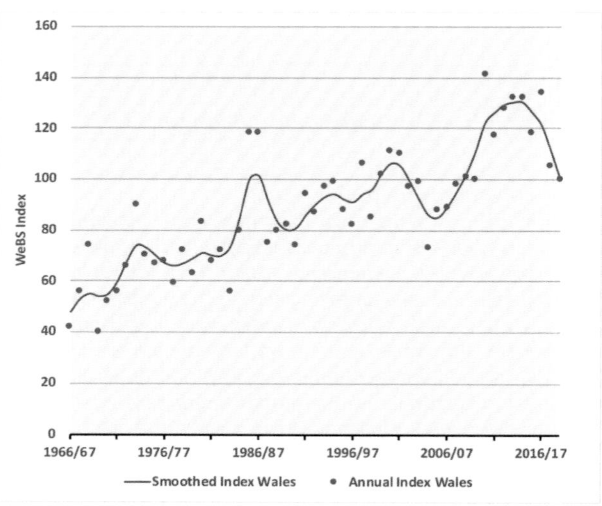

Tufted Duck: Wetland Bird Survey indices, 1966/67 to 2018/19.

Wintering numbers in Wales showed a steady increase from the 1960s to 2014/15, but declined steeply in the second half of the last decade. It is a legal quarry species for wildfowlers; around 5,000 are shot annually in the UK (Aebischer 2019), less than 4% of the winter population. Although it appears to have a secure future here as a breeding bird, there is strong evidence that those nesting in northern and eastern Europe are short-stopping closer to their breeding areas (Lehikoinen *et al.* 2013), so we may expect to see fewer wintering in Wales. Whilst the Tufted Duck does not face any immediate threat, there are many gaps in our knowledge of the breeding population here, and a dedicated survey could provide a more accurate figure for the Welsh breeding population.

Rhion Pritchard

Scaup *Aythya marila* Hwyaden Benddu

Welsh List Category	IUCN Red List			Birds of Conservation Concern			
	Global	Europe	GB	UK	Wales		
A	LC	VU	EN	Red	2002	2010	2016

The Scaup breeds on small, shallow lakes and tundra pools across northern North America and Eurasia, with an almost circumpolar distribution. Except in the very far east, the Eurasian population of the nominate subspecies breeds mainly on the Russian tundra, but there is also a substantial population in Iceland (*HBW*). The majority of European breeders winter in estuaries and sheltered bays in northwest Europe, but some from the northeast winter in the Black and Caspian Seas (*European Atlas* 2020). Around 65% of the European wintering population is found around the coasts of Denmark, Germany and The Netherlands, though the proportion there has reduced significantly since the 1990s and it is thought to be decreasing both globally and within Europe (BirdLife International 2020). It feeds by diving, with molluscs an important food item (*HBW*).

The sole ringing recovery involving Wales shows only local movement during the same winter. Most birds wintering in Welsh waters are presumed to be from the Icelandic population, since the majority of recoveries elsewhere in Britain and Ireland originate there, with a smaller number of birds from farther east, such as Finland and Russia.

Scaup occasionally breed in Scotland, though they may have formerly bred more regularly on the Outer Hebrides and Orkney (*The Birds of Scotland* 2007). There is just one breeding record from Wales: a pair reared five young on Llyn Traffwll, Anglesey, in 1988. Others, including pairs, have summered in potentially suitable breeding habitat, but there has never been substantive indication of breeding. A male Scaup paired with a female Tufted Duck on Llangorse Lake, Breconshire, in 1995 and produced hybrid young.

Eyton (1838) described Scaup in North Wales as "not nearly so common as the other species of the genus, and never killed inland". Forrest (1907) said that it was "not at all uncommon" on saltwater in North Wales, numerous off Point of Ayr, Flintshire, and occasionally found inland. In Montgomeryshire there was a small flock on the Afon Vyrnwy, near its confluence with the River Severn in January 1895. F.T. Fielden was said to have regularly seen small flocks of up to six on the Dyfi Estuary, Ceredigion/Meirionnydd, during October in the 1880s and 1890s, with a few in winter (Roderick and Davis 2010). Mathew (1894) said that it was found all winter in large flocks off the Pembrokeshire coast but warned that it was hardly worth powder and shot for the table, as the flesh was strong and rank. Barker (1905) included it on his list of birds seen in Carmarthenshire, but knew of no specific record and, while it was considered to be the most common *Aythya* species in Glamorgan in the 1890s, Hurford and Lansdown (1995) thought that it was unlikely to have been particularly numerous.

There are few Scaup records from the early part of the 20th century, although Ingram and Salmon (1939) said that it was a regular visitor to the Gwent coast, sometimes in considerable numbers, and Bruce Campbell mentioned it as being present in strength at Ystumllyn, Caernarfonshire, in January 1947, presumably on the sea in Tremadog Bay (North *et al.* 1949). Williams (1977) wrote that the species had occurred in small numbers in Liverpool Bay and the outer Dee Estuary for most of the 20th century, occasionally in flocks of 300–400, and described an influx of 2,000–5,000 on the English side of the estuary in 1949 and 1950. Numbers returned to normal after this date and declined sharply after 1972.

The largest counts from the early 1970s to the mid-1990s were made between Llanddulas, Denbighshire, and Point of Ayr, Flintshire. There were 196 in this area in December 1973, rising to 280 in January 1974 and 400 in February 1986. In winter 1995/96 there were 210 in the Dee Estuary, 271 in the Clwyd Estuary and a Welsh record count of 784 in three flocks between Pensarn and Kinmel Bay, Denbighshire, on 4 February 1996. However, numbers declined by the late 1990s.

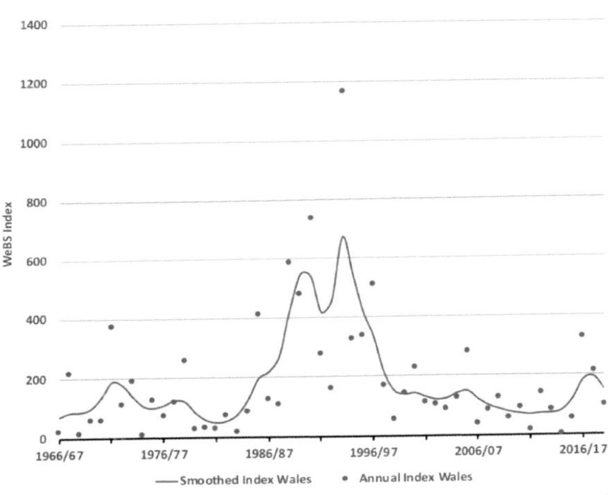

Scaup: Wetland Bird Survey indices, 1966/67 to 2018/19.

Good numbers were also recorded in Cardigan Bay in the late 1980s and early 1990s, with a mobile flock that numbered 90 between Black Rock, Caernarfonshire, and Morfa Harlech, Meirionnydd, on 5 February 1989. The highest count was 160 on the Glaslyn/Dwyryd estuaries, Meirionnydd/Caernarfonshire, in January 1997. Numbers were smaller farther south in Cardigan Bay, with 36 off Ynys-las in January 1970 being the highest count in Ceredigion and 88 off Ragwen Point in February 1991, a county record for Carmarthenshire. Farther east, numbers in the Burry Inlet increased during the 1980s, with peaks of 95 on the Gower side in January 1992 and 106 in winter 1996/97. In the early 1980s a wintering population developed off the Rhymney Estuary, East Glamorgan, peaking at 72 in March 1989. In some years good numbers were also recorded farther up the Severn Estuary, mainly off Peterstone, Gwent, generally peaking in February and March. Numbers here reached 76 in winter 1985/86 and there were 96 on the Welsh side of the Severn Estuary in March 1991.

Numbers probably peaked in early 1996, with *c.*1,245 Scaup off the Welsh coast during January to March. This occurred when numbers off Scotland had already declined sharply, but such a downturn was soon evident in Wales too and there were few large counts after 1996/97. In the 21st century, there have been only four counts in excess of 40 from two sites: Kinmel Bay (60 in January 2001 and 40 in January 2004) and the Burry Inlet (56 in March 2006 and 40 in February–March 2017. A total of 97 in Wales in February 2017 was the highest for many years, during a winter that had seen up to 19 on the Inland Sea, Anglesey, in December 2016. However, the average in the last 20 years has been closer to 50 birds, with as few as eight in 2015.

While seen primarily offshore, the Scaup is a fairly regular visitor to some of the larger lakes and reservoirs. Counts of *c.*50 at Witchett Pool, Carmarthenshire, in January 1974 (Morgan 1984) and 36 in February 1981 were exceptional, but up to eight were annual on Eglwys Nunydd Reservoir, Gower, and Kenfig Pool, East Glamorgan, in the 1980s and early 1990s (Hurford and Lansdown 1995). High counts elsewhere were 12 at Wentwood Reservoir, Gwent, in November 1971; 12 at Eglwys Nunydd Reservoir in February 2009; ten at Cosmeston Lakes, East Glamorgan, in March 2005; and ten on Llyn Trawsfynydd, Meirionnydd, in March 2011.

Scaup is an Amber-listed bird of conservation concern in Wales, owing to its small wintering population (Johnstone and Bladwell 2016). The 54% long-term decline has also triggered a WeBS High Alert for Wales (Woodward *et al.* 2019). Studies in Iceland and Finland have shown considerable fluctuations in breeding success between years (*European Atlas* 1997). It is vulnerable to marine oil pollution and entanglement in fishing gear, which caused annual mortality in the Baltic Sea estimated at 5–10% (Jensen 2009). Around Britain it was formerly found in large numbers near coastal sewage outlets, which for the most part no longer exist as regulations have improved water quality, although Venables *et al.* (2008) noted that the decline in numbers off Peterstone, Gwent, pre-dated the cessation of sewage discharge there. A tiny proportion of the global population winters around Wales, so conservation measures here will have limited impact on the species' global fortunes, but marine protection measures for a range of inshore waterbirds will provide local benefits.

Rhion Pritchard

Lesser Scaup *Aythya affinis* Hwyaden Benddu Fechan

Welsh List Category	IUCN Red List	Rarity recording
	Global	
A	LC	WBRC

Lesser Scaup on Cosmeston Lakes, East Glamorgan, February 2016 © Richard Smith

This Nearctic species breeds from central Alaska through eastern Canada to Hudson Bay and south to Wyoming in the USA. It winters along both coastlines of North America, as well as Central America and the Caribbean (*HBW*). The first British record was in 1987, after which sightings increased so markedly that from 2015 it was no longer considered a rarity, although the average number seen each year in Britain during 2010–18 was only six (White and Kehoe 2020a). However, it remains a rare visitor to Wales, with only eight records:

- male at Penclacwydd, Carmarthenshire, on 4 June 2003;
- 1CY male at Cosmeston Lakes, East Glamorgan, from 26 December 2008; it commuted between there and Cardiff Bay Wetlands until 8 May 2009, returning to the same area annually for the next eight winters until the last sighting on 17 March 2016;
- 1CY female at Eglwys Nunydd Reservoir, Gower, from 1 February to 1 April 2010, which returned the following winter from 16 February to 6 March 2011;
- 1CY male at Cop Hole near Shotton, Flintshire, from 31 January to 4 February 2013;
- male at Bryn Bach country park, Gwent, from 27 March to 27 April 2013;
- subadult male at Llangorse Lake, Breconshire, from 7 October 2014 to 25 February 2015, which had been colour marked with a nasal saddle marked 'VH', at São Jacinto, Portugal, on 20 December 2013, where it remained until 3 February 2014; after visiting Wales it was recorded at Wintersett Reservoir, South Yorkshire, on 4–16 May 2015 and RSPB Loch Leven, Perth & Kinross, on 30–31 July 2015, before being illegally shot at Lough Neagh, Northern Ireland, on 10 September 2016;
- female at Eglwys Nunydd Reservoir, Gower, on 1–3 November 2015 and
- 1CY at Llyn Llygeirian, Anglesey, from 20 October to 31 December 2018.

Jon Green and Robin Sandham

King Eider *Somateria spectabilis* Hwyaden Fwythblu'r Gogledd

Welsh List Category	IUCN Red List		Rarity recording
	Global	Europe	
A	LC	LC	BBRC

The King Eider has a circumpolar breeding distribution, with the closest nesting areas to Wales being in Greenland and northwest Russia. The European population winters along ice-free coasts of the White Sea, northern Norway and Iceland (*HBW*). There were 241 recorded in Britain to 2019, with 4–5 new birds each year, almost exclusively in Scotland (Holt *et al.* 2020). It is an extremely rare visitor to Wales with only two records, both females.

The first was at Black Rock, Caernarfonshire, from 28 January to 5 February 1989, which later moved south as far as Aberdysynni, Meirionnydd, where it summered from 29 May to 8 October 1989 and was seen again from 28 April to 8 May 1990. The second was initially seen off Aberaeron, Ceredigion, on 26–29 June 2017, then farther up the Ceredigion coast at Ynyslas, between 4 July and 22 September. It also spent time on the Meirionnydd side of the estuary at Aberdysynni on 9 July and 26–27 August. It was back at Ynyslas on 9 November 2017 and remained around the Dyfi and Dysynni estuaries for the next 18 months, except for a visit north to the Dwyfor Estuary, Caernarfonshire on 9 June 2018. The last report, with 50 Eiders, was at Aberdysynni on 5 May 2019.

Jon Green and Robin Sandham

Eider *Somateria mollissima* Hwyaden Fwythblu

Welsh List Category	IUCN Red List			Birds of Conservation Concern			
	Global	Europe	GB	UK	Wales		
A	NT	VU	VU	Amber	2002	2010	2016

The Eider has a circumpolar breeding distribution, from well within the Arctic Circle up to 80° North and as far south, in North America, as New England (*HBW*). A southward extension of the breeding range has occurred throughout Europe, including from the Scottish population from the 19th century onwards (*The Birds of Scotland* 2007), which may be a result of improved protection at breeding colonies and rising sea temperatures increasing the abundance of food. It is Europe's most numerous and widespread seaduck, with the highest concentrations in northwest Iceland, where management by farmers, who harvest their downy feathers, led to high breeding densities (*European Atlas* 2020). However, breeding numbers in the large Baltic–Wadden Sea flyway population halved during 2000–09 (Ekroos *et al.* 2012). Eiders winter in the southern part of their breeding range, usually in shallow waters where they can dive for food, mainly shellfish and small crustaceans; the Blue Mussel is a particularly important food item. There is some evidence of passage migration in Wales in spring and autumn, with some parties passing coastal watchpoints, but movements in Wales are not well understood.

The Eider was a rare winter visitor to the Welsh coast in the late 19th century, though local fishermen remembered them in the Burry Inlet, Gower/Carmarthenshire, at least as far back as 1870

(Hurford and Lansdown 1995). In Pembrokeshire, Mathew (1894) had seen adults that had been shot at Stackpole or on the coast between Stackpole and Tenby and mentioned an immature male that had been shot near Pembroke, a young male shot at Dale in January 1891 and that Lord Cawdor's collection contained an adult male thought to have been killed in Carmarthenshire. Forrest (1907) was aware of only one record from North Wales, a female shot near Holyhead, Anglesey, about 1884–85.

Numbers remained low in the first half of the 20th century, but from about 1950 there was a marked increase. Burry Inlet, Gower/Carmarthenshire, is the most important site for wintering Eider in Wales. Numbers reached 54 by 1958, 81 in 1964 and 241 in 1978/79. By the 1980s there was a mean winter peak of 140 birds there and a maximum count of 313 was recorded on 27 April 1990 (Hurford and Lansdown 1995). Numbers have since declined sharply in the Burry Inlet and the five-year mean WeBS count during 2013/14–2018/19 was 66, although 207 off Whiteford, Gower, in February 2019 was the highest count since November 1996. The Meirionnydd coast from Aberdysynni to Llangelynnin also emerged as an important area for non-breeding birds from 1973, with counts in excess of 150 in several years, the highest being 181 off Aberdysynni on 16 November 2018. The only count over 100 elsewhere in Wales was 150 in Tremadog Bay, Caernarfonshire, in May 2002. The WeBS index for Wales showed a rapid decline in the early 1990s, leading to a WeBS High Alert being issued following a 63% long-term and 61% short-term decline (Woodward *et al.* 2019).

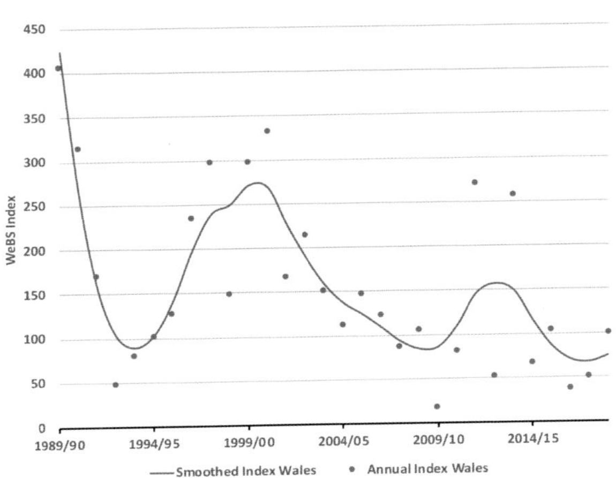

Eider: Wetland Bird Survey indices, 1989/90 to 2018/19.

Inland records are scarce: the only record during 2000–19 was a male and a female on Llyn Tegid, Meirionnydd, on 20 January 2008. A few records precede this in Breconshire, but Eider has never been recorded in Montgomeryshire or Radnorshire.

It is a localised breeding species in Wales. The Eider was suspected of breeding on Caldey, Pembrokeshire, in the early 1980s, and a group of flightless birds in the St Govan's Head/ Barafundle area of that county in July 1983 was thought to be a female with a brood of large young. The first confirmed breeding record in Wales came in May 1997, when a nest with five eggs was found on Puffin Island, Anglesey; a female with chicks seen on the nearby Caernarfonshire coast a few weeks later were thought probably to have been from Puffin Island. A pair probably attempted to breed on Puffin Island in 1998 and the continued presence of Eiders off the east coast suggests annual attempts both there and on the Anglesey mainland. Since 2000, it has bred regularly near Llandygai, Caernarfonshire, where the highest count was in 2014, when at least five females produced young.

The Eider also breeds in Meirionnydd: two nests were found below cliffs south of Llangelynnin in 1998, but the first successful breeding was not confirmed until 2007, when a female with a brood of five was seen at Tonfanau on 8 June. This coast is now the southernmost regular breeding location in Britain and Ireland (*Britain and Ireland Atlas 2007–11*). The *North Wales Atlas 2008–12* confirmed breeding at these three sites—Puffin Island, Llandygai and southern Meirionnydd—and possible breeding at several sites along Llŷn, Caernarfonshire, and around the Mawddach Estuary, Meirionnydd. The total breeding population is probably no more than 20 pairs.

The wintering population has declined across the UK, although to a lesser extent than in Wales, thought to be a combination of smaller European breeding numbers and short-stopping closer to the breeding grounds; numbers wintering in Denmark and Sweden have increased since 2000 (Ekroos *et al.* 2012). The species is Amber-listed in Wales because it is a rare breeder and owing to its Vulnerable status in Europe (Johnstone and Bladwell 2016). The availability of food in fairly shallow water is probably the most important factor governing Eider numbers. In Scotland, increasing development of commercial mussel-farming attracts Eider and brings them into conflict with the operators, leading to licensed shooting in some areas (*The Birds of Scotland* 2007). It seems likely that the number of wintering Eiders in Wales will continue to decline, perhaps to become a scarce visitor. The small Welsh breeding population seems to be holding its own, as is the adjacent Isle of Man breeding population, which has grown from a single pair in 1992, to several hundred birds (Moore 2017). However, long-term climate projections suggest that its breeding range in Britain will contract to northwest Scotland by the end of this century (Huntley *et al.* 2007).

Rhion Pritchard

Sponsored by Sir Richard Williams-Bulkeley Bt

Surf Scoter *Melanitta perspicillata* Môr-hwyaden yr Ewyn

Welsh List Category	IUCN Red List Global	Rarity recording
A	LC	WBRC

The species breeds across Alaska and Canada and winters on both coasts of North America as far south as the USA's Gulf Coast (*HBW*). Records in Britain increased from the 1980s, doubtless a result of improved telescopes and greater interest in winter seawatching, with a mean annual total of 15 during 1980–2018 (White and Kehoe 2020a). It is a scarce visitor to Wales, often occurring with wintering flocks of Common Scoter.

The first Welsh record was a male found dead on the beach at Ginst, Carmarthenshire, on 15 January 1971. It occurs most frequently off the coasts of Carmarthenshire, Pembrokeshire and Denbighshire, generally between November and March. The largest counts in each main wintering area are:

- two off Rhossili, Gower, in March 1991;
- three in Carmarthen Bay off Pendine, Carmarthenshire, and Amroth, Pembrokeshire, in November 1994;
- up to three off Harlech, Meirionnydd, in December 1988 and January 1989, and
- up to seven males and a female between Abergele and Llanddulas, Denbighshire, during winter 2014/15.

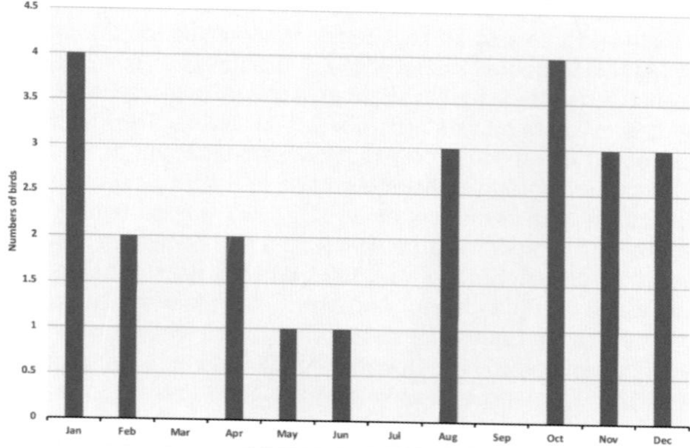

Surf Scoter: Totals by month of arrival, 1971–2019.

Aside from those recorded in wintering areas, seven individuals have passed Strumble Head, Pembrokeshire and it occurs occasionally on inland freshwater, such as one at Eglwys Nunydd Reservoir, Gower, and Kenfig Pool, East Glamorgan, in October/ November 1981. A male on Lake Vyrnwy, Montgomeryshire, on 2 July 2013, is presumed to have been moving overland with migrating Common Scoters. While it has occurred annually off the Welsh coast in recent decades, since many are presumed to be returning individuals, the species may become a true rarity once the current generation has perished.

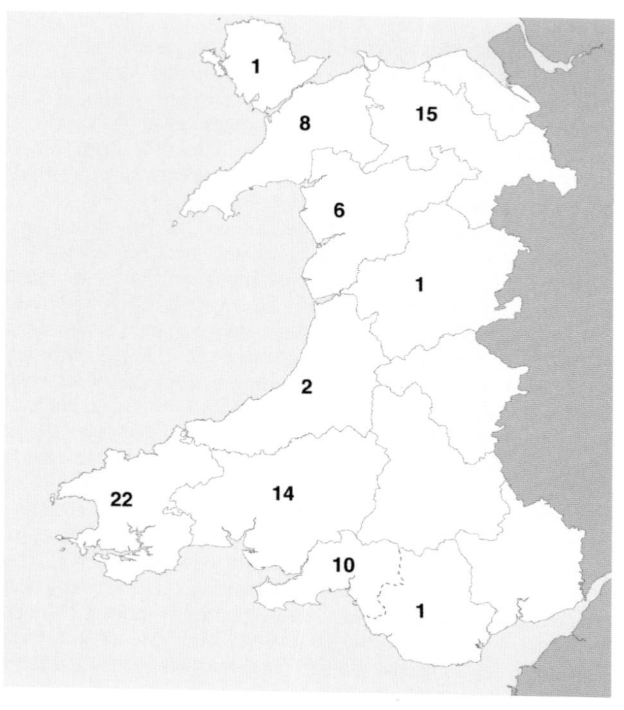

Surf Scoter: Numbers of birds in each recording area, 1971–2019. Some birds were seen in more than one recording area.

Total	Pre-1950	1950–59	1960–69	1970–79	1980–89	1990–99	2000–09	2010–19
74	0	0	0	4	16	15	15	24

Surf Scoter: Individuals recorded in Wales by decade.

Jon Green and Robin Sandham

Velvet Scoter *Melanitta fusca* Môr-hwyaden y Gogledd

Welsh List Category	IUCN Red List			Birds of Conservation Concern			
	Global	Europe	GB	UK	Wales		
					2002	2010	2016
A	VU	VU	VU	Red			

The Velvet Scoter breeds in the far north of Europe and western Siberia, and winters to the south from the Atlantic coast to the Caspian Sea (*HBW*). The largest breeding populations outside Russia are in Sweden and Finland (BirdLife International 2015). The world population has declined by up to 49% over 23 years (three generations), although the rate of decline has apparently slowed (BirdLife International 2020). Large numbers winter in the Baltic Sea, where surveys in 2007–09 estimated the population at *c*.373,000 individuals, down from *c*.933,000 in 1992–93 (Skov *et al.* 2011). The most recent estimate of numbers wintering around Britain is 3,400 (Frost *et al.* 2019a), the majority of which are in Scottish waters. Only small numbers are recorded in Wales. Like Common Scoter, with which it usually associates off the coast in winter, it feeds by diving for crustaceans and molluscs. Groups of Velvet Scoter may form within a larger flock, but unless seen in flight, individuals can be difficult to pick out from Common Scoters at a distance, so it may be under-recorded.

Historic records are few, which almost certainly relates to the absence of suitable optical equipment. In Pembrokeshire, Mathew (1894) mentioned that six were in Goodwick Bay on 16 November 1886, and another was picked up exhausted at Tenby in December 1889. Forrest (1907) regarded it as a rare winter visitor to North Wales. He recorded singles shot in the Penrhyndeudraeth area, Meirionnydd, in 1892, 1896 and 1897, and in the Conwy Estuary near Glan Conwy, Denbighshire, in 1899 and near Deganwy, Caernarfonshire, in 1901. Others were reported in the Dyfi Estuary and Barmouth, Meirionnydd, and near Porthmadog, Caernarfonshire. There was also an unusual inland sighting on Llyn Tegid, Meirionnydd, on 3 November 1906 (Bolam 1913). It was said to be a regular winter visitor to Glamorgan in the late 19th century (Hurford and Lansdown 1995), but there were no records in Gwent during that period.

It remained scarce during the first half of the 20th century. On the north coast, five were off the Great Orme, Caernarfonshire, on 4 August 1919, six in the Conwy Estuary later the same year, and 13 there in April 1929. A count of 19 off Llanddulas, Denbighshire, in late 1939 was exceptional for the period, while an undated record of ten off Aberystwyth, Ceredigion, was probably in the late 1940s (Roderick and Davis 2010). The first record for Flintshire was in the Dee Estuary in 1921 and the species was said to have been recorded in the county in most years until the 1970s (*Birds in Wales* 1994). The first record for Gwent was a female shot at Peterstone in December 1938, but the next county record was not until 1969. A few had been reported in Carmarthenshire at the turn of the 20th century, but Ingram and Salmon (1954) knew of no more recent records. The first record for Anglesey was not until 1951 (Jones and Whalley 2004).

The number of records off the Welsh coast increased from the 1960s. Small numbers were reported with regularity off Llanfairfechan, Caernarfonshire, including 16 on 19 November 1959 and 21 in 1963 (*Birds in Wales* 1994) and 19 were off Rhos Point, Caernarfonshire, on 20 February 1977. Anglesey also recorded elevated numbers, including 31 at Red Wharf Bay on 22 December 1976. Exceptional numbers were off Penmaenmawr, Caernarfonshire, in winter 1991/92, with 70 there in December 1991, 76 in January 1992 and a Welsh record 79 in April 1992. Up to eight were recorded there in every month of 1992 except November. Such numbers have never been matched.

In Cardigan Bay, 14 were off Barmouth in December 1991 and higher numbers were recorded from the late 1990s, including 24 off Morfa Harlech, Meirionnydd, on 30 March 1997, 11 south past Bardsey, Caernarfonshire, on 24 October 2000 and 22 off Morfa Bychan, Caernarfonshire, in March 2002. Numbers in Tremadog Bay have been much lower since 2005, the highest count being five off Criccieth in December 2012. Farther south, *c.*20 were in Borth Bay, Ceredigion, in February–March 1986, and 11–14 were there in November–December 2002. In Pembrokeshire, up to eight were seen most winters in St Brides Bay and off Saundersfoot-Amroth, with 12 off Newgale on 30 December 1988 and 14 passed Strumble Head on 21 October 1991 (Donovan and Rees 1994). Off the south coast, Velvet Scoter was recorded in Carmarthenshire every year during 1987–2007, including 12 off Telpyn Point on 15 February 2004. An estimated 167 were seen in Glamorgan between 1960 and 1992, the majority off Gower, including 17 at Rhossili on 4 January 1970 and 14 in Swansea Bay on 29 April 1979 (Hurford and Lansdown 1995). The species is scarce in Gwent, with just 12 records and no count higher than four.

The species has become more difficult to find since about 2010. The last double-figure count in Tremadog Bay, for example, was ten off Morfa Harlech in January 2012. The exception was the Denbighshire coast between Old Colwyn and Pensarn, where Velvet Scoters associated with large flocks of Common Scoter. There were 30 on 17 February 2010, 25 on 11 March 2013 and at least 18 on 7 February 2015. Counts were lower subsequently, but numbers off the Welsh coast are too small for a trend to be evident.

Inland records are rare, but it has been recorded, usually singly, on lakes in Gwent, Glamorgan, Pembrokeshire, Meirionnydd, Anglesey and Flintshire. The three inland counties, Breconshire, Radnorshire and Montgomeryshire, have yet to record this species. The global decline is not fully understood, but climate change and ocean acidification are predicted to have an impact on their food supply, and it is susceptible to marine oil pollution and entanglement in fishing gear. Declines in Velvet Scoters wintering in the Baltic Sea have coincided with high mortality in gillnet fisheries, in which it is one of the most frequent victims (Dagys 2017). Its Vulnerable status in Europe, as well as its small population size in Wales, led to its Amber-listing in Wales (Johnstone and Bladwell 2016).

Rhion Pritchard

Common Scoter — *Melanitta nigra* — Môr-hwyaden Ddu

Welsh List Category	IUCN Red List			Birds of Conservation Concern			
	Global	Europe	GB	UK	Wales		
A	LC	LC	CR(B)	Red	2002	2010	2016

The Common Scoter breeds from western Greenland and across northern Eurasia as far east as the Lena River in eastern Siberia. It winters in shallow waters, 5–15m deep (Cabot 2009), to the south of its breeding range, some reaching as far south as North Africa (*HBW*). The most southerly breeding populations in the world are in northern Scotland and western Ireland, but both are thought to be declining. An estimated 52 pairs bred in Scotland in 2007, almost half the number found in 1995 (95 pairs) (Underhill *et al.* 1998). There has never been a record of breeding in Wales.

In winter, scoters are concentrated in areas with soft inshore sediments that support a rich and diverse benthic community of short-lived species on which bivalve molluscs feed. The scoters dive for molluscs such as Blue Mussel and Razor Clam and take other invertebrates (Fox 2003). Off North Wales, Common Scoters have occurred when there have been high densities of the bivalves *Lutraria angustior* and *Spisula subtruncata.* It is speculated that they also feed on the Atlantic Jack-knife Clam (*Ensis leei*), an invasive non-native razor clam from the east coast of North America. It occurs in Liverpool Bay and, unlike native razor clams, spends more time partly emerged from the sea bed (Ivor Rees pers. comm.).

The Baltic Sea is of particular importance, holding up to half the total wintering in northwest Europe, estimated at 1.6 million birds (Delany and Scott 2006), but numbers declined by 48% between 1988–93 and 2007–09 (Skov *et al.* 2011). However, it is thought that part of the wintering population has shifted from the Baltic to the North Sea (M. Ellermaa quoted in BirdLife International 2020), so the global population may be unaffected. The departure of birds from Liverpool Bay and Carmarthen Bay in early April, tracked audibly across England, indicates that many of those wintering off the Welsh coast are likely to breed in Russia and Fennoscandia. There is also some evidence that numbers in Liverpool Bay increase in late winter, as birds that wintered in the North Sea use it as a staging post on their return to Iceland. Birds are seen in every month of the year around the Welsh coast, with some large counts even in May and June. These may be immature birds or non-breeding adults, and increased counts in July are likely to signal the arrival of moulting males. Numbers increase in October and decrease again in March.

The comparatively few historic records of this species reflect the lack of suitable optical equipment. In Caernarfonshire, Scoters are mentioned by Pennant (1810) as among the birds killed by a gale and intense frost in February 1776, when bodies were strewn along the coast at Criccieth. Williams (1835) commented that the species could always be seen on the sea between Penmaenmawr and the Great Orme on the north coast. Most late-19th- and early-20th-century authors mentioned it as a common or fairly common winter visitor, but none quantified their descriptions. Common Scoters can be found a long way offshore in places where the water is shallow. Better optics have made it possible to find and identify birds at a greater distance from shore, but only aerial or boat-based surveys provide a true indication of numbers at most sites. The All-Wales Common Scoter Survey found sizeable flocks up to 20km offshore, well beyond the range of shore-based counts, during 2001/02–2003/04. For this reason, and because weather conditions can make birds even fairly close to shore undetectable, Wetland Bird Survey core counts on a pre-arranged date are not a particularly good method of monitoring this species. Numbers are most effectively monitored by aerial surveys, some of which relate to the leasing of the seabed by The Crown Estate for offshore wind generation.

There are three main wintering areas off the coast of Wales: Carmarthen Bay, the northern part of Cardigan Bay and Liverpool Bay, the latter shared with England. Birds apparently tend to remain in the same general area all winter, but can be very mobile within each wintering area.

Carmarthen Bay has long been recognised as an important wintering site for Common Scoter. Reports before the First World War were sporadic, but Bertram Lloyd was the first to notice large numbers there in the 1920s and 1930s, including a flock of at least 1,000 at the mouth of the Burry Inlet on 14 July 1932. In early 1973, the Royal Air Force recorded 8,000 from a stationary boat four miles off Pendine, with an estimated 12,000 more on the radar (Sutcliffe 1975). A boat count in March 1974 estimated 25,000 birds in the bay (Sutcliffe 1975) and there were land-based counts from Cefn Sidan of 16,000 on 3 August and 10,000–12,000 on 14 September 1974. Sample counts showed that 80% were males. Numbers fell in the following years: aerial surveys during 1976–78 recorded a maximum of 10,000 and in 1991 produced estimates of 7,700 in the bay in January and 2,150 in July. Around 11,000 were off Cefn Sidan in August 1992. Around 300 were found stranded on the Pembrokeshire shoreline of the bay following oil spills in 1950 and 1973/74 (Donovan and Rees 1994), but the release of 72,000 tonnes of light crude oil by the *Sea Empress* at the entrance to Milford Haven Sound over seven days in February 1996 caused much greater mortality. About 7,000 oiled birds of all species were collected onshore and an unknown number died at sea. Common Scoter was the most abundant casualty: 4,571, of which 1,818 were dead and 2,753 were picked up alive. Carmarthen Bay contained about 8,000 Scoters when oil arrived there on 22 February, so the majority of those present were contaminated and either washed ashore or died at sea. Further migrants subsequently used the Bay that winter, up to a peak of 15,000, but its suitability as a feeding area was compromised as some bivalve populations on which Common Scoters normally fed were severely reduced and others were probably toxic when ingested (Edwards and White 1999).

Numbers were much reduced the following year and distributed in areas that were potentially suboptimal for foraging. However, counts increased to pre-spill levels within three winters and mapping suggested a return to previously contaminated feeding areas (Banks *et al.* 2008). A ground-based count in February 2005 found 24,460 and an aerial survey in February 2010 produced a record total of 42,515, which was considered likely to include cold-weather movements from elsewhere. Further aerial surveys produced estimates of 32,173 in February 2013 and 36,314 in February 2017 (APEM 2017). Most occur in an area *c.*5km wide running from the northwest of the bay off Amroth and Saundersfoot, Pembrokeshire, to the east side off Pembrey Sands, Carmarthenshire, where the depth is 10m or less (Natural Resources Wales 2018). Carmarthen Bay Special Protection Area (SPA) was designated in 2003 as the first fully marine SPA in the UK, specifically for the protection of Common Scoter (Banks *et al.* 2007a).

One by-product of the *Sea Empress* clean-up was the opportunity to gather data about the effectiveness of rehabilitating and releasing Common Scoters from whose plumage the crude oil could be removed. Of those ringed prior to release, some were subsequently recovered in Pembrokeshire, and others in Cardigan Bay, off North Wales and along the south coast of Devon and Dorset. Several more were recorded on the migration flyway, in The Netherlands and Russia, including two in their breeding area on the River Ob, western Siberia, in 2007 and 2009. The latter bird had been released at Borth, Ceredigion, in March 1996 and holds the British and Irish longevity record for a ringed bird: it was shot 13 years, 2 months and 27 days after being ringed as an adult (Robinson *et al.* 2020).

For many years, Carmarthen Bay was thought to be the most important site in Britain for wintering Common Scoter. However, aerial surveys have established that even larger numbers are found in Liverpool Bay, though only part of the wintering population is in Welsh waters, between the east coast of Anglesey and the Dee Estuary. T.A. Coward recorded a long line, numbering several thousand, flying east from Colwyn Bay towards Abergele on 22 June 1913 (Coward 1914) and several thousand were reported in the Menai Strait on 26 November 1933. Hardy (1941) reported 50,000 in Liverpool Bay in October 1939 and in February 1964 there were 10,000 off Pensarn, Denbighshire.

Aerial surveys have found numbers in Liverpool Bay to be

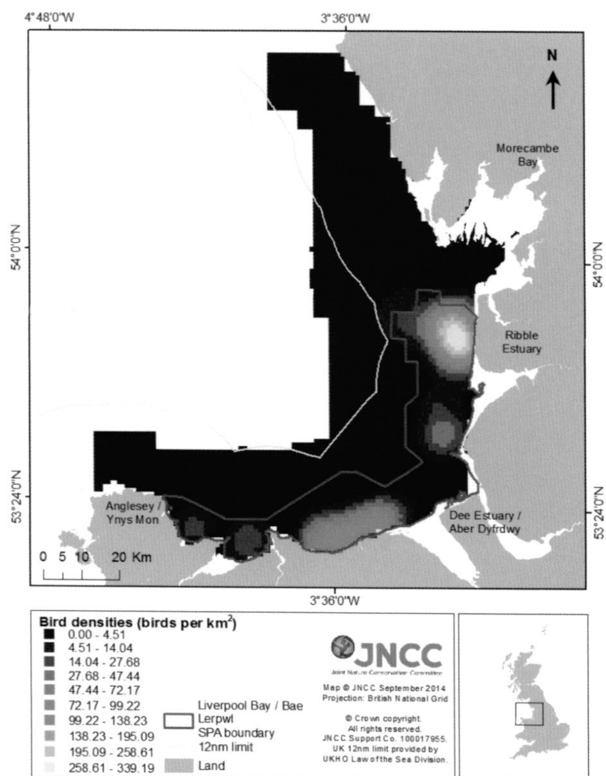

Common Scoter: Estimated mean density from aerial surveys in Liverpool Bay 2004/05 to 2007/08, and 2010/11 (from Lawson *et al.* 2016a; map courtesy of and © JNCC).

far higher than previously suspected. A five-year mean peak of 57,995 during 2004/05–2010/11 (Lawson *et al.* 2016a), 3.4% of the global population, led to the designation of the Liverpool Bay/Bae Lerpwl SPA for the protection of wintering Red-throated Diver, Common Scoter and an assemblage of more than 20,000 waterfowl (Lawson *et al.* 2016a). Further aerial surveys during 2015–20 undertaken by HiDef Aerial Surveying Ltd contracted to Ørsted used cutting-edge species-recognition software to count Common Scoters on high-resolution digital video. The highest estimate from that series of surveys was 289,000 birds in the SPA in February 2015, of which 13% were in Welsh waters. However, a survey in January 2019 estimated there to be 74,000 in the Welsh part of the SPA, 64% of the total. It is unclear whether these represent a real increase or resulted from improved counting techniques. Clearly the scoters move widely within Liverpool Bay, particularly between Shell Flat west of Blackpool and the North Wales coast between Rhos Point, Caernarfonshire, and Point of Ayr, Flintshire. Counts from single points onshore during 2011–17 included several of 20,000, and in Denbighshire alone, between Old Colwyn and Pensarn, 25,000 in March 2012 and February 2013. Clearly Liverpool Bay is a key staging post for this species in midwinter. Mass nocturnal movements recorded across northern England are likely to be birds moving between Liverpool Bay and the North Sea staging posts.

The third major wintering area is the northern part of Cardigan Bay, from Aberystwyth, Ceredigion, to Criccieth, Caernarfonshire. Thousands were recorded off the coast of Meirionnydd in October 1884 (Aplin 1902) and May 1901 (Forrest 1907), while A.W. Boyd noted "fully 2,000" near Harlech on 7 August 1933. Around 1,000 were off Borth, Ceredigion, in October 1944 and *c.*1,200 were there in January 1961. Coordinated shore-based counts during 1991/92–1997/98 included 6,400 between Black Rock and Aberdyfi on 20 November 1991 and 6,500 between Black Rock and Harlech on 11 December that year (Thorpe 2002). Aerial surveys in Cardigan Bay found an estimated 11,771 in winter 2003/04, the majority off Meirionnydd, with concentrations off Morfa Harlech and around Sarn Badrig (Smith *et al.* 2007). Cardigan Bay has been monitored less regularly than the other two areas, but it can hold good numbers in summer, such as 1,000 off Llandanwg on 4 June 2018. Counts in the southern part, south of Aberystwyth, are usually much lower, but there have been up to 350 around Aberporth, Ceredigion (Roderick and Davis 2010).

Smaller numbers can be seen almost anywhere off the Welsh coast. Rafts of up to 400 were recorded off Kenfig Sands, East Glamorgan, prior to the mid-1970s, but numbers farther east are smaller, although a flock of 70 was off the Rhymney Estuary, East Glamorgan, following stormy weather on 5 January 1988 (Hurford and Lansdown 1995). In the early 1990s, 1,000–2,500 wintered in St Brides Bay, Pembrokeshire, where females outnumbered males by approximately 4:1 (Donovan and Rees 1994). Numbers passing coastal headlands often peak in July; a notable year was 2017 when 6,099 passed Strumble Head, Pembrokeshire, between July and October. *Birds in Wales* (1994) described inland records as scarce, but in recent years this species has been found inland regularly in small numbers, particularly on larger lakes and reservoirs, mainly in mid- and South Wales. It was not recorded in Radnorshire, for example, until 1954, but there were 38 records involving 183 birds in the following 60 years (Jennings 2014), while about 95 were recorded in Breconshire during 1969–89 (Peers and Shrubb 1990). Allied to the coastal headland counts, the largest inland counts are often in late June and July, which are probably males undertaking a moult migration from Iceland to Danish and German waters. An exceptional inland count was 90 at Lake Vyrnwy, Montgomeryshire, on 19 June 2014. Other large counts include 20 at Llangorse Lake, Breconshire, on 28 July 2002, 27 at Llyn y Tarw, Montgomeryshire, on 27 July 2005 and 30 in three flocks on the Elan Valley reservoirs, Radnorshire, on 5 July 2016.

Population estimates are now much higher than formerly and Welsh waters are likely to be important in a European context. They increased from 25,000–30,000 in Britain and Ireland during the *Winter Atlas 1981–84* to 40,000–50,000 in Welsh waters alone by 2001–06 (Smith *et al.* 2007). The most recent estimate for Common Scoters in Britain was 135,000 (Frost *et al.* 2019a), but the recent discovery of an average 160,000 in Liverpool Bay alone (HiDef Aerial Surveying Ltd unpublished data) shows that this will need to be revised further. Perhaps 80,000 may winter off Wales in an average year. The counts of several thousand in summer appear to be a more recent change and worthy of further investigation.

The Amber status of the Common Scoter in Wales reflects its localised nature, concentrated in three key areas, and the fact that Wales held 4% of the flyway population (Johnstone and Bladwell 2016), although this proportion may now be much higher. The main threat to the species in its wintering areas is probably from oil pollution. Birds in Liverpool Bay and Carmarthen Bay are in high-risk areas, as illustrated by the *Sea Empress* spill. Marine renewable energy projects also have potential to increase mortality or reduce the availability of suitable feeding areas. Most Common Scoters, including those studied at Gwynt-y-Môr wind farm off the Denbighshire coast, fly too low to be at risk from turbine rotor blades (Cook *et al.* 2012). The bases of turbines can act as artificial reefs (Draget 2014), providing potential feeding opportunities for Common Scoters. Seaducks are particularly sensitive to movements of marine traffic, flying off when boats were up to 2km away in Liverpool Bay (Kaiser *et al.* 2006), so their distribution can be affected by boats servicing energy installations. The impact of tidal lagoons, such as one proposed for the entire Denbighshire coast, is unknown but if constructed through the shallow feeding areas favoured by Common Scoter could be damaging at a flyway level, especially if the lagoons are used subsequently for recreational activities, such as sailing and jet-skiing.

Rhion Pritchard

Sponsored by Carmarthenshire Bird Club

Black Scoter *Melanitta americana* Môr-hwyaden America

Welsh List Category	IUCN Red List	Rarity recording
	Global	
A	NT	BBRC

The Black Scoter breeds in northeast Russia, Alaska and Labrador in North America, and winters around the north Pacific Rim and the eastern seaboard of the USA (*HBW*). It was formerly considered to be a subspecies of Common Scoter (Collinson *et al.* 2006) and is a rare visitor to Britain, with 14 records, typically one record every three years (Holt *et al.* 2020). Only two, both males, have been seen in Wales: the first was off Newgale, Pembrokeshire, from 25 December 1991 to March 1992 and the other was off Llanfairfechan, Caernarfonshire, between 10 March and 8 May 1999. The latter bird returned in January 2001 and then each winter until 2006/07, when it was last seen on 9 April; one off Point Lynas, Anglesey, on 31 January 2003 was considered to be the same bird.

Jon Green and Robin Sandham

Long-tailed Duck *Clangula hyemalis* Hwyaden Gynffonhir

Welsh List Category	IUCN Red List			Birds of Conservation Concern			
	Global	Europe	GB	UK	Wales		
A	VU	VU	NT	Red	2002	2010	2016

A true northern species, the Long-tailed Duck breeds on tundra pools and bogs, rivers and shorelines mostly within the Arctic Circle. The bulk of the European population breeds in high Arctic western Russia, with substantial populations also in Iceland and Norway (*HBW*). Most Icelandic and Greenland breeders probably winter offshore whereas those from Russia and Fennoscandia migrate, mostly to the Baltic Sea, where numbers declined by 65% to 1.5 million birds during 1991–2008 (Skov *et al.* 2011). It dives for crustaceans and molluscs and can feed at least to 60m, deeper than most other ducks. In Britain, the largest numbers are found off northern and eastern Scotland, but it is a scarce visitor to Wales. Most records here occur between October and April.

There are a few records from the 19th century, mostly of single birds, but it was reported that "a good many" were seen in Tremadog Bay, Caernarfonshire, in March–April 1898 and 1899. In Pembrokeshire, Mathew (1894) knew only of two immature birds in a collection that were said to have been shot at Stackpole on an unknown date and at Haverfordwest on 15 June 1843. A few records also came from Ceredigion, where a Captain Cosens of Llanbadarn told Salter that he had once killed three with a single shot on the Dyfi Estuary (Roderick and Davis 2010). Forrest (1907) knew of records from all counties in North Wales, with the possible exception of Denbighshire, and reported that one was shot inland on the River Severn above Welshpool, Montgomeryshire, in December 1874. There was also an inland record of a pair at Llangorse Lake, Breconshire, on 27 October 1893, and one shot on Nanteos Lake, Ceredigion, in winter 1893–94.

There were also few records in the first half of the 20th century,

Female Long-tailed Duck in winter plumage © Richard Stonier

although Farrar (1938) stated that having previously been a great rarity on the Dee Estuary, it had become a regular, though scarce, visitor. The first record in East Glamorgan was one at Roath Park Lake, Cardiff, in December 1905. There was no record in Carmarthenshire until one on Witchett Pool from 20 December 1932 to 13 January 1933. Records became annual, or almost annual, in several coastal counties from the 1960s, which may have been down to better optical equipment. On Anglesey, six were off Newborough on 24 April 1963 and six on Llyn Coron on 30 March 1968. Gwent recorded the species for the first time in 1965, when two were at Wentwood Reservoir on 3–4 April.

There was a further increase in records from the mid-1980s, including 12 off Aberdysynni, Meirionnydd, on 18 January 1984. Late winter 1988/89 produced several high counts, including 28 off Borth, Ceredigion, in January–February 1989, 12 in the Amroth-Saundersfoot area of Pembrokeshire from January to early April, 16 off Llanfairfechan, Caernarfonshire, on 9 April, and ten off Cefn Sidan, Carmarthenshire, on 6 May. Even higher numbers were recorded during winter 1991/92, particularly in Tremadog Bay, Caernarfonshire, where numbers peaked at a Welsh record of 48 in December 1991. At least 12 were in Glamorgan in October and November 1991, the best year on record for the species in the county (Hurford and Lansdown 1995), and 11 were off Penmaenmawr, Caernarfonshire, in the same month. Elevated numbers were recorded for the next few winters, including 27 off Mochras, Meirionnydd, in February 1993 and 11 off Pensarn, Denbighshire, on 18 January 1994, but began to decline from the mid-1990s. It remained a regular visitor to several coastal counties, and Tremadog Bay continued to produce good counts in some winters, with up to 20 there in 1999/2000 and 23 in 2011/12. A county record of 18 flew past Holyhead's Breakwater Country Park, Anglesey, on 4 November 2004. The decline continued at all of the regular sites in the 21st century, though winter 2016/17 produced the highest counts in Wales for many years, peaking in February 2017, with an estimated 43 in Welsh waters, including 17 in Gower.

The Long-tailed Duck has been recorded fairly regularly on freshwater, particularly on Eglwys Nunydd Reservoir in Gower, Llandegfedd Reservoir in Gwent, and on the Anglesey lakes and reservoirs. Some on inland water bodies remain for several months. In 1990, for example, a drake remained all year at Llyn Traffwll, Anglesey, and another remained at Pembroke Mill Ponds, Pembrokeshire, from February 2002 to August 2004. Birds have been recorded in all counties, including those with no coast: there have been ten records in Breconshire but just two in Radnorshire.

Its declining global status and small population size in Wales led to its Red-listing (Johnstone and Bladwell 2016). It is likely that two factors lie behind the reduction in numbers wintering around Wales. The decline in numbers wintering in the Baltic Sea is believed to be a result of a northward shift in winter distribution (Bellebaum *et al.* 2014), but it coincided with heavy mortality caused by gillnet fisheries. The Long-tailed Duck is the most frequently recorded victim in the eastern and south-eastern Baltic Sea (Skov *et al.* 2011), with annual mortality estimates of 1–5% for the total Baltic Sea population (Bellebaum *et al.* 2013). It is quite likely that Long-tailed Duck will go from being a scarce winter visitor to Wales to genuinely rare in most winters.

Rhion Pritchard

Goldeneye *Bucephala clangula* Hwyaden Lygad Aur

Welsh List Category	IUCN Red List			Birds of Conservation Concern			
	Global	Europe	GB	UK	Wales		
A	LC	LC	VU	Amber	2002	2010	2016

The cavity-nesting Goldeneye has an extensive range, breeding over a huge expanse of Alaska, Canada and northern Eurasia on freshwater lakes, pools and rivers surrounded by coniferous forest (*HBW*). The nominate subspecies occurs across Europe, the largest populations of which are in Russia, Finland and Sweden (BirdLife International 2015), but there is a breeding population of around 200 pairs in Britain (Woodward *et al.* 2020), mostly in northern Scotland where nest-boxes are provided (*The Birds of Scotland* 2007). It winters to the south of its breeding range, at sea or on inland lakes. No ringing recoveries involve Wales, but those from elsewhere in Britain and Ireland show that many come from Sweden and Finland, with a few from as far afield as Russia.

There are no records of breeding in Wales, but a pair, one or both of which were injured, was reported to have bred in 1931–32 on the Dee Estuary. This was in the vicinity of Burton Point, just over the border from Flintshire (Hardy 1941). Goldeneye have been recorded in all months and pairs are occasionally seen in suitable habitat as late as June. Nest-boxes have been installed in some areas, but there is no reliable evidence of any having been used by Goldeneye. Subadults seen in late summer sometimes prompt suggestions of breeding but have always been fledged birds that could well have come from farther north.

The Goldeneye is widespread around the Welsh coast in winter, found on both large and small lakes, including in the uplands. It was common off the coast of North Wales in the 19th century, more so off Llŷn (Eyton 1838, Forrest 1907). Around 20–30 wintered in the Dyfi Estuary, Ceredigion/Meirionnydd, in this period (Roderick and Davis 2010), but it was less common in Pembrokeshire (Matthew 1894). Its status and distribution seems to have remained similar until the 1960s, but there are few comparable counts.

Wales only holds a small proportion of the estimated 18,500 birds wintering in Britain (Frost *et al.* 2019a). No site in Wales currently comes close to the threshold for British importance, although Traeth Lafan did reach this threshold during the 1990s. The WeBS index for Wales increased to a peak in the 1990s, driven largely by an increase in breeding populations in countries around the Baltic Sea (*European Atlas* 1997), but subsequently returned to its 1966/67 baseline. WeBS Medium Alerts have been issued for Wales, triggered by a 30% short-term decline and a 45% decline over the long term (Woodward *et al.* 2019). While numbers at some sites have remained similar for many years, there have been significant changes at the majority.

Numbers at Traeth Lafan peaked at 465 on 22 January 1995, a record for Wales. Several other sites recorded counts in excess of 100 between the late 1970s and late 1990s, including 107 at Llyn Alaw, Anglesey, in February 1977; 158 at Trearddur Bay, Anglesey, in January 1978; 180 at the Inland Sea/Alaw Estuary in February 1983; and 100 at the Ogmore Estuary, East Glamorgan, in January

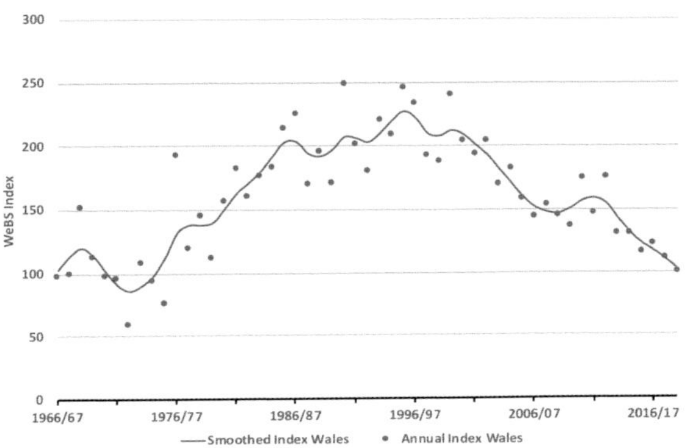

Goldeneye: Wetland Bird Survey indices, 1966/67 to 2018/19.

Male Goldeneye displaying to female © John Hawkins

Site	County	74/75 to 78/79	79/80 to 83/84	84/85 to 88/89	89/90 to 93/94	94/95 to 98/99	99/00 to 03/04	04/05 to 08/09	09/10 to 13/14	14/15 to 18/19
Llangorse Lake	Breconshire	1	2	2	5	34	43	55	71	62
Traeth Lafan	Caernarfonshire/ Anglesey	45	26	99	163	212	132	50	46	55
Inland Sea and Alaw Estuary	Anglesey	67	101	136	77	77	51	53	46	50
RSPB Valley Wetlands inc. Llyn Traffwll	Anglesey	116	21	25	32	35	45	33	44	31

Goldeneye: Five-year average peak counts during Wetland Bird Surveys between 1974/75 and 2018/19 at sites in Wales averaging 70 or more at least once. International importance threshold: 11,400; Great Britain importance threshold: 190.

1987. In Pembrokeshire, numbers formerly increased during cold spells, such as 150 in the county during January–February 1987, including 114 on the Cleddau Estuary (Donovan and Rees 1994), but since the early 2000s, numbers have averaged only single figures and the species has returned to its former relatively scarce status. There has been no record of more than 100 birds at any site in Wales since 113 at Traeth Lafan in January 2004.

The sex ratio of birds wintering in Wales may also have changed over time. Forrest (1907) stated that adult males were "curiously rare" in North Wales, and at Kenfig Pool and Llanishen Reservoir in East Glamorgan during 1921–50, female and first-winter birds outnumbered adult males by a ratio of 9:1 (Hurford and Lansdown 1995). This no longer seems to be true: adult males now make up around half of the birds at Traeth Lafan (pers. obs.) and comprise around one-third of records at Llangorse Lake (Andrew King pers. comm.).

The recent reduction in numbers here is likely to be a combination of a reduction in numbers breeding in Sweden and Finland (*European Atlas* 2020) and an eastward shift in its wintering distribution, as more open water is available around the Baltic (Lehikoinen *et al.* 2013). Assuming this trend continues, it is unlikely that numbers in Wales will return to the level of the 1990s.

Rhion Pritchard

Sponsored by Andrew King

Smew *Mergellus albellus* Lleian Wen

Welsh List Category	IUCN Red List			Birds of Conservation Concern				Rarity recording
	Global	Europe	GB	UK	Wales			
A	LC	LC	CR	Amber	2002	2010	2016	WBRC

This compact sawbill breeds in northeast Europe and in a broad band across subarctic Russia to the Pacific Ocean (*HBW*). It winters to the south of its breeding range and visits Britain, mostly England, in small numbers. It has probably always been scarce in Wales, but declined from the 1990s and no more than four have been recorded in any winter since 2012/13. Most records in Wales are of 'redheads', the colloquial term for the plumage of females and immatures. True estimates of the number of individuals involved in historic records are difficult to gauge since individuals may move between sites and across county boundaries.

During the 19th century, Smew were reported in small numbers, usually associated with hard weather, across Wales. Most of those mentioned by Forrest (1907) in North Wales were on rivers, generally near the sea. The earliest records were both shot, at Llanrwst, Denbighshire, in 1830 and Llanelli Marsh, Carmarthenshire, in 1841. Records in Wales increased from the 1860s, although it remained a scarce visitor. Several were shot on the River Dee in 1861 and one was obtained at Porthmadog market, Caernarfonshire, in January 1881. In East Glamorgan, records during this period include one shot on the River Taff near Cardiff about 1882 and one at Gabalfa in 1892. An influx during severe weather in 1891 brought seven to the River Dee near Saltney Ferry, Flintshire. It was described as "a not uncommon winter visitor" to Pembrokeshire by Mathew (1894), who mentioned birds having been shot at Goodwick and Stackpole Court, and that Bosherston Pools was a favoured site.

Smew remained scarce but appear to have occurred more regularly in the early decades of the 20th century. The number of records increased from the late 1930s, when it became clear that there was a strong southern bias to Welsh records. Small aggregations were recorded more frequently, such as six at Orielton, Pembrokeshire, in 1939 and four on Witchett Pool, Carmarthenshire, on 4 February 1940. Influxes were recorded in South Wales in several winters in the early 1950s and late 1960s. In East Glamorgan, these included three males and ten redheads at Roath Park Lake and at Llanishen and Lisvane reservoirs in winter 1953/54, and two males and 14 redheads in the county in winter 1954/55.

From the middle of the 20th century, the numbers reaching Wales increased. Annual totals fluctuated, being generally higher when winter temperatures were low in eastern Britain or elsewhere in Europe. Totals recorded in each decade were 35 in the 1970s, 75 in the 1980s and 129 in the 1990s. Notable totals include up to 16 in Gwent in winter 1978/79, nine at Ynysmaengwyn, Meirionnydd, on 17 January 1987 and six at Afonwen, Caernarfonshire, on 30 December 1987. Numbers peaked in the second half of the 1990s, with 44 birds in 1996/97 being the largest ever winter total in Wales. The majority were in Gower and East Glamorgan, although there was probably some duplication of records between sites in the two recording areas. Eglwys Nunydd Reservoir, Gower, was a favoured location during this period, with counts of eight in February 1996, December 1997 and December 1998, and 14 in February 1997.

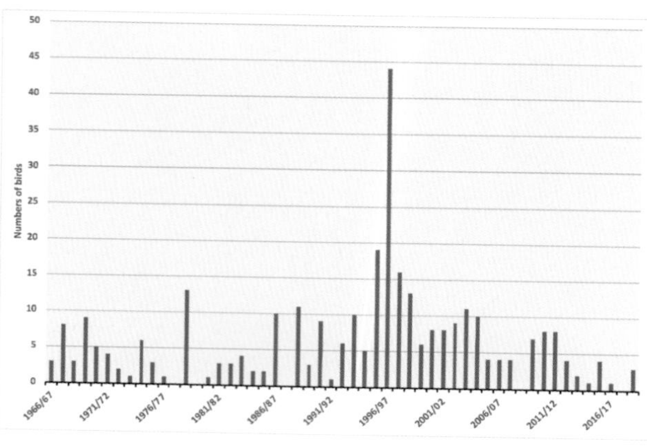

Smew: Minimum numbers recorded each winter, 1966/67–2018/19.

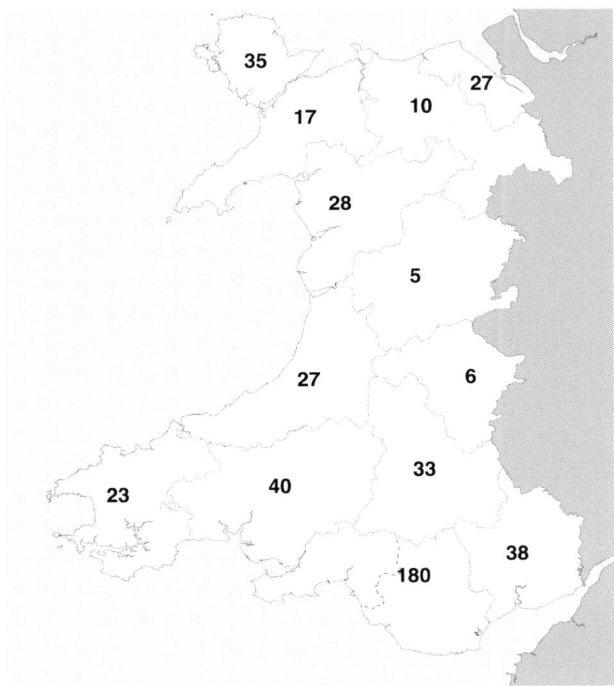

Smew: Numbers of birds in each recording area, 1967–2019. Some birds were seen in more than one recording area. East Glamorgan and Gower were reported together in the *Welsh Bird Report* until 1995.

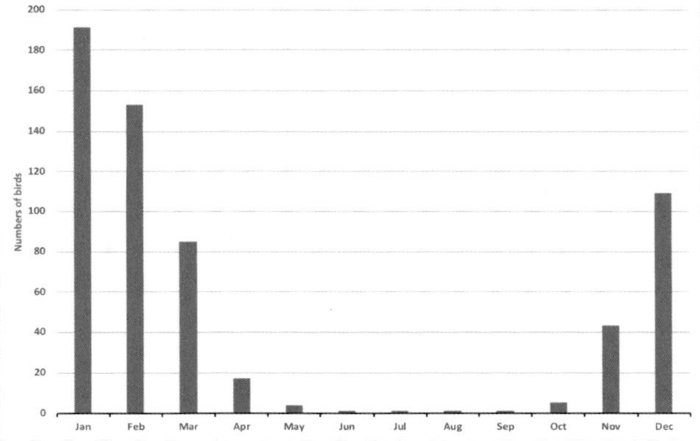

Smew: Totals by month of arrival, 1967–2019.

It is a rare visitor to inland counties. There have been only 10 in Montgomeryshire, the first an immature male shot near Churchstoke, Montgomeryshire, on 5 January 1909 (Forrest 1919). There have been six records in Radnorshire, mostly on the Elan Valley reservoirs, since the first in 1967. It has occurred more frequently in Breconshire, mainly on Llangorse Lake, Talybont Reservoir and the River Usk, with much movement between these sites.

The peak months are January and February, but a few individuals remain for longer, such as a redhead on Eglwys Nunydd Reservoir that was present from 30 November 1966 to 25 April 1967. Most leave by the end of March, but there have been three May records: a male at Kenfig Pool, East Glamorgan, on 10 May 1991, and redheads at Penclacwydd, Carmarthenshire, on 26 May 2001 and at Newport Wetlands Reserve (Uskmouth), Gwent, on 6 May 2012. A male summered at Penclacwydd in 2002.

Wintering Smew have declined markedly in Wales, as elsewhere in Britain in the 21st century: there were 67 from 2000/01 to 2009/10 and 32 from 2010/11 to 2018/19. Its wintering range has rapidly contracted north and east (Pavon-Jordan *et al.* 2015) as a result of milder winters. The small size of its Welsh population led to its Amber-listing (Johnstone and Bladwell 2016). The effect of changing climate, either because of a smaller breeding population or its wintering farther to the east, would suggest that its occurrence in Wales will be far less frequent in future.

Jon Green and Robin Sandham

Hooded Merganser *Lophodytes cucullatus* Hwyaden Gycyllog

Welsh List Category	IUCN Red List	Rarity recording
	Global	
A, E	LC	BBRC

This species breeds in southern Canada and the eastern USA and winters on both the Pacific and Atlantic coasts of North America (*HBW*). It is popular in waterfowl collections and there is a long history of escapes to the wild. A subadult male shot in the Menai Strait, Caernarfonshire/Anglesey, in winter 1830/31 was considered the only acceptable British record until discounted by a review (British Ornithologists" Union 2001). Several other subsequent sightings in Wales were not accepted as being birds of wild origin (see Appendix 1), but from 2000 a number of records were accepted elsewhere in Britain, Iceland and the Azores. The sole Welsh record was among these: a second-calendar-year male at Cei Cydweli, Carmarthenshire, on 7 October 2018. With an 86% increase in the breeding population each decade in the 40 years from 1965/66 (Butcher and Niven 2007), greater numbers of Hooded Mergansers now winter on the eastern seaboard of North America, so it may not be the last.

Jon Green and Robin Sandham

Sponsored by Carmarthenshire Bird Club

Goosander *Mergus merganser* Hwyaden Ddanheddog

Welsh List Category	IUCN Red List			Birds of Conservation Concern			
	Global	Europe	GB	UK	Wales		
					2002	2010	2016
A	LC	LC	LC	Green			

The Goosander breeds in North America and in a broad belt across northern Europe and Russia to the Pacific Ocean, with small populations in central and eastern Europe (*HBW*). Until the 1870s, it occurred only sporadically in Scotland (*Historical Atlas 1875–1900*), but following the first breeding record, in Perthshire in 1871, spread south to breed in Northumberland in 1941. The *Britain and Ireland Atlas 1968–72* estimated the breeding population as 1,000–2,000 pairs. It has since increased its range and breeding density and there are now an estimated 4,800 pairs in the UK (Woodward *et al.* 2019). The breeding range in Scotland contracted between 1968–72 and 1988–91 and the growth in numbers slowed, thought to be due to persecution.

Goosanders in the north of their breeding range are migratory but only partially so farther south (*HBW*). From breeding areas in Wales, some move to estuaries in the Scottish Highlands in June, but most join other European males to moult in fjords in northern Norway. Some 35,000 gather post-moult at Tana Fjord in September. Adult and first-calendar-year birds from Wales move in a NNE–SSW direction in late summer: three ringed in mid-Wales were recovered in Norway, including one at its most northerly point in the Barents Sea. A breeding adult female ringed in 1990 on the River Wye, Radnorshire, was subsequently recaught while moulting at Guardbridge on the River Eden, Fife, in 1992, lending support to the suggestion that early and failed breeding females moult in Scottish estuaries (*BTO Migration Atlas* 2002, Mitchell *et al.* 2008). The increase in breeding numbers in Wales from the 1980s coincided with an increase in moulting females on the rivers Eden, Tay, Tyne and in Montrose Basin and the Moray Firth (Hatton and Marquiss 2004). Late breeding females in Wales are presumed to moult on Welsh rivers, as has been demonstrated in Scotland (Marquiss and Duncan 1994). Welsh breeding birds, along with some Scandinavian breeders, arrive back in Wales—and elsewhere in lowland Britain—between November and January; for example, a duckling ringed in Northumberland in July 1998 was recovered at Llansantffraed, Breconshire, the following December. Winter migrants leave Wales in March.

The breeding colonisation of upland Wales was probably founded by birds from northern England or Scotland. Goosanders were suspected to have bred on the Afon Dyfi and Lake Vyrnwy, both in Montgomeryshire, in 1968, at the latter site in 1969 and confirmed there in 1970, when a brood of ducklings was seen. The first breeding records in other eastern counties came rapidly: on the Elan Valley reservoirs, Radnorshire, in 1972, on the River Usk, Gwent, in 1975, in Carmarthenshire in 1980, in Meirionnydd in 1982, and in Caernarfonshire and Gower in 1987. A possible first breeding record on Anglesey was a brood seen at Traeth Dulas in August 2016.

Goosanders now occur in the breeding season on most rivers, even narrow tributaries, and on some lakes or reservoirs, from the River Monnow in Gwent, across to the Tawe and Neath in Gower and the Teifi in Ceredigion, and north to the Dee, Meirionnydd/Denbighshire, and Conwy, Caernarfonshire/Denbighshire. The *Britain and Ireland Atlas 1988–91* recorded a much lower density in Wales than in Scotland or England. Mitchell *et al.* (2008) found that there was consistently one brood every 6km and mean brood size was 6.3 ducklings on an upland section of the River Wye during 1990–2000. Broods typically gather in a crèche with a small number of adults, one of the largest recorded involving 53 young on the River Wye at Newbridge, Radnorshire, in June 1984.

There were an estimated ten pairs in Wales by 1977, 100 pairs by 1985 (Tyler 1985) and about 150 pairs by the early 1990s (Griffin 1990, Underhill 1993). The BTO Sawbill survey estimated a Welsh population of 883 birds (516 females) based on a territorial pair for every 12.5km of rivers in 1987. Of this, 86% were in an area broadly covered by Gwent, East Glamorgan, Gower, Carmarthenshire, Breconshire, Radnorshire and Montgomeryshire (Gregory *et al.* 1997). No Wales-wide surveys have been undertaken subsequently, but the Waterways Breeding Bird Survey index showed a 32% increase across the UK during 1995–2018. Records in county bird reports are far from comprehensive, typically averaging 40 pairs or territories in 2014–18. They are best monitored in Gwent, where surveys estimated up to 12 pairs on the River Monnow and

Female Goosander © Kev Joynes

13 pairs on the River Usk in 2014 (Tyler 2010, Tyler 2015). However, in neighbouring East Glamorgan, there have been no breeding records since 2013. The Welsh population in 2016 was estimated to be 685 (565–855) pairs (Hughes *et al.* 2020).

Goosanders have long occurred as wintering birds on Welsh rivers, such as the Wye, and on lakes in Snowdonia, especially during severe winters. Seven were off Bardsey, Caernarfonshire, and nine at Tal-y-cafn, Caernarfonshire/Denbighshire, in January 1963, for example (Jones and Dare 1976). The wintering range in Britain and Ireland expanded by 87% between the *Winter Atlas 1981–84* and the *Britain and Ireland Atlas* 2007–2011, mostly into Wales, Ireland and central and southern England. Wintering birds come to Wales from Scotland, northern England, Fennoscandia and Russia, but hard-weather movements may bring others that would otherwise winter in western Europe.

Birds occur on all Welsh rivers in winter, especially on the lower stretches and on many lakes and reservoirs, but it is evident that sites change over time in the extent to which they are favoured. As the number of Red-breasted Mergansers declined at Traeth Lafan, from more than 500 in the 1990s, to fewer than 100 in recent years, the number of Goosanders increased. The site held numbers of British importance between July and September during 2012/13–2015/16, including a record of 242 in August 2013 that is almost certainly a Welsh record count, but substantially fewer in subsequent years. Wintering birds in Wales are supposedly from the breeding population but there is some immigration from elsewhere.

Attempts have been made to assess wintering numbers of Goosanders on some rivers. Simultaneous counts on the upper River Wye, Breconshire, on four mornings in February–March annually during 2004–12, found that numbers were highly variable. The highest count was 95 along 56km between the Elan confluence south of Rhayader and Hay-on-Wye on 22 March 2006, equating to 1.7 birds/km (Andrew King *in litt.*). Coordinated surveys of roosts at inland waters in Gwent recorded 100–150 Goosanders in midwinter during 2011–2013 (Richard Clarke *in litt.*). The wintering population on the River Usk, Gwent, was 102–104 in March 2013, *c.*2.5 birds/km (Tyler 2013).

Flocks of more than 50 may be present during the day on lakes and reservoirs. For example, in Radnorshire up to 110 were on Llanbwchllyn in January 2018 and 50 on Llyn Gwyn in January 1989, and there were 65 on Llangorse Lake, Breconshire, in January 2019. The variable trends in numbers recorded at sites

Site	County	74/75 to 78/79	79/80 to 83/84	84/85 to 88/89	89/90 to 93/94	94/95 to 98/99	99/00 to 03/04	04/05 to 08/09	09/10 to 13/14	14/15 to 18/19
Traeth Lafan	Caernarfonshire/ Anglesey	0	0	1	2	2	5	7	118	102
Llangorse Lake	Breconshire	1	0	1	1	14	18	25	35	27
Dyfi Estuary	Ceredigion/ Meirionnydd	0	0	0	0	2	34	6	16	11
Talybont Reservoir	Breconshire	19	26	36	30	23	36	15	15	7
NE Glamorgan Moorland	East Glamorgan					16	30		13	6

Goosander: Five-year average peak counts during Wetland Bird Surveys between 1974/75 and 2018/19 at sites in Wales averaging 30 or more at least once. International importance threshold: 2,100; Great Britain importance threshold: 150.

in the table suggest that water bodies change in their suitability or attractiveness over relatively short time periods. In Pembrokeshire, for example, Goosander was not even recorded annually by the Wetland Bird Survey until the early 2000s, but it is now the fourth most important site in Wales.

The British wintering population in the early 1980s was believed to be 8,000–9,000 individuals, of which around 2,000 were in Wales (*Britain and Ireland Winter Atlas 1981–84*), and more recently was estimated at 15,000 (Frost *et al.* 2019a). Wintering numbers in Britain have fallen since the turn of the 21st century, which is likely to be related to short-stopping of the Fennoscandian population closer to their breeding areas. The WeBS index for Wales doubled between 1990/91 and 2015/16, although it covers lakes and estuaries better than stretches of river. The Welsh population probably numbers about 1,000 individuals in winter. A more robust estimate of population size in each major Welsh river catchment is planned in 2021, as part of a programme to assess the impact of Goosanders on Atlantic Salmon, particularly on six Special Areas of Conservation for which Atlantic Salmon is a qualifying feature. Salmon populations are assessed as Vulnerable in Europe.

No recent data have been published on the number of Goosanders shot under licence to protect fisheries in Wales. In 2011–13, six Goosanders were killed under five licences issued by Welsh Government, but the level of illegal killing is unknown. Many Goosanders are very wary of humans, presumably because of shooting, but in some places they have become habituated to people, even congregating with Mallard for food hand-outs at some sites, such as Llyn Padarn, Caernarfonshire, and the River Monnow in Monmouth, Gwent. There is anecdotal but unquantified evidence that anglers' nets and night-lines catch Goosanders.

Good fish populations are crucial for piscivores such as Goosander, so any pollution from slurry or runoff from fertilisers threatens the whole ecosystem. Close dialogue and good data on the impacts of all factors affecting Salmon, not just predation, are needed to assess any licence applications to kill Goosanders. Any increase in the numbers shot will need to be monitored closely to understand the effect on Wales' breeding population, which is expected to decline as Wales and England are predicted to be climatically unsuitable for Goosander by the end of this century (Huntley *et al.* 2007).

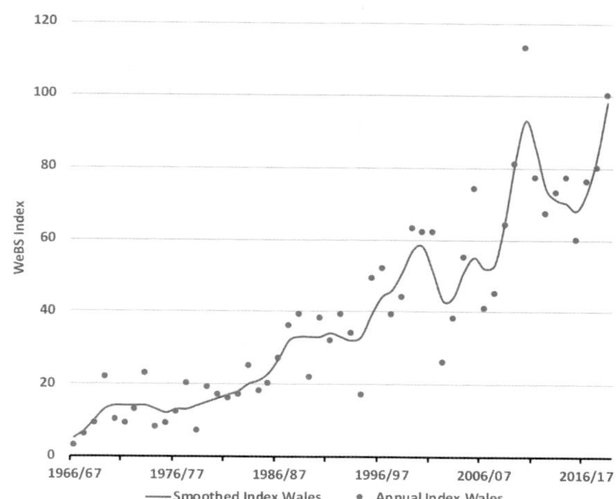

Goosander: Wetland Bird Survey indices, 1966/67 to 2018/19.

Steph Tyler

Sponsored by Kate Gibbs

Red-breasted Merganser *Mergus serrator* Hwyaden Frongoch

Welsh List Category	IUCN Red List			Birds of Conservation Concern				Rarity recording
	Global	Europe	GB	UK	Wales			
A	LC	NT	VU	Green	2002	2010	2016	RBBP

The Red-breasted Merganser breeds over most of northern North America, around the coast of southern Greenland, and across northern Eurasia. It usually breeds on deeper lakes and small rivers, often in wooded country and in sheltered marine habitats, and winters mostly at sea to the south of its breeding range. Males tend to remain closer to nesting areas than females or immature birds (*HBW*). There have been very few ringing recoveries involving Wales: a second-calendar-year male ringed on Holy Island, Northumberland, in August 1990 was found long dead on the Cefni Estuary, Anglesey, in May 1991 and an adult female ringed on the German island of Walfisch in the Baltic Sea in July 1998 was found dead at Porthmadog, Caernarfonshire, in March 2010. The global population is stable, but the European population is thought to be decreasing (BirdLife International 2020).

In Britain, breeding was largely confined to north and west Scotland, until range expansion began in the mid-19th century. Suggested reasons for this expansion include a greater abundance of small fish as commercial marine fish stocks were depleted, and a reduction in predators as a result of more intensive game preservation (*The Birds of Scotland* 2007). The UK population is estimated to be 1,565 pairs (Humphreys *et al.* 2016). The species was mentioned, as "Red-breasted goosander", in the *Swansea Guide* of 1802, but it appears to have been a scarce visitor to Wales during the 19th century and the early years of the 20th century. In North Wales, it occurred off the coast of Meirionnydd, particularly around Penrhyndeudraeth, in some numbers and every winter in small flocks on the Dyfi Estuary. It was less numerous on the north coast and scarce inland, although there were records from the upper Dee and from the River Severn above Welshpool, Montgomeryshire (Forrest 1907). It was an occasional winter visitor to Pembrokeshire and Carmarthenshire (Mathew 1894, Barker 1905) and to Glamorgan in the second half of the 19th century, but there were no records in the latter county between 1890 and 1927 (Hurford and Lansdown 1995), nor in Gwent until 1920.

Numbers began to increase in the Dyfi Estuary from the 1920s, with just over 100 reported off Aberdyfi, Meirionnydd, in February 1938. The first confirmed breeding record in Wales was at Traeth Dulas, Anglesey, when 14 ducklings were seen in the bay on 30 June 1953. A pair bred there the following year and by 1957 breeding had been recorded at several sites on the island. A survey in 1966 confirmed breeding at seven sites on Anglesey and it was suspected at nine others. The first breeding record on the Welsh mainland was in 1957, when two pairs bred in Meirionnydd, one near Traeth Bach south of Porthmadog and one east of Aberdyfi. By 1962 breeding had been confirmed in four estuaries in the county and by 1973, around 15–20 pairs were known (Lovegrove *et al.* 1980). A similar spread was evident in Caernarfonshire, where the species was first recorded breeding in 1958. The county's population expanded to about 25 pairs by 1973 (Jones and Dare 1976). The Dyfi Estuary became one of the most important breeding areas in Wales. Breeding was suspected at RSPB Ynys-hir, Ceredigion, in 1968 though not confirmed until 1973, but colonisation must have been rapid, as 115 juveniles were counted on the Dyfi in 1976. Red-breasted Mergansers probably bred on the Gower side of the Burry Inlet in 1970, when two flightless juveniles were reported (Hurford and Lansdown 1995).

A survey of breeding sawbills on selected Welsh rivers in 1981 found 27–43 breeding pairs of Red-breasted Merganser on nine rivers, of which 9–20 were on the Afon Dyfi and 7–9 on the River Dee (RSPB 1981). A repeat survey in 1985 found 24–41 pairs,

Male Red-breasted Merganser © Norman West

including 7–22 pairs on the Dyfi and nine pairs on the Dee (Tyler 1985). However, another survey in 1990, funded by the National Rivers Authority in response to complaints from anglers, found just 22 adults and three juveniles (Griffin 1990). The *Britain and Ireland Atlas 1968–72* revealed widespread breeding in northwest Wales—on Anglesey, and in Caernarfonshire and Meirionnydd—but few occupied squares elsewhere. By the 1988–91 *Atlas*, they had spread east and south into southern Denbighshire and north Ceredigion. Breeding was recorded in Breconshire for the first time in 1993, when pairs bred at Talybont Reservoir and at Newbridge-on-Wye. The first breeding record for Pembrokeshire came in 1995, on the Gann Estuary.

The earliest breeding records in Wales were mostly in estuaries, but a range of inland sites was used as the population increased. There were *c.*12 pairs on the Ogwen river system, Caernarfonshire, in 1973 (Jones and Dare 1976) and six pairs bred on Llyn Trawsfynydd, Meirionnydd, in 1988. They also bred on small coastal islets on Anglesey: 12 pairs bred or attempted to breed on Ynys Feurig in 1994, but numbers declined to 2–6 pairs during 2010–19; and on The Skerries 1–4 pairs bred annually during 1993–99, but nesting has been sporadic since. Breeding has also occurred on Puffin Island on several occasions in the 21st century (Steve Dodd pers. comm.). By the mid-1990s, western Wales and northwest Scotland had the highest density of breeding Red-breasted Mergansers in Britain, with a mean of 0.24 birds/km of river in summer, excluding young. It estimated that *c.*800 pairs bred in riverine habitats in Britain, of which around 22% were in Wales, suggesting a Welsh riverine population of around 175 pairs (Gregory *et al.* 1997). This was higher than previous estimates of *c.*100 pairs (Lovegrove *et al.* 1980) and *c.*150 pairs (*Birds in Wales* 1994) for all habitats in Wales.

That appears to have been the peak of the Welsh breeding population, and from around the turn of the 21st century many observers formed the impression that it was in decline. The *Britain and Ireland Atlas 2007–11* confirmed widespread losses in south and east Wales since 1988–91, as the range contracted back to Anglesey, Caernarfonshire and Meirionnydd. The number of pairs breeding inland declined even in this core area: numbers on the Afon Ogwen in Caernarfonshire, for example, dropped from 12 pairs in 1973 to 8–10 pairs in 1992 and only 2–3 pairs in 2005–07. In the latter survey just two broods were found in 2005, both on the lower part of the river, but none in the other two years (Gibbs *et al.* 2011). The *North Wales Atlas 2008–12*, which covered most of the core breeding area, found the species in 6% of tetrads, but breeding was confirmed in only 17 (1%) and considered probable in a further 54 (3%). There has been a tendency in Snowdonia for pairs to occur on lakes and rivers in early spring, but to leave by late April, apparently without making a breeding attempt (pers. obs.). The Welsh breeding population is now likely to consist of fewer than 100 pairs.

Little information is available on the origins of Red-breasted Mergansers found in Wales outside the breeding season. Many are presumed to be birds breeding in Wales, but some may come from elsewhere. The WeBS index in Wales declined from the early years of the 21st century, to its lowest level since the early 1970s by 2018/19. Medium Alerts have been issued for Wales for the short (-27%), medium (-37%) and long (-34%) term (Woodward *et al.* 2019). Declines of a similar scale are evident across Britain.

The largest congregation in Wales is usually on Traeth Lafan/Conwy Bay Special Protection Area (SPA), Caernarfonshire, where a moulting flock is present between June and October. This peaked at 553 in 1994, but numbers have declined and it is no longer of British importance for the species. WeBS Medium Alerts have been issued for the medium (ten years) and long term (25 years) at the site. At most other sites, the highest numbers are between October and April. The Inland Sea and Alaw Estuary consistently hold good numbers, although the peak of 140 was back in October 1982. There were 122 off Rhos Point, Caernarfonshire, in January 1995, but more recent counts have been around 20. There was a marked increase in the Dee Estuary from the 1960s, including an exceptional flock of 250 on the English side in February–March 1965 (Williams 1977), but counts have been considerably lower since around 2005. The highest count on the Welsh side was 50 in September 1996, but during 2010–19 there were just 20–30 in the whole estuary, and only 1–2 in Wales.

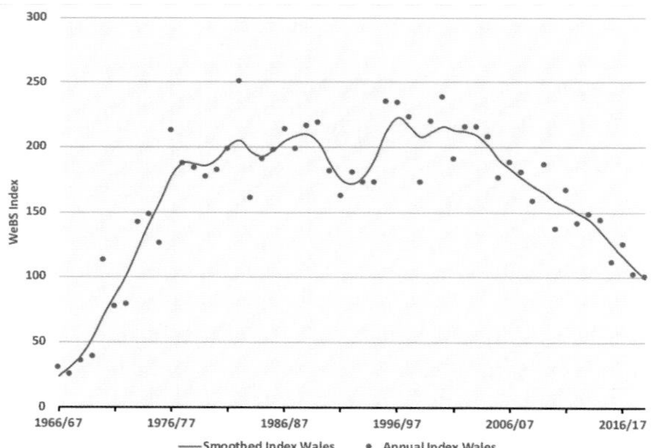

Red-breasted Merganser: Wetland Bird Survey indices, 1966/67 to 2018/19.

Site	County	74/75 to 78/79	79/80 to 83/84	84/85 to 88/89	89/90 to 93/94	94/95 to 98/99	99/00 to 03/04	04/05 to 08/09	09/10 to 13/14	14/15 to 18/19
Traeth Lafan	Caernarfonshire/ Anglesey	15	28	160	68	162	234	133	136	66
Inland Sea and Alaw Estuary	Anglesey	44	80	69	74	72	72	83	80	64
Dee Estuary: Welsh counties (whole estuary)	Flintshire	na (27)	na (27)	na (24)	10 (51)	26 (54)	22 (54)	14 (44)	4 (29)	2 (27)
Traeth Bach	Meirionnydd	41	50	17	25	27	21	23	27	24
Dyfi Estuary	Ceredigion/ Meirionnydd	77	51	66	79	80	48	27	21	16
Mawddach Estuary	Meirionnydd	58	23	35	43	43	46	32	27	16

Red-breasted Merganser: Five-year average peak counts during Wetland Bird Surveys between 1974/75 and 2018/19 at sites in Wales averaging 50 or more at least once. International importance threshold: 860; Great Britain importance threshold: 100.

On the Cardigan Bay coast, several counts over 100 were recorded in the Glaslyn/Dwyryd and Mawddach estuaries from the mid-1970s to the late 1990s, and the Dyfi Estuary held good numbers each winter and spring in the 1980s and 1990s, peaking at 166 in April 1989. Coordinated counts in the late 1990s recorded several totals over 200, including 294 between Clarach in Ceredigion and Pwllheli in Caernarfonshire, in January 1998, of which 111 were off Criccieth, Caernarfonshire. Numbers declined at all these sites during the early 21st century. Farther south, the Cleddau Estuary, Pembrokeshire, and the Burry Inlet, Gower/Carmarthenshire, sometimes held over 40 in the 1970s and 1980s, and 50 were in the Gwendraeth Estuary, Carmarthenshire, in November 1975 and January 2011 (Lloyd *et al.* 2015).

The picture is one of decline in both breeding and wintering birds in Wales in the 21st century. Barnes (1997), commenting on its decline as an inland breeder in Caernarfonshire, suggested that this was at least partly caused by competition from Goosander, whose numbers increased on many river systems at the same time as Red-breasted Merganser numbers began to decline. However, the *Britain and Ireland Atlas 2007–11* showed that the range contraction in Wales between 1988–91 and 2007–11 was matched by declines elsewhere, including in western Ireland, where Goosander does not occur regularly. No recent data is available on the number of Red-breasted Mergansers killed under licence in Wales but 300 were killed in England during 2013–19 (Natural England 2020). The *Britain and Ireland Atlas 2007–11* suggested that predation by American Mink and shooting on game fishing rivers, in response to perceived predation of fish, were potential factors in the decline.

It is an Amber-listed species of conservation concern in Wales, owing to a 29% decline in its breeding range between 1968–72 and 2008–11, a 32% decline in its wintering population and its Near Threatened status in Europe (Johnstone and Bladwell 2016). All of Britain, except the far northwest of Scotland, is projected to become climatically unsuitable for Red-breasted Merganser by the end of this century (Huntley *et al.* 2007), but it is not yet clear whether climate is driving the changes already recorded. There is a great deal that we do not know about the Welsh population, which is on the southern edge of its European range. Dedicated surveys to get a more accurate estimate of the number of breeding pairs would be a good start.

Rhion Pritchard

Ruddy Duck *Oxyura jamaicensis* Hwyaden Goch

Welsh List Category	IUCN Red List	Rarity recording
	Global	
C6	LC	RBBP

Probably no species has appeared on, and then disappeared from, the list of birds breeding in Wales as rapidly as the Ruddy Duck. Native to North America, from Canada to Mexico, this attractive species became popular in wildfowl collections. Inevitably, some birds escaped, the first two from the Wildfowl Trust (now WWT) at Slimbridge, Gloucestershire, in winter 1952/53. Up to 20 escaped from there in 1957 and the first wild brood was seen in England in 1960 (*Birds in England* 2005). The first record in Wales appears to be of two females at Lake Vyrnwy, Montgomeryshire, on 16 June 1960. The first breeding was also in Montgomeryshire, at Lymore, where a pair summered in 1974 and bred in 1975. Up to five pairs bred there annually until at least 1980. There was a brood at another site in the county in 1981. A pair appeared at Llyn Llywenan, Anglesey, in 1977, and breeding was confirmed there in 1978. By the mid-1980s breeding had been recorded at another six sites on the island, and the Anglesey population was estimated at 25–35 pairs (RSPB 1986). By 1991 breeding was confirmed at 13 sites on the island.

The first breeding record for Carmarthenshire came in 1985, when a pair bred at Witchett's Pool, then in 1986 a pair bred in East Glamorgan at Kenfig. Breeding was suspected at a site in Llŷn, Caernarfonshire, (RSPB 1988a) and by the early 1990s it had been recorded at two sites there: Afonwen pools and Llyn Glasfryn. The first breeding in Radnorshire was a pair at Llyn Heilyn in 1987, and breeding was confirmed there in five of the next 11 years, but at no other site in the county. The first confirmed breeding record for Flintshire was in 1990, with a brood on Llyn Helyg, and in Gwent in 2001, when Ruddy Ducks bred at Newport Wetlands Reserve (Goldcliff). It has been recorded in all counties, but there are no breeding records in Breconshire, Meirionnydd, Ceredigion or Pembrokeshire.

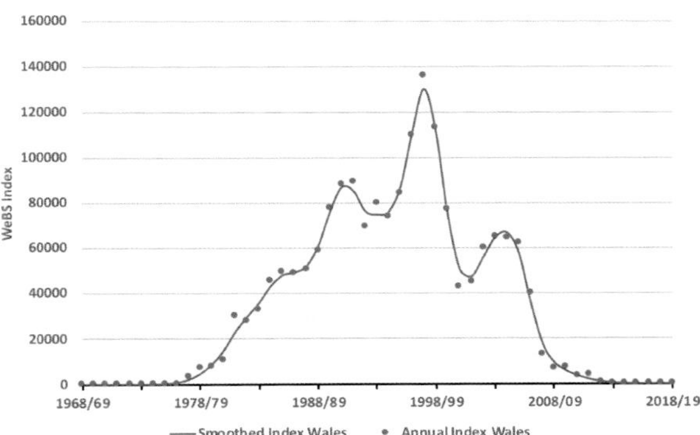

Ruddy Duck: Wetland Bird Survey indices, 1968/69 to 2018/19.

Outside the breeding season, Ruddy Ducks congregated on larger lakes and reservoirs. Anglesey held the largest numbers in Wales, with the first three-figure count in 1985, when 113 were on Llyn Traffwll on 10 October. By the late 1990s, numbers had increased, with a Welsh record 225 at Llyn Alaw on 1 February 1998. Numbers in other counties were much lower, though 30 were at Hanmer Mere, Denbighshire, on 3 October 2000.

By the mid-1990s, a problem was becoming apparent. The Ruddy Duck was dispersing to Spain and interbreeding with the globally threatened White-headed Duck. There was real concern that Ruddy Duck would extend its range to the main White-headed Duck population in Turkey and Central Asia. A Ruddy Duck Control Trial was established by the UK Government to determine the feasibility, cost and access requirements necessary to reduce the population to fewer than 175 individuals in ten years. Three regions were selected for the trial, including Anglesey, where the aim was to reduce the breeding population by at least 70% within three years. In 1999, 80 waterbodies were visited to count all the Ruddy Ducks on the island at the start of the breeding season before the trial, and the population was estimated to be 200 birds. Between April 1999 and May 2002, 515 Ruddy Ducks were shot at 16 sites on Anglesey: 366 during breeding seasons and 149 in post-breeding/winter periods, with considerable immigration evident in most months of the year. The breeding population was reduced by over 70% in the first year and by *c.*93% within 16 months (Defra 2002). The cull was extended to cover all relevant sites where permission could be obtained and by 2009 the UK population was thought to be 300–400 birds, compared to about 6,000 in January 2000 (Henderson 2009). By 2019 it had been reduced to *c.*20 individuals (Niall Henderson *in. litt.*).

The effects of the cull were quickly apparent on Anglesey, where there were 159 on Llyn Alaw on 13 January 2000 but much lower numbers the following winter. A count of 88 at Llyn Alaw in January 2007 was the last anywhere in Wales in excess of 20. Two broods at Shotton Steelworks, Flintshire, in 2008 proved to be the last confirmed breeding record in Wales. Five birds were at Kenfig, East Glamorgan, in January 2011, but thereafter records were confined to a few at Shotton and on some of the Anglesey lakes. There were no records of the species in Wales in 2016, for the first time since 1973, and none in 2017 or 2019, although a single male was at Cosmeston Lakes, East Glamorgan, on 25 September 2018. The UK Government continues to monitor the presence of Ruddy Ducks to assess whether further intervention is necessary.

Rhion Pritchard

Common Nighthawk *Chordeiles minor* Cudylldroellwr

Welsh List Category	IUCN Red List	Rarity recording
	Global	
A	LC	BBRC

The Common Nighthawk breeds throughout temperate North America, south to Panama and the Caribbean. It winters in South America as far south as central Argentina (*HBW*). Some migrate over the west Atlantic, where it is regular, on passage, on Bermuda and the Lesser Antilles. There have been a total of 24 records in Britain to the end of 2019 (Holt *et al.* 2020). The only one of these in Wales is of a dead bird found on the tideline at Mwnt, Ceredigion, on 28 October 1999. Its skin is preserved in the National Museum of Wales, Cardiff.

Jon Green and Robin Sandham

Nightjar *Caprimulgus europaeus* Troellwr Mawr

Welsh List Category	IUCN Red List			Birds of Conservation Concern			
	Global	Europe	GB	UK	Wales		
A	LC	LC	LC	Amber	2002	2010	2016

The Nightjar has an extensive breeding range, over most of Europe except the far north, and extending east, across temperate Asia as far as Mongolia and northwest China. There is also a breeding population in northwest Africa. All populations winter in Africa south of the Sahara (*HBW*). The species is entirely crepuscular and nocturnal.

During the 19th century it appears to have been common and widespread throughout most of Wales, heavily associated with bracken-covered slopes, heather-clad crags, open woodland, sand dunes and, to a lesser extent, heathland and bogs. From the early 20th century, the population underwent a contraction of range and a strong decline. The reasons are uncertain, but habitat loss, climate change and afforestation have been suggested. From about 1980, the population increased again as Nightjars colonised newly planted and clear-felled conifer plantations, often at considerable altitude. This trend has continued through the last three decades. The species is now once again locally frequent in some areas and is characteristic of open areas and cleared patches within conifer plantations in Wales.

Dillwyn (1848) said that the Nightjar was plentiful in the woodlands around Penllergaer, Gower. At the end of the 19th century and in the early years of the 20th century, it was widespread and apparently common in Wales. On Anglesey, it was said to be abundant throughout the island—"where ever we went we heard them calling" (Thomas Coward, Diaries, quoted in *Birds in Wales* 1994). Salter (1900) frequently recorded the species on the outskirts of Aberystwyth, Ceredigion, and stated that it was fairly numerous around oakwoods and on bracken-covered hillsides. This association with bracken-covered slopes was widespread and is reflected in the alternative name of 'Fern Owl'. Matthew (1894) recorded it breeding in many parts of Pembrokeshire and even believed it to be "not uncommon in September turnip fields", although this is hard to believe and may have referred to passage birds. Bolam (1913) reported the Nightjar as common on the hazel-covered hillsides near Dolgellau, Meirionnydd. Several were said to have bred on the bogs of Cors Fochno and Cors Caron, Ceredigion (Roderick and Davis 2010). During this period, the species also nested in coastal dune systems along the Welsh

Male Nightjar in displlay flight © John Hawkins

coast including Ynyslas, Ceredigion (1934), Gwbert, Ceredigion (1927), Merthyr Mawr and Kenfig, East Glamorgan, and on the Gower Peninsula (*Birds in Wales* 1994).

The decline appears to have started earliest, and most severely, in the west of Britain (*Historical Atlas 1875–1900*). Despite their widespread distribution at the time, Walpole-Bond (1904) remarked that Nightjars were becoming less plentiful in mid-Wales. Numbers certainly declined markedly in several counties from the 1920s and 1930s onwards—for example, on the Gower Peninsula, where in 1927 the species was said to be much scarcer than formerly (Hurford and Lansdown 1995). Declines also began in Ceredigion in this period, and by the 1950s and 1960s the species was said to be thinly distributed there (Roderick and Davis 2010). Nightjars were scarce in Breconshire by 1940. Ingram and Salmon (1955) said that they had been getting scarcer in Radnorshire for some years and were now quite rare. In Montgomeryshire, the species decreased during the 1940s. There was no published record in the county between 1954 and 1962 (Holt and Williams 2008).

Farther south, Lockley *et al.* (1949) stated that the species was still fairly common and widely distributed in Pembrokeshire. In Carmarthenshire, Ingram and Salmon (1954) said that although it had been a frequent and widespread breeder in the county 20–30 years previously, it had progressively declined and was now found in extremely few areas. In the north, the *Cambrian Bird Report* 1956 reported that Nightjars were still present at many sites on Anglesey, including sand dunes, but numbers then declined steadily on the island.

The first attempt to estimate Nightjar numbers came in 1957 (Stafford 1962), as a result of concern about these apparent large-scale declines. Data were collected from BTO members and Regional Representatives on the status of the Nightjar in their areas, but coverage was poor and in Wales; the information obtained was often based on the knowledge of just one naturalist per county. Nightjars were reported as uncommon, or very uncommon, in all counties of Wales except Meirionnydd, where it was reported breeding and "fairly numerously on bracken-covered slopes towards the coast". Populations were reported as declining in seven counties: Glamorgan, Radnorshire, Carmarthenshire, Pembrokeshire, Caernarfonshire, Denbighshire and Flintshire.

There was a further decline in the 1960s. *Birds in Wales* (1994) suggested that the Nightjar probably never disappeared completely as a breeding species in Breconshire, as had been reported by others, and stated that, during the late 1960s and early 1970s, it was still found at seven sites in the county. In Meirionnydd, the main reduction in numbers occurred during 1925–40, although the species was still regarded as widespread. It remained fairly common in Caernarfonshire during the 1960s where a local Welsh name was *adar naw o'r gloch* (nine o'clock birds) owing to its habit of churring at dusk (*Birds in Wales* 1994). On Anglesey, several were heard around Mynydd Bodafon in July 1964, but numbers declined rapidly during the 1960s. In 1969 one displaying bird at Mynydd Bodafon was the only one recorded on the island. The *Britain and Ireland Atlas 1968–72* indicated that Nightjar was still fairly widespread across North and South Wales, but there were very few records in mid-Wales.

Gwent appeared to be the main stronghold for this species, and an increase in records was noted there from 1965. The main site at that time was the forest restocks of Wentwood, where there were up to eight 'pairs' in 1970, although numbers declined in subsequent years (Venables *et al.* 2008). Parslow (1973) stated that there were no recent breeding records for Carmarthenshire, Breconshire or Radnorshire. The species had almost disappeared from Anglesey by the 1970s, although there was a churring male at Cors Goch in May 1976. There have been a few records since, including one in Newborough Forest in May 1999 and two displaying there in June 2014. In Pembrokeshire, Nightjars could be found at only eight sites by 1971. Churring birds were heard in five localities in 1984–88 (Donovan and Rees 1994), but there have been no indications of breeding since, and the species is now considered to be an occasional passage migrant in the county (Annie Haycock *in litt.*).

Alarm at the continued decline of the species in Britain resulted in the first attempt at a survey by the BTO in 1981 (Gribble 1983). The results, which included the first detailed census to be undertaken in Wales, suggested that the population here had diminished to a mere 57 singing males, almost half of which were in Gwent at 36 occupied sites. The organiser, however, stated that while most previously occupied sites were visited, coverage in Wales and Scotland was poor. Some important areas in Glamorgan were missed completely. During the 1980s it became apparent that clearfelled areas and young plantations were providing ideal breeding habitat for Nightjars, especially as so much of the Welsh forest estate was then being felled and replanted. In 1990, a sample survey organised by the Forestry Commission identified 36 territorial males in seven counties. Better monitoring following this census revealed at least 67 churring males in 1991 (*Welsh Bird Report*).

Gribble (1983) thought that changed climatic conditions, particularly late, cold springs and wetter summers, were the major cause of the decline. Cold springs delay first-laying dates, so reducing the number of second broods initiated (Berry and Bibby 1981), while wet weather reduces the availability of moths and other insects. However, this seems at odds with the recent increase in numbers in upland plantations in Wales, at considerable altitude, where it could not be much colder and wetter. Agricultural intensification, industrialisation, land-use change and increased human disturbance have all been postulated as possible causes of decline. While these may explain a reduction in overall numbers, as formerly suitable sites were destroyed, none of them explain the loss of birds from seemingly unchanged, little visited, habitats. The use of DDT and other organochlorine pesticides, including in areas away from the nesting locations, has not been demonstrated as causal in population decline in Nightjars, but may have played a part. The decline in some areas began long before widespread use of pesticides, so this could not have been the sole cause.

In 1992, a second census of Nightjars in Britain confirmed that the recovery was continuing (Morris *et al.* 1994). In Wales, 193 churring males were recorded at 107 sites, Meirionnydd being the most important county with 41 males at 19 sites. Anglesey was the only county where no birds were found, although suitable sites still existed. Since coverage in some counties was incomplete, the population at the time is likely to have been in excess of 200 males.

The third, and most recent, UK census, undertaken in 2004 (Conway *et al.* 2007), produced a population estimate of 280 males in Wales, an increase of *c.*24% since 1992. This was largely due to increases in Carmarthenshire, Ceredigion and Glamorgan, but it also found a population decline and range contraction in North Wales. Around 80% of churring males in Wales were recorded

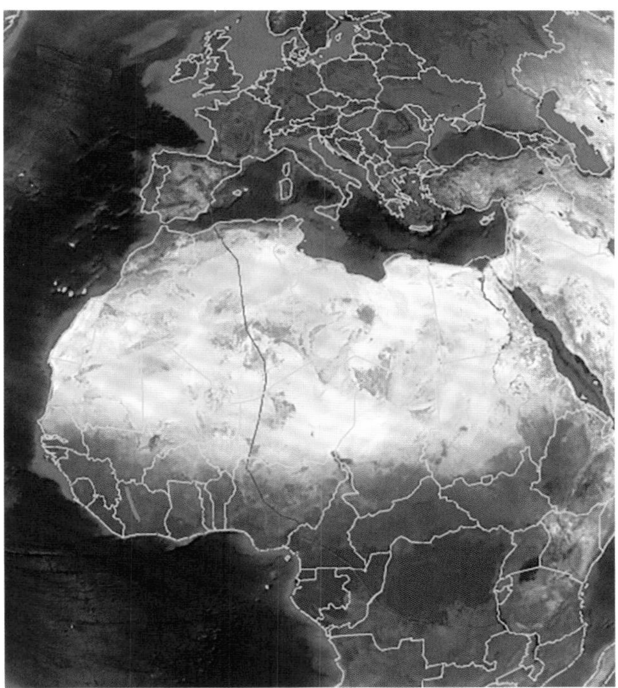

Nightjar: Data from a GPS data logger fitted to a Nightjar near Llyn Brenig, Denbighshire, which moved to the Democratic Republic of Congo side of Lake Tanganyika where the battery ran out. Map courtesy of Google Earth.

in forest plantations, including a higher proportion on unplanted areas than in 1992.

The *Britain and Ireland Atlas 2007–2011* recorded Nightjars in 20% of 10-km squares in Wales, compared to 15% in the 1988–91 *Atlas,* with losses in the north and west and gains clustered around core breeding areas, particularly in East Glamorgan, but also in Gwent, Gower and Caernarfonshire.

Surveys can greatly under-represent the true number of birds present. On cool or windy nights—neither uncommon in upland Wales—the window of opportunity for detecting a churring male can be as little as 20 minutes, before it goes off to feed (pers. obs.). As observers willing to go out into remote areas after dark are in short supply, most large surveys will underestimate the number of birds present. Planning applications and consent agreements for wind farms have led to several large blocks of commercial conifer forest in Wales being surveyed for Nightjars in an intensive and sustained manner. These have shown that the density of Nightjars in upland forestry can be relatively high. Surveys in an 18km² area of Brechfa Forest, Carmarthenshire, during 2012–19 revealed a mean density of 0.82 males/km² (Paddy Jenks *in litt.*), while similar surveys in an area of *c.*36km² in Clocaenog Forest, Denbighshire, in 2014–17 gave equivalent figures of 0.58 males/km² (D. Watson *in litt.*). Surveys at the Pen y Cymoedd wind farm, East Glamorgan, recorded an average of 0.38 males/km², in an area of approximately 46km² (pers. obs.).

In recent years Nightjars have returned to several sites in north Ceredigion, including former, un-afforested, peatland haunts on Cors Fochno. Following a year or two when birds were heard churring, a nest was discovered there on a dry, bracken-covered peat baulk in 2013 with another one in 2016 on a bare, shaded area under gorse. An *ad hoc* survey of apparently suitable habitat at Cors Fochno and Morfa Borth in 2018 found 12–14 churring males (Mike Bailey pers. comm.). In 2019 a Nightjar was heard churring on an area of Cors Caron from which birch scrub had recently been removed (own data). Monitoring at wind-farm projects has included the location and regular checking of nesting attempts, which has produced a robust data set of over 200 nest records during 2014–19. These show that the species can be remarkably successful at breeding in these 'upland' habitats, with average nesting success of 63% and mean productivity of 1.16 chicks per nesting attempt (own data). High nesting success is almost certainly the result of low levels of predation by Fox, Badger and other ground-based predators that appear to avoid thicket-stage forest and tangled piles of brash.

Surprisingly perhaps, Nightjars are more often encountered on the offshore islands of Wales on spring rather than autumn passage, although numbers in both seasons have declined. In Pembrokeshire, Lockley *et al.* (1949) said that they bred occasionally on Skomer, Ramsey and Caldey, but they are now no more than a scarce passage migrant. Birds were formerly recorded annually on spring passage at Skokholm, but there were only five records between 1960 and 2004 (Thompson 2008). Skomer however has 12 modern records, involving 13 individuals. On Bardsey, there have been 33 Nightjar records since 1953, including 19 in spring and 13 in autumn, with a single July record. The last encountered was in September 2006. The maximum recorded there was three together on 3 June 1910 (*Birds in Wales* 1994).

Wind-farm ecological assessments have led to an increasing number of Nightjars being ringed in Wales: 392 in 2015–18, about 25% of the current Britain and Ireland annual total. The Welsh longevity record stands at eight years and eight days, for an adult male ringed on the Rhigos Mountain, East Glamorgan, on 5 June 2009 and re-trapped at the same location on 13 June 2017.

There have been very few recoveries at any distance from the ringing sites, and none from the presumed wintering quarters in central Africa. Three juveniles ringed in Wales have been recovered while on passage in southern England and France. The farthest distance covered was by one ringed on 20 August 2013 at Ynyshir, East Glamorgan, and killed by a cat in Saint-Sulpice, Tarn, France, 32 days later and 953km SSE. While many juvenile and nestling Nightjars ringed in Wales have returned to their natal area in subsequent summers, there is also recent evidence for some natal dispersal (own data). Two chicks ringed at opposite ends of Wales were recaptured at distant breeding sites the following spring: a nestling ringed in Brechfa Forest, Carmarthenshire, on 26 July 2018 was recaught on 19 May 2019 on Cors Fochno, Ceredigion, 61km to the north, and a nestling ringed in Clocaenog Forest, Denbighshire, on 4 August 2018 was recaptured at Thorne Moors, South Yorkshire, on 13 June 2019, 183km to the ENE. Ongoing studies involving the fitting of small GPS tags have just started to provide fascinating insights into the exact wintering locations and migration routes of Welsh Nightjars. Our first retrieved tag, from a female breeding in Clocaenog Forest in 2019, showed that she left her territory on 7 September and crossed into Africa on 5 October. On 15 October she set off across the Sahara taking just nine days to complete the 2,500+ km journey, at one point covering over 600km in a single 24-hour period at an altitude of 4–5km. She finally reached her wintering location in the Democratic Republic of the Congo, near Lake Tanganyika, on 23 November (own data). DNA studies underway at York University should add useful data on the degree of genetic mixing with populations breeding outside Wales.

The current breeding population estimate for Britain, of 4,600 (3,700–5,500) churring males (Woodward *et al.* 2020), is derived from the results of the 2004 survey (Conway *et al.* 2007). The estimate of 280 churring males in Wales in that survey seems rather low, representing just 6.1 % of the GB population. Using the national forestry inventory (Forestry Commission 2017) there is estimated to be an area of at least 764km² of potentially suitable woodland habitat in Wales (i.e. felled areas, ground-prepared areas, low-density woodland and young trees). Combining this with the mean density estimates from Brechfa, Clocaenog and Pen y Cymoedd, we suggest a revised population estimate of 406 churring males for commercial forestry areas alone, with the total in Wales being almost certainly in excess of 500 churring males.

The Nightjar's Amber conservation status in Wales is because of a historical decline, followed by a recovery (Johnstone and Bladwell 2016). Climate modelling suggested that Wales will become more suitable for Nightjars during the 21st century (Huntley *et al.* 2007), but the effects of climate change in its wintering quarters are unknown. There have been marked declines in the abundance of many large moth species in Britain in recent years (Conrad *et al.* 2006). The species is now almost entirely dependent on open areas within coniferous woodland plantations for nesting, so its future in Wales will largely depend on forestry management ensuring that suitable breeding sites remain available.

Tony Cross and Mike Shewring

Sponsored by John Woodruff

Chimney Swift *Chaetura pelagica* Coblyn Simdde

Welsh List Category	IUCN Red List	Rarity recording
	Global	
A	VU	BBRC

This species breeds in eastern North America and winters in north-western South America (*HBW*). There has only been one accepted record in Wales, at Penmon, Anglesey, on 2 November 2005. This was part of an influx into Britain, with 16 recorded between 29 October and 9 November, brought over the Atlantic by the remains of Hurricane Wilma.

Jon Green and Robin Sandham

Alpine Swift *Tachymarptis melba* Gwennol Ddu'r Alpau

Welsh List Category	IUCN Red List		Rarity recording
	Global	Europe	
A	LC	LC	WBRC

This large swift breeds discontinuously around the Mediterranean, Turkey and east as far as the Himalayas. A separate population occurs in southern Africa. Its winter range is uncertain but is assumed to be in the Afro-tropics or western India, where the separation of the local populations from northern migrants is not possible (*HBW*). In total, 660 were recorded in Britain during 1958–2018, with an average of 12–15 annually over the last 40 years (White and Kehoe 2020a), but in Wales this species remains a scarce visitor, mainly in spring.

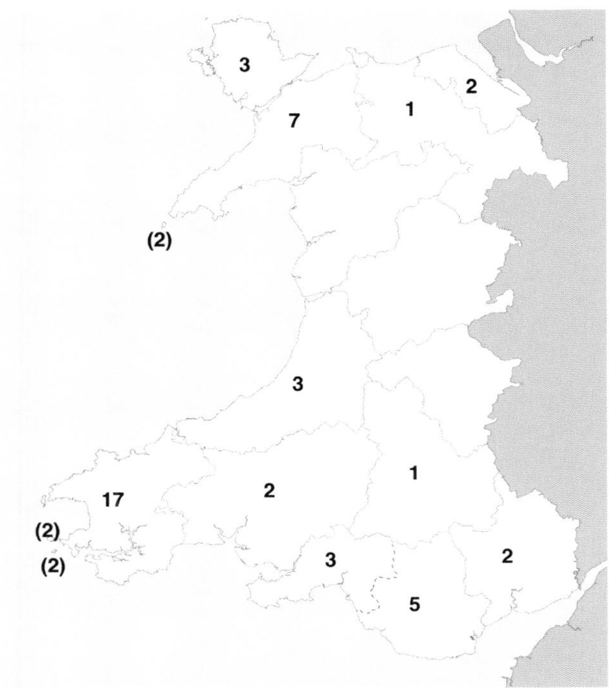

Alpine Swift: Numbers of birds in each recording area 1908–2019. One bird was seen in more than one recording area.

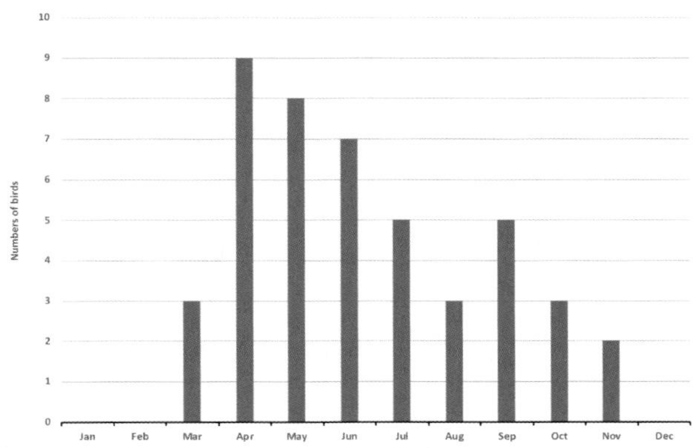

Alpine Swift: Totals by month of arrival, 1908–2019.

The first Welsh record was of two birds seen at Angle Bay, Pembrokeshire, on 20 November 1908, one of which was shot. Two others were seen at St Davids, Pembrokeshire, on 14 September 1934, and singles at Aberdaron, Caernarfonshire, on 27 June 1954 and at Llanfrynach Pool, Breconshire, on 30 June 1956. A further 39 birds have been recorded in Wales since, on average one every other year. There was an increase in records in the 1980s and 1990s, to about one a year, with two recorded in 1983, 1987, 1988 and 1996 and four in 1999. Since the turn of the 21st century, there have been fewer records, with only three between 2010 and 2019.

Most records are of over-shooting spring birds, mainly in the southern counties. The earliest was at Borth, Ceredigion, on 1 March 1987. There have only been 11 autumn records, from August onwards, with particularly late ones being at the Great Orme, Caernarfonshire, on 5–7 October 2015, South Stack, Anglesey, on 8 October 1960, Bardsey, Caernarfonshire, on 12 October 1962, and two at Angle Bay on 20 November 1908.

Jon Green and Robin Sandham

Total	Pre-1950	1950–59	1960–69	1970–79	1980–89	1990–99	2000–09	2010–19
45	4	2	6	5	10	9	6	3

Alpine Swift: Individuals recorded in Wales by decade.

Swift *Apus apus* Gwennol Ddu

Welsh List Category	IUCN Red List			Birds of Conservation Concern			
	Global	Europe	GB	UK	Wales		
					2002	2010	2016
A	LC	LC	EN	Amber			

Seeing the first Swifts of the season brings joy to many people across Wales. Their screaming cries evoke summer like no other sound. Swifts can feed, drink, preen, bathe, mate and sleep on the wing, so much of their life remains high up and out of sight.

Swifts breed throughout Europe, and east across temperate Asia as far as Mongolia and northern China. They are long-distance migrants, with almost all wintering in Africa south of the equator, reaching as far as South Africa (*HBW*). Swifts have few significant predators, and are among the longest-lived birds of their size, but have relatively low productivity, usually laying only 2–3 eggs each year. Several birds ringed as adults in Wales have survived for more than ten years. Welsh-ringed Swifts have been recovered during southward migration in Spain and Morocco. Geolocators have been used to track Swifts, although none have yet been fitted to birds nesting in Wales. Swifts have been found to use different routes on spring migration to autumn, with different stopover sites in southern Europe and North and West Africa. They have important wintering areas in the Congo, but also farther south and east, in Zimbabwe, Tanzania and Mozambique (Åkesson *et al.* 2012, Appleton 2012). Evans *et al.* (2012a) suggested that Swifts range over a greater variety of winter habitat than other migrants.

Relatively brief annual visitors, most Swifts return to Wales in early May and are gone by early August. The first birds usually arrive in Wales in mid-April, but there have been a few records

Swifts © Gary Eisenhauer

in March, such as six at Orielton, Pembrokeshire, on 21 March 1969. The last are usually recorded in September, but there have been several records as late as early November, including one at Plas Gogerddan, Ceredigion, on 9 November 1988. Unseasonal records include one over the Marloes area, Pembrokeshire on 1 January 2020, one at Mwnt, Ceredigion, on 11 January 1998 and one at Ynyslas, Ceredigion, on 8 February 1992.

Arrival date and the length of the breeding season can vary according to foraging conditions. Unseasonably cool weather in June 2019 may have accounted for reports of broods fledging in late August and even the first week of September (pers. obs.). June temperature fluctuations affect breeding success, although variations in July temperatures may have greater long-term impact as these seem to influence adult survival (Thomson *et al.* 1996).

At one time, all Swifts would have nested in natural sites. Historically, Swifts were recorded nesting in coastal cliffs in East Glamorgan, Gower, Pembrokeshire, Ceredigion, Caernarfonshire and Anglesey. They still nest in such sites at Stackpole, St Govan's Head, Cemmaes Head and Dinas, Pembrokeshire, and at Fall Bay, Gower (Paddy Jenks, Bob Haycock and Alan Rosney pers. comm.), although numbers at St Govan's Head and Stackpole have been lower in the last few years (Bob Haycock pers. comm.). Cliff nesting was also suspected at Trefor, Caernarfonshire in 2019 (Adrienne Stratford *in litt.*) and was formerly occasionally reported inland, such as at Betws-y-coed, Caernarfonshire (North *et al.* 1949).

Expansion of urban habitat may have allowed an increase in Swift populations in Wales, although it is not known when the transition from nesting in holes in cliffs and trees occurred. Swifts are now closely associated with humans, breeding in villages, towns and cities, with an apparent bias in favour of settlements with older, taller buildings. Examples of apparent strongholds include the Conwy Valley, Caernarfonshire/Denbighshire, with its network of villages with stone-built architecture, surrounded by riverine and woodland habitat. Historic buildings such as Conwy Castle may have hosted breeding Swifts for centuries. Williams (1835) noted that "The castle and town walls furnish commodious nestling [*sic*] places for great numbers of these birds". Forrest (1907) described

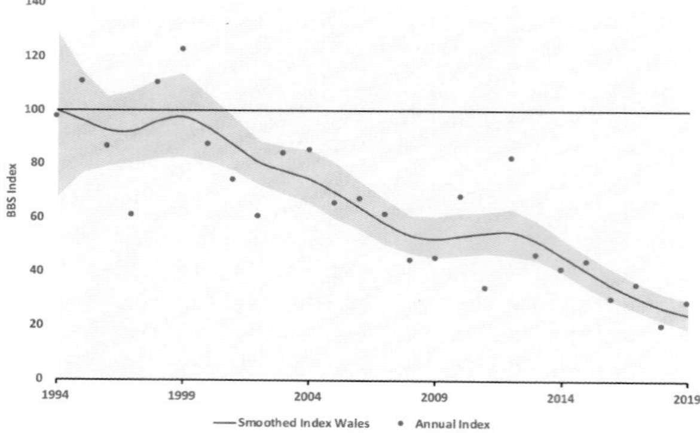

Swift: Breeding Bird Survey indices, 1994–2019. The red line shows the smoothed index, 1995–2018. The shaded area indicates the 85% confidence limits.

the species as "more or less common" in all parts of North Wales. It was said to have been an increasingly common and widespread visitor to Glamorgan in the 1890s, but the population there declined between 1925 and 1966, particularly in the Cardiff suburbs (Hurford and Lansdown 1995). Ingram and Salmon (1954) thought that it was then possibly increasing in Carmarthenshire.

Fieldwork for the *Britain and Ireland Atlas 1968–72*, and for those in 1988–91 and 2007–11, found Swifts in over 90% of 10-km squares in Wales. Relative abundance in Wales in 2007–11 was low compared to most of England and had decreased almost everywhere since 1988–91. Wales held several of the 10-km squares that showed the greatest decline in relative abundance in Britain in this period (*Britain and Ireland Atlas 2007–11*).

The BBS showed a 72% decrease in Swift numbers in Wales between 1995 and 2018. After Greenfinch, this was the second-steepest decline of any species (Harris *et al.* 2020) and an even steeper decline than in England (-58%) or Scotland (-52%). Serious declines may have been under way before 1994, but no wide-scale monitoring scheme could track their fortunes. Rooftop-height screaming parties are a reliable sign of breeding, but locating nest-holes and establishing the numbers of pairs or fledglings is a challenge. BBS data record the wide-ranging foraging behaviour of this species but not its nesting distribution. The RSPB is currently developing a standardised Swift survey methodology that, if widely adopted, could provide meaningful population data.

Gwent, East Glamorgan and Gower are the only counties where counts of over 1,000 have been recorded. Such counts have been made in every month between April and August. There were 1,000 over Eglwys Nunydd Reservoir, Gower, as early as 30 April in 2004, and a remarkable *c.*10,000 birds there on 27 August 1985. The largest spring count recorded in Wales was over 3,000, also at Eglwys Nunydd Reservoir, on 18 May 2000, though "many thousands" were recorded flying over Brynmawr, Gwent, on 10 May 1993. Large counts in June and July could include non-breeding subadult birds that return in the latter half of summer to form pairs and prospect for nest sites, while large congregations of Swifts in late summer around wetlands presumably consist of birds feeding up prior to southward migration. In East Glamorgan, 2,500 were recorded over East Aberthaw ash tip on 20 June 2007 and "several thousand" over Cardiff Bay on 17 July 2001. Counts farther north and west are not as high, though 700 flew south over Bardsey, Caernarfonshire, on 6 June 1969.

At least two-thirds of Swifts in Wales nest in buildings over 100 years old (Bladwell 2014). It is thought that the loss of traditional sites through renovation or demolition of older buildings is a major factor in Swift decline. *Birds in Wales* (1994) linked decreases in local populations to civic rebuilding programmes in the 1970s and 1980s. More recent focus on energy efficiency has caused many of the cavities used by Swifts to be filled in, and as Swifts are strongly site-faithful, direct displacement through nesting habitat loss is clearly an issue.

Recent conservation work in Wales has focused on addressing the loss of nest sites. The provision of purpose-built Swift accommodation may prove significant in the future. North Wales Wildlife Trust has installed over 350 nest-boxes, while in 2019, a bespoke, 90-nest-space Swift Tower was erected in Cardiff Bay, with a solar-powered call-attraction system (Rosney 2019). There are active groups championing Swifts in several towns and cities. There are examples of Swifts occupying both external boxes and swift bricks—integral nest-spaces incorporated into a wall—in most Welsh counties. Increased awareness of the Swift's nesting requirements among homeowners, builders and consultants could help address displacement by protecting existing cavities during renovation, or the installation of nest-boxes as mitigation. Mainstream promotion of Swift bricks in new-build homes could also address the absence of cavities in modern buildings.

Some local examples of Swift decline in Wales do not, however, appear related to loss of nest sites. Landscape-scale insect loss is a possible factor: there has been a decline of more than 75% in total flying insect biomass in Germany over a period of 27 years (Hallmann *et al.* 2017). If this is reflected here, it would be remarkable if Welsh Swifts had *not* been adversely affected. Despite their extreme mobility, there could be an incremental effect on productivity, if Swifts have to travel farther in search of isolated insect hotspots, such as wetlands or deciduous woodland.

The Swift is currently Amber-listed in Wales, on the basis of a moderate breeding-population decline over 25 years (Johnstone and Bladwell 2016). Further population declines since this assessment mean that it is likely to qualify for the Red List in the next assessment. The breeding population in Wales in 2008–11 is estimated at 4,100 pairs (3,000–5,250 pairs) (Hughes *et al.* 2020). There remain various gaps in our understanding of Swift ecology, such as whether there is a threshold colony size below which breeding success suffers. The focus so far has been on factors affecting Swifts in Europe, but they spend nine months each year elsewhere, and the possible effects of insecticide use, climate change and habitat loss in Africa have been little explored, although studies of Swifts nesting in northern Italy found that significant variation in overwinter survival between years was best explained by the El Niño–Southern Oscillation (Boano *et. al.* 2020). Alongside better monitoring of nest sites, research into the causes of decline is needed urgently if wholesale disappearance of Swifts from Welsh towns and villages is to be averted.

Ben Stammers

Sponsored by Swift Conservation Trust and North Wales Wildlife Trust

Pallid Swift *Apus pallidus* Coblyn Gwelw

Welsh List Category	IUCN Red List		Rarity recording
	Global	Europe	
A	LC	LC	BBRC

The Pallid Swift breeds in southern Europe and North Africa, and eastward through the Middle East to Pakistan (*HBW*). The southernmost populations are resident, while others are medium-distance migrants. Most winter in the Sahel area of Africa, but some remain in southern Europe in winter. Out of a total of 125 individuals recorded in Britain to the end of 2019 (Holt *et al.* 2020), only five were in Wales:

- Strumble Head, Pembrokeshire, on 12–13 November 1984;
- Porthgain, Pembrokeshire, on 5–7 October 2001;
- Cefn Sidan, Carmarthenshire, on 20 March 2010;
- 1CY at Skokholm, Pembrokeshire, on 4 November 2018 and
- 1 CY at Newport Wetlands, Gwent, on 10 November 2018.

The two in November 2018 were part of an influx into Britain, following strong southerly winds as ex-hurricane Oscar moved past.

Jon Green and Robin Sandham

Little Swift *Apus affinis* Gwennol Ddu Fach

Welsh List Category	IUCN Red List		Rarity recording
	Global	Europe	
A	LC	VU	BBRC

This species breeds locally in southern Spain and Morocco and throughout sub-Saharan Africa. It also breeds locally through the Middle East and is widespread in India. It is largely resident, but some populations in the Middle East are migratory. There have been 25 individuals in Britain up to the end of 2019 (Holt *et al.* 2020), including two in Wales. The first was on Skokholm, Pembrokeshire from 31 May to 1 June 1981, and the other at Ysgyryd Fawr, Gwent, on 6 June 2004.

A bird at Llanrwst, Denbighshire, on 6 November 1973 was initially accepted by BBRC, but has now been found not proven upon review by BOURC.

Jon Green and Robin Sandham

Great Bustard *Otis tarda* Ceiliog y Waun

Welsh List Category	IUCN Red List			Rarity recording
	Global	Europe	GB	
B	VU	LC	RE	BBRC

The Great Bustard breeds in Iberia and Morocco, and from Germany east to Ukraine, with other populations in Turkey, Central Asia, Mongolia and northeastern China. The species is resident in Iberia, but some eastern populations are migratory, moving south in winter (*HBW*). It formerly bred in parts of England, where it became extinct around 1838 (*Birds in England* 2005) and also bred in Scotland. It has recently been reintroduced to Wiltshire.

Remains of Great Bustard were found at Port Eynon Cave, Gower, dating from the Mesolithic era. This is one of only five records of the species in Britain (Yalden and Albarella 2009). There is only one confirmed record in Wales in historic times, a female shot at Glan-rhedw, near Llanelli, Carmarthenshire, in Christmas week 1890.

Earlier publications listed two other records, which are now considered no longer acceptable following a review. One was of a bird said to have been shot before 1830 at Llanrwst, Denbighshire, and the other of one shot at Pontardawe, Gower, on 20 December 1902.

Jon Green and Robin Sandham

Little Bustard *Tetrax tetrax* Ceiliog Gwaun Bychan

Welsh List Category	IUCN Red List		Rarity recording
	Global	Europe	
A	NT	VU	BBRC

The Little Bustard breeds from southern Europe and northwest Africa east to Central Asia. A total of 212 individuals have been recorded in Britain (Holt *et al.* 2020) but numbers have declined since the 19th century (Parkin and Knox 2010). All six individuals recorded in Wales were shot, three of them in the same area of Carmarthenshire:

- Llanbabo, Anglesey, on 9 December 1884;
- Gileston, East Glamorgan, on 19 November 1885;
- female at Laugharne, Carmarthenshire, on 19 November 1901;
- Laugharne, Carmarthenshire, on 5 February 1914;
- male at Laugharne, Carmarthenshire, on 9 September 1938 and
- 1CY at St Davids Airfield, Pembrokeshire, on 23 November 1968.

The skin of the St Davids individual, examined at Liverpool Museum, resembled a first-winter male from the eastern part of the range of this species.

Jon Green and Robin Sandham

Great Spotted Cuckoo *Clamator glandarius* Cog Frech

Welsh List Category	IUCN Red List		Rarity recording
	Global	Europe	
A	LC	LC	BBRC

This species breeds in Iberia and around the north Mediterranean coast, with scattered populations east to Iran and extensively through sub-Saharan Africa. European birds winter mainly in sub-Saharan Africa, though a few remain in Iberia. There have been 54 records in Britain, on average about one a year over the last 30 years (Holt *et al.* 2020). Most records are of over-shooting migrants in early spring, as were the five in Wales:

- male found dead at Plas Penhelig, Aberdyfi, Meirionnydd, on 1 April 1956;
- Newborough, Anglesey, on 3–15 April 1960;
- 2CY at Trefeiddan, St Davids, Pembrokeshire, on 10 March 2009;
- 2CY at Penally, Pembrokeshire, on 11–23 March 2014 and
- 2CY at Cwm Cadlan, Penderyn, East Glamorgan, on 15–17 April 2015.

Climate modelling indicates that Great Spotted Cuckoos will move north, to breed as close to Britain as Normandy, during the 21st century (Huntley *et al.* 2007), so it may occur in Wales with greater frequency.

Jon Green and Robin Sandham

Yellow-billed Cuckoo *Coccyzus americanus* Cog Bigfelen

Welsh List Category	IUCN Red List	Rarity recording
	Global	
A	LC	BBRC

This species breeds in the eastern USA, Caribbean and Central America. Northern populations are migratory, wintering in South America. There have been 69 records in Britain, about one every two years on average over the last 30 years (Holt *et al.* 2020). There have only been four records in Wales:

- shot at Stackpole, Pembrokeshire in autumn 1832, the first for Britain;
- found dead at Clarach, Aberystwyth, Ceredigion on 26 October 1870;
- found dead at Craig y Don, Menai Bridge, Anglesey, on 10 November 1899 and
- Porthclais, Pembrokeshire on 30 October 1994.

Jon Green and Robin Sandham

Cuckoo *Cuculus canorus* Cog

Welsh List Category	IUCN Red List			Birds of Conservation Concern			
	Global	Europe	GB	UK	Wales		
A	LC	LC	VU	Red	2002	2010	2016

No April would be complete without the song of the male Cuckoo proclaiming its return from its African wintering grounds. For many, the song of the Cuckoo is the first sign that spring has truly arrived. The Cuckoo breeds over most of Europe and Asia north of the tropics, as well as northwest Africa. It is migratory except in the far south of its breeding range. Most winter in Africa south of the Sahara, but some do so in southern Asia (*HBW*). The first Cuckoos usually arrive back in Wales in mid-April, although there are sometimes records in March. Most adults have left by the end of July, but young birds often stay later. There have been records into October or even early November. Unseasonal records include one at Little Milford, Pembrokeshire, on 21–22 December 1954, one heard and seen at Bettws Newydd, Gwent, on 14–15 and 22–26 January 1989, and one at Llanmadoc, Gower, on 10 February 1989.

The Cuckoo is a brood parasite, laying its eggs in the nests of various host species. The Cuckoo chick hatches before the other eggs in the nest and ejects them, leaving it as the sole occupant of the nest. Individual female Cuckoos specialise in a particular host species, a specialisation that is passed on to their female offspring. Over 50 species have been recorded as being hosts to Cuckoo in Britain, but five of these account for 90% of the parasitised nests: Reed Warbler, Meadow Pipit, Dunnock, Robin and Pied Wagtail (Davies 2000). The BTO Nest Record Scheme

Male Cuckoo being mobbed by a Meadow Pipit © Steve Wilce

holds records of 65 nests parasitised by Cuckoo in Wales, dated between 1939 and 2018. The Meadow Pipit was the host in 51% of records, followed by Dunnock in 25%. In eastern England the Reed Warbler is a major host, but the Cuckoos that specialise in that species do not seem to have followed the Reed Warblers that have colonised Wales over the past 40 years. Only 3% of records in Wales had Reed Warbler as the host, though Reed Warblers have been recorded as the main host for Cuckoos at some sites, for example Cors Dyfi, Montgomeryshire (Montgomeryshire Bird Report). Although the sample is small, there appears to have been a change in recent years, with the Meadow Pipit acting as host in 83% of the 23 records since 1990.

The Cuckoo features prominently in early Welsh poetry, notably in the ninth-century poem *Claf Abercuawg* ('The Sick Man of Abercuawg'), where the sufferer is haunted by the singing of the birds outside his sickroom. Late-19th-century and early-20th-century writers described it as a common breeding bird in Wales, including in at least some parts of the lowlands, such as the sand dunes at Abersoch, Caernarfonshire, and in western Anglesey (Aplin 1900, Forrest 1907). By the 1950s there were suggestions of a decline: Ingram and Salmon (1957) said that in Breconshire "we do not think there are as many Cuckoos as there used to be twenty or so years ago". Lockley (1961) stated that it was "less plentiful than formerly" in Pembrokeshire and, in Glamorgan, Heathcote *et al.* (1967) thought that there were far fewer than 30–40 years previously.

The *Britain and Ireland Atlas 1968–72* showed the Cuckoo to be present in 97% of 10-km squares in Wales, with probable or confirmed breeding in most. By the 1988–91 *Atlas*, it occurred in 91% of 10-km squares with losses evident in Pembrokeshire, although there was little change elsewhere. The 2007–11 *Atlas* showed further losses, particularly in coastal Denbighshire and Flintshire, with both gains and losses in Pembrokeshire. Overall, the species was found in 84% of squares. Relative abundance was high in the uplands, but lower in Anglesey and the southwest. In most areas, relative abundance has declined since 1988–91.

Tetrad atlases show evidence of decline where results can be compared. In Gwent, Cuckoos were present in 87% of tetrads during the first county atlas in 1981–85, but in only 67% in 1998–2003, with the decline most marked on farmland (Venables *et al.* 2008). In East Glamorgan, Cuckoos were present in 58% of tetrads in 1984–89, but in 2008–11 they were present in only 23% of tetrads and recorded with breeding codes in just 14% of tetrads. Pembrokeshire showed a similar pattern: 45% of tetrads occupied in 1984–88, but only 16% in 2003–07, with the decline mostly in lowland areas. Cuckoos were recorded in 50% of tetrads in Breconshire visited in 1988–90, but it was noted in the *Welsh Bird Report* 2011 that in this county "the concentration of birds into upland and common-land areas is increasingly noticeable". The *North Wales Atlas 2008–12* found Cuckoos in 33% of tetrads, with most of the occupied tetrads in semi-upland and upland areas. Few tetrads were occupied in Flintshire or Anglesey, but birds were present in 43% of tetrads in Caernarfonshire and 46% in Meirionnydd.

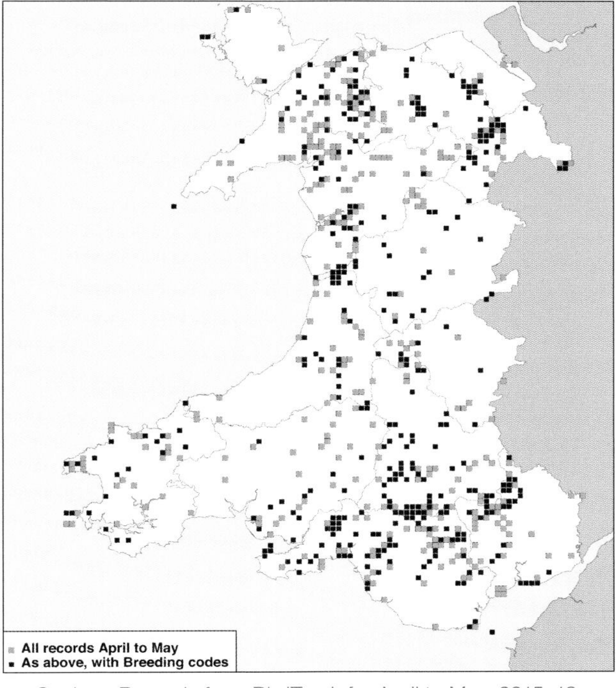

Cuckoo: Records from BirdTrack for April to May, 2015–19, showing the locations of records and those with breeding codes.

The BBS shows a marked difference in the fortunes of the Cuckoo in different parts of Britain. Over the 23-year period 1995–2018, there was a decline of 29% in Wales. This compares to a decline of 71% in England and an increase of 54% in Scotland over the same period (Harris *et al.* 2020).

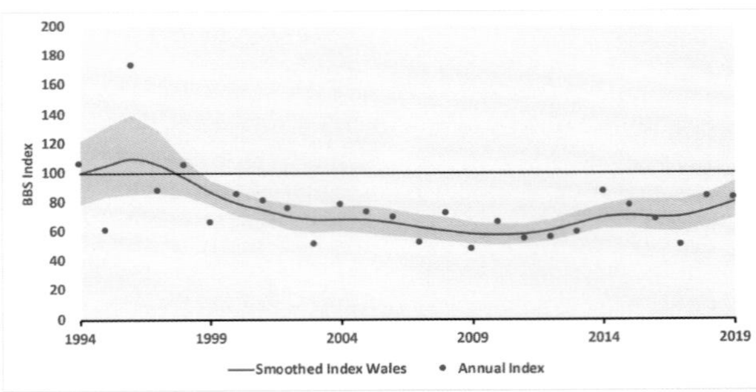

Cuckoo: Breeding Bird Survey indices, 1994–2019. The red line shows the smoothed index, 1995–2018. The shaded area indicates the 85% confidence limits.

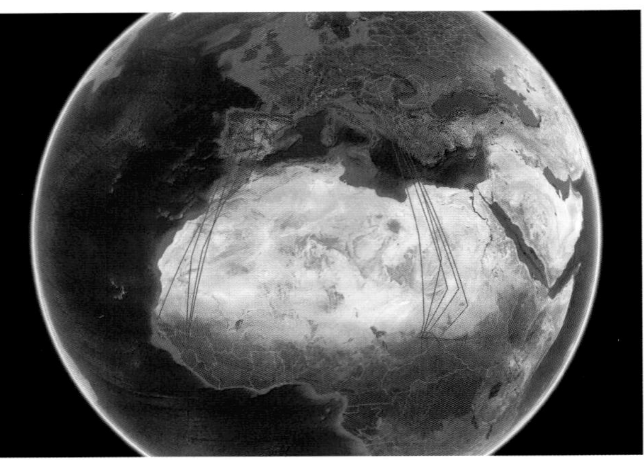

Cuckoo: The movements over four years of the Cuckoo, David, that was ringed in Wales. Map courtesy of Google Earth. Cuckoo data reproduced with permission of BTO (www.bto.org).

Prior to 2011, little was known about the migration routes and wintering areas used by Cuckoos nesting in Britain, except that birds were known to go south via the Po Valley area of Italy. By 2011, satellite tracking devices were small enough to be safely fitted onto males, and the BTO began a Cuckoo-tracking project. Initially five were tagged in eastern England, followed by a further five in mid-Wales, in May 2012. This showed that Cuckoos breeding in Britain used two distinct routes to reach the same wintering grounds in central Africa: some headed southwest via Spain, while others headed southeast via Italy or the Balkans. An individual Cuckoo consistently used the same route in different years. Mortality prior to completion of the Sahara crossing, which is the major ecological barrier for birds on both routes, was higher for birds using the western route through Spain, even though this route is shorter (Hewson *et al.* 2016). Several tracked birds perished when they reached Spain in a drought and were unable to obtain sufficient sustenance for onward migration. Comparison of tracked Cuckoos from nine widely spaced localities in Britain, including one in the uplands of central Wales near Tregaron, Ceredigion, found a strong correlation between population trends in each area and the proportion of Cuckoos following each migration route. This suggests that recent problems on the western migration route through Spain may have contributed to the population decline (Hewson *et al.* 2016). All the birds from the Tregaron study area used the eastern route to reach their wintering grounds, as did those from upland areas in Scotland, whereas birds from England made much greater use of the western route. More research is needed, and the samples are small, but this could be one explanation why Cuckoos are doing better in Wales and Scotland than in England.

The map shows the movements of one of the Welsh birds, given the name 'David', who took a more easterly route through Slovenia, with a brief stop in Montenegro, before setting off across the Mediterranean to Libya, finally arriving in the Democratic Republic of the Congo, where all the British Cuckoos over-winter. The following January, return migration began with a dogleg to West Africa and then northwards across the Sahara into southern Spain, before returning to the same region of mid-Wales where he was ringed.

Mortality on migration is probably not the only reason for the decline of the Cuckoo. Caterpillars of various moth species are important prey, and there have been marked declines in the abundance of many large moth species in Britain in recent years (Conrad *et al.* 2006). Declines in prey abundance in agriculturally 'improved' habitats, but not in semi-natural habitats, may help to explain the loss of Cuckoos from farmland (Denerley 2014). The amount of scrub, on which caterpillars can feed, may be an important factor. Changes in global weather patterns due to climate change could also be important. BBS data indicate that the Cuckoo's main host species in Wales, the Meadow Pipit, has also declined in recent years.

The Cuckoo population of Wales is estimated to be about 1,900 pairs (1,000–2,750 pairs) (Hughes *et al.* 2020). Its Red conservation status in Wales is because of the severe long-term decline in the UK population, rather than the more moderate decline recorded in Wales over the past 25 years (Johnstone and Bladwell 2016). Longer-term changes in Wales cannot be assessed. The Cuckoo may be faring reasonably well in the Welsh uplands at present, but further research into factors affecting the population here is needed.

Kelvin Jones

Pallas's Sandgrouse *Syrrhaptes paradoxus* **Iâr Diffeithwch**

Welsh List Category	IUCN Red List		Rarity recording
	Global	Europe	
B	LC	EN	BBRC

This species lives in Kazakhstan, Mongolia and northern China, makes relatively short-distance northward movements in the breeding season and moves south in winter. It is capable of occasional eruptions across all the Palearctic, although the reasons are not fully understood (*HBW*). Two major eruptions to Europe, in 1863 and 1888–89, resulted in several Welsh records, with a total of 20 individuals:

- three near Tremadog, Caernarfonshire, on 9 July 1859, one of which was shot;
- one shot near Haverfordwest, Pembrokeshire,on 8 February 1864;
- 16 at Llanrhidian, Gower, in May or June 1888, of which two were shot and
- a female shot at Ambleston, Pembrokeshire, on 28 May 1888.

Previously published records of one in Pembrokeshire on 8 February 1864; and a pair near Fonmon, East Glamorgan, in 1888 are now considered no longer acceptable following a review of historical records.

Jon Green and Robin Sandham

Rock Dove / Feral Pigeon *Columba livia* Colomen y Graig/Colomen Ddôf

Welsh List Category	IUCN Red List			Birds of Conservation Concern			
	Global	Europe	GB	UK	Wales		
B, C4	LC	LC	LC	Green	2002	2010	2016

Though very familiar to most people, particularly those living in urban areas, the Feral Pigeon seems of little interest to most bird-watchers and is therefore greatly under-recorded. The Rock Dove, the wild form of the species, has greater respectability, but no longer occurs in Wales. The Rock Dove's natural range extends from the Atlantic and southern coasts of Europe south to northern Africa and east to India and Mongolia. It has spread in urban areas, interbreeding with domesticated Pigeons, and is now wide-spread to the north of its natural range, and has been introduced to Australasia and several Pacific Islands. The degree of mixing of wild and domesticated forms is unknown in many areas (*HBW*). In Britain, the wild Rock Dove is now confined to north and west Scotland, particularly the Inner and Outer Hebrides, which are thought to be among the very few places in the world where intact gene pools may still exist (*The Birds of Scotland* 2007). Wild Rock Doves usually breed on cliffs, while most Feral Pigeons breed on buildings, largely in urban areas, but also in rural areas, although some breed in cliffs like their wild ancestors. Birds breeding in urban areas will move into more rural areas to eat grain and seeds.

The pigeon has been domesticated for thousands of years, with the first evidence from Mesopotamia. Dovecotes, or *columbaria,* are well attested in ancient Rome, by writers such as Pliny. Domestic Pigeons may have been introduced into Britain by the Romans. The species was prized as food in medieval times, and dovecotes were found at monasteries and manor houses. *Colomendy*, the Welsh for 'dovecote' is found in many place names. The best-known structure in Wales is at Culver Hole, Gower, which dates from about the 14th century. It has an 18m high wall with 30 tiers of nesting boxes, giving space for hundreds of pigeons.

Culver Hole Dovecote © Bob Haycock

Many of the birds nesting on the cliffs of Wales might not have been truly wild birds even by the late 18th century. Thomas Pennant (1810), visiting the Little Orme, Caernarfonshire, in 1773, commented on the "Rock Pigeons, abundance of which regularly breed here, in preference to the dove-houses, which they constantly quit at their laying-time". Forrest (1907) referred to the difficulty of distinguishing between genuinely wild Rock Doves and dovecote pigeons that had reverted to a wild state. He considered that in North Wales the Rock Dove was confined to the cliffs of Caernarfonshire and Anglesey and was rare, even there. Mathew (1894) reported a few pairs on the cliffs of Pembrokeshire, including on Ramsey, but stated that he personally had never seen the species in the county. Lockley *et al.* (1949) reported several records of apparent Rock Doves in Pembrokeshire in 1947, including six pairs breeding in rocks near St Davids Head, but thought the species was perhaps extinct as a pure race in the county. Ingram and Salmon (1954) said that the species had been unknown in Carmarthenshire for very many years. Wild Rock Doves were said to nest in good numbers on the Gower peninsula in the late 19th century, with a pair thought to have bred on Sully Island, East Glamorgan, in 1904. One or two pairs were reported on the south Gower cliffs in 1925. Small flocks of what were reputed to be wild Rock Doves were said to be present in the Clyne area of Gower, up to about 1944 (Hurford and Lansdown 1995). The authors of *Birds in Wales* (1994) considered that "it seems likely that the last truly wild Rock Doves in Wales were probably to be found on the more remote headlands of Pembrokeshire and, offshore, on Ramsey, during the 1930s and 1940s." Subsequent records are almost certainly all of Feral Pigeon, domesticated pigeons that have reverted to living in the wild, or of racing pigeons that have failed to find their way back to their lofts. Some of these can resemble wild Rock Doves. Lloyd *et al.* (2015) commented that birds showing the superficial characteristics of pure Rock Doves continued to breed on the cliffs at Llansteffan and Pendine in Carmarthenshire.

The *Britain and Ireland Atlas 1968–72* showed the Feral Pigeon present mainly in the lowlands of Wales, with the majority of the occupied 10-km squares in southern Wales, particularly East Glamorgan and Pembrokeshire. The 1988–91 *Atlas* detected a spread into the countryside in England since the 1970s. This was also evident in some parts of Wales, particularly Anglesey. Surprisingly, northern Anglesey showed the highest relative abundance in 1988–91, with high densities also in coastal areas of the northeast and around the Pembrokeshire coast. Some other areas, notably Llŷn, showed extensive losses since 1968–72. The *Britain and Ireland Atlas 2007–11* showed relative abundance to be highest in coastal areas in the southeast, around the main urban centres, and also in Pembrokeshire and on Anglesey. Several areas along the south coast showed an increase in relative abundance since 1988–91. All three Britain and Ireland atlases showed the species to be largely absent from higher ground. The *Gwent Atlas 1981–85* recorded the species in 38% of tetrads, while the 1998–2003 *Atlas* found it in 31% of tetrads. There was a similar pattern in East Glamorgan, where birds were found in 38% of tetrads in 1984–89, while in 2008–11 the species was present in 34% of tetrads but only recorded with breeding codes in 20% of tetrads. The *West Glamorgan Atlas 1984–89* found birds in 35% of tetrads. There was a notable increase in distribution in Pembrokeshire, where the 1984–88 *Atlas* found the species in 30% of tetrads and the 2003–07 *Atlas* recorded it in 47% of tetrads. Graham Rees commented, in the *Pembrokeshire Atlas 2003–07,* that a diminished interest in maintaining lofts by pigeon racers, leading to the release of birds, may have been a contributory factor. *The North Wales Atlas 2008–12* recorded birds in 16% of tetrads, largely concentrated along the north coast and on Anglesey. The species was scarce in Meirionnydd, where it was found in only 3% of tetrads.

The BBS index for Wales shows a fairly steady increase, in contrast to that for England, which has shown a sharp decline since

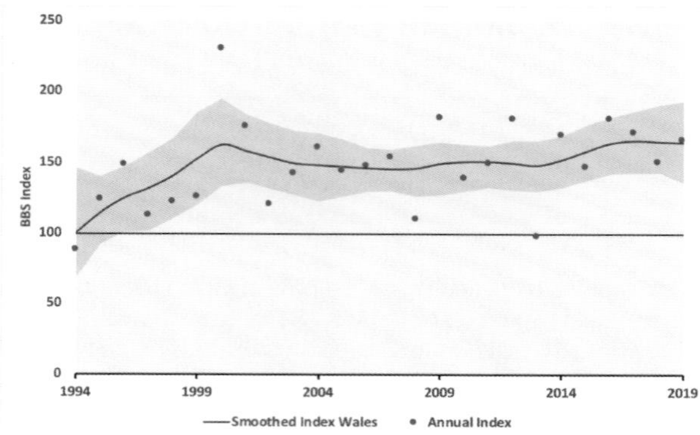

Rock Dove / Feral Pigeon: Breeding Bird Survey indices, 1994–2019. The red line shows the smoothed index, 1995–2018. The shaded area indicates the 85% confidence limits.

about 2005. Records in county bird reports include flocks of 250 or more, with the majority of large counts coming from the cities and towns of Gwent, East Glamorgan and Gower. Up to 300 have been reported in Newport, Gwent. Lloyd *et al.* (2015) stated that several hundred were in Carmarthen town centre in 2002. Venables *et al.* (2008) commented that Feral Pigeons had increasingly visited suburban gardens in Cardiff, to feed from bird tables and beneath feeders, but that this behaviour had not been reported from Gwent. The breeding population in Wales is estimated at 41,000 pairs (33,500–49,000 pairs) (Hughes *et al.* 2020).

The Feral Pigeon is a controversial species in many urban areas, loved and fed by some, but regarded as a pest and a potential health risk by others: "rats with wings" is one epithet sometimes applied to these birds. There have been local culls in some areas of Wales. Numbers of breeding Feral Pigeons are limited by the availability of nest sites, leading to the formation of a reservoir of non-breeding birds. Efforts to reduce numbers by culling could simply result in some of these birds entering the breeding population (*Britain and Ireland Atlas 1988–91*). Feare (1990) noted that a permanent reduction in numbers could only be achieved by removing the available food supply. The future status of the Feral Pigeon in Wales is likely to depend on human attitudes towards it.

Rhion Pritchard

Stock Dove — *Columba oenas* — Colomen Wyllt

Welsh List Category	IUCN Red List			Birds of Conservation Concern			
	Global	Europe	GB	UK	Wales		
A	LC	LC	LC	Amber	2002	2010	2016

The Stock Dove breeds over most of Europe and parts of North Africa and the Middle East, from Ireland eastward as far as northeast China. Birds in the west and south of the range are resident, while those breeding in the north and east are migratory, moving farther south in winter (*HBW*). It is mainly a lowland species, but breeds above 500m in some parts of its range. It seems to prefer a narrow margin between forest and open country, where there are cavities for breeding that are relatively close to fields with weed or crop seeds and near water for drinking. It will use rock crevices, holes in buildings or other structures, such as nest-boxes (*BWP*). The few ringing recoveries involving Wales show no long-distance movements.

In Britain, the Stock Dove was confined to south and east England in the early 19th century. It had spread north and west by 1900, as a result of the increase in arable farming (Parslow 1973). The *Historical Atlas 1875–1900* indicated that Stock Dove was common across North Wales and East Glamorgan/Gower, but uncommon in mid- and West Wales. In northeast Wales, Dobie (1894) called it a common resident, that "ordinarily nests in rabbit holes" or in ivy as well as in holes in trees. Forrest's (1907) account of Stock Dove suggested that it was often confused with Rock Dove, but was common throughout North Wales, more numerous around the coast. It was said to be a common and increasing breeding resident in Glamorgan at the end of the 19th century and thought still to be increasing along the south Gower coast in the 1920s (Hurford and Lansdown 1995).

The proportion of arable land in Wales declined dramatically during the 20th century, from almost a quarter of the land area in the late 1800s to just 3% in 2010 (Blackstock *et al.* 2010), but the effect of this change on Stock Dove numbers in Wales is unclear. Salter (1930) stated that it was much scarcer in northern Ceredigion than formerly, but it was thought to have increased in Ceredigion from the late 1940s, possibly as a result of increased cultivation during the Second World War (Roderick and Davis 2010). The population declined dramatically in England during the 1950s, to a low point in the early 1960s, after which it recovered gradually (O'Connor and Mead 1984). The population in areas of most intensive agriculture dropped to about 10% of its 1950 level, mainly a result of increased egg failure rates caused by sub-lethal doses of organochlorine pesticides. Once bans were imposed on the use of these pesticides, numbers recovered fairly rapidly, helped by the fact that the Stock Dove can lay up to five clutches of, usually, two eggs in a year.

Although O'Connor and Mead had little data from Wales, they concluded that the decline noted in England was not evident in Wales. The lower proportion of arable land in most of Wales meant that the use of organochlorine pesticides would have been much lower, particularly in the west. In Radnorshire, Ingram and Salmon (1955) considered that numbers had decreased compared to 40–50 years previously. In Pembrokeshire, numbers increased from the 1950s, until at least the late 1970s, particularly on the islands and coastal cliffs. This was thought to be linked to the low population of Peregrines in this period. With the recovery of the Peregrine population during the 1980s, Stock Dove numbers declined (*Pembrokeshire Atlas 2003–07*). This was most noticeable on the islands; for example, breeding began on Skokholm in 1967, and by 1975 the population had increased to 62 pairs. This was followed

Stock Dove © Tate Lloyd

by an equally rapid decline, with ten pairs in 1980 and none in 1984 (Betts 1992). A similar trend was evident on Skomer.

The *Britain and Ireland Atlas 1968–72* showed the species occurred in 85% of the 10-km squares in Wales, being present in all except in areas of mountain and some lowlands in Gwent and on Anglesey. Occupation during the *Britain and Ireland Winter Atlas 1981–84* was only 63% of 10-km squares, with presence in most 10-km squares in the Marches but more sparse distribution in western parts. This suggested that either there was some movement from west to east in winter, or that the population in western Wales was very sparse.

The *Britain and Ireland Atlas 1988–91* showed that 74% of 10-km squares were occupied, with extensive losses in the west since 1968–72, particularly in Pembrokeshire and Caernarfonshire, though there were some gains in Carmarthenshire. Relative abundance across Wales in 1988–91 was low compared to most of England. The *Britain and Ireland Atlas 2007–11* showed no net change in the number of occupied squares in the breeding season since 1988–91, with gains on Anglesey and Llŷn but losses in south Caernarfonshire, Meirionnydd and Ceredigion; fewer squares were occupied in winter than in the breeding season in 2007–11. Relative abundance was again low compared to England except in the eastern lowlands, and showed little overall change compared to 1988–91.

There was little change in breeding distribution in Gwent between atlases in 1981–85 and 1998–2003, when Stock Doves were present in 60% of tetrads. The *East Glamorgan Atlas 1984–89* showed breeding presence in 36% of tetrads, mainly in the southern two-thirds of the county but by 2008–11 there were records in just 25% of tetrads and records with breeding codes in only 19%. In Gower, the *West Glamorgan Atlas 1984–89* showed presence in 25% of tetrads. The *Pembrokeshire Atlas 1984–88* found the species in 32% of tetrads, but in only 20% of tetrads by 2003–07. The *North Wales Atlas 2008–12* showed a reduction of 18% in 10-km squares with breeding records since 1968–72. At the tetrad level, the species was recorded in 23% of tetrads in 2008–12, the majority in eastern Flintshire and Denbighshire.

The BBS index showed a fairly steady increase in Wales until about 2014, with a 54% increase between 1995 and 2018 (Harris *et al.* 2020). The reasons for the increase are unclear, though Jennings (2014) noted that numbers had increased in Radnorshire in line with an increase in autumn-sown grain crops and Oil-seed Rape.

There is some evidence of autumn passage. There were 200 at Peterstone Wentlooge, Gwent, on 29 August 2015 and a Welsh record of 1,000 west over Newport Wetlands Reserve, Gwent, on 27 October 2017, the latter associated with a huge movement of Woodpigeons. To put the latter record into perspective, a record of 108 birds over Uskmouth/Goldcliff between 29 October and 13 November 2005 was described as "unprecedented coastal movements" (Venables *et al.* 2008). Other large counts are mainly of feeding flocks or roosts outside the breeding season. There were 200 at Sealand, Flintshire, on 7 November 1903 and about 400 there in February 1905 (Forrest 1907). More recent counts include 300 at Llanover, Gwent, in March 1972, *c.*400 at Sker, Glamorgan, on two dates in December 1971 and 272 roosting at Margam Moors/Morfa Pinewoods, Gower, on 18 January 2003. The record of 400 at Pontiets, Carmarthenshire, in December 1980, published in *Birds in Wales* (1994), is not now thought to be reliable (John Lloyd *in litt.*). In mid-Wales, there were 200 at Dolydd Hafren, Montgomeryshire, in early winter 2003 and again on 2 March 2014. Farther west there were 250 at RSPB Ynys-hir, Ceredigion, on 3 November 1990. Numbers tend to be lower in North Wales, where the largest recent counts were 140, at Holt, Denbighshire, on 14 March 2008 and at Connah's Quay nature reserve, Flintshire, on 4 February 2019.

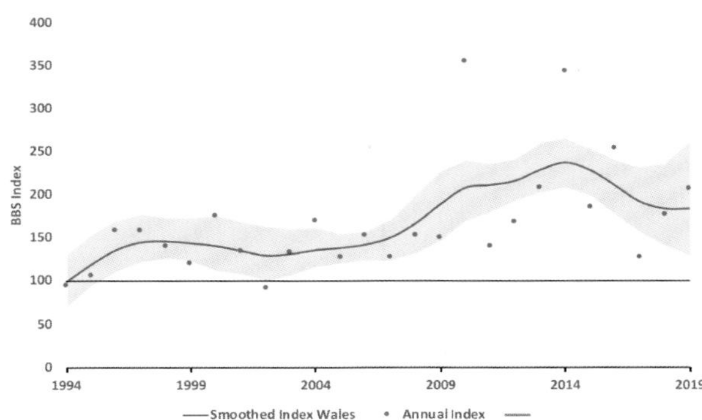

Stock Dove: Breeding Bird Survey indices, 1994–2019. The red line shows the smoothed index, 1995–2018. The shaded area indicates the 85% confidence limits.

The Welsh population is estimated to be about 21,500 territories (Hughes *et al.* 2020). A shortage of old trees with suitable holes for nesting may limit numbers in some areas (Jim Dustow *in litt.*), but there appear to be no major concerns about the status or future prospects of the Stock Dove in Wales at present.

Ian M. Spence

Woodpigeon *Columba palumbus* Ysguthan

Welsh List Category	IUCN Red List			Birds of Conservation Concern			
	Global	Europe	GB	UK	Wales		
					2002	2010	2016
A	LC	LC	LC	Green			

The Woodpigeon, one of the commonest species in Wales, is widespread across Europe to western Siberia and south to North Africa and Pakistan. In the western part of that range it is largely resident, but it is migratory in eastern Europe, Scandinavia and western Russia, wintering as far south as North Africa, Greece and Turkey. It typically inhabits deciduous and coniferous woodland (*HBW*) although now it is closely associated with arable farmland (*European Atlas* 1997). In Britain, the highest densities are in areas of arable land, particularly the eastern lowlands of England (*Britain and Ireland Atlas* 2007–11). Wales now has predominantly pastoral agriculture, with arable land accounting for only 3% of the land area. Extensive areas of arable farmland are limited to the southern lowlands and the eastern lowlands along the border with England (Blackstock *et al.* 2010).

Birds in Wales (1994) suggested that the Woodpigeon probably ranked second to Carrion Crow as the most disliked bird species among Welsh farmers. It may legally be shot in Wales under a General Licence to prevent damage to crops or foodstuff for livestock. Most recoveries of birds ringed in Britain are of birds shot, mainly by farmers (*BTO Migration Atlas* 2002). Relatively few Woodpigeons have been ringed in Wales, but two ringed on Bardsey, Caernarfonshire, moved to France. The large flocks encountered in autumn provide visible evidence of passage, although away from the southeast such flocks may just be aggregations of birds that bred relatively locally.

Woodpigeons were probably more confined to afforested areas in the past. Agriculture provided greater opportunities for feeding, especially in winter on crops and grass/clover leys. Cultivation of Oil-seed Rape is not widespread in Wales, but Woodpigeons have found adequate opportunities to feed in suburban and urban environments too. Living in close association with humans, even in cities, may reduce predation risk (*European Atlas* 1997).

The *Historical Atlas 1875–1900* suggested that Woodpigeons were common across all of Wales in that period, but particularly abundant in Ceredigion and Breconshire. Mathew (1894) said they were found in all wooded parts of Pembrokeshire but could not be considered abundant. Forrest (1907) reported that Woodpigeons had "greatly increased" around the end of the 19th century and that they were most abundant in the eastern counties of North Wales, where they were considered pests by farmers. He also reported large flocks in winter but provided no counts.

Woodpigeons were found in over 97% of 10-km squares in all three Britain and Ireland atlases, in 1968–72, 1988–91 and 2007–11, although relative abundance was low in Wales compared to most of England. Relative abundance increased in Wales between 1988–91 and 2007–11, with the highest densities along the southern coastal and eastern lowlands, particularly in Flintshire and eastern Denbighshire.

At tetrad level, the *Gwent Atlas 1981–85* recorded Woodpigeons in 97% of tetrads, while the 1998–2003 *Atlas* found them in 98% of tetrads. The *East Glamorgan Atlas 1984–89* recorded them in 86% of tetrads, while by the 2008–11 *Atlas* this had increased to 90% of tetrads. The *West Glamorgan Atlas 1984–89* found them in 86% of tetrads in Gower, and they were recorded in 72% of tetrads in Breconshire visited between 1988 and 1990, during fieldwork for the *Britain and Ireland Atlas* 1988–91. Woodpigeons were present in 85% of tetrads in the *Pembrokeshire Atlas 1984–88* increasing to 88% of tetrads in the *Pembrokeshire Atlas 2003–07.* In the *North Wales Atlas 2008–12* they were found in 62% of tetrads, being absent from the highest land where there are few or no trees. In Montgomeryshire, they were almost absent in the breeding season at RSPB Lake Vyrnwy, until about 2010 when there was a marked increase (Jim Dustow *in litt.*).

Late autumn movements are evident, particularly in southeast Wales, although there were few records of very large numbers before about 2010. A count of *c.*5,800, over Cathays, Cardiff, on 6 November 1989, was considered exceptional (Hurford and Lansdown 1995), as was 17,000 over Peterstone, Gwent, on 6 November 1999. In the last decade, greater effort has been made by birdwatchers to monitor visible migration, especially on the Gwent coast. Here notable one-day counts include 98,470 near Peterstone Wentlooge on 6 November 2011, 102,800 there on 12 November 2015, 156,350 over Pen-y-lan on 10 November 2013 (the second-largest movement ever recorded in Britain at the time) and 225,000 over Newport Wetlands Reserve on 27 October 2017. Counts from East Glamorgan include 116,600 over Cardiff Bay on 5 November 2014 and 130,000 over Grangemoor Park on 2 November 2016. Counts farther west and north are usually smaller, but there were 12,000 over Pencarnan Camp, Pembrokeshire, on 6 November 2017 with a flock estimated at 42,000 over Llansantffraid-ym-Mechain, Montgomeryshire, on 13 November 2013. In Radnorshire, 3,500 flew down the Marteg Valley in two hours on 9 November 1996 while 4,400 moved southeast over Radnor Forest in three hours on 29 October 2011 (Jennings 2014). In 2013, Plant (2014) considered that birds from a wide area of central England had gathered in a 'holding area' around Herefordshire and the Wye/Severn valleys in the previous days, until favourable conditions led to a mass movement on 10 November, tracking in a southwesterly direction along the Welsh side of the Severn Estuary. He suggested that the birds dispersed around South Wales for the winter, as there was no record of large flocks crossing the Severn Estuary or the Irish Sea.

Large flocks in late autumn and winter included 3,000 in an unharvested barley field at Ysgyryd Fach, Gwent, in winter 1992/93 (Venables *et al.* 2008), 1,600–2,000 at Llanarthne, Carmarthenshire, on 9 February 1985 and 1,700 at Angle Bay, Pembrokeshire, on 14 February 1975 (Donovan and Rees 1994). In Ceredigion, 2,000 were recorded at RSPB Ynys-hir on 4 November 1990, while in Radnorshire 1,000 were at Glasbury on 21 November 1981. In Meirionnydd, a county record of 2,000 were in oakwoods near Corris in winter 1985, during a good year for acorn mast (Pritchard 2012); a roost of over 700 at Penrhos was the highest count in Anglesey (Jones and Whalley 2004). In Caernarfonshire a flock of 1,000 in January 1977 was considered to be unusual, though in Barnes (1997), Peter Schofield wrote that over 2,000 had been seen in Coedydd Aber (no date). Around 5,000 roosted at Conwy, Caernarfonshire, in December 2016.

The latest estimate of the population in Great Britain is just over five million pairs (Woodward *et al.* 2020). The Welsh population is estimated at 295,000 pairs (280,000–315,000 pairs) (Hughes *et al.* 2020).

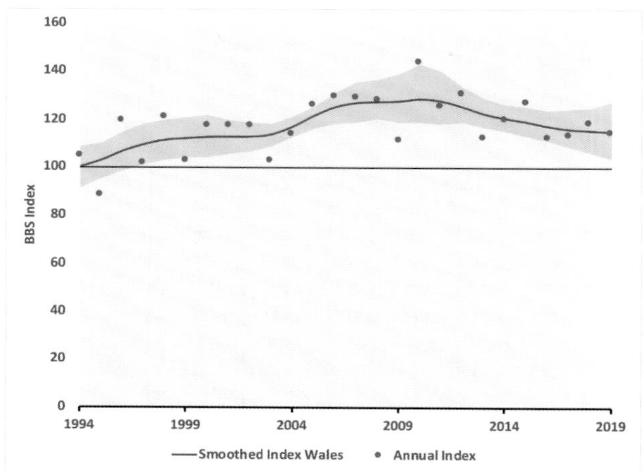

Woodpigeon: Breeding Bird Survey indices, 1994–2019. The red line shows the smoothed index, 1995–2018. The shaded area indicates the 85% confidence limits.

The BBS index for Wales has dipped slightly in the last few years, after a steady rise between 1994 and 2011, but overall, there was an increase of 12% between 1995 and 2018 (Harris *et al.* 2020). There appear to be no major concerns about the population at present. Considerable numbers are shot each year, an estimated 1.9 million across the UK (Aebischer 2019), but there is no evidence that this is having a serious impact on the population.

Ian M. Spence

Turtle Dove *Streptopelia turtur* Turtur

Welsh List Category	IUCN Red List			Birds of Conservation Concern			
	Global	Europe	GB	UK	Wales		
A	VU	VU	CR	Red	2002	2010	2016

The Turtle Dove breeds over most of Europe apart from Ireland and Scandinavia, and its range extends eastward through Central Asia as far as northwest China. Most winter in the Sahel area south of the Sahara (*HBW*). There have been few ringing recoveries involving Wales, though birds ringed on passage on the Welsh islands have been recovered in Ireland, France and Spain. Birds breeding in England move SSW through France and Iberia in autumn, mainly as nocturnal migrants, with juveniles migrating farther to the west than adults. Juveniles form the bulk of recoveries of ringed birds that were shot in southwest France, northwest Spain and Portugal (*BTO Migration Atlas* 2002).

For breeding, the Turtle Dove prefers fairly dry, sunny, sheltered lowlands, mainly below 350m, with cover and access to water. It avoids large forests but uses the edges of open woodland and copses, hedges and trees by riverbanks, preferably near arable land with crop and weed seeds readily available. Fumitories, plants of tilled or disturbed arable land, are often mentioned as a favourite food. It seems to avoid nesting close to human habitation, even farms (*BWP*). In winter it is highly gregarious and roosts in *Acacia* woodland.

Since the 1970s, and especially since 1985, there has been a serious decline across most of its European breeding range.

Turtle Dove © Ben Porter

The formerly large population in European Russia has crashed by more than 90% since 1980 (BirdLife International 2020). The second *European Atlas* (2020) found range contraction mainly on the northern edge of the range, in Britain and Fennoscandia. Elsewhere, the range was largely unchanged, despite a decline of over 30% in population size since 2020.

In Britain, CBC/BBS data indicate a 98% decline in population between 1967 and 2017. If this trend continues, it is expected the species will be close to extinction in Britain within the next two decades (Massimino *et al.* 2019). Habitat destruction, especially the removal of hedges, widespread use of herbicides that have reduced food sources, hunting pressure on migration and, in its wintering range, poor rainfall, deforestation, scrub clearance and agricultural changes have all contributed to the decline (Browne and Aebischer 2005). The species is hunted on its wintering grounds, as well as on migration (Jim Dustow *in litt.*).

Wales was always on the edge of the Turtle Dove's breeding range. It was uncommon but regular in Glamorgan in the late 1890s and increased in the early 20th century. In the early 1920s up to 12 pairs bred in a small area of farmland on the outskirts of Cardiff (Hurford and Lansdown 1995). Farther west, Mathew (1894) said that a few pairs nested in south Pembrokeshire. There were breeding records in north Ceredigion around the turn of the 20th century, although it was declining there by the 1920s (Roderick and Davis 2010). The Turtle Dove was described as "tolerably plentiful in the Vale of Clwyd and at Glyn Ceiriog" and common around Colwyn Bay and Abergele, Denbighshire (Dobie 1894). Forrest (1907) wrote that it was "gradually increasing and spreading westwards" along the north coast as far as Bangor, Caernarfonshire. In 1919 he quoted correspondents as saying that it was increasing in the Llandudno area, and several pairs nested there in 1921 (Pritchard 2017). Forrest also reported that it had recently arrived as a breeding species at Bala, Meirionnydd, but there was no record of breeding on Anglesey in this period, where it was only recorded as a passage migrant.

Birds in Wales (1994) reported that breeding records in Wales were notably fewer by 1950, although there may have been a modest recovery from the mid-1960s to about 1978. Breeding was recorded in several western counties in this period, including the first confirmed breeding on Anglesey in 1969. The species was apparently doing well in the Wye Valley on the Breconshire/Radnorshire border in the 1950s and 1960s, with ten pairs between Glasbury and Llanigon in 1964. The *Britain and Ireland Atlas 1968–72* noted a possible slight increase in Wales, following a decline there in the 1950s and early 1960s. There were few records of confirmed breeding in Denbighshire and Flintshire. The western side of a line down the Cambrian Mountains was virtually devoid of breeding records, but it was more common in Montgomeryshire and eastern Gwent. Its distribution in Britain closely matched the distribution of fumitory spp. The *Britain and Ireland Atlas 1988–91* showed an almost total collapse of the population in Wales, with fewer than ten records of confirmed breeding, all in the eastern half of the country. During the period covered by the *Britain and Ireland Atlas 2007–11* there were just a handful of records in Wales and only one record of confirmed breeding.

Tetrad atlases show the speed and scale of loss. The *Gwent Atlas 1981–85* found Turtle Doves in 18% of tetrads, and the county population was estimated at up to 50 pairs. By 1998–2003, only 2% of tetrads were occupied, a 93% reduction in distribution, and the county population was 0–3 pairs. The species had already virtually disappeared from East Glamorgan by 1984–89, when there was possible breeding in just two tetrads, while the 2008–11 *Atlas* showed presence in one tetrad.

Breeding ceased in Breconshire in 1964, in Carmarthenshire by the late 1960s, in Gower in 1967, in Meirionnydd in 1969, in East Glamorgan in 1972 and there has been no record of breeding in Caernarfonshire since 1974. In 2005, four or five pairs were still hanging on in the Trellech area of Gwent, where grain was being put out for them (Stephanie Tyler *in litt.*). The following year none returned, and there have been no breeding records in Gwent since. Single pairs bred in Denbighshire in 2009 and in a young conifer plantation near Presteigne, Radnorshire, in 2011 (Jennings 2014), but there has been no record of confirmed breeding in Wales since.

The species is now recorded in Wales mainly as a passage migrant in April–June and October, but numbers are much lower than formerly. Records at two island observatories show a gradual increase in 1954–74 and then a decline, with no records in some years since 2000. Similar declines are evident elsewhere, and it seems unimaginable now that 266 were recorded at Lavernock Point, East Glamorgan, between 4 May and 11 June 1965. There were six on Bardsey, Caernarfonshire, on 29 May 2002, but in the decade 2008–17, the total number of records in Wales was 15–20 in most years. 2018 saw a new low, with only five records in Wales.

Of the ten winter (December–February) records in Wales, eight have been since 2010, including two birds at a garden feeding station at Llandegley, Radnorshire, in November and December 2010, with records of single birds in Gwent, East Glamorgan, Gower, Pembrokeshire (three) and Anglesey.

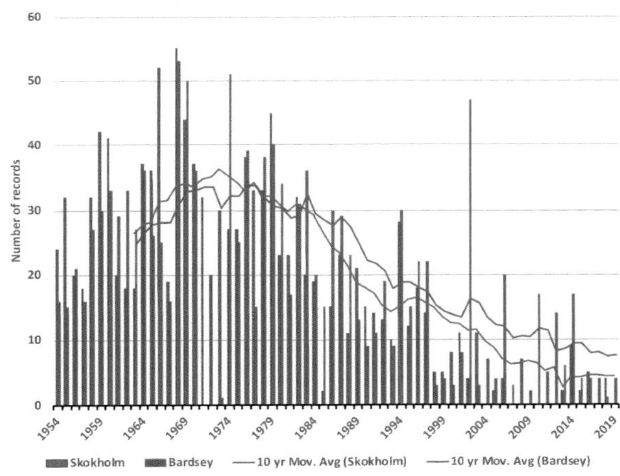

Turtle Dove: Numbers of birds recorded at Skokholm and Bardsey, 1954–2019, with the ten-year moving averages.

The species is Red-listed in Wales under several criteria, including its Vulnerable conservation status, both globally and in Europe, and severe decline in numbers in the UK (Johnstone and Bladwell 2016). In Wales, the reduction in the extent of arable cultivation means that many specialist arable plants are now rare or extinct here (Blackstock *et al.* 2010). The decline throughout Europe suggests that conditions in the wintering grounds and hunting pressure in Europe are having a significant effect. There seems to be little political will to stop hunting in autumn and spring in southern Europe at present. The future prospects for Turtle Doves are bleak, both in Wales and elsewhere in Europe.

Ian M. Spence

Collared Dove *Streptopelia decaocto* Turtur Dorchog

Welsh List Category	IUCN Red List			Birds of Conservation Concern			
	Global	Europe	GB	UK	Wales		
A	LC	LC	NT	Green	2002	2010	2016

The current range of the Collared Dove extends across much of the Palearctic, from Ireland eastward through Europe, and across Asia as far as the Korean Peninsula (*HBW*). In the early 1930s the nearest Collared Doves to Wales were in Hungary, but over the following two decades its range extended about 1,900km northwest (Fisher 1953). It arrived in Britain in 1955 and within ten years outnumbered Turtle Dove. The reasons for the rapid expansion from its core area, in India, are not known but it involved birds that dispersed long distances and were very productive in the areas where they arrived (*European Atlas* 1997).

The first records for Wales were in 1959, all in Caernarfonshire: one on Bardsey on 6 May, two there on the next two days and another at Llandudno on 23–24 November. Collared Doves were recorded in Anglesey and Ceredigion in 1960. In 1961 breeding was confirmed in Pembrokeshire, with four pairs at St Davids. The species may also have bred in Ceredigion that year, although Roderick and Davis (2010) state that there is some uncertainty about this record. Collared Doves quickly spread in Wales, and by the mid-1960s they were present in all Welsh counties except Radnorshire, Breconshire and Carmarthenshire (Hudson 1965). Breeding was confirmed in Radnorshire in 1966, in Carmarthenshire in 1968 and finally in Breconshire in 1969.

Collared Doves favour places where grain is easily available, such as mills, docks, farms and other places with livestock feed. They are associated with mixed habitats, often in suburban environments, villages and large clusters of buildings, avoiding open countryside without human habitation, and city centres, where Feral Pigeons would provide strong competition (*BWP*). Collared Doves can breed in any month of the year and numbers can increase very rapidly if conditions are favourable.

Clearly, Collared Doves have, or had, the capacity to undertake fairly long movements, hence their arrival in Britain, but since becoming established along the eastern Atlantic seaboard they have tended to move shorter distances. Prior to 1980 there were many movements of well over 100km but this is now a rare occurrence (*BTO Migration Atlas* 2002). One recaught at Penrhos, Anglesey, in July 1967 had been ringed as an adult near Ramsgate in Kent, 463km away, in January the same year. Another, killed by a cat at Old Colwyn, Denbighshire, in September 1989, had been ringed 436km away on Sark, Channel Islands, in April 1986. Other ringing recoveries show that, at least until the 1980s, there was immigration into Wales from England. *Birds in Wales* (1994) recorded westerly movements at some places in spring, and in some years, flocks of up to 60 were seen flying west out to sea from RSPB South Stack, Anglesey (Jones and Whalley 2004).

The *Britain and Ireland Atlas 1968–72* showed evidence of breeding across lowland Wales, in 71% of 10-km squares. This had changed little by the *Britain and Ireland Winter Atlas 1981–84*. The *Britain and Ireland Atlas 1988–91* showed a wider distribution, with birds found in 82% of squares, and a spread into higher-altitude areas. Abundance across most of Wales remained lower than elsewhere in Britain, although higher along the coastal lowlands of East Glamorgan, Anglesey, Denbighshire and Flintshire. The *Britain and Ireland Atlas 2007–11* showed a further expansion in breeding distribution, with Collared Doves found in 95% of squares and absent only at the highest altitudes. There was also increased relative abundance across much of North and South Wales compared to 1988–91. The winter distribution in 2008–11 was, not surprisingly, very similar to its contemporary breeding range, but abundance was much lower relative to England at the same latitude.

The *Gwent Atlas 1981–85* found at least possible breeding in 66% of tetrads, with birds absent from the higher-altitude areas in the west and some areas in the east with a lower human population. By 1998–2003, evidence of breeding was recorded in 79% of tetrads, with most of the previous blank areas having been filled. In East Glamorgan, birds were found in 51% of tetrads in 1984–89, while in 2008–11 Collared Doves were present in 59% of tetrads, but only recorded with breeding codes in 52%. In Gower, the *West Glamorgan Atlas 1984–89* recorded evidence of breeding in 47% of tetrads. In Pembrokeshire, the proportion of occupied tetrads increased from 54% in 1984–88 to 62% in 2003–07, although the species was absent from Mynydd Preseli during both periods. The species was recorded in only 12% of tetrads in Breconshire visited during 1988–90, during fieldwork for the *Britain and Ireland Atlas* 1988–91. The *North Wales Atlas 2008–12* showed, at the 10-km level, a 33.3% increase in distribution since 1968–72, with 96% of squares occupied, but at the tetrad level the picture was very different, with Collared Dove recorded in only 49% of tetrads, with most records in the lowlands. This species was recorded in only 22% of tetrads in Meirionnydd, primarily along the coast.

There is evidence from several counties that flocks are smaller now than in the 1970s. Flock sizes in Gwent were highest during 1973–82, when groups of 150–200 occurred, and flocks of 110–120 still occurred in 1994–96, but subsequently flocks were usually of fewer than 50 (Venables *et al.* 2008). In Carmarthenshire, the expansion of poultry farms in the Pembrey/Pinged/Cydweli area in the 1970s attracted regular flocks of over 200 Collared Doves, with a record count of 400 at Pembrey airfield in September 1974. Since the closure of those farms, the highest counts in the county have been at Penclacwydd, where there were 87 in September 1992 and 51 in August 2002, but numbers have been considerably lower there in recent years. Elsewhere, there were 200 at Valley, Anglesey, in November and December 1966, 175 at Cogan Hall Farm, East Glamorgan, in November 1979, 200 at Porthliski, Pembrokeshire, in 1973 and at least 200 at Carreg Farm, Aberdaron, Caernarfonshire, in May 1978. There have been fewer large counts this century, although exceptional numbers were recorded at The Range, Anglesey, on 30 May 2003, with 315 in one flock and over 500 there during the day. There has been only one subsequent count of over 100 in Wales: 103 west of Wylfa, Anglesey, in October 2012. These spring and autumn counts on the coast suggest passage movements, but nothing is known about these.

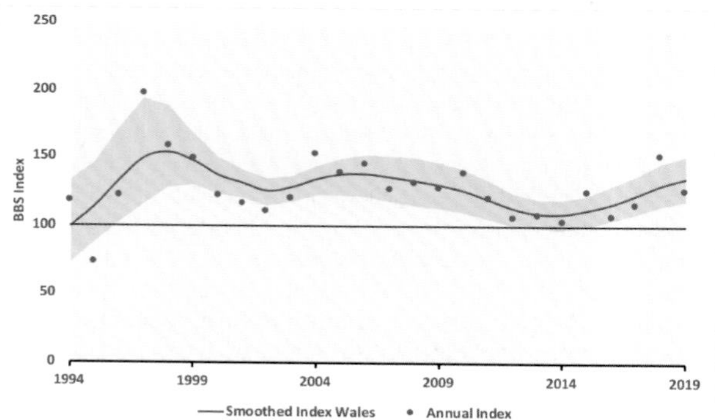

Collared Dove: Breeding Bird Survey indices, 1994–2019. The red line shows the smoothed index, 1995–2018. The shaded area indicates the 85% confidence limits.

The CBC/BBS index for the UK showed a rapid rise during the 1960s and 1970s, then a fall in the early 1980s, followed by a further increase until about 2005 (Massimino *et al.* 2019). The BBS index for Wales showed a 13% increase between 1995 and 2018 (Harris *et al.* 2020), but there have been fluctuations. These fluctuations are similar to the trend for Woodpigeon, which may suggest that common factors have been operating. The population in Wales is estimated to be 87,000 pairs (78,000 to 95,500 pairs) (Hughes *et al.* 2020).

There appear to be no immediate threats to the population, though this species is prone to be infected by *Trichomonosis*. This could result in short-term, local population changes. Large flocks have become scarce in Wales. Winter flocks feeding around farms are smaller than formerly, perhaps because grain storage areas are netted to prevent access by Starlings and Collared Doves (Bob Haycock pers. comm.). This may have reduced winter food availability. However, the Collared Dove is likely to remain an element of our avifauna well into the future.

Ian M. Spence

Sponsored by Peter Crabb

Water Rail *Rallus aquaticus* Rhegen Dŵr

Welsh List Category	IUCN Red List			Birds of Conservation Concern			
	Global	Europe	GB	UK	Wales		
A	LC	LC	LC	Green	2002	2010	2016

With the exception of the northernmost areas of Scandinavia and Russia, the Water Rail breeds across Europe, into North Africa and east as far as western China (*HBW*). The breeding population in western Europe is mainly sedentary. Its numbers are swollen considerably in the winter months by birds migrating from eastern and northern Europe. Across its range, numbers are decreasing. It resides in well-vegetated wetland, both freshwater and saline: reedbeds, swamps, marshes, lakes and ponds, ditches or stretches of slow-moving water, but is more abundant at sites that contain the most wet reed *Phragmites* spp. (Jenkins and Ormerod 2002). At all times of year, it is highly secretive and favours areas of wetland where it can remain largely hidden away from view. This species has an 'announcement call' common throughout the breeding season, with peaks in April and early June, but the courtship call is only made pre-laying. Although vocalisations are more common in the evening, calls peak just before sunrise and before sunset (Polak 2005). The species can be rather silent all year when alone, but they are extremely vocal at sites holding many birds, which makes assessing numbers on small sites and at low population densities difficult (Tony Fox *in litt.*).

Wide-scale drainage of wetlands over the last two centuries undoubtedly had a drastic effect on the breeding and wintering Water Rail populations in Wales, but pockets of suitable habitat remain in all counties. The breeding population is thinly distributed across the country, but the species is absent from upland areas such as Snowdonia, the Cambrian Mountains and Mynydd Preseli. It is more commonly recorded as a passage migrant and winter visitor and there are appreciably more records from wetland sites along the north and south coasts.

There have been only a small number of ringing recoveries involving Wales. Despite appearing to be weak fliers, Water Rails can move long distances on migration. Birds ringed in Germany, Poland and The Netherlands have been recovered in Wales. One of the longest recorded distances involved a bird ringed at Swidwie, Poland, in July 1971 and recovered near Cardigan in December 1973—a distance of 1,285km (Robinson *et al.* 2020). One ringed in Pandy near Abergavenny, Gwent, in January 1976 was found in Glasgow, Scotland, in June 1986. This was the third-longest movement (447km) within Britain and Ireland. This bird also exceeded the longevity record for Britain and Ireland by well over a year, but is not acknowledged as the finding date is not accurate enough (Jerry Lewis *in litt.*). Other ringing records have illustrated the site fidelity of the species: for example, an adult ringed at Cosmeston Lakes Country Park, East Glamorgan, in 2010, was seen there again in 2015.

The earliest record of Water Rail in Wales comes from the Roman period: remains were found in drains at Caerleon, Gwent, dating from the 1st and early 3rd centuries. The species was described as a resident in small numbers, breeding in suitable localities in Gwent in 1937 (Ingram and Salmon 1939). However, the first conclusive record of breeding in the county was not until 1971, when a pair bred at Abergavenny. In Glamorgan, the Water Rail was considered to be a common, regularly breeding resident in the late 1890s, when it was widely distributed in suitable habitat and described as plentiful in the immediate neighbourhood of Cardiff and, in Gower, breeding at Oxwich Marsh and Crymlyn Bog. It is surprising, then, that there were no confirmed breeding records in Glamorgan between 1910, when a pair bred on Sully Island,

Water Rail © Norman West

East Glamorgan, and 1969 when a pair was found at Ynystawe, Gower (Hurford and Lansdown 1995). Forrest (1907) could find no confirmation that it had bred in North Wales, although several had been shot during the breeding season, particularly on the River Clwyd, Denbighshire/Flintshire, and on Anglesey. Mathew (1894) regarded the species as resident in Pembrokeshire, where among the earliest records was a pair breeding on Skokholm in 1929 and again in 1931. In mid-Wales, Cors Caron and Cors Fochno, Ceredigion, and Llangorse Lake, Breconshire, were responsible for some of the earliest historical Welsh records of breeding. There has been no published record of confirmed breeding in Meirionnydd, but the species may be more common in the county than records suggest (Jim Dustow *in litt.*).

The lack of records suggests that the Water Rail is a scarce breeder in Wales, but it is very difficult to assess its true status. It is one of our most elusive birds. The inaccessibility of its preferred habitat means that it is frequently overlooked during the breeding season. Insufficient numbers are recorded using standard survey techniques to produce regular breeding-population estimates or trends. Wintering birds can remain present in suitable habitat in early spring, further complicating our ability to estimate breeding populations.

The number of 10-km squares occupied in Wales in the *Britain and Ireland Atlas 2007–11* was very similar to that in 1968–72, although it was noted that its secretive and skulking behaviour meant that apparent changes could reflect chance patterns of detection in different periods. This behaviour means that few pairs are confirmed breeding each year: between 2009 and 2018 there were, on average, fewer than ten confirmed breeding records annually across Wales.

Birds in Wales (1994) estimated the Welsh breeding population at about 100–150 breeding pairs, an estimate updated by *Birds in Wales* (2002) to 100–200 pairs. Francis *et al.* (2020) reported 53–79 pairs in Wales in 2016–17, out of a total of 2,500–3,900 territories in the UK. However, they noted that dedicated call-playback or acoustic surveys at individual sites had always led to substantially higher population estimates than previously published figures, suggesting that the actual UK population could approach 5,000 pairs. Hughes *et al.* (2020) estimated the population in Wales to be 215 territories (110–315 territories).

Studies have shown that using playback of Water Rail vocalisations to stimulate a response is the most effective way to estimate their abundance (Jenkins and Ormerod 2002). This technique found 53 birds at Newport Wetlands Reserve, Gwent, in 2005, resulting in an estimate of 15 breeding pairs (Clarke 2006). On Anglesey, playback surveys estimated an annual average of 12 breeding pairs at RSPB Valley Wetlands during 2007–13 (max. 22 in 2008) and eight breeding pairs at RSPB Cors Ddyga during 2006–17 (max. 17 in 2010). These figures are higher than would be estimated without playback, but away from wetland nature reserves, such surveys are rarely undertaken. A 1996 breeding-season survey in southeast Carmarthenshire found 66 individuals, including 18 pairs and 30 singles, in 16 out of 40 sites surveyed. The limiting factor was the dryness of reedbeds rather than the size of the site, with birds found in some very small sites (Lloyd *et al.* 2015). Sightings of pre-fledged chicks on Puffin Island, Anglesey, in two years since 2000 (Steve Dodd pers. comm.), a location with no 'typical' nesting habitat, illustrate that Water Rails can breed in the most unlikely places. There has never been a specific survey of Water Rails in any Welsh county. However, combining Gwent's 1998–2003 tetrad atlas and playback response surveys at Newport Wetlands Reserve produced a breeding population estimate of *c.*30 pairs in the county (Venables *et al.* 2008).

The number of Water Rails present in Wales in winter is considerably higher than in the breeding season. During fieldwork for the *Britain and Ireland Atlas 2007–11,* Water Rails were found in more than twice as many Welsh 10-km squares in winter as in the breeding season. Relative abundance was highest in coastal areas, including Anglesey and Gower. The WeBS index for Wales

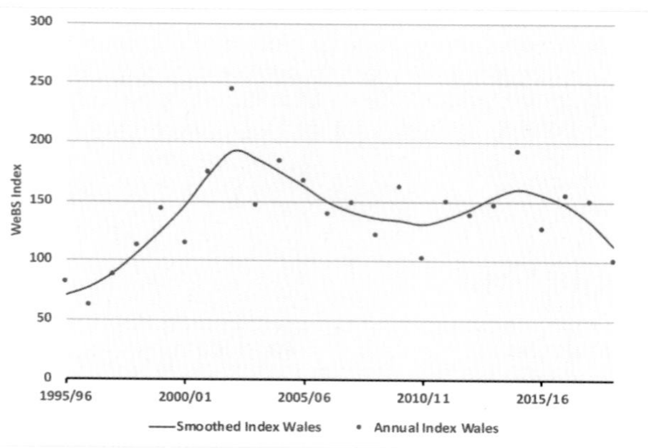

Water Rail: Wetland Bird Survey indices, 1995/96 to 2018/19.

increased between 1995/96 and 2001/02, and has been generally stable since, though these counts do not cover this species well. The highest number recorded on a WeBS count in Wales was 50 at RSPB Cors Ddyga, Anglesey, in November 2011.

The use of playback to estimate numbers has been employed more extensively during the winter, resulting in some significant counts. These included 65 at Penclacwydd, Carmarthenshire, in December 2008, 32 at Cosmeston Lakes in December 2013, 89 at Newport Wetlands Reserve in January 2014 and 24 at RSPB Conwy, Denbighshire, in January 2018. Further evidence that the wintering population is underestimated comes from a study of a small (0.5ha) inland reedbed in mid-Wales, where birds were regularly trapped during autumn and winter 1993/94. It found evidence of territories overlapping and an absolute population density of 14 birds per hectare. This was much higher than densities recorded at other sites where birds were counted only indirectly (Jenkins *et al.* 1995). Harsh winter weather often forces these usually solitary birds to gather in larger concentrations. A count of 200 at Rosehill Marsh, Pembrokeshire/Ceredigion—now part of the Teifi Marshes SSSI—early in 1981 was exceptional. Prolonged spells of cold weather impact the resident breeding population; at Newport Wetlands Reserve in 2011, only two pairs remained following severe weather during the previous winter.

It is not possible to establish the precise timing of the first arrivals of wintering birds in Wales, but at Newport Wetlands, Water Rail are heard more frequently from late August and early September, and numbers increase significantly in October and November, with peak counts between November and January (Clarke 2017). This corresponds broadly with the timing of known passage movement of birds from north and central Europe (*BWP*).

Water Rails are recorded regularly on passage, especially in autumn, on the larger offshore islands. Many were formerly attracted to lighthouses during poor weather and often perished: 50 were killed on Bardsey, Caernarfonshire, on a single night in November 1909, and in 1953 the casualties were consumed by the island's residents (James 2015). The notorious white lights, which caused so many deaths, have now been replaced by red bulbs on Bardsey and Skokholm, Pembrokeshire. These islands continue to record good numbers of birds: Skokholm recorded maximum counts of 15 in October/November between 2014 and 2016, whilst 16 on Skomer in October 2002 is the highest offshore count this century.

Protecting our current wetlands and creating new areas of suitable habitat, especially with a secure water supply, to mitigate for any potential threats that emerge as the climate changes, are key to the Water Rail's future. Addressing the gaps in our knowledge of their breeding ecology and producing an accurate assessment of their breeding population, distribution and abundance are also essential. A nationwide survey of this enigmatic species is long overdue.

Daniel Jenkins-Jones

Sponsored by Richard Clarke

Corncrake *Crex crex* Rhegen yr Ŷd

Welsh List Category	IUCN Red List			Birds of Conservation Concern				Rarity recording
	Global	Europe	GB	UK	Wales			
					2002	2010	2016	
A	LC	LC	LC	Red				WBRC, RBBP

The Corncrake breeds over a wide area of Europe and temperate Asia, from Ireland to northwest China and central Siberia. It is almost always a long-distance migrant, wintering in eastern Africa, south of the equator, though some reach as far south as South Africa. It is primarily a grassland species, breeding mainly in dense herbage at least 50cm tall (*HBW*), and will breed in drier areas of fens. It is far less dependent on wet areas than other members of its family, hence the old name of 'Land Rail'. Very steep long-term declines were recorded in its European breeding population during the 20th century (*HBW*). The population in Europe may now have stabilised. Monitoring since 2002 in Russia, which holds the vast majority of the global population, suggests that numbers are stable or even increasing (BirdLife International 2020). In Britain and Ireland, the bulk of the breeding population is now in the Hebrides and parts of western Ireland.

At one time, the Corncrake was a common and familiar breeding bird in Wales. Owen (1603) recorded that it bred in Pembrokeshire. Pennant (1810) stated that "they are in greatest plenty in Anglesea, where they appear about the 20th of April, supposed to pass over from Ireland, where they abound". Selby (1833) said that they were "very plentiful throughout Wales". The Land Rail was considered a quarry species for shooters, and frequently appeared in the game books of Welsh estates in the 19th century. Numbers shot on the Gogerddan Estate, Ceredigion, peaked in the 1850s, with 90 killed between 1851 and 1860. Most were killed on Partridge shoots in September (Roderick and Davis 2010).

Towards the end of the 19th century, the Corncrake was still common in most of Wales, although it was considered to be decreasing in Glamorgan (Hurford and Lansdown 1995). Anglesey and Llŷn appeared to be the strongholds: Aplin (1900) commented that "The Corn-Crake is very abundant in Lleyn, almost every field with suitable covers holding one". Coward and Oldham (1904), discussing the birds of northern Anglesey, commented that "The Corn-Crake abounds, as it does everywhere on Anglesea". Forrest (1907) regarded it as common in all the counties of North Wales, including on higher ground, such as up to 300m on the Dolwyddelan moors, Caernarfonshire. In Pembrokeshire, Mathew (1894) described it as "numerous in most parts of the county".

Corncrakes were becoming rarer in southeast and eastern England by the late 19th century (*Birds in England* 2005). Concerns about the decline led to an inquiry in 1913–14, summarised by Alexander (1914). He noted that the information from Wales was "very meagre" but reported that there had been none in southwest Glamorgan for ten years and none but an occasional pair in eastern Breconshire for 20 years, although an increase had recently been recorded in eastern Radnorshire. In Gwent, a gradual decrease was reported, but the species was said to be still not rare, while numbers were said to be fairly well maintained in Flintshire and Denbighshire. Farther west, in Carmarthenshire, Ceredigion and Meirionnydd, it was said to be decreasing, but still fairly common, while on Anglesey and Llŷn it was "still evidently a really abundant species". There were no responses from Pembrokeshire.

Another enquiry in 1938, reported by Norris (1945), concluded that the Corncrake was now found only in isolated instances in southeast Wales, recorded at very few sites in Gwent and also very scarce in Breconshire and Radnorshire, with no confirmed breeding records from these counties. It still bred in other Welsh counties, but there had been considerable decreases, even in Anglesey and Llŷn. No information was received from Montgomeryshire in this enquiry, but *Birds in Wales* (1994) noted that it was recorded at 20 widely scattered localities in the county in 1947. Norris (1947) reported that most calling birds were in hay fields which, until about 1850 and in most places much later, were mown almost entirely by hand. This gave Corncrake chicks plenty of time to move out of the way. By 1914 mowing machines were used on estates and larger farms in Wales and their use increased further between the wars, increasingly drawn by tractors rather than horses. A spiral pattern of mowing, starting at the edges of a field and ending in the middle meant that many chicks were killed when the final cuts were made. The date of mowing was also earlier than formerly, increasing the danger of eggs or young being destroyed. Nicholson (1951) thought that the number killed in collision with the increasing number of overhead wires could also affect the population.

Numbers fluctuated considerably from year to year, and despite the general pattern of decline, there were some years when much larger numbers were seen. One such year was 1947 when exceptional numbers were recorded in Carmarthenshire (Lloyd *et al.* 2015) and Montgomeryshire (Holt and Williams 2008). An inquiry in 1952 (Norris 1960) found that the decline had continued, although the species was still thought to breed in small numbers in the western counties of Wales. The *Britain and Ireland Atlas 1968–72* found the species still fairly widespread in the west, with confirmed breeding in 56 of the 10-km squares in Wales, mostly in southwest Anglesey, Llŷn, southern Meirionnydd and Ceredigion. By the *Britain and Ireland Atlas 1988–91*, the Corncrake was almost gone from Wales, with only a single record on Anglesey. The last records of confirmed or suspected breeding in each county are summarised below.

Gwent: there was no regular breeding after the 1938 survey, though there were sporadic breeding attempts, the last in 1982 when birds were considered to have probably bred. A singing male was recorded in 1984.

Glamorgan: regular breeding ceased in 1945. There was sporadic breeding until 1963 and calling birds were heard in 1973, 1978 and 1980 (Hurford and Lansdown 1995).

Carmarthenshire: three or four pairs bred annually in the early 1950s (Ingram and Salmon 1954), with occasional summering birds in the 1960s. There was a breeding record from Cwmann in 1969 (Lloyd *et al.* 2015) and likely breeding attempts at two sites in 1979, including one where a bird had called in 1978. The last calling bird was in 2003.

Pembrokeshire: in May 1962, 4–5 clutches were revealed during silage cutting at Thomas Chapel. The farmers reared three young that were later released at Dale airfield (Donovan and Rees 1994). The *Britain and Ireland Atlas 1968–72* shows confirmed breeding in the Tenby area, and occasional summer calling birds have been recorded since then, the last in 1993.

Ceredigion: Corncrakes were last proved to breed in 1972, although there were a handful of summer records in the following years.

Breconshire: the last confirmed breeding was in 1965, but calling birds were recorded annually between 1963 and 1968, and one summered in 1978.

Radnorshire: there were no positive breeding records in the 1938–39 inquiry, but it bred at three localities just inside the border with Herefordshire in 1945 (Jennings 2014). There were summering records in 1955, 1958, 1966, 1970 and 1982.

Montgomeryshire: Corncrakes last bred in 1973, although calling birds were heard at one site until 1980.

Meirionnydd: the last breeding was recorded in 1973, with probable breeding at the same site in 1974. There was a calling bird in the same area in 1988 and a calling bird at Llyn Barfog in 2013.

Caernarfonshire: the last confirmed breeding was in 1976, but calling birds were noted regularly up to 1991. A pair probably bred at Llanfairfechan in 1996.

Anglesey: at least three pairs bred in the southwest of the island in 1972 (Jones and Whalley 2004). One pair bred in the 1980s and

Total	1967–69	1970–79	1980–89	1990–99	2000–09	2010–19
167	26	75	8	33	9	16

Corncrake: Individuals recorded in Wales by decade since 1967.

another in 1991 and 1992, but all four chicks in the latter year were taken by a Kestrel. A pair probably bred in 1994 and calling birds were heard in 2001, 2011 and 2012.

Denbighshire: Corncrakes bred until 1961. One pair bred in 1992, with another calling nearby in the same year.

Flintshire: the last breeding recorded was in 1959, although calling birds were noted until 1968.

Since 1967, when the *Welsh Bird Report* began, a total of approximately 167 breeding and passage birds have been recorded in Wales. Passage peaks are in April/May, as breeding birds make their way north to Scotland, and in September as adults and juveniles disperse south to Africa.

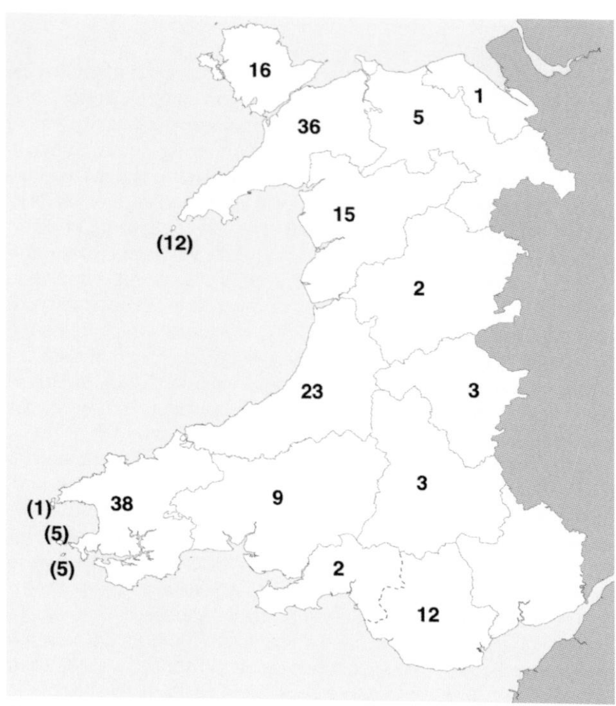

Corncrake: Numbers of birds in each recording area, 1967–2019.

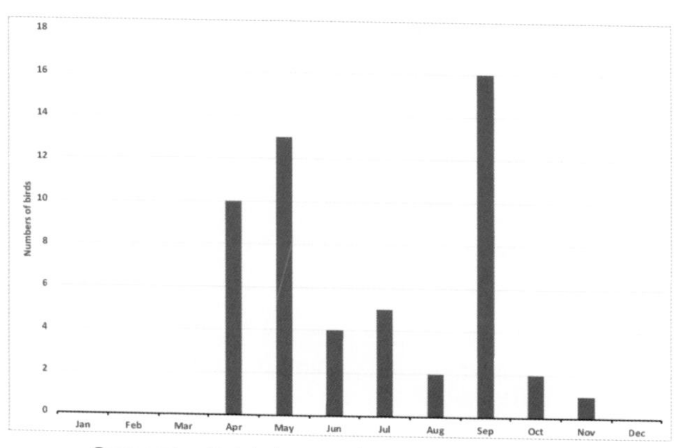

Corncrake: Totals by month of arrival, 1967–2019.

The number of passage migrants recorded in Wales depends partly on the fortunes of the breeding population in the key breeding areas on the Scottish islands. After a period of sustained growth thanks to concerted effort by conservation groups and crofters from around 2000, numbers there have declined since 2015. This is thought to be the result of changes in the agri-environment scheme (Green 2020b). Reintroduction trials in England have had limited success, with 15 singing males at the main site, the Nene Washes, Cambridgeshire, in 2017 (Holling *et al.* 2019). A similar reintroduction scheme might be possible in Wales if appropriate land management could be achieved on a sufficient scale.

Jon Green, Robin Sandham and Rhion Pritchard

Sponsored by Home Front Museum, Llandudno

Sora Rail *Porzana carolina* Rhegen Sora

Welsh List Category	IUCN Red List	Rarity recording
	Global	
A	LC	BBRC

The Sora Rail breeds widely but locally across North America and winters from the southern USA to northern South America. Out of a total of 20 individuals recorded in Britain (Holt *et al.* 2018), three were in Wales. The first, captured at Cardiff Docks, East Glamorgan, in October 1888, was the second record for Britain. In early January 1932 a dead bird was found at Aberdyfi Golf Course, Meirionnydd, and sent to the National Museum of Wales, where it was originally identified as Baillon's Crake and later re-identified as a Spotted Crake. It was only identified as a Sora Rail when it was taken to the 2006 BTO Conference to be used in an identification quiz. The only other record was one trapped and ringed on Bardsey, Caernarfonshire, on 5 August 1981.

Jon Green and Robin Sandham

Spotted Crake *Porzana porzana* Rhegen Fraith

Welsh List category	IUCN Red List			Birds of Conservation Concern				Rarity recording
	Global	Europe	GB	UK	Wales			
A	LC	LC	EN	Amber	2002	2010	2016	WBRC, RBBP

The Spotted Crake breeds locally across western Europe, and more abundantly across eastern Europe and temperate Asia as far as Kazakhstan and northwest China. It is found in freshwater wetlands with muddy edges or shallow water and a dense cover of sedges and rushes. European breeding birds winter throughout Africa. Numbers decreased over most of its European range during the 20th century (*HBW*). It is found almost annually on passage in Wales, and occasionally stays to breed, although confirmation of breeding is very difficult.

Spotted Crake in the Teifi Marshes, Pembrokeshire, July/August 2018 © Tommy Evans

Spotted Crakes probably bred in many parts of Wales up to the middle of the 19th century, but this decreased as suitable habitat was lost to drainage. In Gower, Dillwyn (1848) stated that it was "not very unfrequently found in rushy meadows" in the Swansea area. It was a regular summer visitor to Crymlyn Bog in the 19th century, but breeding was thought to have ceased before 1900 (Hurford and Lansdown 1995).

Most records of confirmed or possible breeding around the turn of the 20th century came from mid-Wales. Phillips (1899) described the Spotted Crake as a regular, but very local, visitor to Breconshire, found every year at Mynydd Illtud and confirmed breeding at Traeth Mawr. The best site in the county was Onllwyn Bog, where Phillips thought that many bred: on one day he flushed six and shot four, of which three were birds of the year. The local station-master told him that there were two broods in part of the bog that year. The species was also found at Rhosgoch Bog, Radnorshire, in the 19th century. In Ceredigion, several birds were shot around the Dyfi Estuary in the late 19th century and the early years of the 20th century. The Borth collector, F.T. Fielden, reported to Forrest (1907) that two young birds had been shot at Ynys-hir, and he thought that the species had probably bred there. It may also have bred at Cors Fochno (Roderick and Davis 2010). Forrest (1907) stated that no nest had ever been found in North Wales, although the authors of *Birds in Wales* (1994) considered that the Spotted Crake probably bred on the Dee Marshes, Flintshire, in the 18th and 19th centuries.

No more records suggested breeding until the 1920s and 1930s, when there was evidence of fairly regular breeding in some areas. E.P. Chance found a broken eggshell at Rhosgoch Bog in 1923, and returned to take a clutch of nine eggs from a nest there the following year (*Birds in Wales* 1994). Two nests were found at Traeth Mawr in 1928, although one was empty. A nest was again found at Rhosgoch in 1929 and young seen there in 1939. Hardy (1941) recorded nesting on the Welsh part of the Dee Marshes in 1920 and 1926. The species was also reported at Puddington Bog, the area now known as RSPB Burton Mere Wetlands, in the 1930s, although this may not have been on the Welsh side of the border. Ingram and Salmon (1939) thought that a few pairs had bred "some years ago" in Gwent.

Since 1967, birds have been recorded in the breeding season in every Welsh county, except Montgomeryshire, Meirionnydd and Denbighshire, including some that held territory over several weeks. In Breconshire, two males and a female were heard calling at Mynydd Illtud in 1979, and a nest was found there. Calling birds were then recorded at this site in several summers up to 1986, with a nest found in 1983. Three calling males were present there in June–July 1993. In Carmarthenshire, a juvenile described as "capable of flight" was ringed in Llangennech reedbeds on 13 August 1988 (Lloyd *et al.* 2015). Some other sites have held birds in the breeding season fairly regularly. For example, calling males were heard at Cors Caron, Ceredigion, in May or June in six of the years between 1967 and 2006. 1993 was a particularly good year: as well as the three calling males at Mynydd Illtud, there were also at least two singing at Ystumllyn, Caernarfonshire, in June. A calling bird was heard at Colwyn Bridge, Hundred House, Radnorshire, from 20 May to 15 June, and single calling birds were recorded in April in Gwent, Carmarthenshire and Pembrokeshire. Calling males have been heard at a number of sites since then, including records in successive years at Trefeiddan, Pembrokeshire, in June 2003 and again in May 2004. Single males were heard at a site in western Breconshire in March and April 2007, then again in May–June 2013 and May 2014.

Elsewhere in Britain there was an increase in breeding Spotted Crakes during the latter decades of the 20th century, perhaps influenced by greater recording effort, but with evidence of a decrease after 2001 (Stroud *et al.* 2012). There have been two dedicated surveys: two singing males were found in Wales in 1999 (Gilbert 2002), but none of the 28 singing males in Britain in 2012 were in Wales (Schmitt *et al.* 2015).

Spotted Crakes are also recorded in Wales as passage migrants. There were many records of birds shot in the late 19th and early 20th centuries, mainly in autumn. A minimum of 123 have been recorded in Wales since 1967. The decline in the last 20 years is a trend evident across Britain, with a decline from an average of 60 a year in 1990–99 to just 26 in 2010–18 (White and Kehoe 2020a).

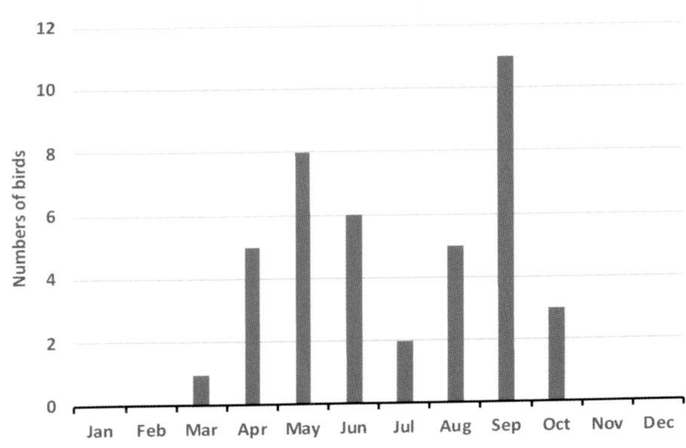

Spotted Crake: Totals by month of arrival, 1967–2019.

The spring peak, mainly in May, includes some long-staying birds which could be breeding, while others were only recorded for a day or two and were presumably on passage. The autumn

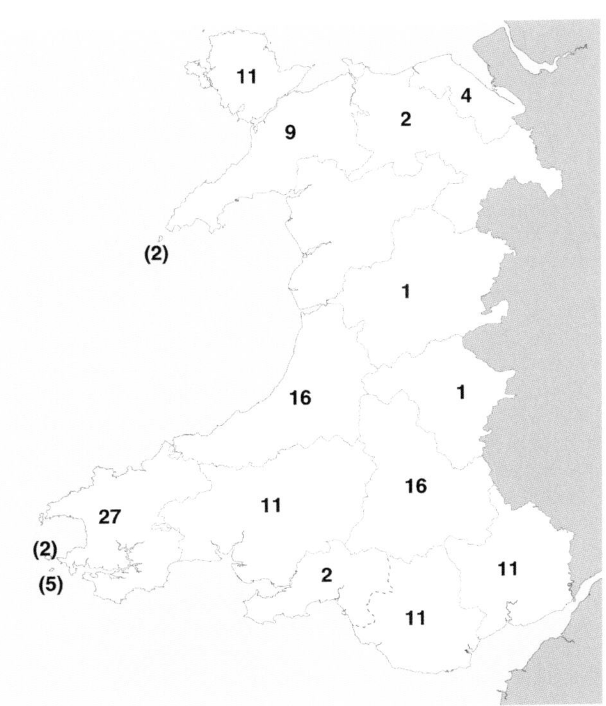

Spotted Crake: Numbers of birds in each recording area 1967–2019. One bird was seen in more than one recording area.

Total	1967–69	1970–79	1980–89	1990–99	2000–09	2010–19
121	12	31	21	26	17	14

Spotted Crake: Individuals recorded in Wales by decade since 1967.

peak, mainly in September, is of birds on passage. There have been a few winter records, including one shot at the Dyfi Estuary, Ceredigion/Meirionnydd, in the winter of 1902/03, one at Trewellwell, Pembrokeshire, in January 1910 and one at Goodwick, Pembrokeshire, in winter 1945. Six birds have been recorded in winter since 1967: one at Pembroke in January 1970; one at Malltraeth Cob, Anglesey, which remained from September to December 1971; one at Shotton, Flintshire, in January 1974; one at Magor Marsh, Gwent, in December 1981; one at Kenfig, East Glamorgan, in January 1985 and one at Pembrey, Carmarthenshire, in January 1987. Some sites have recorded birds on several occasions: since 1967, there have been eight records at Kenfig, seven at Teifi Marshes, Pembrokeshire/Ceredigion, and six at Shotton.

The creation of new wetland areas, for example in Gwent and Anglesey, has increased the area of potentially suitable habitat for Spotted Crake in Wales. However, Stroud *et al.* (2012) commented that across Britain some key sites may have become less suitable for Spotted Crake, as a result of successional vegetation change. They noted the lack of prioritised conservation actions for this species, compared to other rare wetland species. Huntley *et al.* (2007) forecast that climatic change will result in the European range of this species heading northeast, away from Britain, which suggests a reduced likelihood of breeding.

Jon Green, Robin Sandham and Rhion Pritchard

Moorhen *Gallinula chloropus* **Iâr Ddwr**

Welsh List Category	IUCN Red List			Birds of Conservation Concern			
	Global	Europe	GB	UK	Wales		
					2002	2010	2016
A	LC	LC	VU	Green			

The Moorhen has a very extensive range, across Europe, Central and southern Asia, and in suitable habitat south from sub-Saharan Africa to Indonesia. It is found in a variety of natural and artificial wetlands, including ditches and marshes, as well as lakes, ponds and rivers. It is resident or dispersive in western and southern Europe but partially migratory or migratory farther east (*HBW*). Ringing recoveries suggest that most birds breeding in Wales are sedentary. For example, an adult ringed on Bardsey, Caernarfonshire, in September 1984 was recaptured in April 1994, still on the island. A few recoveries indicate that some birds which breed farther east winter in Wales: three ringed as nestlings, or in their first year, in The Netherlands were later found in Wales in winter, and a first-calendar-year bird, ringed in Denmark in September 1961, was shot near Brecon in November that year.

The Moorhen tends to avoid the highest ground, although a nest with eggs was found at 538m on Llyn Du, Ceredigion, in 1927. *Birds in Wales* (1994) suggested that the drainage of lowland wetlands in the 18th and 19th centuries caused local decreases, although the Moorhen is quick to adapt to new wetland habitats, and this may have kept the population stable (*Historical Atlas 1875–1900*). Mathew (1894) described the species as "numerous everywhere" in Pembrokeshire. It was described as abundant in Glamorgan at the end of the 19th century (Hurford and Lansdown 1995). Forrest (1907) stated that it was "so common and generally distributed in North Wales that details are unnecessary".

This species suffered severely in the harsh winter of 1962/63 (Dobinson and Richards 1964), but apart from weather-related declines and recoveries, there is little indication of population trends in this period. Lloyd *et al.* (2015) noted a contraction in range in Carmarthenshire in the second half of the 20th century, with the species confined to the coastal marshes and ponds and a limited number of protected sites away from the coastal strip. Donovan and Rees (1994) considered that its range in Pembrokeshire had retracted since the 1970s, as a result of changing land use, with loss of wetlands and pollution of water by agro-chemicals and farm slurry.

The *Britain and Ireland Atlas 1968–72* found the Moorhen present in 89% of 10-km squares in Wales, absent only from the highest ground. By the *Britain and Ireland Atlas 1988–91*, the species had been lost from some squares around the upland margins and the 2007–11 *Atlas* showed a similar trend, although losses were balanced by gains in the coastal lowlands. The occupation rate was 77% of Welsh squares in both the later atlases. Relative abundance in most of Wales in 2007–11 was very low compared to much of England, although a little higher in Anglesey, Flintshire and along the southern coastal lowlands. Relative abundance had decreased over most of North Wales since 1988–91, with a more mixed picture farther south. The reason for the difference between these two areas is unclear.

The *Gwent Atlas 1981–85* recorded Moorhens in 47% of tetrads, and this was almost unchanged in the 1998–2003 *Atlas*, with birds found in 48% of tetrads. The two East Glamorgan atlases also showed little change, with birds recorded in 26% of tetrads in both 1984–89 and 2008–11, although birds were only recorded with breeding codes in 21% of tetrads in 2008–11. Moorhens were present in 32% of tetrads in Pembrokeshire in 1984–88, increasing slightly to 35% of tetrads in 2003–07. In Gower, the *West Glamorgan Atlas 1984–89* found them in 31% of tetrads. They were present in only 8% of tetrads in Breconshire visited between 1988 and 1990 during fieldwork for the *Britain and Ireland Atlas 1988–91*, although the authors thought that they may have been under-recorded. The *North Wales Atlas 2008–12* found Moorhens in 21% of tetrads. The great majority of occupied tetrads were in the lowlands, particularly in eastern Denbighshire, Flintshire and Anglesey. In Meirionnydd, Moorhens were recorded in only 5% of tetrads.

No breeding population trend is available for Wales, but those for the UK suggest a 20–25% decline during 1995–2018 (Harris *et al.* 2020). The longer-term index has returned to its 1967 baseline, following a rapid increase up to 1971, as numbers recovered from heavy losses in winter 1962/63. Cold winters in 2009/10 and 2010/11 presumably caused another sharp decline, but the index shows no recovery in numbers despite mainly mild winters since.

Breeding Moorhens are not regularly monitored using standardised methods at most sites in Wales. Surveys in 1977–78 produced mean densities of 6.77 territories/10km on the Wye, 19.38 on the Severn and 20.67 on the Vyrnwy (Round and Moss 1984). A survey of Afon Teifi, Ceredigion, in 1981 found 49 territories, with another six on the main tributaries (Roderick and Davis 2010). Unlike the Coot, the Moorhen is a regular or fairly regular breeder on a number of the Welsh islands. Lockley *et al.* (1949) recorded that it bred on Skokholm, Skomer, Ramsey and Caldey. On Bardsey, a survey in 1986 confirmed 21 pairs breeding with another two possibly breeding. By 1990 this population had declined to just two pairs. It recovered to 13 pairs in 2018 but was down to seven pairs in 2019.

Estimating the Moorhen population in Wales is difficult. The population of Gwent was estimated at around 1,000 pairs in the *Gwent Atlas 1981–85*. Venables *et al.* (2008) considered that the estimate of 325–380 pairs in the county, derived from BBS data and sample tetrads, was too low, and considered that the earlier estimate of 1,000 pairs might still be correct. The Pembrokeshire population was estimated at around 340 pairs in 2007 (Rees *et al.* 2009), with estimates of about 50–100 pairs in Breconshire (Peers and Shrubb 1990) and 330–340 pairs in Caernarfonshire in 2008–11 (Pritchard 2017). *Birds in Wales* (1994) estimated the Welsh population at 2,500–3,000 pairs, but the published population estimates for individual counties suggest a higher total. The most recent estimate (Hughes *et al.* 2020) is 14,500 territories (13,500–15,500 territories) in Wales.

In winter, Moorhens do not gather in flocks on larger water bodies as Coots do, so counts are lower and a smaller proportion of the wintering population is recorded by the Wetland Bird Survey (WeBS). There is considerable variation in counts between years at some sites. The WeBS index for Wales appears to have stabilised after a period of decline.

At Penclacwydd, Carmarthenshire, wintering numbers built up steadily after the centre was established in the second half of the 1980s (Lloyd *et al.* 2015), with 281 there in November 1997 and 220 in March 2001, but numbers have been lower since. Good numbers have been recorded on the Monmouth-Brecon Canal, where there were 170 in March 2016. There were 91 on a farm pond near Wrexham, Denbighshire, in November 1972. 141 were recorded on all freshwater sites in Pembrokeshire in January 2010.

Long-term population trends for Europe show a modest decline (PECBMS 2020). Population trends in Wales are uncertain, but there has certainly been a retraction in range in recent years, with the birds having vanished from some formerly regular breeding sites at higher altitudes. *Birds in Wales* (1994) stated that it was absent from sections of the River Monnow, Gwent, and reported declines in recent years on parts of the River Wye, Gwent/Radnorshire, and Afon Teifi, Ceredigion. These declines had been attributed to American Mink, although there was no firm evidence to support this. Predation by Mink has been suggested

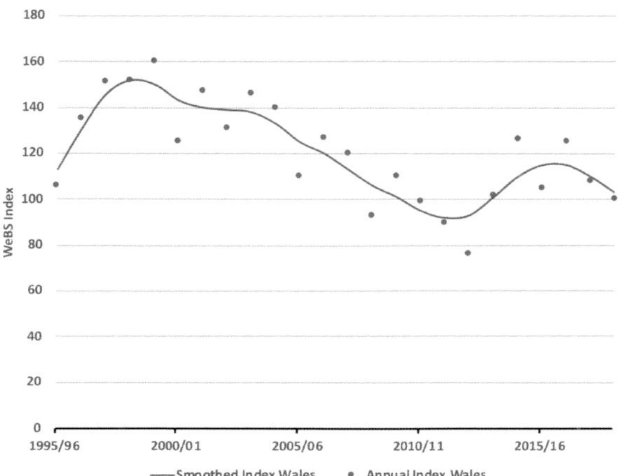

Moorhen: Wetland Bird Survey indices, 1995/96 to 2018/19.

as a reason for declines elsewhere in Wales. The effect of Mink on the Moorhen population has not been studied in Wales. However, they were shown to have a significant effect on the breeding Moorhen population on the upper Thames (Ferreras and Macdonald 1999) and the populations on Lewis and Harris were wiped out as a result of releases of Mink (*The Birds of Scotland* 2007). Otters are also potential predators, although they can benefit Moorhens by predating Mink. The clearing of vegetation along the banks of water bodies reduces the breeding success of Moorhens (Taylor 1984). Improved monitoring of this species in Wales using a standard method would be valuable to detect any population changes.

Rhion Pritchard

Site	**County**	**94/95 to 98/99**	**99/00 to 03/04**	**04/05 to 08/09**	**09/10 to 13/14**	**14/15 to 18/19**
Severn Estuary: Welsh side (whole estuary)	Gwent/ East Glamorgan	10 (349)	25 (582)	89 (606)	57 (299)	63 (345)
Burry Inlet	Gower/Carmarthenshire	178	146	55	17	78

Moorhen: Five-year average peak counts during Wetland Bird Surveys between 1994/95 and 2018/19 at sites in Wales averaging 150 or more at least once. International importance threshold: 20,000; Great Britain importance threshold: 3,000.

Coot *Fulica atra* Cwtiar

Welsh List Category	**IUCN Red List**			**Birds of Conservation Concern**			
	Global	**Europe**	**GB**	**UK**	**Wales**		
A	LC	NT	NT	Green	**2002**	**2010**	**2016**

The extensive range of the Coot stretches across Eurasia and south to the Indian subcontinent, with other subspecies found in New Guinea and Australasia. Those breeding in the north of its range are mainly migratory, moving west or south, but birds are present all year in warm and temperate regions, although not necessarily resident (*HBW*). There is evidence from ringing recoveries that birds from farther east move to Britain in winter, and that some birds which breed in Britain and Ireland undertake regular movements to wintering grounds some distance away (*BTO Migration Atlas* 2002). An adult ringed at the Orielton Duck Decoy, Pembrokeshire, in January 1941 was shot at Prekuln, Latvia, in August 1942, while a bird ringed at Haarlemmermeer, The Netherlands, on 6 December 1969 was found dead at Welshpool, Montgomeryshire, 14 days later. There can also be long-distance movements within Britain: one ringed near Aberdeen in January 1987 was shot near Bodedern, Anglesey, three weeks later.

Coots breed mainly on still or slow-flowing waters, preferably fairly shallow and with marginal, emergent, floating or submerged vegetation (*HBW*). Suitable water bodies are fairly common in some lowland areas of Wales, but scarce in the uplands. Breeding at altitudes much above 350m is unusual in Wales now, although breeding was recorded on Llyn Brenig, Denbighshire, at 376m in 2013. The species apparently bred at higher altitudes in the past. For example, in Ceredigion it bred until the 1970s on Llyn Du, at 540m, and was also recorded on two other lakes above 400m in the county (Roderick and Davis 2010).

The Coot appears to have been present in Wales for a long time. Bones thought to date from the Pleistocene or Early Holocene period were found in Gop's Cave near Dyserth, Flintshire, and remains were found in Roman-era drains at Caerwent, Gwent. The species has expanded its range in Europe since the late 19th century. Although the population fluctuates, it is thought to have generally increased since then, helped by eutrophication, new reservoirs and lakes, and adaptation to urban areas (*HBW*). Mathew (1894) said that in Pembrokeshire it was confined to the few large ponds in the county where there was cover for nests, while it was locally common in Glamorgan in the late 1890s (Hurford and Lansdown 1995). In North Wales, Forrest (1907)

Adult Coot feeding its chick © Kev Joynes

said that it occurred throughout Anglesey, but was distributed somewhat irregularly elsewhere; at some inland localities, it had established itself as a breeding species fairly recently. Up to 150 pairs bred on Llangorse Lake, Breconshire, in 1902 (Peers and Shrubb 1990). Ingram and Salmon (1954) suggested that it was not common in Carmarthenshire, mentioning a few pairs breeding at Witchett Pool and a pair on Bishop's Pond, Abergwili, but not on other inland pools.

The three Britain and Ireland atlases recorded Coot in around 55% of 10-km squares in Wales. In 1968–72 it was present in most squares in eastern Wales, but with a patchier distribution in the west. It was found in many squares in Anglesey, Llŷn and Carmarthenshire, but was largely absent from higher ground and from most of Pembrokeshire. Fieldwork for the *Britain and Ireland Atlas 1988–91* found roughly equal numbers of losses and gains, the latter mainly in the west and in Gwent. The *Britain and Ireland Atlas 2007–11* showed gains in the south, particularly in East Glamorgan, but extensive losses in Caernarfonshire, Meirionnydd and Ceredigion, particularly around the upland margins. Abundance was relatively low in most of Wales, except for the eastern lowlands, Anglesey, East Glamorgan and Gwent, and had decreased since 1988–91 everywhere apart from Anglesey and the southeast.

Tetrad atlases provide some further detail. The *Gwent Atlas 1981–85* found birds in 10% of tetrads, but by the *Gwent Atlas 1998–2003* this had increased to 21%, with an increase from the mid-1980s at a number of previously unoccupied sites (Venables *et al.* 2008). By 2000, at least 30 pairs bred at Newport Wetlands Reserve. In 2008 the county population was estimated at 130–160 pairs, compared to about 20 pairs prior to 1977. A similar range extension was evident in neighbouring East Glamorgan, where Coots were found in only 5% of tetrads in the 1984–89 *Atlas* but in 18% of them in 2008–11, although only with breeding codes in 16%. The increase was much less dramatic in Pembrokeshire, where Coots were recorded in 6% of tetrads in 1984–88 and in 7% in 2003–07, and despite the slight range extension, numbers were considered to have decreased from an estimated 75 pairs in 1988 to no more than 52 pairs in 2007 (Rees *et al.* 2009). In Gower, the *West Glamorgan Atlas 1984–89* found birds in 11% of tetrads. The species was present in 8% of tetrads in Breconshire visited between 1988 and 1990 during fieldwork for the *Britain and Ireland Atlas 1988–91*. The *North Wales Atlas 2008–12* found Coots in 14% of tetrads, the great majority in Flintshire, eastern Denbighshire and Anglesey. The species was very scarce above 250m and in some lowland areas. It was recorded in just seven tetrads on Llŷn, Caernarfonshire, with breeding confirmed in only two of these.

There appears to be considerable regional variation in the fortunes of the Coot as a breeding bird in Wales. The population on Llangorse Lake, Breconshire, was estimated at 110 pairs in 1979, and while Peers and Shrubb (1990) suspected that this estimate was too high, they reported a sharp decline there in the late 1980s—declining to an estimated 30 pairs in 1992 (Francis 1992). In Radnorshire, Jennings (2014) considered it to have been a much rarer breeding bird in the 1950s and 1960s than in the 21st century: a fairly complete survey found 35 pairs at 20 sites in 2000, and a thorough survey of the county in 2010 found 36 pairs. Numbers are thought to have increased in Carmarthenshire from the late 1950s. The population is now thought to be in excess of 60 pairs in most years (Lloyd *et al.* 2015). Anglesey is the main stronghold in Wales, with up to 50 pairs breeding on Llyn Alaw in 1993 and 45 pairs at RSPB Valley Wetlands in 2012. Population trends for the whole island are unknown. The species has probably benefited from the creation of new wetlands, but breeding has ceased at Cemlyn, where there were 11 pairs in 2010, but none in 2015. No trends can be calculated for Wales, but those for the UK as a whole show a 19% decline during the ten-year period 2008–2018 (Harris *et al.* 2020), following a period of increase.

Birds in Wales (1994) considered that the total Welsh breeding population was *c.*1,500–2,000 pairs. Published estimates of county populations suggest that the present population is likely to be towards the lower end of that range, as there has been a marked decline in some western counties in recent years. The most recent estimate for Wales (Hughes *et al.* 2020) is 1,600 pairs (1,250–2,050 pairs). A decline was evident in Caernarfonshire from about 2007. In the next few years birds were lost from several lakes where they had previously bred regularly in small numbers (pers. obs.). The *North Wales Atlas 2008–12* recorded Coots in only 24

Site	County	84/85 to 88/89	89/90 to 93/94	94/95 to 98/99	99/00 to 03/04	04/05 to 08/09	09/10 to 13/14	14/15 to 18/19
Severn estuary: Welsh side (whole estuary)	Gwent/East Glamorgan	0 (217)	0 (120)	0 (242)	210 (601)	403 (779)	369 (679)	443 (689)
Llangorse Lake	Breconshire	282	273	596	841	611	436	439
Llyn Alaw	Anglesey	256	567	401	530	247	121	250
RSPB Valley Wetlands	Anglesey	130	259	272	464	296	281	155
Cemlyn Bay and Lagoon	Anglesey	472	372	536	121	99	60	37

Coot: Five-year average peak counts during Wetland Bird Surveys between 1984/85 and 2018/19 at sites in Wales averaging 450 or more at least once. International importance threshold: 15,500; Great Britain importance threshold: 2,000.

tetrads (6%) in Caernarfonshire, while in Meirionnydd, there was a more dramatic decline since 1988–91, with the species absent from most western areas. In the latter county, the species was found in only eight tetrads (2%), and breeding was confirmed only in three, all on small ponds. In Ceredigion, Roderick and Davis (2010) said that the proliferation of farm ponds in recent years had increased the availability of suitable habitat, but that breeding numbers at some sites had decreased markedly in very recent years.

The numbers recorded in winter in Wales are larger than can be accounted for by birds breeding here, and can increase further in cold weather. Lockley *et al.* (1949) stated that the Coot was a winter visitor in small numbers to Pembrokeshire, except in severe weather, when thousands appeared in the Cleddau Estuary, for example in 1933/34 and 1946/47. The species appears to have been more frequently found in estuaries and sheltered bays in the first half of the 20th century, than it is now.

Counts of over 1,000 have been recorded at a few sites, including 1,100 at Llandegfedd Reservoir, Gwent, in December 1974, 1,200 on Llyn Maelog, Anglesey, in October 1976, 1,159 on Llyn Alaw in September 1997 and 1,046 there in September 1999. The only subsequent count over 1,000 was a site record of 1,075 at Llangorse Lake in November 2003. The Wetland Bird Survey (WeBS) index for Wales shows a fairly steady decline since about 2000/01, resulting in a medium-term Medium WeBS alert (-27%) being issued for Wales (Woodward *et al.* 2019).

Coots are fairly scarce visitors to the Welsh islands, mainly in July and August, although there have been breeding attempts on Skomer and probably on Skokholm, Pembrokeshire. An adult and a juvenile were on the sea off Grassholm, Pembrokeshire, on 6 August 1933, and an adult was among a raft of Manx Shearwaters 4km southwest of Grassholm on 21 August 2002.

Although the global population of the Coot is considered to be increasing, the population in Europe is thought to be decreasing (BirdLife International 2020). This has led to it being Amber-listed in Wales (Johnstone and Bladwell 2016). There is evidence that the breeding population is declining in some western areas of Wales, but the overall picture is unclear. Predation by American Mink (Jennings 2014) and Otter (*Cambrian Bird Report* 2007) has been suggested as a cause of declines in some areas. Disturbance from water sports may have affected breeding or wintering numbers on some lakes and reservoirs, such as at Llangorse Lake (Francis 1992). Coot as a breeding species now needs to be monitored more closely in Wales than it has been in most areas in the past.

Rhion Pritchard

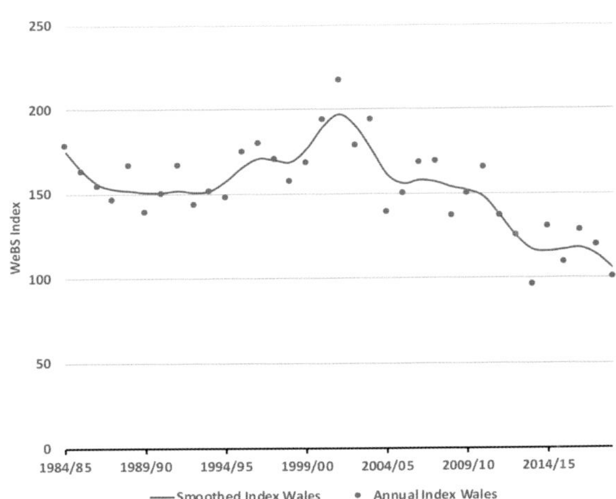

Coot: Wetland Bird Survey indices, 1984/85 to 2018/19.

Sponsored by Fledgemore Nest Recording Group

Baillon's Crake *Zapornia pusilla* Rhegen Baillon

Welsh List Category	IUCN Red List		Rarity recording
	Global	Europe	
A	LC	LC	BBRC, RBBP

Widely but thinly distributed through southern Europe, and from Ukraine to the Pacific Ocean, this species is migratory, wintering to the south as far as Australasia. It bred in Britain in the 19th century (*Historical Atlas 1875–1900*), but only 8–10 individuals have been recorded in Wales. The first was a male, caught by a dog in a ditch at Llangwstenin, on the Caernarfonshire/Denbighshire border, on 6 November 1905. Up to the turn of the 21st century there were two further records, one at Llantwit Major, East Glamorgan, on 8 February 1976 and a second-calendar-year male found dead on Bardsey, Caernarfonshire, on 16 June 1990.

A survey of Spotted Crakes in 2012 found at least four, possibly six, calling Baillon's Crakes at RSPB Cors Ddyga, Anglesey, between 22 May and 14 July, with another bird at RSPB Ynys-hir, Ceredigion, on 25–28 June. Up to 11 males were recorded in Britain that year, an influx thought to have been the result of drought conditions in the Iberian Peninsula. Considering the number of males present, it is quite possible that the species bred in Wales that year. Though none were recorded in Wales in 2019, four were recorded in Britain and unprecedented numbers in France and The Netherlands. Holt *et al.* (2020) stated that it was "tempting to speculate that Baillon's Crake is on the verge of recolonising Britain—reflecting the situation in other parts of northwest Europe".

Jon Green and Robin Sandham

Little Crake *Zapornia parva* Rhegen Fach

Welsh List Category	IUCN Red List		Rarity recording
	Global	Europe	
A	LC	LC	BBRC

The Little Crake breeds locally in western and central Europe and is more widespread through eastern Europe and Central Asia. It winters in sub-Saharan Africa, Iran and Pakistan, though wintering sites are poorly known (*HBW*). A total of 102 individuals have been recorded in Britain to the end of 2019 (Holt *et al.* 2020). The majority were before 1950. Only seven of these records were in Wales:

- Afon Afan, Gower, in 1839, captured by hand;
- found dead at Ynyslas, Ceredigion, in 1894;
- between New Quay and Aberaeron, Ceredigion, on 1 January 1949;
- male at Diles Lake, Llangennith Moors, Gower, on 19 and 22 January 1967;
- male at Broken Bank Tip pools, Shotton, Flintshire, on 18–23 April 1987;
- male at Upper Lliedi Reservoir, Carmarthenshire, from 30 April to 2 May 1987 and
- 1CY at RSPB Conwy, Denbighshire, on 27 September 2000.

Jon Green and Robin Sandham

Crane *Grus grus* Garan

Welsh List Category	IUCN Red List			Birds of Conservation Concern	Rarity recording
	Global	Europe	GB	UK	
A	LC	LC	VU	Amber	WBRC, RBBP

Total	Pre-1950	1950–59	1960–69	1970–79	1980–89	1990–99	2000–09	2010–19
54	1	1	2	6	9	6	15	13

Crane: Individuals recorded in Wales by decade since 1893 (excluding birds thought to be from the re-establishment project).

The Crane breeds from Fennoscandia and northeastern Germany east through temperate Asia to the Russian far east and Mongolia. There are some smaller breeding populations farther south, such as in Turkey and the Caucasus (*HBW*). Most populations are migratory, wintering to the south or west of the breeding range. Many birds from the west European flyway use wintering grounds in Spain and France, although increasing numbers winter to the north and east, while a large part of the eastern European population winters in East Africa. Cranes breed in a wide variety of shallow wetlands. The population in Europe is thought to be increasing (BirdLife International 2020).

Analysis of bird footprints preserved in intertidal mud on the Severn Estuary at Goldcliff East, Gwent, dating from the Late Mesolithic period around 7,000 years ago found that 46% were those of Crane (Barr 2018). Remains have been found at a number of archaeological sites in Wales, including a Bronze Age record at Caldicot, Gwent. There are several records at Roman-era sites, including Pentre Farm, Flintshire, and the legionary fortress at Caerleon, Gwent, with one early-Norman-era record at Hen Domen, Montgomeryshire (Yalden and Albarella 2009). In some versions of the Welsh laws, dating to the 13th century, Crane is included among the "three notable birds" particularly esteemed as a quarry for falconers (Charles-Edwards *et al.* 2000). There are also several references in poetry to indicate that the Crane was valued for eating in medieval times (Williams 2014). *Garan* appears in a number of place names in Wales, though these have to be interpreted with caution as *Garan* was used for Grey Heron in some areas. Several English or Old Norse place names in Pembrokeshire and the Gower also refer to this species (Yalden and Albarella 2009). Owen (1603) recorded that the Crane bred in the bogs of Pembrokeshire, but there were no subsequent records of breeding. The species appears to have become extinct in Britain around this time. It is commonly assumed that it had ceased to breed in England by 1600 (*Birds in England* 2005). Hunting for food and the drainage of wetlands were probably major factors in its extinction.

The Crane has been a scarce visitor to Wales since it ceased to breed. There is only one record from the 19th century, of one captured at a farm near Solva, Pembrokeshire, in April 1893. At the beginning of the 20th century, one was shot at Rhosneigr, Anglesey, in May 1908. The species was not recorded again in Wales until 1959 when one was at Marloes Mere, Pembrokeshire, on 30 October. This was followed by one between Cann Office and Pont Llogel, Montgomeryshire, on 3 May 1961. The only other record in the 1960s was at Cemlyn, Anglesey, in September 1966. There has been an increase in records since the 1970s, coinciding with the recovery of the European breeding population from its low point in the 1960s and the return of the species to England as a breeding bird from the early 1980s.

Birds from the Fennoscandian breeding population follow a route through France to and from their wintering areas. Many of those seen in Wales are likely to have diverged from this flyway. Most records in Wales have been in spring, particularly in April, and in autumn. However, there have been several winter records, including four near Park Wood, Gwent, and nearby at Minorca Farm, East Glamorgan, on 13–20 January 2001. Several individuals were seen in more than one county; these birds are counted once in the table but separately on the county map. Two such birds, first recorded at Fenn's Moss, Denbighshire, in May 2009, might be the same returning, and possibly prospecting, birds that were seen visiting mid- and North Wales each April for the next three years.

Crane: Numbers of birds in each recording area 1893–2019. Some birds were seen in more than one recording area.

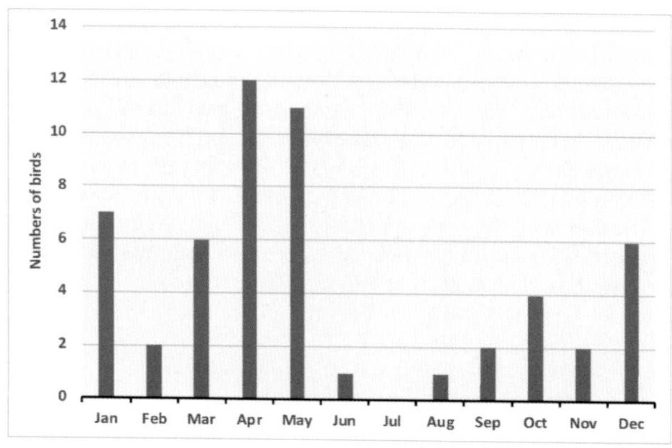

Crane: Totals by month of arrival, 1893–2019.

A small breeding population became established in the Norfolk Broads from 1981. A re-establishment scheme in southwest England, the Great Crane Project, began in 2010, with hand-reared birds released on the Somerset Levels. From 2014, individuals from this project started visiting Wales. Many reports of Cranes, thought to be from this project, have been received in southeast Wales in recent years, including six over Leckwith, East Glamorgan, on 25 September 2017. Released birds are marked with colour rings and are not included in the totals presented here.

A pair, both birds released in Somerset, bred successfully on the Gwent Levels in 2016, raising one young; this was the first breeding record in Wales for over 400 years. In both 2017 and 2018 this pair hatched two chicks, but none survived to fledge. The pair also hatched young in 2019, but these were later predated. A non-breeding pair was also present in East Glamorgan in 2016 and 2017, and bred in the Vale of Glamorgan in 2018, but the single chick died soon after hatching. Two females were present in 2019, but the male from 2016–18 paired up with another female at Slimbridge, Gloucestershire.

The British population was thought to have increased to over 200 individuals in 2019, including 56 breeding pairs (RSPB). Extensive wetland areas are needed to establish a viable breeding population. The creation of new wetlands in some parts of Wales may offer an opportunity for this species to become a regular breeder in Wales once more.

Rhion Pritchard, Jon Green and Robin Sandham

Little Grebe *Tachybaptus ruficollis* Gwyach Fach

Welsh List Category	IUCN Red List			Birds of Conservation Concern			
	Global	Europe	GB	UK	Wales		
A	LC	LC	LC	Green	2002	2010	2016

The Little Grebe breeds across Eurasia, from the Atlantic to the Pacific, and over much of sub-Saharan Africa (*HBW*). When breeding, it favours base-rich water bodies with dense emergent and submergent vegetation, making this shy species difficult to detect visually. The oligotrophic nature of many Welsh ponds, lakes and slow-running streams, especially in the uplands, means that a lack of water bodies with emergent vegetation limits breeding opportunities. It is consequently a rather uncommon and patchily distributed breeding species in Wales, although it can be ubiquitous on lowland farm ponds and oxbow lakes that are subject to little human disturbance.

Where suitable vegetation is present, it will breed on medium-sized lakes such as Llyn Eiddwen, Ceredigion, (10.5ha), on larger lakes such as Llangorse Lake, Breconshire, (153ha) and even on large upland reservoirs such as Nantymoch in Ceredigion (212ha), and Lake Vyrnwy (4.5km^2) and Llyn Clywedog (2.5km^2), both in Montgomeryshire. Breeding has been recorded up to 375m in Caernarfonshire, 390m in Meirionnydd (*Birds in Wales* 1994) and 420m in Ceredigion (Roderick and Davis 2010), but is more common below 250m.

Little is known of the species' abundance prior to the early 20th century. Mathew (1894) said that it was the only grebe breeding in Pembrokeshire, while Bertram Lloyd found a pair breeding at Llanbed Pools, Mathry, in 1936, commenting that it was the first breeding he had seen in Pembrokeshire (Bertram Lloyd Diaries 1925–39). *Birds in Wales* (1994) considered Little Grebe to have declined in Wales during the 1930s, attributing this to the severe winter of 1928/29. Hard winters in 1939/40, 1947/48 and 1962/63 were implicated in further major population reductions, especially the last of these, after which Little Grebes ceased to breed in North Wales for two years (Marchant *et al.* 1990). Although it had re-established in all of North Wales by the *Britain and Ireland Atlas 1968–72*, it remained absent from Pembrokeshire and Carmarthenshire, save for a handful of sites on the south coast.

The number of 10-km squares occupied by Little Grebe in the *Britain and Ireland Atlas 2007–11* increased by 57% over those in 1968–72. Increases were most evident in the valleys of East Glamorgan and Pembrokeshire. Tetrad-level atlases in these two counties emphasised these changes in the breeding season: from 3% of tetrads occupied in Pembrokeshire in 1984–88 to 11% in 2003–07, and from 1% in East Glamorgan in 1984–89 to 12% in 2008–11. Relative abundance during the breeding season is highest in Anglesey and sites close to the south coast.

Little Grebe and chick © Kev Joynes

In the mid-1960s, the Welsh breeding population was estimated to be no more than 50 pairs, but it increased during the 1980s, especially in Anglesey, Caernarfonshire and Carmarthenshire. This was attributed to the creation of new farm ponds and a series of mild winters. *Birds in Wales* (1994) estimated the population at 120–150 pairs. Although this species is by no means exhaustively reported, *Welsh Bird Reports* (2009–15) suggest a total of about 133 pairs in Wales. Up to 11 pairs bred at RSPB Cors Ddyga, Anglesey, in 2018. The *North Wales Atlas 2008–12* recorded confirmed or probable breeding in 92 tetrads in North Wales. Comparing this figure with annual observations recorded in county bird reports for this area suggests that the reports underestimate the true numbers by *c.*60%. Applying the same assumption to the other regions suggests a Welsh breeding population of 330 pairs. The latest population estimate is between 240 and 640 pairs (Hughes *et al.* 2020).

In winter, a large proportion of the Welsh population is thought to move to coastal areas, congregating in muddy estuarine habitats, although solid evidence for this is lacking. For example, numbers on the Anglesey lakes in autumn fell by 70% during midwinter, as counts simultaneously increased at coastal estuarine sites (Vinicombe 1982). In intertidal habitats, Little Grebes are thought to feed on crustaceans and small fish dependent on the tide (Fox 1994), but the species almost never resorts to exposed open maritime coasts with hard substrates. Little Grebes were shot at Aberleri, south of the present Ynyslas dunes, in Ceredigion, as long ago as October 1805 (Roderick and Davis 2010), confirming the species' long-term winter presence in such brackish tidal habitat.

The proportion of 10-km squares occupied by Little Grebe in winter increased from 48% in the *Britain and Ireland Winter Atlas 1981–84* to 62% in 2007–11. Peak winter counts include 91 on the Cleddau Estuary in 1996/97, 74 on the Welsh side of the Dee Estuary in 2016/17 and 69 on the Welsh side of the Severn Estuary in 2005/06. Other high counts are 78 at Shotwick Boating Lake, Flintshire, on 11 November 2018 with 87 there on 15 September 2019. Following construction of the Taff-Ely barrage in 2001, Cardiff Bay has regularly supported 35–60 Little Grebes in winter. Although peak counts are derived across many years, they confirm that, in combination, estuarine habitats support a substantial number of birds in winter. Inland freshwater water bodies also regularly hold ten or more outside the breeding season. For example, there were 28 at Hindwell Pool, Radnorshire, from 18 December 2008 to January 2009 (Jennings 2014). In Breconshire, the breeding population is believed to winter on slack sections of the major rivers, the Usk and Wye, in the county, and at Talybont Reservoir, where 20–30 were counted each winter between 2015/16 and 2019/20 (Andrew King *in litt.*).

Despite there being few recoveries of Welsh-ringed Little Grebes, the recapture of a wintering bird at Aberleri on the Dyfi Estuary, Ceredigion, six years after it was ringed there, suggests that birds exhibit a degree of site fidelity between winters (Robinson *et al.* 2020). Daily observations at Aberleri recorded the first arrivals in the last week of September, although birds were constantly

Site	County	89/90 to 93/94	94/95 to 98/99	99/00 to 03/04	04/05 to 08/09	09/10 to 13/14	14/15 to 18/19
Severn Estuary: Welsh side (whole estuary)	Gwent/East Glamorgan	0 (9)	1 (21)	18 (41)	54 (81)	41 (69)	54 (76)
Dee Estuary: Welsh side (whole estuary)	Flintshire	3 (11)	16 (19)	23 (30)	10 (15)	11 (12)	53 (55)
Llanishen and Lisvane Reservoirs	East Glamorgan	4	9	17	20	15	31
Cleddau Estuary	Pembrokeshire	29	69	55	51	44	27
Cemlyn Bay and Lagoon	Anglesey	22	33	15	16	12	9

Little Grebe: Five-year average peak counts during Wetland Bird Surveys between 1989/90 and 2018/19 at sites in Wales averaging 30 or more at least once. International importance threshold: 4,700; Great Britain importance threshold: 150.

present only from the second week of October. Numbers peaked in late November and again in late February, with lower numbers in late December in three consecutive winters. None was seen after mid-March, by which time birds had reappeared on freshwater breeding ponds around the estuary (Fox 1994). Wetland Bird Surveys (WeBS) indicate that this pattern is typical, with numbers lowest on estuaries and large lakes during April–June and peaking in November–January, before dropping again as birds disperse back to smaller, uncounted breeding sites, where the first eggs are laid from early April (Moss and Moss 1993).

Numbers recorded on WeBS counts in Wales doubled between the early 1990s and early 2000s, followed by a stabilisation in numbers that was interrupted for several years, following the severe winter of 2009/10 (Frost *et al.* 2019b). This trend was reflected across Europe during 1990–2015 (EBCC 2019). Monthly counts at the most important sites in Wales regularly aggregate over 500 birds. Since 184 WeBS sites have registered Little Grebes with an average of ten birds or fewer in 2014/15–2018/19, it is reasonable to assume 1,000 or more Little Grebes winter here. Such a value corresponds to an estimate of a breeding population of 330 pairs, producing an average of one fledged young during the preceding nesting season.

These breeding estimates and winter counts are likely to overlook birds at smaller sites, and it is not known how many birds from outside Wales swell winter numbers. Further expansion of the breeding population may be possible, if mild winters and the construction of small farm ponds continue, but the species may be reaching its capacity in Wales, given the limited availability of nutrient-rich small ponds. Recent increased numbers wintering at some freshwater and estuarine sites suggest that some could support more Little Grebes. However, we lack sufficient knowledge of the species' movements, its feeding ecology or the carrying capacity of its winter and breeding quarters to understand the current situation. Although susceptible to the effects of severe winters, these apparently only affect the population for a year or two, if followed by mild conditions. The immediate future of this species does not seem in jeopardy as long as suitable breeding and wintering habitat continues to be available in Wales.

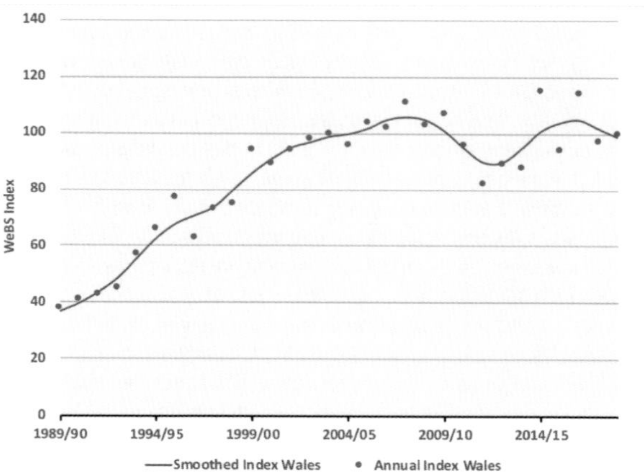

Little Grebe: Wetland Bird Survey indices, 1989/90 to 2018/19.

Tony Fox

Pied-billed Grebe *Podilymbus podiceps* Gwyach Ylfinfraith

Welsh List Category	IUCN Red List Global	Rarity recording
A	LC	BBRC

This grebe breeds throughout the Americas, from central Canada to southern South America. Northern populations are migratory, wintering from the southern USA to Central America. There have been only four records in Wales, out of a total of 45 in Britain up to the end of 2019 (Holt *et al.* 2020). All the records were of single birds:

- Aber Ogwen, Caernarfonshire, from 13 November to 30 December 1984;
- Kenfig Pool, East Glamorgan, from 31 January to 23 April 1987—what was presumed to be the same bird returned there in the autumn from 31 October to 1 April 1988;
- Llangorse Lake, Breconshire, from 15 January to 7 February 1999 and
- Cosmeston Lakes, East Glamorgan, from 31 January to 31 March 1999.

Jon Green and Robin Sandham

Red-necked Grebe *Podiceps grisegena* Gwyach Yddfgoch

Welsh List Category	IUCN Red List			Birds of Conservation Concern				Rarity recording
	Global	Europe	GB	UK	Wales			
A	LC	LC	CR	Red	2002	2010	2016	WBRC

The nominate subspecies breeds across central and northeastern Europe and is mainly a winter visitor to Britain. Another subspecies, *P.g. holbollii,* breeds in northeast Asia and northern North America. This is the scarcest of the five grebes that regularly winter around Wales, with 409 individuals recorded since 1950. Numbers peaked in the 1980s, but have decreased dramatically since. It prefers sheltered localities and there can sometimes be hard-weather influxes from the Continent.

Red-necked Grebes were described as occasional visitors to Wales in the 19th century. The first Welsh record was one shot at Newton Pool, Gower, on 6 March 1833. Another was shot, on an unknown date, on the River Ely, Cardiff, East Glamorgan, before the turn of the century. The first for Anglesey was at Aberffraw in 1887 with one "obtained" on the Dee Estuary, Flintshire in 1894. Two were shot on Afon Dyfi, Ceredigion in October 1899, shortly after two had been seen upriver in Montgomeryshire. In Pembrokeshire, Lockley *et al.* (1949) stated that Tracy had several times killed this species on the millpond at Pembroke and that one was seen at Tenby in spring 1898.

There were no further Welsh records until two at Llanishen and Lisvane reservoirs in Cardiff, East Glamorgan, in 1922, one from 5 February to 2 April, the other from 12 March to 2 April. Others were seen there from 31 January to 14 February 1937, on 2 February 1941, and two from 18 February to 19 March 1956, one of which remained to 5 April. Three were seen in the Menai Strait, Caernarfonshire/Anglesey each winter during 1925–29, with others reported on Anglesey at Menai Bridge in February 1949, December 1950 and January 1954, and at Llanddwyn in 1950. There were singles in Pembrokeshire at Crickmarren Pond on 1 March 1917, Solva on 10 September 1920 and St Ann's Head on 21 April 1938. In Ceredigion one was on the Dyfi Estuary on 2 December 1933 and another was shot on the Afon Rheidol at Llanbadarn in early February 1947.

There was an increase in records from 1960, peaking in the 1980s, but subsequently declining. By far the greatest number of Red-necked Grebes were seen in Anglesey, Caernarfonshire, Pembrokeshire and Glamorgan. Most records were 1–2 birds, but four were at Holyhead, Anglesey, in March 1979. Three were recorded displaying at Llandanwg, Meirionnydd, on 12 March 1987.

Red-necked Grebes are rare inland. There have been eight records at Llandegfedd Reservoir, Gwent, including birds in summer plumage on 14 May 1995, 14–24 April 2006 and 7 April 2007. There have been a total of five records in Breconshire, four of these at Talybont Reservoir, on 26 October 1958, 28 October 1967 (with what was probably the same bird at Llangorse Lake on 1 November), 17 February 1975 and 18 December 1977. There was

	1950–59	1960–69	1970–79	1980–89	1990–99	2000–09	2010–19	1950–2019
Gwent	0	0	2	2	3	3	2	12
Glamorgan	-	12	10	30	17	-	-	69
East Glamorgan	2	-	-	-	-	5	1	8
Gower	-	-	-	-	-	2	5	7
Carmarthenshire	-	4	7	3	3	3	0	20
Pembrokeshire	0	3	12	25	29**	12	3	84
Ceredigion	0	0	4	4	4	1	0	13
Breconshire	1	1	2	0	0	0	1	5
Radnorshire	-	-	-	-	-	1	0	1
Montgomery	-	-	-	-	-	1	0	1
Meirionnydd	-	-	1	4	2	3	2*	12
Caernarfonshire	-	1	3	27	19	13	2*	65
Anglesey	3	9	20	36	18	6	4	96
Denbighshire	-	-	1	2	3	1	2	9
Flintshire	-	-	1	0	2	3	1	7
Total	**6**	**30**	**63**	**133**	**100**	**54**	**23**	**409***

Red-necked Grebe: Individuals recorded in Wales by decade since 1950.

* Includes one on the Afon Glaslyn, seen in Caernarfonshire and Meirionnydd in January 2016.

** Includes 14 in February 1996, with inland records at Llys-y-fran Reservoir, Penberi Reservoir and Heathfield Gravel Pits.

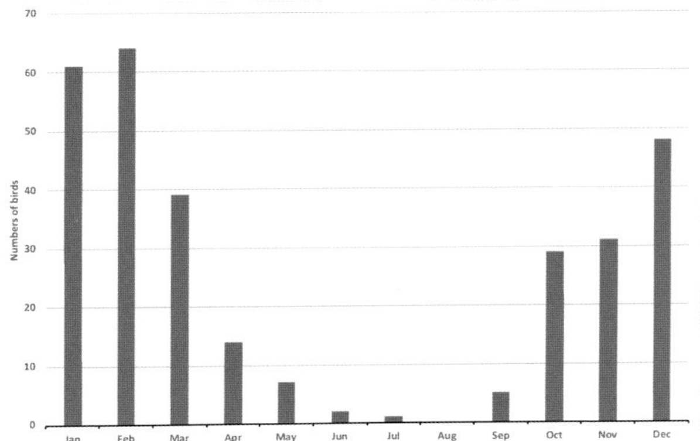

Red-necked Grebe: Totals by month of arrival, 1950 to 2019.

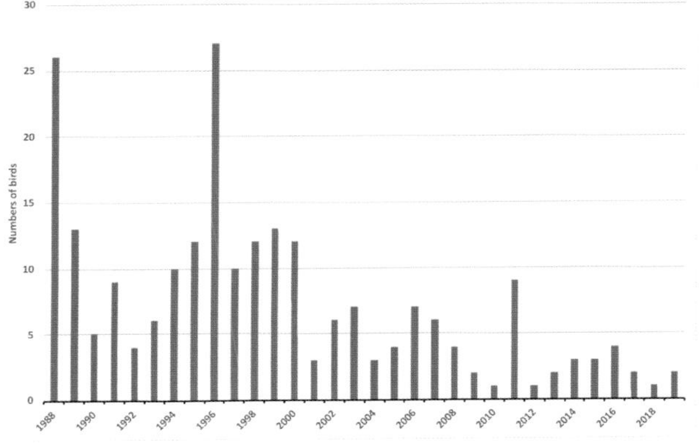

Red-necked Grebe: Numbers seen in Wales each year, 1988 to 2019.

one at Llangorse Lake from 16 February to 4 March 2019. The only record in Radnorshire was a first-calendar-year bird, retrieved from a small stream at Evenjobb on 7 November 2006; it was put in a garden pond, ate several Goldfish and flew off. There is a single Montgomeryshire record, at Morfa Dyfi on 3 March 2000. Inland records in Meirionnydd were one on Llyn Tegid on 23 January 1979 and one on Llyn Trawsfynydd, from 27 October 1992 to 21 March 1993 and in February and March 1996. In Caernarfonshire, there were single birds on Llyn Padarn from 24 February to 2 March 1997 and on 1 November 2003.

Most Red-necked Grebes occur in winter but there have been several summer records. On Anglesey, there was one in breeding plumage on Llyn Maelog in August 1976. There were singles at Llyn Alaw from April to June 1986, at Llyn Penrhyn in April 1989 and possibly the same bird at Llyn Traffwll in July 1989. In Flintshire, a summer-plumage bird was at Shotwick Lake on 27 April 2000. One at Llyn Helyg from 9 May to 6 June 2000 started to build a nest; what is presumed to be the same bird returned on 7–30 June 2001 and 8–10 May 2002.

Jon Green and Robin Sandham

Great Crested Grebe *Podiceps cristatus* Gwyach Fawr Gopog

Welsh List Category	IUCN Red List			Birds of Conservation Concern			
	Global	Europe	GB	UK	Wales		
					2002	2010	2016
A	LC	LC	VU	Green			

This grebe has a wide distribution. The nominate subspecies breeds over much of temperate Europe and Asia, with other subspecies found in Australasia and Africa south of the Sahara. Its preferred breeding sites are shallow eutrophic lowland lakes and reservoirs with emergent vegetation, thus many Welsh lakes are not particularly suitable breeding habitat. It feeds on small fish, crustaceans and other small prey, caught by diving. The birds breeding in western Europe do not usually move very far (*HBW*). In winter, some remain at breeding sites, but most move to the coast, mainly to sheltered bays and estuaries. There is a considerable influx on the Welsh coast in autumn, the wintering population in Wales being much greater than during the breeding season. Most of the birds moving into Wales in winter probably breed in England, as illustrated by the only two ringing recoveries involving Wales: one ringed at Borth, Ceredigion, in January 1992 was found dead at Rothley, Leicestershire, the following June, and one ringed in June 1976 near Wakefield, West Yorkshire, was found dead in March 1986 at Rhyl, Flintshire.

Breeding occurs widely in Wales, though the Great Crested Grebe is more abundant in the east and on Anglesey than elsewhere. It breeds mainly in the lowlands, but it now sometimes nests at a considerably higher altitude than the 200m mentioned in *Birds in Wales* (1994). In recent years a pair has bred regularly on Llyn Idwal, Caernarfonshire, at 375m. A pair held territory at 404m on Llyn Arenig Fawr, Meirionnydd, in June and July 2019, and displaying birds have been noted at still higher altitudes, up to 550m in the Carneddau, Caernarfonshire.

Around the middle of the 19th century, large numbers of Great Crested Grebes were killed to protect fisheries and for their skins, which have a covering of short, dense, waterproof feathers and were used as a substitute for fur. The campaign to protect the birds led to the formation of the RSPB in 1889. Dillwyn (1848) knew of only two records around Swansea, Gower: one killed in winter 1837/38 and the other undated, while Davies (1858) noted that this species had been frequently killed in the Llandeilo area of Carmarthenshire. By 1860, it had apparently been wiped out as a breeding species in Wales (*Birds in Wales* 1994). A close season from hunting was declared for the species in 1880, which protected it from March to July, and numbers slowly recovered. Llangorse Lake, Breconshire, was colonised in 1882, and there were two pairs on Llyn Maelog, Anglesey, by 1885. The species spread in the 1890s, with three pairs on Llyn Penrhyn, Anglesey, in 1892, and a pair bred at Sant-y-nyll pond near St Fagan's, East Glamorgan, in 1894. By 1902, there were 20 breeding pairs at Llangorse Lake (*Birds in Wales* 1994). Forrest (1907) reported that in North Wales a few pairs bred near the English border, such as at Hanmer Mere, Denbighshire, and on Anglesey. The birds were protected at Hensol Castle lake, East Glamorgan, where numbers rose to eight pairs in 1922. The first breeding record for Caernarfonshire was in 1926, and in Radnorshire some years before this (Jennings 2014).

An enquiry into Great Crested Grebe numbers in 1931 found 21–23 pairs in Wales, including 8–10 at Llangorse Lake, with three pairs on Hanmer Mere. Montgomeryshire had three pairs, Glamorgan and Meirionnydd two pairs each, and there were single pairs in Anglesey and Denbighshire (Harrisson and Hollom 1932). The presence of only one pair on Anglesey suggested a decline there as well as at Hensol Lake. Numbers at Llangorse Lake were back to 20 pairs in 1934. The slow spread continued, with breeding confirmed in Ceredigion for the first time in 1949. A census in 1965, which counted adults rather than pairs, found 67–71 birds in Wales, the largest numbers in Anglesey, where there were 24–25, of which 17 were on Llyn Traffwll. Montgomeryshire held 16, including eight on Llyn Ebyr. Breconshire had 12, of which eight were at Llangorse Lake. Birds were also recorded in Flintshire, Denbighshire, Radnorshire and Ceredigion, but none in Glamorgan or Meirionnydd, where the species had been breeding in 1931. Breeding was not confirmed in Carmarthenshire until 1967, and in Gwent not until 1971.

Another census in 1975 found that numbers in Britain had increased by about 50% since 1965, although coverage in Wales was less complete, with no data from Flintshire and fewer sites visited in several other counties. Nonetheless, 134 adults were counted, and the Wales total was estimated at 163, with birds recorded in all counties except Pembrokeshire (Hughes *et al.* 1979). Anglesey, with an estimated 51 birds, remained the stronghold, with 39 in Breconshire, 16 in Montgomeryshire and 12 in Denbighshire. An RSPB survey of Anglesey in 1986 estimated the population at 20–25 pairs.

The *Britain and Ireland Atlas 1968–72* found the Great Crested Grebe to be sparsely distributed in Wales, being almost absent from the southwest. There was a 47% increase in distribution by the 1988–91 *Atlas*, particularly in Anglesey, northern Ceredigion

County	Estimate (pairs)	Source
Gwent	10–15	Darryl Spittle *in litt.*
East Glamorgan	10–12	Phil Bristow *in litt.*
Gower	2–3	Rob Taylor *in litt.*
Carmarthenshire	6	Rob Hunt *in litt.*
Pembrokeshire	1–4	Bob Haycock *in litt.*
Ceredigion	1–4	Roderick and Davis 2010
Breconshire	8–10	Andrew King *in litt.*
Radnorshire	2–4	Pete Jennings *in litt.*
Montgomeryshire	c.14	Holt and Williams 2008
Meirionnydd	10–12	Jim Dustow *in litt.*
Caernarfonshire	10–12	Pritchard 2017
Anglesey	15–20	Nigel Brown *in litt.*
Denbighshire	17	Ian Spence *in litt.*
Flintshire	8	Ian Spence *in litt.*
Total	**114–141**	

Great Crested Grebe: Estimated total of breeding pairs (including unsuccessful breeding) by county based on information from County Recorders and others.

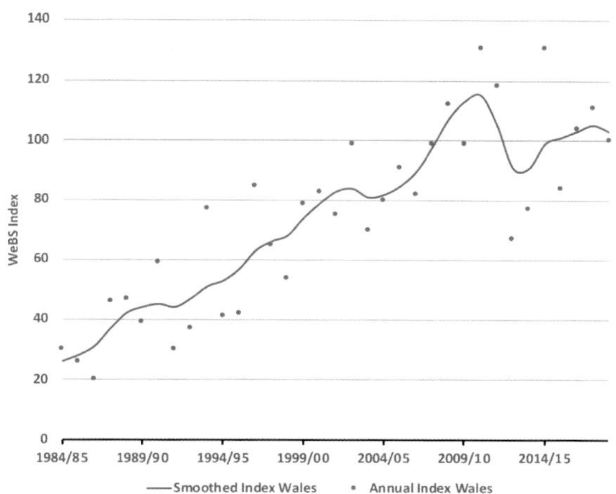

Great Crested Grebe: Wetland Bird Survey indices, 1984/85 to 2018/19.

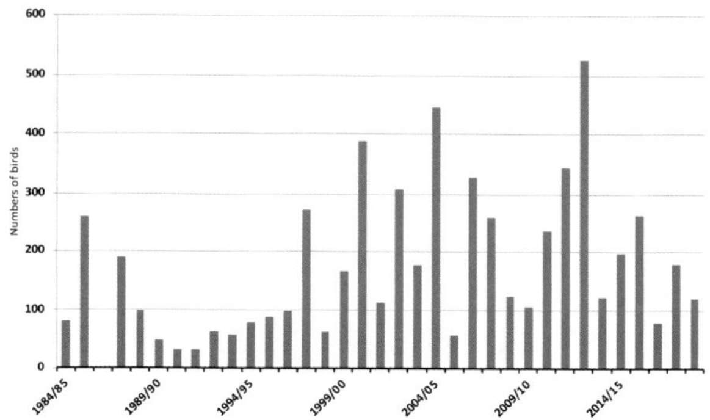

Great Crested Grebe: Occurrence at Traeth Lafan, Caernarfonshire, 1984/85 to 2018/19.

and lowland Glamorgan. There was a further 27% increase by 2007–11, despite some losses in mid-Wales. Relative abundance in Wales was lower than in parts of England and Ireland, except in Anglesey and to some extent southeast Wales and parts of Caernarfonshire, and was lower in Wales than in 1988–91. Breeding atlases based on fieldwork over several years, particularly atlases at tetrad level, may over-estimate the number of pairs, which may breed at a site only in some years.

Birds in Wales (1994) estimated the Welsh breeding population at 100 pairs, by which time birds had bred in all Welsh counties except Pembrokeshire. However, breeding was confirmed there in 1996. The most recent population estimate is that about 270 pairs (210–370 pairs) breed in Wales (Hughes *et al.* 2020), though the total from estimates by county recorders and others is lower. A reduction at some lakes and reservoirs has been attributed to disturbance by water sports. Breeding success appears to be low at some sites: 11 pairs on Llangorse Lake in 2002 fledged only one young between them, and breeding success remained low in subsequent years. Two broods fledged there in 2019, the best performance for some years (Andrew King *in litt.*).

Considerable numbers of Great Crested Grebes use sheltered coastal waters in Wales to moult or winter. The Wetland Bird Survey (WeBS) index for Wales showed a steady increase between 1984/85 and 2010/11.

Two sites in Wales have been of British importance, holding an average of 190 or more: Swansea Bay and Traeth Lafan, although the latter ceased to qualify in 2017/18. Numbers at Traeth Lafan usually peak during July to October: larger counts there include a Welsh record of 536 in October 1976, 508 on 30 August 1994 and 446 in August 2004. It was previously thought that birds at Traeth Lafan moulted there and then moved to winter along the Merseyside coast (Dare and Schofield 1976), but based on the decline in post-moulting birds at Traeth Lafan, coincident with increases in Cardigan Bay, Fox and Roderick (1990) suggested that the latter was the ultimate wintering area for birds moulting at Traeth Lafan. Numbers at Traeth Lafan are lower from November to January, but there is a second peak in February and March, such as 510 on 3 March 2013, as birds move back from Cardigan Bay before dispersing to breeding sites. Small numbers can remain at Traeth Lafan even during the breeding season, although 104 on 5 May 1973 was unusual. There can also be large counts farther east, between Llanddulas, Denbighshire, and Point of Ayr, Flintshire, such as over 400 off Pensarn, Denbighshire, on 19 February 2011. The Dee Estuary is of British importance, but numbers on the Welsh side are usually small, although there were 252 around Bagillt Bank, Flintshire, in September 1997.

Over 500 Great Crested Grebes were recorded in Cardigan Bay in 1981. Counts between 1991/92 and 1997/98 found large numbers there, with a mean winter count of 254 and a peak of 412 in January 1997 (Thorpe 2002). Numbers in Cardigan Bay usually peak between December and February, but there have been no co-ordinated counts there in recent years and it is poorly covered by the Wetland Bird Survey. There were 195 off Morfa Harlech, Meirionnydd, on 29 November 2014.

Fox and Roderick (1990) noted that wintering Great Crested Grebes were much scarcer off the south coast from Pembrokeshire eastwards, though they were regular in small numbers in the Burry Inlet, Gower/Carmarthenshire. This has subsequently changed, with numbers increasing rapidly in Swansea Bay from the late 1990s, peaking at 440 in 2010/11 and at least 440 again in December 2019. There was also a marked increase in the Severn Estuary from 2001/02. On freshwater, Eglwys Nunydd Reservoir, Gower, sometimes holds impressive numbers, such as 156 in January 2007. In total, Wales probably holds about 1,000 birds in winter, of an estimated 17,000 wintering in Britain (Frost *et al.* 2019a).

Great Crested Grebe seems to be faring reasonably well as a breeding bird in Wales, but we lack detailed information and a dedicated survey to establish the number of breeding pairs. There appear to be no serious conservation concerns at present, although human disturbance may prevent breeding on some otherwise-suitable lakes. There is some indication that southern Britain will become climatically less suitable later this century (Huntley *et al.* 2007). Concentrations of wintering birds are vulnerable to oil pollution and developments in the inshore marine environment.

Rhion Pritchard

Site	County	84/85 to 88/89	89/90 to 93/94	94/95 to 98/99	99/00 to 03/04	04/05 to 08/09	09/10 to 13/14	14/15 to 18/19
Swansea Bay	Gower	4	4	58	120	149	299	217
Traeth Lafan	Caernarfonshire/ Anglesey	176	46	120	230	243	267	167
Severn Estuary: Welsh side (whole estuary)	Gwent/East Glamorgan	1 (2)	1 (3)	1 (7)	15 (19)	35 (39)	35 (40)	65 (67)
Inland Sea and Alaw Estuary	Anglesey	28	15	8	25	17	36	56
Llandegfedd Reservoir	Gwent	9	15	45	-	37	39	50

Great Crested Grebe: Five-year average peak counts during Wetland Bird Surveys between 1984/85 and 2018/19 at sites in Wales averaging 50 or more at least once. International importance threshold: 6,300; Great Britain importance threshold: 170.

Slavonian Grebe *Podiceps auritus* Gwyach Gorniog

Welsh List Category	IUCN Red List			Birds of Conservation Concern				Rarity recording
	Global	Europe	GB	UK	Wales			
A	VU	NT	CR	Red	2002	2010	2016	RBBP

The Slavonian Grebe has a wide distribution, breeding across much of Canada (where it is known as Horned Grebe) and temperate Eurasia, from Iceland to Kamchatka. It winters to the south of its breeding range, mainly at sea. It has extended its breeding range in Europe since 1900, but numbers are thought to have decreased within its core range. The largest populations in Europe are in Finland and Sweden (BirdLife International 2015). There is a small breeding population in Scotland, but there has never been any suggestion of nesting in Wales. An absence of ringing recoveries involving Wales means that the origin of birds wintering in Welsh coastal waters is unknown.

The first records in Wales were from the 19th century. One was shot at Marshfield, Gwent, in the late 19th century (Venables *et al.* 2008), while in Gower, Dillwyn (1848) recorded two shot at Penrice, one in February 1830 and the other in January 1847, with a few other records in the 1880s and 1890s. There seem to be no early records for Carmarthenshire, but in Pembrokeshire, Mathew (1894) quoted Dix as saying that a few occurred every winter. In Ceredigion, F.T. Fielden recorded singles in the Dyfi Estuary in September 1886, in 1887 and in winter 1891/92 (Roderick and Davis 2010). Forrest (1907) commented that in North Wales, its range "appears to be almost confined to the Merioneth coast, where it occurs frequently in winter". He quoted Rawlings as saying that this species was more common than Great Crested Grebe off Barmouth, with many in some winters. It was also said to be not uncommon at Penrhyndeudraeth. Records elsewhere were few, though there was one on the River Severn at Welshpool, Montgomeryshire, in December 1900.

The species was recorded fairly regularly during the first half of the 20th century, including a number inland, but there were few large counts, an exception being nine grebes, thought to be this species, in the Gwendraeth Estuary, Carmarthenshire, on 3 January 1950 (Ingram and Salmon 1954). Numbers increased in the second half of the century, and by 1959 it occurred annually in small numbers in the Burry Inlet ('North Gower coast' in the table), where the maximum count was 17 at Whiteford in March 1989. In Pembrokeshire, it became a regular winter visitor from the 1960s, most frequently in St Brides and Angle Bays, where

Site	County	89/90 to 93/94	94/95 to 98/99	99/00 to 03/04	04/05 to 08/09	09/10 to 13/14	14/15 to 18/19
Traeth Lafan	Caernarfonshire/Anglesey	10	4	10	8	7	7
North Gower coast	Gower	5	8	10	9	5	5
Tremadog Bay	Meirionnydd/Caernarfonshire	14	13	11	4	5	5
Beddmanarch Bay/Inland Sea	Anglesey	2	5	7	5	6	4

Slavonian Grebe: Five-year average peak counts during Wetland Bird Surveys between 1989/90 and 2018/19 at sites in Wales averaging seven or more at least once. International importance threshold: 50; Great Britain importance threshold: 9.

Slavonian Grebe in winter plumage © Tate Lloyd

up to seven were recorded (Donovan and Rees 1994) and the county annual total peaked at 16 in 1997. It became almost an annual visitor to Ceredigion from 1977, mainly off Borth and the Dyfi Estuary (Roderick and Davis 2010), but there have been no large counts. Larger numbers have been recorded in Tremadog Bay, Meirionnydd/Caernarfonshire, with a maximum of six in the 1970s. Numbers increased from about 1984, including a Welsh record count of 21 off Llandanwg, Meirionnydd, on 12 March 1987. Counts between Mochras, Meirionnydd, and Black Rock, Caernarfonshire, in the 1990s produced another count of 21 in January 1994.

On the north coast, Traeth Lafan has proved to be an important site, with birds usually recorded on the Caernarfonshire side. Numbers increased from the 1980s, peaking at 14 in March 2002 and 15 in April 2003. Numbers farther east, in Conwy Bay, off Rhos Point and off Llanddulas in Denbighshire, are usually smaller. On Anglesey, Llanddwyn Bay emerged as a site with regular records from the 1950s, and in the 1960s and 1970s it was regarded as the best site for the species in North Wales, with counts peaking at 13 in late 1974, but there have been few records in recent years. Also, in Anglesey, Beddmanarch Bay and Inland Sea had regular records from the 1980s onwards, with a maximum of nine recorded (Jones and Whalley 2004). The species is scarce in the Dee Estuary, Flintshire.

The Slavonian Grebe is not always well covered by WeBS counts, as an accurate count of birds any distance offshore requires a calm day, which seldom coincides with core count dates in winter.

Numbers have declined at most sites in recent years, and the only double-figure counts recorded in Wales since 2006/07 were on Traeth Lafan: ten in 2010/11 and 11 in 2014/15. Evans (2000) gives a mean of 40–43 birds wintering in Wales in 1986/87–1992/93, with a maximum of 63–67. In recent years around 40–50 individuals have been recorded in Wales in an average winter, though some are probably missed; an estimated 920 winter in Britain (Frost *et al.* 2019a).

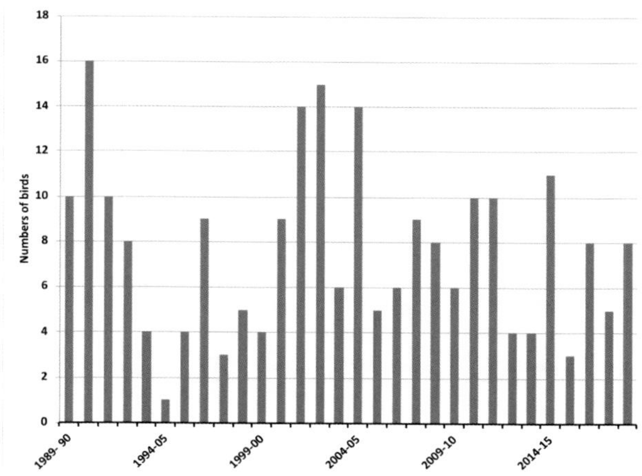

Slavonian Grebe: Occurrence at Traeth Lafan, Caernarfonshire, 1989/90 to 2018/19.

Considering that this is not a numerous species in Wales, it is found inland with surprising frequency, although there is only one record from Radnorshire, a bird on Llandrindod Lake on 11–14 February 1996 (Jennings 2014). The first inland record for Meirionnydd was on Llyn Tegid in January 2018. Llandegfedd Reservoir, Gwent, is a favoured inland site, with 23 records involving 29 birds up to 2019 (Darryl Spittle *in litt.*). There have also been several records from Llangorse Lake, Breconshire, including a pair in breeding plumage in summer 1932 (Peers and Shrubb 1990). Most inland records are of 1–2 birds, but Jones and Whalley (2004) mentioned an undated record of four at Llyn Alaw, Anglesey. Most are recorded between October and early April, peaking in January, but there have been several records in May and one in the Pembroke River on 30 July 1974.

Rhion Pritchard

Black-necked Grebe *Podiceps nigricollis* Gwyach Yddfddu

Welsh List Category	IUCN Red List			Birds of Conservation Concern				Rarity recording
	Global	Europe	GB	UK	Wales			
A	LC	LC	EN	Amber	2002	2010	2016	RBBP

The nominate subspecies of this small grebe breeds in most countries in Europe, though it is scarce in Fennoscandia and around the Mediterranean. Its range extends east to Central Asia. The largest European populations are in Russia and Ukraine (BirdLife International 2015). Other subspecies breed in North America, where it is known as the Eared Grebe, and in southern Africa. It breeds mainly on small, shallow, highly eutrophic water bodies with lush vegetation and stretches of open water (*HBW*), a habitat that is very localised in western Europe.

The second *European Atlas* (2020) found a complex pattern of losses and gains since 1997. Losses dominated in central and eastern Europe, with gains in the west. A westward expansion may be the result of invasions of birds from the east, following desiccation of lakes in steppe areas around the Caspian Sea. The British population has slowly increased to around 55 breeding pairs (Holling *et al.* 2019), the majority at two sites in northern England. Most populations are migratory, wintering to the south of the breeding range. There are no ringing recoveries to indicate the breeding areas of the birds that winter in Wales.

Suitable breeding habitat for this species is scarce in Wales, but it was at one time a regular breeder on Anglesey. One was shot on Llyn Maelog on 1 August 1892, following which Banks (1892) said that he considered the species a resident on the island. He mentioned having seen a male in breeding plumage in May 1892 at an un-named site near Holyhead, which may have been Valley Wetlands, and considered it likely that a female was on eggs nearby. Forrest (1907) was sceptical and considered that the record "doubtless refers to the Great Crested Grebe", although he also reported E. Gosling as stating that "the Eared Grebe was formerly found on Llyn Maelog, but has disappeared for many years, having been shot down". Forrest was, however, unaware of a breeding record on Anglesey in 1904, the first confirmed nest in Britain. On 8 June, Oldham and Cummings found at least five pairs feeding young on Llyn Llywenan, with a sixth pair present (Aplin 1904). At least eight pairs were present the following year, and four pairs in 1910. Breeding occurred at least in some years up to 1934, when a single brood fledged. The only subsequent record of breeding was in 1957, when two young were fed at the site. Birds have been seen subsequently on several Anglesey lakes in the breeding season, including a pair present for at least a month on Llyn Penrhyn in 1990 and a pair that displayed there from 29 March to 8 April 1997. There has been no indication of breeding elsewhere in Wales, although three were at a lake near Builth Wells, Breconshire, in August–September 1932 and a pair was at the same site in July 1934. Three birds were present at Shotwick Boating Lake, Flintshire, on 6–8 May 2019 and display was noted.

In winter, this species is found offshore in sheltered bays and estuaries, and on lakes. Offshore records are usually between September and April, but there can be records in any month. In the late 19th century, the species was apparently more common in North Wales than farther south. Caton Haigh, quoted in Forrest (1907), reported it to be fairly common around Penrhyndeudraeth,

Meirionnydd, during 1890–1906. He saw eight between 1 February and 25 March 1890, of which he shot six. In subsequent years, it was recorded regularly around Tremadog Bay, Meirionnydd/Caernarfonshire, but numbers declined there in the 1920s and 1930s (Pritchard 2012). Two were killed on Hanmer Mere, Denbighshire, in 1864, while in Flintshire it was reported to be regular on the Dee Marshes at the turn of the 20th century (*Birds in Wales* 1994). It was said to be a regular visitor to the Menai Strait in the few years before 1954, but the first certain record for Caernarfonshire was in 1956. Numbers off the north coast increased in the late 1970s and early 1980s, with 24 between the Great Orme and Penmon, Anglesey, on 20 February 1977, eight off Llanfairfechan on 14 February 1982 and seven off Ynys Llanddwyn, Anglesey, on 7 March 1982.

There was no confirmed record from Wales south of the Dyfi Estuary until 1905, when one was on the River Wye near Builth Wells, Breconshire/Radnorshire. The first record for East Glamorgan came in 1921, for Gwent in 1923, for Pembrokeshire in 1925 and for Ceredigion in 1929. The first record for Carmarthenshire was 1938, with the second for the county not until 1968 (Lloyd *et al.* 2015). Up to six were seen in Glamorgan in every winter bar two from 1921/22 up to the Second World War (Hurford and Lansdown 1995). The increase in numbers in Wales might have been linked to a dramatic increase in the number breeding at Lough Funshinagh in Co. Roscommon, Republic of Ireland. Breeding was first proven there in 1929. The population increased to *c.*300 pairs by 1932 before declining (*Birds in Ireland* 1989). Black-necked Grebes were first noted in the Burry Inlet in winter 1953/54 when seven were off Whiteford Point, Gower. There were annual records of up to 12 there during 1956–66, after which there was a gradual decline and the species became irregular after 1985. Farther east, birds were almost annual in the Kenfig/Eglwys Nunydd/Sker area of Gower/East Glamorgan, recorded in 20 of the 26 years during 1967–92, including a county record of 17 off Kenfig Sands, East Glamorgan, on 6 December 1967. In the landlocked counties, the majority of records have come from Breconshire, where there were about 24 records up to 1990, mostly in August and September (Peers and Shrubb 1990), but there have been only six records in the 30 years since (Andrew King *in litt.*).

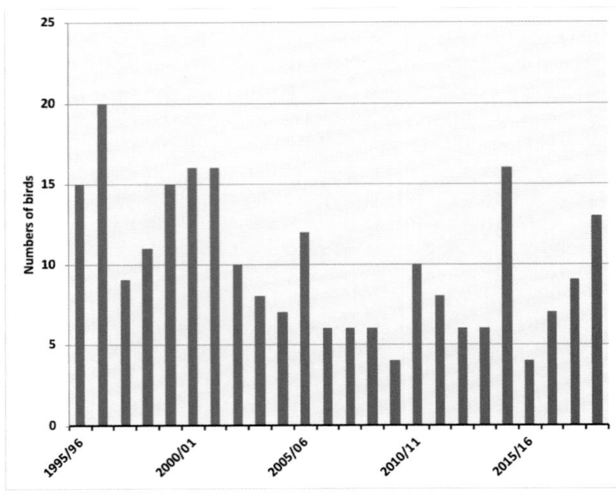

Black-necked Grebe: Occurrence in Wales, 1995/96 to 2018/19.

Numbers recorded in Wales each year vary considerably, but are now much lower than during 1970–99, despite an increase in numbers breeding in England. The most important site for this species in Wales is Tremadog Bay, around the outer part of the Glaslyn/Dwyryd Estuary complex. This is now probably the only site in Wales where Black-necked Grebes are almost annual. Four there on 8 November 2017 was the first count of more than three at any site in Wales since winter 1997/98. Up to five were off Beaumaris, Anglesey, in late April 2018. There are also regular records in the Burry Inlet, on Llandegfedd Reservoir in Gwent, in Cardiff Bay, around Kenfig and Eglwys Nunydd and off Traeth Lafan, Caernarfonshire. On average, about ten winter in Wales out of an estimated 120 in Britain (Frost *et al.* 2019a). The breeding population is modelled to shift eastwards during the 21st century, with none predicted to nest in Britain by 2100 (Huntley *et al.* 2007). The implications for numbers wintering off Wales are unclear, but if these are European birds and there are suitable sheltered bays farther east, this could result in far fewer visiting our coasts.

Rhion Pritchard

Stone-curlew *Burhinus oedicnemus* Rhedwr y Moelydd

Welsh List Category	IUCN Red List			Birds of Conservation Concern	Rarity recording
	Global	Europe	GB	UK	
A	LC	LC	VU	Amber	WBRC, RBBP

Total	Pre-1950	1950–59	1960–69	1970–79	1980–89	1990–99	2000–09	2010–19
48	8	1	1	6	7	11	4	10

Stone-curlew: Individuals recorded in Wales by decade.

Also known as the Eurasian Thick-knee, this species breeds on dry grassland, steppe and arable land with patches of open ground. Its breeding range extends from southern and eastern England east to Central Asia and south to North Africa. Northern populations are migratory, with birds from the northern part of Europe wintering around the Mediterranean and the Sahara Desert. There was a 33% decline in Europe between 1998 and 2016 (PECBMS 2020). Populations are restricted to small areas of suitable habitat, often widely scattered (*HBW*). The decline in numbers in England has been arrested by conservation efforts, and numbers there have increased since the mid-1990s (*Birds in England* 2005). It has never bred in Wales but is a scarce visitor.

The first record for Wales was one shot near Aberystwyth, Ceredigion, in winter 1840/41. The bird was preserved, and L.W. Dillwyn saw it for sale on a visit to Aberystwyth in 1842. Stone-curlews have since been recorded in every Welsh county, mainly in April–August, with one very late record on 15 December 2001

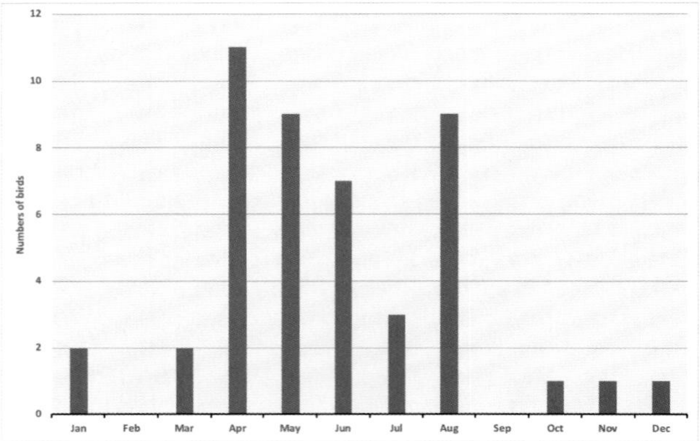

Stone-curlew: Totals by month of arrival where this is known, 1840–2019.

at Llanddulas, Denbighshire. There has been a slight increase in records over the last 50 years, with a peak in the 1990s, principally as a result of five records in 1991. In the 2010s, there were three records in 2011 and two each in 2014 and 2019.

A colour-ringed first-calendar-year bird at Rumney Great Wharf, East Glamorgan, and Wentlooge, Gwent, in August 1990 had been ringed on Salisbury Plain, Wiltshire, on 13 June of that year.

Jon Green and Robin Sandham

Stone-curlew: Numbers of birds in each recording area 1840–2019. Some birds were seen in more than one recording area.

Oystercatcher *Haematopus ostralegus* Pioden Fôr

Welsh List Category	IUCN Red List			Birds of Conservation Concern			
	Global	Europe	GB	UK	Wales		
					2002	2010	2016
A	NT	VU	LC	Amber			

One of our best-known waders, the nominate subspecies of Oystercatcher breeds mainly along the coasts of northern and western Europe. Other subspecies breed across eastern Europe and Asia, as far as the Pacific coast. The largest populations of the nominate subspecies breed in Scotland and The Netherlands (*European Atlas* 2020). The European breeding population increased strongly between the 1960s and the 1990s, aided by a successful adaptation to breeding inland. Subsequently, it declined substantially, with particularly strong decreases in the Wadden Sea area (Van de Pol *et al.* 2014). Some move as far south as West Africa in winter, but many remain around the coasts of western Europe. Numbers wintering on the Welsh coast and estuaries are far larger than the breeding population. Large numbers of Oystercatchers have been ringed in Wales, particularly on Traeth Lafan, Caernarfonshire, by the SCAN Ringing Group. Recoveries indicate that many of the birds wintering in Wales breed in Scotland, the Faroe Islands, Norway and Iceland, although some come from farther afield. Recoveries of birds ringed as nestlings in Wales show a southward movement, with the farthest south being a bird ringed as a nestling on Skokholm, Pembrokeshire, in June 1969 that was subsequently shot south of Casablanca, Morocco, in October 1972. Others winter locally, for example several chicks ringed on Puffin Island, Anglesey, have been recaptured wintering less than 1km away at Penmon. Oystercatchers are long-lived: one ringed as an adult at Wig, Caernarfonshire, in June 1982 was found dead at the same site in April 2018, almost 36 years later.

Although inland breeding is widespread in Scotland and in northern and eastern England, most of the Welsh population breeds on or near the coast. The *Britain and Ireland Atlas 2007–11* showed that relative abundance was low in Wales compared to those areas and showed no pronounced change since the 1988–91 *Atlas*, although there was a marked increase in inland breeding in eastern Wales. No population trends are available for the Welsh breeding population, but the BBS trend for the UK showed an increase until the late 1990s, then a decline.

Breeding numbers in the late 19th and early 20th centuries are difficult to estimate from the available sources. Forrest (1907) described Oystercatcher as common along the coast of North Wales, but said that it only bred as far east as Point of Ayr, Flintshire, and was very rare inland. In Pembrokeshire, Mathew (1894) mentioned breeding at only three coastal sites, but Lockley *et al.* (1949) estimated that 120 pairs bred around the mainland coast and stated that it bred on all the islands. It was described as common along seashores in Carmarthenshire (Barker 1905), while Glamorgan had at least 66 pairs at the end of the 19th century (Hurford and Lansdown 1995).

The largest number of breeding birds is along the coasts of Pembrokeshire, Anglesey and Caernarfonshire. In Pembrokeshire, a 1984–88 survey estimated about 300 breeding pairs, of which at least 50% were on the islands, but the *Pembrokeshire Atlas 2003–07* suggested a 13% decline on the mainland since then (Rees *et al.* 2009). A 1966 survey on Anglesey indicated 140–160 breeding pairs, while a 1986 survey of the island suggested 200–250 pairs in 23 1-km squares. However, by 1997 Oystercatchers bred in only ten of Anglesey's 73 1-km squares (Jones and Whalley 2004). In Caernarfonshire, the population was estimated at 90–120 pairs in 1976 (Jones and Dare 1976) and 120–170 pairs in the 1990s (Barnes 1997). Numbers there have probably remained stable, but there has been no recent survey.

The Ceredigion population was estimated at 20–30 pairs in 1966 and 50–60 pairs in 2010 (Roderick and Davis 2010), while in Meirionnydd, 40 pairs were counted between the Dwyryd and Dyfi estuaries in 1966. There has been no full survey there since. On the south coast, breeding numbers are small, with only sporadic breeding in Carmarthenshire, while in Glamorgan the estimated total was down to 6–8 pairs by 1966, as a result of disturbance and loss of habitat (Hurford and Lansdown 1995). On the Gwent coast, numbers fell from an estimated 15 pairs in 1981–85, to about ten pairs by 2008 (Venables *et al.* 2008). Along the north coast, *Birds in Wales* (1994) reported no breeding on the Denbighshire coast and estimated 40–50 pairs for the entire Dee Estuary in 1992. The *North Wales Atlas 2008–12* recorded 1–2

Nesting Oystercatcher © Ben Porter

pairs present in Denbighshire and confirmed or probable breeding in 20 tetrads on the Flintshire side of the Dee Estuary, suggesting that the population there remains fairly stable. The highest breeding densities are on the islands. For example, in 2015 there were 98 pairs on Bardsey (55 pairs/km²), 73 territories on Skomer (25 territories/km²), 55 on Skokholm (52 territories/km²) and 20 on Ramsey (8 territories/km²). These are the best-monitored populations in Wales, and while numbers fluctuate considerably from year to year, there is no evident trend. Nesting has also been recorded on The Skerries, Anglesey; on Grassholm and Caldey, Pembrokeshire; and on Flat Holm, East Glamorgan.

Inland breeding may have begun in Scotland in the early 19th century, and by the early 21st century had spread to all but the most upland areas (*The Birds of Scotland* 2007). In Wales, the first record of inland breeding was in 1947, when a pair bred at Llyn Trawsfynydd, Meirionnydd, with numbers there peaking at 12 pairs in 1986. In 1974, a nest was found along the River Severn near Caersws, Montgomeryshire, since when there has been a gradual increase along the larger rivers. A survey of river shoals in Wales in 1991, primarily for Little Ringed Plovers, found 25–29 pairs of Oystercatcher along the rivers Severn, Dee, Vyrnwy, Dwyryd and Dyfi, the last holding ten pairs (Tyler 1992b). There has been a

Oystercatcher: Recovery locations of birds ringed in Wales are shown by red circles. Ringing locations of birds that were recovered in Wales are shown by blue triangles. Small brown dots show ringing or recovery locations in Wales.

Site	County	79/80 to 83/84	84/85 to 88/89	89/90 to 93/94	94/95 to 98/99	99/00 to 03/04	04/05 to 08/09	09/10 to 13/14	14/15 to 18/19
Dee Estuary: Welsh side (whole estuary)	Flintshire	na (32,338)	na (29,577)	9,035 (33,322)	13,943 (27,061)	11,580 (22,433)	9,989 (20,804)	16,634 (25,775)	11,521 (23,484)
Burry Inlet	Gower/Carmarthenshire	16,003	18,423	12,391	16,406	15,125	14,059	11,935	16,660
Carmarthen Bay	Carmarthenshire/ Pembrokeshire	825	2,710	5,132	5,018	5,534	8,467	11,936	9,484
Traeth Lafan	Caernarfonshire/ Anglesey	1,585	4,631	3,766	4,574	7,284	6,525	6,225	6,392
Swansea Bay	Gower	2,786	2,302	2,600	2,692	3,373	3,888	2,560	1,882

Oystercatcher: Five-year average peak counts during Wetland Bird Surveys between 1979/80 and 2018/19 at sites in Wales averaging 2,900 or more at least once. International importance threshold: 8,200; Great Britain importance threshold: 2,900.

further spread since then: several pairs now breed at Llyn Brenig and Fenn's Moss in Denbighshire and Oystercatchers have spread along the upper Dee. Breeding occurs fairly regularly at sites such as Dolydd Hafren, Montgomeryshire. In Radnorshire, breeding was first proven along the Wye Valley in 1990 and there have been 1–3 pairs annually since then, between Boughrood and Bronydd (Jennings 2014). There has been annual breeding at Llangorse Lake, Breconshire, since 2010, with occasional breeding by the River Usk, Breconshire, totalling up to five pairs (Andrew King *in litt.*).

Birds in Wales (1994) estimated the Welsh breeding population on the coast at 700–900 pairs. No figure was given for the population breeding inland, but records suggested perhaps another 50 pairs. The figure is probably not greatly different now, perhaps a little lower, with declines in some coastal areas only partly offset by a modest increase in numbers breeding inland.

In winter, Oystercatchers prefer larger estuaries and other coastal areas with soft sediment and good numbers of cockles, mussels and other bivalves, although small numbers are found on rocky shores. Individuals specialise in either cockles or mussels, and changes in the population can be driven by changes in the number of either group. Currently, three sites in Wales have five-year average counts in excess of the threshold for international importance: the Welsh side of the Dee Estuary, the Burry Inlet and Carmarthen Bay. One further site, Traeth Lafan, exceeds the threshold for British importance.

There is comparatively little information available on the numbers of wintering Oystercatchers in Wales before national estuarine bird counts began in 1969/70. Numbers on Traeth Lafan were low in the mid-1960s, with peak counts between 2,500 and 3,000 birds, but there was a large increase in the winter of 1968/69 (Dare 2007). The WeBS population index for Wales shows a fairly steady increase from the 1970s and has been stable since about 1994, but a long-term Medium WeBS Alert (-28%) has been issued for the Dee Estuary SPA (Woodward *et al.* 2019). In the Burry Inlet, the cockle population declined during 1997–2004, before an abrupt 'crash' in stocks during 2004–10. There has been some recovery since then, but stocks of larger cockles are still very low. Body condition and survival of wintering adult Oystercatchers were reduced in the year following the cockle crash (2004/05 to 2005/06), but both subsequently recovered (Bowgen *et al.* in prep.). Counts at other sites are considerably smaller, usually below 1,000, although the Dyfi Estuary, Ceredigion/Meirionnydd, and Conwy Estuary, Caernarfonshire/Denbighshire, had counts in excess of 1,000 birds in the 1990s. Considerable numbers also winter on open coasts. The Non-estuarine Waterbird Survey (NeWS) found a total of 15,010 along the coasts of Wales in 1984/85, 9,012 in 1997/98, 10,687 in 2006/07 and 10,783 in 2015/16. The total number wintering in Wales is probably around 60,000 birds, representing about 20% of the 290,000 estimated for Britain (Frost *et al.* 2019a) and perhaps 6–7% of the flyway population. The 70,000 quoted in *Birds in Wales* (1994) included the English side of the Dee Estuary.

The prominent part played by cockles and mussels in the Oystercatcher's diet sometimes brings it into conflict with commercial interests. The best-known example in Wales is on the Burry Inlet, Gower/Carmarthenshire, where a decline in cockles landed in the 1950s led the South Wales Sea Fisheries Committee to ask the Ministry of Agriculture, Fisheries and Food to initiate a research programme that started in 1958. It found a high mortality of second-winter cockles, which was attributed to Oystercatcher predation (Hancock and Urquhart 1965). A huge spatfall in 1963 led to high cockle numbers for a while, and the Oystercatcher population began to increase significantly as a result of the plentiful food supply. By the early 1970s, cockle numbers had returned to pre-1963 levels, while the average number of wintering Oystercatchers had increased from *c.*8,000 to *c.*14,000. In 1973 the fisheries committee obtained permission to shoot about 11,000 Oystercatchers to protect the cockle fishery, and a total of *c.*10,000 were shot in the autumns of 1973 and 1974. Following the cull, cockle numbers not only failed to increase, but in fact declined sharply for unknown reasons. A new analysis in 1977 concluded that predation by Oystercatchers was not a sufficient explanation of the mortality of second-winter cockles, when these were at medium or high density (Horwood and Goss-Custard 1977).

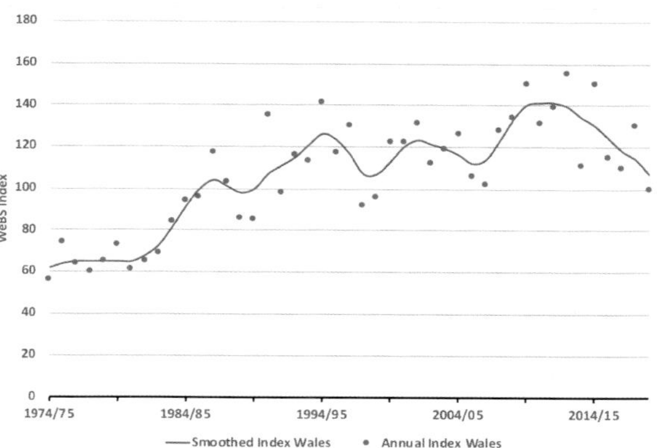

Oystercatcher: Wetland Bird Survey indices, 1974/75 to 2018/19.

Conflicts of this kind are now more likely be resolved by non-lethal methods. The adoption of new methods by growers reduced predation on the mussel beds at Traeth Lafan after 2001 (Atkinson *et al.* 2005). One non-lethal method has been the development of the "Bird-Food Model" adopted by regulating agencies in the UK, including NRW. Surveys of cockle stocks are used to determine the proportion of the stock that is required by Oystercatchers, and then the stock that can be safely harvested. This system is used on the Dee Estuary and Burry Inlet (David Parker *in litt.*). With good management, cocklers, mussel growers and Oystercatchers can co-exist. There do not, otherwise, appear to be any major threats to the population wintering in Wales. Breeding numbers are currently declining in Scotland but increasing in England (Appleton 2016a). A dedicated survey of the breeding population in Wales would provide valuable information.

Rhion Pritchard

Sponsored by the Jenkins family

Black-winged Stilt *Himantopus himantopus* Hirgoes Adeinddu

Welsh List Category	IUCN Red List		Rarity recording
	Global	Europe	
A	LC	LC	WBRC

The nominate subspecies is found in central and southern Europe, Central and southern Asia and Africa. It is a scarce but increasing migrant to Britain and has bred on several occasions in England. While the number of records has increased dramatically in England, from an average of six a year in the 1990s to 26 in the 2010s, there has been no such increase in Wales. There have been 20 individuals recorded, including one breeding attempt.

The first Welsh record was one shot on Anglesey in 1793. There were no further records until a pair attempted to breed at Caldicot Moor, Gwent, in 1952, but they were disturbed by the flow of people interested in seeing them. Most records have been single birds, but there were two on Bardsey, Caernarfonshire, on 29 April 1984. Three birds at Cemlyn, Anglesey, on 10–21 April 1993 then relocated to RSPB Inner Marsh Farm (now Burton Mere Wetlands), where they were seen on both sides of the Flintshire/Wirral border between 22 April and 4 May, and there was a pair at Newport Wetlands, Gwent, on 26 April 2013. Of the 13 dated records, six have been in April." to "Of the 17 individuals whose dates are known, ten arrived in April.

Jon Green and Robin Sandham

Total	Pre-1950	1950–59	1960–69	1970–79	1980–89	1990–99	2000–09	2010–19
20	1	2	2	0	4	6	2	3

Black-winged Stilt: Individuals recorded in Wales by decade.

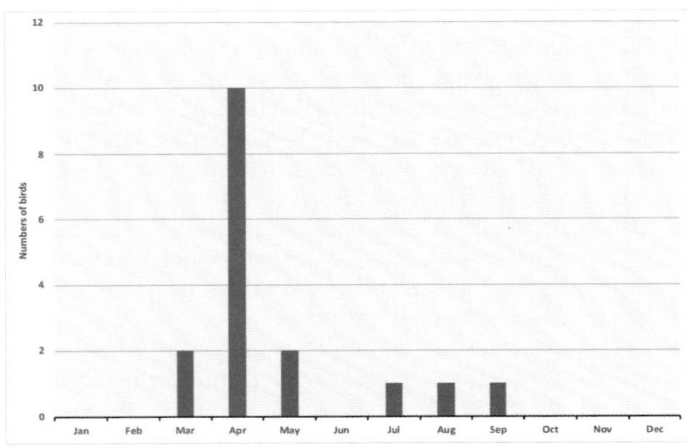

Black-winged Stilt: Totals by month of arrival, 1793–2019.

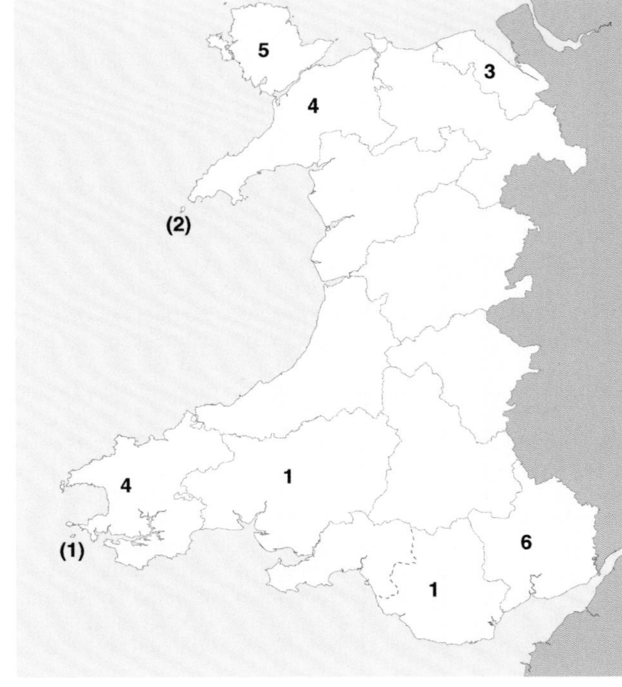

Black-winged Stilt: Numbers of birds in each recording area 1793–2019. Some birds were seen in more than one recording area.

Avocet *Recurvirostra avosetta* Cambig

Welsh List category	IUCN Red List			Birds of Conservation Concern				Rarity recording
	Global	Europe	GB	UK	Wales			
A	LC	LC	LC	Amber	2002	2010	2016	RBBP

The Avocet breeds locally across temperate Europe and Asia, from Britain and the Iberian Peninsula as far east as northern China. It also breeds in eastern and southern Africa, with scattered populations elsewhere. It nests in flat, open areas near shallow brackish lakes and lagoons (*HBW*). The most important characteristics of breeding habitats are water levels that gradually decrease during the summer to expose additional feeding areas, and high salt concentrations that prevent the development of excessive emergent and shoreline vegetation (Johnsgard 1981). Northern populations are migratory, moving to tropical and warm temperate regions south of their breeding range in winter.

The species is thought to have been extirpated from Britain by the middle of the 19th century by the drainage of fens and taking of eggs for food (Parslow 1973). Following breeding in Norfolk in 1941, a growing population was established in Suffolk in the late 1940s (Cadbury and Olney 1978). The breeding population has increased and spread, and in 2016 over 2,000 breeding pairs were reported in Britain for the first time, the largest numbers being in eastern England (Holling *et al.* 2018).

There is an interesting Roman record for Wales. Remains were found in food refuse dating to the late 1st century at the baths of the Second Augustan Legion's base at Caerleon, Gwent. This is one of only two archaeological records of Avocet in Britain (Yalden and Albarella 2009). The species is mentioned in the *Swansea Guide*

Avocet © Tate Lloyd

(Anon. 1802) in a list of the birds found in Swansea and district, but without further comment. Dillwyn (1848) expressed doubt about this record, but stated that he had seen one in the area in 1804 and also mentioned an undated record of one shot near Blackpill. The species remained a rare visitor to Wales for the next 200 years; Mathew (1894) mentioned three shot in Pembrokeshire, including one near Tenby in 1883, while Forrest (1907) knew of only two records in North Wales.

Following its return to eastern England as a breeding species, there was a modest increase in records of Avocet in Wales, and some slightly larger counts; for example four at Little Milford, Pembrokeshire, on 29 January 1955, five at Shotton, Flintshire, on 27 March and 5 April 1960, and six at Dowlais Top, East Glamorgan, on 3 May 1974. When *Birds in Wales* (1994) was published, Avocet was a scarce albeit increasing visitor to Wales, with the greatest concentration of records between March and May, when wintering birds from farther south were returning to their breeding grounds. Most records were of singles, but larger flocks were recorded, including 17 flying up-channel off Cardiff foreshore, East Glamorgan, on 6 May 1991 (Hurford and Lansdown 1995). Most records came from coastal counties, particularly along the south coast from Pembrokeshire to Gwent. At that point there had been no records from Breconshire, Radnorshire, Montgomeryshire or Denbighshire. During 1992–2000, 83 individuals were recorded, the greatest number (38) in Gwent. The total in this period included 26 winter records, 18 of these in Gwent. The highest count away from Gwent was a flock of 12 at Lake Vyrnwy, Montgomeryshire on 22 March 1995, the first county record. Two were recorded in Denbighshire in 1996.

Following the construction of Goldcliff Lagoons at Newport Wetlands Reserve, Gwent, in 1999, a pair of Avocets displayed there in 2002 and successful breeding followed in 2003, the first breeding record for Wales (Venables *et al.* 2008). Breeding success at Newport Wetlands was generally good for the first few years, but has been poor in most years since, which is typical of Avocet colonies. The best years in terms of young fledged per pair were 2008, when seven pairs fledged at least 12 young, and 2014 when 38 pairs fledged 40 young. In 2015, 40 pairs failed to fledge any young, owing to heavy predation, while 39 pairs produced over 80 chicks in 2018, but only three survived the depredations of gulls, corvids and raptors. In 2019, a record 49 pairs bred but not a single chick was fledged (Tom Dalrymple *in litt.*). There was a record count of 134 on the foreshore at Newport Wetlands Reserve (Goldcliff) on 1 April 2018. The site often has an influx of birds in September, with most departing by early November. Ringing recoveries and colour-ring sightings show that some of those at Newport Wetlands in the breeding season had fledged from eastern England. One found dead at Goldcliff in May 2015 had been ringed as a nestling in Vendée, France, in July 2013. Avocets now seem to be well-established at Newport Wetlands Reserve, despite their low breeding success.

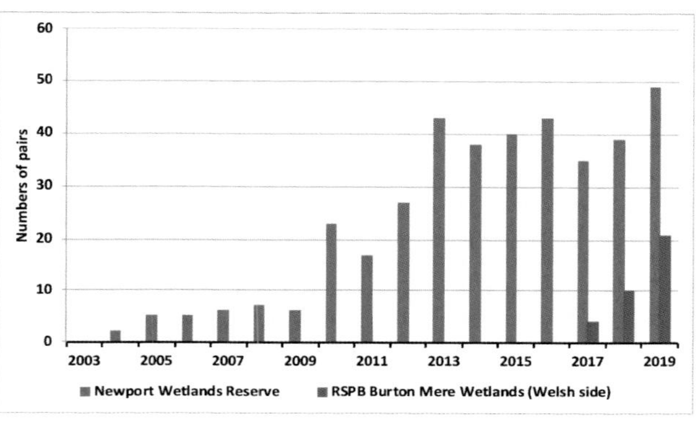

Avocet: Numbers of pairs breeding, Newport Wetlands, Gwent, and the Flintshire part of RSPB Burton Mere Wetlands, 2003–19.

The first breeding in North Wales was recorded in 2017, when four pairs bred on the wet grassland area in the Flintshire part of RSPB Burton Mere Wetlands. Numbers here increased to ten pairs in 2018 and 21 pairs in 2019 (Colin Wells *in litt.*). Elsewhere, a pair was seen copulating at Penclacwydd, Carmarthenshire, in April 2012 but did not stay. Apart from Burton Mere, records are fairly frequent at other sites on the Dee Estuary. The species has now been recorded in all Welsh counties, but it remains a scarce visitor to most. The first record for Radnorshire was of three on a backwater of the Wye between Glasbury and Llowes on 6 July 2010 (Jennings 2014). The first for Breconshire was at Llangasty on 21 March 2016. It remains to be seen whether other breeding colonies will be established in Wales, but suitable habitat is limited.

Rhion Pritchard

Lapwing *Vanellus vanellus* Cornchwiglen

Welsh List Category	IUCN Red List			Birds of Conservation Concern			
	Global	Europe	GB	UK	Wales		
					2002	2010	2016
A	NT	VU	EN (B)	Red	2002	2010	2016

Lapwings are one of our most distinctive farmland birds, their historical familiarity in the British countryside illustrated by the wealth of country names for this species, such as 'Peewit'. They prefer to breed in open habitats in loose colonies, which they defend vigorously. Their global range extends in a broad band across the mid-latitudes of Europe and Asia, reaching as far south as Spain (*HBW*). They generally winter to the south or west of their breeding areas, although these can overlap in Europe, particularly west of the 3°C January isotherm. Historically, they would have bred in natural well-grazed grasslands and steppe, presumably near wet areas and shallow pools for the chicks to feed by. In Britain, they initially inhabited a range of peripheral extensively grazed farmed habitats, then took advantage of the expansion of agriculture, particularly the conversion of land to damp pasture, hay meadows and spring-sown crops (Shrubb 2007).

In the 20th century, Lapwings extended their range northwards, creeping into the Arctic Circle in western Russia and Fennoscandia. Population changes tended to follow changes in agricultural practices, more than factors such as long-term climate amelioration or the egg harvesting that was widespread in parts of Britain until the mid-20th century. Range expansion seemed to be continuing in Asia (Shrubb 2007) but the historic increases have reversed across much of western Europe, largely due to accelerated intensification of grassland and arable farming. Declines have occurred in Norway and Sweden, but in Finland the population is increasing, after a period of decline (Appleton 2019).

Lapwings were once a characteristic feature of open farmland throughout lowland Wales. Mathew (1894) said that it was one of the commonest birds in Pembrokeshire, nesting in most districts of the county. It was reportedly a very abundant breeding species in Glamorgan in the mid- to late 19th century (Hurford and Lansdown 1995). Forrest (1907) described it as numerous in all parts of North Wales, breeding in small numbers up to 600m in the hills. Declines began in some counties after the First World War, in north Ceredigion and Breconshire from at least the 1920s (Salter 1930, Ingram and Salmon 1957). Lockley *et al.* (1949) described it as a common resident in Pembrokeshire, but said that it was "somewhat local in distribution". Condry (1955) reported that it used to be a common breeder in central and northern Wales at around 430m, but that by the early 1950s it was very scarce at this altitude and most of the population bred below 300m.

The *Britain and Ireland Atlas 1968–72* showed the Lapwing to be widespread in Wales, with confirmed breeding in 92% of 10-km squares, the only gaps being in the southwest. By the 1988–91 *Atlas*, occupation was down to 75% of squares, with significant losses evident, particularly in Pembrokeshire, where the species had been lost from most 10-km squares since 1968–72. Losses elsewhere in Wales were mainly on the higher ground. Further extensive losses were clear from the 2007–11 *Atlas*, particularly from southern Meirionnydd, through Ceredigion and Carmarthenshire. Occupation had fallen to below 50% of the 10-km squares in Wales, and in many of the occupied squares breeding was not confirmed. Wales had some of the lowest breeding densities in Britain, with a decline in relative abundance in most of the country since 1988–91. Densities were highest towards the north and east of Wales, at levels comparable with southern England. The best areas—Anglesey and the Dee Estuary—had densities comparable with much of northern England and lowland Scotland, although still well short of those in prime areas in the Pennines and the North York Moors.

Similar declines in range are shown by tetrad atlases. Birds were present in 70% of tetrads in the *Gwent Atlas 1981–85*, but in only 40% by 1998–2003. The *East Glamorgan Atlas 1984–89* found Lapwings in 29% of tetrads, but by 2008–11 this was down to 11%. In Pembrokeshire, where the decline in range began far earlier, the species was present in 6% of tetrads in 1984–88 and in only 0.8% in 2003–07, a decline of 86%; Lapwing is now only occasionally reported as a breeding bird in the county. The *North Wales Atlas 2008–12* recorded Lapwings in 18% of tetrads and confirmed breeding in just 7%.

In the early 1970s the Welsh population was estimated at 15,000 breeding pairs (*Birds in Wales* 1994), a figure that had halved by 1987 to 7,448 pairs (Shrubb *et al.* 1991). The decline in Wales continued at a faster rate than in any other part of Britain. In 1998 the last reliable estimate from a national survey was 1,689 pairs (Wilson *et al.* 2001). In 2006, an all-Wales survey

Lapwing in winter sunshine © Michael Steciuk

	2015	2016	2017	2018	2019
Cors Ddyga, Anglesey	36	41	76	71	71
Dee Estuary, Flintshire	39	42	51	51	57
Ynys-hir, Ceredigion	48	43	24	40	38
Morfa Dinlle, Caernarfonshire	51	50	37	41	30
Valley Wetlands, Anglesey	11	8	4	9	13

Lapwing: Annual totals of nesting pairs on RSPB nature reserves in Wales, 2015–19.

was planned, but by then Lapwings had become too few and too scattered to achieve a reliable figure and there is little evidence of anything other than further declines. On commercial farmland, the Lapwings' acrobatic displays and "peewit" calls are now confined to a few scattered sites or sporadic breeding attempts.

Lapwings are too scarce to generate a BBS index for Wales. Based on average counts from well-monitored nature reserves and the density of Lapwings found in tetrad atlases upscaled to the rest of Wales, it is thought that the population in 2008–11 was between 830 and 1,360 pairs, of which 220 pairs (16–26%) were on nature reserves. Repeating that methodology, and assuming that the distribution and number of records reported from outside nature reserves has contracted evenly across Wales since, 690–970 pairs bred in Wales in 2015–19, of which 265 pairs (27–38%) were on nature reserves. Woodward *et al.* (2020) estimated the total population in Great Britain at 96,500 pairs.

The declines are closely linked to changes in farming policy and practice. Lapwings can nest on pasture or on bare arable land, but tend to have higher productivity on the former. Land drainage has reduced nesting opportunities and particularly chick-rearing areas within the farmed landscape, while the move away from spring-sown crops means that the vegetation in autumn-sown arable fields is too long by March, when Lapwings start to nest. Farms have become increasingly specialised into either arable or pastoral operations, reducing the options for Lapwings, that would, for example, have nested on bare arable to avoid trampling of eggs by livestock. Adults would then have taken their young to pasture, especially wet patches with more invertebrates. Intensification of pasture management, with reseeding and greater use of fertilisers, has resulted in fields of rye-grass leys that support high densities of sheep. Nesting Lapwings are conspicuous to predators in such fields, where the high stocking rates can also result in trampling or the desertion of nests (Shrubb 2007).

Lapwings nesting on nature reserves have fared better, thanks to targeted management, and in the 21st century, they comprise an increasing proportion of the total population. Without dedicated nature-reserve management, the Lapwing population would already be even lower than it is. Lapwings can still breed successfully in extensive lowland wet grassland, but even on reserves it is difficult to keep productivity high enough to sustain these colonies, so the necessary management has intensified. Many key reserves now require electric fences or the control of Foxes to maintain a population. A study during 1996–2003, around the Dyfi Estuary and the Ouse Washes, Cambridgeshire, showed that nest losses to predators were lower where Lapwings nested at a higher density (Macdonald and Bolton 2008), perhaps because they are more effective at mobbing predators and/or because Lapwings select places with lower densities of mammals. At RSPB Morfa Dinlle, Caernarfonshire, a previously successful colony has declined owing to predation, despite the electric fences, and so yet more measures will be required to maintain it. Morfa Madryn LNR, Caernarfonshire, a coastal grassland owned by two local authorities, with an enthusiastic tenant farmer, has managed to sustain a colony of up to 44 pairs of Lapwings since 1998.

In order to maintain a stable population of Lapwings, an annual average of 0.7 chicks fledged/pair is required, and on average this has been achieved across the five sites listed in all five years. Only RSPB Cors Ddyga has exceeded this figure every year, peaking at 2.44 young fledged/pair in 2015. In contrast, the target was not reached at RSPB Morfa Dinlle in any of these five years, although the species had been very successful there during 1996–2014. This illustrates the difficulty of sustaining success at an individual site. Lapwings cannot necessarily depend on a small number of sites in the long term. Elsewhere, at least 27 chicks were fledged from 23 nests at WWT Penclacwydd, Carmarthenshire.

Lapwing movements outside the breeding season are complex. The genetic mixing resulting from this may explain why there are no subspecies or regional variations. While 61% of British Lapwings were recovered within 10km of where they were ringed as chicks, 11% were found more than 100km away in a subsequent breeding season (Thompson *et al.* 1994). A small proportion mix their genes over much larger distances: British-hatched birds have been recorded breeding in Russia, probably after joining flocks travelling east in spring. Having travelled to a new breeding area, they will repeat the journey each year (Appleton 2016b). The oldest Lapwing known to the BTO ringing scheme is a bird from Wales, ringed as a nestling on Skokholm on 24 May 1966, and found dead, on Skokholm, 21 years, one month and 15 days later.

Shrubb (2007) grouped movements of Lapwings into three types: post-breeding dispersal and early summer movements, migration to and from the wintering grounds, and winter cold-weather movements. Lapwings in Britain and Ireland are thought to be partial migrants: some winter close to their breeding sites, including those that move from upland breeding areas to nearby coasts. Summer movements can start as early as late May, with failed and non-breeders dispersing locally. Breeding adults and juveniles disperse in June and July, joining flocks of moulting birds. Ringing recoveries from Wales show that local birds are joined by birds from northern England, Scotland and farther afield: birds ringed as nestlings in The Netherlands, Belgium, Germany and Fennoscandia have been recovered in Wales, and an adult ringed at Shotton, Flintshire, in February 1973 was found dead near Grodno in Belarus, 1,791km away, in April 1980. Numbers peak in December/January in Wales, slightly later than peaks in England and Scotland, suggesting that significant movements within Britain continue during the winter. Some birds breeding in Wales move south for the winter: nestlings ringed here have been recovered as far south as Spain and Portugal in winter. Significantly more juveniles move to the Continent than adults (Imboden 1974).

Cold weather in winter can produce large-scale movements, as birds are forced south and west. 6,900 flew west over Morfa Conwy, Caernarfonshire, in under two hours on 24 December 1963 and 8,900 flew over Kenfig Dunes, East Glamorgan, in three hours on 27 December 1965. There were 15,000 in fields at Magor and Caldicot, Gwent, in December 1970 and flocks of 10,000 were on the Taf Estuary, Carmarthenshire, in January 1956, December 1994 and February 1995. Counts in the 21st century have been

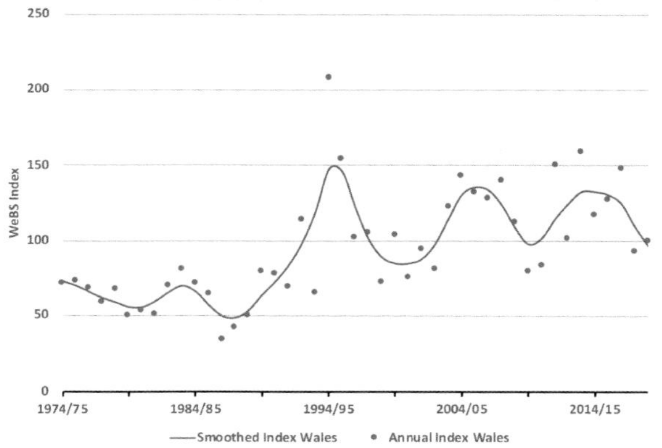

Lapwing: Wetland Bird Survey indices, 1974/75 to 2018/19.

Site	County	74/75 to 78/79	79/80 to 83/84	84/85 to 88/89	89/90 to 93/94	94/95 to 98/99	99/2000 to 03/04	04/05 to 08/09	09/10 to 13/14	14/15 to 18/19
Severn Estuary: Welsh side (whole estuary)	Gwent/East Glamorgan	1,227 (8,205)	1,495 (4,257)	1,623 (4,003)	650 (4,700)	321 (12,104)	395 (11,032)	2,059 (12,725)	1,762 (10,805)	1,107 (11,383)
Dee Estuary: Welsh side (whole estuary)	Flintshire	na (7,219)	na (4,972)	na 6,179)	2,818 (6,983)	4,331 (9,135)	2,560 (7,615)	3,126 (7,112)	2,806 (6,224)	4,321 (8,289)
Cleddau Estuary	Pembrokeshire	2,223	999	952	1,237	2,858	2,185	2,943	3,034	4,553
Carmarthen Bay	Carmarthenshire/ Pembrokeshire	1,375	773	660	3,171	5,824	1,767	4,067	2,304	2,452
Burry Inlet	Gower/ Carmarthenshire	2,757	2,828	1,796	2,184	3,459	2,836	3,045	2,201	1,785

Lapwing: Five-year average peak counts during Wetland Bird Surveys between 1974/75 and 2018/19 at sites in Wales averaging 2,000 or more at least once. International importance threshold: 20,000; Great Britain importance threshold: 6,200.

more modest, perhaps reflecting the lower populations in Europe and perhaps owing to milder winters.

WeBS counts provide the best data for wintering numbers, but may not reflect the full picture for Lapwing, because many flocks occur on farmland which is not counted. Indices for England, Scotland and Wales show that winter populations increased in the late 1980s and early 1990s, followed by a period of stability until around 2005, since when numbers in England and Scotland have decreased, whereas Wales has shown no strong trend. The Dee and Severn estuaries hold wintering numbers of British importance, although no site wholly in Wales reaches that threshold. Eleven WeBS sites in Wales have averaged more than 1,000 during 2014/15–2018/19.

The positive response to targeted management at several nature reserves in Wales offers hope for the future. If this species is to survive as a breeding species in Wales, away from nature reserves, the key factors that have caused its decline need to be addressed. Agri-environment schemes have achieved little for Lapwings in Wales. Much will therefore depend on the future support offered by government, the targeting of actions to places that still hold Lapwings and the responses to that from individual farmers.

Ian Hawkins and Julian Hughes

Sponsored by Tony Prater

Sociable Plover *Vanellus gregarius* Cornchwiglen Heidiol

Welsh List Category	IUCN Red List		Rarity recording
	Global	Europe	
A	CR	CR	BBRC

The Critically Endangered status of this species is due to a very rapid decline and range contraction, although recent fieldwork has shown the population to be larger than previously thought. It breeds on dry steppe and stubble in northern and central Kazakhstan and south-central Russia, and winters in Sudan, Pakistan and northwest India. Small numbers winter on the Arabian Peninsula (*HBW*). There are regular records of vagrants in Europe and it has been suggested that a small number of individuals may over-winter in Iberia. The single record in Wales was one on the River Neath, Gower, on 20–21 October 1984.

Jon Green and Robin Sandham

Golden Plover *Pluvialis apricaria* Cwtiad Aur

Welsh List Category	IUCN Red List			Birds of Conservation Concern			
	Global	Europe	GB	UK	Wales		
A	LC	LC	LC	Green	2002	2010	2016

The global distribution of Golden Plover generally follows the tundra zone of dwarf shrub and sub-alpine heath and mire from eastern Greenland across to northwest Siberia (*HBW*). The small population breeding in Wales belongs to the southern subspecies *P.a. apricaria*, while many of the wintering birds are thought to be of the northern subspecies *P.a. altifrons*. In Britain, Golden Plovers breed towards the centre of major upland blocks, with flat or gently sloping plateaux, but they may fly up to 8km from the nest site to feed on semi-improved grassland (Ratcliffe 1976, Byrkjedal and Thompson 1998). During the 19th century, their main breeding regions in Britain and Ireland were in Scotland, the west of Ireland, northern England and Wales. The main range is now limited to the uplands of north and west Britain and Ireland, with no breeding records in southwest England since the early 2000s (Helen Booker *in. litt.*), leaving the Welsh population as the southernmost breeding birds in the UK and probably Europe.

Early records suggested that Golden Plover was a more widespread breeder in Wales in the 19th and early 20th century. In North Wales, Borrer (1891) found four nests on the summit of Cadair Berwyn, on the Meirionnydd/Denbighshire border, in 1837 and Forrest (1907) stated that it was numerous on moors around Bala and Corwen, Meirionnydd. He found several pairs breeding near the summit of Aran Fawddwy at over 900m. It bred at several other sites in this county and in Denbighshire. In Caernarfonshire,

Golden Plover in winter plumage © Ben Porter

breeding was reported around Dolwyddelan and Moel Siabod, where there are no recent records of breeding, while a few nested in Montgomeryshire (Forrest 1907). There has been one breeding record on Anglesey, on heathland above South Stack in 1959, with no breeding records from Flintshire.

Salter found pairs in suitable breeding habitat at several sites in Ceredigion around the turn of the 20th century (Roderick and Davis 2010). In Breconshire, the Golden Plover bred at several sites. In the hills between Llanafanfawr and Nantgwyllt, part of the Elenydd uplands, "it breeds in some numbers" but "sparingly" on Mynydd Epynt (Phillips 1899), where there have been no recent records of breeding. There were only a few early records of breeding in Radnorshire, with few records from the Elan/Claerwen catchment in this period, although this area was later recognised as important for the species. In South Wales, it bred at several sites in northwest Gwent in the late 19th century (Venables *et al.* 2008), but until recently there had only been one record of possible breeding in Glamorgan, in the second half of the 19th century (Hurford and Lansdown 1995). Ingram and Salmon (1954) thought that a pair or two bred on high moorland in Carmarthenshire west of the upper Tywi valley and that, in about 1894, a few bred in the eastern part of the county, which probably referred to the Mynydd Du area. Pairs bred in this area in 1978 and bred there again, successfully, in 2012 (Rob Hunt *in litt.*). Breeding has never been confirmed in Pembrokeshire.

The *Britain and Ireland Atlas 1968–72* showed records of breeding or possible breeding in 11% of 10-km squares, confined to the central upland spine of Wales, with no records in the west, indicating a marked decline in breeding range when compared to the historical records. This atlas noted that its range and numbers had diminished throughout its breeding range in central and southern Europe. The 1988–91 *Atlas* showed a net loss compared to 1968–72. This trend continued in 2007–11, to 8% of squares in Wales, although the range has remained largely unchanged. Relative abundance was very low in Wales compared to parts of northern England and Scotland.

Ratcliffe (1976) estimated the breeding population by applying known densities for particular habitats and areas to all 10-km squares in which nesting was certain or probable during the 1968–72 *Atlas*. His estimate of 900 pairs in Wales was thought "almost certainly very considerably too high" by the authors of *Birds in Wales* (1994). Surveys of the major upland areas by the RSPB in 1975–78 found 213–224 pairs, including 74 in the Elan/Claerwen (Elenydd) area, 46 pairs on Mynydd Hiraethog, Denbighshire, and 25–35 pairs on Llanbrynmair Moor, Montgomeryshire. Although breeding Golden Plovers are difficult to census accurately, it was considered that the Welsh breeding population in the late 1970s was no more than 250–300 pairs (*Birds in Wales* 1994). During an NCC survey in 1984, Golden Plovers still nested in some of the areas mentioned by Forrest (1907) around Aran Fawddwy and Moel Sych (pers. obs.).

By 2000, the Welsh population was thought to be fewer than 80 pairs. By 2008 it had reduced to *c.*36 pairs, concentrated at seven sites (Johnstone *et al.* 2008). Only 3.2% of the UK's breeding Golden Plover now reside in Wales. Golden Plover breeding populations have always been patchy but declines have occurred in all areas. In the Elenydd area of mid-Wales, declines of *c.*88% were observed between 1982 and 2007, with only 11 pairs remaining. Detailed monitoring was undertaken in the former core breeding area of Abergwesyn Common in the southern Elenydd, Breconshire, as part of a joint National Trust/CCW project that introduced cutting and amended grazing management of *Molinia*-dominant grasslands in 2010 and 2011. A post-treatment survey covering 5km^2 found a decline from three pairs in 2011 and 2012 to a lone male by 2018 (Andrew King *in litt.*). A less severe decline was noted over a similar period at another managed site at Trumau, Radnorshire, administered jointly by Elan Valley Trust and NRW. There was encroachment by Purple Moor Grass over most of the study area. Along with drying out, this was considered to be one of the damaging effects of climate change, affecting what was once prime plover habitat. Surveys over the remainder (104 whole or part 1-km squares) of the National Trust-managed holding of southern Elenydd in Breconshire in 2016 and 2017 failed to find any Golden Plovers. This pattern is reflected on the Pumlumon SSSI, where breeding pairs declined from 12–13 in 1984 to three in 2011 (Crump and Green 2012). The near-complete loss of breeding Golden Plover from south Elenydd by 2018 was offset in part by 15–16 occupied territories in montane areas of the Black Mountains, Brecon Beacons and Mynydd Du that appear not to be severely challenged by *Molinia* encroachment (Richards and King 2019). One pair fledged at least two young at Waun Wen in the north of East Glamorgan in 2019, only the second ever breeding record in Glamorgan, and the first since the 19th century. This was thought to be almost certainly the most southerly breeding record in Britain (Eastern Glamorgan Bird Report 2019). The *North Wales Atlas 2008–12* found Golden Plovers in only 1.7% of tetrads, mainly on the Migneint, Mynydd Hiraethog and the Berwyn range. It was estimated that there were 10–12 breeding pairs in North Wales, the majority on the Migneint.

While breeding numbers have declined drastically, wintering Golden Plover are found in good numbers in coastal fields and estuaries all around Wales. Co-ordinated counts throughout Britain in 1977–78 produced a peak count in Wales of 18,000 in January 1977 (Fuller and Lloyd 1981), including flocks of 1,000 or more

Site	County	74/75 to 78/79	79/80 to 83/84	84/85 to 88/89	89/90 to 93/94	94/95 to 98/99	99/00 to 03/04	04/05 to 08/09	09/10 to 13/14	14/15 to 18/19
Cleddau Estuary	Pembrokeshire	572	395	815	305	735	1,556	3,459	1,710	2,864
Carmarthen Bay	Carmarthenshire/ Pembrokeshire	1,388	225	1,040	3,836	7,646	3,228	7,814	2,199	2,740
Burry Inlet	Gower/Carmarthenshire	3,612	2,477	1,564	1,489	1,392	1,770	1,708	1,455	1,830
Dyfi Estuary	Ceredigion/Meirionnydd	16		165	434	701	851	1,980	971	1,318

Golden Plover: Five-year average peak counts during Wetland Bird Surveys between 1974/75 and 2018/19 at sites in Wales averaging 1,500 or more at least once. International importance threshold: 9,300; Great Britain importance threshold: 4,000.

in Glamorgan, Carmarthenshire, Pembrokeshire, Caernarfonshire and Anglesey. Counts vary between years, which is possibly weather related, with fewer birds in the west of Britain during mild winters. WeBS counts suggest that about 10,000 birds wintered in Wales in 2017/18, though not all sites are covered by this survey.

Two intensive studies have suggested that overwinter survival is important in regulating breeding populations (Parr 1992, Yalden and Pearce-Higgins 1997). Wintering birds are highly mobile and locations are very weather dependent. Several avifaunas noted declines during the 1970s and 1980s (*Birds in Wales* 1994, Peers and Shrubb 1990), but with subsequent increases, for example in Radnorshire (Jennings 2014). Numbers declined from about 2005/06. A Medium WeBS Alert has been issued for Wales following a 39% decline in ten years, although numbers have increased again in the short term (Woodward *et al.* 2019).

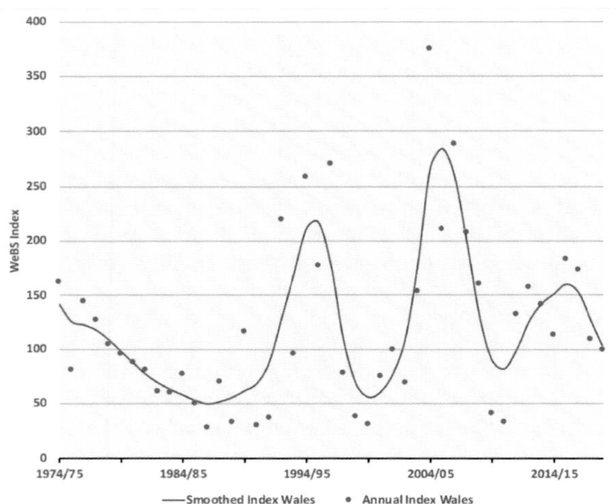

Golden Plover: Wetland Bird Survey indices, 1974/75 to 2018/19.

Peak counts at the main sites include 12,700 in Carmarthen Bay in January 2007, 8,630 at the Cleddau Estuary in December 2005 and 6,450 in the Burry Inlet in March 1977. Golden Plovers at Malltraeth are not counted as regularly as at other sites, but there were 2,600 at Malltraeth Cob and Pools in December 2006 and 2,600 at Malltraeth Sands in January 2014. The Severn Estuary can hold large numbers, but few of these are in Wales. For example, out of 7,109 on the estuary in February 2014, just one was in Wales. There were 10,000 at Llandow, East Glamorgan, on 24 November 1984. Not all the important coastal sites are estuaries: the Gupton/Freshwater West area of the Castlemartin Peninsula, Pembrokeshire, held over 3,000 in January 2013.

Previous publications have concentrated on the locations of coastal flocks as key sites, but it is increasingly evident that inland sites are also very important and can be regularly used for many years. Jennings (2014) noted that although inland flocks had declined in Radnorshire in the 1970s and 1980s, records increased after 2000, with flocks of up to 300 at a number of sites. Surveys of mid-Wales for proposed wind-farm developments have increased monitoring in areas that are not usually visited. Several sites on the border of Montgomeryshire and Shropshire each hold a few hundred birds, although no cumulative assessment has taken place.

Ringing shows a high turnover of individuals at some inland sites. At one site on the Montgomeryshire/Shropshire border, which regularly holds *c.*300 birds, over 1,000 have been ringed between January 2012 and the end of the winter of 2019/20 (Tony Cross pers. comm.). This increases the importance of such sites for the conservation of the species. Two birds fitted with satellite tags at this site displayed very different strategies in response to weather. During a cold spell one moved to the coast at Pendine, Carmarthenshire, returned to the site as the weather improved and then moved back to Pendine in the next cold episode. The second flew to Spain, near Madrid, and then to Portugal where contact was lost. Such ringing information is not available for coastal flocks, but it seems likely that turnover of individuals is also high on these sites.

The varied nature of movements is confirmed by other ringing records, which show no discernible pattern to dispersion or movements. The *BTO Migration Atlas* (2002) stated that some British breeders remain in Britain, while others move to France, Iberia and Morocco, with large movements in response to periods of cold weather. There is no information about where Welsh breeding birds winter. Of 1,400 ringed at wintering sites in Shropshire and mid-Wales, recoveries have included birds in East Yorkshire, Tyne and Wear, Shetland, Iceland and The Netherlands. Of five birds ringed outside Britain and recovered in Wales, three were from Iceland, the others from Belgium and The Netherlands. A Golden Plover ringed at Llaithdu, Montgomeryshire, in February 2012 was found in Norway in May 2014 and was therefore almost certainly breeding there.

There is a considerable passage of birds through Wales in spring and autumn. Jones and Whalley (2004) noted a strong autumn passage on Anglesey from July, peaking in September, with a less-distinct spring passage from March to May. In contrast, Peers and Shrubb (1990) reported a "strong spring passage" through Breconshire, mainly in April. Several authors have noted that passage birds in summer plumage appear to be of the darker, northern race, *P.a. altifrons*. Passage may be through the uplands, as well as along the coast. Large numbers include 2,500 at Mynydd Llanfihangel Rhos-y-corn, Carmarthenshire, on 10 October 2003.

As a breeding species, the Golden Plover has declined significantly in Wales, but the reasons are unclear. Johnstone *et al.* (2008) considered that increases in vegetation height and density were a possible cause, but work in central Wales has shown that although vegetation height and density have fluctuated in favoured breeding areas, this does not seem to have changed significantly since 1984 (Crump 2014). Large areas of previously good habitat have undoubtedly been lost to afforestation and drainage. For example, Llanbrynmair Moor held 40 pairs in 1968, but the entire site was lost to forestry in the 1980s. Afforestation also provides habitat for predators such as Foxes and Carrion Crows, adding to pressure on adjacent nesting habitat (Parr 1992). The increased numbers of generalist predators are a major reason for decline of other wader species such as Curlew (Johnstone *et al.* 2012), and therefore are likely to have an impact on Golden Plover. Moorland drainage also has an impact on breeding success. Re-wetting boggy areas increases significantly the numbers of, for example, Cranefly larvae, which are an important food source (Pearce-Higgins *et al.* 2005). Climate change has added further pressure, causing a mismatch between season and prey availability (Pearce-Higgins *et al.* 2005). Nethersole-Thompson and Nethersole-Thompson (1986) considered earthworms to be an important part of the adult diet, and it is possible that intensification of semi-improved pasture has led to declines in such prey.

The Golden Plover is Red-listed in Wales as a result of severe breeding population decline over 25 years and the longer term, and a decline in breeding range (Johnstone and Bladwell 2016). This is regarded as a priority species in both Wales and Europe, yet the Golden Plover is still listed on Schedule 2 of the Wildlife and Countryside Act 1981 as one of the "birds which may be killed or taken". A bird ringed in mid-Wales in winter was later recovered shot in Cornwall (Tony Cross pers. comm.).

The future of the Golden Plover as a breeding bird in Wales looks uncertain, although habitat improvement at some sites may help if carried out at a sufficient scale. The long-term impact of climate change adds to the uncertainty. The breeding distribution of Golden Plover in Britain is modelled to retract to northern Scotland by the end of this century (Huntley *et al.* 2007). Wales remains an important wintering area for this species and regularly used sites, including seemingly small inland sites, need to be protected.

Heather Crump and Mick Green

Sponsored by Ecology Matters Trust

Pacific Golden Plover *Pluvialis fulva* Corgwtiad Aur y Môr Tawel

Welsh List Category	IUCN Red List	Rarity recording
	Global	
A	LC	BBRC

This plover breeds in northeast Russia and western Alaska and is a long-distance migrant, wintering in northeast Africa, southern Asia and Australasia (*HBW*). Out of a total of 104 individuals in Britain up to the end of 2019 (Holt *et al.* 2020), averaging three each year over the last 30 years, only one was recorded in Wales: at Oakenholt, Flintshire, on 2–4 August 1990. An adult American or Pacific Golden Plover had been seen at the same location on 23 July 1989. There was another record of either American or Pacific Golden Plover at Cefn Sidan, Carmarthenshire, on 26 July 1987.

Jon Green and Robin Sandham

American Golden Plover *Pluvialis dominica* Corgwtiad Aur America

Welsh List Category	IUCN Red List	Rarity recording
	Global	
A	LC	WBRC

A long-distance migrant, the American Golden Plover breeds on coastal tundra across northern Alaska and Canada to Baffin Island. It migrates over the western Atlantic to wintering grounds in northern Argentina and Uruguay. The number of records in Britain increased towards the end of the 20th century, from an average of three each year in the 1970s to 25 each year in the 2010s. There have been only 21 records in Wales out of over 600 in Britain.

Total	Pre-1950	1950–59	1960–69	1970–79	1980–89	1990–99	2000–09	2010–19
21	0	0	0	0	2	0	9	10

American Golden Plover: Individuals recorded in Wales by decade.

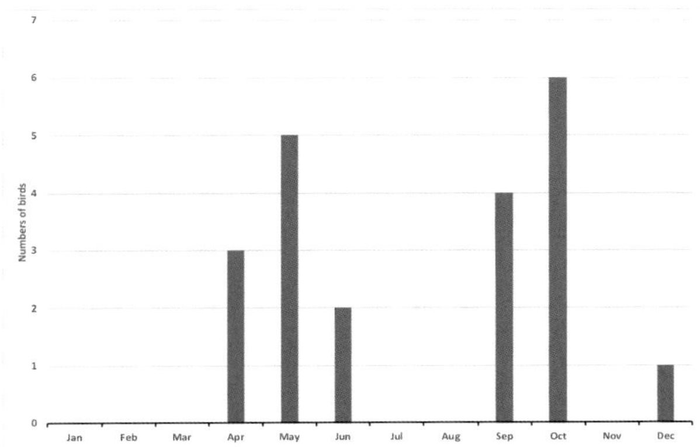

American Golden Plover: Totals by month of arrival, 1981–2019.

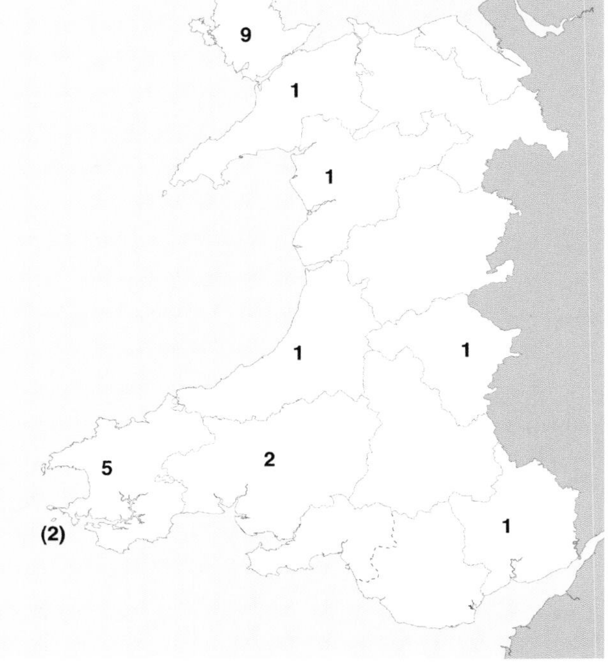

American Golden Plover: Numbers of birds in each recording area 1981–2019.

The first Welsh record was in 1981 when a first-calendar-year bird was on Skokholm, Pembrokeshire, on 26 September. This was followed by an adult at Cemlyn, Anglesey, on 3 April 1983, but there were no further records until 2004, when one was found on the uplands of Garreg Lwyd, Carmarthenshire, on 3–4 May. Records have since increased significantly, possibly partly because of returning individuals, and birdwatchers may also now be more inclined to scan through flocks of Golden Plover to search for this species. Five of the records have been in the Cemlyn area. The only winter record was one at Llanbadarn Fynydd, Radnorshire, on 4 December 2016.

Jon Green and Robin Sandham

Grey Plover *Pluvialis squatarola* Cwtiad Llwyd

Welsh List Category	IUCN Red List			Birds of Conservation Concern			
	Global	Europe	GB	UK	Wales		
A	LC	LC	VU	Amber	2002	2010	2016

The Grey Plover breeds in Arctic Russia, Canada and Alaska, and winters to the south, with some found as far south as Australia (*HBW*). All the birds wintering in, or passing through, Britain and Ireland are presumed to be from the population breeding in Russia (*BTO Migration Atlas* 2002). The few ringing recoveries involving Wales give no clue to the breeding grounds of birds found here. The only two recoveries from outside Britain were both ringed at Ynyslas, Ceredigion, and later recovered in France. Two ringed in Norfolk were later recovered in Pembrokeshire. In winter, this species prefers larger estuaries and other coastal areas with soft sediment.

Birds in Wales (1994) stated that there had been a marked increase in numbers in Wales since the early 1970s, in line with a massive increase in the British wintering population during this period. However, this was followed by a steady decrease that has continued up to the present. WeBS annual population trends for Wales show a steady decline since the early 1990s. A WeBS long-term High Alert has been issued for Wales, following a 57% decline, together with a medium-term Medium Alert (-34%). Long- and medium-term High Alerts have been issued for the Burry Inlet SPA, as a result of 75% long-term and 61% medium-term declines, as well as a short-term Medium Alert (-41%). A long-term Medium Alert has also been issued for the Dee Estuary SPA (Woodward *et al.* 2019). Trends for England and Scotland show a similar decline since the mid-1990s.

There were 350 at Cefn Sidan/Tywyn Point, Carmarthenshire, on 2 February 1980. Numbers are now much lower everywhere in Wales. Away from the main sites, counts in winter 2015/16 produced an estimate of 175 on non-estuarine coasts in Wales, compared to 357 in the winter of 1984/85. The few records of Grey Plover away from the coast are mainly of singles, and most come from some of the larger lakes such as Llangorse Lake, Breconshire. There were four on the River Wye in Radnorshire, near Glasbury, on 4 November 1996.

The reasons for the decline in numbers in Britain are set within the context of a world population that is thought to be decreasing (BirdLife International 2020) and modelling predicting a future retraction in its breeding range away from Europe (Huntley *et al.* 2007). The Grey Plover is known to be more susceptible to cold weather than most waders (*BTO Migration Atlas* 2002) but, with a few exceptions, winters during the period of decline have been fairly mild. Several wader species have shifted their winter distribution farther east or northeast in Europe, a trend known as 'short-stopping'. In The Netherlands, the recent trend for Grey Plover is slightly increasing and the flyway trend is stable, so the recent decline in Britain is likely to be related to shifts in wintering distribution (Frost *et al.* 2017). The main conservation effort for this species needs to be on its breeding grounds. The majority of the main wintering sites in Wales are currently protected areas, and it is important that this level of protection is maintained.

Rhion Pritchard

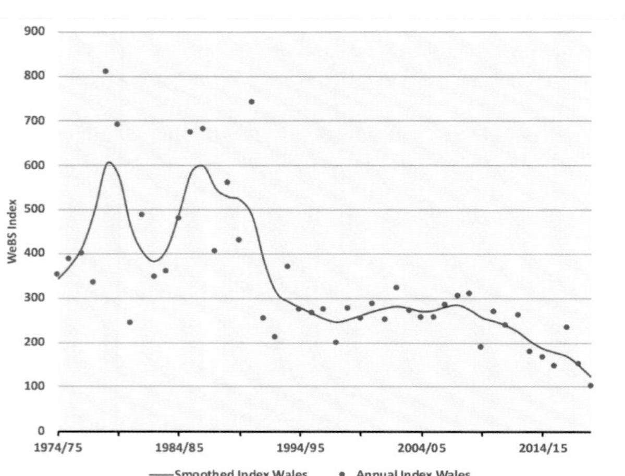

Grey Plover: Wetland Bird Survey indices, 1974/75 to 2018/19.

Site	County	79/80 to 83/84	84/85 to 88/89	89/90 to 93/94	94/95 to 98/99	99/00 to 03/04	04/05 to 08/09	09/10 to 13/14	14/15 to 18/19
Dee Estuary: Welsh side (whole estuary)	Flintshire	na (1,437)	na (1,744)	142 (1,963)	118 (2,816)	139 (1,213)	113 (1,158)	61 (1,077)	61 (973)
Severn Estuary: Welsh side (whole estuary)	Gwent/East Glamorgan	539 (612)	349 (673)	324 (849)	126 (458)	83 (484)	163 (416)	136 (331)	125 (334)
Burry Inlet	Gower/ Carmarthenshire	694	689	484	366	455	474	517	169
Inland Sea and Alaw estuary	Anglesey	48	33	31	46	42	83	210	123
Traeth Melynog	Anglesey		5	0	0	400	62	103	77
Carmarthen Bay	Carmarthenshire/ Pembrokeshire	9	23	48	68	33	48	204	15

Grey Plover: Five-year average peak counts during Wetland Bird Surveys between 1979/80 and 2018/19 at sites in Wales averaging 200 or more at least once. International importance threshold: 2,000; Great Britain importance threshold: 330.

Ringed Plover *Charadrius hiaticula* Cwtiad Torchog

Welsh List Category	IUCN Red List			Birds of Conservation Concern			
	Global	Europe	GB	UK	Wales		
A	LC	LC	VU(NB)	Red	2002	2010	2016

The Ringed Plover breeds mainly in the far north. There are three subspecies: the one breeding in Wales is the nominate *C.h. hiaticula*, which breeds from southern Fennoscandia south to Brittany; *C.h. psammodromus* breeds in Iceland, Greenland and northeastern Canada; while *C.h. tundrae* breeds on coasts and tundra, from northern Scandinavia eastward across Arctic Russia (IOC).

Some *hiaticula* Ringed Plovers in western Europe may winter close to their breeding grounds, but birds breeding in other areas move farther south in winter. Ringed Plovers that breed farthest north usually migrate farthest to the south, many to Africa south of the Sahara, with some reaching as far as South Africa. Birds breeding in the south of the breeding range move much shorter distances (*HBW*). Ringed Plovers usually breed on the coast, on sandy or shingle beaches, but they will also breed alongside lakes, rivers and reservoirs. The fairly small Welsh breeding population is joined by much larger numbers of passage migrants in spring and autumn, with smaller but still considerable numbers wintering here. Recoveries and sightings of ringed birds suggest a mixed origin for the passage migrants and wintering birds, including individuals breeding in The Netherlands, Germany, Denmark, Norway and Sweden.

Ringed Plover: Recovery locations of birds ringed in Wales are shown by red circles. Ringing locations of birds that were recovered in Wales are shown by blue triangles. Small brown dots show ringing or recovery locations in Wales.

There is no detailed information about the fortunes of the Ringed Plover in Wales in the 19th century, but it would appear to have been more common and widespread as a breeding bird than it is now. Even in Pembrokeshire, where there were few nests in later years, Mathew (1894) said that it "nests commonly at many places on the coast." By the 1970s, it was reported that the species was decreasing in many coastal areas as a result of increasing human disturbance.

Survey	Pairs found	Estimated total pairs	Reference
1973–74	186	-	Prater 1976a
1984	167	224	Prater 1989
2007	214	254	Conway *et al.* 2008

Ringed Plover: Number of pairs nesting in Wales from dedicated surveys. No counts were received from Anglesey, Denbighshire or Flintshire in 1984. The estimate of total number of pairs used data from other surveys for these counties.

The most detailed information on the number of pairs nesting in Wales came from three surveys, in 1973–74, 1984 and 2007. In all three surveys, the highest number of pairs was in Caernarfonshire: 67 pairs in 1973–74, 61 in 1984 and 79 in 2007. The 2007 figure included 15 pairs in a single tetrad on the north coast of eastern Llŷn. Anglesey also had a good number, with 40 pairs recorded in 1973–74. The island was not surveyed in 1984, but a survey in 1982 recorded 49 pairs and this figure was used in the estimate. The 2007 census suggested a considerable decline on Anglesey, with only 17 pairs recorded, although some parts of the coast were not surveyed (Gibbs *et al.* 2008). In Meirionnydd, numbers increased from 19 pairs in 1973–74 to 36 pairs in 1984, but decreased to 24 pairs in 2007. Farther south, the 1973–74 survey found no records between the Dyfi Estuary and Carmarthen Bay. In 1984, numbers in Carmarthenshire had increased from 13 to 27 pairs, including 14 pairs on Pendine Sands (Lloyd *et al.* 2015). East Glamorgan/Gower had the highest number of breeding pairs on the south coast, with 34 pairs in 1973–74, increasing to 42 in 1984, but decreasing to 28 pairs in 2007. There were no records in Gwent in the first survey and only one pair in the second, but there were four pairs in 2007. The establishment of Newport Wetlands Reserve in 1999 provided suitable undisturbed habitat and led to an increase in breeding in the county (Venables *et al.* 2008). In northeast Wales, the 1973–74 survey found six pairs in Flintshire and two in Denbighshire. This area was not surveyed in 1984, but in the 2007 survey there were 46 pairs in these two counties. The 1973–74 and 2007 surveys found no inland breeding, but three pairs were found in 1984 (Conway *et al.* 2008). The *Britain and Ireland Atlas 2007–11* showed that, although inland breeding is widespread in Scotland and England, there are still very few inland pairs in Wales, with only two pairs confirmed breeding away from coastal areas.

First-calendar-year Ringed Plover in autumn © Ben Porter

Site	County	74/75 to 78/79	79/80 to 83/84	84/85 to 88/89	89/90 to 93/94	94/95 to 98/99	99/00 to 03/04	04/05 to 08/09	09/10 to 13/14	14/15 to 18/19
Dee Estuary: Welsh side (whole estuary)	Flintshire	na (1,620)	na (449)	na (673)	225 (588)	208 (440)	256 (535)	251 (514)	174 (784)	291 (1,215)
Severn Estuary: Welsh side (whole estuary)	Gwent/East Glamorgan	676 (2,244)	129 (626)	234 (895)	288 (1,096)	152 (789)	98 (622)	90 (1,096)	185 (1,151)	153 (1,034)
Swansea Bay	Gower	466	460	442	269	252	306	259	209	190
Dyfi Estuary	Ceredigion/ Meirionnydd	354		97	87	132	200	155	27	31

Ringed Plover: Five-year average peak counts during Wetland Bird Surveys between 1974/75 and 2018/19 at sites in Wales averaging 200 or more at least once. International importance threshold: 540; Great Britain importance threshold: 420.

On Welsh coastal sites surveyed both in 1984 and 2007, there was a decline of 6%, but there were much greater declines elsewhere: 43% in England and 38% in Scotland (Conway *et al.* 2008). An estimated 5,291 pairs of Ringed Plovers bred in Britain in 2007 compared to 8,483 pairs in 1984. The proportion of the population breeding in Wales had increased, and in 2007 was 5.1% of the total for Britain (Conway *et al.* 2008). The most pronounced decline in Britain as a whole was at inland sites, a decline of 83% (Conway *et al.* 2019). Published records since 2007 do not suggest that there has been a great change in Wales, although breeding appears to have ceased in Pembrokeshire, where no breeding was confirmed during the *Britain and Ireland Atlas 2007–11*.

Larger numbers are recorded in Wales on passage and in winter. Peak May and August counts are thought to be mainly of the subspecies *C.h. psammodromus* on passage (Steve Dodd *in litt.*). The Severn and Dee estuaries are of international importance for Ringed Plover, but most birds are on the English sides, and no site in Wales currently reaches even the threshold for British importance. The Dee Estuary is primarily of importance for passage birds (David Parker *in litt.*).

There were some larger WeBS counts on the Welsh side of the Severn estuary in the early 1970s: a peak of 1,436 in winter 1973/74 and 1,227 in August 1974. Numbers on the entire Dee Estuary peaked at 3,732 in winter 1973/74, but it is not known how many of these were on the Welsh side. There have been two exceptional counts of *c.*1,000 on the Tywi Estuary at Ferryside, Carmarthenshire, on 21 December 1970 and 22 February 1987.

A long-term WeBS High Alert has been issued for Wales, following a decline of 62% (Woodward *et al.* 2019). WeBS counts provide an accurate picture of the number of wintering birds, but a count on one fixed date a month does not give a full picture of the number of birds passing through on migration. Much larger counts have been recorded in spring and autumn. Some sites are of far greater importance at migration times than in winter, particularly the Dyfi Estuary, where there were 1,500 on 7 May 1971. Numbers at this site declined by the 1980s, but there were *c.*1,500 there in May 2010. Most Ringed Plovers are usually on the Ceredigion side of the estuary. On autumn passage, there were 1,150 in the Dyfi Estuary on 25 August 1971 and 940 at Borth/Ynyslas, Ceredigion, in August 2017. Elsewhere, there were 800 on the south side of the Burry Inlet in September 2011 and 787 there in September 2012. Blackpill foreshore, Gower, is always a good site in autumn: the mean autumn peak there during 1967–84 was 520, but this fell to 380 in the late 1980s and 290 in the early 1990s (Hurford and Lansdown 1995). The peak count there was at least 726 in August 1969. On the Dee Estuary, there were 680 at Mostyn Dock, Flintshire, on 25 August 2001.

Large winter counts included 890 on the Dyfi Estuary in December 2012 and 560 at Blackpill in December 2008, although such numbers at a single site in winter are unusual. Overall, the Welsh estuaries probably hold an average of about 900 each winter. Birds are also found on non-estuarine coasts: the 2015/16 Non-estuarine Waterbird Survey (NeWS) found 641 birds in Wales. The total winter population in Wales is probably *c.*1,600, out of an estimated 42,000 wintering in Britain (Frost *et al.* 2019a).

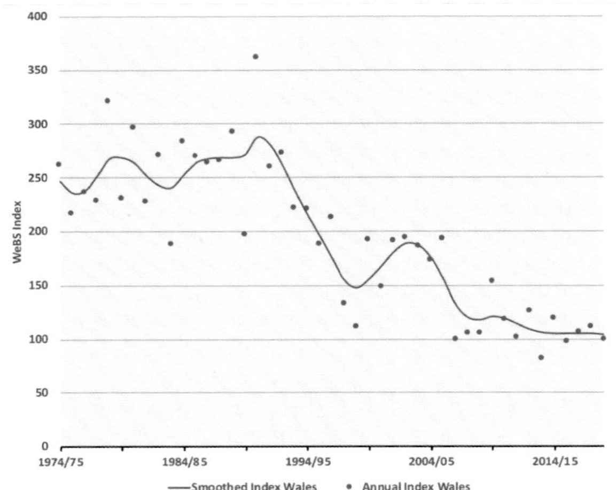

Ringed Plover: Wetland Bird Survey indices, 1974/75 to 2018/19.

Wales held around 5% of the UK breeding population of Ringed Plover in 2007 (Conway *et al.* 2008). The decline in breeding numbers between 1984 and 2007 is likely to be due to habitat change in inland areas and to human disturbance in coastal habitats (Conway *et al.* 2019). Since most of the breeding population in Wales is coastal, the future of this species will depend largely on provision of undisturbed coastal areas, which may require roping or fencing of suitable areas of shingle or requiring dogs to be kept on leads. If this is not done, the current moderate decline could accelerate as human disturbance increases. The area of Europe suitable for breeding Ringed Plovers is modelled to reduce during the 21st century, such that the northern half of Wales may be on the southern edge of the breeding range of this species by 2100 (Huntley *et al.* 2007). A better understanding of why it has declined to a greater extent in coastal areas of England and Scotland than in Wales could also be of value in maintaining breeding numbers in Wales.

Tundra Ringed Plover *C.h. psammodromus*
Cwtiad Torchog y Twndra

This subspecies winters in southern Europe and West Africa, and several ringing recoveries show that it is found in Wales on passage. One ringed at Afon Wen, Caernarfonshire, in April 1985 was hit by a car in Iceland in June 1988 and one, ringed as a nestling in Iceland in June 2010, was caught by a ringer at Ynyslas, Ceredigion, in April 2018. Some come from farther afield: an adult ringed at Collister Pill, Gwent, in May 1973 was shot on Ellesmere Island, Arctic Canada, in July 1979. Peak May and August counts of Ringed Plovers in Wales are thought to be mainly *psammodromus* on passage. There is at present no evidence that *C.h. tundrae* occurs in Wales (Steve Dodd *in litt.*).

Rhion Pritchard

Little Ringed Plover *Charadrius dubius* Cwtiad Torchog Bach

Welsh List Category	IUCN Red List			Birds of Conservation Concern				Rarity recording
	Global	Europe	GB	UK	Wales			
A	LC	LC	LC	Green	2002	2010	2016	RBBP

The Little Ringed Plover is widely distributed across Eurasia and into northwest Africa. It winters to the south, in equatorial Africa and southern Indonesia (*HBW*). It mainly breeds inland on bare flats of sand or shingle with little vegetation, but usually in the vicinity of standing or slow-flowing water. It is also found in temporary, artificial habitat, such as quarries, gravel pits and other rough land (*HBW*). Few have been ringed in Wales, but one, ringed as a nestling on the Afon Tywi, Carmarthenshire, in June 1992, was trapped in Djenne, Mali, in April 1999.

Little Ringed Plover first bred in Britain in 1938 and for a second time in 1944, after which it became annual, mainly in England, where a small breeding population became established (*BWP*). The first record of the species in Wales was at the Gann Estuary, Pembrokeshire, on 16 May 1949, but the first record of breeding was not until 1970, when a pair bred in Flintshire. In 1971, single pairs bred successfully at four sites across Flintshire and Denbighshire. A pair bred at a gravel pit in Denbighshire in 1972, but it was drained in 1973 and there were no known breeding attempts in Wales that year.

In 1974, breeding was recorded on a river in Wales for the first time, when a pair bred on a gravel shoal on the River Severn in Montgomeryshire (Holt and Williams 2008). A small breeding population became established, with seven pairs along this part of the Severn in 1991 (Tyler 1992b). A pair on the River Wye below Glasbury in May 1977 was the first record for Radnorshire. 1–2 pairs bred in this area until 1990 and 3–5 pairs were found along the lower Wye in the 1990s, increasing to eight pairs in 2002, after which numbers decreased. Breeding success was low, in part because of human disturbance, including canoeists landing on shingle banks (Jennings 2014). The first breeding in Gwent was in 1984 at reservoirs with low water levels: a pair at Garnlydan Reservoir and three at Llandegfedd Reservoir. Birds returned to Garnlydan in 1985, and a chick from there was recaught as a breeding female on the Afon Tywi in 1991. In subsequent years, the Garnlydan birds relocated to a nearby small quarry in Breconshire, where they first bred in 1986. Breeding was first confirmed on the River Usk gravel banks in Gwent in 1987. This became the main site for this species in the county until Newport Wetlands Reserve was established in 1999 (Jerry Lewis *in litt.*). Up to four pairs have been confirmed breeding at the Newport Wetlands Reserve in recent years (Darryl Spittle *in litt.*). In East Glamorgan and Gower, many pairs use industrial or post-industrial sites, including 11 pairs at Llanilid, East Glamorgan, in 2006.

Female Little Ringed Plover © John Hawkins

In Carmarthenshire, a pair was seen on a shingle bank on the Afon Tywi near Llandeilo in 1984, although breeding was not confirmed. None was seen in 1985, but breeding was confirmed there in 1986 and numbers built up rapidly to six pairs in 1987, 14 in 1989, 27 in 1990 and 39 in 1991. By this time breeding also occurred on some of the larger tributaries of the Tywi, but overall numbers stabilised, with 45 pairs recorded in 1998. The next count, in 2004, found 62 pairs, of which 14 were on tributaries such as Afon Cothi. The catchment accounted for more than 10% of the UK breeding population at that time. In 2007 there were *c.*76 pairs nesting on shingle banks on the Afon Tywi and its tributaries. There may have been a small decline in numbers on the Tywi since: a survey between Carmarthen and Llandovery in 2018 found only 21 nests, compared to 59 pairs in this section in 2007. However, in 2019 there were 26 pairs between Carmarthen and Rhosmaen (Julian Friese *in litt.*), 24% more than in 2018, in only *c.*60% of the length of river surveyed. Projecting these figures for the whole of the Carmarthen to Llandovery stretch of the Tywi would suggest a total of 43–47 pairs (Rob Hunt *in litt.*). Periods of high water can lead to high failure rates. For example, 40 pairs were found in 1996, but only one flying young was seen (Lloyd *et al.* 2015).

In Ceredigion, the first breeding record was on the Afon Rheidol, near Llanbadarn Fawr, in 1990 and the second on the Rheidol in 1999, with annual breeding there until 2003. A few pairs have nested at other sites in the county (Roderick and Davis 2010). It has been a scarce but regular breeder in Meirionnydd since the first known breeding attempt, at Llyn Trawsfynydd in May 1977 (Pritchard 2012). Elsewhere in North Wales, breeding is confined to Flintshire and Denbighshire. During 2014–18 there were annual records of probable or confirmed breeding in Flintshire, with confirmed breeding in four of the five years, and several records of probable breeding in Denbighshire, with confirmed breeding in, or near, RSPB Conwy in 2018. Up to three pairs nested annually at this site during 1992–98, and 1–2 pairs have nested periodically in subsequent years (Hughes and Money 2016). Breeding has never been confirmed in Caernarfonshire, although a pair was thought to have bred at Morfa Madryn in 1966 (Pritchard 2017). There has never been a record of even possible breeding on Anglesey.

The *Britain and Ireland Atlas 1988–91* showed an increase in occupied 10-km squares in Wales to 20, from three in 1968–72, although there were none in the north away from the Dee Estuary. By the *Britain and Ireland Atlas 2007–11*, the number of occupied squares had increased to 47, in all parts of Wales. Relative abundance was higher in Carmarthenshire than elsewhere in Wales.

The first arrivals in Wales each spring are usually in a three-week window during March, the earliest during 2000–19 being on 2 March 2018, in Carmarthenshire. Departure is from August to October, the latest during that period being in Flintshire on 7 November 2017.

A survey of Welsh rivers in 1991 found 61 pairs, more than half of these on the Afon Tywi (Tyler 1992b). A BTO survey in 2007 found 144 pairs in Wales, out of 901 pairs found in Britain. This included 77 pairs in Carmarthenshire, 23 in Montgomeryshire, 11 in Breconshire and ten pairs in East Glamorgan (Holling *et al.* 2010). It is likely that the current population in Wales is in the order of 150–200 pairs.

A high proportion of the Welsh population breeds on shingle banks in rivers, including quite fast-flowing rivers such as Afon Tywi. Use of this habitat for breeding appears to be confined to South and mid-Wales, with no records of its use north of the Afon Dyfi. Shingle banks are commonly used for breeding on the European mainland, but less frequently in England, where sand and gravel pits are the most commonly used habitat (*Britain and Ireland* Atlas *2007–11*). Shingle banks can be unstable, regularly refreshed in winter floods, but summer floods can lead to complete breeding failure. Trampling of nests by cattle and disturbance by humans can also be problems (Lloyd and Friese 2013). Nesting on river shoals on the River Usk in Breconshire peaked at seven pairs in 2007–10, but this has now been abandoned and records in the county are all from post-industrial sites and lake or reservoir shores. This is thought to be due to the shoals being comprised of substrates of a larger size, through the scouring effect of more severe winter floods having removed all substrate equivalent in size to Little Ringed Plover eggs (Andrew King *in litt.*). This could become a problem elsewhere, as more frequent flood events are anticipated to be a result of climate change. Other areas of suitable breeding habitat are relatively restricted in Wales, with some only temporarily available, such as land cleared for building. While Little Ringed Plover is not a bird of conservation concern in Wales, its prospects for further expansion are probably not good.

Ian M. Spence

Killdeer *Charadrius vociferus* Cwtiad Torchog Mawr

Welsh List Category	IUCN Red List	Rarity recording
	Global	
A	LC	BBRC

The Killdeer breeds in the Caribbean and in North America, as far south as northern Mexico, and winters in Central America and northern South America (*HBW*). It is rarely recorded in Britain, on average once each year, with 59 individuals recorded to the end of 2019, the last in 2018 (Holt *et al.* 2019). It is more likely to occur in Britain during the winter than other Nearctic waders and is prone to long-distance hard-weather movements (Parkin and Knox 2010). There have been three records in Wales, all of single birds:

- Pwll, Llanelli, Carmarthenshire, from 5 February to 16 March 1978;
- Bardsey, Caernarfonshire, on 17–20 March 1982 and
- Holyhead, Anglesey, from 30 December 1993 to 2 January 1994.

Jon Green and Robin Sandham

Killdeer near Holyhead, Anglesey, in December 1993
© Steve Young

Kentish Plover *Charadrius alexandrinus* Cwtiad Caint

Welsh List Category	IUCN Red List		Rarity recording
	Global	Europe	
A	LC	LC	WBRC

This species formerly bred in Britain but now occurs only as a scarce migrant. The breeding range is extensive, from the Mediterranean east to Japan. It usually nests on sand, by the sea or saline lakes and lagoons (Parkin and Knox 2010). Numbers in Britain have declined from an average of 36 a year during the 1990s to just 12 in the 2010s, with most records in the spring. The first Welsh record was of two pairs at Sker, East Glamorgan, in May 1888, but there is no evidence that they bred. The next Welsh record was not until after the Second World War, at Point of Ayr, Flintshire, on 4 November 1951. A further 43 have been seen in Wales since then, almost annually since 1970.

The majority of records are in South and mid-Wales, with Ynyslas, Ceredigion, hosting 11 of these, usually coinciding with large movements of Ringed Plovers in April and May. A female at Sker, East Glamorgan, on 18 April 2016 had been ringed in France the previous year. Modelling suggested that southwest Britain, including southern Wales, will become climatically suitable for

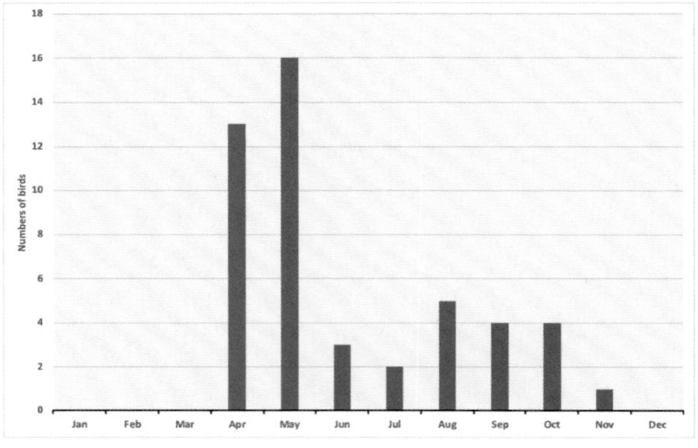

Kentish Plover: Totals by month of arrival, 1888–2019.

Total	Pre-1950	1950–59	1960–69	1970–79	1980–89	1990–99	2000–09	2010–19
48	4	3	3	10	8	5	9	6

Kentish Plover: Individuals recorded in Wales by decade.

breeding Kentish Plovers later this century (Huntley *et al.* 2007), but the availability of habitat is likely to be a major constraint without human intervention.

Jon Green and Robin Sandham

Kentish Plover: Numbers of birds in each recording area 1888–2019. One bird was seen in more than one recording area.

Greater Sand Plover *Charadrius leschenaultii* Cwtiad Tywod Mawr

Welsh List Category	IUCN Red List		Rarity recording
	Global	Europe	
A	LC	VU	BBRC

Scattered populations of this plover breed in desert or semi-desert areas from Turkey east to Mongolia. The westernmost populations winter in eastern Africa (*HBW*). Seventeen individuals were recorded in Britain up to the end of 2019, with the last in 2018 (Holt *et al.* 2019). Only one of these was in Wales, at St Brides Wentlooge, Gwent, on 16 May 1988.

Jon Green and Robin Sandham

Dotterel *Charadrius morinellus* Hutan y Mynydd

Welsh List Category	IUCN Red List			Birds of Conservation Concern				Rarity recording
	Global	Europe	GB	UK	Wales			
A	LC	LC	EN	Red	2002	2010	2016	RBBP

The Dotterel breeds on the Arctic tundra of northern Europe and Siberia, extending into western Alaska, and on Eurasian mountains that offer similar conditions, such as in Scotland (*HBW*). It winters in North Africa and the Middle East, but most recoveries of British-ringed birds are from Morocco (*BTO Migration Atlas* 2002). The Scottish breeding population was estimated at 423 breeding males in 2011 (Hayhow *et al.* 2015), while a few also breed in northern England, at least in some years (*Birds in England* 2005). In a reversal of the usual pattern among birds, the male Dotterel incubates the eggs and looks after the chicks. It is thought that 80–90% of females breeding in Scotland lay a clutch, then relocate to Scandinavia or elsewhere to seek a new mate and lay another clutch (*The Birds of Scotland* 2007). Birds moving through Wales in spring and autumn are thought to be mainly from the Scottish population, although there are few ringing records. A juvenile, predated on Foel Fras, Caernarfonshire, in September 1990, had been ringed as a nestling in Moray, Scotland, in June 1990, while a colour-ringed bird on Pumlumon, Ceredigion, in May 1995 had been ringed as a chick in the Scottish Highlands in 1987 (Roderick and Davis 2010).

Aplin (1903) reported that he found Dotterel on Cadair Idris in Meirionnydd in early May in two successive years and considered them to be on passage. However, he added "it is, of course, just possible that the Dotterel may breed on some of the tops. The ground looks suitable." There are no reports to suggest even possible breeding in Wales before the second half of the 20th century, when breeding was recorded in the Carneddau range, Caernarfonshire. Barnes (1997) described the history of breeding Dotterel in Caernarfonshire as "poorly documented and shrouded in secrecy", and this remains true. There was a report of a possible breeding pair in the 1950s but no confirmed record until a nest was found "in Snowdonia" in 1967 (Ratcliffe 1990), evidently in the Carneddau. A pair probably nested there in 1968 and a nest with four eggs was found in 1969. There were further nesting records in the Carneddau during the early 1970s. Schofield (1975) referred to breeding pairs at three sites in the northern Carneddau. Breeding in this area probably ceased by 1975 or soon afterwards, though a few unconfirmed reports suggested breeding in subsequent years. A bird at one upland site in Meirionnydd, in early June 2019, was reported to have an injured wing. This could have been distraction display, indicating a nesting attempt.

The earliest record of a trip of Dotterel in Wales seems to have been a group on Carnedd Llywelyn, Caernarfonshire, about 1820 (Price 1875). There were a few records in the late 19th century, mostly of small numbers and mainly in the north. Mathew (1894) said that the species was very rarely seen in Pembrokeshire, and Barker (1905) did not mention it in Carmarthenshire, where the first record is of two at Bannau Sir Gar in May 1923. Forrest (1907) recorded a party of 15 in a field near Chester, on the Flintshire side of the border, on 8 May 1903, but otherwise records remained

Female Dotterel in breeding plumage © Jerry Moore

scanty until the second half of the 20th century. As the number of sightings increased, it became clear that Dotterel were found in a few traditional locations, both upland and coastal, where they can be very regular. They were recorded on spring passage on the Mynydd Du range, Carmarthenshire, every year between 1987 and 2011, except in 2001 when restrictions imposed as a result of foot-and-mouth disease made access impossible, and birds were recorded on the Carneddau range, Caernarfonshire, in all but four of the 35 years from 1985 to 2019. Some observers make a point of visiting these favoured sites annually during migration, to look for Dotterel, and birds stopping at other upland sites could well be missed.

Green (1992) examined records of Dotterel in Wales during 1980–89. There were 62 records involving 317 birds, from every county except Flintshire. Of these, 77% were in the uplands and the other 23% were on or close to the coast. Caernarfonshire had the largest share with 83 (26%), followed by Ceredigion with 70 (22%), Anglesey with 56 (18%) and Carmarthenshire with 38 (12%). Green considered that observations of Dotterels in Wales had increased in the 1980s, but this was partly because the number of observers actively looking for Dotterels had increased. The highest count recorded in Wales is 45 on Valley Airfield, Anglesey, on 6–12 May 1980.

Analysis of records during 1990–2017 showed a continued increase (Pritchard 2019): 431 records involving an estimated 1,594 birds, which came from every county. The highest numbers of birds were in Caernarfonshire (42%), Carmarthenshire (22%), Pembrokeshire (8%) and Ceredigion (7%). The proportion of records on Anglesey was much lower than in 1980–89. Across Wales, 81% were recorded in spring and 19% in autumn. One on Garreg Lwyd, Carmarthenshire, on 4 April 2008 appears to be the earliest record, ever, for Wales. Spring passage during 1990–2017 peaked in the two weeks between 30 April and 13 May, although a few birds were recorded in June. There was one July record, of three birds at RSPB Ynys-hir, Ceredigion, on 19 July. The main passage in autumn was more extended than in spring, from mid-August to late October. There were two November records, and one in a wintering Golden Plover flock at Castlemartin, Pembrokeshire, on 13 December 1998 is the only winter record for Wales. The habitat of passage birds was similar to that used during the 1980s: 74% in the uplands, 26% on or near the coast. Most of the highest counts were in the uplands in spring, notably 40 between Foel Fras and Drum in the Carneddau, Caernarfonshire, on 8 May 1993 and 39 between Garreg Lwyd and Bannau Sir Gar, Carmarthenshire, on 7 May 1996. A trip of 35 on the Nantlle Ridge, Caernarfonshire, on 16 October 1999 was an exceptional count for return passage.

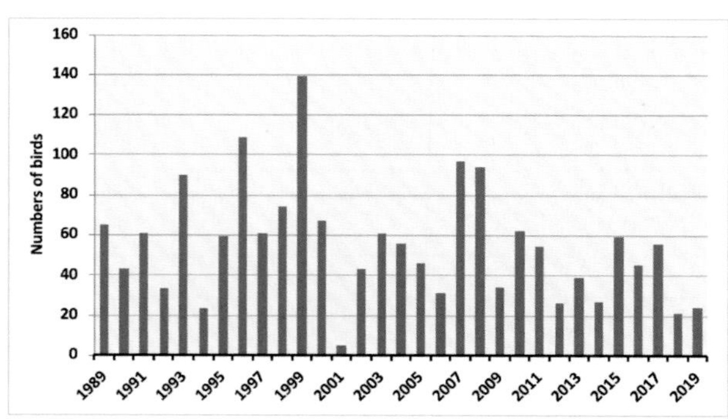

Dotterel: Estimated numbers of birds on passage in Wales, 1989–2019.

Passage numbers peaked in the late 1990s, with an estimated 140 individuals in 1999, the year with the highest total. The average number of birds per trip is consistently higher in the uplands, but a comparison by decade showed that upland trips in spring have reduced in size, from 6.8 in 1990–99 to 3.9 in 2010–17. Dotterel appear to show a consistent preference for certain upland areas: 31.4% of all records during 1990–2017 were in the Carneddau range, and 20.8% in the Mynydd Du range. Foel Fras (944m) in the Carneddau alone held 18.1% of records and Garreg Lwyd, Mynydd Du (616m) hosted 15.6%. Where further details of the habitat were provided, upland records were from high, windswept ridges and summits with short vegetation, while coastal records were mainly from headlands and coastal heath, although there were several records on farmed grassland and at airfields. The coastal sites with most birds recorded were the Great Orme, Caernarfonshire (3.9%), and The Range, RSPB South Stack, Anglesey (3.4%). Bardsey, Skomer, Skokholm and Ramsey all recorded birds in this period.

The likelihood of Dotterel breeding in Wales again is diminishing. Numbers breeding in Britain declined by 57% between 1987 and 2011 (Hayhow *et al.* 2015) and no area of Britain is expected to be climatically suitable for the species later this century (Huntley *et al.* 2007). Since the number recorded on passage in Wales probably depends largely on the fortunes of the Scottish breeding population, lower numbers recorded in the last 20 years reflect the decline in this population. Some Welsh summits may be important refuelling stops for Scottish birds in spring, and it is important that the habitat there is maintained in good condition and not over-grazed.

Rhion Pritchard

Sponsored by Rhion Pritchard

Upland Sandpiper *Bartramia longicauda* Pibydd Cynffonhir

Welsh List Category	IUCN Red List	Rarity recording
	Global	
A	LC	BBRC

This species has an extensive breeding range across North America and winters in South America (*HBW*). It is a rare trans-Atlantic vagrant to Britain with 47 records to the end of 2019, on average one every other year over the past 30 years (Holt *et al.* 2020). There have only been three records in Wales, all in autumn and all in Pembrokeshire:

- Skokholm on 18 October 1960;
- Skomer on 19–20 October 1961 and
- Dale airfield on 1 September 1975.

Jon Green and Robin Sandham

Whimbrel *Numenius phaeopus* Coegylfinir

Welsh List category	IUCN Red List			Birds of Conservation Concern				Rarity recording
	Global	Europe	GB	UK	Wales			
A	LC	LC	CR	Red	2002	2010	2016	RBBP

A widespread species in global terms, the Whimbrel breeds in the far north, from eastern Greenland to eastern Siberia, on subarctic moorland and birch forest and tundra near the tree line. In winter it is found throughout the tropics and subtropics (*HBW*). The birds that pass through Wales are thought to be mainly of the subspecies *N.p. islandicus*, which breeds in Scotland and Iceland. Several birds ringed or colour-ringed in Wales have been recovered in Africa, in Morocco, Senegal, The Gambia and Guinea-Bissau. Two ringed as nestlings in Iceland have been recovered in Wales, and Whimbrels colour-ringed in Wales have been seen in Iceland during the breeding season. One ringed on Skokholm, Pembrokeshire, in August 1967 was found dead in northern Finland in June 1969.

The Whimbrel is usually recorded in Wales on spring and autumn passage, with the highest numbers at most sites in spring. Most birds breeding in Iceland can fly directly to their wintering grounds in Africa in autumn, but they stop in Europe more frequently on spring migration (Gunnarsson and Guðmundsson 2016). Passage is mainly along the coast, though some are recorded flying overland, frequently at night. Spring passage is sometimes recorded from mid-March, and peaks in late April and early May. The status of Whimbrel in Wales does not seem to have changed greatly in the past 150 years, although there are few counts from the 19th century. The area around the upper Severn Estuary, including the Gwent Levels and Somerset Levels, was shown to be the most important area in Britain for spring Whimbrel passage, holding about 2,000 in the period 1973–77. It is perhaps the last major feeding area for birds moving northwest to breed in Iceland (Ferns *et al.* 1979). During peak passage in May, 74% of Whimbrels recorded in Britain during 1972–75 were around the Severn Estuary, feeding on grassland during the day and roosting at night on either side of the estuary, the main site in Wales being Collister Pill, Gwent, where 1,000 roosted on 30 April 1976. Other large counts in Gwent include 400–500 at Peterstone in April 1983 and at Undy in April 1984. *Birds in Wales* (1994) gave a total of 739 scattered across the Gwent Levels on 5 May 1991, although Venables *et al.* (2008) gave a total of 559 for a co-ordinated count of the Gwent coast on that date. The latter also noted that the largest concentrations were now only one-third the size of those during the mid-1970s to mid-1980s, the decline thought to be a result of changes in land use on the Levels that had reduced the area of damp pasture available. Other sites in Britain now have larger counts than the Severn Estuary in spring, notably Barnacre Reservoir, Lancashire.

The Burry Inlet has increased in importance in spring. Hurford and Lansdown (1995) noted that numbers on spring passage had increased from 15 in the 1960s to 70–100 in the 1970s and 1980s; there were 366 there in May 2000, and a Carmarthenshire record of 387 at Cefn Padrig in May 2010. A maximum of 840 was counted in Swansea Bay, Gower, between Blackpill and

Whimbrel on passage during spring © Ben Porter

Site	County	74/75 to 78/79	79/80 to 83/84	84/85 to 88/89	89/90 to 93/94	94/95 to 98/99	99/00 to 03/04	04/05 to 08/09	09/10 to 13/14	14/15 to 18/19
Severn Estuary: Welsh counties (whole estuary)	Gwent/East Glamorgan	21 (1,002)	12 (50)	40 (131)	54 (111)	82 (140)	64 (138)	106 (163)	114 (226)	112 (245)
Burry Inlet	Gower/ Carmarthenshire	150	92	160	92	215	136	133	52	131

Whimbrel: Five-year average peak counts during Wetland Bird Surveys between 1974/75 and 2018/19 at sites in Wales averaging 100 or more at least once. International importance threshold: 6,700; Great Britain importance threshold: 1.

West Cross on 28 April 2003, while in Pembrokeshire there were over 200 at Whitesands on 6 May 2001. In Ceredigion, large counts include 226 at Ynyslas on 26 April 1984 and over 200 at Llanrhystud on 4 May 2015, while in Montgomeryshire there were 156 at the William Condry Reserve on the Dyfi Estuary on 5 May 2005. Numbers in North Wales tend to be lower, with the highest count being 159 on Bardsey, Caernarfonshire, on 4 May 2003. Forrest (1907) commented that the Dee Estuary seemed to be off the main migration route of Whimbrel, and this still appears to be true. Counts on the Welsh side of the Dee seldom exceed 40, although there was a count of 45 in May 2016.

Autumn passage involves smaller numbers than in spring, as in other areas of western Britain, whereas on the east coast of England and Scotland, numbers are greater during southbound migration, involving birds from Fennoscandia and further east (Graham Appleton *in litt.*). There have, however, been a few large counts, including 307 on the Burry Inlet in July 2014. Numbers passing Strumble Head, Pembrokeshire, in autumn vary considerably, but can be high: 680 passed between 16 July and 13 September 1998, with a peak of 370 on 5 August, while in 2013 a remarkable 1,498 passed on 18 August. This species is fairly regular on some lakes and reservoirs away from the coast, usually only in small numbers, but *c.*130 were at Llyn Alaw, Anglesey, on 6 May 2005.

A few Whimbrels spend the winter in Wales, a habit that has been recorded on Bardsey, Caernarfonshire, where 1–4 over-wintered fairly regularly during 1953–79 and 1996–2019, with five there on 6 February 2005. The largest number at a single site in winter was six at Burry Port, Carmarthenshire, on 7 February 1971. On average, 3–5 have been recorded in Wales in recent winters. The highest totals were in winter 2011/12 when ten were present, including four in the Burry Inlet and three on Bardsey, and winter 2018/19 when up to ten individuals were again present. The typical wintering population in Britain is estimated at 38 (Frost *et al.* 2019a). One on Bardsey, in the first winter period of 2013, had been colour-ringed on the island the previous summer.

In Britain, breeding birds are mainly confined to Shetland, with a few pairs in Orkney, the Outer Hebrides and in the far north of the Scottish mainland (*The Birds of Scotland* 2007). Williams (1835) commented, in the *History of Aberconwy*, that the Whimbrel, though not as common as the Curlew, also bred in the mountains above Conwy, Caernarfonshire; this seems unlikely to be correct. The first confirmed record of breeding in Wales was very unexpected: a pair was thought to have held territory on Ynys Gwylan Fawr, off the coast of Llŷn, Caernarfonshire, in 1999, with breeding confirmed there in 2000, amid a Great Black-backed Gull colony. Three young were seen, of which two were ringed.

More than half the world's curlew species have an unfavourable conservation status, ranging from Near Threatened to Critically Endangered. The Whimbrel is an exception and is classified as of Least Concern. The European population is considered to be stable (BirdLife International 2020), but the area of suitable climate for breeding is predicted to reduce in Iceland during this century (Huntley *et al.* 2007). We can expect Whimbrel to continue to be a regular sight on the Welsh coast in spring and autumn, but in the long term, its ability to take advantage of breeding habitat farther north in Greenland may be critical.

Rhion Pritchard

Hudsonian Whimbrel *Numenius hudsonicus* Coegylfinir yr Hudson

Welsh List Category	IUCN Red List	Rarity recording
	Global	
A	Not yet assessed	BBRC

Hudsonian Whimbrel was until recently regarded as the North American subspecies of Whimbrel but is now regarded as a separate species (IOC List 10.2). It breeds in Alaska and northern Canada, as far east as Hudson Bay. A total of 11 individuals had been recorded in Britain up to the end of 2019 (Holt *et al.* 2020). There is one record for Wales, at Newport Wetlands Reserve (Goldcliff), Gwent, on 6–7 May 2000, with what was presumed to be the same bird there again on 3–4 May 2002. This was the third record for Britain.

Jon Green and Robin Sandham

Little Whimbrel *Numenius minutus* Coegylfinir Bach

Welsh List Category	IUCN Red List	Rarity recording
	Global	
A	LC	BBRC

A long-distance migrant, the Little Whimbrel, also known as Little Curlew, breeds in the far northeast of Siberia and winters in Australia. It was formerly considered rare, but the breeding range is more extensive than previously thought (*HBW*). The only Welsh record was an adult at Sker Point, East Glamorgan, from 30 August to 6 September 1982. This was the first recorded in Britain and only the second in the Western Palearctic. There has been one subsequent record in Britain, in Norfolk in August 1985, and a total of only eight records in Europe.

Jon Green and Robin Sandham

Little Whimbrel at Sker Point, East Glamorgan, August – 6 September 1982 © Richard Smith

Curlew *Numenius arquata* Gylfinir

Welsh List Category	IUCN Red List			Birds of Conservation Concern			
	Global	Europe	GB	UK	Wales		
A	NT	VU	EN	Red	2002	2010	2016

The Curlew breeds across western and central Europe, Central Asia and Siberia east to northwestern China (*HBW*). It nests in a range of habitats, from uplands to lowland meadows, and is very site faithful. As our largest wader, with a characteristic and loud call, the Curlew has always been a well-known species. This is especially true in Wales, where it has at least 30 regional names, and its own Saint, St Beuno, who gave it protection (Colwell 2018).

The *Historical Atlas 1875–1900* described it as abundant in all the North Wales counties and Breconshire, and common elsewhere, except Glamorgan and Pembrokeshire. In the latter county, Mathew (1894) stated that he had little doubt that Curlews nested at Mynydd Preseli and he thought that they occasionally nested on Skomer. The species apparently did not breed in Glamorgan in the late 19th century, although it was said to have formerly bred in the hills above Penllergaer, near the northern border of the county (Hurford and Lansdown 1995). In North Wales, Forrest (1907) described Curlew as "common, breeding on all the moorlands and on some lowland bogs". He noted that "throughout the summer months the long rippling whistle or the plaintive call of the curlew constantly greet the wayfarer on the moorlands of North Wales. This is indeed one of the most characteristic birds of the Welsh Uplands throughout the district.'

There appears to have been an increase in the population across Europe around the start of the 20th century (*BWP*), described as "dramatic" in Britain and Ireland by Sharrock (1976). Curlew extended its breeding distribution from moorlands into lowland meadows, although this seems to have halted by the 1950s, with the first sign of declines soon afterwards. In Wales, E. Cambridge Phillips had noted a spread from the hills to adjacent enclosed fields for breeding in Carmarthenshire as early as 1880 (Ingram and Salmon 1954), and from the 1920s there was a considerable increase in the Curlew's breeding range. East Glamorgan was colonised in the 1920s, and pairs were widely distributed in Glamorgan by the end of the Second World War (Hurford and Lansdown 1995). In Carmarthenshire, Ingram and Salmon (1954) said that its distribution had "changed remarkably" in the past 25–30 years, as it bred in the fields of the wider river valleys from the 1920s, adding "now these are the main breeding areas, the higher elevations in the hills being largely deserted". Lockley *et al.* (1949) said that before the early 1930s it was not a common breeding bird in Pembrokeshire, but it had greatly increased and now nested throughout the county, except on the Castlemartin Peninsula, which was colonised in about 1950 (Lockley 1961). A similar pattern was noted in Breconshire (Ingram and Salmon 1957), and in Montgomeryshire, where they formerly nested only high in the hills, but by 1950 bred on the lower ground (Holt and Williams 2008). In Ceredigion, William Condry noted, in his diary for June 1949, that "vast numbers" bred in the Tregaron area; in 1953 he surveyed Cors Caron and estimated under 100 pairs on the bog (Roderick and Davis 2010). Farther north, Curlew bred abundantly in the Snowdonia National Park, on many types of rough ground according to Condry (1966), who reported that more birds were breeding on lower ground than earlier in the 20th century, and fewer on the hills.

A decline was noticeable in some areas by the 1950s. Jennings (2014) reported that the Curlew was a widespread breeder on upland moorland in Radnorshire in the 19th and 20th centuries. It was described as "very numerous" in the Elan Valley in 1896, but had decreased noticeably by 1957. In Glamorgan, a reduction in breeding range was noted between 1957 and 1966 (Hurford and Lansdown 1995). Good numbers still bred in many lowland areas in the 1980s: an RSPB survey of Llŷn in 1986 estimated 200–300 breeding pairs, 50% on marshy grassland and 30% on unimproved grassland. A survey on Anglesey found 11 pairs confirmed breeding, 247 pairs probably breeding and 161 pairs possibly breeding in Anglesey (Jones and Whalley 2004). *Birds in*

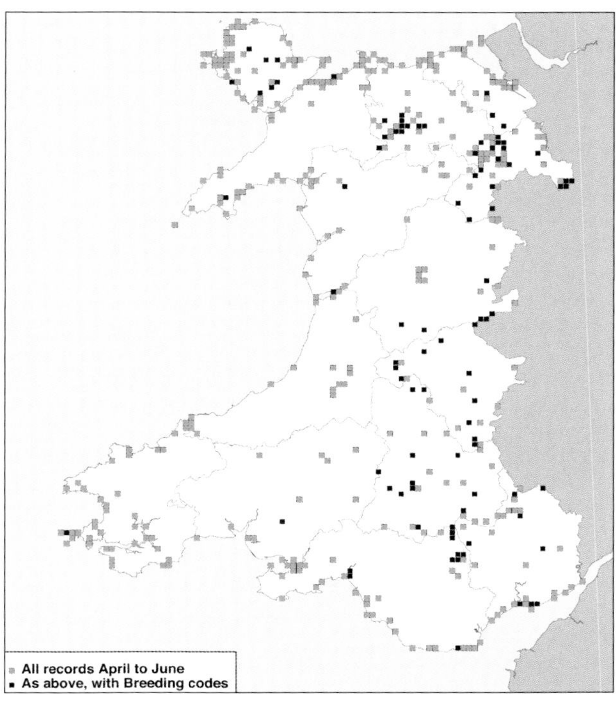

Curlew: Records from BirdTrack for April to June, 2015–19, showing the locations of records and those with breeding codes.

Curlew © Alan Saunders

Wales (1994) concluded that while the Curlew was still "common and widespread", the trend in the previous three decades had been one of decline and diminution of range. The authors considered that, although difficult to gauge with accuracy, the population around 1990 was *c.*2,000 pairs.

Despite evidence of declines in some areas, the *Britain and Ireland Atlas 1968–72* found the Curlew present in 91% of 10-km squares in Wales, with confirmed breeding in most of them. The 1988–91 *Atlas* showed a reduction to 79% of squares occupied, a few gains along the coast in the northeast far outweighed by losses in South Wales, particularly in Pembrokeshire, where the species had been lost from the great majority of squares since 1968–72. The 2007–11 *Atlas* showed 55% of squares were occupied, with further declines particularly in Carmarthenshire, Ceredigion and Meirionnydd. This was part of a pattern of decline in the west of Britain and Ireland, with declines in western Scotland and southwest England, and a particularly severe decline in Ireland. The atlas stated that "the loss of breeding Curlew from Ireland and parts of western Britain is one of the key findings over the last 40 years.'

County tetrad atlases provide further details of this decline. The *Gwent Atlas 1981–85* found Curlew in 49% of tetrads, but the 1998–2003 *Atlas* recorded the species in only 34%, the decline largely in the lowlands (Venables *et al.* 2008). In Pembrokeshire, where the decline had started earlier, it was recorded in only 2% of tetrads in the 1984–88 *Atlas*, and by 2003–07 there were records from only two tetrads (0.4%) with breeding confirmed only on Skomer. Elsewhere, Curlew was found in 19% of tetrads visited in Breconshire in 1988–90, with most records in the north of the county (Peers and Shrubb 1990). The *North Wales Atlas 2008–12* found the species in 23% of tetrads, mainly concentrated around the upland areas east of Afon Conwy, particularly Mynydd Hiraethog and the Berwyn. Occupied squares were sparsely distributed in the west.

BBS data showed a 69% decline in Wales between 1995 and 2018. This compares with declines of 31% in England and 59% in Scotland in the same period (Harris *et al.* 2020).

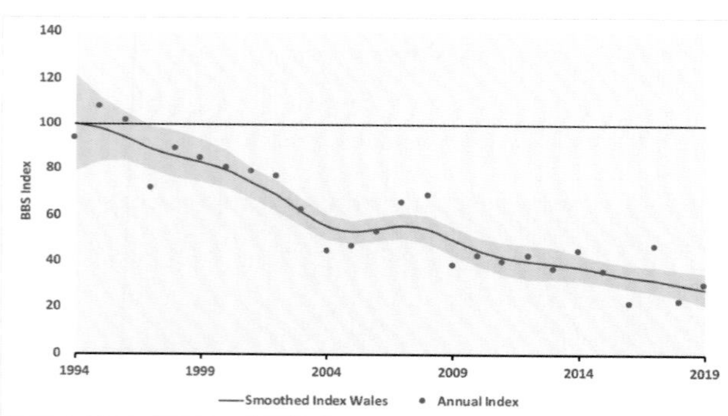

Curlew: Breeding Bird Survey indices, 1994–2019. The red line shows the smoothed index, 1995–2018. The shaded area indicates the 85% confidence limits.

Several surveys have confirmed the scale of the decline. Two surveys of lowland wet grassland in England and Wales, in 1982 and 2002, found a 75% decline in Wales, a reduction even greater than for Lapwing (Wilson *et al.* 2005). A survey of 170 randomly selected 1-km squares in Wales in 1993 produced an estimate of 10,763 pairs (O'Brien *et al.* 1998); these data were reanalysed by Johnstone *et al.* (2007a) using an improved and validated method to produce a revised estimate of 5,713 pairs. A repeat survey in 2006, when 331 squares were visited, resulted in an estimate of 1,099 pairs, a decline of 73% since 1993.

Declines have been reported in all parts of Wales. In the Radnorshire part of the Elan Valley, 18 pairs were recorded in 1976, 8–10 in 1987 and only 2–4 since 2004 (Jennings 2014). Holt and Williams (2008) catalogued a drastic decline across Montgomeryshire, including at RSPB Lake Vyrnwy, where the population of 32 pairs in 1980 had declined to two pairs by 2002. Surveys elsewhere in the county, mostly assessing sites for proposed wind farms, showed similar declines.

Site	Date of previous survey	Pairs	Pairs 2012
Trannon	1995	13	1–2
Mynydd yr Hendre	2005	5–6	1–2
Nant yr Eira	2006	10	2

Curlew: Declines in number of pairs at three sites in Montgomeryshire (from Green 2012).

The three sites in the table show differences in the period of decline. Nant yr Eira declined rapidly after 2006, to two pairs by 2012. Mynydd yr Hendre also declined over 2–3 years between 2005 and 2010 (pers. obs.). However, the Trannon population declined earlier, from 13 pairs in 1995 to 2–3 pairs by 2006 (Whitfield *et al.* 2010), 1–2 pairs by 2012 and none in 2019 (Green 2019a). Trannon Moor wind farm was built in 1996, but there is no obvious direct link to the decline in Curlew, which occurred elsewhere in the county at the same time (Whitfield *et al.* 2010).

Current estimates and predictions of Wales' Curlew breeding population are dire. Taylor *et al.* (2020) modelled scenarios based on a population estimate of 2,584 pairs in Wales in 2009. BBS data suggest that the breeding population in Wales is declining at a rate of about 5% a year, and based on this, the breeding population in 2019 was estimated at 1,101–1,578 pairs. The breeding range appears to be declining even more rapidly than the population, with the Curlew now almost entirely restricted to North and mid-Wales. On the basis of the mean value of all modelled scenarios, it was predicted that Curlews would become extinct as a breeding species in Wales in 2033.

Wales remains an important wintering area for Curlew, mostly around coastal areas but also at some inland sites. The numbers wintering in Wales in the period 2007–11 are estimated at 12,200–17,400 individuals. Numbers wintering in Britain have remained fairly constant over the last 30 years, but the *Britain and Ireland Atlas 2007–11* showed a noticeable easterly shift since the *Britain and Ireland Winter Atlas 1981–84*, which is possibly climate related, with fewer harsh winters.

Wintering birds in Wales appear to come mainly from Scandinavia and northern England: of five foreign-ringed birds caught on the North Wales coast during 1973–2012, four were from Finland, while birds caught in Wales that were ringed in Britain had come from north and east of the capture site. The most distant recovery of a Welsh-ringed bird was an adult ringed at Llanfairfechan, Caernarfonshire, in October 1982 that was found dead, south of Arkhangelsk, Russia, in May 1989, 2,628km to the northeast. It seems likely that Welsh breeding birds move south and west to winter.

A long-term Medium WeBS Alert has been issued for Wales. A long-term High Alert has been issued for the Burry Inlet SPA, together with a medium-term Medium Alert. Long- and medium-term Medium Alerts have also been issued for the Dee Estuary SPA (Woodward *et al.* 2019).

No site in Wales is of international importance. Traeth Lafan and the Cleddau Estuary are of British importance, as are the Dee and Severn estuaries as a whole. Other sites have held good numbers at times, such as 1,680 at the Inland Sea and Alaw Estuary, Anglesey, in February 1980. Sites that regularly have several hundred Curlews in winter include the Dyfi Estuary, Ceredigion/Meirionnydd, the Conwy Estuary, Caernarfonshire/Denbighshire, and the Inland Sea, Anglesey. Small groups are also found on open coast. Inland sites regularly used on passage include Dolydd Hafren, Montgomeryshire, and around Glasbury, Breconshire/Radnorshire.

Ringing shows that Curlews wintering in Wales are getting older. In 2010, the average age of birds recaught in winter was six years. By 2020, the average age was ten years, probably a result of poor breeding success and poor recent recruitment into the breeding population (Taylor 2019). Ringing highlights the site-faithfulness of wintering flocks: an adult ringed at Llanfairfechan, Caernarfonshire, in April 1982 was found dead at the same site in March 2003, nearly 23 years later. A Welsh breeding bird ringed in Lledrod, Ceredigion, was recorded in three successive winters in France. Other breeding birds from the Shropshire/Wales border have been recorded in winter in Devon, Cornwall and Ireland (Mid-Wales Ringing Group).

Curlew: Recovery locations of birds ringed in Wales are shown by red circles. Ringing locations of birds that were recovered in Wales are shown by blue triangles. Small brown dots show ringing or recovery locations in Wales.

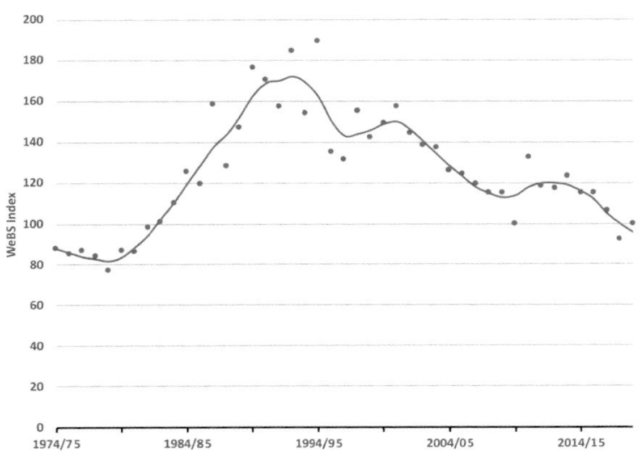

Curlew: Wetland Bird Survey indices, 1974/75 to 2018/19.

There are evident spring and autumn passages through Wales. Venables *et al.* (2008) reported large numbers on autumn passage (July–October) on the Gwent coast, with smaller numbers in spring. Jones and Whalley (2004) also noted a strong autumn passage on Anglesey from July, peaking in September, but with a less-distinct spring passage between March and May.

The main reason for the decline of the breeding population in Wales appears to be a lack of chick production. Curlews need to fledge 0.48–0.62 chicks per pair each year for population stability (Lindley 2018). In one recent study in Shropshire, on the border with Montgomeryshire, no chicks survived to fledging from 13 pairs (Tony Cross pers. comm.). The two key reasons for the low or nil productivity appear to be predation of eggs and chicks, and destruction of nests by silage operations. With the move from hay production to baled silage that is cut in May, it appears that the Curlew's move to the lowlands was not a good evolutionary choice. Studies at Lake Vyrnwy indicate that predators are mainly Fox, Badger and Carrion Crow, although sheep can also predate nests (Fisher and Walker 2015). There is no evidence that adult survival has declined (Johnstone *et al.* 2007). Maintaining an optimum level of grazing may be important in the Welsh uplands; a study on Mynydd Hiraethog (Johnstone *et al.* 2017) found that Curlew density was highest in squares with lower vegetation density. It suggested that habitat conditions for Curlew here could be improved by targeted increases in grazing.

Initial results from studies in North Wales show that Curlew do not keep to distinct territories but move over large areas of the landscape, overlapping with other pairs and favouring different feeding and nesting areas (Bladwell *et al.* 2018). Interventions are currently being trialled in North Wales. In Shropshire, interventions being trialled include the use of electric fencing around nests to

Site	County	74/75 to 78/79	79/80 to 83/84	84/85 to 88/89	89/90 to 93/94	94/95 to 98/99	99/00 to 03/04	04/05 to 08/09	09/10 to 13/14	14/15 to 18/19
Dee Estuary: Welsh side (whole estuary)	Flintshire	na (4,982)	na (3,293)	na (4,763)	1,485 (4,674)	1,673 (5,118)	1,645 (4,690)	1,291 (5,224)	1,558 (3,778)	1,554 (3,715)
Severn Estuary: Welsh side (whole estuary)	Gwent/East Glamorgan	482 (2,703)	780 (1,915)	1,271 (3,523)	998 (3,538)	821 (3,110)	867 (2,399)	1,210 (2,863)	1,179 (3,667)	1,023 (3,398)
Traeth Lafan	Caernarfonshire/ Anglesey	283	689	2,471	1,255	1,111	2,381	2,108	2,143	1,921
Burry Inlet	Gower/ Carmarthenshire	1,871	1,900	2,784	2,163	1,961	2,385	1,733	1,309	1,353
Cleddau Estuary	Pembrokeshire	693	714	1,399	1,653	1,534	1,213	1,433	1,684	954

Curlew: Five-year average peak counts during Wetland Bird Surveys between 1974/75 and 2018/19 at sites in Wales averaging 1,200 or more at least once. International importance threshold: 7,600; Great Britain importance threshold: 1,200.

reduce predation, persuading farmers to delay silage cutting when nests have been located and 'headstarting', whereby eggs are removed under licence, raised in captivity and then released into appropriate habitat once fully fledged. Early results show some success and that this could be an emergency option, but it is very resource intensive (Curlew Country 2020).

The Curlew is Red-listed in Wales for several reasons, including severe declines in the breeding population and range, and its Vulnerable status in Europe (Johnstone and Bladwell 2016). As Curlews are very site faithful on both breeding and wintering grounds, identifying and protecting these sites are vital to protect this iconic wader. It is clear that intensive intervention will be needed to maintain Curlew as a breeding species in Wales. Such intervention is almost certainly essential over the next decade or more, if extinction is to be avoided, but a sustainable future for the Curlew as a breeding bird in Wales will depend on land-management policies that support farmers undertaking management for this species. Its long-term future may be limited to fewer places in southern Britain as its breeding range is predicted to contract north during the 21st century (Huntley *et al.* 2007).

Mick Green

Sponsored by Rachel Taylor

Bar-tailed Godwit *Limosa lapponica* Rhostog Gynffonfraith

Welsh List Category	IUCN Red List			Birds of Conservation Concern			
	Global	Europe	GB	UK	Wales		
A	NT	LC	LC	Amber	2002	2010	2016

The Bar-tailed Godwit breeds in Arctic and subarctic habitats in a discontinuous belt from northern Norway through Siberia to western Alaska (*HBW*). The population breeding from Norway to western Siberia is thought to winter mainly in western Europe, while birds breeding farther east in Siberia, around the Taimyr and Yamal Peninsulas, spend the winter in West Africa, passing through western Europe on migration (*BTO Migration Atlas* 2002). Breeding trends for the population wintering in western Europe are unknown, but substantial declines in the populations that breed in eastern Siberia and winter from India to Australasia led to the species' Near Threatened global status (Birdlife International 2020).

This species is less likely to be found inland in Britain than Black-tailed Godwit, but uses a wider range of coastal habitats including sandy shores, as well as muddy estuaries. There have been few ringing recoveries to cast light on the origins of birds wintering in or passing through Wales, but three birds ringed in Norway have been recovered in Wales, all between October and January.

In Pembrokeshire, Mathew (1894) described the Bar-tailed Godwit as an autumn visitor, also seen occasionally in spring, and Lockley *et al.* (1949) described it in the same terms. Forrest (1907) also said that it was mainly an autumn visitor in North Wales, with smaller numbers seen in spring. In Glamorgan, it was described in the late 1890s as common in autumn and winter, but numbers there declined and by 1925 it was considered to be an infrequent spring and autumn passage visitor (Hurford and Lansdown 1995). By the 1950s larger numbers of wintering Bar-tailed Godwits were being recorded, with *c.*50 present in Swansea Bay, Gower, throughout winter 1953/54. In the Burry Inlet, Gower/Carmarthenshire, the average number of wintering birds rose from 55 in the early 1960s to over 500 by the 1970s. The maximum counts were 1,090 in February 1982 and a flock of *c.*2,000 seen flying south over the Burry Inlet towards Gower on 28 December 1974.

Male Bar-tailed Godwit in breeding plumage © John Marsh

No site wholly in Wales is of British importance for the species, although the whole Dee Estuary qualifies. This site has seen some remarkable fluctuations in numbers. From a peak of 11,149 in February 1977, counts declined to just 32 in winter 1984/85, thought to be the result of recreational disturbance at high-tide roosts that caused the birds to relocate to the Alt Estuary, Lancashire (Mitchell *et al.* 1988). Numbers on the Dee recovered to reach 8,430 in January 1997, fell to 232 in winter 1999/2000, rose again to 12,130 in January 2002 and fell to 140 in winter 2015/16. *Birds in England* (2005) mentioned a count of 18,775 there in February 1992. A WeBS long-term High Alert (-50%) and a short-term High Alert (-51%) have been issued for Wales. A long-term

Site	County	74/75 to 78/79	79/80 to 83/84	84/85 to 88/89	89/90 to 93/94	94/95 to 98/99	99/00 to 03/04	04/05 to 08/09	09/10 to 13/14	14/15 to 18/19
Dee Estuary: Welsh side (whole estuary)	Flintshire	na (7,337)	na (2,498)	na (290)	20 985	88 (5,751)	206 (2,938)	35 (986)	116 (667)	63 (749)
Burry Inlet	Gower/ Carmarthenshire	628	583	433	349	255	160	84	132	314
Inland Sea and Alaw Estuary	Anglesey	276	342	315	225	156	149	71	216	109
Carmarthen Bay	Carmarthenshire/ Pembrokeshire	89	8	164	227	138	225	101	81	20
Swansea Bay	Gower	302	193	143	88	24	15	8	27	12

Bar-tailed Godwit: Five-year average peak counts during Wetland Bird Surveys between 1974/75 and 2018/19 at sites in Wales averaging 150 or more at least once. International importance threshold: 1,500; Great Britain importance threshold: 500.

High Alert and a short-term Medium Alert have been issued for the Dee Estuary SPA (Woodward *et al.* 2019). This species is scarce away from estuaries, with only 12 recorded by the WeBS Non-estuarine Waterbird Survey (NeWS) in 2015/16.

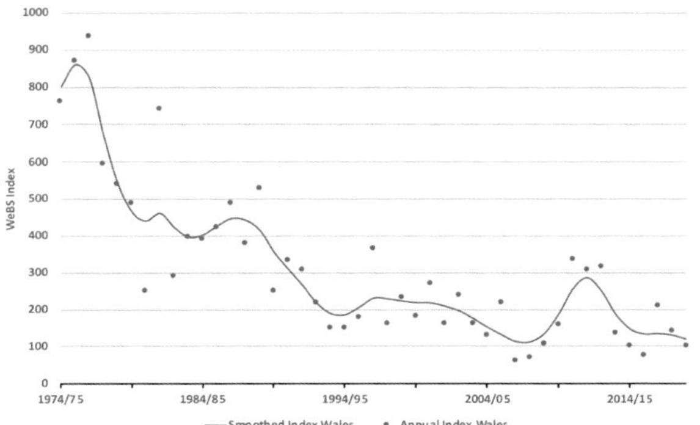

Bar-tailed Godwit: Wetland Bird Survey indices, 1974/75 to 2018/19.

Spring passage, in late April and May, has been more obvious than autumn passage in recent years. There was an exceptional count of 870 on the Dyfi Estuary, Ceredigion, on 4 May 1990, with 600 there and another 300 at nearby Borth on 30 April 2007. Autumn passage starts in late July and continues into October. The WeBS monthly index for Wales shows a pronounced peak in August, yet there have been few large counts from this month, other than 631 in the Burry Inlet in August 2016. Passage off Strumble Head, Pembrokeshire, usually peaks in September. This species is scarce inland, but not unknown. For example, there were ten records in Radnorshire between 1971 and 2012, with the highest count of six in a flooded field near Clyro on 11 May 2012 (Jennings 2014), and Breconshire had *c.*19 records up to 2017. Most inland records are during spring migration, though there have been a few in autumn.

There do not appear to be any pressing conservation issues affecting the species in Wales, which holds only about 1% of the birds wintering in Britain. The WeBS index for England shows no real trend in numbers since the 1970s, while in Scotland numbers increased until the mid-1990s then declined, returning to the levels recorded in the 1970s. WeBS counts showed a consistent increase in eastern Britain and decreases in the west (Austin *et al.* 2000). The decline in the index for Wales reflects this. The trend may be a result of milder winters making it unnecessary for birds to move west. The Bar-tailed Godwit's breeding areas are likely to move east and be squeezed against the north Siberian coast as a result of climate change (Huntley *et al.* 2007), potentially further reducing the number likely to occur in Wales.

Rhion Pritchard

Black-tailed Godwit *Limosa limosa* Rhostog Gynffonddu

Welsh List Category	IUCN Red List			Birds of Conservation Concern			
	Global	Europe	GB	UK	Wales		
					2002	2010	2016
A	NT	VU	EN(B)	Red			

This species has a wide breeding distribution in Europe and Asia, from Iceland to China. It winters to the south of its breeding range. Of the three subspecies, two breed in Britain. The nominate subspecies *L.l. limosa* breeds from western Europe, including parts of eastern England, to Russia, as far east as the Yenisei River, with by far the largest numbers in The Netherlands. However, this population is Near Threatened (BirdLife International 2020), with losses averaging 5% per year at the western edge of its range. Birds of this subspecies mainly winter in freshwater habitats in Africa south of the Sahara (*European Atlas* 1997). Most of the birds that winter in Wales or move through on passage are of the subspecies *L.l. islandica*, which breeds in Iceland, the

Black-tailed Godwits © Bob Garrett

Site	County	79/80 to 83/84	84/85 to 88/89	89/90 to 93/94	94/95 to 98/99	99/00 to 03/04	04/05 to 08/09	09/10 to 13/14	14/15 to 18/19
Dee Estuary: Welsh side (whole estuary)	Flintshire	na (961)	na (505)	na (1,721)	1,372 (1,747)	3,089 (3,529)	4,008 (4,949)	4,824 (5,857)	2,914 (6,183)
Burry Inlet	Gower/ Carmarthenshire	36	42	79	192	105	480	422	1,370
Severn Estuary: Welsh side (whole estuary)	Gwent/East Glamorgan	3 (48)	6 (57)	15 (32)	39 (117)	89 (239)	263 (410)	444 (591)	335 (863)
Cleddau Estuary	Pembrokeshire	3	3	1	13	18	17	73	104
Carmarthen Bay	Carmarthenshire/ Pembrokeshire	4	4	8	8	75	169	158	96

Black-tailed Godwit: Five-year average peak counts during Wetland Bird Surveys between 1979/80 and 2018/19 at sites in Wales averaging 100 or more at least once. International importance threshold: 1,100; Great Britain importance threshold: 390.

Faroes and neighbouring areas, including a few pairs in Orkney and Shetland. These winter mainly in estuarine habitats around Britain and France, some as far south as Spain and Portugal. This subspecies has shown a substantial increase in numbers for over a century, probably as a result of warming conditions in Iceland (Gill *et al.* 2019). Colour-ringing has shown that some individuals colour-ringed in eastern or southern England during the late-summer moult have spent time in Wales later the same year, particularly in the Dee Estuary. Birds sometimes return to the same stop-over sites in subsequent years. Several birds, colour-ringed as nestlings in Iceland, have been seen in Wales, and one ringed in winter in Wales was recovered in the breeding season in Iceland.

Birds in Wales (1994) mentioned three records of possible breeding, including probable breeding in southwest Anglesey during fieldwork for the *Britain and Ireland Atlas 1968–72*, but no further details were available. The other two records appear to have more substance. In May 1938, a pair at Morfa Pools, Glamorgan, behaved like a pair of Redshanks with young, but the nature of the site made it impossible to confirm breeding, and the birds had gone when the site was visited again in June (Hurford and Lansdown 1995). The third record was in Caernarfonshire, at Afonwen, where an egg-collector claimed to have found a nest in the late 1940s. Dixon (2012) found that C.H. Gowland, an egg-collector and dealer from Wirral, had described what must have been the same event in a self-published periodical, *Birdland.* Gowland saw what appeared to be a pair at a marshy site in Llŷn on 29 May 1950 and found a nest with hatching eggs on 9 June. By 19 June, all four eggs had hatched and the chicks had left the nest. Gowland did not name the site, but his description of marshland and a small lake close to a shingle beach makes it likely to have been Afonwen. This provided good evidence that the species has bred in Wales. There has been no recent record indicating breeding, but a bird was seen in display flight at Newport Wetlands Reserve (Goldcliff), Gwent, in 2003 (Venables *et al.* 2008).

As a passage and wintering species, Black-tailed Godwit was quite a scarce bird in Wales in the late 19th and early 20th centuries. Mathew (1894) described it as an occasional and rare autumn visitor to Pembrokeshire. The first record in Glamorgan was in 1821, but the second did not come until September 1916 (Hurford and Lansdown 1995). In North Wales, Forrest (1907) described it as a "somewhat rare" autumn visitor to the estuaries of the Dee, Dyfi and Mawddach, although he noted that one had been 'obtained' at Llyn Maelog, Anglesey, in May 1886. The area occupied by breeding birds in Iceland expanded considerably during the 20th century (Appleton 2015), and numbers in Wales began to increase around the 1930s. By the 1960s, large numbers were reported at some sites during migration: 500 in Bangor harbour, Caernarfonshire, on 14 September 1966 and 500 in the Burry Inlet, Gower/Carmarthenshire, on 7 September 1968, for example. Flocks of 100 or more were recorded in the Dee Estuary by this time, and a large wintering flock built up there from the late 1960s. Counts specifically for the Welsh side are not available for this period, but by the 1990s the majority of the birds on the Dee tended to be on the Welsh side of the estuary, mainly feeding on mudflats near Flint, particularly Bagillt Bank, and roosting at RSPB Oakenholt Marsh at high tide. More recently, the highest numbers have been on the Wirral side of the estuary, but the birds use the estuary as one site, depending on tides, feeding areas and the availability of disturbance-free roost sites (David Parker *in litt.*). The Dee Estuary as a whole is the third most important site for the species in Britain.

Wetland Bird Survey (WeBS) data from the main sites show a dramatic increase in numbers at several sites from the late 1980s, thought to be linked to increases in the Icelandic breeding population.

The Welsh side of the Dee Estuary and the Burry Inlet are currently of international importance for this species, as is the entire Severn Estuary. Good numbers have sometimes been recorded at other sites: for example, 451 at Angle harbour, Pembrokeshire, in December 2018; 129 at Malltraeth Cob pool, Anglesey, in January 2005; and 120 at Porthmadog, Caernarfonshire, in November 2010. Short-term WeBS High Alerts were issued for Wales and specifically for the Burry Inlet, following a 56% decline in numbers between 2005/06 and 2009/10, but numbers at this site subsequently recovered, with a record count of 1,849 in December 2016. The WeBS index for Wales declined sharply from 2013/14, following a period of rapid increase, and a short-term Medium WeBS Alert (-27%) has been issued (Woodward *et al.* 2019).

Numbers now usually peak in November, although good numbers are also recorded on autumn passage between August and October and on spring passage between late March and June. Inland, the species is reported fairly regularly in small numbers: there were 16 records in Radnorshire up to 2013, for example, with a maximum count of eight (Jennings 2014). The largest inland counts appear to be at least 25 at Llangorse Lake, Breconshire, on 12 August 2003 and 51 at Llyn Cefni, Anglesey, in September 2018.

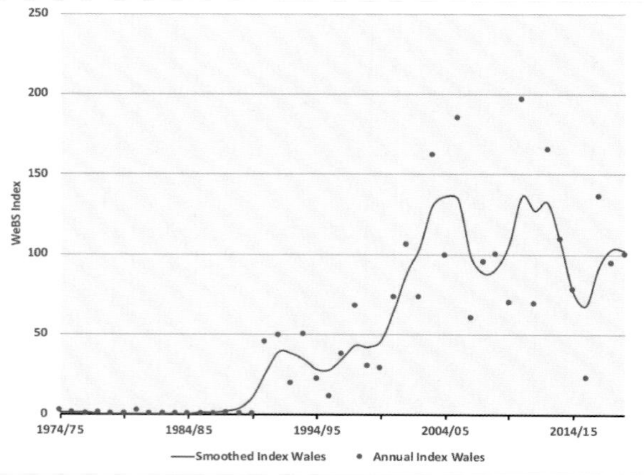

Black-tailed Godwit: Wetland Bird Survey indices, 1974/75 to 2018/19.

There has been no indication of even possible breeding in Wales in recent years. The fortunes of the Black-tailed Godwit as a passage and wintering species in Wales depend largely on the success of the Icelandic breeding population, for which some sites in Wales are of considerable importance in winter. The long-term outlook for this population is uncertain, however, as the area of Iceland that is climatically suitable is set to reduce during the 21st century (Huntley *et al.* 2007). Much may depend on whether habitat in Greenland becomes more suitable for nesting.

L.l. limosa

The occurrence of the nominate subspecies in Wales was confirmed by a bird ringed as a chick in Zwolle, The Netherlands, in 2010 that was present at Penclacwydd, Carmarthenshire, from 1 June to 1 August 2011. Any records of Black-tailed Godwits breeding in Wales would probably also involve birds of the subspecies *L.l. limosa*.

Rhion Pritchard

Turnstone *Arenaria interpres* Cwtiad Traeth

Welsh List Category	IUCN Red List			Birds of Conservation Concern			
	Global	Europe	GB	UK	Wales		
A	LC	LC	VU	Amber	2002	2010	2016

Mainly a High Arctic breeder, this wader has a circumpolar distribution, nesting on tundra in the far north of North America, Greenland, Europe and Asia, usually near water (*HBW*). It breeds farther north than almost any other land bird, but in Europe it was confirmed breeding as far south as Denmark during the latest atlas period, 2014–2017 (*Dansk Ornitologisk Forening* 2020). The global population is thought to be decreasing, although little information is available from many of its Arctic breeding grounds (BirdLife International 2020). It winters to the south of its breeding range and, unlike many wader species, is often found on rocky shores, though it also feeds on the strandline of sandy beaches. The majority of the birds wintering in Britain are thought to belong to the population breeding in Greenland and Arctic Canada. Several ringing recoveries show birds present in the breeding season in Greenland and on Ellesmere Island in Arctic Canada and wintering in Wales. Several other recoveries are likely to be birds breeding in those areas, using Iceland as a staging post. Several birds ringed in their first calendar-year in Norway, Sweden or Finland have been found in Wales, but mainly in autumn and spring: one ringed at Nidingen Bird Observatory in southern Sweden in late August 1983 was recorded at Rhos-on-sea, on the Caernarfonshire/Denbighshire border, in three different winters. Some birds from the Canada/Greenland population are known to stop over in southwest Norway in autumn before wintering in Britain (*BTO Migration Atlas* 2002). Some individuals ringed in Wales during migration have been recovered farther south, including one in The Gambia. Many wintering birds apparently return to the same area every winter: of 243 caught at Rhos Point, Caernarfonshire, by SCAN Ringing Group in 1980–82, 187 (77%) were re-traps of birds ringed there previously (Dodd and Moss 1983). Two ringed at Rhos-on-sea were recaught at the same site just over 20 years later, making them the two oldest Turnstones known to the BTO ringing scheme. Turnstones are also recorded in Wales on autumn and spring migration, with records from July to May, and even a few in June.

Turnstone: Recovery locations of birds ringed in Wales are shown by red circles. Ringing locations of birds that were recovered in Wales are shown by blue triangles. Small brown dots show ringing or recovery locations in Wales.

Site	County	74/75 to 78/79	79/80 to 83/84	84/85 to 88/89	89/90 to 93/94	94/95 to 98/99	99/00 to 03/04	04/05 to 08/09	09/10 to 13/14	14/15 to 18/19
Dee Estuary: Welsh side (whole estuary)	Flintshire	na (409)	na (193)	na (802)	52 (875)	72 (927)	60 (594)	37 (375)	60 (259)	69 (229)
Traeth Lafan	Caernarfonshire/ Anglesey	275	161	129	74	37	68	179	200	179
Colwyn Bay and North Clwyd coast	Denbighshire/Flintshire			217	305	144	83	95	79	146
Swansea Bay	East Glamorgan/Gower	221	259	382	119	50	15	42	53	54
Burry Inlet	Gower/Carmarthenshire	744	720	741	444	249	206	240	10	32

Turnstone: Five-year average peak counts during Wetland Bird Surveys between 1974/75 and 2018/19 at sites in Wales averaging 200 or more at least once. International importance threshold: 1,400; Great Britain importance theshold: 400.

The species may have been less common in the 19th century. Eyton (1838) said that it was "common on the Anglesea coast", but Mathew (1894) described it as "an autumn visitor to the coast; rather rare" in Pembrokeshire, and Salter (1900) said that it was only a passage migrant in the Aberystwyth area of Ceredigion. Forrest (1907) described it as a regular visitor to North Wales in autumn and spring, usually in small parties. Numbers increased in the second half of the 20th century, with 700 recorded at Whiteford, Gower, on 1 September 1963. They peaked in the 1970s, with 600 at Undy in Gwent in May 1971 and 710 at Rhos Point, Caernarfonshire, on 10 November 1979, of which 660 were caught by SCAN Ringing Group (Dodd and Moss 1983). The largest count in Wales was on spring migration: 1,346 at Whiteford in the Burry Inlet on 6 May 1978.

During the mid-1950s the power station at Burry Port, Carmarthenshire, began to discharge warm water from the cooling systems into the Burry Inlet. This led to a marked increase in the food supply and passage numbers along the rocky foreshore increased dramatically in the 1960s and the early 1970s, with over 1,000 there in September 1971, August 1972 and September 1974. Even after the closure of the power station in 1984 large numbers were recorded, including 1,100 on 17 September 1989. However, changes in the marine processes resulted in the formation of a sandy beach instead of the previous rocky shoreline, and there has been only one three-figure count at this site in the last six years (Rob Hunt *in litt.*). WeBS High Alerts for the long term (-89%) and the medium term (-78%) have been issued for the Burry Inlet SPA (Woodward *et al.* 2019).

Male Turnstone in breeding plumage © Ben Porter

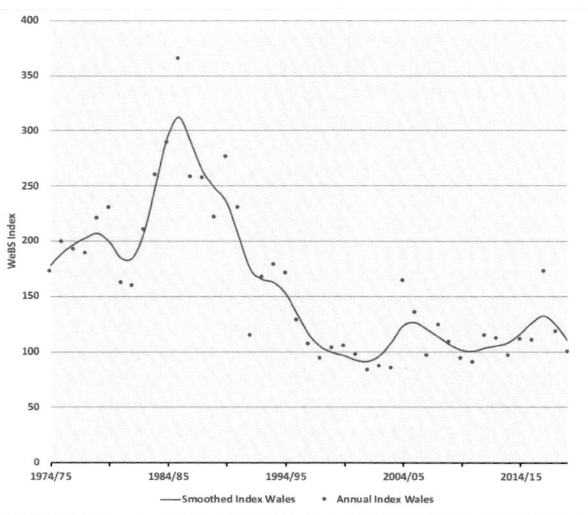

Turnstone: Wetland Bird Survey indices, 1974/75 to 2018/19.

No site in Wales now reaches the threshold of 480 for British importance, although the whole Severn Estuary qualifies. Sites in North Wales do not show the sharp declines recorded at southern sites. Large counts in recent years include over 300 at Kinmel Bay, Denbighshire, on 20 August 2014 with 338 there on 18 October 2015 and *c.*275 in the lower Conwy estuary, Caernarfonshire, in December 2016.

Many Turnstones are found along open coasts, most of which are not included in the WeBS counts. The 2015/16 Non-estuarine Waterbird Survey (NeWS) found 1,008 in Wales, of which 405 were on Anglesey. This was a considerable increase on 766 in 2006/07 and 641 in 1997/98, but much lower than the 1,989 recorded in 1984/85 (Austin *et al.* 2017). Around 2,000 Turnstones probably winter in Wales. Away from the coast, this species is scarce, with only 1–2 records each year during spring or autumn migration. 20 at Rhaslas Pond, East Glamorgan, on 15 May 2013 was exceptional.

This species does not appear to face any specific conservation problems in Wales, particularly since the population is dispersed along the Welsh coast, rather than being concentrated at a few sites. Its future in Wales will depend mainly on its breeding success farther north. The Fennoscandian breeding population is expected to move north during the 21st century (Huntley *et al.* 2007), but no equivalent modelling has been undertaken in Canada, where Welsh wintering Turnstones nest.

Rhion Pritchard

Sponsored by SCAN Ringing Group

Knot *Calidris canutus* Pibydd yr Aber

Welsh List Category	IUCN Red List			Birds of Conservation Concern			
	Global	Europe	GB	UK	Wales		
A	NT	LC	LC	Amber	2002	2010	2016

The Knot breeds on the Arctic tundra. There are five subspecies, but almost all those wintering in Britain and Ireland are of the subspecies *C.c. islandica*, which breeds in northern Greenland and islands in the Canadian Arctic (*BTO Migration Atlas* 2002). Birds of the nominate subspecies, *C.c. canutus*, which breed mainly around the Taymyr Peninsula in Siberia and winter in Africa, have been recorded on the east coast of England, but not confirmed in Wales. In winter the Knot is mainly a bird of muddy estuaries, and is one of the most gregarious wader species, so is well covered by WeBS counts. There have been many ringing recoveries involving Wales, including two birds ringed at Point of Ayr, Flintshire, which were later recovered in Greenland. There have been many more recoveries involving Iceland, which is a very important staging post, particularly in spring. Norway also appears to be used as a staging area. There have also been a good number of recoveries in The Netherlands and it appears that following the breeding season, birds congregate at a few moulting areas, particularly the Wadden Sea (The Netherlands/Germany), but also at some sites in Britain including the Dee Estuary. Following the moult, they disperse to their wintering grounds (*BTO Migration Atlas* 2002).

Knot in winter plumage © Ben Porter

Numbers may have been lower in the 19th and early 20th centuries. The species was described by Dillwyn (1848) as uncommon in the Swansea area, and by the end of the 19th century was said to winter in small numbers on the Cardiff mudflats (Hurford and Lansdown 1995). Salter (1900) described it as an autumn visitor to the Dyfi estuary, Ceredigion. Forrest (1907) said that it was "by no means uncommon" in North Wales in autumn and spring, with flocks reaching 200 or more. He recorded *c.*3,000 on the Dee Estuary on 31 December 1905 and quoted Rawlings as saying there were "vast flocks" around Barmouth, Meirionnydd, in some winters. Forrest knew of no records on Anglesey, where the first record was not until 1912. Elsewhere numbers varied between years, with none in some years. Flocks of thousands became normal on the Dee Estuary from about the 1930s. Large numbers were seen in the Burry Inlet, Gower/Carmarthenshire, in September 1948, and both that site and the Taff/Ely estuary, East Glamorgan, had sizeable wintering populations in the mid-1960s, with flocks of up to 3,000 not uncommon in the Burry Inlet. The Taff/Ely estuary held *c.*4,500 in November 1970, while in Gwent there were over 10,000 at Peterstone in February 1969.

Knot tend to be associated with the larger estuaries and bays in Wales and are less frequently seen at smaller sites. In Pembrokeshire, for example, up to 20, but occasionally 100, are recorded in winter, but seldom stay long (Annie Haycock *in litt.*). Large numbers on less-favoured estuaries are often the result of cold-weather movements, though sometimes, for example during the winter of 2019/20, they are the result of an excellent breeding season, with flocks containing large numbers of juveniles (Steve Dodd *in litt.*).

Good numbers have been recorded in certain years at some sites that in other years may hold very few, or none. On Traeth Lafan, Caernarfonshire, there were 2,000 on 11 January 1969 and 2,028 in February 1991, and there were 2,000 in the Cefni Estuary, Anglesey, in January 1970. Numbers are usually lower in Cardigan Bay, the highest count being 300 in the Dyfi Estuary on 16 January 2006.

Knot: Recovery locations of birds ringed in Wales are shown by red circles. Ringing locations of birds that were recovered in Wales are shown by blue triangles. Small brown dots show ringing or recovery locations in Wales.

Site	County	74/75 to 78/79	79/80 to 83/84	84/85 to 88/89	89/90 to 93/94	94/95 to 98/99	99/00 to 03/04	04/05 to 08/09	09/10 to 13/14	14/15 to 18/19
Dee Estuary: Welsh side (whole estuary)	Flintshire	na (39,632)	na (30,789)	na (17,178)	1,101 (26,414)	764 (17,906)	881 (26,397)	2,011 (14,777)	9,562 (25,590)	1,215 (15,737)
Burry Inlet	Gower/ Carmarthenshire	2,969	4,348	4,642	1,309	4,501	3,532	5,332	1,830	5,529
Severn Estuary: Welsh side (whole estuary)	Gwent/East Glamorgan	5,676 (5,737)	3,603 (3,649)	1,261 (2,403)	1,818 (3,063)	687 (1,878)	154 (1,805)	2,396 (3,300)	1,836 (1,908)	225 (2,133)
Carmarthen Bay	Carmarthenshire/ Pembrokeshire	118	64	73	1,338	404	395	3,175	2,828	1,899

Knot: Five-year average peak counts during Wetland Bird Surveys between 1974/75 and 2018/19 at sites in Wales averaging 2,600 or more at least once. International importance threshold: 5,300; Great Britain importance threshold: 2,600.

WeBS High Alerts have been issued for Wales for the medium term (-50%) and the short-term (-55%). The Burry Inlet is still of British importance, but no longer holds internationally important numbers. A medium-term High Alert has been issued for the Burry Inlet SPA, with Medium Alerts for the long term and short term (Woodward *et al.* 2019).

On the Dee Estuary, numbers declined by 79%, from a peak count of 48,000 in 1979/80 to 10,150 in 1985/86. A similar decline was seen in Bar-tailed Godwit numbers. It was thought that disturbance at their high-tide roosts had led the birds to move to the Alt Estuary (Mitchell *et al.* 1988). Numbers on the Dee subsequently recovered. The average counts on the Welsh side qualify as internationally important in their own right in some years: in December 2012, the peak count for that winter on the Welsh side was 38,101, out of 50,266 for the whole estuary. However, the 2015/16 peak winter count for the whole Dee was only 8,000 (also in December), with just 700 on the Welsh side. These large concentrations of birds use the whole estuary to feed and roost, depending on tides, disturbance and quality of feeding grounds. There is some evidence of a ten-year cycle on the Dee, although the numbers are highly variable (David Parker *in litt.*). The WeBS index for Wales shows a great deal of annual fluctuation, largely as a result of the variations on the Dee. Long-, medium- and short-term Medium Alerts have been issued for the Dee Estuary SPA (Woodward *et al.* 2019). Knots are scarce inland: there have been just two records from Radnorshire, and 15 records of parties of 1–15 birds over the last 60 years in Breconshire (Andrew King *in litt.*).

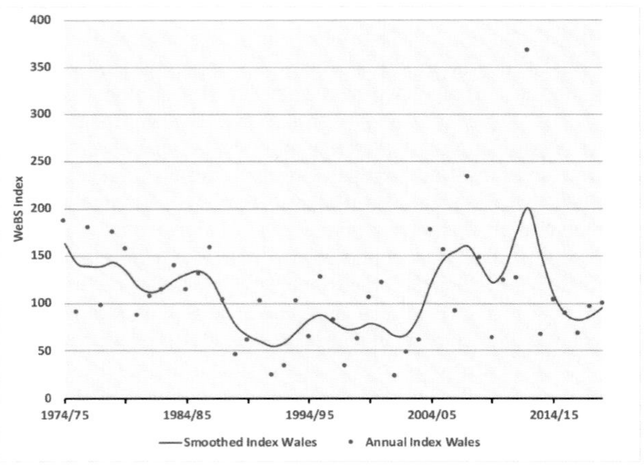

Knot: Wetland Bird Survey indices, 1974/75 to 2018/19.

Van Gils *et al.* (2016) found that earlier snowmelt in the Arctic breeding range of *C.c. canutus* led to a timing mismatch of hatching with peak food abundance, which inhibited the growth rate of chicks. This ultimately reduced survival on their wintering grounds in Africa, resulting in an overall population decline. It is not known whether the same is true of the *islandica* subspecies, which showed a moderate decrease in population size during 2003–12 (BirdLife International 2020).

Rhion Pritchard

Ruff *Calidris pugnax* Pibydd Torchog

Welsh List Category	IUCN Red List			Birds of Conservation Concern			
	Global	Europe	GB	UK	Wales		
A	LC	LC	CR(B)	Red	2002	2010	2016

Ruffs breed in northern Europe and Asia, from Britain to the Bering Strait, mainly above 60°N, on tundra, moorland bordered by pine forest, or farther south on wet low-lying grassland. The bulk of the population breeds in Russia, with most of the European population outside Russia in Fennoscandia (*HBW*). The Russian population is fluctuating, but there have been substantial declines in several other countries, such as Sweden and Finland, with a moderate decline in the European breeding population overall (BirdLife International 2015). In winter, most migrate to Africa south of the Sahara, but some winter in northwest Europe. Few ringing recoveries involve Wales, but one, ringed at Shotton, Flintshire, in August 1962 was found dead in Sweden in May 1966. A bird ringed in The Netherlands, in July 1967, was found dead at Haverfordwest, Pembrokeshire, in November the same year. Another bird ringed at Shotton in August 1962 was shot in Morocco in September the same year.

In Wales the Ruff is mainly a passage migrant, but a few birds winter, preferring areas of soft mud on the edges of freshwater lakes and pools and wet, short-grazed grassland. Although this was once a widespread breeding species in England (*Birds in England* 2005) and small numbers still breed in the East Anglian fens, there has never been a record of confirmed breeding in Wales. A male and two females, at a site in southwest Anglesey in 1970, were thought by the observer to have been probably breeding (*Cambrian Bird Report*), but no further details have been found. Birds on passage in May will sometimes display. For example, a male with developed ruff and wattles displayed to five reeves at Brechfa Pool, Breconshire, on 15 May 2004.

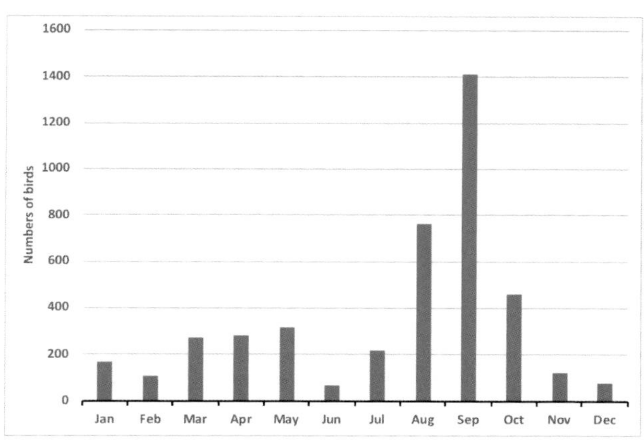

Ruff: Monthly totals in Wales, 1992–2019.

The species seems to have been genuinely rare in Wales in the late 19th and early 20th centuries. There were only four records in Gwent prior to 1963, and although one was recorded in Glamorgan in 1819, the next was not until 1930 (Hurford and Lansdown 1995). Mathew (1894) knew of only a single Pembrokeshire record, one shot "many years ago" at Cuffern, while the first Carmarthenshire record was not until 1931. In North Wales, Forrest (1907) described it as an occasional visitor in spring and autumn, mainly to the Dee and Dyfi estuaries.

By the 1970s there was a substantial increase in numbers on passage, including 35 at Shotton on the Dee Estuary, Flintshire, in autumn 1974 and 30 at Llyn Alaw, Anglesey, in August 1977 and August 1978. There are usually higher numbers in autumn than spring, but the highest counts in Wales have been on north-bound passage. Spring migration was particularly strong in 1987, including 70 at Morfa Coedbach, Cydweli, Carmarthenshire, on 18 April and 46 at Marloes Mere, Pembrokeshire, in late April. A count of at least 85 at Cors Fochno/Dyfi Estuary, Ceredigion, on 1 May 1994, part of an influx into Britain, remains a record for Wales. Spring passage starts earlier than for most waders, with birds moving through from mid-March, for example 12 at William Condry Reserve, Montgomeryshire, on 19 March 2004. The highest total on autumn passage was 54 at RSPB Cors Ddyga, Anglesey, in October 2017.

Wintering Ruff were very scarce in Wales before the 1960s, though there was a record in Glamorgan in 1937. During 1960–65, 1–3 were recorded annually in Wales each winter (Prater 1973), part of a trend of increasing winter numbers in Britain. By 1966–71 the winter average in Wales was 26, a very small proportion of the estimated 1,200 wintering in Britain and Ireland. Shotton Marshes, Flintshire, produced the largest number of records: in 1972, up to 22 wintered in Flintshire and 11 in Glamorgan. Wintering numbers have been lower subsequently, averaging 7–8 birds in Wales. The highest winter count in recent years was nine at Castlemartin Corse, Pembrokeshire, on 17 January 2011.

The Ruff looks likely to continue to be a regular passage migrant in Wales, and may benefit from the creation of new wetland sites. However, this species is not doing well in Europe. The second *European Atlas* (2020) showed massive range losses in the southern and central parts of its range. Strong declines in Estonia, Finland and Sweden suggest a steep decline in Baltic breeding numbers since the 1980s. If the European breeding population continues to decline, we may see fewer birds passing through Wales.

Rhion Pritchard

Broad-billed Sandpiper *Calidris falcinellus* **Pibydd Llydanbig**

Welsh List Category	IUCN Red List		Rarity recording
	Global	Europe	
A	LC	LC	BBRC

The Broad-billed Sandpiper breeds in northern Fennoscandia and Arctic Russia, and migrates through the eastern Mediterranean, Black and Caspian seas to winter around the Arabian Peninsula and Indian Ocean coasts (*HBW*). This is a regular visitor to Britain, with 272 individuals recorded to the end of 2019 (Holt *et al.* 2020). The numbers recorded in Britain have declined steadily since the 1990s, with an average of 4–5 each year over the past 30 years. Most of the birds recorded in Britain have been in May (Parkin and Knox 2010). It is a rare vagrant to Wales with only ten records of 11 individuals. The first Welsh record was in autumn, but all the subsequent records have been in spring:

- two at Shotton, Flintshire, on 22 September 1960;
- Peterstone Wentlooge, Gwent, on 7 May 1979;
- Malltraeth, Anglesey, on 4–6 June 1984;
- Peterstone Sluice Farm, Gwent, on 15 May 1988;
- RSPB Conwy, Denbighshire, on 24 May 1999;
- Newport Wetlands Reserve (Goldcliff), Gwent, on 6–7 May 2008;
- Newport Wetlands Reserve (Goldcliff), Gwent, on 22–23 April 2016;
- Newport Wetlands Reserve (Goldcliff), Gwent on 18–21 June 2016;
- Gronant and Talacre, Flintshire, on 11 and 22 June 2016 also at Kinmel Bay, Denbighshire, on 13–14 June 2016 and
- Llanfwrog and Alaw Estuary, Anglesey, on 1–2 May 2017.

Its European breeding range is expected to be substantially reduced as a result of climate change this century (Huntley *et al.* 2007) and, notwithstanding that the origin of Broad-billed Sandpipers visiting Wales is not known, it may become an even rarer visitor in future.

Jon Green and Robin Sandham

Sharp-tailed Sandpiper *Calidris acuminata* **Pibydd Cynffonfain**

Welsh List Category	IUCN Red List	Rarity recording
	Global	
A	LC	BBRC

This wader breeds in northeast Siberia and winters in Australasia (*HBW*). It is a rare vagrant to Wales with only three records. The first two, both first-calendar-year birds, were found on the same day, 14 October 1973: one at Shotwick, Flintshire, which remained until 25 October, and the other at Morfa Harlech, Meirionnydd, which remained until 15 October. The only other record was an adult at Morfa Dinlle and Foryd Bay, Caernarfonshire, on 25–28 August 1996.

Jon Green and Robin Sandham

Stilt Sandpiper *Calidris himantopus* Pibydd Hirgoes

Welsh List Category	IUCN Red List	Rarity recording
	Global	
A	LC	BBRC

This species breeds mainly in subarctic North America, from northeast Alaska to Hudson Bay, Canada. It migrates overland through the interior USA to winter in central South America (*HBW*). 39 individuals were recorded in Britain up to the end of 2019, on average about one every two years over the last 30 years (Holt *et al.* 2020). Only one of these was in Wales, a second-calendar-year bird at RSPB Conwy, Denbighshire, on 11–13 July 2006.

Jon Green and Robin Sandham

Curlew Sandpiper *Calidris ferruginea* Pibydd Cambig

Welsh List Category	IUCN Red List		Birds of Conservation Concern			
	Global	Europe	UK	Wales		
A	NT	VU	Amber	2002	2010	2016

Curlew Sandpipers breed on tundra along the Arctic coast of Siberia, and winter to the south, mainly along the coasts of sub-Saharan Africa, but also in India and as far south as Australia and New Zealand (*HBW*). They occur in Wales during migration. In spring, a few adults stop here on their way to their breeding grounds, but the largest numbers are seen in autumn. Those seen in August are mainly adults; the males leave the breeding grounds in July, followed later by the females. Larger numbers occur in September, almost all juveniles. There have been few ringing recoveries involving Wales, though one ringed at Angle Bay, Pembrokeshire, in September 1996 was recaught in Spain nearly 15 years later in August 2011, while a first-calendar-year bird ringed in Norway in late August 1991, was found dead near Tywyn, Meirionnydd, just over two weeks later. A first-calendar-year bird at Cei Cydweli, Carmarthenshire, on 17 September 2018 had been colour-ringed on the Vistula Estuary, Poland, 13 days earlier.

The Wadden Sea, off the coast of Denmark, Germany and The Netherlands, is an important staging area for some birds breeding in the western part of their range. Fairly large numbers reach the eastern coasts of Britain, but Wales lies outside the main migration route so numbers are usually lower. In some years, however, sizeable flocks are seen in autumn, dependent on weather conditions over Scandinavia during migration, and/or the number of young fledged that year. That in turn depends on the number of lemmings present on the Curlew Sandpiper's Arctic breeding grounds. If lemmings are numerous, predators such as the Arctic Fox concentrate almost entirely on them. When lemmings are few, the predators turn to wader eggs and chicks (*BTO Migration Atlas* 2002).

There is one archaeological record of Curlew Sandpiper from Wales: bones were found at Bacon Hole Cave, Gower, dating to the Ipswichian interglacial period, 75,000 to 128,000 years ago. Stringer (1977) considered that this could indicate that the species was breeding in Wales in conditions similar to those in the Arctic today. Mathew (1894) described it as an autumn visitor to Pembrokeshire, and Lockley *et al.* (1949) could only list four subsequent records in that county. Small numbers were said to winter regularly in the Cardiff estuaries in the late 1890s, but it was only recorded twice in Glamorgan over the following 50 years (Hurford and Lansdown 1995). In North Wales, Forrest (1907) described it as a passage migrant, not uncommon in autumn on the Dee marshes, but he quoted larger numbers on the Dyfi Estuary, Ceredigion. Salter said that it occurred there in varying numbers: 24 were said to have been shot in an afternoon, though a record he quoted from Fielden, of a flock of 20–30 on 10 February 1902, seems unlikely.

Numbers began to increase from the 1950s. There were 20 at Whiteford Sands, Gower, on 28 September 1952 and *c.*50 were attracted to the lighthouse on Bardsey, Caernarfonshire, on the night of 2/3 September 1954. Large numbers were recorded on the Dee Estuary near Shotton Steelworks in 1960, with about 80 on 22 September. There was an influx into Britain in autumn 1969 and, although numbers in Wales were lower than farther east, around 200 were present during 9–15 September (*Birds in Wales* 1994), including at least 70 at Malltraeth, Anglesey, on 10 September. Autumn 1988 also saw good numbers, with Malltraeth again recording the highest count: 66 on the Cob pool and the Cefni Estuary on 18 September.

First-calendar-year Curlew Sandpiper in autumn © Colin Richards

The late 1990s had several years of above-average occurrence. Autumn 1996 produced high counts including 44 at Malltraeth Cob pool and the Cefni Estuary on 22 September, 79 at Penclacwydd, Carmarthenshire, on 30 September, and a total of at least 112 recorded in Pembrokeshire. In autumn 1998, some 220 were recorded in Wales between 22 July and 27 November, the highest count being 58 at Penclacwydd and Penrhyngwyn, Carmarthenshire, on 12 September. There was an exceptional count of 100 on the Dyfi Estuary on 28 August 1999, but counts in the subsequent two decades have been lower. The only counts over 30 were 36 at Malltraeth Cob pool and the Cefni Estuary on 11 September 2005, and 31 at Cei Cydweli, Carmarthenshire, on 9 October 2017. Monthly WeBS counts do not cover this species well, as most stay at a site only for a few days, although 48 were recorded on a WeBS count in the Burry Inlet, Gower/Carmarthenshire, in September 1996.

Spring records, between late March and mid-June, are fewer and usually involve only 1–2 birds; the count of 13 at RSPB Ynys-hir, Ceredigion, on 18 May 1992 was exceptional. Winter records are very unusual in Wales. Between 1992 and 2019 there were just ten records, all singles, during December to February, although records in November and March may have been either wintering birds or passage migrants. Three of the ten records were on the Dee Estuary, Flintshire, and two at RSPB Conwy, Denbighshire. The species is also rare inland: one in the company of three Pectoral Sandpipers at Llangorse Lake, Breconshire, on 30 September 2008 was the first in the county since 1976, and there have been only two records in Radnorshire (Jennings 2014).

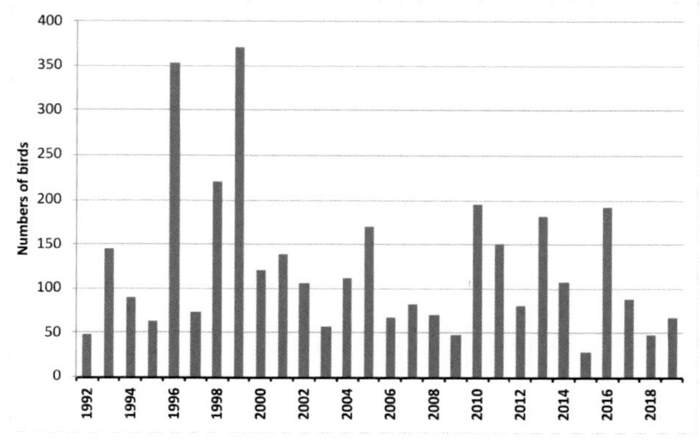

Curlew Sandpiper: Approximate annual totals in Wales, 1992–2019.

Rhion Pritchard

Temminck's Stint *Calidris temminckii* Pibydd Temminck

Welsh List Category	IUCN Red List		Rarity recording
	Global	Europe	
A	LC	LC	WBRC

The Temminck's Stint breeds from northern Fennoscandia east across northern Siberia, and winters in sub-Saharan Africa and southern Asia north of the equator (*HBW*). The numbers recorded in Britain have been stable for the last 30 years, at around 100 a year, while in Wales it is a scarce but regular passage migrant. Its European breeding range is expected to be substantially reduced as a result of climate change this century (Huntley *et al.* 2007) and, although the origin of the Temminck's Stints visiting Wales is not known, it may become an even rarer visitor in future.

Total	Pre-1950	1950–59	1960–69	1970–79	1980–89	1990–99	2000–09	2010–19
60	3	1	4	3	9	10	15	15

Temminck's Stint: Individuals recorded in Wales by decade.

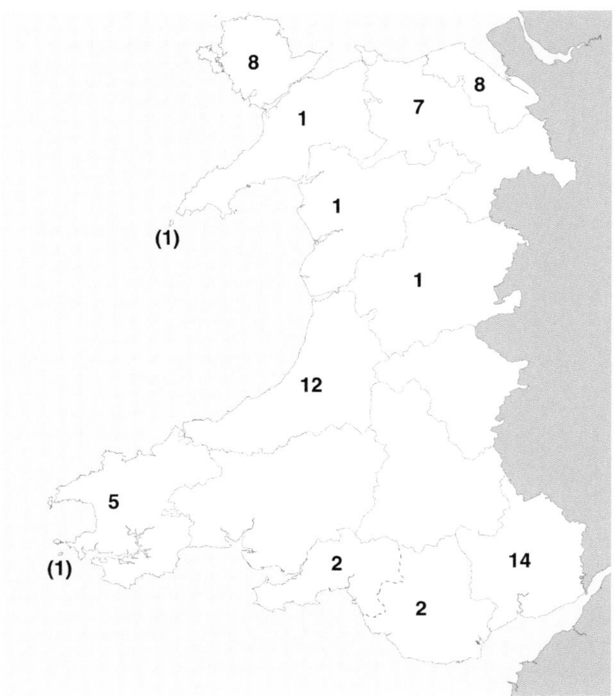

Temminck's Stint: Numbers of birds in each recording area 1945–2019. One bird was seen in more than one recording area.

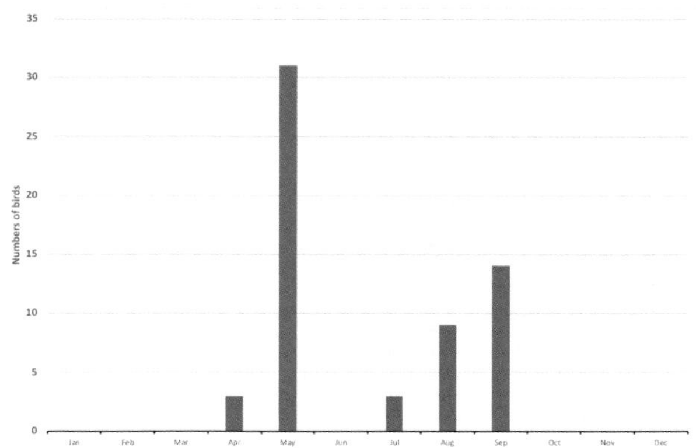

Temminck's Stint: Totals by month of arrival, 1945–2019.

The first Welsh record was two at Malltraeth, Anglesey, on 4–8 September 1945. Records are typically in May, and half have been at just seven sites. There have been ten at Newport Wetlands Reserve (Goldcliff), Gwent, including three on 13–14 May 2017, and seven at Malltraeth, Anglesey, but none since 1970. Elsewhere, there have been five at Cors Caron, Ceredigion, including three together on 12 May 2010, and five at RSPB Conwy, Denbighshire. In Flintshire, there have been four at RSPB Inner Marsh Farm (now RSPB Burton Mere Wetlands), all in May 1993.

Besides those at Cors Caron, there were four other inland records: one at Llanfihangel Gobion, Gwent, on 18 May 1987, two at Borras Pools, Denbighshire, on 16 May 1987, one at Dolydd Hafren, Montgomeryshire, in May 1993 and two at Ynys-y-fro Reservoirs, Gwent, on 18 May 2019.

Jon Green and Robin Sandham

Sanderling *Calidris alba* Pibydd y Tywod

Welsh List Category	IUCN Red List			Birds of Conservation Concern			
	Global	Europe	GB	UK	Wales		
					2002	2010	2016
A	LC	LC	LC	Amber			

Like many waders, the Sanderling breeds on the High Arctic tundra, and has a circumpolar distribution, with the main breeding areas in the Canadian Arctic, Greenland and Siberia. It winters to the south of its breeding range and is one of the world's most widely distributed waders outside the breeding season, reaching as far south as New Zealand and Tierra del Fuego (*HBW*). Unlike many similar waders, it prefers sandy beaches to muddy estuaries when wintering or on passage. Colour-ringing has shown that Sanderling wintering in Britain are mainly from Greenland, not from Siberia (Graham Appleton *in litt.*). Ringing recoveries suggest that birds from the Siberian population also visit Wales. One bird ringed at Point of Ayr, Flintshire, in May 1972, was recovered in July 1974 at Murmansk in the far north of Russia. Two ringed in southern Norway in autumn were later recovered in winter in Wales, but these could have been birds breeding in Arctic Canada and using Norway as a staging post. Interchange with Iceland, a stopping point for Greenland birds, and possibly some from Canada, is more certain, from three adults ringed in Iceland and one ringed in Wales in winter.

Sanderling may have been less common as a wintering species in many parts of Wales in the late 19th and early 20th centuries. In Pembrokeshire, Mathew (1894) and Lockley *et al.* (1949) stated that it was mainly an autumn passage migrant with a few on spring passage. Barker (1905) described it as "occasional" in Carmarthenshire, while it was said to be an uncommon winter visitor to Glamorgan in the late 1890s (Hurford and Lansdown 1995). Salter (1900) reported that large flocks visited the Dyfi Estuary, Ceredigion/Meirionnydd, in spring and August, but knew of only one winter record there, while Ingram *et al.* (1966) said it was only rarely found in winter on the Ceredigion coast. In North Wales it was principally a passage migrant in autumn and spring, with the largest numbers on spring passage in May and early June, although a few wintered (Forrest 1907). During 1969–74, the mean wintering population in Britain was just over 10,300, of which just 423 were in Wales: 228 in Glamorgan, 120 in Carmarthenshire and 75 in Flintshire. It was noted that there were none in some areas, such as the Dyfi Estuary, in winter, even though it held good numbers on passage (Prater and Davies 1978).

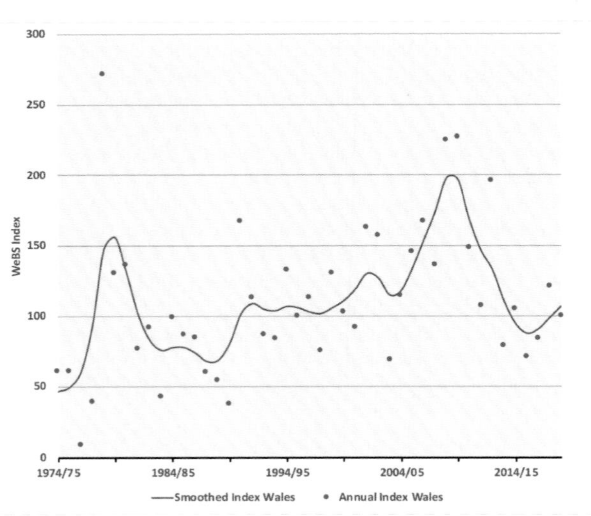

Sanderling: Wetland Bird Survey indices, 1974/75 to 2018/19.

Sanderling: Recovery locations of birds ringed in Wales are shown by red circles. Ringing locations of birds that were recovered in Wales are shown by blue triangles. Small brown dots show ringing or recovery locations in Wales.

Passage numbers were higher in the 1960s and 1970s, but declined from the 1980s. In the Burry Inlet, Gower/Carmarthenshire, mean annual counts of both wintering and passage birds declined from 450 in the 1960s and 1970s to 240 in the 1980s and 40 in the early 1990s. A count of 683 at Whiteford in May 1970 was a record for Gower, and for Glamorgan as a whole. In Swansea Bay, Gower, winter peaks at Blackpill remained stable during the 1980s and early 1990s, but numbers on passage had dropped markedly compared to the late 1960s (Hurford and Lansdown 1995). High numbers were recorded on the Dee Estuary, Flintshire/Wirral, in the 1970s, particularly during migration, peaking at 12,501 in August 1976, but numbers have subsequently been much lower. In the Dyfi Estuary, there were spring passage counts of 380 on 25 May 1971 and 400 on 17 May 1994, but numbers have since fallen there too (Roderick and Davis 2010). There have, nonetheless, still been some large counts, such as over 700 near Gronant, Flintshire, on 1 June 2016 and 880 in the Burry Inlet in August 2010. There were 1,050 at Cefn Sidan/Tywyn Point, Carmarthenshire, in September 2015 and 1,093 there in February 2019.

Birds in Wales (1994) identified three areas of particular importance for Sanderling in Wales: Cefn Sidan beach in Carmarthen Bay, Swansea Bay, especially near Blackpill and the Ogmore Estuary, and in the north, the coast between Llanddulas, Denbighshire, and Point of Ayr, Flintshire. Cefn Sidan was the second most important wintering site in Britain after the Ribble estuary, holding 1,000 or more at its peak.

There was a 113% increase in numbers in Wales between 1984/85 and 2009/10, but there has been a decline since then, triggering WeBS Medium Alerts for the medium term (-45%) and short term (-42%) (Woodward *et al.* 2019). Carmarthen Bay and Swansea Bay are currently of British importance. The highest count in Carmarthen Bay was 2,343 in November 2006, while numbers in Swansea Bay appear to have stabilised, following a 76% decline between 1984/85 and 2009/10 in Blackpill SSSI. Aberafan Sands, Gower, is currently of British importance as a result of a count of 360 in March 2017. The Severn and Dee estuaries as a whole are also of British importance. A lack of regular counts between Llanddulas and the Dee Estuary makes it difficult to be sure of the trend there, but there have been some good counts in recent years, including over 600 near Towyn on 16 February 2014. There

Site	County	74/75 to 78/79	79/80 to 83/84	84/85 to 88/89	89/90 to 93/94	94/95 to 98/99	99/00 to 03/04	04/05 to 08/09	09/10 to 13/14	14/15 to 18/19
Carmarthen Bay	Carmarthenshire/ Pembrokeshire	167	2	57	549	494	1,105	1,531	1,727	766
Dee Estuary: Welsh counties (whole estuary)	Flintshire	na (8,974)	na (970)	na (472)	486 (959)	493 (633)	156 (417)	104 (791)	82 (948)	143 (752)
Swansea Bay	East Glamorgan/ Gower	223	397	255	217	347	300	376	374	441
Burry Inlet	Gower/ Carmarthenshire	198	184	125	61	3	70	16	20	214
Afan Estuary & Port Talbot harbour	Gower	-	-	-	85	205	186	84	146	17

Sanderling: Five-year average peak counts during Wetland Bird Surveys between 1974/75 and 2018/19 at sites in Wales averaging 200 or more at least once. International importance threshold: 2,000; Great Britain importance threshold: 200.

has been a marked increase in wintering birds since the late 1990s on beaches around Tremadog Bay, including 300 at Morfa Bychan, Caernarfonshire, in February 2005. Jones and Whalley (2004) reported that the midwinter total on Anglesey was normally below 100, but the Non-estuarine Waterbird Survey (NeWS) found 215 around the island in winter 2015/16. NeWS totals for Wales were 583 in 1984/95, 436 in 1997/98, 105 in 2006/07 and 583 again in 2015/16 (Austin *et al.* 2017).

There are few inland records, averaging about one record each year in Wales in recent times. These are usually in spring or autumn at large reservoirs or lakes, with most records in May and usually singles, but 12 were at Llyn Tegid, Meirionnydd, in February 2012.

Future prospects for this species in Wales are difficult to assess, partly because there is no reliable information on population trends in the main breeding areas. The sandy beaches preferred by Sanderlings in winter and on passage are also popular with humans, so ensuring undisturbed areas where the birds can feed is important.

Rhion Pritchard

Summer-plumaged Sanderling © Steve Stansfield

Dunlin *Calidris alpina* Pibydd Mawn

Welsh List Category	IUCN Red List			Birds of Conservation Concern			
	Global	Europe	GB	UK	Wales		
A	LC	LC	EN(NB)	Amber	2002	2010	2016

Most Dunlin breed in the far north on Arctic tundra, with a circumpolar distribution from Alaska to eastern Siberia, although some birds breed farther south. Most move south in winter, usually remaining in the northern hemisphere. There are several subspecies, each of which has its own wintering areas and migration routes. The subspecies that breeds in Britain is *C.a. schinzii*, which breeds from southeast Greenland to southern Norway, with the largest number in Iceland. These birds winter mainly in Africa, particularly Mauritania (Hardy and Minton 1980). Two other subspecies can occur in Wales outside the breeding season: *C.a. alpina* breeds from northern Scandinavia east to eastern Siberia, and winters from Britain and Ireland to India, while *C.a. arctica* breeds in northeast Greenland and passes through Britain to winter, mainly in northwest Africa (*HBW*).

In Wales, Dunlin breed mainly in the uplands, although there have been some instances of coastal breeding. The few pairs of Dunlin that breed in Wales are among the most southerly in the world, although an estimated 24 pairs breed on Dartmoor, Devon (Freshney 2020). Dunlin remains were found in Bacon Hole, Gower, dating from the last interglacial period, about 75,000 to 128,000 years ago, and were thought to indicate possible breeding in the area (Stringer 1977). There has been no breeding record in Gower in historic times, nor in East Glamorgan, Carmarthenshire, Pembrokeshire or Anglesey. Elsewhere, breeding Dunlin were more widespread in the 19th century and early 20th century than now. Forrest (1907) quoted Brockholes as saying that several pairs nested on the Dee Marshes in 1871, but did not state whether this was on the Welsh or English side. Forrest commented "Inland a few pairs nest in scattered localities of Denbigh and Merioneth, but not nearly so many as might be expected". It is a very scarce breeder in Meirionnydd now, but Forrest mentioned nesting above Penmaenpool, above Llandderfel, near Trawsfynydd and on Arenig. In 1906, Bolam found at least three pairs breeding on the moors above Llanuwchllyn, with 1–2 pairs at several other sites in that area. This species also bred at several sites in Denbighshire and Ceredigion in this period. It appears to have ceased nesting on Cors Caron, Ceredigion, in the 1950s (Roderick and Davis 2010).

The three Britain and Ireland atlases show little change in the number of 10-km squares with breeding codes for Dunlin in Wales: 14 squares in 1968–72, 16 squares in 1988–91 and 13 squares in 2007–11. The main breeding area is the Elenydd in mid-Wales, an area of upland between Pumlumon in the north and Mynydd Epynt in the south, shared between Ceredigion, Radnorshire and Breconshire, forming part of the Elenydd-Mallaen SPA. There is also regular breeding in the Black Mountains in Breconshire. The last confirmed breeding record for Gwent was in 1973 on moorland in the north of the county. Three records of probable breeding in the *Gwent Atlas 1981–85* and one in the 1998–2003 *Atlas* involved birds that bred on the other side of the county boundary in Breconshire. 3–4 pairs bred in that county just north of the Gwent border at the time (Andrew King *in litt.*). There has been no recent breeding record from Llanbrynmair Moors, Montgomeryshire, where much land was converted to conifer plantation in the1980s. A few pairs may still breed in upland North Wales, including Mynydd Hiraethog, Denbighshire, where Davis (1982a) found 4–5 pairs. Other areas in the north where small numbers have been reported breeding are the Migneint, shared between Denbighshire, Meirionnydd and Caernarfonshire, and the Berwyn, shared between Meirionnydd, Denbighshire and Montgomeryshire. One or two displaying/song-flighting birds have been recorded at Llyn Trawsfynydd, Meirionnydd, in several years since 1991, including most years since 2015 (Reg Thorpe *in litt.*). The *North Wales Atlas 2008–12* recorded birds in six 10-km squares, including one record of confirmed breeding on the

Male Dunlin in breeding habitat in upland Wales © Colin Richards

Year	Reference	No of territories
1975–76	Davies *et al.* 1978	>30
1982	NCC 1982	29
1990	Leversley *et al.* 1990	23
1995	Young *et al.* 1996	24
2007	Johnstone *et al.* 2008	17

Dunlin: Results for a common area within the Elenydd SSSI covered by five surveys (Johnstone and Dyda 2010).

Berwyn on the Meirionnydd/Denbighshire border, with probable breeding on the Caernarfonshire part of the Migneint and on Mynydd Hiraethog.

There have been a number of surveys within the Elenydd SSSI since 1975, but comparing the results is difficult because the surveys covered different areas and some used different methods. Johnstone and Dyda (2010) dealt with this by considering only surveys that were based on either transects or Brown and Shepherd (1993), which had been found to give similar results to transect surveys, and by identifying a common area that had been covered by all the surveys. They found that Dunlin abundance had declined by 41% over the period.

On a larger area of the Elenydd, the 1982 NCC survey found 41 territories. Another survey, using a different method, in 1991 found 37–40 pairs (Hack 1991, Jennings 1991). In the smaller Trumau and Esgair Garthen area of the Elenydd, the number of pairs decreased from six in 1982 to two in 2012 (Crump and Green 2016). The numbers breeding on the Elenydd appear to fluctuate considerably. Hack (1991) suggested that this might be linked to the wetness of the habitat and therefore to the level of rainfall. On Abergwesyn Common, to the south of Esgair Garthen on the Breconshire Elenydd, what had been prime Dunlin breeding bogs in 2014 suffered from drying out and rapid vegetational change, such that by 2018 the bog habitat had been replaced in its entirety by tussock-forming *Molinia*, that forced abandonment by breeding Dunlin. On a wider scale, walk-through surveying over 109 whole or part 1-km squares across south and southeast Elenydd, in 2016 and 2017, found occupation only in a single 1-km square, where breeding by a pair was confirmed (Andrew King *in litt.*).

Breeding on the coast is unusual in Wales, but was proven in Meirionnydd in 1953 and in Flintshire in 1961. More recently, at Morfa Madryn, Caernarfonshire, 1–3 pairs were noted in the breeding season every year between 1996 and 2007: a chick was seen in 1996 and incubating birds recorded in 1998 and 2003. *Birds in Wales* (1994) suggested that the breeding population in Wales was 50–70 pairs. It may now be around 40–50 pairs, although data is lacking away from the Elenydd, as Dunlin breeding areas are usually remote and little-visited.

Large numbers of Dunlin have been ringed in Wales, mainly by the SCAN ringing group on Traeth Lafan, Caernarfonshire/Anglesey. Many ringed in other countries have been recovered in

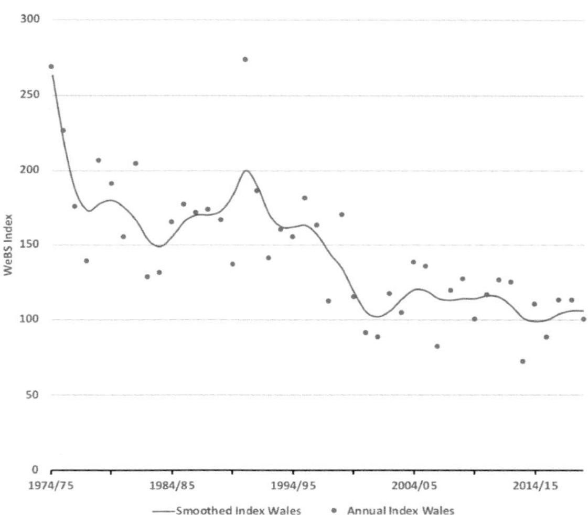

Dunlin: Wetland Bird Survey indices, 1974/75 to 2018/19.

Dunlin: Recovery locations of undetermined subspecies ringed in Wales are shown by red circles. Ringing locations of undetermined subspecies that were recovered in Wales are shown by blue triangles. Small brown dots show ringing or recovery locations in Wales. Pink circles and triangle relate to records of *C.a. alpina* and purple circles and triangle relate to *C.a. schinzii*.

Site	County	74/75 to 78/79	79/80 to 83/84	84/85 to 88/89	89/90 to 93/94	94/95 to 98/99	99/00 to 03/04	04/05 to 08/09	09/10 to 13/14	14/15 to 18/19
Severn Estuary: Welsh side (whole estuary)	Gwent/East Glamorgan	17,687 (46,797)	23,695 (41,020)	26,098 (44,446)	21,541 (47,323)	12,453 (40,029)	3,638 (23,312)	7,701 (20,696)	9,364 (25,281)	7,120 (30,204)
Dee Estuary: Welsh side (whole estuary)	Flintshire	na (38,757)	na (24,541)	na (14,558)	4,250 (21,670)	5,704 (31,122)	4,538 (31,752)	4,297 (15,434)	3,257 (13,502)	6,905 (19,657)
Burry Inlet	Gower/ Carmarthenshire	5,909	4,662	5,619	6,367	8,576	7,286	6,382	7,898	5,198
Cleddau Estuary	Pembrokeshire	1,790	2,083	3,335	4,037	6,071	3,745	3,703	3,383	3,384
Traeth Lafan	Caernarfonshire/ Anglesey	5,367	2,335	4,596	3,643	2,956	4,232	3,571	1,809	2,653
Carmarthen Bay	Carmarthenshire/ Pembrokeshire	2,297	313	726	2,749	1,429	1,954	2,326	2,186	1,143

Dunlin: Five-year average peak counts during Wetland Bird Surveys between 1974/75 and 2018/19 at sites in Wales averaging 2,000 or more at least once. International importance threshold: 13,300; Great Britain importance threshold: 3,400.

Wales. The majority of Dunlin wintering in Wales evidently come from the northeast. Of those ringed outside Britain and Ireland, the largest number of recoveries in Wales are birds ringed as adults in Sweden and Norway, probably on migration from farther east. Most of those wintering in Wales probably breed in Siberia, with some in northern Finland. Several birds ringed in Wales were subsequently found in Russia in the breeding season, including an adult ringed near Beaumaris, Anglesey, in December 1977 and shot in July 1978, 4,033km away in northern Siberia. However, birds passing through Wales on migration apparently come from the northwest. A number of birds ringed in Iceland have been found in Wales, but only in spring and autum. Several individuals recaught on autumn passage recently were *C.a. schinzii*, ringed in southern Europe (Steve Dodd *in litt.*).

The large number of Dunlin recorded at some sites in Wales during November to March are thought to be mainly *C.a. alpina* (e.g. Eades 1974). Adults typically moult at sites around the North Sea, traditionally The Wash, but in recent years probably on the Wadden Sea. Some of these move west to Wales in November, staying until February, when they probably move back to the Wadden Sea, but there are few captures to confirm this (Steve Dodd *in litt.*). Clark (1983) considered that some, fewer than 5%, of those wintering on the Severn Estuary were *C.a. schinzii.* The Severn and Dee estuaries are of international importance for Dunlin, and the Severn Estuary is currently the third most important site in Britain. The majority of the birds are usually on the English side of these two estuaries, although in February 1980, 32,000 out of 34,050 birds in the Severn Estuary were on the Welsh side.

Numbers wintering in Wales declined by 44% during 1991–2016, which has led to a WeBS long-term Medium Alert for Wales being issued (Woodward *et al.* 2019), although numbers appear to have stabilised since about 2000/01. Long-term and short-term Medium Alerts have been issued for the Burry Inlet SPA and a long-term Medium Alert for the Severn Estuary.

The Burry Inlet is of British importance although numbers vary considerably from year to year, with high counts including 15,436 in February 1989, 14,548 in February 1997 and 15,811 in December 2012. Counts of 10,000 or more have been recorded at some other sites, for example, 10,000 on Traeth Lafan in January 1973.

Almost all wintering birds have usually left by the end of March. Spring passage begins in April and peaks in May. This probably mainly involves birds of the subspecies *C.a. schinzii* that breed in Iceland, although there are also likely to be some birds of the North Greenland subspecies *C.a. arctica* (*BTO Migration Atlas* 2002). The Dyfi Estuary, Ceredigion/Meirionnydd, is unusual in often holding larger numbers during migration than in winter, including 4,100 on spring migration in 1972, 3,151 in May 2010 and 6,000 at Ynyslas alone on 6 May 2000. The Alaw Estuary, Anglesey, also often holds larger counts during spring migration than in winter, with more than 1,000 in May in some years. Good numbers are recorded in May at the main wintering sites, the Severn and Dee estuaries and the Burry Inlet, including 6,235 at the latter site in May 1978. Return passage is from late July to September, with high counts including 5,673 on the south side of the Burry Inlet in August 2002 and *c.*5,000 in Malltraeth Bay, Anglesey, on 10 May 1958. Most passage is along the coasts, but smaller numbers are seen inland.

The Dunlin is Red-listed in Wales for several reasons, including severe declines in both its breeding and non-breeding populations over 25 years and the longer term (Johnstone and Bladwell 2016). The Welsh breeding population is small and limited to a few areas, and although it may have been fairly stable in recent years, must be regarded as vulnerable. Johnstone and Dyda (2010) suggested a variety of possible reasons for the decline of this species and Golden Plover on the Elenydd, many of which could also apply elsewhere. These include changes in vegetation density and height, possibly as a result of changes in grazing, and changes in food availability, as a result of climate change or more-intensive grass management. Increased disturbance and increased nest predation are also mentioned as possibilities. Some breeding areas, such as the upper Tywi and Camddwr valleys in Ceredigion, have been lost to afforestation (Roderick and Davis 2010). The outlook for Dunlin as a breeding species in Wales is not good. Models of climate suitability suggest that its range in Britain will be limited to northern Scotland by the end of this century (Huntley *et al.* 2007).

In winter and on passage, Dunlins tend to concentrate in large numbers at a few sites. Ringing recoveries show that individuals often return to the same area of larger sites in subsequent winters. Loss of habitat at these sites, including from sea-level rise, would have a serious effect. The Taff/Ely estuaries held *c.*10,000 birds in December 1979, but all suitable feeding habitat there was lost with the construction of the Cardiff Bay Barrage in 1999 (David Parker *in litt.*). Wintering declines in Wales are probably partly due to short-stopping of Siberian *C.a. alpina* birds, as a result of milder winters (Steve Dodd *in litt.*). Wales is important as a stop-over site for Icelandic birds of the subspecies *C.a. schinzii* (*HBW*).

Greenland Dunlin *C.a. arctica* Pibydd Mawn yr Ynys Las

This subspecies is known to pass through Britain on migration, moving mainly through western Britain in spring (*BTO Migration Atlas 2002*). Its presence in Wales is confirmed by a bird ringed as a nestling in east Greenland in July 2016, and caught by a ringer at Ynyslas, Ceredigion, while on passage in August 2018. The site at which it was ringed is within the breeding range of *C.a. arctica* (Hardy and Minton 1980). Wales could be of importance as a stop-over site for this subspecies, whose global population is estimated at only about 15,000 birds (*HBW*).

Rhion Pritchard

Purple Sandpiper *Calidris maritima* Pibydd Du

Welsh List Category	IUCN Red List			Birds of Conservation Concern			
	Global	Europe	GB	UK	Wales		
A	LC	LC	CR(B)	Amber	2002	2010	2016

This is mainly an Arctic breeder, nesting in the tundra of the Canadian Arctic, Greenland, Iceland and east to the Taymyr Peninsula in Siberia. It also breeds in the mountains of Norway, with a very small number breeding in northern Scotland (*HBW*). It has more northerly wintering populations than any other wader, and some winter as far north as Iceland and northern Norway, though others move as far south as Portugal (*European Atlas* 2020). Populations breeding in Iceland and west Greenland are thought to be largely resident, so most birds wintering in Wales presumably breed in Norway or Canada, although there is no direct evidence of this. Those breeding in Canada have longer bills and wings than those breeding in Norway (*BTO Migration Atlas* 2002).

The Purple Sandpiper prefers rocky shores to the muddy estuaries favoured by most wader species. This species is, therefore, not well covered by core Wetland Bird Surveys. In Wales it has a preference for islands and for certain traditional sites on the mainland. There is evidence of a spring passage, with higher numbers recorded in May at several sites.

One is mentioned as having been shot at Laugharne, Carmarthenshire, in 1792 (Hurford and Lansdown 1995). Mathew (1894) described it as "an autumn visitor; not uncommon" in Pembrokeshire, never encountered in large flocks, though Lockley *et al.* (1949) said that it was found all year round in the county, in small numbers at midsummer, more numerous in winter. Aplin (1905) quoted the lighthouse keeper on St Tudwal's Islands, Caernarfonshire, as saying that a great many were on the larger island in winter 1902/03, so tame that they could almost be caught by hand. Forrest (1907) described it as not uncommon from winter to early summer on western coasts of North Wales, but rare on the north coast. He mentioned no large counts.

The wintering population in Wales in 1968–74 was estimated to be 229 (Atkinson *et al.* 1978). The largest numbers were in Caernarfonshire (70), Glamorgan (67), Anglesey (50) and Denbighshire (20). Numbers appear to have been generally higher in the 1970s than now. On the Gower Peninsula, counts of over 40 were recorded regularly at Crabart, with a peak of 71 on 6 January 1978, and there was a maximum of 62 at Port Eynon Point on 25 December 1977. Farther east in Glamorgan, there were 40 at Black Rocks, Porthcawl, on 8 March 1977. In Pembrokeshire, Donovan and Rees (1994) estimated around 40 on the mainland coast. There was a similar number on Grassholm, Skokholm and Skomer combined, with about the same number on the Smalls, where up to 47 had been counted. In Ceredigion, numbers wintering at Aberystwyth were thought to be low at the end of the 19th century. They increased during the 1920s and 1930s, then fell in the 1970s, before again increasing in the late 1980s to over 20 birds, before declining through the 1990s (Roderick and Davis 2010).

The best-monitored site in Wales is probably Bardsey, Caernarfonshire. Purple Sandpipers were said to be uncommon there in the 1950s, with spring maxima of just ten in some years and often just a few singles in autumn (Roberts 1985). Numbers increased in the 1960s, with up to 65 in April and May 1968. They increased further in the 1990s, reaching 108 in February 2003, a record for the county and probably for Wales. There were 100 on the island in March 2005, but numbers have since declined. 2015 saw the lowest counts since the mid-1980s, with a peak of just 26 in March. However, there were 60 on the island in May 2019.

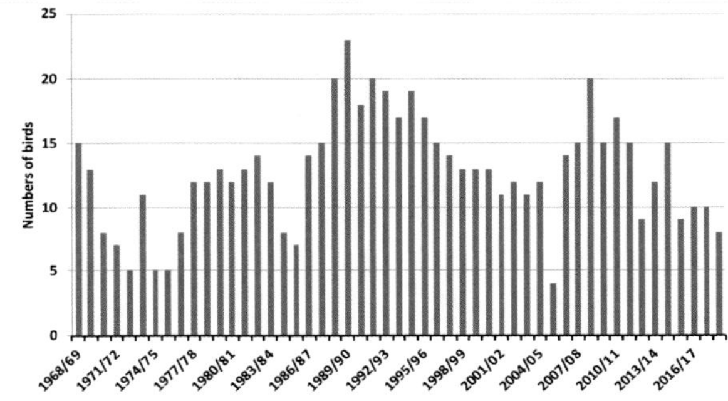

Purple Sandpiper: Numbers seen at Aberystwyth each winter, 1968/69 to 2018/19.

Purple Sandpiper in winter plumage © Ben Porter

1984/85	1997/98	2006/07	2015/16
162	72	87	41

Purple Sandpiper: Non-estuarine Waterbird Survey (NeWS) total counts for Wales.

In Caernarfonshire, the best mainland site is Rhos Point, where numbers peaked at 73 on 10 November 1979. They then declined, coincident with construction of a rock breakwater to prevent coastal flooding, although it is possible that some moved to roost out of sight on the seaward side of the breakwater. Reduced numbers were also recorded at some other sites, such as Bardsey, in the 1980s. Double-figure counts are recorded at Rhos Point most winters, but far fewer than in the 1970s. A count of 23 in December 2011 is the highest in recent years. On Anglesey, a survey during 1968–74 suggested an average winter maximum population of 50. There were 71 on Puffin Island in December 1972. High counts continued in the 1980s at sites such as Trearddur Bay, where 94 were counted in January 1981, and Ynys Llanddwyn, with 50 in April 1986. Numbers on Anglesey fell in the 1990s, with about 75–125 wintering (Jones and Whalley 2004).

The Non-estuarine Waterbird Survey is an important means of monitoring this species and indicates a substantial decline in the population, although it does not cover all sites where the birds are found. A WeBS medium-term High Alert (-51%), and a short-term Medium Alert (-42%), have been issued for Wales (Woodward *et al.* 2019).

Purple Sandpipers are scarce inland. Radnorshire, for example, had just four records—all singles—up to 2013 (Jennings 2014). The only recent inland records were singles in the Elan Valley, around the Breconshire/Radnorshire border, in 2011 and 2018. The species has apparently never been recorded in Montgomeryshire.

Current numbers wintering in Wales are difficult to estimate, but on the basis of published records, the total in winter 2018/19 is unlikely to have been more than 155–170 birds. However, many smaller islands are seldom counted in winter and almost certainly hold birds. For example, at least 25 were recorded on Puffin Island, Anglesey, on 19 January 2019.

The wintering population in Wales is in decline, but factors here are unlikely to be responsible. The species is fairly widely dispersed along the Welsh coast, and its favoured habitat is not very likely to be affected by development or human disturbance. The population trend for the birds breeding in Arctic Canada is unknown, but the Purple Sandpiper is quite likely to be affected by climate change, and it may be that this population is now less likely to fly as far south as Wales to winter.

Rhion Pritchard

Baird's Sandpiper *Calidris bairdii* Pibydd Baird

Welsh List Category	IUCN Red List		Rarity recording
	Global	Europe	
A	LC	LC	BBRC

Breeding in extreme northeast Siberia, across northern Alaska and Arctic Canada as far as northern Baffin Island and northwest Greenland, this species migrates through the North American interior to winter on the high Andean grasslands of South America, from southern Ecuador to Tierra del Fuego (*HBW*). There have been 291 individuals in Britain to the end of 2019, on average about seven a year over the past 30 years (Holt *et al.* 2020). Baird's Sandpiper is a rare vagrant to Wales although the number of records appears to be increasing.

Total	Pre-1950	1950–59	1960–69	1970–79	1980–89	1990–99	2000–09	2010–19
23	0	0	1	5	3	3	5	6

Baird's Sandpiper: Individuals recorded in Wales by decade.

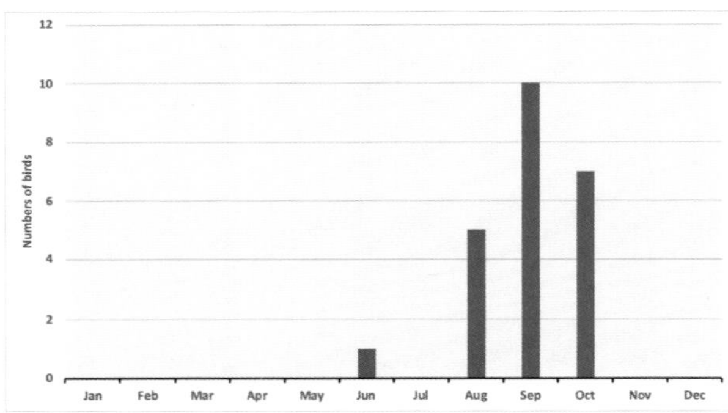

Baird's Sandpiper: Totals by month of arrival, 1967–2019.

The first Welsh record was at the Gann Estuary, Dale, Pembrokeshire on 2–10 October 1967. There has only been one spring record, also at the Gann Estuary, on 11–12 June 2018. The 22 autumn records were all between 7 August and 22 October. The only record of more than one bird was of two together at the Gann Estuary on 13–16 September 2018.

Jon Green and Robin Sandham

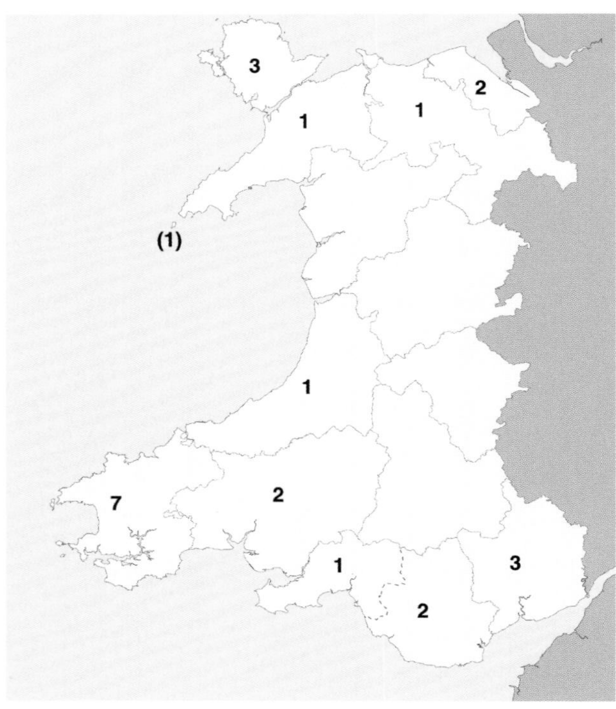

Baird's Sandpiper: Numbers of birds in each recording area 1967–2019.

Little Stint *Calidris minuta* Pibydd Bach

Welsh List Category	IUCN Red List		Birds of Conservation Concern			
	Global	Europe	UK	Wales		
A	LC	LC	Green	2002	2010	2016

The smallest wader seen regularly in Wales, the Little Stint breeds on coastal tundra and islands in the far north, from northern Norway east across Siberia as far as the River Lena (*HBW*). The bulk of the breeding population is in Russia. Global numbers are thought to be increasing (BirdLife International 2020). Most birds from the western parts of its range are believed to winter in Africa south of the Sahara. Wales is on the western edge of the Little Stint's migration route, and numbers recorded here are usually small. Most records are during autumn migration: adults from late July, then birds hatched that summer, whose numbers usually peak in September. Spring passage, in late April and May, is usually smaller.

The number recorded in Wales varies greatly from year to year, in a similar manner to the Curlew Sandpiper that breeds in the same areas. High numbers are recorded in autumn in some years, presumably when a successful breeding season combines with weather patterns that push migrating birds farther west than usual. A good year for Curlew Sandpipers is often a good year for Little Stints, such as 1996 and 1998, although this is not always true: 1999 was an exceptional year for Curlew Sandpipers, but produced only average numbers of Little Stints. There have been few ringing recoveries involving Wales, although birds ringed on the Dee Estuary have later been recovered in southern France and in Morocco.

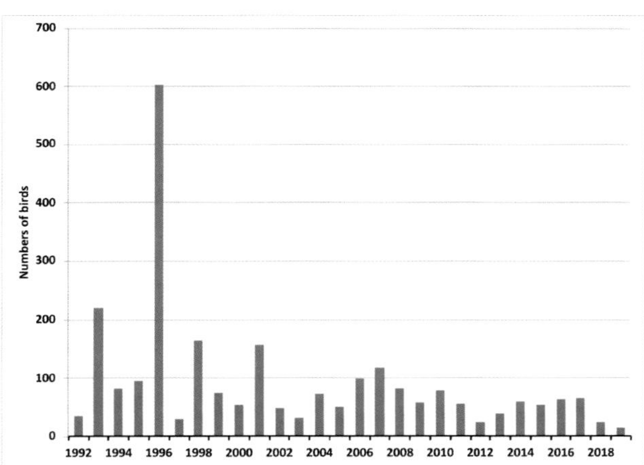

Little Stint: Numbers seen each year, July to November, 1992–2019.

The status of this species in Wales seems to have changed little since the late 19th century, when Mathew (1894) described it as an occasional visitor to Pembrokeshire. Forrest (1907) described it as "by no means common" in North Wales, being most frequent in autumn on the Dee and Dyfi estuaries, although he thought that birds might pass unnoticed. The increase in records since that period probably owes much to the increase in the number of observers and better optical equipment. In an average autumn, 50–100 are recorded in Wales, but there have been much higher totals in some years. In 1960, an influx into Britain in September produced a count of about 350—along with 80 Curlew Sandpipers—at Shotton Pools, Flintshire. This remains a record for Wales and probably for Britain. Above-average numbers were recorded at other sites in Wales that September, including 32 at Malltraeth Cob pool, Anglesey. Autumn 1993 produced a fairly large influx, with at least 220 birds recorded in Wales, including 53 at the Dyfi Estuary, Ceredigion, on 21 September. There was a larger influx in autumn 1996, when over 600 were recorded between 21 July and December (Green 2002), with the main arrival in late September. The largest count that year was at Gronant, Flintshire, where numbers peaked at 102 on 23 September. There were smaller flocks of 37 at Ogmore, East Glamorgan, on 21 September, 34 at Angle Bay, Pembrokeshire, on 25 September and a combined total of 43 at Malltraeth Cob pool and the Cefni Estuary, Anglesey, on 19 September. There were also some large inland counts in 1996: 25 at Llyn Alaw, Anglesey, on 21 September and 36 at Llandegfedd Reservoir, Gwent, on 23 September. 1998 also saw an autumn influx, with a minimum of 164 birds in Wales, including at least 49 in Flintshire. Since then, 2007 has been the only year with over 100 individuals on autumn passage: 117 that year included 26 at RSPB Ynys-hir, Ceredigion, on 2 October.

Numbers on spring passage are much lower. Inland records are fairly scarce, but not as rare, in relation to the total number of Little Stints recorded in Wales, as is sometimes suggested, especially during influx years, such as 1996. In winter, the species is scarce, with most records being of 1–3 birds, although up to five wintered at Shotton following the influx of 1996. In 1969/70 there was something of a winter influx into Britain, which included eight at Berges Island, Gower, on 24 January 1970. The species may be wintering in southern Britain more frequently than in the past; for example, the *Britain and Ireland Atlas 2007–11* showed a 78% increase in the number of occupied squares in winter since the 1981–84 *Winter Atlas*, mostly in southern England.

Rhion Pritchard

First-calendar-year Little Stints © John Hawkins

Least Sandpiper *Calidris minutilla* Pibydd Bychan

Welsh List Category	IUCN Red List Global	Rarity recording
A	LC	BBRC

The smallest of the North American waders, the Least Sandpiper breeds on subarctic tundra and in the far northern boreal forests from Alaska to Newfoundland, breeding farther south than any other North American *Calidris* species (*HBW*). It migrates across the Great Plains and the eastern seaboard of the USA, to winter from the southernmost United States to tropical South America. In Britain, 41 individuals had been recorded up to the end of 2019, with the last in 2017 (Holt *et al.* 2018), about one every two years on average. There has only been one record in Wales, at Aberthaw, East Glamorgan, on 2 September 1972.

Jon Green and Robin Sandham

White-rumped Sandpiper *Calidris fuscicollis* Pibydd Tinwyn

Welsh List Category	IUCN Red List Global	Rarity recording
A	LC	WBRC

This wader breeds in the High Arctic of Alaska and northern Canada on marshy tundra, and winters in southern South America (*HBW*). It is one of the most frequently recorded Nearctic waders in Britain, with a total of 620 individuals recorded between 1958 and 2018 (White and Kehoe 2020a). Parkin and Knox (2010) noted that over 97% of records in Britain and Ireland had been between July and November. After a significant increase in numbers recorded during the late 20th century and early 21st century, annual totals have shown a marked decline in recent years (White and Kehoe 2020a).

It remains very rare in Wales, although it may be under-recorded. There have been 14 records of 15 individuals:

- Blackpill, Gower, on 25 September 1957;
- South Dock, Swansea, Gower, on 19 March 1970;
- Pembrey Airfield, Carmarthenshire, on 12–26 October 1974;
- Pembrey, Carmarthenshire, on 25–27 September 1975;
- the Gann Estuary, Pembrokeshire, on 20 September 1977;
- adult at Shotton, Flintshire, from 30 July to 8 August 1984;
- adult on the Teifi Estuary, Pembrokeshire and Ceredigion, 5 August 1991;
- adult at Newport Wetlands Reserve (Goldcliff), Gwent, on 11–17 August 1995;
- adult at Newport Wetlands Reserve (Goldcliff), Gwent, on 14–21 September 1999;
- adult at Newport Wetlands Reserve (Goldcliff), Gwent, on 2 July 2003;
- 1CY at Kenfig, East Glamorgan, from 29 October to 8 November 2007;
- juvenile at Newport Wetlands Reserve (Goldcliff), Gwent, on 26 October 2008;
- Llanrhidian, Gower, on 20–30 September 2011 and
- two at the Broadwater, Meirionnydd, on 5 November 2011, one remaining until 7 November.

Jon Green and Robin Sandham

Buff-breasted Sandpiper *Calidris subruficollis* Pibydd Bronllwyd

Welsh List Category	IUCN Red List Global	Rarity recording
A	NT	WBRC

The Buff-breasted Sandpiper breeds in the High Arctic of Alaska and northwest Canada, and in the far east of Siberia, migrating overland to winter in northern Argentina and Uruguay (*HBW*). There has been a large increase in the numbers recorded in Britain over the last 50 years, increasing from an average of 15 a year in the period 1990–99 to 41 a year in 2010–19 (White and Kehoe 2020a). Numbers have also increased in Wales, where it may be the commonest trans-Atlantic wader. The Pectoral Sandpiper is often assumed to be the commonest North American wader in Wales, but many of the Pectoral Sandpipers reaching Britain may breed in Siberia (Lees and Gilroy 2004).

The first Welsh record was on Bardsey, Caernarfonshire, on 21 September 1968. Since 1973 it has occurred almost annually in Wales. A series of Atlantic low-pressure systems in late August and September 2011 brought a remarkable influx to Britain and Ireland, with at least 11 in Wales, of which seven were in Pembrokeshire: two at Stackpole, three at Dale airfield and two others nearby at Kete, St Ann's Head. Elsewhere, there was one at Weobley, Gower, two at RSPB South Stack, Anglesey, and one at RSPB Conwy, Denbighshire.

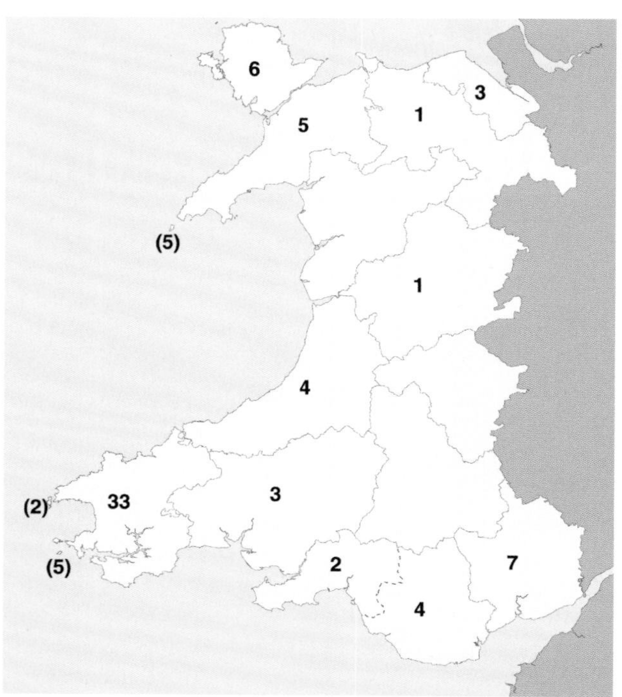

Buff-breasted Sandpiper: Numbers of birds in each recording area 1968–2019.

Total	Pre-1950	1950–59	1960–69	1970–79	1980–89	1990–99	2000–09	2010–19
69	0	0	1	10	20	7	10	21

Buff-breasted Sandpiper: Individuals recorded in Wales by decade.

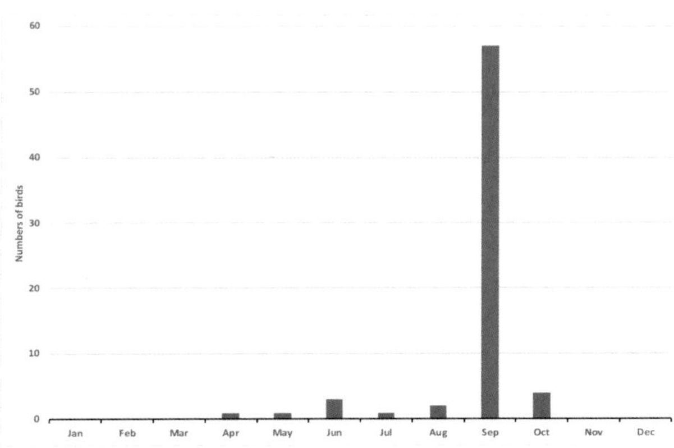

Of the 33 individuals recorded in Pembrokeshire, 19 have been at Dale airfield, including five together on 20–22 September 1984 and three on 23–24 September 2011 and 22–24 September 2018. Four of the six on Anglesey have been in the Cemlyn area. Most records are in autumn, mainly in September, as one would expect from a storm-driven migrant, though there have been five spring records.

Jon Green and Robin Sandham

Buff-breasted Sandpiper: Totals by month of arrival, 1968–2019.

Pectoral Sandpiper *Calidris melanotos* Pibydd Cain

Welsh List Category	IUCN Red List	Rarity recording
	Global	
A	LC	WBRC

The Pectoral Sandpiper breeds in northern Siberia and northern North America, and winters in southern South America, southeast Australia and New Zealand (*HBW*). Many of the birds seen in Britain in late summer, primarily adults, are likely to originate in western Siberia, where their range overlaps with Little Stint and Curlew Sandpiper (Lees and Gilroy 2004). First-calendar-year birds arrive in Britain in September and October, and may be from either the North American or Siberian populations. The species winters in southern South America and Australasia, although Lees and Gilroy (2004) have speculated that some from both populations pass regularly through Europe to undiscovered wintering grounds

Total	Pre-1950	1950–59	1960–69	1970–79	1980–89	1990–99	2000–09	2010–19
142	0	2	15	22	23	14	34	32

Pectoral Sandpiper: Individuals recorded in Wales by decade.

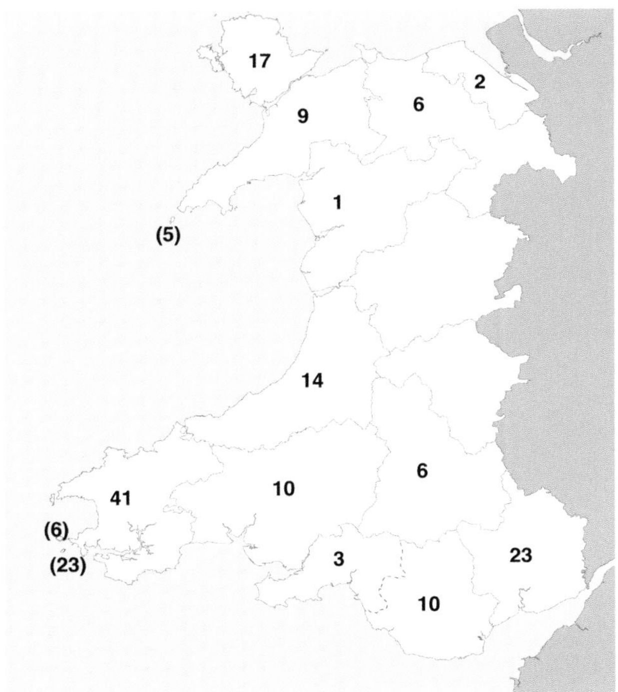

Pectoral Sandpiper: Numbers of birds in each recording area 1958–2019.

Pectoral Sandpiper at St Brides, Gwent, 21–22 September 2011 © Richard Smith

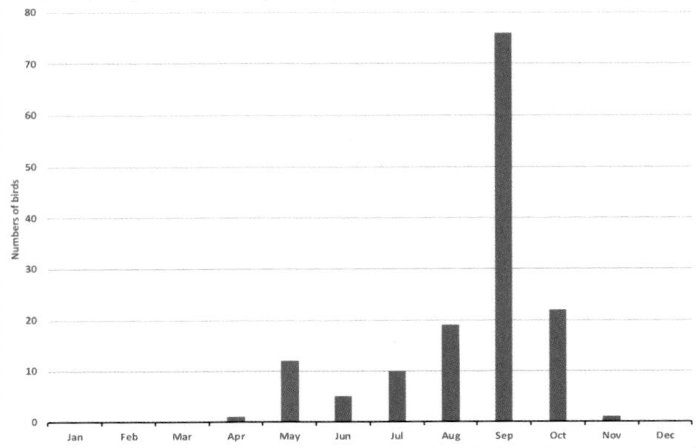

Pectoral Sandpiper: Totals by month of arrival, 1958–2019.

in Africa. A pair may have bred in northeast Scotland in 2004.

An annual mean of 141 was recorded in Britain between 2010 and 2018, compared to a mean of 45 in 1970–79 (White and Kehoe 2020a). This is an increase of over 200%, but the increase in Wales was only about 50% over the same period. The first Welsh record was on Skokholm, Pembrokeshire from 23 September to 12 October 1958. There have been almost annual records since then. Of the 41 recorded in Pembrokeshire, 23 were on Skokholm. In Gwent, 16 of the 23 records were at Newport Wetlands Reserve (Goldcliff). Most records are coastal, but in Breconshire a group of three was at Llangorse Lake between 30 September and 3 October 2008. There were singles at Talybont Reservoir in 1959, at Llangorse in 2003 and at Brechfa Pool in 2013.

There is a distinct autumn bias to records, with September totalling nearly three times the number of records of any other month. The species has been recorded in almost every county in Britain and Ireland (Parkin and Knox 2010), but Radnorshire and Montgomeryshire are two of the exceptions.

Jon Green and Robin Sandham

Semipalmated Sandpiper *Calidris pusilla* Pibydd Llwyd

Welsh List Category	IUCN Red List Global	Rarity recording
A	NT	BBRC

This wader breeds on Arctic and subarctic tundra in Alaska and northern Canada, and winters in the Caribbean and on the coasts of tropical South America (*HBW*). A total of 162 individuals was recorded in Britain to the end of 2019, about four or five a year on average over the last 30 years (Holt *et al.* 2020), including ten in Wales. Most of these were first-calendar-year birds, and all were in the southern half of Wales. All birds were 1CY unless stated otherwise:

- adult on Skokholm, Pembrokeshire, on 20–21 July 1964, which was ringed on the latter date;
- Ogmore Estuary, East Glamorgan, on 6–17 September 1990;
- Ogmore Estuary, East Glamorgan, on 3–4 September 2001;
- Newport Wetlands Reserve (Goldcliff), Gwent, on 6–9 September 2006;
- the Gann Estuary, Dale, Pembrokeshire, on 14–27 October 2007;
- Ynyslas, Ceredigion, on 20–22 September 2012;
- 2CY+ at the Gann Estuary, Dale, Pembrokeshire, on 3–5 September 2013;
- 2CY+ at the Gann Estuary, Dale, Pembrokeshire, on 4–8 August 2016;
- 2CY+ at Sully Island, Swanbridge Bay, East Glamorgan, on 12–13 August 2017 and
- Newport Wetlands Reserve (Goldcliff), Gwent, on 1–5 September 2019.

Jon Green and Robin Sandham

Long-billed Dowitcher *Limnodromus scolopaceus* Gïach Gylfinhir

Welsh List Category	IUCN Red List Global	Rarity recording
A	LC	BBRC

The Long-billed Dowitcher breeds primarily in Arctic Siberia, where its range is expanding west to the Lena River Delta. Its range in North America is restricted to the coastal tundra of western and northern Alaska, and east to the Mackenzie River in Canada. It winters from the coastal southern USA to northern Central America (*HBW*). Some cross to the east coast of Canada and then follow the Atlantic coast south to their wintering grounds. These probably account for many of the records in Europe (Parkin and Knox 2010). The number of records in Britain has increased dramatically over the last 30 years: 276 individuals to the end of 2019 (Holt *et al.* 2020), averaging five each year over the past 30 years. The majority have been recorded in September and October. Although it is still a rare visitor to Wales, the number of records per decade has increased. The first record was one at Nefyn, Caernarfonshire, on 12 October 1963.

The peak month for arrivals is October. Most records are at coastal sites, but there have been two records away from the coast, at Llangorse Lake, Breconshire, on 18–21 October 1997 and at Rhaslas Pond near Merthyr Tydfil, East Glamorgan, on 8–14 October 2013. Some individuals are believed to have made repeat visits in different years. In Ceredigion, one at Cors Fochno, on 23–28 April 2011, was thought to be the same bird as the one at RSPB Ynys-hir from 19 October to 25 November 2010. One first-calendar-year bird at Connah's Quay nature reserve, Flintshire, on 19–21 October 2009, was thought to have returned the following autumn, from 28 August to 11 October 2010. An adult at Penclacwydd, Carmarthenshire, from 28 July to 1 August 2013 could have been the individual that was at the same site on 28–29 October 2012.

Long-billed Dowitcher at Rhaslas Pond, East Glamorgan, October 2013 © Richard Smith

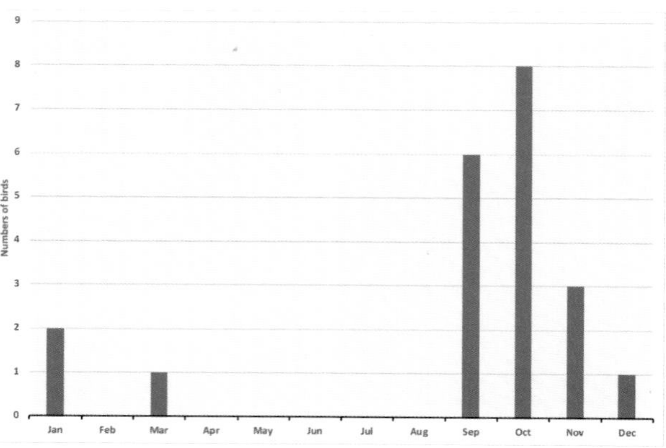

Long-billed Dowitcher: Totals by month of arrival, 1963–2019.

Total	Pre-1950	1950–59	1960–69	1970–79	1980–89	1990–99	2000–09	2010–19
21	0	0	1	1	3	3	6	7

Long-billed Dowitcher: Individuals recorded in Wales by decade.

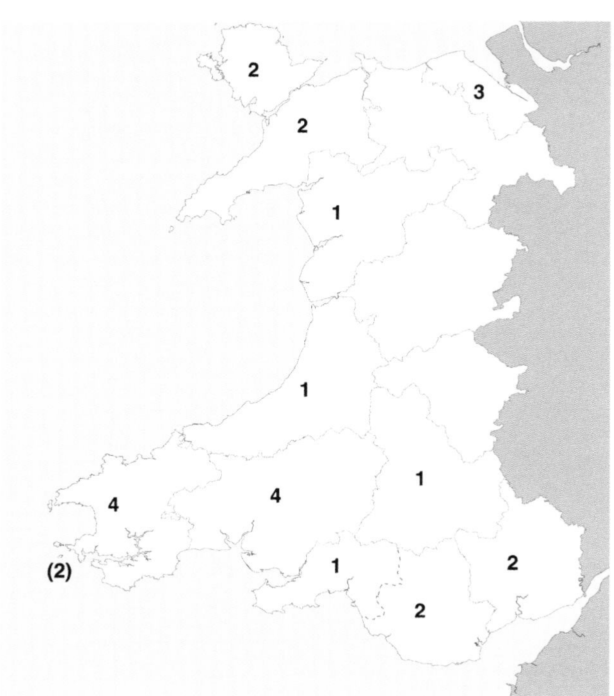

Additionally, there are five records of Dowitchers that were not specifically identified. Short-billed Dowitcher has not yet been recorded in Wales.

Jon Green and Robin Sandham

Long-billed Dowitcher: Numbers of birds in each recording area 1963–2019. Some birds were seen in more than one recording area.

Woodcock *Scolopax rusticola* Cyffylog

Welsh List Category	IUCN Red List			Birds of Conservation Concern			
	Global	Europe	GB	UK	Wales		
					2002	2010	2016
A	LC	LC	VU(B)	Red			

The Woodcock breeds over a wide area of temperate Europe and Asia, from the Azores to Japan. The population breeding in Britain, Ireland and France is mainly resident, but other populations are mainly migratory. The birds breeding in Fennoscandia and western Russia winter in western and southern Europe and North Africa, including Wales (*HBW*). This species is unique in Britain as the only wading bird that favours extensive woodland—broadleaved, coniferous and mixed—for shade, humidity and soft humus, apt to retain moisture (*BWP*). In Wales it has been found breeding in all these woodland types, from the mature broadleaved woodlands of the Wye Valley, Gwent, to the upland coniferous plantations of Snowdonia. In the breeding season, it is usually only discovered at dusk when the male's characteristic 'roding' display is seen and heard; often the only indication of its presence. Nests are extremely difficult to find and rarely located (Campbell and Ferguson-Lees 1972), as the well-camouflaged female sits tightly, not flushing until almost stepped upon.

Woodcock © Tony Cross

A national enquiry in 1934–35 produced a great deal of information about the history of the Woodcock in Britain, documented by Alexander (1945, 1946a, -b, -c, -d, 1947). This showed that breeding numbers had increased rapidly in the first half of the 19th century, thought to be the result of the expansion of plantations and protection from human persecution. The species was well-established in Gwent by the mid-19th century. During 1885–1935, Woodcock nests were reported from all Welsh counties except Anglesey as part of a general increase in breeding range, although regular breeding was confined to Glamorgan, eastern Gwent, northeast Breconshire and east Radnorshire (Alexander 1945). However, it appeared to be less common in Wales than in most of Britain: Alexander commented that it "appears to breed in all suitable localities in the British Isles, except in Wales and the south-western peninsula (Somerset, Devon and Cornwall)." Breeding subsequently increased in Denbighshire, where Harrop (1961) recorded it as a regular breeder in large areas of young fir plantation. Breeding has not been confirmed in Pembrokeshire in the years since 1935. Jones and Whalley (2004) noted that breeding had been claimed at one estate on Anglesey in 1969. They mentioned one other record of roding with two records of birds present in the breeding season. However, they stated that there was so little solid data that it was safer to conclude that the past and present breeding status of the Woodcock on the island was not known, although it was likely to have bred occasionally. *Birds in Wales* (1994) stated that the stronghold was the border counties from Gwent north to Denbighshire.

There has been a marked decline in breeding Woodcock across Britain and Ireland since the 1970s, with breeding evidence found in 50% fewer 10-km squares in the *Britain and Ireland Atlas 2007–11* than in the 1968–72 *Atlas* period. The reduction was even greater in Wales, from 46% of 10-km squares in 1968–72 to 26% in the 1988–91 *Atlas* and 11% in 2007–11, with a decrease in relative abundance between 1988–91 and 2007–11. In North Wales, the number of 10-km squares with probable or confirmed breeding fell from 30 in 1968–72, to 22 by 1988–91 and only six, none confirmed, in the *North Wales Atlas* 2008–12. Despite the difficulties of estimating the size of the breeding population, there is no doubt that there has been a large reduction in breeding Woodcocks in Wales (Hoodless and Hirons 2007). The reasons may include recreational disturbance, declining woodland management and the maturing of plantations (Fuller *et al.* 2005, Hoodless and Hirons 2007). Hoodless *et al.* (2009) suggested a breeding population of around 2,000 males in Wales.

The roding male is polygynous, mating with up to four females and, while attending a female during the egg-laying period, another male can temporarily take over its roding route, with an observer none the wiser to the change (*BWP*). There is practically no information available for Wales relating to nests or density of breeding females. Unpublished work by Steve Roberts and Jerry Lewis in the 1980s and 1990s provided some indication of potential breeding abundance. Using trained spaniels and setters in the Wye Valley and Forest of Dean, in Gwent and Gloucestershire, as well as Wentwood, Trelleck and Ysgyryd Fach, in Gwent, they found a minimum of 13 nests with eggs and 24 broods of chicks during 1983–93. They flushed many more single Woodcock or pairs in suitable breeding habitat, that they considered were either off-nest or about to lay. During 1995–2001, with less effort and fewer dogs, a further 12 broods were found. The majority of those found in the Forest of Dean came from just *c.*10km^2 of woodland near Monmouth, through which the national boundary between Wales and England runs. Although many nests/broods were in Gloucestershire, the contiguous nature of suitable habitat on both sides led Roberts and Lewis to consider Woodcock to be a fairly abundant breeder in that area of Wales.

While the Woodcock's status as a breeding bird in Wales is in steep decline, this is not reflected in its status in winter with the arrival of an estimated half-million migrants, making it, arguably, the commonest wader in Wales (Tony Cross and the mid-Wales Ringing Group pers. comm.). Owen (1603) claimed that it arrived earlier, was present in greater numbers in winter and stayed longer in Pembrokeshire than anywhere else in England and Wales. He said that it occurred in "almost incredible numbers", outnumbering all other gamebirds together. It was taken by nets on "cockroads" in woods at dawn and dusk, and it was not unusual to catch 100–120 from one wood in a day.

The Woodcock remains a legal quarry species in Wales. It can be shot between October and January, but most are shot incidentally during Pheasant shoots, rather than being specifically targeted. As an indication of abundance, records from shooting estates must therefore be regarded with some caution. Even if flushed to guns, Woodcock do not often afford a safe shot, and birds flushed in double figures at a shooting estate in Gwent produced no Woodcock in the bag at the end of the day (own data). There is also some evidence that Pheasant coverts containing many thousands of birds are not particularly attractive to Woodcock (McKelvie 1986). Up to 1935, at nine shooting estates in South Wales, the average bag for the season was 11 Woodcock. For 11 estates in North Wales, the average season's bag was 21 Woodcock (*Birds in Wales* 1994). Perhaps a better indication of Woodcock winter numbers is the 40-year record, kept for an area of approximately 2km^2 of ideal Woodcock habitat near Tregaron, Ceredigion. During 1979–2019 the average number of Woodcock shot to one gun with one dog was 33 a year. Annual bag totals ranged from 15 to 58, with numbers considered stable over the period. Frequently, very large numbers are present: in December 2014, 18 were flushed from a 300m^2 block of mixed larch and silver birch (own data). The valleys of West Wales, after a 'fall' of Woodcock, can produce quite startling numbers when worked with dogs for shooting parties. Over 50 in a single day, flushed from gorse- and blackthorn-covered cwms above Newgale, Pembrokeshire, is not unusual in December and January (own data). Woodcock need soft ground to probe for earthworms with their long beaks, favouring West Wales, which is most likely to remain ice-free in winter. However, in the severe frosts of January–February 1963, 100 were found dead on Bardsey, Caernarfonshire, and over 200 on Skokholm, Pembrokeshire (*Birds in Wales* 1994).

A considerable number of Woodcock have been ringed in Wales, many now caught by 'dazzling': using a powerful beam and net, especially on windy nights. In Ceredigion, as many as 50 have been seen, with 27 caught, by this method in a single night (Tony Cross pers. comm.). There were 57 at Cors Caron, Ceredigion, on 21 January 2019. In an area of mid-Wales stretching across Ceredigion, Radnorshire and Montgomeryshire, over 500 birds have been caught in a single winter by the Mid-Wales Ringing Group. There have been a large number of Woodcock recoveries, mainly birds shot in Britain, Fennoscandia and Russia. Birds ringed as nestlings in Norway, Sweden, France and England have been recovered in Wales, as have birds ringed as adults in The Netherlands, Germany, Sweden, Finland, Latvia, Belarus and Russia. The longest distance recorded of a Woodcock ringed abroad and recovered in Wales in winter is 2,306km. This was a bird ringed in its first calendar-year, at a site south of St Petersburg, Russia, in October 1995, and shot at Llangain, Carmarthenshire, in January 1999. The farthest migration recorded was one ringed in Trawsgoed, Ceredigion, on 4 January 2017, that was shot on 13 May 2018, 6,267km away in Krasnoyarsk, Russia, well to the east of the Urals. Within Britain, one was shot at Mt. Sion, Pembrokeshire, on 20 November 2004, having been ringed the previous month, 428km away at Hartlepool, Cleveland. Individuals are remarkably site-faithful. Some have been recaught in the same small patch of field, down to within 50m, over six or seven winters. The oldest recorded Woodcock ringed in Wales was caught as an adult on 13 January 2008 at Llanilar, Ceredigion, recaught at the same locality on 19 January 2009 and again on 4 February 2016, by which time it was at least eight years and 22 days old.

The global population of the Woodcock is considered to be stable (BirdLife International 2020), but it has been given Red conservation status in Wales because of severe declines in both breeding numbers and breeding range (Johnson and Bladwell 2016). The large increase in Pheasant releases into the wild could have an effect on Woodcock, for example by competition for food

Woodcock: Recovery locations of birds ringed in Wales are shown by red circles. Ringing locations of birds that were recovered in Wales are shown by blue triangles. Small brown dots show ringing or recovery locations in Wales.

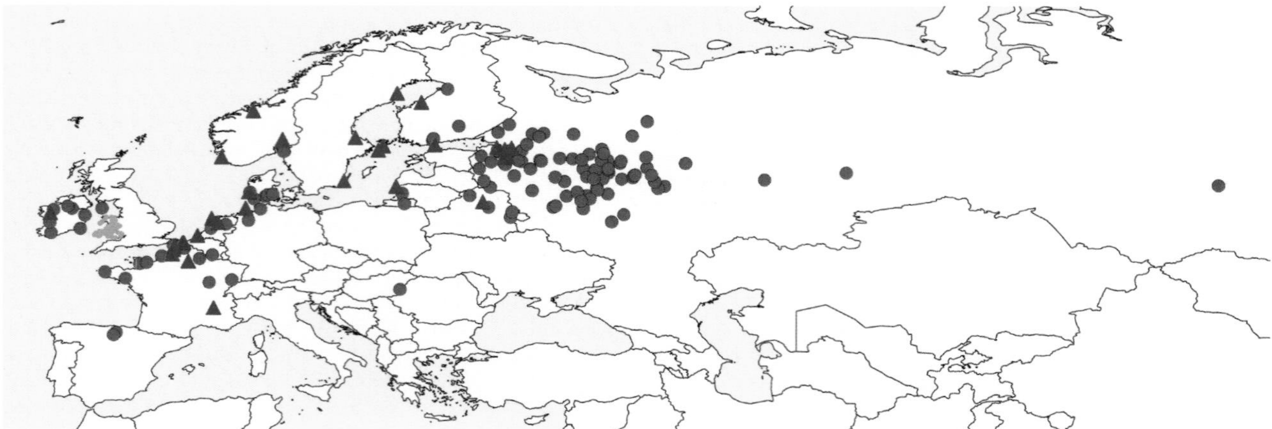

along with the increase in corvid numbers often linked to such releases (David Parker *in litt.*). The most significant threat to the species globally is fragmentation of woodland (*HBW*), together with changes in woodland structure. The area of Europe that is climatically suitable for Woodcock is expected to reduce this century, with very little of southern Britain appropriate by 2100 (Huntley *et al.* 2007). The planting of more large tracts of woodland would probably benefit breeding Woodcock in Wales.

Stephen Roberts

Sponsored by Joan and Brian Iddon

Jack Snipe *Lymnocryptes minimus* Gïach Bach

Welsh List Category	IUCN Red List			Birds of Conservation Concern			
	Global	Europe	GB	UK	Wales		
A	LC	LC	LC	Green	2002	2010	2016

A breeding bird of the forest tundra and northern taiga, Jack Snipe breeds from northeast Fennoscandia and east across Siberia, with a few isolated populations farther south. All populations are migratory, wintering in western Europe, around the Mediterranean, in Africa south of the Sahara and across southern Asia (*HBW*). In winter it is found around brackish or fresh water, usually in fairly short vegetation. It is mainly nocturnal and crepuscular, and roosts during the day. It is well camouflaged and only flushes if closely approached, so is greatly under-recorded. An estimated 100,000 winter in Britain (Frost *et al.* 2019a).

Jack Snipe have been recorded in Wales from 18 August to 12 May, with one summer record, near Llanbedrog, Caernarfonshire, on 18 June 1996. Some of the birds recorded in autumn and spring are passage migrants. A good number of Jack Snipe have been ringed in Wales, including 396 in 2015–18. The Mid-Wales Ringing Group alone caught 141 birds in 2017, including 40 re-traps. The only ringing recovery from a likely breeding area was one ringed at Llanerfyl, Montgomeryshire, in December 2016 and found dead in Udorsky District, northern Russia, in May 2017. Ringing has shown some individuals to be site-faithful, both at wintering sites and at stopover sites on migration. One ringed on Bardsey, Caernarfonshire, in March 2013 was recaught there in March 2014 and again in October 2015.

This species seems to have been more common in Wales in the 19th century. It was described as being "scattered over the county in fair abundance" in Glamorgan in the late 1890s (Hurford and Lansdown 1995), while Mathew (1894) described it as fairly numerous in Pembrokeshire, sometimes in flocks of upwards of a dozen, and Barker (1905) said that it was fairly common in Carmarthenshire. In Breconshire, Phillips (1899) described it as "very common" and mentioned that a shooting party "had moved from thirty to forty Jack Snipe in one day" at a bog near Trecastle in December 1889. An analysis of shooting records in Breconshire, over the period 1890–1950, showed that one Jack Snipe was shot for every three Common Snipe (Massey 1976a). Forrest (1907) described it as a fairly common winter visitor to most areas of North Wales, especially in the west, but not nearly as plentiful as Common Snipe.

Most 19th- and early-20th-century records come from shooting, using dogs to flush the birds. Following the ban on shooting this species, it may have been under-recorded in recent years. Numbers recorded on Wetland Bird Surveys are usually low, such that it is not possible to calculate an annual index for Wales. The highest WeBS total at a site entirely in Wales seems to be 13 at Foryd, Caernarfonshire, in November 2008. Hard weather can bring influxes, such as 20 on Bardsey in January 1963. More recently, there were 25 on the Burry Inlet, Gower/Carmarthenshire, in February 1999 and 30 at Gronant, Flintshire, on 5 December 1999. There have been no subsequent counts of 20 or more, but double-figure counts appear to be more common than formerly. The *Britain and Ireland Atlas 2007–11* showed the species to be widely distributed in coastal and lowland areas of Wales in winter, with almost twice the number of occupied 10-km squares than recorded in the 1981–84 *Winter Atlas*. However, there is insufficient information available to discern long-term trends.

The future of this species in Wales will depend largely on the fortunes of the breeding population in northern Europe, whose range is expected to contract as a result of climate change during this century (Huntley *et al.* 2007). A contraction in range has already been noted in some areas, but the European population is thought to be stable (BirdLife International 2015). Around 5% of the European population are shot each year (*HBW*), although Jack Snipe has had full legal protection in Wales (and in England and Scotland) since 1981 and cannot legally be shot.

Rhion Pritchard

Jack Snipe © Ben Porter

Great Snipe *Gallinago media* Gïach Mawr

Welsh List Category	IUCN Red List		Rarity recording
	Global	Europe	
A	NT	LC	BBRC

The Great Snipe is a scarce and local breeder in Fennoscandia and across Russia, with a smaller fragmented population in eastern Europe. Most of the European population is in Russia (BirdLife International 2015). It winters in sub-Saharan Africa. Many populations have declined and become fragmented, mainly because of loss of wetland habitat, but also possibly over-hunting (Parkin and Knox 2010). Intensification of land use was the main driver of the decline in the 20th century, but in recent decades, overgrowing of breeding habitat, following the abandonment of agriculture, has contributed to negative trends (*European Atlas* 2020). This species appears to have been much more frequent in Britain in the period before the First World War than subsequently. There was a total of 169 records in Britain between 1950 and the end of 2019 (Holt *et al.* 2020).

The earliest acceptable records in Wales were in Breconshire: one shot on Mynydd Eppynt about 1876 and one shot near Cray about 1876 or 1877. In all, 9 birds were shot between 1876 and 1947. Three were in Ceredigion, two in Breconshire, two in Denbighshire and one each in Caernarfonshire and Pembrokeshire.

There have been only two records since then, both of first-calendar-year birds. One was on Bardsey, Caernarfonshire, on 10 October 2003 and the other was trapped and ringed at Llanbadarn Fynydd, Radnorshire, on 9 October 2014.

Jon Green and Robin Sandham

Snipe *Gallinago gallinago* Gïach Cyffredin

Welsh List Category	IUCN Red List			Birds of Conservation Concern			
	Global	Europe	GB	UK	Wales		
A	LC	LC	NT(NB)	Amber	2002	2010	2016

As a breeding bird, the Snipe has a very wide distribution, covering much of Europe and Asia, with isolated populations as far south as Kashmir. It winters to the south of its range, in Africa, India and Southeast Asia, although the British breeding population is thought to be largely resident (*HBW*). It is found in areas with a combination of grassy cover and moist earth, such as open marshland, wet grassland and the edges of lakes and rivers. Snipe breeding in Wales are of the nominate subspecies *G.g. gallinago*; another subspecies, *G.g. faeroeensis*, breeds in Iceland, the Faroe Islands, Orkney and Shetland. Ringing recoveries show that birds ringed in several European countries, particularly Germany, The Netherlands, Denmark and Sweden, move to Wales in winter. Few of these were ringed as nestlings, so some may have been on migration from breeding grounds farther east. Birds thought to show plumage characteristics of the subspecies *G.g. faeroeensis* have been recorded in Wales, and while there have yet to be any recoveries that prove Snipe breeding in Iceland move to Wales, one-quarter of foreign-ringed Snipe recovered in Ireland were ringed in Iceland (Appleton 2016c).

Comments from authors writing in the late 19th and early 20th centuries suggest that the Snipe was then a much commoner breeding bird in Wales than it is now. Forrest (1907) described it as "common on bogs almost everywhere" in North Wales, most numerous in the uplands, but he also noted that several observers considered it to be decreasing. The *Britain and Ireland Atlas 1968–72* showed Snipe to be widely distributed in Wales in the

Snipe: Recovery locations of birds ringed in Wales are shown by red circles. Ringing locations of birds that were recovered in Wales are shown by blue triangles. Small brown dots show ringing or recovery locations in Wales.

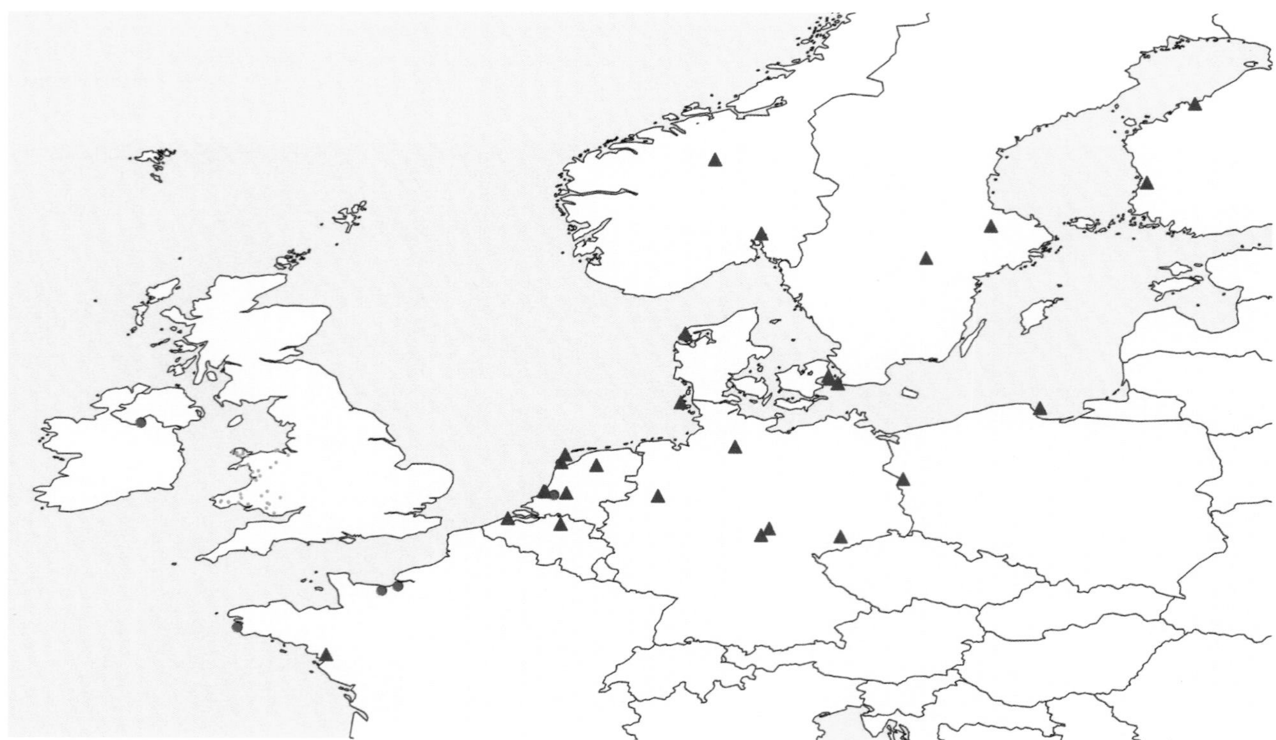

Snipe © John Hawkins

breeding season, occupying 71% of 10-km squares, although few squares were occupied in Pembrokeshire, adjacent areas of Ceredigion and Carmarthenshire, or on Anglesey. The 1988–91 *Atlas* showed a similar pattern of distribution, but with evidence of a contraction in range, to 46% of squares. The *Britain and Ireland Atlas 2007–11* showed a further decline, to 39% of 10-km squares in Wales, with the great majority of breeding Snipe in the uplands and few records from lowland areas. Even in the uplands, relative abundance was low compared with Scotland, northern England and the west of Ireland, and had declined since 1988–91.

In the lowlands, a survey of wet grassland in England and Wales in 2002 found a reduction of 61% in numbers of Snipe, compared to the previous survey in 1982, with birds becoming highly aggregated into a small number of suitable sites. Breeding Snipe were very scarce in Wales in both surveys: 12 pairs in 115km^2 (0.10 pairs/km^2) in 2002, an increase of one pair on 1982 (Wilson *et al.* 2005). There is very little information available on trends in the uplands of Wales, other than from the breeding atlases. A resurvey, in 2000 and 2002, of study areas that had been previously surveyed between 1980 and 1991, suggested a decline in North Wales, but Snipe were again only recorded in very small numbers and thus the changes were not significant (Sim *et al.* 2005). A similar pattern was found in the Berwyn SPA between 1983–85 and 2002.

The most important area for breeding Snipe in Wales is probably the Elenydd, which is split between Radnorshire, Breconshire and Ceredigion. A survey of part of this area by RSPB/CCW in 1995 found 113 territories, of which 69 were in Ceredigion. Walk-through surveys of 109 whole or part 1-km squares, covering the entire National Trust-managed upland in the south and southeast Elenydd, in 2016–17, found only 11 squares (10.1%) occupied (Andrew King *in litt.*). Jennings (2014) estimated that at least 100 pairs bred in the whole of Radnorshire, and the total in Breconshire is likely to be similar. In Ceredigion, Roderick and Davis (2010) quoted an estimate of at least 75 pairs on Cors Caron, with smaller numbers at several other sites in the county. Farther south, numbers are smaller. In Pembrokeshire, where Mathew (1894) said that it still nested all over the county in suitable places, birds were recorded in just nine tetrads in the *Pembrokeshire Atlas 1984–88*, with no confirmed breeding, and Donovan and Rees (1994) estimated a population of about ten pairs in the county. By the *Pembrokeshire Atlas 2003–07*, Snipe were only recorded in three tetrads that contained potentially suitable breeding habitat. In East Glamorgan, Gower and Carmarthenshire the species breeds at scattered locations. The East Glamorgan atlas recorded Snipe with breeding codes in nine tetrads in 2008–11 compared to 38 (9% of all tetrads) in 1984–89. The population in Gwent showed a sharp decline in the lowlands between the 1981–85 and 1998–2003 Gwent atlases, from 51 to 18 occupied tetrads, with a lesser decline in the uplands, from 30 to 23 occupied tetrads. Venables *et al.* (2008) estimated the county population at 20–30 pairs.

In North Wales, the species is most widespread in Denbighshire, on Mynydd Hiraethog, where Bain (1987) found 14–21 pairs, and on the eastern part of the Migneint. In Caernarfonshire, the RSPB survey of Llŷn in 1986 estimated 30–60 pairs, but apart from a few pairs on lowland bogs, the county's population is now largely confined to Snowdonia and estimated at little more than 40 pairs (Pritchard 2017). An RSPB survey estimated 28–31 displaying birds at the Mawddach Estuary, Meirionnydd, in 1994, with 33 drumming/chipping birds there in 2018. A survey in 1986 estimated 18 pairs on bog and rough pasture near Trawsfynydd, Meirionnydd. The *North Wales Atlas 2008–12* recorded Snipe in only 18 tetrads in Meirionnydd, suggesting a significant decline. There were only scattered pairs in Flintshire and Anglesey. *Birds in Wales* (1994) estimated the breeding population in Wales to be *c.*300–500 pairs in an average year; probably an underestimate at that time. The latest population estimate for Wales (Hughes *et al.* 2020) is 1,100 pairs (820–1,400 pairs).

No BBS trend can be calculated for Wales. The UK index increased by 27% between 1995 and 2018 (Harris *et al.* 2020), but the evidence available suggests that this may not be true of Wales. Snipe is poorly recorded by the BBS or atlases as it requires specific techniques to monitor drumming birds at dawn and dusk (Wilson *et al.* 2005).

In winter, large numbers of Snipe move into Wales. It seems always to have been a common species in winter, but Mathew (1894) said that it was far less plentiful in Pembrokeshire than it had been 50–60 years previously, when one wildfowler had shot 120–140 in a day on the Trecwm estate, though Mathew noted that there were still remote areas in the county where 40–60 birds could be shot in a day. The number of Snipe reported shot in most estate game books in the late 19th and early 20th centuries is not large, with few having recorded more than 150 for a season. A total of 244 was shot on the Ashburnham Estate, Carmarthenshire, in the 1911/12 shooting season (Lloyd *et al.* 2015).

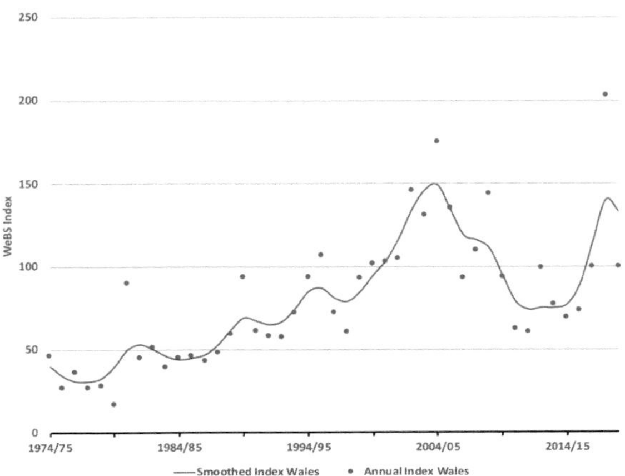

Snipe: Wetland Bird Survey indices, 1974/75 to 2018/19.

WeBS counts probably greatly underestimate the numbers of such a skulking species. The Dee and Severn estuaries can hold good numbers, notably 1,293 on the Dee Estuary in 1972/73, but the majority of birds are usually on the English side of both estuaries. Large counts are sometimes the result of cold weather driving birds westward. In Carmarthenshire, there were an estimated 1,100 at Cydweli Flats on 16 February 1974, and in Pembrokeshire 400–500 at Dale airfield on 4 January 2010. There were 650 at RSPB Ynys-hir, Ceredigion, on 27 November 2008, and there was a cold-weather influx of 500 onto Bardsey, Caernarfonshire, in January 1963. On Anglesey, RSPB Cors Ddyga often holds large numbers, including 570 on November 2004, 573 in October 2007 and 500 on 25 November 2017.

The future of the Snipe as a breeding bird in Wales is uncertain, although the population may now be stable, following a major decline during the second half of the 20th century. Declines on lowland wet grassland are attributed to drainage causing the drying out of sites and changes in the intensity of grazing, including increases in stocking densities and an earlier start to the grazing season (Wilson *et al.* 2005). Drainage is likely to have had an effect on breeding Snipe in some upland areas too, but there has been little research on this. The conservation of existing wetland areas, and restoration of blanket bog and river valley mires, will be key to maintaining the breeding population. However, this might not be enough on its own. Smart *et al.* (2008) noted that Snipe had continued to decline at many lowland wetland reserves, despite management to improve soil conditions, suggesting that other

Site	County	74/75 to 78/79	79/80 to 83/84	84/85 to 88/89	89/90 to 93/94	94/95 to 98/99	99/00 to 03/04	04/05 to 08/09	09/10 to 13/14	14/15 to 18/19
RSPB Cors Ddyga	Anglesey				4	58	62	397	177	236
Castlemartin Corse	Pembrokeshire						125	55	100	218
Cors Crugyll	Anglesey					44	153	86	151	174
Cleddau Estuary	Pembrokeshire	40	93	156	187	173	230	145	99	117

Snipe: Five-year average peak counts during Wetland Bird Surveys between 1974/75 and 2018/19 at sites in Wales averaging 150 or more at least once. International importance threshold: 20,000; Great Britain importance threshold: 10,000.

factors, such as food supply, were important. The area of Britain that is climatically suitable for Snipe is expected to reduce from the south during the 21st century (Huntley *et al.* 2007), adding to the pressure on its status as a breeding species in Wales. However, fewer harsh winters could help this species, as breeding numbers drop after cold winters. For example, BBS data showed a 40% UK-wide decline between 2010 and 2011.

The Snipe is a legal quarry species in Britain, and most ringing recoveries are from birds that have been shot (*BTO Migration Atlas* 2002). Birds breeding in southern Britain are more likely to be found in mainland Europe in winter than those breeding in northern Britain, so hunting in France and elsewhere could also affect the Welsh breeding population. The wintering population in Wales is probably quite a small proportion of the estimated million birds wintering in Britain (Frost *et al.* 2019a), but Wales could be of greater importance during periods of harsh weather farther east.

Rhion Pritchard

Terek Sandpiper *Xenus cinereus* Pibydd Lludlwyd

Welsh List Category	IUCN Red List		Rarity recording
	Global	Europe	
A	LC	LC	BBRC

This wader breeds at northern latitudes from Finland east through Siberia and winters mainly around the Indian Ocean coasts (*HBW*). There has been a total of 92 individuals in Britain to the end of 2019, with an average of about two a year during the last 30 years (Holt *et al.* 2020), most in southeast England. There have only been two records in Wales. The first was a bird at RSPB Conwy, Denbighshire, from 29 April to 3 May 1999, which was presumed to have been the same individual seen at Frodsham, Cheshire on 26–27 April. The other record was at Cemlyn, Anglesey, on 21–23 June 2005.

Jon Green and Robin Sandham

Wilson's Phalarope *Phalaropus tricolor* Llydandroed Wilson

Welsh List Category	IUCN Red List	Rarity recording
	Global	
A	LC	BBRC

This species breeds from interior western Canada south to California and east to the Great Lakes area. Most birds winter in western and southern South America (*HBW*). In Britain, there were 247 records to the end of 2019, with the last in 2017 (Holt *et al.* 2018), an average of 3–4 each year over the last 30 years. Numbers increased from the 1970s to a peak during 1977–99, when there was an average of *c.*11 a year. This coincided with increased autumn migration through eastern North America, perhaps due to drought conditions in Quebec and Ontario (Parkin and Knox 2010). Numbers in Britain have since declined, with none in Wales since 1998.

There have been 12 records of 13 individuals in Wales:

- female at Malltraeth, Anglesey, on 15–16 June 1958;
- Shotton, Flintshire, from 30 August to 4 September 1959;
- Bettisfield Pools, Denbighshire, from 11 October to 1 November 1964;
- Pembrey, Carmarthenshire, on 17 November 1974;
- Llyn Heilyn, Radnorshire, on 6 September 1975;
- Connah's Quay, Flintshire, on 5 September 1982;
- Glan Conwy, Denbighshire, from 4 October to 8 December 1989;
- Point of Ayr, Flintshire, on 6 September 1991;
- two, an adult and a 1CY, at Glan Conwy, Denbighshire, on 28 September 1991;
- Penclacwydd, Carmarthenshire, on 29–30 September 1991;
- Point of Ayr, Flintshire, from 24 September to 6 October 1997, and then at Morfa Madryn, Caernarfonshire, from 21 October to 3 November 1997 and
- Broadwater, Tywyn, Meirionnydd, on 27–28 August 1998.

Jon Green and Robin Sandham

Red-necked Phalarope *Phalaropus lobatus* Llydandroed Gyddfgoch

Welsh List Category	IUCN Red List			Birds of Conservation Concern	Rarity recording
	Global	Europe	GB	UK	
A	LC	LC	EN	Red	WBRC

The Red-necked Phalarope has a circumpolar distribution around the shores of the Arctic Ocean and south as far as Labrador and northern Scotland. It breeds on coastal moorland, bogs and at or near lakes and pools with marshy margins (*HBW*). It is pelagic during the non-breeding season, wintering in tropical waters. A female fitted with a geolocator on Fetlar, Shetland, was found to winter in the southeast Pacific Ocean, between the Galapagos Islands and the South American coast (Smith *et al.* 2014). The numbers of passage migrants reported in Britain increased during 2010–19. In Wales there has been an increase in records since 2000.

Total	Pre-1950	1950–59	1960–69	1970–79	1980–89	1990–99	2000–09	2010–19
67	12	10	9	4	5	3	12	12

Red-necked Phalarope: Individuals recorded in Wales by decade.

The first Welsh record was at Valley, Anglesey, on 5 October 1893. The majority of records have been in autumn, peaking in September, but since 2000, there has been a stronger bias than previously to spring records, with ten of the 24 seen in spring, all but one in June. The site with most records is Kenfig Pool, East Glamorgan, where seven individuals have been recorded, including three on 30 September 1956.

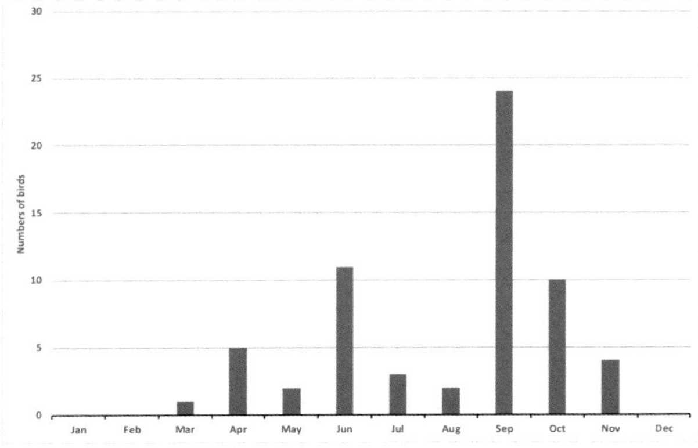

Red-necked Phalarope: Totals by month of arrival, 1893–2019.

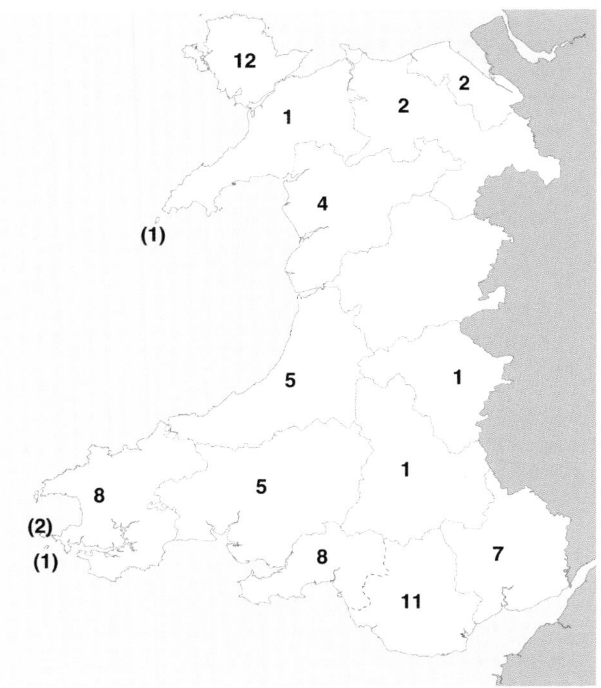

Red-necked Phalarope: Numbers of birds in each recording area, 1893–2019.

The majority of records have been coastal, but a dead bird was found at Cantref Reservoir, Breconshire, on 17 September 1950. Others were seen at Llandegfedd Reservoir, Gwent, on 12–16 September 2004 and Llyn Heilyn, Radnorshire, from 30 April to 2 May 2012.

There is no indication whether those seen in Wales are from Scotland, Iceland or even Fennoscandia, but it may become a rare visitor over time, as the range in all three areas is expected to be reduced substantially as a result of climate change, with none predicted to breed in Scotland by the end of the century (Huntley *et al.* 2007).

Jon Green and Robin Sandham

Grey Phalarope *Phalaropus fulicarius* Llydandroed Llwyd

Welsh List Category	IUCN Red List	
	Global	Europe
A	LC	LC

First-calendar-year Grey Phalarope © Ben Porter

Known in North America as Red Phalarope, this wader breeds mainly on marshy tundra near the coasts of the Arctic Ocean and has a circumpolar distribution, including small populations as close to Wales as Iceland. It is the most pelagic of the phalaropes, wintering at sea in areas of upwelling ocean currents, off the coast of western and southwestern Africa and western South America (*HBW*). The largest numbers recorded in Britain and Ireland are usually off southwestern Ireland, England and Wales, mainly on autumn migration. These are probably birds from northeast North America, Greenland and Iceland that are migrating southeast across the North Atlantic, to winter off western Africa (*BTO Migration Atlas* 2002), although there are no ringing recoveries to confirm this. The few winter records are probably birds wintering in the North Atlantic that have been blown inshore by storms. The global population trend for the species is not known (BirdLife International 2020).

Records in Wales are usually on the coast, but it is also sometimes recorded inland, and there have been records from every county. Numbers vary from year to year, with large influxes following strong westerly winds. Forrest (1907) recorded such an influx in autumn 1891, when 17 were shot on the Dyfi Estuary, Ceredigion/Meirionnydd/Montgomeryshire, in rough weather during 15–24 October, with several more shot inland near Bala, Meirionnydd, in December that year. There was a notable influx in early autumn 1960, with six in Glamorgan and very large numbers passing the Smalls, off Pembrokeshire, including 227 on 29 September. This was part of an extraordinary influx that included totals of over 1,000 on both mainland Cornwall and the Isles of Scilly.

The increased interest in sea-watching, particularly from Strumble Head, Pembrokeshire, and Bardsey, Caernarfonshire, resulted in Grey Phalarope being recorded far more frequently from the 1990s onwards.

A significant influx was evident in October 2001, including 11 in East Glamorgan, six in Gower and 18 in Carmarthenshire. This included seven at Llanelli beach on 4 October. Another

	1990–99	2000–09	2010–19
Gwent	3	3	11
East Glamorgan	4	33	22
Gower	3	25	24
Carmarthenshire	4	35	12
Pembrokeshire, excluding Strumble Head	4	35	16
Strumble Head	70	71+	109
Ceredigion	11	30	23
Breconshire	1	2	0
Radnorshire	0	1	1
Montgomeryshire	0	1	0
Meirionnydd	0	9	13
Caernarfonshire, excluding Bardsey	5	18	20
Bardsey	10	17	28
Anglesey	1	16	7
Denbighshire	1	1	6
Flintshire	1	0	1
Total	**118**	**297**	**293**

Grey Phalarope: Individuals recorded in Wales by decade since 1990.

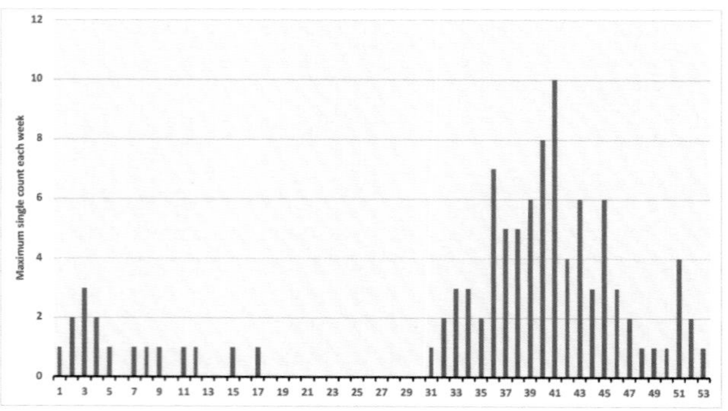

Grey Phalarope: Maximum single counts, each week, of birds passing around Wales, 1955–2019. Week 3 begins *c.*16 January, week 33 begins *c.*13 August and week 51 begins *c.*17 December. Records are from BirdTrack.

influx, in September 2018, followed a deep depression that hit the Western Approaches and resulted in a total of 36 recorded in Wales, including four at Lisvane Reservoir, East Glamorgan, on 21 September and an inland record of one at Claerwen Reservoir, Radnorshire, on 19 September.

The highest count in recent years was 20 past the Smalls, Pembrokeshire, on 17 September 2008. The site with the greatest number of records is Strumble Head, Pembrokeshire, where observations since 1983 have produced an average of nine a year, the highest being 35 in 2001. Autumn passage, linked to storms off Strumble, can occur from late July to mid-December, peaking between mid-September and early October. Totals there have increased, particularly since 2000, and include day totals of ten on 8 October 2001 and 7 October 2011. Most of this increase can be attributed to increased observer effort and experience.

Most records have been from the coast, but birds are sometimes found on freshwater, particularly at large reservoirs. Gwent, for example, has had ten records on freshwater, including five at Llandegfedd Reservoir and three at Ynys-y-fro Reservoirs. In the inland counties, there have been three records in Breconshire and four in Radnorshire. The only record in Montgomeryshire was one at Llyn Coed y Dinas on 19 December 2000.

Jon Green

Common Sandpiper *Actitis hypoleucos* Pibydd Dorlan

Welsh List Category	IUCN Red List			Birds of Conservation Concern			
	Global	Europe	GB	UK	Wales		
A	LC	LC	VU	Amber	2002	2010	2016

The Common Sandpiper breeds over large areas of Europe, Russia and Central Asia, in a very broad range of latitudes and climatic zones, from Mediterranean to montane and steppe habitats and from the coast to high mountains. It winters over a wide area, in Africa, southern Asia and Australasia, in a wide variety of habitats, though it usually avoids large coastal mudflats (*HBW*). In Britain the Common Sandpiper is typically found on upland streams and lakes with rocky or shingle shores.

The species appears to have been common over most of Wales in the late 19th and early 20th centuries, with pairs on most inland streams in Glamorgan in the late 1890s (Hurford and Lansdown 1995), and on the upper reaches of the Western Cleddau and near Maenclochog, Pembrokeshire (Mathew 1894). In North Wales, Forrest (1907) said that it was generally distributed and common in all suitable places, except on the Llŷn Peninsula. It bred from sea level to high in the uplands; a pair nesting at *c.*760m by Ffynnon Caseg in the Carneddau, Caernarfonshire, in 1916 (Forrest 1919).

Lockley *et al.* (1949), despite noting Mathew's comments in 1894, said that there was no evidence that Common Sandpipers had bred in Pembrokeshire. Ingram and Salmon (1954) said that Common Sandpipers had been very scarce in Carmarthenshire for several years, but that there had been a partial recovery in 1953. The same authors in 1955 said that the species was much reduced in Radnorshire, compared with 30 years previously, and in 1957 they noted a sudden and sharp decline in Breconshire in the 1950s. Parslow (1973) considered that the species had decreased in some parts of England, Scotland and Wales since 1950 or earlier, although perhaps only locally, and Nethersole-Thompson and Nethersole-Thompson (1986) stated that "between the 1930s and 1960s numbers fell and the range contracted in SW England, Wales and the English Lake District, east Scotland and Orkney.'

The *Britain and Ireland Atlas 1968–72* showed the Common Sandpiper to be a widespread breeder in Wales, occurring in 57% of 10-km squares. However, it was absent, or largely absent, from Anglesey, Llŷn, Pembrokeshire, adjacent areas of south Ceredigion and western Carmarthenshire, and the southeast coast. The 1988–91 *Atlas* showed a slight net loss in distribution but included gains in Anglesey and Llŷn. By the 2007–11 *Atlas*, the gains in these two areas had been reversed and there were extensive losses throughout eastern Wales; overall, the proportion of occupied squares had fallen to 44%.

Tetrad atlases showed declines in the two southern counties where comparisons are possible. In Gwent, Common Sandpipers were reported in 15% of tetrads in the 1981–85 *Atlas* and in 12% in 1998–2003, though there were more records of confirmed or probable breeding in the later period (Venables *et al.* 2008). The Rivers Usk and Monnow are the strongholds in the county, but fewer were found on the Monnow in the second atlas. In East Glamorgan, they were recorded with breeding codes in 3% of tetrads in the 1984–89 *Atlas* but in only 2% in 2008–11. The *West Glamorgan Atlas 1984–89* found at least possible breeding in 10% of tetrads, when the population was estimated at 30 pairs. No breeding was recorded in Pembrokeshire in either the 1984–88 or the 2003–07 tetrad atlases, although a pair bred at Clarydale near Haverfordwest in 1995, the first and only record of breeding in the county since Mathew's time.

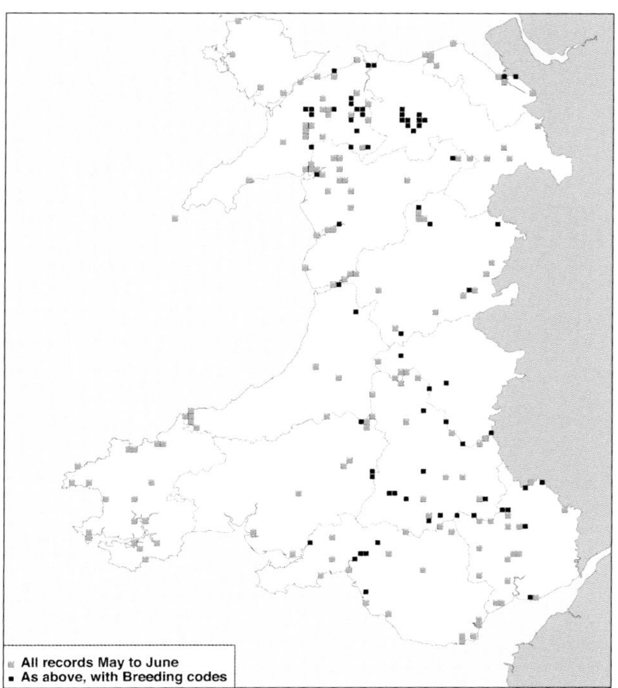

Common Sandpiper: Records from BirdTrack for May to June, 2015–19, showing the locations of records and those with breeding codes.

The *North Wales Atlas 2008–12* found the species in 10% of tetrads, the great majority on higher ground in the west. Aside from a very few records of possible breeding on the coast, which may have been passage migrants, the species was absent from Anglesey and Llŷn. It was also almost absent from Flintshire, where there were records of possible breeding in just four tetrads. Common Sandpipers were recorded in 14% of tetrads in Meirionnydd and 16% in Caernarfonshire. It was noted that while there may have been only a slight contraction in range across North Wales there had been a probable decline in numbers.

Surveys of rivers and lakes provide further evidence of a decline in the population. In Gwent, for example, there were up to 15 pairs on the River Monnow between Pandy and Monmouth in the 1980s, but only six pairs were found in 2002 (Tyler 2002). During 2010–19, no more than three pairs were found on the Monnow, with only two of these between Pandy and Monmouth (Steph Tyler *in. litt.*). There was a decline in some parts of Radnorshire from the mid-1980s, thought to be due to increased recreational disturbance and Mink predation. Up to 30 pairs were in the Radnorshire part of the Elan Valley in 1997, where subsequent whole or partial surveys have shown a decline to 21 in 2008 and 19 pairs in 2011 (Jennings 2014). An incomplete survey of the River Wye and major tributaries in 2003 found 67 pairs, but an exact repeat of this survey in 2010 found only 44 pairs. On the Pumlumon SSSI, across the Ceredigion/ Montgomeryshire border, numbers declined from ten pairs in 1984 to only one in 2011 (Crump and Green 2012). In southern Ceredigion, on the western edge of the Elenydd, similar declines were evident between 1983 surveys and a visit in 2012 (pers. obs.).

In Caernarfonshire, Pritchard (2017) noted that "though distribution does not seem to have changed, indications are that birds are not as numerous as formerly." In that county, a survey of the Afon Ogwen in 1974–75 found nine pairs, but another survey in 2005–07 found only 4–5 pairs (Gibbs *et al.* 2011). At Llyn Conwy there were eight pairs in the 1980s, but only four pairs in 2014. There is little information on the fortunes of this species on Anglesey: confirmed breeding was recorded in both the 1968–72 and 1988–91 Britain and Ireland atlases, but not in the 2007–11 *Atlas*.

Breeding densities vary considerably. For example, in Montgomeryshire, a 1978 survey showed average densities of 2.77 territories/10km on the River Severn, 3.0 territories/10km on the Afon Vyrnwy and 4.86 territories/10km on the River Wye (Round and Moss 1984). In Meirionnydd, there were 19 pairs and another six individuals along the 11.4km shoreline of Llyn Celyn in 1971, and 19 pairs along 14km of Llyn Trawsfynydd shoreline in 1986 (Pritchard 2012). It was estimated that there were 21 pairs on Llyn Tegid, Meirionnydd, in 1991.

Common Sandpipers are recorded in too few squares in Wales, by BBS and WBBS, for a population trend to be calculated, but there is good evidence for a decline in numbers. There was a 40% decline in England, and a 24% decline in Scotland, during 1995–2018 (Harris *et al.* 2020), reflecting the trend in Europe, where there was a 36% decline between 1980 and 2016 (PECBMS 2020). *Birds in Wales* (1994) estimated the size of the breeding population in Wales at *c.*1,000 pairs. The latest estimate (Hughes *et al.* 2020) is just 490 pairs (395–565 pairs).

There is considerable passage of Common Sandpipers through Wales. Birds on spring passage, which starts in April, can be confused with breeding pairs. Southbound migration starts from late June and usually peaks in July, although it can continue into the autumn. Jennings (2014) considered that in Radnorshire peak autumn passage is at the end of July and early August. There are a few records of juveniles in late September and October. In Ceredigion, numbers on estuaries increase from mid-July and in

Displaying Common Sandpiper © Tony Pope

the 1970s and 1980s 25–40 were recorded at Glandyfi and RSPB Ynys-hir on the Afon Dyfi, including a Welsh record count of 82 in July 1981, although numbers subsequently declined (Roderick and Davis 2010). Other large counts on return passage include 41 at the Ogmore Estuary, East Glamorgan, on 8 July 1989 and 45 on the Clwyd Estuary, Denbighshire/Flintshire, on 19 July 1992. Numbers on spring passage are usually lower, but there were 37 at Eglwys Nunydd Reservoir, Gower, on 21 April 1983. Birds on migration can pass through coastal sites very quickly, and therefore this species is not well monitored by monthly WeBS counts.

A small number over-winter in Wales, mainly on sheltered coasts and estuaries, often at traditional sites, which include the lower Usk in Gwent, the Afan Estuary in Gower, the Cleddau Estuary in Pembrokeshire, the Teifi and Dyfi estuaries, Tal-y-cafn on the Conwy Estuary and Church Island, Anglesey, on the Menai Strait. There are also inland records, most frequently on the Wye at Glasbury. *Birds in Wales* (1994) reported that in most years no more than 5–10 birds wintered, and while numbers vary, they have certainly increased since then. The largest winter counts are often in February, notably 16 on the River Wye near Wyndcliff, Gwent, on 4 February 1996, 14 on the Teifi Estuary, Pembrokeshire/Ceredigion, on 28 February 2008 and 11 at Magor Pill, Gwent, on 13 February 2017.

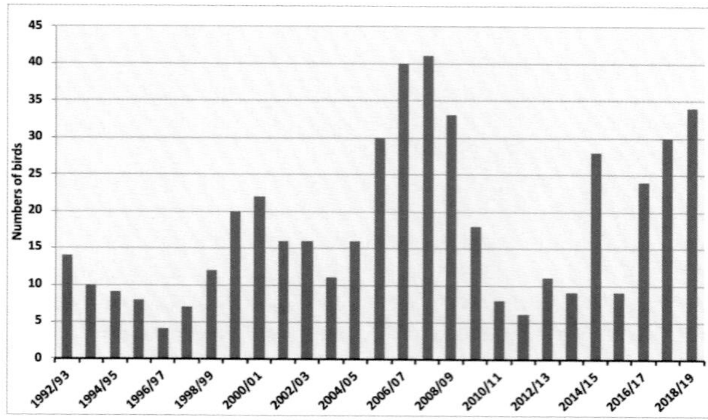

Common Sandpiper: Numbers overwintering in Wales, 1993/94 to 2018/19.

There are few ringing recoveries involving Wales. Two adults ringed in Wales in July, one at Shotton, Flintshire, in 1981 and the other in the Elan Valley, Radnorshire, in 1993, were later recovered in Morocco on spring migration. One, ringed on Bardsey, Caernarfonshire, in July 1957, was shot in Nantes, France, in July 1959, and one ringed as a nestling at Lüneburg, Germany, on 13 May 1963 was recovered at Burry Port, Carmarthenshire, on 7 September the same year. Recoveries from birds ringed across Europe indicate that there is a SSW movement in autumn, with over-wintering in West Africa (Dougall *et al.* 2010).

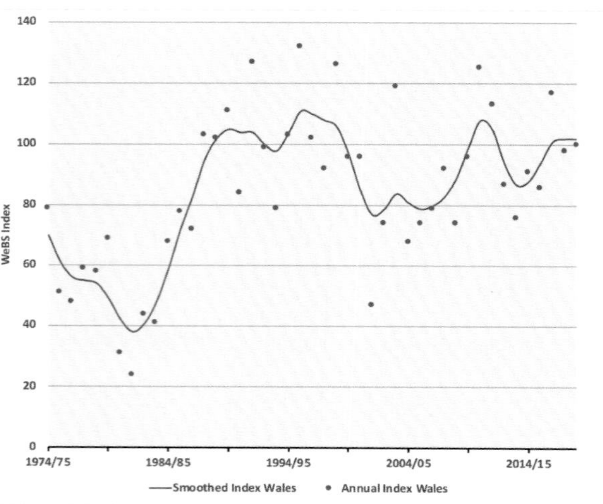

Common Sandpiper: Wetland Bird Survey indices, 1974/75 to 2018/19.

The reasons for the decline of the breeding population are unclear, but potentially include climate change, disturbance, water quality and habitat change, both here and on the wintering grounds, although none of these are proven (Dougall *et al.* 2010). Summers *et al.* (2019a) considered that the increasing frequency and severity of droughts in Iberia and Morocco, both important refuelling stops on migration, were another potential hazard. Common Sandpipers are susceptible to recreational disturbance, but this might not a serious problem on many of the more remote Welsh upland rivers and pools on which declines have occurred. Climate change could affect the populations of some freshwater prey species, since many of the common insect larvae in such watercourses prefer colder water (Durance and Ormerod 2007). Climate change and habitat loss, affecting birds on migration or on the wintering grounds, could also be a factor. In the long term, the species' British range is expected to contract northwards out of Wales and much of England by 2100 (Huntley *et al.* 2007).

The Common Sandpiper is Red-listed in Wales, because of a severe population decline over 25 years and over the longer term (Johnstone and Bladwell 2016). While it remains widespread across much of Wales, the declines are worrying and there have been few surveys and little research. The causes of decline are uncertain, so all parts of the life cycle of the species need to be investigated. This would involve improved monitoring of upland rivers and lakes, along with research on the migration routes and wintering areas. Modern tracking technology should be able to help us understand more about this endearing species' movements and habitat needs, and take action to stem declines.

Mick Green

Sponsored by Rhion Pritchard

Spotted Sandpiper *Actitis macularius* Pibydd Brych

Welsh List Category	IUCN Red List Global	Rarity recording
A	LC	BBRC

Closely related to the Common Sandpiper, this species breeds in North America from western Alaska east to Newfoundland and south to California, Texas and North Carolina. Some birds winter in the coastal USA but most winter in Central and South America (*HBW*). In Britain, its occurrence has increased, with 225 individuals recorded up to the end of 2019 (Holt *et al.* 2020), an average of about five each year over the last 30 years. There has been no such increase in Wales, with only 12 records, on average two per decade. Two of these birds over-wintered.

All records have been of single birds:

- Whitland, Carmarthenshire, on 15–18 May 1960;
- Oxwich, Gower, from 27 August to 4 September 1973;
- Aberthaw, East Glamorgan, on 24–25 August 1974;
- RSPB Ynys-hir, Ceredigion, and Pennal, Dyfi, Meirionnydd, from 9 October to 20 November 1975;
- Bardsey, Caernarfonshire, on 16 September 1977;
- RSPB Ynys-hir, Ceredigion, on 7–12 August 1979;
- Bosherston, Pembrokeshire, from 5 October to 2 December 1980;
- Peterstone, Gwent, from 26 October 1980 to 25 April 1981;
- Porth Colmon, Caernarfonshire, on 29 September 1981;
- 1CY at Lisvane Reservoir, East Glamorgan, from 20 October 2007 to 28 April 2008;
- adult at Malltraeth, Anglesey, on 2 June 2009 and
- adult on Skokholm, Pembrokeshire, on 12 July 2016.

Jon Green and Robin Sandham

Green Sandpiper *Tringa ochropus* Pibydd Gwyrdd

Welsh List Category	IUCN Red List			Birds of Conservation Concern			
	Global	Europe	GB	UK	Wales		
A	LC	LC	EN	Amber	2002	2010	2016

This wader breeds in sub-boreal areas from Fennoscandia and eastern Europe, through Russia to the Pacific Ocean. Most birds winter to the south of the breeding range (*HBW*). In Britain, a few pairs breed in the Highlands of Scotland (*The Birds of Scotland* 2007). Breeding has never been confirmed in Wales. It was thought to have bred at Penllergaer, Gower, in the latter half of the 19th century, but no nest was ever found (Hurford and Lansdown 1995). The status of this species in Wales in the late 19th century and early 20th century was probably much the same as today. It was described as fairly common in Pembrokeshire in autumn and sometimes found there in winter (Mathew 1894), but "distinctly rare" in western parts of North Wales and never found on Anglesey (Forrest 1907). Phillips (1899) stated that the species was found every spring and autumn on the Usk near Talybont, Breconshire, and had often been killed there. He thought that a pair might have bred there in 1898.

Green Sandpipers winter mainly in the Mediterranean basin and in Africa, north and south of the Sahara. A few move as far south as Botswana and South Africa. Some spend the winter in Britain, at the very northern edge of its wintering range, widely spread across southern England and the Midlands (*Britain and Ireland Atlas 2007–11*). Smith *et al.* (1984) suggested a wintering population of 600 in Britain and Ireland, but the most recent estimate was only 290 birds in Britain, though it was noted that this was a minimum (Frost *et al.* 2019a). Limited data from ringing suggests that these birds originate from the Fennoscandian breeding population. Only two ringing recoveries involve Wales: one ringed on the River Wye near Llanigon, Breconshire/Radnorshire, in August 1964 was shot 3km to the northeast at Cusop, Herefordshire, in September 1965, and a first-calendar-year bird ringed in Ceredigion in July 2004, was recaught at Land's End, Cornwall, eleven days later.

Favoured haunts in Wales are coastal ditches, drainage channels and pools, including saline lagoons. They also occur around the edges of lakes and reservoirs, and bogs such as Cors Caron, Ceredigion. Rivers and streams are regular haunts, where birds feed on shoals and muddy areas and in oxbow pools in floodplains. Sewage works are much used in England, but the only site in Wales where this habitat is regularly used is at Penrhyndeudraeth, Meirionnydd. Flooded fields may also be used, with nine noted on this habitat near Denbigh in January 2017.

In Wales, the *Britain and Ireland Atlas 2007–11* found wintering birds were most widespread in southern areas, with scattered records elsewhere. Green Sandpipers were present in almost twice as many 10-km squares as in the 1981–84 *Atlas*, particularly in Pembrokeshire and Anglesey. The East Glamorgan tetrad atlas, covering the winters 2007/08 to 2010/11, found Green Sandpipers in 22 tetrads (5%). The numbers wintering in Wales vary considerably from year to year, but there is some evidence of a decline. WeBS Medium Alerts have been issued for Wales over the long term (-48%) and the medium term (-30%) (Woodward *et al.* 2019). In most years, 20–30 are recorded in January and February, but there were up to 42 birds present in January 2001 and *c.*37 in January 2017. Many other birds will have gone unseen, especially given the species' use of shallow field ditches. The highest winter count recorded in Wales was 12 at Pennard Pill, Gower, on 21 February 1982. The wintering population in Wales probably does not exceed 100–150.

The Green Sandpiper also occurs as a passage migrant. It is difficult to be certain whether birds seen in March and April are on passage or over-wintering, as those doing the latter do not leave until early April or sometimes early May. There are fewer records in May than in other months. It is one of the earliest waders to return on autumn passage, from late June until October, with some into early November. Peak WeBS counts in Britain on passage are 400–600 in August. Autumn passage in Wales is much more marked than spring passage. For example, out of a total of 366 individuals recorded in 2006, 217 (59%) were recorded between July and September (97 in July, 84 in August, 36 in September). Peers and Shrubb (1990) reported that 64% of records in Breconshire during 1969–89 were between July and September. Green Sandpipers are usually seen in ones or twos but larger numbers are sometimes recorded on autumn passage. The highest recent spring count was 12 on the River Usk at Gobion, Gwent, on 6 April 2016. The same section of the Usk, at Llanvihangel Gobion/The Bryn, held 25 in August 1987 and 26 on 28 July 1984. In August 1984, 39 were seen along the Gwent section of the River Usk (Venables *et al.* 2008) and there were 20 at Llandegfedd Reservoir in July 2003. The largest count during 2000–19 was 20 at Llanrhidian, Gower, on 5 August 2013. In mid-Wales, there were 18 at Cors Caron, Ceredigion, in late July and August 2004, and 14 were recorded at Dolydd Hafren, Montgomeryshire, in July in 2003 and 2006. Occasionally birds remain all year, as at Llanishen Reservoir in East Glamorgan, when two remained from October 1931 to September 1932. Occasionally one summers, as at Teifi Marshes, Pembrokeshire/Ceredigion, in 1994.

At least 200–300 birds must pass through Wales every late summer and autumn. Numbers fluctuate from year to year, with no clear trend apparent. There is no evidence that passage birds choose different habitats in Wales to wintering birds, as Smith *et al.* (1984) found in Hertfordshire, where wintering birds preferred watercress beds. That study found them to be more solitary in winter, even territorial, and found a strong fidelity to passage and wintering sites. In Wales there are many sites at which Green Sandpipers occur every year, whether on passage or in winter. The protection of these sites is the key to the future of the species in Wales.

Steph Tyler

Grey-tailed Tattler *Tringa brevipes* Pibydd Cynffonlwyd

Welsh List Category	IUCN Red List	Rarity recording
	Global	
A	NT	BBRC

There has only been one record in Wales of this species, which breeds mainly in northeast Siberia and winters in Indonesia and Australasia (*HBW*). This record, the first for Britain, was on the Dyfi Estuary at RSPB Ynys-hir, Ceredigion, from 13 October to 17 November 1981. The bird was also seen on the Meirionnydd side of the estuary during this period. The only other British record was in Moray, Scotland, in 1994.

Jon Green and Robin Sandham

Lesser Yellowlegs *Tringa flavipes* Melyngoes Bach

Welsh List Category	IUCN Red List	Rarity recording
	Global	
A	LC	WBRC

Lesser Yellowlegs: Castle Pond, Pembroke, September 2019 © Bob Haycock

The Lesser Yellowlegs breeds in Alaska and Canada, as far east as Hudson Bay, and winters in Central and South America (*HBW*). It is among the commonest of the American waders recorded in Britain, with 406 individuals recorded to the end of 2018 (Holt *et al.* 2019), an average of nine each year over the last 30 years. Numbers have increased dramatically over the last 70 years and almost doubled in the last 30 years. A similar trend has been evident in Wales, where some of the 26 individuals recorded have over-wintered. The first Welsh record was at Oxwich, Gower, on 8–15 September 1953.

Total	Pre-1950	1950–59	1960–69	1970–79	1980–89	1990–99	2000–09	2010–19
26	0	1	3	2	2	2	6	10

Lesser Yellowlegs: Individuals recorded in Wales by decade.

Some birds have stayed for long periods: a first-calendar-year bird was at Laugharne, Carmarthenshire, from 12 November 2000 to 18 April 2001. An adult at Fishguard Harbour, Pembrokeshire, on 13 December 2001 and then at Pembroke from 15 December until 17 March 2002 may have been the returning Laugharne individual. A first-calendar-year bird was at the Gann Estuary, Pembrokeshire, from 19 October to 16 November 2003, then at Newgale on 5–6 December, before returning to the Gann until 11 April 2004, and an adult was at WWT Penclacwydd, Carmarthenshire, from 7 July to 29 October 2013.

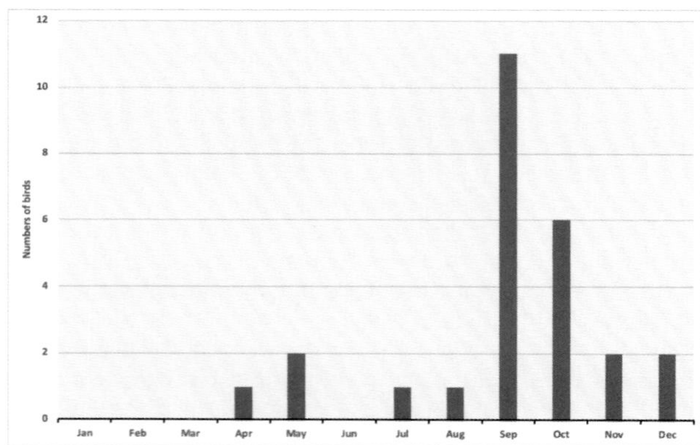

Lesser Yellowlegs: Totals by month of arrival, 1953–2019.

Lesser Yellowlegs: Numbers of birds in each recording area 1953–2019. Some birds were seen in more than one recording area.

There is a distinct autumn peak of arrivals in September and October, which would be expected with a trans-Atlantic migrant. In Pembrokeshire, three of the individuals recorded were at the Gann Estuary and two on Skomer, one of which later moved to Marloes Mere on the mainland. One was seen on both sides of the Burry Inlet, in Carmarthenshire and Gower, in 2017.

Jon Green and Robin Sandham

Redshank *Tringa totanus* Pibydd Coesgoch

Welsh List Category	IUCN Red List			Birds of Conservation Concern			
	Global	Europe	GB	UK	Wales		
					2002	2010	2016
A	LC	LC	VU(B)	Amber			

Rather like the Curlew, anyone looking at large flocks of Redshank on a Welsh mudflat in winter might think that the species was doing well in Wales. However, the European breeding population is greatly reduced and the breeding population in Wales has fallen dramatically. The subspecies breeding in Wales is *T.t. totanus*, which breeds across much of Europe and into North Africa. On the coast, it breeds on saltmarsh, wet grassland and grassy marshes (*HBW*) and inland on lowland wet grassland and upland rough pasture habitats in Britain (Hale 1988). At one time it bred at altitudes well over 300m in Wales: a pair probably bred at an altitude of 480m near Llyn y Gwaith, Carmarthenshire, in 1977 (Lloyd *et al.* 2015) and another pair bred at 500m in Breconshire in the early 1990s (Andrew King *in litt.*).

There is little information about the status of the species in Wales before the late 19th century. Ray (1662) described a bird breeding at Aberavon in Glamorgan that was probably a Redshank, although he did not name it as such. Several later authors listed the Redshank's presence but gave no further details, although there was a reference to breeding in Williams' *Llandudno Guide* of 1861. As a breeding bird, the Redshank spread west and south in Britain between 1865 and 1925 (Parslow 1973). Phillips (1899) considered the species to be "quite a rarity" in Breconshire, but thought that it might have bred at Llangorse Lake, where an adult and a young bird had been seen in August 1890. The first confirmed breeding record for Glamorgan was at Kenfig dunes in 1911, with confirmed breeding at Morfa Moors soon afterwards, and about 12 pairs were present in that area in 1925 (Hurford and Lansdown 1995). In Carmarthenshire, the first confirmed breeding was not until 1929, although the species was suspected of breeding prior to 1926 (Ingram and Salmon 1954). Redshank was apparently in decline in Pembrokeshire in the 1890s. Mathew (1894) noted "Sir Hugh Owen has told us that it had become rare at the time he was accustomed to shoot wild fowl about Milford Haven in his punt, where, fifty or sixty years ago, it was probably a common nesting species." Breeding was suspected near St Davids in 1932 and 1948, but the first confirmed record for the county was not until 1955 (Donovan and Rees 1994). In Ceredigion, Salter suspected breeding on the Afon Leri on Cors Fochno in summer 1904. It had increased in the county by the 1920s, with Oldham reporting many nesting at the head of the Dyfi Estuary in 1922 (Roderick and Davis 2010). Breeding was confirmed in Breconshire in 1913 (Massey 1976a), while breeding in Radnorshire was first recorded at Radnor Forest in 1914 (Jennings 2014). The Redshank bred or probably bred at scattered locations in all North Wales counties, except Montgomeryshire, at the turn of the 20th century (Forrest 1907), and in 1919, Forrest commented that "as a breeding species the Redshank has greatly increased in recent years in Anglesey and Carnarvonshire." Most of the breeding records Forrest mentioned were coastal, but there were a few inland records, on Arenig, Meirionnydd, in 1897, with two pairs near Bala in 1905.

Redshank in winter plumage © Fausto Riccioni

The Redshank Enquiry of 1939–40 (Thomas 1942) found that the species bred in all counties of North Wales, "perhaps least plentifully in Montgomery and most plentifully in Merioneth and Anglesey." It was reported that Redshanks bred at 4–6 sites in Glamorgan and in Radnorshire and at one each in Breconshire and Carmarthenshire, but not in Pembrokeshire. Norris (1960), reporting on a national enquiry in 1952, commented that "Caernarvonshire, one of the first parts of Wales to be colonised, still shows an area of higher density than any other part of Wales or south-west England in the survey."

Fieldwork for the *Britain and Ireland Atlas 1968–72* recorded Redshank in 31% of the 10-km squares in Wales, with widespread confirmed breeding, including inland, particularly in Ceredigion and adjoining areas of Radnorshire and Breconshire. *Birds in Wales* (1994) stated that a decrease in range and numbers was evident in Wales from at least the 1960s, which the authors considered may have been at least partly triggered by mortality during the severe winter of 1962/63. A decline in numbers nesting in lowland wet grassland was evident by the early 1980s. A survey in 1982–83 found 53 pairs of Redshank in this habitat in Wales, of which 31 were in Gwent, and just six pairs in Caernarfonshire, Anglesey and Meirionnydd combined (Smith 1983). A survey of saltmarsh sites in 1985 included 13 sites in Wales: three in Gwent, five in Glamorgan and four in Carmarthenshire. RSPB Ynys-hir in Ceredigion held 99–102 breeding pairs of Redshank in 678ha (6.78km^2) (Allport *et al.* 1986). An RSPB survey of Llŷn in 1986 produced no breeding records of this species.

A decline was evident in all habitats by the early 1990s. The *Britain and Ireland Atlas 1988–91* showed more losses than gains compared to the 1968–72 *Atlas*, with losses predominating, particularly in inland Ceredigion. There were some gains inland, but a dedicated survey in 1991 found only 20 inland pairs, including six pairs each in Gwent and Ceredigion. There were 156–163 pairs at 13 sites on the coast, including 55 pairs on the Flintshire part of the Dee Estuary, at Burton Marsh and White Sands (Colin Wells *in litt.*). Elsewhere, the largest numbers were on the Dyfi Estuary, Ceredigion/Meirionnydd, which had 20 pairs, and the Loughor estuary, Gower/Carmarthenshire, with 16–18 pairs. Overall, there was a 58% decrease in the breeding population in Wales between 1985 and 1991 (Griffin *et al.* 1991). A sample survey of saltmarshes by the RSPB in 1996, repeating the 1985 survey, found that average densities on vegetated saltmarsh, the only parts where Redshank will breed, had declined in Wales from 40.98 pairs/km^2 in 1985 to 26.43 in 1996 (Malpas *et al.* 2013). The highest concentrations in 1996 were at two sites in Carmarthenshire: 17 pairs in 92.1ha at Trostre and 9–10 pairs at the Gwendraeth Estuary. Combining the figures from this survey with counts at the Dyfi Estuary, where 87–92 pairs bred in 1996, gives a minimum of 126–135 pairs breeding on the coast of Wales in 1996, although many areas of coast were not covered.

A repeat survey of waders in lowland wet grassland in 2002 found only 19 pairs in Wales, at a density of 0.17 pairs/km^2, compared to 53 pairs in 1982, a decline of 62% (Wilson *et al.* 2005). The *Britain and Ireland Atlas 2007–11* showed that breeding Redshank had almost vanished from inland Wales and there had also been extensive losses on the coast, particularly in Carmarthenshire. The number of occupied 10-km squares in Wales had fallen to 27, just 10% of the total. There were a few records of probable breeding inland in southern Wales, but records south of the Dyfi Estuary were largely confined to the Glamorgan and Gwent coasts. The *East Glamorgan Atlas 2008–11* found Redshank in 14 tetrads compared to 12 in 1984–89, but most of those present in 2008–11 were not considered to be breeding. Breeding was confirmed in nine tetrads in 1984–89, but not at all in 2008–11. The first *Gwent Atlas* in 1981–85 recorded the species in 44 tetrads, but it was only found in 18 in the 1998–2003 *Atlas*, with losses both inland and on the coast. The number of tetrads with confirmed breeding had halved from 12 to six (Venables *et al.* 2008), and these were confined to Wentlooge Level, Newport Wetlands Reserve

(Goldcliff) and the Nedern Wetlands near Caldicot. Subsequent surveys of Newport Wetlands Reserve recorded 14 pairs in 2008, 17 that fledged 11 young in 2012, and 33 territories in 2014. The *North Wales Atlas 2008–12* confirmed breeding in only seven tetrads, of which three were in the Meirionnydd part of the Dyfi Estuary and three on the Dee Estuary, Flintshire, with the other on the Meirionnydd coast. It recorded probable breeding pairs in another 12 tetrads, several on the Meirionnydd coast and the west coast of Anglesey, but there was no indication of inland breeding.

Most of the saltmarsh sites surveyed in 1985 and 1996 were resurveyed in 2011 (Malpas *et al.* 2013). Average densities on vegetated parts of saltmarsh had declined further in Wales, from 26.43 pairs/km² in 1996 to 22.51 pairs/km² in 2011. The authors anticipated that the rate of decline would lead to the loss of breeding Redshank from the majority of British saltmarshes within 25 years. They concluded that declines were likely to be driven by a lack of suitable nesting habitat, and that conservation management schemes implemented since 1996 appeared not to be delivering the grazing pressures and associated habitat conditions required by this species. There was insufficient grazing pressure and conservation management data for saltmarsh sites in Wales to draw meaningful conclusions about the likely drivers of the Redshank declines there. Their population estimate for saltmarsh in Wales, extrapolating from densities on the surveyed sites to the total area of saltmarsh in Wales, was 1,519 pairs in 2011 (Malpas *et al.* 2013).

Combining estimated totals for farmland and upland in 1993 (O'Brien *et al.* 1998) and on saltmarsh in 1996 (Brindley *et al.* 1998) suggests a population of around 1,686 pairs in Wales in the 1990s. Subsequent declines in many parts of Wales have been partly offset by increases on some nature reserves. This is particularly true of the Dyfi Estuary, where management to benefit breeding Redshank has produced good results. The Dyfi Estuary SSSI, which includes RSPB Ynys-hir and Cors Fochno, is now by far the most important site for breeding Redshank in Wales, holding up to 70–80 pairs. At RSPB Burton Mere Wetlands, due to the creation of wet grassland and the introduction of a predator exclusion fence, there is now a significant breeding population on the Flintshire part of the reserve. Redshank first started nesting within the fenced area in 2012, with eight pairs in Flintshire, and numbers increased to 27 pairs in Flintshire in 2019 (Colin Wells *in litt.*).

Published records suggest that the total number of breeding pairs is likely to be much lower now than in the 1990s. In northern Wales, the coastline between the Dyfi and RSPB Burton Mere Wetlands probably holds only about ten pairs. There may be about 25 pairs at coastal sites in Gwent, another ten pairs at coastal sites in Glamorgan and Carmarthenshire, and perhaps a very few pairs remaining at inland sites. A total of about 150 pairs would be the most optimistic interpretation of these figures.

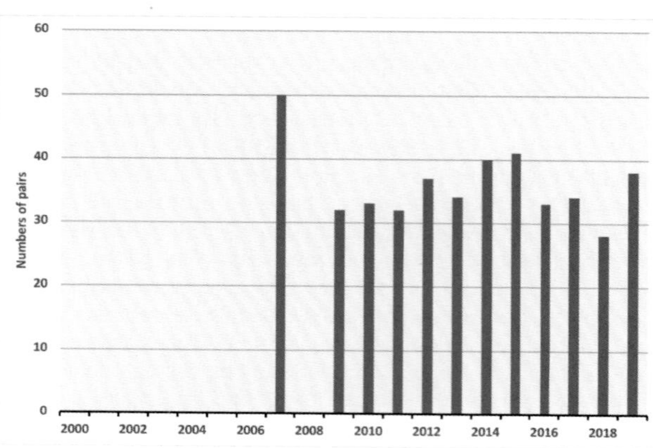

Redshank: Numbers of pairs breeding at RSPB Ynys-hir, 2000–19.

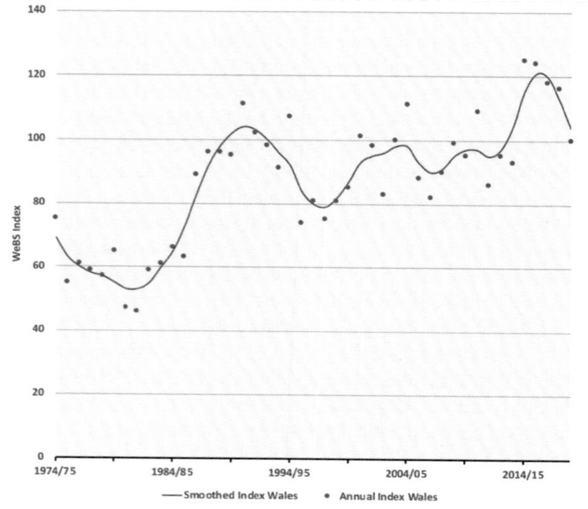

Redshank: Wetland Bird Survey indices, 1974/75 to 2018/19.

Redshank: Recovery locations of birds ringed in Wales are shown by red circles. Ringing locations of birds that were recovered in Wales are shown by blue triangles. Small brown dots show ringing or recovery locations in Wales.

Site	County	74/75 to 78/79	79/80 to 83/84	84/85 to 88/89	89/90 to 93/94	94/95 to 98/99	99/00 to 03/04	04/05 to 08/09	09/10 to 13/14	14/15 to 18/19
Dee Estuary: Welsh side (whole estuary)	Flintshire	na (8,186)	na (4,332)	na (7,686)	2,267 (8,070)	3,697 (9,089)	3,228 (9,846)	4,511 (10,831)	3,453 (9,450)	4,734 (9,325)
Severn Estuary: Welsh side (whole estuary)	Gwent, Glamorgan	1,077 (2,538)	1,226 (1,980)	1,859 (2,781)	1,850 (2,373)	1,334 (2,123)	1,022 (1,864)	1,316 (2,353)	1,354 (3,268)	1,789 (5,791)
Traeth Lafan	Caernarfonshire/ Anglesey	175	452	704	353	428	1,292	1,492	1,480	1,318
Conwy Estuary	Caernarfonshire/ Denbighshire	510	353	677	546	524	285	340	437	979
Burry Inlet	Gower/ Carmarthenshire	1,928	1,031	1,051	975	1,182	775	618	608	701
Cleddau Estuary	Pembrokeshire	685	418	852	1,093	639	709	659	644	687

Redshank: Five-year average peak counts during Wetland Bird Surveys between 1974/75 and 2018/19 at sites in Wales averaging 940 or more at least once. International importance threshold: 2,400; Great Britain importance threshold: 940.

Wintering birds in Wales apparently include both *T.t. robusta*, from Iceland and the Faroes, and *T.t. totanus* breeding in Scotland and northern England. Redshank are very site-faithful in winter, some returning to the same wintering sites for many years. An adult ringed in Bangor harbour in November 1994 was found dead at the same site in November 2010, nearly 16 years later.

Counts outside the breeding season show no great change over the last 40 years, with decreases at some sites balanced by increases at others. The Wetland Bird Survey index for Wales shows an increase since the 1970s.

The Dee Estuary, as a whole, is the most important site in Britain for Redshank, and both it and the Severn Estuary are of international importance. The Welsh side of the Dee is of international importance on its own, while the Welsh side of the Severn Estuary and Traeth Lafan are of British importance. An important site for wintering Redshanks was lost when the Cardiff Bay Barrage was closed in November 1999, removing the intertidal habitat of the Taff and Ely estuaries, East Glamorgan. The site became freshwater and the number of Redshank present dropped substantially. Most displaced birds appeared to move to Rhymney, East Glamorgan, but these struggled to maintain their body condition in the first winter following closure. A colour-ringing study showed that winter survival of adult Redshanks from Cardiff Bay fell after their displacement, from 84.6% in the two years prior to barrage-closure, to 77.8% in the three following years. The survival rate of Rhymney birds did not change. Evidently birds that tried to find space in a new area were severely disadvantaged (Burton *et al.* 2003b).

The decline of the Western Palearctic population of Redshank—for example, the 54% decline in Europe between 1980 and 2016 (PECBMS 2020)—is thought to be due to loss of winter and breeding habitat, agricultural intensification, wetland drainage and encroachment of *Spartina anglica* on mudflats, among other factors (*HBW*). The Redshank has almost vanished as a breeding species from inland Wales. The decline in Radnorshire was thought to be the result of increased grazing pressure and predation (Jennings 2014). This is likely to be also true elsewhere. Grazing pressure is probably also an important factor on saltmarshes. Norris *et al.* (1998) noted that Wales was among the areas where grazing pressure on saltmarshes had increased most in Britain since 1985. Sharps *et al.* (2015) found that on the Ribble Estuary in northwest England even light conservation grazing could reduce Redshank nest survival rates to near zero.

Where land is managed to favour breeding Redshank, numbers have increased, for example at RSPB Ynys-hir on the Dyfi. This species, like the Lapwing, now breeds mainly on nature reserves in Wales, and its future may depend largely on such dedicated management. It is unclear to what extent this can mitigate against changes expected in the climate suitability of Wales for Redshank. Its range in Britain is predicted to contract to Scotland and northern England by the end of this century (Huntley *et al.* 2007). A dedicated survey aimed at producing a reliable estimate of the current breeding population in Wales would be very valuable. It is also important that wintering sites are protected. Schemes that inundate mudflats, such as some proposed to produce energy from the tide, could seriously affect wintering numbers in Wales.

Icelandic Redshank *T.t. robusta* Pibydd Coesgoch Gwlad yr Iâ

A good number of ringing recoveries linking Iceland and Wales prove that birds of the subspecies *T.t. robusta* spend the winter in Wales. This subspecies has different breeding plumage characteristics compared to *T.t. totanus.* In non-breeding plumage, most birds can be separated, in the hand, by measurements, *robusta* being larger. Two catches on Traeth Lafan, Caernarfonshire, in the early 1980s, by the SCAN ringing group, showed a British:Icelandic ratio of 56:44 (Johnson 1985). The proportion of Icelandic birds there had increased by the late 1990s to form up to 70% of the wintering population and has since remained constant (Margrave 2018).

Rhion Pritchard

Sponsored by Jim Marshall

Marsh Sandpiper *Tringa stagnatilis* Pibydd Cors

Welsh List Category	IUCN Red List		Rarity recording
	Global	Europe	
A	LC	LC	BBRC

This species breeds in eastern Ukraine and western Russia, and east through the forest steppe of Siberia to Mongolia and northeast China (*HBW*). Breeding has been recorded occasionally in Finland and the Baltic states. It winters throughout sub-Saharan Africa and South Asia. A total of 149 individuals had been recorded in Britain to the end of 2019 (Holt *et al.* 2020), on average about three each year over the last 30 years, including just five in Wales:

- Malltraeth Cob pool, Anglesey, from 30 June to 2 July 1977;
- RSPB Oakenholt Marsh, Flintshire, on 7 May 1990;
- Penclacwydd, Carmarthenshire, on 19 May 1990;
- Afon Clwyd near Rhuddlan, in both Denbighshire and Flintshire, on 9–27 August 1994 and
- RSPB Conwy, Denbighshire, on 14 June 1996.

Jon Green and Robin Sandham

Wood Sandpiper *Tringa glareola* Pibydd Graean

Welsh List category	IUCN Red List			Birds of Conservation Concern
	Global	Europe	GB	UK
A	LC	LC	EN	Amber

A passage migrant in Wales, the Wood Sandpiper breeds in northern Europe and Asia, from Norway to eastern Siberia. It breeds mainly in marshes and swamps, usually surrounded by trees, or in vegetation close to large lakes (*HBW*). The global population is thought to be stable (BirdLife International 2020). The largest populations in Europe are in Russia, Finland and Sweden (BirdLife International 2015). Some breed farther south, including in northern Scotland; 27 pairs bred there in 2017 (Holling *et al.* 2019). It winters mainly in the tropics. There are no ringing recoveries involving Wales, but birds passing through Wales are almost at the western limit of their global range, and probably breed in Fennoscandia and winter in West Africa (*BTO Migration Atlas* 2002).

It is difficult to be certain whether numbers passing through Wales have changed since the late 19th century. Early records are few and far between. In Pembrokeshire, for example, Mathew (1894) recorded one in spring 1886, but the next county record was not until 1955. Forrest (1907) knew of only three records for northern Wales, all of which were shot: in Montgomeryshire in 1873, at Glandyfi on the Ceredigion side of the Dyfi Estuary in 1880 and on the Caernarfonshire side of the Glaslyn Estuary in May 1898. The next record for Caernarfonshire was not until 1968, and there were no records from Gwent or Anglesey until 1952. The first records in Glamorgan were in 1962, Breconshire in 1964, and Carmarthenshire and Meirionnydd in 1976. It has now been recorded in all counties, and in recent years has been annual, or almost annual, in several. There may have been a real increase, but more birdwatchers, better optical equipment and a better understanding of the differences between this species and Green Sandpiper could be partly responsible. An influx into Britain, in autumn 1952, included four at Malltraeth Cob pool, Anglesey, on 26–28 August (Nisbet 1956).

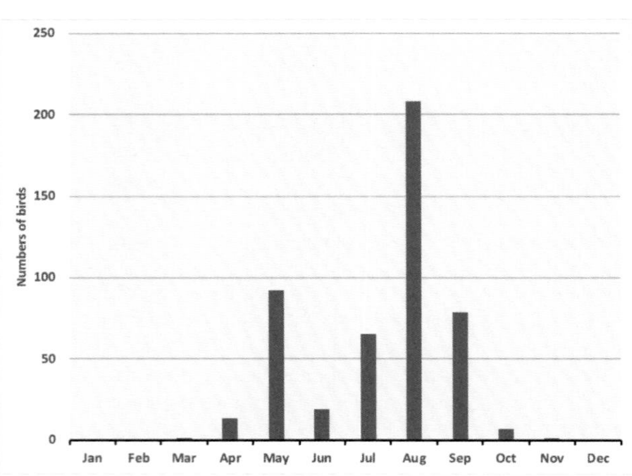

Wood Sandpiper: Totals recorded by month, 1992–2019.

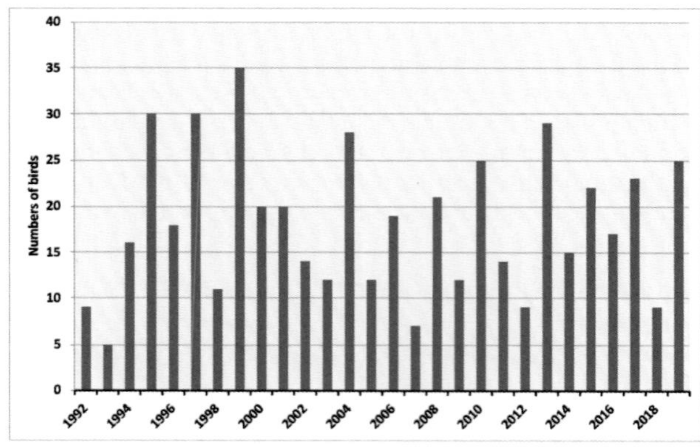

Wood Sandpiper: Numbers recorded in Wales, 1992–2019.

There has been no marked trend over the last 20 years in Wales. The highest annual total was 35 in 1999, of which six were in spring and 29 in autumn, including ten in Flintshire. The highest count in Wales was 12 at Newport Wetlands Reserve (Goldcliff), Gwent, on 10 August 2004. Almost all records were between mid-April and mid-October, although one was at Pont Marquis, Anglesey, on 27 November 2017. This species seems never to have over-wintered in Wales.

Inland records are not particularly uncommon, but notable were four at Eglwys Nunydd Reservoir, Gower, on 16 and 18 August 1964. A pair was reported displaying beside a bog pool at Cors Caron, Ceredigion, on 21 May 1973, but they were not found subsequently (Roderick and Davis 2010) and there has never been a record suggesting breeding in Wales.

Rhion Pritchard

Spotted Redshank *Tringa erythropus* Pibydd Coesgoch Mannog

Welsh List Category	IUCN Red List			Birds of Conservation Concern			
	Global	Europe	GB	UK	Wales		
A	LC	LC	EN	Amber	2002	2010	2016

A breeding bird of the northern tundra and swampy pine or birch forest near the tree line, the Spotted Redshank nests from northern Fennoscandia to eastern Siberia. It winters to the south, in western and southern Europe, Africa south of the Sahara and east to Southeast Asia (*HBW*). In winter, when it is primarily a bird of muddy estuaries, Wales is on the northwest fringe of its range; those that winter or pass through on migration are likely to be from Fennoscandia, although there are no ringing recoveries to confirm this.

This species seems to have been rare in Wales in the 19th century. In Glamorgan, one was "obtained" at St Fagans in September 1876, with one nearby at Sant-y-Nyll in 1895 and five in the Porthcawl area, between August and October 1898 (Hurford and Lansdown 1995). Mathew (1894) described it as a rare autumn visitor to Pembrokeshire, with several shot in autumn seen in a taxidermist's shop at Pembroke. Forrest (1907) knew of only four records from North Wales. One, probably of this species, was killed on the Dee Marsh near Burton in 1864, so was possibly not in Wales. One was shot at Hanmer Mere, Denbighshire vice-county, about 1880 and two were shot in September 1899, one on the Dyfi Estuary, Meirionnydd, and the other on marshes below Llanrwst, Caernarfonshire/Denbighshire.

There were few records during the first half of the 20th century, but this species began to be reported more frequently from the mid-1950s. One on the Dyfi Estuary on 2 September 1951 was the first for Ceredigion, but after 1955 it was found there annually (Roderick and Davis 2010). Two records in Glamorgan in October 1959 were the first for over 60 years, but thereafter the species became regular. The highest numbers are on autumn passage, from July to early November, peaking between August and October. Numbers in Gower peaked at 33 at Llanrhidian on 16 August 1970, while the county record for Carmarthenshire is 28, at the Gwendraeth Estuary on 25 August 1981 and at Penclacwydd in September 1994. There were 27 at Hook, Pembrokeshire, on 12 October 1978, while on the Dyfi Estuary numbers reached 25 on 26 September 1969 and 26 on 22 September 1972, although counts were much lower from the mid-1980s onwards. The Dee Estuary is a particularly good site for this species, with a peak of 180 at RSPB Oakenholt Marsh, Flintshire, in August/September 1977 and regular counts of 60–199 there in autumn up to the mid-1990s. Numbers at many sites have been lower since the mid-1990s than during the previous 30 years, as evident elsewhere in Britain. There were 40 at Connah's Quay nature reserve/RSPB Oakenholt Marsh on 24 September 1992 and *c.*75 there during June to November 1993, but these numbers have not been matched anywhere in Wales since. Counts of 20 or more have been confined to the Dee Estuary in recent years, and even there, only in some years: there were 26 at Connah's Quay nature reserve on 1 November 2015 and 24 there on 29 September 2019. Spring passage is from late March to May, with a few birds remaining into early June. Numbers are usually considerably lower than in autumn, but there were *c.*75 at Connah's Quay nature reserve/RSPB Oakenholt Marsh, between March and May 1993. The decline is clear in Carmarthenshire, where the mean annual maximum count was 15.8 during 1991–2000, but there has been no double figure count since 2010 (Rob Hunt *in litt.*).

The species is also found away from the coast fairly regularly in autumn and sometimes in spring, usually in small numbers, although there was a county record of 18 at Llyn Cefni, Anglesey, on 4 September 1976. The first record for Breconshire was one at Llangorse Lake on 3 April 1963. There had been 22 more records

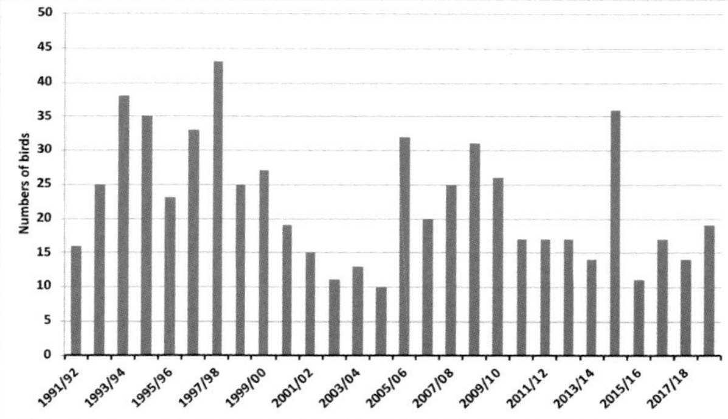

Spotted Redshank: Numbers in Wales, December to February, 1991/92 to 2018/19.

by 1989, all but three in autumn (Peers and Shrubb 1990). The first in Radnorshire was in August 1968, and there were 16 records by 2011, all but one between 8 July and 1 October (Jennings 2014). The first record for Montgomeryshire was not until April 1974.

Numbers are lower in winter and have declined in many counties. In a typical year, 15–20 are recorded in Wales in winter, quite a high proportion of the estimated 67 wintering in Britain (Frost *et al.* 2019a). Up to 17 regularly wintered in the Burry Inlet during 1967–87 (Hurford and Lansdown 1995). More recently, there were eight at Penclacwydd in January 2010, but double-figure counts are now confined to the Dee Estuary, where there were 23 at Connah's Quay nature reserve on 5 February 2015. Overwintering birds are usually found at the coast; one at Pwll Patti, Radnorshire, on 8 February 1996 was exceptional.

The area that is climatically suitable for Spotted Redshanks to breed is expected to reduce greatly this century, with potential loss from Europe outside northern Russia (Huntley *et al.* 2007), which is likely to further reduce the numbers recorded in Wales.

Rhion Pritchard

Greenshank *Tringa nebularia* Pibydd Coeswyrdd

Welsh List Category	IUCN Red List			Birds of Conservation Concern			
	Global	Europe	GB	UK	Wales		
A	LC	LC	LC	Amber	2002	2010	2016

The Greenshank breeds over a large area of northern Europe and Asia, from northwest Scotland to eastern Siberia and Kamchatka, mainly in marshlands in the taiga region. It winters farther south over an even wider area, from western Europe to Africa and Australia (*HBW*). Around 1,000 pairs breed in Scotland (*The Birds of Scotland* 2007). Some birds pass through Wales on migration, while others winter here. *Birds in Wales* (1994) considered that most Greenshanks wintering in Wales were birds that bred in Scotland, but that those seen on passage could include birds breeding in Fennoscandia and Russia. The departure times of most birds wintering in Wales does suggest that these breed in Scotland, where breeding birds are on territory by late March (*BTO Migration Atlas* 2002), while those passing through Wales in May are presumably birds that go on to breed in Fennoscandia. Few ringing recoveries throw light on this, but one recaught at Aberffraw, Anglesey, in November 2016 had been ringed as an adult in the Scottish Highlands in June 2012, presumably at a breeding site. A bird tagged on the breeding grounds on the north coast of Scotland was later recaptured at Bangor, Caernarfonshire (Summers *et al.* 2020). Ringing recoveries also suggest that some birds return to the same areas every winter.

In 1848, Dillwyn described the species as not common in the Swansea area, and it was regarded as uncommon in Glamorgan, as a whole, at the end of the 19th century (Hurford and Lansdown 1995). Mathew (1894) said that it was "not very uncommon" in Pembrokeshire, while Barker (1905) knew of only a few records in Carmarthenshire. It seems to have been rather more common in North Wales at this time. Forrest (1907) described it as not uncommon on coasts and estuaries in autumn and spring, with fewer in winter. Groups of up to five were mentioned, and it was said that a few were around the Dee Estuary every winter and early spring. By the 1950s it was a regular autumn visitor to all coastal counties. Most published records from this period suggest only small numbers on passage, although there were 20 at the Dyfi Estuary on 28 August 1955.

WeBS counts at the main sites show considerable fluctuations, but peak counts are almost always on autumn passage, between August and October. Numbers at most sites peak in September or October, as birds move through on return passage, presumably heading for West Africa or the western Mediterranean. Large counts include 84 in the Burry Inlet in September 2014 and 87 at Penclacwydd, Carmarthenshire, (part of the Burry Inlet WeBS count) in September 2016. There were 60 at the Gann Estuary (part of the Cleddau WeBS count) in September 1969. In the north there were *c.*50 at Traeth Lafan on 10 October 1976. The Dyfi Estuary can also hold good numbers on passage, including 48 on 28 August 1983. Passage birds are seen away from the coast in smaller numbers, although there have been double-figure counts, such as 12 at Llyn Cefni, Anglesey, in September 1976, 12 at Glasbury, Breconshire/Radnorshire, on 17 August 1982 and 20 at Llandegfedd Reservoir, Gwent, on 27 August 1997. Spring passage, in April and May, involves much smaller numbers.

The WeBS index shows a steady increase in the number wintering in Wales. WeBS counts suggest that Wales holds

Greenshanks © John Hawkins

Site	County	74/75 to 78/79	79/80 to 83/84	84/85 to 88/89	89/90 to 93/94	94/95 to 98/99	99/00 to 03/04	04/05 to 08/09	09/10 to 13/14	14/15 to 18/19
Burry Inlet	Gower/ Carmarthenshire	23	25	34	26	40	51	14	4	66
Dee Estuary: Welsh side (whole estuary)	Flintshire	na (34)	na (10)	na (49)	16 (37)	29 (76)	26 (79)	23 (46)	18 (35)	17 (33)
Cleddau Estuary	Pembrokeshire	20	21	34	18	43	35	30	31	32
Severn Estuary: Welsh side (whole estuary)	Gwent/East Glamorgan	6 (22)	1 (12)	1 (37)	3 (19)	2 (23)	9 (22)	6 (15)	11 (20)	9 (25)
Foryd Bay	Caernarfonshire	5	6	10	14	19	20	17	12	15
Traeth Lafan	Caernarfonshire/ Anglesey	0	7	16	13	7	12	12	8	15
Carmarthen Bay	Carmarthenshire/ Pembrokeshire	43	4	11	11	4	1	13	12	13
Dyfi Estuary	Ceredigion/ Meirionnydd	9	-	10	5	7	11	9	5	9
Inland Sea and Alaw Estuary	Anglesey	13	6	12	9	10	8	5	7	9

Greenshank: Five-year average peak counts during Wetland Bird Surveys between 1974/75 and 2018/19 at sites in Wales averaging 8 or more at least once. International importance threshold: 3,300; Great Britain importance threshold: 8.

100–120 birds in winter, a significant proportion of the population wintering in Britain, which is estimated at 810 (Frost *et al.* 2019a). Regular wintering is largely confined to a few sites.

Wales is of some importance as a wintering area for Greenshank, and most of the sites at which birds congregate in winter are protected. However, the long-term outlook is not so positive, with Greenshank expected to be lost as a breeding species in Scotland by the end of this century (Huntley *et al.* 2007).

Rhion Pritchard

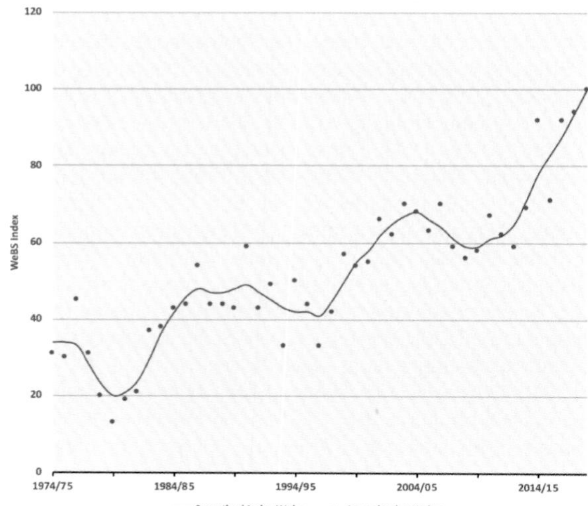

Greenshank: Wetland Bird Survey indices, 1974/75 to 2018/19.

Greater Yellowlegs *Tringa melanoleuca* Melyngoes Mawr

Welsh List Category	IUCN Red List	Rarity recording
	Global	
A	LC	BBRC

The Greater Yellowlegs breeds from southern Alaska across subarctic Canada, east to Labrador and Newfoundland. It migrates through interior North America and along coasts to winter in the Caribbean, Central America and South America (*HBW*). It is a rare vagrant to Europe, with just 34 records in Britain to the end of 2019, on average about one every two years over the last 30 years (Holt *et al.* 2020). There have been two records in Wales, both in Flintshire: at Shotton on 23 July 1961 and at Shotwick Lake on 24 November 1983.

Jon Green and Robin Sandham

Cream-coloured Courser *Cursorius cursor* Rhedwr Twyni

Welsh List Category	IUCN Red List		Rarity recording
	Global	Europe	
A	LC	NT	BBRC

A bird of arid habitats, the Cream-coloured Courser breeds in the Canary Islands, northern and central Africa and Central Asia, as far east as Pakistan and northwest India (*HBW*). Populations in North Africa are migratory, wintering on the southern edge of the Sahara. There have been four records in Wales. The first was shot somewhere in North Wales in 1793. Montagu (1802) wrote "We are assured by Mr Dickinson that a specimen of this very rare bird was shot in 1793 by Mr George Kingstone of Queen's College, Oxford; a very accurate ornithologist. The bird was preserved in the collection of the late Professor Sibthorp." Another was shot at Ynyslas, Ceredigion, on 2 October 1886. The two other records were both seen alive: at Cefn Sidan, Carmarthenshire, on 23 October 1968 and a well-watched individual at Bradnor Hill Golf Course in Herefordshire, which flew into Radnorshire on its departure on 23 May 2012.

Jon Green and Robin Sandham

Collared Pratincole *Glareola pratincola* Cwtiad-wennol Dorchog

Welsh List Category	IUCN Red List		Rarity recording
	Global	Europe	
A	LC	LC	BBRC

This species breeds locally around the Mediterranean basin, the Black Sea and in Central Asia, with other populations in Africa. The largest European populations are in Spain and Turkey (BirdLife International 2015). It winters in sub-Saharan Africa. To the end of 2019 there had been 98 records in Britain, about one a year on average over the last 30 years (Holt *et al.* 2020). Five of these were in Wales:

- Penclawdd, Gower, on 27 May 1973;
- Rhosneigr, Anglesey, on 6 June 1983;
- WWT Penclacwydd, Carmarthenshire, from 14 June to 13 July 2005;
- Skokholm, Pembrokeshire, on 1 May 2014 and
- WWT Penclacwydd, Carmarthenshire, on 3–4 July 2019.

A pratincole at Bosherston, Pembrokeshire, on 13 April 1981 was not specifically identified.

Jon Green and Robin Sandham

Black-winged Pratincole *Glareola nordmanni* Cwtiad-wennol Adeinddu

Welsh List Category	IUCN Red List		Rarity recording
	Global	Europe	
A	NT	VU	BBRC

The Black-winged Pratincole breeds from Romania east through Ukraine to southwest Russia and northern Kazakhstan. It winters mainly in South Africa, Namibia and Botswana, though some winter in West Africa (*HBW*). It is a rare vagrant to Britain, where there had been 40 records up to the end of 2019 (Holt *et al.* 2020). Four of these were in Wales:

- RSPB Inner Marsh Farm (now RSPB Burton Mere Wetlands), Flintshire, on 2 June 1988;
- Newport Wetlands Reserve, Gwent, on 25 June 2001;
- Mona Airfield, Anglesey, on 4–20 July 2001 and
- RSPB Burton Mere Wetlands on 3–4 May 2012.

Jon Green and Robin Sandham

Kittiwake *Rissa tridactyla* Gwylan Goesddu

Welsh List Category	IUCN Red List			Birds of Conservation Concern			
	Global	Europe	GB	UK	Wales		
A	VU	VU	CR	Red	2002	2010	2016

The Kittiwake has an Arctic circumpolar distribution, nesting mainly at high latitudes, but in Europe as far south as Galicia, Spain (*HBW*). It breeds mainly where the sea summer temperature is below 10°C. Iceland holds 31% of the Kittiwakes breeding in Europe, while the UK holds the second-largest population, with 20% of the European total. The populations in Britain and Ireland, where the sea temperature is higher than 10°C, are among the most southerly in Europe, although small numbers breed in France (BirdLife International 2015). Breeding is usually on coastal cliffs, often in large colonies, although this species will also breed on human-made structures, such as on Mumbles Pier, Gower.

Kittiwakes were heavily persecuted during most of the 19th century, being shot both for sport and for the millinery trade, but a measure of protection towards the end of the century led to a rapid increase in numbers (*Birds in Wales* 1994). Williams (1835) referred to Kittiwakes nesting on the Ormes in Caernarfonshire, while in Pembrokeshire "great numbers" nested on Ramsey, Skomer and Grassholm, and in places on the mainland cliffs (Mathew 1894). Aplin (1903) described large numbers at the Carreg y Llam colony in Caernarfonshire: "a vast crowd—thousands perhaps—of Kittiwakes, which come out like a snowstorm when they are alarmed." Breeding was first recorded on Bardsey in the 1920s (Coulson 1963). Lockley *et al.* (1949) recorded *c.*1,900 pairs on Skomer and 115 pairs on Grassholm in 1946, but there were few counts at most sites before the Operation Seafarer census in 1969–70. Since then, most large Kittiwake colonies have been regularly counted, although pre-1986 data are spasmodic for some. Counts tend to demonstrate similar trends throughout Wales, with an accelerating decrease in the last decade, although a few colonies seem to be more stable.

There are no regular counts at St Tudwal's Islands, Caernarfonshire, where there were 363 nests in 2014. There used to be more colonies, but several mainland sites have disappeared completely or have greatly reduced numbers. Colonies that have disappeared include three in Pembrokeshire: Stackpole Head, with over 250 pairs in 1978, but extinct in 1999; Elegug Stacks,

Colony	Pre-1980	1980–89	1990–99	2000–09	2010–19
Mumbles Pier, Gower		Colony established	54 in 1994 increased to 92 in 2000	No annual data but 110 pairs in 2009; stable	Peak at 167 pairs in 2011, 90 pairs in 2018*, 141 in 2019
Worms Head, Gower	First recorded in 1943	140 in 1986 decreased to 104 in 1990	114 in 1991, 94 in 1999	119 in 2001, 20 pairs in 2010	Further decrease, 11 pairs in 2018
Caldey/St Margaret's, Pembrokeshire	120 in 1949, but 435 in 1967 then a roughly stable 250 pairs around 1979	Increased to 406 pairs in 1983 then decreased to 196 pairs in 1989	200 pairs in 1990, decreased to 39 pairs in 1999	6 pairs in 2000, rapid increase to 331 pairs in 2006; 315 in 2009	265 pairs in 2010 and stable to 2019 when 225 pairs
Elegug/ Castlemartin, Pembrokeshire	Peak 375 pairs in 1978	Increased to 542 pairs in 1985	486 pairs in 1990, 147 pairs in 1999	Decreased throughout, to 28 pairs in 2009	Decline to a single pair in 2019 (none in 2020)
Skomer Island, Pembrokeshire	1,731 pairs in 1961; slow but variable increase to 2,296 pairs in 1979	2,522 pairs in 1988, decreased to 2,302 pairs in 1989	2,423 pairs in 1990, irregular slow decline to 2,156 pairs in 1999	2,257 pairs in 2000, declined slowly to 2,046 pairs in 2009	1,922 pairs in 2010, rapid decline to 1,236 pairs in 2018, 1,451 pairs in 2019
Ramsey Island, Pembrokeshire	476 pairs in 1969 and 473 pairs in 1973 suggest stability	Two counts in 1985 and 1987 when 306 pairs suggested a decrease	361 pairs in 1992, peaked at 489 in 1995, 471 pairs in 1999	459 pairs in 2000, then declined to 156 in 2008 but 225 pairs in 2009	191 pairs in 2010 then decreased to 83 pairs in 2019
Newquay Head, Ceredigion	*c.*55 nests in 1969, 216 in 1979	175 nests in 1987	477 nests in 1996	375 nests in 2000	280 nests in 2012, 332 in 2018
Bardsey, Caernarfonshire	90–100 pairs in 1955/56 then steadily increased	Peak of 318 pairs in 1989	Decreased in mid-1990s to 175 pairs in 1998 but 243 pairs in 1999	278 pairs in 2000, increased to peak at 365 in 2005 then decreased to 208 pairs in 2009	128 pairs in 2010, then 188 in 2011 before decrease to 90 pairs in 2018, 121 in 2019
Carreg y Llam, Caernarfonshire	Irregular counts; 778 pairs in 1970, 566 pairs in 1979	Decreased to 435 pairs in 1989	473 pairs in 1990, 586 pairs in 1997	919 pairs in 2006 then declined to 552 pairs in 2009	845 pairs in 2015, 317 pairs in 2018, but 627 pairs in 2019
Great Orme, Caernarfonshire	750 pairs in 1959, increased to 884 pairs in 1979	Increased to 1,494 pairs in 1988	744 pairs in 1990, 1,448 pairs in 1996 then declined to 652 pairs in 1999	1,147 pairs in 2000 then decreased	516 pairs in 2010, 314 in 2013 and 854 pairs in 2019
Little Orme, Caernarfonshire		Counts peaked at 924 pairs in 1987	Counts peaked at 856 pairs in 1994, 661 pairs in 1999	582 pairs in 2000, then decreased to 296 pairs in 2010	488 pairs in 2011, 324 pairs in 2019
Puffin Island, Anglesey	110 pairs in 1959	Increased to 310 pairs in 1986	370 pairs in 1990, 658 pairs in 1997, then 595 in 1999	571 pairs in 2001, 727 pairs in 2005, 535 pairs in 2009	413 pairs in 2010, 313 pairs in 2019

Kittiwake: Counts at the larger colonies in Wales.

* Collapse of some artificial nest platforms.

Adult and juvenile Kittiwakes © Ben Porter

where there were 586 in 1990 but just one pair in 2019 and none in 2020; and Saddle Head, where there were 101 pairs in 1987 but none in 1999. Other former colonies include Devil's Truck, Gower, with 297 pairs in 1987 but extinct by the early 2000s, and Freshwater Bay, Anglesey, where there were 247 pairs in 2001 but none in 2019. There were 141 pairs at Penmon, Anglesey, in 1959, but this colony was lost as a result of quarrying in the 1970s (Ivor Rees *in litt.*). The colony at South Stack, Anglesey, appears to be heading for extinction; there were 250 pairs in 1993 but none in 2001 and 2002, and only small numbers since. Many smaller colonies were abandoned as the population declined. Most adults are site-faithful, but they may change colony as a result of disturbance, cliff falls, predation or nest infestation by ticks. On Caldey Island, Pembrokeshire, a small colony was established on an offshore stack in 2001. It reached 84 pairs in 2005 but all the nests were lost to Carrion Crow predation that year. The colony was deserted and never subsequently re-established, the displaced birds moving to nearby St Margaret's Island.

The estimated population of Kittiwakes in Wales was 3,745 pairs in 1959 (Coulson 1963) and 5,250 in 1969 (Coulson 1983). In 2010, when almost every colony was counted in a single year, the all-Wales count was 5,150 pairs. The total counted in 2015–19, as part of the ongoing national Seabirds Count, indicates a decline of 49.1% since the peak in the mid-1980s.

Operation Seafarer 1969–70	SCR Census 1985–88	Seabird 2000 1998–2002	Seabirds Count Provisional 2015–19
6,891	8,771	7,293	4,527

Kittiwake: Total count of apparently occupied nests (AON) recorded in Wales during UK seabird censuses.

Kittiwakes forage on small fish and crustaceans on or very near the ocean surface. By the early 21st century it became evident that they were struggling to find adequate food to raise their chicks. Fledging rates in Wales were generally low, on average around 0.6 chicks fledged per pair. Low breeding success is most likely to be the primary driver of population decrease (Coulson 2019). The decline in the population in Wales has prompted research into feeding areas and food supplied to chicks, colony attendance, adult survival and winter distribution.

One aspect of the species' biology that has been well studied in Wales is breeding success on Skomer. Over 30 years, the average fledging rate has been just 0.62 chicks/pair and productivity has been over 0.80 in only five years, the threshold calculated as required to sustain populations. Also, on Skomer a study of a colour-ringed population showed that during 1978–2017, annual survival of breeding adults averaged 85%, with cyclic fluctuations. There may be a long-term decline in the survival rate of adults, but in 2015–16 it was 96% (Stubbings *et al.* 2018).

In winter Kittiwakes are truly pelagic; two birds ringed as nestlings on Puffin Island, Anglesey, in July 1963 and July 1964, were both recovered in Newfoundland, Canada, in October 1964. A bird found dead at Pwllheli, Caernarfonshire, in March 1961 had been ringed as a nestling the previous July, 2,780km to the northeast on Kharlov Island near Murmansk, Russia, while another found dead at Port Talbot, Gower, in March 1957 had been ringed as a nestling in western Greenland in July 1955, 3,187km to the northwest. Oceanic records are rare so pioneering tracking studies in 2009/10 provided comparative wintering locations of Kittiwakes from many European colonies (Frederiksen *et al.* 2012). Surprisingly, although over 70% of birds tracked from 19 North Atlantic colonies wintered in the central and west Atlantic, the seven tracked from Skomer all stayed within 500km of the colony. Similar results came from birds tracked from Rathlin, Northern Ireland, but most birds from eastern Atlantic colonies moved farther west and the majority wintered more than 3,000km from their breeding colonies. Some from the Isle of May, in the Firth of Forth, wintered in the Western Approaches/Irish Sea along with the Welsh birds.

Kittiwakes are present in Welsh waters throughout the year, and may be attracted to the coast in some numbers by an abundance of small fish, for instance over 1,000 off Borth, Ceredigion, on 29 November 1988, or when southwesterly storms result in numbers 'wrecked' along beaches (Tony Fox *in litt.*). Pennant (1810) mentioned a great wreck of seabirds on the coast between Morfa Bychan and Criccieth, Caernarfonshire, in 1776, apparently in February, which was said to have included "many thousands" of Kittiwakes. There was another wreck in January and February 1957, when many birds were found dead on the Cardigan Bay coast and the south coast, including 91 in Glamorgan (McCartan 1958). Post-breeding dispersal is recorded every year around the coast of Wales with large flocks seen in September and October passing islands and headlands, including 18,000 off Bardsey, Caernarfonshire, on 26 October 2002. In Pembrokeshire, Strumble Head recorded day counts of up to 30,000 in November 1985. The number of Kittiwakes feeding in the shallow upwelling of Broad Sound, between Skomer and Skokholm, peaks in November, sometimes numbering several thousand, including 2,820 in

Kittiwake: Recovery locations of birds ringed in Wales are shown by red circles. Ringing locations of birds that were recovered in Wales are shown by blue triangles. Small brown dots show ringing or recovery locations in Wales.

2016. This is at a time when birds from most other colonies are already in the western Atlantic and supports the tracking data that shows that Welsh and Northern Irish birds do not move very far. Kittiwakes are rare inland; for example, there were 15 records in Radnorshire up to 2014 (Jennings 2014).

The Kittiwake is Red-listed in Wales because of the severe decline in the breeding population over the past 25 years and because of its Vulnerable status in Europe (Johnstone and Bladwell 2016). Even so, the species' decline may be less severe in Wales than elsewhere in Britain. There have been dramatic declines farther north: more than 87% in Orkney and Shetland and 96% on St Kilda. As a result, the Welsh population may now be more than 10% of the total British population. It may be that climate change is causing shifts in the distributions of planktonic copepods, both on the wintering grounds and around the breeding colonies. Further research may provide more information, but how this could influence decision-makers to manage the marine environment to better conserve Kittiwakes around Wales remains to be seen. The outlook for Kittiwakes in Wales does not look promising.

Steve Sutcliffe

Sponsored by Neil and Eldeg Lukes

Ivory Gull *Pagophila eburnea* Gwylan Ifori

Welsh List Category	IUCN Red List		Rarity recording
	Global	Europe	
A	NT	LC	BBRC

Ivory Gull at Black Rock Sands, Caernarfonshire, February 2002
© Steve Young

This Arctic circumpolar gull breeds in Greenland, Spitsbergen, Arctic Canada and Russia. It winters over sea ice on the southern edge of the Arctic Circle, with the Davis Strait between Canada and Greenland being of particular importance. It is a rare vagrant to Britain, usually to Scotland. There were 142 records in Britain up to the end of 2019, an average of one or two a year over the last 30 years (Holt *et al.* 2020). The first bird recorded in Wales, at Burry Port, Carmarthenshire, in October 1988, was attracted to a discarded fishing net full of rotting fish. All four records were of single birds:

- 1CY at Burry Port, Carmarthenshire, on 10–12 October 1988;
- 2CY at Gileston, East Glamorgan, on 2 January 1998;
- adult at Black Rock, Caernarfonshire, on 9–28 February 2002, also seen at Morfa Harlech, Meirionnydd, and
- 1CY at Blackpill, Gower, from 29 November to 5 December 2002.

A number of older records, from East Glamorgan, Gower and Carmarthenshire, are not now considered sufficiently well documented.

Jon Green and Robin Sandham

Sabine's Gull *Xema sabini* Gwylan Sabine

Welsh List Category	IUCN Red List	
	Global	Europe
A	LC	LC

First-calendar-year Sabine's Gull © Ben Porter

This gull breeds on the Arctic tundra and has a circumpolar distribution. It winters far to the south, mainly in the southern hemisphere, off western South America and southwest Africa. Birds seen in European waters in autumn are thought to breed in eastern Canada and Greenland and to be on migration to southwest Africa (*HBW*).

In Wales it occurs in coastal waters mainly as a result of Atlantic storms, as birds move south for the winter. There were 19th-century records from Pembrokeshire in 1839 and 1892, Flintshire in 1884 and Ceredigion in 1891 and 1896. The latter included six shot, with 2–3 others seen during a gale in late September. During 1900–79 there were only 29 records, but there has since been an increase. By the end of 1989, a total of 281 individuals had been recorded in Wales, including 124 in Pembrokeshire, 44 in Caernarfonshire and 25 on Anglesey. That total included at least 37 in September 1983, 20 of which were in Glamorgan (*Birds in Wales* 1994). Since 1990, a further 885 Sabine's Gulls have been recorded in Wales.

	1990–99	2000–09	2010–19	Total 1990–2019
Gwent	3	3	3	9
East Glamorgan	5	4	15	24
Gower	2	0	2	4
Carmarthenshire	2	0	0	2
Pembrokeshire Strumble Head	94	174	222	490
elsewhere	3	12	21	36
Ceredigion	5	12	9	26
Meirionnydd	2	0	3	5
Caernarfonshire Bardsey	11	47	61	119
elsewhere	1	12	32	45
Anglesey Point Lynas	6	18	27	51
elsewhere	12	39	14	65
Flintshire	2	6	1	9
Decadal total	**148**	**327**	**410**	**885**

Sabine's Gull: Individuals recorded in Wales by decade since 1990.

The vast majority of records are from Strumble Head, where annual totals have varied from none in 1993 to 86 in 2011, depending upon the occurrence of suitable winds during the passage period. In most years the highest daily count is fewer than five birds, but there were 12 on 13 September 1997, 11 on 16 September 2010, a record 18 on 18 September 2011 and 15 on 11 September 2017. Adults usually pass earlier in the season than juveniles, mainly between mid-August and mid-September, while later records are usually almost entirely juveniles (Green 2005a). The well-defined peak is in early to mid-September. An increase in records at Strumble during 1990–2019 probably reflects increased observer effort and experience and prevalence of suitable winds at peak times of passage. There were totals of 86 in 2011 and 52 in 2017 at Strumble, while 28 were recorded passing Bardsey, Caernarfonshire, in 2011 and 32 in 2017.

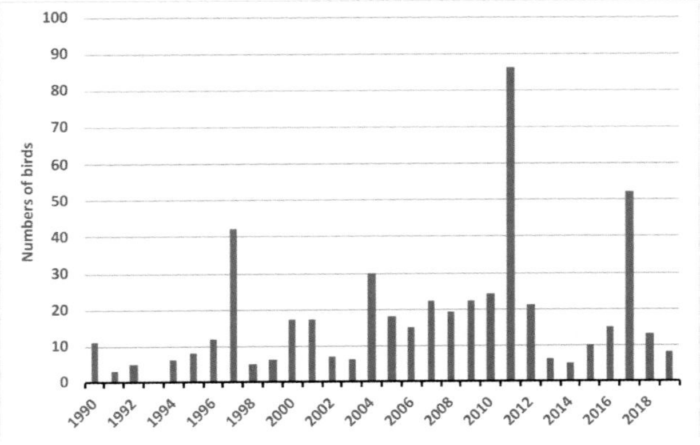

Sabine's Gull: Numbers recorded at Strumble Head each year, 1990–2019.

Winter records are rare, as most birds have left Welsh waters and the Western Approaches by mid-December, but a few individuals have occurred following storms: at Strumble Head on 3 January 1999, at the Rhymney Estuary, East Glamorgan, on 26 February 1990, at Aberdysynni, Meirionnydd, on 8 January 2005 and at Pwll, Carmarthenshire, on 9 January 2008. There have also been a handful of individuals in spring and early summer: at Goldcliff Pill, Gwent, on 9–11 April 1994, off Bardsey on 11 May 2004, at New Quay, Ceredigion, on 28 June 2008, at Bracelet Bay, Gower, on 12 April 2011, at Cemlyn, Anglesey, on 25 May 2011 and at Porthcawl, East Glamorgan, on 24 May 2013.

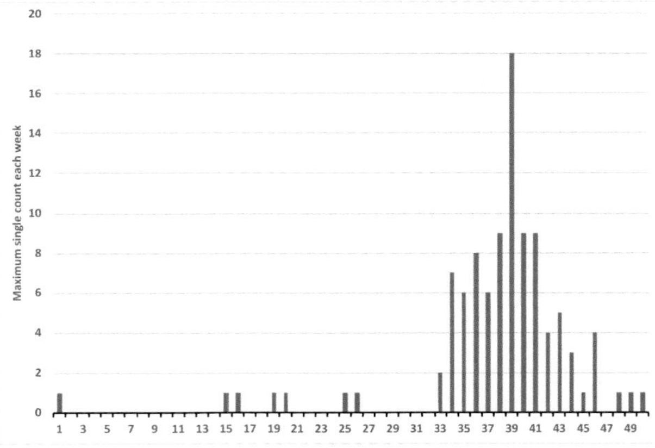

Sabine's Gull: Maximum single counts, each week, of birds passing around Wales, 1966–2019. Week 34 begins *c.*20 August and week 46 begins *c.*12 November. Records are from BirdTrack.

The global population trend for this species is stable (BirdLife International 2020) and the small European population (*c.*10% of global range) is thought to be increasing (BirdLife International 2015). The increase in records off the Welsh coast may be more apparent than real, a result of increased interest in sea-watching and improved optical equipment. There are no particular conservation issues affecting the species in Wales.

Jon Green

Bonaparte's Gull *Chroicocephalus philadelphia* Gwylan Bonaparte

Welsh List Category	IUCN Red List	Rarity recording
	Global	
A	LC	BBRC

Bonaparte's Gull breeds widely across northern North America. It winters locally on ice-free rivers and lakes in the northern USA, and south along both coasts of the USA to Mexico and the Caribbean (*HBW*). A total of 265 individuals had been recorded in Britain by the end of 2019 (Holt *et al.* 2020). Numbers have increased steadily since the 1980s, to an average of seven a year over the last 30 years, reflecting increasing observer awareness and a healthy source population (Holt *et al.* 2019).

Total	Pre-1950	1950–59	1960–69	1970–79	1980–89	1990–99	2000–09	2010–19
26	1	0	0	0	2	5	7	11

Bonaparte's Gull: Individuals recorded in Wales by decade.

A total of 26 individuals have been recorded in Wales. The first Welsh record was in Pembrokeshire in spring 1888, but the next was not until almost a century later, an adult off Bardsey, Caernarfonshire on 6–7 November 1984. This was followed by a second-calendar-year bird at Kenfig in March 1986, which was then at Rhymney until May of that year, both in East Glamorgan.

The monthly distribution of records shows that most are seen in winter or early spring. The map below shows the number of individuals recorded in each recording area, with assumed returning birds only counted once. At least ten individuals have been recorded in East Glamorgan, where birds were seen almost annually in Cardiff Bay and nearby estuaries between 2004 and 2019. These included two in March 2014, March 2015, April 2016 and April 2019. Records elsewhere included two second-calendar-year birds in Beddmanarch Bay, Anglesey, on 16 May 1998. The only inland record was from Breconshire, at Llangorse Lake on 7 April 2017.

Jon Green and Robin Sandham

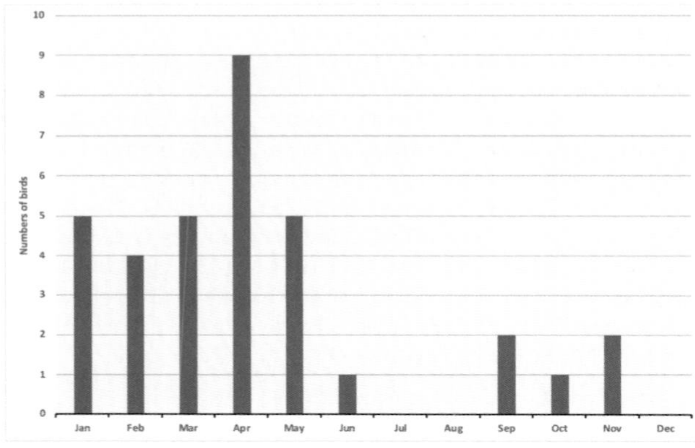

Bonaparte's Gull: Totals by month of arrival 1984–2019.

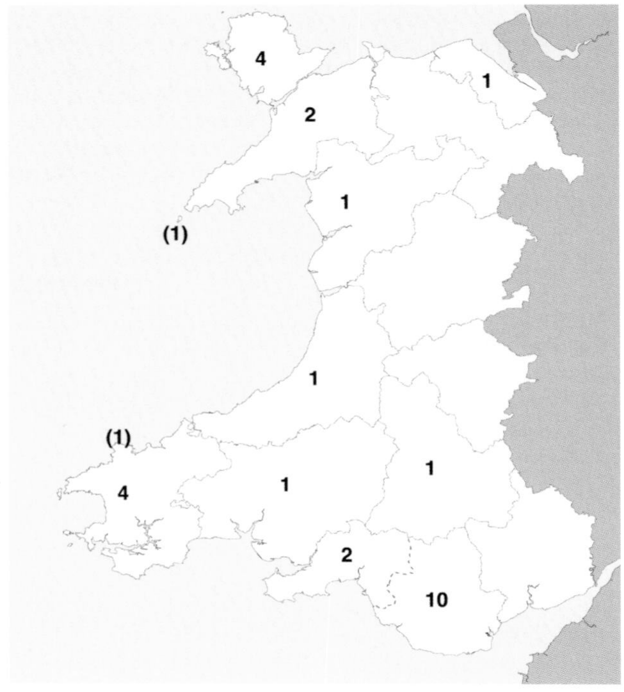

Bonaparte's Gull: Numbers of birds in each recording area 1888–2019. One bird was seen in more than one recording area.

Black-headed Gull *Chroicocephalus ridibundus* Gwylan Benddu

Welsh List Category	IUCN Red List			Birds of Conservation Concern			
	Global	Europe	GB	UK	Wales		
A	LC	LC	VU (NB)	Amber	2002	2010	2016

The Black-headed Gull breeds from southern Greenland and Iceland east across most of temperate Eurasia to the Kamchatka Peninsula, with a small but increasing presence in Newfoundland since 1977. In Europe, small numbers breed as far south as the Mediterranean. It usually breeds near calm, shallow water, on the coast or inland. In winter, its distribution tends to be more coastal, although some winter inland. Northern populations are migratory, while those farther south are resident or dispersive. Its diet is largely aquatic and terrestrial insects, earthworms and marine invertebrates (*HBW*). Ringing suggests that many of the birds that winter in Wales are from mainland Europe, the highest number of records coming from Poland, Finland, Denmark and The Netherlands. Some are from farther afield, such as an adult ringed at Criccieth, Caernarfonshire, in December 1985, that was found dead at Pskov, Russia, just south of St Petersburg, in August 1993. MacKinnon and Coulson (1987) estimated that 71%

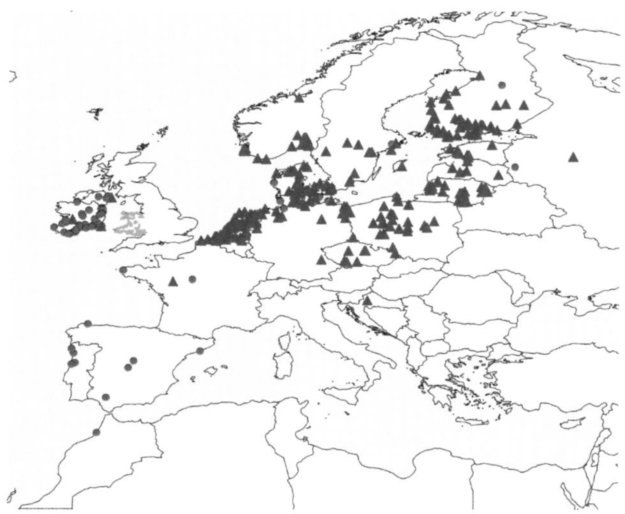

Black-headed Gull: Recovery locations of birds ringed in Wales are shown by red circles. Ringing locations of birds that were recovered in Wales are shown by blue triangles. Small brown dots show ringing or recovery locations in Wales.

of all Black-headed Gulls in England and Wales during the winter are of continental origin, although the influence of this population is lower to the north and west: 68% in southern and mid-Wales and 40% in northern Wales and northwest England. Recoveries of birds ringed as nestlings in Wales show that many winter in Britain but some move southwest, with many Welsh-hatched birds recorded in Ireland, France, Spain and Portugal in autumn and winter. There have also been recoveries in The Netherlands.

There are few early records of this species breeding in Wales. It was mentioned by Ray (1662) as breeding on a small island between Caldey, Pembrokeshire, and the mainland, where he noted "In one part of this island the puits [an old name for Black-headed Gull] and gulls, and sea-swallows nests be so thick that a man can scarce walk but he must needs set his foot upon them". Forrest (1907) misquoted Thomas Pennant as saying that this species bred on Llyn Conwy and Llyn Llydaw, both in Caernarfonshire, in the late 18th century, and this was repeated by several subsequent authors. However, Pennant actually wrote that "black back Gulls" nested on these two lakes (Pennant 1810).

The species seems to have declined greatly in Britain during the 19th century, because of drainage and uncontrolled egg-taking by humans for food. A colony of Black-headed Gulls noted by Martin (1864) on an unnamed mountain tarn near Beddgelert, Caernarfonshire, may have been at Llyn yr Adar, Meirionnydd (Pritchard 2012). The Black-headed Gull's fortunes were helped by the passing of the Sea Birds Preservation Act 1869 (*Historical Atlas 1875–1900*) and a widespread increase began about the end of that century (Parslow 1973). A parallel increase and expansion in range in Europe is thought to have been due to improved survival rates, as Black-headed Gulls began to exploit food sources linked to human activities (*European Atlas* 1997). Several colonies were founded in Wales in the late 19th century or the early years of the 20th century. The earliest confirmed breeding in Meirionnydd was at Llyn Mynyllod where two pairs founded a colony in 1880. This colony grew rapidly with protection by the landowners, the Robertson family of Palé. Forrest (1907) considered this to be the largest colony in North Wales, with 2,000 there in 1904. Its subsequent history is not known. It had been abandoned by 1938, although it was later reoccupied. The first confirmed breeding on Anglesey was around 1895, followed by Glamorgan in 1899, Radnorshire about 1900, Breconshire in 1908, Montgomeryshire at Llanllugan (Llyn Hir) in 1910, and Ceredigion before 1923. Some of these colonies quickly grew to a considerable size.

Most colonies in Wales are either on small islands in lakes or in marshy areas, which provide some protection from ground predators. There have been a number of colonies at altitudes over 500m, including a small colony in 1977 at Llyn Rhuddnant, Ceredigion, at 555m. Birds are often very mobile, with colonies being abandoned and new ones formed, but few have been counted regularly. The fortunes of this species in Wales can only

Operation Seafarer 1969–70	SCR Census 1985–88	Seabird 2000 1998–2002	Seabirds Count Provisional 2015–19
800*	3,010	1,986	827+

Black-headed Gull: Total number of apparently occupied nests (AON) recorded in Wales during UK seabird censuses.

* Coastal colonies only.

+ Coverage not yet complete (76% of Seabird 2000 sites covered).

be gauged from a series of national surveys. The first, in 1938, found 34 occupied colonies in Wales, with another seven considered "doubtful" (Hollom 1940). The majority were in North and mid-Wales, with none in Gwent, Glamorgan or Pembrokeshire. Seven were coastal. Most of the remaining 27 were on moorland or moorland pools, on average over 300m above sea level. Two colonies with over 1,000 pairs were in Montgomeryshire, at Llanllugan and Llyn y Tarw, both north of Newtown.

A survey in 1958 seemed to show a decrease, but coverage was said to be incomplete in Denbighshire and Caernarfonshire, and possibly in Flintshire and Anglesey (Gribble 1962), whereas coverage in Wales had been good in 1938. The majority of the 33 colonies counted were at altitudes of 300–450m. There were no colonies of over 1,000 pairs. The largest colonies were again in Montgomeryshire, with 700–750 pairs at Llyn y Tarw and at least 800 at ponds near Llangadfan. A 1973 census (Gribble 1976) had much better coverage in Wales than the 1958 survey and found 64 colonies, with a large increase in numbers. Two colonies held over 1,000 pairs: Llyn Trawsfynydd, Meirionnydd, with 1,116 nests and Cors Caron, Ceredigion, with 1,000–1,200 pairs.

Seabird censuses also provided information for this species, but Operation Seafarer in 1969–70 counted only coastal colonies. The Seabird Colony Register (SCR) census of 1985–88 (Lloyd *et al.* 1991) surveyed inland as well as coastal colonies, but coverage was incomplete. Far better coverage was achieved during Seabird 2000 in 1998–2002 (Mitchell *et al.* 2004). It showed a decline of around 75% in Wales since the dedicated Black-headed Gull survey of 1973. Numbers were lower almost everywhere in Wales except Anglesey, which then held the two largest colonies in Wales: 440 nests at Cemlyn and 400 at Llyn Alaw. The large colony at Llyn Trawsfynydd had been deserted, and other large colonies held much-reduced numbers.

There was no known breeding in Gwent until 1982, when six pairs attempted to breed at Garnlydan Reservoir, but there were

Vice-county	1938+	1958+	1973+	Seabird 2000*
Gower	0	50	0	75
Carmarthenshire	60	0	0	0
Pembrokeshire	0	98	0	0
Ceredigion	715	1,000–1,050	1,739–2,054	213
Breconshire	80	70–80	152–162	82
Radnorshire	342	250–300	251–265	170
Montgomeryshire	3050	1,690–1,750	1,710–1,883	158
Meirionnydd	315–365	300–355	1,434–1,514	0
Caernarfonshire	90–330	230	481–698	46
Anglesey	305	306–506	703–849	840
Denbighshire	590–840	50–55	1,181–1,442	244
Flintshire	10–20	400	305	158
Total	**5,557–6,107**	**4,444–4,874**	**7,956–9,172**	**1,986**

Black-headed Gull: Total number of pairs or occupied nests by county.

* Apparently occupied nests

+ Pairs

no breeding records there during Seabird 2000 (the figures given for Gwent in *Birds in Wales* (2002) are an error and relate to other counties). Venables *et al.* (2008) reported that there were breeding attempts at Garnlydan and two other upland sites in only six more years after 1982, with none known to have been successful. One pair bred at Newport Wetlands Reserve (Goldcliff) in 2005, again unsuccessfully.

Since the Seabird 2000 counts there has been no record of any site in Wales reaching 500 nests or pairs. The largest colonies in recent years have included Penclacwydd, Carmarthenshire, with 360 pairs in 2016, 418 pairs in 2018 and 399 in 2019, and Llyn Coed-y-dinas, Montgomeryshire, with about 300 pairs in 2015 and 200 pairs in 2019. In North Wales, the largest colonies are at Cemlyn, Anglesey, with an estimated 450 nests in 2015, down to 200 nests in 2019, and Shotton Steelworks, Flintshire, with 370 nests in 2016, which decreased to 82 in 2018 and 118 in 2019. Two other significant colonies are on the border with England. At Fenn's Moss, Denbighshire, around 200 pairs were recorded during 2008–12 (*North Wales Atlas 2008–12*), although some may not have been in Wales. A colony at RSPB Burton Mere Wetlands reached 963 nests in 2019, of which 216 were in Flintshire (Graham Jones *in litt.*). The lack of regular counts at many inland colonies makes it difficult to estimate the current population, but it is unlikely to be more than 1,500 pairs. In Breconshire, only the Brechfa Pool colony persisted into the current century (up to 50 pairs), but breeding became sporadic and was last noted in 2017. Aggressive colonisation of the pool by non-native New Zealand Pigmyweed since 2010 is considered to be the primary reason for loss (Andrew King *in litt.*). Aside from Llyn Coed-y-Dinas and Fenn's Moss, all the remaining colonies of any size in Wales are coastal. The numbers nesting inland are just a small fraction of those recorded in 1973.

Colonies that have held more than 1,000 pairs at times include:

Cors Caron, Ceredigion

The site is also known as Cors Goch Glan Teifi. Several locations have been used in the bog, which is now a National Nature Reserve. Occupied since 1923, F.C.R. Jourdain's diaries recorded a colony of several thousand pairs in May 1934 (Roderick and Davis 2010). Published counts indicated that the colony fluctuated between none and 250 pairs during 1932–68, increased to 500 pairs in 1969, and 1,000–1,200 pairs in 1973. It then declined to 200–300 nests in the 1980s and early 1990s; there were 350 nests in 1995, but subsequent counts were much lower, with just one pair in 2005.

Rhosgoch Bog, Radnorshire

50–60 birds and several nests were found here in 1911. In 1932, F.C.R. Jourdain, who was searching for Spotted Crakes, counted 5,000 birds. There were 300 pairs in 1934, about 500 pairs in 1948 and over 2,000 pairs in 1954, but numbers were down to 30–40 pairs in 1958 and 1959. Numbers increased again to about 500 pairs in the early 1960s. There were ten pairs in 1983 and breeding has been only occasional since (Jennings 2014).

Llyn Hir, Montgomeryshire

This site has also been reported as Llanllugan and Tregwynt Gull Pool (Holt and Williams 2008), although separate counts are sometimes given for a small pool near Tregwynt as 'Tregwynt Gull Pool'. The colony at Llyn Hir was established in 1910, the first in Montgomeryshire. It held 20 pairs in 1911, which grew to 1,000 pairs in 1936 and higher in 1937 and 1938. The 1958 survey found up to 350 pairs, with 700 pairs in 1968 and 400 pairs in 1973, but no nests were found there in 2000.

Llyn y Tarw, Montgomeryshire

The colony began soon after 1910, reached 1,000–1,500 pairs by 1926 and *c.*2,500 pairs in 1927. Oldham reported 3,000–4,000 there in 1939 (*Birds in Wales* 1994), although it is not clear whether he was referring to pairs or individuals. During the Second World War, eggs were collected for shipment to London for food. By 1958 numbers were down to 700–750 pairs, with 650 pairs in 1968 and 600 in 1973. Seabird 2000 found the site deserted.

Llyn Trawsfynydd, Meirionnydd

The date this colony was established is not known. The report of the 1938 survey noted that no nests were found there, and that birds had been "discouraged" over several years. At least 400 pairs bred at this site in the early 1950s, and by 1969 there were 1,260 pairs (Jones 1986). The 1973 survey found that the colony held 1,116 pairs, making it one of the two largest colonies in Wales at the time (Gribble 1976). The SCR Census in 1985–88 found 1,330 pairs, almost half the 3,010 pairs recorded in Wales during those years. However, by the time of the Seabird 2000 survey in 1998–2002, the colony had been abandoned (Mitchell *et al.* 2004). There has been no record of breeding there since.

Llyn Llywenan, Anglesey

This colony probably began in 1895, and about 500 birds were present in 1903. There were *c.*400 pairs in about 1918, 300 pairs in 1936 and a peak of 2,000 pairs in 1952. Numbers were down to 300 pairs by 1958 and 360–445 nests in 1973. About 800 birds were present in 1986, but the colony is thought to have disappeared in the late 1980s (Nigel Brown *in litt.*).

Outside the breeding season, numbers in autumn and winter are much higher than can be accounted for by birds breeding in Wales. Gull counts are optional in the Wetland Bird Survey (WeBS), so it may underestimate numbers at some sites.

Numbers recorded on the WeBS counts peak in July and August, then fall by *c.*50% by the winter months. Some sites have held over 10,000 in July and August, when many birds pass through Wales after breeding, presumably on their way farther south. Large counts in these months include 11,155 birds along the south shore of the Burry Inlet on 23 August 1990 and 12,000 at the Rhymney Estuary, East Glamorgan, in July 1990. Large numbers have been recorded in late autumn off Bardsey, Caernarfonshire, in some years, among flocks of several gull species. There were 12,000 Black-headed Gulls there on 28 October 2003 and 15,000 on 8 October 2005.

In late autumn and winter, there were 28,500 at the Rhymney Estuary on 30 November 1985, *c.*24,000 there in February 1987 and 25,000 at Rumney Great Wharf in November 2005. High numbers have also been recorded in spring in this area, notably *c.*17,000 at the Rhymney Estuary on 8 April 1973. Inland roosts have included 15,000–20,000 birds at Llandegfedd Reservoir, Gwent, in winters 1993/94 and 1994/95, and up to 15,000 at Llangorse Lake, Breconshire (Peers and Shrubb 1990). Numbers in Wales have been lower since 2005, however, with no counts over 10,000 at any site. The highest recent count was 9,746 off Bardsey in November 2016.

Site	County	94/95 to 98/99	99/00 to 03/04	04/05 to 08/09	09/10 to 13/14	14/15 to 18/19
Burry Inlet	Gower/Carmarthenshire	6,993	7,345	7,507	6,242	4,986
Severn Estuary: Welsh side (whole estuary)	Gwent/Glamorgan	1,084 (5,605)	1,932 (7,529)	1,261 (7,453)	3,469 (8,668)	4,627 (12,170)
Dee Estuary: Welsh side (whole estuary)	Flintshire	9 (2,759)	1,025 (6,035)	459 (5,528)	1,408 (10,214)	3,114 (9,486)
Traeth Lafan	Caernarfonshire/ Anglesey	176	860	1,966	1,474	2,591
Cleddau Estuary	Pembrokeshire	3,741	3,598	2,640	1,639	1,715
Swansea Bay	Gower/East Glamorgan	2,503	2,595	3,462	1,603	986

Black-headed Gull: Five-year average peak counts during Wetland Bird Surveys between 1994/95 and 2018/19 at sites in Wales averaging 2,000 or more at least once. International importance threshold: 20,000; Great Britain importance threshold: 22,000.

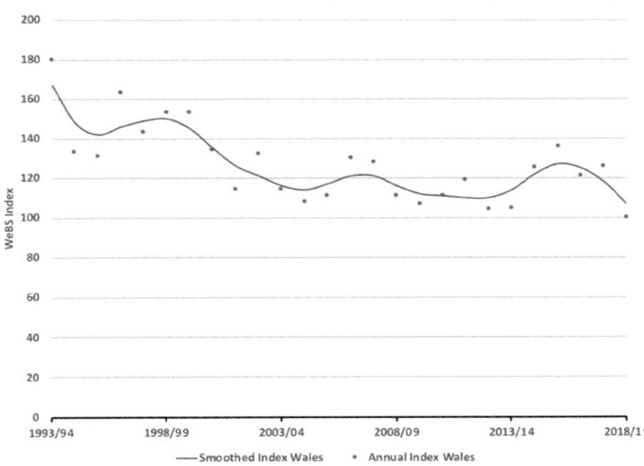

Black-headed Gull: WeBS indices for Wales to 2018/19.

Winter gull roost surveys were carried out every ten years from 1953 onwards, initially mainly in England. Coverage for Wales was incomplete until 1983 (Banks *et al.* 2009), when 15,213 Black-headed Gulls were recorded at inland sites and 101,227 at coastal sites. The 1993 survey recorded 85,729 birds in Wales: 16,993 inland and 68,736 on the coast (Burton *et al.* 2003a). The survey in 2003/04–2005/06 found a total of 100,836 Black-headed Gulls in Wales, 5% of the estimated total for Britain (Banks *et al.* 2007b).

The future of the Black-headed Gull in Wales is uncertain. Several reasons have been suggested for the desertion or drastic decline of colonies. These include the growth of scrub, predation by species ranging from Red Kite to American Mink and Otter, drainage, acidification, invasive non-native waterweeds, human disturbance, agricultural intensification of surrounding land and nest-space competition from increasing numbers of Canada Geese. Any conservation action for the species as a breeding bird in Wales would have to start by understanding the causes of its decline. This highlights the importance of ensuring that all the inland sites are counted regularly. Maintaining breeding Black-headed Gulls could also benefit other species. For example, Sandwich Terns at Wales' only colony, at Cemlyn on Anglesey, nest among Black-headed Gulls. Sandwich Terns are thought to deliberately settle to nest near Black-headed Gulls and to derive a degree of protection as a result (Fuchs 1977).

Rhion Pritchard

Sponsored by WWBIC – West Wales Biodiversity Information Centre

Little Gull *Hydrocoloeus minutus* Gwylan Fechan

Welsh List Category	IUCN Red List		Birds of Conservation Concern			
	Global	Europe	UK	Wales		
A	LC	NT	Green	N/A 2002	2010	2016

The Little Gull breeds in freshwater boreal forest wetlands in northern Fennoscandia, the Baltic States and western Russia, which is where most birds seen in Wales probably originate, although there are no ringing recoveries to confirm this. The breeding range extends east across Russia to the Pacific Ocean and small numbers have nested in North America since the 1960s. The species winters south and west of its breeding range, most likely far offshore but its precise distribution, diet and habitat use in winter remain poorly known (*HBW*).

Madden and Ruttledge (1993) confirmed that easterly winter storms in the Irish Sea 'wrecked' Little Gulls along steep boulder storm beaches in Co. Wicklow in the 1970s, as westerly storms did in Ceredigion (Fox 1986b). Only after Ruttledge's death did midwinter wind-turbine impact assessments, undertaken during 2001–11, finally locate several hundred Little Gulls dip-feeding above the surface of the Irish Sea. These were associated with shoals of small fish during November to March, especially over the Arklow and Kish Banks off Co. Wicklow, sometimes affiliated with feeding Razorbills (MRG 2012).

Although these Little Gulls may continue out into the Western Approaches and Atlantic Ocean during each winter, it seems likely that a modest wintering population remains in the Irish Sea. These, potentially, represent the source of midwinter storm-blown birds in Cardigan Bay, which typically feed in the breakers and spume along steep pebble beaches for a few hours after severe westerly or southwesterly storms, from Pwllheli and Criccieth, Caernarfonshire, to Aberdysynni, Meirionnydd, and especially at Tanybwlch, Aberystwyth and Clarach, Ceredigion. Aerial surveys have revealed wintering aggregations in Liverpool Bay that were previously almost unknown, numbering up to an estimated 333 wintering birds, mostly off Blackpool, Lancashire (Lawson *et al.* 2016a). Greater numbers occur off the east coast of England: over 2,000 off The Wash, Lincolnshire/Norfolk, in early winter (Lawson *et al.* 2016b) and over 250 off the Thames Estuary, Essex/Kent, (Lawson *et al.* 2016c) suggesting that further offshore, unseen winter aggregations await discovery around British coasts.

The erratic occurrence of the Little Gull makes it difficult to judge its true status, abundance or any rate of change in Wales. Fewer than ten records were reported in the 19th century, all associated with stormy weather, but the species has likely been long overlooked in all seasons—and may still be—because of the small numbers involved, its discreet behaviour and because the species rarely remains long in one area. Numbers were considered to have increased considerably in Britain and Ireland during the 1960s and 1970s (Hutchinson and Neath 1978), which was thought to be a result of increases in the breeding population east of the Baltic. The species has increased in number and expanded its breeding range in Fennoscandia and the Baltic States, especially in Finland, during the last three decades (Valkama *et al.* 2011). Successful breeding was recorded in Scotland in 2016, but breeding has never been suspected in Wales.

Although the Little Gull can occur almost anywhere in Wales, it is mainly a coastal species. Most are reported singly, and rarely are more than 3–4 seen together, which are usually the result of aggregations wrecked by midwinter storms. *En route* to its pelagic wanderings in winter, it seems likely that birds pass overland across Britain from the Angus coastline southward, crossing northern England to funnel into the Mersey Estuary and thence out into the Irish Sea and the Atlantic Ocean (Hutchinson and Neath 1978). Hence, the species is a passage commuter through and past Wales during autumn, and in reverse in spring. During April/May and again in late July to November, Little Gulls pass coastal watch points, including Strumble Head, Pembrokeshire, and Bardsey, Caernarfonshire. In autumn, this species is not uncommon on the North Wales coast, as well as migrating through brackish coastal lagoons, such as the Glaslyn Estuary, Meirionnydd/Caernarfonshire. Although the species is regularly reported on passage from some freshwater bodies close to the coast, such as Kenfig Pool in East Glamorgan, Eglwys Nunydd Reservoir in Gower and Llandegfedd Reservoir in Gwent, during April/May and August to October, inland records are relatively scarce.

There were notable spring passages in the period 1973–75. In April 1974, the single biggest count was 74 at Kenfig Pool, but it was thought that at least 201 individuals visited the site during the month (Hurford and Lansdown 1995). There were 73 at Eglwys Nunydd Reservoir in May 1974. Large counts off coastal

Little Gull in winter plumage © Ben Porter

headlands and islands in autumn include 75 past Strumble Head on 10 November 1985 and 76 there on 13 November 1987; 81 past Aberaeron, Ceredigion, on 6 February 2002; and at least 75 adults past Bardsey on 30 September 2005. A total of 230 passed Point of Ayr, Flintshire, between 13 February and 3 April 1997, including 62 that flew west on 18 February, and there were 65 at Black Rock Sands, Caernarfonshire, on 7 March 1999. Numbers have been lower since 2005, with the peak counts of 36 past Bardsey on 22 October 2010 and up to 35 at Tanybwlch/Aberystwyth in February 2014. In recent years, the total reported in Wales each year has rarely exceeded 100–120 birds, exceptionally more than 150.

The transient and ephemeral nature of the Little Gull in Wales makes it difficult to judge its conservation status. Observations suggest that, although scarce and variable, the species is probably no less common than it was 30 years ago. Its western European status appears favourable, but modelling suggests that the Baltic Sea will no longer have a suitable climate for the species by the end of this century (Huntley *et al.* 2007). The second *European Atlas* shows significant range changes since the 1997 *European Atlas*, with large-scale losses in eastern Europe but a massive range expansion further north, in Sweden, northern Norway and northen Finland (*European Atlas* 2020). Its pelagic wintering habitat and the transitory nature of its annual migratory passage through Wales mean that this species is unlikely to respond to nature conservation management.

Tony Fox

Ross's Gull *Rhodostethia rosea* Gwylan Ross

Welsh List Category	IUCN Red List		Rarity recording
	Global	Europe	
A	LC	EN	BBRC

This attractive gull is a locally common breeder on the tundra of northeast Siberia, and a local breeder in Greenland and Canada (*HBW*), from where it is assumed visitors to Britain originate. A total of 102 were recorded in Britain to the end of 2019, on average about two a year over the last 30 years, with numbers stable or decreasing (Holt *et al.* 2020). There have been five records in Wales, thought to involve four individuals:

- adult in Fishguard Harbour, Pembrokeshire, on 15–16 February 1981;
- adult at Aberystwyth, Ceredigion, on 30 December 1994, with what was presumed to be the same bird at Porthmadog and Morfa Bychan, Caernarfonshire, on 14–16 and 22 January 1995;
- adult at Blackpill, Gower, on 10 February 2002 and
- 2CY at Aberavon, Gower, on 8–16 February 2014.

Jon Green and Robin Sandham

Laughing Gull *Leucophaeus atricilla* Gwylan Chwerthinog

Welsh List Category	IUCN Red List	Rarity recording
	Global	
A	LC	BBRC

Laughing Gull at Porthmadog, Caernarfonshire, winter 2005/06
© Steve Young

Laughing Gull: Numbers of birds in each recording area 1978–2019. One bird was seen in more than one recording area.

The Laughing Gull breeds in eastern North America and the Caribbean, and winters in Central and northern South America (*HBW*). A total of 205 were recorded in Britain to the end of 2019, an average of about five a year over the last 30 years, with influxes in 2005 and possibly 2006 (Holt *et al.* 2020). Most of the 19 individuals recorded in Wales arrived in November 2005, following a fast-moving low-pressure system in the Atlantic, the tail-end of Hurricane Wilma.

Total	Pre-1950	1950–59	1960–69	1970–79	1980–89	1990–99	2000–09	2010–19
19	0	0	0	1	2	1	14	1

Laughing Gull: Individuals recorded in Wales by decade.

The first record was at Nant y Moch Reservoir, Ceredigion, on 19–20 May 1978. There were first-calendar-year birds at Colwyn Bay, Denbighshire, on 27–28 December 1981 and at Newton Point, East Glamorgan, on 12 September 1988, followed by a second-calendar-year bird at Point of Ayr, Flintshire, on 20–22 December 1991.

There were then no further records until the tail end of a hurricane in November 2005 brought at least 13 individuals to Wales, out of a total of 56 recorded in Britain. Six of these were in Pembrokeshire, two each in Carmarthenshire and Gower, with singles in East Glamorgan, Ceredigion and Porthmadog on the Meirionnydd/Caernarfonshire border. Eight were first-calendar-year birds, one a second-calendar-year bird and four adults. The first to arrive were two first-calendar-year birds on 4 November. Some remained into 2006. The longest-staying bird was an adult at Porthmadog, which remained from 14 November 2005 to 4 April 2006.

There have been two subsequent records, both in Pembrokeshire: an adult on Skomer on 21 May 2009 and a third-calendar-year bird on Skokholm on 10 and 26 June 2019.

Jon Green and Robin Sandham

Franklin's Gull *Leucophaeus pipixcan* Gwylan Franklin

Welsh List Category	IUCN Red List	Rarity recording
	Global	
A	LC	BBRC

The Franklin's Gull breeds locally throughout the interior provinces of temperate Canada and the northern USA and winters along the Pacific coast of South America (*HBW*). A total of 81 individuals had been recorded in Britain to the end of 2019, about two or three a year over the last 30 years (Holt *et al.* 2020).

There have been only four records in Wales:

- adult roosting at Eglwys Nunydd Reservoir, Gower, from 28 October to 1 November 1998;
- adult at Blackpill, Gower, on 2–6 July 1999;
- 1CY at Pentre Davis, Carmarthenshire, on 6–7 November 2005, which arrived at the same time as an influx of Laughing Gulls and
- 2CY at Penclacwydd, Carmarthenshire, on 18–28 April 2014, which was also seen on the other side of the Burry Inlet, at Crofty, Gower.

A previously published record of this species at Aber Dysynni, Meirionnydd, on 22 March 1986 was later deemed not proven by the BBRC following a review.

Jon Green and Robin Sandham

Mediterranean Gull *Ichthyaetus melanocephalus*

Gwylan Môr y Canoldir

Welsh List Category	IUCN Red List			Birds of Conservation Concern				Rarity recording
	Global	Europe	GB	UK	Wales			
A	LC	LC	LC	Amber	2002	2010	2016	RBBP

Until the 1950s, this species was confined mainly to the areas around the Black Sea coast of Ukraine, with small numbers in the Balkans and Turkey. There had been only 4–5 British records prior to 1940. It has undergone a huge expansion in its range and now breeds across western Europe and east to Azerbaijan, wintering around the coasts of Europe and North Africa from Denmark south to Mauritania, with some around the Black Sea (*HBW*). The first breeding record in Britain was in Hampshire in 1968, and now up to 1,500 pairs nest in the UK (Holling *et al.* 2019). The population around the Black Sea shows strong fluctuations in numbers in approximately 20-year cycles. The expansion into western and central Europe in the 1950s and 1960s coincided with peak numbers in the Black Sea area (Zielińska *et al.* 2007).

The first record for Wales was one at the Ogmore Estuary, East Glamorgan, on 4 April 1964, and was quickly followed by one off Bardsey, Caernarfonshire, on 26 June the same year. A first for Pembrokeshire was recorded in 1968, with records in Anglesey and Gower in 1970, Denbighshire in 1971, Flintshire and Gwent in 1975, Carmarthenshire in 1976, Ceredigion in 1980 and Meirionnydd in 1981. The three inland counties only recorded their first more recently: Breconshire in 1994, Montgomeryshire in 2003 and Radnorshire in 2009.

From the mid-1990s, a series of breeding attempts were made at Cemlyn on Anglesey. A bird unsuccessfully tried to establish itself within a colony of Black-headed Gulls in April 1995, a second-calendar-year bird displayed to a Black-headed Gull on 31 March 1996 and a pair displayed there in 1998, but no breeding took place. A lone male spent six years trying to pair up with Black-headed Gulls between 2009 and 2015, and in the last year was seen to feed Black-headed Gull chicks, having probably ousted the male from the pair. The first successful breeding in Wales took place at Cemlyn in 2010, when a pair raised two chicks. A pair again nested in 2011, but no chicks fledged. In 2013 three pairs laid clutches, and two young fledged from two nests. In 2014 at least five pairs were present, but only two young were raised. In 2015 there were four pairs, but only one pair bred, raising a single chick. In 2016, several birds were present, but only one pair nested and this nest was abandoned prior to hatching. In 2017, 19 were present at Cemlyn in late April, but no pairs were seen beyond that date, although a recently fledged juvenile was seen on 29 July. In 2018 and 2019 mating was observed but the birds did not stay to breed.

Elsewhere, an adult and a second-calendar-year female spent a week in the Black-headed Gull colony at Brechfa Pool, Breconshire, in May 2005, with courtship, copulation and preliminaries to nest-building noted, but they did not persist (Andrew King *in litt.*). At Penclacwydd, Carmarthenshire, a pair was seen mating on 29 April 2009, but no breeding took place. This was the start of a series of breeding attempts there over the following years, with the first successful breeding in 2014, when a pair fledged a single chick. In 2015 two pairs nested there, but both failed, the eggs predated by a Moorhen in one nest, while the chick died shortly after hatching in the other. In 2016, five pairs attempted breeding at Penclacwydd but no chicks fledged, owing to predation by Foxes. A number of pairs have bred at RSPB Burton Mere Wetlands in recent years, but usually on the English part of the reserve. In 2019, 11 pairs nested in the Black-headed Gull colony there, of which three pairs nested in the Flintshire part of the reserve (Colin Wells *in litt.*). In the inland counties, a pair built a nest at Llyn Coed-y-Dinas, Montgomeryshire, in May 2003 and again in 2016, but did not breed. A single bird was seen in the Black-headed Gull colony at Brechfa Pool, Breconshire, in 2018.

Mediterranean Gull © Dan Rouse

Outside the breeding season there has been a large increase in the numbers of Mediterranean Gulls seen around the coast. In 1984, a count of 12 at Blackpill, Gower, on 11 July was considered notable. From about 2005, there have been increasingly large late summer and autumn counts along the Welsh coast. The highest counts have been on the Ceredigion coast between Aberystwyth and New Quay, with 421 between Llanon and Llanrhystud in 2013 and 565 there in 2014, 720 in the Llanon/Llanrhystud/Llansantffraed area in August 2018 and 620 at Llanrhystud on 25 August 2019. There were 500 at Cei Bach, Ceredigion, on 4

		1994/95 to 1998/99	1999/00 to 2003/04	2004/05 to 2008/09	2009/10 to 2013/14	2014/15 to 2018/19
Swansea Bay	Gower	8	14	24	111	120
Mumbles Head	Gower				32	88

Mediterranean Gull: Five-year average peak counts during Wetland Bird Surveys between 1994/95 and 2018/19 at sites in Wales averaging 40 birds at least once. International importance threshold: 2,400; British importance threshold: 40.

Mediterranean Gull: Ringing locations of birds that were recovered in Wales are shown by blue triangles. Small brown dots show recovery locations in Wales.

October 2019. Large increases have also been recorded at other coastal sites, including 436 at Blackpill, Gower, in July 2019. 407 off Skokholm in November 2018 was a county record for Pembrokeshire. In 2015 it was estimated that a total of 1,500 were present during post-breeding dispersal across Carmarthenshire, Gower, Pembrokeshire and Ceredigion. Numbers in North Wales tend to be lower, but at least 170 were off Bardsey, Caernarfonshire, on 29 October 2017 and there were 83 at the Inland Sea, Anglesey, in July 2019. The species remains scarce inland.

The large counts on the Ceredigion coast appear to be birds dispersing from breeding colonies on the near continent prior to wintering around the Mediterranean. One, colour-ringed as a nestling near Antwerp, Belgium, on 16 August 2008, was seen on the Ceredigion coast at Llanon or Aberystwyth in July/August, in most years during 2010–18. During the winter months it was seen in Lisbon, Portugal, in October 2012 and February 2015 and at Gijon, Spain, in January 2016, and near Antwerp again in the breeding season in 2013, 2014 and 2018. A colour-ringed juvenile, seen at Burry Port Harbour, Carmarthenshire, in July 2019, had been ringed as a nestling at Rehbach Gravel Pit, Leipzig, eastern Germany, in June 2019 (Rob Taylor *in litt.*).

Not all birds leave Wales in winter. Smaller numbers can be seen around all coasts. One individual, colour-ringed as a chick in Poland in June 2015, was seen at Cardiff Bay, East Glamorgan, in March 2016 and at Connah's Quay nature reserve, Flintshire, in February 2019. Other colour-marked birds seen in Wales had been ringed in France, The Netherlands, Germany, Czechia, Hungary and elsewhere in Britain. A Czech-ringed bird returned to Fishguard Harbour, Pembrokeshire, for seven consecutive years during 2001–07 and another ringed bird was noted to be at least ten years old.

Birds showing characteristics of hybrid Mediterranean x Black-headed Gulls are seen periodically, and may be under-reported. Examples include birds at the Ogmore Estuary, East Glamorgan, on 5 and 26 September 2003, at Traeth Dulas, Anglesey, on 28 January 2010 and at Point Lynas, Anglesey, on 9 December 2017. A mixed pairing fledged two hybrid young at a site in south Anglesey in 2019.

The Mediterranean Gull has established a toehold in Wales, but it remains to be seen whether it will become a regular breeding species. The population in Europe is now thought to be decreasing (BirdLife International 2015), with predation and disturbance at colonies thought to be the main threats (*HBW*).

David Winnard

Sponsored by Arfon Williams

Common Gull *Larus canus* Gwylan Gweunydd

Welsh List Category	IUCN Red List			Birds of Conservation Concern			
	Global	Europe	GB	UK	Wales		
					2002	2010	2016
A	LC	LC	LC	Amber			

The Common Gull breeds from Iceland and across much of northern Europe and Siberia to the Pacific and also in Alaska and western Canada. Birds winter to the south of their breeding range (*HBW*). It breeds in western Ireland and in much of Scotland, usually in wide inland valleys and on coasts. Sporadic breeding farther south in Britain is thought to stem from continental winter visitors that remain to breed, though nothing is known of the origin of birds that formed a small colony on Anglesey during the 1960s and 1970s. Elsewhere in Wales, there have been sporadic reports of breeding Common Gulls, but these almost certainly involve summering birds that were not attempting to breed. It is otherwise a winter visitor. Recoveries and resightings of ringed birds show that birds nesting in Fennoscandia, Estonia, Poland, Denmark, Germany and The Netherlands winter in Wales, as well as birds from the Scottish breeding population. The majority of ringed birds recorded in Wales have been from Norway.

Williams (1835) stated that this species bred on the Great and Little Ormes. This statement was repeated in the *Llandudno Guide* later in the century, but Forrest (1907) considered this to be an error. Forrest knew of no breeding records in North Wales. Adults and juveniles seen together in Caernarfonshire, around Criccieth in June 1908 and Conwy Sands on 28 June 1913, were thought not to be Welsh bred (Forrest 1919). It was not until 1960 that the first proven breeding took place, at Flagstaff Quarry near Penmon, Anglesey. Between 1960 and 1963, fewer than half a dozen pairs nested at this site, scattered among a colony of Herring Gulls. Breeding continued, perhaps annually, until the late 1960s when only three pairs were present. During the early 1970s, the number of ground-nesting birds decreased markedly at the site, although breeding Common Gulls were again recorded in 1980 and probably in 1985, but not subsequently. There were probably never more than six pairs in any one year (Jones and Whalley 2004). The *Britain and Ireland Atlas 1968–72* also reported Common Gulls, with breeding codes, at three other sites in Anglesey and north Caernarfonshire, but these almost certainly involved first-calendar-year birds from the Penmon area.

It is only in winter that the Common Gull becomes numerous and widespread in southern Britain. In Wales, its winter distribution is mainly coastal, although sizeable flocks roost on some inland reservoirs. Birds arrive from mid-July to November, with return movements between late February and early April. Very small numbers of first-calendar-year birds remain into May and June at a few established summering sites along the North Wales coast.

The number of wintering birds increased gradually at the beginning of the 20th century and by 1925 large flocks were said to be common in Glamorgan, both inland and along the coast (Hurford and Lansdown 1995). Flocks were often seen feeding on estuaries and coastal fields throughout South Wales, although there was little subsequent information on numbers from any Welsh county until a survey of coastal gull roosts in 1956–57. Coverage in Wales was poor, with records only collated from four counties. The largest number recorded in Wales was 2,000 at three roosts at the confluence of Afon Taf and Afon Tywi, Carmarthenshire (Hickling 1960). In winter 1963/64, large numbers were observed flying over Abergavenny, Gwent, down the River Usk and across the Bristol Channel to roost at Avonmouth.

A survey of inland roosts in England and Wales in 1973 reported 2,960 Common Gulls at five roosts across Wales (Hickling 1977). High counts on the coast in the 1970s included *c.*5,000 at Mumbles Head, Gower, in February 1973. In 1983, another midwinter gull roost survey found 25,447 at coastal sites in Wales, but only 87 inland (Bowes *et al.* 1984). 10,177 were counted on the Welsh side of the Dee estuary in Flintshire. There were also 6,000 in Ceredigion (Roderick and Davis 2010), and sizeable roosts were counted along the South Wales coast. Numbers increased at several sites during this period, notably at Blackpill, Gower, where the mean winter peak increased from *c.*1,115 to *c.*3,080 through the late 1960s and 1970s (Hurford and Lansdown 1995). There were 3,100 there in February 1986, but numbers declined steadily through the 1980s and early 1990s. In Ceredigion, there were 4,000 at Ynys-las in December 1982 and 3,100 at Borth in December 1984. Large November influxes into Borth Bay, when shoals of fish came close inshore, included an estimated *c.*12,000 in the Dyfi Estuary on 5 November 1983, and what was described as a "huge influx" on 26 November 1988 (Roderick and Davis 2010).

The Burry Inlet recorded a maximum of 5,085 in autumn 2000 and Blackpill regularly hosted flocks of over 2,000 annually during 1991–2001, with a maximum of 3,150 in January 1996, but numbers have declined at both sites since about 2010. There is little indication of a decline in Carmarthen Bay, which recorded a peak of 4,796 in February 2016. Llandegfedd Reservoir, Gwent, holds sizeable numbers in winter, and Venables *et al.* (2008) noted that numbers had increased dramatically since 1977. An estimated 13,000 birds there in winter 1997/98 was a Welsh record count, and there were three other maxima of 9,000–10,000.

Numbers tend to be lower in North Wales, but the Flintshire coast and the Dee Estuary see a build up from December into January/February, including 3,500 at Bagillt in February 1995, 3,662 at RSPB Oakenholt Marsh on 1 January 2019 and *c.*3,500 at Gronant on 18 February 2019. At Rhuddlan, Flintshire, close to the Clwyd Estuary, counts included 5,000 in January 2007 and 8,000 in December 2012, with several thousand in subsequent winters. Several inland sites in Denbighshire also hold good numbers, notably 3,500 at Pulford in February 2007. Other large flocks in the depths of winter included 3,000 at Nant-y-Caws landfill site, Carmarthenshire, in January 1998 and 3,000 at Nicholaston Pill in Oxwich Bay, Gower, on 1 January 2012.

Common Gulls are regularly recorded on passage in all counties, but the largest numbers are along the north and south coasts. Vernon (1969) described a northward and westward movement of birds through Wales in spring. Large flocks in Radnorshire during February are thought to be a result of pre-migratory flights from the Severn Estuary roosts. These birds are believed to follow a route east across the English Midlands and on to The Netherlands, Germany and Denmark. Large numbers recorded along the North Wales coast in March and early April are believed to pass through Cheshire and Lancashire and move northeast to Scotland and Scandinavia.

Common Gull: Recovery locations of birds ringed in Wales are shown by red circles. Ringing locations of birds that were recovered in Wales are shown by blue triangles. Small brown dots show ringing or recovery locations in Wales.

Site	County	94/95 to 98/99	99/00 to 03/04	04/05 to 08/09	09/10 to 13/14	14/15 to 18/19
Dee Estuary: Welsh side	Flintshire	0	1,045	530	829	1,586
(whole estuary)		(349)	(3,010)	(997)	(1,437)	(1,748)
Carmarthen Bay	Carmarthenshire	597	1,874	1,368	1,786	1,723
Traeth Lafan	Caernarfonshire/Anglesey	74	357	701	432	379
Burry Inlet	Gower/Carmarthenshire	1,263	3,288	1,710	917	304
Swansea Bay	Gower	1,125	767	840	401	277

Common Gull: Five-year average peak counts during Wetland Bird Surveys between 1994/95 and 2018/19 at sites in Wales averaging 700 or more at least once. International importance threshold: 16,400; Great Britain importance threshold: 7,000.

Large flocks are recorded immediately prior to the main eastward movement in winter and early spring, including at sites away from the main winter areas, such as 6,400 at Bagillt Bank, Flintshire, in March 2003. As numbers decreased, a count of 1,000 at the Gwendraeth Estuary, Carmarthenshire, in March 2017 was notable. Inland, Gresford Flash, Denbighshire, recorded 1,800 in February 2009. In autumn, some adults arrive in late July and throughout August, sometimes accompanied by juveniles, although the main westward movement occurs in October. Several thousand have been recorded at Point of Ayr, Flintshire, usually in August, with a maximum of 3,000 on 20 August 1989. Llanrhidian, Gower, on the Burry Inlet, also sees a build-up in late summer and autumn, including 2,336 in August 2007.

The WeBS indices for Wales have shown a worryingly rapid decline since 2010/11. Indices in England and Scotland are falling too. This species is Red-listed in Wales because of the large decline in the wintering population over the past 25 years (Johnstone and Bladwell 2016). The reasons for the decline are unclear.

Since the early 1990s, flock sizes in South Wales appear to have decreased, while those on the North Wales coast have maintained their numbers, and in some cases, increased. It is possible that some birds are now wintering further east or further north than formerly. The population in Europe is thought to be decreasing overall, though most of the main populations are stable (BirdLife International 2015). It is anticipated that in Europe the climate suitable for breeding Common Gulls will move northwards, away from the Baltic Sea, during the 21st century (Huntley *et al.* 2007). It is already possible that milder winters enable the species to winter closer to their breeding grounds.

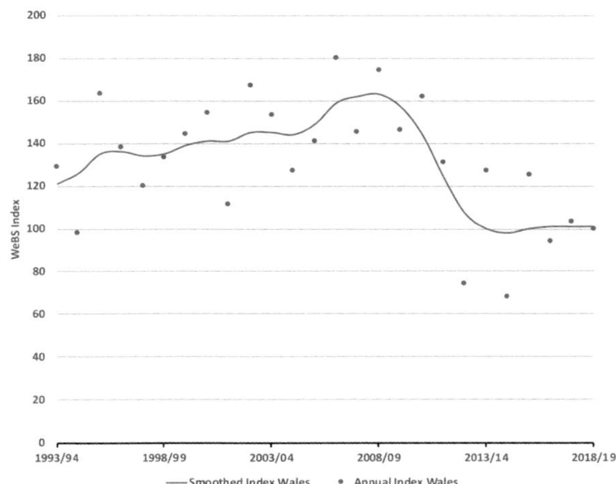

Common Gull: Wetland Bird Survey indices, 1993/94 to 2018/19.

Marc Hughes

Sponsored by HiDef Aerial Surveying Ltd

Ring-billed Gull *Larus delawarensis* Gwylan Fodrwybig

Welsh List Category	IUCN Red List Global	Rarity recording
A	LC	WBRC

Total	Pre-1950	1950–59	1960–69	1970–79	1980–89	1990–99	2000–09	2010–19
421	0	0	0	24	99	194	92	12

Ring-billed Gull: Approximate number of individuals recorded in Wales by decade.

This gull breeds in the northern USA and southern Canada, primarily inland. Most birds winter to the south of their breeding range, in Central America and the Caribbean (*HBW*). The first records in Britain were at Blackpill, Gower, on 14–31 March and 7 April 1973. The number of records increased rapidly over the following two decades, so that it was no longer considered a rarity in Britain by 1987. The number of new individuals recorded in Wales increased until 1992, but has declined subsequently, particularly in the decade since 2010.

It is very difficult to calculate the total number of individuals recorded in Wales. At sites such as Blackpill, where up to a dozen individuals have been observed annually, it is assumed that many birds return in subsequent years, clouding the true number of birds involved. An estimated 421 individuals have been recorded in Wales and over 2,000 in Britain.

The map of records in each recording area shows that the majority have been in southern Wales, particularly Pembrokeshire, East Glamorgan and Gower. Blackpill held the monopoly for this species in Britain for several years. More birds have been recorded in this area than at any other site in Wales, with a minimum of 188 individuals observed. A further 16 have been recorded elsewhere on Gower with good numbers also in East Glamorgan. It is quite possible that these birds also visited Blackpill. In Pembrokeshire, one returned to winter annually at Llys-y-frân Reservoir between 2006 and at least 2019. The only inland record was at Llangorse, Breconshire, on 20 February 1988.

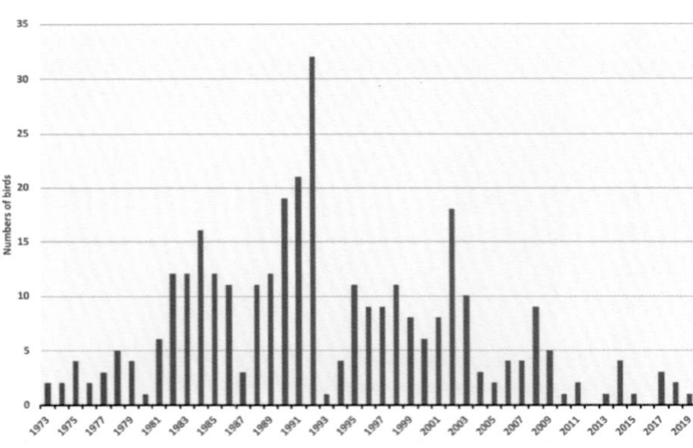

Ring-billed Gull: Approximate numbers recorded in Wales, 1973–2019.

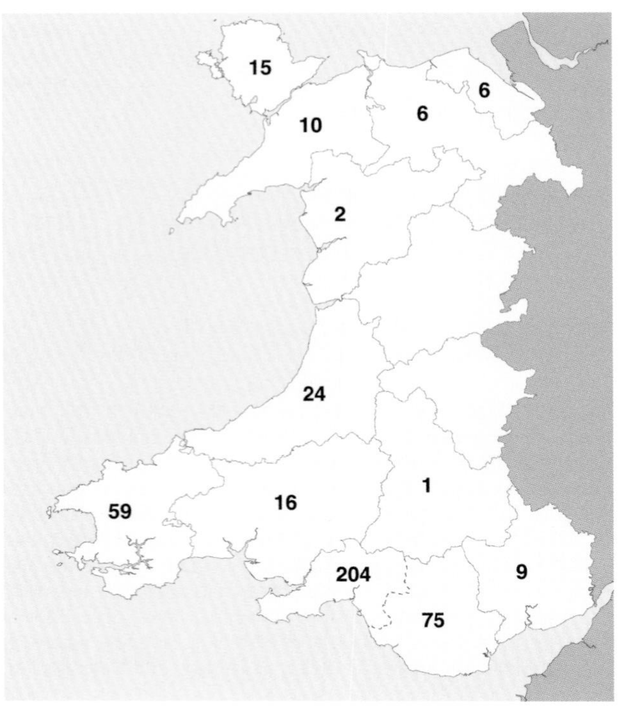

Ring-billed Gull: Approximate numbers of birds in each recording area 1973–2019. Some birds were seen in more than one recording area.

Far fewer have been recorded in mid- and North Wales. The first record for the north was not until 16 November 1987 at Aberdysynni, Meirionnydd, followed by one at Flint Marsh, Flintshire, on 26 October 1989.

The reduction in the number of records in recent years is not confined to Wales. Jones (2019) noted that, since the turn of the 21st century, the number of records had decreased in Britain, Ireland, France and Spain. The reasons for this were unclear, but he thought that it might be linked to localised declines in the breeding population, for example in the Great Lakes area, where the population peaked in 1989–90 then declined.

Jon Green and Robin Sandham

Great Black-backed Gull *Larus marinus* Gwylan Gefnddu Fawr

Welsh List Category	IUCN Red List			Birds of Conservation Concern			
	Global	Europe	GB	UK	Wales		
A	LC	LC	LC	Amber	2002	2010	2016

The Great Black-backed Gull breeds in eastern North America, Greenland and Iceland, across northern Europe as far east as the White Sea and south to western France (*HBW*). The largest population is in Norway, which holds 43% of the breeding pairs in Europe. Another 14% of the European population nests in the UK (BirdLife International 2015). The UK population is mostly in Scotland, with only 2.5% in Wales. This species is more inclined to nest in isolated pairs, on islets or rock outcrops, than our other gull species. It is the most maritime of the large gulls, although a few may be encountered inland wherever a food supply attracts large numbers of gulls, such as landfill sites. It does not generally follow the plough, but will forage around farms, often alongside Ravens. Consequently, this species can come into conflict with, for example, farmers of free-range chickens. It also appears inland at lambing time, meaning that it is greatly disliked by many sheep farmers.

Pennant (1810) said that "black back Gulls" nested on Llyn Conwy and Llyn Llydaw in Caernarfonshire in 1773. The description in his *British Zoology* (Pennant 1812) left little doubt that the Great Black-backed Gull was meant here. Pennant said that a man who tried to swim to the nests on Llyn Conwy was violently attacked by the adults, receiving "dreadful bruises... on all the upper part of his body" and nearly drowned. The species may well have been more common in Pennant's time, as it was heavily persecuted during the second half of the 19th century, with eggs being taken and adults shot. Martin (1864) stated that nests could be found at various places in the mountains of North Wales and at Aber Ogwen, Caernarfonshire. However, Forrest (1907) noted that "some of the inland resorts have been so harried of late years by egg-collectors that they are now deserted." A few pairs nested on the Pembrokeshire islands (Mathew 1894), and it was said to breed in very small numbers on the Gower Peninsula in the late 1890s (Hurford and Lansdown 1995). Lockley *et al.* (1949) noted a great increase in Pembrokeshire since Mathew's day. It then bred on all the islands and along the whole coast, with at least 150 pairs on Skomer and Middleholm in 1946.

A survey in 1956 produced an estimate of 718–748 pairs breeding in Wales, of which the majority (490–520 pairs) were in Pembrokeshire. Anglesey had an estimated 150 pairs and Caernarfonshire 55 pairs (Davis 1958). Much of the increase was thought to have been after 1930, and continued steadily thereafter, but already in the 1950s, control measures in the form of nest destruction were being instituted on Skokholm, Pembrokeshire, to reduce predation levels on Manx Shearwaters. There were 72 pairs there in 1949, but this had been reduced to five pairs in 1962, and nest destruction kept the population below ten pairs until 1978. Control measures on Skomer during 1960–76, involving shooting and trapping adults and nest destruction, resulted in 350 being killed. The population was reduced from a peak of 283 nests in 1961, to 25 pairs in 1984 (Perrins and Smith 2000). Numbers at St Margaret's Island increased from 24 pairs in 1949 to 147 in 1976, and it is tempting to suggest that some birds were displaced from Skomer and Skokholm. From the late 1970s and into the mid-1980s, some Great Black-backed Gull colonies were affected by a botulism outbreak that, by 1988, had reduced the population on St Margaret's to 55 pairs, and probably contributed to the decline on Skomer. However, the species was generally less susceptible than Herring Gulls, probably because only a few birds fed at landfill sites. The Seabird Colony Register census in 1985–88 found a 55% decline in Wales, compared to the Operation Seafarer counts in 1969–70. Only slight declines were recorded in Scotland and England over the same period. The reduction in numbers caused

	County	1960–69	1970–79	1980–89	1990–99	2000–09	2010–19
Skomer	Pembrokeshire	260 pairs, reduced to 160 pairs	133 pairs, reduced to 85 pairs in 1978	95 pairs, reduced to 27 pairs in 1989	41 pairs, increased to 69 pairs in 1994, 65 pairs in 1999	61 pairs, increased to 114 pairs in 2006, 106 pairs in 2009	118 pairs, 120 pairs in 2018, 108 pairs in 2019
Skokholm	Pembrokeshire	5–12 pairs	7–16 pairs	10–16 pairs	16 pairs, increased to 49 pairs	53 pairs, increased to 69 pairs	73 pairs, increased to 93 pairs, 86 pairs in 2019
Middleholm	Pembrokeshire	100 pairs in 1969	No data	23 pairs in 1985	34 pairs in 1993, 69 pairs in 1997, decreased to 34 pairs	14 pairs, increased to 64 pairs in 2004, 26 pairs in 2009	No reliable data
Caldey and St Margaret's	Pembrokeshire	80 pairs in 1962, increased to 169 pairs in 1969	125–156 pairs (max. in 1974), 151 pairs in 1979	114 pairs, decreased to 58 pairs in 1989	85 pairs in 1998; incom-plete data	51–80 pairs but all from boat	38 pairs in 2019; incomplete data
Ramsey and the Bishops	Pembrokeshire	99 pairs in 1969	No data	14 pairs in 1987	17 pairs in 1994, increased to 40 pairs in 1997	24 pairs in 2001, 40 pairs in 2007, 25 pairs in 2008	10 pairs in 2013, increased to 22+ pairs in 2018.
The Gwylans	Caernarfonshire	About 50 pairs	No data	No data	5 pairs in 1990, 54 pairs in 1999	61 pairs in 2001, decreased to 19 pairs in 2009	35 pairs, increased to 92 pairs in 2015, 50 pairs in 2019
The Skerries	Anglesey	No data	No data	16 pairs in 1986, then 5–7 pairs	9–20 pairs	14 pairs in 2000, increased to 40 pairs in 2009	39 pairs in 2010, decreased to 26 pairs in 2019
Denny Island	Gwent	25–30 pairs	30–39 pairs	40–60 pairs	20–30 pairs	20–30 pairs	16–30 pairs

Great Black-backed Gull: Number of pairs at key breeding sites.

by botulism led to the cessation of control measures on Skomer and Skokholm, and numbers increased steadily there until 2014, since when they have stabilised.

Aside from the larger colonies included in the table above, smaller colonies include Grassholm, Pembrokeshire, where 67 pairs bred in 1934 and 55 pairs in 1973 (Williams 1978) but only seven pairs in 2018 (Greg Morgan pers. comm.). Strumble Head, Pembrokeshire, held 26 pairs in 2018. In Ceredigion there were 30–40 pairs on Cardigan Island in 1966, 48 nests in 1972 but only ten pairs in 2011. On the St Tudwal's Islands, Caernarfonshire, there were 34 nests in 2014 and 20 pairs in 2016. On Puffin Island, Anglesey, 15–20 pairs bred in 1952, 85 pairs in 1977 and 107 pairs in 2017.

Operation Seafarer 1969–70	SCR Census 1985–88	Seabird 2000 1998–2002	Seabirds Count Provisional 2015–19
905	289	425	504

Great Black-backed Gull: Total counts of apparently occupied nests (AON) recorded in Wales during UK seabird censuses.

This species seems much less willing to nest on buildings than Herring Gull and Lesser Black-backed Gull. The first record of roof nesting in Wales appears to have been a pair in Aberystwyth, Ceredigion, in 1998. Roof-nesting pairs have been recorded in several towns and cities since, including up to eight pairs on buildings in Cardiff. A few pairs breed inland, usually on islands in lakes, including on Llyn Conwy where Pennant recorded them in the 18th century.

In winter, Great Black-backed Gulls disperse widely, with subadults tending to travel farther from natal colonies than adults (Brown and Eagle 2019). Large numbers, mostly adults, roost on Grassholm, Pembrokeshire, in late autumn and they are often seen in large numbers behind trawlers in the Celtic Sea. Once the food glut of fledgling Manx Shearwaters has left, this species is the last of the gulls to leave Skomer and Skokholm in late September and October, although some individuals remain through the winter, or at least close by.

Great Black-backed Gull: Numbers of breeding pairs on Skomer, Skokholm and Caldey/St Margaret's, 1960–2019.

The *BTO Migration Atlas* (2002) showed a predominantly southerly and coastal movement from Welsh breeding sites. Birds ringed in Wales as nestlings have been recovered in France, Spain and Portugal. More recent colour-ringing studies in South Wales suggest a more maritime movement, across the Bristol Channel and Western Approaches. Those from Denny Island disperse initially to north Somerset, before heading into Cornwall, where observations strongly suggest wintering site-faithfulness within and between years (Clarke 2016). The same seems true of Skokholm-ringed birds that have visited the same location over several winters. Most

Great Black-backed Gulls from Skokholm seem to winter locally, but first-calendar-year birds, in particular, winter in Cornwall, although a few have been found along the south coast of England, in France and Ireland. Large numbers can be found at Burry Inlet, Gower/ Carmarthenshire, and at some other coastal sites in autumn, which are presumed to be relatively local breeders. Less information is available on the movements of birds from the North Wales colonies. Large numbers from north European colonies move to Britain for the winter, and it is likely that some reach Wales. A bird ringed as a nestling on Great Ainov Island, Russia, in June 1933 was recovered at Leckwith, East Glamorgan, in January 1934.

Large autumn counts include *c.*1,000 roosting on Grassholm on 13 October 1982 and a flock of *c.*800, many oiled, on Newgale Beach, Pembrokeshire, in autumn 1978, following oil spillage when the *Christos Bitas* oil tanker ran aground on rocks off Milford Haven. More recently, there were 675 on the south side of the Burry Inlet in September 1996 and 617 in Carmarthen Bay in November 1999. The 2003/04–2005/06 Winter Gull Roost Survey produced an estimate of 4,365 individuals in Wales, very similar to previous counts in Wales (Banks *et al.* 2009). WeBS counts suggest that numbers in Wales are fairly stable.

The Great Black-backed Gull is Red-listed in Wales, as a result of the long-term population decline (Johnstone and Bladwell 2016).

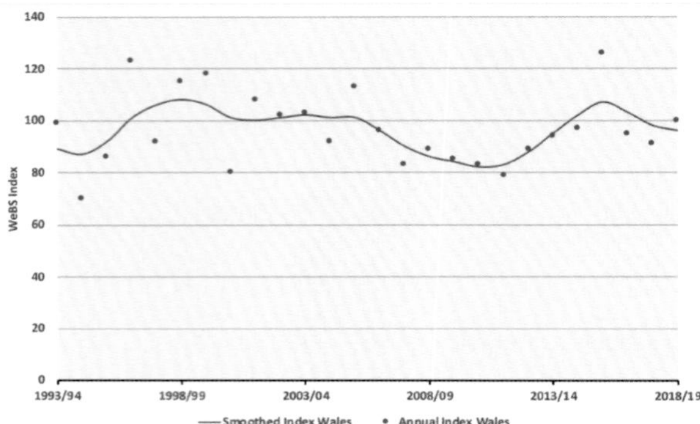

Great Black-backed Gull: Wetland Bird Survey indices, 1993/94 to 2018/19.

However, it has been doing quite well in Wales since the 1980s, in contrast to Scotland, where there has been a prolonged decline since about 2000 (JNCC 2020b). There do not appear to be any major threats at present, but this species is likely to be affected by any reduction in the availability of fishery discards (Bicknell *et al.* 2013).

Great Black-backed Gull: Recovery locations of birds ringed in Wales are shown by red circles. Ringing locations of birds that were recovered in Wales are shown by blue triangles. Small brown dots show ringing or recovery locations in Wales.

Steve Sutcliffe

Sponsored by Steve Sutcliffe

Glaucous-winged Gull *Larus glaucescens* Gwylan Adeinlas

Welsh List Category	IUCN Red List Global	Rarity recording
A	LC	BBRC

This North Pacific species breeds from the Kamchatka Peninsula in Russia, through the Aleutian Islands and coastal North America, to the northeast USA. It winters offshore in this area, extending south to Japan and Mexico (*HBW*). It is an extremely rare visitor to Britain, with only three records, the last in 2016 (Holt *et al.* 2018). One of these was in Wales. This was a bird originally found and colour-ringed on a refuse tip in Gloucestershire on 15–16 December 2006, the first record for Britain. Later that winter it was found at Ferryside, Carmarthenshire, on 2–5 March 2007.

Jon Green and Robin Sandham

Sponsored by Carmarthenshire Bird Club

Glaucous Gull *Larus hyperboreus* Gwylan y Gogledd

Welsh List Category	IUCN Red List			Birds of Conservation Concern
	Global	Europe	GB	UK
A	LC	LC	VU	Amber

The Glaucous Gull is very much a northern species, breeding in the Arctic with a circumpolar distribution. The European population breeds in western Iceland, Svalbard and northern coasts of European Russia. It is partly migratory, wintering from the edge of the sea ice as far south as northern Europe, Japan and the northern USA. Britain and Ireland are close to the southern edge of its usual winter range in Europe (*HBW*). The origin of those wintering in Wales is uncertain. Dean (1984) suggested that most birds recorded in Britain were of the subspecies *L.h. leuceretes* from eastern Greenland. However, a colour-ringed second-calendar-year bird in Holyhead, Anglesey, from late January to May 2017 had been ringed in 2016 as a chick on Svalbard, where the breeding birds are of the nominate form.

This does not seem to have been a common species in Wales at the end of the 19th century, although one or two were recorded in several coastal counties during this period. Forrest (1907) reported one inland record near Llanwddyn, Montgomeryshire, a bird shot in the winter of 1874. There were no records from Anglesey until 1943 or Flintshire until 1959. The first record for Carmarthenshire was not until 1974. The first record for Radnorshire was an adult at Craig Goch Reservoir in October 1993 (Jennings 2014) with the first for Breconshire at Pontsticill Reservoir on 2 February 2013.

Records remained few until the 1960s, when the species apparently became more frequent. There were only five records in Glamorgan up to 1963. After this, there were records in most years, with an estimated 83 individuals between 1963 and 1992, including 12 in 1984 and in 1991 (Hurford and Lansdown 1995). In Gwent, the first record was in 1893, but the second not until 1981 (Venables *et al.* 2008). The highest number of individuals seen together in Wales appears to be four at Cwrt-y-carne, Gower, on 19 February 1985 and 30 December 1986 (Hurford and Lansdown 1995).

The species has been recorded annually in Wales at least since 1989, but numbers vary considerably from year to year. An influx in the early months of 1984 involved about 26 individuals, and coincided with a spell of strong northwesterly winds and severe cold. However, cold weather on its own does not always produce more records: 35 were recorded in both 2014, during stormy weather, and 2017, despite the weather being generally mild. In 2010, which saw severe cold in both winter periods, only two individuals were recorded in Wales, the lowest total for many years. The largest numbers recorded in Britain are usually in Orkney and Shetland, where influxes often follow prolonged northerly or northwesterly gales, thought to displace birds feeding around fishing vessels in Icelandic and Norwegian waters (*The Birds of Scotland* 2007). By far the greatest numbers are recorded in the early part of the year, between January and March, with numbers decreasing as birds move north. There have, however, been a handful of long-staying and summering individuals in Wales: at Blackpill and Swansea, Gower, from March 1992 to 11 April 1993; at Llanstadwell, Pembrokeshire, from 20 February 1993 to 15 January 1995; and birds at Llanddulas, Denbighshire, from 7 April to 12 July 2008, 23 December 2017 to 8 August 2018, and during July and August 2019.

	1990–99	2000–09	2010–19	Total
Gwent	4	1	1	6
East Glamorgan	18	16	13	47
Gower	46	17	11	74
Carmarthenshire	5	3	8	16
Pembrokeshire	29	27	39	95
Ceredigion	23	6	25	54
Breconshire			2	2
Radnorshire	1			1
Montgomeryshire		2	2	4
Meirionnydd		2	6	8
Caernarfonshire	6	13	34	53
Anglesey	13	17	25	55
Denbighshire/ Flintshire	26			26
Denbighshire		5	10	15
Flintshire	13	3	5	21
Total	**184**	**112**	**181**	**477**

Glaucous Gull: Approximate number of individuals recorded in Wales by decade since 1990. Many individuals wander from one recording area to another, so the true number cannot be established.

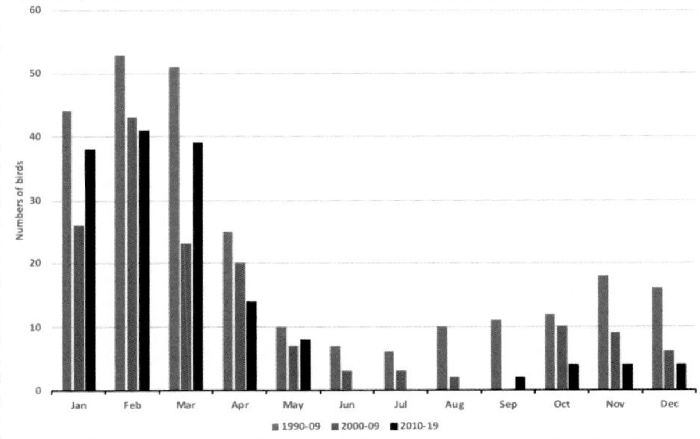

Glaucous Gull: Monthly distribution of records in Wales, 1990–2019.

There is no clear trend for numbers in Wales over the last 30 years and it is not certain what factors cause the annual variations. The global population trend for the Glaucous Gull is thought to be stable (BirdLife International 2020), but the long-term outlook for this Arctic-nesting species is uncertain, given the changing climate. Its breeding range is predicted to contract northwards during the 21st century (Huntley *et al.* 2007). Ultimately, both population size and its distance from Wales will determine the frequency of its future occurrence.

Jon Green

Iceland Gull *Larus glaucoides* Gwylan yr Arctig

Welsh List Category	IUCN Red List			Birds of Conservation Concern
	Global	Europe	GB	UK
A	LC	LC	VU	Amber

The Iceland Gull is an Arctic species, breeding in northeast Canada and Greenland. The closest breeding population to Wales is the nominate *L.g. glaucoides* that breeds on the south and east coast of Greenland. Despite its English name, it does not breed in Iceland, but does winter in Icelandic waters, and northwestern parts of Europe (*HBW*). This subspecies accounts for almost all the records in Britain and Ireland, but small numbers of two other subspecies, Kumlien's Gull *L.g. kumlieni* and Thayer's Gull *L.g thayeri*, have been recorded.

There are very few dated records of Iceland Gull in Wales before the 1950s. The species appears to have been very scarce in the late 19th century, with only three dated records: in East Glamorgan, Ceredigion and Meirionnydd. However, Salter (1895) described the species as "not at all infrequent" in Ceredigion and said that "after a very severe gale some years since Iceland Gulls were plentiful off Aberystwyth." The 1950s saw an increase in records, including the first records in Pembrokeshire, Caernarfonshire, Anglesey and Flintshire. The species was recorded more regularly by the 1970s, but only in small numbers and less frequently than Glaucous Gull. There was a substantial influx in the early months of 1983 and again in early 1984, when over 40 occurred in Wales. There were also good numbers in 2007, with 61 records, including 29 in Pembrokeshire alone, and in 2008, with the arrival of 43, of which five (all first-calendar-year birds) roosted at Llys-y-frân Reservoir, Pembrokeshire, during March.

	1990–99	2000–09	2010–19	Total
Gwent	7	3	12	22
East Glamorgan	26	21	24	71
Gower	39	17	17	73
Carmarthenshire	8	4	6	18
Pembrokeshire	20	62	74	156
Ceredigion	9	9	41	59
Breconshire	0	3	12	15
Montgomeryshire	0	0	1	1
Meirionnydd	0	3	2	5
Caernarfonshire	10	5	28	43
Anglesey	4	7	15	26
Denbighshire/ Flintshire	1			1
Denbighshire	0	3	14	17
Flintshire	7	3	11	21
Total	**131**	**140**	**257**	**528**

Iceland Gull: Approximate number of individuals recorded in Wales by decade since 1990.

There was an unprecedented influx into northwestern Europe during January–March 2012, which included over 150 birds on Shetland and even larger numbers on the Faroe Islands. This was linked to storm-force northwesterly winds in the last week of 2011, together with severe weather on the east coast of North America (Fray *et al.* 2012). An estimated 94 individuals were recorded in Wales in this period, including a total of 22 in Ceredigion and three together at Oxwich, Gower, on 2–12 January.

A few Iceland Gulls appear in November and December, but the majority occur between January and March, with a few remaining into April and May. Adults form only a small proportion of those present; first-calendar-year birds are the most numerous. Inland records are rare: Breconshire's first was at Pontsticill Reservoir on 12 February 2007, increasing to three, all first-calendar-year birds, on 18 February, with at least one remaining until 26 March. Montgomeryshire recorded its first at Dolydd Hafren on 15 February 2009 and there were two in the county in early 2012. Iceland Gull has not yet been recorded in Radnorshire.

Allowing for annual variation, the number of Iceland Gulls recorded in Wales has increased in recent years. This may be partly through increased observer interest in gulls, but it is notable

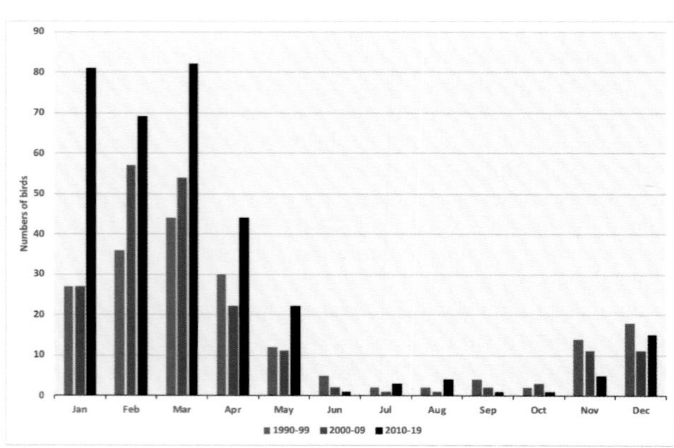

Iceland Gull: Monthly distribution of records in Wales, 1990–2019.

that the position prior to the late 1990s has been reversed, so that numbers of Iceland Gull recorded in Wales now exceed Glaucous Gull numbers in most years. The global population trend for this species is stable (BirdLife International 2020), so the change in occurrence may not be related to an increase in population, but to the availability of food resources outside the breeding season. The decline of the Icelandic fishing industry, leading to a reduction in discards, has been suggested as a reason why birds are arriving in Britain in increased numbers (*BTO Migration Atlas* 2002). In the longer term, Arctic breeding species may be disproportionately affected by climate change, ultimately influencing the numbers that occur at the edge of their wintering range.

Kumlien's Gull *L.g. kumlieni Gwylan* Kumlien

There remains some debate about this subspecies. Some authorities regard it as a hybrid between *L.g. thayeri* and the nominate *L.g. glaucoides* (McGowan and Kitchener 2001). It breeds mainly on Baffin Island, Canada, and winters on the Atlantic coast of North America and around the Great Lakes (*HBW*). Records are assessed by WBRC.

Nine individuals of this subspecies have been recorded in Wales:

- 2CY at Llys-y-frân Reservoir, Pembrokeshire, on 1–5 February 1998;
- adult at Blackpill, Gower, between 15 December 1998 and 2 March 1999 (this bird returned to the same site from 9 September 1999 to 20 February 2000 and 24 November to 11 December 2000);
- adult at Cosmeston Lakes Country Park, East Glamorgan, on 23 November 2012;
- adult at Newgale, Pembrokeshire, 10 January to 22 February 2014;
- another adult at Newgale on 15–22 February 2014;
- 2CY at Blackpill, Gower, on 8 February and 2 March 2014;
- adult at Aberavon, Gower, on 16 February 2014;
- adult at Ogmore, East Glamorgan, on 11 and 17 March 2014 and
- adult at Llys-y-frân Reservoir on 12 January 2019.

Five of the records were part of an influx of Kumlien's Gulls in early 2014.

Thayer's Gull *L.g. thayeri* Gwylan Thayer

As with *L.g. kumlieni*, there is some controversy over the taxonomy of this form, with some authors treating it as a separate species or a subspecies of American Herring Gull. It breeds in the Canadian Arctic and northwest Greenland, and winters on the Pacific coast of North America (*HBW*). Records of this subspecies are assessed by BBRC. A total of seven individuals had been recorded in Britain to the end of 2019, with the last in 2016 (Holt *et al.* 2018). The only Welsh record was a second-calendar-year bird at Burry Holm, Gower, on 4–19 January 2014. This was the sixth record for Britain.

Marc Hughes and Jon Green

Herring Gull *Larus argentatus* Gwylan Penwaig

Welsh List Category	IUCN Red List			Birds of Conservation Concern			
	Global	Europe	GB	UK	Wales		
					2002	2010	2016
A	LC	NT	EN (NB)	Red			

The Herring Gull is the most ubiquitous and best known of our large gulls, breeding around almost the entire Welsh coastline, especially on offshore islands and stacks and increasingly on buildings around the coast and inland. It has undergone extraordinary and well-documented changes in population in the last century, largely as a result of its close relationship with human activities.

This species breeds from Iceland across northwest and northern Europe and as far east as the Kola Peninsula in Russia. There are two subspecies: *L.a. argenteus* breeds across much of western Europe, including Wales, while the nominate *L.a. argentatus* breeds in Fennoscandia and east to Russia. Birds from eastern parts of the range move south in winter, while other populations are resident (*HBW*). Recoveries and colour-ring sightings mostly involve birds in their first two years of life, but there have been much older birds. A colour-ringed bird seen at Clydach, Gower, in October 2013 had been ringed as a second-calendar-year bird at Goring Beach, West Sussex, in December 1980. At well over 32 years old, this is the oldest Herring Gull currently known to the BTO ringing scheme. Birds ringed as nestlings in Wales have been recovered in many European countries, ranging from Iceland and Norway to Spain and Portugal, and one ringed as a nestling at Llyn Trawsfynydd, Meirionnydd, in June 2011 was found sick near Casablanca, Morocco, in March 2012. Birds ringed as nestlings or juveniles, in countries ranging from Italy to Russia, have been recovered in Wales.

Early authors described the species as common, but did not give counts at breeding colonies. It was noted to be a common breeder at South Stack, Anglesey, in 1808 (Jones and Whalley 2004). The statement by Eyton (1838) that the species "breeds plentifully along the whole line of the Welsh coast" may have referred to North Wales only. Mathew (1894) stated that this species nested in great numbers on the islands and many of the cliffs on the Pembrokeshire coast, and that since the Sea Birds Preservation Act 1869 it had greatly increased in numbers. Forrest (1907) noted that in North Wales the Herring Gull did not breed in Meirionnydd, Denbighshire or Flintshire, but bred at many sites in Caernarfonshire and Anglesey.

Some early counts are available from the Gower Peninsula, where a survey in 1921 found about 600 pairs. Numbers there fluctuated greatly over the following three decades, but there were still 568 pairs present in 1958. In East Glamorgan, a small colony was found at Southerndown in 1912, and cliff-nesting birds along the coast of the Vale of Glamorgan increased steadily, to peak at 906 in 1974 (Hurford and Lansdown 1995). In Caernarfonshire, a survey in May 1939 found about 1,000 pairs on the Great Orme (Hardy 1941), while numbers increased steadily in Pembrokeshire. By 1946 there were 300 pairs on Newport Head, 300 pairs on Skokholm and 650 pairs on Skomer (Lockley *et al.* 1949). Breeding numbers increased substantially across Wales between 1930 and 1970, although numbers on Bardsey were kept at a low level, because most eggs were collected for human consumption until about 1964 (Harris 1970). Egg-collecting for human consumption was recorded in several parts of Wales, particularly during, and immediately after, the Second World War. This was banned by the Wildlife and Countryside Act 1981.

Operation Seafarer 1969–70	SCR Census 1985–88	Seabird 2000 1998–2002	Seabirds Count Provisional 2015–19
48,576	11,089	13,974	7,988*

Herring Gull: Total counts of apparently occupied nests (AON) recorded in Wales during UK seabird censuses.

* Provisional figures do not include all roof-nesting pairs.

Herring Gull: Recovery locations of birds ringed in Wales are shown by red circles. Ringing locations of birds that were recovered in Wales are shown by blue triangles. Small brown dots show ringing or recovery locations in Wales.

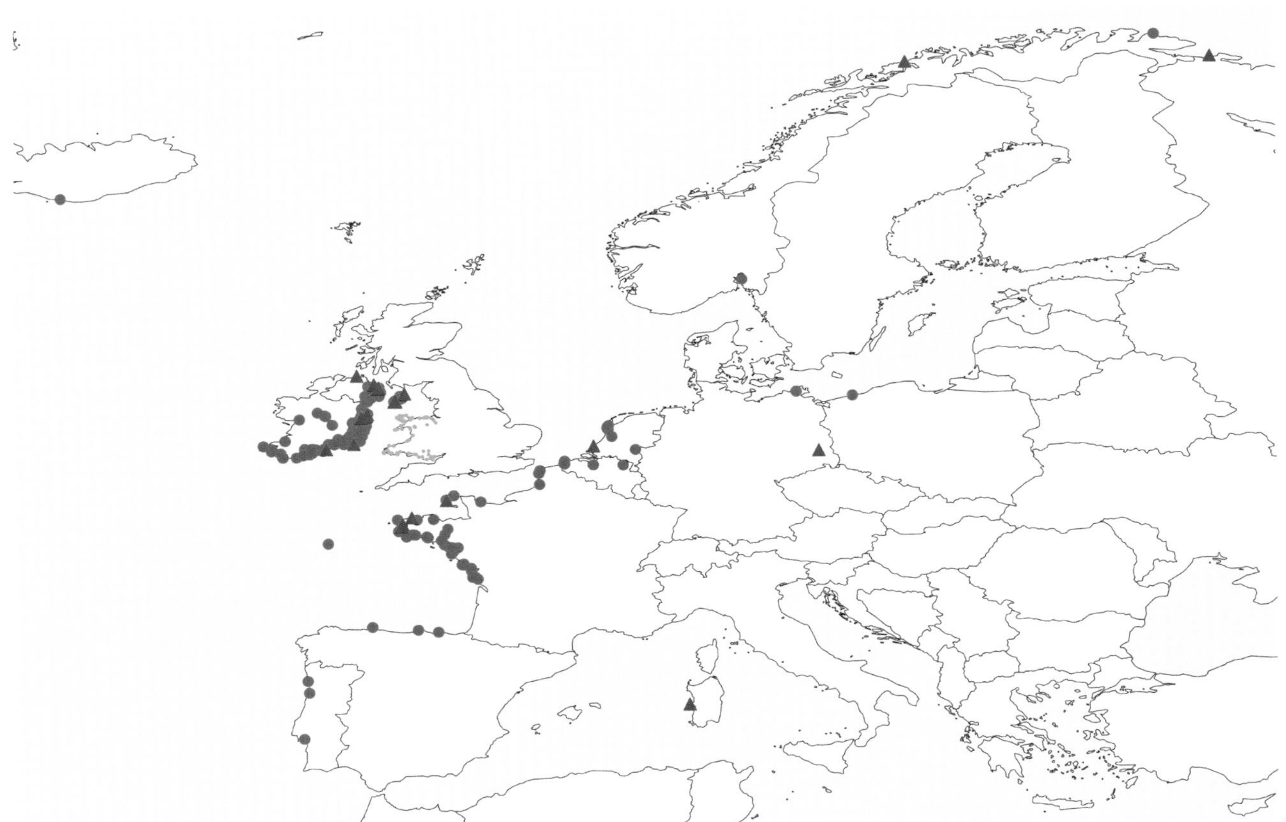

There was a population surge between the 1950s and the late 1970s as birds foraged on discards from fishing boats, at fish docks and at landfill sites. There followed a dramatic decrease in the 1980s, as a result of botulism that was almost certainly linked to landfill sites and which almost exclusively killed adult breeding birds (Sutcliffe 1986). The SCR Census in 1985–87 showed a drop of 77% in breeding numbers in Wales since 1969–70. It was estimated that the breeding population in Glamorgan declined by at least 90% between 1974 and 1992 as a result of botulism, which had been noted in the area since the introduction of black plastic bags for refuse collection (Hurford and Lansdown 1995). The largest decline at any colony in Wales was on Puffin Island, Anglesey, where there were *c.*15,000 pairs in 1969–70, but only 900 pairs in 1985–87. The decline of some colonies was linked to other factors, such as land-use changes. The most notable examples in Wales are two colonies on Anglesey: Newborough Warren, where 4,780 pairs bred in 1969–70, and Bodorgan Head, which was estimated to have 7,000 pairs in 1986 but only 1,110 in 1987, 384 pairs in 2008 and just three pairs in 2016, although 50 pairs were there in 2019. The disappearance of the Newborough Warren colony was thought to be caused by the arrival of Foxes on Anglesey in the late 1960s, canopy closure of the conifers planted on the dunes and the outbreak of botulism (Jones and Whalley 2004). On Anglesey, before the arrival of Foxes, the gulls nested on the slopes above the cliffs as well as on the cliffs themselves at some sites (Ivor Rees *in litt.*).

As management of refuse waste changed, with the closure of smaller landfill sites and the use of nets and soil to cover the waste, mortality from botulism decreased. The result has been a stabilisation of numbers at much lower levels since the early 2000s. Only the large colony on Caldey Island has partially recovered from the botulism crash, but even this has stabilised at around 1,928 pairs, 50% of its peak in 1975.

Colony	Pre-1970	1970–79	1980–89	1990–99	2000–09	2010–19
Caldey Island, Pembrokeshire	No counts before 3,556 pairs in 1970	3,857 pairs in 1975	2,337 pairs in 1980, many dead adults on nests; 675 pairs in 1988, 763 pairs in 1989	736 pairs in 1990, increased to 1,638 pairs in 1999	2,134 pairs in 2000, 2,096 pairs in 2008	2,258 pairs in 2010, 1,438 pairs in 2017 but 1,832 in 2019
Skomer, Pembrokeshire	600 pairs in 1946, 700 pairs in 1961, increased to 2,200 pairs in 1969	Stable; max. 2,940 pairs in 1979	2,350 in 1980, 1,409 in 1981, 979 pairs in 1982 and 430 pairs in 1989	430 pairs in 1990, 568 in 1994, 299 in 1998, 374 pairs in 1999	367 pairs in 2000, increased to 444 in 2009	431 pairs in 2010, decreased to 297 pairs in 2017 and 2019
Skokholm, Pembrokeshire	250–300 pairs during 1928–46, increased to 1,350 pairs in 1969	Stable; max. 1,250 in 1978	750 pairs in 1981, decreased to 253 pairs in 1989	254 pairs in 1990, 450 pairs in 1993, decreased to 330 pairs in 1999	Stable; 309 pairs in 2000, 353 pairs in 2009	Stable; max. 320 pairs in 2018, but 288 pairs in 2019
Ramsey and the Bishops and Clerks, Pembrokeshire	No counts	No counts	232 pairs in 1987, of which 114 on Ramsey	383 pairs on Ramsey in 1999	315 pairs in 2000, 277 pairs in 2008	221 pairs in 2013, 306 pairs in 2018
Flat Holm, East Glamorgan	75 pairs in 1958, increased to 920 pairs in 1969	4,055 pairs in 1974, decreased to 1,302 pairs in 1980	450 pairs in 1986, 70 pairs in 1989	234 pairs in 1990, 341 pairs in 1995, decreased to 243 pairs in 2000	285 pairs in 2001 but 323 pairs in 2008	326 pairs in 2011, est. 611 pairs in 2012, 319 pairs in 2017, but 7 pairs in 2019
The Skerries, Anglesey			458 pairs in 1986, decreased to 300 pairs in 1989	348 pairs in 1990, increased to 892 pairs in 1993	943 pairs in 2000, 1,011 pairs in 2009	1,123 pairs in 2010, 529 pairs in 2017, 665 pairs in 2019
Puffin Island, Anglesey	Estimated at 15,500 pairs in 1969	Rabbits disappeared and vegetation changed	919 pairs in 1986	Brown Rats eradicated from the island	400 pairs in 2001	326 pairs in 2010, 492 pairs in 2012, 472 pairs in 2017
St Tudwal's Islands, Caernarfonshire			212 pairs in 1986		793 pairs in 2013, 383 pairs in 2014	518 pairs in 2016
Bardsey, Caernarfonshire	150–200 pairs from early 1950s to 1969 (probably estimates)	250 pairs in 1973, increased to 419 pairs in 1978, 387 in 1979	450 pairs in 1981, down to 95 pairs in 1984, then 410 pairs in 1989	400 pairs in 1990, 275 pairs in 1991, 508 pairs in 1999	663 pairs in 2001, 172 pairs in 2007	323–417 pairs through decade, 345 pairs in 2019
The Gwylans, Caernarfonshire	No counts	No counts	110 pairs in 1986	270 pairs in 1991, increased to 320 pairs in 1999	235 pairs in 2000, 489 pairs in 2006, then 51 pairs in 2007 and 134 pairs in 2009	115 pairs in 2000, 143 pairs in 2017, 89 pairs in 2019

Herring Gull: Counts at the top ten historic and recent colonies in Wales.

Other colonies over 200 pairs include Castlemartin coast, Pembrokeshire (210 pairs in 2018), Greenscar, Pembrokeshire (217 pairs in 2018), Strumble Head islands, Pembrokeshire (329 pairs in 2018), Newport/Poppit coastline, Pembrokeshire (324 pairs in 2018) and Point Lynas to Trwyn Du, Anglesey (330 pairs in 2016). Away from the coast, birds breed at several lakes, notably at Llyn Trawsfynydd, Meirionnydd. There were up to 44 pairs at Llyn Brenig, Denbighshire, in 2016, and 55 nests at Llyn Elsi,

Location	County	Date last counted and numbers	Comments
Cardiff	East Glamorgan	2017 – 866 pairs	A detailed survey by Peter Rock
Prestatyn area	Flintshire	2019 – 470 pairs	Aerial survey of six 1-km squares (Woodward *et al.* 2020)
Rhyl area	Flintshire	2019 – 402 pairs	Aerial survey of six 1-km squares (Woodward *et al.* 2020)
Llyn Trawsfynydd: islands in the lake	Meirionnydd	2018 – 408 pairs	133 pairs reported in 1986, apparently still increasing slowly
Colwyn Bay area (Llandudno Junction to Old Colwyn)	Denbighshire	2019 – 385 AON	Aerial survey of twenty 1-km squares (Woodward *et al.* 2020)
Llandudno area	Caernarfonshire	2019 – 282 pairs	Aerial survey of four 1-km squares (Woodward *et al.* 2020)
Griffiths Crossing, Caernarfon	Caernarfonshire	2014 – 270 pairs	On an old industrial building
Caernarfon (town)	Caernarfonshire	2019 – 102 pairs	Reduction from *c.*250 pairs in 2015
Newport	Gwent	2017 – 225 pairs	Detailed surveys in 2007 and 2017 (Clarke 2018b)
North Dock, Llanelli	Carmarthenshire	2014 – 195 pairs	This is apparently the only complete count available

Herring Gull: Counts at the top ten rooftop and inland breeding sites in Wales.

Caernarfonshire, in 2014. Smaller numbers breed, at least occasionally, on several other upland lakes, up to an altitude of at least 680m on Ffynnon Lloer in the Carneddau, Caernarfonshire. Most inland colonies are not counted regularly enough to have a reliable estimate of trends.

Herring Gulls have taken advantage of the rooftops and human food supplies to nest in towns and cities, although there have been few counts, so accurately gauging the population is difficult. Rooftop nesting in Wales seems to have been first recorded in Merthyr Tydfil, East Glamorgan. *Birds in Wales* (1994) stated that it dated back to the 1940s there, although Hurford and Landsown (1995) dated it to 1958. The habit spread during the 1960s and was recorded in several towns in South and North Wales by the end of the decade. There were an estimated 1,826 pairs nesting inland, mainly on roofs, in 1999–2000 and *c.*2,960 were counted at 45 sites during 2014–19. The location with the largest count is Cardiff, East Glamorgan, where 866 pairs were counted in 2017 (Peter Rock pers. comm.). Numbers there have fluctuated considerably: there were 425 pairs in 1975 but none during Seabird 2000. Flat factory roofs have been occupied for many years in Gower, East Glamorgan and Gwent. In Gwent, a county-wide breeding survey in 2017 recorded 361 apparently occupied nests, an 88% increase since a comparable survey a decade earlier (Clarke 2018b). Small numbers now nest on inland rooftops, for example, at Welshpool and Newtown, Montgomeryshire. This trend seems set to continue.

An aerial survey in 2019 of a total of 100 selected 1-km squares along the North Wales coast east of the Conwy Estuary, along the Flintshire side of the Dee Estuary and south to the Wrexham area in Denbighshire, found a total of 1,763 apparently occupied nests. The highest totals were along the north coast, particularly in the areas around Prestatyn and Rhyl, Flintshire (Woodward *et al.* 2020). The total number of Herring Gull nests calculated from the aerial survey was approximately three times the number found by ground-level counts of the same 1-km squares made as part of this study. This suggests that aerial surveys in other parts of Wales would probably also reveal higher totals, and that roof-nesting birds could now make up around 50% of the total Welsh population.

Rooftop breeding can bring gulls into conflict with people, because although most birds forage away from town centres (Coulson 2019), some have learned to specialise on rubbish and other street discards and occasionally take food directly from people. Various methods have been used in some towns and cities to control gulls, including the replacement of eggs, the removal of nests, the use of spikes or netting to prevent breeding and the use of acoustic deterrents.

In general, adults remain within 45km of their colonies throughout the year (*BTO Migration Atlas* 2002), whereas immature birds tend to travel more widely. Many forage in fields within a few miles of the coast and roost on offshore stacks and islands. Winter counts of known roosts almost certainly underestimate the true number of gulls in winter.

	1983 (Bowes *et al.* 1984)	1993 (Burton *et al.* 2002)
Inland	2,267	2,112
Coastal	31,220	27,065
Total	**33,487**	**29,177**

Herring Gull: Winter counts in Wales.

The 2003/04–2005/06 Winter Gull Roost Survey combined counts from key sites with estimates for the numbers of birds wintering elsewhere, derived from stratified sampling (Banks *et al.* 2007b). This produced an estimate of 93,613 Herring Gulls in Wales, 13% of the estimated total for Great Britain.

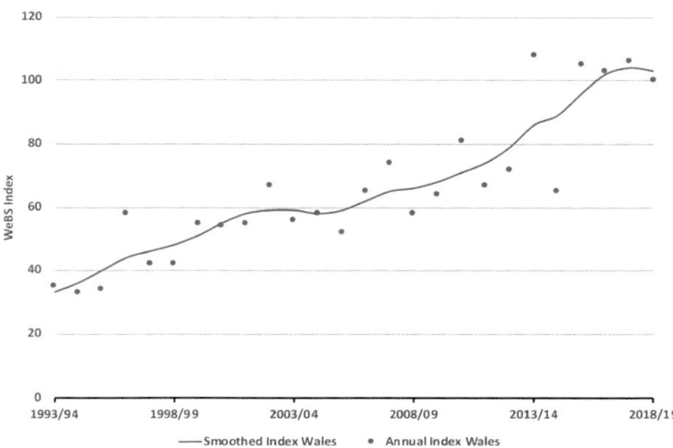

Herring Gull: Wetland Bird Survey indices, 1993/94 to 2018/19.

Adult and juvenile Herring Gulls © Ben Porter

Site	County	94/95 to 98/99	99/00 to 03/04	04/05 to 08/09	09/10 to 13/14	14/15 to 18/19
Dee Estuary: Welsh side (whole estuary)	Flintshire	na (1,416)	1,307 (2,369)	1,655 (2,432)	2,300 (5,540)	5,628 (8,328)
Burry Inlet	Gower/Carmarthenshire	1,408	2,791	2,619	3,618	3,049
Carmarthen Bay	Carmarthenshire	1,472	1,600	1,983	1,635	2,897
Pontsticill Reservoir	East Glamorgan/Breconshire	31	216	1,024	3,256	2,869

Herring Gull: Five-year average peak counts during Wetland Bird Surveys between 1994/95 and 2018/19 at sites in Wales averaging 2,500 or more at least once. International importance threshold: 10,200; Great Britain importance threshold: 7,300.

The WeBS indices for Wales have shown a steady increase since the 1990s. Some very large gatherings have been recorded in Wales, for example, 10,000 at Point of Ayr, Flintshire, in August 2018, 12,000–14,000 feeding on a shellfish wreck at Red Wharf Bay, Anglesey, on 26 November 2016, *c.*12,500 at Kenfig Sands, East Glamorgan, on 14 July 1968 and 15,000 at Cefn Sidan, Carmarthenshire, on 26 July 1994.

The Red-listed conservation status of the Herring Gull in Wales is because of the severe long-term decline in both the breeding and the non-breeding population, together with its Near Threatened status in Europe (Johnstone and Bladwell 2016). The increase in the roof-nesting population in Wales has partly counterbalanced the decline in birds nesting at natural sites on the coast, where their nesting productivity has declined. In 1994, productivity was 1.46 chicks/pair, but during 1998–2015 it was typically below one chick/pair. The reasons for the continued decline in productivity are largely unknown (JNCC 2020b). Much less is known about the productivity of roof-nesting pairs in Wales. Research on Skomer has shown changes in adult survival rates over the past 50 years, with lower rates more frequent since 1995. Botulism is now less of a risk than formerly, but this species appears to be particularly vulnerable to risk of collision with offshore wind turbines (Bradbury *et al.* 2014). It appears that the declines recorded in Wales may continue unless the causes of decline are investigated and addressed.

Scandinavian Herring Gull *L.a. argentatus* Gwylan Penwaig Llychlyn

The nominate subspecies is a regular winter visitor to Wales, though it is probably greatly under-recorded, as not all bird-watchers are interested in searching through gull flocks to look for it. Though this subspecies has been recorded in the majority of coastal counties, most records come from East Glamorgan and Gower. It was estimated that 12 of this subspecies frequented the Rhymney Estuary, East Glamorgan, in 1992 (Hurford and Lansdown 1995).

Steve Sutcliffe

Sponsored by HiDef Aerial Surveying Ltd

Caspian Gull *Larus cachinnans* Gwylan Bontaidd

Welsh List Category	IUCN Red List			Birds of Conservation Concern	Rarity recording
	Global	Europe	GB	UK	
A	LC	LC	VU	Amber	WBRC

The Caspian Gull was formerly regarded as a subspecies of Herring Gull, but is now treated as a separate species. Its breeding stronghold is Central Asia, from where it winters in the Black Sea and around the Arabian Peninsula. However, in the second half of the 20th century, this species spread west to breed on lakes in eastern Europe and, since the 1990s, in Germany (*HBW*). As the breeding population spread west, so did its non-breeding range, and although the first record for Britain was only in 1995, it quickly became a scarce but regular autumn and winter visitor to central and southeastern England (*Birds in England* 2005). A BTO survey of winter gull roosts between 2003/04 and 2005/06 found a total of 110 individuals in Britain (Parkin and Knox 2010). Despite the increase in the number of records in England, and in western Europe generally, there was no record in Wales until birds were seen at two sites in December 2014. Ten have now been recorded in Wales, and it seems set to become an annual winter visitor.

- 1CY at Gresford Flash, Denbighshire, on 2–6 December 2014;
- 1CY at Cosmeston Lakes, East Glamorgan, on 8 December 2014;
- 2CY at RSPB Conwy, Denbighshire, on 13 May 2017;
- 2CY at Llanrhystud, Ceredigion, on 19–28 March 2017, then at Nevern Estuary, Pembrokeshire, on 12 July 2017, which had been colour-ringed at Brandenburg, Germany, on 6 June 2016;
- 1CY at Llanrhystud, Ceredigion, on 12 August 2017;
- 2CY at Cardiff Bay, East Glamorgan, from 16 January to 8 April 2018;
- 2CY at Ginst Point, Carmarthenshire, on 2 April 2018;
- 2CY at Penclacwydd, Carmarthenshire, on 20 April 2018;
- 1CY at Bettws, Gwent, on 14–21 December 2018, then, as a 2CY, at Cardiff Bay, East Glamorgan, on 16 January 2019 and
- 2CY at Cardiff Bay, East Glamorgan, on 3 January and 27 February 2019, different from the above.

Jon Green and Robin Sandham

Yellow-legged Gull *Larus michahellis* Gwylan Goesfelen

Welsh List Category	IUCN Red List			Birds of Conservation Concern
	Global	Europe	GB	UK
A	LC	LC	EB(B)	Amber

Formerly considered to be a subspecies of Herring Gull, the Yellow-legged Gull is now recognised as a distinct species (Sangster *et al.* 2005) and considered more closely related to the Lesser Black-backed Gull. It occurs mainly around the coasts of the Mediterranean, but colonies are found on the Atlantic coast of Europe as far north as Brittany and on the southern shore of the Black Sea (*HBW*). The second *European Atlas* shows a considerable range expansion since the 1997 *Atlas*, with an expansion

Adult Yellow-legged Gull © Alan Saunders

inland into central Europe (*European Atlas* 2020). After breeding it undertakes a northward moult migration, bringing a large influx to southeast England between July and September. Smaller numbers make it to Wales and northern Britain. A few pairs now breed in southern England, sometimes in mixed pairings with Herring Gulls or Lesser Black-backed Gulls.

Records prior to 2005 often referred to birds "showing the characteristics" of Yellow-legged Gull. The first published record for Wales seems to be one at Malltraeth, Anglesey, in April and May 1978. In December 1984, two were at Llandegfedd Reservoir, Gwent, and one at Llanishen Reservoir, East Glamorgan. A rapid increase in records followed in Glamorgan, and by 1992 it was estimated that more than 40 individuals had frequented the county (Hurford and Lansdown 1995). It was also recorded almost annually in Gwent (Venables *et al.* 2008). The first record for Pembrokeshire came in 1985 and in Caernarfonshire in 1986. Records increased from the early 1990s, although the numbers involved are difficult to estimate, as birds may have moved between sites. The graph is approximate but shows the general trend.

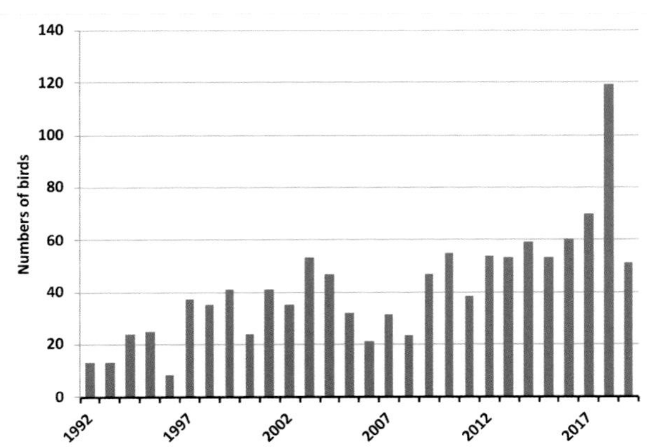

Yellow-legged Gull: Numbers of individuals in Wales each year, 1992–2019.

Yellow-legged Gulls are recorded in all months of the year, with no clear pattern evident, although the highest counts are usually in autumn and winter. The largest numbers are still in Glamorgan, where there have been good numbers in both East Glamorgan and Gower, including 12 in Cardiff Bay in January 2017. The species is regularly recorded in other counties of South Wales, but usually in small numbers. Most records are coastal, but there are also regular inland records, particularly in Breconshire, where five adults roosted at Talybont Reservoir in October 2012. In Montgomeryshire, by contrast, the first record was not until 2009, and this species remains scarce in North Wales. Two adults, presumably a pair, gathered nesting material at RSPB Conwy, Denbighshire, in 1998, but there has been no record of confirmed breeding in Wales. A number of hybrids between this species and Lesser Black-backed Gull have also been reported, although it should be noted that hybrids between Lesser Black-backed and Herring gulls also exist and can look very similar (Donovan and Rees 1994).

Modelling suggests that the breeding range will extend to southern England by the end of this century (Huntley *et al.* 2007). Since colonisation seems already to be underway, it would seem likely that nesting in Wales will occur eventually.

Rhion Pritchard

Lesser Black-backed Gull *Larus fuscus* Gwylan Gefnddu Fach

Welsh List Category	IUCN Red List			Birds of Conservation Concern			
	Global	Europe	GB	UK	Wales		
A	LC	LC	DD	Amber	2002	2010	2016

The Lesser Black-backed Gull breeds from Greenland across northern Europe to western Siberia and as far south as the Iberian Peninsula. The subspecies that nests in Wales is *L.f. graellsii*, which breeds in Greenland, Iceland, the Faroe Islands and coastal western Europe. The subspecies *L.f. intermedius* breeds along the southern and eastern coasts of the North Sea and occurs in Wales in winter. The nominate *L.f. fuscus* breeds throughout the rest of Fennoscandia (IOC) and appears to be a rare visitor to Britain. Most populations are migratory, with those that breed farthest north moving the greatest distances south in winter, as far as equatorial Africa (*HBW*).

In Wales this species breeds primarily on islands free from predators, though there are also inland colonies, and increasing numbers nest on buildings, particularly in Cardiff. Away from the major colonies listed, scattered groups and single pairs breed around the southern and northern coasts of Wales but few, if any, around Cardigan Bay between Bardsey and Cardigan Island. Over 94% of Wales' breeding Lesser Black-backed Gulls are in the southern counties.

The Lesser Black-backed Gull was noted by Ray (1662) on the south coast of Llŷn, Caernarfonshire, in May 1662, but there is no detailed information about its status in Wales until the late 19th century. Mathew (1894) described it as a common resident in Pembrokeshire, nesting on the various islands, but mentioned only colonies of 20–30 pairs. The only breeding birds in Glamorgan in the late 19th century were small numbers on the Gower Peninsula (Hurford and Lansdown 1995). Forrest (1907) said that it bred at only a few sites in North Wales, with the largest colony on Puffin Island, Anglesey, but that none were known to breed along the coasts of Denbighshire and Flintshire, and it was seldom seen inland. An inland colony at Cors Caron, Ceredigion, held 57 nests in 1892, but died out in the late 1920s or early 1930s (Roderick and Davis 2010). By 1946 there had been a big increase in Pembrokeshire, with about 800 pairs breeding on Skokholm, and 1,000 pairs on Skomer (Lockley *et al.* 1949). Breeding was first confirmed in Carmarthenshire in 1954 (Lloyd *et al.* 2015). By the mid-1950s, breeding numbers were said to be much lower across Wales, due to persistent egg-collecting (*Birds in Wales* 1994). A rapid increase followed even though egg-collecting for food continued at a low rate on the Pembrokeshire islands until 1973.

There was a marked increase in Wales in the 1970s and 1980s. The main colonies on Skomer and Skokholm saw massive increases from about 1961 with a peak in 1993. On Skomer there were just 900 pairs in 1961 but an estimated 20,200 pairs in 1993, while counts on Skokholm were 350 and 4,652 pairs over the same period. The Flat Holm colony was 1,100 pairs in 1969, peaking at 4,055 pairs in 1975. On Cardigan Island a small colony of 350 pairs in 1986 rapidly increased to a peak of 4,700 in 1994. Other smaller colonies on Ramsey and the Bishops, on Caldey, and at a number of very small island and coastal sites showed similar trends, although counts were generally more spasmodic and some sites were only counted for the first time during the mid- to late 1990s.

Increases in North Wales were more moderate. On Bardsey there were 199 pairs in 1985 and a peak of 652 pairs in 2003. On Anglesey, a colony on The Skerries more than doubled in size from 484 pairs in 1986, to 1,169 pairs in 1993, while RSPB South Stack, which recorded only two pairs in 1986, held 175 in 1989.

Breeding at Newborough Warren began about 1951. It peaked at around 2,000 pairs in 1969, then declined to zero by 1982, probably for the same reasons as the decline in Herring Gull numbers there: the arrival of Foxes on Anglesey in the late 1960s, canopy closure of the conifers planted on the dunes and the outbreak of botulism (Jones and Whalley 2004). A colony at Bodorgan Head was reported to have 2,000 pairs in 1986 but only 740 in 1987, 386 in 1989 and just three pairs in 2018.

Colony	1970–79	1980–89	1990–99	2000–09	2010–19
Flat Holm, East Glamorgan	Rapid increase from 455 pairs in 1960 to 4,055 pairs in 1974	Decline from mid-1970s to 2,379 pairs in 1980 then low of 1,133 in 1987 but 1,397 in 1989	Sustained increase from 1,403 pairs in 1990 to 3,336 in 1999	3,309 in 2000, increasing to peak 4,298 pairs in 2009	4,137 in 2010 then steady decrease to 2,055 pairs in 2019; decline from peak of 52%
Skomer, Pembrokeshire	4,106 to 9,600 – increase 133%	13,030 to 17,500 – increase 34.3%	12,800 to 20,200 then to 12,028 – increase 57.8% then decline 41%	13,253 to 10,219 – decline 23%	10,249 to 5,216 – decline 49.1%
Skokholm, Pembrokeshire	2,500 to 3,000 (estimates)	3,654 to 2,595 (4 years data) – probable decline	2,605 to 4,652 to 2,894 – increase 78.5% then decline 37.8%	2,419 to 2,396 – decline 1%	2,486 in 2010, 1,008 in 2019 – decline 59.1%
Caldey, Pembrokeshire	104 single count	45 to 113 – increase 151%	149 to 507 – increase 340%	550 to 735 then to 484 – overall decrease 12%	509 to 676 to 536 – overall increase 5.3%
Cardigan Island, Ceredigion	No data	350 to 800 – increase 129%	1,300 to 4,700 to 2,763 – increase 362% then decline 41.2%	2,468 to 950 – decline 41.5%	803 to 323 – decline 59.8%
Bardsey, Caernarfonshire	No data	199 in 1986, 260 in 1989	235 to 478 – increase 203%	594 in 2000, peak 652 then 176 in 2009	251 in 2010, 164 in 2019 – decline 34.7%
The Skerries, Anglesey	No data	484 to 451	425 to 1,169 (4 years data) – increase 275%	747 in 2000, 522 in 2005 – decline 30%	472 to 92 – decline 80.5%; 115 in 2019
Puffin Island, Anglesey		90 pairs in 1986	No data	100 pairs in 2001	579 in 2012 and 526 in 2017

Lesser Black-backed Gull: Number of pairs and current numbers at key sites in Wales.

Region, city or town	County	Historic count	Most recent count	Notes
Newport	Gwent	293 in 2007	285 in 2017	
Chepstow	Gwent	21 in 2007	17 in 2017	
Monmouth	Gwent	None in 2007	79 in 2017	
Ebbw Vale	Gwent	29 in 2002	3 in 2017	
Brynmawr	Gwent	None in 2000; 39 in 2007	36 in 2017	
Cardiff	East Glamorgan	2,431 in 2006	2,357 in 2017	
Barry	East Glamorgan	676 in 2005	No recent data	
Bridgend	East Glamorgan	587 in 2005*	No recent data	
Tredegar	East Glamorgan	40 in 2007	103 in 2017	
Trethomas	East Glamorgan	14 in 2007	80 in 2017	
Maesteg	East Glamorgan	52 in 2004	37 in 2017	
Port Talbot	Gower	50 in 2002	No recent data	
Swansea	Gower	50 in 2002	No recent data	
Pembroke Dock	Pembrokeshire		27	
Caernarfon	Caernarfonshire	11 in 2000	17 in 2019	78 at a nearby industrial site in 2014
Wrexham area	Denbighshire		32 in 2019	Aerial survey of four 1-km squares (Woodward *et al.* 2020)
Prestatyn area	Flintshire	18 in 2002	123 in 2019	Aerial survey of six 1-km squares (Woodward *et al.* 2020)
Rhyl area	Flintshire	6 in 2002	87 in 2019	Aerial survey of six 1-km squares (Woodward *et al.* 2020)
Connah's Quay area	Flintshire		71 in 2019	Aerial survey of 17 1-km squares (Woodward *et al.* 2020)

Lesser Black-backed Gull: Rooftop nesting birds by region. Gwent data supplied by Richard Clarke.

* May include Herring Gull.

Roof nesting was recorded at Beaumaris, Anglesey, sometime after 1976, but even in the late 1980s, only nine birds were recorded nesting on buildings in Caernarfonshire. Seabird 2000 (Mitchell *et al.* 2004) attempted to quantify the numbers of roof-nesting birds in Britain for the first time and found that 12% of all nesting attempts were on buildings. However, in Wales only around 2% nested on buildings, with the largest concentrations in Cardiff, Newport and Pembroke Dock.

The Welsh population peaked in the early 1990s, since when there has been a sharp fall in numbers almost everywhere, with a shift from natural nest sites to buildings. The fall has almost certainly been the result of very poor breeding success in coastal colonies since 1987. On Skomer, productivity has been below 0.20 fledged chicks/pair in many years, and over 0.60 in only a few years. This inevitably led to poor recruitment into the breeding population and a subsequent decline in numbers. Anecdotal observations suggest a switch from maritime food supplies, including discards from fishing vessels, to farmland since the 1970s, presumably driven by a decline in the availability of food at sea. A study involving tracking birds from Skokholm in 2014 found that some foraged exclusively inland throughout the breeding season while others switched to offshore foraging after chicks hatched. However, the same switch did not occur in 2015, when most continued to forage inland (Brown and Eagle 2014, 2016). Observations in the 1980s showed that the birds fed primarily on earthworms. This may have reduced breeding success, for example in the dry weather of 1989 and 1990, when breeding success was effectively zero (Thompson 2008).

Alongside poor breeding success, adult survival on Skomer reduced from an average of over 90% in the 1970s and early 1980s to 88% in 2018, but in some years it was below 80%. Survival has improved since 2010, with 93% in 2019 (Wilkie *et al.* 2019). The sample size in the study area is reducing in line with the overall population, so results may be less robust than in the 1980s.

The colonies on Skomer and Skokholm have declined by 75% and 79% respectively since 1993, and in 2019 together totalled just 6,226 pairs. On Flat Holm, following a 72% decline to 1,133 pairs in 1987, the colony started to increase again, and reached a new peak of 4,298 pairs in 2009, before declining again to 2,060 pairs in 2019. Cardigan Island has seen a 93% reduction from

Operation Seafarer 1969–70	SCR Census 1985–88	Seabird 2000 1998–2002	Seabirds Count Provisional 2015–19
11,529	20,043	20,722	Coastal – 10,190 Inland rooftops – >3,244*

Lesser Black-backed Gull: Total counts of apparently occupied nests (AON) recorded in Wales during UK seabird censuses.

* Rooftop nest counts not yet complete.

4,700 pairs in 1993 to just 323 in 2019. The colonies on Caldey have reduced to a smaller extent, from 735 pairs in 2006 to 536 pairs in 2019. Decline rates seem to have increased since 2000. In North Wales, numbers on Bardsey fell from a peak of 652 pairs in 2003 to 164 in 2019, and at RSPB South Stack from 175 in 1989 to just six pairs in 2019. Similarly, The Skerries peaked at 1,169 pairs in 1993, but fell to only 92 pairs in 2018, although there were 115 pairs in 2019. Intriguingly, the Puffin Island population showed a different trend, with 90 pairs in 1986 and 100 in 2001 but 579 pairs in 2011 and 526 in 2017.

Whilst most colonies on natural sites are declining, those on rooftops seem to have continued to increase, except for the large Cardiff colonies that have been stable since at least 2006. A greater interest in rooftop nesting birds may have resulted in improved counts, but surveys almost certainly undercount the number of pairs (Coulson and Coulson 2015). There are now many more colonies, including some in counties without previous nesting, such as Welshpool, Montgomeryshire, (Mike Haigh *in litt.*) and at Barmouth and Aberdyfi, Meirionnydd, in 2018.

An aerial survey in 2019 of a total of 100 selected 1-km squares along the North Wales coast east of the Conwy Estuary, along the Flintshire side of the Dee Estuary and south to the Wrexham area in Denbighshire found a total of 368 apparently occupied nests (Woodward *et al.* 2020). The total number of Lesser Black-backed Gull nests calculated from the aerial survey was approximately six times the number found by ground-level counts of the same 1-km squares made as part of this study. This suggests that aerial surveys in other parts of Wales would probably also reveal much higher totals than those recorded by counts from ground level.

There are also some colonies on inland lakes, notably at Llyn Trawsfynydd, Meirionnydd, where there were about 15 pairs in 1969 increasing to 115 pairs in 1986, with 79 nests there in 2018. There were 12 nests at Llyn Elsi, Caernarfonshire, in 2017.

Until the 1960s, it appears that few individuals spent the winter in Wales, though authors such as Mathew (1894) had described the Lesser Black-backed Gull as "resident". Increasing numbers fed on inland fields and roosted on some reservoirs in winter, at least until 1993, but may have declined thereafter (Banks *et al.* 2009). This may simply reflect changes in the breeding population, but in other parts of the UK, the number of inland wintering birds is still increasing. At Llys-y-frân Reservoir, Pembrokeshire, up to 10,000 were recorded in winter 1988, whilst the maximum in 2018 was *c.*4,000 (Berry and Green 2018). Up to 2,000 were counted at Llangorse Lake, Breconshire, in November 2017 and again in January 2019.

It is not known whether birds wintering in Wales are from Welsh breeding colonies or from farther afield. There is certainly a gap between departure from breeding sites, after the chicks have fledged in late July to late August, and the establishment of the main reservoir roosts in late autumn. This suggests that many wintering birds are not Welsh breeders, but come from colonies to the north. Sightings of birds colour-ringed in Cardiff have been made around the Severn Estuary in winter, which suggests that at least some are not travelling far from their colonies. Colour-ringing has shown a degree of winter fidelity by some individuals over many years. More recently, GPS tagging of adults at Skokholm has confirmed this (Brown and Eagle 2018a).

Many breeding birds and their offspring migrate to Spain, Portugal and Morocco. Ringing, mainly of chicks, at the major colonies has produced many recoveries, the vast majority on the Iberian Peninsula (*BTO Migration Atlas* 2002). There are a few ringing recoveries from farther afield in Senegal, the Canary Islands, Iceland, the Faroe Islands and Denmark, but these are exceptional.

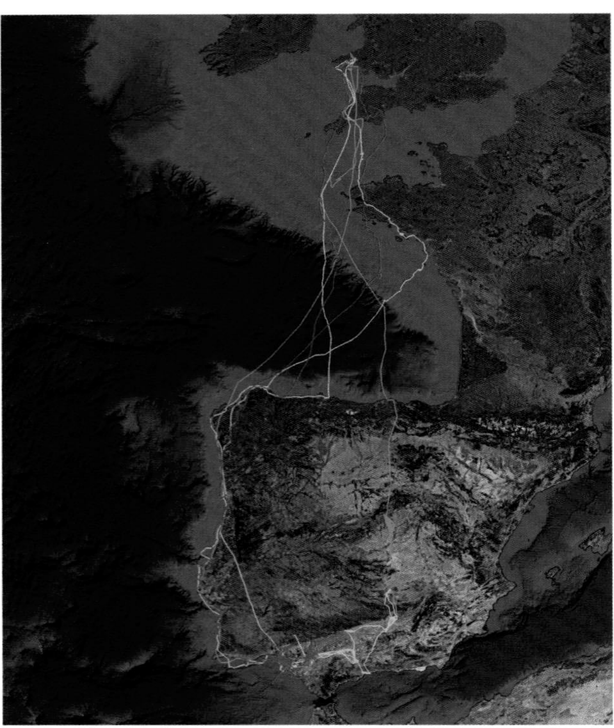

Lesser Black-backed Gull: Comparing the winter ranges of one bird over successive winters. Red tracks are those made between the 2014 and 2015 breeding seasons, yellow tracks are those made between 2015 and 2016 and green tracks are those made between 2016 and 2017. Map courtesy of Google Earth. Lesser Black-backed Gull data reproduced with permission of BTO (www.bto.org).

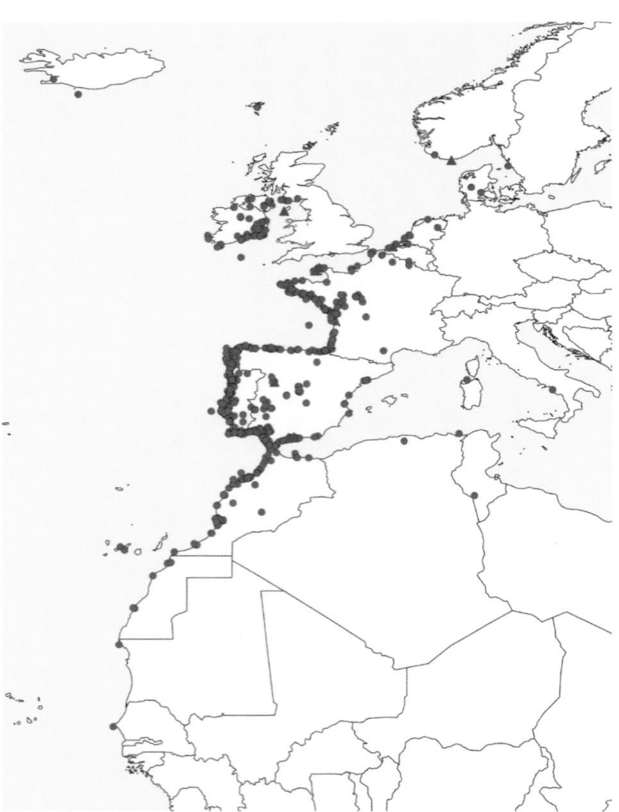

Lesser Black-backed Gull: Recovery locations of birds ringed in Wales are shown by red circles. Ringing locations of birds that were recovered in Wales are shown by blue triangles. Small brown dots show ringing or recovery locations in Wales.

The fortunes of the Lesser Black-backed Gull have changed dramatically in Wales over the past 60 years. The reasons for the rapid increase from the 1960s to the 1980s probably include an increase in the availability of domestic refuse and discards from fisheries. The subsequent decline is likely to have been driven by a reversal in these factors and a decrease in adult survival rates (JNCC 2020b). A scampi fishery developed on the Smalls Grounds, 65km west of Pembrokeshire, from the 1950s. Large numbers of unwanted fish were initially discarded, but availability reduced following an increase in legal net mesh size in 1986. The increased

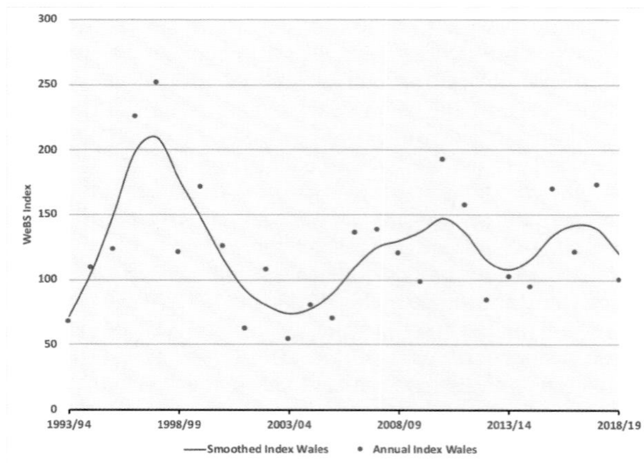

Lesser Black-backed Gull: Wetland Bird Survey indices, 1993/94 to 2018/19.

tendency to nest on buildings may be because such sites are closer to human food waste: rooftop-nesting gulls are more successful in raising chicks than those in natural sites (Rock 2005).

Wales is of considerable importance for Lesser Black-backed Gulls. The Seabird 2000 census estimated the UK population at just under 112,000 pairs, 38.4% of the global population of the species and 62.6% of the subspecies *L.f. graellsii* (Mitchell *et al.* 2004). The Welsh population made up 18.5% of the UK total, and thus 6.6–7.8% of the world population. The JNCC Seabird Monitoring Programme suggests a large decline in the UK breeding population since then, but urban sites are not well covered, and only a full census will provide a reliable estimate of trends (JNCC 2020a). In 2019 the total population of Lesser Black-backed Gulls in Wales was at least 13,500 pairs, of which 32% were nesting on buildings. This population is almost certainly undercounted. It is likely that Wales still holds a similar proportion of the wider population to that recorded in 1998–2002. The species has Amber conservation status in Wales, on the basis of localised breeding and the international importance of the Welsh breeding population (Johnstone and Bladwell 2016). Prospects for the species in Wales do not look good. The Welsh colonies are likely to continue to decline, maybe back to the 1940s levels, which were less than 25% of even today's depleted numbers.

Continental Lesser Black-backed Gull *L.f. intermedius* Gwylan Gefnddu Fach y Cyfandir

L.f. intermedius, which breeds in southern Scandinavia and Germany, was first confirmed in Glamorgan in 1984, when one was at Cosmeston Lakes on 3 May and 16 November. However, Hurford and Lansdown (1995) considered that many earlier records in the county of birds said to show characteristics of the northern subspecies *L.f. fuscus* were probably in fact *L.f. intermedius*. Venables *et al.* (2008) expressed the same opinion of early records in Gwent, and this is likely to be true of other counties. Birds said to show characteristics of *L.f. fuscus* had been reported fairly regularly in South Wales since 1929, and in increased numbers from 1970, including 35 at Blackpill, Gower, on 2 April 1985. There were records of *L.f. intermedius* in several other counties from the mid-1980s onwards, including 30 at the Gann Estuary, Pembrokeshire, on 18 January 1987. A flock of 110 at Glasbury, Radnorshire, on 6 October 1994, was considered to be made up of 85 *intermedius* and 25 *graellsii* (Jennings 2014). There have been few published records in recent years, probably owing to a lack of interest on the part of many observers.

In light of current knowledge about the identification and migration movements of *L.f. fuscus*, none of the records claimed to be of this subspecies in Wales is now regarded as acceptable. It is genuinely rare in Britain (Phil Bristow *in litt.*).

Steve Sutcliffe

Gull-billed Tern *Gelochelidon nilotica* Môr-wennol Ylfinbraff

Welsh List Category	IUCN Red List		Rarity recording
	Global	Europe	
A	LC	LC	BBRC

This large tern has a broad global but discontinuous population, from the coasts of North and South America, northern Africa and southern China, to the steppes of Central Asia. In Europe, it breeds primarily around the Mediterranean and the Black Sea, particularly in Spain, but there are small colonies on the North Sea coasts of Germany and Denmark. European birds winter in coastal West Africa, south to the Gulf of Guinea (*HBW*). There was a substantial decline in the European population between 1970 and 1990 (Tucker and Heath 1994), but numbers were thought to be increasing in 2015 (BirdLife International 2015). A total of 377 were recorded in Britain to the end of 2019, about three or four a year over the last 30 years (Holt *et al.* 2020).

The first Welsh record was at Shotton, Flintshire, on 1 August 1960. Since then a further 19 individuals have been recorded in Wales. Most records have been in southern Wales, between April and October, but mostly in May and July. Four individuals have been recorded in the Burry Inlet, Gower/Carmarthenshire, four on the Dyfi Estuary, Ceredigion/Meirionnydd, and two passing Strumble Head, Pembrokeshire. Only five have been recorded in North Wales.

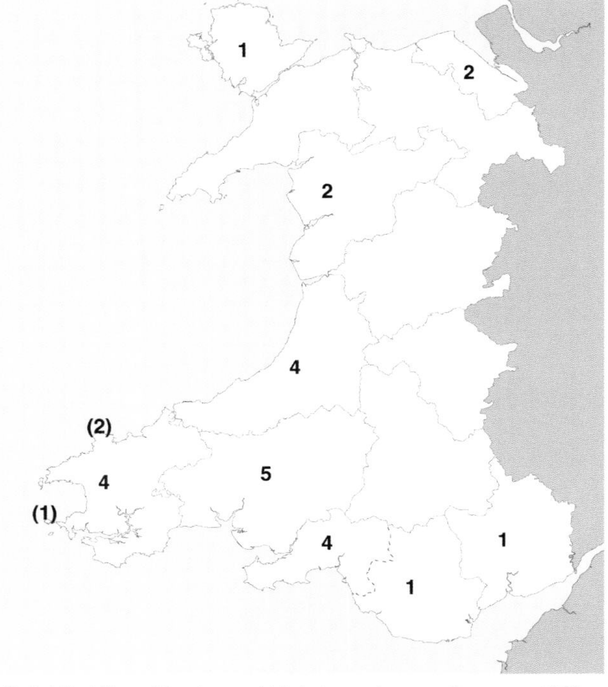

Gull-billed Tern: Numbers of birds in each recording area 1960–2019. Some birds were seen in more than one recording area.

Welsh Total	Pre-1950	1950–59	1960–69	1970–79	1980–89	1990–99	2000–09	2010–19
20	0	0	2	3	3	2	3	7

Gull-billed Tern: Individuals recorded in Wales by decade.

The second *European Atlas* (2020) showed evidence of range expansion since the 1997 *Atlas*, particularly in Iberia and along the northern shore of the Black Sea, though the picture is more complex in eastern Europe. If this expansion continues, the Gull-billed Tern could well become a more regular visitor to Wales in future.

Jon Green and Robin Sandham

Caspian Tern *Hydroprogne caspia* Môr-wennol Gawraidd

Welsh List Category	IUCN Red List		Rarity recording
	Global	Europe	
A	LC	LC	BBRC

This global seabird breeds in five continents, but in Europe is limited to the north coast of the Black Sea and the coasts of the northern Baltic Sea. The latter population winters in West Africa, as far south as the Gulf of Guinea (*HBW*). A total of 323 were recorded in Britain up to the end of 2019, about five a year on average over the last 30 years (Holt *et al.* 2020). There have been just 11 in Wales, all between 11 April and 6 August, except for one record off Bardsey in October:

- Llangorse Lake, Breconshire, on 20 July 1962;
- Blackpill, Gower, on 19–21 August 1973;
- Dyfi Estuary, Ceredigion/Meirionnydd, on 8–28 May 1974;
- Cemlyn, Anglesey, on 26 May 1980;
- Cemlyn, Anglesey, on 5 August 1988;
- Kenfig, East Glamorgan, on 11 April 1989;
- Skomer, Pembrokeshire, on 28 May 1994;
- Kenfig, East Glamorgan, on 6 August 1997, and Eglwys Nunydd Reservoir, Gower, on 6–8 August;
- Bardsey, Caernarfonshire, on 28 May and 1 June 1998;
- Bardsey, Caernarfonshire, on 12 October 2005 and
- Kenfig Sands and Pool, East Glamorgan, and Crymlyn Burrows, Gower, on 25 June and 24 July 2017, also at Penclacwydd, Carmarthenshire, on 26–29 June, 3–6 and 24 July 2017.

Jon Green and Robin Sandham

American Royal Tern *Thalasseus maximus* Môr-wennol Fawr America

Welsh List Category	IUCN Red List	Rarity recording
	Global	
A	LC	BBRC

The American Royal Tern breeds in the coastal USA, the Caribbean and the Atlantic coast of South America. It winters more widely across those regions (*HBW*). There have been two records in Wales. The first, at Kenfig Pool, East Glamorgan, on 24 November 1979, had been ringed in North Carolina as a nestling and was the first record for Britain. In 2018, one was seen at Traeth Dulas, Anglesey, on 10 December and at nearby Traeth Lligwy the following day; it then appeared on the Gann Estuary, Pembrokeshire, on 22 December and was subsequently seen in the Scilly Isles. Also ringed, this bird was thought to be the same individual that frequented the Channel Islands between February 2017 and August 2018, with a brief visit to Pagham Harbour, West Sussex, in June 2018. It was seen again from the Channel Islands in 2019. A previously published record of one at The Mumbles, Gower, in 1987 is not now considered acceptable.

One other bird seen in Wales was accepted by the BBRC as a Royal Tern. This individual, which showed some characters associated with the West African form *albididorsalis*, was seen at several sites in Caernarfonshire in 2009, first at Porth Ceiriad, Abersoch and Black Rock Sands on 15 June, then at Llandudno on 20 June. Subsequent analysis of DNA showed that the West African birds were more closely related to Lesser Crested Tern than to the American population of Royal Tern. It is now recognised as a species: West African Crested Tern, *T. albididorsalis*, which breeds along the west coast of Africa from Western Sahara to the Gulf of Guinea and outside the breeding season disperses north as far as Morocco and south to Namibia (IOC List 10.2). The BOU is currently assessing whether there is sufficient evidence to confirm that this species has occurred in Britain.

Jon Green and Robin Sandham

Lesser Crested Tern *Thalasseus bengalensis* Môr-wennol Gribog Fach

Welsh List Category	IUCN Red List	Rarity recording
	Global	
A	LC	BBRC

This species breeds on the coasts of North Africa, the Indian subcontinent, Southeast Asia and Australia (*HBW*). The only record for Wales, which was also the first British record, was at Cymyran Bay, Anglesey, on 13 July 1982. Since then, only another eight individuals have been recorded in Britain, the last in 2005 (Holt *et al.* 2020). The species is closely related to Sandwich Tern and can pair with it to produce hybrid young. Most records in Britain, like the first one at Cymyran Bay, have been in Sandwich Tern colonies.

Jon Green and Robin Sandham

Sandwich Tern *Thalasseus sandvicensis* Môr-wennol Bigddu

Welsh List Category	IUCN Red List			Birds of Conservation Concern			
	Global	Europe	GB	UK	Wales		
A	LC	LC	LC	Amber	2002	2010	2016

Unique among the family, with its shaggy crest and yellow-tipped black bill, the Sandwich Tern is our largest regularly occurring tern species and one of the most gregarious. It breeds in colonies around the coast of Europe, from the Baltic Sea to Ukraine and in the Caspian Sea (*HBW*), usually in association with Black-headed Gulls or other tern species (Cabot and Nisbet 2013). Most populations winter to the south of their breeding range. West European birds move south along the west coast of Africa, to winter mainly in the tropics (*HBW*). It exhibits the most erratic population trends and distribution of any seabird breeding in Britain and Ireland (Mitchell *et al.* 2004).

Birds were breeding around the Irish Sea in the mid-19th century, at Walney Island, Cumbria, from at least 1843 and in Ireland from 1850 (Cabot and Nisbet 2013). There appears, however, to be no dated record of this species in Wales during the 19th century, although it was included in a list of birds for the Cardigan area (James 1899). Mathew (1894) knew of no records in Pembrokeshire, but thought that the species must occur occasionally, as it was seen in north Devon. Forrest (1907) knew of only one record in North Wales, a pair on The Skerries, Anglesey, in 1902 that did not stay to breed. By the time of Forrest's update volume in 1919, the position had changed in North Wales. A breeding colony of 40 nests had been found by Cummings and Oldham on a stack off Ynys Llanddwyn, Anglesey, in 1915. The ship pilots there reported that the birds had been present in 1914 (Jones and Whalley 2004). Forrest stated that Sandwich Terns were noted visiting the Conwy Estuary, Caernarfonshire/Denbighshire, in 1913 and were present in numbers at the western end of the Menai Strait in June 1918. He had seen them as far east as Prestatyn, Flintshire. It remained a rare bird farther south. A corpse was found at Cors Caron, Ceredigion, in June 1922, and the first record for Pembrokeshire was four at Newport in June 1928. The first Carmarthenshire record followed in August 1929.

Numbers breeding at Ynys Llanddwyn dwindled after 1915, and in 1924 and 1925 there was only one pair. There were only occasional breeding attempts on Anglesey over the next four decades, including at Newborough, Rhoscolyn Beacon, Inland Sea, Cemlyn Lagoon and The Skerries. In 1956 there were said to be no recent breeding records on the island (Jones and Whalley 2004). In 1970, 20 pairs nested at Ynys Feurig, off the west coast of Anglesey, increasing to 100 pairs in 1971 and 200 pairs in 1972. Numbers varied through the remainder of the 1970s and early 1980s, peaking at 241 pairs in 1983. Within a few years they had ceased breeding there and few attempts have been made there since, such is the fickle nature of the species.

Colonies became temporarily established at another couple of sites during that period, but were not sustained. One pair bred at Cemlyn lagoon in 1972. A successful season was reported in 1976, but 1977 was a disastrous season because of flooding. The site has been occupied every summer since 1984. The North Wales Wildlife Trust employs seasonal wardens to keep people, dogs and predators away from the two nesting islands in the lagoon. Numbers have fluctuated at this colony, with peaks of 1,563 pairs in 2004 and 2,650 pairs in 2015. In 2015, the colony held about 20% of the British population and 2% of the world population of the species (Chris Wynne pers. comm.). Disastrous breeding seasons in 2007, 2008 and 2017, due to predation and desertion of nests,

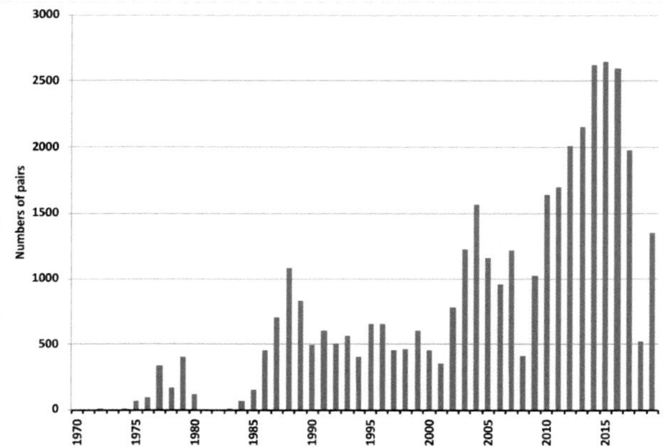

Sandwich Tern: Numbers of pairs breeding at Cemlyn, Anglesey, 1970–2019 (data from the NWWT and RSPB).

led to a sharp drop in numbers in subsequent years. It can take a few years for the numbers to build up again. Disturbance and predation of eggs by an Otter in 2017 led to complete breeding failure and only 519 pairs returned to breed the following year. Electric fencing is now installed each year around the perimeter of the islands and numbers increased to an estimated 1,200–1,500 pairs in 2019.

In 2018, following the crash at Cemlyn, an influx was noted at RSPB Hodbarrow, Cumbria, closely matching the number that had deserted Cemlyn. These are long-lived birds. The oldest individual known to the BTO ringing scheme was caught alive 30 years and nine months after it had been ringed as a nestling. Breeding failure or desertion events are part of their ecology. Terns can recover from such failure in a single year, providing there are sufficient alternative sites in the region. Tracking studies show that even breeding adults routinely prospect other sites immediately before or after a breeding attempt (Fijn *et al.* 2011).

It is estimated that an average productivity of 1.10 fledged chicks/pair is required to maintain a stable population (JNCC 2020b). In 2018, average UK productivity was 0.54. The productivity at Cemlyn has been higher than at most other UK colonies for many years, exceeding one fledgling/pair in three years: 1997, 2003 and 2006. Since 2009, productivity there averaged 0.73 fledglings/pair but dropped to *c.*0.5 when the colony was at its most dense in 2014–15. Nonetheless, 1,450 and 1,200 young fledged in these years, despite poor weather conditions during

Sandwich Tern © Ben Porter

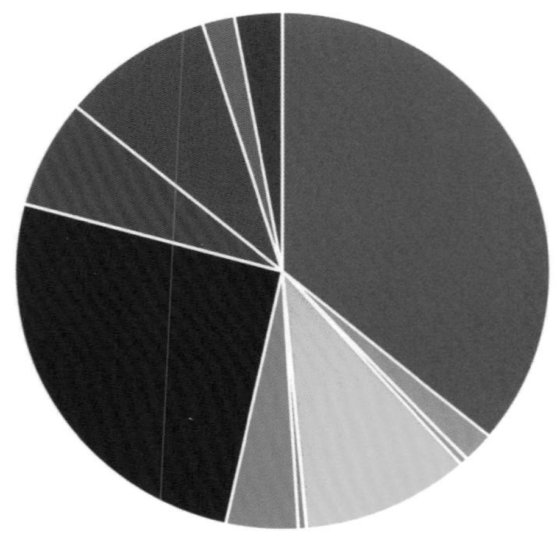

Sandwich Tern: The colonies or sites where terns later seen around North Wales were colour-ringed. The percentages of the total of 206 sightings are shown for each ringing site.

Sandwich Tern: Recovery locations of birds ringed in Wales are shown by red circles. Ringing locations of birds that were recovered in Wales are shown by blue triangles. Small brown dots show ringing or recovery locations in Wales.

late May in the latter year. Cold, wet weather can frequently lead to the loss of early nests (Ratcliffe *et al.* 2000).

This is typically the first tern species to appear off the Welsh coast each spring, with the earliest records usually in March, although the bulk of birds arrive in April. Particularly early records in recent years include 15 February 2018 at Port Eynon Point, Gower, and 20 February 2014 at Holyhead Harbour, Anglesey. Over-wintering has been suspected in some cases. Inland records are scarce, but some occur in spring and autumn, for example five at RSPB Lake Vyrnwy, Montgomeryshire, on 22 September 2018.

Significant post-breeding gatherings are evident at several sites along the North Wales coast, including at Gronant in Flintshire, the Clwyd Estuary, Denbighshire/Flintshire, and Glan-y-môr Elias, Caernarfonshire, all of which have hosted 500–1,000 birds during August in recent years. Many are in family groups, with youngsters still begging for food. The shallow waters between Sarn Cynfelin, Ceredigion, and Aberdysynni, Meirionnydd, are a very important post-breeding nursery area for this species, where many hundreds of adults forage for food to take to fledged chicks along the coast. There are regularly more than 500 in the mouth of the Afon Dyfi, and the highest count was 1,100 there on 14 August 1981 (Roderick and Davis 2010). The true numbers are probably far greater, underlining this area as being of vital importance as a pre-migration feeding area immediately after fledging (Tony Fox *in litt.*).

The latest records in recent years have generally been in October, but birds at Strumble Head, Pembrokeshire, on 19 December 2014, at Kenfig Pool, East Glamorgan, on 10 December 2015 and at Fishguard harbour, Pembrokeshire, on 11–15 December 2019 were particularly late. Sandwich Terns sometimes winter off the south coast of England. This may occur increasingly with the trend for warmer winters.

Birds ringed at colonies on Anglesey have been recovered in a number of West African countries. Two were recovered farther south, in Angola. Birds ringed outside the breeding season at Ynys-las, Ceredigion, have been recovered as far south as Namibia and South Africa. The use of inscribed colour rings at most UK colonies since 2014 has greatly increased the number of resightings, although this only started at Cemlyn in 2019. Colour-ringing has provided new information on post-breeding movements and long-distance migration. Post-breeding observations along the north coast of Wales in autumn 2019 resulted in the resighting of one-third of the young fledged at Cemlyn that year, indicating that a significant number use this coast as a post-breeding area, to feed up before migrating. Other birds seen here originated at colonies in Ireland, Cumbria, Northumberland, northeast Scotland and Rudkøbing, Denmark, 955km to the northeast.

The Sandwich Tern is Amber-listed in Wales owing to its highly localised breeding (Johnstone and Bladwell 2016) and Amber-listed in the UK because of a moderate breeding population decline over the long term (Eaton *et al.* 2015). The UK population was estimated at 12,500 pairs in the Seabird 2000 census (Mitchell *et al.* 2004). In the years since then, the JNCC Seabird Monitoring Programme has reported a 13% increase from counts of sample colonies. Even if that rising trend continues, the limited number of breeding sites will keep this species firmly on the Amber list for the foreseeable future. The European population is about 79,900–148,000 pairs, with no discernible trend (BirdLife International 2020). The concentration of the Welsh breeding population at a single site makes it very vulnerable. Effective wardening is vital to protect against predation and disturbance, but there are other important factors, such as food supply in warming waters and overwinter survival.

Henry Cook

Sponsored by Sue Stolton and Nigel Dudley

Elegant Tern *Thalasseus elegans* Môr-wennol Gain

Welsh List Category	IUCN Red List	Rarity recording
	Global	
A	NT	BBRC

Up to 95% of the global population of this Pacific species is restricted to Isla Rasa in the Gulf of California, although a few other islands in the USA and Mexico are used (BirdLife International 2020). Outside the breeding season birds disperse along the Pacific coast as far south as Chile (*HBW*). Birds thought to be Elegant Terns have been reported in Europe since 1974, initially in France and then in Spain, where a pair nested on the Valencia coast from 2009–18 and two pairs in 2019. DNA analysis showed that these were Elegant Terns and not hybrids (Dies *et al.* 2019b).

Several birds resembling this species were recorded in Britain in 2002, but the possibility that these may have been hybrids meant that Elegant Tern was not accepted onto the British List until January 2018. This followed the publication in 2017 of results of DNA analysis on similar birds in France and Spain. Three individuals were thought to have been present in Britain in 2002. The first, at Dawlish Warren, Devon, on 18 May 2002, was confirmed as the first record for Britain. The only Welsh record was an adult at Porthmadog, on both sides of the Meirionnydd/Caernarfonshire border, on 23–26 July 2002. This is considered to be the same bird as a second individual recorded in Devon, seen near Dawlish on 8 July and at Dawlish and Torbay on 18–19 July (Stoddart and Batty 2019).

Jon Green and Robin Sandham

Little Tern *Sternula albifrons* Môr-wennol Fach

Welsh List Category	IUCN Red List			Birds of Conservation Concern				Rarity recording
	Global	Europe	GB	UK	Wales			
A	LC	LC	VU	Amber	2002	2010	2016	RBBP

Little Tern on nest © Michael Steciuk

The 'little sea-swallow', to translate the Welsh name literally, is widely distributed, breeding in Asia, Africa, Australia and widely across Europe, although only locally in the west. European and Central Asian breeders move south in winter, those from western Europe mainly to the coast of West Africa (*HBW*). The Little Tern usually arrives back on our shores in the latter half of April, nesting in a shallow scrape on the sliver of habitat between the foredune and the tidally washed shore—a narrow and vulnerable strip of shingle and sand a few metres wide. In Britain and Ireland, this species breeds only on the coast, although in other parts of western Europe it often nests far inland on shingle banks in rivers and occasionally in quarries. Nest sites in eastern Europe include meadows in Belarus and flat roofs of buildings in Latvia and Finland (Norman 2020). It does not forage far from the breeding site, so access is necessary to shallow, sheltered feeding areas, where it can easily locate small fish, crustaceans and other invertebrates (Mitchell *et al.* 2004). Within ten days of hatching, the young reach full adult weight (Norman 1992). This is one of the fastest-known growth rates of any non-passerine. The chicks fledge after three weeks, after which birds gather on the shoreline for a few days, before leaving *en masse* by mid-August.

The breeding population in Britain and Ireland is thought to have increased around the turn of the 20th century and peaked in the 1930s, followed by a decline (Cabot and Nisbet 2013). In Wales, the species is mentioned in the *Swansea Guide* of 1802. Nesting occurred at the entrance of the Menai Strait near Traeth Lafan, Caernarfonshire, in 1865 and at Point of Ayr, Flintshire, in 1866 (Smith 1866). In the late 19th and early 20th century, it was a common breeding species in northern Wales, although there were only two colonies farther south. These had been long established in East Glamorgan, at Sker Point and The Leys near Aberthaw (Hurford and Lansdown 1995). Around 50 pairs bred at Sker Point in the mid-1880s, putting it among the largest colonies in Wales at this time. This species has never been recorded breeding in the other southern coastal counties, but there were many colonies from Ynyslas, Ceredigion, northwards to the Dee Estuary. Forrest (1907) described it as common on the coast of North Wales, especially in the west, with many colonies ranging from about ten to 50–60 pairs. Forrest (1919) stated that the colony at Point of Ayr had increased greatly. In 1915, S.G. Cummings thought it to be the largest colony in North Wales and in 1916 it held 200 pairs (Charles Oldham diaries, quoted in *Birds in Wales* 1994). There was said to be another large colony about three miles away towards Prestatyn, probably in the Gronant area.

A decline was already underway in Wales by the 1930s but tracing the fortunes of individual colonies is difficult. Both East Glamorgan colonies had been abandoned by 1930, although a pair bred successfully at Sker Point in 1936. Many smaller colonies between the Dyfi Estuary and the Dee Estuary had disappeared by the late 1960s. The Ynyslas colony was lost when the area became a firing range during the Second World War, and the species has not bred in Ceredigion since (Roderick and Davis 2010). In Meirionnydd, a decrease in numbers was reported in the 1920s and 1930s, and by the 1950s there was only one colony, at Aberdysynni. In the early 1970s, fewer than ten pairs bred there annually, although one pair bred at a second site in 1962. The RSPB set up a protection scheme at Aberdysynni. Numbers peaked in 1977 with 36 pairs, but in many years eggs and young were lost to predators, including Kestrel, Stoat and even Oystercatcher. In 1985 the colony numbered nine pairs. Three pairs attempted to breed in 1987 but had failed by late May. *Birds in Wales* (1994) stated that the colony remained active until 1988.

Only two colonies remained in Caernarfonshire by the late 1970s, one in Tremadog Bay around Aber Dwyfor near Criccieth, and one at Morfa Dinlle, opposite a colony at Abermenai Point

Operation Seafarer 1969–70	SCR Census 1985–88	Seabird 2000 1998–2002	Seabirds Count Provisional 2015–19
28	55	75	171

Little Tern: Number of apparently occupied nests (AON) recorded in Wales during UK seabird censuses.

on Anglesey. At Aber Dwyfor, the North Wales Naturalists' Trust arranged wardening from 1973. A full-time RSPB warden guarded the colony from 1976, but there was heavy predation and breeding success was low. The last record there was of three pairs at Penychain in 1986. At Morfa Dinlle, there was one pair in 1986, 15 in 1987 and three in 1988, but there has been no subsequent record of breeding in the county. On Anglesey, the breeding population had dwindled to 12 pairs at four sites by 1967. Jones and Whalley (2004) thought that the species bred for the last time on Anglesey at Abermenai in 1978, although an adult, with a very young juvenile, was seen there in August 1997. Thus, by 1989, only one regular colony remained in Wales, at Gronant, Flintshire.

The factors causing the declines included habitat loss, human disturbance and predation. Most of the favoured beaches around Wales are now either stabilised, developed or subjected to too much recreational disturbance to sustain breeding attempts. While terns have always been mobile, human use of the coast means that they have little in the way of alternative options. The impact of human activities, on top of predation, has tipped the natural balance against the terns. A range of predators has been recorded taking eggs or chicks, including Hobby and American Mink, but the main predators are Fox, Carrion Crow and Kestrel. The colonial breeding of Little Tern means that a Fox can clear out a large proportion of the eggs in a night and a persistent Kestrel can take many chicks over several weeks. Severe predation events at Gronant caused by Foxes occurred in 1986, 1990 and 2001, when all the eggs or chicks were taken, and Kestrels took 33 chicks in 2015 (*Gronant Little Tern Report 2015*).

Aside from nests among the *machair* of the Outer Hebrides, the Little Tern only remains as a breeder in Britain thanks to intensive conservation effort. Many colonies have declined despite this. The UK breeding population was estimated at well over 3,000 pairs in the mid-1970s, but declined to under 2,000 pairs by 2005. This is attributed to low breeding success, leading to low recruitment into the breeding population (Cabot and Nisbet 2013). There were indications of a partial recovery between 2005 and 2012, but then a further decline until at least 2018, with a 40% decline since 1986 (JNCC 2020b). Woodward *et al.* (2020) estimated the population in Britain at 1,450 pairs in 2013–17 (there were none in Northern Ireland).

The Gronant colony almost died out in the 1960s, with only a couple of pairs in 1967, when volunteers from Clwyd Ornithological Society deterred predators. A protective wardening scheme has enabled a steady increase since the 1970s. The RSPB managed the site from 1975 to 2004, deploying electric fences, 24-hour wardening and using anti-predator devices. Numbers grew steadily from 15 to 87 pairs. Since 2005, the colony has been managed by Denbighshire County Council, during which time it has grown to be one of the largest in the UK, peaking at 171 pairs in 2018, representing 10% of the UK population. The most productive year was 2010, when 216 young fledged. The Gronant Dunes and Talacre Warren area is designated as an SSSI and the Dee Estuary and Liverpool Bay are designated as SPAs, with breeding Little Tern as a qualifying feature. Protection measures are now quite sophisticated, but still require a large number of staff and volunteers to achieve good outcomes. For example, there were 2,042 volunteer hours in one recent season (Cook *et al.* 2018). There are few records of breeding elsewhere in Wales in recent years. A pair attempted to breed at Horton's Nose, Rhyl, Denbighshire, in 2014, and since 2015 a pair has bred at RSPB Point of Ayr, Flintshire, rising to three pairs in 2019 and six pairs in 2020.

A significant post-breeding build-up is evident at Gronant or off Point of Ayr in most years, usually in late July. In 2016 this flock contained 813 birds on 20 July. Colour-ring resightings show that birds from all the Irish Sea colonies feed up there, prior to migration, including Irish, Isle of Man and Cumbrian breeders. Occasionally, birds ringed on the east coast of England have been seen too. Elsewhere in Wales, the Little Tern is fairly scarce, although late April to mid-May and the latter half of August can see a small passage, chiefly along northern and western coasts. With the loss of all the colonies, except Gronant, since 1988, counts away from the northeast are lower than formerly, but include, for example, 23 at Penclacwydd, Carmarthenshire, on 6 May 1989. In 1992, 15 passed Skomer, Pembrokeshire, on 3 September, followed by 19 the next day and 23 on 7 September. 26 passed the Little Orme, Caernarfonshire, on 12 August 2010 and in the Burry Inlet there were 23 at Wernffrwd, Gower, on 15 August 2012. Birds are regularly recorded passing Strumble Head in August and September, with a highest day count of 16 on 7 September 2010. A few birds remain in Welsh waters into October and the species has been recorded as late as the first week of November in Pembrokeshire and Ceredigion. It is infrequent on inland water bodies in Wales, except at Llangorse Lake, Breconshire, which hosts birds fairly regularly, including 12 on 21 October 1967 and ten, all juveniles, on 15 September 2006.

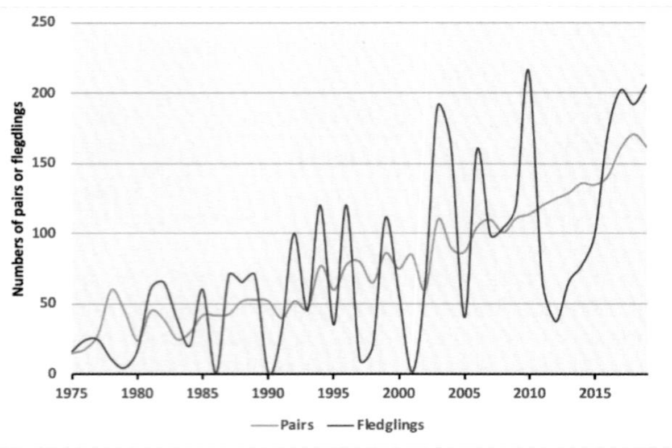

Little Tern: Numbers of pairs and fledglings at Gronant, Flintshire, 1975–2019.

Little Tern: Recovery locations of birds ringed in Wales, taken from published Ringing Reports 1980–2018, are shown by red circles. Ringing locations of birds that were recovered in Wales are shown by blue triangles. The small brown dot shows the ringing or recovery location in Wales.

The Merseyside Ringing Group has ringed chicks at Gronant since 1983, and colour-ringed chicks and adults from 2015, through which the age structure of the colony is being established. One bird ringed as a nestling in 1993 and colour-ringed in 2018 was resighted at Gronant in 2019. At 26 years old, it is the oldest Little Tern known to any European ringing scheme. Ringed chicks from Gronant have been found visiting other British colonies in their second calendar-year, and several birds have bred successfully in their third calendar-year. Most Little Terns spend their first summer in Africa and do not usually breed until their fourth calendar-year.

Wintering records of birds ringed in Wales have come from the coasts of Mauritania, The Gambia and Senegal, while two reached Guinea-Bissau. A bird ringed at Gronant in July 1995 was in a roost of several tern species on the Belgian coast on 21 April 2003, hinting that on spring migration some birds may cross England to reach North Wales. Research is underway using geolocators to track birds from Gronant.

The Little Tern is Red-listed in Wales because of the reduction in its breeding range and its dependence on a single site (Johnstone and Bladwell 2016). At present, its future as a breeding bird in Wales depends on the Gronant colony. Although it is currently doing well, a single colony is vulnerable to chance events. A storm surge hit the colony on 14 June 2018, washing away 89 active nests, over half the total, although many birds relaid. A focus of future conservation effort is to reduce the risk, by increasing the distribution of the species. Several suitable sites along the northeast coast of Wales have been fenced off in the hope that the birds will expand their breeding range. During 2014–18, the Little Tern was the subject of an EU LIFE+ Nature Programme, which led to increased colony protection at 29 sites in the UK, including Gronant. This project reduced the rate of its decline (Wilson *et al.* 2019). Future threats include rising sea levels and changing food stocks induced by climate change. Proposed tidal lagoon developments in North Wales could also present a threat. The tenacity of the Little Tern in the face of long odds, with tireless support from conservationists, might just be enough to ensure its future.

Henry Cook and David Norman

Sponsored by North Wales Little Tern Group

Bridled Tern *Onychoprion anaethetus* Môr-wennol Ffrwynog

Welsh List Category	IUCN Red List	Rarity recording
	Global	
A	LC	BBRC

The Bridled Tern breeds on tropical and subtropical islands in the Pacific and Indian Oceans and the Caribbean. It is pelagic outside the breeding season, and ranges across tropical seas and oceans (*HBW*). Out of a total of 25 birds recorded in Britain to the end of 2019 (Holt *et al.* 2020), two were in Wales. A dead bird was found at Three Cliffs Bay, Gower, on 11 September 1954, and one was at the tern colony at Cemlyn, Anglesey, on 1–23 July 1988.

Jon Green and Robin Sandham

Sooty Tern *Onychoprion fuscatus* Môr-wennol Fraith

Welsh List Category	IUCN Red List	Rarity recording
	Global	
A	LC	BBRC

This tern breeds on islands around the Equator in the Pacific, Indian and Atlantic oceans, including the Caribbean (*HBW*). Like the similar Bridled Tern, it is pelagic outside the breeding season, ranging across tropical seas and oceans. There have been 22 records in Britain to the end of 2019, half of them before 1950 (Holt *et al.* 2020). Two of these were in Wales. The first was a bird knocked over by a boy with a stick at Barmouth, Meirionnydd, on 17 August 1909. The second was recorded at several locations in 2005. It was found on Anglesey, initially on Ynys Feurig on 5 July, then on The Skerries on 7–10 July and Cemlyn on 10–26 July. It was later seen passing Strumble Head, Pembrokeshire, on 23 August, as it passed out into the Western Approaches.

Jon Green and Robin Sandham

Roseate Tern *Sterna dougallii* Môr-wennol Wridog

Welsh List Category	IUCN Red List			Birds of Conservation Concern				Rarity recording
	Global	Europe	GB	UK	Wales			
A	LC	LC	EN	Red	2002	2010	2016	RBBP

The Roseate Tern is perhaps the most attractive of all the terns breeding in Wales, presenting a study of subtle plumage characteristics in the breeding season: a suffusion of pink on the underparts, a solid black bill and a light grey mantle. Combined with its current scarcity in Wales, it evokes much excitement for any observer. Globally, it is widespread, but it has an unusual distribution. It is primarily a tropical species, with large colonies in the Pacific and Indian Oceans, and the Caribbean Sea. In addition, there are five small populations in temperate seas, including one in western Europe that is divided into two groups: one in Britain, Ireland and northwest France; the other in the Azores (Cabot and Nisbet 2013). The European population is estimated to be 2,300–2,900 pairs and is increasing after historic declines (BirdLife International 2015). Roseate Terns breed on rocky islands, but prefer sheltered nest sites, often with a covering of Sea Beet or Tree Mallow.

This is the rarest tern species breeding in Wales and has always been localised, but it was formerly much more numerous. The population in Britain was thought to be at a very low ebb, perhaps even on the verge of extinction, by the end of the 19th century. This was a result of shooting, for sport and for plumes to decorate ladies' hats, together with the depredations of egg-collectors (*Birds in England* 2005). There are few records of Roseate Terns being shot in Wales, but egg-collecting was a major problem

(Forrest 1907). The story of the rise and fall of colonies is often shrouded in secrecy, since site details and numbers were kept vague, to protect the birds from egg-collectors.

In the late 19th century and the early 20th century, Wales was the stronghold for this species in Britain and Ireland (Williams 2000a). Almost all records of breeding in Wales have been on Anglesey, but there were records of breeding in Pembrokeshire in the late 19th century. Mathew (1894) said that Roseate Terns formerly bred on Skokholm Stack, and E. Lort Phillips visited a small colony breeding on Grassholm in 1883 (Donovan and Rees 1994). The only other record of possible breeding away from Anglesey was in Flintshire where, in July 1916, Charles Oldham recorded about six pairs among a colony of Common Terns at Point of Ayr, although he was unable to identify their nests (Oldham Diaries, quoted in *Birds in Wales* 1994).

On Anglesey, Forrest (1907) stated that the species had bred on The Skerries for many years, a location that had previously been kept secret. Forrest felt that it was by then so widely known among egg-collectors that further secrecy was pointless. The nests were scattered among the much larger number of breeding Arctic Terns. The lighthouse keeper described the colony as "very large" in 1892, although it is unclear whether this referred to the Roseate Terns only or to the whole tern colony. Nine clutches of eggs were said to have been taken in 1896 and preserved at Oxford University Museum. S.G. Cummings reported "not many pairs" present in 1902. Forrest (1919) mentioned a small breeding colony there, but breeding appears to have ceased by the early 1920s.

Another colony, on small rocky stacks off Ynys Llanddwyn, is thought to have been established about 1900, and by 1915 was said to number hundreds of pairs. Forrest (1919) stated that there were about 300 pairs, and thought that the colony, which was "rigidly protected", must have been one of the largest in Europe. The birds there nested under Tree Mallow (Coward 1922). Elizabeth and William Jones, who lived in the pilots' cottages on the island, were hired by the RSPB to act as watchers to protect the birds but, despite all efforts, numbers declined, partly because of competition from nesting Herring Gulls and Cormorants. There were 100 pairs in 1924, but only one pair in 1925. Smaller colonies were noted at several sites on Anglesey in the early 20th century, such as 20 pairs at Ynys Dulas in 1912. *Birds in Wales* (1994) quoted "a good many" at Rhoscolyn Beacon in 1916 and nesting "in some numbers" on a reef in Cymyran Bay and on Ynys Dulas in 1922.

The Skerries were recolonised by 1925, and in 1928 there were thought to be at least 75 nests there, with possibly up to 300 pairs (A. Whitaker Diaries, Edward Grey Institute, quoted in *Birds in Wales* 1994). There were at least 150 pairs in 1933 and "several hundred" in 1935, before numbers declined to 75 pairs in 1947, 40–50 pairs in 1957 and 30 pairs in 1961, when breeding failed because of predation by rats and gulls. Small colonies were recorded at other sites around Anglesey in the 1950s and early 1960s, including 60 pairs at East Mouse (Ynys Amlwch) in 1957, at Abermenai and on Ynys Gorad Goch in the Menai Strait. Small numbers also bred once more on a small stack off Ynys Llanddwyn. From 1959, the main site was at Ynys Feurig, although there were a few pairs at other Anglesey tern colonies in most years. The number of pairs breeding in northwest Europe fell rapidly, by an average of 14%/year during 1968–82 and by 1.4%/year during 1982–91, but then increased by 4%/year from 1991 to 2001 (Cabot and Nisbet 2013). The pattern in Wales was different. Between 1969 and 1986 the numbers breeding on Anglesey fluctuated, with a peak of 251 pairs at three colonies in 1974 and a low point of 130 pairs at two colonies in 1981. There were still 208 pairs at three colonies in 1986, but in 1987 there was a large decline, to 77 pairs at four colonies. The Ynys Feurig colony suffered from predation by Brown Rats and Peregrines, and in 1987 a Fox got onto the island in late May and killed 12 Roseate Terns. The population on Ynys Feurig dropped from 200 pairs in 1986 to 40 pairs in 1987, and gull predation led to a complete failure of the colony in 1990. By 1991 there were just five pairs at two colonies on Anglesey.

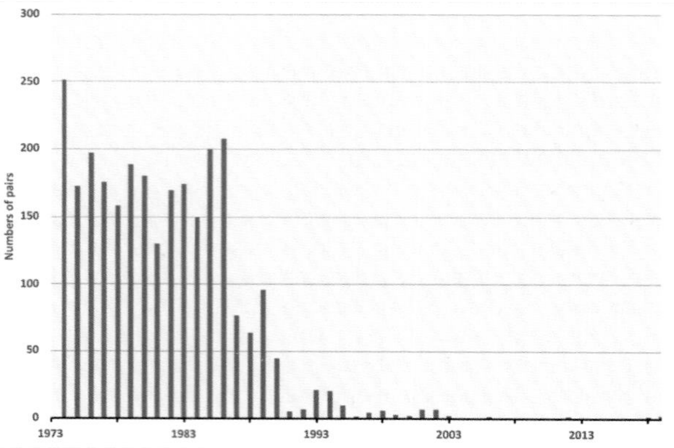

Roseate Tern: Numbers of pairs that bred in Wales, 1974–2019.

The largest colony in northwest Europe is at Rockabill in Dublin Bay, Republic of Ireland. In 1990, ring numbers were read on 241 birds at this colony and 65 of these proved to have been ringed at Ynys Feurig. Evidently, at least some of the birds that formerly bred on Anglesey had moved across the Irish Sea, so the decline on Anglesey did not reflect a decline in the whole Irish Sea population at this time. A few pairs continued to breed annually on Anglesey until 2003, since when breeding has been spasmodic. One or two birds have been present in the main Anglesey tern colonies every year, and in several years Roseate Terns have paired with Common Terns, with some hybrid chicks fledged. Single pairs of Roseate Terns fledged chicks in 2006, 2012 and 2018, and in 2019 two pairs each fledged two chicks on The Skerries.

Away from breeding colonies, the Roseate Tern has been recorded in all Welsh counties except Radnorshire, but it is scarce everywhere. Not surprisingly, the largest counts were on Anglesey before the collapse of the breeding population: a flock of 60 at Llugwy prior to 1955 and 60 at Penmon in July 1962. More unexpected was a record of 50 off The Range (Penrhosfeilw Common) on 21 May 2004. The largest numbers off the coast of mainland Wales have been at Strumble Head, Pembrokeshire, including counts of 11 on 29 July 1989 and 12 on 8 September 2009. Those passing Strumble in autumn are presumed to be from the Irish breeding population.

Roseate Terns breeding in western Europe spend the winter in a fairly restricted area around the Gulf of Guinea, West Africa. There are recoveries of Welsh-ringed birds from several countries in that area, but by far the largest number are from Ghana. One reason for the decline of the European population from the late 1960s is high mortality on the wintering grounds. Large numbers of terns are caught for sport and for food in Ghana, mainly by young boys, and Roseate and Common Terns are thought to be particularly susceptible. Despite projects to raise awareness of conservation issues and legislation, trapping continues (Quartey *et al.* 2018), although perhaps on a smaller scale than formerly.

The Roseate Tern's Red-listed status in Wales is the result of the huge decline in the breeding population. The prospects for its return as a regular breeding bird in Wales are uncertain. Numbers at the two largest colonies in northwest Europe, Rockabill and Lady's Island Lake, both on Ireland's east coast, have increased steadily in recent years. In 2018 there was a record 1,633 pairs on Rockabill (Newton 2018) and 227 pairs at Lady's Island, although breeding productivity on Rockabill has declined in recent years. Most birds seen on Anglesey originate from Ireland: of 17 ringed birds recorded at The Skerries, 12 were from Rockabill, four from Lady's Island Lake and one from elsewhere on Anglesey. There is hope that continuing efforts to maintain suitable nesting conditions at Anglesey's tern colonies could be rewarded by the return of the Roseate Tern as a regular breeding species in Wales.

Henry Cook and Rhion Pritchard

Common Tern *Sterna hirundo* Môr-wennol Gyffredin

Welsh List Category	IUCN Red List			Birds of Conservation Concern			
	Global	Europe	GB	UK	Wales		
A	LC	LC	NT	Amber	2002	2010	2016

The Common Tern breeds widely across much of Europe, temperate Asia and North America. In winter, birds move south, with the majority of the European population wintering off western and southern Africa. As a general rule, the northernmost breeders winter farthest south (*HBW*).

The main breeding areas for Common Tern in Wales are offshore islands and coastal lagoons on Anglesey and at Shotton steelworks, Flintshire, where it breeds on artificial islands. It tolerates breeding relatively close to human activity, provided access is controlled to reduce the risk of disturbance. At Shotton, for example, the nests are within a busy industrial site. Many coastal areas in Wales, as elsewhere, are prone to disturbance, hence offshore islands have become important. The Common Tern is not restricted to coastal areas, but while there has been an expansion of inland breeding colonies in England since the 1970s, the few in Wales have been short-lived.

Common Terns do not normally breed until three years old (*BWP*). Most second-calendar-year birds stay in or near their winter quarters, but a few return to Europe, some indistinguishable from adults by plumage. Many third-calendar-year birds visit breeding colonies, often arriving late in the season. Ringing records suggest considerable interchange between natal and breeding colonies. These may be local movements within the Irish Sea, but some birds disperse more widely: chicks ringed at Shotton in 1987 and at Ynys Gorad Goch in the Menai Strait in 1988 were recovered in the Scottish Highlands in July 1990 and July 1994 respectively, while another chick ringed at Shotton in 1986 was found dead at a colony in Cork harbour, Republic of Ireland, in 1993. In contrast, adults are highly faithful to their breeding colony (*BWP*) although prior to wardening, colonies could be mobile from year to year within a local area, if incidences of predation or disturbance occurred.

Ray (1662) visited St Margaret's Island, Pembrokeshire, and mentioned the nests of "sea-swallows", which along with Black-headed Gull nests were "so thick that a man can scarce walk but he must needs set his foot upon them." Sage (1956) considered that these were almost certainly Common Terns. Ray also mentioned seeing the species on the coast of Llŷn, Caernarfonshire. Common Terns were more widespread in Wales in the 19th century than they are today. Evidence of breeding in Glamorgan is ambiguous. Doddridge-Knight recorded the species as abundant around Sker/Kenfig in the 1830s (*Birds in Wales* 1994), but these sightings may have related to passage birds, as recorded in the area today. Dillwyn (1848) stated that Common Terns bred regularly in the county, but did not mention the site. Hurford and Lansdown (1995) suggested that this colony was probably in the west of the county. A small colony was recorded on Skokholm, Pembrokeshire, at the end of the 19th century and remained until 1916. Forrest (1919) stated that several pairs nested on the south side of the Dyfi Estuary near Ynyslas, Ceredigion, in July 1914, but Roderick and Davis (2010) considered this to be an error. In Meirionnydd, 2–3 pairs nested near Tywyn during 1911–14, with another nest in 1927. In Caernarfonshire, a colony at Morfa Dinlle comprised 40 nests in 1902 and remained active until at least 1932. Barnes (1997) considered that it became extinct in the late 1960s or early 1970s.

Forrest (1907) reported several colonies on Anglesey in the early years of the 20th century. Common Terns outnumbered Arctic Terns at Newborough Warren, where there was a colony of up to 120–130 pairs in sand dunes towards Abermenai Point, from at least 1902 to 1974. Nearby Ynys Llanddwyn was another site where Common Tern outnumbered Arctic Tern. Forrest (1907) reported that in 1903, on a rocky islet and neighbouring small stack, "so thickly was the surface covered with nests that it was impossible to avoid treading on some of them." There were reports of egg-harvesting in the 1950s by local farmers for human consumption and supplementing animal feed (R.P. Cockbain and G. Thomason pers. comm.). The Llanddwyn colony remained active until at least 1958.

On Rhoscolyn Beacon a few pairs of Common Terns nested among more numerous Arctic Terns in 1916, but by 1925 the 56 pairs outnumbered the Arctics. Nesting also occurred on the Inland Sea where 180 chicks were ringed in 1961 (Merseyside Naturalists' Association 1962) but in most years, numbers fluctuated between none and 20 breeding pairs. On The Skerries, Common Terns were greatly outnumbered by Arctic Terns; 50 pairs were estimated to breed there in 1933. On the east coast of Anglesey, Forrest (1907) noted colonies on Ynys Moelfre and Ynys Dulas. In 1903, the latter held "large numbers" and sporadic breeding was reported until 1952. Terns also nested on a shingle ridge at Traeth Dulas from 1923 to at least 1953. In the early 1950s, up to 55 pairs nested on coastal heath at Traeth Llugwy. From 1953, until at least 1963, up to 30 pairs nested at Penmon limeworks tip. Ynys Gorad Goch, an islet in the Menai Strait, regularly held Common Terns: 80+ pairs in 1957 (Merseyside Naturalists' Association 1958) and 175 in 1975, but numbers declined to fewer than 20 after 1998. Two neighbouring islets, Ynys Benlas and Ynys Welltog, occasionally held small colonies. At the former, 30 chicks fledged in 2010 and the latter had 19 pairs in 2012. Breeding now appears to have ceased there.

In Flintshire, a colony at Point of Ayr was recorded in 1898, reached a peak of *c.*150 pairs in 1916 and remained until the early 1950s, moving closer to Gronant in some years, with an isolated nesting occurrence in 1965. Farther up the Dee Estuary, a colony was recorded at Connah's Quay in 1915 and on the rapidly accreting saltmarsh between Burton Point and the main river channel, on the border with England, in 1918. Fifty pairs were present there in 1934 (Farrar 1938) and a small colony, varying between ten and 80 pairs, maintained a tenuous foothold up to 1970. From the 1970s, Merseyside Ringing Group created three artificial nesting islands totalling 1,000m^2. This developed into the largest colony in Wales, peaking at 762 pairs in 2007.

Operation Seafarer 1969–70	SCR Census 1985–88	Seabird 2000 1998–2002	Seabirds Count Provisional 2015–19
292	514	674	858

Common Tern: Number of apparently occupied nests (AON) recorded in Wales during UK seabird censuses.

The Common Tern breeds at inland sites in many parts of its range, usually at sea level, although in Scotland up to 300m, and at even higher altitudes elsewhere. However, inland breeding in Wales has been very limited. Up to ten pairs nested at Llyn Trawsfynydd, Meirionnydd, from 1968 to the early 1970s. A colony was established on islets at Llyn Alaw, Anglesey, shortly after the reservoir was filled in 1966. This colony peaked at 141 pairs in 1969 and fluctuated between 20 and 100 until breeding ceased in 1993. Both sites are within 13km of the nearest coastline.

The Welsh breeding population represented 5% of the Britain and Ireland population in 2000 and increased steadily up to 2007. It declined sharply in 2009, with the collapse of the Shotton colony, when birds returned to the Dee Estuary but did not breed. Lack of food was identified as the most probable cause. This was related to exceptional flooding in summer 2007 and winter 2007/08, that scoured the channel leading to the estuary, destroying spawning and nursery grounds for fish and sandeels. Common Terns continued to return to the Dee Estuary each year and, after a false start in 2013, when nests were predated by a Fox, 225 pairs fledged 445 chicks in 2014 (Coffey 2015). An average of 373 pairs nested there during 2015–19.

Three large colonies on Anglesey, including The Skerries and Cemlyn Lagoon, account for most of the remaining Common Tern nests in Wales. The Skerries has been wardened by the RSPB since the late 1980s, and Common Terns nested there from 1991. The colony grew in parallel with a larger Arctic Tern colony, to 386 pairs in 2017, although numbers fell back in the subsequent two years, and the colony was abandoned in 2020, when it was not wardened owing to the coronavirus pandemic. Cemlyn lagoon held small numbers of Common Terns from the mid-1960s. It regularly attracted 80–100 pairs during 1975–91 and 100–194 pairs during 2006–14. Numbers fell in 2017 because of predation and disturbance by Otters. Although electric fencing was installed, only 11 pairs nested in 2018 and 25 in 2019. Another private site has maintained a colony of 100–190 breeding pairs since 2002. Cumulatively, by 2014 the Common Tern population, at the three main Anglesey colonies and Shotton, had recovered to the level recorded in the Seabird 2000 survey, and increased by 22% during 2015–19.

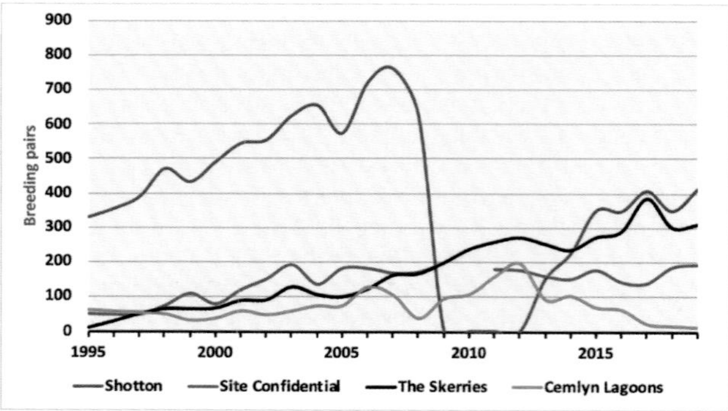

Common Tern: Totals of breeding pairs at the four main colonies in Wales, 1995–2019.

Post-fledging dispersal starts as early as July. A few birds travel quickly to West Africa: a chick ringed at Shotton on 25 June 1972 was found dead in Liberia on 2 September the same year. However, for the first two months, most birds stay close to their natal site (*BTO Migration Atlas* 2002). Welsh birds begin to move south in August, but Common Terns from colonies elsewhere in Britain and Europe may also move on passage across Wales. For example, a chick ringed in Poland on 17 June 2012 was at Ynyslas, Ceredigion, on 21 August the same year. Counts of passage birds are hampered by the difficulty of distinguishing Common and Arctic Terns at a distance. Many large movements have been recorded as "Commic Terns", but in recent years observers have become more aware of the differences between the two species. Passage is observed from many coastal locations, occasionally with spectacular numbers: 740 on 15 September 2010 off Bardsey, Caernarfonshire, was a record for the island, and there were remarkable numbers in the Burry Inlet in August 2012, peaking on 17 August, when 1,707 roosted at Llanrhidian Pill and another 300 at Pwll-y-froga, both in Gower. A record day count of 1,028 passed Strumble Head, Pembrokeshire, on 19 September 2019.

Birds from Norway and the Baltic Sea move south later, in September and October. Several records of Common Terns in November are likely to be from those populations. There have also been a few December records: one at Aberporth, Ceredigion, on 4 December 1973, one at Berges Island, Gower, on 26–28 December 1978 and two in 2006, one off the Great Orme, Caernarfonshire, from 26 November to 2 December, and one off Carmel Head, Anglesey, on 16–17 December. Away from the breeding colonies, counts in spring are usually lower than in autumn. However, good numbers have been recorded, for example on 8 May 2019, 500+ passed Sker Point, East Glamorgan, and there were 1,500 at Ynyslas, Ceredigion.

Some birds move overland within Wales. The species is regularly recorded at Llangorse Lake, Breconshire, with an exceptional 125 there on 4 August 2011. Counts at Llandegfedd Reservoir, Gwent, include 90 that flew south on 8 September 2009. Other inland locations have recorded smaller numbers, and these may include both British and overseas birds. The remains of one, ringed as a chick in The Netherlands on 24 June 1998, were found three months later at Abertillery, Gwent, having been taken by a bird of prey. An adult ringed in Namibia in 2006 was found dead in Newtown, Montgomeryshire, in mid-September 2010.

Away from Britain, migration follows the Atlantic coast of southwest Europe and northwest Africa. Ringing recoveries of British and Irish birds show that their main wintering area is the west coast of Africa and the Gulf of Guinea, from Senegal to Ghana. Nearly all ringed birds recovered in West Africa have been caught by humans, either boys trapping them for fun or food, or by fishermen as 'bycatch'. Significant educational programmes by the RSPB and local BirdLife partners have reduced the deliberate trapping, though some still continues. Fishermen have also been encouraged to try to remove birds caught in their nets, and some are even reporting ringed birds and releasing them alive. Welsh birds have benefited from this changed practice, notably a chick from Shotton that was reported off Guinea-Bissau in its first December and then eight months later was found 2,000km farther east off Togo. It was accidentally caught, then released, by fishermen on both occasions.

Common Terns breeding in Fennoscandia and the Baltic states perform a leapfrog migration to southern Africa, with some passing through Britain on passage. This surely accounts for a bird caught and ringed in Wales as an adult at Point of Ayr, Flintshire, on 6 September 1975 and found five years later, east of the Cape, South Africa, in the Indian Ocean. Occasionally, a small number of British or Irish birds, mostly individuals in their first calendar-year, appear to get caught up with their Fennoscandian conspecifics and spend the winter in southern Africa. Ringers in Namibia in 2005 and 2007 caught birds from the Shotton colony, and a chick ringed at Shotton in 1986 is one of only two British-ringed chicks to have been found in South Africa east of the Cape. Welsh-ringed birds have been recorded in 16 African and five European countries. The three individuals mentioned above are the only foreign-ringed Common Terns recorded in Wales.

The Amber conservation status of the Common Tern in Wales is because of a moderate long-term range decline and its highly localised breeding distribution (Johnstone and Bladwell 2016). Common Terns at Welsh colonies are among the most productive in the UK, regularly fledging more than one chick per pair (JNCC 2020b). Liverpool Bay is a particularly important spawning area for sandeels and other fish eaten by terns. These stocks may be vulnerable to climate change (Green 2017), but this species has a broader diet than many other tern species and so is less affected by changes in prey availability.

The Common Tern is among the longest-lived birds. The biggest threats come from hazards that jeopardise the survival of adults or prevent them from nesting successfully. The principal risks, including marine pollution and overfishing, are encountered on migration and in the winter quarters. Breeding colonies are vulnerable to depredation by mammals, particularly Brown Rat, Otter, European Mink and Fox, and large gulls. They are also vulnerable to competition for space from other nesting birds, including gulls and other tern species, and weather events such as storms, particularly if they coincide with high tides. Fluctuations in annual productivity reflect these pressures. The limited number of breeding colonies in Wales remains a concern. The temporary collapse of the Shotton colony during 2009–13 and the reduction of the Cemlyn colony, from 194 pairs in 2012 to ten pairs in 2019, demonstrate their vulnerability. Formerly, birds would respond by moving to new sites, but the lack of beaches free from disturbance, and with suitable adjacent habitat, provides birds with few other options.

Peter Coffey

Sponsored by Merseyside Ringing Group

Arctic Tern *Sterna paradisaea* Môr-wennol y Gogledd

Welsh List Category	IUCN Red List			Birds of Conservation Concern			
	Global	Europe	GB	UK	Wales		
A	LC	LC	VU	Amber	2002	2010	2016

The Arctic Tern, as its name suggests, is very much a northern species. It has a circumpolar breeding range, with the bulk of the population breeding north of the Arctic Circle. Many colonies in the High Arctic are inaccessible to counters, so the global breeding total is poorly known. This species winters far to the south, many birds reaching the edge of the Antarctic pack ice (*HBW*). Birds nesting in The Netherlands have been shown to cover an average annual round-trip distance of about 90,000km (Fijn *et al.* 2013). Another study found a British bird covering a total of 96,000km, flying down to the pack ice of the Weddell Sea, Antarctica. It can spend a third of the annual cycle there, feeding on krill (Redfern and Bevan 2019). This is the longest known annual migration of any animal on earth (Alerstam *et al.* 2019) and this species spends more time in daylight than any other animal (*BWP*). Birds ringed as nestlings on Anglesey have been recovered in a range of West African countries. Two have been recovered in South Africa, while one, ringed at Valley, Anglesey, in June 1966, was picked up dead in New South Wales, Australia, 17,314km away, in December the same year.

The Arctic Terns breeding in Wales are close to the southern edge of the breeding range of this species. This is the commonest breeding tern species in Wales, but colonies are restricted to just a few sites on the coast of Anglesey. The Arctic Tern favours low-lying, sparsely vegetated, rocky islets for nesting, safe from the threat of ground predators. The first birds arrive back in Welsh waters from mid-April, but most do not arrive until May.

There are very few records of breeding in Wales away from Anglesey. Up to 12 pairs bred with Common Terns at Morfa Dinlle, Caernarfonshire, between 1915 and 1950, and a couple of pairs have occasionally attempted to breed within the Common Tern colony at Shotton, Flintshire. On Anglesey, monitoring over many decades has highlighted the boom-and-bust nature of tern colonies. Eyton (1838) stated that the species bred on The Skerries, off northern Anglesey, and this remains the main breeding site for Arctic Terns in Wales. At the turn of the 20th century, there were two colonies on Anglesey, on The Skerries and at Rhoscolyn (Forrest 1907). Numbers at The Skerries have varied hugely, with up to 10,000 pairs in 1908 (Jones and Whalley 2004), at the time perhaps the largest colony in Britain. In 1952, eggs were laid but all the terns deserted the colony. There were no records of breeding there between 1962 and 1979, when a few pairs recolonised. There were 200 pairs in 1981 and about 150 pairs in 1983, when the RSPB established annual wardening on the islets, which are owned by Trinity House. Numbers increased steadily to be of British importance, with a recent peak of 3,833 pairs in 2014, attaining a level not seen for a century and making this the largest colony in Britain. Numbers subsequently dipped, and in 2020, when there were no wardens on site owing to the Covid-19 pandemic, the colony was deserted. Disturbance by Peregrines is thought to have been the main reason for this.

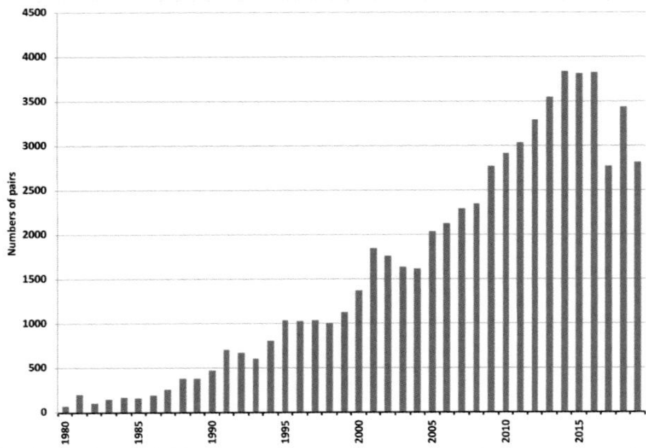

Arctic Tern: Numbers of pairs breeding on The Skerries, Anglesey, 1980–2019.

Arctic Tern © Bob Garrett

There were 200–300 pairs in the Rhoscolyn colony between 1884 and 1910, but numbers had declined to "a few" by 1925. At several other sites, Arctic Terns have bred among Common Tern colonies. For example, up to 75–100 pairs nested at Llanddwyn Island during 1909–24. Occasional pairs have attempted to breed with limited success, at sites such as Dulas in 1909 and 1953, Abermenai/Newborough Warren in 1915 and 1951, and Ynys Gorad Goch in the Menai Strait in 1959.

There have been three other major colonies on Anglesey, at least two of which are still in existence. The largest colony in the 1970s was in the Inland Sea, between mainland Anglesey and Holy Island, but numbers there had declined considerably by 1977. The very close proximity of a refuse tip led to heavy predation by rats, gulls and other species, and it is thought that many of the birds previously breeding there had moved to Cemlyn (Williams 2000a). There were six apparently occupied nests (AON) at this site in 2012, but none have been reported since. Cemlyn Lagoon, a site managed by North Wales Wildlife Trust, held over 100 pairs in 1972 and 255 pairs in 1982, but fortunes have fluctuated since, with only 2–3 pairs in the late 1990s. This number rose to 64 pairs in 2008, then fell again to 25 pairs in 2019. At Cemlyn, the species has to compete with Black-headed Gulls and Sandwich Terns in a mixed colony. Competition for space pushes them out to the edges of the islands, where they are vulnerable to bad weather and predation. The other main site, in southern Anglesey, is privately owned and is wardened to protect the terns from people and predators. The site has had problems with Brown Rats in the past, with some adults being taken, and in 1978 all the eggs and chicks were predated. Nonetheless, numbers increased from an average 215 pairs in the 1990s to an average of 475 pairs in the 21st century with a recent peak of 635 pairs in 2010.

Together, The Skerries, Cemlyn Lagoon and the privately owned site form the Anglesey Terns Special Protection Area, designated under European legislation and affording protection from future developments. Designating all the tern colonies in a single protection area acknowledges that the species operates in a meta-population: multiple colonies that interact over time with breeding adults moving between colonies. This is true of the Irish Sea population as a whole. Declines at one site can result in increases at another. Annual variations at a single site can be considerable. Longer-term population-wide trends are more important for assessing the species' fortunes.

Operation Seafarer 1969–70	SCR Census 1985–88	Seabird 2000 1998–2002	Seabirds Count Provisional 2015–19
436	732	1,705	3,994

Arctic Tern: Total counts of apparently occupied nests (AON) recorded in Wales during UK seabird censuses.

With the main risks to breeding productivity being food availability and predation pressure, research has been undertaken to understand the role of these factors. Food provisioning to chicks has been studied at a couple of the Anglesey colonies, including The Skerries where, during 1997–99, sandeels made up 90% of the diet, with clupeids (young Sprat and Herring) making up most of the rest (Newton and Crowe 2000). Elsewhere on Anglesey, Sandeels represented a smaller percentage of the food brought in for chicks. Small crustacea and even squid will also be provided on occasion, but are not the preferred prey, as they are much less energy rich in terms of protein and fat (Massias and Becker 1990). Breeding birds from The Skerries have been tracked flying an average distance of 8.1km to collect food (Perrow *et al.* 2010), but climate change means that food stocks could decline or shift to other locations, potentially requiring longer flights or more time spent fishing.

Novel techniques are being tested at various colonies to improve the fledging rate. One example, on the Farne Islands, was that canes were placed at angles to deter gulls from landing near nests and eating eggs or chicks. This reduced the number of predation attempts by half (Boothby *et al.* 2019). Another method used a laser to deter crows and gulls from flying into the colony, with varying degrees of success, depending on the location and target.

Passage birds are sometimes seen in good numbers on the Welsh coast. An exceptional spring passage was noted in Glamorgan between 21 April and 6 May 1947, when parties of up to 45 were seen almost daily over every stretch of water from the Rhymney Valley to Port Talbot (Hurford and Lansdown 1995). More recently, 1,100 passed Sker Point, East Glamorgan, in 30 minutes on 26 April 1995, a Welsh record away from the breeding colonies. There were *c.*700 off Goldcliff, Gwent, and over 150 at the Rhymney Estuary, East Glamorgan, on 3 May 2004, and 220 at Ynyslas, Ceredigion, on 3 May 2011. Departure from breeding colonies takes place largely during August, once the young have fledged. Arctic Terns are frequently seen from headlands during late summer and early autumn, but distinguishing them from the very similar Common Tern at long range is difficult. Many records are aggregated as 'Commic Tern', but in recent years there have been more counts specifically identified as Arctic Tern. 2018 was an excellent year for autumn passage, with 470 past Point Lynas, Anglesey, on both 29 August and 16 September, 440 past Cemlyn on 14 September and an island record of 765 past Bardsey, Caernarfonshire, on 19 September. Tardy individuals can occur into November. There have been a few winter records: a juvenile off Bardsey in 2009 remained from 25 November into December, and a second-calendar-year bird was on Sandy Water Park, Carmarthenshire, on 5 January 1992 following southwesterly gales. The species is fairly regular inland on passage, particularly at Llangorse Lake, Breconshire, but usually only in small numbers. Large counts in the inland counties include 17 at Llangorse on 24 April 2001 and 13 on Llyn Heilyn, Radnorshire, on 25 April 2012.

The Red-listed conservation status of this species in Wales is because of a long-term 60% reduction in breeding range and the high degree of localised breeding (Johnstone and Bladwell 2016). The last UK population estimate was 53,380 pairs in 2000. When the final results of the Seabirds Count 2015–21 census are published, it is anticipated that they will show that the Welsh population is now of greater importance in a UK context than in 2000, as both the Scottish and English populations have experienced declines in recent years, linked to food availability (Green 2017, JNCC 2020b). The European population is estimated to be over 564,000 pairs but is predicted to decrease by 25% in the next three tern generations (BirdLife International 2015). The availability of food is a crucial factor for the Welsh population. This could be affected by climate change, both around Anglesey and on its Antarctic wintering grounds. Modelling suggested that suitable climatic conditions for Arctic Terns will contract north during the 21st century. Within Britain, only the far north of Scotland may accommodate the species (Huntley *et al.* 2007). In 2016, a total of 820 adults and young birds died from botulism on The Skerries late in the season. Although the Arctic Tern has been doing well in Wales in recent years, the breeding population remains vulnerable.

Henry Cook

Forster's Tern *Sterna forsteri* Môr-wennol Forster

Welsh List Category	IUCN Red List Global	Rarity recording
A	LC	BBRC

The Forster's Tern breeds in the interior of North America and winters around the Gulf of Mexico, and the Caribbean and Pacific coasts of Central America (*HBW*). Only 21 individuals were recorded in Britain up to the end of 2019, with the last in 2016 (Holt *et al.* 2017). Of these, five, or possibly six, birds were recorded in Wales. Once birds have crossed the Atlantic, they sometimes return to the same sites over several years, making it difficult to determine the number of individuals involved.

- 2CY at Point of Ayr, Flintshire, from 3 July to 6 August 1984 and later at Penmon, Anglesey, from 30 September to 20 October. There was no record in 1985, but an adult in 1986 was thought to be the same bird as in 1984. It was recorded at Holyhead, Anglesey, from 19 January to 6 February, at Point of Ayr, Flintshire, from 27 September to 7 October, at Abergele, Denbighshire, on 7 October and at Penmon, Anglesey, from 22 October to 23 November.
- One was at Gronant, Flintshire, from 20 August 1987, and another joined it there on 31 August; both birds remained there until 20 September. One of these was presumably the bird from 1986, the other a new arrival. Single birds, thought to be one or other of these two, were seen off the Great Orme, Caernarfonshire, on 23 August and at Penmon, Anglesey, on 17 October, then at Holyhead from 10 January to 26 February 1988, at Penmon on 15–16 October and back at Holyhead from 21 January to 1 March 1989.
- Fishguard Harbour, Pembrokeshire, on 10–11 January 1994;
- Bangor Harbour, Caernarfonshire, on 20–24 January 1995 and at Foryd, Caernarfonshire, on 3–9 February, which may have been one of the birds seen during 1987–89;
- 2CY at Bangor Harbour and the Menai Strait, Caernarfonshire/Anglesey on 2–18 December 2000 and
- The Gann Estuary, Pembrokeshire, on 4 December 2000.

Jon Green and Robin Sandham

Whiskered Tern *Chlidonias hybrida* Corswennol Farfog

Welsh List Category	IUCN Red List		Rarity recording
	Global	Europe	
A	LC	LC	BBRC

The Whiskered Tern is widely distributed across Eurasia and southern Africa, but in Europe breeds in small, scattered colonies, as close to Britain as southern Brittany (*HBW*). The Volga/Ural River complex holds most of the European population, but it has undergone considerable population expansion since the 1960s. The eastern and western European populations are genetically distinct, the latter wintering in tropical West Africa (Dayton *et al.* 2017). A total of 227 were recorded in Britain up to the end of 2019, about four or five a year on average over the past 30 years (Holt *et al.* 2020). It is much more regular in southern and eastern England than elsewhere in Britain (Parkin and Knox 2010).

There have been 13 records of 14 birds in Wales, all between 21 April and 30 September. There is a distinct easterly spread of records in Wales, and at least half the records have been on freshwater.

- Llanbwchllyn, Radnorshire on 21–22 April 1956;
- adult at Glandyfi, Ceredigion, on 11 July 1965;
- two adults at Blackpill, Gower, on 13 May 1974;
- adult at Eglwys Nunydd Reservoir, Gower, on 7 September 1974;
- adult at Shotton, Flintshire, on 5 July 1976;
- Llyn Traffwll, Anglesey, on 19 June 1993, then at Cemlyn, Anglesey, on 20–23 June;
- adult at Llandegfedd Reservoir, Gwent, on 15–18 July 1994;
- adult at Llyn Maelog, Anglesey, on 3 August 2002;
- Kenfig, East Glamorgan, on 4 May 2008;
- RSPB Conwy, Denbighshire, on 9 May 2008, and the same bird at RSPB Burton Mere Wetlands on 9–20 May, also feeding over the adjacent Shotwick Fields, Flintshire;
- Llyn Login, Breconshire, on 23 May 2008, then at Llangorse Lake, Breconshire, on 27–28 May;
- 1CY at RSPB Burton Mere, Flintshire, on 18 September 2010 and
- 1CY at Eglwys Nunydd Reservoir, Gower, on 25–30 September 2010.

Jon Green and Robin Sandham

White-winged Black Tern *Chlidonias leucopterus* Corswennol Adeinwen

Welsh List Category	IUCN Red List		Rarity recording
	Global	Europe	
A	LC	LC	WBRC

This species breeds mainly inland, on freshwater lakes and marshes in eastern Europe and Siberia. The Western Palearctic population winters in Africa, mostly south of the Sahara (*HBW*). Just over 1,000 were recorded in Britain between 1958 and 2018 (White and Kehoe 2020a), with numbers stable over the last 30 years, an average of 18 a year. Wales, however, records very few of these marsh terns, with an average of about six per decade over the last 70 years.

Total	Pre-1950	1950–59	1960–69	1970–79	1980–89	1990–99	2000–09	2010–19
44	3	1	4	14	3	9	4	6

White-winged Black Tern: Individuals recorded in Wales by decade.

The first Welsh record was of two adults, one of which was shot, at Cardiff, East Glamorgan, on 25 May 1891. This was followed by one, also shot, at Llangorse Lake, Breconshire, on 23 August 1910. Since then, a further 41 individuals have been recorded in Wales.

An adult at the Gwendraeth Estuary, Carmarthenshire, on 8 March 1958 was the earliest record in Wales. The latest was a first-calendar-year bird at Llys-y-frân Reservoir, Pembrokeshire, on 16–28 October 1976. The county map shows that most

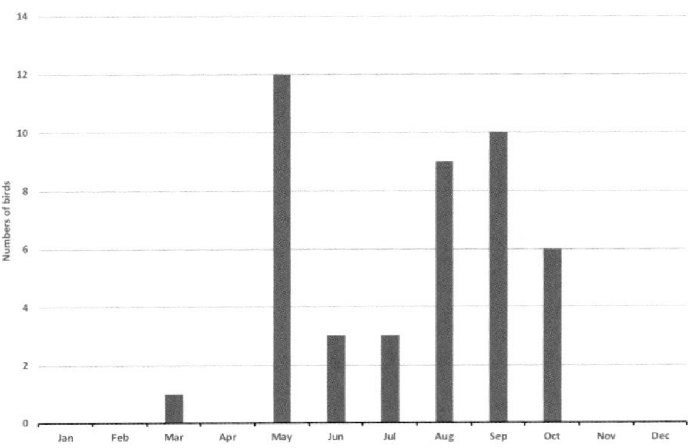

White-winged Black Tern: Totals by month of arrival 1891–2019.

White-winged Black Tern: Numbers of birds in each recording area 1891–2019. One bird was seen in more than one recording area.

records are from South Wales: six have been recorded at Kenfig, East Glamorgan, including three in summer plumage on 18 May 1992, and all five of the Gower records were at Eglwys Nunydd Reservoir. Three of the seven recorded in Pembrokeshire were seen during sea-watches off Strumble Head, while the three records in Breconshire were all at Llangorse Lake. In North Wales, the Anglesey records come from various lakes and pools, while in Flintshire there were three at Shotton and two each at Shotwick and RSPB Burton Mere.

Jon Green and Robin Sandham

Black Tern *Chlidonias niger* Corswennol Ddu

Welsh List Category	IUCN Red List			Birds of Conservation Concern			
	Global	Europe	GB	UK	Wales		
A	LC	LC	RE	Green	2002	2010	2016

The nominate form of Black Tern breeds locally over much of Europe and Central Asia and winters mainly on tropical West African coasts and rivers (*HBW*). The subspecies *C.n. surinamensis* breeds in North America and winters off the coasts of Central and South America. There are no records of this form in Wales, but with several records in England in the 21st century, the possibility should not be discounted.

The global population of this species is thought to be declining, although the European population trend is unknown (BirdLife International 2020). Results from the second *European Atlas* indicate substantial losses in breeding range in western Europe since the 1997 *Atlas*, though there appears to be no substantial change in the core breeding area further east in Belarus, Ukraine and Russia (*European Atlas* 2020). The Black Tern breeds on well-vegetated inland pools and lakes, lowland peat bogs, marshes and sometimes rice fields (*HBW*). It has never been recorded breeding in Wales, probably because of the shortage of suitable breeding habitat, but it bred in considerable numbers in eastern England until the middle of the 19th century. There have been only a few instances of subsequent breeding in Britain, although there was an unexpected breeding record in Ireland in 1967 (Cottier and Lea 1969).

In Wales, the Black Tern is an uncommon passage migrant, usually in spring and autumn. Numbers are normally highest in the post-breeding period, peaking in late August, when the population is bolstered by juveniles and records can be quite widespread. Most records come from the coast, where Black Terns are often seen roosting with other terns, or seen on active migration during sea-watches. Numbers vary considerably from year to year, largely depending on weather conditions. Flocks can occur on inland water bodies, peaking in May, when birds are in breeding plumage. These spring inland records may be associated with easterly winds and thunderstorms.

Although this species was mentioned in the *Swansea Guide* (Anon. 1802), there were few records from the 19th century or the early 20th century. Phillips (1899) stated that two had been shot at Llangorse Lake, Breconshire, in 1889. Mathew (1894) said that the Black Tern was "seen occasionally on its passage in the autumn" in Pembrokeshire, but quoted only two undated records. Lockley *et al.* (1949) were only able to add one more record from the county, in 1904. Forrest (1907) described it as a "somewhat rare passing migrant" in North Wales, although he thought it likely to often pass unnoticed. The first Caernarfonshire record was in 1909, but 47 years passed before the next in 1956. By the 1950s, Black Terns were seen more frequently. Records in Glamorgan were sporadic in the first half of the 20th century but from 1954 it was recorded in the county every year (Heathcote *et al.* 1967).

Small flocks are now an annual occurrence, often appearing at a number of sites concurrently in spring between April and June. There are influxes in some years, with counts of 20–40 recorded, particularly in the south. The largest spring count on record in Wales is of 286 off Black Rock, Gwent, on 2 May 1990. 70 were at the Rhymney Estuary, East Glamorgan, on the same date, both part of a massive movement of Black Terns in southern Britain.

Autumn passage is between late July and early November. The largest numbers are usually recorded passing headlands, particularly Strumble Head, Pembrokeshire. A remarkable 536 passed there on 31 August 2005, with 74 recorded inland on Llangorse Lake on the same date. Other large counts past Strumble Head include 177 on 18 August 2012 and 105 on 23 August 2015. Counts at other sea-watching points are usually lower, but there were 70 past Point Lynas, Anglesey, on 20 September 2000 and 69 at Point of Ayr, Flintshire, on 27 August 1997. There are almost annual records at Llangorse Lake, with large counts including 75 on 11 August 1969 and at least 100 on 30 August 2008. Eglwys Nunydd Reservoir, Gower, held 77 birds on 6 September 1974. A late bird was seen at RSPB Ynys-hir, Ceredigion, on 24 November 1994. There was an exceptionally late bird on the Dyfi Estuary, Ceredigion/Meirionnydd, during a gale on 17 December 1989.

We do not know where the birds passing through Wales breed. This makes it difficult to assess future prospects for the Black Tern as a passage migrant. Up to 80,000 birds congregate at the IJsselmeer, The Netherlands, in July and August, undergoing a partial moult before moving south through the English Channel from late August. These birds are thought to be drawn from the whole Western Palearctic population (Cabot and Nisbet 2013). The birds seen in Wales may be birds deflected by weather conditions from the main migration route.

Henry Cook

South Polar Skua/Brown Skua *Stercorarius maccormicki/antarcticus* Sgiwen Pegwn y De/Sgiwen Frown

Both these species have a circumpolar distribution, Brown Skua nesting on sub-Antarctic islands and South Polar Skua on the Antarctic continental shelf. Both are highly pelagic outside the breeding season in the Southern Oceans (*HBW*). South Polar Skuas have been satellite tracked regularly into the North Atlantic and there are sight records off the Azores and Canary Islands, as well as one shot in the Faroe Islands in 1889 (Mitchell 2018).

An exhausted skua found in a Port Talbot supermarket car park, Gower, on 1 February 2002, was taken into care at Kenfig, East Glamorgan, rehabilitated and released on 16 February. DNA samples, taken from feathers dropped while in captivity, were initially thought to indicate that this bird, and another recorded on the Isles of Scilly in January 2002, were Brown Skua *S. antarcticus*. However, further work suggested that although both birds belonged to the "southern" skua group, it was not possible to assign them with certainty to either *S. antarcticus* or *S. maccormicki*. The "southern" skua group also includes Chilean Skua *S. chilensis*, but this could be excluded on plumage characteristics (Votier *et al.* 2007). The Gower record is currently accepted by BBRC as *Stercorarius antarcticus/chilensis/maccormicki*. It may later be possible to identify this bird to species.

Jon Green and Robin Sandham

Great Skua *Stercorarius skua* Sgiwen Fawr

Welsh List Category	IUCN Red List			Birds of Conservation Concern			
	Global	Europe	GB	UK	Wales		
A	LC	LC	LC	Amber	2002	2010	2016

The Great Skua has a more restricted range and a smaller world population than our other skua species. About half the global population breeds in northern Scotland, with other populations in Iceland, Norway, the Faroe Islands and islands in the Barents Sea. Most winter around the Bay of Biscay, although young birds move farther south and west. Numbers have increased greatly since 1900, when the populations in Scotland and the Faroe Islands were close to extinction (*HBW*). The latest estimate of the global population is 16,000–17,000 breeding pairs and it is considered to be stable (BirdLife International 2020). Declines in Shetland, thought to relate to sandeels becoming less available in the North Sea (Furness *et al.* 2018), have been offset by increases elsewhere. Ringing recoveries in Wales are of birds ringed as nestlings in Scotland (ten, of which eight were from Shetland), Iceland (four) and the Faroe Islands (two).

As with the other skuas, there were few records during the late 19th century but, unlike the others, this species was probably a genuinely rare bird off the Welsh coasts in that period. In Pembrokeshire, Mathew (1894) quoted Sir Hugh Owen as saying that it was "always to be seen in Goodwick Bay in a good herring season", and there was a record of one shot in Solva harbour in 1894. Other records came from Glamorgan, where the species occurred off the Gower coast on occasions, with singles seen in Cardiff and Barry in 1883. In Ceredigion, Salter noted in his diaries that the local taxidermist, Hutchings, had only ever received one specimen, and reported another "probably seen at Aberystwyth in 1905" (Roderick and Davis 2010). Forrest (1907) was aware of a few records in North Wales, but regarded all of them as doubtful. There were few sightings in the early part of the 20th century. One off Holyhead in July 1903 was the first for Anglesey. There were several Caernarfonshire records in the following decade; R.W. Jones recorded one in the Conwy Estuary in August 1912 and two off the Great Orme in September 1915. He felt sure that the species occurred more frequently than was supposed. Off the coast of Anglesey, there were two in the Menai Strait in November 1912 and further autumn records of singles off Beaumaris in 1927 and Holyhead in 1931, and two off Penmon in August 1935. Off Glamorgan, there was one in 1915, then no records until 1926, with a total of seven records between that year and 1954. The first record for Carmarthenshire was one found on the beach at Pendine in January 1939. The following year saw the first dated record for Ceredigion, one shot at Aberystwyth harbour in November. There had been a few early doubtful records off Meirionnydd, but the first confirmed record was not until 1951. After the 1894 bird, there was no further record in Pembrokeshire until 1955. The first documented record for Gwent was one at Peterstone in September 1953 (Venables *et al.* 2008).

The population in Scotland had increased and extended its distribution by the 1960s (*The Birds of Scotland* 2007). Numbers seen off the Welsh coast increased concurrently. A total of 17 past Bardsey, Caernarfonshire, in autumn 1967 was described as "unprecedented", while 13 past the island on 2 October 1970 was noted as "remarkable" (Jones and Dare 1976). Numbers increased further through the 1970s and 1980s, with greater interest in sea-watching. Strumble Head, Pembrokeshire, emerged as the main site for the species in Wales. Records of this species there were as numerous as those of Arctic Skua (Donovan and Rees 1994).

Passage past Strumble Head starts in mid-summer, increases through August to peak in September, then drops off in October. Annual totals vary depending on the occurrence of autumn storms, from as few as 53 in 1993 to 503 in 1983, that included

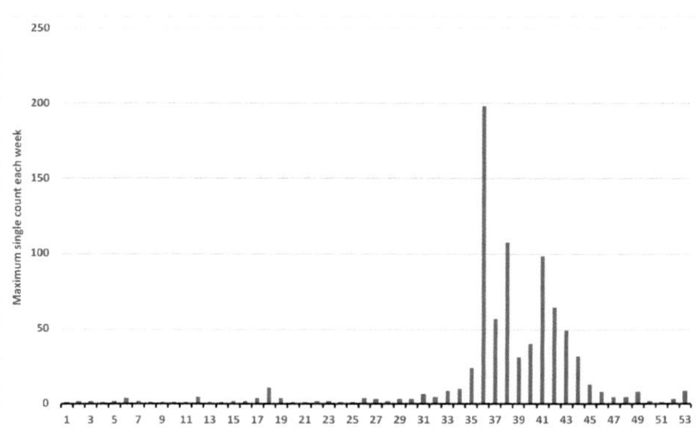

Great Skua: Maximum single counts, each week, of birds passing around Wales, 1955–2019. Week 14 begins *c.*2 April, week 36 begins *c.*3 September and week 45 begins *c.*5 November. Records are from BirdTrack.

a record day total of 198 on 3 September. Greater numbers were logged in the 1980s than in the subsequent two decades, helped by totals of 501 in 1983 and 472 in 1985. There was an increase in 2010–19, including 481 in 2011. The peak during mid-September in this period was caused by large counts of 86 on 15 September 2010 and three days with over 40 in September 2011.

Bardsey had its best ever autumn in 2017, with 265 birds recorded, including a record day total of 57 on 11 September. 85 passed Point Lynas, Anglesey, in autumn the same year. Numbers peak in September and October, although some passage continues into November. The bulk of records are from Pembrokeshire, Caernarfonshire and Anglesey, but all coastal counties have recorded some birds, with the largest counts including 10–12 past Point of Ayr, Flintshire, in autumn 1997 and up to ten off Goldcliff Point, Gwent, on 18 November 2015.

There are regular December records and a smaller number in January and February, which are presumably birds wintering in the Celtic and Irish Seas. Gales in December 1997 resulted in an unusual number of records. There were good numbers in the southeast in similar weather in December 2006, including 16 off Cardiff Heliport, East Glamorgan, on 8 December. Spring passage, from late March to mid-June, is generally on a much smaller scale than in autumn, although some years are the exception. In 2006, for example, 21 were recorded in Wales between 27 March and 22 June. There were 14 off The Range, Anglesey, on 5 May 2005 and 11 there on 28 April 2019.

Inland records are very rare, as this species seems to avoid migrating over land. One picked up near Bettws Cedewain, Montgomeryshire, on 23 May 1973 had been ringed in Iceland. Fresh remains of a Great Skua were found at Llandegfedd Reservoir, Gwent, on 16 October 1977. The first record for Breconshire was one on Llangorse Lake on 3–5 March 2003. The only Radnorshire record was one picked up exhausted at Llandegley in December 1989. After eating two tins of sardines and half a pound of Sprats, it was released on the Ceredigion coast two days later (Jennings 2014).

The decrease in the Scottish population, and the increase on islands to the north, is in line with the climate modelling by Huntley *et al.* (2007) that forecasts the species' loss from Scotland by the end of this century. Providing that the Icelandic population is able to increase in parallel, there may not be a noticeable difference in passage numbers recorded off the Welsh coast.

Rhion Pritchard

Pomarine Skua *Stercorarius pomarinus* Sgiwen Frech

Welsh List Category	IUCN Red List		Birds of Conservation Concern			
	Global	Europe	UK	Wales		
				2002	2010	2016
A	LC	LC	Green			

The Pomarine Skua breeds in the far north, on the Arctic tundra of North America and Russia. It is a trans-equatorial migrant, mostly wintering between the Tropics of Cancer and Capricorn, and along the coasts of Australia and Argentina (*HBW*). The world population, estimated at 400,000 pairs, is thought to be stable (BirdLife International 2020), although the number of pairs breeding in any one year depends on the abundance of lemmings. Fox and Aspinall (1987) observed that the occurrence of inland Pomarine Skuas in Britain, particularly of juveniles, coincided with good lemming years in Arctic Russia, and that a greater number of Pomarine Skuas were recorded wintering off the British coast in the winters following good lemming years. There are no ringing recoveries to indicate the breeding areas of the birds seen off the Welsh coast.

Pomarine Skuas are usually recorded in Wales as passage migrants in spring and autumn, although in smaller numbers than along North Sea coasts of Britain. Identification uncertainties make it difficult to be confident about its status prior to the 1960s. In Pembrokeshire, Mathew (1894) stated that it was "by far the commonest of the family upon our coasts and a few are to be seen every autumn", but Lockley *et al.* (1949) knew of only four having been recorded in the county, and thought that Mathew had confused this species with Arctic Skua. Barker (1905) knew of only one record from Carmarthenshire. Forrest (1907) regarded it as an occasional visitor to the coasts of North Wales, but Salter thought that many of the skuas seen after autumn gales in Ceredigion at the end of the 19th century were this species (Roderick and Davis 2010). There was a record of nine off Criccieth, Caernarfonshire, in late May 1925 following gales.

During 1967–75 fewer than three a year were recorded in Wales on average, but the numbers recorded annually increased subsequently, with the growing interest in sea-watching, although the totals each year depend on weather conditions. The highest numbers are seen when strong winds push birds closer inshore.

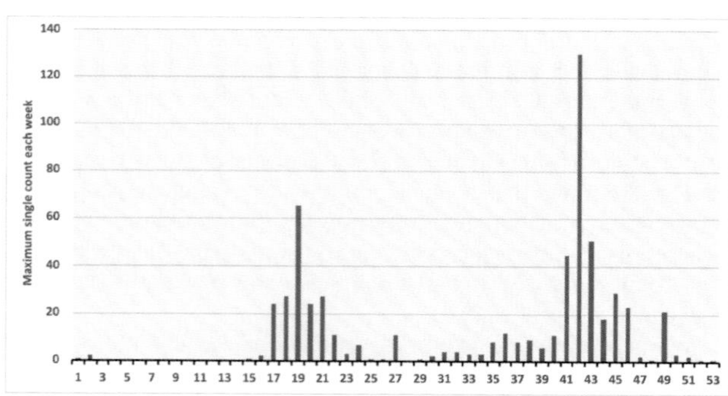

Pomarine Skua: Maximum single counts, each week, of birds passing around Wales, 1966–2019. Week 19 begins *c.*7 May, week 41 begins *c.*8 October and week 49 begins *c.*3 December. Records are from BirdTrack.

From the late 1970s it became clear that good numbers passed Welsh headlands in autumn in some years. The main site for the species in autumn is Strumble Head, Pembrokeshire. Donovan and Rees (1994) suggested that the path of birds flying south through the Irish Sea is blocked by the north coast of Pembrokeshire, which they follow until able to head out to sea again. Annual numbers at Strumble Head reflect the occurrence of autumn storms, from just seven individuals in 2002 to 306 in 1991. This included 97 on 17 October and 130 the following day, a record for Wales. Occasional individuals pass as early as late July, but most are between the end of August and early November, peaking in October–November, although a few have been seen to the end of December. Other sea-watching points where good numbers have been recorded in autumn are Bardsey, Caernarfonshire, which recorded 36 on 17 October 1991 and autumn totals of 82 in 2011 and 75 in 2017, and Point Lynas, Anglesey, where 35 passed between 28 September and 4 October 2009.

Spring passage peaks from late April to early June, but the sites with the highest counts are not those that record the greatest numbers in autumn. Exceptional numbers passed South Stack, Anglesey, in spring 1983, with 125 between 8 and 20 May, and nearby, 27 passed The Range on 28 April 2019. Another important site is around Criccieth, Caernarfonshire, as strong southwesterly winds between late April and early June can push large numbers into Tremadog Bay. Counts there include an estimated 65 on 24 May 2002, with *c.*30 present the following day, 30 on 4 May 2004 and at least 35, possibly 65, on 10 May 2013. An accurate count can be difficult there, as birds circle in the bay.

Other coastal counties record smaller numbers, but there have been some notable records, including 11 off Peterstone, Gwent, on 9 May 1997, 17 off Mumbles, Gower, on 12 June 1993, ten off Pembrey, Carmarthenshire, on 23 May 1994 and 11 off Aberdysynni, Meirionnydd, on 20 April 1987. Inland records are not common. Forrest (1907) mentioned an immature bird shot at Llanidloes, Montgomeryshire, in October 1895. Jennings (2014) referred to a bird exhibited in 1895, said to have been from Radnorshire, just on the border with Herefordshire. Both were stated to be in the collection of Mr A. Gwynne Vaughan, and it is uncertain whether they referred to different individuals. The first record for Breconshire was in 2007, when one was at Mynydd Llangatwg on 5 November, having been seen earlier at Garnlydan Reservoir, Gwent.

After Great Skua, Pomarine Skua is the second most recorded skua in Britain during the winter, although the numbers involved are small. Passage may continue into December in some years, but during 2000–19 there were only 14 Welsh records in January or February, perhaps involving only 11 individuals. There are also a few records in March, which may be early spring migrants. It is unlikely that more than a handful winter anywhere near Welsh waters.

Since the number of birds recorded depends so much on suitable weather conditions, breeding success earlier that year and the intensity of sea-watching, it is not possible to identify trends for Wales and there are no obvious conservation issues.

Rhion Pritchard

Arctic Skua *Stercorarius parasiticus* Sgiwen y Gogledd

Welsh List Category	IUCN Red List			Birds of Conservation Concern			
	Global	Europe	GB	UK	Wales		
A	LC	LC	CR	Red	2002	2010	2016

The Arctic Skua breeds mainly on coastal tundra in the Arctic. It has a circumpolar distribution, including about 2,000 pairs in northern and western Scotland, most in Shetland and Orkney (*HBW*). The world population is estimated to be around 200,000–300,000 pairs and is thought to be stable (BirdLife International 2020). Almost the entire population winters in the southern hemisphere, around the coasts of South America, southern Africa and Australasia. The Arctic Skua is more inclined to migrate over land than our other skua species (*HBW*). There appear to be no ringing recoveries involving Wales, but birds migrating along the coasts of Britain are thought to include both birds breeding in Scotland and elsewhere in northern Europe (Furness 1987).

There were comparatively few Welsh records in the late 19th and early 20th centuries. Mathew (1894) said that it was scarcer on the southwest coasts of Pembrokeshire than Pomarine Skua, but that "a season rarely passes without one or two being noticed". The only specific record mentioned was one shot at Goodwick, with no date given. In Carmarthenshire, Barker (1905) recorded one shot at Ferryside, about 1880. The first dated records for East Glamorgan were two shot on the mudflats of east Cardiff in 1895 (Hurford and Lansdown 1995). There were several early records from Ceredigion, where it was considered an occasional autumn visitor in this period. In North Wales, in contrast to Pembrokeshire, Forrest (1907) said that although by no means common it was seen more frequently in the west than any other skua. It had not been recorded on the north coast, except at the Dee Estuary, where an adult and an immature were shot in 1903. He knew of two inland records from Montgomeryshire. Another inland county, Breconshire, had its first record in 1903. There was no great change in the number of records during the first half of the 20th century, although 44 off the Great Orme, Caernarfonshire, on 19 September 1919, was notable. Lockley *et al.* (1949) said that the species was occasionally seen in Pembrokeshire off Skokholm and Grassholm, usually in August and September, but sometimes in May and June. The first record for Gwent was a bird found at Abergavenny after gales in the autumn of 1946, while the first dated record in Gower was one at Whiteford Point on 25 May 1958. Numbers recorded in Wales began to increase from the 1960s, particularly on autumn passage. In 1963, hurricane-force winds resulted in unprecedented numbers in Glamorgan, with 80 past Lavernock Point between 25 September and 21 October, including 26 on 7 October. A total of 38 passed Bardsey, Caernarfonshire, on 13–15 September 1966. Increased numbers were recorded off the island from about 1967, at least partly due to better identification skills (Roberts 1985). Records off Pembrokeshire never involved more than four birds until ten passed Strumble Head on 11 September 1971, followed by 50 on 8 September 1974 (Donovan and Rees 1994). Strumble has proved to be the best site in Wales for this species in autumn. Between 1980 and 2006, 67% of records were adults. The remainder were predominantly first-calendar-year birds, although some second- and third-calendar-year birds were identified (Rees 2010a). Passage past Strumble Head ran from July to December, peaking in late August and early September.

As with other species seen off headlands, the numbers recorded depend largely on wind strength and direction. The largest cumulative counts from Strumble Head were 436 in autumn 2006 and 661 in autumn 2011, with the lowest annual total of this period recorded in 1993, when there were 69. Large day totals included 103 on 3 September 1983, 84 on 6 September 2008 and 106 on 29 September 2019. Decadal means have increased slightly from 267 during 1980–89 to 305 in 2010–19.

Bardsey, Caernarfonshire, and Point Lynas, Anglesey, are also good sites for this species in autumn. Numbers have increased off Bardsey in recent years, with a site season record of 214 in autumn 2011, and a record day count of 50 on 29 September 2015. Both records were broken in 2017, with an autumn total of 455 and a day count of 70 on 11 September. At Point Lynas, 116 passed in autumn 1987 and 140 in autumn 1997, with a site record day total of 37 on 8 September 2001. In September and October 2011, an excellent year for autumn passage, 60 passed Ramsey, Pembrokeshire, and 62 passed Mwnt, Ceredigion. A county record of 106 off Ceredigion in autumn 2017 included 35 past Mwnt on 11 September.

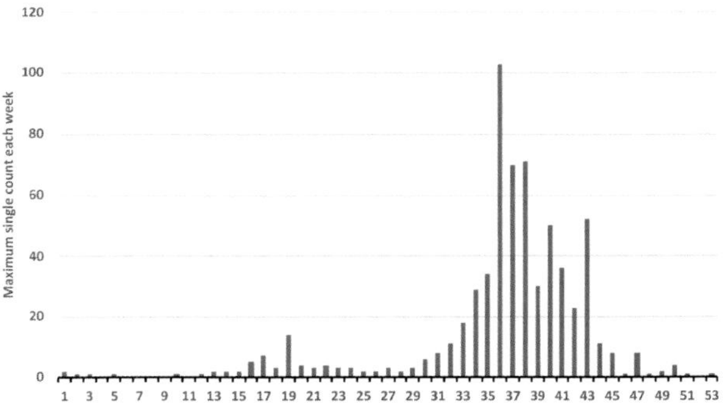

Arctic Skua: Maximum single counts, each week, of birds passing around Wales, 1939–2019. Week 33 begins *c.*13 August and week 47 begins *c.*19 November. Records are from BirdTrack.

Spring passage generally involves far fewer birds than autumn, although some sites, such as South Stack, Anglesey, sometimes have good numbers. In 1983, 30 passed north there, between 2 and 24 May, peaking at 16 on 21 May. However, around the mouth of the Severn Estuary, Gwent, spring records are more frequent than autumn (Venables *et al.* 2008). 27 were recorded in Gwent in May and June 2013, including 14 past Goldcliff Point on 11 May. Numbers recorded in Wales, on spring passage in the 21st century, have ranged from just eight in 2001, to at least 85 in 2007. The main spring passage is from late April to mid-June, peaking in May. Winter records are scarce, with birds recorded in December to February in just eight years during 2001–17. The few recorded in March may have been wintering rather than on passage. Eight were off the Welsh coast in December 2006, but no other year has produced more than three.

Inland records are more frequent than for other skua species, but not common. There was a remarkable record of a flock of 17 flying southwest over Ynysyfro Reservoir, Gwent, on 29 August 1992. Rather surprisingly, the species has still not been recorded in Radnorshire. Some overland movements have been recorded between Tremadog Bay and the north coast of Caernarfonshire, as well as across Anglesey from the Cefni Estuary to Red Wharf Bay (Nigel Brown *in litt.*).

The breeding population in Scotland is not doing well. The Arctic Skua declined more than any other seabird in the UK during 1986–2018, including a large decrease in Shetland. This is thought to be linked to the recent sandeel scarcity around Shetland, and the consequent decline in breeding numbers of several seabird species parasitised by Arctic Skuas (JNCC 2020b). If these declines continue, as forecast by climate modelling (Huntley *et al.* 2007), this species may become a less common sight in Wales.

Rhion Pritchard

Long-tailed Skua *Stercorarius longicaudus* Sgiwen Lostfain

Welsh List Category	IUCN Red List		Birds of Conservation Concern			
	Global	Europe	UK	Wales		
				2002	2010	2016
A	LC	LC	Green			

The smallest, most graceful and also the least common of the skuas regularly seen off the Welsh coast, the Long-tailed Skua is a passage migrant in spring and autumn as it moves between its breeding grounds on the Arctic tundra and wintering areas far to the south. It breeds across northernmost Eurasia and North America and winters at sea, mainly in sub-Antarctic areas. It is more pelagic than our other skua species, and seldom comes within sight of land outside the breeding season (*HBW*). The population is thought to be stable (BirdLife International 2020).

Until the increased interest in sea-watching from the mid-1970s, this species was thought to be a rare visitor to the Welsh coastline. Forrest (1907) knew of only one record in North Wales, an immature recorded by Dobie as having been shot at Foryd in the Clwyd Estuary, Denbighshire/Flintshire, in 1869. Mathew (1894) knew of only one record in Pembrokeshire, another immature shot at Tenby in autumn 1889 or 1890, but also mentioned one shot in January 1892, at Rhymney, East Glamorgan.

Up to 1991, there were 91 dated records in Wales, involving 180 individuals. There has been a great increase in the number recorded since: 725 between 1992 and 2019, of which 30 were in spring and 695 in autumn. Autumn passage can start in July, the earliest being one past Point Lynas, Anglesey, on 14 July 2005. A few stragglers are recorded in December, with the latest passing Point Lynas on 15 December 1999. The highest Welsh total in a single autumn was 83 in 2011, which included 40 past Strumble Head, Pembrokeshire, and 15 past Bardsey, Caernarfonshire. Other high autumn totals for Wales were 55 in 1999 and 49 in 2011. Strumble Head typically provides about half the total autumn records for Wales, with the highest autumn total there of 73 in 1991, including 18 on 15 September, which is the highest day count in Wales. Bardsey and Point Lynas, Anglesey, are also regular sites for records of this species in autumn. Most autumn records are juveniles. Passage peaks in September, with the numbers recorded depending on the occurrence of autumn storms.

Spring passage is much smaller, with birds only recorded in ten of the 28 years between 1992 and 2019. The only year with more than four spring records was 2013, when there were 12 in Wales, including six past The Range, Anglesey, on 12 May. The earliest in any year during this period was one past The Range on 28 April 2019. Nearly all the records in Wales have been on the coast, but an immature bird was found dead at Cwmdu, Breconshire, in August 1981.

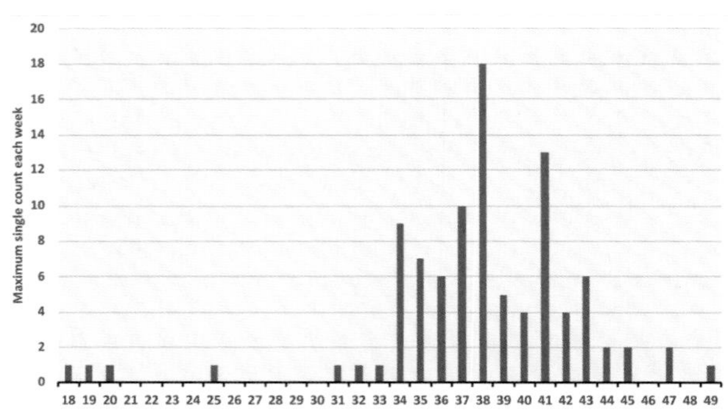

Long-tailed Skua: Maximum single counts, each week, of birds passing around Wales, 1957–2019. Week 19 begins *c.*7 May, week 34 begins *c.*20 August and week 45 begins *c.*5 November. Records are from BirdTrack.

As with many other birds mainly seen from coastal headlands, the numbers recorded depend to a great extent on weather conditions, although since a large proportion are juveniles, breeding success in that year will also have an influence. The increase in records is likely to be due to a combination of greater interest in sea-watching, improved telescopes and greater awareness of features to look for in identifying the species. There are currently no conservation issues for this species within Wales, but its occurrence here may depend on its fortunes in Scandinavia, where the area that is climatically suitable for breeding is predicted to reduce substantially during the 21st century (Huntley *et al.* 2007).

Rhion Pritchard

Little Auk *Alle alle* Carfil Bach

Welsh List Category	IUCN Red List			Birds of Conservation Concern
	Global	Europe	GB	UK
A	LC	LC	DD	Green

The Little Auk is very much a northern species, breeding mainly in the high Arctic, where it has a circumpolar distribution, from northeastern Canada to the Severnaya Zemlya archipelago in Russia. It usually breeds in crevices of coastal scree, often in large colonies. In winter, it is mainly pelagic and most birds do not move very far south, often feeding around the edge of the pack ice. It is thought to be the most abundant auk in the Atlantic, with the world population estimated at around 12 million pairs, including 10 million in northwest Greenland (*HBW*).

Wales is close to the southern limit of its winter distribution. In the late 19th and early 20th centuries it was a scarce visitor to coastal counties, with many records being of birds found dead or moribund after gales. Many seen alive were promptly shot. Hurford and Lansdown (1995) stated that 33 of the 49 recorded in Glamorgan were known to have died. Forrest (1907) knew of a few records from each county of North Wales, but regarded it as only an accidental visitor driven in by bad weather. Mathew (1894) said that it was found in very limited numbers in Pembrokeshire, and Lockley *et al.* (1949) thought that its status in that county had not changed. There were five records of singles in Glamorgan in the 19th century, the first being one found dead near Swansea in 1805. Three of these five records were in East Glamorgan and two in Gower.

Little Auks are particularly susceptible to 'wrecks', the term used when seabirds are washed up on beaches following prolonged storms, although this is far less common in the Irish Sea than in the North Sea. There was a wreck in 1877, when Forrest (1907) quoted Rawlings as stating that many were washed ashore dead at Barmouth, Meirionnydd. In February 1912, many Little Auks appeared on the coasts of Flintshire and Denbighshire, although Forrest (1919) only mentioned one at Llandudno, Caernarfonshire, that month. Another wreck in February 1950 affected mainly southwest Britain (Sergeant 1952), when ten were recorded in Glamorgan, eight in Meirionnydd, six around Aberystwyth, Ceredigion, 5–10 in Anglesey and 20–50 in Pembrokeshire, including the remains of at least 13 found on Skokholm. There was no wreck as such in 1988, but southwesterly gales in January and February produced an unusual number of records. There was another influx in winter 1990/91, when 50 were recorded in Wales, many found dead on beaches in January 1991.

The increased interest in sea-watching from the mid-1970s led to more sightings of live birds. Most records are from Pembrokeshire, Caernarfonshire and Anglesey, although birds are recorded in most years in Glamorgan. While most are seen between late September and March, a few have been recorded as late as May and June. By far the largest day count in Wales was on 2 October 1981, when severe northwesterly gales pushed 176 south past Bardsey, Caernarfonshire, with more in the following few days, including 17 on 4 October (Roberts 1985). Strong NNW winds brought 22 past Bardsey on 31 October 2001. In Pembrokeshire, a count of 21 past Strumble Head, on 20 November 1983, was exceptional, but the species has been almost annual in the county since, mainly off Strumble. Numbers recorded at Strumble usually peak in October, either indicating that this is the principal month of passage, or perhaps because of more intensive sea-watching at this time than in winter. Analysis of records in Ceredigion between 1980 and 2004 showed the peak was in January (Roderick and Davis 2010).

Inland records usually result from gales. Birds have been recorded inland in all Welsh counties, though Radnorshire did not record its first Little Auk until one was picked up alive in a Llandrindod Wells garden in December 1991. They have been found in other unusual places. In Gwent, one was found on a factory roof, at Risca, in December 1968 and one in a bramble bush at Nash in October 1977; both flew off strongly when released at Goldcliff.

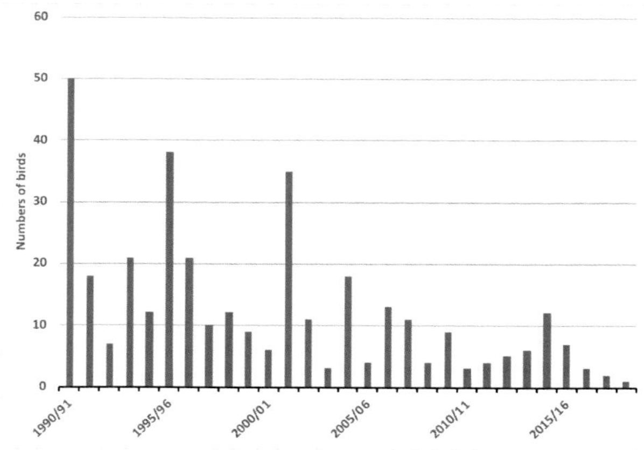

Little Auk: Numbers of birds recorded in Wales, 1990/91 to 2018/19.

The number of Little Auks seen in Wales has declined in recent years, with few winters now producing double-figure counts. This may be due to changing weather conditions or a result of fewer birds wintering as far south as Wales. Welsh waters may not be particularly suitable for wintering Little Auks. They feed largely on marine copeopods, which are larger in colder water (Leinaas *et al.* 2016). The smaller copeopods in our waters may be rather too small, even for Little Auks.

Rhion Pritchard

Common Guillemot *Uria aalge* Gwylog

Welsh List Category	IUCN Red List			Birds of Conservation Concern			
	Global	Europe	GB	UK	Wales		
A	LC	NT	LC	Amber	2002	2010	2016

The Guillemot is among the most abundant seabirds in the northern hemisphere, occurring in both the North Atlantic and Pacific Oceans (Mitchell *et al.* 2004). It breeds colonially, often in large numbers, in a range of habitats, from narrow cliff ledges to low-lying flat islands, usually at very high densities. Guillemots breed in cool temperate waters and generally winter south of their breeding range. In Wales, Guillemots breed from Anglesey in the north to Gower in the south, often in the company of Razorbills and, sometimes, Puffins. The breeding population is of the subspecies *U.a. albionis*. The nominate subspecies *U.a. aalge* breeds in Scotland and occurs in the Irish Sea during the winter months (IOC).

The discovery of incompletely ossified bones, probably dating from the Mesolithic period, in Port Eynon Cave, Gower, indicates that birds were breeding here about 6,000 to 9,000 years ago (Harrison 1987). Ray (1662) recorded "guillems" breeding on Puffin Island, Anglesey. Guillemot numbers in Wales were apparently high in the late 18th and early 19th centuries. At Elegug Stacks, Pembrokeshire, in 1792 Sir Richard Colt Hoare reported "an astonishing quantity of eligugs [guillemots]… The number of birds is so great that these two rocks are so completely covered with them that in parts no rock whatever is to be seen" (Hoare 1983). Numbers at some sites were suppressed by harvesting of eggs for human consumption. Williams (1835) described Guillemots as very numerous on the Great and Little Ormes, Caernarfonshire, where great quantities of eggs were gathered annually, and sold at about half a crown a dozen. He stated that the birds "are not nearly so numerous now, as they were formerly, owing to the incessant removal of their eggs. About thirty years ago, two men were killed, by being precipitated down the rock, which deterred the gatherers from their usual employment, for several seasons, and the birds increased in a short time to an immense number.'

The subsequent history of the Guillemot in Wales is a story of mixed fortunes. From photographs taken between the late 1800s and the 1950s, at various Welsh colonies, changes in Guillemot numbers can be assessed with some confidence. This is based on the accumulated knowledge of the population biology of Guillemots on Skomer, the largest colony in Wales. The number of birds here probably remained stable between the late 1800s and the 1930s (Birkhead 2016). Despite Ronald Lockley's well-deserved reputation as a pioneer of seabird biology, quantification was not his *forté* and his estimates of Guillemot numbers should be treated with caution. He stated that there were "at least 5,000 pairs" on Skomer in 1934 (Buxton and Lockley 1950), but high-resolution photographs he took that year, of just one large colony, indicate that there must have been *c.*100,000 birds on the island (Birkhead 2016).

By the end of the Second World War, Lockley noted that the Guillemot population on Skomer was much reduced. It is now clear that a decline of *c.*95% occurred largely as the result of oil pollution created as ships were sunk during the war. Lockley's daughter, growing up with her parents on the adjacent island of Skokholm, mentioned the frequency with which oiled Guillemots were encountered (Lockley 2013), and indeed Guillemots seem to be more susceptible to oiling than other species. As well as wartime sinkings, tank washing not far from the coast continued at least into the 1950s (Ivor Rees *in litt.*). The decline in seabirds on Skomer, and at colonies in southern England, from the 1940s to the 1970s was of such a scale that, despite no reliable census figures, it was obvious that a massive reduction in the numbers of Guillemots (and Razorbills and Puffins) had occurred (*BWP*).

The colony on Skomer was the first to be accurately censused, from 1962, when numbers were low. There were further declines resulting from ongoing chronic oiling, but also as a result of the *Torrey Canyon* oil-tanker disaster in 1967, when tens of thousands of birds died. After the *Torrey Canyon*, there were several other notorious oil events that caused losses from Welsh colonies. During the so-called Irish Seabird Wreck in 1969, at least 15,000 Guillemots were found dead. At the time it was thought that the cause was toxic chemicals, but it now appears that it was the result of severe weather preventing birds from obtaining sufficient food, when they were already stressed by the energetic demand of moulting. Similar auk 'wrecks' have been recorded subsequently.

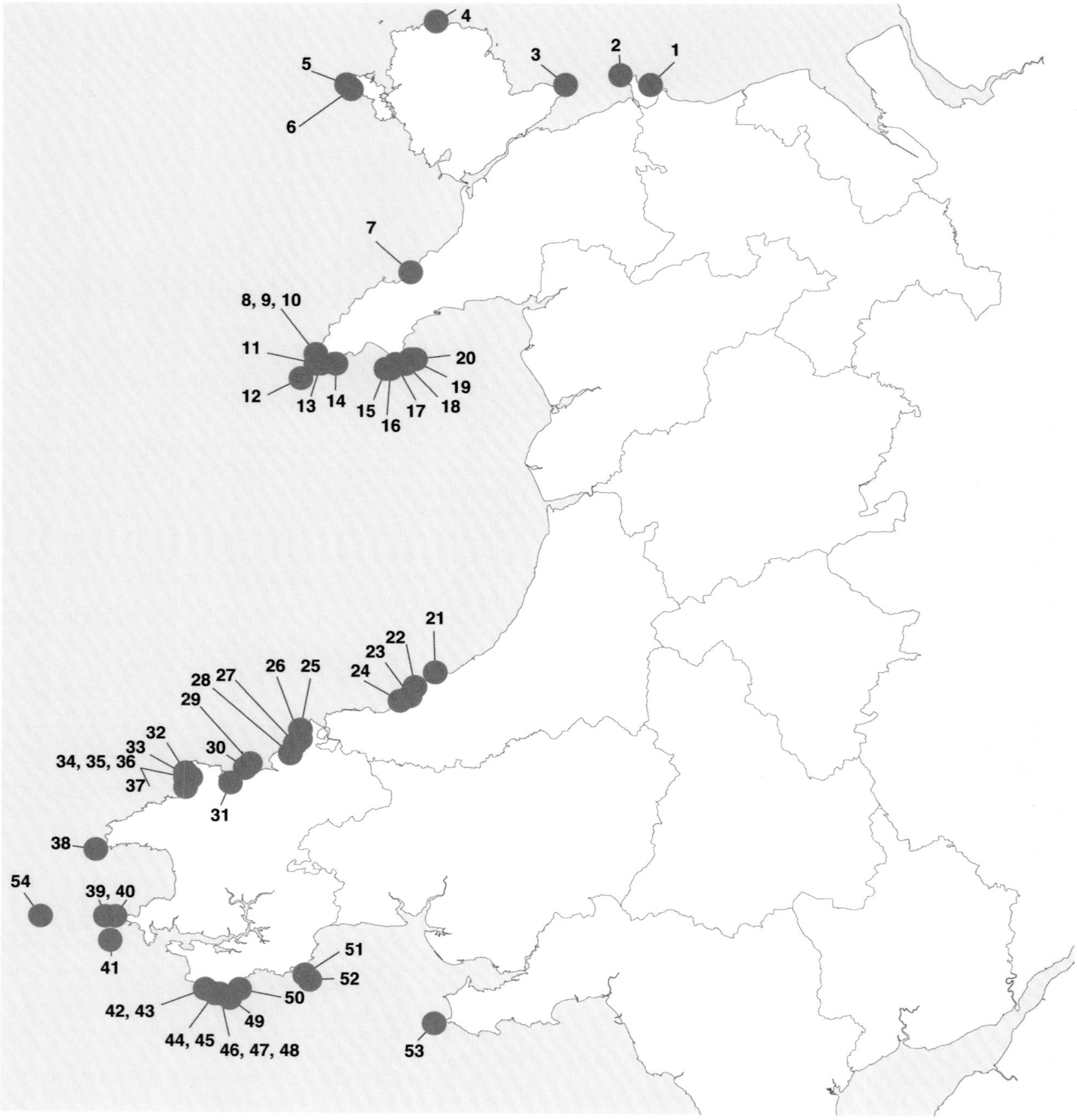

Guillemot: Locations of all currently known colonies around the Welsh coast. See the key, below:

The number of Guillemots on Skomer reached its lowest level (2,634 individuals) in 1972 and numbers remained around that level until about 1980. From 1973, several study plots were established there, to obtain a mean and standard deviation of numbers, thus providing a more reliable estimate than those from single whole-island counts. From about 1980, numbers on Skomer increased at around 5% per annum. A detailed population model suggested that this increase was driven almost entirely by the Skomer population itself, rather than by recruitment from other colonies (Meade *et al.* 2013). There is a small amount of colony interchange: a Guillemot ringed as a chick on Stora Karlso in the Baltic Sea moved to Skomer to breed. Most colony interchanges, however, are of non-breeding birds (i.e. younger than seven years old). The Guillemot breeding population on Skomer reached 28,798 individuals in 2019, a post-war record.

The increase in Guillemot numbers on Skomer since 1980 is reflected elsewhere in Wales. Between the Seabird Colony Register (1985–88) and Seabird 2000 (1998–2002), numbers in Gower increased by 124%, in Pembrokeshire by 127%, in Ceredigion by 62%, in Caernarfonshire by 19% and on Anglesey by 98% (Mitchell 2005), showing a lower rate of increase in the north.

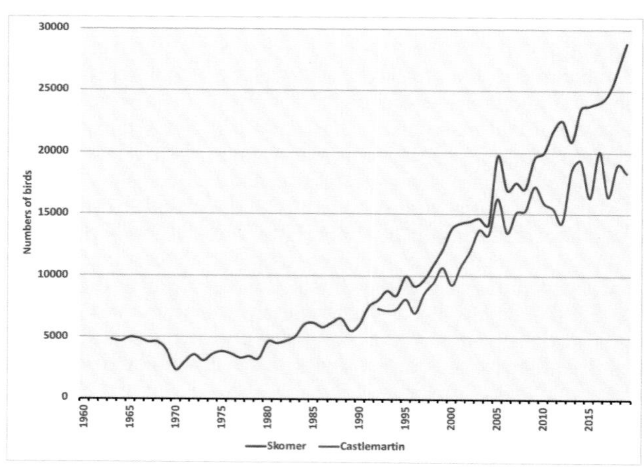

Guillemot: Annual counts of the populations on Skomer and at Castlemartin, Pembrokeshire, 1963–2019.

New site no.	Colony	Vice-county	Grid ref.	2000 count	2018/19 count	% difference
1	Little Orme	Caernarfonshire	SH816827	603	573	4.9-
2	Great Orme	Caernarfonshire	SH756842	1,512	2,027	34.1+
3	Puffin Island	Anglesey	SH647820	1,821	3,606	98.0+
4	Middle Mouse	Anglesey	SH382959	2,464	5,550	125.2+
5	RSPB reserve	Anglesey	SH205821	3,528	5,243	48.3+
6	Abraham's Bosom	Anglesey	SH215813	165	315	84.8+
7	Carreg y Llam	Caernarfonshire	SH333438	7,980	11,000	37.8+
8	Maen Du	Caernarfonshire	SH140263			
9	Maen Du	Caernarfonshire	SH140263	11	40	264.0+
10	Braich Anelog	Caernarfonshire	SH146265			
11	Porth Felen	Caernarfonshire	SH147246			
12	Bardsey	Caernarfonshire	SH117216	848	1,112	31.1+
13	Parwyd	Caernarfonshire	SH155242			
14	Ynysoedd Gwylan	Caernarfonshire	SH183244	18	59	228.0+
15	Penymynydd	Caernarfonshire	SH286237	15	0	
16	Trwyn Cilan	Caernarfonshire	SH291235	3,000	1,632	45.6-
17	Muriau	Caernarfonshire	SH303242			
18	Porth Ceiriad	Caernarfonshire	SH320243			
19	St Tudwal's Island W	Caernarfonshire	SH334253	231	352	52.3+
20	St Tudwal's Island E	Caernarfonshire	SH341258	731	1,139	55.8+
21	New Quay Head	Ceredigion	SN381604	4,235	5,418	27.9+
22	Penmoelciliau	Ceredigion	SN348572	234	174	25.6-
23	Lochtyn 1	Ceredigion	SN332556	477	477	0
24	Llangrannog 1	Ceredigion	SN310543	270	229	15.2-
25	Cemaes Head	Pembrokeshire	SN119485	140	51	65.6-
26	Ceibwr2	Pembrokeshire	SN110460	0	83	
27	Ceibwr 1	Pembrokeshire	SN100450	279	24	91.4-
28	Carreg Bica	Pembrokeshire	SN09064393			
29	Needle Rock	Pembrokeshire	SN016410	172	291	69.2+
30	Dinas Head	Pembrokeshire	SN000408	38	73	92.1+
31	Fishguard to Pwllgwaelod 3 (Fishguard Needle rock)	Pembrokeshire	SM97563797	82	101	23.1+
32	Pen Brush	Pembrokeshire	SM880396	14	361	255.0+
33	Penbwchdy N	Pembrokeshire	SM890382	67	40	40.3-
34	Ynys y Dinas	Pembrokeshire	SM88633866	0	1	
35	Ynys Ddu	Pembrokeshire	SM88603887	56	354	532.0+
36	Ynys Melyn	Pembrokeshire	SM88523859	0	109	
37	Carreg Dandy	Pembrokeshire	SM882360	0	74	
38	Ramsey	Pembrokeshire	SM700235	3,031	4,403	45.3+
39	Skomer	Pembrokeshire	SM725095	13,852	24,788	79.0+
40	Middleholm	Pembrokeshire	SM748094	323	302	6.5-
41	Skokholm	Pembrokeshire	SM736049	996	4,038	305.4+
42	Green Bridge of Wales to Flimston Bay (incuding Elegug Stacks)	Pembrokeshire	SR926944	6,817	14,432	111.7+
43	The Wash to Green Bridge of Wales	Pembrokeshire	SR923944	763	1,001	31.2+
44	Mewsford Arches	Pembrokeshire	SR943938	433	689	59.1+
45	Mewsford Arches to Crickmail Point	Pembrokeshire	SR945938	56	127	126.8+
46	Crickmail Point to The Castle	Pembrokeshire	SR950936	179	379	111.7+
47	The Castle	Pembrokeshire	SR955934	11	18	63.6+
48	The Castle to Saddle Head	Pembrokeshire	SR957931	13	3	76.9-
49	St Govan's Chapel to New Quay	Pembrokeshire	SR974926	6	3	50-
50	Saddle Point to Griffith Lorts Hole (Stackpole Head)	Pembrokeshire	SR993942	970	1,695	74.7+
51	St Margaret's Island	Pembrokeshire	SS123973	501	1,806	260.5+
52	Caldey Island – West Beacon	Pembrokeshire	SS133965	9	57	533.3+
53	Worms Head – Worm North	Gower	SS383878	190	169	11.0-
54	Grassholm	Pembrokeshire	SM598092	995	2,462	147.4+

Guillemot: Key to map of colonies in Wales.

Operation Seafarer 1969–70	SCR Census 1985–88	Seabird 2000 1998–2002	Seabirds Count Provisional 2015–19
17,238	32,126	57,961	96,802

Guillemot: Total count of individuals recorded in Wales during UK seabird censuses.

The largest counts in North Wales have been at Carreg y Llam on Llŷn, Caernarfonshire, although numbers fluctuate. There were 2,750 adults on land (AOL) in 1968–69, increasing to a peak of 17,230 individuals in 2009. A decline followed, although numbers recovered to 9,676 AOL in 2015 and an estimated 11,000 in 2019. Large numbers are also found on North Stack and South Stack on Anglesey, where there were 9,690 birds in 2017.

Recoveries of Guillemots ringed mainly on Skomer, but also elsewhere in the Irish Sea, show that at the end of the breeding season, in late June and early July, they move south. Younger birds, birds of the year and those up to six years old disperse more widely than breeding adults (Votier *et al.* 2008). Tracking of 25 adults using geolocators in 2009–2013 provided a more nuanced view of movements. At the end of the breeding season, birds moved south towards the Bay of Biscay, a swimming migration with males accompanying their (still flightless) young. During moult the adults lose all their primaries simultaneously and are flightless for about six weeks. After several weeks in Biscay, having completed their moult, Guillemots fly north to Scottish waters, with some reaching the vicinity of the Faroe Islands, where they spend several weeks feeding, before flying south again to the English Channel and Biscay (own data).

These movements almost certainly account for the tens of thousands sometimes seen flying north or south during sea-watches between late September and early December. At Strumble Head, Pembrokeshire, sample counts, taken for 15 minutes in each hour, produced estimated totals of 24,000 on 17 October 1983, 26,000 on 13 November 1987 and 35,000 on 20 October 1984 (Donovan and Rees 1994). Guillemots are scarce in the Severn Estuary off Gwent, with none seen in some years, though a total of eight birds was recorded there in 2000.

By October, adult Guillemots remain within commuting distance of their breeding colonies, making irregular visits, usually in the morning, but returning to roost and feed on the water overnight. There have been recoveries of birds ringed as chicks in Wales over an area from Scandinavia to the Mediterranean, but the bulk of those outside Britain and Ireland have been in France. The majority of such recoveries have been birds in their first or second-calendar-year (Meade *et al.* 2013).

Since 1972, the Guillemots on Skomer have been the subject of a long-term study of their numbers, ecology, behaviour and population biology. Annually since the mid-1980s, the study has monitored adult and immature survival, breeding success, the timing of breeding and chick diet. It consequently provides the basis for much of the information presented here (Birkhead 2009, 2014).

The Skomer study shows that the combination of high annual adult survival (around 95%) and high (*c.*50%) survival to breeding age (*circa* seven years) is largely responsible for driving the increase in the population. Breeding success on Skomer is high, with 70–80% of pairs successfully fledging a chick, but this has a more limited effect on the population increase. The ultimate reason for the increase in numbers is not known. There may have been a shift and/or increase in food supplies (chicks on Skomer are fed predominantly on Sprats (Riordan and Birkhead 2018)) but almost nothing is known about the abundance or availability of this or other prey species. One explanation for the increase in numbers

Guillemot: Recovery locations of birds ringed in Wales are shown by red circles. Ringing locations of birds that were recovered in Wales are shown by blue triangles. Small brown dots show ringing or recovery locations in Wales.

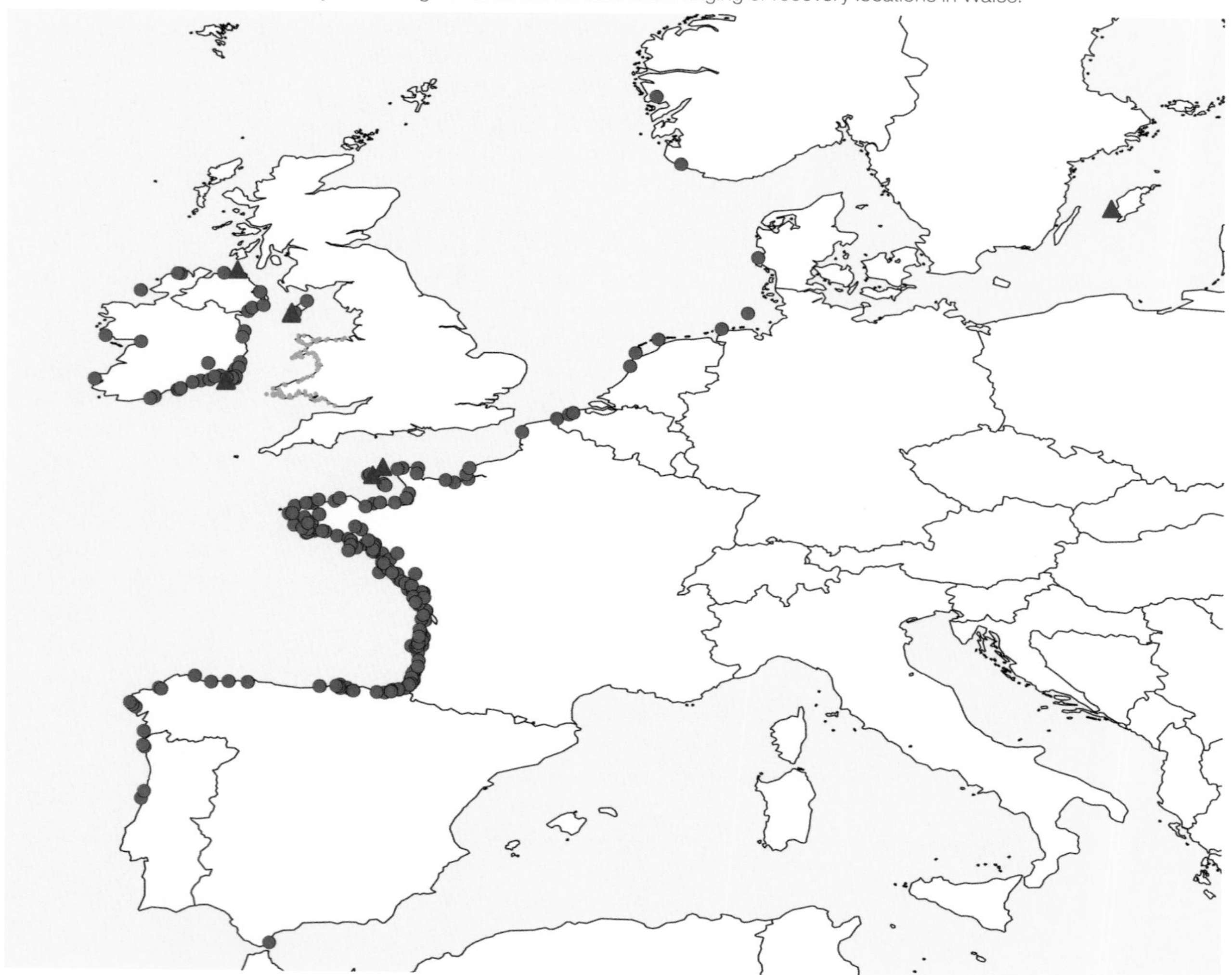

since 1980 is that it is simply a 'recovery' from the devastating effects of oil-induced mortality during the Second World War. At the current rate of population increase, it will be several more decades before the Skomer population reaches its pre-1930s level.

The Amber conservation status of the Guillemot in Wales is based on the international importance of the breeding population and its Near-Threatened conservation status in Europe (Johnstone and Bladwell 2016). In 1998–2002, Wales held approximately 4% of the breeding Guillemots in Britain and Ireland (Mitchell *et al.* 2004). Numbers have since increased by 6% annually in Wales. The substantial increase in the Welsh population, in the second half of the 20th century and the early 21st century, is in contrast to those in Scotland and farther north, where massive declines have occurred in the last 20 years (Dunn 2019). Threats, including further oiling incidents and storm-related 'wrecks', as occurred in 1969 and 2014, may disrupt this increase as the frequency and intensity of storms increase, as a result of climate change (Morley *et al.* 2016). Climate change may be implicated in increasingly earlier breeding (Votier *et al.* 2009) and could affect food availability in Welsh waters. In terms of chick diet, Guillemots in Welsh colonies are less reliant on sandeels than those in Scotland, feeding instead mostly on Sprats and gadids. During 2013–17, the proportion of gadids in the Skomer chicks' diet increased from 6% to 18%, suggesting a change in the availability of different prey species (Riordan and Birkhead 2018). However, in 2018–19 the proportion of gadids declined and that of Sprats increased. In 2018 clupeids formed 78% of the diet and gadids 12%. In 2019 the proportions were 78% clupeids and 8% gadids (own data). The breeding range of Guillemots is expected to contract northwards during the 21st century. Southern Britain, including Pembrokeshire, may prove to be unsuitable in 80 years' time (Huntley *et al.* 2007).

Northern Guillemot *U.a. aalge* Gwylog y Gogledd

A number of birds ringed as chicks or adults within the breeding range of the nominate subspecies *U.a. aalge*, in Scotland north of 55° 38" (*The Birds of Scotland* 2007) and in southern Norway, have been found in Wales, most often outside the breeding season.

Tim Birkhead

Sponsored by Aaron Davies and Anon. in memory of the late Peter Hope Jones

Razorbill *Alca torda* Llurs

Welsh List Category	IUCN Red List			Birds of Conservation Concern			
	Global	Europe	GB	UK	Wales		
					2002	2010	2016
A	NT	NT	LC	Amber			

The Razorbill breeds in eastern North America, Greenland, and northwest and northern Europe as far east as the White Sea. It winters at sea, with northern populations moving south or southwest (*HBW*). Welsh breeding birds are of the subspecies *A.t. islandica*, occurring in Brittany, Britain, Ireland, the Faroe Islands and Iceland (IOC). This species is a resident breeder and partial migrant in Wales, nesting at suitable coastal sites from Worms Head, Gower, to the Little Orme, Caernarfonshire, but has not been recorded breeding in Carmarthenshire or Meirionnydd. The largest colonies are mainly on the islands, particularly Skomer and Skokholm in Pembrokeshire, and Bardsey, Caernarfonshire, with smaller numbers on mainland cliffs in the southwest and northwest. Outside the breeding season some birds remain in Welsh waters, while others move south. Despite some individuals being conspicuous on cliffs, Razorbills have proven very difficult to census, so that while their general distribution is known, reliable estimates of numbers are not.

During the breeding season, incubating birds, fitted with miniature radios on Skomer, flew on average 140km from the colony on trips averaging 16 hours. However, when bringing food to chicks they returned more frequently, being away only around seven hours, with an average maximum distance of 50km (Shoji *et al.* 2014).

The chicks leave the breeding site on average about 19 days after hatching (Harris and Wanless 1989). At this point, when they are about one-third grown and still flightless, they leap dramatically into the sea at dusk and swim away from the colony with the male parent. Little is known about the next period of their lives, including how long the chicks remain flightless, but at this stage dispersal is restricted to swimming and the movement of currents. During this period, moulting (and now flightless) adults disperse. They are able to fly again by September, as are their chicks. A small number of Welsh-ringed birds have been recovered in the English Channel, with others in the North Sea as far north as Norway. However, these are in the minority: by far the largest number of ringing recoveries of Welsh birds are of birds that headed south, along the French and Spanish coasts and into the Mediterranean. Some reach as far east as Italy and a small number have been recovered along the North African coast. This pattern is strongly influenced by the age of the birds. On average, first-calendar-year birds move

Razorbill © Ben Porter

Razorbill: Recovery locations of birds ringed in Wales are shown by red circles. Ringing locations of birds that were recovered in Wales are shown by blue triangles. Small brown dots show ringing or recovery locations in Wales.

farther than older ones, and account for the large majority of those entering the Mediterranean. The timing of movements is also age-related. Birds of breeding age return to colonies before the third-calendar-year and non-breeding fourth-calendar-year birds. Second-calendar-year birds are the last to return, and many of these may not return to the colonies at all. Those that do rarely join the breeders on the cliffs.

Razorbills generally do not start to breed until four years or older. They lay a single egg each year. although they may replace this if lost, especially early in the season. Breeding success (chicks fledged/egg laid) was 0.53 on Skomer between 1995 and 2018, but varies with nest site. Those in burrows and under crevices are more successful than those on ledges. The annual survival rate of breeders was about 90% on Skomer during 1970–2018. Two birds, both ringed as nestlings, vie for the title of oldest Welsh Razorbill. One was ringed on Bardsey in 1962 and recaptured there in June 2004, a week short of being 42 years old. The other was ringed on Skokholm in June 1970 and found dead in southern France, possibly caught in a fishing net, in December 2013, 43 years and six months later, although the exact date of death is not known.

Historically, we know little of the status of the Razorbill in Wales, although the presence of incompletely ossified and seemingly juvenile bones at Bacon Hole, Gower, suggests that this cave was used for breeding during the Ipswichian interglacial period, from 115,000 to 130,000 years ago. Razorbill bones from this period were also found at Minchin Hole nearby (Harrison 1987). Ray (1662) recorded the species breeding on Puffin Island, Anglesey. The adult birds and their eggs were important sources of food. Lewis (1833) recorded Razorbills breeding on the north side of the Great Orme, Caernarfonshire, and stated that "the eggs of the razor-bill are esteemed a delicacy, and the sale of them, generally at two shilling per dozen, affords a livelihood to several families employed during the season in procuring them." This later became less important, but local consumption of seabird eggs probably increased in wartime and may have affected numbers.

Towards the end of the 19th century, the Razorbill population suffered further pressures. As the opening of the railways allowed easier access to the coast, hunting seabirds became a pastime: they were heavily shot for sport and their eggs collected, for food and for decoration. From 1869, the hunting of seabirds during the breeding season was banned in the UK with the introduction of the Sea Birds Preservation Act. The very general descriptions of distribution in Pembrokeshire, for example by Mathew (1894), and in Caernarfonshire and Anglesey, by Forrest (1907), suggested it occurred in the same places that it is seen today, and broadly the same as those reported in the first seabird survey of Britain and Ireland in 1969–70 (Cramp *et al.* 1974). About 70% of the Welsh population is concentrated on six islands, or regions of cliffs. In order of importance, these are: Skomer, Skokholm, Bardsey and Ramsey (all islands), Green Bridge of Wales to Flimston Bay (Elegug Stacks) in Pembrokeshire, and RSPB South Stack, Anglesey. Four other sites—Carreg y Llam on Llŷn, Caernarfonshire; Middleholm, Pembrokeshire; and Ynys Badrig (Middle Mouse) and Puffin Island on Anglesey—account for a further 10% of the total. Almost all of the remaining 20% are spread in small, scattered groups along the coasts of Pembrokeshire, Ceredigion, Caernarfonshire and Anglesey. While the numbers recorded may fluctuate quite markedly between years, most of the sites have been occupied since at least the end of the Second World War.

For several reasons, it is difficult to compare current and past records of Razorbill numbers.

1. The nest site varies from a Puffin-like burrow at the cliff top, through those deep under boulders to those on open ledges, sometimes among Guillemots.
2. At most sites the numbers visible bear little relation to the number of breeding pairs. Many non-breeders are more conspicuous in the colony than breeders. Almost no

second-calendar-year birds return to the colony, but many third-calendar-year and older non-breeders do return (Lloyd and Perrins 1970). The presence of birds may indicate the existence of a colony, but is a poor guide to breeding numbers.
3. As with Puffins and Guillemots, prior to the start of incubation, Razorbill attendance at colonies has a strikingly cyclic pattern: about every fourth day numbers reach a high level, then collapse and gradually build to the next peak, four days later.
4. Over time, counters have recorded counts of eggs, the number of individuals visible, the number of pairs, the number of apparently occupied sites (AOS) or even just numbers, with no further information. The first survey of Britain and Ireland (Cramp *et al.* 1974) noted that "there is no information, even in the vaguest terms, concerning the past size of most colonies." As a result, most older counts cannot be compared easily with current ones. Most estimates are now based on the mean number of individuals present, over an average of ten counts taken between 0700 and 1500 GMT, in fine weather during 1–20 June. However, some colonies can only be counted from the sea, and more than one count a year may not be possible.

Thus, little is known about numbers of Razorbills prior to the Second World War, although descriptions indicated that both Puffins and Guillemots were much more plentiful before the war than afterwards. It seems safe to assume that trends in Razorbill numbers closely matched those of these two species, especially Guillemot. The widely accepted explanation for the declines is the high mortality from oil pollution arising from the torpedoing of ships, especially tankers bringing oil to the UK, during the Second World War. Notwithstanding this, it seems likely that the numbers in the early post-war years were higher than they were later, as were those of Guillemot. If so, Razorbill numbers may have declined until around 1970.

Lloyd *et al.* (1991) stated that "most Welsh Razorbill colonies either hardly changed in size from 1969/70 to 1985–87, or declined a little." The Welsh total increased by 33% between 1985–87 and 1998–2002, and by a further 68% by 2015–19. In the 1998–2002 census (Mitchell *et al.* 2004), 80% of Razorbills were in Pembrokeshire/Ceredigion and almost all the rest in Caernarfonshire and Anglesey. Most sites seem to broadly match the long-term population-level increases, though numbers on Cardigan Island, Ceredigion, and Bardsey, seem to have increased more markedly. Razorbills first bred on Cardigan Island in 1983, increasing to 19 pairs in 1990 and 90 birds in 2019.

Operation Seafarer 1969–70	SCR Census 1985–88	Seabird 2000 1998–2002	Seabirds Count Provisional 2015–19
9,316	9,501	12,638	21,233

Razorbill: Total counts of individuals recorded in Wales during UK seabird censuses.

Bardsey records until 2009 are hard to interpret. The island was recorded as having at least 20 pairs in 1905 and 1913 (Ticehurst 1919), with numbers unchanged in 1930 (Wilson 1930). However, numbers had increased very considerably by 1952, when Norris (1953) estimated numbers to be "many times more than the 'at least twenty pairs' recorded by Ticehurst." Numbers from 1953–1993, recorded as "eggs/yng", vary wildly. From 1994, counts were also recorded as "AOL" (adults on ledges). Counts of AOL up to 2008 varied between 89 and 2,010 with no apparent trend and no apparent relationship between the "eggs/yng" and the AOL figures in the same year. Since 2009, the AOLs have been counted consistently and the AOLs and the "eggs/yng" counts bear a constant relationship. In the period 2009–2019, the mean number of AOLs was 1,859, with a slight increase over this period.

In contrast, the population on Ramsey has remained stable: 1,599 birds were recorded in 2019, similar to the 25-year average of 1,467. This population is thought to be constrained by limited suitable breeding habitat: "most nest in crevices on steep cliffs that would have been out of bounds to rats" (Greg Morgan pers. comm.).

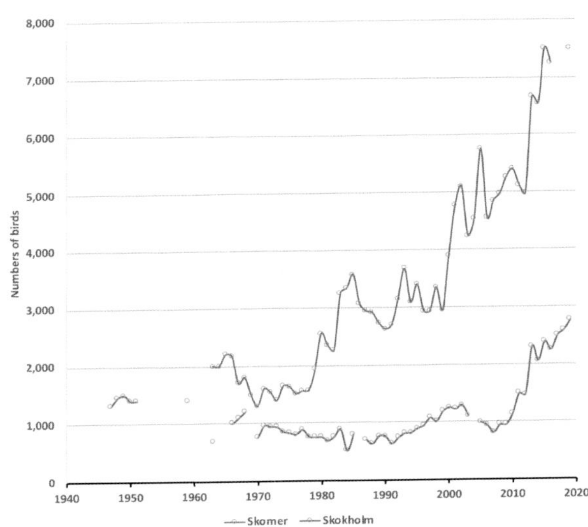

Razorbill: Numbers on Skomer and Skokholm, 1947–2019.

At the two largest colonies, Razorbills on Skokholm have been counted almost every year from shortly after the Second World War and on Skomer since 1960. Numbers on Skokholm declined during 1950–68 and then crashed to 750 in 1970, in line with a major summer mortality in 1969 (see below). After 1970, there was a small recovery, but numbers were largely stable from then until about 1991. Thereafter, numbers rose steadily to 1,285 in 2002, declined to 812 in 2007, then rose to 2,755 in 2019. On Skomer, the pattern of change differed from that on Skokholm, with a decline of about 30% between 1968 and 1970 and a marked increase between 1978 and 1983. Numbers then remained stable until 1999, since when they have risen steadily, with 4,847 in 2007 and 7,500 in 2019. In 2020 the Razorbill count on Skokholm increased by a massive 27.7% to 3,517. There was no count on Skomer in 2020.

The differences in the recent rates of increase on the two islands is striking. During 2007–19 Skokholm numbers rose by 240%, while the comparable increase for Skomer is just 55%. It is difficult to believe that the rate of increase on Skokholm could be achieved by a species that lays only one egg each year and takes four years or more to attain breeding age, unless the numbers were augmented by immigration. It seems reasonable to assume that a significant part of the Skokholm increase must be due to immigration from Skomer, the nearest part of which is only 4km away.

Sizeable passage is sometimes observed at headlands, mainly in autumn. At Strumble Head, Pembrokeshire, an estimated 20,000 passed in 11 hours on 25 October 1992 and 12,000 passed in seven hours on 2 October 2000. Farther north, large numbers were recorded on 22 October 2014, when 12,100 passed Cemlyn, Anglesey, and an island record 14,637 passed Bardsey. On the same date, about 5,000 were off Mwnt, Ceredigion. Most cannot have been Welsh birds, so must have come from colonies farther north.

As with many other Welsh seabirds, human activities have had deleterious effects on numbers. Hunting was a serious cause of mortality overseas for many years. Winter shooting of auks, particularly in Norway (Lloyd 1974), was believed to be an important cause of overwinter mortality in British and Irish auks. Shooting was banned in Norway in 1979 and is illegal across the European Union. Relatively few Welsh birds move to Norway or Denmark, so these changes probably benefited Scottish and North Sea populations more than Welsh ones. The importance of hunting on historic numbers is difficult to quantify, since ringed birds that have been shot are more likely to be reported than is the case with most causes of death, except entanglement in fishing gear.

Rats play a major part in the distribution of almost all seabirds, and their introduction to islands can be devastating. The largest numbers of Razorbills occur on rat-free islands, whereas there are fewer on mainland Wales, except on steep cliffs, where Brown Rats presumably seldom venture. Lundy, an island in the Bristol Channel regarded as part of Devon, illustrates the effect of rats: in

1939 there were estimated to be 10,500 pairs of Razorbill (Perry 1940), but rats subsequently reached the island and numbers rapidly declined. Only 841 individuals were reported in 2004. Brown and Black rats were eliminated from Lundy by 2006 and numbers subsequently more than doubled. Brown Rats were eliminated from Puffin Island, Anglesey, in 1998–99 and Razorbill numbers increased dramatically in the next 12 years, but subsequently declined. Brown Rats were eradicated from Ramsey during winter 1999/2000, but Razorbill numbers have remained stable there.

The Razorbill is the second commonest species, after Guillemot, to die in most major oil spills. These include the *Torrey Canyon* in Cornwall in 1967, *Amoco Cadiz* in Brittany, France, in 1978, *Sea Empress* in Milford Haven, Pembrokeshire, in 1996, *Erika* in Biscay, France, in 1999–2000 and *Prestige* in Galicia, Spain, in 2002. All of these spills are likely to have affected Welsh-nesting Razorbills. Fishing, especially with modern nylon monofilament nets, is also a cause of Razorbill mortality. Mortality due to fishing nets was particularly serious off the west and south coasts of Ireland, from at least the 1960s (Lloyd *et al.* 1991).

Prolonged periods of strong winds during the winter may also increase mortality, especially of young birds. A prolonged series of gales in 2013/14 resulted in a large 'wreck' of seabirds along the English Channel, Bay of Biscay and Spanish coasts. A minimum of 55,000 seabirds, mostly dead, were washed ashore. About 7% of these were Razorbills (Morley *et al.* 2016). On Anglesey, Razorbills made up more than 90% of dead birds found, and from examining beak grooves, most were of breeding age (Ivor Rees *in litt.*). Rings recovered from Razorbills that died in this wreck showed that some were from Welsh colonies. It is predicted that climate change may lead to an increase in the frequency and intensity of such gales. Food supply may also be an important factor, but little is known about this. The Razorbill is heavily dependent on sandeels, the stock of which has been greatly reduced in the North Sea by commercial fishing. There is no commercial fishery off Wales, but higher sea temperatures reduce sandeel recruitment (Wanless *et al.* 2018).

Other factors affect Razorbill numbers and breeding success. Usually these are unexplained, often probably not even noticed. Shortly after the end of the 1969 breeding season, an unusually high number of dead auks, including Razorbills—adults and fledglings—was reported around the Irish Sea (Holdgate 1971), followed by a decline in breeding numbers in 1970. Possible explanations included toxic chemical pollution, but no firm conclusion was reached. In 2008, breeding on Bardsey was seriously reduced. The number of adults on land (eggs/young in brackets) recorded was 2,000 (334) in 2007, 800 (38) in 2008 and 1,464 (295) in 2009. Almost the whole season's breeding was lost. Many breeding birds did not even attend the colonies, although they returned the following year. Guillemot breeding was similarly reduced (Steve Stansfield pers. comm.). The effect was local, in as much as breeding numbers on Skomer and Skokholm in 2008 did not decline.

The Razorbill has Amber-listed conservation status in Wales because of its Near Threatened status in Europe and because of the high degree of localisation of the breeding population (Johnstone and Bladwell 2016). It is also Amber-listed in the UK (Eaton *et al.* 2015) because of the vulnerability of the major breeding colonies and the international importance of the population. There have been large declines in Iceland, which holds the largest population of the *A.t. islandica* subspecies (Berglund and Hentati-Sundberg 2015). The Welsh breeding population seems to be doing well at present, but factors such as food shortages and an increased number of storms, as a result of climate change, are potential threats. The breeding range of Razorbill is forecast to contract away from southern Britain, including Pembrokeshire, during the 21st century (Huntley *et al.* 2007).

Chris Perrins

Sponsored by Pembrokeshire Coast National Park Authority

Black Guillemot — *Cepphus grylle* — Gwylog Ddu

Welsh List Category	IUCN Red List			Birds of Conservation Concern			
	Global	Europe	GB	UK	Wales		
A	LC	LC	LC	Amber	2002	2010	2016

The Black Guillemot is a circumpolar species, concentrated mainly around the North Atlantic, and Barents and Baltic seas, with smaller numbers around the Chukchi Sea in northern Alaska and northeast Siberia (*HBW*). Only small numbers breed in Wales, which is close to the southern limit of the range of the subspecies *C.g. arcticus* in Europe. All populations are thought to winter offshore from their breeding colonies, travelling much shorter distances than other auks, although it is conceivable that birds of the subspecies *C.g. islandicus*, from Iceland, or *C.g. mandtii*, from Svalbard and northeast Canada (IOC), could occur in Welsh waters during the winter. Approximately half of the UK's population of 19,000 pairs breeds in Orkney and Shetland, with most of the remainder on the coasts of north and west Scotland and Northern Ireland. It breeds away from large seabird cliff colonies, preferring small rocky islets and low-lying, indented stretches of rocky coast, which makes it more difficult to survey.

The distribution of the Black Guillemot is predominantly determined by the availability of suitable nest cavities, in rock crevices and under boulders, that are safe from land predators, such as Brown Rat, American Mink and Stoat. The species feeds in relatively shallow waters. On Mousa, Shetland, 70–80% of fish fed to chicks were Lesser Sandeel and Butterfish (Ewins 1990).

Pennant (1810) visited the Great Orme and the Little Orme in Caernarfonshire in the 1770s, and noted that a few Black Guillemots bred; he also mentioned that the species bred on Anglesey at Ynys Llanddwyn. Williams (1835) and Price (1875) stated that it bred on the western side of the Little Orme. It also bred in Pembrokeshire: Montagu (1802) wrote "we have seen it rarely on the coast of Wales near Tenbeigh [*sic*] where a few breed annually." However, by the end of the 19th century, breeding appears to have ceased in Wales. Mathew (1894) knew of no recent record in Pembrokeshire and commented that the species had deserted Anglesey. Forrest (1907) knew of only three records in North Wales, including a bird at Ynys Llanddwyn in June 1901, although a pair possibly bred on Ynys Moelfre, Anglesey, in 1912. There was some evidence of breeding activity around Penmon, Anglesey, in the early 1950s, but this was not proven. A pair was regularly seen at St Tudwal's Islands, Caernarfonshire, in the mid-1950s. On the Great Orme, Paul Whalley watched birds carrying material in 1953, but asserted that they did not breed (*Birds in Wales* 1994). The first confirmed breeding in Wales for many years was in 1962, when a bird sat on an addled egg near Valley, Anglesey (Jones and Whalley 2004). By the mid-1960s at least two pairs were breeding at Fedw Fawr, Anglesey.

The *Britain and Ireland Atlas 1968–72* recorded probable breeding at five sites around Anglesey. This was the most southerly breeding in Britain at the time, although birds bred farther south in Ireland. Between seabird censuses in 1969–70 and 1985–91, the range expanded in Britain and Ireland, in particular, with the colonisation of new sites around the Irish Sea, including artificial structures such as harbour walls, jetties and piers. The Seabird Colony Register counts in 1985–88 produced an estimate of 26 individuals in Wales, mainly around Anglesey.

Operation Seafarer 1969–70	SCR Census 1985–88	Seabird 2000 1998–2002	Seabirds Count Provisional 2015–19
5	26	28	19

Black Guillemot: Total number of individual pre-breeding birds recorded in Wales during UK seabird censuses.

Black Guillemot in breeding plumage © Kev Joynes

The key strongholds in Wales remained in the north, particularly on Anglesey. A small colony below the cliffs at Fedw Fawr has probably been present since the 1960s. The number of breeding pairs is difficult to estimate at this site, but the highest counts of adults recorded there were 24 on 30 March 2002 and 21 on 18 April 2004. Following Brown Rat eradication from nearby Puffin Island in 1998–2000, 1–2 pairs have been seen there each year, with one chick ringed in 2002. At Holyhead Harbour, nest-boxes were installed to increase breeding opportunities, and up to 14 adults have been present during the breeding season. Elsewhere on Anglesey, breeding has been recorded at Point Lynas/Porth Eilian, where up to three pairs have bred since 1997, and at Amlwch and Benllech. Breeding was proved in Beddmanarch Bay in the 1970s, but this site is thought not to have been used subsequently.

In Caernarfonshire, two well-grown chicks were found in a burrow among the Puffin colony on Ynys Gwylan Fawr in June 2000, the first confirmed record of breeding there. There have been occasional records of birds around the Great Orme during the breeding season in many years since 2003, including two pairs in 2006. In July 2018, 14 birds there included four young birds. The site is difficult to survey, but it appears that a few pairs breed. In the early 1980s two pairs entered crevices below Pen Cilan, displaying agitation and showing every sign of breeding. The species has never been recorded breeding on Bardsey.

In Pembrokeshire, Lockley *et al.* (1949) recorded singles at the Teifi Estuary, St Davids Head and Jack Sound in June/July in five years during 1914–33. A pair was in Ramsey Sound during the summers of 1968 and 1969. One was seen next to a hole at Stackpole Head in 1987 (Donovan and Rees 1994) and birds were seen in suitable breeding habitat in 2003 and 2007. Breeding was confirmed in 2008 in Fishguard harbour, the most southerly nest in Wales, with a chick seen each year between 2009 and 2011. Access difficulties have meant that only the presence of birds, assumed to be breeding, has been recorded since (Annie Haycock *in litt.*).

Data collected in 2015–19, as part of the ongoing Seabirds Count census, show a reduction to 19 pairs, although counts of 19 adults at Fedw Fawr in April 2018, and again on 1 May 2019, suggest that this might be an underestimate.

During autumn and winter, birds disperse locally and moult into white and mottled plumage. Winter observations of Black Guillemots, although not numerous, have increased since the 1970s. Birds are seen regularly at many locations around Anglesey, often in the wake of storms. They are relatively scarce off other counties, even adjacent Caernarfonshire, although there was an exceptional count of *c.*30 off Nefyn in March 1972 and up to 14 around Bangor Pier and off the Great Orme in early 2019. Meirionnydd has produced only a handful of winter records, mainly as tideline corpses. In Ceredigion only three records were known up to 1974, but there have been 19 subsequently, some probably involving the same individuals. In Pembrokeshire, 1–2 are now seen in most years, usually between September and March. Black Guillemots are rare on the south coast, east of Pembrokeshire. There has been only one Carmarthenshire record, at Burry Port on 28 October 1973, although there have been more regular records off Gower, usually singles but with four in 1990. One off Goldcliff Point in 2015 was the first record for Gwent. On the north coast, birds are seldom seen farther east than Rhos-on-sea, Caernarfonshire, although singles were at Point of Ayr, Flintshire, in July 1990 and March 1992. Many of those around the Welsh coast in winter may be from the Irish breeding population, rather than from North Wales. The three ringing recoveries in Wales all involved birds ringed as nestlings in Ireland. One found dead at Pwllheli, Caernarfonshire, following a storm in October 2002, had been ringed at Rockabill, near Dublin, the previous July, while another ringed at Rockabill in July 2006 was found dead in Holyhead harbour in July 2012. One ringed at Bray Head, Co. Wicklow, in July 1987 was found long dead at Aberffraw, Anglesey, in July 2002.

The breeding population in Wales is small and, because it remains a rare breeder, it has an Amber conservation status (Johnstone and Bladwell 2016). Many Welsh colonies are not well-monitored because they are difficult to access. No systematic data on productivity in Wales is available, and future prospects cannot be assessed. The population may remain stable in the short term, but by the end of this century, climate modelling indicates that Wales will no longer be within its breeding range (Huntley *et al.* 2007).

Robin Sandham

Sponsored by Cambrian Ornithological Society

Puffin *Fratercula arctica* Pâl

Welsh List Category	IUCN Red List			Birds of Conservation Concern			
	Global	Europe	GB	UK	Wales		
A	VU	EN	LC	Red	2002	2010	2016

This charismatic seabird breeds around the North Atlantic, from northeast Canada to Novaya Zemlya in Russia (*HBW*). Like other burrow-nesting seabirds, it is very vulnerable to terrestrial predators such as rats, and so usually breeds on offshore islands or steep coastal cliffs. The largest populations are in Iceland and Norway. About 600,000 pairs, around 9% of the world population, breed in Britain and Ireland (Harris and Wanless 2011). Of these, around 27,000 pairs breed in Wales.

Puffins were much more numerous in Wales in the past, with particularly large colonies in Pembrokeshire and Anglesey, totalling several hundred thousand birds. Numbers had plummeted by the early 20th century, with some colonies greatly reduced (e.g. Skomer, Pembrokeshire) and others almost completely deserted (Puffin Island, Anglesey). The exact causes of these declines are difficult to pinpoint, but include rats and human exploitation, along with likely changes in the marine environment affecting food availability. Nonetheless, the two current largest colonies, on Skomer and Skokholm in Pembrokeshire, have recovered steadily since the 1990s, although they remain far below their previous peaks.

Puffins are migratory. Ringing recoveries of Welsh birds reveal movements to the Bay of Biscay and the Irish Sea, with a few recoveries farther afield in Norway, the Mediterranean Sea and even the Canary Islands. Advances in tracking technology have revealed their wintering movements in greater detail and shown that patterns differ between colonies (Fayet *et al.* 2017). Puffins from Skomer have a dispersive migration. Neighbours, breeding a few metres apart in summer, may spend winter thousands of kilometres apart. Some remain in the Irish and Celtic Seas, but others head out to the mid- or west North Atlantic, and some even visit the Mediterranean in late winter. Puffins can be vulnerable to bad weather, including outside the breeding season. Extreme storms in the North Atlantic in winter 2013/14 led to mass mortality, with over 29,000 found wrecked, mainly on the French coast. About 7% of recovered ringed birds came from Wales (Morley *et al.* 2016), and adult survival on Skomer in 2014 dropped sharply from approximately 90% to 75%, suggesting that Welsh Puffins had been substantially affected.

In Wales, Puffins appear near their breeding colonies in mid- to late March. They usually arrive, *en masse*, on land in April, and after a period of short visits, settle and start laying their eggs. The breeding season usually lasts until mid-July, after which fewer birds are seen, and most are gone by August. Puffins are particularly sensitive to handling and tracking. It is only with the development of miniature ultra-light GPS loggers, in the late 2010s, that it became possible to track their feeding movements around the colonies whilst breeding. Emerging results suggest that Puffins take multiple short trips 5–10km from the colony, interspersed with much longer ones (Fayet *et al.* 2021). On Skomer, birds took long trips westwards into the Celtic Sea, often covering over 100km on a single trip.

Operation Seafarer 1969–70	SCR Census 1985–88	Seabird 2000 1998–2002	Seabirds Count Provisional 2015–19
4,255	11,116	10,328	27,831

Puffin: Total number of apparently occupied burrows (AOB) or pairs recorded in Wales during UK seabird censuses.

Seabird censuses over the last 50 years have shown an 11-fold increase in occupied Puffin burrows in Wales, most of which occurred between 1998–2002 and 2015–19. Puffin numbers can be measured in different units, and this account reports the units of the original sources. These include number of individuals,

Puffin: Recovery locations of birds ringed in Wales are shown by red circles. Ringing locations of birds that were recovered in Wales are shown by blue triangles. Small brown dots show ringing or recovery locations in Wales.

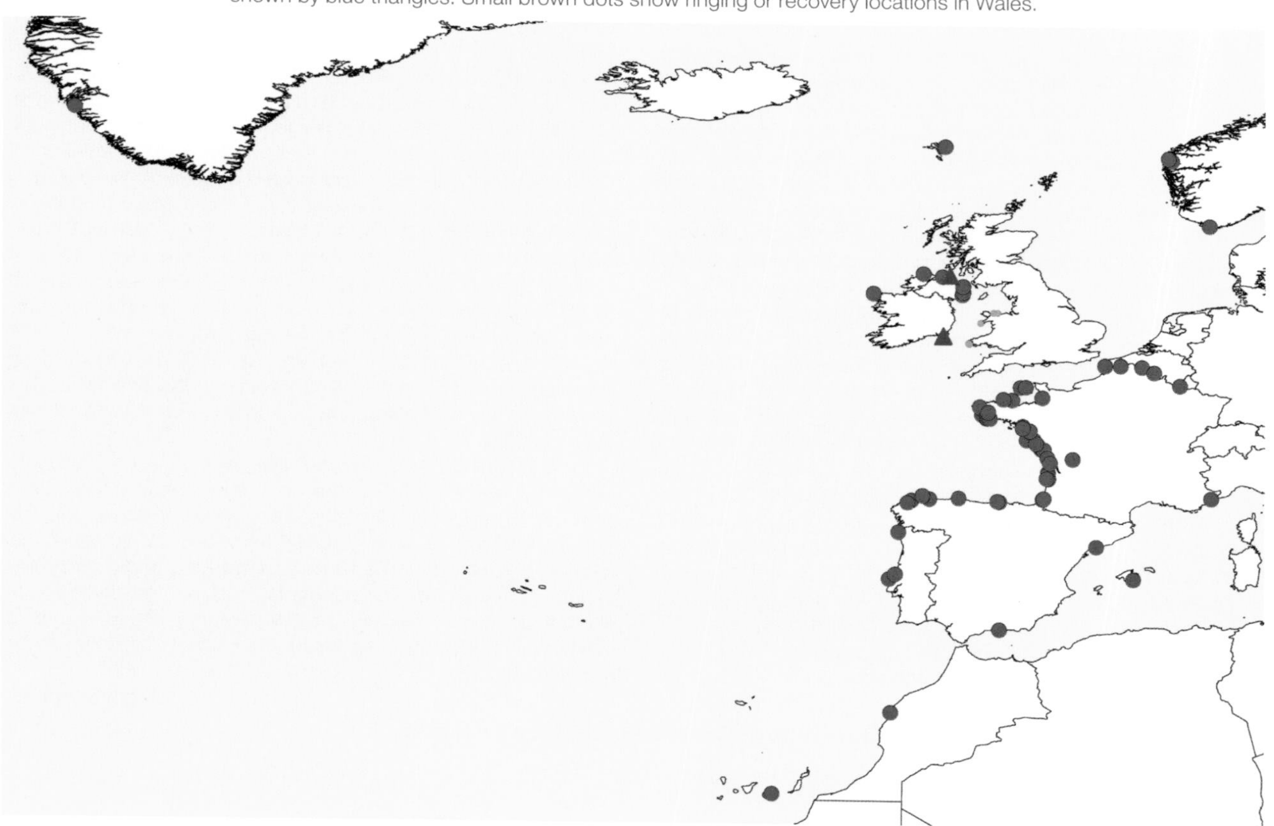

Puffin © Ben Porter

number of pairs (half the number of individuals) and number of apparently occupied burrows (AOB). AOB is a common metric to record underground nesting species, being approximately equivalent to number of pairs. The history of Puffin colonies in each recording area is summarised below.

Gower

Very small numbers of Puffins have bred at Worms Head. In 1848 the species was said to nest there annually. Numbers are thought to have declined towards the end of the 19th century, because of the depredations of rats (Hurford and Landsdown 1995). The maximum count in the last 50 years has been eight birds, with seven there in June 2017.

Pembrokeshire

Pembrokeshire has long been, and continues to be, the Puffin stronghold in Wales, although numbers now are far lower than in the past. Until the end of the 19th century, key colonies included Grassholm and the sister islands of Skomer and Skokholm. The last two are now the major Puffin colonies in Wales, with smaller colonies found on other islands.

Caldey

There were said to be "many" on Caldey in 1804 (Donovan 1805). Dix (1869) noted that Puffins bred in great numbers there, and that the boys and men of Tenby slaughtered them wholesale on Whit Monday, a traditional date for this activity. Walpole-Bond (1904) noted "a small colony" in 1902. Sage (1956) stated that there were "a few pairs" in 1927, but no breeding in 1949. No breeding has been recorded since.

St Margaret's Island

"A small colony" was present in 1903 (Walpole-Bond 1904), around ten pairs in 1952–55 (Sage 1956) and 4–10 pairs in 1983. Up to five occupied nests were recorded in 1999–2009 and 1–4 birds during 2014–19. The presence of rats on the island is most likely responsible for the small number of pairs, restricted to breeding in cliff crevices.

Grassholm

Mathew (1894) wrote that he believed that the number of Puffins on Grassholm was equal to all other birds in Pembrokeshire added together. In 1890, J.J. Neale estimated over half a million Puffins there and other authors estimated several hundred thousand, but the most credible figure is around 200,000 birds (Williams 1978), which would have been one of the largest colonies in Britain. The colony collapsed at the turn of the 20th century, partly due to birds destroying the habitat, although other reasons must have played a part. Donovan and Rees (1994) stated that the ground became so honeycombed with burrows that it collapsed and the birds largely forsook Grassholm, probably by 1920. Local opinion was that many moved to breed on Skokholm. There were 130–200 pairs in 1928–34, but Lockley *et al.* (1949) said that there were "scarcely 50 pairs" in 1946. There were 1–2 breeding pairs in 1973, and a few birds have been seen on the sea close to the island since. Much of the island is now occupied by the large Gannet colony.

Skokholm

Numbers increased on Skokholm in the first part of the 20th century, partly as a result of birds abandoning nearby Grassholm. Numbers decreased sharply after the Second World War, possibly as a result of oil pollution, but increased again after 2000. There were *c.*20,000 pairs in 1930–40 (Harris and Wanless 2011), 6,000 pairs in 1955 and 2,500 pairs in 1971. About 3,000 birds were present in 1994–2000, 5,000 birds in 2007, 8,700 birds in 2018 and 7,447 in 2019.

Skomer

Skomer is the major Puffin colony in Wales since the collapse of Grassholm in the early 20th century. As on Skokholm, the breeding population plummeted around the 1950s and is now

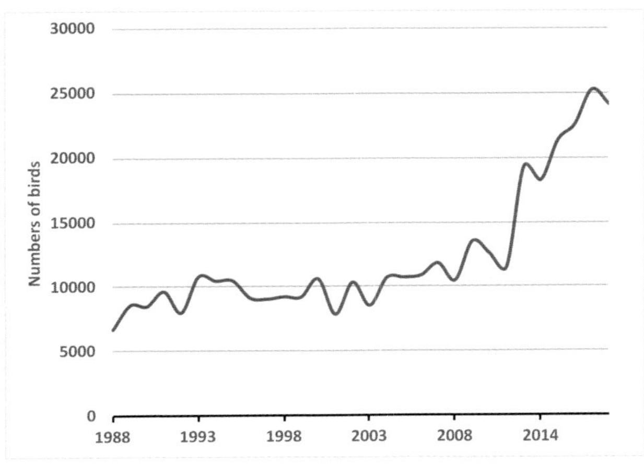

Puffin: Maximum spring counts of individuals on Skomer, 1988–2019.

growing again. Matthew (1894) described the numbers there as "marvellous" and noted "there is scarcely a yard of ground free from them." Lockley *et al.* (1949) said that the total number on Skomer and Skokholm was "probably not less than one hundred thousand pairs". About 6,000 pairs were estimated there during 1963–71, and *c.*5,200–6,500 pairs in 1975. The Seabird Colony Register counts in 1985–88 estimated *c.*5,000–6,000 pairs. The *Skomer Seabird Report* reported around 10,000 birds between 1990 and 2004. Counts from 2004 also include Middleholm. There were *c.*10,500 birds in 2004–08 and 11,500 birds in 2012, followed by a rapid increase to 24,108 birds in 2019.

Middleholm

Middleholm is a very small island next to Skomer and in recent years its counts have been amalgamated with those of Skomer. While records in the 1930s estimated *c.*1,000 pairs, numbers had dropped to *c.*80 pairs by 1966 and subsequently grew to *c.*200 pairs in 1983 (Lockley and Saunders 1967).

Ramsey

Large numbers of Puffins allegedly bred on Ramsey in the 16th century, but the reliability of those accounts is difficult to establish. The arrival of Brown Rats, presumably after a shipwreck in the 19th century, is likely to have been the key driver of the Puffins' disappearance from the island. By the late 1890s, the birds were restricted to a small area of the island (Mathew 1894) and eventually all breeding activity stopped. Despite a successful rat eradication in 2000, the Puffins have yet to return. In 2009 the RSPB began deploying decoys to try to attract birds back to the island. Birds have been seen on land, including 80 in 2018, but no breeding activity has yet been reported.

Ynys Bery (off Ramsey)

A single account from 1810 (Fenton 1811) reported large numbers of Puffins on Ynys Bery, but there have been no more reports of birds breeding on the islet.

Bishops and Clerks

A few pairs were reported to breed by Lockley (1953). 40–50 pairs were estimated on North Bishop in the 1970s and *c.*30 pairs in 1992. 30–60 individuals were reported in 2010.

Pembrokeshire mainland

A few pairs of Puffins breed along the south Pembrokeshire coast. A handful are also seen on the Castlemartin Peninsula each year, mainly near Stackpole Head in cliff crevices. A photograph from the 1930s showed at least 30 Puffins on land. It suggested that they nested in burrows along the cliff edge. At that time potential predators were heavily controlled by Cawdor Estate keepers and coastal access was restricted. Today the Puffins only occupy cliff crevices, in competition with Razorbills (Bob Haycock *in litt.*). Ten pairs were recorded in 1983, and 1–5 occupied crevices confirmed in more recent years (2012–19).

Ceredigion

Cardigan Island

Dix (1869) stated that Puffins were breeding there, and the colony held 20–30 pairs in 1924. It met the same fate as Ramsey: following a shipwreck in 1934, rats landed on the island and the colony disappeared. The rats were eradicated in 1969 and attempts were made to attract the Puffins back by placing decoy Puffins on the grassy slopes, but the birds have not returned (Roderick and Davis 2010). There were reports of a few birds on mainland Ceredigion around New Quay Head in the mid-20th century but breeding was never confirmed.

Caernarfonshire

Colonies on the south coast of Llŷn collapsed in the 20th century, most likely because of rats.

Bardsey

Ray and Willughby visited Bardsey in 1662 and recorded the "Prestholm puffin" breeding there, but Aplin (1902) thought that the birds they had seen were probably breeding on nearby Ynysoedd Gwylan. Breeding was suspected on Bardsey during the first half of the 20th century, although never proven, with the last evidence of possible breeding in 1953. Birds were seen coming ashore in the 1990s, but there was no evidence of breeding until 2000. At least two pairs bred in 2002 and 2003, and the colony grew to an estimated ten pairs by 2005, 75 AOB in 2015 and 143 AOB in 2019.

St Tudwal's Islands

Pennant (1810) mentioned Puffins there in 1773, and Bingley (1804), who visited in 1798, recorded that a considerable sum of money was annually made of these islands as Puffin-warrens, with birds harvested for food. Forrest (1907) stated that this was the largest colony in North Wales, and that "the Puffins here are so numerous that they cannot be estimated, but they probably number some hundreds of thousands", although this was probably an over-estimate. Ticehurst said that there were "some thousands of pairs" there in 1922, but in 1935 James Fisher said that the colony was much reduced, though still large. He counted 10–20,000 burrows, but by no means all were in use. There were no rats there in 1935, but a visit in 1951 found much evidence of their presence on St Tudwal's Island East, and no sign of breeding Puffins. The rats were said to have reached the island in 1949. There was no evidence of rats on St Tudwal's Island West, but this island had also been deserted by the Puffins (Thearle *et al.* 1953).

Ynysoedd Gwylan (Gwylan Islands)

Puffins have long been reported to breed on these two small islands, and while numbers decreased during the 20th century, they do not seem to have plummeted so much as in other Welsh colonies. Aplin (1901) recorded "a good many" burrows there in 1900. In 1910 he estimated as many as 1,000 pairs on Ynys Gwylan Fawr with small numbers on Ynys Gwylan Bach. He also mentioned a few pairs on the mainland opposite the islands, between Trwyn y Penrhyn and Ogof Lwyd. J.M. Harrop estimated 450–500 pairs on Ynys Gwylan Fawr in 1958. An estimated 400 pairs bred on the larger island in 1966. There were 15–20 pairs on Ynys Gwylan Fach in 1961, although there was no evidence of breeding there in 1968. The first accurate count of apparently occupied burrows on both islands was in 1976, when 777 were counted on Ynys Gwylan Fawr and 52 on Ynys Gwylan Bach. There were 457 AOBs in 1998, rising to 524 in 1999. There was then a large increase to 1,113 apparently occupied burrows in 2000, although numbers dropped again after 2001. Parts of the colony are overgrown with Tree Mallow, which makes a full count difficult, but in 2014 there were 871 AOB on Ynys Gwylan Fawr, the highest total since 2001. There were 619 AOB in 2019.

Great and Little Ormes

Pennant (1810) mentioned a few Puffins on the Great Orme in the 1770s, but said that the Little Orme "swarmed" in season with Puffins. Forrest (1919) stated that a few pairs bred on both headlands. Jones and Dare (1976) stated that occasional pairs probably bred on the Great Orme in the 1940s and early 1950s. There was thought to be possible breeding on the Little Orme in 2001, and in 2005, a pair was present in June and a young bird seen with an adult in July.

Anglesey

Anglesey has had three main breeding sites for Puffins: The Skerries, South Stack cliffs and Puffin Island. South Stack and Puffin Island, like many others, declined greatly in the 20th century and now host a handful of pairs. The population on The Skerries has fluctuated around a few hundred pairs since the 1990s, but it is unclear how this compares to earlier numbers.

The Skerries

Birds were reported present in the 1770s but did not breed there in 1839 (Jones and Whalley 2004). There were about 20 pairs in 1911 (Jones and Whalley 2004) and 10–50 pairs in 1951 (Whalley 1954). By the end of the 1990s there were 250–300 pairs, 170–340 nests were recorded in 2001–09 and 250–580 nests in 2010–18. There were 602 AOB in 2019.

South Stack Cliffs

Coward (1902) recorded "large numbers" here. There were 10–15 pairs in 1950. RSPB reports indicated 120–170 pairs in 1971–82 and 30–110 pairs in 1983–92. 60 birds were reported in 2000, 20–30 birds in 2006–09 and 7–16 birds in 2010–19.

Puffin Island (Priestholm or Ynys Seiriol)

Ray (1662) visited the island and recorded Puffins breeding. Pennant (1810) visited in about 1784 and commented on "the myriads which annually resort to Priestholm". He said that the young birds were pickled and preserved by spices. Bingley (1804) visited in 1798 and stated that "upward of fifty acres of land were literally covered with Puffins". He estimated their numbers at "upwards of 50,000". In the early 19th century, the very large population on Puffin Island was still heavily exploited by locals, but in 1817 rats were said to have reached the island after a shipwreck. Williams (1835) stated that Puffins were found there "in great numbers" and Hicklin (1858) stated that they were present in large numbers. However, R.E. Williams of Beaumaris, in 1866, stated that there was scarcely a single bird to be seen, and Price (1875) considered that the Puffins had gone. There is evidence that the rat population of the island was controlled and perhaps eliminated in the late 19th century (Arnold 2004). Forrest (1907) stated that the island was entirely deserted by Puffins for several years in the 1880s, but that birds returned in considerable numbers. He estimated fewer than 50 pairs in 1903, but "at least two thousand" in 1907. Coward (1922) doubted that the island had ever been completely deserted. Puffin numbers fluctuated between the 1920s and the 1960s, with up to 500 pairs estimated to be present in some years. Rats may have recolonised the island sometime after about 1960 (Arnold 2004). Puffin numbers were high in the early 1970s, with a maximum count of 450 pairs during 1970–74, but declined steadily over the next 25 years, with a maximum count of just 18 individuals during 1995–99. Counts of occupied burrows in this period showed the same trend (Arnold 2004). Rats were eradicated from the island in 1998–2000, but this did not lead to any great increase in the population, with counts during 2011–19 varying between five and 29 birds.

In Europe, where most of the world population breeds and where declines have been most pronounced, the species was classified as Endangered in 2015. The Puffin's Red-listed conservation status in Wales is because of its global and European conservation status and also because of the high degree of localisation of the breeding population (Johnstone and Bladwell 2016). Despite recent increases, Puffin numbers in Wales are much reduced compared to historical levels and represent less than 3% of the global population. Nonetheless, Puffin populations, on Skomer and Skokholm in particular, are potentially important for the species' conservation. Unlike many colonies in Scotland, Iceland and Norway, where breeding success and population size have declined substantially in recent times, some since the 1970s, these populations are growing, and adult survival and breeding success are high. Why this is the case remains unclear, but it is likely linked to greater food availability during the breeding season (Fayet *et al.* 2021). Modelling suggests that the breeding range of Puffins will contract during the 21st century, but that around Britain the impact will be far greater in the North Sea than on the west coast (Huntley *et al.* 2007). Studying these populations, and comparing their ecology to declining colonies, may help us to understand why Puffins are declining elsewhere; as the populations have grown, their importance for the species increases. Long-running programmes carefully monitoring Puffins on these islands, as well as research investigating their ecology, must continue in the future. Biosecurity measures, to ensure rats never invade these islands, are also crucial to safeguard these populations.

Annette Fayet

Sponsored by Ian Beggs

Red-throated Diver *Gavia stellata* Trochydd Gyddfgoch

Welsh List Category	IUCN Red List			Birds of Conservation Concern			
	Global	Europe	GB	UK	Wales		
					2002	2010	2016
A	LC	LC	LC	Green			

The smallest of the three divers regularly seen off the coasts of Wales, this species has a circumpolar distribution, extending across Canada, northern Europe and Russia (*HBW*). The European population trend is unknown, but the Scottish breeding population of about 1,000–1,600 pairs is thought to be decreasing (BirdLife International 2015). It winters off coasts to the south of its breeding range; in Europe as far south as Spain (*HBW*). The Red-throated Diver has never been recorded breeding in Wales, but good numbers spend the winter at some sites here. Birds are usually recorded between September and May, but have been seen in all months of the year. Only three ringed birds have been recovered in Wales, of which two had been ringed in Sweden and one in Orkney, all as nestlings.

This has always been the commonest diver species along the Welsh coast, although it occurs inland less frequently than Great Northern Diver. Mathew (1894) and Lockley *et al.* (1949) reported that it was a common winter visitor off Pembrokeshire, especially in Milford Haven. Forrest (1907) described it as common on the west coast of North Wales from autumn to May, but less frequent on the north coast, although he quoted a record of about 50 between Llanddulas and Old Colwyn, Denbighshire, on 31 March 1907.

Cardigan Bay began to emerge as an important site for this species in the late 1970s, with *c.*100 off Wallog, Ceredigion, on 10 January 1977. Numbers increased in the next few years, with 557 off Borth, Ceredigion, on 2 December 1990 (Fox and Roderick 1990). Co-ordinated counts in the northern part of the Bay, between 1991/92 and 1997/98, produced some impressive numbers, with a mean winter count of 666 (Thorpe 2002). The largest concentrations were usually south of the Mawddach estuary, off Wallog, Borth and Aberdysynni. These were associated with shallow water over Sarn Cynfelin in Ceredigion and Sarn y Bwch in Meirionnydd, with smaller numbers around Sarn Badrig, also in Meirionnydd (Thorpe 2002). A count of 1,916 between Wallog and Llanbedrog, Caernarfonshire, in January 1997 is a record for Wales.

Outside this survey, there was a concentration of 900 in Borth Bay on 14 December 1995. Numbers in Cardigan Bay usually peaked during January, then fell sharply in February. Counts off Borth were lower during 1998–2006 (Roderick and Davis 2010), although 732 were there at the end of 2001, and remained low in the following decade. It seems likely that there are fewer birds in Cardigan Bay now than in the 1990s, but in the absence of further co-ordinated counts this cannot be confirmed.

Good numbers have been recorded off Llanfairfechan, Caernarfonshire, including 150 in March 1993, 202 on 5 April 2003 and 298 in March 2012. Large numbers have also been recorded in Caernarfon Bay, between Fort Belan and Trefor, including 193 on 27 March 2012 and 286 on 19 February 2013. There can also be significant numbers farther east in Colwyn Bay, such as 131 off Rhos Point, Caernarfonshire, in March 2016, and 189 off

1991/92	1992/93	1993/94	1994/95	1995/96	1996/97	1997/98
427	390	740	252	404	1916	536

Red-throated Diver: Peak counts in northern Cardigan Bay in winter between 1991/92 and 1997/98.

Site	County	94/95 to 98/99	99/00 to 03/04	04/05 to 08/09	09/10 to 13/14	14/15 to 18/19
Caernarfon Bay	Caernarfonshire	2	8	15	169	78
Traeth Lafan	Caernarfonshire/Anglesey	1	87	17	94	70
Colwyn Bay and north Clwyd coast	Denbighshire	3	15	2	11	35

Red-throated Diver: Five-year average peak counts during Wetland Bird Surveys between 1994/95 and 2018/19 at sites in Wales averaging 30 or more at least once. International importance threshold: 3,000; Great Britain importance threshold: 210.

Llanddulas, Denbighshire, in February 2009. Numbers around the Dee Estuary are generally low.

Off the south coast, numbers are usually lower, but there was an exceptional count of 200 at Amroth, Pembrokeshire, from late February to early March 1993. 70 were in Carmarthen Bay in December 1980, 66 flew east at Cefn Sidan, Carmarthenshire, on 23 January 1992 and there were 85 between Pendine and Telpyn Point on 21 February 2000. Off the Glamorgan coast, numbers were low prior to a dramatic increase during 1980–92 (Hurford and Lansdown 1995). The largest counts there were off Gower, including 80 in Rhossili Bay on 26 December 1992. Farther east, the maximum counts were lower, such as a minimum of 35 off Sker, East Glamorgan, on 30 January 1989 and 40 at Nash Point on 2–7 December 2016. The species is scarcer farther up the Severn Estuary. Up to the end of 2019 there had been 14 records of Red-throated Diver in Gwent, involving 20 birds (Darryl Spittle *in litt.*).

Counts at the main sites off the Welsh coast vary greatly from year to year. This species is not well covered by the Wetland Bird Survey (WeBS), as many birds are usually a long way offshore and accurate counts require calm conditions and good visibility. The WeBS index for Wales shows a decline in recent years.

The highest counts in Caernarfon Bay and off Traeth Lafan are usually in spring. The movements of Red-throated Divers around the Welsh coast are not well understood, but the increase in numbers off the north coast of Caernarfonshire, in March and early April, coincides with a reduction in Cardigan Bay from February (Thorpe 2002, Roderick and Davis 2010). Birds are recorded passing coastal watch points in autumn and winter, sometimes in considerable numbers. Off Point Lynas, Anglesey, 90 passed in two and a half hours on 27 November 2012 and 122 passed in three and a half hours on 21 November 2015. Counts off Strumble Head, Pembrokeshire, are usually lower, although a total of 207 passed there between 4 September and 31 December 2000, including 66 on 14 December. Inland, there are fairly regular records, although fewer than for the Great Northern Diver. Red-throated Divers have been recorded in all counties, but there have been none in Radnorshire since 1910 (Jennings 2014).

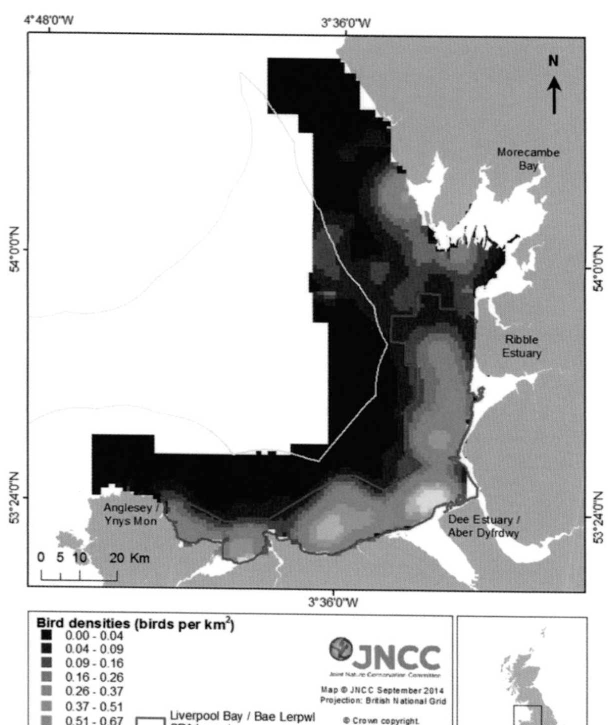

Red-throated Diver: Estimated mean density from aerial surveys in Liverpool Bay 2004/05 to 2007/08, and 2010/11 (from Lawson *et al.* 2016a; map courtesy of and © JNCC).

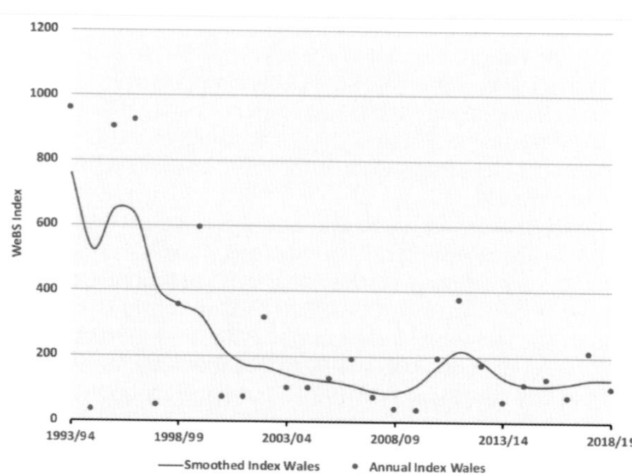

Red-throated Diver: Wetland Bird Survey indices, 1993/94 to 2018/19.

O'Brien *et al.* (2008) estimated the wintering population in Britain, primarily from aerial surveys carried out in 2001–06. They estimated 249 off the southern coast of Wales, 612 in Cardigan Bay and 1,061 in Liverpool Bay, which included the coast of northwest England as well as North Wales. This would suggest approximately 1,300–1,400 birds around the Welsh coast, about 8% of the estimated total of 17,000 wintering in British waters. Oil pollution may be the main threat to the conservation of this species here, as illustrated by the 81 Red-throated Divers known to have been casualties of the *Sea Empress* oil spill in Pembrokeshire, in February 1996. This was documented as being potentially "a very high proportion of the local wintering population at the time of the spill" (*Sea Empress* Environmental Evaluation Committee 1998). It was thought that many of the 20 casualties examined were not from the Scottish/Irish breeding population (Weir *et al.* 1997).

Rhion Pritchard

Black-throated Diver *Gavia arctica* Trochydd Gyddfddu

Welsh List Category	IUCN Red List			Birds of Conservation Concern			
	Global	Europe	GB	UK	Wales		
A	LC	LC	VU	Amber	2002	2010	2016

This diver breeds across northern Europe and Asia to western Alaska. It winters to the south, in Europe as far south as Spain and the northern Mediterranean (*HBW*). The largest European populations are in Russia and Finland, with a smaller population of about 190–250 pairs in Scotland (BirdLife International 2015). There have been no ringing recoveries to indicate the origin of the birds wintering off the coasts of Wales. Some are likely to be from the Scottish population, but there may also be birds from Fennoscandia.

The Black-throated Diver is the least common of the three regular divers seen in Wales. There is one archaeological record: bones found in Port Eynon Cave, Gower, dating from the Mesolithic era, approximately 6,000 to 12,000 years ago (Yalden and Albarella 2009). Mathew (1894) thought that this species must occur off Pembrokeshire, but knew of no definite record. Barker (1905) noted that it had occurred near the mouth of the Gwendraeth Estuary in Carmarthenshire. Forrest (1907) said that this was by far the rarest of the divers off North Wales, though he quoted Professor Salter as saying that immature birds were "not infrequent" in the Dyfi Estuary, Ceredigion/Meirionnydd. The first record for Gwent was one found dead in a Newport garden on 5 February 1915. The first dated record in Glamorgan was in 1924, with ten more between 1925 and 1967, some of them inland (Hurford and Lansdown 1995).

The Black-throated Diver was described as "almost as frequent as Great Northern Diver in winter" in Pembrokeshire after the Second World War (Lockley *et al.* 1949), but the first reliable record for Ceredigion was not until 1961 (Roderick and Davis 2010). Although it is fairly regular off the Carmarthenshire coast, numbers are usually low, with none recorded between 2007 and 2015. There have been 15 records in Gwent, involving 18 birds (Darryl Spittle *in litt.*). There was an early inland record in Radnorshire, one shot on the River Wye at Glasbury "many years before 1882" (Jennings 2014).

Numbers vary from year to year, with no trend apparent in recent decades. Most records are between late August and May, with numbers peaking in December and January. There is something of an influx in some years, notably in 1979, but also in 1986 and 1987. Birds are normally seen in ones and twos, but there have been some larger counts: up to ten in Holyhead harbour, Anglesey, in March 1979; up to six in St Brides Bay, Pembrokeshire, from January to March 1986; and eight in Red Wharf Bay, Anglesey, on 22 February 1987. A small spring passage is sometimes evident, for example up to seven off Aberystwyth,

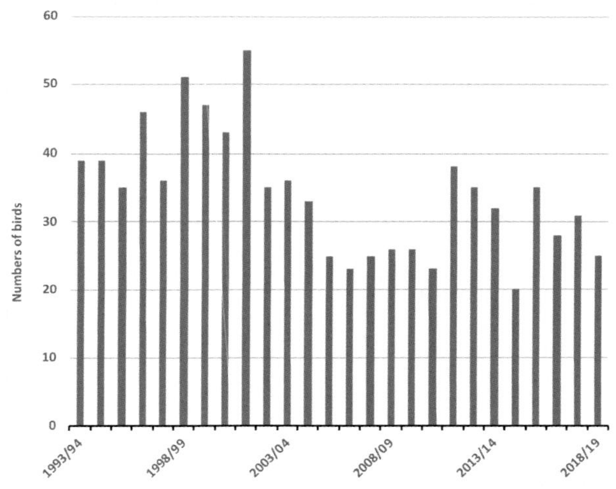

Black-throated Diver: Estimated number of individuals recorded in Wales during each winter period, 1993/94 to 2018/19.

Ceredigion, in spring 1995. Autumn passage is usually also small, although a total of 20 passed Strumble Head, Pembrokeshire, in late 2015, including a Welsh day record of 15 on 11 December.

Although there are fewer inland records than for Great Northern Diver, this species has occurred on many lakes and reservoirs. There were 13 records totalling 15 birds from Llandegfedd Reservoir, Gwent, during 1975–2002, and there have been several records from Eglwys Nunydd Reservoir, Gower. Most inland records are singles, but there were three on Llyn Trawsfynydd, Meirionnydd, on 28 March 1979 and at Llandegfedd Reservoir on 7 January 1985.

The Welsh wintering population may be no more than around 30–40 birds, of no great significance in a British or European context: the British winter population is estimated at 560 (Frost *et al.* 2019a). Threats include oil spills and birds being caught in fishing nets. By the end of this century, it is expected that Scotland will cease to be climatically suitable for breeding Black-throated Divers (Huntley *et al.* 2007), while Fennoscandian birds may stay closer to their breeding grounds through the winter. These outcomes would probably lead to it becoming a far rarer visitor to Wales.

Rhion Pritchard

Pacific Diver *Gavia pacifica* Trochydd y Môr Tawel

Welsh List Category	IUCN Red List	Rarity recording
	Global	
A	LC	BBRC

The Pacific Diver breeds across Canada and Alaska, and in northeast Siberia, and winters in the Pacific Ocean, as far south as northern Mexico (*HBW*). Up to the end of 2019 there were nine records in Britain (Holt *et al.* 2020). Only one has been recorded in Wales, although this bird returned to spend part of three winters at the same site, Llys-y-frân Reservoir, Pembrokeshire. Its first occurrence, from 2 February to 20 March 2007, was only a few days after the first record for Britain, in North Yorkshire. The third British record was in Cornwall, also in February–March 2007. The Pembrokeshire bird returned to Llys-y-frân from 16 January to 11 February 2008 and on 25–26 February 2009.

Jon Green and Robin Sandham

Pacific Diver on Llys-y-fran Reservoir, Pembrokeshire, January - February 2008 © Richard Stonier

Great Northern Diver *Gavia immer* Trochydd Mawr

Welsh List Category	IUCN Red List			Birds of Conservation Concern			
	Global	Europe	GB	UK	Wales		
					2002	2010	2016
A	LC	VU	LC	Amber			

The Great Northern Diver breeds in northern North America and Greenland, with smaller numbers in Iceland and Svalbard (*HBW*). The species has been suspected of breeding occasionally in Scotland, although breeding has been confirmed there only once, in 1970. Birds that winter along the coasts of western Europe, as far south as Spain, are thought to be mainly from the population that breeds on Iceland, but probably include some from Greenland and Canada (*HBW*). An estimated 4,300 winter around Britain (Frost *et al.* 2019a), but Welsh coastal waters hold only a small proportion of these. The species is usually recorded off the coasts of Wales from late August or September onwards, with some staying into late May and even into June. This is the largest of the three divers regularly found in Wales. It can feed in deeper water and is often found farther offshore than the other two species, making an accurate count difficult if conditions are less than ideal.

One was reported shot on the Tawe Estuary, Gower, on 23 November 1838, with several others said to have been killed in Glamorgan prior to 1848. According to A.R. Martin in 1864, the species was found occasionally in the Menai Strait around Bangor, Caernarfonshire, although adults were rare. Forrest (1907) reported that it visited western coasts of North Wales almost every winter, but was comparatively scarce along the north coast, although he mentioned ten in Porth Lleidiog near Aberffraw, Anglesey, following southwesterly gales in winter 1886/87. In the south, it was said to be numerous at times in Milford Haven, Pembrokeshire (Mathew 1894). Inland, three were on Llangorse Lake, Breconshire, in winter 1881.

There were fewer records in the first half of the 20th century, although Lockley *et al.* (1949) said that the species was sometimes numerous in Pembrokeshire. Numbers apparently increased from the late 1950s, with 7–8 in Conwy Bay, Caernarfonshire, in March 1956 and eight in the Dee Estuary, Flintshire/Wirral, in December 1961. It was said to be by far the commonest diver in Glamorgan during 1967–79 (Hurford and Lansdown 1995), with at least 15 in 1976, including seven in Oxwich Bay, Gower, on 11 January.

The tidal race off Strumble Head, Pembrokeshire, held 43 Great Northern Divers on 6 January 1991, following prolonged gales (Donovan and Rees 1994). Eight were among the casualties of the *Sea Empress* oil spill off Milford Haven in February 1996, which the final report on the environmental impact suggested "may have been a very high proportion of the local wintering population at the time of the spill" (*Sea Empress* Environmental Evaluation Committee 1998). Co-ordinated counts in the northern part of Cardigan Bay, between 1991/92 and 1997/98, produced only small numbers, between none and five in each year (Thorpe 2002).

Records increased in the early years of the 21st century, particularly as it became evident that Caernarfon Bay, Caernarfonshire, held large numbers in late winter and early spring. Regular counts showed that it consistently held numbers above the threshold of international importance. The only other sites of international importance in Britain are around Orkney, Shetland and the Hebrides. It seems likely that 30–40 Great Northern Divers winter at this site, with numbers augmented in late winter and spring by birds that have wintered farther south, possibly using the bay as a moulting area. Adult Great Northern Divers moult between February and April, shedding all their primaries simultaneously and becoming temporarily flightless (Appleby *et al.* 1986). The highest counts in Caernarfon Bay are usually in March or early April. There were 87 there in early April 2015, 110 on 13 March 2016 and 89 on 7 April 2018. Unlike Red-throated Diver, no large numbers are known to winter in Cardigan Bay, so these birds must have come from farther afield.

Other than birds passing coastal watch points, there are

Great Northern Diver in winter plumage © Peter Howlett

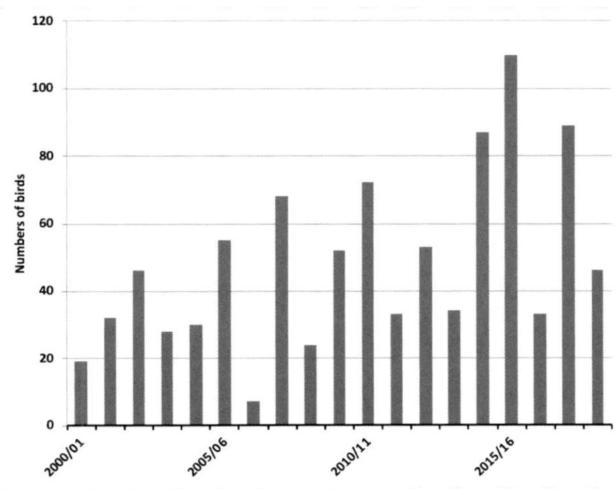

Great Northern Diver: Maximum WeBS counts in Caernarfon Bay, 2000/01 to 2018/19. International importance threshold: 50; British importance: 43.

few double-figure counts elsewhere. The highest counts are usually in March and April. Off the north coast, there were ten off Llanfairfechan, Caernarfonshire, on 1 April 2000, 20 off Traeth Lafan, Caernarfonshire, in March 2006 and 13 off Point of Ayr, Flintshire, on 18 April 2001. In northern Cardigan Bay, there were 13 off Traeth Bach, Meirionnydd, on 27 March 2011 and 30 off Abersoch, Caernarfonshire, on 27 March 2012. The Inland Sea and Alaw Estuary, Anglesey, held up to nine birds in winter 2013/14.

In the south, the Cleddau Estuary, Pembrokeshire, usually holds the highest numbers. A maximum of seven wintered there in 2000/01 and again in 2002/03. The 2015/16 Non-estuarine Waterbird Survey (NeWS) found a total of 34 off Pembrokeshire and Carmarthenshire, including 15 between St Brides Haven and Little Haven, Pembrokeshire, on 1 January 2016. Passage birds are seen regularly from Strumble Head in Pembrokeshire, Bardsey in Caernarfonshire and Point Lynas in Anglesey, mainly in autumn. *Birds in Wales* (1994) reported up to 26 a day past Strumble Head during autumn passage, but subsequent counts have been lower, the best years being 1999, when 71 were recorded between 11 September and 27 December, and 2015, when there were 23 on 11 December.

This species is found on freshwater more frequently than the other diver species, and it appears to favour certain lakes and reservoirs. Of a total of 30 records in Gwent, 18 have been at Llandegfedd Reservoir (Darryl Spittle *in litt.*). During 1992–2000, seven were recorded at Eglwys Nunydd Reservoir, Gower, while Llyn Tegid, Meirionnydd, has held birds regularly in recent years, including one or two in seven of the nine winters between 2007/08 and 2015/16. Other counties have very few freshwater records, despite having apparently suitable water bodies. For example, Anglesey has only had one, on Llyn Coron in January 1986, and Ceredigion also only a single record, one on Nant y Moch Reservoir, from 27 November to 4 December 1990.

The Welsh wintering Great Northern Diver population is probably no more than about 150 birds, with more moving through in spring and autumn. Although some have been caught and drowned in fishing nets, there appear to be few systemic conservation threats in Welsh waters. The species is very vulnerable to oil spills (Camphuysen 1989). This poses a risk, particularly in view of the concentration of birds in Caernarfon Bay.

Rhion Pritchard

Sponsored by Andy Jones

White-billed Diver *Gavia adamsii* Trochydd Pigwyn

Welsh List Category	IUCN Red List		Rarity recording
	Global	Europe	
A	NT	VU	WBRC

This diver breeds in northern Canada and northern Russia, mostly east of the Ural Mountains. It winters in the Pacific, and off the coasts of Greenland and Fennoscandia (*HBW*). Over 650 have been recorded in Britain, with increased occurrence since 2000; an average of four a year in 1970–79 increased to 18 in 2000–09 and 32 in 2010–18 (White and Kehoe 2020a). This included a record total of 72 individuals in 2018. The increase has been put down to greater recording effort, including the discovery and monitoring of spring staging areas off the Outer Hebrides and the coasts of Moray and Nairn in Scotland, which have produced most of the British records. This species remains a very rare visitor to Wales, with only three records. The first was one in Holyhead harbour, Anglesey, from 24 February to 19 May 1991. One passed Strumble Head, Pembrokeshire, on 27 September 1999, and another was recorded there on 1 September 2011.

Jon Green and Robin Sandham

Wilson's Petrel *Oceanites oceanicus* Pedryn Drycin Wilson

Welsh List Category	IUCN Red List	Rarity recording
	Global	
A	LC	WBRC

Regarded as one of the world's most abundant seabirds, with a population of 4–10 million breeding pairs (Brooke 2004), Wilson's Petrel breeds on islands in the southern hemisphere. However, outside the breeding season it occurs in all the world's oceans, from the coast of Antarctica to the Arctic waters of the Atlantic (*HBW*). In Britain, 986 were recorded between 1958 and 2018, mainly from dedicated pelagic trips, between June and September, off the Isles of Scilly (White and Kehoe 2020a). Smaller numbers have been recorded annually from sea-watches during strong onshore winds in late summer and early autumn.

It is a rare bird off the Welsh coast with a total of 13 individuals recorded, all either off Strumble Head, Pembrokeshire, or from pelagic trips into the Celtic Deep off Pembrokeshire:

- from the Rosslare to Fishguard ferry, a few kilometres off Strumble Head, on 12 September 1980;
- past Strumble Head on 3 September 1986;
- past Strumble Head on 6 September 1990;
- past Strumble Head on 5 September 1997;
- past Strumble Head on 11 September 1998;
- in the Celtic Deep off Pembrokeshire on 16 August 1999;
- three in the Celtic Deep off Pembrokeshire on 25 August 1999;
- two in the Celtic Deep off Pembrokeshire on 6 August 2000—one on 7 August was presumed to be one of these two;
- past Strumble Head on 1 August 2009 and
- in the Celtic Deep off Pembrokeshire on 5 August 2009.

Jon Green and Robin Sandham

Black-browed Albatross *Thalassarche melanophris* Albatros Aelddu

Welsh List Category	IUCN Red List	Rarity recording
	Global	
A	LC	BBRC

This huge seabird breeds on islands in the southern Atlantic and Indian oceans. Outside the breeding season, it disperses north to the cold-water areas of the Benguela and Humboldt currents, off the coasts of southwest Africa and western South America, reaching as far as the Tropic of Capricorn, around Namibia and northern Chile (*HBW*). It is the albatross species most frequently found in the eastern North Atlantic, with 34 records in Britain to the end of 2019 (Holt *et al.* 2020). Once in the northern hemisphere, some birds return to seabird colonies annually, notably one seen almost annually at Hermaness, Shetland, during 1972–95. It has occurred three times in Welsh waters. The first record was off Skokholm, Pembrokeshire, on 19 August 1990. One passed South Stack, Anglesey, on 12 February 2005, and the third was off Bardsey, Caernarfonshire, on 16 October 2016.

Additionally, an albatross species was seen off Bardsey on 1 May 1976, but was not identified to species.

Jon Green and Robin Sandham

Storm Petrel *Hydrobates pelagicus* Pedryn Drycin

Welsh List Category	IUCN Red List			Birds of Conservation Concern			
	Global	Europe	GB	UK	Wales		
A	LC	LC	LC	Amber	2002	2010	2016

The Storm Petrel, weighing less than 25g (the weight of two two-pound coins), is the smallest species of seabird nesting in Britain and Ireland. Outside the breeding season it is truly pelagic, spending the majority of its life at sea. It feeds mainly in flight, picking up small fish, squid and crustaceans from the sea surface, although it will dive to depths of up to 2m (*HBW*). Its vulnerability to mammalian predators on land means that virtually all breeding colonies are on remote, largely rat-free and rocky islands, scattered predominantly on the fringe of the eastern North Atlantic. Its global range is virtually restricted to Europe, although colonies are found on islands within the territorial waters of Morocco and Algeria (*HBW*). The only crude estimate of its global population is 300,000–680,000 breeding pairs, with the biggest concentrations found in the Faroe Islands and Iceland. The only census of the species in Britain, in 1998–2002, suggested a population of 25,650 apparently occupied sites (AOS), over 80% of which were at 59 colonies in Scotland, with small populations in western Wales and on the Isles of Scilly. There is a much larger population, estimated at 99,000 AOS, in the Republic of Ireland (Mitchell *et al.* 2004).

Outside the breeding season, Storm Petrels winter in the South Atlantic, predominantly off the coast of Namibia and South Africa. Non-breeding birds are thought to remain in that area until they return north, in their third or fourth calendar-year, to explore European coasts for potential nest sites in future years (*BTO Migration Atlas* 2002). There is a greater tendency for birds ringed around the Irish Sea to be recovered in the Bay of Biscay, while those ringed in Scotland are more likely to be recovered off southern Africa. This difference remains unexplained. One of the most surprising recoveries involved a bird ringed on Skomer, Pembrokeshire, in 1993, that was found dead following a storm in northwest Switzerland, on 28 December 1999. Of the Storm Petrels ringed at Strumble Head, Pembrokeshire, or caught there having been ringed previously elsewhere, during 1983–2003, most exchanges were with Ireland and the west coast of Britain. However, a small number involved Orkney and the North Sea coasts of England and Scotland (Spence 2015). Ringing also shows the longevity of these tiny birds: one ringed as an adult on Skokholm, Pembrokeshire, in 1974, was 37 years and 11 days old when recaught by a ringer on Alderney, Channel Islands, in 2011. This was only a year younger than the oldest recorded in the Britain and Ireland scheme (Robinson *et al.* 2020).

Storm Petrel: Recovery locations of birds ringed in Wales are shown by red circles. Ringing locations of birds that were recovered in Wales are shown by blue triangles. Small brown dots show ringing or recovery locations in Wales.

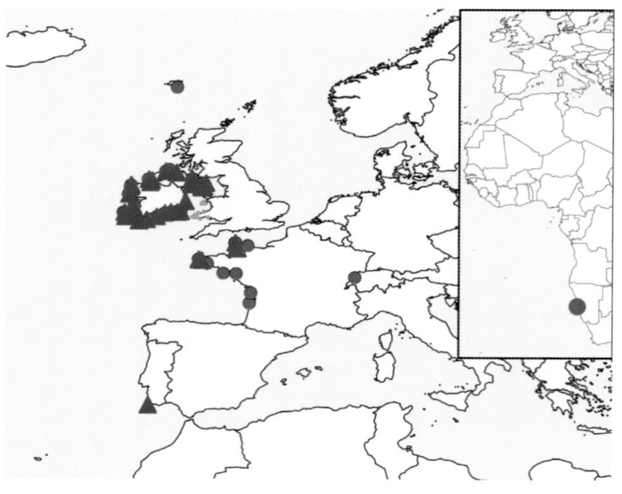

Their nocturnal habits and secretive nesting locations made it difficult for early naturalists to confirm whether Storm Petrels bred in Wales. The only confirmed nesting site at the turn of the 20th century was on Skomer, where they nested in a tumble-down wall (Mathew 1894). 30 pairs were believed to nest there in 1946 (Lockley *et al.* 1949). Their presence was discovered on Skokholm in 1931. An estimate made in 1939 suggested that 600 pairs nested there, with others on Stack Rocks (Lockley *et al.* 1949). They were suspected to breed in Aberdaron Bay, Caernarfonshire, in 1906 (Forrest 1907), but the first confirmed nesting in the county was not until a nest was found on Bardsey in 1926. For the remainder of the 20th century Bardsey's population was estimated to be around ten pairs. Three pairs were reported to have bred at a site on Anglesey in 1966 (Jones and Whalley 2004).

The accurate monitoring of Storm Petrels presents a considerable challenge as they nest in deep rocky crevices, stone walls and sometimes in burrows in soft soils in inaccessible or fragile habitats. In addition, they are nocturnal and sensitive to disturbance. As well as the three sites known prior to the Second World War, *Birds in Wales* (1994) reported contemporary breeding by small numbers on Ramsey, Middleholm and North Bishop, all in Pembrokeshire. Elsewhere in the county, it reported historic nesting on Green Scar, Stack Rocks and St Margaret's, but that the status of the species there at that time was unknown. The Seabird 2000 survey, the first to definitively describe the status of the Storm Petrel in Wales, recorded breeding on five islands in Pembrokeshire (Carreg Rhoson, North Bishop, Ramsey, Skokholm and Skomer) and Bardsey in Caernarfonshire (Mitchell *et al.* 2004).

Operation Seafarer 1969–70	Seabird 2000 1998–2002	Seabirds Count Provisional 2015–19
5,100–7,100	2,663–2,962	2,486

Storm Petrel: Total count of apparently occupied sites (AOS) recorded in Wales during seabird censuses. These counts are not comparable owing to different count methods.

The largest Storm Petrel breeding colony in Wales, and the fourth largest in Britain in 1998–2002, is on Skokholm, where 2,450 AOS counted during Seabird 2000 constituted *c.*3% of the British and Irish population (Mitchell *et al.* 2004). Skokholm, Skomer and Middleholm together form the Skomer, Skokholm and the Seas off Pembrokeshire/Sgomer, Sgogwm a Moroedd Penfro Special Protection Area (SPA) for which Storm Petrel is a classified feature. Censuses were attempted on Skokholm in 2003 (Thompson 2008) and Skomer in 2003–04 (Brown 2006), but these and previous estimates (Scott 1970, Vaughan and Gibbons 1996, Mitchell *et al.* 2004) are not comparable owing to differences in survey methods and reliability. As part of the fourth British seabird census, Seabirds Count (2015–21), NRW commissioned a repeat census of Skomer and Skokholm. Owing to concerns that previous methods may have provided inaccurate population estimates, this census used a playback method, along with the development of habitat-specific correction factors. This method is considered to be more reliable, as a result of the extent of survey coverage and the fact that it included the use of infra-red imaging, to find previously unknown breeding sites within inaccessible rock falls. Previous estimates are now considered likely to have over-estimated the population on the islands. A revised estimate of 2,130 AOS for the two islands combined, and hence the SPA, was probably not greatly different to the number present in 1998–2002 (Wood *et al.* 2017). Recalculated whole-island counts for Skokholm are 1,937 AOS in 2001, 1,810 AOS in 2003 and 1,925 AOS in 2016 (Matty Murphy *in litt.*). The largest colony in Wales outside Pembrokeshire is on Bardsey, where numbers have increased in the 21st century. A survey in 2017 produced an estimate of about 175 pairs there.

Breeding colony	County	1998–2002	2014–2018
Skokholm	Pembrokeshire	2,450	1,910
Skomer	Pembrokeshire	110	220
Bardsey	Caernarfonshire	35	175
Carreg Rhosen	Pembrokeshire	51	82
North Bishop	Pembrokeshire	57	81
Grassholm	Pembrokeshire		11
Ramsey	Pembrokeshire	102	7
Total		**2,805**	**2,486**

Storm Petrel: Number of apparently occupied sites (AOS) recorded by surveys in Wales. Counts are not comparable between periods owing to different count methods (see main text).

Away from potential nesting colonies, Storm Petrels are recorded in small numbers around the Welsh coast, including the Bristol Channel. Now, they are seen from April to October, but during the 19th and the first half of the 20th centuries they were recorded mainly as storm-blown waifs, usually in October. Generally, numbers were small, although the species was described as "numerous" following a storm in Glamorgan in 1828 (per Hurford and Lansdown 1995). They were occasionally found some distance inland, such as one donated to Shrewsbury Museum that was labelled as having been taken alive at Glansevern, Montgomeryshire, in 1843, and one found in a turnip field at Rhug, Meirionnydd, in 1870 (Forrest 1907). The only two records from Radnorshire were both prior to 1900 (Jennings 2014).

The Storm Petrels regularly recorded from boats in the southern Irish and Celtic seas were presumably making feeding forays from colonies in Wales, Ireland and the Isles of Scilly. Passage in small numbers is noted from coastal headlands, but this species is far less prone to being 'wrecked' by bad weather than is Leach's Petrel. Daytime counts of more than 100 away from breeding sites are rare. Examples include: 117 past Mumbles Head, Gower, on 4 July 1990, and 100 past Strumble Head on 3 September 1983. Large numbers can be attracted to night-time mist-netting, away from established breeding colonies, such as at Strumble Head and Wooltack Point, Pembrokeshire, but this does not necessarily suggest breeding. There is some evidence that Storm Petrels are attracted to inshore waters at night by the presence of intertidal benthic crustaceans, but concentrations of these are probably highly erratic and may shift between years. As a result, Storm Petrel numbers vary and trends are impossible to discern.

The availability of suitable nesting habitat may well determine the size and distribution of Storm Petrel colonies in Wales. In response to the collapse of dry-stone walls used for nesting on Skokholm, the Wildlife Trust of South and West Wales built a herringbone stone wall that contains over 100 Storm Petrel nest-boxes, with accessible hatches to facilitate ringing. The wall, appropriately named 'The Petrel Station', should replace some of the nesting sites lost, but Welsh Storm Petrel numbers may also be limited by factors such as predation. On Skomer and Skokholm, Little Owls specialise on predating Storm Petrels and may have contributed to their decline. Short-eared Owls have also been recorded predating Storm Petrels on Skokholm. Maintaining strict biosecurity on their nesting islands, particularly to keep them free of rats, is of utmost importance, but wider environmental influences may play a role in their future too. There is evidence that the breeding productivity of Leach's Petrels on the Atlantic coast of the USA may be negatively influenced by increasing temperatures (Mauck *et al.* 2017). It is unclear whether increases in temperature, particularly of the sea surface, will play a part in the breeding success or adult survival of Storm Petrels breeding around the North Atlantic fringe.

Patrick Lindley

Sponsored by Skokholm Bird Observatory

Leach's Petrel *Oceanodroma leucorhoa* Pedryn Drycin Leach

Welsh List Category	IUCN Red List			Birds of Conservation Concern			
	Global	Europe	GB	UK	Wales		
A	VU	LC	LC	Amber	2002	2010	2016

Leach's Petrel is a truly pelagic seabird, spending much of its life in the North Pacific and North Atlantic oceans, extending as far as southwest Africa, as it follows plankton associated with the Benguela Current. It breeds on remote islands around both ocean basins, including off the USA, Canada, Iceland, Norway, the Faroe Islands and the UK (*HBW*). Its diet comprises mainly small fish, squid and crustaceans, taken on the wing from the surface of the sea. The population is around seven million pairs, but has declined by more than 30% over three Leach's Petrel generations, including at its largest colonies in Canada and its largest British colony, St Kilda (BirdLife International 2020). The British population was estimated to be 48,000 pairs in 1998–2002 (Mitchell *et al.* 2004). Birds seen off the Welsh coast, primarily in autumn, are assumed to be largely from the Scottish breeding population, but there are no ringing recoveries from Wales. The majority winter in the tropical Atlantic off the west coast of Africa and the northeast coast of Brazil (*HBW*).

There is no evidence that Leach's Petrel has ever attempted to breed in Wales. The nearest colonies are on St Kilda and the Flannan Islands off western Scotland, but individuals have been found among Storm Petrel colonies in Pembrokeshire: on St Margaret's Island in May 1902 and on Skokholm in 1966, 1976, 1977 (two individuals), 1978, 1980 and 1989.

Birds in Wales (1994) stated that early records of Leach's Petrel in Wales dated from the middle of the 19th century. There were sporadic reports from all coasts of Wales in the 20th century, but most records come from the north coast of Wales, where birds are pushed into Liverpool Bay during northwesterly winds in September, with smaller numbers to early December. These birds move west along the North Wales coast and around the top of Anglesey to continue south through the Irish Sea. Such a movement is likely to have been involved in the occurrence of 8–9 being shot close inshore at the mouth of the Dee Estuary, Flintshire, in 1899 (Forrest 1907).

As sea-watching became more popular in the late 20th century, the timing of Leach's Petrels on southward passage became easier to predict from records at headlands and islands such as Bardsey, Caernarfonshire. Numbers vary depending on the extent to which birds are pushed into the North Channel of the Irish Sea, rather than maintaining their usual passage down the west coast of Ireland.

Date	Selected site counts	'Wrecks' in Wales
October–November 1952		*c.*900
October 1978	181 Point Lynas, Anglesey	
20 October 1983	51 Prestatyn, Flintshire	
17–24 December 1989		191
7 September 1990	72 Point of Ayr, Flintshire	
19–21 September 1990	62 Point of Ayr, Flintshire	
7–17 September 1997	488 Point of Ayr, Flintshire	
19 August–16 September 2001	260 Strumble Head, Pembrokeshire, including 122 on 16 September	
3–16 September 2001	563 Point of Ayr, Flintshire, including 305 on 13 September	
6 February 2002	314 Aberaeron, Ceredigion	
5–9 December 2006		1,239
15–16 September 2010	110 Point of Ayr, Flintshire	
15–17 September 2010	537 Rhos Point, Caernarfonshire, including 240 on 16 September	

Leach's Petrel: Number of individuals recorded during major passage periods and wrecks in Wales.

There can also be "wrecks" of Leach's Petrels following prolonged storms at sea, often later in the winter than typical passage. A large wreck in 1891 produced records in several coastal counties and as far inland as Radnorshire. In autumn 1952, 6,700 were found across Britain, of which around 900 were recorded in Wales, including 29 inland. Large counts of living birds included "hundreds" along the coast between Criccieth, Caernarfonshire, and Penrhyndeudraeth, Meirionnydd, on 25 October 1952 and about 200 offshore at Aberystwyth, Ceredigion, on the same date. Large numbers were also found dead on beaches, including 100–120 at the Tywi estuary, Carmarthenshire, on 28 October (Boyd 1954).

The 1989 wreck was focused in South Wales, with 22 in Gwent, 23 in East Glamorgan, 106 in Gower, 17 in Carmarthenshire, 31 in Ceredigion (including one inland at Tregaron) and singles in Breconshire and Radnorshire, but only two in Pembrokeshire (Powell 1990). Records in December 2006 came from every coastal county, including 73 in Gwent, 509 in East Glamorgan (including 93 off Porthcawl on 9 December), 129 in Carmarthenshire, 239 in Pembrokeshire (including 150 at Freshwater West on 5 December) and 124 in Caernarfonshire (including 62 off Criccieth/Borth-y-gest on 7 December).

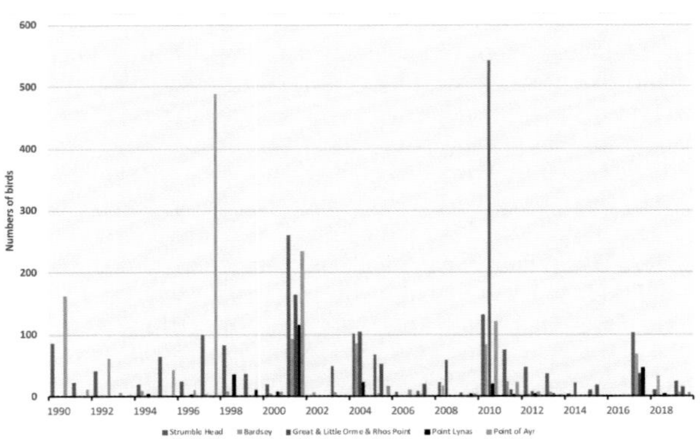

Leach's Petrel: Annual totals counted passing the main sea-watching sites: Strumble Head, Pembrokeshire; Bardsey, Caernarfonshire; Great Orme, Little Orme and Rhos Point, Caernarfonshire; Point Lynas, Anglesey and Point of Ayr, Flintshire.

Leach's Petrel is very sensitive to climatic change, which may explain the marked decline in the size of breeding colonies. The single most important predictor of hatching success at a colony in Nova Scotia, Canada, was annual global mean temperature, more so even than regional sea surface temperatures in either the breeding season or winter (Mauck *et al.* 2017). Thus, it seems likely that the global number of Leach's Petrel will diminish and its occurrence in Wales will become much less frequent.

Jon Green

Fulmar *Fulmarus glacialis* Aderyn Drycin y Graig

Welsh List Category	IUCN Red List			Birds of Conservation Concern			
	Global	Europe	GB	UK	Wales		
A	LC	EN	LC	Amber	2002	2010	2016

The Fulmar is essentially a northern species, breeding on steep cliffs around the Atlantic Ocean, from northern Greenland southward as far as northern France. It also breeds in the North Pacific (*HBW*). It feeds on small fish and fish offal, squid and zooplankton, particularly amphipods, mainly by seizing them from the sea surface. A colonial breeder, it nests on narrow ledges or in hollows, on or near the coast.

This is a long-lived species, and most do not reach sexual maturity until they are about nine years old (*HBW*). Dunnet and Ollason (1978) calculated that Fulmars can live for more than half a century; the oldest known in Britain is a bird ringed as an adult on the Isle of Canna in Scotland. It was caught by a ringer on Sanda Island, also in Scotland, almost 42 years later, so it may well have been over 50 years old. The oldest recorded in Wales

Fulmar © Ben Porter

was a bird ringed as an adult on Great Saltee, Republic of Ireland, in July 1980. It was found dead at Harlech, Meirionnydd, following a storm in March 2014, almost 34 years later. One ringed as a nestling in Iceland in July 2013 was found dead at Port Eynon, Gower, in December the same year, while a nestling ringed on Skokholm, Pembrokeshire, in August 1974 was found long dead in Pas-de-Calais, France, in July 1980. Several birds ringed as nestlings in Shetland and Orkney have been found dead on Welsh beaches. Breeding Fulmars can move great distances in search of food, as illustrated by a bird fitted with a GPS tag and geolocator at its breeding site on Orkney in 2012, that travelled as far as the mid-Atlantic Ridge and covered over 6,200km in 14.9 days, although such a journey was not thought to be typical (Edwards *et al.* 2013).

The world population is estimated at about seven million pairs and thought to be increasing, but European-breeding birds, which number 3.38–3.5 million pairs, have declined by more than 40% since the 1980s (BirdLife International 2020). Fulmars sitting on ledges are quite easy to count, but not all these are necessarily breeding; some may be immature birds.

The spread of the Fulmar population in the North Atlantic is thought to have begun in Iceland prior to 1753 (Fisher 1966). Before the mid-19th century, breeding birds in Britain were probably restricted to St Kilda, where they had been an important item of food for the human population since at least the Iron Age (*The Birds of Scotland* 2007). There is no confirmed record of breeding elsewhere in Britain or Ireland at that time, although Fulmar bones have been found at archaeological sites as far south as the Isle of Man and possibly Guernsey, though not in Wales (Yalden and Albarella 2009). Fulmars colonised the Faroe Islands in the mid-19th century, and it may be that the initial extension in range in Britain, which began in 1878 when Foula, Shetland, was colonised, involved birds from the Faroes. Other colonies in Shetland were established by the start of the 20th century. A southward spread began that continued until about the mid-1980s, with birds eventually breeding as far south as France.

The reasons for the range expansion were the subject of much debate in the 1950s and 1960s. Thompson (2006) reviewed the main hypotheses put forward in this period. One was that it was caused by increases in food availability, following the expansion of whaling and commercial fisheries (Fisher 1952). Other suggestions were that the birds were responding to natural changes in food availability, coinciding with a period of warming in the temperate North Atlantic, or that the spread resulted from a genetically or culturally based change in dispersal behaviour. Thompson concluded that Fulmar populations responded to multiple drivers, and considered that an earlier hypothesis, that the spread had been caused by reduced predation by humans, had been neglected. Fulmars were killed in large numbers by island communities around Iceland, the Faroe Islands and St Kilda, for their oil, down and meat, but the numbers killed had reduced dramatically by the end of the 1930s.

In the late 19th century and early 20th century, the Fulmar was very scarce in Wales. In the north, Forrest (1907) described it as a rare occasional visitor, chiefly to the Meirionnydd coast. He knew of only one record from the north coast, an immature bird shot near Rhyl, Flintshire, about 1860. In Ceredigion, one was shot off Aberystwyth in January 1892. Hutchins, the well-known Aberystwyth taxidermist, was reported to have received five or six in 1894, but it is not known where they were obtained. All other records in Ceredigion, up to the Second World War, were tideline corpses (Roderick and Davis 2010). Farther south, there were even fewer early records. Mathew (1894) knew of only one in Pembrokeshire, a bird caught on a cod line in Tenby Bay in December 1890. The only early record for Carmarthenshire seems to be one in Jefferies' taxidermist shop in Carmarthen in 1926, which was said to have been obtained in the county, but without further details (Ingram and Salmon 1954). The first record for Glamorgan was not until 1929, when one was found dead at Kenfig Sands in December. The species was not recorded in Gwent until 1967.

By 1930, the southward spread of breeding birds had reached the Isle of Man and the Saltee Islands, Co. Wexford, Ireland. Lockley *et al.* (1949) noted that in Pembrokeshire, Fulmars appeared in 1930 at Flimston Stack Rocks and landed on cliffs there in 1931, after which they were present in the breeding season at many cliffs in the county. Saunders (1976) dated the first egg found in Wales, near Flimston, to about 1940. In April 1937, two were around the Great Orme cliffs, Caernarfonshire. Birds were seen around South Stack, Anglesey, from 1938–39, with two on the cliff in June 1940.

Prospecting continued at various sites in several coastal counties, with at least five sites occupied in Ceredigion by 1944 (Fisher 1952). Breeding was confirmed in Caernarfonshire in May 1945, when an egg was found on the Great Orme. In 1947, at least two young hatched at New Quay Head, Ceredigion, and three downy young were reported at Trwyn Dinmor, Anglesey, the same year. In Pembrokeshire, breeding was confirmed on Skomer and near Mathry in 1949 (Fursdon 1950). Breeding probably occurred in Gower in 1955 and was confirmed in 1956. A survey in 1982 found a total of 50 apparently occupied sites (AOS). Breeding was strongly suspected in East Glamorgan in 1977, but not confirmed until 1988, when single pairs bred at Llantwit Major and Nash Point (Hurford and Lansdown 1995). This proved to be the limit of the Fulmar's range expansion in Wales, leaving Gwent as the only coastal county not to have recorded breeding Fulmars.

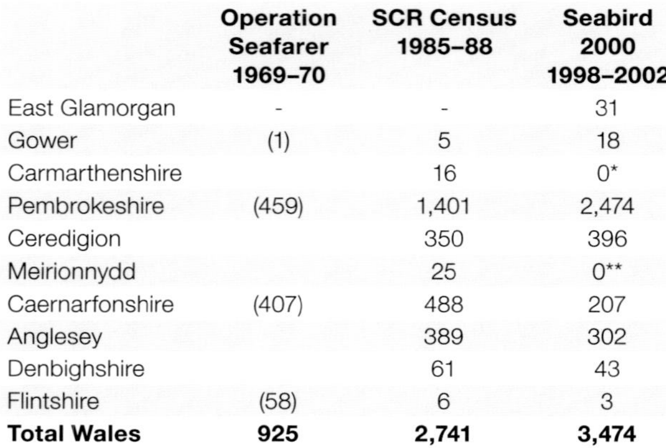

	Operation Seafarer 1969–70	SCR Census 1985–88	Seabird 2000 1998–2002
East Glamorgan	-	-	31
Gower	(1)	5	18
Carmarthenshire		16	0*
Pembrokeshire	(459)	1,401	2,474
Ceredigion		350	396
Meirionnydd		25	0**
Caernarfonshire	(407)	488	207
Anglesey		389	302
Denbighshire		61	43
Flintshire	(58)	6	3
Total Wales	**925**	**2,741**	**3,474**

Fulmar: Number of apparently occupied sites (AOS) recorded in Wales (Mitchell *et al.* 2004, Mitchell 2005). A breakdown of Operation Seafarer numbers (Cramp *et al.* 1974) by county is not available. Figures given are for the administrative counties of West Glamorgan, Dyfed, Gwynedd and Clwyd.

* The colonies at Telpyn Point and Gilman Point, Carmarthenshire, were not surveyed in 1998–2002.

** Friog Cliffs, the only breeding site in Meirionnydd, was not surveyed in 1998–2002.

There was a rapid increase between 1969–70 and 1985–88, but by 1998–2002 the rate of increase had slowed. It was apparent that numbers had decreased in North Wales, particularly in Caernarfonshire, though there was a continued increase farther south. The largest colony in Wales is on Skomer, which peaked at 742 AOS in 1990. The most recent count here was 578 AOS in 2018, compared to 675 AOS in the previous whole-island count in 2016 (Wilkie *et al.* 2019). Other large colonies in Pembrokeshire include Ramsey, which peaked at 321 AOS in 1997 and again in 2017, and Skokholm, where the maximum count was 217 AOS in 2018.

Colonies in North Wales tend to be smaller. The only northern colony ever to hold over 100 AOS was the Great Orme, which held 101 AOS in 1987, but numbers there declined thereafter and the highest count since 2000 was 33 AOS in 2018. Most colonies are on coastal cliffs, but several disused quarries near the coast have also been used, such as at Llanddulas, Denbighshire, and at Prestatyn, Flintshire. Breeding has also been recorded at Dyserth quarry, Flintshire, about 5km from the sea. The north–south difference noted during Seabird 2000 is still evident, with colonies in North Wales showing a more pronounced decline than those in the south. The provisional total for Wales from the ongoing Seabirds Count is 2,193 AOS (Matty Murphy *in litt.*), a 37% decrease since the Seabird 2000 count.

The abundance index of Fulmars in Wales, based on Seabird Monitoring Programme data, increased until the mid-1990s. The decline in abundance for the UK and for Scotland, which started in the mid- to late 1990s, was not reflected in Wales until 2005–09, although the index since 2010 has been fairly stable.

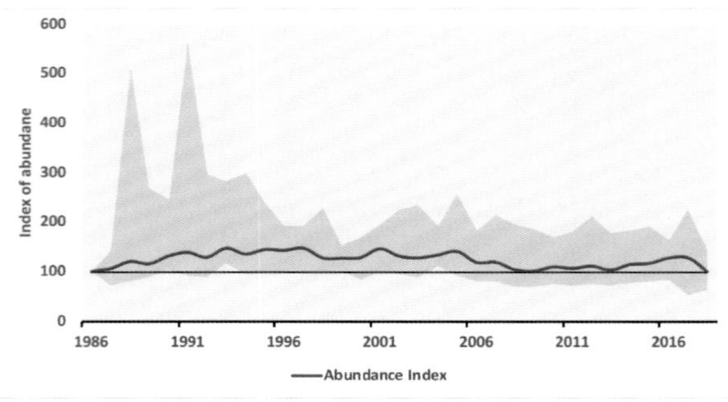

Fulmar: The line shows the abundance index calculated from the Seabird Monitoring Programme database. The shaded area shows the 97.5% confidence limits.

Nesting productivity in Wales has, on average, been higher than in Scotland and England over the same period, but has declined since 2006, to be closer to that recorded in Scotland. The reasons for this are unknown (JNCC 2016). Fulmar numbers declined by 31% in the UK between 1998–2002 and 2015 (JNCC 2016). This was probably caused by a combination of declines in the volume of discards by fishing fleets and the abundance of natural prey such as sandeels in the North Sea, and of certain species of zooplankton in the North Atlantic, particularly the copepod *Calanus finimarchius* (Mitchell *et al.* 2004). Climate change is likely to have contributed to these declines, but the effect is clearly more nuanced than a straightforward contraction in range away from the south, as predicted by Huntley *et al.* (2007). The diet of Fulmars varies considerably, both between colonies and between years at the same colony, and the diet of those breeding in Wales has been barely studied.

Fulmars often do not leave nesting colonies until September and can be back as soon as late November, with most back on ledges by February. Large counts of passage birds have been recorded off Welsh headlands, notably off Strumble Head, Pembrokeshire. Counts there include 5,500 on 10 February 1988, 1,300 in three and a quarter hours on 8 February 2000 and 2,000 in three hours on 20 February 2002. Off Point Lynas, Anglesey, 600 per hour were recorded on 21 August 1988. Birds can sometimes be blown inland by storms, such as one over Llangorse Lake, Breconshire, in July 1986. One was chased by a Peregrine in the Tanat Valley, Montgomeryshire, in June 1976. Radnorshire is the only Welsh county not to have recorded the species. There have been two inland records in Gwent: one at Llanvihangel Ystern Llewern on 7 September 1967 and one over Llangybi on 19 August 1994 (Darryl Spittle *in litt.*).

Fulmars breeding in Wales are pale morph birds, but the population breeding in the Arctic includes a high proportion of dark birds, often called 'blue' morph. Dark morph birds are recorded fairly regularly off the Welsh coast, although generally not at breeding colonies. An exception was one at Prestatyn Quarry, Flintshire, throughout the 1992 breeding season.

The Fulmar's Amber conservation status in Wales is a result of its Endangered status in Europe (Johnstone and Bladwell 2016), and with more colonies showing a downward trend than increase during 2015–19, there is evidence of a decline in breeding numbers in Wales. At present, this decline appears less severe than in some other parts of Europe, but the recent reduction in nest productivity may be a portent of a future decline in breeding numbers of this long-lived species. A potential problem is the high incidence of plastic fragments, often expanded polystyrene, found in Fulmar stomachs: 58% of Fulmars washed up on North Sea coasts contained more than 0.1g of plastic, and the same is likely to apply to the Irish Sea (Ivor Rees *in litt.*). Large numbers of Fulmars may be caught and killed by long-lining fleets in the Norwegian Sea and in the North Atlantic (JNCC 2020b). The effect of these factors on the Welsh population is uncertain, but there is now something of a question mark concerning the future of this species in Wales.

Rhion Pritchard

Zino's/Fea's/Desertas Petrel *Pterodroma madeira/feae/deserta*

Pedryn Madeira/Cabo Verde/Desertas

The *Pterodroma* species are collectively known as gadfly petrels because of their speedy weaving flight. The *feae*-complex comprises three species that nest on islands in the eastern Atlantic Ocean. Outside the breeding season they are pelagic, but the extent to which they visit western European waters is unclear. Fea's Petrel breeds on the Cape Verde Islands, Zino's Petrel in the central mountains of Madeira, and Desertas Petrel on Bugio and Deserta Grande, about 25km southeast of Madeira (Ramos *et al.* 2016).

Identification to species level, particularly from a land-based watch point, is extremely difficult (Shirihai *et al.* 2010), although it may prove possible in future with sufficiently good photographs. There were 82 records in British waters up to the end of 2019, all of which were accepted by BBRC as either Fea's/Desertas Petrel or Fea's/Zino's/Desertas Petrel (Holt *et al.* 2020). There have been five records in Wales, all of single birds, but none was identified to species level. Three of these were off Bardsey, Caernarfonshire, on 10 September 1994, 15 September 2013 and 20 August 2019. The other two were off Pembrokeshire: off Strumble Head on 4 October 1996 and off Grassholm on 11 July 2010.

Jon Green and Robin Sandham

Cory's Shearwater *Calonectris borealis* Aderyn Drycin Cory

Welsh List Category	IUCN Red List		Rarity recording
	Global	Europe	
A	LC	LC	WBRC

This large shearwater breeds on the Berlengas archipelago off the coast of Portugal, Madeira, the Azores and the Canary Islands. Outside the breeding season it roams widely around the Atlantic Ocean, and is most frequently seen in late summer off the British and Irish coasts (*HBW*). Increasingly large numbers have been recorded in Britain, mainly from southwest England, with average annual totals increasing from 17 during 1960–69 to over 1,000 during 2010–18 (White and Kehoe 2020a). Far fewer are seen off the Welsh coast, and numbers have fallen since a peak in 1990–99.

The remains of a Cory's Shearwater were found at Bacon Hole Cave, Gower, dating from about 125,000 years ago, during the Ipswichian interglacial period. This is the only archaeological record of the species in Britain (Yalden and Albarella 2009). It is not included in the table.

Total	Pre-1950	1950–59	1960–69	1970–79	1980–89	1990–99	2000–09	2010–19
116	0	0	1	3	29	41	25	17

Cory's Shearwater: Individuals recorded in Wales by decade.

The first modern Welsh record was seen from a ferry off Pembrokeshire on 25 July 1965, while the first from the mainland was off Sker Point, East Glamorgan, on 27 July 1975. Most records have occurred in late summer and autumn, but there was one off South Bishop, Pembrokeshire, on 22 February 1976. A previously published record of one off Telpyn Point, Carmarthenshire, on 29 February 1992 is now considered no longer acceptable.

Of the total of 116 records in Wales, 78 have come from Pembrokeshire. Most of these were seen from Strumble Head, where a total of 65 birds have been seen, including seven on 15 August 1999 and five on 1 September 2000. Elsewhere in the county, six individuals have been recorded off Skokholm, with smaller numbers off St Govan's Head, Porthgain, St Davids Head, Grassholm and South Bishop. This total does not include other individuals recorded at sea, from the Rosslare–Fishguard Ferry and on pelagic trips to the Celtic Deep off Pembrokeshire.

Numbers elsewhere in Wales are much smaller, although birds have been recorded from all coastal counties except Denbighshire and Flintshire. The nine off Anglesey is the highest total away from Pembrokeshire and includes three off South Stack. Four of the five seen from Caernarfonshire were off Bardsey. Other records were from Gwent (1), East Glamorgan (4), Gower (4), Carmarthenshire (3), Ceredigion (2) and Meirionnydd (1).

Most records (68) are in August and September. The latest in the year was one off Strumble Head on 29 November 1999, concluding a record season of 23 in Wales, coincident with a large passage of Great Shearwaters off Ireland and southwest England (Nightingale and Elkins 2000). The previous year, 1998, had seen the largest ever passage of Cory's Shearwaters off Cornwall (Nightingale 1999), but only three were seen in Wales that year, all off Strumble. The accepted explanation for records off Pembrokeshire was that southwesterly gales pushed birds from the Western Approaches into the Irish Sea. Birds were seen relocating back south out to the Atlantic in subsequent days (Green 2005a). However, since 2010, Cory's Shearwaters have been seen even in the absence of such gales, and it may be that some birds have joined the feeding flocks of Manx Shearwaters in Cardigan Bay.

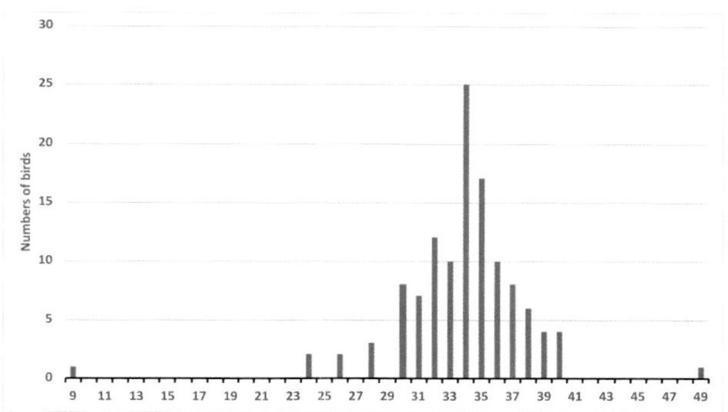

Cory's Shearwater: Maximum single counts each week, of birds passing around Wales, 1965–2019. Week 30 begins *c.*23 July and week 40 begins *c.*1 October. Records are from the WBRC.

Jon Green and Robin Sandham

Sooty Shearwater *Ardenna grisea* Aderyn Drycin Du

Welsh List Category	IUCN Red List	Birds of Conservation Concern			
	Global	UK	Wales		
A	NT	Green	2002	2010	2016

This medium-sized shearwater breeds on small islands in the South Atlantic and South Pacific, but outside the breeding season, it undertakes extensive pelagic movements as far north as Iceland. Those that occur off the European coast will have bred in the Falkland Islands and southern Patagonia during the austral summer, moved up the east coast of the Americas to reach subarctic waters in June, and then crossed the Atlantic from west to east as the northern summer progresses. Birds feed off northwest Europe during August and September *en route* back to their breeding colonies, to which they return in November (*HBW*). Small numbers occur off Wales each autumn, the totals depending in part on the strength of southwesterly winds pushing them into the Irish Sea.

There are few historical records, but they do include the only record from an inland county, in Radnorshire, where one was shot upstream of Glasbury "a few years before 1899" (Jennings 2014). Remarkably, this may be the first record in Wales, since the species is mentioned neither by Forrest (1907) from North Wales nor Matthew (1894) in Pembrokeshire.

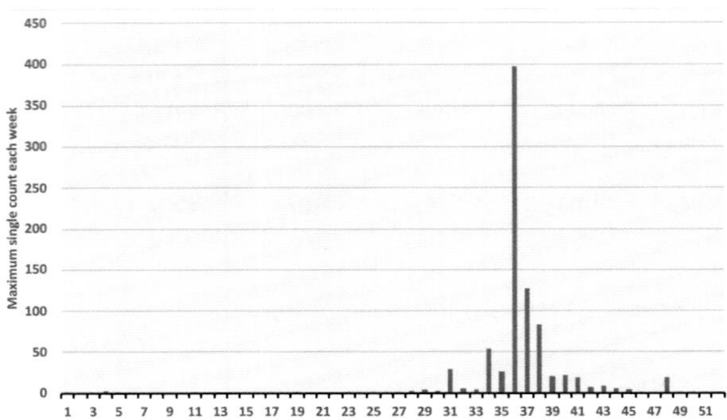

Sooty Shearwater: Maximum single counts, each week, of birds passing around Wales, 1960–2019. Week 34 begins *c.*20 August and week 44 begins *c.*29 October. Records are from BirdTrack.

	1990–99	2000–09	2010–19
East Glamorgan	6	0	2
Gower	4	7	3
Carmarthenshire	6	0	0
Pembrokeshire			
Strumble Head	636	656	623
elsewhere	156	90	58
Ceredigion	8	9	21
Meirionnydd	0	1	0
Caernarfonshire			
Bardsey	180	71	160
elsewhere	3	9	11
Anglesey	30	50	17
Denbighshire	2	0	0
Flintshire	4	1	0
Total	**1,035**	**894**	**895**

Sooty Shearwater: Numbers recorded in Wales by decade since 1990.

The greatest numbers are seen from Pembrokeshire, where Lockley *et al.* (1949) regarded this species as fairly regular during early September and Saunders (1976) assessed it as annual in extremely small numbers. From the 1950s until the 1970s, there were a handful of records each autumn, predominantly from Skokholm, but regular sea-watching, from Strumble Head since the early 1980s, showed that greater numbers occur. Strumble, on the northwest coast of Pembrokeshire, is an ideal location to observe larger shearwaters as they move out of the Irish Sea into the Southwestern Approaches, especially in the days following southwesterly gales. Passage off Strumble starts in late July, increases in late August, peaks in the first ten days of September and tails off into October. There are occasional records outside this period, even into December.

Annual totals at Strumble vary, with 50–100 in most years but the averages for each decade are quite similar. During 1983–87, totals averaged 81 per year, but with peaks of 237 on 1 September 1985 and 397 on 3 September 1983, which remain the highest day counts recorded in Wales. During the 1990s, annual totals off Strumble ranged from just 14 in 1993 to 119 in 1994, compared to fewer than ten each year off Skokholm. 2011 was the best recent year for the species off Strumble Head, with 277 counted, including 87 on 7 September.

The species was first recorded in Caernarfonshire in 1965, off Bardsey, where it is now seen annually in small numbers, except for 1992 and 1993 when there were no records. Annual numbers typically range from a single in 1995 to 55 in 1994, but 208 in September 1980 was exceptional. Sooty Shearwater is much scarcer from the Caernarfonshire mainland. The best year was 2004 when nine were seen. The first record from Anglesey was off Cemlyn in 1970. There have been only sporadic records from the island, except at RSPB South Stack and Point Lynas, where the species is seen annually in small numbers.

Several Sooty Shearwaters were recorded from Worms Head, Gower, in August 1963 and 1964, and one in August 1969, but there were no further records in the recording area until 1980. There were 45 records during 1980–92, all from Port Eynon Point (Hurford and Landsdown 1995). The species is scarce in East Glamorgan: the first record was at Lavernock Point on 25 July 1964, and there have been just a handful of subsequent records, all from Lavernock, Porthcawl and Sker Point. The first Meirionnydd record was on 22 August 1983, and there were only two further records during that decade, both at Aberdysynni in 1988, on 22 July and 12 August. There have been no records from Gwent, or from any inland county since the 19th century Radnorshire record.

The Sooty Shearwater has a world population of 19–23 million birds. This population is in moderately rapid decline, thought be as a result of the impact of pelagic fisheries, harvesting of the young (in Australia and New Zealand) and climate change (BirdLife International 2020). Numbers seen from Wales have not changed dramatically during the last three decades, but if the global trend continues, it could become a far rarer sighting here.

Jon Green

Great Shearwater *Ardenna gravis* Aderyn Drycin Mawr

Welsh List Category	IUCN Red List	Birds of Conservation Concern	Rarity recording
	Global	UK	
A	LC	Green	WBRC

The Great Shearwater breeds in colonies on islands in the South Atlantic, and is pelagic outside the breeding season, completing a clockwise migration around the Atlantic Ocean. The birds move north up the American coast and return south along the eastern side of the Atlantic. They are, therefore, seen in British and Irish waters mainly during late summer and early autumn (*HBW*). This species is a scarce passage migrant off the Welsh coast, although there have been more than twice as many records of Great Shearwater as of Cory's Shearwater.

The first Great Shearwater recorded in Wales was in 1957, when one was seen in Jack Sound, Pembrokeshire, on 15 August. A previously published record of one off Caernarfonshire in 1912 is now considered to be no longer acceptable. Most occur in late August and early September. The latest bird was one found dead on the beach at Llanon, Ceredigion, on 19 November 1967.

Total	Pre-1950	1950–59	1960–69	1970–79	1980–89	1990–99	2000–09	2010–19
289	0	1	3	5	52	110	61	57

Great Shearwater: Individuals recorded in Wales by decade.

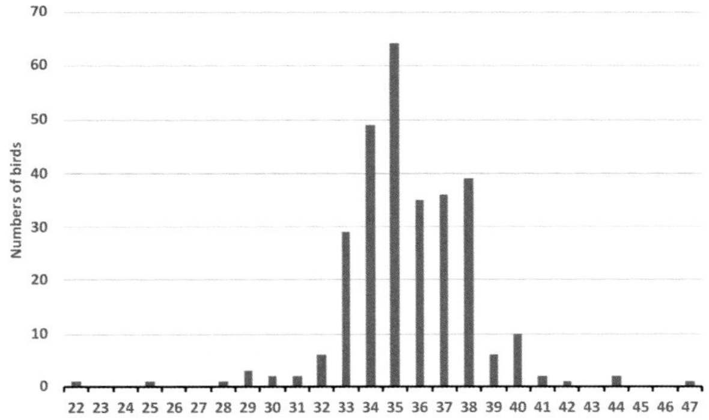

Great Shearwater: Numbers, each week, of birds passing around Wales, 1957–2019. Week 33 begins *c*.13 August and week 38 begins *c*.17 September. Records are from the WBRC.

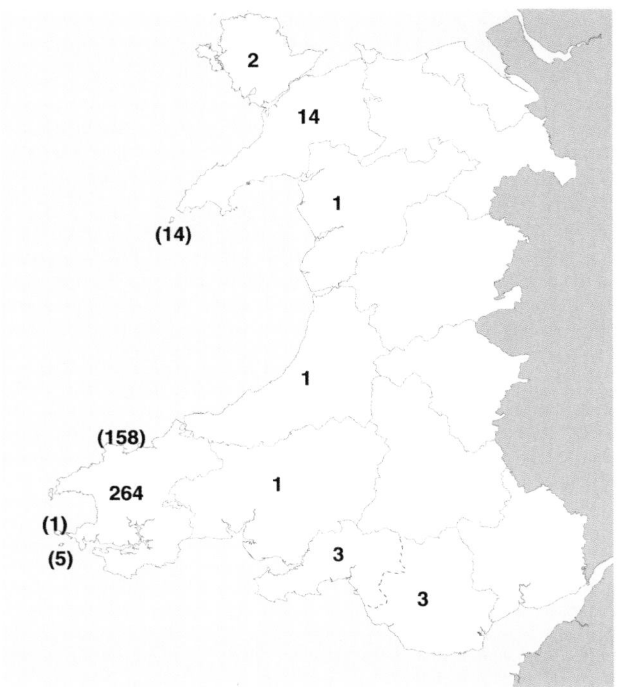

Great Shearwater: Numbers of birds in each recording area 1957–2019.

The vast majority of Welsh records have come from Pembrokeshire, most of these being birds passing Strumble Head on their way out of the Irish Sea, following southwesterly gales in autumn (Green 2005a). The number of records at Strumble Head in any given year depends on the weather, with none in some years, but many birds pass in a single day in ideal conditions. The highest one-day count was 25 on 31 August 2002. Other large numbers were 14 on 10 September 1993 and ten on 30 August 2014. Good numbers have also been seen at sea off Pembrokeshire. This includes birds seen in Pembrokeshire waters from the Rosslare to Fishguard ferry, with a highest count of 23 on 15 September 1981. Other records have come from pelagic trips to the Celtic Deep, which produced a high count of 20 on 15 August 1999. Birds have also been seen off Tenby, Skomer, Skokholm, South Bishop and the Smalls. The highest annual totals in Pembrokeshire were 50 in 1999, when there was also a large movement off Ireland and southwest England (Nightingale and Elkins 2000), and 43 in 2004.

Birds have been recorded in all coastal counties except Gwent, Denbighshire and Flintshire. A total of 14 individuals have been recorded off Bardsey, six of which were in 2018, including three on 16 August. Two were seen off Nash Point, East Glamorgan, on 3 August 1986. There have been three records of singles off Gower, and Carmarthenshire, Ceredigion and Meirionnydd have one record apiece, the last two found dead. Rather surprisingly, there have been only two records of singles off Anglesey. Comparing the totals recorded for Pembrokeshire, Caernarfonshire and Anglesey shows how markedly the number of records declines further north in the Irish Sea.

Jon Green and Robin Sandham

Manx Shearwater *Puffinus puffinus* Aderyn Drycin Manaw

Welsh List Category	IUCN Red List			Birds of Conservation Concern			
	Global	Europe	GB	UK	Wales		
A	LC	LC	LC	Amber	2002	2010	2016

Manx Shearwaters are summer visitors to breeding colonies in the North Atlantic, with the bulk of the population nesting in Wales, Scotland and Ireland, a small number in Iceland and Norway, and a tiny number in northeast USA and Canada (*HBW*). The most recent published estimate is a European population of 342,000–393,000 pairs (BirdLife International 2015) but as a result of surveys in Wales in 2018 this will soon be revised upwards (Perrins *et al.* 2020). It is clear that Wales holds well over half the global population. Manx Shearwaters catch their prey, mainly small fish such as sardines and sandeels, only during daylight hours, typically diving up to 10m, but sometimes up to 55m.

In Wales, the Manx Shearwater now breeds only on the predator-free islands of Skomer, Skokholm, Ramsey and Middleholm in Pembrokeshire, and Bardsey, Caernarfonshire. In Pembrokeshire, the species bred on Caldey until the 1880s, and perhaps on St Margaret's Island in 1893 (Mathew 1894), while in Caernarfonshire it bred on St Tudwal's Islands prior to the arrival of Brown Rats in the 1950s (Barnes 1997). There is some evidence of breeding on mainland cliffs on the south coast of Llŷn, Caernarfonshire (Coward 1895), where the species was thought still to breed in the 1970s (Jones and Dare 1976). There is no written evidence that the Manx Shearwater bred on Cardigan Island, Ceredigion, but unfledged chicks were translocated from Skokholm to artificial burrows on Cardigan Island in 1980–84, and an egg was found in 1984 (Roderick and Davis 2010). This species is occasionally heard calling around mainland cliffs in the breeding counties, and less frequently in other counties such as Ceredigion, and at several locations in north Anglesey. Eight pairs were reported as attempting to breed at a site on Anglesey in 1966 (Jones and Whalley 2004) and, in 1990, near Rhossili, Gower. Breeding has not been confirmed at these locations, but is difficult to prove.

On the islands, Manx Shearwaters were taken, sometimes in considerable numbers, for use as fertiliser or as bait in lobster pots. Mathew (1894) noted than on Skomer "Mr. Vaughan Davies informed us that one year he ploughed cartloads of the poor 'Cockles' into the ground for manure". Some were also taken for human consumption. Morris (1857) reported the sale of shearwaters from Wales in Leadenhall Market, London. Rabbit-warreners on Skomer and Skokholm killed shearwaters and Puffins because they considered that they disturbed does in the breeding season, seriously reducing the annual yield of rabbits. Captain Davies, owner of Skomer, campaigned against the passing of the Sea Birds Preservation Act 1869, and later for its withdrawal from areas with rabbit warrens, on the grounds that the birds had become more numerous and the rabbit crop had halved. Barrington (1888) wrote that Davies offered his farm boys a small reward for the destruction of the birds and "he told me they brought him I think it was twenty-four dozen shearwaters in a few hours, striking them with sticks as they fluttered along the ground.'

Early studies by Lockley (1930) on Skokholm established the very long incubation and fledging periods of this species. He showed its homing ability by taking incubating birds, releasing them elsewhere and noting the time that they took to return to the nest. These early studies were followed by more detailed ones by Geoffrey Matthews (1953), and more detailed studies of the breeding biology by Harris (1966) and Brooke (1990).

Breeding adults start to leave in late August, and most have left the colonies by mid-September, before their young, which consequently are not fed for the last ten days or so of the pre-fledging period. They are not commonly seen from land, except near colonies or at a few headlands, such as Strumble Head, Pembrokeshire, where large numbers may be observed heading to and from their feeding grounds. They can be seen almost anywhere in Welsh waters, with large concentrations in late afternoon and evening near colonies. A small number, mainly newly fledged juveniles, are sometimes blown ashore, occasionally far inland, by strong gales in autumn.

Manx Shearwaters are long-lived, with several birds ringed on Bardsey living at least another 40 years. One, the oldest known in the world, was re-trapped on Bardsey in May 2008, 50 years, 11 months and 21 days after it had first been ringed there. As it was an adult when first ringed in May 1957, it must already have been at least two years old at that time.

Geolocators have produced a more detailed picture of the movements of breeding adults. The birds return to the same burrow each year, and so can be recaptured easily to download

Manx Shearwater at night on Bardsey Island © Ben Porter

Manx Shearwater: Recovery locations of birds ringed in Wales are shown by red circles. Ringing locations of birds that were recovered in Wales are shown by blue triangles. Small brown dots show ringing or recovery locations in Wales.

the tag data. Southbound migration takes them down the east Atlantic coast to The Gambia (15°N), from where they cross to the most eastern point of Brazil. They continue south to the area off the River Plate, Uruguay/Argentina, where they spend the austral summer in an area of upwelling and an abundance of marine food over the Patagonian Shelf (Guilford *et al.* 2009). Ringing recoveries away from breeding colonies are limited, but there has been a steady trickle of birds, many of which were subadults, washed up in South America. One exceptional recovery was of a fledgling ringed on Skokholm on 9 September 1960 that was found dead near Venus Bay, Australia, on 22 November 1961.

In February, the birds head north following the South American coast, then northwest towards the Caribbean Sea. They gradually turn clockwise in a wide arc, until they reach about 50°N from where they head east to the breeding colonies (Guilford 2019). These journeys take 14–44 days, but only a few individuals achieve the fastest times. A fledgling ringed on Skokholm was recovered dead in Brazil, just 16 days after ringing. Most stop-over to rest for some days. Data from immersion loggers suggest they were feeding (Guilford *et al.* 2009).

First breeding has been recorded at age four, but 6–7 years is more usual. Once birds have returned to their burrows and found their mate or acquired a new one, most females leave the nesting area for about a fortnight. This is a period referred to as the 'Exodus'. They head out to the Atlantic Shelf in search of the food needed to make the single large egg, that is 15% of her body weight. After laying, females usually return to sea almost immediately, leaving their mate to undertake the first incubation. The pair take it in turns, usually of about a week each, while the off-duty bird wanders widely. A favoured area is the Irish Sea Front Special Protection Area, a food-rich area about 36km northwest of Anglesey, that hosts the third largest marine aggregation of breeding Manx Shearwaters in the UK (Kober *et al.* 2012). From Skomer, this is a return journey of some 400km. Once hatched, the chick is fed every 2–3 days, so parents forage closer to the colony. During 1995–2018, breeding success averaged 0.6 fledged chicks/burrow on Skomer. The annual survival rate of breeders was about 86% during the period 1977–2018. Porter and Stansfield (2018) noted that GPS tracking of birds breeding on Bardsey also showed the Irish Sea Front area to be important, and speculated that these birds might benefit from having a shorter journey to this area compared to those breeding on Skomer and Skokholm.

The Manx Shearwater's habit of breeding in burrows and being nocturnal on land makes it difficult to census. Lockley (1930) guessed that there might be 10,000 birds on Skokholm, an estimate which he subsequently altered to 10,000 pairs (1942). Conder and Keighley (1950) estimated there were 10,000–15,000 pairs there, based on ringing re-traps. Independent attempts agreed an estimate of *c.*35,000 pairs (Harris 1966, Perrins 1968). Skomer proved more difficult to census because of its larger size.

Since 1998, the colonies on Skokholm and Skomer have been censused using a tape-playback technique, providing more reliable estimates of apparently occupied burrows (AOBs). The Seabird 2000 census (1998–2002) estimated a total Welsh population of 168,133 AOBs, including 147,978 on the two main Pembrokeshire islands: 46,184 on Skokholm and 101,794 on Skomer (Smith *et al.* 2001). Perrins *et al.* (2020) re-examined the 1998 census and discovered an inconsistency in the methods, recalculating that year's estimates as 362,027 AOBs for Skomer and Skokholm combined, 2.4 times higher than the original and only a little lower than the next census more than a decade later.

Censuses on Skokholm found 63,536 AOBs in 2012–13 (Perrins *et al.* 2017) and 90,000 AOBs (67,000–110,000) in 2018 (Perrins *et al.* 2020), while counts on Skomer found 316,070 pairs in 2011 (Perrins *et al.* 2012) and 350,000 pairs (256,000–443,000) in 2018 (Perrins *et al.* 2020). The most recent estimates for nearby islands are 16,548 AOBs on Middleholm in 2018 (Perrins *et al.* 2020) and 4,796 AOBs on Ramsey in 2016 (G. Morgan pers. comm.). The total of 900,000 breeding adults, together with an unknown number of non-breeding subadults, makes it likely that more than 1.5 million Manx Shearwaters are present in the Skomer, Skokholm and the Seas off Pembrokeshire Special Protection Area during the breeding season.

Together with at least 20,675 AOBs on Bardsey in 2014–16, there were an estimated 487,471 pairs in Wales during the 2015–19 part of the ongoing Seabirds Count. There are 126,000 pairs in Scotland (*The Birds of Scotland* 2007), *c.*200,000 pairs in Ireland (which are the most difficult to census) and about 50,000 pairs elsewhere, so Wales holds well over half of the world population. The large confidence intervals mean that trends cannot be determined, but sample plots, surveyed annually on Skomer and Skokholm, indicate that the population is at least stable and may be increasing (Brown and Eagle 2018a, Stubbings *et al.* 2018).

Away from breeding sites, Manx Shearwaters are seen from coastal watch points between March and early October. Some large movements seen from headlands in summer are probably non-breeders. These were formerly thought to be heading for the Bay of Biscay to feed on sardine stocks (Perrins and Brooke 1976), but in recent decades such movements have been evident in the

Bristol Channel, usually in June, which suggests good feeding conditions there. There are several counts of more than 7,000 from Gower, including 10,000 off Port Eynon on 16 June 1985. County record day counts farther east include 11,250 off Porthcawl, East Glamorgan, on 22 June 2012 and 3,400 off Peterstone Wentlooge, Gwent, on 22 June 2013. Congregations occur regularly off Borth and the Dyfi Estuary, Ceredigion, in late July and August, with flocks of up to 50,000 there in 2010, 30,000 in 2011 and 20,000 in 2013, 2015 and 2019. Such feeding flocks sometimes occur farther north in the Bay, such as 4,800 off Aberdysynni, Meirionnydd, in July 1984 and 10,000 off Tywyn, Meirionnydd, in July 2011. The highest count from Bardsey was 29,868 on 16 August 2018.

The largest counts are, not surprisingly, from Strumble Head on the north coast of Pembrokeshire, where they are often too numerous for accurate counts to be recorded, and frequently greater than 10,000 birds daily in August and September. Counts are generally highest early in the morning and after 5pm, as birds leave and later return to the nesting colonies. The largest count on a single day is probably 140,000 in two hours on 9 July 2010. Although the south coast of Pembrokeshire is less well watched, numbers passing St Govan's Head can also be considerable, such as 6,000 an hour on 29 July 2018. To the east, counts in Carmarthen Bay are generally smaller although Cefn Sidan has recorded 2,920 on 16 May 1992, 3,600 on 1 August 1993 and 1,600 on 16 August 2015.

Numbers off the north coast tend to be smaller than in Cardigan Bay. The largest counts on Anglesey were 10,977 off Penrhyn Mawr in three hours on 11 August 1982, 12,630 off RSPB South Stack, Anglesey, in four hours on 6 June 2017, 10,000 off Cemlyn on 23 June 2013 and 5,000 an hour past Point Lynas on 12 August 2010. Smaller numbers are seen at sites farther east, with the highest counts being 3,000 off Rhos Point, Caernarfonshire, on 23 June 2013 and an exceptional 11,000 off Point of Ayr, Flintshire, on 2 August 2009.

Inland records tend to result from autumn gales and often involve juveniles; for example, 35 were recorded in Radnorshire between 1926 and 2011 (nine in the latter year alone), including several that had been ringed on Skokholm. Manx Shearwaters are sometimes seen during December and January, occurring off any coastal county. These are usually singles, but sometimes small groups, such as seven off Ramsey on 11 December 2014, and occasionally even calling at breeding sites, such as one on Bardsey on 16 November 2019.

As with many other seabird species, successful breeding only occurs in rat-free areas. Skomer and Skokholm were both farmed historically and it was fortunate that this did not result in the arrival of rats. Rats did reach Ramsey, which lost most of its shearwaters, down to 896 AOBs in 1998. A successful rat eradication programme was undertaken there by the RSPB in winter 1999/2000, resulting in a five-fold increase in Manx Shearwater numbers by 2016. Rat eradication was also achieved on Lundy, just outside Welsh waters. The number of birds in the Welsh section of the Bristol Channel may thus be expected to rise.

Two natural factors may affect numbers. Great Black-backed Gulls are major predators of shearwaters and on Skokholm an average 1,954 full-grown birds and 1,100 fledglings were killed each year during 2017–19 (R. Brown pers. comm.). These are minimum figures. Statistically, the 'full-grown' birds are likely to be subadults, since they spend much more time on the surface than breeders. On both islands an unexplained disease, *Puffinosis*, causes the deaths of fledglings. Brooke (1990) thought that it might lead to the deaths of *c.*4% of the young. At sea, a number of mortality factors have been identified, but the effects cannot be quantified. Most have changed over time. These include oiling, especially from major tanker spills, hunting and fishing. In the past there were recoveries of several ringed birds from nets used by sardine fishermen in the Bay of Biscay (Perrins and Brooke 1976), implying that sometimes considerable numbers of birds may have been killed, but there have been no such reports in recent years.

In the longer term, there is concern that fish stocks are moving northwards as a result of climate change and that this may affect the breeding success and ultimately the distribution of the species. Climate models show that Brittany and southwest Britain, including Pembrokeshire, may not be suitable for Manx Shearwaters by the end of this century (Huntley *et al.* 2007). Studies of breeding success, adult survival, foraging range, food and ingested plastics will all be valuable in monitoring the status of the bird species for which Wales is of greatest importance.

Chris Perrins and Steve Sutcliffe

Sponsored by Pembrokeshire Bird Group

Balearic Shearwater *Puffinus mauretanicus* Aderyn Drycin Balearig

Welsh List Category	IUCN Red List			Birds of Conservation Concern			
	Global	Europe	GB	UK	Wales		
A	CR	CR	VU	Red	2002 N/A	2010	2016

This medium-sized shearwater nests in burrows and caves on Mallorca, Menorca and Ibiza. It was considered a subspecies of Manx Shearwater until 1991, then as 'Mediterranean Shearwater' until 2000, when that form was split into Balearic and Yelkouan Shearwater (IOC). Most birds winter locally in the western Mediterranean, but an unknown proportion enter the Atlantic Ocean after breeding and feed off the British and Irish coasts. It has undergone a rapid decline, of more than 90% in three generations of this species. The world population is estimated to be 3,142 pairs (BirdLife International 2020). Longline fisheries are thought to be the biggest factor in the Balearic Shearwater's decline, but predation, by introduced mammals at nesting colonies, is also a problem.

The similarity of this species to Manx Shearwater, and its inclusion within the same species complex for many years, means that it has been difficult to know the true status of Balearic Shearwater in Wales until relatively recently. It was not mentioned from North Wales by Forrest (1907) nor from Pembrokeshire by Mathew (1894), but Donovan and Rees (1994) reported that one was collected in Pembrokeshire in September 1900 and is in a museum at Harvard, USA. It became apparent that it occurred regularly in small numbers off Pembrokeshire, following the first sighting of a live bird in 1955. Several were seen from Worms Head, Gower, in August 1964, after which records from the county occurred in 1968, 1973 and in four years during the 1980s.

The first record from Caernarfonshire was from Bardsey on 23 October 1965, and from the north coast, off the Great Orme on 15 September 1985. The first dated Ceredigion record was off Ynyslas on 2 May 1972 and the first Anglesey record from Point Lynas in November 1980. The first record from East Glamorgan, where it remains a rare bird, was off Lavernock Point on 30 September 1982, while the first Meirionnydd record flew in from Ceredigion, across the mouth of the Dyfi, on 7 November 1997.

Birds in Wales (1994) reported that Balearic Shearwaters were being seen with increasing frequency, particularly off Strumble Head, Pembrokeshire, mainly in September and October, but with records in every month except February. It would appear that the increase in records was genuine, as numbers reported in Britain and Ireland grew from an average of 300 each year in the 1980s to 2,000 a year in the late 1990s. Wynn and Yésou (2007) reported that this coincided with a reduction in numbers remaining in the Bay of Biscay and highlighted the importance of the Southwestern Approaches and adjacent coastal waters for the species.

Birds in Wales (2002) noted that it was a regular occurrence in the daily logs off Strumble Head, with peak daily counts of up to 36 birds. Annual totals there vary, reflecting the availability of

	1990–99	2000–09	2010–19
East Glamorgan	2	4	2
Gower	12	276	28
Carmarthenshire	0	2	1
Pembrokeshire			
Strumble Head	1,351	2,154	963
elsewhere	97	82	98
Ceredigion	12	154	102
Meirionnydd	1	16	8
Caernarfonshire			
Bardsey	146	410	149
elsewhere	33	12	40
Anglesey	79	347	196
Flintshire	28	0	1
Total	**1,761**	**3,457**	**1,588**

Balearic Shearwater: Individuals recorded in Wales by decade since 1990.

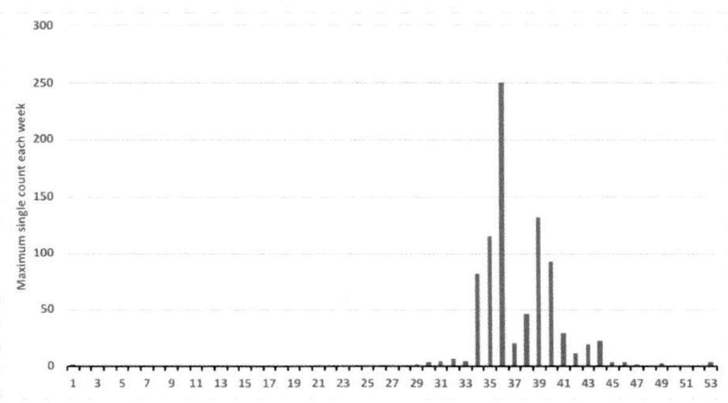

Balearic Shearwater: Maximum single counts, each week, of birds passing around Wales, 1960–2019. Week 34 begins *c.*20 August and week 44 begins *c.*29 October. Records are from BirdTrack.

food, summer temperatures and autumn gales. 2006 saw the greatest number, with 889 recorded, but a decline in the last decade is evident, reflecting its global status. Passage usually starts in August, increasing towards the end of the month, peaking in September and then decreasing in October, with very few in November and occasional records into December.

Smaller numbers are recorded off the other Welsh sea-watching hotspots, including Bardsey, RSPB South Stack and Point Lynas, but the collective annual totals away from Strumble Head are generally less than 20. Exceptions include 32 off Bardsey in 1997, 91 there in 1999 and 297 in 2006, and 225 from Point Lynas in 2003. The Gower total was boosted by an unusual flock of at least 250 off Rhossili/Port Eynon, during late August and early September 2003. There have been no records from Denbighshire, Gwent or the inland counties.

The species' Critically Endangered global status means that Balearic Shearwater is likely to become rarer in Welsh waters, unless a solution can quickly be found to mortality from fishing gear, especially artisanal long-lining in the Mediterranean, and rodent predation at its nesting burrows. It is also possible that its autumn feeding grounds, having moved already from Biscay to the British coast, will move even farther north with a changing climate.

Jon Green

Barolo/Boyd's/Audubon's Shearwater *Puffinus baroli/boydi/lherminieri*

Aderyn Drycin Barolo/Boyd/Audubon

The taxonomy of this closely related group of species is difficult. At one time, all the North Atlantic 'small' shearwaters were considered to be subspecies of Little Shearwater *Puffinus assimilis*. Molecular analysis of this group by Austin *et al.* (2004) found that the birds breeding in the North Atlantic were clearly distinct from those in the southern hemisphere. From 2005, these were treated as two separate species, *P. lherminieri* (Audubon's Shearwater) *and P. baroli* (Macaronesian Shearwater), following Sangster *et al.* (2005). Macaronesian Shearwater is itself now treated as two distinct species, *P. baroli* (Barolo Shearwater) and *P. boydi* (Boyd's Shearwater). Barolo Shearwater breeds on the Azores, Madeira and the Canary Islands, while Boyd's Shearwater breeds on the Cape Verde Islands. Audubon's Shearwater breeds mainly in the West Indies and Caribbean (IOC). Their respective ranges outside the breeding season are uncertain.

Records in Britain are assessed by BBRC. There has been a total of 68 records in Britain up to the end of 2019 (Holt *et al.* 2020). The classification and status of this group of species were under review by the BBRC at the time of publication.

The previously published record of a bird off Aberdaron, Caernarfonshire, on two dates in May 1951 is now no longer deemed acceptable. The first Welsh record was of a male captured ashore on Skomer, Pembrokeshire, in late June 1981. This bird is highly likely to have been a Barolo Shearwater, based on sonagrams of the call. The bird remained on the island until late July and was observed visiting a burrow. The same bird returned to Skomer in late June 1982 and again stayed until late July (James and Alexander 1984, James 1986). The acceptance of this individual as Barolo Shearwater will depend on the outcome of the BBRC review.

All other records were observed on sea-watches. A total of seven individuals have been observed passing Strumble Head, Pembrokeshire, six past South Stack, Anglesey, and one past Bardsey, Caernarfonshire. A remarkable four birds passed South Stack on 12 September 1987, with two more there on 15 September the same year. Of the total of 15 birds recorded in Wales, 14 were seen between 1981 and 1989. The only record since then was one off Strumble Head on 12 September 1997.

Jon Green, Robin Sandham and Rhion Pritchard

Black Stork *Ciconia nigra* Ciconia Du

Welsh List Category	IUCN Red List		Rarity recording
	Global	Europe	
A	LC	LC	BBRC

The Black Stork breeds in southern Africa, temperate Asia and locally in continental Europe. Those from northern latitudes winter to the south, European birds generally in the Sahel. Within its range it is usually seen in pairs or small flocks in marshy areas, by rivers or inland waters (*HBW*). A total of 280 were recorded in Britain to the end of 2019, about seven a year on average over the last 30 years (Holt *et al.* 2020).

A large piece of bird bone found in Lynx Cave, near Bryn Alun, Denbighshire, in 1991 was eventually identified by the Natural History Museum as being from a Black Stork. The bone, which showed cut marks, was dated to 2,945 years ago, in the mid- to late Bronze Age (Blore 2016). This is one of only two finds of

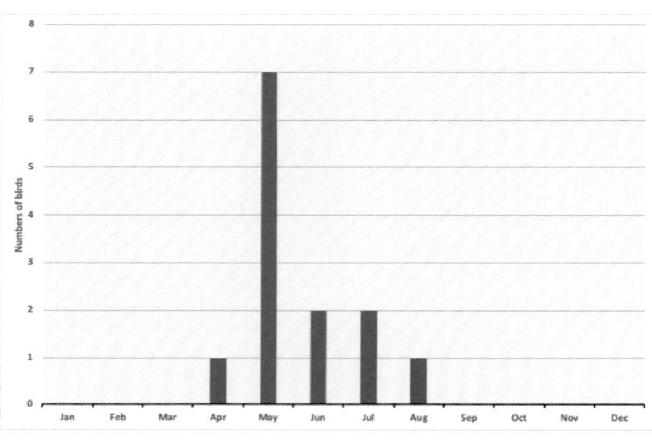

Black Stork: Totals by month of arrival, 1989–2019.

Black Stork remains in Britain (Yalden and Albarella 2009). More recently, this species has been a rare vagrant to Wales with 13 individuals recorded, all since 1989.

Records were mainly in spring, and most from South Wales:

- Cwm Eigiau, Caernarfonshire, on 3 May 1989, also seen nearby at Foel Fras on 7 May;
- Upper Teme Valley, Radnorshire, from 30 August to 6 September 1990;
- Skokholm, Pembrokeshire, on 27–28 April 1991 and on Skomer, Pembrokeshire on 29 April;
- Carmel Head, Anglesey, on 22 June 1991;
- Skomer on 29 July 1991 and then at Marloes, Pembrokeshire, on 2 August;
- adult on the Alaw Estuary, Anglesey, from 31 July to 31 August 2007;
- Machynys, Llanelli, Carmarthenshire, on 1 June 2009;
- Twyn-yr-Odyn, Llanfilo, Breconshire, on 8 May 2011;
- Maesteg, East Glamorgan, on 22 May 2011;
- Tregaron, Ceredigion, on 27 May 2012;
- Gwytherin, Denbighshire, on 6–9 May 2013;
- over Conwy, Caernarfonshire, and Colwyn Bay, Denbighshire, on 20 May 2014 and
- over Skomer, Pembrokeshire on 25 May 2015. This bird moved east across South Wales the following day, being seen at Afan Argoed Park in Gower, Nantyffyllon in East Glamorgan and Coed Morgan, Abergavenny in Gwent.

Jon Green and Robin Sandham

White Stork *Ciconia ciconia* Ciconia Gwyn

Welsh List Category	IUCN Red List		Rarity recording
	Global	Europe	
A, E	LC	LC	WBRC

The White Stork breeds across much of continental Europe, the Middle East and Central Asia (*HBW*). The largest populations in Europe are in Poland and Spain (BirdLife International 2015). Most populations migrate south in winter, with the majority of European birds wintering in the Sahel area of Africa or farther south (*HBW*). This species is doing well in Europe, with a 46% increase in the population between 1980 and 2016, though numbers have been stable over the last decade (PECBMS 2020). The new *European Atlas* shows range expansion in several parts of Europe since the 1997 *Atlas* (*European Atlas* 2020).

A pair nesting on St Giles High Kirk in Edinburgh in 1416 was the only documented breeding record in Britain until 2020 and it is uncertain whether this species ever bred regularly here. White Stork footprints were found preserved in intertidal mud at Goldcliff East, Gwent, on the Severn Estuary, dating from the Late Mesolithic period, around 7,000 years ago. The small size of the footprints suggested they might have been made by a juvenile (Barr 2018). There has never been a confirmed record of breeding in Wales.

The numbers recorded in Britain increased in the second half of the 20th century, from an average of three records a year during 1960–69 to an average of 21 during 1990–99. This increase has continued into the 21st century. However, separating genuine wild migrants from wandering White Storks released or escaped from collections, and more recently from rehabilitation and re-establishment projects in southern England and Europe, has become increasingly difficult. The figures presented here are of birds thought to be of wild origin. White Storks were formerly very rare visitors to Wales, with only three individuals recorded before 1971. There has been an increase in records, particularly in the last decade.

Total	Pre-1950	1950–59	1960–69	1970–79	1980–89	1990–99	2000–09	2010–19
74	3	0	0	12	14	8	12	25

White Stork: Individuals recorded in Wales by decade.

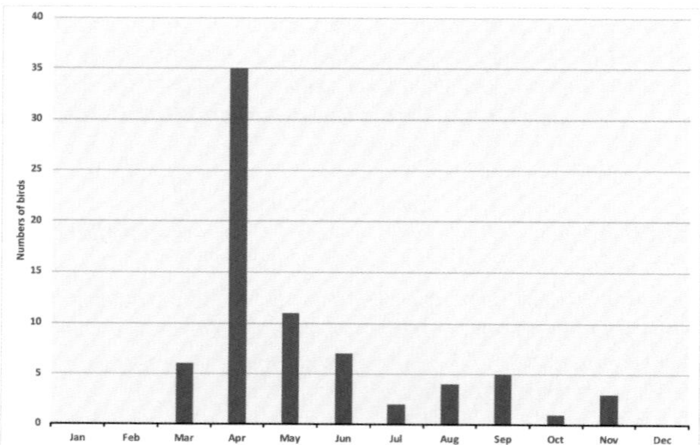

White Stork: Totals by month of arrival, 1900–2019.

The first Welsh record was of three at Bishopston, Gower, on 19 May 1900, one of which was shot. The next Welsh record was not until 1971, when one was in Ceredigion, at Bronant on 9 April and Tregaron on 18 April. A further 70 individuals have been recorded in Wales since, with records in most years since 1971.

The majority of records relate to singles, but two together were seen at Roath Park Lake, East Glamorgan, on 19 November 1982 and two were at Rogerstone, High Cross, Gwent, and then Moylegrove, Pembrokeshire, on 9 April 2010, and subsequently at Wylfa, Anglesey, and Llyn Brân, Denbighshire, on 10–11 April. It is thought that these two individuals passed through Wales again in the autumn, on their way south, and were seen at Brawdy, Pembrokeshire, on 1 September. Larger groups were three at Bishopston, Gower, in May 1900, four at Dingestow, Gwent, on 5–9 June 2012 and six at Bagillt, Flintshire, on 22 April 2013.

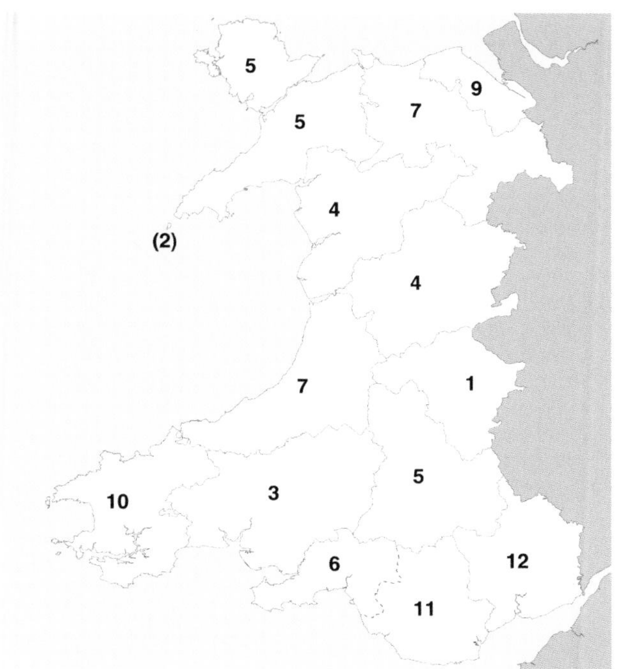

An injured bird found at Pendine, Carmarthenshire, on 24 September 1975 had been ringed as a nestling in Oldeborg, Germany, on 17 July the same year. A colour-ringed bird, seen at Halton, Ruabon, Denbighshire, in March 1978, had been ringed as a nestling in Denmark in June 1976.

A re-establishment scheme in West Sussex led to successful breeding at Knepp Castle Estate in 2020, the first recorded in Britain for 600 years. If a breeding population can be established, it is possible that the White Stork could one day breed in Wales.

Jon Green and Robin Sandham

White Stork: Numbers of birds in each recording area 1900–2019. Some birds were seen in more than one recording area.

Frigatebird species *Fregata spp.* Rhywogaeth Aderyn Ffrigad

There are five species of Frigatebird, all of which breed on tropical and subtropical islands, although only two species occur regularly in the Atlantic Ocean and are on the British List: Magnificent Frigatebird (*F. magnificens*), which breeds in the Caribbean and South America, and Ascension Frigatebird (*F. aquila*), which breeds on Ascension Island in the South Atlantic (*HBW*). Records in Britain are assessed by BBRC. Identification is not straightforward. Neither of the two Welsh records was identified to species level, but both were accepted by BBRC as Frigatebird species.

A female over Skomer, Pembrokeshire, on 14 June 1995 was thought to be probably a Magnificent Frigatebird, but identification was uncertain. A little over ten years later, on 6 November 2005, one was seen over Flat Holm, East Glamorgan, in the wake of Hurricane Wilma, one of the most ferocious storms ever recorded in the Atlantic and responsible for reports of Magnificent Frigatebird along the eastern seaboard of North America, as far north as Nova Scotia, in the last week of October 2005. The day after the Flat Holm sighting, a farmer found a moribund Magnificent Frigatebird in a field near Whitchurch, Shropshire. It was taken to Chester Zoo, but died on 9 November, and was later accepted as a first for Britain. The Welsh record could not be identified to species level, but the BBRC considered it "very likely" to have been the same individual, working its way up the River Severn (Bradbury *et al.* 2008).

Jon Green and Robin Sandham

Gannet *Morus bassanus* Hugan

Welsh List Category	IUCN Red List			Birds of Conservation Concern			
	Global	Europe	GB	UK	Wales		
A	LC	LC	LC	Amber	2002	2010	2016

The Gannet is the largest breeding seabird in the North Atlantic, with two widely separated breeding populations, one in eastern Canada and the northeastern USA, the other in northwest Europe from France in the south to Iceland and Norway (*HBW*). There is no evidence of recruitment from either breeding population into the other, although Gannets ringed at Canadian colonies have been found in European waters. There is a single breeding colony in Wales, on Grassholm in Pembrokeshire, but both during and outside the breeding season, Gannets are recorded around almost the entire Welsh coast. The global population is estimated to be over 527,000 pairs (Murray *et al.* 2015).

The *BTO Migration Atlas* (2002) suggested that birds from all east Atlantic colonies intermingle outside the breeding season to the southwest of Britain, as far south as Senegal, but Deakin *et al.* (2019) showed that the core wintering range of most tagged adults from Grassholm is in the Canary Current Large Marine Ecosystem, off the coast of northwest Africa. Most immature Gannets remain in this area, prior to first breeding at 4–5 years of age. Most recoveries of birds ringed on Grassholm are in that general direction. Less expected was the recovery of a nestling ringed in July 1934 that was found dead in the Baltic Sea, off Germany, in March 1936. Gannets are long-lived. A nestling ringed on Grassholm in September 1961 was found dead in south Devon in January 1998; at over 36 years old, this was only one year younger than the oldest ringed in Britain and Ireland.

Gannets eat a wide range of species and sizes of prey, but favour surface-shoaling pelagic fish such as Mackerel and Herring; they also consume fisheries discards. Breeding Gannets will travel a substantial distance from their nest site to good foraging areas. Satellite tracking shows that Gannets from 12 colonies in the UK each foraged in mutually exclusive areas. Those from Grassholm feed predominantly south and southwest of the colony, off Cornwall and in St George's Channel. The sizes of foraging ranges are determined by density-dependent competition, rather than territorial behaviour (Wakefield *et al.* 2013).

Since the 1980s, the numbers and distribution of Gannets at breeding colonies in Britain and Ireland have been assessed using aerial photographs to identify the number of apparently occupied sites (AOS) holding a single bird or a pair of Gannets, irrespective of whether nest material or chicks are visible. Murray *et al.* (2015) estimated that the UK held 293,200 AOS in 2013–15, representing

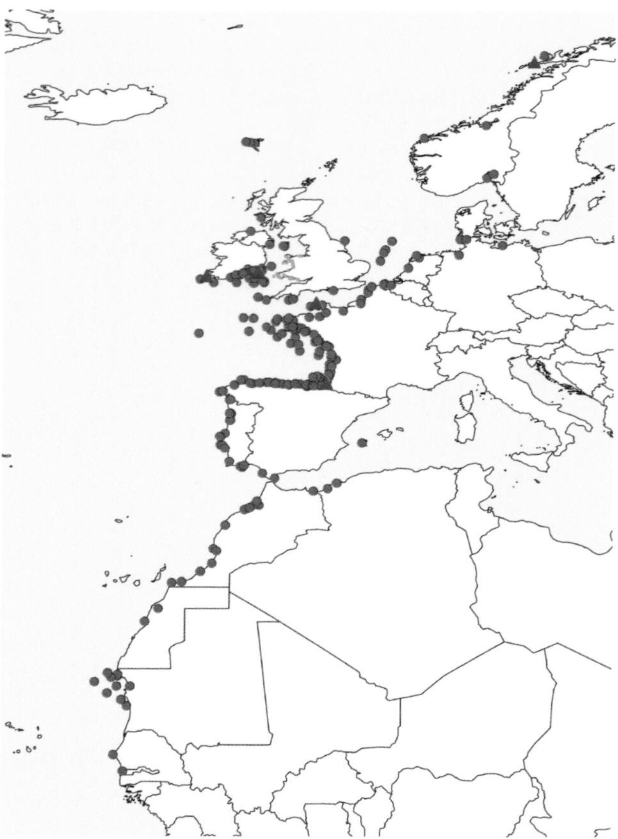

Gannet: Recovery locations of birds ringed in Wales are shown by red circles. Ringing locations of birds that were recovered in Wales are shown by blue triangles. Small brown dots show ringing or recovery locations in Wales.

70% of the biogeographical and 55% of the world population. The UK holds 18 gannetries, most on remote offshore islands, some of which have been occupied for centuries (Fisher and Vevers 1944).

Grassholm is the only breeding colony ever recorded in Wales. Williams (1978) believed that this colony became established shortly before 1860, when 20 pairs were present, although Gurney (1913) suggested that birds may have been present as early as 1820. Mathew (1894) reported that colonisation there was coincident with the decline of a colony on Lundy, Devon, which was finally declared extinct in 1909 as a result of human exploitation (Fisher and Vevers 1944). Egg-collecting and shooting was also a problem on Grassholm during the early phase of the colony's establishment in the late 19th century. *Birds in Wales* (1994) provided a synopsis of breeding counts there between 1883 and 1987, although the count unit varied between nests and pairs. In 1964 the first aerial survey of Grassholm produced an estimate of 15,500 breeding pairs, and the colony increased at a rate of 2.6% per year, to a peak of 39,292 AOS in 2009.

The most recent census in 2015 estimated 36,011 AOS, an apparent decrease of 8.4%, although Murray *et al.* (2015) suggested that the 2009 count mis-classified some non-breeders within the total, leading to an over-estimate of its size. If correct, the Gannet breeding population in Wales is probably stable, and comparisons of sample areas in 2009 and 2015 show no obvious differences in colony density. Grassholm is probably the fourth largest gannetry in the world, after Bass Rock and St Kilda in Scotland and Bonaventure Island in Quebec. It holds 12% of the UK breeding population, 8.6% of the biogeographic population and 6.8% of the global population. Grassholm was classified as a Special Protection Area (SPA) in 1986.

Adults, possibly prospecting birds, have been recorded occasionally lingering around mainland sea cliffs, such as the Great Orme, Caernarfonshire, where birds were seen on ledges in 1908 and 1941 and even copulating in the latter year. They were also recorded on St Margaret's Island, Pembrokeshire, where one sat on a nest annually during 1998–2003 and laid an egg in 2004. In 2019 (and again in 2020), up to 200 adults displayed and built nests on Ynys Badrig (Middle Mouse), off the north coast of Anglesey, but no breeding was recorded.

Breeding birds move back towards their colonies from December, usually occupying nest sites from March, and leave between September and November. They can occur off all coasts of Wales in any month, although they were a rare sighting in the 19th and the first half of the 20th centuries, proportionate to the smaller breeding populations in Britain and Ireland. Tracking studies from colonies in Norway and Iceland showed that adults from those populations also move through British and Irish waters. A bird ringed as a nestling on the Lofoten Islands, northern Norway, in July 1985 was found sick at Rhossili, Gower, in October the same year.

Off the North Wales coast, the largest numbers tend to occur in autumn following strong winds. Record high day counts include 2,100 off Bardsey, Caernarfonshire, on 17 October 1991, 1,720 off Cemlyn, Anglesey, on 7 October 1988, 700 past the Little Orme, Caernarfonshire, on 29 August 2010 and 250 off Point of Ayr, Flintshire, on 5 September 2009. Recoveries of ringed birds in North Wales include birds from colonies as far afield as Ailsa Craig in Ayrshire, Bass Rock in Lothian, Hermaness on Shetland, Les Etacs in Alderney and Great Saltee Island in Co. Wexford, Ireland, as well as Grassholm. Maximum counts from coastal watch points in Cardigan Bay are generally smaller, under 300 a day, although there were 500 feeding in Tremadog Bay, Caernarfonshire, on 15 September 1986 and 760 at Borth, Ceredigion, on 16 July 2013.

Large numbers pass Strumble Head, Pembrokeshire, peaking in August, although their abundance means that numbers tend to be less well recorded. Counts here include 2,980 on 31 August 2019. Other county maxima from southern Wales include 1,400 off Cefn Sidan, Carmarthenshire, on 1 August 1993, 600 from Port Eynon, Gower, on 24 July 1986, 300 past Sker Point, East Glamorgan, on 25 July 1971 and at least 51 at Goldcliff Point, Gwent, on 28 April 2003. Gannets are occasionally recorded, sometimes dead or moribund, at inland sites following autumn or winter gales. Although Gannets have been found in all landlocked counties, this species is rarer inland than some other seabirds; there were, for example, just three in Radnorshire in the 20th century (Jennings 2014).

Gannets can be locally vulnerable to oil discharge and 'ghost' fishing gear. Polypropylene and Courlene netting are frequently used as nest material on Grassholm, where 49% of nests were estimated to contain this material (Nelson 1978). Some estimates put the total weight of this material used at around 20 tonnes. A small proportion of chicks become trapped in the netting each year. Some are cut free in the autumn by the RSPB. This problem has not, to date, affected nest productivity, which, since 2002, has shown no significant variation, with an average of 0.74 fledged chicks/nest (JNCC 2020b).

The improved conservation status of Gannet in Wales contrasts markedly with that of several other seabirds whose populations are declining and for which breeding failures are becoming more frequent. Ironically, while Gannets seem less sensitive to climate change than many other northwest European seabirds, the species may be more vulnerable to technologies established to mitigate against climate change. Cook *et al.* (2012) identified it as being vulnerable to mortality as a result of collision with offshore wind turbines, owing to its flight elevation and plunge-dive height overlapping with the height of the rotor sweep (Langston 2010). A further increase in offshore wind farms in Welsh waters may negatively influence future breeding populations, so it is essential that decadal aerial surveys of Grassholm continue.

1969–70[1]	SCR Census 1985–88[1]	2003–04[2]	2009[2]	2015[2]
16,128	28,545	32,095	39,292	36,011

Gannet: Total count of apparently occupied sites (AOS) recorded in Wales during UK seabird[1] and Gannet[2] censuses.

Patrick Lindley

Sponsored by Lisa Morgan and Dave Astins

Cormorant *Phalacrocorax carbo* Mulfran

Welsh List Category	IUCN Red List			Birds of Conservation Concern			
	Global	Europe	GB	UK	Wales		
A	LC	LC	NT	Green	2002	2010	2016

Cormorant © Peter Howlett

The Cormorant has a wide distribution, breeding in all continents except South America and Antarctica (HBW). Five subspecies are recognised by the IOC List, of which two occur in Wales. The nominate form breeds around the North Atlantic, including the rocky coasts of Britain. The birds breeding in northern latitudes are migratory, but in most of Britain, Ireland and France the species is resident (*HBW*). Mitchell *et al.* (2004) estimated the global population of *P.c. carbo* to be 53,000 pairs, of which approximately 23% breed in Britain and Ireland, including 3.1% in Wales.

In Wales, Cormorants breed mostly on extensive sea cliffs with large ledges or platforms that are relatively open in approach. They tend to form larger colonies than Shags. Numbers at colonies vary annually, possibly in response to food availability or weather conditions in early spring. Ringing recoveries of Cormorants, mainly birds ringed as nestlings at St Margaret's Island, Pembrokeshire, and Puffin Island, Anglesey, show extensive dispersal to the south and east, whereas northerly movements are more modest. Substantial numbers spend at least their first winter in southwest Europe. Of 708 recoveries of Cormorants ringed as chicks on Puffin Island, 19% were found outside Britain and Ireland, mostly in the Bay of Biscay (13% in France, 3.5% in Spain), and one went as far as Italy. In contrast, recoveries of birds ringed as chicks on Denny Island, Gwent, have mostly been at inland waters across Britain (R.M. Clarke *in litt.*). In 1986, a Norwegian-hatched bird was recorded in Glamorgan in its first winter, but 90% of ringing recoveries in that county were of birds ringed at Welsh colonies (Hurford and Landsdown 1995). Breeding adults do not lend themselves to close approach, which restricts opportunities to fit (and read) colour rings, making it difficult to monitor adult survival, and no information is available on colony fidelity. However, colour-ringing of chicks has shown that many individuals spend a substantial period of time at a single site outside the breeding season and there is evidence that many return to the same site in subsequent years (pers. obs.).

The breeding distribution of the Cormorant in Wales is primarily around the western sea cliffs of Pembrokeshire, Ceredigion, Caernarfonshire and Anglesey. There have been no known breeding records in Flintshire, Denbighshire, East Glamorgan or the landlocked counties. One of the best-known colonies is Craig yr Aderyn (Bird Rock), Meirionnydd, with 30–70 nests, some 7km inland from the sea, which has presumably been occupied continuously since the sea lapped at its base. This colony was mentioned by Edward Llwyd in 1695 and Thomas Pennant in 1778. Aside from Craig yr Aderyn, there are no records of inland nesting in Wales, although Forrest (1907) thought that there was occasionally breeding at Llyn Gwynant or Llyn Peris, Caernarfonshire. Tree nesting has not been confirmed in Wales, although Mathew (1894) claimed this nesting habitat in Pembrokeshire. The colony on Puffin Island was documented as long ago as 1662 (Ray and Willughby 1676). Bertram Lloyd estimated 100 pairs on St Margaret's Island in 1930. Cormorants nested in Gower through the 19th century, and in small numbers at Worms Head during the 1950s, occasionally up to 1976 and in 2018. Up to 11 pairs bred at Thurba Head until 1971. In Gwent, a significant colony has been on Denny Island since at least 1999. The mean number of apparently occupied nests between 2000 and 2011 was 62.8 with a peak in 2004 of 94 (Clarke 2011b). The colony continues to be active with 55 nests in 2019 (Richard M. Clarke *in litt.*). There has been no confirmed breeding in Carmarthenshire, although it was suspected at Telpyn Point in 1989 (Lloyd *et al.* 2015).

Operation Seafarer 1969–70	SCR Census 1985–88	Seabird 2000 1998–2002	Seabirds Count Provisional 2015–19
1,468	1,668	1,634	1,491

Cormorant: Number of apparently occupied nests (AON) recorded in Wales during UK seabird censuses.

The breeding population of Cormorants in Wales has been relatively stable for the last five decades, against a backdrop of a decline of nests at Scottish colonies since 2000 and increased breeding at inland colonies in England. However, the most recent counts in 2015–19, as part of the ongoing Seabirds Count, show declines at a number of colonies, including the Little Orme, the largest in Britain during 1998–2002, but now substantially smaller than the peak of 452 nests in 2003. It is unclear whether this is related to an increase on Puffin Island, 16km to the west, during the same period.

Site	County	1985–88	1998–2002	2015–19
Little Orme	Caernarfonshire	198	428	158
Puffin Island	Anglesey	370	353	695
Tandinas Quarry to Fedw Fawr	Anglesey	35	125	139
Bodorgan Head	Anglesey	0	138	28
St Tudwal's West	Caernarfonshire	nc	112	0
Craig yr Aderyn	Meirionnydd	57	65	55
Penderi 2	Ceredigion	140	69	48
St Margaret's Island	Pembrokeshire	238	137–198*	140

Cormorant: Numbers of apparently occupied nests (AON) at the main breeding colonies in Wales during the last three censuses.

* Minimum and maximum counts 1998–2002 from SMP database.

The *BTO Migration Atlas* (2002) showed that 59% of ringing recoveries have been birds deliberately taken by humans, including accidental by-catch in pelagic fishing nets, but also birds shot. Since 1981, Cormorants have been legally protected, but can be controlled under licence to protect fisheries. No consistent data are available on the number of licences issued or birds killed, but data released by the Welsh Government to the Welsh Ornithological Society showed that licences were issued

Site	County	89/90 to 93/94	94/95 to 98/99	90/00 to 03/04	04/05 to 08/09	09/10 to 13/14	13/14 to 18/19
Dee Estuary: Welsh side (whole estuary)	Flintshire	160 (338)	240 (420)	370 (697)	382 (940)	416 (1,340)	543 (1,664)
Denbighshire coast	Denbighshire	13	28	52	55	49	784
Clwyd Estuary	Denbighshire/Flintshire	36	134	188	214	61	250
Severn Estuary: Welsh side (whole estuary)	Gwent	39 (79)	42 (73)	39 (70)	68 (96)	54 (99)	91 (159)
Carmarthen Bay	Carmarthenshire	237	82	74	163	152	70
Dysynni Estuary	Meirionnydd	127	167	124	47	116	62

Cormorant: Five-year average peak counts during Wetland Bird Surveys between 1989/90 and 2018/19 at sites in Wales averaging 150 or more at least once. International importance threshold: 1,200; Great Britain importance threshold: 620.

to 31 individuals during 2011–13. Only 13 birds were reported to have been killed under these licences, although there is no way to verify these figures. Far greater numbers are killed each winter under licence in England: 13,797 during 2014–18 (Natural England 2020). However, no assessment has been made of the impact of shooting, in either country, on the Welsh breeding population.

Outside the breeding season Cormorants use coastal, estuarine and inland waters, and have become more widespread. Over 60 10-km squares in Wales gained wintering Cormorant records during the *Britain and Ireland Atlas 2007–11*, compared to 1981–84, and just five lost them. The WeBS index for Wales has grown steadily since the species was included in counts in 1987, with a more rapid increase during 2012–16. Nonetheless, away from the Dee Estuary, numbers are modest and have not shown a substantial change in the last 20 years.

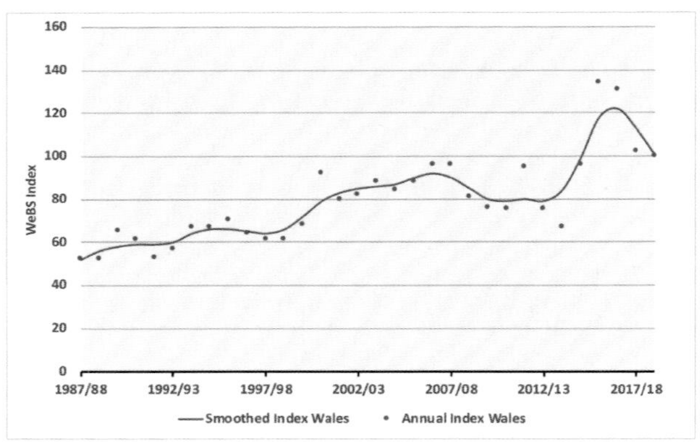

Cormorant: Wetland Bird Survey indices, 1987/88 to 2018/19.

Large roosts occur at several Welsh estuaries, such as 377 at Morfa Harlech, Meirionnydd, in September 2003 and again in September 2004, 283 at Tywyn Point, Carmarthenshire, in August 1974 and 470 at Gronant, Flintshire, in November 2015. Roosts have also been recorded on artificial structures, such as 248 at a pylon roost in the Dwyryd Estuary, Meirionnydd, in September 1987. As numbers breeding on the Little Orme diminished, so have counts there in late summer, from 1,306 in August 2015 to 451 in August 2019. Cormorants also roost on Denny Island, where 200 birds were seen flying to roost in 1994.

Inland, Cormorants occur in modest numbers at a limited number of sites outside the breeding season. Llandegfedd Reservoir, Gwent, can hold good numbers, for example 138 in September 2018. In Radnorshire, there was a marked increase in records from the late 1970s to the early 1990s, peaking at 35–45, mostly on the Elan reservoirs and the River Wye, after which numbers declined. In Breconshire, only Llangorse Lake consistently holds Cormorants in any number, where the mean count has remained around 60, from the mid-1980s to 2019. Ringing shows that most Cormorants seen in mid-Wales originate from Welsh coastal colonies, although some have come from as far away as southwest Scotland and northern France. These birds are also likely to be *P.c. carbo*. In Gwent, Llandegfedd Reservoir consistently holds birds with numbers peaking at 50–130 in the autumn. The most significant roost site in Gwent during the 1980s was at Piercefield on the river Wye, which held over 100 birds during winter months between 1986 and 1988.

The Cormorant is Amber-listed in Wales because of the international importance of the non-breeding population (Johnstone and Bladwell 2016). Collection of productivity data at colonies in Wales has been sporadic, but the information available suggests a steady decline in the number of chicks fledged per pair since about 2004 (JNCC 2020b). While the species is not currently under threat in Wales, regular monitoring of breeding numbers and productivity at the Welsh colonies is important.

Continental Cormorant *P.c. sinensis*

Mulfran y Cyfandir

The 'continental' subspecies *P.c. sinensis* breeds from central Europe, east to India and China (IOC). In the 1980s, it extended its range westward, initially as a winter visitor to Britain and then colonising England, mainly breeding at inland lakes and reservoirs. By 2005, around 2,100 pairs were breeding at these sites in England (Newson *et al.* 2007). There has never been a record of *sinensis* breeding in Wales, but there have been non-breeding records.

Identification of Continental Cormorants is not straightforward, and the only diagnostic feature is the angle of the gular patch (Newson *et al.* 2004). Prior to this realisation, Cormorants with white crown feathering or pale breast feathers were sometimes considered to be of this subspecies. A total of 41 published records during 1981–90 were reported in *Birds in Wales* (1994) and another 18 reported during 1992–2000 in *Birds in Wales* (2002).

Although *P.c. sinensis* is likely to be a regular visitor to Wales, particularly along the north and south coasts and in the border counties, its status is not clearly established. The following records are the only ones that have been accepted by the Welsh Birds Rarities Committee:

- RSPB Conwy, Denbighshire, on 12 October 2004;
- Llyn Parc Mawr, Abergele, Denbighshire, on 19 February 2017 and
- Kinmel Bay, Denbighshire, on 9 September 2018.

Steve Dodd

Sponsored by Pembrokeshire Ringing Group

Shag *Phalacrocorax aristotelis* Mulfran Werdd

Welsh List Category	IUCN Red List			Birds of Conservation Concern			
	Global	Europe	GB	UK	Wales		
					2002	2010	2016
A	LC	LC	EN	Red			

The Shag is a marine species strongly associated with rocky coasts, and largely avoiding sandy and estuarine shores. The breeding range of the nominate subspecies extends from Iceland and northern Fennoscandia to Portugal, with Britain and Ireland accounting for about 40% of the global population. Other subspecies are found in the Mediterranean/Black Sea and in northwest Africa (*HBW*). Wales holds about 3% of the British population (Mitchell 2005).

This species nests under large boulders, in caves, or failing that, under an overhanging rock. The normal breeding season is from March until July, with the peak number of chicks in nests through June. However, abnormal breeding cycles are frequent, with first eggs recorded as early as 20 November (1966) and 21 January (1967) at Ynys yr Adar off Llanddwyn. Winter nests are vulnerable to poor weather. The majority of birds breed at, or very close to, their natal colony, but a bird ringed as a nestling at Puffin Island, Anglesey, in 1997, bred at the Calf of Man between 2008 and 2018, while birds ringed as nestlings at the Ynysoedd Gwylan, Bodorgan and Pen-y-parc colonies were all later found breeding on Puffin Island.

There is little information available on the numbers breeding on the coasts of Wales in the 19th century and the early 20th century. Shag colonies were noted in Pembrokeshire, on Ramsey, Caldey, St Margaret's Island and Elegug Stacks, by Matthew (1894), and at multiple sites in Anglesey and Caernarfonshire by Forrest (1907). The species bred annually in small numbers in south Gower until 1925. A single pair bred here in six of the years between 1953 and the 1970s, and up to five pairs during the 1980s, all on Worms Head or nearby (Hurford and Lansdown 1995).

The Operation Seafarer counts in 1969–70 found that the largest numbers in Wales were in Caernarfonshire, followed by Anglesey and Pembrokeshire (Mitchell 2005). The highest number of nests was on Puffin Island, Anglesey. Shag numbers increased almost everywhere in Britain and Ireland between the Operation Seafarer counts and the SCR census in 1985–88. Shags probably benefited from increased protection for Cormorant in the Republic of Ireland and the UK since 1976 and 1981 respectively, as undoubtedly many Shags will have been shot by mistake. However, better coverage of previously inaccessible coastlines was also thought to have contributed to the apparent increase (JNCC 2020b). Breeding numbers in Wales increased by almost 43% between Operation Seafarer and the SCR census, and almost doubled on Ynys Gwylan Fawr, Caernarfonshire, the largest colony in Wales in 1985–88.

Numbers declined by 27% in Britain and Ireland between the SCR census and the Seabird 2000 counts in 1999–2000, but increased by 16% in Wales during the same period. In the years following the Seabird 2000 counts, numbers in Wales have remained stable or showed a small increase, although other colonies in the Irish Sea, such as the large colony at Lambay Island, County Dublin, showed declines (JNCC 2020b). However, counts between 2015 and 2019, for the ongoing Seabirds Count, suggest that Shag populations are now declining at a similar rate in Wales to those in Scotland and England, and are now below the level recorded in the Operation Seafarer survey in 1969–70. This is a matter of urgent conservation concern.

Operation Seafarer 1969–70	SCR Census 1985–88	Seabird 2000 1998–2002	Seabirds Count Provisional 2015–19
550	785	914	502

Shag: Total number of apparently occupied nests (AON) recorded in Wales during UK seabird censuses.

Shag © Ben Porter

The principal colonies in Wales are in Caernarfonshire, from west of Abersoch around Pen Llŷn to Carreg y Llam, on the Great and Little Ormes, and on islands off Anglesey. The number of occupied nests has declined at all these colonies and on the Pembrokeshire islands. A handful of pairs nest at the most southerly colony in Wales, on Worms Head, Gower. Around 60 pairs nest in four loose colonies in Ceredigion, while two nests at Friog Cliffs in May 2008 were the first and sole breeding records in Meirionnydd. The largest colony is now on Puffin Island, where numbers increased following eradication of Brown Rats in 1998–2000. Caution is required because boat-based surveys do not always adequately reflect the number of nesting Shag, many of which may not be visible if they are nesting under boulders or deep in caves. For example, land-based surveys of Puffin Island recorded 401 nests in 2010 and 425 in 2017, far more than the numbers seen from a boat.

At the end of the breeding season, birds remain in the vicinity of the colony or disperse locally along adjacent rocky coasts. Ringing recoveries show that the majority of Shags from Welsh colonies remain within the Irish Sea. During 1982–2016, there were 473 recoveries of Shag ringed by SCAN Ringing Group, mainly as chicks on Puffin Island, Anglesey. Of these, 19% were found on Anglesey, 56% in Caernarfonshire/Meirionnydd and 9% in Ceredigion/Pembrokeshire. The majority of the remainder were found between southwestern Scotland and Cornwall. Shags are vulnerable to bad weather and recoveries are often associated with periods of stormy weather. Despite this, substantial 'wrecks' or population crashes occur rarely in Wales, compared to the North Sea. The only winter of serious mortality between 2000 and 2019 was 2012/13, when the survival rate of adults was dramatically below the mean of 78%.

The species is not particularly well covered by WeBS counts in Wales, as most birds are found away from the regularly counted WeBS sites. Some of the largest counts outside the breeding

Site	County	1969–70	1985–88	1998–2000	2015–19
Grassholm	Pembrokeshire	27	7	8	8
Skomer	Pembrokeshire	20	3	1	6
Middleholm	Pembrokeshire	22	nc	29	11
St Margaret's Island	Pembrokeshire	18	18	4	2
Cardigan Island	Ceredigion	6	26	12	4
Ynys Gwylan Fawr	Caernarfonshire	65	125	133	66
Bardsey	Caernarfonshire	28	23	44	
Carreg y Llam	Caernarfonshire	5	27	15	7
St Tudwal's	Caernarfonshire	11	123	114	80
Bodorgan Head	Anglesey	43	33	45	14
Ynys yr Adar	Anglesey	51	28	35	19
Puffin Island	Anglesey	70	113	178	122

Shag: Counts of apparently occupied nests (AON) at the main colonies during the four UK seabird censuses.

season have been close to breeding colonies, for example 271 in the sound between Anglesey and Puffin Island on 20 November 2006. Co-ordinated counts of north Cardigan Bay in the 1990s produced a peak count of 427 individuals in December 1992 (Thorpe 2002). A survey of the North Wales coast from Traeth Lafan, Caernarfonshire, to Mostyn, Flintshire, in January 1998 found a total of 233 birds, of which 166 were off Penmaenmawr, Caernarfonshire. Sightings are scarcer on this coast east of the Little Orme, although they are annual in small numbers between Rhos-on-Sea and Llanddulas, Denbighshire. Off the south coast, there were 134 off Oxwich Point, Gower, on 28 December 2011. Up the Bristol Channel, two seen from the Newport Wetlands Reserve (Goldcliff) in August 2014 were the first in Gwent since 2005, although there was another record in 2015. Sightings inland are by no means annual in Wales, though birds are sometimes found on large lakes and reservoirs.

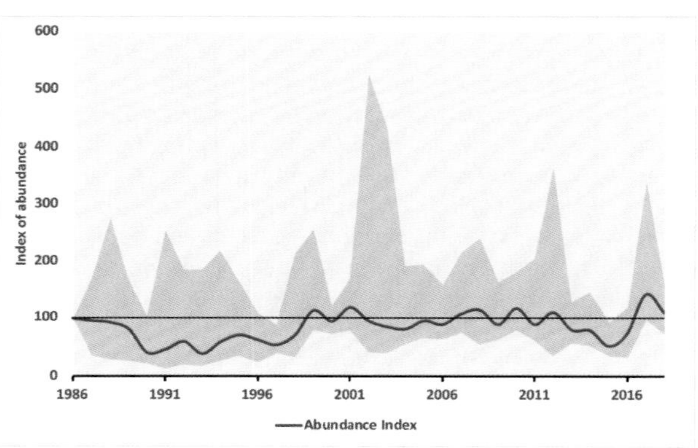

Shag: The line shows the abundance index calculated from the Seabird Monitoring Programme (SMP) database. The shaded area shows the 97.5% confidence limits.

Future prospects for the Shag in Wales are uncertain. Since about 1992, Shag numbers have declined in Scotland and England, particularly on the east coast of Britain, on a similar scale to other seabirds that rely largely on sandeels and other small fish. As a result, the species is now Red-listed in the UK. The Shag is Amber-listed in Wales because of its status in the UK and Europe (Johnstone and Bladwell 2016), not because of factors specific to Wales. This species is susceptible to predation by invasive mammals, so ensuring that islands remain rodent-free is essential. It is also locally vulnerable to oil pollution and discarded fishing gear, increasingly so if the population falls as a result of other factors. The reasons for the decline in breeding numbers on the east coast are thought to include mortality during periods of prolonged gales. Such 'wrecks' could become more frequent in the Irish Sea as a result of a predicted increase in the number and severity of storms due to climate change (Frederiksen *et al.* 2008). Such storms are a key cause of mortality of both immature and adult birds, and could adversely affect Shag populations in Wales in future.

Food supply is also important—particularly the availability of sandeels, which are an important food item for the Shag (Harris and Wanless 1991). Increases in sea surface temperature are thought to have led to a decline in sandeel abundance (Dye *et al.* 2013). Trends in prey availability in Welsh waters are poorly known at present, but high productivity figures suggest that this is not a problem in the breeding season. Productivity in Wales averaged just under two chicks/pair between 1986 and 2018, compared to 1.25 chicks/pair in Scotland (JNCC 2020b). The effects of climate change will ultimately determine the fortunes of the species in Wales. Colonies will need to be closely monitored to establish whether or not the recent reduction in breeding numbers in Wales is the start of a long-term trend.

Steve Dodd

Sponsored by Maggie and Ian Wright

Glossy Ibis *Plegadis falcinellus* **Ibis Du**

Welsh List Category	IUCN Red List		Rarity recording
	Global	Europe	
A	LC	LC	WBRC

Total	Pre-1950	1950–59	1960–69	1970–79	1980–89	1990–99	2000–09	2010–19
129	12	1	0	1	1	1	27	86

Glossy Ibis: Individuals recorded in Wales by decade.

The Glossy Ibis is a migrant from warm temperate and tropical zones, where it is associated with a wide variety of inland wetland habitats and, to a lesser extent, coastal lagoons and estuaries. It is the most widespread ibis species, with breeding populations around the Caribbean Sea, southern Europe, Central Asia, Africa and Australia (*HBW*). The number of records in Britain increased dramatically from an average of nine each year in 2000–09 to 76 in 2010–18 (White and Kehoe 2020a). The increase has been linked to a rise in numbers in Spain and to drought issues in its usual wintering areas. Although still a

Glossy Ibis at Borth, Ceredigion, February 2012 © Kev Joynes

scarce bird in Wales, since 2009 the Glossy Ibis has become a more regular visitor.

The first record in Wales was one, from a group of five, shot near Beaumaris, Anglesey, in September 1806. Early records suggest that other birds suffered a similar fate or died from exhaustion, including the first three records in Carmarthenshire (between 1858 and 1910), two in Pembrokeshire (in 1834 and 1917), one in Gwent (in 1902) and another on Anglesey (in 1945). During the following 60 years up to 2005, there were only six further records in Wales: singles in Flintshire and Caernarfonshire, two in Pembrokeshire and two in East Glamorgan, one of which was also seen in Gower.

The autumn of 2009 saw a remarkable influx of Glossy Ibises into Britain, including an estimated 25 in Wales. Among the first to arrive was one in the lower Gwendraeth Fawr Estuary near Cydweli, Carmarthenshire, on 3 September. The following day, 25 were discovered nearby at Pembrey, Carmarthenshire, where up to ten remained until 6 September, including four that had been colour-ringed as nestlings at Doñana National Park, Spain, in May 2009. These birds were thought to have moved north later in the month, there being 12 on the Alaw Estuary, Anglesey, on 17 September, of which ten flew south over Bardsey, Caernarfonshire, on 18 September. Nine were seen later that day on the Teifi Estuary, Ceredigion/Pembrokeshire. Single birds were seen at a few other locations up to 13 October. Numbers arriving in the autumns of 2010 and 2011 were smaller, involving singles at several wetland sites on Anglesey and in Meirionnydd, Carmarthenshire, East Glamorgan and Gwent.

There was another influx in 2012, this time in winter, involving at least 43 individuals. Their arrival coincided with drought conditions in Spain and other parts of southern Europe that possibly contributed to their dispersal away from regular wintering areas (Hudson *et al.* 2013). The first sightings were of two at Westfield Pill, Pembrokeshire, on 12 January that were thought to have moved to the St Davids peninsula by the following day. Singles were seen at Sageston, Pembrokeshire, on 14 January and at Newport Wetlands Reserve, Gwent, on 14–31 January and in East Glamorgan at Sully Moors and Cosmeston Lakes on 16 and 17 January. Further sightings were at Flat Holm and Cosmeston Lakes on 6 February and near the Kenson River, Llancadle, on 22–24 February, involving records of one or two birds. Larger numbers were then reported from Pembrokeshire, including 23 in the Carew and Sageston areas between 28 January and 14 February, and up to four in the Marloes area, between 18 January and 9 June, one of which remained until 20 August 2013. An adult at Borth, Ceredigion, on 3–21 February 2012 had been ringed at Doñana National Park on 7 May 2007. One at Montgomery on 24–25 February and seen there again on 28–30 April, was the first

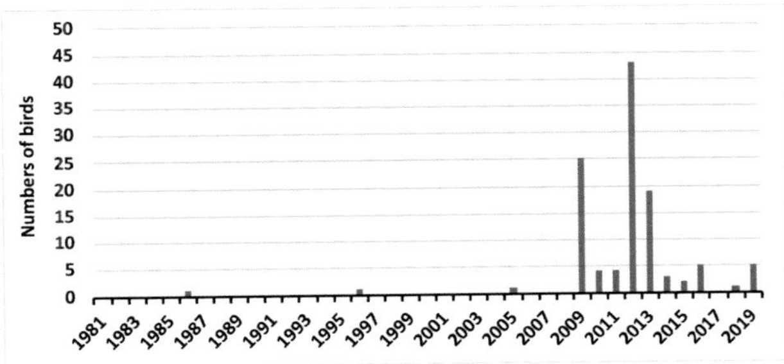

Glossy Ibis: Estimated numbers in Wales, 1981–2019.

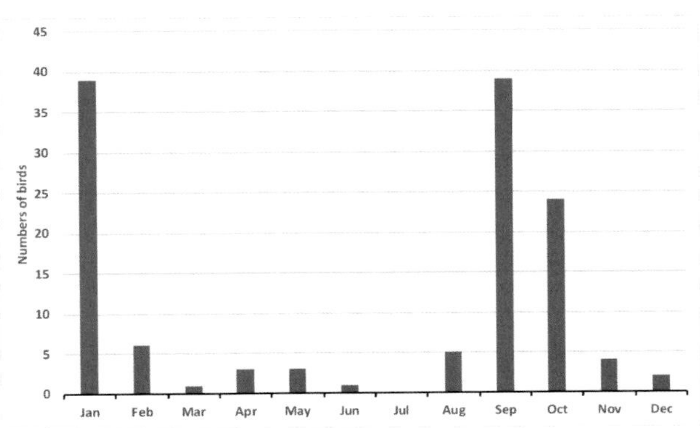

Glossy Ibis: Totals by month of arrival (when known), 1806–2019.

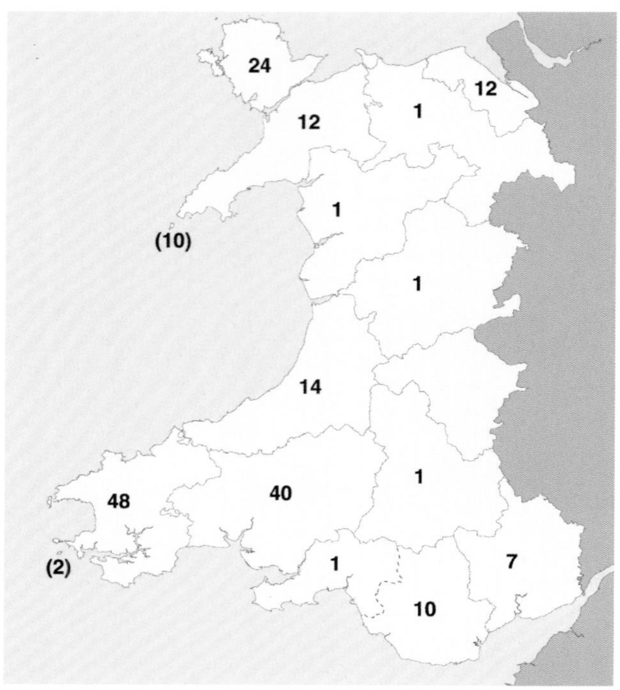

Glossy Ibis: Numbers of birds in each recording area, 1806–2019. Some birds were seen in more than one recording area.

record for Montgomeryshire. The final record of the year was of a single in Carmarthenshire, at Penclacwydd on 28 August.

In 2013, there was another autumn arrival, involving at least 19 individuals. These included ten at Bagillt Bank, Flintshire, on 1 October and five at Machynys, Carmarthenshire, on 4 October that moved to Penclacwydd where they remained from 6–26 October. There were, however, only *c.*16 recorded during 2014–19 and none in 2017. These included a long-staying individual at Newport Wetlands Reserve, Gwent, from 14 January 2019 into 2020. One seen flying southeast over Llangorse Lake, Breconshire, about ten days earlier (the first county record) may have been this bird.

By the end of 2019, the Glossy Ibis had been recorded in all Welsh counties except Radnorshire. The majority of sightings were in coastal counties, with the highest numbers in southwest and northwest Wales. Its European distribution contracted markedly in historical times owing to human activities (Huntley *et al.* 2007). Numbers had started to increase after a long period of decline during the second half of the 20th century (*European Atlas* 1997). Since then its range has also expanded considerably in westerly and northerly directions (*European Atlas* 2020). It is likely that its recovery, combined with a changing climate, will result in increased occurrences in Britain. Pairs built nests in Lincolnshire in 2014 and Somerset in 2016. Influxes in spring are associated with mild winters and smaller numbers breeding in Coto Doñana (Ausden *et al.* 2019). It seems quite feasible that Glossy Ibis could breed in Wales in the coming decades.

Jon Green and Robin Sandham

Spoonbill *Platalea leucorodia* Llwybig

Welsh List Category	IUCN Red List			Birds of Conservation Concern				Rarity recording
	Global	Europe	GB	UK	Wales			
A	LC	LC	EN	Amber	2002 N/A	2010	2016	RBBP

The Spoonbill has a wide but patchy distribution in Europe, Asia and along the coast of Africa north of the Equator. It winters in seasonal wetlands within that range (*HBW*). The largest populations in Europe (outside Russia) are in Spain, Hungary, Ukraine and The Netherlands (Birdlife International 2015). The species prefers shallow wetlands, including deltas, river floodplains and marshes, where it nests in reedbeds, on islands or in trees.

The only evidence that Spoonbills once bred in Wales is a brief reference in Owen (1603) quoted in Donovan and Rees (1994): "On highe trees the heronflewes, the floveler". The name 'Shoveler' was used for the Spoonbill, and sometimes for the duck, in this period. The duck was also sometimes called the 'spoonbill', which adds to the confusion and means that older records must be treated with caution. However, Owen's reference to breeding in high trees indicates that he meant *P. leucorodia*. The birds mentioned by Dillwyn (1848), on the authority of the *Swansea Guide*, as being "sometimes caught in the decoys at Stouthall" [Gower], probably half a century previously, must surely refer to the duck, although it is listed as *P. leucorodia*. There is more evidence for breeding in eastern England up to the late 17th century (*Birds in England* 2005). As in many parts of its range, drainage of wetlands was probably a major factor in its disappearance.

Records since the early 1900s have most likely involved birds breeding in mainland Europe. Mathew (1894) noted that the species was common in The Netherlands, but thought that these birds would not wander west as far as Wales and considered that those recorded in Pembrokeshire were probably from southern Spain. Colour-ringing has subsequently shown that many seen in Wales originate from The Netherlands, where the population has increased greatly over the last 30 years and is now well over 2,000 breeding pairs (Oudman *et al.* 2017). These birds winter mainly in France, Iberia and south to Mauritania.

The first record after George Owen's time was one shot near Aberystwyth, Ceredigion, in 1838 (Roderick and Davis 2010). In Pembrokeshire, Mathew (1894) noted that in 1854 and 1855, as many as 11 Spoonbills were shot on the shores of Milford Haven. In 1885, five or seven were shot in one day near Mullock Bridge. A flock of seven was on Goodwick Sands in 1856 and 2–3 were seen around Pembroke almost every year. On the Dyfi Estuary, Ceredigion/Meirionnydd, there was a flock of 14 on 16 May 1893, below Glandyfi Castle (Forrest 1907).

The species became much less common in Wales during the first half of the 20th century. *Birds in Wales* (1994) documented only about ten occurrences in this period. These included records in Carmarthenshire at Laugharne in 1912 or 1913, at Cydweli Flats in 1919 and at Laugharne on 22 June 1931 (Ingram and Salmon 1954). The lower numbers may have been the result of declines in the population in Europe. *Birds in Wales* (1994) recorded a marked increase in the number of records in the second half of the 20th century (52 between 1950 and 1991) and the number of sightings continued to increase between 1992 and 2019. Sightings of colour-ringed birds show how mobile they can be. In some years it is difficult to ascertain the numbers involved.

The largest count recorded in Wales after 1893 was over a century later, when 11 first-calendar-year birds were on saltmarsh at Uzmaston, Pembrokeshire, on 18 September 2005. It is presumed that these birds, joined by one other, accounted for a flock of 12 on the Nevern Estuary, Pembrokeshire, that afternoon. Two of these, which had been colour-ringed as nestlings in The Netherlands, had been seen at Newport Wetlands Reserve (Goldcliff), Gwent, prior to their arrival in Pembrokeshire. By October they had moved to Co. Mayo, Republic of Ireland. Most recent arrivals have been in April–June or September–October. The great majority of these have been coastal, such as eight at Goldcliff Lagoons, Newport

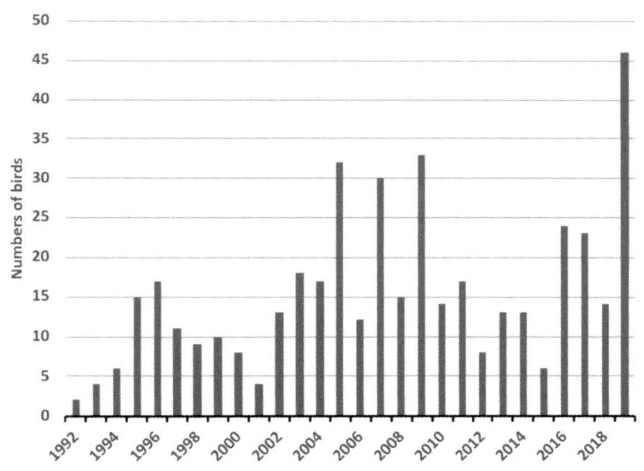

Spoonbill: Estimated numbers of birds in Wales, 1992–2019.

Wetlands Reserve, on 22 and 23 September 2019 and 11 at Penclacwydd, Carmarthenshire, on 28 September 2019. There have also been a few on inland lakes and marshes.

The Spoonbill's breeding range, although still patchy, has expanded considerably since the *European Atlas* (1997). New colonisations have taken place within its existing range, but additional countries have also been occupied, including Britain. Some losses have occurred as well, mainly in southeast Europe (*European Atlas* 2020). Following initial breeding attempts in northwest England, at Frodsham Marsh, Cheshire, in 1996 (Norman 2008) and in eastern England in 1997, the first there for perhaps 350 years (*Birds in England* 2005), breeding occurred in Lancashire in 1999 and in Galloway in 2008. It also occurred in Norfolk from 2010, where a small colony has become established at Holkham and 28 pairs bred, fledging 48 young, in 2018 (Eaton *et al.* 2020). A pair built a nest on the Wirral side of the Flintshire boundary at RSPB Burton Mere Wetlands in 2019. The female appeared to incubate for a short period, but no chicks hatched. The adults, plus a third male, fed primarily on the Welsh side of the Dee Estuary. Small numbers summered in South Wales in consecutive years in the 2010s. It seems likely that Spoonbill will eventually breed again in Wales, after an absence of over 400 years.

Rhion Pritchard

Bittern *Botaurus stellaris* Aderyn y Bwn

Welsh List Category	IUCN Red List			Birds of Conservation Concern				Rarity recording
	Global	Europe	GB	UK	Wales			
A	LC	LC	VU	Amber	2002	2010	2016	RBBP

Wales is at the western edge of the Bittern's breeding range, which extends over a wide area of temperate Europe and Asia as far east as northern Japan. The bulk of the European breeding population is in Russia, Ukraine, Poland and Belarus (*HBW*). These birds migrate to southern Europe and the Sahel zone in Africa during the winter. Bitterns breed in large freshwater *Phragmites* reedbeds. They feed on fish, amphibians, small mammals and young birds: European Eel is a particularly important food source. This is a very cryptic species, but the 'booming' calls made by males in late winter and spring reveal their presence and are used to estimate the number of territories. In winter, birds move into Wales from the east, particularly during freezing weather. There have been two ringing recoveries in Wales, both in East Glamorgan. One was ringed as a nestling in Belgium and found dead at Kenfig Pool in October 1957, and another found dead at St Mellons, Cardiff, in March 2009 had been ringed in Finland the previous year.

Before large-scale drainage of wetlands in the late 18th and early 19th centuries, the Bittern was likely to have been a familiar bird in Wales. It was sufficiently widespread for perhaps the world's only nursery rhyme about a Bittern, *Deryn y Bwn o'r bannau*, to have been written in Welsh. In medieval Welsh law, the Bittern was named as one of three notable quarry birds in falconry, along with Grey Heron and Curlew (said by some) or Crane. If a bird, flown by the king's falconer, brought down one of these, the falconer was to be honoured by the king in specified ways (Jenkins 2000). Remains dating to the early Norman period were found at Hen Domen castle, Montgomeryshire (Yalden and Albarella 2009). The Bittern was also mentioned in Welsh poetry as a quarry for falconers. For example, a poem by Guto'r Glyn in the 15th century mentioned Bittern on Anglesey (Centre for Advanced Welsh and Celtic Studies 2013). Owen (1603) recorded that it bred in Pembrokeshire.

The Bittern was said to breed at Cwm Maelog and Crymlyn Bog, both in Gower, in 1802 (Hurford and Lansdown 1995), although by 1848 Dillwyn reported that the latter site had been drained and the Bittern had become rare. Later authors knew of no record of confirmed breeding in the county. Mathew (1894) commented that in former days the Bittern "was, doubtless, a common Pembrokeshire bird" and Salter (1900) implied that the species had formerly bred at Cors Fochno and Cors Caron, both in Ceredigion. Forrest (1907) knew of no record of breeding in North Wales, although he thought it extremely probable that it had bred

The unusual sight of a Bittern swimming across a lake © Bob Garrett

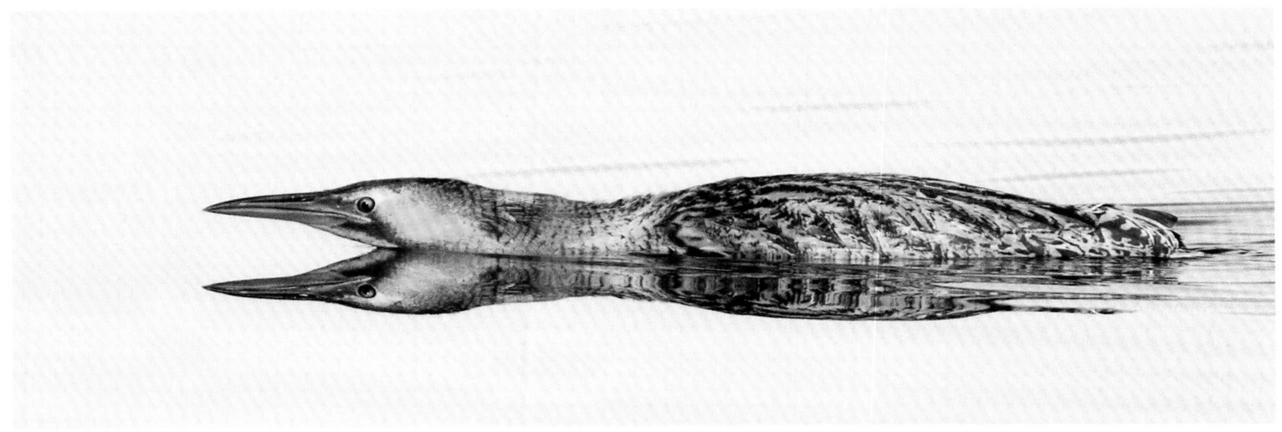

in the past. He was shown a female Bittern that had been shot on Anglesey on 13 August 1899. He said that "it was just possible that this hen bird had nested on Anglesey, for the plumage was in poor condition and seemed to indicate that it had recently been sitting". In 1919, he quoted H.S. Davenport as stating that there was evidence that a Bittern had nested in a marsh between Bala and Dolgellau, Meirionnydd, in the 1880s, having been heard booming repeatedly in the evenings. Lockley *et al.* (1949) quoted a record of two at Dowrog, Pembrokeshire, on 26 June 1909, a date that could have suggested possible breeding. It is likely, however, that the Bittern no longer bred regularly in Wales by the mid-19th century and had ceased to breed in Britain by about 1886. They began to recolonise the Norfolk Broads in the early 1900s and, by the mid-1950s, the British population had reached around 80 pairs (*Birds in England* 2005).

Booming was recorded on Anglesey in 1947. Two were heard there in 1955 and it was then heard on the island every year until 1966, when males were at five sites. Breeding was confirmed at Llyn Traffwll (Valley Wetlands) in 1968. By 1971 at least ten pairs were thought to have bred on Anglesey, but numbers gradually declined in the following years (Jones and Whalley 2004). At Oxwich Marsh, Gower, birds were heard booming between April and June in nine years during 1969–79. A pair was present in 1973, but there was no evidence of breeding. The last breeding record in Wales during this period was at Valley Wetlands, Anglesey, in 1984, by which time numbers breeding in England were also declining rapidly. By 1997 there were just 11 booming males in Britain (*Birds in England* 2005).

In the 1990s, the RSPB began work on Anglesey. It was aimed at restoring a breeding population of the Bittern in Wales, by creating suitable large wet reedbeds, as part of a strategy to reduce its dependency on East Anglian coastal reedbeds at risk from sea-level rise. The RSPB Malltraeth Marsh reserve (now Cors Ddyga) was established in 1994, converting pasture fields to reedbed. In 2016, a pair nested and reared young there; the first breeding in Wales for 32 years. Four young fledged from three nests on Anglesey in 2017 and there were two nests at RSPB Cors Ddyga in both 2018 and 2019, although the outcome of these was uncertain.

In recent years, several sites, including others on Anglesey, have recorded birds booming in late winter. Others have been recorded at Llangorse Lake, Breconshire; at Newport Wetlands Reserve, Gwent; at Aberleri, Ceredigion and at sites in East Glamorgan. In 2019, almost 200 booming males were heard at around 90 sites across Britain, including eight males in Wales, with continuing signs of expansion on Anglesey (RSPB 2019d).

In addition to breeding at RSPB Cors Ddyga, there were two booming males at RSPB Valley Wetlands, plus a third partial territory at Cerrig Bach/Treflesg. One nest was confirmed, and a first-calendar-year bird was found dead at Cors Crugyll in late summer 2019. Bitterns were present in the breeding season and heard booming at Newport Wetlands Reserve, Gwent, between 2016 and 2019, but there was no evidence of breeding success. However, in 2020, 2–3 young fledged there from two separate nests (Dalrymple 2020). This was the first confirmed breeding in South Wales for at least 200 years.

As a winter visitor, the Bittern has occurred regularly in several counties. Mathew (1894) said that it was very rare in Pembrokeshire but had been more common 50 years previously. He quoted a taxidermist named Mr Tracy as saying that, in about 1842, he had 13 Bitterns to set up, all killed the same week during severe weather. Forrest (1907) stated that "in 1876, no less than ten were obtained on the Dysynni and Dovey Estuaries, and some years earlier Mr Henry Franklin shot five at Towyn [Meirionnydd] early on a winter morning". Forrest stated that records of Bitterns in all the counties of North Wales were so numerous that it was not necessary to give further details. Hutchins, a taxidermist at Aberystwyth, Ceredigion, received 12 in the winter of 1925. Since then, wintering numbers have been fairly low in most years, although influxes can be caused by severe weather to the east, such as in early 1982, when at least 27 were reported in southeast and southwest Wales. A cold period in January 2010 resulted in records of at least 42 in Wales, including 12 on Anglesey. At least 32 were seen across Wales during another cold spell in December that year, including eight in East Glamorgan. Bitterns were recorded in almost all counties, but the largest numbers were at Kenfig, East Glamorgan, where five were thought to be present in early 2002, in late 2004 and again in early 2011.

When *Birds of Conservation Concern in Wales 3* was published (Johnstone and Bladwell 2016), the Bittern was classed as a bird that had formerly bred in Wales, but was then only a winter visitor, and so was consequently given Amber status. Its return as a breeding species makes it of higher conservation concern. Breeding birds occur on nature reserves where reedbeds are managed on a rotational basis. This requires a long-term commitment to habitat management and the maintenance of water levels to prevent the accumulation of dead vegetation and the encroachment of scrub. The wetlands also require a good supply of fish. Bitterns have doubtless had to switch prey away from European Eel, which itself is Critically Endangered following a decline of over 90% since the 1980s.

Rhion Pritchard

Sponsored by Graham Williams

American Bittern *Botaurus lentiginosus* Aderyn Bwn America

Welsh List Category	IUCN Red List	Rarity recording
	Global	
A	LC	BBRC

The American Bittern breeds in wetlands across southern Canada and the northern USA, and winters in the southern USA, Mexico and islands in the northern Caribbean (*HBW*). This species is an extremely rare trans-Atlantic visitor to Britain. There have been 39 records in Britain, with the last in 2018 (Holt *et al.* 2019). Seven of these were in Wales, all single birds. Of the four pre-1950 records, the first three were shot:

- an unknown location, Anglesey, about December 1851;
- St Davids, Pembrokeshire, in October 1872;
- male at Dale, Pembrokeshire, on 11 December 1905;
- Ramsey, Pembrokeshire, on 19 October 1946;
- Bardsey, Caernarfonshire, on 12–15 September 1962, which was taken into captivity but later died;
- Magor, Gwent, between 29 October 1981 and 7 January 1982 and
- the remains of an individual at St Davids Airfield, Pembrokeshire, on 30 November 2008.

American Bittern at Magor Marsh, Gwent, winter 1981/82
© Richard Smith

Jon Green and Robin Sandham

Little Bittern *Ixobrychus minutus* Aderyn Bwn Lleiaf

Welsh List Category	IUCN Red List		Rarity recording
	Global	Europe	
A	LC	LC	BBRC

Total	Pre-1950	1950–59	1960–69	1970–79	1980–89	1990–99	2000–09	2010–19
35	15	0	2	9	1	4	2	2

Little Bittern: Individuals recorded in Wales by decade.

The Little Bittern is widely distributed as a summer migrant at wetlands across western Asia and continental Europe south of Fennoscandia. These populations winter in sub-Saharan Africa, where they join a widespread resident population (*HBW*). Over 500 have been recorded in Britain, with an average of about 3–4 each year over the last 30 years (Holt *et al.* 2020). In the last decade, up to six pairs have held territory in England, primarily on the Somerset Levels, where breeding was confirmed in 2017, although there was an absence of records there in 2018 (Eaton *et al.* 2020). It is a very rare species in Wales, with 36 records up to 2019. Most have been in South Wales and were overshooting spring migrants.

Of 16 pre-1950 records in Wales, including the first at Talacre, Flintshire, in 1836, half were shot, mostly in the late 19th and early 20th centuries up to 1929. These included records from Glamorgan, Carmarthenshire, Pembrokeshire, Ceredigion, Breconshire, Meirionnydd, Anglesey and Flintshire. All were single birds, except for two shot near Arenig Fach, Meirionnydd, in 1867 or 1868. There were no records between 7 May 1929, when one was at Connah's Quay, Flintshire, and 1964, when a dead male was found at Broad Haven, Pembrokeshire, on 26 April. Another bird, a female, was seen at Oxwich Marsh, Gower, on 4 July of that year.

Nine individuals were recorded in the 1970s, five of which were part of a spring influx in 1970, involving East Glamorgan, Gower, Pembrokeshire and Flintshire. Others in that decade were all in South Wales. Three were at Oxwich Marsh, including a male in June and August 1976, a female/immature in September 1977 and another male between 25 May and 30 June 1979. A first-calendar-year male was found dead at Kenfig river mouth in October 1977, on the Gower/East Glamorgan boundary. The only 1980s record was of a male at Dowrog Common, Pembrokeshire, on 3 May 1983. There were eight further records, all single birds, between 1993 and 2019. Apart from one by a garden pond in Welshpool, Montgomeryshire, on 29 October 2004, all the others were in April or May. Most of these were in Pembrokeshire, where four birds were found, two of which were dead and another that subsequently died. The others, all live birds, were in Gwent and on the Pembrokeshire/Ceredigion border. The most recent one was at Gowerton, Gower, on 13 May 2019.

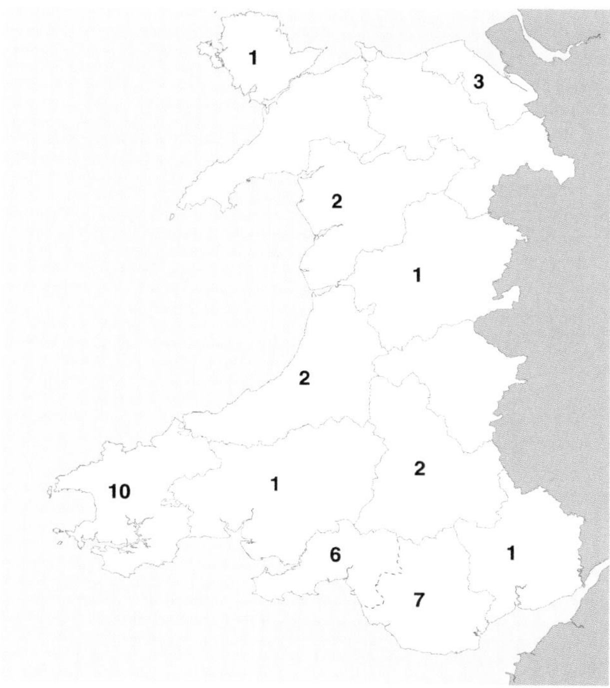

Little Bittern: Numbers of birds in each recording area, 1836–2019. One was seen on both sides of the Ceredigion/Pembrokeshire border.

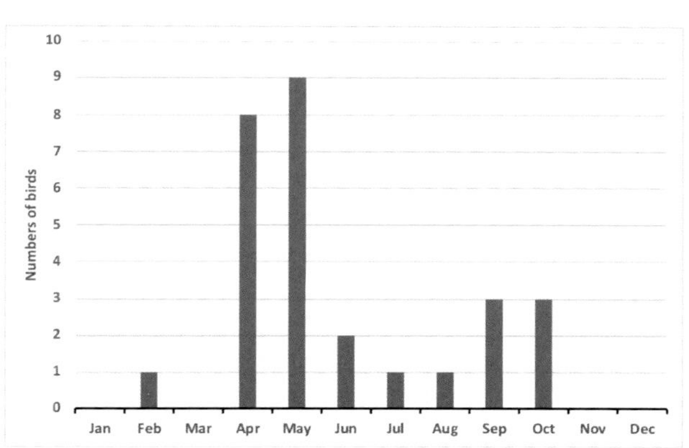

Little Bittern: Totals by month of arrival (when known), 1836–2019.

This is a species that breeds widely in northern France. Its spread into southwest England is in line with predictions for its distribution by the end of this century (Huntley *et al.* 2007), and with suitable management of reedbeds in southeast Wales, it is likely to occur here more frequently, with breeding quite possible.

Jon Green and Robin Sandham

Night-heron *Nycticorax nycticorax* Crëyr Nos

Welsh List Category	IUCN Red List		Rarity recording
	Global	Europe	
A	LC	LC	WBRC

Total	Pre-1950	1950–59	1960–69	1970–79	1980–89	1990–99	2000–09	2010–19
51	10	0	2	2	4	18	7	8

Night-heron: Individuals recorded in Wales by decade.

The Night-heron (or Black-crowned Night Heron) has an almost global distribution. In Europe it occurs as a summer visitor to wetlands across continental Europe to the south of Britain and Fennoscandia. The European population winters in Africa south of the Sahara (*HBW*). It is a rare migrant to Britain, with typically fewer than 20 records annually, although subject to occasional influxes in spring (White and Kehoe 2020a). The first confirmed breeding record in the UK was in Somerset in 2017, following suspected breeding there in the previous two years (Holling *et al.* 2019).

Most of the birds recorded in Wales have been spring overshoots to coastal counties, with the largest numbers in Pembrokeshire. The first record was of one at St. Asaph, Flintshire in 1810. Between then and 1919, there were a further six records, from Pembrokeshire, Carmarthenshire and Anglesey, involving nine birds, of which half were shot. There were no more reports until 1960, when one was present on the Menai Strait, Anglesey/Caernarfonshire between 7 February and 11 March. It might have originated at Edinburgh Zoo but was accepted by the BBRC as probably being a wild bird. Between then and 1989 there were only seven further birds, two in North Wales and five in South Wales.

There was a spring influx into southwest Britain in 1990, with a dozen or perhaps more, mostly adults, reaching South Wales. At least seven were recorded in Pembrokeshire, including two that died on Skomer, one in March and the other in April, and two at Teifi Marshes that were also seen in Ceredigion, between 19 March and 29 April. In Gower, there was a single at Oxwich Marsh between 21 April and 7 May and a second bird there on 6 May.

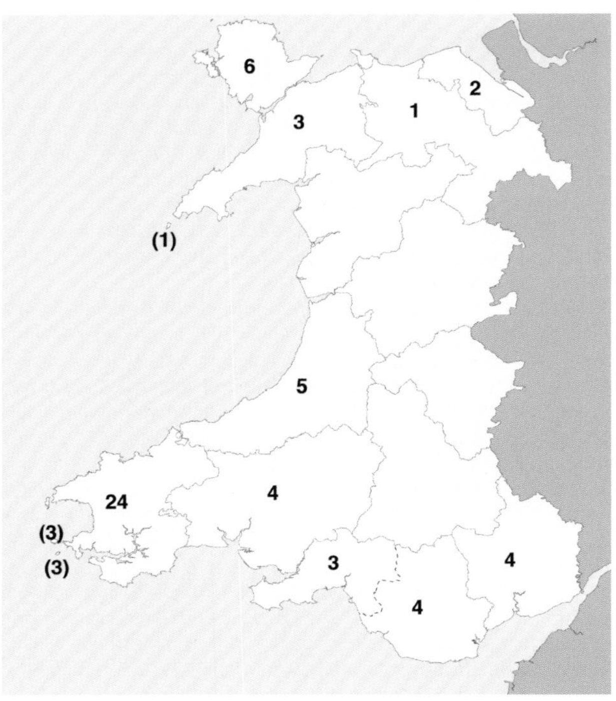

Night-heron: Numbers of birds in each recording area, 1810–2019. Some birds were seen in more than one recording area.

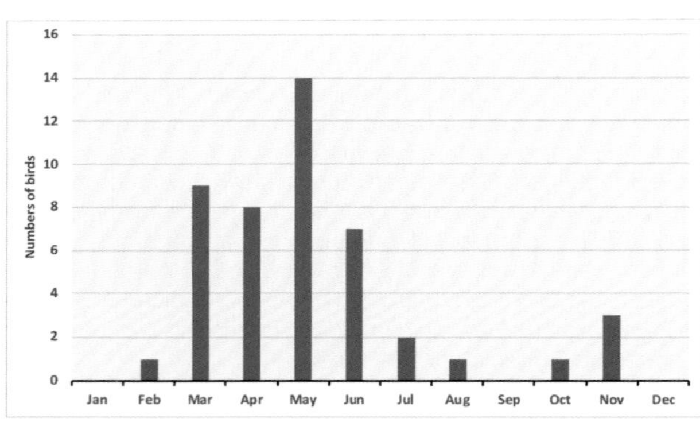

Night-heron: Totals by month of arrival (when known), 1810–2019.

An adult or second-calendar-year bird was in the Tywi Valley near Ffairfach, Carmarthenshire, on 16–21 July. Other records during the 1990s included one at Ddol Uchaf, Caerwys, Flintshire in June 1993; two at Woodstock/Morgans Pool, Newport, Gwent, in May 1994 and one at Llandegfedd Reservoir, Gwent, in July 1997; singles on Anglesey at Llyn Cefni in June 1994 and at Pentre Berw village in April 1998. Between 2000 and 2019 single, mostly adult, birds were seen in Carmarthenshire, Ceredigion, Caernarfonshire and Anglesey; and three were in East Glamorgan and seven in Pembrokeshire, including two adults and an immature that flew over Skokholm on 30 May 2019.

The Night-heron is a species that is predicted to breed in southern Britain regularly during the 21st century (Huntley *et al.* 2007). If colonisation in Somerset becomes firmly established it seems likely that it will occur more frequently in Wales and may well breed in reedbeds in southern Wales.

Jon Green and Robin Sandham

Green Heron *Butorides virescens* Crëyr Gwyrdd

Welsh List Category	IUCN Red List	Rarity recording
	Global	
A, E	LC	BBRC

The Green Heron breeds just into Canada and across the United States, avoiding drier areas of the west, south to Florida and the Bahama Islands; through mainland Mexico to South Panama, Greater Antilles, Tobago and islands off the North Venezuelan coast. However, only the US population is migratory. It is an

Green Heron in Red Wharf Bay, Anglesey, November 2005 © Steve Young

extremely rare trans-Atlantic vagrant, with eight records in Britain, two of them in Wales, the last being in 2018 (Holt *et al.* 2019). The first Welsh record was a first-calendar-year bird at Red Wharf Bay, Anglesey, between 30 October and 20 November 2005. The only other record, an adult, also stayed for several weeks, at Llanmill, Pembrokeshire, between 28 April and 7 May 2018.

Jon Green and Robin Sandham

Squacco Heron *Ardeola ralloides* Crëyr Melyn

Welsh List Category	IUCN Red List		Rarity recording
	Global	Europe	
A	LC	LC	BBRC

The Squacco Heron occurs locally across southern Europe, the Middle East and Central Asia. It winters in Africa south of the Sahara, where it joins a resident population (*HBW*). There were 176 in Britain up to the end of 2019 (Holt *et al.* 2020). A total of 13 have been recorded in Wales, where it is a very rare vagrant, the last of these in 2018. All records to date have been of single birds, with only one between 1875 and 1988:

- shot at Furnace, near Bodnant on the Afon Conwy, Denbighshire, on 11 July 1828;
- shot on the Radnorshire side of the River Wye, near Hay-on-Wye, on 3 May 1867;
- shot on an unknown date at Garthmyl, Montgomeryshire, in about 1875;
- Nottage Court, near Porthcawl, East Glamorgan, on 17–30 May 1954;
- Cemlyn, Anglesey, on 11 June 1988;
- Kenfig Pool, East Glamorgan, on 28 June 1994;
- Aberleri, Ceredigion, on 3–5 June 2003;
- Newport Wetlands Reserve (Uskmouth), Gwent, on 11 June 2003, and at Lamby Lake, East Glamorgan, on 24–25 June;
- 1CY at Angle Bay, Pembrokeshire, between 3 October and 15 November 2010;
- 2CY at Tyn Llan, Cemlyn, Anglesey, on 30 June 2015;

Squacco Heron at Angle, Pembrokeshire, October 2010 © Richard Crossen

- 2CY+ at The Watermill, Ogmore, East Glamorgan, on 29 May 2016;
- 1CY+ in Pembrokeshire between 9 October and 12 November 2016, at Westfield Pill, Llangwm, Narberth and Saundersfoot and
- 3CY+ at Llangwm, Pembrokeshire, on 3 July 2018.

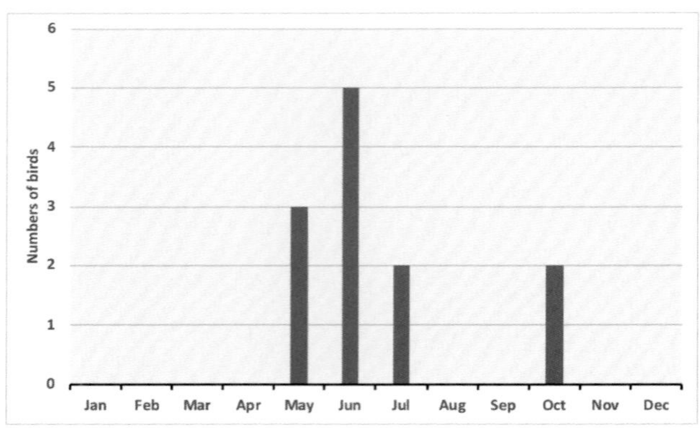

Squacco Heron: Totals by month of arrival (when known), 1828–2019.

Squacco Heron populations in Europe have fluctuated markedly, but in recent years, there have been population increases in Spain, Italy and southern France (*European Atlas* 2020). During this century, they are expected to spread as breeding species into northern Europe, as far as the Channel coast of France (Huntley *et al.* 2007). However, they may also be affected by conditions in sub-Saharan Africa. As such, population fluctuations and shifts appear to be an integral aspect of the biology of this widespread and successful species (Heron Conservation undated).

Jon Green and Robin Sandham

Cattle Egret *Bubulcus ibis* Crëyr Gwartheg

Welsh List Category	IUCN Red List		Rarity recording
	Global	Europe	
A, E	LC	LC	RBBP

The Cattle Egret has a widespread global distribution (*HBW*) and is common and widespread in Europe, with the largest populations in Spain and Portugal. Populations there are believed to have decreased, but smaller populations in France and in Italy appear to be expanding (BirdLife International 2015). Northern populations disperse outside the breeding season, mostly into Africa. During the first 20 years of this century, its status has changed from a rarity in Britain to a scarce but regular visitor and breeding species. In 2017, 10–15 pairs bred in Britain, including 6–7 pairs in Somerset and 1–3 pairs on the Wirral part of RSPB Burton Mere Wetlands, just across the Flintshire border. There were fewer breeding attempts in the UK in 2018, those closest to Wales being in Somerset where three pairs bred successfully (Eaton *et al.* 2020). In 2019, a pair nested again at RSPB Burton Mere Wetlands, near the Flintshire border (Graham Jones *in litt.*).

Total	Pre-1950	1950–59	1960–69	1970–79	1980–89	1990–99	2000–09	2010–19
143	0	0	0	0	5	3	25	110

Cattle Egret: Individuals recorded in Wales by decade.

A small influx in winter 1980/81 brought one or more to Pembrokeshire, including the first Welsh record at Haroldston Chins, Pembrokeshire, on 11 December 1980. It, or another, was at Redstone Cross on 13 December and at Stackpole between 16 December and 17 January 1981. Others were at Abergwyngregyn, Caernarfonshire, between 25 January and 13 March 1981; at Llyn Alaw, Anglesey, between 15 January and 24 May 1981; at Crundale, Pembrokeshire, on 5–14 April 1981, and one was found dead at Llandenny, Gwent, on 1 March 1981.

There were just three further records in the following decade: in Caernarfonshire (at Aberdaron on 2–4 May 1992 and at Foryd Bay on 19 April 1998), and on Skomer, Pembrokeshire, on 30 April 1996. The next, in 2007, was a single bird at Kenfig, East Glamorgan, on 5–6 November and then at Eglwys Nunydd Reservoir, Gower, on 6–7 November. An influx into Britain in 2008 included 18 in Wales, with singles in Gwent, East Glamorgan, Gower, Ceredigion and Anglesey, four in Carmarthenshire and nine in Pembrokeshire including a group of five at Trefasser on 15–29 December. There were five more records in Wales in 2009 and 2010, but only nine during 2011–15.

There was a considerable increase in the number of records in 2016, including seven that over-wintered at Pont Marquis, Anglesey, between December 2016 and April 2017. At least five were at RSPB Burton Mere Wetlands, Flintshire/Wirral, three in East Glamorgan and there were singles in Gwent, Carmarthenshire, Pembrokeshire and Ceredigion. The number of records continued to rise, making it increasingly difficult to track individuals and, therefore, the numbers involved. At least 28 were recorded in 2017. These included six at Ynyslas, Ceredigion, between 6 May and 23 June that were also seen in Montgomeryshire and in Meirionnydd in that period. Thirty-three in 2018 included the first record from Breconshire, at Llangorse Lake, on 11–23 October. Of at least 16 recorded in the following year, the majority were in spring. These included a flock of four that remained in the Porthcawl area between 31 December 2018 and 6 April, and four

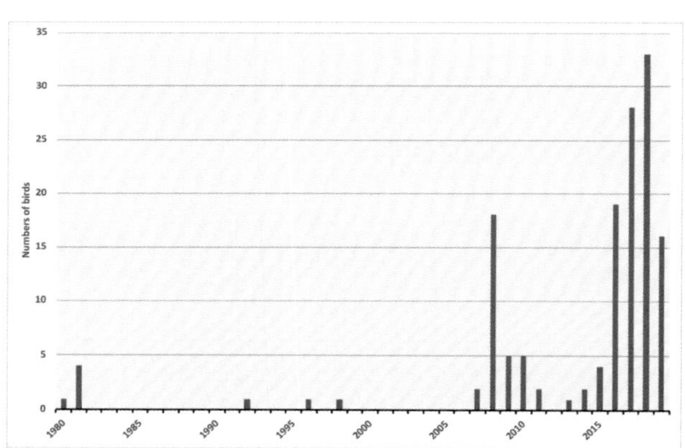

Cattle Egret: Estimated numbers, 1980–2019.

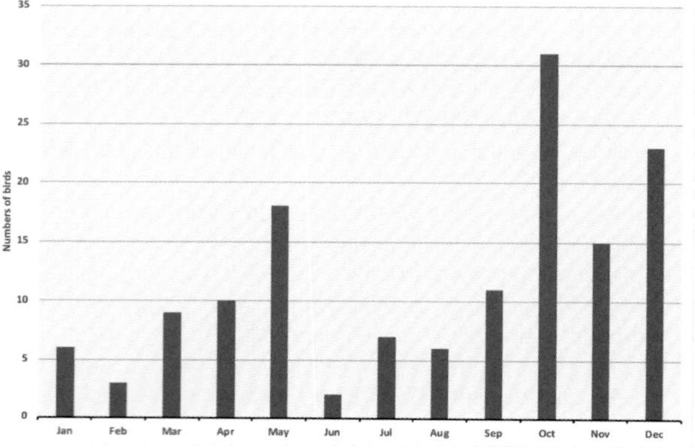

Cattle Egret: Totals by month of arrival, 1980–2019.

at Trefasser, Pembrokeshire, on 17 May. There were long-staying individuals at the Teifi Estuary at St Dogmaels, Pembrokeshire/Ceredigion, between January and 31 March and at Penclacwydd, Carmarthenshire, between 4 August and 5 October 2019.

Cattle Egrets tend to arrive in autumn, peaking in October, and again in December, with fewer seen in late winter. There has been a slight increase in records in spring, peaking in May. There were possible breeding attempts at Little Egret colonies in Gwent and Meirionnydd in 2017 (Holling *et al.* 2019). It seems likely to be only a matter of time before Cattle Egrets start to breed in Wales. Further influxes from Spain could be a source of future colonisation, but an extended cold spell in winter could set that back substantially.

Jon Green and Robin Sandham

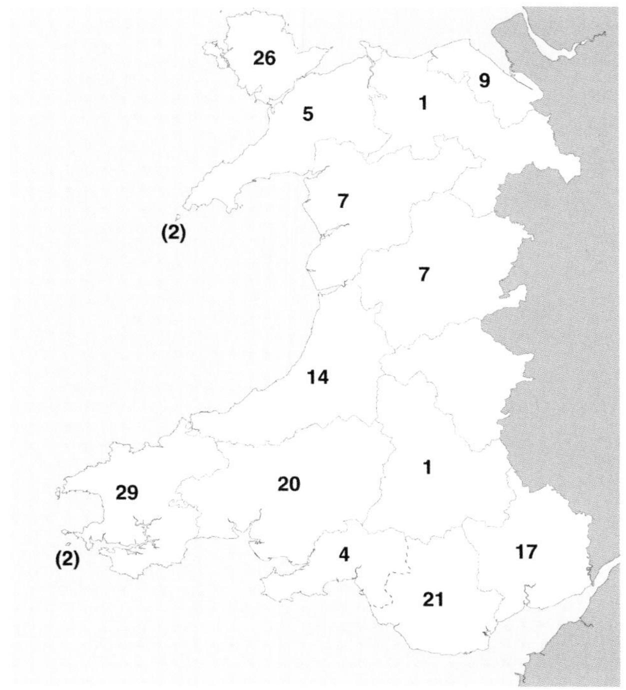

Cattle Egret: Minimum numbers of birds in each recording area, 1980–2019.

Grey Heron *Ardea cinerea* Crëyr Glas

Welsh List Category	IUCN Red List			Birds of Conservation Concern			
	Global	Europe	GB	UK	Wales		
A	LC	LC	NT (B)	Green	2002	2010	2016

The Grey Heron has a wide distribution in Europe, Asia and Africa (*HBW*). It feeds mainly on small fish caught in shallow water, although it also takes frogs, reptiles, birds and small mammals. Although this species usually breeds colonially, single nests are also encountered. Nests are usually built in tall trees close to water, with conifers often preferred as they provide more cover early in the year when the Grey Heron starts to breed. Historically, it nested on cliffs at South Stack, Anglesey (Pennant 1810, Eyton 1838), on the Great Orme, Caernarfonshire (Williams 1864), at Porth Wen, Anglesey, and at Nefyn and Porth Meudwy, Caernarfonshire (Forrest 1907). No cliff-nesting occurs today, although it only ceased in Pembrokeshire, at Linney Head and St Brides Bay, as recently as 1974 (Donovan and Rees 1994). Most heronries are in the lowlands, but there are records of nests at altitudes over 350m in Wales. For example, they have been recorded near Llyn Brenig, Denbighshire, at around 380m (*North Wales Atlas 2008–12*) and in Cwm Ystwyth, Ceredigion, at 370m (Roderick and Davis 2010). Birds nesting in the northern part of their range usually move south in winter, but those in Britain and Ireland are mainly resident (*BTO Migration Atlas* 2002). Birds ringed as nestlings in Norway, Germany, The Netherlands, Belgium, France and Ireland have been recovered in Wales. Recoveries of nestlings ringed in Wales suggest that most do not move any great distance, although one ringed in Carmarthenshire in 1967 was found dead near Bedford nearly 12 years later and several have been recovered in Ireland.

In the Middle Ages, the Grey Heron was prized both as quarry for falconers and as food. In medieval Welsh laws, it was named as one of the three notable quarry birds in falconry. If a bird flown by the King's falconer brought one down, the falconer was to be honoured by the King (Jenkins 2000). It may be that the bird's importance in falconry is one reason that so many heronries are in the grounds of mansions. This species was probably more common before large-scale drainage reduced the area of wetlands, but early records gave few indications of the numbers involved. There was said to be "an extensive cranery" at Peniarth-uchaf, Llanegryn, Meirionnydd, in 1768 (*Birds in Wales* 1994). A heronry at Margam, Gower, supported some 71 pairs in 1848 (Hurford and Lansdown 1995)—a total that appears to be the highest ever recorded at a single colony in Wales. Forrest (1907) said that the Grey Heron appeared to be more numerous on the western estuaries and on Anglesey than elsewhere in Wales. He knew of no heronry in Montgomeryshire. Although there was a single nest at Marrington, which is just in Shropshire close to the Welsh border, he did not mention any other details. In North Wales, the largest colony mentioned by Forrest was 25 nests at Pennant, Tal-y-cafn, Denbighshire, with several other heronries holding around 20 nests.

Since 1928, the BTO Heronries Census has arranged annual counts of heronries. Around two-thirds of those in England and Wales are currently counted each year, with more complete censuses carried out in 1928, 1954, 1964, 1985, 2003 and 2018. This represents the longest-running monitoring dataset for any breeding bird in the world (BTO Heronries Census undated). An index, compiled from the annual counts in Wales, shows a gradual

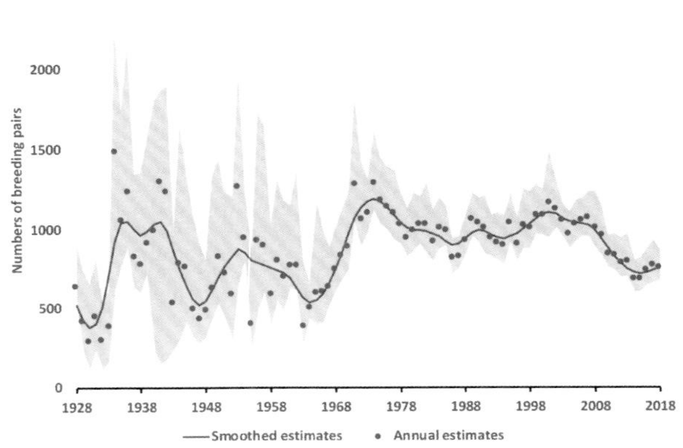

Grey Heron: Estimated population size in Wales (blue dots), 1928–2018, with smoothed trend in red. The shaded area indicates the 85% confidence limits.

	1928		1954		1964		1985		2003		2018	
Recording area	**C**	**N**	**C**	**N**	**C**	**N**	**C**	**N**	**C**	**N**	**C**	**N**
Gwent	3	27–28	3	50	5	14	5	90	9	121	8	67
East Glamorgan	2	23	1	16	1	3	1	44	6	74	5	50
Gower	1	6	1	10	1	9	9	51	11	62	6	21
Carmarthenshire	5	39–40	3	81–84	3	17	15	172	10	85	7	28
Pembrokeshire	4	22–26	2	43	3	16	4	43	4	32	8	39
Ceredigion	3	30–32	5	40–47	5	37	9	108	12	92	9	66
Breconshire	6	46–49	7	72	2	15	6	37	6	80	4	84
Radnorshire	3	8	2	25	1	7	3	19	2	23	4	32
Montgomeryshire	3	22–23	2	30	0	0	6	51	5	53	3	45
Meirionnydd	3	18–22	2	7	3	7	7	53	3	13	2	9
Caernarfonshire	7	47	6	69	8	63	7	43	8	82	11	64
Anglesey	2	9	4	28–29	6	20	8	26	4	31		
Denbighshire	3	30	3	19	0	0	2	16	5	39	3	20
Total	**45**	**327–343**	**41**	**490–501**	**38**	**208**	**82**	**753**	**85**	**787**	**70**	**525**

Grey Heron: Estimated number of colonies (C) and occupied nests (N) in Wales, in full census years, 1928–2018.

rise, but with considerable fluctuations over 90 years. The low counts recorded in 1948 and 1964 can be attributed to severe weather in preceding winters, which greatly reduced Grey Heron numbers for a few years. By the late 1980s, non-lethal methods of preventing Heron predation on commercial fisheries were also in use (Carss and Marquiss 1992), which probably helped improve adult survival rates.

It is likely that coverage was by no means complete in the early years; however, the 1954 survey probably covered most sites in Wales. Numbers were fairly stable between the mid-1970s and about 2007, but subsequently returned to the fluctuating levels seen in the 20th century. The peak count in Wales was in the full survey of 2003, when a minimum of 787 nests was recorded involving all counties except Flintshire. The absence of results for that county is very noticeable, although there have been a few small heronries there in the past and the *North Wales Atlas 2008–12* confirmed breeding in at least two tetrads. 2007 was another good year, with a minimum of 763 nests counted in Wales. The highest totals that year were in Glamorgan (154), Carmarthenshire (134) and Gwent (115), but only 408 nests were recorded in 2014. Not all heronries are counted every year. An apparent decline could be due to a few large heronries no longer being counted, but it is suspected that the decline is real: 47 heronries counted in both 2007 and 2018 held 33% fewer nests, down from 493 to 332.

Many of the heronries mentioned in accounts of a century ago are still occupied today, although some may have been deserted for years in the intervening period. Some have been abandoned and new heronries founded elsewhere. If a heronry is abandoned because nesting trees are felled, either by humans or in a gale, a number of small heronries may form, the birds later coming together in a new large heronry, sometimes at the original site, after an interval of many years. Most heronries in Wales are fairly small, with few containing more than 40 nests. The largest heronry recorded in Wales by the BTO Heronries Census was 60 nests at Allt y Gaer, Dryslwyn, Carmarthenshire, in 1974. No heronry in North Wales has ever reached 40 pairs, although one at Coed Benarth, Conwy, had 39 nests in 1974. Only one has reached this level in South Wales since the beginning of the 21st century: near Talybont-on-Usk, Breconshire, with 44 nests in 2002.

Outside the breeding season, this species is not well monitored by the Wetland Bird Survey (WeBS), as it can be found in a wide range of habitats and only a small proportion of the population occurs at count sites. The WeBS index for Wales suggests that numbers have declined a little since 2000, but has shown no significant change over the last 20 years, with the highest numbers usually recorded in September and October. The sites with the highest five-year average counts during 2014/15–2018/19 were the Conwy Estuary, Caernarfonshire/Denbighshire (35); the Welsh side of the Dee Estuary, Flintshire (32 of 72 on the whole Dee); and Traeth Lafan, Caernarfonshire (25). The Severn Estuary held an average of 75 Grey Herons during this period, of which only 19 were on the Welsh side. There have been some larger counts, including 66 at Aber Ogwen, Caernarfonshire, on 8 November 1987 and 72 on the north Burry Inlet, Carmarthenshire, in August 1994.

The Grey Heron is on the Amber List of birds of conservation concern in Wales, because of a moderate breeding range decline over 25 years (Johnstone and Bladwell 2016). Declines evident in the BTO Heronries Census in recent years give cause for concern. During 2003–07 the Welsh population must have been around 850–900 pairs, as even in those years some colonies were not counted, and some individual nests or small groups were probably

Heronry	Recording area	Max count	Year
Wern Wood, Allt y Gaer, Dryslwyn	Carmarthenshire	60	1974
Whitson Court	Gwent	51	1998
Neuadd-Fawr, Cilycwm, Llandovery	Carmarthenshire	50	1882
Rookery Wood, Pant-y-goitre	Gwent	49	1991
Coed Llwyn Rhyddid, Hensol Castle	East Glamorgan	44	1985
Crafty Wood & Picton Castle Wood, Slebech	Pembrokeshire	44	1974
Ashford Wood, Talybont-on-Usk	Breconshire	44	2002

Grey Heron: Heronries in Wales that at one time held 40 or more nests.

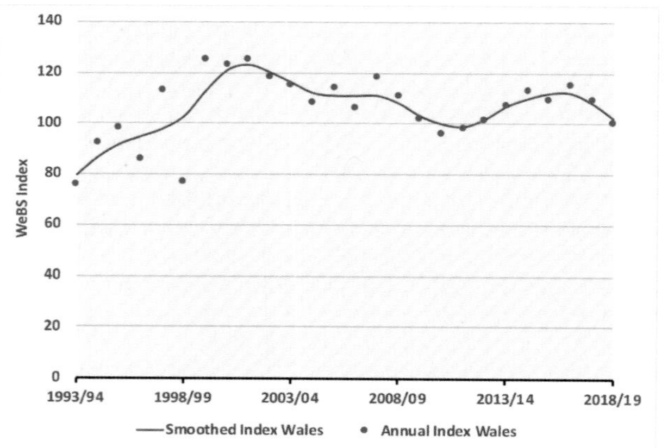

Grey Heron: Wetland Bird Survey indices, 1993/94 to 2018/19.

missed. The population, allowing for colonies not counted, is probably now 550–600 pairs. Cold periods in the winters of 2009/10 and 2010/11 may have increased mortality, but although winters since then have been quite mild, there has been little sign of recovery. Climate modelling indicates that inland areas of Wales and England will become unsuitable for Grey Heron during the 21st century (Huntley *et al.* 2007).

Rhion Pritchard

Sponsored by BIS – Biodiversity Information Service for Powys and Brecon Beacons National Park

Purple Heron *Ardea purpurea* Crëyr Porffor

Welsh List Category	IUCN Red List		Rarity recording
	Global	Europe	
A	LC	LC	WBRC, RBBP

The Purple Heron is a widespread wetland species in Africa and Asia, and in Europe it is a localised summer visitor (*HBW*). This species has bred occasionally in southern England in recent years and has possibly done so in Wales. There has been a steady increase in the number of records in Britain, from an annual mean of seven in the 1960s to 25 in 2000–2018 (White and Kehoe 2020a). The Purple Heron remains a rare migrant to Wales, although the total recorded during 2010–19 was more than double that in the two preceding decades, and closer to totals last seen in 1970–89. Most records have been from South Wales with peak numbers occurring in April and May, as overshooting spring migrants.

Total	Pre-1950	1950–59	1960–69	1970–79	1980–89	1990–99	2000–09	2010–19
72	5	0	3	14	15	8	8	19

Purple Heron: Individuals recorded in Wales by decade.

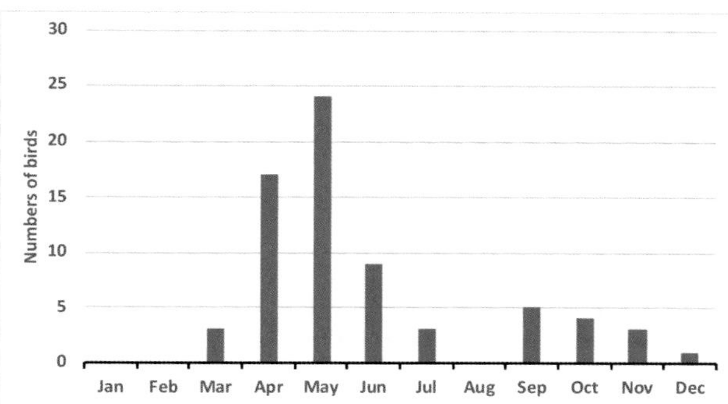

Purple Heron: Totals by month of arrival, 1932–2019.

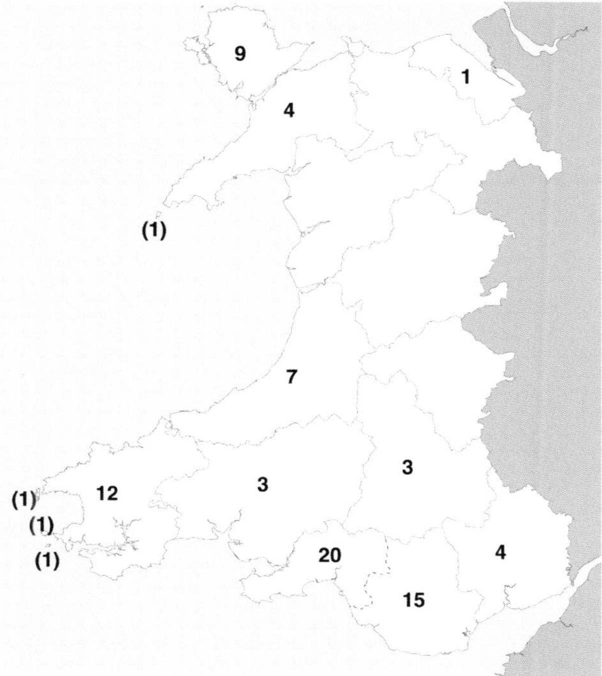

Purple Heron: Numbers of birds in each recording area, pre-1882–2019. Some birds were seen in more than one recording area.

The earliest record in Wales was of three, one of which was shot at Talybont-on-Usk, Breconshire, sometime before 1882. There were two further records before 1950, both in Pembrokeshire, at Haverfordwest in April 1932 and at Hubberston in April 1945. There were no further records in Wales, until one at Shotton, Flintshire, between 24 September and 15 October 1961 and two at Eglwys Nunydd Reservoir, Gower, in 1965 when there was an adult in June and a first-calendar-year bird there in October.

During the 1970s and 1980s there was a distinct increase in the occurrence of Purple Herons in Wales, which included 14 in Gower (some of which were seen in East Glamorgan), five in East Glamorgan (all at Kenfig), three in Ceredigion, two in Carmarthenshire, two in Pembrokeshire, two on Anglesey and one in Denbighshire. There was a cluster of records at Oxwich Marsh, Gower, where 12 individuals were found between 1971 and 1984, including two adults in 1981. Quite possibly a breeding attempt took place but, due to the inaccessible nature of the reedbed, there was no firm evidence. There were fewer records during 1990–2009, but numbers increased from 2010. There have been long-staying birds: one at Kenfig saltmarsh and Margam Moors, East Glamorgan/Gower, between 25 April and 10 July 2013; two, possibly three, subadult birds at RSPB Cors Ddyga, Anglesey, between 28 July and 1 October 2016, where two were also present on 14–20 May 2019.

There has been no obvious change at the outer boundaries of its distribution across Europe since the *European Atlas* (1997). However, within its range, the number of occupied sites has increased. This might partly be the result of improved monitoring, but the pattern is also visible in the well-studied areas (*European Atlas* 2020). Like other herons, the southern European population is expected to move north into suitable habitat during the 21st century and is predicted to nest in southern England by the end of this century (Huntley *et al.* 2007). It seems likely, therefore, that it will be encountered more frequently in Wales.

Jon Green and Robin Sandham

Great White Egret *Ardea alba* Crëyr Mawr Gwyn

Welsh List Category	IUCN Red List		Rarity recording
	Global	Europe	
A	LC	LC	RBBP

The Great White Egret has a global range across all six populated continents. In Europe, it breeds in France, The Netherlands, and across central and eastern countries (*HBW*). The European population, of the nominate subspecies *A.a. alba*, winters mostly in the northern Mediterranean. Lawicki (2014) reported on its population increase and expansion in Europe since 1980, and the subsequent large increase in the wintering populations in western and central Europe. Several thousand have been recorded over-wintering in France and The Netherlands since 2000, for example. Since 2012, when a pair bred in the Somerset Levels (Anderson *et al.* 2013), it has bred at several sites in England. There were 12–18 pairs in 2018 (Eaton *et al.* 2020) and 15 pairs in Somerset alone in 2020 (BirdGuides 2020b). A pair probably bred in a reedbed on the Wirral side of the Dee Estuary in 2017 (Holling *et al.* 2019). A pair bred on the Wirral part of RSPB Burton Mere Wetlands in 2019 and three pairs nested there in 2020. It was a rare bird in Wales until the turn of the 21st century, but is now a regular visitor and a potential breeding species.

Total	Pre-1950	1950–59	1960–69	1970–79	1980–89	1990–99	2000–09	2010–19
520+	0	0	0	0	4	7	23	486+

Great White Egret: Individuals recorded in Wales by decade.

Great White Egret © Alan Saunders

The first Welsh record was at Gronant, Flintshire, on 29 May 1981, followed by others at Penmaenpool in Meirionnydd in 1982, Bangor in Caernarfonshire in 1984, St Davids in Pembrokeshire in 1988 and on Bardsey, Caernarfonshire, in 1990. There were no further records until 1995, after which there were a further five by the end of the century. Numbers increased in 2000–09 and by 2014 it occurred regularly in Wales. Up to six were noted at the Conwy Estuary, Caernarfonshire, in 2016, and at the Dee Estuary, at White Sands/Connah's Quay, Flintshire, in 2017. There were an estimated 95 individuals in Wales in the latter year, with larger groups including up to eight at Staffal Haegr, Gower. Many roamed widely making it difficult to assess the total number involved. This pattern continued through 2018 when the estimated total for the year increased to 138, including at least ten roosting at Staffal Haegr in November. There was a slight drop to around 100 in 2019. These included at least ten at White Sands/Connah's Quay, presumed to be from nearby RSPB Burton Mere Wetlands. Smaller groups of around 4–6 birds were seen at several other locations, including: Pembrey Harbour, Carmarthenshire, in

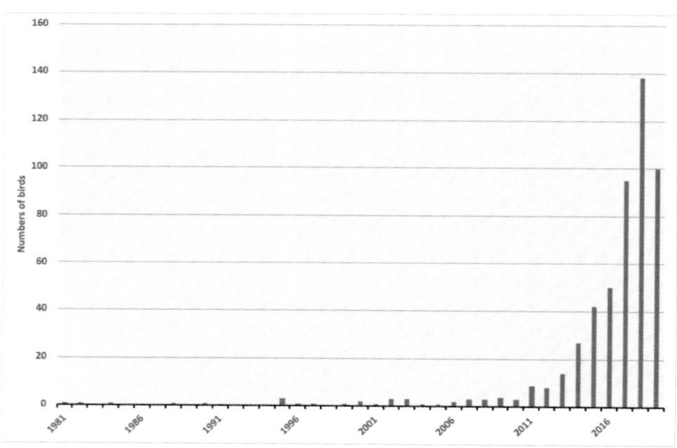

Great White Egret: Estimated numbers, 1981–2019.

October 2018; Llangorse Lake, Breconshire, in February 2018 and between February and March 2019; Llanrhystud, Ceredigion in October 2019; and RSPB Cors Ddyga, Anglesey in August 2019. Three flying over Bardsey, Caernarfonshire, on 1 November 2019 was unusual.

Resightings of one that was colour-ringed as a nestling at Brière, Brittany, France, in May 2009 and seen in northwest England later that year include at St Mellons, East Glamorgan, from 22 January 2010 and also in Gwent and Gloucestershire over the winter. It moved to Somerset in late April 2010, where it was present in most years up to 2015.

The Great White Egret now occurs in Wales throughout the year, and with expanding Dutch and English populations benefiting from landscape-scale wetland creation, it is only a matter of time before this species is recorded breeding in Wales.

Jon Green and Robin Sandham

Little Egret *Egretta garzetta* Crëyr Bach

Welsh List Category	IUCN Red List			Birds of Conservation Concern				Rarity recording
	Global	Europe	GB	UK	Wales			
A	LC	LC	LC	Green	2002	2010	2016	RBBP

Although a widespread species globally across four continents (*HBW*), the Little Egret's distribution in western Europe was mainly confined to the Mediterranean for much of the 20th century. A northward spread led to breeding in Brittany in 1960, which may have been the springboard for colonising Britain. The first successful nest was at Brownsea Island, Dorset, in 1996 (Lock and Cook 1998) and two nests were built in Wales that year. The Little Egret has a greater affinity with saltwater than other heron species but is not confined to coasts. It often breeds communally with other heron species. The nests are, initially, smaller than those of Grey Heron and tend to be lower and in denser cover, which can make an accurate count difficult.

Thomas Pennant said that he had been sent the feathers of a bird shot on Anglesey before 1768, that he suspected was a Little Egret and included the species in his *British Zoology* (Pennant 1812) on the strength of that record. He speculated that the species might once have been common in Britain but had been exterminated by humans and he knew of no other record in Britain. Early records in Wales came from Pembrokeshire, where individuals were recorded at Goodwick in November 1909 and Dale on 25 May 1938, although it was thought that the latter might have been an escaped bird (Lockley *et al.* 1949). Another was at Dale in May 1949. There were only about 12 records in Britain before 1952 (Lock and Cook 1998), but there were several records that year, including one in Carmarthenshire at Laugharne on 6 June. *Birds in Wales* (1994) reported about 70 records of 78 individuals in Wales up to 1991, most after 1970, with 75% during April to early July. Most were in southern counties: 75% in Glamorgan, Carmarthenshire and Pembrokeshire. There was an influx into Britain in 1989, with up to six in the Burry Inlet, Gower/Carmarthenshire, and a steady increase followed during the 1990s.

The first record of potential breeding activity in Wales was in 1995, when birds displayed at a heronry in Pembrokeshire. There was no attempt at nesting that year, but in 1996, two pairs built nests and one may have incubated eggs at this site, while a single bird collected nest material in Carmarthenshire. Numbers of Little Egrets rose rapidly in South Wales during the late 1990s, with numbers usually higher in autumn than in spring. The first successful breeding in Wales in 2001 was not unexpected: a pair reared four young at Whitson, Gwent, that year, with ten nestlings

Little Egret © Alan Saunders

recorded there the following year (Venables *et al.* 2008). From 2006 to 2009 the colony increased, reaching a peak of 22 AON in 2008. Numbers then tailed off and although there was a resurgence in 2013 to 20 AONs, this level was not sustained. Numbers continued to reduce until the colony was eventually deserted in 2018 with birds thought to have relocated to Magor Marsh (Clarke 2018a).

2002 saw the first successful breeding in North Wales, on the Anglesey side of the Menai Strait, where two pairs fledged one and four young respectively. Successful breeding soon followed in other counties: Pembrokeshire in 2003, Ceredigion in 2004, Gower in 2005 and Carmarthenshire in 2006. The Menai Strait colony moved to the Caernarfonshire side near Llandygai in 2006, and by 2010 was the largest colony in Wales, with an estimated 50–60 nests. 97 nestlings were colour-ringed there that year. Numbers at this colony subsequently decreased following the severe winter of 2010/11. Predation of eggs and young by Ravens was also observed at the colony (Brenchley *et al.* 2013).

By 2017, 16 colonies held 147 confirmed, probable or possible breeding pairs in nine counties. Breeding was recorded in four locations in Carmarthenshire, and three in Gwent and on Anglesey, but Caernarfonshire had the highest number of pairs, at 60 (Holling *et al.* 2019). Breeding has now been confirmed in all coastal counties, and since 2016, in Radnorshire.

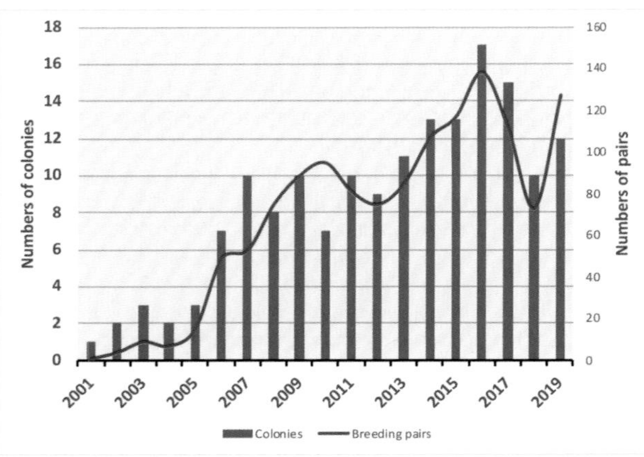

Little Egret: Numbers of breeding pairs and colonies in Wales, 2001–19 (data from BTO Heronries Census, RBBP reports and additional records).

From the late 1990s there was a rapid increase in numbers outside the breeding season, and in 2000 there were an estimated 304 individuals (*Birds in Wales* 2002). The Wetland Bird Survey (WeBS) trend for this species mirrors, to a large extent, the pattern shown by the breeding population. Numbers usually peak between July and October. While some birds move on, others remain over winter. Several sites in Wales are currently, or have been, of British importance for Little Egrets.

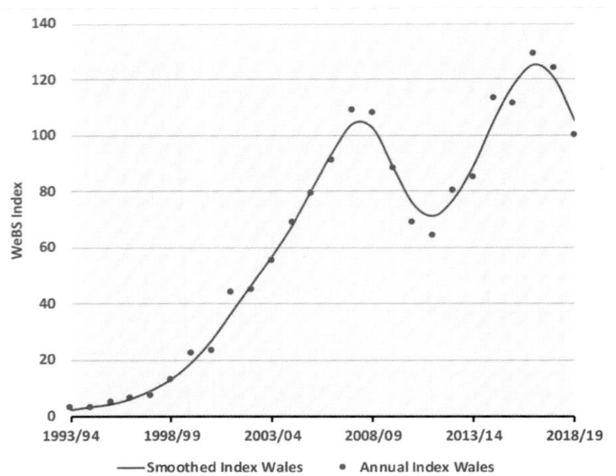

Little Egret: Wetland Bird Survey indices, 1993/94 to 2018/19.

Colour-ringing has shown that in their first calendar-year Little Egrets disperse widely, some within weeks of fledging. For example, birds from the Whitson colony in Gwent, ringed between 2002 and 2009, have shown wide-scale dispersal to almost all points of the compass, with resightings in England: at Stockton-on Tees, to the northeast; in Buckinghamshire to the east; at Radipole, Dorset, to the south; at Minehead, Somerset, to the southwest and at Leighton Moss, Lancashire, to the north (Clarke 2018a). Others, from a heronry in Caernarfonshire, have moved north as far as southern Scotland, and one from the same colony wintered in the Canary Islands (Brenchley *et al.* 2013). Another bird colour-ringed in Kent as a nestling, in May 2009, was resighted at Connah's Quay Nature Reserve, Flintshire, on 31 July 2019 and at over ten years of age is the oldest reported in Wales so far (Robinson *et al.* 2020).

Numbers levelled out in some parts of Wales in the 2010s, both during and outside the breeding season, which may have been at least partly caused by the cold winters of 2009/10 and 2010/11. The Little Egret is vulnerable to severe winter weather, but climate change makes milder winters and warmer, wetter summers more likely. Its breeding range is expanding far more quickly than expected by modelling (Huntley *et al.* 2007) and its prospects in Wales look positive.

Site	**Recording area**	**1994/95 to 1998/99**	**1999/00 to 2003/04**	**2004/05 to 2008/09**	**2009/10 to 2013/14**	**2014/15 to 2018/19**
Dee Estuary: Welsh side (whole estuary)	Flintshire	1 (2)	7 (16)	29 (162)	84 (292)	133 (305)
Severn Estuary: Welsh side (whole estuary)	Gwent/East Glamorgan	0 (4)	19 (34)	52 (91)	47 (111)	55 (228)
Traeth Lafan	Caernarfonshire/ Anglesey	1	29	110	125	139
Burry Inlet	Gower/Carmarthenshire	18	92	102	74	102
Carmarthen Bay	Carmarthenshire	3	11	61	56	56
Cleddau Estuary	Pembrokeshire	16	38	89	44	55

Little Egret: Five-year average peak counts during Wetland Bird Surveys between 1994/95 and 2018/19 at sites in Wales averaging 50 or more at least once. International importance threshold: 1,100; Great Britain importance threshold: 110.

Rhion Pritchard

Sponsored by North Wales Wildlife Trust (Arfon Group)

Osprey *Pandion haliaetus* Gwalch Pysgod

Welsh List Category	IUCN Red List			Birds of Conservation Concern				Rarity recording
	Global	Europe	GB	UK	Wales			
A	LC	LC	NT	Amber	2002	2010	2016	RBBP

The Osprey is a truly global raptor, one of the few species to be found on every continent except Antarctica. Birds breeding in the tropics are resident, but elsewhere it is mostly migratory (*HBW*). Those that breed in western Europe, where the population is estimated at 8,400–12,300 pairs and thought to be increasing (BirdLife International 2015), spend the winter in coastal West Africa, although an increasing number of birds over-winter on the Iberian Peninsula (Mackrill 2019). It feeds almost exclusively on live fish.

The species was heavily persecuted in Britain during the 18th and 19th centuries, owing to its habit of taking fish from stew ponds, which were used to store live fish reared for human consumption. In Britain, by the end of the 19th century, just a handful of pairs remained in Scotland, and the last known pair bred near Loch Loyne, Wester Ross, in 1916 (*The Birds of Scotland* 2007). Breeding was only confirmed again when a pair raised two young at Loch Garten in 1954, but it is thought that occasional nesting occurred in Scotland between those dates (Dennis 2008). Thanks to the visionary conservation efforts of George Waterston and the RSPB, the Scottish population began to recover slowly. By 2000 there were 147 pairs, and by 2017, 258 pairs (Holling *et al.* 2019). Ospreys now also breed in England and Wales, as well as being seen on migration in spring and autumn. These passage birds are mostly from the Scottish breeding population, although, in 1998 and 1999, a female that had been ringed as a nestling at Lake Muritz, eastern Germany, in June 1996, summered in mid-Wales. Passage birds are usually noted between mid-March and the end of May, and between August and October. There are just two Welsh records of Osprey in winter: at Llandegfedd Reservoir, Gwent, on 3 December 1994. Possibly the same bird was at Llanrhidian Marsh, Gower, on 19 January 1995.

There is no evidence that Ospreys bred in Wales prior to 2004, although it is quite likely that they did. Elias (2004) found ten recorded Welsh names linked to the Osprey, but noted that many of these were also used for White-tailed Eagle. The Osprey appears on the coat of arms of Swansea, which was granted in 1316, but there is no documentary evidence that the original emblem included an Osprey (swansea.gov.uk/coatofarms). A reference to several "fishey-hawkes" breeding close together at the Dyfi Estuary, Ceredigion/Meirionnydd, in 1604, suggested Osprey (Evans 2014).

There appears to be no definitive record of the species in Wales before the 19th century. Williams (1835) mentioned one shot by the gamekeeper in Gloddaeth Wood, Caernarfonshire, in 1828, though Forrest (1907) noted that this bird was described in the *Faunula Grustensis* as "a large Eagle". Between this date and the end of the 19th century, there were another 17 records in Wales: in Glamorgan, Breconshire, Radnorshire, Montgomeryshire, Meirionnydd and Denbighshire, several of which were birds shot or trapped. Most records were of singles, although three were at Aberthaw, Glamorgan, in 1841 and two at Ogmore-by-Sea in the same county in 1887 (Hurford and Lansdown 1995). *Birds in Wales* (1994) stated that Ospreys appeared to have been fairly regular visitors to Llangorse Lake, Breconshire, although Massey (1976a) could trace only three 19th-century records in the county. There were few records of Osprey in Wales in the first half of the 20th century. In Pembrokeshire for example, Lockley *et al.* (1949) knew of only two occurrences, on the Pembroke River in 1904 and again in September 1931. One was also seen near Twm Siôn Cati's cave, now on RSPB Dinas nature reserve, in the upper Tywi Valley, Carmarthenshire, on 12 June 1904 (Ingram and Salmon 1954). There were no records in Glamorgan in 1928–52. A record in Breconshire, probably at Llangorse Lake, on 8 May 1938, was the only county record between 1886 and 1960 (Massey 1976a). The first confirmed Caernarfonshire record was one killed at Waunfawr in October 1937, but the next was not until one at Aber Ogwen in August 1973. The first documented record for Gwent was in 1965.

Records in Wales became slightly more numerous from the 1950s. Up to the late 1970s, the Osprey was regarded as a scarce passage migrant, with usually one, or at most two records, each year. The number of records increased from the 1980s, with birds occasionally staying for long periods. For example, one was at Llyn Brianne on the borders of Carmarthenshire, Ceredigion and Breconshire between June and August 1986 and another at the

Ospreys 'Aran' and 'Mrs G' at the Glaslyn Osprey Project, 2020 © Glaslyn Osprey Project

Burry Inlet, Gower/Carmarthenshire, between 23 June and 10 July 1992. Long-staying individuals were also recorded at Llyn Brenig in Denbighshire, the Dyfi Estuary in Ceredigion/Meirionnydd, and the Elan Valley lakes, Breconshire/Radnorshire. In early April 1986, a male was recorded displaying, with a fish, on several occasions at the upper Lliedi Reservoir, Carmarthenshire.

The *Welsh Bird Report 1990* listed 36 records, although some may have involved the same individual seen at multiple sites, and the number of records continued to increase during the decade. From 1996, at least one Osprey summered in Wales each year, including two that were recorded carrying fish in Meirionnydd that year (*Birds in Wales* 2002). Breeding was confirmed in Wales in 2004, when two pairs bred (Anon. 2006). One pair was on the Meirionnydd side of the Afon Glaslyn, near Ynysfor, the male having been colour-ringed as a chick near Aviemore, Highland, and translocated to Rutland Water as part of a re-establishment project in 1998. The female was unringed. The two nestlings died when the nest was damaged during a gale on 30 June. A second pair nested just south of Welshpool, Montgomeryshire, and fledged one chick; the male had also been translocated to Rutland Water, in 1997, and the female hatched at Loch Tummel, Perthshire, in 2001. Neither bird returned to the Montgomeryshire nest the following year, but both Glaslyn adults returned and fledged two young in 2005. These were the first of 26 young raised by this pair between 2005 and 2014. The male did not return in 2015, but the original female found a new, unringed mate and, up to 2019, they raised a further 12 young.

A nesting platform was erected at Cors Dyfi by Montgomeryshire Wildlife Trust in autumn 2007. It was occupied by a male Osprey in April 2008, which was joined by a female later in the season. A male Osprey was present in 2009 and 2010, but although females passed through, none of them stayed. In 2011 the Dyfi male was joined by a colour-ringed, Rutland-hatched female and the pair successfully raised three chicks. All three chicks were satellite tagged and their journeys to Africa were followed by the BBC *Springwatch* programme. The pair raised another chick in 2012. In 2013 the female failed to return. The male found a new mate, a colour-ringed female from Rutland Water that had intruded at the nest the previous year. They raised two female chicks together in 2013 and went on to fledge a further ten chicks during 2014–17. Another Rutland-hatched female joined the male in 2018 when the previous female failed to return. They raised three chicks in 2018, with a further three in 2019.

By 2012 there were three active breeding Osprey nests in Wales, the third being a pair on a nest platform at another location in Meirionnydd. The adults were both unringed and successfully fledged one chick. This pair has bred annually since, raising a total of 17 chicks, although failed due to disturbance in 2018. In 2014, an unringed male that had been observed pair-bonding with a ringed female at the Dyfi nest went on to pair bond with another unringed female on a nest near Llyn Clywedog, Montgomeryshire. This pair fledged two chicks together and a further three chicks in 2015. In 2016, a new unringed male replaced the original male and between 2016 and 2019 they raised a further ten chicks. In 2017, a male and a female, both of which had previously been seen at other Osprey nests in Wales, took up residence on a nest platform at Llyn Brenig, Denbighshire, but too late in the season for breeding to take place. Both returned in 2018 and in 2019, successfully fledging single chicks each year.

Apart from the nests noted here, there have been cases of polygyny. In 2016, other females were recorded on nest platforms in proximity to the Dyfi and Glaslyn nests, mated with the resident males. Both females laid eggs, but neither nest was successful, as in both cases the males favoured their primary nests over the secondary ones. The secondary Glaslyn nest was on the Caernarfonshire side of the Afon Glaslyn, and the breeding attempt there was the first in the county. A similar pattern was observed at the Glaslyn in 2017. The second platform at Cors Dyfi had been removed under licence, to reduce the likelihood of polygyny. This strategy worked; the female moved to Llyn Brenig and bred for the first time at eight years old.

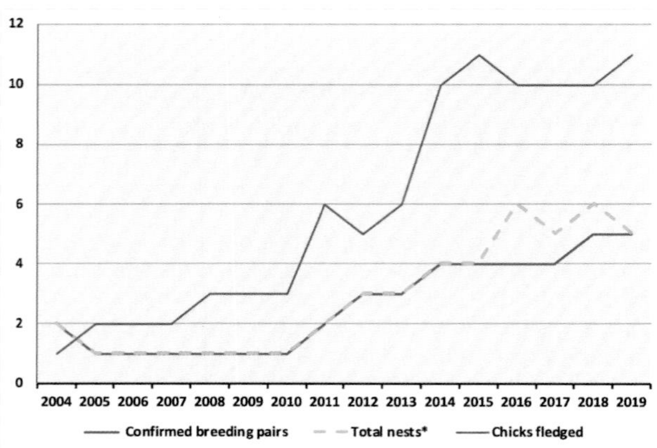

Osprey: Numbers of confirmed breeding pairs, totals of nests (*including confirmed and probable breeding) and chicks fledged, 2004–19.

Viewing facilities for the public have been established at some Welsh Osprey nests. At the Glaslyn, the RSPB set up a viewing point at Pont Croesor and from 2005 cameras provided close-up views of the nest. Bywyd Gwyllt Glaslyn Wildlife took over the stewardship of the Glaslyn Ospreys in 2014. Montgomeryshire Wildlife Trust set up the Dyfi Osprey Project in 2009 and cameras were added to the nest platform. Viewing facilities have also been set up at Llyn Brenig. These centres have proved very popular with the public. For example, nearly 75,000 people visited the Glaslyn visitor centre in 2005 and by 2014 over half a million people had visited the two Osprey projects at the Glaslyn and the Dyfi (Evans 2014).

By the end of 2019, 95 chicks are known to have fledged from Welsh nests. Of these, a minimum of 16 are known to have survived to breeding age. At least nine Glaslyn-hatched Ospreys are known to have returned to Britain. Of these, six subsequently bred successfully: one in Dumfries and Galloway, two in Northumberland, two in Cumbria and one in Perthshire, this last one being the only female among this cohort. Four Dyfi-hatched birds have returned to Britain and a fifth had a failed breeding attempt in Denmark in 2018. This bird, colour-ringed Blue W1 as a nestling in July 2015 and initially thought to be a female, was resighted in its fourth calendar-year in West Jutland, Denmark, on 13 May 2018 (Robinson *et al.* 2020). However, DNA analysis has since confirmed that it was in fact a male (Dyfi Osprey project 2017 and 2018). Two chicks from the second Meirionnydd nest were subsequently resighted in Britain as adults. One of them bred in Kielder Forest, Northumberland, in 2019, but the chicks did not survive. To date, no Welsh-hatched Osprey has nested in Wales, although several have visited established nests.

Four chicks from the Dyfi nest were satellite tagged. All of them reached wintering grounds in Senegal and Gambia. Only one Glaslyn bird has been recorded in winter quarters, a colour-ringed female in Senegal in October 2017. Another was photographed near Zurich, Switzerland, on 26 August 2010, more than 1,000km away and just four days after leaving Glaslyn. This was somewhat east of the normal migration route and was not reported again.

There is plenty of suitable habitat for Ospreys in Wales. Nest platforms have been set up at a number of likely sites to encourage breeding. Birds have summered at sites including the Cleddau Estuary, Pembrokeshire, in 2010 and 2011 and at Sylen Lakes, Carmarthenshire, in 2019. A further increase in the number of breeding pairs can be expected, and the species may well spread to other areas of Wales such as Anglesey and Ceredigion.

Heather Corfield

Sponsored by Bywyd Gwyllt Glaslyn Wildlife (Gwennan Williams)

Honey-buzzard *Pernis apivorus* Boda Mêl

Welsh List Category	IUCN Red List			Birds of Conservation Concern				Rarity recording
	Global	Europe	GB	UK	Wales			
A	LC	LC	EN	Amber	2002	2010	2016	RBBP

The Honey-buzzard is unique among raptors, with its primary diet of wasps *Vespidae* and their larvae, and its spectacular 'butterfly' display flight. The species is found across Europe south of the Arctic Circle and the west Siberian plain, mainly in temperate zones (*HBW*). It can occur at altitudes up to nearly 2,000m in the Alps and Pyrenees, where the tree line is much higher than in Britain. It winters in Africa south of the Sahara, those from Britain probably doing so in equatorial west Africa (*HBW*).

BirdLife International (2015) estimated the Honey-buzzard breeding population in Europe to be in the range of 118,000–171,000 breeding pairs. However, it is a rare breeding summer visitor to the UK, with around 24–39 pairs reported in 2018, which was a decline from 47 recorded in 2017 (Eaton *et al.* 2020). Speculation that low numbers in the UK may be down to the cool, oceanic climate and low wasp numbers due to agriculture (Brown 1976) have not been substantiated by more recent research (Roberts and Law 2014). The Honey-buzzard arrives in Wales around the second week of May and leaves any time from the end of August to the end of September. One adult tends to leave earlier, while the remaining bird continues provisioning the fledged young. Both adults usually leave before their young, which will commence their migration alone. In Wales, it breeds at altitudes up to 500m in coniferous and deciduous woodland, most doing so very successfully in commercial forestry plantations.

The Honey-buzzard has probably always been a rare bird in Wales. There were only four records in North Wales in the 19th century from 1827 and only six in South Wales from 1851, none of which were of confirmed breeding. The situation changed little through most of the 20th century. Witherby *et al.* (1940) stated that Honey-buzzards bred near the Welsh border between 1928 and 1932, but did not state whether the nest was in England or Wales. F.C.R. Jourdain also mentioned this pair, in his diaries (quoted in *Birds in Wales* 1994), writing that one of the adults shot and

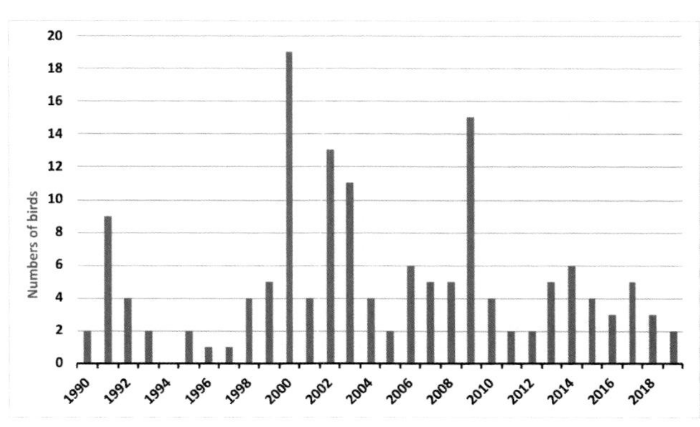

Honey-buzzard: Estimated numbers of passage birds in Wales, 1990–2019.

displayed in a taxidermist's shop in Shrewsbury, was "obtained locally", suggesting the nest was in Shropshire. This record carries more significance in light of recent breeding discoveries on the Shropshire/Radnorshire border.

Records through the 20th century provided little evidence of breeding, although a bird "summered" in the Doethie and Pysgotwr valleys, Ceredigion in 1968. There was a significant upturn in sightings during 1980–91, with birds recorded every year except 1984. In 1981 and 1991, seven and nine sightings were reported respectively, the majority from Caernarfonshire, Glamorgan and the Pembrokeshire islands (*Birds in Wales* 1994).

During the three decades between 1990 and 2019, *c.*142 passage records of Honey-buzzard have been reported across

Honey Buzzard © John Hawkins

Wales, excluding those from confirmed breeding sites. The majority were from East Glamorgan/Gower, Pembrokeshire and Caernarfonshire, and most were in autumn, including three over the Honddu Valley near Upper Chapel, Breconshire, on 29 September 2000. Numbers reported each year have generally been in low single figures, although during 2000–09, there were double-figure totals in three years: 19 in 2000, part of a large movement through the UK that autumn; 11 in 2003; and 15 in 2009. One of those recorded in 2009 had been satellite tagged in Moray in Scotland that year. It roosted on Gower Peninsula on 12 September, having flown from Fife, 521km to the NNE, during the day.

In 1991, displaying birds were recorded at two sites in Meirionnydd. One pair built a nest in a Sitka Spruce but did not lay eggs. A pair returned to the same nest in 1992, but a breeding attempt failed during incubation (*Birds in Wales* 1994). Intriguingly, in September that year, a recently fledged 'Buzzard' was rescued from the roadside, about 500m from the failed Honey-buzzard nest, and taken to a falconry centre. To everyone's surprise, the following spring it morphed into a Honey-buzzard; the failed pair having almost certainly re-laid in a new nest. Honey-buzzards have bred annually in Wales ever since, including, in subsequent years, that pair in a Western Hemlock close to where the chick was found. By 1997, there was a widely scattered population in Meirionnydd. Three occupied nests were found, a fourth pair probably bred and pairs were present at another five sites in the county. Although consistent monitoring of this population effectively petered out after 2010, when two chicks were ringed, casual observations indicated birds continued to display and bring food to some of these sites.

Since 1993, 1–3 pairs have attempted to breed in Radnorshire (Jennings 2014). There is evidence that the Honey-buzzard will maintain long historical links to nesting areas. For example, a pair confirmed to be breeding in Radnorshire in 2005, at a nest with two chicks, had been monitored for many years breeding in Shropshire. This site was close to the area where Jourdain referred to the bird shot in 1932. In 2007, a pair bred successfully in Montgomeryshire. However, none of these sites is currently monitored sufficiently to determine their present status.

Honey-buzzards were observed in Gwent during summer 1999 and almost certainly bred. That location was still occupied in 2003 when an unfinished nest was found at the end of May, but the pair subsequently moved and was not re-found. A bird was seen carrying food in Carmarthenshire in 2009 and a pair confirmed. Although the nest was not found, birds were at least present until 2011. In Gower, Honey-buzzards were first confirmed breeding in 1999 at a nest with two chicks, increasing to three active nests by 2001 and four by 2006. Mirroring natural fluctuations across Britain, this population had fallen to two pairs by 2013 and only one in 2015. This pair has continued to breed successfully. Breeding Honey-buzzard numbers increased in 2018 and 2019, with a new male setting up territory in Gower, hopefully heralding a future increase in the number of breeding pairs. Nine different territories have been recorded in Gower (own data). This population has been the subject of a long-term collaborative study involving Natural Resources Wales, Neath Port Talbot Council, and Cardiff and Newcastle universities (Parnell 1997, Roberts 2009). This was the first study of its kind involving Honey-buzzards in the UK. It comprised a range of methods, including colour-ringing, nest-camera recording, satellite tagging and faecal analysis. The results have placed Wales at the forefront of Honey-buzzard research in the UK.

Colour-ringing and nest-cameras provided the first confirmed instance of a British Honey-buzzard returning to breed in the UK: a male chick ringed in Meirionnydd in 2002 bred 150km away in Gower in 2006, mated with a female that had been ringed as a chick in Shropshire in June 2000, 95km away. The female bred in Gower until, at the age of more than 12 years, she was found long dead on a beach in Cornwall in June 2013, during spring migration. By contrast, two males, colour-ringed in Gower, moved only 4km from their natal site to breed. One of these males, ringed in 2008, was still breeding successfully in 2019.

In 2008, two breeding adults were satellite tagged. A male from Gower migrated to Guinea, West Africa, by September and then moved through Liberia to Ivory Coast by November. The other, a female from Meirionnydd, left on 28 August and reached

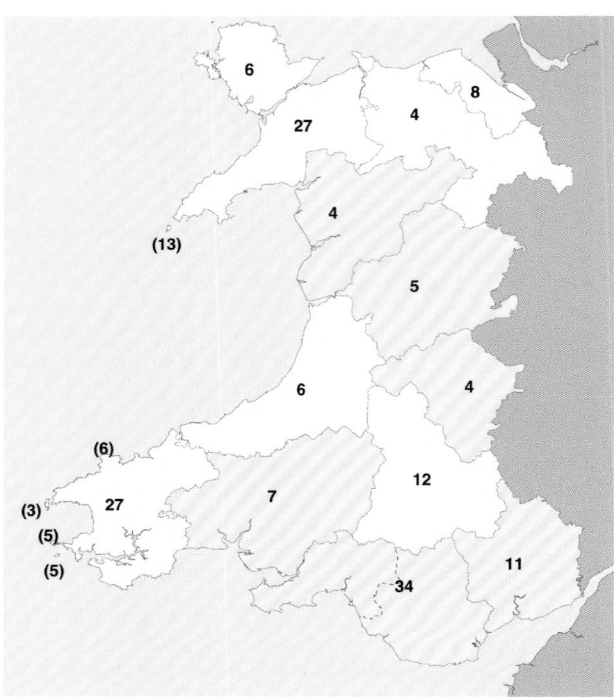

Honey-buzzard: Counties in Wales with breeding records (shaded areas) and the numbers of passage migrant records in each recording area, 1990–2019.

Ivory Coast on 6 October. She returned to Meirionnydd in spring 2009, crossing through Spain and France during 15–18 May, and returned to Ivory Coast that autumn. On the return journey north in spring 2010, the satellite signal was lost over the Solent, Hampshire, on 28 May and she was presumed to have died. Notable ringing recoveries from Wales include: a chick ringed in Meirionnydd in July 1995 that was found injured in Winchester, Hampshire, on the late date of 19 October 1995; a chick ringed in 2000 that was found having been shot dead in Ghana in 2001, 5,212km from its natal area; and a chick ringed in 2005 that was also shot, in Liberia in 2014 (Robinson *et al.* 2020).

Between 1991 and 2019 there have been 84 confirmed nests in Wales. These included 'summer' nests, built by non-breeding birds, that may be used for breeding in the following year, in a minimum of 14 territories. From these, 93 chicks are known to have fledged, most of which were colour-ringed. Four were subsequently resighted as adults breeding in Wales. One chick that was colour-ringed in Shropshire later bred in Wales (own data).

Most nests are found in coniferous trees, mainly Sitka Spruce, Western Hemlock and Douglas Fir, while those in broadleaved woodland are usually in birch or oak. This reflects the tree species available rather than selection by the birds. As commercial forest plantations have matured and been subject to rotational felling, the structure created may have proved more attractive to Honey-buzzards. Woodland cover averages 50% in the 5km^2 area around the nest. They generally lay two eggs and are adapted to feed their chicks on protein-rich wasp larvae, excavated from the ground or extracted from dense vegetation. However, when Honey-buzzards return to Wales in May, wasp nests are still quite small, and so they require a ready supply of Common Frogs to rapidly attain breeding condition after an arduous migration. Despite the generally wetter climate of Wales, food availability appears to be good, and the climate seems to have little impact upon breeding success. Nest-camera observations and faecal analysis show that Welsh woodlands appear to harbour abundant supplies of the social wasp species on which they thrive. The birds mainly forage within woodland and will travel considerable distances from one area to another: one female travelled 14km to forage (Roberts and Law 2014).

Population estimates for Honey-buzzards have always proved difficult, as this is not a well-monitored species, but there are indications in Europe and elsewhere that their population is decreasing (BirdLife International 2015). There is an estimate for Scotland of 50 pairs (*The Birds of Scotland* 2007) and for the UK, as a whole, in the low hundreds (Roberts and Law 2014). Breeding Honey-buzzards are very secretive. It can take many weeks of observation to locate and prove breeding pairs. Although they will often re-occupy the same nest each year, they can suddenly move distances of *c.*3–5km. A pair that nested at a site in Gower in 2015 had, after successive movements, moved nearly 7km away by 2019. Observers can often fail to relocate birds that have moved and conclude they have not returned. A shortage of dedicated fieldworkers and the difficulties of surveying many of the remote upland forests of Wales, riven with deep valleys, make the likelihood of under-recording high.

The Honey-buzzard can show great affinity to nesting areas. At two Meirionnydd sites not visited effectively since 2010, birds were seen in 2019 and were likely still breeding. If all the Honey-buzzard breeding territories used since 1991 are totalled, this may be a better estimate of the current breeding population than the confirmed annual figure. On this basis, the Welsh breeding population is likely to be 14–20 pairs, distributed across Gwent, East Glamorgan, Gower, Carmarthenshire, Radnorshire, Montgomeryshire and Meirionnydd.

The Honey-buzzard has enjoyed remarkable success in Britain, which has been reflected in Wales (Roberts and Law 2014). It is quite likely that it is benefiting from climate change. The southern half of Wales is expected to be climatically suitable for the species by the end of this century (Huntley *et al.* 2007). However, its continued success and further potential increase here will depend upon sympathetic management of the forest habitat it requires. Large-scale wind-farm developments close to breeding Honey-buzzards could pose a high risk of collisions and fatalities. Long-term forest management plans should take account of its requirements, whether present or potentially so. Scalloping rides, to increase the amount of internal forest edge, facilitates the Honey-buzzard's ability to locate worker wasps returning to their nests. Thinning of trees provides more access for foraging, while creation (and maintenance) of ponds helps to maintain and increase Common Frog populations. The wholesale rapid felling of all larch plantations in response to the disease *Phytophthora* is of particular concern. These open, airy plantations provide ideal foraging areas for the species and are also the preferred nesting habitat of Goshawk. Goshawks were known to have killed Honey-buzzard chicks on three occasions in Gower and have been recorded killing adult birds in Sussex (Roberts and Law 2014). With no mature larch, Goshawks will be forced to move into mature spruce or Douglas Fir, closer to Honey-buzzards, with perhaps fatal consequences.

Stephen Roberts

Sponsored by Iolo Williams

Golden Eagle *Aquila chrysaetos* Eryr Euraid

Welsh List Category	IUCN Red List			Birds of Conservation Concern	Rarity recording
	Global	Europe	GB	UK	
B, E	LC	LC	NT	Green	WBRC

The Golden Eagle has a Holarctic range, extending across vast areas of the Northern Hemisphere from Europe and North Africa to countries bordering the North Pacific. In northern Europe it occurs through the forests and montane habitats of Fennoscandia to the peatlands of the Baltic lowlands (*HBW*), and in the Scottish Highlands and islands, where the population in 2015 was estimated to be 508 territorial pairs, about 15% higher than in 2003 (Hayhow *et al.* 2017a).

There are many references to 'eagles' in early Welsh literature, although Golden Eagles and White-tailed Eagles were not distinguished and, since some of the earliest Welsh-language poetry comes from southern Scotland, it cannot be taken as evidence for their presence in Wales. However, some references were certainly located in Wales. For example, an eagle appeared in the *Mabinogion*. *Eryri*, the Welsh name for Snowdonia, had been said by several authors to mean 'the place of eagles', but the current expert view by Owen and Morgan (2007) is that it does not refer to the bird, and is more akin to 'the raised land'. There are, however, several place names in Wales that do refer to 'eagle', but again this could be either species. Evans *et al.* (2012b) identified 12 place names in Wales that they considered to be a reference to Golden Eagle, rather than White-tailed Eagle. These were mainly in North Wales, but some were in mid-Wales and as far south as Pembrokeshire and East Glamorgan. Pritchard (2020) found a total of 51 place names in Wales containing the element *eryr*, even after eliminating those that may not have referred to the birds. Using the same criteria as Evans *et al.* (2012b), 22 probably referred to White-tailed Eagle and 19 to Golden Eagle, while the remaining ten could refer to either species.

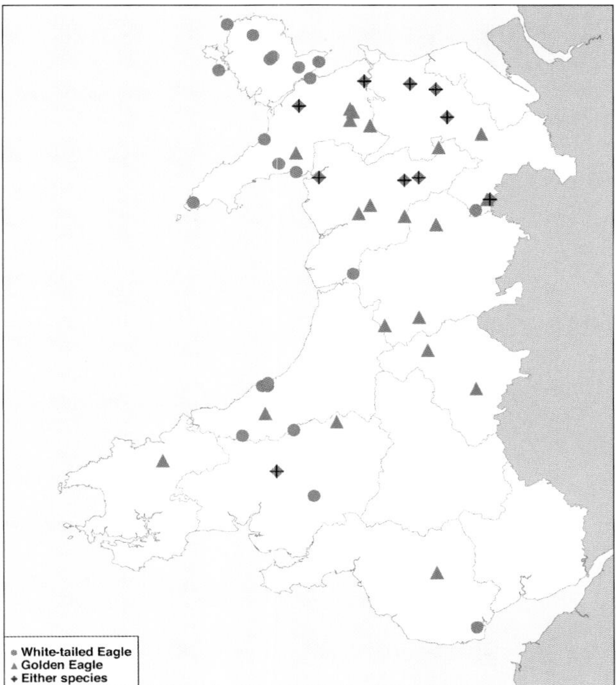

Golden Eagle: Place names in Wales containing the element 'Eryr' (map from Pritchard 2020, amended in the light of new information from Dewi Lewis *in litt.*).

Some accounts mentioned eagles in South Wales but there is no mention of breeding, and these could well be birds that had wandered from the north. However, there is reasonably good evidence of breeding in North Wales. John Leland visited Castell Dinas Brân near Llangollen, Denbighshire, about 1536, and said: "here in the rok side that the castelle stondith on bredith every yere an egle. And the egle doth sorely assaut hym that distroith the nest, goyng doun in one basket, and having another over his hedde to defend the sore stripe of the egle" (Leland 1769).

Specific evidence that Golden Eagles once bred in Wales comes in the fossil record, with remains discovered in Cat Hole Cave, Gower, dated to the Devensian period about 20,000 years ago (Yalden 2007). *Birds in Wales* (1994) mentioned remains found at Coygan Cave, Carmarthenshire. Research by Eagle Reintroduction Wales used observational records from ornithological literature, place-name records, and archaeological and paleontological records to model the predicted Golden Eagle and

White-tailed Eagle distribution in Wales (Williams *et al.* 2020). This produced compelling evidence that both species were historically widespread in Wales and that the core historic range of Golden Eagle was weighted towards North Wales, particularly centred on the upland areas of Snowdonia.

A number of early accounts suggested that the Golden Eagle bred in Snowdonia at one time. Botanist Thomas Johnson climbed Carnedd Llywelyn, Caernarfonshire, in 1639 to collect plants, but found that the boy acting as his guide refused to take him to some of the crags, as he was afraid of the eagles that he said nested there (Johnson 1641). Willughby (1678) said that the species "doth not only come hither to prey, but also many times builds and breeds with us yearly (they say) upon the high rocks of Snowdon in Carnarvanshire". An account of Snowdonia compiled for Edward Llwyd, probably by William Rowlands, in 1693 may be the most reliable source. This stated that "there is scarcely anyone of the older inhabitants who had not seen the eagles; an eagle has been seen within the past three years, and people believe that there is still one surviving" (Emery 1985). Llwyd himself visited Snowdonia several times between 1688 and his death in 1709. In his revision of the sections on Wales in the second edition of Camden's *Britannia* (1722), he noted the possible derivation of the Welsh name for Snowdonia, *Eryri*, from *Eryr*, and that it was "generally understood by the Inhabitants to be so call'd from the Eagles that formerly bred here too plentifully, and do yet haunt these Rocks some years, though not above three or four at a time, and that commonly one summer in five or six; coming hither, as is suppos'd, out of Ireland".

There seems to be reasonable evidence that Golden Eagle bred in Snowdonia during the 17th century, although none of the authors cited had personally seen an eagle. After this period, the evidence is less convincing. Pennant, in the late 18th century, said that Golden Eagles sometimes migrated into Caernarfonshire but rarely bred. However, this may have been information gleaned from 17th-century publications. The Rev. John Evans in 1812 stated that "to the present time, a few of these ravenous birds are found skulking from human ken amid the lofty clefts of the precipitous heights".

Edward Williams, in his Welsh dictionary published in 1826, mentioned that several eagles had been seen in Wales and England "in this age". He told the tale of the killing of an eagle, which he said he had seen himself, at Llansanwyr near Cowbridge, East Glamorgan, in 1776. The eagle was shot in the act of killing a lamb and its wings were damaged. It then attacked the shooter's dog, which was only rescued with great difficulty by beating the eagle with the gun. He did not specify the species, but Phillips (1891) considered that it was probably a Golden Eagle. Williams (1826) referred to a place In North Wales called *Creigiau'r eryrau* (Eagles' Crags), where the birds used to breed. Under *Eryri* (Snowdonia) he said that the name was taken by some to mean "the crags of the eagles", because formerly eagles bred frequently there, and stated that they were still found in that area.

Accounts of breeding in the 19th century seem unlikely to be correct, and it may be that regular breeding even in Snowdonia had ceased by the early 18th century. There was a record of an eagle killed at Penpont, Breconshire, in about 1859, but Forrest (1907) commented that he had been unable to find any reliable records from the 19th century in North Wales, with birds reported as Golden Eagles probably being immature White-tailed Eagles or Common Buzzards. Most records from the early 20th century onwards could have been birds that had escaped from captivity (see Appendix 1), although one at Cwm Bychan, above Llanbedr, Meirionnydd, on 2 July 1909 and another in the same county at Llanuwchllyn in September 1923 are considered to have occurred naturally.

A report on the feasibility of re-establishment in Wales indicated that this was possible, but considered that the prospects of success were not as good for Golden Eagle as for White-tailed Eagle (Marquiss 2005). Williams *et al.* (2020) cited cases where collation of records and reconstruction of historical ranges for regionally extinct species aided the successful regional restoration of a species, such as the White-tailed Eagle in Britain. They considered the extent to which the Welsh landscape has changed since Golden Eagle and White-tailed Eagle last bred in Wales over 150 years ago and suggested that additional analysis was needed, should re-establishment be thought a realistic possibility. Apart from the likely reaction of sheep farmers, the main problem would be the level of human disturbance in most areas of Snowdonia. Obtaining public support and minimising the risk to other species of conservation concern would also be essential. However, re-establishment is underway in Ireland and southern Scotland and important information has been assembled about its historic distribution in Wales as a possible first step.

Rhion Pritchard

Sponsored by Iolo Williams

Sparrowhawk *Accipiter nisus* Gwalch Glas

Welsh List Category	IUCN Red List			Birds of Conservation Concern			
	Global	Europe	GB	UK	Wales		
A	LC	LC	NT	Green	2002	2010	2016

The Sparrowhawk's range extends across Europe and temperate Asia (*HBW*). A bird of coniferous, deciduous or mixed forests, it particularly favours a mosaic of woodland and more open areas, where it can feed almost exclusively on small and medium-sized birds. In Britain and Ireland, the population is non-migratory and relatively sedentary, but in winter their numbers are augmented by migrants from northern Europe (*BTO Migration Atlas* 2002). Those that breed in Fennoscandia, for example, winter to the south and ringing has also shown that some from that area either winter in Wales or pass through on migration. A first-calendar-year bird ringed at Utsira, Norway, in October 2009 was found dead at Dwygyfylchi, Caernarfonshire, in March 2010, while two ringed, presumably on migration, at Helgoland, Germany, were later recovered in Wales. One ringed there in April 1944 was killed when it hit a window in Rhyl, Flintshire, in March 1945, and one ringed there in October 2011 was found dead near Rhayader, Radnorshire, in February 2012. Another long-distance movement involved one ringed on Bardsey, Caernarfonshire, in November 1986, that was recaptured on Fair Isle, Shetland, in May 1987, presumably on migration to Scandinavia. Although the majority of birds ringed as nestlings in Wales disperse over fairly short distances, several have been recovered in Ireland.

There are few early records of the Sparrowhawk in Wales, but bones thought to date from the Mesolithic era, 6,000 to 9,000 years ago, were found at Port Eynon Cave, Gower (Yalden and Albarella 2009). In the historic era, Sparrowhawks were used for falconry, but were considered of lower status than Goshawk or Peregrine. The Domesday Book (1082) has references to "*airae accipitru*" (hawk's eyries) and includes four in North Wales, but Yalden and Albarella (2009) considered that these were probably Goshawks. The Sparrowhawk is listed in the *Swansea Guide* (Anon. 1802) as being "found in the woods and rocks about Margam and Penrice".

The development of gamebird shooting in the second half of the 19th century led to intensive persecution of the Sparrowhawk by gamekeepers. On lands belonging to Penrhyn Estate between Dolwyddelan and Penmachno in Caernarfonshire, for example, 738 were killed between 1874 and 1902. Nonetheless, it remained

Male Sparrowhawk with a Goldfinch © Kev Joynes

a quite common species in Wales in the late 19th and early 20th centuries. In Pembrokeshire, Mathew (1894) described it as "a common resident; numerous in the wilder unpreserved parts of the county". The species was described as a common, regularly breeding resident in Glamorgan at the end of the 19th century, despite intense persecution by gamekeepers (Hurford and Lansdown 1995). Barker (1905) described it as fairly common in Carmarthenshire, while Forrest (1907) noted that Sparrowhawk probably outnumbered Kestrel in the eastern half of North Wales, particularly in Montgomeryshire, although the reverse was true in the west. He described it as comparatively scarce in Llŷn and on Anglesey, due to a lack of trees, and quoted Aplin as stating that it was not common in western Meirionnydd. As with several other species targeted by gamekeepers, it is likely that a reduction in persecution during the two world wars led to increases. Lockley *et al.* (1949) recorded it as a common resident in Pembrokeshire.

Widespread afforestation in the Welsh uplands during the 20th century, particularly after the Second World War, provided breeding sites in areas where there had previously been none. However, the population declined in many parts of Wales in the late 1950s and the 1960s, linked to organochlorine pesticides, such as Cyclodiene used in seed dressings (Walker and Newton 1998), that came into widespread agricultural use in the late 1940s. These caused birds to lay eggs with thinner shells, making them prone to breakage and reducing breeding success. Chemicals such as aldrin and dieldrin, which were in widespread use after 1955, were more toxic and killed adult birds (Newton and Haas 1984). There may have been a rather earlier decline in Sparrowhawk numbers in some areas. Ingram and Salmon (1954) stated that in Carmarthenshire and elsewhere in South Wales, the Sparrowhawk was "by no means as numerous as it used to be, in spite of the reduced amount of game-preserving". Some parts of Wales were apparently more affected than others. An investigation in some counties in England and Wales, after the 1960 breeding season, indicated that Sparrowhawk numbers had been badly affected in many areas, including Montgomeryshire, but no evidence of any decline in numbers was reported in Pembrokeshire and Flintshire (Cramp 1963). Massey (1976a) noted no decline in Breconshire in the 1960s, but in Gwent the species was said to have been "reduced almost to rarity" by the early 1960s (Humphreys 1963) and declines were also reported in Glamorgan (Hurford and Lansdown 1995). With less arable farming than farther east in Britain and thus lower use of pesticide dressings on cereal grain, the effect of these chemicals was probably less in western counties of Wales than in other areas. It was considered that there was no evidence of declines in Ceredigion (Roderick and Davis 2010), Meirionnydd (Jones 1974a) or Caernarfonshire (Jones and Dare 1976). Restrictions on the use of these chemicals were introduced from the 1960s and the Sparrowhawk population began to recover. With the exception of the southern coastal lowlands, Newton and Haas (1984) placed Wales in Zone 1, where Sparrowhawks survived in greatest numbers through the height of the organochlorine era around 1960. The population decline was judged to be less than 50% in this zone and recovery was effectively complete before 1970.

The Welsh population may, therefore, have largely recovered by the time of the *Britain and Ireland Atlas 1968–72*, which found Sparrowhawks present in almost every 10-km square in Wales, except for a few squares in the uplands of mid-Wales. The 1988–91 *Atlas* showed a mixed picture, with some gains in distribution, but more losses scattered through parts of South, mid- and North Wales. However, the 2007–11 *Atlas* showed net gains, particularly in the northwest, with just a few losses, mainly in mid- and southwest Wales. Relative abundance during 2007–11 was lower than in much of England and eastern Ireland, with decreases evident in Pembrokeshire and Carmarthenshire.

Tetrad atlases in Wales provided more detail about change. However, Sparrowhawks were probably under-recorded by bird atlases, as they can be very secretive, often only seen when they display above breeding woodlands in spring or visit gardens to hunt at bird feeders. The *Gwent Atlas 1981–85* found them in *c.*71% of tetrads, which had increased to *c.*75% of tetrads in 1998–2003. In East Glamorgan the species was found in 41% of tetrads in both the 1984–89 and the 2008–11 atlases, although breeding evidence was recorded in only 24% of tetrads in the latter period. There was a marked reduction in Pembrokeshire, where birds were found in *c.*55% of tetrads in 1984–88, but in only *c.*40% in 2003–07, although there was virtually no change in the total number of tetrads with confirmed or probable breeding. In Gower, the Sparrowhawk was found in *c.*49% of tetrads surveyed for the *West Glamorgan Atlas 1984–89*, but was present in only 14% of tetrads in Breconshire during 1988–90 (Peers and Shrubb 1990). The *North Wales Atlas 2008–12* found the species in 28% of tetrads, with few records from the largely treeless uplands; most breeding records being below 250m. It was recorded in only *c.*16% of tetrads in Meirionnydd.

The Sparrowhawk is an occasional breeder on some Welsh islands, having nested on Caldey and Ramsey in Pembrokeshire (Donovan and Rees 1994) and on Bardsey, Caernarfonshire. Influxes are sometimes evident on the islands. For example on Bardsey in 2003, at least 25 individuals were recorded in spring, and minima of 20 passed through the island in August, 18 in September and ten in October.

Population trends for Wales are uncertain as Sparrowhawks are recorded in too few BBS squares for a trend to be calculated. In England, the population increased steadily until the early 1990s and was stable until about 2008, but subsequently declined by 21% during 2013–18 (Massimino *et al.* 2019). There was also evidence of a decrease in parts of Scotland (*The Birds of Scotland* 2007). The population in Wales may have been affected by the increase in Goshawk numbers. Newton (1986) noted that where Goshawks were present, Sparrowhawks tended to concentrate in denser woods, avoiding otherwise suitable but more open woodlands. Lloyd *et al.* (2015) noted this in Carmarthenshire, but said that they remained common in areas with high Goshawk numbers and that the Sparrowhawk's colonisation of upland conifer blocks meant that the county's population was probably higher than at any recent time. Jennings (2014) stated that in Radnorshire numbers had been much reduced in some plantations by Goshawks, with numbers in the Elan Valley 80% lower than in the early 1990s.

A survey of 71km^2 around Tregaron, Ceredigion, found a density of 31 pairs/100km^2 in an area with 6% woodland cover (Newton *et al.* 1977), while a survey of 150km^2 around Llanwrtyd Wells, Breconshire, in 1986–90, found a minimum of *c.*11 pairs/100km^2 (Peers and Shrubb 1990). A higher density was found in a coniferous woodland in Denbighshire in 1991, where 23 occupied territories were in 55km^2, a density of *c.*42 territories/100km^2, despite the presence of Goshawks (Patrick Lindley *in litt.*). Venables *et al.* (2008) estimated the population in Gwent to be at the upper end of a range of 430–610 pairs, Rees *et al.* (2009) estimated a minimum of 400 pairs in Pembrokeshire, but the true figure was thought likely to be higher, while Peers and Shrubb (1990) considered the Breconshire population to be *c.*200 pairs or fewer. Roderick and Davis (2010) estimated "well over 800 pairs" in Ceredigion, but the Caernarfonshire population was estimated at only 155–163 pairs (Pritchard 2017). The most recent estimate of the population in Wales, up to 2018, is 2,950 (2,700–3,200) pairs (Hughes *et al.* 2020).

There are still instances of illegal killing of Sparrowhawks in Wales, usually birds that are shot, but the level appears to have no significant effect on the population. Predation by Goshawks is thought to have reduced numbers in some areas and the recent clear-felling of some upland conifer plantations will have reduced the availability of nesting sites. However, Government proposals for increased woodland cover may ultimately provide more opportunities for the species. The main factor affecting the size of the Sparrowhawk population in Wales will probably be the availability of sufficient numbers of small and medium-sized birds as a food source.

Rhion Pritchard

Sponsored by Andrew Dale

Goshawk *Accipiter gentilis* Gwalch Marth

Welsh List Category	IUCN Red List			Birds of Conservation Concern				Rarity recording
	Global	Europe	GB	UK	Wales			
B, C3, E	LC	LC	NT	Green	2002	2010	2016	RBBP

The Goshawk has a widespread Northern Hemisphere distribution. In Europe, the population is scattered in some western countries, including Britain and Ireland (*HBW*). Those in central and northern Europe are of the nominate form *A.g. gentilis*, as are those now found in Wales (IOC). Through a combination of deforestation and persecution, the Goshawk was exterminated in Britain by the late 19th century. By the 1960s, it was becoming a regular breeding species in Britain again, with an estimated 60 known nests by 1979–80 (Marquiss and Newton 1982). This expanding breeding population originated entirely from released birds or falconers' escapes, rather than from immigration. An estimated 20 were released each year in Britain in the 1970s and 30–40 each year in the early 1980s (Kenward *et al.* 1981). These secretive and unofficial releases meant that there is only limited data specifically for Wales. Studies in eastern Gwent/Forest of Dean (Anon. 1989, 1990) confirmed that some of the original colonists were escaped/released falconers' birds that were still wearing leather anklets or bells. The birds were large and pale, almost certainly originating from Scandinavia. The first bird known to have been released in Meirionnydd was a three-year-old German-bred female in October 1967. Subsequent releases during the late 1960s and early 1970s, mainly in the Coed y Brenin area, involved birds from Germany or from Norway (Squires *et al.* 2009).

The Goshawk, a large and powerful raptor, is renowned as an aggressive hunter. Looking much like a huge Sparrowhawk, this is a forest bird, and despite its size and relative abundance, is surprisingly difficult to see. Many of its activities are hidden within the forest, but on fine days in February and March, displaying birds fly above the forest canopy, and within the forest emitting what Kenneth Richmond (1958) aptly described as "wild Viking cries". These far-carrying calls are often the first indication of breeding activity. Eggs are laid in early April in a very large nest, usually high in a larch or Douglas Fir. Young are fed largely on Woodpigeons, corvids and Grey Squirrels found within the forest (Anon. 1989, 1990; Lindley and Jenkins 1991).

Ringing recoveries involving birds born in Wales confirmed that many are sedentary, but some range widely. Of 42 recoveries of birds ringed in Gwent, 29 involved movements of less than 9km (Venables *et al.* 2008). However, a bird ringed in Wales in May 1985 was recovered 97km away in Gloucester the following October and a bird ringed near Llandovery, Carmarthenshire, in June 1995 was found dead in Co. Wexford, Republic of Ireland, in September that year, 177km to the WNW. The oldest Goshawk recorded in Wales was ringed near Llandovery on 5 June 1994 and found dead at Llandeilo, Carmarthenshire, 13 years, 11 months and 28 days later.

The history of the Goshawk in Wales is clouded, as one of the English names of the Peregrine, 'Goose Hawk', was often abbreviated to Goshawk (*Birds in Wales* 1994). Remains of this species were found in excavations at Hen Domen Castle, Montgomeryshire, dating from the early Norman period around the early 12th century. The bones had a faint green stain, thought to be the remains of a falconer's ring (Browne 1988). Poets, such as Lewis Glyn Cothi (1447–86) from Carmarthenshire, mentioned the Goshawk, which was popular for falconry at the time. Quoting Linnard (1982), *Birds in Wales* (1994) mentions that in 1484, a warrant was issued to

Goshawk with young © John Hawkins

take for falconry "such goshawks, tarcells, fawcons, laneretts and other hawks as can be gotten in the Principality of Wales", which also suggested that the Goshawk was widespread at that time.

By the late 18th century, through loss of woodland habitat, persecution and removal of chicks for falconry, the Goshawk had become rare in Wales, eventually becoming extinct as a breeding bird around 1800, when the last recorded nest was found at the head of the Neath Valley (now Gower) (Dillwyn 1848). Aside from a few scattered records in the early 19th century, some of which were probably falconry escapees, breeding was not confirmed in Wales again until 1969, when a fledged youngster was seen with an adult at Pembrey, Carmarthenshire. This pair continued to breed until at least 1974. A rapid increase followed across the coniferous forests of Wales, with breeding confirmed in Gwent, Radnorshire and Breconshire by the end of the 1970s. Over the next ten years, Goshawks continued to colonise the mature forests and woodlands in upland and lowland Wales (*Birds in Wales* 1994). In 50 years, it had gone from almost absent to a relatively common breeding raptor in all the large forested areas of Wales and, increasingly, also in smaller blocks, as little as 2.3ha in size. The *Britain and Ireland Atlas 2007–11* highlighted Wales as a major stronghold of the species, estimated by Hughes *et al.* (2020) to be 50% of the total population in Britain in 2016.

High densities and excellent breeding success were recorded in eastern Gwent/Forest of Dean, Gloucestershire. Many pairs, if unmolested, laid five eggs and fledged five young (Anon. 1989, 1990). These early findings developed into an ongoing longer-term study of the species in Gwent. From the first breeding pair in 1979, the number of confirmed pairs rose to four by 1985, 11 pairs by 1990, 27 pairs by 1995, 30 pairs by 2000 and 36 by 2003 (Venables *et al.* 2008). This reflected the rapid increase and expansion in breeding pairs across Wales. As the population expanded, pairs were found around the margins of the main forest blocks, in less densely wooded areas. The average amount of woodland in a radius of 750m (177ha), centred around 40 nest sites, was 67% (maximum 89%) but one territory was in the farmed landscape with just 16% woodland cover (Jerry Lewis *in litt.*). By 2019, there were 65 confirmed breeding pairs in Gwent, and probably another ten that are no longer monitored. Territory spacing would suggest a further five pairs likely in the Wye Valley. Taking potentially suitable habitat into account that is not surveyed, the breeding population in Gwent could be close to 100 pairs (Jerry Lewis pers. comm.).

Territorial pairs nest approximately 1.5km apart, but have been found as close as 700m apart in prime woodland (Venables *et al.* 2008). Territories that once held a single pair with an alternative nest now hold two pairs, with the second pair occupying the original pair's alternative site. As the population increased, nearest neighbour distances have decreased and, as a result of competition, clutch and brood sizes have reduced, with 3–4 eggs and 2–3 chicks now more usual (Jerry Lewis pers. comm.). Illustrating the density at which Goshawk can breed, a survey in Coed y Cymoedd forest district, Gower, found five nests. The nearest nests were 500m apart, although with different aspects, and the farthest was 4km from its neighbour.

In Breconshire, breeding was first suspected by Herman Ostroznick Snr (pers. comm.) in 1975 and the first nest was found in 1976. Surveys indicated that, by 1990, the colonisation of Breconshire and surrounding counties was complete. Infilling of territories meant that there were at least 54 pairs in Breconshire (Mike Coleman pers. comm.), although the true total was likely to be 60 pairs. In North Wales, no breeding Goshawks were recorded in the *Britain and Ireland Atlas 1968–72* and probable/confirmed breeding was recorded in just eight 10-km squares in 1988–91. Between 1992 and 2006, Squires *et al.* (2009) gathered information from the nest sites of 30 pairs within three conifer plantation study sites in mid- and North Wales and recorded an overall breeding density of 14 pairs/100km^2. Other surveyors found 15–20 pairs in some forests in Montgomeryshire, Meirionnydd and Denbighshire. Judging by the number of forests that had not been surveyed, this suggested that the true number of pairs in mid- and North Wales was probably double that recorded by Squires *et al.* (Iolo Williams *in litt.*). The *North Wales Atlas 2008–12*, which recorded species at tetrad level, found breeding Goshawks in 64 tetrads and acknowledged that some pairs were missed. A comparison of its winter distribution and abundance, between the *Britain and Ireland Atlas* 2007–11 and the 1981–84 *Winter Atlas*, showed significant gains in distribution at 10-km square level across all Welsh counties, and relatively high abundance in South and mid-Wales in 2007–11.

Birds in Wales (1994) reported a rapid growth in the population and suggested there were about 150 pairs, and Toyne (1994) considered the population in Wales to be 200 pairs by the early 1990s. The most recent estimate, by Hughes *et al.* (2020), suggested there were 310 (260–350) pairs in 2018. Many of the remoter forested parts of Wales are likely to have been under-recorded by atlas fieldwork, so this may still be an underestimate. RBBP reports suggest that the southwest of England and Wales held the largest concentrations in 2018 (Eaton *et al.* 2020).

The powerful Goshawk is a known predator of other scarce species in Wales, with proven instances of them killing Honey-buzzard adults, chicks in the nest and recently fledged young (Roberts and Law 2014). They have also been known to kill Buzzards, Merlins, Barn Owls (own data), Kestrels, Black Grouse (Kenward 2006) and Hen Harrier (Iolo Williams pers. comm.). A study of Goshawk diet in Wales recorded similar species, plus Sparrowhawk and Tawny Owl (Toyne 1996). The impact of this predation on the conservation status of these species is not understood, but the spread of the Goshawk into maturing upper reaches of upland plantations coincides with the decline of the Merlin that formerly resided there.

Young Goshawks, in some cases perhaps forced to seek territories outside timber production forests, can be drawn to woodlands managed for gamebird shooting, where they can cause damage to shooting interests, particularly at Pheasant release pens. There is little quantitative data on the impact of Goshawk predation on Pheasants or Red-legged Partridges. In a survey of gamekeepers in 1995, only 8% said that the species was a problem around Pheasant release pens, but this probably reflected the species' limited British range at that time. Total losses to all raptors were less than 5% of release stock for 90% of the shoots surveyed (Harradine *et al.* 1997). Toyne (1996) found no evidence of gamebirds in more than 2,200 Goshawk prey items examined. This was expected, as the land use in his Welsh study area was mainly sheep farming and forestry, and virtually free of partridge and Pheasant rearing. In Wales, the main prey species taken during the breeding season were pigeons (*c.*31%), corvids (*c.*36%) and mammals, mostly Grey Squirrels and Rabbits (*c.*13%).

The Goshawk has full legal protection, but still suffers illegal persecution. At just one Pheasant release pen in Gwent, 13 were killed over a ten-year period (Venables *et al.* 2008). Disturbance, due to inappropriately sited leisure activities, such as cycle routes or motor rallies, has caused individual pairs to fail or move. The main threats stem from forest operations that can remove active nests or cause failure through disturbance. Large-scale clear-felling of mature timber, notably larch to control the disease *Phytophthora*, which it often favours for nesting, could prove detrimental. Some wind-farm developments have been sited within forests, where the risk to Goshawk and other woodland species through bird strikes and disturbance is not known. Studies show (e.g. Toyne 1997) that with appropriate knowledge of the locations of Goshawk populations, it should be possible to minimise disturbance from forestry management and similar activities, so the overall long-term prospects for this species in Wales seem good.

Stephen Roberts

Sponsored by Wrexham Birdwatchers Society

Marsh Harrier *Circus aeruginosus* Boda'r Gwerni

Welsh List Category	IUCN Red List			Birds of Conservation Concern				Rarity recording
	Global	Europe	GB	UK	Wales			
A	LC	LC	NT	Amber	2002	2010	2016	RBBP

The Marsh Harrier, the largest of the European harriers, has a wide breeding distribution across the lowlands of Europe, the Middle East and Central Asia. Most Marsh Harriers nesting in Britain overwinter around the Mediterranean and northwest Africa (*HBW*). Like some other scarce wetland species, including Bittern and Bearded Tit, it is a reedbed specialist, usually seen quartering low, gliding with the v-shaped wing profile characteristic of harriers. The Marsh Harrier breeds almost exclusively in large reedbeds. In early spring it may be seen sky-dancing high over the reeds. In late spring acrobatic aerial food passes are made, as the hunting male provides for the nesting female, and in summer those same food passes can be seen between adults and recently fledged juveniles. The Marsh Harrier is often polygynous: a male may provide for up to three nesting females, although the third nest seldom succeeds.

In the UK, breeding numbers have fluctuated considerably. The Marsh Harrier was once widespread and locally common, reflecting the amount of wetland available before large-scale drainage and agricultural mechanisation. Numbers decreased through the 19th century, as land was drained and persecution increased. By 1900, the Marsh Harrier had become extinct in the UK, although sporadic breeding attempts started again in 1911. These became more frequent and the population started to recover, until the 1950s, when a second crash resulted from indirect poisoning by the widespread application of organochlorine pesticides. By 1971, the British population had shrunk to just a single pair, in Suffolk. From that low point, the population increased steadily, thanks to concerted efforts by conservationists, legal protection and the reduction in use of organochlorine pesticides, following voluntary moratoria from 1961 and their eventual ban in 1984. The Marsh Harrier has benefited from considerable investment in revitalising existing reedbeds and the creation of new ones since the 1990s, to help the recovery of the Bittern population. A population census of the British Marsh Harrier population in 2005 estimated 363–429 nests (Holling *et al.* 2008), none of which were in Wales. Its numbers and range have continued to increase, although no formal census has taken place.

The Marsh Harrier was a common resident across much of lowland Wales before drainage in the late 18th and early 19th centuries. Historic accounts from Carmarthenshire, Glamorgan, Caernarfonshire and Anglesey are cited in *Birds in Wales* (1994), including an observation made by George Montagu (1802) who found it to be "the most common of the falcon tribe about the sandy flats on the coast of Carmarthenshire", which was a possible indication of the wealth of wetland habitat available at that time. In Glamorgan, Dillwyn (1848) thought a pair may have bred at Crymlyn Bog as late as 1882, and Forrest (1907) mentioned west Anglesey and a wetland near Llandudno, Caernarfonshire, as former breeding sites from at least a generation previously. Salter (1900) regarded the species as a former resident of Ceredigion, where it had been "numerous" on Cors Fochno and Cors Caron before the middle of the 19th century. It might have nested at the former location about 1866.

The Marsh Harrier appears to have bred in Wales only sporadically during the 20th century. There was an unconfirmed report of breeding on Anglesey in 1935 and a nest was found at Llyn Tai Hirion, now part of RSPB Cors Ddyga, on Anglesey in 1945 that fledged one chick (Colling and Brown 1946). There was another possible nesting attempt in Caernarfonshire in the same year. Birds summered on Anglesey in 1969 and 1970, a pair fledged three young at Llyn Bodgylched in 1973 and breeding may have occurred there in 1974 but was not confirmed. The next confirmed breeding attempt was a report of two young at an unspecified location on Anglesey in 1991, and breeding was reported to have occurred here again in 1992.

In recent years more sustained breeding attempts have been made in both North and South Wales, including summering pairs in Pembrokeshire in 2010, 2011 and 2016. A pair nested at RSPB Cors Ddyga, Anglesey, in 2015, but the attempt failed. Both sexes were present throughout the breeding season at Newport Wetlands Reserve, Gwent, in these years too. In 2016, two females nested at Cors Ddyga with a polygynous male and fledged four young. A pair nested at Newport Wetlands Reserve, but failed to rear any chicks (Dalrymple 2020), and a pair held territory on the Sennybridge Ranges, Breconshire, and may have bred. In 2017, a male with three females nested at Cors Ddyga, fledging five young from two nests. A pair also fledged a chick at Newport Wetlands. In 2018, four young fledged from two of three nests at Cors Ddyga and two young fledged from a nest at Newport Wetlands. In 2019, there were two males and three nesting females at Cors Ddyga, but only one nest succeeded, fledging two young. Also, three young fledged from a nest at Newport Wetlands, a pair raised three young on the Gower part of Kenfig saltmarsh, and there was an unsuccessful breeding attempt at a second location on Anglesey.

Before the recent recolonisation, most records of Marsh Harriers in Wales were passage birds, typically in April and May and from August into September. Records during spring passage have remained at similar levels over the last 40 years, but autumn passage has increased recently. The *Britain and Ireland Winter Atlas 1981–84* estimated that as few as ten wintered, but that number has increased greatly over the last 30–40 years. The *Britain and Ireland Atlas 2007–11* found birds wintering in a broad swathe across central and southern England, and also in isolated locations farther north into Scotland. In Wales during this period, wintering birds were found in several places along the south coast, and farther north on the Dyfi, Dee and Glaslyn estuaries, and on Anglesey.

Years	Average sightings per month in spring		Average sightings per month in autumn	
	April	May	August	September
1988–91*	4	9	2	4
1995–97	9	10	11	5
2005–07	5	5	7	3
2015–17**	10	9	17	15

Marsh Harrier: Average sightings per month during spring and autumn in Wales, in selected periods, between 1988–91 and 2015–17.

* Averaged from data in Birds in Wales (1994).

** The 2015–17 sightings exclude records from Newport Wetlands Reserve and RSPB Cors Ddyga.

Winter sightings have increased from 1–2 each month in a small number of localities, in the mid-1990s and the mid-2000s, to 6–10, from a similar number of localities, in the mid-2010s, not including RSPB Cors Ddyga and Newport Wetlands Reserve, where 3–4 Marsh Harriers are present throughout the year. Three to four birds have been recorded wintering at several other locations, such as Castlemartin Corse, Pembrokeshire, and the Dee Estuary, Flintshire/Wirral. *Birds in Wales* (1994) commented that most winter records were of females, but recognisable adult and second-calendar-year males are now frequently seen. There have been several recoveries in Wales of Marsh Harriers ringed in Norfolk. One such young male, wing-tagged as a nestling in Norfolk, was at Newport Wetlands between June and November 2012, before crossing the Severn Estuary to Somerset and then returning to Norfolk in February 2013. Surprisingly, it was resighted back in Wales at Marloes Mere, Pembrokeshire, on 4 April that year.

Marsh Harrier: Recoveries in Wales of bird ringed in Norfolk. Ringing sites are marked with small blue circles and single journey recovery sites by larger blue circles. One harrier (red circles) made movements to and from Wales, ending in Pembrokeshire.

The Welsh Marsh Harrier population is underpinned by high numbers in England, but it has a year-round presence at each end of Wales, where large wetland nature reserves are actively managed for reedbed species, particularly Bittern. Although it is likely to remain localised, owing to the limited availability of suitable wetlands, the prospect of the Marsh Harrier becoming a permanent fixture in the Welsh avifauna has never been higher.

Ian Hawkins

Sponsored by Iolo Williams

Hen Harrier *Circus cyaneus* Boda Tinwyn

Welsh List Category	IUCN Red List			Birds of Conservation Concern				Rarity recording
	Global	Europe	GB	UK	Wales			
A	LC	NT	VU	Red	2002	2010	2016	RBBP

The buoyant ease of a hunting Hen Harrier, dallying here and tacking there, exudes an elegance equalled by few other raptors. There can be few springtime sights more exhilarating than a male, with his gull-grey plumes and jet-black primaries, undertaking a skydiving display. As a breeding species, the Hen Harrier is widely distributed across northern Eurasia (*HBW*). In Britain it breeds primarily over 300m, on heather-covered uplands. In common with Asian counterparts, north European Hen Harrier populations, and some from Britain, migrate to more southerly areas in winter. At this time, unless there are plentiful vole populations on the uplands, most Hen Harriers move to lower ground in Wales, such as around the estuaries of the Dee, Flintshire/Wirral, and the Dyfi, Ceredigion/Meirionnydd, the coastal plain of Pembrokeshire, or to lowland mires and bogs such as Cors Caron, Ceredigion. These sites provide vast areas of suitable hunting ground, with prey in the form of voles and passerines such as Meadow Pipits, many of which have also moved to lower ground from the moors, and wintering populations of Starlings, thrushes and finches.

It was reported during the first half of the 19th century that Hen Harriers were fairly common on the Berwyn (within Denbighshire, Meirionnydd and Montgomeryshire) but no species could survive the scale of slaughter at that time. Eyton (1838) observed them near Corwen, Meirionnydd, and reported "it is remarkable with what regularity they return to the same beat at the same time for many days together, which propensity often tends to their destruction". This account clearly referred to the annual return of birds from their wintering grounds to their traditional upland breeding sites. Around Swansea, Gower, Dillwyn (1848) described the Hen Harrier as formerly abundant, but it underwent a decline that was reflected across Britain. Mathew (1894) reported that the species had previously bred on the moors of Pembrokeshire. This was discounted by Lockley *et al.* (1949) as probably referring to Montagu's Harrier, although an egg in the collection of A.D. Griffiths of Brighton was reputed to be of a Hen Harrier taken in Pembrokeshire. Evidence in Ceredigion, where sporadic breeding probably occurred through the latter part of the 19th century, came from a pair killed at Troed-yr-Aur in 1868. Watson (1977) documented the disappearance of the breeding Hen Harrier throughout mainland Britain by the end of the 19th century. Even on the extensive moorlands of North Wales, the species had become very scarce. Dobie (1894) and Bolam (1913) confirmed that the species had only "maintained a foothold" on the Rhug and Palé estates of Berwyn until the middle of the century.

By the turn of the 20th century, it was a very scarce breeding species in North Wales (Forrest 1907), but there continued to be sporadic reports. A pair probably nested near Nefyn, Caernarfonshire, in 1902, and a pair attempted to nest in Cwm Bychan, Meirionnydd, in 1909. The last breeding pair in Radnorshire nested near Rhayader in 1903. Other reports included a pair in Carmarthenshire at Newcastle Emlyn in 1865, although Lloyd *et al.* (2015) suggested that this had been confused with Montagu's Harrier. A pair bred near Nantgwyllt, on the Breconshire side of the Elan Valley, for several years in the 1860s and were subsequently trapped. Although identified as Hen Harriers at that time, several naturalists have subsequently doubted this identification (*Birds in Wales* 1994). By 1910, Hen Harriers had ceased to breed regularly in Wales, with only sporadic reports, including a pair that nested in Caernarfonshire sometime between 1910 and 1940. A pair that reared two young on Anglesey in 1924 (Aspden 1939) was the first confirmed breeding in the county; they returned to breed until 1926.

During this era of intense persecution, the Hen Harrier was all but eradicated from mainland Britain. It remained only on the Western Isles and Orkney where, significantly, there was no driven grouse-shooting and, in the latter case, plentiful supplies of Orkney Vole. A pair bred on the Scottish mainland in 1939 and, after the Second World War, extensive afforestation in the Scottish uplands provided, at least temporarily, ideal conditions for the species to nest unmolested (*Birds in Wales* 1994). In Wales, a pair bred, again, on the Berwyn in 1958, although the eggs disappeared that year. There were repeated breeding attempts at the same site in the following two years and the population reached five pairs on the Berwyn in 1962, while a pair bred on Mynydd Hiraethog, Denbighshire, in 1963.

Over the next 15 years, the Welsh population increased. In 1977 it was estimated to be 27–30 pairs, mainly on the upland blocks of the Berwyn and Migneint-Arenig-Dduallt in Meirionnydd/Denbighshire (both of these now being Special Protection Areas, with Hen Harrier a qualifying feature) and on Mynydd Hiraethog. In contrast with Scotland, where recolonisation was more reliant on

Male Hen Harrier © Keith Offord

young conifer plantations, these open moorlands have remained key strongholds for the Welsh population, although numbers have fluctuated over the years, in line with vole density. Breeding attempts have occasionally been recorded in other locations. A pair bred in Ceredigion in 1972, for example, fledging young in the north Tywi Forest. However, most of the hill land in the county, outside of restocked forest areas, is considered unsuitable, due to sheep grazing pressures (Roderick and Davis 2010). In Radnorshire, Jennings (2014) mentioned that breeding was likely to have taken place in the 1970s and that 1–3 pairs bred regularly between 1983 and 2001, with up to eight pairs annually since. He commented on a lack of rank, ungrazed and unburned heather moorland being a major limitation on breeding numbers. Peers and Shrubb (1990) were aware of single pairs breeding several times in north Breconshire, but gave no dates. Venables *et al.* (2008) mentioned a pair nesting along the Gwent/Breconshire border in 1975, plus two other suggestive breeding records in Gwent on Mynydd Garnclochdy, where males were present in June in 1997 and 1998. Three territorial pairs were also present in Snowdonia in 2019.

During 1988–94, a study by RSPB and the NCC (CCW from 1991) of breeding pairs and wing-tagged first-calendar-year birds showed that the population within Wales was stable, with 20–28 occupied territories located annually and an annual average of 25 territories. Nest failures were caused by natural phenomena, including bad weather and predation, but some losses were due to illegal persecution by gamekeepers (Williams 1999a).

Five national surveys of Hen Harriers have been carried out, the first in 1988–89 (Bibby and Etheridge 1993) and the most recent in 2016 (Wotton *et al.* 2018). Each has been aimed principally at enhancing and expanding on the annual monitoring of Hen Harriers that takes place around the UK to assess accurately the number of pairs breeding in different regions. An assessment of the number of territorial pairs in Wales in 1994, published in that years *Welsh Bird Report*, was considered by Sim *et al.* (2001) (writing about the results of the second national survey in 1998) to be the most reliable first population estimate for Wales. The 1994 estimate is referred to here. These surveys showed that numbers of breeding pairs of Hen Harriers in Wales gradually increased from 27 territorial pairs in 1994 to 57 in 2010. The reduction to 35 territorial pairs in 2016 has been explained by a series of colder springs, which adversely affected productivity in preceding years and delayed the onset of breeding that year (Wotton *et al.* 2018, Wotton 2019).

Given that there can be variation in breeding populations from one year to the next, choice of year for national surveys may provide results with misleading trends. This is supported by data from a survey area covering the northern two-thirds of Berwyn, which sustains a significant proportion of the Welsh Hen Harrier population. Since 1975, this area has been the subject of consistent, comprehensive annual monitoring. Here the population has been relatively stable, in contrast to the sharp decline suggested by the most recent national survey between 2010 and 2016. Results from 2010–19 are presented in the table.

A review of breeding data from 1980 to 2000 on the Berwyn and Migneint-Arenig-Dduallt SPAs showed a relationship between the two areas: a steady decline in the number of territorial pairs and nesting attempts on Berwyn was balanced by an increase in numbers and productivity on Migneint-Dduallt (Offord 2002). Overall, the combined population of 15 territorial pairs across both areas was reasonably stable. Clutch size, which can be an indicator of pre-laying food availability, declined in the two SPAs between the late 1980s and early 1990s, and the late 1990s and early 2000s, but this had little influence on productivity (Whitfield *et al.* 2008) or the apparent population expansion since the late 1990s.

While persecution associated with management for shooting of driven Red Grouse continues to significantly affect Hen Harriers in northern England and parts of Scotland, there has been relatively little evidence of this occurring in Wales, where the low population of Red Grouse has made driven grouse-shooting unviable. Whitfield and Fielding (2009) concluded that its absence provided a substantial explanation for the improved fortune of the Welsh Hen Harrier population after 1997, although there has been some evidence of human persecution on one North Wales moor, where two satellite-tagged Hen Harriers 'disappeared' in 2018 and a Raven was found poisoned in 2019.

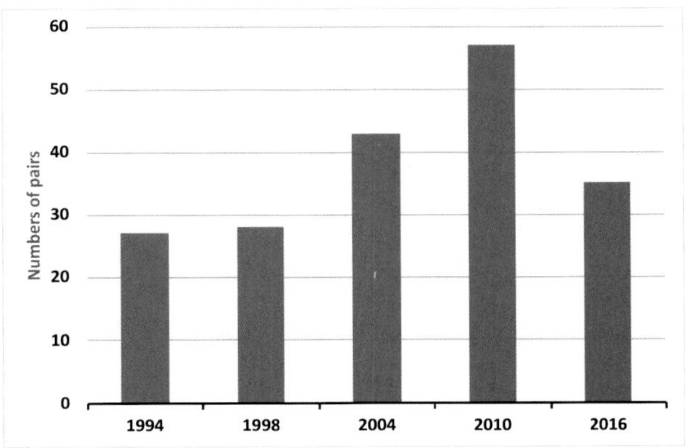

Hen Harrier: Numbers of territorial pairs, from five national surveys in Wales, 1994–2016.

Year	Territorial pairs	Young fledged	Productivity
2010	7	24	3.42
2011	6	20	3.33
2012	9	22	2.44
2013	8	22	2.75
2014	9	22	2.44
2015	12	25	2.08
2016	9	4	0.44
2017	15	20	1.33
2018	11	27	2.46
2019	9	24	2.66

Hen Harrier: Annual breeding performance in the northern two-thirds of Berwyn, 2010–19.

Natural predation accounts regularly for nest failures, the most likely suspect being Fox. The control of Foxes by gamekeepers might be expected to benefit the Hen Harrier and other ground-nesting species, but studies in Scotland found no clear evidence of this (Green and Etheridge 1999). Bad weather can have a considerable effect on the survival of young. On Berwyn in 2016, for example, persistent wet weather coincided with what had been a promising season: with 34 eggs having been laid, chicks hatched at eight nests, but only four young fledged. Many young, some well-feathered, were found dead near their nests. In contrast, in 2018, in the same survey area, 27 chicks fledged from 54 eggs laid in 11 nests, reflecting much better weather that season (own data).

In autumn, Hen Harriers can travel considerable distances from their breeding areas. They are regularly recorded on islands such as Bardsey, Caernarfonshire, and the Pembrokeshire islands of Ramsey, Skomer and Skokholm, where 1–3 are recorded most years. There were few winter records in Wales in the first half of the 20th century, but surveys of winter roosts in Britain and Ireland in 1983/84 and 1984/85 provided records from six roost sites in Wales (Clarke and Watson 1990). While most winter records from Wales are of 1–2 birds, communal roosts have been recorded at a number of sites. These include Malltraeth on Anglesey; the Dee Estuary, Flintshire/Wirral; Cors Caron and Cors Fochno, Ceredigion; Dowrog-Tretio Commons, Pembrokeshire, where up to 14 have been recorded (Donovan and Rees 1994); Rhosgoch Bog, Radnorshire, which held up to 12 in winter 1986/87 (Jennings 2014); and Llanrhidian Marsh, Gower. Occasional influxes also occur, such as at Newborough

Warren, Anglesey, during the severe 1962/63 winter, which may have been birds from continental Europe. However, wing-tagging confirmed that both Welsh and Scottish birds often over-winter on the Welsh coast.

Our understanding of the movements and dispersal range of both breeding and wintering birds has been greatly increased by ringing, wing-tagging and more recently by satellite tagging. Wing-tagging studies in Scotland, for example, have shown that the distance moved by the Hen Harrier between its natal area and the place it first bred is greater than any subsequent movement between successive breeding sites. Once established at a breeding site, there is considerable fidelity, the median distance moved between nest sites being only 0.7km (Etheridge *et al.* 1997). Ringing recoveries show that many Welsh-ringed birds moved south and east within Britain following the breeding season, although some have moved farther. A chick ringed in Montgomeryshire in 1980 was recovered in Monte Aldura, Spain, 1,053km away the following October. Another ringed in the county in 1993 was recaught by a ringer at Pancas, Portugal, in November that year, 1,615km from its nest site.

In North Wales, 155 nestlings—70 females and 85 males—were wing-tagged during 1990–95. Although the number of resightings was relatively small, first-calendar-year birds, particularly females, wintered mainly in East Anglia or remained in North Wales. However, two males were reported in France and Portugal in November of their first year. All bar one of 23 tagged birds returned to breed in North Wales, indicating the very substantial level of fidelity of this species to its natal grounds (Brian Etheridge pers. comm.). Some moved between sites within Wales, such as a tagged female that bred on Berwyn SPA in 1999, having fledged from Mynydd Hiraethog in 1995 (own data). Most Welsh females make their first breeding attempt in their first year and most males in their second year. This is in line with studies elsewhere, such as in Scotland (Etheridge *et al.* 1997).

Satellite and GPS tracking has provided greater insight into the movements of Hen Harriers, including the way they use the landscape and the fate of fledged young. During 2017–19, 12 chicks that fledged in North and mid-Wales were fitted with tags: 11 of the birds died before the following breeding season. Seven of these were judged to have died of natural causes while the cause of death of another was uncertain. Of the other four, one was almost certainly killed illegally and another two disappeared in highly suspicious circumstances: two in Denbighshire and one in Devon. The remaining bird, a female nicknamed 'Bomber' that was tagged as a chick in 2019, spent her first winter in northern Spain and returned to Wales in 2020, where she paired and nested in Caernarfonshire (RSPB data).

A small number of Hen Harriers tagged outside Wales visited during this period. Two from Scotland and one from northern England spent only a short time here, *en route* to and from their wintering quarters in France, Devon and Spain respectively, but two females emigrated to Wales after fledging on the Isle of Man. Both of these died in Wales. One that had hatched in 2018 died of uncertain causes on Anglesey in October 2018, a month after crossing the Irish Sea. The other hatched in 2016 and moved to Berwyn, Denbighshire, where she paired with a male, but did not nest in 2017. She remained in the area through the following winter, but died in suspicious circumstances in February 2018 (RSPB data). The sample size, although small, suggests that illegal persecution contributes to low survival rates of Hen Harriers in Wales in their first year.

The Hen Harrier has experienced a clear range loss in central Europe, leading to the separation of western and eastern populations. There have been population declines in many European countries in both the west (e.g. in Britain and Ireland and The Netherlands) and in the east. In contrast to the widespread losses, its distribution in France has remained mostly unchanged and it has expanded its range southwards in Spain (*European Atlas* 2020). In Wales, although it has shown a recovery following historic declines, it remains a scarce breeding species. In the absence of human interference and with the continued availability of suitable habitat to feed, roost and nest, recent productivity figures suggest that this should be sufficient to maintain a population here and that it is not reliant on the species' fortunes in other breeding areas. However, this also assumes that habitat quality is maintained and that other variables, such as food availability, levels of predation, persecution or weather-related issues, do not pose increased threats in future.

Keith Offord

Sponsored by Nature of Snowdonia (Mike Raine)

Pallid Harrier *Circus macrourus* Boda Llwydwyn

Welsh List Category	IUCN Red List		Rarity recording
	Global	Europe	
A	NT	NT	BBRC

The Pallid Harrier breeds mainly in southeast Europe and Central Asia. It winters mainly in sub-Saharan Africa and South Asia (*HBW*) but breeding records in Finland are now regular (BirdLife International 2020). Breeding was confirmed for the first time in The Netherlands in 2017 and the female from that pair was discovered breeding in Spain in 2019 (Holling *et al.* 2019). This north-westward expansion has resulted in several records each year in Britain, including two mixed pairings of male Pallid Harrier with female Hen Harrier in Scotland in 1993 and 1995, and territorial males in northern England in 2017. However, the Pallid Harrier is an extremely rare vagrant to Wales, with only two records up to the end of 2019:

- 2CY male on Skomer, Pembrokeshire, between 20 April and 1 May 2013 and
- 2CY+ female at Connah's Quay nature reserve, Flintshire, on 30 September 2017.

On current trends, there will be more.

Jon Green and Robin Sandham

Montagu's Harrier *Circus pygargus* Boda Montagu

Welsh List Category	IUCN Red List			Birds of Conservation Concern	Rarity recording
	Global	Europe	GB	UK	
A	LC	LC	CR	Amber	WBRC

The Montagu's Harrier breeds in northwest Africa, across continental Europe, Turkey and Kazakhstan to northwest China. Those from Europe winter in Africa, south of the Sahara (*HBW*). It formerly bred in Wales but is now a rare passage migrant.

The Welsh breeding population numbered 6–10 pairs in the 1950s, but dwindled, and by the mid-1960s it was lost as a breeding species here (*Birds in Wales* 1994). Unfortunately, the earliest evidence of its breeding distribution in Wales is clouded by possible confusion with Hen Harrier. Although it had been confirmed as a separate species from the Hen Harrier at the beginning of the 19th century (Montagu 1802), some later ornithologists still regarded breeding reports of both species from the 19th century with a degree of suspicion.

The earliest breeding evidence in Wales was from Leweston, Pembrokeshire, in 1854. However, the Montagu's Harrier was largely unknown to Mathew (1894), who stated: "it must doubtless have been both shot and trapped occasionally without recognition". The first breeding evidence from Ceredigion was near Aberaeron in 1892, which failed when the female was shot. In Breconshire, Cambridge Phillips (1899) did not know of the species, but there was also confusion with Hen Harrier: a pair trapped at a nest on the Breconshire side of the Wye near Nantgwyllt, Radnorshire, were later identified as Montagu's Harriers (*Birds in Wales* 1994). Breeding occurred on moorland near Bala, Meirionnydd in 1900, where four eggs were laid, but both adults were shot by a keeper. This was the only breeding record known to Forrest (1907). He was also aware of possible breeding in Caernarfonshire at around this time (Forrest 1919).

Breeding records in the 20th century are rather sparse until the 1940s–1960s, when it was confirmed or suspected in several counties. Ingram and Salmon (1954), for example, said that it was reported to have bred in Carmarthenshire, but they found no evidence of a definite nest. A pair reared three young at Morfa Harlech, Meirionnydd, in 1951 and there was an unconfirmed report of a pair having bred in Caernarfonshire that year, although this may have referred to the Meirionnydd pair (*Birds in Wales* 1994). A maximum of five pairs was considered to have bred in Pembrokeshire, where two nests, both serviced by a single male, fledged six chicks on Dowrog Common, near St Davids, in 1954. Breeding continued occasionally there and at two other Pembrokeshire sites until 1962. They were seen in suitable areas throughout the 1960s and an attempt at breeding might have been made in 1968 (Donovan and Rees 1994). In Ceredigion, a pair managed to rear three young at Cors Caron in 1953 and, although absent in 1954, birds were reported there again in 1955. A pair possibly bred at Gernos in 1955 and breeding was suspected in the Dyfi area during the 1950s. A pair nested at Ynys-Eidiol, Ceredigion, in 1963, but there were no further confirmed breeding records in the county. In Carmarthenshire, breeding was suspected at Pembrey Burrows in 1966 and 1967. A female was seen carrying nesting material or food into undergrowth at Wharley Point a year later and a pair summered at Mynydd Llanfihangel Rhos-y-corn in 1969 (*Birds in Wales* 1994).

By far the best documented breeding evidence came from Newborough Warren and the adjacent Cefni saltmarsh, Anglesey. Breeding occurred between 1945 and 1964, with a peak population of five pairs in 1957. However, despite wardening efforts, the species suffered greatly from egg-collectors. Ninety eggs (from 30 clutches) were known to have been laid during 1949–60. These produced 31 flying young, but 32 (36%) of the eggs were taken by collectors. From 1961, eggs were removed under licence and hatched in an incubator. From this treatment, 19 eggs produced ten flying young, a 53% success rate (Jones and Whalley 2004). Despite these valiant efforts, the species last bred on Anglesey in 1964, when two young were reared from four eggs. This was also the last known successful breeding attempt recorded in Wales. By

Total	1967–69	1970–79	1980–89	1990–99	2000–09	2010–19
71	22	21	12	7	6	3

Montagu's Harrier: Individuals recorded in Wales by decade (excluding birds holding territory) since 1967.

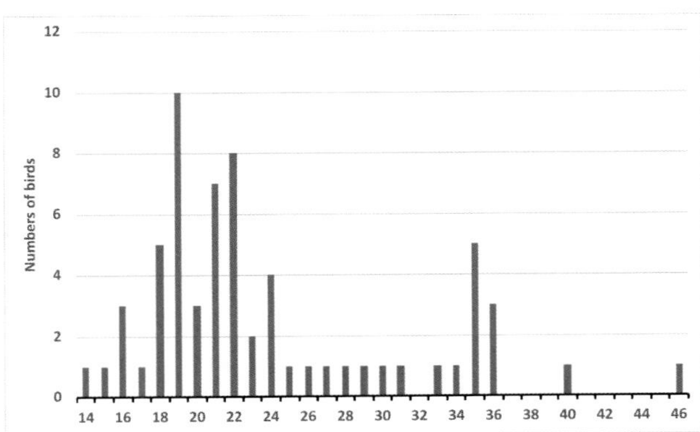

Montagu's Harrier: Totals by week of arrival (when known, n=64), 1967–2019. Week 18 begins *c.*30 April and week 35 begins *c.*27 August.

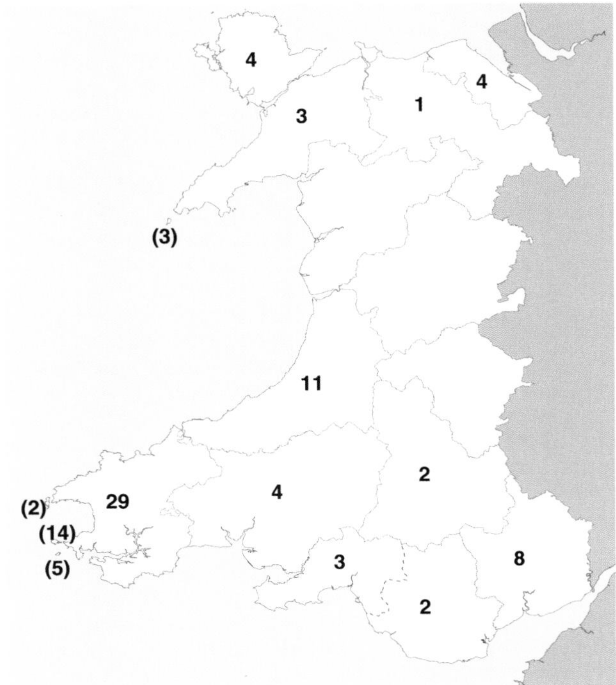

Montagu's Harrier: Minimum numbers of birds in each recording area, 1967–2019. Some birds were seen in more than one recording area.

the mid-1960s, the habitat at Newborough Warren was no longer ideal for this ground-nesting species, due to plantation forestry growth (*Birds in Wales* 1994).

Since then, there have been a few other records of long-staying or possible summering adults, but no suggestion of breeding. A melanistic female was at Cors Caron, Ceredigion, between 30 May and 21 July 1967; a different female was there over a similar period in 1968 and single males were present briefly in July in both years (Roderick and Davis 2010). A pair was present at Wentwood Forest, Gwent, from 19–31 May 1969 and a single was there on 21 June 1969 (Venables *et al.* 2008). A male was present between 24 June and 22 July on the Dee marshes, Flintshire, in 1980. Birds visited former breeding haunts in Pembrokeshire in the 1980s and a displaying male was present for several weeks at a moorland site in Denbighshire from early May to late July 1993.

The Montagu's Harrier is otherwise a rare passage migrant to Wales, having occurred in all counties except Montgomeryshire. The number of records decreased following the demise of the Welsh breeding population. It is mainly a late spring migrant, with a peak in the third week of May and smaller numbers in autumn.

The records have a southerly bias, most of them being in Pembrokeshire, including a first-calendar-year bird that unexpectedly over-wintered on Skomer between 1 October 1988 and 25 March 1989. The last decade has seen the fewest number of records in Wales since, probably, the 1920s. Climate modelling indicates that southern Britain will become suitable for the species later this century (Huntley *et al.* 2007), but with no sites in England now regularly used for breeding, it is likely to remain a rare visitor for the foreseeable future.

Bob Haycock and Jon Green

Red Kite *Milvus milvus* Barcud Coch

Welsh List Category	IUCN Red List			Birds of Conservation Concern			
	Global	Europe	GB	UK	Wales		
					2002	2010	2016
A, C5	LC	NT	LC	Green			

For several generations of birdwatchers, the Red Kite was synonymous with Wales. It bred nowhere else in Britain for almost a century, being confined to just one or two remote valleys in the heart of mid-Wales by years of deliberate and sustained persecution elsewhere. This is almost exclusively a European species, although there are two small and declining populations in North Africa. Its range extends from southern Sweden to Spain and Portugal, and east as far as Ukraine and Bulgaria. The populations in northern and central Europe are mostly migratory, wintering mainly in southern France and the Iberian Peninsula (*HBW*). The global population is estimated at 25,200–33,400 pairs (BirdLife International 2015), with the largest populations in Germany and eastern France, Britain and Spain.

The Red Kite is essentially a bird of open wooded country, although historically the species was found in cities, living in close association with humans. It can take a wide variety of food items but depends on carrion and small- to medium-sized mammals and birds (*HBW*). Sheep carrion is particularly important in mid-Wales (Davis and Davis 1981). The Welsh population is essentially sedentary, but young birds can move considerable distances. For example, one ringed as a nestling in mid-Wales in June 1992 was seen on a gas rig off the Norfolk coast, 387km away, on 19 July that year. Movements into Wales of birds ringed elsewhere have also been occasionally recorded, such as one ringed as a nestling in Schleswig-Holstein, Germany, in June 1971, that was found dead near Rhayader, Radnorshire, in July the following year. Birds from re-establishment schemes in England, Scotland and Ireland have also been observed in Wales. For example, one ringed as a nestling at the Black Isle, Scotland, in July 1997, was seen at Rhayader, 592km away, in November that year.

Red Kite © Alan Saunders

Red Kite remains found at Bacon Hole, Gower, date back as far as the Ipswichian interglacial period, between 115,000 and 130,000 years ago. There were Roman-era remains from Caerwent, Gwent (Yalden and Albarella 2009). Little information is available about the species in Wales in the medieval period, compared to some other raptors, as it was of no interest to falconers. There are, however, many references to Red Kite in Welsh poetry of the period. The bird was clearly a familiar species, although its image was consistently negative, probably because it was a scavenger (Williams 2014). Place names containing the element *barcud* or *beri* are widespread in Wales, particularly in Denbighshire and Meirionnydd.

The decline of the Red Kite in Wales is not well chronicled. It is thought that the species was already in decline in Britain in the early 18th century (Cross and Davis 2005). Although its presence was recorded in various parts of Wales by 18th-century authors, they gave no information on its abundance. It had probably gone from Flintshire and much of Denbighshire before 1800 (Cross and Davis 2005) and the first half of the 19th century saw a more widespread decline. The Red Kite is conspicuous and slow flying, and thus an easy target for shooting; and as a carrion feeder, it is vulnerable to poisoning. Eyton (1838) stated that the Kite had formerly been common in Wales, but was becoming rare due to persecution by gamekeepers. Dillwyn (1848) stated that in the Swansea area it was "not infrequently seen, but every year becomes less common" and it had probably disappeared from many parts of Wales by that time. Forrest (1907) considered that the Red Kite was scarce in North Wales by the 1830s and had ceased to breed there by 1850, although it may still have bred in southern Meirionnydd in the 1850s. Forrest quoted an old gamekeeper who remembered it breeding at Dinas Mawddwy about 1855, and noted that Salter had been told that old people remembered the Kite breeding on Craig yr Aderyn. It was also disappearing from southeast Wales. The last breeding pair in Glamorgan was near Cardiff in

1853 (Hurford and Lansdown 1995), although the last breeding in Gwent was not until the 1870s at Nantyderry (Humphreys 1963). In Pembrokeshire, Mathew (1894) stated that he had heard from old people who remembered the Kite as quite a common bird when they were young, but it was long gone as a breeding bird by his time.

In the Llandeilo area of Carmarthenshire, Davies (1858) said that it was "plentiful of former years, now seldom or ever seen". As a breeding species, Red Kite gradually disappeared from most of Ceredigion during the second half of the 19th century, although it may have hung on in the upper Teifi Valley until about 1900 (Roderick and Davis 2010). In Breconshire, Phillips (1899) thought that it had been nearly extinct in 1875 but had increased by 1889 when several birds were killed. The population was still under heavy pressure: young birds were taken for sale as pets, and birds were killed as pests or fell victim to poison put out for corvids and Foxes. From the 1890s, systematic commercial egg-collection became a major problem for the few remaining pairs. Their scarcity meant that a clutch was worth £4 or £5 (equivalent to £500–£620 in 2020) and very few young are known to have fledged around the turn of the century (Lloyd *et al.* 2015).

By about 1900 the Welsh breeding population was concentrated in two core areas. These were the upper Wye Valley, Breconshire/Radnorshire, and particularly the upper Tywi Valley and that of its tributary, Afon Cothi, mainly in Carmarthenshire, but with a small part in Ceredigion; although a nest was recorded on the Meirionnydd side of the lower Dyfi Estuary *c.*1905–06. The Red Kite had ceased to breed in England and Scotland by this time. It had disappeared from Ireland even earlier, by the end of the 18th century (Price and Robinson 2008), leaving the tiny Welsh population as the only one remaining in Britain and Ireland. Around 1910, probably no more than 10–12 pairs remained (Cross and Davis 2005).

The fascinating story of attempts to conserve the Welsh Red Kite population, based largely on information collected by J.H. Salter and later by H. Morrey Salmon, has been told by Lovegrove (1990), *Birds in Wales* (1994), Carter (2001) and Cross and Davis (2005). The first attempt to preserve Red Kites in Wales was made in north Breconshire in the 1880s, where landowners were inspired by Cambridge Phillips to protect the few remaining pairs. In 1903, J.H. Salter wrote to the British Ornithologists' Club (BOC) calling for the establishment of a protection scheme in the upper Tywi Valley. This involved guarding the nests, with local people paid to keep a 24-hour watch at several sites. By 1923 the RSPB had taken over from the BOC and guarding of nests continued until 1939. It is thought that the population continued to decline until the 1930s, largely because of the killing of birds that wandered outside the protected area and the plundering of outlying nests for eggs. The population in northern Carmarthenshire probably increased during 1905–25 and remained quite stable thereafter, despite low breeding success. In 1931–35, although up to 11 pairs are known to have existed in Wales, only two pairs are thought to have fledged young, both in the upper Tywi Valley (Cross and Davis 2005). A study of female bloodlines, DNA that passes unaltered from mother to daughter, concluded that at one point the whole of the Welsh population had just a single matrilineage (May *et al.* 1993). This means that at some time, probably in the early 1930s, all pairs had females that were direct descendants of a single female.

Away from this area, nesting was recorded at Llanbwchllyn, Radnorshire, in 1927, and breeding persisted in Breconshire until about 1930 (Peers and Shrubb 1990). In 1927–28 and in 1934–35, the Liverpool egg dealer C.H. Gowland had Red Kite eggs from Spain placed in the nests of Buzzards in the Builth Wells area, Breconshire/Radnorshire, and around Rhayader, Radnorshire. A good many young are said to have fledged, but there is no indication that these birds joined the native stock (Davis 1993), possibly owing to the Kites imprinting on their adoptive Buzzard parents. Sightings of these Gowland birds petered out during the Second World War.

By the end of the 1930s, there were indications that the only remaining population in Wales, in the upper Tywi Valley, was expanding slightly, with successful breeding confirmed in the Tregaron area of Ceredigion in 1939. Despite the ending of the nest-watcher scheme in 1939, a reduced level of persecution during the Second World War probably benefited Red Kites, and the population increased slowly in the years that followed. A new Kite Committee was established in 1950, when there were about a dozen breeding pairs. The breeding range extended north into Ceredigion, with a few birds returning to the upper Wye Valley (Cross and Davis 2005), and breeding resumed in Breconshire in the 1950s (Peers and Shrubb 1990). The Kite Committee, supported by the Nature Conservancy (later Nature Conservancy Council and Countryside Council for Wales, CCW) and the RSPB, attempted to locate all pairs every year. Breeding productivity remained low. This was probably the result of the reduction in the Rabbit population, following the arrival of myxomatosis in 1955 and the severe winters of 1961/62 and particularly 1962/63. DNA studies showed that a female from outside Wales, probably of German origin and joining the population in the 1970s, extended the gene pool (Davis 1993). Egg-collecting remained a problem, however. During 1960–89, 72 clutches were known to have been taken, equating to 8.5% of all nesting attempts, which slowed the annual rate of population growth by 0.6% (Bibby *et al.* 1990).

Numbers rose slowly until the mid-1980s, then more rapidly, as new pairs occupied more productive lowland sites (Davis 1993). In 1993, the Welsh population exceeded 100 breeding pairs, probably for the first time since the early 19th century. This milestone prompted the withdrawal of funding by the RSPB and CCW for annual monitoring of the entire Welsh population. As a result, the Welsh Kite Trust was established in 1996 to raise funds for continued monitoring, conservation and research. Egg-collecting continued to be a problem, with a minimum of 29 clutches taken during 1990–2019, of which 23 were in 1990–99, but none are known to have been taken since 2010 (RSPB data). By 1999 the population had increased to more than 200 pairs, but it was becoming increasingly difficult to cover all potential sites, meaning the number of pairs was probably higher. In 2000, the first survey of Red Kites in the UK produced a total of *c.*430 breeding pairs, of which *c.*259 were in Wales (Cross 2001). By 2005, the population in Wales alone was thought to be *c.*500 territorial pairs and by 2008, this total was exceeded (Welsh Kite Trust data). The breeding range had also expanded, with Kites returning to breed in Meirionnydd by 1963 and successfully in Radnorshire in 1976. The first modern records of breeding in Glamorgan and Caernarfonshire were in 1996. Young fledged in Pembrokeshire in 2003, the first successful breeding known there for over 150 years (Cross 2004). In the first decade of the 21st century, the Welsh population spread into England, with nests in Herefordshire and Shropshire that both had at least one wing-tagged adult of Welsh ancestry. English-released birds have also been seen in Wales and, although it seems possible, it is not known if any of these have been recruited into the Welsh breeding population.

The Welsh Kite Trust's last full attempt to monitor the entire Welsh population in a single year was in 2010, when 578 apparently occupied sites were recorded. Reduced monitoring in 2011 found 528 apparently occupied sites and 502 nesting pairs. Mid- and West Wales remained the stronghold, with 161 pairs in Ceredigion, 113 in Radnorshire, 76 in Carmarthenshire, 58 in Breconshire, 51 in Montgomeryshire and 13 in Pembrokeshire. Numbers had increased in South Wales, with 13 in Gower, six in Gwent and six in East Glamorgan, but its spread into North Wales was slower. By 2011 there were four occupied territories in Meirionnydd and one in Caernarfonshire. Although no occupied territories were found in Denbighshire in 2011, breeding had been recorded there in the early years of the 21st century. The *North Wales Atlas 2008–12* confirmed breeding in only 11 tetrads, mainly in Meirionnydd and southern Denbighshire. This undoubtedly underestimated the total, and since 2012 there has been a marked increase in records in Caernarfonshire and Denbighshire, particularly in Llŷn and the Conwy Valley.

After decades of secrecy to protect and conceal the precise locations of breeding sites, the improving fortunes of the Red Kite and growth of public interest in wildlife and conservation resulted in it becoming an important tourist attraction and an emblem of Powys Council and Mid-Wales Tourism. The first Kite feeding station was established at Tregaron in the early 1980s. Keen to benefit from the increased interest in Kites, the RSPB initiated the Kite Country project and started a joint venture with Gigrin Farm, just outside Rhayader, Radnorshire. The feeding station initially attracted just two Kites, but numbers grew rapidly. Gigrin soon became an independent venture and Red Kite feeding stations were set up by organisations and individuals at several other locations. Many regularly attracted more than 100 birds, but Gigrin maintained the largest gatherings. As part of a co-ordinated Europe-wide Red Kite roost count there in January 2010, during a spell of cold weather, a panoramic series of photographs confirmed the attendance of over 750 Kites in the air simultaneously (Tony Cross own data).

While the highest counts of Red Kites are at the feeding stations, there have been instances in recent years of large gatherings in late spring and early summer in coastal areas of Wales. A total of around 20 at Uwchmynydd, on Llŷn, Caernarfonshire, on 30 May 2009, was considered remarkable (Pritchard 2017) as was 66 over Ramsey, Pembrokeshire, on 3 June 2011. These were thought to be part of an influx into southwest Britain, which included up to 80 in Cornwall. A few years later, 51 were recorded over Ramsey, on 6 June 2015. Other coastal parts of southern and western Britain have also seen large and possibly increasing numbers of birds, such as 306 near Marazion, Cornwall, on 10 May 2020 (*Cornwall Birds* undated). Ian Carter (*in. litt.*) has suggested that these late spring/early summer flocks outside breeding areas are mostly immature birds of British origin that are making exploratory movements. Having reached the end of a land-mass, their social nature means that when they meet up by chance, they tend to stick together. Higher numbers are now seen in coastal areas in April and May, extending into June, than was the case before the re-establishment schemes, when spring peaks occurred in March and April, with a few in May.

Re-establishment programmes in England and Scotland, started by the Nature Conservancy Council and the RSPB in 1989, involved 93 young Kites, mainly from Spain, Sweden and eastern Germany, being released at each of two sites, one in southern England and another in northern Scotland over five years. By 2016 the total population in Britain had risen to an estimated 4,350 pairs (Woodward *et al.* 2020). As a result of the Red Kite population doing well in Wales, in 2007 the Welsh Kite Trust joined forces with the Golden Eagle Trust and RSPB in the first stage of a five-year programme to re-establish kites in both the Republic of Ireland (Wicklow) and Northern Ireland. Under licence from the Countryside Council for Wales, between 2007–11, a total of 250 kite chicks were collected, at about four weeks of age, from within the core breeding area. This has resulted in small breeding populations becoming established in both areas.

The populations in the core European breeding areas in Germany, France and Spain are thought to have declined since 1990, but this has been partly offset by increases in Sweden, Poland and Switzerland (BirdLife International 2015). Britain may now have the second-largest breeding population in the world, after Germany. Data from the *Britain and Ireland Atlas 2007–11* indicated that Wales held just over 40% of the British population. The Red Kite is sufficiently widespread within Wales to be monitored annually here by the BBS. Between 1995 and 2019 the population index increased by more than 400% (Harris *et al.* 2020).

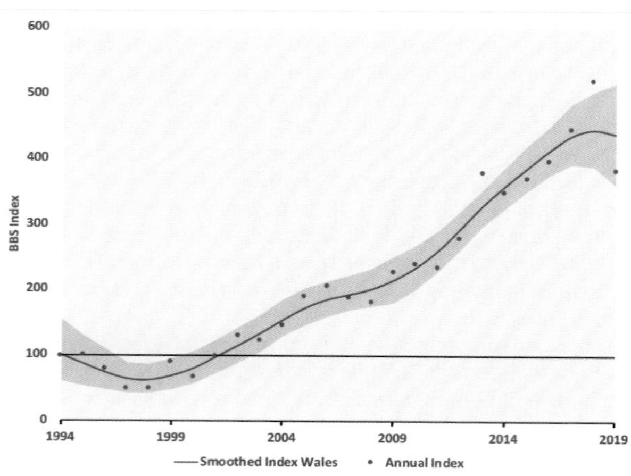

Red Kite: Breeding Bird Survey indices, 1994–2019. The red line shows the smoothed index, 1995–2018. The shaded area indicates the 85% confidence limits.

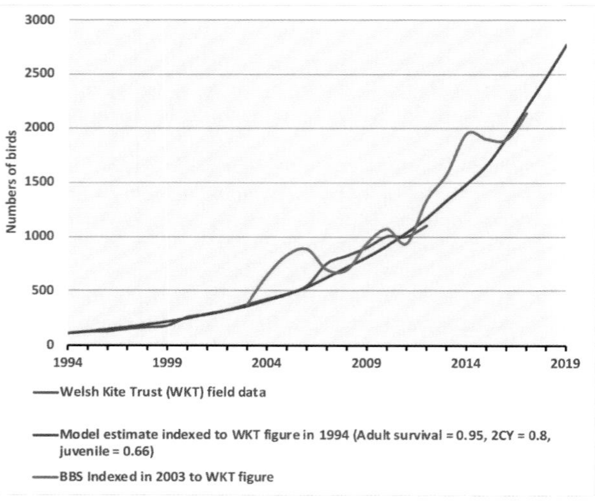

Red Kite: Population estimates, 1994–2019.

An estimate by the Welsh Kite Trust, based on field data, BBS trends and modelling using annual productivity from a large sample of monitored nests and survival data, analysed by Newton *et al.* (1989), suggested that by 2019 the Red Kite population could have been as high as 2,500 pairs (Welsh Kite Trust 2020).

Because Red Kites will scavenge dead animals such as rats, they are susceptible to the effects of rodenticides, including second-generation anticoagulant rodenticides (SGARs), which can be toxic to all mammals and birds. The Wildlife Incident Investigation Scheme (WIIS) and the Predatory Bird Monitoring Scheme (PBMS) have shown that some mortalities result from this secondary exposure. Analyses of 77 Red Kites found dead in Britain in 2017 and 2018 found detectable liver residues of at least one SGAR in all but one of these. Comparison with the previous two years showed that the proportion exposed to SGARs was always 90% or more. The percentage of kites examined that were diagnosed as birds in which SGARs were implicated as a contributory cause of death did not differ significantly between individual years nor show a significant trend across the years; the overall average across the four years was 22% (Walker *et al.* 2019).

Davis *et al.* (2001) showed that *c.*57% of mortality in Red Kites between 1950 and 1999 occurred during February to May, and particularly among adults at this time. They suggested that this seasonal bias probably arose because of increased presence of poisoned baits laid in the lambing season. Poisoning accounted for at least 38% of known causes of death. The annual average number of confirmed persecution incidents in Wales involving Red Kite was only 11% lower in 2010–19 than in 1990–99, but, occurring during a period of substantial growth, this represents a much smaller proportion of the population. However, the number of poisoning incidents, a crime to which Red Kites are susceptible by nature, was higher in the latter decade than the previous two (Hughes *et al.* in prep). Illegal poisoning is still reported but annual survival is typically around 95% for territorial adults (Newton *et al.* 1989).

Approximately 17% of the Red Kite's European range—and with the exception of a small population in Morocco, therefore its global range—is in Wales (*European Atlas* 2020). This makes it one of a small number of bird species for which Wales has a truly global responsibility. The future prospects of the Red Kite in Wales are good, although it remains on the Amber List of Birds of Conservation Concern in Wales, because of its Near Threatened status in Europe and the international importance of the Welsh breeding population (Johnstone and Bladwell 2016). Reduced sheep stocking rates and a decline in the Rabbit population could limit further increases in the core breeding area, but there remains much scope for further expansion in South Wales and particularly in North Wales. Breeding has still not been confirmed on Anglesey or in Flintshire, but this is only a matter of time.

Tony Cross and Rhion Pritchard

Sponsored by Gwyn Harrison

Black Kite *Milvus migrans* Barcud Du

Welsh List Category	IUCN Red List		Rarity recording
	Global	Europe	
A	LC	LC	WBRC

The Black Kite breeds throughout continental Europe, excluding the maritime northwest and Scandinavia. The largest breeding populations are in Spain, France and Germany. To the east, it breeds through European Russia, Central Asia and the Pacific coast. Birds from Europe winter in sub-Saharan Africa (*HBW*). It was extremely rare prior to the 1980s, but there has been an increase in the number of records in Britain since then. An average of 23 were reported each year in 2010–18 (White and Kehoe 2020a), with a spring bias. However, it is still a rare migrant to Wales, with a total of 24 records up to the end of 2019.

Total	Pre-1950	1950–59	1960–69	1970–79	1980–89	1990–99	2000–09	2010–19
24	0	0	0	3	3	3	5	10

Black Kite: Individuals recorded in Wales by decade.

The first Welsh record was at Llangorse Lake, Breconshire, on 19 October 1976. It was followed by two in 1979, at Overton and Paviland, Gower, on 17–18 April and at Cathays, East Glamorgan, on 19 May. In 1985, one was seen at Lake Vyrnwy, Montgomeryshire, and at Dyfi Forest near Corris, Meirionnydd, on 26 April. One was also seen over Lady Park Wood, Gwent, on 15 May that year. There was one other record in that decade, at Llansadwrn, Anglesey, on 24 April 1987.

There were only three records in the 1990s and five during the next decade, involving several counties. The majority were seen in spring, although the first record for Radnorshire, an immature bird, somewhat bucked this trend. It turned up at the Red Kite feeding station at Gigrin Farm, near Rhayader, Radnorshire, and over-wintered there between at least 3 January and 15 March 2010.

The number of records doubled during 2010–19, with reports in every year except 2013, 2015 and 2017. One at Parc Slip, East Glamorgan, on 28 September 2011 was the only autumn record in the period. One at Llanrhystud, Ceredigion, on 2 May 2012 was a first county record. Some individuals were probably seen in more than one county, although there were no individually recognisable plumage features or tags to confirm this. One seen from Bardsey, Caernarfonshire, on 10 May 2016 was presumed to have been the same one seen at The Range, Anglesey, on the same day. Another at Bardsey on 13 May 2018 was probably the bird seen over Skomer, Pembrokeshire, the next day. One at Holyhead Mountain and at Carmel Head, Anglesey, on 6–7 April 2019 had probably moved to Dinas Dinlle, Caernarfonshire, where it was present on 7 April 2019; sixteen days later, one was also seen at Goldcliff, Gwent, on 23 April 2019.

Up to 2019, most Black Kites had occurred in April and May, with similar numbers in North Wales and South Wales and only a few in autumn. Although the overall balance of both distribution and abundance appear to be stable, there are indications of expansion in some areas including along the northern borders of its range (*European Atlas* 2020). On current trends, it seems likely that the status of Black Kite in Wales is slowly changing. It is a potential future breeding species here. Huntley *et al.* (2007) have shown that the climate in southern England will be suitable for the species later this century.

Jon Green and Robin Sandham

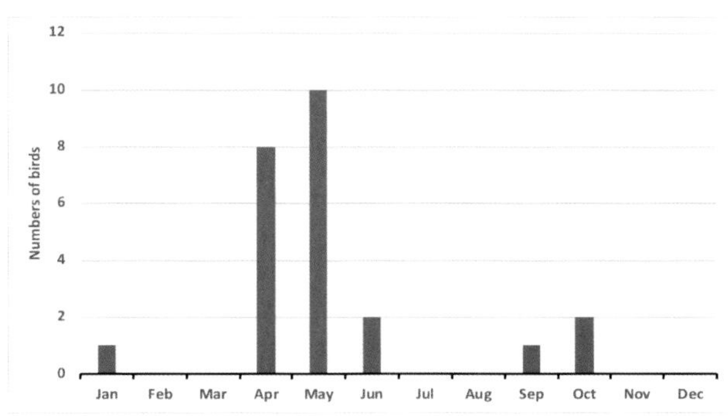

Black Kite: Totals by month of arrival, 1976–2019.

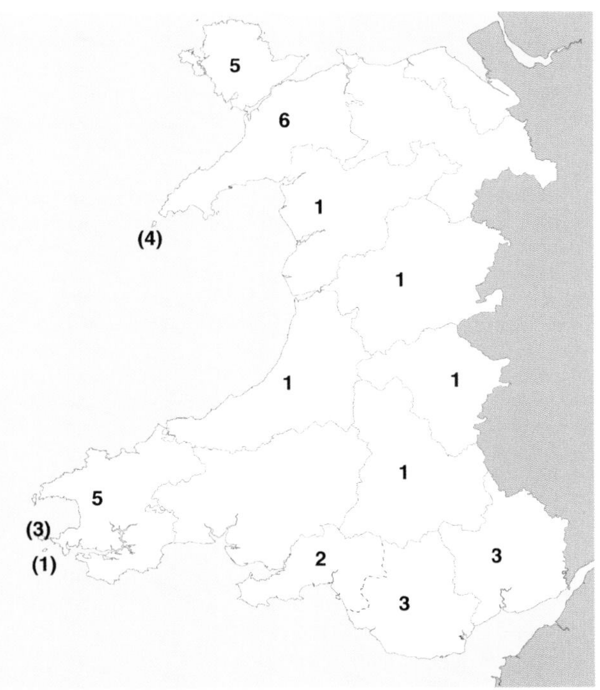

Black Kite: Numbers of birds in each recording area, 1976–2019. Some birds were seen in more than one recording area.

Black Kite © Tony Cross

White-tailed Eagle *Haliaeetus albicilla* Eryr Môr

Welsh List Category	IUCN Red List			Birds of Conservation Concern	Rarity recording
	Global	Europe	GB	UK	
A	LC	LC	EN	Red	WBRC

This large eagle has a wide distribution in northern Europe and Asia, from Greenland to the Pacific Ocean (*HBW*). Unlike Golden Eagle, this species is usually associated with lowland areas with coasts and wetlands, where fish form an important part of its diet. Persecution led to its extermination from many parts of Europe from 1800; its European strongholds are in Norway and Russia (BirdLife International 2015). It became extinct in Britain early in the 20th century, where the last breeding pair was on Skye in 1916 (Love 1988). Attempts at re-establishment started on Fair Isle in 1959, but it was a re-establishment programme that started in 1975 on Rhum that proved successful. Further releases in East Scotland followed, and now *c.*123 pairs breed in Scotland (Holling *et al.* 2019). Adults in the western part of its range are usually sedentary, but subadults roam widely and those seen in Wales to date are likely to be from the Scottish population.

There is more archaeological evidence for the presence of White-tailed Eagle in Wales than for Golden Eagle, with bones found at Cat Hole, Gower, from the Devensian period about 20,000 years ago; at Port Eynon Cave, also in Gower, from the Mesolithic, 6,000–9,000 years ago; and two Roman-era records, from Caerleon, Gwent, and Segontium, Caernarfonshire. Eagle remains from Pembrokeshire and Denbighshire were not assigned to species (Yalden 2007, Yalden and Albarella 2009), although those found in Pembrokeshire were later identified as White-tailed by Eastham (2016).

Place names and early references in literature do not distinguish between the two species of eagle. Evans *et al.* (2012b) identified 18 place names in Wales that contained the word *eryr*. Two were thought not to refer to the birds and the others were attributed to Golden Eagle or White-tailed Eagle according to altitude and presence or absence of nearby wetland. Using the same criteria as Evans *et al.* (2012b), Pritchard (2020) found that out of a total of 51 place names containing the element *eryr*, 22 of these probably referred to White-tailed Eagle.

Many references to eagles in early Welsh literature could refer to either species of eagle. One of the best-known references is in *Canu Heledd*, dating probably from the 9th or 10th century, where two eagles were described waiting to feed on the corpses of the slain, following a battle near Pengwern in the kingdom of Powys. Both the location and the description of the birds suggested that these were White-tailed Eagles. Research by the Eagle Reintroduction Wales (ERW) project (referred to in the Golden Eagle account) suggested that the core historic range of the White-tailed Eagle was larger than that of the Golden Eagle and identified two core areas: one, centred on Gwynedd in northwest Wales (including Anglesey, parts of Meirionnydd and Caernarfonshire), accounted for *c.*36% of historic distribution records. The other, centred on southern and southeast Wales, included parts of Dyfed, mainly the Ceredigion coast, and parts of Glamorgan, predominantly the Kenfig area and Gower coast, and accounted for almost 43% of historic distribution records (Williams *et al.* 2020).

There is no direct record of the species breeding in Wales, although there was a record of possible breeding in Caernarfonshire. Lovegrove (2007) recorded that Hale was told by a well-known egg-collector of men climbing to a nest near Nefyn around 1880. Lovegrove commented that "although the evidence is circumstantial and the date a very late one, it was in all probability Sea Eagles". In the 19th and early 20th centuries, this species was recorded more regularly in Wales than Golden Eagle. In the Swansea area, Dillwyn (1848) commented that "formerly a year rarely passed without one or more of these Eagles being seen in the neighbourhood; but they now appear to be becoming more and more rare". *Birds in Wales* (1994) mentioned 17 records between 1818 and 1910; many of these birds had been shot, and their identification confirmed from mounted specimens. The records covered all counties except Radnorshire and Gwent. The last of these records, at Abersoch, Caernarfonshire, on 29 November or October 1910, was said to have been in the area for a fortnight before it was shot in the wing and captured alive; the bird was kept in an aviary at Wrexham, Denbighshire. That was the last White-tailed Eagle recorded in Wales, until three records in the modern era. These were a subadult on Skomer, Pembrokeshire, on 10 and 11 November 1993, a first-calendar-year bird at Cors Caron, Ceredigion, on 12 November 1997 and one over Dolwyddelan, Caernarfonshire, on 28 March 2008.

If the Scottish population continues to increase and to extend its distribution, it is possible that birds might eventually recolonise Wales. However, even within northwest Scotland, their spread has been slow, so this would be a very long-term prospect. Re-establishment has been suggested for Wales, like that undertaken in Scotland and, from 2019, on the Isle of Wight in England. A report on the feasibility of re-establishment in Wales (Marquiss 2005) indicated that there would be a good likelihood of success, with better prospects for this species than for Golden Eagle. Research to reconstruct the historic range of both the White-tailed Eagle and the Golden Eagle in Wales has provided important information about the past distribution of both species (Williams *et al.* 2020). These data, together with additional analysis of the modern landscape, will provide important information needed to assess the possibility of re-establishing one or both species in Wales. Parts of South Wales and Llŷn, for example, would certainly provide suitable habitat for White-tailed Eagle, although it would be essential to build public support for the project first and to minimise the risk of potential conflicts with other species of conservation concern in Wales.

Rhion Pritchard

Rough-legged Buzzard *Buteo lagopus* Boda Bacsiog

Welsh List Category	IUCN Red List		Rarity recording
	Global	Europe	
A, E	LC	LC	WBRC

The Rough-legged Buzzard has a circumpolar distribution (*HBW*), breeding in Arctic and subarctic regions and wintering in lower latitudes. In Europe it breeds in Fennoscandia and northwest Russia and winters from the Baltic Sea southward (*HBW*). In some years, more westerly movements result in influxes into Britain, mainly into eastern counties of England. There is great variability in the numbers recorded each year, generally reflecting the harshness of the winter on the continent and the vole cycle to the east. Although there are records from all Welsh counties, it is a rare autumn and winter vagrant to Wales. Only the nominate form has been recorded in Wales, but one of the North American subspecies *B.l. sanctijohannis* (IOC) has been recorded in Ireland (Mullarney and Murphy 2005).

Total	19th century	1900–49	1950–59	1960–69	1970–79	1980–89	1990–99	2000–09	2010–19
75	21	3	4	12	13	12	4	3	3

Rough-legged Buzzard: Individuals recorded in Wales by decade.

Most of the early records in Wales refer to individuals either shot or trapped, including the earliest ones from Glamorgan in Gower, at Margam in 1840 and Penllergaer in 1843. Birds that probably suffered a similar fate later in the 19th century included a few others in Glamorgan, Pembrokeshire, Ceredigion and Caernarfonshire, and several in Meirionnydd—six of which had been killed on the Rhug estate's grouse moors near Corwen, before 1891. Some authors have commented on possible confusion with Common Buzzard, which can show considerable plumage variation. Mathew (1894), for example, referring to Pembrokeshire, mentioned that "when shot or trapped it might very well be confounded by people who were not well up in birds with the Common Buzzard, from which it is always to be easily separated by its feathered tarsi".

There were just three records of single birds in the first half of the 20th century: in Ceredigion in October 1910, in Breconshire in 1921 (month unknown) and in Pembrokeshire in September 1931. There were no further records until the 1950s, with four singles between 1952 and 1957: three in Montgomeryshire and one in Meirionnydd. The number of reports increased during the next three decades, with about a dozen in each period. These included overwintering birds at Cors Caron, Ceredigion, in 1961/62, on Skomer, Pembrokeshire, between 22 October 1962 and 2 March 1963, and at Sennybridge, Breconshire, between November 1968 and April 1969. Records in the 1970s included single birds at Montgomery and Skomer on 30 September 1971; on Bardsey, Caernarfonshire, between 4 November 1974 and 27 March 1975; at Cors Fochno, Ceredigion, on 31 October 1979 and, what was presumed to have been the same bird there on 20–21 November 1979. There were records from eight counties in the 1980s, including a subadult found thin and exhausted near Castlemartin, Pembrokeshire, in early October 1988. After being rehabilitated, it was released at Dowrog Common, Pembrokeshire, later that month.

Records were fewer in the 1990s, but included a long-staying individual in Montgomeryshire and Radnorshire between 4 March and 23 April 1995. The number of records continued to fall, with only six during 2000–19, including birds at Lligwy, Anglesey, on 24 January 2002, on Bardsey on 31 October 2002 and from Pen-rhiw-fawr, Gower, on 8 January 2003. It was almost ten years before another was seen, at Cemaes Head, Anglesey, on 12 December 2012, and another five before the next, at Fenn's Moss, Denbighshire, on 25 November 2017. The last of the decade was at Pen-y-pass, Caernarfonshire, on 20 November 2018.

Most records were passage birds in late autumn, particularly those recorded in Pembrokeshire, although winter records and early spring records have been boosted by a handful of longer-staying overwintering birds. The average numbers seen annually in Britain increased between 1990–99 and 2010–18 (White and Kehoe 2020a), although these are biased by influx winters. It remains a rare bird in Wales and would seem set to remain so.

Jon Green and Robin Sandham

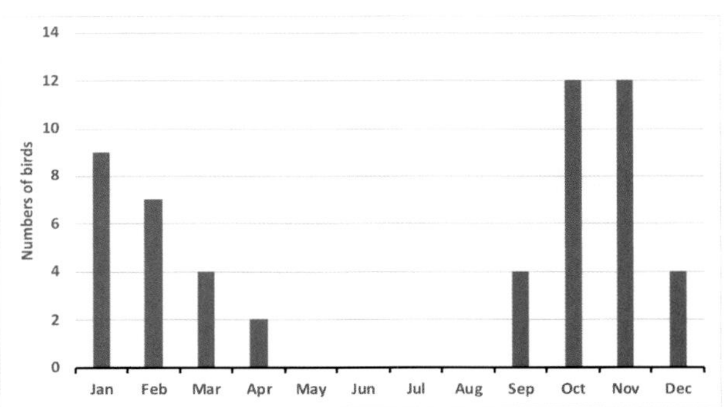

Rough-legged Buzzard: Totals by month of arrival (when known), 1893–2019.

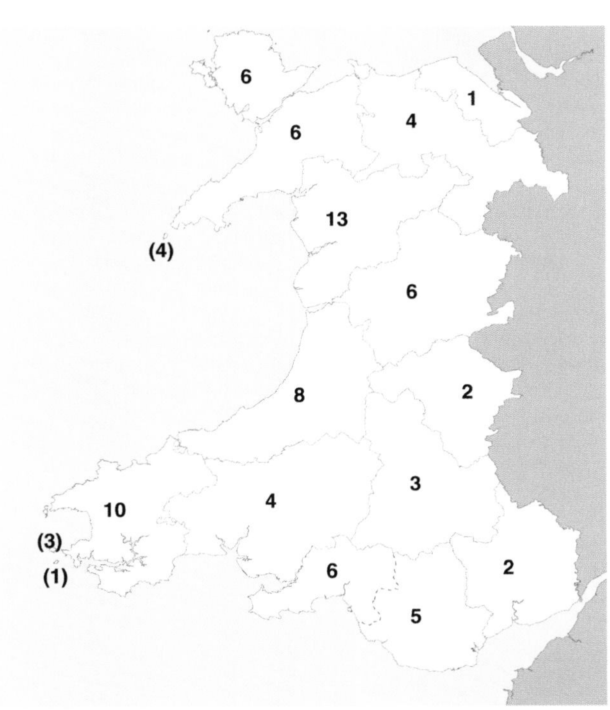

Rough-legged Buzzard: Numbers of birds in each recording area, c.1840–2019. One bird was seen in more than one recording area.

Buzzard *Buteo buteo* Bwncath

Welsh List Category	IUCN Red List			Birds of Conservation Concern			
	Global	Europe	GB	UK	Wales		
					2002	2010	2016
A	LC	LC	LC	Green			

The Buzzard is found across Europe and, east of the Urals, through the West Siberian Plain as far as northwest China. Eastern and northern populations are migratory, whereas territorial birds in Britain and the rest of western Europe are sedentary (*HBW*). This is a versatile species that breeds from sea level to upland areas. It takes a wide variety of prey: small mammals, birds (particularly corvids), frogs, reptiles and invertebrates such as earthworms and beetles, as well as carrion (Dare 2015). Nests in Wales are most commonly built in trees, while others are on mountain and sea cliffs or in disused quarries.

Adult Buzzards in Wales hold territory throughout the year while first-calendar-year birds typically disperse only short distances. One, ringed as a nestling at Trawsfynydd, Meirionnydd, in July 1986 was found sick at Blaenau Ffestiniog, 10km away in December 2016 and released again in January 2017. This 30 year old is the oldest Buzzard known in Britain and Ireland. Some young Buzzards will move considerably farther, however. A female nestling ringed near Lampeter, Ceredigion, in June 1977 was found dead 412km to the northeast at Bamburgh, Northumberland, in May 1990, and one seen at Pentrefelin, near Criccieth, Caernarfonshire, in January and February 1995, had been wing-tagged in Dumfries, southwest Scotland, in spring 1994.

The Buzzard is thought to have been very common in Britain during the Middle Ages (*Birds in Wales* 1994), but there is surprisingly little evidence of its being in Wales apart from remains dating from between the 12th and 14th centuries, found during excavations at Loughor Castle, Gower (Lewis 1993). There is no evidence in parish accounts of persecution in Wales before the advent of large-scale gamebird preservation that began in the early 19th century (Lovegrove 2007). Moore (1957) considered that the species bred in every county in Wales in 1800, and it was described by Eyton (1838) as common in North Wales. It was noted in *The Field* as common in the Beddgelert area of Caernarfonshire in 1859 and was said to be common in the Elan Valley, Radnorshire, in 1869 (Jennings 2014). Persecution by gamekeepers drastically reduced numbers during the second half of the 19th century, and by 1865 there was little evidence of breeding in Flintshire, in Denbighshire and on Anglesey (Moore 1957). It is interesting that 'only' 135 Buzzards were reported killed on Penrhyn Estate, mainly between Penmachno and Dolwyddelan, Caernarfonshire, between 1874 and 1902, compared to 1,988 Kestrels and 735 Sparrowhawks (Lovegrove 2007).

More information is available from the last years of the 19th century and the early 20th century. Mathew (1894) said that in Pembrokeshire "there may be possibly still some half dozen nesting stations of the bird on the islands, and on the cliffs along the coast". J.H. Salter's diaries show the Buzzard to have been scarce in most of Ceredigion at the end of the 19th century. It could only have been described as being at all common in the Tywi and Camddwr valleys, in the southeast of the county, on the border with Breconshire and Carmarthenshire, where there were no gamekeepers (Roderick and Davis 2010). In Glamorgan, the species was said to be rare at the end of the 19th century but could still be found in some numbers in more remote hill country (Hurford and Lansdown 1995). In North Wales, Forrest (1907) commented that the species had ceased to breed in the eastern counties, although he noted a record of breeding in Denbighshire in 1904. He said that Buzzards were still found in fair numbers farther west and could be described as common in some areas, although "almost unknown on Anglesey". He also stated that while bygone writers described it as nesting in trees, by this period it had become almost entirely a cliff-nester.

The Buzzard population of Wales was at a low ebb in the first decade of the 20th century, but was not extirpated, as it was from land to the east. Walpole-Bond (1914) described Wales as the "head-quarters" of the species in Britain and added "there it still deserves to be called common". He thought that the population in Wales was around 250 pairs, of an estimated 450 pairs in Britain, and that Buzzards nested in every county except Anglesey and possibly Flintshire. He had visited over 60 different nests in Breconshire, Radnorshire, Ceredigion and Carmarthenshire, particularly in the last two. He knew the location of about another 50 and said that in the best areas it was not unusual to encounter 20 or more birds in a day's walking.

The species benefited from greatly reduced persecution, as a result of the First World War when many gamekeepers left to join the forces. The increase in Buzzard numbers seems to have started even while the war was still raging. In 1915, Tomkinson was able to find ten nests in a day in the Abergwesyn area of Breconshire (*Birds in Wales* 1994) and by the 1920s there were clear indications of an increase in a number of counties. When Salter returned to the Aberystwyth area of Ceredigion in 1923, he recorded many more breeding birds than in the late 19th and early 20th centuries, but Kennedy Orton said that only a few pairs bred in Snowdonia and that these were strictly confined to the higher mountains (Carr and Lister 1925). By the 1930s, the Buzzard was recorded as being increasingly common in the Barmouth, Dolgellau and Aberdyfi areas of Meirionnydd (Pritchard 2012). Ingram and Salmon (1939) knew of two breeding records in Gwent, but considered it doubtful that the species bred in the county when they wrote, although it was often seen in autumn and winter. By the onset of the Second World War in 1939, the Buzzard population had recovered to its highest level for over 150 years (*Birds in Wales* 1994). Campbell (1946) thought that the Buzzard was then spreading in Gwent, as elsewhere along the Welsh border. By 1949, Lockley *et al.* estimated that not fewer than 120 pairs bred in Pembrokeshire, on the coast, islands and inland, compared to Mathew's half a dozen pairs in the 1890s. In Montgomeryshire, two egg-collectors found 18 nests "in a short spell" in the Llanbrynmair area in 1951 (Holt and Williams 2008), but the species remained absent from Llŷn, Caernarfonshire, in the early 1950s (Bark Jones 1954).

Buzzard © John Hawkins

Prospects for the Buzzard in Britain were boosted by the passing of the Protection of Birds Act in 1954, but persecution was soon replaced by food as the main factor affecting their recovery. In 1953, the potent Rabbit viral disease myxomatosis was deliberately introduced into southern England, quickly leading to rapid, large and widespread declines in Rabbit populations and to a temporary abundance of food for Buzzards and other scavengers, followed by scarcity. A survey of Buzzards in 1954 (Moore 1957), aimed at recording their status in Britain and Ireland immediately before the myxomatosis outbreak started to spread, confirmed that Wales, apart from Anglesey and Flintshire, was one of the most important areas for Buzzards. The highest densities were in central and southwest Wales. The highest recorded in Britain was on Skomer, Pembrokeshire, where there were 7–8 pairs on this small (2.92km^2) but food-rich island (Davis and Saunders 1965).

Views about the effects of myxomatosis on Buzzards in Wales varied. Roderick and Davis (2010) said that although Rabbits in Ceredigion had been wiped out, this had no obvious effect on local Buzzard numbers. Condry (1966) thought that in Snowdonia there had been no great effect, because Buzzards there had never depended heavily on Rabbits. In Radnorshire, Buzzard numbers crashed in the mid-1950s but were back to 75% of pre-myxomatosis levels by 1959 (Jennings 2014). In Pembrokeshire, the sudden Rabbit scarcity initially caused widespread breeding failures. In 1955, there were no successful nests along 30km of coastline on the St Davids Peninsula or Ramsey, and the Skomer population declined to two pairs by 1956 (Donovan and Rees 1994). Another problem that became apparent during the 1960s was the lethal effect of poisoning by toxic organochlorine pesticides (aldrin, dieldrin, heptachlor) on raptor numbers. The varied diet of Buzzards meant that they were less affected than specialist bird-eating species, such as Sparrowhawk and Peregrine. However, the use of dieldrin in sheep dips meant that upland Buzzards, for whom sheep carrion was an important item of diet, were at risk (Dare 2015). Despite this, there were some signs of a range expansion during the 1960s, including three pairs breeding on Anglesey in 1967.

The *Britain and Ireland Atlas 1968–72* showed that Buzzards were present in almost every 10-km square in most of Wales, although in only a few squares in Flintshire, eastern Denbighshire and Anglesey. By 1988–91, there had been some gains in these northern areas, but the species was still largely absent from western Anglesey, its spread probably still restricted by persecution. Elliott and Avery (1991), reviewing incidents in Britain in 1975–89, found that Buzzards were about twice as likely to be reported dead through persecution on the edge of their range compared with the middle. The *Britain and Ireland Atlas 2007–11* showed that the remaining gaps in the Buzzard's distribution in Wales had been filled and it had become one of the most widespread species in Wales.

Tetrad atlases provided more detail on this upward trend. In Gwent, breeding evidence was found in 82% of tetrads in 1981–85, and this had increased to *c.*97% by 1998–2003. In East Glamorgan, Buzzards were found in 59% of tetrads in 1984–89 and 84% in 2008–11, although there were fewer with breeding evidence codes (69%). There was a slight reduction in Pembrokeshire, where breeding evidence was recorded in *c.*85% of tetrads in 1984–88 and in 81% in 2003–07 (*Pembrokeshire Avifauna* 2020). Buzzards were found in *c.*77% of tetrads in the *West Glamorgan Atlas 1984–89* and in 73% of those surveyed in Breconshire during 1988–90 (Peers and Shrubb 1990). In the three counties with repeat surveys, their occupation had risen on average by 12%, to 87% of tetrads.

The *North Wales Atlas 2008–12* found the Buzzard to be among the most widespread species, occurring in *c.*86% of tetrads, including most of those in Flintshire, eastern Denbighshire and Anglesey. In a study area in eastern Denbighshire, the sharpest increase occurred after the early 1990s (Roberts and Jones 2009). The Buzzard now breeds almost everywhere in Wales, except in urban areas and those scattered rural districts where there are no suitable nest sites. The Buzzard breeds along sea cliffs on several Pembrokeshire islands, with the largest number usually on Skomer, where there were five pairs in 2017, but it has never been recorded breeding on Bardsey, Caernarfonshire.

A BTO survey in 1983 showed that Buzzard density was higher in Wales than anywhere else in Britain, with a mean of 1.45 soaring birds per tetrad (Taylor *et al.* 1988). Its Welsh distribution in the *Britain and Ireland Atlas 1988–91* was very similar to that in 1954 (Moore 1957). It showed high relative abundance in central and southwest Wales, somewhat less in much of the northwest and the southeast, and lowest abundance in Flintshire, eastern Denbighshire and Anglesey. The only other areas of Britain with comparable abundance were southwest England and parts of western Scotland. By the 2007–11 *Atlas*, however, relative abundance had increased on Anglesey and in northeast Wales. Away from these areas, there were no great changes.

The most reliable information on population densities comes from dedicated surveys. In a 475km^2 area in the southwest Cambrian Mountains, mostly in Ceredigion, Newton *et al.* (1982b) found densities of 41 territories/100km^2 on farmland and 24 territories/100km^2 on hill sheepwalks with conifer plantations. These densities were, at the time, among the highest recorded in Britain, although breeding success was low, with an average of 0.6 young fledged per pair. In north Breconshire, Shrubb (2000) noted 228 soaring Buzzards during spring 1999, in four 10-km squares, an increase of 115% on the number reported in 1983. Snowdonia has been particularly well surveyed. Dare (2015) estimated 30–35 pairs present during the 1950s, but by 1977–84 a further survey found 96 pairs (one per 9.7 km^2 of total terrain) in a 926km^2 study area, mainly in Caernarfonshire and west Denbighshire, but also including a small part of northwest Meirionnydd (Dare and Barry 1990, Dare 1995). Numbers doubled in the same area, to 194 pairs, in 2003–04 and the density had increased from 10.4 pairs/100km^2 to 21 pairs/100km^2 (Driver and Dare 2009). The largest increase occurred in the Carneddau range, from 29 pairs in 1977–84 to 78 pairs in 2003–04. Two studies on smaller areas of the Arfon coastal plain, Caernarfonshire, in 1998–2001 and 2014–15, found densities of 45–55 territories/100km^2 (Pritchard 2017).

The annual BBS population index for Wales showed that numbers fluctuated, but declined by 8% between 1994 and 2019, compared to a large increase in the UK as a whole (Harris *et al.* 2020). This is somewhat unexpected, considering the evidence for range expansion and an increase in areas where the Buzzard was still a scarce bird in the early 1990s. Buzzards and other raptor species are less detectable in the air in bad weather but, given that the BBS is not undertaken if conditions are unsuitable, this would not explain the wide variance in the two BBS trends for this species, meaning the reason for it is not known.

First-calendar-year Buzzards, dispersing from natal sites, tend to wander the countryside and to gather in autumn and winter, wherever invertebrate and carrion food is plentiful. In Wales, a notably large gathering included 57 feeding in a stubble field near Newcastle Emlyn, Ceredigion, in December 1999 (James 2000). Impressive movements of nomadic immature birds have also been seen, including: 185 north over Bow Street, Ceredigion, in two hours on 21 September 2001; 58 flying west over Treginnis, Pembrokeshire, on 5 October 2005; and 52 (plus eight Red Kites and six Ravens) on one thermal near Llandysul, Montgomeryshire, in September 2018.

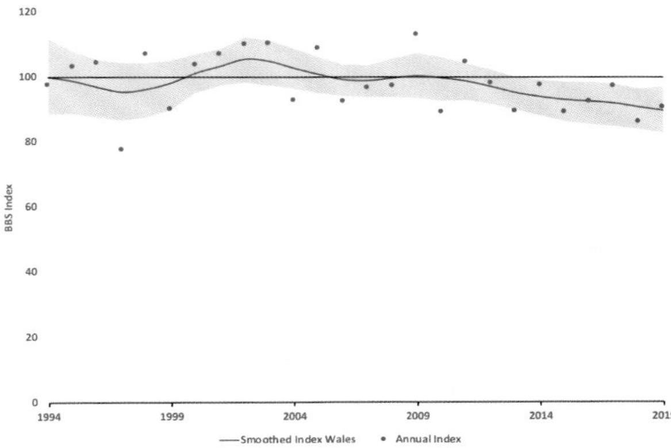

Buzzard: Breeding Bird Survey indices, 1994–2019. The red line shows the smoothed index, 1995–2018. The shaded area indicates the 85% confidence limits.

Birds in Wales (1994) estimated the Buzzard population to be 3,600–4,000 pairs, some 25% of the total British population at that time. The population in 2018 was estimated to be 9,850–13,500 pairs (Hughes *et al.* 2020). It does not appear to face any particular threat in Wales, although a reduction in sheep stocking densities and the wintering of sheep on low ground could reduce numbers in the uplands, where sheep carrion is an important food. This has already been noted in Snowdonia over the last five years (Julian Driver pers. comm.). Persecution is no longer a major threat to the species, at a population level, in Wales. The annual average number of confirmed persecution incidents in Wales involving Buzzard declined by 50% between 1975–87 and 2000–09, during a period of population growth and then stability, although the number of poisoning incidents, a crime to which Buzzards are by their nature susceptible, was higher in the latter decade than the previous two (Hughes *et al.* in prep).

Peter Dare and Rhion Pritchard

Barn Owl *Tyto alba* Tylluan Wen

Welsh List Category	IUCN Red List			Birds of Conservation Concern			
	Global	Europe	GB	UK	Wales		
A, E	LC	LC	LC	Green	2002	2010	2016

The Barn Owl has a worldwide distribution, occurring on all human-inhabited continents. In southern and western Europe, including Wales, the nominate subspecies *T.a. alba* is widespread (*HBW*). The subspecies *T.a. guttata* breeds in central Europe (IOC) and occurs occasionally in Britain. A previously published record of this subspecies in Gwent in 1908 currently does not meet the criteria set out by the BOURC (Harrop 2011) and is now considered no longer acceptable following a review of historical records (Green 2020a).

In the Western Palearctic, the Barn Owl breeds in areas with oceanic or moderately continental climates, where winter snow is light and typically lasts fewer than 40 days. Thus, it favours lowland districts, although a nest in Denbighshire has been successful over more than a decade at 390m (own data). The Barn Owl occupies open habitats, with some trees, especially farmland where there is some rough vegetation, and roadside verges where mice and other small mammals can be hunted during low flight. It nests and roosts primarily in old buildings with suitable access holes and enough room to nest or a shady recess for daytime roosting. It sometimes uses holes in old trees and readily accepts large nest-boxes placed on buildings or trees (*HBW*).

The Barn Owl prefers lowland areas that have the most dependable food supplies in hard winters. Forrest (1907) reported that at one nest site the species fed mainly on House Sparrows, and on Anglesey, the head of a Dunlin was found in a regurgitated pellet. An analysis of pellets from two roosts in Carmarthenshire (2004) and West Gower (2007) by Facey and Vafidis (2009) has provided a more recent insight into typical prey species with varying ecologies. Field Voles predominated at both roosts, although other species taken included Common Shrew, mice *Apodemus* species, Bank Vole, juvenile Brown Rat and unidentified birds. Breeding success has been shown to be related to vole abundance and the vagaries of the weather: snow, rain and wind decrease hunting effectiveness (Shawyer 1987, Formaggia 2002). Reductions in the availability of nesting sites, as old buildings have been removed or renovated, and the storage of grain in closed silos, reducing the spilt grain available to rodents, have probably contributed to declines in the Barn Owl population across Europe.

Some road developments have also proved hazardous by removing habitat and contributing to increased mortality, because of the speed of traffic (*European Atlas* 1997). The extent to which this is an issue across Wales is unknown, but Project Splatter, for example, has accumulated records of more than 50 Barn Owl road casualties from various parts of Wales between 2005 and 2019 (Sarah Perkins *in litt.*). A study by Birdwatch Ireland showed that individual tagged Barn Owls spent more time than expected hunting small mammals along motorway verges that support a similar abundance and greater diversity of small mammal species than the surrounding countryside (Lusby 2018). An ongoing study of the A55 trunk road across Anglesey showed that larger numbers of owls were killed in the early years of the road development. This was possibly due to the temporary creation of 'perfect' vole habitat—long grass, including new tree planting plots—in a landscape of closely grazed pasture. Since 2009, reported numbers of owl deaths along the route have declined, whereas the Barn Owl population trend, based on the number of newly ringed birds by members of the Anglesey Barn Owl project, has been one of increase over the same time period. The decline in the mortality trend may be attributable, at least in part, to mitigation measures. Some grassland has been allowed to develop into patches of gorse scrub and plantation areas are being allowed to mature, thus reducing the amount of desirable feeding habitat alongside the road (Jill Jackson *in litt.*).

Barn Owl © Bob Garrett

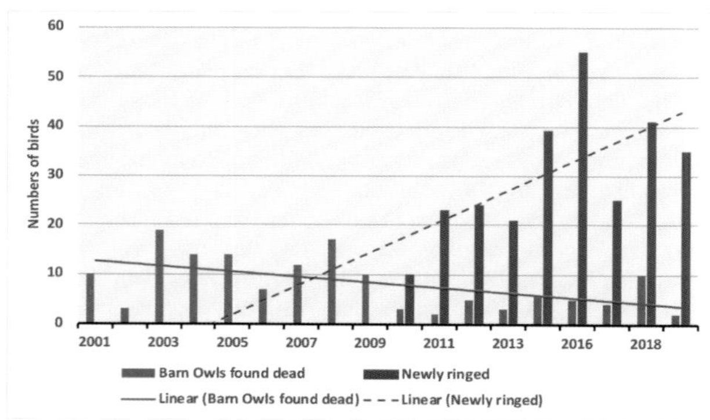

Barn Owl: Numbers found dead on the A55 on Anglesey, 2001–19. Total = 146. The overall trend is shown by the blue descending line. The linear trend for the numbers ringed on Anglesey is shown by the red, dashed line.

Barn Owls have been studied intensively through ringing across much of Britain. Around 90% of recoveries are of birds that were ringed as nestlings, usually in nest-boxes. Breeding adults in Britain are largely sedentary. Young Barn Owls do not disperse very far, with a median dispersal distance of 12km from the nest, while only *c.*4% of movements were over 100km. In mainland Europe, dispersal distances are greater, often more than 50km, with 9–22% moving more than 100km (*BTO Migration Atlas* 2002). Most recoveries in Wales have been less than 20km from their ringing site, but there are quite a few exceptions of longer movements into and out of Wales. These include: a bird from Glenurquhart, Highland, ringed in July 2007, that was found 624km away at St Ishmael's, Pembrokeshire, in December 2007; one that was ringed at Newcastle, Gwent, in December 1997 and found near Whitby, North Yorkshire, in January 1998; and one ringed at Jubbega, The Netherlands, in June 2015 that was found at Cosmeston Lakes, East Glamorgan, in February 2016, 657 km to the west.

Barn Owl: Recovery locations of birds ringed in Wales are shown by red circles. Ringing locations of birds that were recovered in Wales are shown by blue triangles. Small brown dots show ringing or recovery locations in Wales. All movements shown are greater than 50km.

The Barn Owl appears to have been widespread in the late 19th century, abundant in lowland counties but far scarcer in upland areas (*Historical Atlas 1875–1900*). Dobie (1894) described it as fairly common and "generally distributed" in northeast Wales, while Forrest (1907) stated that it was resident and common in lowland areas, but avoided "elevated districts" of North Wales. However, Mathew (1894) reported it as being "far from common" in Pembrokeshire, but he knew of two locations that he termed as 'Owleries', where large numbers were driven out or killed. He stated the following: "Although the Barn Owl is generally a solitary recluse, we have, in our experience, met with two instances of its living in society in such numbers that the association might fairly be termed an 'Owlery'. One of these had its location in some old cottages, just below a beautiful Henry VII church tower. The roofs of the cottages all communicated, and were tenanted by such a number of Barn Owls that at last the cottagers rose up against them, being annoyed by the smell and the noises proceeding from the birds, and we were informed that between forty and fifty were either driven out or destroyed".

During the 19th century, particularly in Breconshire, Glamorgan and Gwent, there was much persecution by gamekeepers. In Radnorshire, Jennings (2014) reported that there was a decline in population from the middle of the 19th century, possibly due to persecution to provide specimens for taxidermy. In Glamorgan, Hurford and Lansdown (1995) reported that the population declined because of "wanton destruction" by gamekeepers and hunters, but numbers increased during the First World War, when gamekeeper pressure reduced. Persecution was less of an issue in North Wales, where gamekeeping was focused on upland grouse moors (*Birds in Wales* 1994).

Numbers decreased by up to 33% during the 1920s, with weather extremes and severe winters being implicated (Blaker, quoted in *Birds in Wales* 1994). A study in 1932 (Blaker, quoted in Shawyer 1998) made an estimate of 1,416 pairs breeding in Wales, with Pembrokeshire (210) and Carmarthenshire (187) the two most populous counties, a positive contrast to the species' status in these counties a few decades previously. By the 1950s, Barn Owls were scarce in many counties, this being attributed largely to more intensive agricultural practices. Massey (1976a) wrote that, in Breconshire, Barn Owls were uncommon or rare at the start of the 20th century and that while they remained uncommon in the 1970s, they were widespread, particularly in river valleys. Post-war increases were adversely affected by the severe winters of 1946/47, 1961/62 and 1962/63, although in the winter of 1949/50 there was "an extraordinary influx" in the south of Glamorgan (Hurford and Lansdown 1995). The *Britain and Ireland Atlas 1968–72* showed that breeding occurred over most of Wales, except the highest upland areas, but its distribution had decreased by the *Britain and Ireland Winter Atlas 1981–84*, particularly in Caernarfonshire, Meirionnydd, Ceredigion and East Glamorgan, although the species can easily be overlooked during atlas fieldwork. It was found to be scarce and local during surveys in Glamorgan, but there was a moderate increase in sightings in the early 1990s (Hurford and Lansdown 1995). An estimate of the numbers of Barn Owls breeding in Britain in 1982–85, estimated there were just 462 pairs in Wales, much of the change attributed to climatic fluctuations. Among factors that negatively affected the population were: snow cover lasting for 20 days or longer; above average summer rainfall, which reduces nest productivity; and markedly lower than average rainfall, which decreases vole numbers, particularly during the breeding season (Shawyer 1987).

Birds in Wales (2002) showed that the Wales population declined further following the 1982–85 survey. The *Britain and Ireland Atlas 2007–11* showed further losses in distribution across Wales, although patchy areas of greater abundance were evident across northern parts of Wales, particularly in southern Anglesey. The *North Wales Atlas 2008–12* showed clusters of breeding records in lowland areas, but no evidence of breeding in *c.*90% of tetrads. The *Gwent Atlas 1981–85* showed breeding evidence in *c.*18% of tetrads, reduced to *c.*13% in the 1998–2003 *Atlas*. The *East Glamorgan Atlas 1984–89* showed breeding evidence in *c.*2% of tetrads. This had increased to 5% in 2008–11. In Gower, the *West Glamorgan Atlas 1984–89* showed evidence of possible, probable and confirmed breeding in 8% of tetrads and the *Pembrokeshire Atlas 1984–88* showed evidence in 23%, which reduced slightly to *c.*21% in 2003–07 (*Pembrokeshire Avifauna* 2020).

Recording area	Estimate (pairs)	Year(s)	Source
Gwent	25–50	1998–2003	Gwent Atlas 1998–2003
East Glamorgan	34–68	2008–11	*East Glamorgan Atlas 2008–11*
Gower	8	2019	Rob Taylor (*in litt.*)
Carmarthenshire	50	2014	Lloyd *et al.* (2015)
Pembrokeshire	*c.*50–100	2003–20	Jenks (2021)
Ceredigion	140	2010	Roderick and Davis (2010)
Breconshire	25–30	2019	Andrew King (*in litt.*)
Radnorshire	19–25	2016	Peter Jennings (*in litt.*)
Montgomeryshire	*c.*85	2019	Montgomery Barn Owl Group (2019)
Meirionnydd	37	2008–12	Pritchard (2012)
Caernarfonshire	30+	2017	Pritchard (2017)
Anglesey	25–35	2008–12	*North Wales Atlas 2008–12*
Denbighshire	45–55	2008–12	*North Wales Atlas 2008–12*
Flintshire	10–15	2008–12	*North Wales Atlas 2008–12*
Total	**583–728**		

Barn Owl: Most recent population estimates in Wales.

There has not been a recent survey of the numbers of Barn Owls in Wales and their largely nocturnal behaviour makes a reliable population assessment difficult. The best recent estimates are tabulated, based on available information from county avifauna and other sources. This suggested a Welsh Barn Owl breeding population of 583–728 pairs, out of an estimated UK population of 4,000–14,000 pairs (Woodward *et al.* 2020). In Denbighshire and Flintshire, they were badly affected by periods of heavy snow in February 2009 and March 2013, which resulted in a population reduction of *c.*30% across these two counties that, by 2019, had not yet fully recovered (own data).

In addition to impacts from weather, Barn Owls are affected by loss of nest sites and adverse changes to habitat and food supply, including potential effects of second-generation anticoagulant rodenticides (SGARs). Analyses of Barn Owl livers from samples across Britain in 2018 showed that 87% had detectable residues of one or more SGAR, and that there were few differences in liver SGAR accumulation, compared with baseline years of 2006 and 2012 (Shore *et al.* 2019). Although these rodenticide chemical traces were not as high as those found in Kestrel samples, for example, their potential impacts on owl behaviour and on overall survival are unknown.

Information from the *European Atlas* (2020) suggests that the breeding range of the Barn Owl had not changed since the *European Atlas* (1997), although an increase in population size was observed in regions where intensive nest-box projects had started some years previously, such as in Britain and Denmark. Barn Owl study groups in Wales have been doing their best to help its conservation, such as by placing nest-boxes in suitable places in farmland, and their efforts are likely to be needed into the future. The Barn Owl Trust advises that nest-boxes should not be erected within at least 1km of a major road, unless there are continuous screens on both sides, to minimise collision risk (Barn Owl Trust 2020).

Although climate change has been predicted to leave all of Wales suitable for Barn Owl through to the end of the 21st century (Huntley *et al.* 2007), there are likely to be some adverse effects. There may be less snow in winter, but increased rainfall during the winters and summers may make hunting conditions more difficult. There is no immediate threat, but the future does not look entirely rosy for the Barn Owl.

Ian M. Spence

Sponsored by Cymdeithas Ted Breeze Jones in memory of Dewi Jones, 'Stiniog

Scops Owl *Otus scops* Tylluan Sgops

Welsh List Category	IUCN Red List		Rarity recording
	Global	Europe	
A	LC	LC	BBRC

The Scops Owl breeds across southern and eastern Europe, the Middle East and southern Russia as far east as Mongolia, and winters in Africa south of the Sahara (*HBW*). It is a rare migrant to Britain. 86 had been recorded to the end of 2019 (Holt *et al.* 2020), but with only two from Wales, both in Pembrokeshire, it is an extremely rare vagrant here:

- one was caught near Pembroke in spring 1868 and
- the other was caught and ringed on Skokholm on 25 April 1955.

The west European population is expected to spread north, potentially breeding in southeast England later this century (Huntley *et al.* 2007), so we may expect to see an increase in records.

Jon Green and Robin Sandham

Snowy Owl *Bubo scandiacus* Tylluan yr Eira

Welsh List Category	IUCN Red List		Rarity recording
	Global	Europe	
A	VU	LC	BBRC

The Snowy Owl is found in the Arctic regions of Eurasia and North America and winters to the south of its range, usually no closer to Britain than Iceland and southern Scandinavia (*HBW*). There were, on average, about three records each year over the last 30 years in Britain, the vast majority in Scotland, although irruptions farther south occur in some years. A total of 431 birds had been recorded in Britain up to 2019 (Holt *et al.* 2020), but only nine of these records, all involving single birds, were in Wales. Two records previously published in *Birds in Wales* (1994), in Carmarthenshire in winter 1902/03 and in Gwent in winter 1915/16, are now considered no longer acceptable following a review of historical records (Green 2020a).

- Bwlch, Breconshire, on 7 January 1947;
- Bishton, Gwent, on 23 December 1953 (recorded as 1952 by Venables *et al.* 2008);
- Valley, Anglesey, on 27 March 1959;
- immature bird at Penarth Moors, East Glamorgan, on 28 March 1972;
- a dead female at Mynachdy, Anglesey, on 3 May 1972;
- Abergavenny/Rhaglan area, Gwent, on 28 January 1976;
- Bardsey, Caernarfonshire, on 13 April 2001 and
- Twmbarlwm, Gwent, on 22 March 2018 that, based on documentation submitted, could not be linked to another individual seen at multiple locations (see below).

A remarkable series of sightings in 2018 included what was presumed to be a single second-calendar-year female across three counties, part of a small influx of at least 11 individuals into Britain that year. It was found at St Davids Head, Pembrokeshire, on 30 March where it stayed until at least 6 April. Several weeks later it was seen briefly on Skomer on 30 May. The next encounter was on Anglesey, near Amlwch, on 15–17 June and then at RSPB South Stack on 7–8 July. Three months later it was at Craig Ddrwg, Rhinogau, Meirionnydd, on 7–11 October. The records were the first for Pembrokeshire and for Meirionnydd.

Jon Green and Robin Sandham

Tawny Owl *Strix aluco* Tylluan Frech

Welsh List Category	IUCN Red List			Birds of Conservation Concern			
	Global	Europe	GB	UK	Wales		
A	LC	LC	NT	Amber	2002	2010	2016

The Tawny Owl has an extensive range across Europe, except Ireland and Iceland, and just into western Asia and northwest Africa (*HBW*). It occurs mainly in the lowlands, although in Scotland has been found up to 560m, and in the Alps, up to 1,600m. In Wales, it inhabits deciduous, mixed and coniferous woodland, farmland and parks with trees, large gardens and churchyards (*HBW*), but is largely absent from very exposed open landscapes devoid of trees, up to about 400m. There has been one notable exception. A nest was found on a rock ledge in the Carneddau, Caernarfonshire, at 560m (Driver 2010). Habitats likely to support breeding Tawny Owls occur over 26% of Wales' land area (Blackstock *et al.* 2010). Tawny Owl feeds on voles and mice using a 'sit and wait' technique. A dependence on small mammals, during winter and early spring, means that the Tawny Owl can be affected by vole population cycles, which causes breeding populations to fluctuate. It prefers to nest in holes in trees, but will nest in crevices in buildings, crags and sometimes the stick nests of other species (*European Atlas* 1997). It will use nest-boxes where natural sites are limited and has occasionally been known to nest on the ground, such as at the base of a tree (Iolo Williams *in litt.*).

The *Historical Atlas 1875–1900* suggested that the Tawny Owl was common across much of Wales, but less so in Gwent, Flintshire, Denbighshire, Caernarfonshire and on Anglesey. The species was heavily persecuted in the 19th century, because it predated gamebirds, young Rabbits and Brown Hares that were then of economic importance. It was also taken for the taxidermy trade, when owl exhibits were popular. Forrest (1907) said that the species was "resident and generally distributed in the wooded lowlands; most numerous in Montgomeryshire and Merioneth", but less so in Llŷn and on Anglesey. He stated that it "eats birds habitually" and was known to take young Pheasants from coops where they were being raised for shooting. Mathew (1894) had a different view of its diet, based on his observations in Pembrokeshire, saying that "although we had so many Tawny Owls in our plantation, we never missed any of our young Pheasants, and are certain that the Owls never molested them, confining themselves almost exclusively to the rats and mice". By the 1890s, although the Tawny Owl was considered a common and regularly breeding resident in Glamorgan, numbers were suppressed because of trapping. However, this moderated after the First World War and the species recovered, becoming numerous and widespread from the 1920s (Hurford and Lansdown 1995). In Ceredigion its numbers also increased substantially in the 1920s and 1930s, due to less persecution (Ingram *et al.* 1966).

The *Britain and Ireland Atlas 1968–72* showed evidence of breeding in 92% of 10-km squares in Wales. There was a slight reduction noticed in distribution in the *Britain and Ireland Atlas 1988–91*, with no records from 18% of squares and the highest relative abundance in Denbighshire, Flintshire, Radnorshire and Pembrokeshire. It had made a partial recovery, to occupy 87% of squares by the *Britain and Ireland Atlas 2007–11*. Relative breeding abundance was highest in Montgomeryshire and southwest Carmarthenshire. Winter distribution was similar to that in the breeding season, except on higher ground. However, unless evening visits are made, atlas surveys tend to underestimate the distribution and abundance of crepuscular and nocturnal species, so it is difficult to determine whether these changes are due to under-recording or are genuine.

Taking into account the acknowledged difficulties in surveying this and other owl species, at tetrad level the *North Wales Atlas 2008–12* found its breeding distribution to be patchy, but related to the distribution of areas of mature woodland, with breeding evidence in *c.*23% of tetrads. In the southern counties of Wales, the *Pembrokeshire Atlas 1984–88* found breeding evidence in *c.*42% of tetrads, although this had reduced to *c.*33% in 2003–07 (*Pembrokeshire Avifauna* 2020). The *East Glamorgan Atlas 1984–89* showed breeding evidence in 35% of tetrads but in only 13% in 2008–11. In Gower, the *West Glamorgan Atlas 1984–89*

Tawny Owl © John Hawkins

found breeding evidence in *c.*57% of tetrads. The *Gwent Atlas 1981–85* found evidence of breeding in a much higher proportion of tetrads, *c.*76%, although this had reduced to *c.*71% in 1998–2003.

In Carmarthenshire, Lloyd *et al.* (2015) thought that the species was more widespread in the early 2000s than previously, owing to mature coniferous forest growing on formerly open moorland. Using tape play-back, they reported 1–3 males/km^2 in mixed deciduous woodland and farmland. Roderick and Davis (2010) thought the Tawny Owl was widespread in Ceredigion for similar reasons. In Meirionnydd, the species was found in *c.*13% of tetrads, mainly in lowlands and foothills (Pritchard 2012). Five tetrads near Bangor, Caernarfonshire, each held an average of just over three pairs. Extrapolating this across the county's 96 occupied tetrads, Pritchard (2017) suggested a population of about 300 pairs in Caernarfonshire. In Montgomeryshire, Holt and Williams (2008) considered the Tawny Owl to be the commonest owl while in Radnorshire, where it was found in all 10-km squares, it was relatively scarce on treeless moorland (Jennings 2014).

The Tawny Owl is sedentary, with no movements of over 20km made by the relatively small number of birds ringed in Wales. A study in the mainly coniferous Clocaenog Forest, Denbighshire, found 3–9 active nests in nest-boxes across three 10-km squares and no movements over 5km during 1988–95 (Spence and Lloyd 1996). However, there have been examples of longer-distance movements into Wales, including one that was hit by a car at Pembrey Forest, Carmarthenshire, in November 1987. It had been ringed as a nestling at Camore Woods in the Scottish Highlands, 688km to the north, in May that year. Another, ringed as a nestling near Pewsey, Wiltshire, in May 1998, was found near Cardigan, Ceredigion, in January 2000, 218km WNW of its natal area (Robinson *et al.* 2020).

The BTO Tawny Owl Point Surveys (TOPS) assessed the proportion of tetrads in which Tawny Owls were recorded in 2018. This was based on visits to a series of selected tetrads that were also surveyed in 1989 and 2005 (Massimino and Hanmer 2019). This showed that there had been a decline in the proportion of tetrads across Britain with positive records in 2018 (53%), compared with the two previous surveys (65% in 2005 and 62% in 1989). The proportion of tetrads with positive records across BTO Welsh regions in 2018 was variable: at 20–40%, this was lowest in the southwest (Gower, Carmarthenshire, Pembrokeshire, Ceredigion) and in Montgomeryshire and on Anglesey; and at 60–80%, highest in parts of the southeast (Breconshire, part of Glamorgan and Gwent) and in North Wales (Caernarfonshire and Denbighshire).

Their nocturnal behaviour creates difficulties for assessing population levels, but an average of *c.*3 pairs/tetrad was estimated in a sample of five tetrads in Caernarfonshire in 2014–15 (Pritchard 2017) and 2–3 pairs/tetrad, in a sample of five tetrads with optimal woodland habitat, in Pembrokeshire during 2003–07 (Rees *et al.* 2009). Applying an average estimate of 2–3 pairs/tetrad to the whole of Wales, taking into consideration the amount of likely suitable habitat within *c.*4,900 'whole' tetrads (Spence 2016), gives an estimated population of 2,550–3,800 pairs. Judging by the TOPS, the population level might currently be at the lower end, but this species, like the other resident owl species, would benefit from a more dedicated survey across Wales.

There is no evidence that persecution has had a significant impact on its current population status, but it is not known if second-generation rodenticides could be having a localised impact. Its population is likely to be affected by food availability. An increase in tree cover in Wales could benefit the species, by providing additional nest-site opportunities and also creating more suitable habitat for its prey species. Apart from the fluctuations in vole numbers, the population of Tawny Owls is unlikely to suffer any major adverse influences, at least in the medium term. However, a study of Tawny Owls and prey cycles in Northumberland (Millon *et al.* 2014) has shown that a long-term decline and depression of density (dampening) of vole population cycles could have a deleterious impact on this and other vole-eating predators.

Ian M. Spence

Sponsored by Clare Ryland

Little Owl *Athene noctua* Tylluan Fach

Welsh List Category	IUCN Red List	
	Global	Europe
C1(1)	LC	LC

Although the Little Owl has a very wide distribution, extending throughout temperate Europe and Asia and locally in the Middle East and Africa north of the equator (*HBW*), the species is not native to Britain. While mainly nocturnal and crepuscular, it can be seen out in daylight more frequently than most of our owl species. It can be found in a variety of open habitats but tends to avoid dense woodland and seems to be more abundant in mixed farmland, where small parcels of farmland are divided by hedgerow and woodland-edge habitats (Toms 2014). It feeds on small mammals and birds, although earthworms and insects are an important part of its diet. Little Owls are not normally found at altitudes above 300m in Britain (Lever 2009). It nests in cavities and is mainly sedentary: ringing recoveries in Wales show only short-distance movements, usually less than 10km. A first-calendar-year female, ringed on Bardsey in October 1978, was recaught on the island ten years later in October 1988. Some longer-distance travellers may relate to pioneer colonists, such as one of unknown age ringed at Rhydycroesau, Denbighshire, in December 1981, that was recovered 60km to the southwest at Pantydwr, Radnorshire, in the following year (*BTO Migration Atlas* 2002).

The first recorded introduction of Little Owls into Britain was in 1842 or 1843, when some young birds purchased in Rome, Italy, were released on an estate in Yorkshire. These birds were not seen again. The first successful introduction may have been in Kent between 1874 and 1880, which led to an expanding population (Lever 2009). This was followed by other introductions, notably in Northamptonshire by Lord Lilford in 1888. Lever noted that away from one attempted introduction in the New Forest, Hampshire, in the 1860s, about 20 records before 1874 were probably vagrants from continental Europe. In Wales, there were a number of early records of 'Little Owl' or 'Little Night Owl' that may not have been this species. Pennant (1812), when visiting the Overton area of Denbighshire, stated that "I must not leave this neighbourhood without observing that the little owl, that rare English species, has been shot in some adjacent woods". He also mentioned in his *British Zoology* (Pennant 1812), that the Little Owl (*Strix passerina*) "is sometimes found in Yorkshire, Flintshire, and also near London". The reference to Flintshire is presumably the Overton bird. Forrest (1907) thought that this bird was more likely to have been a Tengmalm's Owl, which is not on the Welsh species list. Williams (1864) stated that "The Little Night Owl can only be considered an occasional visitor to this country… It would appear to have been seen at Gloddaeth". Another 'Little Owl' was mentioned by Dobie (1894) as having been caught in a wood near Gresford, Denbighshire, before 1870. This was identified as a Tengmalm's Owl, but Forrest thought it more likely to have been a Little Owl, as it was said to have been young and unable to fly when first caught.

The first dated record of Little Owl in Wales appears to be one killed at Merthyr Mawr, East Glamorgan, in the 1860s (Hurford and Lansdown 1995). Toms (2014) stated that some early records in Britain may well have been genuine immigrants, and the Merthyr Mawr bird might be one such. Others were seen at Caswell, Gower, in about 1880, at Ffrwdgrech, Breconshire, in 1890, and one was 'obtained' on Anglesey by a Pheasant-shooting party in winter 1899/1900. There were further records at Chepstow, Gwent, in

Little Owl © Ben Porter

1901, at Cilycwm, Carmarthenshire, in 1903, with a pair seen at Penmon, Anglesey, in 1909. In Radnorshire, Davies (1912) included Little Owl in a list of uncommon birds "known to have bred within the county", but the first confirmed breeding in Wales did not come until 1914, when a pair bred at Chepstow. The species is thought to have increased in Britain during the First World War, perhaps because many gamekeepers were away in the armed forces. In 1916 a pair nested at Nottage Court, East Glamorgan, and there was another breeding record in the county in 1918, when a pair bred in a Rabbit burrow at Merthyr Mawr. 1918 also saw the first confirmed breeding record in Radnorshire, when a pair bred at Knighton, and possible breeding in Ceredigion, although there is some confusion about early records in this county (Roderick and Davis 2010).

The species was found breeding at several sites in Pembrokeshire in 1920. It was also proved breeding in Meirionnydd and Breconshire in 1922, and in Carmarthenshire the following year. *Birds in Wales* (1994) stated that, by the early 1920s, it had become established as a breeding species in all counties except Caernarfonshire, where a pair bred in 1930, although Lever (2009) gave the date of the first confirmed breeding in Denbighshire as 1931. However, Lever contradicted himself by stating that by the 1920s, it occurred in every county in Britain south of the River Humber, except Caernarfonshire and north Denbighshire, implying that nesting had already occurred in the south of the county. Jones and Whalley (2004) stated that the first conclusive evidence of breeding on Anglesey was not until 1946. In most counties, the species seems to have been fairly widespread by the 1930s, but at the end of that decade it probably reached a peak. Numbers declined thereafter and it was remarked upon, in several counties, that the species suffered severely during the harsh winter of 1946/47, with local extinctions in some parts.

Lockley *et al.* (1949) recorded that the Little Owl was a common resident in Pembrokeshire, where it was the most abundant owl in 1927, but less numerous by 1946. Bruce Campbell (in North *et al.* 1949) referred to the species breeding in abandoned hill farms in Snowdonia. There was a recovery in the 1950s, although Roderick and Davis (2010) stated that a decline apparent in Ceredigion by 1954/55 had become very marked by the early 1960s. The severe winter of 1962/63 caused further losses in many areas. Jones and Dare (1976) said that there was some evidence of a decrease in Caernarfonshire in the previous 10–20 years.

The *Britain and Ireland Atlas 1968–72* gave the first complete picture of the distribution of the Little Owl in Wales, where it was recorded in 140 10-km squares. Most squares in Gwent and East Glamorgan were occupied, as were many on Llŷn and along the north coast, but elsewhere the species was sparsely distributed. The 1988–91 *Atlas* showed a small net loss, although there were gains in Ceredigion. The 2007–11 *Atlas* recorded it in 59 fewer 10-km squares compared with the 1968–72 *Atlas* and relative abundance was low. A survey of Anglesey in the late 1980s and early 1990s found up to 60 pairs (Jones and Whalley 2004). A Little Owl nest-box study in Montgomeryshire recorded the species at 20 different sites in 2019. Some of these sites were occupied by single birds and some by pairs, with confirmed breeding recorded at six sites. The remaining Little Owls left in Montgomeryshire were very much in "isolated pockets", in three separate areas of the county (Chris Griffiths *in litt.*).

Where tetrad atlases for the same area can be compared, there was a substantial drop in the population after the 1980s. In Gwent, breeding evidence was recorded in *c.*70% of tetrads in 1981–85 but in only 37% during 1998–2005. There was an even greater decline in East Glamorgan, where breeding evidence was found in 30% of tetrads in 1984–89, but in only *c.*2% during 2008–11. In Pembrokeshire, where the population is much smaller, breeding evidence declined from *c.*7% of tetrads in 1984–88 to just over 1% in 2003–07 (*Pembrokeshire Avifauna* 2020). Elsewhere, it was found in *c.*19% of tetrads in the *West Glamorgan 1984–89 Atlas* and in 7% in North Wales during 2008–12, concentrated in eastern Denbighshire, southeast Flintshire, Anglesey and Llŷn, with few records elsewhere. Brenchley *et al.* (2013) noted that individuals that had previously been conspicuous at known sites had become more elusive, so a change in behaviour may have contributed to the reduced recording rate during atlas fieldwork. Reports from several counties indicated that the decline continued, with birds lost from traditional breeding locations and it now appears to be very scarce in a number of counties. BirdTrack records since 2015 suggest there are large gaps in its distribution, but that there are still possibly stronger populations on Anglesey, in parts of northeast Wales and southeast Wales.

Little Owls breed on several of the Welsh islands. On Bardsey, Caernarfonshire, where the species was first recorded in 1952, seven territories in 2016 was the highest number since 1989, when the island's population peaked at eight territories. In Pembrokeshire, 2–4 pairs breed on Ramsey in most years, with five pairs there in 2013, while numbers on Skomer have declined from five pairs in 2006 to one pair between 2013 and 2016, and none has bred there since 2017. On Skokholm, there were small influxes of 5–10 between September and November in the early 1930s. One pair attempted to breed there up to 1954 but was discouraged as they were feeding on Storm Petrels. Nests were destroyed and adults caught and translocated, first to Marloes, on the nearby mainland, and when this proved not to be far enough, to Bath and London (Thompson 2008). A bird there, in March 2014, was the first on the island for 18 years. Little Owls have also been recorded as predators of Storm Petrels on Skomer (Green *et al.* 2005), although no Storm Petrel remains were found in pellets in a recent study (Compton *et al.* 2016).

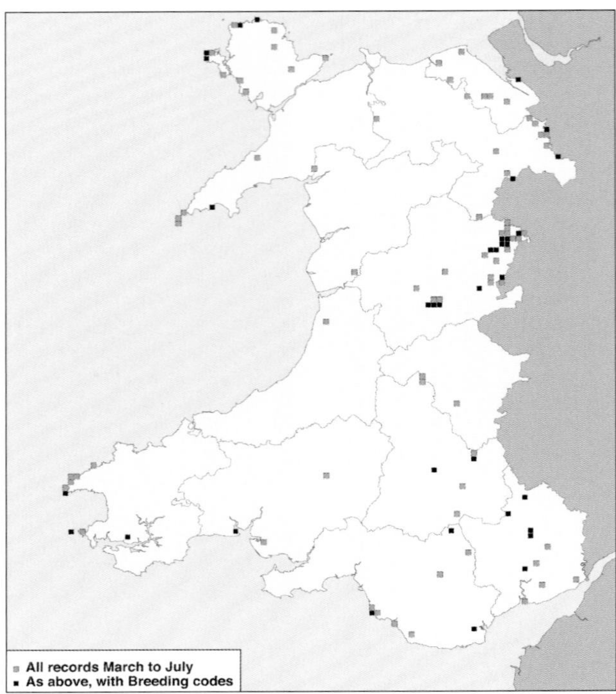

Little Owl: Records from BirdTrack and *Welsh Bird Reports* for March to July, 2015–19, showing the locations of records and those with breeding codes.

The Little Owl is found in too few BBS squares in Wales for a trend to be calculated, but CBC/BBS data show a steady decline across Britain since about 1985, measured at 62% during 1995–2019 (Harris *et al.* 2020). This is clearly a species in decline, but as it is a non-native species, there has been little research here to determine the causes. There is also evidence of a continuing population decline across its natural range in northern Europe and it has disappeared from parts of central Europe (*European Atlas* 2020). Work in other countries may provide clues to the reasons. In The Netherlands, Le Gouar *et al.* (2011) found that reduced survival of first-calendar-year birds was driving the decline, while in Denmark a fall in the number of young fledged/pair was apparently caused by lack of food. The provision of extra food led to more young being fledged (Thorup *et al.* 2010).

Birds in Wales (1994) estimated the Welsh population at around 2,000 pairs while *Birds in Wales* (2002) stated that the published records for 1992–2000 suggested a breeding population of 300 pairs, and that even this might be an over-estimate. However, the population in Gwent was estimated at 250–390 pairs during the second county atlas in 1998–2003 (Venables *et al.* 2008). Population estimates in the other Welsh counties are much smaller, however, and in view of the indications of continued decline, a current Welsh population of 250–350 pairs seems reasonable. A dedicated survey using tape-playback would be required for a more accurate estimate.

Although the Little Owl is a naturalised species in Wales, it seems to have filled a vacant ecological niche and, although its varied diet has included Storm Petrels on seabird islands such as Skomer, it has caused few other problems. Otherwise, it has been an interesting and generally welcomed addition to our avifauna. However, if the decline observed over the last 30 years continues, the Little Owl could well be lost from Wales.

Rhion Pritchard

Sponsored by Chris Griffiths

Long-eared Owl *Asio otus* Tylluan Gorniog

Welsh List Category	IUCN Red List			Birds of Conservation Concern				Rarity recording
	Global	Europe	GB	UK	Wales			
A	LC	LC	LC	Green	2002	2010	2016	RBBP

The Long-eared Owl is found in temperate habitats across the Northern Hemisphere (*HBW*) and is the most nocturnal and elusive of the owls that breed in Wales. This species is equally at home in scattered thorn bushes in the warm, sunny plains of Spain as in the wild, wet uplands of Wales. Its preferred habitat, here, is upland forest adjacent to rough pasture, moorland and young restocked plantation, a not uncommon habitat in Wales usually found in remote areas. It nests at, or close to, the edge of woodland where it can lay eggs in the old nests of Carrion Crow or Magpie, or on old squirrel dreys (Toms 2014). It forages over the adjacent open moorland after dark for Field Voles, Wood Mice and small birds, quartering the ground in a similar manner to Short-eared Owls. In some areas they favour the more open structure of Lodgepole Pine to nest (Hatch 2006), but will readily use Sitka Spruce and European Larch.

Long-eared Owl © John Hawkins

In the breeding season the territorial call of the male, a repetitive, single-note, soft 'hoot', although described as audible at up to a kilometre (Scott 1997, Toms 2014), can be difficult to hear over 200m away (Hatch 2006), particularly in windy mountain localities. Males can start hooting at dusk or some considerable time after dark, making the timing of visits to locate birds difficult (Hatch 2006). If a female is present prior to, or during, laying she can often be heard giving a soft sighing call—described as like blowing a comb and paper—from close to or on the nest, but only audible at about 50m. During display both birds perform distinctive and quite far-carrying wing-claps, the wings meeting below the body (Scott 1997). Vocal activity falls away considerably once eggs are laid, which can be any time from early March, so attempts to locate Long-eared Owls in late March and early April, by listening for calls, may already have missed the optimal time.

Most Long-eared Owls are discovered once the young have fledged. At this time, after dark, the hungry juveniles call persistently, a penetrating 'squeaking gate' call, likened to an unoiled hinge, that can be heard up to a kilometre away, for a month or two after fledging (Scott 1997). However, this accounts only for successful nests. Owing to the large variation in laying and fledging dates, several visits through May and June may be required to ascertain presence. Consequently, the likelihood of under-recording is high and any attempts at population estimates must bear this in mind. The *Britain and Ireland Atlas 1968–72* gave an estimate of 3,000–10,000 pairs while the 1988–91 *Atlas* estimated a population of 2,200–7,200 pairs across Britain and Ireland. *Birds in Wales* (1994) gave an estimate, albeit tentative, of probably fewer than 30 pairs in Wales, which is thought to be much too low in light of more recent knowledge.

During the second half of the 19th century, the Long-eared Owl was considered by many authors to be widespread and locally common in Wales. For example, Dobie (1894) stated that it was widespread around Colwyn Bay, Denbighshire, while Forrest (1907) recorded pairs nesting in all counties of North Wales, except Anglesey. Salter (1900) considered it a fairly numerous resident in parts of Ceredigion, while Barker (1905) reported it as not uncommon in Carmarthenshire. Mathew (1894) regarded it as a scarce winter visitor to Pembrokeshire and reported that "bird-stuffers" received a few occasionally. In later years, Lockley *et al.* (1949) considered it a scarce resident and noted breeding near Granant, at Solva, near St Dogmaels, and at Dowrog, near St Davids. The last breeding record was in 1935 (Donovan and Rees 1994).

In South Wales, the Long-eared Owl population declined between 1900 and 1950, coinciding with the recovery of Tawny Owls in Wales from historic persecution. Like the Tawny Owl, the Long-eared Owl suffered from pole-trapping by gamekeepers and from egg-collecting in Pembrokeshire, Denbighshire and Caernarfonshire. The late Gordon Ireson located nine nests in Radnorshire between 1952 and 1965. *Birds in Wales* (1994) stated that it was surprising, considering the proliferation of coniferous plantation throughout much of upland Wales, that breeding numbers of Long-eared Owl had not increased. Maturing upland conifer plantations may well have proved attractive and ideal habitat for the Long-eared Owl. However, because of the remoteness of the areas concerned, and difficulties associated with locating this species due to its erratic and very nocturnal behaviour, this could have led to under-recording.

Lending weight to this notion are illuminating studies in the upland coniferous plantations of Gwent (Hatch 2006). Here, Long-eared Owls bred near Abergavenny and Chepstow around 1900 (Ingram and Salmon 1939), but there was no further confirmed breeding evidence until 1992, when three young were raised at a nest on the edge of a remote forestry plantation in the Black Mountains. In 1993, two nests were at this location and breeding was recorded at two other sites. The Black Mountains Forest held two pairs in 1995 and three in 1998 (Venables *et al.* 2008). By undertaking a concerted effort to locate Long-eared Owls, Hatch found seven breeding pairs within a radius of 8km where no breeding Long-eared Owls had been previously recorded and, in 2005, found 11 breeding pairs in a wider area of Gwent, all in upland coniferous plantation above 350m. He considered that with so much similar habitat available, there could readily be up to 30 pairs in the county. This habitat is replicated over many parts of Wales.

The *North Wales Atlas 2008–12* confirmed breeding in only three widely dispersed tetrads during 2008–12, with breeding possible in another five squares, across an area of 6,307km^2. The usual daytime atlas survey fieldwork methods do not cover highly nocturnal species particularly well, and with many large upland coniferous plantations bordering moorland and rough grassland, it is likely that the true population of this secretive owl in North Wales will have been under-recorded.

Many breeding pairs of Long-eared Owls are sedentary. Hatch (2006) found birds at their upland breeding sites in the harshest of winter weather, when many Long-eared Owls also visit Britain from continental Europe, particularly from Fennoscandia. Large numbers of these rarely reach Wales, although influxes were recorded on Bardsey in 1975 and 1989, with up to six in a single day, such as on 6 December 1989. A substantial winter roost of up to 12 was present on the Wentlooge Levels, East Glamorgan, in 1990. The roost moved to the Rhymney Estuary in 1991/92, when 18 were present (*Birds in Wales* 1994). During January to March 1998, at least 12 roosted in a tall hawthorn hedge adjacent to set-aside at Clytha, Gwent, where it was reported that birds had been present in previous years. Six were ringed and one was found dead at Usk, Gwent, two months later (Venables *et al.* 2008).

The farthest distance travelled by a Welsh Long-eared Owl was recorded when one ringed as a nestling in Blaenafon, Gwent, on 12 May 2005, was found unwell in Oxford on 14 May 2006, 124km to the east of its natal area. A Long-eared Owl ringed in Hanko, Finland, on 25 September 1997 was found dead at Chepstow, Gwent, on 7 February 1999.

Recording area	*Total Pairs 2014–18
Gwent	11
East Glamorgan	2
Ceredigion	2
Breconshire	1
Radnorshire	8
Montgomeryshire	3
Meirionnydd	1
Anglesey	1
Denbighshire	2
Flintshire	1
Total	**32**

Long-eared Owl: Population estimates in Wales, 2014–18, from *Welsh Bird Reports* and RBBP data.
* The highest value for the total number of confirmed, probable and possible breeding pairs, in any of the five years.

The Long-eared Owl is likely to be one of the UK's most under-recorded breeding species (Eaton *et al.* 2020). The data presented here is therefore almost certainly an underestimate of the population. A paucity of records in county bird reports is likely due to a combination of difficulties associated with observing this nocturnal species and low observer effort in many areas. Gwent (where effort has increased in recent years) was the only county to record a breeding population in double figures, with 11 pairs in 2014. A maximum of only 15 pairs were reported in Wales in 2017, of which four were confirmed to have bred. In 2018, confirmed breeding was reported from three sites in Gwent, from five in Radnorshire and from single sites in Breconshire, Ceredigion and Denbighshire. Breeding also occurred on Anglesey, where a family group in a plantation near Holyhead in May was the first confirmed breeding on the island since 1978. Hatch (2006) considered that Long-eared Owls are likely to be present in all mature upland coniferous forest in Wales adjacent to suitable hunting habitat. With so much un-surveyed potential habitat, a figure of 100+ pairs would seem possible, but a Wales-wide survey is needed to confirm that.

Although secretive and elusive, Long-eared Owls are quite tolerant of disturbance. They sit very tightly on nests and, whilst roosting, close to the trunk of a tree. They are difficult to see and reluctant to fly. At some regularly successful nest sites in Gwent there is considerable disturbance from dog-walkers, vehicles, and people camping and lighting fires. Of greater concern are forestry operations that can result in nesting trees being felled, because forest planners were unaware of the owls' presence. This happened at one site in Gwent but now there is liaison between Natural Resources Wales forest planners and fieldworkers. All prospective felling of forest edge adjacent to suitable hunting habitat for Long-eared Owl should include an environmental impact assessment for this species. The Goshawk population in Wales, burgeoning as forests mature, has moved closer to breeding Long-eared Owls and may yet prove detrimental. Goshawks are predators of owls and other birds of prey (Kenward 2006). The maintenance of good hunting areas is essential for the owls to be able to obtain small mammals. A reduction in sheep numbers in areas such as in the Black Mountains, Breconshire, is commensurate with the improvement in the quality of the heather and vegetation, and should benefit prey for owls and raptors.

Stephen Roberts

Short-eared Owl *Asio flammeus* Tylluan Glustiog

Welsh List Category	IUCN Red List			Birds of Conservation Concern				Rarity recording
	Global	Europe	GB	UK	Wales			
A	LC	LC	EN	Amber	2002	2010	2016	RBBP

The Short-eared Owl has an extensive circumpolar range, occurring on all continents except Antarctica and Australasia (*HBW*). This is a ground-nesting species, favouring a range of open habitats, including bogs and marshes, wet or dry grassland, moorland, damp woodland and forest clear-fell, but generally avoiding cultivated land in the breeding season (*European Atlas* 1997). BirdLife International (2015) has estimated a population of 55,000–186,000 pairs in Europe, the majority (88%) in Russia, which accounts for *c.*9–16% of the global population. Toms (2014) suggested a British breeding population of 750–3,500 pairs, with the bulk of the population in Scotland and northern England. Across Europe, its numbers fluctuate, including within the UK, but there are trends of decreasing breeding populations, particularly evident in some eastern and northern European countries (*European Atlas* 2020). However, Short-eared Owls are extremely difficult to census, resulting in relatively broad population estimates (Calladine *et al.* 2007). They are probably greatly under-recorded in Wales. This is because much of their breeding habitat is in remote areas, although small numbers do breed on offshore islands and lowland coastal sites. Also, populations and breeding success vary annually, linked to the abundance of prey species such as Field Vole, while the species' nomadic nature may not be synchronised in response to vole abundance.

During the 19th century the Short-eared Owl was already a scarce breeding bird in Wales, confined mainly to the uplands, coastal marshes and dunes of Anglesey (*Birds in Wales* 1994). It nested in Ceredigion from about 1874, according to a letter sent to *The Field* (cited by Roderick and Davis 2010), and Forrest (1907) mentioned several nests on both the Ceredigion and Montgomeryshire side of Pumlumon in the late 19th century. In Pembrokeshire, Mathew (1894) considered it to be common in the winter but noted that eggs had been taken from Skomer; half-grown young were seen there in 1895 (Benoit *et al.* 1958). Breeding was also recorded from Gower, near Penclawdd and in dunes at Margam burrows, during the 19th century (Williams 1989). Breeding in Wales may have become more frequent in the latter part of the 20th century, with local increases in upland areas probably a response to the expansion in afforestation after the Second World War. The early stages of conifer growth may have led to higher densities of preferred prey such as voles. *Birds in Wales* (1994) estimated 20–23 breeding pairs during 1988–92, although this was thought to be the minimum population. *Birds in Wales* (2002) highlighted the significance of four upland areas (Abergwesyn, Elenydd, Llanbrynmair and Berwyn) and Skomer. An unprecedented minimum of 12–13 pairs reared 32 young on Skomer in 1993, significantly contributing to a minimum population estimate of at least 18 pairs in Wales.

Between 1968–72 and 2008–11, the breeding range contracted northwards, a trend consistent with climatic modelling by Huntley *et al.* (2007), which suggested that the Welsh climate will not be suitable for Short-eared Owl later this century. It is increasingly restricted to the uplands of northern England, mainland Scotland, the Outer Hebrides and Orkney. In Wales, the Short-eared Owl no longer nests in coastal areas, apart from the Pembrokeshire islands of Skomer and Ramsey, and is a scarce breeding species elsewhere. On Skokholm, Pembrokeshire, a pair bred successfully in 2017—the first confirmed breeding record for the island. Storm Petrels formed an important part of their diet, with the remains of more than 90 individuals recorded that year on this vole-free island (Brown and Eagle 2018b). The timing of breeding of Short-eared Owls in Britain, including Wales, varies, probably in relation to vole abundance, and in some years, birds may not breed, even if holding a territory. Short-eared Owl population densities in an area can thus vary substantially between years (Calladine *et al.* 2007). Since 2000, monitoring illustrated the cyclical nature of their breeding populations, although observer effort also varies. The 18 pairs in 2012 followed a large autumn influx to Wales the previous year when, for example, 24 were recorded on Bardsey in October, including a single flock of 15. The uplands of Radnorshire, Montgomeryshire, Meirionnydd and Denbighshire, and the Pembrokeshire islands, hold almost the entire Welsh population, which in 2013–17 was estimated to be 23 breeding pairs (Holling *et al.* 2019). County estimates suggested a population of around 20 pairs in 2018. The total was less than half of this in 2019, although the decline was in line with general population fluctuations recorded over the years.

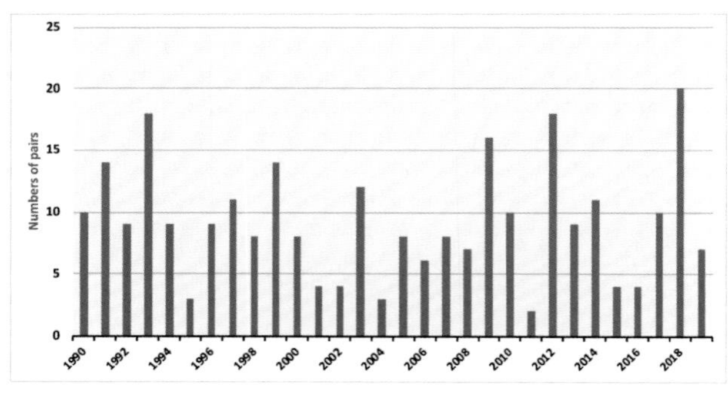

Short-eared Owl: Minimum numbers of pairs in Wales, each breeding season, 1990–2019.

Short-eared Owl © Ben Porter

Most winter Short-eared Owl records are in lowlands habitats, such as estuaries, coastal dunes and inland marshes, with influxes to Britain from Fennoscandia and Iceland boosting resident populations. Examples include at least 40 on Ramsey, Pembrokeshire, during a cold spell in January 1982 (Donovan and Rees 1994) and up to 20 in west Anglesey in December 2019, which was the highest there for many years and was not obviously weather-related. Some coastal sites in Flintshire, Anglesey, Pembrokeshire, East Glamorgan and Gwent are occupied every winter, but birds are recorded in most counties annually, demonstrating their highly nomadic nature in response to local vole availability. Ringing recoveries in South Wales indicate the origins of some that had been ringed as nestlings in Scotland, Derbyshire and Norfolk. The most distant movement was of one, ringed as an adult in Aberdeenshire in September 1988, that was killed by a car near Cardiff, East Glamorgan, one month later, 648km south of its ringing location.

The wintering movements of British-bred birds can be extensive, with several having migrated from England and Scotland to North Africa. Satellite tracking by the BTO has provided evidence of one female that nested in Scotland and then in Norway in the same season, illustrating its ability to follow the food resource to maximise breeding success. Such research may help us to better understand the extent of this species' breeding and non-breeding movements. There is a need for more accurate population estimates for the species nationally and within designated Special Protection Areas.

Understanding the fluctuating numbers and movements within and between breeding seasons is challenging because of the species' nomadic behaviour. Owing to declines in the size of the breeding population and its breeding range, the Short-eared Owl remains on the Red List of species of conservation concern in Wales (Johnstone and Bladwell 2016). There is an urgent need to understand the integration of Welsh, British and other European populations, since the future of this species does not, currently, look very secure.

Patrick Lindley

Sponsored by Iolo Williams

Hoopoe *Upupa epops* Copog

Welsh List Category	IUCN Red List	
	Global	Europe
A	LC	LC

The Hoopoe breeds widely throughout warm and temperate regions of Eurasia and Africa, except in the desert regions. Some European birds over-winter in the Mediterranean basin and North Africa, but most migrate to sub-Saharan Africa (*HBW*). The first Welsh record was of one shot at Penyrhiw, Fishguard, Pembrokeshire, in 1811 (Fenton 1811). It has been a scarce migrant to Wales throughout the last two centuries. There has been no indication of any change in the frequency of records, either in Wales or across Britain, for which the annual total averaged 121 each year during 2010–18 (White and Kehoe 2020a). More than one-third of Welsh records have come from Pembrokeshire. The majority are overshooting migrant birds in April and May, although Hoopoes have been recorded during every month of the year. There have been a few December records during the last three decades and a much earlier record of one that over-wintered on Tenby Golf Course, Pembrokeshire, between 14 December 1963 and March 1964.

Recording area	First record	Pre-1967	1967–89	1990–99	2000–09	2010–19	Total
Gwent	Peterstone Wentlooge, in early July 1934	8	11	6	5	3	33
East Glamorgan	Southerndown, August 1836	17	16	4	1	4	42
Gower	Burrows Lodge, Swansea, 7 April 1840	17	12	3	3	5	40
Carmarthenshire	Ferryside, before 1909	11	17	9	8	6	51
Pembrokeshire (excl. Skomer & Skokholm)	Penyrhiw, Fishguard, 1811	40	49	52	15	30	186
Skomer	spring 1957	4	4	4	3	3	18
Skokholm	3 May 1928	22	7	10	2	5	46
Ceredigion	Abermâd, about 1889	8	18	8	10	9	53
Breconshire	Llangorse, before 1899	13	8	2	2	2	27
Radnorshire	near Knighton, about 1901	13	1	4	3	5	26
Montgomeryshire	Aberhafesp, about 1870	4	5	4	2	3	18
Meirionnydd	Garthgell, Dolgellau, April 1901	5	14	1	1	0	21
Caernarfonshire (excl. Bardsey)	Llithfaen, 10 August 1951	7	11	7	6	7	38
Bardsey	23 May 1957	8	18	2	7	3	38
Anglesey	Llyn Maelog, 1898	11	8	2	10	5	36
Denbighshire	Coed Coch, about 1878	2	2	4	2	6	16
Flintshire	Nannerch, 1812	5	2	2	1	2	12
Total		**195**	**203**	**124**	**81**	**98**	**701**

Hoopoe: Locations and dates of first records, and approximate totals of individuals in each recording area in Wales, 1811–2019, from county avifauna and WBRC records.

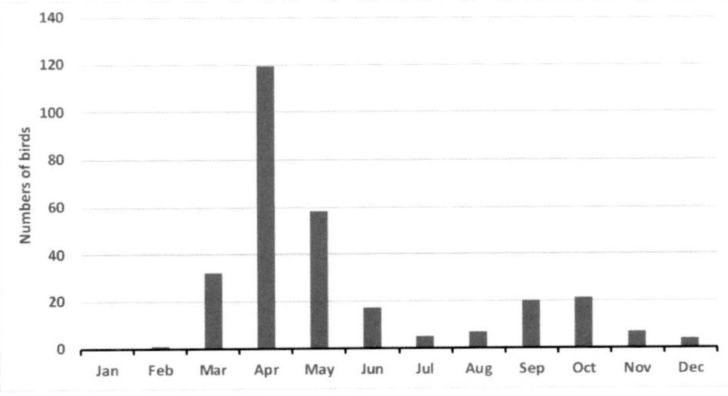

Hoopoe: Totals by month of arrival, 1990–2019.

Hoopoes have bred sporadically in England but only once in Wales. A pair bred on a farm between Pontrobert and Dolanog, Montgomeryshire, in 1996. They nested in an old stone building and reared three young that were fed on worms from the farm manure heap (Holt 1997). A bird returned the following year, but did not stay more than a few days (Holt and Williams 2008).

We should expect to hear more of its distinctive call in Wales: the population in Europe increased by 244% between 1980 and 2016 (PECBMS 2020) and it is likely that climate change will cause the Hoopoe's breeding range to move northward in the coming decades. The climate in the southern half of Wales will be suitable for the species to breed by the end of this century (Huntley *et al.* 2007).

Jon Green and Robin Sandham

Roller *Coracias garrulus* Rholydd

Welsh List Category	IUCN Red List		Rarity recording
	Global	Europe	
A	LC	LC	BBRC

The Roller breeds around the Mediterranean, central Europe, the Middle East and Central Asia, and winters in tropical and southern Africa (*HBW*). A total of 301 had been recorded in Britain up to 2019 (Holt *et al.* 2020), with an average of one or two a year over the last 30 years. It is a very rare bird in Wales, with only 11 records of single birds, three of which were in the 19th century and only four in the last 30 years.

- shot at Holywell, Flintshire, in 1857 (date unknown, possibly in September or August);
- Abergele, Denbighshire, on 19 October 1874;
- Nant-y-glyn Valley, near Colwyn Bay, Denbighshire, on 7 October 1897;
- Penygarreg Reservoir, Radnorshire, on 12 July 1962;
- Pentrebach, near Llandovery, Carmarthenshire, on 4–6 July 1965;
- Tal-y-bont, Ceredigion, on 7–8 June 1970;
- Michaelston-y-Fedw, Castleton, Gwent, on 30 September 1987;
- Panteg, near St Nicholas, Pembrokeshire, on 2–11 August 1991;
- 1CY Skokholm, Pembrokeshire, on 26 October 2001;
- St Davids, Pembrokeshire, on 1–16 July 2005 and
- 2CY at Usk Reservoir, Breconshire/Carmarthenshire, on 29–31 July 2007, which moved to Bryn Common, Gower, on 6–7 August 2007.

The areas of Europe suitable for Roller are expected to shift northwards during this century, and by 2100 may be as close as Brittany, France (Huntley *et al.* 2007). If land management is sympathetic, it may occur more frequently in Wales as an over-shooting spring migrant.

Jon Green and Robin Sandham

Kingfisher *Alcedo atthis* Glas y Dorlan

Welsh List Category	IUCN Red List			Birds of Conservation Concern			
	Global	Europe	GB	UK	Wales		
A	LC	VU	LC	Amber	2002	2010	2016

The Kingfisher is widespread throughout Europe, northwest Africa, and temperate and tropical parts of Asia and Australasia, as far south as Papua New Guinea. Several subspecies are recognised. In Wales we have *A.a. ispida* (IOC) which occurs in Britain and Ireland and over much of Europe, excluding the Mediterranean basin, where it is replaced by the nominate form *A.a. athis*. Birds breeding in the northern part of the range move south in winter, but those farther south are generally resident (*HBW*). First-calendar-year birds disperse after fledging and, although most do not move far, some have been recorded moving considerable distances (*BTO Migration Atlas* 2002). A first-calendar-year bird ringed at Mullock, Pembrokeshire, on 14 August 1993, was caught by a ringer 967km away at Irun on the north coast of Spain, 19 days later, the longest Kingfisher movement recorded from Britain. Another first-calendar-year bird ringed near Newtown, Montgomeryshire, in August 1974, was found dead near Calvados, France, 430km away, in October the same year. One ringed as a nestling on Île d'Ouessant, France, in July 1970 was found dead at Abersoch, Caernarfonshire, in March 1972, having moved 487km.

Kingfishers depend on open water, which needs to be clear and preferably slow flowing or still. They feed primarily on small fish, up to about 125mm long, but also on aquatic invertebrates (*HBW*). Nest locations in earth banks mean that sandy or clay soil is necessary for the excavation of nest-burrows. In Wales, they are scarce at altitudes over 250m, but have been recorded nesting at up to 305m in north Montgomeryshire. The highest densities recorded in Wales are on large rivers, such as the Usk, Wye, Severn and Dee. The species is scarcer where rivers are small, such as on Anglesey, or are fast flowing, such as in Caernarfonshire. In winter, many move to the coast or to large lakes.

The *Swansea Guide* (Anon. 1802) noted the presence of the "King's Fisher". Mathew (1894) stated that the species was a common resident in Pembrokeshire. Forrest (1907) described it as generally distributed in the lowlands of North Wales, most common in Montgomeryshire, but said that it was principally a winter visitor to Anglesey and Llŷn. In Glamorgan, Hurford and Lansdown (1995) considered that the species was a common, regularly breeding resident until about 1925, when a gradual decline began due to pollution, habitat destruction and disturbance.

Kingfishers can be severely affected by very cold weather, particularly in harsh winters such as those of 1946/47, 1961/62 and 1962/63. Lockley *et al.* (1949) said that in Pembrokeshire many had died in 1946/47 and that Kingfishers had been subsequently very scarce. The winter of 1962/63 may have caused even higher mortality. On average, numbers were 85% lower in 1963 than the previous year, although the severity of the decline varied by area. In Breconshire, there were just two breeding pairs in 1963, compared to an estimated 46 the previous year. However, the species can recover quickly from such losses. Survival was apparently better in southwest Wales and elsewhere some recovery was already evident by 1964. By 1969 the population was thought to be much the same as before 1962/63 (Smith 1969).

The effect of cold winters needs to be considered when comparing the results of atlases. The *Britain and Ireland Atlas*

Kingfisher © Jerry Moore

1968–72 showed the Kingfisher to be widespread in Wales, absent only from the highest ground and northern Anglesey. Around 25% fewer 10-km squares were occupied in the 1988–91 *Atlas*, the losses most evident in northwest Wales and the southern coastal lowlands. This was perhaps a result of the very cold winter of 1981/82 followed by several other cold winters in the mid-1980s (Kington 2010). That is illustrated by the *Britain and Ireland Winter Atlas 1981–84*, which found Kingfisher in just 36% of 10-km squares, compared to 66% in the breeding seasons of 1968–72 and 49% in 1988–91. The 2007–11 *Atlas* showed gains in the northwest and south that were offset by losses in mid-Wales. 53% of Welsh squares showed some evidence of breeding, which was 20% fewer than in 1968–72, and there were fewer records of confirmed breeding. Relative abundance was highest in the east and south of Wales and lowest in the northwest. In winters 2007–11, the highest relative abundance was along the south coast.

Tetrad atlases showed considerable regional variation within Wales. In the *Gwent Atlas 1981–85*, breeding evidence for Kingfishers was found in *c.*25% of tetrads, while in the 1998–2003 *Atlas* they were found in *c.*30%. In East Glamorgan, they were found in 15% of tetrads in 1984–89. The proportion was similar in 2008–11, but there was breeding evidence in just 9% of tetrads. There was also a drop in the number of occupied tetrads in Pembrokeshire, from *c.*10% in 1984–88 to *c.*7% in 2003–07 (*Pembrokeshire Avifauna* 2020). Kingfishers were recorded in 9% of tetrads in the *West Glamorgan Atlas 1984–89* and in 6% of those visited in Breconshire in 1988–90 (Peers and Shrubb 1990). The *North Wales Atlas 2008–12* found Kingfishers in *c.*5% of tetrads, the majority in the east, along the River Dee and its tributaries.

In the breeding season, Kingfishers are monitored by the UK BBS and by the Waterways Breeding Bird Survey (WBBS), but there are too few sites for an index to be calculated for Wales. Information from WBBS and its predecessor, the Waterways Bird Survey (WBS), shows a dramatic decline in 1975–82, before a steady recovery, then another decline from 2007 and a slow recovery subsequently.

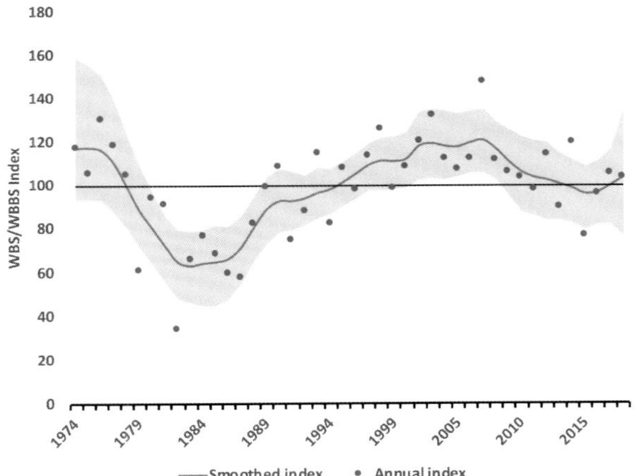

Kingfisher: UK Waterways Bird Survey / Waterways Breeding Bird Survey indices, 1974–2018. The shaded area shows the 85% confidence limits.

The Kingfisher is also included in the Wetland Bird Survey (WeBS), which has assessed numbers of non-breeding birds since the early 1990s. The annual index in Wales shows that Kingfisher numbers fluctuate, but have remained relatively stable overall, although the numbers involved are small. The highest WeBS count at a site in Wales was seven at the Cleddau Estuary, Pembrokeshire, in December 1995.

Across the past century, while there have been widespread records of Kingfishers in suitable habitat during the breeding season, records of confirmed breeding are scarce. The presence of pairs is still the best indicator of the distribution of the population. Aside from weather-related fluctuations in the population,

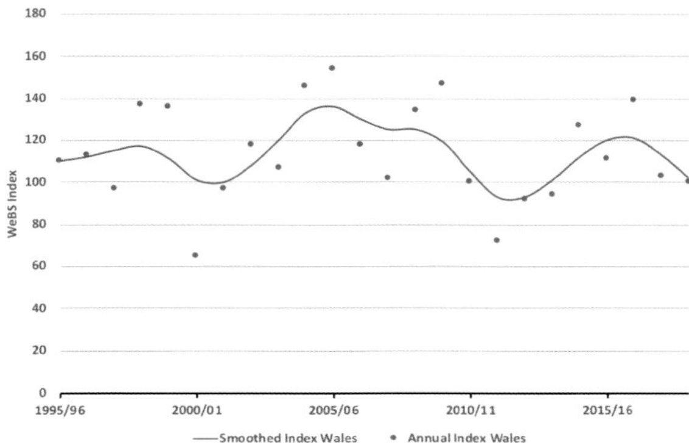

Kingfisher: Wetland Bird Survey indices, 1995/96 to 2018/19.

there are indications of genuine long-term changes in some areas. For example, there has been a marked decline in Meirionnydd, south of the Afon Mawddach (*North Wales Atlas 2008–12*), but on Anglesey, there appears to have been an increase. *Birds in Wales* (1994) described the species as absent from the island. Jones and Whalley (2004) said that it bred only sporadically there, with very few records of confirmed breeding over the previous century, which they suggested could be because of the absence of the Common Minnow, an important prey item elsewhere. However, Kingfishers were confirmed breeding in four tetrads on Anglesey in the *North Wales Atlas 2008–12*. Although it was apparently unknown to Jones and Whalley, the Common Minnow seems to have been present on the island for some time. There is a commonly held view that it was accidently introduced by anglers using it as live bait to catch trout. Records held by Cofnod suggest that the Minnow has been on Anglesey since at least the 1960s. Its current distribution seems very patchy and to follow the larger lakes and reservoirs on the island (Huw Jones, NRW *in litt.*).

Any reasonably accurate estimate of the size of the population in Wales is dependent on surveys. Round and Moss (1984) reported on surveys in 1977–78 that found 48 territories on the Wye (mean density of 2.14 territories/10km of river), 29 territories on the Severn (2.23/10km) and 16 territories on the Vyrnwy (2.67/10km). These figures probably represent typical densities in good habitat at a time of high numbers. Two 10km sections of the River Severn in Montgomeryshire held six territories in 1978, but none in 1982, following a cold winter (Crosby 1982). On smaller and faster-flowing rivers, the Kingfisher is likely to be much scarcer or absent: a survey of the entire Afon Ogwen, Caernarfonshire, in 2005–07, found no territories (Gibbs *et al.* 2011).

Birds in Wales (1994) suggested that *c.*400 breeding pairs were probably the maximum in a good year in Wales. The most recent estimate, by Hughes *et al.* (2020), suggested a range of 445–970 pairs in 2018, but they acknowledge difficulties in reliably estimating the population of this species. Winter records have only been described in detail in *Welsh Bird Reports* in the last decade, with most records coming from Gower, Carmarthenshire, Caernarfonshire and Anglesey.

Population trends over most of the Kingfisher's extremely large range are unknown, but there has been a decline in Europe, thought mainly to be a result of chemical and biological river pollution (BirdLife International 2020). The Kingfisher is Amber-listed in Wales because of its Vulnerable status in Europe (Johnstone and Bladwell 2016). Freezing winter conditions currently outweigh other threats to Kingfisher survival in Wales, since periods of extreme low temperatures reduce survival both directly from the cold and through ice cover making it difficult to find food. High, silt-laden water levels from prolonged rainfall events can impact feeding and flood nests, affecting productivity and subsequently recruitment into the local population. Other threats to Kingfishers include pollution and disturbance of the nest by both humans and machinery working on riverbanks. An all-Wales survey is long overdue.

Katharine Bowgen

Bee-eater *Merops apiaster* Gwybedog Gwenyn

Welsh List Category	IUCN Red List		Rarity recording
	Global	Europe	
A	LC	LC	WBRC

The Bee-eater breeds in southern Europe, northwest Africa, the Middle East and Central Asia. European birds winter in the Sahel zone in West Africa. Another population is resident in southern Africa (*HBW*). The species is a scarce migrant visitor to Britain, usually as a spring overshoot, although some have occasionally remained to breed in England. The number of records in Britain has increased, with the annual mean in 2010–18, 73 records, almost double that of 1990–99 (White and Kehoe 2020a). Sightings in Wales have mirrored this trend, although the species remains a scarce visitor here.

Total	Pre-1950	1950–59	1960–69	1970–79	1980–89	1990–99	2000–09	2010–19
67	5	3	1	8	11	5	14	20

Bee-eater: Individuals recorded in Wales by decade.

Bee-eater on Llŷn, Caernarfonshire © Ben Porter

The first Welsh record was of one killed at Johnston, Pembrokeshire, in about 1854. There was one other 19th-century record from that county: three at Milford Haven on 13 May 1896. It was 50 years before the next Welsh record, at Moelfryn above the Wye Valley, near Rhayader, Radnorshire, on 3–4 August 1949. There were only five records during the next three decades: one at Borth y Gest, Caernarfonshire, in 1955; two at Dale Fort, Pembrokeshire, in May 1958; one near St Athan, East Glamorgan, in July 1965; a party of seven near Mumbles, Gower, in June 1973; and one at Aberporth, Ceredigion, in May 1978.

Having increased in the 1980s, numbers dipped in the 1990s, but have gradually increased since. There were records in most years from 2010 onwards. The majority were in May and June, involving coastal counties, with only a few in autumn. Most records have been of single birds but, in addition to the party of seven in Gower in 1973, other groups have included: four at RSPB South Stack, Anglesey, on 11 July 1987; five at Llanrhaeadr ym Mochnant, Montgomeryshire/Denbighshire, on 27 May 2007; and three at St Davids Head, Pembrokeshire, on 17 May 2014.

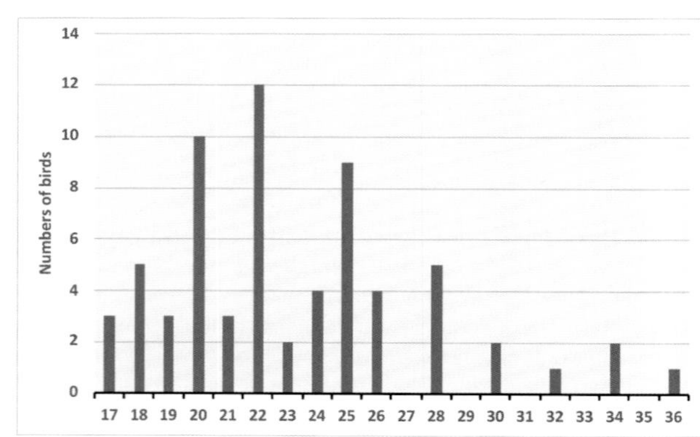

Bee-eater: Totals by week of arrival, 1896–2019. Week 18 begins *c.*30 April and week 28 begins *c.*9 July.

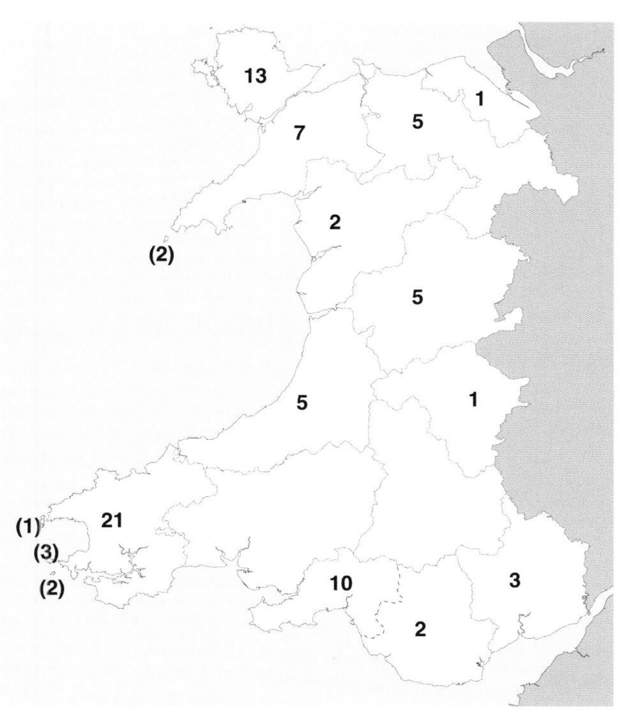

Of *c.*14 records in Pembrokeshire, about half of these were around the Dale–Marloes and Skomer–Skokholm areas. Four of the five in Ceredigion were at Aberporth–Blaenporth, while on Anglesey five birds were recorded at RSPB South Stack and three in the Cemlyn area.

The distribution of the Bee-eater has changed greatly since the *European Atlas* (1997). At the northern edge it has expanded strongly, while in some southern countries it seems to be locally in contraction. These overall changes in distribution are consistent with known population trends (*European Atlas* 2020). It seems likely that the species' occurrence in Wales will increase if its European breeding range moves north to the English Channel coast, as projected by Huntley *et al.* (2007). Breeding in Wales is also an increased possibility: there are plenty of soft coastal cliffs and sandy riverbanks to provide nest sites, but success will also require a good supply of large flying insects.

Jon Green and Robin Sandham

Sponsored by Teulu Owsianka Roberts

Bee-eater: Numbers of birds in each recording area, *c.*1854–2019. Some birds are suspected to have occurred in more than one recording area.

Wryneck — *Jynx torquilla* — Pengam

Welsh List Category	IUCN Red List		
	Global	Europe	GB
A	LC	LC	RE

Europe's only regular long-distance migratory woodpecker, the Wryneck breeds in most of continental Europe and temperate Asia as far east as Japan. Although some in the south of Europe remain throughout the year, most European birds winter in the Sahel zone of Africa (*HBW*). This is a bird of open forest, orchards, wooded pasture and unimproved meadowland with scattered trees, where it forages mainly on ground-dwelling ants taken directly from their nests. It breeds in tree cavities (Coudrain *et al.* 2010). The Wryneck is a scarce but regular migrant in Wales. Birds passing through are likely to be from the Scandinavian population, but there have been no ringing recoveries involving Wales to confirm this.

Wryneck © Elfyn Lewis

The species bred in Wales in the 19th century, but the limited number of documented breeding reports suggest that it may never have been common here. There were no confirmed breeding records from Anglesey, Caernarfonshire, Meirionnydd or Ceredigion, and just one in Carmarthenshire, at Ammanford in 1904. In Pembrokeshire, Mathew (1894) was informed of birds being present at Picton Castle during the summer months. He considered that, if the report was correct, they probably nested there. Monk (1963) mentioned two 19th-century Pembrokeshire breeding records, but gave no further details. Farther east in Wales, the species may at one time have been more numerous. In Glamorgan, a clutch of nine eggs was taken from a nest at Lower Penarth in 1890 and a pair bred at Mayals, Gower, in 1906 (Hurford and Lansdown 1995). That was apparently the last confirmed breeding record in Wales, although Monk (1963) mentioned a possible breeding record in Glamorgan in 1913. In Gwent, Ingram and Salmon (1939) quoted four breeding records from around 1900. Breconshire may have had a sizeable breeding population at one time, and the species was said to be fairly regular in the Usk and Wye valleys up to about 1910–20. A nest was said to have been taken in the county sometime before 1882 and a pair nested at Llansantffraed in 1903. Monk (1963) thought that probably more than one pair bred annually in southwest Breconshire until at least 1900 and stated that a pair was reported in the Usk Valley in 1926, although no nest was found. Davies (1912) said that the species was known to have bred in Radnorshire, and Forrest (1907) quoted a record of a pair nesting near Llandinam, Montgomeryshire, about 1860. In North Wales, Monk (1963) said that in the mid-19th century it was occasionally recorded breeding, especially in Denbighshire and Flintshire, but the only firm record of breeding in North Wales in Forrest (1907) was of eggs found at Plas Heaton, Denbighshire, about 1866. The species was said to have been heard frequently at this site between 1846 and 1869. There appears to be no confirmed record of breeding in Flintshire.

A Wryneck survey by the BTO in 1954–58 found no evidence of breeding in Wales (Monk 1963) and there has been no serious suggestion of breeding in Wales since the early 20th century. A census in 1964–66 identified two records of possible breeding in South Wales and one record of "just possibly breeding" in North Wales, all in 1964, but no further details were provided. There were no records in 1965 or 1966 (Peal 1968). A bird held territory in a garden at Llanfachraeth, Meirionnydd, in June 1988, and there have been a few other records in June. More recently, there was an intriguing record of one calling at RSPB Dinas, Carmarthenshire, on 1–2 May 2010. The loss of Wryneck as a breeding species in Wales was part of a trend across Britain, from at least 1840. During the 20th century there was a huge decline in numbers and contraction in range, with breeding restricted to southeast England by the 1950s, and by 1973, Parslow considered that the species was near to extinction in Britain. A small breeding population became established in the Scottish Highlands, at Strathspey, where three nests were found in 1969. This population peaked at five nests in 1977 but declined subsequently (*The Birds of Scotland* 2007). There has been no confirmed breeding in Britain since 2002.

On passage, Wrynecks seem to have been recorded fairly regularly in several counties in the late 19th century, but then became much scarcer. As illustration, the first 20th-century record in Ceredigion was not until 1976 and there were none in Meirionnydd during 1900–88 or in Gwent between *c.*1900 and 1964. Numbers were very small in other counties between the 1930s and 1960s, but then increased. In the 31 years between 1969 and 1999, a total of 211–213 were recorded in Wales (*Birds in Wales* 1994, 2002). The increasing trend has continued, with at least 345 birds recorded in the 20 years between 2000 and 2019, averaging about 17 per annum. This increase was evident across the whole of Britain, where the annual mean rose from 259 in 1990–99 to 377 in 2010–18 (White and Kehoe 2020a).

The larger number of records may be partly down to more observers visiting areas where migrants can be expected, but there does seem to be a genuine increase. Between 2000 and 2019, birds were recorded in all counties of Wales, except Montgomeryshire. The largest totals were in Pembrokeshire (*c.*48%) and Caernarfonshire (*c.*29%), which predominate because

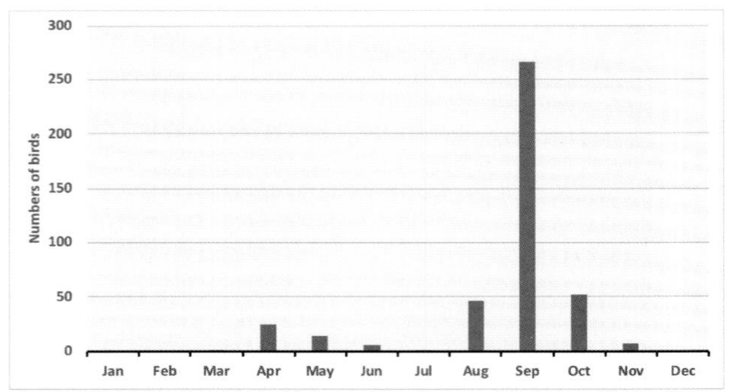

Wryneck: Totals by month of arrival, 1990–2019.

many records come from the islands: 56% of those recorded in Wales in this period were found on four islands: Bardsey in Caernarfonshire (*c.*26%) and three Pembrokeshire islands—Skomer (*c.*16%), Skokholm (*c.*8%) and Ramsey (*c.*6%).

Wrynecks were recorded in all months between March and November, although there was only one March record, a bird at St Brides, Pembrokeshire, on 11–14 March 2014. There were two July records of single birds in gardens: near Pant y Dwr, Radnorshire, on 1–16 July 2001, which died after hitting a window, and at Llandysul, Ceredigion, on 18 July 2017. The latest record, in 2000–19, was on Skokholm on 12 November 2014. Outside this period, one was found dead at St Davids, Pembrokeshire, on 31 January 1965 and one wintered around Manorbier/Penally, also in Pembrokeshire, from 23 December 1965 to 1 February 1966. Some recent years have seen good numbers of Wrynecks, notably 2016 when *c.*47 were recorded, all but three in autumn, including at least 22 on Bardsey. That autumn, at least four were on Skomer on 1 September and four on Bardsey on 14 September. The record day total for Wales seems to be five on Bardsey on 11 September 2005.

The European population declined 64% between 1980 and 2016, although this seems to have levelled off since about 2000 (PECBMS 2020). Suggested reasons include pesticides and agricultural improvement reducing food supplies, in particular ants; habitat changes, including the loss of open ground near potential nest sites; and climatic changes leading to increased rainfall in the breeding season (*HBW*). The area of Europe that is climatically suitable for Wryneck is expected to shift northwards during the 21st century, to include southern and eastern England (Huntley *et al.* 2007), but given the decline in Green Woodpecker in Wales, which has a similar diet, the prospect of the species returning here to breed seems remote.

Rhion Pritchard

Lesser Spotted Woodpecker *Dryobates minor* Cnocell Fraith Fach

Welsh List Category	IUCN Red List			Birds of Conservation Concern				Rarity recording
	Global	Europe	GB	UK	Wales			
A	LC	LC	EN	Red	2002	2010	2016	RBBP

This is the smallest and scarcest of our resident woodpeckers, a species that many Welsh birdwatchers seldom see. The Lesser Spotted Woodpecker has a large range, covering almost all of Europe and temperate Asia, as far as the Kamchatka Peninsula and northern Japan (*HBW*). Birds breeding in Wales are of the subspecies *D.m. comminutus* (IOC), which is confined to Britain. The species is not found in Ireland or Scotland, so Wales is at the western edge of its range. A resident species, the Lesser Spotted Woodpecker is found mainly in open deciduous woodland or parkland, particularly mature oak woods, orchards and riverine wet woodland. The presence of dead, rotting wood is important, providing both a good invertebrate food supply and nest sites. The species is undoubtedly under-recorded, being inconspicuous when it feeds in the upper branches of tall trees. Late February to early April is the period when it calls and drums, making the species more detectable, but once trees are in full leaf, the birds can be difficult to see. This is very much a lowland species, not usually recorded in Wales above about 150–180m (*Birds in Wales* 1994).

The earliest records in Wales included three shot at Ynysygerwn and Ynystawe, Gower, in 1840 and 1841 (Hurford and Lansdown 1995). In the late 19th and early 20th centuries, the species appears to have been very scarce in the western part of Wales, but more widely distributed in border counties, although not common even there. In Pembrokeshire, Mathew (1894) described it as "a rare occasional visitor" and was aware of records only from Goodwick. It was described as scarce in Ceredigion at the end of the 19th century, known only from mature woodlands in the grounds of large country houses (Roderick and Davis 2010). Forrest (1907) described this species as largely absent from the western part

Lesser Spotted Woodpecker © John Hawkins

of North Wales, with no authentic records from Caernarfonshire or Anglesey, although it was quite common in some parts in the east. It was apparently extending its range westward at that time, notably along the upper Dee Valley. The westward spread continued, with the first records for both Caernarfonshire and Anglesey coming in 1926. By the 1940s it was regarded as fairly common in Snowdonia (North *et al.* 1949). In 1966, Condry described it as rather scarce in Snowdonia, but noted that it had increased in the previous few years. In Pembrokeshire, Lockley *et al.* (1949) described it as scarce, and knew of it in only a few localities.

The Lesser Spotted Woodpecker is not sufficiently widespread for a population trend for Wales to be calculated. The trend for Britain as a whole shows a peak in the late 1970s and early 1980s, followed by a steady decline, with a loss of 60% of the population during 1968–99 (Massimino *et al.* 2019). However, since 2000 the species has become too rare for trends to be calculated from the BBS. The *Britain and Ireland Atlas 1968–72* showed it to be widely distributed in Wales, although absent from higher ground and from much of Anglesey and Pembrokeshire. The spread of Dutch Elm disease from 1969 had a considerable effect. The population initially increased due to the availability of dead wood invertebrates, but then decreased as dead mature trees were felled. Elms are used for nesting by this species to a greater extent than the other two woodpecker species (Osborne 1982). The *Britain and Ireland Atlas 1988–91* showed only a small net loss in the number of 10-km squares occupied, compared to 1968–72, with losses in Llŷn, Gower and East Glamorgan balanced by gains in Carmarthenshire and Montgomeryshire. The 2007–11 *Atlas* showed more losses than gains, occupying one-third fewer squares in Wales, with losses particularly in Carmarthenshire and Montgomeryshire, although there were some gains in Pembrokeshire in 2007–11. While densities appeared to be low, which may have reduced its detectability, especially as atlas fieldwork was mostly undertaken outside the period of peak activity, it still occurred in 27% of 10-km squares in Wales in 2007–11.

Recording area	**1980s survey**		**Most recent survey**	
Gwent	1981–85	112 (28%)	1998–2003	60 (15%)
East Glamorgan	1984–89	*17 (4%)	2008–11	*15 (4%)
West Glamorgan (Gower)	1984–89	19 (8%)	na	na
Pembrokeshire	1984–88	20 (4%)	2003–07	19 (4%)
Breconshire	1988–90	31 (7%)	na	na
North Wales (Flintshire, Denbighshire, Caernarfonshire, Anglesey, Meirionnydd)	na	na	2008–12	32 (2%)

Lesser Spotted Woodpecker: Number and percentage of tetrad records from county/regional atlas surveys in Wales.

* Sixteen tetrads with breeding evidence in 1984–89 and two with breeding evidence in 2008–11.

The results of surveys at tetrad level showed that only in Gwent has the species been recorded in more than a small percentage of tetrads in recent decades. There was a 46% contraction in range there between the 1981–85 and 1998–2003 atlases. The *Britain and Ireland Atlas 2007–11* showed a comparatively high relative density in eastern Carmarthenshire but, in common with several other counties, the 2011 *Carmarthenshire Bird Report* mentioned it to be in serious decline. In Ceredigion, it was described as widespread but scarce (Roderick and Davis 2010). By 2010 the population in Radnorshire was estimated at just 15–20 pairs, a decline that began in the early 1990s (Jennings 2014). The species is probably still found in all Welsh counties but appears to be

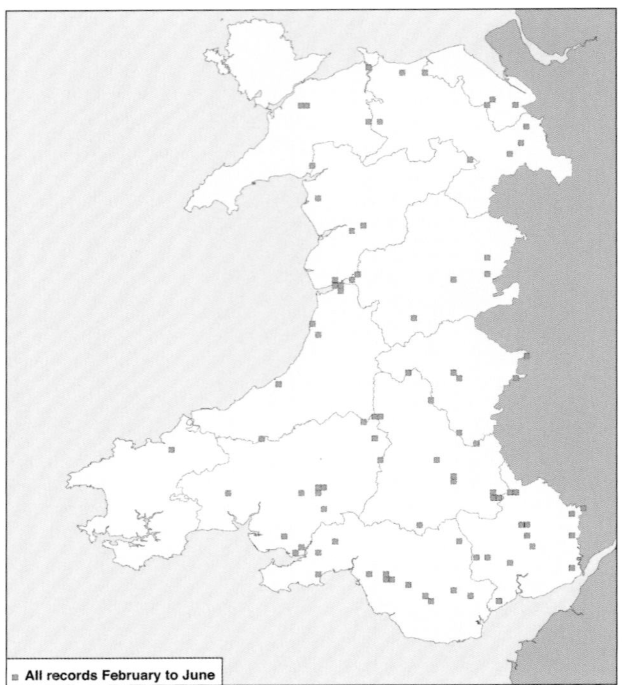

Lesser Spotted Woodpecker: Records from BirdTrack and *Welsh Bird Reports* for February to June, 2015–19.

extremely scarce on Anglesey. It was not recorded there during fieldwork for the *Britain and Ireland Atlas 2007–11* or in the *North Wales Atlas 2008–12*, and it is uncertain whether there is a regular breeding population on the island.

Breeding-season records of Lesser Spotted Woodpeckers from *Welsh Bird Reports* and BirdTrack, during 2015–19 provide the most recent indication of its distribution across Wales. Although these records have not been collected in a systematic way, they show remaining concentrations in parts of South and southeast Wales in particular.

In the absence of a dedicated Wales-wide survey, it is not possible to determine a reasonably accurate breeding population estimate for this elusive and declining species. *Birds in Wales* (1994) suggested 400–600 pairs while *Birds in Wales* (2002) suggested 300 pairs. The most recent assessment by Hughes *et al.* (2020) considered it to be in the range of 105–320 pairs in 2018.

Several suggestions have been made for decline in the population. Charman *et al.* (2012) found that breeding success was lower in England than previously reported and lower than elsewhere in Europe. This was thought to be due to food shortages and poor weather in the breeding season and possibly the pre-breeding season (Smith and Charman 2012). The lack of dead wood in woodlands has been suggested as a cause, as has nest predation by Great Spotted Woodpecker and competition with this species for nest-holes (Fuller *et al.* 2005). Data from the Repeat Woodland Bird Survey (RWBS) showed that the Lesser Spotted Woodpecker population had decreased more heavily in woods with relatively high numbers of squirrel dreys, suggesting that predation by squirrels could be an important factor (Amar *et al.* 2006). The decline in this species is happening on a wide scale, having reduced by 80% in Europe between 1980 and 2016 (PECBMS 2020). The Lesser Spotted Woodpecker is Red-listed in Wales for several reasons, including declines in both the breeding population and in breeding range (Johnstone and Bladwell 2016). Its future in Wales is now very uncertain, although appropriate woodland management, to ensure a plentiful supply of dead wood of deciduous trees, could help to retain it as a breeding species here.

Rhion Pritchard

Sponsored by Enid Griffiths

Great Spotted Woodpecker *Dendrocopos major* Cnocell Fraith Fawr

Welsh List Category	IUCN Red List			Birds of Conservation Concern			
	Global	Europe	GB	UK	Wales		
A	LC	LC	LC	Green	2002	2010	2016

The Great Spotted Woodpecker has an extensive distribution in Europe and Asia, extending from the Atlantic to the Pacific coast, with a small population breeding in northwest Africa (*HBW*). This medium-sized woodpecker is found in all types of woodland, broadleaved and coniferous. Those in Britain are resident and of the endemic subspecies, *D.m. anglicus* (IOC), although birds of the nominate form, which breed in Fennoscandia, show eruptive movements in some years and can reach Britain (*BTO Migration Atlas* 2002, Smith 2010). Ringing recoveries from Wales suggest that most birds move very little, although a first-calendar-year bird ringed at Leominster, Herefordshire, in 2016, was recovered at Wrexham, Denbighshire, a few months later, 92.5km to the NNW. The Great Spotted Woodpecker is now found in all areas of Wales, except at the highest altitudes, although in several counties the species is now found at higher elevations than formerly, for example, in Gwent (Venables *et al.* 2008). *Birds in Wales* (1994) reported that it bred up to 365m in Wales, but more recently pairs have been found nesting at altitudes of slightly over 400m, such as in Breconshire (Andrew King *in litt.*) and in Caernarfonshire (pers. obs.).

The Great Spotted Woodpecker seems not to have been a common bird in many parts of Wales in the late 19th and early 20th centuries. At the end of the 19th century, the species was said to be an uncommon, but regularly breeding, resident in Glamorgan, where it was thought to be increasing (Hurford and Lansdown 1995). Rather surprisingly, it was not recorded in Gwent until 1926 (Venables *et al.* 2008). In Pembrokeshire, Mathew (1894) described it as "a rare occasional visitor" and knew of a few records but had never seen one in the county. He said that it was more common in neighbouring Carmarthenshire. In Ceredigion it was said to be localised in the 19th century, mainly associated with mature woodland on the estates of large country houses (Roderick and Davis 2010). Phillips (1899) described the species as rare in Breconshire but noted that it was increasing there. Forrest (1907) stated that it was less numerous than the Green Woodpecker but was found in all wooded areas of North Wales. It was nowhere common, although perhaps more numerous than was generally supposed. The species was said to have increased in some parts of Caernarfonshire and Meirionnydd around that time but was unknown on Anglesey.

Numbers increased steadily in many counties during the 20th century, in many places becoming more common than Green Woodpecker, an observation made by Salter in 1932 for Ceredigion (Roderick and Davis 2010). The first record for Anglesey came from Penmon on 3 April 1912, but there was only one more record on the island before 1954, which was of one shot at Bodorgan in 1924 or 1925. Lockley *et al.* (1949) said that in Pembrokeshire it was formerly less common than Green Woodpecker, but had increased during recent years, while Ingram and Salmon reported it to be increasing in Carmarthenshire in 1954. In Radnorshire, Ingram and Salmon (1955) said that the Great Spotted Woodpecker was "possibly not more numerous than the Green Woodpecker". In Meirionnydd, Jones (1974a) described it as a widespread but rather uncommon breeder, more confined to closed-canopy woodlands than the Green Woodpecker.

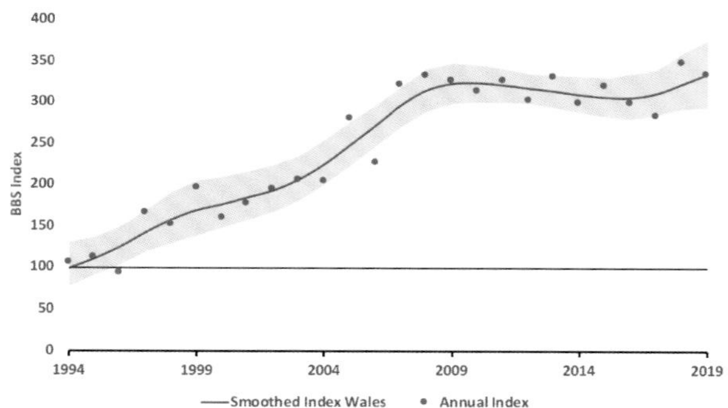

Great Spotted Woodpecker: Breeding Bird Survey indices, 1994–2019. The red line shows the smoothed index, 1995–2018. The shaded area indicates the 85% confidence limits.

The *Britain and Ireland Atlas 1968–72* showed Great Spotted Woodpecker to be found everywhere in Wales, except for northern Anglesey, and even in 1980 the species was still absent from much of the island (Jones and Whalley 2004). On Bardsey, Caernarfonshire, this species was recorded less frequently than Green Woodpecker up to 1985, with only eight records of singles, but there have been many more records in recent years. An estimated 15 birds passed through the island in autumn 2008, including seven on 12 October. The rise in the number of records probably reflects increasing numbers on Llŷn (Pritchard 2017), although it is possible that birds of the nominate Fennoscandian subspecies *D.m. major* were involved in some years. A bird on Bardsey on 10 November 1975, for example, was thought to show characteristics of this form (Pritchard 2017). *Birds in Wales* (1994) noted that this species now outnumbered Green Woodpecker everywhere, except in Monmouthshire (Gwent) and probably West Glamorgan (Gower). It now outnumbers the Green Woodpecker everywhere, even in the latter's remaining Welsh stronghold in Gwent, where the Great Spotted Woodpecker population was estimated by the *Gwent Atlas 1998–2003* at 670–1,100 pairs, compared to 420–770 pairs of Green Woodpecker (Venables *et al.* 2008).

The *Britain and Ireland Atlas 2007–11* showed an increase from 87% to 94% occupation of 10-km squares in Wales, compared to 1968–72, with gains in western areas, notably northern Anglesey. Relative abundance in 2007–11 had increased almost everywhere in Wales compared to the 1988–91 *Atlas* but was highest in eastern Wales. At a local level, the *Gwent Atlas 1998–2003* recorded breeding evidence in 81% of tetrads. The proportion of tetrads recorded as occupied in the *East Glamorgan Atlas 2008–11* was 60%, 52% with breeding evidence. The *Pembrokeshire Atlas 2003–07* found breeding evidence in *c.*52% of tetrads compared to 36% in 1984–88 (*Pembrokeshire Avifauna* 2020). The *North Wales Breeding Atlas 2008–12* found evidence in 58% of tetrads, although it was mainly absent from the higher ground and from parts of Anglesey with fewer trees.

The BBS index for Great Spotted Woodpecker in Wales increased by 189% between 1995 and 2019, with no indication yet of levelling off (Harris *et al.* 2020). Formerly it was mainly a bird of heavily wooded areas, but it now appears to have moved into some areas with comparatively few trees. As numbers have increased, so has the species' use of gardens in Wales (BTO Garden Birdwatch 2020), with an increase from 1995 that very much mirrors the pattern indicated by the wider countryside BBS trend.

The most recent estimate of 21,500 (19,500–23,500) pairs in Wales in 2018 (Hughes *et al.* 2020) accords with the considerable increase in numbers recorded by surveys over the last 25 years.

Evidence of Great Spotted Woodpecker movements in autumn includes birds seen flying out to sea from several west-facing headlands. The species has long been absent from Ireland, but the island has recently been recolonised, with populations in the north and in Co. Wicklow, where breeding was first confirmed in 2009. DNA analysis indicates that these birds originated from the British population (McDevitt *et al.* 2011), so it seems likely that Wicklow was colonised from Wales.

There are likely to be several reasons why this species is doing so well in Wales, including the maturing of the conifer forests planted after the Second World War and the fact that this species has learned to take advantage of garden feeders. The decrease in the Starling population may also have helped, reducing competition for nest-holes (Smith 2005). This species is likely to continue doing well in Wales, provided that woodland management ensures plenty of mature trees and dead wood.

Rhion Pritchard

Sponsored by Shelagh Hourahane

Green Woodpecker *Picus viridis* Cnocell Werdd

Welsh List Category	IUCN Red List			Birds of Conservation Concern			
	Global	Europe	GB	UK	Wales		
					2002	2010	2016
A	LC	LC	LC	Green			

This species, the largest of our resident woodpeckers, is found across most of Europe (except Iberia, where there is a separate species, *P. sharpei*) and east to Iran. Over most of its range, the population is thought to be stable or increasing, but declines have been recorded in some European countries (*HBW*). The Green Woodpecker is a bird of forest edges, clearings, copses, parks and orchards, usually near mature deciduous trees. It feeds mainly on ants, chiefly meadow-dwelling species and particularly the Yellow Meadow Ant. In Wales, it breeds mainly in the lowlands, with few breeding records at altitudes over 300m, although it may move higher to feed. The few ringing recoveries involving Wales show movements of only a few kilometres.

The Green Woodpecker was considered a common and regularly breeding resident in Glamorgan in the late 1890s (Hurford and Lansdown 1995) and was described by Barker (1905) as not uncommon in Carmarthenshire. In Pembrokeshire, Mathew (1894) described it as a common resident, and the only common woodpecker in the county, although in neighbouring Ceredigion it was said to be a rather scarce, albeit widespread, resident breeder at the end of the 19th century. Forrest (1907) described it as the most numerous and widely distributed of the woodpeckers in North Wales, found everywhere there were trees and even sometimes where there were none. It may have been extending its range in this period. On Anglesey, where it was the only woodpecker, it was said to have been rarely seen in the Holyhead area up to 1890 but was then increasing. The population probably remained at roughly the same level for most of the 20th century, except for the effects of unusually cold winters such as 1946/47 and 1962/63, which made ground feeding difficult. Lockley *et al.* (1949) said that there had been a considerable decrease in Pembrokeshire in the late 1940s. Similar decreases were noted after the winter 1962/63 (Dobinson and Richards 1964), but numbers seemed to have recovered quite quickly.

Within Wales, there has been an east–west divergence in the fortunes of this species, which is evident in the results of the three breeding birds of Britain and Ireland atlases. The 1968–72

Green Woodpecker © Tom Wright

Atlas recorded the species in 90% of 10-km squares in Wales. South Wales had the joint highest concentration of confirmed breeding records in Britain, along with the area of England south of the Severn and Thames valleys. The 1988–91 *Atlas* showed a pattern of losses in western counties of Wales, particularly Anglesey, Ceredigion and Pembrokeshire. This trend continued in the 2007–11 *Atlas*, with losses in these three counties and in Meirionnydd. In total, 62% of 10-km squares were occupied in Wales. In eastern Wales there was little change in breeding distribution, although relative abundance decreased in many areas between 1988–91 and 2007–11. Scotland and Cornwall also showed a contraction in breeding distribution since the 1968–72 *Atlas* and a decline in relative abundance between the 1988–91 and 2007–11 atlases.

Results from county atlases at tetrad level illustrated the regional differences in the number and the percentage of tetrads in which Green Woodpecker was recorded.

Recording area	**1980s survey**		**Most recent survey**	
Gwent	1981–85	326 (83%)	1998–2003	333 (85%)
East Glamorgan	1984–89	255 (63%)	2008–11	249 (61%)
West Glamorgan	1984–89	183 (78%)	na	na
Pembrokeshire	1984–88	139 (29%)	2003–07	32 (7%)
North Wales (Flintshire, Denbighshire, Caernarfonshire, Anglesey, Meirionnydd)	na	na	2008–12	220 (12%)

Green Woodpecker: The number and percentage of tetrad records from county/regional atlas surveys in Wales.

The BBS index for Wales showed a 29% decrease in the population between 1995 and 2019, in contrast to the trend in England, which showed a 23% increase for the same period, although there has been a gradual decline there since about 2009 (Harris *et al.* 2020).

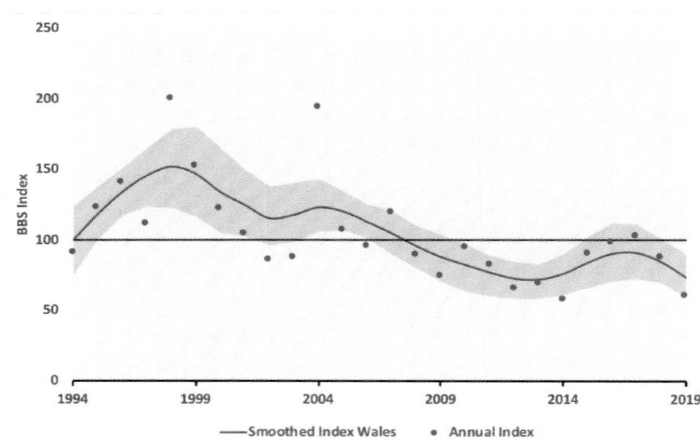

Green Woodpecker: Breeding Bird Survey indices, 1994–2019. The red line shows the smoothed index, 1995–2018. The shaded area indicates the 85% confidence limits.

The one exception to the retreat from the west is Llŷn, Caernarfonshire, where the *North Wales Atlas 2008–12* showed the Green Woodpecker to be widespread. It sometimes visits the larger Welsh islands, but records have become less frequent than formerly. For example, on Bardsey there were 15 records up to 1985, but there has been none since. BirdTrack records from 2015–19 suggest that, elsewhere in the west, the species has declined. By 2019 it had almost disappeared from Anglesey and Pembrokeshire, and from large areas of Meirionnydd and Ceredigion, and had become very scarce in western Denbighshire.

The loss of ant-rich permanent pasture may be one of

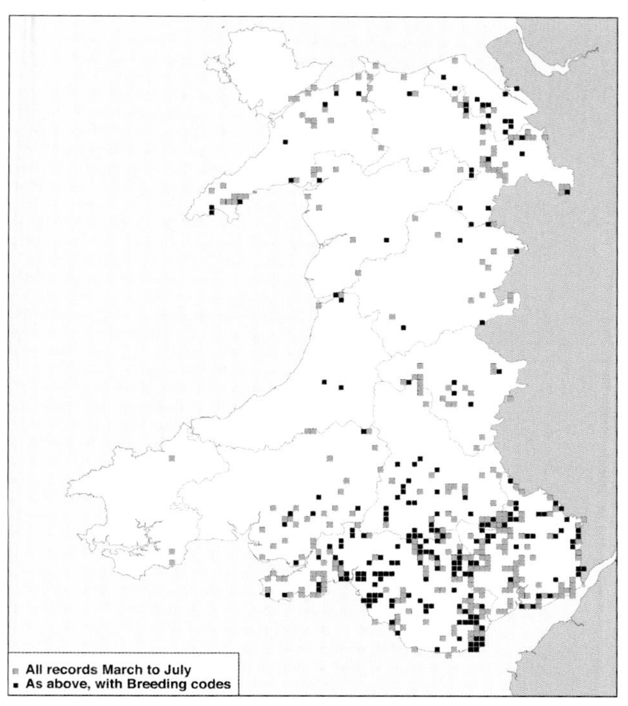

Green Woodpecker: Records from BirdTrack for March to July, 2015–19, showing the locations of records and those with breeding codes.

the reasons for this decline. Lloyd *et al.* (2015) noted that in Carmarthenshire there was evidence that the ploughing of old *ffridd* pasture, during the 1970s and 1980s, resulted in the loss of some pairs, but that as the 'improved' pasture reverted to its former state in the 1990s, Green Woodpeckers returned. Lloyd *et al.* also stated that Green Woodpeckers had colonised old colliery sites, with their mixture of trees and free-draining ant-rich grasslands. 2015–19 BirdTrack records indicate that it is still quite widespread in southern and eastern counties of Wales. Gwent probably still has the largest Green Woodpecker population, where the most recent estimate was 420–770 pairs in 1998–2003 (Venables *et al.* 2008). The latest assessment of the Welsh population in 2018 estimates it to be 3,350 (3,000–3,750) pairs (Hughes *et al.* 2020).

The Green Woodpecker has Amber conservation status in Wales because of its population decline. Temporary declines could be down to cold winters, with frozen ground making feeding difficult, but this does not explain the difference between east and west, which is likely to be linked to food availability, at least partly due to the ploughing and re-seeding of permanent pasture where ants were numerous. A decline in grazing by Rabbits due to the arrival of Rabbit Viral Haemorrhagic Disease (RVHD) during the 1990s, cooler wetter summers and mild wet winters could also be significant factors. Research to understand the reason for the declines in western Wales would be beneficial. It remains to be seen whether the population reductions will spread farther east.

Rhion Pritchard

Sponsored by Clwyd Ornithological Society

Kestrel *Falco tinnunculus* Cudyll Coch

Welsh List Category	IUCN Red List			Birds of Conservation Concern			
	Global	Europe	GB	UK	Wales		
					2002	2010	2016
A	LC	LC	VU	Amber			

The Kestrel is the most numerous and widespread falcon in the Old World, with a range extending from the Atlantic to Pacific coasts of Eurasia and from the Arctic Circle in Norway to South Africa (*HBW*). It is an adaptable species, using a wide range of open habitats, but in Wales is usually associated with rough grassland. Suitable foraging habitats include grass moorland/heath, coastal heath, unmanaged brownfield sites, fallow arable fields and margins, very young forestry plantations or any other grassy habitat with plenty of voles, its favoured prey. The Kestrel can also breed in urban areas, where it predominantly hunts for birds rather than voles, and even in vole-rich habitats it will also take other prey including birds, lizards and insects, especially beetles.

The Welsh name for Kestrel translates as 'red hawk', referring to the rich chestnut colour of the upperparts of the adult male, which is noticeable even at great distance. It has earned the colloquial names of 'windhover' and 'motorway hawk', due to its habit of conspicuously hovering over unmanaged roadside verges as it hunts for small rodents. Its ability to remain stationary in the air is a feat that most observers find impressive. There is no better place to witness such aerial mastery than on the coastal cliffs of Wales, where the updrafts are strong enough to make flapping unnecessary. The Kestrel will often hold its position with its tail raised and held pencil thin, and wings reduced to downcurved, narrow scimitars, as if frozen mid-stoop. Then, with the slightest flick of feathers, the hovering bird will sweep away gracefully to a new stationary vantage point.

In Wales, Kestrels have been recorded using a variety of nest sites, including cliffs, buildings, tree cavities, the stick nests of other species and nest-boxes. Cliff ledges and crevices are probably their nest site of choice, although in their absence, birds readily take to nest-boxes, especially when placed in very open situations, away from Buzzard nests. The use of stick nests is much more common in eastern Welsh counties (Shrubb 1993a), where perhaps there is less exposure to extreme weather and fewer other options.

Most adults, especially males, are territorial and remain on their patch throughout the year. Ringing of nestlings in Wales has shown that young birds can disperse long distances: five have been recorded in France and four in Spain, as far as 1,635km from the ringing site. Most were found in the autumn/winter period following fledging. However, some were found in the second winter, and even four years later in the case of one in France, implying either that they remain away from their natal area or, perhaps more likely, return to favoured wintering areas each year. More typical movements are within mainland Britain, with many Welsh-born birds leaving their natal area during their first autumn, and individuals hatched elsewhere arriving in Wales. There have also been several exchanges across the Irish Sea—in both directions—and nine from northern Europe with examples from Denmark (1), Norway (1), Sweden (3) and Finland (4). The longest recorded movement relating to Wales was a female ringed as a nestling in Finland in 2014 and found dead at Llandyfriog, Ceredigion, in October that year, 2,264km to the southwest.

During autumn, Kestrels occasionally congregate for brief periods, typically in small groups of 4–8 (pers. obs.), although 30 on Skomer in September 1986 was exceptional. Such gatherings are probably associated with a mass hatching of insects. On 1 October 2000, seven were together at Freshwater West, Pembrokeshire, walking through grassland and gorging on

Kestrel: Recovery locations of birds ringed in Wales are shown by red circles. Ringing locations of birds that were recovered in Wales are shown by blue triangles. Small brown dots show ringing or recovery locations in Wales.

hatching craneflies. The timing of these gatherings implies that they may involve dispersing or migrating birds on passage.

Being both small and familiar is a good recipe for being ignored by naturalists, and the Kestrel did not always attract the attention received by larger, rarer species, so charting its fortunes in Wales is open to interpretation. The first detailed references appear in Pembrokeshire, where Mathew (1894) wrote "The most numerous of all our Hawks, to be met with all over the county, nesting in woods, in old ruins, and in many places on the cliffs all round the coast". At the time, many raptor populations would have been severely reduced by persecution. Therefore, the accolade of being the most numerous did not necessarily mean much, but nonetheless, the description implies that the species was far more numerous than it is at present.

Although the Kestrel population seems not to have been affected to the extent of most other raptor species during the persecution era, the species was nonetheless targeted. Forrest (1907) recorded that 1,988 Kestrels were killed on Penrhyn Estate in the uplands of North Wales, between 1874 and 1902, a higher total than for any other raptor. This equates to an average of 71 per annum, a figure only possible if the local population was far more substantial than at present. Whether this level of persecution caused a population reduction is not possible to ascertain, and although it was widely believed to have done so in some parts of the UK, observers across Wales consistently listed the Kestrel as having been common around the beginning of the 20th century (*Birds in Wales* 1994). By the 1920s, the intensity of raptor persecution had begun to recede, and if that had suppressed Kestrel numbers, the population would have probably bounced back quickly thanks to high productivity. Given that farming practices did not change dramatically until later in the century, it is a reasonable assumption that suitable habitat would have been widespread on farmland across Wales. Additional habitat may have been added in the form of upland afforestation from the 1950s, which, during the early stages of growth, supported high populations of voles. However, much of the planting took place on wet heath and grass moorland, habitats that also supported breeding Kestrels, and as the canopy closed and became unsuitable, in the long term the plantations may have caused habitat loss.

The UK Kestrel population declined during the early 1960s, due to the use of organochlorine pesticides, with the decrease being most apparent in the arable areas of eastern England. Although in Wales there was evidence of a decline in the south, numbers apparently increased in the north (Prestt 1965), so Welsh Kestrels may have come through this period relatively unscathed. Any negative impacts were probably short-lived. The *Britain and Ireland Atlas 1968–72* showed 97% occupancy of 10-km squares and, by the 1970s, most observers agreed that the Welsh Kestrel population was at a new peak, with breeding noted in many towns. Around this time, the first detailed estimates of population and breeding density were made, showing that even at their apparent peak, Kestrels bred at a relatively low density in Wales, compared to many other parts of their UK range (Dare 1986b, Shrubb 1993b). Competition with Buzzards was suggested as a possible reason. A population estimate for Wales, based on fieldwork undertaken in 1982–86, was 800–1,000 pairs (*Birds in Wales* 1994), while a WOS survey in 1997–2002 estimated 530–850 pairs (Shrubb 2003a). This latter survey gives the most robust recent population estimate, because it was based on dedicated fieldwork from a range of habitat types across Wales. It also took into account the variation in site occupancy rates between years, which ranged from 46% to 74% (mean 59%). This is particularly relevant when interpreting distribution maps with a sampling period covering several seasons, as the registered distribution may imply a higher number of territorial birds than are present in any single year. There is no BBS index for Kestrel in Wales, because it occurs in too few monitored 1-km squares, but since 2007, there has been a 26% decrease at the UK level. The decline has been much greater in western regions: 49% in southwest England and 59% in Scotland, for example. It can be assumed that the Welsh population has declined by a similar extent. This assumption is backed up by county bird reports that have recorded decreases, and from local tetrad-atlas surveys.

The *Gwent Atlas 1998–2003* recorded a decline of 21% in the number of tetrads with breeding evidence in 1998–2003 when compared with the *Gwent Atlas 1981–85*. The *Pembrokeshire Atlas 2003–07* recorded a 38% decline in tetrad occupancy since the *Pembrokeshire Atlas 1984–88*. Comprehensive surveys also showed that the Kestrel population had declined in Pembrokeshire to *c.*31 breeding territories in 2008, from an estimated 50 pairs in 1984–88 (Rees *et al.* 2009), but was down to 11 pairs by 2016 (Welsh Kite Trust website). A decline of a similar scale was evident in East Glamorgan, where the combined total of tetrads with probable and confirmed breeding records decreased from 113 to 33 between 1984–89 and 2008–11; a decline of 71% (*East Glamorgan Atlas 2008–11*). Based on population change rates measured in Pembrokeshire and in East Glamorgan, it is likely that by 2020 the Welsh population had dropped to 265–475 breeding pairs. If decline rates have been less severe in North Wales than they have in South Wales, then this estimate may be slightly pessimistic, but the population is considered to be much lower than the estimate of 1,750 pairs in 2016 (Hughes *et al.* 2020).

Despite the obvious and continuing decline, the *Britain and Ireland Atlas 2007–11* showed that the Kestrel population remained reasonably widespread, occurring in 98% of 10-km squares in winter and 81% in the breeding season. Although the 10km resolution tends to mask the much bleaker picture already described, it indicates where changes have occurred across Wales. The 1988–91 *Atlas* showed gaps in breeding distribution since the 1968–72 *Atlas*, particularly affecting Pembrokeshire. By the 2007–11 *Atlas*, declines in breeding distribution had extended eastwards and northwards to other parts of Wales. There had also been a widespread decline in relative breeding abundance across almost all areas of Wales since 1988–91. Despite observer biases, the BirdTrack map of Kestrel records between 2015–19 possibly illustrates better its much patchier distribution, with extensive gaps, especially in areas dominated by improved farmland. The key areas are currently the coastal strip, unfenced grass moorland (much of which is under conservation management), and lowland farmland in North and southeast Wales.

A combination of factors is likely to be driving the decline. As well as the reduction in habitat due to less mixed and arable farming, habitat quality is also likely to have been reduced by increased sheep grazing densities during the 1980s and 1990s. This probably reduced vole numbers across large swathes of Wales (Villar *et al.* 2014). Most enclosed sheep and dairy pastures are now unsuitable for Kestrels. A radio-tracking study of breeding adults in Pembrokeshire showed that they completely ignored such habitat for foraging (own data). The trend towards farm specialisation, with associated loss of fallow land, stubble and hay meadows, together with more intensive grass management, has left many parts of Wales with no suitable habitat.

However, habitat change does not explain all the declines. There are areas of apparently suitable habitat that are not occupied or only used for breeding sporadically. Breeding productivity at 70 nests across Wales during 2008–16 was the same as that recorded during earlier times, when the population was much higher (Welsh Kite Trust 2016). This was sufficient to maintain a population, which suggests that increased mortality must have been driving the decline. During their first few months of independence, young Kestrels feed primarily on large insects such as beetles, craneflies and grasshoppers, whose populations

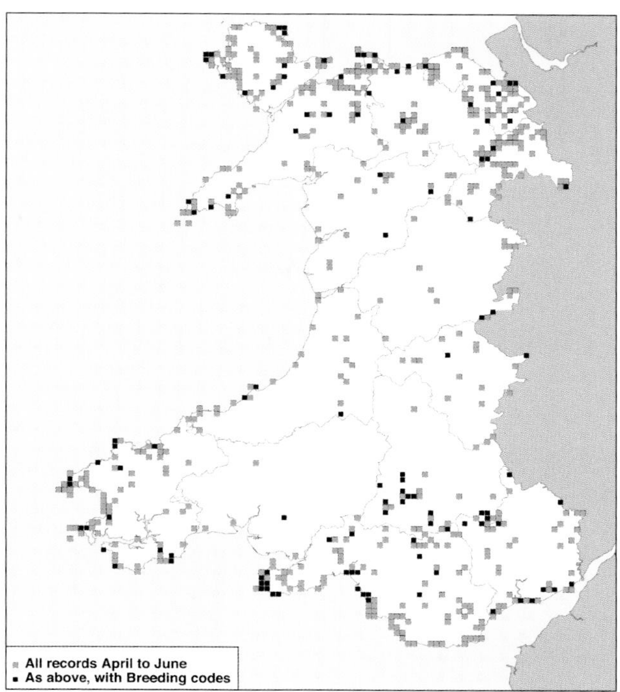

Kestrel: Records from BirdTrack for April to June, 2015–19, showing the locations of records and those with breeding codes.

have declined significantly in farmland (Shortall *et al.* 2009). Therefore, newly independent young may be struggling to find enough food. They are also potentially vulnerable to predation at this stage, especially by Peregrine and Goshawk. The latter species is increasing in number and expanding its range, with a stronghold in Wales, and was shown to cause a Kestrel population decline in a study in northern England (Petty *et al.* 2003). Shrubb (2004) thought it highly likely that an increasing population of Goshawks was having an impact on Kestrel populations in parts of Wales. He also noted that when a large forest block along the northern ridge of Mynydd Epynt, Breconshire, was removed in 2002, taking out two Goshawk sites, two pairs of Kestrels moved in during 2003. They bred successfully, occupying territories last used a decade or more previously. Adults and first-calendar-year Kestrels struggle to find food in prolonged periods of wet or cold weather. There have been several reports of emaciated individuals during such conditions (own data). A study by Newton *et al.* (1999) showed that most Kestrels died from collision or starvation. The trend to wetter autumns may be having an effect, as wet weather and its associated poor visibility restrict their ability to hover and therefore to hunt. Another potential cause of mortality is poisoning from rodenticides, particularly more potent second-generation chemicals, traces of which were detected in 76% of adult and 50% of first-year Kestrels analysed between 1997 and 2005, in a study of UK birds (Shore *et al.* 2018). Rodenticide traces were recorded on average at twice the concentration of that found in Barn Owls and a further study (Roos *et al.* 2021) has provided evidence of a population-limiting effect of SGARs on Kestrels in the UK.

Kestrel populations have declined across much of their European range by an estimated 24% between 1980 and 2016 (PECBMS 2020). Their future as a breeding species in Wales is far from secure. Due to severe breeding population declines, it is on the Red List of species of conservation concern (Johnstone and Bladwell 2016). Targeted research to establish the causes of the population decline is urgently required before effective conservation measures can be selected and applied.

Paddy Jenks

Red-footed Falcon *Falco vespertinus* Cudyll Troedgoch

Welsh List Category	IUCN Red List		Rarity recording
	Global	Europe	
A	NT	NT	WBRC

This rare falcon breeds in warmer eastern Europe, western and central Siberia, and winters in the sub-tropical savannah and grasslands of southern Africa (*HBW*). The number of records in Britain increased dramatically in the late 20th century, although it had settled back to an average of 17 records each year during 2000–18, most in eastern Britain (White and Kehoe 2020a). The number seen in Wales has not shown such an increase and it remains a very rare vagrant here, with 21 records up to 2019.

Total	Pre-1950	1950–59	1960–69	1970–79	1980–89	1990–99	2000–09	2010–19
21	3	2	0	6	4	2	1	3

Red-footed Falcon: Individuals recorded in Wales by decade.

Adult male Red-footed Falcon at Strumble Head, Pembrokeshire, May 2017 © Richard Stonier

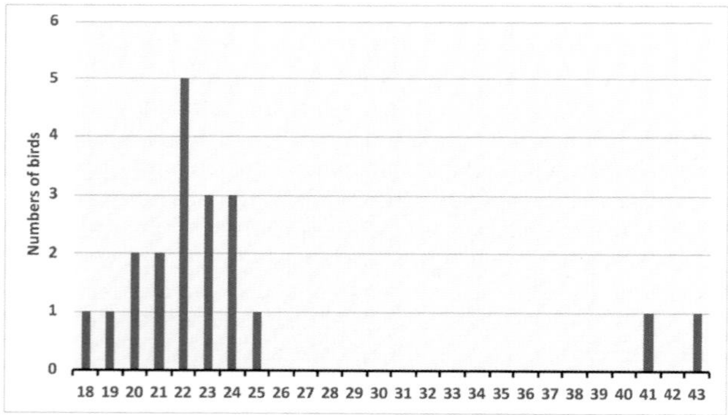

Red-footed Falcon: Totals by week of arrival (when known, n=20), 1903–2019. Week 18 begins *c.*30 April, week 25 begins *c.*18 June and week 41 begins *c.*8 October.

The first recorded Red-footed Falcon in Wales was shot at Wrexham, Denbighshire, in May 1868. In Pembrokeshire, Mathew (1894) quoted two early occurrences, near Cuffern on 5 May 1887 and an undated "example obtained" at Tregwnt, but the evidence presented was found to be unconvincing by later reviews. Two were at St Fagans, East Glamorgan, on 1 June 1903, one of which was shot. Fifty years elapsed before the next Welsh records: males at Clyne Common, Gower, on 23 May 1957 and on Ramsey, Pembrokeshire, on 24 May 1959.

There were no further Welsh records until 1972, when there was a female near Solva, Pembrokeshire, on 6–16 October and a first-calendar-year bird at Sker Point, East Glamorgan, on 16–17 October. These were the only autumn records. All other records, all single birds, have been in early summer, the majority in May and a smaller number in June, and mainly in southern and western coastal counties. Between 1972 and 1989 there were ten records in total, but only six during the last three decades between 1992 and 2017, three of which were in the latter year: a male at the Alaw Estuary, Anglesey, on 6–7 May, a male at Strumble Head, Pembrokeshire, on 11–13 May and a female in the hills above Tregaron, Ceredigion, on 12–13 June.

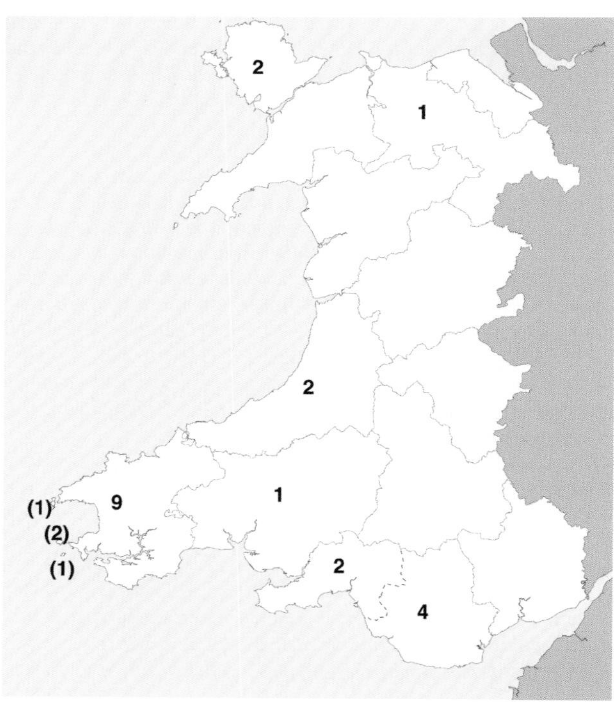

Red-footed Falcon: Numbers of birds in each recording area, 1868–2017.

This species has undergone significant declines in population and distribution across much of its European range (*European Atlas* 2020) and so is likely to remain a rare visitor to Wales.

Jon Green and Robin Sandham

Merlin *Falco columbarius* Cudyll Bach

Welsh List Category	IUCN Red List			Birds of Conservation Concern				Rarity recording
	Global	Europe	GB	UK	Wales			
A	LC	LC	EN	Red	2002	2010	2016	RBBP

Nothing quite matches a Merlin, locked onto its quarry like a heat-seeking missile, zig-zagging around the sky and teasing the eye of the observer in its dazzling and relentless pursuit. Its diminutive size conceals a fearless attitude, as it thinks nothing of terrorising any hapless Buzzard that crosses its path. It is found across the Northern Hemisphere, breeding in North America, northern Europe and subarctic Asia and wintering to the south. Some European birds migrate as far as North Africa (*HBW*). It is an upland breeding species in Britain, although there are historic records of pairs breeding on coastal dunes and dry lowland bogs. Its small size and secretive behaviour make it difficult to locate, although it regularly returns to the same area to breed. The subspecies breeding in Wales is *F.c. aesalon* and, outside the breeding season, it is possible that the population here is supplemented by small numbers of larger Icelandic-breeding birds of the subspecies *F.c. subaesalon* (IOC). There have been ringing recoveries of this subspecies concentrated in Ireland and the west side of Britain, but so far none in Wales. A few birds ringed in autumn in Caernarfonshire showed characteristics of Icelandic birds. However, measurements of breeding birds in northern Britain show an overlap with the size of Icelandic birds (*BTO Migration Atlas* 2002) and this subspecies has yet to be confirmed in Wales.

The Merlin was previously much more widespread in Wales than it is today. *Birds in Wales* (1994) described it as being common on several of the North and mid-Wales grouse moors up until the First World War, and as "a ubiquitous breeder, widespread throughout Wales, even in lowland areas" during the 19th century. Forrest (1907) considered it to be common on Anglesey, where Coward and Oldham (1905) described it as being numerous on the north coast. At that time, it nested on Anglesey at a number of heath-clad headlands, on sand dunes and at inland locations. Breeding sites near the coast were also known in Caernarfonshire on Llŷn and in sand dunes in Meirionnydd. Salter (1895) seems to have been unaware of any nest sites in Ceredigion but was told of breeding at Cors Caron and Cors Fochno. Mathew (1894) knew of nests in heathland near the coast and on Mynydd Preseli in north Pembrokeshire. It was believed to have nested on the dunes in Carmarthenshire in 1879. In East Glamorgan, breeding occurred near the mouth of the Ogmore Estuary at the end of the 19th century. At least six pairs were known to have bred in the dune systems there and in Gower prior to the First World War (*Birds in Wales* 1994).

Despite the egg-collecting, coastal sites were less vulnerable to human interference than those inland, where this harmless falcon was taken for falconry or pointlessly persecuted by gamekeepers.

Female Merlin © Tony Pope

228 were killed, for example, on the Penrhyn Estate around Betws y Coed, Caernarfonshire, between 1874 and 1902. Across the grouse moors of North Wales, especially on Mynydd Hiraethog, Denbighshire, collections of eggs, such as that by the particularly avid collector W.M. Congreave, indicated that Merlins were numerous in the early years of the 20th century. This was probably the case on the Berwyn, Montgomeryshire/Denbighshire, and the heather uplands of Snowdonia. Farther south, in Breconshire for example, they were recorded as being uncommon. The removal of a nest with four eggs from the heather on Grouse Hill in 1888 (Phillips 1899) did not help that situation.

Dune nesting in Carmarthenshire stopped during the 1920s, when many of these areas were afforested (Lloyd *et al.* 2015). By the 1920s, the Glamorgan population was beginning to decline. Disturbance by increasing numbers of day-visitors and by military training activities during the Second World War ended the last breeding attempts along its coast by the mid-1940s (*Birds in Wales* 1994). In Ceredigion, breeding had occurred at Ynyslas sand dunes until the area became a Second World War bombing range (Roderick and Davis 2010). By 1939, only three pairs remained in the Gwent uplands and a decline was noted in Pembrokeshire around this time. By the 1950s, the Carmarthenshire breeding population was reduced to a few pairs on the northern moors. In Ceredigion, numbers almost halved during 1940–50, although there was a slight recovery in the 1960s, with a pair breeding on coastal cliffs in 1952 and 1961. In the Abergwesyn area of Breconshire, O.H. Wells found several breeding pairs, including six in 1942, most of which were using old Carrion Crow nests, or in one case, an old Buzzard nest. Merlins still bred on Anglesey in the 1950s, at sites such as Newborough Warren, Holyhead Mountain, Carmel Head and Porth Wen, and at coastal sites on the mainland coast of North Wales. However, by the 1960s, the inland, upland birds became the core breeding population, with few records elsewhere. This is indicated by the *Britain and Ireland Atlas 1968–72*, when confirmed breeding was recorded in approximately 30 10-km squares within the upland regions of southeast, central, North and northeast Wales, plus lowland Anglesey. There were sporadic reports of breeding in Ceredigion and Carmarthenshire and in other southern counties in the 1960s and 1970s. These included a pair that nested unsuccessfully at a coastal site in Carmarthenshire in 1971 (Lloyd *et al.* 2015) and in Pembrokeshire, where breeding was suspected on Ramsey, until the mid-1970s.

Although numbers had also fallen sharply in the uplands of southern Powys, RSPB surveys in the late 1970s found at least 12 pairs in Breconshire and up to ten pairs in Radnorshire. Here, Ingram and Salmon (1955) reported its status as "a breeding bird in relatively small numbers… [T]hirty or so years ago its numbers were appreciably greater than they are today". Jennings (2014) remarked that the decline in Radnorshire had continued, due mainly to huge losses of heather moorland in the 1960s, which were ploughed up for sheep grazing and to a lesser extent for forestry plantations. Analysis by the Institute of Terrestrial Ecology (now the UK Centre for Ecology and Hydrology) of an egg from Radnorshire in the 1970s found that it contained the highest concentration of polychlorinated biphenyls (PCBs) (24.3 parts per million) ever discovered in a Merlin egg at that time (*Birds in Wales* 1994). There was already a body of evidence linking these and other organochlorine pesticides to the reproductive failure of Merlins, as well as Peregrines and Sparrowhawks, during the 1960s (Newton 1973, Newton *et al.* 1982b).

In the 1970s, decline was evident in North Wales. Only one pair remained in the Clwydian Range, Denbighshire/Flintshire, and numbers decreased even in key areas, such as Ruabon Moor, Denbighshire. The Welsh breeding population between 1968 and 1975 was estimated to be at least 150 pairs (Williams 1981), with the majority in North Wales, nesting in heather, and a smaller number in the south, predominantly using trees. This figure may have been an over-estimate: a survey in 1983–84 suggested a population closer to *c.*40 pairs in Wales, most of which were on the Berwyn and Migneint moors, Meirionnydd/Denbighshire (Bibby and Natrass 1986).

The decline was further illustrated by long-term monitoring on Ruabon Moor, from eight pairs in the mid-1970s to a single pair in 1982 (Roberts and Green 1983). Pesticides, persecution, fire and poor weather may all have contributed to this loss, with no single factor being solely responsible. An additional 15 years of data showed an increase in fledging success, despite smaller mean egg size, clutch size and an adverse ratio of adult to immature birds. From a density of 7–8 pairs/30km^2 in the late 1970s, the population had declined to 1–3 pairs/30km^2 between 1982 and 1997. It was suggested that the comparatively high density in the mid-1970s was a result of rich supply of food, with flocks of Starlings and House Sparrows supplementing moorland birds in June and July, plentiful high-quality nesting sites and a near absence of Peregrines in the study area (Roberts and Jones 1999a).

In the mid-1980s, a shift towards the use of old corvid nests in mature plantations adjacent to open moorland was observed in the Cambrian Mountains (Parr 1991). The move into plantation edge habitat around this time was also observed in other areas, such as in Gwent where there was an indication of a slight recovery in the breeding population (Venables *et al.* 2008). It was estimated that a much higher proportion of pairs in mid-Wales favoured this habitat, perhaps because of the relative lack of heather-covered upland in that area. Due to the difficulties of locating Merlins in forestry, and because several tree-nesting pairs were not detected, an estimate of fewer than 50 pairs in the early 1980s was probably too low (*Birds in Wales* 1994). The 1988–91 *Atlas* showed an increase in the number of upland 10-km squares with confirmed breeding records since the previous *Atlas*, with some gains in North Wales, but losses in South Wales, in northeast Wales and on Anglesey, where it was no longer breeding. Where there had been signs of a recovery, this was thought partly attributable to the recent use of forest margins (Gibbons *et al.* 1993). Breeding associated with forestry also occurred elsewhere, such as in Northumberland, but it faded away when Goshawks became established. In contrast, in Ireland, where Goshawks were still rare at the time of the *Britain and Ireland Atlas 2007–11*, almost the entire Merlin population was thought to be nesting on the edges of forests (Ian Newton *in litt.*). The Merlin nested in conifers on Mynydd Hiraethog for a while, and its demise there could also be linked to an increased number of Goshawks (Brenchley *et al.* 2013).

A survey of Merlins in 1993–94 found 81 pairs in Wales, an apparently marked increase on the *c.*40 recorded in 1983–84 (Rebecca and Bainbridge 1998). The next survey, in 2008, estimated a breeding population of 94 pairs, a further 16% increase, although this was not statistically significant (Ewing *et al.* 2011). However, in areas of Wales that were more comprehensively monitored, numbers declined by 38%. It is likely that the Welsh Merlin population was closer to the lower end of the confidence interval of 62–130 breeding pairs.

The 2007–11 *Atlas* recorded further losses at the 10-km square level in the southern upland regions of Wales. Apart from in the northeast, there were more gains than losses in North Wales, although relative breeding abundance had declined across most of its breeding range by 2007–11, compared to 1988–91. At a county level, the *Gwent Atlas 1981–85* suggested that the Black Mountains were one of South Wales' strongholds for the Merlin, although it was never common. The *Gwent Atlas 1998–2003* recorded it in 25 tetrads (a few more than in the first *Atlas*). However, the number of tetrads with probable and confirmed breeding records had declined by 67%, and the population in 1998–2003 was thought unlikely to be above three pairs (Venables *et al.* 2008). The *North Wales Atlas 2008–12* recorded the Merlin in 90 tetrads, with probable or confirmed breeding in a third of these. However, it was found in 20% fewer 10-km squares than in the *Britain and Ireland Atlas 1968–72* and was now confined to upland areas of at least 400m altitude. Although the number of 10-km squares with breeding attempts in North Wales had hardly changed in 40 years, the actual number of breeding pairs had dramatically declined (*North Wales Atlas 2008–12*).

There has been no subsequent national Merlin population survey, but volunteers, licensed by Natural Resources Wales, have continued to record nesting attempts and productivity. The number of breeding pairs monitored in Wales averaged only 12–14 during 2010–15, most being in the Berwyn and Migneint-Arenig-Dduallt Special Protection Areas (in Meirionnydd/Denbighshire), where Merlin is a qualifying feature. Perhaps of greater significance was the low productivity reported in 2010–15, with a mean of only 2.7 eggs and 1.74 fledged young per successful confirmed nest. While territorial pairs regularly appear on breeding sites, a proportion disappear, often before the middle of May. Of three pairs on Berwyn annually since 2010, for example, it has not been unusual for two pairs to fail, often after egg-laying. In the same area, numbers peaked at 11 territorial pairs in 2002, when they fledged 33 young, but only two territorial pairs fledged three young in 2019. The productivity of the few remaining pairs has fallen by over 50% (own data). There has been further evidence of a declining Merlin breeding population elsewhere in Wales. The number of pairs breeding in disused corvid nests, mainly on moorland forest edge at sites in mid- and South Wales, had decreased from around 14 pairs between 1992 and 2000 to only four pairs by 2011. Predation of recent fledglings by Goshawk was observed and Merlin feather remains were also found at Peregrine nest sites (Haffield 2012). A survey in 2018 of the Merlin population on the mid-Wales Cambrian Mountains, comprising Pumlumon SSSI and Elenydd-Mallaen SPA, found five breeding pairs and two lone males. All but one pair were nesting in conifer forestry plantations adjacent to grass moorland. There was a steep decline, from 11 pairs breeding within the Elenydd-Mallaen SPA in 2003, to just three pairs in 2018. The Merlin population, which is also a designated SPA feature, was no longer in favourable condition (Green 2019b).

Recording area	***Total Pairs 2014–18**
Gwent	2
East Glamorgan	1
Carmarthenshire	2
Ceredigion	1
Breconshire	8
Radnorshire	8
Montgomeryshire	4
Meirionnydd	13
Caernarfonshire	3
Denbighshire	4
Total	**46**

Merlin: Population estimate in Wales, 2014–18, from RBBP data.

* The highest value for the total number of confirmed, probable and possible breeding pairs, in any of the five years. The data presented here are not based on a standardised sample but, in the absence of a more recent national survey, provide an indication of the population decline in Wales since the last national survey.

been recovered in an arc from Cornwall to Liverpool; most of these movements have been in a southerly direction. There have been some records from farther south, such as one found on a ship in the Atlantic, south of Ireland in October 1991, 683km southwest from where it was ringed as a nestling in Breconshire the previous July.

Merlin: Recovery locations of birds ringed in Wales are shown by red circles. Ringing locations of birds that were recovered in Wales are shown by blue triangles. Small brown dots show ringing or recovery locations in Wales. Only movements over 50km are shown.

Outside the breeding season, Merlins occur at a low density in all Welsh counties, with the highest numbers of passage birds observed in March/April and from September to November. In mild winters, some will remain on the uplands, but the majority move to the lowlands, where the *Britain and Ireland Atlas 2007–11* showed quite a wide winter distribution and highest relative abundance. Coastal marshes and inland wetland sites are favoured, but they can turn up anywhere there are flocks of wintering passerines, such as finches, thrushes and Starlings, large roosts of which can attract several Merlins. It is speculated that their habit of hunting in association with Hen Harriers enables them to opportunistically take advantage of the passerines flushed by a passing harrier. They also regularly turn up on islands such as Bardsey, Caernarfonshire, and the Pembrokeshire islands, where numerous migrant passerines can be present, and sometimes remain for long periods.

Merlins ringed in Britain tend to show a southerly movement away from their breeding areas, but most remain in Britain (*BTO Migration Atlas* 2002). There are several recoveries of birds in Wales that were ringed in Scotland and of birds ringed in Wales that have

Another, recovered at Brizambourg in western France in December 1983, had been ringed as a nestling on the Migneint, Meirionnydd, in June that year, 827km to the north. One ringed in Breconshire, in June 2007, moved an even greater distance of 1,022km, being recovered four months later near Castro Urdiales in northern Spain.

Due to severe long-term declines in the breeding population and breeding range, Merlin is now included on the Red List of species of conservation concern (Johnstone and Bladwell 2016) and its future in Wales is uncertain. The parts of Britain that are suitable for Merlin are expected to contract north to the Scottish Highlands during this century (Huntley *et al.* 2007). This may explain the steady declines recorded in southwest England and South Wales. Although the habitat appears to look unchanged in many places in Wales, diminishing availability of invertebrates could have a dramatic impact on prey species such as Meadow Pipit. Declines have continued (Wales Raptor Study Group data), with no part of Wales showing significant recovery. It would be necessary to investigate all factors that might explain this worrying trend, along with another UK breeding survey.

Keith Offord

Sponsored by Steven Davies

Hobby *Falco subbuteo* Hebog yr Ehedydd

Welsh List Category	IUCN Red List			Birds of Conservation Concern				Rarity recording
	Global	Europe	GB	UK	Wales			
A	LC	LC	LC	Green	2002	2010	2016	RBBP

With its long scimitar wings, striking plumage and manoeuvrable flight, the Hobby is arguably the most agile and delightful falcon to grace our shores. It is a summer visitor, usually arriving in Wales from Africa, around the beginning of May, and leaving towards the end of September, following its principal late summer prey, hirundines. The adults are a joy to watch in flight, as they undertake high-speed aerial pursuits of dragonflies to feed themselves, or pursue hirundines to feed their young.

The breeding range of the Hobby spans the breadth of Eurasia, as far as the Arctic Circle in Fennoscandia (*HBW*), from which the entire population retreats to southeast Africa during the northern winter, mainly to Zambia, Zimbabwe and Mozambique (Chapman 1999). Europe forms *c.*30% of its global range, where an estimated population size of 92,100–147,000 pairs appears to be stable (BirdLife International 2015), although the population trend appears to be decreasing (BirdLife International 2020).

Hobbies occupy disused nests of Carrion Crows and Rooks or old squirrel dreys. They lay their eggs around the middle of June. Chicks hatch around the middle of July and fledge in mid-August (Chapman 1999). They are typically a farmland bird in Wales, as they are across much of Britain now, although they were more restricted to heathland in the past. They are usually very site-faithful, returning to the same area, and sometimes the same tree, over many years. Both parents will vigorously defend their nest against predators such as Carrion Crows. As a result, productivity is often high in Wales. They usually fledge 2–3 chicks (own data), although a small proportion of complete broods have disappeared, due to suspected predation by Ravens, Goshawks or Buzzards. An adult found dead on a nest showed signs of having been killed overnight by a Tawny Owl (Roberts and Lewis 1990).

Most early records of breeding were in South Wales, although Walpole-Bond (1903) recorded it breeding at two locations in mid-Wales at the turn of the 20th century, as well as at Llanmaes, East Glamorgan, in 1891 and near Margam, Gower, in 1897. It bred at Wentwood, Gwent, in 1910 and may have bred at St Fagans, Glamorgan, in 1913 (*Birds in Wales* 1994). Forrest (1907) considered that the Hobby did not breed in North Wales, although he recorded that in Meirionnydd one was caught alive at Bala on the unusually late date of 15 November 1873 and a pair was 'taken' near Tywyn. He also mentioned one having been shot near Betws-y-coed, Caernarfonshire, about 1880. A pair was reported to have nested in a solitary tree on moorland above Dolgellau, Meirionnydd, in 1912, a location more typical of a Merlin (*Birds in Wales* 1994). In the 1930s, it was reported to have bred in three southern Welsh counties, and in 1962 and 1963 in Radnorshire (Jennings 2014).

In the early 1970s there were an estimated 75–100 pairs of Hobby in Britain, mainly in southern counties of England (Bijleveld 1974). The *Britain and Ireland Atlas 1968–72* showed that it was restricted to dry heath with scattered pines, in areas such as the New Forest, Hampshire, and the heaths of Surrey and Sussex. It was recorded in just two 10-km squares in Wales, both in the South. Since that time, the Hobby's population increase and range expansion in Britain, and in Wales in particular, has been remarkable. Fuller *et al.* (1985) first highlighted that it was breeding regularly on farmland in the south Midlands of England, away from southern heathlands. By the time of the *Britain and Ireland Atlas 1988–91*, the population estimate for Britain had risen steeply to 500–900 pairs alongside a seven-fold increase in the number of occupied 10-km squares. It was still largely absent from the west, but its distribution had significantly increased to 22 of Wales' 10-km squares, mostly in southeast Wales.

A pair of Hobbies attempted to breed in Pembrokeshire in 1989 and may have occurred in the county since, but confirmed evidence has been lacking. In 1990–92, breeding occurred in Gwent and Radnorshire, and during 1992–2000 also in Breconshire, Montgomeryshire, Denbighshire, Flintshire and probably elsewhere in Wales, reflecting the increases across Britain. *Welsh Bird Reports* have reported increased numbers since the early 1990s: from 27 records in 1990 to up to *c.*200, mainly passage sightings, by 2019. Gwent and East Glamorgan had the highest proportion of records, during the period 1990–2019, accounting for 43% of the total.

The *Britain and Ireland Atlas 2007–11* found breeding evidence in 32% of 10-km squares in Wales, across many counties and with gains in the southwest and north, although it was still a scarce breeding species in both. At tetrad level, there was a staggering 295% increase in Gwent between 1981–85 and 1998–2003, with an estimated population of 20–28 pairs in 2005 (Venables *et al.* 2008). East Glamorgan also recorded an expansion in distribution

Adult and juvenile Hobbies © John Hawkins

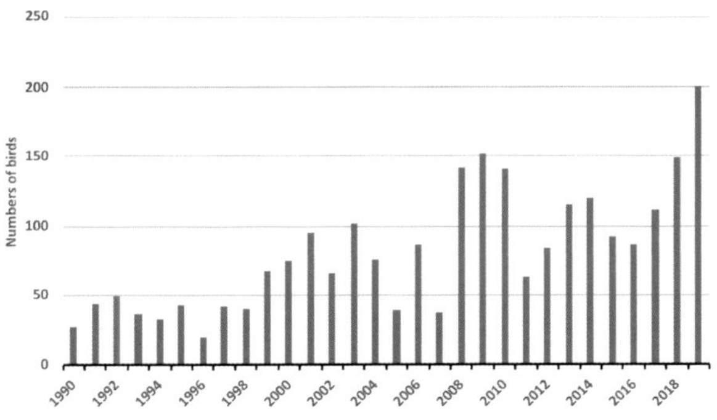

Hobby: Estimated number of passage birds, 1990–2019.

at tetrad level. It was present with no breeding evidence in only two tetrads, less than 1% of the county total, in 1988–91. By 2008–11, it was present in 40 tetrads (10%) with possible breeding recorded in eight of these. The *North Wales Atlas 2008–12* showed the northward expansion with confirmed breeding in three tetrads, probable breeding in 11 and possible breeding in 13, as far west as Llŷn.

The Hobby's fortunes in Gwent, being perhaps the best documented in Wales, provide an indication of the pattern of likely expansion across the rest of the country. Although the Hobby is often described as noisy and demonstrative (Hardey *et al.* 2009), it can be extremely elusive and secretive. Confirmed breeding can be hard to establish unless one is in the vicinity of the nest when food is provisioned and calling can be heard (pers. obs.). In the early 1980s, a concerted effort to locate breeding birds in Gwent, based on reports of sightings, resulted in the discovery of a nest in 1985. This was thought to be the first known nest there in recent times, although records of fledged young came to light subsequently for 1966, 1980, 1982 and 1984 (Roberts and Lewis 1985). In the case of the successful 1966 nest, this was found in a Scots Pine, by a gamekeeper on the Clytha Estate near Raglan. The gamekeeper, Richard Hester, an excellent naturalist, intriguingly stated that, although no nest was found in subsequent years, Hobbies were seen in the area annually and he believed that they continued to breed undetected. Since 1985, the Hobby has been discovered breeding at a minimum of 22 sites in Gwent. Nests have usually been in hedgerow oaks, some located on the fringes or at the centre of larger woods and occasionally around clear-fells in forestry plantations. Although site-faithful, some pairs having occupied territories for over 20 years, they have been known to move short distances, perhaps 1km, between years. In the most intensively monitored areas of Gwent, around the central Usk Valley, Gobion, Llanarth and Raglan, three occupied nests were found, only 2–3km apart, highlighting the potential density of this elusive falcon.

Hobbies generally return to near their natal area to breed. Ringing evidence suggests that there is an average dispersal distance from natal site to breeding site of 25km, based on recoveries of mature adults that had been ringed as nestlings (*BTO Migration Atlas* 2002). Two ringed as chicks in Gwent fitted this general pattern. One ringed in 1992 was found injured, in the breeding season, three years later near Hereford. The other, ringed in 2000, was recaught at Llangorse Lake in Breconshire, the following summer (Venables *et al.* 2008). Ringing has also provided some insights into movements and longevity. The most distant recovery of a Welsh-ringed Hobby within Britain is of one colour-ringed as a nestling in Radnorshire, in July 2010, that was seen alive at Minsmere, Suffolk, in June 2011 and then at Wareham, Dorset, in June 2013. Another ringed as a chick near Glasbury, Radnorshire, in August 2010, was found dead 978km away in Pyrénées-Atlantiques, France, in October that year. One ringed as a nestling in Leicestershire on 7 August 1988 was found dead 207km away, near Bridgend, Gower, 11 years, nine months and two days later.

There have been a few Hobby sightings in March, but these are unusual. The average first arrival date in Gwent is around 25 April. While normally seen singly, up to five birds have been observed hunting over tracts of heather moorland in the Black Mountains, Breconshire, in late spring, where they appear to hunt large day-flying moths, such as Fox Moth, and, later, dragonflies (Andrew King *in litt.*). Although most depart by late September, records in October are not unusual, some of which are family parties still at, or close to, their breeding site. There have also been a few November sightings during the last 30 years, such as one near St Davids, Pembrokeshire, on 28 November 2007. Hobbies on passage or dispersing from breeding areas can sometimes be seen over water, such as at Llangorse Lake, where they hawk for insects or pursue pre-roosting hirundines.

It has been suggested that the expansion of the Hobby population in the southern Midlands and eastern England could be linked to improved survival of immature birds resulting from an increased availability of insect prey, and particularly Migrant Hawker dragonflies in late summer (Clarke *et al.* 1996). The expansion of Hobby distribution in Wales could also be similarly linked to the spread in range of this dragonfly species across most lowland areas of Wales since the 1990s (Cham *et al.* 2014). The most recent population estimate for Britain was 2,050 pairs (Woodward *et al.* 2020)—a massive increase in 40 years. Reports suggest that its range is still expanding northwards in England and Scotland (Eaton *et al.* 2020). However, the BBS index trend at the UK level has indicated a decline of *c.*23% in the population during the ten-year period 2008–2018 (Harris *et al.* 2020). The Hobby is reported from too few BBS squares in Wales to determine the trend here.

By 2019, the Hobby had bred, or probably bred, in every Welsh county except Anglesey. The most recent population estimate suggests that in 2018 there were around 205 (160–265) pairs in Wales (Hughes *et al.* 2020). The area of Britain of suitable climate is expected to include all of Wales, except the far northwest, by the end of this century (Huntley *et al.* 2007). If the trend continues, the future for Hobby in Wales seems good, provided agricultural intensification does not reduce populations of large insects, hirundines and other small passerine prey species. In the New Forest, southern England, previously considered a stronghold, numbers declined significantly with the arrival and rapid expansion of breeding Goshawks (Andy Page pers. comm.). Hopefully, the Hobby's preference in Wales for open farmland may preclude this potential threat.

Stephen Roberts

Sponsored in remembrance of David Reading Thomas and Roland Pugh of Montgomery Barn Owl Group

Gyr Falcon *Falco rusticolus* Hebog y Gogledd

Welsh List Category	IUCN Red List		Rarity recording
	Global	Europe	
A, E	LC	LC	BBRC

The Gyr Falcon is the world's largest falcon with a circumpolar breeding range. In Europe, the species is most numerous in Iceland and Norway, with smaller populations breeding in other parts of Fennoscandia. It also breeds in Arctic Siberia, Alaska, northern Canada and Greenland (*HBW*). European birds are mostly resident, but high-arctic breeders from northern Canada and Greenland are migratory, occasionally wintering in northern Europe. There were 372 records in Britain to 2019 (Holt *et al.* 2020) with, on average, about three a year during the last 30 years. In Wales, however, the Gyr Falcon is extremely rare, with 12 records of single birds, eight of which were before 1950. There have been only four since then, of birds that were considered to be of wild origin, and only one since 1982. An early Pembrokeshire record, at Stackpole *c.*1850, mentioned by Mathew (1894) and cited elsewhere (e.g. Donovan and Rees 1994) does not meet the criteria for acceptance by the BOURC (Harrop 2011). Other reports not listed here are considered to have escaped from captivity (see Appendix 1).

Of 12 confirmed records, seven were white-morph birds:

- adult male (white-morph) shot at Llanbedr, Rhuthun, Denbighshire, on 1 March 1876;
- adult female (white-morph) trapped at Buckland, near Talybont-on-Usk, Breconshire, in the winter of 1893;
- white-morph shot near Brecon between 1893 and 1899;
- white-morph at Pentrefoelas, Denbighshire, pre-1909;
- white-morph shot at Trelech, Carmarthenshire, on 5 February 1913;
- immature (white-morph) trapped at Boncath, Pembrokeshire, on 26 March 1921;
- greyish-brown bird sketched by C.F. Tunnicliffe on the Cefni Estuary shore at Malltraeth, Anglesey, on 3 December 1947
- immature female (white-morph) trapped at Rhoscolyn, Anglesey, on 11 January 1949;
- Cefn Coch, Abergwesyn, Breconshire, on 23 December 1963;
- adult female (grey-morph) shot at Pen-lôn, Newborough, Anglesey, on 31 March 1972;
- probable immature at Connah's Quay, Flintshire, on 12 November 1982 and
- white-morph at RSPB South Stack, Anglesey, on 8 March 2002.

As an Arctic breeding species, the Gyr Falcon is particularly vulnerable to global climate change and the area of suitable climate in the breeding season is expected to reduce in Iceland and Fennoscandia (Huntley *et al.* 2007). If those seen in Wales originate from migratory populations in Greenland or even Canada, it is unclear what effect this will have on its future occurrence here.

Jon Green and Robin Sandham

Peregrine *Falco peregrinus* Hebog Tramor

Welsh List Category	IUCN Red List			Birds of Conservation Concern				Rarity recording
	Global	Europe	GB	UK	Wales			
A	LC	LC	LC	Green	2002	2010	2016	RBBP

Few birds in Wales stir the heart and evoke wild places as much as the imperious Peregrine. To watch an adult leaving its lofty perch to embark on a hunting flight, which may culminate in an audible 'whoosh' to accompany its fearsome stoop onto prey, is an unforgettable experience. The Peregrine is one of the most widespread species in the world, occurring on all continents and absent only from the high Arctic, Antarctica and some desert regions. Within Europe it breeds in Britain, Ireland and in countries with a Mediterranean coast, and locally in central and eastern Europe (*HBW*). Breeding populations in western Europe are largely sedentary.

The majority of Peregrines in Wales breed on precipitous coastal cliffs and natural crag and quarry faces in the uplands. In the 21st century there has been a shift into lowland Wales, including the use of artificial structures in urban areas. The Welsh Peregrine population appears to be rather sedentary, with most recoveries of ringed birds showing local dispersal within Wales or from neighbouring English counties. A small number of nestlings ringed in Wales have been recovered, usually as first- or second-calendar-year birds in Scotland and northern England. One from a foster-nest on Skokholm, Pembrokeshire, was found dead in Finistère, France, about 11 months later, on 5 January 1994 (*BTO Migration Atlas* 2002). Evidence of movements into Wales from elsewhere have included: an adult found dead in Powys (vice-county unknown), in April 1989, five years after being ringed as a nestling in Co. Wicklow, Ireland; a female, colour-ringed as a nestling in Gloucestershire, in 2011, that bred in south Pembrokeshire in 2016; and a highly productive breeding female,

Peregrine © Tony Cross

colour-ringed as a nestling near Craven Arms, Shropshire, in 2006, that bred annually between 2009 and 2020, at a Breconshire territory.

Peregrine nest sites (eyries) can be occupied by generations of birds. Ancestral sites on cliff ledges are often devoid of vegetation but include decomposing bones that are incorporated into the scrape. Tree-nesting has aided the expansion of Peregrines in England, but there have been very few instances in Wales. As in parts of England, Peregrines have adopted nest sites on buildings in some places (South Wales Peregrine Monitoring Group, unpublished data) and this, plus occasional use of tree sites, has aided their population expansion. They have also nested on other structures in Wales, such as pylons, and there have been a few instances of ground-nesting. Old Raven nests are also commonly used. Although Ravens tend to nest earlier than Peregrines, there can be competition between both species for nest sites. Egg and chick predation by Ravens have been recorded occasionally, and at coastal sites, gulls have also been suspected of depredating nestlings.

There is an archaeological record dating from the Mesolithic era (6,000 to 9,000 years ago) at Port Eynon Cave, Gower (Yalden and Albarella 2009), but it is probably reasonable to surmise that Peregrines were less numerous historically than currently, owing to greater tree cover across Wales. Human activities created more open landscapes that probably led to marked increases (Ratcliffe 1980). The oldest written records relate to their use in falconry and for hunting. Giraldus Cambrensis (1191) described how Henry II was impressed by the speed and aggression shown by Peregrines when flying his own hawks in Pembrokeshire and from 1171 birds were taken from the cliffs for his own falconry. Over subsequent centuries, chicks were plundered from several traditional breeding eyries on the coasts of West and North Wales, by those in pursuit of falconry interests.

During 1200–1900, despite the medieval preoccupation with falconry, there is little information to indicate the numbers and distribution of Peregrines in Wales. Repeated references to 'cliffs' as occupied sites imply that the coast probably held most pairs, although mountain crags were also in use. Real expansion inland came with quarrying in the 19th and 20th centuries (Ratcliffe 1980). Forrest (1907) stated that Caernarfonshire was the county holding the greatest numbers in North Wales, while 5–6 pairs usually bred on Anglesey, but he believed there were many more inland in Snowdonia. In Ceredigion, only single coastal and inland sites were known to Salter (1900).

Persecution by game-preservation interests had probably prevented expansion in numbers in the 19th century (*Birds in Wales* 1994). In Caernarfonshire, however, only 16 were shot on the Penrhyn Estate around Betws-y-coed between 1874 and 1902. This compared, for example, with 228 Merlin and 1,988 Kestrels (Forrest 1907), suggesting that Peregrine was not a common bird in that area. The increased popularity of pigeon racing was associated with increases in the Peregrine population in central Wales from the turn of the 20th century, by providing a huge increase in biomass of optimum-sized prey. It also fuelled a great deal of antipathy from pigeon fanciers, whose persecution probably hampered the increase, especially into the industrial valleys of South Wales (Dixon *et al.* 2006).

Tracing breeding history at traditional breeding sites provides a fascinating insight into the scale of the onslaught suffered by these birds (Richards *et al.* 2012). The South Wales Peregrine Monitoring Group (SWPMG) monitors up to 100 sites annually and has compiled site histories from museum egg collections and collectors' diaries. One such history traces the occupation of a Breconshire nest-ledge frequently targeted, especially by two men working together, who reputedly accounted for removing 48 clutches across eight sites in the Brecon Beacons prior to the Second World War. None bred at this site at the height of poisoning by organochlorine pesticides during 1961–71, but breeding resumed there from 1974, with chicks successfully reared from 1978.

Peregrine numbers were reduced during 1940–45, under the Destruction of Peregrine Falcons Order 1940 that specifically sanctioned killing in coastal Pembrokeshire, Denbighshire, Caernarfonshire and Anglesey. This was mainly to protect message-carrying homing pigeons, despatched from downed

Period	No. of years with data	Years of egg theft	Breeding success	No. of chicks fledged
1910–1919	5	5	No	0
1920–1929	6	6	No	0
1930–1939	10	9	No	0
1940–1949		No information available		
1950–1959	4	2	No	0
1960–1969	10	0	No	0
1970–1979	7	1	Yes	3
1980–1989	10	3	Yes	7+
1990–1999	10	0	Yes	12+
2000–2009	10	0	Yes	23
2010–2019	10	0	Yes	12

Peregrine: Breeding success and losses to egg-collectors at a Breconshire eyrie, 1910–2019; compiled from museum egg collections and from collectors' diaries by the SWPMG.

RAF aircraft patrolling coastal regions of Britain (Ratcliffe 1980). Diaries held by the National Museum of Wales described the scale of control measures in Pembrokeshire. With up to 30 pairs, it was the Welsh county with the largest population at that time. Although 87% control was achieved around the coasts of southwest England, it was much less complete in Wales. Lockley *et al.* (1949) considered that 18 of the Pembrokeshire pairs remained in 1947. The degree to which the Destruction Order was carried out in other Welsh counties is largely unknown.

After the Order was rescinded in 1946, the Welsh Peregrine population began to recover, but was soon subject to a spectacular crash through contamination by organochlorine pesticides. Pigeon fanciers and 'homing pigeon' keepers in South Wales mining communities noticed increased losses of their birds during this short-lived recovery and in 1960 petitioned the Home Office for the removal of legal protection for the species. This prompted the Nature Conservancy, the government's statutory advisory body on nature conservation, to commission the BTO to census the UK Peregrine population. It quickly highlighted an absence and low breeding success of Peregrines in many areas, thereby negating demands to remove protection and stimulating research into the causes of the population crash. Of 149 known territories in Wales, only eight, seven inland and one coastal, were successful in 1961 and just two in 1962; 90 territories were empty and another 17 were occupied, but breeding was either not proven or was unsuccessful (Ratcliffe 1963).

Increased adult mortality resulted from the widespread use of dieldrin, aldrin and heptachlor pesticides in cereal seed dressings between 1956 and 1962. Granivorous birds, including finches, buntings and pigeons, species common in the Peregrine's diet, had accumulated these toxins. Harmful effects were first suspected in the late 1940s, but toxicity in adult falcons peaked in 1961–62. Although these pesticides were progressively banned during 1962–84, their impact on raptors was long lasting, because breeding success was reduced through thinning of eggshells (Ratcliffe 1970). Distance from arable farming was no sanctuary, as lethal toxic residues were found in adults and eggs even in remote mountainous areas, but there was generally a decrease in their impact from south to north in Britain. Ratcliffe (1980) calculated that the population crash linked to pesticides caused a 56% reduction in UK Peregrine numbers, compared to the estimated population in the 1930s. Desertion of approximately 90% of Welsh territories, when combined with low breeding success, depressed the population for a considerable time. In Wales, the *Britain and Ireland Atlas 1968–72*, for example, recorded Peregrine breeding in northern Snowdonia and on the north Pembrokeshire coast, but only at a handful of other sites in mid-Wales.

The Peregrine recovery in Wales, tracked by decadal surveys since 1961, and by Britain and Ireland breeding atlases, shows how the species spread back across Wales. However, even as recently as the 2007–11 *Atlas*, it remained absent from north

Carmarthenshire, inland Anglesey, where there is a lack of suitable nest sites, and parts of northeast Wales. During the 1990s, Peregrine numbers peaked in inland counties and new nest sites were adopted on smaller natural crags and disused quarry faces for the first time. Some appeared to have been selected due to their proximity to an abundance of medium-sized prey, such as colonies of Rook or Black-headed Gull (SWPMG unpublished data).

The 2002 Peregrine survey (Banks *et al.* 2010), for the first time, divided Wales into 'north' and 'south', while maintaining the distinction between coastal and inland territories. This approach had the unfortunate consequence of missing territories in parts of North and mid-Wales, particularly in Montgomeryshire, part of which fell into North Wales, with the remainder in South Wales. Their estimate was about 283 pairs (88 coastal and 195 inland) divided between North Wales (104) and South Wales (179). The Wales Raptor Study Group estimated the Welsh Peregrine population to be 321 occupied territories: 89 coastal and a much higher number inland, 232 (Thorpe and Young 2004). Dixon *et al.* (2008) were able to show that the 2002 survey results, reported by Banks *et al.*, had underestimated the actual breeding population in Wales. They quantified the differences, particularly where these affected inland breeding territories, revealing a difference of 15%.

The most recent survey, in 2014, estimated a UK Peregrine population of 1,769 pairs. It assessed the number in Wales to be about 280 pairs, but did not report coastal and inland territories separately (Wilson *et al.* 2018). The population in Wales was described as stable, whereas there had been a significant increase in England, a slight increase in Northern Ireland and a decline in Scotland since 2002. In an assessment of the 2014 survey results from Wales, Williams (2018) gave a total of 279 nesting ranges (territories) occupied by a pair or a single adult (85 coastal and 194 inland). In accordance with previous censuses in Wales, Williams used data from site-based surveys and showed the results at county level, as these were directly comparable with results from Thorpe and Young (2004).

The surveys clearly showed the dramatic decline in the Peregrine population in the 1960s and 1970s, due to the impacts of the organochlorine pesticide era, and its subsequent recovery from 1981. The population reached a peak significantly above the pre-Second World War estimate by 2002 but, by 2014, had declined by about 13%.

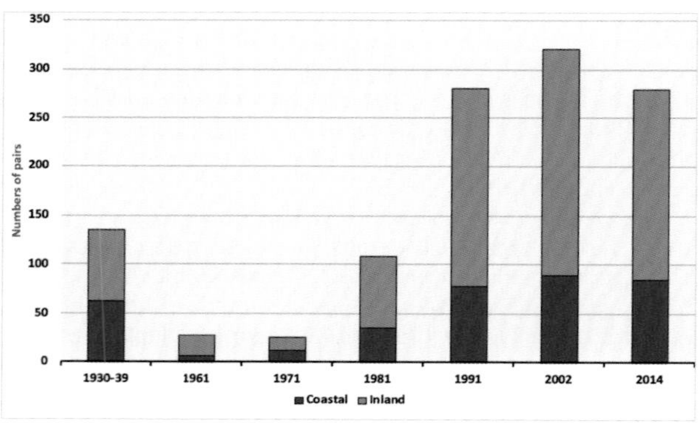

Peregrine: Estimated breeding population in Wales, 1930–2014 (from Ratcliffe 1963, 1972 and 1984; Williams 1991; Thorpe and Young 2004; and Williams 2018).

This trend is supported by regional evidence, for example, from detailed monitoring of the breeding population in Pembrokeshire and in South and mid-Wales. In Pembrokeshire, the population peaked in 2004, but had fallen by 31% by 2019, while numbers in South and mid-Wales fell by 20% during 2000–19. The *North Wales Atlas 2008–12* also showed an absence from some previously occupied 10-km squares, compared with the *Britain and Ireland Atlas 1988–91*.

Occupancy at monitored eyries across Wales in 2014 was higher in quarries than at coastal crags, which in turn had higher occupancy levels than natural inland sites. Occupancy rates were higher in working quarries than disused ones and at larger natural inland sites than small crags. In some industrial areas of

Peregrine: Annual territory occupancy in South and mid-Wales* and Pembrokeshire, 1969–2019, from data provided by South Wales Peregrine Monitoring Group (SWPMG) and Pembrokeshire Bird Group. No data were collected in 2001 owing to access restrictions during an outbreak of foot-and-mouth disease.

* The South and mid-Wales region includes Breconshire in its entirety and parts of Gwent, Glamorgan, Carmarthenshire, Ceredigion and Radnorshire, where the SWPMG monitors up to 100 sites annually.

Wales, buildings, chimneys and pylons have also been adopted as breeding sites. Such sites are used for nesting in coastal areas either side of Newport, Gwent, for example, and these now account for most of the few pairs nesting in that part of the county (Jerry Lewis *in litt.*).

Winter distribution increased between the 1981–84 and 2007–11 Britain and Ireland winter atlases, in line with the increased breeding population. Assuming the pair bond is maintained with no loss of a partner, presence at a nest site over winter depends on availability of food (Ratcliffe 1980). Many Welsh adult Peregrines can be found within their breeding territories throughout the year, but the 2007–11 *Atlas* showed that Peregrines move away from some of the highest-altitude areas outside the breeding season. Observations suggested that the smaller male has a better chance of survival, especially in the uplands, than the female with her greater demand for larger prey. Consequently, while the male will more commonly defend an inland nest site over winter, the female leaves for long periods and operates over a much larger range. Coastal pairs are more likely to remain together in the breeding range at or close to their favoured nest site.

Peregrines take living prey, normally in flight. Adults may also cache food temporarily in the breeding season. Prey remains collected by the SWPMG since 1985 included 107 identifiable prey species within a study area incorporating coastal and inland breeding sites. The dominant prey species in summer were domestic pigeons and corvids, while in winter the main prey were Jackdaw, Woodcock, Starling and thrush species (Dixon *et al.* 2018). Ratcliffe (1980) found that the diet of coastal pairs included, in addition, auks, gulls and terns. Peregrines can hunt when light levels are low, which enables them to take night migrants, such as Woodcock and thrushes. Artificial light sources in towns and cities have further aided their colonisation of urban areas (Dixon and Richards 2005).

With declines in abundance of medium-size wild prey species over the last 30 years, such as breeding Golden Plover (-88%) and Lapwing (-77%) (Johnstone *et al.* 2012), especially around upland Peregrine breeding eyries, the role of pigeon species in the diet has become more important (Roberts and Jones 2004, Stirling-Aird 2015). The notion that Peregrines might selectively specialise on racing pigeons is difficult to confirm, but these are likely to be particularly vulnerable, because of their mass, flight behaviour and movement through unfamiliar habitats during races. Racing pigeons, with a typical mass of 435g, are closer to the upper range considered to be the optimal prey mass for Peregrines (Ratcliffe 1993). Being selectively bred to home accurately and rapidly, domestic pigeons are unlikely to exhibit the range of predator-avoidance strategies found in wild birds, or to effectively find refuges from attack. In a UK-wide study, at least

70% of racing pigeons predated by Peregrines had already adopted a feral existence or had strayed significantly from their racing or training routes (Shawyer *et al.* 1998). Recent declines in Peregrine numbers may be attributed to a reduction in food availability in these regions. Changes to racing-pigeon race routes, to avoid birds flying across parts of Wales where Peregrines occur, together with a decline in the popularity of pigeon racing could have led to a reduction in the availability of this favoured prey source (Dixon *et al.* 2010).

Egg-collecting and destruction of adults and chicks probably suppressed the distribution and breeding success of Peregrines across Wales for many years, but perhaps to a lesser extent than in other parts of Britain. In South and mid-Wales in 2000–19, there were 46 instances where breeding attempts failed due to illegal disturbance, most of which were intentional (SWPMG unpublished data). Only five of these incidents resulted in prosecutions, four through the activities of one persistent egg thief, and the other a poisoning case. Across Wales, the annual average number of confirmed persecution incidents in Wales involving Peregrine was 73% lower in 2010–19 than 1990–99, mostly reflecting a reduction in South Wales (Hughes *et al.* in prep). In the last three decades, it was the Schedule 1 species most frequently targeted by egg-collectors, with a minimum 40 clutches known to have been taken in Wales during 1990–99, but only five during 2000–07, and none since (RSPB data).

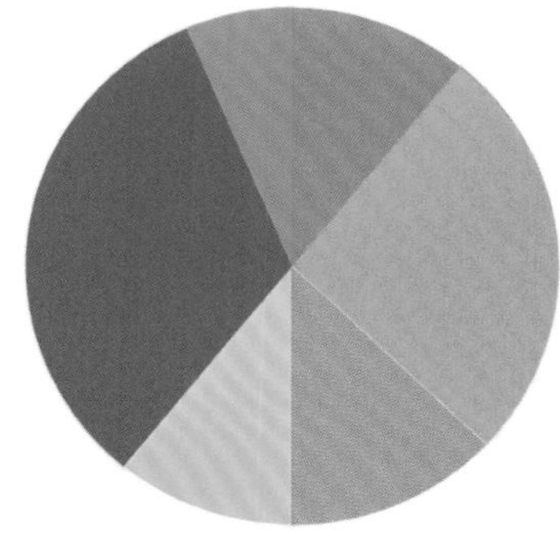

Peregrine: Causes of nest failure (n=46) in South and mid-Wales, 2000–19 (compiled by Colin Richards from South Wales Peregrine Monitoring Group annual reports).

If recent population trends continue, the future for Peregrines in Wales could be one of a continued reduction in site occupancy and numbers of breeding pairs. In some areas this may in part be due to a continuing redistribution from the uplands to coastal, agricultural and urban districts, although numbers may be thinning out in some coastal areas too. Adults may be more resilient to the effects of climate change, but eggs and chicks will be vulnerable to the predicted increase in intensity of summer rainfall, and projections indicate that South and West Wales may no longer be climatically suitable for Peregrines by the end of this century (Huntley *et al.* 2007). Declining prey resources could also be a factor in some areas. Maintaining the decadal surveys initiated in the 1960s will be important for tracking future changes.

Andrew King

Sponsored by Cambrian Ornithological Society

Red-backed Shrike *Lanius collurio* Cigydd Cefngoch

Welsh List Category	IUCN Red List			Birds of Conservation Concern	Rarity recording
	Global	Europe	GB	UK	
A	LC	LC	CR	Red	WBRC

The Red-backed Shrike breeds throughout most of mainland Europe apart from southern Spain and northern Fennoscandia. It also breeds in Turkey, Caucasia and east through Russia, as far as the Altai Mountains. All populations winter in eastern and southern Africa (*HBW*). It prefers dry terrain with scattered bushes or low trees as a perch from which to hunt its prey, which is mostly insects but includes voles and small birds. There was a dramatic decline in western and northern parts of its breeding range during 1970–90 (Harris and Franklin 2000), although the European population is now thought to be stable and there has been some increase in range in Spain, France and the Low Countries since the 1990s (BirdLife International 2020, *European Atlas* 2020).

It was once a widespread breeding bird in Wales, described as very common by Eyton (1838), referring mainly to North Wales. He reported it to be particularly abundant around Barmouth, Meirionnydd, and near Capel Curig, Caernarfonshire, where grasshoppers seemed to be its chief food and "some dozens of them may be seen on the side of the hill above the lakes, which is thinly covered with scattered hawthorn bushes, and abounds with their prey". Around Llandeilo, Carmarthenshire, it was "said to be common" (Davies 1858). It appears to have remained widespread in Wales at the end of the 19th century, when it was thought to be increasing in the lowlands of the southeast, particularly numerous around Cardiff, Merthyr Mawr and in the Vale of Glamorgan near Cowbridge, but was absent from west Gower and scarce in east Gower (Hurford and Lansdown 1995).

Barker (1905) said that it was a not uncommon summer visitor in Carmarthenshire but may have been less abundant in Pembrokeshire. Mathew (1894) had never seen one in the north of the county. In mid-Wales, it was "fairly well distributed" in the lowlands of northern Ceredigion, common in Breconshire and described as quite common around Knighton in Radnorshire and along the border with Herefordshire (Salter quoted in Roderick and Davis 2010, Phillips 1899, Jennings 2014). Forrest (1907) quoted correspondents from Montgomeryshire who reported that it bred every year around Welshpool, but was rare around Montgomery, Newtown and Llanerfyl. In North Wales, he stated that it could hardly be described as common, but was plentiful in coastal Meirionnydd, where there had been a marked increase around Penrhyndeudraeth, and widespread in Caernarfonshire. It was said to be scarce in north and east Anglesey, Flintshire and Denbighshire, although in the last county it was increasing around Colwyn Bay. However, by 1919, Forrest implied that it had become more localised in North Wales and was common only in one or two localities. The last confirmed breeding record on Anglesey was in 1917 (Jones and Whalley 2004).

There were indications of decline elsewhere by the 1920s. In Glamorgan, this began soon after 1910, and after about 1925 this decline led to a contraction in breeding range. A record near Pembroke in 1924 was the last of breeding in Pembrokeshire (Lockley *et al.* 1949). Salter (1930) wrote that he had not seen one in northern Ceredigion since 1924, though 1–2 were reported to him annually. In Breconshire, several pairs were around Builth Wells and Brecon in the 1920s, but it was decreasing in both areas. By 1925 it was described as quite rare on the Radnorshire/Herefordshire border (Jennings 2014) and had decreased around Barmouth, Meirionnydd (Peakall 1962). The decline continued through the 1930s, though a few pairs probably still bred in all Welsh counties, except Anglesey and Pembrokeshire. Its status in Flintshire was unclear: Hardy (1941) stated that Red-backed Shrike

Total	1967–69	1970–79	1980–89	1990–99	2000–09	2010–19
158	14	23	22	33	37	29

Red-backed Shrike: Individuals recorded in Wales by decade since 1967.

had bred at Mold and Bodfari, but he gave no dates and may have been referring to old records quoted by Forrest.

The Red-backed Shrike was lost as a breeding bird from many counties during the 1940s. The *Cambrian Bird Report 1956* stated that there were confirmed records for Denbighshire in 1937–39, but there appeared to have been no subsequent breeding record in the county. The last nesting suspected in Ceredigion was at Llanfarian and Nanteos in 1942, and in Radnorshire around Llanelwedd-Llandrindod in 1944. A pair nested successfully at Llanfoist, Gwent, in 1944–46, but this was the last breeding record in the county for many years. The last successful breeding in East Glamorgan was at Tongwynlais in 1948, although two pairs were at Rhoose in 1948 and single pairs near Llandough in 1949 and 1950. Its stronghold in Montgomeryshire in the 1940s was Llanymynech Hill on the Shropshire border, where eight pairs fledged 25 young in 1943, but by 1948 there were only three pairs. The last successful breeding there was in 1950 (Holt and Williams 2008).

By the 1950s, Red-backed Shrike was a rare summer visitor to Wales, with regular breeding only in Carmarthenshire and even there it was very scarce and local. Ingram and Salmon (1954) mentioned breeding at six sites but gave no dates. The last nest recorded in Breconshire was at Abergwesyn early that decade (Peers and Shrubb 1990), in Meirionnydd in 1951 (Peakall 1962) and in Caernarfonshire in the Aber Valley in 1952, from which the egg was taken by a collector (*Cambrian Bird Report 1956*). These were the last breeding records in Wales for 30 years. A census in 1971 found only 81 pairs, all in southern and eastern England (Bibby 1973). This decline was attributed to the effect of climatic change on food supply, but changes of land use and agricultural practice were also likely to have been factors, with egg-collecting important once it had become rare. Regular breeding in England ceased by the end of the 1980s (*Birds in England* 2005).

The next breeding in Wales was in 1981, when a pair bred successfully at The Kymin, Gwent. A pair was found at Tumble, Carmarthenshire, on 8 July that remained to 5 August. A pair summered at the latter site in 1982–85 and a male was present in 1986, and although breeding was not proven there, it was strongly suspected (Rob Hunt *in litt.*). The only other nesting records were in Gwent, where a pair probably bred in 2004, two young fledged in 2005 and four young in 2006. The pair returned in 2007, but two nesting attempts failed during persistent rain and the birds did not return in 2008 (Roberts and Lewis 2012).

Since the 1950s, Red-backed Shrike has been primarily a scarce visitor to Wales, almost annually since records were collected nationally in 1967. These are most likely to be from the breeding population in Fennoscandia. The majority are seen in spring, with peak passage during the last week of May and the first week of June, although the earliest was at St Davids Head, Pembrokeshire, on 30 April 1977. Passage is more prolonged in autumn, from the last week of August to the end of September, although a few occur much later, such as in Newtown, Montgomeryshire, on 4 November 1992 and at Porthcawl Golf Course, East Glamorgan, on 15–29 November 2019. Most occurred on the islands: 20 on Bardsey, Caernarfonshire; 23 on Skomer, Pembrokeshire and 12 on Skokholm, Pembrokeshire. It was rare on the mainland, except in Pembrokeshire where it was recorded annually. Aside from the 1992 Montgomeryshire record, the only other inland-county records were in Breconshire, at Buckland on 29 July 1968, at Llangattock in August 1965 and June 1966, at Storey Arms in May 1974; in Radnorshire, in the Begwyns in August 1995, and in Montgomeryshire, at Carno in June 1996 and the aforementioned record in November 1992. The average numbers visiting Britain each year declined by 30% between the 1980s and 2010–18 (White and Kehoe 2020b), although the European population increased by 32% during 2007–16 (PECBMS 2020).

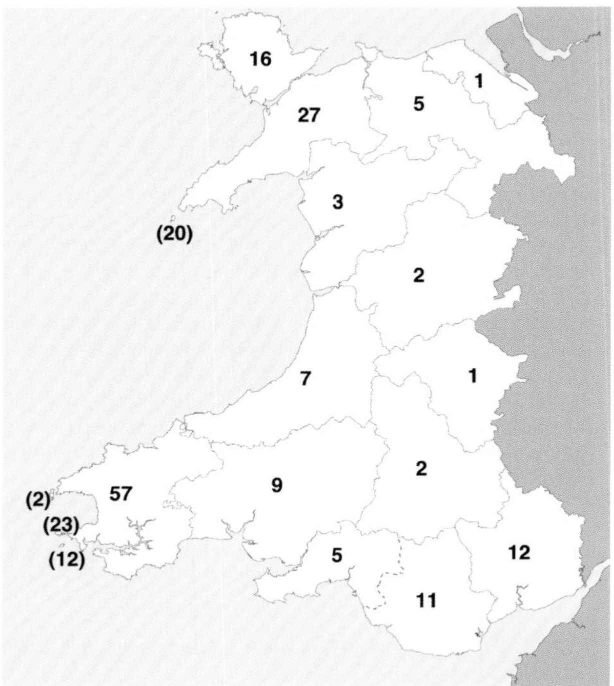

Red-backed Shrike: Numbers of birds in each recording area, 1967–2019.

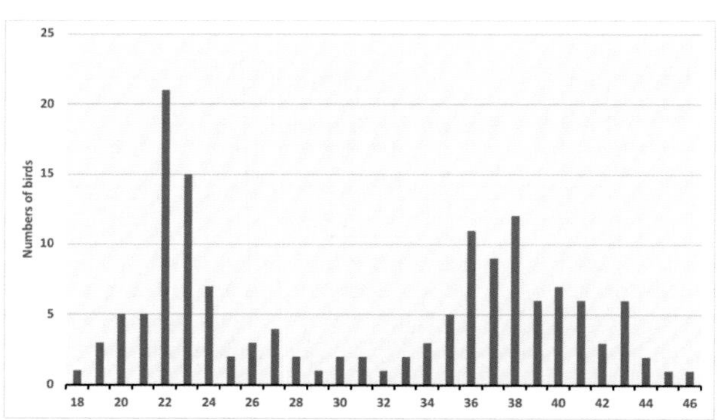

Red-backed Shrike: Totals by week of arrival (when known, n=152), 1967–2019. Week 22 begins *c.*28 May and week 43 begins *c.*22 October.

The prospects of Red-backed Shrike returning as a regular breeding species in Wales seem slim. Modelling suggests that southeast England may become climatically suitable by the end of the 21st century (Huntley *et al.* 2007), but whether land-use changes can cater for its needs is less certain. Some optimism may be gained from The Netherlands, where the population increased from 100 pairs in 1985 to 400–490 pairs in 2019, primarily where polder floodplains were converted back to semi-natural habitats from Rye-grass agriculture, from the late 1990s (Sovon 2020).

Jon Green, Robin Sandham and Rhion Pritchard

Turkestan Shrike *Lanius phoenicuroides* Cigydd Tyrecstan

Welsh List Category	IUCN Red List	Rarity recording
	Global	
A	LC	BBRC

The Turkestan Shrike, also known as Red-tailed Shrike, breeds in Central Asia and winters in the Sahel zone and east of Africa, in Arabia and around the Persian Gulf. There have been just seven records in Britain (Holt *et al.* 2020), of which the sole Welsh sighting was the second, a female at Cemlyn, Anglesey, from 2 July to 8 August 1998.

Daurian/Turkestan Shrike *Lanius isabellinus/phoenicuroides* Cigydd Dawria/Tyrcestan

Formerly considered one species, known as Isabelline Shrike, these two are now treated separately (IOC). Daurian Shrike breeds in northern China and southern Russia and winters in eastern Africa, the Sahel zone and Arabia, but also in northern India (*HBW*). While 107 records had been accepted as either one form or the other in Britain to the end of 2019 (Holt *et al.* 2020), only adults were identified specifically, some based on their plumage but many through DNA analysis. None recorded in Wales has been confirmed as Daurian Shrike, but there have been eight records that were either Daurian or Turkestan Shrike:

- Holyhead, Anglesey, on 25 October 1985;
- Nine Wells, Pembrokeshire, on 27 October 1995;
- 1CY at Bardsey, Caernarfonshire, on 25–26 October 1996;
- 1CY at Llanbedrog, Caernarfonshire, on 27–29 November 2003;
- 1CY at Great Orme, Caernarfonshire, on 22 October 2006;
- 1CY at Porth Clais, Pembrokeshire, from 28 October to 7 November 2011;
- 1CY at Breakwater Country Park, Anglesey, on 15–16 October 2017 and
- 1CY at Cosmeston Lakes, East Glamorgan, on 17 November 2018.

Jon Green and Robin Sandham

Lesser Grey Shrike *Lanius minor* Cigydd Glas

Welsh List Category	IUCN Red List		Rarity recording
	Global	Europe	
A	LC	LC	BBRC

The Lesser Grey Shrike breeds in southeastern Europe and east to Central Asia, with smaller populations in the northern Mediterranean, as far west as southern France and northeast Spain. It winters in southern Africa (*HBW*). The European population has declined steeply since at least 1990 (PECBMS 2020), particularly in the west of its range from where most British vagrants are likely to originate. It is now extinct, or nearly so, as a breeding species in several countries in which it was formerly quite abundant, including France. Climate fluctuations, particularly cooler and wetter summers, and the impact of increased use of insecticides and fertilisers on large invertebrates are the main causes of poor breeding success (BirdLife International 2020). In addition, there has been a series of chronic droughts in the Kalahari Desert of southern Africa, its main wintering grounds, since the 1970s. These factors make it unlikely that this species will occur too many more times in Wales, even though the area of Europe that is climatically suitable for Lesser Grey Shrike should theoretically enable it to spread north (Huntley *et al.* 2007).

To the end of 2019, there were 201 records in Britain, typically 2–3 each year during 1990–2019 (Holt *et al.* 2020), but it is a very rare vagrant in Wales, with only ten records:

- Holyhead Mountain, Anglesey, on 26 May 1961;
- Shotton, Flintshire, on 21–22 September 1961;
- Pen-y-groes, Caernarfonshire, on 8 June 1967;
- Skomer, Pembrokeshire, on 18 September 1974;
- Ferryside, Carmarthenshire, on 13 October 1975;
- Fan Pool, Montgomeryshire, on 16 May 1982;
- Abersoch, Caernarfonshire, on 17–18 October, then at Aberdaron, Caernarfonshire, from 20 October to 15 November 1986;
- St Davids, Pembrokeshire, on 24 June and then Skomer, Pembrokeshire, on 2–4 July 1993;
- Lower Caernhedryn, Pembrokeshire, on 22 September 1998 and
- 2CY at St Justinians, Pembrokeshire, on 4–9 July 2011.

Jon Green and Robin Sandham

Great Grey Shrike *Lanius excubitor* Cigydd Mawr

Welsh List Category	IUCN Red List	
	Global	Europe
A	LC	VU

The Great Grey Shrike breeds over large parts of Europe, Asia and North Africa (IOC). In Europe, the largest populations are in Fennoscandia and European Russia, where most are partial migrants, wintering south or west of their breeding range (BirdLife International 2015). Those that occur in Wales are probably from the population breeding in Fennoscandia, most of which winter in Poland and Germany. Its main requirement is an open area with perches from which it can watch for prey, such as small birds and mammals. When prey is abundant, birds will store some for later, typically on a thorn bush or barbed-wire fence. One such 'larder' in the Elan Valley, Radnorshire, contained 14 Wrens, six Blue Tits and four Short-tailed Voles (Jennings 2014).

Great Grey Shrikes wintering in Wales use wetland areas, commons with gorse or scrub, *ffridd* and commercial forestry, both recently planted and clear-felled areas. The availability of the latter in Wales increased from the 1990s as trees planted following the Second World War were felled. There is evidence that some individuals return to the same wintering site in successive winters

Great Grey Shrike at Garwnant, East Glamorgan © Tim Collier

(*BTO Migration Atlas* 2002). The sole ringing recovery involving Wales was one caught by a ringer at Cross Inn, Ceredigion, in March 2015, which had been ringed at Gibraltar Point, Lincolnshire, in the previous October.

As well as long-staying wintering individuals, there are passage migrants in autumn and spring and transient individuals in Britain during the winter (Fraser and Ryan 1995). The first arrivals usually reach Wales in mid-October. Records span the winter until early May, although one was in Nant Ffrancon, Caernarfonshire, on 17 June 1949. During 1990–2019, the latest spring record was at Lake Vyrnwy, Montgomeryshire, on 3 June 2014, and the earliest in autumn was at Abergavenny, Gwent, on 3 October 2008.

There were only about 11 records in Wales before 1900, such as one near Capel Curig, Caernarfonshire, which Eyton (1838) reported was in May, but did not give the year, and one shot near Builth Wells, Radnorshire, on 10 March 1865 (Jennings 2014). However, W.A. Rogers, whose father was a taxidermist at St Asaph, Flintshire, referencing one shot in the parish in December 1897, stated that about a dozen had been shot previously in the Denbigh and St Asaph districts (Forrest 1907). Great Grey Shrike remained a very scarce visitor in the first half of the 20th century, with only about eight dated Welsh records during 1900–49, although Hutchings, a taxidermist at Aberystwyth in Ceredigion, received about six specimens up to 1927 (Salter, cited in Roderick and Davis 2010). Numbers increased from the 1960s, and birds began to be recorded in successive winters at some sites. For example, one was around Cors Caron, Ceredigion, each winter during 1966–77.

There were 13 occupied 10-km squares in Wales over the four winters of the *Britain and Ireland Winter Atlas 1981–84*, mainly on the fringes of upland areas in South and mid-Wales, with few in the north. In the early 1990s, *c.*35 individuals regularly held territory each winter in Britain. An average of 40 was recorded for short periods of time at various sites (Fraser and Ryan 1995). At this time, no more than 5–10 wintered in Wales each year (*Birds in Wales* 1994), but records increased from the turn of the 21st century. Over the four winters of the *Britain and Ireland Atlas 2007–11* there were 35 occupied 10-km squares in Wales, widely distributed across mainland Wales. Increased occupancy since 1981–84 was also evident in southwest England, but there were fewer in eastern England, suggesting a westward shift in wintering distribution.

Great Grey Shrikes have been recorded, regularly, at some favoured sites. These include Wentwood in Gwent, Garwnant Forest in East Glamorgan, Mynydd Margam in Gower, Brechfa Forest in Carmarthenshire and the Llys-y-frân Reservoir/Ty Rhyg/Pantmaenog area of Mynydd Preseli, Pembrokeshire. In mid-Wales, favoured sites have been Cross Inn Forest and the upper Teifi Valley around Tregaron, Cors Caron and Llanddewi Brefi in Ceredigion, Glasfynydd Forest/Usk Reservoir on the Breconshire/Carmarthenshire border, and Lake Vyrnwy and the Dyfnant Forest in Montgomeryshire. In North Wales, there have been fairly regular records in the area around Bala and Cynwyd in eastern Meirionnydd since 2010, but the species has been most frequently recorded in Denbighshire, particularly in Clocaenog Forest, where birds were present in 15 of the 16 winters during 2003/04–2018/19, with two and possibly three present in some years. Other areas in Denbighshire with regular records have been Ruabon Moor/World's End/Coed Llandegla and Fenn's Moss. All but one of the records on Anglesey were passage birds, and 15 modern records in Flintshire between 1973 and 2013 were all recorded on a single date and mostly on the coast, so were most probably passage migrants.

Few have been recorded on the islands or on coastal headlands during migration. The last confirmed records from Welsh islands were in 2002, when there was one on Skomer on 10 October and one on Bardsey on 31 October. Some appeared to hold territories in November and December but were not recorded subsequently. These birds may have moved farther south over the coldest months of the winter, but very little is known about their movements. Some birds first appear in March and hold territory for a few weeks before moving on.

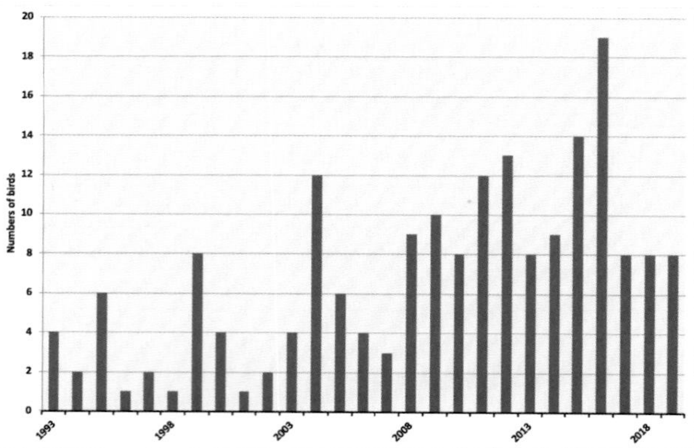

Great Grey Shrike: Estimated number of individuals in Wales in January–February during 1993–2019.

Estimating the number wintering in Wales is complicated by the potential mobility of birds at widely separated sites. A winter territory can be large: one at Mynydd Illtud, Breconshire, in 2008/09 was estimated to cover 5km². In 2009–18 there was an average of 11 Great Grey Shrikes in Wales in January and February each year, although up to 20 were present in 2016. These have contributed to an overall increase in the numbers recorded annually in Britain, averaging 233 in 2010–18, compared to 128 each year during 1986–2009 (White and Kehoe 2020b). There is no obvious explanation for the recent increase in its occurrence in Wales. These birds are assumed to originate from the large breeding population in northern Europe, which is thought to be stable (BirdLife International 2020), in contrast to the serious decline in western, central and eastern Europe since the mid-20th century. The proportion of its small British population wintering in Wales increased from 5% in the *Britain and Ireland Winter Atlas 1981–84* to 12% in 2007–11. Providing its north European population remains stable and suitable habitat is available in Wales, it should continue to be a regular winter visitor.

Rhion Pritchard and Jon Green

Woodchat Shrike *Lanius senator* Cigydd Pengoch

Welsh List Category	IUCN Red List		Rarity recording
	Global	Europe	
A	LC	LC	WBRC

This species breeds across much of Iberia and around the Mediterranean basin east to Iran, and winters in Africa south of the Sahara (*HBW*). The numbers recorded in Britain have gradually increased since the 1960s, to an average of 28 each year during 2010–18 (White and Kehoe 2020b). It is a scarce visitor to Wales. The first Welsh record was at Penally, Pembrokeshire, on 4 May 1923 and only five others were recorded prior to 1950. The numbers recorded in each subsequent decade were similar, at 10–15, except in the 1960s, which included six in 1968, and in the 2010s, including eight in 2011.

Total	Pre-1950	1950–59	1960–69	1970–79	1980–89	1990–99	2000–09	2010–19
124	6	11	24	12	12	13	14	32

Woodchat Shrike: Individuals recorded in Wales by decade.

The majority of records have been in the southern half of Wales. The earliest records in each South Wales county were at Singleton Park, Gower, in May 1947, at Sully, East Glamorgan, in May 1957, and at Peterstone, Gwent, in October/November 1983. One appeared to over-summer in southeast Gwent in 1993, having initially been seen at Magor Marsh on 14–18 May, then *c.*2km away at Magor Pill on 8 June and *c.*2.5km NE at Caldicot Moor on 8–16 August. Of the five records in Ceredigion, the first was at Glandyfi in August 1954. Of 61 in Pembrokeshire, the majority have been on the islands, but curiously, while Skomer hosted six prior to 1990 and eight since, Skokholm held 19 prior to 1990 and only three subsequently. In mainland Pembrokeshire, nine were around St Davids, and three each at Dale and Martin's Haven/Deer Park.

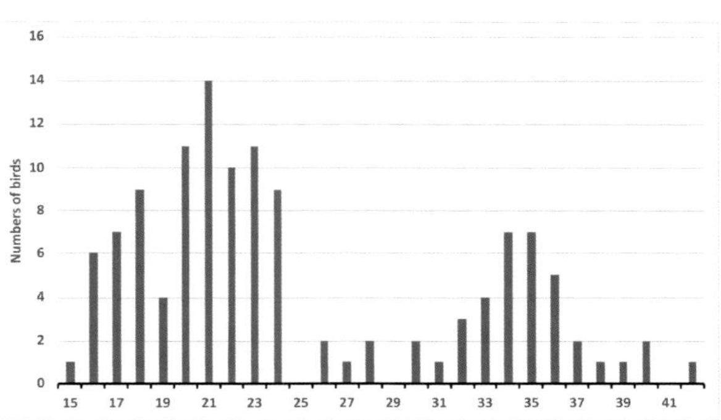

Woodchat Shrike: Totals by week of arrival, 1923–2019. Week 18 begins *c.*30 April and week 36 begins *c.*3 September.

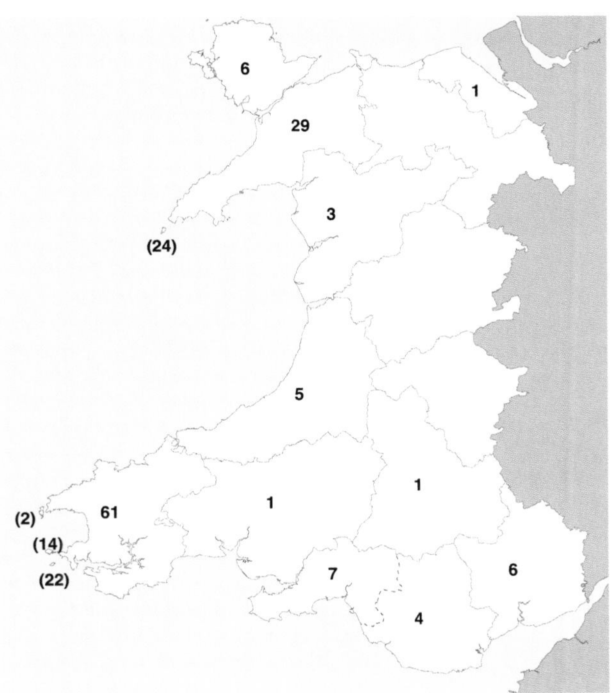

Woodchat Shrike: Numbers of birds in each recording area, 1923–2019.

The species has always been scarcer in North Wales. The majority of records there have been on Bardsey, Caernarfonshire, where the first was in August 1956. There were 18 during 1956–1990 and six since. The first record on Anglesey was at Porth Dafarch in June 1928 and the first in Meirionnydd at Tonfanau in June 1976 but there were no records in mainland Caernarfonshire, until one was found on the Great Orme in May 2006. The only Flintshire record was one caught at Shotton in August 1970. Cemlyn hosted four of Anglesey's 16 records, in August 1981, July 2012, May 2016 and May 2019.

Almost all Welsh records were close to the coast, with the exception of those at Trawsfynydd, Meirionnydd, in May 1986, at Llanwrtyd Wells, Breconshire, in August 1996 and at Tumble, Carmarthenshire, in June 2002, the only record from that county. The majority were in spring, overshooting the Mediterranean as they migrated north. It is a relatively late migrant to its breeding areas; hence the peak occurred in the last week of May and first week of June. A second-calendar-year male at Cosmeston Lakes, East Glamorgan, on 18–19 May 2002, was considered to be possibly the same, based on plumage similarities, as one at Aberthaw, 15km to the west, on 10 August 2002. Fewer were recorded in autumn, with a peak during the last week of August.

Jon Green and Robin Sandham

Red-eyed Vireo *Vireo olivaceus* Fireo Llygatgoch

Welsh List Category	IUCN Red List	Rarity recording
	Global	
A	LC	BBRC

This trans-Atlantic vagrant breeds from Canada south through the central and eastern USA and winters in northern South America (*HBW*). Up to the end of 2019, there were 156 records in Britain, typically 3–4 annually during 1990–2019 (Holt *et al.* 2020). There have been seven Welsh records, most of which were seen on just a single day.

- Skokholm, Pembrokeshire, on 14 October 1967;
- Aberdaron, Caernarfonshire, on 25–26 September 1975;
- Bardsey, Caernarfonshire, on 15 October 1985;
- Porth Clais, Pembrokeshire, on 18–19 October 1995;
- Bardsey, Caernarfonshire, found dead on 19 September 1998;
- South Stack, Anglesey, on 4 October 2001 and
- Skokholm, Pembrokeshire, on 12 October 2019.

Jon Green and Robin Sandham

Golden Oriole *Oriolus oriolus* Euryn

Welsh List Category	IUCN Red List			Birds of Conservation Concern	Rarity recording
	Global	Europe	GB	UK	
A	LC	LC	CR	Red	WBRC

This is a scarce visitor to Wales, usually as a spring overshoot, from the Mediterranean, though it breeds across Europe as far north as southern Sweden and Finland, east across Russia to western Mongolia, and south to Turkey and northwest Iran. It winters in equatorial and southern Africa (*HBW*). There was a small breeding population in the East Anglian Fens from the 19th century, which peaked at 20–30 pairs during 1989–98 (*Birds in England* 2005), but dwindled during the early 21st century. The last confirmed breeding in Britain was in 2009 (Holling *et al.* 2017).

The yellow-green rump of a Golden Oriole can lead to identification confusion with Green Woodpecker. An early dispute was reported by Giraldus Cambrensis in the 12th-century *Itinerarium Cambriae*, when the party argued about whether a bird heard between Caernarfon and Bangor, Caernarfonshire, was an oriole or a woodpecker. Golden Oriole has been recorded in all Welsh counties except Montgomeryshire, the majority on the westernmost islands and peninsulas. Most occurred in May and June, as birds overshot their continental breeding areas. There are few records after the start of July. Records in Britain increased towards the end of the 20th century, but it has occurred less frequently since 2000. Annual means increased from 34 in 1960–69 to 132 in 1990–99, but the annual average has since fallen to 77 in 2010–18 (White and Kehoe 2020b).

Male Golden Oriole © Tony White

The only suggestions of breeding in Wales were a pair reported nesting at Penarth, East Glamorgan, in a year between 1858 and 1863, a pair at Coedarhydyglyn, East Glamorgan, in 1883 and 1886, and one that summered at Hendrefoilan, Gower, in 1885 (Hurford and Lansdown 1995). These reports did not occur during a period of regular records of migrants in Wales, although five males were in Breconshire during 1889–91 (three in the last year). There were just four other Welsh records in the second half of the 19th century: at Goodwick, Pembrokeshire, and Rhuthun, Denbighshire, in April 1870, a pair at Llanilar, Ceredigion, before 1895 and a single at Cwm Woods, Ceredigion, in May 1895. It remained an extremely rare visitor to Wales, with only eight further records prior to 1967.

During 1967–82, males and females were seen at Wentwood, Gwent, and although there was no indication of breeding, it cannot be ruled out, as proof of nesting is difficult to obtain. The first 20th-century record in East Glamorgan was on 1 May 1971 at Kenfig, a site where a further four occurred subsequently, although there have been only two 21st-century records in the county. The first 20th-century record in Gower was one at Crawley Woods in May 1976, and there has been none in the county since one at Oxwich in May 1984, and none in Carmarthenshire since two at Dinefwr Park in June 2002.

Over half of Pembrokeshire's records were on the islands, including three on Skomer during 14–20 May 2006 and at least four there on 11–12 May 2015. RSPB Ynys-hir has hosted eight of the 22 records in Ceredigion, the last in May 2005. The vast majority of Caernarfonshire records were on Bardsey, where the first was seen on 25 April 1945. Numbers on the island peaked during 1990–99, when 14 were recorded, including five different birds during May 1994, and in 2000–09, when 17 were recorded. However, there were only nine in 2010–19. All the Anglesey records have been since 1984, when the first was at Penmon on 29 April. Four counties have recorded one each in the modern era: at Coed Glanrafon, Meirionnydd, on 19 May 1983, Erwood, Radnorshire, on 7 April 1990, Llangynidr, Breconshire, on 24 May 1992 and Connah's Quay, Flintshire, on 4 July 1996.

Total	Pre-1950	1950–59	1960–69	1970–79	1980–89	1990–99	2000–09	2010–19
219	22	4	10	37	27	51	43	25

Golden Oriole: Individuals recorded in Wales by decade.

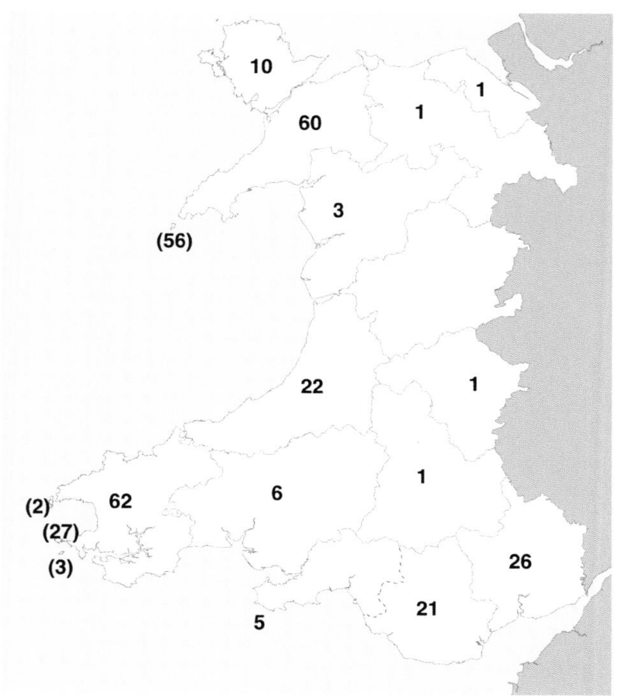

Golden Oriole: numbers of birds in each recording area, c.1858–2019.

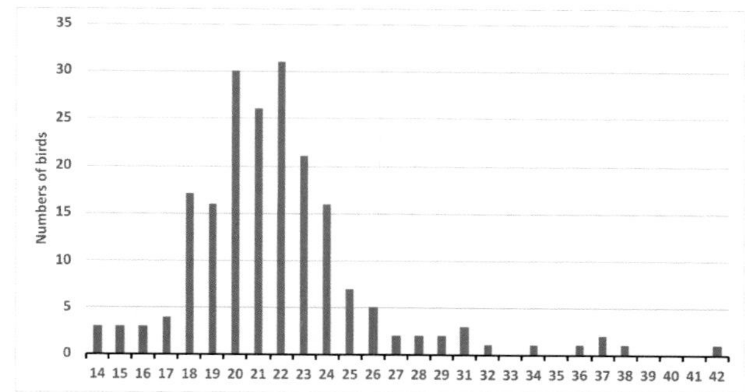

Golden Oriole: Totals by week of arrival (when known, n=196), 1870–2019. Week 18 begins *c.*30 April and week 26 begins *c.*25 June.

The authors of *Birds in Wales* (1994) expressed a hope that Golden Oriole would soon be added to the list of the nation's breeding species, but while the European population increased by 16% during 1982–2016 (PECBMS 2020), the trends in Britain make this highly unlikely in the short term.

Jon Green and Robin Sandham

Jay *Garrulus glandarius* Sgrech y Coed

Welsh List Category	IUCN Red List			Birds of Conservation Concern			
	Global	Europe	GB	UK	Wales		
A	LC	LC	LC	Green	2002	2010	2016

The Jay is found over almost all of Europe, northwest Africa and the Middle East, its range extending east across temperate Asia as far as Japan (*HBW*), although the subspecies *G.g. rufitergum* occurs only in Britain and northwest France (IOC). It is mainly a bird of woodland, usually of deciduous trees, but can be found in conifers, parkland and large gardens. Jays in Britain are mostly resident, with 98% of ringing recoveries within 50km of the ringing site (*HBW*). Pairs are highly site-faithful. Recoveries involving Wales showed no long-distance movements, although one ringed near Talsarnau, Meirionnydd, in June 1971, was shot in northeast Montgomeryshire in April 1973, 56km to the east.

Jays are omnivorous, primarily eating caterpillars and beetles gleaned from the tree canopy in spring and summer, but also birds' eggs when the opportunity arises. In autumn, nuts and seeds predominate and a single Jay can cache 3,000 acorns in a month for future consumption (*HBW*). Their scatter-hoarding behaviour can play an important role in tree regeneration (Pesendorfer *et al.* 2016). One in Radnorshire was observed repeatedly flying a distance of more than 2.5km to cache acorns (Jennings 2014).

Its distribution in Wales is closely linked to the extent of woodland and it has been reported breeding at up to 400m in several counties and up to 450m in Caernarfonshire. It was presumably very common before extensive forest clearance began in the Iron Age. It probably hit a low point in the late 19th century, when forest cover was at its lowest and persecution by gamekeepers was at its most intense. On Penrhyn Estate, around Penmachno and Dolwyddelan, Caernarfonshire, 306 were killed between 1874 and 1902, primarily on moorland where Jay densities would have been low. The number killed on lowland Pheasant-shooting estates was probably far higher. Contemporary naturalists suggested that it was common in Welsh woodlands but scarcer in counties such as Pembrokeshire (Mathew 1894) and almost unknown on Anglesey, despite being common in northern Caernarfonshire (Forrest 1907).

The Jay benefited from the reduction in gamekeeping during and after the First World War. Although there are few indications of population size between the wars, it most likely increased, although in Pembrokeshire it was still described as "resident, but not numerous" (Lockley *et al.* 1949). It surely benefited from the planting of conifers in much of Wales, spreading uphill in Ceredigion in the late 1940s (Roderick and Davis 2010), although it decreased considerably in Radnorshire and was not common anywhere in the county (Ingram and Salmon 1955). On Anglesey, Jay reached woodland skirting the Menai Strait by the early 1950s, but it remained scarce elsewhere on the island (Jones and Whalley 2004). The number of Jays killed by gamekeepers and farmers probably continued to fall through this period and certainly did so later in the 20th century: the National Gamebag Census showed that the number killed on estates managed for shooting in the UK fell 27% during 1966–2016. In the last year an estimated 8,500–17,000 were killed (Aebischer 2019), but there are no data specifically for Wales.

The Britain and Ireland atlases of 1968–72 and 2007–11 recorded Jays in around 90% of 10-km squares in Wales, though fewer during 1988–91. Its relative abundance in Wales was higher in 2007–11 than elsewhere in Britain and had increased since 1988–91, except in treeless upland areas and a few western coastal districts. Its distribution on Anglesey barely changed between the 1950s and 1988–91. The *North Wales Atlas 2008–12* showed an expansion through the east of the island, but it remained largely absent in the west. Where local atlases have been undertaken, the proportion of tetrads occupied varied between 43% across North Wales and Pembrokeshire, and 84% in Gwent. Where comparisons with earlier atlases are possible, only small increases in tetrad occupation are evident. While its range has not increased, other than on Anglesey, it has increased in abundance. The Repeat Woodland Bird Survey showed an increase of 54% in Welsh woodlands between 1965–72 and 2003–04 (Amar *et al.* 2006). During 2004–18, Jay numbers increased by 24% in Wales

Jay © Alan Saunders

in contrast to a stable population in England (Harris *et al.* 2020). There may be local variations, however: Jennings (2014) reported that following a large increase in Radnorshire during 1950–90, there had been a noticeable decline in some areas, which he suggested was a result of predation by Goshawks. The Welsh breeding population in 2018 was estimated to be 35,500 territories (Hughes *et al.* 2020).

There is no proof from ringing that other subspecies of Jay, *G.g. hibernicus* from Ireland or the nominate *G.g. glandarius* from northern Europe, have occurred in Wales, but irruptions of the latter occur from Scandinavia, when the acorn mast fails (Selås 2017). Such movements were recorded in eastern Britain in 1965, 1983 and 1997, the second of which saw a major influx into southern counties of Wales. The largest counts in 1983 were in Pembrokeshire, where 200 passed south over Martin's Haven on 6 October, 32 were on Skomer on 6 October and 38 on 17 October, and 127 flew west at Strumble Head on 19 October. In East Glamorgan, 118 flew NNW over Kenfig Pool on 6 October and a flock of at least 60 moved north there the following day. Large counts elsewhere that autumn included 45 heading west over Monmouth, Gwent, in late September and 50 north over RSPB Ynys-hir, Ceredigion, on 12 October. Numbers were lower in North Wales, the maximum day count on Bardsey, Caernarfonshire, was 19 on 29 October (John and Roskell 1985).

Birds in Wales (1994) mentioned other autumn influxes in 1923, 1935, 1947, 1975, 1977 and 1981. Subsequent movements were evident in 1996, 2005, 2009 and 2014, but none surpassed 1983 in volume. Counts in the 21st century included 50 heading west at Mewslade, Gower, on 10 October 2005 and 46 over Newport Wetlands, Gwent, on 4 October 2009. A flock of 50 at Bettws Bledrws, Ceredigion, on 11 November 2012 was also notable, but there were no unusually large counts elsewhere in Wales that year.

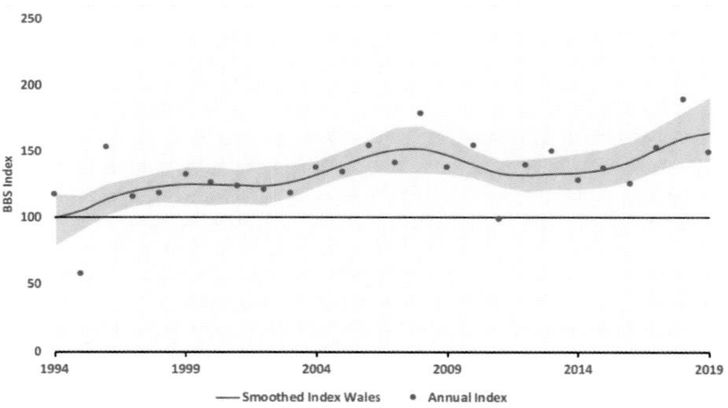

Jay: Breeding Bird Survey indices, 1994–2019. The red line shows the smoothed index, 1995–2018. The shaded area indicates the 85% confidence limits.

The species can legally be killed under General Licence for the purpose of conserving wild birds, but it was removed from licences in Wales to prevent damage to agriculture and public health and safety in 2019. There are no data on numbers killed in Wales. Studies have shown Jay to be the principal predator of Wood Warbler nests in Wales (Mallord *et al.* 2012a) and of Spotted Flycatcher nests in southern England (Stevens *et al.* 2008). However, neither study implicated the Jay as the cause of decline. Evidence of their impact on population levels of breeding songbirds is weak (Newson *et al.* 2019). The Jay does not appear to face any major conservation threat in Wales at present and recent levels of control do not appear to have a significant effect on the population.

Rhion Pritchard

Magpie *Pica pica* Pioden

Welsh List Category	IUCN Red List			Birds of Conservation Concern			
	Global	Europe	GB	UK	Wales		
A	LC	LC	LC	Green	2002	2010	2016

The Magpie is found over almost the whole of Europe and in a wide band across temperate Asia as far east as the Kamchatka Peninsula and south to Vietnam. It is a bird of open country with scattered trees, avoiding extensive woodlands, but since the late 20th century has become increasingly common in urban areas. It is an omnivore but mainly a carnivorous scavenger (*HBW*). Unlike Carrion Crow, it is seldom recorded at altitudes over about 400m in Wales. Those that breed in the most northerly part of its range move south in winter, but otherwise it is remarkably sedentary (Birkhead 1991) and site-faithful. Magpies move only short distances in Wales, with one exception: a ring fitted on a nestling on Skomer, Pembrokeshire, in May 1992 was found at Aberystwyth, Ceredigion, 112km away, in June 2009. This is the longest recovery distance recorded in Wales, although it is not known that this is where the Magpie perished.

The inclusion of eggs and young birds in the Magpie's diet led to it being heavily persecuted by gamekeepers from the 19th century. On Cawdor Estate at Stackpole, Pembrokeshire, 113 were killed in 1821 (per Donovan and Rees 1994) and 334 were killed on Penrhyn Estate between Penmachno and Dolwyddelan, Caernarfonshire, during 1874–1902. Dix (1866) said that at a site in the far south of Wales, two or three barrows full of dead Magpies were collected the morning after poison had been laid near a roost. In Pembrokeshire he reported that it was "very common but so readily destroyed that I fear it will soon be a rarity; still, as there are large tracts of country without a gamekeeper, it has a chance for the present". A generation later, however, the Magpie was described as "very numerous about all the wild and unpreserved districts of the county" (Mathew 1894). Contemporary authors said that it was common everywhere in Wales, except where numbers had been reduced by humans.

The Protection of Animals Act 1911 made it illegal to use poison in the open to kill wild birds. This, along with the reduction in gamekeeping during and following the First World War, enabled numbers to increase. It also benefited from afforestation of the Welsh uplands in the 20th century, which provided nest sites where there were none previously. In Ceredigion, it had spread into upland plantations by the 1940s (Roderick and Davis 2010).

Magpies were found in the breeding season in at least 98% of 10-km squares in Wales, in all three Britain and Ireland atlases, in 1968–72, 1988–91 and 2007–11. In the most recent atlas period, densities were highest in the southeast, Pembrokeshire and Anglesey, and lowest in the uplands, where they were lower than in 1988–91. In 2018 the breeding population in Wales was estimated to be 79,500 territories (Hughes *et al.* 2020).

Atlases at tetrad level suggested a contraction in distribution of between 3% and 8% since the 1980s, but all those undertaken in the southern half of Wales recorded Magpie in more than 80% of tetrads. The *North Wales Atlas 2008–12* found Magpie in 77% of tetrads, being least widespread in Meirionnydd, where it was recorded in 58% of tetrads and absent from some low ground, as well as the considerable areas of the county above 300m. Magpies also breed on several Welsh islands. Bardsey, Caernarfonshire, was colonised in 1952 and numbers there increased from the early 1970s, peaking at 15 territorial pairs in 2016. Six pairs bred on Skomer, Pembrokeshire, in 2017, but it is a rare visitor to nearby Skokholm, where there were only about 18 records up to 2019, of which 60% were in March–June.

Magpie numbers increased steadily across England until the early 1990s, substantially during 1964–1993 (Gregory and Marchant 1996), after which the population stabilised. Although there are no Welsh data for this period, commentaries in bird reports described a simultaneous increase. Sheep numbers in Wales increased considerably in the 1970s and 1980s, providing

A Magpie atop a Welsh Mountain ram © Ben Porter

valuable food sources of carrion and stock-feed. Roadkill from increased volumes of traffic, including a huge increase in numbers of Pheasants released for shooting, are likely to have provided a new food source.

Outside the breeding season, Magpies are frequently seen in parties of up to 30, with larger numbers sometimes observed at roosts. Lockley (1957) stated that winter roosts of 100 were common in Pembrokeshire, and Donovan and Rees (1994) mentioned a roost of 112 at Plumstone on 21 February 1988. A roost at Rhosgoch Bog, Radnorshire, numbered 150 in February 1982. One at Ynysyfro Reservoir, Gwent, contained up to 131 in winter 1987/88, and 133 were at Sker Farm, East Glamorgan, on 8 January 1989. The highest roost counts recorded in Wales were in an osier bed at Abersoch, Caernarfonshire, where 172 roosted on 9 December 1970 and 177 roosted in 1977 (*Birds in Wales* 1994). There have been no three-figure counts in recent years, but 92 fed on discarded takeaway food at Cross Hands, Carmarthenshire, on 22 September 2017 and there were 76 at Aber Ogwen, Caernarfonshire, on 13 February 2019.

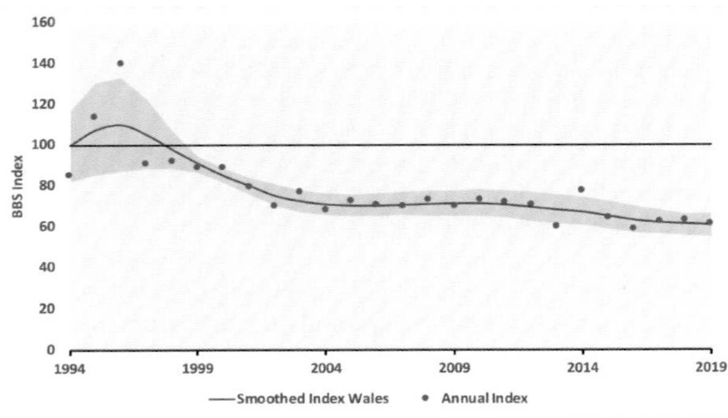

Magpie: Breeding Bird Survey indices, 1994–2019. The red line shows the smoothed index, 1995–2018. The shaded area indicates the 85% confidence limits.

The BBS index for Magpie in Wales declined by 43% during 1995–2018 (Harris *et al.* 2020), in contrast to a stable population across the UK as a whole. No analysis has been done to suggest possible reasons for this, though the reduction in sheep numbers in many areas may have reduced the food supply for rural Magpies. The species may be killed under General Licences to prevent serious damage to agriculture and to conserve wild birds, although numbers killed by gamekeepers in Britain have been lower than any other common corvid, except Jay (Aebischer 2019). The widespread adoption of the Larsen trap for predator control since 1990 was responsible for a large increase in the number of Magpies killed on shooting estates in the UK, but the trend is less clear in Wales (Aebischer and Harradine 2007). Jennings (2014) observed a noticeable decline in Radnorshire from the late 1980s, attributed to predation of adults by Goshawks and of young by Red Kites and Buzzards. The Magpie's predation of eggs and chicks has led some to implicate it in declines of songbird species, but several major reviews have found no evidence to support this. The most recent concluded that "while there is some evidence that Magpies can reduce the local productivity and abundance of prey species where they occur at high density, analyses of large datasets at a UK-scale provide little evidence of magpies on avian prey populations" (Newson *et al.* 2019).

Its noisy predation of the eggs and chicks of other birds means that some will not mourn the decline in the number of breeding Magpies that is evident in Wales. If it is related to the reduced availability of carrion from livestock, it may well be considered by many to be an acceptable outcome of more sustainable land use, but it would be valuable to properly understand this dynamic change.

Rhion Pritchard

Nutcracker *Nucifraga caryocatactes* Malwr Cnau

Welsh List Category	IUCN Red List		Rarity recording
	Global	Europe	
A	LC	LC	BBRC

This is a widespread bird of forests, particularly in mountainous areas, from northern and central Europe east across Asia (*HBW*). Irruptions occur as a result of a shortage of seeds of the Siberian Stone Pine, which sporadically results in arrivals in Britain, the largest and most recent of which involved at least 339 individuals in 1968. That irruption is thought to have involved *N.c. macrorhynchos* birds from western Siberia, rather than those of the subspecies *N.c. caryocatactes*, which breed in Europe (Slack 2009). There were also major European irruptions in 1911, 1933 and 1991, but none of these resulted in records in Wales. There have been eight records, involving nine birds, in Wales, of which five were during 1968:

- Mostyn, Flintshire, on 5 October 1753, the first British record;
- one killed in Glamorgan *c.*1915 and now at Taunton Castle Museum, Somerset.
- Newport, Gwent, on 19 October 1954, the only British record during an irruption across western Europe;
- Llangattock, Breconshire, in autumn 1957;
- two at Henllan, Carmarthenshire, on 26–31 August 1968;
- Beddgelert, Caernarfonshire, on 27 August 1968;
- Whiteford, Gower, on 30 August 1968 and
- Llanhilleth, Gwent, on 12 November 1968.

One reportedly killed near Swansea, Gower, in the early 1800s (*Birds in Wales* 1994) is no longer considered an acceptable record (Naylor 2021).

Jon Green and Robin Sandham

Chough *Pyrrhocorax pyrrhocorax* Brân Goesgoch

Welsh List Category	IUCN Red List			Birds of Conservation Concern				Rarity recording
	Global	Europe	GB	UK	Wales			
A	LC	LC	VU	Green	2002	2010	2016	RBBP

The Chough has a broad distribution from northwest Africa through southern Europe, the Middle East and Central Asia, with isolated populations in Ethiopia and northwest Europe, including Wales (*HBW*). The nominate form *P.p. pyrrhocorax* occurs only in Brittany, Cornwall, Ireland, the Inner Hebrides of Scotland, the Isle of Man and Wales, where it is largely restricted to the rugged, wild and weather-beaten coastal fringes. Here, they cavort in the updrafts of the seemingly never-ending onshore Atlantic winds that buffet the precipitous sea cliffs and breed in sheltered crevices. Smaller numbers cling to a more precarious existence in some of the disused slate quarries and mines in North Wales away from the coast, the only place in Britain where it nests inland, but numbers there are diminishing rapidly.

Breeding pairs occupy a territory that contains at least one suitable nest site within commuting distance of an adequate food supply, their 'home range' (Bignal *et al.* 1997). The edge of one pair's territory may overlap with those of neighbouring territories, especially if the food supply is good. Foraging habitats include

Choughs © Ben Porter

short, open maritime grassland and grassy heath, soft cliffs, bare soil/vegetation interfaces, earth/stone banks (*cloddiau*), sand dunes and coastal fields grazed by livestock. Using their slightly down-curved bill, Choughs probe for soil invertebrates, including the larvae of flies, especially Cranefly, and of beetles, such as Welsh Chafer. They also seek out ants and bees during spring and summer. Their diet in autumn can be supplemented by spilt grain in coastal arable stubbles (Meyer 1991) and Bilberries in upland areas such as Snowdonia. In some coastal areas, such as on Bardsey, Caernarfonshire, invertebrates associated with the strandline can be locally important. Birds have been recorded taking supplementary food, such as that provided on a rocky knoll at Church Bay, Anglesey, and at RSPB South Stack, Anglesey, some have learned to access sunflower seeds from a hanging feeder.

Chough remains found in southern Wales were dated to the Late Glacial period and the Mesolithic era at Paviland Cave, Gower (11,000 years ago), at Hoyle's Mouth near Tenby, Pembrokeshire, (10,000 years ago) and at Port Eynon Cave, Gower (6,300–11,000 years ago) (Yalden and Albarella 2009, Eastham 2016). In the historic period, the species was found in Flintshire, Caernarfonshire and Anglesey during the 1770s (Pennant 1810), but rarely away from the coast (Montagu 1802). Chough was described as rather numerous in Aberconwy, Caernarfonshire, (Williams 1835) and as common on all the headlands of the Welsh coast (Eyton 1838).

A review of Choughs in Britain during 1780–1982 suggested that the British breeding population fell from 509–737 pairs in the 1750s to 105–127 pairs in the 1940s, with the steepest declines occurring between 1820 and 1900 (Owen 1988). That is a fair reflection of its history in Wales. It was common on the Pembrokeshire coast in the mid-19th century, but much scarcer there by the turn of the 20th century (Mathew 1894). In North Wales it became scarcer from *c.*1865, with many sites deserted by 1900, although considerable numbers remained in Meirionnydd and Caernarfonshire (Forrest 1907). Choughs ceased to breed at Dunraven, East Glamorgan, before 1860, on the Gower coast about 1895 (Hurford and Lansdown 1995) and at Telpyn Point, Carmarthenshire, around or shortly after 1920 (Lloyd *et al.* 2015). Less is known about the inland populations. Choughs were abundant around mines and quarries. Such was the affinity of miners for the birds that chicks were sometimes removed from nests, reared as pets and shown at local 'cage-bird' shows. However, Choughs were said to have ceased nesting at Llanberis slate quarries, Caernarfonshire, in 1894 (Forrest 1907). It is unlikely that Choughs ever bred in Flintshire, despite Pennant's comments, and although they were said to have nested at Ysgyryd Fawr, Gwent, around 1880 (per Humphreys 1963), this must also be regarded as doubtful. There are no breeding records from Breconshire or Radnorshire.

Nest plundering for egg-collections was rife in the decades around the turn of the 20th century. In Pembrokeshire, Mathew wrote in 1894 that "There can be no doubt that 50 years ago the Chough was a common bird on the coast all the way round from Tenby to St Davids Head, and on towards Cardiganshire about Dinas etc. It is now rapidly becoming scarce, and if it were not for its sagacity in building in holes and crannies of inaccessible cliffs, it would long ago have been exterminated, as all its eggs would have been taken to meet the demands of collectors". Forrest (1907) attributed the initial declines in North Wales to natural causes, particularly competition from Jackdaws, but said that egg-collecting rapidly became an important factor.

Breeding may have ceased on Anglesey in the 1930s. It was stated in 1933 that Choughs had not been seen on the island for some years (Jones and Whalley 2004). It was lost as a breeding bird on the Great and Little Ormes, Caernarfonshire, in the 1940s (North *et al.* 1949), but in Pembrokeshire it was described as "probably no scarcer now than it was fifty years ago", with an estimated 35 breeding pairs (Lockley *et al.* 1949). By the 1950s there were a few signs of recovery. Ingram *et al.* (1966) reported that Choughs had been slowly increasing along the Ceredigion coast for the previous decade. A pair may have bred on Puffin Island, Anglesey, in 1953, and breeding was confirmed on Anglesey from 1959. Little is known about the small population in Montgomeryshire before the 1950s (Holt and Williams 2008).

A survey estimated 99–104 breeding pairs of Chough in Wales in 1963 (Rolfe 1966). This was almost certainly an underestimate, as many areas were not covered well, including central Wales and Pembrokeshire. Furthermore, it was undertaken immediately after the severe weather of winter 1963, which may have killed many birds. Numbers increased gradually in Pembrokeshire, North Wales and Ceredigion in the late 1960s and 1970s (Donovan 1972, Harrop 1970, Roderick 1978). By the next national survey in 1982, the Welsh population had increased to *c.*140 pairs (Bullock *et al.* 1983). There continued to be real growth into the early 21st century, although some of the increased population figures, charted in 1992, 2002 and 2014, are almost certainly a result of improved coverage, generated by knowledge of historic and traditional nest sites. In 2014, Wales held 213–258 pairs, which was *c.*55% of the UK and Isle of Man population (Hayhow *et al.* 2018b). The most recent survey estimated that *c.*800–1,000 pairs bred in the Republic of Ireland (Gray *et al.* 2003).

Given their natural charisma and the importance of the Welsh population, Choughs have attracted much interest. There has been almost complete monitoring of the population for over 25 years in mid- and North Wales (through the Cross and Stratford Welsh Chough Project), at Bardsey Bird and Field Observatory and in Pembrokeshire (Hodges and Haycock 2020). This effort, alongside national surveys, provides a detailed account of the status of Choughs in Wales over 60 years. Around the coast, there have been some welcome increases, such as the breeding

	1963[1]	1982[2]	1992[3]	1995–99[6]	2000–04[6]	2002[4]	2005–09[6]	2010–14[6]	2014[5]	2015–19[6]
Anglesey	2	8	13 (16)	29 (30)	37 (40)	37 (39)	36 (42)	38 (41)	35 (40)	39 (41)
Caernarfonshire	42	51	56 (66)	65 (74)	81 (96)	81 (101)	92 (101)	87 (94)	78 (92)	89 (93)
Denbighshire	1	1	0 (1)	0	2 (2)	2	2 (3)	1 (2)	2 (2)	2 (2)
Meirionnydd	7	12	8 (9)	16 (17)	17 (19)	19	19 (21)	19 (20)	17 (18)	14 (14)
Ceredigion	9	16	15 (21)	16 (18)	25 (29)	25 (29)	29 (34)	25 (29)	21 (27)	20 (29)
Montgomeryshire	5 (7)	2	4 (5)	2 (2)	1 (1)	1	0 (1)	0	0	0
Pembrokeshire	33 (36)	49–52	53 (58)	46 (50)	54 (59)	46 (67)	65 (71)	59 (67)	56 (71)	70 (78)
East Glamorgan/ Gower	0	0	1			1 (4)			4 (8)	
Carmarthenshire	0	0	0	0	0	0	0	0	0	1(1)
Wales total	**99–104**	**139–142**	**150 (177)**	**174 (191)**	**217 (246)**	**212 (262)**	**243 (273)**	**229 (253)**	**213 (258)**	**235 (258)**

Chough: Breeding pairs during single-year surveys: (1) Rolfe 1966, (2) Bullock *et al.* 1983, (3) Green and Williams 1992, (4) Thorpe and Johnstone 2003, (5) Hayhow *et al.* 2018b; and average in five-year periods since 1995 using data from Cross and Stratford Welsh Chough Project and Hodges and Haycock (6). Figures are confirmed pairs; those in parentheses are the total confirmed, probable and possible pairs.

population on Anglesey more than doubling from 16 pairs in 1992 to 41 in 2019, and an increase to at least seven nesting pairs on the Great and Little Ormes, in northeast Caernarfonshire. Numbers in Pembrokeshire have increased by 25% since 1992. Up to eight pairs now breed on Gower following the return of the first pair in 1990, and 1–2 pairs nest on the Vale of Glamorgan coast. The total Welsh population appears to be holding up well, but these figures hide some very worrying local trends.

Since the mid-1990s, almost all of the dozen or so inland breeding sites in Montgomeryshire and Ceredigion, first identified in the 1960s, have been lost. The last active nest of seven known in Montgomeryshire was near Llangynog in 2008. Of six inland Ceredigion sites active in the 1980s or 1990s, only the most northerly one was still occupied in 2019 (Cross and Stratford data), having been occupied continuously since at least 1979 (Roderick and Davis 2010). A similar decline has occurred more recently in southern Snowdonia and around Blaenau Ffestiniog, Meirionnydd, although some pairs do still hang on in these areas. By 2019, the number of occupied inland territories was 72% lower than in 1994. The coastal population was 12% lower, with the majority of the decline occurring in 2014–19 (Cross *et al.* 2020).

Prior to their final abandonment, several inland sites were occupied for a year or more by single birds (Cross and Stratford data). This was also noted by Montagu (1802): "a pair of these birds had for many years bred in the ruins of Crow Castle [Castell Dinas Brân], in the Vale of Llangollen, Denbighshire. By accident one of them was killed, and the other continued to haunt the same place for two or three years without finding another mate." An increased incidence of widowed birds, pairing up from adjacent inland sites, instead of each recruiting new partners, has been observed in Snowdonia over the last decade, usually leading to abandonment of one nest site (Cross and Stratford data). Extra sites on the Ceredigion and North Wales coasts have been created by the provision of nest-boxes since 1991. These have helped to offset inland losses and maintain the overall population, but do not compensate for the reduction in range and abandonment of sites that were used for many decades.

Choughs are gregarious and may flock at all times of the year. Williamson (1959), writing about the Isle of Man in the 1930s, referred to the flock as "the hub of Chough social life". The largest gatherings usually form in late summer or autumn, when family parties join subadult non-breeders at communal feeding areas and roosting sites. Colour-ringing shows that breeding pairs fly to these areas with their recently fledged young, which are 'deposited' in the communal flock, before the adults return to their breeding territory. Such movements can be as far as 80km, from Bardsey to Conwy Mountain in Caernarfonshire, and from the Ceredigion coast to the Nantlle Valley in Caernarfonshire. This must present a substantial risk to the young birds, so the benefits of being in a social group must be considerable. Several sites in Caernarfonshire are regular gathering points from nest sites as far away as mid-Ceredigion, north Anglesey, Bardsey and Llŷn, although one in Llanberis Pass has fallen into disuse in recent years. The largest feeding flocks in North Wales, all in Caernarfonshire, were *c.*90 at Uwchmynydd in Septembers 2002 and 2020, 70–80 on Conwy Mountain in 2011, 67 at Nant Peris in 1996 and 65 in the Carneddau in 2000 and 2017. In North and mid-Wales these are often in the Bilberry areas of the uplands or on semi-improved pastures with good invertebrate populations. In Pembrokeshire, feeding flocks of up to 70 occur on the coastal grasslands and slopes. The largest roost in Wales was at Craig-yr-Aderyn, Meirionnydd, where up to 80 roosted in 1991, but in recent years the largest have been in Caernarfonshire, where several roosts of more than 60 birds have been recorded (Cross and Stratford data).

National surveys showed a reduction in the proportion of non-breeders in the population, from 30% in 1963 and 1982 (Rolfe 1966, Bullock *et al.* 1983) to 20% in 1992 (Green and Williams 1992), though this increased to 28% in 2014 (Hayhow *et al.* 2018b). In Pembrokeshire, estimates of the non-breeding population made in spring, during 1992–2019, averaged 25% of the total, with no obvious change through that period (Hodges and Haycock 2020). While adult Choughs are fairly sedentary, pre-breeding birds (up to four years old) disperse widely and some move a long distance. Since June 1991, the Cross and Stratford Welsh Chough Project has colour-ringed almost 6,000 birds in North and mid-Wales, most as nestlings. Of these, only 13 are known to have moved outside Wales: two to Yorkshire, one to Lancashire and ten to the Isle of Man, eight of which were probably in a single exodus in 2004. Two of that eight, both males, subsequently returned to nest in Wales and one female joined the Isle of Man breeding population, as did a male that hatched in Wales in 1997. No Isle of Man reared birds have yet been proven to nest in Wales. In 2019/20, two first-calendar-year siblings, from a nest in north Anglesey, spent six months near Settle in the Yorkshire Dales National Park, 153km to the northeast, but one was found dead there in March 2020. The longest confirmed movement was a bird ringed as a nestling in 2016, coincidentally at the same north Anglesey nest site as the two that later went to Yorkshire. The 2016 nestling was at Lock's Common, East Glamorgan, in November that year, 224km to the southeast, but had returned to Anglesey by the following spring.

Colour-ringing in Pembrokeshire and Gower has shown similarly low levels of dispersal, although there is some movement between South Wales and Somerset, which currently has no breeding Chough population. A chick hatched in Gower in 2004 was at Sand Point, near Weston-super-Mare, in April 2006. It returned to Somerset in March 2007, among a flock of five, including a sibling that was at Brean Down, and displayed to a bird that had hatched from the same Gower site in 2003. That displaying pair had been at Porthcawl, East Glamorgan, just six days previously. The 2004 siblings subsequently returned to South Wales, but the 2003 bird was not recorded again.

After wandering widely in the first year or two, Choughs generally return to their natal area to establish a nesting territory. From a sample of 411 Welsh nesting adults (200 males and 211 females), the mean distance between hatching site and their first nesting site was 19.6km. Typical of most bird species (Kingma *et al.* 2017), females averaged nearly twice the distance of males, 25.5km to 13.3km respectively (Cross and Stratford 2015). The difference was also evident in Pembrokeshire (Haycock 2019b). This strategy may reduce the likelihood of in-breeding, although that does occur occasionally. Choughs are fairly long-lived, averaging about seven years, but the current longevity record for one in Britain and Ireland is held by a female ringed as a nestling between Borth and Aberystwyth, Ceredigion, on 27 May 1996. She reared three chicks near Llanrhystud, Ceredigion, in 2019 and was aged 23 years and 11 days when the nest was visited on 7 June 2019 (Robinson *et al.* 2020).

The reasons for the severe reduction of the inland population during the last three decades are not fully understood. It may be due to a combination of factors, including reduced livestock grazing, afforestation and predation by a recovered Peregrine population or increasing numbers of Goshawk, both of which will hunt Chough. It may be coincidental, but the peak occupancy of isolated inland sites was during the lowest ebb in the Peregrine population, during 1960–80. Adult survival and chick productivity appear not to have changed during 1991–2019, but there has been a substantial reduction in survival of first-time breeders being recruited into the breeding population at around 2–4 years old (Cross *et al.* 2020). Lower post-fledging survival may have been a factor in the decline in the non-breeding population that occurred during the 1980s and 1990s.

Concerns have been raised since the 1980s about the serious impact of agricultural change on Chough feeding areas. Bullock *et al.* (1985) cited the negative effect of the removal of sheep grazing on Ramsey, Pembrokeshire, in 1968–82, and the positive effect of a steady increase in stocking density on Bardsey, Caernarfonshire, in 1958–82. Rabbits can also help to create good conditions for foraging Chough, but their numbers have also fallen sharply, by 58% in Wales between 1996 and 2018 (Harris *et al.* 2020).

A reduction in, or removal of, livestock can result in pastures and coastal slopes becoming rank and overgrown, reducing the availability of soil invertebrates to Choughs. Where grazing of coastal slopes is abandoned, scrub encroachment can render favoured feeding areas totally inaccessible. Arable farming, which is more frequent in Pembrokeshire than other parts of its Welsh range, can reduce the number and species composition of soil invertebrates exploited as food, but can also provide additional feeding opportunities in autumn and winter.

Welsh agri-environment schemes (Tir Cymen, Tir Gofal and Glastir) mostly focused on reducing or removing livestock from coastal grasslands and upland pastures, frequently with the aim of increasing areas of heather; but they have often resulted in impenetrable gorse or other scrub. Other changes in farming practice, such as ploughing and reseeding of permanent pasture and changes in the breeds of sheep and cattle favoured, have resulted in previously frequented areas now being in unfavourable condition for feeding Choughs. This is particularly true of inland sites in North Wales: pairs feeding young in Blaenau Ffestiniog now frequently spend in excess of two hours away from the nest on foraging trips, whereas 30–60 minutes would have been typical in the past (Cross and Stratford data).

Weather events can affect Chough survival. Very cold, snowy winters and prolonged dry conditions make it difficult for the birds to feed, although these problems tend to be short-term. Choughs could, however, face potentially longer-term challenges from climate change. More regular periods of increased storminess and more intense rainfall could affect soil invertebrate populations and cause

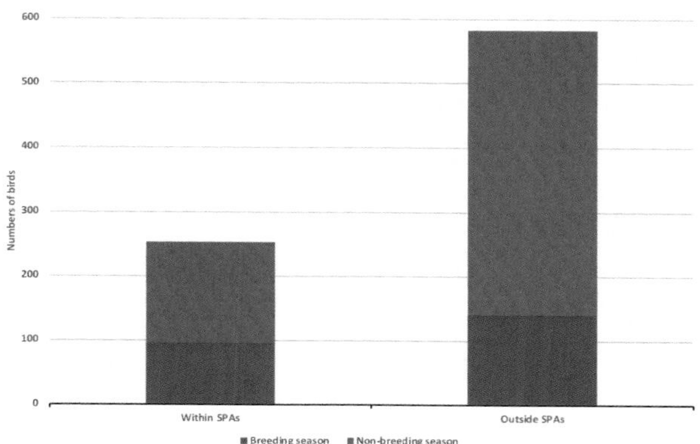

Chough: Number of (a) breeding pairs and (b) non-breeding individuals in Special Protection Areas in Wales in 2014–19, based on Hayhow *et al.* (2014) and Cross *et al.* (2020).

the loss of nest sites through coastal erosion. Storms in April can affect the nesting success of west-facing nests, especially those in sea caves. Despite legislation that prohibits egg collecting, including the introduction of custodial sentences in 2001, the relative scarcity of Choughs and their strong tendency to use traditional nest sites made them targets for egg-collectors in particular problem areas. There were 28 known nest thefts in 1990–99 and 11 during 2000–08, but none since, and additionally there were two incidents of disturbance: two adults shot and one nest destroyed intentionally during 1994–2003 (RSPB data). One nest in a quarry building at Penmaenmawr, Caernarfonshire, was targeted annually from at least 1992 to 2002, with repeat clutches stolen in some years, until quarry staff fitted grills to the doors and windows in 2003. A nest in a building in Penrhyn Quarry, Caernarfonshire, was also targeted almost annually, and the pair abandoned the site after 2009.

Chough should be a high conservation priority in Wales given the country's relative importance to the UK population and the signs of decline and range loss. The establishment of Special Protection Areas (SPAs) is important, but only if land management within these areas delivers their needs, particularly for short sward grassland and an abundance of soil invertebrates. While 40% of the Welsh Chough population breeds within the designated SPAs, the proportion of the non-breeding population protected by SPAs is only 26%. The importance of pre-breeding flocks to the species' ecology makes this an important gap to plug. Schemes by the Pembrokeshire Coast National Park Authority and Llŷn AONB are aimed at tackling the decline of traditional management on coastal slopes and cliff tops that should benefit Chough, but these efforts need to become mainstream in agricultural policies and funding across the species' range. Increasing its resilience will also be important in the face of climate change. In order to assess such changes, the wealth of information collected by long-term surveillance and colour-ringing studies is essential in monitoring the health of the population in Wales.

Tony Cross and Adrienne Stratford
with additional information by Jane Hodges
and Bob Haycock

Sponsored by Bardsey Bird and Field Observatory

Jackdaw *Coloeus monedula* Jac-y-do

Welsh List Category	IUCN Red List			Birds of Conservation Concern			
	Global	Europe	GB	UK	Wales		
A	LC	LC	LC	Green	2002	2010	2016

The Jackdaw is found over most of Europe except for northern Fennoscandia. Its range extends in temperate Asia as far east as Kashmir and western China, as well as to northwest Africa (*HBW*). Jackdaws breeding in Wales are of the subspecies *C.m. spermologus*, but a small number of the nominate form *C.m. monedula* ('Nordic Jackdaw') also occur (IOC). It is a bird of open country, preferably with scattered trees, and breeds semi-colonially in natural and artificial cavities. It has been recorded nesting in burrows on Skomer in Pembrokeshire, Bardsey in Caernarfonshire and sometimes on the mainland. It frequently nests in chimneys and can become very tame in urban areas. It may well have been associated with castles and walled towns from the medieval period. It is mainly a lowland species, scarce as a breeding bird above *c.*300m in Wales, and is mainly a resident (*HBW*). There have been no ringing recoveries to suggest movements between Wales and continental Europe, but several show that some Jackdaws breeding in Wales spend the winter in Ireland: several ringed at North Slob, Co. Wexford, in winter were later recovered in Wales, most during the breeding season, and a nestling ringed on Bardsey in June 1960 was found dead in Co. Meath in February 1964.

The earliest records of Jackdaws in Wales come from parish records and relate to payments made to kill them, such as one shilling for killing 14 Jackdaws at Llanfair Dyffryn Clwyd, Denbighshire, in 1757 (Matheson 1932). Late-19th-century and early-20th-century authors described it as common in rural counties such as Pembrokeshire (e.g. Mathew 1894) and in urban towns in Glamorgan (per Hurford and Lansdown 1995). Forrest (1907) commented that while Jackdaws were common almost everywhere in North Wales, "to see it literally in thousands we must visit its rocky haunts along the coasts of Merioneth and Carnarvon".

Data are scant for changes in the 20th century, but "in England and Wales, although little numerical evidence is available, the local literature is almost unanimous in recording increased numbers during the 20th century" (Parslow 1973). The *Britain and Ireland Atlas 1968–72* showed Jackdaws to be present in over 98% of 10-km squares in Wales, and subsequent atlases showed little change. In 2007–11 the highest relative abundance was along the north and south coasts, particularly Pembrokeshire and Anglesey, and in the valleys of eastern Wales. It is scarcer in upland areas. However, relative abundance was lower than it had been in the 1988–91 *Atlas*, particularly in North and mid-Wales, the only part of Britain or Ireland to show such a change. Some tetrad atlases have also shown a small decline in breeding distribution. In East Glamorgan, Jackdaws were recorded in 88% of tetrads in 1984–89, but in only 79% in 2008–11, while in Pembrokeshire they were in 94% of tetrads during 1984–88, but only in 89% in 2003–07 (*Pembrokeshire Avifauna* 2020). The decline was less marked in Gwent, with birds in 92% of tetrads in 1981–85 and 90% in 1998–2003. Elsewhere, its distribution was more sparse, being found in only 65% of tetrads in the *North Wales Atlas 2008–12*. Although recorded in almost every tetrad in Flintshire and Anglesey, it occupied only 34% of tetrads in Meirionnydd, and was generally absent from high mountains and moorlands.

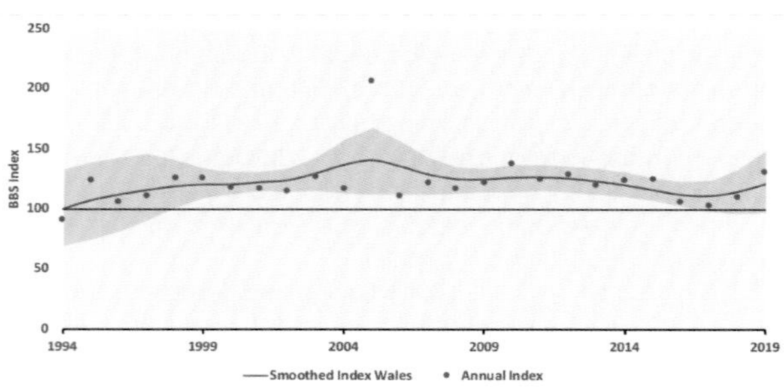

Jackdaw: Breeding Bird Survey indices, 1994–2019. The red line shows the smoothed index, 1995–2018. The shaded area indicates the 85% confidence limits.

The population increased across the UK during 1967–2017 (Massimino *et al.* 2019). In Wales the BBS index was stable during 1995–2018, 7% above its baseline at the end of that period (Harris *et al.* 2020). The Welsh breeding population in 2018 was estimated to be 155,000 (135,000–175,000) territories (Hughes *et al.* 2020).

There is little evidence that Jackdaws significantly affect livestock or crops (Newson *et al.* 2019), but they can be killed under General Licence to prevent serious agricultural damage or to conserve wild birds. It is not known how many are killed under these licences, but the estimated numbers killed by gamekeepers across the UK increased from 75,000 in 2012 to 91,000 in 2016, a far greater total than Magpies (Aebischer 2019). There was no clear trend in the numbers killed in Wales during 1960–2006 (Aebischer and Harradine 2007).

The best-studied Jackdaw populations in Wales are on the Pembrokeshire islands, where their fortunes have fluctuated. They bred in burrows on Skomer as early as 1860 (Lockley *et al.* 1949), nested in "fair numbers" on Caldey (Wintle 1924) and were an established breeding species on Ramsey in 1927 (per Donovan and Rees 1994). Numbers on Skomer increased from 20 pairs during 1946–58 to 200–250 pairs in 1961, after which control measures were deployed until 1971, as their predation was considered to be damaging the auk colonies (Saunders 1963, Gilham 2004). Jackdaws have been recorded predating Manx Shearwater eggs from burrows on Skomer (Annie Haycock *in litt.*) and on Skokholm (Brooke 1990). A census in 1991 recorded 248 pairs, but a substantial decrease followed: 53 pairs bred in 2015 and only 21 pairs in 2018. Skokholm was colonised in 1965, possibly by birds from Skomer. Numbers peaked at 60 pairs in 1975 and 1978, but declined to as few as six pairs during 1989–96 and were estimated at 15–20 pairs during 2011–17. Studies on the islands showed very low productivity, even during periods of population growth, indicating that birds were being recruited from the mainland. The decline that followed was suggested to be a result of there no longer being a surplus of birds on the mainland (Rees *et al.* 2009). On Bardsey, Caernarfonshire, the number of breeding pairs remained consistently at 30–50 pairs during 1953–83 (Roberts 1985) and at around 20–30 pairs during the 1990s, but there was a sudden drop from 23 pairs in 2004 to just one pair in 2005, and Jackdaws have not bred on Bardsey since. The reason was unclear, as the species was still doing well in Llŷn.

Autumn movements are sometimes observed on the coast and islands, such as 1,850 that flew southwest from Bardsey in October 1975, only to return later and cross to the mainland. Movements on Bardsey during 16–27 October 2003, involved flocks flying towards Ireland, peaking at 600 on 17 October. There were counts of 603 there on 3 October 2019 and 611 on 30 October 2019.

The largest counts of Jackdaws are associated with communal roosts, sometimes in the company of other corvids. Many of the largest gatherings have been reported from East Glamorgan, including 5,000–10,000 over Cosmeston Lakes in December 2005, *c.*6,000 at Danygraig in January 1991 and *c.*5,000 at Craig Llanishen in November 1965, Groesfaen in October 1991 and Llwynypia in January 1994. Flocks have been much smaller in recent years, but there were 3,000 over Cwm Bach on 20 October 2019. Large counts elsewhere included *c.*10,000 in Pembrey Forest, Carmarthenshire, in November 1976, up to 5,000 at Crymlyn Bog, Gower, at both ends of 1996, and a roost of at least 3,000 in Penrhyn Park, Caernarfonshire, in January 1966. Large numbers, such as 2,500 at Maelienydd Common, Radnorshire, in August 1995, were formerly recorded in late summer on high ground in several counties. Often in the company of Rooks and Carrion Crows, the birds fed on grassland, possibly on Cranefly larvae. While flocks are still seen, the numbers involved are now far smaller.

Despite evidence of local declines, such as on the islands, the Jackdaw is a very adaptable species and should continue to do well, although the availability of suitable nest sites may limit numbers in some areas.

'Nordic' Jackdaw *C. monedula monedula*
Jac-y-do Llychlyn

The nominate form breeds mainly in Fennoscandia, where it is migratory in the northernmost part of its range. Movements to western Europe occur during severe winter weather. There are also areas of intergrade between it and *C.m. spermologus* in eastern Europe. Individuals showing characteristics of *C.m. monedula* have been assessed by the WBRC only since 2010. During 2010–19, 21 were recorded, of which ten occurred in Pembrokeshire, including one on a housing estate in Haverfordwest in 2010–13 and two at Stackpole in 2017. Aside from singles at Reynoldston, Gower, in February 2014, Aberystwyth, Ceredigion, in October 2014 and at Rhossili, Gower, in December 2018 and January 2019, the remainder were in North Wales. There were three at Llandudno Junction, Caernarfonshire, in January 2013, singles on Anglesey at Pentraeth in February 2013 and Llanddona in October 2013, at Deganwy Golf Course, Caernarfonshire, in December 2015 and St Asaph, Flintshire, in December 2016. It is not certain whether these records relate to genuine Fennoscandian birds or whether the plumage characteristics they show are within normal variation for British Jackdaws.

Rhion Pritchard and Jon Green

Rook *Corvus frugilegus* Ydfran

Welsh List Category	IUCN Red List			Birds of Conservation Concern			
	Global	Europe	GB	UK	Wales		
A	LC	LC	NT	Green	2002	2010	2016

The Rook breeds across most of Europe, but is absent from much of Fennoscandia, the Iberian Peninsula and Italy. Its range extends east through temperate Asia as far as China. It is resident in the western and southern parts of its range but migratory in the north and east (*HBW*). It is mainly a bird of lowland mixed farmland with tall trees, which it uses for nesting and roosting. In Wales, most rookeries are in the lowlands below 250m; the highest recorded was at 350m at Bryngwyn, Radnorshire, in 1980 (Jennings 2014). Many tree species are used for rookeries, but oaks and Ash are the most frequently adopted in Wales. The Rook is less of a carrion feeder than most other crows, with earthworms, small invertebrates and grain being important food items.

Rooks have long been regarded as an agricultural pest where grain is grown, as they will eat newly sown seed and damage emerging plants. Parish records from the 19th century show payments made to destroy Rooks, usually by shooting the young before they had fully fledged (Lovegrove 2007). Authors around the turn of the 20th century agreed that Rooks were abundant in many parts of Wales, although Forrest (1907) considered them to be comparatively scarce in Llŷn, western Meirionnydd and Anglesey. Few figures are available from this period, but surveys in central Wales showed a 30% increase in nests between the First World War and the 1960s, although the average colony size reduced by c.65% (Chater *et al.* 1974). Concerns about the impact of Rooks on the wartime supply of grain prompted the Ministry of Agriculture, Fisheries and Food to commission a national survey in 1944/46 (Fisher 1948). When the first legislation was passed in Britain to protect birds in 1954, Rook was included on a schedule of birds that could be killed at any time of the year to protect crops.

	Rookeries	Nests	Nests/rookery
1913–16	16	1,550	96.9
1959–67	67	1,833	27.4

Rook: Survey of north Ceredigion and neighbouring districts in Montgomeryshire (Walton 1928, Chater *et al.* 1974).

Serious declines were recorded in north European countries in the mid-20th century as a result of farming changes, pesticides and persecution (Brenchley 1986). Surveys showed a 60% decline in Wales between 1944/46 and 1975, despite incomplete coverage in the earlier period. This was a greater reduction than elsewhere in Britain (Sage and Whittington 1985). *Birds in Wales* (1994) suggested that numbers had increased up to the 1960s, but that there was a steep decline in the year immediately preceding the 1975 survey, which estimated the Welsh population to be 38,916 nests in 1,546 rookeries (Sage and Vernon 1978). A sample survey in 1980 suggested that numbers increased after 1975, and a survey in 1996 recorded a 36% increase in Wales during the previous decade (Marchant and Gregory 1999). There has been no dedicated survey since 1996, but the BBS index for Wales fell by 58% during 1995–2018, accelerating after 2010. This rate of decline is greater than in any other UK nation (Harris *et al.* 2020). The Welsh breeding population in 2018 was estimated to be 35,500 (28,500–37,500) pairs (Hughes *et al.* 2020), around 9% lower than in 1975. Densities are low in Wales compared to much of the rest of the UK, averaging 1.9 nests/km^2 in 1975 (Sage and Whittington 1985). Abundance in Wales recorded by the *Britain and Ireland Atlas 2007–11* remained low, although densities were higher on Anglesey, in Pembrokeshire and through eastern counties.

1944–46	1975	1996	2018
98,260	38,916	53,140	35,500

Rook: Welsh population estimates from surveys (Fisher 1948, Sage and Whittington 1985, Marchant and Gregory 1999) and from Hughes *et al.* (2020).

Local surveys told a similar story of an increase and subsequent decline in the last quarter of the 20th century. In north Ceredigion, the number of nests increased by 30% between 1975 and the mid-1980s, after which the population remained fairly stable to 1994, with a density of 3.0–4.2 nests/km^2 (Chater 1996). In Gwent, there was a 30% increase during 1980–85 (Tyler *et al.* 1987) and in Gower numbers were stable during the 1990s, but halved from 1,104 to 539 between 2000 and 2018.

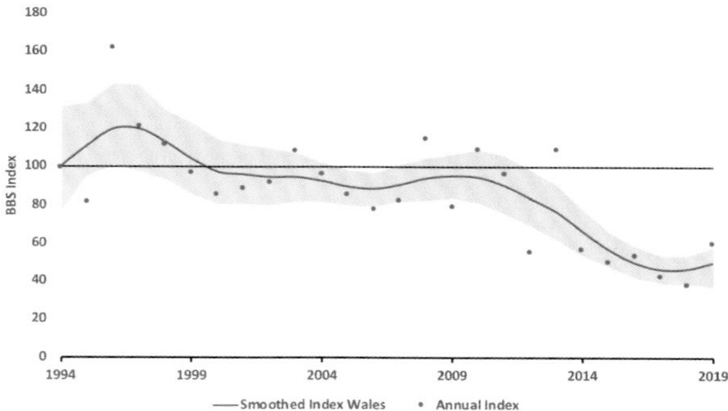

Rook: Breeding Bird Survey indices, 1994–2019. The red line shows the smoothed index, 1995–2018. The shaded area indicates the 85% confidence limits.

Local atlases in Gwent found little change in the number of tetrads containing rookeries, which remained around 34% of the total, between 1981–85 and 1998–2003, whereas in East Glamorgan breeding was confirmed in just 13% of tetrads in 2008–11, a decline of 38% since 1984–89. In Pembrokeshire, the proportion of tetrads with confirmed breeding fell slightly, from 47% to 44%, between 1984–88 and 2003–07. A subsequent survey in 2013/14 found 9% fewer rookeries than in 2003–07 and estimated a county population of 7,550 nests (*Pembrokeshire Avifauna* 2020). Nesting occurred in 27% of tetrads in North Wales in 2008–12, and while widespread on Anglesey, in Denbighshire and Flintshire, it was much scarcer in Snowdonia.

Average rookery size reduced across Britain between the mid-1940s and 1975. Most in Wales were fairly small in the latter year, a mean of 25.2 nests (Sage and Vernon 1978). More recent surveys suggested only a slight further reduction: 27 nests/rookery in Pembrokeshire in 2013–14 and an average of 19 in Gower in 2014–18.

Large rookeries are now largely consigned to history. Among the largest on record were 280–300 nests at Mabws, Ceredigion, in 1913–16 (Walton 1928), 292 in Tudweiliog and 204 at Botwnnog in 1966, both in Caernarfonshire, and 200 at Aberllolwyn, Ceredigion, in 1992 (Chater 1996), though many others must have been uncounted. A rookery at Stouthall, Gower, held 219 nests in 1966, which peaked at 278 nests in 1984, then declined in size during the 1990s. It shrank more rapidly from 2006 and ceased to exist after 2011. Initially this was the result of felling unsafe Beech trees, but the final loss coincided with the development of new colonies at nearby locations. There has been no published record of any rookery holding more than 200 nests in Wales since Stouthall in 1990, though one at Carreg near Aberdaron, Caernarfonshire, contained 188 nests during the *North Wales Atlas 2008–12* and one at Dale, Pembrokeshire, held 178 nests in 2014. Some of the largest rookeries in North Wales are on Llŷn, Caernarfonshire, where the species had been described as comparatively scarce in 1907.

Outside the breeding season, most Rooks remain close to their breeding colonies, usually feeding in flocks of up to 200, often with other corvids, on improved pasture, up to an altitude of *c.*600m. A flock of 1,000 feeding at Maeliennydd Common, Radnorshire, in August 1995 seems exceptional in Wales for recent times. They gather in large roosts, also in the company of other corvids, and while most counts in recent years have involved fewer than 200, up to 1,000 roosted in western Newport, Gwent, between October and December 2011. There has been some evidence of autumn movements, including large-scale departures to the west over South Bishop, Pembrokeshire, in November 1975, where the largest single day count was 350 on 6 November (Donovan and Rees 1994). Counts over Bardsey, Caernarfonshire, also have peaked in October but usually number only 20–30 (Roberts 1985). These seem likely to have involved birds moving to Ireland, but there is only one ringing exchange, a first-calendar-year bird ringed at North Slob, Co. Wexford, in December 1975, that was found dead at Brynhalen near Corwen, Meirionnydd, in April 1980. There is good historic evidence of winter movements to eastern Britain from Fennoscandia and Russia, but the only longer-distance movement from areas to the east or north of Wales was a nestling ringed in Yeadon, North Yorkshire, in May 1935, which was recovered at Nant Fawr, Denbighshire, in January 1936.

Overall, the Rook seems to have been doing well in Wales until around the 1960s, after which there was a sharp decline prior to 1975, followed by a gradual recovery until about 1997, since when it has declined again. Its decline has received less attention than it merits. The decline prior to 1975 was attributed to changes in agricultural practice (Sage and Whittington 1985), such as the ploughing and reseeding of fields. It can feed in permanent pasture for much of the year, because the grasses grow less vigorously than in fertilised grass grown for silage (Anne Brenchley *in litt.*), although ploughing provides a temporary boon in exposed invertebrates (Chater 1996).

Rook numbers and densities are highest in areas of mixed farming (Brenchley 1986), so the increase in specialised livestock farming in Wales may well be a major factor. Until 2019, the species could be killed under a General Licence in Wales to prevent damage to crops, although it was subsequently excluded, at least in part because of the ongoing decline in numbers (Natural Resources Wales 2019). *Breconshire Birds* (2014) suggested that a significant reduction in the size of several formerly large rookeries was probably due to shooting under General Licence, although predation by Goshawk was implicated in the collapse of one large rookery near Talybont on Usk, Breconshire. No data are available on the numbers shot in Wales, but an estimated 88,000 were killed across the UK in 2016 (Aebischer 2019). The rapid decline in Rooks in Wales will qualify it for Red-listing. Understanding the reasons for the population changes would be beneficial.

Rhion Pritchard

Sponsored by Anne Brenchley

Carrion Crow *Corvus corone* **Brân Dyddyn**

Welsh List Category	IUCN Red List			Birds of Conservation Concern			
	Global	Europe	GB	UK	Wales		
					2002	2010	2016
A	LC	LC	LC	Green			

With the Hooded Crow again considered a separate species, the Carrion Crow has a disjunct distribution. The nominate subspecies occurs only in western Europe, while *C.c. orientalis* occurs in Asia, east of the Caspian Sea (IOC). It is a bird of woodland and open country, preferably with at least a few scattered trees for nest sites. It is largely resident in Wales, with the only ringing recovery of more than a few kilometres being a nestling ringed on Bardsey, Caernarfonshire, that was later recovered in north Pembrokeshire, 91km to the south.

The Carrion Crow is very adaptable, mainly nesting in trees but sometimes using rock ledges, including sea cliffs. On the treeless Epynt military training area, Breconshire, derelict tanks and vehicles, positioned as bombing targets, have been used as nest sites (Andrew King *in litt.*). It can nest up to quite a high altitude. Driver (2005) found a nest at 650m in the Carneddau, Caernarfonshire. Although carrion is an important part of the diet, it eats a variety of foods, including the eggs and young of other birds, such as ground-nesting species and colonial seabirds. Their impact varies according to the resilience of each species' population. Early records of persecution from parish records include five shillings paid for killing crows at Caerwys, Flintshire, in 1676, and three shillings for destroying 240 chicks at Llanuwchllyn, Meirionnydd, in 1712 (Matheson 1932). It was heavily persecuted by gamekeepers in the 19th and early 20th centuries, owing to its consumption of the eggs and small young of Pheasant and Red Grouse: 1,538 were killed on Penrhyn Estate between Penmachno and Dolwyddelan, Caernarfonshire, during 1874–1902; 134 on Cawdor Estate at Stackpole, Pembrokeshire, in 1821. Some 150 were counted on a gibbet at Slebech, Pembrokeshire, in 1930 (per Donovan and Rees 1994). Sheep farmers also killed considerable numbers, but these were rarely documented. Nonetheless, Carrion Crow seems always to have been common in Pembrokeshire and North Wales (Mathew 1894, Forrest 1907). The latter author described it as "astonishingly numerous" around Penrhyndeudraeth and Llanbedr, Meirionnydd. He stated that he had seen about a dozen nests in one small wood, each in a separate tree. Numbers increased following a ban on the use of poison in the open by the Protection of Animals Act 1911, and a reduction in gamekeeping during and after the First World War.

Afforestation of the Welsh uplands provided secure nest sites in areas where there had previously been none, and after the Second World War, Carrion Crows benefited from increasing numbers of sheep. Large-scale persecution continued in some areas. It remained a scarce and local breeding bird on the Gower Peninsula into the early 1960s, for example (Hurford and Lansdown 1995). The *Cambrian Bird Report 1956* reported a rapid increase on Anglesey, in Caernarfonshire and Denbighshire, but noted that a single funnel trap in Denbighshire caught 120 in just one month. It also spread into urban areas following the Second World War (Prestt 1965).

All three Britain and Ireland atlases found the Carrion Crow in virtually every 10-km square in Wales during the breeding season, and as would be expected from a sedentary species, the results of winter atlases were similar. Relative abundance in 2007–11 was highest in eastern Wales and on Anglesey, and lowest on elevated ground in the west. It had changed little since 1988–91. Tetrad atlases confirm that Carrion Crow is one of the most widely distributed birds, present in 99% of tetrads in Gwent in 1981–85 and 1998–2003. In East Glamorgan it was found in *c.*90% of tetrads during the 1984–89 and 2008–11 breeding seasons, although it was recorded with a breeding code in only 80% in the latter period. In Gower it occupied 94% of tetrads in the *West Glamorgan Atlas 1984–89*. It was found in 96% of tetrads in Pembrokeshire in 1984–88 and in 92% in 2003–07 (*Pembrokeshire Avifauna* 2020). Elsewhere, it was present in 97% of tetrads visited in Breconshire in 1988–90, and in 93% in the *North Wales Atlas 2008–12*.

Across the UK, Carrion Crows increased consistently from the 1960s, with numbers doubling between 1964 and 1977 (Gregory and Marchant 1996). This is likely to have been also true for Wales. The BBS index for Wales has remained relatively stable since 1994, whereas numbers continued to increase in England (Harris *et al.* 2020). The Welsh breeding population in 2018 was estimated to be 160,000 territories (Hughes *et al.* 2020).

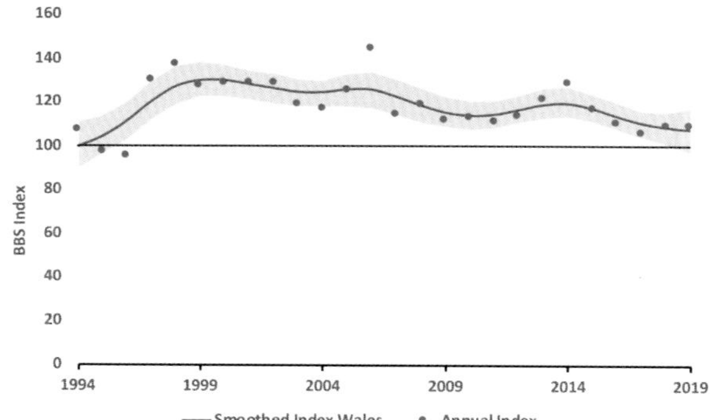

Carrion Crow: Breeding Bird Survey indices, 1994–2019. The red line shows the smoothed index, 1995–2018. The shaded area indicates the 85% confidence limits.

Studies showed wide variation in nesting densities according to habitat and altitude. The population on Skomer, Pembrokeshire, (1.2km²) reached an exceptional 26 nests in 1965, which at 21.7/km² is the highest density recorded in Wales. By 2018, there were ten nests on the island, a decline that Donovan and Rees (1994) attributed to increased competition for food and nest sites from Magpies, although 8.3 nests/km² is still double that of mainland Pembrokeshire. Carrion Crows also breed on the other Welsh islands, although numbers have varied. They declined on Skokholm, Pembrokeshire, from 12 pairs in the 1960s to none, for several years, in the 1980s and 1990s, but subsequently recovered. They were temporarily eradicated by trapping from Ramsey, Pembrokeshire, during the 1980s, but seven pairs bred in 2013 (Morgan 2013). 5–10 pairs usually breed on Bardsey, Caernarfonshire.

In many parts of Wales, the breeding population is at the carrying capacity of the habitat, and there are large numbers of subadults and adults that have failed to find a breeding territory. *Birds in Wales* (1994) estimated that these could form as much as 50% of the population. Large counts are probably such non-breeders, often with other corvids, at plentiful food sources such as, in Carmarthenshire, 350 on the tideline at Laugharne in February 1966 and the same number at Cei Cydweli on 10 July 2018. Large numbers are drawn to fields immediately after grass has been cut for silage. Roosts in Gwent have included *c.*1,000 near Llandegfedd Reservoir in winter

Study area	County	Nests/km²	Land management
Skomer	Pembrokeshire	21.7	Mixed grass/ maritime heath, seabird colony
Newbridge-on-Wye	Radnorshire	4.88	Sheep pasture, 150–300m asl(1)
Carneddau	Caernarfonshire	7.52	Sheep pasture, 150–350m asl(2)
Aber/ Llanfairfechan	Caernarfonshire	4.0	Coastal plain/ wooded hillsides(2)
Foel Fras	Caernarfonshire	0.46	Rough grassland, >350m asl(2)

Carrion Crow: Breeding densities in selected studies in Wales. (1) RSPB 1981 (cited in Jennings 2014); (2) Driver 2005.

1994/95, 1,500–2,000 at Goytre in December 1994 and at least 1,000 at Llanwern in February 1997. East Glamorgan also recorded gatherings over 1,000 at Talbot Green in October 2006 and at Pant yr Awel in February 2013. Numbers in North Wales are usually lower, but 600 were counted from RSPB Conwy, Denbighshire, in November 2006.

Carrion Crows can be killed legally under General Licences to prevent serious damage to agriculture and for the conservation of wild birds, which may reduce numbers locally. The number of Carrion Crows killed in Wales, recorded by the National Gamebag Census, declined slowly during 1960–2004 (Aebischer and Harradine 2007), but no subsequent data have been published. There is no assessment of numbers killed by sheep farmers, although this control is probably on a smaller scale now than formerly. Carrion Crow can be one of the main predators of eggs and young of birds of high conservation concern, particularly those that nest on or under the ground. A pair on Bardsey in 2002 took at least 80 Manx Shearwater eggs from accessible nest chambers in May and June, but predation has not prevented continued substantial growth in that seabird's population. Analyses of extensive monitoring data provided little evidence that Carrion Crows affected songbird populations in the UK (Newson *et al.* 2019).

High densities of Carrion Crow are a factor in the decline of Curlew and Lapwing in Wales, although Ausden *et al.* (2009) found that they were far less important than mammals, particularly Fox, as predators of Lapwing nests. An experiment on moorland in northern England showed that reducing the abundance of Carrion Crows and mammalian predators led to a more than threefold increase in Curlew, Lapwing and Golden Plover breeding success, but it could not distinguish the relative contribution made by each (Fletcher *et al.* 2010). Larsen cage traps are effective in reducing the number of territorial Carrion Crows, but this is very short-term, as birds from the plentiful non-breeding population will move into the vacant territory (Ausden *et al.* 2009). Populations of generalist predators are relatively high in the UK and, while control of Carrion Crows may provide a short-term solution for wader conservation in tandem with habitat restoration, resolving the factors that lead to such high densities of Carrion Crows will be a more sustainable solution.

Rhion Pritchard

Hooded Crow *Corvus cornix* Brân Lwyd

Welsh List Category	IUCN Red List		Birds of Conservation Concern			
	Global	GB	UK	Wales		
A	LC	NT	Green	2002 N/A	2010	2016

The Hooded Crow was for many years regarded as a subspecies of the Carrion Crow, but is now treated as a separate species, although it can mate with Carrion Crow and hybrids are frequently recorded in Wales. The nominate subspecies is found in northern and eastern Europe, and other subspecies extend east in Siberia as far as the River Yenisei, and south as far as Turkey, Iraq and part of Iran (IOC). It is mainly a bird of open country with at least scattered trees, and is largely resident, although some northern populations move south in winter. Large numbers from Fennoscandia formerly wintered in western Europe, as far as eastern England

Hooded Crow © Steve Stansfield

and northern France, but fewer have done so since the 1930s, perhaps because they now remain in towns and villages in their breeding areas in winter (*HBW*). The nearest breeding populations to Wales are in Ireland, Isle of Man and northwest Scotland, but there have been no ringing recoveries to indicate the origins of birds seen in Wales.

The species may have been fairly common along the coast of Glamorgan in the 19th century. In Gower, Dillwyn (1848) referred to a pair, called by the old name of Royston Crow, wintering at the mouth of the Afon Tawe in March 1809. In East Glamorgan it was said to be "common on shore" at Newton Nottage and in at least two winters at East Moors, Cardiff (Doddridge-Knight 1853). Five were recorded at St Athan in 1884 and seven in 1885. Hooded Crow was also a not infrequent winter visitor to the Dee Estuary, Flintshire/Wirral, in the first half of the 19th century (Dobie 1894). Numbers in all areas declined in the second half of the century and by the 1890s it was an occasional winter visitor (Mathew 1894, Forrest 1907, Hurford and Lansdown 1995). Forrest (1907) mentioned breeding near Llanrwst, Denbighshire, in 1900, a hybrid brood near Barmouth, Meirionnydd, and in his 1919 volume, a pair that nested in the Carneddau, Caernarfonshire, in 1917.

Hooded Crows continued to be recorded fairly regularly in Wales during the 20th century, but as some can be mobile and stay for several months or even years, it is difficult to assess the number of individuals involved or to discern any trend. The increased number of records since Hooded Crow was considered to be a separate species in 2002 (Parkin *et al.* 2003) may reflect the increased interest of observers rather than an increase in numbers. There are records from all months of the year, although usually fewer during July to September, and it has been recorded in every county, being almost annual in Caernarfonshire, Pembrokeshire and Anglesey. Records of more than one together are not unusual, usually on western coasts. Five at Frongoch, Montgomeryshire, on 18 March 1951 were notable for their inland location, but a group of 35 at RSPB South Stack, Anglesey, on 2 February 2019 was exceptional, and almost certainly the largest single count in Wales. Bardsey, Caernarfonshire, has had several records of groups of up to four, and during April–June 1964, ten individuals were thought to be present.

The only confirmed nesting of Hooded Crows in Wales, since the 1917 record, was at RSPB South Stack, Anglesey, in 1996 (Bagguley *et al.* 2000). It is usually unclear whether hybrids seen in Wales have fledged locally, but there have been a number of records of Hooded Crow nesting with a Carrion Crow mate. It occurred annually on Anglesey during 2007–10, for example. Other records of adult Hooded Crows in company with hybrids in late summer/early autumn suggested local breeding. Sightings are generally in western counties, but a Hooded Crow at Parc Lettis, Gwent, present through winter 2019/20, fledged three hybrid young the following spring.

There has been a decline in the range of Hooded Crow in Scotland and numbers have fallen by over 50% (Harris *et al.* 2020), where Carrion Crow may have a competitive advantage (*The Birds of Scotland* 2007). Numbers in Ireland increased during 1998–2008 (Crowe *et al.* 2010). While it is unlikely to establish itself as a regular breeding species in Wales in the near future, it should continue to be a regular visitor.

Rhion Pritchard

Raven *Corvus corax* Cigfran

Welsh List Category	IUCN Red List			Birds of Conservation Concern			
	Global	Europe	GB	UK	Wales		
A	LC	LC	LC	Green	2002	2010	2016

A remarkably adaptable species, the Raven is found over most of the northern hemisphere, from the high Arctic as far south as Nicaragua, North Africa and northwest India. It uses a wide variety of habitats, although it tends to avoid urban areas over much of its range. It is mainly an opportunistic scavenger, historically associated with large mammals such as Wolves (*HBW*), and is quick to benefit from the activities of humans. In Wales, Ravens nest from close to sea level to around 800m. In Snowdonia 76% of nests were at 100–500m, but a nest was recorded at 790m on Pen yr Ole Wen in the Carneddau, Caernarfonshire (Driver 2007). They nest mainly on formidable cliffs along the coast and in the mountains, with some in disused quarries. Tree nesting is also fairly common at lower altitudes and apparently increasing, and some pairs nest on human-made structures. Breeding birds are usually sedentary, but non-breeders range widely: one ringed as a nestling at Mynydd Parys, Anglesey, in April 1960, was found dead in March 1962 at North Berwick, East Lothian, 318km to the northeast. A bird ringed as a nestling near Morfa Bychan, Caernarfonshire, in April 1976 was found dead in December 1977 near Dundalk, Ireland, 178km to the northwest.

There are many references to Ravens in the earliest Welsh poetry, largely related to their feeding on the corpses of the fallen after a battle. It was thus regarded as a bird of ill omen, foretelling battle and death. A notable warrior would be described as a "feeder of Ravens", as his exploits on the battlefield provided carrion for them (Williams 2014). In the Welsh uplands, many place names include *Cigfran* as an element, including several crags that still have nesting Ravens today.

The species has long been regarded as a pest by sheep farmers who blame it for killing new-born lambs. There are many records of Ravens being killed during the 18th century, the payments recorded in churchwardens' accounts. In Llanfor parish, near Bala, Meirionnydd, 1,656 Ravens were killed during 1720–58, at 4 pennies a head (Jones 1974b). With the development of large-scale game shooting in the 19th century, the gamekeeper became another enemy of the Raven. A total of 464 were killed between 1874 and 1902 on Lord Penrhyn's estate between Penmachno and Dolwyddelan, Caernarfonshire. Forrest (1907) stated that it was "cordially detested" by Welsh farmers. Bolam (1913) described persecution in the Llanuwchllyn area of Meirionnydd and noted that poisoned bait was the main threat to Ravens. Egg-collectors also targeted their nests.

With so many enemies, it is not surprising that Raven numbers were much reduced in Wales by the end of the 19th century, but it was considered to be "more common in Wales than anywhere else in Britain, probably because it was less persecuted" (Walpole-Bond 1914). In Glamorgan, Ravens nested in Cathays Park, Cardiff, until the 1860s, but the species was considered to be uncommon and decreasing in the county by the 1890s (Hurford and Lansdown 1995). It was still found in sufficient numbers to justify considering it to be one of the characteristic birds of Pembrokeshire by Mathew (1894), who wrote that he scarcely visited the coast without seeing one and suggested there were *c.*20 breeding pairs in the county. Forrest (1907) stated that it was found in fair numbers in the western part of North Wales, but that "considering the pertinacity with which it is pursued by farmers and shepherds, and the equal pertinacity with which the eggs are raided by collectors, and the young by men who sell them as pets, it is wonderful that there are any left at all!" Ingram and Salmon (1957) estimated about 20 pairs in Breconshire in 1914.

The tide began to turn for the Raven in the second decade of the 20th century. The Protection of Animals Act 1911 banned the use of poison in the open, although the legislation did not end the practice. The First World War resulted in greatly reduced persecution, as many gamekeepers joined the armed forces, and by 1919 Raven numbers had increased in parts of Caernarfonshire and Meirionnydd. The recovery continued in the upland core between the two World Wars, but there was little indication of an increase in their range. Tree-nesting resumed during this period, a habit that had almost vanished in the early years of the 20th century. It was recorded in Ceredigion from 1927 (Ingram *et al.* 1966). Numbers increased to at least 80 pairs in Pembrokeshire, of which about 60 were on coastal and island cliffs (Lockley *et al.* 1949). A study in Montgomeryshire in 1949–53 also recorded an increase since 1935 (Simson 1966).

The *Britain and Ireland Atlas 1968–72* confirmed that Ravens occupied 92% of 10-km squares in Wales, with a few gaps, such as in parts of Flintshire and eastern Denbighshire. The 1988–91 *Atlas* showed no great change in distribution, but by the 2007–11 *Atlas*, it had spread into 98% of Welsh squares, with confirmed breeding even in easternmost Denbighshire, as it expanded east across Britain. Despite increases elsewhere, the relative abundance of Ravens in Wales was higher than in any other part of Britain or Ireland and was higher throughout the nation than in 1988–91. The greatest increase occurred in the eastern part of mid-Wales.

The size and distribution of the Welsh breeding population was assessed as 1,200–1,275 pairs in the early 1990s (Ratcliffe 1997). Eight counties each held more than 100 breeding pairs: Ceredigion (165–190), Breconshire (160), Pembrokeshire (140), Montgomeryshire (120–140), Caernarfonshire (115), Glamorgan (110), Carmarthenshire (100) and Meirionnydd (100). Gwent held 75 pairs, Denbighshire 50–55 and Flintshire 10–12. Tetrad atlases showed a marked increase in Raven distribution in southeast Wales: the *Gwent Atlas 1981–85* recorded Ravens in 51% of tetrads and the 1998–2003 *Atlas* recorded them in 71%. In East Glamorgan, they were recorded in 28% of tetrads in 1984–89 and 57% in 2008–11, although there was evidence of breeding in only 32% in the latter period. Its recovery was illustrated by the return of Ravens to Cardiff city centre, including breeding on the clock tower of City Hall. In Pembrokeshire, the proportion of tetrads occupied in 1984–88 and 2003–07 was similar, at 28–29%, but a greater proportion of these were inland in the later period.

	1950–57(1)	1978–85(2)	1998–2005(3)
Snowdonia	37–46	97–100	169–180
Migneint/Hiraethog		20	28–31

Raven: Population estimates in mountain ranges in Snowdonia (Caernarfonshire and north Meirionnydd) and Migneint/Mynydd Hiraethog, primarily in west Denbighshire, from (1) Ratcliffe 1997; (2) Dare 1986a; (3) Driver 2006.

Consecutive studies in Snowdonia, from the 1950s to the present day, showed a 130% increase between the 1950s and 1980s and a further 69% increase by the early years of the 21st century (Allin 1968, Ratcliffe 1997, Dare 1986a, Driver 2006). Increasing sheep stocks and greatly reduced persecution were thought to have resulted in substantially higher densities of nesting pairs than elsewhere in Wales, although they were similar to those in the Cambrian Mountains in 1975–79 (Newton *et al.* 1982b). Densities were far lower on Anglesey and Llŷn, Caernarfonshire, where most pairs nested on sea cliffs: 55–69 territories on Anglesey and 43–50 on Llŷn (Driver 2006).

Region	Pairs/100km²	Period
Snowdonia	10.5	1978–85(1)
	18.3–19.4	1998–2005(2)
Migneint/Mynydd Hiraethog	4.2	1978–85(1)
	5.9–6.5	1998–2005(2)
Anglesey	7.66–10.03	2002–05(2)
Llŷn	4.9–5.7	2002–05(2)
Cambrian Mountains	15.2	1975–79(3)
Brecon Beacons	11.3	1993–95(4)
	7.7	2014–16(5)
Llynfi catchment, Breconshire	18.7	2014–16(5)

Raven: Density of territorial pairs in Wales, from: (1) Dare 1986a; (2) Driver 2006; (3) Newton *et al.* 1982b; (4) Dixon 1997; (5) Dixon 2016

Increases in the Raven population were evident in other areas of Wales. A population on the fringes of a grouse moor in eastern Denbighshire increased from one to around ten pairs during 1988–97, thought to be a result of diminished persecution at nesting sites (Roberts and Jones 1999b). By contrast, the Gower Peninsula population showed a small increase from the late 1930s, when a few pairs began to nest inland in trees, but has remained relatively stable (Dixon *et al.* 2007). More recently, increases have occurred on lower ground, such as an area of enclosed farmland in Breconshire, centred on Llangorse/Talgarth/Talachddu, where 28 breeding territories in 2014–16 all contained nests in trees, evenly spaced across the primarily beef and sheep pastoral study area (Dixon 2016). Most Welsh islands support breeding Ravens, even tiny Grassholm, Pembrokeshire (11ha), and The Skerries, Anglesey (17ha), which are probably the smallest Raven territories in Wales. Seabird colonies there ensure that sufficient food is available during the breeding season, although birds may also visit the mainland to forage.

The BBS index for Raven in Wales increased until 2005,

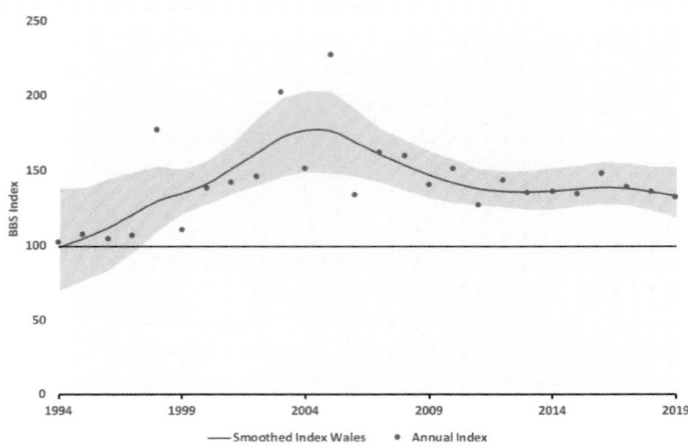

Raven: Breeding Bird Survey indices, 1994–2019. The red line shows the smoothed index, 1995–2018. The shaded area indicates the 85% confidence limits.

followed by a modest decline, but by 2018 was 29% higher than in 1995 (Harris *et al.* 2020). In Snowdonia and the Brecon Beacons, there has been a decline in Raven numbers and breeding success, probably due to reductions in the availability of sheep carrion. A survey of upland sheepwalk in the Brecon Beacons showed a 30% reduction in the density of Raven territories between 1993–95 and 2014–16 (Dixon 2016). Sheep are no longer over-wintered on high ground, lambing is now confined to sheltered low ground and legislation requires more effective removal of carcasses. The Raven population on Anglesey fell to 28–30 pairs in 2018, with significant loss on the northwest coast, from 13 pairs in 2000 to only one in 2018, and at inland sites, from 19 to ten pairs over the same period. This coincided with a reduction in sheep numbers and many farms switching to rearing beef cattle on the island (Driver *in litt.*). The Welsh breeding population in 2018 was estimated to be 2,150 pairs (Hughes *et al.* 2020).

Raven © John Hawkins

Non-breeding Ravens form flocks comprised mainly of immature birds up to three years of age, but also older birds, often paired, that are unable to find territories (*Birds in Wales* 1994). Large counts are sometimes made at good food sources, such as 150 on Mynydd Hiraethog, Denbighshire, in May 1992 and 162 at a chicken waste dump near Knill, Radnorshire/Herefordshire, in February 1996. The numbers involved may be much higher than any individual count. At a landfill site at Tylwch near Llanidloes, Montgomeryshire, where about 100 were present at any one time in autumn 1997, resightings of colour-ringed birds suggested that over 1,000 used the site over two months (*Birds in Wales* 2002).

The largest counts have been at roosts, which probably also contain mostly juvenile and other non-breeding birds, although making accurate counts of Ravens coming to roost is difficult. A roost monitored in Newborough Forest, Anglesey, from 1986, increased to over 1,000 in winter 1996/97 and 1,871 at its peak in January 1997. It was considered to be the second-largest Raven roost in the world (Jones and Whalley 2004). At its peak, there was circumstantial evidence that it included Ravens from south Snowdonia, and perhaps even farther afield (Nigel Brown *in litt.*). This roost did not fall below 500 birds during 1995–2000, even in summer months. Numbers remained high for some years, such as 1,200 in December 2001, but were smaller subsequently, synchronous with a decline on the island and possibly because new roosts formed in eastern Anglesey. The highest recent count at Newborough was *c.*500 in January 2018. Studies there made a significant contribution to understanding the use of roosts by birds to gather information, and showed that birds using the same geographical area during the day roosted together in the same part of the roost at night (Wright *et al.* 2003). Another large roost, at Blaencanaid in East Glamorgan, held 577 Ravens in June 2016. Unlike Newborough, which peaks in January, the Blaencanaid roost usually peaks between May and September. There is often an autumn influx onto some islands, coinciding with the fledging period of Manx Shearwaters (*Skomer Island Bird Report* 2005). There were 133 on Skomer on 20 September 2005, for example. Flocks of non-breeding Ravens have declined noticeably, although the closure of open landfill refuse dumps has probably had a greater impact on their numbers around North Wales (Julian Driver pers. comm.).

Ravens in Wales, unlike some other corvid species, cannot be killed except under specific licence from Natural Resources Wales. Several licences were issued in 2016 that resulted in 49 being killed, including 23 at a site in Denbighshire to 'prevent damage to livestock', in this case to young Pheasants (Sheldon 2018). No Ravens were legally killed in Wales during 2017–19 (Sheldon 2019). However, Ravens are the targets of illegal persecution, and in particular of poison baits that can result in multiple deaths and also kill scavenging raptors. There were 84 Ravens killed in confirmed persecution incidents during 1990–2019, including 28 in a single poisoning incident in Ceredigion in 1999. The number of confirmed incidents that resulted in Raven deaths was higher in 2010–19 than in the previous two decades and included two that were shot. Almost one quarter of all the incidents during 1990–2019 were in Breconshire (RSPB data). Concerns have been expressed about the impact of increasing Raven numbers on declining ground-nesting birds, such as Curlew, although a study at sites across Britain, including the Berwyn and Migneint in Wales, found no significant effects (Amar *et al.* 2010).

It seems that the Raven's recovery over the last century is complete and that numbers are at carrying capacity for the available number of nest sites and available food.

Peter Dare and Rhion Pritchard

Sponsored by The Mountain Training Trust (Plas y Brenin)

Waxwing *Bombycilla garrulus* Cynffon Sidan

Welsh List Category	IUCN Red List		
	Global	Europe	GB
A	LC	LC	LC

Waxwing has an almost circumpolar breeding distribution, from northern Fennoscandia to the Kamchatka Peninsula, and from Alaska to Hudson Bay. It breeds in coniferous forests, often near water, and most individuals move south in winter (*HBW*). Britain is on the edge of its usual wintering range, with large numbers wintering in some years, but very few in others. Berries, particularly Rowan, form the most important part of its winter diet, and high numbers in Britain may follow a poor berry crop in its usual wintering areas.

Birds recorded in Wales are mainly from Fennoscandia, although some may be from Russia. Most colour-ringed Waxwings seen here had been ringed in eastern Britain earlier in the same winter, but there have been a few recoveries of birds ringed in Norway. One colour-ringed at Newtown, Montgomeryshire, on 1 December 2012, was seen at Lewes, East Sussex, later that month and then near Geneva, Switzerland, in early March 2013.

The first record in Wales was a bird killed in Denbighshire in December 1788 (Pennant 1812), about which Forrest (1919) wrote that "The ancient specimen mentioned by Pennant as killed at Garthmeilio in 1788 is still extant and in the possession of Mr. R.D. Roberts of St Asaph". There were few records in Wales during the 19th century, although William Davies (Gwilym Teilo) included the species in a "list of winter migrators" in a book about Llandeilo, Carmarthenshire, published in 1858. From the same county, Mathew (1894) mentioned a record by Dix in 1869 that had been shot near Llandysul, Ceredigion, "a few years since". Flocks at Roath, East Glamorgan, in 1859 and Llanwrtyd Wells, Breconshire, (undated) were mentioned by Phillips (1899). In Ceredigion, there were one or two Waxwings in the Penglais collection at the end of the 19th century, said to have been collected locally (Roderick and Davis 2010). Forrest (1907) mentioned six near St Asaph, Flintshire, in December 1898.

Birds in England (2005) listed 37 winters between 1679/80 and 2000/01 in which irruptions occurred into Britain and Ireland. There have been further irruptions in 2004/05, 2008/09, 2010/11, 2012/13 and 2016/17. Waxwings are usually first recorded in northeast Scotland in October, from where they gradually disperse south and west. As poor berry crops in northern Europe can initiate an irruption, so birds are more likely to disperse if crops in Scotland are poor (*The Birds of Scotland* 2007). In many irruptions, particularly earlier ones, few or no birds were recorded in Wales, and in the first half of the 20th century, only small numbers were seen. Winter 1946/47 saw a major influx into Britain, but fewer than 30 individuals were recorded in Wales, and while a large-scale arrival in 1965/66 produced a count of 52 at Llysfaen, Denbighshire, numbers in Wales were low compared to eastern Britain. A very large influx in 1988/89 included about 100 in Wales, with records in all counties except Carmarthenshire, Pembrokeshire and Meirionnydd. An influx in winter 1995/96 produced 157 in Wales between 27 January and 10 March 1996, recorded in all counties except Flintshire. It included 41 in Glamorgan, 31 in Ceredigion and 26 in Radnorshire (*Birds in Wales* 2002). Good numbers were also recorded in Ireland and northwest England that winter, and it has been suggested that at least some could have come from North America (*Birds in England* 2005). January 2001 saw reasonable numbers in Wales, the biggest flock being 31 at Wrexham Industrial Estate, Denbighshire, that remained into early February.

The winter 2004/05 influx produced numbers never recorded previously in many European countries, such as a single flock of 3,000 at Lutry, Switzerland (Fouarge and Vandevondele 2005). Several flocks of over 1,000 were recorded in Scotland, including 1,800 at Kincorth, Aberdeenshire, on 21 November, with a minimum of 15,000 thought to be involved (*The Birds of Scotland* 2007). Numbers in Wales were lower but still greater than anything recorded previously, particularly in eastern counties. In Gwent, a flock of 270 was near Chepstow on 26 January 2005, and an estimated 720 were in the county on 29 January, mostly around Cwmbran. In Cardiff, East Glamorgan, a 'super-flock' formed during January, expanding and contracting in size as birds searched for berry-bearing trees, with the largest count a Welsh record of 650. A flock of 212 was recorded in Swansea, Gower, on 23 January and in mid-Wales there were an estimated 200 around Welshpool, Montgomeryshire. The highest counts in North Wales were 150 at Rhuthun, Denbighshire, on 19 December 2004 and 150 at Mold, Flintshire, on 14 January 2005. Numbers in western counties were much lower, although one on Bardsey, Caernarfonshire, on 23 October 2004, was only the second record for the island. A characteristic of this influx was that birds lingered until May in several countries (Fouarge and Vandevondele 2005), including four near Pont Croesor, Meirionnydd, on 1 May 2005.

An influx in winter 2008/09 produced only moderate numbers in Wales, the highest count being 25 at Usk, Gwent, in early January. Larger numbers arrived during winter 2010/11, mainly in November and early December 2010. There were several flocks of over 200 in East Glamorgan, including 250 at Heath, Cardiff, on 13 December. Elsewhere, 110 were near Knighton, Radnorshire, and 150 in Rhuthun on 24 November. Another large influx occurred in 2012/13, the highest numbers again in November and early December, including a maximum of 300 in the Cardiff area. There were also flocks of *c.*200 in Newtown in Montgomeryshire, Llandudno in Caernarfonshire, Denbigh and Wrexham in Denbighshire, and Rhuddlan in Flintshire. None was recorded in Wales in winters 2013/14–2015/16, but good numbers were again recorded in 2016/17, when the highest counts were in Denbighshire, including 190 in Denbigh on 7 January and up to 160 at Kinmel Bay during 2–12 January.

Large influxes into Wales have occurred more frequently in the last 30 years than historically, most likely because of a 683% increase in the European breeding population during 1988–2016 (PECBMS 2020). Most birders, and even shoppers at retail parks with planted Rowan trees, will continue to welcome such visits.

First-calendar-year Waxwing © John Hawkins

Rhion Pritchard

Cedar Waxwing *Bombycilla cedrorum* Cynffon Sidan y Cedrwydd

Welsh List Category	IUCN Red List	Rarity recording
	Global	
A	LC	BBRC

This is a very rare trans-Atlantic vagrant, which breeds throughout northern states of the USA and southern provinces of Canada, where the population is increasing. Many eastern Cedar Waxwings winter in the southeastern USA, but some travel as far south as Costa Rica and Panama. There were seven British records between 1985 and 2016, including the sole Welsh record (Holt *et al.* 2017). Its remains were found at Treginnis, St Davids, in Pembrokeshire on 26 June 2015.

Jon Green and Robin Sandham

Coal Tit *Periparus ater* Titw Penddu

Welsh List Category	IUCN Red List			Birds of Conservation Concern			
	Global	Europe	GB	UK	Wales		
					2002	2010	2016
A	LC	LC	LC	Green			

The Coal Tit is widely distributed, found over all of Europe, except the far north, extending across temperate Asia as far as the Kamchatka Peninsula and Japan (*HBW*). The subspecies *P.a. britannicus* breeds only in Britain and northeast Ireland, while small numbers of the nominate *P.a. ater* occur in Britain from continental Europe (IOC). The Coal Tit's fine bill and acrobatic ability make it well adapted to feeding in conifers (Snow 1954); it particularly favours Spruce. In the south and western part of its range, including Wales, it is resident, but northern populations move south, sometimes in large numbers if the seed crop fails (*HBW*). In Wales it benefited from the planting of conifers, particularly after the Second World War, but it is not a conifer specialist like Crossbill, and is found in a wide variety of habitats, including deciduous woodland, parks and gardens. It will use garden feeders and nest-boxes if unable to find natural holes, although it is subordinate to Blue and Great Tits if there is competition (Perrins 1979). Ringing recoveries involving Wales show few long-range movements. The greatest distance covered was by a bird ringed in its first calendar-year at Llangorse Lake, Breconshire, in August 2015, which was killed by a cat at Burton on Trent, Staffordshire, 148km to the northeast, in May 2016. A first-calendar-year bird, ringed on Bardsey in August 1985, was caught by a ringer at Little Sutton, Wirral, in February 1986, 136km to the northeast.

There was far less coniferous forest in Wales in the late 19th and early 20th century, so it can be assumed that Coal Tits were less widespread, but it seems to have been fairly common nonetheless. Mathew (1894) described it as a common resident in Pembrokeshire and noted that one of his correspondents, Mr Dix, considered it more common than he had ever seen in England. It was similarly described in Glamorgan (Hurford and Lansdown 1995), in northern Ceredigion (Salter 1900) and Carmarthenshire (Barker 1905), while in North Wales, Forrest (1907) said that it was most common in the west, where winter flocks of 50–60 were reported at Aberglaslyn, Caernarfonshire. It was said to be "exceedingly abundant" around Penrhyndeudraeth, Meirionnydd. There is little information about its fortunes between 1900 and the late 1960s, although in the mid-1930s the Gower population was said to be recovering after a distinct decline (Hurford and Lansdown 1995). It is likely that the species increased significantly in Wales during the 1950s and 1960s, as conifer plantations matured, but this is not well documented. It also meant that Coal Tit could extend its breeding range uphill. It will breed at a higher altitude than other tit species, at over 600m in Radnorshire (Jennings 2014) and Caernarfonshire (pers. obs.).

The *Britain and Ireland Atlas 1968–72* showed the Coal Tit to be present in 90% of 10-km squares in Wales, with the most obvious gap in northern Anglesey. There was no great change in distribution in the 1988–91 *Atlas*. There was a small expansion in 2007–11, with the gap in northern Anglesey no longer evident at 10-km square level. Abundance in Wales in 2007–11 was generally higher than in England but lower than in Ireland, with highest densities in semi-upland and upland forests. Relative abundance had increased in twice as many 10-km squares than it had decreased, since 1988–91.

Local tetrad atlases showed that Coal Tit is less widespread than Blue and Great Tit but was found in most tetrads in the majority of counties surveyed. In Gwent, while relative abundance had decreased, its distribution increased during 1981–85 and 1998–2003. In the latter period it occurred in 82% of tetrads, but was absent from the coastal levels and the treeless moorlands of the northeast (Venables *et al.* 2008). In Gower it was found in at least 70% of tetrads in 1984–89. In East Glamorgan there was little change in the number of tetrads in which there was evidence of breeding in 1984–89 and 2008–11, at 61% and 63% respectively. It is less widespread in Pembrokeshire, where it was found in 43% of tetrads in 2003–07, only marginally more than in 1984–88. Rees *et al.* (2009) considered that clear-felling of conifer plantations must have reduced Coal Tit numbers in that county. It was less widespread in Breconshire, where it was found in 59% of tetrads visited during 1988–90. The *North Wales Atlas 2008–12* recorded it in 56% of tetrads, absent primarily in the treeless uplands and sparse in north and west Anglesey.

In many coniferous plantations of Wales, it is the commonest tit species and one of the most numerous of all bird species, but there can also be good numbers in deciduous woodland, particularly Sessile Oak. It was the fifth most abundant bird in 74 native oak woods in Wales (Bibby *et al.* 1989). The number of territories remained very stable in Welsh broadleaved woodlands between 1965–72 and 2003–04 (Amar *et al.* 2006). Breeding densities equated to 27 pairs/km^2 in oak woodland at Abergwyngregyn, Caernarfonshire, in the early 1970s (Gibbs and Wiggington 1973) and 35.7 territories/km^2 in deciduous woodland at RSPB Ynys-hir,

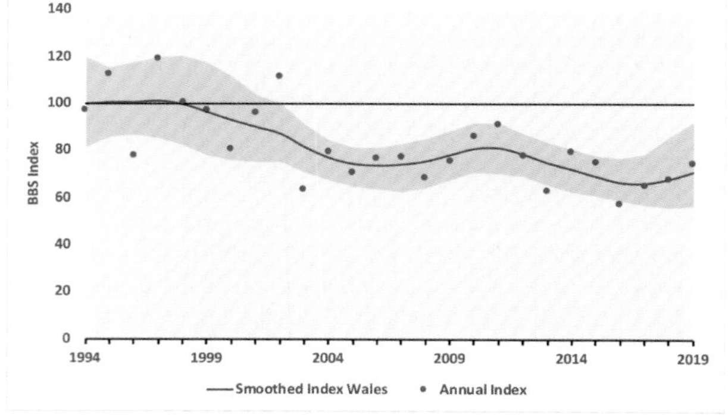

Coal Tit: Breeding Bird Survey indices, 1994–2019. The red line shows the smoothed index, 1995–2018. The shaded area indicates the 85% confidence limits.

Ceredigion, in 2011 (RSPB). In two mature oak woods near Maentwrog, Meirionnydd, Coal Tit was only slightly less common than Blue Tit in winter (Jones 1975).

The long-term CBC/BBS index for the UK showed a substantial increase during 1966–76, followed by a fluctuating but generally stable trend until 2011, a different pattern from the indices for Blue and Great Tit (Massimino *et al.* 2019). The BBS index for Wales declined by 32% in Wales during 1995–2018, a loss that is similar to indices for Blue and Great Tit but very different to the trends for Coal Tits in other parts of the UK (Harris *et al.* 2020). The Welsh breeding population in 2018 was estimated to be 75,000 territories (Hughes *et al.* 2020).

Cold winters can affect numbers. At RSPB Ynys-hir, it was said to be the only tit species markedly affected by a cold spell in winter 1981/82 (Roderick and Davis 2010). In Radnorshire, Jennings (2014) stated that there were losses during very cold winter weather, but usually only following years with a poor Beech-mast crop, such as in 1962/63. The Coal Tit's habit of caching food may help it to survive periods of severe weather, but it has also taken advantage of supplementary food in gardens. The reporting rate in the BTO Garden Birdwatch in Wales is variable, with the annual peak ranging between 53% and 86% during 1995–2019 and consistently in October–December.

Some autumn passage is recorded in most years along the coast, with larger numbers in some years, when birds from elsewhere in Europe may be involved. With Blue and Great Tit, this species formed part of a major irruption in autumn 1957, when numbers on Bardsey, Caernarfonshire, peaked at *c.*120 on 8 October, although numbers on Skokholm were lower, with no more than ten on any one day (Cramp *et al.* 1960). An influx in autumn 1991 produced a count of 100 on Bardsey on 8 October, while 170 were on Ramsey, Pembrokeshire, in autumn 1997. Other autumn passage counts included 50 on the Great Orme, Caernarfonshire, on 28 September 2001; 50 off the sea onto Ramsey on 30 September 2015 and 47 on Bardsey on 6 October 2019. Large counts in winter included 250 at Pontneddfechan, Gower/Breconshire, on 26 December 1983 and 200 in Llantrisant Forest, East Glamorgan, on 30 December 2007.

The Coal Tit clearly benefited from the planting of coniferous woodland in Wales, particularly in the period after the Second World War. Many of these have now matured, and clear-felling of some these may reduce Coal Tit numbers in the short term, but it should continue to be a common species.

Continental Coal Tit *P. ater ater*
Titw Penddu y Cyfandir

The nominate form breeds across much of Europe, east to western Siberia and central Turkey. It is mainly sedentary but, in some years, migrates in large numbers to the southwest. Continental Coal Tits are a regular feature of autumn on the east coast of England, especially in Kent. Although several birds in Wales have showed some characteristics of this subspecies, the only two identified with certainty, and thus accepted by the WBRC, were on Bardsey, Caernarfonshire, on 14 May 2018. This was part of an unseasonal influx, when there were records along the south coast of England from Kent to Dorset, a few days prior to the records on Bardsey and two at Red Rocks, Wirral, on the same day.

Rhion Pritchard

Marsh Tit *Poecile palustris* Titw'r Wern

Welsh List Category	IUCN Red List			Birds of Conservation Concern			
	Global	Europe	GB	UK	Wales		
					2002	2010	2016
A	LC	LC	LC	Red			

Marsh Tit is a widespread Palearctic species with distinct populations in Europe and Southeast Asia and generally a more southerly distribution than Willow Tit. The subspecies *P.p. dresseri* occurs only in Britain and western France (IOC) and inhabits broadleaved woodlands as far north as southeast Scotland. It rarely breeds in secondary habitats such as gardens and hedgerows, though may use these habitats in winter. It nests in tree cavities, although rarely in old woodpecker holes, and only occasionally uses nest-boxes (Broughton and Hinsley 2015). It requires a complex, multi-layered woodland structure, with a dense canopy and a dense understorey/shrub layer, especially at 2–4m above the woodland floor (Hinsley *et al.* 2007). Breeding densities, calculated from the BTO Common Birds Census, estimated 0.4 pairs/km^2 in farmland and 4.4 pairs/km^2 in woodland, but these figures were drawn from across Britain in the 1980s, when the breeding population was much larger than today.

The historical status of the species is difficult to interpret. Even after Marsh Tit and Willow Tit were differentiated as species, it was not appreciated that Willow Tit occurred in Britain until 1897. Birds recorded as Marsh Tit before this date may thus have been wrongly identified. During the early 20th century, Marsh Tit spread north in Britain and bred in Scotland from 1921. It may have reached the maximum extent of its distribution in the early 1950s, when it made a temporary incursion into Anglesey for a few years, although there were only a dozen records and no proof of breeding after 1954. The *Britain and Ireland Atlas 1968–72* provided a clear picture of the species' distribution in Wales for the first time, showing that it occurred across much of eastern Wales, except on open moorland. Occupation in West Wales and the northeast was more patchy and it was absent from Ceredigion, Anglesey, and much of Meirionnydd and Caernarfonshire.

By the *Britain and Ireland Atlas 1988–91*, Marsh Tits had withdrawn from much of Gower and East Glamorgan and from many parts of North and mid-Wales. The contraction in the south and northwest continued to 2007–11, although there were some local range increases in the northeast by the latter period. The highest levels of abundance in the 2007–11 *Atlas* were in a band from Pembrokeshire, through Carmarthenshire and into Radnorshire and Breconshire, but the relative abundance of breeding Marsh Tits was lower throughout Wales in the later period.

The scale of loss in the south is evident from atlases in East Glamorgan, where the proportion of occupied tetrads fell from 22% in 1984–89 to 6% in 2008–11, and there was breeding evidence in only one-third of the latter. In Gwent, it contracted from 45% of tetrads in 1981–85 to 33% in 1998–2003. In Pembrokeshire, the range loss was less severe, with the proportion of occupied tetrads falling from 36% in 1984–88 to 30% in 2003–07. The only other recent local atlas, in North Wales during 2008–12, found Marsh Tits in just 3% of tetrads, with few records west of the Conwy Valley. Carmarthenshire was considered to be a stronghold, with little evidence of change in some parts of the county, illustrated by counts of 40 pairs in a single 10-km square in 1993 (*Birds in Wales* 2002) and 19–20 pairs in the mid-Tywi Valley in 2001 (Rob Hunt *in litt.*). They also remained in reasonable numbers in southeast Wales, from the Wye Valley to Wentwood Forest, with reporting rates little changed since the 1980s. In south Ceredigion, Morris (2017) found 7–12 territories in the 10-km square SN65, which is bisected by the Afon Teifi, during 2012–17.

The Marsh Tit population decreased by 78% in England during 1967–2018 (Massimino *et al.* 2019), with periods of steep decline in the early 1970s and the early 21st century, although there is some evidence that the rate slowed during 2012–17. There are insufficient data to generate a population index for Wales, but it is unlikely to differ significantly. The Welsh breeding population in 2016 was estimated to be 4,850 (4,400–5,250) pairs (Hughes *et al.* 2020), around half the estimate made in *Birds in Wales* (2002). Marsh Tit is probably very much under-recorded in county reports, and although it occurred in all counties during 2010–19, most records were south of a line from Llanrhystud in Ceredigion to Llanigon in Breconshire.

Once a pair has established within a woodland, it remains there throughout the year, although its home range in winter is larger than the 5–6ha required during the breeding season. Broughton *et al.* (2015) found that winter territories averaged 31ha in a Cambridgeshire woodland, with extensive overlap of neighbouring territories, because they need large areas of woodland to thrive. Social groups are fluid and, although they mix with other tits, do not form stable winter flocks. Typically, only 1–2 Marsh Tits are seen at a time. Larger counts from Welsh woodlands are now largely historic. Site counts from Gwent included 18 in Coed-y-Prior in November 1980, 12 near Tintern in November 1997 and 12 at Redbrook in December 2006. They also visit garden bird tables more readily outside the breeding season.

The winter distribution of Marsh Tit in the *Britain and Ireland Atlas 2007–11* was similar to that of the breeding atlas in that period, 61% of 10-km squares. It included records in two 10-km squares on Anglesey that may have been overlooked during the summer. The reduction in occupation since the *Britain and Ireland Winter Atlas 1981–84* was small, less than 7% and mainly in north Ceredigion, Meirionnydd and coastal East Glamorgan. There was an increase in the number of squares occupied in central Ceredigion and Pembrokeshire.

Marsh Tits rarely move any great distance in Britain: 85% of ringed birds were recovered within 5km of the ringing site and only 1% travelled farther than 20km (*BTO Migration Atlas* 2002). In Wales, only two ringed birds have been recovered more than 10km from their natal site. One ringed as a nestling near Rhydymwyn, Flintshire, on 14 May 2011, was caught 10km away at Shotton, Flintshire, on 19 June 2011. Another, ringed at Meusydd Brwyn, Denbighshire, on 20 May 1972, was caught 12km away at Henfryn Hall, Flintshire, on 26 February 1978. Increased numbers were reported along the coast and inland in 1957, a year of large-scale tit irruption (Cramp *et al.* 1960). There have been offshore records in Pembrokeshire: on Skomer in November 1961 and October 1989, and on Skokholm in October 1988.

There is little doubt that Marsh Tits have seriously declined in Wales in the last 30 years, although there are no adequate recent data to quantify the scale of this and they remain more widespread than Willow Tits. On the face of it, habitat conditions should better suit Marsh Tits, as trees have matured and the understorey in most woodlands has become more dense. It may be that the farmed landscape between these woods is not suitable for this sedentary species, so locally poor breeding success over just a couple of years can result in loss with no recruitment feasible from elsewhere. Future expansion of valley woodland provides a potential lifeline for Marsh Tits, but the management of these woods will determine whether they can take advantage of any land-use changes. There are some suggestions that competition from Blue Tit and Great Tit may be a factor in the Marsh Tit's decline (Richard Broughton *in litt.*). That will be a more difficult challenge to overcome even if the habitat issues can be resolved.

Steph Tyler

Sponsored by Stephanie Tyler

Willow Tit — *Poecile montanus* — Titw'r Helyg

Welsh List Category	IUCN Red List			Birds of Conservation Concern				Rarity recording
	Global	Europe	GB	UK	Wales			
A	LC	LC	EN	Red	2002	2010	2016	RBBP

Willow Tit is resident across Europe, except in the far northwest and Iberia, and across temperate Russia to the Bering Sea, although it avoids the far north (*HBW*). The subspecies *P.m. kleinschmidti* is endemic to Britain (IOC), where its distribution is concentrated in southern Wales, the Midlands and northern England, with outlying populations in southwest Scotland, Devon/Cornwall and Hampshire/Wiltshire. The *Britain and Ireland Atlas 2007–11* mapped substantial losses from southern and eastern England since 1968–72. Within Wales it resides at low densities, with relatively large territories.

Globally, Willow Tit is a specialist of mixed conifer and birch bog woodlands, but in Britain it occupies a variety of wooded and scrubby habitats, preferring young, damp woodland with some standing rotten timber into which, unlike Marsh Tit, it excavates a new nest cavity each year (Perrins 1979, Siriwardena 2004). In Wales, Willow Tits also inhabit conifer plantations, especially where there are pockets of scrubby deciduous woodland. It can also occur on higher, more exposed ground with few mature trees, and in dense overgrown hedgerows close to willow scrub. It occurs at 350m in Clocaenog Forest, Denbighshire, and above Lake Vyrnwy, Montgomeryshire, and was recorded at up to 450m in Radnor Forest (*Birds in Wales* 1994). Outside the breeding season it is seen in a wider range of habitats, including at garden feeders. In Wales, most recoveries of the small number of ringed Willow Tits have been very local, but a juvenile ringed near Penmaenpool, Meirionnydd, in July 1995 was recovered at Waunfawr, Caernarfonshire, in December that year, 44km to the northwest and the other side of the Rhinogau.

Willow Tit was not proven to occur in Britain until 1897, and for several subsequent decades its distinct identification from Marsh Tit was not universally accepted (Perrins 1979). As a result, its historical distribution and status are uncertain, and even atlases may not be wholly accurate. In most Welsh counties, first sightings occurred between the World Wars, but records did not increase until after 1945. It is difficult to judge whether this was an increase in abundance or in observer knowledge and effort.

In Pembrokeshire, there were only sporadic records prior to the 1960s, by which time it was widely but thinly distributed in the county (Donovan 1965), while in Carmarthenshire it was unknown by Barker (1905). Little was known about its distribution by Ingram and Salmon (1954), who quoted six records around Rhandirmwyn during 1934–50. Records in Gower, East Glamorgan and Gwent increased during the 1960s, although it was localised in all three counties (Hurford and Lansdown 1995, Venables *et al.* 2008). The *Britain and Ireland Atlas 1968–72* showed that these areas, along with Breconshire and Radnorshire, formed the core distribution in Wales. The number of 10-km squares occupied in Wales changed little in the *Britain and Ireland Atlas 1988–91*, when losses in southern counties and Meirionnydd were offset by in-filling in Carmarthenshire and parts of mid-Wales. This was, however, a period of expansion on Anglesey, particularly around Llyn Cefni and Cors Erddreiniog, where it bred during 1985–99. It seems always to have been scarce in Meirionnydd, Caernarfonshire and in north and west Denbighshire, but records also peaked there during the 1980s and 1990s, with isolated pockets on Llŷn and in the Conwy Valley that were maintained to the turn of the century. It was thinly distributed in river valleys through mid-Wales in the 1980s, and in swamp and carr habitats in Ceredigion, although it was always scarce south of Aberaeron (Roderick and Davies 2010). In most parts of Wales, it was always probably less abundant than Marsh Tit, which favours mature broadleaved woodland.

The *Britain and Ireland Atlas 2007–11* showed considerable erosion in distribution, mainly around the periphery of Wales but to a lesser extent than in England. Nonetheless, Wales held 22% of the British population in that period. Local atlases that have been repeated showed large reductions in the number of occupied tetrads: from 40% in Gwent in 1981–85 to 16% in 1998–2003, and from 22% in Pembrokeshire in 1984–88 to 10% in 2003–07. In

Willow Tit © Norman West

East Glamorgan, the number of territories with breeding evidence fell from 20% in 1984–89 to just 3% in 2008–11, when it was found in just a few sites in the Cynon and Taff valleys. In North Wales, its range contracted from 37% of 10-km squares in 1968–72, to 19% in 2008–12, but in the latter period these records came from less than 2% of tetrads surveyed. There were none in Meirionnydd or Anglesey, and only three in Caernarfonshire, although it has possibly been overlooked in some mainland forestry plantations (Pritchard 2012, 2017). Northeast Wales was the one area where Willow Tits expanded between 1988–91 and 2007–11, mainly in forestry plantations in Denbighshire. In Ceredigion, the losses were mainly in their former stronghold in the north. In the south of that county, Morris (2017) found 10–13 pairs in 25 tetrads straddling the Afon Teifi on the Carmarthenshire border, where the species is spatially separate from Marsh Tit and used conifer stands.

The Willow Tit is the second-fastest-declining species in the UK after the Turtle Dove. The UK BBS index fell by 82% during 1995–2018 (Harris *et al.* 2020). Although there are no Wales-specific figures, there is no reason to believe the scale of decline is very different to that in England. Breeding was recorded in all counties, except Anglesey and Caernarfonshire, during 2010–19.

Willow Tits are most readily detected in early spring, just before the breeding season, when they become more vocal. Most comprehensive surveys are undertaken from mid-February to mid-April, when birds are more likely to respond to a playback of their song. A survey was organised in 2019–21 based on: (1) non-random 'core' coverage (e.g. Lake Vyrnwy, Montgomeryshire), (2) a stratified random sample of tetrads from occupied 10-km squares, either known to have held Willow Tits or containing potentially suitable habitat, and (3) historical or recently occupied sites in counties with few recent records. In the first two seasons of fieldwork, Willow Tits were found in only 56% of tetrads that had been occupied previously. From these 279 tetrads surveyed, a provisional population of 1,541 territories (1,177–1,919) has been estimated, which will be refined following the final year of survey (Simon Wotton *in litt.*). The core range is now north Pembrokeshire, south Ceredigion (particularly around Cors Caron), north Carmarthenshire, Breconshire north of the Brecon Beacons, and Montgomeryshire. In the latter county, there remain strong populations in Coed Hafren, the Dyfnant Forest and at RSPB Lake Vyrnwy, where there were *c.*30 territories in 2019. It is considered that there are "at least 100 pairs of Willow Tit in Montgomeryshire and possibly as many as 300" (Mike Haigh *in litt.*). The provisional Welsh total is 20% of the total British population (Simon Wotton *in litt.*).

Several hypotheses have been put forward to explain the cause of the decline. The Willow Tit's preference for birch and conifer bog woodlands and wet scrub gives it an advantage over other tit species, with which it would otherwise compete. Numbers had remained stable in wet woodlands but had declined in deciduous woods (Siriwardena 2004). Sites still holding the species tend to be wetter than those that have lost it (Lewis *et al.* 2007, 2009a, 2009b). However, a reduction in soil moisture, as a result of drainage and increasingly dry summers, may reduce the quality of this habitat (Vanhingsbergh *et al.* 2003) and potentially make it more suitable for competitor tit species. Reduced management has increased the amount of dead wood available. It is not thought that nest sites are a limiting factor, as they require only one suitable dead stem within 5–10ha of woodland (Richard Broughton *in litt.*).

Deterioration in woodland quality, through canopy closure and increased browsing by deer, has been evident in English study areas (Perrins 2003, Fuller *et al.* 2005, Newson *et al.* 2011). However, a comparison of the distribution of deer and Willow Tits in Gwent suggested that browsing was unlikely to be the cause of the 60% decline in the county between 1981–85 and 1998–2003 (Venables *et al.* 2008). Deer densities remain low in Wales compared to England.

Nest eviction by Blue Tits causes 23% of first breeding attempts by Willow Tit to fail, but only 3% of second attempts did so. The survival rate of hatched chicks was high, but late broods may be less likely to be recruited into the breeding population (Broughton and Hinsley 2015, Parry and Broughton 2019). Predation of eggs and chicks, including by Great Spotted Woodpecker, can result in the failure of 20–35% of nests (Richard Broughton *in litt.*), but studies found no difference in the density of predators or competitors at sites that had retained or lost Willow Tits (Lewis *et al.* 2007, 2009a, 2009b, Siriwardena 2004). Competition for food appears not to be an issue, at least during the nestling stage, but more study is required to understand fully its potential implications.

Trials of management solutions for Willow Tits are under way on former colliery sites in northern England, but it is unclear how applicable these, or the research undertaken to date, are to Wales, which may now hold a greater proportion of the remaining British population than historically. The prospects for this species are not bright, especially as it occurs at low density in a patchwork landscape managed by many different private owners. The result of the recent survey provides an opportunity to focus conservation effort in its core range and for changing land-management policies to support the management of its peripheral habitats.

Wayne Morris

Blue Tit *Cyanistes caeruleus* Titw Tomos Las

Welsh List Category	IUCN Red List			Birds of Conservation Concern			
	Global	Europe	GB	UK	Wales		
A	LC	LC	LC	Green	2002	2010	2016

Although it may be unremarked upon by many birdwatchers, the Blue Tit is one of the best-known and best-loved birds in Wales, recognised by almost everyone as a common visitor to garden feeders, even in the most urban areas. It is found throughout Europe, except in higher latitudes, and its distribution extends to parts of the Middle East as far east as Iran (*HBW*). The subspecies *C.c. obscurus* occurs only in Britain and Ireland (IOC). While birds seen in Wales in autumn may include those of the continental nominate subspecies, there have been no recoveries here of birds ringed outside Britain. An adult male shot near Erwood, Radnorshire, on 29 October 1939, was considered at the time to be of the nominate subspecies and reported as such in *Birds in Wales* (1994). The skin is preserved at the National Museum of Wales, but on re-examination, Jennings (2014) stated that "it is probably safer to say that it shows some of the characteristics of that subspecies".

Its preferred habitat is lowland and sub-montane deciduous woodland, although it is also commonly found in parkland, hedgerows, urban and suburban areas (*HBW*). It is not particularly common in coniferous forest, where Coal Tit is much more numerous, and is infrequent in *ffridd* habitats. It was present in 21 of 118 study sites in central Wales surveyed in 1985–87, although it was the most widespread of the tits (Fuller *et al.* 2006). It seldom breeds above 400m in Wales, although there are a few records of breeding up to around 500m. Populations in Britain and Ireland are mostly resident. The longest distance covered by a British-hatched Blue Tit involving Wales was a nestling ringed in North Yorkshire in June 2000 that was recovered at Uskmouth, Gwent, in June 2003, 358km to the southwest. One ringed on Bardsey, Caernarfonshire, on 4 October 2003, was re-trapped at Cape Clear Bird Observatory on the southern tip of Ireland, 353km to the southwest, just 19 days later. It is likely that this was only part of a longer migration.

Blue Tit has probably been one of the most abundant birds in Wales for a very long time. It was considered by Forrest (1907) to be the most populous of its clan in North Wales, found in all suitable habitat, and contemporary authors made similar comments elsewhere in Wales. All three Britain and Ireland atlases confirmed Blue Tits breeding in almost every 10-km square in Wales. In 2007–11, abundance was relatively high in Pembrokeshire and the wide river valleys of eastern Wales, but much lower in the hills. It had increased or remained stable since 1988–91 in all but one 10-km square. Tetrad atlases invariably showed Blue Tit to be among the most widespread of species, absent only from the highest ground and occurring in more than 80% of tetrads in counties surveyed in the 21st century. The county with the most widespread distribution was Gwent, with occurrence in 99% of tetrads in 1998–2003. A decrease was noted only in East Glamorgan, where it was found in 87% of tetrads in 1984–89 and 83% in 2008–11, with breeding evidence in 79%.

Densities can be high in suitable habitat. There were 115 territories in 112ha of woodland at RSPB Ynys-hir, Ceredigion, in 2006, a density of 103 territories/km^2 (Roderick and Davis 2010). By contrast, in upland conifer plantations the mean density was 19 individuals/km^2, making it only the 15th most numerous species (Bibby *et al.* 1989). In restocked conifer plantations there were only five individuals/km^2 (Bibby *et al.* 1985). Densities in Welsh broadleaved woodlands increased by 148% between 1965–72 and 2003–04 (Amar *et al.* 2006).

The long-term CBC/BBS index for the UK showed a fairly steady increase between 1966 and 2011, after which it had declined to 1990 levels by 2018 (Massimino *et al.* 2019). In Wales, the BBS index peaked in 2005. It had declined by 23% by 2018, returning to a level similar to its 1995 baseline (Harris *et al.* 2020). The Welsh breeding population in 2018 was estimated to be 415,000 territories (Hughes *et al.* 2020).

Blue Tit © John Freeman

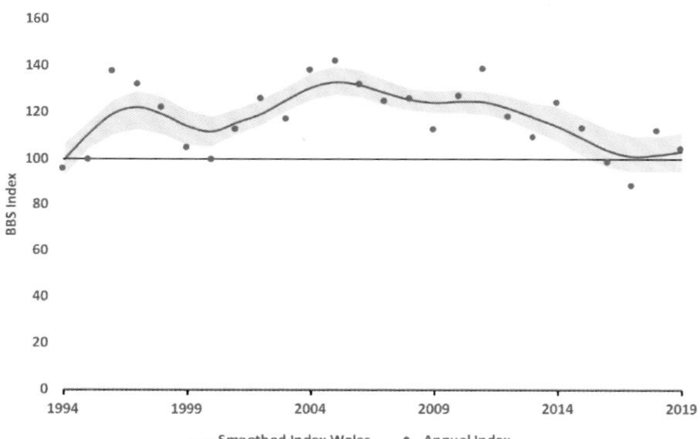

Blue Tit: Breeding Bird Survey indices, 1994–2019. The red line shows the smoothed index, 1995–2018. The shaded area indicates the 85% confidence limits.

Provision of food in gardens may have been a factor in the long-term increase in the population (Plummer *et al.* 2019), and nest-boxes may lower the risk of predation of eggs or chicks. Even small gardens may have much larger populations than expected. For example, ringing showed that over 390 visited one Swansea garden in winter 1974/75 (Hurford and Lansdown 1995), 315 were caught in a garden at Nant Glas, Radnorshire, in 2001 (Jennings 2014) and 166 were ringed in a Monmouth garden, Gwent, during ten days in 1979 (Venables *et al.* 2008). In a small garden in Pwllglas, Denbighshire—a village surrounded by woodland—705 were ringed between 18 September 1987 and 8 April 1988.

Some coastal sites and islands record fairly large numbers in some autumns, usually peaking in late September and October. There are regular southwesterly movements of part of the Scandinavian population and there can be irruptions from continental Europe in some years, notably in autumn 1957 when, although numbers were lower than in eastern Britain, there were some large counts in Wales. Bardsey, Caernarfonshire, recorded a peak of 75 on 8 October and there were 400 on Skokholm, Pembrokeshire, in October, including 50 on 16 October. Large numbers were seen along the coasts of Ceredigion, on 13 October, and Pembrokeshire, on 14 October. In the latter county, about 300 were recorded along a quarter mile (*c.*400m) stretch of coast at Marloes on 19 October (Cramp *et al.* 1960). There was another irruption in 1959, but only small numbers reached the Welsh islands, although they were said to be in every gulley on the south coast of Pembrokeshire on 13 October (Cramp 1963). Up to 50 were on Skokholm in October 1964 and 85 on Skomer, Pembrokeshire, in October 1991. On Bardsey, the best years were 1981, when there were 120 on 6 October and further counts of 120 on 14 October, and 1988, on 9 October. Movements on the mainland included 120 southwest along the Gwent coast on 3 October 1988 and 150 at Lavernock, East Glamorgan, on 9 October 1977 and 2 October 1988. It is possible that birds seen during irruptions may have been of the subspecies *C.c. caeruleus* from northern Europe, but it is uncertain how frequently this subspecies reaches Wales.

In winter, the species is less territorial than Great Tit and tends to be more mobile, forming flocks, although they usually do not move far (*BTO Migration Atlas* 2002). The Blue Tit's winter range is very similar to its breeding distribution. *Birds in Wales* (1994) noted that winter numbers in some areas, such as parts of Ceredigion, recorded during the *Britain and Ireland Winter Atlas 1981–84,* were surprisingly low considering the apparent suitability of much of the countryside. The 2007–11 *Atlas*, however, showed relatively high winter abundance in most of the county. A study of the winter bird population in Coed Cymerau, Meirionnydd, between 1965/66 and 1968/69, found Blue Tit to be the most common species, comprising 30% of the total bird population (Jones 1975). Large counts in winter include *c.*150 at Welsh St Donats, East Glamorgan, on 7 March 1964. In Breconshire, at least 185 fed in flocks around Cwm Gwenffrwd and Llangenny in November 2010, and 160–180 were in a mixed flock at Felinfach on 1 January 2012.

The Blue Tit has adapted well to the opportunities provided in gardens, occurring weekly in 95–99% of gardens in Wales participating in the BTO Garden Birdwatch, although the reporting rate has slowly declined during 1995–2019. Heavy use of garden bird feeders can increase the transmission of disease within a population and between species, but Avian Pox is thought unlikely to be behind the decline in Blue Tit since 2011 (Lawson *et al.* 2018). As a single-brooded species, Blue Tit can be susceptible to cool, wet weather in June, when nestlings need food. This may have influenced a decline in nest productivity across the UK since the late 1990s: nest-box schemes generally recorded poor productivity during 2012–16. Adult survival has slowly increased since 1990, perhaps a result of milder winters, but warmer springs have created a trophic mismatch, whereby chicks hatch later than the peak abundance of caterpillars. There is no significant difference in the extent of the mismatch between southern and northern Britain (Burgess *et al.* 2018). Expansion of broadleaved woodland in Wales should benefit Blue Tits, but the phenological mismatch in chick food may well prove to be the biggest driver of its future population level.

Rhion Pritchard

Great Tit *Parus major* Titw Mawr

Welsh List Category	IUCN Red List			Birds of Conservation Concern			
	Global	Europe	GB	UK	Wales		
					2002	2010	2016
A	LC	LC	LC	Green			

The Great Tit is widely held to be the most studied bird species in the world (Gosler 1993). It is the most widespread tit species, with a distribution covering almost all of Europe, northwest Africa and large parts of Asia as far as western China (*HBW*), although the subspecies *P.m. newtoni* breeds only in Britain and Ireland (IOC). It is a cavity nester, mainly in tree holes, and takes more readily to nest-boxes than other tit species (Perrins 1979). The Great Tit's preferred habitat is open deciduous woodland, but it also uses hedgerows, parkland and gardens, including those in urban areas. It is not common in coniferous woodland, but the presence of a few mature deciduous trees in conifer plantations can considerably increase numbers (Bibby *et al.* 1985). It was much less abundant than Blue Tit in coniferous woodland in Breconshire (Peers and Shrubb 1990) and occurred in only 17 of 118 *ffridd* study sites in central Wales surveyed in 1985–87 (Fuller *et al.* 2006). It breeds at altitudes up to about 500m in Wales.

Populations are largely resident: ringing recoveries involving Wales show that most birds move only very short distances, and none has involved birds from outside Britain. An adult ringed at Northwold, Norfolk, in March 2000, was caught by a ringer 393km away at Pwllcrochan, Pembrokeshire, in December the same year, while one ringed near Dwyran, Anglesey, in July 2007, was caught by a ringer at Ardeonaig, Stirling, 371km to the north, in June 2008. Established males remain on the breeding territory throughout the year if conditions permit (Gosler 1993), although females and young birds are more likely to join winter flocks. Great Tits were recorded in *c.*80% of Welsh gardens participating in the BTO Garden Birdwatch during 1995–2019. The peak generally occurred in winter, although the difference between seasons has become less marked during 2012–19.

The Great Tit seems always to have been abundant in Wales, although where earlier authors compared this to Blue Tit, the latter was said to be more numerous. It was somewhat scarce in north and northwest Anglesey (Forrest 1907). The three Britain and Ireland atlases, in 1968–72, 1988–91 and 2007–11, confirmed breeding in almost all 10-km squares in Wales. The few gains between the first and most recent periods were in coastal areas. Relative abundance in Wales in 2007–11 was higher in most areas than in 1988–91. It was at a similar level to Ireland, but lower than much of England. The highest densities in Wales were in the east and along the south coast. As expected from a generally sedentary species, relative abundance in the breeding season and in winter are very similar.

Local atlases showed its widespread distribution. It occurred in more than 98% of tetrads in Gwent in 1998–2003, 92% in Gower in 1984–89 and 84% in East Glamorgan in 2008–11. This had changed little over time where a comparison was available. In Pembrokeshire, however, its occurrence increased from 81% of tetrads in 1984–88 to 86% in 2003–07 (*Pembrokeshire Avifauna* 2020). Presence was lower in Breconshire, where it was found in 70% of tetrads visited in 1988–90, and in North Wales where it was in 79% of tetrads during 2008–12, being absent mainly from higher ground. In suitable habitat, numbers can be high; for example, there were 89 territories in 112ha of woodland at RSPB Ynys-hir, Ceredigion, in 2006, a density of 79.5 territories/km^2. It does not breed regularly on the Welsh islands, but nested on Bardsey, Caernarfonshire, in five years between 1984 and 2015.

Passage movements south and west are evident on the coast in autumn, mainly in late September and October, although they are outnumbered by Blue Tits in bird-observatory logs from Skokholm, Pembrokeshire and Bardsey. An irruption of tits from continental Europe along the coasts of Ceredigion on 13 October and of Pembrokeshire on 14 October 1957 included large numbers of Great Tit, which were thus likely to have been of the nominate subspecies (Cramp *et al.* 1960). The peak day count on Skokholm was 25 on 9 October, with 200 recorded that month. Good numbers have been recorded on Bardsey in some autumns, with a peak day count of at least 80 on 7 October 2010. A smaller late spring passage has been observed at some sites, such as Bardsey and South Stack, Anglesey (Jones and Whalley 2004). Movements of birds ringed around Llysdinam, Breconshire, suggest that at least some first-calendar-year Great Tits move towards the South Wales coast in autumn and return in spring (Robertson and Slater 1997). Outside passage periods, counts are generally smaller than for Blue Tit, although a mixed flock of over 240 tits in the Elan Valley, Radnorshire, in November 1989, contained at least 100 Great Tits.

The population increased in Wales during 1995–2011, continuing a fairly steady increase across the UK from at least 1966 (Massimino *et al.* 2019). This increase built on a 109% increase recorded in Welsh woodlands between 1965–72 and 2003–04 (Amar *et al.* 2006). This could have been at least partly due to

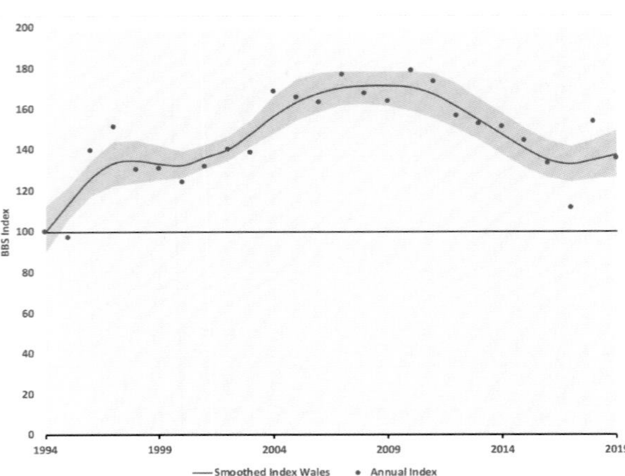

Great Tit: Breeding Bird Survey indices, 1994–2019. The red line shows the smoothed index, 1995–2018. The shaded area indicates the 85% confidence limits.

human factors, such as the provision of nest-boxes and particularly the provision of food in gardens (Plummer *et al.* 2019), which can be very important in winter for birds forced out of woods by food shortage (Gosler 1993). However, from a peak around 2007–11, the BBS index for Wales fell by 21% during 2012–18 (Harris *et al.* 2020). This is likely to be the result of successive poor breeding seasons, judging by low fledging rates at many nest-box schemes in Wales. Ringing at Constant Effort Sites across the UK showed a reduction of 20% in the proportion of juveniles to adults during 2011–16 (Massimino *et al.* 2019), although productivity improved greatly in 2017–18. The Welsh breeding population in 2018 was estimated to be 255,000 territories (Hughes *et al.* 2020).

It is too soon to determine if the recent decline in Great Tit abundance in Wales is of any significance in the long term. A potential problem for Great Tit, as for some other woodland birds, is a mismatch between the times of peak caterpillar availability and the hatching of eggs (Burgess *et al.* 2018). Great Tits are laying earlier: the mean laying date in the UK advanced by ten days during 1967–2017 (Massimino *et al.* 2019). A mismatch has been observed at some Welsh sites, such as Coed y Felin NWWT reserve, Flintshire (Ian Spence pers. obs.), and this is likely to increase as a result of climate change. Great Tit is the member of the tit family that is most susceptible to a strain of Avian Pox that was first reported in southeast England in 2006 and across Wales by 2019, with a seasonal peak in late summer. It is not thought to have a population-level impact, but continued surveillance is essential. This is a very adaptable species and a potential beneficiary of any increase in broadleaved woodland planting in the coming decades, but the phenological mismatch in availability of chick food could change its status relatively quickly.

Rhion Pritchard

Penduline Tit *Remiz pendulinus* Titw Pendil

Welsh List Category	IUCN Red List		Rarity recording
	Global	Europe	
A	LC	LC	WBRC

This bird of riparian woodland breeds locally around the Mediterranean basin and more abundantly through eastern Europe and east to Kazakhstan (*HBW*). It was formerly a very rare visitor to Britain, with very few accepted records prior to 1990. Numbers, mainly in eastern and southern England, then increased rapidly, following expansion of its breeding range in Denmark, Belgium and The Netherlands, but there have subsequently been losses at the western edge of its range (*European Atlas* 2020). Most records in Britain have been during autumn and winter as juveniles dispersed, but the numbers have been small and stable, at an average of just ten each year during 1990–2018 (White and Kehoe 2020b). Some have predicted Penduline Tit as a future breeding species, but any colonisation of Wales will need movements in far greater numbers.

It is a very rare visitor to Wales, with just five records, all since 1981:

- Bardsey, Caernarfonshire, on 9–13 May 1981;
- Llyn Rhos Ddu, Anglesey, on 21 October 1992;
- a male and one other at Kenfig, East Glamorgan, from 11 November 1996 to 9 March 1997;
- Cardiff Bay, East Glamorgan, on 14 October 2009 and
- Newport Wetlands Reserve (Uskmouth), Gwent, on 7–9 November 2013.

Jon Green and Robin Sandham

Bearded Tit *Panurus biarmicus* Titw Barfog

Welsh List Category	IUCN Red List			Birds of Conservation Concern				Rarity recording
	Global	Europe	GB	UK	Wales			
A	LC	LC	LC	Green	2002 N/A	2010	2016	RBBP, WBRC

Bearded Tit is primarily an Asian species, occurring from the Black Sea east through Central Asia, Mongolia and China. The nominate form, found in Europe, has a more scattered distribution, from Finland in the north to the Mediterranean coast, that is restricted by its need for large reedbeds (*HBW*). Its distribution in Britain is centred in southern and eastern England, with populations at several sites in northern England and Scotland, as well as South Wales (Eaton *et al.* 2020). It is largely resident, but prone to irruption, which can result in its finding new sites far from the natal area, although this does not always lead to colonisation. The precise trigger for irruptions is not clear, but it involves adults as well as first-calendar-year birds. Autumn movements in Britain are always westward (*BTO Migration Atlas* 2002).

It was rarely recorded in Wales until the 1960s. It did not feature in Forrest's 1919 review of North Wales. It was mentioned by Mathew (1894) as having occurred only once in Pembrokeshire, at Sealyham in 1860, and once in Carmarthenshire, at Castellgorfod in November 1891, although Ingram and Salmon (1954) considered the latter record unsatisfactory. There seem to be no records in the 20th century before the 1960s, a period in which the availability of quality reedbed reached its low point and the 1946/47 winter had reduced the British population to 2–5 pairs, all in Norfolk (Marchant *et al.* 1990).

A series of irruptions from the Low Countries and eastern England in the 1960s and early 1970s (Axell 1966, O'Sullivan 1976) resulted in the first records in Gwent, where three were at Newport Docks from 29 January to 6 March 1966, and in East Glamorgan, where two were at Kenfig on 17 October 1965. However, the most substantial influxes were in North Wales. At least 16 were at Shotton Steelworks, Flintshire, in autumn 1965, of which 13 were caught by Merseyside Ringing Group, including a male and female that had been ringed as juveniles with consecutively numbered rings the previous summer in Oost Flevoland, The Netherlands. None was seen there after November 1965 (Rob Cockbain pers. comm.). This first wave of irruptions made particular headway on Anglesey, which had the most suitable habitat at the time. A minimum of 40 were at Valley Wetlands and Cors Ddyga in November and December 1965, leading to single pairs breeding at Valley in 1967, at Llyn Bodgylched in 1968 and at Cors Ddyga in 1969.

There were further arrivals into Wales in the autumns of 1971–73. Six were ringed at Shotton Steelworks in October 1971, but all moved on rapidly, with one being recovered near Hanover, Germany, in January 1972. Another five were ringed there in October 1972, at least some of which stayed until January 1973. One was recovered in Gelderland, The Netherlands, the following

Male Bearded Tit © Norman West

August. Two pairs were present for a short period in April 1973, but there was no evidence of breeding and no more were seen until October 1973, when another ten were ringed. These were the last of a minimum of 38 at the Flintshire site during 1965–73 (Rob Cockbain pers. comm.).

The 1972 irruption also saw the first accepted records for Carmarthenshire, with one at Ashburnham, Pembrey, at the end of October, two at Trostre, Llanelli, on 16 November and four there on 3 December. The first record for Ceredigion was of two at Cors Fochno on 17 February 1973. Six Bearded Tits at Brynsiencyn in November and four at Llyn Bodgylched from January to late August 1973 were the first on Anglesey since 1969. Several were present at the latter site in 1974, although breeding was not proven. The 1972 irruption did, however, lead to the establishment of a small breeding population at Oxwich Marsh, Gower, which had greater longevity than that on Anglesey. The first Bearded Tit record there was on 17 October 1972 and at least ten were present in November. It was thought to have bred every year during 1973–1986, peaking at three pairs in 1978 and 1979, but there was just a single pair in 1984–86. The largest counts, which included fledged young, were 18 on 30 September 1984, 15 in October/November 1983 and 12 on 30 June 1976. The last sighting was on 9 April 1989.

After the demise of the Oxwich breeding population, Bearded Tit was a sporadic visitor to Wales, usually in late autumn, for almost 20 years. A pair may have bred at Llangorse Lake, Breconshire, in 1977. There were 12 there in January 1978 and birds remained until September 1980. The majority of sightings during the 1980s and 1990s were in Gwent, where the first had only been recorded in December 1974, and in East Glamorgan, where there were several arrivals of small groups during 1970–99.

There was an increase in records at Uskmouth, soon to become part of Newport Wetlands Reserve, Gwent. Up to four were present during 2001–05, with nesting confirmed in 2005, when 6–9 juveniles were recorded. They have bred annually there since: four pairs in 2008–10, five in 2011–13 and six in 2014–19. Peak autumn counts included 20 on 21 September 2009, 29 on 10 November 2013 and 31 on 29 September 2015. Individuals from this nucleus have been reported at other sites in Gwent and East Glamorgan, including a pair at Cosmeston Lakes in October 2010 that had been ringed at Uskmouth. There have been repeated records at Cardiff Bay Wetlands Reserve, East Glamorgan, including four in March/April 2007. The only other recent breeding record was at RSPB Conwy, Denbighshire. A female and two males there, in April 2010, comprised the first county record, from which a pair went on to fledge three young, with adults or young birds seen until immediately prior to a period of freezing weather in December that year.

Away from southeast Wales, Bearded Tit remains rare. Records prior to 2000 were listed in *Birds in Wales* (1994) and *Birds in Wales* (2002). These included four at Witchett Pool, Carmarthenshire, in November 1988 and November 1989, a female that summered at RSPB Ynys-hir, Ceredigion, in 1993–95, and perhaps the same individual there in March 1997. Singles at Pwllheli in August 1986 and August 1987 constitute the only Caernarfonshire records, and there was just one further record from Anglesey, at Llyn Penrhyn on 11 January 1981.

Away from Gwent and East Glamorgan, records in Wales are assessed by the WBRC. Those since 2000 were:

- pair at Tre'r Ddol, Ceredigion, on 2 May 2001;
- RSPB Ynys-hir, Ceredigion, on 7–20 May 2002;
- Teifi Marshes, Pembrokeshire/Ceredigion, on 15 February 2003 and 23 November 2003;
- Pentrosfa Mire, Radnorshire, on 31 October 2004, the only county record;
- female and two males at RSPB Conwy, Denbighshire, in April 2010, the only county record, a pair from which went on to breed successfully;
- pair at Llangorse Lake, Breconshire, from 26 February to 10 March 2012 and
- Dolydd Hafren, Montgomeryshire, from 31 January to 15 February 2017, which is the only county record.

A major programme of reedbed restoration and creation since the mid-1990s, prompted by the very real risk of the extinction of Bittern, has helped the Bearded Tit population to grow to around 700 pairs across Britain (Eaton *et al.* 2020). Cold winters are a major cause of mortality (Campbell *et al.* 1996) and deep flooding in spring can cause nesting attempts to fail. Bearded Tits generally prefer to nest in drier reedbeds, with a layer of leaf-litter at the base. It seems better able to withstand short periods of freezing winter conditions than in the past, at least where it is well-established. In the short term, there is no reason that it should not thrive in southeast Wales. In the longer term, its climate space is modelled to move from southeast to southwest Britain, including South Wales (Huntley *et al.* 2007), so its future will depend on the creation of new, well-managed reedbeds in this area. Its irruptive behaviour, however, means that if sufficient numbers were to get to Anglesey, where there is more reedbed than in the 1970s, its prospects for sustaining a breeding population may be better. Bearded Tits bred within sight of the Flintshire border at RSPB Burton Mere Wetlands in 2019 and 2020, so that possibility has certainly increased.

Julian Hughes and Jon Green

Woodlark *Lullula arborea* Ehedydd Coed

Welsh List Category	IUCN Red List			Birds of Conservation Concern	Rarity recording
	Global	Europe	GB	UK	
A	LC	LC	VU	Green	RBBP

The Woodlark's breeding range is from southern Britain to northwest Africa, and across temperate Europe as far east as the Ural Mountains and northern Iran. The nominate subspecies occurs across the northern part of its range, including Britain (IOC). It is typically resident in the west and south of its range, but migratory in the north and east (*HBW*). It favours open habitats with low-intensity management, which in Britain tends to be lowland heath, young forestry plantations and recently felled woodland (*HBW*). The Woodlark underwent several major changes in numbers and distribution across Britain during the 20th century, fluctuating between 100 and 400 pairs, and became concentrated in five discrete areas of heathland in southern England (Langston *et al.* 2007). The population increased to over 3,000 pairs in 2009 (Conway *et al.* 2009). Breeding now extends as far north as Derbyshire and Yorkshire (Eaton *et al.* 2020). However, aside from a handful of breeding records in Gwent early in the 21st century, it is a scarce passage migrant and winter visitor to Wales.

During the mid-19th century, Woodlark was a widespread breeding species in many parts of Wales and England, although it was always more common in the southern counties of the latter (*Historical Atlas 1875–1900*). Its precise status was confounded by its Welsh name being very similar to Tree Pipit (Corhedydd y Coed), leading Forrest (1907) to reject many North Wales records for publication. In Gower, it occurred very locally around Swansea and on the southern half of the peninsula in the first half of the 19th century (Hurford and Lansdown 1995). It was described as plentiful in Denbighshire and Flintshire around 1830, but had gone from these counties by 1854, according to a Liverpool taxidermist quoted by Dobie (1894). Similarly, in Pembrokeshire, Woodlark was said to be "very generally distributed, and a constant resident" (Dix 1866), but within 30 years Mathew (1894) lamented that "the local birdcatchers used to obtain 36s. a dozen for fresh caught Woodlarks, hens and cocks taken together, so it is no wonder that they sought after them persistently, and have nearly obliterated this sweet songster from our county list".

By the turn of the 20th century, in the northern half of Wales it was resident only locally, in parts of Montgomeryshire, Denbighshire and Flintshire, sporadic in winter in Meirionnydd and there were no records from Anglesey or Caernarfonshire (Forrest 1907). Elsewhere in Wales, it was a localised breeding species in all counties, but most authors considered that its numbers had declined during the previous 50 years. There seemed to have been a resurgence early in the 20th century, at least in Carmarthenshire, Pembrokeshire and Ceredigion, perhaps a result of clear-felling of woodland during the First World War (Salter 1930).

Severe winter weather in 1928/29 resulted in a population crash, from which it took around a decade to recover. By the 1940s and 1950s, it was said to be plentiful in rough open country up to c.275m in Ceredigion, but absent from moorland in the north of the county and more localised in the south (Condry 1950). It was widely distributed on heath and moorland in Pembrokeshire, particularly in the foothills of Mynydd Preseli (Lockley *et al.* 1949). Numbers also increased in Carmarthenshire during the second quarter of the 20th century, with regular nesting in the coastal lowlands and at many inland sites, including into the hills up the Tywi Valley (Lloyd *et al.* 2015). It was more localised in the far south. It bred in very small numbers near Margam, Gower, until at least 1925, and in the Vale of Glamorgan, around Dinas Powys and St Fagans, until at least 1936, but subsequent records in Glamorgan were sporadic, with most being of birds on passage. A reasonable population was present in Breconshire during 1920–45, but the 1947 winter caused another serious crash, and it was slow to return to its former haunts. In the 1950s it bred locally in Radnorshire and in Meirionnydd, primarily on the coastal plain around the Dyfi and Mawddach estuaries. It occurred more irregularly in Caernarfonshire, and although reported to have nested in the east of the county between the 1930s and 1950s, confusion with Tree Pipit cannot be ruled out.

While the 1928/29 winter set the population back, and the 1947 winter seemed to have only local effects, the 1962/63 winter was more terminal for its breeding status in Wales. However, it seemed to have been in decline in most areas for the preceding decade, so the freezing conditions hit an already vulnerable population. It had already ceased to breed in East Glamorgan, with the last pair near Merthyr Tydfil in 1955–59. In Gower, breeding records were only sporadic, such as a pair at Fforest-fach in May 1958 and the last at Penllergaer in 1967. It was said to be quite common in eastern Montgomeryshire during 1950–65, but was not recorded in the county after one at Aberhafesp in June 1965. The only breeding record in Pembrokeshire after winter 1962/63 was at Amroth in 1965, though a few pairs remained in Breconshire, including three near Hay-on-Wye in 1966.

Woodlarks managed to cling on as a breeding species in a few places, mainly in mid-Wales, and were recorded in 20 of the 10-km squares in Wales during the *Britain and Ireland Atlas 1968–72*. These included five territorial males in Ceredigion in 1967–69, several pairs in Montgomeryshire, and *circa* ten pairs around the Dyfi (Jones 1974a) and a pair at Dyffryn Ardudwy, Meirionnydd, in 1966–68. In Breconshire, several pairs were in the Irfon Valley in 1964–1970, and breeding was confirmed at Ystradgynlais, Defynnog, Aber Car and Glan Wye during 1969–70. The last proven breeding in Carmarthenshire was at Cynwyl Elfed in 1972, although singing males were at Tumble and Cilycwm until 1975 and there were occasional records at former sites into the 1980s. A pair bred in Gwent near Pen-y-fan Pond in 1974 and singing birds were heard there until 1977. Records from the Llanelwedd area of Radnorshire suggested a small population may have remained until the late 1970s, with a pair seen in May 1979 and a singing male in 1980 (Jennings 2014). Singing males were also heard in Ceredigion, near Cellan in April 1980 and at Llangeitho in 1980 and 1981 (Roderick and Davis 2010).

During the 1970s and 1980s, Woodlark became a genuinely scarce visitor to Wales and during 1991–2012 all records were assessed by the WBRC. There were four isolated records in Wales during the period covered by the *Britain and Ireland Winter Atlas 1981–84,* but none during that of the 1988–91 breeding atlas, when the English population reached a low point. A survey in 1986 recorded c.250 pairs in England, its decline attributed to severe winters, climatic change and habitat suitability (Sitters *et al.* 1996), although *Birds in Wales* (1994) judged that there was no shortage of habitat in Wales. The English population made a substantial recovery to 1,426–1,552 territories in 1997, when over 85% of territories recorded were on heathland or in planted forests (Wotton and Gillings 2000). Neither survey included Wales, nor is there reason to believe that Woodlarks bred here during that period, although one was seen at a site in Gwent in May 1987 that was later used by a breeding pair. The *Britain and Ireland Atlas 2007–11* recorded a substantial expansion to the north and east of its range in England since 1988–91, as far as Nottinghamshire and East Yorkshire. There was a series of breeding records at a site in Gwent around this time: a pair fledged four young in 2006, a male held territory in 2010 and a pair fledged three young in 2011. No breeding has occurred since, as work on heathland restoration in Gwent stalled and succession to birch scrub rapidly covered the nesting areas (Jerry Lewis *in litt.*).

Aside from those breeding records, Woodlark occurs only as a passage migrant and occasional winter visitor in Wales. Following a low point in records in the 1980s, numbers picked up a little during 1990–2019, especially during severe weather farther east. Southern coastal counties have recorded the majority of records, with 46 in Gwent, 74 in East Glamorgan (of which 23 were in October and November 2019) and 88 in Gower during this period. Double-figure counts have been recorded at several locations, including: 12 at Pitton, Gower, in November 2003 and counts of ten at Dingestow, Gwent, in October 1994, Scurlage, Gower, in November 1994, Llanarth, East Glamorgan, in January 2009 and Lavernock Point, East Glamorgan, in October 2019. It is scarcer farther west: five at Broad Oak in October 1997 was the only Carmarthenshire record in the last 30 years and there were only 18 in Pembrokeshire, including four on Skomer, two on Skokholm and four at Strumble Head.

Woodlark is scarcer in mid- and North Wales, although it has become more frequent on Bardsey, Caernarfonshire. It was recorded on the island on just four occasions during 1956–1990, but on 28 occasions during 1990–2019, including groups of five in December 2010 and four in October/November 2014. During the latter period, 30 have been recorded on the Creuddyn Peninsula, Caernarfonshire, including ten on Bodafon Fields in February 2003 and 14 at Penrhyn Bay in January 2010. Cold weather in January 2010 also brought two to nearby RSPB Conwy, the only Denbighshire record in that period. It remains rare on Anglesey: one at Newborough Warren on 26 May 1951 was the first and there have been only three subsequent records, in October 1987, May 2006 and November 2008.

Medium-term climatic change has been suggested as the cause of the previous fluctuations in the Woodlark's fortunes (Sitters *et al.* 1996). Modelling suggests that Britain, south of a line from the Dyfi to the Humber, will be climatically suitable for the species by the end of the 21st century (Huntley *et al.* 2007). The European population increased by 41% during 1980–2016 (PECBMS 2020), perhaps driven by the ability of pairs to make more nesting attempts in warmer years (Wright *et al.* 2009). Woodlarks spread in England between 1988–91 and 2007–11, in a north and east direction, with a reduction in range in southwest England. Where it is doing well, it continues to be associated with lowland heathland and clear-fell conifer plantations, which tend to be on dry, sandy soil in the lowlands, unlike most of those in Wales, away from the west coast. Its history in Wales would suggest that predictions about this species are challenging and much will depend on the nature of future climatic change and the distribution of available habitat.

Jon Green and Julian Hughes

Sponsored by Geoff Gibbs and Arthur Chater

Skylark *Alauda arvensis* Ehedydd

Welsh List Category	IUCN Red List			Birds of Conservation Concern			
	Global	Europe	GB	UK	Wales		
A	LC	LC	LC	Red	2002	2010	2016

The Skylark is found across the Palearctic, except for the far north of Fennoscandia and Russia. It is resident in western and southeast Europe, whereas it is a summer visitor across most of its range. Those breeding in Scandinavia winter in the Mediterranean basin (*HBW*). It favours open areas that are relatively dry, covered with grasses or other low vegetation, such as farmland, sand dunes, saltmarsh, rough pasture and moorland, up to 930m in the Carneddau, Caernarfonshire (Rhion Pritchard, pers. comm.). It is able to tolerate taller and denser plant cover than other larks, requiring vegetation 20–50cm high in which to nest, but avoiding vertical features such as trees, tall hedges and rock faces (*BWP*). Only a small number of ringing recoveries involve Wales and most movements were of less than 10km (Robinson *et al.* 2020).

In the 19th century, Skylark was popular as a cage bird and as a delicacy. The *Historical Atlas 1875–1900* did not state if these practices took place in Wales, but it does not appear to have affected their abundance. At the turn of the 20th century, all nature writers in Wales described it as abundant in open country during the breeding season. There was little to quantify these statements until the *Britain and Ireland Atlas 1968–72*, when Skylarks bred in over 98% of 10-km squares in Wales. Its range had changed little in the 1988–91 and 2007–11 atlases, with abundance highest in the uplands, but having fallen in many coastal areas between the two periods.

Tetrad atlases showed the patchy distribution of Skylarks. In the *North Wales Atlas 2008–12*, for example, it occurred in 95% of 10-km squares, little changed from 1968–72, but in only 49% of tetrads. The highest densities were in moorland, even mountainous, districts, and on arable farmland and coastal dunes, but they were absent from lowland areas dominated by pasture for dairy cattle and sheep, such as the Vale of Clwyd, Denbighshire. In counties where tetrad atlases have been repeated, all showed a reduction in range: from 85% of tetrads in the *Gwent Atlas 1981–85* to 71% by 1998–2003, from 74% of tetrads in the *East Glamorgan Atlas 1984–89* to 64% (and only 57% showed breeding evidence) in 2008–11, and in Pembrokeshire from 88% of tetrads in 1984–88 to 58% in 2003–07 (*Pembrokeshire Avifauna* 2020).

Birds in Wales (1994) attributed the Skylark's slow decline after the Second World War to intensification of agriculture and afforestation of upland grass that would have held breeding birds. Increased numbers of sheep and changes to grassland management, such as ploughing and reseeding, resulted in pasture that was too short to provide cover for nests. Skylarks were more successful in arable or mixed farming regimes but found at much lower densities in Wales than elsewhere in Britain (O'Connor and Shrubb 1986). The population in England declined by 63% during 1967–2017, much of which occurred on farmland during the late 1970s and 1980s (Massimino *et al.* 2019). In Wales, the trend prior to 1995 is unknown, but the BBS index had declined by 11% by 2018 (Harris *et al.* 2020). The Welsh breeding population in 2018 was estimated to be 115,000 territories (Hughes *et al.* 2020).

Skylark © Ben Porter

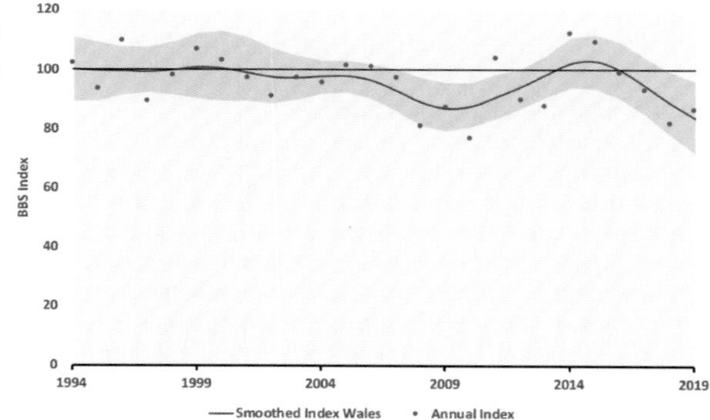

Skylark: Breeding Bird Survey indices, 1994–2019. The red line shows the smoothed index, 1995–2018. The shaded area indicates the 85% confidence limits.

Skylarks move away from upland areas in winter and, in addition, birds migrate from elsewhere through Wales. Dobie (1894) quoted Brockholes' belief that it was a partial migrant, and Salter (1900) reported large passage down the coast of Ceredigion during hard weather (Roderick and Davis 2010). In North Wales, Forrest (1907) reported that it was seen in "considerable flocks" on stubble and as a migrant on coasts from mid-September to mid-February, but particularly during October. It has long been suspected that some Skylarks from upland Britain winter in Ireland, which has less snow cover. There is little ringing evidence to substantiate this, but relatively few Skylarks are ringed and the recovery rate is low. Visible migration over Bardsey, Caernarfonshire, and Skomer, Pembrokeshire, confirmed that Forrest's observations remain valid, although the numbers involved now are lower, probably because of milder winters and a smaller British breeding population. The highest counts have generally been in South and West Wales. Notable autumn movements over Skomer during 2000–19 included day counts of 600 on 4 November 2002, 412 on 30/31 October 2003 and 640 on 20 October 2016. The highest autumn count on Bardsey during this period was 788 on 15 October 2019. On the mainland, highest autumn counts were 500 at Llanon, Ceredigion, on 19 October 2002 and 370 at Norton Farm, East Glamorgan, on 28 September 2013.

The *Britain and Ireland Winter Atlas 1981–84* showed that mountainous areas were devoid of Skylarks, but there were large counts from coastal locations. The 2007–11 *Atlas* also recorded Skylarks over much of Wales in winter away from high ground and that they were relatively scarce across much of mid- and North Wales. Abundance was highest on Anglesey, in Pembrokeshire and in parts of coastal South Wales. Severe winter weather, particularly a covering of snow, can cause westward movements of Skylarks into Wales. One exceptional winter flock numbered 4,000 in the Camlad Valley, Montgomeryshire, in February 1969, but most recent counts in Wales are of fewer than 1,000. The largest in recent times were 3,000 at White Sands, Flintshire, on 19 December 2010, 2,000 near Hawarden, Flintshire, on 1 January 2002, and 1,400 at Bardsey, Caernarfonshire, on 19 December 2010, which most likely indicated departure to Ireland.

The decline of Skylark has been well researched, focusing on those nesting in arable habitats in eastern England, which still hold the highest densities. This showed that the switch from spring to autumn sowing of cereals, which are too dense and tall by the time Skylarks are searching for a territory, has been a significant driver of decline. The effect is likely to have been far smaller on Skylarks in Wales during the late 20th century, because a much smaller proportion of the population nests in arable. However, the loss of arable crops and particularly winter stubbles from North and West Wales, as farm businesses became more specialised and mixed farming less widespread, may have reduced over-winter survival, not only of Welsh breeding Skylarks, but those from the English and Scottish uplands, especially during severe winter weather. It is quite likely that the intensification of dairy and sheep farming in Wales has reduced the availability of suitable nesting habitat for Skylark, as suggested by the *North Wales Atlas 2008–12*, but no studies have tested this. The Skylark's destiny in Wales is very much linked to the types of agricultural production supported by government policy, farmers' willingness to create suitable nesting and foraging habitat, and the extent to which further parts of Wales are afforested.

Ian M. Spence

Sponsored by LERC Wales (Local Environmental Records Centres Wales)

Crested Lark *Galerida cristata* Ehedydd Copog

Welsh List Category	IUCN Red List		Rarity recording
	Global	Europe	
A	LC	LC	BBRC

This species occurs across temperate Eurasia but also in Africa almost to the equator, although avoiding central desert areas. It is non-migratory throughout almost its entire range and is rare in Britain despite its abundance as close as Pas-de-Calais in northeast France (*HBW*). There have been 24 British records, the most recent in 2013, including the only Welsh record, which was one on Bardsey, Caernarfonshire, on 5–6 June 1982.

Jon Green and Robin Sandham

Shore Lark *Eremophila alpestris* Ehedydd Traeth

Welsh List Category	IUCN Red List			Birds of Conservation Concern	Rarity recording
	Global	Europe	GB	UK	
A	LC	LC	EN	Amber	WBRC

Also known as Horned Lark, this Holarctic species breeds across the northern hemisphere, on tundra and alpine moors above the tree limit, in the Western Palearctic, as far south as the High Atlas Mountains in Morocco. It is a winter visitor to Britain, particularly in eastern counties where birds are mainly of the subspecies *E.a. flava* that breed in Fennoscandia and Russia and usually winter around North Sea and Baltic. The population has declined dramatically since the 1950s, owing to overgrazing of lichens by Reindeer (*HBW*). In Wales, the Shore Lark is mainly a scarce passage migrant in autumn and has wintered in only 13 of the last 60 years. There have only been counts of ten or more in two winters.

Although small numbers over-wintered on the Dee Estuary

Total	Pre-1950	1950–59	1960–69	1970–79	1980–89	1990–99	2000–09	2010–19
290	1	1	146	29	21	75	9	9

Shore Lark: Individuals recorded in Wales by decade.

Shore Lark © Ben Porter

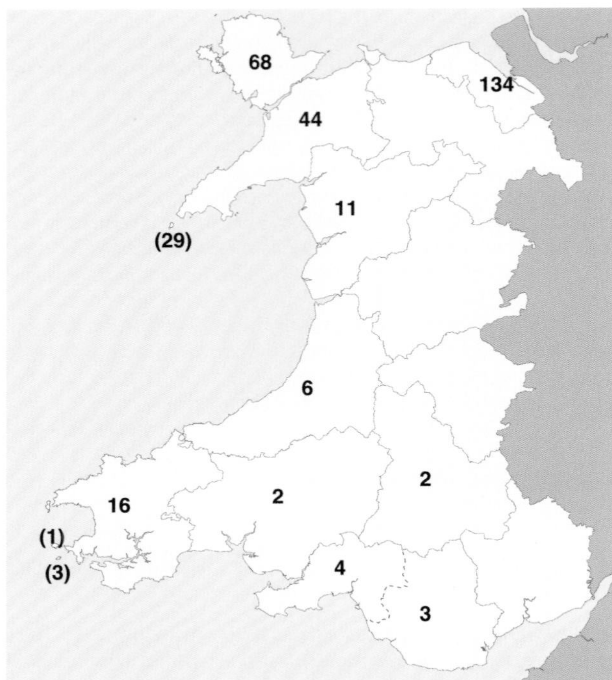

Shore Lark: Numbers of birds in each recording area, 1905–2019.

at the turn of the 20th century, on Hilbre Island on the English side (Forrest 1907), the first documented record in Wales was in Aberystwyth, Ceredigion, on 17 October 1905. The second was not until 1957, when one was on Skokholm, Pembrokeshire, on 23–24 April. Shore Lark has since been reported almost annually in Wales, generally on passage in small numbers.

The majority of records are from North Wales, although curiously there have been none in Denbighshire. The first record on Anglesey was at Malltraeth on 30 October 1960, with another there in February 1963. A cold-weather influx in November and December 1966 brought 48 to Red Wharf Bay, Anglesey, and up to 70 at Gronant and the Dee Training Wall at Shotton, Flintshire. The first record for Bardsey, Caernarfonshire, was at the start of the same winter, on 13 October 1966, the flock increasing to ten, which remained until mid-November.

It is much rarer farther south. The first records for several counties occurred during the 1960s and 1970s: at Morfa Harlech, Meirionnydd, on 3 January 1965, on the Rhymney Estuary, East Glamorgan, in November 1972 (at the time within the Gwent recording area) and at Berges Island, Gower, from 6 January to 17 February 1979. There have been only two records in Carmarthenshire: at Llangennech Marsh in November 1964 and at Burry Port from 17 February to 21 April 1979.

There were 21 records in North Wales in the 1970s, including up to five at Cemlyn, Anglesey, in 1972/73 and three at Point of Ayr, Flintshire, in December 1978. The first record in mainland Caernarfonshire was at Braich-y-pwll on 19 March 1982. During the early 1980s and late 1990s, Shore Lark occurred annually in Flintshire, primarily at Gronant and Point of Ayr, with 1–4 birds during winters 1981/82–1983/84, up to seven in 1994/95 and four in 1997/98. In winter 1998/99, up to 35 were present there, numbers peaking in January. Winter 1998/99 also saw small numbers in other counties: three at Castlemartin, Pembrokeshire, two at Cemlyn, Anglesey, and one on the Great Orme, Caernarfonshire. The only two winters with significant numbers in Wales, 1966/67 and 1998/99, coincided with increased numbers wintering in southeast England, mainly in Norfolk, Sussex and Essex. The influxes followed a summer of high breeding productivity and coincided with severe weather in their continental wintering area and persistent easterly winds during migration (*Birds in England* 2005).

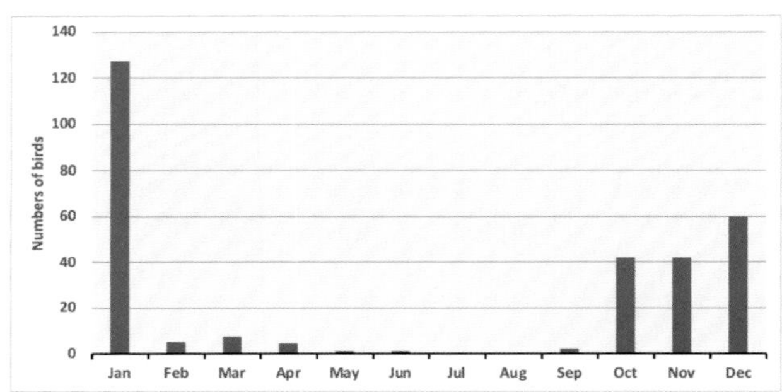

Shore Lark: Totals by month of arrival, 1905–2019.

It is otherwise a bird of autumn passage, recorded between mid-September and mid-November. Spring passage is recorded less frequently, but there have been three on Bardsey, of which two stayed for several weeks in 1970 and the other was on 30–31 May 2015. One on Skokholm on 4 June 1990 was also unseasonal.

Shore Lark records have become more sporadic since the turn of the 21st century, with only 11 birds in North Wales: four on Anglesey, of which only one wintered, at Abermenai Point in December 1999 and January 2000, and only one further wintering record, at Gronant, where two were present during 2009/10. The only two records in Breconshire were during this period, on Mynydd Epynt in October/November 2001 and at Llyn y Fan Fawr in March 2016. There was one at Whiteford, Gower, in 2002/03 one at Newton Point, East Glamorgan, in November 2019, and four in Pembrokeshire, including two at Marloes Mere in January/February 2005. It seems likely that the trend for reduced occurrence will continue, particularly as the suitable climate range of the Fennoscandian breeding population is expected to diminish through the remainder of this century (Huntley *et al.* 2007) and milder winters may reduce their need to migrate southwest.

Jon Green and Robin Sandham

Short-toed Lark *Calandrella brachydactyla* Ehedydd Llwyd

Welsh List Category	IUCN Red List		Rarity recording
	Global	Europe	
A	LC	LC	WBRC

This migratory lark breeds in southern Europe, North Africa, Central Asia and parts of the Middle East, on open dry plains and arable fields. The European breeding population winters in North Africa (*HBW*). Numbers in Britain increased at the end of the 20th century, from annual means of nine in 1958–89 to 22 in 1990–2018 (White and Kehoe 2020b). Numbers recorded in Wales have broadly followed the same pattern since 2000, but the species remains a rare visitor. The majority of records have been spring overshoots, with a peak in May and smaller numbers in autumn.

Total	Pre-1950	1950–59	1960–69	1970–79	1980–89	1990–99	2000–09	2010–19
37	0	2	3	3	3	5	13	8

Short-toed Lark: Individuals recorded in Wales by decade.

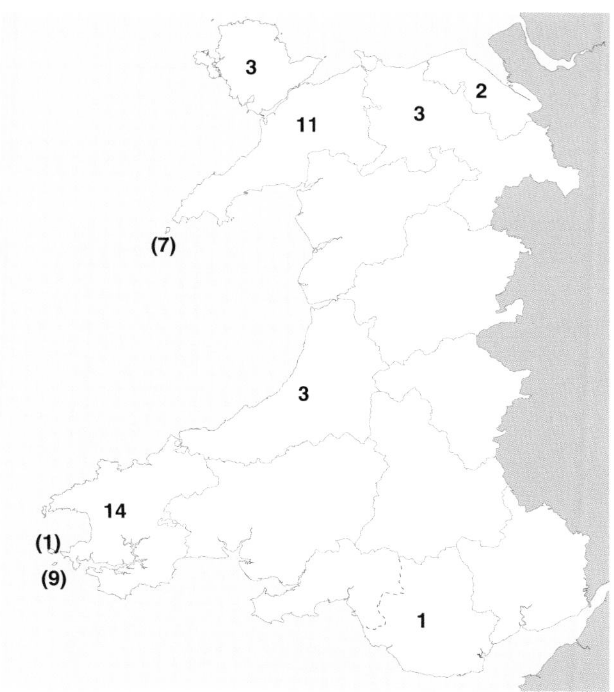

Short-toed Lark: Numbers of birds in each recording area, 1952–2019.

The first Welsh record was on Skokholm, Pembrokeshire, on 9–13 April 1952 and the next five were all from the same island. The first on Bardsey, Caernarfonshire, was in October 1975 and the first Welsh mainland record was at Llanfairfechan, Caernarfonshire, during October/November 1976.

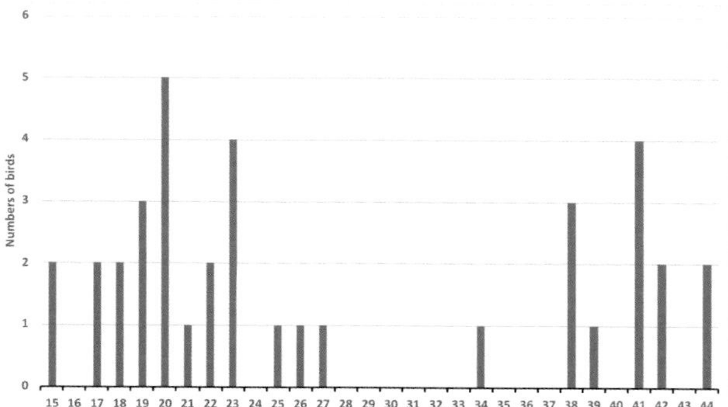

Short-toed Lark: Totals by week of arrival, 1952–2019. Week 19 begins *c.*7 May and week 41 begins *c.*8 October.

The majority of records have been in Pembrokeshire, including the only record involving more than one bird: two were together on Skokholm in October 1995. All records have been from coastal sites, although it is perhaps surprising that more than 10% of records have been from northeast Wales. The only record from the southeast was one at Sker-Kenfig, East Glamorgan, in May 2008.

Jon Green and Robin Sandham

Black Lark *Melanocorypha yeltoniensis* Ehedydd Du

Welsh List Category	IUCN Red List		Rarity recording
	Global	Europe	
A	LC	CR	BBRC

This species breeds only in Central Asia, on the steppe grasslands of northern Kazakhstan and southwest Russia, making only a short-distance migration to winter around the Black Sea and Caspian Sea (*HBW*). It is an extremely rare vagrant to Britain and there has been only one record in Wales, a male at RSPB South Stack, Anglesey, on 1–8 June 2003. At the time, this was thought to be the first for Britain and attracted large crowds during its stay. However, its appearance prompted a record from Spurn, East Yorkshire, on 27 April 1984 to be unearthed and accepted as the first British record. There has been only one further record in Britain. The Anglesey bird was one of the most unexpected additions to the Welsh List, and it seems unlikely that there will be another.

Jon Green and Robin Sandham

Black Lark at RSPB South Stack, Anglesey, June 2003 © Steve Young

Sand Martin *Riparia riparia* Gwennol y Glennydd

Welsh List Category	IUCN Red List			Birds of Conservation Concern			
	Global	Europe	GB	UK	Wales		
					2002	2010	2016
A	LC	LC	LC	Green			

Sand Martin collecting nesting material © Tony Pope

The Sand Martin breeds across Eurasia and much of North America, wintering to the south around the Tropics. Those from Europe spend the winter immediately south of the Sahara in the Sahel zone of West Africa (*HBW*). It is a highly gregarious species, generally associated with wetlands for feeding and roosting. Ringing recoveries, including many from Welsh locations, show an autumn migration route that follows the French and Iberian Atlantic coast to the Senegal Delta. The spring migration route is farther to the east, across the central Mediterranean (*BTO Migration Atlas* 2002). Ringing recoveries in Wales include birds from colonies in northern Scotland and Ireland. It typically nests colonially on exposed vertical faces in natural riverbanks, in sand and gravel pits and in coastal soft cliffs and dunes, but human constructions, such as drainage pipes or purpose-built nest walls, are sometimes used.

Around the turn of the 20th century, Sand Martin was described as common in lowland river valleys throughout Wales, including the Dyfi, Dysynni and Dee (Forrest 1907), the Ystwyth and Rheidol valleys in Ceredigion (Roderick and Davis 2010), and the Usk Valley, Breconshire (Phillips 1899). As well as nesting in riverbanks, Mathew (1894) described colonies in gravel pits and old quarries in Pembrokeshire. One was established in a sand/gravel quarry face at Sarnau, near Bala, Meirionnydd, from 1895 (Forrest 1907). Small colonies were also known in boulder-clay cliffs along the Ceredigion coast in this period (Roderick and Davis 2010). In East Glamorgan, riverbank 'improvements' led to the loss of several large colonies along the Afonau Rhymney, Taf, Ely and Ogmore by 1925 and those in sand dunes in the county ceased to exist during 1920–1966 (Hurford and Lansdown 1995). Similar accounts of colony losses to insensitive river management or natural erosion were reported from other counties, including Gwent and Ceredigion, although two colonies remained in coastal sand dunes at Laugharne, Carmarthenshire, in 1952 (Ingram and Salmon 1954). Despite these losses, there were large numbers in some areas, such as 1,500 nests along the River Usk in Gwent in 1965. In Montgomeryshire, a survey of the Afon Vyrnwy found 585 Apparently Occupied Burrows (AOB), and another 1,271 AOB on the River Severn, downstream to the English border in 1969 (*Birds in Wales* 1994).

The *Britain and Ireland Atlas 1968–72* showed Sand Martins to be widespread across much of Wales, although breeding

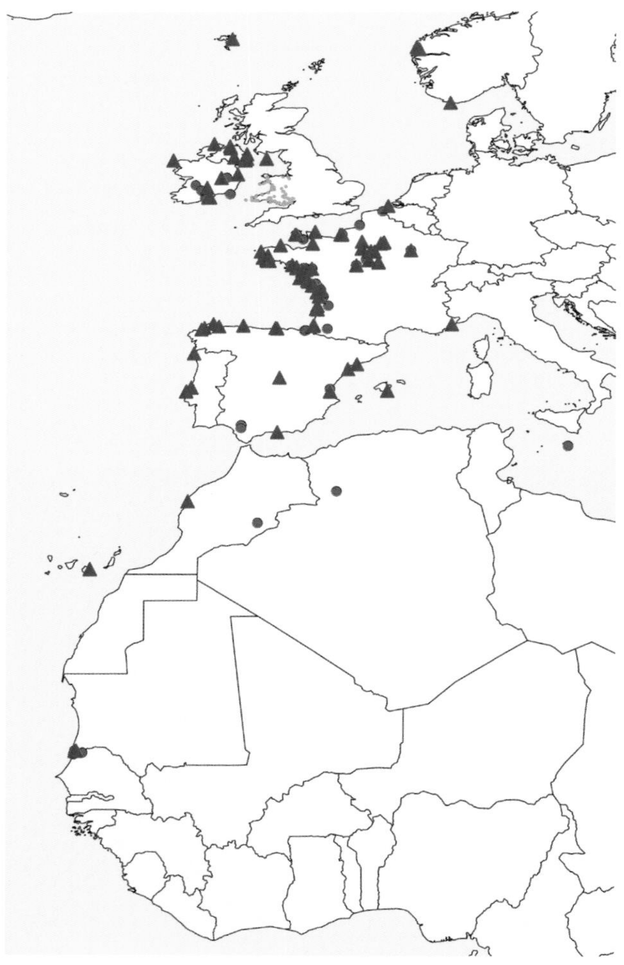

Sand Martin: Recovery locations of birds ringed in Wales are shown by red circles. Ringing locations of birds that were recovered in Wales are shown by blue triangles. Small brown dots show ringing or recovery locations in Wales.

evidence was patchy in the southwest and northwest. However, during this period there was a well-documented crash in the population, caused by drought in the Sahel (Cowley 1979). The population recovered in the mid-1970s, but there was another temporary decline in 1984–86, owing to a lack of rainfall in the Sahel (Cowley and Siriwardena 2005). The *Britain and Ireland Atlas 1988–91* showed a patchier breeding distribution in Wales than in 1968–72, and although good concentrations remained in some areas, there were more losses at the 10-km square level than gains. By 2007–11, the distribution had largely recovered from the earlier crashes and relative abundance had increased in many areas, particularly in North Wales.

Local monitoring confirmed Sand Martin to be concentrated in the major river valleys of eastern and South Wales. Colonies are generally smaller and more scattered in the west and absent in the uplands, although breeding has been recorded up to 320m at Mynydd Epynt, Breconshire, 330m in Ceredigion, 340m in Meirionnydd and at 375m at Llyn Aled Isaf, Denbighshire. In Gwent, it is concentrated along the Wye, Usk and Monnow valleys, particularly the latter two, and while its distribution changed little between 1981–85 and 1998–2003, numbers were reduced. In Gower, the *West Glamorgan Atlas 1984–89* found that the Afon Loughor on the Carmarthenshire border was by far the most densely populated watercourse in the county, while a survey of the Afon Tywi, Cothi and their tributaries in Carmarthenshire recorded 2,915 AOB in 1995 (Lloyd *et al.* 2015). In the late 1980s, *c.*400 pairs were estimated to nest in Breconshire, most along the Usk and Wye valleys (Peers and Shrubb 1990). Following earlier declines in Radnorshire, Jennings (2014) considered that numbers had increased from 1995 and peaked at 650–700 AOB in the county in 2008–13, including 225 AOB on the River Wye between Boughrood and Cabalva in 2010. Subsequent bankside erosion and summer floods led to lower numbers, and a county estimate closer to 500 AOB in 2019 (Pete Jennings *in litt.*). In Montgomeryshire, a colony at Dolydd Hafren nature reserve fluctuated between 70 and 200 pairs during 1992–2005, but was completely washed out by high river levels in summer 2006 (Holt and Williams 2008).

In Pembrokeshire, Donovan and Rees (1994) mentioned colonies in riverbanks, such as the Western Cleddau, in sand quarries at Dale and Gupton (which no longer hold colonies now that active quarrying has ceased) and soft coastal cliffs such as Abermawr and Freshwater West. They estimated the county breeding population to be around 70–100 pairs based on the *Pembrokeshire Atlas 1984–88* and this had not changed significantly by 2003–07. In Ceredigion, Roderick and Davis (2010) wrote of breeding colonies along all major rivers and some tributaries. One of the largest, near Capel Bangor on the Afon Rheidol, held 160 nests in 2003. They also noted that nesting continued in coastal boulder-clay cliffs, some holes just a metre or so above the beach. There was little change in distribution in Meirionnydd between 1968–72 and 1988–91, and only a small increase by 2007–11. There are currently at least two colonies larger than 100 AOBs along the coast, at Mochras sand dunes and in soft cliffs at Tonfanau. In Caernarfonshire, colonies along riverbanks are small and the large ones are on sea cliffs along Llŷn, but it is absent from central Snowdonia and the coast east of Caernarfon. On Anglesey, at least four colonies were known in the 1960s, the largest having 120 AOB in 1967. A 1986 survey by the RSPB recorded colonies in three 1-km squares, but the *North Wales Atlas 2008–12* confirmed Sand Martins nesting in 18 tetrads, mostly in the west where they had been uncommon in Forrest's time. During the 21st century, there have consistently been around half a dozen colonies across Denbighshire and Flintshire.

As well as nesting in sand and gravel quarries, other artificial sites occupied include eroding ash cliffs at a former power station near Burry Port, Carmarthenshire, where there were *c.*100 AOB in 2016, and drainage holes in a concrete wall under the A5 trunk road on Anglesey. This was occasionally flooded, so the North Wales Trunk Roads Agency constructed a nesting wall above the flood level, which was occupied from 2011. They have also occupied holes in an artificial nest-bank at Port Talbot, Gower, where 81 AOB were recorded in 2012, but only 1–2 in 2018 and none in 2019.

Most early arrivals in Wales are reported during the first week of March, although some occasionally arrive in February. During 2000–19, the earliest arrival dates were typically on 1–9 March in the southern half of Wales and on 7–14 March in the northern half, but there were February records in six of these years. There was a notably early arrival of hirundines in 2019, with February records in seven counties, including Gower, where one at Cwm Ivy on 12 February was the earliest ever in Wales. Numbers build during April, with large counts occasionally reported, such as a minimum of 5,000 on Ramsey, Pembrokeshire, on 3 April 2006, 3,000 at Cardiff Bay, East Glamorgan, on 7 April 2008 and 2,000 held up by adverse weather conditions at Llangorse Lake, Breconshire, on 6 May 2015.

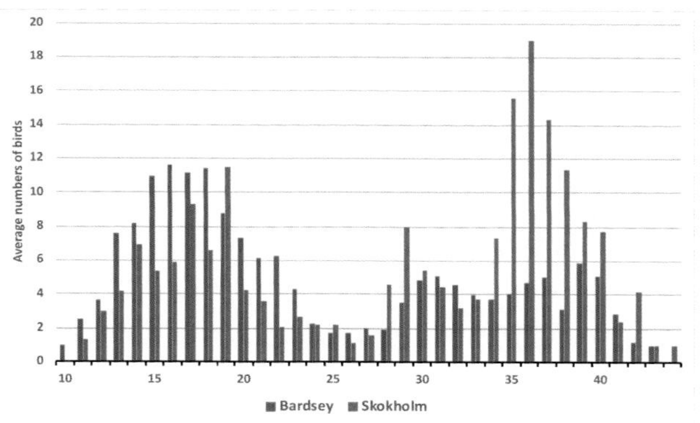

Sand Martin: Average numbers seen, by week number, at Bardsey and Skokholm, 1953–2019. Week 14 begins *c.*8 April and week 39 begins *c.*2 September.

Daily counts at Skokholm, Pembrokeshire, and Bardsey, Caernarfonshire, peak between mid-April and mid-May. Weekly totals on Bardsey are higher than on Skokholm in spring, but the reverse is true in autumn, when passage is more protracted, from mid-July to the third week of October. The largest spring counts on the islands were 400 on Bardsey on 17 April 1967 and on Skokholm on 8 May 1989, while the highest autumn count was 500 on Skokholm on 8/9 September 1967.

On the mainland, post-breeding flocks build during July with hundreds frequently reported over wetlands, including at reedbed roosts. The largest counts include *c.*10,000 at Llangorse Lake on 2 August 1979 and 7,000 at Lavernock Point, East Glamorgan, on 28 September 2018, but peak numbers are typically lower, such as *c.*2,000 at Llangorse Lake on 6 August 2004, and at Eglwys Nunydd Reservoir, Gower, on 29 August 2002 and 24 August 2004. Most Sand Martins have left by mid-October, but small numbers were seen in the first half of November in East Glamorgan, Gower and Pembrokeshire during 2000–19, and the latest record was in Gwent on 4 December 2005.

There are insufficient data to produce a population trend for Wales, but the UK Waterways Breeding Bird Survey index was stable during 1998–2013 and then increased by 40% in 2014–18 (Harris *et al.* 2020). Prospects for Sand Martins in Wales are influenced by several factors, primarily relating to climate. Arrival dates in the UK advanced by over three weeks between the 1960s and the 2000s (Newson *et al.* 2016), but laying dates have not changed, so it is unclear whether this has an effect on the population. The success of nesting in riverbanks is affected by summer flooding that is predicted to become more frequent. Drier winter conditions during June to October in the Sahel could reduce annual survival, although more recent wintering conditions have been less extreme (Masoero *et al.* 2016). Colonies in quarry faces and sand dunes will always be ephemeral, but there may be more opportunities to create artificial nesting walls at locations with good freshwater foraging.

Bob Haycock

Swallow *Hirundo rustica* Gwennol

Welsh List Category	IUCN Red List			Birds of Conservation Concern			
	Global	Europe	GB	UK	Wales		
					2002	2010	2016
A	LC	LC	LC	Green			

The Swallow breeds throughout Europe, North America and Asia north of the Himalayas, and generally south of the Arctic Circle. It is semi-colonial, but single breeding pairs are not uncommon. Being aerial feeders, Swallows are dependent on a supply of small insects and the presence of large domestic animals often influences the choice of breeding sites (*HBW*). Permanently open agricultural buildings are preferred as nest sites, but there are examples in Wales of nesting in abandoned cottages, castles, disused military bunkers, bird hides on nature reserves, and the entrances of old mineral mines and natural caves. Breeding is mostly in lowland areas, but occurs at 420m at Lake Vyrnwy, Montgomeryshire. It has been recorded at isolated farms up to 440m in Ceredigion (Roderick and Davis 2010) and 460m in Meirionnydd (Pritchard 2012). At migration time, in autumn especially, large flocks form to roost communally overnight in reedbeds, often with Sand Martins.

Swallow: Recovery locations of birds ringed in Wales are shown by red circles. Ringing locations of birds that were recovered in Wales are shown by blue triangles. Small brown dots show ringing or recovery locations in Wales.

Swallows from Britain and Ireland migrate to southern Africa for the winter. Some of the earliest recoveries in South Africa were from Wales. For example, birds ringed at Laugharne, Carmarthenshire, in the 1920s, were recovered in the Transvaal, East Griqualand and Cape Province (Lloyd *et al.* 2015). More recent ringing recoveries include two ringed as chicks in Montgomeryshire in 2014 that were found in South Africa in 2015. One from Llanfechain was recaught at Umzumbe, KwaZulu Natal, among a roost of two million Swallows, 9,835km to the south. Another ringed on Anglesey in July 2000 was trapped in the Democratic Republic of Congo in January 2001. This was either already on return passage or had over-wintered well to the north of the usual area for this species. Swallows migrating along the Welsh coast include movements to and from Ireland: a juvenile ringed in the Vale of Glamorgan in August 1965 was recaptured in Co. Tipperary in May 1966, and another caught on Flat Holm, East Glamorgan, in September 1996 had been ringed on Inch Island, Co. Donegal, the previous month.

Authors at the turn of the 20th century reported Swallow to be common and widespread, although Hurford and Lansdown (1995) suggested that numbers in Glamorgan were already decreasing. Salter (1930) and W. Miall Jones (per Roderick and Davis 2010) considered that numbers declined in Ceredigion during the 1920s, and Ingram and Salmon (1954) believed it had decreased appreciably during the second quarter of the 20th century in Carmarthenshire. In Gwent, numbers were apparently decreasing in 1937 and 1963, when county avifaunas were published (Venables *et al.* 2008).

The Britain and Ireland atlases showed that it bred in almost every 10-km square in Wales in 1968–72, 1988–91 and 2007–11. In the latter period, abundance was relatively high in the southwest and northwest, with fewer along the central spine of upland Wales. It was lower around Carmarthen Bay, East Glamorgan and in south Meirionnydd relative to the rest of Britain than in 1988–91. Local county atlases confirmed their wide distribution, but with differing trends. It increased its presence, from 94% to 98% of tetrads, in Gwent between 1981–85 and 1998–2003, but reduced from 73% of tetrads to 69% in East Glamorgan, between 1984–89 and 2008–11. It remained stable in Pembrokeshire, where 89% of tetrads were occupied in 1984–88 and 2003–07. In North Wales, 84% of tetrads were occupied in 2008–12.

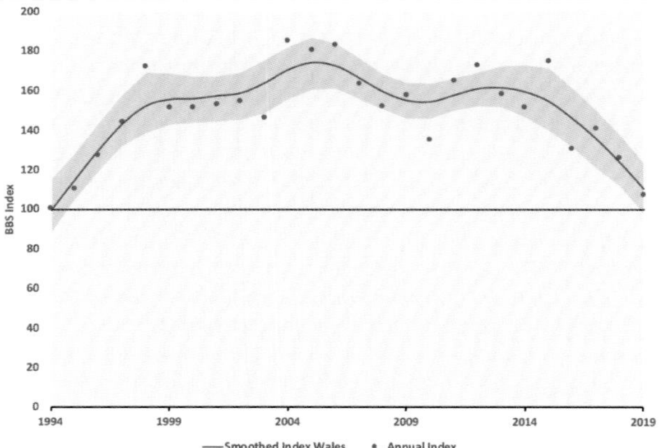

Swallow: Breeding Bird Survey indices, 1994–2019. The red line shows the smoothed index, 1995–2018. The shaded area indicates the 85% confidence limits.

The BBS index for Wales increased significantly to 2005, but declined by 22% during 2008–18, in tandem with even steeper declines in Northern Ireland and England. In 2018, the index for Wales was 7% above the 1995 baseline (Harris *et al.* 2020) and the Welsh breeding population was estimated to be 82,000 territories (Hughes *et al.* 2020). However, there have been increasing reports of Swallows failing to return to regular nesting locations in recent years, including farms that have been occupied for at least 40 years (Mike Haigh *in litt.*).

The earliest records in Wales each year are usually in the first two weeks of March, although they are slightly later in landlocked counties. The first can arrive even earlier. During 2000–19 there were February records from six counties, including four that registered their earliest ever arrivals in 2019. There were winter records from all coastal counties, except Carmarthenshire and Meirionnydd, in 2000–19: 30 in December, four in January and ten in February, but none in landlocked counties. One-third of these winter records were from Pembrokeshire. They include four at the refinery in Rhoscrowther, from December 2004 to the third week of January 2005, although at least one perished; three at Amroth on 26–27 January 2005; and one at the Rhoscrowther refinery on 5 January 2009, where a second bird was found dead.

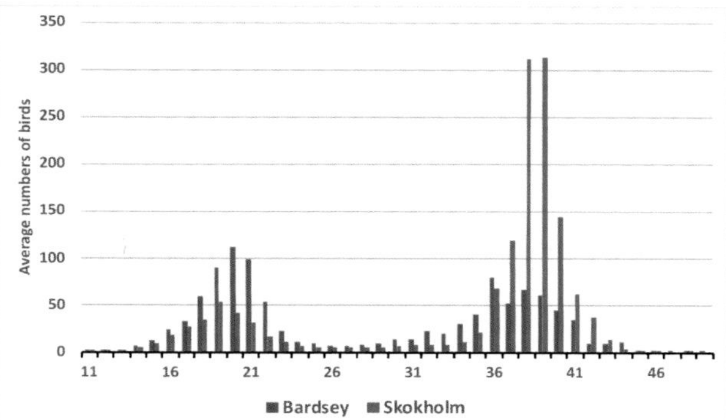

Swallow: Average numbers seen, by week number, at Bardsey and Skokholm, 1953–2019. Week 20 begins *c.*13 May and week 38 begins *c.*16 September.

Numbers increase during April and May, when large counts are sometimes reported. An exceptional movement was 12,600 north through New Quay, Ceredigion, in three hours on 15 May 2010. Average weekly counts from bird observatories on Skokholm, Pembrokeshire, and Bardsey, Caernarfonshire, showed that movements peak in May, slightly larger and about a week later on Bardsey than Skokholm.

Following breeding, numbers build in August, with overnight roosts usually in reedbeds, such as 50,000 at RSPB Cors Ddyga, Anglesey, in 2004. Roosts have also been recorded in crops, such as 3,000 in a Maize field in Denbighshire on 3 September 2012. In Breconshire, small groups have roosted in regenerating mixed woodland, such as 30 at Mynydd Wysg on 20 July 2013 and 35 at Llaneglwys Forest on 24 August 2016. The importance of these stop-over roost sites was illustrated at Ffrwd Fen, Carmarthenshire, where ringing on successive nights in autumn 1990 revealed an almost 100% turnover each night (Lloyd *et al.* 2015). Peak passage occurs in September, with very large movements in some years, such as 50,000 over Flat Holm on 21 September 2002, which was the largest ever count in East Glamorgan. In neighbouring Gwent, *c.*20,000 moved down the Usk Valley near Abergavenny on 23 September 2007. Several thousand over Llangorse Lake, Breconshire, the same day, were probably part of this movement. Average weekly counts on Skokholm in autumn are far higher than on Bardsey, probably because they include a proportion of the Irish breeding population, which moves south via Wales and southern England (*BTO Migration Atlas* 2002). The peak in mid-September on Skokholm is more than five times the size of its spring peak, significantly influenced by a few large counts. The maximum autumn count on Bardsey was *c.*5,000 on 4 September 2012, while an estimated 13,000 passed Skokholm on 18 September 2017, the same day that *c.*17,000 passed Skomer, Pembrokeshire, and a similar number flew east over Rhossili Bay, Gower. Although most have left by the end of October, small numbers remain into November and nine Welsh counties have now recorded Swallow in December.

Swallow numbers fluctuate in relation to rainfall in the early austral summer in South Africa and in the western Sahel, prior to northbound migration (Robinson *et al.* 2008, 2014). Lower rainfall can affect over-winter survival and may carry over to affect the following breeding season (Saino *et al.* 2012). Arrival dates in the UK advanced by 15 days between the 1960s and 2000s (Newson *et al.* 2016) and laying dates advanced by 12 days. Between 1967 and 2017 there was a 65% increase in the failure rate of nests containing chicks (Massimino *et al.* 2019). There is evidence that hot, dry summers increase chick mortality and reduce post-fledging survival (Turner 2009). The effect of agricultural intensification or air pollution on insect abundance in both breeding and wintering areas are also likely to be important drivers. Locally, Lloyd *et al.* (2015) commented that the conversion and modernisation of farm buildings, and the replacement of traditional roof slates with corrugated sheets, may reduce the availability of nest sites. The declines recorded in recent years, and the changes in their breeding metrics related to food and weather, throw some doubt on the future abundance of this delightful herald of spring.

Bob Haycock

Sponsored by Gareth Jenkins

Swallows mating © Ben Porter

Crag Martin *Ptyonoprogne rupestris* Gwennol y Clogwyn

Welsh List Category	IUCN Red List		Rarity recording
	Global	Europe	
A	LC	LC	BBRC

Crag Martins nest in caves and cavities on cliff faces, both coastal and inland, in southern Europe, North Africa and Central Asia. Much of the population is sedentary, but those from the north of their breeding range in Europe spend the winter in West Africa (*HBW*). It is a rare visitor to Britain, with 12 records to the end of 2019 (Holt *et al.* 2020), of which the second was the only one to have been seen in Wales, at Llanfairfechan, Caernarfonshire, on 3 September 1989. While primarily a sedentary Mediterranean species, the breeding range has extended to central Europe since the 1980s, and those in southern Germany move south outside the breeding season (Holt *et al.* 2020). It is expected to occupy the whole of France by the end of this century (Huntley *et al.* 2007) and thus we might expect it to occur with greater frequency in Britain, including in Wales.

Jon Green and Robin Sandham

House Martin *Delichon urbicum* Gwennol y Bondo

Welsh List Category	IUCN Red List			Birds of Conservation Concern			
	Global	Europe	GB	UK	Wales		
A	LC	LC	VU	Amber	2002	2010	2016

The House Martin breeds in Europe, western Siberia and Central Asia, favouring temperate, steppe and Mediterranean climate zones, where it is tolerant of both maritime and continental conditions, providing there is an abundance of aerial insects. It winters across sub-Saharan Africa (*HBW*), but far less is known about its typical wintering areas and roosting preferences than those of Swallows and Sand Martins. There has been only one recovery of a House Martin ringed in Britain and Ireland from south of the Sahara (Robinson *et al.* 2020). One ringed in Wales in September 1989 and recovered in Genoa, Italy, in May 1991, accords with other ringing evidence of a broad front of passage from eastern Spain to Italy.

Traditional breeding sites include inland rock faces and sea cliffs, but these are now widely used only in southern Europe. Since the 19th century, most have nested on artificial structures, typically under the eaves of buildings in villages, towns and cities (*HBW*). House Martins nested on the Menai suspension bridge, Caernarfonshire/Anglesey, in 1907, in the 1950s and 1960s, and in 1984, but it is not clear whether this use was continuous. There were also 150 nests annually on Knucklas Viaduct near Knighton, Radnorshire, until 2014, when rail contractors reportedly cleared the nests and the site was abandoned. Artificial nest-cups are also used, such as on Ramsey, Pembrokeshire, where a small colony was established in 2013, having been previously absent as a breeding species on the island. In Wales, nesting has been recorded up to 410m in Ceredigion, Meirionnydd and Caernarfonshire, and up to 430m in the Brecon Beacons. Cliff colonies were known around 1902–03 on the north and east coasts of Anglesey, and considerable numbers were said to breed around the two Orme headlands, Caernarfonshire, in the same period (Forrest 1907). Nesting continued on the Great Orme cliffs in the 1950s (Pritchard 2017). There were sea-cliff nests between Gwbert and Tresaith, Ceredigion, at least to the 1960s (Roderick and Davis 2010) but not since, and on inland limestone crags at c.400m on Craig-y-Cilau, Breconshire, until the 1950s. Up to six pairs bred on cliffs above the Afon Elan, Radnorshire, in 1984–90.

Sea-cliff colonies in the Vale of Glamorgan increased to 250 nests during the 1960s, but fell back during the 1970s, and now only small numbers, usually around half a dozen, nest there. On Anglesey, nests on cliffs were recorded at Traeth Lligwy in 1967, at Point Lynas and Traeth Bychan in the 1970s (Jones and Whalley 2004) and near Moelfre early in the 21st century (Brenchley *et al.* 2013). Sea-cliff nesting continues in Pembrokeshire at Ceibwr, Nolton Haven and between Great Furzenip and Greenala.

Forrest (1907) considered House Martin to be common in the lowlands of North Wales, except in Llŷn and parts of Anglesey, but less abundant than Swallow. It nested in the centre of Aberystwyth at the end of the 19th century, but had gone from there and from coastal villages in Ceredigion by the late 1920s. Salter (1930) put this down to a series of poor summers, although it bred there again by the 1960s. In that decade, it was considered to be more numerous than ever in Glamorgan, despite low numbers in the cities (Hurford and Lansdown 1995). It was described as well-distributed in Gwent in 1963, but less on the coastal levels than elsewhere (Venables *et al.* 2008).

The Britain and Ireland atlases showed House Martins bred or probably bred in over 98% of 10-km squares in Wales, in 1968–72, 1988–91 and 2007–11. The most recent atlas found that abundance was highest along eastern border counties and lowest in upland areas. Although there had been changes in relative abundance since 1988–91, there was no obvious geographic pattern to these. However, House Martins range widely away from their nests, so records of confirmed or probable breeding can provide a more accurate assessment. The *North Wales Atlas 2008–12* recorded probable or confirmed breeding in 78% of tetrads, with higher occupancy in Flintshire and Denbighshire than the three other counties, perhaps due to more mixed farming and the locally drier climate providing more insect food. The number of tetrads with confirmed or probable breeding increased in Gwent and Pembrokeshire between their first and second tetrad atlases: from 78% to 85% of tetrads in Gwent between 1981–85 and 1998–2003 and from 43% to 51% in Pembrokeshire between 1984–88 and 2003–07. However, the proportion of tetrads with confirmed or probable nesting in East Glamorgan reduced from 53% in 1984–89 to 33% in 2008–11.

Surveys in the older parts of Cardiff, East Glamorgan, found 68 nests in 32 streets in 1976 and 67 nests in 22 streets in 1980. Although the population was stable, breeding sites were lost as old houses were demolished (Hurford and Lansdown 1995). A 1981 survey found 522 occupied nests in an area of 314 km² in Gower, numbers similar to a 1963 survey (Tallack 1982). In Ceredigion, Roderick and Davis (2010) referred to the decline of one of the largest and oldest colonies at the Hafod Arms Hotel, Devil's Bridge, known to Salter in 1893. It held 90 nests in the late 1960s, 31 by 1976, just six by 1998 and none by 2004. Colonies can exist for many years and then switch to other locations, so the demise of one colony may be balanced by another becoming established, making it difficult to assess local trends. The BBS index for Wales increased strongly up to 2000 but was 19% lower in 2018 than 1995. The decline is even greater in England, but numbers have increased in Scotland and Northern Ireland (Harris *et al.* 2020).

Early arrivals are usually reported in the middle two weeks of March. Although they are generally later in some North Wales counties, two of the three Welsh counties with February records in 2000–19 were Anglesey (19 February 2019) and Denbighshire (28 February 2015). Gwent's earliest arrival was also in February 2019, at Newport Wetlands on the 24th. Daily counts at bird observatories

House Martin © Ben Porter

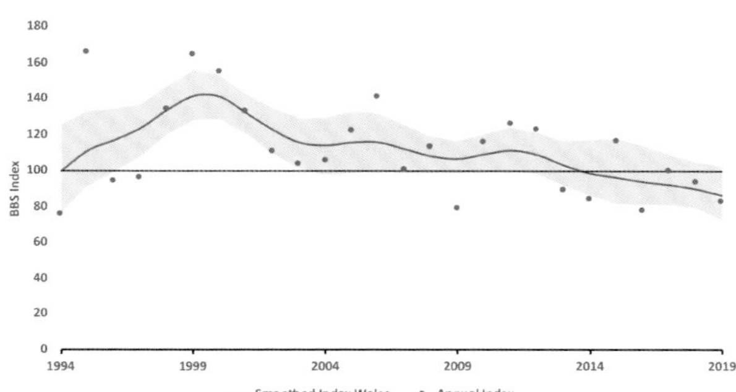

House Martin: Breeding Bird Survey indices, 1994–2019. The red line shows the smoothed index, 1995–2018. The shaded area indicates the 85% confidence limits.

on Skokholm, Pembrokeshire, and Bardsey, Caernarfonshire, show steadily increasing passage through April and May, with greater numbers over Bardsey, that peak in mid- to late May. The highest day count on Bardsey was 639 on 20 May 2014, while on the mainland, 2,000 were at Eglwys Nunydd Reservoir, Gower, on 11 May 1995.

Numbers on autumn passage build from August, but House Martins can continue to feed nestlings through September. Average weekly counts show a protracted migration, especially on Bardsey, where there are peaks in early September and the

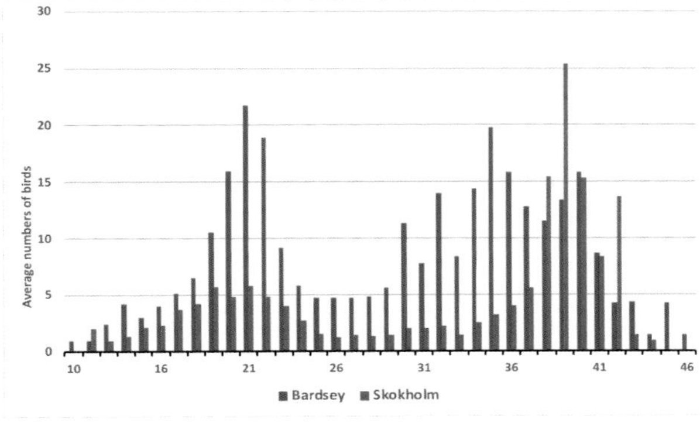

House Martin: Average numbers seen, by week number, at Bardsey and Skokholm, 1953–2019. Week 21 begins *c.*20 May and week 39 begins *c.*23 September.

second week of October. On Skokholm, numbers peak in late September and high counts there included 450 on 23 September 1959 and at least 700 on 28 September 2014. Four-figure counts have been reported from a few counties, including *c.*3,100 at Peterstone Wentlooge, Gwent, on 5 September 2015. An eastward movement of 42,000 in three hours at Black Rock, Gwent, on 22 September 1986 was exceptional. Small numbers have been recorded in October in most counties, with seven having records from the second half of November during 2000–19. A few may have attempted to over-winter, such as individuals seen on 16 December 1980 in Gwent, on 12 February 1981 and 5 February 1989 in Carmarthenshire. The first ever January records in Wales have been noted since 2000, with one at Hakin, Pembrokeshire, on 29 January 2002, two at New Quay, Ceredigion, on 16 January 2007 and one at Tenby, Pembrokeshire, from 29 December 2015 to 1 February 2016. There were also records in the first half of February during 2000–19: at Staple Point, Gower, and Laugharne, Carmarthenshire, on 8 February 2004.

House Martins are a difficult species for which to make a population estimate. A BTO survey in 2015–16 provisionally estimated 70,000 pairs in Wales, very similar to an estimate made by Hughes *et al.* (2020). Each surveyed 1-km square typically held 0.5–3 colonies across much of Wales. Densities were lowest (<1 colony/1-km square) on Anglesey, Meirionnydd and Montgomeryshire and highest (>3 colonies/1-km square) in Carmarthenshire and East Glamorgan (BTO 2016a). The same survey indicated that nesting started later, and breeding performance was lower in the west than in eastern Britain, which appears to be related to spring rainfall (Kettel *et al.* 2020).

Despite full legal protection, there continue to be annual reports of nests being destroyed by householders who object to droppings on walls and pathways. In some instances, loosely hanging wires have been placed below weatherboarding or under the eaves as a deterrent. Nests built on plastic soffits were less likely to be multi-brooded and less likely to be successful, compared with other materials (Kettel *et al.* 2020). Arrival dates in the UK advanced by 16 days between the 1960s and 2000s (Newson *et al.* 2016), but it is not known if this has affected breeding productivity. Winter weather and food resources may be more limiting than conditions during the breeding season. Annual survival rates of nesting birds were positively correlated with maximum monthly rainfall in West Africa during 1994–2004, but declines in survival rate did not correspond well to the period of population decline (Robinson *et al.* 2008). The prospects for House Martin in Wales, as for many trans-Saharan migrants, will probably depend on how climatic factors affect insect food supply, breeding success and survival.

Bob Haycock

Red-rumped Swallow *Cecropsis daurica* Gwennol Dingoch

Welsh List Category	IUCN Red List		Rarity recording
	Global	Europe	
A	LC	LC	WBRC

The subspecies *C.d. rufula* breeds on coastal and inland cliffs, and occasionally on human-built stone structures, around the Mediterranean basin and through the Balkans and Middle East, and winters in Africa south of the Sahara. Other subspecies breed in Africa, and South and East Asia (*HBW*). Records in Britain increased as the breeding population spread north from Spain into southern France from the 1960s, but it is a rare visitor to Wales, occurring mainly as an overshooting spring migrant in April and May, with far fewer in autumn. Unusually for a migrant of Mediterranean origin, Welsh records are equally divided between the north and south of Wales.

The first Welsh record was at Eglwys Nunydd Reservoir, Gower, on 15 August 1973, followed by singles in Pembrokeshire in 1977: at Dale on 5 March and Bosherston on 30 June. Seven in the 1980s included a group of five, possibly up to seven, at Point of Ayr, Flintshire, on 26–28 October 1987, part of a remarkable influx of 64 birds to Britain in late October and early November. Ten singles were recorded in each of the next two decades, from Skokholm (2), Marloes Mere and Bosherston (2) in Pembrokeshire, Mynydd Rhiw and Bardsey in Caernarfonshire, Bull Bay and RSPB South Stack on Anglesey, and Shotwick, Flintshire.

Total	Pre-1950	1950–59	1960–69	1970–79	1980–89	1990–99	2000–09	2010–19
32	0	0	0	3	7	4	6	12

Red-rumped Swallow: Individuals recorded in Wales by decade.

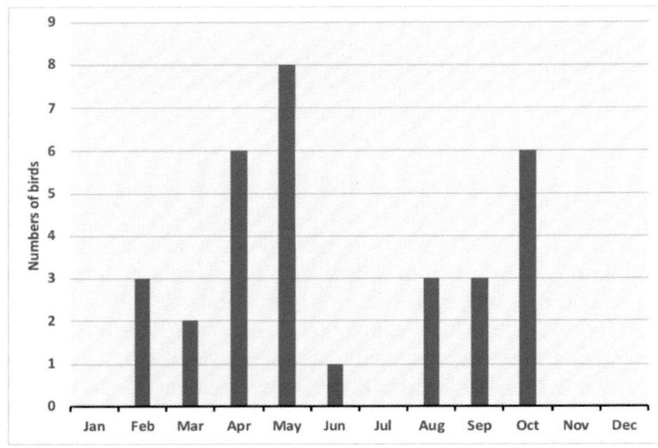

Red-rumped Swallow: Totals by month of arrival, 1973–2019.

There was a marked increase across Wales during 2010–19, with the first county records including in Breconshire, at Llanwrthwl in April 2010, Gwent at Newport Wetlands Reserve in May 2012, Denbighshire at RSPB Conwy in September 2012 and East Glamorgan at Kenfig Rivermouth in April 2019. There were two together at Poppit, Pembrokeshire, in May 2014. The increased trend may well continue, given higher breeding densities in northern Spain since the 1990s (*European Atlas* 2020) and with its breeding range predicted to move northwards through France during the 21st century (Huntley *et al.* 2007).

Jon Green and Robin Sandham

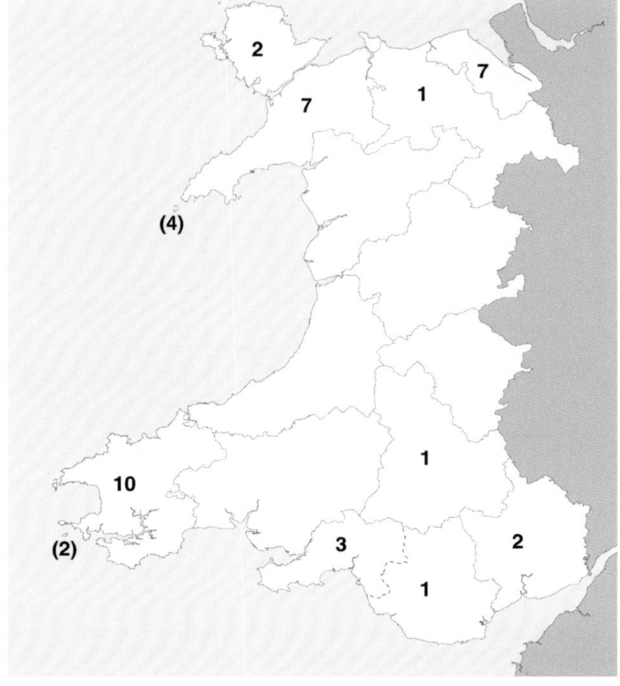

Red-rumped Swallow: Numbers of birds in each recording area, 1973–2019. Some birds were seen in more than one recording area.

Cetti's Warbler *Cettia cetti* Telor Cetti

Welsh List Category	IUCN Red List			Birds of Conservation Concern			
	Global	Europe	GB	UK	Wales		
A	LC	LC	LC	Green	2002	2010	2016

A bird more often heard than seen, Cetti's Warbler was until fairly recently a species with which Welsh birdwatchers would be familiar from trips to the Mediterranean, although its range extends eastward in an arc, south of the Alps through the Middle East to northwest China. Those in Britain are the most northerly bush-warblers in the world (*HBW*). It is a bird of swampy lowlands, preferring reeds with sparse bushes. It began to spread north in France around the 1920s (Bonham and Robertson 1975) and was first recorded in Britain in 1961. It probably bred in southern England in 1972 and did so annually from 1973. The harsh winters of 1978/79 and 1984/85 caused a withdrawal from the northern part of its range in England and Wales (*HBW*) but it can recover quickly from losses.

The first record in Wales was on Bardsey, Caernarfonshire, on 26–30 October 1973. Others followed over the next few years: at Rhosneigr, Anglesey, in December 1976, at Oxwich Marsh, Gower, in July 1977 and at Bosherston Pools, Pembrokeshire, from January to March 1979. The first confirmed nesting in Wales was

Cetti's Warbler © Norman West

at Oxwich Marsh in 1985, when the site held at least ten singing males. In 1987, breeding was confirmed in Pembrokeshire, where the species was recorded at seven sites south of the Cleddau Estuary. The first Carmarthenshire record was at Ffrwd Fen in 1986, where breeding was confirmed in 1989. The population at Pembrokeshire's Teifi Marshes had spread to the Ceredigion side of the river by 1991. In mid-Wales, the first record in Breconshire was at Llangorse Lake in 1992 and in Radnorshire near Glasbury sewage works in 2003. It has subsequently bred in small numbers in both counties. The first record in Montgomeryshire was in 2009, and there have been several subsequent records, including possible breeding at Cors Dyfi.

It was a considerable period before Cetti's Warbler reached North Wales as a breeding species, but birds were recorded at RSPB Cors Ddyga and RSPB Valley Wetlands on Anglesey in 1997. There were five territories at the latter site the following year, including confirmed breeding at Llyn Cerrig-bach. A small breeding population became established at a site in Caernarfonshire in 2005 and held 3–4 territories in 2019. Meirionnydd recorded its first bird in 2007, although it remains a scarce bird in the county, and the first in Flintshire was in 2009, when four were ringed at Shotton Steelworks. The first record for Denbighshire was in 2012, with singles at RSPB Conwy in January and November. It has occurred there almost annually outside the breeding season since, and is on the brink of becoming established as a breeding species.

The size of the Welsh population varies as it can be badly hit by a cold winter. The population crashed at Teifi Marshes in 1996, for example. At Newport Wetlands Reserve, Gwent, counts of singing males were down 75% following winter 2010/11 and 33% after a short cold snap in February/March 2018. Paradoxically, a consistent 3–5 breeding territories at Llangorse Lake, Breconshire, since 2002 seemed unaffected by cold weather in those two winters, but deep and prolonged floodwater for 5–6 months in winter 2019/20 reduced the population to 1–2 territories (Andrew King *in litt.*).

Notwithstanding the effects of cold winters, Newport Wetlands Reserve is the species' stronghold in Wales, particularly Uskmouth, since 1994. There were 51 singing males there in 2005 and 70 in 2016. Another important site in Gwent is Llanwern, where there were 20 singing males in 2012. The population in Gwent was a minimum of 100 pairs in 2016–17. It now occupies many sites along the south coast of Wales: at least 20 in East Glamorgan, 12 in Gower and 15 in Carmarthenshire, where Penclacwydd is the primary site. It is less abundant in Pembrokeshire and Ceredigion, but there are good populations at Castlemartin, Teifi Marshes and the Dyfi Valley. It is scarcer inland, but there were five territorial pairs at Llangorse Lake in 2019. In the north it remains more localised, centred on RSPB Cors Ddyga and RSPB Valley Wetlands, which each hold around ten territories. There are small breeding populations in Flintshire, at Shotton and Gronant, and at a site on Llŷn, Caernarfonshire. There are few other regular breeding sites in the north and, although records of singing males over successive weeks in the main river valleys suggest probable breeding, it has yet to be confirmed. There are probably 300–600 territories in Wales.

Ringing recoveries show that adult Cetti's Warblers in Wales are site-faithful, even if they wander locally during the winter. For example, a female trapped at Uskmouth in August 1998 was recaptured there several times over the following seven years and one ringed at Llangorse Lake in July 2010 was recaught there in May 2019, almost nine years later. Young birds will move considerable distances, however. First-calendar-year birds caught at Kilpaison Marsh, Pembrokeshire, have included birds ringed as juveniles near Peterborough in Cambridgeshire and near Tring in Hertfordshire, while birds ringed at Kilpaison in autumn were recaught subsequently at Brandon Marsh, Warwickshire, and at Attenborough NR, Nottinghamshire, the following summer.

Robinson *et al.* (2007) found that as the population increased, over-winter survival became increasingly dependent on winter temperatures. Milder winters resulting from climate change should help to secure the future of this species in Wales and it is expected to extend its range to southern Scotland during this century (Huntley *et al.* 2007). It has benefited greatly from the creation of wetlands such as on the Gwent Levels and RSPB Cors Ddyga, but at present seems to be more specific in its habitat needs in Britain than around the Mediterranean. Unless this changes as densities increase, it may ultimately limit the species' distribution in Wales.

Rhion Pritchard

Long-tailed Tit *Aegithalos caudatus* Titw Cynffonhir

Welsh List Category	IUCN Red List			Birds of Conservation Concern			
	Global	Europe	GB	UK	Wales		
					2002	2010	2016
A	LC	LC	LC	Green			

The Long-tailed Tit is widely distributed over most of Europe except the far north, and across temperate Asia to the Pacific Ocean (*HBW*). The subspecies *A.c. rosaceus* is endemic to Britain and Ireland (IOC). Populations of the white-headed nominate *A.c. caudatus* sometimes show irruptive behaviour from Scandinavia, and birds showing characteristics of this form have been recorded in eastern England (*Birds in England* 2005), but there has been no confirmed record in Wales.

It is found in woodland with a well-developed shrub layer, particularly around the edges, in thick hedgerows and in parks and gardens with the requisite cover (Perrins 1979). Occurrence in gardens in Wales has increased in recent decades. It was reported by BTO Garden Birdwatch participants from 15% of gardens in the late 1990s but in at least 35% during 2010–19, increasing to almost 50% during severe weather, such as in February 2018. It is largely confined to the lowlands, seldom occurring above 280m, although there have been records up to 460m in Radnorshire (Jennings 2014).

Severe winters can cause heavy losses, even wiping out the population in some areas. This is not simply a matter of low temperatures. Winter losses are most severe when tree branches are coated in ice, which prevents foraging. Rainfall is also important: wet springs reduce adult survival (Gullett *et al.* 2014). Those in Britain and Ireland are very sedentary, the mean distance between ringing and recovery sites being just 2km (*BTO Migration Atlas* 2002). However, some ringed birds have moved considerably farther: one caught at Fenn's Moss, Denbighshire, in November 2013 had been ringed as an adult at Low Hauxley, Northumberland, 277km to the northeast, in October 2012, and an adult ringed on Bardsey, Caernarfonshire, in October 1973 was caught by a ringer 217km to the southeast at Chew Valley Lake, Somerset, in January 1976.

There is little information about the Long-tailed Tit's fortunes in Wales before the late 19th century. Authors at the turn of the 20th century did not give a detailed account of its status other than to report that it was widespread and abundant in southern Wales, while in North Wales, it was absent from mountainous and moorland districts and scarce on Llŷn and Anglesey. Forrest (1907) also reported that it had been extinguished around Conwy, Caernarfonshire, and Palé, Meirionnydd, by an intense frost in February 1895. At Palé none was seen for several years, but it had bounced back to its previous numbers by 1902–03. Population trends in the 20th century have been largely governed by weather conditions. It suffered severely in winter 1939/40, when ice formation on tree branches was a particular feature (Dobinson and Richards 1964), and in 1946/47, although it remained fairly common in Pembrokeshire (Lockley *et al.* 1949) and some recovery was evident in Carmarthenshire by 1949–50 (Ingram and Salmon 1954). Long-tailed Tits probably did not suffer as severely in 1962/63 as in some previous winters (Dobinson and Richards 1964), but it took the Gwent population several years to recover from its effects (Venables *et al.* 2008).

The Britain and Ireland atlases of 1968–72, 1988–91 and 2007–11 found the species in at least 90% of 10-km squares in Wales, except for the highest upland areas and a few less-wooded areas in the west. The only real change was that birds had spread into northern Anglesey by the latest period. Relative abundance in 2007–11 was highest along the south coast and in some eastern areas but generally low compared to most of England, and lower in Carmarthenshire and south Ceredigion than in 1988–91. It had become more abundant and widespread in Gwent, where it increased from 80% of tetrads in 1981–85 to 88% in 1998–2003, having filled gaps in the western valleys. It remained absent only from the treeless moorlands of the far west. There was also a marked increase in East Glamorgan, where there was breeding evidence in 47% of tetrads in 1984–89 and in 54% in 2008–11. The increase was less pronounced in Pembrokeshire, where it occurred in 41% of tetrads in 1984–88 and 46% in 2003–07. The *North Wales Atlas 2008–12* found it in 40% of tetrads, largely absent from the uplands. It also remained absent from much of western and central Anglesey. Jennings (2014) considered that the species underwent a huge increase in Radnorshire from the late 1980s, and that it was probably three or four times more common there than 25 years previously.

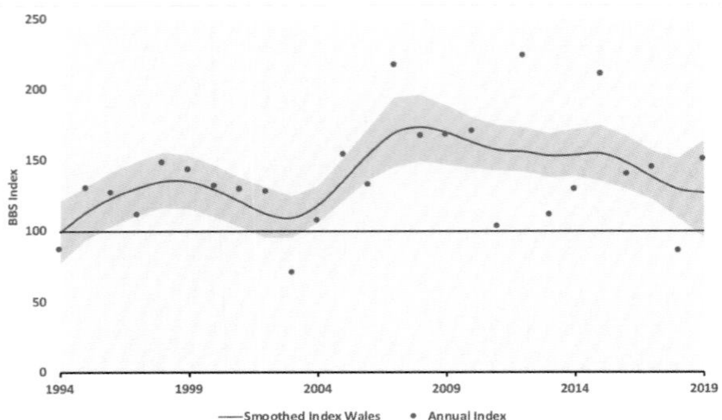

Long-tailed Tit: Breeding Bird Survey indices, 1994–2019. The red line shows the smoothed index, 1995–2018. The shaded area indicates the 85% confidence limits.

Numbers in Welsh woodlands fell by 41% between 1981–88 and 2003–04, having been stable during the 15 years prior to 1981–88 (Amar *et al.* 2006), whereas across a wider variety of habitats, the BBS index for Wales increased by *c.*70% during 1995–2008. However, by 2018 the index was only 14% above its 1995 baseline (Harris *et al.* 2020). The annual indices were much more variable during 2007–19 than in the previous decade, which may be a combination of several winters of cold weather and a number of wet springs. The Welsh breeding population in 2018 was estimated to be 28,000 territories (Hughes *et al.* 2020).

Outside the breeding season, Long-tailed Tits congregate in flocks based on family parties, and associated helpers, and will sometimes join with other small birds. It is not usually found in great numbers on the Welsh islands, but there are occasional influxes. For example, there was a record arrival on Bardsey in late October and early November 2008, with at least 60 present on 1 November. However, 120 there on a day in 1957, quoted by Barnes (1997), was an error and should have referred to Coal Tit. In Pembrokeshire, a flock of 12 was seen heading out from Skomer towards Grassholm on 3 November 2008 and there were 44 on Skomer on 27 October 2017. Ramsey recorded this species in November 2008 for the first time since 1995. Winter flocks usually number up to 30 but can be considerably larger, such as 64 at Llangynwyd, East Glamorgan, in autumn 1990 and at least 70 at Garden Village, Wrexham, Denbighshire, on 25 January 2009.

Long-tailed Tit does not appear to face any particular threat in Wales and may have benefited from climate change (Pearce-Higgins and Crick 2019), especially warmer springs and autumns. The average laying date has moved from 19 April in 1967 to 3 April in 2017, but there is no evidence of a population effect to date. Although mainly insectivorous, supplementary food supply in gardens may reduce winter mortality and it seems likely that the occurrence of cold winters will continue to be the main determinant of the Long-tailed Tit's fortunes in Wales.

Rhion Pritchard

Sponsored by Caroline de Carle

Wood Warbler *Phylloscopus sibilatrix* Telor Coed

Welsh List Category	IUCN Red List			Birds of Conservation Concern			
	Global	Europe	GB	UK	Wales		
A	LC	LC	VU	Red	2002	2010	2016

Along with Pied Flycatcher and Redstart, the Wood Warbler is one of the characteristic species of the Sessile Oak woodlands of Wales. It breeds across Europe, except for Iberia, Greece and the far north, and east to the west Siberian plain (*HBW*). It is scarce in Ireland, so Wales and western Scotland are at the western edge of its range (*European Atlas* 2020). It is usually found in deciduous forest, including Beech and mixed oak, but it also occurs in mixed stands of spruce and pine (*HBW*). It nests on the ground and so requires sufficient cover to hide the nest, but otherwise needs a closed canopy with sparse undergrowth. Branches less than 2.5m above the ground are important as perches when bringing food to the young (Tiedemann 1971). In Wales, the Wood Warbler is most abundant below 300m, which is the upper limit for deciduous forest in most areas, but it has been recorded breeding in conifers at 420m in the Tywi Forest, Ceredigion, and up to 500m in Meirionnydd.

It winters in the forests of sub-Saharan Africa (*HBW*). The only recovery of a Welsh-ringed Wood Warbler outside Britain was a nestling that was ringed in June 2007 at Glasbury, Radnorshire, and shot near Pietrastornina in Campania, southern Italy, in October that year. Most arrive in Wales in the second half of April, though earlier arrivals occur, such as one at Cwmere near Eglwysfach, Ceredigion, on 26 March 2003. There were also records in the first few days of April in Caernarfonshire in 2005 and East Glamorgan in 2017. Most have left by early August, but there have been September records in several counties and a few in October, the latest being on Ramsey, Pembrokeshire, on 21–25 October 1997.

Its distinctive song resulted in several early references to its presence. It was said to be numerous in Coed Benarth near Conwy, Caernarfonshire, in the early 19th century, for example (Williams 1835). At the turn of the 20th century, Wood Warblers appear to have been distributed to a similar extent as now. It was said to be uncommon in Glamorgan in the late 1890s, although regular in old deciduous woodland on the northern fringes of the county (Hurford and Lansdown 1995). It was a scarce and very local species in Pembrokeshire, which Mathew (1894) doubted bred to the west of Mynydd Preseli, though he quoted Dix as saying that it was numerous in almost any oak or Beech woodland along the Ceredigion border. Salter (1895) described it as "numerous in all suitable localities, its range extending to the highest beech and oak plantations in the valleys" of Ceredigion and abundant in the upper Tywi Valley, Carmarthenshire. In North Wales it was numerous in wooded valleys up to at least 300m. Around Llanberis and some other western districts "it swarms in all the woods to such an extent that its shivering song becomes almost wearisome, while its call note resounds through the woods incessantly" (Forrest 1907).

Ringing recoveries showed that birds can change their breeding site between years, illustrated by an adult male, colour-ringed in Macclesfield Forest, Cheshire, in June 2007, which was at Coed Bwlch-glas in the Elan Valley, Breconshire, in May 2009. Such movements between years mean that populations can fluctuate locally to an extent that is not representative of the wider picture (*European Atlas* 1997). However, changes in range in Wales are evident around the periphery, in southern Glamorgan, Anglesey and Llŷn.

The *Britain and Ireland Atlas 1968–72* showed that Wood Warblers were widespread in Wales, although absent from most of Anglesey, Llŷn, western Pembrokeshire and the Gower Peninsula. The 1988–91 *Atlas* showed a fairly even balance of gains and losses. However, the 2007–11 *Atlas* revealed a marked decline, particularly in northeast Wales and southern Ceredigion. Abundance was relatively high in most of Wales, matched only by some areas of northwest Scotland, but lower everywhere in Wales than in 1988–91. Tetrad atlases confirmed the decline in range around the turn of the 21st century. The *Gwent Atlas 1981–85* recorded Wood Warbler in 39% of tetrads but only 31% in 1998–2003, and its range in Pembrokeshire reduced from 11% of tetrads in 1984–88 to 5% in 2003–07. The *North Wales Atlas 2008–12* recorded it in 14% of tetrads, centred on Meirionnydd, where it occurred in 28% of tetrads. However, there were no confirmed or probable breeding records in Flintshire, where Birch *et al.* (1968) had described Wood Warbler as "local, though fairly widespread". On Anglesey, it was said to be common along the Menai Strait in the 1950s (Jones and Whalley 2004) and recorded in 11 1-km squares on the island in 1986 (RSPB), but there were no confirmed breeding records in 1988–91. A survey of Llŷn, Caernarfonshire, by the RSPB in 1986, found it at 12–14 sites, but it is now very scarce on most of the peninsula.

The population changes in Glamorgan were perhaps the best documented in the 20th century. Its distribution expanded to several sites in the south during 1910–25, from which it contracted after 1927. Breeding increased again following an influx into the Rhymney Valley in 1947 and it had become more widespread by the end of the 1970s and increasingly abundant by the early 1990s

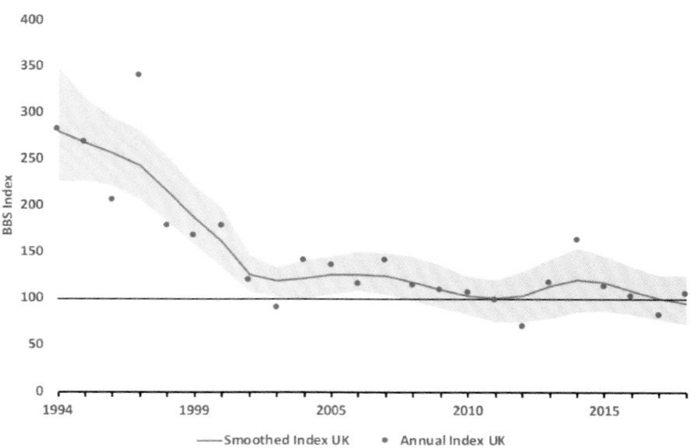

Wood Warbler: UK Breeding Bird index, 1994–2018. The grey line shows the smoothed index, 1995–2018. The shaded area indicates the 85% confidence limits.

Location	Woodland type	Density (per km²)	Year
Coed Penrhyn Mawr, Ceredigion	Ungrazed Sessile Oak	29–79.7 territories	1995–2006
RSPB Ynys-hir, Ceredigion	Sessile Oak	45.5 territories	2006
RSPB Ynys-hir, Ceredigion	Sessile Oak	47.3 territories	2011
Cwm Clettwr, Ceredigion	Sessile Oak/Western Hemlock	92 territories	1988
Garth Wood, Radnorshire	Not stated	93.3 singing males	1984–85
Elan/Claerwen valleys, Radnorshire	Broadleaved/coniferous	41.1 singing males	1991
Elan/Claerwen valleys, Radnorshire	Broadleaved/coniferous	26.3 singing males	2010
Wentwood, Gwent	Larch plantation	10 nests	2010

Wood Warbler: Estimated density of singing males at selected woodlands in Wales.

Wood Warbler © Kev Joynes

(Hurford and Lansdown 1995). The species was found in 24% of tetrads in East Glamorgan in 1984–89, but then contracted to just 13% in 2008–11, with breeding evidence in less than 10%. In Ceredigion, by contrast, Roderick and Davis (2010) considered that it was more common than in the 20th century, because oak woodlands were no longer coppiced (although this is at odds with the species' need for an open understorey), and that it had become more common in conifer plantations, particularly larch.

The increase evident in Glamorgan during 1970–90 was also apparent in Woodland Bird Surveys that recorded a 99% increase between 1965–72 and 2003–04, but populations had been even higher in 1981–88 and fell by 25% during the latter 15 years (Amar *et al.* 2006, Hewson and Noble 2009). Trends can only be calculated at the UK level, but since the estimated 2,850 (2,650–3,100) singing males in Wales comprise 44% of the UK population (Hughes *et al.* 2020), the 66% decline during 1995–2018 may be a fairly accurate guide to the trend here (Harris *et al.* 2020). The European population declined by 44% between 1980 and 2016 (PECBMS 2020), but the magnitude of this varied regionally (Vickery *et al.* 2014).

While the standard count unit is the number of singing males, this is not necessarily an accurate guide to the number of territories, as some males fail to find a mate, while others may be polygynous and sing in secondary territories in an attempt to attract another female (Temrin *et al.* 1984). There was a median of 29 males/occupied 10-km square in Wales in the 1980s. This was the highest number recorded anywhere in Britain, but there is wide variation according to woodland type; some sites held over 90 territories/km^2 (Bibby 1989).

Wood Warblers from Britain take a route through the central Mediterranean via Italy when they migrate to Africa but use a more western route in spring (*BTO Migration Atlas* 2002). Numbers recorded on the islands and at coastal watchpoints have been small. There have been no day counts of more than three on Bardsey in spring and no more than four in autumn. A count of ten on migration at Llandegfedd Reservoir, Gwent, on 7 August 1969 was exceptional.

Habitat changes at its breeding sites may have been partly responsible for the decline but the main causes probably lie elsewhere. Mallord *et al.* (2016) found that it was unlikely that the structure of upland oak woods in western and northern Britain had caused population changes in four Afro-Palearctic migrants, including Wood Warbler. Although nest predation—the primary predator being Jay—can be high, it is not a cause of decline (Mallord *et al.* 2012a). Caterpillars are an important food for nestlings, but although there is a mismatch between hatching dates and caterpillar availability, this does not appear to affect reproductive fitness but it may affect adult survival rates (Whytock *et al.* 2015). Woodland management can also cause significant local changes. For example, in Wentwood, Gwent, during 2004–12, a mature larch plantation contained an average of around ten nests/km^2, but numbers subsequently declined as forestry compartments were thinned, allowing ground vegetation to increase. That population was lost when the trees were prematurely felled to combat the spread of the larch disease *Phytophthora* (Jerry Lewis *in litt.*). In Meirionnydd, the population in larch forestry is considered to have increased in recent years (Jim Dustow *in litt.*).

The main causes of decline appear to be on the wintering grounds or along migration routes. Regional changes in climate or land use in the humid tropics may be driving declines in Wood Warbler, among other long-distance migrants. It may be that peaks of food abundance at critical stop-over sites are no longer aligned (Ockendon *et al.* 2012). Climate change may be important in Britain too. Modelling suggests that large parts of Wales may become climatically unsuitable for Wood Warblers by the end of this century (Huntley *et al.* 2007), making it more important that Welsh woodlands continue to offer prime habitat in order to increase their resilience to change. Wood Warblers are associated with structural features of woodland that relate to past management. Habitat quality could be restored through the introduction of a moderate grazing regime (Mallord *et al.* 2012b). Future land-management policies should be sufficiently flexible to encourage the right levels of temporary grazing to achieve good conditions for Wood Warblers and Pied Flycatchers. This is being demonstrated by the Celtic Rainforest Wales project in Meirionnydd and Ceredigion. Timber harvesting during summer can damage Wood Warbler nesting habitat as cut trees are hauled to the road. Such activities should not take place in forests where the species nests. Welsh woodlands are a special place for Wood Warbler, but there are substantial challenges ahead for these long-distance migrants.

Rhion Pritchard

Sponsored by John Young

Western Bonelli's Warbler *Phylloscopus bonelli* Telor Bonelli y Gorllewin

Welsh List Category	IUCN Red List		Rarity recording
	Global	Europe	
A	LC	LC	BBRC

This very rare visitor breeds in western Europe as far east as Austria and as far north as the France/Belgium border, and locally in the Atlas Mountains of North Africa. It winters along the southern edge of the Sahara, from Senegal and south Mauritania to western Chad (*HBW*). There were 155 records in Britain to the end of 2019, typically 3–4 each year, with the number of records having increased since 2000 (Holt *et al.* 2020).

The first British record was one caught on Skokholm, Pembrokeshire, on 31 August 1948, which later died. Almost two-thirds of the Welsh total were recorded on Bardsey, Caernarfonshire. The first two were found there on separate dates in 1959. There have been three records on the Pembrokeshire islands: one on Ramsey and two on Skokholm. Two of the three Welsh mainland records were, typically, in autumn, at St Davids Head, Pembrokeshire, on 3 October 2011 and at Pwll y Cyw, Aberdaron, Caernarfonshire, from 31 August to 5 September 2019. The sole record in spring was a male that sang at Gwastedyn, Radnorshire, on 17–18 May 2006.

The climate in southern and eastern England is expected to become more suitable for Western Bonelli's Warbler during the 21st century (Huntley *et al.* 2007). Almost half of the Welsh records were during 2010–19, so its occurrence appears to be increasing as projected.

Jon Green and Robin Sandham

Total	Pre-1950	1950–59	1960–69	1970–79	1980–89	1990–99	2000–09	2010–19
15	1	2	2	0	1	0	2	7

Western Bonelli's Warbler: Individuals recorded in Wales by decade.

Eastern/Western Bonelli's Warbler *Phylloscopus orientalis/bonelli* Telor Bonelli y Dwyrain/Gorllewin

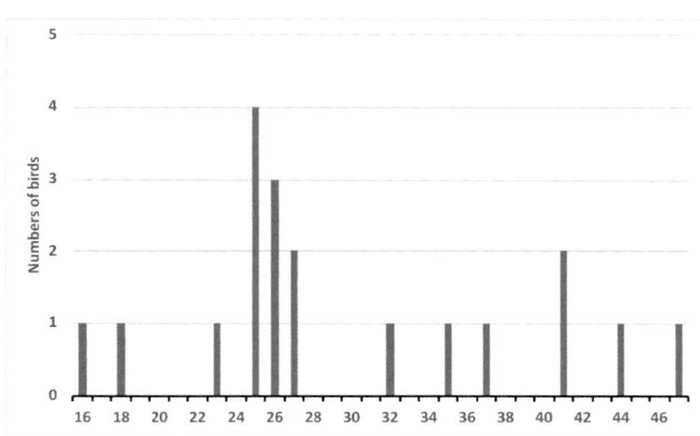

Bonelli's Warbler spp.: Totals by week of arrival, 1948–2019. Week 25 begins *c.*18 June and week 41 begins *c.*8 October.

Western Bonelli's Warbler breeds from Austria and Italy westwards and Eastern Bonelli's Warbler in the Balkans and Turkey. Both breed in open woodland with understorey, and winter in the Sahel zone of sub-Saharan Africa. Each can only be specifically identified by its call, and five seen in Wales remained silent and so were accepted as Bonelli's Warbler spp. These were at:

- Lavernock, East Glamorgan, on 30 August 1963;
- Llaniestyn, Caernarfonshire, on 17 September 1968;
- Skokholm, Pembrokeshire, on 31 August 1991;
- Skomer, Pembrokeshire, on 31 August 2017 and
- Skomer, Pembrokeshire, on 16 September 2018.

There were 84 undetermined Bonelli's Warbler spp. in Britain to 2019 (Holt *et al.* 2020).

Jon Green and Robin Sandham

Hume's Warbler *Phylloscopus humei* Telor Hume

Welsh List Category	IUCN Red List	Rarity recording
	Global	
A	LC	BBRC

This very rare vagrant to Wales breeds in mountain forest in Mongolia, western China and the Himalayas, where it largely replaces Yellow-browed Warbler, of which it was formerly treated as a subspecies. Hume's Warbler migrates a relatively short distance to India and Southeast Asia (*HBW*). Those that are found in Britain are frequently seen from November onwards. Just three of the 170 British records up to 2019 (Holt *et al.* 2020) occurred in Wales:

- Strumble Head, Pembrokeshire, on 20 November 1993;
- Caernarfon, Caernarfonshire, from 14 December 2003 to 9 January 2004 and
- Penrhyn Bay, Caernarfonshire, on 18 November 2007.

Jon Green and Robin Sandham

Yellow-browed Warbler *Phylloscopus inornatus* Telor Aelfelyn

Welsh List Category	IUCN Red List	
	Global	Europe
A	LC	LC

Synonymous, in Britain at least, with an autumn day on the coast, the high-pitched *'soo-eeet'* call of a Yellow-browed Warbler has become a far more frequent occurrence in the last 20 years, but still triggers excitement in keen birdwatchers. It breeds in broadleaved woodland and the margins of conifer forests on the eastern edge of Europe, in the western foothills of the Ural Mountains, and across Russia to the Pacific Ocean. A handful of singing males have been recorded in spring in Finland in recent years (per Lintutiedotus, the Finnish bird recording app). Most winter in Southeast Asia, but in recent decades it has become the most frequent Siberian migrant to occur in western Europe in autumn. The reasons for its change in status are not known, but older theories of reverse migration have been superseded by the concept that the species is forging a new westward migration route in the same way that some Blackcaps in central Europe developed a new wintering area in Britain in the 20th century.

There was a steady increase in its occurrence in Britain from the early 21st century: the average annual total increased from 72 in 1968–79 to 325 in 1980–99 and 1,276 in 2000–17 (White and Kehoe 2019). The majority are thought to be first-calendar-year birds (Gilroy and Lees 2003). Stable isotope analysis from feathers of birds caught in Norway suggests that they were from the westernmost or southern part of the breeding range (de Jong *et al.* 2019). Winter records in Iberia, the Macaronesian islands and North Africa have also increased and it may be that these birds return to Siberia in spring (de Juana 2008). Although this has yet to be proven, the resighting of one over-wintering in November 2018 in the same Cork Oaks in Andalucía, Spain, in which it had been ringed the previous winter is intriguing (Tonkin and Perea 2019).

While several thousand can occur on the east coast of Britain in a good autumn, numbers arriving in Wales are more modest, but it is now considered a regular migrant and has occasionally over-wintered. Arrivals start in late September and peak in the second week of October, with a few discovered in November and December. Whether the latter are late arrivals or just the late discovery of earlier arrivals is uncertain.

The first Welsh record was on Skokholm, Pembrokeshire, on 2–3 October 1959. There was an average of only one each year in Wales over the following two decades. A huge increase occurred during 1985–89 and the reporting rate really picked up in the 21st century, specifically in 2007, 2013 and 2014. Those years were, however, dwarfed by autumn 2016, when at least 4,500 were recorded across Britain, of which *c.*200 occurred in Wales (White and Kehoe 2018). This total included 74 in Caernarfonshire (of which a minimum of 60 were on Bardsey), 30 in Gower, 27 in Pembrokeshire, 26 in East Glamorgan and 13 on Anglesey. Peak day counts that year included 14 on Bardsey and six at Mewslade/Middleton, Gower, on 9 October. Of 16 ringed at Oxwich Marsh, Gower, that autumn, nine were on 8 October. A smaller influx in 2018 included 17 in Pembrokeshire and 19 in Caernarfonshire, of which 13 were on Bardsey, including nine on 26 September. The vast majority pass south. There has yet to be a spring record in Wales, but ten seen in January/February are presumed to have over-wintered. The first of these was at Wentlooge Industrial Park, East Glamorgan, from 6 December 1992 to at least 17 January 1993, since when there have been four in Pembrokeshire, three on Anglesey and two in East Glamorgan. Numbers wintering in southwest England have increased, with 50 there in 2018–19.

Yellow-browed Warbler © Ben Porter

Recording area	First record	1960–69	1970–79	1980–89	1990–99	2000–09	2010–19	Total
Gwent	Magor, 14 October 1988			1	1	3	2	7
East Glamorgan	Kenfig Pool, 17 October 1981			7	5	18	55	85
Gower	Cwm Ivy, 1 November 1981			1	1	12	57	71
Carmarthenshire	Pembrey, 4–5 October 1997				1	2	8	11
Pembrokeshire*	The Smalls, 19 November 1983			12	11	31	73	127
Skomer	10 October 1986			1	2	17	33	53
Skokholm	2–3 October 1959	1	5	4	4	3	16	33
Ceredigion	Aberporth, 23 October 1988			1		1	14	16
Breconshire	Llangammarch, 8 October 1986			2		2	3	7
Radnorshire	Claerwen, 17 October 1992				1	1	2	4
Montgomeryshire	Newtown, 12–16 October 2015						3	3
Meirionnydd	Dolgellau, 14 October 1996				1		1	2
Caernarfonshire*	Aberdaron, 20 October 1985			10	2	16	43	71
Bardsey	26 September 1960	8	5	54	9	52	180	308
Anglesey	Penmon, 19 October 1985			6	2	14	40	62
Denbighshire	Clocaenog, 22 October 1994				2	2	8	12
Flintshire	Point of Ayr, 29 October 1987			1	2		7	10
Total		**9**	**10**	**100**	**44**	**174**	**545**	**882**

Yellow-browed Warbler: First records and total numbers seen by decade (*totals for Pembrokeshire and Caernarfonshire are away from the islands listed).

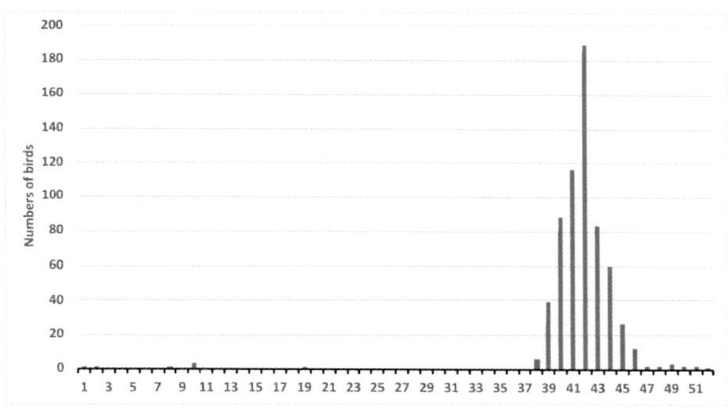

Yellow-browed Warbler: Totals by week of arrival, 1959–2019. Week 39 begins *c.*24 September and week 46 begins *c.*12 November. Records are from BirdTrack.

Over one-third of records in Wales are from Bardsey, while the lion's share of records in the south are from Pembrokeshire, where a greater proportion of records are on the mainland compared to Caernarfonshire. Fewer observers may account for smaller numbers recorded in neighbouring Carmarthenshire and Ceredigion, whereas more have been seen on the coasts of Gower and East Glamorgan. It remains a rare bird inland, with only 14 records in Breconshire, Radnorshire and Montgomeryshire combined.

Our understanding of westward migration of Yellow-browed Warblers is in flux, and its breeding range may be extending west too: birds summered at several sites in Finland in 2020 (Alex Lees *in litt.*). If Britain is now a staging post on a regular passage migration, it may become an increasing wintering species on the mild coastal fringe of Wales.

Jon Green, Robin Sandham and Julian Hughes

Sponsored by Jim Dustow

Pallas's Warbler *Phylloscopus proregulus* Telor Pallas

Welsh List Category	IUCN Red List	Rarity recording
	Global	
A	LC	WBRC

This is a scarce migrant to Wales, which breeds in the coniferous taiga forests of southern Siberia east of the Ural Mountains and in mixed forest as far south as Mongolia. It winters in southeast China and adjacent countries (*HBW*). Not surprisingly, the majority of British records are from the east coast of England and Scotland. Far fewer make it to the west. The first record was on Bardsey, Caernarfonshire, on 7 November 1975, since when the numbers recorded in each decade have increased in line with the rest of Britain, although the average number seen annually in Britain was significantly lower in 2010–18 than in the previous decade (White and Kehoe 2020b). Reasons for the increased occurrence during 1994–2004 are not clear, but are thought not to be linked to the number of observers nor any extensive change in the range of its breeding population. It may have been a result of reverse migration, by which first-calendar-year birds headed in the opposite direction to their expected route and found themselves in Fennoscandia, combined with anti-cyclonic weather conditions that enabled them to cross the North Sea (Thorup 1998).

Pallas's Warbler on Bardsey, Caernarfonshire © Ben Porter

Total	Pre-1950	1950–59	1960–69	1970–79	1980–89	1990–99	2000–09	2010–19
48	0	0	0	1	7	11	11	18

Pallas's Warbler: Individuals recorded in Wales by decade.

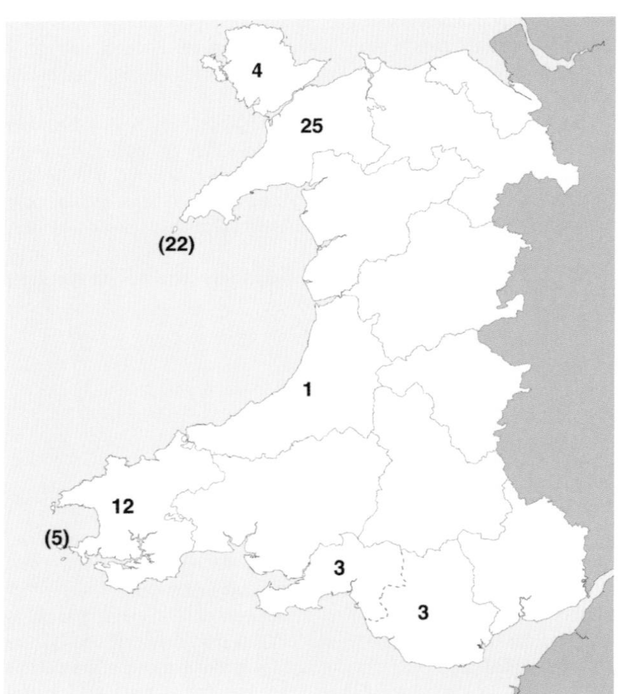

Pallas's Warbler: Numbers of birds in each recording area, 1975–2019.

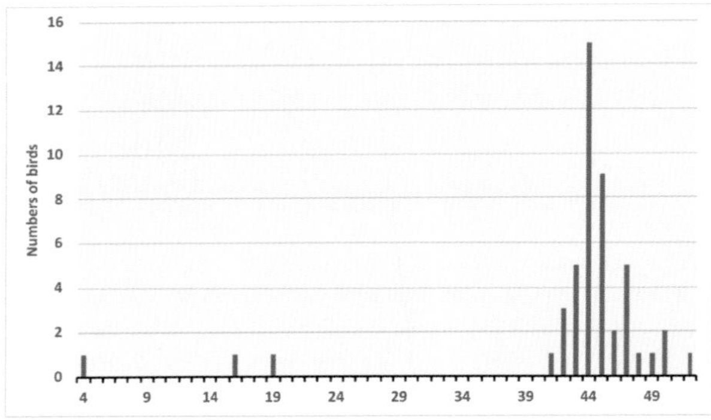

Pallas's Warbler: Totals by week of arrival, 1975–2019. Week 16 begins *c.*16 April and week 44 begins *c.*29 October.

Almost half of the Welsh total, 22 individuals, occurred on Bardsey, Caernarfonshire, presumably as migration halted when they reached the sea. Five have been seen on Skomer, but none on neighbouring Skokholm, while seven were recorded on mainland Pembrokeshire. The majority have been in late autumn, the last week of October being the peak period of arrival, associated with high pressure stretching across from northern Europe. Three have turned up in December and one in January, which presumably over-wintered. There have also been two records in spring, both on Bardsey in 2017, on 18 April and 7 May. The latter had been ringed at Spurn, East Yorkshire, in October 2016, which suggested that it had over-wintered somewhere along the Atlantic seaboard and was attempting a return towards its breeding area.

Jon Green and Robin Sandham

Radde's Warbler *Phylloscopus schwarzi* Telor Radde

Welsh List Category	IUCN Red List	Rarity recording
	Global	
A	LC	WBRC

Radde's Warbler breeds in shrubby, open deciduous woodland, often close to water, in southern Siberia along the Mongolian border and east to the Korean Peninsula. It migrates along the west Pacific flyway to winter in Vietnam and Thailand (*HBW*). There were 444 records in Britain up to the end of 2018, the recording rate having increased during the second half of the 20th century. An average of 13 were recorded in Britain each year during 2000–18 (White and Kehoe 2020b). While reverse migration, of first-calendar-year birds heading in the 'wrong' direction, is a possible explanation, it is unclear why this would have increased in frequency. However, it is a very rare vagrant to Wales, with only eleven records. All were in October and all on the islands, except for one on the Great Orme, Caernarfonshire:

- Skokholm, Pembrokeshire, on 22 October 1968;
- Bardsey, Caernarfonshire, on 29 October 1987;
- Bardsey, Caernarfonshire, on 18 October 1990;
- Bardsey, Caernarfonshire, on 29–30 October 2006;
- Bardsey, Caernarfonshire, on 30–31 October 2006;
- Skomer, Pembrokeshire, on 17 October 2007;
- Great Orme, Caernarfonshire, on 28 September 2008;
- Bardsey, Caernarfonshire, on 14 October 2012;
- Bardsey, Caernarfonshire, on 13 October 2016;
- Skomer, Pembrokeshire, on 20 October 2017 and
- Skokholm, Pembrokeshire, on 26 October 2017.

Jon Green and Robin Sandham

Dusky Warbler *Phylloscopus fuscatus* Telor Tywyll

Welsh List Category	IUCN Red List	Rarity recording
	Global	
A	LC	WBRC

This very rare vagrant to Wales breeds in taiga bogs and wet meadows in central and eastern Siberia, south to central China, and winters in Southeast Asia (*HBW*). The number of records elsewhere in Britain has increased markedly, from an average of 13 each year in 1990–99 to 27 in 2010–18, bringing the total to 590 (White and Kehoe 2020b). There have been only six records in Wales:

- Bardsey, Caernarfonshire, on 30 October 1982;
- Bardsey, Caernarfonshire, on 7 November 1987;
- Strumble Head, Pembrokeshire, on 15 October 1988;
- Point of Ayr, Flintshire, on 11–12 November 1997;
- Porth Clais, Pembrokeshire, on 10–13 November 2003 and
- Rhostryfan, Caernarfonshire, on 16–17 October 2013.

Jon Green and Robin Sandham

Willow Warbler *Phylloscopus trochilus* Telor Helyg

Welsh List Category	IUCN Red List			Birds of Conservation Concern			
	Global	Europe	GB	UK	Wales		
A	LC	LC	LC	Amber	2002	2010	2016

Willow Warbler is a summer migrant to northern Europe, from central France north and east across Russia and Siberia, just reaching the Pacific coast north of the Kamchatka Peninsula. It winters south of the Sahara and the Horn of Africa. The nominate *P.t. trochilus* breeds in Britain, in deciduous woodland and conifer plantations, mixed forests in which birch predominates, and in willow and birch scrub (*HBW*). Records of the subspecies *P.t. acredula*, which breeds across most of Fennoscandia and as far east as central Siberia (IOC), were included in *Birds in Wales* (1994, 2002), but it was subsequently removed from the Welsh List because there were no records with sufficient evidence to prove its occurrence.

Accounts from the late 19th century and through much of the 20th century described Willow Warbler as abundant across Wales, although in Pembrokeshire it was considered "a not very numerous summer visitor" by Mathew (1894), who lived a few miles south of Fishguard. Most authors regarded it as the most populous warbler species in Wales, and it can occur at high densities, in both native woodland and pre-thicket conifer plantations. Around farmland, O'Connor and Shrubb (1986) showed that the mean density of 19.2 pairs/km² was higher than elsewhere in the UK. That figure was, however, lower than densities of 33–61 pairs/km² on farmland in southern Wales and 37–156 pairs/km² in woodlands in North Wales and Breconshire, although most of those studies were over only 1–2 years (*Birds in Wales* 1994).

The *Britain and Ireland Atlas 1968–72* and those for 1988–91 and 2007–11 showed occupation in at least 96% of 10-km squares in Wales, with no real change between the surveys. In the 2007–11 *Atlas* abundance was relatively high in Wales, particularly on higher ground, and far greater than almost anywhere in England at the same latitude. The Welsh breeding population in 2018 was estimated to be 265,000 territories (Hughes *et al.* 2020). On the face of it, there seemed to be little cause for concern.

However, the Repeat Woodland Bird Survey showed a 68% reduction in the density of territories in Wales between 1965–72 and 2003–04, much of which occurred during the 1980s (Amar *et al.* 2006). Ringing on BTO Constant Effort Sites has shown a steep and consistent decline over the same period and declines

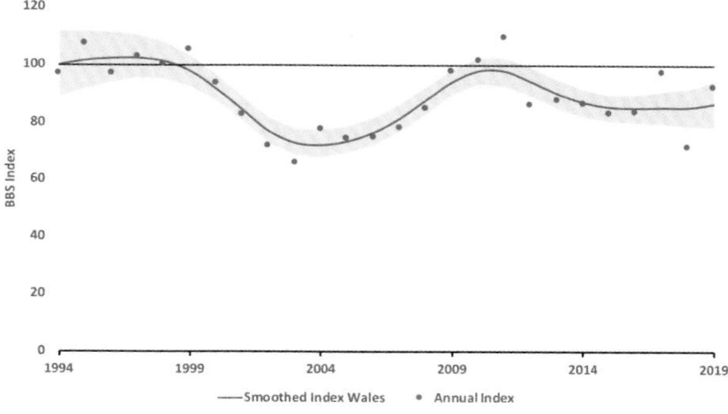

Willow Warbler: Breeding Bird Survey indices, 1994–2019. The red line shows the smoothed index, 1995–2018. The shaded area indicates the 85% confidence limits.

were noticed in parts of southern Wales from the 1990s, such as in Radnorshire (Jennings 2014). A change was apparent in tetrad atlases repeated in Gwent, East Glamorgan and Pembrokeshire in the early 21st century. Over 95% of tetrads held Willow Warblers in the *Gwent Atlas 1981–85* and although the occupancy rate was only marginally lower in 1998–2003, at 92%, the average density of singing birds had fallen dramatically. Although different methods were used to make the estimates in the two periods, Venables *et al.* (2008) were confident that the population had reduced dramatically. In East Glamorgan, breeding evidence was recorded in 83% of tetrads in 1984–89 but only 72% in 2008–11, with a substantial reduction in the southern coastal plain that was only partially offset by increases in the more elevated north. In Pembrokeshire, 89% of tetrads showed breeding evidence in 1984–88, but this had reduced to 75% by 2003–07, with gaps having appeared throughout the county (*Pembrokeshire Avifauna* 2020).

Farther north, it remained fairly numerous and widespread at the time of the *North Wales Atlas 2008–12*, which reported evidence of Willow Warblers breeding in 84% of tetrads and absence only from the highest moorland and mountains. The BBS index for Wales has fluctuated but was 16% lower in 2018 than in 1995 (Harris *et al.* 2020). However, this almost certainly hides a greater reduction in the south, especially in the lowlands. The five English regions that lie south and east of Wales have seen an average decline of 70% in the same period. Estimates from the 1988–91 *Atlas* suggested that Willow Warblers outnumbered Chiffchaff by almost 3:1 in Wales (*Birds in Wales* 2002). The most recent population estimate for Wales shows that the populations of the two species are now very similar (Hughes *et al.* 2020). The

Willow Warbler: Recovery locations of birds ringed in Wales are shown by red circles. Ringing locations of birds that were recovered in Wales are shown by blue triangles. Small brown dots show ringing or recovery locations in Wales.

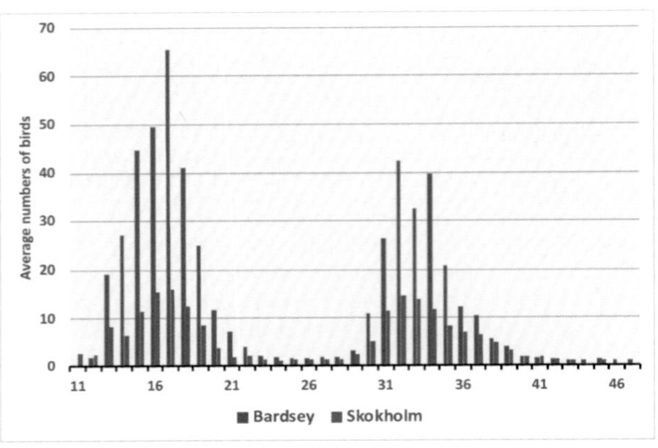

Willow Warbler: Average numbers seen, by week number, at Bardsey and Skokholm, 1953–2019. Week 17 begins *c.*22 April and week 32 begins *c.*5 August.

Willow Warbler population will soon be overtaken, in size, by that of Blackcap.

The main wintering area for British Willow Warblers is in countries on the Gulf of Guinea in West Africa (*BTO Migration Atlas* 2002). The current Britain and Ireland longevity record is one ringed as a juvenile at Bronbannog, Clocaenog Forest in Denbighshire on 25 July 1999 and caught by another ringer at Hilton of Fern in Angus, 409km to the north, on 13 July 2010 (Robinson *et al.* 2020). Ten years 11 months and 18 days is a remarkable age for a small bird that weighs 9g, a fraction more than a £1 coin, and had flown to West Africa and back 11 times, a total of at least 154,000km!

The first arrivals in Wales in the 20th century were almost always in April, but during 2000–19 every county recorded March arrivals, the earliest being on 29 February 2004 at Bridgend, East Glamorgan. Spring migration peaks in the last week of April. After breeding, most Willow Warblers leave in August, the autumn migration peaking on the islands in the middle of the month but continuing through September. Large falls of *Phylloscopus* warblers occur occasionally, such as on 7 August 1975, when *c.*5,000 were counted on Bardsey, Caernarfonshire, presumed to be all Willow Warblers. Most Welsh counties except Radnorshire and Flintshire recorded Willow Warbler in October during 2000–19,

Willow Warbler © John Freeman

and several also in November, the latest being on Bardsey on 19 November 2016. Such late records are unusual but not unprecedented: there was one at Margam, Gower, on 15 November 1935 (Hurford and Lansdown 1995). There has also been a handful of winter records: *Birds in Wales* (1994) referred to one in East Glamorgan in each of the three decades prior to publication, though Hurford and Lansdown (1995) doubted their veracity. One sang at Stackpole, Pembrokeshire, on 9 February 1991 and there were a further four records between December and mid-February during 2000–19.

Willow Warbler is on the Red List of conservation concern in Wales (Johnstone and Bladwell 2016), based on the decline in its breeding population in woodlands (Amar *et al.* 2006). In southern Britain, including South and mid-Wales, this has been driven by reduced breeding success, a metric that has remained stable in northwest Britain, including North Wales (Morrison *et al.* 2015). This suggested that either habitat condition or climate in Europe is the primary factor, and although little research has been undertaken to determine the mechanism, it is speculated that climate change is driving reduced breeding success. The population declined by 33% across Europe during 1980–2016 (PECBMS 2020) including in Finland, which holds 10% of the population (*European Atlas* 2020). Modelling suggested that by the end of the 21st century, Willow Warblers will have withdrawn northwards and be absent from much of mainland Europe (Huntley *et al.* 2007). We can expect a further significant reduction in its Welsh population, and it may rapidly become only a passage migrant in South and West Wales.

Ian M. Spence

Sponsored by Ian M. Spence

Chiffchaff *Phylloscopus collybita* Siff-siaff

Welsh List Category	IUCN Red List			Birds of Conservation Concern			
	Global	Europe	GB	UK	Wales		
A	LC	LC	LC	Green	2002	2010	2016

The Chiffchaff is largely a summer visitor to most of Europe, although resident around the Mediterranean and western Turkey. It winters in northern Africa and the Sahel, and the Arabian Peninsula (*HBW*). The nominate subspecies, *P.c. collybita*, is a summer visitor to Wales. Two other subspecies that breed to the east of Britain—*P.c. abietinus* and *P.c. tristis*—are occasional visitors (IOC). It prefers fairly open and mature woodland, both deciduous and coniferous, with a medium to tall understorey, and also breeds in parks, gardens or hedges with tall trees. It occurs within and at the edge of woodland, except where Willow Warbler breeding density is high (*BWP*).

Authors at the turn of the 20th century wrote that Chiffchaff was common across Wales, although Forrest (1907) believed it to be less numerous than in English counties adjacent to North Wales and he reported it was scarcer around Bala in Meirionnydd, on Llŷn in Caernarfonshire and in north Anglesey, where there were few mature woodlands. Chiffchaff was considered more numerous than Willow Warbler in Pembrokeshire, and Mathew (1894) estimated that there were 11 for every Willow Warbler. Only in Radnorshire does it appear to have been relatively scarce (Ingram and Salmon 1955). Jennings (2014) speculated that this may have been a result of overgrazing by livestock, but reported that numbers had subsequently increased, benefitting from ground cover in stock-free areas.

Chiffchaff: Recovery locations of birds ringed in Wales are shown by red circles. Ringing locations of birds that were recovered in Wales are shown by blue triangles. Small brown dots show ringing or recovery locations in Wales.

The *Britain and Ireland Atlas 1968–72* showed that Chiffchaff bred across most of Wales, except in north Meirionnydd, a distribution that was largely unchanged in 1988–91, when it was found in around 94% of 10-km squares. The areas of highest abundance in the second period were along the north mainland coast, the southern tip of Llŷn and much of Pembrokeshire, western Ceredigion and Carmarthenshire. The *Britain and Ireland Atlas 2007–11* showed that occupation had increased to 98% of squares and relative abundance had changed little. Comparable tetrad atlases undertaken in three counties also showed a slight increase in distribution over time. The number of occupied tetrads increased from 85% in the *Gwent Atlas 1981–85* to 96% in 1998–2003, which was thought also to reflect an increase in abundance. The East Glamorgan atlases showed a smaller increase, with breeding evidence in 70% of tetrads in 1984–89 and 73% in 2008–11. In Pembrokeshire, the increase was from 80% of tetrads in 1984–88 to 87% in 2003–07 (*Pembrokeshire Avifauna* 2020). The *North Wales Atlas 2008–12* showed an increase of 7% in distribution at the 10-km level since 1968–72, and 75% of tetrads were occupied.

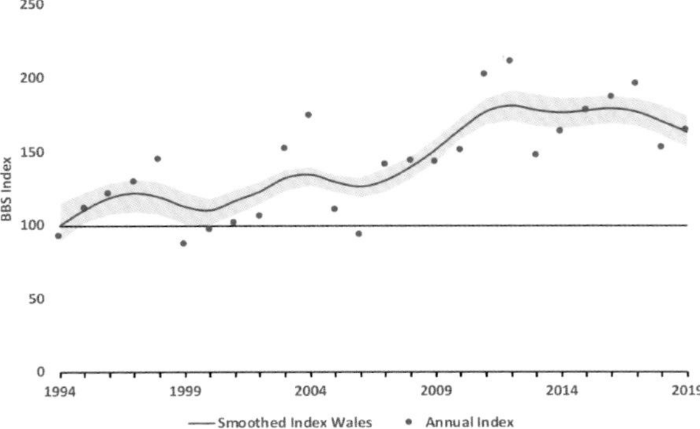

Chiffchaff: Breeding Bird Survey indices, 1994–2019. The red line shows the smoothed index, 1995–2018. The shaded area indicates the 85% confidence limits.

County	Details	Source
Gwent	Since about 1968, almost annual	Venables *et al.* (2008)
Glamorgan	December 1947 and annual since 1967	Hurford and Lansdown (1995)
Pembrokeshire	Small numbers annual from late 19th century	Mathew (1894)
Ceredigion	Aberporth, 22 January 1974	Roderick and Davis (2010)
Breconshire	Tretower, 10 January 1967	Massey (1976a)
Meirionnydd	Aberdyfi, winter 1931–32	Pritchard (2017)
Caernarfonshire	Llandudno, 15 February until mid-March 1913	Forrest (1919)

Chiffchaff: First records of overwintering birds in selected counties.

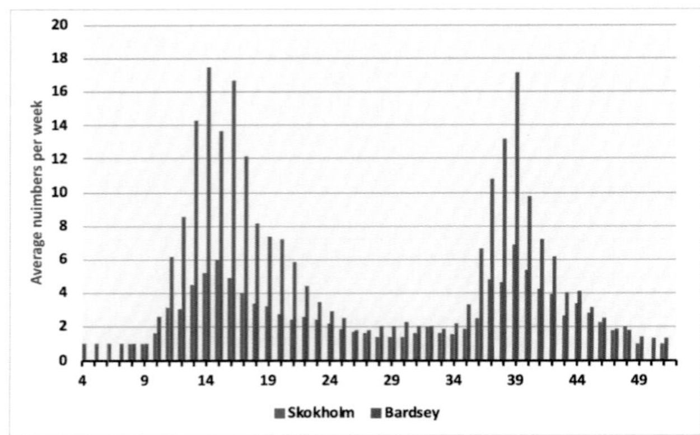

Chiffchaff: Average counts by week number recorded by bird observatories, 1953–2019. Week 14 starts *c.*1 April and week 39 starts *c.*23 September.

Long-term monitoring through the CBC/BBS shows that by 2000, across the UK, Chiffchaffs had recovered from a substantial fall in the early 1970s, a result of drought in their Sahel wintering area, and the index doubled its turn-of-the-century level by 2018 (Massimino *et al.* 2019). Surveys of Welsh broadleaved woodland found that the density of territories increased by 76% between 1981/88 and 2003–04 (Amar *et al.* 2006). The BBS index for Wales increased by 64% during 1995–2018, much of the growth prior to 2012 (Harris *et al.* 2020). Reduced frequency of drought in sub-Saharan Africa since the mid-1980s has probably aided survival. The Welsh breeding population in 2018 was estimated to be 245,000 territories (Hughes *et al.* 2020).

Migration on the islands peaks in the second half of September, several weeks later than that of Willow Warblers. Chiffchaffs move from their natal areas first towards the south and southeast of England and then to northern France and the Low Countries, before heading southwest to the Atlantic coasts of Spain, Portugal and Morocco. Passage in Wales continues through October into early November, when it merges with those that winter in Britain. Some cross the Sahara to winter in West Africa, from Mauretania to Guinea-Bissau.

Wintering in Wales occurred with increasing frequency during the 20th century, particularly towards the southwest, where the climate is more temperate. It had been reported over winter in Pembrokeshire from at least the late 19th century (Mathew 1894). The *Britain and Ireland Winter Atlas 1981–84* showed a small number of records, mostly singles, on Anglesey, south Llŷn and along the south coast. By 2007–11, wintering Chiffchaff were more numerous and widespread, with records in the lowlands of all coastal counties. Only five ringed birds have been recovered in Wales in winter, but they demonstrate a variety of migration strategies: one remained at St Margaret's, Pembrokeshire, from August to November 1976; two were ringed elsewhere in Britain/Isle of Man in spring and found in Wales the following winter; one was ringed in Belgium in early November and recovered in Denbighshire 22 days later; and one was ringed in Cornwall in October and found in Gower the following January.

The dates of birds over-wintering in Wales and the first arrivals on spring passage may overlap in late February and early March. One on Bardsey, Caernarfonshire, on 25 February 2019, for example, had evidence of pollen around the base of the bill, an indicator that it had probably wintered overseas. Monitoring there and at Skokholm, Pembrokeshire, shows that migration picks up in the second week of March, and peaks in the last week of the month and the first week of April. Movements are earlier and on a larger scale on Bardsey than on Skokholm. Birds make a two-step migration via southeast England. The average date of arrival in the UK advanced by two weeks between 1962–66 and 2002–11 (Newson *et al.* 2016). The average laying date has similarly advanced, by 11 days since 1967 (Massimino *et al.* 2019).

The Chiffchaff breeding population is doing well, driven by increased over-winter survival (Johnston *et al.* 2016). Modelling suggests that the climatic range of Chiffchaff in Britain and Ireland is unlikely to change during the 21st century (Huntley *et al.* 2007), although changes in woodland management could cause local reductions, especially where grazing is introduced to benefit woodland specialists such as Pied Flycatcher and Redstart.

Siberian Chiffchaff *P. collybita tristis*

Siff-siaff Siberia

This subspecies breeds across Russia, as far west as the White Sea, and south to northern Mongolia. It winters across Arabia and east to Myanmar (IOC). Records were monitored by the WBRC from 2007–18 to provide an understanding of the frequency of its occurrence in Wales. It is now evident that Siberian Chiffchaff occurs regularly on autumn passage and it has been suggested that southern Britain is becoming an alternative wintering area for part of the population. An average of 248 was recorded each year in Britain during 2010–18 (White and Kehoe 2020b).

Total	1995–99	2000–09	2010–19
89	10	20	59

Siberian Chiffchaff: Individuals recorded in Wales by decade.

Siberian Chiffchaffs are late autumn migrants, most arriving in late October and November, but some have remained over winter, hence the discovery of individuals in January and February. There were 89 Welsh records during 1995–2019, most found along the north and south coasts of Wales with just a handful inland. Over one-third of the records during 2007–19 were from Bardsey. A small number of spring migrants have also been recorded, mainly on Bardsey, which are presumed to be wintering individuals moving north.

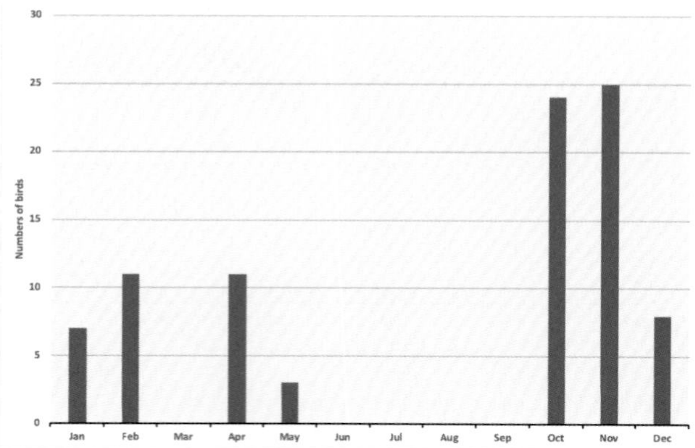

Siberian Chiffchaff: Totals by month of arrival, 1995–2019.

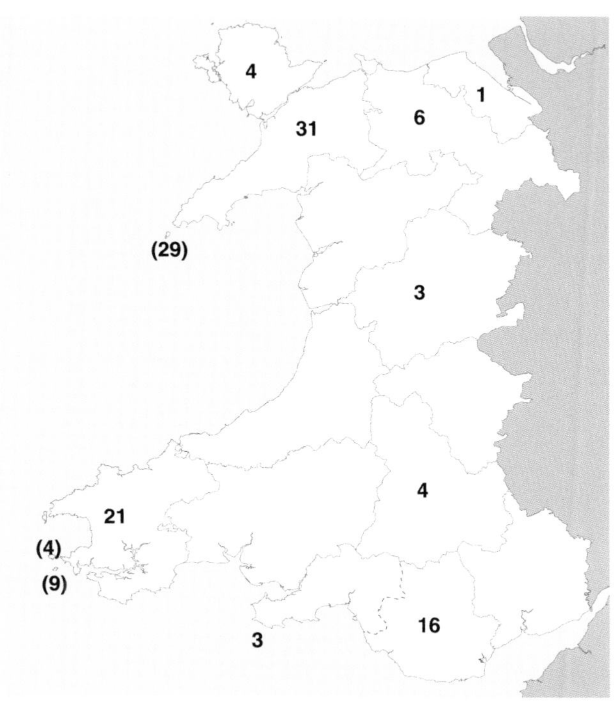

Siberian Chiffchaff: Numbers of birds in each recording area, 1995–2019.

Scandinavian Chiffchaff *P. collybita abietinus*
Siff-siaff Llychlyn

This subspecies breeds across northern Fennoscandia, Belarus and Russia as far as the Ural Mountains and winters around the eastern Mediterranean, East Africa and the Middle East (IOC). However, identification is far from easy (e.g. Dean and Svensson 2005) and there are few ringing records. From the dates involved, a bird ringed on Bardsey on 19 April 1984 and subsequently recaught in Finland on 27 June that year could only really have been a Scandinavian Chiffchaff. Mitochondrial DNA analyses of 'eastern-plumaged' Chiffchaffs caught by ringers in winter and during passage migration in The Netherlands and Britain (de Knijff *et al.* 2012, Collinson *et al.* 2018) found that relatively few could be confirmed as *P.c. abietinus*. Collinson *et al.* (2018) suggested that birds of this form are simply 'lost' among the passage of nominate birds, from which they may not stand out sufficiently. As a result, the status of Scandinavian Chiffchaff in Wales remains unclear.

Ian M. Spence, Jon Green and Robin Sandham

Sponsored by Ian M. Spence

Iberian Chiffchaff — *Phylloscopus ibericus* — Siff-siaff Iberia

Welsh List Category	IUCN Red List		Rarity recording
	Global	Europe	
A	LC	LC	RBBP, BBRC

The breeding range of Iberian Chiffchaff is limited to Portugal, northern Spain west of the Pyrenees and the very southwest of France, with pockets in southern Spain and along the coast of northwest Africa, where its preferred habitat is riverine woodland (Copete 2008). Its wintering range is poorly known, but isotope analysis shows that it is likely to be in sub-Saharan Africa and similar to that of Willow Warbler (de la Hera *et al.* 2020). There have been 80 British records, on average six each year (Holt *et al.* 2020), with heightened awareness by birders since its treatment as a separate species by the British Ornithologists' Union (BOURC) in 1998. Most records in Britain involved singing males that held territory for several weeks. Based on recent trends, we might expect more in the future. There have been four records in Wales, including the only breeding record in Britain:

- Wentwood, Gwent, from 10 May to 18 June 2010;
- Pwll, Llanelli, Carmarthenshire, from 17 April to 8 July 2013;
- a pair bred, fledging at least seven young, in Gower in 2015, birds being present from 15 May to 5 July and
- Gwydir Forest, Caernarfonshire, on 16–18 April 2019.

Jon Green and Robin Sandham

Greenish Warbler — *Phylloscopus trochiloides* — Telor Gwyrdd

Welsh List Category	IUCN Red List		Rarity recording
	Global	Europe	
A	LC	LC	WBRC

In Europe, Greenish Warblers breed in a variety of woodland types in countries to the south and east of the Baltic Sea, extending east through Russia as far as central China. It winters in South Asia, east to northern Thailand (*HBW*). Of the 754 records in Britain up to the end of 2018, almost 30% were during 2010–18, an average of 25 each year (White and Kehoe 2020b). Numbers have increased steadily since the 1980s, with breeding reported more regularly in Fennoscandia and Germany, but the recent atlas of Russian Breeding Birds reported that the expansion there during the 20th century has now stopped (Kalyakin and Volzit 2020).

Total	Pre-1950	1950–59	1960–69	1970–79	1980–89	1990–99	2000–09	2010–19
34	0	2	2	1	2	6	1	20

Greenish Warbler: Individuals recorded in Wales by decade.

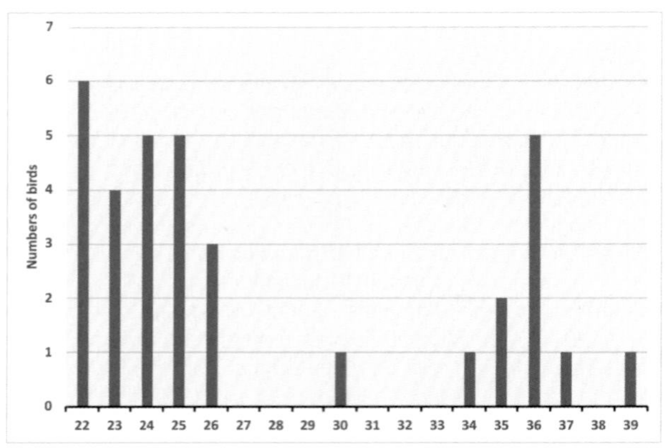

Greenish Warbler: Totals by week of arrival, 1954–2019. Week 22 begins *c.*28 May and week 36 begins *c.*3 September.

It is a rare visitor to Wales, with 34 records, the first of which was on Bardsey, Caernarfonshire, on 16–17 June 1954. All were singles, except two together on Bardsey in May 2016. More were seen during 2010–19 than in all the previous decades combined. Of the total, 23 (67%) were recorded in spring, but it is a relatively late migrant with the peak during the first week of June. All other records were from the last week of August and September, except for one on Skomer on 23 July 2015. Almost all Welsh records have come from the islands: 19 on Bardsey, and in Pembrokeshire seven on Skokholm and six on Skomer. There have been only two mainland records: on the Great Orme, Caernarfonshire, on 29 August 2012 and at Martin's Haven, Pembrokeshire, on 24 September 2017.

Jon Green and Robin Sandham

Arctic Warbler — *Phylloscopus borealis* — Telor yr Arctig

Welsh List Category	IUCN Red List		Rarity recording
	Global	Europe	
A	LC	LC	WBRC

This species breeds in the birch forests of northern Fennoscandia across northern Asia and into western Alaska, and winters in Indonesia (*HBW*). Of 410 records in Britain, almost two-thirds were on Shetland or Orkney (Holt *et al.* 2019). There has been just a single record in Wales, on Bardsey, Caernarfonshire, on 13 September 1968. Its breeding range in Europe is expected to contract eastwards during the 21st century, almost entirely out of Scandinavia (Huntley *et al.* 2007). Given its southeast autumn migration, it is likely to remain a very rare visitor to Wales.

Jon Green and Robin Sandham

Great Reed Warbler — *Acrocephalus arundinaceus* — Telor Cyrs Mawr

Welsh List Category	IUCN Red List		Rarity recording
	Global	Europe	
A	LC	LC	BBRC

The breeding range of this reedbed-dwelling summer visitor is across much of continental Europe and Fennoscandia, extending east to Mongolia. It winters in tropical and southern Africa (*HBW*). There were 308 Great Reed Warblers recorded in Britain to 2019, around six each year during 1990–2019 (Holt *et al.* 2020). However, it is a very rare spring migrant in Wales, with 11 records:

- Skokholm, Pembrokeshire, on 13–14 May 1967;
- Skokholm, Pembrokeshire, on 11 May 1970;
- Bardsey, Caernarfonshire, on 28–29 May 1976;
- Penmaenpool, Meirionnydd, on 30 July 1978;
- Bardsey, Caernarfonshire, on 8 June 1979;
- Llangynog, Montgomeryshire, on 15 June 1988;
- Bardsey, Caernarfonshire, on 22 May 1991;
- Skomer, Pembrokeshire, on 21 May 1998;
- Skomer, Pembrokeshire, on 16 May 2002;
- RSPB Conwy, Denbighshire, on 10–18 June 2005 and
- RSPB Cors Ddyga, Anglesey, on 15 June 2010.

Its range is patchy in western Europe, but it does breed up to the Channel coast in northeast France and is a potential breeding species in Britain in the 21st century (Huntley *et al.* 2007), although its occurrence in Wales will be constrained by the extent of suitable nesting habitat.

Jon Green and Robin Sandham

Aquatic Warbler — *Acrocephalus paludicola* — Telor Dŵr

Welsh List Category	IUCN Red List		Birds of Conservation Concern	Rarity recording
	Global	Europe	UK	
A	VU	VU	Red	WBRC

The global population of this species is in serious decline as a result of the destruction of the wet sedge meadows in which it breeds, which has been under way since the second half of the 19th century. It breeds locally in eastern Europe and Central Asia and undertakes an anti-clockwise migration circuit, which brings it west from its breeding areas in early autumn, and on to winter in north and tropical Africa (*HBW*). Around ten are seen annually in Britain, but numbers have fallen around 80% since 1990 (Holt *et al.* 2017). It is a scarce migrant in Wales.

Total	Pre-1950	1950–59	1960–69	1970–79	1980–89	1990–99	2000–09	2010–19
84	1	0	2	13	15	35	15	4

Aquatic Warbler: Individuals recorded in Wales by decade.

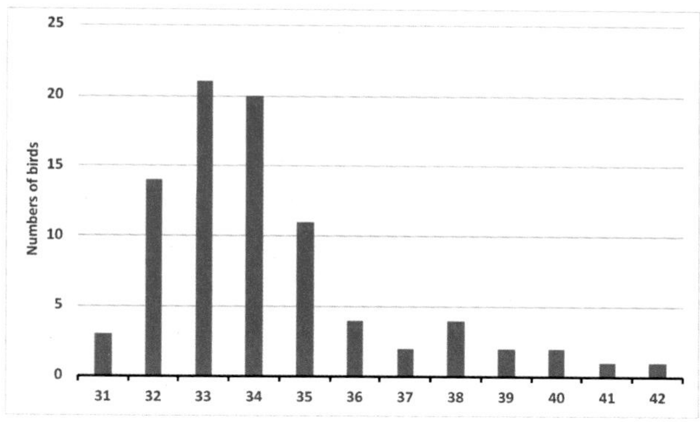

Aquatic Warbler: Totals by week of arrival, 1949–2019. Week 32 begins *c.*6 August and week 38 begins *c.*17 September.

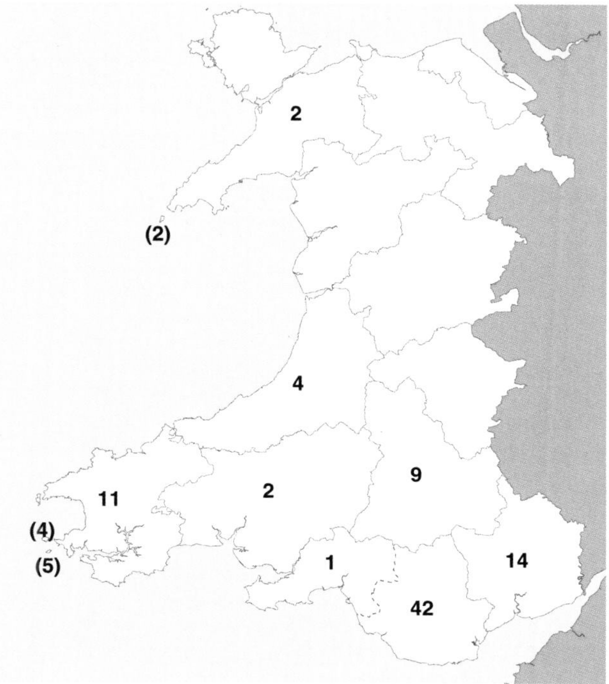

Aquatic Warbler: Numbers of birds in each recording area, 1949–2019.

Aquatic Warbler ringed at Llangorse Lake, Breconshire, August 2019 © Carlton Parry

The first Welsh record was on Skokholm, Pembrokeshire, on 5 September 1949. There were only two further records before the 1970s, which saw the start of an increase in occurrence that peaked in the 1990s. The increase reflected a more sustained ringing effort in reedbeds, but that effort has remained fairly constant in the last 30 years, so the significant decline across Britain reflected continued declines in the population, at least at the western end of its breeding range. The numbers in Wales in 2010–19 were only 8% of the average recorded during 1990–2009.

The majority of records resulted from ringing activity in reedbeds during August, in particular at Kenfig, East Glamorgan, where 39 have been caught, including four on 31 August 1989. There have been 12 at Newport Wetlands Reserve (Uskmouth), Gwent, ten at Llangorse Lake, Breconshire, and five at Teifi Marshes, Pembrokeshire/Ceredigion. Although the first was found on Skokholm, the islands accounted for very few, in contrast to most other scarce warblers. A significant reversal of its fortunes in eastern Europe is necessary for the future of Aquatic Warbler. Hope comes from Lithuania where a conservation programme, including translocation of birds from Belarus, resulted in an increase from 50 birds in 2013 to 316 in 2020 (Griniene 2020).

Jon Green and Robin Sandham

Sponsored by Carlton Parry and Llangorse Ringing Group

Sedge Warbler *Acrocephalus schoenobaenus* Telor Hesg

Welsh List Category	IUCN Red List			Birds of Conservation Concern			
	Global	Europe	GB	UK	Wales		
A	LC	LC	LC	Green	2002	2010	2016

The Sedge Warbler is a summer visitor to lowland Europe north of the Mediterranean and east to Central Asia and the West Siberian Plain. It winters in Africa from the Sahel region to northern South Africa (*HBW*). It inhabits low vegetation, frequently near water and usually in the lowlands, but at up to 330m near Bronbannog in Clocaenog Forest, Denbighshire. It is more widespread than Reed Warbler, nesting low to the ground in the drier margins of reedbeds or other dense vegetation, often with willows. It can attain high breeding densities (e.g. 300–600 pairs/km^2) in prime damp habitat (*European Atlas* 1997). The western European population declined from about 1970, adversely affected by severe droughts in the Sahel that reduced over-winter survival (Peach *et al.* 1991). Its recovery in Britain stalled following further drought in the early 1980s. Although it had recovered most of its losses by 2000, numbers in Britain have since fallen again and declined across Europe by 29% during 1980–2016 (PECBMS 2020).

Prior to extensive drainage of wetlands, it seems likely that Sedge Warbler would have been widespread in lowland Wales. Authors around the turn of the 20th century described it as abundant in suitable habitat, but that was inevitably localised in North and mid-Wales, except around wetlands on Anglesey and in Ceredigion at Cors Fochno and Cors Caron (Dobie 1894, Salter 1895, Forrest 1907). In Pembrokeshire, Mathew (1894) considered that it was nearly as abundant as Chiffchaff, the most common warbler species in the county at that time. In Glamorgan, it had declined by the mid-1920s (per Hurford and Lansdown 1995).

The *Britain and Ireland Atlas 1968–72* showed that Sedge Warblers bred across the Welsh lowlands, occurring in 66% of 10-km squares. The 1988–91 *Atlas* showed a reduction in the number of occupied squares, to 57%, following drought in the Sahel, and that abundance was relatively low away from hotspots on the northeast coast and in southwest Pembrokeshire. The

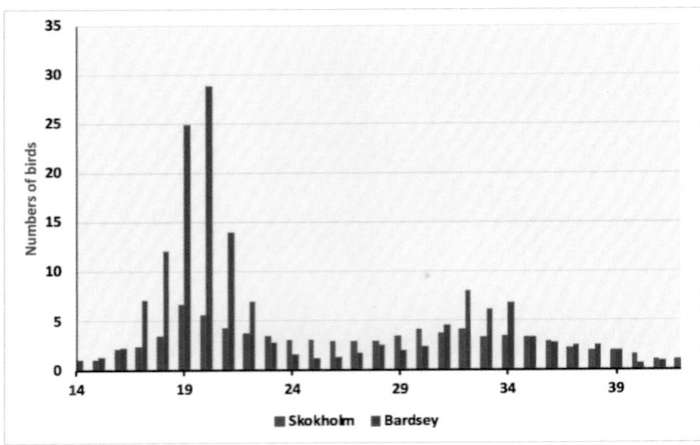

Sedge Warbler: Average counts by week number recorded by bird observatories, 1953–2019. Week 20 starts *c.*13 May and week 34 starts *c.*19 August.

range had expanded once again by the *Britain and Ireland Atlas 2007–11* to occupy 79% of 10-km squares, with most of the gains in the valleys of mid-Wales. The areas of relatively high abundance, on a par with the highest densities elsewhere in Britain, were on Anglesey, Carmarthenshire's Tywi Valley and northwest Pembrokeshire.

Tetrad atlases reflected the availability of suitable habitat, generally following the soft coastal fringe and river valleys. The area occupied in Gwent increased from 11% of tetrads in 1981–85 to 15% in 1998–2003, while the proportion with breeding evidence in the late 1980s and the first decade of the 21st century was unchanged in East Glamorgan and Pembrokeshire, at around 6% and 29% of tetrads respectively. In Gower, 20% of tetrads were occupied during the *West Glamorgan Atlas 1984–89*, mainly in the Neath-Port Talbot area. It was present in 21% of tetrads in the *North Wales Atlas 2008–12*, being most widespread on Anglesey and western Llŷn, Caernarfonshire.

The main period of spring arrival is in late April and peaks in mid-May but can be far earlier. During 2000–19, all counties had records in the first half of April, some even in March in the southern half of Wales, such as on the Afon Ogwr, East Glamorgan, on 18 March 2016, and by the upper Dyfi Estuary, Montgomeryshire, on 19 March 2015. Post-breeding Sedge Warblers undertake two movements between late July and September. The first is a relatively short distance to reedbeds in southern Wales, England, or northwestern France where they fatten up, often doubling in weight, before a non-stop flight to the Sahel (*BTO Migration Atlas* 2002). Autumn passage on the island bird observatories is protracted, with a much flatter peak than in spring and continuing through September.

Irish-breeding Sedge Warblers move along the western fringe of Wales in spring, before crossing the Irish Sea, but their autumn passage is on a broader front through South Wales, with only two ringing recoveries in North Wales, both on Bardsey. Ringers in South Wales report high numbers in autumn, such as 101 ringed at Newport Wetlands Reserve (Uskmouth), Gwent, on 5 August 2016, 82 on 8 August 2016 and 71 on 4 August 2016 (Richard M. Clarke *in litt.*). Small numbers continue on passage through September, and there were records in the second half of October in Gwent and East Glamorgan during 2000–19. One at Malltraeth Cob, Anglesey, on 19 November 2005, was exceptionally late. The few recoveries of Sedge Warblers from Wales in sub-Saharan Africa have come from Senegal, but these were mostly ringing recoveries and it is likely that they also winter elsewhere in West Africa.

Sedge Warbler: Recovery locations of birds ringed in Wales are shown by red circles. Ringing locations of birds that were recovered in Wales are shown by blue triangles. Small brown dots show ringing or recovery locations in Wales.

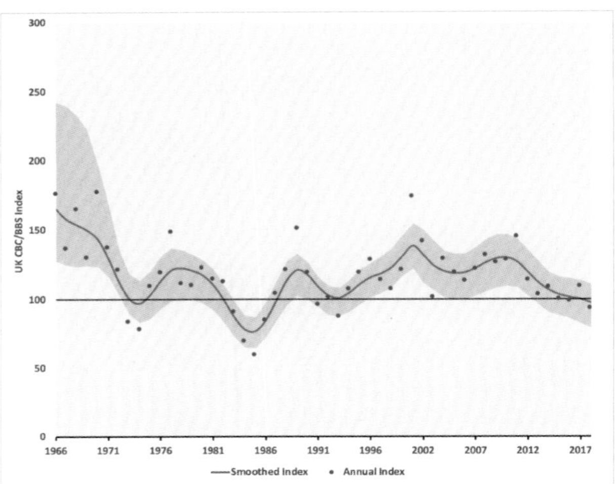

Sedge Warbler: Common Birds Census and Breeding Bird Survey indices for the UK, 1966–2019 (Massimino *et al.* 2019). The grey line shows the smoothed index 1967–2018 and the shaded area indicates the 85% confidence limits.

There are no data on breeding trends in Wales, but the BBS showed that the UK population declined by 24% during 2008–18 (Harris *et al.* 2020). The Welsh breeding population in 2018 was estimated to be 19,500 (17,500–21,500) territories (Hughes *et al.* 2020), although this is probably on the high side. Most breeding parameters monitored have changed little in the last 30–50 years, but the average egg-laying date is now ten days earlier than in the late 1960s. The ratio of juveniles to adults recorded by ringers at Constant Effort Sites halved during 1984–2017 (Massimino *et al.* 2019). Modelling suggested that by the late 21st century, the climate in France, much of Germany and parts of southeast England will no longer be suitable for Sedge Warblers (Huntley *et al.* 2007), so in the longer term, South Wales may also become less suitable. However, this modelling did not account for the frequency of severe drought in the Sahel, which has the strongest influence on the population. The species' fate in Wales will depend largely on that. In the meantime, the protection of marginal scrub and creation of damp and wet features on Welsh lowland farms, as well as large-scale wetland restoration, would provide suitable foraging habitat to increase the resilience of Sedge Warblers.

Ian M. Spence

Sponsored by Goldcliff Ringing Group

Paddyfield Warbler *Acrocephalus agricola* Telor Padi

Welsh List Category	IUCN Red List		Rarity recording
	Global	Europe	
A	LC	LC	BBRC

The western edge of the breeding range of this reedbed dweller is around the northern edge of the Black Sea, although it has bred as close as The Netherlands, but the major part of the population is farther east through Central Asia to western China. Western populations winter in southeast Iran, Pakistan and northwest India (*HBW*). There were 118 British records to the end of 2019, with an increase to 3–4 a year during 1990–2019. The majority of records, around 70%, have been in autumn, but late spring records are becoming less remarkable, and breeding has now been confirmed in Finland and The Netherlands (Holt *et al.* 2020). However, it is a very rare vagrant to Wales, there having been just four records. The first was caught and ringed at Llangorse Lake, Breconshire, on 11 September 2004, after which all the records have been on Bardsey, Caernarfonshire: on 11 October 2008, 17 September 2009 and 7 June 2013.

Paddyfield Warbler on Bardsey, Caernarfonshire, October 2008 © Steve Stansfield

Jon Green and Robin Sandham

Sponsored by Jerry Lewis and Peter Jenkins

Blyth's Reed Warbler *Acrocephalus dumetorum* Telor Cyrs Blyth

Welsh List Category	IUCN Red List		Rarity recording
	Global	Europe	
A	LC	LC	WBRC

It is not long ago that Blyth's Reed Warbler was a mega-rarity in Britain. There were no records in Britain during 1928–79, but a surge in the number of records from the late 20th century resulted in this species no longer being considered a British rarity, although it remains rare in Wales. It breeds in scrub within forests or along watercourses from northeast Europe through Belarus and Russia to northwest Mongolia. It winters in India, Bangladesh and Myanmar (*HBW*). The population has expanded westwards in the last century, with breeding recorded in Estonia from 1938, where there is now a population of over 60,000 pairs, and there has been a similar rate of increase in Finland since the first breeding record in 1947 (BirdLife International 2015). Since first breeding in Sweden in 1970, it is now widespread, especially in the south, and has extended its range along the southern coast of Norway since 2000 (*European Atlas* 2020).

Of the 261 recorded in Britain, 62% were during 2010–18, averaging 18 each year during that decade (White and Kehoe 2020b). Its increased occurrence in western Europe is a result of birds overshooting its enlarged breeding range during spring migration. Autumn records in late September/October, which account for almost half the Welsh total, are believed to have originated farther east in its range, as the Baltic population departs by the end of August (Hudson *et al.* 2008). It remains a rare vagrant in Wales, with nine records:

- Bardsey, Caernarfonshire, on 13 October 2001;
- Skokholm, Pembrokeshire, on 27 September 2013;
- Skomer, Pembrokeshire, on 26 May 2014;
- Bardsey, Caernarfonshire, on 7 June 2014;
- Bardsey, Caernarfonshire, from 30 September to 1 October 2016;
- Bardsey, Caernarfonshire, on 25–27 May 2018;
- Bardsey, Caernarfonshire, on 6 June 2018;
- Cemlyn, Anglesey, on 15 June 2019 and
- Porth Eilian, Anglesey, on 27 October 2019.

On the evidence of the recent trend, it will undoubtedly be recorded more frequently.

Jon Green and Robin Sandham

Reed Warbler *Acrocephalus scirpaceus* Telor Cyrs

Welsh List Category	IUCN Red List			Birds of Conservation Concern			
	Global	Europe	GB	UK	Wales		
A	LC	LC	LC	Green	2002	2010	2016

The Reed Warbler breeds through much of Europe and Central Asia, although it is patchily distributed around the Mediterranean basin and in Turkey and the Middle East. It winters in sub-Saharan Africa, mainly south of the Sahel (*HBW*). Its distribution is limited by its need to nest in mature reeds that have strong stems taller than 1m. Reedbeds can be small or in narrow margins of lakes or watercourses, but those sheltered from wave action are favoured and in Wales such habitat is below 200m. Its range in Europe expanded north from the 1940s and into Ireland in 1981, where a population of 50–150 pairs occurs on the east and south coasts, and in Northern Ireland (*Birds in Ireland* 1989, IRBBP 2020). Its range has continued to extend northwards in Britain, Ireland and Finland, but it has declined in some other countries (*European Atlas* 2020).

Around the turn of the 20th century, it was a localised breeding species in Wales and appears to have become even more sporadic until the 1950s, in some areas until the 1970s. In the north, the main population was in east Denbighshire, close to its regional stronghold on the Shropshire meres. Forrest (1907) reported that the numerous small lakes in North Wales did not have reedbeds and that reeds along embankments or ditches were not used, hence Reed Warbler was unknown on Anglesey and scarce in Caernarfonshire. In Meirionnydd, it was found locally around Fairbourne and probably bred at Arthog before 1907

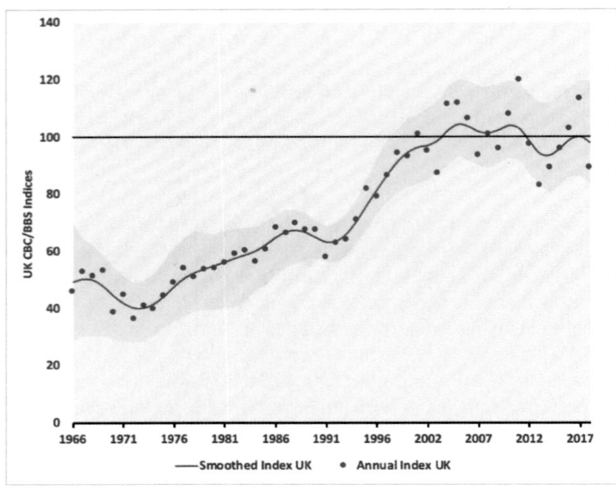

Reed Warbler: Common Birds Census and Breeding Bird Survey indices for the UK, 1966–2018 (Massimino *et al.* 2019). The grey line shows the smoothed index 1967–2017 and the shaded area indicates the 85% confidence limits.

(Jones 1974a) but not again in the county until the mid-1970s. In Montgomeryshire, a pair bred in 1900, but the next nesting record was not until 1955 (Holt and Williams 2008), while in Breconshire it has bred at Llangorse Lake since at least the late 19th century (Phillips 1899). It appears always to have been rare in Radnorshire, but bred at Rhosgoch Bog in 1942–63, until habitat succession made the area unsuitable, and sporadically at Llanbwchllyn from the early years of the 20th century and annually after 1983 (Jennings 2014).

In Gwent, it was scarce and local in the 1930s and early 1960s (Venables *et al.* 2008). In Glamorgan, it was known at just two sites by the 1950s (Heathcote *et al.* 1967), both of which were subsequently destroyed by development. It was rare in Gower until the 1950s, when dramatic increases were recorded at Oxwich Marsh as the reedbed became more extensive (Hurford and Lansdown 1995). In Carmarthenshire, it may have been in the lower Gwendraeth Valley in the 1950s, but the first dated record was not until 1971, when two pairs bred at Trostre (Lloyd *et al.* 2015). Reed Warbler was not recorded in Pembrokeshire by Mathew (1894), and there were only four records between the first at The Smalls lighthouse on 17 October 1908 and a review by Lockley *et al.* (1949). There were claims of nesting at Goodwick Moor in 1934 or 1935 and possibly at Tenby Marsh in 1931, but the first confirmed breeding in Pembrokeshire was at Nevern in 1975 (Donovan and Rees 1994). In Ceredigion, a nest was reportedly found at Clarach Bog in the 1920s, but the next was not until 1974 near RSPB Ynys-hir (Roderick and Davis 2010). There were records on Anglesey from the 1950s. The first nest was found in 1968 (Jones and Whalley 2004), but the first credited records in Caernarfonshire were not until the early 1970s.

The *Britain and Ireland Atlas 1968–72* showed its very limited distribution in Wales, occupying just 32 of the 10-km squares (12% of the total) and presumably present in only a few tetrads in each. There was breeding evidence around the Conwy Estuary and western Anglesey in the north, a few sites in Carmarthenshire, Gower, the coastal strip of Glamorgan and Gwent and a handful of sites in the South Wales valleys. The atlas period saw colonisation of a few places farther inland, such as Lymore, Montgomeryshire. Occupation had almost doubled to 59 squares (22%) by the *Britain and Ireland Atlas 1988–91*, with considerable expansion around the Dee Estuary, Flintshire, the coast of Llŷn and Conwy in Caernarfonshire, much of Anglesey, the coast of Meirionnydd, Teifi Marshes, Pembrokeshire/Ceredigion, and along the south coast. Abundance was relatively low in Wales, except in parts of Gower and the coast of Gwent. The restoration, creation and improved management of reedbeds for wildlife aided the expansion of Reed Warblers, such as at RSPB Conwy in Denbighshire, RSPB Cors Ddyga and Valley Wetlands on Anglesey, Llyn Coed-y-Dinas and Dolydd Hafren in Montgomeryshire and Newport Wetlands, Gwent. The *Britain and Ireland Atlas 2007–11* showed it had expanded further to 108 squares (41%), with breeding records in every county at low density but greater abundance in eastern Flintshire and Denbighshire, Anglesey and along the coast of South Wales.

The localised, habitat-restricted nature of Reed Warblers is evident at the tetrad scale of atlases. In the *North Wales Atlas 2008–12*, it was recorded in just 5% of tetrads, scattered along the coast and across Anglesey, with an inland record at Llyn Tegid, Meirionnydd. There were no records from Fenn's or Bettisfield Mosses on the Denbighshire/Shropshire border, in contrast to Forrest's account from a century previously. In the south, the breeding evidence was more widespread in the most recent atlases: 12% of tetrads in Gwent in 1998–2003, 8% in East Glamorgan in 2008–11, 12% in Gower in the *West Glamorgan Atlas 1984–89* and 6% in Pembrokeshire in 2003–07. In the three counties where these surveys repeated earlier atlases in the 1980s, there had been a significant increase in range, including a spread beyond the coast and up the river systems, notably in East Glamorgan.

The distribution of Reed Warbler now includes all the reedbeds in Wales listed by Tyler (1995), primarily on Anglesey and around the coast. The only consistent and significant breeding population inland is 70–90 territories at Llangorse Lake, Breconshire. In the last 50 years, Reed Warbler has gone from being a passage migrant, encountered infrequently in Wales, to a widespread breeder, albeit limited by the availability of quality reedbed. The breeding population in 2018 was estimated to be 11,000 (8,700–13,500) pairs (Hughes *et al.* 2020), although this may be too high. There is no population trend data specifically for Wales and even the trend at a UK level is uncertain: the CBC/BBS index shows that, despite a fall in the early 1970s following drought in the Sahel, the Reed Warbler population doubled between 1967 and 2018, but ringers at BTO Constant Effort Sites found a 27% decline in the number of adults during 1984–2017 (Massimino *et al.* 2019).

Reed Warblers generally arrive from mid-April and go directly to potential breeding sites, but migration can continue through much of May. The average laying date is 10 June, although that has moved forward 11 days since the late 1960s. During 2000–19,

Reed Warbler: Recovery locations of birds ringed in Wales are shown by red circles. Ringing locations of birds that were recovered in Wales are shown by blue triangles. Small brown dots show ringing or recovery locations in Wales.

the Reed Warbler occurred in every county during the first half of April, and much earlier in 2015, when the first was on 20 March at Parc Slip nature reserve, East Glamorgan, and the following day at Newport Wetlands Reserve (Uskmouth), Gwent. Passage movement at the island bird observatories is negligible, unsurprising given the relative lack of breeding numbers to the north and west of Wales until quite recently. Southward passage begins in late July and continues into late September in most years and through October in some. During 2000–19, there were records in the second half of October in five counties and November records in two, including at Oxwich Marsh on 19 November 2016. The initial movement away from Wales is to the north or southwest coast of France, from where birds move to the coast of Portugal and on to Morocco and Senegal. During the winter, birds move to other countries in West Africa (*BTO Migration Atlas* 2002).

The Reed Warbler was a 20th-century success story, expanding its range despite the loss of many wetland reedbeds, especially to industrial development in the coastal lowlands of South Wales. It is, however, very dependent on nature conservation management to maintain reedbeds and create new ones. It benefited significantly from wetland restoration to support Bittern in Wales, such as at Newport Wetlands and RSPB Cors Ddyga, Anglesey. Its breeding success has improved over recent decades. There is some evidence that its expanded distribution is a result of a warming climate, with a longer breeding season that enables more pairs to raise a second brood (Eglington *et al.* 2015). Experiments in Cardiff Bay Wetlands and Cosmeston Lakes, East Glamorgan, showed that supplementary food resulted in an earlier laying date, in more time incubating and a quicker response to the simulated presence of a predator. This suggested a mechanism by which the improvements in breeding productivity may have occurred (Vafidis *et al.* 2016, 2018). Providing its reedbed habitats are maintained and no factors adversely affect the species on migration or in Africa, the outlook is good.

Ian M. Spence

Sponsored by Llangorse Ringing Group

Marsh Warbler *Acrocephalus palustris* Telor Gwerni

Welsh List Category	IUCN Red List			Birds of Conservation Concern	Rarity recording
	Global	Europe	GB	UK	
A	LC	LC	CR	Red	WBRC

A small population of Marsh Warbler breeds in Britain, generally in eastern counties from Kent to Shetland (Eaton *et al.* 2020), but the main breeding grounds are in eastern Europe where it nests in tall, herbaceous vegetation, usually near water (*HBW*). Its European population declined by 19% during 2007–16 (PECBMS 2020). The numbers recorded in Britain have remained stable since 1990, with typically 50 records each year (White and Kehoe 2020b), but there was a 70% decrease in breeding numbers in 1993–2018 (Eaton *et al.* 2020).

Total	Pre-1950	1950–59	1960–69	1970–79	1980–89	1990–99	2000–09	2010–19
28	0	1	0	2	6	4	6	9

Marsh Warbler: Individuals recorded in Wales by decade.

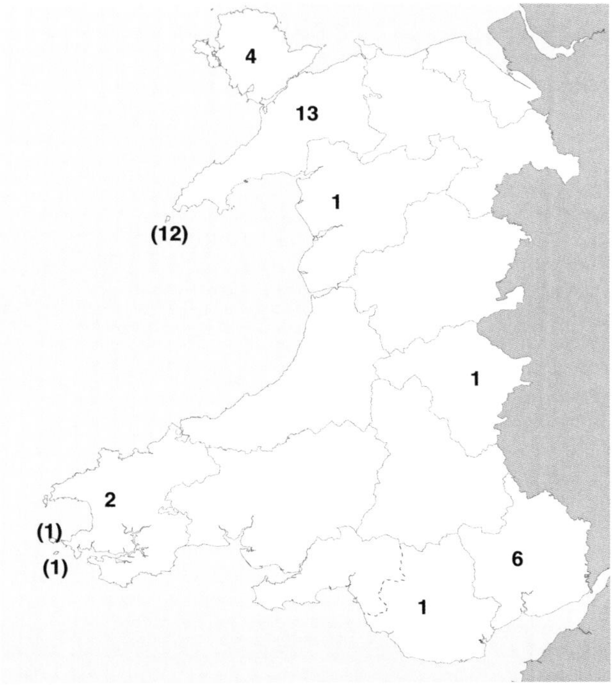

Marsh Warbler: Numbers of birds in each recording area, 1959–2019.

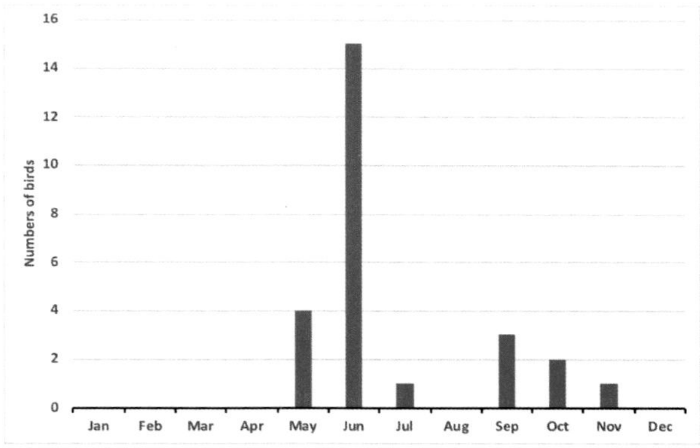

Marsh Warbler: Totals by month of arrival, 1959–2019.

It is a rare visitor to Wales, although the recording rate has increased a little since 2000. The first Welsh record was on Bardsey, Caernarfonshire, on 30 June 1959, closely followed by another there on 2 November that year. A pair that fledged five young at Llanwern, Gwent, in 1972 is the sole Welsh breeding record. An earlier claim of nesting at St Fagans, East Glamorgan, in 1914, has been discredited (Heathcote *et al.* 1967). Of the other 26 Welsh records, 18 involved overshooting spring migrants, with a distinct peak in June. Almost half the total are from Bardsey, where 12 have been recorded. That there have been only single records from the Pembrokeshire islands of Skokholm, in June 2017, and Skomer, in November 2009, is rather surprising. Excluding the five young that fledged in 1972, Gwent has had four records, including a territorial male at Peterstone from 22 June to 10 July 1996. Singing males have also been recorded near Glasbury, Radnorshire, in June–July 1984 and at Llyn Cwellyn, Caernarfonshire, in June 1995. Climate modelling suggested a reduction at the western edge of its range during the 21st century (Huntley *et al.* 2007), so we may well see fewer records in Britain, particularly in the west.

Jon Green and Robin Sandham

Booted Warbler *Iduna caligata* Telor Bacsiog

Welsh List Category	IUCN Red List		Rarity recording
	Global	Europe	
A	LC	LC	BBRC

This species nests in thick vegetation on the arid, open steppes of western Russia and northern Kazakhstan, although a small population has spread into eastern Finland since 2000. It winters in India (*HBW*). There were 172 records in Britain to the end of 2019, an average of five each year during 1990–2019 (Holt *et al.* 2020). Of these, just six have occurred in Wales, all except one on the islands:

- 1CY caught and ringed on Skokholm, Pembrokeshire, on 25–28 September 1993;
- Skomer, Pembrokeshire, on 14–15 September 2000;
- Bardsey, Caernarfonshire, on 30 August 2005;
- 1CY on Ramsey, Pembrokeshire, on 23 September 2013;
- 1CY on Skokholm, Pembrokeshire, on 25 September 2013 and
- Great Orme, Caernarfonshire, on 3–9 October 2016.

Two other records, on Bardsey on 25–26 September 1998 and at Rhossili, Gower, on 25–30 September 2017, were considered to be either Booted or Sykes's Warbler but could not be identified with certainty. Sykes's Warbler *I. rama* was previously treated as a subspecies of Booted Warbler but is now recognised as a separate species. It breeds to the south of Booted Warbler, from central Kazakhstan to the Persian Gulf and winters in India and Sri Lanka, but has not been recorded in Wales.

Jon Green and Robin Sandham

Melodious Warbler *Hippolais polyglotta* Telor Pêr

Welsh List Category	IUCN Red List		Rarity recording
	Global	Europe	
A	LC	LC	WBRC

The Melodious Warbler breeds in open deciduous woodland and scrubby habitats in northwest Africa and western Europe, and winters in Africa, south of the Sahara (*HBW*). In Britain, it occurs more frequently in the south and west and is a scarce migrant to Wales, with almost twice as many records as Icterine Warbler, which tends to occur farther east and north.

Total	Pre-1950	1950–59	1960–69	1970–79	1980–89	1990–99	2000–09	2010–19
222	0	16	51	30	39	27	26	33

Melodious Warbler: Individuals recorded in Wales by decade.

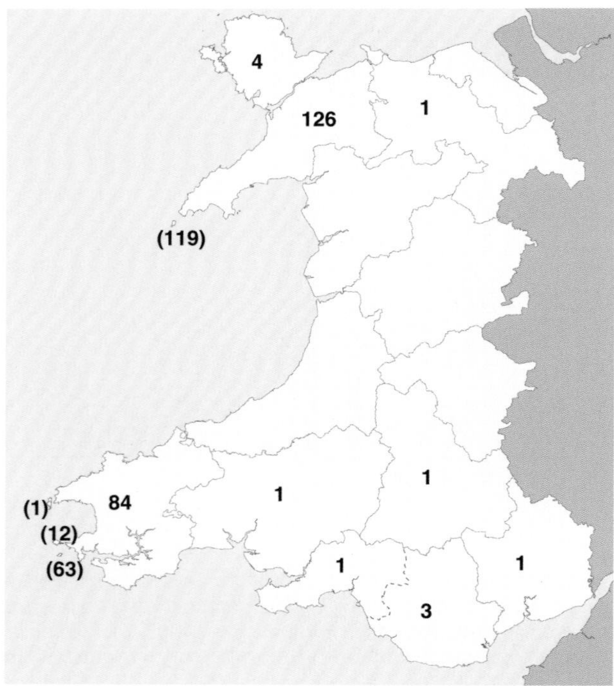

Melodious Warbler: Numbers of birds in each recording area, 1954–2019.

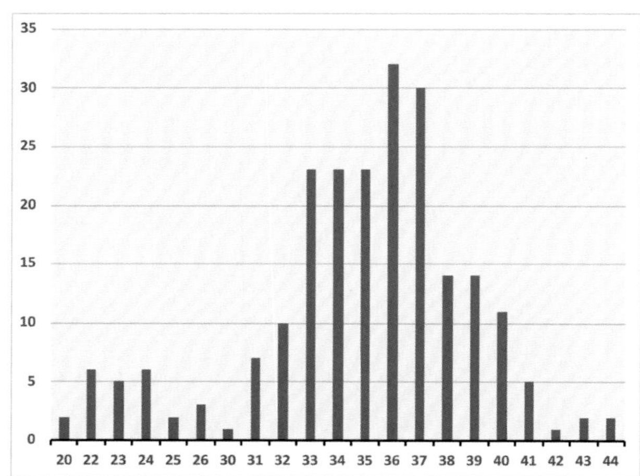

Melodious Warbler: Totals by week of arrival, 1954–2019. Week 33 begins *c.*13 August and week 40 begins *c.*1 October.

The first Welsh record was on Bardsey, Caernarfonshire, on 27 August 1954. The island has now claimed more than half of Welsh records and the Pembrokeshire islands, primarily Skokholm, have recorded one-third of the total. Mainland records are much rarer and almost exclusively coastal. The only truly inland records were at Llangernyw, Denbighshire, in September 1980, and a singing male at Heol Senni, Breconshire, on 21–23 June 2015.

There has been a small peak of spring overshooting migrants at the end of May and into mid-June but the majority were seen in autumn, from August to October. The autumn peak has been in the last week of August and the first week of September. Populations have increased across its breeding range in the last two decades and there has been a marked increase in numbers breeding in France and the Low Countries since the 1990s (BirdLife International 2015). Climate modelling suggested that it will expand further its breeding range from northern France into southern Britain during the 21st century (Huntley *et al.* 2007). There is no sign of that occurring yet, however, and there has been a steady decline in the number of records in Britain since the 1980s (White and Kehoe 2020b).

Jon Green and Robin Sandham

Icterine Warbler *Hippolais icterina* Telor Aur

Welsh List Category	IUCN Red List		Rarity recording
	Global	Europe	
A	LC	LC	WBRC

This species replaces Melodious Warbler in northern and eastern Europe, and consequently occurs less frequently in Wales. It breeds in woodland from northeast France to western Siberia and north to Fennoscandia, and winters in Africa south of the Sahara. Although its range has contracted in the westernmost part of its range in recent years (*HBW*), pairs bred in Scotland in 2002 and 2009. It is a scarce migrant to Wales, where records increased during the second half of the 20th century but have fallen since a peak in the 1990s. These trends mirror those for the rest of Britain, with an annual mean of 139 during 1990–99 but 78 in 2010–18 (White and Kehoe 2020b).

Icterine Warbler on Bardsey, Caernarfonshire, August 2011
© Steve Stansfield

Total	Pre-1950	1950–59	1960–69	1970–79	1980–89	1990–99	2000–09	2010–19
120	0	2	20	14	26	30	16	12

Icterine Warbler: Individuals recorded in Wales by decade.

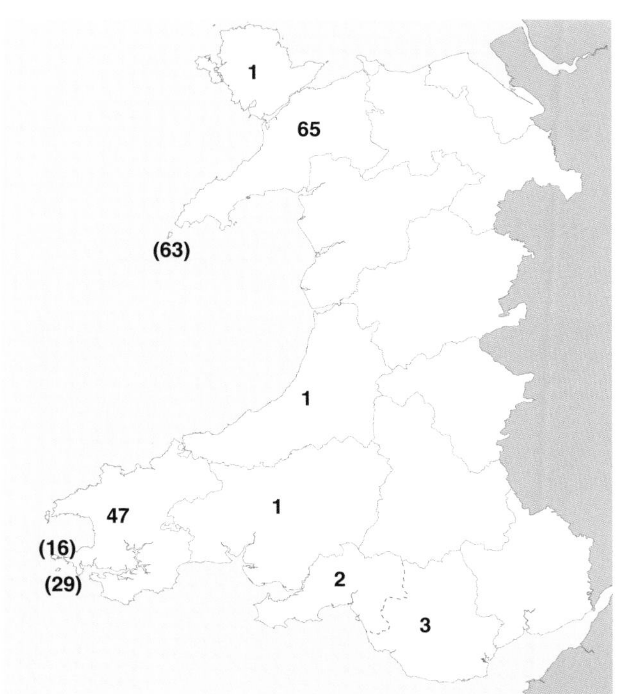

Icterine Warbler: Numbers of birds in each recording area, 1955–2019.

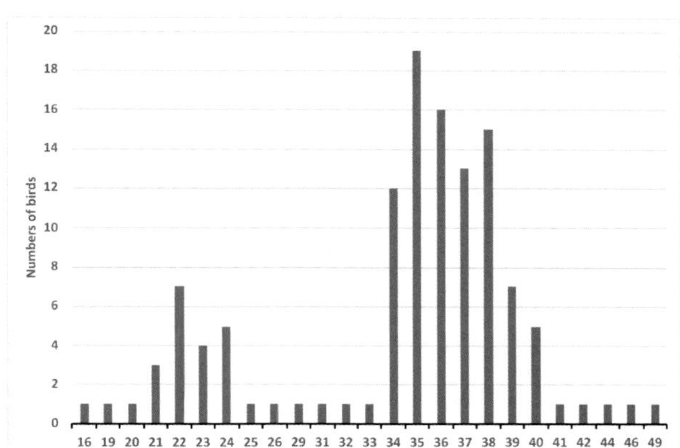

Icterine Warbler: Totals by week of arrival (when known, n=120), 1955–2019. Week 22 begins *c.*28 May and week 39 begins *c.*24 September.

The first Welsh record was on Skokholm, Pembrokeshire, on 31 August 1955. The vast majority have been on the islands: 53% on Bardsey, Caernarfonshire, 24% on Skokholm and 13% on Skomer, Pembrokeshire. In addition, one was ringed on Flat Holm, East Glamorgan, in August 1995 and there was one on The Skerries, Anglesey, in June 1998.

There have been just 10 mainland records: in East Glamorgan at Lavernock Point in September 1978 and Cosmeston Lakes in November 1982; in Gower at Port Eynon in September 1987 and at Mewslade in September 2006; in Carmarthenshire at Abergwili on the very late dates of 2 and 6 December 1996; in Pembrokeshire at Martin's Haven in June 1986 and St Ann's Head in September 1997; in Ceredigion at Cors Fochno in September 1990; and in Caernarfonshire at Llanfairfechan in May 1977 and Uwchmynydd May 2008. Those at Mewslade and Taliesin were ringed on site.

A small number occurred as spring overshooting migrants, usually at the end of May or early June, but the majority have been seen in the autumn, with the peak during the last week of August. The European breeding population declined by 49% during 1980–2016 (PECBMS 2020), which may explain the reduced numbers in Britain since 2000, including in Wales.

Jon Green and Robin Sandham

Lanceolated Warbler *Locustella lanceolata* Troellwr Bach Rhesog

Welsh List Category	IUCN Red List		Rarity recording
	Global	Europe	
A	LC	LC	BBRC

This species breeds in swampy forests through temperate Russia to the Pacific coast, and winters in Southeast Asia (*HBW*). Its range has expanded west into Finland since the 1990s, but numbers and its distribution there are uncertain (*European Atlas* 2020). There were 163 British records to the end of 2019, about four each year during 1990–2019 (Holt *et al.* 2020). Of these just three were in Wales and all were on Bardsey, Caernarfonshire, during the 1990s: on 18 October 1990, a moribund bird on 8 October 1994 and a first-calendar-year bird ringed on 27 September 1997.

Jon Green and Robin Sandham

River Warbler *Locustella fluviatilis* Telor Afon

Welsh List Category	IUCN Red List		Rarity recording
	Global	Europe	
A	LC	LC	BBRC

This bird of swamp and wet forests breeds in central and eastern Europe, as close to Britain as The Netherlands. Its range extends from southern Fennoscandia to the Adriatic and east to the west Siberian plain. It migrates through the Middle East and northeast Africa to winter in East Africa (*HBW*). It is a rare visitor to Britain, with 48 records to 2019 (Holt *et al.* 2020), and an extremely rare vagrant to Wales with just one record: an individual found moribund on Bardsey, Caernarfonshire, on 17 September 1969. The Central European population declined by 74% during 1982–2016 (PECBMS 2020), but its range has extended north to the higher latitudes of the Gulf of Bothnia in Fennoscandia (*European Atlas* 2020), as projected by Huntley *et al.* (2007), so it may occur more frequently as a drift migrant to Britain.

Jon Green and Robin Sandham

Savi's Warbler *Locustella luscinioides* Telor Savi

Welsh List Category	IUCN Red List			Birds of Conservation Concern	Rarity recording
	Global	Europe	GB	UK	
A	LC	LC	CR	Red	RBBP, BBRC

This warbler, of extensive reedbeds, breeds locally across Central Asia and Europe, including in very small numbers in Britain, and winters in tropical Africa (*HBW*). During 1950–2019, 824 were recorded in Britain (Holt *et al.* 2020). Numbers peaked during 1970–84, when up to 30 singing males were recorded in a year (Slack 2009). There have been nine records in Wales:

- Skokholm, Pembrokeshire, on 31 October 1968, which was the latest ever autumn migrant in Britain;
- Dowrog, Pembrokeshire, on 18 June 1983;
- Oxwich Marsh, Gower, on 13–20 May 1987;
- Llangorse Lake, Breconshire, on 4–7 July 1994;
- RSPB Cors Ddyga, Anglesey, on 8–11 June 1999;
- Newport Wetlands Reserve (Uskmouth), Gwent, from 24 May to 8 June 2014;
- Newport Wetlands Reserve (Uskmouth), Gwent, from 29 May to 7 June 2016;
- Newport Wetlands Reserve (Uskmouth), Gwent, on 25–30 May 2019 and
- pair bred at RSPB Cors Ddyga, Anglesey, in 2019 (birds present from 14 June to 25 July).

The number of records declined in the 1990s, reflecting a reduction in populations in western Europe that was thought to be linked to conditions in their Sahel wintering area (Hudson *et al.* 2010). It was, therefore, a considerable surprise when a pair bred at RSPB Cors Ddyga in 2019, fledging at least one young. There had been an average of only six singing males in Britain during 2014–18, and the Anglesey brood was the first confirmed breeding in Britain since 2010 (Eaton *et al.* 2020). Climate modelling suggested that its European breeding range could extend into South Wales later this century (Huntley *et al.* 2007), although this does not account for changes in its wintering area nor the limited number of large reedbeds in the region. The frequency of singing males at Uskmouth, Gwent, may be a precursor to future breeding there.

Jon Green and Robin Sandham

Savi's Warbler at RSPB Cors Ddyga, Anglesey, June–July 2019 © Martin Jones

Sponsored by Cofnod – North Wales Environmental Information Service

Grasshopper Warbler *Locustella naevia* Troellwr Bach

Welsh List Category	IUCN Red List			Birds of Conservation Concern			
	Global	Europe	GB	UK	Wales		
A	LC	LC	LC	Red	2002	2010	2016

The Grasshopper Warbler is a summer migrant to Europe, broadly north of the Alps and across southern Russia to western Mongolia (*HBW*). It spread into Denmark and southern Norway in the late 20th century, but the rate of northerly expansion has since slowed (*European Atlas* 2020). Its sub-Saharan wintering areas are relatively restricted and discontinuous, with British birds most likely to winter in Senegal (*HBW*). It is not especially site-faithful in either its breeding or wintering areas, a flexible strategy that helps it to adapt to changing conditions in the marginal habitats that it favours (*BTO Migration Atlas* 2002).

It breeds in upland bog and damp moorland, rough grass, downland scrub and pre-canopy forestry plantations up to *c.*500m but avoids mature forests. It conceals its nest on the ground among shrubby cover, usually near water, or at least damp ground (*BWP*). Only small numbers have been ringed in Wales, so few movements have been recorded. The only recovery from outside Britain and Ireland was one ringed at Ynyslas, Ceredigion, on 19 April 1985, which was found in Algeria on 5 February 1986 and may have been a migrant heading north. A first-calendar-year bird ringed near Zandvoort, The Netherlands, on 13 September 2009, was killed by a cat in Caerphilly, East Glamorgan, just three days later, indicating that continental breeding birds sometimes migrate through Wales (Robinson *et al.* 2020).

The *Historical Atlas 1875–1900* suggested that North and mid-Wales were strongholds for the species in Britain and that it was abundant or common in most counties, though scarce and local in Radnorshire, Breconshire and Pembrokeshire. Contemporary writers indicated that its distribution was restricted by the availability of suitable habitat, such as rushy meadows and rough grassland. In North Wales, it was common around Wrexham and Nant-y-Ffrith (Dobie 1894) and "fairly common on hills inland, less numerous on flats by the coast" (Forrest 1907). It remained scarce in mid-Wales through the mid-20th century, being described as scarce and localised in Radnorshire (Ingram and Salmon 1955) and rare in Breconshire until the early 1950s (Massey 1976a). It is difficult to gauge trends but a decline in the early 1970s was linked to drought conditions in the Sahel where our birds winter (Marchant *et al.* 1990).

The *Britain and Ireland Atlas 1968–72* showed that it was widespread in Wales, occurring in 81% of 10-km squares and absent only from mountainous areas. By 1988–91 its range had contracted substantially, occurring in 51% of squares, in line with the scale of loss elsewhere in Britain. Its disappearance from many areas was attributed to loss and deterioration of habitat. Many young conifer plantations that held Grasshopper Warblers in the earlier period had matured and ceased to be suitable, while the drainage of moorland and rough pasture had reduced its 'traditional' habitat. It was suspected that other factors outside the breeding season led to fluctuations in population levels (*Britain and Ireland Atlas 1988–91*), but there is insufficient data on winter survival to be sure. Records in the *Britain and Ireland Atlas 2007–11* show that it largely recovered its 1968–72 range in Wales, except in the eastern lowlands, but a far smaller proportion of squares included breeding records. This recovery was in marked contrast to much of southern England, from which Grasshopper Warbler

A singing male Grasshopper Warbler © John Hawkins

was lost entirely. Relative abundance was high in the western half of Wales away from the coast. There had been a westward shift in the areas holding the highest densities since 1988–91, to this part of Wales, western Ireland and western Scotland.

Its localised nature is evident in tetrad atlases, but its fortunes have varied where these surveys have been repeated. The *Gwent Atlas 1981–85* recorded it in 8% of tetrads, mainly in the east, whereas it occurred in 6% of tetrads in 1998–2003, mainly in the west, a change that was attributed to the phases of planting, maturing and felling of plantations (Venables *et al.* 2008). In East Glamorgan breeding evidence was found in less than 2% of tetrads in 1984–89 but over 9% in 2008–11, again relating to forestry management cycles. Occurrence in Pembrokeshire changed little between 1984–88 and 2003–07, when it was in 19% and 20% of tetrads respectively (*Pembrokeshire Avifauna* 2020). The *North Wales Atlas 2008–12* showed occupation in 22% of tetrads with particularly strong presence in Meirionnydd, where occupation of 10-km squares had increased by 35% since 1968–72 to almost all the squares in the county. In Caernarfonshire, it occupied 82% of 10-km squares in 2007–11, compared to just 36% in 1988–91.

Fluctuating numbers make accurate population estimates difficult, and county estimates made at different times are barely comparable. In 1986, an RSPB survey of Anglesey recorded 132 singing males. In 1998–2003, the Gwent population was estimated at less than 35 pairs (Venables *et al.* 2008), while the *Pembrokeshire Atlas 2003–07* estimated 480–670 pairs. Pritchard (2012) estimated there could be 200–320 pairs in Meirionnydd, but it remained uncommon in Radnorshire, where Jennings (2014) estimated the population at 50–100 pairs. The Welsh breeding population in 2018 was estimated to be 2,000 (1,800–2,200) territories (Hughes *et al.* 2020).

The first Grasshopper Warblers generally arrive in early April. During 2000–19 every county had records before 11 April. The earliest was at Llanrhystud, Ceredigion, on 29 March 2011. Spring passage at the island bird observatories peaks from the third week of April to early May. While males announce their arrival with a far-carrying 'reel', the species' skulking nature later in the season means that most leave from mid-August without being noticed, except where ringing is undertaken regularly, such as at Oxwich Marsh in Gower, Newport Wetlands in Gwent and bird observatories. As in spring, the autumn peak on Bardsey, Caernarfonshire, has been larger than at Skokholm, Pembrokeshire, and has been far more protracted, extending from late July to mid-September. Most have left Wales by the end of September, but there have been October records in several counties. During 2000–19, the latest date recorded was on 31 October 2004 at Redberth, Pembrokeshire.

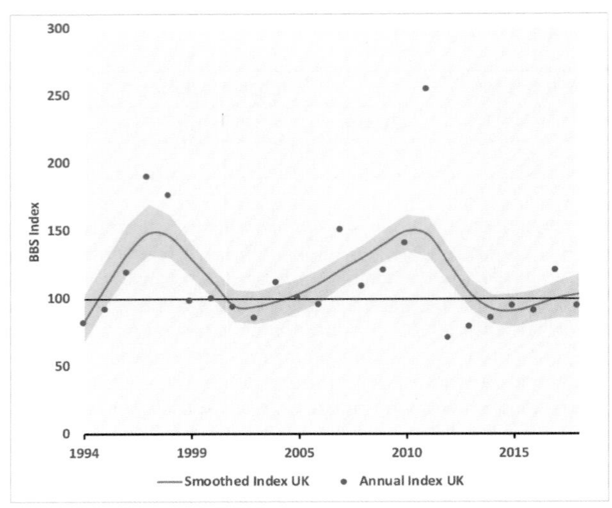

Grasshopper Warbler: UK Breeding Bird Survey indices, 1994–2018 (Massimino *et al.* 2019). The shaded area indicates the 85% confidence limits.

The BBS index shows some evidence of 12-year population cycles in the UK. The Grasshopper Warbler occurs on too few BBS squares to generate an index in Wales and while that for England shows a long-term decline, the most recent atlas suggested that any change may be less severe in Wales. A large part of its Welsh population may now be associated with rotational forestry management, which by its nature is ephemeral and so its fortunes will be tied to future felling policy, scrub and grazing management in the uplands. Active intervention to increase the heterogeneity of habitat structure would help to secure the Grasshopper Warbler in places from which it would otherwise be lost (Gilbert 2012) and would benefit other species such as Black Grouse and Cuckoo. Climate change may affect a large part of its breeding range in Europe, where the population has already declined by 73% since 1980 (PECBMS 2020), though modelling suggests that the climate of the Welsh uplands will continue to be suitable (Huntley *et al.* 2007). As a long-distance migrant, it is also subject to changes in its wintering area and on migration. It is a species for which we need to understand much more about its ecology and resilience to habitat and climate change.

Ian M. Spence

Blackcap *Sylvia atricapilla* Telor Penddu

Welsh List Category	IUCN Red List			Birds of Conservation Concern			
	Global	Europe	GB	UK	Wales		
A	LC	LC	LC	Green	2002	2010	2016

The Blackcap's range extends from the Macaronesian islands through Europe to the Ural Mountains and south to the Black Sea basin. Those that breed in the north and east of their range are largely migratory, wintering in northwest and sub-Saharan Africa (*HBW*). In Wales, as elsewhere in Britain and Ireland, the breeding population migrates to the south and is replaced by a far smaller wintering population from Europe, extending from northern Spain to Poland (Delmore *et al.* 2020). The Blackcap breeds primarily in broadleaved woodland, requiring trees as song posts and shrubby undergrowth in which to nest, and occurs at lower density in coniferous woodland, scrub, and suburban parks and gardens (*BWP*).

Prior to the species' increased abundance in Wales, the first Blackcaps arrived from mid-April and left by mid-September (*Birds in Wales* 1994). It is now more difficult to distinguish migrants from wintering birds, especially in spring, but the reporting rate for BirdTrack in Wales and passage through Skokholm, Pembrokeshire, and Bardsey, Caernarfonshire, shows a significant increase from around 20 March. This is 4–5 weeks earlier than described by Forrest (1907) for arrivals in North Wales, who cited one at Llandderfel, Meirionnydd, on 17 April 1877, as the earliest record. Larger numbers occurred on the islands in some years, such as 300 attracted to the lighthouse on Bardsey on 2 May 2005 and 226 there on 24 April 2013 (Pritchard 2017).

A marked autumn passage is evident through the second half of September. In Gwent, for example, 51 were ringed at Newport Wetlands Reserve (Uskmouth) on 15 September 2018, 65 on 21 September 2014 and 72 on 3 October 2015 (Richard M. Clarke *in litt.*). Autumn migration through the island bird observatories is protracted through October and into November, with smaller numbers than in spring. Ringing recoveries during September and October are mainly from western France, Iberia and North Africa. Some cross the Sahara to winter in the Sahel (*BTO Migration Atlas* 2002), although the only sub-Saharan recovery from Wales was a bird ringed as a juvenile in Penallt, Gwent, in June 1993 and, rather

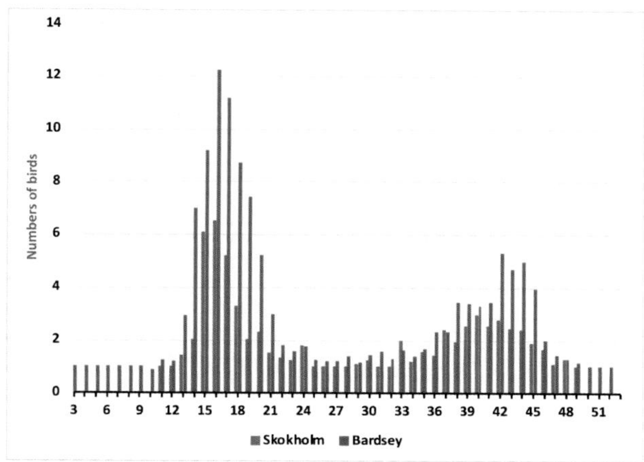

Blackcap: Average counts by week number recorded by bird observatories, 1953–2019. Week 17 starts *c*.17 April and week 42 starts *c*.14 October.

curiously, trapped alive in Ivory Coast in June 2000. This is the most southerly recovery of any Blackcap in the Britain and Ireland ringing scheme (Robinson *et al.* 2020).

Around the turn of the 20th century, the Blackcap appears to have been abundant in areas of quality woodland but scarce or absent in parts of North Wales (Dobie 1894, Forrest 1907). In Glamorgan, it was common in the 1890s but declined from the 1930s, before recovering after the Second World War (Hurford and Lansdown 1995). In Pembrokeshire, having been uncommon in the late 19th century (Mathew 1894), it was described as fairly common by the 1940s (Lockley *et al.* 1949). Ingram and Salmon, quoted in Jennings (2014), considered the Blackcap to be quite scarce and thinly distributed in Radnorshire during the 1950s but numbers increased considerably. It was described as fairly common in Breconshire by Massey (1976a), although less numerous in the north of the county.

Blackcap: Recovery locations of birds ringed in Wales are shown by red circles. Ringing locations of birds that were recovered in Wales are shown by blue or black triangles (see text for explanation). Small brown dots show ringing or recovery locations in Wales.

The *Britain and Ireland Atlas 1968–72* confirmed breeding from almost all of Wales, suggesting an expansion in range during the middle of the 20th century. A further increase was recorded by the atlases of 1988–91 and 2007–11, and by the latter period it occupied 96% of 10-km squares, absent only from a few upland areas. The range has expanded in southern Wales during the last 40 years. In Gwent, occupation of tetrads increased from 85% during 1981–85 to 93% in 1998–2003, and in East Glamorgan from 67% in 1984–89 to 70% with breeding evidence in 2008–11. Reductions in the south of the latter recording area were more than compensated for by expansion in the northern valleys. In Pembrokeshire, the number of occupied tetrads also increased between 1984–88 and 2003–07, from 71% to 81% (*Pembrokeshire Avifauna* 2020). Abundance in 2007–11 was relatively low across North and mid-Wales compared to South and West Wales and much of England. During the *North Wales Atlas 2008–12*, the Blackcap was found in 68% of tetrads, being absent from the treeless mountains and moorland.

Birds in Wales (1994) considered the Blackcap to be less of a woodland bird than in England because of overgrazing of semi-natural habitats and that, consequently, it was more likely to nest in large gardens, wooded watersides and thick hedgerows. Nonetheless, territory density in Welsh broadleaved woodlands increased by 57% between 1965–72 and 2003–04 (Amar *et al.* 2006). Since 1995, the BBS index for Wales has increased by 143%, a trend consistent across northern and western parts of the UK (Harris *et al.* 2020). The increase is reflected in monitoring during passage migration at bird observatories: on Skokholm there has been a six-fold increase in the number of 'bird-days' logged since 2011, particularly in spring. The Welsh breeding population in 2018 was estimated to be 265,000 territories (Hughes *et al.* 2020), making it the most abundant summer migrant passerine jointly with Willow Warbler.

Blackcaps make use of gardens in winter, particularly in suburban areas. Their occurrence has become more strongly associated with the provision of food over time and is also linked to warmer temperatures (Plummer *et al.* 2015). It was reported very occasionally as a winter visitor in the first half of the 20th century, such as a "pugnacious" female in a garden at Porthmadog, Caernarfonshire, in January 1906 (Forrest 1919). It was encountered more frequently from the early 1960s and became annual from later that decade in Gwent and Glamorgan (Venables *et al.* 2008, Hurford and Lansdown 1995), although winter records were still the exception in Breconshire (Massey 1976a) and did not feature in Pembrokeshire until the 1980s. It occurred in 30% of 10-km squares in Wales, scattered around the lowlands, during the four winters of the *Britain and Ireland Winter Atlas 1981–84*. *Birds in Wales* (2002) estimated that 50–150 Blackcaps were present each winter in Wales during 1992–2000.

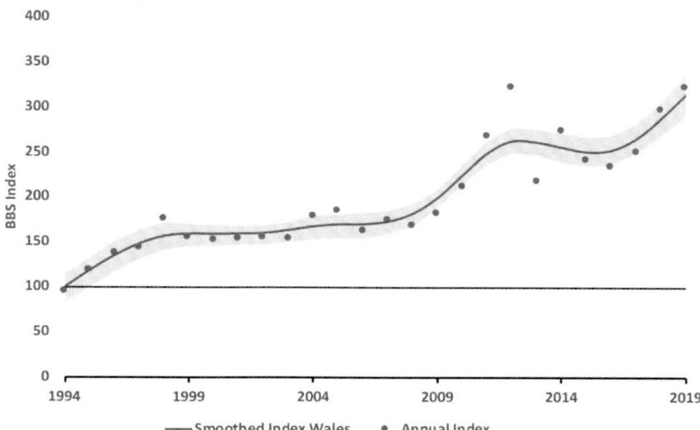

Blackcap: Breeding Bird Survey indices, 1994–2019. The red line shows the smoothed index, 1995–2018. The shaded area indicates the 85% confidence limits.

The *Britain and Ireland Atlas 2007–11* showed winter distribution had increased to over 48% of 10-km squares, with increases particularly in South and West Wales. It was recorded in 10–15% of gardens in Wales participating in the BTO Garden Birdwatch during 1995–2019, although this may be biased to gardens that provide more supplementary food than the average. At least 2,000 Blackcaps winter here in some years, and the total may be far higher (Hughes and Carter, in prep.).

Although, only 1–2 may be seen in a garden at once, the turnover may be far higher: in winter 1984/85, 11 were ringed in a garden in Bangor, Caernarfonshire (Pritchard 2017). Blackcaps wintering in Britain are from west-central Europe, a migration strategy that developed from the 1960s, with a few pioneering individuals that found adequate amounts of food available in gardens to sustain themselves, even through harsh weather (Berthold and Helbig 1992). Blackcaps arrive in Britain from this area from mid-September, with the bulk of arrivals in mid-October (Ozarowska 2012). A dozen Blackcaps ringed in Europe between 31 August and 30 November (shown as black triangles on the map) were recovered in Wales during November to March, of which three were in one garden in Rhos-on-Sea, Denbighshire. Those ringed in the Low Countries may have originated farther east or southeast.

The Blackcap is doing very well in Wales and its breeding distribution is expected to shift only slightly north during the remainder of this century (Huntley *et al.* 2007). It is a species that has already demonstrated its ability to respond to a changing environment. Although no Blackcap, ringed in Wales during the breeding season, has yet been shown to stay for the winter, it is possible that part of the population in Wales may become resident in the future.

Ian M. Spence

Garden Warbler *Sylvia borin* Telor yr Ardd

Welsh List Category	IUCN Red List			Birds of Conservation Concern			
	Global	Europe	GB	UK	Wales		
					2002	2010	2016
A	LC	LC	LC	Green			

The Garden Warbler breeds across Europe, except in the far north of Fennoscandia and around the Mediterranean, and across temperate Russia as far as the west Siberian plain. It winters in sub-Saharan Africa, most British birds around the Gulf of Guinea (*HBW*). It nests in woodland glades and clearings with scrub and in habitats with a dense understorey, including young conifer plantations and even Rhododendron thickets. High densities occur in Britain in Sessile Oak woodland (*HBW*). It cannot tolerate disturbance so, despite its name, rarely breeds in gardens (*BWP*). As most winter south of the Sahel, it avoided declines on the scale of other migrants such as Whitethroat in the 1960s and 1980s. Nonetheless, the population in Britain halved during 1968–75; while it recovered by the late 1980s, it has since undergone a more gradual decline (Massimino *et al.* 2019). That change is Europe-wide, the species having declined by 28% in Europe during 1980–2016 (PECBMS 2020).

Forrest (1907) reported that the Garden Warbler was likely to be overlooked by "all but the practised ornithologist," so it is difficult to draw reliable conclusions about its status from earlier works. At the turn of the 20th century, it was considered much less common than the Blackcap in North Wales, rare on Anglesey and Caernarfonshire except for the Conwy Valley but more numerous in Montgomeryshire and Meirionnydd (Forrest 1907). In counties to the south, however, the Garden Warbler was more common than the Blackcap through much of the century (Lockley *et al.* 1949, Ingram and Salmon 1955, Ingram *et al.* 1966, Massey 1976a, Jennings 2014). All described it as numerous and widespread, but its distribution could be patchy because of its habitat requirements.

Garden Warbler © John Hawkins

The *Britain and Ireland Atlas 1968–72* showed breeding records in 86% of 10-km squares in Wales but with minimal breeding evidence from Anglesey in particular. There was no net change in the 1988–91 *Atlas* and a contraction in northwest Wales proved to be short term and was not evident in the 2007–11 *Atlas*, when almost 92% of squares were occupied in Wales. Densities in Pembrokeshire, mid-Wales and the South Wales uplands were among the highest in Britain. Tetrad atlases showed differing trends, but this may simply be a factor of the different time periods in which they were undertaken. It occurred in 50% of tetrads in the *Gwent Atlas 1981–85* and increased to 55% by 1998–2003, whereas in the *East Glamorgan Atlas,* it declined considerably from 40% of tetrads with breeding codes in 1984–89 to 27% in 2008–11, having gone from much of the coastal plain. In Gower, the *West Glamorgan Atlas 1984–89* reported breeding evidence from 37% of tetrads but almost total absence from the Gower Peninsula. In Pembrokeshire there was no change in overall occupation, at *c.*40% of tetrads, between 1984–88 and 2003–07, but it had almost gone from south of the Cleddau Estuary in the later survey (*Pembrokeshire Avifauna* 2020). The *North Wales Atlas 2008–12* showed a 10% increase in occupation of 10-km squares since 1968–72. It occurred in 39% of tetrads, with few records on Anglesey and along the border with England. The southern part of Denbighshire, the Conwy Valley in Caernarfonshire/Denbighshire, and the valleys of the Glaslyn, Dwyryd, Gain and Wnion in Meirionnydd held the highest concentrations in the region.

The Blackcap population, which inhabits a similar habitat to that of the Garden Warbler and with which its song is easily confused, increased by 143% in Wales during 1995–2018, whereas the BBS index for Garden Warbler declined by 18% over the same period. In many counties, it is likely that the Blackcap is now more populous: in Caernarfonshire, Pritchard (2017) suggested there were two Blackcaps for every Garden Warbler, and in Ceredigion, the estimate was almost 9:1 (Roderick and Davis 2010). The Welsh breeding population in 2018 was estimated to be 39,500 territories (Hughes *et al.* 2020). Compared with the population estimate of 265,000 for Blackcap, the ratio is about 7:1.

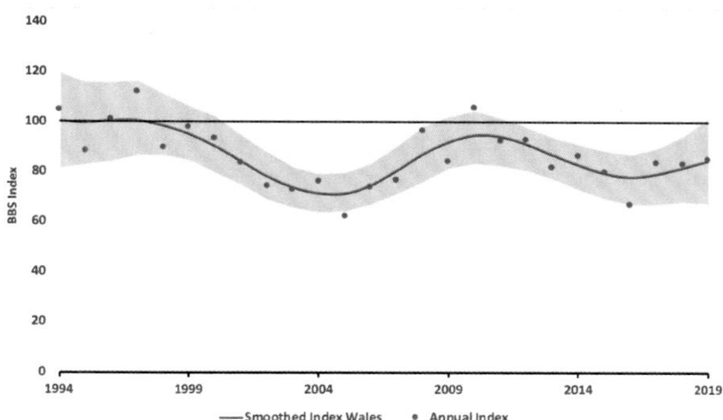

Garden Warbler: Breeding Bird Survey indices, 1994–2019. The red line shows the smoothed index, 1995–2018. The shaded area indicates the 85% confidence limits.

Garden Warbler: Recovery locations of birds ringed in Wales are shown by red circles. Ringing locations of birds that were recovered in Wales are shown by blue triangles. Small brown dots show ringing or recovery locations in Wales.

No Garden Warbler ringed in Wales has been recovered in West Africa, where the majority of the British population spends the winter, but several have been recovered in western France and Morocco, most on autumn passage. There have also been several Garden Warblers ringed in The Netherlands, Belgium and Norway, and recaught by ringers or found dead in Wales in autumn of the same year, but all of these were prior to 1980. More recently, a first-calendar-year Garden Warbler ringed at Llangorse Lake, Breconshire, on 21 July 2018 was recovered in Norway 11 days later. Given the ringing date, it had probably hatched near Llangorse and dispersed northeast to investigate a potential breeding area for the future, before migrating south (Lewis and Haycock 2019). Such movements across the North Sea occur in both directions in autumn (*BTO Migration Atlas* 2002).

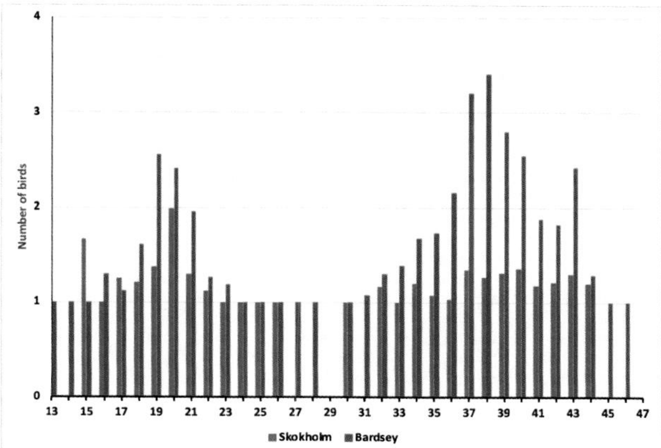

Garden Warbler: Average counts by week number recorded by bird observatories, 1953–2019. Week 18 starts *c.* 30 April and week 37 starts *c.* 9 September.

The first arrivals in spring are during the first half of April, with every county having records before 16 April during 2000–19. The earliest in Wales during this period was on the exceptionally early date of 24 March 2004 at Plas Tan y Bwlch, Meirionnydd. The Garden Warbler is a less-frequent passage migrant at the island bird observatories than most other warblers, reflecting the small population in Ireland and restricted range in Scotland. The peak for spring records on Skokholm, Pembrokeshire, and Bardsey, Caernarfonshire, is in mid-May. Most depart in September, but some occur into October most years in coastal counties, six of which had November records during 2000–19. The latest record in this period was at St Mary's Well Bay, East Glamorgan, on 27 November 2019. There have been a handful of winter records, mostly in Pembrokeshire: at Pembroke on 28 February 1976, at Pwllcrochan in February 1992 and at Hook on 7 December 2001. The only other winter record was at Kenfig, East Glamorgan, on 19 February 2000.

The Garden Warbler appears to be undergoing a slow, long-term decline, driven by reduced productivity or juvenile survival (Johnston *et al.* 2016). Ringing at Constant Effort Sites showed that nest failure rates have increased while there was a decline in adult survival, the number of fledglings/breeding attempts and post-fledging productivity (BTO 2020). The causes are not understood, but analysis of Woodland Bird Surveys undertaken in the 1980s and repeated in 2003–04 showed that a reduction in understorey cover and increased temperatures in May were correlated with a 42% decline, and that the probability of occupancy was lower at more southerly latitudes (Mustin *et al.* 2014). Modelling suggested a significant contraction from the south of the Garden Warbler's European range by the end of the 21st century (Huntley *et al.* 2007), and while there is no imminent risk of losing the species in Wales, future land management should account for its needs as part of creating resilient and diverse woodlands.

Ian M. Spence

Barred Warbler *Curruca nisoria* Telor Rhesog

Welsh List Category	IUCN Red List		Rarity recording
	Global	Europe	
A	LC	LC	WBRC

The Barred Warbler nests in scrub and woodland across the farmed landscape in Europe, east of a line from Denmark to the Adriatic, through southern Russia and Central Asia to Mongolia. The southern edge of Fennoscandia is the northern limit of the breeding range. It winters in a fairly discrete area of East Africa, centred on Kenya (*HBW*). Little is known about the population trend across most of its breeding range, but numbers have increased in Germany since the 1980s (BirdLife International 2015). It is a scarce autumn migrant to Wales, although the number of records has increased during the last 30 years. Such a pattern has been evident across Britain, where the annual number of records was 84% higher in 2010–18 than in 1968–89 (White and Kehoe 2020b). It is believed that those found here are part of extended post-juvenile dispersal.

Total	Pre-1950	1950–59	1960–69	1970–79	1980–89	1990–99	2000–09	2010–19
111	1	2	20	6	13	18	22	29

Barred Warbler: Individuals recorded in Wales by decade.

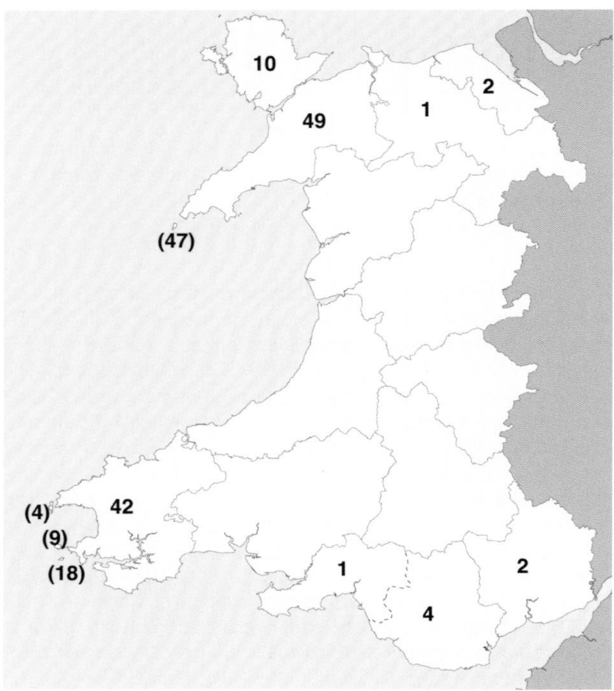

Barred Warbler: Numbers of birds in each recording area, 1910–2019.

The first Welsh record was one found dead on The Skerries, Anglesey, in September 1910. It is an exclusively autumn visitor, between mid-August and the end of November, with a peak in late September and early October. Bardsey, Caernarfonshire, has had 42% of the Welsh total, and Skokholm, Pembrokeshire, 16%. Away from these counties and Anglesey, it is a very rare bird, with only a handful of records along the south and north coasts: in Gwent, at Newport Wetlands Reserve in August/September 1994 and October 2015; in East Glamorgan, at Nash Point in September 1993 and Kenfig in September 1979, 1994 and 2004; at Mewslade, Gower, in October 2016; at Rhos-on-Sea, Denbighshire, on 18 and 24 November 2016; and in Flintshire, at Shotton in August 1986 and at Shotwick Broken Bank in October 2010. The bird in Rhos-on-Sea was caught and ringed, and is one of the latest ever records in Britain.

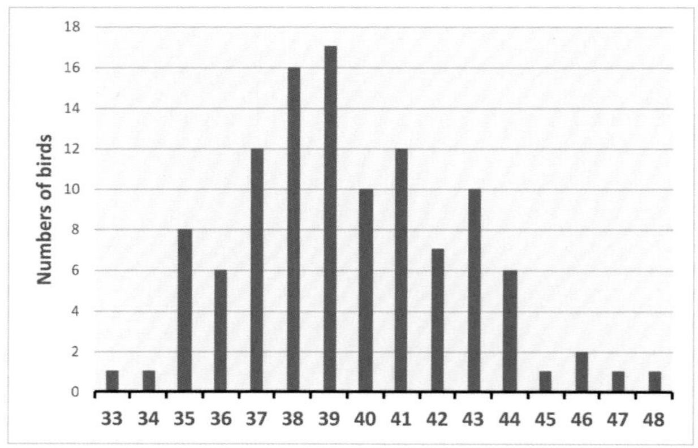

Barred Warbler: Totals by week of arrival, 1910–2019. Week 36 begins *c.*3 September and week 44 begins *c.*29 October.

The breeding range of Barred Warbler is projected to extend into Fennoscandia during this century (Huntley *et al.* 2007), which may result in increased numbers of records in Britain.

Jon Green and Robin Sandham

Lesser Whitethroat *Curruca curruca* Llwydfron Fach

Welsh List Category	IUCN Red List			Birds of Conservation Concern			
	Global	Europe	GB	UK	Wales		
A	LC	LC	LC	Green	2002	2010	2016

Wales is at the western edge of the Lesser Whitethroat's extensive range, which covers most of temperate Eurasia as far as Mongolia. Birds breeding in Wales are the nominate form that breeds across Europe, except in Iberia and the far north, Central Asia and much of Siberia (IOC). It breeds in a variety of habitats where there is low, dense cover, such as scrub, hedgerows and plantations with small trees (*HBW*). It is very scarce at altitudes over 200m in Wales, presumably because of a lack of its favoured Hawthorn and Blackthorn scrub, but one sang at 310m at Llanfihangel Nant Melan, Radnorshire, in 1991 and a pair probably bred at 400m near Arenig Fawr, Meirionnydd, in 2011. European breeding birds winter in northeast Africa, mainly from Chad eastward (*HBW*), and so take an easterly route through the Mediterranean, illustrated by two recoveries of Welsh-ringed birds: an adult ringed at Usk, Gwent, in May 1973, that was shot in Lebanon in May 1974, and a first-calendar-year bird ringed at Llangorse Lake, Breconshire, in August 1986, that was trapped on the Greek island of Khios in October the same year.

There were few records of Lesser Whitethroat in Wales until the mid-20th century. Hurford and Lansdown (1995) stated that it was listed in the Swansea Guide (Anon. 1802) as "Pettychaps". However, this term was used generically for warblers and the scientific name given, *Motacilla hippolais*, was sometimes used for Chiffchaff. It was said to be uncommon in Glamorgan in the late 1890s and was not recorded on the Gower Peninsula until 1927 (Hurford and Lansdown 1995). Barker (1905) knew of no specific records in Carmarthenshire, simply noting that it was "on Mr Davidson's list", although Ingram and Salmon (1954) stated that Barker added a manuscript note in 1909 that it had been seen near Abergwili. Mathew (1894) had never seen the species in Pembrokeshire and said that "we only admit it doubtfully" on the basis that Mr Mathias of Haverfordwest claimed to have found a nest at Lamphey around 1832. He also stated that it did not appear to visit Ceredigion, and Salter (1900) did not mention it as occurring around Aberystwyth, although he recorded one in 1903 and then almost annually after he returned to the area to live at Llanbadarn from 1923 (Roderick and Davis 2010). Very little was published on its status in North Wales during the 19th century, although nesting was proved around Llyn Tegid, Meirionnydd, in 1867. A record at Abersoch, Caernarfonshire, in May 1893, was based on a nest with eggs found by Thomas Coward, but an egg from the clutch was later re-identified as being from a Common Whitethroat (Aplin 1902). Forrest (1907) concluded that it was found in "fair numbers" in Flintshire, Denbighshire, eastern Montgomeryshire and eastern Meirionnydd, but that it was unknown in Caernarfonshire west of the Conwy Valley and was very rare in western Meirionnydd and Anglesey. By 1919, however, Forrest was able to quote several additional records from the latter.

Lesser Whitethroat remained a very localised breeding species through the first few decades of the 20th century, primarily in Gwent (Ingram and Salmon 1939) and in Pembrokeshire, where it bred from the 1930s (Lockley *et al.* 1949), and later on Anglesey (Alexander and Lack 1944). The first complete picture of its distribution came from the *Britain and Ireland Atlas 1968–72*, when it was very sparsely distributed except along the eastern border, and absent from the uplands and much of the western lowlands. There were no records from Llŷn in Caernarfonshire, or Meirionnydd, and few on Anglesey, or in Ceredigion and Pembrokeshire. Parslow (1973) stated that there was no evidence of widespread changes in numbers or range and Simms (1985) commented "it is generally believed, without real evidence to support it, that the species has moved farther west and north during the present century".

However, it certainly had increased on Anglesey by the 1980s: 4–5 pairs were at Cors Erddreiniog and six at Cors Ddyga in 1985, and in 1986 it was recorded in 75 1-km squares on the island (RSPB). A survey in 1986 noted that it had recently colonised Llŷn and estimated the population there to be 40–60 pairs. The *Britain and Ireland Atlas 1988–91* showed extensive gains not only in northwest Wales but also in Gwent, Carmarthenshire and Ceredigion. The 2007–11 *Atlas* showed a further net increase, to 58% of 10-km squares, with substantial gains in Pembrokeshire and Caernarfonshire, although losses predominated in Ceredigion. Abundance was relatively low compared to much of southern England, but the areas of highest density within Wales included several from which it had been very rare or absent a century previously: Anglesey, Llŷn and Pembrokeshire. In Caernarfonshire, for example, Lesser Whitethroat was recorded in three 10-km squares in 1968–72, but in 17 in 2007–11, the third-highest increase of any species in the county (Pritchard 2017).

Tetrad atlases have indicated a decline in the east in the last 30 years. It was found in 15% fewer tetrads in Gwent in 1998–2003 than in 1981–85, when it had been recorded in 33% of the county's tetrads. It had spread in the north, but decreased in the south, which Venables *et al.* (2008) attributed, at least in part, to the loss of old Hawthorn hedges on the coastal levels. A reduction also occurred in East Glamorgan, where there was breeding evidence in 27% of tetrads in 1984–89 but only 15% in 2008–11, whereas it increased slightly in Pembrokeshire, from 14% of tetrads in 1984–88 to 16% in 2003–07 (*Pembrokeshire Avifauna* 2020). Although there are no comparisons from other counties, the *West Glamorgan Atlas 1984–89* found it in only 14% of tetrads in Gower and the *North Wales Atlas 2008–12* found it in 16%. In the latter, records were concentrated in Flintshire, eastern Denbighshire, the lower Clwyd and Conwy valleys, Anglesey and Llŷn, and it was almost absent from upland habitats. In Meirionnydd, it occurred in only 3% of tetrads, mainly in the coastal lowlands and not at all in the east, where Forrest (1907) had said that it was found in fair numbers.

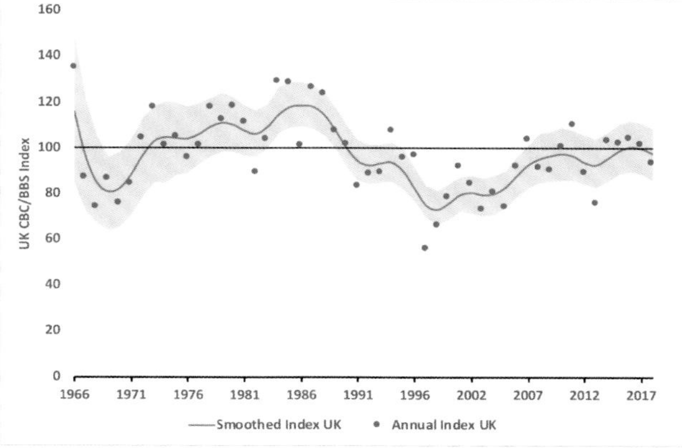

Lesser Whitethroat: Common Birds Census/Breeding Bird Survey indices for the UK, 1966–2018 (Massimino *et al.* 2019). The shaded area indicates the 85% confidence limits.

The Lesser Whitethroat is recorded in too few squares in Wales to calculate a population trend. The CBC/BBS index for the UK during 1967–2017, and that for Europe during 1980–2016, show considerable fluctuations but no overall trend (Massimino *et al.* 2019, PECBMS 2020). Declines during the 1990s were probably mostly due to pressures during migration and in winter (Fuller *et al.* 2005), although a different migration route and wintering area to most other migrants that nest in Britain meant that it was not affected by lack of rainfall in the Sahel. The Welsh breeding population in 2018 was estimated to be 6,950 (6,150–7,800) territories (Hughes *et al.* 2020), but even the lower estimate is thought likely to be too high.

Lesser Whitethroats usually arrive on their breeding grounds in late April, although during 2000–19 all counties except Breconshire recorded the earliest arrivals during the first half of the month, and there were March records at Collister Pill, Gwent, on 20 March 2012, at Llyn Llywenan, Anglesey, on 23 March 2003 and at Milford Haven golf course, Pembrokeshire, on 31 March 2002. The species is not usually recorded in large numbers on migration, and double-figure counts are unusual. There were ten on Bardsey, Caernarfonshire, on 13 May 1980, 11 at Lavernock, East Glamorgan, on 15 August 1965 and 12 at Pant Norton, East Glamorgan, on 17 August 2007.

Besides those above, there were eight other records from six counties in December to March during 2000–19, including three in East Glamorgan. Most winter records were seen on a single date, but four were seen over a longer period: at Griffithstown, Gwent, on 11–15 January 2010; at Skewen, Gower, from 17 December 2010 to 16 April 2011; at Heath, East Glamorgan, from 27 January to 10 February 2014 and at Llanelli, Carmarthenshire, from December 2015 to 14 January 2016. Over-wintering was unusual before 2010, but one at Llandudno, Caernarfonshire, remained from 23 December 1991 to 11 April 1992. Records of *C.c. blythi* (below) tend to occur in late autumn/early winter, but in the cases above there was insufficient information or photographs to suggest its identification to subspecies.

Although the Lesser Whitethroat is now a widespread species in Wales, there appear to be few areas where it can be described as common. It is little studied here, making it difficult to prescribe management improvements, but preserving thick hedges and thorny scrub is clearly important, and hard trimming of hedgerows can result in the loss of valuable habitat. Furthermore, modelling suggested that the suitable climate for Lesser Whitethroats will shift north and east during the 21st century, with scope for substantial loss from France, southern Wales and England (Huntley *et al.* 2007).

Siberian Lesser Whitethroat *C. curruca blythi* Llwydfron Fach Siberia

There have been several records of this subspecies, which breeds in north central Siberia and Kazakhstan (IOC). Records in Britain and Ireland increased from the 1980s, a trend matched by increased westward vagrancy of other Siberian passerines (Baker 1988). All bar one of the five records accepted by the WBRC were in the last week of October, or later:

- Bardsey, Caernarfonshire, on 16 May 2014, confirmed by DNA analysis and the first confirmed spring record in Britain;
- Skokholm, Pembrokeshire, on 3–5 October 2014, its identity confirmed by DNA analysis;
- Llandwrog, Caernarfonshire, from late October 2017 to January 2018;
- Nant Gwynant, Caernarfonshire, from 27 November to 24 December 2017 and
- Bardsey from 20 October to 18 November 2018 (ringed on 23rd), confirmed by DNA analysis.

It may be that *C.c. blythi* occurs more frequently in Wales and that any Lesser Whitethroat present in late autumn or winter might be of this subspecies. Another subspecies, *C.c. halimodendri*, which breeds in Central Asia and western Mongolia, has been recorded elsewhere in Britain, but not in Wales.

Rhion Pritchard, Jon Green and Robin Sandham

Sponsored by Giles Pepler

Western Orphean Warbler *Curruca hortensis* Telor Llygad Arian y Gorllewin

Welsh List Category	IUCN Red List		Rarity recording
	Global	Europe	
A	LC	LC	BBRC

This species breeds in woodland, Olive groves and in scrubby riparian vegetation in northwest Africa, Iberia, southern France, Switzerland and Italy, and winters in sub-Saharan Africa from Senegal to Chad (*HBW*). It is a rare vagrant to Britain, with only six records to 2019 (Holt *et al.* 2020). Two of those were in Wales: one in a garden near St Brides, Pembrokeshire, from 10 November to 5 December 2013 and one ringed on Bardsey, Caernarfonshire, on 19 May 2019. These may prove to be the vanguard of an increasing number of records. The European population increased by 84% during 2006–16 (PECBMS 2020) and its breeding range is expected to extend north to the Channel coast by the end of this century (Huntley *et al.* 2007).

Jon Green and Robin Sandham

Western Orphean Warbler at St Brides, Pembrokeshire, November 2013 © Richard Stonier

Rüppell's Warbler *Curruca ruppeli* Telor Rüppell

Welsh List Category	IUCN Red List		Rarity recording
	Global	Europe	
A	LC	LC	BBRC

This warbler breeds in dense thorn bushes on the dry mountain slopes of southern Greece and Turkey and is a short-distant migrant that winters in northeast Africa, on both sides of the Sahara Desert (*HBW*). It is thus a rare vagrant to countries outside the eastern Mediterranean, although it has been recorded as far north as the Faroe Islands (Mitchell 2018). There had been four previous British records when one was identified at Porth Meudwy, Caernarfonshire, on 21 June 1995, and there have been no subsequent British records.

Jon Green and Robin Sandham

Sardinian Warbler *Curruca melanocephala* Telor Sardinia

Welsh List Category	IUCN Red List		Rarity recording
	Global	Europe	
A	LC	LC	BBRC

This species breeds in dense vegetation in woodland and farmland and even in gardens, around the Mediterranean. Most are sedentary, but some migrate to North Africa and to sub-Saharan Africa (*HBW*). Of 82 records in Britain to the end of 2019 (Holt *et al.* 2020), only three occurred in Wales: a male on Skokholm, Pembrokeshire, on 28 October 1968, and two in Caernarfonshire in 1994—a singing male on the Great Orme on 17–20 May and a male on Bardsey on 2–7 June. Its occurrence here may increase during the 21st century as the area of suitable climate is projected to extend north to the Channel coast (Huntley *et al.* 2007), although the trend in British records has been downward since 2010.

Jon Green and Robin Sandham

Eastern Subalpine Warbler *Curruca cantillans* Telor Brongoch y Dwyrain

Welsh List Category	IUCN Red List		Rarity recording
	Global	Europe	
A	LC	LC	BBRC

There are two subspecies: the nominate form that breeds in Sicily and southern Italy, and *C.c. albistriata* that breeds in the very northeast of Italy, the Balkans and western Turkey (Zuccon 2020). Both are presumed to winter in the eastern part of the Sahel. There were 104 records in Britain to 2019 (Holt *et al.* 2020). The first Welsh record was a female on Bardsey, Caernarfonshire, on 3 May 1984, followed by a second-calendar-year female there in 1985. Of the 18 Welsh records, all were in spring, between 13 April and 29 May. Of these, 11 were on Bardsey, and in Pembrokeshire three were on Skokholm and two on Skomer. The others were also in Pembrokeshire: on Ramsey on 12 May 1993 and the only Welsh mainland record, at St Davids Head on 13 April 2014.

Jon Green and Robin Sandham

Eastern Subalpine Warbler, Bardsey, Caernarfonshire, May 2016 © Steve Stansfield

Western Subalpine Warbler *Curruca iberiae* Telor Brongoch y Gorllewin

Welsh List Category	IUCN Red List		Rarity recording
	Global	Europe	
A	LC	LC	BBRC

This species breeds around the western Mediterranean, from Tunisia to the extreme northwest of Italy, and winters in the Sahel zone in Africa (Zuccon 2020). There were 21 British records to the end of 2019 (Holt *et al.* 2020), of which four were in Wales:

- 2CY male on Bardsey, Caernarfonshire, on 18 April 2009;
- Bardsey, Caernarfonshire, on 29 April 2011;
- male at Uwchmynydd, Caernarfonshire, on 19–30 April 2013 and
- 2CY female ringed on Skomer, Pembrokeshire, on 20 May 2018.

Unlike Eastern Subalpine Warbler, the breeding range of this species is expected to extend north as far as the English Channel by 2070–2100 (Huntley *et al.* 2007), so it may well occur more frequently in Wales in the coming decades.

Jon Green and Robin Sandham

Subalpine Warbler species *Curruca cantillans/iberiae* Rhywogaeth Telor Brongoch

The group of species formerly described as 'Subalpine Warbler' are now treated as three species: Western and Eastern Subalpine Warbler, which are presented above, and Moltoni's Warbler *Curruca subalpina*, which breeds in northern Italy, Corsica, Sardinia and the Balearic Islands (Zuccon 2020). All three winter in the Sahel, from Senegal to Sudan, but the precise areas are not yet known for each species.

Total	Pre-1950	1950–59	1960–69	1970–79	1980–89	1990–99	2000–09	2010–19
56	0	1	0	3	9	17	16	10

'Subalpine Warbler': Individuals recorded in Wales but not assigned to species, by decade.

Records of those that were identified specifically are presented above, but the remainder are reported here as 'Subalpine Warbler sp.', of which there were 872 accepted records to the end of 2019 (Holt *et al.* 2020). The BBRC is reviewing all records supported by photographs to see whether species identification can be determined.

The first Welsh record of 'Subalpine Warbler sp.' was a first-calendar-year female on Skokholm, Pembrokeshire, on 1 October 1953 and the second was also on Skokholm, in May 1970. There was a marked increase in records from the 1980s, in Wales as across Britain, which may well reflect the 176% increase in its European population during 1989–2016 (PECBMS 2020). Most records have been on islands: 20 on Bardsey, Caernarfonshire; and in Pembrokeshire, 12 on Skokholm, 12 on Skomer and four on Ramsey. There have been just a handful of mainland records: at Uskmouth, Gwent, on 26–28 April 2009; St Davids Head, Pembrokeshire, on 23–30 April 2013; Talybont, Meirionnydd, on 22 April 2000; and five in Caernarfonshire, at Porth Meudwy on 30 May 1985, Penmaenmawr on 2 May 2000, Llandudno on 14–17 May 2000, and on the Great Orme on 8–9 June 2013 and 14 May 2017.

Jon Green and Robin Sandham

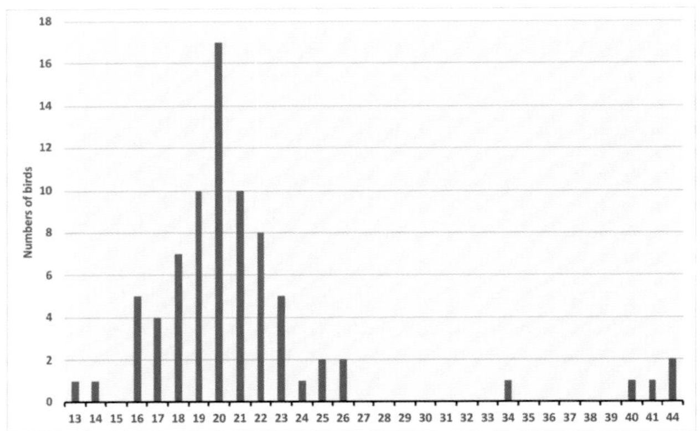

Subalpine Warbler spp.: Totals by week of arrival, 1953–2019. Week 16 begins *c.*16 April and week 26 begins *c.*25 June. Includes those identified as Eastern or Western Subalpine Warbler and those not assigned to species.

Whitethroat *Curruca communis* Llwydfron

Welsh List Category	IUCN Red List			Birds of Conservation Concern			
	Global	Europe	GB	UK	Wales		
A	LC	LC	LC	Green	2002	2010	2016

The Whitethroat breeds across most of Europe, except for northern Fennoscandia, and across temperate Asia to western Mongolia (*HBW*). It nests in hedgerows and scrub in open country with a mosaic of scattered bushes adjacent to grassy patches. In Wales it is mainly a lowland species but has been recorded at higher altitudes in scrub around young conifer plantations. It nested in long, dense heather at 420m on a moor managed for grouse shooting in northeast Wales (Roberts 1983). It is, however, scarce in *ffridd:* the Whitethroat was found at only three of 118 sites surveyed in the mid-1980s (Fuller *et al.* 2006).

The Whitethroat is migratory, the western European population wintering mainly in the Sahel zone of West Africa, south of the Sahara. Ringing recoveries from the Welsh islands suggested that some pass through Wales, to and from Ireland and Scotland, while recoveries in France, Portugal and Spain indicate the main migration route. Two ringed on Bardsey, Caernarfonshire, have been recovered in Africa: one ringed in May 1987 was recovered in Morocco in October 1990 and another, ringed in May 1991, was caught by a ringer in Senegal in April 1992. One ringed on Bardsey on 24 May 1989 showed some characteristics of the eastern subspecies *C.c. icterops*, which breeds from Turkey east to Turkmenistan (IOC).

Authors around the turn of the 20th century agreed that it was common in Wales. Forrest (1907) regarded the Whitethroat as the second commonest warbler in North Wales after Willow Warbler and stated that it was especially numerous in Llŷn, while in Pembrokeshire Mathew (1894) said that it was the third most common bird after Chiffchaff and Spotted Flycatcher. It remained common for the first six decades of the 20th century, with one-day passage counts as large as *c.*500 at Skokholm, Pembrokeshire, on 22 May 1959, 1,500 on Bardsey on 14 May 1967 and 2,500 there on 30 August 1968.

Numbers are greatly affected by a lack of rainfall in the Sahel region during June to October. Drought there during 1968–1975, particularly in 1968–69, led to a reduction of 50–90% in the population in western and central Europe (*HBW*). Only a small fraction of those that departed in autumn 1968 returned to breed in spring 1969. The decrease in Britain and Ireland was estimated to have been 77% (Winstanley *et al.* 1974). By 2017, it had only recovered around 15% of that loss (Massimino *et al.* 2019). Local evidence of the crash comes from surveys on Gower that found 78 pairs in 1968 but only 22 in 1969 and six in 1972, and on Bardsey, where 15 pairs bred in 1960, ten pairs in 1968 and none in 1969.

Fieldwork for the *Britain and Ireland Atlas 1968–72* started the year before the crash and found Whitethroats in almost every 10-km square, except in the uplands of mid-Wales. A gradual recovery followed after 1975, interrupted by another crash in 1984, so it was little surprise that the 1988–91 *Atlas* showed losses in several upland squares. The *Britain and Ireland Atlas 2007–11* showed a full range recovery, to 97% of 10-km squares in Wales, but relative abundance remained fairly low compared to eastern England. Within Wales, densities were highest in Pembrokeshire, Llŷn, Caernarfonshire, and Anglesey. A survey of Anglesey in 1986 found a minimum of 3,198 singing males. This had increased by 59% in 1997 (Jones and Whalley 2004).

Male Whitethroat © Ben Porter

A small increase in the last 30 years was evident in southern counties. In Gwent, occupation increased from 70% of tetrads in 1981–85, to 73% in 1998–2003, and in East Glamorgan, the number of tetrads with breeding evidence was 54% in 1984–89 and 55% in 2008–11, although it was recorded in 62% of tetrads in the latter period. However, tetrad occupation in Pembrokeshire remained the same between 1984–88 and 2003–07 at 85% (*Pembrokeshire Avifauna* 2020). Elsewhere, the Whitethroat was present in 64% of tetrads in Gower in 1984–89 and 54% in North Wales in 2008–12, but only 12% of tetrads surveyed in Breconshire in 1988–90, where the county population was estimated to be only *c.*100 pairs (Peers and Shrubb 1990).

The Whitethroat population trend is cyclical, as is evident in the BBS index for Wales which was 28% lower in 2018 than in 1995. However, the population across Britain increased over that period, driven by strong growth in Scotland (Harris *et al.* 2020). Its cyclical nature is evident on Skomer, where 20–26 pairs bred in 1990 and 1991, but there were only three pairs in 2005, the lowest total since the 1970s. By 2014, there were 26 territories, then 14 in 2016 and 2017. The Welsh breeding population in 2018 was estimated to be 79,500 territories (Hughes *et al.* 2020).

Whitethroats usually arrive, in Wales, in mid-April, but it has been recorded in all counties during the first eight days of the month except in Ceredigion and Montgomeryshire. During 2000–19 there were March records from Flintshire, East Glamorgan and in Gwent, where one was at Newport Wetlands Reserve on 3 March 2017. Numbers passing through the Welsh islands have never recovered to levels prior to the population crash. Day counts of 250 on Bardsey in May 1994 and 2004 were the highest since 1968, but totals have been much lower in most years. A few remain into late October or early November, such as on Bardsey on 5 November 2014 and Collister Pill, Gwent, on 12 November 2017. There have been a handful of winter records: at Clarach, Ceredigion, on 9 December 1987; at a bird table in Haverfordwest, Pembrokeshire, from 20 December 1990 to April 1991; at RAF St Athan, East Glamorgan, on 24 February 1992; and at Cwm Ivy Woods, Gower, on 23–27 November and 14 December 2003.

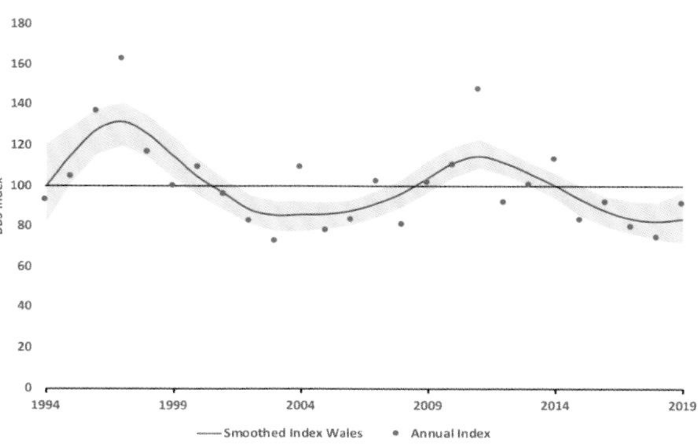

Whitethroat: Breeding Bird Survey indices, 1994–2019. The red line shows the smoothed index, 1995–2018. The shaded area indicates the 85% confidence limits.

Fluctuations in Whitethroat populations result mainly from their reduced over-winter survival when low rainfall in the Sahel reduces the abundance of invertebrates (Baillie and Peach 1992). The density of trees in the Sahel is also important (Stevens *et al.* 2010), and in particular the availability of *Salvadora* trees, the fruits of which assist fattening of Whitethroats before their return migration (Wilson and Cresswell 2006). The prospects for this species in Wales depend to a great extent on conditions in the Sahel, where there has been more rainfall in recent years, but desertification remains a problem and most climate models have predicted that this region will be drier in the 21st century.

Rhion Pritchard

Marmora's Warbler *Curruca sarda* Telor Marmora

Welsh List Category	IUCN Red List		Rarity recording
	Global	Europe	
A	LC	LC	BBRC

This bird of rocky coastal scrub and maquis breeds only on Corsica, Sardinia and small islands off the west coast of Italy and northern Tunisia. It is a short-distance migrant to northwest Africa, principally to Tunisia, where it spends the winter. It is thus a rare vagrant to Britain, with only seven records, of which one was in Wales: a male that held territory on moorland at Blorenge, Gwent, on 3–15 June 2010.

Jon Green and Robin Sandham

Dartford Warbler *Curruca undata* Telor Dartford

Welsh List Category	IUCN Red List			Birds of Conservation Concern				Rarity recording
	Global	Europe	GB	UK	Wales			
A	NT	NT	VU	Amber	2002	2010	2016	RBBP

The Dartford Warbler has a fairly restricted global distribution, from southern Britain to the Iberian Peninsula and Italy, and the coastal strip of northwest Africa. Within that range are three subspecies, of which *C.u. dartfordiensis* occurs only in Britain and near the Atlantic coast of Europe (IOC). Spain holds about 75% of the global population. The highest breeding densities in England were found where there are 1–2m high gorse bushes or thickets scattered among tall heather. Its breeding productivity is directly related to the abundance of Gorse (Bibby 1979). Adults are resident, only leaving their breeding territories in very severe winter weather, to which they are vulnerable (*HBW*). The population in southern England was reduced to just 11 pairs after winter 1962/63 but recovered strongly as a result of fairly mild winters. By 1994, its range had expanded into several new areas and the population numbered at least 1,600 pairs (Gibbons and Wotton 1996).

An early claim of a Dartford Warbler near Tremadog, Caernarfonshire, in May 1932, was considered doubtful by subsequent authors. The first accepted Welsh record was at Langland, Gower, in December 1969. There had been only six records up to 1991, all in Gwent, Gower or Pembrokeshire, and the authors of *Birds in Wales* (1994) expressed surprise that there had been even that many, since the population in southern England was small and considered to be non-migratory. The number of records increased during 1992–2000, when 14 were recorded in Wales (excluding breeding pairs), all from the same three counties (*Birds in Wales* 2002). The first record of breeding was in 1998, when two broods fledged at a site in northwest Gwent and a third brood was lost to predation (Williams 2000). Following a spell of cold weather in April 1999, no further birds were seen at this site. In 2000, three pairs bred in East Glamorgan and there was another non-breeding pair in the county. Two of these breeding pairs were in the Vale of Glamorgan, where numbers rose to nine pairs in 2008 but crashed after the cold winter of 2008/09 (Roberts 2013). The first breeding in Pembrokeshire was recorded in 2003, with six pairs at one site. A national sample survey in 2006 found 41 territories in Wales (see table), including one at 431m on the Brecon Beacons, Gwent, and the population was estimated to be 72 breeding pairs (Wotton and Conway 2008).

Male Dartford Warbler © Richard Stonier

The species has since had mixed fortunes, driven largely by the severity of winter weather. Four pairs held territory at two sites in Carmarthenshire in 2005 and 2006, but breeding was not confirmed, and it is considered that "severe gorse clearance" prevented establishment (Lloyd *et al.* 2015). 2008 produced the first records for Ceredigion, Radnorshire and inland Carmarthenshire. A bird was in suitable habitat on Anglesey in the breeding season, but there was no evidence of a mate (Steve Culley *in litt.*). Cold winters from 2008/09 to 2010/11 reduced the population in many areas, with few records in Gwent after 2006 and a retraction of range in East Glamorgan, with no confirmed breeding after 2011, although a juvenile was ringed at Kenfig in August 2016. By contrast, breeding in North Wales was first confirmed in 2011, when a male was seen carrying food at a site on the southern coast of Llŷn, Caernarfonshire. This may have resulted from birds moving from breeding areas during harsh weather several months previously. Dartford Warbler bred at this site annually to at least 2020, with up to three pairs in some years (Eddie Urbanski pers. comm.). It bred at a second site in Caernarfonshire, in the northeast of the county, during 2012–17. The main breeding areas in Wales have been the Gower Peninsula and Pembrokeshire. The Pembrokeshire population was reduced by salt-burn of gorse during winter 2013/14 but recovered to 23 pairs in 2017. The Gower population, in contrast, decreased from 20 pairs in 2015 to five in 2017. Severe weather in late winter 2018 caused numbers to crash by 39–70% in core areas in southern England (Eaton *et al.* 2020). In Wales, only eight pairs were recorded during the breeding season that followed: two in Gower, five in Pembrokeshire and one in Llŷn.

Small numbers have been recorded in winter or on passage, including in counties where they have yet to breed. Of the 20 Welsh records prior to 2001, 14 were between September and April. Since 2001, there were October records on the Gwent coast in 2009 and 2013, in Carmarthenshire in 2014, and one on Ramsey, Pembrokeshire, in 2012, that was only the second record for the island. In addition, three October records on Skomer, during 2011–15, were the only island sightings since the first in 1971 and one was near Trefechan, Ceredigion, in November 2014. Individuals stayed for several weeks or months in Gwent in winters 2003/04 and 2010/11, and there were inland records in the county in November and December 2003 and January 2006. Two first-calendar-year birds were near Painscastle, Radnorshire, in November/December 2008, and one was at Dunraven, East Glamorgan, in January 2017. Birds have also been seen in March/April, such as a male at Morfa Bychan, Carmarthenshire, in 2003, one at World's End, Denbighshire, in 2012, and in East Glamorgan one was at Sker in 2015 and two on the Heritage Coast in 2018.

County	2006	2015–19
Gwent	4	0
Gower/East Glamorgan	12	10
Carmarthenshire	10	<1
Pembrokeshire	15	9
Caernarfonshire	0	<2
Total	**41**	**<22**

Dartford Warbler: Territories recorded by county in 2006 survey (Wotton and Conway 2008) and average number reported during 2015–19.

The Dartford Warbler is declining in Spain, France and Italy (BirdLife International 2015). While the Welsh population is small, it is pioneering, at the northwestern edge of its range. There is plenty of maritime heath available as potential habitat, and if it is able to breed up to 400m, as it has done in South Wales and on Exmoor, Somerset, it could occupy substantial areas of inland moor. The very localised nature of the population makes it vulnerable to human disturbance, fires and cold winter weather until it becomes more firmly established, but management to ensure the availability of suitable stands of gorse should enable it to build on its foothold in Wales. The climate of most of southern Britain, including much of Wales, is expected to suit Dartford Warbler by the end of the 21st century (Huntley *et al.* 2007).

Rhion Pritchard

Sponsored by Gower Ornithological Society

Firecrest *Regulus ignicapilla* Dryw Penfflamgoch

Welsh List Category	IUCN Red List			Birds of Conservation Concern			
	Global	Europe	GB	UK	Wales		
A	LC	LC	LC	Green	2002	2010	2016

Weighing just 5–7g, the Firecrest shares the status of being our smallest native bird species with the closely related Goldcrest, and is one of our most striking birds. It breeds over most of Europe south of Fennoscandia, and in pockets extending to Georgia and northwest Africa. Most populations are resident, but those in central Europe winter to the south and west (*HBW*). It prefers coniferous forest but is also found in mixed and deciduous forest (*BWP*). The sites holding the highest number of breeding birds in Britain have been mature plantations of Norway Spruce (Gilbert *et al.* 1998). The UK population is now thought to exceed 6,000 pairs (Holling *et al.* 2019).

The Firecrest was a rare vagrant to Wales until the 1960s. The earliest record was of one said to have been shot at Pwllheli, Caernarfonshire, on 24 March 1878 (Forrest 1907). There were only four other records in the 19th century: two in Glamorgan and one each in Carmarthenshire and Breconshire. It continued to be very scarce in the first half of the 20th century. The first confirmed record of breeding in Britain was in Hampshire in 1962 (*Birds in England* 2005), part of a northwestern extension of its range. In the 1960s, small numbers started to be recorded regularly in Wales, mainly on autumn passage on Bardsey, Caernarfonshire, and Skokholm, Pembrokeshire. An influx was recorded in 1974, when birds wintered in several counties and one sang at Llandegfedd Reservoir, Gwent, in March. In that county, Wentwood Forest became the hub of a pioneering breeding population. At least three sang there in May/June 1975 and territorial birds were regular in subsequent years. A nest was built in 1977, but successful breeding was not confirmed until 1987, when at least four juveniles were seen. Breeding peaked at Wentwood in 1989, when there were 21 singing males and 11 pairs confirmed breeding, with a minimum of 75 young fledged. The Gwent breeding population crashed, probably due to cold weather, in February 1991, and none was found at any site that year (Venables *et al.* 2008). In 1982, six territories were found at RSPB Lake Vyrnwy, Montgomeryshire, from which at least four pairs produced 14 young, and 3–6 pairs were recorded there during the remainder of the 1980s (Holt and Williams 2008). The species had further extended its breeding range by the end of the 1980s, with singing birds recorded at other sites in Gwent, such as three in the Trelleck area in 1987, the first breeding in North Wales at Nercwys, Flintshire, in 1987, and singing males in Breconshire and Radnorshire in 1990. Breeding was not confirmed again in Wales until 1999 when a pair with young was seen at Nercwys, having been present through the previous breeding season.

It is difficult to ascertain the true extent of the breeding distribution in Wales: it is a tiny species that nests high in conifers, so is easily overlooked, and requires considerable fieldwork effort to monitor. Territories or singing males in Wales were reported every year during 2000–19 except in 2005, most regularly from Gwent and Radnorshire. The highest total was 47 in 2015, thanks mainly to a survey in Gwent that found 45 territories: 22 in Wentwood Forest and a further 23 in the Wye Valley and around Abergavenny. Breeding is also regular in Radnorshire, where four pairs bred between Evenjobb and Presteigne in 2008 and seven singing birds were found at four sites in 2009, with young fledged by at least three pairs (Jennings 2014). Territories or singing males were also recorded during 2000–18 in East Glamorgan, Carmarthenshire, Ceredigion, Breconshire, Montgomeryshire, Caernarfonshire and Denbighshire. The first successful breeding in East Glamorgan occurred in 2013, and at least possible breeding has now been

Male Firecrest © Bob Garrett

recorded in every county in Wales, except Pembrokeshire and Anglesey.

The Firecrest is, otherwise, an uncommon passage migrant, principally at coastal sites and on the islands in October and November, and to a lesser degree between March and May. The largest counts have been from Bardsey, Caernarfonshire, in autumn, notably 15 on 14 October 2001 and 16 on 10 October 2009. Smaller numbers have wintered in Wales since the early 1970s, usually in sheltered scrub or woodland edge at coastal sites, and around sewage treatment works, perhaps because of the availability of invertebrates. They have returned annually, or almost so, to a few favoured sites, such as Bosherston, Pembrokeshire, where there were five in December 2009, and RSPB Conwy, Denbighshire, which held five in December 2015. In most winters, 15–30 Firecrests probably winter in Wales, although they are likely to be under-recorded.

The number of 10-km squares occupied in Wales in winter increased from 38 (15%) in the four years of the *Britain and Ireland Winter Atlas 1981–84* to 60 (23%) in 2007–11, with a particular increase in the southwest. The *Britain and Ireland Atlas 2007–11* showed a concentration of winter records along the north and south coasts of Wales. These records probably included some late passage migrants. The origins of Firecrests passing through Wales on migration or wintering here are uncertain, and nothing is known about where birds that have bred in Wales spend the winter. There have been few ringing recoveries. The only one from outside Britain was a bird that had been ringed in Belgium in September 2014 and was recaught at Teifi Marshes, Pembrokeshire/Ceredigion, in November 2014. A first-calendar-year male ringed at Portland Bill, Dorset, on 4 October 1988, was caught by a ringer on Bardsey 18 days later.

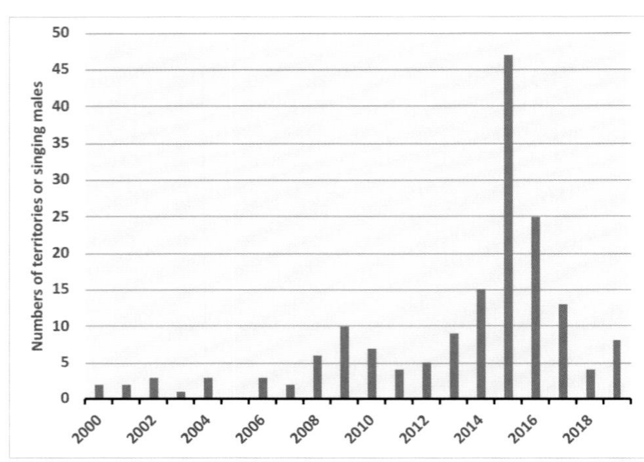

Firecrest: Number of territories or singing males in Wales, 2000–19.

The Firecrest appears to have recovered its breeding range in Wales since the 1991 setback and continues to increase. It is Amber-listed in Wales because of its small breeding population (Johnstone and Bladwell 2016), but there is much suitable habitat available and it may well spread farther. Climate modelling suggested its breeding range will extend across southern Britain, including much of Wales, by the end of the 21st century (Huntley *et al.* 2007), but the frequency of severe winter weather will determine that pace.

Chris R. Jones

Goldcrest *Regulus regulus* Dryw Eurben

Welsh List Category	IUCN Red List			Birds of Conservation Concern			
	Global	Europe	GB	UK	Wales		
					2002	2010	2016
A	LC	LC	LC	Green			

The Goldcrest's persistent high-pitched call is evocative of conifer forests and misty autumn hedgerows. It breeds in almost every part of Europe, although limited to higher altitudes in the south, and its range extends east to the Himalayas and parts of China and Japan. Most populations are migratory to some extent, but only the most northerly parts of the breeding range are completely abandoned in winter. Many birds breeding in northern Europe winter in the south of the continent, with a few going as far as northwest Africa (*HBW*).

This insectivore is widespread in Wales throughout the year, predominantly nesting in coniferous plantations, where it can occur at high densities. It breeds from sea level, such as those on the planted dunes of Newborough, Anglesey, to some of the highest upland plantations, for example, up to 625m in Radnor Forest, Radnorshire (Jennings 2014). Goldcrests can, however, be readily encountered almost anywhere that provides groups of conifers, either exotic or native, and is frequently found in churchyard Yews, large gardens and urban parks, especially outside the breeding season. One was recorded at an altitude of 816m on Yr Wyddfa, Caernarfonshire. The only parts of Wales from which breeding Goldcrests are normally absent are urban and industrial centres, treeless uplands, and parts of Anglesey and coastal stretches.

The Goldcrest appears to have been less numerous as a breeding bird in the 19th century and the early 20th century than subsequently. Dillwyn (1848) described it as "not uncommon, particularly in winter" in the Swansea area of Gower. By the late 19th century, it was locally common in Glamorgan around conifers, while in most parts of Pembrokeshire and North Wales, it was described as a common resident, with increased numbers in winter (Mathew 1894, Forrest 1907). The planting of conifers, following the First World War, increased the area of suitable habitat for Goldcrests and by the middle of the century there were large numbers in Radnor Forest, where conifers had been planted in the mid-1920s (Ingram and Salmon 1955). More extensive planting after the Second World War led to a further increase and although the species was hit hard by severe winters in 1946/47 and 1962/63, Roderick and Davis (2010) reported that it recovered quickly after these winters in Ceredigion.

The *Britain and Ireland Atlas 1968–72* showed it to be present in 93% of 10-km squares in Wales, and despite a slight fall in occupancy in 1988–91, by 2007–11 it occupied 95% of squares. Where the results can be compared, county tetrad atlases also suggested an expansion since the 1980s. The *Gwent Atlas 1981–85* recorded Goldcrests in 77% of tetrads, a rate that had increased to 84% in 1998–2003. The number of tetrads with breeding evidence in East Glamorgan was similar in 1984–89 and 2008–11 at around 53% to 52%, while in Pembrokeshire, it increased from 52% of tetrads in 1984–88 to 68% in 2003–07 (*Pembrokeshire Avifauna* 2020). The cold winter of 1979 was thought to have caused high mortality and the population was lower in Glamorgan and Pembrokeshire in the 1980s (Hurford and Lansdown 1995, Rees *et al.* 2009).

Numbers recorded by the Repeat Woodland Bird Survey in Wales in 2003/04 were 236% higher than in 1965–72 (Amar *et al.* 2006). The population was stable during the early years of the BBS in 1995–2001, but this was a high point from which it declined subsequently. There were noticeable drops following the cold weather in winters 2009/10, 2010/11 and March 2018, and by 2019 the BBS index was 54% below its 1995 level (Harris *et al.* 2020). Other parts of the UK showed similar temporary dips but not the underlying decline and there is no obvious explanation

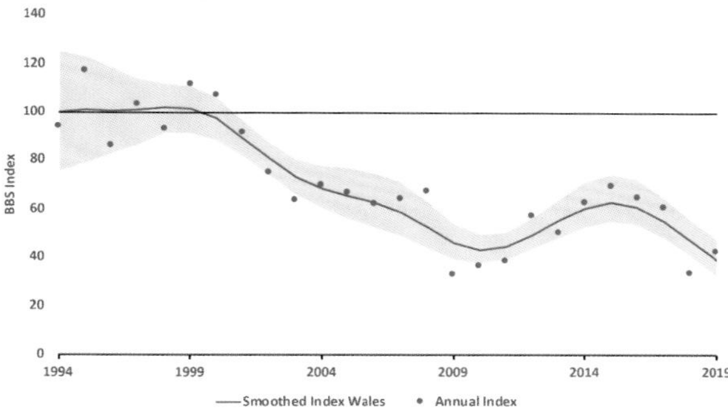

Goldcrest: Breeding Bird Survey indices, 1994–2019. The red line shows the smoothed index, 1995–2018. The shaded area indicates the 85% confidence limits.

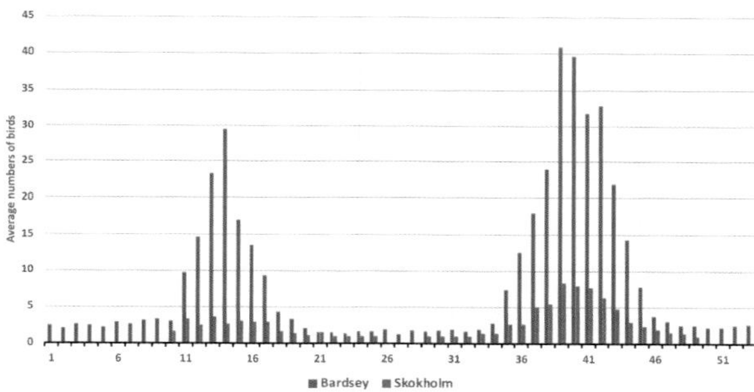

Goldcrest: Average counts by week number recorded by bird observatories, 1953–2019. Week 14 begins *c.*1 April and week 38 begins *c.*16 September.

for the difference. The Welsh breeding population in 2018 was estimated to be 85,500 territories (Hughes *et al.* 2020).

Autumn sees an influx of Goldcrests into Wales, with passage sometimes apparent inland but more discernible on offshore islands and coastal headlands. Over 27,000 were ringed on Bardsey, Caernarfonshire, during 1953–2008 (Archer *et al.* 2010), but there have been comparatively few recoveries outside Britain. Most recoveries involved those ringed at island bird observatories and show that many pass through Wales but do not stay to winter: an adult female ringed near Kaliningrad, Russia, was recaught on Bardsey on 26 October 1975, a journey of 1,695km in just 22 days, while a juvenile female ringed on Bardsey in September 1994 was killed by a cat in Switzerland 47 days later. Within Britain, there is a broadly southeasterly movement in autumn and a northwesterly movement in spring (*BTO Migration Atlas* 2002). Little information is available on the movements of the Welsh breeding population, but most Goldcrests nesting here probably do not move very far.

Bardsey usually records the highest numbers of migrants, with peaks in autumn and spring passage, in mid-October and mid-April respectively. Peak autumn counts there include 1,000 on 19 September 1988, 1,200 on 26 September 1989 and 1,300 on 3 October the same year. An exceptional influx in spring 1998 started on 25 February and continued until late March, peaking on 29 March, when 2,169 were counted. On the Pembrokeshire islands passage is usually far more modest, with daily autumn totals over 50 on Skokholm or Skomer being unusual, though 250–300 were forced down by rain on Skokholm on 8 October 1959 (Betts 1972) and 50 on Ramsey on 15 October 2000 was notable. Spring daily maxima are rarely in double figures in the south.

Our wintering population presumably consists of resident birds, supplemented by an influx of birds from farther north. Loose winter flocks of Goldcrests are often encountered in association with Coal Tit, Blue Tit, Great Tit, Long-tailed Tit and Treecreeper. There was a small increase in distribution between the *Britain and Ireland Winter Atlas 1981–84* and the 2007–11 *Atlas*, primarily in central Wales, and relative abundance in winter was similar to the breeding season in the latter period.

The Goldcrest is Amber-listed in Wales because of a moderate breeding population decline over 25 years (Johnstone and Bladwell 2016), although it may qualify for Red-listing in the next review. In the short term, Goldcrest numbers are governed by the severity of winters. Mortality can be very high in protracted cold periods, especially when frost or lying snow prevent access to the supply of minute insects in the axils of conifer needles. The population generally recovers from such events within a couple of years, but this may be slower following very severe or consecutive cold winters.

There is little evidence available on the causes of longer-term population change. Goldcrest numbers are likely to be influenced by the availability of mature coniferous forestry plantations. During 2001–16, there was a reduction of 18,000ha (10%) in the area of coniferous forest in Wales (Warren-Thomas and Henderson 2017). The decline in Goldcrest numbers since about 2000 may reflect the move away from even-aged stands of single conifer species. The future prospects for this species in Wales may depend largely on forestry policies.

Chris R. Jones

Sponsored by Sîan Stacey

Wren *Troglodytes troglodytes* Dryw

Welsh List Category	IUCN Red List			Birds of Conservation Concern			
	Global	Europe	GB	UK	Wales		
A	LC	LC	LC	Green	2002	2010	2016

The Wren is widely distributed over most of Europe and parts of Asia as far east as Japan and is the only member of the large genus *Troglodytes* found outside the Americas. Those that breed in Wales are of the subspecies *T.t. indigenus*, which is found over most of Britain and Ireland, but sedentary endemic subspecies occur on Shetland, Fair Isle and St Kilda, with another on the remaining Outer Hebrides (IOC). Populations in northern Europe are highly migratory (*HBW*) and ringing recoveries suggest that some move south, both from and into Wales, in autumn and winter. A nestling ringed at Tremadog, Caernarfonshire, in May 2005 was killed by a cat at Bognor Regis, West Sussex, 337km to the southeast in September 2005, while a first-calendar-year bird ringed near Ravenglass, Cumbria, in October 1977, was found dead near Llandigwynnet, Pembrokeshire, during cold weather in February 1978, a distance of 309km. Some Wrens breeding in the uplands move to lower altitudes in winter and some remote uplands may be completely deserted (*Britain and Ireland Atlas 2007–11*), although its ability to hunt spiders and other invertebrates in deep recesses beneath boulders and scree may render it almost invisible in winter.

A Wren typically lives around two years, but the oldest known to the BTO ringing scheme was a juvenile ringed on Bardsey on 22 July 1997 and last recorded there on 28 October 2004, aged seven years, three months and six days. Of 84 ringed on the island that were later found dead, only one had not remained on Bardsey. It was killed by a cat at Llanbedrog, Caernarfonshire, 24km to the northeast.

It seems always to have been a common and well-known bird in Wales, at one time having considerable cultural significance, perhaps dating back to a very early period. 'Hunting the Wren' was

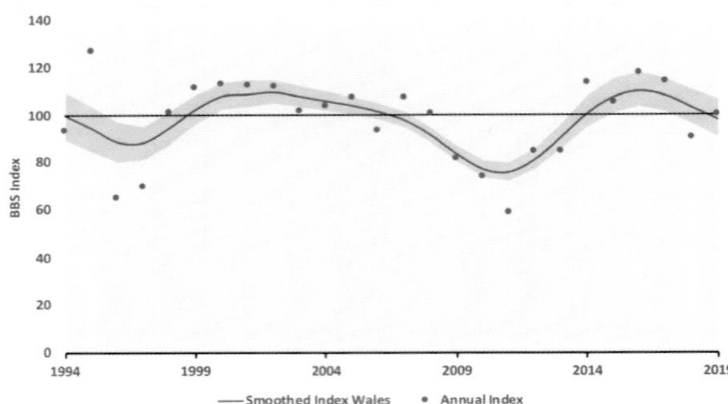

Wren: Breeding Bird Survey indices, 1994–2019. The red line shows the smoothed index, 1995–2018. The shaded area indicates the 85% confidence limits.

a tradition in several countries but particularly prevalent in Ireland and Wales between St Stephen's Day (26 December) and Twelfth Night (6 January). It involved young people catching a Wren and carrying it to neighbouring houses in a cage or wooden box to wish the householder good fortune. In Wales, the tradition was particularly associated with Pembrokeshire, and is thought to have disappeared around 1890 (Gwyndaf 1995).

Wren is one of the most widespread species in Wales, breeding in most habitats from sea level to around 700m, but probably most numerous in deciduous woodland. It was found in at least 98% of 10-km squares in Wales in all three Britain and Ireland atlases, in 1968–72, 1988–91 and 2007–11. In the last period, relative abundance within Wales was highest on Anglesey, in Pembrokeshire and along the south coast. Local atlases confirmed its status as among the most widely distributed of birds, occurring in over 98% of tetrads in both Gwent in 1998–2003 and in Gower by the *West Glamorgan Atlas 1984–89*. It was found in 95% of tetrads in Pembrokeshire in 2003–07, in 93% in the *North Wales Atlas 2008–12*, in 90% visited in Breconshire in 1988–90 and in 88% in East Glamorgan, with breeding evidence in 84%, in 2008–11.

Wrens suffer severely in very cold winters, probably more than any other species (Parslow 1973). This is the main factor influencing population size (Peach *et al.* 1995). Losses were very heavy during the exceptionally severe winter of 1962/63, when the breeding population in Glamorgan was virtually eradicated (Hurford and Lansdown 1995) and numbers were reduced by 90% in Radnorshire (Jennings 2014). On Bardsey the number of territories fell from 24 in 1962 to only four in 1963. However, numbers can recover quickly. They were thought to have recovered to pre-1963 levels in Glamorgan by 1966. Winter 1981/82 reduced breeding numbers in Gwent by 60–70% and at RSPB Ynys-hir, Ceredigion, by 79%. The short-term effects of cold conditions in 1995/96, 2009/10, 2010/11 and 2017/18 are evident on the BBS index, but the population in Wales nonetheless increased by 9% during 1995–2018 (Harris *et al.* 2020). That period of population growth was preceded by a 73% increase in numbers in Welsh woodlands between 1965–72 and 2003–04 (Amar *et al.* 2006). Relative density increased strongly on Anglesey and the lowlands of mainland North Wales between 1994–96 and 2007–09, with similar increases in northern England and mainland Scotland. Farther south, there was little change (Massimino *et al.* 2019).

In good habitat, and following a run of mild winters, it can occur at high density, more than 250 territories/km² in high-quality broadleaved woodland (e.g. Marshman 1979). Population density in coniferous woodland is closely correlated with the volume of trimmed branches or other ground cover present (Moss 1978). Densities are lowest on moorland and mountains. One was reported breeding 50m below the summit of Yr Wyddfa, Caernarfonshire, at about 1,035m (Pritchard 2017), although males will construct an unlined nest for roosting and one at 830m on the same mountain was a 'cock's nest' of this kind (Ratcliffe 1990). Ratcliffe also noted that little was known about Wrens in the uplands.

The islands have some of the best monitored populations. On Bardsey, numbers grew to 246 territories (137/km²) in 2017, having increased very quickly from 34 to 122 territories during 1997–2001, but then crashed by 45% in 2018, following cold weather in early March that year. Wrens also breed on the Pembrokeshire islands, for example 92 territories (32/km²) on Skomer in 2007 and 60

Fledgling Wren © Ben Porter

Study area	Territories/km²	Year(s)	Habitat
RSPB Ynys-hir, Ceredigion(1)	144	2006	All woodland types
RSPB Coed Penrhyn Mawr(1)	255	1985–92	Oak woodland
Aber Valley, Caernarfonshire(2)	107	1972	Sessile Oak woodland
Elan Valley, Radnorshire(3)	167	1995–2010	Mostly ungrazed broadleaved woodland
Ystrad/Llwynypia, East Glamorgan(4)	61	1975	Mixed habitat above 150m
Ruabon Moor, Denbighshire(5)	10.7	2003	Moorland

Wren: Territory density at selected locations in Wales. From: (1) RSPB, per Roderick and Davis 2010, (2) Gibbs and Wiggington 1973, (3) Jennings 2014, (4) Marshman 1979, (5) *Welsh Bird Report* 2003.

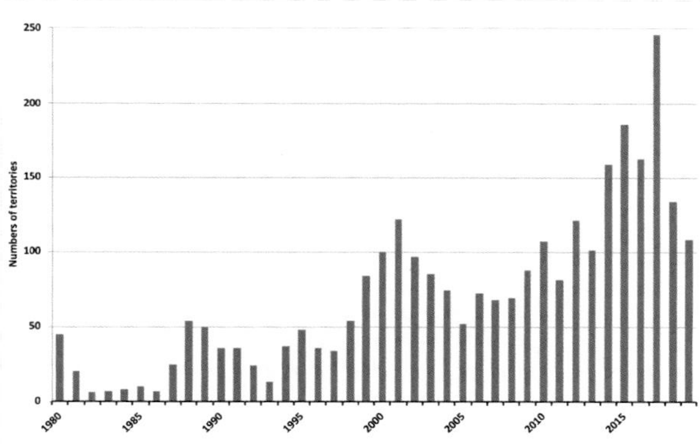

Wren: Territories on Bardsey, Caernarfonshire, 1980–2019 (from Bardsey Bird & Field Observatory reports).

pairs (23/km²) on Ramsey in 2014. On Skokholm, where breeding was first recorded in 1988, numbers since 2011 have been much higher than formerly. There were 57 singing males (54/km²) in 2014, but prior to 2011 the previous peak had been 19 territories in 1994. There were 37 singing males (86/km²) on Puffin Island, Anglesey, in 2000.

There is a small autumn passage in coastal areas and islands in some years, although the Wren is such a ubiquitous resident that this is difficult to detect. Prior to the establishment of a breeding population on Skokholm in 1988, it was a common winter visitor with a substantial arrival each October. On Grassholm, Pembrokeshire, Wrens appear regularly between late August and November, and although humans are rarely present to witness this, 25 were seen on 27 September 1972. Several were recorded at a gas platform 39km WNW of St Davids Head, Pembrokeshire, in October 2005. Communal roosts in cold weather can involve a considerable number of birds: up to 100 roosted in a nest-box at Tudweiliog, Caernarfonshire, in early 1997, and at least 60 in an old House Martin nest at Doldowlodd, Radnorshire, in January 2001.

Severe winter weather formerly led to considerable annual fluctuations in the size of the population, but such conditions are becoming less frequent as a result of climate change, which is benefiting the Wren (Pearce-Higgins and Crick 2019). The Welsh breeding population in 2018 was estimated to be 1.3 million territories (Hughes *et al.* 2020), making it the most numerous species in Wales, and there seems no reason to think that its status will change.

Rhion Pritchard

Nuthatch *Sitta europaea* Delor Cnau

Welsh List Category	IUCN Red List			Birds of Conservation Concern			
	Global	Europe	GB	UK	Wales		
A	LC	LC	LC	Green	2002	2010	2016

The Nuthatch has an extensive Palearctic distribution, through much of Europe, except northern Fennoscandia and Ireland, and through temperate Asia to Japan and the Kamchatka Peninsula. It is a bird of mature forest or parkland with large, old trees where it can find nesting cavities. Even small woodlands are used provided they are linked by mature hedgerows, small copses or corridors of streamside trees (*Birds in Wales* 1994). It is rarely found above 300m, although one at 365m at Ystrad/Llwynpia, East Glamorgan, was about 800m from the nearest tree (Hurford and Lansdown 1995). Adults remain on territory all year, and although juvenile dispersal is usually local, a few travel an exceptional distance: one ringed as a nestling near Crickhowell, Breconshire, in May 2010, was recaught at Southam, Warwickshire, in January 2011, 134km to the northeast.

In the late 19th century, Nuthatch was common only in lowland areas of Wales close to the English border, such as in eastern Radnorshire (Jennings 2014). It was scarce in Flintshire and Denbighshire in the 1890s (Dobie 1894), although a decade later it was common around Bangor-on-Dee in Denbighshire and along the Severn Valley, Montgomeryshire, a distribution that was described as "peculiar and almost unaccountable" by Forrest (1907). There was a single record in the northwest, at Caernarfon in 1902. The first dated record for Ceredigion was in 1899, although it was said to be seen occasionally at Gogerddan prior to 1894. It remained scarce in the county until the 1920s (Roderick and Davis 2010), as it was in Breconshire at that time (Peers and Shrubb 1990). Farther south, there were records from Glamorgan from 1802, and Dillwyn (1848) said that it was "not infrequent" at Penllergaer, Gower, but very few were documented later in the 19th century (Hurford and Lansdown 1995). Davies (1858) described Nuthatch as rare and local in Carmarthenshire, and Mathew (1894) said that although it was not uncommon in Carmarthenshire, he did not think it was resident in Pembrokeshire. He knew of only one record from the county, which had been shot at Slebech on 7 September 1893.

The European range of Nuthatch expanded north and west during the 20th century. This was evident in Wales, apparently unaffected by extensive felling of woodland during the First World War. The first Nuthatch record on Anglesey was near Llangoed in 1910. The population there spread, albeit slowly, and it remained confined to the south of the island for many years. It spread west in Radnorshire during 1920–40 (Jennings 2014) and rapidly through Ceredigion from 1920. By 1950 it was found in all deciduous woodland in northeast Ceredigion but was less common in the uplands (Condry 1950, Roderick and Davis 2010). It was fairly common around Aberystwyth in 1946–56 (Peach and Miles 1961).

Nuthatch © John Hawkins

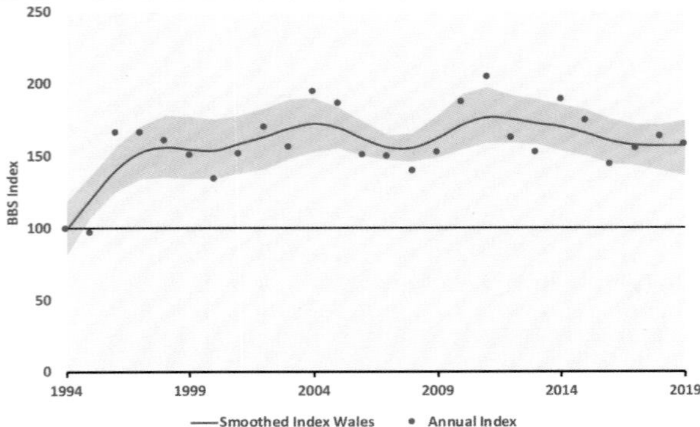

Nuthatch: Breeding Bird Survey indices, 1994–2019. The red line shows the smoothed index, 1995–2018. The shaded area indicates the 85% confidence limits.

It spread west from Gwent through East Glamorgan and Gower during the 1930s, and by the 1960s was numerous and widely distributed (Hurford and Lansdown 1995). By the late 1940s to early 1950s, it was increasing in Carmarthenshire (Ingram and Salmon 1954), common in Pembrokeshire (Lockley *et al.* 1949) and locally common in Caernarfonshire (Whalley 1954).

The first complete picture of distribution was given by the *Britain and Ireland Atlas 1968–72*, which showed it to be found in 87% of 10-km squares in Wales, absent in the lowlands only from north and west Anglesey. It was not greatly different by 1988–91, but by 2007–11 much of Anglesey had been colonised and the species occurred in 90% of 10-km squares in Wales, with absence only in the uplands of mid-Wales and from some coastal squares. Abundance was relatively high compared to the rest of Britain and had increased almost everywhere since 1988–91. The highest densities were in the southern half of lowland Wales, including in Pembrokeshire, where it had been unknown a century previously.

Where tetrad atlases have been repeated, all show some expansion in range since the 1980s. The Nuthatch was found in 76% of tetrads in Gwent (1998–2003), with gains mainly in the upper parts of the western valleys (Venables *et al.* 2008). The increase was marginal in East Glamorgan (2008–11), where there was breeding evidence in 51% of tetrads, and in Pembrokeshire (2003–07), where the equivalent figure was 44%. Elsewhere, it was found in 63% of tetrads in Gower (1984–89), 52% of those visited in Breconshire (1988–90) and 48% of tetrads in North Wales (2008–12). In all cases it remained absent from the highest ground. It was widespread in the lowlands, except on Anglesey, which still showed large gaps in the north and west, where the few patches of large deciduous trees are isolated. It is not surprising that such a sedentary species is rarely recorded on the islands. Bardsey, Caernarfonshire, did not record its first Nuthatch until 1979, and there have been only five further records, all in September or October. Skomer first recorded the species in 1986.

The Nuthatch population increased by 36% in a suite of Welsh woodlands surveyed in 1965–72 and in 2003–04 (Amar *et al.* 2006). The BBS showed a continued increase in the more recent period, the index for Wales increasing by 30% during 1995–2018 (Harris *et al.* 2020). The Welsh breeding population in 2018 was estimated to be 60,000 territories (Hughes *et al.* 2020).

The expansion into western Wales was matched by a northward expansion through England, and from the early 1990s it spread rapidly in Scotland (*The Birds of Scotland* 2007), in line with climate modelling (Huntley *et al.* 2007). Mild winters have benefited Nuthatch and the increased availability of food in gardens may also have helped, as this species is a regular visitor to peanut feeders. It occurs in 40% of gardens in Wales participating in the BTO Garden Birdwatch. Additionally, the average brood size has increased from 3.7 juveniles/breeding attempt in 1967 to 5.5 juveniles in 2017 (Massimino *et al.* 2019). The population in Wales appears to have occupied almost all suitable habitat. Its future here depends largely on the availability of large, mature, deciduous trees. There do not appear to be any serious threats, and it should remain a common bird.

Rhion Pritchard

Sponsored by Teulu Owsianka Roberts

Treecreeper *Certhia familiaris* Dringwr Bach

Welsh List Category	IUCN Red List			Birds of Conservation Concern			
	Global	Europe	GB	UK	Wales		
					2002	2010	2016
A	LC	LC	LC	Green			

The Treecreeper is found over most of Europe and across temperate Asia as far east as Japan. The subspecies *C.f. britannica* breeds only in Britain and Ireland (IOC). It is a woodland bird, favouring mature trees with plenty of cracks and crevices for feeding, roosting and breeding. The nest is usually in such a crevice or behind loose bark. It will sometimes use specially designed nest-boxes but does not take to them readily. In Britain it is more common in deciduous and mixed woodland than in conifers. It is also found in well-wooded farmland, parks and large gardens. A lack of suitable nest sites, rather than food supply, may limit its population in conifer plantations (*Britain and Ireland Atlas 1968–72*). It is mainly a lowland species, seldom breeding above 300m in Wales, perhaps because of a lack of suitable nest sites, although possible breeding has been recorded up to around

430m in Meirionnydd (Pritchard 2012). The few ringing recoveries involving Wales show that it is very sedentary, with the majority of movements being 5km or less.

There does not appear to have been any great change in the status of the Treecreeper in Wales. It was described as a common resident in the 1890s in Glamorgan, Pembrokeshire and Carmarthenshire, and in North Wales below 300m (Hurford and Lansdown 1995, Mathew 1894, Barker 1905, Forrest 1907). Its status in Ceredigion at this time was unclear. Salter described it as "numerous" but later retracted the statement, writing that he thought it was rare until the mid-1920s (Roderick and Davis 2010). The planting of conifers after the Second World War provided fewer benefits to the Treecreeper than for some other woodland birds.

This is a rather unobtrusive species, whose song and calls can easily be missed among the bird sounds in a woodland in spring. Treecreepers sing early in the year and may have stopped before woodland surveys begin and so can be under-recorded. It was found in 90% of 10-km squares in Wales during the *Britain and Ireland Atlas 1968–72*, but fewer in the 1988–91 *Atlas*, although it made a partial recovery by the 2007–11 *Atlas*, to 88% of squares. Its distribution in the last period was, not surprisingly, almost identical in the breeding season and winter. Abundance in 2007–11 was relatively low on Anglesey, but elsewhere was as high as anywhere in Britain and Ireland. In northern and southern Wales, relative abundance was lower than in 1988–91, but it was higher in mid-Wales.

Tetrad atlases showed considerable variation in the proportion of squares occupied, related to the availability of suitable habitat. In Gwent, the species was recorded in 79% of tetrads in 1981–85 and in 75% in 1998–2003, whereas it was less widespread in East Glamorgan, occupying 51% in 1984–89 and only 37% in 2008–11, with breeding evidence in just 29%. In Gower it was found in 58% of tetrads in the *West Glamorgan Atlas 1984–89* and in 41% of those visited in Breconshire in 1988–90 (Peers and Shrubb 1990). In Pembrokeshire, it occurred in 44% of tetrads in 1984–88 and in 40% in 2003–07. It is far less widespread in northern counties. The *North Wales Atlas 2008–12* found it in only 31% of tetrads, absent from most of north and western Anglesey, where tree cover is limited, as well as from the treeless uplands.

The highest densities are usually in deciduous woodland. The small number of woodland Common Birds Census plots surveyed in Wales recorded a maximum density of *c.*11.3 pairs/km^2. A survey of RSPB Ynys-hir, Ceredigion, in 2006 found a density of 31.3 territories/km^2, and in the Radnorshire part of the Elan and Claerwen valleys there averaged 60.3 singing birds/km^2 in 1996. It is far less numerous in conifer plantations, which offer few suitable nest sites. Bibby *et al.* (1989) recorded Treecreeper only four times in two visits to each of 253 survey points in plantations in North Wales. However, the Charles Ackers Redwood Grove at Leighton, Montgomeryshire, held 30 pairs/km^2 in 1970 (Williamson 1971). Treecreepers can hollow out roost sites in the soft bark of *Sequoia* and *Sequoiadendron* (Wellingtonia) trees. At least 11 roosted in such trees around Llandrindod Wells Museum, Radnorshire, in January 1998.

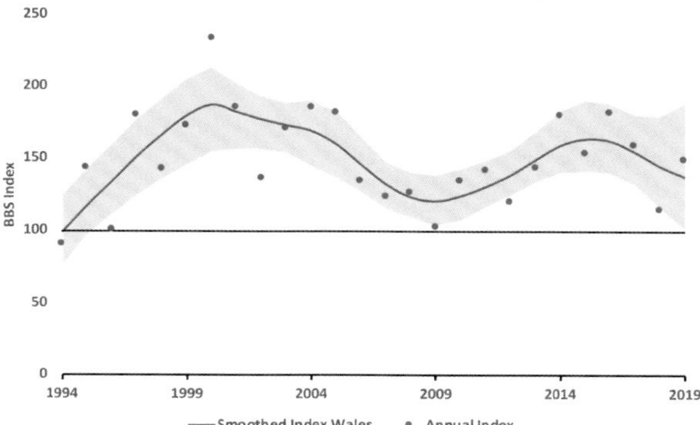

Treecreeper: Breeding Bird Survey indices, 1994–2019. The red line shows the smoothed index, 1995–2018. The shaded area indicates the 85% confidence limits.

Treecreeper © Kev Joynes

Treecreepers can be affected by both cold and wet winter weather (Peach *et al.* 1995). The population increased substantially in the UK between 1966 and 1973, when it may still have been recovering from the severe 1962/63 winter. It decreased through the remainder of the 1970s and has been relatively stable since (Massimino *et al.* 2019). The only broad-scale data from Wales during the second half of the 20th century showed a 92% increase in the same woodlands surveyed in 1965–72 and 2003–04, although numbers fell slightly in the second half of that 30-year period (Amar *et al.* 2006). The BBS index for Wales shows considerable variation between years but increased 22% during 1995–2018 (Harris *et al.* 2020). In Radnorshire, Jennings (2014) considered that the severe cold of winter 2009/10 reduced numbers by an average of 15% across six broadleaved woodlands. The Welsh breeding population in 2018 was estimated to be 37,500 territories (Hughes *et al.* 2020).

There is little evidence of movements outside the breeding season, although a few Treecreepers do sometimes reach the Welsh islands, most frequently in September. These, and a few apparent 'migrants', sometimes seen around mainland coasts, are likely to be young birds making a fairly local post-natal dispersal. Treecreepers have been recorded almost annually on Bardsey, Caernarfonshire, usually singly, but four were there on 8 September 1961. The nearest probable breeding site is 15km away. Treecreepers are less frequent on the other islands but have been found on most, at least once or twice, including The Skerries, Anglesey. One on Grassholm, Pembrokeshire, on 17 June 1957, was found clinging to the only wooden post on the island, and another was on the island on 13 October 2001. In winter, Treecreepers sometimes join roving tit and Goldcrest flocks, seen at up to 550m in the conifers of Radnor Forest. There were ten at Coed y Wenallt, East Glamorgan, on 1 March 1985 and at least 13 with a large tit flock in the Elan Valley, Radnorshire, in November 1989. There were 45 roosting in a nest-box at Llaneilian, Anglesey, on 11 February 1996.

The Treecreeper appears to be doing well in Wales, and it faces no immediate conservation threat, although, in the longer term, climate modelling indicated that most of southern and central Wales will be unsuitable for Treecreeper by the end of this century (Huntley *et al.* 2007).

Rhion Pritchard

Grey Catbird *Dumetella carolinensis* Cathaderyn Llwyd

Welsh List Category	IUCN Red List	Rarity recording
	Global	
A	LC	BBRC

This rare trans-Atlantic vagrant has occurred just twice in Britain, the first at South Stack, Anglesey, on 4–6 October 2001. It was the first American passerine recorded on Anglesey, although it beat a Red-eyed Vireo to that accolade by just a few hours. Along much of the Atlantic coast of the USA it is a resident species, but those that breed in New England and the rest of North America migrate to the Gulf Coast, Caribbean and the west coast of Central America in autumn. Their preferred habitat is thorny scrub and woodland edge (*HBW*). There has been only one other British record, from Cornwall in October 2018 (Holt *et al.* 2019).

Jon Green and Robin Sandham

Rose-coloured Starling *Pastor roseus* Drudwen Wridog

Welsh List Category	IUCN Red List		Rarity recording
	Global	Europe	
A	LC	LC	WBRC

This nomadic breeder nests in steppe and semi-desert habitat from southeast Europe through Central Asia and south to Afghanistan, and all populations winter in India (*HBW*). The species' breeding range had been slowly spreading west into the Balkans, but there has been little significant change in the last 30 years (*European Atlas* 2020). Populations are prone to irrupt in summer after several good breeding years, generally to the west, perhaps related to the availability of food. A wide-scale irruption in spring 2020 led to the first breeding in France. There is no clear reason as to why irruptions have occurred more frequently, nor from where birds seen in Britain originate.

Total	Pre-1950	1950–59	1960–69	1970–79	1980–89	1990–99	2000–09	2010–19
132	13	1	3	5	4	17	42	47

Rose-coloured Starling: Individuals recorded in Wales by decade.

Rose-coloured Starlings have been recorded in all Welsh counties except Montgomeryshire. There were five records from Wales in the 19th century, all of which were shot. The first was at Holyhead, Anglesey, in 1835, followed by birds at Magor in Gwent and Singleton Park in Gower in 1836. The others were at Point of Ayr, Flintshire, in 1861; near Bangor, Caernarfonshire, in 1867; and an adult shot at Cynghordy, Carmarthenshire, on an unknown date (Phillips 1899). There were very few records in the first half of the 20th century, and the recording rate did not increase until the 1990s. The number of records in Britain accelerated in the first decade of the 21st century: three times as many were recorded each year in 2000–18 than in 1990–99 (White and Kehoe 2020b).

Rose-coloured Starling at Rhos-on-Sea, Caernarfonshire, June 2012 © Bob Garrett

The vast majority of records have been adults in the first three weeks of June, as spring overshoots from southeast Europe. A smaller number, predominantly first-calendar-year birds, occurred in late summer and autumn, although these are less likely to be reported by the general public than the more distinctive pink-and-black adults. Some autumn arrivals have over-wintered in Wales, often using garden feeding stations: at Horton, Gower, in 2001/02; at Cydweli, Carmarthenshire, in 2006/07; in Pembrokeshire, at St Davids in 1986/87 and Pembroke Dock in January 1996; at Llandudno, Caernarfonshire, in 2018/19; and on Anglesey, at Holyhead in 2011/12 and Cemlyn in 2019/20.

Most records have been of singles, although there are five records of two together: at Mostyn, Flintshire, on 18 August 1928; two adults at St Davids, Pembrokeshire, on 24–25 August 2002; two first-calendar-year birds at Ynyslas, Ceredigion, on 25–26 September 2016; two adults at Trearddur Bay, Anglesey, on 27 May 2018; and two adults at Ynyslas, Ceredigion, on 6 June 2018. There have been fewer than a dozen truly inland records since the

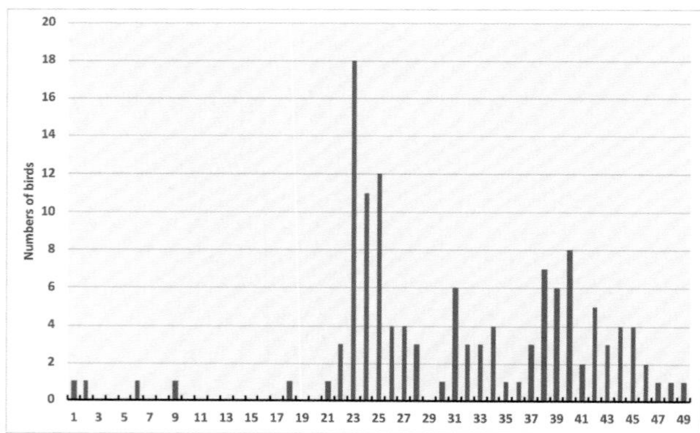

Rose-coloured Starling: Totals by week of arrival (when known, n=127), 1928–2019. Week 23 begins *c.*4 June and week 45 begins *c.*5 November.

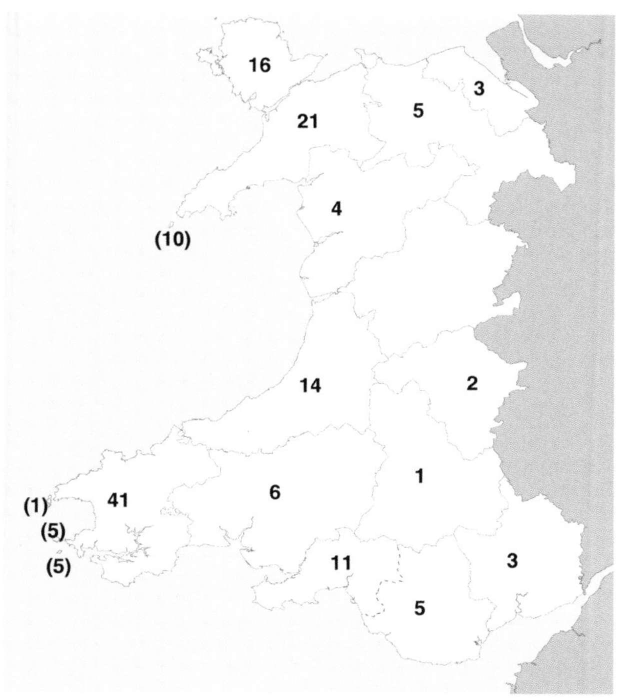

Rose-coloured Starling: Numbers of birds in each recording area, 1836–2019.

one shot in Carmarthenshire in the 19th century: at Monmouth, Gwent, in September 1937; Ystumtuen, Ceredigion, in July 1975; Llansantffraed-Cwmdeuddwr, Radnorshire, in September 1994; Aberdare, East Glamorgan, in August 2000; Rhuthun, Denbighshire, in June 2002; Llangefni, Anglesey, in June 2012; Llandysul, Carmarthenshire/Ceredigion, in October 2013; Llanyre, Radnorshire, in October 2014; Brecon, Breconshire, in June 2018; Ammanford, Carmarthenshire, in September 2018 and Acton Park, Denbighshire, in June 2019.

Rose-coloured Starling seems to be undergoing a significant change in its population and distribution, and the number of records looks set to increase, based on recent trends. Historically, it has been greatly affected by insecticide spraying of locusts (*European Atlas* 1997). Locust swarms may increase in Europe as a result of climate change, and if insecticides were to be the response, this could stem the long-term trend.

Jon Green and Robin Sandham

Starling — *Sturnus vulgaris* — Drudwen

Welsh List Category	IUCN Red List			Birds of Conservation Concern			
	Global	Europe	GB	UK	Wales		
A	LC	LC	VU	Red	2002	2010	2016

The Starling is resident across much of Europe and east through Siberia to northern Mongolia. Populations that breed in Fennoscandia and western Russia winter to the west and south as far as the Mediterranean (*HBW*). Large numbers of these occur in Wales.

As a breeding species, its status has changed from widespread to localised in lowland Wales during the last 30 years. It occurs in close proximity to humans where buildings offer nesting opportunities in summer and surplus food around farms in winter. As well as urban and suburban concentrations, it breeds in woodlands, inland and coastal cliffs, quarries and farmsteads, and even in earth banks or stone walls, such as on Bardsey, Caernarfonshire. It occasionally uses the holes of other species such as Rabbit or Sand Martin (*BWP*). Birds feed on soil invertebrates and in rural landscapes rely on Cranefly larvae to feed nestlings.

The Starling has not always been numerous throughout Wales, although it has been present at least since the Ice Age: fossilised remains were found at Bacon Hole cave, Gower, from an interglacial period 115,000–130,000 years ago (Stringer 1977). In the early 19th century, it was virtually unknown as a breeding bird and occurred only as a numerous passage migrant and abundant winter visitor, although it was recorded as breeding in Glamorgan in 1802 (Hurford and Lansdown 1995). There was a north and westward expansion across the UK between *c.*1830 and 1880 (Parslow 1973), although the mountain core of Wales formed a genuine barrier to colonisation. Only a few pairs bred in Pembrokeshire by the 1850s (Mathew 1894) and in northwest Wales it was unknown as a breeding species prior to 1880. The spread in North Wales was presumably from east to west, as the first breeding record on Anglesey was not until 1886, but by 1902 it nested in Rabbit burrows at Newborough Warren and in 1903 in cliffs on the island's east coast (Jones and Whalley 2004). Forrest (1907) reported that it was abundant and increasing in all parts of North Wales. By the 1890s it was also common and increasing in southeast Wales, and rapidly becoming established in Pembrokeshire, although it remained a very scarce nesting species in much of West Wales, especially Ceredigion, even in the late 1940s (North *et al.* 1949).

By the 1960s, the Starling probably reached its numerical peak in Wales. The *Britain and Ireland Atlas 1968–72* confirmed breeding in 99% of 10-km squares in Wales and it was considered to be the most numerous breeding resident in Glamorgan (Hurford and Lansdown 1995). Subsequent atlases showed increasing gaps, and by 2007–11 it occurred in 89% of 10-km squares during the breeding season, losses having occurred primarily in upland areas. Breeding abundance was low in Wales relative to the rest of Britain and Ireland in 1988–91 and 2007–11.

The highest densities are around the conurbations of southeast Wales but even there, tetrad atlases showed a marked decline. In East Glamorgan, the number of occupied tetrads with breeding evidence fell from 84% in 1984–89 to 51% in 2008–11, with losses primarily in the valleys and hills of the north. The contraction was smaller in Gwent, from 96% of occupied tetrads in 1981–85 to 90% in 1998–2003. Severe declines were also evident in rural districts: 39% of tetrads were occupied in Pembrokeshire in 1984–88 but only 12% in 2003–07. The *North Wales Atlas 2008–12* recorded Starling in 33% of tetrads, but whereas there were breeding records in most of Flintshire and east Denbighshire, it was absent from much of Meirionnydd and Caernarfonshire, away from the coastal towns and inland slate villages. The Welsh breeding population in 2018 was estimated to be 90,000 (79,000–100,000) pairs (Hughes *et al.* 2020).

The CBC/BBS index for England charted a sharp decline of 89% during 1967–2018 but primarily after 1980 (Massimino *et al.* 2019). There are no comparable data for Wales, but it is likely that it underwent a similar reduction here. This has continued throughout the period for which a BBS index has been generated for Wales: it declined by 68% during 1995–2009, since when it has been stable

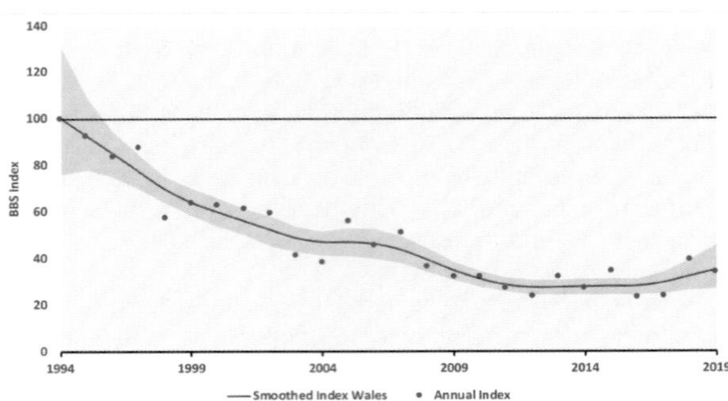

Starling: Breeding Bird Survey indices, 1994–2019. The red line shows the smoothed index, 1995–2018. The shaded area indicates the 85% confidence limits.

(Harris *et al.* 2020). The decline can be traced back to the early 1970s in some counties, such as Breconshire (Peers and Shrubb 1990). In Pembrokeshire, there were no more than 50 nesting pairs in 1948 (Saunders 1976) and *c.*2,000 pairs in 1988 but just 200 pairs by 2009 (Rees *et al.* 2009).

The decline is also evident in the BTO Garden Birdwatch, which showed a reduction in the recording rate in participating Welsh gardens in May/June, from *c.*50% in the late 1990s to 30% in 2011–19. Records from the larger islands show an increase in the 1960s that was reversed to extirpation. On Bardsey, Caernarfonshire, Starlings ceased to breed in the early 1950s but recolonised in 1962 and increased rapidly to 40–60 pairs by 1969. The population declined to 25 pairs in 1984, increased to 80–90 pairs in 1989, then fell rapidly, and it ceased to breed after 1997. On Skokholm, Pembrokeshire, breeding first occurred in 1946. Numbers rose to over 50 pairs by 1969 but declined to 6–8 pairs in the early 1990s, and it ceased to breed in 2006.

Post-breeding flocks of resident Starlings are common in many parts of Wales from mid-June and can involve large numbers feeding on invertebrate-rich improved grassland and on Bilberries around the moorland fringe. Flocks increase in size as summer progresses and can result in spectacular movements, as they are joined by others from elsewhere in Britain and Europe. Very large numbers of migrants pass through Wales in autumn. The first arrive in late September and passage continues into November. These are most prominent along the North Wales coast and to a lesser degree along the Glamorgan coast. When birds were attracted to lighthouses, Starlings were one of the principal casualties, with many thousands killed or grounded on Bardsey during nights of poor visibility. Southward passage on Bardsey can be 5,000–10,000 birds/day, with maxima of 17,000 on 5 November 1978 and 22 November 2006. Numbers passing Skokholm are generally lower, but up to 10,000 passed southwest over the island on 6 November 1970.

A key feature of Starling social activity is their overnight roosts, preceded by spectacular murmurations. Vernon (1963) showed how temporary summer and autumn roosts of resident Starlings are deserted for larger communal roosts, which are dominated by winter immigrants. He detailed the distribution of roosts in Glamorgan and Gwent during 1948–63 and found that feeding flocks made long evening flights to the main roosts, often well over 30km. Gower birds have been known to cross Carmarthen Bay to roost at Pembrey Forest, Carmarthenshire, for example, and birds have been watched leaving the East Glamorgan coast, such as Lavernock Point, to cross the Bristol Channel to roost in Somerset, and return the following morning.

In most winters a few roosts in Wales exceed 50,000 Starlings, but the favoured locations are not the same every year. Most have been in woodland or reedbeds, but the latter frequently collapse under the sheer weight of birds. In Pembrokeshire, the roost at Slebech switched to a crop of *Miscanthus* (Elephant Grass) in 2015, after the reeds were flattened. Probably the best known and easiest roost to watch is on Aberystwyth Pier, Ceredigion, where up to 75,000 Starlings may be present. The largest roosts in recent decades have been in northern and western counties, including several estimated to hold in excess of one million birds. Smaller numbers roost in the south and inland, although reedbeds at Llangorse Lake, Breconshire, hosted 150,000 in 2017. Night-time roosts break into feeding parties up to 100 strong, although larger concentrations occur where food is abundant, such as sewage treatment works and, formerly, landfill sites.

Hard-weather movements can produce spectacular numbers resting or feeding on headland fields, perhaps the largest being 500,000 that flew west over Red Wharf Bay, Anglesey, on 24 January 1998. The return movement in spring, generally in March, is equally well-marked, although numbers are not so large. Aside from changes in estimated numbers at roosts, it is difficult to gauge trends in the winter population, which are confounded by severe weather events. The BTO Garden Birdwatch reporting rate from Welsh gardens in winter fell from an average 69% of gardens in 1995/96–1999/2000 to 60% in 2000/01–2009/10 and 54% in 2010/11–2019/20, even though the last period included higher rates during cold weather in winters 2009/10 and 2010/11.

The majority of winter visitors originate from the Baltic states, Finland and north European Russia. The farthest recorded movement was one ringed at Lydstep, Pembrokeshire, on 9 January 1986 and shot six months later at Shchyolkovo, Russia, 2,832km to the east. Several other movements of *c.*2,000km have been recorded, and recoveries from Ireland show that some birds ringed in Wales were on passage farther west. Starlings that bred in Wales have been recovered in other parts of Britain and Ireland, but not from farther afield.

Starling flocks gather over Aberystwyth Pier, prior to roosting © Jerry Moore

Location	County	Estimate	Year
Newport Wetlands	Gwent	200,000	2017
Cwm Ogwr	East Glamorgan	1.5 million	1960s
Merthyr Mawr	East Glamorgan	400,000	1971
Margam	Gower	300,000	1967
Pencader	Carmarthenshire	1 million	2001
Carmarthen	Carmarthenshire	300,000	2011
Plumstone Mountain	Pembrokeshire	1 million	2009, 2010, 2011
Pickle Wood/Slebech	Pembrokeshire	300,000	2006
Llanfihangel	Radnorshire	400,000	2009
Llyn Traffwll	Anglesey	1 million	1983
Llyn Garreg Lwyd	Anglesey	600,000	2003
Mynydd Parys	Anglesey	300,000	2018
RSPB Cors Ddyga	Anglesey	250,000	1998
Holyhead (pre-roost)	Anglesey	250,000	2014
RSPB Conwy	Denbighshire	250,000	1998

Starling: Peak counts at selected roosts holding in excess of 200,000 at least once.

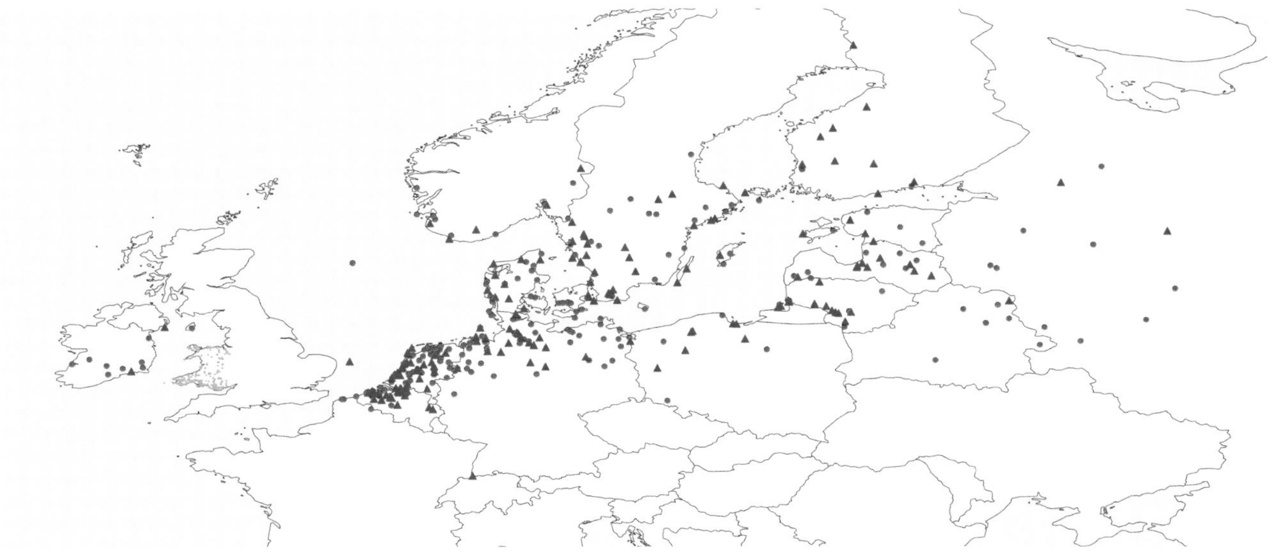

Starling: Recovery locations of birds ringed in Wales are shown by red circles. Ringing locations of birds that were recovered in Wales are shown by blue triangles. Small brown dots show ringing or recovery locations in Wales.

Declines in Britain have been mirrored in other countries; the European population declined by 70% during 1980–2016 (PECBMS 2020). There is good evidence that changes in first-year survival rates have driven the declines (Versluijs *et al.* 2016). Changes in livestock farming are likely to have been a major factor, related to changes in food resources: a reduction in cattle, which create a different sward structure to sheep, explains much of the decline across Europe (Heldbjerg *et al.* 2019). This has likely reduced foraging opportunities for Starlings, exacerbated by the use of insecticides on grassland (Vickery *et al.* 2001). Drying out of grasslands through drainage (e.g. Newton 2004) or prolonged hot summer weather is also thought to have reduced the quality of foraging conditions. Reversing these changes on a large scale will be challenging and will only follow structural changes in farming to encourage soil and nature conservation. With a stable breeding population in Wales during the last decade, it may have found a new equilibrium, but the Starling's future in some parts is by no means secure.

Annie Haycock

Grey-cheeked Thrush *Catharus minimus* Brych Bochlwyd

Welsh List Category	IUCN Red List Global	Rarity recording
A	LC	BBRC

This extremely rare trans-Atlantic vagrant breeds in conifer and shrubby deciduous forests in northern North America and northeast Siberia, and winters in northern South America (*HBW*). There were 64 British records up to 2018 (Holt *et al.* 2019), but none in 2019. Three of these were in Wales, all on Bardsey, Caernarfonshire: on 10 October 1961, 31 October 1968 and 20 October 1971, the last of which was found dead.

Jon Green and Robin Sandham

Swainson's Thrush *Catharus ustulatus* Brych Swainson

Welsh List Category	IUCN Red List	Rarity recording
	Global	
A	LC	BBRC

This extremely rare vagrant to Wales breeds in dense coniferous forest across northern North America and migrates to winter in the northern Andes of South America (*HBW*). Migration routes vary between eastern and western populations but are on a broad front, extending over the west coast of Mexico, as well as most of the continental United States. There were 42 records in Britain up to 2019 (Holt *et al.* 2020), of which two were in Wales and both on Skokholm, Pembrokeshire. The first, on 14–19 October 1967, was also the first British record. The second, on 2–10 June 2015, was the first British record in spring. It had perhaps crossed the Atlantic the previous autumn, although a spring record in Co. Mayo, Ireland, in May 1956, was considered to have made the crossing during its spring migration (Slack 2009).

Jon Green and Robin Sandham

Ring Ouzel *Turdus torquatus* Mwyalchen y Mynydd

Welsh List Category	IUCN Red List			Birds of Conservation Concern			
	Global	Europe	GB	UK	Wales		
A	LC	LC	VU	Red	2002	2010	2016

The Ring Ouzel has a restricted global range, limited to mountainous areas in Europe and Caucasia. The British population of the nominate subspecies, *T.t. torquatus*, occurs in upland areas from Dartmoor through Wales, the Pennines and Cumbria to Scotland. It migrates to spend the early part of the winter in southern Spain, particularly the Sierra Nevada. Around the turn of the year, these birds move to northwest Africa, where they mix with those from Scandinavia and birds from central and southern Europe of the subspecies *T. t. alpestris*. Extensive Juniper scrub in the Atlas Mountains provides the berries that dominate the winter diet.

In Britain and Ireland, breeding Ring Ouzels are mainly found at altitudes over 250m (*HBW*), although historically in Wales and currently in northwest Scotland, they can breed close to sea level in suitable habitat. Steep-sided upland valleys, crags and gullies are favoured for breeding, particularly where heather, in which to nest, is abundant, although Bracken is also used. Disused quarries and old mine workings are also frequently used as nest sites, typically where vegetation has colonised the rock face. Breeding sites are strongly traditional and some crags or gullies in core population areas have held breeding Ring Ouzels for at least 40 years (Smith 2014). Breeding has also been recorded in old farm or quarry buildings and very occasionally in trees (Sim *et al.* 2010). Young are fed invertebrates, mainly earthworms, Cranefly and beetle larvae, and adult beetles. From July until birds start to migrate in September, the diet switches to berries such as Bilberry, Crowberry and Rowan (*BWP*).

The Ring Ouzel is one of our earliest returning migrants, the first typically reaching Wales in mid-March, with migration continuing until early May, depending on weather conditions. Passage is evident on the Pembrokeshire islands, on Bardsey in Caernarfonshire, and on headlands such as the Great Orme, Caernarfonshire. Most sightings involve small numbers, so 33 at Amlwch, Anglesey, on 12 April 1988 and 20 on the Great Orme on 10 April 2015 were exceptional. Early arriving birds were recorded on Carmel Head, Anglesey, on 13 February 1996, at Maesteg golf course, East Glamorgan, on 27 February 2008 and on Skomer, Pembrokeshire, on 29 February 1984. Five on the Aran Mountains, Meirionnydd, in a snowstorm on 19 February 1905, was very unusual (Bolam 1913).

The Ring Ouzel has bred in every Welsh county except Anglesey, exploiting the extensive hill and mountain terrain. Naturalists John Ray and Francis Willughby in the 17th century and Pennant (1810), referring to the 1770s, considered the species to be frequent in Snowdonia. It was described as abundant on the Berwyn in northeast Wales by Forrest (1907). It was "fairly common on heathery areas" in Breconshire, with regular nesting on Mynydd Epynt (Phillips 1899). Walpole-Bond (1903) considered it to be more abundant in that county than elsewhere in mid-Wales. A small breeding population was centred on the Rhondda Valley at the turn of the 20th century, but it was never common in East Glamorgan (Hurford and Lansdown 1996). In Gower, it was recorded in the Swansea and Neath valleys in the first half of the 19th century (Anon 1802, Dillwyn 1848) but did not breed regularly. Ring Ouzel was described as a frequent breeder in Mynydd Preseli, Pembrokeshire (Mathew 1894).

It has been in a long and steady decline in Britain, probably since the early 20th century (*Britain and Ireland Atlas 1988–91*). By the 1950s, Ring Ouzel distribution in Radnorshire and Carmarthenshire, where the species was relatively well studied, was already much reduced (Jennings 2014, Lloyd *et al.* 2015). Declines appear to have accelerated in the latter decades of the 20th century, with breeding lost or much reduced in most areas. It was still relatively common in Breconshire in the early 1970s and occurred regularly in the Brecon Beacons, but 5–10 occupied territories in a good year was a substantial reduction on former levels (Massey 1976a). It was lost from areas such as Gwenffrwd/Dinas, Carmarthenshire, in the early 1990s, and an RSPB survey of Mynydd Du, Carmarthenshire/Breconshire, in 1996 found only ten territories across 150km^2, a 40% decline since 1978.

In the south, numbers declined in East Glamorgan and Gwent, until the last breeding in each county was recorded in 1990 and 1997 respectively. In Pembrokeshire, the last regular breeding was in 1971, except for a successful nest on Mynydd Crwn in 1995. The only regular breeding site in Carmarthenshire in the 21st century was on the Bannau Sir Gaer/Llyn y Fan Fach escarpment, where breeding is no more than "probably" annual (Lloyd *et al.* 2015). It is now only sporadic in Ceredigion and former haunts such the upper Ystwyth and Tywi valleys are quiet in most years. A survey of the Elan and Claerwen valleys in Radnorshire in 1997 failed to find any breeding birds, the area having held 16 pairs in 1987 (Jennings 2014). Montgomeryshire still held 30–40 pairs in the late 1980s (Tyler and Green 1994), but breeding had ceased on Llanbrynmair Moor by 1994 (Harris and Williams 1995) and on Pumlumon—which extends into Ceredigion—by 2006 (Green 2007). Elsewhere in Montgomeryshire, significant declines occurred on the Berwyn, around Llangynog and Lake Vyrnwy, with no confirmed breeding at the latter site since 2008 (Gethin Elias pers. comm.). Breeding distribution is greatly reduced in Denbighshire, with only a handful of pairs remaining on Mynydd Hiraethog and the Clwydian Hills. Ruabon Moor and the extensive Eglwyseg escarpment held 19 pairs in 1981 but just one in 1995 (Hurford 1996). Breeding there is now irregular and generally confined to quarries in the Horseshoe Pass (John Lawton Roberts pers. comm.).

The higher elevation ranges in Snowdonia seem to hold more robust populations, although little historic data exists to assess trends. There were 23–25 territories on Cadair Idris, Meirionnydd, in 2008, which was similar to 1999 (Smith 2011). The population in

Juvenile and adult Ring Ouzel © Bob Garrett

the rugged, heather-clad Rhinogau, Meirionnydd, has halved since the late 1960s, to 20 territories in 2009. The majority of historic breeding sites are still used, in a shifting mosaic of occupancy from year to year, suggesting that the habitat remains suitable (Smith 2014). A first complete survey of the nearby Aran range in 2014 located 14 pairs (Smith 2018). In northern Snowdonia, Caernarfonshire, 164 territories were located in 2009–10, the main concentrations on the highest parts of the Carneddau, Glyderau and Snowdon ranges and particularly on the steep, craggy heather slopes above the Nant Ffrancon and Llanberis passes (Driver 2011). Monitoring in those three ranges during 2009–16 suggested a relatively stable population of *c.*100 pairs, although shifting occupation of nest sites was noted there too (Driver 2017). Elsewhere in Caernarfonshire, a few pairs nest on Llŷn, in the hills above Trefor. It seems that the highest elevations in North Wales still provide a favourable breeding environment that has been lost across most areas farther south.

The Ring Ouzel is a Red-listed bird of conservation concern in Wales, based on a 50% decline in range (Johnstone and Bladwell 2016). The *Britain and Ireland Atlas 1968–72* found it in 37% of 10-km squares in Wales. That proportion fell to 24% in 1988–91 and 18% in 2007–11, as the Ring Ouzel's range contracted to a core in Snowdonia, with a smaller population in the western part of the Brecon Beacons. Two stratified sample surveys revealed a decline of 29% in Britain between 1999 and 2012. Somewhat unexpectedly, the 2012 survey recorded a 39% increase in Wales, to 444–547 pairs, although there were limitations in the earlier survey methodology and a direct comparison suggested a non-significant 11% decline (Wotton *et al.* 2016).

Most Ring Ouzels depart during September and October, although it was recorded as late as November in every county except Montgomeryshire and Denbighshire during 2000–19. Most autumn sightings involve small numbers, although in favoured habitat where food sources are good, large numbers gather. Autumn 2005 was particularly notable, with flocks of 52 at Craig Walter, East Glamorgan, on 22 October, and at least 110 in the Elan Valley, Radnorshire, on the same day, including 69 that roosted at Cnwch Wood, Breconshire. Other notable counts included at least 30 below Craig Goch Dam in the Elan Valley in September 1992, up to 27 in Blaen-y-cwm, East Glamorgan, from 19 September to 7 October 1967 and 26 on Bardsey on 14 October 2014.

Occasional sightings occur in December, but it is unclear whether these are late migrants or have chosen to over-winter. During 2000–19 there were records between 1 December and mid-February in nine counties. Most were singles, but five were in the Nant Ffrancon Valley, Caernarfonshire, between 16 January and 16 February 2015 and a male was in nearby Dinorwig Quarry, Llanberis, on 31 January 2015. Three were seen on Aberdyfi beach, Meirionnydd, in January 1945, of which two were seen alive and a third was found dead. Increasing numbers may be choosing to remain here for winter, although the majority favour the milder climate of lowland England (*Britain and Ireland Atlas 2007–11*).

The reasons for its widespread and severe decline in Britain are probably complex and yet to be fully understood. Undoubtedly, large-scale afforestation and agricultural intensification of moorland across upland Wales caused the loss of favourable habitat, but the widespread nature of declines suggests that large-scale factors are responsible. Threats during migration, including hunting pressure, require more research, as does the loss and degradation of Juniper scrub in northwest Africa (Ryall and Briggs 2006). There is evidence that increasing summer temperatures may be responsible for the declines (Beale *et al.* 2006). The northward shift already detected in its range is expected to result in Wales and northern England being climatically unsuitable for Ring Ouzel by the end of the 21st century (Huntley *et al.* 2007). Research in Scotland suggested that poor survival rates of first-calendar-year birds may be driving declines, with reduced frequency of double broods and poor productivity in the early stages of the breeding season (Sim *et al.* 2011), but the exact nature of the mechanisms causing these changes is yet to be determined.

Continued monitoring of core populations in Wales is important as these now mark the southern edge of its breeding range in Britain, with only a small relict population to the south, on Dartmoor, Devon. This would help to detect any further declines and range contraction in the key higher-altitude areas. At a time of changes in upland agriculture and land use, habitat management trials are necessary to maintain and enhance the mosaics that Ring Ouzels require. Only through a combination of research and conservation action will we better understand what is required to keep this charismatic bird in its Welsh haunts.

David Smith

Sponsored by Ecology Matters Trust

Blackbird *Turdus merula* Mwyalchen

Welsh List Category	IUCN Red List			Birds of Conservation Concern			
	Global	Europe	GB	UK	Wales		
					2002	2010	2016
A	LC	LC	LC	Green			

The Blackbird is a favourite garden bird, well-known even to people with no particular interest in birds. Found throughout Europe, except the far north, and east to Central Asia, it makes use of a wide range of habitats. It may originally have been a bird of open forest and woodland edge (*HBW*). It only began to use gardens and parks in Britain in the mid-19th century (Parslow 1973). During 1995–2019, Blackbird was consistently recorded in at least 95% of gardens in Wales that participated in the BTO Garden Birdwatch. The peak counts were always during the winter, but it occurred in at least 60% of gardens, even in September, the time of year when visits to gardens are at their lowest.

This species seems always to have been common but was historically less numerous than Song Thrush in most areas (e.g. Mathew 1894, Forrest 1907). Numbers increased as it spread into new habitats, and it is now one of the most widespread bird species, breeding on most of the islands and up to a considerable altitude, aided by conifer afforestation during the 20th century. For example, birds nested at 620m in Radnorshire (Jennings 2014). In the breeding season, it was found in almost every 10-km square in Wales in all three *Britain and Ireland Atlases*, although its abundance in most of Wales is fairly low relative to England and Ireland. In 2007–11, the highest densities were along the south coast from Gwent to Pembrokeshire, which may be linked to the higher human population. The BBS shows Blackbirds to be more common in villages and towns than other habitats. The *Gwent Atlas 1998–2003* found breeding evidence in over 99% of tetrads, the *East Glamorgan Atlas 2008–11* in 86% and the *Pembrokeshire Atlas 2003–07* in 93%. It occurred in 88% of tetrads in the *North Wales Atlas 2008–12*, including almost all those below 400m. Winter distribution is very similar to that of the breeding season. Most males will defend their territory throughout the year, although some of those breeding on higher ground move to nearby lowlands.

Blackbirds in Britain have a complex pattern of movements, with some highly sedentary and others migratory (Snow 1987). Most Blackbirds in Wales seem not to move far. Four ringed in Wales as young birds and recaptured over ten years later had all remained at the original ringing site. In autumn, there is an influx from eastern and northern Europe, but the estimated total number in Britain during the winter (6.7–11.4 million birds, Hayhow *et al.* 2017b) is lower than the breeding population and that year's juveniles. Ringed birds from Fennoscandia, Germany, the Low Countries and northeast France have been found in Wales, some presumably *en route* from farther east when ringed. Some of those winter in Wales but many pass through to Ireland or go on to winter in southwest Europe. Peak passage is usually in October and early November. The highest counts included 2,000 on Bardsey, Caernarfonshire, on 22 October 1962, 1,000 on Skokholm, Pembrokeshire, on 18 October 1964 and at least 1,000 around Aberdaron, Caernarfonshire, on 29 October 2004. On 12 November 1966, there was a massive arrival of Blackbirds in southwest Anglesey, with an estimated 20,000 in less than

Blackbird: Recovery locations of birds ringed in Wales are shown by red circles. Ringing locations of birds that were recovered in Wales are shown by blue triangles. Small brown dots show ringing or recovery locations in Wales.

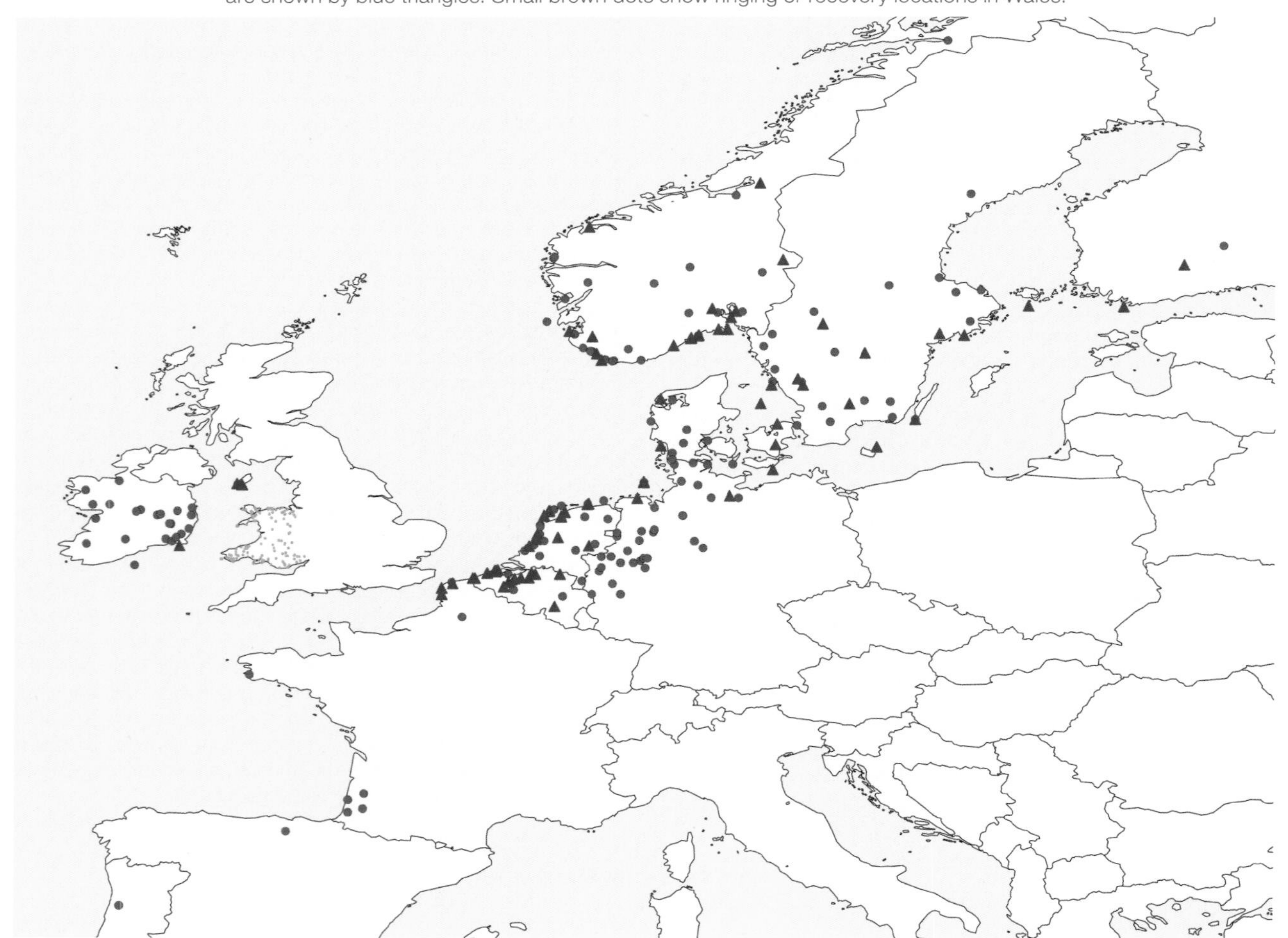

8km between Brynsiencyn and Newborough. Spring passage is usually less obvious, and counts are lower. There may also be cold-weather movements west and south in winter, such as 500 on Bardsey in January 1963.

The Welsh breeding population in 2018 was estimated to be 665,000 (635,000–695,000) pairs (Hughes *et al.* 2020), making it the fourth most abundant passerine. The highest published population density in Wales is probably an equivalent of 119 pairs/km² in mature woodland at RSPB Ynys-hir, Ceredigion, in 1985–91, but densities of up to 296 pairs/km² have been recorded in suburban habitats in western Europe (Mason 2003).

The BBS shows that the species is doing well in Wales, the index increasing by 43% during 1995–2018, and has been very stable during the second half of that period (Harris *et al.* 2020). An increase, albeit smaller, across the UK over that period represents a recovery following a major decline in England during 1970–95. No trend information is available for Wales prior to 1995. The decline was thought to be linked to the intensification of arable farming, which reduced adult survival (Siriwardena *et al.* 1998). The authors of *Birds in Wales* (1994) believed that this probably did not apply to Wales, where agriculture is dominated by grassland, although the Common Birds Census recorded reductions in woodland too. A run of colder winters between 1978/79 and 1986/87 (Kington 2010) may also have contributed.

There is evidence from elsewhere in Europe that climate change has enabled Blackbird to start nesting earlier, thus extending the breeding season (Najmanová and Adamík 2009), but there is no evidence of this in the UK to date. There is also evidence of it breeding at higher altitude than formerly. Nesting now occurs at 430–500m in Breconshire, where it was absent in

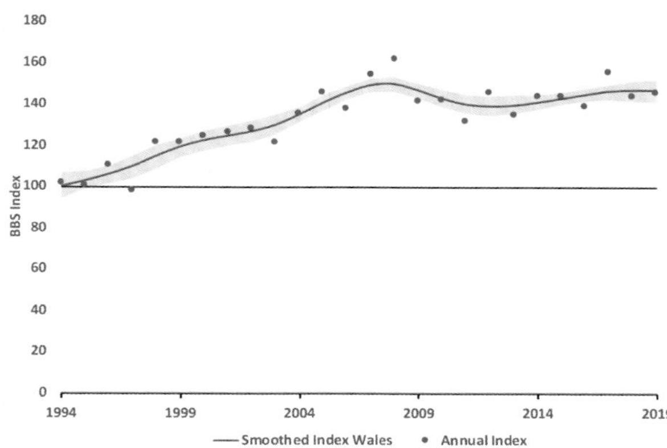

Blackbird: Breeding Bird Survey indices, 1994–2019. The red line shows the smoothed index, 1995–2018. The shaded area indicates the 85% confidence limits.

the 1990s (Andrew King *in litt.*). Winter food resources are a major factor in Blackbird survival. It is likely that soil invertebrates are more abundant in the moister west of Britain, with more woodland and pasture, than in the arable-dominated east. By contrast, there is no apparent difference in nest productivity between the two areas (Robinson *et al.* 2012). Blackbirds are now doing well across Wales, and there seems every likelihood that they will continue to do so.

Rhion Pritchard

Sponsored by Richard Groves

Eyebrowed Thrush *Turdus obscurus* Brych Aeliog

Welsh List Category	IUCN Red List	Rarity recording
	Global	
A	LC	BBRC

This extremely rare vagrant breeds in dense conifer forest and taiga in Russia east of the Ural Mountains and winters in Southeast Asia (*HBW*). There were 23 British records up to 2018 (Holt *et al.* 2019) but none in 2019. Only one was in Wales: on Bardsey, Caernarfonshire, on 12 October 1999.

Jon Green and Robin Sandham

Eyebrowed Thrush on Bardsey, Caernarfonshire, October 1999 © Steve Stansfield

Black-throated Thrush *Turdus atrogularis* Brych Gyddfddu

Welsh List Category	IUCN Red List	Rarity recording
	Global	
A	LC	BBRC

This species breeds around clearings in conifer and mixed forests on both sides of the Ural Mountains in Russia and south to Mongolia. This is the same breeding area as that of the easternmost Redwings that winter in Europe. Black-throated Thrush, however, has a shorter migration, wintering from eastern Arabia to northeast India (*HBW*). There were 86 British records up to 2019, an average 2–3 each year during 1990–2019, some of which remained over winter (Holt *et al.* 2020). Of these, two were in Wales: a male in a Swansea garden, Gower, from 29 December 2005 to 15 March 2006, and a first-calendar-year bird at St Asaph, Denbighshire/Flintshire, on 16–31 December 2016.

Jon Green and Robin Sandham

Dusky Thrush *Turdus eunomus* Brych Tywyll Asia

Welsh List Category	IUCN Red List	Rarity recording
	Global	
A	LC	BBRC

Dusky Thrush breeds in open woodland in northern Russia east of the Ural Mountains, as far north as the edge of the tundra. It winters in southeast China, the Korean Peninsula and Japan (*HBW*). It is a rare vagrant in Britain, having occurred only 16 times up to 2019 (Holt *et al.* 2020). One of these was in Wales: on Skomer, Pembrokeshire, on 3–5 December 1987.

Jon Green and Robin Sandham

Fieldfare *Turdus pilaris* Socan Eira

Welsh List Category	IUCN Red List			Birds of Conservation Concern			
	Global	Europe	GB	UK	Wales		
A	LC	LC	LC	Red	2002	2010	2016

The Fieldfare breeds in Iceland, over much of northern and central Europe, and east across Siberia (*HBW*) and has bred sporadically in Scotland since 1967 (*The Birds of Scotland* 2007). Migratory northern populations winter in southern and western Europe, including in Wales, although the extent of movements can vary between years. In Norway, a significant proportion of the breeding population can remain in country, providing that autumn and winter temperatures do not fall too low and there is a good supply of Rowan berries (*HBW*). The largest breeding populations in Europe outside Russia are in Fennoscandia (BirdLife International 2015). Several ringed as nestlings in Norway and Finland have been recovered in Wales, and one ringed in Wales was found in Sweden in the breeding season.

An estimated 15.1 million Fieldfares are in Britain in mid-winter (Hayhow *et al.* 2017b), making it by far the most abundant thrush species at that time of year. The earliest arrivals in Wales are in mid- to late September in most years, with greater numbers usually between mid-October and mid-November. Many have moved on by the end of December, presumably to western France or northwest Iberia, although they may remain longer if plenty of berries are available. The variable numbers make it difficult to discern trends. It was described as common in the lowlands, particularly inland, at the turn of the 20th century (Mathew 1894, Forrest 1907) and generally more common in severe winters than in mild ones. It was found throughout Wales during winter in the four years of the *Britain and Ireland Atlas 2007–11*, but abundance in Wales was low relative to England and Ireland.

In some years, large numbers have been recorded passing over, or stopping for short periods, in October or early November. Many records simply referred to 'thousands' of birds, but there were 20,000 at Tal-y-cafn, Caernarfonshire/Denbighshire, on 8 November 1986 and 10,000 southwest over Cwm Ystwyth, Ceredigion, on 11 November 1997. Particularly large numbers were recorded in late October 2009, including what Jennings (2014) described as "probably the most extraordinary passage of migrant birds ever seen in Radnorshire," peaking on 28 October when at least 40,000 were in the Elan Valley and 20,000 flew south over Radnor Forest in one hour. Huge numbers were also recorded over Breconshire, where passage peaked during 22–30 October. No counts were made, but there were reports of 'deafening' levels of calls. In the same county, 10,000 flew over The Cwm, Llanwrthwl, on 21 October 2013 and there were 30,000 there on 28 October 2016.

Influxes can occur later in the winter during hard weather, such as in Gwent and Glamorgan in advance of severe conditions in 1962/63. Despite the severity of that winter, Fieldfares were thought to have suffered lower mortality in Britain than in some previous winters, such as 1916/17 (Dobinson and Richards 1964), perhaps because many moved farther south ahead of the worst conditions. It is a less frequent visitor to Welsh gardens than Redwing, typically reported in fewer than 5% of those participating in the BTO Garden Birdwatch during 1995–2019. Higher reporting rates tend to occur in late winter, such as in February

Fieldfare © Kev Joynes

1996 when Fieldfares were recorded in 17% of participating gardens. Thousands were reported in frozen riverside fields between Caerleon and Usk, Gwent, on 18 February 1969 and several thousand were on the Gwent Levels in winter 1973. Flocks of up to 4,000 were recorded on the Gwent Levels in December 1985, and it was suggested that the total was as high as 15,000 (Venables *et al.* 2008).

Large flocks congregate at good food sources, sometimes accompanied by Redwings. There were 6,000 with Redwings at Llandenny, Gwent, in January 1977, *c.*5,000 at Merthyr Mawr, East Glamorgan, on 16 January 1987, attracted to Sea Buckthorn berries, and 5,000 at Llantilio Crossenny, Gwent, on 15 January 2017. Other large counts included at least 5,000 at Llanrhos, Caernarfonshire, in December 1981 and *c.*3,500 over Kenfig Pool, East Glamorgan, on 3 February 1991. Large numbers have also sometimes been recorded at roosts. For example, *c.*4,000 roosted in Sitka Spruce at Pantmaenog Forest, Pembrokeshire, in February 1984.

Numbers on return passage are usually lower, either because Fieldfares that have over-wintered in southwest Europe take a more direct route across continental Europe to their breeding sites or because they make the journey in a single flight, unless bad weather intervenes (*BTO Migration Atlas* 2002). Return passage usually begins in February. Most have left by the end of April, although some remain into May. A flock of 6,000 over Cardiff Bay, on 2 March 2018, was unusual for spring but associated with cold weather. The Fieldfare has been recorded in Wales in every month of the year and there are records of courtship behaviour and of birds giving anxiety calls in suitable breeding habitat, but there have been no confirmed records of breeding in Wales. Some appear to have summered here: two were in moult at Crymlyn Bog, Gower, on 2 August 1983 and a female trapped on Bardsey, Caernarfonshire, on 23 May 2015, had a fully vascularised brood patch, suggesting that it had attempted to breed somewhere in Britain, perhaps even in Wales (*Bardsey's Wildlife 2015*).

The European breeding population increased by 18% during 1980–2016 (PECBMS 2020) and it should continue to be a regular winter visitor to Wales.

Rhion Pritchard

Redwing *Turdus iliacus* Coch Dan Adain

Welsh List Category	IUCN Red List			Birds of Conservation Concern			
	Global	Europe	GB	UK	Wales		
A	NT	NT	LC	Red	2002	2010	2016

The Redwing breeds in northern Eurasia, the nominate form from Fennoscandia to northeast Siberia, and *T.i. coburni* in Iceland and the Faroe Islands (IOC). Both subspecies occur in Wales in winter, although the vast majority are from northeast Europe. The largest breeding population in Europe is in Finland. While it is essentially a northern species, from about 1930 it spread south to the Baltic states and to northern Scotland, where the population peaked in the 1970s and 1980s (*European Atlas* 1997, *The Birds of Scotland* 2007). The number of pairs reported in Scotland fluctuated between ten and 40 during 2000–18 (Eaton *et al.* 2020). Almost all populations are migratory, wintering in western and southern Europe, North Africa and the Middle East. Population size can vary greatly, with reductions caused by cold summers on the breeding grounds or severe conditions in its usual wintering areas (*HBW*). The population in Europe has declined by 16% since 1980 (PECBMS 2020) but trends in Asian Russia are not known.

The Redwing seems always to have been a common passage migrant and winter visitor to Wales and was described as such by late-19th- and early-20th-century authors (Mathew 1894, Forrest 1907). Ringing recoveries indicated that most birds wintering in Wales came from Finland, Norway and Sweden, but there have been recoveries of birds that bred farther east. One ringed near the Dysynni Estuary, Meirionnydd, in January 1971 was killed by a cat near the Ural Mountains in Bashkortostan, Russia, 3,869km away, in July 1972, and an adult ringed near St Petersburg, Russia, on 4 October 2018 was recaught at Abermule, Montgomeryshire, 28 days later. Numbers fluctuate, with influxes sometimes during cold weather. This makes it difficult to identify a trend, although Jennings (2014) considered that numbers had increased markedly in Radnorshire since the 1980s. They are usually most numerous on lower ground and avoid the uplands.

An estimated 8.6 million Redwings are in Britain in mid-winter, only about half the number of Fieldfares (Hayhow *et al.* 2017b), but in Wales it is generally the more abundant species. The first are usually recorded in September, with the main arrival in October and early November. Redwings often migrate at night. There are many reports of what appeared to be huge flocks passing over in darkness in autumn, with numbers often noted as 'many thousands'. It has sometimes been recorded in large numbers on Bardsey, Caernarfonshire, where it was the species most frequently attracted to the lighthouse. Some of the largest counts there were in 2002, when there were at least 10,000 on 3/4 November. This was followed by the largest lighthouse attraction recorded on the island, involving at least 30,000, possibly 40,000, on 9/10 November. Numbers on other Welsh islands are usually smaller. On the mainland, there was a very large arrival into Radnorshire overnight on 1 November 1992, with at least 20,000–30,000 in fields alongside the A44 near Llandegley and Penybont the following day. Other large autumn counts included 12,000 through Lavernock, East Glamorgan, on 5 November 2007 and 10,000 over Caerau near Maesteg, East Glamorgan, on 16 October 2005.

At first, they usually feed on berries. When these have been stripped, they turn to ground feeding, usually in pasture and often in the company of Fieldfares. Redwings can be very mobile, moving on as local food supplies are exhausted. Large hard-weather movements have been recorded. The highest winter counts in Wales included *c.*12,000 between St Brides Wentlooge

Redwing: Recovery locations of birds ringed in Wales are shown by red circles. Ringing locations of birds that were recovered in Wales are shown by blue triangles. Small brown dots show ringing or recovery locations in Wales.

Redwing feasting on Rowan berries © Kev Joynes

and Peterstone, Gwent, on 19 February 1968 and a roost of 10,000 at Gwysaney, Flintshire, in December 1992. A westward movement over Llangennith, Gower, on 8 December 1967 involved around 26,000, while 17,000 passed over Pontypool, Gwent, on 18 January 2013.

Numbers fall as many move on to Ireland or to France, Spain and Portugal. In some counties, such as Meirionnydd, few over-winter in most years. Birds do not always winter in the same area each year. Around 64% of recoveries of Redwings ringed in Wales were from countries to the south and west, in a different autumn/winter to the one in which they were ringed. These included three in Italy. Some of those ringed in Wales were recovered soon after ringing to the southeast: both Belgian recoveries and two of three recoveries from eastern England in the same autumn had been ringed in Wales (three in Caernarfonshire) less than two weeks previously.

The Redwing sometimes suffers heavy mortality in very cold weather. In East Glamorgan, thousands were said to have died under hedgerows around Cardiff during early 1947 (Hurford and Lansdown 1995) and thousands perished in Pembrokeshire that winter (Lockley *et al.* 1949). Mortality appears to have been lower in 1962/63, as most birds moved on ahead of the cold weather (*Birds in Wales* 1994). Visits to gardens are not common but more frequent than those by Fieldfare. It was recorded in more than 5% of Welsh gardens participating in the BTO Garden Birdwatch in 15 years during 1995–2019. The highest reporting rates were during cold weather in December 1996, January 2010 and December 2010.

Numbers on spring passage are usually lower, but there were 5,000 on Bardsey on 15 March 1991, following a lighthouse attraction. Those that over-winter normally leave in late March and April, but Redwing has been recorded in Wales in all months of the year. A few sometimes remain into May or June, and although some are heard singing in most years, there has been no confirmed record of breeding in Wales. A claim of nesting at Maentwrog, Meirionnydd, in 1855 is unreliable: the eggs were later identified as those of a Blackbird (Forrest 1907).

There do not appear to be any particular conservation issues affecting the Redwing in Wales, although its breeding range is expected to contract northwards within Fennoscandia and Iceland during this century (Huntley *et al.* 2007). It is likely to continue to be a common winter visitor, although unlikely ever to become established as a breeding bird.

Icelandic Redwing *T. iliacus coburni* Coch Dan Adain Gwlad yr Iâ

Redwings of the subspecies *T.i. coburni* are longer-winged, darker and more heavily streaked than the nominate subspecies. It winters in Scotland, western France and Iberia, but Ireland is the most important wintering area (Milwright 2002). A number of Redwings examined in the hand on the Welsh islands show characteristics of *coburni* and there have been a few records of such birds from the mainland. On Bardsey, birds showing characteristics of the Icelandic subspecies are relatively scarce in autumn, but there are more records between March and May. The sole ringing recovery between the two countries was of an adult ringed at Keeston, Pembrokeshire, in November 2018 and taken by a cat at Mosfellsbaer, Iceland, in July 2020.

Rhion Pritchard

Song Thrush *Turdus philomelos* Bronfraith

Welsh List Category	IUCN Red List			Birds of Conservation Concern			
	Global	Europe	GB	UK	Wales		
A	LC	LC	LC	Red	2002	2010	2016

The Song Thrush breeds across Europe, except in the far south, east to central Siberia (*HBW*). The subspecies that breeds in Wales, *T.p. clarkei*, is restricted to western Europe (IOC). However, birds of the nominate *T.p. philomelos*, from central, eastern and northern Europe, are highly migratory. They occur in Wales in autumn, and probably in winter and on spring passage (*HBW*). Most ringing exchanges involving Wales are along the southern coast of the North Sea.

It requires access to moist ground to forage for soil invertebrates, and within Wales is most abundant in lowland woodland, both deciduous and coniferous, with a shrub understorey. It is much less common on bare moorlands and mountains, but conifer plantations have enabled an extension of its distribution to higher altitudes, to 610m in Radnorshire. Jennings (2014) considered that most of those breeding above 250m in that county leave the area for the winter and return in February. The same is likely to be true in all upland counties, but such movements may be less frequent now than historically. Over the last century or so it has adapted to small, wooded areas, such as parks and gardens with trees and shrubs, even in urban areas (*BWP*). In the early 20th century, it nested on rocky ledges in North Wales, including on the treeless Puffin Island, Anglesey (Forrest 1907).

It becomes more obvious at dawn and dusk in spring, when its strident repetition of short phrases emanates from woodland edges, hedgerows and gardens, where it is able to dive into undergrowth when disturbed. Authors around the turn of the 20th century reported Song Thrush to be abundant in all parts of Wales, including on moors and mountains during the breeding season, prior even to widespread conifer afforestation (Forrest 1907). O.V. Aplin informed Forrest that in winter Song Thrushes were more numerous in southeast Anglesey than anywhere else in Wales, because of the arrival of migrants. Following a severe winter in 1880 "there was scarcely a Thrush left in North Pembrokeshire" but the following autumn "the woods and copses became again replenished by immigrants" (Mathew 1894).

In most of Wales, its population is presumed to have remained abundant and stable through the first half of the 20th century, although Lockley *et al.* (1949) reported that it was common, but not numerous, in Pembrokeshire in the 1940s. A substantial decline, of around 60%, was recorded across the UK during 1969–98 (Massimino *et al.* 2019) and local observations in Wales reflected this. In Gwent, the ratio of Blackbirds to Song Thrushes was estimated at 4:1 in 1946/47 and closer to 7:1 by the early 2000s, the decline occurring from the 1960s (Venables *et al.* 2008). A sharp decline was noted in Glamorgan in the same period, particularly during 1967–79 (Hurford and Lansdown 1995).

Almost every 10-km square in Wales held Song Thrush during the *Britain and Ireland Atlas 1968–72*, with breeding confirmed in most. Its distribution diminished only slightly by 1988–91, but breeding was confirmed in fewer squares, which aligned with the population decreases reported. By the 2007–11 *Atlas* it had recovered these losses, to occupy 97% of squares, and it was more abundant across much of Wales relative to England than in 1988–91. The highest densities were in a band across South Wales from lowland Gwent to Pembrokeshire. At the tetrad level, differences between lowland and upland, and between north and south, were more obvious. The *North Wales Atlas 2008–12* showed breeding in 75% of tetrads, absent primarily from upland pasture, moorland and mountains. Atlases undertaken in southeast Wales showed little change over time: in the most recent surveys it occurred in 96% of tetrads in Gwent in 1998–2003 and in 83% in East Glamorgan in 2008–11. However, it was absent from higher elevations and there was breeding evidence in only 75% of tetrads. In Gower, the *West Glamorgan Atlas 1984–89* reported breeding in 87% of tetrads and the *Pembrokeshire Atlas 2003–07* in 82%.

The breeding population in Welsh woodlands increased by 69% between 1965–72 and 2003–04. The rate accelerated in the

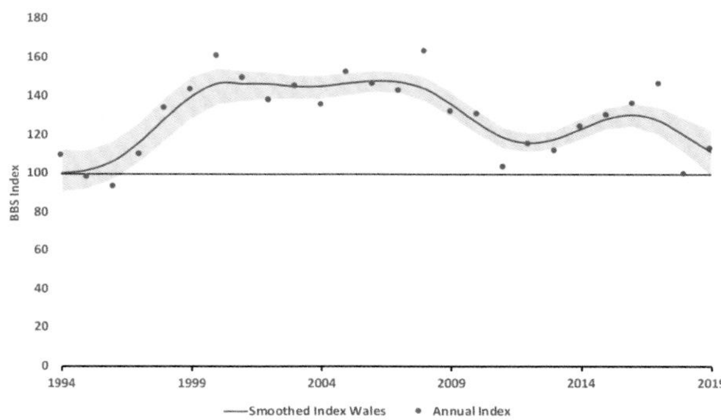

Song Thrush: Breeding Bird Survey indices, 1994–2019. The red line shows the smoothed index, 1995–2018. The shaded area indicates the 85% confidence limits.

second half of that 30-year period (Amar *et al.* 2006). The BBS index for Wales also increased, by 46%, during 1994–2000. The index fell by 17% during 2008–18 (Harris *et al.* 2020), leading to an overall increase of 18% between 1994 and 2019. The Welsh breeding population in 2018 was estimated to be 160,000 territories (Hughes *et al.* 2020), about one-quarter the size of the Blackbird population.

Some Welsh breeding birds move south in autumn, although it is not clear what proportion of the population is involved. It is assumed that many of the large movements witnessed from the second week of October involve birds of the nominate subspecies. The largest autumn movement through Bardsey, Caernarfonshire, involved an astonishing 4,500 on 15 October 2002, when flocks headed north along the island in groups of 40–50 all morning (Steve Stansfield, pers. comm.). Most other passage counts over Bardsey have involved fewer than 500 in a day, although there were 750 on 27 October 1979. Most go on to winter in Iberia, some perhaps via Ireland, which holds higher winter densities of Song Thrush than much of Britain. One ringed on Bardsey, Caernarfonshire, on 30 October 1987, was found dead at Varkaus, Finland, on 8 May 1988, presumably having returned to its breeding area.

Cold weather in Europe can produce winter concentrations along the west coast, such as 2,000 over Bardsey on 1–2 January 1963 and 1,000 over Skokholm, Pembrokeshire, on 16 February 1929. In 1935, thousands were on Skokholm in "hard east winds" (Betts 1992). Recent cold-weather movements have been much smaller, such as 30 at Aberdyfi, Meirionnydd, on 1 January and 55 on Bardsey on 2 January 2009. In 2010 there were 158 at Rhossili, Gower, on 11 December and 96 on Bardsey in late December. Spring passage through Wales is far lighter, perhaps indicating that birds return to eastern Europe across the continent. Skokholm does, however, have higher counts during a short period in mid-March than in autumn. One ringed on Bardsey on 1 May 1977 was recovered on the west coast of Italy on 7 November 1977. It was probably also a Scandinavian breeder that had taken different routes in successive winters.

The Song Thrush is widespread in winter, occurring in 95% of 10-km squares in the *Britain and Ireland Winter Atlas 1981–84* and in 99% in 2007–11, but the numbers present in mid-winter are lower than the breeding population estimate for Britain (Hayhow *et al.* 2017b). Some of the highest concentrations in Britain were on Anglesey, Llŷn, Pembrokeshire and the coastal plain of South Wales. Abundance was relatively low in mountainous areas. The winter population includes a mix of Welsh breeding birds and migrants from farther north in Britain and northeast Europe. In Breconshire, the highest densities on farmland in winter have been found feeding on molluscs and other invertebrates under the canopy of forage rape, fodder beet and swedes (Andrew

Song Thrush: Recovery locations of birds ringed in Wales are shown by red circles. Ringing locations of birds that were recovered in Wales are shown by blue triangles. Small brown dots show ringing or recovery locations in Wales.

King *in litt.*). It is a frequent visitor to gardens in winter, although the peak reporting rate in the BTO Garden Birdwatch was 35% lower in Wales during 2012–19 than in 1995–2009. Cold weather, such as in January and December 2010, can bring more birds into gardens, and in those months, 56% of Welsh gardens reported Song Thrush.

The Song Thrush is Amber-listed in Wales (Johnstone and Bladwell 2016) on account of the decline in its UK breeding population. Decreases during the 1970s and 1980s were linked to increased mortality during the post-fledging period and first year (Robinson *et al.* 2014). Declines in Wales may have been less severe than elsewhere in Britain, as it is less reliant on arable farmland, which has been the focus of most ecological research. There was no evidence of the effects of predation by other birds or by Grey Squirrels in England (Newson *et al.* 2010) and no reason to think that the situation would be any different in Wales. The moist climate and greater woodland coverage in Wales would seem to favour Song Thrushes. Drainage of damp ground and the depletion of woodland shrub layers may have played a role in the wider decline (Fuller *et al.* 2005), but the extent to which that has occurred in Wales is not known. It may have benefited from climate change (Pearce-Higgins and Crick 2019) but can struggle during long periods of drought and cold, when ground conditions reduce access to soil invertebrates, reducing survival (Robinson *et al.* 2004b). Wales is the only UK nation where Song Thrush declined during 2008–18 and this will need to be watched in case the rate increases. Otherwise, it would seem that the Song Thrush population here is relatively secure.

Geoff Gibbs and Ian M. Spence

Mistle Thrush *Turdus viscivorus* Brych Coed

Welsh List Category	IUCN Red List			Birds of Conservation Concern			
	Global	Europe	GB	UK	Wales		
					2002	2010	2016
A	LC	LC	VU	Red	2002	2010	2016

The Mistle Thrush is now a relatively common bird throughout Wales, although less numerous than Blackbird or Song Thrush. It breeds across most of Europe and northwest Africa, east to central Siberia and is migratory over its northern and eastern range (*HBW*). It nests in open, mature woodland but requires short-turf grassland nearby in which to feed and moves locally, to use berry-bearing trees in autumn. It nests early in the year, with eggs usually laid from late February, although sometimes earlier. Fledged young were seen at Cardiff Bay, East Glamorgan, on 20 January 2011.

Its song is very evocative, uttered from an elevated perch and is often held to herald an imminent gale, hence its old English name, 'Storm Cock'. It will fiercely defend its nest against predators such as Grey Squirrel and Magpie, accompanied by a football-rattle alarm call. Snow and Snow (1988) described how pairs, or individuals, guard a winter food source such as a well-berried Holly or a

Mistle Thrush © Ben Porter

tree with Mistletoe. Other thrushes are chased away, but a solitary Mistle Thrush is frequently overwhelmed by flocks of Fieldfares or Redwings. Often birds that defended a food source over the winter remain to nest nearby.

Mistle Thrush was virtually unknown in north and west Britain prior to 1800 and is thought to have colonised Wales and Ireland during the first half of the 19th century (*Historical Atlas 1875–1900*). By the early 20th century, it was common throughout most of Wales, usually less abundant than Blackbird or Song Thrush, although authors in Gwent suggested that it was locally more common than Song Thrush, at least during 1937–77 (Venables *et al.* 2008). In North Wales it was described as less common than in neighbouring English counties (Dobie 1894, Forrest 1907). There were areas where it remained relatively scarce even much later in the century, such as in Ceredigion (Ingram *et al.* 1966) and parts of Pembrokeshire (Donovan and Rees 1994).

By 1970, its distribution "would have astonished 18th century ornithologists, for this species was then apparently largely restricted to S England and Wales" (*Britain and Ireland Atlas 1968–72*). It was widespread in both that and subsequent Britain and Ireland atlases, occurring in at least 90% of 10-km squares. In 2007–11, abundance was relatively high across the southern half of Wales, a little lower in the northern half but on a par with much of England. Tetrad atlases revealed more localised changes. In Gwent and Pembrokeshire, the number of tetrads with breeding evidence increased: in Gwent from 89% in 1981–85 to 92% in 1998–2003 and in Pembrokeshire from 45% in 1984–88 to 56% in 2003–07, with gains primarily in the middle of the county (*Pembrokeshire Avifauna* 2020). By contrast, the *East Glamorgan Atlas 1984–89* recorded breeding evidence in 63% of tetrads but this fell to 59% in 2008–11. The *North Wales Atlas 2008–12* found breeding evidence in 63% of tetrads, with greater presence in upland areas than Song Thrush, but it was absent from parts of central Anglesey, easternmost Denbighshire and parts of Llŷn, Caernarfonshire.

The BBS index for Wales was relatively stable during 1995–2018 (Harris *et al.* 2020). It was also stable in Welsh broadleaved woodlands between 1965–72 and 2003–04. Numbers reduced only 9% in that period, mostly prior to 1981–88 (Amar *et al.* 2006). The Welsh breeding population in 2018 was estimated to be 26,000 territories (Hughes *et al.* 2020).

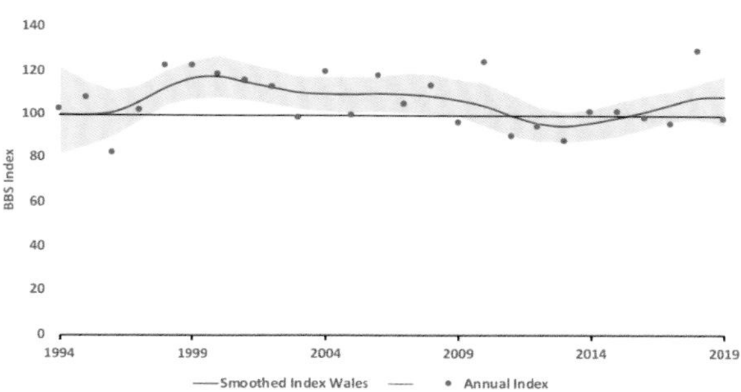

Mistle Thrush: Breeding Bird Survey indices, 1994–2019. The red line shows the smoothed index, 1995–2018. The shaded area indicates the 85% confidence limits.

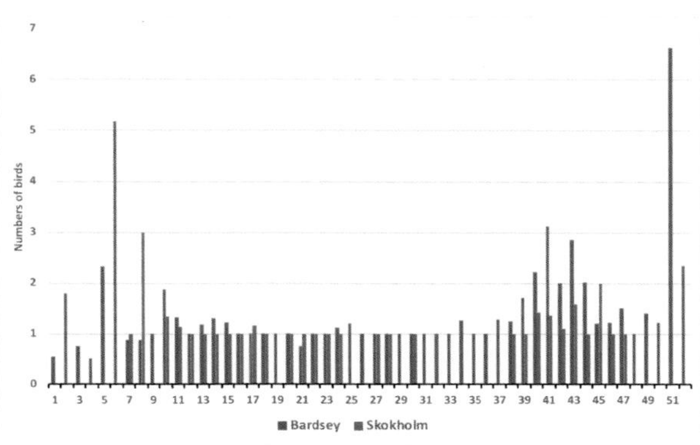

Mistle Thrush: Average counts by week number recorded by bird observatories, 1953–2019. Week 6 starts *c.*4 February and week 42 starts *c.*14 October.

Family parties flock together at the end of the breeding season. There is visual evidence of movements from lowland breeding territories uphill in late summer to forage on Bilberries, before birds return to the valleys to feed on berries with other thrushes (e.g. Massey 1976a). Some flocks can be quite large, such as up to 150 at Moel y Ci, Caernarfonshire, in mid-August 2006.

Mathew (1894) stated that, in Pembrokeshire, flocks would gather in July and disappear soon afterwards, assumed to have migrated south. There is no evidence now of large-scale migration over the Welsh coastal bird observatories, but small-scale movements along the Welsh coast in October are likely to include Fennoscandian breeding birds (*BTO Migration Atlas* 2002). The number present in Britain in mid-winter (422,000–715,000) is higher than the estimated breeding population (Hayhow *et al.* 2017b). Groups of up to 30 have been recorded in Gwent (Venables *et al.* 2008) and up to 50 on Bardsey, Caernarfonshire. More exceptional autumn movements include *c*.500 over Ramsey, Pembrokeshire, on 12 October 1992, 100 on Skomer, Pembrokeshire, in October 1983, and 77 over the Great Orme, Caernarfonshire, on 13 October 2001. Ringing sheds no light on these movements: relatively few Mistle Thrushes have been ringed in Wales and none has been recovered from outside Britain and Ireland, nor have there been local recoveries of foreign-ringed birds in Wales. Occasional records of larger numbers in spring, such as 100 at Abercynafon, Breconshire, on 1 April 2013, at a time when local birds would be feeding young, may be on return passage to a more northerly location.

Forrest (1907) stated that Mistle Thrush had been found nesting in rock cavities at about 460m. That altitudinal limit remains generally true today, but he also wrote that these birds moved to lowland areas in winter, as it was less tolerant of cold weather than other thrushes. Such movements occurred in Breconshire in the 1970s (e.g. Massey 1976a) but may be less frequent now. At the 10-km square level, winter distribution in Wales was slightly greater than during the breeding season during the *Britain and Ireland Atlas 2007–11*, as it occurred in at least 96% of squares. Abundance was relatively high across the lowlands and it was absent only from the most-elevated areas. It may be that Mistle Thrush is benefiting from milder winters in the uplands, although some local movement into the valleys almost certainly still occurs.

It is Amber-listed in Wales because of its Red List status in the UK (Johnstone and Bladwell 2016). The CBC/BBS index for Mistle Thrush in the UK fell by more than half between the late 1970s and 2010, driven mainly by a change in the English population. The greatest declines occurred in built-up areas and on mixed farmland, during 1995–2011 (Massimino *et al.* 2019). A reduction in survival of juveniles is suspected to be the cause of the decline, but no research has been undertaken that can explain this. With the Welsh population very stable, there seems no reason to believe that the Mistle Thrush's rattling call will not continue to be a feature of the Welsh countryside.

Geoff Gibbs and Ian M. Spence

American Robin *Turdus migratorius* Robin America

Welsh List Category	IUCN Red List	Rarity recording
	Global	
A	LC	BBRC

This extremely rare vagrant to Wales breeds widely across North America, although only those in the northern half are migratory, some of which winter around the Gulf of Mexico (*HBW*). There were 29 records in Britain up to 2018 (Holt *et al.* 2019), but none in 2019. It has occurred in Wales just once, on Bardsey, Caernarfonshire, on 11–12 November 2003.

Jon Green and Robin Sandham

American Robin on Bardsey, Caernarfonshire, November 2003
© Alan Clewes

Spotted Flycatcher *Muscicapa striata* Gwybedog Mannog

Welsh List Category	IUCN Red List			Birds of Conservation Concern			
	Global	Europe	GB	UK	Wales		
A	LC	LC	LC	Red	2002	2010	2016

The Spotted Flycatcher breeds in northwest Africa, Europe and across western Siberia to Lake Baikal. It is a long-distance migrant, wintering in Africa south of the equator (*HBW*). In Wales, the preferred habitats for this species include mature conifer and deciduous woodland with clearings and glades, large gardens, parks and churchyards, where high perches, from which to hunt insect prey, are in plentiful supply. It is often found near open water. Most Spotted Flycatchers breed at elevations below 400m, although nests were found in Radnor Forest at 500m (Jennings 2014) and at 460m in Breconshire (Massey 1976a). Crevices, ledges and creepers on house and garden walls are popular nest sites, and it will use open-fronted nest-boxes.

Spotted Flycatchers are one of the latest breeding migrants to return in spring. Whilst the earliest arrival dates for most counties are in the second half of April (the earliest record in the years 2000–18 was on 5 April 2007 in Breconshire), good numbers seldom appear until the second week of May and peak passage is in the last two weeks of May. The last autumn sightings are usually in October, though one stayed until 10 November 2010 at Monk Haven, Pembrokeshire. On Bardsey, Caernarfonshire, and Skokholm, Pembrokeshire, the main autumn passage runs from early August to the end of September, with a few stragglers in October. Much higher numbers were recorded on Bardsey.

At the end of the 19th century and the beginning of the 20th

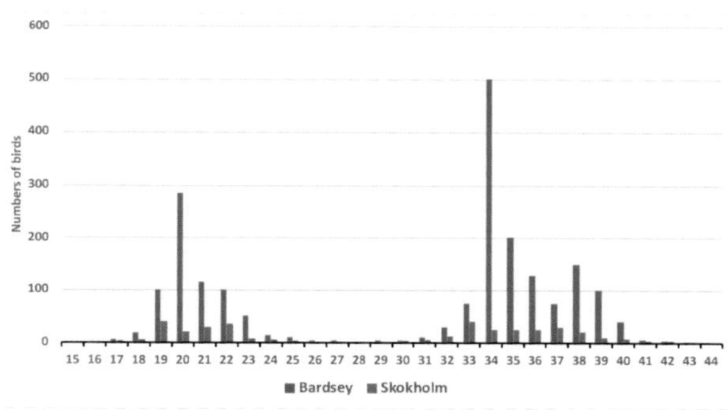

Spotted Flycatcher: Maximum counts by week number, at Bardsey, 1953–2019, and at Skokholm, 1928–2019. Week 20 starts *c.*13 May and week 34 starts *c.*19 August.

century, the Spotted Flycatcher was considered to be common or very common over much of the Welsh lowlands in North Wales (Forrest 1907), Pembrokeshire (Mathew 1894), Breconshire (Phillips 1899) and Glamorgan (Heathcote *et al.* 1967). Up to the 1950s, its status appeared to change very little. From 1937 to at least 1963 it was well distributed and quite numerous in Gwent (Venables *et al.* 2008). In Carmarthenshire, Ingram and Salmon (1954) described it as fairly numerous across the county, and in Ceredigion, Roderick and Davis (2010) reported that in the 1940s William Condry thought it was abundant in the north, particularly in the lowlands. Holt and Williams (2008) wrote that in 1951 an observer found 127 nests in the area around Llanymynech, Montgomeryshire, close to the border with Shropshire, something that would be thought impossible nowadays.

Changes in the fortune of the Spotted Flycatcher began to be reported in the second half of the 20th century. In Radnorshire, a decline occurred after about 1951 (Jennings 2014), and in Carmarthenshire, Lloyd *et al.* (2015) wrote that it was more numerous in the 1950s than in later decades. The *Britain and Ireland Atlas 1968–72* confirmed breeding in almost all 10-km squares in Wales, which showed that the Spotted Flycatcher still had a quite widespread distribution across lowland Wales but it was not until the next *Atlas* that any abundance information could be shown. The *Britain and Ireland Atlas 1988–91* revealed a slight reduction in range, most notably in the uplands and coastal locations, but otherwise the Spotted Flycatcher remained widely distributed. Relatively high abundance was shown across west Montgomeryshire, Radnorshire, northeast Carmarthenshire and southeast Pembrokeshire. In Gwent it was reported still to be widespread in the 1980s, though some absences were noted and a decline was clearly underway from the 1990s (Venables *et al.* 2008). During the 1980s, in Glamorgan, its distribution became patchy, with absence from upland areas and those of intensive farming (Hurford and Lansdown 1995). In Caernarfonshire, it was said to have been locally common and widely distributed, though scarce on Llŷn in 1976 (Jones and Dare 1976), but by the 1990s the population was in sharp decline (Barnes 1997). Jones and Whalley (2004) reported that it was "sparse" on Anglesey in 1956 and very few remained by the 1990s.

By the early 21st century, the Spotted Flycatcher was no longer a common breeding bird in Wales, although the *Britain and Ireland Atlas 2007–11* showed little change in distribution and relative abundance. However, there was higher abundance in northeast Meirionnydd and southern Denbighshire than in 1988–91. It was even considered to have increased in some parts of Radnorshire since 2001 (Jennings 2014) but in most other parts of Wales further declines were noted. Donovan and Rees (1994) wrote that in Pembrokeshire, during 1984–88, it was widespread. The population was estimated to be 900 pairs based on an average density of four pairs per tetrad, but 20 years later the *Pembrokeshire Atlas 2003–07* estimated a reduction to 500 pairs. In Ceredigion, Roderick and Davis (2010) noted that the Spotted Flycatcher was less numerous than it had been in the 1960s and 1970s. In Meirionnydd, it was reported to be in sharp decline in 2007 (Pritchard 2012) and it had disappeared as a breeding species in southeast Carmarthenshire by 2019 (Rob Hunt *in litt.*).

At the tetrad level, the picture is one of decline in those counties where a comparison can be made. The *Gwent Atlas 1981–85* showed evidence of breeding in 70% of tetrads, but this had reduced to 55% by 1998–2003. The *East Glamorgan Atlas 1984–89* showed evidence of breeding in 31% of tetrads with a marked decline to just 10% by 2008–11, whereas the *Pembrokeshire Atlas 1984–88* had records in 47% of tetrads with a smaller decrease, to 43%, in 2003–07 (*Pembrokeshire Avifauna* 2020). Other local atlases recorded breeding evidence in about one-quarter to one-third of tetrads: 36% in Gower (*West Glamorgan Atlas 1984–89*), 38% of the tetrads surveyed in Breconshire in 1988–90 (Peers and Shrubb 1990) and just 25% in the *North Wales Atlas 2008–12*.

There seems to have been no substantial change in the Spotted Flycatcher's status in Wales up to the 1960s, but a major decline has taken place subsequently. There are no adequate

Adult Spotted Flycatcher feeding its brood © John Hawkins

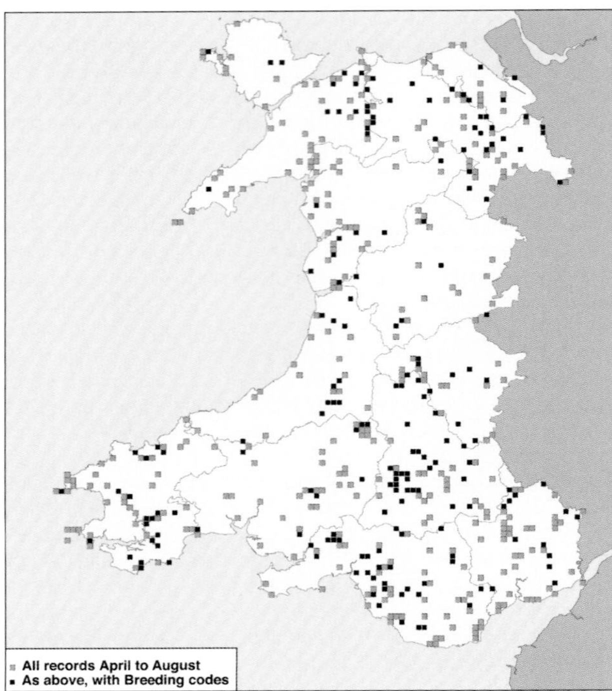

Spotted Flycatcher: Records from BirdTrack for April to August, 2015–19, showing the locations of records and those with breeding codes.

monitoring data for Wales but there is no reason to believe that the trend differs greatly from England, where the CBC/BBS indices show a 93% decline during 1967–2017 (Massimino *et al.* 2019).

The Welsh breeding population in 2018 was estimated to be 5,700 (5,300–6,050) territories (Hughes *et al.* 2020).

In the winter, Spotted Flycatchers move to Africa. Two ringed on Skokholm, one an adult ringed on 14 May 1964 and the second a first-calendar-year bird ringed on 9 August 1968, were recovered in Nigeria in October 1966 and 1969, respectively. They provide an indication of where Welsh breeding birds spend the winter. Ringing recoveries also suggest that some Irish breeding birds pass along the western coast of Wales: six ringed on Bardsey (three in spring and three in autumn) were recovered in Ireland during the breeding season. An exceptional long-distance recovery was a juvenile ringed on Bardsey on 30 August 1968 that was found dead in Cape Province, South Africa, on 7 March 1969. It is now thought that this might have been a passage bird breeding in eastern Europe, as these do winter further south than British breeders (*BTO Migration Atlas* 2002).

The decline in Wales, which is clearly part of a wider picture throughout much of Britain and Ireland and across Europe (PECBMS 2020), gives serious cause for concern. The only area where there has been no contraction of range is northern Scotland (*Britain and Ireland Atlas 2007–11*). Modelling of the likely effect of climate change during the 21st century indicated that Wales should remain within the appropriate climate window for this species (Huntley *et al.* 2007). Reduced survival of birds in their first year after fledging, either immediately after fledging or before the first breeding attempt, is thought to be the main demographic factor behind the decline (Freeman and Crick 2003). Lower survival may be a result of deterioration in woodland quality for young birds in the immediate post-fledging period, or from conditions in winter or on migration. This is one of several long-distance migrants wintering in the humid zone of West Africa that have undergone steep population declines. These declines may have a common cause, most likely related to climate change (Pearce-Higgins and Crick 2019). The sighting of a Spotted Flycatcher in Wales could well become an even less common occurrence in the future.

Geoff Gibbs and Ian M. Spence

Sponsored by Alan Williams

Robin — *Erithacus rubecula* — Robin Goch

Welsh List Category	IUCN Red List			Birds of Conservation Concern			
	Global	Europe	GB	UK	Wales		
A	LC	LC	LC	Green	2002	2010	2016

The Robin is our second most common species in Wales, the Wren being the commonest, but is probably our most well recognised and loved bird (Hughes *et al.* 2020). The subspecies *E.r. melophilus* breeds only in Britain and Ireland, but others breed across Europe (notably the nominate subspecies) and into western Asia, as well as in northwest Africa (*HBW*). It is a summer visitor to Fennoscandia and eastern Europe. Many of these pass through Britain *en route* to southern Europe in autumn (*BTO Migration Atlas* 2002). Increased numbers are evident in Wales, from late August to the end of October, either movements of young birds from local areas or arrivals from overseas. Some nominate birds (*E.r. rubecula*) winter in Britain. Robins are found in all types of woodland, hedgerows, parks, gardens and other suburban habitats, and occur in the uplands where moist woodland is present (*BWP*). The European population increased by 38% between 1980 and 2016 (PECBMS 2020), with some fluctuations caused by severe winters, though numbers can recover within 5–6 years (*European Atlas* 1997).

A review of the avifaunas written since 1894 indicates that the Robin has been common and widespread in Wales during the last couple of centuries. Afforestation has enabled it to breed at higher altitudes than formerly, at least up to 500m. The three atlases of Britain and Ireland, in 1968–72, 1988–91 and 2007–11, showed that Robins bred in nearly every 10-km square in Wales. The 1988–91 *Atlas* showed relatively high abundance in lowland areas, particularly in southwest and southeast Wales, and this remained the case in 2007–11.

Local tetrad atlases indicated that Robin distribution hardly changed between the 1980s and 20 years later, with evidence of breeding in close to 99% of tetrads in both atlases for Gwent (1981–85 and 1998–2003) and 91% in both Pembrokeshire surveys (1984–88 and 2003–07) (*Pembrokeshire Avifauna* 2020). The *East Glamorgan Atlas 1984–89* reported evidence of breeding in 87% of tetrads, but that reduced to 82% in the 2008–11 *Atlas*. In Gower, the *West Glamorgan Atlas 1984–89* showed breeding in 97% of tetrads. The *North Wales Atlas 2008–12* showed evidence of breeding in 86% of tetrads, with absence mainly in upland tetrads without trees or shrubs. The highest occupied tetrad was at 600m, in a forestry plantation on the northern fringe of the Dyfi Forest, Meirionnydd.

The winter distribution of the Robin is very similar to that in the breeding season. No change is apparent between the *Britain and Ireland Winter Atlas 1981–84* and the 2007–11 *Atlas*. The highest winter relative abundance in 2007–11 was at coastal locations and in the southern counties, most notably on Anglesey and in Pembrokeshire, Carmarthenshire and Gwent.

British Robins tend to be fairly sedentary. First-calendar-year birds do move out of their natal areas a few weeks after fledging, but most do not move far and usually breed within a few kilometres of their natal area (*BTO Migration Atlas* 2002). There are, however, a few examples of Welsh-bred birds making long-distance movements, with recoveries in southwest France and Spain. Large numbers pass through Britain in the autumn (from August

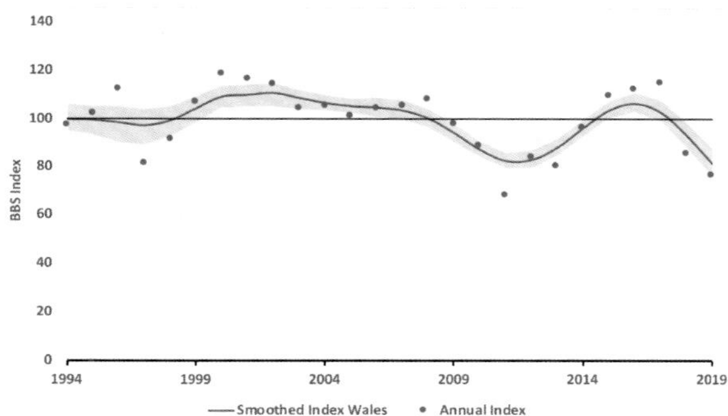

Robin: Breeding Bird Survey indices, 1994–2019. The red line shows the smoothed index, 1995–2018. The shaded area indicates the 85% confidence limits.

to November, peaking in October), arriving from Fennoscandia, the Baltic states and Poland (*BTO Migration Atlas* 2002). Many large counts come from the islands, such as 150 on Skomer, Pembrokeshire, on 30 September 2000 and 300 there following a southeasterly gale on 29–30 October 2004, and 120 on Bardsey, Caernarfonshire, attracted to the beam from the lighthouse on the night of 3–4 September 2002.

Breeding numbers fall after a prolonged period of cold weather. For example, at RSPB Ynys-hir, Ceredigion, the number of territories on the CBC monitoring plot (12ha) dropped from 23 in 1981 to five in 1982, following a cold winter. The BBS index for Wales was relatively stable between 1994 and 2008 but fell for a few years following cold weather in the 2009/10 and 2010/11 winters. There was an increase for five years, followed by a substantial decrease during 2018/19, but there is every reason to suppose that rapid recovery will occur, because the Robin is one of the most resilient members of the Welsh avifauna. The Welsh breeding population in 2018 was estimated to be 905,000 territories (Hughes *et al.* 2020). The European Atlas (2020) suggests that the Robin could be a beneficiary of climate change and associated milder winters. The Robin is likely to remain one of the commonest species in Wales.

Continental Robin *E.r. rubecula* Robin Goch y Cyfandir

Definitive evidence of the presence of Continental Robins in Wales is hard to find. There are clearly records of increased numbers in winter and this subspecies has been accepted onto the Welsh List. This is because there are several records of birds ringed in Europe in the spring and summer months that have been subsequently recovered or controlled in Wales, on passage or in winter. These include a first-calendar-year bird ringed on 28 August 1989 at Kabil, Parnu Region, Estonia, which was found long dead at Llanstadwell beach, Dale, Pembrokeshire, 2,002km west of the ringing site. Another first-calendar-year bird ringed on 29 August 2009, at Grimstadyatnet, Haried, Norway was found freshly dead on 19 March 2012 at Llanelli, Carmarthenshire, perhaps having wintered there.

Geoff Gibbs and Ian M. Spence

Sponsored by Heather Coats

Bluethroat *Luscinia svecica* Bronlas Smotyn

Welsh List Category	IUCN Red List		Rarity recording
	Global	Europe	
A	LC	LC	WBRC

The Bluethroat breeds in northern and western Europe and winters in North Africa and South Asia (*HBW*). There are two subspecies: the nominate *Luscinia s. svecica* (Red-spotted) and *L.s. cyanecula* (White-spotted). Both have been recorded in Wales, although only males can be safely identified sub-specifically. There have been 51 records of Bluethroat in Wales: 16 Red-spotted males, five White-spotted males and 32 others that could not be identified to subspecies, including one male with a completely blue throat on Skokholm on 27 May 2017. Within Britain, the number of records has been falling since a peak in the 1980s, when the annual mean was 186 records, to 69 during the 2010s (White and Kehoe 2020b). The number of Welsh records similarly increased towards the end of the 20th century but peaked in the 1990s and has declined since.

The Bluethroat has been recorded in all counties except Carmarthenshire. Most records have been from coastal locations, notably the islands. There have been only seven inland records, five of which were of Red-spotted males: at Llanbedrog, Caernarfonshire, in May 1976; at Beulah, Breconshire, in May 1978; at Lake Vyrnwy, Montgomeryshire, April 1985; at Knighton, Radnorshire, in May 1989; and one, which occurred at Arenig Fawr, Meirionnydd, on 10–17 June 2007, that appeared to hold territory. The sixth record was of a first-calendar-year female, which cannot be identified to subspecies, found dead at Ceiriog Valley, Denbighshire, in September 2002 with another first-calendar-year individual seen in fields bordering the River Usk near Llanfair Kilgeddin, Gwent, on 5–6 October 2017.

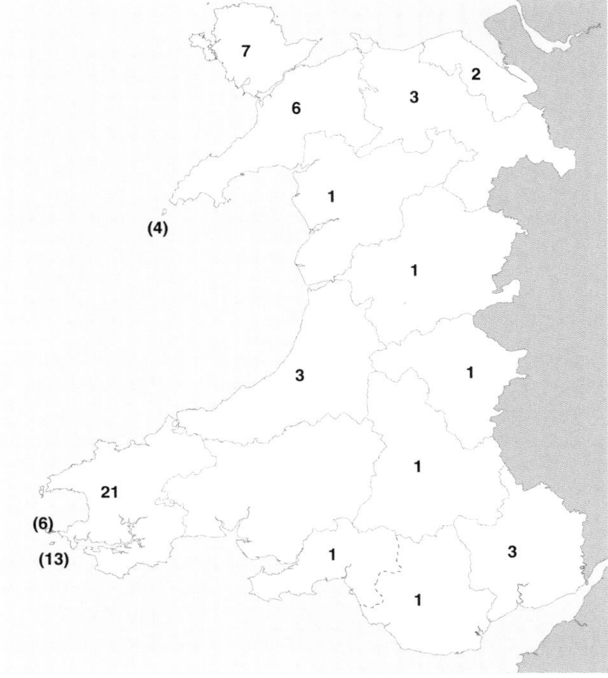

Bluethroat: Numbers of birds in each recording area, 1943–2019.

Peak passage periods are in late May/early June in spring and, in autumn, from mid-September to mid-October.

Total	Pre-1950	1950–59	1960–69	1970–79	1980–89	1990–99	2000–09	2010–19
51	1	2	8	9	6	10	9	6

Bluethroat: Individuals recorded in Wales by decade.

Red-spotted Bluethroat *Luscinia s. svecica*
Bronlas Smotyn Coch

This subspecies breeds in northern Europe, northern Asia and Alaska, and winters in North Africa and South Asia (IOC). The breeding range of this subspecies has been contracting in Fennoscandia and the eastern Baltic (*European Atlas* 2020). The majority of Bluethroat records in Wales, identified to subspecies, were Red-spotted.

White-spotted Bluethroat *Luscinia s. cyanecula*
Bronlas Smotyn Gwyn

This subspecies breeds in the temperate zones, locally in western and central Europe and more widely in northeast Europe, and winters in North Africa (IOC). The population has had a notable range expansion in central Europe in the past twenty years (*European Atlas* 2020). The first Welsh record of Bluethroat was of a white-spotted male on Skomer, Pembrokeshire, on 9 May 1946. The white-spotted form is predicted to move north with climate change, and could breed in southwest Britain, including Wales, by 2070–2100 (Huntley *et al.* 2007).

- Skomer, Pembrokeshire, on 9 May 1946;
- Skokholm, Pembrokeshire, on 20 October 1968:

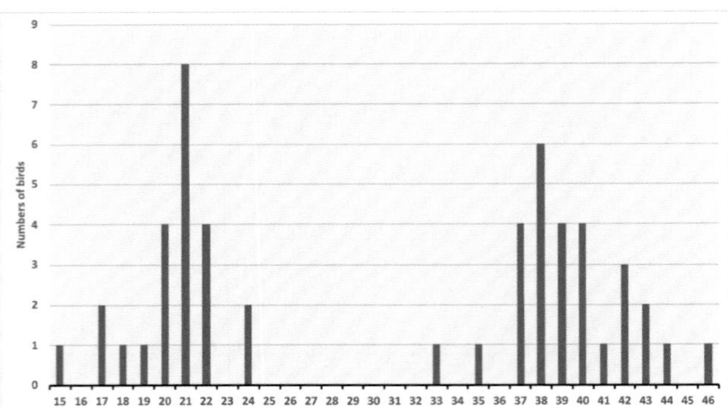

Bluethroat: Totals by week of arrival, 1946–2019. Week 21 begins *c.*21 May and week 38 begins *c.*17 September.

- Cors Caron, Ceredigion, on 26 May 2001;
- Skomer, Pembrokeshire, on 10–11 April 2008 and
- Bardsey, Caernarfonshire on 26 May 2018.

Jon Green and Robin Sandham

Thrush Nightingale *Luscinia luscinia* Eos Fraith

Welsh List Category	IUCN Red List		Rarity recording
	Global	Europe	
A	LC	LC	BBRC

The Thrush Nightingale breeds in northern and eastern Europe from Fennoscandia, Poland and Ukraine eastwards into central Russia, and winters in southeast Africa (*HBW*). A total of 226 have been recorded in Britain, with on average about five a year (Holt *et al.* 2020). It is an extremely rare vagrant to Wales, with just one record of a dead first-calendar-year bird, found on Bardsey, Caernarfonshire, on 20 September 1976.

Jon Green and Robin Sandham

Nightingale *Luscinia megarhynchos* Eos

Welsh List Category	IUCN Red List			Birds of Conservation Concern	Rarity recording
	Global	Europe	GB	UK	
A	LC	LC	VU	Red	WBRC

The Nightingale breeds in a broad band across much of central and western Europe, south to northwest Africa, and east in a scattered distribution to northwest China. In Britain it is now restricted to southern and eastern England, south of a line from the Severn to the Humber. It occupies a variety of habitats, including scrub and open woodland with thickets and dense patches of vegetation. Nightingales are migratory, wintering in tropical Africa south of the Sahara (*HBW*).

Although the Nightingale seems never to have been common in most of Wales, it has had considerable cultural significance and has often been mentioned in Welsh literature. The earliest reference was in a poem by Hywel ab Owain Gwynedd (died 1170), in which he praised the beauties of his father's realm of Gwynedd and included among these a Nightingale singing "at the confluence of the rivers". On the other hand, in his account of his journey through Wales with the Archbishop of Canterbury in 1188, Giraldus Cambrensis related that when the exhausted archbishop stopped to rest after some particularly steep hills, a member of the party commented that the Nightingale was never heard in that country. The archbishop smiled and said "The Nightingale followed wise counsel, and never came into Wales; but we, unwise counsel, have penetrated and gone through it".

The Nightingale was one of the four most frequently mentioned bird species in Welsh poetry of the 14th century (Williams 2014). The others—Song Thrush, Blackbird and Cuckoo—were presumably as common in Wales then as now, but it is unlikely that this applied to the Nightingale. Many place names in Wales include the element *Eos*, but interpreting these is complicated by the habit of giving the sobriquet '*Eos*' to gifted poets, musicians or singers. Most places that included *Eos* in the northern part of Wales were buildings and may have referred to people rather than to the bird. In southern Wales, such place names referred mainly to natural features, such as groves, and could indicate places where Nightingales bred (Brown 2008).

The population in the late 19th century and early 20th century seems to have been mainly in Gwent and Glamorgan. The Nightingale was an annual summer visitor to southeast Glamorgan, sometimes as far west as Pyle (Hurford and Lansdown 1995). A questionnaire survey of the distribution of Nightingale in Britain (Ticehurst and Jourdain 1911) found that in Gwent it was most numerous in the Wye Valley, being seen also in the Monnow and Usk valleys. In Glamorgan, the species was said to have increased in numbers largely in the southeast of the county during the previous ten years, and it appeared to be slowly spreading west along the coast. It was particularly numerous in the Vale of Glamorgan. It was estimated in 1910 that there were 20–30 within a seven-mile (11km) radius of the centre of Cardiff. In Breconshire, Phillips (1899) described it as "sparingly scattered over the

Total	1967–69	1970–79	1980–89	1990–99	2000–09	2010–19
123	21	39	21	26	8	8

Nightingale: Individuals recorded in Wales by decade, 1967–2019.

county", but mainly to the east of Brecon. A pair nested at Talgarth in the east of the county, a few years prior to 1899, and a pair was present in this area on and off up to 1907. In Radnorshire, nesting was reported at Glasbury in 1903 and near Llandrindod in 1912 (Jennings 2014).

Farther west, there were few records, although a nest with eggs was found 8km north of Llandovery, Carmarthenshire, in 1898 (Ingram and Salmon 1954). There were a few records of singing males in Ceredigion, but it was evidently rare. It is said that hundreds of people went to listen to one at Rhydyfuwch, near Cardigan, on 20 May 1886 (Roderick and Davis 2010). Mathew (1894) did not record the species in Pembrokeshire, and Forrest (1907) said that "North Wales lies outside the normal range of the Nightingale, and it never visits regularly any part of it". He was aware of a few records from the Severn Valley, Montgomeryshire, and of a considerable number of records in southern Flintshire and eastern Denbighshire, though none of confirmed breeding. There was one doubtful record from Caernarfonshire, and none from Meirionnydd and Anglesey.

The species was apparently still locally numerous in Glamorgan and Gwent in the 1920s and 1930s. In 1926, breeding was recorded in Radnorshire, at Glasbury in the Wye Valley, and in Breconshire near Glangrwyney in the Usk Valley. Hurford and Lansdown (1995) said that Nightingales declined rapidly in Glamorgan after 1936, with the only regularly visited localities being Llandough and Leckwith, where singing individuals were last heard in 1961 and 1962 respectively. There were occasional records of breeding outside the core area in the 1940s, including a nest of four eggs that was found at Brechfa, Carmarthenshire, in 1945: two were taken by a boy, one later finding its way to the National Museum of Wales, but the others were said to have hatched and the young fledged (Ingram and Salmon 1954). In Montgomeryshire, a pair nested on Llanymynech Hill in 1941. There were occasional records in the county up to 1963, including three singing at different localities in 1954 and 1959 (Holt and Williams 2008). Nightingales were evidently scarce in Gwent by 1961, when a male in song at Risca in May was said to have attracted a large audience (Venables *et al.* 2008). The last record of confirmed breeding in Gwent was at Lord's Grove, Monmouth, in 1975. A census in 1976 found just one singing male in Wales, in Gwent (Hudson 1979), and a similar survey in 1980 (Davis 1982) again found just one singing bird, this time in Dyfed [Pembrokeshire/Ceredigion], but details of this record have not been located and cannot be verified. The last definitive record of confirmed breeding in Wales was in East Glamorgan in 1981, when a pair fledged five young at Llanishen (Hurford and Lansdown 1995). However, singing birds, some long-staying, continued to be recorded in Gwent and Glamorgan, when both tetrad atlases in the 1980s recorded probable breeding in four locations. Even as late as the turn of the 21st century, probable breeding was recorded in one tetrad in the *Gwent Atlas 1998–2003*. There were also some records of long-staying singing males outside the core area. One sang for a month at Bodfari, Flintshire, in 1981 and one for two weeks at Middletown, Montgomeryshire, in 1983.

The Nightingale is now recorded as a scarce migrant in Wales, though there have been no ringing recoveries to indicate the likely origins of passage birds. Since 1967 at least 123 individuals, including a few breeding birds, have been recorded in Wales.

The majority of records have been in spring, either prospecting breeders or overshooting migrants, with a peak during the first three weeks in May. By far the most have been seen in East Glamorgan and Gwent, usually spring records of singing birds. In the 21st century, a good proportion of records were of passage migrants from Bardsey, Caernarfonshire, and in Pembrokeshire from Skokholm and Skomer. A smaller number were recorded in autumn, the latest being one at Kenfig, East Glamorgan, on 6 November 1993.

The Nightingales that formerly bred in parts of Wales were on the extreme northwestern edge of their breeding range. The Welsh climate may never have been ideally suited to a species whose core populations are in areas with a warmer or more continental climate. In the maritime areas of western Europe, it usually breeds south of the 19°C July isotherm (*European Atlas* 1997). In the last few decades, the European population has been stable (PECBMS 2020). Wilson *et al.* (2002) used climate-change scenarios to predict that in the period 2010–39 there could be small potential range extensions in Britain, including westwards into South Wales. However, in recent years the breeding range in England has continued to contract towards the south and east. The decline in the number of records is presumably linked to this. Several factors have been suggested as possible causes, including conditions on the wintering grounds. Holt *et al.* (2012) considered that reduced woodland management in recent decades and browsing by increased deer populations had caused a deterioration of habitat quality in woodland. At present, there seems little prospect of the Nightingale becoming a regular breeding bird in Wales again.

Rhion Pritchard, Jon Green and Robin Sandham

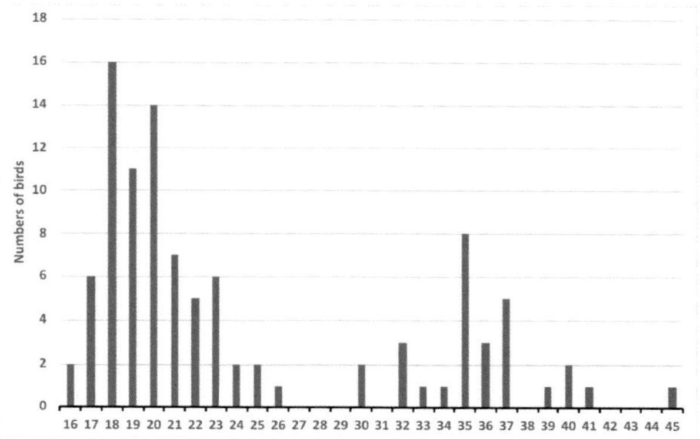

Nightingale: Totals by week of arrival (when known, n=100), 1967–2019. Week 18 begins *c.*30 April and week 35 begins *c.*27 August.

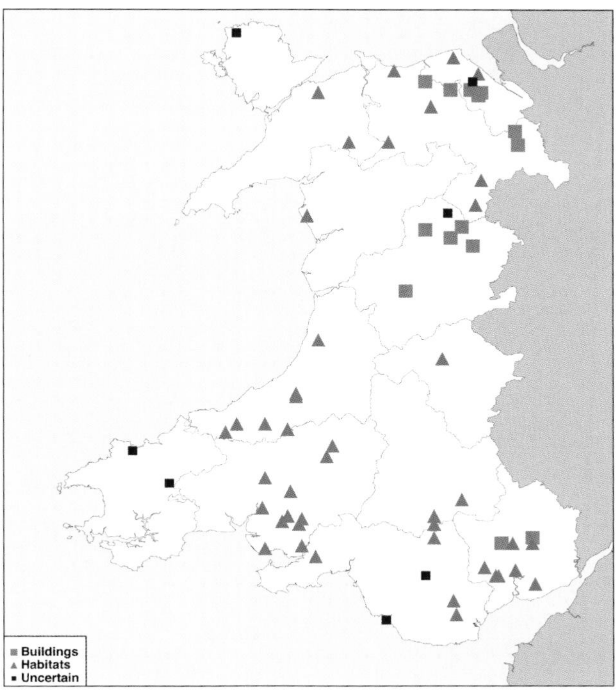

Nightingale: Map showing the distribution of place names containing "Eos" in Wales (Pritchard 2020).

White-throated Robin *Irania gutturalis* Robin Gyddfwyn

Welsh List Category	IUCN Red List		Rarity recording
	Global	Europe	
A	LC	LC	BBRC

The White-throated Robin breeds in Turkey, the Middle East and southern Central Asia. It winters in the Rift Valley of East Africa (*HBW*). This is an extremely rare vagrant, with one record in Wales: a female on Skokholm, Pembrokeshire, on 27–30 May 1990. This was the first record for Britain, although there had been one on the Calf of Man in June 1983.

Jon Green and Robin Sandham

Red-flanked Bluetail *Tarsiger cyanurus* Cynffonlas Ystlysgoch

Welsh List Category	IUCN Red List		Rarity recording
	Global	Europe	
A	LC	LC	WBRC

The Red-flanked Bluetail's main breeding range extends through the cool temperate forests of Russia to northern Japan and northeast China, with a small but expanding population in Fennoscandia. It winters in Southeast Asia (*HBW*). This species was once extremely rare in Britain, but numbers increased towards the end of the 20th century, possibly linked to the westward expansion of its range.

By the end of 2018 there had been 191 individuals in Britain (White and Kehoe 2020b). In Wales it remains a very rare vagrant, with five records up to 2019. The first four were all found at coastal localities in October, but the fifth was at an inland forested area:

- male on Bardsey, Caernarfonshire, on 1 October 2007;
- female on Skomer, Pembrokeshire, on 25 October 2010;
- 1CY/female on Bardsey, Caernarfonshire, on 25 October 2015;
- 1CY/female on the Great Orme, Caernarfonshire, on 9–10 October 2016 and
- 1CY/female at Wern Ddu, Caerphilly, East Glamorgan, from 26 January to 21 February 2017.

Jon Green and Robin Sandham

Red-flanked Bluetail on Bardsey, Caernarfonshire, 1 October 2007 © Steve Stansfield

Red-breasted Flycatcher *Ficedula parva* Gwybedog Brongoch

Welsh List Category	IUCN Red List		Rarity recording
	Global	Europe	
A	LC	LC	WBRC

The Red-breasted Flycatcher breeds in northern and eastern Europe as far as the Ural Mountains, and in the Caucasus, and winters in South Asia (*HBW*). It is a forest species in the west of its range, found in both mixed and deciduous stands, at altitudes higher than most flycatchers (*BWP*). The number of British records has increased steadily over the last 60 years, particularly since 2012, largely fuelled by an increase in autumn records. The average autumn totals are around 80, while spring annual totals have typically been 12 or fewer. It is a scarce passage migrant to Wales. Although Welsh totals each decade showed little trend, the weekly distribution of records is in line with the rest of Britain (White and Kehoe 2020b).

Total	Pre-1950	1950–59	1960–69	1970–79	1980–89	1990–99	2000–09	2010–19
177	2	9	24	29	34	21	24	34

Red-breasted Flycatcher: Individuals recorded in Wales by decade.

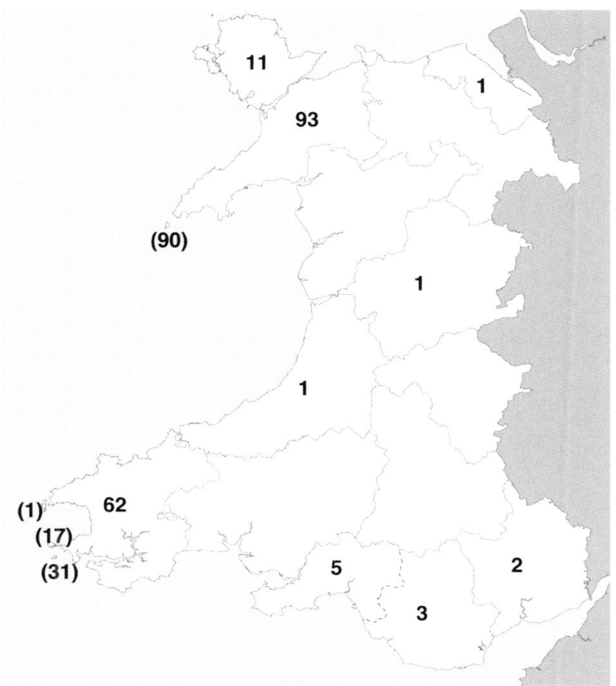

Red-breasted Flycatcher: Totals by week of arrival 1943–2019. Week 22 begins *c*.28 May and week 40 begins *c*.1 October.

The first Welsh record was a female at Llanishen, East Glamorgan, on 12 September 1943. A further 176 individuals have been recorded in Wales, of which just over half were on Bardsey, Caernarfonshire, and a quarter in Pembrokeshire, mainly on Skokholm and Skomer. Almost all of the remaining records were at mainland coastal locations. The only record from a landlocked county was in Montgomeryshire at Powis Castle, on the unusual date of 3–4 February 1983.

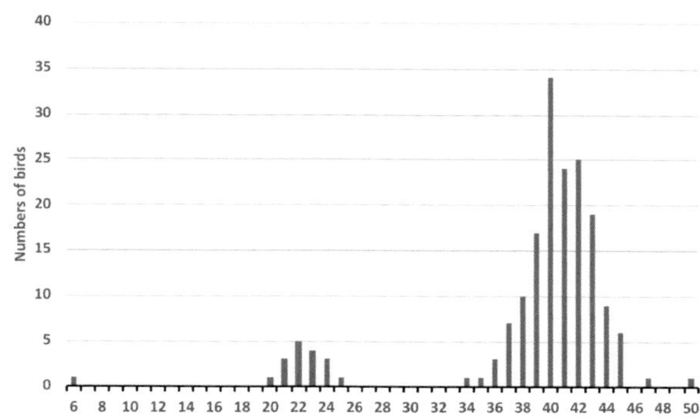

Red-breasted Flycatcher: Totals by week of arrival 1943–2019. Week 22 begins *c*.28 May and week 40 begins *c*.1 October.

A few Red-breasted Flycatchers arrive in spring, but the vast majority are from late September to the end of October. Most Welsh records relate to singles, but Bardsey has recorded three individuals, present from mid-September to late October 1968, two on 24 September 1970, another two on 4 October 1973 and five in October 1975.

Jon Green and Robin Sandham

Pied Flycatcher *Ficedula hypoleuca* Gwybedog Brith

Welsh List Category	IUCN Red List			Birds of Conservation Concern			
	Global	Europe	GB	UK	Wales		
A	LC	LC	VU	Red	2002	2010	2016

This charismatic black-and-white flycatcher breeds in the mountains of northwest Africa and northern Spain, in south and west France, Britain and east across central and northern Europe to central Siberia (*HBW*). In Britain, its preferred breeding habitat is mature deciduous and mixed woodlands, but it will also use riparian Alder and Beech woods (*BWP*). The best woodland habitat has an open canopy, with little or no understorey, often the result of stock-grazing. Wales is the stronghold for this species within Britain, with 69–76% of the population (Hughes *et al.* 2020). Here it is one of the most characteristic species of the Celtic 'rain-forest', woodland dominated by Sessile Oak. The Pied Flycatcher is a long-distance migrant. During the winter months it mainly lives in west and west-central Africa (Moreau 1972, Ouwehand *et al.* 2015). While it has been extensively studied during the breeding season, studies of the migration routes and the majority of its life cycle in winter are only now being undertaken (Vickery *et al.* 2014, Marra *et al.* 2015, Ouwehand *et al.* 2015).

It was probably Swainson (1893), using correspondence via *The Field* magazine, and Forrest (1907) who attempted the first comprehensive assessment of the species' presence in Wales. Their results, and additional observations by others (Campbell 1954), indicated that by the end of the 19th century Pied Flycatchers

Male Pied Flycatcher © Kev Joynes

had been recorded in all Welsh counties except Anglesey, with breeding in Caernarfonshire, Denbighshire, Meirionnydd, Montgomeryshire, Radnorshire and Breconshire. By the mid-20th century, the breeding range had increased southwards, although breeding was still unrecorded in Pembrokeshire, Glamorgan and Gwent (Campbell 1954, 1955). By 1962, up to a total of 50 pairs bred in Flintshire, Gower, East Glamorgan and Gwent, and all reported increasing trends. Other counties had more than 100 breeding pairs, with increasing counts in Meirionnydd, west Montgomeryshire and Breconshire (Campbell 1965).

By the time of the *Britain and Ireland Atlas 1968–72*, the range of the Pied Flycatcher had expanded a little more, including a pair on Anglesey (Jones and Whalley 2004). Although suspected in previous years, it was not until 1978, at Ffynone, that breeding was confirmed in Pembrokeshire (Donovan and Rees 1994), but by the *Pembrokeshire Atlas 1984–88*, at least 100 breeding pairs were recorded. The *Britain and Ireland Atlas 1988–91* described the Pied Flycatcher as "common in Wales," with a preference for upland valleys and hillsides dominated by mature Sessile Oaks. Between these two atlas periods there had been a 35% range expansion, mainly towards the coast in Ceredigion and into Pembrokeshire. The availability of nest-boxes, doubling breeding densities where these have been provided, may have been a factor in this expansion, which was probably the peak of the Pied Flycatcher's distribution in Wales. By the time of the 2007–11 *Atlas*, the range had contracted by 27% compared to 1968–72, such that all the previous gains had been lost. There were losses in Pembrokeshire, west Carmarthenshire and the Severn Valley, and noticeable decreases in abundance in Carmarthenshire, south Meirionnydd, north Ceredigion, Radnorshire and Breconshire. A clear decline in relative abundance was evident. An increase in Montgomeryshire is considered to have been an artefact of under-reporting in 1988–91 (Mike Haigh pers. comm.). Recent data indicated a further range contraction away from the northwest and the Marches (Gillings *et al.* 2015 and Simon Gillings pers. comm.), as well as a loss of breeding abundance throughout this reduced range as documented in county bird reports.

Some of the local tetrad atlases showed this decline in range in further detail. Though there was no great change in Gwent, where the species was found in 25% of tetrads in the 1981–85 *Atlas* and in 24% in the 1998–2003 *Atlas,* a much greater decline was evident in East Glamorgan. There, Pied Flycatchers were recorded in 14% of tetrads in 1984–89 but only 5% in 2008–11, with breeding evidence in only 3%. There was also a decline in Pembrokeshire, from 7% of tetrads in 1984–88 to 4% in 2003–07 (*Pembrokeshire Avifauna* 2020). The *North Wales Atlas 2008–12* recorded this species in 23% of tetrads. It was absent from Anglesey but recorded in 39% of tetrads in Meirionnydd, where more suitable habitat is found.

Analysis of Welsh records, submitted to BirdTrack and county bird reports during 2000–18, indicated that Pied Flycatchers typically arrive in Wales early in April, with an average arrival date of 15–17 April. March arrivals on the coast are not unusual, but there have been occasional early records from breeding sites, such as at Farchynys, Meirionnydd, on 19 March 2018; Plas Tan y Bwlch, Meirionnydd, on 24 March 2004; and RSPB Ynys-hir, Ceredigion, on 26 March 2003. Older males arrive first, followed by females a few days later, and younger males last (Potti 1998, Both *et al.* 2016). Ringing illustrated the faithfulness of this species to its natal site (Kern *et al.* 2014), with males and females returning to woodlands where breeding was successful.

Pied Flycatchers often select nest-boxes over natural sites (Lundberg and Alatalo 1992) and this, together with their tolerance of minor disturbance, has permitted extensive breeding studies. These studies have provided valuable detailed insights into their biology, particularly with evidence that climate change is resulting in a mismatch between the peak of caterpillar abundance and chick feeding demands (Burgess *et al.* 2018). If the mismatch is large, flycatchers in predominantly oak woodlands fare worse than those breeding in mixed habitats (Burger *et al.* 2012), indicating that these birds will feed on whatever is available and are not tied to oak-borne invertebrates (Burgess *et al.* 2018). However, declines in other aerial insects (Shortall *et al.* 2009), moths (Fox 2013) and butterflies (Fox *et al.* 2015) may also be having a negative influence.

In Wales, first-egg laying dates have become earlier: on average by 5.2 days +/-3.7 (from 66 nest-box schemes for which data were available to the author), compared to ten days across Britain as a whole (Massimino *et al.* 2019). Clutch size and fledging rate varied across sites both between and within years, with no consistent increase or decrease, but there is evidence of increased failure rates at the chick stage (Massimino *et al.* 2019). Once juveniles have fledged in the second half of June, they and their parents move into the tree canopy, often out of sight, or disperse away from their natal woodlands and, as a result, are less often recorded from late June onwards. Humphrey (1973), who recaught 12 juveniles travelling along hedges up to 5km away from their natal woodland in the Tywi Valley, Carmarthenshire, showed there is a period of local dispersal before birds started to move southeast in July. Migration starts in mid-August, with departure typically between 26 August and 3 October (BirdTrack data). At some breeding sites, departure is now up to seven days later than previously, from 66 nest-box schemes for which data were made available to the author. The last seen dates in Wales are 18 October 2006 from Bardsey and 26 October from Skomer, although these most probably will be birds on passage from breeding areas further north.

Recoveries of birds ringed in Wales show an autumn migration route down the west coast of mainland Europe, through The Netherlands, Belgium, France, Portugal and Spain, into North Africa and finally to Liberia and Ghana. Once in Africa, birds move slightly east during the winter (Ouwehand *et al.* 2015, Malcolm Burgess pers. comm.), causing some to take a more easterly route on their spring return. Welsh-ringed birds have been recovered in Algeria, Tunisia, Italy, Austria and Switzerland on return passage.

Hughes *et al.* (2020) estimated the Welsh breeding population in 2018 to be in a range between 15,000 and 19,000 territories. Unlike previous estimates, this took account of the density recorded in Wales in 2008–11. Given what is known about the reductions across much of its range, it is likely that previous figures underestimated the population size and so the decline is far greater than it may appear.

Pied Flycatcher breeding sites are not well represented in BBS squares in Wales. No Welsh index can be calculated, but there has been a UK decline of 45% between 1995 and 2018 (Harris *et al.* 2020). In Europe, the species has also been declining (29% between 1980 and 2017), but the rate of decline has slowed to 1% since 2008 (PECBMS 2020).

To try and determine the reasons for the decline, initial research focused on nest-box studies. Although increased temperatures are a major driver of change, first-egg dates showed correlation to early spring temperatures (Ockendon *et al.* 2013). Other weather events, increasing in frequency, frosts, stormy days and intense spring precipitation (pers. obs., Osborn and Maraun 2008), have played a part. Habitat limitations may confound any move to higher latitudes or altitudes, as is occurring with other species (Huntley *et al.* 2007). There is also evidence of increased competition with Great Tits for nesting cavities, which can result in the death of young, typically male, Pied Flycatchers (Samplonius and Both 2019).

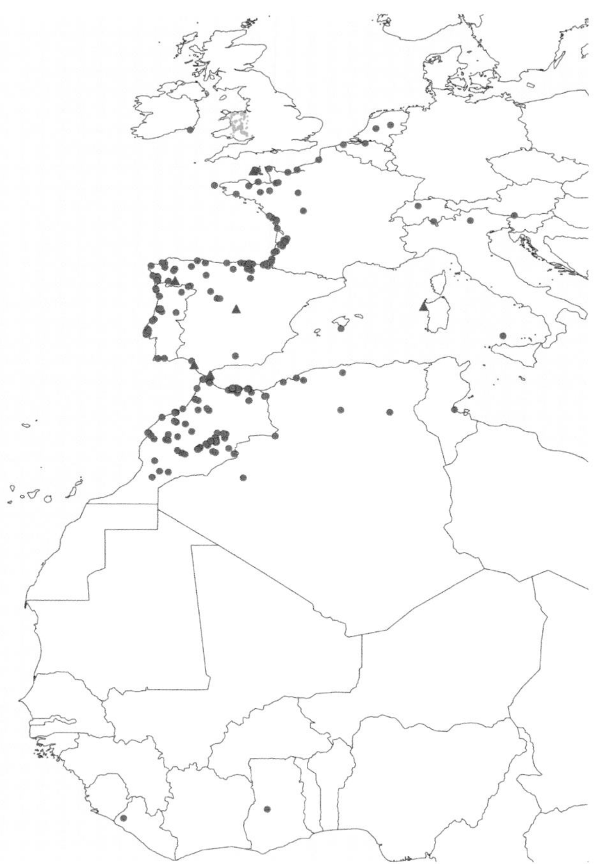

Pied Flycatcher: Recovery locations of birds ringed in Wales are shown by red circles. Ringing locations of birds that were recovered in Wales are shown by blue triangles. Small brown dots show ringing or recovery locations in Wales.

The reasons for the decrease in the range and the decline in abundance of the Pied Flycatcher in Wales are complex and may not be associated only with the breeding part of its life cycle. However, it is important to continue to monitor sites comprehensively with consistent data collection, recording successes and failures, and documenting productivity with respect to the caterpillar peak and abundance, in order to report an accurate picture of populations. Further attention should be given to investigating the diet of Pied Flycatcher, which has been scant to date (Tyrell 2017, Smith 2019a). Given concerns about caterpillar peak mismatches and the apparent improved breeding success in mixed woodland, dietary studies are important, especially in the light of future woodland management and any existing or future nest-box schemes within them. It would be tragic if a sighting of this iconic Welsh species were to become a rare occurrence.

Bob Harris

Sponsored by Mark Wilson

Collared Flycatcher — *Ficedula albicollis* — Gwybedog Torchog

Welsh List Category	IUCN Red List		Rarity recording
	Global	Europe	
A	LC	LC	BBRC

The Collared Flycatcher breeds in eastern and southeastern Europe into western Russia, and winters in central southern Africa (*HBW*). This is an extremely rare vagrant with a total of 55 British records up to 2019—on average, about one a year (Holt *et al.* 2020). The only Welsh record was of an adult male on Bardsey, Caernarfonshire on 10 May 1957.

Jon Green and Robin Sandham

Black Redstart *Phoenicurus ochruros* Tingoch Du

Welsh List Category	IUCN Red List			Birds of Conservation Concern				Rarity recording
	Global	Europe	GB	UK	Wales			
A	LC	LC	EN	Red	2002	2010	2016	RBBP

The Black Redstart is found over most of Europe and parts of Asia as far east as China, with a small population in northwest Africa. It is generally found in sparsely vegetated rocky areas, often at high altitudes. In the European Alps, density increases with elevation up to about 2,400m, while in the Himalayas it is found up to 5,200m. It is resident around the Mediterranean and through much of France but is migratory elsewhere (*HBW*). In Wales, it is mainly a passage migrant and winter visitor. The few ringing recoveries involving Wales provide little information on the origin of these birds. Langslow (1977) considered that most birds recorded at observatories breed well to the east of Britain and are migrating to and from their wintering grounds around the western Mediterranean. Those that move into Britain in autumn are thought to have originated from the western part of continental Europe, including The Netherlands, Belgium and France, while spring birds may include a higher proportion from central Europe (*BTO Migration Atlas* 2002). Some appear to use the same wintering site for several years. A bird ringed at Aberystwyth, Ceredigion, in December 2010, was seen there again in January 2012 and in December 2015. Wales holds the British longevity record for this species. An adult male ringed on 5 December 2010 at Aberystwyth, Ceredigion, was seen alive exactly five years and five days later at the same location on 7 December 2015.

Since the mid-19th century, the breeding range has expanded in northwest Europe, reaching Sweden, Norway and Britain, where it first bred in 1845. However, it is thought that populations around the periphery have lower breeding success and may not be self-sustaining (*HBW*). The British breeding population is almost entirely in England, where it is found largely in urban areas and derelict industrial sites. There was a rapid increase in the early 1940s in London, where birds bred on bombed sites and docks. Black Redstarts later spread to the West Midlands and northern England, but the population remained small, with a peak of 119 singing males at 92 localities in 1986 (*Birds in England* 2005). There has been a reduction of 45% in breeding numbers over the past 25 years, with the five-year mean in recent years a little over 50 pairs (Holling *et al.* 2019).

There have been a few breeding records in Wales, many of them in quarries. An apparently unmated male sang in Cardiff city centre, East Glamorgan, from 24 June to 8 July 1948, but the first confirmed breeding record was not until 1981, when a pair fledged three young in Cardiff city centre. In 1984, a pair that had over-wintered at Point of Ayr colliery, Flintshire, remained to breed there, rearing two broods. A pair fledged four young at Tremorfa, Cardiff, in 1988, and a juvenile in the middle of Knighton, Radnorshire, in July 1998 was thought to be of very local origin and possibly not

Male Black Redstart © Norman West

fully independent (Jennings 2014). In July 2006, a juvenile at RSPB Lake Vyrnwy, Montgomeryshire, was thought to have hatched locally. There have been more records of confirmed or probable breeding in recent years. A pair raised two broods in disused quarries in the eastern Brecon Beacons, Breconshire, in 2010 and again in 2011. A singing male there in 2012 appeared to be unpaired, but breeding was confirmed again in 2013 and a singing male has been present at this site in several subsequent years to

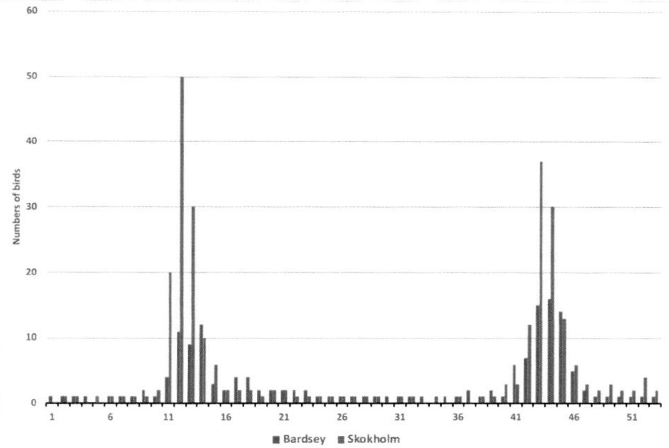

Black Redstart: Maximum counts by week number recorded by bird observatories, 1953–2019. Week 12 starts *c.*19 March and week 42 starts *c.*14 October.

2019 (Andrew King *in litt.*). In Radnorshire, a pair bred successfully in 2010 and 2011 and probably also attempted to breed in 2012, but no birds were seen in 2013 (Jennings 2014). A recently fledged juvenile at a site in Ceredigion in 2012 was believed to have fledged locally, and there was a possible breeding pair there the following year. A pair bred at a quarry in Denbighshire in 2012–15, fledging at least one and three young in the first two years respectively. A pair fledged four young at a disused quarry near Llanberis, Caernarfonshire, in 2016. There have also been a number of records of singing males in May and June. In Meirionnydd, for example, a singing immature male held territory for three weeks in May and June 2012.

There have been far more records of birds on passage, with smaller numbers wintering. Black Redstarts seem to have been scarce in Wales in the 19th and early 20th centuries. Mathew (1894) described it as "a winter visitor, not common" in Pembrokeshire but it was said to be a regular winter visitor to Pembroke. He quoted Mr Tracy as mentioning two birds killed in autumn 1847, one at Tenby and the other by Tracy himself, with an 'air cane' gun loaded with small shot, on the water trough of his neighbour's house in Pembroke. A female was shot at Cydweli, Carmarthenshire, in March 1864, with two seen in Carmarthen in December 1875. One killed in Cardiff in November 1884 was the only 19th-century record in Glamorgan. The first for Ceredigion was one at Borth in January 1887. Forrest (1907) listed only five records in North Wales, all between 1886 and about 1889: one in Flintshire, three in Meirionnydd and one on Traeth Mawr on the Meirionnydd/Caernarfonshire border, on 5 April 1888, that was thought to be on migration.

Numbers appeared to increase during the first half of the 20th century. Forrest (1919) said that "it seems probable that this species regularly passes through North Wales on migration in very small numbers". R.W. Jones recorded six at Llandudno in January and February 1921 (Pritchard 2017). It was scarce on Anglesey, where the first record in 1912 was followed by another in 1914, but there was only one further record there during the next 40 years. Ingram and Salmon (1939) described it as a rare winter visitor to Gwent, and gave only two dated records, although they thought that it was probably overlooked. Lockley *et al.* (1949) said that it was a regular visitor to Pembrokeshire, chiefly from September to November and from March to April, but also seen from December to February. Ingram and Salmon (1954) stated that it was an almost annual winter visitor to the Carmarthenshire coast. Records increased from the 1960s.

As spring passage can be from mid-March to early June, and autumn passage is usually from mid-September to late November, it is sometimes difficult to distinguish passage birds from wintering or potentially breeding individuals. Records analysed by month in Breconshire (Peers and Shrubb 1990), Ceredigion (Roderick and Davis 2010) and Meirionnydd (Pritchard 2012) showed the largest number of records in November. The great majority of passage and wintering birds have been recorded near the coast. Pembrokeshire usually recorded the largest numbers, but the species is annual, or almost annual, in most coastal counties and can occur in considerable numbers in some years.

Numbers also peak on the island observatories in late autumn, with large counts in some years. There were up to 40 on Skokholm, Pembrokeshire in October 1968 (Thompson 2008), and up to 15 on Bardsey, Caernarfonshire in the same month. Autumn 1982 also saw higher than average totals, with 111 along the Pembrokeshire coast during October, including 54 at Castlemartin on 24 October (Donovan and Rees 1994) and up to 14 on Bardsey. In 1996 there were up to eight on the Great Orme, Caernarfonshire, in September and 14 on Bardsey on 25 October, and in 1997 there were 28 on Skokholm on 23 October, many of which roosted in the island's outside toilet (Thompson 2008). Since the turn of this century, passage numbers have tended to be lower, although there was an influx into Pembrokeshire in late October 2004, with 15 at Dale on 28 October and 13 on Skomer the following day. There were 13 on Bardsey on 21 October 2007 and 12 on Skomer on 28 October 2009. Wintering numbers (December to February) although usually much lower, are often from regular locations that are occupied in most years. There were six at Rover Way, Cardiff, on 22 December 1986. Counts in spring are lower, but there were 50 on Skokholm in March 1948 and 30 there in March 1949.

Records of passage or wintering birds, more than a few kilometres from the coast, are less frequent, but include some seen in the uplands, including one on the summit of Pumlumon, Ceredigion, in May 1989. There were only two records in Breconshire before 1961, but Peers and Shrubb (1990) reported 19 subsequently, eight of them in November, while Jennings (2014) listed 25 records in Radnorshire following the first record in 1923. The first record for Montgomeryshire was not until 1975.

It remains to be seen whether the increased number of breeding records since 2010 will be maintained, but the preferred breeding habitat here seems to be disused quarries, which are certainly not in short supply in Wales. Although Wales is on the northwestern edge of its breeding range, and despite the suggestion that peripheral populations have lower breeding success, the pairs here in recent years seem to have fledged good numbers of young. This may help to extend its breeding range northwards, as indicated by climate modelling in Huntley *et al.* (2007), which forecasts that all of Wales, except Snowdonia, will be suitable for the Black Redstart by the end of this century.

Rhion Pritchard

Sponsored by Alan Williams

Redstart *Phoenicurus phoenicurus* Tingoch

Welsh List Category	IUCN Red List			Birds of Conservation Concern			
	Global	Europe	GB	UK	Wales		
A	LC	LC	LC	Amber	2002	2010	2016

Wales is of great importance for the Redstart, holding about 37% of the UK breeding population (Hughes *et al.* 2020). It breeds over most of Europe and in temperate Asia as far east as central Siberia, with a small population in northwest Africa. It is a long-distance migrant, most crossing the Sahara to winter in the Sahel and savannah to the south. It is mainly a bird of open woodland, particularly clearings and margins, preferably with semi-open undergrowth or herbage (*HBW*). In Wales, it generally avoids coastal areas, except on migration, and while the majority of breeding Redstarts are found at 100–450m, a few pairs can be found up to 500m (pers. obs.). Redstarts nest in cavities, often in trees, although dry-stone walls and old buildings are much used in Wales, as are nest-boxes. Along with Pied Flycatcher and Wood Warbler, this is one of the characteristic species of Welsh Sessile Oak woodland, but can be found in a wider range of wooded habitats where nest sites are available. Bibby *et al.* (1989) found it at 47 of 253 survey points in upland conifer forests in North Wales, and Fuller *et al.* (2006) recorded it in 50 of 118 *ffridd* study sites in central Wales in 1985–87. Redstarts show a preference for Bracken-dominated *ffridd* over other types of habitat (Conway and Fuller 2010).

Redstarts usually arrive in Wales in early to mid-April, but there was a record as early as 12 March, at Kenfig, East Glamorgan, in 1991. Most counties had March records during 2000–18, and Newson *et al.* (2016) found that their arrival in the UK had become significantly earlier between 1962–66 and 2002–11. Most birds have left by the end of September, but some remain into October and there were November records from six counties during 2000–18. There were also a few December records in the 20th century, including one at Holyhead, Anglesey, on 17 December 1978 and another at Pennard, Gower, on 31 October 1995, which remained to 26 December. A male over-wintered by the upper Dyfi Estuary, Montgomeryshire, from 7 January to 16 March 2015, and was singing on 20 February. Recoveries of birds ringed in Wales illustrated a migration route through France, Spain and Portugal and into Morocco, but there have been no recoveries from farther south in Africa (*BTO Migration Atlas* 2002).

Forrest (1907) considered that "the distribution of the Redstart in North Wales is peculiar and at first sight puzzling", being rare on Anglesey and absent from Llŷn. He concluded that it was "found commonly in all the well-wooded lowland valleys and hardly anywhere else". His correspondents suggested that it was particularly numerous in parts of Montgomeryshire. Salter (1900) considered it rather scarce in northern Ceredigion, but it increased in the 1920s and was described as locally common by the late 1940s, although scarce in the lowlands, except around the Dyfi Estuary (Roderick and Davis 2010). Mathew (1894) stated that Redstart was abundant in south Ceredigion but extremely rare in Pembrokeshire, and it remained so into the 20th century, with Lockley *et al.* (1949) reporting several nesting pairs in the centre of the county in 1928. Barker (1905) said that it was not common around Carmarthen, but he understood from Dr Salter that it was very abundant in the upper Tywi Valley. Ingram and Salmon (1954) reported that it was more numerous there than farther south in Carmarthenshire. In Gower, Dillwyn (1848) said that it was not uncommon in the Swansea area. By the end of the 19th century, it was thought to be increasing in Glamorgan, although it was restricted to estates with a good supply of old

Male Redstart © Kev Joynes

trees with nesting holes. Its range expanded in the county during the 1930s, an increase that continued until the mid-1960s (Hurford and Lansdown 1995). In Radnorshire, Ingram and Salmon (1955) described it as fairly widespread, although numbers fluctuated.

Numbers crashed across Europe in 1969, one of several migrants affected by drought in the Sahel region of Africa (Marchant *et al.* 1990). There are no data for Wales in this period, but the UK CBC index showed a sharp decline that continued into the early 1970s (Massimino *et al.* 2019). This was followed by a rather uneven recovery, although numbers in Wales may have recovered more rapidly than in some other areas of Britain. In northwest Wales, the *Cambrian Bird Report* stated that breeding numbers were still low in 1974, but had recovered in most districts by 1980. In Radnorshire, Jennings (2014) considered that numbers had recovered by the mid-1980s. In parts of Ceredigion, it was considered possibly more abundant by 1988–91 than before the population crash, with an expansion in the south (Roderick and Davis 2010). The recovery in this period was associated with a significant improvement in the number of fledglings per breeding attempt and progressively earlier laying dates (Massimino *et al.* 2019).

The *Britain and Ireland Atlas 1968–72* gave the first complete picture of Redstart distribution. Although it was found over much of Wales, there were few records from Anglesey, Llŷn or the south coast, and it was absent from large areas of Pembrokeshire. The general distribution of the Redstart across Wales changed very little in the subsequent two atlases, in 1988–91 and 2007–11, unlike in England and Scotland during this period, where a contraction in range was evident. However, there were changes in the relative abundance between these two atlases. The 1988–91 *Atlas* showed that Redstarts were most abundant in the core range but less so on the margins, and a similar pattern was evident in the 2007–11 *Atlas*. Relative abundance in 2007–11 was higher in Wales than elsewhere in Britain, but more 10-km squares showed a decrease in relative abundance than showed increases.

Where county tetrad atlases have been repeated, they showed some differences in trend. In Gwent evidence of breeding was recorded in 51% of tetrads in 1981–85, but only 44% in 1998–2003, with a greater number of tetrads occupied in the western valleys in the later period, but a pronounced decline in the centre of the county and the Wye Valley. In East Glamorgan it showed a slight decrease, from 29% of tetrads to 24% between 1984–89 and 2008–11, but there were few breeding records from coastal areas in either county. Pembrokeshire atlases recorded Redstart in 8% of tetrads in 1984–88 and 11% in 2003–07. Fieldwork for the first atlas recorded a previously unknown population in the Mynydd Preseli region, which had expanded by 2003–07. The *West Glamorgan Atlas 1984–89* recorded Redstart in 39% of tetrads, mainly in the uplands. It confirmed only one breeding pair on the Gower Peninsula, while the species occurred in 79% of tetrads visited in Breconshire during 1998–90 and was especially numerous in the north and west. The *North Wales Atlas 2008–12* recorded it in 46% of tetrads but with great variation between counties. There was only one record, of possible breeding, on

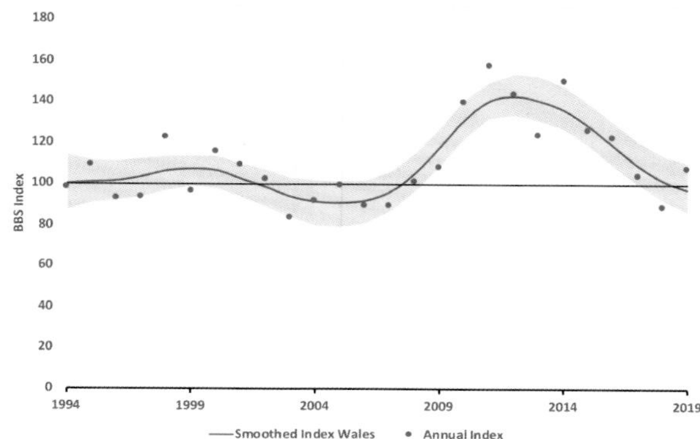

Redstart: Breeding Bird Survey indices, 1994–2019. The red line shows the smoothed index, 1995–2018. The shaded area indicates the 85% confidence limits.

Anglesey and one in western Llŷn, Caernarfonshire, but there were breeding records in 67% of tetrads in Meirionnydd (Pritchard 2012).

Redstarts can occur at high densities. For example, an RSPB survey found 40 singing males/km² near Rhayader, Radnorshire, in 1979, and 48 pairs were found in 82ha of broadleaved woodland in the Radnorshire part of the Elan Woodlands SAC in 2010 (Jennings 2014), a density of almost 59 pairs/km². At one of the few lowland strongholds, RSPB Ynys-hir in Ceredigion, there were 56 territories in 112ha of woodland in 2011, a density of 50 territories/km².

Sites surveyed in Wales by the RSPB as part of the Repeat Woodland Bird Survey (RWBS) showed a decline of 51% between the mid-1980s and 2003–04 (Amar *et al.* 2006). This had not been evident from national monitoring schemes, but was matched in other parts of Britain (Hewson *et al.* 2007). The BBS index for Wales increased sharply during 2005–12, followed by an equally sharp fall and by 2019 was much the same as in 1995 (Harris *et al.* 2020). The Welsh breeding population in 2018 was estimated to be 50,000 (36,500–65,000) territories (Hughes *et al.* 2020).

In contrast, the European population increased by 30% during 2007–16 (PECBMS 2020). Changes in the structure of upland oak woods in western and northern Britain are considered unlikely to have been the main driver of change in the Redstart population (Mallord *et al.* 2016). However, in northwest Switzerland, clutch sizes were greater where there was limited or little ground vegetation (Martinez 2012). Introducing the right levels of grazing into Welsh woodlands could, therefore, benefit Redstarts. The recovery of the population in the 1970s was driven by increased nesting productivity and earlier laying dates. These have moved by 14 days over 50 years, which is consistent with climate change (Massimino *et al.* 2019). However, conditions on its wintering grounds and along its migration route are also likely to play an important role in the future status of the Redstart in Wales.

Rhion Pritchard

Sponsored by Peter Rathbone

Moussier's Redstart — *Phoenicurus moussieri* — Tingoch Moussier

Welsh List Category	IUCN Red List	Rarity recording
	Global	
A	LC	BBRC

The Moussier's Redstart breeds in Morocco, northern Algeria and Tunisia. It is mostly resident but makes altitudinal and short-distance movements outside the breeding season (*HBW*). This is an exceptionally rare vagrant, with only one northern European record, of a male at Dinas Head, Pembrokeshire, on 24 April 1988.

Jon Green and Robin Sandham

Rock Thrush *Monticola saxatalis* Brych Craig Cyffredin

Welsh List Category	IUCN Red List		Rarity recording
	Global	Europe	
A	LC	LC	BBRC

The Rock Thrush has a scattered breeding distribution across southern Europe, the Middle East, Central Asia to northern China and Mongolia. It winters in East Africa (*HBW*). There have been 30 British records of this extremely rare vagrant to the end of 2017 (Holt *et al.* 2018). Three of those were from Wales:

- male at Ynyslas, Ceredigion, on 21 June 1981;
- female at Llyn Alaw Dam, Anglesey, on 4–6 June 1986 and
- 1CY male at Pwll Du Quarry, Gwent, from 12 October to 3 November 2017.

It may occur with increasing frequency during the 21st century, as the area that is climatically suitable for Rock Thrush has been modelled to extend north to southern England (Huntley *et al.* 2007).

Jon Green and Robin Sandham

Sponsored by John Marsh

Moulting male Rock Thrush at Pwll-du, Gwent, October-November 2017 © Richard Smith

Blue Rock Thrush *Monticola solitarius* Brych Craig Glas

Welsh List Category	IUCN Red List		Rarity recording
	Global	Europe	
A	LC	LC	BBRC

The Blue Rock Thrush is found all year round in much of Mediterranean Europe and southern Turkey, and its breeding range as a summer migrant extends east to Japan and Indonesia (*HBW*). There were nine British records of this extremely rare vagrant up to 2019 (Holt *et al.* 2020), of which two were in Wales. A male was at Moel-y-Gest, Caernarfonshire, on 4 June 1987 and another male was seen in the Elan Valley, Radnorshire, on 11 April 2007.

It may occur with increasing frequency during the 21st century, as the area that is climatically suitable for Blue Rock Thrush has been modelled to extend north to southern England (Huntley *et al.* 2007).

Jon Green and Robin Sandham

Whinchat *Saxicola rubetra* Crec yr Eithin

Welsh List Category	IUCN Red List			Birds of Conservation Concern			
	Global	Europe	GB	UK	Wales		
A	LC	LC	NT	Red	2002	2010	2016

The Whinchat breeds across much of central and northern Europe, including Fennoscandia and western Russia, and spends the winter in a wide band of tropical and sub-tropical Africa, from Senegal to Ethiopia and south as far as Tanzania and Malawi (*HBW*). Some range losses occurred in southern and western Europe between the *European Atlas* (1997) and *European Atlas* (2020). The Whinchat is typically found breeding in agriculturally unimproved upland habitats, mainly *ffridd*, usually at 300–450m, but occasionally higher. These upland margins of Wales hold an important proportion of the British Whinchat population. After breeding, Whinchats from Britain, *en route* to their wintering grounds, move down the west coast of France to southern Spain and Portugal, where they stop to feed up. Two ringed as nestlings in Gwent were recovered later in the same year, on 15 August in southwest France and on 15 September in the Algarve, Portugal.

It is one of the later spring migrants to arrive on its breeding grounds, usually in the third week of April, after which a steady influx continues into the first week of May. Peak arrivals on Bardsey, Caernarfonshire, and Skokholm, Pembrokeshire, occur between late April and mid-May. Earlier arrivals have been recorded, with some mid–late March dates in the southern counties, but 1 March 1946 on Skomer, Pembrokeshire, and 2 March 2004 at Eglwys Nunydd, Gower, were exceptional.

During the breeding season, the Whinchat is associated with a mosaic of habitats often dominated by Bracken and heather with scattered trees and shrubs (Henderson *et al.* 2017). It prefers the damper features of these habitats, where an insect-rich food supply, particularly craneflies, is available. The Whinchat has a rich far-reaching song that was once a typical sound of upland *ffridd* through summer. It sings and forages from prominent features, often returning to favoured perches around its territory. The nests are well-concealed on the ground among grassy tussocks. As a long-distance migrant, the Whinchat has a relatively short breeding window and so is usually single-brooded, relying on predictable prey abundance. It is, therefore, quite susceptible to disruption in the timing of peak prey availability (Visser *et al.* 2015). These marginal or transitional habitats have a restricted distribution and are regularly under threat from planting, drainage and overgrazing (Fuller *et al.* 2006, Border *et al.* 2016). In many upland locations, the Whinchat and Stonechat breed in the same general areas where there is transitional habitat, particularly where damp bog merges with heather or gorse, but the contrasting ecological requirements between the two species offer a likely explanation for their divergent population trends (Jiguet *et al.* 2007). A survey of breeding Whinchats on Mynydd Preseli, Pembrokeshire, found that the two species never shared the same territory for breeding and had different habitat preferences (Jenks *et al.* 2012). Stonechats occurred in extensive gorse, especially with dense three- to four-year-old patches, while Whinchats preferred a mosaic of Bracken, stream and bushes. Where Whinchat territories did include gorse, these tended to be mature and leggy bushes, unlike the more rounded structure that Stonechats seemed to prefer. There was no evidence that Stonechats presented a displacement threat to Whinchats.

In North Wales, Forrest (1907), regarded it as a "rather common bird… numerous on the belt of flats from Pwllheli to Tywyn [Caernarfonshire/Meirionnydd]… [while] inland its favourite haunts seem to be railways and the skirts of moors". This certainly does not hold true today, as the Whinchat is absent from the whole of the coastal lowland between Pwllheli and Tywyn (Rhion Pritchard pers. obs). In Glamorgan, it was common and widespread at the turn of the 20th century, diminishing gradually over the next 25 years. The decline continued until the 1950s, when areas on the flanks of the coalfield began to be reoccupied, aided by the expansion of forestry plantations, the development of rough uncultivated vegetation on the spoil tips and the abandonment of many small-holdings in the hills (Heathcote *et al.* 1967). In Pembrokeshire, Mathew (1894) said that Whinchat was "a summer visitor. Although Mr Dix wrote that the Whinchat was decidedly rare in his district, which was the north-eastern corner of the county immediately adjoining Cardiganshire, we have found that it is pretty generally distributed". In the mid-20th century, the Whinchat was regarded as a scarce summer resident in Pembrokeshire, with breeding confined to the north and east of the county (Lockley *et al.* 1949, Saunders 1976). In Gwent, Ingram and Salmon (1939) and Humphreys (1963) described Whinchat as breeding regularly on the coastal levels, railway embankments and other lowland habitats, as well as in the hills.

From the mid-20th century, a general decline in distribution and numbers in Britain was evident, most notably in the lowlands and primarily due to land-use changes. In Wales, the Whinchat's range progressively decreased between 1968–72 and 2007–11, as the three atlas surveys showed a definite retreat to higher elevations. The *Britain and Ireland Atlas 1968–72* showed 70% of 10-km squares occupied, decreasing to 67% in 1988–91 and only 48% in 2007–11. In 1968–72, there were still Whinchats in parts of lowland Llŷn, Caernarfonshire, coastal Flintshire, Ceredigion, Pembrokeshire and Gwent, but many had gone by 1988–91. On Anglesey, RSPB surveys in 1986 found pairs in only 16 of 762 1-km squares (0.02/km^2). In Breconshire, the Whinchat is most common in the north and west. During 1988–90, it was present in 39% of tetrads visited (Peers and Shrubb 1990). In Pembrokeshire, it was present in 9% of tetrads in the 1984–88 *Atlas*, declining to just 3% by 2003–07. Those that remained were on Mynydd Preseli, save for probable breeding on the St Davids Peninsula. From the 2003–07 results, a maximum county population of 25 pairs was estimated. Adverse conditions on migration and changes in agricultural practice were cited as causes of the decline (*Pembrokeshire Avifauna* 2020). The *East Glamorgan Atlas 1984–89* showed that 17% of tetrads were occupied, but the 2008–11 *Atlas* indicated this had reduced to 11%. In Gower, the *West Glamorgan Atlas 1984–89* showed the Whinchat to be uncommon and confined to the uplands, found in 14% of tetrads. By the 21st century it had almost disappeared as a breeding bird, occurring primarily as a

Male Whinchat © Bob Garrett

scarce passage migrant. In Gwent, between 1981–85 and 1998–2003, there was a decline in the percentage of occupied tetrads from 23% to 19%. At present the Whinchat predominantly resides in the hills of the north and west of the county. It benefited in the short term from conifer planting in the uplands, which provided good habitat in the early stages of growth, but the population declined as the trees grew larger (Venables *et al.* 2008). The *North Wales Atlas 2008–12* recorded Whinchat in only 16% of tetrads, the main strongholds being the Berwyn, Mynydd Hiraethog and Ruabon Moor in Denbighshire, the Migneint, Caernarfonshire/Denbighshire/Meirionnydd, and eastern slopes of the Carneddau, Caernarfonshire.

Comparisons between the 1988–91 and 2007–11 Britain and Ireland atlases also showed that relative abundance had decreased markedly, most notably in Caernarfonshire and Breconshire.

In good upland habitat in Wales, detailed breeding studies, from the 1980s to the early 21st century, indicated that densities of 14–30 pairs/km^2 were encountered, compared to the lower average densities calculated from the results of local atlases. RSPB surveys on Moel Famau, Denbighshire, in 1986 found 118 pairs on 861ha (maximum density 13.7 pairs/km^2), almost all of which were clustered on the westward-facing slopes. In Gwent, there were estimates of 77 pairs above the 220m contour on the Henllys to Penyrheol ridge in 1980, and 20 pairs on the slopes of Mynydd Llwyd at Cwm Lickey in 1985 (Venables *et al.* 2008). The *Gwent Atlas 1981–85* suggested an average of 1.25 pairs/km^2 with an estimated population of *c.*450 pairs, thought to be high at the time, but quite possibly an underestimate, given the results of later studies. Detailed studies in 1999–2001 produced average population densities of 30.3 pairs/km^2 on The Blorenge, 38.5 pairs/km^2 on Mynydd-y-garn-Fawr and 18.2 pairs/km^2 on Mynydd Garnclochdy. There was a mean annual total of 106 pairs on these three sites, plus Coity Mountain (Smith 2002). These data contributed to the Gwent population estimate of *c.*550 pairs (Steve Smith pers. data), but this figure was subject to considerable annual variation, as wet summers had depressed breeding success in Gwent during Smith's study. A comparison between the two Gwent atlases of 1981–85 and 1998–2003 gave the impression that the Whinchat population had increased, a result contrary to the national trend. However, the second population estimate was thought to be more accurate, being based on high-quality density data (over 300 nests monitored), and a comparison may not be a true reflection of the situation at the time. Sadly, the high breeding densities of 1998–2003 have not been repeated (Steph Tyler and Jerry Lewis pers. comm.), and the Gwent population has declined in the 21st century.

In Breconshire, Peers and Shrubb (1990) suggested an average of three pairs in occupied tetrads (0.75 pair/km^2) between 1988–91, implying a total of about 600 pairs for the county. In Pembrokeshire, a detailed survey in 2012 by Jenks *et al.* (2012) found no evidence of Whinchats on the St Davids Peninsula, but mapped 29 pairs on Mynydd Preseli, now their only breeding locality in the county. This total was not dissimilar to the earlier estimate, in 2007, of 25 pairs (*Pembrokeshire Atlas 2003–07*). In Carmarthenshire, it is more or less confined to the hills in the north and, while 16 pairs were at RSPB Gwenffrwd/Dinas in 1998 and 27 pairs at Mynydd Du in 1996, a reduction in numbers was noted at most sites in the 21st century (Lloyd *et al.* 2015).

The UK BBS index for 1995–2018 declined by 53%, but it was fairly stable during 2009–18 (Harris *et al.* 2020). This pattern was quite similar in southern and western Europe, where the greatest decline appeared to occur between 1980 and 2007, but there was a lower rate of decline (29%) from 2007 to 2018 (*European Atlas* 2020, PECBMS 2020). In 2013, Henderson *et al.* (2017) estimated the Welsh population to be 2,825 breeding pairs (a range of 1,046–6,027).

Whinchats move away from their breeding grounds as early as the end of June, after which individuals and family parties are recorded at lowland and coastal habitats. Movement out of the uplands continues through August, although a few may linger even into October. Small groups occur at coastal sites from early July to mid-October, with stragglers into November. The peak autumn counts from the bird observatories occur from the last week of August through to the first days in October. In the 21st century, whilst November records have come from all the southern counties, with the latest being on 24 November 2011 at Southerndown in East Glamorgan, true winter records are very rare. Possible overwintering birds were recorded at Whiteford, Gower, on 25 December 1966 (Heathcote *et al.* 1967), at Newport Wetlands Reserve, Gwent, on 2 December 2006 and at Sker, East Glamorgan, on 5–7 December 2016.

Several potential factors are involved in the decline of the Whinchat. Blackburn and Cresswell (2016) found that selection of habitat or territory quality in winter makes little difference to survival. In fact, the survival rates for all age and sex classes on the wintering grounds were high. This implies that most mortality occurs primarily outside the winter period and is likely to be during the post-fledging stage or on the first migration. Changes in both the suitability and availability of breeding habitat have resulted in a reduction in range, away from the lowlands across all of Wales. Drainage and drying of upland boggy habitats, through land management and climate warming, can significantly reduce the abundance of important prey species. Regular burning of Bracken

▪ All records April to July
▪ As above, with Breeding codes

Whinchat: Records from BirdTrack for April to July, 2015–19, showing the locations of records and those with breeding codes.

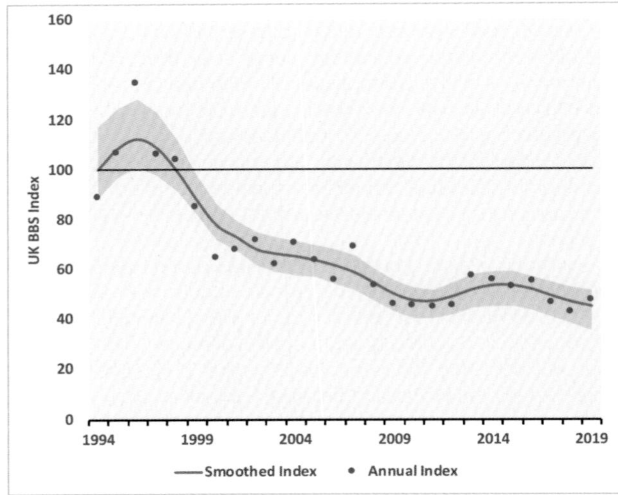

Whinchat: Breeding Bird Survey indices for the UK, 1994–2018. The shaded area indicates the 85% confidence limits.

and shrub mosaics, which are favoured by the Whinchat, could reduce the availability of suitable nesting habitat and encourage gorse to dominate, which is more suited to Stonechats. The squeeze in availability of suitable upland habitat since the late 20th century is now thought to be one of the main reasons for population declines (Calladine and Bray 2012). Climate modelling suggested that the westernmost parts of the breeding range will become less suitable for Whinchat during the 21st century, and that little, if any, of Wales will be occupied by 2100 (Huntley *et al.* 2007). In the near future, therefore, we can expect Whinchat populations to continue to decline in Wales.

Anne Brenchley and Robin Sandham

Stonechat *Saxicola rubicola* Clochdar y Cerrig

Welsh List Category	IUCN Red List			Birds of Conservation Concern			
	Global	Europe	GB	UK	Wales		
A	LC	LC	LC	Green	2002	2010	2016

The Stonechat breeds in Europe, North Africa, Turkey and the Caucasus (*HBW*). The subspecies *S.r. hibernans* is confined to Britain and Ireland, western France and Portugal (IOC). The nominate *S.r. rubicola* is found across the rest of its range and, while it is likely to have occurred in Britain, this has not been confirmed by ringing recoveries. Stonechats are partial migrants; northerly breeding populations migrate to the Mediterranean Basin for the winter. In Wales, it is abundant in habitats where gorse and Bracken are found, such as coastal heath and dunes, higher altitude *ffridd*, moorland and rank pasture. In the upland margins, it occupies a similar geographic range to Whinchat, but in general it breeds in a wider variety of habitats, preferring the drier ones (Henderson *et al.* 2017). Stonechats feed predominantly on the adult and larval stages of invertebrates. They nest on the ground or slightly elevated in vegetation, usually less than 30cm off the ground.

In the 19th and early 20th centuries, the Stonechat was considered to be widespread and common in coastal habitats across Wales and little had changed by the 21st century. In North Wales, Forrest (1907) noted it was "most numerous near the coasts, especially in Anglesey, where it is extremely common", while in Pembrokeshire it was numerous where 'furze' or gorse was to be found, from the coast to the hills (Mathew 1894). In Gwent, Venables *et al.* (2008) quoting Ingram and Salmon (1939) said the "earliest records were birds taken at Marshfield in November 1926 and at Wentwood Forest in July 1930" and described Stonechat as "a resident breeding species that was more numerous and widely distributed than the Whinchat". In that county, it was most abundant in the north and northwest and localised in the southern lowlands.

Male Stonechat © Ben Porter

Mid-20th-century authors reported that it was less common in the hills than on the coast in Ceredigion (Ingram *et al.* 1966) and Caernarfonshire (Jones and Dare 1976). The timing of these observations is critical because Stonechat is particularly susceptible to long periods of cold weather, when temperatures are regularly below zero. Population crashes occurred during several winters in the 19th and 20th centuries (*Birds in Wales* 1994). In Pembrokeshire, Donovan and Rees (1994), citing Lockley *et al.* (1949) and Donovan (1963), said the effects of the severe "winter of 1947 almost wiped them out" and "that of 1962 also left very few survivors". On Bardsey, Caernarfonshire, the highest recorded breeding population was about 15 pairs in 1960, but the hard winter of 1962/63 eliminated them all (Barnes 1997). After a population crash, coastal populations can recover quite rapidly, but it can take several years to recolonise former breeding areas in the uplands. Its recovery is aided by the fact that it is an early breeder, starting as early as late March and often rearing three broods in a year. A pair at Cwmbrandy, near Fishguard, Pembrokeshire, was suspected of rearing four broods in 1962 (Donovan and Rees 1994).

The effects of the 1962/63 winter were still evident by the time of the *Britain and Ireland Atlas 1968–72*, when it was mainly restricted to coastal areas and absent from much of inland Wales. In Breconshire, for example, where its preferred habitat was young conifer plantations and heather, the winters of 1962/63 and 1981/82 took a major toll on populations. It had failed to recover from the latter winter by the time of the *Britain and Ireland Atlas 1988–91*, when pairs were found in only three 10-km squares in the county and singing males were present in two others. In other upland areas, some recolonisation had occurred, notably in North Wales, by the 1988–91 *Atlas*. By the end of the 20th century, as the occurrence of severe winter weather became less frequent, dramatic fluctuations in Stonechat numbers were less evident. In Pembrokeshire, the population expanded its distribution between the local atlas periods of 1984–88 and 2003–07 from 26% to 34% of tetrads and occurred at a higher density in many more localities than previously. A population of 650–700 pairs in 2007 was suggested, equating to an average of one pair/km² of suitable habitat (*Pembrokeshire Atlas 2003–07*). Breeding also occurs on the islands of Skomer, Skokholm and Ramsey, albeit this is sporadic because Stonechats disappear after severe winters and recolonise once mainland populations are at a high level.

In Gwent, following the 1962/63 winter, there were a few records from coastal locations but none inland until 1975 when breeding was confirmed in the western valleys. In the late 1970s the Stonechat was still considered to be mainly a winter visitor. Further severe winters in 1978/79 and 1981/82 reduced the population again, so that during the *Gwent Atlas 1981–85* only 8% of tetrads were occupied and the population was estimated to be 20–30

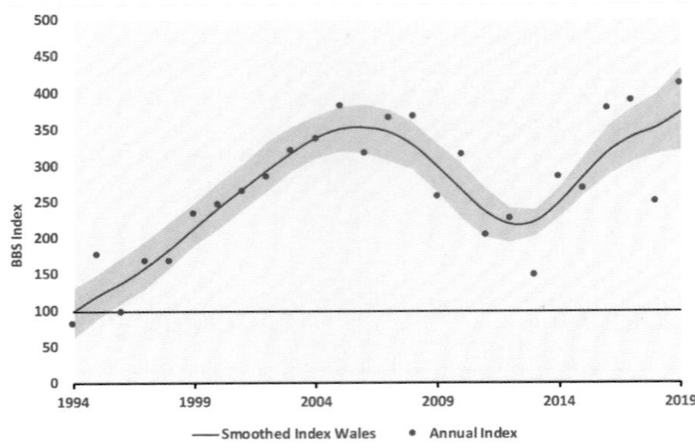

Stonechat: Breeding Bird Survey indices, 1994–2019. The red line shows the smoothed index, 1995–2018. The shaded area indicates the 85% confidence limits.

pairs. In Gower, the *West Glamorgan Atlas 1984–89* showed 34% tetrad occupancy, with a strong presence at coastal locations, but it was scarcer inland. In East Glamorgan there were significant gains in the number of occupied tetrads between the 1984–89 and 2008–11 atlases, with a noticeable increase on higher ground. On the southwest coast, however, the range contracted away from the more urbanised areas of Porthcawl and Ogmore-on-Sea.

At the turn of the 21st century, the range and numbers of many Welsh Stonechat populations had clearly increased. The *Gwent Atlas 1998–2003* showed a striking expansion with 19% of tetrads being occupied. It also showed an extension in range in the central and southern part of the county and estimated a population of 110–210 pairs, most likely towards the top end of that range (Venables *et al.* 2008). The *North Wales Atlas 2008–12* recorded tetrad occupancy of 23%. Stonechats were found to be widespread around the coasts of Anglesey, Llŷn in Caernarfonshire, and Meirionnydd but absent from other lowland areas.

In 2013, the BBS index for Wales fell to its lowest level for 15 years following several cold winters during 2009–12, but then increased by 93% in 2014, relative to the previous year, and by 2017 it had recovered its loss completely. The BBS index may under-represent change in remote parts of Wales where the species has increased most, but overall, the Welsh population increased by 191% during 1995–2018 (Harris *et al.* 2020).

In general, the Stonechat is resident within its breeding territory, provided there is no prolonged period of harsh winter weather. Birds breeding in the hills move to lower elevations and the coast in some winters. The *Britain and Ireland Winter Atlas 1984–88* clearly shows a strong low-altitude/coastal distribution. However, the winter distribution in the *Britain and Ireland Atlas 2007–11* showed the Stonechat to be a very widespread species, found at all but the highest elevations. The differences between these two winter surveys are likely to be due to the timing of surveys relative to the effects of hard winters and a run of milder winters in the years prior to the second *Atlas*. Ringing recoveries indicated a southerly movement in autumn. Some breeding birds from northern England and Scotland over-winter on the south coast of England, the Iberian Peninsula and even North Africa (*BTO Migration Atlas* 2002). Eight recoveries of Stonechats ringed in Wales show a similar pattern, but it is uncertain to what extent this is a common occurrence. One female, ringed in Huesca, Spain, in November 1999, bred in Blaenavon, Gwent, in July 2000 (Venables *et al.* 2008). Interestingly, Wales holds the longevity record for Stonechat, as a first-calendar-year male was ringed on Bardsey, Caernarfonshire, on 1 October 1987 and was recaptured on 16 September 1992, four years, 11 months and 15 days later. It is quite possible that this bird was resident on the island in the intervening period.

Whilst Stonechat habitat is susceptible to wild fires and to some loss along the coastal strip and margins of the uplands, due to intensification of farmland, the species is doing well in Wales. Results from the Welsh Chat survey 2012/13 gave a population estimate of 12,082 pairs (a range of 5,173–22,926) (Henderson *et al.* 2017). The Stonechat appears well-adapted to benefit from the short-term climate changes of mild wet winters and periods of dry summer weather, which improve survival rates and breeding success (Jiguet *et al.* 2009). It seems likely that the Stonechat will continue to do well in Wales.

Anne Brenchley and Robin Sandham

Sponsored by Geoff Gibbs

Siberian/Stejneger's Stonechat *Saxicola maurus/stejnegeri*

Clochdar y Cerrig Siberia/Stejneger

Welsh List Category	IUCN Red List		Rarity recording
	Global	Europe	
A	LC	LC	BBRC

These two species are not, currently, identifiable to species level in the field outside the breeding season. Siberian Stonechat breeds in European Russia, Turkey, the Caucasus and east to China, and winters in northeast Africa, the Middle East and India (IOC). Stejneger's Stonechat breeds farther to the east than *S. maurus*, in eastern Siberia, Mongolia and Japan, and winters in Southeast Asia (IOC). Records to species level in Britain have only been accepted with photographic and DNA evidence: 17 of Siberian Stonechat and eight of Stejneger's Stonechat by 2019 (Holt *et al.* 2020). A further 413 records of Siberian/Stejneger's Stonechat occurred in Britain up to 2018, about ten a year on average (Holt *et. al* 2020). Twelve of these records, all of single individuals, came from Wales:

- 1CY at Bardsey, Caernarfonshire, on 25–27 October 1983;
- Llanfairfechan, Caernarfonshire, on 6–7 November 1983;
- 1CY/female at Strumble Head, Pembrokeshire, on 12 October 1986;
- South Stack, Anglesey, on 10–13 October 1987;
- South Stack, Anglesey, on 15–18 October 1987;
- 1CY/female on Skokholm, Pembrokeshire, on 11–15 October 1991;
- 1CY/female at Strumble Head, Pembrokeshire, on 6 October 1997;
- 1CY on Bardsey, Caernarfonshire, on 29 October 1997;
- 1CY on Bardsey, Caernarfonshire, on 14 October 2000;
- Bardsey, Caernarfonshire, on 24–29 September 2013;
- 1CY on Skomer, Pembrokeshire, on 2 November 2017 and
- 1CY female on Bardsey, Caernarfonshire, on 16–18 October 2018.

Jon Green and Robin Sandham

Wheatear *Oenanthe oenanthe* Tinwen y Garn

Welsh List Category	IUCN Red List			Birds of Conservation Concern			
	Global	Europe	GB	UK	Wales		
A	LC	LC	LC	Green	2002	2010	2016

The first view of a Wheatear is usually of its white rump as it flies away. Despite the male's bold plumage, it is remarkably well camouflaged against a rocky background. The breeding range of the nominate subspecies, *O.o. oenanthe*, is very extensive, covering almost all of Eurasia and Alaska (*HBW*). The subspecies *O.o. leucorhoa* breeds in northeast Canada, Greenland and Iceland (IOC). Wheatears breed on open ground where there is a mixture of bare rocky areas and a short herb layer, nesting in crevices or old animal burrows. All populations are migratory, wintering in Africa south of the Sahara (*HBW*). No birds breeding in Wales have been recovered on the wintering grounds, but 17 ringed in Wales have been recovered *en route*, in Morocco, France, Spain and Portugal. One ringed on Skokholm, Pembrokeshire, on 16 August 1949, was found dead 942km away on the southwest coast of France, two days later. More recently, one colour-ringed in its first-calendar-year on Skokholm on 16 July 2019 was identified as a male at Banc d'Arguin in southwest France on 4 April 2020, and had returned to Skokholm by 16 April to become part of a breeding pair.

Wheatears are one of the earliest migrant passerines to arrive in Wales in spring, with returning breeders seen from early March. In recent years, a few have been suspected of over-wintering. There were two December records (1–12 December 2011 at Pwll, Carmarthenshire, and 14 December 2002 at Rhossili, Gower) and three January records: at Corwen, Meirionydd, on 17–19 January 2018, one in Bangor, Caernarfonshire on 22 January 2000 and 4–5 near Trecastle, Breconshire, on 24 January 1966. The few February records may have been early migrants. One at Pembrey Harbour, Carmarthenshire, on 28 February 2019, was considered to be a migrant (Rob Hunt *in litt.*). By early April, Wheatears begin settling into pairs. In Wales, Wheatears breed up to an altitude of *c.*1,000m in the Carneddau, Caernarfonshire (Pritchard 2017). In autumn, birds leave their Welsh breeding grounds between mid-August and mid-September, though migration along the Welsh coast can continue through October and into November.

The earliest documented record of the Wheatear in Wales is probably by Thomas Pennant, who recorded them near Llyn Glas on Yr Wyddfa, Caernarfonshire, in 1778. Wheatear was described as common around the coast of Pembrokeshire and very numerous on Mynydd Preseli (Mathew 1894) and on the hills and moors of Breconshire "but in no great numbers" (Phillips 1899). It was said to be common in the hills and along the coasts of Glamorgan in the late 1890s (Hurford and Lansdown 1995). In North Wales, Forrest (1907) said that Wheatear was most numerous near the sea, though comparatively scarce in northern Anglesey and along the coasts of Denbighshire and Flintshire.

Coastal populations declined during the 20th century. Myxomatosis, which arrived in Wales in 1954, is likely to have been a major factor, as it greatly reduced the number of Rabbits whose grazing maintained the close-cropped vegetation required by Wheatears for foraging. Increased human disturbance in coastal areas also probably played a part. Heathcote *et al.* (1967) stated that the Wheatear was quite numerous in Glamorgan in the 19th century, up to the first quarter of the 20th century. During this time there was a strong coastal population along the cliff tops and in sand dunes between Ogmore and Swansea and in Gower. Declines in Glamorgan were noted from the mid-20th century onwards, coincident with the arrival of Myxomatosis and the increase in visitor disturbance, particularly at the coast (Hurford and Lansdown 1995). A marked decline was evident in Breconshire during the 1940s and 1950s (Massey 1976) and Wheatear was lost from the coast of Meirionnydd in this period (Jones 1974a). There may also have been a decline inland in parts of Wales. The authors of *Birds in Wales* 1994 commented that "the afforestation of uplands has clearly had a substantial effect locally on distribution and numbers", though they noted that much of the afforestation had been on ground not suitable for Wheatear, such as heather moor.

Fieldwork for the *Britain and Ireland Atlas 1968–72* found the Wheatear to be widespread in Wales, recorded in 82% of 10-km squares during the breeding season, but largely absent from inland Anglesey, southern Ceredigion and much of Carmarthenshire. The 1988–91 *Atlas* showed some losses, particularly in Pembrokeshire and along the south coast, but the 2007–11 *Atlas* showed gains in these areas with a very similar picture to that in 1968–72. Relative abundance in 2007–11 was highest in the uplands, particularly in Snowdonia and in Mynydd Du, Carmarthenshire, but in most areas was lower than in 1988–91.

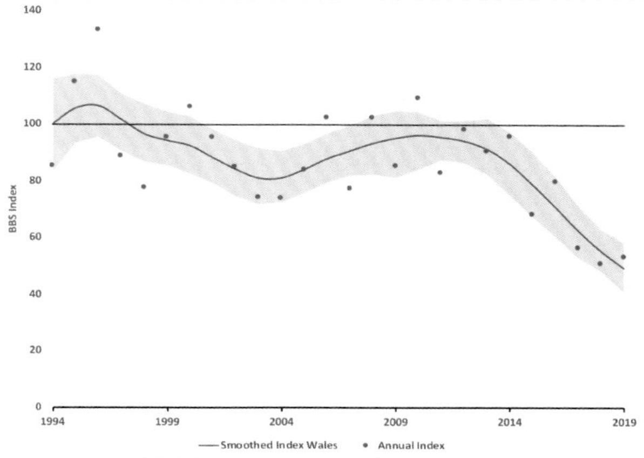

Wheatear: Breeding Bird Survey indices, 1994–2019. The red line shows the smoothed index, 1995–2018. The shaded area indicates the 85% confidence limits.

Tetrad atlases gave a more detailed picture of distribution, though the difficulty of distinguishing migrants from breeding birds, particularly in coastal areas, means that results must be interpreted with care. The *Gwent Atlas 1981–85* recorded the species in 31% of tetrads, almost all in the uplands of the north and west. The 1998–2003 *Atlas* showed a similar pattern, though there was a small decline, with Wheatears found breeding in 28% of tetrads. Venables *et al.* (2008) stated that coastal breeding had never been recorded in Gwent. Both the East Glamorgan Atlases, in 1984–89 and 2008–11, found the species mainly in the uplands, with little evidence of change between the two. In Pembrokeshire, the distribution of this species was essentially the same in 1984–88 and 2003–07, with birds found mainly on Mynydd Preseli and on the coasts and islands. The *North Wales Atlas 2008–12* found Wheatear breeding in 42% of tetrads, mainly in the uplands and in some coastal areas. Throughout Wales, the Wheatear seemed to be largely absent from the lowlands except along the coast, with few birds breeding at altitudes below 300m.

The BBS indices for Wales shows that the Wheatear is not doing well, with a 48% decline between 1995 and 2018 (Harris *et al.* 2020), most of it since 2014. A similar trend is evident in England and Scotland. As a result of the Welsh Chat survey in 2012 and 2013, Henderson *et al.* (2017) estimated the Welsh breeding population to be 13,759 pairs (a range of 7,153–25,861). It was noted that Wheatears were difficult to survey accurately using a multi-species approach because of problems in timing the fieldwork and distinguishing breeding birds from passage migrants.

Wheatears breed on all the larger Welsh islands, with the greatest number of breeding pairs usually on Ramsey, Pembrokeshire, where there were an exceptional 115 territories in 2009, a population density of 42.9 pairs/km^2, but a normal average is 70 pairs, a density of 26 pairs/km^2 (Greg Morgan pers. comm.).

Outside the breeding season, large numbers of migrants have sometimes been recorded in coastal areas and on the islands. There was a fall of an estimated 1,200–1,500 birds in a field south of Dinas Dinlle, Caernarfonshire, in strong easterly winds on 19 April 2008. The largest passage total recorded on Bardsey was 561 on 30 March 1998. In autumn, there were *c.*300 on The Smalls, Pembrokeshire, on 19 September 1982.

Future prospects for the Wheatear in Wales depend largely on the continued availability of suitable habitat. It feeds mainly in short turf, maintained by grazing or climate, and is usually absent from areas of taller vegetation. The intensity of grazing is therefore of great importance. A good example can be seen on Bardsey, Caernarfonshire, where changes to grazing from 1997, as a result of the agri-environment scheme, led to the loss of close-cropped grassland (Archer *et al.* 2010) and a large reduction in the breeding population. Later, the intensity of grazing was increased on Bardsey, mainly to benefit Chough. As a result, Wheatear numbers recovered, to a record 38 territorial pairs in 2017. Any reduction in sheep numbers in the uplands of Wales could be bad news for the Wheatear. A study in Cumbria (Douglas *et al.* 2017) found that where sheep numbers were reduced, sward length increased and Wheatear numbers declined. There has already been a moderate decline in sheep numbers in some upland areas of Wales, such as the Carneddau in Caernarfonshire (Pritchard 2017). Rabbit grazing is important in maintaining short turf in many coastal areas, but there has been a decline in the Rabbit population. The 48% decline, shown in BBS data (Harris *et al.* 2020) is likely to be partly due to Rabbit Viral Haemorrhagic Disease (RVHD), which was first reported in Britain in the early 1990s. Wet weather in late spring and early summer could also affect Wheatear breeding success.

Male Wheatear © Kev Joynes

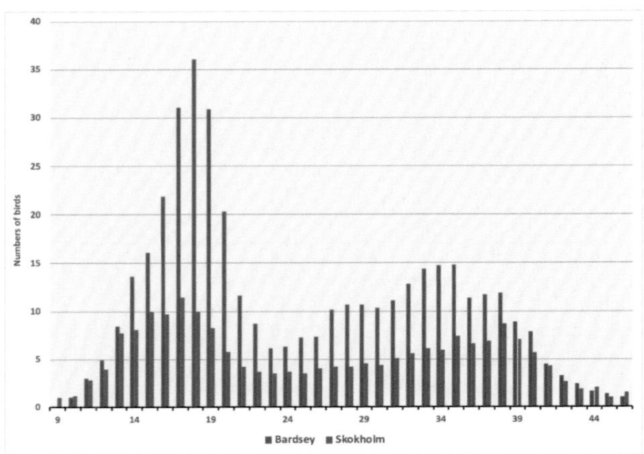

Wheatear: Average counts by week number recorded by bird observatories, 1953–2019. Week 18 starts *c.*29 April and week 35 starts *c.*26 August.

The Wheatear has undergone a major decline of 72% across Europe since 1980 (PECBMS 2020) and in the years between the two European atlases the range of this species has decreased, notably in France and Germany (*European Atlas* 2020). More work is needed to understand the decline in the Welsh breeding population, including identifying the wintering sites and migration pathways of our breeding birds. The recent decline means that the Wheatear is very unlikely to keep its Green conservation status in Wales at the next assessment, and may well be a candidate for Red-listing.

Greenland Wheatear *O.o. leucorhoa* Tinwen y Garn yr Ynys Las

This subspecies breeds in northeast Canada, Greenland and Iceland and winters in West Africa (IOC). One fitted with a light-level geolocator crossed *c.*3,400km of ocean from Baffin Island to reach western Scotland in no more than four days, possibly via Greenland (Bairlein *et al.* 2012). Larger and brighter than the nominate subspecies that breeds in Wales, Greenland Wheatears pass through in late April and May, by which time local breeding is well advanced, and again on return migration in autumn between late August and mid-October. Two ringed in Greenland have been found dead in Wales, one hit by a car at Wenvoe, East Glamorgan, on 12 May 1969 and one that had hit a building in Colwyn Bay, Denbighshire, on 15 May 1981. One ringed in Iceland has been recovered in Wales, and another ringed in Wales was recovered in Iceland.

Greenland Wheatears have been recorded in Wales since the late 19th century. They have often been found in mixed flocks with migrants of the nominate subspecies, making accurate counts difficult, particularly in autumn. A fall of 200 Wheatears on Bardsey on 4 May 2005 were all thought to be of this subspecies. There was a large spring passage on Bardsey in 2012, with birds recorded from 26 April to mid-May. The largest counts were 403 on 28 April and 454 on 2 May. On the mainland there were 90 at Llanrhystud, Ceredigion, on 29 April 2007. A count of 1,500 on Skokholm on 28 April 1938 were also likely to have been mainly of this subspecies in view of the date.

Ian Beggs

Sponsored by Glamorgan Bird Club

Isabelline Wheatear *Oenanthe isabellina* Tinwen Felynllwyd

Welsh List Category	IUCN Red List		Rarity recording
	Global	Europe	
A	LC	LC	BBRC

The Isabelline Wheatear breeds in extreme southeast Europe, the Middle East, Central Asia, northern China and Mongolia, and winters in the Nile Valley and sub-Saharan Africa and Arabia, east to northern India (*HBW*). There have been 51 British records to the end of 2019, with on average one or two a year (Holt *et al.* 2020). It is a very rare vagrant to Wales, with seven records, all since 1997:

- 1CY on Bardsey, Caernarfonshire, on 20–21 September 1997;
- 1CY on Skokholm, Pembrokeshire, on 24–26 September 1997;
- Bardsey, Caernarfonshire, on 16 October 2002;
- Mynachdy, Anglesey, on 22–23 September 2006;
- 1CY male at Wernffrwd, Gower, on 5–10 November 2011;
- 1CY at Martin's Haven, Pembrokeshire, from 30 September to 10 October 2013 and
- 1CY at Bardsey, Caernarfonshire, on 10–16 September 2019.

Jon Green and Robin Sandham

Sponsored by West Coast Birdwatching

Isabelline Wheatear on Bardsey, Caernarfonshire, September 2019 © Steve Stansfield

Desert Wheatear *Oenanthe deserti* Tinwen y Diffeithwch

Welsh List Category	IUCN Red List		Rarity recording
	Global	Europe	
A	LC	NT	BBRC

The Desert Wheatear breeds in North Africa, locally in the Middle East, and east to China and southern Mongolia. It winters in North Africa, Arabia and east to northwest India (*HBW*). There were 161 British records up to 2019, an average of 4–5 records a year (Holt *et al.* 2020). There have been ten records in Wales:

- 1CY male at Penclawdd, Gower, on 21–22 November 1989;
- 1CY male at Peterstone, Gwent, on 16–20 December 1996;
- female at Skokholm, Pembrokeshire, on 12 December 1997;
- female at Skokholm, Pembrokeshire, on 13 November 2003;
- male at Towyn, Denbighshire, on 20 November 2007;
- male at North Stack, Anglesey, on 26 November 2011;
- 1CY male at Skomer, Pembrokeshire, on 15 November 2011;
- female near Rhyl, Flintshire, on 23–26 November 2012;
- 1CY male at Skomer, Pembrokeshire, on 28 October 2014 and
- 1CY male at Pen y Cil, Aberdaron, Caernarfonshire, on 6 November 2019.

Jon Green and Robin Sandham

Male Desert Wheatear at Towyn, Merionethshire, 20 November 2007 © Steve Young

Western Black-eared Wheatear *Oenanthe hispanica* Tinwen Glustddu'r Gorllewin

Welsh List Category	IUCN Red List		Rarity recording
	Global	Europe	
A	LC	LC	BBRC

In 2020, the Black-eared Wheatear, formerly made up of two subspecies, the nominate *hispanica* and *melanoleuca,* was split into two species (IOC). The Western Black-eared Wheatear, which breeds in northwest Africa, the Iberian Peninsula, southern France and northern Italy, is the species most likely to occur in Britain. The Eastern Black-eared Wheatear (*O. melanoleuca*) breeds from southern Italy to Greece and southwest Asia from Turkey to the southern Caucasus, south to Israel and SW Iran (*HBW*). Up to the end of 2019 there were 61 British records of the two species, but the majority were not attributed to one or the other (Holt *et al.* 2019).

The only Welsh record of a definite Western Black-eared Wheatear was a male on Bardsey, Caernarfonshire, on 18 April 1970.

Two other records, which were not at the time assigned to subspecies and are now described as 'Black-eared Wheatear sp.', were of male individuals seen on Skomer, Pembrokeshire, on 4 May 1990 and on Bardsey, Caernarfonshire, on 6 May 1992.

Jon Green and Robin Sandham

Pied Wheatear *Oenanthe pleschanka* Tinwen Fraith

Welsh List Category	IUCN Red List		Rarity recording
	Global	Europe	
A	LC	LC	BBRC

The Pied Wheatear breeds in the stony habitats of southeast Europe and east across China to Mongolia, and winters in Yemen and northeast Africa (*HBW*). This is an extremely rare vagrant to Britain, with 86 records to the end of 2019, about two-to-three per year on average (Holt *et al.* 2020). There have been three records in Wales:

- female on Skokholm, Pembrokeshire, on 27–29 October 1968;
- male at Ramsey, Pembrokeshire, on 25 October 1993 and
- male on Bardsey, Caernarfonshire, on 13 October 2013.

Jon Green and Robin Sandham

Dipper *Cinclus cinclus* Bronwen y Dŵr

Welsh List Category	IUCN Red List			Birds of Conservation Concern			
	Global	Europe	GB	UK	Wales		
					2002	2010	2016
A	LC	LC	NT	Amber			

Dippers occur on fast-flowing streams and rivers from Fennoscandia in the north to the Atlas Mountains of Morocco in the south, and from Ireland in the west to central China in the east (*HBW*). They are found throughout northern and western Britain, with breeding and wintering ranges coinciding closely. The Dippers resident in Wales are of the subspecies *C.c. gularis* and are largely sedentary as adults. Migratory 'Black-bellied Dippers', *C.c. cinclus* from northern and western mainland Europe and *C.c. aquaticus* from central and southern Europe (IOC), have occurred as occasional immigrants in Britain, mostly in eastern counties. Birds of this subspecies have occurred in Gloucestershire, Devon and Ireland and so may have been overlooked in Wales.

Dippers breed throughout Wales from mountain streams at about 600m, like those in Snowdonia and the upper tributaries of the Afon Irfon, Breconshire and Afon Tywi, Carmarthenshire, to near sea level. For example, they breed on a tributary of the River Wye at Tintern, Gwent, and streams that drain steeply into Cardigan Bay (Tyler and Ormerod 1985, 1994). Most pairs, however, breed below 300m. In North Wales, owing to a lack of suitable watercourses, birds are absent from the Great Orme and western Llŷn, Caernarfonshire. They are scarce on Anglesey, although breeding has occurred annually, mainly on the Afon Cefni but occasionally on the Afonydd Cadnant and Braint (Jones and Whalley 2004, Nigel Brown pers. comm.).

The Dipper has probably always been fairly common on suitable rivers and streams across Wales. This was certainly the case in the late 19th and early 20th centuries in North Wales (Forrest 1907) and Pembrokeshire (Mathew 1894). Dippers did not apparently suffer from persecution by anglers in Wales, as they did in Scotland. In one Scottish district in the mid-19th century, 548 Dippers were killed in three years because of their perceived impact on salmonid fish eggs and fry (Holloway 1996). However, river pollution, such as from heavy metals on the Afon Rheidol and Afon Ysytwyth in Ceredigion, meant that Dippers were either absent or scarce on those rivers (Salter 1895). Between the 1920s and the 1960s, little changed, although the distribution of the Dipper in Glamorgan contracted away from areas of increasing population, thought to be due mainly to an increase in disturbance and river pollution (Heathcote *et al.* 1967). There was little change in the range of the Dipper between the periods of the Britain and Ireland atlases (1968–72, 1988–91 and 2007–11), but there was a decline in relative abundance in many parts of Wales between 1988–91 and 2007–11. This was most notable in the uplands of Breconshire, Caernarfonshire and Meirionnydd in particular. In southern Wales, in the three counties where a comparison between tetrad atlases was possible, a more positive trend was shown. In Gwent, tetrad occupancy in 1981–85 and 1998–2003 remained the same, at 35% (Venables *et al.* 2008). In Pembrokeshire, tetrad occupancy declined slightly, from 14% in 1984–88 to 11% in 2003–07 (*Pembrokeshire Avifauna* 2020). In the East Glamorgan atlases for 1984–89 and 2008–11, there was an increase in tetrad occupancy, from 14% to 26%. In Gower, Thomas (1992b) reported a decline during 1900–70, followed by an increase during 1970–90, when Dippers were found mostly in the north and east of the county and in reasonable numbers on the Afonydd Afan, Neath and Tawe.

Breeding densities vary according to the nature of the water-course and its invertebrate productivity. Highest densities of 2–3 pairs/km are on narrow, fast-flowing rocky streams with plentiful exposed rocks and riffles, overlying base-rich rocks as in the Black Mountains in Gwent. They favour watercourses lined with broadleaved trees, which, unlike conifers, contribute leaves for herbivorous aquatic invertebrates and also caterpillars and other terrestrial invertebrates that are particularly important for Dippers when they are moulting and unable to swim or dive.

Dippers will, however, nest on very narrow streams tumbling down tree-less hillsides on moorland. They are scarcer on acidic streams because there are fewer caddisfly larvae and mayfly nymphs in such streams. In particular, numbers declined in the 20th century on some rivers overlying base-poor rocks that were subject to conifer planting and consequent acidification, such as on the Afon Irfon, Breconshire, where the population fell by 80% between the early 1960s and mid-1980s (Ormerod *et al.* 1985, 1986, and Tyler and Ormerod 1994). On more lowland sections of rivers, Dippers are restricted to weirs and side tributaries. In Gwent there are, in most years, at least 53 territories on the River Monnow and two of its tributaries, the majority on the upper narrower rocky

Pair of Dippers with nesting material © John Hawkins

reaches where territories are 300–500m long and adjacent to each other. Only nine are on the main river about 25km below Pandy, from where the river meanders and has sandy banks and gravel or pebble shoals and is less suitable for Dippers (Tyler and Ormerod 1985, Tyler 2004, Tyler and Burge 2011).

Sulphur deposition, from acid rain, has been much reduced in recent decades, with an apparent recovery in stream invertebrates. Likewise, the quality of rivers in the coalfield valleys of South Wales has improved greatly since 1971, when *c.*60% of rivers draining the coalfield were described as grossly polluted or of doubtful quality (Tyler and Ormerod 1994). There has been a marked recovery of invertebrate, fish and Dipper populations. There are now healthy populations on the Afonydd Taff, Ebbw, Sirhowy and Rhymney in Gwent and Glamorgan, for example (Tyler and Ormerod 1994). However, on 29km of the Grwyne Fawr and Grwyne Fechan in Gwent and Breconshire, there has been a marked decrease in breeding abundance (Jerry Lewis pers. comm.). In the 1980s, there were on average 31 pairs/year (10.7 pairs/10km, max. 39 pairs); in the 1990s, the average was 27 pairs (9.3 pairs/10km) and in the 2000s an average of 25 pairs (8.6 pairs/10km), with a low count in one year of only 17 pairs (5.9 pairs/10km). A long-running study on some tributaries of the lower River Wye, Gwent, showed a decline of about 30% in the 1990s, with losses mainly in sub-optimal habitats (i.e. slow-flowing lowland sections with few riffles and exposed rocks), and a reduction in clutch and brood size (Tyler and Burge 2011). There has been some recovery since, although densities on the favoured reaches have, however, changed little in over 40 years. In Caernarfonshire, a survey of 16km of the Afon Ogwen in 1974/75, and repeated in 2005–07, found a small increase, from 11–14 territories (6.9–8.75/10km) to 14–15 (8.75–9.4/10km) (Gibbs *et al.* 2011).

In parts of the range, as in the Black Mountains, Gwent, densities in winter are similar to the breeding season, with many pairs remaining on territory. That year's young soon find a vacant stretch of river. Counts in winter have been few, but eight were noted on 3km of the Afon Afan, Gower, in January 2009. Some movements occurred from the highest altitude streams to lower levels in severe winters when hill streams froze, although this rarely happens now. Birds may then occur on lowland rivers, at the coast or on estuaries. For example, in Caernarfonshire, wintering Dippers are often found on the shore between Llanfairfechan and Aber Ogwen and by tidal waters at Conwy, Foryd Bay and Aberdaron.

There have been 18 recoveries of Dippers ringed in Wales that moved more than 50km. The longest movements were all ringed as nestlings and found in a subsequent season: one on the Afon Leri, near Talybont, Ceredigion, in May 2017, that was caught 63km away on the Afon Lugg at Monaughty, Radnorshire, in October 2017, and which subsequently bred 4km upstream on the Lugg at Llangunllo, Radnorshire, in 2018 and 2019. One ringed as a nestling at Cynghordy, Carmarthenshire, on 27 April 2016, was caught 77km away at The Grove, Craven Arms in Shropshire on 10 October 2016. A male ringed as a nestling at Horderley, Shropshire, on 11 April 2011, was caught 95km away at Pontllanfraith, East Glamorgan, on 19 June 2013. This movement was unusual as males usually establish a breeding territory only a few kilometres from their natal site, whereas young females often disperse to new river systems in the autumn.

The Dipper lives on average for three years, but many records of Britain's oldest Dippers come from Wales, mainly because of the long-term studies conducted in Gwent. Several have lived to their seventh year, but the two longest-lived birds were one ringed at Troed-y-rhiw, Rhuddlan, Ceredigion, on 11 June 1994, that was found dead in Carmarthen on 24 October 2002, eight years and four months later, and a female at Llangefni, Anglesey, that was last caught as an eight-and-a-half year old.

Johnstone *et al.* (2010a) and Johnstone and Bladwell (2016) placed Dipper on the Amber-list in Wales based on a moderate decline in the breeding population of 25–50%. Apart from the declines noted above on the eastern fringe of Wales, especially in the Black Mountains and Wye tributaries in Breconshire and Gwent, other populations appear to be stable or have increased. The Welsh breeding population in 2018 was estimated to be in the range of 1,150–4,050 territories (Hughes *et al.* 2020).

Any activity that causes a reduction in invertebrate prey, especially of caddis and mayflies, will affect Dippers. There may be a possible impact on Dippers from a resurgence of conifer planting schemes, but many forest districts now have standard practice riparian management along watercourses, where conifers up to 20m from the river are felled away from the river to prevent water acidification. This may well benefit Dipper greatly where forestry abuts rivers suitable for this species.

Agriculture can also have a major impact on Dipper numbers, especially nitrogen and phosphorus pollution from silage or slurry effluent spread on improved and arable fields that is washed downstream, along with contamination by pesticides, other chemicals and heavy metals (Tyler and Ormerod 1994). Abstraction of water for agriculture can reduce river flows, which adversely affects stream invertebrates. Problems are also caused by sedimentation from soil run-off, where fields are ploughed adjacent to watercourses or by livestock poaching fields by watercourses. Heavy and prolonged rainfall may badly affect Dippers, as a rapid increase in water flow brings soil and silt into rivers, smothering invertebrates on the bed. This reduces feeding opportunities, because they cannot forage for long in deep water and are unable to see their prey in silt-laden water. We need to know more about how climate change, especially increased amounts and intensity of rainfall, affects Dipper populations. We also need to know the impact of a wide range of agricultural contaminants and pollutants from industrial and domestic use, including micro-plastics that have been found to occur in half of all river insects in South Wales rivers and are ingested by Dippers (D'Souza *et al.* 2020).

Generally, a reduction of sulphur pollution has helped alleviate the localised acidification issues but agricultural payments are needed to benefit the environment. These could help to reduce the use of herbicides and insecticides and result in fewer nitrates and phosphates entering watercourses. The water quality from sewage works and domestic septic tanks also needs to be improved and river flows must be maintained. The long-term outlook is not rosy for Dipper in Wales: the climate of South and mid-Wales is not expected to be suitable for the species by the end of this century (Huntley *et al.* 2007). However, up to 2020, there has been no indication that Dipper populations in Wales are anything other than healthy.

Steph Tyler

Sponsored by Gwent Ornithological Society

House Sparrow *Passer domesticus* Aderyn y Tô

Welsh List Category	IUCN Red List			Birds of Conservation Concern			
	Global	Europe	GB	UK	Wales		
					2002	2010	2016
A	LC	LC	LC	Red			

Possibly one of the most familiar of all birds to non-birdwatchers, the House Sparrow is closely associated with human habitation. It has an extensive range, covering most of Eurasia and parts of North Africa. It has also been introduced to many other parts of the world, including the Americas and Australia, making it one of the world's most abundant passerine species (*HBW*). This is a very sedentary species. The British longevity record was of a bird ringed at Pontypool, Gwent, on 11 April 1966, as an adult male, found dead at exactly the same location on 23 April 1978, 12 years and 12 days later. There have been no long-distance ringing recoveries involving Wales. It is often absent from isolated houses and farms, particularly upland farms, though it can be found where there is a good supply of winter feed for livestock.

The species may well have been more abundant in the late 19th and early 20th centuries, when there were more cereals grown in Wales and hence more spilt grain. Forrest (1907) described it as the most abundant of birds in North Wales, numerous almost everywhere in the lowlands and up to a moderate elevation on the mountains and moors, though perhaps least numerous on Anglesey. It seems to have been less abundant in Pembrokeshire. Mathew (1894) stated that it was a common resident in Pembrokeshire, but not as abundant as in England and rather scarce in the "mountain" districts. He said, "The absence of cornlands, and the sparsely inhabited country, in which isolated mountain farms are far apart, would account for the comparative scarcity of the House Sparrow, in most places a far too abundant pest", while Lockley *et al.* (1949) described it as "not numerous, and... absent from whole villages, though present in small numbers in all large villages and towns". The species was said to be undergoing a marked increase in Glamorgan in the late 1890s, thought be to be linked to urban spread and development (Hurford and Lansdown 1995).

The House Sparrow underwent a large-scale population reduction in England, where the CBC showed a precipitous decline between 1976 and *c.*1994, but the BBS subsequently showed a less marked decline. Too few CBC plots were covered in Wales to give a reliable indication of Welsh population trends prior to 1995, since when the BBS index has increased by 92% (Harris *et al.* 2020), so it is not known whether this represents a recovery following a previous decline. The Welsh breeding population, in 2018, was estimated to be 800,000 pairs (725,000–870,000), making this species one of the most common in Wales (Hughes *et al.* 2020).

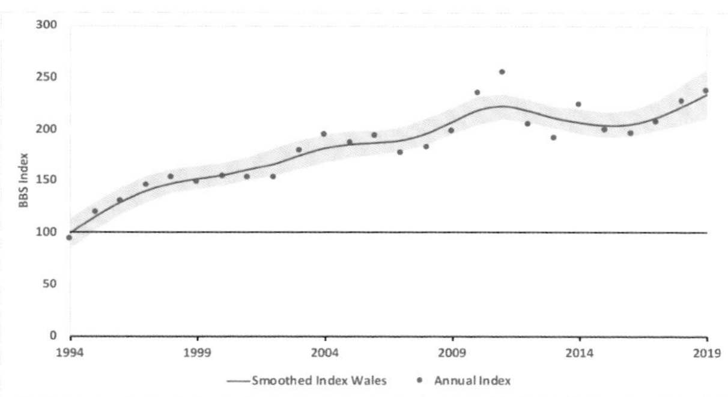

House Sparrow: Breeding Bird Survey indices, 1994–2019. The red line shows the smoothed index, 1995–2018. The shaded area indicates the 85% confidence limits.

The *Britain and Ireland Atlas 1968–72* showed the House Sparrow to be present in almost every 10-km square in Wales. A few small gaps appeared in the uplands and on the coast in the 1988–91 *Atlas*, but the 2007–11 *Atlas* once more confirmed breeding in almost every square. Relative abundance was highest in the lowlands, particularly in Gwent, East Glamorgan, Pembrokeshire and Anglesey. Most parts of Wales showed an increase in relative abundance compared to 1988–91. Where comparison has been possible between tetrad atlases, no great change was evident. The species was recorded with breeding evidence in 76% of tetrads in East Glamorgan in 1984–89 and in 79% in 2008–11, though there were fewer tetrads with confirmed breeding. The first *Gwent Atlas* in 1981–85 recorded it in 93% of tetrads, and the second in 1998–2003 showed almost no change. In Pembrokeshire, it was recorded in 78% of tetrads in 1984–88 and in 84% in 2003–07. In West Glamorgan, it was recorded in 81% of tetrads in 1984–89. It was less widely distributed in counties where there are fewer human settlements. For example, it was recorded in 45% of tetrads visited in Breconshire in 1988–90. The *North Wales Atlas 2008–12* found it in 71% of tetrads in the region, including almost every lowland tetrad, but it was absent from higher ground, possibly related to reduced human habitation. Some county-atlas authors commented that the population in urban areas had decreased while it had increased in suburban areas.

The largest flocks have usually been recorded between July and October. Most counties have recorded maxima of 150–500, but the largest counts came from the southeast, including 1,500–2,000 that roosted near Cwmbran Boating Lake, Gwent, in 1985 (Venables *et al.* 2008) and *c.*1,000 near Llandough Hospital, East Glamorgan, in October 1963 (Heathcote *et al.* 1967). Whilst such large numbers have not been recorded in recent years, flocks of 100+ have been regularly reported in the 21st century. These included: 300 at Abercastle, Pembrokeshire, on 1 August 2010; 300+ at Warren Road, Breconshire, on 24 August 2004; 300+ at Rhuddlan, Flintshire, on 14 August 2015; and 500 around Cemlyn, Anglesey, on 18 August 2010. The species no longer breeds on any of the islands, although several have fairly regular records of small numbers on passage. Bardsey was said to have been first colonised about 1785. It held up to 25 breeding pairs during 1953–1969, but only four pairs bred in 1970 and it has not bred there since. The House Sparrow is now a scarce visitor to Bardsey, with none in some years. During 1982–92, birds were only recorded in 1986, for example. Lockley *et al.* (1949) stated that the species had bred on Caldey and Ramsey, but was casual on the other Pembrokeshire islands, though it bred on Skomer in 1965 and 1966, and was resident on Caldey until about 1985 (Donovan and Rees 1994). There seems to be no record of breeding on Skokholm, but the species was recorded there in 47 years between 1928 and 2018.

In Wales, species found in remote areas are often better studied than species found on the doorsteps of the majority of the human population. Certainly, there is still much we do not know about the House Sparrow in Wales, and no paper has been published on this species in *Birds in Wales* or its predecessors. The declines in England are thought to be mainly driven on farmland by changes in survival rates, due to lack of food resources as a result of agricultural intensification, and in towns by poor breeding performance (Woodward *et al.* 2018). Its Amber-listing in Wales is because of its UK Red List status as a bird of conservation concern, rather than any factors within Wales. While the House Sparrow appears to be doing well in Wales at present, a detailed study of the species here would be very valuable.

Rhion Pritchard

Sponsored by SEWBReC – South East Wales Biodiversity Records Centre

Spanish Sparrow *Passer hispaniolensis* Golfan Sbaen

Welsh List Category	IUCN Red List		Rarity recording
	Global	Europe	
A	LC	LC	BBRC

The nominate subspecies, *P.h. hispaniolensis*, breeds from the Atlantic islands, east across North Africa, Iberia, Corsica and Sardinia, and from the Balkans and Romania to Israel (*HBW*). The subspecies *P.h. transcaspicus* is found in Iran eastwards to Kazakhstan and Afghanistan. Eastern populations are mostly migratory, wintering from Egypt to India, and those in the west of the range are largely sedentary (IOC). There have been ten British records of this extremely rare vagrant up to the end of 2019, the most recent of which was in 2013 (Hudson *et al.* 2014). None of these sightings have been sub-specifically identified. There has been only one record in Wales at Martin's Haven, Pembrokeshire, on 18 May 1993.

Jon Green and Robin Sandham

Tree Sparrow *Passer montanus* Golfan Mynydd

Welsh List Category	IUCN Red List			Birds of Conservation Concern			
	Global	Europe	GB	UK	Wales		
A	LC	LC	VU	Red	2002	2010	2016

The Tree Sparrow is largely resident across all but high Arctic Eurasia, extending into the southern hemisphere in Indonesia (*HBW*). The nominate subspecies is found across Europe (IOC). There its range has expanded since the 1990s, with a notable ongoing northern shift across Fennoscandia, whilst a contraction in range in Britain and western France was noted (*European Atlas* 2020). In Wales it is normally a bird of lowlands, generally below 250m, but elsewhere in Europe can be found up to 1,500m. It is patchy in its distribution, with inexplicable colonisation and desertion of colonies. It nests in holes in buildings, coastal cliffs with ivy, trees and in urban areas where nest-boxes are provided (*BWP*).

Its populations have fluctuated widely. In Britain and Ireland, the population increased from the mid-1950s to 1960 and remained stable for about 20 years. Summers-Smith (1995) showed this increase graphically and suggested that it may have been linked with cyclical increases of populations on mainland Europe. The European population then declined rapidly, a decrease of 65% between 1980 and 2017, although since 2007, the population has been relatively stable to increasing slightly (*European Atlas* 2020, PECBMS 2020). In general, the Welsh population went into a decline from the 1920s until the mid-1950s, followed by a modest recovery during the 1960s and 1970s and then a further decline that has brought it to the verge of extinction in most areas.

Forrest (1907) reported that it was overlooked because of its similarity to House Sparrow, and he had only a few records in North Wales. On Anglesey, there was a breeding colony at Penmon Priory at that time and a few records from other parts of the island. From Caernarfonshire, Forrest received a few records, including breeding at Deganwy, while it seems to have been more common in Flintshire around Sandycroft, Sealand and between Hope and Mold. There were records from several places in Denbighshire, but it was considered rare in Montgomeryshire and absent from Meirionnydd. In the same period, Mathew (1894) did not list Tree Sparrow as present in Pembrokeshire. In Ceredigion, Roderick and Davis (2010) reported that a few were present at the end of the 19th century. Four were at Clarach on 20 December 1901, but there were no further records until the 1960s. In Glamorgan, after a couple of records at the end of the 19th century, there were none until 1936 (Hurford and Lansdown 1995).

By the mid-20th century, the Tree Sparrow was recorded more widely. In Montgomeryshire, Holt and Williams (2008) reported that during 1947–82 it was reasonably widespread in the east but scarce in the west. By 1944, Hurford and Lansdown (1995) considered it to be resident in Glamorgan though only occasional in Swansea. There were no further records until 1958, after which it became re-established. Slowly, it spread to Gower and the area south of the coalfields. In Radnorshire, it was "scarce, local and irregularly distributed" and declined during the 1930s (Ingram and Salmon 1955), but the population had increased by the 1970s (Jennings 2014). The Tree Sparrow was recorded as a common breeding resident in south and east Breconshire but uncommon in the north and west (Massey 1976).

The Britain and Ireland Atlas 1968–72 showed that the species bred in all counties, albeit much less widespread in the western counties. The *Britain and Ireland Atlas 1988–91* showed declines in the west, but confirmed breeding along the valley of the Afon Tywi, Carmarthenshire. While it was relatively abundant in Breconshire, Gwent and Radnorshire, plus the English counties east of the border, its abundance across much of western Wales was relatively low. By the time of the *Britain and Ireland Atlas 2007–11*, the breeding distribution was very much reduced in all areas and it was absent as a breeding species from western counties, except for three 10-km squares with confirmed breeding in the Tywi Valley.

Venables *et al.* (2008) reported a dramatic change in the fortunes of Tree Sparrow in Gwent. Ferns *et al.* (1977) had described the species as "fairly common in all areas apart from the industrial valleys". In the late 1970s there were records of flocks of 200–300, including 200 at Magor in August 1976. The *Gwent Atlas*

Tree Sparrow © John Hawkins

1981–85 recorded breeding in 52% of tetrads, but by the *Gwent Atlas 1998–2003*, there was evidence of breeding in just 9% of them. A reduction in the number of pollarded willows and orchards was cited as having led to the decline. The *East Glamorgan Atlas 1984–89* found evidence of breeding in 12% of tetrads but, by 2008–11, this was the case in less than 1% of them. In Gower, the *West Glamorgan Atlas 1984–89* recorded breeding in 7% of tetrads. In Pembrokeshire, evidence of breeding was recorded in less than 1% of tetrads in 2003–07, compared with just over 1% in 1984–88 (*Pembrokeshire Avifauna* 2020). In Radnorshire, Tree Sparrow populations declined from the mid-1990s before a noticeable, but slight and gradual, increase from around 2005 to 2019 (Jennings 2014, Pete Jennings *in litt.*). In Ceredigion, a pair that bred at RSPB Ynys-hir in 1967 was the first county breeding record and whilst there have been a few breeding records since, sightings are generally sporadic and mostly in winter. Counts of up to 15 at Ynys-las during 1977–81 were notable.

In Carmarthenshire, Lloyd *et al.* (2015) reported that outside the Tywi floodplain, the Tree Sparrow's range had contracted by 2008 and that the largest count in recent decades had been 14 at Ginst Point on 12 December 1995. Within the floodplain, a count of 30 was "notable" in August 1961. In 1987 a flock of 39 was at Dryslwyn on 25 September. A nest-box project started in 1999 had, by 2005, resulted in 85 boxes being occupied. It was thought that the population in the valley was about 100–200 pairs, at the time considered to be about 50% of the Welsh population (Lloyd *et al.* 2009).

In Meirionnydd, Pritchard (2012) reported that the Tree Sparrow was an uncommon breeding resident, with *c.*30 at Aberdysynni in late October and early November 1967 with several breeding records in the late 1960s and early 1970s. The last record of that period was at Dyfi Bridge on 14 June 1988. A pair probably bred in the northeast of the county during 2008–11, since when there have been few records. The *North Wales Atlas 2008–12* recorded evidence of breeding in only 3% of tetrads, with no records west of Llansannan, Denbighshire.

The Tree Sparrow was never an abundant or regular breeding species in Caernarfonshire (Pritchard 2017) and was in decline by the time of the *Britain and Ireland Atlas 1968–72*. It was last confirmed to breed in the county at Abergwyngregyn in 1973, although a few isolated pairs probably bred in the 1980s. Some "strong" colonies on Anglesey continued through the 20th century, before declining in the 1990s (Jones and Whalley 2004). However, small autumn and winter flocks have been recorded: 25 were at Newborough Warren on 13 January and 5 March 1996, 30 were at the Braint Estuary on 3 January 2000 and, in autumn 2018, 15 fed with finches and buntings on an arable field at Cemlyn (Nigel Brown *in litt.*). A few pairs probably do still breed occasionally on Anglesey where at least two pairs bred at Bryngwran in 2017–19.

Tree Sparrows have become reliant on nest-boxes to maintain relatively large colonies but nonetheless can desert apparently successful colonies for no clear reason. An example was near Bodelwyddan, Denbighshire, where a colony of up to 15 pairs in nest-boxes was suddenly abandoned in September 2008, there having been no evidence of predation at either the boxes or feeding stations (Ian Spence pers. obs.). This type of event has been referred to by Alexander and Lack (1944) and quoted by Summers-Smith (1995), but with no further discussion or explanation.

While Tree Sparrow populations fluctuated, but generally declined, across the whole of the UK for much of the latter part of the 20th century, there has been a remarkable increase in the 21st century, driven largely by its fortunes in Scotland, where they have increased by 426% since 1995 (Harris *et al.* 2020). Sadly, the situation in Wales is very different. There are too few Tree Sparrows for their populations to be recorded via BBS, so recent records come from BirdTrack, local ringing projects and bird reports. As fewer than 50 10-km squares were occupied during the *Britain and Ireland Atlas 2007–11*, Hughes *et al.* (2020) were not able to provide a population estimate, but information from the counties in 2020 suggested that only 200–325 pairs were now breeding in Wales. Breeding was no longer known to occur in Anglesey, Caernarfonshire, Meirionnydd, Flintshire, Pembrokeshire, Ceredigion, Gower, East Glamorgan and Gwent. Breconshire, Carmarthenshire and Denbighshire only had a few pairs known to still breed. The bulk of the Welsh population resided in Radnorshire, where between 150 and 250 pairs were thought to be present. Most of these birds were in the east, between Radnor Forest and the English border, but a few were in the south along the Wye valley, downstream from Boughrood and the English border. Also a few birds have started to spread back into the northwest of the county (Pete Jennings pers. comm.).

During 2009–18, counts were generally low, with fewer than 25 from most counties, with three exceptions. In Carmarthenshire, the nest-box study continued, but the number of pairs declined, although there was a reduction in effort to check the boxes and counts of over 40 were regular during winter 2011. The colony was of at least 100 pairs that produced well over 300 young each year between 2000 and 2008 (Lloyd *et al.* 2009). Although this colony is still in existence, it is no longer being studied at the same level of intensity (John Lloyd pers. comm.). It was thought that some of the birds may have used old Ash trees, but the threat of Ash Dieback may be another problem for the future (John Lloyd pers. comm.). No double-figure counts of Tree Sparrows have been recorded in the Tywi Valley, Carmarthenshire, since 2013 (Rob Hunt *in litt.*). In Flintshire, a colony in nest-boxes at Sandycroft, which was monitored from 2004 to 2016, went into decline from 2013, when just three broods were ringed (Ian Spence pers. obs.). During 2000–14, two breeding colonies in Denbighshire each produced over 100 fledglings over several years (pers. obs.) However, these three colonies declined for reasons that are not all understood, though one was disturbed by human intervention.

In Breconshire, prior to 2010 and colony desertion/extinction, several nest-box colonies were found to have dead incubating adults/females on eggs during the breeding season. The reasons were unknown. In 2019, 6–8 pairs were rediscovered breeding in central Breconshire, on a small-holding with nest-boxes, old stone barns that offered natural breeding sites, year-round supplementary food and an acreage of cover-crops planted for a Pheasant shoot. This is now the only known colony in the county (Andrew King *in litt.*).

In winter, its distribution in Wales was little changed from the *Britain and Ireland Winter Atlas 1981–84*, with greatest numbers in the east, on the border with England. In winter, during 2007–11, there were a few records to the west of the breeding range, on Anglesey, Llŷn, southwestern Ceredigion and north Pembrokeshire.

During periods when Tree Sparrow breeding populations were larger, there was some evidence of passage at coastal sites. Donovan and Rees (1994) said that it was recorded on passage in Pembrokeshire, particularly on the islands and coastal stubbles, during March to June and August to November. In Gwent, the largest passage flock was 130 at Goldcliff on 6 October 1968, but groups of 10–60 were more usual, moving SSW. Small numbers continued to occur on passage in the 21st century, such as an exceptional 11 on the Great Orme, Caernarfonshire, on 15 October 2007, but counts at the bird observatories have not exceeded five. In winter, larger numbers were recorded: in Gwent there were 200 at Newport rubbish tip in December 1975 and 300 at Monmouth from November 1976 to mid-June 1977. In the 1980s, counts were smaller, with 150 near Tredegar Park in February 1980 and 70 there in December 1983, and with 220 at Llandewi Skirrid in January and February 1989. In Caernarfonshire, a flock of *c.*200 was at Glanwydden on 26 November 1975. Towards the close of the 20th century, such large flocks became uncommon. In Pembrokeshire, Donovan and Rees (1994) stated that 1–2 was typical, with numbers occasionally up to ten, but that 32 at Pengawse in February 1988 and 40 at Ratford Bridge on 27 January 1990 were unusual. During 2000–18, the maximum flock sizes suggested decline, with occasional hints at improvement. Bardsey recorded them more regularly than Skokholm, though generally in small numbers, usually less than five.

Most Tree Sparrows that breed in Wales are sedentary, but there has been evidence of movements in all counties. The mean distance moved by ringed Tree Sparrows is less than 1km, but 23% of movements are greater than 20km (*BTO Migration Atlas* 2002). Lloyd *et al.* (2009) reported a skeleton found in a box at Dryslwyn, Carmarthenshire, on 5 January 2006, that had been ringed as a nestling at Broomhill Flash, South Yorkshire, in 2000, a movement of 281km. The origins of non-breeding parties are a mystery.

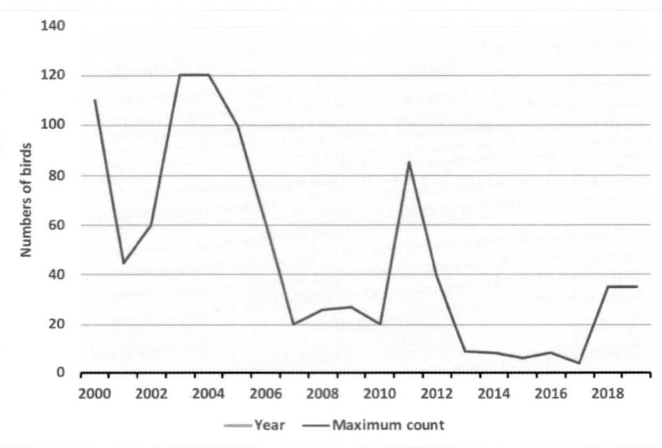

Tree Sparrow: Maximum flock sizes recorded in Wales each year, from *Welsh Bird Reports*, 2000–18.

The conservation of Tree Sparrows will probably depend on better targeted agri-environment schemes that support land-owners to provide nesting sites and ensure adequate feeding opportunities: for example, a range of seeds in winter and a variety of invertebrates, many of which have strong aquatic associations, for feeding nestlings (Lloyd *et al.* 2009). They are likely to remain on the Red List of conservation concern in Wales for the foreseeable future, but the previous fluctuations and their recovery in parts of England give some hope that they can return, if the right habitat conditions are provided.

Ian M. Spence

Sponsored by Betty and Bob Lee

Alpine Accentor *Prunella collaris* Llwyd Mynydd

Welsh List Category	IUCN Red List		Rarity recording
	Global	Europe	
A	LC	LC	BBRC

The Alpine Accentor is largely resident in its discontinuous habitat of the major mountain ranges of northwest Africa, central and southern Europe and Asia (*HBW*). In winter, most European birds descend below the snowline near breeding areas, but some disperse to lowlands. This is an extremely rare vagrant to Britain, with 39 records up to 2019 (last recorded in 2016, Holt *et al.* 2017), on average one every two or three years.

The first published account was of one on the slopes of Yr Wyddfa, Caernarfonshire, on 20 August 1870 (Saunders 1889). The timing and location of this record is at odds with the non-breeding patterns of occurrence of this species in Britain and northern Europe, as an extremely rare autumn or spring migrant. Also, none of the British records since 1950 occurred away from the coastal lowlands (Slack 2009). For this reason, the record is no longer thought to be acceptable.

The only Welsh record was of one found on the lighthouse steps at Strumble Head, Pembrokeshire, by a visiting birder, on 30 October 1997. It is possible that the number of records in Wales will increase, as its suitable climatic range is forecast to extend north to southern England by 2070–2100 (Huntley *et al.* 2007).

Jon Green and Robin Sandham

Dunnock *Prunella modularis* Llwyd y Gwrych

Welsh List Category	IUCN Red List			Birds of Conservation Concern			
	Global	Europe	GB	UK	Wales		
A	LC	LC	LC	Amber	2002	2010	2016

The Dunnock's breeding range covers most of Europe away from the Mediterranean and extends east to the Caspian Sea (*HBW*). The subspecies *P.m. occidentalis* occurs only in western France and Britain, except the Outer Hebrides (IOC). It is found in a variety of habitats: around woodland edges, young conifer plantations, scrub and gardens, and in gorse and heather in the uplands. Open scrub is perhaps the preferred habitat. The Dunnock's Welsh name, like its alternative English name of 'Hedge Sparrow', indicates its liking for hedgerows. The Dunnock is predominantly a bird of the lowlands and semi-uplands in Wales, with few breeding above 450m, though some have bred considerably higher: *Birds in Wales* (1994) stated that it had been found in rank heather at 530m in the Rhiniogau, Meirionnydd, and they have been recorded at 600m in Meirionnydd (Pritchard 2012) and 610m in Radnorshire (Jennings 2014). There have been three records—including two singing birds—near the summit of Tryfan in the Glyderau, Caernarfonshire, at an altitude of 917m. The population in the east of its range is largely migratory, but in Britain, Dunnocks are sedentary. The *BTO Migration Atlas* (2002) commented that in Britain and Ireland, bird-watchers were unlikely to see a Dunnock that was more than 1km away from where it hatched and most recoveries of birds ringed in Wales have been close to the original ringing site. Birds breeding at higher altitudes elsewhere in Europe move to lower ground in response to cold weather (*BWP*), and this probably happens in Wales. However, Dunnocks ringed elsewhere showed evidence of long-distance movement; for example, a first-calendar-year bird ringed at Creeting St Mary, Suffolk, in September 2009, was caught by a ringer at Oxwich Marsh, Gower, 364km to the west, in March 2014.

The Dunnock seems always to have been common in Wales. Mathew (1894) said that it was a common resident in Pembrokeshire and in North Wales. Forrest (1907) considered it to be so common everywhere that further details were unnecessary. Other authors from this period made similar comments. The species seems to have benefited from afforestation in the Welsh uplands, and in several counties, it was recorded that it had spread into young conifer plantations. Peers and Shrubb (1990) commented that in northwest Breconshire, apart from villages and farmyards and nearby hedgerows, it was only found in young conifer plantations. However, once the canopy closed, Dunnocks become much less common in conifer plantations.

The species was recorded in almost every 10-km square in Wales in the *Britain and Ireland Atlas 2007–11*, as it had been in the two previous atlases, in 1968–72 and 1988–91. Relative abundance in Wales in 2007–11 was generally lower than in England and Ireland, and was highest in Pembrokeshire and Anglesey, with little difference between the breeding season and winter. Tetrad atlases showed it to be among the most widespread species in most counties. It was found with evidence of breeding in 96% of tetrads in Gwent in 1981–85 and in 97% in 1998–2003, in 77% of tetrads in East Glamorgan in 1984–89 and in 82% in 2008–11. In Gower, the *West Glamorgan Atlas 1984–89* recorded it in 90% of tetrads, while in Pembrokeshire there was evidence of breeding in 91% of them in atlas surveys in 1984–88 and 2003–07. The *North Wales Atlas 2008–12* found it in 79% of tetrads. Only in Breconshire was it recorded in fewer than half the tetrads, the species being present in only 45% of tetrads visited during fieldwork for the 1988–91 *Breeding Atlas*, although the authors considered it under-recorded. In all these atlases, most of the tetrads from which Dunnock was absent were on the highest ground in the uplands.

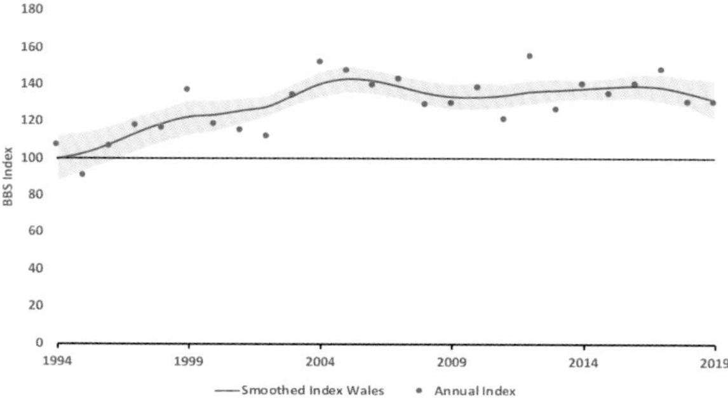

Dunnock: Breeding Bird Survey indices, 1994–2019. The red line shows the smoothed index, 1995–2018. The shaded area indicates the 85% confidence limits.

The BBS index for Wales shows a fairly steady increase from 1994, with a 32% increase during 1995–2018 (Harris *et al.* 2020). The BBS map of change in relative density between 1994–96 and 2007–09 indicates that the increase in the UK index had occurred mainly in Northern Ireland, in northwest Britain and Wales (BTO 2020). Dunnock populations can be quite severely affected by cold winters. On Bardsey, Caernarfonshire, for example, numbers dropped from 14 territories in 1981 to four in 1982. A sharp decline was noted in Glamorgan in 1963, but numbers had recovered by 1966 (Hurford and Lansdown 1995). Surveys suggested that the Dunnock is not particularly numerous in mature deciduous forest, particularly if there is little ground cover. A survey of woodland at RSPB Ynys-hir, Ceredigion, in 2006, found 21 territories in 112ha (18.8 pairs/km^2), with 18 territories in the same area in 2011.

There may be post-juvenile eruptions in early autumn in some years (*BTO Migration Atlas* 2002). A small autumn passage is evident on several islands and coastal watch points in some years, peaking in September and October, although the presence of residents almost everywhere makes influxes hard to detect, unless they are fairly large. On Skomer, Pembrokeshire, there was a large influx in autumn 2002, with 200 there in mid-October, although some would have been residents. A total of 47 was counted on Puffin Island, Anglesey, in November 1999. Dunnocks do not usually congregate in winter, but fairly high numbers can occur where there is a good food source. For example, 40 were in a field near Glasbury, Breconshire, in February 2010, feeding on the seeds of Fat-hen.

Estimating population size for this species is somewhat complicated by the Dunnock's notorious breeding strategy. Although it may be the archetypal 'little brown job' with its understated brown and grey plumage, the Dunnock makes up for it by having a varied sex life. Some form pairs in the conventional manner, but there are also associations of two males and a female, two females and a male, two of each sex or other combinations. This is described as a "variable mating system" (Davies and Lundberg 1984). Measuring population in terms of 'pairs' is thus an over-simplification. Hughes *et al.* (2020) did, however, estimate the Welsh breeding population to be 310,000 territories in 2018.

The Dunnock appears to be doing well in Wales at present. There is scope for more research on its movements, since the sedentary habits suggested by ringing contrast with the evidence of autumn movements.

Continental Dunnock *P.m. modularis* Llwyd y Gwrych y Cyfandir

A first-calendar-year bird ringed at Dolembreux, Belgium, in September 1986 was killed by a cat at Abercraf, Breconshire, 667km to the west, in February 1989. This bird is likely to have been of the nominate subspecies *P.m. modularis*, whose distribution covers much of north and central Europe (IOC).

Rhion Pritchard

Sponsored by S. De Clancy

Western Yellow Wagtail *Motacilla flava* Siglen Felen y Gorllewin

Welsh List Category	IUCN Red List			Birds of Conservation Concern			
	Global	Europe	GB	UK	Wales		
A	LC	LC	NT	Red	2002	2010	2016

The Western Yellow Wagtail is a summer visitor to Europe, western and central Siberia, wintering in tropical and subtropical areas to the south (*HBW*). In Britain it breeds mainly in England and eastern Wales, with small numbers in Scotland. There are ten subspecies, of which five have been recorded in Wales, but only one, *M.f. flavissima*, known now by the IOC as 'British Yellow Wagtail', breeds here regularly. In most parts of its range, the Western Yellow Wagtail prefers wet grassland habitats (*BWP*), but in Wales, due to the paucity of lowland wet grassland, the breeding habitat is usually land farmed for spring-sown root crops and cereals or pasture, mainly in the lowland river valleys of the counties bordering England.

Western Yellow Wagtails from southwest England, and probably Wales too, migrate south to Portugal in one movement (*BTO Migration Atlas* 2002). From there they move to spend the winter in West Africa, south of the Sahara. For their return, they move north up the western edge of the Sahara, through the Atlas Mountains in Morocco and then central Spain. There have been very few recoveries in spring of British-ringed Western Yellow Wagtails in France, so it is not clear if they overfly France to arrive at the south coast of England.

Forrest (1907) wrote that it was found "irregularly" in most counties of North Wales, favouring low ground in fields of root crops, marshy land and wet meadows, especially near the sea, such as near Tywyn, Meirionnydd. It was common at Sealand, Flintshire. Forrest stated that it was rarely found above 300m, but family parties were seen near Cerrigydrudion, Denbighshire, on 15 July 1901, and a nest with eggs at Palé, Meirionnydd, on 15 July 1903. Breeding was sporadic in Meirionnydd until the 1930s, after which only one breeding record (in 2008) was confirmed (Pritchard 2012). It occurred on Anglesey and up the west coast, mainly as a passage migrant. Roderick and Davis (2010) quoted one or two nesting records in Ceredigion in 1897. They cited Salter (1930), who had suggested that a series of wet summers in the 1920s may have been a factor in its demise as a breeding species, at least in Ceredigion. It was only considered as a passage migrant in Pembrokeshire by Mathew (1894) and Lockley *et al.* (1949). In Glamorgan at the start of the 20th century, Western Yellow Wagtails

Western Yellow Wagtail feeding around Welsh mountain sheep © Ben Porter

were thought to be common by some, but this was probably in error as it was the least numerous wagtail and sightings of large numbers at pre-migration roosts were misleading (Heathcote *et al.* 1967). In Carmarthenshire, it was a scarce summer visitor to just a few breeding locations, mainly in the Tywi Valley (Ingram and Salmon 1954). In Radnorshire, Ingram and Salmon (1955) said that it was rare and had not bred in the county for 30 years after 1925, but it is likely that a small but regular breeding population at Glasbury, on the Brecon/Radnor border was overlooked (Jennings 2014). In Ceredigion, there had been breeding records up to about 1927, but thereafter it occurred only as a passage migrant (Ingram *et al.* 1966). In Flintshire, it was fairly common in the lowlands in the east, around Sealand (Birch *et al.* 1968). The Western Yellow Wagtail was a scarce visitor to Breconshire, with a few breeding localities, particularly near Llangorse Lake, where passage birds were also recorded, and by the Usk below Brecon. It was also found at a couple of places near Glasbury in the Wye Valley (Massey 1976).

By the mid-20th century, the Western Yellow Wagtail was still considered to be reasonably common in suitable habitat across many parts of eastern Wales, such as the small breeding population in meadows along the Severn and Camlad rivers, Montgomeryshire, in the 1940s and 1950s (Holt and Williams 2008). In Gwent, it was most numerous on the coastal flats and river valleys in 1963 (Venables *et al.* 2008). In Radnorshire, up to 20 pairs bred near Glasbury in the 1960s. A roost count of up to 100 in July 1963 was counted by James Cobb, who also ringed 40 of these birds, most of which were thought to have bred locally. However, of the 120 birds ringed, some were in August and almost certainly these birds included migrants (Jennings 2014).

As farming became more intensive after the Second World War, a widespread decline in populations of Western Yellow Wagtail was noted from the 1960s. The *Britain and Ireland Atlas 1968–72* showed evidence of breeding in 29% of 10-km squares in Wales, with confirmed breeding mostly concentrated in the eastern border counties from Flintshire south to Gwent, with a few scattered records in the west in Caernarfonshire, Carmarthenshire and Gower. In 1967, in Montgomeryshire, it was considered to be fairly common in the east of the county, but by the 2000s there were only a few breeding pairs left (Holt and Williams 2008). In Radnorshire in the 1980s, *c.*35 pairs were still breeding in the Wye catchment in the southernmost part of the county and this population remained fairly stable into the 1990s (Jennings 2014). Breeding in the Tywi Valley, Carmarthenshire, had ceased by the mid-1970s. The last known breeding record in the county was at Penclacwydd in 1987 (Lloyd *et al.* 2015). In Gwent, there had been little change prior to 1977, but a widespread decline followed (Venables *et al.* 2008). The only two breeding attempts in Pembrokeshire, at Talbenny in 1977 and Treginnis in 1983, involved a male of the Blue-headed subspecies, *M.f. flava*, and only the former was successful (Donovan and Rees 1994). A pair nested in 1981 at Tre'r Ddol, Ceredigion (Roderick and Davis 2010). In Caernarfonshire, Pritchard (2017) reported that there had been very few breeding records. The *Britain and Ireland Atlas 1988–91* showed a contraction in range, to 19% of 10-km squares, especially from the westernmost part of its range in South Wales (Carmarthenshire). Its distribution was largely unchanged in mid-Wales, but it occurred at low densities. Tetrad atlases highlighted this contraction in more detail. The *Gwent Atlas 1981–85* found evidence of breeding in 29% of tetrads, which had reduced to 15% by 1998–2003. The *East Glamorgan Atlas* found records of breeding in 3% of tetrads in 1984–89 and in fewer than 1% of them in 2008–11. In Gower, the *West Glamorgan Atlas 1984–89* recorded breeding in just 2% of tetrads.

By the 21st century, the Western Yellow Wagtail had all but disappeared as a breeding species in Wales. The *Britain and Ireland Atlas 2007–11* found that just 7% of 10-km squares were occupied. There were a few records of possible breeding in the southeast and one in Carmarthenshire. Birds were present elsewhere in the south and along the north coast, without evidence of breeding. There was one notable record of confirmed breeding at Abersoch, Caernarfonshire, where a pair nested in a potato field, where only migrants would normally be expected. The other four tetrads with confirmed breeding records in the *North Wales Atlas 2008–12* were in the northeast, on the eastern boundaries of Flintshire and Denbighshire. Breeding was still recorded in the Wye Valley near Glasbury, Radnorshire, between 2000 and 2013, with *c.*12–16 pairs found near Boughrood (Jennings 2014). At least 15 pairs still breed in the area, probably due to the mixed agricultural system of cereals, Oil-seed Rape and potatoes, combined with stock-grazed grassland, providing good habitat for feeding and breeding (Pete Jennings *in litt.*). In 2019, 3–4 pairs were still thought to breed in the Usk Valley, Gwent.

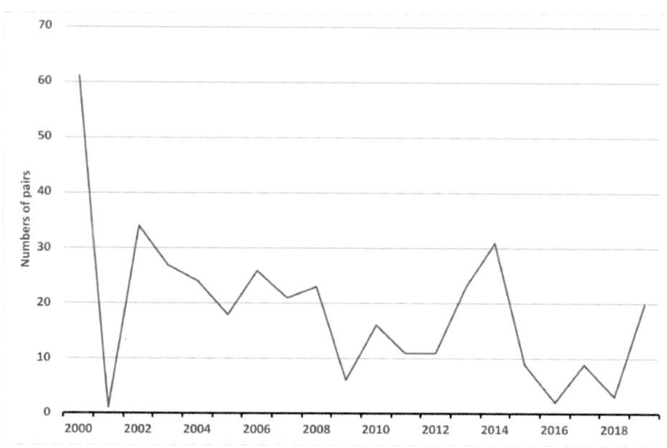

Western Yellow Wagtail: The minimum numbers of pairs breeding in Wales, 2000–18 (NB restricted fieldwork in 2001 because of foot-and-mouth disease).

The general contraction in range and numbers of the Welsh population is reflected in the BBS index for the UK, which declined by 42% during 1995–2018. It occurs on an insufficient number of BBS squares in Wales to generate a trend (Harris *et al.* 2020). *Birds in Wales* (1994) estimated a population of 180–190 breeding pairs, confined to the eastern counties, and noted that the coastal belt of Glamorgan and Gower had been abandoned in the 1980s. *Birds in Wales* (2002) revised the estimated number of breeding pairs down to 100, including 30 pairs at Shotwick Fields, Flintshire. During 2000–18, the decline in numbers continued, as did the number of counties with breeding activity. This decline is not restricted to Wales, as is seen in the UK CBC/BBS data (Harris *et al.* 2020). It has not been possible to calculate a more recent population estimate (Hughes *et al.* 2020), but it is almost certain that the population is less than it was in 2002, possibly as low as 20 pairs.

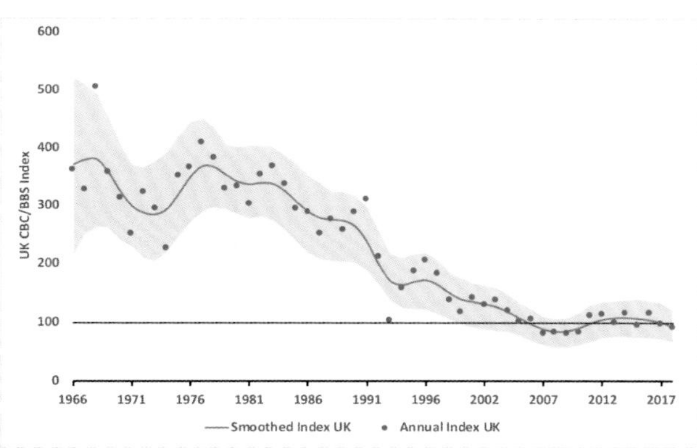

Western Yellow Wagtail: Common Birds Census and Breeding Bird Survey indices for the UK, 1966–2018 (Massimino 2020). The shaded area indicates the 85% confidence limits.

In most counties, the Western Yellow Wagtail is seen only as a passage migrant in small numbers. Spring migration in the south and west of Wales is typically noted at coastal sites, during the second and third weeks of April, with some birds still passing through in May. Maximum counts have declined, though not as markedly as breeding numbers.

Autumn migration typically occurs between mid-August and October, and even up until the 1980s, large parties of over 200 were recorded in Glamorgan, such as those at Aberthaw on 24 August 1976 and 31 August 1987, and the 170 at Kenfig Pool, Gower, on 28 August 1989. It was still seen on passage, in Gwent between 1981 and 1985, with larger counts in autumn, but these counts were, then, usually less than 200. Counts in subsequent years have been small, rarely of more than 50 (Hurford and Lansdown 1995). In 2019, the maximum count in Gwent was of 30 individuals at Newport Wetlands on 25 August. It is still a regular

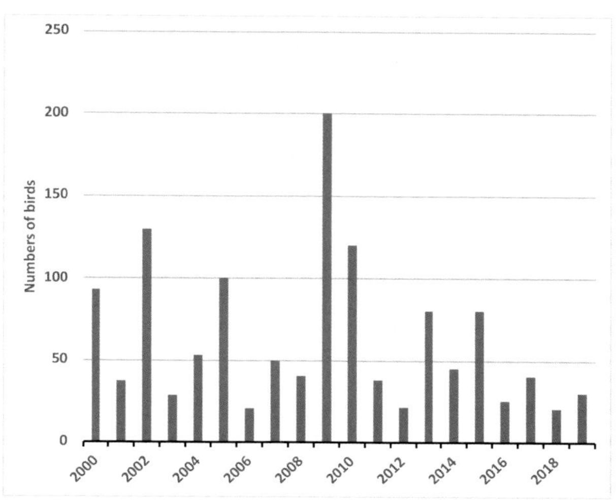

Western Yellow Wagtail: Maximum counts, 2000–19.

passage migrant on Skokholm, Pembrokeshire, but whereas historic counts included 150 there, on 25 August 1952, numbers are now quite small. Since 1987 it has been relatively scarce in Carmarthenshire, with some small groups and a maximum count of ten at Banc-y-Lord on 8 September 2010 (Lloyd *et al.* 2015). Apart from on Bardsey, Caernarfonshire, in most of North Wales it is now a scarce passage migrant. In Meirionnydd, most records were of 1–2, but there have been just five records of singles in the county since 1996. On Bardsey, it was a regular passage migrant in autumn, with counts of 10–15 in most years prior to 1985 (Roberts 1985), but counts have been lower in recent years (Pritchard 2017).

Chamberlain and Fuller (2000, 2001) showed that range contractions were greatest in regions, such as Wales, dominated by pastoral agriculture. Farmland drainage, the conversion of pasture to arable land, changes in grazing and cutting regimes and the loss of insects associated with cattle are likely to have reduced the quality of grasslands as a nesting and foraging habitat. Climate modelling suggested that parts of western Britain will become less suitable for breeding Western Yellow Wagtail during the 21st century, but did not distinguish between subspecies (Huntley *et al.* 2007). That the Western Yellow Wagtail has already been lost from much of Wales strongly points to habitat factors being responsible. The prospects for Western Yellow Wagtail as a breeding species in Wales are not promising.

The identification of the subspecies of the Yellow Wagtail group can be extremely difficult, due to the frequency of intergrades showing mixed characteristics to a greater or lesser extent. Due to this, a hard line is usually adopted in accepting records of the rarer subspecies. In general, only males, of Grey-headed, Black-headed, Ashy-headed and Iberian, are accepted and even those showing only minor differences from the norm are found unacceptable. For example, what was thought to be a male Black-headed Wagtail on the River Clwyd, Denbighshire, on 8 May 2016 was accepted by BBRC as most likely to be an intergrade between *M.f. feldegg* and *M.f. flava/bema* as it showed only a trace of a pale supercilium and greyish tones to upperparts, both non-feldegg characteristics.

Blue-headed Wagtail *M.f. flava* Siglen Benlas

This subspecies breeds in north and central Europe, south of Fennoscandia, and east to the Urals, and winters in sub-Saharan Africa (IOC). A nest was found at Llangorse Lake, Breconshire, in May 1896, reported in Massey (1976). Forrest (1907, 1919) suggested that this subspecies went unnoticed because of its similarity to *M.f. flavissima*. Two specimens were shot near Penrhyndeudraeth, Meirionnydd, in April 1897 and 1898, and a few at Llandudno, Caernarfonshire, in 1912, 1913 and probably 1914. A male was seen at Witchett, Carmarthenshire, on 12–13 April 1952 (Ingram and Salmon 1954) and in Breconshire. Other records were at Llansantffraed in spring 1899, near Glasbury on 21 July 1963 and in August 1965, and at Llangorse Lake on 26 April 1974 (Massey 1976).

Birds in Wales (1994) stated that there were 30 records of 36 individuals during 1978–87. In Montgomeryshire, three individuals were recorded: a male was present at Garthmyl in 1968, 1970 and 1971, which in the last two years bred with a female of the subspecies *flavissima*, and there was one at the Gaer from 26 July to 19 August 1982 and another on 19 April 1993 (Holt and Williams 2008). *Birds in Wales* (2002) reported records involving 53 individuals during 1992–2000, with most in Caernarfonshire (20) and Pembrokeshire (16) along with 2–4 in each of Flintshire, Anglesey, Ceredigion, Carmarthenshire, Glamorgan, Breconshire and Gwent. During 2000–18, it was recorded on Bardsey, Caernarfonshire, in spring and autumn every year, but was not recorded in 2019. Elsewhere, records were low and infrequent. Most sightings were in the northwest in spring, at Bardsey or on Anglesey, mainly at Cemlyn Bay.

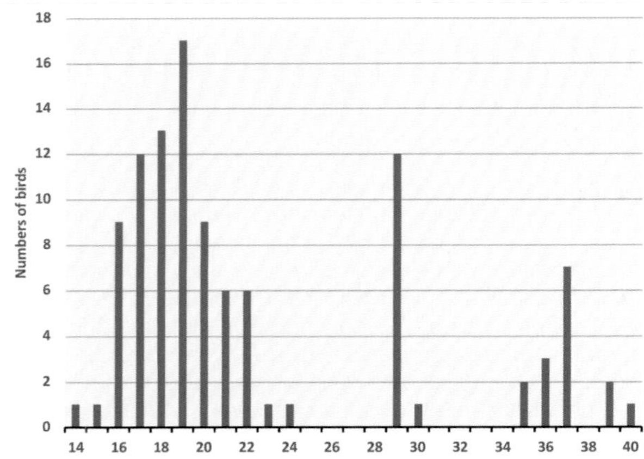

Blue-headed Wagtail, *M.f. flava*: The arrival dates of *flava*, by week number, recorded in Wales, 2000–19. Week 20 starts *c.*13 May and week 38 starts *c.*16 September.

Grey-headed Wagtail *M.f. thunbergi*
Siglen Benlwyd

This subspecies breeds in Fennoscandia, European Russia and northwest Siberia, and winters in sub-Saharan Africa, South and Southeast Asia (IOC). It is assessed by WBRC. The first Welsh record was on Bardsey, Caernarfonshire, on 29–31 May 1975. There have been a further 16 records in Wales, all arriving in May. Of the 17 individuals, a total of four have been on Bardsey (1975, 1976, 1987 and 1993); four on Skokholm, Pembrokeshire (1978, 1987, 1989 and 1990); two at Cemlyn, Anglesey, in 2011 and 2016; two at RSPB Conwy, Denbighshire, in 1999 and 2004; and singles at Goldcliff, Gwent, in 1994; Strumble Head (1983) and Flimston (1998) in Pembrokeshire; Aberdysynni, Meirionnydd, in 2017; and the Little Orme, Caernarfonshire, in 1980.

Black-headed Wagtail *M.f. feldegg*
Siglen Benddu

This subspecies breeds in southeast Europe from the Balkans to south Ukraine, in Turkey, the eastern Mediterranean area, the Arabian Peninsula to Iran and north Afghanistan. It winters in Africa from Nigeria east to Sudan and Uganda (IOC). Historic sources did not mention this subspecies, which is now assessed by BBRC. There have been four records in Wales since 1950, all males in spring:

- Skomer, Pembrokeshire, on 7 May 1986;
- RSPB Conwy, Denbighshire, on 8–9 May 1998;
- Cemlyn, Anglesey, on 6 May 2011 and
- Marloes Mere, Pembrokeshire, on 12–13 May 2013.

'Southern' Yellow Wagtail (Iberian x Ashy-headed Wagtail) *M.f. iberiae* x *cinereocapilla* Siglen Felen Iberia x Siglen Benllwyd yr Eidal

Subspecies *iberiae* breeds across the Iberian Peninsula, the Balearic Islands and Morocco, and sporadically in southwest and southern France, east to the Camargue. It is presumed to winter with other *flava* wagtails in sub-Saharan Africa. Subspecies *cinereocapilla* breeds in Sardinia, Italy, Sicily, southwest Slovenia, northwest Croatia and southwest Morocco. This intergrades with *iberiae* in southern France and northwest Spain. It winters in west-central Africa from Mali to Nigeria, east to Lake Chad. Both subspecies are assessed by BBRC.

These subspecies both occasionally overshoot on spring migration and turn up in Britain, but intergrades are frequent. The intergrades between them are called 'Southern Wagtails'. There is one record of Southern Yellow Wagtail (*M.f. iberiae/cinereocapilla*) at RSPB Conwy, Denbighshire, on 28–30 April 2008.

Ian M. Spence, Jon Green and Rob Sandham

Sponsored by Richard Arnold

Eastern Yellow Wagtail *Motacilla tschutschensis* Siglen Felen y Dwyrain

Welsh List Category	IUCN Red List	Rarity recording
	Global	
A	LC	BBRC

The Eastern Yellow Wagtail has a large range from Siberia and Transbaikalia, south to Mongolia and Manchuria, east to Kamchatka, north west to the Aleutian Islands, south to northern Japan (northern Hokkaido), as well as extreme northwest Alaska in the USA (*HBW*). The species winters in southeastern Asia as far as the Philippines, south to Indonesia and northern Australia. There have been 16 records in Britain to the end of 2019 (Holt *et al.* 2020). This extremely rare vagrant has only occurred in Wales once and that was a first-calendar-year bird at Cemlyn, Anglesey, on 25–27 September 2019.

Jon Green and Robin Sandham

Citrine Wagtail *Motacilla citreola* Siglen Sitraidd

Welsh List Category	IUCN Red List		Rarity recording
	Global	Europe	
A	LC	LC	WBRC

The Citrine Wagtail, which can be semi-colonial, breeds in Finland, Turkey and eastern Europe, across much of Russia, western China and Mongolia (*HBW*). Two subspecies occur in Europe: the nominate *M.c. citreola* and *M.c. werae* (Alstrom and Mild 2003). The nominate subspecies occurs in European northeast Russia and further north and eastwards, and winters in Southeast Asia. In the rest of Europe, the subspecies *M.c. werae* breeds in Finland and eastern Europe, as far east as southwest China, and winters in East Africa. In the past fifty years or so the range of *M.c. werae* has expanded westwards and it is now a common breeding bird in southern Finland, Belarus, Poland, the Baltic States and Ukraine

Citrine Wagtail on Bardsey Caernarfonshire, May 2014
© Ben Porter

(*European Atlas* 2020). This change in range goes some way to explaining why this species was a great rarity in Britain until the 1990s, but thereafter the number of records increased dramatically, and it is no longer considered a British rarity. Up to 2018 there had been 349 records in Britain, with an average of 14 a year during 2010–18 (White and Kehoe 2020b). This is a rare migrant in Wales, with the first record for Wales being in 2000 (Brown 2001) and another 11 records from 2008 onwards:

- 1CY at Skomer, Pembrokeshire, on 28 September 2000;
- 2CY male at RSPB Conwy, Denbighshire, on 30 April 2008;
- 2CY male at Brecon, Breconshire, on 5 June 2008;
- male at RSPB Conwy, Denbighshire, on 15–16 May 2011 then at Cemlyn, Anglesey, on 17 May 2011;
- 1CY at Bardsey, Caernarfonshire, on 10–11 October 2012;
- 1CY at RSPB Conwy, Denbighshire and Caernarfonshire, on 22 August 2013;
- 2CY female on Bardsey, Caernarfonshire, on 25–26 May 2014;
- 1CY on Bardsey, Caernarfonshire, on 8–11 August 2014;
- 1CY on Bardsey, Caernarfonshire, on 17 August 2016;
- female at Morfa Madryn, Caernarfonshire, on 6–7 May 2017;
- male at Llwyn-onn Reservoir, East Glamorgan, on 27 April 2019 and
- 2CY male at Llangorse Lake, Breconshire, on 8 May 2019.

Jon Green and Robin Sandham

Sponsored by Keith Noble

Grey Wagtail *Motacilla cinerea* Siglen Lwyd

Welsh List Category	IUCN Red List			Birds of Conservation Concern			
	Global	Europe	GB	UK	Wales		
A	LC	LC	NT	Red	2002	2010	2016

The Grey Wagtail breeds in hills and mountains discontinuously across temperate Eurasia and in North Africa. Those in southern and western Europe, including in Wales, are resident, whereas those in Fennoscandia and from the Baltic eastwards migrate to the south (*HBW*). It is widespread in Britain and Ireland but most abundant in the north and west, as it favours small, wooded, fast-flowing streams and rivers from high mountains to sea level. In Wales, some undertake a limited seasonal movement from the uplands to lower ground in winter, where they use a wider variety of habitats, such as sewage farms, lakes, garden ponds and coastal locations. In lowland Wales it breeds along rivers, as well as lake outflows and along canals. Nests are usually sited on riverside cliffs, in crevices or ledges in walls and under bridges but occasionally, it nests in buildings up to 1km away from water (Tyler 1972).

Grey Wagtails are widespread in every county, but breeding densities are far higher in the uplands, where it nests up to 500m and occasionally higher. It favours smaller watercourses bordered by broadleaved trees, rather than large rivers or those bordered by moorland or conifers (Tyler and Ormerod 1991). In optimal conditions, territories can be less than 250m apart: two active nests found in May 2007, on a minor tributary of the River Wye in Gwent, were only 65m apart (Tyler and Medland 2008). Unlike Dippers, the breeding biology and abundance of Grey Wagtails were unaffected by the acidification of rivers such as the Afon Irfon in the 1970s to 1990s. Their diet is catholic and includes a wide range of terrestrial invertebrates, as well as adult aquatic prey (Ormerod and Tyler 1991, Tyler and Ormerod 1991). Wales holds the longevity record for Grey Wagtail. A nestling ringed on 9 June 1982 at Cwn Cleisfer,

Female Grey Wagtail © Bob Garrett

Crickhowell, Breconshire, was recaught at Grwyne Fawr Reservoir, Crickhowell, on 10 June 1989, seven years and one day later.

At the end of the 19th century, the Grey Wagtail was widespread and common in suitable riverine habitats throughout Wales, except on Anglesey (Forrest 1907, Mathew 1894, Phillips 1899). Authors noted that, in winter, birds seemed to desert upland watercourses and move to lower elevations. Little had changed by the mid-20th century. The Grey Wagtail was still to be commonly found along many watercourses in all counties, although Massey (1976) made a particular note that it was badly affected by the 1962/63 winter in Breconshire.

The *Britain and Ireland Atlas 1968–72* provided the first picture of breeding distribution across the whole of Wales. It showed that it was absent only around the coasts of Anglesey, southwest Pembrokeshire and western Gower. Twenty years later, the range had contracted slightly, especially in Llŷn and west Pembrokeshire (*Britain and Ireland Atlas 1988–91*), but by the 2007–11 *Atlas*, only western Anglesey and southwest Pembrokeshire had no breeding Grey Wagtails. Over these 40 years, there was a small net increase in occupied squares from 85% in 1968–72 to 87% in 2007–11. Relative abundance in 1988–91 and 2007–11 was highest in the uplands, compared to many other parts of its Welsh range. Jones and Dare (1976) reported densities of 1.6 pairs/km along Afon Ogwen, Caernarfonshire. In 1974, the BTO Waterways Survey (Marchant and Hyde 1979) recorded 1.14 pairs/km of natural waterways in North Wales.

Local tetrad atlases and studies provided a more detailed picture. There has been little change in most counties in the last 30–40 years, although localised increases were reported in the western valleys of Gwent (Venables *et al.* 2008) and in east Pembrokeshire.

Watercourse	Year	No. pairs	Pairs/10km	Reference
River Wye, Montgomeryshire, Radnorshire, Breconshire, Gwent and Herefordshire	1976	74 pairs in 223km	3.3	RSPB 1977
River Severn, Montgomeryshire	1977	31 pairs in 107km	2.9	RSPB 1978
Afon Vyrnwy, Montgomeryshire	1977	28 pairs in 67km	4.2	RSPB 1978
Afon Irfon, Breconshire	1990	19 pairs in 20km	9.5	Tyler and Ormerod (1991)
North Wales		114 pairs in 100km	11.4	*Britain and Ireland Atlas 1968–72*
Grwyne Fawr and Grwyne Fechan, Breconshire and Gwent	1992 and 1995	21 pairs in 29km in 1992 39 pairs in 29km in 1995	7.2–13.4	Jerry Lewis *in litt.*
Afon Honddu, Gwent	1990s and 2000s	9 pairs in 7km	12.9	Tyler (own data)
Angidy brook, Gwent	1990s	7–8 pairs in 3km	23.3–26.7	Tyler (2003)
Afon Eglwyseg, Llangollen, Denbighshire	2002	5–8 pairs in 2km	25–40	Roberts and Jones (2003)
Afon Ogwen, Caernarfonshire	2005–2007	12–14 pairs in 16km	7.5–8.75	Gibbs *et al.* 2011

Grey Wagtail: Breeding abundances on various watercourses in Wales.

There were also losses in parts of the Western Cleddau catchment and south and east of Haverfordwest (*Pembrokeshire Atlas 2003–07*). Recent county population estimates include 430–610 pairs in Gwent, where 66% of tetrads were occupied (Venables *et al.* 2008), and 450 pairs in Pembrokeshire, where 23% of tetrads were occupied (*Pembrokeshire Avifauna* 2020). In the East Glamorgan atlases, the Grey Wagtail was widespread in the breeding season and was found in 38% of tetrads in 1984–89 and 36% in 2008–11. The *West Glamorgan Atlas 1984–89* recorded it in 50% of tetrads, the stronghold being upland areas in the north and east of Gower. The *North Wales Atlas 2008–12* showed that Grey Wagtail distribution was widespread, with 33% of tetrads occupied. Most suitable watercourses were occupied, but upland areas were particularly favoured. It should be noted that despite few breeding pairs being recorded on Anglesey during 2008–12, Nigel Brown (*in litt.*) reported that the Grey Wagtail had increased on the island since 2000 and that it is found breeding on all the main rivers.

In the 21st century, changes in UK populations became evident. There are no BBS indices for Grey Wagtails in Wales because the sample sizes in many years have been too small. However, the Waterways Breeding Bird Survey (WBBS) index for the UK recorded a 17% decline during 1999–2018 (Harris *et al.* 2020), most of this occurring in 2002–12. The reason for this is unclear and is not thought to be due to a decrease in aquatic invertebrates, which have recovered in numbers since 2012, probably due to improved river management and a reduction in pollution, which has been shown to increase freshwater invertebrate populations (Outhwaite *et al.* 2020). In Montgomeryshire and Radnorshire, where there has been a growing problem of chicken manure being washed into rivers in the past decade, the impact of this pollution on the Grey Wagtails has yet to be quantified. In Wales, the main factor affecting populations was thought to be low late-winter temperatures, but the species soon recovered from temporary declines (Tyler 1972) and this is still the case (pers. obs.). Milder winters in recent years may have benefited the Grey Wagtail and information, from the rivers in south Wales that are studied in most detail, suggests that numbers are holding up well and there is no evidence of any declines in occupancy (pers. obs.). The Welsh breeding population in 2018 was estimated to be 5,950 territories (Hughes *et al.* 2020).

Grey Wagtail: Recovery locations of birds ringed in Wales are shown by red circles. Ringing locations of birds that were recovered in Wales are shown by blue triangles. Small brown dots show ringing or recovery locations in Wales. All movements shown are greater than 50km.

Sharrock (1964) first drew attention to coastal passage of Grey Wagtails in Britain. In Wales, all coastal counties experience some passage. On the south coast this is evident from mid-August to October, occasionally later, but peaks in the first two weeks of September. Usually up to ten have been recorded daily at any site but 20–40 are not uncommon. For example, counts in Gwent have included up to 45/hour at Black Rock and 20/hour west at Collister Pill (Venables *et al.* 2008). Some may be young Welsh birds dispersing or moving to southern and southwest England, but it may also be that birds from farther north in Britain or elsewhere in Europe move through these areas. There is a southerly movement from Bardsey, Caernarfonshire, where passage starts from early June and peaks in late August, with the highest day count of 61 on 1 September 2007. A similar pattern occurs on Skokholm but with far lower numbers.

In winter, Grey Wagtails occur in larger numbers in lowland areas of North and South Wales, as Welsh breeding birds are joined by others from Scotland and northern England, especially during cold weather. The British winter population is boosted by birds from Europe (Tyler 1979, Robinson *et al.* 2020) and some evidently reach Wales: one ringed in Denmark on 22 August 1975 was killed by a cat at Ponthir, Gwent, on 20 February 1977 and another ringed in the Channel Islands on 3 October 1980 was found near Chepstow on 12 October 1980. Some British breeding birds move to France (Tyler 1979): from Wales a juvenile ringed at Tintern, Gwent, on 27 June 1987, was found in France on 9 January 1988. Unlike Pied Wagtails, Grey Wagtails do not usually gather in large roosts in winter but are usually found singly, in pairs or family groups of fewer than ten birds. However, roosts of migrating birds do occur, such as 34 on 2 September and 31 on 19 September 2019 at Rhydymwyn, Flintshire (Ian Spence pers. obs.).

Dispersal of young birds from other parts of Britain may also result in movements in Wales. For example, a Grey Wagtail ringed as a nestling in Devon in 2011 was caught near Kentchurch on the River Monnow, Gwent/Herefordshire, on 18 October 2011 and a first-year female ringed near Brecon on 5 August 1986 moved to Wiltshire, where it was recaught on 11 January 1987.

More information on movements, further detailed studies on breeding biology and repeat surveys of rivers for which there are historical data, will help us to understand what factors determine the fortunes of the Grey Wagtail. If the health of our watercourses improves, then the Grey Wagtail should remain one of the commonly encountered and characteristic birds associated with Welsh rivers and streams.

Steph Tyler

Sponsored by John Lloyd

White/Pied Wagtail *Motacilla alba* Siglen Wen/Fraith

Welsh List Category	IUCN Red List			Birds of Conservation Concern			
	Global	Europe	GB	UK	Wales		
A	LC	LC	LC	Green	2002	2010	2016

This species has a very extensive distribution, over virtually the whole of Europe and Asia outside the tropics, and western Alaska (*HBW*). Within this range, nine subspecies are recognised (IOC). The birds that breed in Wales are *M.a. yarrellii*, the familiar Pied Wagtail that breeds in Britain and Ireland and in coastal parts of northwest continental Europe. The nominate subspecies, *M.a. alba*, the White Wagtail, is a familiar passage visitor to Wales.

The Pied Wagtail is a partial migrant. Some birds move within Britain, while others move farther south in winter to the western Mediterranean (*HBW*). A number of birds ringed in Wales as juveniles have been recovered further south in winter, in France, Spain and Portugal. It may be that, among the breeding population in southern Britain, young birds are more likely to move south, while adults are more sedentary. Adults breeding further north in Britain, in Scotland and northern England, are more likely to move south (*BTO Migration Atlas* 2002). Some Scottish-ringed birds have been recovered in Wales on spring and autumn passage.

The Pied Wagtail breeds in a variety of habitats in Wales, from coastal areas and the centres of towns to the shores of mountain lakes in Snowdonia, up to an altitude of at least 678m. While it is often found around the edges of rivers and lakes, it is much less restricted to this type of habitat than Grey Wagtail. Good numbers are often around farms with livestock and in areas of sheep-grazed grassland with dry-stone walls, a habitat type that is very common in Wales. It is generally absent or scarce in dense woodland, moorland and some intensively farmed areas. All the late-19th-century and early-20th-century authors described it as a common breeding bird. There seems to have been little change over the following century, except for a reduction in numbers following severe winters. Like several other small insectivorous passerines, the Pied Wagtail suffered heavy mortality in the winters of 1946/47 (Ticehurst and Hartley 1948) and 1962/63 (Dobinson and Richards 1964).

The *Britain and Ireland Atlas 1968–72* showed the species to be present in virtually every 10-km square in Wales, and no significant change in distribution was shown in the 1988–91 and 2007–11 atlases. Relative abundance in the breeding season was highest in semi-upland (200–400m) areas, and lower around the coast and in the eastern lowlands, as well as on the very highest ground.

Tetrad atlases confirmed that this is a very widespread species. Birds were recorded in 90% of tetrads in the *Gwent Atlas 1981–85* and in 94% in 1998–2003. Pied Wagtails occupied 65% of tetrads in East Glamorgan in 1984–89 and 69% in 2008–11, while in Pembrokeshire, birds were found in 60% of tetrads in 1984–88 and in 75% in 2003–07 (*Pembrokeshire Avifauna* 2020). Pied Wagtails were found in 79% of tetrads during fieldwork for the *North Wales Atlas 2008–12*.

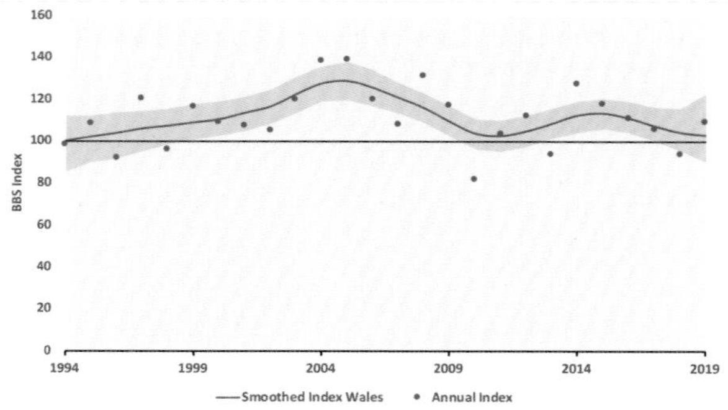

Pied Wagtail: Breeding Bird Survey indices, 1994–2019. The red line shows the smoothed index, 1995–2018. The shaded area indicates the 85% confidence limits.

The BBS index for Wales showed some fluctuation but no pronounced trend, with just a 2% increase between 1995 and 2018 (Harris *et al.* 2020). Pied Wagtails regularly breed on most of the larger Welsh islands, and on some smaller islands such as The Skerries, Anglesey. Bardsey, Caernarfonshire, held a record 14 pairs in 2015. There is not a great deal of information available on population density in Wales, although surveys commissioned by the Nature Conservancy Council in 1977/78 found this to be the most widely distributed riparian species on three Welsh rivers. There were 11 pairs/10km on the River Wye, eight pairs/10km on the River Severn and seven pairs/10km on the Afon Vyrnwy (Round and Moss 1984).

Birds that breed on higher ground are often absent from there in winter, moving to lower ground or farther afield. Some birds apparently move into urban areas, but there was no evidence of any large-scale hard-weather movements, even during the very severe 1962/63 winter (Dobinson and Richards 1964). Almost all the birds that breed on Bardsey depart for the winter and return from early March (*North Wales Atlas 2008–12*).

Good numbers of Pied Wagtails can sometimes be seen on passage, particularly in early autumn, when they are often in mixed flocks with White Wagtails. There were 704 at Eglwys Nunydd Reservoir, Gower, on 30 August 2001. The largest counts of Pied Wagtails have usually been at winter roosts, often in urban areas. These can sometimes hold over 600 birds. There was a roost of 600–1,000 in gorse at Martletwy, Pembrokeshire, between January and March 1935, up to 600 roosted at Llyn Coron, Anglesey, in late December 1964 and *c.*1,400 at Treforest Trading Estate, East Glamorgan, on 20 December 1974. Ysbyty Gwynedd in Bangor, Caernarfonshire, held a roost of over 600 on 9 October 2000 and there were 750 at a roost in Cardiff city centre, East Glamorgan, on 5 December 2012. Some of the largest counts were at a roost in Abergavenny, Gwent: 600 on 3 January 2005, 1,000 on 16 December 2007 and 2,000 on 20 February 2008. Outside the winter months such gatherings are unusual, but 1,000, likely to have been on passage, were at the Gulf Refinery, Pembroke, on 18 May 1998.

The *Britain and Ireland Atlas 2007–11* found relative abundance in Wales in winter to be highest around the south coast, and very low in the uplands. This was in marked contrast to the pattern in the breeding season, when relative abundance was highest in upland and semi-upland areas. The Welsh breeding population in 2018 was estimated to be 56,000 (a range of 49,000–63,500) territories (Hughes *et al.* 2020). This species is doing well in Wales, and no particular conservation issues have been identified.

White Wagtail *M.a. alba* Siglen Wen

The White Wagtail breeds in Iceland and the Faroe Islands, through most of Europe and east to the Urals, the Middle East, the Caucasus and Central Asia. It is more migratory than Pied Wagtail, with northern and eastern populations wintering well to the south, some reaching the Sahel and East Africa (IOC). A few pairs breed in Scotland in most years, mainly in the Northern Isles, and there are also records of mixed *alba/yarrellii* pairs (*The Birds of Scotland* 2007). There have been several reports of birds showing characteristics of this subspecies breeding in Wales, although the reliability of some older records is uncertain. Between 1900 and 1930, birds thought to be White Wagtail were reported breeding in Glamorgan, Meirionnydd, Caernarfonshire and possibly Carmarthenshire, although Ingram and Salmon (1954) considered the Carmarthenshire record to be doubtful. A female showing plumage characteristics of White Wagtail mated with a male Pied Wagtail and bred successfully near Minffordd, Penrhyndeudraeth, Meirionnydd, in 1970.

White Wagtail is otherwise seen on passage in Wales. An adult female ringed on Bardsey, Caernarfonshire, on 8 October 1981, and recovered in Morocco on 3 November the same year, is likely to have been *M.a. alba*. The majority of the birds passing through Wales are probably from the Icelandic population, as illustrated by ringing recoveries, such as one ringed as a nestling at Austur Skaftafells, Iceland, in July 1970 and found dead at Rhuthun, Denbighshire, on 1 September the same year. Two birds, ringed while on passage through Wales, were recovered during the breeding season in Iceland. However, there may also be birds that breed in east Greenland and the Faroe Islands, and Donovan and Rees (1994) suggested that some could have come from Scandinavia. Spring passage is usually between late March and mid-May, but birds have been recorded between early March and late June. Return passage is mainly from late August to mid-October, but a few have been recorded well into November.

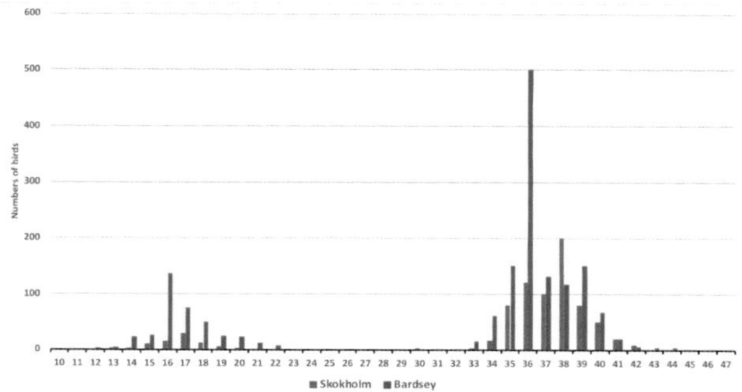

White Wagtail: Maximum counts by week number at Bardsey and Skokholm. Week 16 starts *c.*15 April and week 36 starts *c.*2 September.

This passage is not confined to the coast. Large counts have been recorded on the shores of inland lakes and reservoirs, such as 100–150 at Llyn Trawsfynydd, Meirionnydd, on 26 April 1958, and a mixed flock of 134 birds, including both Pied and White Wagtails, feeding at Talybont Reservoir, Breconshire, on 26 October 2014 (Andrew King *in litt.*).

Spring passage can involve considerable numbers of birds. For example, there were 150 on the Meirionnydd/Caernarfonshire border at Porthmadog Cob on 29 April 1997, 150 at Gronant, Flintshire, on 20 April 2009, over 180 at Llyn Maelog, Anglesey, on 22 April 2002, *c.*400 at Eglwys Nunydd Reservoir, Gower, on 25 April 1977 and over 500 at Cemlyn, Anglesey, in May 1957. High counts of birds roosting at the Teifi Marshes, Ceredigion/Pembrokeshire, included 215 on 16 April 2011, 300 on 14 April 2015 and 200 between 10 and 13 April 2019. There were 200 at Newport, Pembrokeshire, on 14 April 2019. Birds on autumn passage present greater identification difficulties, as they often form mixed flocks with Pied Wagtails, and the two subspecies are more difficult to tell apart than in spring, particularly those that have not completed their moult out of juvenile plumage. The largest numbers have been recorded on the islands: 750 on Bardsey, on 2 September 1987 (twice the previous day record for the island), 500 of which were considered to be White Wagtails. Thompson (2008) stated that up to 500 White/Pied Wagtails had been recorded on Skokholm, from late August to mid-September 1953, but the proportion of White Wagtails remains uncertain. However, 200 White Wagtails were reported on 15 September 1988. On the mainland, 200 were at Pencarnan, St Davids, Pembrokeshire, on 4 September 2014.

Masked Wagtail *M.a. personata* Siglen Fygydog

The Masked Wagtail breeds in northern Iran, southwest Siberia, Mongolia, northwest China and the western Himalayas (*HBW*). Records are assessed by BBRC. A bird of this subspecies was at Camrose, Pembrokeshire, from 29 November to 26 December 2016. This was the first record for Britain (Holt *et al.* 2018).

Rhion Pritchard

Richard's Pipit *Anthus richardi* Corhedydd Richard

Welsh List Category	IUCN Red List
	Global
A	LC

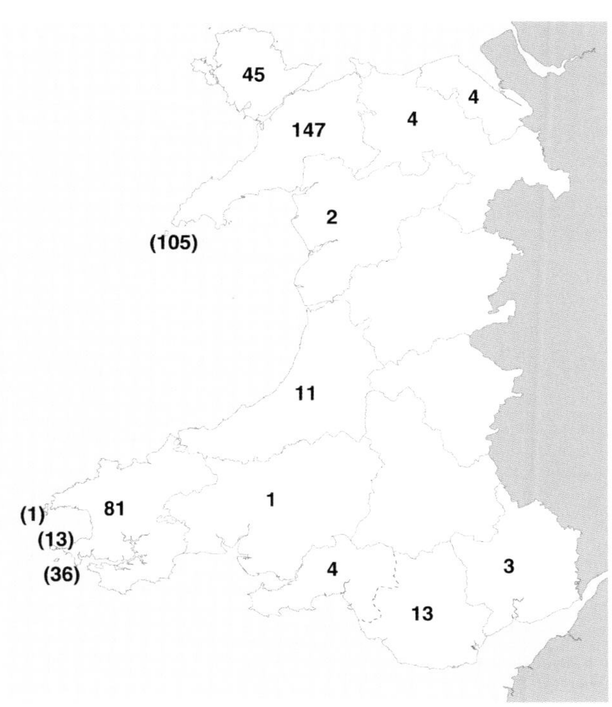

Richard's Pipit: Numbers of birds from each recording area, 1921–2019.

The Richard's Pipit breeds in central and eastern Siberia, Mongolia and eastern China, and winters in South and Southeast Asia (*HBW*). It is a scarce autumn migrant to Wales, with a total of 313 individuals recorded. This number has increased dramatically since the 1950s. The annual number of British records has remained essentially unchanged for almost 30 years, with an average of 118 individuals recorded each year (White and Kehoe 2020b).

The first Welsh record was on the Great Orme, Caernarfonshire, on 26 December 1921. Almost one-third of all Welsh records have been from Bardsey. The adjacent mainland of Caernarfonshire has also recorded good numbers, as has Anglesey. In contrast, very few have been recorded in the other North Wales counties. In the south, the majority have been in Pembrokeshire, with 31 on the

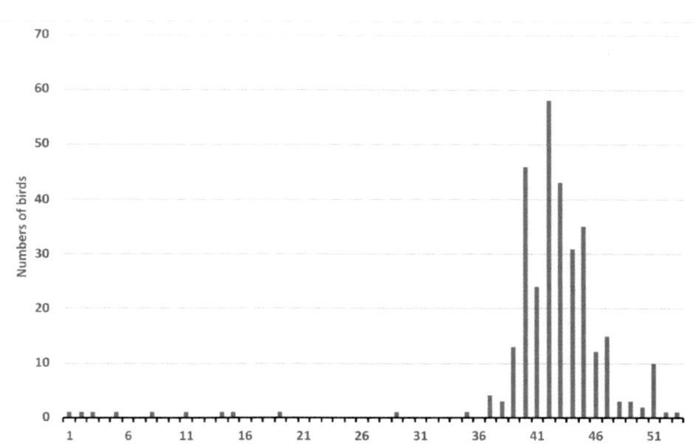

Richard's Pipit: Totals by week of arrival 1921–2019. Week 37 begins *c.*10 September and week 47 begins *c.*19 November.

Total	Pre-1950	1950–59	1960–69	1970–79	1980–89	1990–99	2000–09	2010–19
315	1	3	36	22	20	35	108	90

Richard's Pipit: Individuals recorded in Wales by decade.

mainland and a further 50 on the islands. Elsewhere, there have been a handful of records from Gwent and Gower, and several more in East Glamorgan and Ceredigion.

Most have been recorded in the autumn, from the end of September to late November, with a peak in the last week of October. A few have been discovered in December and into January. Just three have been recorded in spring.

Jon Green

Blyth's Pipit *Anthus godlewskii* Corhedydd Blyth

Welsh List Category	IUCN Red List	Rarity recording
	Global	
A	LC	BBRC

The Blyth's Pipit breeds in Mongolia and southernmost Siberia, and winters in South Asia (*HBW*). It is a rare vagrant in Britain, with 30 records up to 2019 (Holt *et al.* 2020), two of which were in Wales: on Bardsey, Caernarfonshire, on 16–17 October 2005 and at St Davids Head, Pembrokeshire, on 18 November 2014.

Jon Green and Robin Sandham

Tawny Pipit *Anthus campestris* Corhedydd Melyn

Welsh List Category	IUCN Red List		Rarity recording
	Global	Europe	
A	LC	LC	BBRC

The Tawny Pipit breeds in the drier habitats of the western Mediterranean and eastern Europe, North Africa, Turkey and Central Asia. Most European birds winter in Arabia or sub-Saharan Africa (*HBW*). This is a very rare migrant to Wales, with 20 records, all during 1961–2008. The European population declined dramatically in the mid-1990s and recovered slightly in the latter part of that decade (PECBMS 2020). This population decline was thought to be due to the loss of open scrubland breeding habitat, which has come about mainly as a result of afforestation and agricultural intensification (*European Atlas* 1997, Birdlife International 2020). Although numbers in Europe are now relatively stable, the population remains at a much lower level than in the early 1990s. However, the range of the Tawny Pipit in NW Europe has contracted since the 1990s. It is now almost completely absent from The Netherlands, Germany, Denmark and Sweden (*European Atlas* 2020). This may explain why there has been a recent decline in the number of British records: from an annual average of 36 records between 1980 and 1989, to 11 records between 2000–09 (Fraser and Rogers 2006, White and Kehoe 2014b).

Total	1960–69	1970–79	1980–89	1990–99	2000–09	2010–19
20	2	6	4	3	5	0

Tawny Pipit: Individuals recorded in Wales by decade, 1961–2019.

The first Welsh record was on Skokholm, Pembrokeshire, on 19 September 1961. The Tawny Pipit has been recorded in Pembrokeshire only on the islands, with four each on Skomer and Skokholm, and two on Ramsey. Only one has been recorded on Bardsey, Caernarfonshire. Elsewhere, there have been three records from southern counties, Gwent (1) and East Glamorgan (2), and seven records from North Wales (Anglesey, Caernarfonshire and Flintshire). There have been far fewer records of this species than of its larger relative, Richard's Pipit, the occurrence of which, in Wales, appears to be increasing. Only five records occurred in spring with more in autumn, most in September and October; nine and three respectively. The most recent record was at The Range, Anglesey on 10 May 2008. Sixteen records involved single individuals, but in 1975 two individuals turned up on Skomer, Pembrokeshire on 8 October and two more birds were seen at Cemlyn, Anglesey in late September 2006.

The decline in its occurrence in Britain and in its European breeding population is contrary to the northward extension of its distribution modelled by Huntley *et al.* (2007), which suggested that it could breed in southern England later this century. This suggests that the habitat drivers are currently more important than climatic suitability.

Jon Green and Robin Sandham

Meadow Pipit *Anthus pratensis* Corhedydd y Waun

Welsh List Category	IUCN Red List			Birds of Conservation Concern			
	Global	Europe	GB	UK	Wales		
A	NT	NT	LC	Amber	2002	2010	2016

The Meadow Pipit breeds from southeast Greenland to westernmost Siberia, and south to the latitude of central France (*HBW*). It is a bird of open country, including rough pasture, moorland and bog, preferring denser and longer vegetation than the Skylark. It favours areas with grass tussocks under which it can hide the nest and patches of shorter vegetation for feeding. A mosaic of heather, bog and grassland is thought to be the optimum habitat in the British uplands. In western Europe it is a resident or partial migrant, but is fully migratory elsewhere, some birds reaching Mauritania in West Africa (*HBW*). In Wales, it is most common in the uplands, where it breeds up to at least 950m in the Carneddau, Caernarfonshire. Most birds move off the hills in winter, some wintering locally on lower ground, but ringing provided evidence of a southerly migration for many. Welsh ringing recoveries came from France, Portugal and Spain. Birds ringed as adults on Welsh islands may have been passing through, having fledged farther

Meadow Pipit © Ben Porter

north, but several ringed as nestlings in Wales have also been recovered in these countries. The *Britain and Ireland Winter Atlas 1981–84* suggested that 80% of the British population emigrates in winter, with birds moving in from farther north to replace them. However, the *BTO Migration Atlas* (2002) found no evidence that overseas Meadow Pipits winter in Britain or Ireland, and ringing showed little evidence of birds from Scotland or northern England wintering in Wales. Those passing through Wales in autumn and spring may include birds breeding in Iceland and the Faroe Islands, and perhaps Scandinavia.

Accounts from the late 19th and early 20th centuries described the Meadow Pipit as a common resident in various parts of Wales but gave few details. Forrest (1907) said that it was one of the most abundant species in North Wales and added, "nowhere is it more numerous than in the mountainous parts of Carnarvon and Merioneth". He also said that it "literally swarms in summer" on the Dolwyddelan moors, Caernarfonshire. Breeding numbers were thought to have declined considerably in Glamorgan between 1925 and 1966, as a result of large areas of suitable habitat being lost (Hurford and Lansdown 1995). *Birds in Wales* (1994) suggested that, compared to the early years of the 20th century, the population in Wales might be less than 50% of its earlier level, with the loss of moorland to forestry among the main factors. The *Britain and Ireland Atlas 1968–72* showed the species breeding in almost all the 10-km squares in Wales, though it was absent from much of inland Anglesey. Neither the 1988–91 *Atlas* nor the 2007–11 *Atlas* showed any great change in distribution, and the latter indicated high relative abundance in the Welsh uplands. Atlases at the tetrad level, however, showed declines that were not apparent at the 10-km square level. A marked decline is evident in all the counties where the results of two tetrad atlases can be compared. In Gwent, the species was found with evidence of breeding in 57% of tetrads in the 1981–85 *Atlas*, but in only 47% in 1998–2003. A similar pattern occurred in East Glamorgan, where birds were present in 65% of tetrads in 1984–89, almost all with breeding codes. In the 2008–11 *Atlas*, birds were found in 52% of tetrads, and recorded with breeding codes in only 45%. In Pembrokeshire, the species was found in 45% of tetrads in the 1984–88 *Atlas* and in 39% in 2003–07 (*Pembrokeshire Avifauna* 2020). Elsewhere, in Breconshire, the species was found in 76% of tetrads visited in 1988–90 (Peers and Shrubb 1990). The *North Wales Atlas 2008–12* recorded it in 62% of tetrads, but in a higher percentage in upland counties: 77% in Caernarfonshire and 83% in Meirionnydd. The BBS index for Wales shows a moderate

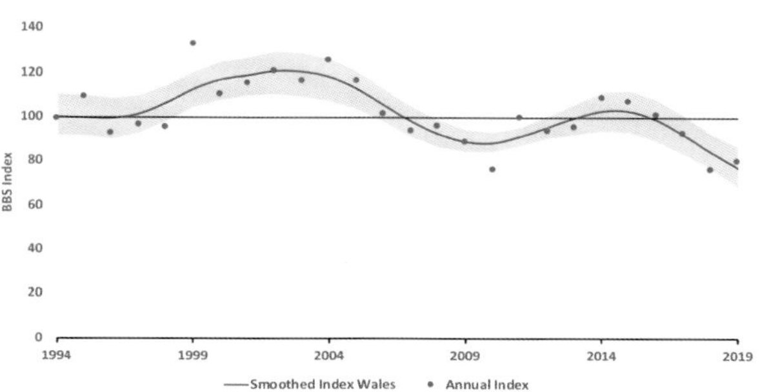

Meadow Pipit: Breeding Bird Survey indices, 1994–2019. The red line shows the smoothed index, 1995–2018. The shaded area indicates the 85% confidence limits.

decline of 15% between 1995 and 2018 (Harris *et al.* 2020).

A survey in 1972–75 found that on sheepwalk at 300–350m altitude at Nant y Benglog, between Llyn Ogwen and Capel Curig, Caernarfonshire, breeding season densities varied between 33 and 74 pairs/km², with an average of 48 pairs/km² (Seel and Walton 1979). Counts across *c.*66km² in Breconshire in 2002 found an average of 11 pairs/km². Surveys at RSPB Lake Vyrnwy, Montgomeryshire, between 2005 and 2013, found numbers increasing until they peaked in 2012, at 71.6 pairs/km² (RSPB Lake Vyrnwy Reserve records unpubl.). These surveys ceased in 2014.

The species does well on the Welsh islands: for example, a minimum of 118 males held territory on Bardsey, Caernarfonshire, (65.9 males/km²) in 2015 and 115 pairs on Skomer, Pembrokeshire, (39.4 pairs/km²) in 2016.

The size of spring and autumn passage varies from year to year but can be very large in autumn. *Birds in Wales* (1994) mentioned a record of 4,000 over Llangynidr, Breconshire, in September, but did not give the year. Many of the largest counts have come from the Gwent coast, including 3,500 moving southwest in 90 minutes on 5 October 1985, 3,400 past Peterstone Wentlooge on 21 September 2013 and 5,243 there on 20 September 2014. In East Glamorgan, there was a westerly movement of *c.*4,000 birds at Lavernock Point on 5 October 1976 and counts at Nash Point included *c.*2,000 on 21 September 1997 and 24 September 2005. About 2,000 passed in several waves over Cynghordy, Carmarthenshire, on 29 September 2010, while autumn passage peaks on the islands included 1,800 over Bardsey on 12 September

Year	2005	2006	2007	2008	2009	2010	2011	2012	2013
Pairs/km²	21.9	20.2	38.8	47.9	50.6	30.8	58.7	71.6	48.4

Meadow Pipit: Average number of pairs of Meadow Pipit per square kilometre at RSPB Lake Vyrnwy, 2005–2013.

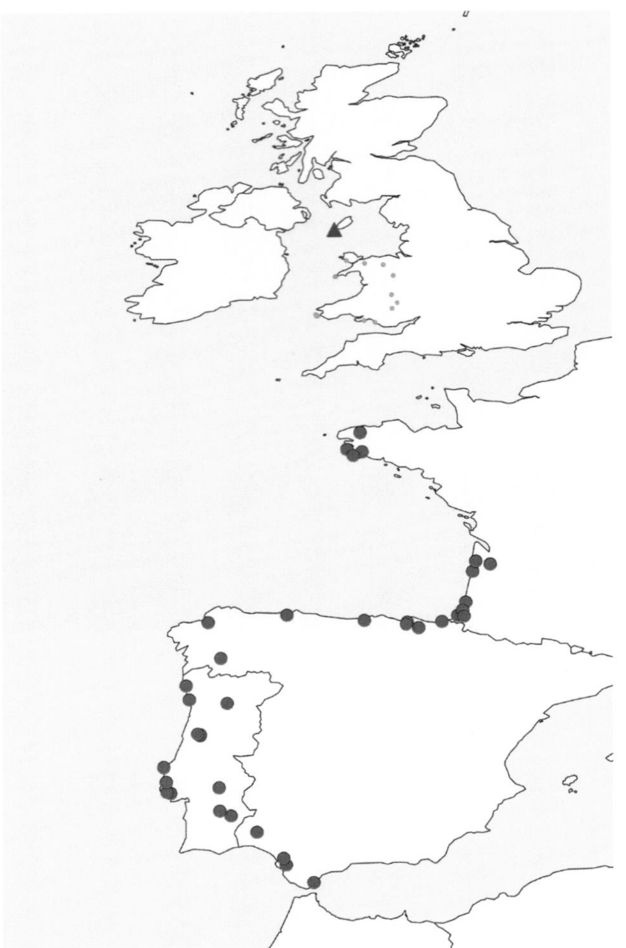

Meadow Pipit: Recovery locations of birds ringed in Wales are shown by red circles. Ringing locations of birds that were recovered in Wales are shown by blue triangles. Small brown dots show ringing or recovery locations in Wales.

1965 and 1,353 past Skokholm, Pembrokeshire, on 21 September 2014. Numbers on spring passage, usually between early March and early May, have not been as high, but *c.*1,000 were counted on Bardsey on 22 March 1995. There can also be winter cold-weather movements, as birds move south or west, notably 1,500 over Bardsey on 1 January 1963.

The Welsh breeding population in 2018 was estimated to be 170,000 (150,000–195,000) territories (Hughes *et al.* 2020). The Meadow Pipit is Amber-listed in Wales because of a moderate breeding population decline in the longer term and because of its Near Threatened status in Europe (Johnstone and Bladwell 2016). A number of factors may have contributed to the decline of the Meadow Pipit in Wales. Afforestation during the 20th century led to considerable areas of suitable habitat being lost, together with changes in agriculture. Reseeded pasture that is intensively grazed by cattle or sheep is unsuitable for this species, leaving fewer invertebrates, no cover for nests and exposing them to trampling. Heavy grazing can make the habitat less suitable for Meadow Pipits so this species could have benefited from a reduction in sheep numbers in Wales in recent years. The Meadow Pipit is of considerable importance for the ecology of the Welsh uplands. It is the main host species for the Cuckoo in Wales and an important prey item for species such as Merlin and Hen Harrier. Climate modelling strongly suggests that the distribution for this species will shift northwards and reduce in extent by the end of the 21st century, and that the uplands of Britain and Ireland may become the southern edge of its range in western Europe (Huntley *et al.* 2007). In addition to this, the increasing incidence of wet summers, which are climate-change induced, could have a negative impact on breeding success and further studies may be required to examine this possibility. Maintaining a healthy population of this species in its upland strongholds in Wales should be a key conservation priority.

Rhion Pritchard

Sponsored by John Lloyd

Tree Pipit — *Anthus trivialis* — Corhedydd y Coed

Welsh List Category	IUCN Red List			Birds of Conservation Concern			
	Global	Europe	GB	UK	Wales		
A	LC	LC	LC	Red	2002	2010	2016

The Tree Pipit breeds across most of temperate Eurasia, as far east as China. It is a long-distance migrant, with the European population wintering in Africa, south of the Sahara (*HBW*). The overall European range of the Tree Pipit has changed very little between the *European Atlas* (1997) and the *European Atlas* (2020), but small losses have occurred in the lowland agricultural landscapes of central and western Europe. The European breeding population of the Tree Pipit has, however, declined by 55% between 1980 and 2016 (PECBMS 2020). It prefers open habitat, with scattered trees or bushes, open woodland and woodland edges (*HBW*). It is unique among European pipits in its affinity for trees and bushes, though it nests and feeds on the ground. In Wales, the Tree Pipit prefers open habitats, but not intensively managed grassland, with scattered trees that are used as song posts. Much of the most suitable habitat, such as *ffridd* and newly planted conifer plantations, is found in the semi-uplands (200–450m). In general, the species is largely absent from the lowlands and less common at higher altitudes. However, it breeds in considerable numbers up to at least 560m at the edge of Beddgelert Forest, Caernarfonshire, (Pritchard 2017) and singing males have been recorded up to 580m at Radnor Forest (Jennings 2014). There have been few ringing recoveries involving Wales. A first-calendar-year bird ringed on Skokholm, Pembrokeshire, in August 1956 was recovered in Portugal in September 1957, and one ringed in Dorset in August 2011 hit a window at Rhandirmwyn, Carmarthenshire, in May 2015.

Forrest (1907) described the Tree Pipit as common in suitable habitat in North Wales, though scarce on Anglesey, where it was largely confined to the area around the Menai Strait, and not numerous on the Llŷn Peninsula, Caernarfonshire. It was said to be common and widely distributed in Glamorgan in the late 1890s (Hurford and Lansdown 1995). Mathew (1894) described it as a common summer visitor to Pembrokeshire and quoted Dix as stating that it was "generally distributed", but by no means numerous in the county. In northern Ceredigion, Salter (1930) considered it to be declining in coastal areas in the late 1920s.

The *Britain and Ireland Atlas 1968–72* showed the species to be widely distributed in mainland Wales, with the exception of a few areas of the southwest and Anglesey. Jones and Whalley (2004) could find no record of proven breeding on the island. The 1988–91 *Atlas* showed losses, particularly in coastal areas of Gwent, Pembrokeshire and Llŷn, but some gains in inland Carmarthenshire. A survey of *ffridd* sites, mainly in central Wales, during 1985–87 (Fuller *et al.* 2006) found the Tree Pipit to be the second most widespread and third most abundant species in 94 of 118 sites, showing that this can be a common species in

Tree Pipit © Tom Wright

suitable habitat. The 2007–11 *Atlas* showed a balance of gains and losses since 1988–91, but losses predominated in coastal Denbighshire and Flintshire. Relative abundance in Wales was among the highest in Britain but was lower, in most areas, than in 1988–91.

Tetrad atlases showed a similar picture. In Gwent, the 1981–85 *Atlas* found the Tree Pipit, with evidence of breeding, in 41% of tetrads, but the second Gwent *Atlas*, in 1998–2003, found it in only 31%. In East Glamorgan, the species occurred in 30% of tetrads in the 1984–89 *Atlas*. In the 2008–11 *Atlas*, it was found in 29% of tetrads, but breeding was only confirmed in 3%, compared to

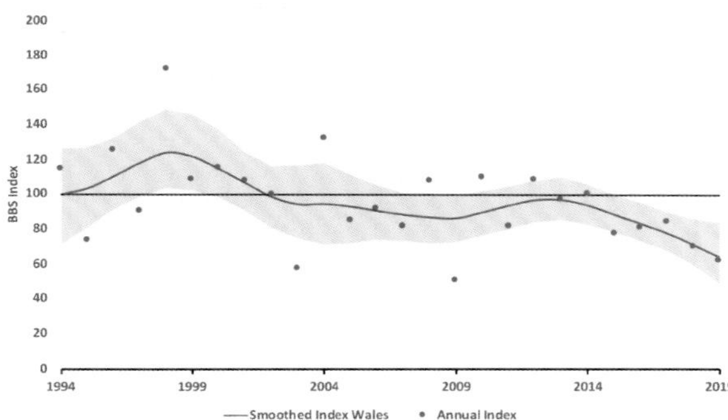

Tree Pipit: Breeding Bird Survey indices, 1994–2019. The red line shows the smoothed index, 1995–2018. The shaded area indicates the 85% confidence limits.

13% in 1984–89. There was a marked decline in Pembrokeshire, where it was recorded in 18% of tetrads in 1984–88 but only in 6% in the second *Atlas*, in 2003–07 (*Pembrokeshire Avifauna* 2020). The species was present in 62% of tetrads visited in Breconshire during fieldwork in 1988–90 (Peers and Shrubb 1990). The *North Wales Atlas 2008–12* found it in 23% of tetrads. There was a great deal of variation between counties, with 47% of tetrads in Meirionnydd having records with breeding codes but only 20% in Caernarfonshire. They were almost absent from Llŷn with no birds recorded with breeding codes on Anglesey.

Despite a small increase in the 1990s, the BBS index showed a 31% decline in Wales between 1995 and 2018 (Harris *et al.* 2020). Over the same period, there was a 56% decline in England, but an increase of 80% in Scotland.

A few birds have been recorded in mid- to late March, with the earliest arrival date being 16 March 1966 on Skokholm, Pembrokeshire, but most arrive around mid-April. The return migration peaks in late August and September, with a few stragglers leaving in mid-October. The latest record since 2000 was at Mynydd Epynt, Breconshire, on 18 October 2001. Singles on Bardsey on 6 November 2011 and at Nash Point, East Glamorgan, on 1–7 December 2016 were exceptional. Numbers recorded on spring migration in Wales have not usually been very large, suggesting that they fly over Wales on their journey north. There

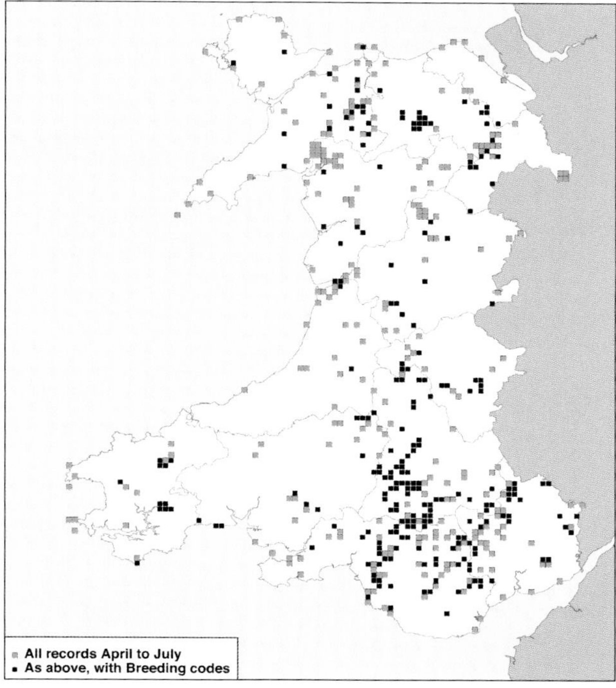

Tree Pipit: Records from BirdTrack for April to July, 2015–19, showing the locations of records and those with breeding codes.

have sometimes been large day counts in autumn, such as 186 at Lavernock Point, East Glamorgan, on 21 August 1971 and 120 south over Bardsey on 17 August 1977, but nothing of this order has been recorded in recent years. There were 44 at Machen, Gwent, on 8 September 2007. A count of 30 on Bardsey on 25 August 2019 was the highest there since August 1990. There have been few large spring totals. For example, the highest spring count on Bardsey is only 28, on 23 April 1998, but there was a record of over 100 sheltering in a gale in the Vron Valley, Radnorshire, on 26 April 1959 (Jennings 2014).

The Welsh breeding population in 2018 was estimated to be 16,500 (10,500–22,500) territories (Hughes *et al.* 2020). The Tree Pipit is Amber-listed in Wales, on the basis of a moderate population decline over the long term. Fuller *et al.* (2005) suggested that the declines could be linked to changes in the age structure of woodlands and a reduction in management of broadleaved woodland, resulting in increased shading and fewer open spaces. Declines could also be linked to rainfall patterns on the Tree Pipit's wintering grounds in the humid zone of West Africa (Ockendon *et al.* 2014), as well as problems encountered on migration, particularly in crossing the Sahara. The Tree Pipit faces a rather uncertain future in Wales.

Rhion Pritchard

Olive-backed Pipit *Anthus hodgsoni* Corhedydd Cefnwyrdd

Welsh List Category	IUCN Red List		Rarity recording
	Global	Europe	
A	LC	LC	WBRC

The Olive-backed Pipit (also known as the 'Indian Tree Pipit') breeds from just west of the Ural Mountains through Siberia and south to China. It winters in South and Southeast Asia (*HBW*). The first British occurrence of this species was in Wales in 1948. Sightings in Britain increased dramatically from 1950, with 610 records by 2018 (White and Kehoe 2020b). This was an average of 13 a year during 1990–2009 but increasing to 31 per year since then. The majority are seen in October. The Olive-backed Pipit, once considered to be a rare vagrant, is now considered to be a scarce migrant in Britain (White and Kehoe 2019). However, it is still a very rare visitor in Wales, where there have been only five records:

- Skokholm, Pembrokeshire, on 14–18 April 1948;
- Skomer, Pembrokeshire, on 24 April 2001;
- Skomer, Pembrokeshire, on 22–23 October 2002;
- Bardsey, Caernarfonshire, on 20 October 2003 and
- St Brides, Pembrokeshire, on 22–25 October 2017.

Jon Green and Robin Sandham

Pechora Pipit *Anthus gustavi* Corhedydd y Pechora

Welsh List Category	IUCN Red List		Rarity recording
	Global	Europe	
A	LC	VU	BBRC

The Pechora Pipit breeds in the scrub-tundra and taiga of subarctic Eurasia, from west of the Ural Mountains to the Pacific Ocean, and winters in the Philippines and central Indonesia (*HBW*). The Pechora Pipit is an extremely rare vagrant to Britain, and most of the 113 records (Holt *et al.* 2020) come from Shetland, the most northerly part of Britain. There has been only one Welsh record, a bird found at Goodwick Moor, Pembrokeshire, on 19–23 November 2007.

Jon Green and Robin Sandham

Red-throated Pipit *Anthus cervinus* Corhedydd Gyddfgoch

Welsh List Category	IUCN Red List		Rarity recording
	Global	Europe	
A	LC	LC	BBRC

The Red-throated Pipit breeds in Eurasia and western Alaska around the Arctic Circle, and winters to the south, birds from Fennoscandia and Russia migrating to the Sahel zone of Africa (*HBW*). Numbers recorded in Britain have fallen since the turn of the 21st century. There were 573 records in Britain to 2019 (Holt *et al.* 2020), but only 14 of these were in Wales.

Total	1970–79	1980–89	1990–99	2000–09	2010–19
14	1	2	4	6	1

Red-throated Pipit: Individuals recorded in Wales by decade since 1970.

Most were autumn birds, with a total of four in September and seven in October. Three were seen in spring, one in April and two in May.

The first Welsh record of this rare migrant was on Skokholm, Pembrokeshire on 13 October 1970. Twelve of these records were from the islands: four each on Skokholm and Skomer, Pembrokeshire, and four on Bardsey, Caernarfonshire. The only mainland records were at Kenfig, East Glamorgan, on 3 May 1992 and a first-calendar-year bird at Colwyn Bay, Denbighshire, on 10 October 2018.

Jon Green and Robin Sandham

Buff-bellied Pipit *Anthus rubescens* Corhedydd Melynllwyd

Welsh List Category	IUCN Red List	Rarity recording
	Global	
A	LC	BBRC

There are two subspecies of Buff-bellied Pipit. The North American subspecies *A.r. rubescens* breeds in the high Arctic of north and northwest Canada and Alaska to western Greenland. It winters in west and south USA, Mexico and Central America (IOC). The Asian subspecies *A.r. japonicus* breeds in central and eastern Siberia, and winters in Southeast Asia and Japan (IOC). This is an extremely rare migrant to Britain, with 48 records to the end of 2019 (Holt *et al.* 2020), all of them being assigned to the *rubescens* subspecies, about 1–2 a year in the last six years since it was added to the British List. The only Welsh record (a 1CY) occurred on Bardsey, Caernarfonshire, on 29–30 September 2019.

Jon Green and Robin Sandham

Water Pipit *Anthus spinoletta* Corhedydd y Dŵr

Welsh List Category	IUCN Red List			Birds of Conservation Concern			
	Global	Europe	GB	UK	Wales		
A	LC	LC	VU	Amber	2002	2010	2016

The Water Pipit was formerly regarded as a subspecies of the Rock Pipit (*A. petrosus*), and interest in its status increased when it was split off as a separate species by the BOU in 1986. The species has an extensive but discontinuous distribution, breeding through central and southern Europe, Turkey, the Caucasus and east to Mongolia and central China (*HBW*). It winters on lower ground in western and southern Europe, as far north as Britain and south as far as North Africa (*HBW*). It breeds on alpine pastures and high mountain meadows, and winters in wet areas, on the coast or inland. Most populations are altitudinal migrants, moving to nearby lowlands in winter, or are short-distance migrants. The largest breeding populations in Europe are in the Alps and the Caucasus (*European Atlas* 2020). The birds found in Wales are presumed to originate from a population that breeds in the Alps or the Pyrenees, which are of the nominate subspecies, *A.s. spinoletta*. One ringed at Shotton Pools, Flintshire, on 7 November 1964, was re-trapped there in October 1965 (Johnson 1970). Ringing recoveries in other countries also suggested that birds may return to the same sites, sometimes for several winters.

Forrest (1907) mentioned several Water Pipits shot by G.H. Caton Haigh on the Caernarfonshire side of the Afon Glaslyn near Porthmadog: singles in April and December 1897 and another in February 1898. The next records were not until the 1930s, when there were singles on Skokholm, Pembrokeshire, in October 1933 and July 1934 (Lockley *et al.* 1949). There was one in Menai Bridge harbour, Anglesey, in February 1950. However, the 1960s saw an increase in records, with the first one for East Glamorgan in April 1961, when three were at Llanishen Reservoir, followed by records in Gwent (1967), Ceredigion (1968), Breconshire (1969) and Meirionnydd (1970). Birch *et al.* (1968) said that it was fairly regular at Shotton Pools, with up to six recorded at a time. There were nine at Abergavenny sewage works, Gwent, in November 1969. Numbers increased further in the 1970s, including five in Breconshire in March 1975.

Most records in Wales have been between October and April, with a suggestion of some passage migration in spring, because most records occur between January and March. Published records suggest a total of about 60 individuals wintering in Wales in recent years. The majority of records occur at coastal locations while records from inland counties such as Breconshire, Montgomeryshire and Radnorshire are rare.

The largest counts were all in the southern half of Wales: at least 11 at Cors Caron, Ceredigion, in November 2004; 13 on the Dyfi Estuary around the Clettwr confluence, Ceredigion, in March 2004; 13 at Aberthaw, East Glamorgan, in February 2012; 14 at Ffrwd Fen, Carmarthenshire, in March 1999; and at least 16 at Rhymney Great Wharf, East Glamorgan, in February 2014. Gwent has regularly had larger counts than any other county, including up to 25 at Sluice Farm, from January to April 2005, 22–35 at Peterstone Wentlooge in early 2006, 30 there on 23 January 2011 and a remarkable count of 50 on 23 March 2019. Numbers in North Wales are lower, with no double-figure counts, although nine were counted at Shotwick Fields and the Welsh section of RSPB Burton Mere Wetlands, Flintshire, on 15 January 2019. There were five at RSPB Cors Ddyga, Anglesey, on 4 December 2018.

The wintering numbers present in Wales are small and no change in the pattern of occurrence is apparent, so there do not appear to be any particular conservation issues at present. However, the area of climatic suitability for Water Pipit in Europe is predicted to move north (Huntley *et al.* 2007), but it is highly limited by its need for mountainous habitat. It nests principally above 1,400m and frequently near glaciers, so peaks in Wales are unlikely to provide an adequate alternative.

Rhion Pritchard

Rock Pipit *Anthus petrosus* Corhedydd y Graig

Welsh List Category	IUCN Red List			Birds of Conservation Concern			
	Global	Europe	GB	UK	Wales		
A	LC	LC	LC	Green	2002	2010	2016

The Rock Pipit has quite a restricted world range, confined to coasts of Europe from northwest France to northwest Russia. Those breeding in Wales are of the nominate subspecies, *A.p. petrosus*, and are resident on rocky coasts and islands only in Britain, Ireland and France, though it will sometimes breed a few hundred metres from the shore. In winter it can be found on a wider variety of coasts, for example feeding on the tideline on sandy beaches. The other subspecies, *A.p. littoralis*, occurs in Wales as a passage or winter visitor and breeds in Fennoscandia (*HBW*). Evans (1966) stated that of over 2,000 *petrosus* ringed on Skokholm, Pembrokeshire, none had produced a long-distance recovery. *Birds in Wales* (1994) stated that no Welsh Rock Pipit was known to have moved even 10km from the ringing site. Colour-ringing on Bardsey, Caernarfonshire, has also shown that most birds are sedentary, but a first-calendar-year bird ringed there in September 2008 was seen 180km away at Walney Island, Cumbria, in February 2015 and 2016. Wales holds the longevity record for Rock Pipit. A first-calendar-year individual was colour-ringed on Bardsey on 26 September 2008, and resighted at the same location on 26 January 2018, nine years, four months and zero days later.

It was said to be a common species along rocky shores in Glamorgan in 1848 (Hurford and Lansdown 1995). Authors such as Mathew (1894) and Forrest (1907) described the species as a common resident on rocky shores in Pembrokeshire and North Wales respectively. There is little evidence of significant population changes, although this is a difficult species to census accurately. The best-monitored populations are on the islands, where there are considerable fluctuations but no clear trends. Bardsey averages around 30 breeding pairs along its 8.2km coastline, although numbers fell to 13 pairs in 1963, following the preceding cold winter, and had risen to 56 by 1990. A minimum of 50 territories in 2014 was the highest total since that year. A new island record was set in spring 2015, with 59 singing males. The population on Skokholm, Pembrokeshire, peaked at 67 pairs in 1959. There were 53 occupied territories there in 1992, 14 in 1999, 32 in 2013 and 61 in 2017. On Skomer, there were 51 territories in 1993, 19 in 2016 and 27 in 2017. Ramsey, Pembrokeshire, usually has 20–30 territories, although there were 48 territories in 2002.

The *Britain and Ireland Atlas 2007–11* found the Rock Pipit breeding all along the Welsh coast, except for Flintshire and eastern Denbighshire, which are without rocks or cliffs. Compared to the 1988–91 *Atlas*, there were gains along the coasts of Gwent, East Glamorgan, Gower, Pembrokeshire and Caernarfonshire. In counties where two tetrad atlases could be compared, such as East Glamorgan and Pembrokeshire, the species occupied virtually the same proportion of tetrads in the first and subsequent atlases. In Gwent, however, the species was not recorded at all in the 1981–85 *Atlas*, but by 1998–2003 a small breeding population, estimated by Venables *et al.* (2008) at 2–3 pairs, had become established, including on Denny Island. Elsewhere, the *North Wales Atlas 2008–12* found the species in 7% of tetrads, mainly along the coasts of Anglesey and Llŷn.

In areas where there is no suitable natural habitat, Rock Pipits will sometimes use human-made structures such as sea walls for nesting.

The peak counts on the Welsh islands have usually been in autumn, but most of the birds recorded may be the resident breeding population and their young, rather than migrants. The highest count made in Wales was 165 on Skokholm on 21 September 2014. The highest count on Bardsey outside the breeding season was 70 on a beach in October 1988. Inland records are more scarce, but birds have been recorded around the shores of large lakes and reservoirs in several counties outside the breeding season. Peers and Shrubb (1990) found 18 records at Llangorse Lake, Breconshire, since 1969.

The available data suggested an average of 7–10 pairs per tetrad in suitable habitat. The largest population is probably in Pembrokeshire, where there were estimated to be around 900 pairs in 2003–07 (Rees *et al.* 2009). The Welsh population could be about 2,000 pairs, but a dedicated survey of this species would be needed to obtain a reliable estimate. Future coastal developments might create local concerns in some areas, but currently there are no known threats to what appears to be a fairly stable population in Wales.

Scandinavian Rock Pipit *A.p. littoralis* Corhedydd y Graig Llychlyn

While the nominate subspecies is largely resident, northern and eastern populations of *littoralis* are migratory, breeding in Fennoscandia and wintering in western Europe and as far south as Morocco (*HBW*). Identifying *littoralis* is fairly easy in spring, when birds are coming into breeding plumage. In autumn and winter, many Rock Pipits cannot be reliably identified to subspecies. Svensson (1992) considered that *littoralis* is "very similar to nominate, and apparently inseparable in autumn". In England, birds of this subspecies are said to winter commonly on soft coastlines, including sand dunes, harbours and, particularly, saltmarshes (*Birds in England* 2005). British Rock Pipits are not commonly found in these habitats and it may be that Welsh birds, wintering on saltmarshes in areas well away from suitable breeding habitat, are likely to be *littoralis*.

Birds showing characteristics of this subspecies have been recorded fairly regularly in spring and autumn since the 1990s. A nestling ringed in Sweden in June 1983 was hit by a car at Rhosneigr, Anglesey, in October the same year, and one on Bardsey in March 2006 had been ringed in Finland. Spring passage is between March and May, usually peaking in April, while autumn passage is from September to November. There appear to be influxes in some years, notably in autumn 2004 when 20, thought to be of this subspecies, were on Bardsey on 12 October. Some of these birds were ringed and subsequently over-wintered on the island, eventually moulting and proving to be indeed *littoralis*.

An average of about seven individuals a year, recorded in Wales over the 15 years since 2004, probably did not represent the real numbers of this subspecies found here. Identification difficulties in autumn and winter, and a lack of interest on the part of many birdwatchers, mean that the true status of *littoralis* in Wales is uncertain.

Rhion Pritchard

Chaffinch *Fringilla coelebs* **Ji-binc**

Welsh List Category	IUCN Red List			Birds of Conservation Concern			
	Global	Europe	GB	UK	Wales		
A	LC	LC	LC	Green	2002	2010	2016

The Chaffinch is the most common breeding finch, and the fifth most common breeding passerine, in Wales (Hughes *et al.* 2020). It breeds across Europe (except for the very far north), southwest Siberia and in North Africa (*HBW*). The birds breeding in Wales are of the subspecies *F.c. gengleri*, which is restricted to Britain and Ireland. It is generally smaller and more brightly coloured than the nominate *F.c. coelebs* that breeds over most of the rest of Europe (IOC). Most populations are resident, but northern populations of *F.c. coelebs* are strongly migratory, moving south and southwest in autumn (*HBW*).

In Wales, the Chaffinch is found throughout the year in a wide variety of locations from sea level, including some of the offshore islands, to altitudes of about 600m. These include both rural and urban habitats, generally with some tree cover. In winter, adult Chaffinches feed almost exclusively on seeds, taken on the ground, but in the breeding season they forage for invertebrates to sustain their young.

The Chaffinch has been common throughout Wales for at least 150 years. In the late 19th century, it was described as the commonest bird species in Breconshire (Phillips 1899), one of the commonest in Glamorgan in the late 1890s (Hurford and Lansdown 1995) and common in Pembrokeshire (Matthew 1894). Forrest (1907) described it as abundant in North Wales, but noted that there were comparatively few on Anglesey. It is now common and widespread on the island. Jones and Whalley (2004) attributed the increase by the 1940s to "changes linked to the alterations in farming practices on the island", with more arable land (and seed) available to the birds during the Second World War. Elsewhere in Wales, there appears to have been little change in distribution or numbers during the 20th century. The severe winter of 1962/63 reduced numbers. Hurford and Lansdown (1995) stated that in Glamorgan numbers had still not fully recovered by 1966. The use of organochlorine seed dressings on farmland also caused a decline in Britain in the 1960s, although it is uncertain to what extent the Chaffinch population of Wales was affected. Afforestation of the uplands, following the Second World War, allowed it to spread into areas where there had previously been no suitable nest sites.

The three breeding atlases of Britain and Ireland, in 1968–72, 1988–91 and 2007–11, showed Chaffinch to have occurred in almost every 10-km square in Wales. Relative abundance in 2007–11 was highest in Pembrokeshire, and showed little change since the 1988–91 *Atlas*, although there were increases in a few 10-km squares. Local atlases recorded it in over 90% of tetrads in most counties surveyed, for example, in 99% of tetrads in Gwent in both 1981–85 and 1998–2003.

Finch Trichomonosis, caused by the protozoal parasite *Trichomonas gallinae*, was first recognised in Britain in 2005. There was significant mortality of both Chaffinch and Greenfinch in 2006, with Wales among the parts of Britain most affected (Robinson *et al.* 2010). Chaffinches make much use of garden feeders, which are thought to have facilitated the spread of the disease. Although initially Chaffinch populations were regionally diminished by the epidemic, Lawson *et al.* (2012b) stated that this was not apparent in subsequent years, at least, to about 2010.

The BBS index for Wales declined by 38% during 1995–2018 (Harris *et al.* 2020), but had been relatively stable until about

Female Chaffinch © Ben Porter

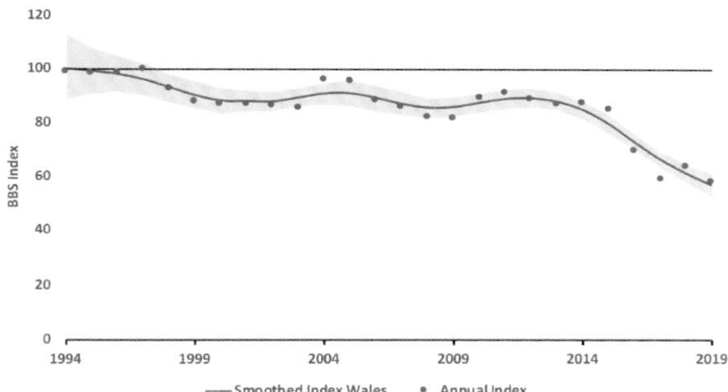

Chaffinch: Breeding Bird Survey indices, 1994–2019. The red line shows the smoothed index, 1995–2018. The shaded area indicates the 85% confidence limits.

2015, after which there was a steep decline. Similar declines have been recorded in England and Scotland, but are not mirrored across Europe, where the population has been stable since 1980 (PECBMS).

Breeding densities can be very high in good habitat: in an 18.8ha oakwood at RSPB Ynys-hir, Ceredigion, they ranged from 15 to 31 territories between 1995 and 2000, densities of 123 to 255 territories/km² (Roderick and Davis 2010). An area of 112ha of woodland at RSPB Ynys-hir held 119 territories/km² in 2006 and 87 territories/km² in 2011. A Norway Spruce plantation at 350m in Cwm Einion, Ceredigion, held 190 territories/km² in 1984–87 (Roderick and Davis 2010).

Most Chaffinches breeding in Wales, regardless of age and sex, do not move very far, although there may be some limited post-breeding juvenile dispersal up to 50km. There is some local movement, generally less than 10km. This is often from higher ground to lower elevations, usually in response to fluctuating food availability and weather conditions (*BTO Migration Atlas* 2002).

The flocks passing through Wales in spring and autumn are likely to include both birds breeding further north in Britain and birds of the nominate subspecies *F.c. coelebs* breeding in continental Europe. Spring passage, in March and early April, has been noted at most coastal locations in Wales but usually only in small numbers. Autumn passage is on a much larger scale. Lockley *et al.* (1949) stated that "tens of thousands" were seen on the northern and western coasts of Pembrokeshire in October, coming over Cardigan Bay from the northwest. Many thousands passed Strumble Head in autumn 1986. Many more were heard that were too high to see, but considerably fewer have occurred in Pembrokeshire since the 1980s (Rees (2010b).

Bardsey, Caernarfonshire, regularly records hundreds of Chaffinches migrating past the island in October: an estimated 33,000 moved south between 19 and 31 October 1966, including 20,000 on 19 October. 9,349 headed south there on 31 October 2015 and 10,000 passed Llanon/Llanrhystud, Ceredigion, the same day. Large numbers are also recorded moving west along the north coast of Wales. 6,500 passed the Great Orme, Caernarfonshire, on 4 November 2007. That autumn also saw counts of 6,950 over Carmel Head, Anglesey, on 21 October and over 6,000 at Uwchmynydd, Caernarfonshire, on 29 October. The largest counts on the south coast were *c.*8,500 on 18 October 1985 and over 7,000 on 2 November 2015 at Lavernock Point, East Glamorgan.

There has not been enough information from ringing recoveries to estimate what proportion of the birds passing through Wales in autumn originated from outside Britain and how many from farther north in Britain. Equally, we do not know where the majority of these birds spent the winter. The authors of *Birds in Wales* (1994) stated that many moved on to Ireland, and that they could on occasions be found in "immense numbers" on headlands in Pembrokeshire, Llŷn and Anglesey, awaiting the right conditions to make the sea crossing to Ireland. Although only a few birds ringed in Wales have been recovered in Ireland, and most movements reported from Bardsey have been of birds moving south, rather than out to sea towards Ireland, there are other indications of such movements. In 2019, visible migration was observed from South Stack, Anglesey. On 20 October, 770 birds were seen moving south and west, with another 614 moving in the same direction on 29 October.

Fieldwork for the *Britain and Ireland Atlas 2007–11* found that relative abundance in Wales in winter was highest in Pembrokeshire. Winter flocks can reach several hundred birds, and there have been a few records of larger numbers. There were 3,000 at Hubberston, Pembrokeshire, on 5 January 2006 and

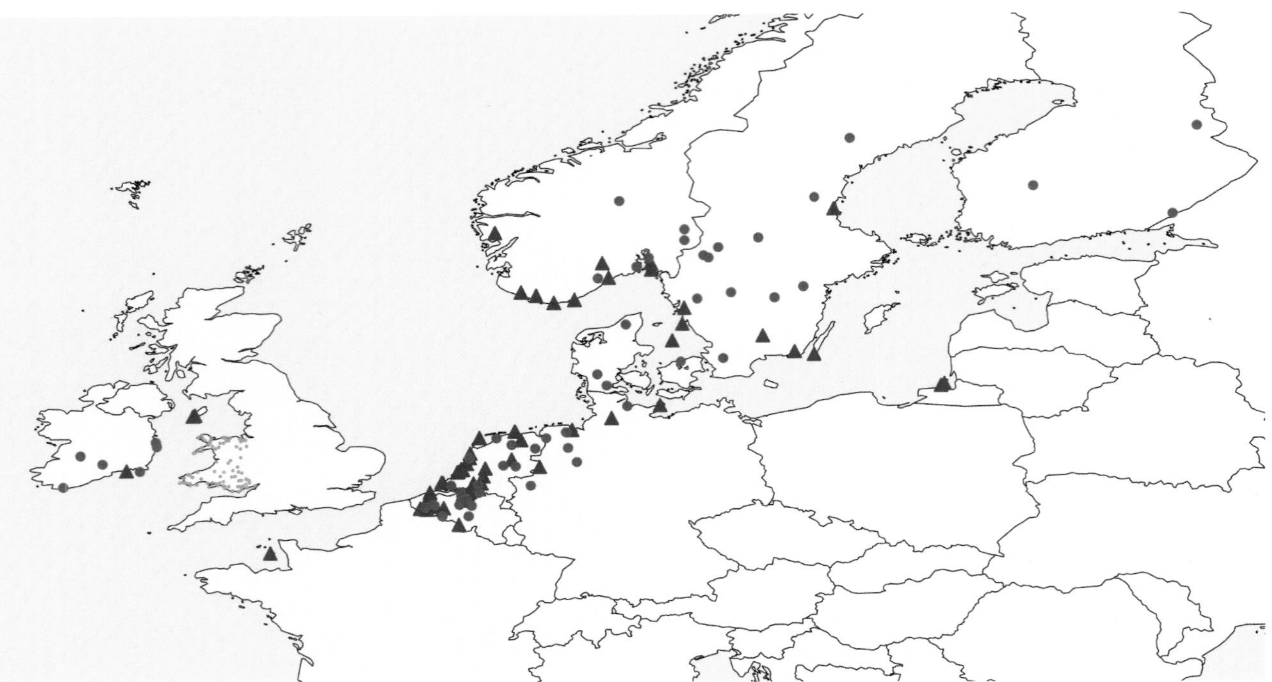

Chaffinch: Recovery locations of birds ringed in Wales are shown by red circles. Ringing locations of birds that were recovered in Wales are shown by blue triangles. Small brown dots show ringing or recovery locations in Wales.

3,000 at Lampeter Vale, Pembrokeshire, on 13 January 2009. Inland, there were 2,500 at Llyn Heilyn, Radnorshire, in December 1978 (Jennings 2014) and an exceptional flock of 5,000 feeding in root crops at Heol Senni, Breconshire, in January 1980 (Peers and Shrubb 1990). BTO Garden Birdwatch data showed that the peak use of gardens in Wales is between November and March, with a moderate decline in both peak counts and the reporting rate since 2014.

The recent decline in the breeding population could qualify Chaffinch for Amber-listing in Wales in the next assessment of its conservation status. It is thought that decreases in adult survival have caused the downturn, but the reasons for this are unclear. The evidence for attributing the decreases to Trichomonosis is less strong for Chaffinch than for Greenfinch (Massimino *et al.* 2019). More research is needed into the factors responsible for a worrying decline. Despite this concern, the Welsh breeding population in 2018 was estimated to be 470,000 territories (Hughes *et al.* 2020) and the Chaffinch is still one of our most widespread and common birds. If the recent decline does not slow, however, this situation could easily change for the worse.

Continental Chaffinch *F.c. coelebs* Ji-binc y Cyfandir

Large numbers of Chaffinches pass through Wales in autumn, while others move in to spend the winter here. Ringing recoveries showed that autumn movements and winter flocks include birds breeding in Europe, and therefore of the nominate subspecies *F.c. coelebs.* Records of this subspecies are assessed by WBRC. These birds come mainly from Scandinavia with a few from Finland and Russia. There have also been recoveries in Wales of birds ringed in Belgium and The Netherlands. These were mainly ringed in October and may have been birds breeding in Scandinavia. Most Scandinavian Chaffinches migrating to Britain for the winter follow a route through Denmark and northern Germany to The Netherlands, Belgium and northeast France, before crossing the English Channel (*BTO Migration Atlas* 2002). Some birds originate from farther east: three ringed in Wales have been recovered in Finland, including one ringed at Llangernyw, Denbighshire, in December 1975, and killed by a cat near Oulu in northern Finland, 2,236km away, in early September 1978. Two ringed in Russia have been recovered in Wales: one ringed near Kaliningrad in September 1958 was caught by a ringer at Margam, Gower, in January 1962. More recently, one ringed in Rybachy on 1 October 2019 was caught at Bagillt, Flintshire, on 3 November the same year, having covered 1,658km in 33 days (Giles Pepler *in litt.*).

Whilst it is clear that this subspecies is found in Wales in winter, the proportion of these birds in the Welsh winter population is not known. The *Britain and Ireland Winter Atlas 1981–84* estimated that, of 30 million Chaffinches present in midwinter, 10–20 million were immigrants. Newton (1967b) found that, in Oxfordshire, continental Chaffinches behaved differently from local breeding birds, which tended to remain near their breeding areas in winter and mostly fed in small groups. Continental birds were found in large flocks in open fields and formed large communal roosts at night. However, large feeding and roosting flocks in northern Scotland were found to be mainly local birds with very few, if any, continental birds present (Swann 1988). In Wales, Lloyd (1929–1939) considered that small groups in Pembrokeshire were local birds and larger flocks in open country were from the continent.

Nigel Brown and Rhion Pritchard

Sponsored by Ian and Jane Hemming

Brambling *Fringilla montifringilla* Pinc y Mynydd

Welsh List Category	IUCN Red List			Birds of Conservation Concern			
	Global	Europe	GB	UK	Wales		
A	LC	LC	LC	Green	2002	2010	2016

The range of the Brambling extends in a broad band across Fennoscandia and Russia, below the Arctic Circle, to the Pacific, with occasional breeding in Scotland. Though a few birds are resident in Scandinavia, most of the population is migratory, wintering to the south of the breeding range. Birds from Scandinavia are thought to winter mainly in Britain, Belgium and France (*HBW*). In years of high population, very large numbers may be recorded farther south in Europe, such as a flock of about five million birds in Slovenia, in winter 2018/19 (Zagoršek 2019). Birds concentrate where there is a widespread, rich beechmast crop, keeping movements to the minimum, until the time that severe weather forces migration further south and west (Jenni 1987). Ringing recoveries involving Wales were of birds on migration or at their wintering sites, including some birds ringed in Wales that were recovered in Norway in May. Ringing recoveries elsewhere showed that while many birds return to the same area in subsequent winters, others will winter in different areas. Some birds ringed in Britain or Belgium in one winter were recovered in Turkey and the Balkans in a later winter (Newton 2007). A female ringed on Bardsey, Caernarfonshire, in October 1965 and recovered at Modena, Italy, in October 1967, is another such an example.

A very early autumn bird was recorded on Bardsey on 22 August 1993, but Bramblings usually arrive in mid-October. Most have departed by mid-April, but one or two have lingered into early May in some years. One was reported in the *Cambrian Bird Report* to have summered at Conwy, Caernarfonshire, in 1953. There was a singing male on Ruabon Moor, Denbighshire, on 21 June 1993, and one was still visiting a feeder at Pontsaeson, Ceredigion, on 13 June 2006. A number of these late birds were recorded singing, but there has never been any evidence of breeding.

There is scant indication of the Brambling's status in Wales during the 19th century. Williams (1830) recorded the species in the Llanrwst, Denbighshire, area. The information given to Forrest (1907) by correspondents in North Wales also suggested that it was generally fairly scarce, with flocks in some winters, and at Palè, Meirionnydd, flocks almost every winter. Tracy (1850) said that it was very common in Pembrokeshire in some winters, but Mathew (1894) considered it to be a rare bird, at least in the north of the county. In the late 1890s, large flocks were seen in Glamorgan, during severe winter weather (Hurford and Lansdown 1995). Phillips (1899) stated that it was not common in Breconshire, although winter flocks sometimes occurred. In the 20th century, this species was recorded in Glamorgan in fewer than 20 winters between 1900 and 1955, but after the winter of 1961/62 it was recorded annually (Hurford and Lansdown 1995). Lockley *et al.* (1949) noted that although it was seldom recorded in Pembrokeshire before 1930, by 1949 it was a regular winter visitor. The *Britain and Ireland Winter Atlas 1981–84* found it to be widely distributed in the lowlands of Wales, but absent from higher ground. The 2007–11 *Atlas* showed some gains in semi-upland areas.

The species is now recorded in Wales every winter, but numbers vary greatly from year to year, and no trend is detectable. Extensive snow cover farther north and east in Europe makes beechmast inaccessible and forces movements to the south and southwest (Jenni 1987). Prior to 2007, the highest counts recorded in Wales were 1,500 in a weedy failed swede crop near Rhayader, Radnorshire, on 7 December 1994, and 1,500 at Pwllpeiran, Ceredigion, on 27 December 1995. The 1995 count was part of an influx that also included over 800 roosting at Llyn Fanod, Ceredigion, in December 1995, 800 at Dinefwr Park, Carmarthenshire, on 7 January 1996 and 800 feeding on beechmast at Gregynog Hall, Montgomeryshire, on 16 March 1996. Numbers were much lower in the northern counties that winter; the highest count was 100 at Ogwen Bank, Caernarfonshire, in February 1996.

Male Brambling © Bob Garrett

Brambling: Recovery locations of birds ringed in Wales are shown by red circles. Ringing locations of birds that were recovered in Wales are shown by blue triangles. Small brown dots show ringing or recovery locations in Wales.

Winter 2007/08 saw the largest flocks ever recorded in Wales. Numbers were highest in Gwent, where there were flocks of 1,000 at several sites, including 1,500–2,000 at Hafodyrynys on 14 January 2008, over 2,000 in Carno Forest on the Breconshire border on 19 December 2007 and a remarkable 10,000 on Mynydd Manmoel on 18 December 2007. That winter also saw *c.*1,000 Bramblings at Coed-y-Wenallt and Fforest Ganol, East Glamorgan, on 11 December. In Ceredigion, there were *c.*2,000 west of Pontrhydygroes on 25 November 2007 and 4,000 at The Arch, Ystwyth Forest, on 12 January 2008. Numbers were again much lower farther north. Since that winter, the highest counts have been 600 at Moel Garegog/Llandegla Forest, Denbighshire, on 9 April 2011, 1,200 at Caemeirch, Ceredigion, on 1 December 2013 and 1,000 feeding on beechmast at Buckland Wood, Bwlch, Breconshire, on 8 November 2018. Ceredigion is the only western county to have recorded such large numbers in winter, but there have been good passage counts elsewhere, such as a fall of 800 on Skokholm, Pembrokeshire, on 22 October 1966 and 500 over Bardsey on 30 October 1992. In winter 2018/19, the Mid Wales Ringing Group ringed nearly 1,500 Brambling at three main locations: one at Pontrhydygroes, Ceredigion, and two at Dolfra, Montgomeryshire (Cross 2019).

The numbers of Brambling recorded in Wales depend largely on the availability of beechmast farther north and east. Increasingly frequent milder winters could reduce snow cover in these areas and make beechmast more accessible there, resulting in fewer large influxes into Wales.

Rhion Pritchard

Hawfinch *Coccothraustes coccothraustes* Gylfinbraff

Welsh List Category	IUCN Red List			Birds of Conservation Concern				Rarity recording
	Global	Europe	GB	UK	Wales			
A	LC	LC	EN	Red	2002	2010	2016	RBBP

The Hawfinch was, until recently, one of the least known and most under-recorded species in Wales. It breeds over most of Europe, apart from Ireland and northern and central Fennoscandia, in northwest Africa and across temperate Asia. Eastern populations are largely migratory, while some birds in central and western Europe move west or south in winter (HBW). It is found mainly in areas of extensive woodland with a high proportion of broadleaved trees, where it feeds on the large fruits of species such as Hawthorn, Holly, Hornbeam, Lime, Rowan, Wild Cherry and Yew. Other large trees such as Ash, Beech, Field Maple and Sycamore are also exploited, and elms are an important food source in spring. It can also be found in parkland, including the grounds of country houses such as Powis Castle, Montgomeryshire, which can be occupied over many years. Churchyards, with their ancient Yew trees, are also frequented.

Alexander and Lack (1944) suggested that the species only began to breed in England in the early 19th century, although records of damage to pea crops in East Anglia, some 200 years earlier, may indicate an earlier date. It was breeding in southeast England in the 1820s and 1830s, then apparently extended its

Male Hawfinch © John Hawkins

range and increased in numbers throughout the remainder of the 19th century (*Birds in England* 2005). The earliest mention of the species in Wales was in 1802, when it was included, without further comment, in a list of the birds of the area in the *Swansea Guide*. Dillwyn (1848) said that he had "grounds for suspecting" that one had been killed at Stouthall, Reynoldston, Gower, half a century previously. Possibly both Dillwyn and the *Swansea Guide* were referring to the same record. Mathew (1894) mentioned one at Llanstinan [Llanstinian], Pembrokeshire, in the spring of 1854 and two were shot at Ynys-hir, Ceredigion, about 1868 (Roderick and Davis 2010). Phillips (1899) mentioned a large flock at Tregunter, Breconshire, during a severe winter about 1864. By the late 19th century and early 20th century, the species was a scarce breeder in parts of Wales. Breeding was confirmed at Llandaff, East Glamorgan, in 1899, and it was said to have bred on occasions in this period at Merthyr Mawr (Hurford and Lansdown 1995). Phillips (1899) stated that the species was increasing in Breconshire, and had bred near Brecon some years previously, as well as near Crickhowell. A party of four above Llanelwedd, Radnorshire, in June 1910 was probably a family group (Jennings 2014).

Forrest (1907) said that, a quarter of a century previously, the Hawfinch had been practically unknown in North Wales. By 1907 it was fairly common in the Severn Valley in Montgomeryshire and in the Vale of Clwyd in Denbighshire, and "pretty well distributed" throughout the eastern half of North Wales, including eastern Meirionnydd. Forrest considered that it was gradually increasing and extending its range westward. In 1919, Forrest noted that the westward expansion was continuing, and that it then nested regularly around Bangor, Caernarfonshire, although it was still unknown on Llŷn and there were only a few records from Anglesey. The westward spread may have continued during the 1920s and 1930s. Two pairs were said to have bred near Pumpsaint, Carmarthenshire, in 1926 (Ingram and Salmon 1954). The only confirmed record of breeding on Anglesey was at Porthamel in 1930 (Jones and Whalley 2004). A pair seen in May 1938 at Nanteos, Ceredigion, was suspected of breeding (Roderick and Davis 2010). Nests were said to have been robbed annually by egg-collectors in the Abergele area of Denbighshire from the 1930s to the 1950s. Eggs were also said to have been collected regularly near Bala, Meirionnydd, in the 1950s (*Birds of Wales* 1994). There were surprisingly few early records from Gwent, where Ingram and Salmon (1939) described it as a local breeder in the central and southern parts of the county.

The first overall picture of the distribution of the Hawfinch in Wales was given by the *Britain and Ireland Atlas 1968–72*, which showed 25 10-km squares were occupied in Wales, mostly in the eastern half of the country. The 1988–91 *Atlas* showed 20 occupied 10-km squares, with gains particularly evident in Gwent and losses in Breconshire and Montgomeryshire. The *Britain and Ireland Atlas 2007–11* also showed local gains and losses in Wales, but the number of occupied 10-km squares was almost identical to 1988–91. There were widespread losses in the former strongholds in southeast England, contributing to a 76% reduction in the number of occupied 10-km squares in Britain between 1968–72 and 2007–11, most since 1988–91. The species was least likely to be lost from 10-km squares where broadleaved woodland cover was high (Kirby *et al.* 2015). Many seed-eating bird species declined over this period, but the factors specifically affecting Hawfinches remain unclear. The loss of English Elm from much of eastern Britain in the 1970s may have contributed to the decline there, but with its more westerly distribution, Wych Elm will have helped to sustain the Welsh population of Hawfinch.

Gwent is a stronghold for this species, with birds present in 22 tetrads and breeding confirmed in seven of these, during fieldwork for the *Gwent Atlas 1981–85*. It was widely distributed in the Wye Valley and there was a small colony near Usk. The *Gwent Atlas 1998–2003* found birds present in 18 tetrads and confirmed breeding in only two, with losses in central Gwent including the Usk colony, which was attributed to management work removing cherry trees (pers. obs.). Breeding is now recorded almost annually in the Wye Valley, usually with 4–10 sites or nests/year, although 21 nests were found in 2014, by radio-tracking females back to their nests. Elsewhere in the county, breeding has been recorded in small numbers at Slade or Minnetts Woods and Dingestow. Three feeding sites were established in the Wye Valley in 2008, and up to 2019, 1,085 individuals were ringed, with an additional 510 birds (part of the same population) ringed in the adjacent Forest of Dean, much of which is in Gloucestershire (Jerry Lewis and Adrian Thomas own data). It was estimated, from the ratio of colour-ringed to unringed birds, that 60–80 were visiting one supplementary feeding area on 2 May 2016 (Martin Peacock and Dave Potter *in litt.*). A Re-trapping Adults for Survival project generated an estimate for the whole Wye Valley and Forest of Dean population of 650 pairs (Lewis 2018), of which some 250–300 pairs are likely to be in Gwent.

Elsewhere in South Wales, a family group at Wenvoe in 2009 was the first confirmation of breeding in East Glamorgan in the 21st century. In 2019, a pair of Hawfinch bred in Nolton Churchyard, Bridgend. Previous records included two adults and a juvenile seen at Castell Coch in 1997. A female was recorded nest building in Cardiff in 1999 and there were other records of pairs, including two pairs at Tongwynlais in 2003. Successful breeding was reported from three sites in Gower in 2017. In Carmarthenshire, records since 1970 have been concentrated in the Tywi Valley (Lloyd *et al.* 2015). One, at Pen-y-banc, Llandeilo, in 1995, attended a Greenfinch nest and assisted in feeding the chicks. Pairs with juveniles were recorded at Cynghordy in 1996 and 2013, and a pair was at Pen-y-banc in 1999. A successful pair at RSPB Dinas reserve in 2019 was the first breeding record for this site (Rob Hunt *in litt.*). Breeding has never been confirmed in Pembrokeshire.

In mid-Wales, three fledglings seen in the north of Breconshire in 2007 was the first breeding record since 1979, and there were two pairs at The Cwm, Llanwrthwl, in May–June 2017. In Radnorshire, there were three pairs, in the Elan Valley, Presteigne and Penybont, in 2008, five pairs bred in 2009, two family groups were present in 2010, and breeding was confirmed in 2012. Jennings (2014) estimated that at least 10–15 pairs bred in Radnorshire. There have been no recent records of confirmed breeding in Ceredigion or Montgomeryshire, although it was suggested in the 2011 *Ceredigion Bird Report* that small numbers may have bred in the Devil's Bridge area. In Montgomeryshire, a pair was reported at Guilsfield in 1996. The 2017 *Montgomeryshire Bird Report* suggested breeding was possible, as birds had been sighted in suitable habitat at Dyfnant Forest on 3 June and, for the third consecutive year, one had visited a Dolanog garden on 4–16 June.

Meirionnydd holds the largest breeding population in North Wales. Monitoring between 1999 and 2003 identified 25–30 territories in the Dolgellau area, with an additional six pairs at another wood in the Mawddach valley, from which Smith (2004) estimated a population of *c.*50 pairs. Twelve nests were found by radio-tracking females there in 2012. Birds have been reported during the breeding season at several other sites. There may be a small population in the Maentwrog area, where an adult and juvenile were seen near Pont Croesor, on the border with Caernarfonshire, in 2004. In Caernarfonshire, there may be a small population in the Conwy Valley, where two recently fledged juveniles were at Llanbedr-y-cennin in June 2000 and a pair were at a nest nearby in May 2003. On Anglesey, possible breeding occurred in 2003, which was also the first record for the county since 1990. There has been no recent evidence of breeding in Denbighshire or Flintshire. These counties, in the early 20th century, were the strongholds of this species in North Wales.

The decline of *c.*40% in the British breeding population, between the mid-1980s and the late 1990s (Langston *et al.* 2002), was not evident in Wales, but the increased numbers now recorded in Wales are due to dedicated monitoring, rather than to an increased population. Smith (2004) considered the Meirionnydd population around Dolgellau to have been previously overlooked, and increased numbers in the Wye Valley, Gwent, are believed to be the result of more interest in monitoring the species (Jerry Lewis and Adrian Thomas *in litt.*). Ringing studies at feeding stations in Gwent, and subsequently in Meirionnydd, East Glamorgan and Caernarfonshire, have started to provide a greater insight into the species' true status.

The Wye Valley and Dolgellau area are now the strongholds for Hawfinch in Britain. 70–80% of the numbers ringed in Britain during the last decade have been in Wales. Recoveries show that Hawfinches are highly mobile in late winter and early spring. 'Gwent' birds range throughout the whole of the Wye Valley and Forest of Dean. Some movements are associated with poor mast years of Beech and Hornbeam but are still not fully understood (own data). Their mobility is confirmed by radio-tracking and GPS studies in the Wye Valley and Meirionnydd. One female visited a feeding site while nesting some 5km away and another ranged over 10km^2 during five days in late July (Will Kirby *in litt.*). Movements between feeding sites in Gwent, Meirionnydd and Fforest Ganol, East Glamorgan, are regular. However, some individuals caught in Dolgellau had previously been ringed in Somerset and Cumbria. The longevity record for Hawfinch in Britain is held by a Wye Valley bird, ringed in May 2008 and recaught in May 2018, ten years and four days later. Birds older than seven years are regularly caught, and it is not unusual for many years to elapse between captures, illustrating the species' wide-ranging habits.

Several foreign exchanges have also been recorded since 2010, principally with southern Norway: five to or from the Dolgellau area, one from the Conwy Valley and one to the Wye Valley and Forest of Dean. Only one Swedish ringed bird has been found in Britain, also caught in the Wye Valley and Forest of Dean (Robinson *et al.* 2020). All these were caught in Wales between 15 March and 19 April. The presence of Scandinavian birds in Wales during the winter is probably annual, although numbers vary.

Hawfinches are more widespread in Wales outside the breeding season and have been recorded in all counties in Wales in recent years. There have been regular records from Fforest Ganol, where there were 25 in January 2014, and up to 20 in January–February 2015. A ringing study commenced there in 2018, and in two years, 61 were ringed (Richard Facey *in litt.*). In Breconshire, up to 14 fed in Hornbeam trees on the outskirts of Brecon in April 2013. Birds were present at 11 sites in Radnorshire in 2014, mainly in autumn and winter. Birds were seen at Powis Castle, Montgomeryshire, including up to 15 in January–February 2015 and 15 in December 2016. At a nearby feeding site, 26 were ringed in 2018, following an influx in 2017, but all had left the site by mid-April and no ringed birds have been seen locally since (Tony Cross and Paul Roughley *in litt.*).

The importance of the Dolgellau area of Meirionnydd was first appreciated from studying roost sites, with two near Dolgellau, and three more near Afon Mawddach in December 1996. Counts included 40 at each of two roosts in December 1999, 63 and 17 in two roosts in August 2000 and 61 in August 2001. There were 56 at a roost in Coed y Brenin in January 2001, 52 there in October 2001 and 61 in December 2002. More recently the Coed y Brenin roost count was 69 on 30 November and 65 on 4 December 2019. Churchyards in the Dolgellau area yielded regular counts, including over 30 in April 2002, 33 in January 2011 and 30 in December 2013. A ringing study was started at two sites in 2011, and by 2019, 1,034 individuals had been ringed, with over 6,000 resightings of colour rings (Tony Cross and Dave Smith *in litt.*). Sightings of birds ringed there have now been received from Bala, Llanuwchllyn, Maentwrog and Mallwyd (Jim Dustow *in litt.*). All this evidence now suggests a much higher breeding population than the *c.*50 pairs estimated by Smith (2004), which is at least comparable to that for Gwent.

There are regular winter records from the Conwy Valley, Caernarfonshire. At Llanbedr-y-cennin, the highest counts were in January, including 35 in 2014, 27 in 2016, 31 in 2017 and 48 in 2018. At Caerhun, there were 20 in February 2014. A study in the Conwy Valley involved ringing some 40 birds during 2016–19

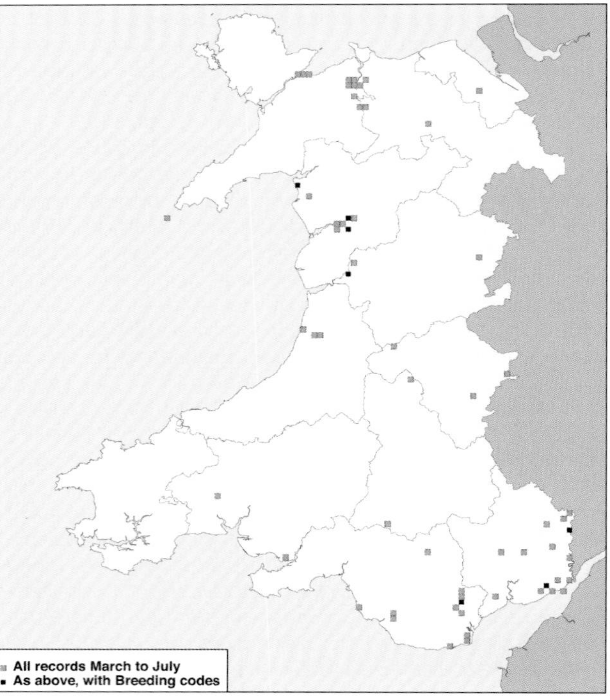

Hawfinch: Records from BirdTrack for March to July, 2015–19, showing the locations of records and those with breeding codes.

(Kelvin Jones *in litt.*). There were up to 14 in the Vale of Clwyd, Denbighshire, in February 2014, and up to 11 there between January and April 2015. Records have been less than annual from Pembrokeshire, Carmarthenshire, Ceredigion, Anglesey and Flintshire.

In autumn 2017, unprecedented numbers of continental Hawfinches arrived in Britain, with flocks of up to 600 in southern England. The highest counts recorded in Wales included 38 on the Pontypool-Blaenafon ridge at the Folly, Gwent, on 1 November 2017 and 20–50 at Dingestow, Gwent, during January–March 2018. There were up to 40 at Fforest Ganol in January and 50 there in March, 40 at Bleddfa, Radnorshire, in early 2018, a county record 48 at Llanbedr-y-cennin, Caernarfonshire, in January 2018 and 23 near Rhuthun, Denbighshire, in January 2018. Several of these sites have been used regularly in winter, making it difficult to distinguish local birds from immigrants.

There are more records in Wales now than at any time in recorded history, although this is likely to be largely due to improved monitoring. The Welsh population is probably in the order of 700 pairs, centred mainly in four areas—with broadly similarly sized populations in the Wye Valley of eastern Gwent, and the Dolgellau area of Meirionnydd, and smaller populations in the Cardiff and Caerphilly area of East Glamorgan, and the Conwy Valley in Caernarfonshire. Small numbers have also been recorded breeding in other counties. Woodward *et al.* (2020) estimated the British population at 500–1,000 pairs. There is considerable uncertainty about this population estimate, given the number of birds now being ringed in Wales, which form the main component of the British population. The Hawfinch seems to be doing well in Wales at present, but its future prospects remain unclear. Woodland management is likely to be a critical factor in maintaining the population.

Jerry Lewis

Sponsored by John Harrop

Bullfinch *Pyrrhula pyrrhula* Coch y Berllan

Welsh List Category	IUCN Red List			Birds of Conservation Concern			
	Global	Europe	GB	UK	Wales		
					2002	2010	2016
A	LC	LC	LC	Amber			

The Bullfinch breeds across Europe, except around the Mediterranean, and the southern half of Siberia except for the far north, to Japan (*HBW*). The IOC recognises ten subspecies, which differ in intensity of plumage colour and size, but separating them is not always easy. The subspecies *P.p. pileata* is restricted to Britain and Ireland, and is smaller, less bright and generally darker than its northern European counterpart, *P.p. pyrrhula*.

Breeding habitat includes dense woodland of all types, scrub, thick hedgerows, parks and gardens with mature shrubs and trees, primarily in rural areas, but occasionally in towns and cities. It is mainly a lowland species in Wales, being formerly scarce above *c.*300m. However, afforestation in the uplands has created suitable habitat at higher altitudes, particularly in young conifer plantations, so this altitudinal limit has now changed. It was present up to 550m in Radnor Forest, Radnorshire (Jennings 2014). The young are fed on invertebrates, but adults feed on seeds and fruits of fruit trees and other woody members of the Rose family, as well as some seed-bearing herbaceous plants, such as Hogweed. They can be a problem for commercial fruit-growers by eating fruit-tree buds or tips and young fruit. Large numbers were killed by trapping or shooting in parts of England in the past, but Lovegrove (2007) found no evidence of organised Bullfinch control in Wales, in historical or modern times. However, over 30 were said to have been shot in a garden near Caerleon, Gwent, during February 1930 (Venables *et al.* 2008).

The Bullfinch has probably been a fairly widespread and moderately common breeder in Wales over the last 150 years. However, it exhibits little or no overt territorial behaviour, so may easily go undetected. It was reported as being a common and regular breeder in Glamorgan in the 1890s (Hurford and Lansdown 1995). Mathew (1894) described it as a common resident in Pembrokeshire and Phillips (1899) said that it was "very common" in Breconshire. Forrest (1907) described it as common in wooded country throughout North Wales, most numerous in Montgomeryshire and parts of Denbighshire and Flintshire, but less common in Llŷn, Caernarfonshire, and north and west Anglesey.

Ingram and Salmon (1954) said that it was widely distributed but not numerous in Carmarthenshire. In 1955 the same authors described the species as "very thinly distributed" in Radnorshire. Parslow (1973) considered that the population throughout Britain and Ireland had increased considerably from about 1955. Newton (1967a) considered that this was linked to a behavioural change, involving a move to more open areas, a change that could have been facilitated by a reduction in Sparrowhawk numbers at that time. There is not a great deal of evidence from Wales, but Humphreys (1963) described it as a common resident in Gwent and suggested that numbers had increased.

The species' distribution has probably remained unchanged throughout Wales since the early 20th century. The *Britain and Ireland Atlas 1968–72* found the species present in 92% of 10-km squares in Wales, with 87% in 1988–91 and 94% in 2007–11, which indicated no significant change overall. Relative abundance in 2007–11 was highest along the south coast and eastern lowlands. At tetrad level, the *Gwent Atlas 1981–85* found the species present in 85% of tetrads but the 1998–2003 *Atlas* in only 73%. There was no significant change in East Glamorgan, where Bullfinch was present in 55% of tetrads in 1984–89 and 57% in 2008–11. However, in Pembrokeshire it was found in 71% of tetrads in 1984–88 but only in 65% in 2003–07 (*Pembrokeshire Avifauna* 2020). The *North Wales Atlas 2008–12* found the species in 42% of tetrads, but it was absent from higher ground in the west and present in only 26% of tetrads in Meirionnydd. It was also very sparsely distributed in some lowland areas in the west, particularly north and west Anglesey.

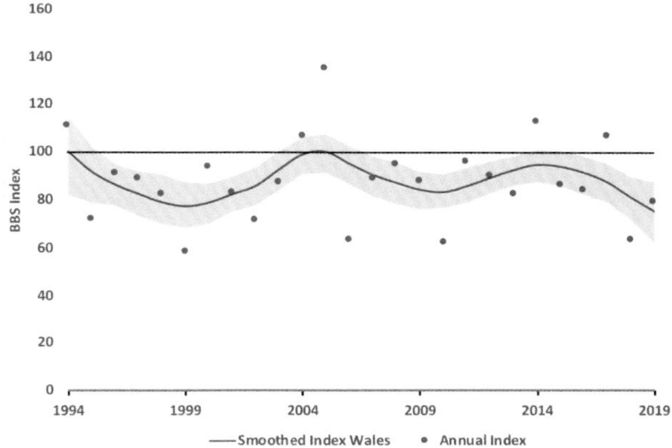

Bullfinch: Breeding Bird Survey indices, 1994–2019. The red line shows the smoothed index, 1995–2018. The shaded area indicates the 85% confidence limits.

There was a large decline in the Bullfinch population in England from the early 1970s, and CBC/BBS data showed a 58.8% decline in the UK population during 1975–2000 (Massimino *et al.* 2019). Trends in Wales before 1995 are not known, although Jennings (2014) commented that any declines since the early 1980s had not been evident in Radnorshire. The BBS indices for Wales showed a decline of 12% during 1995–2018 (Harris *et al.* 2020), but this masked some fluctuations.

The reasons for the decline in the early part of this period are unclear, although changes in adult survival might be important, and agricultural intensification is thought to have played a part (Massimino *et al.* 2019). The recovery of the Sparrowhawk population may have limited the use of more open areas for feeding. Where Sparrowhawks are present, the food available to Bullfinches is determined not solely by its abundance, but also by its close proximity to shrubby cover (Marquiss 2007).

The rather scant CBC data available for Wales suggested an average population of 7.5 pairs/km² on farmland in Gwent (*Birds in Wales* 1994). In conifer plantations in North Wales, densities of five pairs/km² were recorded (Bibby *et al.* 1985), while densities in restocked plantations were estimated at 21 pairs/km² (Bibby *et al.* 1989). The Sessile Oak woodlands of the west usually have little suitable breeding habitat. For example, a survey in 2006 found only one territory in 112ha (*c.*0.9 pairs/km²) of woodland at RSPB Ynys-hir, Ceredigion (Roderick and Davis 2010).

Breeding pairs are generally solitary but once the breeding season is over, they form loose aggregations, exhibiting small-scale movements, generally less than 5km, in search of food, often in hedgerows, gardens or orchards. Groups are usually no more than a dozen, but 30 have been recorded in winter in Gwent, Carmarthenshire and Pembrokeshire. Ringing recoveries suggested that most do not move far, although a bird shot at Itton, Chepstow, Gwent, in February 1978, had been ringed at Langstone Harbour, Hampshire, 153km away, in January 1977. Autumn movements are sometimes recorded, but the numbers involved are not usually very large. Hurford and Lansdown (1995) mentioned autumn movements involving up to 35 at coastal locations in Glamorgan, and 23 flew south over Bardsey, Caernarfonshire, on 12 October 2008. The species is recorded almost annually on Bardsey, but seldom more than 2–3 birds. It is scarcer on Skokholm and Skomer, Pembrokeshire. There can also be dispersal flights in winter, such as a movement of 30 in several small groups over a heavily snow-covered Palé Moor, Meirionnydd, on 5 January 2010. An influx into Gwent was reported in winter 1986/87, with many small flocks of up to 25, mainly males (Venables *et al.* 2008).

The breeding population in Wales in 2018 was estimated to be 29,500 territories (Hughes *et al.* 2020). The Bullfinch remains on the Red List because of a severe long-term decline in the UK breeding population (Johnstone and Bladwell 2016). The Welsh population is probably at a considerably lower level than in the 1960s and has shown a moderate decline in recent years. More research to discover the reasons for this would help to ensure the future of this attractive finch in Wales.

Male Bullfinch © Bob Garrett

Northern Bullfinch *P.p. pyrrhula* Coch y Berllan y Gogledd

The nominate subspecies, which breeds from northern and eastern Europe to Siberia, is partly migratory and prone to eruptive movements in winter, mainly in a southerly and/or westerly direction (*HBW*). Movements are much more marked in years of widespread failure of relevant tree-seed crops (Newton 1993). The presence of larger and more brightly coloured immigrants in Wales has been noted since the early 20th century (Forrest 1907); hence records of this subspecies are now assessed by WBRC. There have only been two confirmed records in Wales, a male on Bardsey on 11–16 December 2004 and a female at RSPB Lake Vyrnwy, Montgomeryshire, on 15 February 2005. These followed an influx into Britain in autumn 2004, with the largest numbers in Orkney and Shetland (Parkin and Knox 2010).

Nigel Brown and Rhion Pritchard

Sponsored by Emily Jenkins and Joshua Robertson

Common Rosefinch *Carpodacus erythrinus* Llinos Goch

Welsh List Category	IUCN Red List		Rarity recording
	Global	Europe	
A	LC	LC	WBRC

The Common Rosefinch is a widespread summer visitor across much of northern and eastern Europe, Asiatic Russia and China (*HBW*). It is a spring and autumn migrant to Britain, where there has been a large increase in records over the past 50 years.This is a little surprising, as overall its European population is on the decrease (Birdlife International 2020) although its range has moved slightly north and west (*European Atlas* 2020). It is now regularly seen each year, with an annual mean of 147 individuals recorded over the last three decades (White and Kehoe 2020b). While it is still considered to be a scarce passage migrant to Wales, 161 individuals have now been recorded and the recent increase in numbers mirrors the pattern across Britain.

Total	Pre-1950	1950–59	1960–69	1970–79	1980–89	1990–99	2000–09	2010–19
161	2	0	3	7	23	41	41	44

Common Rosefinch: Individuals recorded in Wales by decade.

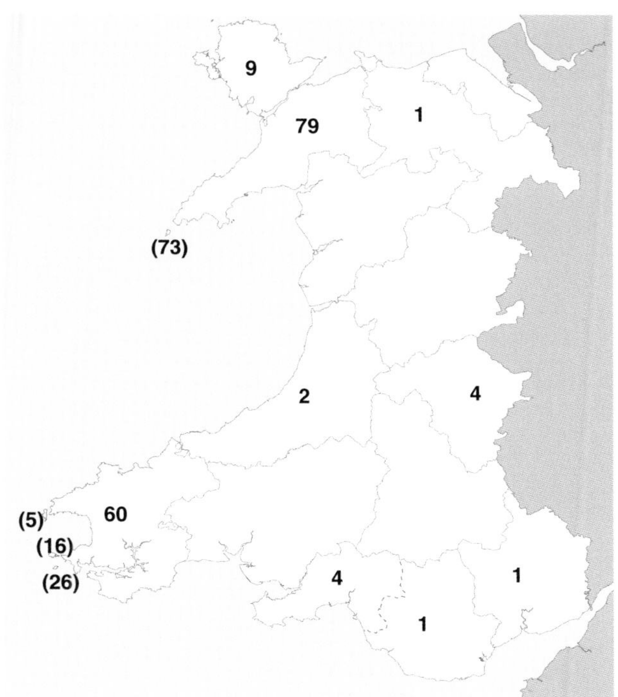

Common Rosefinch: Numbers of birds in each recording area, 1949–2019.

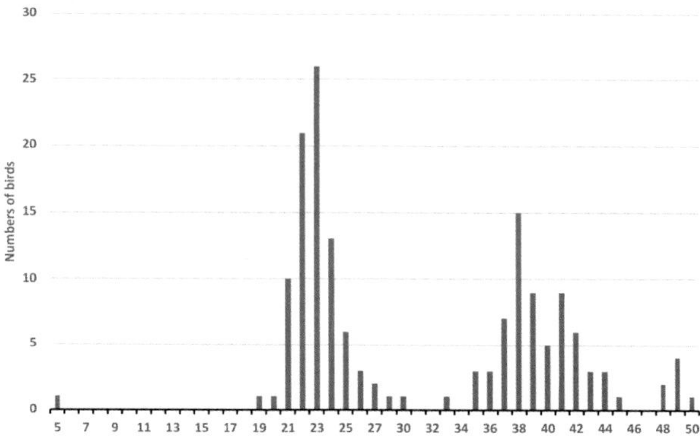

Common Rosefinch: Totals by week of arrival (when known, n=158) 1949–2019. Week 22 begins *c.*28 May and week 38 begins *c.*17 September.

The first Welsh record was of a male shot near Glascwm, Radnorshire, in 1875 (*Birds in Wales* 1994). Nearly half of all Welsh records came from Bardsey, Caernarfonshire, with only six records on the mainland of that county. Over one-third of records were from Pembrokeshire, mostly from the islands but 13 on the mainland.

Elsewhere, most records were from coastal locations in Gwent, East Glamorgan, Gower, Ceredigion, Caernarfonshire and Denbighshire. The two inland records from Radnorshire, the first in 1875 and then another at Llangunllo in May 2013, were unusual.

The distribution of records coincides with the expected British spring and autumn migration pattern, with peaks at the end of May/early June and in mid-September. The species' range is expected to move northward, almost entirely out of central and eastern Europe by 2070–2100 (Huntley *et al.* 2007), and it remains to be seen what effect this will have on its occurrence in Britain, although most autumn records are assumed to be from Fennoscandia.

Jon Green and Robin Sandham

Greenfinch — *Chloris chloris* — Llinos Werdd

Welsh List Category	IUCN Red List			Birds of Conservation Concern			
	Global	Europe	GB	UK	Wales		
					2002	2010	2016
A	LC	LC	EN	Green			

The Greenfinch breeds throughout most of Europe, apart from the far north, and its range extends to northwest Africa and east through western Asia as far as Kazakhstan. The subspecies that breeds in Wales is *C.c. harrisoni*, which is endemic to Britain and Ireland, but the nominate subspecies *C.c. chloris*, which breeds in northern Scotland and the rest of northern and central Europe, may also occur in Wales in winter (IOC). Most populations are resident, but birds breeding in the north and east of the range migrate south or west in winter (*HBW*). Greenfinches feed mainly on seeds, berries and fruits, but some insects are also taken. In winter they can form large feeding flocks on farmland, often in the company of other finches, buntings and larks.

Ringing recoveries suggest that there is a southwesterly movement within Britain in winter. A female ringed near Cresselly, Pembrokeshire, in December 1993 was recovered 765km away near Dounreay, Caithness, in northern Scotland in May 1998. Other recoveries suggested a movement from Scotland and northern England into Wales in winter. Several ringed in Wales have been recovered in Ireland, and birds ringed as adults, in France (February 1965) and the Channel Islands (March 1996) in winter, have later been recovered in Wales (March 1968 and October 1997, respectively). Most winter visitors to Britain from overseas are from Norway (Wernham *et al.* 2002), but there have been no ringing recoveries to prove that these birds reach Wales. It is quite possible some of these movements involve the nominate subspecies.

Around the turn of the 20th century, the Greenfinch appears to have been a common species in Wales, although authors from the period gave few details. Greenfinch populations seem to have fluctuated locally, but otherwise probably remained fairly stable through the 20th century. The first assessment of its distribution in Wales was provided by the *Britain and Ireland Atlas 1968–72*. The Greenfinch was found in almost all 10-km squares in Wales, apart from some upland squares and a gap in northern Anglesey. No huge changes were evident in the 1988–91 and 2007–11 atlases, but occupancy had increased to 97% of squares by the latter period. The *Britain and Ireland Atlas 1988–91* showed that relative abundance was highest at coastal locations, particularly in East Glamorgan, Anglesey and Denbighshire/Flintshire. The 2007–11 *Atlas* suggested that had not changed, but in general, numbers over most of Wales had increased since 1988–91, but were low compared to England and eastern Ireland.

Occupation of tetrads by Greenfinches in southern Wales increased between the 1980s and the early 20th century. The *Gwent Atlas 1981–85* found Greenfinches, with evidence of breeding in 81% of tetrads, with the gaps mainly on the higher ground in the northwest, while the *Gwent Atlas 1998–2003* found it present in 92% of tetrads. The increase, mainly in the northwest, was attributed to improved habitat, with birds in the later period occupying mature conifer plantations. Greenfinches had become more common in gardens, even in urban areas, but scarcer on farmland, owing to changes in agricultural practices (Venables *et al.* 2008), a trend also noted in other counties. An increase was also apparent in East Glamorgan, where birds showed evidence of breeding in 62% of tetrads in the 1984–89 *Atlas* but in 69% in 2008–11, with an expansion onto higher ground also evident there. In Pembrokeshire, the 1984–88 *Atlas* found the species in 55% of tetrads while in 2003–07 it was present in 78% (*Pembrokeshire Avifauna* 2020). The *North Wales Atlas 2008–12* found Greenfinches in 58% of tetrads, with most lowland areas occupied, but in few tetrads on higher ground. Only 35% of tetrads in Meirionnydd were occupied, for example, with very few tetrads at an altitude above 300m holding this species.

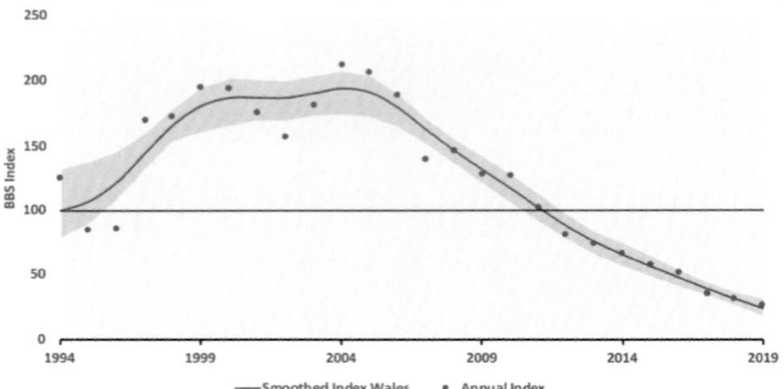

Greenfinch: Breeding Bird Survey indices, 1994–2019. The red line shows the smoothed index, 1995–2018. The shaded area indicates the 85% confidence limits.

The BBS index for Wales increased between 1994 and 2005, unlike that for several other primarily seed-eating species. *Birds in Wales* (1994) noted that their former habit of congregating at stockyards had been replaced by exploitation of garden feeding stations. This may have fuelled the increase. Population density increased between 1994–96 and 2007–09 in large parts of lowland Wales, with an increase of over 75% in many areas (Harris *et al.* 2020). However, the BBS index declined dramatically from 2007, by 78% during 2008–18. In 2005, Britain saw the first cases of Trichomonosis in wild finches, a disease caused by a protozoan parasite, *Trichomonas gallinae*. It had not previously been reported in finches but soon proved fatal for Greenfinch and Chaffinch (Robinson *et al.* 2010). The disease spreads rapidly from saliva via contaminated food and water, wherever infected individuals mix with conspecifics: at garden feeders, in feeding flocks, roosts and through the semi-colonial behaviour of Greenfinch during the breeding season, with a pronounced seasonal peak of mortality in August and September (Lawson *et al.* 2018).

The initial centres of disease outbreak in Britain were in central and western parts of England and Wales. The first case of Trichomonosis in Wales was confirmed in Greenfinch in October 2005 in Glamorgan and, by 2006, reported incidences rose quickly. Between April and September 2006, 50% of all reports of garden bird mortality involved Trichomonosis, 80% of which involved Greenfinches and to a lesser extent Chaffinches (Robinson *et al.* 2010). This led to a huge reduction in the Greenfinch population. Incidence of the disease has fluctuated in more recent years, with some local signs of recovery reported by county recorders, for example, in Radnorshire. Finches appear to be particularly susceptible to this disease, and others such as salmonellosis and colibacillosis, although the reasons for this are unclear (Robinson *et al.* 2010). Congregation around garden feeders is likely to aid the spread of the disease, but this does not explain why other species such as the tits are not similarly affected by Trichomonosis.

Passage has been recorded in spring and autumn on the Welsh islands and on the coasts. Spring passage has usually been light, but larger numbers have been recorded in autumn, particularly on Bardsey, Caernarfonshire, including as many as 1,100 on 27 October 1975 and 1,500 on 23 October 2007. Numbers on the Pembrokeshire islands have usually been lower. Large numbers have also been recorded at coastal sites on the mainland. For example, at least 5,000 passed the Great Orme, Caernarfonshire, between 15 September and 6 November 2005, with a peak of at least 400 on 22 October, and there were over 500 there on 14 October 2007.

Feeding flocks have often been found in open cultivated farmland habitats outside the breeding season. Large counts may be feeding flocks or at roosts: published records do not always distinguish the two. A finch flock of *c.*1,500 birds at Sker, East Glamorgan, in January 1964, mostly comprised Greenfinches, and there were 1,500 at Hubberston, Pembrokeshire, on 5 January 2006. The highest counts in mid-Wales have been in Radnorshire, where *c.*500 were at David's Well on 1 November 1953 and over 500 were at Newchurch on 17 December 1971. Counts in North Wales have tended to be lower.

There still appeared to be good numbers on passage and in winter in 2009, including 600 on Bardsey on 24 October, 390 at Newport Wetlands, Gwent, in November and several other counts in excess of 200. There have been fewer large counts since the breeding population declined across Britain, although 450 were at Dolydd Hafren, Montgomeryshire, on 13 January 2012, and over 300 at a roost in Llandudno, Caernarfonshire, during winter 2014/15.

The Welsh breeding population in 2018 was estimated to be 38,000 (a range of 35,500–40,500) territories. Whilst the Greenfinch is still a relatively common species, the population is still declining due to reduced adult survival, of which the impact of Trichomonosis is still considered to be a major factor (Lawson *et al.* 2018, Massimino *et al.* 2019). Garden feeders can be a major source of pathogen infection and transmission. Good hygiene management to prevent this is important (Lawson *et al.* 2018). The population in Wales is still declining and future prospects for the Greenfinch remain a concern. The current Amber-listing of the Greenfinch in Wales is based on a moderate breeding population decline over 25 years (Johnstone and Bladwell 2016), but the rate of decline should qualify the species for Red-listing at a future assessment.

Nigel Brown and Rhion Pritchard

Sponsored by Giles Pepler

Twite — *Linaria flavirostris* — Llinos y Mynydd

Welsh List Category	IUCN Red List			Birds of Conservation Concern			
	Global	Europe	GB	UK	Wales		
					2002	2010	2016
A	LC	LC	NT	Red			

The Twite has a discontinuous distribution, from Ireland east to western China. There are two main population groups, one breeding from eastern Turkey to central China and south to the Himalayas, the other in Norway and extreme northwest Russia, Britain and Ireland (*HBW*). Of the nine subspecies, the one that breeds in Wales is *L.f. pipilans*, which is restricted to Britain and Ireland. Those breeding in Fennoscandia are of the nominate subspecies, *L.f. flavirostris*.

Twite breed on lower montane and submontane plateaux, moorland, barren hillsides and areas with stunted vegetation. They are mainly seed eaters and are very dependent on a year-round supply of small seeds. The nest is usually on the ground or low down in bushes or other vegetation (*HBW*). Most of the birds breeding in Scotland and Ireland nest near the northern and western coasts, but farther south in Britain they breed only in the uplands. In the South Pennines, Twite have been recorded breeding at altitudes of over 500m (Orford 1973). There is little information available on the altitude of nest sites in Wales, although they are likely to breed in Snowdonia at an altitude of at least 500m. Many upland populations move to lower ground in winter, and birds breeding in northern Europe move south and southwest, to winter in coastal lowlands and estuaries around the North Sea and central Europe (*HBW*).

The highest densities of breeding Twite in Britain are in Shetland, Orkney, Caithness and the Hebrides. The upland populations farther south, in the Southern Uplands of Scotland, the Pennines and Snowdonia, are at a much lower density. There has been a 19% range contraction in Britain since 1968–72, and an 80% range contraction in Ireland (*Britain and Ireland Atlas 2007–11*). A survey in 2013 produced an estimate of 7,831 pairs in the UK, with Scotland holding 98% of the population (Wilkinson *et al.* 2018).

Eyton (1838) described the Twite as "common in North Wales, where it breeds". Forrest (1907), however, noted that although many parts of North Wales seemed to have suitable breeding habitat, "the most diligent search by several competent ornithologists has failed to reveal a single authentic instance of the nest being discovered". Forrest concluded that Twite was a winter visitor in fair numbers to certain hilly districts, but that it rarely, if ever, remained to breed. Bolam (1913) recorded finding two nests in the Bala area of Meirionnydd in May 1905, but Forrest (1919) thought that Bolam was mistaken as to the species. One of the nests was on a railway embankment, which Forrest noted was "a most unlikely spot for a Twite's nest". T.A. Coward watched a pair of Twite on Mynydd Hiraethog, Denbighshire, in June 1905.

Elsewhere in Wales, the only suggestion of possible breeding was a comment by one of A.G. More's correspondents that Twite bred sparingly on the Black Mountains, on the western border of Herefordshire with Gwent and Breconshire (More 1865). The Twite was an occasional winter visitor in Glamorgan in the late 19th century (Hurford and Lansdown 1995), but it was unknown in many counties of Wales in this period. It was not recorded in Pembrokeshire until 1925 (Donovan and Rees 1994), in Ceredigion until 1943 (Roderick and Davis 2010), in Anglesey until 1949 (Jones and Whalley 2004) and in Carmarthenshire until 1962 (Lloyd *et al.* 2015). In Breconshire, Phillips (1899) described it as "fairly distributed in the winter throughout the county", but Massey (1976) thought that this was an error, probably through confusion with Linnet, though there was a definite record of two obtained from the county in 1897.

Breeding was confirmed in 1944 on Tryfan, in the Glyderau, Caernarfonshire, with adults seen carrying food. This is adjacent to the present core breeding area around Nant Ffrancon. Breeding was also proved near Llanberis from 1953 onwards. A record of a nest on saltmarsh at Shotton, Flintshire, in May 1967 (*Birds in Wales* 1994), seems to be the only record of coastal nesting in Wales. Orford (1973) was unaware of any breeding population in Wales, but from 1973 until the late 1980s, there was an increasing number of breeding records from several areas in North Wales. These included the area surrounding Nant Ffrancon in Caernarfonshire and sites in Mynydd Hiraethog, near Cerrigydrudion and Bylchau (Jones and Roberts 1983). Four pairs were present at one site on Mynydd Hiraethog in 1978, and nesting was also confirmed near Llangollen, Denbighshire, in 1990 (*Birds in Wales 1994*). A small breeding population was found in some upland areas of Meirionnydd in this period, with breeding confirmed on the Arenig range and probable in the Trawsfynydd area. In 1989, three pairs were reported to have bred at two sites in the county (Pritchard 2012). Breeding has never been confirmed farther south in Wales, but two pairs found on the north side of Pumlumon, on the Ceredigion and Montgomeryshire border, in 1988 were presumed to be nesting (*Birds in Wales* 1994). The *Britain and Ireland Atlas 1988–91* showed a small but significant expansion in Wales since 1968–72, with confirmed breeding in seven 10-km squares.

A decline in the breeding population was evident in the 1990s. In 1995, it was recorded that the birds were absent from one traditional breeding site in Meirionnydd. As part of a UK survey in 1999, a complete census of known and suspected breeding sites in Wales was attempted and found just five pairs. However, coverage was incomplete, and it was noted that in August and September that year, a flock of 100–200 Twite was recorded in Nant Ffrancon. These were thought to be probably local, post-breeding birds, perhaps representing a population of 20–40 pairs, suggesting that the breeding population may have been considerably larger than the estimate from this survey (Langston *et al.* 2006). The 1999 survey found none at three traditional sites on the Migneint, and it is likely that the breeding population in Meirionnydd had died out by that year (Pritchard 2012).

The breeding range had contracted markedly by the early 21st century. A survey in 2002 produced an estimate of 26–33

Twite in winter © Andy Davis

breeding pairs in Wales, most of them adjacent to Nant Ffrancon. Another survey in 2008 estimated 14–17 breeding pairs in Wales (Johnstone *et al.* 2011). The *North Wales Atlas 2008–12* confirmed breeding in just three tetrads, all in the area around Nant Ffrancon. Most of the eight tetrads with probable breeding were also in Snowdonia, although there was probable breeding in one tetrad on the Berwyn, Meirionnydd, and one near Ruabon Mountain, Denbighshire. It is possible that a few pairs may have bred outside the core area; for example, there was a record of a pair in apparently suitable breeding habitat in the Elan Valley, Radnorshire, in June 2002 (Jennings 2014).

Breeding Twite need a supply of suitable seed close to the nest sites, and this depends largely on livestock management, with traditional hay meadows also important. Cattle were formerly out-wintered in fields and fed on hay, which provided a source of seeds for Twite in the spring. The ending of this practice may have reduced food availability (Johnstone *et al.* 2011). Specific action to help Twite in Snowdonia was stepped up from 2015 when the RSPB and the BTO worked with local farmers, who were tenants of The National Trust, in Nant Ffrancon. This targeted six farms with funding from the Glastir Advanced agri-environmental scheme. The actions involved the provision of feeding stations and seed pasture with favoured plants (such as Common and Sheep's Sorrel, Cat's Ear and hawkbits), and the topping of rushes to prevent them from dominating the herbaceous swards. Farmers withdrew grazing from these fields in rotation for 8–10 weeks during May to September, so the plants could set seed, providing a continuous food supply for the birds during the breeding season (Pierce 2016). Nyjer seed continues to be provided and may be especially important in April and early May, when natural food is in short supply. Historically, Twite may have eaten seed spilt from hay given to livestock in early spring, but most stock is now wintered indoors or away from the valley, due to Glastir restrictions. There have been no further surveys, but flocks of up to 40 have been counted in August during 2016–18 (Rhian Pierce pers. comm).

Birds remain around the breeding area in late summer, and sometimes into early winter. Post-breeding flocks in Nant Ffrancon included 80–100 in August 1997 and 100–200 in August/September 1999, but such flocks declined in size between 2000 and 2008 (Johnstone *et al.* 2011). There were *c.*150 at Llyn Ogwen, at the head of Nant Ffrancon, in December 1990 and flocks of 50–100 were seen regularly in Nant Ffrancon during the winter months of 1989–93, feeding on seeds of Soft Rush (Pritchard 2017). Other post-breeding and winter flocks in the hills included a flock of 25 near Esgairgeiliog (Ceinws), Montgomeryshire, in August 1983 (Holt and Williams 2008) and a remarkable 120 at Moel yr Iwrch, near Pentrefoelas, Denbighshire, on 11 December 2014.

In autumn and winter, Twite have been recorded in small numbers on the coasts of Wales, with the largest numbers on the North Wales coast. Farther south, records of wintering Twite have been much less regular. There used to be a wintering flock at Llanfairfechan, Caernarfonshire, during the 1990s, peaking at 70 in November 1995, but numbers gradually declined after that year and the birds no longer regularly occur there. The most reliable site to find Twite in winter during the last 20 years has been the Dee Estuary, Flintshire. At Connah's Quay Nature Reserve, Flintshire, maximum counts during the 1980s were in the order of 12–25. During the 1990s there were up to 87, although counts were not undertaken every year in either decade. A peak count of 100 was recorded in the period 2000–09. There was a large increase in maximum numbers in 2014–16, peaking at 200 in January 2015, although after 2016 the maximum dropped to about 100 in 2018 and 63 in February 2019.

Ringing has provided a good deal of information on the origins of these birds. Some birds, ringed in the breeding season in the Nant Ffrancon area, have been recorded in winter at Connah's Quay and other coastal sites in Wales. However, others have moved east in winter, with records from as far afield as Titchwell, Norfolk. These birds were using the same wintering areas as birds that bred in the South Pennines, which have been found to winter predominantly on the east and southeast coasts of England (Raine *et al.* 2006).

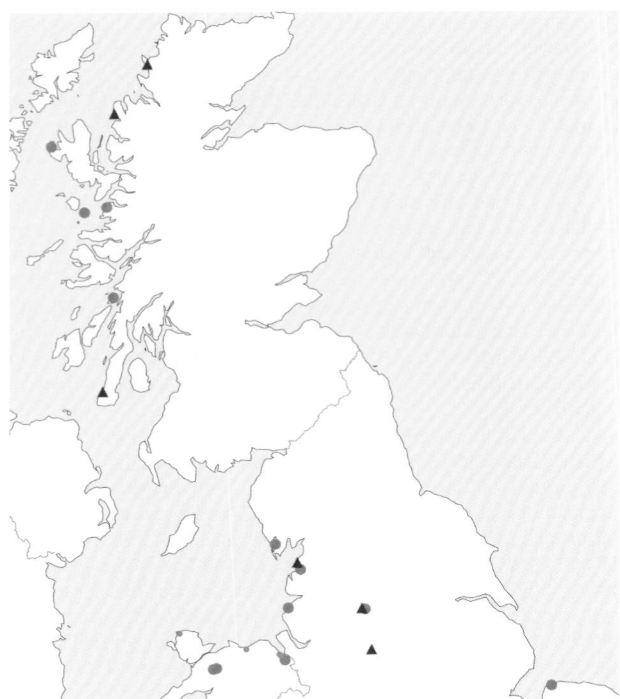

Twite: Recovery locations of birds ringed in Wales are shown by red circles. Ringing locations of birds that were recovered in Wales are shown by blue triangles. Small brown dots show ringing or recovery locations in Wales. All movements shown are greater than 50km.

Many of the Twite found on the Welsh coast in winter, from November through to early April, appear to have been from the western Scottish population, particularly from the islands of Argyll. However, since about 2016, Scottish birds appear to have largely stopped coming to North Wales in winter. This coincided with the creation of feeding habitat by the RSPB on Islay, Argyll, which appeared to have led breeding birds in the Inner Hebrides to stay more locally for the winter. Flocks of over 1,000 birds have been recorded on Islay, for example in November 2016 (RSPB 2017). In the last three or four years, even the Islay site has no longer attracted large numbers of Twite. It is not known whether this is because of a decline in the population or the birds wintering elsewhere, at an unknown site or sites.

There does not appear to be a great deal of interchange between the breeding populations within Britain (Dunning *et al.* 2020), but one bird ringed post-breeding in the South Pennines in 2004 was sighted on Yr Wyddfa, Caernarfonshire, on 30 July 2005, presumably breeding nearby (Raine *et al.* 2006). There has been no indication of interchange between the Welsh and Scottish breeding populations and none between the British populations and the continent.

The Twite is Amber-listed in Wales, because of a moderate decline in the breeding population over 25 years and its rarity as a breeder and a wintering species (Johnstone and Bladwell 2016). A similar decline has been recorded in the Pennines: the 2013 survey found a decline of 72% in England since 1999 and this survey produced an estimate of just 16 pairs breeding in Wales (Wilkinson *et al.* 2018). Any population as small as this is inevitably in danger of extinction. The future of the Twite as a breeding bird in Wales remains uncertain, and its long-term prospects are threatened by climate change. The area of Britain suitable for breeding Twite is expected to contract entirely to Scotland by 2070–2100 (Huntley *et al.* 2007).

Ian M. Spence and Rhion Pritchard

Sponsored by Ian M. Spence and Kelvin Jones

Linnet *Linaria cannabina* Llinos

Welsh List Category	IUCN Red List			Birds of Conservation Concern			
	Global	Europe	GB	UK	Wales		
A	LC	LC	NT	Red	2002	2010	2016

The Linnet is found over most of Europe except in central and northern Fennoscandia, and its range extends to northwest Africa and northwest China. Some populations are resident, but those that breed in the north and east of its range move south and southwest in winter (*HBW*). Linnets breeding in Wales are of the nominate subspecies, *L.c. cannabina*. It is probable that birds of the endemic Scottish subspecies, *L.c. autochthona*, are present in Wales in winter. A juvenile ringed on South Uist in the Outer Hebrides in August 2017 was recaught at Mwnt, Ceredigion, in February and again in March 2018.

The British breeding population is partially migratory. Some spend the winter in Britain, while others move south or southwest to winter in France, Spain or Morocco (*BTO Migration Atlas* 2002). There have been few ringing recoveries involving Wales, where most were within 10km of the ringing site. The only two recovered from outside Britain and Ireland were ringed on Bardsey, Caernarfonshire, in the summers of 1959 and 1961 and recovered on the Atlantic coast of France, in autumn 1959 and winter 1964 respectively. A year-round ringing project, that started on the Ceredigion coast in 2016, indicated that there are very few year-round residents there. The population changes over almost completely in late March and April and in September and October.

Linnets feed mainly on small to medium-sized seeds. Unlike most finches, the young are fed entirely on seeds, both ripe and unripe (Newton 1967b), which are masticated into a paste by the adults. They usually nest in loose colonies of 4–6 pairs, rarely up to 20 pairs, on rough common land, heathland, hedgerows and patches of scrub. They are particularly common on rough ground above cliffs around the Welsh coast, and on *ffridd* on the upland margins, where they occur in higher densities than in other habitats during the breeding season (Fuller *et al.* 2006). They also breed in higher areas in young conifer plantations or where gorse or thick heather is plentiful. The planting of conifers in upland areas may have enabled them to breed higher up on the hills, for example, up to 580m in Radnor Forest, Radnorshire (Jennings 2014). They are largely absent from urban and suburban areas in the breeding season, and do not usually visit garden feeders.

A decline in Linnet numbers was evident in some parts of Britain in the later part of the 19th century. There is no evidence for such a decline in Wales, where it was abundant and widely distributed in Glamorgan and in Pembrokeshire in the late 1890s (Hurford and Lansdown 1995, Mathew 1894). Phillips (1899) thought it was less numerous in Breconshire than in England, though fairly common, and it was "one of the commonest and most generally distributed birds in North Wales" (Forrest 1907).

Little information exists on population trends in Wales during the 20th century. There was a severe decline in England in the late 1960s, and again between the mid-1970s and mid-1980s, after which the population stabilised (Massimino *et al.* 2019). This was due to the reduced availability of food in the breeding season on arable farmland, the result of agricultural intensification reducing the supply of weed seeds (Moorcroft and Wilson 2000). Changes in arable farming would have had less effect in Wales, but the authors of *Birds in Wales* (1994) considered that the breeding population in Wales was then probably as low as it had ever been. They considered that scrub clearance, hill 'improvement' (involving the ploughing and reseeding of grassland) and the intensity of grazing by sheep were the main factors for decline in Wales.

Fieldwork for the *Britain and Ireland Atlas 1968–72* confirmed breeding in almost every 10-km square in Wales, and little change was evident in the 1988–91 and 2007–11 atlases. Relative abundance increased in some areas between the 1988–91 and

Male Linnet (right) feeding juvenile (left) © Bob Garrett

the 2007–11 atlases, but overall, more 10-km squares showed a relative decline. Tetrad atlases, where a comparison was possible, did not suggest any large changes in distribution. The *Gwent Atlas 1981–85* found Linnets with evidence of breeding in 78% of tetrads. By the second *Atlas* in 1998–2003 this had increased slightly to 83% of tetrads. In the *East Glamorgan Atlas 1984–89*, the species was recorded with breeding evidence in 62% of tetrads, and in 51% in the 2008–11 *Atlas*. Linnets were present in 72% of tetrads in Pembrokeshire in 1984–88 and in 78% in 2003–07 (*Pembrokeshire Avifauna* 2020). The *North Wales Atlas 2008–12* found Linnets in 56% of tetrads. They were absent from the higher ground and found in only 39% of tetrads in Meirionnydd.

The BBS index showed there was a 25% decline in Wales between 1995 and 2018, with a recent drop since 2015 (Harris *et al.* 2020). In Europe, populations fluctuated widely between 1980 and 2000, with an overall decline of 49%. Since then, the population has broadly stabilised, with a slight increase since 2015 (PECBMS 2020).

Post-breeding flocks have in the past reached 1,000 or more at sites where a good supply of seed is available, and could include birds on passage. Autumn passage is mainly between August and October, and spring passage in March and April. Ferns *et al.* (1977) suggested that autumn passage in southeast Wales was generally west along the Gwent coast and on to Lavernock Point in East Glamorgan, before birds crossed the Bristol Channel around the islands of Steep Holm and Flat Holm. Large counts in this area included over 2,000 southwest in two hours at Goldcliff, Gwent, on 6 October 1968, *c.*3,000 over Lavernock Point on 8 October 1976 and a remarkable count of *c.*10,500 birds there on 18 October 1985. More recently, there were 2,000 at Lavernock Point on 17 October 2018 and 940 on 14 October 2019. Similar, but smaller, movements along the north coast included 1,350 past Gronant, Flintshire, on 17 September 2005. The highest day total on Bardsey was 1,650 on 28 October 1975. Numbers on spring passage there were usually lower. The highest count was 360 on 3 April 2012. Fewer were usually recorded passing the Pembrokeshire islands, with counts rarely exceeding double figures on Skokholm, although there were up to 220 there in October 1959 (Betts 1992), 250 in October 1967 and 239 on 7 October 2016.

The *Britain and Ireland Winter Atlas 1981–84* showed that Linnets were largely absent from the uplands in winter, and the same pattern was evident in the 2007–11 *Atlas*. Relative abundance in the winter periods of 2007–11 was highest in coastal areas of Gower, Pembrokeshire, south Ceredigion, Llŷn in

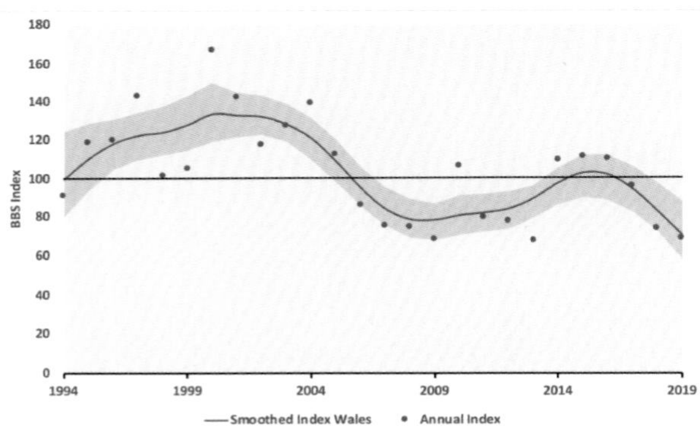

Linnet: Breeding Bird Survey indices, 1994–2019. The red line shows the smoothed index, 1995–2018. The shaded area indicates the 85% confidence limits.

Caernarfonshire, Anglesey, and in the east along the border with England. Winter flocks often foraged in weedy fields or stubble, although saltmarshes and dunes are also used, particularly when fields inland are frozen. Numbers wintering in Glamorgan declined steadily between the early 1960s and early 1990s (Hurford and Lansdown 1995), whereas flock size in Anglesey was much higher in 1980–90 than in earlier years (Jones and Whalley 2004). There have not been enough ringing recoveries in Wales to indicate what proportion of these winter flocks were non-migrant residents and what proportion incomers from farther north and east. The largest winter flocks in the 21st century included two in Pembrokeshire: 1,000 at Abercastle on 31 December 2008 and 1,200 at Gupton Farm on 16 December 2012. Elsewhere, many counties still reported winter flocks of 100 or more on a regular basis, including 200+ at Dolydd Hafren, Montgomeryshire, on 4 January 2015 and 100+ there in 2017.

The increasing availability of Oil-seed Rape was probably an important factor in arresting the decline of Linnets in England (Moorcroft *et al.* 2006). Nestling condition and growth rate have been higher with proximity to an Oil-seed Rape field (Bradbury *et al.* 2003). Weed-rich stubble fields are also important in winter (Moorcroft *et al.* 2002). There is little arable farming in many areas of Wales, but on the sandy soils around the coast of Pembrokeshire, arable land in National Trust ownership and/or management has been shown to support relatively large numbers of wintering birds given the right management prescriptions. At Gupton Farm, south Pembrokeshire, large numbers of Linnets were counted in the winter of 2012/13, using fields where Maize and spring Barley had been grown the previous summer. Surveys of six mid- and North Pembrokeshire farms, which probably represent 'more typical' coastal arable farmland, started in 2018/19 and in just two winters, numbers in excess of 500 were counted. Here, the NT were attempting to improve conditions for wintering bird populations. Tenants have been encouraged to retain rough margins and/or to plant marginal ruderal bird-crop strips and so forth (National Trust Winter Farmland Bird Surveys in Pembrokeshire unpubl. data). It is too soon to show whether the benefits for the Linnet can be sustained. Elsewhere, agri-environment schemes (AES) also have the potential to benefit this species, as recorded in *Breconshire Bird Reports* between 2016 and 2019. Flock sizes, usually varying between 30 and 100, were seen feeding on weed seeds found in fields of root vegetables grown for sheep. Mustard seed, grown within a Glastir wild-bird cover crop in Pennorth, attracted 200 on 5 January 2016. In 2017, 200 were seen feeding in a weedy field adjoining Priory Groves on 11 February and in mid-September a flock feeding on Oil-seed Rape stubble peaked at 250. The following year 220 were seen coming to a field of organically grown swedes near Newbridge-on-Wye on 7 February. In 2019, 150 fed in a weedy field at Ffrwdgrech on 19 January.

The Linnet's Red-listing in Wales is because of a severe breeding population decline over 25 years and in the longer term, together with its Red-listed status in the UK (Johnstone and Bladwell 2016). The Welsh breeding population in 2018 was estimated to be 47,500 territories (Hughes *et al.* 2020). Jennings (2014) estimated that numbers in Radnorshire were probably only 25% of their level in the 1960s, although there is less evidence of a major decline farther west. Arable land, managed with conservation aims and/or assisted by appropriate AES options, could be highly beneficial to seed-eating birds such as the Linnet. The future prospects for this species in Wales will depend largely on such schemes continuing for many more years and further studies into the ecology of this species in Wales would be very valuable.

Chris M. Jones

Sponsored by Chris M. Jones

Common Redpoll *Acanthis flammea* Llinos Bengoch

Welsh List Category	IUCN Red List			Birds of Conservation Concern	Rarity Recording
	Global	Europe	GB	UK	
A	LC	LC	VU	Amber	BBRC

The Common Redpoll has a circumpolar distribution, generally breeding farther south than the Arctic Redpoll. Redpoll taxonomy remains contentious, but the IOC recognises two subspecies, *A.f. flammea* (known as 'Mealy Redpoll'), which breeds in Fennoscandia, Siberia, Alaska and most of northern Canada, and *A.f. rostrata*, which breeds in northeast Canada, Greenland and Iceland. A few pairs of Common Redpoll breed in northern Scotland, including as many as 14–15 pairs in the Outer Hebrides in 2004. Most of those breeding in Scotland are thought to resemble *A.f. rostrata*, but the possibility of *A.f. flammea* cannot be excluded (*The Birds of Scotland* 2007).

Total	1967–69	1970–79	1980–89	1990–99	2000–09	2010–19
260	1	1	82	112	44	20

Common Redpoll: Individuals recorded in Wales by decade since 1967.

Many populations of Common Redpoll are migratory, although some, such as the birds breeding in Iceland, are largely resident. Most of the Mealy Redpolls breeding in northern Europe move south or southeast in autumn, many wintering in central Russia. Smaller numbers from northern Norway and Sweden move southwest (*HBW*). Widespread birch-seed failure, when the population is at a high level, can trigger large-scale movements (Riddington *et al.* 2000).

The Lesser Redpoll and Mealy Redpoll were considered to be separate species in the early 20th century, but were later regarded as subspecies of Common Redpoll. In 2001, the British Ornithologist's Union Records Committee split Lesser Redpoll (*A. cabaret)* from Common Redpoll, but the IOC did not accept the Lesser Redpoll as a separate species until 2017. All records of Common Redpoll in Wales are considered to refer to Mealy-type Redpolls (*A.f. flammea*). There has been no authenticated record of *A.f. rostrata*.

There was only one 19th-century record in Wales, a specimen taken in the Cardiff area, East Glamorgan, sometime during the latter half of the 19th century (Hurford and Lansdown 1995). There were several records in the early years of the 20th century. Forrest (1907) was informed of a party of six or eight near Menai Bridge, Anglesey, in November and December 1900, of which two were shot, and one of these, an adult male, preserved. Two more males were shot at the same site on 24 December 1904. J. Walpole-Bond saw several with Lesser Redpolls near Builth Wells, Breconshire, on 10 February 1903 (Ingram and Salmon 1957).

There were no further acceptable Welsh records until individuals were reported at Cors y Sarnau, Meirionnydd, in 1967 and 1970. Numbers increased from 1980 onwards, reflecting increased observer interest in splitting Common Redpolls from Lesser Redpolls and influxes into western Europe.

The majority of records have come from Bardsey, Caernarfonshire, where there would appear to be a small spring passage, April to May, in most years, with larger numbers including 15 in May 1995, 29 in May 1999 and 15 in May 2000. Spring passage has also been noted on the Great Orme, Caernarfonshire, in 2005, 2007, 2017 (six individuals) and 2018, on Skomer, Pembrokeshire, in 2002 and Carmel Head, Anglesey, in 2018. The only double-figure counts outside this period included 60 on Bardsey on 1 October 1981 and 60 at Marford Quarry, Denbighshire, on 10 March 1996.

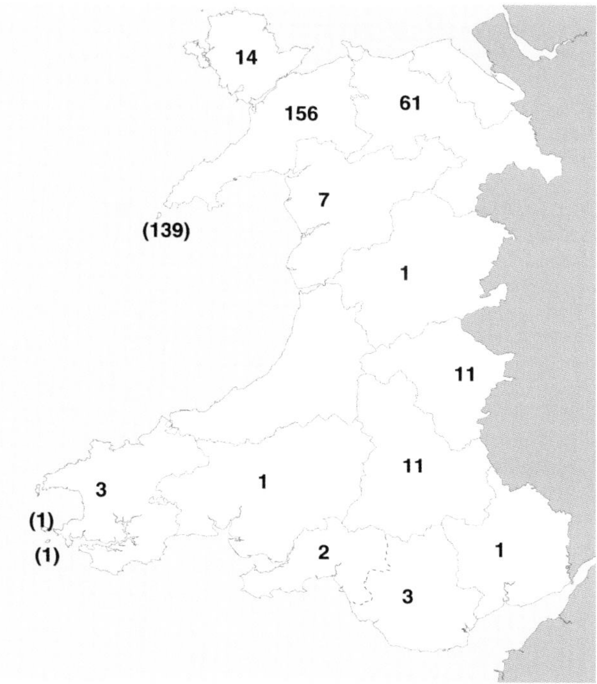

Common Redpoll: Numbers of birds in each recording area.

Away from North Wales, this species is rare and usually seen during the winter months. Interestingly, most records in the southern half of Wales were from Radnorshire, including a group of six at Franksbridge on 2 January 1988 and five in the Elan Valley, Radnorshire, on 3 February 1988.

Jon Green and Robin Sandham

Lesser Redpoll *Acanthis cabaret* Llinos Bengoch Fechan

Welsh List Category	IUCN Red List			Birds of Conservation Concern			
	Global	Europe	GB	UK	Wales		
					2002	2010	2016
A	LC	LC	LC	Red			

The Lesser Redpoll is confined to Europe, where it breeds in Britain and Ireland, Fennoscandia, the Alps and countries adjacent to the southern North Sea (*European Atlas* 1997). In the mid-1800s it was confined largely to northern Britain, Ireland and the Alps. From about 1950, its range expanded considerably on the continent, with populations established from France to southern Sweden and in central Europe (Knox *et al.* 2001). This was formerly considered to be a subspecies of the Common Redpoll (*A. flammea*), but it has been treated as a separate species by the BOU since 2001 (BOURC 2001), although the IOC did not follow suit until 2017. Older works referred simply to 'Redpoll', although any birds present in Wales in the breeding season were assumed to have been Lesser Redpoll. However, the taxonomy of redpolls is complex. Recent work in Europe failed to differentiate between the species, treating all as subspecies of *A. flammea* (Lifjeld 2015). The *European Atlas* (2020) has shown the range of the Lesser Redpoll to have moved away from southeast England, northern France and the eastern Alps, but increased northwards, which may be in line with climate change predictions.

It feeds on very small seeds, particularly birch and Alder seeds, and also on invertebrates in the breeding season. In Wales it occurs in young conifer plantations, birch woodland, and willow and birch carr in boggy areas, and breeds to an altitude of *c.*610m, although most of the population nests below 450m.

In some of its range, the population is largely migratory, in others sedentary (*HBW*). It appears that at least some birds breeding in Wales move south in autumn, with a number of Welsh-ringed birds recovered on the south coast of England in autumn. Others moved into Wales from farther north. A number of birds ringed in northern Scotland have been recovered in Wales outside the breeding season. The longest distance recorded is one ringed as an adult in Ordiequish Forest, Moray, in March 2015 and recaught on Bardsey, Caernarfonshire, 545km away in April 2016. Several birds ringed in Wales were recovered in Belgium, including two ringed in mid-April and recovered there in November. Birds ringed in Belgium and France in autumn have been recovered in Wales in the breeding season.

Lesser Redpolls seem to be prone to population fluctuations, both long term, as a result of changes in habitat, and short term, as a result of changes in the availability of suitable small seeds (Newton 1972). They were regarded as uncommon as a breeding species in Glamorgan in the late 1890s (Hurford and Lansdown 1995). Mathew (1894) described them as resident in small numbers and common winter visitors in Pembrokeshire, while they were said to be very scarce, and probably localised, breeding birds in Ceredigion at the end of the 19th century (Roderick and Davis 2010). Forrest (1907) observed that they were increasing all over North Wales and were common in the eastern part but scarcer on the west coast. Ingram *et al.* (1966) said that they were still scarce and local in Ceredigion until the 1920s. Lockley *et al.* (1949) considered that they remained scarce residents in the eastern half of Pembrokeshire.

Numbers increased in many parts of Wales owing to the expansion of upland forestry from the 1950s. Condry (1960) said that redpolls, which had been scarce and local across northern and central Wales until the 1920s, bred in young plantations up to 600m by the late 1950s. Parslow (1973) considered that the species had benefited more than any other from afforestation in Wales. There has been a severe decline in the UK since the 1970s, particularly in England, where the CBC/BBS index showed a precipitous decrease between 1978 and 1995. Fuller *et al.* (2005) considered that a reduction in planting of conifers, and hence a decrease in the extent of suitable young growth, may have been important in southern Britain. There was also a reduction in the area of scrub, particularly birch, within forests by the late 1970s (Locke 1987), which could have reduced food availability. In northern Britain and Wales, the extent of population change was unclear, because the CBC index was not representative of habitats there. Local evidence suggested that there was no significant decline in Wales. Big increases in the Gwent breeding population were recorded in the late 1960s and early 1970s (Venables *et al.* 2008). Local surveys in Glamorgan in the 1980s and early 1990s indicated a stable population, possibly increasing in some areas (Hurford and Lansdown 1995). *Birds in Wales* (1994) stated that numbers may have declined after the 1960s, as the forests matured, but increased again in the late 1980s and early 1990s. The UK population of the Lesser Redpoll has been faring well in recent years. The BBS index has risen by 31% from 1994 to 2018. Only a short-term trend is available for Wales, but this showed no overall change during 2008–18 (Harris *et al.* 2020).

The *Britain and Ireland Atlas 1968–72* showed the species to be widely distributed across Wales, but sparse in the southwest, particularly Pembrokeshire, and in Gwent. The 1988–91 *Atlas* showed gains in these areas. This pattern continued in the 2007–11 *Atlas*, where it was found in 74% of 10-km squares. There were more gains than losses in Wales, contrasting with large-scale losses in England, between 1968–72 and 2007–11. Abundance was relatively high in the Welsh uplands in 2007–11, higher than in 1988–91. Tetrad atlases provided a more detailed picture of its distribution. It was recorded with breeding evidence in 20% of tetrads in the *Gwent Atlas 1981–85* and in 15% in the *Gwent Atlas 1998–2003*. Of the tetrads occupied, 70% in the first *Atlas* contained conifer plantations, but this had fallen to 44% in the second *Atlas*. It was thought the distribution of Lesser Redpolls had contracted, as trees matured and the canopy closed, exacerbated by the removal of birch from plantations (Venables *et al.* 2008). In East Glamorgan, it was present with breeding evidence in 17% of tetrads in the 1984–89 *Atlas* and in 17% in 2008–11, while in West Glamorgan it was present in 28% of tetrads in 1984–89 but absent from the Gower Peninsula. In Pembrokeshire, it occurred in only 6% of tetrads in 1984–88, increasing slightly to 8% in 2003–07 (*Pembrokeshire Avifauna* 2020). In Breconshire it bred in 24% of tetrads visited in 1988–90 (Peers and Shrubb 1990). The *North Wales Atlas 2008–12* found it in 37% of tetrads and more widely distributed in western counties than in the east, occurring in 48% of tetrads in Meirionnydd and 44% in Caernarfonshire. Occupied tetrads were mainly in semi-upland and upland areas, coinciding with forestry plantations. Few tetrads were occupied in coastal areas. There have been few studies of population densities in Wales, but it can be abundant in young forestry plantations before the canopy closes. Harris and Williams (1995) found Lesser Redpoll to be the most abundant passerine in

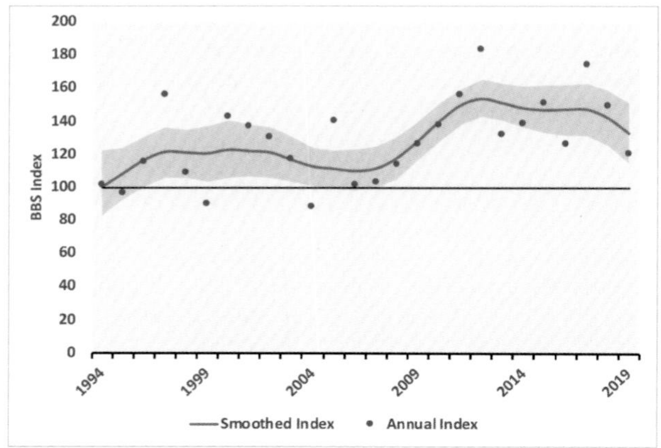

Lesser Redpoll: Breeding Bird Survey indices for the UK, 1966–2018. The shaded area indicates the 85% confidence limits.

plantations on Llanbrynmair Moor, Montgomeryshire, in 1994, with a density of 52 pairs/km^2.

Outside the breeding season, Lesser Redpolls gather in flocks, sometimes in the company of other finches, usually Siskins but also Linnets and Goldfinches. Flocks feed largely on birch and Alder seeds, although weed seeds are also important. Winter distribution increased by 45% between the *Britain and Ireland Winter Atlas 1981–84* and the 2007–11 *Atlas*, from 50% of 10-km squares to 72%. Counts of over 300 have been recorded in several counties, such as *c.*320 at RSPB Ynys-hir, Ceredigion, in August 1997. Spring and autumn passage has been observed along the coast and on the islands, revealing fairly small numbers in most years, but at least 150 passed the Great Orme, Caernarfonshire, on 6 April 2007. Autumn numbers were usually lower, seldom exceeding 50, though 200 at Cwmgwdi, Breconshire, on 21 October 2011, were thought to be migrants.

This species was formerly Red-listed in Wales. This was based on the same listing in the UK, following a long-term population decline, but as it appeared to be doing well in Wales, this listing was subsequently revised to Amber (Johnstone and Bladwell 2016). Rees *et al.* (2009) considered that extensive felling of conifer plantations in Pembrokeshire was likely to have had a detrimental effect on this species. However, in situations where extensive clear-felling of mature conifers leads to either the restocking of conifers or the regeneration of birch forest, Lesser Redpoll numbers quickly increase again, as happened in Pembrokeshire (Paddy Jenks *in litt.*) and in the Aber Valley, Caernarfonshire (pers. obs.). The Lesser Redpoll's medium-term prospects in Wales may depend largely on the future of woodland policy. The Welsh Government is committed to increasing tree planting and to diversify woodland by moving away from even-aged, single-species stands (Welsh Government 2018a), but it remains to be seen what effect this will have on bird species that rely on conifers. However, its longer-term prospects are less certain, as southern and central Wales are expected to be no longer climatically suitable for Lesser Redpoll later by the end of this century (Huntley *et al.* 2007).

Rhion Pritchard

Sponsored by Heather Coats

Arctic Redpoll *Acanthis hornemanni* **Llinos Bengoch Coue**

Welsh List Category	IUCN Red List		Rarity recording
	Global	Europe	
A	LC	LC	WBRC

Although the taxonomy of all the redpolls remains uncertain, two subspecies of Arctic Redpoll are recognised: Coues's Arctic Redpoll, *A.h. exilipes*, which breeds in northern Eurasia, Alaska and northwest Canada, and Hornemann's Arctic Redpoll, *A.h. hornemanni*, which breeds in northeast Canada and Greenland (IOC). No Welsh records have been identified to subspecies, but it is presumed that they were *A.h. exilipes*, which nests in northern Europe and has a range that partially overlaps with that of the 'Mealy' Common Redpoll (*A.f. flammea*). Occurrence in Britain has increased dramatically, boosted by influxes in 1990/91 and 1995/96, when 76 and 459 were recorded respectively (Holt *et al.* 2019, White and Kehoe 2019, White and Kehoe 2020b). However, this remains a very rare winter visitor to Wales with only six recorded individuals:

- Bardsey, Caernarfonshire, on 3–4 May 1987;
- Goodwick, Pembrokeshire, from 6 February to 17 March 1996;
- Great Orme, Caernarfonshire, on 11 April 1996;
- two at Marford, Denbighshire, on 6–11 March 1996 and
- Llanfairfechan, Caernarfonshire, on 15 and 23 March 2002.

Jon Green and Robin Sandham

Crossbill *Loxia curvirostra* **Gylfingroes**

Welsh List Category	IUCN Red List			Birds of Conservation Concern			
	Global	Europe	GB	UK	Wales		
A	LC	LC	LC	Green	2002	2010	2016

The Crossbill has a very extensive distribution, with its core range in the great northern forests of Eurasia and North America, but extending into Central America, North Africa and the Himalayas (*HBW*). This species is able to nest at different times of year, depending on food availability. In the northern part of its range, the seeds and buds of Norway Spruce are of particular importance, but it also feeds on the seeds of other conifer species. In Scotland, nesting can start as early as August and continue through the winter, with birds feeding on the seeds of Sitka Spruce, while in spring it can switch to Scots Pine seeds (Nethersole-Thompson 1975). Birds tend to disperse in summer. Eurasian birds are periodically eruptive, when poor spruce and pine seed crops coincide with high population levels. In these years, large numbers disperse from their core range. There were 78 eruptions between 1800 and 2000 (*HBW*). In those years, large numbers moved into Britain, usually from midsummer, some from over 4,000km away in northern Russia. Recoveries of birds ringed mainly in Germany confirmed that birds do not always return to their natal breeding grounds in the same calendar-year, but breed at new sites for a year or two, before at least some return to their original area in a later year, when a new Norway Spruce crop becomes available (Newton 2006).

Conifers were comparatively scarce in Wales in the 19th and early 20th centuries. This was reflected in the scarcity of Crossbills. Dillwyn (1848) recorded a flock in Glamorgan in winter 1806. Good numbers were recorded in some counties during a few irruption years, such as 1830, 1865, 1867–68 and 1898, although most reports simply mention "large numbers" or "large flocks", without providing counts. Matthew (1894) knew of only two records in Pembrokeshire, and there were apparently only three records from Anglesey prior to 1907. The first breeding record in Wales seems to have been at Llanyblodwel, Montgomeryshire, in 1880, when a pair nested in a fir tree (Forrest 1907). In Caernarfonshire, 2–3 pairs nested in larches at Penmaenmawr in 1890 or 1891 (Pentland 1899). Breeding was suspected in Gwent in 1898, when a female shot at Michaelston y Fedw on 28 June had a bare breast, apparently a brood patch (Venables *et al.* 2008). Breeding was also suggested at Margam, Gower, the same year. A pair with one young was reported in Carmarthenshire in July 1904 (Barker 1905), while the first breeding in Radnorshire was recorded in 1910 (Jennings 2014).

There appears to have been no change in the status of this species in Wales during the first half of the 20th century, with the majority of records confined to irruption years such as 1930, but

Male Crossbill © Tony Pope

there was an increase in records from the 1950s. In Pembrokeshire, Donovan and Rees (1994) mentioned one record in 1929, then records in 24 subsequent years, most during 1950–92. Following the Second World War, there was extensive planting of conifers, particularly Sitka Spruce, in Wales, and as these matured and began to produce seeds, the prospects for Crossbills in Wales improved. An irruption in July 1962 led to increased numbers in several counties, with breeding or probable breeding in 1963 and 1964, including the first confirmed record for Glamorgan.

The Crossbill is stimulated to breed by the abundance of food, regardless of the season (*BWP*). In the upland forestry plantations of South Wales, Dixon and Haffield (2013) found that the breeding season began early in the year, with a median laying date of 13 February. Few nests were initiated after April, when the availability of Sitka Spruce seeds declined. This unusual breeding season, together with the highly variable numbers from year to year, means that the Crossbill is not well covered by fieldwork for breeding atlases, and breeding birds may easily be missed. However, the general pattern is clear. The *Britain and Ireland Atlas 1968–72* found very few Crossbills in Wales, with no records at all in North Wales and those that were present being mainly in Gwent. Crossbills were more numerous during the period of the 1988–91 *Atlas*, being present in 28% of the 10-km squares in Wales. Birds were widespread in most counties, although few were recorded in Pembrokeshire or much of Ceredigion. The species had been confirmed breeding in all Welsh counties by 1991, when it occurred in Pembrokeshire (Rees *et al.* 2009). The 2007–11 *Atlas* showed a further increase in all counties except Anglesey, with birds found in 48% of 10-km squares and many records of confirmed breeding.

Atlases at tetrad level also suggested an increase where two atlases could be compared. The first *Gwent Atlas* in 1981–85 recorded the species in 6% of tetrads, while the second *Atlas* in 1998–2003 found it in 9%. There was a greater increase in East Glamorgan, where the species was found with evidence of breeding in only 2% of tetrads in 1984–89. It was present in 16% of tetrads in 2008–11, although recorded with breeding codes only in 11%. This species is far less widespread in Pembrokeshire, where it was not recorded in the 1984–88 *Atlas*, but was found in four tetrads (less than 1%) in 2003–07, although with no confirmed breeding (*Pembrokeshire Avifauna* 2020). The *North Wales Atlas 2008–12* recorded the presence of Crossbill in 10% of tetrads. As elsewhere, records were concentrated in the large blocks of forestry plantations.

There was a particularly large irruption in 1990, when birds, apparently from northern Russia, reached much of western Europe. By winter 1990/91 there were thought to be at least half a million in Scotland, possibly over a million (Jardine 1992). Numbers reaching Wales were smaller, but some large flocks were seen: for example, over 250 at Abbeycwmhir, Radnorshire, from August onwards; over 600 in the Gwaelod Plantation in the Elan Valley, Radnorshire, in October; 200 in Pentraeth Forest, Anglesey, in October and "hundreds" in Clocaenog Forest, Denbighshire, by the end of the year. Following this irruption, higher numbers bred the following year, with for example about 200 pairs in Radnorshire (Jennings 2014).

A large irruption into Britain in 2001 produced no high counts in Wales, but another in 2002 produced good numbers, including over 1,000 in Clocaenog Forest in January 2003. This appeared

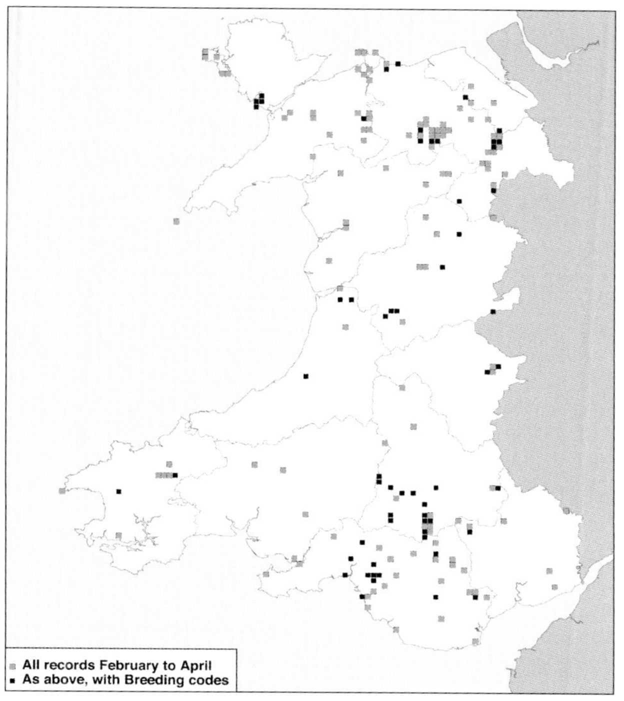

Crossbill: Records from BirdTrack for February to April, 2015–19, showing the locations of records and those with breeding codes.

to be the highest count recorded in Wales, but was matched by 1,000 at the same site on 30 December 2007. Other notable counts during the last decade included 250 in Radnor Forest in 2011, along with passage birds at coastal watch points, such as 180 at Lavernock, East Glamorgan, on 30 September 2015, 195 there on 30 October 2018, and 290 at Dunraven, East Glamorgan, on 29 September 2015.

The number of pairs breeding in Wales is highly variable, so it is difficult to give any meaningful population estimate. Breeding numbers are much higher in the year or two following an irruption, but there does now appear to be a core breeding population in the larger coniferous forests, even in years with no significant immigration. Many of the commercial conifer forests in Wales, planted in the 1950s and 1960s, are now a suitable age for harvesting. The future of Crossbills, here, depends largely on commercial forestry management ensuring a good supply of mature, 'old-growth' trees. In the longer term its distribution may well be governed by climate, affecting the ability to grow suitable trees in Wales and also leading to a projected northward shift. The northeast may be the only part of Wales suitable by 2070–2100 (Huntley *et al.* 2007).

Rhion Pritchard

Two-barred Crossbill *Loxia leucoptera* Gylfingroes Adeinwyn

Welsh List Category	IUCN Red List		Rarity recording
	Global	Europe	
A	LC	LC	BBRC

The Two-barred Crossbill breeds in Fennoscandia and Russia, east to the Pacific Ocean, and across northern North America (*HBW*). In the Palearctic, the subspecies *L.l. bifasciata* is prone to occasional irruptions, especially in the summer months, and on these occasions, birds can sometimes reach as far as Britain (*HBW*). The last influx, in 2019, involved 219 individuals. There have been 604 records of this extremely rare vagrant in Britain, up to the end of 2019 (Holt *et al.* 2020). Despite this, there have been only three records in Wales, comprising six birds:

- one found dead at Llandrindod Wells, Radnorshire, in November 1912;
- one at Llanfihangel Glyn Myfyr, Denbighshire, on 3–26 March 1991 and
- four at Nercwys Mountain, Flintshire, on 17 February 2014.

Jon Green and Robin Sandham

Goldfinch *Carduelis carduelis* Nico

Welsh List Category	IUCN Red List			Birds of Conservation Concern			
	Global	Europe	GB	UK	Wales		
A	LC	LC	LC	Green	2002	2010	2016

The Goldfinch is widespread in Europe, as far north as 65 degrees (the southern half of Fennoscandia), and east through European Russia and western Siberia, and extends to North Africa and the Middle East (*HBW*). It is more common in the south of its range, where food availability is highest for much of the year (*European Atlas* 1997). In winter, most of its eastern range is vacated and enormous numbers concentrate in the Mediterranean basin, where the mild, damp winters ensure the almost continuous growth and seeding of the herbaceous plants they need for food (*European Atlas* 1997).

Since the 1980s the Goldfinch has become a regular, colourful visitor at garden bird-feeders in Britain, with a special liking for Nyjer seeds and sunflower hearts. The number of bird feeders in gardens doubled between 1973 and 2010, and the proportion of gardens, where Goldfinches were observed using feeders, rose from less than 20% to over 80% (Plummer *et al.* 2019). It is widespread and common across lowland Wales, although infrequently recorded above 350m. Preferred habitats are farmland, woodland edge, waste ground, gardens and parkland, where there are plentiful thistles and other plants of the daisy family to provide abundant seeds.

In the 19th and early 20th centuries, it was commonly kept as a cage bird, with thousands caught across the country each year for this trade. Forrest (1907) claimed that one bird-catcher in the Llandudno area caught as many as 3,000 in one year. Mathew (1894) remarked that Goldfinches remained abundant in Pembrokeshire "despite great numbers being taken in the autumn by bird-catchers, for example, thirty-three dozen were caught at one time at Fishguard". However, in Breconshire they were "not much troubled with professional bird-catchers… but only by a few amateurs, so that I think this may in some way account for their numbers" (Phillips 1899). Inevitably numbers declined, although not to the extent seen in England (Witherby and Ticehurst 1907–08), but the species soon recovered once legal protection was introduced in the early 20th century.

The *Britain and Ireland Atlas 1968–72* showed that the Goldfinch bred in 94% of 10-km squares in Wales, with absence or unconfirmed breeding in a few 10-km squares in the uplands. During the 1988–91 *Atlas*, all but three squares were occupied (96%) and by the 2007–11

Atlas, there was 99% occupancy. This pattern was reflected in local atlases, which showed increases between the 1980s and the early 2000s. In East Glamorgan, tetrad evidence of breeding increased from 67% in 1984–89 to 69% in 2008–11. In Pembrokeshire there was an increase in tetrad occupancy from 62% in 1984–88 to 81% in 2003–07 (*Pembrokeshire Avifauna* 2020). In Gwent the increase was from 85% to 93% between 1981–85 and 1998–2003 (Venables *et al.* 2008). The *North Wales Atlas 2008–12* showed an increase from 95% of 10-km squares in 1968–72 to 99%. The relative abundance maps in the *Britain and Ireland Atlas 2007–11* showed the greatest densities in the lowlands of eastern Wales, Pembrokeshire and Anglesey, and in sheltered valleys elsewhere.

Trends from the BBS and CBC also showed an increase in Goldfinches across the UK. In Wales, the BBS index increased by 104% between 1995 and 2018, though the rate of increase has slowed since 2002 (Harris *et al.* 2020). This increase was reflected in the reporting rates, as well as in numbers reported, in the BTO Garden BirdWatch (from 20% to 70%) since the mid-1990s.

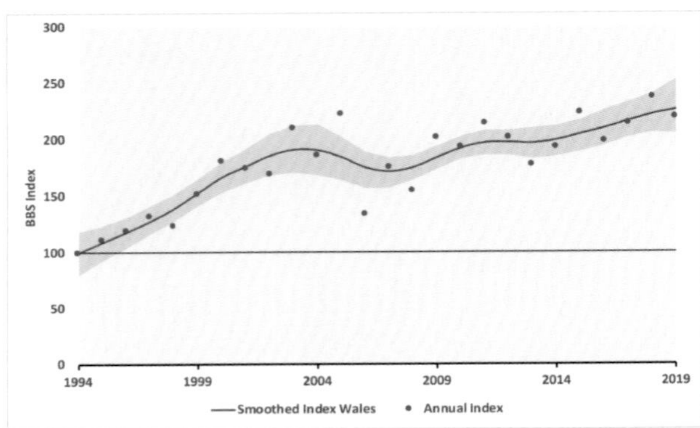

Goldfinch: Breeding Bird Survey indices, 1994–2019. The red line shows the smoothed index, 1995–2018. The shaded area indicates the 85% confidence limits.

Goldfinches form roving flocks after fledging, sometimes in association with Linnets. These very mobile flocks increase in size from August onwards. They move to wherever suitable food (e.g. seeds of thistles and knapweeds) is available, then move on when it is depleted. In the 20th century, flocks of dozens of birds were frequently reported, but only a few reached 50 or more. Since 1995, flocks of 50–100 have been recorded in most Welsh counties, and flocks of over 100 have become more common. By October there is evidence of migration. Flocks of over 1,000 are regular along the Rhymney Estuary, and over 1,000 passed east over Lavernock Point, East Glamorgan, on 8 October 1976. In 2016, counts there reached 2,000 and 3,000 on 31 October and 1 November respectively. In October 2019, flocks of between 200 and 500 were regularly seen feeding on linseed and sunflower seeds in fields at The Vile, Rhossili, Gower. Passage has also been observed along the North Wales coast, especially at Rhyl in Flintshire, RSPB Conwy in Denbighshire, and on Anglesey, but the numbers involved have been much lower, usually 100–200 per day. Autumn passage has also been quite marked at the bird observatories, with over 250 recorded on some days at Bardsey, Caernarfonshire, where the highest count was 439 on 31 October 2015. 100–150 were on Skokholm, Pembrokeshire, reaching a maximum of 285 on 4 October 2013.

The *Britain and Ireland Winter Atlas 1981–84* and the 2007–11 *Atlas* both showed similar distributions in winter and summer, although the winter density was far lower in Wales than in England. Some are thought to migrate. However, the extent and direction of the migration seems to depend on food supply and weather conditions, rather than the birds being hard-wired to a particular route or destination (*BTO Migration Atlas* 2002). Many Welsh breeding birds migrate to Spain, but ringing data indicated there is also an interchange with Ireland. Few birds came from farther north to winter in Wales, in contrast to southern England, where birds arrived from northern Britain, possibly en route to wintering grounds in western France and parts of Iberia. Those that remained in Wales were susceptible to severe weather conditions. Winters such as 1895, 1962/63 and 1980/81 took a considerable toll on Goldfinches (*Birds in Wales* 1994). However, there is little evidence that the more recent cold winters of 2009/10 and 2010/11 had any effect, possibly because of the ready availability of supplementary food in gardens.

Spring migration is most noticeable on the islands. At the bird observatories on Bardsey, Caernarfonshire, and Skokholm, Pembrokeshire, there is a marked increase in the numbers seen between the last week of March and mid-May. However, while numbers on Bardsey have exceeded 150 on several occasions, there are generally fewer than 20 in a day on Skokholm.

Ringing recoveries showed that Goldfinches ringed in Welsh counties have been found across most parts of Britain, while a similar number ringed elsewhere have turned up in Wales. Of 80 recoveries, there were 31 movements to England, four to Scotland and 18 to continental Europe. Another 19 moved to Ireland, all of which were ringed between October and March, but were found at any time of year, though mostly in March–April. These included birds found 11 and 39 days after ringing, suggesting a spring movement westward. One ringed in Northern Ireland, in April 2002, was found in Wales the following March. Of those birds moving into Wales, only six were from the west. 55 of the other 70 were from England. The spread of ringing and recovery dates does not suggest any consistent direction or timing of movements. The greatest distance travelled was by a bird ringed near Talybont-on-Usk, Breconshire, on 21 August 1976, that was trapped 15 months later on 7 November 1977, some 1,229km away in Portugal (place name unknown). The oldest known Goldfinch in Wales was ringed as an adult near Abergavenny, Gwent, on 30 December 2007, and was last caught there on 17 January 2016 at the age of 8 years, zero months and 18 days.

Goldfinch: Recovery locations of birds ringed in Wales are shown by red circles. Ringing locations of birds that were recovered in Wales are shown by blue triangles. Small brown dots show ringing or recovery locations in Wales.

The Goldfinch has been doing very well, expanding its range from 60 degrees north to 65 degrees in Europe and occasionally farther (*HBW*). This northwards expansion, which is consistent with climate-change modelling, is anticipated to go as far as northern Sweden by 2070–2100 (Huntley *et al.* 2007). The European population, which went through a major increase in the late 1980s and early 1990s, is relatively stable now, with a slight increase of 9% between 2007–2018 (PECBMS 2020). The Goldfinch has also increased in abundance in Wales, over the last two decades especially. The Welsh breeding population in 2018 was estimated to be 180,000 (160,000–195,000) territories (Hughes *et al.* 2020). If garden bird-feeding continues to supplement or replace any losses of the natural food supply, there is no reason why these increases should not continue.

Annie Haycock

Serin *Serinus serinus* Llinos Frech

Welsh List Category	IUCN Red List		Rarity recording
	Global	Europe	
A	LC	LC	WBRC

The Serin is resident in southern Europe and France, and in the breeding season extends its range into the rest of continental Europe west of Russia (*HBW*). Numbers recorded in Britain each decade increased during the second half of the 20th century, peaking in the 1990s, but have since declined, with an average of 39 recorded each year during the 2010s (White and Kehoe 2020b). It occurs infrequently in Wales, apart from the ten recorded in the 2000s, repeating the British pattern.

Total	Pre-1950	1950–59	1960–69	1970–79	1980–89	1990–99	2000–09	2010–19
34	4	2	1	6	3	5	10	3

Serin: Individuals recorded in Wales by decade.

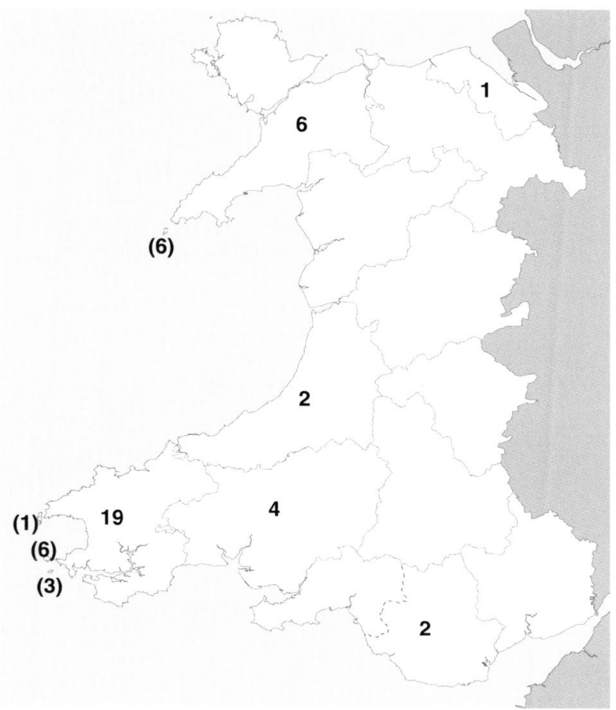

Serin: Numbers of birds from each recording area, 1933–2019.

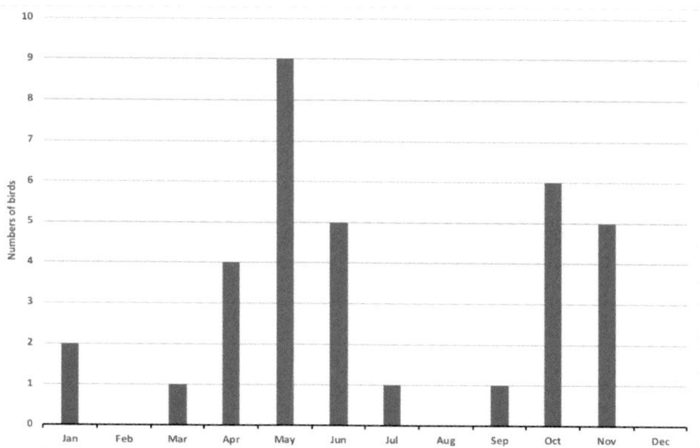

Serin: Occurrence of arrival dates by month, 1933–2019.

The first Welsh record was of two at Marloes, Pembrokeshire, on 21 October 1933. The majority of records have come from that county: a total of 19, of which six have been on Skomer, three on Skokholm, one on Ramsey and nine on the mainland. Another six occurred on Bardsey, Caernarfonshire. The other records were from East Glamorgan: a male at Kenfig in March 1972 and one at Nash Point on 24 November 2019. There have been two records in Carmarthenshire: two at Cydweli in January 1955 and a pair at Burry Port in April 1987. There were also singles at RSPB Ynys-hir, Ceredigion, in October 1975 and June 2002 and at Point of Ayr, Flintshire, in June 1990.

The Serin could extend its breeding range north from the French coast to the whole of Britain south of a line from the Dyfi to the Humber Estuary by the end of this century (Huntley *et al.* 2007), so we may expect many more than have been recorded to date.

Jon Green and Robin Sandham

Siskin *Spinus spinus* Pila Gwyrdd

Welsh List Category	IUCN Red List			Birds of Conservation Concern			
	Global	Europe	GB	UK	Wales		
A	LC	LC	LC	Green	2002	2010	2016

The breeding range of the Siskin extends over much of central and northeast Europe, locally at altitude in western and southern Europe, and discontinuously east to the Pacific Ocean. Whilst many western populations are mainly resident, those in the east move south in winter (*HBW*). It breeds mainly in coniferous and mixed forests, and feeds on seeds, buds and fruit. In Wales, breeding has been recorded up to 530m in Meirionnydd (Pritchard 2012). Conifer seeds are an important food source in spring, when the cones open (*HBW*), although other seeds, including dock, dandelion and thistle, become important later in the year.

In Britain, the Siskin, at one time, bred only in the Caledonian pinewoods of Scotland (Newton 1972). The earliest record in Wales dates from 1802, when the species was listed in the *Swansea Guide*. It was mainly a winter visitor to Wales in the late 19th century and early 20th century. The earliest record of possible breeding was in 1872, when W.J. Kerr saw a flock of about 20 on 6 August and noted that most were birds of the year (Kerr 1873). The location was not given, but Kerr lived in western Denbighshire, near the border with Meirionnydd. Forrest (1907) mentioned a few instances of possible breeding in North Wales, the most definite being a young bird, not fully fledged, found dead near Colwyn Bay, Denbighshire, in June 1899. There was a record of nesting near

Male Siskin in displaying song-flight © John Hawkins

Llandygai, Caernarfonshire, the same year, but Forrest considered this less reliable. He commented that the Siskin "probably breeds in North Wales more frequently than is supposed, though still not often". It was, however, a regular winter visitor to northern counties. A flock of 100 was on Alders at Palé, Meirionnydd, on 7 December 1902, for example. Farther south, it was a scarce winter visitor in the 1890s, described as rare in Pembrokeshire, occasional in small flocks in Breconshire, and regular but uncommon in Glamorgan (Mathew 1894, Phillips 1899, Hurford and Lansdown 1995).

There was little indication of any change in the status of the Siskin in Wales over the following 50 years, although the authors of *Birds in Wales* (1994) considered that breeding may have been continuous since Forrest's time. Ingram and Salmon (1954) regarded it as an occasional winter visitor in Carmarthenshire. The first confirmed breeding in the county was in 1971 at RSPB Dinas. Large-scale planting of conifers in the years after the Second World War created a great deal of suitable habitat, as trees matured and began to bear cones. Condry (1966) considered that Siskins had probably bred in the Gwydyr Forest, Caernarfonshire, since the mid-1940s. Breeding was confirmed at Lake Vyrnwy, Montgomeryshire, in 1957 (Holt and Williams 2008), in the Elan Valley, Radnorshire and at Corris, Meirionnydd, in 1967. About 40 singing males were recorded in the Gwydyr Forest in April 1968.

The *Britain and Ireland Atlas 1968–72* found Siskins present in 12% of 10-km squares in Wales. It confirmed breeding in several counties, but primarily in Meirionnydd and Caernarfonshire. As conifer plantations matured, the Siskin spread rapidly. By the *Britain and Ireland Atlas 1988–91*, it occurred in 50% of squares, with confirmed breeding in all counties except Pembrokeshire and Flintshire. Fieldwork for the 2007–11 *Atlas* recorded Siskin in 84% of 10-km squares in Wales and confirmed breeding in the majority of these across all counties.

Tetrad atlases also showed the dramatic expansion of this species. It was fairly scarce in Gwent at the time of the 1981–85 *Atlas*, found with evidence of breeding in 10% of tetrads, but by 1998–2003 it occurred in 28% of them. Siskins were found with breeding evidence in only 7% of tetrads in the *East Glamorgan Atlas 1984–89*, but in 26% in 2008–11. The *Pembrokeshire Atlas 1984–88* found the species in only 1% of tetrads, but this had increased to 15% by 2003–07. The *North Wales Atlas 2008–12* recorded Siskins in 40% of tetrads, including 58% of those in Meirionnydd.

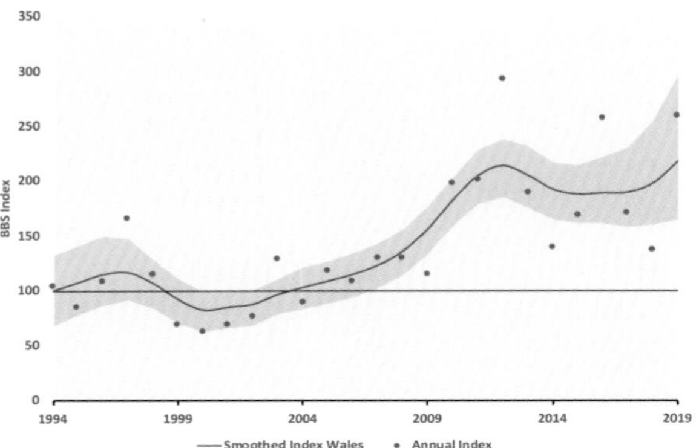

Siskin: Breeding Bird Survey indices, 1994–2019. The red line shows the smoothed index, 1995–2018. The shaded area indicates the 85% confidence limits.

The BBS index for Wales increased by 84% between 1995 and 2018 (Harris *et al.* 2020), the rapid growth being possibly partly a result of birds moving into Wales from farther north in winter and remaining to breed. Breeding densities in Wales range from 47 individuals/km^2 (Bibby *et al.* 1985) to the equivalent of 165 pairs/km^2 (Currie and Bamford 1982) and may depend on the seed crop. Some move south after breeding in Wales, many to southern England but others fly farther south: one ringed in Guipuzcoa, northern Spain, on 15 April 2009 was recaught near Glyn Ceiriog, Denbighshire, 14 days later.

The usual autumn migration season for Siskins is mid-October and early November, soon after the normal seed-shedding period of birch, which is one of their main foods in late summer and autumn (Newton 1972). Eruptions are recorded in some years, possibly when the local breeding population has reached a high level, resulting in unusually large counts on passage and in winter. Skokholm, Pembrokeshire, recorded 1,200 in fog on 26 October 1988, while in East Glamorgan there have been several large counts at Lavernock, including *c.*1,200 flying east on 18 October 1985, and 1,000/hour east over Dunraven on 22 October 1997. On the north coast of Anglesey, 1,056 passed over Carmel Head in three hours on 21 October 2007 and 1,500 moved east,

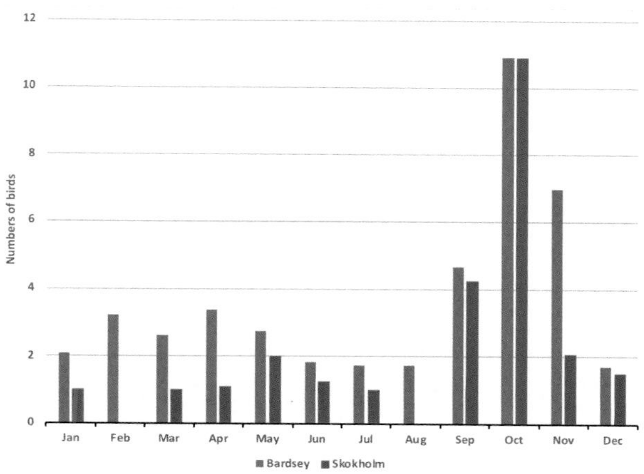

Siskin: Mean counts, by month, at Bardsey and Skokholm, 1954–2019.

past Cemlyn, on 29 October 2014. Further east, 911 were seen on migration at Point of Ayr, Flintshire, on 21 October 2019. The numbers recorded on spring passage have usually been much lower, possibly because some individuals make their northerly journey at speed. This was illustrated by one ringed at Bagillt, Flintshire, at 16:20hrs on 17 April 2019 and recaught at Peebles in southern Scotland at 19:20hrs on 18 April, a movement of 267km in 27 hours (Giles Pepler *in litt.*).

In winter, Siskin numbers increase in Wales. Ringing recoveries indicate that the winter population is boosted by breeding birds from northern Britain and Europe, notably Norway and Sweden. A few come from farther east: one ringed in February 2013 at Mynydd Llandygai, Caernarfonshire, was recaught 2,360km away near St Petersburg, Russia, in July the same year. The areas of highest relative abundance in winter are different to those in the breeding season, as birds often vacate the coniferous forest and the uplands of North and mid-Wales once cone seed supplies are depleted. Many birds move into the lowlands, notably in the south and east (*Britain and Ireland Atlas 2007–11*).

Siskins historically fed mainly on Alder, birch and larch seeds in winter, often in mixed flocks with Lesser Redpoll. In recent years, they have become a common sight at garden feeders, a habit first recorded in 1963 (Spencer and Gush 1973). Nyjer and sunflower hearts can maintain a population through winter and have become a supplemental food source for newly fledged young from early May. BTO Garden Birdwatch data, between 1995 and 2019, showed that the species is recorded in over 40% of gardens in Wales in some years, with peaks of 46% in March 1995 and 45% in April 2013. There has been considerable variation from year to year, but no trend is evident. However, there is a very pronounced seasonal pattern, with the peak use of gardens in March and April, when there is a mix of birds: those returning north and fuelling up, and local birds gaining weight for the breeding season. The lowest level of garden use is in October and November. McKenzie *et al.* (2007) found that Siskins were present in gardens in Britain more often in years of poor Sitka Spruce cone crops.

Flocks of several hundred are recorded quite frequently in late winter and early spring, with larger flocks in some years when there has been an influx. The largest counts have been in Denbighshire, including at least 995 at Brynhyfryd, Clocaenog Forest, on 23 February 2008, 1,000 near Pendinas Reservoir in Llandegla Forest on 22 March 2007 and an exceptional 3,000 at Clocaenog on 18 January 1997.

The Siskin population is doing well in Wales, both as a breeding and wintering bird. The Welsh breeding population in 2018 was estimated to be 58,500 territories (Hughes *et al.* 2020). An ongoing study in Pembrokeshire indicated that adult survival has increased, perhaps aided by garden feeding and changing patterns of migration and wintering (own data). Its medium-term prospects in Wales depend to a large extent on the area of mature coniferous forest available. There has been a reduction of 18,000ha in the area of coniferous forest in Wales since 2001. Most new planting is of broadleaved trees, although 129,000ha of stocked conifer woodlands remained in 2012 (Warren-Thomas and Henderson 2017). In the long term, its future will depend on Welsh forestry policy and climate change. The area that is climatically suitable for Siskin is expected to retract northwards during the 21st century, and by 2100 only a small part of North Wales may accommodate the species (Huntley *et al.* 2007).

Richard Dobbins

Sponsored by Giles Pepler

Lapland Bunting *Calcarius lapponicus* Bras y Gogledd

Welsh List Category	IUCN Red List			Birds of Conservation Concern			
	Global	Europe	GB	UK	Wales		
A	LC	LC	VU	Amber	2002	2010	2016

The Lapland Bunting is a circumpolar breeder of Arctic dwarf scrub tundra that winters to the south of the high-latitude coniferous forest regions of Eurasia and North America, often along coasts, but also in steppes, prairies and on agricultural land (*HBW*). Lapland Buntings passing through Britain, especially the east, are thought to be of the nominate subspecies from breeding areas in Fennoscandia and western Russia. However, several authors have argued that small numbers of the subspecies *C.l. subcalcaratus* from eastern Greenland reach Europe via Iceland, occurring from southwest Norway south to northwest France, including Britain and Ireland (e.g. Williamson and Davis 1956, Jacobsen 1963, Yésou 1983, Schekkerman 1989, Fox *et al.* 1992, Petersen 1998).

The Lapland Bunting is a scarce but regular autumn migrant in Wales with 60% of records in September and October. Variable numbers have remained to winter in some years, rarely more than 100, but there is no marked return spring passage, with only 6% of records in March–May. Sightings primarily came from beach strandlines, saltmarsh and coastal grazing land, but this may reflect the fact that these habitats attract many birdwatchers, since the species can be observed very far inland, especially on migration. It can be found gleaning grain in stubble fields, so habitat does not seem to limit its numbers. It seems more likely that this super-abundant Arctic nesting passerine simply does not pass through Wales in large numbers on a regular basis.

Surprisingly for such a scarce species, there has been a recovery in Wales of a French-ringed Lapland Bunting. It was ringed as a first-calendar-year male in Saintes, Charente-Maritime, northwest France on 14 October 2007 and found freshly dead on 11 December 2007 near Llanrwst, Denbighshire (Coiffait *et al.* 2008).

The species was first reported in Wales from Skokholm, Pembrokeshire, in September 1936 (Lockley *et al.* 1949). It has been more or less regular there, on neighbouring Skomer and on Bardsey, Caernarfonshire, every year since, although it has now also been regularly reported from mainland migration points on the Great Orme, Caernarfonshire, and Strumble Head, Pembrokeshire. The species occurred annually in reasonable numbers on Anglesey and Gower, but its relative scarcity makes it

Lapland Bunting © Ben Porter

difficult to assess any trend in Wales. Annual numbers have varied greatly, but between 2009 and 2019, an average year reported *c.*50 individuals, with most reports being of singles, and occasionally flocks of up to ten. However, there was an exceptional influx in autumn 2010 when records in the *Welsh Bird Report* suggested that over 300 individuals were present in Wales in autumn and early winter that year. This unusual occurrence was mirrored throughout Britain, following the arrival of hundreds of Lapland Buntings in the Northern and Western Isles in the last week of August. Their arrival coincided with prolonged and persistent easterly winds, which suggested these were of Fennoscandian and European Russian origin, rather than Greenland birds that perhaps comprise the normal, smaller numbers. Unprecedented numbers occurred in Wales from the last few days of August 2010, with 26–50 remaining above Common Cliff, Gower, in November–December, 36 in January 2011 and 12 remaining to 9 March. Large numbers were also reported from Pembrokeshire, such as 20 at Dale Airfield, 16 on Skomer and eight at Martin's Haven in October, but these seemingly did not over-winter. Up to 37 were on Bardsey in early October, where birds persisted until the following spring. There were remarkable numbers on Anglesey, including 54 west of Wylfa and 70 around Hen Borth throughout October. These numbers were very much greater than anything witnessed in normal years, the previous largest count having been 30 in November–December 1962 on Bardsey and 26 at Strumble Head in October 1987 and September 1988 (*Birds in Wales* 1994). It is not clear whether the 2010 influx was caused solely by weather patterns at the time of migration or by exceptional breeding success. In west Greenland, periodic plague outbreaks of larvae of the Great Brocade moth (Lund *et al.* 2017) have been associated with years of elevated Lapland Bunting reproductive success (Fox *et al.* 1987). Since this moth is said to be widely distributed across the Arctic (Dahl *et al.* 2017), this may be the case in other parts of its breeding range, explaining its periodic eruptive occurrence to Wales and elsewhere. Capture and measurement of wing lengths by ringers could potentially help differentiate birds of Greenland origin from those of continental Europe (Schekkerman 1989, Francis *et al.* 1991).

The Lapland Bunting's distribution in Europe is expected to contract as a result of a changing climate by the end of this century (Huntley *et al.* 2007). It is assumed that there will be similar changes in Greenland, but it is unclear how this will affect its abundance or its migration strategy.

Tony Fox

Snow Bunting *Plectrophenax nivalis* Bras yr Eira

Welsh List Category	IUCN Red List			Birds of Conservation Concern			
	Global	Europe	GB	UK	Wales		
A	LC	LC	EN	Amber	2002	2010	2016

The Snow Bunting, which is an Arctic species with a circumpolar distribution, is the most northerly occurring passerine in the world. It breeds in treeless, rocky areas, often near snow, where it is regularly found around human settlements. It also breeds on high ground farther south, where conditions are broadly similar. Most populations are migratory, wintering to the south of its breeding range; the females moving farther south than males. In winter it is found on open fields, shingle beaches and saltmarshes (*HBW*). About 60 pairs breed in Scotland (Hayhow *et al.* 2018a), usually above 900m where snow remains into spring and summer (*The Birds of Scotland* 2007). These are mainly of the subspecies *P.n. insulae*, whose core breeding population is in Iceland (IOC). The

Welsh list refers to this subspecies as Icelandic Snow Bunting and it is also thought that 70–85% of the wintering birds in northern Scotland are of this subspecies (Banks *et al.* 1991). The great majority of Snow Buntings that winter in Wales on the coast and in the mountains, but seldom in between, are also believed to be *insulae*. The nominate subspecies, *P.n. nivalis*, which breeds in North America, Greenland and Fennoscandia (IOC), also winters in Britain.

There does not seem to have been any great change in the status of this bird in Wales over the last century. Forrest (1907) said that it was found on the north coast almost annually, usually in flocks of 5–20, and there were also records of flocks on the high tops. He quoted L.F. Lort as saying there were "immense numbers" at Llanllugan, Montgomeryshire, in 1881. Salter (1895) said that it was numerous in Ceredigion and the areas bordering it after snowfalls in January 1892 and January 1893, but only listed a few records on the coast between Aberystwyth and Borth, Ceredigion. There were only two records in Ceredigion between 1900 and 1940 (Roderick and Davis 2010). This species may have been less frequent on the south coast. Mathew (1894) described it as "rather rare" in Pembrokeshire and it was said to be an occasional visitor to Glamorgan in the late 1890s, recorded in only six winters during 1900–40, with no further records until 1959 (Hurford and Lansdown 1995).

Most records have been between mid-October and late March, but birds have been recorded on Bardsey, Caernarfonshire, as early as 4 September (in 2002) and as late as 22 June (in 1970). There is an autumn passage, particularly on the islands and headlands, and a lighter spring passage. Numbers recorded in Wales increased from the late 1960s, possibly because of increasing numbers of observers. Birch *et al.* (1968) noted that it had become a regular visitor to the Flintshire coast, usually between Point of Ayr and Gronant, where flocks of up to 50 had been reported. There was an influx in autumn 1996, with 160 recorded at Strumble Head, Pembrokeshire, between 24 October and 8 December, and 53 at Gronant on 18 November.

Large counts in winter included *c.*80 at Glandyfi, Ceredigion, on 6–9 February 1951 and *c.*50 at Whiteford Point, Gower, on 1 January 1960. There were good counts along the north coast in early 2000, including 45 at Pensarn, Denbighshire, on 14 January and 34 at Gronant, Flintshire, from 27 January to 28 February. There were 26 at Horton's Nose, Denbighshire, on 10 December 2012, but there has been no double-figure count in winter anywhere on the Welsh coast since 2013. The only double-figure count in autumn was a record of ten on the Great Orme, Caernarfonshire, on 23 October 2019.

There have been fewer records from the high tops, probably because few birdwatchers visit in winter, but flocks tend to be larger than on the coast. These included *c.*60 on Aran Fawddwy, Meirionnydd, on 3 January 1977, over 40 on Foel Grach in the Carneddau, Caernarfonshire, on 17 February 2001, and 18 on Fan Fawr, Mynydd Du, Carmarthenshire, on 8 February 2015, also seen over the border in Breconshire. A flock of 22–24 on Bwlch Giedd, south of Fan Brycheiniog, on 5 March 2014, was a record count for Breconshire.

Numbers are thought to have declined significantly in North America over the last 40 years, but the population trend in Iceland is not known (BirdLife International 2020). The breeding population in Iceland and elsewhere could decline, if climate changes lead to the current breeding areas becoming unsuitable. Even if the population remains stable, the numbers wintering in Wales could continue to decrease, if birds are able to winter further north, nearer their breeding grounds.

P.n. nivalis

The nominate subspecies is widespread: Nearctic birds winter across North America, while the Fennoscandian population migrates south as far as central and eastern Europe in winter, with some wintering on the east coast of Britain (*HBW*). There have been five records of a total of ten individuals in Wales:

- male and four female/1CY birds at Soldier's Point, Anglesey, on 25 November 2012;
- male and a female/1CY at Cemlyn, Anglesey, on 9 March 2014;
- female, trapped and ringed on Skokholm, Pembrokeshire, on 19 October 2014;
- female/1CY at Cemlyn, Anglesey, on 16 November 2014 and
- female at Cemlyn, Anglesey, on 28 November 2016.

The WBRC believes that this subspecies occurs frequently in Wales, but it is rarely documented owing to the difficulty of identification in the field.

Rhion Pritchard

Male Snow Bunting in winter plumage © Tony Pope

Corn Bunting *Emberiza calandra* Bras yr Ŷd

Welsh List Category	IUCN Red List			Birds of Conservation Concern				Rarity recording
	Global	Europe	GB	UK	Wales			
A	LC	LC	NT	Red	2002	2010	2016	WBRC

The Corn Bunting, the largest European bunting, is resident or a summer visitor across Europe (except Scandinavia), in North Africa and parts of the Middle East (*HBW*). It breeds mainly in drier lowland habitats with gentle slopes, rather than level terrain, and appears to like coastal areas. It needs perches to act as song posts. Its presence is strongly associated with arable farmland, particularly Barley. Since the 1950s the Corn Bunting has declined in numbers across much of its range, and by 81% across Europe in 1980–2016 (PECBMS 2020). The contraction of range and population decline has been confirmed by fieldwork in Europe. In Britain, the Netherlands and Switzerland, this species has almost disappeared (*European Atlas* 2020).

Dobie (1894) wrote that Corn Bunting was not abundant in northeast Wales but occurred between the Rivers Mersey and Conwy. It was found along the coast and was common near Holyhead, Anglesey. Mathew (1894) found it to be resident and local in Pembrokeshire, within five miles of the coast and plentiful near St Davids, Pembroke and Tenby. Forrest (1907) described its distribution in North Wales in a striking manner: "if a line were drawn all round the coast at a distance of one mile inland, the narrow strip of country included between this line and the sea would be found to contain about 90 per cent. of all the Corn-Buntings in North Wales, whilst the individuals on the west coast would outnumber those in the north at least two to one." It was common from Meirionnydd round the coast, northwards, to Denbighshire, where it was "fairly common", particularly during the breeding season. Corn Buntings were harder to find during the winter months and it was rarely seen inland. The association with the coast is likely to have been because farms there included some arable that provided food throughout the year (Stamp 1962, Shrubb 1997).

In Britain, the population decline began to be evident from the 1950s, but it was not until the *Britain and Ireland Atlas 1968–72* that the distribution was mapped. In Wales, the 1968–72 *Atlas* recorded that the Corn Bunting occupied 14 10-km squares, with breeding evidence provided for only 5% of these. The species was confined to eastern Flintshire and Denbighshire. The *Britain and Ireland Atlas 1988–91* found no confirmed breeding in Wales and presence in only two tetrads, while the *Britain and Ireland Atlas 2007–11* showed considerable contraction of range eastwards into England, with no records in Wales.

Between 1970 and 1990 the UK CBC/BBS index dropped by 88.6%. Between 2000 and 2018 the UK Breeding Bird Survey (BBS) showed an increase of 2.7%, although Wales has not contributed to this slight increase (Massimino *et al.* 2019). The 70 years of decline in Wales are likely to have been due to changes in agricultural practice, including a reduction in the cropped area of, particularly, Barley and changes from harvesting hay to silage. A reduction in winter food supplies, due to the loss of spring tillage, increased pesticide use and modernisation of harvesting and storage of grain, is particularly evident across the northern parts of its European range (*European Atlas* 2020).

The sad demise of the Corn Bunting has been documented in most counties. Ingram and Salmon (1954) and Lloyd *et al.* (2015) reported that it was extinct as a breeding species in Carmarthenshire by the 1950s. Ingram and Salmon (1955) reported just one, undated, record of an unspecified number at Llandrindod Wells, Radnorshire. In Breconshire there were occasional records of singles until the late 1960s, with the last at Pennorth in July 1978 (Andrew King *in litt.*). Donovan and Rees (1994) reported that the Corn Bunting had been considered to be local in Pembrokeshire as far back as 1894, and Lockley *et al.* (1949) reported it was found near the coast, in the St Davids area, the Dale Peninsula and near Castlemartin. Breeding there probably ceased soon after 1963, although there were occasional records up to 1981. There was no evidence of breeding in the *Pembrokeshire Atlas 1984–88*. Ingram *et al.* (1966) and Roderick and Davis (2010) said that at the start of the 20th century it occurred locally near the coast in Ceredigion, being numerous in the north of the county in 1914, but a decline had started by the 1920s. They had disappeared from Aberystwyth by 1934, although there was a sighting in 1957. There were six other records during 1969–95. There were just 13 records in Gwent up to 1990, with breeding last recorded in 1970. Heathcote *et al.* (1967) wrote that this former breeding resident of Glamorgan was then rarely seen, with the last two records in 1964. There were no records in the *West Glamorgan Atlas 1984–89*. In Meirionnydd, they vanished from their coastal haunts during the 1920s. A last record in the first half of the 20th century was in 1935. Since then, there have been just three subsequent sightings: in 1979, 1983 and 1990 (Pritchard 2012). In Caernarfonshire, the period from the 1920s to the 1950s saw the population collapse. Subsequently, the only records have been a few on Bardsey in 1950–56, May 1994 and April 2010 (Pritchard 2017). In winter, singles were recorded from several counties up to 1994, with one record in Gwent in 1998, though this record is not included in Venables *et al.* 2008. There were 26 at Shotwick, Flintshire, on 26 January 1996. In northeast Wales, by the period of the early 2000s up to about 2007, there were records of just one or two from two areas: Bettisfield, Denbighshire, and the Shotwick area, Flintshire. There were no records in 2008, only one in 2009 (a few miles from the Shropshire border near Llanrhaeadr-ym-Mochnant on 19 December), one at Shotwick in June 2011 and one there in July 2015. Since then, there have been no further records.

By the time of *Birds in Wales* (1994), the total Welsh breeding population of the Corn Bunting was confined to the 15 pairs in the Denbighshire and Flintshire area. *Birds in Wales* (2002) reported that this had declined to 12 pairs in 1998 and four pairs in 1999. Other than occasional sightings of individuals, no reports of breeding have occurred up to 2019. Therefore, an exciting report of a male carrying food for young, seen on the border with England at Bettisfield, Denbighshire, in June 2020 gave cause for hope. This was the area most recently vacated by Corn Buntings but it was not known in which country the nest was located.

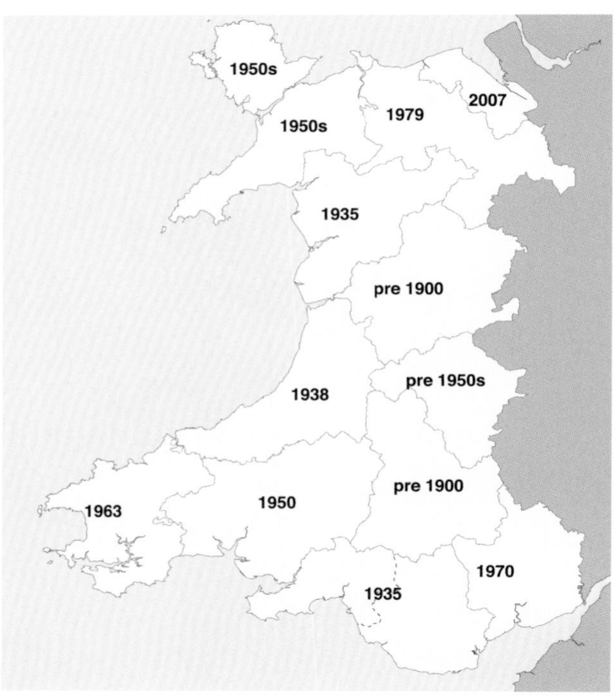

Corn Bunting: Last year of recorded breeding in each county.

Unfortunately, breeding Corn Buntings have now withdrawn from the Shropshire side of the border near Bettisfield, mainly further east (Smith 2019b). Therefore, it is unlikely that the Shropshire population will be the source of recolonisation of Wales in the near future. Johnstone *et al.* (2010a) considered that a species should not breed for 20 years before it is called 'extinct'. Unfortunately, the farmland management options prescribed in Johnstone *et al.* (2010b) have not been deployed and so Corn Bunting looks set to be declared a former breeding species in Wales.

Ian M. Spence

Yellowhammer *Emberiza citrinella* **Bras Melyn**

Welsh List Category	IUCN Red List			Birds of Conservation Concern			
	Global	Europe	GB	UK	Wales		
A	LC	LC	LC	Red	2002	2010	2016

The Yellowhammer is found in Europe away from the Mediterranean and below the Artic Circle, and across western and central Siberia (*HBW*). Breeding birds in Wales are of the subspecies *E.c. caliginosa*, whose distribution is restricted to Ireland and most of mainland Britain. The other subspecies found in Britain is the nominate, *E.c. citronella*, which is found in southeast England and much of the rest of Europe (IOC). The Yellowhammer lives in open areas that are dry and sunny with scrub such as gorse, broom, Hawthorn and Juniper, and farm hedgerows. In Wales, it is associated with lowland mixed farmland and *ffridd*. In Breconshire, such *ffridd* on common-land margins is generally between 350 and 450m (Andrew King *in litt.*). It nested up to 460m in Radnorshire (Ingram and Salmon 1955). It needs song posts, so also occurs on the edge of new forestry plantations, with adjacent open areas in which to feed. On the Pembrokeshire coast, it is found where there is mixed deciduous woodland/scrub adjacent to coastal heath and grassland. It requires a supply of arthropods to feed its young, though at other times of the year it eats seeds of weeds or spilt grain. In winter, it feeds on seeds and invertebrates in weedy field margins and stubble. Whilst most Yellowhammers are reasonably sedentary, those that breed on higher slopes move down to lowland areas in winter, forming foraging flocks where food resources are more abundant. Several birds, colour-ringed in the winter of 2014/15 in the Glaslyn Valley, Caernarfonshire, were later resighted in the breeding season in *ffridd* above the valley (Steve Dodd pers. obs).

At the end of the 19th century, the Yellowhammer was common and widespread in Pembrokeshire (Mathew 1894) and abundant in northeast Wales (Dobie 1894). Until the mid- to late 20th century, many avifaunas continued to describe it as widespread and common. It was especially numerous in western Llŷn, Caernarfonshire, (Jones and Dare 1976) in areas with gorse, and flocks were seen in winter, usually in the lowlands.

The *Britain and Ireland Atlas 1968–72* showed that Yellowhammers were breeding in almost every 10-km square in Wales and there was no marked change in range when the 1988–91 *Atlas* showed breeding was still evident in 86% of 10-km squares. Indications of a major UK decline started to be apparent by the late 1980s into the 1990s (Siriwardena *et al.* 1998), although the first mention of population decline in Wales was from Ceredigion where it was noticed from the early 1980s (Roderick and Davis 2010). In Caernarfonshire, Barnes (1997) noted a decline in the 1990s linked to a decline in mixed farming. Subsequently, lower numbers were noted in every county, but they were increasingly noted at feeding stations in gardens during the winter (Barnes 1997, Venables *et al.* 2008, Jennings 2014). By 1999, on Anglesey, the species was restricted to the northeast corner (Jones and Whalley 2004). In Gower, the *West Glamorgan Atlas 1984–89* found breeding in only 43% of tetrads. The *Gwent Atlas 1981–85* provided evidence of breeding in 77% of tetrads but the 1998–2003 *Atlas* showed a decline to 49% of tetrads.

The *Britain and Ireland Atlas 2007–11* showed breeding evidence from 64% of 10-km squares and low relative breeding abundance across Wales, with western Llŷn and Pembrokeshire showing lower relative abundance than during 1988–91. Relative abundance in winters 2007–11 was considerably lower than in 1981–84. The *North Wales Atlas 2008–12* showed evidence of breeding in 75% of 10-km squares but in only 17% of tetrads, emphasising their restricted distribution. In East Glamorgan, the 1984–89 *Atlas* had evidence of breeding in 35% of tetrads with a decline to just 11% in the 2008–11 *Atlas*. In Pembrokeshire, there was evidence of breeding in 80% of tetrads during the 1984–88 *Atlas* but only 38% in 2003–07 (*Pembrokeshire Avifauna* 2020). The later results confirmed that the Yellowhammer, whilst still widespread, could no longer be considered common but was patchily distributed in Wales.

Shrubb (2003b) documented many of the changes in modern agriculture after the Second World War. A reduction in cereal growing and other changes in arable land management, along with the increased use of pesticides and herbicides, are strongly implicated in the decline of farmland birds. Traditionally, many farmers would have grown a field of spring Barley (often unsprayed) for over-winter livestock feed, leaving weedy stubbles through the winter. The cost of machinery has made this less attractive. In the more extensive cereal-growing areas, in the lowlands of eastern Wales, there has been a shift towards intensive livestock pastoral systems, such as dairy with associated Maize, neither of which provide suitable feeding habitat for Yellowhammers. Farmyards tended to have less spilt grain, and more winter feed for stock was in the form of processed pellets, rather than grain. Yellowhammers have a close association with the arable component of mixed farms, plus the presence of wide, weed-rich field margins and thick, varied hedgerows (Bradbury *et al.* 2000, Robinson *et al.* 2001). Chamberlain and Fuller (2000) commented on the loss of arable in western Britain, with the increased specialisation of farming systems; but in Wales, the amount of arable land has fluctuated by about 20% (from 200,000ha and 250,000ha) between 1983 and 2016 (Historical Statistics 1974–1996, Statistics for Wales 2018) with no overall downward trend.

The conversion of rough grazing land, particularly in the upland margins adjacent to *ffridd*, to more intensive pasture suitable for sheep, along with the subsequent increase in sheep numbers (e.g. from 3.9 million to 9.5 million between 1945 and 2019—Welsh Government 2019a), has been another contributory factor in the decline of the Yellowhammer. Hopkins (2001) noted that modern pasture management promotes the use of vigorous grass varieties, which outcompete broadleaved plants that are then prevented from setting seed. In intensive sheep pasture areas, hedges can become much thinner due to sheep grazing, unless protected by double fencing. Thin, undercut hedges are unsuitable habitat for nesting. Also, the grass margins adjacent to these hedges, when over-grazed, no longer have an abundance of large invertebrates or useful winter seed resources.

The majority of Yellowhammers in sheep-dominated landscapes find refuge in more semi-natural habitat, such as *ffridd* (Fuller *et al.* 2006), with Bracken, gorse, low shrubs and young trees. This habitat is getting squeezed as farmers seek to maximise the yield on land that is economically marginal. *Ffridd*, which is generally outside designated areas such as SSSIs, and is not supported by agri-environmental schemes (AES) management options (Johnstone *et al.* 2019), continues to be lost to agricultural intensification or used to establish new woodlands (Anon. undated).

Since 1992, Welsh AES have sought to address declines

Male Yellowhammer © Bob Haycock

in biodiversity on farmland in Wales. The Yellowhammer was among a suite of key species that should have benefited from schemes that included scrub conservation, hedgerow management, maintenance and enhancement of semi-natural grassland habitats, and provision of seed resources, such as unsprayed root crops and over-winter stubbles, but *ffridd* is not included in any of these (see above). MacDonald *et al.* (2019) found significant effects on numbers of breeding Yellowhammers on farms that were in AES compared to farms that were not. Winter surveys of AES arable options showed the highly beneficial effects of the presence of Wild Bird Cover, stubbles and unsprayed root crops, compared to non-AES arable land. However, the voluntary nature of the schemes meant that the targeting necessary to meet their objectives was lacking. AES hedgerow prescriptions provided significant benefits for Yellowhammers, but the presence of arable was a more significant driver of whether birds remained on the land (Dadam and Siriwardena 2019).

Further evidence of decline is shown by the BBS index for Wales, a decline of 64% between 1995 and 2018 (Harris *et al.* 2020).

Research in England has shown that much of the Yellowhammer's decline is driven by reduced over-winter survival, as a result of low seed availability in late winter (Siriwardena *et al.* 2008). Recent work in North Wales highlighted the potential for intensively managed Rye-grass to be left to set seed as a food resource for buntings, particularly late in winter, in addition to the planting of conventional wild-bird seed mixes. However, Yellowhammers continued to decline in the study area, suggesting other factors were also at work, probably related to key features of breeding habitat (Johnstone *et al.* 2019).

The *Britain and Ireland Winter Atlas 1981–84* showed relatively low abundance across most of Wales, with absence from the areas of highest land. By the 2007–11 *Atlas*, relative abundance in winter had decreased even further across all of Wales, with most birds found in coastal areas, notably Pembrokeshire and southern East Glamorgan as well as the lowlands along the English border. Even there, relative abundance was much lower than in the east of England and Scotland. There appears to have been little change in winter flock sizes, which have usually been between 10 and 30, and more rarely over 100, but the frequency of reporting has decreased in the 21st century. Since 2000, there have been just seven records of flocks exceeding 100 in Wales: over 100 were at Ewenny Moor, Glamorgan, in January 2000; 100 fed on a sacrificial crop at Dolydd Hafren, Montgomeryshire, on 31 January 2006; 230 at Llanilar, Ceredigion, on 5 December 2001 and 104 at Brynteg, Anglesey, in March 2012. A further three records were in Breconshire, on two farms in AES, within 5km of each other near Llangorse, where wild-bird crops were grown. These crops were based on unharvested spring cereals, for 5–15 years, rotated around the farms. The mixed farm with the longest history attracted a flock of over 100 in three consecutive winters: 130 in January 2017, and 100 in January and December 2018 (Andrew King *in litt.*). Relatively few Yellowhammers have been ringed in Wales since 2000 and most recoveries have been less than 5km from the site of ringing. However, over six winters since March 2015, 650 individuals were ringed by the Llangorse Ringing Group. Resightings of these birds showed little movement between fields with different crops and only one recovery from an adjacent garden, emphasising their sedentary nature. The greatest movement of a Yellowhammer in Wales involved one ringed on its breeding site on *ffridd* above Prenteg, Caernarfonshire, on 17 May 2015, that moved to a garden in Penmachno, Caernarfonshire, on 3 December 2015, a distance of 23km (Robinson *et al.* 2020). There seems to be a small, but distinct, passage of Yellowhammers, at Bardsey and Skokholm, in the spring and autumn, but the origins of these birds are unknown.

The published estimate for the Welsh breeding population in 2018 was 31,500 (29,000–33,500) territories (Hughes *et al.* 2020) but this is now considered too high and the population is probably closer to 22,500. In Europe, there has been very little change in the range of the Yellowhammer between the *European Atlas* (1997) and the most recent fieldwork (*European Atlas* 2020), with just some localised losses in central Italy, coastal Netherlands, northwest Scotland and Ireland. However, Yellowhammer numbers have declined between 1980 and 2016 by 47% (PECBMS 2020).

The Yellowhammer is Red-listed in Wales (Johnstone and Bladwell 2016), owing to a severe population decline and range contraction over the past 25 years. Considerable effort will be

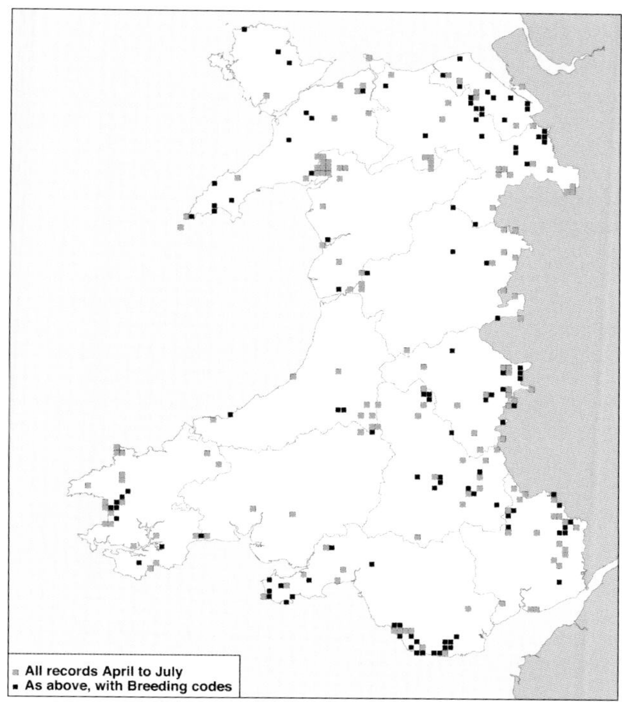

Yellowhammer: Records from BirdTrack for April to July, 2015–19, showing the locations of records and those with breeding codes.

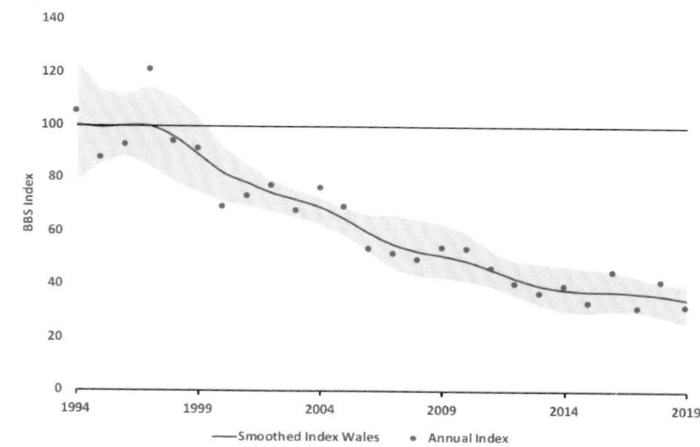

Yellowhammer: Breeding Bird Survey indices, 1994–2019. The red line shows the smoothed index, 1995–2018. The shaded area indicates the 85% confidence limits.

required to ensure that Yellowhammers can find suitable feeding habitat throughout the year. As Wales adopts a new approach to support for farmers and land management, it will be essential that the needs of Yellowhammers are targeted both during and outside the breeding season, that *ffridd* habitats are maintained and that the lessons are learned from previous agri-environment schemes.

Steve Dodd, Ian M. Spence and Anne Brenchley

Sponsored by Bob and Annie Haycock

Pine Bunting *Emberiza leucocephalos* Bras Pinwydd

Welsh List Category	IUCN Red List		Rarity recording
	Global	Europe	
A	LC	VU	BBRC

The Pine Bunting's breeding range extends across Russia east of the Ural Mountains, and it winters in the mountains of eastern Kazakhstan, the Himalayas, northeast China and the Korean Peninsula (*HBW*). This is an extremely rare vagrant to Britain with 64 records, the last being in 2018 (Holt *et al.* 2019), of which two were in Wales. A male was seen on Skokholm, Pembrokeshire, on 28 April 2000 and another male was recorded on Bardsey, Caernarfonshire, on 30 April 2001.

Jon Green and Robin Sandham

Rock Bunting *Emberiza cia* Bras y Graig

Welsh List Category	IUCN Red List		Rarity recording
	Global	Europe	
A	LC	LC	BBRC

The Rock Bunting breeds in northwest Africa, southern and central Europe, the Middle East, Central Asia and the Himalayas (*HBW*). Most European birds are resident or altitudinal migrants. There have been six records of Rock Bunting in Britain, the most recent being in 2011 (Hudson *et al.* 2013). The only record in Wales was a male on Bardsey, Caernarfonshire, on 1 June 1967, which was the fifth to be recorded in Britain.

Jon Green and Robin Sandham

Ortolan Bunting *Emberiza hortulana* Bras Gerddi

Welsh List Category	IUCN Red List		Rarity recording
	Global	Europe	
A	LC	LC	WBRC

The Ortolan Bunting has a patchy breeding distribution across Europe, the Middle East, Central Asia and western Siberia. It winters locally in West Africa and the Sahel zone, returning in late April or early May (*HBW*). It is a rare migrant to Wales, with 182 individuals recorded. Decadal totals in the table show that numbers increased slightly towards the end of the 20th century, but have declined since. This mirrors the British trend, where annual means peaked at 72 in the 1990s, but declined to 31 in the 2010s (White and Kehoe 2020b). Although the Ortolan Bunting is classified by the International Union for Conservation of Nature (IUCN) as being of Least Concern, its European population declined by 90% between 1980 and 2016 (PECBMS 2020) and it is Red-listed in many European countries. White and Kehoe (2019) highlighted that the French breeding population had "suffered a dramatic southward retraction" in the 21st century. The whole breeding population in northern Europe had declined due to "the familiar suite of factors afflicting birds of farmland and open country: agricultural intensification, habitat loss and fragmentation". Also "continued illegal hunting in France, resulting in 15,000–30,000 birds being killed annually, significantly increases the risk of localised extinction in some of the northern and western parts of the species' breeding range". Hunting is still considered to be one of the main threats to this species in Europe (*European Atlas* 2020).

Total	Pre-1950	1950–59	1960–69	1970–79	1980–89	1990–99	2000–09	2010–19
182	15	12	40	15	25	32	21	22

Ortolan Bunting: Individuals recorded in Wales by decade.

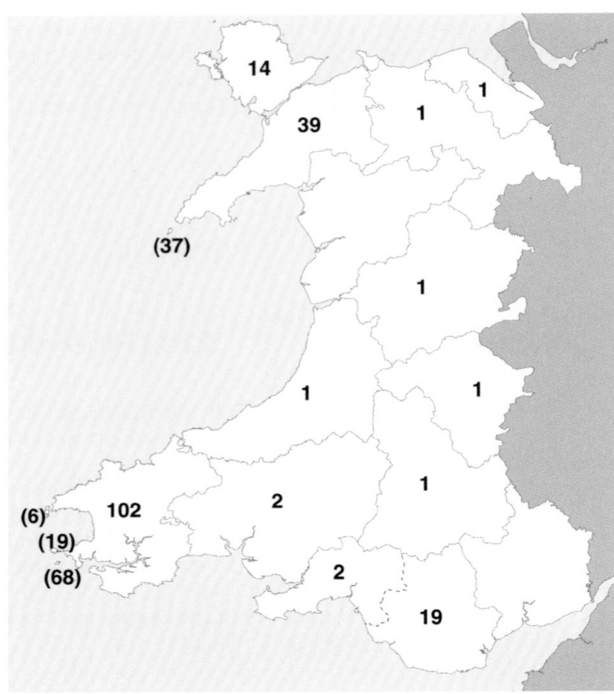

Ortolan Bunting: Numbers of birds in each recording area, 1907–2019.

The first Welsh record was of one killed at the lighthouse on Bardsey, Caernarfonshire in June 1913. Individuals have been recorded in all Welsh counties except Meirionnydd, Montgomeryshire and Gwent. Over half of the Welsh records have come from Pembrokeshire (102), of which Skokholm has recorded by far the most individuals (68), almost twice as many as Bardsey, Caernarfonshire, with 19 on Skomer, six on Ramsey, two on other islands and only seven on the mainland. Of the 19 in East Glamorgan, 12 were at Lavernock Point, including a flock of seven on 12–13 September 1992. The inland record in Breconshire and Radnorshire referred to the same bird at Glasbury on 4 September 1993.

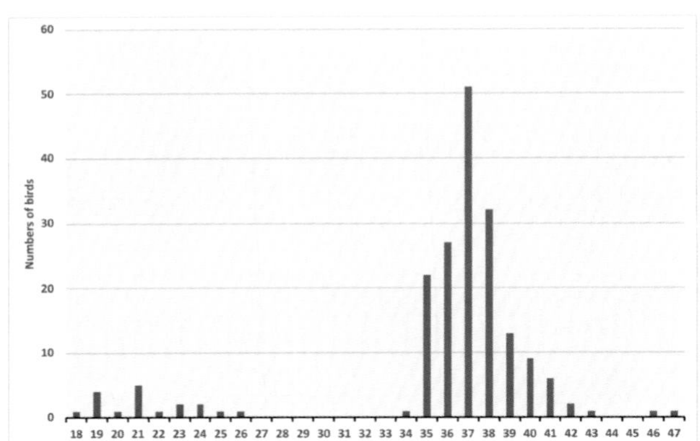

Ortolan Bunting: Totals by week of arrival 1907–2019. Week 19 begins *c.*7 May and week 37 begins *c.*10 September.

The Ortolan Bunting is mainly an autumn migrant in Wales, with a distinct peak of records in September. There are very few spring records. Most records relate to single birds or twos, but larger groups, all reported in September, included five on Skokholm in 1956, five at Brynsiecyn, Anglesey, in 1968 and seven at Lavernock Point, East Glamorgan, in 1992. Climate modelling suggested potential breeding by Ortolan Bunting in southern England by the end of this century (Huntley *et al.* 2007), but the scale of its decline and contraction away from the edge of its current range suggests that habitat factors are over-riding any benefits of climate change.

Jon Green and Robin Sandham

Cretzschmar's Bunting *Emberiza caesia* Bras Cretzschmar

Welsh List Category	IUCN Red List		Rarity recording
	Global	Europe	
A	LC	LC	BBRC

The Cretzschmar's Bunting breeds in arid habitats around the eastern Mediterranean and winters in the Nile Valley and the coast of Sudan and Eritrea (*HBW*). There have been seven individuals recorded in Britain up to the end of 2019 (Holt *et al.* 2020) and only one record of this extremely rare vagrant in Wales. This was a second-calendar-year male on Bardsey, Caernarfonshire, on 10–20 June 2015. This sighting, one of the rarest birds Bardsey has seen in recent times, caused major national interest. Its appearance was sporadic, until a regular pattern was noted. It was seen around the lighthouse compound from 14 June until it disappeared on 20 June.

Jon Green and Robin Sandham

Male Cretzschmar's Bunting on Bardsey, Caernarfonshire, June 2015 © Steve Stansfield

Cirl Bunting *Emberiza cirlus* Bras Ffrainc

Welsh List Category	IUCN Red List			Birds of Conservation Concern	Rarity recording
	Global	Europe	GB	UK	
A	LC	LC	LC	Red	WBRC

The Cirl Bunting is mainly a species of western and southern Europe, and North Africa, with a small population in southwest England being the most northerly. It is usually found in bushes and small woodlands in open country, forest edges and farmland. Birds breeding in the north of the range in mainland Europe are partially migratory, but birds breeding farther south are resident (*HBW*). Ringing has shown that birds breeding in Devon wintered within a few kilometres of their breeding sites.

The species was first recorded in Britain in 1800, in Devon, and it may have been a fairly recent colonist at that time (*Birds of England* 2005). Later that century, it was mainly found in southern England, but "of late years it appears to have been increasing its range and pushing its way into Wales" (Aplin 1892). Aplin quoted C.G. Beale as stating that between 1875 and *c.*1890 it was common in Glyn Ceiriog, Denbighshire, although it appeared to be very locally distributed. It also bred in small numbers in Breconshire, Glamorgan and Ceredigion. The first record in Breconshire was in 1888, although Phillips (1899) said that it was found at several sites in the county and may have been over-looked previously, owing to its resemblance to Yellowhammer. Salter recorded several pairs in the Aberystwyth and Llanbadarn areas of Ceredigion, between 1893 and 1899, with others up the Rheidol and Clarach valleys. S.G. Cummings recorded several singing around New Quay in July 1901 (Roderick and Davis 2010). Ceredigion appeared to be the only western county in Wales with a breeding population at that time. Mathew (1894) described it as a rare occasional visitor to Pembrokeshire, and Aplin knew of only one record from Meirionnydd. It was regarded as a very local breeder in Gwent at the turn of the 20th century (Venables *et al.* 2008). It bred at Solva, Pembrokeshire, in 1895, where up to three pairs nested annually until 1910 (Donovan and Rees 1994). A nest was found in Radnorshire near Builth Wells in 1902 (Jennings 2014). Forrest (1907) said that in North Wales the species was practically confined to Denbighshire and Flintshire, where it had been recorded at many sites, and around the Conwy Estuary on the border with Caernarfonshire. There had been no records from Caernarfonshire until the 1890s, but in June 1899 two were singing as far west as near Llanbedrog in Llŷn (Aplin 1900). Aplin was uncertain whether the species was extending its range in North Wales or had previously been overlooked. Forrest also knew of several records in Montgomeryshire, including a pair near the confluence of the Severn and the Vyrnwy in May 1900, although a nest he mentioned at Marrington in 1901 was probably just over the border in Shropshire. Cummings (1908) said that in some parts of North Wales the Cirl Bunting might be described as common. He had heard nine singing in a comparatively small area at Llanddulas, Denbighshire, and in Flintshire it occurred all along the valley between Mold and Bodfari. He failed to find the species on Anglesey and considered that it showed a decided preference for limestone districts. Elsewhere, there was no firm evidence of breeding in Meirionnydd or in Carmarthenshire, although one was recorded near Bishop's Pond on 1 May 1905 and several singing males were recorded up to 1912. The species apparently disappeared from Breconshire in this period, with no record in the county after 1915 (Peers and Shrubb 1990).

There was no record of breeding in Gwent after 1925 (Venables *et al.* 2008) and regular breeding may have ceased in Caernarfonshire soon afterwards (Pritchard 2017). When Salter returned to Aberystwyth in 1923, he found that the population in that area had declined. Several pairs were found along the south coast of the Gower Peninsula in June 1925 and a pair nested near Clyro, Radnorshire, in 1928. There were fewer records in the 1930s, although breeding was recorded at Abergele, Denbighshire, in 1931 and probably 1932 (Jones and Roberts 1983). A pair, with fledged young, was seen near Port Eynon, Gower, in July 1936 and a small number were still around Aberystwyth and Llanbadarn in Ceredigion up to 1940. The last known breeding pair in Ceredigion was near Aberystwyth sometime during 1939–44 (Roderick and Davis 2010). Hardy (1941) stated that Cirl Bunting occurred at Chirk and Ceiriog in Denbighshire, and quoted H. Pollit as reporting that they bred near Rhyl, Flintshire. The *Cambrian Bird Report 1956* stated that the species bred in the Aber Valley, Caernarfonshire, in 1952, although no further details were given. A pair at Moelfre, Anglesey, in 1954, was only the second record for the island, and the first since 1878 or 1879 (Jones and Whalley 2004). A pair was at Michaelchurch-on-Arrow, Radnorshire, in June 1960. Singles were recorded in the breeding season on Anglesey in 1960, in Pembrokeshire in 1961 and in Denbighshire in 1963. The last record of breeding in Wales appeared to be at Llanarth, Gwent, where a male was in song in July–September 1968 and two young were seen (Venables *et al.* 2008), with a male in song there throughout the summer of 1969. A male was at Magor, Gwent, in June 1970.

Cirl Buntings continued to be recorded outside the breeding season in several counties. An influx into South Wales in winter 1981/82 included seven at Llanybri, Carmarthenshire, on 27–28 December 1981 and five males at Burry Green, Gower, on 24 January 1982. There has been only one subsequent record: a female on Bardsey, Caernarfonshire, on 23 April 2004.

The Cirl Bunting's breeding range in England contracted markedly in the 1970s and 1980s and a survey in 1989 found only 118 pairs, all but a handful in Devon (Evans 1992). Elsewhere, it had become a rarity. This stimulated research and conservation action that, since the early 1990s, has led to a gradual increase. A survey in 2016 estimated the population at 1,079 territories, with birds starting to recolonise former areas in north Devon and north Cornwall (Jeffs *et al.* 2018). This has yet to lead to an increase in records in Wales, but if the population in Devon and adjacent areas continues to do well, there must be the possibility of more records across the Bristol Channel into southeast Wales. However, Cirl Bunting seems unlikely to return as a breeding bird in Wales without substantial changes in farming, even though almost all of Wales is expected to be climatically suitable for the species by the end of this century (Huntley *et al.* 2007).

Rhion Pritchard, Jon Green and Robin Sandham

Little Bunting *Emberiza pusilla* Bras Bychan

Welsh List Category	IUCN Red List		Rarity recording
	Global	Europe	
A	LC	LC	WBRC

Little Bunting on Bardsey, Caernarfonshire, October 2016 © Steve Stansfield

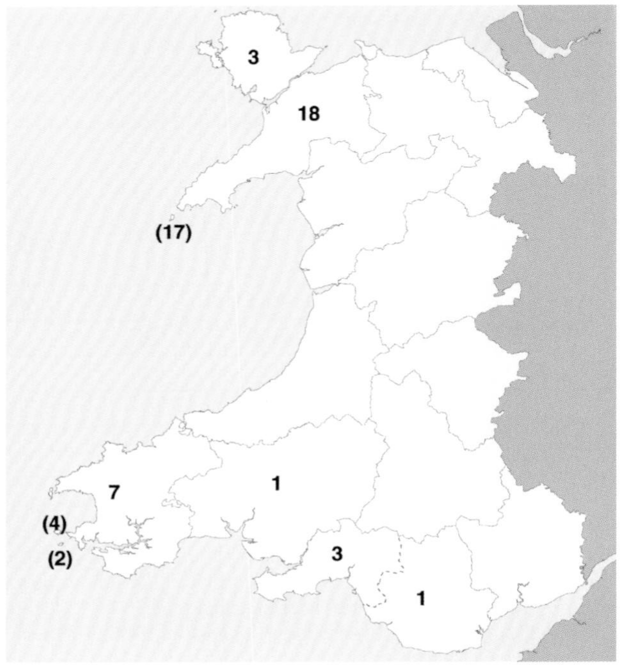

Little Bunting: Numbers of birds in each recording area, 1957–2019.

Total	1957–59	1960–69	1970–79	1980–89	1990–99	2000–09	2010–19
33	3	2	0	7	4	10	7

Little Bunting: Individuals recorded in Wales by decade, 1957–2019.

The Little Bunting has a broadly similar breeding range to that of Rustic Bunting, although slightly farther north through Fennoscandia and across Russia to the Pacific coast. It winters in southern China and adjacent territories (*HBW*). There has been a dramatic increase in the number of records in Britain over the last 40 years, with annual means of 24 in the 1980s increasing to 74 in the 2010s (White and Kehoe 2020b). There was a slight southerly shift in its range, in Fennoscandia, during this time period, which may explain part of the increase in numbers being recorded in the UK (*European Atlas* 2020). This is a rare migrant to Wales, with a slight increase in the number of records since the 1980s.

The first Welsh record was at Llanddeusant, Anglesey, on 8 January 1957, closely followed by one at Oxwich, Gower, on 28 September that year.

Little Buntings have been seen in Wales in most winter, early spring and autumn months, with the majority in October. Over half the Welsh Little Buntings have been recorded on Bardsey, Caernarfonshire, (17) with very few on the Pembrokeshire islands.

After the first two records for Wales in 1957, all but one of the next 17 individuals recorded up to 2003 were on the islands: Bardsey, The Skerries, Skomer and Skokholm. The only exception was one at Ffairfach, Carmarthenshire, in December 1998. Since 2004, out of 14 individuals recorded, six were on the mainland or on Anglesey. One was at the Braint Estuary, Anglesey, from January to April 2004, one at Mewslade, Gower, in November 2005 and one at the Brewery Fields, Bangor, Caernarfonshire, in December 2008. There was one at Fforest Farm, East Glamorgan, from February to April 2015, then two records in 2016: one at St Ann's Head, Pembrokeshire, in April and one ringed at Oxwich, Gower, in October.

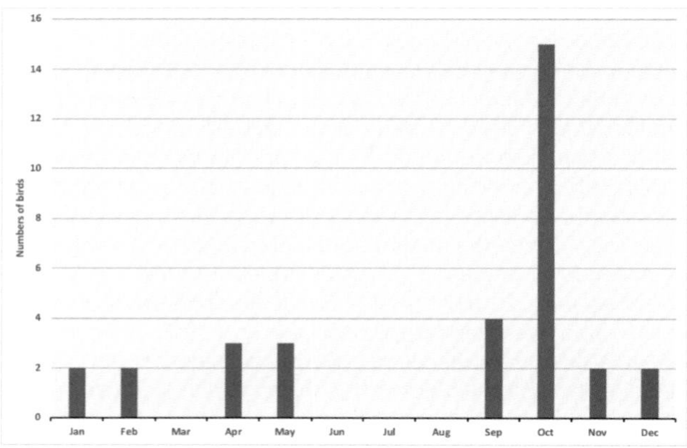

Little Bunting: Totals by month of arrival 1957–2019.

Jon Green and Robin Sandham

Rustic Bunting *Emberiza rustica* Bras Gwledig

Welsh List Category	IUCN Red List		Rarity recording
	Global	Europe	
A	VU	VU	WBRC

The Rustic Bunting breeds in Fennoscandia and across temperate Russia, and winters in eastern China, Japan and the Korean Peninsula (*HBW*). There have been 586 individuals recorded in Britain up to 2019. The annual occurrence of this species appears to be stable or decreasing slightly, with an average of ten a year (Holt *et al.* 2020). The majority of records were from Shetland. It is a very rare vagrant to Wales with only ten records:

- female on Skokholm, Pembrokeshire, on 8 June 1953;
- male on Skokholm, Pembrokeshire, on 6–7 June 1975;
- male on Bardsey, Caernarfonshire, on 29 March 1981;
- male on Skomer, Pembrokeshire, on 31 May 1987;
- Arthog, Meirionnydd, on 5–10 April 1991;
- male on Skokholm, Pembrokeshire, on 20 May 1993;
- Skokholm, Pembrokeshire, on 23 May 1994;
- 1CY male at Cors Caron, Ceredigion, on 11 November 1997;
- male found dead at Abergele, Denbighshire, on 7 April 2000 and
- 2CY male on Bardsey, Caernarfonshire, on 19 May 2008.

Jon Green and Robin Sandham

Yellow-breasted Bunting *Emberiza aureola* Bras Bronfelyn

Welsh List Category	IUCN Red List		Rarity recording
	Global	Europe	
A	CR	CR	BBRC

The Yellow-breasted Bunting breeds across European and Asiatic Russia, and in northern Mongolia and China. It winters in Southeast Asia and the Himalayas (*HBW*). There have been 239 individuals of this extremely rare Asian vagrant recorded in Britain, mainly on the Isles of Shetland, up to 2019. Numbers increased during the latter half of the 20th century, but have since dropped significantly to an average of one record a year (Holt *et al.* 2019). The collapse in the global population in the 21st century (Birdlife International 2015) is largely attributable to roosting flocks being caught outside the breeding season, cooked and sold as 'sparrows' or 'rice-birds'. This practice was formerly restricted to a small area of southern China, but has become more widespread and popular owing to increased affluence in China (Slack 2009). The sole Welsh record was of a female on Bardsey, Caernarfonshire, on 4–5 September 1973.

Jon Green and Robin Sandham

Black-headed Bunting *Emberiza melanocephala* Bras Penddu

Welsh List Category	IUCN Red List		Rarity recording
	Global	Europe	
A & E	LC	LC	BBRC

The Black-headed Bunting breeds in southeastern Europe, Turkey and parts of the Middle East, and in winter migrates to central west India (*HBW*). There have been 229 individuals of this rare vagrant recorded in Britain up to 2019 (Holt *et al.* 2020), of which 32 were in Wales. Numbers appear to be decreasing, with an average of about five a year since 1990 (Holt *et al.* 2019).

The first Welsh record was on Bardsey, Caernarfonshire, on 27–29 May 1963. The majority of records refer to overshooting males in late May and June, with a peak in the last week of May. Only three have been recorded in autumn, two of these on Bardsey: one on 2 October 1988 and a female on 19–24 September 1992. There was then a first-calendar-year bird on Skomer, Caernarfonshire, on 5 September 2004. Thirty-two individuals were recorded in Wales, of which two-thirds have been seen on the islands.

Mainland records were mostly from coastal locations. In Pembrokeshire, these records included individuals at Marloes on 28–30 May 1990, Lower Town Fishguard on 18–20 May 1995, Freshwater West on 16 May 2001, and in 2015 at Saundersfoot on 10 June and St Justinians on 29 June. Elsewhere, coastal records came from Aberystwyth, Ceredigion, on about 30 May 1987; at Carmel Head, Anglesey, on 18 May 1986 and Cemlyn, Anglesey, on 12 June 1993; a male at Aberdaron, Caernarfonshire, on 2–5 June 1995; and at RSPB South Stack, Anglesey, on 18–20 May

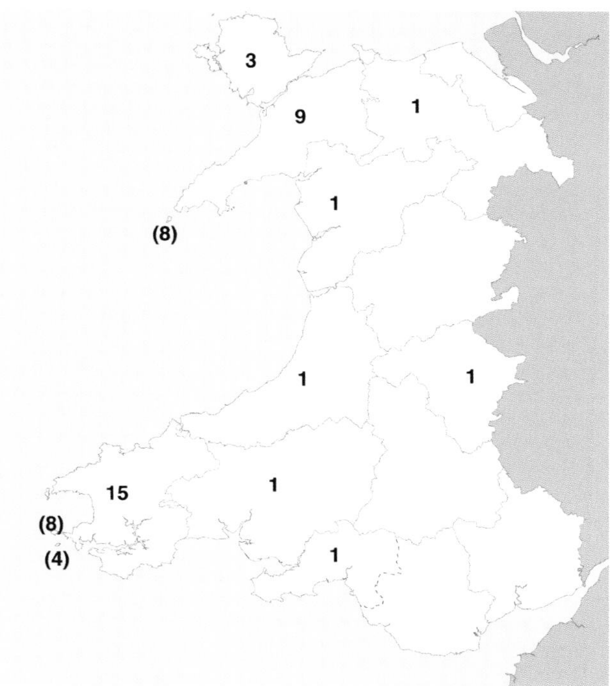

Black-headed Bunting: Numbers of birds in each recording area.

Total	1963–69	1970–79	1980–89	1990–99	2000–09	2010–19
32	5	0	5	9	8	5

Black-headed Bunting: Individuals recorded in Wales by decade.

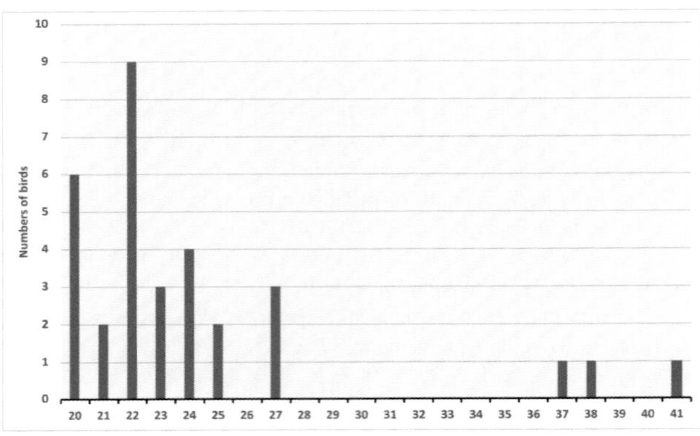

Black-headed Bunting: Totals by week of arrival 1963–2019. Week 22 begins *c*.28 May and week 37 begins *c*.9 September.

2001. In 2012, a male was seen at Godre'r-graig, Gower, on 17 June and the same individual (identified from photographs) was later seen at Porthyrhyd, Carmarthenshire, on 18–24 June.

The two inland records, at Rhayader, Radnorshire, on 3–12 June 1994 and at Llanarmon-yn-ial, Denbighshire, on 9–10 June 1992, were unusual.

Jon Green and Robin Sandham

Reed Bunting *Emberiza schoeniclus* Bras Cyrs

Welsh List Category	IUCN Red List			Birds of Conservation Concern			
	Global	Europe	GB	UK	Wales		
A	LC	LC	LC	Amber	2002	2010	2016

Despite its English and Welsh names, the Reed Bunting is not particularly associated with reedbeds, although it will breed in reed-dominated vegetation. The largest populations are usually found in wet areas with rough vegetation. It is a widespread species, breeding over almost all of Europe away from the Mediterranean and much of Asia to the Pacific (*HBW*). North European populations move south in winter, while even residents tend to move to drier habitats, such as scrub and farmland, in winter (*European Atlas* 1997). In Britain, ringing recoveries have shown that the breeding population is fairly sedentary (*BTO Migration Atlas* 2002), although one first-calendar-year female that was ringed at Pwllcrochan, Pembrokeshire, in September 2002 was recaught at Insh in the Scottish Highlands in May 2003, a distance of 604km. This may well have been a Scottish-bred bird that moved south to over-winter. No birds ringed outside Britain have been recovered in Wales.

The species was probably much more common before large-scale drainage of wetlands began in the 18th century, but there is no information on its distribution before the late 19th century. It was recorded as locally distributed and not very common in Glamorgan in the late 1890s (Hurford and Lansdown 1995) and as locally common in Breconshire (Phillips 1899). In Pembrokeshire, Mathew (1894) and Lockley *et al.* (1949) regarded the Reed Bunting as a scarce resident, breeding in boggy places. Ingram and Salmon (1954) reported that it was found in relatively small numbers in Carmarthenshire, chiefly in wetland areas, while in 1955 they described it as very local in Radnorshire. Forrest (1907) described the species as widely distributed in bogs and marshes throughout North Wales, but seldom found at altitudes above 150m.

The authors of *Birds in Wales* (1994) considered that the status of the Reed Bunting had probably not changed radically in historic times prior to 1950, after which it showed a marked increase in Britain, spreading into sub-optimal breeding areas. Expansion into young conifer plantations and rough farmland hedgerows was noted in some counties. The presence of young conifer plantations in the uplands helped the species to breed at higher altitudes than formerly, up to 460m in Meirionnydd, Caernarfonshire and Radnorshire. The severe winter of 1962/63 greatly reduced the population across Britain, however, with a recovery occurring only in the late 1960s, followed by a precipitous decline during 1970–83.

The *Britain and Ireland Atlas 1968–72* showed that the Reed Bunting was found in 87% of 10-km squares in Wales, although there were a few gaps, particularly in southern Ceredigion and western Carmarthenshire. Fieldwork for the 1988–91 *Atlas* showed more losses than gains, with breeding in 78% of 10-km squares, particularly in the uplands of mid-Wales. By 20 years later, the *Britain and Ireland Atlas 2007–11* showed that many of these losses had been reversed (to 88%), and a very similar picture to that of 1968–72 was produced. *Birds in Wales* (1994) observed that the distribution of the species in Wales was strangely uneven, even allowing for the effect of hill areas, with unexplained absences from the lower Wye Valley in Gwent, south Ceredigion and areas of Denbighshire and Flintshire. In the 2007–11 *Atlas*, relative abundance in the breeding season was generally low in Wales, compared to Ireland and parts of England, but within Wales was highest on Anglesey.

The results of tetrad-level atlases showed a mixed pattern. In Gwent, there was little difference between the *Gwent Atlas 1981–85* (28% of occupied tetrads) and the *Gwent Atlas 1998–2003* (25% of occupied tetrads). There was a slight increase in East Glamorgan, however, where the Reed Bunting was recorded with evidence of breeding in 26% of tetrads in 1984–89 and in 29% in the 2008–11 *Atlas.* An increase was also evident in Pembrokeshire, where the species was found in 21% of tetrads in the 1984–88 *Atlas* and in 28% in 2003–07 (*Pembrokeshire Avifauna* 2020). In Gower, the *West Glamorgan Atlas 1984–89* found the Reed Bunting was present in 43% of tetrads, while the *North Wales Atlas 2008–12* found it in 28% of them.

The BBS index for Wales showed a 34% increase between 1995 and 2018 (Harris *et al.* 2020), with a degree of stability since 2010. The Welsh breeding population in 2018 was estimated to be 17,500 territories (Hughes *et al.* 2020).

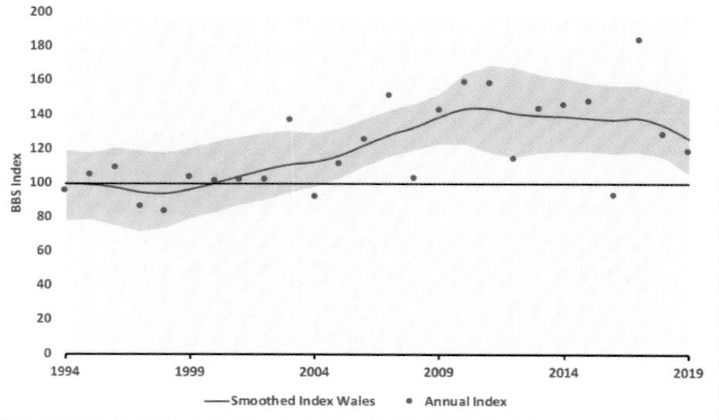

Reed Bunting: Breeding Bird Survey indices, 1994–2019. The red line shows the smoothed index, 1995–2018. The shaded area indicates the 85% confidence limits.

RSPB surveys in 1986 found minima of 384 pairs on Llŷn and 680 singing males or pairs at 450 sites on Anglesey. The largest populations tended to be in large wetlands. For example, Oxwich NNR, Gower, held 66 pairs in 1977, but only 25 pairs in 1992 (*West Glamorgan Atlas 1984–89*). Cors Caron, Ceredigion, held 71 territorial males in 1999 and 77 in 2004 (Roderick and Davis 2010). There were at least 43 territories at RSPB Cors Ddyga, Anglesey, in 2012, an increase compared to ten pairs in 2000. Jennings (2014) stated that it had increased very noticeably in Radnorshire since the late 1980s and was moving into new areas and drier habitats.

Flocks can form in winter where there are good food sources. *Birds in Wales* (1994) noted that flocks up to 100 had been recorded, such as 100 at Eglwys Nunydd Reservoir, Gower, in October 1966, 200 in the Peterstone area of Gwent in Octobers 1967 and 1984, and 250–300 there in February–March 1984 (Venables *et al.* 2008). There were 140 at Clyne Common, Gower, on 26 January 1991 (Hurford and Lansdown 1995). Flocks of over 100 have been recorded more frequently in recent years, with several counts of over 200, including 222 at Blaencaerau Farm, Maesteg, East Glamorgan, on 12 March 2004 and 250 at Cors Caron on 20 January 2013. The highest counts recorded in Wales were at a roost at Mynydd Garn-goch, Gower, in December 2001, which peaked at 525 on 15 December. These flocks were likely to contain both local breeders and some wintering birds from farther north. There is evidence of a modest autumn passage on the islands and headlands, although numbers are variable. Bardsey recorded 70 on 29 September 1964 and 64 on 30 October 2017, but numbers elsewhere tended to be lower.

This is another species that seems to have benefited from winter seed grown for birds under agri-environment schemes. For example, in Breconshire there was a maximum count of up to 140 in January 2016 (Andrew King *in litt.*). The continuation of such beneficial land-management schemes could be vital to the future of this species. In Europe, the second *European Atlas* (2020) has identified that the Reed Bunting has been lost from the southern edge of its range around the Mediterranean, especially in Spain and Italy. It is suggested that this may be a response to climate change. The species may be faring well in Wales at present, even though it is Amber-listed, on the basis of a moderate long-term decline in the UK breeding population, but the potential impact of climate change cannot be discounted in the future.

Rhion Pritchard

Sponsored by Teifi Ringing Group

Dark-eyed Junco *Junco hyemalis* Jwnco Penddu

Welsh List Category	IUCN Red List	Rarity recording
	Global	
A	LC	BBRC

The Dark-eyed Junco is a common breeding bird of coniferous and mixed forests across much of Canada and Alaska south of the Arctic Circle, and more locally in the USA. In winter, it is found across the USA and northern Mexico (*HBW*). Junco taxonomy is complex, as there are 15 subspecies that form two large groups and 3–5 monotypic ones (IOC). Only birds of the 'Slate-coloured Junco' group (*J.h. hyemalis*, *J.h. carolinensis* and *J.h. cismontanus*) have reached Europe. There were 49 records in Britain to 2019, about one a year on average over the last 30 years (Holt *et al.* 2020). There have been three Welsh records:

- Bardsey, Caernarfonshire, from 25 April to 3 May 1975;
- Skomer, Pembrokeshire, on 7–8 May 2017 and
- in a Dolgellau garden, Meirionnydd, on 12 May 2017.

Jon Green and Robin Sandham

White-throated Sparrow *Zonotrichia albicollis* Bras Gyddfwyn

Welsh List Category	IUCN Red List	Rarity recording
	Global	
A	LC	BBRC

The White-throated Sparrow breeds across much of Canada, excluding the far north and west, and spends the winter in the southern USA (*HBW*). There were 50 records of this extremely rare trans-Atlantic vagrant in Britain, with the last record in 2018 (Holt *et al.* 2019), with an average of one a year over the last 30 years. There have been two records in Wales, both on Bardsey, Caernarfonshire: a first-calendar-year bird from 15 October to 7 November 1970 and one on 11 June 2010.

Jon Green and Robin Sandham

White-throated Sparrow on Bardsey, Caernarfonshire, June 2010 © Steve Stansfield

Song Sparrow *Melospiza melodia* Bras Persain

Welsh List Category	IUCN Red List	Rarity recording
	Global	
A	LC	BBRC

The Song Sparrow occurs throughout most of North America. Birds breeding in Canada and the north central USA migrate south in autumn to the southern states of the USA and northern Mexico (*HBW*). This extremely rare trans-Atlantic vagrant has only occurred in Britain on eight occasions (Holt *et al.* 2019), only one of which was in Wales: on Bardsey, Caernarfonshire, on 5–8 May 1970.

Jon Green and Robin Sandham

Bobolink *Dolichonyx oryzivorus* Bobolinc

Welsh List Category	IUCN Red List	Rarity recording
	Global	
A	LC	BBRC

The Bobolink is a long-distance migrant that makes a 20,000km round trip every year, one of the longest migrations of any songbird in the New World. From its northern breeding grounds in the northern and central states of the USA and southern Canada, it flies in groups through Florida and across the Gulf of Mexico to its wintering grounds in South America around the borders of Brazil, Bolivia, Paraguay and into Argentina (*HBW*). This very rare trans-Atlantic vagrant was recorded in Britain on 33 occasions to 2019 (Holt *et al.* 2020), with three records in Wales:

- Skokholm, Pembrokeshire, on 13–14 October 1999 (Thompson 2000);
- Eglwys Nunydd Reservoir, Gower, on 20 September 2010 and
- Skomer, Pembrokeshire, on 8 and 11 October 2010.

Jon Green and Robin Sandham

Baltimore Oriole *Icterus galbula* Euryn y Gogledd

Welsh List Category	IUCN Red List	Rarity recording
	Global	
A	LC	BBRC

The Baltimore Oriole is a North American long-distance migrant. From early April to late May, flocks arrive in eastern and central states of the USA and south-central Canada. They start to leave as early as July for wintering grounds in Florida, the Caribbean, Central America, and the northern tip of South America (*HBW*). There have been 26 records of this extremely rare trans-Atlantic vagrant in Britain up to 2019 (Holt *et al.* 2019), of which there were three Welsh records, all in Pembrokeshire:

- probable female (although there is some uncertainty over the age/sex) on Skokholm on 5–10 October 1967;
- male at Hook on 6–7 May 1970 and
- 2CY female that over-wintered at Roch, from 2 January to 24 April 1989.

Jon Green and Robin Sandham

Brown-headed Cowbird *Molothrus ater* Aderyn Gwartheg Penfrown

Welsh List Category	IUCN Red List	Rarity recording
	Global	
A	LC	BBRC

The Brown-headed Cowbird is widely distributed across the USA, eastern Canada and northern Mexico, expanding as forests are cleared for agriculture (Parkin and Knox 2010). Northern populations winter in Mexico (*HBW*). There are three records of this extremely rare vagrant in Britain, of which one was in Wales, at Angle, Pembrokeshire, on 14–15 July 2009.

Jon Green and Robin Sandham

Black-and-white Warbler *Mniotilta varia* Telor Brith

Welsh List Category	IUCN Red List	Rarity recording
	Global	
A	LC	BBRC

The Black-and-white Warbler breeds in central and eastern North America. Some migrate to Florida, Baja California and the Caribbean. Others go as far as northern South America to spend the winter (*HBW*). There has been a single record of this extremely rare vagrant in Wales, on Skomer, Pembrokeshire, on 10 September 1980. This was the fifth record for Britain. There were a further ten British records to the end of 2019 (Holt *et al.* 2020).

Jon Green and Robin Sandham

Common Yellowthroat *Geothlypis trichas* Aderyn Gyddf-felyn Cyffredin

Welsh List Category	IUCN Red List	Rarity recording
	Global	
A	LC	BBRC

The Common Yellowthroat is a common breeding bird across the lower 48 states of the USA and southern Canada. It winters in the Caribbean and Central America (*HBW*). There have been only ten records in Britain up to 2019, of which two were in Wales: a female on Bardsey, Caernarfonshire, on 27 September 1996 and a second-calendar-year male at Rhiwderin, Newport, Gwent, from 10 February to 31 March 2012. The latter record was a remarkable find amongst the farmland of southeast Wales and was the most recent occasion on which this species was recorded in Britain (Hudson *et al.* 2013).

Jon Green and Robin Sandham

2CY Common Yellowthroat at Rhiwderin, Gwent, February–March 2012 © John Marsh

Blackburnian Warbler *Setophaga fusca* Telor Blackburn

Welsh List Category	IUCN Red List	Rarity recording
	Global	
A	LC	BBRC

The Blackburnian Warbler breeds in southeast Canada and New England, and migrates across the Gulf of Mexico to winter in Columbia and the Andes of Peru (*HBW*). There are only two records of this extremely rare vagrant in Britain, the first of which was on Skomer, Pembrokeshire, 5 October 1961. The other British record was on Fair Isle on 7 October 1988.

Jon Green and Robin Sandham

Yellow Warbler *Setophaga aestiva* Telor Melyn

Welsh List Category	IUCN Red List	Rarity recording
	Global	
A	LC	BBRC

The Yellow Warbler is a common breeding bird in much of North America and migrates south to winter in southern Central America, Columbia and Venezuela (*HBW*). There have been six records of this extremely rare trans-Atlantic vagrant in Britain, the most recent in 2017 (Holt *et al.* 2018). The first British record was in Wales on Bardsey, Caernarfonshire, on 29 August 1964, where it was caught and ringed. Unfortunately the bird died the following day (Evans 1965).

Jon Green and Robin Sandham

Yellow Warbler on Bardsey, Caernarfonshire, 29–30 August 1964 © Hugh Miles

Blackpoll Warbler *Setophaga striata* Telor Penddu America

Welsh List Category	IUCN Red List	Rarity recording
	Global	
A	LC	BBRC

The Blackpoll Warbler breeds across much of Canada and Alaska, and winters in northern South America (*HBW*). There have been 48 records in Britain, the most recent being in 2017 (Holt *et al.* 2018). Three of these records were in Wales:

- Bardsey, Caernarfonshire, on 22–23 October 1968;
- Bardsey, Caernarfonshire, on 7–9 October 1976 and
- Marloes Mere, Pembrokeshire, on 7 October 2008.

The first British record was on the Isles of Scilly on 12–25 October 1968, just ten days before the first Bardsey record. The 1976 Bardsey bird was one of ten that arrived in Britain that autumn.

Jon Green and Robin Sandham

Myrtle Warbler *Setophaga coronata* Telor Tinfelyn

Welsh List Category	IUCN Red List	Rarity recording
	Global	
A	LC	BBRC

The Myrtle Warbler breeds across Alaska and Canada, and in the northwestern states of the USA (*HBW*). Britain has recorded this rare trans-Atlantic vagrant on 24 occasions to the end of 2019, on average one every two years over the last 30 years (Holt *et al.* 2020). There have been three Welsh records, all from the Pembrokeshire islands:

- 1CY on Ramsey, Pembrokeshire, from 31 October to 4 November 1994;
- 2CY male on Skokholm, Pembrokeshire, on 18 June 2017 and
- 2CY male on Ramsey, Pembrokeshire, from 31 May to 1 June 2019.

Jon Green and Robin Sandham

2CY male Myrtle Warbler on Ramsey, Pembrokeshire, October-November 1994 © Steve Young

Summer Tanager *Piranga rubra* Tanagr Haf

Welsh List Category	IUCN Red List	Rarity recording
	Global	
A	LC	BBRC

The Summer Tanager breeds in the southern states of the USA and northern Mexico, and migrates south to central and northern South America in winter (*HBW*). The only record in Britain of this extremely rare trans-Atlantic vagrant occurred in Wales. This was of a first-calendar-year male on Bardsey, Caernarfonshire, from 11 to 25 September 1957.

Jon Green and Robin Sandham

Sponsored by Bardsey Bird and Field Observatory

Rose-breasted Grosbeak *Pheucticus ludovicianus* **Tewbig Brongoch**

Welsh List Category	IUCN Red List	Rarity recording
	Global	
A	LC	BBRC

The Rose-breasted Grosbeak breeds in the northeastern states of the USA and central Canada, and winters in Central America and the Caribbean (*HBW*). There were 30 records of this extremely rare trans-Atlantic vagrant in Britain to 2019, on average one every two years in the last 30 years (Holt *et al.* 2020). There have been four records in Wales:

- female on Skokholm, Pembrokeshire, on 5 October 1967;
- 1CY male on Bardsey, Caernarfonshire, on 14 October 1983;
- 1CY female on Skomer, Pembrokeshire, from 29 September to 11 October 1988 and
- 2CY at St Davids Head, Pembrokeshire, on 21 April 2016.

Jon Green and Robin Sandham

Indigo Bunting *Passerina cyanea* **Bras Goleulas**

Welsh List Category	IUCN Red List	Rarity recording
	Global	
A	LC	BBRC

The Indigo Bunting breeds throughout the eastern states of the USA and in Utah, Arizona and New Mexico. It winters in Central America and the Caribbean (*HBW*). This extremely rare trans-Atlantic vagrant has been recorded in Britain on two occasions, both in Wales. The first was a first-calendar-year male on RSPB Ramsey, Pembrokeshire, on 18–26 October 1996. The other record was a male photographed on a bird table at Llansadwrn, Anglesey, on 20 May 2013.

Jon Green and Robin Sandham

Appendix 1. Probable and presumed escapes

Records that have occurred as suspected or known escapes are not included in the species accounts. This brief review of known, or suspected, escapes is inevitably incomplete, but only recently have most birders and county recorders recognised the importance of reporting and recording systematically these sightings; indeed, many still go unrecorded. In the case of passerines in particular, many of the records come from the island bird observatories, where recording is more systematic, or from garden feeders where escaped cage birds seeking food to survive are more likely to be seen and identified.

Neither category D nor E species form part of the British or Welsh Lists. Several species were formerly assigned to Category C, but in reviewing the records for this book, it was not clear that Welsh populations of two of these had ever been self-sustaining (Golden Pheasant and Lady Amherst's Pheasant) nor is it evident that Ring-necked Parakeet has yet become so.

Category D

Category D is a holding category for species about which there is doubt of their origin in a wild state. It is not intended to be a permanent assignment for any species, and those records are reviewed periodically with a view to assign them to either Category A or Category E.

Bald Eagle *Haliaeetus leucocephalus* Eryr Moel
(North America)
An adult fishing in Llyn Coron, Anglesey, on 17 October 1978 was placed in Category D with the comment by the BOURC that genuine vagrancy was unlikely but that the escape likelihood was also slight. Two records in Ireland have been accepted as genuine vagrants.

Category D and E

Red-headed Bunting *Emberiza bruniceps* Bras Pengoch
(Central Asia)
This species was formerly recorded quite regularly in Wales but its frequency in the cage-bird trade has always counted against its acceptance. The bulk of records in Wales—as across the rest of Britain—occurred during the 1960s and 1970s, and records tailed off in the 1990s, after the Indian Government banned their export.

Five records remain in category D: on Skokholm, Pembrokeshire, in May 1959 and September 1961; on Bardsey, Caernarfonshire, in July and August 1959; and on The Skerries, Anglesey, in September 1961. The remaining records, including 16 on Bardsey up to 1992 and others in Ceredigion (4), Meirionnydd (1), Pembrokeshire (3) and Glamorgan (1), are in Category E. The only records since the start of the 21st century were both in Pembrokeshire: a first-calendar-year bird on Skomer on 5 September 2004 and an adult male at Marloes Mere on 28 November 2012.

Category E

Category E contains records about which there is reasonable doubt of occurrence in a natural state, although in some cases, noted where relevant, a species is also on Category A, B or C of the Welsh List. Some Category E species have been deliberately introduced or are wild birds that were transported here with human assistance. It is illegal to release non-native birds into the wild under Section 14 of the Wildlife and Countryside Act 1981, and to release certain species of native and naturalised birds, listed on Schedule 9 of the Act.

Helmeted Guineafowl *Numida meleagris* Iâr Gini Helmog
(Africa)
As well as being reared for their meat, birds are also released by gamekeepers to act as alarm birds and to encourage Pheasants to roost in trees (reported to Rhion Pritchard by a gamekeeper in Meirionnydd). Records associated with shoots come from a number of counties, the highest counts being 15 at Flemingston, East Glamorgan, in March 2019, and seven at Lleweni Watermeadows, Denbighshire, in November 2018.

Northern Bobwhite *Colinus virginianus* Sofliar Virginia
(North America and Mexico)
Forrest (1907) reported that one had been shot near Bala, Meirionnydd, in 1898, and two shot near Llyn Elsi, Caernarfonshire. He was unaware of any being released in North Wales, but other accounts show that Bobwhite Quails, as they were then known, were introduced to several English estates in the 19th century, and again in East Anglia and the Isles of Scilly in the 1950s and 1960s (Lever 1979).

Wild Turkey *Meleagris gallopavo* Twrci Gwyllt
(North America)
An unknown number were released by a Pheasant shoot at Lake Vyrnwy, Montgomeryshire, in the 1990s, apparently to deter predators around release pens. Some of these bred, but all the birds subsequently disappeared (Mike Walker pers. comm.). One near Stackpole Warren, Pembrokeshire, on 22 April 2010 was some distance from any known Turkey farms.

Chukar Partridge *Alectoris chukar* Petrisen Siwcar
(southeast Europe and Asia)
These were released for shooting in the UK from 1970, until the practice was banned in 1992 (because of hybridisation with the equally non-native Red-legged Partridge, (see page 53). It is not known how abundant they became in Wales but records continued for a few years after the ban. Four at Tainant, on the eastern edge of Ruabon Moor, Denbighshire, in April 2007, if correctly identified, must have been illegal releases. Chukar is on Schedule 9 of the Wildlife and Countryside Act 1981, making it illegal to release birds into the wild.

Red-legged Partridge *Alectoris rufa* Petrisen Goesgoch
(Europe)
A small number of released birds go on to breed in Wales (see page 53), although it is not clear whether this population is self-sustaining. Around 3% of the 10 million birds released in the UK are in Wales, based on those reported on the DEFRA poultry register (Bicknell *et al.* 2010). It is estimated that 480,000 Red-legged Partridges were released for shooting in Wales in 2016.

Grey Partridge *Perdix perdix* Petrisen
(Eurasia)
The native population has declined dramatically in recent decades (see page 54), and it is believed that some of those recorded now have been released for shooting.

Swinhoe's Pheasant *Lophura swinhoii* Ffesant Swinhoe
(Taiwan)
One was seen in Aberystwyth, Ceredigion, on 2 April 1999.

Reeves's Pheasant *Syrmaticus reevesii* Ffesant Reeves
(China)
Individuals were recorded at six locations in Wales on BirdTrack during 2000–19 and from five 1-km squares during 2007–19 (aderyn.lercwales.org.uk), usually associated with releases of Common Pheasants for shooting. Woodland south of Tywyn, Meirionnydd, provided the greatest number of records. Reeve's Pheasant is on Schedule 9 of the Wildlife and Countryside Act 1981, making it illegal to release birds into the wild.

(Common) **Pheasant *Phasianus colchicus* Ffesant**
(Asia)
It is estimated that around 2.54 million Pheasants are released and 0.75 million are shot each year in Wales (see page 56).

Golden Pheasant *Chrysolophus pictus* Ffesant Euraid
(China)
This highly decorative pheasant was released at various locations in England and Scotland from 1725. There are no records of releases in Wales until the 20th century, and the only county in which an attempt was made to establish a population was Anglesey. One population became established on Bodorgan Estate in southwest Anglesey, following a release of birds in the early 1960s, either on the estate (*Birds in Wales* 1994) or at nearby Trefeilir (Jones and Whalley 2004). The other population was in the southeast of the

island, at Pen-y-Parc Woods above Beaumaris, which originated from a pair given as an anniversary gift in 1963 that escaped after a heavy gale destroyed their enclosure at Baron Hill (Lever 2009). At both sites, birds were in dense woodland with very restricted access, which made them difficult to monitor. The population at these two sites was estimated at *c.*30–35 birds (*Birds in Wales* 1994), but it was thought to be extinct at both after 1998. Elsewhere on Anglesey, birds possibly bred at Plas Newydd in 1991. It is now believed that both the small populations on Anglesey were reliant on supplementary food and not self-sustaining. Foxes, thought to have been absent from Anglesey until the late 1960s or early 1970s, may have played a part in their disappearance.

Released Golden Pheasants have been recorded at several sites elsewhere, such as a male and two females at Hirnant, Montgomeryshire, from August to October 1976. Birds were said to have been released at Glynllifon Estate, Caernarfonshire, in the 1970s or 1980s and at Halkyn, Flintshire, in the late 1970s but did not become established. Golden Pheasant is on Schedule 9 of the Wildlife and Countryside Act 1981, making it illegal to release birds into the wild.

Lady Amherst's Pheasant *Chrysolophus amherstiae* Ffesant Amherst

(southeast Asia)

One of the most spectacular gamebirds, Lady Amherst's Pheasant became popular in ornamental collections in Britain during the second half of the 19th century, and from around the turn of the 20th century there were attempts to introduce it into the wild. A population became established at Woburn, Bedfordshire, following releases *c.*1890 (*Birds in England* 2005). Birds from there were introduced to the grounds of Halkyn Castle, Flintshire, *c.*1950. The area was intensively 'keepered for Pheasants and the birds prospered, spreading onto the neighbouring Gwysaney Estate. The population peaked in the early 1980s, estimated by local ornithologists at around 40, but said by local landowners to have been as high as 150 (*Birds in Wales* 1994). Birds were introduced to Plas Isaf, Flintshire, by accident *c.*1980, when the owner brought young pheasants from Halkyn Castle and some of them proved to be Lady Amherst's Pheasants. That population survived for a while, with six recorded in 1987, but no records thereafter.

In the 1980s and early 1990s, the old cemetery at Halkyn became a popular site with birdwatchers who wanted to see this species. The introduction there of about 30 Golden Pheasants in the late 1970s caused some identification problems, and hybridisation occurred between these and the Lady Amherst's Pheasants. By the mid-1980s, the area was no longer intensively 'keepered, and adults, young and eggs were lost to predators, mainly Foxes and domestic cats. *Birds in Wales* (1994) estimated the population to be four males and two females, and while the birds may have been present in 1995, there were no subsequent records. Nightingale (2005) considered that this population died out about 1998. These populations were never self-supporting, since the birds were provided with supplementary food. This species is reported elsewhere occasionally, presumably from collections or having been released by shooting estates. Two males were in a woodland at Bryncrug, Meirionnydd, on 11 October 2007 for example, and a male visited a garden in Ammanford, Carmarthenshire, in September 2008. Lady Amherst's Pheasant is on Schedule 9 of the Wildlife and Countryside Act 1981, making it illegal to release birds into the wild.

Indian Peafowl *Pavo cristatus* Paun

(South Asia)

Cheke (2019) assessed the distribution of the species in Britain, finding records in 23 1-km squares in Wales in the *Britain and Ireland Atlas 2007–11* and via an internet search in 2018. There was some evidence of breeding in Gower and Meirionnydd, and proven breeding at Tremeirchion, Flintshire, in 2007. Breeding was also recorded at Faenol Fawr, Denbighshire, in 2017 but this was in the grounds of a country house hotel and so presumed not to be 'wild'. As well as being kept in ornamental gardens, some rural farms keep Indian Peafowl, presumably as an alarm bird. Several have occupied a residential area in central Bangor, Caernarfonshire, since at least 2019.

Fulvous Whistling Duck *Dendrocygna bicolor* Hwyaden Chwibanog Fechan

(Central and South America, Africa)

Two were at Teifi Marshes, Pembrokeshire, on 2 December 2017.

Wandering Whistling Duck *Dendrocygna arcuata* Hwyaden Chwibanog Grwydrol

(Indonesia and Australia)

Three were in the Cefni Estuary, Anglesey, for one night (date not recorded) *c.*2000.

Red-breasted Goose *Branta ruficollis* Gŵydd Frongoch (Russia)

There are two records of wild birds recorded in Wales (see page 59), but all records since 1950 are considered to have been escapes from captivity. These include two on the Alaw Estuary, Anglesey, during winter 1975/76, and singles at Shotwick Fields, Flintshire, in November 1991, RSPB Ynys-hir, Ceredigion, on 13–15 October 1999, Llyn Alaw in winter 1999, on the Dyfi Estuary on 25–26 June 2005, RSPB Oakenholt Marsh, Flintshire, on 14 July 2005 and Gresford Flash, Denbighshire, on 5 February 2007. Records have increased since 2012, with one at Penberi Reservoir, Pembrokeshire, in June and at Dolydd Hafren, Montgomeryshire, from June to December that year. One frequented the Dyfi Estuary, Ceredigion, between October 2016 and June 2019, associating with the naturalised flock of Barnacle Geese each autumn and winter. In 2018, what was presumed to be the same bird spent time at Tanygrisiau Reservoir, Meirionnydd, and Llyn Brenig, Denbighshire, in June, before returning to RSPB Ynys-hir in August; the same bird was at Cors Caron, Ceredigion, on 1 January 2019. Another was at various sites around Parc Bryn Bach in Gwent, Caerphilly in East Glamorgan and Llangorse Lake in Breconshire during winter 2017/18.

Lesser Canada Goose *Branta canadensis parvipes/moffitti* Gŵydd Fach Canada

(*parvipes* breeds in northern Canada and winters in southern USA; *moffitti* breeds in the Rocky Mountains and the Pacific coast of North America)

The Canada Goose is the most numerous goose species in the world, and there is a possibility that vagrants from North America could reach Wales. The introduced birds in Britain (see page 59) are physically large, apparently descended from one or more of the larger races of the species. There have been reports in winter of smaller races of Canada Goose in Wales. One identified as *B.c. parvipes* was at Ffordd Fawr, Breconshire, from January to April 2010 and on 10 February 2011. Birds thought to be either *B.c. parvipes* or *B.c. moffitti* were recorded on Skomer, Pembrokeshire, and at RSPB Burton Mere Wetlands on the Flintshire/Wirral border in 1998, and at Bronydd, Radnorshire, in January 2010. Canada Goose is on Schedule 9 of the Wildlife and Countryside Act 1981, making it illegal to release birds into the wild.

Barnacle Goose *Branta leucopsis* Gŵydd Wyran

(Greenland, Europe)

As well as a naturalised breeding and wintering population, and the possibility of birds of Svalbard and Greenland origin, in Wales, it is believed that small numbers of feral origin also occur. Barnacle Goose is on Schedule 9 of the Wildlife and Countryside Act 1981, making it illegal to release birds into the wild.

Cackling Goose *Branta hutchinsii* Gŵydd Glegyrog

(North America)

This species was added to the British List only in 2016, with the first accepted record being in Lancashire and North Merseyside in 1976. There are four subspecies, including the nominate Richardson's *B.h. hutchinsii* and Taverner's *B.h. taverneri*, but the latter is difficult to distinguish from 'Lesser Canada Goose' *B. canadensis parvipes*. Although there have been several claims of this species in Wales, only one at Llanishen Reservoir, East Glamorgan, on 22 April 2012 was assessed by the BBRC. It had been in Avon and Somerset through the previous winter and was considered to be of uncertain origin. One reportedly of the subspecies *B.h. taverneri* was on the River Wye at Bronydd/Glasbury, Radnorshire, from 28 October 2009 to 20 March 2012. As well as the timing and location of occurrence being important to acceptance by the BBRC, genuine birds of wild origin are likely to be with 'carrier species' from the same area, such as Barnacle Geese and Greenland White-fronted Geese.

Bar-headed Goose *Anser indicus* Gŵydd Benrhesog
(Asia)
One of the escapes most frequently recorded in Wales, this species has been sighted in most Welsh counties, usually with Canada Geese or Greylag Geese at urban wetland sites. Most records are singles but there have been a number of breeding attempts: eggs were laid in a nest at Llyn Alaw, Anglesey, in the 1980s; a probable family party was reported in Newport, Pembrokeshire, in September 2011; a female laid three infertile eggs at a farm in Llanpumsaint, Carmarthenshire, in 2012 and one was seen at Dyfi Junction, Meirionnydd, with a single hybrid gosling of unknown parentage in 2018. Bar-headed Goose is on Schedule 9 of the Wildlife and Countryside Act 1981, making it illegal to release birds into the wild.

Emperor Goose *Anser canagicus* Gŵydd Frech
(Alaska)
One was at Hook on the Cleddau Estuary, Pembrokeshire, from 25 June 1983 until at least the end of 1984. Others were at sites on Anglesey in 1999, Aberystwyth, Ceredigion, in May 2003, and in the Usk Valley, Gwent, in 2009–10. Emperor Goose is on Schedule 9 of the Wildlife and Countryside Act 1981, making it illegal to release birds into the wild.

Ross's Goose *Anser rossii* Gŵydd Ross
(North America)
This species is on categories D and E of the British List and may well be added to Category A on the basis of small numbers that have been seen in Iceland and Scotland, sometimes with Pink-footed Geese (Holt *et al.* 2020). However, it would seem unlikely that any of the Welsh records will be accepted onto Category A in that review. One roamed the wetlands of Anglesey in 1996–98 but most recent records have come from northeast Wales, with 1–2 on former gravel pits in the Dee Valley, Denbighshire/Flintshire, and occasionally moving as far north as Point of Ayr, Flintshire, during 2011–14.

Snow Goose *Anser caerulescens* Gŵydd yr Eira
(North America)
There is no evidence that this species has occurred as a wild vagrant in Wales although some Scottish records were probably genuine trans-Atlantic visitors. It was removed from consideration by the BBRC in 1963, because of the difficulty of distinguishing wild vagrants in western Scotland from a population that had become established on Mull and Coll in the 1950s. That population remains small: 12 birds in 2017 and 2018, half the number recorded in 2014 (Eaton *et al.* 2020). Colour-ringing of a large proportion of the Coll birds in 2002 showed these to be highly sedentary, not recorded away from the island (*The Birds of Scotland* 2007). There have been small populations elsewhere in the UK: 40 in Norfolk in 2000 and 21 in Yorkshire in 2012, but those in Orkney and Hampshire dwindled to a handful in the same period (Holling *et al.* 2017).

Snow Geese have been kept in captivity for a long time: the sole record in Forrest (1907), at Abergwyngregyn, Caernarfonshire, in 1901 was traced to a collection in Parc Faenol, a few miles to the west. It is one of the most commonly recorded escapees in Wales, with 452 records on BirdTrack in 2000–19, occurring in almost every Welsh county. Up to 18 have occurred in any year, but the 'escape' label means that birdwatchers and authors probably record its presence inconsistently.

A few pairs have bred on islands off West Wales. Two of four pinioned Snow Geese that arrived on Skokholm, Pembrokeshire, in June 2000 mated and fledged four young in 2001, of which three were regularly seen around Marloes and one as far out as Grassholm, Pembrokeshire, in 2004. Two were on Skokholm in 2005–07, of which one paired with a Canada Goose in 2007 and produced five infertile eggs. A female remained in 2008–10, pairing with a Canada Goose in the latter two years and laying eggs, but no young fledged. The only other nesting record in Wales was one that incubated eggs on Cardigan Island, Ceredigion, in June 1983.

Most records have been of 1–2 birds, but there are some larger groups. The largest was eight at Margam Breakwater, Gower, on 3 September 1974 and seven were at Newport Wetlands Reserve, Gwent, on 15 April 2013. Flocks of six were seen over Bardsey, Caernarfonshire, on 26 October 1999, at Llanon, Ceredigion, on 16 October 2009, at Kenfig, East Glamorgan, on 18 November 2015 and on Skomer, Pembrokeshire, on 24 June 2018, one of which was a first-calendar-year bird. Snow Goose is on Schedule 9 of the Wildlife and Countryside Act 1981, making it illegal to release birds into the wild.

Greylag Goose *Anser anser* Gŵydd Lwyd
(Eurasia)
In addition to the growing naturalised population in Wales, some birds of feral origin occur away from farmsteads.

Swan Goose *Anser cygnoides* Alarch Ŵydd
(China)
This species has been domesticated in Britain since at least the mid-18th century, and the number in the "wild" is doubtless under-recorded. One or two have been recorded at a number of sites over a long period, particularly on the Afon Clwyd, Denbighshire; at Newport Wetlands Reserve (Goldcliff), Gwent; and up to 3–5 at urban wetlands in East Glamorgan, including Dare Valley Country Park, Coed Craig Ruperra and Cardiff in 2017–18.

Taiga/Tundra Bean Goose *Anser fabalis/serrirostris* Gŵydd Lafur y Taiga/Twndra
(Asia and northeast Europe)
While small numbers of wild origin have occurred (see page 63), several other records of 'Bean Goose' not assigned to subspecies have occurred but were not accepted as wild: at Maesllyn Pool, near Tregaron, Ceredigion, on 1 December 1989 ('thought to be of feral origin'—Roderick and Davis 2010); and on Anglesey, at Llyn Alaw in March–April 1986, at Llyn Traffwll in September 1986, and at both sites in January 1992.

Single Taiga Bean Geese on Llyn Alaw, Anglesey, on 4–10 January 1985 and from 1 January to 15 July 1998 and a Tundra Bean Goose at Llyn Alaw, Anglesey, from 27 December 1996 into January 1997 were considered to be escapes.

Pink-footed Goose *Anser brachyrhynchus* Gŵydd Droetbinc
(Greenland/Europe)
Flocks of wild birds occur each winter, mainly in northeast Wales, but individuals are recorded elsewhere and at all times of the year, often with Greylag Goose or Canada Goose, which are presumed to be of feral origin.

Lesser White-fronted Goose *Anser erythropus* Gŵydd Dalcenwen Fechan
(Fennoscandia/Russia)
There are two records of wild birds having been recorded in Wales (Category A, see page 69) but all records since 1971 are considered to have been escapes or to have originated from re-establishment attempts in Fennoscandia. One record that pre-dated the releases was a bird described as very tame at Brynmawr, Breconshire, on 19–24 September 1973, and presumed to be an escape. A re-establishment scheme in Swedish Lapland involved the release of captive-bred birds alongside Barnacle Geese from 1981, with continued supplementary releases. These birds migrated with the Barnacle Geese to The Netherlands rather than to parts of eastern Europe, where hunting pressure was high. During 1987–97, around 150 captive-bred birds were also released in Finnish Lapland but mortality was high and none of the released birds made breeding attempts (Marchant and Musgrove 2011). At least one marked bird from the Finnish scheme was recorded in Wales, at Dryslwyn, Carmarthenshire, on 2–10 February 1991, and was also seen at Slimbridge, Gloucestershire.

There were no further records until a series during 2002–10 that suggested several long-staying birds were present in North Wales, all of which were considered to have been escapes. One was seen intermittently on the Conwy Estuary and adjacent lagoons at RSPB Conwy, Denbighshire, from April 2002 to September 2010, with two there in March and July 2006 and four in June 2008. A Lesser White-fronted Goose x Barnacle Goose hybrid was also recorded at RSPB Conwy, a combination that occurred frequently following the Swedish release programme (Tony Fox pers. comm.).

Records in the Dee Estuary, Flintshire, may be the same birds seen at Conwy: singles were on the border at RSPB Burton Mere Wetlands in July 2003 and at Talacre/RSPB Point of Ayr in

February 2009, and there were three at Connah's Quay NR on 15 August 2010 and at RSPB Burton Mere Wetlands on 17 December 2017. One bearing an orange or yellow ring was at Llyn Traffwll, Anglesey, on 30 June 2005 and then at various sites on the island until 2010. Two were at Llyn Traffwll from 27 February to 20 March 2019, initially with 12 European White-fronted Geese and six Pink-footed Geese. The only recent record in the south was of two at Llangennith Moor, Gower, on 8 May 2016.

Black Swan *Cygnus atratus* Alarch Du
(Australia)
This species has been widely kept in ornamental collections, from which individuals have frequently escaped. Forrest (1907) mentioned Black Swans being seen at the mouth of the Dee Estuary, Flintshire, and near Abergele, Denbighshire, in spring 1879. It was recorded in 42 10-km squares between 1923 and 2019, at least once in every county (aderyn.lercwales.org.uk). As a long-lived bird, some can remain in an area for many years. Most records have been 1–2 birds, but multiple counts include five in Roath Park, East Glamorgan, in 1988–89, four at White Sands, Flintshire, in February and October 2016, four at West Aberthaw, East Glamorgan, in October 2018, and four off Skokholm, Pembrokeshire, on 26 March 2019. While around 25 pairs of Black Swans breed annually in England, we have been unable to find any evidence of breeding in Wales. Black Swan is on Schedule 9 of the Wildlife and Countryside Act 1981, making it illegal to release birds into the wild.

Black-necked Swan *Cygnus melancoryphus* Alarch Gyddfddu
(South America)
One was in the Dyfi Estuary, Ceredigion, from 11 November to 2 December 1975.

Egyptian Goose *Alopochen aegyptiaca* Gŵydd yr Aifft
(Africa)
Birds seen in Wales may occur from the naturalised English breeding population (see page 73), and it is included in Category A on the basis of known movements, but some individuals may be of more local, feral stock.

Ruddy Shelduck *Tadorna ferruginea* Hwyaden Goch yr Eithin
(southeast Europe and Asia)
This species is on Category B on the basis of a record in 1892 (see page 76), and all subsequent sightings relate to escapes or movements of birds released in continental Europe. The BOURC has accepted the possibility of wild birds reaching Britain but did not accept any post-1950 records as being wild (Harrop 2002). Large numbers moult in The Netherlands each summer, with counts of over 600 on the Eemmeer (Lensink *et al.* 2013). These are thought to originate mostly from naturalised populations in Germany and Switzerland, and could be the source of at least some modern records in Wales. Since birds here are usually presumed to have escaped, the species is likely to be under-reported.

Aside from one at Newgale, Pembrokeshire, in 1955 (Donovan and Rees 1994), all other Welsh records have occurred since 1974. Aside from singles, these include three at Aberaeron, Ceredigion, for several winters between early November 1974 and 1979, eight at Llyn Alaw, Anglesey, on 26 July 1979, two at Llyn Traffwll, Anglesey, from 29 June to 19 July 1989, and four at RSPB Ynys-hir, Ceredigion, on 28 July 1992.

In 1994 there was an influx of birds, with exceptional numbers recorded in Fennoscandia and smaller numbers in Britain. The largest flock recorded was eight at Point of Ayr, Flintshire, on 24–25 July that moved to Wirral and formed the nucleus of a congregation that reached a peak of 12 birds. In late autumn, some of this flock dispersed west, providing records of singles at Foryd Bay, Caernarfonshire, on 13–16 November; at Broadwater, Meirionnydd, from 21 November to 14 December; near Glan Conwy, Denbighshire, on 17–31 December; and near Holyhead, Anglesey, on 19–31 December. The flock was thought to have included one or two escaped birds that had been in the area for several years, one with a red colour-ring (Vinicombe and Harrop 1999).

Records in the 21st century include two at Celtic Manor golf course, Gwent, throughout 2000; three over Porthcawl, East Glamorgan, on 17 June 2002; four at Newport, Pembrokeshire, on 18 June 2002; two at RSPB Conwy, Denbighshire, on 19 June 2004; two at Wernffrwd, Gower, on 16–17 July 2007 and two on the Gwendraeth Estuary, Carmarthenshire, on 5 August 2007; 2–3 at Newport Wetlands Reserve, Gwent, from July to September 2009; two at Peterstone Wentlooge, Gwent, on 5 October 2013; four at Peterstone Gout, Gwent, on 14 July 2014; and two on the River Clwyd, Flintshire, between September and November 2019. There were two hybrids at Connah's Quay NR, Flintshire, in July 2009, although the parentage was uncertain. Birds occasionally breed in England, such as a pair that probably bred in Wiltshire in 2017 (Holling *et al.* 2019), but there appears to be no record of any breeding attempt in Wales. Ruddy Shelduck is on Schedule 9 of the Wildlife and Countryside Act 1981, making it illegal to release birds into the wild.

South African Shelduck *Tadnorna cana* Hwyaden Goch y Penrhyn
(southern Africa)
These have appeared sporadically in Wales. A pair, one or both possibly of mixed South African/Australian Shelduck *T. tadornoides* parentage, were at several sites in Breconshire and Radnorshire between April 2009 and December 2011. They bred at Brechfa Pool, Breconshire, in 2010 and hatched at least nine chicks, of which six survived to at least November. This was only the third breeding record in Britain. Up to four birds were present in Breconshire through 2011 but there was no proven breeding attempt.

Australian Shelduck *Tadorna tadornoides* Hwyaden Fraith Awstralia
(Australia)
This species occurs occasionally, such as in Caernarfonshire in January 2005 and in Gwent in September 2006 and April 2008. The latter was also recorded on the East Glamorgan side of the River Usk in 2008.

Muscovy Duck *Cairini moschata* Hwyaden Fwsg
(Central and South America)
This is the most commonly recorded non-native species in Wales that has not established a breeding population. There were 610 records on BirdTrack during 2000–19, and it was recorded in 34 10-km squares between 1973 and 2019 (aderyn.lercwales.org.uk). While birds occur widely, many are found close to habitation, suggesting they are at least partly dependent on supplementary food. Sites with regular records include up to six at Fendrod Lake, Gower, in 2006–19; up to four on the Brecon Canal in 2013–14 (a pair hatched young in both years); up to four at Parc Taf-Bargoed, East Glamorgan, in 2017–18; 1–2 in Clydach Vale CP, East Glamorgan, in 2010–14; and one at Brynmawr Ponds, Gwent in 2018–19.

Wood Duck *Aix sponsa* Hwyaden Gribog y Coed
(North America)
One or two are reported in Wales every few years (in 14 1-km squares between 1973 and 2019—aderyn.lercwales.org.uk). The only exceptional count was nine at Festival Lakes, Ebbw Vale, Gwent, in December 2018. A pair, thought to have been deliberately (illegally) released at St Mellons, East Glamorgan, hatched four chicks in 2018, but none fledged (Eaton *et al.* 2020). Wood Duck is on Schedule 9 of the Wildlife and Countryside Act 1981, making it illegal to release birds into the wild.

Mandarin Duck *Aix galericulata* Hwyaden Mandarin
(Asia)
Like Wood Duck, this species is kept in collections and some past records in Wales almost certainly were released birds, although as the naturalised population has increased and spread, it has become impossible to distinguish these unless they are pinioned or marked. Mandarin Duck is on Schedule 9 of the Wildlife and Countryside Act 1981, making it illegal to release birds into the wild.

Maned Duck *Chenonetta jubata* Hwyaden Fyngog
(Australia)
A pair was at Glandyfi, Ceredigion, on 7 March 1997.

Ringed Teal *Callonetta leucophrys* Corhwyaden Dorchog
(South America)
This is occasionally recorded outside collections, such as at RSPB Conwy, Denbighshire, in June 2007, at Cosmeston Lakes CP, East Glamorgan, in 2008 and at Shotwick, Flintshire, in September 2013.

Cinnamon Teal *Spatula cyanoptera* Corhwyaden Winau America
(Americas)
There is one record from ST17, the 10-km square in East Glamorgan covering Cardiff, between 1973 and 1983 (aderyn.lercwales.org.uk).

Blue-winged Teal *Spatula discors* Corhwyaden Asgell-las
(Americas)
While a small number of birds considered to be of wild origin has been recorded (see page 78), the species is also held in collections. An adult male at Pembroke Mill Pond from 21 May to 4 June 2003 bore a plastic ring and was presumed to be an escape.

Chiloé Wigeon *Mareca sibilatrix* Chwiwell Magellan
(South America)
This occurs in Wales periodically and some reside in the same place for a long time, such as two in Roath Park, East Glamorgan, in 1988–2005. Most records during 2007–19 were along the South Wales coast between Carmarthen and Barry, East Glamorgan, although individuals have occurred elsewhere.

Indian Spot-billed Duck *Anas poecilorhyncha* Hwyaden Bigfannog India
(South Asia)
One was in Fishguard Harbour, Pembrokeshire, on 25 August 1999.

Mallard *Anas platyrhynchos* Hwyaden Wyllt
(worldwide)
As well as the native breeding, wintering and passage populations in Wales, an unknown number are released for shooting.

White-cheeked Pintail *Anas bahamensis* Hwyaden y Bahamas
(Caribbean)
One was at Penclacwydd, Carmarthenshire, in October 2018.

Marbled Duck *Marmaronetta angustirostris* Hwyaden Gleisiog
(Spain, northwest Africa, Middle East and Central Asia)
This globally Vulnerable duck has a small population around the Mediterranean and is kept in captivity in Britain. One was at Kenfig Pool, East Glamorgan, on 7 June 1994.

Red-crested Pochard *Netta rufina* Hwyaden Gribgoch
(Eurasia)
There is considerable uncertainty about the origins of birds seen in Wales (see page 90), although it is on Category C of the Welsh List on the basis of a breeding population in England. One at Greenfield Valley, Flintshire, on 22–30 January 1992 was considered to have been an escape and one at Withybush Woods pond, Pembrokeshire, in November–December 1999 certainly was, bearing a metal leg ring and having a badly-pinioned left wing. Red-crested Pochard is on Schedule 9 of the Wildlife and Countryside Act 1981, making it illegal to release birds into the wild.

Barrow's Goldeneye *Bucephala islandica* Hwyaden Lygad Aur Barrow
(Iceland and North America)
Three records of this northern species have been accepted as wild birds in Scotland and one in Northern Ireland, but the origin of a first-calendar-year male at Clarach, Ceredigion, on 9–15 March 1975, was deemed not proven by the BBRC.

Hooded Merganser *Lophodytes cucullatus* Hwyaden Gycyllog
(North America)
Although a bird of wild origin has been accepted in Wales (see page 107), this species is popular in waterfowl collections and there is a long history of escapes to the wild. A first-calendar-year male shot in the Menai Strait, Caernarfonshire/Anglesey, in winter 1830/31 was considered the only acceptable British record until discounted by a review in 1999. Other claims of doubtful origin and therefore not accepted onto the Welsh List include one shot in Glamorgan in 1838, a male in the Menai Strait in March 1911, another there in May 2011, and a colour-ringed female at Malltraeth, Anglesey, from November 2000 to January 2004.

White-cheeked Turaco *Tauraco leucotis* Twraco Bochwyn
(Africa)
Turacos of various species have escaped from captivity and seem capable of living in suburban gardens for several weeks, or even months. Most are not identified specifically, but one at Newport Wetlands Reserve, Gwent, in September 2014 was identified as this species.

African Collared Dove *Streptopelia roseogriseo* Turtur Barbari
(Africa)
The domesticated form, known as Barbary Dove, is kept in captivity by specialist breeders, from which some escape. One was seen with (Eurasian) Collared Doves on Bardsey, Caernarfonshire, on 30 April 1984 and one was near Gelly, Pembrokeshire, in February–April 1999.

Laughing Dove *Spilopelia senegalensis* Turtur Chwerthinog
(Africa, Middle East, Central and South Asia)
Escapes from aviaries occur, including several in Pembrokeshire: on Skomer on 6–18 June 1988, at Haverfordwest from April to June 1992 and at St Brides on 27–30 June 2019.

Diamond Dove *Geopelia cuneata* Colomen Ddeimwnt
(Australia)
Escapes occur sporadically in Wales but sightings are rarely logged with county recorders.

Gallinule spp. *Porphyrio* Corsiar spp.
(several species worldwide)
Several 'Purple Swamphens' were reported in the late 19th and early 20th centuries: one shot at Llandeilo, Carmarthenshire, prior to September 1893 when it was seen in a local taxidermy shop (Ingram and Salmon 1954); one caught at Overton-on-Dee, Denbighshire, in April 1901 and kept in an aviary for more than a year; and one shot in the same location in May 1903. These species have been widely held in captivity and although both Purple Gallinule *P. martinica* (North America) and Western Swamphen *P. porphyrio* (southwest Europe and North Africa) have subsequently been admitted to the British List, the Welsh records pre-date the split into multiple species and there is insufficient evidence to consider them to have been wild birds.

Crane *Grus grus* Garan
(Eurasia)
While there has been a number of records of wild birds (see page 136), a marked increase in sightings in southeast Wales from 2012 is considered to relate to the Great Crane Project in Somerset, where young birds were released during 2010–15. Most records were of 1–2 birds in Gwent and East Glamorgan, and there were breeding attempts in both counties during this period. Crane is on Schedule 9 of the Wildlife and Countryside Act 1981, making it illegal to release birds into the wild except under licence.

Flamingo spp. *Phoenicopterus* Fflamingo spp.
(several species worldwide)
Members of this family have been kept in captivity in Britain for many years, and their tendency to escape seems matched only by the high likelihood of them being reported. Forrest (1907) reported one as early as 1898, shot at Traeth Bach, Meirionnydd, while there were four on the Dysynni Estuary in the same county in 1913. None was identified to species level, and that continued to be the case with most records through the 20th century, the exceptions being the three species below.

Greater Flamingo *Phoenicopetrus roseus* Fflamingo Mawr
(southern Europe and Africa)
This species is widely kept in zoos and collections in Britain. One in the Cleddau Estuary, Pembrokeshire, from 11 May to early June 1969 and two on Bardsey, Caernarfonshire, in April 1974 are the only records we have been able to find from Wales in the published

literature. A flock of seven in The Netherlands in July 2020 included two that had been ringed in Spain, illustrating the potential for wild birds to appear in northwest Europe.

American Flamingo *Phoenicopterus ruber* Fflamingo America
(Caribbean)
One flew past Strumble Head, Pembrokeshire, on 23 June 1990.

Chilean Flamingo *Phoenicopterus chilensis* Fflamingo Chile
(South America)
A flamingo at Llyn Heilyn, Radnorshire, in July 1968 was attributed to this species by Jennings (2014); another was with two probable Greater Flamingos at Cemlyn, Anglesey, on 30 April 1977; one was on the Gann Estuary, Pembrokeshire, in December 1983; and one was on the Afon Cefni, Anglesey, in 1989 that was reportedly killed by Mute Swans. One at several sites in Ceredigion in 1976 could have been the bird involved in several reports of flamingo spp. in this county through the 1970s. Only two have featured in *Welsh Bird Reports* in the 21st century: at Trevalyn Meadows, Denbighshire, in November 2014 and at RSPB Ynys-hir, Ceredigion, from December 2014 into 2015.

Cape Petrel *Daption capense* Pedryn Patrymog
(sub-Antarctic Islands)
One was shot on the Dyfi Estuary, Ceredigion/Meirionnydd, in 1879 and held in the Gogerddan collection (as 'Cape Pigeon', Salter 1895). During the 19th century, birds were brought to and released in European waters by sailors (Haas and Crochet 2009).

Collared Petrel *Pterodroma brevipes* Pedryn Torchog
(Pacific Ocean)
One was claimed to have been shot by a fisherman near Aberystwyth, Ceredigion, in winter 1889 (*Birds in Wales* 1994).

White Stork *Ciconia ciconia* Ciconia Gwyn
(Eurasia)
While White Stork is on category A of the Welsh List (see page 282), the species is also widely held in captivity and since 2016 has been the subject of an introduction attempt in Sussex, with the first young fledged in 2020. It is hard to determine the origin of White Storks seen in Wales and most spring/summer records are assumed to be of wild birds. One at Peterstone Wentlooge and St Brides Wentlooge, Gwent, on 9 November 1996 and at Sker, East Glamorgan, on 10–12 November 1996 was considered to have been a bird that escaped from Bristol Zoo the previous month. One on the Nevern Estuary, Pembrokeshire, on 14 June 2004 was so approachable that it was considered to be an escape.

African Sacred Ibis *Threskiornis aethiopicus* Crymanbig y Deml
(sub-Saharan Africa)
Naturalised populations have become established from zoo escapes in Italy, France and The Netherlands, although extensive culling has been undertaken in the last two countries, and it has not bred in The Netherlands since 2010 (Sovon 2020). There is concern about the impact of depredation by Sacred Ibis on native species, particularly tern species in France, and the GB Non-native Species Secretariat lists it as an Alert species, requesting that any sighting is reported immediately. It lists the first British record as at Pantyffynnon, Carmarthenshire, in 1995 (nonnativespecies.org), which may have been one reported in Ceredigion from 1993–95 by Roderick and Davis (2010). However, that was pre-dated by one in the Glasbury area, Radnorshire, from July 1972 to January 1973 (Jennings 2014). More recent records include one at Trefeiddan, Pembrokeshire, in spring 1989; one on the Teifi Marshes, Pembrokeshire/Ceredigion, throughout 1994, having escaped from a collection at Newcastle Emlyn, Ceredigion/Carmarthenshire; two at Rhuddlan, Flintshire, in July 2012; and one at Llyn Coed-y-Dinas, Montgomeryshire, in August 2017.

Spoonbill *Platalea leucorodia* Llwybig
(Eurasia)
A female of the Mauritanian subspecies *P.l. balsaci*, nicknamed "Maureen" by birdwatchers, was a ship-assisted arrival to Britain that was ringed and released at WWT Slimbridge, Gloucestershire, in 1998. She was on the Inland Sea, Anglesey, from 16 December 1998 to 5 January 1999, at Aber Ogwen, Caernarfonshire, from 9 January to 5 March 1999 (Pritchard 2017) and then on the Dee Estuary, generally on the English side, to at least May 2000.

Green Heron *Butorides virescens* Crëyr Gwyrdd
(North and Central America)
While there are two records on Category A of the Welsh List (see page 294), one on the River Wye below Glasbury on 22–29 January 1991 had a greyish (probably old and discoloured) leg-ring (Jennings 2014).

Cattle Egret *Bubulcus ibis* Crëyr Gwartheg
(Europe, Middle East, Africa, Americas)
Prior to the first Welsh record in 1980 (see page 296) one at Morfa Harlech, Meirionnydd, on 13–26 July 1971 and then, presumably the same, in the Malltraeth/Newborough area of Anglesey, from 28 August to 9 December 1971 was considered by the BBRC to be of captive origin (Pritchard 2012).

Great White Pelican *Pelecanus onocrotalus* Pelican Gwyn
(southeast Europe, Central Asia and sub-Saharan Africa)
There have been several Welsh records of a species that has been kept in collections since at least 1664. There was at least one at RSPB Ynys-hir and Aberystwyth, Ceredigion, in October/November 1974, one at Whiteford, Gower, on 24 September 1989 and another at Hay-on-Wye, Radnorshire, on 23 October 2001. One roamed across North Wales in September and October 2006, being seen at several points in Flintshire, Denbighshire and Caernarfonshire before settling at Malltraeth Cob, Anglesey, for ten days. An unidentified Pelican was at Llyn Trawsfynydd, Meirionnydd, on 3 September 1973 and a 'brown' Pelican was in north Ceredigion in August and September 1974.

Black-winged Kite *Elanus caeruleus* Barcud Ysgwydd-ddu
(southwest Europe and Africa)
One was in the Elan Valley, Breconshire/Radnorshire, in 2010.

Bearded Vulture *Gypaetus barbatus* Fwltur Barfog
(southern Eurasia and Africa)
An immature at Sudbrook, Gwent, on 12 May 2016 was subsequently seen in Devon and Cornwall and, it later emerged, had earlier been in Belgium and Kent. It was considered to have originated from a re-establishment project in the Alps, which at the time of the sighting was not considered to be a self-sustaining population. The BOURC thus placed the record in Category E.

Egyptian Vulture *Neophron percnopterus* Fwltur yr Aifft
(southern Europe, Asia, and Africa)
A locally escaped bird was seen at Taliaris Woods, Carmarthenshire, on 7 January 2012.

Hooded Vulture *Necrosyrtes monachus* Fwltur Cycyllog
(Africa)
One, from a collection in North Devon, was seen over a number of sites in Wales during August 2000.

White-backed Vulture *Gyps africanus* Fwltur Cefnwyn Affrica
(Africa)
One was seen near the upper Tywi Estuary, Carmarthenshire, in 2006.

Griffon Vulture *Gyps fulvus* Griffon
(southwest Europe and North Africa)
One was seen over Llandovery, Carmarthenshire, in 2009.

Cinereous Vulture *Aegypius monachus* Fwltur Du
(southern Eurasia)
Also known as Black Vulture or Monk Vulture, an unringed adult wintered in Breconshire and Radnorshire from October 1977 to 20 February 1978, and ventured to RSPB Gwenffrwd, Carmarthenshire, during early February. After much debate by the BOURC, it was placed in Category D in 1992, and then Category E in 2008 (Vinicombe 1994).

Bateleur *Terathopius ecaudatus* Eryr Cwta
(Africa)
One that escaped from a Llandeilo aviary remained at large in the Llanelltyd area, Meirionnydd, in 2005–06. Another was reported over Cardiff Bay, East Glamorgan, on 25 February 2007.

Booted Eagle *Hieraaetus pennatus* Gwalcheryr Bacsiog
(southwest Europe and North Africa)
One was seen at various sites in South Wales in 2008, including Neath Port Talbot, Gower, and the Rhondda Valley, East Glamorgan.

Golden Eagle *Aquila chrysaetos* Eryr Euraid
(North America and Eurasia)
The historical references to Golden Eagle are discussed on page 307, but most dated records relate to escaped birds. Two kept on Skomer, Pembrokeshire, were released there in 1915: one was shot on the mainland soon afterwards, but the other survived for years, moving to Ramsey in 1928. The following year a female was obtained from London Zoo and released to provide a mate for it, but it drowned on the nearby mainland coast a few months later. The original bird was shot on Ramsey in 1932 by a farmer who said that it was killing sheep. It turned out also to be female.

One that flew over Bardsey, Caernarfonshire, in 1968 was thought to be a bird kept partially free-flying at Pwllheli. Records at Foel, Montgomeryshire, near Trawsfynydd, Meirionnydd, in 1965, near Llanuwchllyn, Meirionnydd, in June 1967, at Coed-y-Prior Common, Gwent, in March 1985 and at Cors Caron, Ceredigion, in January 1990 were also considered escapes. A third-calendar-year bird, initially with jesses and rings visible, turned up at Nant Irfon, Breconshire, in March 2009, and in subsequent years was seen in Ceredigion, Radnorshire, Montgomeryshire and Meirionnydd. It was found dead in Nant Irfon in July 2020.

Goshawk *Accipiter gentilis* Gwalch Marth
(Eurasia and North America)
The current breeding population in Wales originated entirely from falconers' escapes or deliberate releases, which went on to breed (see page 310). An estimated 20 were released each year in Britain in the 1970s and 30–40 each year in the early 1980s. Goshawk is now on Schedule 9 of the Wildlife and Countryside Act 1981, making it illegal to release birds into the wild.

Black Kite *Milvus migrans* Barcud Du
(Eurasia)
In addition to the accepted records on page 321, one with visible jesses flew along the Gower and East Glamorgan coast on 10 September 2019.

White-tailed Eagle *Haliaeetus albicilla* Eryr Môr
(Eurasia)
Records of wild origin, in Category A, are featured on page 322, but birds released in England have been tracked into Wales and this species is also kept in captivity from which escapes occur. One in apparently suitable habitat was a third-calendar-year bird at St Govan's Head, Pembrokeshire, on 16 August 2016 that was recaptured the following day. White-tailed Eagle is on Schedule 9 of the Wildlife and Countryside Act 1981, making it illegal to release birds into the wild except under licence.

Harris's Hawk *Parabuteo unicinctus* Gwalch Harris
(Americas)
One of the most frequent raptor escapes, birds are reported in the wild periodically, particularly in the South Wales valleys. Up to four were around Aberdaron, Caernarfonshire, in 2017–18, from which a pair nested at Plas-yn-Rhiw. Efforts were made by public authorities to catch the birds in December 2018, but it is not certain whether this was achieved. The owner was prosecuted for releasing a non-native species into the wild but the case collapsed in court.

Red-tailed Hawk *Buteo jamaicensis* Bwncath Cynffongoch y Gogledd
(Americas)
This raptor is commonly held in captivity and is an unlikely vagrant. One with jesses was captured at Rhosycaerau, Pembrokeshire, on 23 March 2001 but escaped again the following day.

Rough-legged Buzzard *Buteo lagopus* Boda Bacsiog
(Fennoscandia)
Although this is an occasional winter visitor to Wales (see page 322), escaped birds are not unknown; one at Garreg-ddu Reservoir, Radnorshire, on 10 October 1971 was wearing jesses.

Barn Owl *Tyto alba* Tylluan Wen
(worldwide)
In response to declines in wild populations, a number of projects were undertaken to release captive-bred Barn Owls in Britain, including some in Wales, such as in Pembrokeshire (Bob Haycock pers. comm.). It is uncertain whether any of these birds bred, and follow-up monitoring was limited. Barn Owl is now on Schedule 9 of the Wildlife and Countryside Act 1981, making it illegal to release birds into the wild, although rehabilitated birds of wild origin can be released under licence.

Eagle Owl *Bubo* Eryrdylluan spp.
(Europe, Asia and Africa)
Eagle Owls of various species have been found in the wild at a number of locations, most having originated from known collections or aviaries, but able to hunt successfully and remain at large for several years. Only the Eurasian Eagle Owl (*Bubo bubo*) is a slight possibility as a vagrant but there is no evidence that birds have made the sea-crossing and Eagle Owl (then a 'lumped' species) was removed from the British List in 1996. A review of all occurrences in Britain between 1678 and 1990 rejected all 79 records, including the two in Wales: in Swansea, Gower, around 1836, and in Llanidloes, Montgomeryshire, in 1863 (Melling *et al.* 2008).

Around 3,000–4,000 Eagle Owls are thought to be kept in Britain. Breeding in the wild has occurred in several places in Scotland (since 1984) and England (since 1993), but not in Wales. Welsh records since 2008 have come from a minimum of seven counties, several of which have been known to be wild-living for many months or even years. Some individuals in Wales have been identified as non-European species, such as an east Asian species hit by a train near Porthmadog, Caernarfonshire, in 1999 and the African species Spotted Eagle Owl *B. africanus* on Anglesey in 2014. Eagle Owl is on Schedule 9 of the Wildlife and Countryside Act 1981, making it illegal to release birds into the wild.

Lanner Falcon *Falco biarmicus* Hebog Lanner
(southeast Europe, Arabia and Africa)
Widely kept for falconry, escapes occur sporadically. The only records we can find in Wales are an unringed second-calendar-year bird of the North African subspecies *F.b. erlangeri* on Bardsey, on 22 April 2004, and one at Mynydd Du, Carmarthenshire, in 2007.

Saker Falcon *Falco cherrug* Hebog Saker
(Asia)
A species kept in captivity for falconry, individuals have been recorded on Bardsey, Caernarfonshire, in 1989 and 2013; on Skokholm, Pembrokeshire, wearing jesses in 1998; at Penclacwydd, Carmarthenshire, in 1989; around Maesteg, East Glamorgan, in 2005; at Llangorse Lake, Breconshire, in 2006; at Cemlyn, Anglesey, and Rumney Great Wharf, East Glamorgan, in May 2011; in Swansea and Perriswood, Gower, in 2013; and at Newport Wetlands Reserve, Gwent, in 2016.

Gyr Falcon *Falco rusticolus* Hebog y Gogledd
(Holarctic)
In addition to the records in Category A (page 353), this species is kept by falconers and several records in Wales relate to birds seen with jesses. A falcon breeding centre in Carmarthenshire is reported to have lost several Gyr Falcons, with individuals assumed to be escapees from there or elsewhere noted at several locations: at Dale and Pembroke Dock, Pembrokeshire, in January–February 2009; at Llanhennock, Gwent, on 5 October 2009; at Llanddewi, Gower, on 17–19 December 2009; and at Rhossili, Gower, on 27 March 2010. There was also one at Skomer, Pembrokeshire, on 17–18 August 2014, and what was presumed to be the same bird, on Bardsey, Caernarfonshire from 20–21 August 2014.

Pied Crow *Corvus albus* Brân Fraith
(sub-Saharan Africa)
One at St Justinians, Pembrokeshire, on 3–7 July 2018 had been seen in East Yorkshire, Lincolnshire, Norfolk and Somerset during the previous month and subsequently returned to East Yorkshire until March 2019.

Cockatiel *Nymphicus hollandicus* Cocatïl
(Australia)
These are widely held in captivity as a companion animal. Escapes occur sporadically, usually in built-up areas.

Scarlet Macaw *Ara macao* Macaw Sgarlad
(Central and South America)
An escaped pet was seen around Mynydd Llandygai, Caernarfonshire, in 2002–07.

Plum-headed Parakeet *Psittacula cyanocephala* Paracît Pen Cochlas
(India)
One was on Bardsey, Caernarfonshire, on 8–11 April 1992.

Ring-necked Parakeet *Psittacula krameri* Paracît Torchog
(India and the African Sahel)
Also known as Rose-ringed Parakeet, this species has been widely introduced to at least one part of every continent except Antarctica. Genetic analysis of birds in Europe suggested that they are descended primarily from populations in Pakistan and northern India (Jackson *et al.* 2015). Although there were isolated reports of breeding in England from as early as 1855, the naturalised population was apparently founded in the late 1960s, with suspected breeding in Kent in 1969 and confirmed breeding in Surrey in 1971 (*Birds in England* 2005). London and southeast England were the focal areas of population growth, but it also occurs in several other large conurbations, including Liverpool and Manchester. The species was admitted to Category C of the British List in 1983 but although there have been breeding records in Wales, the population here has yet to become self-sustaining.

Early records in Wales were singles at Penarth, East Glamorgan, in summer 1974, at Rumney, East Glamorgan, on 6 December 1975, and at Peterstone Wentlooge, Gwent, in 1975. A parakeet on Bardsey, Caernarfonshire, on 15 October 1976 was thought to be this species but not positively identified. The first certain record for North Wales was also the first breeding record for Wales, a pair that fledged young at Gresford Flash, Denbighshire, in 1979. Breeding was attempted there again in 1980 but failed when the male was shot. Regular records of 1–3 birds followed over the next decade, with a concentration around Cardiff, East Glamorgan.

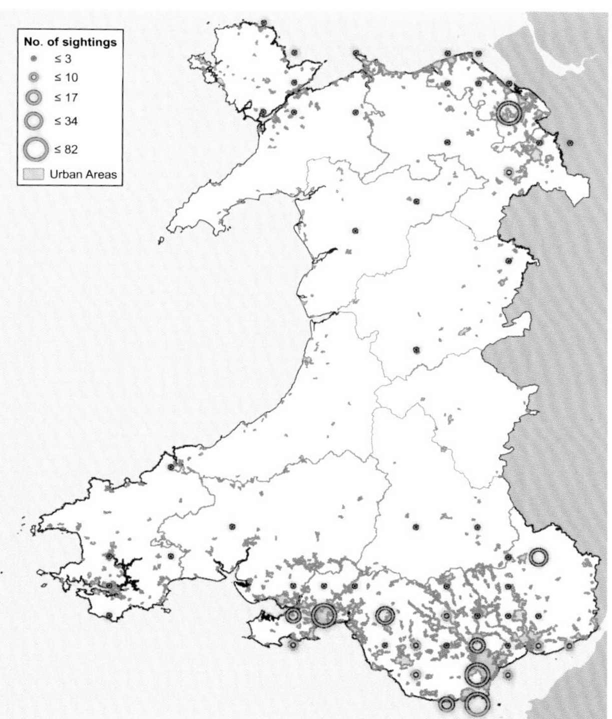

Ring-necked Parakeet: Distribution of 511 records of Ring-necked Parakeet in Wales between 1974 and 2017, sourced from the NBN Atlas. Circles are placed in the centre of the 10-km square in which birds were recorded. Map contains Ordnance Survey data © Crown copyright and database right 2020. Based on Facey and Vafidis (2020).

Birds in Wales (1994) listed 40 records up to 1990 from all Welsh counties except Breconshire and Anglesey, as well as 13 other parakeet records that may not have been this species.

There was no real indication of an increase in Wales during the 1990s and the early years of the 21st century. An average of two birds/year was recorded in 2000–09 and of nine/year in 2010–19 (max. 17 in 2016). Birds have been recorded every year since 2006, the recent growth mainly in urban areas of South and northeast Wales. There is no evidence of a self-sustaining breeding population in Wales (Facey and Vafidis 2020), but 6–7 near Llanerchaeron, Ceredigion, in mid-September 2004 and six at Gwaenysgor, Flintshire, on 6 October 2016, might suggest family groups. While across the border there is a growing, probably self-sustaining population in Liverpool, it has yet to breed in Wirral or Cheshire, and has not become well established in Bristol (Steve Waite *in litt.*, Hugh Pulsford *in litt.*, Mielcarek 2019). Ringing in southeast England suggests the introduced birds are highly sedentary (Robinson *et al.* 2020), so for now, there is insufficient evidence to include the species on Category C in Wales. It may be under-reported by birdwatchers, however, as some may regard any parakeet in Wales as an escape from captivity. Ring-necked Parakeet is on Schedule 9 of the Wildlife and Countryside Act 1981, making it illegal to release birds into the wild.

The Ring-necked Parakeet has now been recorded in all Welsh counties. Based on its distribution outside its native range to date, it may be that most of rural Wales would not be suitable. If the species was to become established outside urban areas, it has the potential to be not only an agricultural pest but also a serious conservation problem, out-competing hole-nesting native species in areas where the supply of suitable nest-holes is limited (*Birds in England* 2005). There is some evidence from Belgium that it may compete for nest-holes with native species, such as Nuthatch (Non-native Species Secretariat 2019). Otherwise, it currently appears to have had little ecosystem impact, but vigilance and improved recording are required.

Budgerigar *Melopsittacus undulatus* Byjerigar
(Australia)
Common as pets, escaped birds are probably not usually reported to county recorders. Most stay around suburban gardens and probably do not survive for long, although breeding in the wild has occurred in several English counties and a colony was established on Tresco, Isles of Scilly, in the 1970s. One that made it to Bardsey, Caernarfonshire, in August 1982 and one at Strumble Head, Pembrokeshire, in September 2000 are the only Welsh records we can find well away from residential areas.

Blue Jay *Cyanocitta cristata* Sgrech Las
(central and eastern North America)
One was seen just south of Aberystwyth, Ceredigion, in 1987.

Northern Mockingbird *Mimus polyglottos* Gwatwarwr y Gogledd
(North and Central America)
One photographed at Worm's Head, Gower, from 24 July to 3 August 1978 was not admitted to the British List by the BOURC, following extensive debate. More than half of Committee members favoured its admittance to Category A, but this was insufficient to reach the required two-thirds majority. Subsequent British records, in Cornwall in 1982 and Essex in 1988, were included on Category A.

Asian Glossy Starling *Aplonis panayensis* Drudwen Loyw Asia
(South Asia and Indonesia)
One visited a garden feeder at Wolf's Castle, Pembrokeshire, on 5–17 July 2005 and was probably taken by a Sparrowhawk.

Brahminy Starling *Sturnia pagodarum* Drudwen Benddu
(South Asia)
One was caught on Bardsey, Caernarfonshire, on 15 September 1956 having been attracted by the lighthouse beam.

Fischer's Starling *Lamprotornis fischeri* Drudwen Fischer
(East Africa)
One visited garden feeders in Llanddewi Brefi, Ceredigion, for five months over winter 1988/89.

Sudan Golden Sparrow *Passer luteus* Golfan Aur
(northeast Africa)
Sold by pet shops as 'Golden Song Sparrow', one visited garden feeders in Rhiw, Caernarfonshire, in June 1995.

Village Weaver *Ploceus cucullatus* Gwehydd Pentref
(Africa)
Following one at Strumble Head, Pembrokeshire, on 5 October 1985, there was a series of records across a wide area of West Wales in late September 2002: one at feeders in a Brongest garden, Ceredigion, on 17 September, one in Pembroke on 21 September and one at Llawhaden, Pembrokeshire, on 26 September.

Red-billed Quelea *Quelea quelea* Cwelia Pig Goch
(equatorial and southern Africa)
The world's most numerous bird is kept in captivity by some specialist breeders. One was caught by ringers on Bardsey, Caernarfonshire, on 11 October 1962.

Yellow-crowned Bishop *Euplectes afer* Esgob Euraid
(Africa)
One was seen around gardens in Swansea, Gower, in 2014.

Zebra Finch *Taeniopygia guttata* Pila Gwellt Rhesog
(Australasia)
A common cage bird, escapes are not uncommon, but it is presumed that most survive for only a short time.

Pin-tailed Whydah *Vidua macroura* Wida Llostfain
(East and southern Africa)
One was at RSPB Cors Ddyga, Anglesey, in 2015.

Chinese Grosbeak *Eophona migratoria* Tewbig Cynffonddu
(southeast Asia)
Formerly known as Yellow-billed Grosbeak, one visited a bird table at Tudweiliog, Caernarfonshire, in June 2002.

Black-headed Greenfinch *Chloris ambigua* Llinos Werdd Yunnan
(southeast Asia)
One was in Broughton, Flintshire, in 2011.

Yellow-fronted Canary *Crithagra mozambica* Caneri Aelfelyn
(sub-Saharan Africa)
Known to aviculturists as a 'Green Singing Finch', one was at Aberarth, Ceredigion, on 5–7 December 2006.

Atlantic Canary *Carduelis canaria* Caneri
(Macaronesia)
Commonly kept as a domestic pet and by specialist breeders, small numbers are recorded in the wild and some appear able to survive on garden feeders, even through cold winters. One at Amlwch, Anglesey, survived from November 1981 to January 1982, for example.

Yellow-throated Bunting *Emberiza elegans* Bras Gyddf-felyn
(southeast Asia)
One at Treffos, Anglesey, in January 1998 had escaped from a local butterfly farm.

Chestnut Bunting *Emberiza rutila* Bras Gwinau
(East Asia)
This species was accepted onto the British List on the basis of a first-calendar-year male on Orkney in October 2015, but the only record from Wales, an immature male on Bardsey on 18–19 June 1986, was considered to be an escaped cage bird because of the timing of its appearance.

Red-winged Blackbird *Agelaius phoeniceus* Tresglen Adeingoch
(North America)
One was seen at Nash Lighthouse, East Glamorgan, on 27 October 1886. Red-winged Blackbird was subsequently admitted to the British List on the basis of a female on Orkney in April 2017, which occurred after the cessation of wild-bird imports to the European Union.

Rusty Blackbird *Euphagus carolinus* Tresglen Winau
(North America)
One was shot near Cardiff, East Glamorgan, on 4 October 1881.

Common Diuca Finch *Diuca diuca* Pila Diwca Cyffredin
(South America)
This is widely kept in aviculture; one was seen in north Cardiff, East Glamorgan, in 2019.

Julian Hughes

Appendix 2. Scientific names of non-bird species mentioned in the text

Common Name	**Species Name(s)**
VERTEBRATES	
Mammals	
American Mink	*Neovison vison*
Arctic Fox	*Vulpes lagopus*
Badger	*Meles meles*
Black Rat	*Rattus rattus*
Brown Rat	*Rattus norvegicus*
deer	Generic term for British deer species including Red (*Cervus elaphus*), Roe (*Capreolus capreolus*), Fallow (*Dama dama*) and Reeve's Muntjac (*Muntiacus reevesi*)
(Domestic) cat	*Felis catus*
European Mink	*Mustela lutreola*
Fox (Red)	*Vulpes vulpes*
Grey Squirrel	*Sciurus carolinensis*
Irish Elk	*Megalocerus giganteus*
lemming	*Lemmus spp.*
mice	Generic term for House (*Mus musculus*), Wood (*Apodemus* sylvaticus) or Yellow-necked mouse (*Apodemus flavicollis*)
Muskrat	*Ondatra zibethicus*
Orkney Vole	*Microtus arvalis orcadensis*
Otter	*Lutra lutra*
Rabbit	*Oryctolagus cuniculus*
rats	Generic term for what is usually Brown Rat (*Rattus norvegicus*) but could include Black Rat (*R. rattus*)
Reindeer	*Rangifer tarandus*
Field Vole (Short-tailed Vole)	*Microtus agrestis*
shrews	*Sorex spp.*
Stoat	*Mustela erminea*
squirrels	*Sciurus spp.* Grey Squirrel (*S. carolinensis*) or Red Squirrel (*S. vulgaris*)
voles	Generic term for Common Vole (*Microtus arvalis*) and/or Field Vole (*M.* agrestis)
Wild Horse	*Equus ferus*
Wolf	*Canis lupus*
Wood Mouse	*Apodemus sylvaticus*
Amphibians	
Common Frog	*Rana temporaria*
frogs	Usually refers to Common Frog (see above)
Reptiles	
lizards	Generic term for Common Lizard (*Zootoca vivipara*) and/or Sand Lizard (*Lacerta agilis*)
Fish	
Atlantic Salmon	*Salmo salar*
Butterfish	*Peprilus triacanthus*
Common Minnow	*Phoxinus phoxinus*
clupeid(s)	Generic term for Herring and Sprat (see below)
European Eel	*Anguilla anguilla*
gadids	Marine fish including cod (*Gadus spp.*), Haddock (*Melanogrammus arglefinus*), Whiting (*Merlangius* merlangius) and pollock (*Pollachius spp.*)
Goldfish	*Carassius auratus*
Grass Carp	*Ctenopharyngodon idella*
Herring (Atlantic)	*Clupea harengus*
Lesser Sandeel	*Ammodytes tobianus*
Mackerel	*Scomber scombrus*
sandeel(s)	Small fish of the genera *Hyperoplus, Gymnammodytes* and *Ammodytes*
Sardine (young pilchard)	*Sardina sp.* (may include other genera)
Sprat	*Sprattus sprattus*
Trout (Brown)	*Salmo trutta*
Wels Catfish	*Silurus glanis*

INVERTEBRATES

Molluscs

Jenkin's Spire Shell/New Zealand Mud Snail	*Potamopyrgus antipodarum*
Atlantic Jack-Knife Clam	*Ensis leei*
Blue Mussel or mussel	*Myrtilis edulis*
Common Cockle or cockle	*Cerastoderma edule*
Zebra Mussel	*Dreissena polymorpha*
Mud Snail (or Laver Spire Shell)	*Peringia ulvae*
Razor Clam	*Siliqua patula*
Squid	Various genera in the Class Cephalopoda
Narrow Otter Shell	*Lutraria angustior* (bivalve clam)
Surf Clam or Trough Shell	*Spisula subtruncata* (bivalve clam)

Crustacea

Krill	*Euphausia sp.*
Lobster	*Nephrops sp.*
Scampi	*Nephrops sp.*
Signal Crayfish	*Pacifastacus leniusculus*
copepods	Group of small planktonic crustaceans
amphipods	Order Amphipoda (many genera—includes shrimps)

Annelids

earthworms	Generic term for soil or grassland segmented worms

ARTHROPODS

Insects

ants	Order Hymenoptera, Family Formicidae
bees	Order Hymenoptera, Family Apidae
beetles	Order Coleoptera
caddisflies	Order Tricoptera
craneflies	Order Diptera, Family Tipulidae
dragonflies	Order Odonata
grasshoppers	Order Orthoptera
Great Brocade moth	*Eurois occulta* Order Lepidoptera
locusts	Order Orthoptera
mayflies	Order Ephemeroptera
Migrant Hawker dragonfly	*Aeshna mixta* Order Odonata
moths	Order Lepidoptera
termites	Order Isoptera
wasps	Order Hymenoptera Family Vespidae
Welsh Chafer	*Hoplia philanthus*
Yellow Meadow Ant	*Lasius flavus*

Arachnids

ticks	Usually refers to Sheep Tick *Ixodes ricinus*

PLANTS

Higher Plants (Angiosperms)

acacia	*Acacia spp.*
Alder	*Alnus glutinosa*
Annual Sea-blite	*Suaeda maritima*
Ash	*Fraxinus excelsior*
Barley	*Hordeum vulgare*
Beech	*Fagus sylvatica*
bilberry	*Vaccinium spp.*
birch	Generic name for *Betula spp.* Either Silver Birch *Betula pendula* or Downy Birch *Betula pubescens*
Blackthorn	*Prunus spinosa*
bramble	*Rubus spp.*
broom	*Cytisus spp.*
Cat's Ear	*Hypochaeris radicata*
clover	*Trifolium spp.*
Common Eelgrass	*Zostera marina*
Common Cord-grass	*Spartina anglica*
Common Glasswort	*Salicornia europaea*

Common Sorrel	*Rumex acetosa*
Cork Oak	*Quercus suber*
Crowberry	*Empetrum nigrum*
dandelion	*Taraxacum spp.*
dock	*Rumex spp.*
Elephant-grass	*Miscanthus x giganteus*
elm	Generic term for English Elm (*Ulmus procera*) or Wych Elm (*Ulmus glabra*)
English Elm	*Ulmus procera*
Fat-hen	*Chenopodium album*
Field Maple	*Acer campestre*
fumitory	*Fumaria spp.*
gorse	*Ulex spp.*
hawkbits	*Leontodon spp.*
Hawthorn	*Crataegus monogyna*
heather	*Calluna vulgaris/Erica spp.*
Himalayan Balsam	*Impatiens glandulifera*
Hogweed	*Heracleum sphondylium*
Holly	*Ilex europaea*
Hornbeam	*Carpinus betulus*
ivy	*Hedera spp.*
Juniper	*Juniper communis*
lime	*Generic name for Large-leaved Lime, Small-leaved Lime or hybrid Tilia spp.*
Maize	*Zea mays*
Mistletoe	*Viscum album*
New Zealand Pigmyweed	*Crassula helmsii*
Oilseed Rape	*Brassica napus*
Olive	*Olea europaea*
Phragmites reedbed/wet reed/reeds	*Phragmites australis*
Purple Moor-grass	*Molinia caerulea*
Rhododendron	*Rhododendron ponticum*
Rowan	*Sorbus aucuparia*
Rye-grass	*Lolium perenne*
Samphire	*Salicornia europaea*
Sea Beet	*Beta maritima*
Seablite	*Suaeda maritima*
Sea Buckthorn	*Hippophae rhamnoides*
Sessile Oak	*Quercus petraea*
Sheep's Sorrel	*Rumex acetosella*
submerged macrophytes	Species of rooted aquatic plants with the vegetative parts predominantly submerged
Sycamore	*Acer pseudoplatanus*
thistles	*Carduus* or *Cirsium spp.*
Tree Mallow	*Malva (Lavatera) arborea*
Wheat	*Triticum aestivum*
Wild Cherry	*Prunus avium*
willow	*Salix spp.*
Wych Elm	*Ulmus glabra*

Higher Plants (Gymnosperms)

Douglas Fir	*Pseudotsuga menziesii*
European Larch	*Larix decidua*
fir	*Abies spp.*
larch	*Larix spp.* European Larch (see above) or Japanese Larch (*L. kaempferi*)
Lodgepole Pine	*Pinus contorta*
Norway Spruce	*Picea abies*
pine	*Pinus spp.*
Scots Pine	*Pinus sylvestris*
Siberian Stone Pine	*Pinus sibirica*
Sitka Spruce	*Picea sitchensis*
spruce	*Picea spp.*
Wellingtonia/Redwood	*Sequoia/Sequoiadendron spp.*
Western Hemlock	*Tsuga hetreophylla*
Yew	*Taxus baccata*

Ferns

Bracken	*Pteridium aquilinum*

Diseases

Avian Pox	A group of avian poxviruses
Botulism (bacterial disease)	*Clostridium botulinum*
Colibacillosis	A systemic infection caused by the avian pathogen *Escherichia coli*
Myxomatosis	A mammalian poxvirus (*Myxoma virus)*
Puffinosis	A viral disease of shearwaters
Phytopthora	Genus of plant-damaging oomycetes (water molds) which particularly affect larch and oak in the UK
Rabbit Viral Haemorrhagic Disease	A mammalian viral disease particularly noted in rabbits
Salmonellosis	A very infectious bacterial disease
Trichomonosis	A disease of particularly doves, pigeons and finches caused by the protozoan parasite *Trichomonas gallinae.*

For the scientific and vernacular names of plants and mammals, the following references have been used (Stace 2010, Crawley *et al.* 2020). Other taxa were checked using the internet.

Bibliography

Aderyn. LERC Wales' Biodiversity Information and Reporting Database. **https://aderyn.lercwales.org.uk/recorder/data**

Aebischer, N.J. 2009. *The GWCT Grey Partridge Recovery Programme: a Species Action Plan in action*. In Cederbaum, S.B. *et al*., (eds.) *Gamebird 2006:* Quail VI and Perdix XII (pp.291–301). Athens, USA: Warnell School of Forestry and Natural Resources.

Aebischer, N.J. 2019. Fifty-year trends in UK hunting bags of birds and mammals, and calibrated estimation of national bag size, using GWCT's National Gamebag Census. *European Journal of Wildlife Research* 65(4): p.64.

Aebischer, N.J., and Harradine, J. 2007. Developing a tool for improving bag data of huntable birds and other bird species in the UK: a report to Defra and the Scottish Executive. GCT and BASC.

Agreement on the Conservation of Albatrosses and Petrels. 2010. ACAP Species assessment: Black-browed Albatross *Thalassarche melanophris*.

AHVLA. 2013. *Great Britain Poultry Register (GBPR) Statistics: 2013*. Animal Health and Veterinary Laboratories Agency.

Åkesson, S., Klaassen, R., Holmgren, J., Fox, J.W., and Hedenström, A. 2012. Migration routes and strategies in a highly aerial migrant, the Common Swift *Apus apus*, revealed by light-level geolocators. *PLoS One* 7(7): e41195.

Alerstam, T., Backman, J., Gronroos, J., Olofsson, P., and Strandberg, R. 2019. Hypotheses and tracking results about the longest migration: the case of the arctic tern. *Ecology and Evolution* 9: pp.9,511–9,531.

Alexander, H.G. 1914. A report on the Land Rail inquiry. *British Birds* 8(4): pp.82–92.

Alexander, W.B. 1945. The Woodcock in the British Isles: Publication of the British Trust for Ornithology, based on a Report on the Inquiry, 1934–35. *Ibis* 87(4): pp.512–550.

Alexander, W.B. 1946a. The Woodcock in the British Isles. *Ibis* 88(1): pp.1–24.

Alexander, W.B. 1946b. The Woodcock in the British Isles. *Ibis* 88(2): pp.159–179.

Alexander, W.B. 1946c. The Woodcock in the British Isles. *Ibis* 88(3): pp.271–286.

Alexander, W.B. 1946d. The Woodcock in the British Isles. *Ibis* 88(4): pp.427–444.

Alexander, W.B. 1947. The Woodcock in the British Isles. *Ibis* 89(1): pp.1–28.

Alexander, W.B., and Lack, D. 1944. Changes in status among British breeding birds. *British Birds* 38(3): pp.42–45; 38(4): pp.62–69; 38(5): pp.82–88.

Allen, D. 2004. *A rapid review of SSSI feature condition in Wales*. Bangor: Countryside Council for Wales.

Allin, E.K. 1968. Breeding notes on Ravens in north Wales. *British Birds* 61(12): pp.541–545.

Allport, G., O'Brien, M., and Cadbury, C.J. 1986. *Survey of redshank and other breeding birds on saltmarshes in Britain 1985*. Sandy: RSPB Report, unpublished.

Alstrom P. and Mild K. 2003. *Pipits and Wagtails of Europe, Asia and North America*. London: Helm.

Amar, A., Hewson, C.M., Thewlis, R.M., Smith, K.W., Fuller, R.J., Lindsell, J.A., Conway, G., Butler, S., and MacDonald, M.A. 2006. *What's happening to our woodland birds: long-term changes in the populations of woodland birds*. Sandy: RSPB Research Report 19; Thetford: BTO Research Report 169.

Amar, A., Redpath, S., Sim, I., and Buchanan, G. 2010. Spatial and temporal associations between recovering populations of common raven *Corvus corax* and British upland wader populations. *Journal of Applied Ecology* 47(2): pp.253–262.

Amat, J.A., and Soriguer, R.C. 1984. Kleptoparasitism of Coots by Gadwalls. *Ornis Scandinavica* 15: pp.188–194.

Anderson, K., Clarke, S., and Lucken, R. 2013. Nesting behaviour of the first breeding Great White Egrets in Britain. *British Birds* 106(5): pp.258–263.

Andersson, Å., and Larsson, T. 2006. Reintroduction of Lesser White-fronted Goose *Anser erythropus* in Swedish Lapland. In Boere, G.C., Galbraith, C.A., and Stroud, D.A., (eds.) *Waterbirds around the world*. Edinburgh: The Stationery Office. pp.635–636.

Animal Health and Veterinary Laboratories Agency. 2013. Great Britain Poultry Register Statistics 2013. **https://assets.publishing.service.gov.uk/government/uploads/system/uploads/attachment_data/file/314715/pub-avian-gbpr13.pdf**

Anon. (undated). *Ffridd – a habitat on the edge*. Cardiff: RSPB and NRW.

Anon. 1802. *The Swansea Guide: containing such information as was deemed useful to the traveller, through the counties of Glamorgan and Monmouth: from the exemplifications of ancient and modern authors*. Z.B. Morris.

Anon. 1954. The New Protection of Birds Act. *British Birds* 47(12): pp.409–413.

Anon. 1989. Goshawk breeding habitat in lowland Britain. *British Birds* 82(2): pp.56–67.

Anon. 1990. Breeding biology of Goshawk in lowland Britain. *British Birds* 83(12): pp.527–540.

Anon. 2006. Osprey: a new breeding bird in Wales. *Welsh Birds* 4(4): pp.396–397.

APEM. 2017. Aerial surveys in Carmarthen Bay SPA get common scoter in focus. **www.apemltd.co.uk/aerial-surveys-in-carmarthen-bay-spa-get-common-scoter-in-focus/**

APHA. 2019. *Livestock Demographic Data Group: Sheep population report Livestock population density maps for Great Britain*. Animal and Plant Health Agency.

APHA. 2020. *Livestock Demographic Data Group: Cattle population report Livestock population density maps for Great Britain*. Animal and Plant Health Agency.

Aplin, O.V. 1892. On the distribution of Cirl Bunting in Great Britain. *Zoologist* 3rd Series 16: pp.121–128, 174–181.

Aplin, O.V. 1900. The birds of Lleyn, West Carnarvonshire. *Zoologist* 4th Series 4: pp.489–505.

Aplin, O.V. 1901. Further notes from Lleyn, West Caernarvonshire. *The Zoologist* 4th Series 5: pp.401- 415.

Aplin, O.V. 1902. The birds of Bardsey Island, with additional notes on the birds of Lleyn. *Zoologist* 4th Series 6: pp.8–17, pp.107–110.

Aplin, O.V. 1903. Letter. *Ibis* 45(1): pp.133–135.

Aplin, O.V. 1904. On the breeding of the Black-necked Grebe (*Podiceps nigricollis*) in the British Islands. *Zoologist* 4th Series 8: pp.417–420.

Aplin, O.V. 1905. Winter notes from Lleyn. *Zoologist* 4th Series 9: pp.41–50, 101–106.

Appleby, R.H., Madge, S.C., and Mullarney, K. 1986. Identification of divers in immature and winter plumages. *British Birds* 79(8): pp.365–391.

Appleton, G. 2012. Swifts start to share their secrets. *BTO News* May–June 2012: pp.16–17.

Appleton, G. 2015. Black-tailed Godwits expand their range in Russia and Iceland. *Wader Tales*. **https://wadertales.wordpress.com/2015/12/18/black-tailed-godwits-expand-their-range-in-russia-and-iceland**

Appleton, G. 2016a. Oystercatchers: from shingle beach to roof-top. *Wader Tales*. **https://wadertales.wordpress.com/2016/06/13/oystercatchers-from-shingle-beach-to-roof-top**

Appleton, G. 2016b. Lapwing moult. *Wader Tales*. **https://wadertales.wordpress.com/2016/07/27/lapwing-moult**

Appleton, G. 2016c. Snipe and Jack Snipe in the UK and Ireland. *Wader Tales*. **https://wadertales.wordpress.com/2016/09/20/snipe-jack-snipe-in-the-united-kingdom**

Appleton, G. 2019. Fennoscandian wader factory. *Wader Tales*. **https://wadertales.wordpress.com/2019/11/26/fennoscandian-wader-factory**

Archer, M., Grantham, M., Howlett, P., and Stansfield, S. 2010. *Bird observatories of Britain and Ireland*. London: T&AD Poyser.

Armstrong, E. 2016. *The farming sector in Wales*. National Assembly for Wales Research Service.

Arnold, R. 2004. A history of the birds of Puffin Island. In Jones, P.H., and Whalley, P., *Birds of Anglesey*. Menter Môn.

Arnold, R. 2006. The Brent Geese (*Branta bernicla*) of eastern Anglesey. *Cambrian Bird Report* 2006: pp.156–159.

Arnold, R., Dixon, S., Korboulewsky, N., and Märell, A. 1997. Common Eiders breeding in North Wales in 1997. *Welsh Birds* 1(6): p.80.

Aspden, W., 1939. Hen Harriers breeding in Anglesey. *British Birds* 32(3): pp.326–328.

Atkinson, N.K., Davies, M., and Prater, A.J. 1978. The winter distribution of Purple Sandpipers in Britain. *Bird Study* 25(4): pp.223–228.

Atkinson, P.W., Clark, N.A., Dodd, S.G., and Moss, D. 2005. *Changes in survival and recruitment of Oystercatchers Haemotopus ostralegus at Traeth Lafan North Wales in relation to shellfish exploitation*. Thetford: BTO Research Report 393.

Ausden, M., Bolton, M., Butcher, N., Hoccom, D.G., Smart, J., and Williams, G. 2009. Predation of breeding waders on lowland wet grassland – is it a problem? *British Wildlife* 21(1): pp.29–38.

Ausden, M., White, G., and Eaton, M. 2013. Breeding Baillon's Crakes in Britain. *British Birds* 106(1): pp.7–16.

Ausden, M., White, G., and Santoro, S. 2019. The Changing Status of the Glossy Ibis *Plegadis falcinellus* in Britain. *SIS Conservation* 1: 116–121. IUCN Stork, Ibis and Spoonbill Specialist Group.

Austin, G., Frost, T., Mellan, H., and Balmer, D. 2017. *Results of the third Non-Estuarine Waterbird Survey, including Population Estimates for Key Waterbird Species*. Thetford: BTO Research Report No. 697.

Austin, G.E., Peachel, I., and Rehfisch, M.M. 2000. Regional trends in coastal wintering waders in Britain. *Bird Study* 47(3): pp.352–371.

Austin, G.E., Rehfisch, M.M., Allan, J.R., and Holloway, S.J. 2007. Population size and differential population growth of introduced Greater Canada Geese *Branta canadensis* and re-established Greylag Geese *Anser anser* across habitats in Great Britain in the year 2000. *Bird Study* 54(3): pp.343–352.

Austin, J.J., Bretagnolle, V., and Pasquet, E. 2004. A global molecular phylogeny of the Small *Puffinus* Shearwaters and implications for systematics of the Little-Audubon's Shearwater complex. *The Auk* 121(3): pp.847–864.

Axell, H.E. 1966. Eruptions of Bearded Tits during 1959–1965. *British Birds* 59(12): pp.513–543.

Baggott, G.K. 1986. The fat contents and flight ranges of four warbler species on migration in north Wales. *Ringing & Migration* 7(1): pp.25–36.

Bagguley, R., Bagguley, M., and Richardson, D. 2000. Hooded Crow nesting in Anglesey. *Welsh Birds* 2(4): p.228.

Baillie, S.R., and Peach, W.J. 1992. Population limitation in Palaearctic-African migrant passerines. *Ibis* 134 Supplement 1: pp.120–132.

Bain, C. 1987. *Breeding wader habitats in an upland area of North Wales (Hiraethog)*. Sandy: RSPB Report.

Bairlein, F., Norris, D.R., Nagel, R., Buite, M., Voigt, C.C., Fox, J.W., Hussell, D.J.T., and Schmalijohann, H. 2012. Cross-hemisphere migration of a 25 g songbird. *Biology Letters* 8(4): pp.505–507.

Baker, H., Stroud, D.A., Aebischer, N.J., Cranswick, P.A., Gregory, R.D., McSorley, C.A., Noble, D.G., and Rehfisch, M.M. 2006. Population estimates of birds in Great Britain and the United Kingdom. *British Birds* 99(1): pp.25–44.

Baker, K. 1988. Identification of Siberian and other forms of Lesser Whitethroat. *British Birds* 81(8): pp.382–390.

Ballance, D.K. 2020. *Avifaunas, atlases & authors: a personal view of local ornithology in the United Kingdom, from the earliest times to 2019.* Wareham: Calluna Books.

Balmer, D.E., Gillings, S., Caffrey, B.J., Swann, R.L., Downie, I.S., and Fuller, R.J. (eds.) 2013. *Bird atlas 2007–11: the breeding and wintering birds of Britain and Ireland.* Thetford: BTO Books.

Banks, A., Bolt, D., Bullock, I., Collier, M., Fairney, N., Hasler, C., Haycock, B., Maclean, I., Roberts, P., Sanderson, B., Schofield, R., Smith, L., Swan, J., Taylor, R., and Whitehead, S. 2007a. *Ground and aerial monitoring for Carmarthen Bay SPA 2004–07.* CCW Marine Monitoring Report No: 48.

Banks, A.N., Burton, N.H.K., Calladine, J.R., and Austin, G.E. 2007b. *Winter gulls in the UK: population estimates from the 2003/04–2005/06 Winter Gull Roost Survey.* Thetford: BTO Research Report No. 456.

Banks, A.N., Burton, N.H.K., Calladine, J.R., and Austin, G.E. 2009. Indexing winter gull numbers in Great Britain using data from the 1953 to 2004 Winter Gull Roost Surveys. *Bird Study* 56(1): pp.103–119.

Banks, A.N., Crick, H.Q.P., Coombes, R., Benn, S., Ratcliffe, D.A., and Humphreys, E.M. 2010. The breeding status of Peregrine Falcons *Falco peregrinus* in the UK and Isle of Man in 2002. *Bird Study* 57(4): pp.421–436.

Banks, A.N., Sanderson, W.G., Hughes, B., Cranswick, P.A., Smith, L.E., Whitehead, S., Musgrove, A.J., Haycock, B., and Fairney, N.P. 2008. The *Sea Empress* oil spill (Wales, UK): Effects on Common Scoter *Melanitta nigra* in Carmarthen Bay and status ten years later. *Marine Pollution Bulletin* 56(5): pp.895–902.

Banks, K.W., Clark, H., Mackay, I.R.K., Mackay, S.G., and Sellers, R.M. 1991. Origins, population structure and movements of Snow Buntings *Plectrophenax nivalis* wintering in Highland Region, Scotland. *Bird Study* 38(1): pp.10–19.

Banks, W. 1892. Eared Grebes on Anglesea. *Zoologist* 3rd Series 16: pp.411–12.

Barclay-Smith, P. 1959. The British contribution to bird protection. *Ibis* 101(1): pp.115–122.

Bark Jones, R. 1954. Nesting birds of the North Wales coast. *Merseyside Naturalists' Association Report 1954*.

Barker, T.W. 1905. *Handbook to the natural history of Carmarthenshire*. W. Spurrell and Son.

Barn Owl Trust. 2020. **https://www.barnowltrust.org.uk/hazards-solutions/barn-owls-major-roads**

Barnes, J. 1997. *The birds of Caernarfonshire.* Cambrian Ornithological Society.

Barr, K. 2018. *Prehistoric avian, mammalian and* H. sapiens *foot-print-tracks from intertidal sediments as evidence of human palaeoecology.* PhD Thesis, University of Reading.

Barrington, R.M. 1888. The Manx Shearwater on Skomer Island. 1888. *Zoologist* 3rd Series 12: pp.367–371.

Batten, L., Bibby, C.J., Clement, P., Elliott, G.D., and Porter, R.F. 1990. *Red data birds in Britain.* London: T&AD Poyser.

Batten, L.A. 1973. Population dynamics of suburban Blackbirds. *Bird Study* 20(4): pp.251–258.

BBC News 1999. Final step for Cardiff Bay Barrage. **http://news.bbc.co.uk/1/hi/wales/503837.stm**

Beale, C.M., Burfield, I.J., Sim, I.M., Rebecca, G.W., Pearce-Higgins, J.W., and Grant, M.C. 2006. Climate change may account for the decline in British ring ouzels *Turdus torquatus*. *Journal of Animal Ecology* 75(3): pp.826–835.

Beekman, J., Koffijberg, K., Wahl, J., Kowalik, C., Hall, C., Devos, K., Clausen, P., Honman, M., Laubek, B., Luigujõe, L., Wieloch, M., Boland, H., Švažas, S., Nilsson, L., Sti, A., Keller, V., Gaudard, C., Degen, A., Shimmings, P., Larsen, B.H., Portolou, D., Langendoen, T., Wood, K.A., and Rees, E.C. 2019. Long-term population trends and shifts in distribution of Bewick's Swans *Cygnus columbianus bewickii* wintering in northwest Europe. *Wildfowl* Special issue 5: pp.73–102.

Bellebaum, J., Kube, J., Schulz, A., Skov, H., and Wendeln, H. 2014. Decline of Long-tailed Duck *Clangula hyemalis* numbers in the Pomeranian Bay revealed by two different survey methods. *Ornis Fennica* 91: pp.129–137.

Bellebaum, J., Schirmeister, B., Sonntag, N., and Garthe, S. 2013. Decreasing but still high: bycatch of sea-birds in gillnet fisheries along the German Baltic coast. *Aquatic Conservation* 23: pp.219–221.

Benoit, P.M., Lockley, R.M., and Miles, P.M. 1958. Island of Skomer – A special supplement. *Nature in Wales* (The Quarterly Journal of the West Wales Field Society). Volume 4(4).

Berglund, P.-A., and Hentati-Sundberg, J. 2015. Arctic seabirds breeding in the African-Eurasian Waterbird Agreement (AEWA) area: status and trends 2014. CAFF's Circumpolar Seabird expert group (CBird), CAFF Assessment Series Report No. 13. CAFF International Secretariat.

Berry, R., and Bibby, C.J. 1981. A breeding study of Nightjars. *British Birds* 74(4): pp.161–169.

Berry, S.E., and Green, J. 2018. Pembrokeshire Bird Report 2018. Published online: **https://drive.google.com/file/d/1aKauJuT6XILAOKnBLDEMBuPakc46dF9M/view**

Berthold, P., and Helbig, A. 1992. The genetics of bird migration: stimulus, timing and direction. *Ibis* 134 Supplement 1: pp.35–40.

Betts, M. 1992. *Birds of Skokholm.* BioLine.

Bibby, C. 1973. The Red-backed Shrike: a vanishing British species. *Bird Study* 20(2): pp.103–110.

Bibby, C., Robinson, P., and Bland, E. 1990. The impact of egg collecting on scarce breeding birds. *RSPB Conservation Review* 4: pp.22–25.

Bibby, C.J. 1979. Breeding biology of the Dartford Warbler *Sylvia undata* in England. *Ibis* 121(1): pp.41–52.

Bibby, C.J. 1989. A survey of breeding Wood Warblers *Phylloscopus sibilatrix* in Britain, 1984–1985. *Bird Study* 36(1): pp.56–72.

Bibby, C.J., and Etheridge, B. 1993. Status of the Hen Harrier *Circus cyaneus* in Scotland in 1988–89. *Bird Study* 40(1): pp.1–11.

Bibby, C.J., and Natrass, M. 1986. Breeding status of the Merlin in Britain. *British Birds* 79(4): pp.170–185.

Bibby, C.J., Aston, N., and Bellamy, P. 1989. The effect of broadleaved trees on birds of upland conifer plantations in North Wales. *Biological Conservation* 49: pp.17–29.

Bibby, C.J., Phillips, B.N., and Seddon, A.J.E. 1985. The birds of restocked conifer plantations in Wales. *Journal of Applied Ecology* 22: pp.215–230.

Bicknell, A.W.J., Oro, D., Camphuysen, K., and Votier, S.C. 2013. Potential consequences of discard reform for seabird communities. *Journal of Applied Ecology* 50: pp.649–658.

Bicknell, J., Smart, J., Hoccom, D., Amar, A., Evans, A., Walton, P., and Knott, J. 2010. *Impacts of non-native gamebird release in the UK: a review.* Sandy: RSPB Research Report Number 40.

Bignal, E.M., Bignal, S., and McCracken, D. 1997. The social life of the Chough. *British Wildlife* 8(6): pp.373–383.

Bijleveld, M. 1974. *Birds of prey in Europe*. London: Macmillan.

Bingley, W. 1800. *A tour round North Wales performed during the summer of 1798.* London: E. Williams.

Bingley, W. 1804. *North Wales; including its scenery, antiquities, customs and some sketches of its natural history.* London: Longman and Rees.

Birch, J.E., Birch, R.R., Birtwell, J.M., Done, C., Stokes, E.J., and Walton, G.F. 1968. *The birds of Flintshire.* Flintshire Ornithological Society.

BirdGuides News. 2020a. Norfolk spoonbill population hits new heights. News release 14 August 2020. **https://www.birdguides.com/news/norfolk-spoonbill-population-hits-new-heights**

BirdGuides News. 2020b. Great Egret enjoys another successful breeding season in Somerset. News release 19 August 2020. **https://www.birdguides.com/news/great-egret-enjoys-another-success-ful-breeding-season-in-somerset**

BirdLife International. 2004. *Birds in Europe: population estimates, trends and conservation status.* Cambridge: BirdLife International.

BirdLife International. 2015. *European Red List of Birds.* Luxembourg Office for Official Publications of the European Communities.

Birdlife International. 2016. Almost half of all Finnish Birds are in danger. **https://www.birdlife.org/europe-and-central-asia/news/almost-half-all-finnish-birds-are-danger**

BirdLife International. 2019. IUCN Red List for birds. Downloaded from: **http://www.birdlife.org**

BirdLife International. 2020. Data Zone. **http://datazone.birdlife.org/home**

Birkhead, T.R. 1991. *The Magpie*. London: T&AD Poyser.

Birkhead, T.R. 2009. Guillemots on Skomer: a long-term view. *Natur Cymru* 30: pp.33–35.

Birkhead, T.R. 2014. Guillemots on Skomer. *Natur Cymru* 52: pp.10–15.

Birkhead, T.R. 2016. Changes in the numbers of Common Guillemots on Skomer since the 1930s. *British Birds* 109(11): pp.651–659.

Birks, H.J.B. 1988. Long-term ecological change in the British uplands. In Usher, M.B., and Thompson, D.B.A., (eds.) *Ecological Change in the Uplands*, pp.37–56. Oxford: Blackwell.

Blackburn, E., and Cresswell, W. 2016. High within-winter and annual survival rates in a declining Afro-Palaearctic migratory bird suggest that wintering conditions do not limit populations. *Ibis* 158(1): pp.92–105.

Blackburn, T.M., and Gaston, K.J. 2018. Abundance, biomass and energy use of native and alien breeding birds in Britain. *Biological Invasions* 20: pp.3,563–3,573.

Blackstock, T.H., Howe, E.A., Stevens, J.P., Burrows, C.A., and Jones, P.S. 2010. *Habitats of Wales: a comprehensive field survey 1979–1997*. Cardiff: University of Wales Press.

Bladwell, S. 2014. Swifts: life on the wing. *Natur Cymru* 49: pp.8–11.

Bladwell, S., Noble, D.G., Taylor, R., Cryer, J., Galliford, H., Hayhow, D.B., Kirby, W., Smith, D., Vanstone, A., and Wotton, S.R. 2018. *The state of birds in Wales 2018*. The RSPB, BTO, NRW and WOS. Cardiff: RSPB Cymru.

Blathwayt, Rev. F.L. 1923. Report to RSPB Watchers Committee.

Blore, J.D. 2016. *Lynx Cave excavations: Zoology*. **http://lynxcave.webs.com/zoology.htm**

Boano, G., Pellegrino, I., Ferri, M., Cucco, M., Minelli, F., and Åkesson, S. 2020. Climate anomalies affect annual survival rates of swifts wintering in sub-Saharan Africa. *Ecology and Evolution* 10(14): pp.7,916–7,928.

Boase, T. 2018. *Mrs Pankhurst's Purple Feather: Fashion, Fury and Feminism – Women's Fight for Change*. Aurum Press.

Bolam, F.C., *et al*. 2021. How many bird and mammal extinctions has recent conservation action prevented? *Conservation Letters*: 14(1): e12762.

Bolam, G. 1913. *Wildlife in Wales*. London: Frank Palmer.

Bonham, P.F., and Robertson, J.C.M. 1975. The spread of Cetti's Warbler in north-west Europe. *British Birds* 68(10): pp.393–408.

Boothby, C., Redfern, C., and Schroeder, J. 2019. An evaluation of canes as management technique to reduce predation by gulls on ground-nesting birds. *Ibis* 161(2): pp.453–458.

Border, J.A., Henderson, I.G., Redhead, J.W., and Hartley, I.R. 2016. Habitat selection by breeding Whinchats *Saxicola rubetra* at territory and landscape scales. *Ibis* 159(1): pp.139–151.

Borrer, W. 1891. *The birds of Sussex*. R.H. Porter.

Both, C., Bijlsma, R.G., and Ouwehand, J. 2016. Repeatability in spring arrival dates in pied flycatchers varies among years and sexes. *Ardea* 104(1): pp.3–21.

BOURC. 2001. British Ornithologists' Union Records Committee: 27th Report (October 2000). *Ibis* 143(1): pp.171–175.

Bowes, A., Lack, P.C., and Fletcher, M.R. 1984. Wintering gulls in Britain, January 1983. *Bird Study* 31(3): pp.161–170.

Bowgen, K.M., Wright, L.J., Calbrade, N.A., Coker, D., Dodd, S.G., Hainsworth, I., Howells, R.J., Hughes, D.S., Jenks, P., Murphy, M.D., Sanderson, W.G., Taylor, R.C., and Burton, N.H.K. In prep. Adaptability mitigates impacts of a crash in cockle stocks on Oystercatcher wintering at a protected site in south Wales, UK.

Boyd, H. 1954. The 'wreck' of Leach's Petrels in the autumn of 1952. *British Birds* 47(5): pp.137–163.

Bradbury, G., Trinder, M., Furness, B., Banks, A.N., Caldow, R.W.G., and Hume, D. 2014. Mapping seabird sensitivity to offshore wind farms. *PLOS One* 9(9).

Bradbury, R., Eaton, M., Bowden, C., and Jordan, M. 2008. Magnificent Frigatebird in Shropshire: new to Britain. *British Birds* 101(6): pp.317–321.

Bradbury, R.B., Kyrkos, A., Morris, A.J., Clark, S.C., Perkins, A.J., and Wilson, J.D. 2000. Habitat associations and breeding success of Yellowhammers on lowland farmland. *Journal of Applied Ecology* 37(5): pp.789–805.

Bradbury, R.B., Wilson, J.D., Moorcroft, D., Morris, A.J., and Perkins, A.J. 2003. Habitat and weather are weak correlates of nestling condition and growth rates of four UK farmland passerines. *Ibis* 145(2): pp.295–306.

Brazil, M. 1991. *The birds of Japan*. London: Helm.

Brecon Beacons National Park. 2019. *A Future with Nature at its Heart: a nature recovery action plan 2019–24*.

Brenchley, A. 1986. The breeding distribution and abundance of the rook *Corvus frugilegus* L. in Great Britain since the 1920s. *Journal of Zoology* 210: pp.261–278.

Brenchley, A., Gibbs, G., Pritchard, R., and Spence, I.M. (eds.) 2013. *The breeding birds of North Wales = Adar Nythu Gogledd Cymru*. Liverpool: Liverpool University Press.

Brides, K., Wood, K., Hearn, R., and Fijen, T.P. 2017. Changes in the sex ratio of the Common Pochard *Aythya ferina* in Europe and North Africa. *Wildfowl* 67: pp.100–112.

Brindley, E., Norris, K., Cook, T., Babbs, S., Forster-Browne, C., and Yaxley, R. 1998. The abundance and conservation status of redshank (*Tringa totanus*) nesting on saltmarshes in Great Britain. *Biological Conservation* 86: pp.289–297.

British Ornithologists' Union (BOU). 2001. Records Committee: 27th Report. *Ibis* 143(1): pp.171–175.

Brooke, M. 1990. *The Manx Shearwater*. London: T&AD Poyser.

Brooke, M. 2004. *Albatrosses and Petrels across the world*. Oxford: Oxford University Press.

Broughton, R.K., Bellamy, P.E., Hill, R.A., and Hinsley, S.A. 2015. Winter Social Organisation of Marsh Tits *Poecile palustris* in Britain. *Acta Ornithologica* 50(1): pp.11–21.

Broughton, R.K., and Hinsley, S.A. 2015. The ecology and conservation of the Marsh Tit in Britain. *British Birds* 108(1): pp.12–29.

Brown, A., and Grice, P. 2005. *Birds in England*. London: T&AD Poyser.

Brown, A.F., and Shepherd, K.B 1993. A method for censusing upland breeding waders. *Bird Study* 40(3): pp.189–195.

Brown, D. 2008. Eosiaid Gwlad y Gan. *Llên Natur* 5.

Brown, J.G. 2001. Citrine Wagtail on Skomer Island – first for Wales. *Welsh Birds* 3(2): p.139.

Brown, J.G. 2006. Census of European storm-petrels *Hydrobates pelagicus* on Skomer Island. *Atlantic Seabirds* 8(1/2): pp.21–30.

Brown, L. 1976. *British birds of prey: a study of Britain's 24 diurnal raptors*. London: HarperCollins.

Brown, R., and Eagle, G. 2014. *Skokholm Bird Observatory Seabird Report 2014*. Online publication. **https://www.welshwildlife.org/wp-content/uploads/2014/07/Skokholm-Seabird-Report-2014.pdf**

Brown, R., and Eagle, G. 2016. *Skokholm Bird Observatory Seabird Report 2015*. Online publication. **https://www.welshwildlife.org/wp-content/uploads/2016/01/Skokholm-Seabird-Report-2015.pdf**

Brown, R., and Eagle, G. 2018a. *Skokholm Bird Observatory Seabird Report 2017*. Online publication: **https://welshwildlife-6aa7.kxcdn.com/wp-content/uploads/2019/03/Skokholm-Seabird-Report-2017.pdf**

Brown, R., and Eagle, G. 2018b. *Skokholm Bird Observatory Annual Report 2017*. pp 96–97. Online publication. **https://welshwildlife-6aa7.kxcdn.com/wp-content/uploads/2018/11/Skokholm-Annual-Report-2017-sml.pdf**

Brown, R., and Eagle, G. 2019. *Skokholm Bird Observatory Seabird Report 2018*. Online publication: **https://welshwildlife-6aa7.kxcdn.com/wp-content/uploads/2019/03/Skokholm-Seabird-Report-2018.pdf**

Browne, S. 1988. The animal bones. In Higham, R.A., and Barker, P. *Hen Domen, Montgomery: a timber castle on the English–Welsh border: a final report*. Exeter: University of Exeter Press: pp.126–134.

Browne, S.J., and Aebischer, N.J. 2005. Studies of West Palearctic birds: Turtle Dove *Streptopelia turtur*. *British Birds* 98(2): pp.58–72.

BTO. 2016a. House Martin Survey 2015 Results. **https://www.bto.org/our-science/projects/house-martin-survey/house-martin-survey-2015/house-martin-survey-2015-results**

BTO. 2016b. Wetland Bird Survey species threshold levels.

BTO. 2020. Maps of population density and trends. **https://www.bto.org/our-science/projects/bbs/latest-results/maps-population-density-and-trends**

BTO Garden Birdwatch. 2020. **https://www.bto.org/our-science/projects/gbw/results/long-term-patterns**

BTO Heronries Census. Undated. **https://www.bto.org/our-science/projects/heronries-census/about-heronries-census**

Bullock, I.D., Drewett, D.R., and Mickleburgh, S.P. 1983. The Chough in Britain and Ireland. *British Birds* 76(9): pp.377–401.

Bullock, I.D., Drewett, D.R., and Mickleburgh, S.P. 1985. The Chough in Wales. *Nature in Wales* 4(1&2): pp.46–57.

Buner, F.D., Browne, S.J., and Aebischer, N.J. 2011. Experimental assessment of release methods for the reestablishment of a red-listed galliform, the grey partridge (*Perdix perdix*). *Biological Conservation* 144: pp.593–601.

Burger, C., Belskii, E., Eeva, T., Laaksonen, T., Magi, M., Mand, R., Qvarnstrom, A., Slagsvold, T., Veen, T., Visser, M.E., Wiebe, K.L., Wiley, C., Wright, J., and Both, C. 2012. Climate change, breeding date and nestling diet: how temperature differentially affects seasonal changes in pied flycatcher diet depending on habitat variation. *Journal of Animal Ecology* 81: pp.926–936.

Burgess, M.D., Smith, K.W., Evans, K.L., Leech, D., Pearce-Higgins, J.W., Branston, C.J., Briggs, K., Clark, J.R., du Feu, C.R., Lewthwaite, K., and Nager, R.G., 2018. Tritrophic phenological match–mismatch in space and time. *Nature ecology & evolution* 2(6): pp.970–975.

Burns, F., Eaton, M.A., Barlow, K.E., Beckmann, B.C., Brereton, T., Brooks, D.R., *et al.* 2016. Agricultural management and climatic change are the major drivers of biodiversity change in the UK. *PLoS ONE* 11(3): e0151595.

Burton, N.H.K., Banks, A.N., Calladine, J.R., and Austin, G.E. 2013. The importance of the United Kingdom for wintering gulls: population estimates and conservation requirements. *Bird Study* 60(1): pp.87–101.

Burton, N.H.K., Musgrove, A.J., Rehfisch, M.M., and Sutcliffe, A. 2002. *Winter gull roosts in the United Kingdom in January 1993 with recommendations for future surveys of wintering gulls.* Thetford: BTO Research Report No. 277.

Burton, N.H.K., Musgrove, A.J., Rehfisch, M.M., Sutcliffe, A., and Waters, R. 2003a. Numbers of wintering gulls in the United Kingdom, Channel Islands and Isle of Man: a review of the 1993 and previous Winter Gull Roost Surveys. *British Birds* 96(8): pp.376–401.

Burton, N.H.K., Rehfisch, M.M., and Clark, N.A. 2003b. *The effect of the Cardiff Bay Barrage on waterbird populations: final report*. Thetford: BTO Research Report No. 343.

Butcher, G., and Niven, D. 2007. *Combining data from the Christmas Bird Count and the Breeding Bird Survey to determine the continental status and trends of North America birds*. National Audubon Society report.

Buxton, J., and Lockley, R.M. 1950. *Island of Skomer: A preliminary survey of the natural history of Skomer Island, Pembrokeshire, undertaken for the West Wales Field Society*. London: Staples Press.

Byrkjedal, I., and Thompson, D. 1998. *Tundra plovers: the Eurasian, Pacific and American Golden Plovers and Grey Plover.* London: T&AD Poyser.

Cabot, D. 2009. *Wildfowl*. London: Collins.

Cabot, D., and Nisbet, I. 2013. *Terns*. London: Collins.

Cadbury, C.J., and Olney, P.J.S. 1978. Avocet population dynamics in England. *British Birds* 71(3): pp.102–121.

Cadman, W.A. 1949. Distribution of Black Grouse in North Wales forests. *British Birds* 42(11): pp.365–367.

Cagnon, C., Lauga, B., Hemery, G., and Mouches, C. 2004. Phylogeographic differentiation of storm petrels (*Hydrobates pelagicus*) based on cytochrome *b* mitochondrial DNA variation. *Marine Biology* 145: pp.1,257–1,264.

Calladine, J., Baines, D., and Warren, P. 2002. Effects of reduced grazing on population density and breeding success of black grouse in northern England. *Journal of Applied Ecology* 39(5): pp.772–780.

Calladine, J., and Bray, J. 2012. The importance of altitude and aspect for breeding Whinchats *Saxicola rubetra* in the uplands: limitations of the uplands as a refuge for a declining, formerly widespread species? *Bird Study* 59(1): pp.43–51.

Calladine, J., Garner, G., and Wernham, C. 2007. *Developing methods for the field survey and monitoring of breeding Short-eared Owls (Asio flammeus) in the UK: an interim report from pilot fieldwork in 2006.* Thetford: BTO Research Report No. 472.

Camden, W. 1722. *Britannia: Or a Chorographical Description of Great Britain and Ireland, Together with the Adjacent Islands.* 2nd ed., revised by Edmund Gibson. London: Mary Matthews.

Campbell, B. 1946. Notes on some Monmouthshire and East Glamorgan birds, 1943–1945. *British Birds* 39(11): pp.322–325.

Campbell, B. 1954. The breeding distribution and habitats of the Pied Flycatcher (*Muscicapa hypoleuca*) in Britain. *Bird Study* 1(3): pp.81–101.

Campbell, B. 1955. The breeding distribution and habitats of the Pied Flycatcher (*Muscicapa hypoleuca*) in Britain. *Bird Study* 2(1): pp.24–32.

Campbell, B. 1960. The Mute Swan Census in England and Wales 1955–56. *Bird Study* 7(4): pp.208–223.

Campbell, B. 1965. The British breeding distribution of the Pied Flycatcher, 1953–62. *Bird Study* 12(4): pp.305–318.

Campbell, B., and Ferguson-Lees, J. 1972. *A field guide to birds' nests*. London: Constable.

Campbell, L., Cayford, J. and Pearson, D. 1996. Bearded Tits in Britain and Ireland. *British Birds* 89(8): pp.335–346.

Camphuysen, C.J. 1989. *Beached bird surveys in the Netherlands 1915–1988. Seabird mortality in the southern North Sea since the early days of oil pollution.* Techn. Rapport Vogelbescherming 1, Werkgroep Noordzee, Amsterdam.

Carr, H.R.C., and Lister, G.A. 1925. *The mountains of Snowdonia*. Crosby Lockwood.

Carss, D.N., and Marquiss, M. 1992. Avian predation at farmed and natural fisheries. *Proceedings of the 22nd Institute of Fisheries Management Annual Study Course*: pp.179–196.

Carter, I. 2001. *The Red Kite*. Chelmsford: Arlequin Press.

Centre for Advanced Welsh and Celtic Studies 2013. Guto's Wales. The life of a poet in fifteenth-century Wales. **http://www.gutorglyn.net/gutoswales/diddordebau-hela-adar.php**

Cham, S., Nelson, B., Parr, A., Prentice, S., Smallshire, D., and Taylor, P. 2014. *Atlas of Dragonflies in Britain and Ireland*. Field Studies Council for Biological Records Centre, NERC Centre for Ecology and Hydrology, Joint Nature Conservation Committee with British Dragonfly Society.

Chamberlain, D.E., and Fuller, R.J. 2000. Local extinctions and changes in species richness of lowland farmland birds in England and Wales in relation to recent changes in agricultural land-use. *Agriculture, Ecosystems & Environment* 78(1): pp.1–17.

Chamberlain, D.E., and Fuller, R.J. 2001. Contrasting patterns of change in the distribution and abundance of farmland birds in relation to farming system in lowland Britain. *Global Ecology and Biogeography* 10(4): pp.399–409.

Chaniotis, P., Cioffi, B., Farmer, R., Cornthwaite, A., Flavell, B., and Carr, H. 2018. Developing an ecologically-coherent and well-managed Marine Protected Area network in the United Kingdom: 10 years of reflection from the Joint Nature Conservation Committee. *Biodiversity* 19(1–2): pp.140–147.

Chapman, A. 1999. *The Hobby*. Chelmsford: Arlequin Press.

Charles-Edwards, T.M., Owen, M.E., and Russell, P. 2000. *The Welsh King and his court*. Cardiff: University of Wales Press.

Charman, E.C., Smith, K.W., Dillon, I.A., Dodd, S., Gruar, D.J., Cristinacce, A., Grice, P.V., and Gregory, R.D. 2012. Drivers of low breeding success in the Lesser Spotted Woodpecker *Dendrocopos minor* in England: testing hypotheses for the decline. *Bird Study* 59(3): pp.255–265.

Charman, E.C., Smith, K.W., Gruar, D.J., Dodd, S., and Grice, P.V. 2010. Characteristics of woods used recently and historically by Lesser Spotted Woodpeckers *Dendrocopos minor* in England. *Ibis* 152(3): pp.543–555.

Chater, A., Chater, M., and Chater, E.H. 1974. Rookeries in Cardiganshire. *Nature in Wales* 14(2): pp.69–75.

Chater, A.O. 1996. Colonies of the Rook in North Ceredigion 1975–1994. *Welsh Birds* 1(3): pp.3–21.

Cheke, A. 2019. A long-standing feral Indian Peafowl population in Oxfordshire, and a brief survey of the species in Britain. *British Birds* 112(6): pp.337–348.

Clark, N.A. 1983. *The ecology of Dunlin (*Calidris alpina L.*) wintering on the Severn Estuary.* Thesis presented for the degree of Doctor of Philosophy. Edinburgh: University of Edinburgh.

Clarke, A., Prince, P.A., and Clarke, R. 1996. The energy content of dragonflies (Odonata) in relation to predation by falcons. *Bird Study* 43(32): pp.300–304.

Clarke, R., and Watson, D. 1990. The Hen Harrier *Circus cyaneus* winter roost survey in Britain and Ireland. *Bird Study* 37(2): pp.84–100.

Clarke, R.M. 2006. Survey to determine the population of breeding pairs of Water Rail at the Uskmouth Lagoons. *Gwent Bird Report* 41: pp.62–65.

Clarke, R.M. 2008. A survey of breeding *Larus* gulls in Gwent during 2007. *Gwent Bird Report 2007*: pp.80–86.

Clarke, R.M. 2011a. Gwent Goosander Winter Roost Survey 2011. *Gwent Bird Report* 46: pp.105–111.

Clarke, R.M. 2011b. The birds of Denny Island. *Birds in Wales* 8(1): pp.29–34.

Clarke, R.M. 2016. Monitoring of Great Black-backed Gulls *Larus marinus* on Denny Island, Monmouthshire. *Gwent Bird Report* 51 (2015): pp.115–121.

Clarke, R.M. 2017. Assessing the population of Water Rail *Rallus aquaticus* wintering at the Uskmouth reedbeds, Newport Wetlands Reserve. *Gwent Bird Report* 53: pp.101–105.

Clarke, R.M. 2018a. Breeding Little Egrets in Gwent. *Gwent Bird Report* 53: pp.101–109.

Clarke, R.M. 2018b. Breeding *Larus* Gulls in Monmouthshire (VC 35). *Gwent Bird Report* 53: pp.115–120.

Cleasby, I.R., Bodey, T.W., Vigfusdottir, F., McDonald, J.L., McElwaine, G., Mackie, K., Colhoun, K., and Bearhop, S. 2017. Climatic conditions produce contrasting influences on demographic traits in a long-distance Arctic migrant. *Journal of Animal Ecology* 86(2): pp.285–295.

Coffey, P. 2015. Common Terns at Shotton – a Welsh success story. *Natur Cymru=Nature of Wales* 54: pp.19–23.

Coiffait, L., Clark, J.A., Robinson, R.A., Blackburn, J.R., Grantham, M.J., Marchant, J.H., Barber, L., De Palacio, D., Griffin, B.M., and Moss, D. 2008. Bird ringing in Britain and Ireland in 2007. *Ringing & Migration* 24: pp.104–144.

Colling, A.W., and Brown, E.B. 1946. The breeding of Marsh and Montagu's Harriers in North Wales in 1945. *British Birds* 39(8): pp.233–243.

Collinson, J.M., Murcia, A., Ladeira, G., Dewars, K., Roberts, F., and Shannon, T. 2018. Siberian and Scandinavian Common Chiffchaffs in Britain and Ireland – a genetic study. *British Birds* 111(7): pp.384–394.

Collinson, M., Parkin, D.T., Knox, A.G., Sangster, G., and Helbig, A.J. 2006. Species limits within the genus *Melanitta*, the scoters. *British Birds* 99(4): pp.183–201.

Colwell, M. 2018. *Curlew moon*. London: Collins.

Compton, E., Daley, L.F., Stubbings, E.M., Büche, B.I., and Wood, M. 2016. Diet, ecology and biosecurity: analysis of owl pellets from Skomer Island. *Birds in Wales* 13(1): pp.57–72.
Conder, P. 1989. *The Wheatear*. London: Helm.
Conder, P., and Keighley, J. 1950. *Skokholm Bird Observatory Report for 1949*.
Condry, W. 1955. The breeding birds of the Welsh mountains. *Nature in Wales* 1(1): pp.25–27.
Condry, W. 1960. The breeding birds of conifer plantations. *Quarterly Journal of Forestry* 54(4): pp.357–362.
Condry, W. 1966. *The Snowdonia National Park*. London: Collins.
Condry, W. 1981. *The Natural History of Wales*. London: Collins.
Condry, W. 1996. *Welsh country essays*. Gomer Press.
Condry, W.M. 1950. The breeding birds of an area of Central Wales. *The Naturalist* Jan. – March: pp.11–16.
Conrad, K.F., Warren, M.S., Fox, R., Parsons, M.S., and Woiwod, I.P. 2006. Rapid declines of common, widespread British moths provide evidence of an insect biodiversity crisis. *Biological Conservation* 132(3): pp.279–291.
Conway, G., Wotton, S., Henderson, I., Eaton, M., Drewitt, A., and Spencer, J. 2009. The status of breeding Woodlarks *Lullula arborea* in Britain in 2006. *Bird Study* 56(3): pp.310–325.
Conway, G., Wotton, S., Henderson, I., Langston, R., Drewitt, A., and Currie, F. 2007. Status and distribution of European Nightjars *Caprimulgus europaeus* in the UK in 2004. *Bird Study* 54(1): pp.98–111.
Conway, G.J., Austin, G.E., Handschuh, M., Drewitt, A.L., and Burton, N.H.K. 2019. Breeding populations of Little Ringed Plover *Charadrius dubius* and Ringed Plover *Charadrius hiaticula* in the United Kingdom in 2007. *Bird Study* 66(1): pp.22–31.
Conway, G.J., Burton, N.H.K., Handschuh, M., and Austin, G.E. 2008. *UK population estimates for the 2007 breeding Little Ringed Plover and Ringed Plover survey*. Thetford: BTO Research Report No. 510.
Conway, G.J., and Fuller, R.J. 2010. *Multi-scale relationships between vegetation and breeding birds in the upland margins (*ffridd*) of North Wales*. Thetford: BTO Research Report No. 566.
Cook, A.S.C.P., Barimore, C., Holt, C.A., Read, W.J., and Austin, G.E. 2013. *Wetland Bird Survey Alerts 2009/2010: Changes in numbers of wintering waterbirds in the Constituent Countries of the United Kingdom, Special Protection Areas (SPAs) and Sites of Special Scientific Interest (SSSIs)*. Thetford: BTO Research Report 641. **http://www.bto.org/volunteer-surveys/webs/publications/webs-annual-report**
Cook, A.S.C.P., Johnston, A., Wright, L.J., and Burton, N.H.K. 2012. *A review of flight heights and avoidance rates of birds in relation to offshore wind farms*. Thetford: BTO Research Report 618.
Cook, H., Dipple, M., MacCormack, F., and Taylor, S. 2018. *Gronant Little Tern (*Sternula albifrons*) Report 2018*. Denbighshire County Council Countryside Services. Unpublished.
Copete, J. 2008. Distribution and identification of Iberian chiffchaff. *British Birds* 101(7): pp.378–379.
Cordeaux, J. 1866. Ornithological notes from North Wales. *Zoologist* 2nd Series 1: pp.436–440.
Cornwall Birds. Undated. **https://www.cbwps.org.uk/cbwpsword/red-kite-map**
Cottier, E.J., and Lea, D. 1969. Black-tailed Godwits, Ruffs and Black Terns breeding on the Ouse Washes. *British Birds* 62(7): pp.259–270.
Coudrain, V., Arlettaz, R., and Schaub, M. 2010. Food or nesting place? Identifying factors limiting Wryneck populations. *Journal of Ornithology* 151(4): pp 867–880.
Coulson, J.C. 1963. The status of the Kittiwake in the British Isles. *Bird Study* 10(3): pp.147–179.
Coulson, J.C. 1983. The changing status of the Kittiwake, *Rissa tridactyla* in the British Isles 1969–1970. *Bird Study* 30(1): pp.9–16.
Coulson, J.C. 2019. *Gulls*. London: Collins.
Coulson, J.C., and Coulson, B.A. 2015. The accuracy of urban nesting gull censuses. *Bird Study* 62(2): pp.170–176.
Countryside Council for Wales. 2006. *Sites of Special Scientific Interest (SSSIs) in Wales: current state of knowledge report for April 2005–March 2006* (unpublished).
Coward, T.A. 1895. Manx Shearwater breeding on the coast of Carnarvonshire. *Zoologist* 3rd Series 19: p.72.
Coward T.A. 1910. *Mammals and birds: The fauna of Cheshire Vol. 1*. London: Witherby.
Coward, T.A. 1913. Common scoters in summer: in Cheshire and North Wales. *British Birds* 7(9): p.118.
Coward, T.A. 1922. *Bird haunts and nature memories*. London: F. Warne.
Coward, T.A. and Oldham, C. 1902. Notes on the birds of Anglesea. *The Zoologist* 4th Series 6: pp.213–230.
Coward, T.A., and Oldham, C. 1904. Notes on the birds of Anglesea. *Zoologist* 4th Series 8: pp.7–29.
Coward, T.A., and Oldham, C. 1905. Notes on the birds of Anglesea. *Zoologist* 4th Series 9: pp.213–230.
Cowley, E. 1979. Sand Martin population trends in Britain, 1965–1978. *Bird Study* 26(2): pp.113–116.
Cowley, E., and Siriwardena, G.M. 2005. Long-term variation in survival rates of Sand Martins *Riparia riparia*: dependence on breeding and wintering ground weather, age and sex, and their population consequences. *Bird Study* 52(3): pp.237–251.
Cramp, S. 1963. Toxic chemicals and birds of prey. *British Birds* 56(4): pp.124–139.
Cramp, S., Bourne, W.R.P., and Sanders, D. 1974. The seabirds of Britain and Ireland. London: Collins.
Cramp, S., and Simmons, K.E.L. 1977–1994. *Handbook of the birds of Europe, the Middle East and North Africa: The birds of the Western Palearctic. Vol. 1–9.* Oxford: Oxford University Press.
Cramp, S., Vettet, A., and Sharrock, J.T.R. 1960. The irruption of tits in autumn 1957. *British Birds* 53(2): pp.49–77; 53(4): pp.176–192.
Crawley, D., Coomber, F., Kubasiewicz, L., Harrower, C., Evans, P., Waggitt, J., Smith, B., and Mathews, F. 2020. *Atlas of the mammals of Great Britain and Northern Ireland*. Exeter: Pelagic Publishing.
Crick, H.Q.P., and Ratcliffe, D.A. 1995. The Peregrine (*Falco peregrinus*) breeding population of the United Kingdom in 1991. *Bird Study* 42(1): pp.1–19.
Crick, H.Q.P., and Sparks, T.H. 1999. Climate change related to egg-laying trends. *Nature* 399: pp.423–424.
Cromie, R., Newth, J., and Strong, E. 2010. *Compliance with the environmental protection (restriction on use of lead shot) (England) Regulations 1999*. DeFRA, Bristol.
Crosby, M.J. 1982. Birds of the River Severn and its tributaries. *Montgomery Bird Report 1981 and 1982*: pp.6–9.
Cross, A.V., and Stratford, A. 2015. Juvenile survival, pre-breeding dispersal and natal fidelity of Red-billed Choughs on the Llŷn peninsula, Gwynedd. *Birds in Wales* 12(1): pp.26–49.
Cross, A.V., Stratford, A., Johnstone, I., Thorpe, R.I.T., Dodd, S., Peach, W., Buchanan, G., and Moorhouse-Gann, R. 2020. *Red-billed Chough Wales research programme*. Natural Resources Wales Evidence Report No. 486.
Cross, T. 2001. Results of the UK Red Kite survey 2000. *Welsh Birds* 3(1): pp.60–61.
Cross, T. 2004. 2003 Red Kite breeding season in Wales. *Welsh Birds* 4(2): p.168.
Cross, T. 2019. A tale of two finches. *Ruffled Feathers*. **https://midwalesringers.blogspot.com/**
Cross, T., and Davis, P. 2005. *The Red Kites of Wales*. Subbuteo Natural History Books.
Crowe, O., Coombes, R.H., Lysaght, L., O'Brien, C., Choudhury, K.R., Walsh, A.J., Wilson, J.H., and O'Halloran, J. 2010. Population trends of widespread breeding birds in the Republic of Ireland 1998–2008. *Bird Study* 57(3): pp.267–280.
Crowe, O., McElwaine, J.G., Boland, H., and Enlander, I.J. 2015. Whooper *Cygnus cygnus* and Bewick's *C. columbianus bewickii* Swans in Ireland: results of the International Swan Census, January 2015. *Irish Birds* 10: pp.151–158.
Crump, H. 2014. *Evaluating causes of Golden Plover (*Pluvialis apricaria*) decline in Mid Wales: a geographic approach.* PhD Thesis, Aberystwyth University.
Crump, H., and Green, M. 2012. Changes in breeding bird abundances in the Plynlimon SSSI 1984–2011. *Birds in Wales* 9(1): pp.9–13.
Crump, H., and Green, M. 2016. Changes in breeding bird abundances in the Elenydd SSSI between 1982 and 2012. *Birds in Wales* 13(1): pp.32–36.
Cummings, S.G. 1908. Notes on the habits and distribution of the Cirl Bunting in North Wales. *British Birds* 1(9): pp.275–279.
Curlew Country 2020. **https://curlewcountry.org/shropshire-welsh-marches-recovery-project**
Currie, F.C., and Bamford, R. 1982. The value to wildlife of retaining small conifer stands beyond normal felling age within forests. *Quarterly Journal of Forestry* 76: pp.153–160.
Cymdeithas Edward Llwyd a Cymdeithas Ted Breeze Jones. 2015. *Adar y Byd*. **http://termau.cymru/**
Dadam, D., and Siriwardena, G.M. 2019. Agri-environment effects on birds in Wales: Tir Gofal benefited woodland and hedgerow species. *Agriculture, Ecosystems & Environment* 284. **https://doi.org/10.1016/j.agee.2019.106587.**
Dagys, M. 2017. *Species Status Report for Velvet Scoter* Melanitta fusca. *Western Siberia & Northern Europe/NW Europe Population*. LIFE Project: Coordinated Efforts for International Species Recovery Euro SAP. Report commissioned by the European Commission Directorate General for the Environment. Lithuanian Ornithological Society, Vilnius, Lithuania.
Dahl, M.B., Priemé, A., Brejnrod, A., Brusvang, P., Lund, M., Nymand, J., Kramshøj, M., Ro-Poulsen, H., and Haugwitz, M.S. 2017. Warming, shading and a moth outbreak reduce tundra carbon sink strength dramatically by changing plant cover and soil microbial activity. *Scientific Reports* 7(1): p.16035.

Dalrymple, T. 2020. Newport Wetlands National Nature Reserve: a review of the first 20 years. *Birds in Wales* 17(1): pp.36–55.

Dandy, J.E. 1969. *Watsonian vice-counties of Great Britain*. London: The Ray Society.

Dansk Ornitologisk Forening. 2020. Atlas III: Dansk Ornitologisk Forenings fugleatlas 2014–17. **dofbasen.dk/atlas**

Dare, P.J. 1986a. Raven *Corvus corax* populations in two upland regions of north Wales. *Bird Study* 33(3): pp.179–189.

Dare, P.J. 1986b. Notes on the Kestrel population of Snowdonia, North Wales. *Naturalist* III: pp.49–54.

Dare, P.J. 1995. Breeding success and territory features of Buzzards in Snowdonia and adjacent uplands of North Wales. *Welsh Birds* 1(2): pp.69–78.

Dare, P.J. 1996. Autumn seabird movements at Point Lynas, Anglesey, during 1976 to 1993. *Welsh Birds* 1(3): pp.29–44.

Dare, P.J. 2007. Oystercatcher *Haematopus ostralegus* at Traeth Lafan, North Wales, in the 1960s: before the advent of commercial cockle fishing. *Welsh Birds* 5(1): pp.65–77.

Dare, P.J. 2015. *The life of Buzzards*. Whittles Publishing.

Dare, P.J., and Barry, J.T. 1990. Population size, density and regularity in nest spacing of Buzzards *Buteo buteo* in two upland regions of North Wales. *Bird Study* 37(1): pp.23–29.

Dare, P.J., and Schofield, P. 1976. *Ecological survey of the Lavan Sands: ornithological survey 1969–74*. Cambrian Ornithological Society.

Davies, A.K. 1988. The distribution and status of the Mandarin Duck *Aix galericulata* in Britain. *Bird Study* 35(3): pp.203–207.

Davies, J. 1974. The end of the great estates and the rise of freehold farming in Wales. *Welsh History Review* 7: pp.186–212.

Davies, L. 1912. *Radnorshire*. Cambridge County Geographies. Cambridge University Press.

Davies, M., Inskip, T., and Hulme, R.A. 1978. *Elenydd surveys (1975–77)*. Sandy: RSPB, unpublished.

Davies, N.B. 1992. *Dunnock behaviour and social evolution*. Oxford: Oxford University Press.

Davies, N.B. 2000. *Cuckoos, Cowbirds and other cheats*. London: T&AD Poyser.

Davies, N.B., and Lundberg, A. 1984. Food distribution and a variable mating system in the Dunnock *Prunella modularis*. *Journal of Animal Ecology* 53: pp.895–912.

Davies, W. 1858. *Llandeilo-Vawr and its neighbourhood: past and present*. Llandeilo: D.W. and G. Jones.

Davis, P. 1993. The Red Kite in Wales: setting the record straight. *British Birds* 86(7): pp.295–298.

Davis, P., Cross, T., and Davis, J. 2001. Movement, settlement, breeding and survival of Red Kites *Milvus milvus* marked in Wales. *Welsh Birds* 3(1): pp.18–43.

Davis, P.E. 1982a. *NCC Breeding Bird Survey (Mynydd Hiraethog)*. NCC.

Davis, P.E., and Davis, J.E. 1981. The food of the Red Kite in Wales. *Bird Study* 28(1): pp.33–40.

Davis, P.G. 1982b. Nightingales in Britain in 1980. *Bird Study* 29(1): pp.73–79.

Davis, T., and Jones, T. 2007. *The birds of Lundy*. Berrynarbor, Devon: Devon Bird-watching and Preservation Society and Lundy Field Society.

Davis, T.A.W. 1958. The breeding distribution of the Great Black-Backed Gull in England and Wales in 1956. *Bird Study* 5(4): pp.191–215.

Davis, T.A.W., and Saunders, D.R. 1965. Buzzards on Skomer Island 1954–1964. *Nature in Wales* 9(3): pp.116–124.

Dayton, J., Ledwoń, M., Paillisson, J.M., Atamas, N., and Szczys, P. 2017. Genetic diversity and population structure of the Eurasian Whiskered Tern (*Chlidonias hybrida hybrida*), a species exhibiting range expansion. *Waterbirds* 40(2): pp.105–117.

de Jong, A., Torniainen, J., Bourski, O.V., Heim, W., and Edenius, L. 2019. Tracing the origin of vagrant Siberian songbirds with stable isotopes: the case of Yellow-browed Warbler (*Abrornis inornatus*) in Fennoscandia. *Ornis Fennica* 96: pp.90–99.

de Juana, E. 2008. Where do Pallas's and Yellow-browed Warblers (*Phylloscopus progregulus* and *Ph. inornatus*) go after visiting northwest Europe in autumn? An Iberian perspective. *Ardeola* 55(2): pp.179–192.

De Knijff, P., van der Spek, V., and Fischer, J. 2012. Genetic identity of grey chiffchaffs trapped in the Netherlands in the autumn of 2009. *Dutch Birding* 34: pp.386–392.

de la Hera, I., Gómez, J., Dillane, E., Unanue, A., Pérez-Rodríguez, A., Pérez-Tris, J., and Torres-Sánchez, M. 2020. Wintering grounds, population size and evolutionary history of a cryptic passerine species from isotopic and genetic data. *Journal of Avian Biology* 51(9).

Deakin, Z., Hamer, K.C., Sherley, R.B., Bearhop, S., Bodey, T.W., Clark, B.L., Grecian, W.J., Gummery, M., Lane, J., Morgan, G., Morgan, L., Phillips, R.A., Wakefield, E.D., and Votier, S.C. 2019. Sex differences in migration and demography of a wide-ranging seabird, the northern Gannet. *Marine Ecology Progress Series* 622: pp.191–201.

Dean, A.R. 1984. Origins and distribution of British Glaucous Gulls. *British Birds* 77(4): pp.165–166.

Dean, A.R., and Svensson, L. 2005. 'Siberian Chiffchaff' revisited. *British Birds* 98(8): pp.396–410.

Defra (Department for Environment Food and Rural Affairs). 2002. *UK Ruddy Duck Control Trial Final Report*. **www.defra.gov.uk**

Defra (Department for Environment Food and Rural Affairs). 2018. Wild bird populations in the UK, 1970 to 2017. **https://www.gov.uk/government/statistics/wild-bird-populations-in-the-uk**

Defra (Department for Environment Food and Rural Affairs). 2020. Wild Bird Populations in the UK, 1970–2019.

del Hoyo, J., Elliott, A., Sargatal, J., and Christie, D.A. (eds.) 1992–2011. *Handbook of the birds of the world*. Vols. 1–16. Barcelona: Lynx Edicions.

Delany, S. 1992. *Pilot survey of breeding Shelduck in Great Britain and Northern Ireland*. WWT unpubl.

Delany, S., and Scott, D. 2006. *Waterbird population estimates*. Wageningen, The Netherlands: Wetlands International.

Delany, S., Greenwood, J.J.D., and Kirby, J. 1992. *National Mute Swan Survey 1990*. Report to the Joint Nature Conservation Committee. JNCC.

Delmore, K.E., Van Doren, B.M., Conway, G.J., Curk, T., Garrido-Garduño, T., Germain, R.R., Hasselmann, T., Dieter Hiemer, D., van der Jeugd, H.P., Justen, H., Ramos, J.S.L., Maggini, I., Meyer, B.S., Phillips, R.J., Remisiewicz, M., Roberts, G.C.M., Sheldon, B.C., Vogl, W., and Liedvogel, M. 2020. Individual variability and versatility in an eco-evolutionary model of avian migration. *Proceedings of the Royal Society B*. 287 (1938): (as yet unpublished).

Denerley, C. 2014. *The impact of land use change on a brood parasite system: Cuckoos, their hosts and prey*. PhD thesis, University of Aberdeen.

Dennis, R. 2008. *A life of Ospreys*. Whittles Publishing.

Dickenson, H., and Howells, R.J. 1962. Divers (Genus *Gavia)* in Wales. *Nature in Wales* 8(2): pp.47–53.

Dies, J.I., Abad, A., and Chardí, M. 2019a. First record of multiple Elegant Tern nests in Spain. Birdguides.com. **https://www.birdguides.com/articles/rare-birds/first-record-of-multiple-elegant-tern-nests-in-spain**

Dies, J.I., Chardi, M., and Abad, A. 2019b. Elegant Terns breeding at L'Albufera de Valencia, Spain. *British Birds* 112(2): pp.110–117.

Dillwyn, L.W. 1848. *Materials for a Fauna and Flora of Swansea and the neighbourhood*. D. Rees.

Dix, T. 1865. Notes on birds in Carmarthenshire. *Zoologist* 23: pp.9,963–9,964.

Dix, T. 1866. A list of birds observed in Pembrokeshire. *Zoologist* 2(1): pp.132–140

Dix, T. 1869. Ornithological notes from Pembrokeshire. *Zoologist* 2(4): pp.1,670–1,681.

Dix, T. 1870. Quails in Pembrokeshire. *Zoologist* Second 5th series: pp.2,394–2,396.

Dixon, A. 1997. Breeding biology of a Raven population in the Brecon Beacons. *Welsh Birds* 1(5): pp.27–38.

Dixon, A. 2011. A list of breeding species in Wales. *Birds in Wales* 8(2): pp.219–224.

Dixon, A. 2012. An historical record of Black-tailed Godwit breeding in Wales. *Birds in Wales* 9(1): pp.50–51.

Dixon, A. 2014. Egg collectors and the Peregrines of south-central Wales. *The Oologist:* 1: pp.2–5 and 24–27.

Dixon, A. 2016. Ravens in Breconshire. *Breconshire Birds*. pp.77–81.

Dixon, A., and Haffield, J.P. 2013. Seed availability and timing of breeding of Common Crossbills *Loxia curvirostra* at Sitka Spruce *Picea sitchensis* dominated forestry plantations. *Ardea* 101(1): pp.33–38.

Dixon, A., Haffield, P., Richards, C., and Trobe, W. 2007. Breeding Ravens on the Gower Peninsula. *Welsh Birds* 5(1): pp.29–36.

Dixon, A., and Richards, C. 2005. Peregrine predation of Woodcocks: evidence of nocturnal hunting. *Birds in Wales* 4(3): pp.221–226.

Dixon, A., Richards, C., Haffield, P., Lawrence, M., Thomas, M., Roberts, G., and Lowe, A. 2006. Peregrine (*Falco peregrinus*) population trends in south-central Wales. *Welsh Birds* 4(5): pp.433–441.

Dixon, A., Richards, C., Haffield, P., Roberts, G., Thomas, M., and Lowe, A. 2008. The National Peregrine Survey 2002. How accurate are the published results for Wales? *Welsh Birds* 5(4): pp.276–283.

Dixon, A., Richards, C., Haffield, P., Thomas, M., Lawrence, M., and Roberts, G. 2010. Population declines of Peregrines (*Falco peregrinus*) in central Wales associated with a reduction in racing pigeon availability. *Birds in Wales* 7: pp.3–12.

Dixon, A., Richards, C., and King, V.A. 2018. Diet of Peregrine Falcons in relation to temporal and spatial variation in racing pigeon availability in Wales. *Ornis Hungarica* 26(2): pp.188–200.

Dobie, W.H. 1894. Birds of West Cheshire, Denbighshire, and Flintshire: Being a list of species occurring in the District of the Chester Society of Natural Science. *Proceedings of the Chester Society of Natural Science and Literature* 4: pp.282–351.

Dobinson, H.M., and Richards, A.J. 1964. The effects of the severe winter of 1962/63 on birds in Britain. *British Birds* 57(10): pp.373–434.
Dockray, J.A. 1910. The Dee as a wildfowl resort. In Coward, T.A., (ed.) *The vertebrate fauna of Cheshire and Liverpool. Volume 2.* Witherby.
Dodd, S., and Moss, D. 1983. The Turnstone population at Rhos-on-sea. *SCAN Ringing Group Report 1981 and 1982*: pp.56–57.
Dodd, S.G. 2017. Colour-ring resightings and flock counts link the increasing Welsh wintering Barnacle Goose *Branta leucopsis* flock to a naturalised breeding population in the Lake District. *Ringing & Migration* 32(1): pp.54–57.
Doddridge-Knight, E. 1853. An account of Newton Nottage, Glamorgan. Chapter 3. *Archaeologia Cambrensis* New Series IV: pp.229–262.
Domesday Book, The. **https://opendomesday.org**
Donald, P.F., Sanderson, F.J., Burfield, I.J., Bierman, S.M., Gregory, R.D., and Waliczky, Z. 2007. International conservation policy delivers benefits for birds in Europe. *Science* 317: pp.810–813.
Donovan, E. 1805. *Descriptive Excursions through South Wales and Monmouthshire in the year 1804 and the four preceding summers*. 2 vols. London: The Author.
Donovan, J., and Rees, G. 1994. *Birds of Pembrokeshire: status and atlas of Pembrokeshire's birds*. Dyfed Wildlife Trust.
Donovan, J.W. 1963. Bird notes. *Nature in Wales* 8: p.205.
Donovan, J.W. 1965. Field notes. *Nature in Wales* 9: p.155.
Donovan, J.W. 1972. The Chough in Pembrokeshire. *Nature in Wales* 13(1): pp.21–23.
Dougall, T.W., Holland, P.K., and Yalden, D.W. 2010. The population biology of Common Sandpipers in Britain. *British Birds* 103(2): pp.100–114.
Douglas, D.J.T., Beresford, A., Selvidge, J., Garnett, S., Buchanan, G.M., Gullett, P., and Grant, M.C. 2017. Changes in upland bird abundances show associations with moorland management. *Bird Study* 64(2): pp.242–254.
Draget, E. 2014. Environmental impacts of offshore wind power production in the North Sea: a literature overview. WWF, Norway.
Driver, J. 2005. A study of breeding Carrion Crows *Corvus corone* in Snowdonia. *Welsh Birds* 4(3): pp.227–235.
Driver, J. 2006. Raven *Corvus corax* population census of northwest Wales, 1998 to 2005. *Welsh Birds* 4(5): pp.442–453.
Driver, J. 2007. Breeding performance of Ravens *Corvus corax* in Snowdonia, 1998–2005. *Welsh Birds* 5(1): pp.55–64.
Driver, J. 2010. Tawny Owl nesting on a high mountain crag in Snowdonia. *Birds in Wales* 7(1): pp.130–131.
Driver, J. 2011. Population census of Ring Ouzels *Turdus torquatus* breeding in Snowdonia 2009–10. *Birds in Wales* 8(1): pp.3–13.
Driver, J. 2017. Ring Ouzels (*Turdus torquatus*) in Snowdonia 2009–16. *Birds in Wales* 14(1): pp.3–9.
Driver, J., and Dare, P. 2009. Population increase of Buzzards in Snowdonia 1977–2007. *Welsh Birds* 6(1): pp.38–48.
D'Souza, J.M., Windsor, F.M., Santillo, D., and Ormerod, S.J. 2020. Food web transfer of plastics to an apex riverine predator. *Global Change Biology* 26(7): pp.3,846–3,857.
Dunn, E. 2019. Seabirds in freefall – the silent cliffs. *Marine Conservation* Spring 2019: pp.20–23. Marine Conservation Society (MCS).
Dunn, P.O., and Møller, A.P. (eds.) 2019. *Effects of climate change on birds*. Oxford: Oxford University Press.
Dunnet, G.M., and Ollason, J.C. 1978. The estimation of survival rate in the fulmar, *Fulmarus glacialis*. *Journal of Animal Ecology* 47: pp.507–520.
Dunning, J., Finch, T., Davison, A., and Durrant, K.L. 2020. Population-specific migratory strategies of Twite *Linaria flavirostris* in Western Europe. *Ibis* 162(2): pp.273–278.
Durance, I., and Ormerod, S.J. 2007. Climate change effects on upland stream macroinvertebrates over a 25-year period. *Global Change Biology* 13: pp.942–957.
Dyda, J., Symes, N., and Lamacraft, D. 2009. *Woodland management for birds: a guide to managing woodland for priority birds in Wales*. Sandy: RSPB and Forestry Commission Wales.
Dye, S.R., Hughes, S.L., Tinker, J., Berry, D.I., Holliday, N.P., Kent, E.C., Kennington, K., Inall, M., Smyth, T., Nolan, G., Lyons, K., Andres, O., and Beszczynska-Möller, A. 2013. Impacts of climate change on temperature (air and sea). *MCCIP Science Review 2013*: pp.1–12.
Dyfi Osprey Project. 2017. **http://www.dyfiosprey-project.com/blog/emyr-mwt/2017/11/26/welsh-ospreys-dna-and-genetics-part-iii**
Dyfi Osprey Project. 2018. **http://www.dyfiospreyproject.com/blog/emyr-mwt/2018/05/25/merin-and-tegid-return**
EAC. 2019. Invasive species – first report of session by the Environmental Audit Committee. London: House of Commons.
Eades, R.A. 1974. Monthly variations in foreign-ringed Dunlins on the Dee estuary. *Bird Study* 21(2): pp.155–157.
East Glamorgan Bird Atlas. **http://www.eastglamorganbirdatlas.org.uk**
Eastham, A. 2016. *Goosey goosey Gander with Jemima Shelduck in attendance: two Stone Age occupation caves in South Pembrokeshire*. Pembrokeshire Historical Society. **http://www.pembrokeshirehistoricalsociety.co.uk/goosey-goosey-gander-jemima-shelduck-attendance-two-stone-age-occupation-caves-south-pembrokeshire/**
Eaton, M., Aebischer, N., Brown, A., Hearn, R., Lock, L., Musgrove, A., Noble, D., Stroud, D., and Gregory, R. 2015. Birds of Conservation Concern 4: the population status of birds in the UK, Channel Islands and Isle of Man. *British Birds* 108(12): pp.708–746.
Eaton, M., and the Rare Breeding Birds Panel. 2020. Rare breeding birds in the UK in 2018. *British Birds* 113(12): pp.737–791.
EBCC. 2017. *Trends of common birds in Europe, 2017 update*. **http://ebcc.birdlife.cz/trends-of-common-birds-in-europe-2017-update**
EBCC. 2019. Population index for Little Grebe (*Tachybptus ruficollis*) 1990–2015. EBCC/Bird Life/RSPB/CSO, Prague. Accessible at: **http://www.birds.cz/pecbm/species.php?ID=&result_set=Publish2017&species%5B70%5D=1**
Edlin, H.L. 1969. *Snowdonia: forest park guide*. HMSO.
Edwards, E.W.J., Quinn, L.R., Wakefield, E.D., Miller, P.I., and Thompson, P.M. 2013. Tracking a northern fulmar from a Scottish nesting site to the Charlie-Gibbs Fracture Zone: Evidence of linkage between coastal breeding seabirds and Mid-Atlantic Ridge feeding sites. *Deep Sea Research II* 98: pp.438–444.
Edwards, M., Atkinson, A., Bresnan, E., Helaouet, P., McQuatters-Gollup, A., Ostle, C., Pitois, S., and Widdicombe, C. 2020. Plankton, jellyfish and climate in the North-East Atlantic. *MCCIP Science Review* 2020: pp.322–353.
Edwards, R., and White, I. 1999. *The* Sea Empress *oil spill: environmental impact and recovery*. International Oil Spill Conference, 1999 Proceedings.
Eglington, S.M., Julliard, R., Gargallo, G., van der Jeugd, H., Pearce-Higgins, J.W., Baillie, S., and Robinson, R.A. 2015. Latitudinal gradients in productivity of European migrant warblers have not shifted northwards during a period of climate change. *Global Ecology & Biogeography* 24: pp.427–436.
Ekroos, J., Fox, A.D., Christensen, T.K., Petersen, I.K., Kilpi, M., Jónsson, J.E., Green, M., Laursen, K., Cervencl, A., de Boer, P., Nilsson, L., Meissner, W., Garthe, S., and Öst, M. 2012. Declines amongst breeding Eider *Somateria mollissima* numbers in the Baltic/Wadden Sea flyway. *Ornis Fennica* 89: pp.1–10.
Elias, T. 2004. Hanes Gwalch y Pysgod yng Nghymru. *Llygad Barcud* Haf 2004: pp.10–14.
Elliott, G.D., and Avery, M.I. 1991. A review of reports of Buzzard persecution 1975–1989. *Bird Study* 38(1): pp.52–56.
Elliott, R.D. 1985. The exclusion of avian predators from aggregations of nesting lapwings (*Vanellus vanellus*). *Animal Behaviour* 33(1): pp.308–314.
Emery, F.V. 1982. Edward Lhuyd and 'A Natural History of Wales'. *Nature in Wales* New Series 1(2): pp.34–38.
Emery, F.V. 1985. Edward Lhuyd and Snowdonia. *Nature in Wales* (New series) 4: pp.3–11.
Environment Systems. 2019. *Tree Suitability Modelling – Planting Opportunities for Sessile Oak and Sitka Spruce in Wales in a Changing Climate*. Report to the Committee on Climate Change.
Etheridge, B., Summers, R.W., and Green, R.E. 1997. The effects of illegal killing and destruction of nests by humans on the population dynamics of the Hen Harrier *Circus cyaneus* in Scotland. *Journal of Applied Ecology* 34: pp.1,081–1,105.
European Commission – Environment. 2019. *Sandwich Tern Sterna sandvicensis*. **https://ec.europa.eu/environment/nature/conservation/wildbirds/threatened/s/sterna_sandvicensis_en.htm**
Evans, A.D. 1992. The numbers and distribution of Cirl Buntings Emberiza cirlus breeding in Britain in 1989. *Bird Study* 39(1): pp.17–22.
Evans, D. 1997. *A history of nature conservation in Britain*. London: Routledge.
Evans, D.M., Redpath, S.M., Elston, D.A., Evans, S.A., Mitchell, R.J., and Dennis, P. 2006. To graze or not to graze? Sheep, voles, forestry and nature conservation in the British uplands. *Journal of Applied Ecology* 43: pp.499–505.
Evans, E. 2014. *Ospreys in Wales: the first ten years*. The author.
Evans, G.H. 1965. Yellow Warbler on Bardsey Island: a bird new to Britain and Ireland. *British Birds* 58(11): pp.457–460.
Evans, J. 1812. *The beauties of North Wales: The beauties of England and Wales Vol. XVII*.
Evans, K.L., Newton, J., Mallord, J.W., and Markman, S. 2012a. Stable isotope analysis provides new information on winter habitat use of declining avian migrants that is relevant to their conservation. *PloS One* 7(4): e34542.
Evans, P.R. 1966. Some results from the ringing of Rock Pipits on Skokholm 1952–65. *Report of the Skokholm Bird Observatory 1966*: pp.22–27.
Evans, R.J. 2000. Wintering Slavonian Grebes in coastal waters of Britain and Ireland. *British Birds* 93(5): pp.218–226.

Evans, R.J., O'Toole, L., and Whitfield, D.P. 2012b. The history of eagles in Britain and Ireland: an ecological review of placename and documentary evidence from the last 1500 years. *Bird Study* 59(3): pp.335–349.

Ewing, S.R., Rebecca, G.W., Heavisides, A., Court, I.R., Lindley, P., Ruddock, M., Cohen, S., and Eaton, M.A. 2011. Breeding status of Merlins *Falco columbarius* in the UK in 2008. *Bird Study* 58: pp.379–389.

Ewins, P.J. 1990. The diet of black guillemots *Cepphus grylle* in Shetland. *Ecography* 13(2): pp.90–97.

Eyton, T.C. 1836. *A history of rarer British birds.* Longman.

Eyton, T.C. 1838. An attempt to ascertain the fauna of Shropshire and North Wales; II Aves. *Annals of Natural History, or Magazine of Zoology, Botany, and Geology* 1(4): pp.285–293.

Facey, R.J., and Vafidis, J.O. 2020. The Ring-necked Parakeet (*Psittacula krameri*) in Wales: a preliminary review of records from 1975 to 2017. *Birds in Wales* 17(1): pp.26–35.

Farmer, R. *History of the RSPB in Wales.* Unpublished sabbatical report.

Farrar, G.B. 1938. *Feathered folk of an estuary.* Country Life.

Fayet, A.L., Clucas, G.V., Anker-Nilssen, T., Syposz, M.M., and Hansen, E.S. 2021. Local prey shortages drive foraging costs and lower breeding success in a declining seabird, the Atlantic puffin. *Journal of Animal Ecology* March 2021. **https://doi.org/10.1111/1365-2656.13442**

Fayet, A.L., Freeman, R., Anker-Nilssen, T., Diamond, A.W., Erikstad, K.E., Fifield, D.A., and Guilford, T. 2017. Ocean-wide drivers of migration strategies and their influence on population breeding performance in a declining seabird. *Current Biology* 27(4): pp.3,871–3,878.

Feare, C.J. 1990. Pigeon control: towards a humane alternative. *Environmental Health* 98: pp.155–156.

Fenton, R. 1811. *A historical tour through Pembrokeshire.* Longman, Hurst, Rees, Orme & Co.

Ferguson-Lees, J., and Williamson, K. 1960. Phalaropes in abundance. *British Birds* 53(11): pp.529–531.

Ferns, P.N., Green, G.H., and Round, P.D. 1979. The significance of the Somerset and Gwent Levels in Britain as feeding areas for migrant Whimbrel *Numenius phaeopus*. *Biological Conservation* 16: pp.7–22.

Ferns, P.N., Hamar, H.W., Humphreys, P.N., Kelsey, F.D., Sarson, E.T., Venables, W.A., and Walker, I.R. 1977. *The birds of Gwent.* Gwent Ornithological Society.

Ferreras, P., and Macdonald, D.W. 1999. The impact of American mink *Mustela vison* on waterbirds in the upper Thames. *Journal of Applied Ecology* 36(5): pp.701–708.

Fieldsports. October 2018. **https://fieldsports-journal.com/**

Fijn, R.C., Hiemstra, D., Phillips, R.A., and van der Winden, J. 2013. Arctic Terns *Sterna paradisaea* from the Netherlands migrate record distances across three oceans to Wilkes Land, East Antarctica. *Ardea.* 101(1): pp.3–12.

Fijn, R.C., Wolf, P., Courtens, W., Poot, M.J.M., and Stienen, E.W.M. 2011. Post-breeding dispersal, migration and wintering of Sandwich Terns *Thalasseus sandvicensis* from the southwestern part of the Netherlands. *Sula* 24: pp.121–135. [In Dutch, English summary and figure captions]

Fisher, G., and Walker, M. 2015. Habitat restoration for curlew *Numenius arquata* at the Lake Vyrnwy reserve, Wales. *Conservation Evidence* 12: pp.48–52.

Fisher, J. 1948. Rook Investigation. *Journal of the Ministry of Agriculture* 55: pp.20–23.

Fisher, J. 1952. *The Fulmar.* London: Collins.

Fisher, J. 1953. The Collared Turtle Dove in Europe. *British Birds* 46(5): pp.153–182.

Fisher, J. 1966. The Fulmar population of Britain and Ireland, 1959. *Bird Study* 13(1): pp.5–76.

Fisher, J., and Vevers, H.G. 1944. The breeding distribution, history and population of the North Atlantic Gannet (*Sula bassana*). *Journal of Animal Ecology* 13(1): pp.49–62.

Fisher, S. 1947 (unpublished MS.). *A summary of the results of the Rook investigation.* Edward Grey Institute of Field Ornithology Report R4.

Fletcher, J. (ed.) 2001. *Where truth abides: diaries of the 4th Duke of Newcastle-under-Lyme.* Bakewell: Country Books.

Fletcher, K., Aebischer, N.J., Baines, D., Foster, R., and Hoodless, A.N. 2010. Changes in breeding success and abundance of ground-nesting moorland birds in relation to the experimental deployment of legal predator control. *Journal of Applied Ecology* 47(2): pp.263–272.

Folliot, B., Guillemain, M., Champagnon, J., and Caizergues, A. 2018. Patterns of spatial distribution and migration phenology of Common Pochards *Aythya ferina* in the Western Palearctic: a ring-recoveries analysis. *Wildlife Biology* 2018(1): pp.1–11.

Forest Research. 2019. Forestry Statistics 2019. **www.forestresearch.gov.uk/tools-and-resources/national-forest-inventory**

Forest Research. 2020. Chalara (*Hymenoscyphus fraxineus*) confirmed infection sites. **https://www.forestresearch.gov.uk/documents/7631/Ashdieback_UK_outbreak_Map4_10_Web_Version_May2020.jpg**

Forestry Commission. 2017. *National Forest Inventory Woodland Wales 2017.* **https://data.gov.uk/dataset/02e2489e-65c9-4fc7-ac67-6774833552f7/national-forest-inventory-woodland-wales-2017**

Formaggia, B. 2002. The Barn Owl *Tyto alba* in Montgomeryshire: a survey 1991–2001. *Welsh Birds* 3(3): pp.171–181.

Forrest, H.E. 1907. *The vertebrate fauna of North Wales.* London: Witherby.

Forrest, H.E. 1919. *A handbook to the vertebrate fauna of North Wales.* London: Witherby.

Forrester, R., and Andrews, I. 2007. *The birds of Scotland.* Scottish Ornithologists Club.

Fouarge, J., and Vandevondele, P. 2005. Synthèse d'une exceptionnelle invasion de Jaseurs boréaux (*Bombycilla garrulus*) en Europe en 2004–2005. *Aves* 42(4): pp.281–312.

Foulkes, I. 1862–64. *Cymru Fu: yn cynwys hanesion, traddodiadau, yn nghyda chwedlau a dammegion Cymreig.* Hughes and Son.

Fox, A.D. 1986a. The breeding Teal (*Anas crecca*) of a coastal raised mire in central West Wales. *Bird Study* 33(1): pp.18–23.

Fox, A.D. 1986b. The Little Gull *Larus minutus* in Ceredigion, West Wales. *Seabird* 9: pp.26–31.

Fox, A.D. 1988. Breeding status of the Gadwall in Britain and Ireland. *British Birds* 81(2): pp.51–66.

Fox, A.D. 1994. Estuarine winter feeding patterns of Little Grebes *Tachybaptus ruficollis* in Central Wales. *Bird Study* 41(1): pp.15–24.

Fox, A.D. 2003. Diet and habitat use of scoters *Melanitta* in the Western Palearctic – a brief overview. *Wildfowl* 54: pp.163–182.

Fox, A.D., and Aspinall, S.J. 1987. Pomarine Skuas in Britain and Ireland in autumn 1985. *British Birds* 80(9): pp.404–421.

Fox, A.D., Balsby, T.J.S., Jørgensen, H.E., Lauridsen, T.L., Jeppesen, E., Søndergaard, M., Fugl, K., Myssen, P., and Clausen, P. 2019a. Effects of lake restoration on breeding abundance of declining common pochard. *Hydrobiologia* 830: pp.33–44.

Fox, A.D., Caizergues, A., Banik, M.V., Dvorak, M., Ellermaa, M., Folliot, B., Green, A.J., Grüneberg, C., Guillemain, M., Håland, A., Hornman, M., Keller, V., Koshelev, A.I., Kostyushin, V.A., Kozulin, A., Ławicki, Ł., Luigujõe, L., Müller, C., Musil, P., Musilová, Z., Nilsson, L., Mischenko, A., Pöysä, H., Šćiban, M., Sjeničić, J., Stīpniece, A., Švažas, S., and Wahl, J. 2016. Recent changes in the abundance of breeding Common Pochard *Aythya ferina* in Europe. *Wildfowl* 66: pp.22–40.

Fox, A.D., and Christensen, T.K. 2018. Could falling female sex ratios among first-winter northwest European duck populations contribute to skewed adult sex ratios and overall population declines? *Ibis* 160(4): pp.929–935.

Fox, A.D., Francis, I.S., Madsen, J., and Stroud, J.M. 1987. The breeding biology of the Lapland Bunting *Calcarius lapponicus* in West Greenland in two contrasting years. *Ibis* 129(S2): pp.541–552.

Fox, A.D., Francis, I.S., McCarthy, J.P., and McKay, C.R. 1992. Body mass dynamics of the Lapland Bunting *Calcarius lapponicus* in West Greenland. *Dansk Ornitologisk Forenings Tidsskrift* 86: pp.155–162.

Fox, A.D., Francis, I., Norris, D., and Walsh, A. 2019b. *Report of the 2018/19 International Census of Greenland White-fronted Geese.* Aarhus: Greenland White-fronted Goose Study and National Parks and Wildlife Service.

Fox, A.D., Gitay, H., Owen, M., Salmon, D.G, and Ogilvie, M.A. 1989. Population dynamics of Icelandic-nesting geese, 1960–1987. *Ornis Scandinavica* 20: pp.289–297.

Fox, A.D., and Leafloor, J.O. (eds.) 2018. *A global audit of the status and trends of Arctic and Northern Hemisphere Goose populations (component 2: population accounts).* Akureyri, Iceland: Conservation of Arctic Flora and Fauna International Secretariat.

Fox, A.D., and Mitchell, C.R. 1988. Migration and seasonal distribution of Gadwall from Britain and Ireland. *Wildfowl* 39: pp.145–152.

Fox, A.D., and Roderick, H.W. 1990. Wintering divers and grebes in Welsh coastal waters. *Welsh Bird Report 1989*: pp.53–60.

Fox, A.D., and Salmon, D.G. 1988. Changes in the non-breeding distribution and habitat of Pochard in Britain. *Biological Conservation* 46: pp.303–316.

Fox, A.D., and Salmon, D.G. 1989. The winter status and distribution of Gadwall in Britain and Ireland. *Bird Study* 36(1): pp.37–44.

Fox, A.D., and Stroud, D.A. 1985. The Greenland White-fronted Goose in Wales. *Nature in Wales* 4(1&2): pp.20–27.

Fox, R. 2013. The decline of moths in Great Britain: a review of possible causes. *Insect Conservation and Diversity* 6: pp.5–19.

Fox, R., Brereton, T.M., Asher, J., August, T.A., Botham, M.S., Bourn, N.A.D., Cruickshanks, K.L., Bulman, C.R., Ellis, S., Harrower, C.A., Middlebrook, I., Noble, D.G., Powney, G.D., Randle, Z., Warren, M.S., and Roy, D.B. 2015 *The state of the UK's butterflies 2015.* Butterfly Conservation and the Centre for Ecology and Hydrology, Wareham, Dorset.

Francis, I. 1992. *The birds of Llangorse Lake: factors affecting population trends of wetland species*. Sandy: RSPB report, unpublished.

Francis, I., and the Rare Breeding Birds Panel. 2020. Water Rails in the United Kingdom – an estimate of the breeding population. *British Birds* 113(2): pp.105–109.

Francis, I.S., Fox, A.D., McCarthy, J., and Mackay, C. 1991. Measurements and moult of the Lapland Bunting *Calcarius lapponicus* in west Greenland. *Ringing & Migration* 12: pp.28–37.

Fransson, T., Jansson, L., Kolehmainen, T., Kroon, C., and Wenninger, T. 2017. *EURING list of longevity records for European birds*. **https://euring.org/files/documents/EURING_longevity_list_20170405.pdf** (Accessed 03/2020).

Fraser, P., and Ryan, J. 1995. Status of the Great Grey Shrike in Britain and Ireland. *British Birds* 88(10): pp.478–484.

Fraser, P.A., and Rogers, M.J. 2006. Report on Scarce Migrant birds in Britain in 2003. *British Birds* 99(2): pp.129–147.

Fray, R., Pennington, M., Riddington, R., Meek, E., Higson, P., Forsyth, A., Leitch, A., Scott, M., Marr, T., Rheinallt, T.A., and Olofson, S. 2012. An unprecedented influx of Iceland Gulls in the northeastern Atlantic in January/February 2012. *British Birds* 105(5): pp.263–272.

Frederiksen, M., Daunt, F., Harris, M.P., and Wanless, S. 2008. The demographic impact of extreme events: stochastic weather drives survival and population dynamics in a long-lived seabird. *Journal of Animal Ecology* 77(5): pp.1,020–1,029.

Frederiksen, M., Moe, B., Daunt, F., Phillips, R.A., Barrett, R.T., Maria, I., Bogdanova, M.I., Boulinier, T., Chardine, J.W., Chastel, O., Chivers, L.S., Christensen-Dalsgaard, S., Clement-Chastel, C., Colhoun, K., Freeman, R., Gaston, A.J., González-Solís, J., Goutte, A., Gremillet, D., Guilford, T., Jensen, G.H., Krasnov, Y., Lorentsen, S.-H., Mallory, M.L., Newell, M., Olsen, B., Shaw, D., Steen, H., Strøm, H., Systad, G.H., Thorarinsson, T.L., and Anker-Nilssen, T. 2012. Multicolony tracking reveals the winter distribution of a pelagic seabird on an ocean basin scale. *Diversity and Distributions* 18: pp.530–542.

Freeman, S.N., and Crick, H.Q.P. 2003. The decline of the Spotted Flycatcher Muscicapa striata in the UK: an integrated population model. *Ibis* 145(3): pp.400–412.

Freshney, F. 2020. Science underpins the Dartmoor Moorland Bird Project. *Devon Birds* 73(1): pp.4–10.

Friggens, N.L., Hester, A.J., Mitchell, R.J., Parker, T.C., Subke, J.A., and Wookey, P.A. 2020. Tree planting in organic soils does not result in net carbon sequestration on decadal timescales. *Global Change Biology* 26: pp.5,178–5,188.

Frost, T., Austin, G., Hearn, R., McAvoy, S., Robinson, A., Stroud, D., Woodward, I., and Wotton, S. 2019a. Population estimates of wintering waterbirds in Great Britain. *British Birds* 112(3): pp.130–145.

Frost, T.M., Austin, G.E., Calbrade, N.A., Mellan, H.J., Hall, C., Hearn, R.D., Stroud, D.A., Wotton, S.R., and Balmer, D.E. 2017. *Waterbirds in the UK 2015/16: The Wetland Bird Survey*. BTO, RSPB and JNCC, in association with WWT. Thetford: British Trust for Ornithology.

Frost, T.M., Austin, G.E., Calbrade, N.A., Mellan, H.J., Hearn, R.D., Robinson, A.E., Stroud, D.A., Wotton, S.R., and Balmer, D.E. 2019b. *Waterbirds in the UK 2017/18: The Wetland Bird Survey*. Thetford: BTO/RSPB/JNCC.

Frost, T.M., Calbrade, N.A., Birtles, G.A., Mellan, H.J., Hall, C., Robinson, A.E., Wotton, S.R., Balmer, D.E., and Austin, G.E. 2020. *Waterbirds in the UK 2018/19: The Wetland Bird Survey*. Thetford: BTO/RSPB/JNCC.

Fuchs, E. 1977. Predation and anti-predator behaviour in a mixed colony of terns *Sterna* sp. and Black-Headed Gulls *Larus ridibundus* with special reference to the Sandwich Tern *Sterna sandvicensis*. *Ornis Scandinavica* 8(1): pp.17–32.

Fuller, R.J., Atkinson, P.W., Garnett, M.C., Conway, G.J., Bibby, C.J., and Johnstone, I.G. 2006. Breeding bird communities in the upland margins (ffridd) of Wales in the mid-1980s. *Bird Study* 53(2): pp.177–186.

Fuller, R.J., Baker, J.K., Morgan, R.A., Scroggs, R., and Wright, M. 1985. Breeding population of the Hobby, *Falco subbuteo* on farmland in southern Midlands of England. *Ibis* 127(4): pp.510–516.

Fuller, R.J., and Gough, S.J. 1999. Changes in sheep numbers in Britain: implications for bird populations. *Biological Conservation* 91: pp.73–89.

Fuller, R.J., Gregory, R.D., Gibbons, D.W., Marcant, J.H., Wilson, J.D., Baillie, S.R., and Carter, N. 1995. Population declines and range contractions among lowland farmland birds in Britain. *Conservation Biology* 9: pp.1,425–1,441.

Fuller, R.J., and Lloyd, D. 1981. The distribution and habitats of wintering Golden Plovers in Britain, 1977–1978. *Bird Study* 28(3): pp.169–185.

Fuller, R.J., Noble, D.G., Smith, K.W., and Vanhinsbergh, D. 2005. Recent declines in populations of woodland birds in Britain: a review of possible causes. *British Birds* 98(3): pp.116–143.

Furness, R.W. 1987. *The Skuas*. London: T&AD Poyser.

Furness, R.W., Boesman, P., and Garcia, E.F.J. 2018. *Great Skua (Catharacta skua)*. Barcelona. **https://www.hbw.com/node/53957**

Fursdon, J. 1950. 12th Annual Report of the West Wales Field Society: pp.14–15.

Gaget, E., *et al*. 2020. Benefits of protected areas for nonbreeding waterbirds adjusting their distributions under climate warming. *Conservation Biology* 27pp. **https://doi.org/10.1111/cobi.13648**

Game Conservancy Trust. 2001. *GCT Annual Review*. Fordingbridge.

Garthwaite, D., Ridley, L., Mace, A., Parrish, G., Barker, I., Rainford, J., and MacArthur, R. 2019. *Pesticide Usage Survey Report no. 284: Arable Crops in the United Kingdom 2018*. Fera Science Ltd.

Gibbons, D., and Wotton, S. 1996. The Dartford Warbler in the United Kingdom in 1994. *British Birds* 89(5): pp.203–212.

Gibbons, D.W., Amar, A., Anderson, G.Q.A., Bolton, M., Bradbury, R.B., Eaton, M.A., Evans, A.D., Grant, M.C., Gregory, R.D., Hilton, G.M., Hirons, G.J.M., Hughes, J., Johnstone, I., Newbery, P., Peach, W.J., Ratcliffe, N., Smith, K.W., Summers, R.W., Walton, P., and Wilson, J.D. 2007. *The predation of wild birds in the UK: a review of its conservation impact and management*. Sandy: RSPB Research Report no 23.

Gibbons, D.W., Avery, M.I., Baillie, S.R., Gregory, R.D., Kirby, J., Porter, R.F., Tucker, G.M., and Williams, G. 1996. Bird Species of Conservation Concern in the United Kingdom, Channel Islands and Isle of Man: revising the Red Data List. *RSPB Conservation Review* 10: pp.7–18.

Gibbons, D.W., Reid, J.B., and Chapman, R.A. 1993. *The New Atlas of Breeding Birds in Britain and Ireland: 1988–1991*. London: T&AD Poyser.

Gibbs, R.G., Small, J., and Schofeld, P. 2011. Breeding birds of the Ogwen river corridor, North Wales in 2005–07. *Birds in Wales* 8(1): pp.23–28.

Gibbs, R.G., White, T., and Anning, D. 2008. Breeding Ringed Plovers in the COS area in 2007. *Cambrian Bird Report 2007*: pp.189–90.

Gibbs, R.G., and Wiggington, M.J. 1973. A breeding bird census in a sessile oak-wood at Aber, Caernarfonshire. *Nature in Wales* 13: pp.158–162.

Gibbs, R.G., and Wood, J.B. 1974. A 1968 survey of Stonechat in Wales. *Nature in Wales* 14(1): pp.7–12.

Gilbert, G. 2002. The status and habitat of Spotted Crakes *Porzana porzana* in Britain in 1999. *Bird Study* 49(1): pp.79–86.

Gilbert, G. 2012. Grasshopper Warbler *Locustella naevia* breeding habitat in Britain. *Bird Study* 59(3): pp.303–314.

Gilbert, G., Gibbons, D.W., and Evans, J. 1998. *Bird monitoring methods*. Sandy: RSPB.

Gilbert, G., MacGillivray, F.S., Robertson, H.L., and Jonsson, N.N. 2019. Adverse effects of routine bovine health treatments containing triclabendazole and synthetic pyrethroids on the abundance of dipteran larvae in bovine faeces. *Scientific Reports* 9: p.4,315.

Gilham, M.E. 2004. *Memories of Welsh islands*. Dinefwr Publishers Ltd.

Gill, F., Donsker, D., and Rasmussen, P. (eds.) 2020. *IOC world bird list* (v. 10.2). **http://www.worldbirdnames.org**

Gill, J.A., Alves, J.A., and Gunnarsson, T.G. 2019. Mechanisms driving phenological and range change in migratory species. *Philosophical Transactions of the Royal Society B* 374(1781): 20180047.

Gillings, S., Balmer, D.E., and Fuller, R.J. 2015. Directionality of recent bird distribution shifts and climate change in Great Britain. *Global Change Biology* 21(6): pp.2,155–2,168.

Gilman, E., Chopin, F., Suuronen, P., and Kuemlangan, B. 2016. *Abandoned, lost and discarded gillnets and trammel nets: Methods to estimate ghost fishing mortality, and the status of regional monitoring and management*. Food and Agriculture Organization of the United Nations. Rome: FAO Fisheries and Aquaculture Technical Paper No. 600.

Gilmore, D. 2006. *The birds of Cardiff, a full review of all the birds recorded in Cardiff UA from the earliest historical records to 2004*. Glamorgan B.C.

Gilroy, J.J., and Lees, A.C. 2003. Vagrancy theories: are autumn vagrants really reverse migrants? *British Birds* 96(9): pp.427–438.

Giraldus Cambrensis. 1191. *Itinerarium Cambriae* (Journey through Wales).

Gooch, S., Baillie, S.R., and Birkhead, T.R. 1991. Magpie *Pica pica* and songbird populations. Retrospective investigation of trends in population density and breeding success. *Journal of Applied Ecology* 28(3): pp.1,068–1,086.

Gosler, A. 1993. *The Great Tit*. London: Hamlyn.

Gray, N., Thomas, G., Trewby, M., and Newton, S.F. 2003. The status and distribution of Choughs *Pyrrhocorax pyrrhocorax* in the Republic of Ireland 2002/03. *Irish Birds* 7: pp.147–156.

Great Britain Non-native Species Secretariat. 2019. **http://www.nonnativespecies.org/factsheet/factsheet.cfm?speciesId=3537** (Retrieved 8 August 2019).

Green, E. 2017. *Tern diet in the UK and Ireland: a review of key prey species and potential impacts of climate change*. Sandy: RSPB Report.

Green, J. 2002. *Birds in Wales 1992–2000*. Welsh Ornithological Society.

Green, J. 2005a. Autumn sea-watching at Strumble Head 1994–2003. *Welsh Birds* 4(3): pp.247–267.

Green, J. 2020a. Scarce and rare birds in Wales 2019. *Birds in Wales* 17(2).

Green, M. 1992. Migration of the Dotterel through Wales. *Welsh Bird Report* 5: pp.70–73.
Green, M. 2005b. Densities of breeding Skylarks *Alauda arvensis* in upland Wales. *Welsh Birds* 4(3): pp.219–220.
Green, M. 2007. Wales Ring Ouzel survey in 2006. *Welsh Birds* 5(1): pp.37–41.
Green, M. 2012. *Montgomeryshire Curlew Survey.* Unpublished Report to the Countryside Council for Wales (CCW).
Green, M. 2019a. *Montgomeryshire Curlew Survey.* Unpublished report to Montgomery Wildlife Trust.
Green, M. 2019b. Population trends of Merlin in the Cambrian Mountains. *Birds in Wales* 16(1): pp.24–27.
Green, M., Cross, A., and Wilson, A. 2005. *Little Owls and Storm Petrels on Skomer Island.* CCW Contract Science Report 674.
Green, M., and Williams, I. 1992. The status of the Chough in Wales in 1992. *Welsh Bird Report* 6: pp.77–84.
Green, R. 2020b. Corn Crake conservation. *British Birds* 113(11): pp.671–685.
Green, R.E., and Etheridge, B. 1999. Breeding success of the hen harrier *Circus cyaneus* in relation to the distribution of grouse moors and the red fox *Vulpes vulpes*. *Journal of Applied Ecology* 36(4): pp.472–483.
Green, R.E., and Pain, D.J. 2016. Possible effects of ingested lead gunshot on populations of ducks wintering in the UK. *Ibis* 158(4): pp.699–710.
Green, R.M.W., Burton, N.H.K., and Cook, A.S.C.P. 2019. *Review of the migratory movements of Shelduck to inform understanding of potential interactions with off-shore windfarms in the southern North Sea.* Thetford: BTO Research Report No. 718.
Green, S., and Walker, E. 1991. *Ice Age hunters: Neanderthals and early modern hunters in Wales.* Cardiff: National Museum of Wales.
Greenwood, J.G. 2007. Earlier laying by Black Guillemots *Cepphus grylle* in Northern Ireland in response to increasing sea-surface temperature. *Bird Study* 54(3): pp.378–379.
Gregory, R.D., Carter, S.P., and Baillie, S.R. 1997. Abundance, distribution and habitat use of breeding Goosanders *Mergus merganser* and Red-breasted Mergansers *Mergus serrator* on British rivers. *Bird Study* 44: pp.1–12.
Gregory, R.D., and Marchant, J.H. 1996. Population trends of Jays, Magpies, Jackdaws and Carrion Crows in the United Kingdom. *Bird Study* 43(1): pp.28–37.
Gribble, F.C. 1962. Census of Black-headed Gull Colonies in England and Wales 1958. *Bird Study* 9(1): pp.56–71; also Letter, *Bird Study* 9(3): p.183.
Gribble, F.C. 1976. A census of Black-headed Gull colonies. *Bird Study* 23(2): pp.135–145.
Gribble, F.C. 1983. Nightjars in Britain and Ireland in 1981. *Bird Study* 30(3): pp.165–176.
Griffin, B. 1990. *Breeding sawbills on Welsh rivers.* Sandy: RSPB unpublished report.
Griffin, B., Saxton, N., and Williams, I. 1991. *Breeding redshanks in Wales, 1991.* Sandy: RSPB unpublished report.
Griniene, R. 2020. Aquatic Warbler's population in Lithuania hits record high. Baltic Environmental Forum. **https://meldine.lt/en/aquatic-warblers-population-in-lithuania-hits-record-high**
Gronant Little Tern (Sternula albifrons) Report 2015. Denbighshire County Council Countryside Services. Unpublished.
Grove, S.J., Hope Jones, P., Malkinson, A.R., Thomas, D.H., and Williams, I. 1988. Black Grouse in Wales, spring 1986. *British Birds* 81(1): pp.2–9.
Guilford, T. 2019. The shearwater's world. *British Birds* 112(1): pp.9–25.
Guilford, T., Meade, J., Willis, J., Phillips, R.A., Boyle, D., Roberts, S., Collett, M., Freeman, R., and Perrins, C.M. 2009. Migration and stopover in a small pelagic seabird, the Manx Shearwater *Puffinus puffinus*: insights from machine learning. *Royal Society Proceedings. Biological sciences* 276: pp.1,215–1,223.
Guillemain, M., and Elmberg, J. 2014. *The Teal.* London: T&AD Poyser.
Guillemain, M., Aubry, P., Folliot, B., and Caizergues, A. 2016. Duck hunting bag estimates for the 2013/14 season in France. *Wildfowl* 66: pp.126–141.
Gullett, P., Evans, K.L., Robinson, R.A., and Hatchwell, B.J. 2014. Climate change and annual survival in a temperate passerine: partitioning seasonal effects and predicting future patterns. *Oikos* 123: pp.389–400.
Gunnarsson, T.G., and Guðmundsson, G.A. 2016. Migration and non-breeding distribution of Icelandic Whimbrels *Numenius phaeopus islandicus* as revealed by ringing recoveries. *Wader Study* 123(1): pp.44–48.
Gurney, J.H. 1913. *The Gannet: a bird with a history.* London: Witherby.
Guyomarc'h, C., Combreau, O., Pugicerver, M., Fontoura, P., Aebischer, N.J., and Wallace, D.I.M. 1998. *Coturnix coturnix* Quail. BWP Update 2: pp.27–46.
GWCT (Game & Wildlife Conservation Trust). 2013. Long-term trends in grey partridge abundance. **www.gwct.org.uk/research__surveys/species_research/birds/grey_partridge_bap_species/218.asp**
GWCT (Game & Wildlife Conservation Trust). 2018. *The Knowledge: Every Gun's Guide to conservation.* Fordingbridge.
GWCT (Game & Wildlife Conservation Trust) website. Undated (a). National Gamebag Census 1961–2011. **https://www.gwct.org.uk/research/long-term-monitoring/national-gamebag-census/bird-bags-summary-trends**
GWCT (Game & Wildlife Conservation Trust) website. Undated (b). *Pheasant.* **https://www.gwct.org.uk/game/research/species/pheasant/**
GWCT (Game & Wildlife Conservation Trust) website. Undated (c). Research & Surveys: Grey partridge *Perdix perdix.* **https://www.gwct.org.uk/research/species/birds/grey-partridge/**
GWCT (Game & Wildlife Conservancy Trust) website. Undated (d). Research & Surveys: Red-legged partridge *Alectoris rufa.* **https://www.gwct.org.uk/research/long-term-monitoring/national-gamebag-census/bird-bags-summary-trends/red-legged-partridge**
Gwyndaf, R. 1995. *Chwedlau Gwerin Cymru.* Amgueddfa Werin Cymru.
Haas, M. and Crochet, P.A. 2009. Western Palearctic list updates: Cape Petrel. *Dutch Birding* 31: pp.24–28.
Hack, P. 1991. *A survey of the Breconshire and South Cardiganshire section of the Elenydd SSI for breeding golden plover and dunlin.* Sandy: RSPB report, unpublished.
Haffield, P. 2012. Merlins in mid and South Wales. *Birds in Wales* 9(1): pp.41–49.
Hagemeijer, W.J., and Blair, M. 1997. *The EBCC Atlas of European Breeding Birds: their distribution and abundance.* London: T&AD Poyser.
Hale, W.G. 1988. *The Redshank.* Shire Natural History Series.
Hall, C.O., Crowe, O., McElwaine, G., Einarsson, Ó., Calbrade, N., and Rees, E. 2016. Population size and breeding success of the Icelandic Whooper Swan *Cygnus cygnus*: results of the 2015 International Census. *Wildfowl* 66: pp.75–97.
Hallmann, C.A., Sorg, M., Jongejans, E., Siepel, H., Hofland, N., and Schwan, H. 2017. More than 75 percent decline over 27 years in total flying insect biomass in protected areas. *PLoS One* 12(10): e0185809.
Hance, J. 2009. *Proving the 'shifting baselines' theory: how humans consistently misperceive nature.* **https://news.mongabay.com/2009/06/proving-the-shifting-baselines-theory-how-humans-consistently-misperceive-nature**
Hancock, D.A., and Urquhart, A.E. 1965. The determination of natural mortality and its causes in an exploited population of cockles (*Cardium edule* L.). *Fishery Investigations* Series II 24(2): pp.1–40.
Hancock, M., Baines, D., Gibbons, D., Etheridge, B., and Shepard, M. 1999. Status of male black grouse *Tetrao tetrix* in Britain in 1999. *Bird Study* 46(1): pp.1–15.
Hardey, J., Crick, H., Wernham, C., Riley, H., Etheridge, B., and Thompson, D. 2009. *Raptors. a field guide for surveying and monitoring.* Edinburgh: The Stationary Office.
Hardy, A.R., and Minton, C.D.T. 1980. Dunlin migration in Britain and Ireland. *Bird Study* 27(2): pp.81–92.
Hardy, E. 1941. *The birds of the Liverpool area.* Merseyside Naturalists' Association Handbooks No. 1.
Harradine, J. 1985. Duck shooting in the United Kingdom. *Wildfowl* 36: pp.81–94.
Harradine, J., Reynolds, N., and Laws, T. 1997. *Raptors and gamebirds – a survey of game managers affected by raptors.* Marford Mill, Rossett: British Association for Shooting and Conservation.
Harrap, S., and Quinn, D. 1996. *Tits, Nuthatches and Treecreepers.* London: Helm.
Harris, A., and Williams, I. 1995. *Ornithological survey of Llanbrynmair Moor: a report to RSPB (Wales) and CCW.*
Harris, M.P. 1959. The status of the Eider (Somateria mollissima) in Wales. *Nature in Wales* 5(4): pp.849–852.
Harris, M.P. 1962 Weights from five hundred birds found dead on Skomer Island in January 1962. *British Birds* 55(3): pp.97–103.
Harris, M.P. 1966. Breeding biology of the Manx Shearwater *Puffinus puffinus. Ibis* 108(1): pp.17–33.
Harris, M.P. 1970. Rates and causes of increases of some British gull populations. *Bird Study* 17(4): pp.325–335.
Harris, M.P., and Wanless, S. 1989. The breeding biology of Razorbills *Alca torda* on the Isle of May. *Bird Study* 36(2): pp.105–114.
Harris, M.P., and Wanless, S. 1991. The importance of the lesser sandeel *Ammodytes marinus* in the diet of the shag *Phalacrocorax aristotelis. Ornis Scandinavica* 22: pp.375–382.
Harris, M.P., and Wanless, S. 2011. *The Puffin.* London: T&AD Poyser.
Harris, S., Morris, P., Wray, S., and Yalden, D. 1995. *A review of British mammals: population estimates and conservation status of British mammals other than cetaceans.* Peterborough: JNCC.

Harris, S.J., Massimino, D., Balmer, D.E., Eaton, M.A., Noble, D.G., Pearce-Higgins, J.W., Woodcock, P., and Gillings, S. 2020. *The Breeding Bird Survey 2019*. Thetford: BTO Research Report 726.

Harris, S.J., Massimino, D., Eaton, M.A., Gillings, S., Noble, D.G., Balmer, D.E., Pearce-Higgins, J.W., and Woodcock, P. 2019. *The Breeding Bird Survey 2018*. Thetford: BTO Research Report 717.

Harris, S.J., Massimino, D., Newson, S.E., Eaton, M.A., Marchant, J.H., Balmer, D.E., Noble, D.G., Gillings, S., Procter, D., and Pearce-Higgins, J.W. 2016. *The Breeding Bird Survey 2015*. Thetford: BTO Research Report 673.

Harris, T., and Franklin, K. 2000. *Shrikes and Bush-shrikes*. London: Helm.

Harrison, C.J.O. 1987. Pleistocene and prehistoric birds of South-west Britain. *Proceedings of the University of Bristol Spelaeological Society* 18(1): pp.81–104.

Harrison, T. 2005. Bird populations in the broadleaved woodland of North Wales, with a focus on Pied Flycatcher *Ficedula hypoleuca*, Redstart *Phoenicurus phoenicurus* and Wood Warbler *Phylloscopus sibilatrix*. *Cambrian Bird Report* 2005: pp.165–173.

Harrisson, T.H., and Hollom, P.A.D. 1932. The Great Crested Grebe enquiry, 1931 (Part 1). *British Birds* 26(3): pp.62–92.

Harrop, A.H.J. 2002. The Ruddy Shelduck in Britain: a review. *British Birds* 95(3): pp.123–128.

Harrop, A.H.J. 2011. The Wiltshire Hawk Owl and criteria for accepting historical records. *British Birds* 104(3): pp.162–163.

Harrop, J.M. 1961. The Woodcock in Denbighshire. *Nature in Wales* 7(3): pp.79–82.

Harrop, J.M. 1970. The Chough in North Wales. *Nature in Wales* 12(2): pp.65–69.

Hatch, C. 2006. Long-eared Owls in the Gwent Uplands. *Gwent Bird Report 2006*. Vol 42. Gwent Ornithological Society.

Hatton, P.L., and Marquiss, M. 2004. The origins of moulting Goosanders on the Eden estuary. *Ringing & Migration* 22 (2): pp.70–74.

Hayashi, E., Hayakawa, M., Satou, T., and Masuda, N. 2002. Attraction of little terns to artificial roof-top breeding sites and their breeding success. *Strix* 23, pp.143–148.

Haycock, A.N. 2019a. *A review of the status of the wetland birds in the Milford Haven Waterway and Daugleddau Estuary*. Unpublished report to the Milford Haven Waterway Environmental Surveillance Group.

Haycock, B., and Hinton, G. 2010. Monitoring Stoneworts *Chara* spp. at Bosherston Lakes. Section 5: 25. Lake and Wetland Case Studies. In Hurford, C., Schneider, M., and Cowx, I., (eds.) *Conservation Monitoring in Freshwater Habitats Practical Guide and Case Studies*. Springer publication.

Haycock, R.J. 2019b. *A review of the Chough breeding population on the Castlemartin Peninsula, including an analysis of colour-ring re-sighting records*. A Report to the MOD Defence Infrastructure Organisation.

Hayhow, D., Johnstone, I., Moore, A., Mucklow, C., Stratford, A., Šúr, M., and Eaton, M. 2018c. Breeding status of Red-billed Choughs *Pyrrhocorax pyrrhocorax* in the UK and Isle of Man in 2014. *Bird Study* 65(4): pp.458–470.

Hayhow, D.B., Benn, S., Stevenson, A., Stirling-Aird, P.K., and Eaton, M.A. 2017a. Status of Golden Eagle *Aquila chrysaetos* in Britain in 2015. *Bird Study* 64(3): pp.281–294.

Hayhow, D.B., Bond, A.L., Douse, A., Eaton, M.A., Frost, T., Grice, P.V., Hall, C., Harris, S.J., Havery, S., Hearn, R.D., Noble, D.G., Oppel, S., Williams, J., Win, I., and Wotton, S. 2017b. *The state of the UK's birds 2016*. Sandy: RSPB, BTO, WWT, DAERA, JNCC, NE, NRW and SNH.

Hayhow, D.B., Eaton, M.A., Bladwell, S., Etheridge, B., Ewing, S., Ruddock, M., Saunders, R., Sharpe, C., Sim, I.M.W., and Stevenson, A. 2013.The status of the Hen Harrier, *Circus cyaneus*, in the UK and Isle of Man in 2010. *Bird Study* 60(4): pp.446–458.

Hayhow, D.B., Eaton, M.A., Stanbury, A.J., Douse, A., and Marquiss, M. 2018a. The first UK survey and population estimate of breeding Snow Bunting *Plectrophenax nivalis*. *Bird Study* 65(1): pp.36–43.

Hayhow, D.B., Ewing, S.R., Baxter, A., Douse, A., Stanbury, A., Whitfield, P., and Eaton, M.A. 2015. Changes in the abundance and distribution of a montane specialist bird, the Dotterel *Charadrius morinellus*, in the UK over 25 years. *Bird Study* 62(4): pp.443–456.

Hayhow, D.B., Johnstone, I., Lindley, P., Stratford, A., and Bladwell, S. 2018b. The breeding status of Red-billed Choughs (*Pyrrhocorax Pyrrhocorax)* in Wales in 2014. *Birds in Wales* 15(1): pp.9–20.

Hayhow, D.B., *et al.* 2016. *State of nature 2016*. The State of Nature partnership.

Hayhow, D.B., *et al.* 2019. *State of nature 2019*. The State of Nature Partnership.

Heathcote, A., Griffin, D., and Salmon, H.M. 1967. *The birds of Glamorgan*. Cardiff Naturalists' Society.

Heavisides, A., Barker, A., and Poxton, I. 2017. Population and breeding biology of Merlins in the Lammermuir Hills. *British Birds* 110(3): pp.138–154.

Heldbjerg, H., *et al.* 2019. Contrasting population trends of Common Starlings (*Sturnus vulgaris*) across Europe. *Ornis Fennica* 96(4): pp.153–168.

Henderson, I. 2009. Progress of the UK Ruddy Duck eradication programme. *British Birds* 102(12): pp.680–690.

Henderson, I., Noble, D., and Jones, K. 2017. Population estimates and habitat associations for Stonechat (*Saxicola rubicola*), Whinchat (*Saxicola rubetra*) and Wheatear (*Oenanthe oenanthe*) in Wales, 2012 and 2013. *Birds in Wales* 14(1): pp.10–19.

Heron Conservation. Undated. **https://www.heronconservation.org/herons-of-the-world/list-of-herons/squacco-heron**

Hewson, C.M., Amar, A., Lindsell, J.A., Thewlis, R.M., Butler, S., Smith, K., and Fuller, R.J. 2007. Recent changes in bird populations in British broadleaved woodland. *Ibis* 149 (Suppl. 2): pp.14–28.

Hewson, C.M., and Noble, D.G. 2009. Population trends of breeding birds in British woodlands over a 32-year period: relationship with food, habitat use and migratory behaviour. *Ibis* 151(3): pp.464–486.

Hewson, C.M., Thorup, K., Pearce-Higgins, J.W., and Atkinson, P.W. 2016. Population decline is linked to migration route in the Common Cuckoo. *Nature Communications* 7: DOI: 10.1038/ncomms12296.

Hicklin, J. 1858. *Handbook to Llandudno and its vicinity*. The Author.

Hickling, R.A.O. 1960. The coastal roosting of Gulls in England and Wales 1955–56. *Bird Study* 7(1): pp.32–54.

Hickling, R.A.O. 1977. Inland wintering of Gulls in England and Wales, 1973. *Bird Study* 24(2): pp.79–88.

Hill, D.A., and Robertson, P.A. 1988. *The Pheasant: Ecology, management and conservation*. London: Blackwell Scientific Publications.

Hinsley, S.A., Carpenter, J.E., Broughton, R.K., Bellamy, P.E., Rothery, P., Amar, A., Hewson, C.M., and Gosler, A.G. 2007. Habitat selection by Marsh Tits *Poecile palustris* in the UK. *Ibis* 149(2): pp.224–233.

Hirons, G., and Thomas, G. 1993. Disturbance in estuaries: RSPB nature reserve experience. *Wader Study Group Bulletin* 68: pp.72–78.

Historical Statistics 1974–1996. 1998. *Digest of Welsh Historical Statistics 1974–1996 Chapter 3 Agriculture*. **https://gov.wales/digest-welsh-historical-statistics**

Hoare, R.C. 1983. *The journeys of Sir Richard Colt Hoare through Wales and England, 1793–1810*. A. Sutton.

Hodges, J.E., and Haycock, R.J. 2020. *Annual surveillance of Choughs in the Pembrokeshire Coast National Park Report No. 2. Analysis of data obtained between 1992 and 2019*. A Report to the Pembrokeshire Coast National Park Authority.

Höglund, J., Piertney, S.B., Alatolo, R.V., Lindell, J., Lundberg, A., and Rintamäki, P.T. 2002. Inbreeding depression and male fitness in black grouse. *Proceedings of the Royal Society of London, Series B – Biological Sciences* 269: pp.711–715.

Holdgate, M. (ed.) 1971. *The seabird wreck in the Irish Sea, autumn 1969*. NERC Publication Series C, No. 4.

Holling, M., and the Rare Breeding Birds Panel. 2008. Rare breeding birds in the United Kingdom in 2005. *British Birds* 101(6): pp.276–316.

Holling, M., and the Rare Breeding Birds Panel. 2010. Rare breeding birds in the United Kingdom in 2007. *British Birds* 103(1): pp.2–52.

Holling, M., and the Rare Breeding Birds Panel. 2013. Rare breeding birds in the United Kingdom in 2011. *British Birds* 106(9): pp.496–554.

Holling, M., and the Rare Breeding Birds Panel. 2016. Rare breeding birds in the United Kingdom in 2014. *British Birds* 109(9): pp.491–545.

Holling, M., and the Rare Breeding Birds Panel. 2017. Non-native breeding birds in the UK. 2012–14. *British Birds* 110(2): pp.92–108.

Holling, M., and the Rare Breeding Birds Panel. 2018. Rare Breeding Birds in the UK. *British Birds* 111(11): pp.635–710.

Holling, M., and the Rare Breeding Birds Panel. 2019. Rare breeding birds in the UK in 2017. *British Birds* 112(12): pp.706–758.

Hollom, P.A.D. 1940. Report on the 1938 survey of Black-headed Gull colonies. *British Birds* 33(8): pp.202–221.

Holloway, S. 1996. *The historical atlas of breeding birds in Britain and Ireland: 1875–1900*. London: T&AD Poyser.

Holt, B. 1997. Hoopoe breeding in Wales in 1996. *Welsh Birds* 1(6): p.79.

Holt, B., and Williams, G. 2008. *The birds of Montgomeryshire*. Published by the authors.

Holt, C., and the Rarities Committee. 2017. Report on rare birds in Great Britain in 2016. *British Birds* 110(10): pp.562–631.

Holt, C., and the Rarities Committee. 2018. Report on rare birds in Great Britain in 2017. *British Birds* 111(10): pp.557–627.

Holt, C.A., French, P., and the Rarities Committee. 2019. Report on rare birds in Great Britain in 2018. *British Birds* 112(10): pp.556–626.

Holt, C., French, P., and the Rarities Committee. 2020. Report on rare birds in Great Britain in 2019. *British Birds* 113(10): pp.585–655.

Holt, C.A., Hewson, C.M., and Fuller, R.J. 2012. The Nightingale in Britain: status, ecology and conservation needs. *British Birds* 105(4): pp.172–187.

Hoodless, A.N., and Hirons, G.J.M. 2007. Habitat selection and foraging behaviour of breeding Eurasion Woodcock: a comparison between contrasting landscapes. *Ibis* 149 (Supp 2): pp.234–249.

Hoodless, A.N., Lang, D., Aebucher, N.J., Fuller, R.J., and Ewald, J.A. 2009. Densities and population estimates of breeding Eurasian Woodcock in Britain in 2003. *Bird Study* 56(1): pp.15–25.

Hopkins, A. 2001. *Grass: its production & utilization* (3rd ed.). Oxford: Blackwell.
Horwood, J.W., and Goss-Custard, J.D. 1977. Predation by the Oystercatcher, *Haematopus ostralegus (L.)*, in relation to the Cockle, *Cerastoderma edule (L.)*, fishery in the Burry Inlet, South Wales. *Journal of Applied Ecology* 14(1): pp.139–158.
Howard, C., Stephens, P.A., Pearce-Higgins, J.W., Gregory, R.D., Butchart, S.H.M., and Willis, S.G. 2020. Disentangling the relative roles of climate and land cover change in driving the long-term population trends of European Migratory Birds. *Diversity and Distributions* 26: pp.1,442–1,455.
Howells, G. 1962. The status of the Red-legged Partridge in Britain. *Game Research Association Annual Report* 2: pp.46–51.
Hudson, N., and the Rarities Committee. 2008. Report on rare birds in Great Britain in 2007. *British Birds* 101(10): pp.516–577.
Hudson, N., and the Rarities Committee. 2010. Report on rare birds in Great Britain in 2009. *British Birds* 103(10): pp.562–638.
Hudson, N., and the Rarities Committee. 2013. Report on rare birds in Great Britain in 2012. *British Birds* 106(10): pp.570–641.
Hudson, N. and the Rarities Committee. 2014. Report on rare birds in Great Britain in 2013. *British Birds* 107(10): pp.578–653.
Hudson, R. 1965. The spread of the Collared Dove in Britain and Ireland. *British Birds* 58(4): pp.105–139.
Hudson, R. 1979. Nightingales in Britain in 1976. *Bird Study* 26(4): pp.204–212.
Hughes, B., Underhill, M., and Delany, S. 1998. Ruddy ducks breeding in the United Kingdom in 1994. *British Birds* 91(8): pp.336–353.
Hughes, J., Mason, H., Bruce, M. and Sharrock, G.H. In prep. Crimes against raptors in Wales 1990–2019. *Birds in Wales*.
Hughes, J. and Carter, T. In prep. Blackcaps wintering in Wales, 1999/2000–2020/21. *Birds in Wales*.
Hughes, J., and Money, S. 2016. RSPB Conwy nature reserve: a review of the first 20 years. *Birds in Wales* 13(1): pp.73–92.
Hughes, J., Spence, I.M., and Gillings, S. 2020. Estimating the sizes of breeding populations of birds in Wales. *Birds in Wales* 17(1): pp.56–67.
Hughes, S.W.M., Bacon, P., and Flegg, J.J.M. 1979. The 1975 census of the Great Crested Grebe in Britain. *Bird Study* 26(4): pp.213–26.
Humphrey, J. 1973. Mist-netting in July. Pied Flycatcher Study Group First report pp.8–9.
Humphreys, E.M., Gillings, S., Musgrove, A., Austin, G., Marchant, J., and Calladine, J. 2016. *An update of the review on the impacts of piscivorous birds on salmonid populations and game fisheries in Scotland.* Inverness: SNH Report No. 884.
Humphreys, P.N. 1963. *The birds of Monmouthshire.* Newport: Newport Museum.
Huntley, B., Green, R.E., Collingham, Y.C., and Willis, S.G. 2007. *A climatic atlas of European breeding birds*. Barcelona: Durham University, the RSPB and Lynx Edicions.
Hurford, C. 1996. The decline of the Ring Ouzel *Turdus torquatus* breeding population in Glamorgan. *Welsh Birds* 1(3): pp.45–51.
Hurford, C., and Lansdown, P. 1995. *Birds of Glamorgan*. National Museum of Wales.
Hutchinson, C. 1989. *Birds in Ireland.* London: T&AD Poyser.
Hutchinson, C.D., and Neath, B. 1978. Little Gulls in Britain and Ireland. *British Birds* 71(12): pp.563–582.
Hutchinson, G.A. 1908. Wild birds of the neighbourhood. *Llandudno Advertiser and List of Visitors* 19 December 1908.
Imboden, C. 1974. Zug, Fremdansiedlung und Brutperiode des Kiebitz *Vanellus vanellus* in Europa. *Der Ornithologische Beobachter* 71: pp.5–134.
Ingram, G.C.S., and Salmon, H.M. 1939. The birds of Monmouthshire. *Transactions of the Cardiff Naturalists' Society* 70: pp.93–127.
Ingram, G.C.S., and Salmon, H.M. 1954. *A hand list of the birds of Carmarthenshire*. West Wales Field Society.
Ingram, G.C.S., and Salmon, H.M. 1955. *A hand list of the birds of Radnorshire*. Herefordshire Ornithological Club.
Ingram, G.C.S., and Salmon, H.M. 1957. The birds of Brecknock. *Brycheiniog* 3: pp.181–255.
Ingram, G.C.S., Salmon, H.M., and Condry, W.M. 1966. *The Birds of Cardiganshire.* West Wales Naturalists' Trust.
Ingrouille, M. 1995. *Historical Ecology of the British Flora*. London: Chapman and Hall.
IOC. 2020. See Gill *et al.* 2020.
IRBBP. 2020. Species Profile: Reed Warbler. Irish Rare Birds Breeding Panel. **http://irbbp.org/species-list/species-profiles/81_reed-warbler**
IUCN. 2020. Nature-based solutions for people and planet. **iucn.org/theme/nature-based-solutions**
Jackson, H., Strubbe, D., Tollington, S., Prys-Jones, R., Matthysen, E., and Groombridge, J.J. 2015. Ancestral origins and invasion pathways in a globally invasive bird correlate with climate and influences from bird trade. *Molecular Ecology* 24: pp.4,269–4,285.
Jacobsen, J.R. 1963. Laplandsværlingens træk og overvintring. *Dansk Ornitologisk Forenings Tidsskrift* 57: pp.181–220. [In Danish]
James, D. 2000. Collective ground feeding of Welsh Buzzards. *Welsh Birds* 2(5): p.292.
James, J. 2015. *A bird observatory is born: the edited diaries of the Bardsey Bird and Field Observatory 1953 to 1955*. Privately published.
James, P.C. 1986. Little Shearwaters in Britain and Ireland. *British Birds* 79(1): pp.28–33.
James, P.C., and Alexander, M. 1984. Madeiran Little Shearwater *Puffinus assimilis baroli* prospecting on Skomer Island, UK. *Ardea* 73: pp.105–106.
James, W.E. 1899. *Guide book to Cardigan and district*. M.M. & W.R. Thomas.
Jardine, D.C. 1992. Crossbills in Scotland 1990 – an invasion year. *Scottish Bird Report* 1990: pp.65–69.
Jeffs, C., Croft, S., Bradbury, A., Grice, P., and Wotton, S. 2018. The UK cirl bunting population exceeds one thousand pairs. *British Birds* 111(3): pp.144–156.
Jenkins, D. 2000. Hawk and hound: hunting in the laws of court. In Charles-Edwards, T.M., Owen, M.E., and Russell, P., (eds.) *The Welsh king and his court.* Cardiff: University of Wales Press, pp.255–280.
Jenkins, R.K.B., Buckton, S.T., and Ormerod, S.J. 1995. Local movements and population density of Water Rails *Rallus aquaticus* in a small inland reedbed. *Bird Study* 42(1): pp.82–87.
Jenkins, R.K.B., and Ormerod, S.J., 2002. Habitat preferences of breeding Water Rail *Rallus aquaticus*. *Bird Study* 49(1): pp.2–10.
Jenks, P. 2021. *A Summary of Barn Owl Surveys in Pembrokeshire 2003–20.* Unpublished Report to the Pembrokeshire Nature Partnership.
Jenks, P., Knight, T., and Hodges, J.E. 2012. A Survey of Breeding Whinchats in Pembrokeshire 2012. *Pembrokeshire Bird Report*: pp.56–58.
Jenni, L. 1987. Mass concentrations of Bramblings *Fringilla montifringilla* in Europe 1900–1983: their dependence upon beech mast and the effect of snow-cover. *Ornis Scandinavica* 18(2): p.84.
Jennings, P. 1991. *A part-survey of the Elenydd uplands for dunlin and golden plover territories*. Sandy: RSPB report, unpublished.
Jennings, P. 2014. *The birds of Radnorshire*. Leominster: Ficedula Books.
Jensen, P.H. 2009. *European Union management plan for Scaup Aythya marila 2009–2011.* European Comission.
Jiguet, F., Gadot, A.-S., Julliard, R., Newson, S.E., and Couvet, D. 2007. Climate envelope, life history traits and the resilience of birds facing global change. *Global Change Biology* 13: pp.1,672–1,684.
Jiguet, F., Gregory, R.D., Devictor, V., Green, R.E., Voříšek, P., Van Strien, A., and Couvert, D. 2009. Population trends of European common birds are predicted by characteristics of their climatic niche. *Global Change Biology* 16: pp.497–505.
JNCC. 2016. *Seabird population trends and causes of change: 1986–2015.* Report (**http://jncc.defra.gov.uk/page-3201**). Joint Nature Conservation Committee. Updated September 2016.
JNCC. 2020a. *Seabird Monitoring Programme.* **https://jncc.gov.uk/our-work**
JNCC. 2020b. Seabird population trends and causes of change: 1986–2018 report (**https://jncc.gov.uk/our-work/smp-report-1986-2018**). Joint Nature Conservation Committee, Peterborough. Updated 10 March 2020.
John, A.W.G., and Roskell, J. 1985. Jay movements in autumn 1983. *British Birds* 78(12): pp.611–637.
Johnsgard, P.A. 1981. *The plovers, sandpipers and snipes of the world.* Lincoln, USA: University of Nebraska Press.
Johnson, C. 1984. County bird recording in Wales. *Nature in Wales* New Series 2(1/2): pp.57–62.
Johnson, G. 1985. The breeding areas of the Lavan Sands Redshank. *SCAN Ringing Group Report 1983–84*: pp.32–34.
Johnson, I.G. 1970. The Water Pipit as a winter visitor to the British Isles. *Bird Study* 17(4): pp.297–319.
Johnson, T. 1641. *Mercurii botanici pars altera sive plantarum gratia suscepti itineris in Cambrian sive Walliam description.* London: Cotes. **https://www.biodiversitylibrary.org/item/188095#page/5/mode/1up**
Johnston, A., Robinson, R.A., Gargallo, G., Julliard, R., van der Jeugd, H., and Baillie, S.R. 2016. Survival of Afro-Palaearctic passerine migrants in western Europe and the impacts of seasonal weather variables. *Ibis* 158(3): pp.465–480.
Johnstone, I., and Bladwell, S. 2016. Birds of Conservation Concern in Wales 3: the population status of birds in Wales. *Birds in Wales* 13(1): pp.3–31.
Johnstone, I., Dodd, S., and Peach, W.J. 2019. Seeded ryegrass fills the late winter "hungry gap" but fails to enhance local population size of seed-eating farmland birds. *Agriculture, Ecosystems & Environment* 285: 106619.
Johnstone, I., and Dyda, J. 2010. Patterns in Golden Plover and Dunlin abundance over 25 years in relation to management of the Elenydd, mid-Wales. *Birds in Wales* 7(1): pp.13–38.

Johnstone, I., Dyda, J., and Lindley, P. 2007a. The population and hatching success of Curlews in Wales in 2006. *Welsh Birds* 5(1): pp.78–87.

Johnstone, I., Elliot, D., Mellenchip, C., and Peach, W.J. 2017. Correlates of distribution and nesting success in a Welsh upland Eurasian Curlew *Numenius arquata* population. *Bird Study* 64(4): pp.535–544.

Johnstone, I., Scott, D., and Webster, L. 2010b. On the brink: the breeding population status of Turtle Doves and Corn Buntings in Wales. *Birds in Wales* 7(1): pp.92–99.

Johnstone, I., Stratford, A., Roberts, D., Lindley, P., Lamacraft, D., and Jones, K. 2011. The status, ecology, movements and conservation of Twite *Carduelis flavirostris* in Wales. *Welsh Birds* 8(1): pp.35–59.

Johnstone, I., Thorpe, R., Moore, A., and Finney, S. 2007b. Breeding status of Choughs *Pyrrhocorax pyrrhocorax* in the UK and Isle of Man in 2002. *Bird Study* 54(1): pp.23–34.

Johnstone, I., Young, A., and Thorpe, R.I. 2010a. The revised population status of birds in Wales. *Birds in Wales* 7(1): pp.39–91.

Johnstone, I.G., Dyda, J., and Lindley, P. 2008. The population status of breeding Golden Plover in Wales in 2007. *Welsh Birds* 5(4): pp.300–310.

Johnstone, I.G., Thorpe, R.I., Taylor, R., and Lamacraft, D. 2012. *The state of birds in Wales 2012*. Cardiff: RSPB Cymru.

Jones, J. 2019. The rise and fall of Ring-billed Gull. *Bird Guides*. **www.birdguides.com/articles/britain-ireland/the-rise-and-fall-of-ring-billed-gull**

Jones, K.H., Spence, I.M., and Stratford, A. 1995. Mute Swans in Gwynedd. *Cambrian Bird Report 1995*: pp.84–87.

Jones, L.P., Turvey, S.T., Massimino, D., and Papworth, S.K. 2020. Investigating the implications of shifting baseline syndrome on conservation. *People and Nature* 2(4): pp.1131–1144.

Jones, P. 1981. The geography of Dutch Elm Disease. *Transactions of the Institute of British Geographers* 6(3): pp.324–336.

Jones, P.H. 1974a. *Birds of Merioneth*. Penrhyndeudraeth: Cambrian Ornithological Society.

Jones, P.H. 1974b. Wildlife records from Merioneth parish documents. *Nature in Wales* 14: pp.35–43.

Jones, P.H. 1975. Winter Bird Populations in a Merioneth Oakwood. *Bird Study* 22(1): pp.25–34.

Jones, P.H. 1979. Ring Ouzel *Turdus torquatus* territories in the Rhinog Hills of Gwynedd. *Nature in Wales* 16(4): pp.267–269.

Jones, P.H. 1989. The chequered history of the Black Grouse in Wales. *Welsh Bird Report 3*: pp.70–78.

Jones, P.H., and Dare, P. 1976. *Birds of Caernarvonshire*. Cambrian Ornithological Society.

Jones, P.H., and Roberts, J.L. 1983. Birds of Denbighshire. *Nature in Wales* New Series 1(2): pp.56–65.

Jones, P.H., and Whalley, P. 2004. *Birds of Anglesey = Adar Môn*. Llangefni: Menter Môn.

Jones, R.W. 1910. Bird-life in Creuddyn in 1909. *Llandudno Advertiser and List of Visitors* 1st January 1910.

Jones, T. 1986. Gwylanod ac adar eraill y dŵr ar Lyn Trawsfynydd. *Y Naturiaethwr* 15: pp.12–16.

Kaiser, M.J., Bullimore, B., Newman, P., Lock, K., and Gilbert, S. 1996. Catches in 'ghost fishing' set nets. *Marine Ecology Progress Series* 145: pp.11–16.

Kaiser, M.J., Elliot, A.J., Galanidi, M., Rees, E.I.S., Caldow, R.W.G., Stillman, R.A., Showler, D.A., and Sutherland, W.J. 2002. *Predicting the displacement of common scoter* Melanitta nigra *from benthic feeding areas due to offshore windfarms*. Final report – Executive Summary. Bangor: Centre for Applied Marine Sciences, School of Ocean Sciences, University of Wales.

Kaiser, M.J., Galanidi, M., Showler, D.A., Elliot, A.J., Caldow, R.W.G., Rees, E.I.S., Stillman, R.A., and Sutherland, W.J. 2006. Distribution and behaviour of Common Scoter *Melanitta nigra* relative to prey resources and environmental parameters. *Ibis* 148 (Suppl 1): pp.110–128.

Kalyakin, M., and Volzit, O.V. (eds.) 2020. *Atlas gnezdyashchikhsya ptits evropeyskoy chasti Rossii* [*Atlas of the Breeding Birds of European Russia*]. Moscow: Fiton XXI.

Keller, I., Korner-Nievergelt, F., and Jenni, L., 2009. Within-winter movements: a common phenomenon in the Common Pochard *Aythya ferina*. *Journal of Ornithology* 150: pp.483–494.

Keller, V., Herrando, S., Voříšek, P., Franch, M., Kipson, M., Milanesi, P., Martí, D., Anton, M., Klvaňová, A., Kalyakin, M.V., Bauer, H.-G., and Foppen, R.P.B. 2020. *European breeding bird atlas 2: distribution, abundance and change*. Barcelona: European Bird Census Council & Lynx Edicions.

Kenward, R. 2006. *The Goshawk*. London: T&AD Poyser.

Kenward, R.E., Marquiss, M., and Newton, I. 1981. What happens to goshawks trained for falconry? *Journal of Wildlife Management* 45: pp.803–806.

Kern, M., Slater, F., and Cowie, R. 2014. Return rates and dispersal distances of Welsh Pied Flycatchers *Ficedula hypoleuca* and factors that influence them. *Ringing & Migration* 29(1): pp.1–9.

Kerr, W.J. 1873. Ornithological notes from North Wales for the summer and autumn of 1872. *Zoologist* 2nd series 8: pp.3,409–3,411.

Kettel, E.F., Woodward, I.D., Balmer, D.E., and Noble, D.G. 2020. Using citizen science to assess drivers of Common House Martin *Delichon urbicum* breeding performance. *Ibis*. **https://doi.org/10.1111/ibi.12888**

Kingma, S.A., Komdeur, J., Burke, T., and Richardson, D.S. 2017. Differential dispersal costs and sex-biased dispersal distance in a cooperatively breeding bird. *Behavioural Ecology* 28(4): pp.1,113–1,121.

Kington, J.A. 2010. *Climate and weather*. London: Collins.

Kirby, W.B., Bellamy, P.E., Stanbury, A.J., Bladon, A.J., Grice, P.V., and Gillings, S. 2015. Breeding season habitat associations and population declines of British Hawfinches *Coccothraustes coccothraustes*. *Bird Study* 62(3): pp.348–357.

Knox, A., Helbig, A.J., Parkin, D.T., and Sangster, G. 2001. The taxonomic status of Lesser Redpoll. *British Birds* 94(6): pp.260–267.

Kober, K., Wilson, L.J., Black, J., O'Brien, S., Allen, S., Win, I., Bingham, C., and Reid, J.B. 2012. *The identification of possible marine SPAs for seabirds in the UK: The application of Stage 1.1–1.4 of the SPA selection guidelines*. Peterborough: JNCC Report No. 461.

Lack, P.C. 1986. *The atlas of wintering birds in Britain and Ireland*. Calton: T&AD Poyser.

Lampila, S.M., Orell, M., Belda, E.J., and Koivula, K. 2016. Importance of adult survival, local recruitment and immigration in a declining boreal forest passerine, the Willow Tit *Parus montanus*. *Oecologia* 148: pp.405–413.

Langslow, D.R. 1977. Movements of Black Redstarts between Britain and Europe as related to occurrences at observatories. *Bird Study* 24(3): pp.169–178.

Langston, R., Gregory, R., and Adams, R. 2002. The status of the Hawfinch in the UK 1975–1999. *British Birds* 95(4): pp.166–173.

Langston, R.H.W. 2010. *Offshore wind farms and birds at sea: Round 3 zones, extensions to Round 1 & Round 2 sites, & Scottish Territorial Waters*. Sandy: RSPB Research Report No. 39.

Langston, R.H.W., Smith, T., Brown, A.F., and Gregory, R.D. 2006. Status of breeding Twite *Carduelis flavirostris* in the UK. *Bird Study* 53(1): pp.55–63.

Langston, R.H.W., Wotton, S.R., Conway, G.J., Wright, L.J., Mallord, J.W., Currie, F.A., Drewitt, A.L., Grice, P.V., Hoccom, D.G., and Symes, N. 2007. Nightjar *Caprimulgus europaeus* and Woodlark *Lullula arborea* – recovering species in Britain? *Ibis* 149(2): pp.250–260.

Lawicki, L. 2014. The Great White Egret in Europe: population increase and range expansion since 1980. *British Birds* 107(1): pp.8–25.

Lawson, B., Lachish, S., Colvile, K.M., Durrant, C., Peck, K.M., Toms, M.P., Sheldon, B.C., and Cunningham, A.A. 2012a. Emergence of a novel avian pox disease in British tit species. *PLoS ONE* 7(11): e40176.

Lawson, B., Robinson, R.A., Colvile, K.M., Peck, K.M., Chantrey, J., Pennycott, T.W., Simpson, V.R., Toms, M.P., and Cunningham, A.A. 2012b. The emergence and spread of finch trichomonosis in the British Isles. *Philosophical Transactions of the Royal Society B* 367: pp.2,852–2,863.

Lawson, B., Robinson, R.A., Toms, M.P., Risely, K., MacDonald, S., and Cunningham, A.A. 2018. Health hazards to wild birds and risk factors associated with anthropogenic food provisioning. *Philosophical Transactions of the Royal Society B* 373: 20170091.

Lawson, J., Kober, K., Win, I., Allcock, Z., Black, J., Reid, J.B., Way, L., and O'Brien, S.H. 2016a. *An assessment of the numbers and distribution of wintering waterbirds and seabirds in Liverpool Bay/Bae Lerpwl area of search*. Peterborough: JNCC Report No 576.

Lawson, J., Kober, K., Win, I., Allcock, Z., Black, J., Reid, J.B., Way, L., and O'Brien, S.H. 2016b. *An assessment of the numbers and distribution of wintering red-throated diver, little gull and common scoter in the Greater Wash*. Peterborough: JNCC Report No 574.

Lawson, J., Kober, K., Win, I., Allcock, Z., Black, J., Reid, J.B., Way, L., and O'Brien, S.H. 2016c. *An assessment of the numbers and distribution of little gull* Hydrocoloeus minutus *and great cormorant* Phalacrocorax carbo *over winter in the Outer Thames Estuary*. Peterborough: JNCC Report No 575.

Lawton, J.H., Brotherton, P.N.M., Brown, V.K., Elphick, C., Fitter, A.H., Forshaw, J., Haddow, R.W., Hilborne, S., Leafe, R.N., Mace, G.M., Southgate, M.P., Sutherland, W.J., Tew, T.E., Varley, J., and Wynne, G.R. 2010. *Making Space for Nature: a review of England's wildlife sites and ecological network*. Report to Defra.

Le Gouar, P.J., Schekkerman, H., van der Jeugd, H.P., Boele, A., van Harxen, R., Fuchs, P., Stroeken, P., and van Noordwijk, A.J. 2011. Long-term trends in survival of a declining population: the case of the Little Owl (*Athene noctua*) in the Netherlands. *Oecologia* 166: pp.369–379.

Leclère, D., *et al*. 2020. Bending the curve of terrestrial biodiversity needs an integrated strategy. *Nature* 585: pp.1–6.

Lees, A., and Gilroy, J. 2004. Pectoral Sandpipers in Europe: vagrancy patterns and the influx of 2003. *British Birds* 97(12): pp.638–646.

Lehikoinen, A., Jaatinen, K., Vähätalo, A., Clausen, P., Crowe, C., Deceuninck, B., Hearn, R., Holt, C.A., Hornman, M., Keller, V., Nilsson, L., Langendoen, T., Tománková, I., Wahl, J., and Fox, A.D. 2013. Rapid climate driven shifts in winter distributions of three common waterbird species. *Global Change Biology* 19: pp.2,071–2,081.

Leinaas, H.P., Jalal, M., Gabrielsen, T.M., and Hessen, D.O. 2016. Inter- and intraspecific variation in body- and genome size in calanoid copepods from temperate and arctic waters. *Ecology and Evolution* 6(16): pp.5,585–5,595.

Leland, J. 1769. *The itinerary of John Leland the antiquary, by Thomas Hearne*. Vol. 4. Oxford: James Fletcher.

Lensink, R., Ottens, G., and van der Have, T. 2013. *Vreemde vogels in de Nederlandse vogelbevolking: een verhaal van vestiging en uitbreiding*. Bureau Waardenburg bv.

Lever, C. 1979. *The naturalised animals of the British Isles*. London: Harper Collins.

Lever, C. 2009. *The naturalized animals of Britain and Ireland*. London: New Holland.

Leversley, P., Williams, I., Griffin, B., and Brooks, D. 1990. *Breeding dunlin and golden plover on the Elan Uplands*. Sandy: RSPB, unpublished report.

Lewis, A.J.G., Amar, A., Charman, E.C., and Stewart, F.R.P. 2009b. The decline of the Willow Tit in Britain. *British Birds* 102(7): pp.386–393.

Lewis, A.J.G., Amar, A., Cordi-Piec, D., and Thewlis, R.M. 2007. Factors influencing Willow Tit *Poecile montanus* site occupancy: a comparison of abandoned and occupied woods. *Ibis* 149(2): pp.205–213.

Lewis, A.J.G., Amar, A., Daniells, L., Charman, E.C., Grice, P., and Smith, K. 2009a. Factors influencing patch occupancy and within-patch habitat use in an apparently stable population of Willow Tits *Poecile montanus kleinschmidti* in Britain. *Bird Study* 56(3): pp.326–337.

Lewis, J. 2018. The Hawfinch population in the Forest of Dean/Wye Valley. *British Birds* 111(3): pp.168–169.

Lewis, J.M. 1993. Excavations at Loughor Castle, West Glamorgan. 1969–73. *Archaeologia Cambrensis* 142: pp.99–181.

Lewis, J.M.S. and Haycock, R.J. 2019. *Llangorse Ringing Group Bulletin No. 41.*

Lewis, S. 1833. *A topographical dictionary of Wales*. Vol 2. S. Lewis and Co.

Lifjeld, J.T. 2015. When taxonomy meets genomics: lessons from a common songbird. *Molecular Ecology* 24: pp.2,901–2,903.

Lindley, P. 2018. Curlew conservation action in Wales. Presentation at the Curlew workshop on 12 September 2018. **www.daera-ni.gov.uk/publications/curlew-conservation-action-wales-curlew-workshop-presentation**

Lindley, P., Johnstone, I., and Thorpe, R. 2003. The status of black grouse in Wales in 2002 and evidence for population recovery. *Welsh Birds* 3(5): pp.318–329.

Lindley, P., Young, A., and Thorpe, R.I. 2006. Status of Hen Harrier *Circus cyaneus* in Wales in 2004. *Welsh Birds* 4(5): pp.415–422.

Lindley, P.J., and Jenkins, Z. 1991. *Status, distribution and prey selection of the Goshawk* Accipiter gentilis *in commercial upland forests in North Wales*. Sandy: Unpublished RSPB Report.

Lindley, P.J., Johnstone, I., Mellenchip, C., and Young, A. 2007. The status of male black grouse *Tetrao tetrix* in Wales in 2005, and comparison with previous surveys. *Welsh Birds* 5(1): pp.42–50.

Linnard, W. 1982. *Welsh woods and forests: history and utilization*. National Museum of Wales.

Lloyd, B. 1929–1939. *Diaries of Bertram Lloyd*. National Museum of Wales.

Lloyd, B. 1934. Birds of Bardsey Island. *North Western Naturalist* 9: pp.331–334.

Lloyd, C. 1974. Movement and survival of British Razorbills. *Bird Study* 21(2): pp.102–116.

Lloyd, C.S., and Perrins, C.M. 1970. Survival and age at first breeding in the Razorbill (*Alca torda*). *Bird-Banding* 48: pp.239–252.

Lloyd, C.S., Tasker, M.L., and Partridge, K. 1991. *The status of seabirds in Britain and Ireland*. London: T&AD Poyser.

Lloyd, J., and Friese, J. 2013. Little Ringed Plovers in southwest Wales. *British Birds* 106(1): pp.30–35.

Lloyd, J.V., Facey, R., and Friese, J. 2009. The Tree Sparrow *Passer montanus* in the Tywi valley: observations on an isolated population. *Welsh Birds* 6(1): pp.3–8.

Lloyd, J.V., Powell, D.M., and Roberts, D.H.V. 2015, unpublished. *Birds of Carmarthenshire.*

Lock, L., and Cook, K. 1998. The Little Egret in Britain: a successful colonist. *British Birds* 91(7): pp.273–280.

Locke, G.M.L. 1987. *Census of woodlands and trees 1979–82.* Forestry Commission Bulletin 63. HMSO.

Lockley, A. 2013. *Island child: my life on Skokholm with R.M. Lockley.* Gwasg Carreg Gwalch.

Lockley, R.M. 1930. On the breeding habits of the Manx Shearwater with special reference to the incubation- and fledging periods. *British Birds* 23(8): pp.202–218.

Lockley, R.M. 1936. Skokholm Bird Observatory. *British Birds* 29(8): pp.222–235.

Lockley, R.M. 1942. *Shearwaters*. London: Dent.

Lockley, R.M. 1953. *Puffins*. London: J.M. Dent.

Lockley, R.M. 1957. *Pembrokeshire*. Robert Hale.

Lockley, R.M. 1961. The birds of the south-western peninsula of Wales. *Nature in Wales* 7(4): pp.124–134.

Lockley, R.M., Ingram, C.S., and Salmon, H.M. 1949. *The birds of Pembrokeshire*. West Wales Field Society.

Lockley, R.M., and Saunders, D.R. 1967. Middleholm, Pembrokeshire. *Nature in Wales* 10: pp.146–151.

Love, J.A. 1988. *The reintroduction of the white-tailed sea eagle to Scotland: 1975–1987.* Research and survey in nature conservation series. Nature Conservancy Council.

Lovegrove, R. 1990. *The Kite's tale: the story of the Red Kite in Wales*. Sandy: RSPB.

Lovegrove, R. 2007. *Silent fields: the long decline of a nation's wildlife.* Oxford: Oxford University Press.

Lovegrove, R., Hume, R.A., and McLean, I. 1980. The status of breeding wildfowl in Wales. *Nature in Wales* 17(1): pp.4–10.

Lovegrove, R., Shrubb, M., and Williams, I. 1995. *Silent fields: the current status of farmland birds in Wales*. Sandy: RSPB.

Lovegrove, R., Williams, G., and Williams, I. 1994. *Birds in Wales*. London: T&AD Poyser.

Loxton, D., and Jones, P.H. 1999. The breeding birds of Bardsey, Skomer, Skokholm and the Calf of Man. *BBFO Report* 38: pp.84–159.

Loxton, D., Kittle, T., and Jones, P.H. 1999. *Atlas of recoveries of birds ringed by Bardsey observatory 1953–1996*. Bardsey Bird and Field Observatory.

Ludwig, G.X., Alatalo, R.V., Helle, P., Lindén, H., Lindström, J., and Siitari, H. 2006. Short- and long-term population dynamical consequences of asymmetric climate change in black grouse. *Proceedings of the Royal Society of London, series B* 273: pp.2,009–2,016.

Lund, M., Raundrup, K., Westergaard-Nielsen, A., López-Blanco, E., Nymand, J., and Aastrup, P. 2017. Larval outbreaks in West Greenland: Instant and subsequent effects on tundra ecosystem productivity and CO2 exchange. *Ambio* 46 (Suppl 1): pp.26–38.

Lundberg, A., and Alatalo, R.V., 1992. *The Pied Flycatcher.* London: T&AD Poyser.

Lusby,J. 2018. **https://www.researchgate.net/publication/340285154_Barn_Owls_and_Major_Roads_IENE_Conference_September_2018**

MacDonald, M.A., Angell, R., Dines, T.D., Dodd, S., Haysom, K.A., Hobson, R., Johnstone, I.G., Matthews, V., Morris, A.J., Parry, R., and Shellswell, C.H. 2019. Have Welsh agri-environment schemes delivered for focal species? Results from a comprehensive monitoring programme. *Journal of Applied Ecology* 56(4): pp.812–823.

MacDonald, M.A., and Bolton, M. 2008. Predation of Lapwing *Vanellus vanellus* nests on lowland wet grassland in England and Wales: effects of nest density, habitat and predator abundance. *Journal of Ornithology* 149: pp.555–563.

MacKinnon, G.E., and Coulson, J.C. 1987. The temporal and geographical distribution of Continental Black-headed Gulls *Larus ridibundus* in the British Isles. *Bird Study* 34(1): pp.1–9.

Mackrill, T. 2019. *RSPB Spotlight Ospreys*. London: Bloomsbury Wildlife.

Madden, B., and Ruttledge, R.F. 1993. Little Gulls in Ireland 1979–1991. *Irish Birds* 5: pp.23–34.

Magee, J.D. 1965. The breeding distribution of the Stonechat in Britain and the causes of its decline. *Bird Study* 12(2): pp.83–89.

Mallord, J.W., Charman, E.C., Cristinacce, A., and Orsman, C.J. 2012b. Habitat associations of Wood Warblers *Phylloscopus sibilatrix* breeding in Welsh oakwoods. *Bird Study* 59(4): pp.403–415.

Mallord, J.W., Orsman, C.J., Cristinacce, A., Butcher, N., Stowe, T.J., and Charman, E.C. 2012a. Mortality of Wood Warbler *Phylloscopus sibilatrix* nests in Welsh Oakwoods: predation rates and the identification of nest predators using miniature nest cameras. *Bird Study* 59(3): pp.286–295.

Mallord, J.W., Smith, K.W., Bellamy, P.E., Charman, E.C., and Gregory, R.D. 2016. Are changes in breeding habitat responsible for recent population changes of long-distance migrant birds? *Bird Study* 63(2): pp.250–261.

Malpas, L.R., Smart, J., Drewitt, A., Sharps, E., and Garbutt, A. 2013. Continued declines of Redshank *Tringa totanus* breeding on saltmarsh in Great Britain: is there a solution to this conservation problem? *Bird Study* 60(3): pp.370–383.

Marchant, J.H., and Gregory, R.D. 1999. Numbers of nesting Rooks *Corvus frugilegus* in the United Kingdom in 1996. *Bird Study* 46(3): pp.258–273.

Marchant, J.H., Hudson, R., Carter, S., and Whittington, P. 1990. *Population trends in British breeding birds*. Tring: British Trust for Ornithology.

Marchant, J.H., and Hyde, P.A., 1979. Population changes for waterways birds, 1974–78. *Bird Study* 26(4): pp.227–238.

Marchant, J.H., and Musgrove, A.J. 2011. *Review of European flyways of the Lesser White-fronted Goose* Anser erythropus. Research Report 595. Thetford: British Trust for Ornithology.

Margrave, A.B. 2018. *How does breeding success vary between British and Icelandic race Redshank over the period 1980 to 2015*? Dissertation, University of Bangor.

Marquiss, M. 2005. *Scoping study for the possible reintroduction of Golden Eagle and White-tailed Sea Eagle to Wales*. Contract Science Report No. 692. CCW.

Marquiss, M. 2007. Seasonal pattern in hawk predation on Common Bullfinches *Pyrrhula pyrrhula*: evidence of an interaction with habitat affecting food availability. *Bird Study* 54(1): pp.1–11.

Marquiss, M., and Duncan, K. 1994. Seasonal switching between habitats and changes in abundance of Goosanders *Mergus merganser* within a Scottish river system. *Wildfowl* 45: pp.198–208.

Marquiss, M., and Newton, I. 1982. The Goshawk in Britain. *British Birds* 75(6): pp.243–260.

Marra, P.P., Cohen, E.B., Loss, S.R., Rutter, J.E., and Tonra, C.M. 2015. A call for full annual cycle research in animal ecology. *Biology Letters* 11(8): 20150552.

Marshman, P. 1979. The Wren: population and song studies. *Glamorgan Bird Report 1979*. Cardiff Naturalists' Society.

Martin, A.R. 1864. *A week's wanderings amongst the most beautiful scenery of North Wales*. New edition. North Wales Chronicle.

Martinez, N. 2012. Sparse vegetation predicts clutch size in Common Redstarts *Phoenicurus phoenicurus*. *Bird Study* 59(3): pp.315–319.

Masoero, G., Tamietti, A., Boano, G., and Caprio, E. 2016. Apparent constant adult survival of a Sand Martin *Riparia riparia* population in relation to climatic variables. *Ardea* 104: pp.253–262.

Mason, C.F. 2003. Some correlates of density in an urban Blackbird *Turdus merula* population. *Bird Study* 50(2): pp.185–188.

Mason, L.R., Bicknell, J.E., Smart, J., and Peach, W. 2020. *The impacts of non-native gamebird release in the UK: an updated evidence review.* Sandy: RSPB Centre for Conservation Science.

Massey, M. 1976a. *Birds of Breconshire: a review of status and distribution.* Brecknock Naturalists' Trust.

Massey, M. 1976b. Winter populations of waterfowl in Brecon Beacons National Park. *Nature in Wales* 15(1): pp.14–21.

Massias, A., and Becker, P.H. 1990. Nutritive value of food and growth in Common Tern *Sterna hirundo* chicks. *Ornis Scandinavica* 21: pp.187–194.

Massimino, D. and Hanmer, H. 2019. Silent Night: Tawny Owl Point Survey Monitoring. BTO News Autumn 2019. pp.10–11. **https://www.bto.org/sites/default/files/tops-article-autumn-2019-bto-news.pdf**

Massimino, D., Woodward, I.D., Hammond, M.J., Harris, S.J., Leech, D.I., Noble, D.G., Walker, R.H., Barimore, C., Dadam, D., Eglington, S.M., Marchant, J.H., Sullivan, M.J.P., Baillie, S.R., and Robinson, R.A. 2019. *BirdTrends 2019: trends in numbers, breeding success and survival for UK breeding birds.* Thetford: BTO Research Report 722.

Matheson, C. 1932. *Changes in the fauna of Wales within historic times*. National Museum of Wales.

Matheson, C. 1960. Additional gamebook records of Partridges in Wales. *British Birds* 53(2): pp.81–84.

Matheson, C. 1963. The Pheasant in Wales. *British Birds* 56(12): pp.452–455.

Mathew, M. 1894. *The birds of Pembrokeshire and its islands*. London: Porter.

Mathew, M.A. 1884. A visit to Skomer Island. *Zoologist*, 3rd series 8: p.95.

Matthews, G.V.T. 1953. Navigation in the Manx Shearwater. *Journal of Experimental Biology* 30: pp.370–396.

Mauck, R.A., Dearborn, D.C., and Huntington, C.E. 2017. Annual global mean temperature explains reproductive success in a marine vertebrate from 1955 to 2010. *Global Change Biology* 24(4): pp.1,599–1,613.

Maxwell, J. 2002. Nest-site competition with Blue Tits and Great Tits as a possible cause of declines in Willow Tit numbers: observations in the Clyde area. *Glasgow Naturalist* 24: pp.47–50.

May, C.A., Wetton, J.H., Davis, P.E., Brookfield, J.F.Y., and Parkin, D.T. 1993. Single-locus profiling reveals loss of variation in inbred populations of the red kite (*Milvus milvus*). *Proceedings of the Royal Society B: Biological Sciences* 251(1332): pp.165–170.

McCartan, L. 1958. The wreck of Kittiwakes in early 1957. *British Birds* 51(7): pp.253–266.

Mcdevitt, A.D., Kajtoch, L., Mazgajski, T.D., Carden, R.F., Coscia, I., Osthoff, C., Coombes, R.H., and Wilson, F. 2011. The origins of Great Spotted Woodpeckers *Dendrocopos major* colonizing Ireland revealed by mitochondrial DNA. *Bird Study* 58(3): pp.361–364.

McGowan, R.Y., and Kitchener, A.C. 2001. Historical and taxonomic review of the Iceland Gull *Larus glaucoides* complex. *British Birds* 94(4): pp.191–195.

McKelvie, C.L. 1986. *The book of the Woodcock*. Debrett.

McKenzie, A.J., Petty, S.J., Toms, M.P., and Furness, R.W. 2007. Importance of Sitka Spruce *Picea sitchensis* seed and garden bird-feeders for Siskins *Carduelis spinus* and Coal Tits *Periparus ater. Bird Study* 54(2): pp.236–247.

McKnight, S.K., and Hepp, G.R. 1998. Diet Selectivity of Gadwalls Wintering in Alabama. *Journal of Wildlife Management* 62: pp.1,533–1,543.

McMahon, B.J., Doyle, S., Gray, A., Kelly, S.B.A., and Redpath, S.M. 2020. European bird declines: Do we need to rethink approaches to the management of abundant generalist predators? *Journal of Applied Ecology* 57(10): pp.1,885–1,890.

Mead, C.J. 1974. The results of ringing auks in Britain and Ireland. *Bird Study* 21(1): pp.45–86.

Mead, C.J. 1984. Sand Martin slump. *BTO News* 133: p.1.

Meade, J., Hatchwell, B.J., Blanchard, J.L., and Birkhead, T.R. 2013. The population increase of Common Guillemots *Uria aalge* on Skomer Island is explained by intrinsic demographic properties. *Journal of Avian Biology* 44: pp.55–61.

Melling, T., Dudley, S., and Doherty, P. 2008. The Eagle Owl in Britain. *British Birds* 101(9): pp.478–490.

Menter Môn Morlais Limited. 2020. *Morlais Project Environmental Statement: Marine Ornithology*. Chapter 11, Vol. 1, pp.60–61.

Merilä, J., Björklund, M., and Baker, A.J. 1997. Historical demography and present day population structure of the Greenfinch, *Carduelis chloris* – an analysis of mtDNA control-region sequences. *Evolution* 51(3): pp.946–995.

Merseyside Naturalists' Association. 1958. *Bird Report 1957–58.*

Merseyside Naturalists' Association. 1962. *Bird Report 1960–62.*

Met Office Hadley Centre. 2019. *UK Climate Projections*. **https://www.metoffice.gov.uk/research/approach/collaboration/ukcp/index**

Meyer, R.M. 1991. *The feeding ecology of the red-billed chough in West Wales, and the feasibility of re-establishment in Cornwall.* PhD Thesis, University of Glasgow.

Mielcarek, R. 2019. *Birds in Avon: an annotated checklist*. Bristol Ornithological Club.

Millon, A., Petty, S.J., Little, B., Giminez, O., Cornulier, T., and Lambin, X. 2014. Dampening prey cycle overrides the impact of climate change on predator population dynamics: a long-term demographic study on tawny owls. *Global Change Biology* 20: pp.1,770–1,781.

Milwright, R.D.P. 2002. Redwing (*Turdus iliacus*) migration and wintering areas as shown by recoveries of birds ringed in the breeding season in Fennoscandia, Poland, the Baltic Republics, Russia, Siberia and Iceland. *Ringing and Migration* 21(1): pp.5–15.

Mitchell, C. 2016. *Status and distribution of Icelandic-breeding geese: results of the 2015 international census.* Slimbridge: Wildfowl & Wetlands Trust.

Mitchell, C., and Hall, C. 2020. *Greenland barnacle geese* Branta leucopsis *in Britain and Ireland: results of the International census, spring 2018*. Scottish Natural Heritage Research Report No. 1,154.

Mitchell, C., Green, M., Jones, R., Lindley, P., and Dodd, S. 2018. Year-round movements of Greenland White-fronted Geese ringed in Wales in winter 2016/17 revealed by telemetry. *Birds in Wales* 15(1): pp.38–48.

Mitchell, C., Hearn, R., and Stroud, D. 2012. The merging of populations of Greylag Geese breeding in Britain. *British Birds* 105(9): pp.498–505.

Mitchell, C., Hughes, B., and Cross, T. 2008. Goosander broods on the River Wye, 1990–2000 and a summary of Welsh ringing returns. *Welsh Birds* 5(4): pp.268–275.

Mitchell, D. 2018. *Birds of Europe, North Africa and the Middle East: an annotated checklist.* Barcelona: Lynx Edicions.

Mitchell, J.R., Moser, M.E., and Kirby, J.S. 1988. Declines in midwinter counts of waders roosting on the Dee estuary. *Bird Study* 35(3): pp.191–198.

Mitchell, P.I. 2005. Breeding seabirds in Wales. *Welsh Birds* 4(3): pp.183–218.

Mitchell, P.I., Newton, S.F., Ratcliffe, N., and Dunn, T.E. (eds.) 2004. *Seabird populations of Britain and Ireland: results of the Seabird 2000 census (1998–2002).* London: T&AD Poyser.

Monk, J.F. 1963. The past and present status of the Wryneck in the British Isles. *Bird Study* 10(2): pp.112–132.

Montagu, G. 1802. *Ornithological dictionary; or alphabetical synopsis of British birds*. J. White.

Moorcroft, D., Whittingham, M.J., Bradbury, R.B., and Wilson, J.D. 2002. The selection of stubble fields by wintering granivorous birds reflects vegetation cover and food abundance. *Journal of Applied Ecology* 39(3): pp.535–547.

Moorcroft, D., and Wilson, J.D. 2000. The ecology of Linnets *Carduelis cannabina* on lowland farmland. In Aebischer, N.J., Evans, A.D., Grice, P.V., and Vickery, J.A., (eds.) *Ecology and conservation of lowland farmland birds* (pp.173–181). Tring: BOU.

Moorcroft, D., Wilson, J.D., and Bradbury, R.B. 2006. Diet of nestling Linnets *Carduelis cannabina* on lowland farmland before and after agricultural intensification. *Bird Study* 53(2): pp.156–162.

Moore, A.S. 2017. The Manx Bird Report for 2014. *Peregrine* 10: p.5.

Moore, N.W. 1957. The past and present status of the Buzzard in the British Isles. *British Birds* 50(5): pp.173–197.

Moore, P.D. 1973. The influence of prehistoric cultures upon the initiation and spread of blanket bog in upland Wales. *Nature* 241: pp.350–353.

Moore, T.J. 1859. Occurrence of the Sand Grouse (*Syrrhaptes paradoxus*) in Wales. *Zoologist* 17: p.6,728.

Moralee, S.J. 2000. *Seabird 2000 in North Wales*. CCW Contract Science Report no. 409.

More, A.G. 1865. On the distribution of birds in Great Britain during the nesting-season. *Ibis* 7(1): pp.1–27; 7(2): pp.119–142; 7(4): pp.425–458.

Moreau, R.E. 1951. The British status of the Quail and some problems of its biology. *British Birds* 44(8): pp.257–276.

Moreau, R.E. 1956. Quail in the British Isles. 1950–53. *British Birds* 49(5): pp.161–66.

Moreau, R.E. 1972. *The Palaearctic-African Bird Migration Systems*. London: Academic Press.

Morgan, G. 2013. Ramsey's breeding birds – 40 years on. **community.rspb.org.uk/placestovisit/ramseyisland/b/ramseyisland-blog/posts/ramsey-39-s-breeding-birds-40-years-on**

Morgan, H.W., and Morgan, R. 2007. *Dictionary of Welsh place names*. Llandysul: Gomer.

Morgan, I. 1984. Birds of the Carmarthenshire coast. *Llanelli Naturalists Newsletter*, March 1984.

Morley, T.I., Fayet, A.L., Jessop, H., Veron, P., Veron, M., Clark, J.A., and Wood, M.J. 2016. The seabird wreck in the Bay of Biscay and South-Western Approaches in 2014: A review of reported mortality. *Seabird* 29: pp.22–38.

Morris, A.J., Burges, D., Fuller, R.J., Evans, A.D., and Smith, K.W. 1994. The status and distribution of Nightjars *Caprimulgus europaeus* in Britain in 1992. A report to the British Trust for Ornithology. *Bird Study* 41(3): pp.181–191.

Morris, F.O. 1857. *A history of British birds*. London: Groombridge and Son.

Morris, I. 2017. Marsh Tits (*Poecile palustris*) and Willow Tits (*P. montanus*) in 10km square SN65 in Ceredigion. *Birds in Wales* 14(1): pp.35–41.

Morrison, C.A., Robinson, R.A., Clark, J.A., Leech, D.I., and Gill, J.A. 2015. Season-long consequences of shifts in timing of breeding for productivity in Willow Warblers, *Phylloscopus trochilus*. *Bird Study* 62(2): pp.161–169.

Morrison, C.A., Robinson, R.A., Clark, J.A., Risely, K., and Gill, J.A. 2013. Recent population declines in Afro-Palaearctic migratory birds: the influence of breeding and non-breeding seasons. *Biodiversity Research* 19(8): pp.1,051–1,058.

Moss, D. 1978. Song-bird populations in forestry plantations. *Quarterly Journal of Forestry* 75: pp.5–14.

Moss, D., and Moss, G.M. 1993. Breeding biology of the Little Grebe *Tachybaptus ruficollis* in Britain and Ireland. *Bird Study* 40(2): pp.107–114.

MRG Consultant Engineers Limited. 2012. *Dublin Array: An offshore wind farm on the Kish and Bray Banks. Environmental Impact Statement January 2012 – Revision 1. Volume 2 of 5 Main Environmental Impact Statement*. Saorgus Energy, Dublin. Accessible at: **http://www.dublinarray.com/downloads/1eis/EIS-Vol-2-Main-Text.pdf [AQ-LS6]**

Mullarney, K., and Murphy, J. 2005. The Rough-legged Hawk in Ireland. *Birding World* 18: pp.503–504.

Murray, S., Harris, M.P., and Wanless, S. 2015. The status of the Gannet in Scotland in 2013–14. *Scottish Birds* 35(1): pp.3–18.

Musgrove, A., Aebischer, N., Eaton, M., Hearn, R., Newson, S., Noble, D., Parsons, M., Risely, K., and Stroud, D. 2013 Population estimates of birds in Great Britain and the United Kingdom. *British Birds* 106(2): pp.64–100.

Mustin, K., Amar, A., and Redpath, S.M. 2014. Colonization and extinction dynamics of a declining migratory bird are influenced by climate and habitat degradation. *Ibis* 156(4): pp.788–798.

Najmanová, L., and Adamík, P. 2009. Effect of climatic change on the duration of the breeding season in three European thrushes. *Bird Study* 56(3): pp.349–356.

National Gamebag Census. **https://www.gwct.org.uk/research/long-term-monitoring/national-gamebag-census/bird-bags-summary-trends**

Natural England. 2020. Natural England Licensing Statistics. **https://naturalengland.blog.gov.uk/2020/01/17/natural-englands-licensing-statistics-for-2018/**

Natural England and Countryside Council for Wales. 2009. The Seven Estuary European Marine Site Regulation 33. **https://naturalresources.wales/media/673887/severn-estuary-sac-spa-and-ramsar-reg-33-advice-from-ne-and-ccw-june-09.pdf**

Natural England and Countryside Council for Wales. 2009. Severn Estuary SAC, SPA and Ramsar Site: Regulation 33 Advice from CCW and Natural England.

Natural England and Countryside Council for Wales. 2010. *The Dee Estuary European Marine Site: comprising: Dee Estuary / Aber Dyfrdwy Special Area of Conservation, The Dee Estuary Special Protection Area, The Dee Estuary Ramsar Site: Natural England & the Countryside Council for Wales" advice given under Regulation 33(2) of the Conservation (Natural Habitats &c.) Regulations 1994*. Natural England and CCW.

Natural History Museum. 2020. UK has 'led the world' in destroying the natural environment. **https://www.nhm.ac.uk/discover/news/2020/september/uk-has-led-the-world-in-destroying-the-natural-environment.html**

Natural Resources Wales (NRW). 2016. *State of Natural Resources Report (SoNaRR): Assessment of the Sustainable Management of Natural Resources*. Technical Report. Natural Resources Wales.

Natural Resources Wales (NRW). 2018. *Carmarthen Bay and Estuaries/ Bae Caerfyrddin ac Aberoedd European Marine Site: advice provided by Natural Resources Wales in fulfilment of Regulation 37 of the Conservation of Habitats and Species Regulations 2017*. NRW.

Natural Resources Wales (NRW). 2019. Extent of Semi-Natural Habitat in Wales (Indicator 43). **https://naturalresources.wales/evidence-and-data/maps/extent-of-semi-natural-habitat-in-wales-indicator-43/?lang=en#:~:text=We%20found%20that%20in%202017,of%20healthy%20ecosystems%20in%20Wales**

Natural Resources Wales (NRW). 2020. *State of Natural Resources Report (SoNaRR)*. Natural Resources Wales.

Naylor, K.A. 2021. Historical Rare Birds. **http://www.historicalrarebirds.info**

NBN Atlas. Undated. Zebra Mussel. **http:/species.nbnatlas.org/species/NBNSYS0000006809[AQ-LS7]**

NCC. 1982. *Birds of the Elan Valley Uplands*. NCC, unpublished.

Nelson, J.B. 1978. *The Sulidae: Gannets and Boobies*. Oxford: Oxford University Press.

Nethersole-Thompson, D. 1975. *Pine Crossbills*. Calton: T&AD Poyser.

Nethersole-Thompson, D., and Nethersole-Thompson, M. 1979. *Greenshanks*. Calton: T&AD Poyser.

Nethersole-Thompson, D., and Nethersole-Thompson, M. 1986. *Breeding Waders, their breeding, haunts and watchers*. Calton: T&AD Poyser.

Newson, S., Hughes, B., Russell, I., Ekins, G., and Sellers, R. 2004. Sub-specific differentiation and distribution of Great Cormorants *Phalacrocorax carbo* in Europe. *Ardea* 92: pp.3–9.

Newson, S.E., Calladine, J., and Wernham, C. 2019. *Literature review of the evidence base for the inclusion of bird species listed on General Licences 1, 2 and 3*. Scottish Natural Heritage Research Report No. 1136.

Newson, S.E., Johnston, A., Renwick, A.R., Baillie, S.R., and Fuller, R.J. 2011. Modelling large-scale relationships between changes in woodland deer and bird populations. *Journal of Applied Ecology* 49(1): pp.278–286.

Newson, S.E., Marchant, J.H., Ekins, G.R., and Sellers, R.M. 2007. The status of inland-breeding Great Cormorants in England. *British Birds* 100(5): pp.289–299.

Newson, S.E., Moran, N.J., Musgrove, A.J., Pearce-Higgins, J.W., Gillings, S., Atkinson, P.W., Miller, R., Grantham, M.J., and Baillie, S.R. 2016. Long-term changes in the migration phenology of UK breeding birds detected by large-scale citizen science recording schemes. *Ibis* 158(3): pp.481–495.

Newson, S.E., Rexstad, E.A., Baillie, S.R., Buckland, S.T., and Aebischer, N.J. 2010. Population change of avian predators and grey squirrels in England: is there evidence for an impact on avian prey populations? *Journal of Applied Ecology* 47(2): pp.244–252.

Newton, I. 1967a. The feeding ecology of the Bullfinch (*Pyrrhula pyrrhula L.*) in southern England. *Journal of Animal Ecology* 36(3): pp.721–744.

Newton, I. 1967b. The adaptive radiation and feeding ecology of some British finches. *Ibis* 109(1): pp.33–98.

Newton, I. 1972. *Finches*. London: Collins.

Newton, I. 1973. Egg breakage and breeding failure in British Merlins. *Bird Study* 20: 241–244.

Newton, I. 1986. *The Sparrowhawk*. London: T&AD Poyser.

Newton, I. 1993. Studies of West Palearctic birds: 192. Bullfinch. *British Birds* 86(12): pp.638–648.

Newton, I. 2004. The recent declines of farmland bird populations in Britain: an appraisal of causal factors and conservation actions. *Ibis* 146(4): pp.579–600.

Newton, I. 2006. Movement patterns of Common Crossbills *Loxia curvirostra* in Europe. *Ibis* 148(4): pp.782–788.

Newton, I. 2007. *The migration ecology of birds*. London: Academic Press.

Newton, I. 2020. *Uplands and Birds*. London: HarperCollins.

Newton, I., Bogan, J., Meek, E., and Little, B. 1982a. Organochlorine compounds and shell-thinning in British Merlins *Falco columbarius*. *Ibis* 124(3): pp.328–335.

Newton, I., Davis, P.E., and Davis, J.E. 1982b. Ravens and Buzzards in relation to sheep-farming and forestry in Wales. *Journal of Animal Ecology* 19: pp.681–706.

Newton, I., Davis, P.E., and Davis, J.E. 1989. Age of first breeding, dispersal and survival of Red Kites *Milvus milvus* in Wales. *Ibis* 131(1): pp.16–21.

Newton, I., and Haas, M.B. 1984. The return of the Sparrowhawk. *British Birds* 77(2): pp.47–70.

Newton, I., Marquiss, M., Weir, D.N., and Moss, D. 1977. Spacing of Sparrowhawk nesting territories. *Journal of Animal Ecology* 46: pp.425–441.

Newton, I., Wyllie, I., and Dale, L. 1999. Trends in the numbers and mortality patterns of sparrowhawks (*Accipiter nisus*) and kestrels (*Falco tinnunculus*) in Britain, as revealed by carcass analyses. *Journal of Zoology* 248: pp.139–147.

Newton, S. 2018. Highs and lows of tern season. Wings Rockabill Summary. *Wings*: pp.22–23.

Newton, S.F., and Crowe, O. 2000. *Roseate Terns – the natural connection.* Martine Ireland/Wales INTERREG Report NO.2. Monkstown, Co. Dublin: IWC-BirdWatch Ireland

NFFN. 2020. *Nature means business: establishing the balance between food production and improving nature.* Nature Friendly Farming Network.

Nicholson, E.M. 1951. *Birds and men.* London: Collins.

Nightingale, B. 1999. The ornithological year 1998 – part one. *British Birds* 92(7): pp.354–361.

Nightingale, B. 2005. The status of Lady Amherst's Pheasant in Britain. *British Birds* 98(1): pp.20–25.

Nightingale, B., and Elkins, N. 2000. The birdwatching year 1999. *British Birds* 93(10): pp.470–487.

Nilsson, L. 2008. Changes in numbers and distribution of wintering waterfowl in Sweden during forty years, 1967–2006. *Ornis Svecica* 18: pp.135–226.

Nilsson, L., Persson, H., and Voslamber, B. 1997. Factors affecting survival of young Greylag Geese *Anser anser* and their recruitment into the breeding population. *Wildfowl* 48: pp.72–87.

Nisbet, I.C.T. 1956. Records of Wood Sandpipers in Britain in the autumn of 1952. *British Birds* 49(2): pp.49–62.

Non-native Species Secretariat 2019. **http://www.nonnativespecies.org/factsheet/factsheet.cfm?speciesId=2886**

Norman, D. 1992. The growth rate of Little Tern *Sterna albifrons* chicks. *Ringing & Migration* 13(2): pp.98–102.

Norman, D. 2008. *Birds in Cheshire and Wirral: A Breeding and Wintering Atlas.* Liverpool: Liverpool University Press.

Norman, D. 2020. Little Tern. In Keller, *European Breeding Bird Atlas 2: Distribution, Abundance and Change.*

Norris, C.A. 1945. Summary of a report on the distribution and status of the Corn-crake (*Crex crex*). *British Birds* 38(8): pp.142–148; 38(9): pp.162–168.

Norris, C.A. 1947. Report on the distribution and status of the corn-crake Part Two: a consideration of the causes of the decline. *British Birds* 40(8): pp.226–244.

Norris, C.A. 1953. The birds of Bardsey Island in 1952. *British Birds* 46(4): pp.131–137.

Norris, C.A. 1960. The breeding distribution of thirty bird species in 1952. *Bird Study* 7(3): pp.129–184.

Norris, K., Brindley, E., Cook, T., Babbs, S., Forster-Brown, C., and Yaxley, R. 1998. Is the density of Redshank *Tringa totanus* nesting on saltmarshes in Great Britain declining due to changes in grazing management? *Journal of Applied Ecology* 35(5): pp.621–634.

North, F.J., Campbell, B., and Scott, R. 1949. *Snowdonia: the National Park of North Wales.* London: Collins.

Nuijten, R.J.M., Wood, K.A., Haitjema, T., Rees, E.C., and Nolet, B.A. 2020. Concurrent shifts in wintering distribution and phenology of migratory swans: individual and generational effects. *Global Change Biology* 26(8): pp.4,263–4,275.

O'Brien, M. 2004. Estimating the number of farmland waders breeding in the United Kingdom. *International Wader Studies* 14: pp.135–139.

O'Brien, M., Green, M., Harris, A., and Williams, I. 1998. The numbers of breeding waders in Wales in 1993. *Welsh Birds* 2(1): pp.35–42.

O'Brien, S.H., Wilson, L.J., Webb, A., and Cranswick, P.A. 2008. Revised estimate of numbers of wintering Red-throated Divers *Gavia stellata* in Great Britain. *Bird Study* 55(2): pp.152–160.

O'Connor, R.J., and Mead, C.J. 1984. The Stock Dove in Britain, 1930–80. *British Birds* 77(5): pp.181–201.

O'Connor, R.J., and Shrubb, M. 1986. *Farming and birds.* Cambridge: Cambridge University Press.

O'Sullivan, J.M. 1976. Bearded Tits in Britain and Ireland, 1966–1974. *British Birds* 69(12): pp.473–489.

Ockendon, N., Hewson, C.M., Johnston, A., and Atkinson, P.W. 2012. Declines in British-breeding populations of Afro-Palaearctic migrant birds are linked to bioclimatic wintering zone in Africa, possibly via constraints on arrival time advancement. *Bird Study* 59(2): pp.111–125.

Ockendon, N., Johnston, A., and Baillie, S.R. 2014. Rainfall on wintering grounds affects population change in many species of Afro-Palaearctic migrants. *Journal of Ornithology* 155: pp.905–917.

Ockendon, N., Leech, D., and Pearce-Higgins, J.W. 2013. Climatic effects on breeding grounds are more important drivers of breeding phenology in migrant birds than carry-over effects from wintering grounds. *Biology Letters* 9(6): 20130669.

Offord, K. 2002. *Review of raptor data on Berwyn and Migneint-Dduallt SPAs.* Bangor: CCW Contract Science Report 479, Countryside Council for Wales.

Ogilvie, M., and the Rare Breeding Birds Panel. 1996. Rare breeding birds in the United Kingdom in 1993. *British Birds* 89(2): pp.61–91.

Ogilvie, M.A. 1981. The Mute Swan in Britain 1978. *Bird Study* 28(2): pp.87–106.

Ogilvie, M.A. 1986. The Mute Swan *Cygnus olor* in Britain 1983. *Bird Study* 33(2): pp.121–137.

Orford, N. 1973. Breeding distribution of the Twite in central Britain. Part 3 – range and distribution of the Twite. *Bird Study* 20(2): pp.121–126.

Ormerod, S.J., Allinson, N., Hudson, D., and Tyler, S.J. 1985. Factors influencing the abundance of breeding Dippers *Cinclus cinclus* in the catchment of the River Wye, mid Wales. *Ibis* 127(3): pp.332–340.

Ormerod, S.J., Boilstone, M.A., and Tyler, S.J. 1986. The distribution of breeding Dippers (*Cinclus cinclus* (L) Aves) in relation to stream acidity in upland Wales. *Freshwater Biology* 16: pp.501–507.

Ormerod, S.J., and Tyler, S.J. 1991. The influence of stream acidification and riparian land-use on the feeding ecology of Grey Wagtails *Motacilla cinerea* in Wales. *Ibis* 133(1): pp.53–61.

Orros, M.E., and Fellowes, M.D.E. 2015. Wild bird feeding in an urban area: intensity, economics and numbers of individuals supported. *Acta Ornithologica* 50: pp.43–58.

Orton, K. 1925. Bird life in the mountains. In Carr, H.R.C., and Lister, G.A. 1925. *The mountains of Snowdonia.* Crosby Lockwood.

Osborn, T., and Maraun, D. 2008. *Factsheet # 15: Changing intensity of rainfall over Britain.* University of East Anglia: Climate Research Unit.

Osborne, P. 1982. Some effects of Dutch elm disease on nesting farmland birds. *Bird Study* 29(1): pp.2–16.

Ottenburghs, J., Honka, J., Müskens, G.J.D.M., and Ellegren, H. 2020. Recent introgression between Taiga Bean Goose and Tundra Bean Goose results in a largely homogenous landscape of genetic differentiation. *Heredity* 125: pp.73–84.

Oudman, T., de Goeij, P., Piersma, T., and Lok, T. 2017. Colony-breeding Eurasian Spoonbills in the Netherlands: local limits to population growth with expansion into new areas. *Ardea* 105(2): pp.113–124.

Outhwaite, C.L., Gregory, R.D., Chandler, R.E., Collen, B., and Isaac, N. 2020. Complex long-term biodiversity change amongst invertebrates, bryophytes and lichens. *Nature Ecology and Evolution* 4: pp.384–392.

Ouwehand, J., Ahola, M.P., Burgess, M., and Han, S. 2015. Light-level geolocators reveal migratory connectivity in European populations of pied flycatchers *Ficedula hypoleuca*. *Journal of Avian Biology* 47(1): pp.69–83.

Owen, D. 1988. Factors affecting the status of the Chough in England and Wales; 1780–1980. In Bignal, E., and Curtis, D.J., (eds.) *Choughs and land-use in Europe.* Scottish Chough Study Group.

Owen, G. 1603. *The description of Pembrokeshire. Part 1* (ed. H. Owen 1892). London.

Owen, H.W. and Morgan, R. 2007. Dictionary of the place-names of Wales. Gomer.

Owen, M., Atkinson-Willes, G.L., and Salmon, D.G. 1986. *Wildfowl in Great Britain, 2nd ed.* Cambridge: Cambridge University Press.

Owen, M., Callaghan, D., and Kirby, J. 2006. *Guidelines on avoidance of introductions of non-native waterbird species.* AEWA Technical Series No. 12. Bonn.

Owen, M., and Mitchell, C. 1988. Movements and migrations of Wigeon *Anas penelope* wintering in Britain and Ireland. *Bird Study* 35(1): pp.47–59.

Ozarowska, A. 2012. The origin of Blackcaps *Sylvia atricapilla* wintering on the British Isles. *Ornis Fennica* 89: pp.254–263.

Papworth, S.K., Rist, J., Coad, L., and Milner-Gulland, E.J. 2009. Evidence for shifting baseline syndrome in conservation. *Conservation Letters* 2: pp.93–100.

Parish, D.M.B., and Sotherton, N.W. 2007. The fate of released captive-reared grey partridges *Perdix perdix*: implications for reintroduction programmes. *Wildlife Biology* 13: pp.140–149.

Parkin, D.T., Collinson, M., Helbig, A.J., Knox, A.G., and Sangster, G. 2003. The taxonomic status of Carrion and Hooded Crows. *British Birds* 96(6): pp.274–290.

Parkin, D.T., and Knox, A.G. 2010. *The status of birds in Britain and Ireland.* London: Helm.

Parnell, I. 1997. *A study of Honey-buzzard* Pernis apivorus *diet using faecal analysis.* Unpublished BSc paper, Cardiff University.

Parr, R. 1992. The decline to extinction of a population of Golden Plover in North-East Scotland. *Ornis Scandinavica* 23(2): pp.152–158.

Parr, S.J. 1991. Occupation of new conifer plantations by Merlins in Wales. *Bird Study* 38(2): pp.103–111.

Parrinder, E.D. 1989. Little Ringed Plovers *Charadrius dubius* in Britain in 1984. *Bird Study* 36(3): pp.147–153.

Parry, M. 2010. *Mapping a sustainable future*. Institute of Welsh Affairs. **https://www.iwa.wales/click/2010/10/mapping-a-sustainable-future-for-man-and-the-environment**

Parry, W., and Broughton, R.K. 2019. Nesting behaviour and breeding success of Willow Tits *Poecile montanus* in north-west England. *Ringing & Migration* 33(2): pp.75–85.

Parslow, J. 1973. *Breeding birds of Britain and Ireland*. Calton: T&AD Poyser.

Pavon-Jordan, D., Fox, A.D., Clausen, P., Dagys, M., Deceuninck, B., Devos, K., Hearn, R.D., Holt, C.A., Hornman, M., Keller, V., Langendoen, T., Ławicki, L., Lorentsen, S.H., Luingujoe, L., Meissner, W., Musil, P., Nilsson, L., Paquet, J.Y., Stipniece, A., Stroud, D.A., Wahl, J., Zanatello, M., and Lehikoinen, A. 2015. Climate-driven changes in winter abundance of a migratory waterbird in relation to EU protected areas. *Diversity and Distributions* 21: pp.571–582.

Peach, W., Baillie, S., and Underhill, L. 1991. Survival of British Sedge Warblers *Acrocephalus schoenobaenus* in relation to west African rainfall. *Ibis* 133(3): pp.300–305.

Peach, W., du Feu, C., and McMeeking, J. 1995. Site tenacity and survival rates of Wrens *Troglodytes troglodytes* and Treecreepers *Certhia familiaris* in a Nottinghamshire wood. *Ibis* 137(4): pp.497–507.

Peach, W.J., Siriwardena, G.M., and Gregory, R.D. 1999. Long-term changes in over-winter survival rates explain the decline of Reed Buntings *Emberiza schoeniclus* in Britain. *Journal of Applied Ecology* 36(5): pp.798–811.

Peach, W.S., and Miles, P.M. 1961. An annotated list of some birds found in the Aberystwyth district, 1946–56. *Nature in Wales* 7(1): pp.11–20.

Peakall, D.B. 1962. The past and present status of the Red-backed Shrike in Great Britain. *Bird Study* 9(4): pp.198–216.

Peal, R.E.F. 1968. The distribution of the Wryneck in the British Isles 1964–1966. *Bird Study* 15(3): pp.111–126.

Pearce-Higgins, J.W., and Crick, H.Q.P. 2019. One-third of English breeding bird species show evidence of population responses to climatic variables over 50 years. *Bird Study* 66(2): pp.159–172.

Pearce-Higgins, J.W., Eglington, S.M., Martay, B., and Chamberlain, D.E. 2015. Drivers of climate change impacts on bird communities. *Journal of Animal Ecology* 84(4): pp.943–954.

Pearce-Higgins, J.W., Grant, M.C., Robinson, M.C., and Haysom, S.L. 2007. The role of forest maturation in causing the decline of black grouse *Tetrao tetrix*. *Ibis* 149(1): pp.143–155.

Pearce-Higgins, J.W., Lindley, P., Johnstone, I., Thorpe, R., Douglas, J.T., and Grant, M.C. 2019. Site-based adaptation reduces the negative effects of weather upon a southern range margin Welsh black grouse *Tetrao tetrix* population that is vulnerable to climate change. *Climatic Change* 153: pp.253–265.

Pearce-Higgins, J.W., Wright, L.J., Grant, M.C., and Douglas, D.J.T. 2016. The role of habitat change in driving black grouse *Tetrao tetrix* population declines across Scotland. *Bird Study* 63(1): pp.66–72.

Pearce-Higgins, J.W., Yalden, D.W., and Whittingham, M.J. 2005. Warmer springs advance the breeding phenology of golden plovers *Pluvialis apricaria* and their prey (Tipulidae). *Oecologia* 143: pp.470–476.

PECBMS. 2020. *Trends of common birds in Europe*. EBCC/BirdLife International. **https://pecbms.info/trends-and-indicators/species-trends/**

Pederson, M.B. 1994. Jack Snipe *Lymnocryptes minimus*. In Tucker, G.M., and Heath, M.F., (eds.) *Birds in Europe: their conservation status. Conservation Series No. 3*. pp.266–267. Cambridge: BirdLife International.

Peers, M., and Shrubb, M. 1990. *Birds of Breconshire*. Brecknock Wildlife Trust.

Peers, M.F. 1985. *The Birds of Radnorshire and Mid-Powys*. Privately published, Llangammarch Wells.

Pembrokeshire Avifauna. 2020. **https://pembsavifauna.co.uk**

Pennant, T. 1810. *Tours in Wales*. London: Wilkie and Robinson.

Pennant, T. 1812. *British Zoology*. New Edition. London: Wilkie and Robinson.

Pentland, G.H. 1899. Crossbill in North Wales. *Zoologist* 4th Series 3: p.182.

Perrins, C. 2003. The status of Marsh and Willow Tits in the UK. *British Birds* 96(9): pp.418–426.

Perrins, C., Padget, O., O'Connell, M., Brown, R., Büche, B., Eagle, G., Roden, J., Stubbings, E. and Wood, M.J. 2020. *A census of breeding Manx Shearwaters Puffinus puffinus on the Pembrokeshire Islands of Skomer, Skokholm and Midland in 2018. Seabird* 32: pp.106–118.

Perrins, C.M. 1968. The numbers of Manx Shearwaters on Skokholm. *Skokholm Bird Observatory report for 1967*: pp.22–29.

Perrins, C.M. 1979. *British Tits*. London: Collins.

Perrins, C.M., and Brooke, M. de L. 1976. Manx Shearwaters in the Bay of Biscay. *Bird Study* 23(4): pp.295–299.

Perrins, C.M., Padget, O., O'Connell, M., Brown, R., Buche, B., Eagle, G., Roden, J., Stubbings, E., and Wood, M.J. 2019. A census of breeding Manx Shearwaters on the Pembrokeshire Islands of Skomer, Skokholm and Midland in 2018. *Seabird* 32: pp.106–118.

Perrins, C.M., and Smith, S.B. 2000. The breeding *Larus* Gulls on Skomer Island National Nature Reserve, Pembrokeshire. *Atlantic Seabirds* 2(3/4): pp.195–210.

Perrins, C.M., Wood, M.J., Garroway, C.J., Boyle, D., Oakes, N., Revera, R., Collins, P., and Taylor, C. 2012. A whole-island census of the Manx Shearwater *Puffinus puffinus* breeding on Skomer Island in 2011. *Seabird* 25: pp.1–13.

Perrins, C.M., Wood, M.J., Gilham, J., Brown, R., Eagle, G., and Taylor, C. 2017. *Manx Shearwater census on Skokholm Island 2012 and 2013*. Report to Wildlife Trust for South and West Wales.

Perrow, M.R. (ed.) 2019. *Wildlife and wind farms – conflicts and solutions. Volume 3: potential effects*. Exeter: Pelagic Publishing.

Perrow, M.R., Gilroy, J.J., Skeate, E.R., and Mackenzie, A. 2010. *Quantifying the relative use of coastal waters by breeding terns: towards effective tools for planning and assessing the ornithological impacts of offshore wind farms*. ECON Ecological Consultancy Ltd. Report to COWRIE Ltd.

Perry, R. 1940. *Lundy: Isle of Puffins*. Lindsay Drummond.

Pesendorfer, M.B., Sillett, T.S., Koenig, W.D., and Morrison, S.A. 2016. Scatter-hoarding corvids as seed dispersers for oaks and pines: a review of a widely distributed mutualism and its utility to habitat restoration. *The Condor* 118(2): pp.215–223.

Petersen, Æ. 1998. *Íslenkir fuglar*. Vaka-Helgafell, Reykjavik. [In Icelandic]

Petty, S.J., Anderson, D.I.K., Daison, M., Little, B., Sherratt, T.N., Thomas, C.J., and Lambin, X. 2003. The decline of Common Kestrels *Falco tinnunculus* in forested area of northern England: the role of predation by Northern Goshawks *Accipiter gentilis*. *Ibis* 145(3): pp.472–483.

Phillips, E.C. 1891. On the Welsh names of birds of prey. *Transactions of the Woolhope Naturalists' Field Club* 1891: pp.254–257.

Phillips, E.C. 1899. *The birds of Breconshire*. Edwin Davies.

Pierce, R. 2016. *Fight for Twite in Snowdonia*. RSPB Guest Blog. Accessed 29 June 2020. **community.rspb.org.uk/getinvolved/wales/b/wales-blog/posts/fight-for-twite-in-snowdonia-guest-blog-by-rhian-pierce**

Plant, A.R. 2014. The autumn passage of Woodpigeon (*Columba palumbus*) through Southeast Wales in 2013. *Gwent Bird Report 2013*: pp.109–118.

Plummer, K.E., Risely, K., Toms, M.P., and Siriwardena, G.M. 2019. The composition of British bird communities is associated with long-term garden bird feeding. *Nature Communications* 10: p.2,088.

Plummer, K.E., Siriwardena, G.M., Conway, G.J., Risely, K., and Toms, M.P. 2015. Is supplementary feeding in gardens a driver of evolutionary change in a migratory bird species? *Global Change Biology* 21(12): pp.4,353–4,363.

Polak, M. 2005. Temporal pattern of vocal activity of the Water Rail *Rallus aquaticus* and the Little Crake *Porzana parva* in the breeding season. *Acta Ornithologica* 20: pp.21–26.

Porter, B., and Stansfield, S. 2018. GPS tracking Manx Shearwaters (*Puffinus puffinus*) from Bardsey's breeding colony. *Birds in Wales* 15(1): pp.21–37.

Potti, J. 1998. Arrival time from spring migration in male pied flycatchers: individual consistency and familial resemblance. *The Condor* 100(4): pp.702–708.

Potts, G.R. 1978. The effects on a partridge population of predator control, insect shortages, different shooting pressures and releasing reared birds. *Game Conservancy Annual Review* 9 (1977): pp.75–83.

Potts, G.R. 2012. *Partridges*. London: Collins.

Powell, M.C. 1990. Leach's Petrels in Wales December 1989. *Welsh Bird Report* 1989: pp.36–37.

Pöysä, H., Lammi, E., Pöysä, S., and Väänänen, V.M. 2019. Collapse of a protector species drives secondary endangerment in waterbird communities. *Biological Conservation* 230: pp.75–81.

Prater, A.J. 1973. The wintering population of Ruffs in Britain and Ireland. *Bird Study* 20(4): pp.245–250.

Prater, A.J. 1976a. Breeding population of the Ringed Plover in Britain. *Bird Study* 23(3): pp.155–161.

Prater, A.J. 1976b. The midwinter estuarine population of waders in Wales, 1971–74. *Nature in Wales* 15(1): pp.2–7.

Prater, A.J. 1981. *Estuary birds of Britain and Ireland*. Calton: T&AD Poyser.

Prater, A.J. 1989. Ringed Plover *Charadrius hiaticula* breeding population of the United Kingdom in 1984. *Bird Study* 36(3): pp.154–160.

Prater, A.J., and Davies, M. 1978. Wintering Sanderlings in Britain. *Bird Study* 25(1): pp.33–38.

Prestt, I. 1965. An enquiry into the recent breeding status of some of the smaller birds of prey and crows in Britain. *Bird Study* 12(3): pp.196–221.

Price, J. 1875. *Llandudno and how to enjoy it*. Llandudno: Woodcock.

Price, R., and Robinson, J.A. 2008. The persecution of Kites and other species in 18th century Co. Antrim. *The Irish Naturalists' Journal* 29: pp.1–6.

Pringle, H., Wilson, M., Calladine, J., and Siriwardena, G.M. 2019. Associations between gamebird releases and general predators. *Journal of Applied Ecology* 56(8): pp.2,102–2,113.

Pritchard, R. 2012. *Adar Meirionnydd = Birds of Meirionnydd*. Cambrian Ornithological Society.

Pritchard, R. 2017. *Birds of Caernarfonshire*. Cambrian Ornithological Society.

Pritchard, R. 2018. Traeth Lafan. *WeBS News* 34: pp.4–5.

Pritchard, R. 2019. The status of the Dotterel (*Charadrius morinellus*) in Wales. *Birds in Wales* 16(1): pp.28–36.

Pritchard, R. 2020. The prehistoric and historic bird fauna of Wales: evidence from archaeology, literature and place names. *Birds in Wales* 17(1): pp.15–25.

Pyman, G.A. 1959. The status of the Red-crested Pochard in the British Isles. *British Birds* 52(2): pp.42–56.

Quartey, J.K., Nuoh, A.A., Emmanuel Taye, E., and Ntiamoa-Baidu, Y. 2018. *Assessment of the wintering population, conservation requirements and priority actions for Roseate tern (*Sterna dougallii*) in Ghana*. Centre for African Wetlands Final Report.

Raine, A.F., Sowter, D.J., Brown, A.F., and Sutherland, W.J. 2006. Migration patterns of two populations of Twite *Carduelis flavirostris* in Britain. *Ringing & Migration* 23(1): pp.45–52.

Raines, J. 1962. *Birds of the Wirral Peninsula 1961: the wild geese of Mersey and Dee*. The Liverpool Ornithologists Club.

Ramos, R., Ramírez, I., Paiva, V., Militão, T., Biscoito, M., Menezes, D., Phillips, R.A., Zino, F., and González-Solís, J. 2016. Global spatial ecology of three closely-related gadfly petrels. *Scientific Reports* 6: 23447.

Ratcliffe, D. 1970. Changes attributable to pesticides in egg breakage frequency and eggshell thickness in some British birds. *Journal of Applied Ecology* 7: pp.67–107.

Ratcliffe, D. 1990. *Bird life of mountain and upland*. Cambridge: Cambridge University Press.

Ratcliffe, D. 1993. *The Peregrine Falcon*. 2nd ed. London: T&AD Poyser.

Ratcliffe, D. 1997. *The Raven: a natural history in Britain and Ireland*. London: T&AD Poyser.

Ratcliffe, D.A. 1963. The status of the Peregrine in Great Britain. *Bird Study* 10(2): pp.56–90.

Ratcliffe, D.A. 1972. The Peregrine population of Great Britain in 1971. *Bird Study* 19(3): pp.117–156.

Ratcliffe, D.A. 1976. Observations on the breeding of the Golden Plover in Great Britain. *Bird Study* 23(2): pp.63–116.

Ratcliffe, D.A. 1980. *The Peregrine Falcon*. London: T&AD Poyser.

Ratcliffe, D.A. 1984. The Peregrine breeding population of the United Kingdom in 1981. *Bird Study* 31(1): pp.1–18.

Ratcliffe, N., Pickerell, G., and Brindley, E. 2000. Population trends of Little and Sandwich Terns *Sterna albifrons* and *S. sandvicensis* in Britain and Ireland from 1969 to 1998. *Atlantic Seabirds* 2: pp.211–226.

Ray, J. 1662. *Itinerary III*. Reprinted in Lankester, E. (ed.) 1846. *Memorials of John Ray, consisting of his life by Dr. Derham; biographical and critical notices by Sir J.E. Smith, and Cuvier and Dupetit Thouars*. With his itineraries, etc. Ray Society, London.

Ray, J. 1713. *Synopsis Methodica Avium*.

Ray, J., and Willughby, F. 1676. *Francisci Willughbeii Ornithologiae*. John Martyn, London.

Rebecca, G.W., and Bainbridge, I.P. 1998. The breeding status of the Merlin *Falco columbarius* in Britain in 1993–94. *Bird Study* 45(2): pp.172–187.

Redfern, C.P.F., and Bevan, R.M. 2019. Use of sea ice by arctic terns *Sterna paradisaea* in Antarctica and impacts of climate change. *Journal of Avian Biology* 51(2).

Rees, E.I.S. 1969. Guillemot kill in North Wales – autumn 1969. *Seabird Report 1969*: pp.7–11.

Rees, G. 2010a. Arctic Skua – 2007 migration. In *Pembrokeshire Avifauna*. 2020. **https://pembsavifauna.co.uk**

Rees, G. 2010b. Chaffinch – 2008 migration. In *Pembrokeshire Avifauna*. 2020. **https://pembsavifauna.co.uk**

Rees, G., Haycock, A., Haycock, B., Hodges, J., Sutcliffe, S., Jenks, P., and Dobbins, R. 2009. *Atlas of breeding birds in Pembrokeshire*. Pembrokeshire Bird Group.

Rehfisch, M.M., Allan, J.R., and Austin, G.E. 2010. *The effect on the environment of Great Britain's naturalized Greater Canada* Branta canadensis *and Egyptian Geese* Alopochen aegyptiacus. In *BOU Proceedings – The Impacts of Non-native Species*.

Reneerkens, J., Benhoussa, A., Boland, H., Collier, M., Grond, K., Gunther, K., Hallgrimsson, G.T., Hansen, J., Meissner, W., Meulenaer, B., Ntiamoa-Baidu, Y., Piersma, T., Poot, M., Roomen, M., Summers, R., Tomkovich, P., and Underhill, L. 2009. Sanderlings using African-Eurasian flyways: a review of current knowledge. *Wader Study Group Bulletin* 116: pp.2–20.

Reynardson, C.T.S. Birch. 1887. *Sports and anecdotes of bygone days: in England, Scotland, Ireland, Italy and the sunny south*. London: Chapman and Hall.

Rhind, P., and Jones, B. 2003. The Vegetation History of Snowdonia since the Late Glacial Period. *Field Studies* 10: pp.539–552.

Rhymer, C.M., Devereux, C.L., Denny, M.J.H., and Whittingham, M.J. 2012. Diet of Starling *Sturnus vulgaris* nestlings on farmland: the importance of Tipulidae larvae. *Bird Study* 59(4): pp.426–436.

Richards, C. 2009. Dotterel migration through Mynydd Du, Carmarthenshire. *Welsh Birds* 6(1): pp.27–37.

Richards, C., Dixon, A., and other members of the South Wales Peregrine Monitoring Group. 2012. Interim summary of South Wales Peregrine Monitoring Group for 2011. *Birds in Wales* 9(1): pp.22–26.

Richards, C., and King, A. 2019. Golden Plover. *Breconshire Birds Annual Report 2018* 7(3): pp.74–80.

Richmond, K. 1958. *Wild Venture*. London: Geoffrey Bles Ltd.

Riddington, R., Votier, S.C., and Steele, J. 2000. The influx of redpolls into Western Europe, 1995/96. *British Birds* 93(2): pp.59–67.

Riddington, R., and Ward, N. 1998. The invasion of Northern Bullfinches to Britain in autumn 1994, with particular reference to the Northern Isles. *Ringing & Migration* 9: pp.48–52.

Ridgill, S.C., and Fox, A.D. 1990. *Cold weather movements of waterfowl in Western Europe*. IWRB special publication 13. International Waterfowl and Wetlands Research Bureau. Slimbridge: The Wildfowl & Wetlands Trust.

Riordan, J., and Birkhead, T.R. 2018. Changes in the diet composition of Common Guillemots *Uria aalge* chicks on Skomer Island, Wales between 1973 and 2017. *Ibis* 160(2): pp.470–474.

Roberts, J.L. 1983. Whitethroats breeding on a Welsh heather moor. *British Birds* 76(10): p.456.

Roberts, J.L. 1998. Wintering Hen Harriers on a moor in North East Wales: Sex ratios, seasonality and possible links with numbers of Red Grouse. *Welsh Birds* 2(1): pp.46–53.

Roberts, J.L. 2010. Bird numbers from September to March on a grouse moor in northeast Wales 1978–2005. *Birds in Wales* 7(1): pp.100–129.

Roberts, J.L. 2014. Short and long term changes in breeding bird abundance on a grouse moor in north-east Wales, 1979–2003. *Birds in Wales* 11(1): pp.3–21.

Roberts, J.L., and Green, D. 1983. Breeding failure and decline of Merlins on a North Wales moor. *Bird Study* 30(3): pp.193–200.

Roberts, J.L., and Jones, M.S. 1999a. Merlins *Falco columbarius* on a NE Wales moor – breeding ecology (1983–1997) and possible determinants of density. *Welsh Birds* 2(3): pp.88–108.

Roberts, J.L., and Jones, M.S. 1999b. Increase of a population of Ravens in north east Wales – its dynamics and possible causation. *Welsh Birds* 2(3): pp.121–130.

Roberts, J.L., and Jones, M.S. 2003. Grey Wagtails (*Motacilla cinerea*) breeding at high density. *Welsh Birds* 3(5): pp.340–342.

Roberts, J.L., and Jones, M.S. 2004. Increase of Peregrines in the NE Wales borders, 1973–2003. *Birds in Wales* 4(1): pp.48–59.

Roberts, J.L., and Jones, M.S. 2009. Spacing and breeding production of Buzzards in north east Wales, 1978–2008. *Welsh Birds* 6(1): pp.9–26.

Roberts, J.W. 2009. *The specialized diet of the European Honey-buzzard* Pernis apivorus *in Britain*. Unpublished BSc thesis, Newcastle University.

Roberts, P. 1985. *The birds of Bardsey*. Bardsey Bird and Field Observatory.

Roberts, P. 2013. Life on the edge – the changing fortunes of Dartford Warblers in Glamorgan. *Birds in Wales* 10(1): pp.3–6.

Roberts, S., and Lewis, J. 1985. The Hobby in Gwent. *Gwent Bird Report* 21: pp.7–14.

Roberts, S., and Lewis, J. 1990. Breeding Hobbies in Gwent – a status review. *Gwent Bird Report* 26: pp.7–11.

Roberts, S., and Lewis, J. 2012. Red-backed Shrike breeding in Wales. *Birds in Wales* 9(1): pp.14–21.

Roberts, S.J., and Law, C.D. 2014. Honey-buzzards in Britain. *British Birds* 107(11): pp.668–691.

Robertson, G.S., Aebischer, N.J., and Baines, D. 2017a. Using harvesting data to examine temporal and regional variation in red grouse abundance in the British uplands. *Wildlife Biology* 4: pp.1–11.

Robertson, J.C., and Slater, F.M. 1997. Seasonal movements of Blue Tits *Parus caeruleus* and Great Tits *P. major* in Mid Wales. *Welsh Birds* 1(5): pp.38–40.

Robertson, P.A., Mill, A.C., Rushton, S.P., McKenzie, A., Sage, R.B., and Aebischer, N.J. 2017b. Pheasant release in Great Britain: long-term and large-scale changes in the survival of a managed bird. *European Journal of Wildlife Research* 63(6): p.100.

Robinson, J.A., Colhoun, K., Gudmundsson, G.A., Boertmann, D., Merne, O.J., O'Briain, M., Portig, A.A., Mackie, K., and Boyd, H. 2004a. *Light-bellied Brent Goose* Branta bernicla hrota *(East Canadian High Arctic population) in Canada, Ireland, Iceland, France, Greenland, Scotland, Wales, England, the Channel Islands and Spain 1960/61 – 1999/2000*. Waterbird Review Series, The Wildfowl & Wetlands Trust/Joint Nature Conservation Committee, Slimbridge.

Robinson, R.A. 2005. *BirdFacts: profiles of birds occurring in Britain & Ireland*. BTO, Thetford. **http://www.bto.org/birdfacts**

Robinson, R.A., Baillie, S.R., and King, R. 2012. Population processes in European Blackbirds *Turdus merula*: a state–space approach. *Journal of Ornithology* 152: pp.419–433.

Robinson, R.A., Balmer, D.E., and Marchant, J.H. 2008. Survival rates of hirundines in relation to British and African rainfall. *Ringing & Migration* 24: pp.1–6.

Robinson, R.A., Freeman, S.N., Balmer, D.E., and Grantham, M.J. 2007. Cetti's Warbler *Cettia cetti:* analysis of an expanding population. *Bird Study* 54(2): pp.230–235.

Robinson, R.A., Green, R.E., Baillie, S.R., Peach, W.J., and Thomson, D.L. 2004b. Demographic mechanisms of the population decline of the song thrush *Turdus philomelos* in Britain. *Journal of Animal Ecology* 73(4): pp.670–682.

Robinson, R.A., Lawson, B., Toms, M.P., Peck, K.M., Kirkwood, J.K., Chantrey, J., Clatworthy, I.R., Evans, A.D., Hughes, L.A., Hutchinson, O.C., and John, S.K. 2010. Emerging infectious disease leads to rapid population declines of common British birds. *PLoS one* 5(8): e12215.

Robinson, R.A., Leech, D.I., and Clark, J.A. 2020. *The Online Demography Report: bird ringing and nest recording in Britain and Ireland in 2019*. Thetford: BTO. **http://www.bto.org/ringing-report**

Robinson, R.A., Morrison, C.A., and Baillie, S.R. 2014. Integrating demographic data: towards a framework for monitoring wildlife populations at large spatial scales. *Methods in Ecology and Evolution* 5: pp.1,361–1,372.

Robinson, R.A., Wilson, J.D., and Crick, H.Q.P. 2001. The importance of arable habitat for farmland birds in grassland landscapes. *Journal of Applied Ecology* 38: pp.1,059–1,069.

Rock, P. 2005. Urban gulls: problems and solutions. *British Birds* 98(7): pp.338–355.

Roderick, H., and Davis, P. 2010. *Birds of Ceredigion*. The Wildlife Trust, South and West Wales.

Roderick, H.W. 1978. The Chough in Ceredigion. *Nature in Wales* 16(2): pp.127–128.

Rolfe, R. 1966. The status of the Chough in the British Isles. *Bird Study* 13(3): pp.221–236.

Roos, S., Campbell, S.T., Hartley, G., Shore, R.F., Walker, L.A. and Wilson, J.D. 2021. Annual abundance of common Kestrels (*Falco tinnunculus*) is negatively associated with second generation anticoagulant rodenticides. *Ecotoxicology* pp.1–15.

Rose, E. 2020. *The UK's Enforcement Gap 2020*. Report to Unchecked UK.

Rosney, A. 2019. The Glamorgan Bird Club Swift Project. *Birds in Wales* 16(1): pp.43–48.

Ross-Smith, V.H., Robinson, R.A., and Clark, J.A. 2015. *Dispersal and Movements of Lesser Black-backed gull in Europe*. Thetford: BTO Research Report No 671.

Round, P.D., and Moss, M. 1984. The waterbird populations of three Welsh rivers. *Bird Study* 31(1): pp.61–68.

Rowe, E., Sawicka, K., Mitchell, Z., Smith, R., Dore, T., Banin, L.F., and Levy, P. 2019. *Trends in critical load and critical level exceedances in the UK*. Centre for Ecology and Hydrology (CEH).

Rowell, H.E., and Spray, C.J. 2004. *The Mute Swan* Cygnus olor *(Britain and Ireland populations) in Britain and Northern Ireland 1960/61–2000/01.* Waterbird Review Series, The Wildfowl & Wetlands Trust/ Joint Nature Conservation Committee.

RSPB. 1909. *RSPB Annual Report 1909*. London.

RSPB. 1977. *Breeding birds of the River Wye, 1976*. Sandy: Unpublished report.

RSPB. 1978. *Breeding birds of the River Severn, 1977.* Sandy: Unpublished report.

RSPB. 1981. *Sawbills in Wales – a short survey of selected rivers*. Sandy: Unpublished report.

RSPB. 1986. *The Anglesey Report*. Sandy: Unpublished.

RSPB. 1988a. *The Lleyn Report.* Sandy: Unpublished.

RSPB. 1988b. *Ornithological studies at Trawsfynydd*. Sandy: Unpublished.

RSPB. 2017. Inner Hebrides blog. Accessed 28 June 2020. **community.rspb.org.uk/placestovisit/islay/b/weblog/posts/twite**

RSPB. 2019a. High nature carbon map. **https://rspb.maps.arcgis.com/apps/Cascade/index.html?appid=2b383eee459f4de18026002ae648f7b7**

RSPB. 2019b. Raptor persecution map hub. **https://www.arcgis.com/apps/opsdashboard/index.html#/0f04dd3b78e544d9a6175b7435ba0f8c**

RSPB. 2019c. How does RSPB South Stack help the local economy? **https://community.rspb.org.uk/placestovisit/southstackcliffs/b/southstackcliffs-blog/posts/how-does-rspb-south-stack-help-the-local-economy**

RSPB 2019d. Booming year for bitterns. RSPB press release, 10 September 2019. **https://www.rspb.org.uk/about-the-rspb/about-us/media-centre/press-releases/booming-bitterns**

Ruttledge, R.F., and Hall Watt, R. 1958. The distribution and status of wild geese in Ireland. *Bird Study* 5(1): pp.22–23.

Ryall, C., and Briggs, K.B. 2006. Some factors affecting foraging and habitat of Ring Ouzels *Turdus torquatus* wintering in the Atlas Mountains of Morocco. *Bulletin of the African Bird Club* 13(1): pp.17–31.

Ryves, B.H. 1948. *Bird life in Cornwall*. London: Collins.

Saffery Champness Chartered Accountants. 2016. *Landed estates annual review.*

Sage, B., and Whittington, P.A. 1985. The 1980 sample survey of rookeries. *Bird Study* 32(2): pp.77–81.

Sage, B.L. 1956. Notes on the birds of Caldey and St. Margaret's islands, Pembrokeshire. *Nature in Wales* 2: pp.333–340.

Sage, B.L., and Vernon, J.D.R. 1978. The 1975 National Survey of rookeries. *Bird Study* 25(2): pp.64–86.

Saino, N., Romano, M., Caprioli, M., Ambrosini, R., Rubolini, D., Scandolara, C., and Romano, A. 2012. A ptilochronological study of carry-over effects of conditions during wintering on breeding performance in the barn swallow *Hirundo rustica*. *Journal of Avian Biology* 43: pp.513–524.

Salmon, H.M., and Ingram, G.C.S. 1957. The birds of Brecknock. *Brycheiniog* 3: pp.182–259.

Salter, J.H. 1895. Observations on birds in Mid-Wales. *Zoologist* 3rd series 19: pp.129–143.

Salter, J.H. 1900. *A list of the birds of Aberystwyth and neighbourhood.* University College of Wales Scientific Society.

Salter, J.H. 1930. Changes in a local avi-fauna. *North Western Naturalist* 5(16): pp.14–17.

Samplonius, J.M., and Both, C. 2019. Climate change may affect fatal competition between two bird species. *Current Biology* 29(2): pp.327–331.

Sandham, R. 2017. *Scarce and rare birds in North Wales: historic records up to and including 2016.* Published privately.

Sangster, G., Collinson, J.M., Helbig, A.J., Knox, A.G., and Parkin, D.T. 2005. Taxonomic recommendations for British birds: third report. *Ibis* 147(4): pp.821–826.

Saunders, D., and Sutcliffe, S. 2020. Great bird reserves: Skokholm. *British Birds* 113(7): pp.377–397.

Saunders, D.R. 1962. The Great Black-backed gull on Skomer. *Nature in Wales* 8(2): pp.59–66.

Saunders, D.R. 1963. Skomer Bird Report for 1962. *Nature in Wales* 8(3): pp.99–104.

Saunders, D.R. 1976. *A brief guide to the birds of Pembrokeshire*. H.G. Walters.

Saunders, D.R. 2019. Orielton duck decoy, Pembrokeshire: a rich history. *Birds in Wales* 16(1): pp.3–16.

Saunders, H. 1889. *An illustrated manual of British birds*. London: Gurney and Jackson.

Schekkerman, H. 1989. Herfsttrek en biometrie van de IJsgors (*Calcarius lapponicus*) te Castricum. *Limosa* 62: pp.29–34. [In Dutch]

Schmitt, S., Eaton, M., and Drewitt, A. 2015. The spotted crake in the UK: results of the 2012 survey. *British Birds* 108(4): pp.220–230.

Schofield, P. 1975. *Ornithological survey of a proposed North Wales express roadway route between Aber and Roewen*. NCC.

Scott, D. 1997. *The Long-eared Owl.* London: The Hawk and Owl Trust.

Scott, D.A. 1970. *The breeding biology of the Storm-petrel* Hydrobates pelagicus. Oxford: University of Oxford.

Scragg, E. 2016. Seas of change. *BTO News* Winter 2016.

Sea Empress Environmental Evaluation Committee. 1998. *The environmental impact of the* Sea Empress *oil spill.* Final Report. The Stationery Office, London.

Seel, D.C., and Walton, K.C. 1979. Numbers of Meadow Pipits on mountain farm grassland in N. Wales. *Ibis* 121(2): pp.147–164.

Selås, V. 2017. Autumn irruptions of Eurasian Jay (*Garrulus glandarius*) in Norway in relation to acorn production and weather. *Ornis Fennica* 94: pp.92–100.

Selby, P.J. 1833. *Illustrations of British ornithology.* W.H. Lizars.

Sergeant, D.E. 1952. Little Auks in Britain, 1948–1951. *British Birds* 45(4): pp.122–133.

Sharps, E., Smart, J., Skov, M.W., Garbutt, A., and Hiddink, J.G. 2015. Light grazing of saltmarshes is a direct and indirect cause of nest failure in Common Redshank *Tringa totanus. Ibis* 157(2): pp.239–249.

Sharrock, J.T.R. 1964. Grey Wagtail passage in Britain. *British Birds* 57(1): pp.10–23.

Sharrock, J.T.R. 1976. *The atlas of breeding birds in Britain and Ireland.* Berkhamsted: T&AD Poyser.

Shawyer, C. 1998. *The Barn Owl.* Chelmsford: Arlequin Press.

Shawyer, C., Clarke, R., and Dixon, N. 1998. *A study into the raptor predation of domestic pigeons* Columba livia. Unpublished contract report from Hawk and Owl Trust to DETR and DoENI. 115 pp, South Wales Peregrine Monitoring Group Annual Reports 2017/2018/2019. Limited circulation only.

Shawyer, C.R. 1987. *The Barn Owl in the British Isles: its past, present and future.* The Hawk Trust.

Sheldon, R. 2018. Welsh Ravens again: some more thoughts on licences. **www.rdsconservation.com/?p=630**
Sheldon, R. 2019. Wales: a safe haven for Ravens? **www.rdsconservation.com/?p=852**
Shirihai, H., Bretagnolle, V., and Zino, F. 2010. Identification of Feae's, Desertas and Zino's petrels at sea. *Birding World* 23: pp.239–275.
Shoji, A., Owen, E., Bolton, M., Dean, B., Kirk, H., Fayet, A.L., Boyle, D., Perrins, C., and Guilford, T. 2014. Flexible foraging strategies in a diving seabird with high flight cost. *Marine Biology* 161(9): pp.2,121–2,129.
Shore, R.F., Potter, E.D., Walker, L.A., Pereira, M.G., Chaplow, J.S., Jaffe, J.E., Sainsbury, A.W., Barnett, E.A., Charman, S., Jones, A., Giela, A., Senior, C., and Sharp, E.A. 2018. The relative importance of different trophic pathways for secondary exposure to anticoagulant rodenticides. *Proceedings of the Vertebrate Pest Conference* 28. Retrieved from: **https://escholarship.org/uc/item/5gv7t7w1**
Shore, R.F., Walker, L.A., Potter, E.D., Chaplow, J.S., Pereira, M.G., Sleep, D., and Hunt, A. 2019. *Second generation anticoagulant rodenticide residues in barn owls 2018.* CEH contract report to the Campaign for Responsible Rodenticide Use (CRRU) UK.
Shortall, C.R., Moore, A., Smith, E., Hall, M.J., Woiwod, I.P., and Harrington, R. 2009. Long-term changes in the abundance of flying insects. *Insect Conservation and Diversity* 2(4): pp.251–260.
Shrubb, M. 1993a. Nest sites in the Kestrel *Falco tinnunculus*. *Bird Study* 40(1): pp.63–73.
Shrubb, M. 1993b. *The Kestrel*. London: Hamlyn.
Shrubb, M. 1997. The impact of changes in farming and other land uses on bird populations in Wales. *Welsh Birds* 1(5): pp.4–26.
Shrubb, M. 2000. An increase in the Buzzard population in North Breconshire. *Welsh Birds* 2(5): pp.251–256.
Shrubb, M. 2003a. The Kestrel in Wales. *Welsh Birds* 3(5): pp.330–339.
Shrubb, M. 2003b. *Birds, scythes and combines. A history of birds and agricultural change*. Cambridge: Cambridge University Press.
Shrubb, M. 2004. The decline of the Kestrel in Wales. *Welsh Birds* 4(1): pp.65–66.
Shrubb, M. 2007. *The Lapwing*. London: T&AD Poyser.
Shrubb, M., Lack, P.C., and Greenwood, J.J.D. 1991. The numbers and distribution of Lapwings *Vanellus vanellus* nesting in England and Wales in 1987. *Bird Study* 38(1): pp.20–37.
Sim, I., Rollie, C., Arthur, D., Benn, S., Booker, H., Fairbrother, V., Green, M., Hutchinson, K., Ludwig, S., Nicoll, M., Poxton, I., Rebecca, G., Smith, L., Stanbury, A., and Wilson, P. 2010. The decline of the Ring Ouzel in Britain. *British Birds* 103(4): pp.229–239.
Sim, I.M.W., Dillon, I.A., Eaton, M.A., Etheridge, B., Lindley, P., Riley, H., Saunders, R., Sharpe, C., and Tickner, M. 2007. Status of the Hen Harrier *Circus cyaneus* in the UK and Isle of Man in 2004, and a comparison with the 1988/89 and 1998 surveys. *Bird Study* 54(2): pp.256–267.
Sim, I.M.W., Eaton, M.A., Setchfield, R.P., Warren, P.K., and Lindley, P. 2008. Abundance of male Black Grouse *Tetrao tetrix* in Britain in 2005, and change since 1995–96. *Bird Study* 55(3): pp.304–313.
Sim, I.M.W., Gibbons, D.W., Bainbridge, I.P., and Mattingley, W.A. 2001. Status of the Hen Harrier *Circus cyaneus* in the UK and the Isle of Man in 1998. *Bird Study* 48(3): pp.341–353.
Sim, I.M.W., Gregory, R.D., Hancock, M.H., and Brown, A.F. 2005. Recent changes in the abundance of British upland breeding birds. *Bird Study* 52(3): pp.261–275.
Sim, I.M.W., Rebecca, G.W., Ludwig, S.C., Grant, M.C., and Reid, J.M. 2011. Characterizing demographic variation and contributions to population growth rate in a declining population. *Journal of Animal Ecology* 80(1): pp.159–170.
Simms, E. 1985. *British warblers*. London: Collins.
Simson, C. 1966. *A bird overhead*. London: Witherby.
Siriwardena, G.M. 2004. Possible roles of habitat, competition and avian nest predation in the decline of the Willow Tit *Parus montanus* in Britain. *Bird Study* 51(3): pp.193–202.
Siriwardena, G.M., Baillie, S.R., Buckland, S.T., Fewster, R.M., Marchant, J.H., and Wilson, J.D. 1998. Trends in the abundance of farmland birds: a quantitative comparison of smoothed Common Birds Census indices. *Journal of Applied Ecology* 35(1): pp.24–43.
Siriwardena, G.M., Calbrade, N.A., and Vickery, J.A. 2008. Farmland birds and late winter food: does seed supply fail to meet demand? *Ibis* 150(3): pp.585–595.
Sitters, H.P., Fuller, R.J., Hoblyn, R.A., Wright, M.T., Cowie, N., and Bowden, C.G.R. 1996. The Woodlark *Lullula arborea* in Britain: population trends, distribution and habitat occupancy. *Bird Study* 43(2): pp.172–187.
Skokholm and Skomer Nature Reserves Report for 1976. WWNT.
Skov, H., Heinänen, S., Žydelis, R., Bellebaum, J., Bzoma, S., Dagys, M., Durinck, J., Garthe, S., Grishanov, G., Hario, M., Kieckbusch, J.J., Kube, J., Kuresoo, A., Larsson, K., Luigujoe, L., Meissner, W., Nehls, H.W., Nilsson, L., Petersen, I.K., Roos, M.M., Pihl, S., Sonntag, N., Stock, A., Stipniece, A., and Wahl, J. 2011. *Waterbird populations and pressures in the Baltic Sea.* Nordic Council of Ministers, Copenhagen.
Slack, R. 2009. *Rare birds where and when: an analysis of status and distribution in Britain and Ireland.* Rare Bird Books.
Slater, F.M. 1990. Population trends of mute swans in Mid Wales and the Lower Wye Valley. *Welsh Bird Report* 4: pp.74–79.
Smart, J., Amar, A., O'Brien, M., Grice, P., and Smith, K. 2008. Changing land management of lowland wet grasslands of the UK: impacts on snipe abundance and habitat quality. *Animal Conservation* 11: pp.339–351.
Smith, D. 2004. Breeding and Roosting Hawfinches (*Coccothraustes coccothraustes*) in Merioneth. *Welsh Birds* 1(5): pp.11–19.
Smith, D. 2011. Status of the Ring Ouzel *Turdus torquatus* on Cadair Idris in 2008. *Birds in Wales* 8(1): pp.14–23.
Smith, D. 2014. The Ring Ouzel *Turdus torquatus* on the Rhinog Mountains: a 40 year perspective. *Birds in Wales* 11(1): pp.22–31.
Smith, D. 2018. The status of the Ring Ouzel (*Turdus torquatus*) on the Aran Mountains. *Birds in Wales* 15(1): pp.55–58.
Smith, H.E. 1866. Nesting of the Little Tern. *Zoologist* 2nd Series 1: p.100.
Smith, J. 2019a. *Impacts of climate change on woodland birds; from individual behaviour to population change.* PhD thesis Cardiff University.
Smith, K.W. 1983. The status and distribution of waders breeding on wet lowland grassland in England and Wales. *Bird Study* 30(3): pp.177–192.
Smith, K.W. 2005. Has the reduction in nest-site competition from Starlings *Sturnus vulgaris* been a factor in the recent increase of Great Spotted Woodpecker *Dendrocopos major* numbers in Britain? *Bird Study* 52(3): 307–313.
Smith, K.W. 2010. Continental great spotted Woodpeckers *Dendrocopos major* in Britain – further analyses of wing-length data. *Ringing & Migration* 25(2): pp.65–70.
Smith, K.W., and Charman, E.C. 2012. The ecology and conservation of the Lesser Spotted Woodpecker. *British Birds* 105(6): pp.294–301.
Smith, K.W., Reed, J.M., and Trevis, B.E. 1984. Habitat use and site fidelity of Green Sandpipers *Tringa ochropus* wintering in Southern England. *Bird Study* 39(3): pp.155–164.
Smith, L. 2019b. *The birds of Shropshire*. Liverpool: Liverpool University Press.
Smith, L.E., Hall, C., Cranswick, P.A., Banks, A.N., Sanderson, W.G., and Whitehead, S. 2007. The status of Common Scoter in Welsh waters and Liverpool Bay 2001–2006. *Welsh Birds* 5(1): pp.4–28.
Smith, M., Bolton, M., Okill, D.J., Summers, R.W., Ellis, P., Liechti, F., and Wilson, J.D. 2014. Geolocator tagging reveals Pacific migration of Red-necked Phalarope *Phalaropus lobatus* breeding in Scotland. *Ibis* 156(4): pp.870–873.
Smith, M.E. 1969. Kingfisher in Wales: effects of severe weather. *Nature in Wales* 11: pp.109–115.
Smith, S., Thompson, G., and Perrins, C.M. 2001. A census of the Manx Shearwater on Skomer, Skokholm and Middleholm, West Wales. *Bird Study* 48(3): pp.330–340.
Smith, S.J. 2002. A study of Whinchats *Saxicola rubetra* on the moorland edge. *Welsh Birds* 3(3): pp.183–190.
Snow, B., and Snow, D. 1988. *Birds and Berries*. Calton: T&AD Poyser.
Snow, D.W. 1954. The habitats of Eurasian tits (*Parus* spp.). *Ibis* 96(4): pp.565–585.
Snow, D.W. 1987. *The Blackbird.* Shire Natural History.
Snow, D.W., and Perrins, C.M. 1998. *The birds of the Western Palearctic*. Oxford: Oxford University Press.
Sovon. 2020. Species: trends and indices. **https://www.sovon.nl/en/content/vogelsoorten**
Spence, I.M. 2015. Some results from catching Storm Petrels at Strumble Head, Pembrokeshire. *Birds in Wales* 12(1): pp.3–11.
Spence, I.M. 2016 The distribution of birds breeding in Wales at tetrad level: what we know now and future prospects. *Birds in Wales* 13(1): pp.37–56.
Spence, I.M., and Lloyd, I.W. 1996. Tawny Owls *Strix aluco* breeding in Clwyd Forest District. *Welsh Birds* 1(3): pp.52–58.
Spencer, R., and Gush, G.H. 1973. Siskins feeding in gardens. *British Birds* 66(3): pp.91–99.
Squires, D., Grasse, J., and Falshaw, C. 2009. Northern Goshawks in mid/north Wales 1992–2006. *Welsh Birds* 6(1): pp.49–65.
Stace, C.A. 2010. *New flora of the British Isles*. 3rd ed. Cambridge: Cambridge University Press.
Stafford, J. 1962. Nightjar Enquiry, 1957–58. *Bird Study* 9(2): pp.104–115.
Stamp, L.D. 1962. *The land of Britain: Its use and misuse* (3rd ed.). London: Longmans, Green and Co Ltd.
Stanbury, A., Brown, A., Eaton, M., Aebischer, N., Gillings, S., Hearn, R., Noble, D., Stroud, D., and Gregory, R. 2017. The risk of extinction for birds in Great Britain. *British Birds* 110(9): pp.502–517.
Staneva, A., and Burfield, I. 2017. *European birds of conservation concern: populations, trends and national responsibilities.* BirdLife International.
Stansfield, S. 2004b. Breeding birds Ynys Enlli Ynysoedd Gwylan in 2004. Report of Bardsey Bird and Field Observatory 48: pp.73–85.

Statistics for Wales. 2018. *Welsh agricultural statistics 2016*. Cardiff: Welsh Government.

Steffen, W., Richardson, K., Rockström, J., Cornell, S.E., Fetzer, I., Bennett, E.M., Biggs, R., Carpenter, S.R., De Vries, W., De Wit, C.A., and Folke, C. 2015. Planetary boundaries: Guiding human development on a changing planet. *Science* 347: no. 6,223.

Stevens, D.K., Anderson, G.Q.A., Grice, P.V., Norris, K., and Butcher, N. 2008. Predators of Spotted Flycatcher *Muscicapa striata* nests in southern England as determined by digital nest-cameras. *Bird Study* 55(2): pp.179–187.

Stevens, M., Sheehan, D., Wilson, J., Buchanan, G., and Cresswell, W. 2010. Changes in Sahelian bird biodiversity and tree density over a five-year period in northern Nigeria. *Bird Study* 57(2): pp.156–174.

Stirling-Aird, P. 2015. *Peregrine Falcon*. Bloomsbury.

Stoddart, A., and Batty, C. 2019. The Elegant Tern in Britain and Europe. *British Birds* 112(2): pp.99–109.

Stoddart, A. and Hudson, N. 2021. From the rarities committee's files: BBRC and newly split species. *British Birds* 114(1): pp.8–17.

Storch, I. 2000. *Grouse status survey and conservation action plan 2000–2004*. WPA/BirdLife/SSC. Grouse Specialist Group. IUCN.

Stott, M., Callion, J., Kinley, I., Raven, C., and Roberts, J. 2004. *The breeding birds of Cumbria – a tetrad atlas 1997–2001*. Cumbria Bird Club.

Stowe, T.J. 1987. *The habitat requirements of some insectivorous birds and the management of sessile oakwoods*. PhD Thesis (Unpublished).

Stringer, C. 1977. Evidence of climatic change and human occupation during the last interglacial at Bacon Hole Cave, Gower. *Gower Journal* 28: pp.36–44.

Stroud, D., Francis, I., and Stroud, R. 2012. Spotted Crakes breeding in Britain and Ireland: a history and evaluation of current status. *British Birds* 105(4): pp.197–220.

Stroud, D.A. 2011. *The legal status of Greenland White-fronted Geese in England and Wales*. JNCC Briefing note, 9pp.

Stroud, D.A., Bainbridge, I.P., Maddock, A., Anthony, S., Baker, H., Buxton, N., Chambers, D., Enlander, I., Hearn, R.D., Jennings, K.R., Mavor, R., Whitehead, S., and Wilson, J.D., on behalf of the UK SPA & Ramsar Scientific Working Group (eds.). 2016. *The status of UK SPAs in the 2000s: the third network review*. 1,108 pp. Peterborough: JNCC. **http://jncc.defra.gov.uk/page-7309**

Stubbings, E.M., Büche, B.I., Miquel Riera, E., Green, R.M., and Wood, M.J. 2015. *Seabird monitoring on Skomer Island in 2015*. Peterborough: JNCC Report.

Stubbings, E.M., Büche, B.I., Riordan, J.A., Baker, B., and Wood, M.J. 2018. *Seabird monitoring on Skomer Island in 2018*. Peterborough: JNCC Report.

Summers, R., Bates, B., de Raad, L., Elkins, N., and Etheridge, B. 2019a. The migrations of British Common Sandpipers. *British Birds* 112(8): pp.431–443.

Summers, R., Christian, N., Etheridge, B., Rae, S., Cleasby, I., and Pálsson, S. 2020. Scottish-breeding Greenshanks *Tringa nebularia* do not migrate far. *Bird Study* 67(1): pp.1–7.

Summers, R.W., de Raad, A.L., Bates, B., Etheridge, B., and Elkins, N. 2019b. Non-breeding areas and timing of migration in relation to weather of Scottish-breeding common sandpipers *Actitis hypoleucos*. *Journal of Avian Biology* 50(1). **https://doi.org/10.1111/jav.01877**

Summers-Smith, J.D. 1995. *The Tree Sparrow*. Guisborough: Published privately.

Sutcliffe, S.J. 1975. Common Scoter in Carmarthen: an oiling incident. *Nature in Wales* 14(4): pp.243–249.

Sutcliffe, S.J. 1986. Changes in the gull populations of SW Wales. *Bird Study* 33(2): pp.91–97.

Svensson, L. 1992. *Identification guide to European Passerines (4th ed.)*. Stockholm: The author.

Swainson, Capt. E.A. 1893. On the distribution and habits of the Pied Flycatcher in Wales. *Zoologist*, 3rd Series, 17: pp.420–424.

Swann, R.L. 1988. Are all large Chaffinch flocks composed of Continentals? *Ringing & Migration* 9(1): pp.1–4.

Tallack, R.E. 1982. The breeding population of the House Martin in Gower. *Gower Birds* 3: p.5.

Tapper, S. 1992. *Game heritage: an ecological review from shooting and gamekeeping records*. Game Conservancy Trust.

Tapper, S. 1999. *A question of balance: game animals and their role in the British countryside*. Game Conservancy Trust.

Taylor, K. 1984. The influence of watercourse management on Moorhen breeding biology. *British Birds* 77(4): pp.141–148.

Taylor, K., Hudson, R., and Horne, G. 1988. Buzzard breeding distribution and abundance in Britain and Northern Ireland in 1983. *Bird Study* 35(2): pp.109–118.

Taylor, R. 2019. Curlews in Wales. In *Call of the Curlew: Curlew Forum Newsletter 8*. **www.curlewcall.org/curlew-forum-newsletter-8-december-2019/**

Taylor, R., Bowgen, K., Burton, N., and Franks, S. 2020. *Understanding Welsh breeding Curlew: from local landscape movements through to population estimations and predictions*. NRW Evidence Report (in prep.).

Taylor, R., Noble, D., and Gillings, S. 2016. *Wales-specific patterns of change in distribution and relative abundance derived from the Bird Atlas 2007–11*. BTO unpublished report.

Taylor, R.H.A. 1999. *Checklist of the birds of Gower*. Gower Ornithological Society.

Taylor, R.H.A. 2008. *Gower birds' checklist*. Gower Ornithological Society.

Temrin, H., Mallner, Y., and Winden, M. 1984. Observations on polyterritoriality and singing behaviour in the Wood Warbler *Phylloscopus sibilatrix*. *Ornis Scandinavica* 15: pp.67–72.

Tew, I. 1990. Quail records in Wales in summer 1989. *Welsh Bird Report 3*: pp.38–39.

Thatcher, M. 1990. Speech at the second World Climate Conference, Geneva. 6 November 1990. Margaret Thatcher Foundation. **https://www.margaretthatcher.org/document/108237**

Thaxter, C.B., Sansom, A., Thewlis, R.M., Calbrade, N.A., Ross-Smith, V.H., Bailey, S., Mellan, H.J., and Austin, G.E. 2010. *Wetland Bird Survey Alerts 2006/2007: Changes in numbers of wintering waterbirds in the Constituent Countries of the United Kingdom, Special Protection Areas (SPAs) and Sites of Special Scientific Interest (SSSIs)*. Thetford: BTO Research Report 556.

Thearle, R.F., Hobbs, J.T., and Fisher, J. 1953. The birds of the St Tudwal's Islands. *British Birds* 46(5): pp.182–188.

Thomas, D. 1992a. *Marine wildlife and net fisheries around Wales*. Unpublished report by the Royal Society for the Protection of Birds and the Countryside Council for Wales.

Thomas, D.K. 1992b. *An atlas of breeding birds in West Glamorgan*. Gower Ornithological Society.

Thomas, J.F. 1942. Report on the Redshank Enquiry 1939–40. *British Birds* 36(1): pp.5–14.

Thompson, G. 2000. Bobolink *Dolichonyx oryzivorus* on Skokholm – First Welsh record. *Welsh Birds* 2(6): p.380.

Thompson, G.V.F. 2008. *The natural history of Skokholm Island*. Trafford Publishing.

Thompson, P.M. 2006. Identifying drivers of change; did fisheries play a role in the spread of North Atlantic fulmars? In Boyd, I.L., Wanless, S., and Camphuysen, C.J., (eds.) *Management of marine ecosystems: monitoring change in upper trophic levels*. Cambridge: Cambridge University Press.

Thompson, P.S., Baines, D., Coulson, J.C., and Longrigg, G. 1994. Age at first breeding, philopatry and breeding site-fidelity in the Lapwing *Vanellus vanellus*. *Ibis* 136(4): pp.474–484.

Thomson, D.L., Douglas-Home, H., Furness, R.W., and Monaghan, P. 1996. Breeding success and survival in the Common Swift *Apus apus*: a long-term study on the effects of weather. *Journal of Zoology* 239: pp.29–38.

Thorpe, R.I. 2002. Numbers of wintering seaducks, divers and grebes in North Cardigan Bay 1991–1998. *Welsh Birds* 3(3): pp.155–170.

Thorpe, R.I., and Johnstone, I. 2003. The status of Chough in Wales 2002. *Welsh Birds* 3(5): pp.354–362.

Thorpe, R.I., and Stratford, A. 2019. The Welsh list: a checklist of the birds of Wales (3rd edition). *Birds in Wales* 16(3): pp.218–219.

Thorpe, R.I., and Young, A. 2002. The population status of birds in Wales: an analysis of conservation concern 2002–07. *Welsh Birds* 3(4): pp.289–302.

Thorpe, R.I., and Young, A. 2004. The breeding population of the Peregrine (*Falco peregrinus*) in Wales in 2002. *Welsh Birds* 4(1): pp.44–47.

Thorup, K. 1998. Vagrancy of Yellow-browed Warbler *Phylloscopus inornatus* and Pallas's Warbler *Ph. proregulus* in north-west Europe: Misorientation on great circles? *Ringing & Migration* 19(1): pp.7–12.

Thorup, K., Sunde, P., Jacobsen, L.B., and Rahbek, C. 2010. Breeding season food limitation drives population decline of the Little Owl *Athene noctua* in Denmark. *Ibis* 152(4): pp.803–814.

Ticehurst, N.F. 1919/1920. The birds of Bardsey island. Part 1. *British Birds* 13(2): pp.42–51; 13(3): pp.66–75; 13(4): pp.101–106; 13(5): pp.129–134; 13(7): pp.182–193 and 13(8): pp.214–216.

Ticehurst, N.F., and Hartley, P.H.T. 1948. Report on the effect of the severe winter of 1946–1947 on bird life. *British Birds* 41(11): pp.322–334.

Ticehurst, N.F., and Jourdain, F.C.R. 1911. On the distribution of the Nightingale during the breeding season in Great Britain. *British Birds* 5(1): pp.2–21.

Tiedemann, G. 1971. Zur ökologie und sieldlunsdichte des Waldlaubsängers (*Phylloscopus sibilatrix*). *Vogelwelt* 92: pp.8–17.

Toms, M. 2014. *Owls*. London: HarperCollins.

Tonkin, S., and Perea, J.M.G. 2019. Ringing recovery of Yellow-browed Warbler in Andalucía confirms overwintering in consecutive winters. *British Birds* 112(11): pp.686–687.

Toulmin Smith, L. (ed.) 1906. *The itinerary in Wales of John Leland in or about 1536–39*. London.

Toyne, E.P. 1994. *Studies on the ecology of the northern goshawk (*Accipiter gentilis*) in Britain.* London: PhD thesis. Dept. Biology, Imperial College of Science, Technology & Medicine.

Toyne, E.P. 1996.The northern goshawk: friend or foe? *The Raptor* 23: pp.45–49.

Toyne, E.P. 1997. Nesting chronology of northern goshwaks (*Accipiter gentilis*) in Wales: implications for forest management. *Forestry* 70(2): pp.121–127.

Tracy, J. 1850. Catalogue of birds taken in Pembrokeshire with observations on their habits, manners &c. *The Zoologist* 8: pp.2639–2642.

Trolliet, B., and Girard, O. 2006. Anatidae numbers and distribution in West Africa in winter. In Boere, G.C., Galbraith, C.A., and Stroud, D.A., (eds.) *Waterbirds around the world.* Edinburgh: The Stationery Office. pp.226–227.

Tucker, G.M., and Heath, M.F. 1994. *Birds in Europe: their conservation status.* Cambridge: BirdLife International.

Tuite, C.H., Owen, M., and Paynter, D. 1983. Interaction between wildfowl and recreation at Llangorse Lake and Talybont Reservoir, South Wales. *Wildfowl* 34: pp.48–63.

Tunnicliffe, C.F. 1949. Letter to RSPB. 27 October 1949. Sandy: RSPB archive.

Tunnicliffe, C.F. 1992. *Shorelands winter diary.* London: Robinson Publishing.

Turner, A. 2009. Climate change: a Swallow's eye view. *British Birds* 102(1): pp.3–16.

Tyler, S. 2003. Birds of the lower Wye and tributaries. *Gwent Bird Report* 39: pp.6–8.

Tyler, S. 2010. An explosion of Mandarin Ducks and a continued spread of Goosanders on the Monnow. *Gwent Bird Report 2009*, 45: pp.77–78.

Tyler, S. 2013. Counts of fish-eating birds and other waterbirds on the River Usk in the winter of 2012/2013. *Gwent Bird Report for 2012*, 48: pp.108–113.

Tyler, S. 2015. Breeding waterbirds on the River Usk in 2014. *Gwent Bird Report* 50: pp.116–123.

Tyler, S., Coleman, J., and Ruston, R. 2015. Waterbirds on the River Monnow and lower River Wye in the 2014 breeding season. *Gwent Bird Report* 50: pp.105–115.

Tyler, S.J. 1972. Breeding Biology of the Grey Wagtail. *Bird Study* 19(2): pp.69–80.

Tyler, S.J. 1979. Movements and mortality of the Grey Wagtail. *Ringing & Migration* 2(3): pp.122–131.

Tyler, S.J. 1985. The wintering and breeding status of Goosanders in Wales. Sandy: Unpublished RSPB Report.

Tyler, S.J., *et al.* 1987. *The Gwent atlas of breeding birds: an atlas of breeding birds in the county of Gwent from 1981–85.* Gwent Ornithological Society.

Tyler, S.J. 1992a. A review of the status of breeding waders in Wales. *Welsh Bird Report 5*: pp.74–86.

Tyler, S.J. 1992b. *Little Ringed Plovers in Wales in 1991.* Sandy: Report to the National Rivers Authority by RSPB.

Tyler, S.J. 1995. Reed-beds in Wales – their extent, distribution and birds and threats facing them. *Welsh Birds* 1(1): pp.25–35.

Tyler, S.J. 2002. Birds of the River Monnow. *Gwent Bird Report* 38: pp.6–13.

Tyler, S.J. 2004. The status of Dippers and Grey Wagtails in south and mid Wales. *Welsh Birds* 4(1): pp.4–10.

Tyler, S.J., and Burge, F. 2011. Dippers on the Monnow and other Wye tributaries in Monmouthshire and Herefordshire. *Gwent Bird Report*, 47: pp.100–108.

Tyler, S.J., and Green, M. 1994. The status and breeding ecology of Ring Ouzels in Wales with reference to soil acidity. *Welsh Bird Report 7*: pp.78–89.

Tyler, S.J., and Medland, B. 2008. Grey Wagtails nesting in close proximity. *Gwent Bird Report for 2007*, 43: p.94.

Tyler, S.J., and Ormerod, S.J. 1985. Aspects of the breeding biology of Dippers in the southern catchment of the River Wye, Wales. *Bird Study* 32(3): pp.164–169.

Tyler, S.J., and Ormerod, S.J. 1991. The influence of stream acidification and riparian land-use on the breeding biology of Grey Wagtails *Motacilla cinerea* in Wales. *Ibis* 133(3): pp.286–292.

Tyler, S.J., and Ormerod, S.J. 1994. *The Dippers*. London: T&AD Poyser.

Tyrell, C. 2017. Molecular analysis of the diet of Pied Flycatchers (*Ficedula hypoleuca*) in the context of climate change, productivity and carry-over effects: a pilot study. *Birds in Wales* 14(1): pp.20–32.

UK Biodiversity Partnership. 2007. *Conserving Biodiversity – the UK Approach.* London: Defra.

Underhill, M.C. 1993. *Numbers and distribution of Cormorants and Goosanders on the River Wye and its main tributaries.* Cardiff: Unpublished report to the National Rivers Authority. 69pp.

Underhill, M.C., Gittings, T., Callaghan, D.A., Hughes, B., Kirby, J.S., and Delany, S. 1998. Status and distribution of breeding Common Scoters *Melanitta nigra nigra* in Britain and Ireland in 1995. *Bird Study* 45(2): pp.146–156.

Väänänen, V.-M., Nummi, P., Rautiainen, A., Asanti, T., Huolman, I., Mikkola-Roos, M., Nurmi, J., Orava, R., and Rusanen, P. 2009. The effect of raccoon dog *Nyctereutes procyonoides* removal on waterfowl breeding success. *Suomen Riista* 53: pp.49–63.

Vafidis, J., Facey, R., Leech, D., and Thomas, R. 2018. Supplemental food alters nest defence and incubation behaviour of an open-nesting wetland songbird. *Journal of Avian Biology* 49(8): **DOI: 10.1111/jav.01672**

Vafidis, J.O., Vaughan, I.P., Jones, T.H., Facey, R.J., Parry, R. and Thomas, R.J. 2016 The effects of supplementary food on the breeding performance of Eurasian reed warblers *Acrocephalus scirpaceus*; implications for climate change impacts. *PloS One* 11: p.e0159933

Valkama, J., Vepsäläinen, V., and Lehikoinen, A. 2011. *The Third Finnish Breeding Bird Atlas.* Finnish Museum of Natural History and Ministry of Environment, Helsinki. Little Gull (*Hydrocoloeus minutus*) account accessible at: **http://atlas3.lintuatlas.fi/results/species/little%20gull**

Van de Pol, M., Atkinson, P.W., Blew, J., Duriez, O.P.M., Ens, B.J., Hälterlein, B., Hötker, H., Laursen, K., Oosterbeek, K.H., Petersen, A., Thorup, O., Tjørve, K., Triplet, P., and Yésou, P. 2014. A global assessment of the conservation status of the nominate subspecies of Eurasian oystercatcher (*Haematopus ostalegus ostralegus*). *International Wader Studies* 20: pp.47–61.

van Gils, J.A., Lisovski, S., Lok, T., Meissner, W., Ożarowska, A., de Fouw, J., Rakhimberdiev, E., Soloviev, M.Y., Piersma, T., and Klaassen, M. 2016. Body shrinkage due to Arctic warming reduces red knot fitness in tropical wintering range. *Science* 352: pp.819–821.

van Kleunen, A., and Lemaire, A.J.J. 2014. A risk assessment of Mandarin Duck (*Aix galericulata*) in the Netherlands. Sovon-report 2014/15. Nijmegen: Sovon Dutch Centre for Field Ornithology.

Vanhingsbergh, D., Fuller, R.J., and Noble, D. 2003. *A review of possible causes of recent changes in populations of woodland birds in Britain.* Thetford: BTO Research Report, no. 245.

Vaughan, D., and Gibbons, D.W. 1996. The status of breeding Storm-petrels *Hydrobates pelagicus* on Skokholm Island in 1995. *Seabird* 20: pp.12–21.

Venables, W.A., Baker, A.D., Clarke, R.M., Jones, C., Lewis, J.M.S., Tyler, S.J., Walker, I.R., and Williams, R.A. 2008. *The birds of Gwent.* London: Helm.

Vernon, D.R. 1963. The distribution of Starling roosts in Glamorgan and Monmouthshire. *Nature in Wales* 8(4): pp.165–174.

Vernon, J.D.R. 1969. Spring migration of the Common Gull in Britain and Ireland. *Bird Study* 16(2): pp.101–107.

Versluijs, M., van Turnhout, C.A.M., Kleijn, D., and van der Jeugd, H.P. 2016. Demographic changes underpinning the population decline of Starlings *Sturnus vulgaris* in The Netherlands. *Ardea* 104: pp.153–165.

Viain, A., Corre, F., Delaporte, P., Joyeux, E., and Bocher, P. 2011. Numbers, diet and feeding methods of common Shelduck *Tadorna tadorna* wintering in the estuarine bays of Aiguillon and Marennes-Oleron, Western France. *Wildfowl* 61: pp.121–141.

Vickery, J.A., Ewing, S.R., Smith, K.W., Pain, D.J., Bairlein, F., Škorpilova, J., and Gregory, R.D. 2014. The decline of Afro-Palearctic migrants and an assessment of potential causes. *Ibis* 156(1): pp.1–22.

Vickery, J.A., Tallowin, J.T., Feber, R.E., Asteraki, E.A., Atkinson, P.W., Fuller, R.J., and Brown, V.K. 2001. The management of lowland neutral grasslands in Britain: effects of agricultural practices on birds and their food resources. *Journal of Applied Ecology* 38(3): pp.647–664.

Village, A. 1990. *The Kestrel.* London: T&AD Poyser.

Villar, N., Cornulier, T., Evans, D., Pakeman, R., Redpath, S., and Lambin, X. 2014. Experimental evidence that livestock grazing intensity affects cyclic vole population regulation processes. *Population Ecology* 56(1): pp.55–61.

Vinicombe, K. 1982. Breeding and population fluctuations of the Little Grebe. *British Birds* 75(5): pp.204–218.

Vinicombe, K. 1994. The Welsh Monk Vulture. *British Birds* 87(12): pp.613–622.

Vinicombe, K.E. 2000. Identification of Ferruginous Duck and its status in Britain and Ireland. *British Birds* 93(1): pp.4–21.

Vinicombe, K.E., and Harrop, A.H.J. 1999. Ruddy Shelducks in Britain and Ireland 1986–1994. *British Birds* 92(5): pp.225–255.

Visser, M.E., Gienapp, P., Husby, A., Morrisey, M., de la Hera, I., Pulido, F., and Both, C. 2015. Effects of spring temperatures on the strength of selection on timing of reproduction in a long-distance migratory bird. *Plos Biology*: 13(4): e1002120.

Votier, S.C., Birkhead, T.R., Oro, D., Trinder, M., Grantham, M.J., Clark, J.A., McCleery, R.H., and Hatchwell, B.J. 2008. Recruitment and survival of immature seabirds in relation to oil spills and climate variability. *Journal of Animal Ecology* 77(5): pp.974–979.

Votier, S.C., Hatchwell, B.J., Beckerman, A., McCleery, R.H., Hunter, F.M., Pellatt, J., Trinder, M., and Birkhead, T.R. 2005. Oil pollution and

climate have wide-scale impacts on seabird demographies. *Ecology Letters* 8: pp.1,157–1,164.
Votier, S.C., Hatchwell, B.J., Mears, M., and Birkhead, T.R. 2009. Changes in the timing of egg-laying of a colonial seabird in relation to population size and environmental conditions. *Marine Ecology Progress Series* 392: pp.225–233.
Votier, S.C., Kennedy, M., Bearhop, S., Newell, R.G., Griffiths, K., Whitaker, H., Ritz, M., and Furness, R. 2007. Supplementary DNA evidence fails to confirm presence of Brown Skuas *Stercorarius antarctica* in Europe: a retraction of Votier *et al.* (2004). *Ibis* 149(3): pp.619–621.
Wakefield, E.D., Bodey, T.W., Bearhop, S., Blackburn, J., Colhoun, K., Davies, R., Dwyer, R.G., Green, J.A., Grémillet, D., Jackson, A.L., Jessopp, M.J., Kane, A., Langston, R.H.W., Lescroël, A., Murray, S., Le Nuz, M., Patrick, S.C., Péron, C., Soanes, L.M., Wanless, S., Votier, S.C., and Hamer, K.C. 2013. Space partitioning without territoriality in Gannets. *Science.* 341 (6,141): pp.68–70.
Waldron, M.J. 2020. The Barnacle Goose in Brecknock – a review. *Breconshire Birds 2020.*
Wales Environment Link. 2020. Environmental NGOs unite in call for a moratorium on intensive poultry units. **https://www.waleslink.org/sites/default/files/a_moratorium_on_intensive_poultry_units_final_sept_20.docx.pdf**
Walford, T. 1818. *The scientific tourist through England, Wales and Scotland.* Vol. II. J. Booth.
Walker, F., and Kenmir, B. 1994. *Grouse moors in Wales, 1994.* CCW unpublished report.
Walker, L.A., Jaffe, J.E., Barnett, E.A., Chaplow, J.S., Charman, S., Giela, A., Hunt, A.G., Jones, A., Pereira, M.G., Potter, E.D., Sainsbury, A.W., Sleep, D., Senior, C., Sharp, E.A., Vyas, D.S., and Shore, R.F. 2019. *Anticoagulant rodenticides in red kites (*Milvus milvus*) in Britain in 2017 and 2018.* Lancaster: Centre for Ecology & Hydrology.
Walpole-Bond, J.A. 1904. *Birdlife in wild Wales* (with annotations by H.M. Salmon). T. Fisher Unwin.
Walpole-Bond, J.A. 1914. *Field-studies of some rarer British birds.* London: Witherby.
Walton, C.L. 1928. Rooks and agriculture in Mid and North Wales. *Welsh Journal of Agriculture* 4: pp.353–356.
Wanless, S., Harris, M.P., Newell, M.A., Speakman, J.R., and Daunt, F. 2018. Community-wide decline in the occurrence of lesser sandeels *Ammodytes marinus* in seabird chick diets at a North Sea colony. *Marine Ecology Progress Series* 600: pp.193–206.
Ward, R.M. 2004. *Dark-bellied Brent Goose* Branta bernicla bernicla *in Britain 1960/61 – 1999/2000.* Slimbridge: Waterbird Review Series, The Wildfowl & Wetlands Trust/Joint Nature Conservation Committee.
Ward, R.M., Cranswick, P.A., Kershaw, M., Austin, G.E., Brown, A.W., Brown, L.M., Coleman, J.C., Chisholm, H.K., and Spray, C.J. 2007. Numbers of Mute Swans *Cygnus olor* in Great Britain: results of the national census in 2002. *Wildfowl* 57: pp.3–20.
Warren, P., and Baines, D. 2012. *Changes in upland bird numbers and distribution in the Berwyn Special Protection Area, North Wales between 1983 and 2012.* Game & Wildlife Conservation Trust.
Warren, P., and Baines, D. 2014. Changes in the abundance and distribution of upland breeding birds in the Berwyn Special Protection Area, North Wales 1983–2002. *Birds in Wales* 11(1): pp.32–42.
Warren-Thomas, E., and Henderson, E. 2017. *Woodlands in Wales: a quick guide.* Research Briefing. National Assembly for Wales.
Watson, A., and Moss, R. 2008. *Grouse: the natural history of British and Irish species.* London: Collins.
Watson, D. 1977. *The Hen Harrier.* Berkhamsted: T&AD Poyser.
Watson, M., Aebischer, N.J., Potts, G.R., and Ewald, J.A. 2007. The relative effects of raptor predation and shooting on overwinter mortality of grey partridges in the United Kingdom. *Journal of Applied Ecology* 44(5): pp.972–982.
Watson, S.J., and Sale, R.G. 2020. A significant decline of breeding Peregrine Falcons in coastal north and west Cornwall, 2015–19. *British Birds* 113(6): pp.354–362.
Weir, D.N., Kitchener, A.C., McGowan, R.Y., Kinder, A., and Zonfrillo, B. 1997. *Origins, population structure, pathology and diet of samples of diver and auk casualties of the* Sea Empress *oil spill.* CCW Contract No. 73-02-69.
Welsh Assembly Government. 2005. *Tir Gofal Full Management Options.* Cardiff: Welsh Assembly Government.
Welsh Government. 2009. *Woodlands for Wales: The Welsh Government's Strategy for Woodlands and Trees.*
Welsh Government. 2018a. *Woodlands for Wales: The Welsh Government's Strategy for Woodlands and Trees.* **https://gov.wales/sites/default/files/publications/2018-06/woodlands-for-wales-strategy_0.pdf**
Welsh Government. 2018b. *Valued and Resilient: The Welsh Government's Priorities for AONBs and National Parks.* Written Statement 27 July 2018. **https://gov.wales/written-statement-valued-and-resilient-welsh-governments-priorities-areas-outstanding-natural#:~:text=I%20am%20committed%20to%20ensuring,of%20the%20people%20of%20Wales.**
Welsh Government. 2019a. *Survey of Agriculture and Horticulture.* Statistics for Wales.
Welsh Government. 2019b. *Farm incomes in Wales, April 2018 to March 2019.* Statistics for Wales.
Welsh Government. 2019c. *Phytophthora ramorum* strategy for Wales.
Welsh Government. 2019d. *Emissions of greenhouse gases by year.* Statistics Wales. **https://statswales.gov.wales/Catalogue/Environment-and-Countryside/Greenhouse-Gas/emissionsofgreenhousegases-by-year**
Welsh Government. 2019e. *Welsh National Marine Plan.* Cardiff.
Welsh Kite Trust. Undated. **https://welshkitetrust.wales**
Welsh Kite Trust. 2016. A survey of breeding Kestrels in Pembrokeshire 2016. **http://welshkitetrust.wales/wp-content/uploads/2017/02/WKT-Pembrokeshire-kestrel-survey-2016.pdf**
Welsh Kite Trust. 2020. **http://welshkitetrust.wales/how-many-kites-are-there-in-wales**
Welsh Ornithological Society. 2021. The Welsh List. **https://birdsin.wales/welsh-list-2021-2/**
Wenzel, M.A., Webster, L.M.I., Blanco, G., Burgess, M.D., Kerbiriou, C., Segelbacher, G., Piertney, S.B., and Reid, J.M. 2012. Pronounced genetic structure and low genetic diversity in European red-billed chough, *Pyrrhocorax pyrrhocorax* populations in *Conservation Genetics* 13: pp.1,213–1,230.
Wernham, C.V., Toms, M.P., Marchant, J.H., Clark, J.A., Siriwardena, G.M., and Baillie, S.R. (eds.) 2002. *The migration atlas: movements of the birds of Britain and Ireland.* London: T&AD Poyser.
Whalley, P.E.S. 1954. List of birds seen in Anglesey and Caernarfonshire with notes on their distribution and status. *North West Naturalist:* pp.604–618.
White, S., and Kehoe, C. 2014a. Report on Scarce Migrants in Britain 2008–2010: Part 1. *British Birds* 107(3): pp.251–281.
White, S., and Kehoe, C. 2014b. Report on scarce migrant birds in 2008–10 Part 2: passerines. *British Birds* 107(5): pp.251–281.
White, S., and Kehoe, C. 2015. Report on scarce migrant birds in 2011–12. Part 2: passerines. *British Birds* 108(4): pp.192–219.
White, S., and Kehoe, C. 2018. Report on scarce migrant birds in Britain in 2016 Part 2: passerines. *British Birds* 111(9): pp.519–542.
White, S., and Kehoe, C. 2019. Report on scarce migrant birds in Britain in 2017 Part 2: passerines. *British Birds* 112(11): pp.639–660.
White, S., and Kehoe, C. 2020a. Report on scarce migrant birds in Britain in 2018: Part 1: non-passerines. *British Birds* 113(8): pp.461–482.
White, S., and Kehoe, C. 2020b. Report on scarce migrant birds in Britain in 2018: Part 2: passerines. *British Birds* 113(9): pp.533–554.
Whitfield, D.P., and Fielding, A.H. 2009. *Hen Harrier population studies in Wales.* CCW Contract Science Report 879, Bangor: Countryside Council for Wales.
Whitfield, D.P., Fielding, A.H., and Whitehead, S. 2008. Long-term increase in the fecundity of Hen Harriers in Wales is explained by reduced human interference and warmer weather. *Animal Conservation* 11: pp.144–152.
Whitfield, D.P., Green, M., and Fielding, A.H. 2010. *Are breeding Eurasian curlew* Numenius arquata *displaced by wind energy developments?* Natural Research Projects.
Whittaker, I. 1941. North Wales ornithological notes. *Northwest Naturalist* 16: p.194.
Whytock, R.C., Davis, D., Whytock, R.T., Burgess, M.D., Minderman, J., and Mallord, J.W. 2015. Wood Warbler *Phylloscopus sibilatrix* nest provisioning rates are correlated with seasonal caterpillar availability in British Oak *Quercus* woodlands. *Bird Study* 62(3): pp.339–347.
Wilkie, N.G., Zbijewska, S.M., Piggott, A.R., Hastie, V., and Wood, M.J. 2019. *Seabird monitoring on Skomer Island in 2019.* JNCC Report.
Wilkinson, N.I., Eaton, M.A., Colhoun, K., and Drewitt, A.L. 2018. The population status of breeding Twite *Linaria flavirostris* in the UK in 2013. *Bird Study* 65(2): pp.174–188.
Williams, A.H. 2014. *Adar yng ngwaith y cywyddwyr.* PhD Thesis, Aberystwyth University.
Williams, E. 1826. *Cyneirflyfr neu eiriadur Cymraeg.* William Williams.
Williams, G. 1978. Notes on the birds of Grassholm. *Nature in Wales* 16(1): pp.2–15.
Williams, G. 2018. The breeding population of the Peregrine (*Falco peregrinus*) in Wales in 2014. *Birds in Wales* 15(1): pp.3–8.
Williams, G.A. 1977. Status and distribution of wildfowl in the Dee estuary. *Nature in Wales* 15(4): pp.166–179.
Williams, G.A. 1981. The Merlin in Wales. *British Birds* 74(5): pp.205–214.
Williams, G.A. 2000a. Breeding terns in Wales, 1975–1999. *Welsh Birds* 2(5): pp.274–279.
Williams, G.A. 2005. Breeding waders in Wales. *Welsh Birds* 4(3): pp.269–270.
Williams, G.A. 2013. Notable Welsh Ornithologists of the Twentieth Century. *Birds in Wales* 10(1): pp.29–39.
Williams, I., Griffin, B., and Holloway, D. 1991. *The Red Grouse population of Wales based on winter counts 1990/91.* Sandy: RSPB Report, unpublished.

Williams, I.T. 1989. Breeding Short-eared Owls in Wales. *Welsh Bird Report 2 (1988)*: pp.64–71.

Williams, I.T. 1991. The breeding population of the Peregrine in Wales in 1991. *Welsh Bird Report 5*: pp.62–69.

Williams, I.T. 1999a. Breeding Hen Harriers in Wales 1988–1994. *Welsh Birds* 2(3): pp.113–120.

Williams, I.T. 1999b. Lapwing nesting on a school roof. *Welsh Birds* 2(3): p.137.

Williams, I.T. 2000b. Dartford Warbler breeding in Wales – first Welsh breeding record. *Welsh Birds* 2(4): p.227.

Williams, I.T. 2000c. Colonisation and extinction of birds in Wales 1945–1999. *Welsh Birds* 2(5): pp.257–263.

Williams, I.T., and Parr, S. 1995. Breeding Merlins in Wales in 1993. *Welsh Birds* 1(1): pp.14–20.

Williams, J. 1830. *Faunula Grustensis: being an outline of the natural contents of the Parish of Llanrwst*. John Jones.

Williams, R. 1835. *The history and antiquities of the town of Aberconwy: with notices of the natural history of the district*. Thomas Gee.

Williams, S., Perkins, S.E., Dennis, R., Byrne, J.P. and Thomas, R.J. 2020. An evidence-based assessment of the past distribution of Golden and White-tailed Eagles across Wales. *Conservation Science and Practice*: e240.

Williams, T. 1864. *The complete guide to Llandudno: its history and natural history*. Llandudno.

Williams, T.S. 1949. The wild geese of Mersey and Dee. *Proceedings of the Chester Society of Natural Science, Literature and Art.*

Williamson, K. 1959. Observations on the Chough. *Peregrine (A Journal of Manx Natural History)* 3: pp.8–14.

Williamson, K. 1971. Breeding birds in Redwood Grove. *Quarterly Journal of Forestry*: pp.109–121.

Williamson, K., and Davis, P. 1956. The autumn 1953 invasion of Lapland Buntings and its source. *British Birds* 49(1): pp.6–25.

Willughby, F. 1678. *The ornithology of Francis Willughby of Middleton in the county of Warwick.* Printed by A.C. for John Martyn.

Wilson, A.M., Henderson, A.C.B., and Fuller, R.J. 2002. Status of the Nightingale *Luscinia megarhynchos* in Britain at the end of the 20th century with particular reference to climate change. *Bird Study* 49(3): pp.193–204.

Wilson, A.M., Vickery, J.A., Brown, A., Langston, R.H.W., Smallshire, D., Wotton, S., and Vanhinsbergh, D. 2005. Changes in the numbers of breeding waders on lowland wet grasslands in England and Wales between 1982 and 2002. *Bird Study* 52(1): pp.55–69.

Wilson, A.M., Vickery, J.A., and Browne, S.J. 2001. Numbers and distribution of Northern Lapwings *Vanellus vanellus* breeding in England and Wales in 1998. *Bird Study* 48(1): pp.2–17.

Wilson, J.M., and Cresswell, W. 2006. How robust are Palearctic migrants to habitat loss and degradation in the Sahel? *Ibis* 148(4): pp.789–800.

Wilson, L.J., Rendell-Read, S., Lock, L., Drewitt, A.L., and Bolton, M. 2019. Effectiveness of a five-year project of intensive, regional-scale, coordinated management for little terns *Sternula albifrons* across the major UK colonies. *Journal of Nature Conservation* 53(2): pp.1–12.

Wilson, M.W., Balmer, D.E., Jones, K., King, V.A., Raw, D., Rollie, C.J., Rooney, E., Ruddock, M., Smith, G.D., Stevenson, A., Stirling-Aird, P.K., Wernham, C.V., Weston, J.M., and Noble, D.G. 2018. The breeding population of Peregrine Falcon *Falco peregrinus* in the United Kingdom, Isle of Man and Channel Islands in 2014. *Bird Study* 65(1): pp.1–19.

Wilson, W. 1930. Some further notes on the birds of Bardsey Island. *British Birds* 24(5): pp.121–123.

Winstanley, D., Spencer, R., and Williamson, K. 1974. Where have all the Whitethroats gone? *Bird Study* 21(1): pp.1–14.

Wintle, W.J. 1924. Some Caldey birds. In *The Quarterly Review of The Benedictines of Caldey*. Caldey: Pax.

Wiseall, C. 2018. *The farming sector in Wales*. National Assembly for Wales Research Service.

Witherby, H.F., Jourdain, F.C.R., Ticehurst, N.F., and Tucker, B.W. 1940–1941. *The Handbook of British Birds*. Witherby. 5 Vols.

Witherby, H.F., and Ticehurst, N.F. 1907–1908. On the more important additions to our knowledge of British birds since *1899. Part II. British Birds* 1(6): Part I: pp.52–56; Part II: pp.81–85; Part III: pp.109–114; Part IV: pp.147–152; Part V: pp.178–184 (for Goldfinch, see p.178).

Wood, K.A., Brown, M.J., Cromie, R.L., Hilton, G.M., Mackenzie, C., Newth, J.L., Pain, D.J., Perrins, C.M., and Rees, E.C. 2019. Regulation of lead fishing weights results in mute swan population recovery. *Biological Conservation* 230: pp.67–74.

Wood, M.J., Taylor, V., Wilson, A., Padget, O., Andrews, H., Büche, B., Cox, N., Green, R., Hooley, T.A., Newman, L., Miquel-Riera, E., Perfect, S., Stubbings, E., Taylor, E., Taylor, J., Moss, J., Eagle, G., and Brown, R. 2017. *Repeat playback census of breeding European Storm-petrels* Hydrobates pelagicus *on the Skokholm and Skomer SPA in 2016.* NRW Evidence Report 190.

Woodburn, M.A., and Robertson, P.A. 1990. Woodland management for Pheasants: economics and conservation effects. In Lumeij, J.T., and Hoogeveen, Y.R., (eds.) *The future of Wild Galliformes in The Netherlands*. pp.185–198. Organisatiecommissie Nederlandse Wilde Hoenders, Amersfoort, The Netherlands.

Woodward, I., Aebischer, N., Burnell, D., Eaton, M., Frost, T., Hall, C., Stroud, D., and Noble, D. 2020. Population estimates of birds in Great Britain and the United Kingdom. *British Birds* 113(2): pp.69–104.

Woodward, I.D., *et al.* 2018. *BirdTrends 2018: Trends in numbers, breeding success and survival for UK breeding birds.* Thetford: British Trust for Ornithology.

Woodward, I.D., Frost, T.M., Hammond, M.J., and Austin, G.E. 2019. *Wetland Bird Survey Alerts 2016/2017: changes in numbers of wintering waterbirds in the Constituent Countries of the United Kingdom, Special Protection Areas (SPAs), Sites of Special Scientific Interest (SSSIs) and Areas of Special Scientific interest (ASSIs).* Thetford: BTO Research Report 721. **www.bto.org/webs-reporting-alerts**

Wotton, S. 2019. Hen Harriers in Wales in 2016. *Birds in Wales* 16(1): pp.17–23.

Wotton, S., and Conway, G. 2008. The Woodlark and Dartford Warbler surveys in Wales in 2006. *Welsh Birds* 5(4): pp.322–327.

Wotton, S., Conway, G., Eaton, M., Henderson, I., and Grice, P. 2009. The status of the Dartford Warbler in the UK and the Channel Islands in 2006. *British Birds* 102(5): pp.230–246.

Wotton, S., and Vanhinsbergh, D. 2005. Changes in the numbers of breeding waders on lowland wet grasslands in England and Wales between 1982 and 2002. *Bird Study* 52(1): pp.55–69.

Wotton, S.R., Bladwell, S., Mattingley, W., Morris, N.G., Raw, D., Ruddock, M., Stevenson, A., and Eaton, M.A. 2018. Status of the Hen Harrier *Circus cyaneus* in the UK and Isle of Man in 2016. *Bird Study* 65(2): pp.145–160.

Wotton, S.R., and Gillings, S. 2000. The status of breeding Woodlarks *Lullula arborea* in Britain in 1997. *Bird Study* 47(2): pp.212–224.

Wotton, S.R., Stanbury, A.J., Douse, A., and Eaton, M.A. 2016. The status of the Ring Ouzel *Turdus torquatus* in the UK in 2012. *Bird Study* 63(2): pp.155–164.

Wright, J., Stone, R.E., and Brown, N. 2003. Communal roosts as structured information centres in the Raven, *Corvus corax*. *Journal of Animal Ecology* 72(6): pp.1,003–1,014.

Wright, L.J., Hoblyn, R.A., Green, R.E., Bowden, C.G.R., Mallord, J.W., Sutherland W.J., and Dolman, P.M. 2009. Importance of climatic and environmental change in the demography of a multi-brooded passerine, the woodlark *Lullula arborea*. *Journal of Animal Ecology* 78(6): pp.1191–202.

WWT/JNCC 2010. Goose News: the newsletter of the goose and swan monitoring programme no. 9.

Wynn, J. 1770. *The History of the Gwydir family.* Hon. Dines Barrington.

Wynn, R.B. and Yésou, P. 2007. The changing status of Balearic Shearwater in northwest European waters. *British Birds* 100(7): pp.392–406.

Wynne, G. (ed.) 1993. *Biodiversity Challenge: an agenda for conservation action in the UK*. Sandy: RSPB.

Wynne-Edwards, V. 1962. *Animal dispersion in relation to social behaviour.* Edinburgh: Oliver & Boyd.

Yalden, D.W. 2007. The older history of the White-tailed Eagle in Britain. *British Birds* 100(8): pp.471–480.

Yalden, D.W., and Albarella, U. 2009. *The history of British birds*. Oxford: Oxford University Press.

Yalden, D.W., and Pearce-Higgins, J.W. 1997. Density dependence and winter weather as factors affecting the size of a population of Golden Plovers *Pluvialis apricaria*. *Bird Study* 44(2): pp.227–234.

Yésou, P. 1983. Le Bruant Lapon *Calcarius lapponicus* en Bretagne. *Alauda* 51: pp.161–178.

Yésou, P., and Clergeau, P. 2005 Sacred Ibis: a new invasive species in Europe. *Birding World* 18(12): pp.517–526.

Young, A. 2001. *Collation of data for selected farmland birds in Wales*. Sandy: RSPB, unpublished report.

Young, A., and Thorpe, R.I. 2002. The population status of birds in Wales: an analysis of conservation concern 2002–2007. *Welsh Birds* 3(4): pp.289–302.

Young, A., Doody, D., and Williams, I. 1996. *Upland breeding Waders – Elenydd 1995*. CCW/RSPB.

Zagoršek, T. 2019. Incredible flock of 5 million bramblings wows Slovenia. **www.birdlife.org/europe-and-central-asia/news/incredible-flock-5-million-bramblings-wows-slovenia**

Zielińska, M., Zieliński, P., Kołodziejczyk, P., Szewczyk, P., and Betleja, J. 2007. Expansion of the Mediterranean Gull *Larus melanocephalus* in Poland. *Journal of Ornithology* 148: pp.543–548.

Zuccon, D., Pons, J.-M., Boano, G., Chiozzi, G., Gamauf, A., Mengoni, C., Nespoli, D., Olioso, G., Pavia, M., Pellegrino, I., Raković, M., Randi, E., Rguibi Idrissi, H., Touihri, M., Unsöld, M., Vitulano, S., and Brambilla, M. 2020. Type specimens matter: New insights on the systematics, taxonomy and nomenclature of the subalpine warbler (*Sylvia cantillans*) complex. *Zoological Journal of the Linnean Society*.

Index to bird species

English species name index

Scientific species name index

Mynegai enwau rhywogaeth Cymraeg

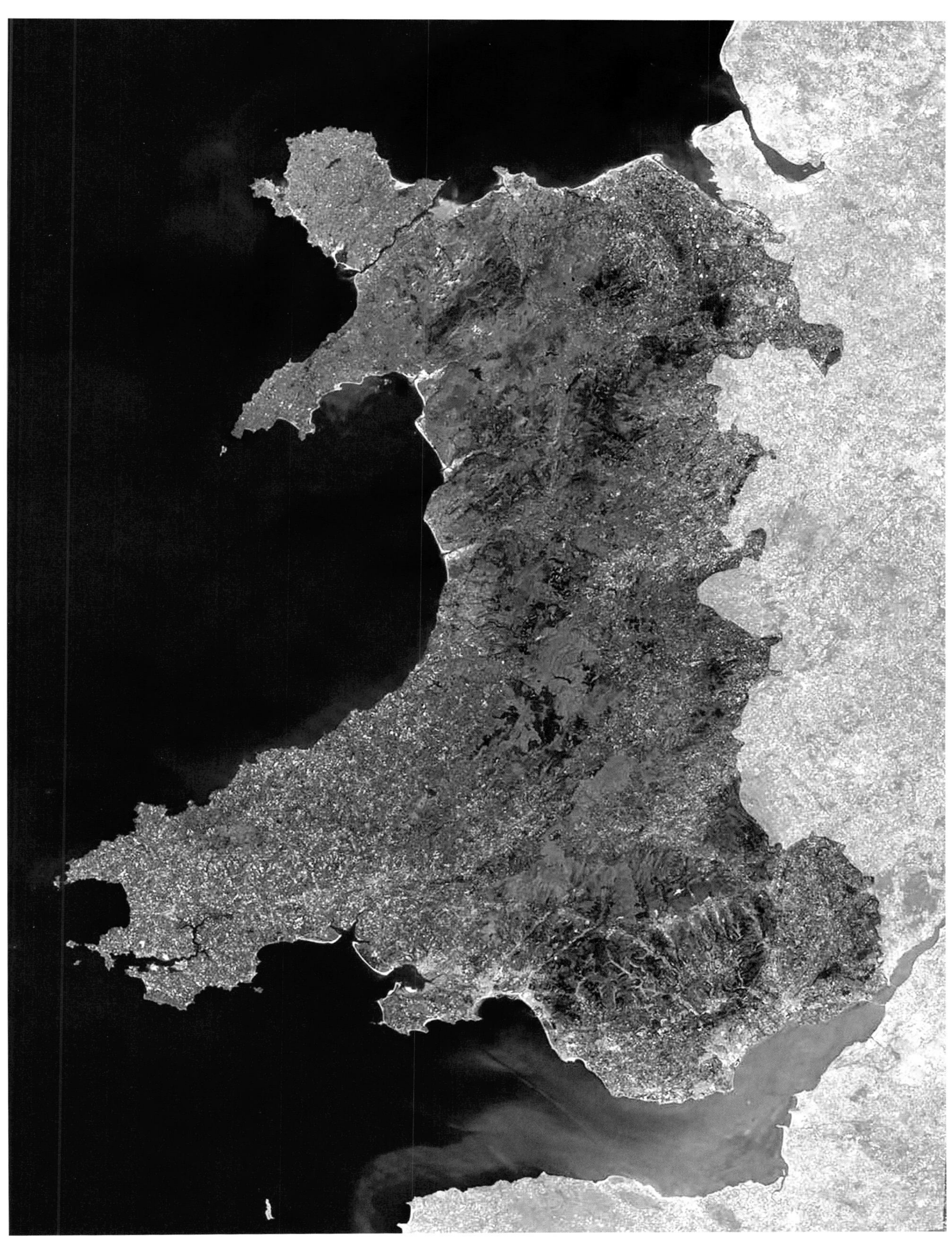